PÁLMASON
TALWANI & ELDHOLM
DEWEY
ROBERTS
MONTADERT, et al.
AUZENDE & OLIVET
ROSS
BIJU-DUVAL et al.
STÖCKLIN
STONELEY
MOORE
LUDWIG
SCHOLL
KIMURA
UYEDA
HINZ
COLEMAN
CLOSS et al.
DELTEIL et al.
CURRAY & MOORE
GINSBURG & JAMES
DRIVER & PARDO
KENT
VEEVERS
SIESSER

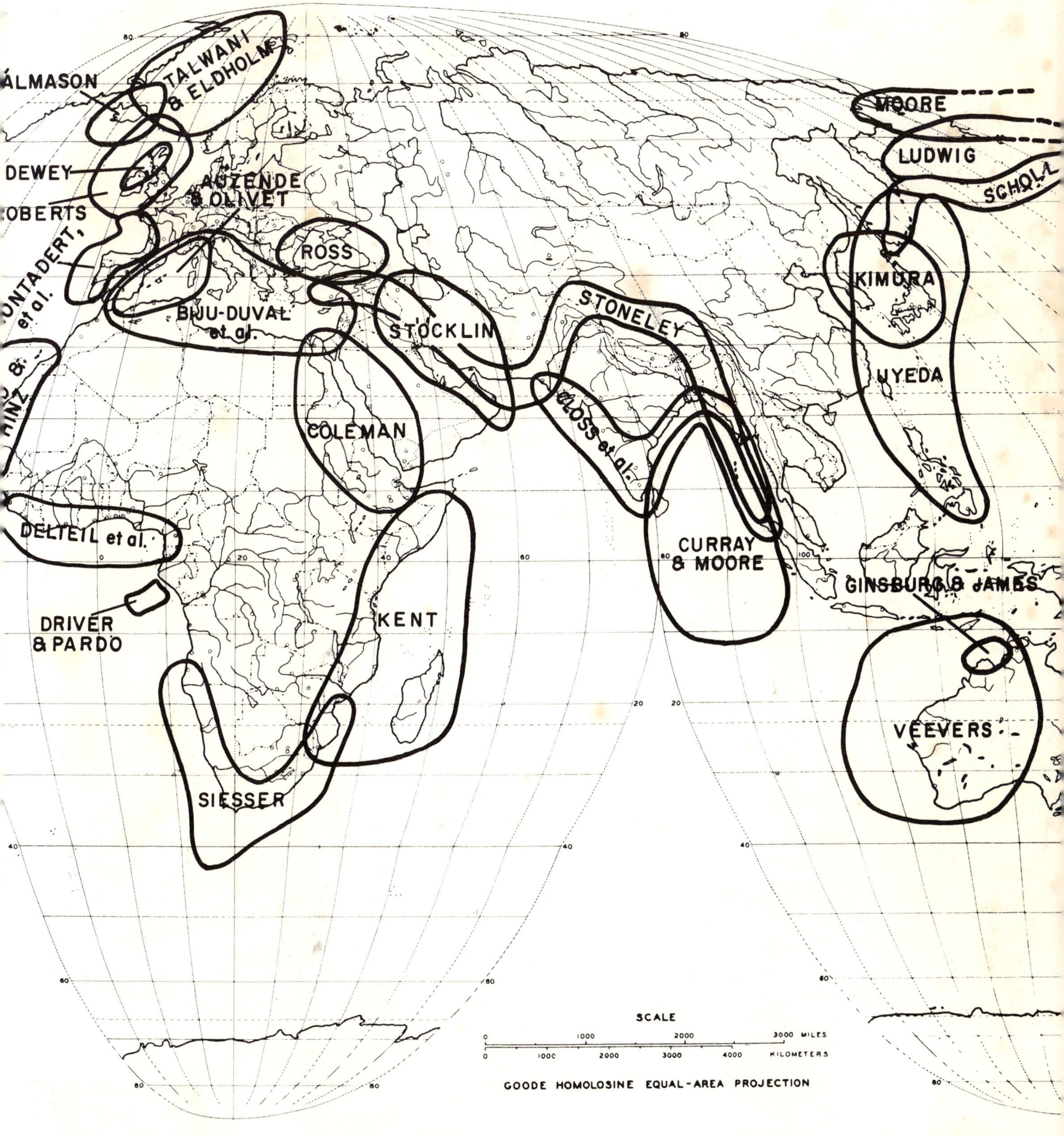

SCALE
0 1000 2000 3000 MILES
0 1000 2000 3000 4000 KILOMETERS
GOODE HOMOLOSINE EQUAL-AREA PROJECTION

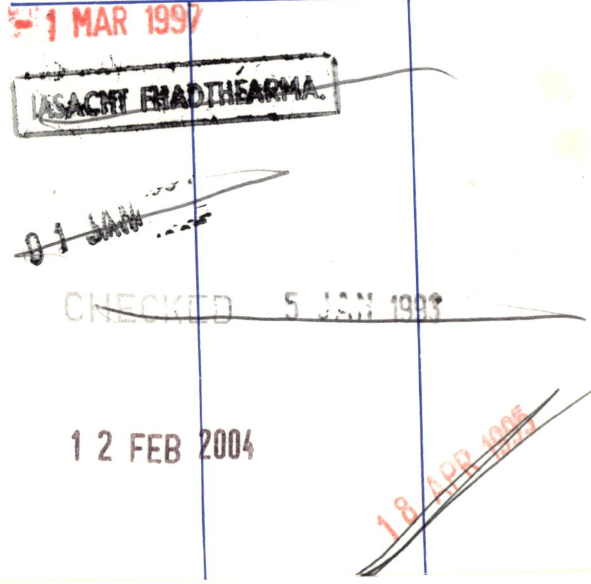

THE GEOLOGY OF CONTINENTAL MARGINS

edited by

Creighton A. Burk

and

Charles L. Drake

Springer-Verlag New York Heidelberg Berlin

Creighton A. Burk
Chief Geologist
Mobil Oil Corporation

Charles L. Drake
Professor of Earth Sciences
Dartmouth College

Second corrected printing.

Library of Congress Cataloging in Publication Data

Burk, Creighton A.
The Geology of Continental Margins.

1. Continental margins. 2. Submarine geology.
I. Drake, Charles L., 1924- joint author.
II. Title.
QE39.B85 551.4'608 74-16250

Printed in the United States of America

ISBN 0-387-06866-X Springer-Verlag New York • Heidelberg • Berlin
ISBN 3-540-06866-X Springer-Verlag Berlin • Heidelberg • New York

Preface

The continental margins of the world constitute the most impressive and largest physiographic feature of the earth's surface, and one of fundamentally great geological significance. Continental margins have been the subject of increasing attention in recent years, an interest focused by a body of new data that has provided new insights into their character. This interest was further stimulated by the realization that, in addition to the abundant living resources, continental margins contain petroleum and mineral resources that are accessible with existing technology. This realization, along with their basic geological importance, has provoked further research into the nature of continental margins throughout the world. A summary of these findings, as related to both recent and ancient continental margins, is the subject of this book.

At various times in the past we had been approached individually to prepare a basic reference to continental margins; we then proposed to do such a volume jointly. However, the stimulus for the present volume eventually arose from a Penrose Conference arranged through the Geological Society of America. This conference was attended by specialists of numerous disciplines and from throughout the world, many of whom insisted that such a volume would be both timely and useful. Consequently, we agreed to undertake the task of assembling this book, with the objectives of making it available as soon and as inexpensively as possible.

The volume received further impetus from the International Geodynamics Project. The U.S. Program noted that "a fundamental step at the outset of the Geodynamics Project is the preparation of syntheses designed with a view to their bearing on important problems of geodynamics." In view of the importance of continental margins to geodynamics, we expect that this volume contributes to this goal.

The objectives of this volume are to provide an overall view of our present state of knowledge of all aspects of continental margins throughout the world, both modern and ancient. We hope that this will be a valuable reference for the researcher, economic explorer, teacher, student, and all others interested in continental margins and their diverse geological aspects, and that it will help guide future research and exploration. All contributors have been invited (some have been blackmailed or threatened). We have acted as sole editors and reviewers, and consequently must accept full responsibility for important contributions omitted and for any significant errors included. All contributors were encouraged to include important references in the bibliographies, whether cited or not. We hope that this will be an added contribution.

During the final stages of editing, we were notified of the death of Maurice Ewing, one of whose last papers appears in the volume, The study of continental margins owes a great debt to Maurice. His interest goes back to the early 1930s when he carried out the first of a series of investigations entitled, "Geophysical Investigations on the Emerged and Submerged Atlantic Coastal Plain." These studies led to cooperative programs on many other margins of the world, including Argentina, Brazil, Japan, Canada, South Africa, Antarctica, Norway, the Caribbean and Mediterranean seas and elsewhere. Prof. Ewing was also instrumental in developing or modifying many geological and geophysical techniques. His drive and abilities made him a leader in this and other fields. He will be sorely missed.

We are particularly grateful to all of the contributors for the rapid response to short deadlines that made it possible for this volume to be produced so quickly. A debt of gratitude is owed to Mobil Oil Corporation, particularly to John D. Moody, for encouragement and support during the inception and editorial stages. The cover was conceived and designed by Yvonne T. Burk. We are pleased to acknowledge particularly the patience and unstinting efforts of the charming and almost vice-free Carol L. Thompson, without which this book would still be in rough draft.

C. A. Burk
Princeton, New Jersey

C. L. Drake
Hanover, New Hampshire

August 1974

Publisher's Note

Based on the advice of the Editors, an index has not been included in this volume. The reader should be able to locate specific information by using the detailed Table of Contents and well-designed chapter subdivisions.

Contributors

John W. Antoine, Decca Survey Systems, Inc., 8204 Westglen Drive, Houston, Texas 77042

F. Aubertin, Compagnie Francais du Petrole, 5 Rue Michel-Ange, Paris, France

Jean-Marie Auzende, Centre Oceanologique de Bretagne, Boite Postale 337, 29.273 Brest Cedex, France

Muawia Barazangi, Cornell University, Department of Geological Sciences, Ithaca, New York 14850

Felix P. Bentz, Mobil Oil Corporation-MEPSI, P.O. Box 900, Dallas, Texas 75221

Wolfgang H. Berger, University of California, San Diego, Scripps Institution of Oceanography, Geological Research Division, P.O. Box 1529, La Jolla, California 92037

B. Biju-Duval, Institut Francais du Petrole, 1 et 4, Avenue de Bois-Preau, 92-Rueil-Malmaison, Paris, France

M. Clark Blake, Jr., U.S. Department of the Interior, U.S. Geological Survey, Geologic Division, Branch of Western Environmental Geology, 345 Middlefield Road, Menlo Park, California 94025

William R. Bryant, Texas A & M University, Oceanographic Department, College Station, Texas 77843

Creighton A. Burk, Mobil Oil Corporation, 221 Nassau Street, Princeton, New Jersey 08540

Carlos W. M. Campos, Petroleo Brasileiro S. A., Caixa Postal 809 XZ-00, Rio de Janeiro, Brazil

M. J. Carr, Dartmouth College, Department of Earth Sciences, Hanover, New Hampshire 03755

James E. Case, U.S. Department of the Interior, U.S. Geological Survey, Office of Marine Geology, P.O. Box 6732, Corpus Christi, Texas 78411

Hans Closs, Bundesanstalf fur Bodenforschung, 3 Hannover, Postfach 230153, Federal Republic of Germany

Robert G. Coleman, Branch of Field Chemistry and Petrology, U.S. Geological Survey, Geologic Division, 345 Middlefield Road, Menlo Park, California 94025

P. Courrier, Societe ELF pour la Recherche et l'Exploitation des Hydrocarbures, 7, Rue Nelaton, Paris, France

Michael J. Cruickshank, U.S. Geological Survey, Conservation Division, 345 Middlefield Road, Menlo Park, California 94025

Joseph R. Curray, Scripps Institution of Oceanography, University of California, Geologic Research Division, P.O. Box 109, La Jolla, California 92037

Ian W. D. Dalziel, Columbia University, Department of Geology, New York, New York 10027

Jean-Raymond Delteil, ELF-ERAP, 7, Rue Nelaton, Paris, France

John F. Dewey, State University of New York at Albany, Department of Geology, 1400 Washington Avenue, Albany, New York 12203

Charles L. Drake, Dartmouth College, Department of Earth Sciences, Hanover, New Hampshire 03755

Edgar S. Driver, Gulf Research and Development Center, Gulf Building, P.O. Drawer 2138, Pittsburgh, Pennsylvania 15230

Kenneth J. Drummond, Mobil Oil Corporation, 221 Nassau Street, Princeton, New Jersey 08540

J. Dubois, Office de la Recherche Scientifique et Technique Outre-Mer, Centre de Noumea, Boite Postale 4, Noumea, New Caledonia

N. Terence Edgar, Scripps Institution of Oceanography, University of California, Deep Sea Drilling Project, P.O. Box 1529, La Jolla, California 92037

Olaf Eldholm, University of Oslo, Department of Geology, Oslo, Norway

W. Gary Ernst, University of California, Department of Geology, Institute of Geophysics and Planetary Physics, Los Angeles, California 90024

Maurice Ewing (deceased), Earth and Planetary Sciences Division, Marine Biomedical, University of Texas Medical Branch, Galveston, Texas 77550

Robert L. Fisher, University of California, San Diego, Geologic Research Division, Scripps Institution of Oceanography, P.O. Box 109, La Jolla, California 92037

Roger Flood, Department of Earth and Planetary Sciences, Massachusetts Institute of Technology, Cambridge, Massachusetts 02139

Catherine Fondeur, Institut Francais du Petrole, 1 et 4, Avenue de Bois-Preau, 92-Rueil-Malmaison, Paris, France

Gerald A. Fowler, University of Wisconsin, Parkside, Division of Science, Kenosha, Wisconsin 53140

S. C. Garde, National Geophysical Research Institute, Hyderabad-500007 (A.P.), India

Robert N. Ginsburg, Comparative Sedimentology Laboratory, Division of Marine Geology and Geophysics, Rosenstiel School of Marine and Atmospheric Science, University of Miami, Fisher Island, Miami Beach, Florida 33139

Alan M. Goodwin, University of Toronto, Precambrian Research Group, Toronto 181, Canada

G. Grau, Institut Francais du Petrole, 1 et 4, Avenue de Bois-Preau, 92-Rueil-Malmaison, Paris, France

R. Guillaume, Societe Nationale des Petroles D'Aquitaine, B.P. No. 65, 64 Pau, France

Trevor Hatherton, Department of Scientific and Industrial Research, Geophysics Division, P.O. Box 8005, Wellington, New Zealand

Charles W. Hatten, Great Basin Petroleum Company, 1011 Gateway West, Century City, Los Angeles, California 90067

James E. Case, U.S. Department of the Interior, U.S. Geological Survey, Office of Marine Geology, P.O. Box 6732, Corpus Christi, Texas 78411

Hans Closs, Bundesanstalf fur Bodenforschung, 3 Hannover, Postfach 230153, Federal Republic of Germany

Robert G. Coleman, Branch of Field Chemistry and Petrology, U.S. Geological Survey, Geologic Division, 345 Middlefield Road, Menlo Park, California 94025

P. Courrier, Societe ELF pour la Recherche et l'Exploitation des Hydrocarbures, 7, Rue Nelaton, Paris, France

Michael J. Cruickshank, U.S. Geological Survey, Conservation Division, 345 Middlefield Road, Menlo Park, California 94025

Joseph R. Curray, Scripps Institution of Oceanography, University of California, Geologic Research Division, P.O. Box 109, La Jolla, California 92037

Ian W. D. Dalziel, Columbia University, Department of Geology, New York, New York 10027

Jean-Raymond Delteil, ELF-ERAP, 7, Rue Nelaton, Paris, France

John F. Dewey, State University of New York at Albany, Department of Geology, 1400 Washington Avenue, Albany, New York 12203

Charles L. Drake, Dartmouth College, Department of Earth Sciences, Hanover, New Hampshire 03755

Edgar S. Driver, Gulf Research and Development Center, Gulf Building, P.O. Drawer 2138, Pittsburgh, Pennsylvania 15230

Michael J. Keen, Faculty of Arts and Sciences, Dalhousie University, Halifax, Nova Scotia, Canada

Peter E. Kent, BP Petroleum Development Ltd., Britannic House, Moor Lane, London EC2Y 9BU, England

Toshio Kimura, Geological Institute, University of Tokyo, Tokyo, Hongo, Japan

Laverne D. Kulm, Department of Oceanography, Oregon State University, Corvallis, Oregon 97331

C. A. Landis, University of Otago, Geology Department, Dunedin, New Zealand

J. Launay, Office de la Recherche Scientifique et Technique Outre-Mer, Centre de Noumea, Boite Postale 4, Noumea, New Caledonia

J. Letouzey, Institut Francais du Petrole, 1 et 4, Avenue de Bois-Preau, 92-Rueil-Malmaison, Paris, France

J. Louis, ELF-ERAP, 7, Rue Nelaton, Paris, France

William J. Ludwig, Lamont-Doherty Geological Observatory of Columbia University, Palisades, New York 10964

Ray G. Martin, U.S. Geological Survey, P.O. Box 6732, Corpus Christi, Texas 78411

Jean Mascle, Centre Oceanologique de Bretagne, Boite Postale 337, 29.273-Brest Cedex, France

John C. Maxwell, Department of Geological Sciences, University of Texas, Austin, P.O. Box 7909, Austin, Texas 78712

Michael A. Mayhew, University of Wisconsin, Department of Geological Sciences, Sabin Hall—Greene Museum, Milwaukee, Wisconsin 53201

Arthur A. Meyerhoff, A.A.P.G. Headquarters, Box 979, Tulsa, Oklahoma 74101

K. Miura, Petroleo Brasileiro S.A., Caixa Postal 809 XZ-00, Rio de Janeiro, Brazil

Lucien Montadert, Institut Francais du Petrole, 1 et 4, Avenue de Bois-Preau, 92-Rueil-Malmaison, Paris, France

David G. Moore, Department of the Navy, Naval Undersea Center, San Diego, California 92132

J. Casey Moore, University of California, Santa Cruz, Division of Natural Sciences, Department of Geological Sciences, Santa Cruz, California 95060

Carlos Mordojovich, Empresa Nacional del Petroleo, Corporacion de Fomento de la Produccion, Casilla 3556, Santiago, Chile

J. F. Mugniot, Compagnie Franciase des Petroles, 5, Rue Michel-Ange, Paris 16e, France

Hari Narain, National Geophysical Research Institute, Hyderabad 500007 (A.P.), India

Jack E. Oliver, Cornell University, Department of Geological Sciences, Ithaca, New York 14850

Jean-Louis Olivet, Centre Oceanologique de Bretagne, Boite Postale 337, 29.273-Brest Cedex, France

Ned A. Ostenso, Ocean Science & Technology Division, Department of the Navy, Arlington, Virginia 22217

Gudmunder Palmason, Icelandic National Energy Authority, Reykjavik, Iceland

Georges Pardo, Gulf Oil Company, U.S.A., P.O. Box 2100, Houston, Texas 77001

Philippe Patriat, Institut de Physique de Globe de Paris, Paris France

F. C. Ponte, Petroleo Brasileiro, S.A., Caixa Postal 809 XZ-00, Rio de Janeiro, Brazil

Thomas Pyle, University of South Florida, 830 1st Street South, St. Petersburg, Florida 33701

Philip D. Rabinowitz, Lamont-Doherty Geological Observatory of Columbia University, Palisades, New York 10964

C. Ravenne, Institut Francais de Petrole, 1 et 4, Avenue de Bois-Preau, 92-Rueil-Malmaison, Paris, France

Vincent Renard, Centre Oceanologique de Bretagne, B.P. 337, 29.273 Brest Cedex, France

David G. Roberts, Institute of Oceanographic Sciences, Natural Environment Research Council, Brook Road, Wormley, Godalming, Surrey GU8 5UB, England

David A. Ross, Woods Hole Oceanographic Institution, Department of Geology & Geophysics, Woods Hole, Massachusetts 02543

J. Sancho, Societe Nationale des Petroles D'Aquitaine, B.P. No. 65, 64 Pau, France

David W. Scholl, U.S. Geological Survey, Pacific-Arctic Branch of Marine Geology, Office of Marine Geology, 345 Middlefield Road, Menlo Park, California 94025

R. A. Scrutton, Grant Institute of Geology, Edinburgh, Scotland

Donald R. Seely, Esso Production Research Company, P.O. Box 2189, Houston, Texas 77001

Eugen Seibold, Geologisch-Palaontologisches, Institat und Museum, Der Universitat Kiel, Olshausenstrasse 40/60, Kiel, FRG, Germany

Robert E. Sheridan, University of Delaware, College of Arts and Sciences, Department of Geology, Newark, Delaware 19711

George G. Shor, Jr., University of California, Scripps Institution of Oceanography, P.O. Box 1529, La Jolla, California 92037

W. G. Siesser, Marine Geology Unit, Department of Geology, University of Cape Town, Cape Town, South Africa

E. S. W. Simpson, Marine Geology Unit, Department of Geology, University of Cape Town, Cape Town, South Africa

R. K. Stevens, Memorial University of Newfoundland, Department of Geology, St. John's, Newfoundland, Canada

Jovan Stocklin, United Nations, Geological Survey Institute, Box 1555, Tehran, Iran

R. E. Stoiber, Dartmouth College, Department of Earth Sciences, Hanover, New Hampshire 03755

R. L. Stoneley, BP Petroleum Development Ltd., Britannic House, Moor Lane, London EC2Y 9BU, England

Donald J. P. Swift, Atlantic Oceanographic and Meteorological Laboratories, 15 Rickenbacker Causeway, Miami, Florida 33149

Manik Talwani, Lamont-Doherty Geological Observatory of Columbia University, Palisades, New York 10964

Carlos Urien, Cabot Petroleum (Argentina) Inc., Avenida Ptd R.S. Pena 852, Buenos Aires, Argentina

Seiya Uyeda, Earthquake Research Institute, The University of Tokyo, Bunkyo-Ku, Tokyo, Japan

Peter R. Vail, Esso Production Research Company, P.O. Box 2189, Houston, Texas 77001

Pierre Valery, ELF-ERAP, 7, Rue Nelaton, Paris, France

John J. Veevers, School of Earth Sciences, Macquarie University, North Ryde, New South Wales, Australia 2113

Roland E. von Huene, U.S. Geological Survey, Department of the Interior, Reston, Virginia 22070

G. G. Walton, Esso Production Research Company, P.O. Box 2189, Houston, Texas 77001

Lewis G. Weeks, Weeks Natural Resources, Inc., 287 Riverside Avenue, Westport, Connecticut 06880

Harold Williams, Memorial University of Newfoundland, Department of Geology, St. John's Newfoundland, Canada

E. Winnock, Institut Francais du Petrole, 1 et 4, Avenue de Bois-Preau, 92-Rueil-Malmaison, Paris, France

Donald U. Wise, Department of Geology, University of Massachusetts, Amherst, Massachusetts 01002

J. Lamar Worzel, Earth & Planetary Sciences Division, Marine Biomedical Institute, University of Texas Medical Branch, Galveston, Texas 77550

J. J. Zambrano, Direccion Nacional de Hidrocarburos, Av. Julio A. Roca 651, 7°P., Buenos Aires, Argentina

Contents

I. GEOLOGICAL SIGNIFICANCE OF CONTINENTAL MARGINS

II. GENERAL BATHYMETRY AND TOPOGRAPHY

III. TRANSITION FROM CONTINENT TO OCEAN

IV. RECENT SEDIMENTATION

V. DEFORMATION AT CONTINENTAL MARGINS

VI. GEOLOGY OF SELECTED MODERN MARGINS: ATLANTIC REGION

VII. GEOLOGY OF SELECTED MODERN MARGINS: PACIFIC REGION

VIII. GEOLOGY OF SELECTED MODERN MARGINS: INDIAN OCEAN REGION

IX. GEOLOGY OF SELECTED SMALL OCEAN BASINS

X. ANCIENT CONTINENTAL MARGINS

XI. IGNEOUS ACTIVITY AND ANCIENT MARGINS

XII. RESOURCES AT CONTINENTAL MARGINS

XIII. CONTINENTAL MARGINS IN PERSPECTIVE

Part I

Geological Significance of Continental Margins

Geological Significance of Continental Margins

C. L. Drake and C. A. Burk

INTRODUCTION

Detailed studies of continental margins do not have a long history. Although surface geological sampling in shallow water dates back to the eighteenth century, systematic investigation of the topography and the surface geology date back only for the last half century or so. Early gravity measurements gave some clues about the deeper character of the margins, but serious geophysical investigations were not carried out until after World War II. Since that time many margins have been studied in greater or lesser detail using geological and geophysical techniques that continued to improve significantly.

In spite of the efforts of the last two decades, many aspects of continental margins still remain unknown and controversial. Continental margins are among the most difficult in which to work since they consist of great variations in water depth and in configuration of the sedimentary and crustal layers, major changes in physical properties, and fundamental changes in composition of the rocks, combined within a single limited region.

Ancient margins are even more difficult to identify and study because typically they have been greatly deformed. Thus it is not uncommon to find wide latitude in possibilities for reasonable interpretation and for differing concepts of origin, history, and development of various continental margins. It is difficult to find areas more thoroughly embedded in the rich organic topsoil of geological speculation.

Geologists of the past worked seaward until the exposures were covered by water; then they turned landward again, where many exposed problems still remained to be solved. Similarly, marine geophysicists of the last few decades developed indirect techniques which allowed them to work successfully in the deep oceans (where the 4-5 km of water served as a useful filter to remove problematic details, and where the structure of the deep-ocean crust proved to be simpler than that on land); but when the study of shallower-water areas became uncomfortably complex, they turned seaward again, where many deep-water problems still remained to be solved. There has thus always existed a "belt of ignorance," extending sinuously across the earth's surface, marking the transition from the exposed continents to the deep oceans—the world's continental margins.

EARLY IDEAS

In his monumental summation of the geology of the world, Suess (1904) noted that "the possibility was recognized [in the 1885 edition] of deducing from the uniform strike of the folds of a mountain-chain a mean general direction or trend-line; such trend-lines were seen to be seldom straight, but consisted of arcs or curves, often violently bent curves of accommodation; the trend-lines of central Europe were observed to possess a certain regular arrangement and to be traceable in part as far as Asia. It was further recognized that the ocean from the mouth of the Ganges to Alaska and to Cape Horn is bordered by folded mountain chains, while in the other hemisphere this is not the case, so that a Pacific and Atlantic type may be distinguished; that the Mediterranean is not a part of the Atlantic Ocean, but the remains of a sea which once crossed the existing continent of Asia, and has since been enlarged by subsidence; that at various times, as for instance during the middle and upper Cretaceous period, extension of the seas occurred much too general and too equable to be explained by the subsidence of continental land, and so on."

Thus Suess recognized almost 90 years ago the fundamental differences between the active (Pacific) and passive (Atlantic) continental margins; he saw the continuity of the circum-Pacific and Alpine-Himalayan belts, which are now known to be the regions in which the deep earthquakes are also found; he realized that it was difficult to explain the nature of marginal seas such as the Mediterranean, except in terms of massive subsidence; and he was aware of the problems associated with major transgressions of the sea, such as that found in the late Cretaceous.

Detailed analysis of these areas was hampered by lack of information from the water-covered areas—indeed, Suess considered that the ocean crust was similar to that of the continents and that "It is to subsidence and collapse that...the largest oceans owe their origins and enlargements."

INACTIVE CONTINENTAL MARGINS

Alfred Wegener (1924) noted that "on the oceans, gravity possesses almost its normal value in spite of the evident defects of mass formed by the great ocean basins."

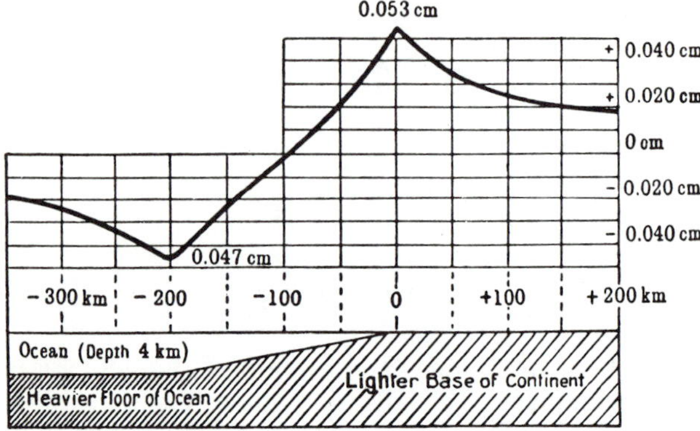

Fig. 1. Variation in gravity across a continental margin (after Helmert, 1909).

This mass defect must be compensated by an excess of mass at depth, and furthering the arguments of Schweydar (1921), Wegener suggested that (continental) crust was completely absent in the ocean basins and that the oceanic crust must be very thin.

Helmert (1909) very early showed the characteristic gravity anomaly on the margins of the continents (Fig. 1). There is an edge-effect across the margin characterized by a gravity high on the outer part of the continental shelf and a low on the continental rise, with values approaching worldwide normal at large distances from the continental edge. Wegener concluded that this confirmed the fundamental difference between continent and ocean and further that there would be a large pressure differential as a result of this variation. This pressure differential would cause the material of the continental platform to press out into the oceanic areas and thus cause the step-faulting found on such continental margins as Atlantic South America and Africa.

The development of the submarine gravity-pendulum apparatus by Vening-Meinesz led to many more accurate determinations of gravity variations in the oceans. Vening-Meinesz (1941) made a study of the variations across continental margins, mostly of the Atlantic (inactive) type, and concluded that the most probable explanation of the data was an abrupt thinning of the sialic layers of the earth's crust at the continental margin. This was in accord with the thinking of this period, which assumed that sial was absent only in the north polar basin and in large parts of the Pacific. It was not until the advent of seismic refraction measurements at sea (Hill and Laughton, 1954; Hill, 1952; Ewing et al., 1950) that it was firmly established that the sialic rocks were not merely thin, but absent from the ocean basins.

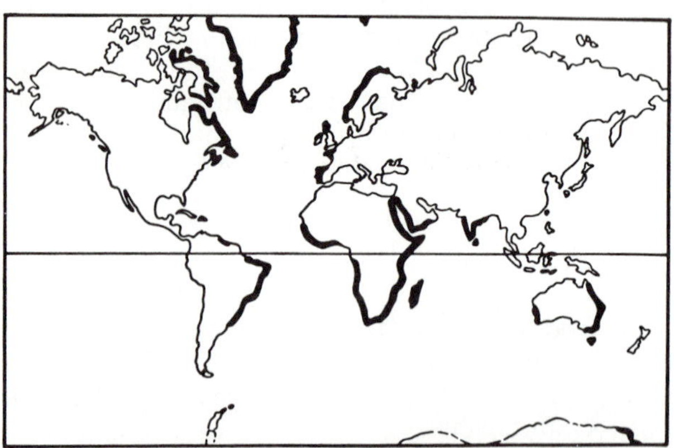

Fig. 2. Fault-line coasts (heavy line) (after Du Toit, 1937).

The literature of continental drift is voluminous, but the implications, recognized early in the development of the concept (Taylor, 1910; Wegener, 1924), suggested major differences between continental margins on the leading edge of proposed drifting-blocks (Pacific type) and those on the trailing edges (Atlantic type). Du Toit (1937) compared the Atlantic-type margins with rift valleys and concluded that certain characteristics, including an uplifted edge, drainage toward the continent, and a steep continental slope, could serve to identify a margin as a fault-line coast (Fig. 2). Du Toit noted that these characteristics could be found along many coasts in the Atlantic and Indian oceans. The areas where these characteristics are absent in the Atlantic are principally those in which the Paleozoic structures parallel the coast (e.g., eastern United States and northwestern Africa), and along which obvious structural truncations are absent.

Umbgrove (1947) rejected the idea, advanced by Wegener and Du Toit, that the continental shelf might have a fault origin, on the basis of seismic refraction measurements by Ewing et al. (1937, 1939) and Bullard and Gaskell (1941). These measurements, extending part way across the shelf, indicated a regular increase in sediment thickness and suggested that the shelf was not structurally controlled but was extended by sedimentary processes. Umbgrove preferred Bourcart's (1938) marginal flexure concept, which proposed that spasmodic warping of the continental border caused submergence of what was formerly the margin of the continent, and simultaneous bowing up of an inland marginal area parallel to the newly formed coastline.

Jesson (1943), on the other hand, cited the faults that bordered the continental margin of northeastern Australia and concluded that the margin and the Great Barrier Reef were underlain by a tilted and subsided fault-block of continental crust (Fig. 3).

Ross Gunn presented an interesting series of papers in the early 1940s in which he considered the influence of a strong lithosphere on gravity anomalies and isostatic compensation. In one of these papers (Gunn, 1944), he carried Bourcart's (1938) marginal flexure idea a step further by making a quantitative study of the behavior of the lithosphere at continental margins under a sedimentary and a fluid load. The purpose was to explain the variations between theoretical and observed gravity anomalies at these margins. He concluded that the

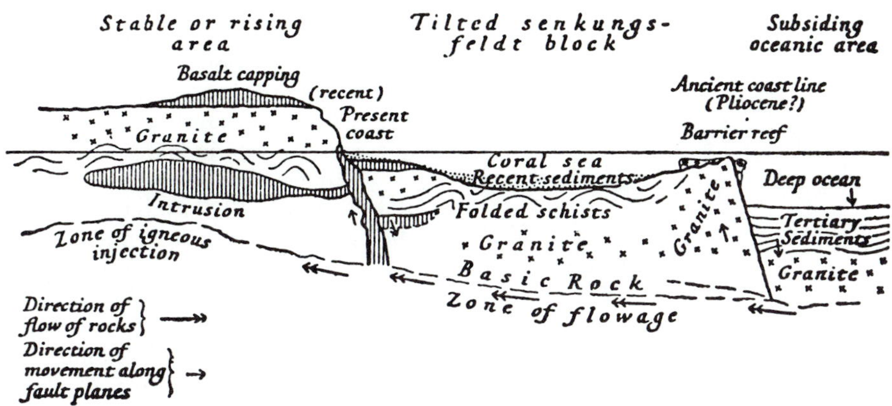

Fig. 3. Continental margin of eastern Australia (after Jesson, 1943).

assumption of a lithosphere with finite strength would better fit the observations and that the deformation would produce a marginal sediment-filled trough with its axis migrating seaward as the load increased.

The distribution of relief on the earth's surface had been summarized as early as 1933 by Kossina, and has not been changed significantly since that time (Fig. 8). Kuenen (1950) used this hypsometric curve as an innovative way to estimate reasonable models for an isostatic transition between continental and oceanic crust (Fig. 4). In this unique approach Kuenen eliminated unreasonable alternatives and arrived at solutions similar to many profiles being published today. It was further noted by Kuenen (1950, p. 168-169) that since the present continental margins "...are comparatively young, Mesozoic or Tertiary, one may well ask where are the accumulations of sediment and the masses planed off by marine erosion that must have developed during the preceding geological ages. ...Should one assume that in some mysterious manner the sialic foundation was chemically altered to sima? Did vast invasions of ultrabasic igneous rock radically change to specific gravity? Was the sial stretched out to a thin stratum by continental drift or by gravitational pull?"

ACTIVE CONTINENTAL MARGINS

The seismically active continental margins and island arcs received early attention because of the structural similarities between these areas and both modern and ancient mountain systems. Speculations about their nature were abundant and emotions ran high when their history was discussed. Taylor (1910), as did Wegener, associated these with the lateral movements of the continents.

There was much early preoccupation with the geometry of island arcs. Sollas (1903) pointed out that the poles of circular volcanic arcs in the western Pacific lay on a great circle. Lake (1931) followed this by suggesting that the arcs coincided with large thrust planes and that the thrusting took place perpendicular to the great circle, passing through the poles. From this he surmised that the Asiatic continent as a whole was being pushed over the floor of the Pacific.

The discovery of deep-focus earthquakes (Turner, 1922) and the finding by Wadati (1935) that the earthquakes beneath Japan fall in a zone that

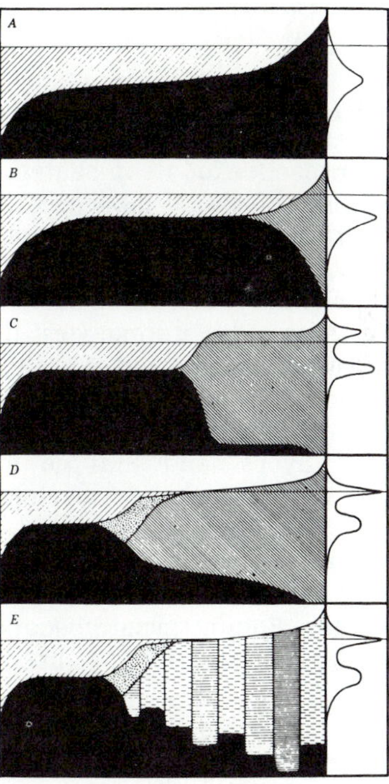

Fig. 4. Hypothetical models of the earth's crust with deduced frequency curves of elevation (after Kuenen, 1950): A. Crust of uniform density, a possibility imagined by Wegener, but excluded on account of isostatic equilibrium. B. Uniform sial of variable thickness. No external forces. Deeps due to lack of isostatic balance and to local heavy masses. C. Uniform sial of uniform thickness outside mountain ranges. No external forces. Postulated by Wegener. D. Uniform sial of variable thickness, cutdown by denudation. Terrace built by deposition. E. Sial of variable density and thickness, cutdown by denudation. Terrace built by deposition.

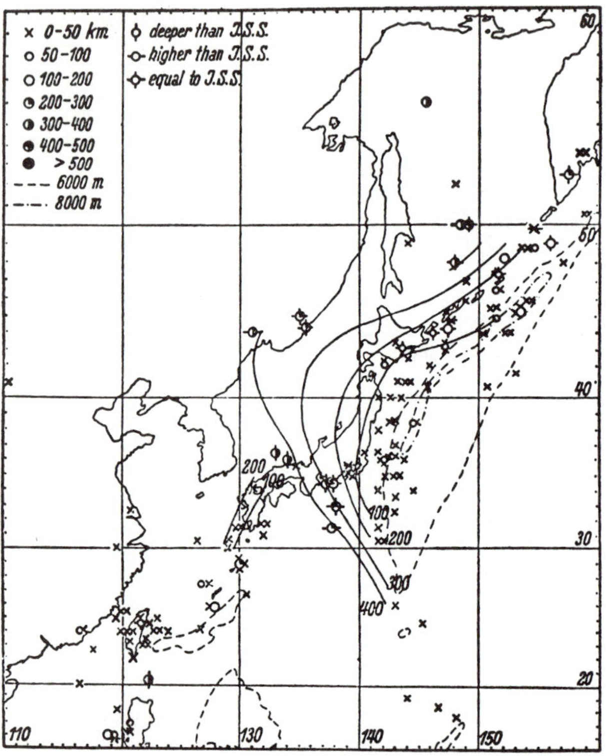

Fig. 5. Distribution of deep earthquakes in the vicinity of Japan (after Wadati, 1935).

dipped from the deep oceanic trench westward beneath Japan and the Asian continent (Fig. 5) implied the presence of deep-seated shear zones along the Pacific border (Umbgrove, 1947).

Gutenberg and Richter (1938) noted that "The true deep-focus shocks appear to be associated with events that originally took place early in the history of the earth...it is natural to suggest that there has been motion over a long period of geological time, by which the uppermost layers surrounding the Pacific basin have been displaced toward its center relative to the lower layers, and that no new zones of faulting or weakness have developed at great depths."

Schwinner (1941) was convinced that there was overthrusting of the continents, rather than underthrusting of the Pacific crust, because the latter case would require the Pacific to become stretched and enlarged on all sides (a problem that can now be resolved by the sea-floor spreading model). Benioff (1954) noted the different attitudes of seismic thrust planes beneath island arcs as compared to those beneath continents and suggested that differences between continents and oceans (arcs) might extend as deep as 300 km.

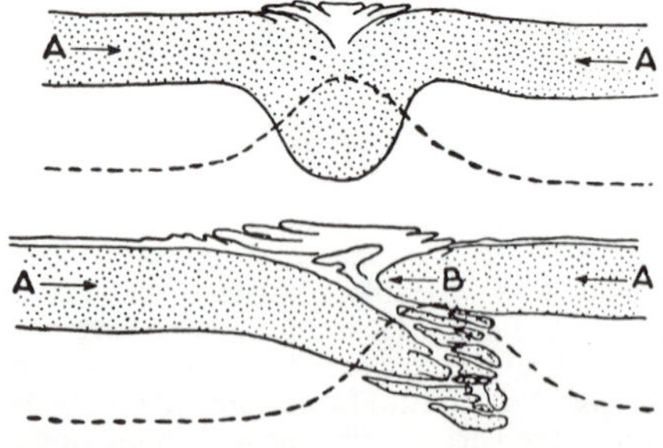

Fig. 6. Symmetrical and asymmetrical types of mountain roots (after Umbgrove, 1947).

During his gravity-measuring cruises to the East Indies in 1929 and 1930, Vening-Meinesz (1933) discovered the large negative gravity anomalies associated with the deep-sea trenches. His explanation was that under horizontal stresses, the earth's crust buckled downward and the deformation of the surface rocks accompanied this downbuckling. Thus the earthquake data suggested a shearing type of deformation (Fig. 6, bottom), while the gravity data were interpreted as downbuckling (Fig. 6, top).

Hess (1938) pointed out the presence of serpentinized peridotites in the axial zones of folded geosynclines, which he regarded as derived from mantle material at a depth of about 60 km, and a conclusive indication of uplift resulting from crustal shortening and thrusting. He considered the occurrence of ultrabasic rocks to be characteristic of ancient-type continental margins.

RECENT STUDIES OF CONTINENTAL MARGINS

The importance of systematic geological and geophysical studies was recognized and supported following World War II, and international cooperation in such studies has grown continuously since that time.

The new geological studies of continental margins were natural extensions of such early work as that of Shepard (1934) and Stetson (1936, 1938), but the advances in geophysics were mostly due to new instruments and techniques applied to the study of the sea floor. The interpretation of the results of these efforts on the continental margins both assisted and were assisted by studies within the continents of the nature and history of ancient continental margins (Kay, 1951).

New mechanisms of sediment transport and deposition on the margins were discovered (Kuenen and Migliorini, 1950; Heezen and Ewing, 1952; Heezen and Drake, 1964; Heezen et al., 1966), and sufficient geophysical data were accumulated on some continental margins that serious comparisons could be made between these and ancient geosynclines (Fig. 6).

Interest in the margins increased not only for basic scientific reasons, but because of their resource potential (Hedberg, 1970; Burk, 1972, 1973) and because of the changes in international agreements that were, and are, taking place with regard to jurisdiction over marine areas that had traditionally lacked formal controls.

The advent of the sea-floor spreading and the plate tectonics model created enormous new interest in the continental margins. Not only did this

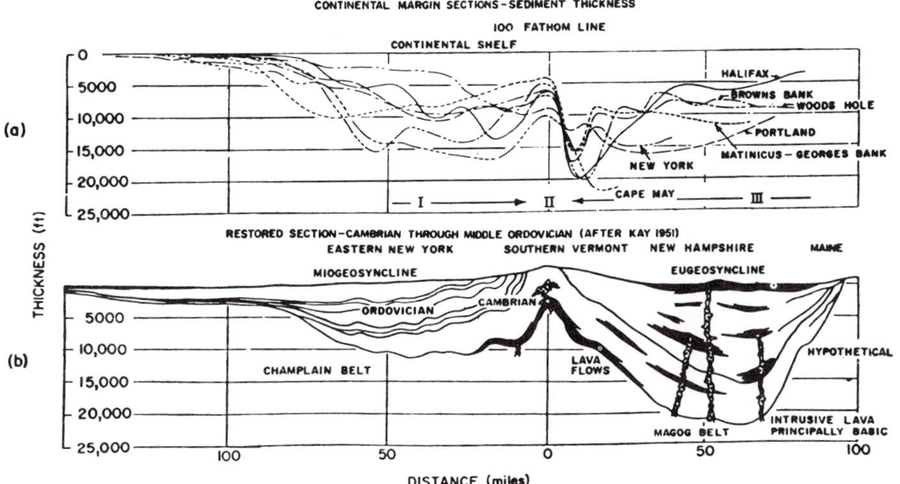

Fig. 7. Comparison of continental margin of eastern North America with Kay's (1951) reconstruction of the Appalachian system in Middle Ordovician time (after Drake et al., 1959).

model revive the concept of continental drift, which had fallen upon hard times, but it created fresh possibilities for determining the relationships between the continental margins and orogenic mountain belts, and provided a framework for possibly relating continental margins to the geological effects of past orogenies and to present tectonic activity.

BATHYMETRY AND DISTRIBUTION OF CONTINENTAL MARGINS

The broad relief of the earth's solid surface has been known for many decades, and recent compilations have modified this distribution only slightly (Fig. 8). The earth consists of two great topographic surfaces—one essentially at sea level, representing the great continental masses of the world, and another at nearly 5 km below sea level, representing the earth's great oceanic basins. These two surfaces are unquestionably the major features of the face of the earth. The boundaries between these surfaces are the world's continental margins—the subject of this volume.

Continental margins represent a relatively small area of the earth's total surface, yet they are of great geological importance, and of enormous linear extent, stretching for a total distance of nearly 350,000 km. Approximately 15% of the world's oceans actually overlie the submerged margins of continental crust. The extreme deviations of the frequency curve (Fig. 8) represent tectonically active areas, inherently out of isostatic equilibrium—high orogenic mountain belts and deep oceanic trenches. The minimum between these two topographic peaks represents the world's continental margins.

Nearly 71% of the earth is covered by water; of this total about 5.3% is less than 200 m deep. However, not all of this very shallow water covers areas that actually represent true continental margins (the transition from continents to oceanic basins). Approximately 8.7 x 10^6 km^2 are actually inland seas or bays ("interior shelves") which cover continental interiors and are thus unrelated to true continental margins (e.g., Baltic Sea, North Sea, Persian Gulf, Hudson Bay, Gulf of Venezuela, Gulf of Paria, Gulf of Carpentaria, etc.) These "interior shelves" are the areas where most marine resources are presently being developed. The boundary between such areas and true continental margins is often difficult to determine, but it should be kept in mind that their geology and resource potential do not necessarily have any relationship to true continental margins (Table 1).

The classical configuration of continental margins was established by Heezen et al. (1959), based on detailed bathymetric summaries in the North Atlantic Ocean. Three significant marine areas were recognized, including the *continental shelf*, *continental slope*, and the *continental rise*, based on major changes in inclination of the bathymetric surface (Fig. 9).

The possibility for variations obviously is end-

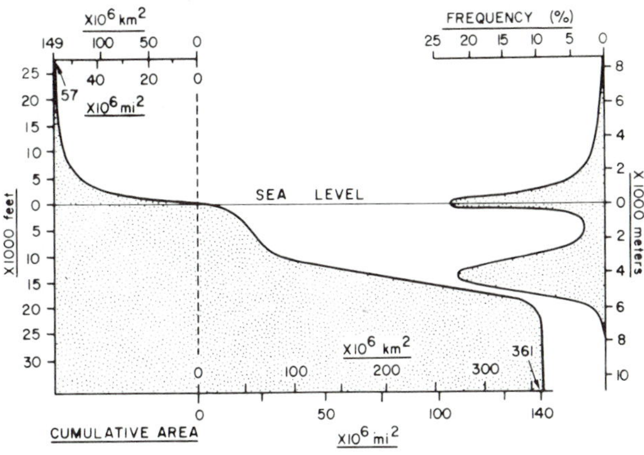

Fig. 8. Cumulative and frequency distribution of the earth's solid surface (modified from various sources).

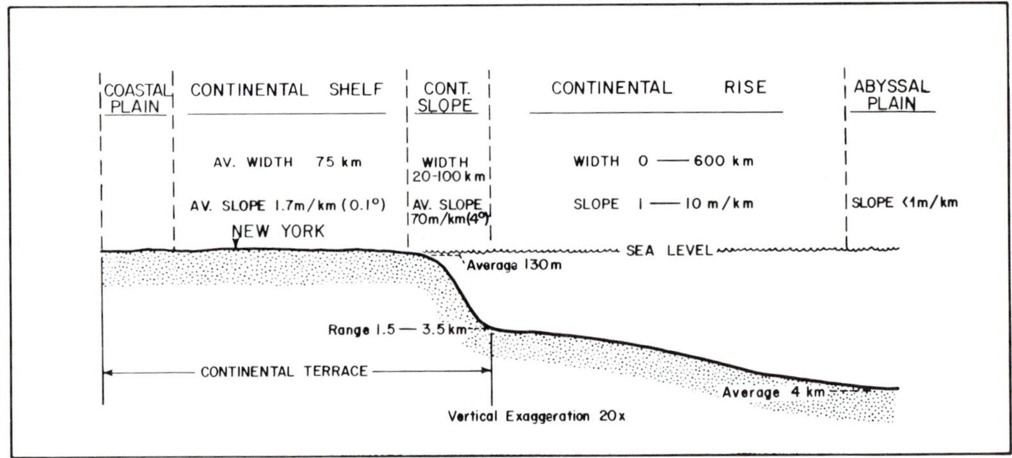

Fig. 9. Idealized diagram of the principal elements of a continental margin.

less. For example, Curray (1966) proposed that "continental terrace" includes the slope, shelf, and emerged coastal plain as well. However, "terrace" has been used in the past at continental margins in a great variety of ways, and the term has not become widely accepted in a formal sense (Fig. 9).

The North Atlantic margins are typically of the inactive type, but Heezen et al. (1959) proposed added terminology based on the excursion into the Atlantic of the active Caribbean arc, as well as on other local variations of the typical Atlantic margin.

Table 1. Areal distribution of the earth's major physiographic provinces.

	AREA Miles^2 (Km^2) × 10^6	PERCENT of Earth
EARTH	197.0 (510.1)	100.0 %
Land	57.2 (148.1)	29.2
Water	139.8 (362.0)	70.8
CONTINENTAL*	78.7 (203.8)	40.0 %
Land	57.2 (148.1)	29.1
Interior shelf	3.4 (8.7)	1.7
Shelf	7.1 (18.4)	3.6
Slope	11.0 (28.7)	5.6
OCEANIC	118.2 (306.2)	60.0 %
Rise	9.6 (25.0)	4.8
Ocean Basins**	108.6 (281.2)	55.2

* Edge of the continents is considered to be at 2 km of water depth

** Includes oceanic ridges, volcanic peaks, abyssal plains and abyssal hills

A *continental borderland* (or plateau) is a deeply submerged and commonly irregular surface, corresponding elsewhere to the shelf (e.g., Blake Plateau, California borderlands). *Marginal escarpments* are recognized where there is no gentle continental slope (e.g., Blake Escarpment). A *trench slope* leads from a continental shelf to a deep oceanic trench, and continues to be a useful bathymetric and geologic term (Fig. 10).

Seismic refraction and gravity data suggest that the true boundary between continental and oceanic crust occurs at a water depth of 2-3 km. Thus, although salt water covers nearly 71% of the world, the earth's surface consists of approximately 40% continental crust and 60% oceanic crust (Table 1). The continental shelves and slopes constitute about 9.2% of the world's surface (about 47 X 10^6 km^2) and contain an area equivalent to all the land in North, Central, and South America; all of Europe; and northern Africa.

The total area of the continental rises is much more difficult to estimate, but the area of these "oceanic margins" may be nearly equal to the total area of the world's continental slopes. It is important to note that rises are essentially absent in the trench-margins of the circum-Pacific.

SMALL OCEAN BASINS AND SMALL CONTINENTS

Small ocean basins, such as the Mediterranean Sea, the Gulf of Mexico, the Caribbean, or even the Caspian and Black seas usually are of a different structural character from the major ones (Neprochnov et al., 1970; De Roever, 1969; Ewing et al., 1971; Ryan et al., 1971).

The character of these small ocean basins is problematical, as related to large continents and oceans, perhaps because they are made up in cases totally of continental margins or perhaps actually because of a different structural history.

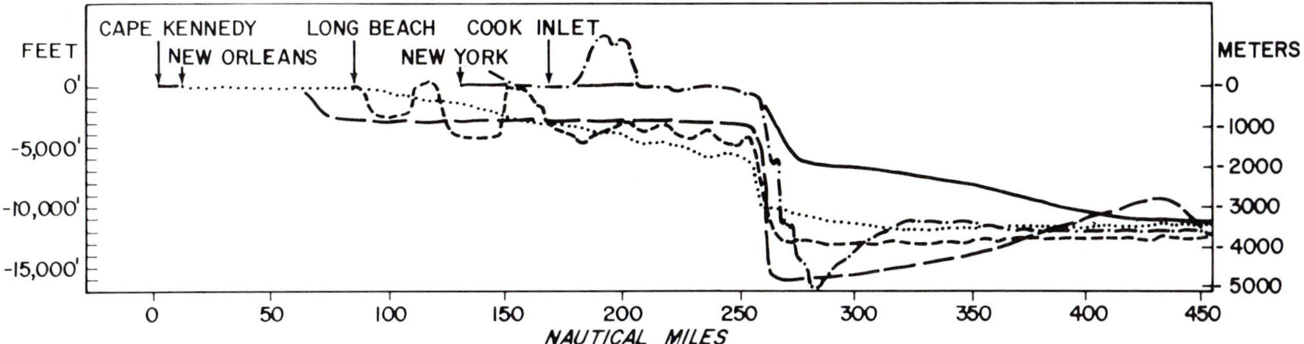

Fig. 10. Selected bathymetric profiles of continental margins of North America.

Small continents, or microcontinents, such as Madagascar or the Seychelles islands, are called such because of the age and nature of the rocks of which they are constructed (Baker, 1963; Baker and Miller, 1963; Dixey, 1960). These are quite different from true oceanic islands of similar dimensions, such as the Azores, associated with the mid-ocean ridge system, or the Hawaiian Islands, associated with the open ocean.

The problems of interpreting ancient continental margins and their history become very much more complex if one must consider the possibility of "continents" the size of the Seychelles and "oceans" of the structural character of the Caspian Sea. We do not know in detail the differences between an insular continental margin and that of a micro-continent, nor do we know the reasons for the differences between the structure of small ocean basins and the major ones. If we do not have type sections for continental crust, oceanic crust, or typical crustal sections for continental margins, our difficulties in interpreting the geological record are greatly magnified.

BIBLIOGRAPHY

Baker, B. H., 1963, Geology and mineral resources of the Seychelles Archipelago: Min. Com. and Ind. Geol. Survey, Kenya Mem. 3, 140 p.

———, and Miller, J. A., 1963, Geology and geochronology of the Seychelles Islands and the structure of the floor of the Arabian Sea: Nature, v. 199, p. 346-348.

Benioff, H., 1954, Seismic evidence for crustal structure and tectonic activity: Geol. Soc. America Spec. Paper 62, p. 61-74.

Bourcart, J., 1938, La Marge continentale: Bull. Soc. Geol. France, v. 8.

Bullard, E. C., and Gaskill, T. F., 1941, Submarine seismic investigations *in* Proceedings of the Royal Society, Series A, v. 177, p. 476-499.

Burk, C. A., 1972, Global tectonics and world resources: Am. Assoc. Petr. Geol. Bull., v. 56, no. 2, p. 196-202.

———, 1973, Mineral resources of the oceans: Australia Bur. Min. Resources Bull., v. 141, p. 115-122.

Curray, J. R., 1966, Continental terrace, *in* Fairbridge, R., eds., The encyclopedia of oceanography: New York, Van Nostrand, Reinhold, p. 207-214.

De Roever, W. P., ed., 1969, Symposium on the problem of oceanization in the western Mediterranean: Verhandel. Ned. Geol. Mijnbouwk. Genoot. Geol. Ser., v. XXVI, 165 p.

Dixey, F., 1960, The geology and geomorphology of Madagascar, and a comparison with eastern Africa: Quart. Jour. Geol. Soc. London, v. 116, p. 255-268.

Drake, C. L., Ewing, M., and Sutton, G. H., 1959, Continental margins and geosynclines: the east coast of North America north of Cape Hatteras, *in* Physics and chemistry of the earty: Elmsford, N.Y., Pergamon, v. 5, p. 110-198.

Du Toit, A., 1937, Our wandering continents: Edinburgh, Oliver and Boyd, 366 p.

Ewing, M., Crary, A. P., and Rutherford, H. N., 1937, Geophysical investigations in the emerged and submerged Atlantic coastal plain, pt. I, Methods and results: Geol. Soc. America Bull., v. 48, p. 753-802.

———, Woollard, G. P., and Vine, A. C., 1939, Geophysical investigations in the emerged and submerged Atlantic coastal plain, pt. III, Barnegat Bay, N.J., section: Geol. Soc. America Bull., v. 50, p. 257-296.

———, Worzel, J. L., Steenland, N. C., and Press, F., 1950, Seismic refraction measurements in the Atlantic Ocean Basin, pt. I: Seismol. Soc. America Bull., v. 40, p. 233-242.

———, Edgar, N. T., and Antoine, J. W., 1971, Structure of the Gulf of Mexico and Caribbean Sea, *in* The Sea, v. 4, pt. II, New York, Wiley-Interscience, p. 321-358.

Gunn, R., 1944, A quantitative study of the lithosphere and gravity anomalies along the Atlantic Coast: Jour. Franklin Inst., v. 237, p. 139-154.

Gutenberg, B., and Richter, C. F., 1938, Depth and geographical distribution of deep focus earthquakes: Geol. Soc. America Bull., v. 49.

Hedberg, H. D., 1970, Continental margins from the viewpoint of the petroleum geologist: Am. Assoc. Petr. Geol. Bull., v. 54, p. 3-43.

Heezen, B. C., and Drake, C. L., 1964, Grand Banks slump: Am. Assoc. Petr. Geol. Bull., v. 48, p. 221-225.

———, and Ewing, M., 1952, Turbidity currents and submarine slumps, and the 1929 Grand Banks earthquake: Am. Jour. Sci., v. 250, p. 849-873.

———, Hollister, C. D., and Ruddiman, W. F., 1966, Shaping of the continental rise by deep geostrophic contour currents: Science, v. 152, p. 502-508.

———, Tharp, M., and Ewing, M., 1959, The floors of the oceans: I. The North Atlantic: Geol. Soc. America

Spec. Paper 65, 122 p.
Helmert, F. R., 1909, Die Tiefe Der Ausgleichfläche Bei Des Prattschen Hypothese Für Das Gleicagewicht Der Erdkruste Und Der Verlauf Der Schwerestörung vom Innern Der Kontinente Und Ozeane Nach Den Küsten: Sitzber. Deut. Kgl. Preusz. Akad. Wiss., v. 18, p. 1192-1198.
Hess, H. H., 1938, A primary peridotite magma: Am. Jour. Sci., v. 35, p. 321-344.
Hill, M. N., 1952, Seismic shooting in an area of the eastern Atlantic: Phil. Trans. Roy. Soc. London, ser. A., v. 244, p. 561-596.
______, and Laughton, A. S., 1954, Seismic investigations in the eastern Atlantic: Proc. Roy. Soc. London Ser. A., v. 222, p. 348-356.
Hill, M. N., and Laughton, A. S., 1954, Seismic investigations in the eastern Atlantic: Proc. Roy. Soc. London ser. A., v. 222, p. 348-356.
Jesson, O., 1943, Die Randschwellen Der Kontinente: Peterm. Geogr. Mitt. Erg., Heft 241.
Kay, M., 1951, North American geosynclines: Geol. Soc. America Mem. 48, 143 p.
Kossina, E., 1933, Die Erdoberflache: Handb. d. Geophysik, Bd. 2, p. 869-954, Berlin.
Kuenen, Ph. H. 1950, Marine Geology: New York, Wiley, 568 p.
______, and Migliorini, C. I., 1950, Turbidity currents as a cause of graded bedding: Jour. Geology, v. 58, p. 91-126.
Lake, P., 1931, Island arcs and mountain building: Geograph. Jour., v. 78.
Neprochnov, Y. P., Kosminskaya, I. P., and Malovitsky, Y. P., 1970, Structure of the crust and upper mantle of the Black and Caspian seas: Tectonophysics, v. 10, p. 517-538.
Ryan, W. B. F., Stanley, D. J., Hersey, J. B., Fahlquist, D. A., and Allen, T. D., 1971, The tectonics and geology of the Mediterranean Sea, *in* The Sea, v. 4, pt. II: New York, Wiley-Interscience, p. 387-492.
Schweydar, W., 1921, Berierkungen zu Wegener's Hypothese der Verschiebung der Kontinente: Zeichr. d. Ges. f. Erdk. zu Berlin, p. 120-125.
Schwinner, R., 1841, Seismik Und Reictouische Geologie Der Jetzizeit: Z. Geophys., v. 17.
Shepard, F. P., 1934, Origin of Georges Bank: Geol. Soc. America Bull., v. 45, p. 281-320.
Sollas, W. J., 1903, The figure of the earth: Quart. Jour. Geol. Soc. London, v. 59.
Stetson, H. C., 1936, Geology and paleontology of Georges Bank canyons: Geol. Soc. America Bull., v. 47, p. 339-366.
______, 1938, Sediments of the continental shelves off the eastern coast of the United States: Woods Hole Oceanogr. Inst. Paper Phys. Ocean. and Met., v. 5, p. 5-48.
Suess, E., 1904, The face of the earth (Das Antlitz der Erde): Engl. ed.: New York, Oxford University Press, Inc., 5 v.
Taylor, F. B., 1910, Bearing of the tertiary mountain belt on the origin of the earth's plan: Geol. Soc. America Bull., v. 21, p. 179-226.
Turner, H. H., 1922, On the arrival of earthquake waves at the Antipodes and on the measurement of the focal depth of an earthquake: Monthly Notices Roy. Astron. Soc., Geophys. Suppl., v. 1, p. 1-13.
Umbgrove, J. H. F., 1947, The pulse of the earth: The Hague, Martinus Nijhoff, 358 p.
Vening-Meinesz, F. A., 1933, The mechanism of mountain formation in geosynclinal belts: Koninkl. Ned. Adad. Wetenschop. Proc., v. 33.
______, 1941, Gravity over the continental edges: Koninkl. Ned. Akad. Wetenschap. Proc., v. 44.
Wadati, K., 1935, On the activity of deep focus earthquakes in the Japanese Islands and neighborhood: Geophys. Mag. (Tokyo), v. 3, p. 305-325.
Wegener, Alfred, 1924, The origin of continents and oceans: London, Methuen Press.

Part II

General Bathymetry and Topography

Atlantic-Type Continental Margins

Bruce C. Heezen

INTRODUCTION

Continental margins can be characterized as of "Atlantic" or "Pacific" type, essentially depending on whether they have experienced a relatively long period of stability, in which case they are called Atlantic type, or whether they have suffered active tectonism during latter geological times, in which case they are assigned to the Pacific type. We are concerned in this chapter with the Atlantic-type continental margins which bound the Arctic and Norwegian seas, the North and South Atlantic oceans, and all of the Indian Ocean (with the exception of the Sunda Arc), and all which circles Antarctica (with the exception of the Scotia Arc). Parts of the marginal basins of the Pacific can also be considered to be Atlantic type; for example, the continental margin of Alaska in the Bering Sea, the Siberian continental margin in the Okhotsk Sea, the Asian continental margin in the Japan Sea, and the Southeast Asian continental margin in the South China Sea. Parts of the Mediterranean margin might also be considered to be Atlantic type. Atlantic-type continental margins have been studied most extensively in the North Atlantic, but a significant assemblage of data provides insight into the structure of the continental margin of west Africa, Argentina, and Kenya.

ATLANTIC-TYPE CONTINENTAL MARGIN

Topographically, the Atlantic-type continental margin is typified by a relatively wide continental shelf which may vary from 30 to over 300-km in width (Fig. 1). It is also characterized by a relatively dissected continental slope indented by submarine canyons (at least in those areas which have been surveyed in sufficient detail to reveal the presence or absence of submarine canyons).

The boundary between the "continental shelf" and the continental slope, known as the "shelf break," varies in depth, often as a direct function of latitude (shelf breaks in the polar areas are often as great as 600 m, whereas those in lower latitudes rarely exceed 100 m). The base of the "continental slope" and its boundary with the relatively more gentle, seaward-sloping "continental rise" is defined as that point where the seaward gradient drops below 1 in 40. The declinity of the continental slope ranges from 1 in 5 to 1 in 25. Continental rise gradients are generally less than 1 in 100, so the sometimes gradational change between continental slope and continental rise although defined by an arbitrary slope angle, is generally not difficult to pick on regional profiles. The continental rise spreads out 80 to more than 500-km as a gentle-relief feature which is delineated by yet-another marked boundary in seaward gradient, and a contrast in microtopography, at the margin of the "abyssal plains of the ocean basin floor." The gradients of the abyssal plains are by definition and by observation less than 1 in 1000. Abyssal plains have relief of less than 1 fathom over distances of 1 or 2 miles. The width of the abyssal plains range from a few km to 300 or 500 km. (Fig. 2).

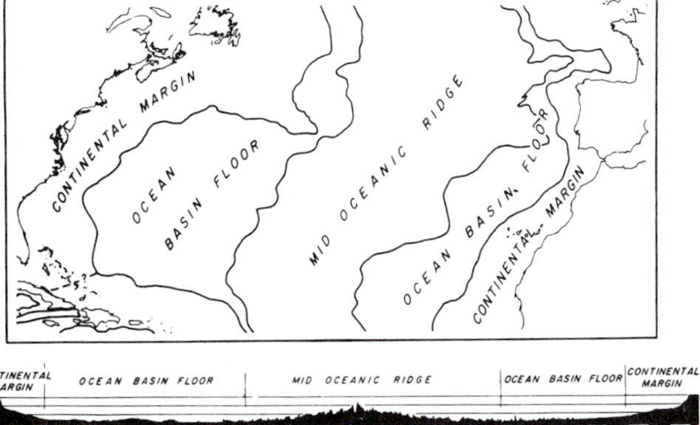

Fig. 1. An Atlantic continental margin is a fundamental part of the threefold division of the entire ocean basin. Here are shown the three major morphologic divisions of the ocean in a trans-Atlantic profile from New England to the Spanish Sahara (after Heezen et al., 1959).

INTERNAL MARGIN COMPOSITION

The Atlantic-type continental margins are characterized topographically by a relatively smooth relief, which is quite obviously a surface expression of relatively large sediment accumulations (Fig. 3). This sediment has accumulated under a relatively stable but incessantly subsiding tectonic environment which has produced little in the way of deformation. Therefore, it is ideally suited to seismic reflection exploration techniques since relatively undisturbed stratified sequences provide acoustical contrast and continuity. Where the seismic refraction technique has been employed to determine the total thickness of sediments at the foot of the continental slope, thicknesses of as much as 10 km have been measured. However, the sediment thicknesses of the continental rise are often so great that most seismic systems are incapable of recording reflections from the areas of greatest sediment accumulation.

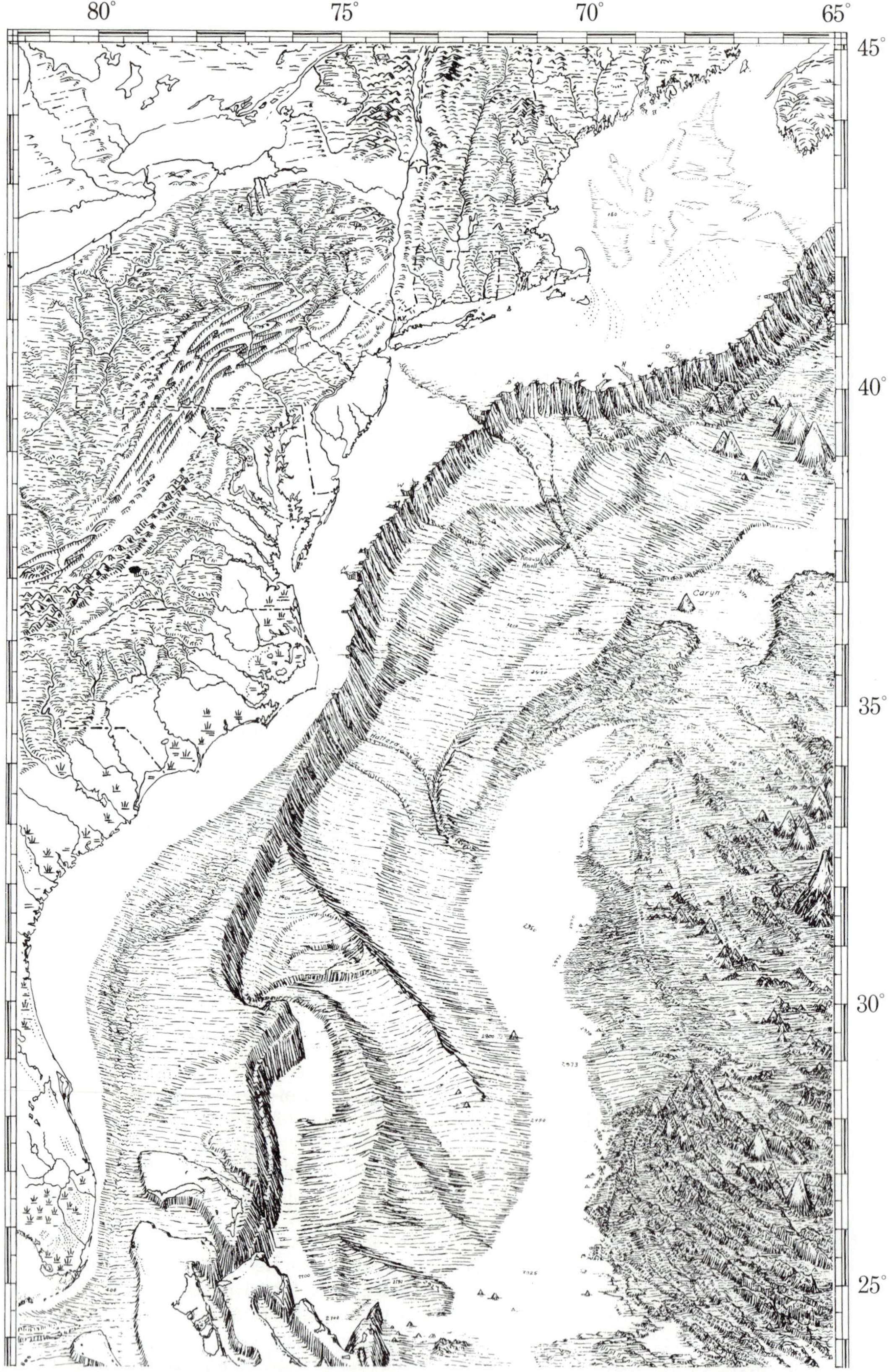

Fig. 2. The continental margin of the eastern United States illustrates the principal characteristics of Atlantic continental margins: the broad, smooth, emerged and submerged continental shelf; the precipitous continental slope dissected by canyons; and the continental rise, a broad apron of sediments sweeping out toward the ocean basin floor. Also shown are a current-eroded marginal plateau (the Blake Plateau) off Georgia and Florida and its seaward escarpment, and the subsiding Bahama Platform. (From Physiographic diagram of the North Atlantic Ocean, Geological Society of America, copyright Bruce C. Heezen and Marie Tharp, 1968.)

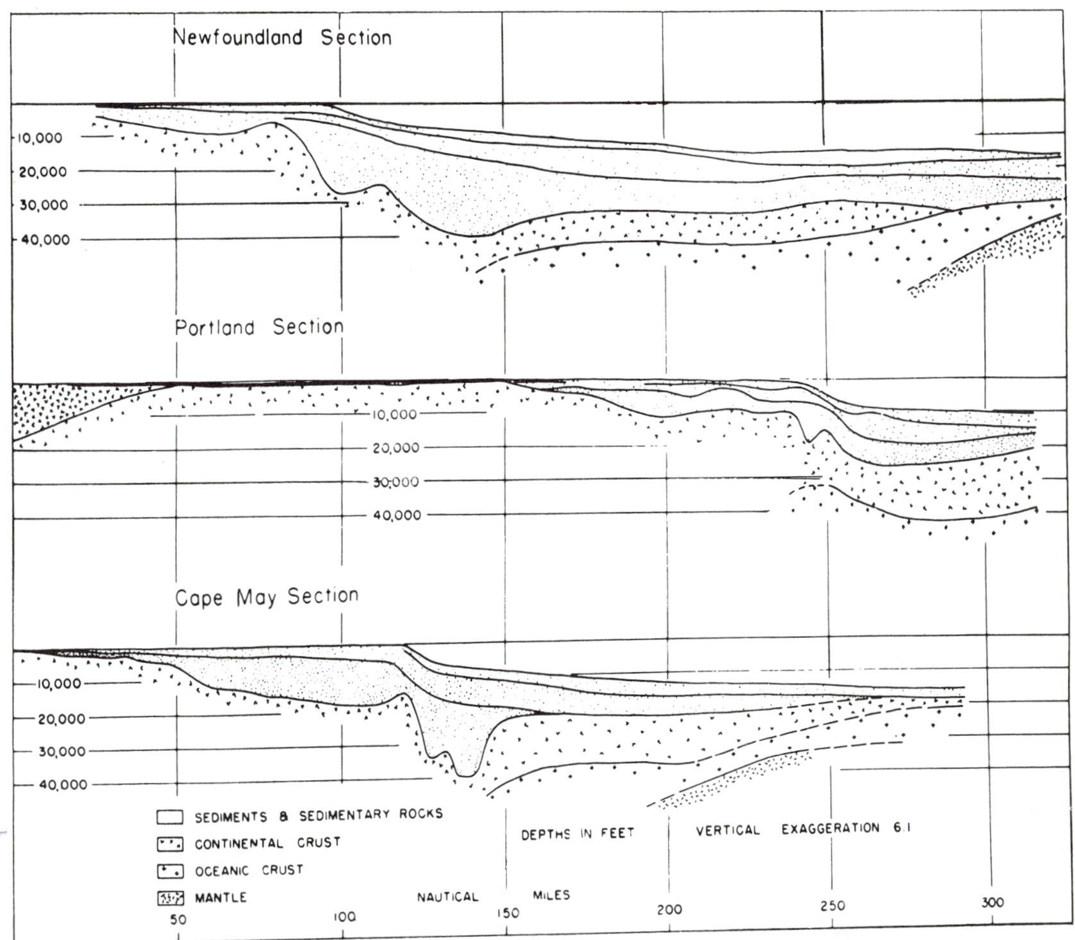

Fig. 3. The continental rise and the continental shelf are sites of thick sediment accumulation. Three structural profiles of the eastern continental margin of North America (after Drake et al., 1959).

Structural processes produce the subsiding or subsided area at the base of the continental slope in which the thick sequence of sediments are deposited, but sedimentary processes create the continental rise sedimentary wedges. Certain minor structural forms, such as diapirs, are seen in various parts of the world on the continental rise. However, the rise is emphatically an area of little or no structural deformation. The abyssal plains which lie seaward of the continental rise are created by the same processes that create the rise. Abyssal plains result from the ponding of turbidity current deposits in the last stages of deposition. Only a small fraction of the clastic sediment that reaches a continental rise continues on to the adjacent abyssal plain. Seismic reflection profiles across continental margins generally indicate a thin sequence of highly stratified beds beneath the abyssal plains and a much thicker landward-expanding sequence of wedges which constitute the continental rise.

CONTINENTAL MARGIN SEDIMENTATION

The continental shelf, an area of gentle relief with overall slopes of about 1 in 100 or less, has been repeatedly emerged and submerged over the geologically recent past as a consequence of rising and falling sea level (Fig. 4). The most recent submergence was completed approximately 5,000 years ago, at the end of the Pleistocene-Holocene transgression. Many of the relief features of the shelf are modified subaerial forms. A thin layer of surface sediments, millimeters to meters in thickness, in equilibrium with the present current regimes, overlays peat, clay, coal, marl, sand, and various Tertiary marine formations in a confusing array.

The wedging, upper sedimentary strata of the continental shelf often crop out on the continental slope, particularly on the walls of submarine canyons. Some submarine canyons are dynamically active features eroding at the present time, whereas others appear to be quite inactive. Canyons provide conduits for the near-bottom transport of sediments from continental shelf to continental rise and on to the abyssal plains. In many areas of the continental slope, deposition is indeed temporary, being intermittently interrupted by collapse of oversteepened slopes. A certain amount of prograding has occurred on the continental slope, but the concept that the continental terrace (continental shelf and continental slope) is a prograded terrace does not conform with the facts as presently known (Fig. 5).

Fig. 4. The continental margins of Europe and North Africa, although similiar to North America (Fig. 2), display marked differences. The sharply indented Iberian continental slope suggests tectonic control. The continental shelf south of France is much narrower than the shelf in the western Atlantic. The Canary Islands have no counterpart to the west. (From Physiographic Diagram of the North Atlantic Ocean, Geological Society of America, copyright Bruce C. Heezen and Marie Tharp, 1968.)

Fig. 5. The continental margin of West Africa possesses a narrow shelf but a wide continental rise. The Congo has built a huge cone in the Angola Basin. The Walfisch Ridge and the Cammeron seamounts interrupt the smooth rise. (From Physiographic Diagram of the South Atlantic Ocean, Geological Society of America, copyright Bruce C. Heezen and Marie Tharp, 1961.)

The continental rise is a large and very significant accumulation of sediment dominated by two important near-bottom sedimentary processes: (1) turbidity currents and (2) contour currents.

Turbidity currents flowing down the submarine canyons produce natural leveed submarine channel systems on the continental rise and contribute significantly to the buildup of the rise sediments (Fig. 6).

Contour currents, the other very important factor in the shaping of the continental rise sediment bodies, are currents involved in the overall circulation of the deep waters of the ocean. The significant role of these currents was revealed by anomalies in the distribution of sediment off the eastern United States. The Blake-Bahama Outer Ridge, a large, 3-km-thick, 600-km-wide accumulation of sediment east of Georgia and the Carolinas, is separated from the continent by a wide, sediment-free limestone plateau and a small segment of ocean basin. The Blake-Bahama Outer Ridge is a sedimentary drift composed of sediments derived mostly from eastern Canada and the northeastern United States and transported by contour currents 1,000-3,000 km to the south, and finally drifted into this peculiar and gigantic spit-like mass.

The margin of the southeastern United States is not directly contributing sediment to the adjacent deeper continental margins. The demonstration of this very large and active drift provided compelling evidence that at the depositional surface the sediment of the continental rise is very mobile and moves hundreds, if not thousands, of kilometers parallel to the rise before reaching its final resting place. Thus, in the creation of the continental rise physiography and stratigraphy, one is compelled to evoke two principal processes: (1) downslope transport of sediment largely by gravity-controlled turbidity currents through submarine canyon systems and (2) the lateral contour-following transport of sediments by geostrophic contour current systems which are involved in the overall circulation of the waters of the world oceans. The two sedimentary processes that appear to be most active on the continental slope are gravity slumping and turbidity current erosion; the processes most significant on the continental shelf are wave and strandline processes which have left their fossil imprint on these large terraces.

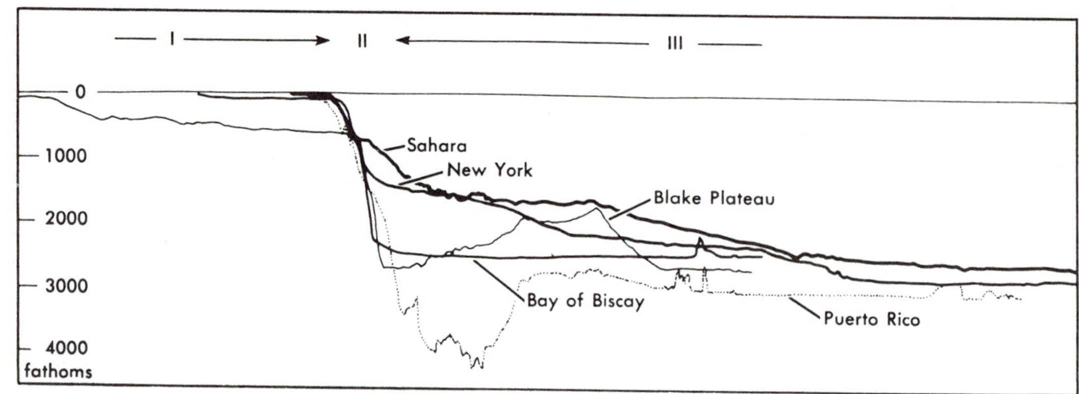

Fig. 6. Categories of continental margin provinces. Representative profiles taken from various parts of the North Atlantic.

POTENTIAL MARGIN RESOURCES

The continental margins of Atlantic-type may be of the greatest fascination for the petroleum geologist and the greatest potential interest to mankind. This is not to say that Pacific-type continental margins do not have their own fascination, particularly in the dramatic activity involved in the grand landscapes which they present, but the Atlantic continental margins are areas of large sediment accumulations and thus may possess great potential for petroleum reserves. From what can be inferred from oceanographic studies concerning the yet-undrilled and unexplored offshore areas, it is conceivable that they may possibly possess orders-of-magnitude more petroleum reserves than all the continental shelves of the world combined, or, conversely, very little at all.

Several problems bear very significantly on the continental rise as a potential petroleum province. First, at the present time there is, of course, no known way to produce oil in such deep water, although the solution of this problem is considered by most within the capabilities of present engineering technology. Second, there are no facilities available at the moment for drilling safely in depths of the continental rise. Such engineering problems will be solved when the need is sufficient. Then we may conclude that the real questions are the same as those posed by any other prospective oil province; are reservoir rocks and traps present? The processes operating on the continental rise and their variation in the past thus become immensely important, as does correct evaluation of the marine geology of the continental rise to a world concerned with the problem of evaluating future energy sources.

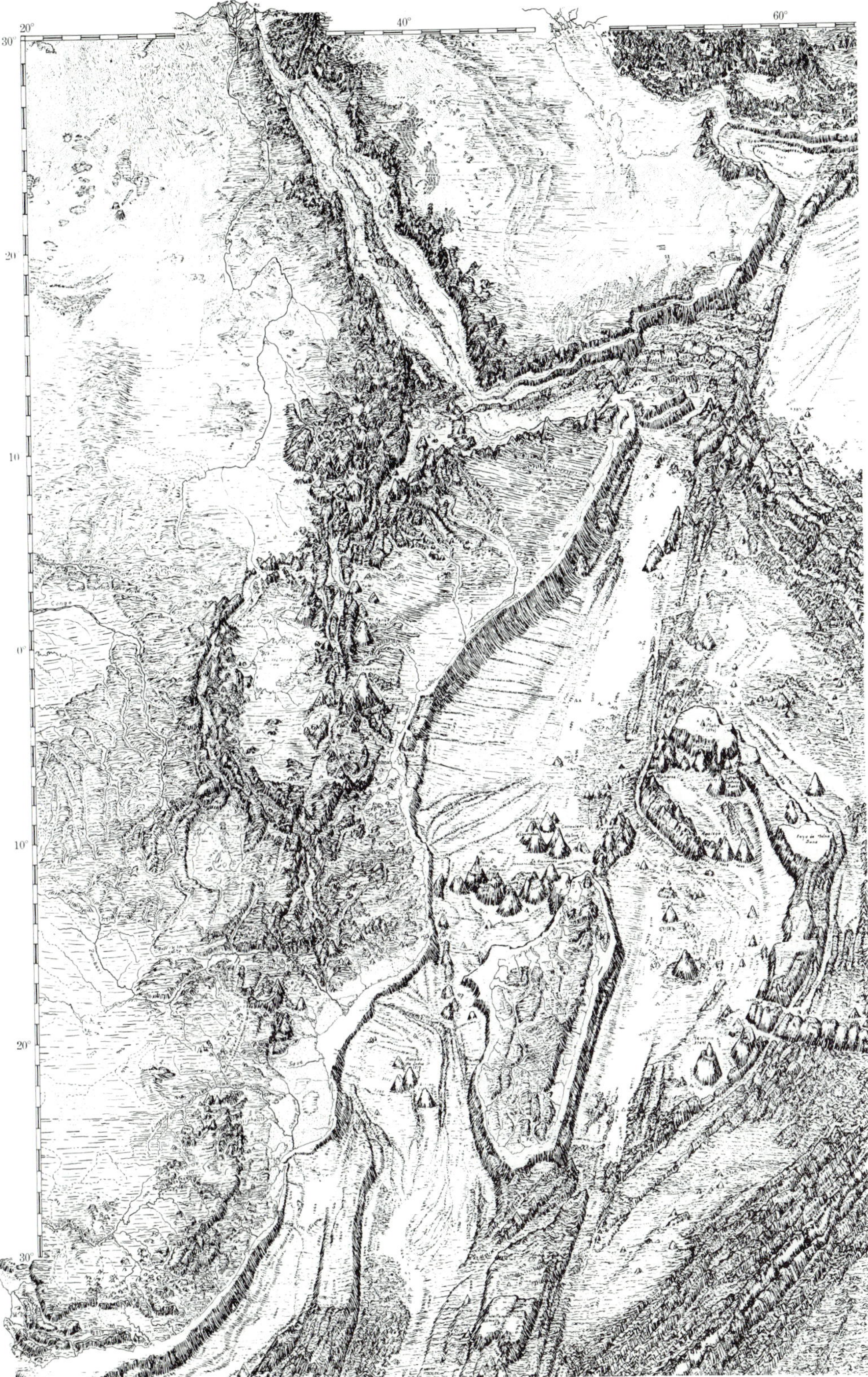

Fig. 7. The eastern margin of Africa can easily be classified as an Atlantic-type margin. It does, however, possess a narrow shelf and introduces a microcontinent into the pattern. The Red Sea presumably represents juvenile Atlantic-type margins in the earliest stages of development. (From Physiographic Diagram of the Indian Ocean, Geological Society of America, copyright Bruce C. Heezen and Marie Tharp, 1964.)

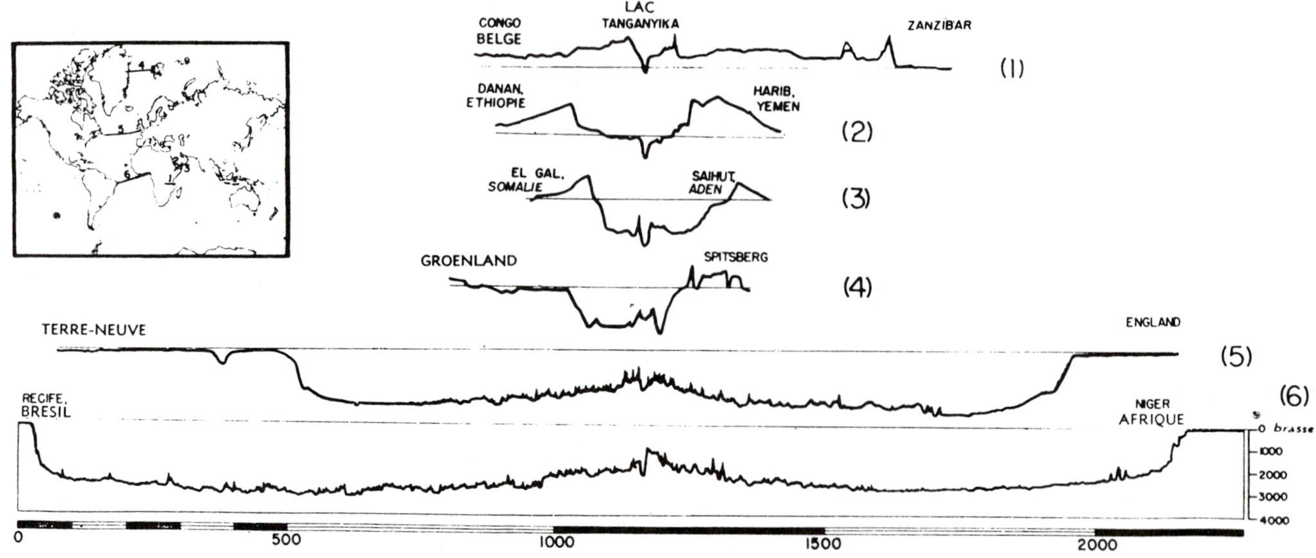

Fig. 8. All continental margins presumably originated in the same manner by the rifting of a continent. This series of topophysiographic profiles represents margin in successive stages of development. This sequence represents a genetic series in which the rift valleys of Africa represent an early stage of development and profiles 5 and 6 represent later stages (after Heezen, et al. 1959).

NORTH ATLANTIC

Having considered the general characteristics of Atlantic-type continental margins, we can now examine briefly some specific continental margins of the Atlantic type. We will consider first the continental margin of eastern United States, as its study has provided many of the generalizations we have cited. The widest continental rises in the North Atlantic occur off New England and the Sahara. These continental margins are also the oldest in the Atlantic, so we might conclude tentatively that the width of a continental rise is roughly a function of its age. Given a uniform rate of deposition along all continental margins, this would obviously be the case. But is it reasonable to assume that deposition rates in all continental margins have been roughly the same? When we consider, for instance, the great differences in sedimentation rates that have resulted from the influence of major rivers, particularly in contrast with the relatively meager seaward dispersal of sediment off, let us say, deserts or limestone areas, should we not expect that any difference in total accumulation relating to initial age would be obscured?

The oldest margins of North America lie between Nova Scotia and the Bahamas. The widest continental rise of the Atlantic occurs in this area; however, regional differences in deposition rate can be seen to have a profound influence. For instance, off the southeastern United States the continental rise is relatively narrower and contains a much smaller total volume of sediment. In fact, the entire rise appears to have been constructed by contour currents, and without their active participation as a major sedimentary process there would be, in all probability, no continental rise off the southeastern United States.

SOUTH ATLANTIC

In comparison to the North Atlantic, the continental margins of the South Atlantic are in general less well-developed sedimentary accumulation centers (Fig. 7). Areas of smooth relief are narrower than in the North Atlantic (Fig. 8). This observation is in keeping with the much later date of the initial rifting of the South Atlantic (late Early Cretaceous), in contrast with Late Triassic or Early Jurassic for the North Atlantic (Fig. 9).

INDIAN OCEAN

The continental margins of the Indian Ocean certainly demonstrate vast differences produced by variations in rate of deposition. In the northern Indian Ocean the great Ganges and Indus cones stretch out from either side of the Indian subcontinent. Huge accumulations of sediment, much of which was laid down in the later parts of geological time, produce a large, very significant, but relatively easily identifiable pattern. The continental margin of east Africa presents one of the really formidable sedimentary accumulations of the world (Fig. 10).

TECTONIC HISTORY OF MARGINS

Atlantic continental margins are considered to be the sites of initial rifting, which gave rise to the present configuration of the continental masses. The sites of Atlantic continental margins seem to have no previous deep-oceanic history prior to the proposed early Mesozoic breakup of Pangea. Of course, there may have been earlier oceans of

Fig. 9. The eastern and western margins of South America clearly demonstrate the difference between the Atlantic and Pacific types of margins: one with broad, smooth expanses of sediment-covered sea floor; and the other with a deep trench and few major sediment bodies. (From Physiographic diagram of the South Atlantic Ocean, Geological Society of America, copyright Bruce C. Heezen and Marie Tharp, 1961.)

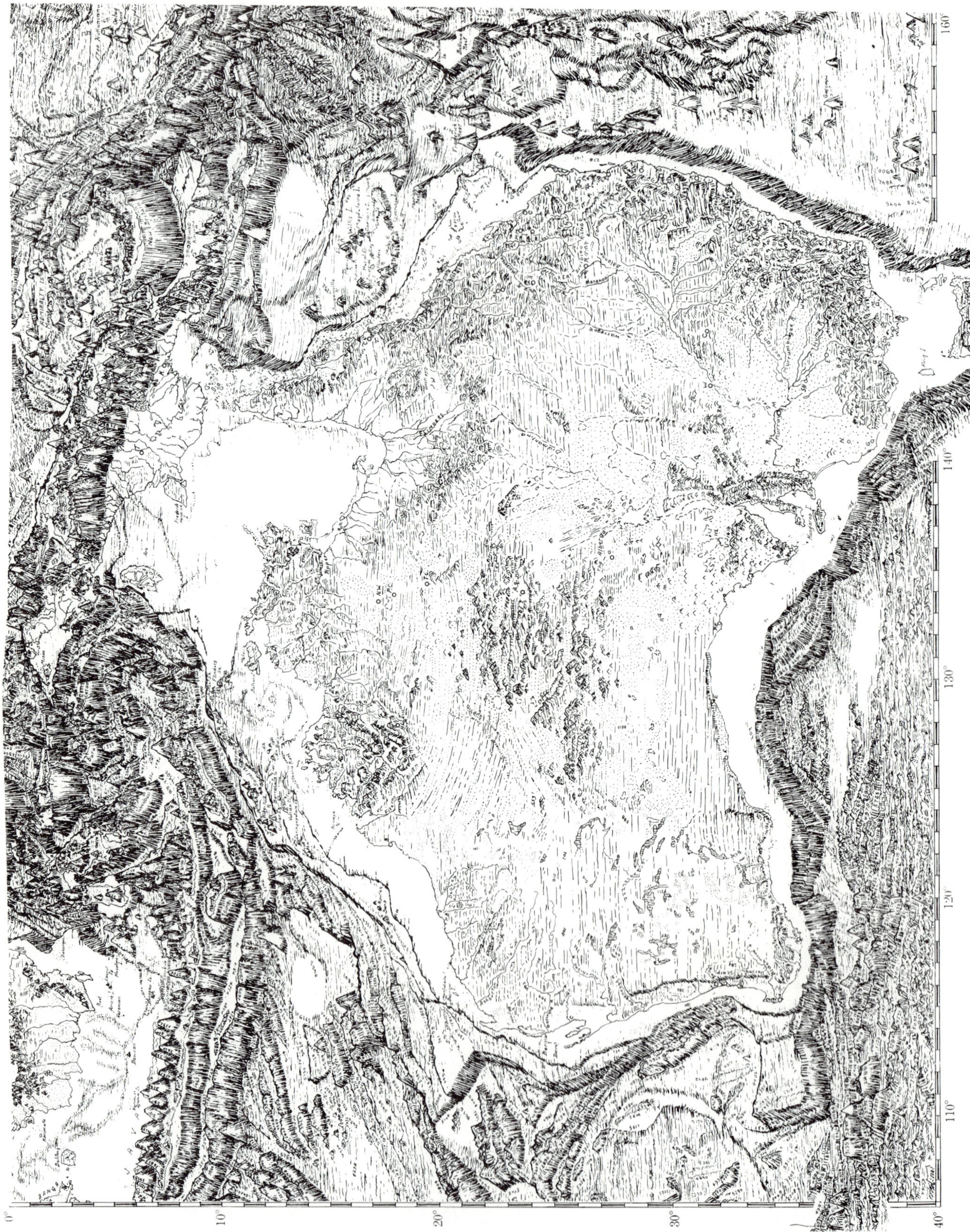

Fig. 10. The west, south, and east coasts of Australia belong to the class of Atlantic-type margins. The Indonesian arc with its Java Trench is of Pacific type (From Physiographic diagram of the western Pacific Ocean, Geological Society of America, copyright Bruce C. Heezen and Marie Tharp, 1971.)

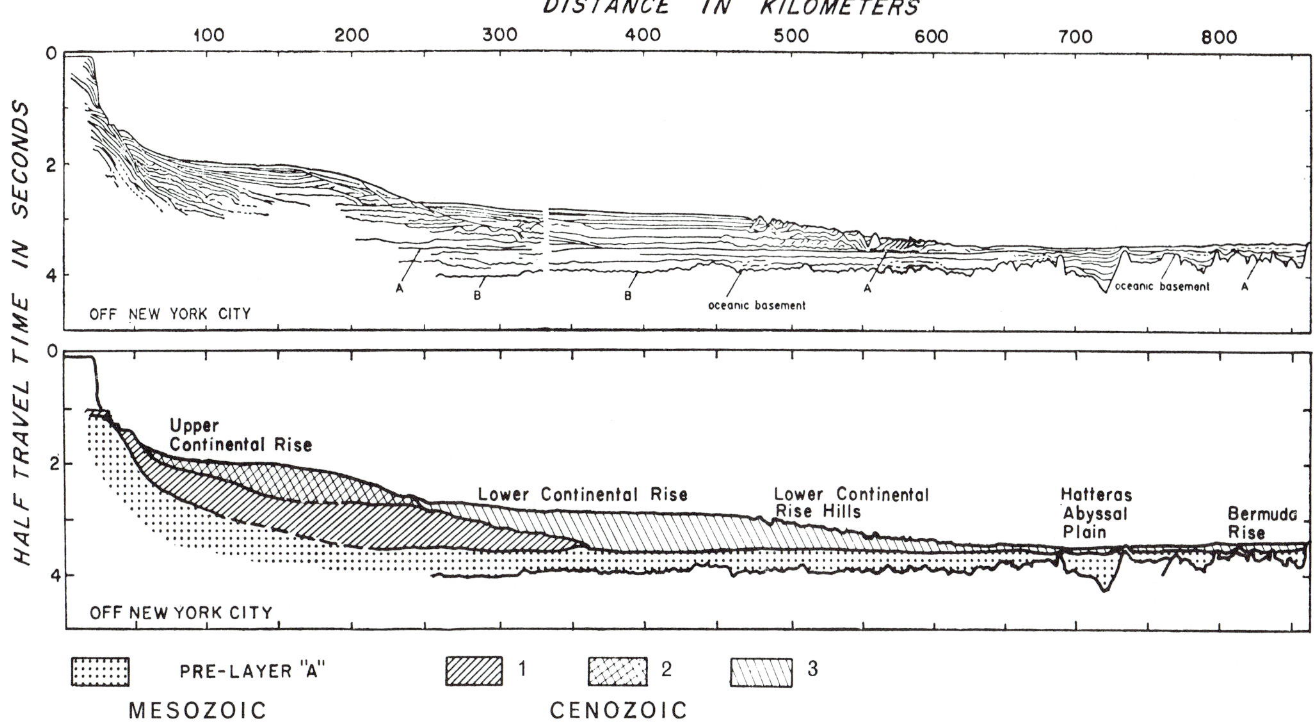

Fig. 11. The continental rise off New York is composed of a seaward-migrating progression of continental rises stacked against the base of the continental slope. Turbidity currents and pelagic sources provide the necessary sediment; the Western Boundary Undercurrent subsequently and penicontemporaneously smoothes and shapes, occasionally eroding or depositing fine-grained hemipelagic continental rise sediment. (From Hollister and Heezen, 1972.)

somewhat similar outlines. Thus the Atlantic continental margins have apparently had a fairly simple history: an initial rift, the opposed drift of the continental blocks, the creation of oceanic crust by normal axial sea floor accreation processes, and the deposition and accumulation of sediments on the new crust (Fig. 8). This appears to capsulate the history of the continental rise.

The continental shelf appears to have subsided with time, following earlier periods of basin filling and truncation. Deep beneath the continental shelf are sedimentary basins which apparently existed as topographic basins until their eventual filling, after which the thin wedges of shelf sediments accumulated on top. Therefore, in certain areas of the eastern United States the geology of the upper kilometer of the sediment gives virtually no hint of the geology to be found in the deeper parts of the sedimentary pile.

This is less true off Argentina, where structural trends near the coast can be seen to continue out under the shelf and to control the depositional style of the continental shelf and continental slope sediments. In the continental margins off North America the deeper structure is obscured by a conformable series of marine beds of Tertiary age which cap the presumably Cretaceous and earlier accumulations, which show apparently great regional variation in the continental margin of the eastern United States.

The continental margin of the eastern United States has been compared to a reconstructed Paleozoic Appalachian geosyncline, the shelf deposits thought to represent a miogeosyncline, and the continental rise deposits thought to represent the eugeosyncline. The flysch and graywacke sediments of the eugeosyncline would then be represented by the deposits of turbidity currents, which were then assigned a large role in the shaping of the continental rise. It now becomes apparent that contour currents have at least as great and probably a greater role in the final depositional pattern of the continental rise than do turbidity currents (Ericson et al., 1961; Heezen et al., 1966; Hollister and Heezen, 1971). Turbidity currents still retain a significant role in seaward dispersal, but now a lesser role is assigned to them in final deposition on this type of continental margin. Whether contour currents could produce sufficiently thick sands to provide the necessary reservoir rocks is an important but unanswered question. Figure 11 shows the sedimentary wedges shaping the lower continental margin off eastern North America.

GENERAL COMMENTS

The world's need for petroleum in the future is so obviously significant that it clearly will be necessary to examine this problem thoroughly throughout the world. It is of utmost concern to determine whether or not significant reserves of petroleum are to be found in the continental rise, as this certainly will affect world policy on the cost, use, and conservation of energy. Any responsible approach must consider how the civilization of our descendants is to exist in a world without fossil fuels. If the continental rise possesses significant quantities of petroleum, the problem certainly must be considered from an entirely different economical approach.

The total area of the world's deeper continental margins may exceed the known area of sedimentary basins on land (which has yielded all the known large energy resources of the world), yet these continental rises and slopes are among the most poorly known parts of the earth.

BIBLIOGRAPHY

Bunce, E. T., Emery, K., Gerard, R., Knott, S., Lidz, L., Saito, T., and Schlee, J., 1965, Ocean drilling on the continental margin: Science, v. 150, p. 709-716.

Drake, C. L., Ewing, M., and Sutton, G. H., 1959, Continental margins and geosynclines: The east coast of North America north of Cape Hatteras, *in* Physics and chemistry of the earth: Elmsford, N.Y., Pergamon Press, v. 3, p. 110-198.

Emery, K. O., Uchupi, E., Phillips, J. D., Bowin, C. O., Bunce, E. T., and Knott, S. T., 1970, Continental rise off eastern North America: Am. Assoc. Petr. Geo. Bull., v. 54, p. 44-108.

Ericson, D. B., Ewing., M., and Heezen, B. C., 1952, Turbidity currents and sediments in the North Atlantic: Am. Assoc. Petr. Geol. Bull., v. 36, p. 489-512.

______, Ewing, M., Wollin, G., and Heezen, B. C., 1961, Atlantic deep-sea sediment cores: Geol. Soc. America Bull., v. 72, p. 193-286.

Ewing, J., Ewing, M., and Leyden, R., 1966, Seismic profiler survey of Blake Plateau: Am. Assoc. Petr. Geol. Bull., v. 50, p. 1948-1971.

Fox, P. J., Harian, A., and Heezen, B. C., 1968, Abyssal anti-dunes: Nature, v. 220, p. 470-472.

Heezen, B. C., and Drake, C. L., 1964, Gravity tectonics, turbidity currents and geosynclinal accumulations in the continental margin of eastern North America, *in* Carey, S. W., ed., Snytaphral tectonics: Univ. Tasmania, p. D1-D10.

______, and Hollister, C. D., 1964, Deep-sea current evidence from abyssal sediments: Marine Geology, v. 1, p. 141-174.

______, Hollister, C. D., and Ruddiman, W. F., 1966, Shaping of the continental rise by deep geostrophic contour currents: Science, v. 152, p. 502-508.

______, and Hollister, C. D., 1971, The face of the deep: New York, Oxford Univ. Press, 650 pp.

______, Tharp, M., Ewing, M., 1959, The floors of the oceans: 1; The North Atlantic: New York, Geol. Soc. America, 122 p.

Hollister, C. D., and Heezen, B. C., 1971, Geological effects of ocean bottom currents, western North Atlantic, *in* Gordon, A. L., Studies in physical oceanography: New York, Gordon & Breach, v. 2, p. 37-66.

______, et. al., 1972, Initial reports of the Deep Sea Drilling Project, v. XI: Washington, D.C., U.S. Govt. Printing Office.

Jones, E. J. W., Ewing, M., Ewing, J. I., and Eittrem, S. L., 1970, Influences of Norwegian Sea overflow water on sedimentation in the northern North Atlantic and Labrador Sea: Jour. Geophys. Res., v. 75, p. 1655-1680.

Knott, S. T., and Hoskins, H., 1968, Evidence of Pleistocene events in the structure of the continental shelf off the northeastern United States: Marine Geology, v. 6, p. 5-43.

McCoy, F. W., Jr., 1968, Bottom currents in the western Atlantic Ocean between Lesser Antilles and the Mid-Atlantic Ridge: Deep-Sea Res., v. 15, p. 179-184.

Needham, H. D., Habib, D., and Heezen, B. C., 1969, Upper Carboniferous palynomorphs as a tracer of red sediment dispersal patterns in the northwest Atlantic: Jour. Geology, v. 77, p. 113-120.

Nesteroff, W. D., and Heezen, B. C., 1960, Les Depots de courants de turbidite, le flysch et leur signification tectonique: Comptes Rendus, v. 250, p. 3690-3692.

Schneider, E. D., Fox, P. J., Hollister, C. D., Needham, D., and Heezen, B. C., 1967, Further evidence for contour currents in the western North Atlantic: Earth and Planetary Sci. Letters, v. 2, p. 351-357.

Wust, G., 1933, Das Bodenwasser und die Gliederum der Atlantischen Tiefsee: Deutsche Atlantische Exped. METEOR 1925-1927, Wiss. Erg., Bd. 6, 1 T: Walter de Gruyter & Co., 520 p.

______, 1936, Schichtung und Zirkulation des Atlantischen Ozeans. Das Bodenwasser und die Stratosphare: Wiss. Erg. Deutsch. Atlant. METEOR 1925-1927, v. 6, p. 1-288.

______, 1957, Quantitative Untersuchungen zur Statik und Dynamik des Atlantischen Ozeans. Stromgeschwindigkeiten und Strommengen in den Tiefen des Atlantischen Ozeans: Deutsche Atlantischen Exped. METEOR, 1925-1927, Wiss. Erg., VI(2), 420 p.

______, 1958, Die Stromgeschwindigkeiten un Strommengen in der Atlantischen Tiefsee: Geol. Rundschau, v. 47, p. 187-195.

Pacific-Type Continental Margins

Robert L. Fisher

INTRODUCTION

In the introduction of "Atlantic-type" continental margins (Heezen, this volume) it is noted that Pacific-type or "tectonically active" margins are not confined to the Pacific ocean but are found in the Atlantic (Antilles and South Antilles) and the Indian oceans (Indonesia-Sunda Trench) as well. Earthquake epicenter plots bear out that observation (Barazangi and Dorman, 1969). Nevertheless, the present report will be concerned almost wholly with Pacific trench occurrences, and even more narrowly with topographic-bathymetric exploration and ancillary observations relating to, or elucidating, morphological characteristics. Later in this volume various authors will treat in detail integrated geological-geophysical observations on specific trenches or island arcs in order to explain the geological history of the Pacific Ocean borders.

Illustrations in this report will be limited to a few areas where the author has made field observations. It will cite several recent sources where well-located and precisely sounded profiles are reproduced, and geological processes become obvious. Finally, several pre-1964 articles, foreshadowing the present plate tectonics-subduction zone paradigm, are noted. Dickinson (1973) has excellently placed the Pacific-margin structures within the post-1964 framework and nomenclature.

GENERAL CHARACTERISTICS

Trench topography, comparisons between trenches, and processes in their morphologic development as shown by precision depth sounding were discussed by Fisher and Hess (1963), who also reviewed evidence of origin, structure, and compressive versus tensional effects, and included a long bibliography for trench geological-geophysical observations prior to the introduction of routine seismic-reflection methods. Reflection profiling such as the detailed work of Ross and Shor (1965), has supported earlier conclusions as to sedimentary processes. Frequently, it has also raised the question of why sedimentary beds or turbidite layers often show little or no contortion along the inner margin of trench floors, where subduction processes should be most effective and where sediments conveyed into the trench from offshore and buried by turbidites would be crumpled and plastered against the nearshore flank.

Very few well-located rocks from the deepest and most seismically active trenches are available, but some dredged samples in trenches free of ice-rafted materials suggests that the lower nearshore walls may contain igneous rocks of the deep oceanic crust, or magmatic cumulates rather than metamorphosed and macerated pelagic sediments scraped off a subducting plate. On the other hand, drilling (to 926 m and 370 m penetration) at two Aleutian Terrace sites, just shoreward of the lower nearshore flank of the moderately deep Aleutian Trench, recovered only horizontally bedded to slightly deformed late Tertiary-Quaternary diatomaceous and silty clay layers and ash beds (Creager et al., 1973; Grow, 1973). Hayes and Ewing (1971) provide an updated review and discussion of reflection profiling and other geophysical measurements in Pacific boundary structures, with emphasis on the Aleutian Arc and northwest Pacific occurrences. Documentation for many of the assertions made below is available in the field studies cited by Hayes and Ewing (1971) and Fisher and Hess (1963).

Bathymetric exploration with precise depth recorders (to a precision of ± 1 m even at the greatest depths) supported by bomb sounding has established the essential similarity of topographic profiles on the Pacific-type margins. The major differences are in the presence or absence of an extensive nearshore bench or terrace and with the amount of fill present (Tayama, 1950; Udintsev, 1955; Brodie and Hatherton, 1958; Nichols and Perry, 1966). Fill usually can be related more closely to weather conditions in source areas, to fluvial transport, or to interception on the nearshore flank than to swallowing or inferred rates of subduction. Even the deepest, steepest, most seismically active V-shaped trenches contain minor fill, a flat floor ½-2 km wide that represents some tens of meters of sediment. Commonly such fill is ponded or terraced along the trench axis, is horizontally bedded in at least its shallower horizons, and intersects the walls at a large angle. Sometimes the shallow, horizontally bedded layers (often turbidites) onlap and overlie with marked disconformity sediment layering continuous with that on the offshore flank. In the deepest trenches the fill is siliceous ooze, volcanic debris and pyroclastics, brown clays, and silts or even sands. Manganese nodules or coating on dredged samples are absent at great depths; dendrites are present on hard fragments from mid-depths and some basalt pebbles or cobbles dredged well outside are encrusted.

Direct measurements of water movements are very rare; one, at depths of 9,400-9,600 m in the

Tonga Trench over a 3-day period gave a value of only several hundredths of a knot along the trench. However, trench-floor photographs occasionally reveal well-developed ripple marks in coarse debris that attest to episodic slumps and turbidity current action, presumably triggered by earthquakes. More commonly the trench-floor photographs reveal little direct evidence of macroscopic animal life—holothurians, worms, starfish, and glass sponges—but extremely abundant tracks, mounds, and discoloration in coarse to fine sediment. The conclusion most easily drawn is that all tracks and disturbances persist for extremely long periods in a very tranquil environment. Photographs of the trench walls in their lower steeper reaches reveal siliceous—and perhaps occasionally carbonate—sediments moving downslope by mass wasting, trickling, and cascading over and around boulders, ledges, and rubble. Off some coasts gullies or canyons convey sediments to the flat floors so that toes, small aprons, or even bajadas appear on well-sounded profiles in detailed surveys.

Mass wasting results in local ponds or terraces on the steep nearshore flanks but also on the characteristically (or ideally) more gently sloping or step-faulted offshore flank. Irregularities damming the sediments can be horsts, massive dikes, or small cones. Reflection profiling reveals that offshore flanks commonly display grabens, with sediment layers conformable to one another and correlatable to the intervening horsts, but all with an overall dip into the trench. Such grabens were predicted on theoretical grounds (Gunn, 1947) and the expected layering early was reported for the Japan Trench (Ludwig et al., 1966). Locally, offshore flanks may be steepened by extensive fault scarps, by peaks or cones or dikes. The earliest direct evidence for "subduction" came with measurement of trenchward tilting of flat-topped Capricorn Guyot outside Tonga Trench (Raitt et al., 1955; Fisher and Revelle, 1955).

From sea-floor spreading considerations, extensive exposures of igneous rock on the gently sloping offshore flanks are rare and expectably mantled by pelagic sediments. Manganese-encrusted pillows have been photographed on the offshore flank of Palau Trench at 4,900 m depth and columnar-jointed basalts and contorted flow basalts have been collected on the deep offshore flank of Tonga Trench at 8,800 m depth (Fig. 1). Locally in the latter area, as well as in the westernmost part of the New Britain Trench, the quality of reflections—both by 12-kHz sounder and airgun—suggests the presence of new flows near the base of the offshore flank; no material appropriate for isotopic dating is yet in hand.

At mid-depths of less than 5,000-6,000 m, nearshore slopes are variable; they may continue irregular and ponded, decrease in slope and become thickly sedimented or even terraced, or remain smoothly precipitous and apparently faulted clear up to the continental shelf (if any) or island shore. Offshore, the trench flank may merge smoothly with a flat ocean basin floor or a low swell or "outer ridge," or it may continue rough, jumbled, steep, and apparently in fault or extrusive contact with the more horizontal basin floor. Examples of all these instances appear in the Melanesian and Micronesian profiles discussed below. Samples from depths shoaler than 4,500 m on trench flanks are predominantly calcareous, composed of *Globigerina* oozes and silts or are volcanic debris; coralline fragments are common in tropical areas. Pumice, too, is abundant to ubiquitous in the shallower nearshore samples from volcanic arcs and serves to trace gross water movements.

EASTERN PACIFIC MARGINS

Except in the northeastern Pacific, the eastern limit of the Pacific Basin lies several thousand kilometers shoreward of the East Pacific Rise, the complexly faulted, seismically active, and very rapidly spreading local representative of the worldwide mid-ocean ridge system and the presumed principal source of oceanic igneous rock impinging against that margin. Several much smaller discordant features—Nazca Ridge, Carnegie Ridge, Cocos Ridge, Tehuantepec Ridge—resembling buttresses to that spreading feature, make a reticulate pattern in the region east and north of the East Pacific Rise-South Chile Ridge. All four ridges, aseismic or only moderately active and not of the spreading variety, meet the continent at very large angles. They mark profound changes in the topographic and geophysical characteristics of the series of trenches that front the shoreline from southern Chile to Ecuador and from southeastern Costa Rica to the mouth of the Gulf of California. Finally, a nearly filled trench or marginal basin, Cedros Deep, lies at the foot of the continental slope off Baja California, north of the juncture of the East Pacific Rise and the Gulf of California. Quiescent and not especially deep, that basin's subfill level is that of the very active and topographically spectacular trenches off Central and South America.

The Peru-Chile Trench is relatively shallow, ranking only fourteenth of the world's trenches in maximum depth, but in length it extends over 45° of latitude, being surpassed only by the less continuous Japan-Kuril and East Indian trench systems. Topographic studies (Zeigler et al., 1957; Fisher and Raitt, 1962) reveal marked differences between the relatively shallow and gently basined or flat-bottomed trench, with wide shelf and sediment-mantled flank, off central Peru and the much deeper, almost sediment-free trench, with a shelf only minor or entirely missing, off northern Chile. Between these regions the Nazca Ridge constricts and deforms the trench (Fisher, 1958, Fig. 8). South of Valparaiso, Chile, the trench is flat-floored and continues as a sedimented deep, with mantled slope and wider shelf, to the vicinity of 40°S; south of that

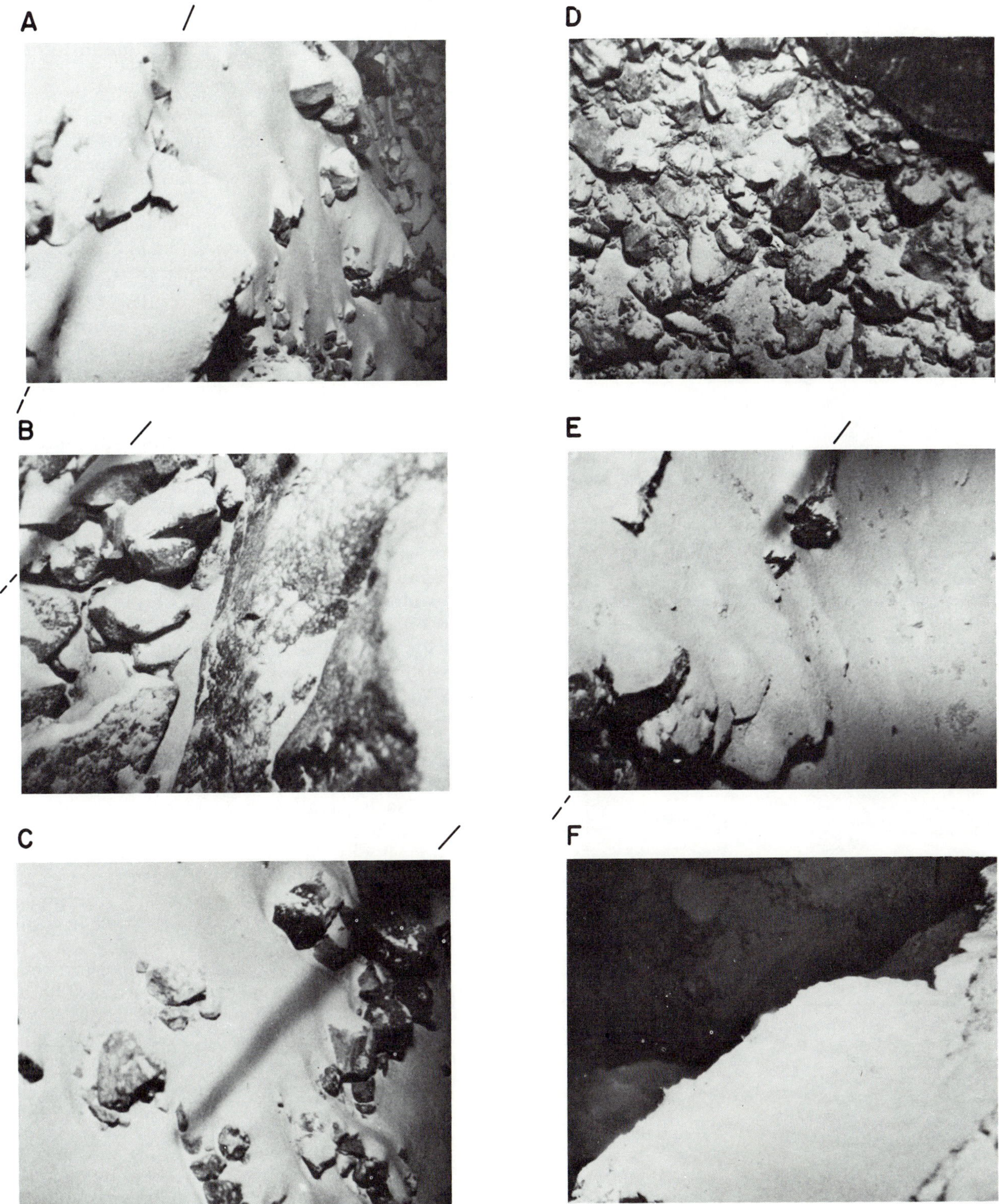

Fig. 1. Bottom photographs on oceanward slope of Tonga Trench, showing rock exposures, angular rubble and siliceous sediments trapped in crevices and shifting downslope by slumping. (Field of view 2-4 m^2.) Small ticks indicate the vertical. (Seven-Tow 61-C, 8,790-8,870 m near 20°34.5'S, 173°17.2'W.)

latitude it is traceable on reflection profiles to at least 44°S (Hayes, 1966; Scholl et al., 1970). Both station seismology (Benioff, 1949) and shipborne refraction profiling (Fisher and Raitt, 1962) indicate that the oceanic crust descends below the continental margin (Fig. 2, in an area where fill is minor to missing).

Reflection profiling (Scholl et al., 1968; Hayes and Ewing, 1971) clearly portrays details of horizontal and undeformed sediment layering in the trench

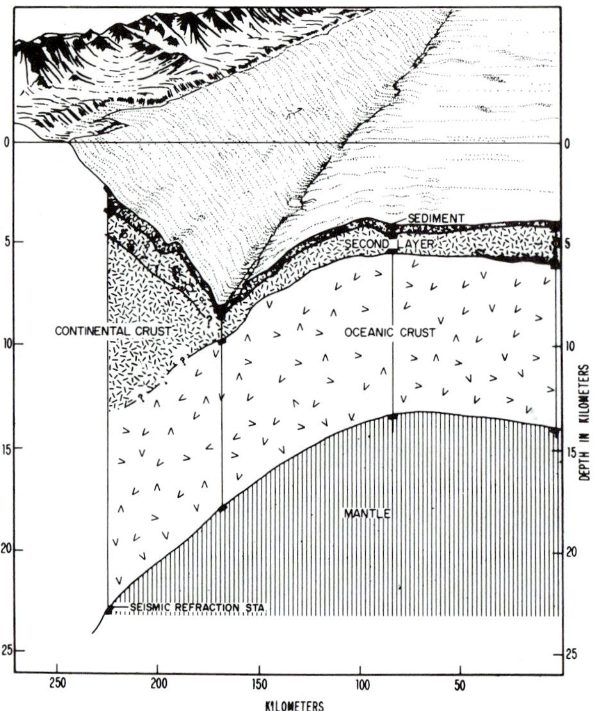

Fig. 2. Crustal structure along section off Antofagasta, northern Chile, from seismic-refraction studies (after Fisher and Raitt, 1962, Fig. 13). (Vertical exaggeration X 10.) Here the trench reaches 8,055 ± 10 m, the deepest point along the eastern Pacific margin.

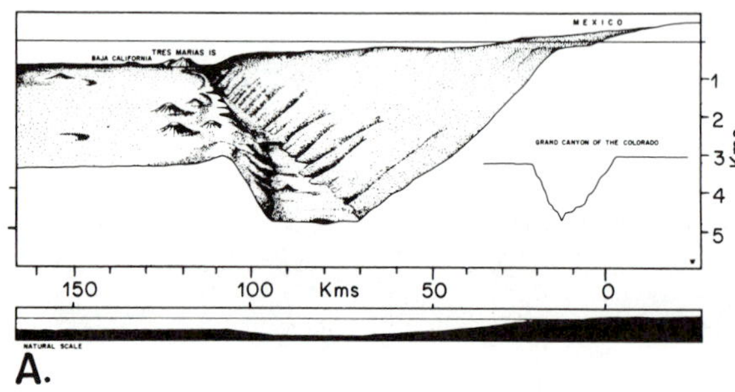

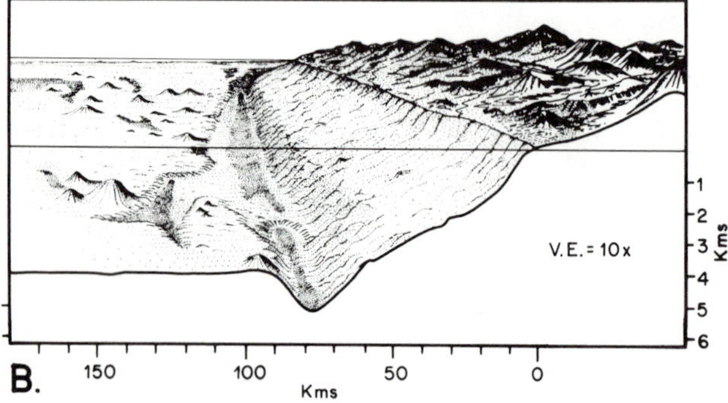

Fig. 3. Perspective section of the Mid-America Trench based on detailed exploration by Scripps prior to 1970. A. Ponded and terraced sediments, knolls, and aprons on trench floor; gullied lower flank. View to northwest, with line of section near Manzanillo, Colima. Grand Canyon, Arizona, drawn to same scale (X10). B. View to northwest, with line of section passing near Acapulco, Guerrero.

bottom, contrary to expectation from simple plate tectonics models in a rapidly subducting (5-10 cm/yr) area. Scholl et al. (1970) assessed quantity of fill, sediment attitude, and paleontologically determined ages of the floor, walls, and upper flank of the Chilean segment with respect to rainfall and sediment sources in the several climatic zones, from the arid north near 23°S to the extremely humid south near 44°S. From analysis of three dozen profiles, they conclude that (1) sediment removal possibly implying spreading occurred at the base of the margin and adjacent sea floor in Late Cretaceous and perhaps earliest Tertiary time; (2) Eocene to Pliocene sediments, of continental derivation, have accumulated on the margin and in minor amount in the trench but evidence for tectonic removal during this time is not conclusive, and (3) late Cenozoic turbidites, perhaps 70,000 km^3 in volume, within the trench suggest that with conventional rates of continental denudation either late Cenozoic underthrusting has not taken place (or has been much slower than implied by magnetic lineation data) or that underthrusting at the "geophysical" rate, 5-10 cm/yr, has not involved removal of a significant volume of sediment from the trench.

No other trench is so favorably situated for such a zonal examination of prolonged denudation-sedimentation budget from an extensive provenance; however, smaller-scale island arc occurrences where trench floor-nearshore flank sediment contortion has been reported include the Antillean Trench (Chase and Bunce, 1969), the Japan Trench (Hilde and Raff, 1970), the "Manila Trench" (Karig, 1973), and the Java Trench (Beck and Lehner, 1974); see also Seely et al. and Kulm and Fowler, this volume.

The Middle America Trench, off the west coasts of Mexico and Central America, strikingly resembles, on a smaller and shoaler (maximum depth 6,660 m) scale, the Peru-Chile Trench. Again the shallower, more arcuate, and less active northern section is separated from the deep, more seismically and volcanically active southern section by an offshore, wholly submerged discordant topographic high, here the Tehuantepec Ridge (Fisher, 1961). Southeast of this ridge, fill along the trench floor is minor, until the V-shaped trench dies out near Cocos Ridge. Northwest of the Gulf of Tehuantepec the trench floor is ponded or terraced (Fig. 3), with clays and silts conveyed across the narrow shelf and down moderately steep flanks by canyons and gullies from the volcanic, metamorphic, and silicic sources.

Seismic refraction measurements (Shor and Fisher, 1961) and reflection profiling (Ross and Shor, 1965) indicate 0.5 -1.5 km of sediments, little deformed, in several of the basins between Islas Tres Marias and Acapulco. Apparently the northwest part of the Mid-America Trench is relatively inactive, in a subduction sense, at present; faulting and volcanism associated with the East Pacific Rise and its cross fractures seaward of the trench are

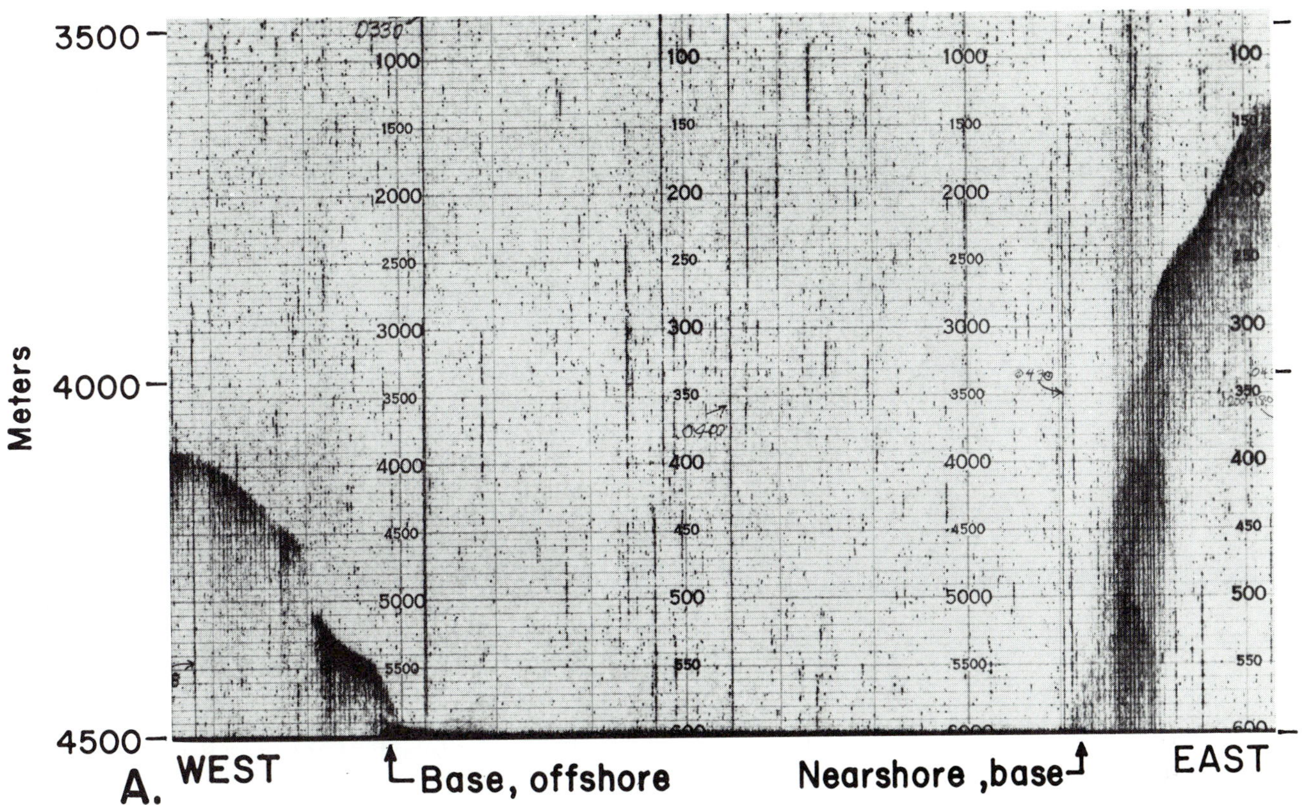

Fig. 4. Cedros Deep, a flat-floored sediment trap at the base of the continental slope off Sebastian Vizcaino Bay, central Baja California. A. Sediment pond (4,490 ± 1 m) in central part of Cedros Deep. Length of section approximately 30 km. (Vertical exaggeration X19.) B. View to south, with Cedros Island at left. Flat sea floor, ponded sediments, partially "inundated" knolls.

more significant. If so, this rapidly sedimenting part of the Mid-America Trench may soon, in terms of geological time, resemble the minor, strikingly flat-floored sediment traps bordering the west coast of central Baja California (Cedros Deep).

Cedros Deep, a spectacularly flat-floored and terraced marginal trough about 250 km long and 60-80 km wide (Figs. 4 and 5A) lies at the foot of the steep and locally rocky continental slope off central Baja California. Within the northwest-trending basin are two principal levels (Fig. 5B) and several minor terraces. Profiling shows that these are sediment terraces, formed behind fault slivers paralleling, and ridges athwart, the trough; the near perfect horizontality, except for minor aprons at the base of both flanks, attests to extremely fluid sediment masses inundating basement and knolls (Figs. 4A and 5A). Early reflection profiling in Cedros Deep (Fisher, 1953) established the customary asymmetrical V-shaped basement cross section, with 1.6 sec maximum two-way travel time to the base of the sediments near the foot of the precipitous eastern flank. Cedros Deep, now in an essentially senile state in a practically aseismic locale, has a basement profile very nearly that of the extremely active Mid-America Trench segment off Guatemala.

More recent profiling, with closely spaced airgun impulses, confirmed the 1.6 sec maximum sedi-

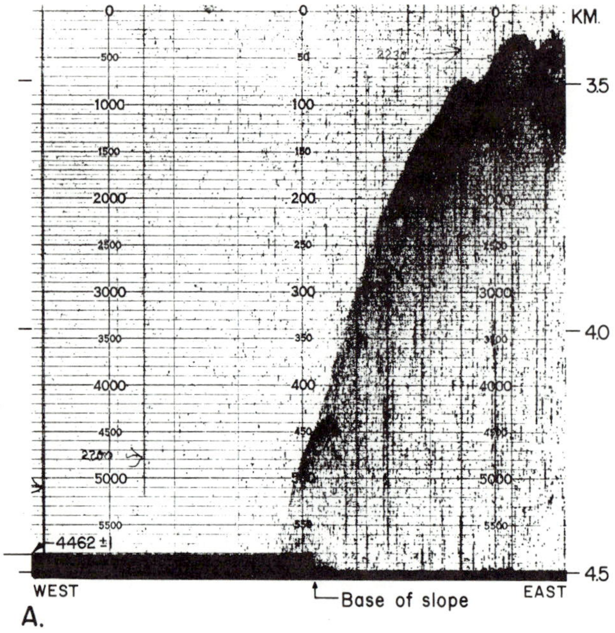

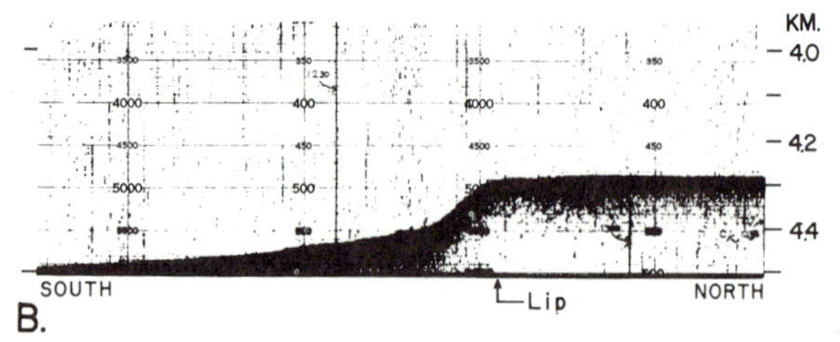

Fig. 5. Cedros Deep. A. Sediment pond (at 4,462 ± 1 m) and lower east flank. Length of section approximately 14 km. B. Longitudinal profile within the eastern deep showing sediment terraced or ponded behind lip and spilling over barrier. Length of section approximately 18 km.

ment thickness but revealed a longitudinally irregular basement profile, and at least the upper 0.6+ sec of the sediment section, to consist of distinctly bedded turbidites, undeformed and essentially perfectly horizontal for more than 65 km along the deepest part of the basin.

The nearest land to Cedros Deep are Islas San Benitos, three tiny islands immediately atop the continental slope, of interest because they consist of Franciscan-like graywackes, chert, carbonates, glaucophane schist, basalt, and serpentinite (Cohen et al., 1963, assigned a Jurassic-Cretaceous age on the basis of that similarity). Those authors aver that the islets' rocks are extensively folded, sheared, and deformed; strike-slip faults, of right-lateral displacement, trend north-northwesterly, approximately parallel to the continental slope. Some linear offsets and reentrants on the slope may be similar fault slivers or slumped blocks. In all this tectonic deformation there is no definitive evidence for overall motion normal to the slope in Tertiary-Quaternary time; Pleistocene terraces are variously tilted.

MICRONESIAN AND MELANESIAN MARGINS

Eleven of the twelve deepest trench areas, all very active seismically, lie in the Western Pacific; Puerto Rico Trench in the Caribbean, number seven in ranking, is the only outsider among this deepest dozen. All are similar in profile, whether on the Pacific basin or "interior" side of the associated islands, with variations within any trench equal, except perhaps in scale, to differences displayed between the trenches. Profiles recorded on a 12-kHz Precision Depth Recorder during a 1962 Scripps investigation of eight west equatorial and southwestern Pacific trenches illustrate such characteristics (Figs. 6-15).

Eleven profiles across the southern part of the Marianas Trench (Figs. 6 and 7) display abrupt and large-scale depth changes along the rugged axis of that feature. This east-trending part of Marianas Trench contains Challenger Deep, the greatest depth in the oceans (Fig. 6, D-D'). On the single Challenger Deep profile shown here the traverse deeper than 8,000 m was made at reduced speed; slopes there are as steep and rough as on the other profiles. It reveals a narrow, flattish, filled, and terraced axial zone that has been explored in detail on later Scripps' cruises. It resembles in scale and setting the very deepest, sedimented part of the Tonga Trench, Horizon Deep, discussed in the next section. Shoaling of the axis by 2,500 m is abrupt in the 75 km between G-G' and H-H'; profiles on Figure 7 suggest that both the Marianas and Yap trenches exist as shoal, but unsedimented clefts on K-K', then reinforce to form a single deeper cleft at their intersection. Most of the profiles display rough and precipitous walls; tilting and normal faulting appears on the deep basin floor south of Challenger Deep.

Yap Trench, intersecting the western end of the Marianas Trench at a large angle, is not long but it displays great relief along its little-sedimented axis (Fig. 8, A-A', B-B') and very steep and smooth flanks particularly on the shoreward side. Near its southern end it becomes irregular, with considerable relief on its seaward side. Palau Trench (Fig. 9), overlapping Yap Trench immediately to the south and west, is much shoaler, and its profiles are exceptional in their displays of superimposed, large-scale fault and volcanic activity: large rugged peaks hilly tracts, and ridges lie oceanward of the axis, pillow basalts were photographed there, the trench bottom is V-shaped to broadly ponded, and one profile (Fig. 9, E-E') reveals a secondary, apparently recent cleft within the strongly reflecting, gently sloping floor.

South of the equator and interior to some major Melanesian island groups, a series of linked trenches intersecting at large angles extends from Huon Gulf off eastern New Guinea to the southern border of the Fiji Plateau (Figs. 10-15). Again the series is extremely variable: resembling Palau

Trench, the western representatives (Figs. 10, 11, and 12A-A', B-B') are rough-walled, disrupted, jumbled to hilly on the flanks, terraced along the axis, with the previously cited evidence for a recent flow at the base of the scarp north of the Papuan Peninsula (Fig. 10, B-B'). Shoal and variable off the central part of the Solomon chain, to the east of Rennell Island the profiles regain regularity in the smoother steep trench flanks, but there is little evidence of sedimentary fill (Fig. 12 D-D', E-E', F-F').

Off the Santa Cruz Islands (Fig. 13) the rough-floored, steep-walled North New Hebrides Trench plunges to depths of well over 9,000 m, then shoals markedly to very moderate depths and a gently sedimented floor off the New Hebrides Islands proper (Fig. 14). These profiles (G-G', H-H_1-H', I-I') off the New Hebrides group resemble in width, shape, and scale the crossings of the complex Sunda Trench off Sumatra and the Nicobar Islands. Off Makelula (Fig. 14, H-H') however, and possibly off Espiritu Santo (F-F'), there appears evidence of an inner, more recent, rough-walled, faulted trench or cleft. Finally, between the Loyalty Islands and the southernmost New Hebrides (Fig. 15) there is a very classical (by joining D-D' to E-E_1-E') shoal trench (7,000$\pm$ m) bearing only minor sediment and associated with a steep-walled 1,000-1,500 m deep rift near Aneityum Island in the western sector of the Fiji Plateau. Hayes and Ewing (1971, Fig. 11) present a similar trench-rift pairing on a section across the North New Hebrides Trench-"Vityaz Trench," by reflection profiling.

DETAILED STUDIES IN TWO SECTORS OF THE TONGA TRENCH

In the foregoing progression from the general to the particular in terms of Pacific-margin or trench topographic characteristics, assertions were made linking morphology to tectonic processes and, by extrapolation, to composition of the walls of the deeper, more spectacular, difficult-to-sample trenches. Support for such conclusions comes in part from shipboard studies in the Tonga Trench by Scripps scientists variously over an 18-year interval. Work since late 1967, employing precision depth recorders, seismic-reflection profiling, deep-sea camera, and pinger-controlled dredging is especially pertinent (Figs. 16-18).

Horizon Deep (Fisher, 1954), the second deepest part of the oceans, displays excellently in scale and geometry the characteristics of such extreme hadal regions as Challenger Deep in the Marianas Trench (Fig. 6, D-D') and Cape Johnson Deep in the Philippine Trench (Fisher and Hess, 1963, Figs. 3 and 4), the third deepest area. In these localities the lowest point is flat, sedimented, and 1/2-2 km wide. Mirror-like, it yields a relatively strong return on a high-frequency echo-sounder trace. Customarily, with normal beam-width echo sounders, the floor of such deep trenches never appears as a first arrival, even when the ship is directly over the axial region. The lower walls appear to be rough, with small knobs or knolls and reentrants, and the echo from the flat-floored basin turns downward, "deepening" as the ship starts climbing the wall. Persistence of these strong, spuriously deep returns accounts for some reports of excessive record depths even in familiar, well-sounded regions. It is obvious that a local depression in a sediment flat or terraced pond is extremely unlikely; note in this regard Figures 4A and 5B, where shoal depths and widely spaced walls permit clear demonstration of this.

The extremely steep to precipitous, occasionally vertical, lower trench flanks on both sides of such faults are due to the presence of hard rock, either as outcrop or talus. Well-positioned sampling of these remote exposures is rare; Bowin et al. (1966) and Petelin (1964) are two of very few references other than the Scripps work noted above. The former paper reports nearly completely serpentinized peridotites (total H_2O of 12-13 wt%, bastite pseudomorphs after orthopyroxene), "talc rock," and altered plagioclase-clinopyroxene basalt from scarps at 7,100-6,300 m on the northern flank of the Puerto Rico Trench. These dredged rocks were recovered in company with Late Cretaceous and early Tertiary sediments. Petelin (1964) reports tholeiitic pillow basalt from 7,600 to 7,300 m in the New Britain Trench, north of H, Figure 10, and basalt (unspecified) from the west side of Tonga Trench at 8,500-7,500 m.

Possibly the most detailed rock recovery work in a very deep trench was carried out in Tonga Trench between 20 and 21°S, a sector where topographic, seismic refraction, and magnetic profiling had been done much earlier (Raitt et al., 1955). Fourteen successful dredge hauls and 15 camera stations were made at depths of 5,000+ to 9,900 m, principally on the nearshore flank, during expeditions in 1967 and 1970. Figure 18, redrawn and amplified from Fisher and Engel (1969, Fig. 2), displays the generalized results.

In the Tonga region overall, and on a composite profile, the deep offshore "subducting" flank yielded diabase, some well-cemented shales, and variolitic basalts that show contortions and pillow structure (Fig. 1). The shallow ($<$5,500 m) nearshore flank is mantled by "sediments"—siltstones, calcareous and coralline debris, pumice, intermediate and silicic volcano-clastics—that subdue irregularities and develop gentle slopes except where intruded or cut by a massive diabase body. Farther downslope the steep scarps yield augite-hornblende gabbro and altered or fresh-jointed alkali olivine basalts. On the deepest, steepest part of the nearshore flank are fresh harzburgites (total H_2O $<$0.10 wt%: C. G. Engel in Fisher and Engel, 1969), some dunite, serpentinized lherzolite, serpentine lenses, and variolitic low-alkali basalts.

Samples from these deeper nearshore hauls often are slickensided, show considerable grain strain, and may be veined by zeolites. Bottom photographs show crushed and sheared platy zones as well as massive outcrops. These rocks, and the outer flank variolitic basalts, occur in the lower parts of the wall, where seismic-refraction work yielded compressional wave velocities of 5.2 ± 0.1 km/sec, taken to be "volcanic basement" (Raitt et al., 1955). Neither the topographic, rock-sampling, nor reflection-profiling studies suggest that pelagic sediments were plastered against the deep near-shore walls here in the Tonga Trench. Neither did the dredging yield rocks derived from metamorphosed or melted deep-sea sediments.

ACKNOWLEDGMENTS

Several illustrators contributed their talents to the scenes reproduced here, in particular Robert Winsett and Howard Taylor. Profiles were prepared by David G. Crouch, T. W. C. Hilde, Marie Z. Jantsch, and Robert M. Beer. Additional drafting was done by Judy Clinton and Judy Morley. Robert J. Mann assisted in bottom photography in Tonga Trench, carrying out shipboard film processing preparatory to selection of dredging sites. Field work was carried out under NSF grants G-19935 and GA-776 and contracts with the Office of Naval Research.

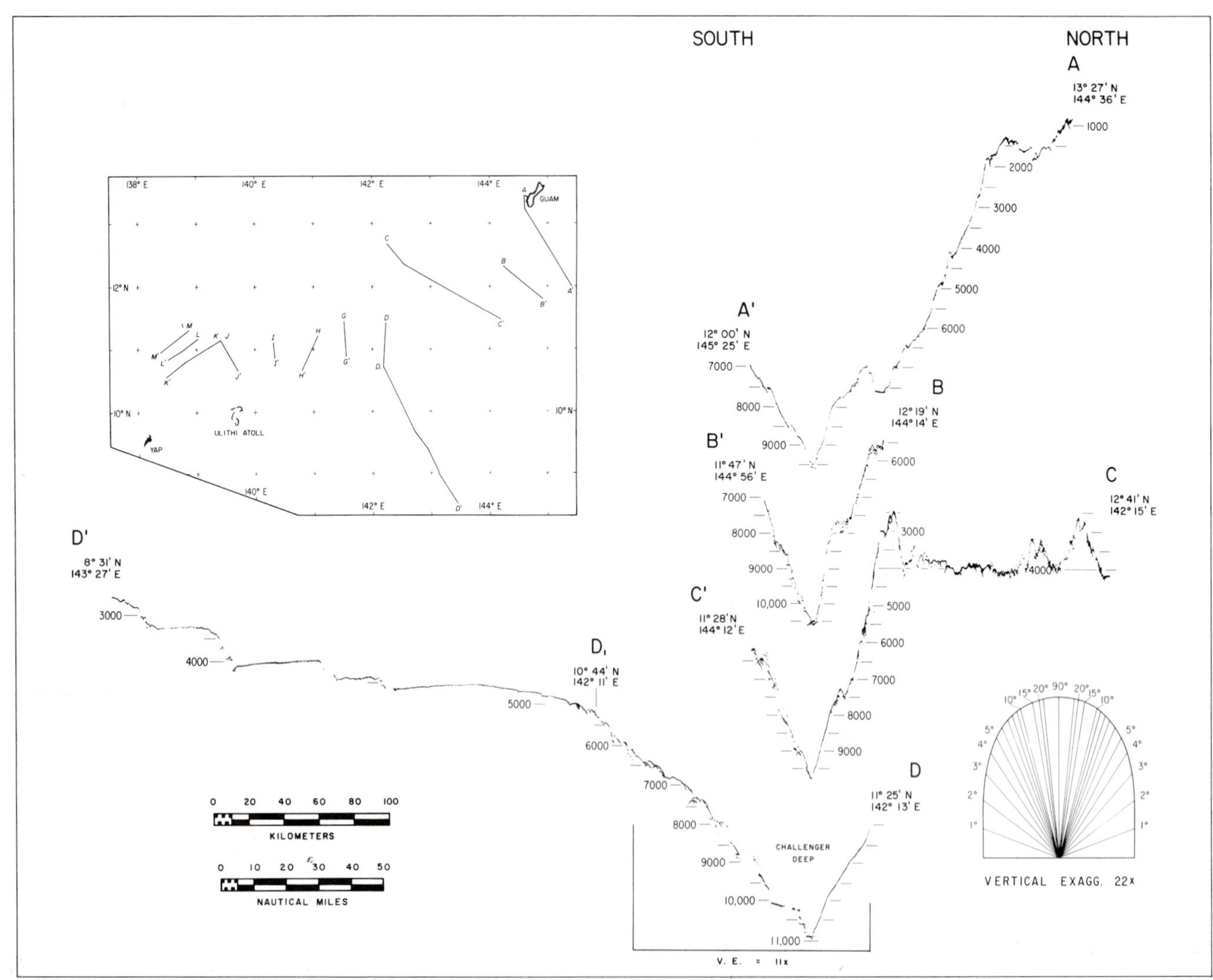

Fig. 6. Precision depth records (PDR) in vicinity of Challenger Deep, Marianas Trench (vertical scale corrected after Matthews, 1939). From Scripps PROA Expedition.

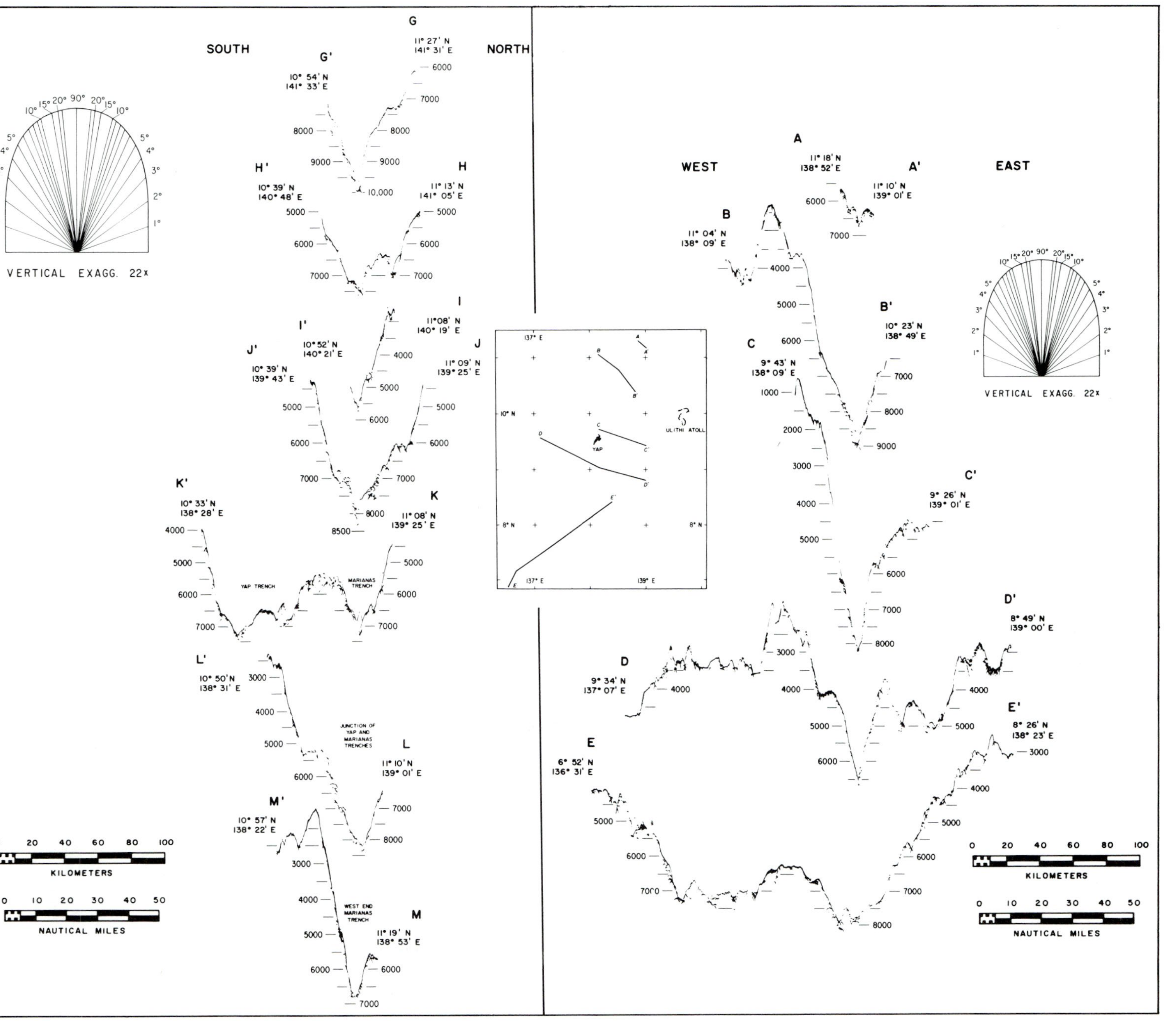

Fig. 7. PDR traverse in vicinity of Marianas Trench. From Scripps PROA Expedition.

Fig. 8. PDR traverse in vicinity of Yap Trench. From Scripps PROA Expedition.

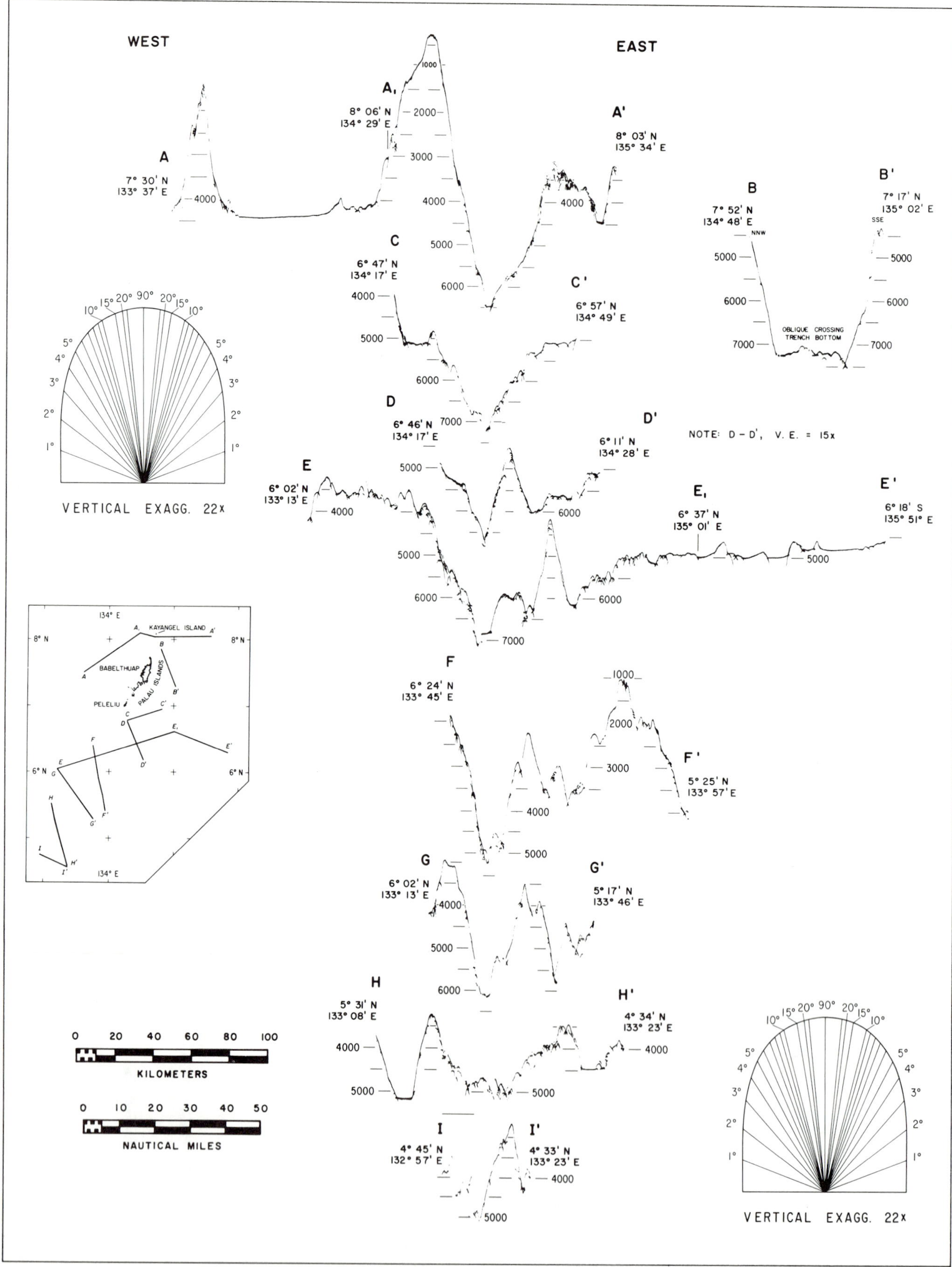

Fig. 9. PDR traverse in vicinity of Palau Trench. From Scripps PROA Expedition.

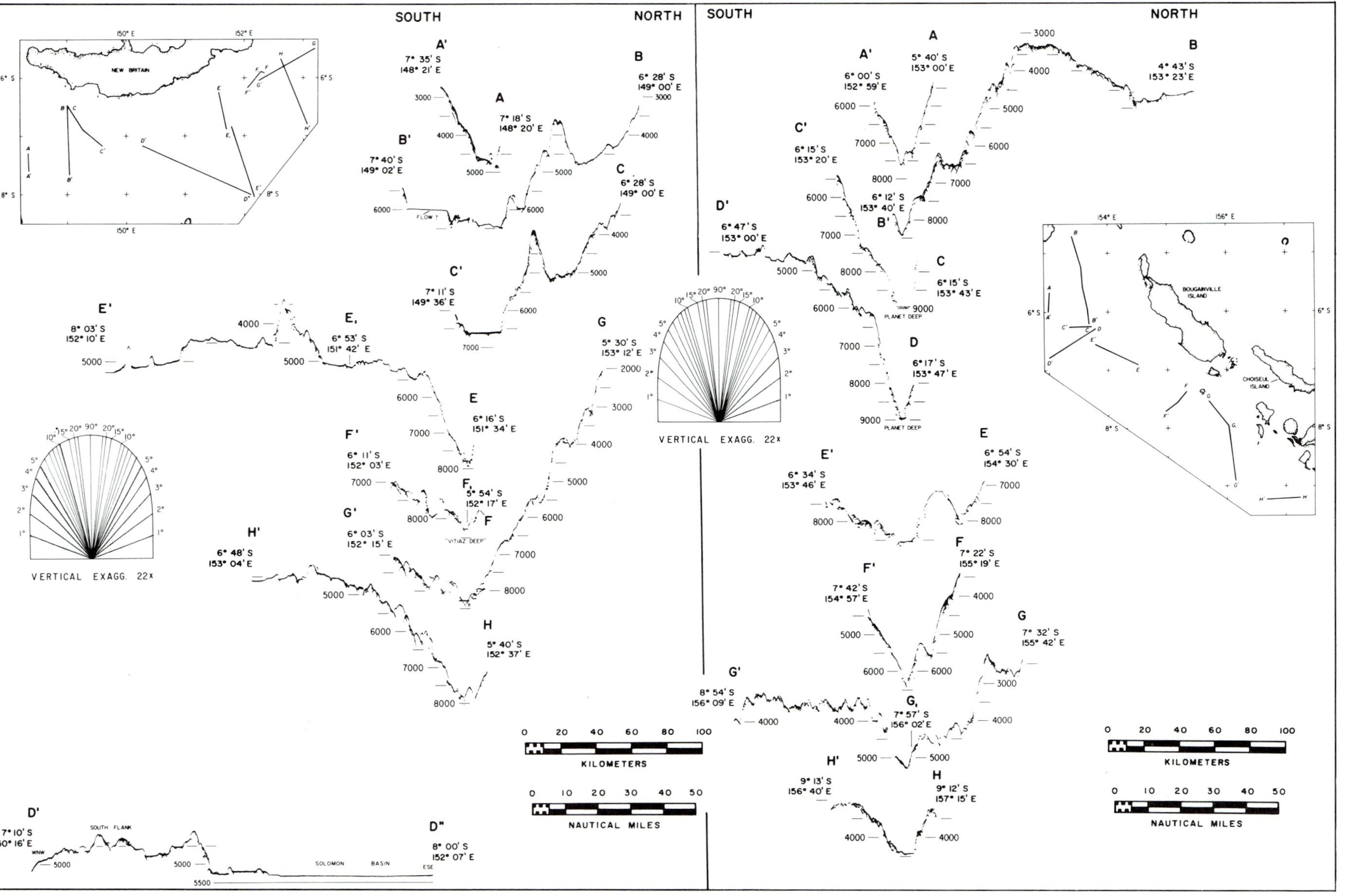

Fig. 10. PDR traverse in vicinity of New Britain Trench. From Scripps PROA Expedition.

Fig. 11. PDR traverse in vicinity of North Solomons Trench. From Scripps PROA Expedition.

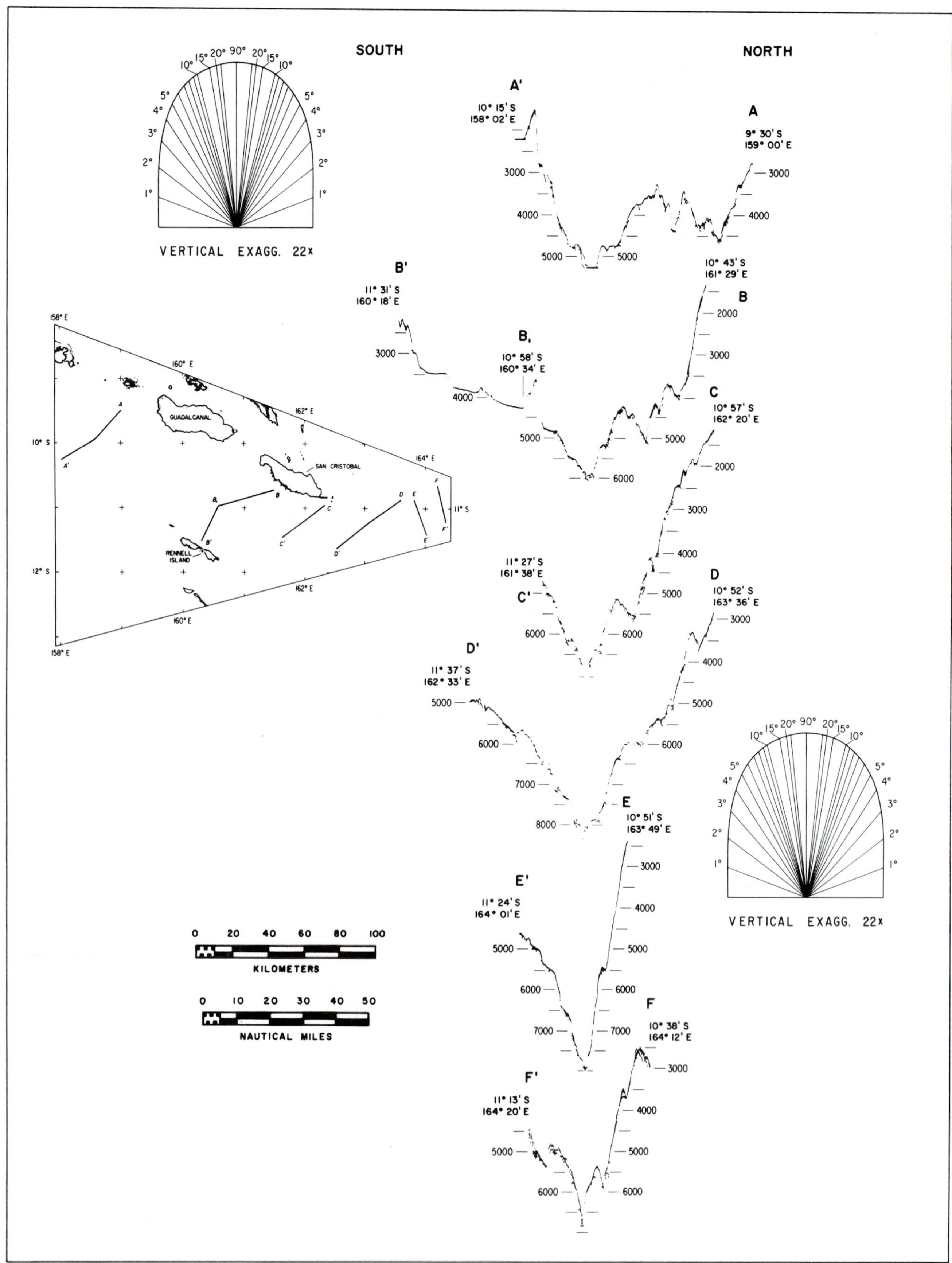

Fig. 12. PDR traverse in vicinity of South Solomons Trench.

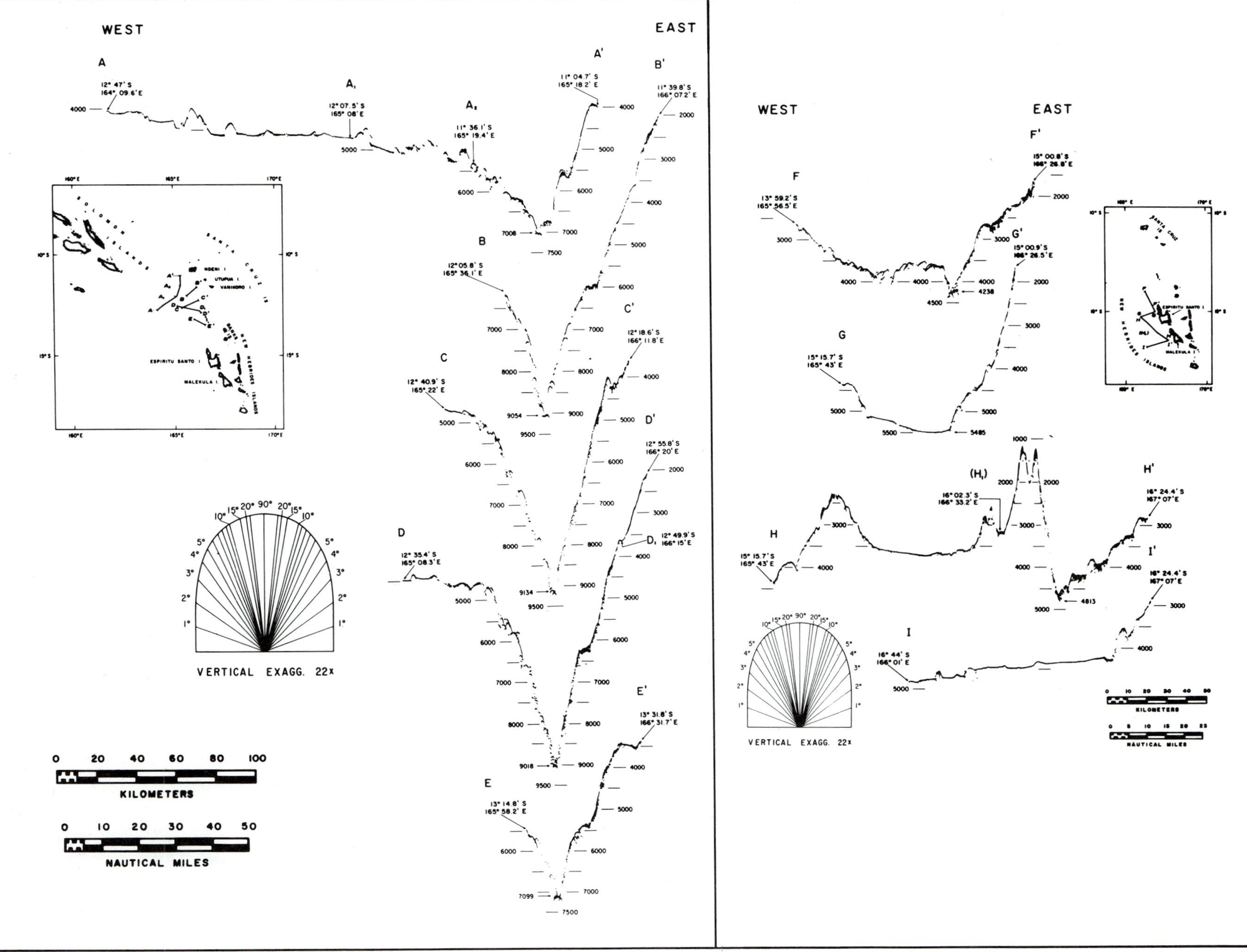

Fig. 13. PDR traverse in vicinity of North New Hebrides Trench, off Santa Cruz Group. From Scripps PROA Expedition.

Fig. 14. PDR traverse in vicinity of North New Hebrides Trench, off New Hebrides Group. From Scripps PROA Expedition.

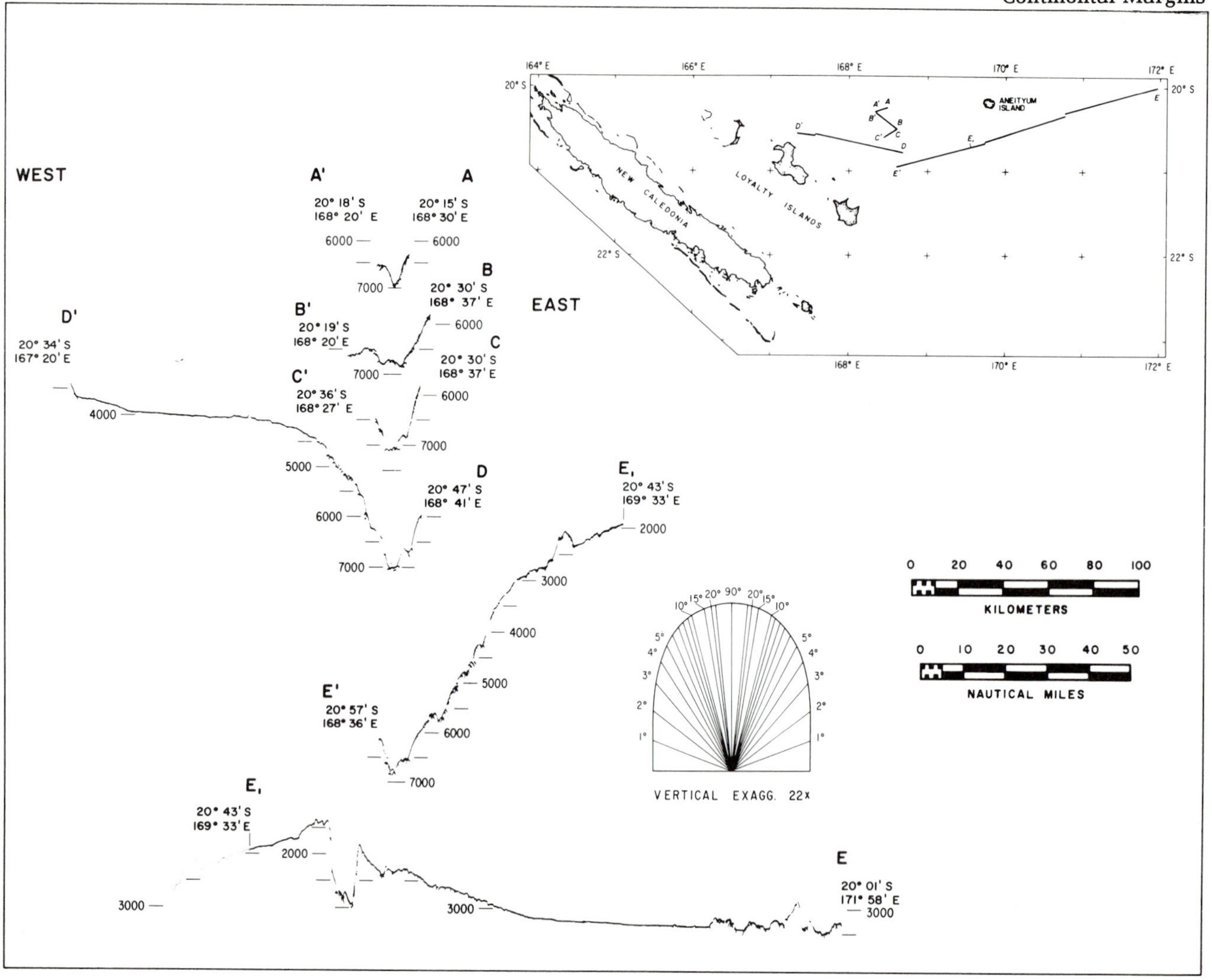

Fig. 15. PDR traverse in vicinity of South New Hebrides Trench. From Scripps PROA Expedition.

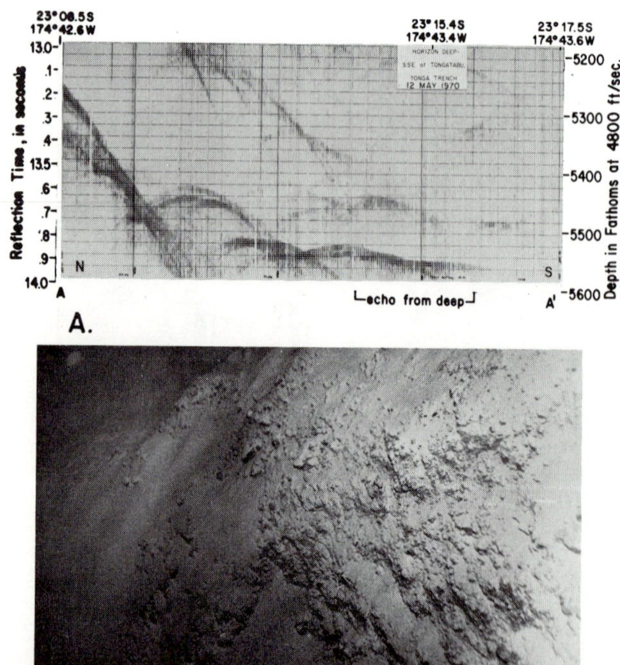

Fig. 16. Horizon Deep, Tonga Trench. A. PDR records show that strongly reflecting sedimented trench floor (13.92 ± sec) does not give any first arrival. Depth approximately 5,566 ± fm (uncorr.), approximately 10,800 m (corr.) B. Unconsolidated siliceous sediment moving down the nearshore flank. Depth approximately 10,400 m; approximately 23°16.2S, 174°46.9W (Seven-Tow 67-C).

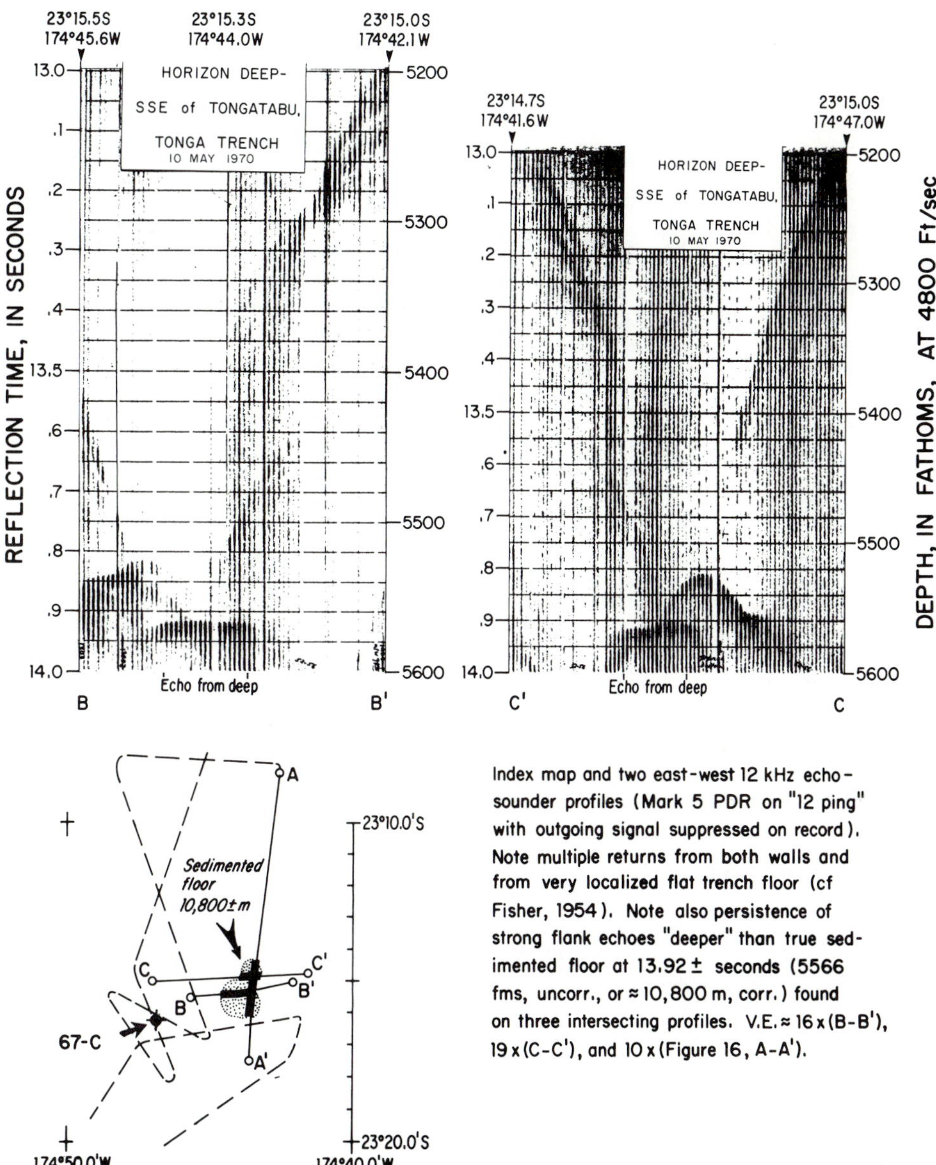

Index map and two east-west 12 kHz echo-sounder profiles (Mark 5 PDR on "12 ping" with outgoing signal suppressed on record). Note multiple returns from both walls and from very localized flat trench floor (cf Fisher, 1954). Note also persistence of strong flank echoes "deeper" than true sedimented floor at 13.92 ± seconds (5566 fms, uncorr., or ≈ 10,800 m, corr.) found on three intersecting profiles. V.E. ≈ 16 x (B-B'), 19 x (C-C'), and 10 x (Figure 16, A-A').

Fig. 17. Horizon Deep, Tonga Trench. Index map and PDR records. (R/V Thomas Washington, Scripps Expedition.)

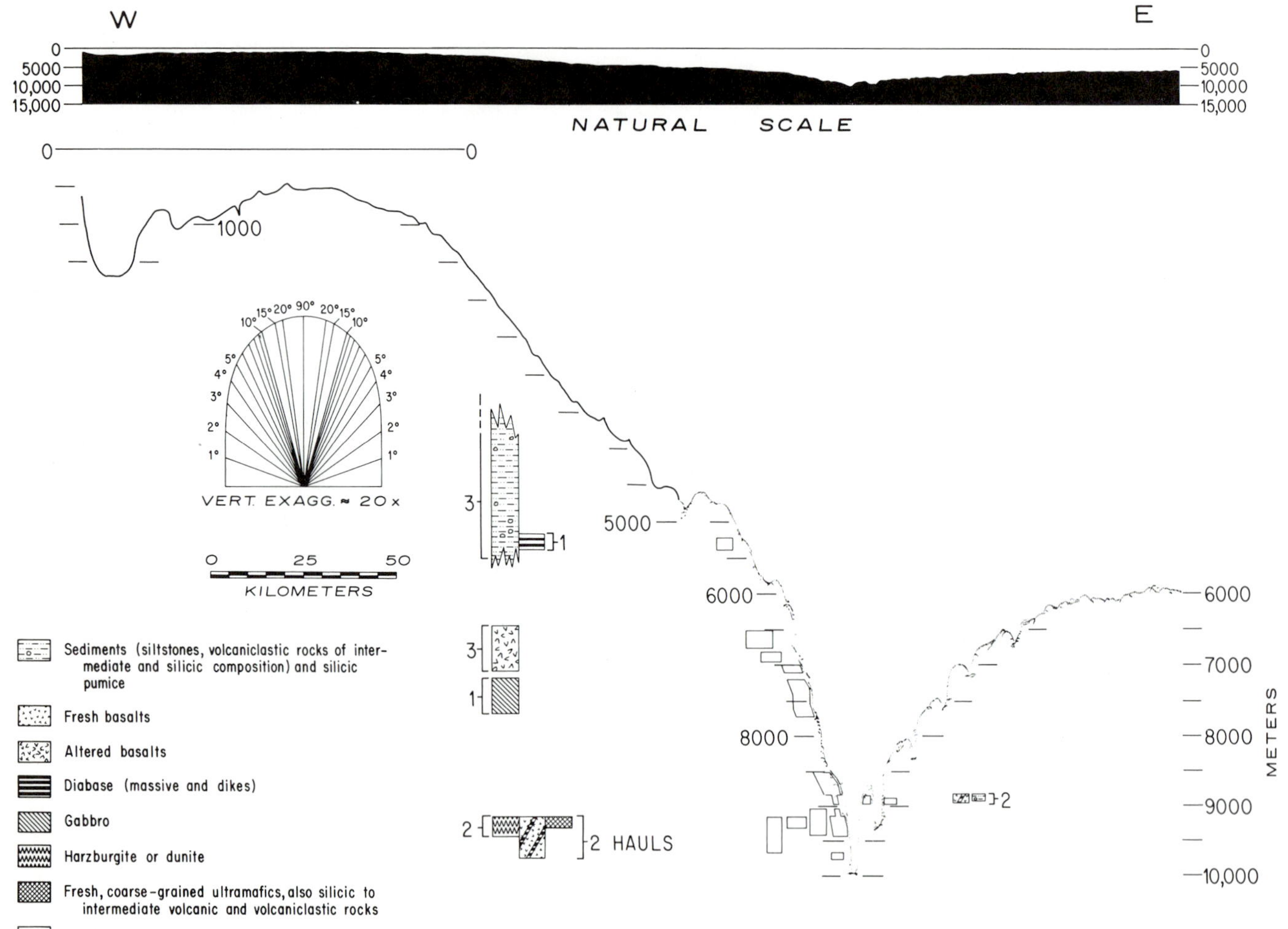

Fig. 18. Composite cross section of Tonga Trench: that part of the profile to the right of the 5,000-m notation was traced from a PDR tape. Dredge hauls made as much as 35 km north or south of this profile have been projected onto it. The rock symbols are keyed, by length, to the depth-range indicators (except for three hauls in the sedimented area shoaler than 5,000 m).

BIBLIOGRAPHY

Barazangi, M., and Dorman, J., 1969, World seismicity maps compiled from ESSA, coast and geodetic survey, epicenter data, 1961-1967: Seismol. Soc. America Bull., v. 59, p. 369-380.

Benioff, H., 1949, The fault origin of oceanic deeps: Geol. Soc. America Bull., v. 60, p. 1837-1866.

———, 1955, Seismic evidence for crustal structure and tectonic activity: Geol. Soc. America Spec. Paper 62, p. 61-73.

Bijlaard, P. P., 1936, Théorie des déformations plastiques et locales par rapport aux anomalies négatives, aux fosses océaniennes, aux géogynclinaux, etc.: I.U.G.G., Rept. 6th Gen. Assembly, Assoc. Geod., Edinburgh.

Bowin, C. O., Nalwalk, A. J., and Hersey, J. B., 1966, Serpentinized peridotite from the north wall of the Puerto Rico Trench: Geol. Soc. America Bull., v. 77, p. 257-270.

Brodie, J. W., and Hatherton, T., 1958, The morphology of Kermadec and Hikurangi trenches: Deep-Sea Res., v. 5, p. 18-28.

Chase, R. L., and Bunce, E. T., 1969, Underthrusting of the eastern margin of the Antilles by the floor of the western North Atlantic Ocean, and origin of the Barbados Ridge: Jour. Geophys. Res., v. 74, p. 1413-1420.

Cohen, L. H., Condie, K. C., Kuest, L. J., MacKenzie, G. S., Meister, F. H., Pushkar, P., and Steuber, A. M., 1963, Geology of the San Benito Islands, Baja California, Mexico: Geol. Soc. America Bull., v. 74, p. 1355-1370.

Creager, J. S., Scholl, D. W., and others, 1973, Initial reports of the Deep Sea Drilling Project, v. 19, Washington, D. C., U.S. Govt. Printing Office, 913 p.

Beck, R. H., and Lehner, P., 1974, Oceans, new frontier in exploration: Am. Assoc. Petr. Geol. Bull., v. 58, p. 376-395.

Dickinson, W. R., 1973, Width of modern arc-trench gaps proportional to past duration of igneous activity in associated magmatic arcs: Jour. Geophys. Res., v. 78, p. 3376-3389.

Ewing, M., and Heezen, B. C., 1955, Puerto Rico Trench topographic and geophysical data: Geol. Soc. America Spec. Paper 62, p. 255-267.

Fisher, R. L., 1953, Sedimentary fill in two Mexican foredeeps (abstr.): Geol. Soc. America Bull., v. 64, p. 1422-1423.

———, 1954, On the sounding of trenches: Deep-Sea Res., v. 2, p. 48-58.

———, 1958, DOWNWIND investigation of the Nasca Ridge: Preliminary Report on Expedition DOWNWIND: Washington, D.C.: IGY World Data Center A, I.G.Y. Gen. Rept. Ser. 2, p. 20-23.

———, 1961, Middle America Trench: topography and structure: Geol. Soc. America Bull., v. 72, p. 703-720.

———, and Engel C. G., 1969, Ultramafic and basaltic rocks dredged from the nearshore flank of the Tonga Trench: Geol. Soc. America Bull., v. 80, p. 1373-1378.

———, and Hess, H. H., 1963, Trenches, *in* Hill, M. N., ed., The sea, v. 3, The earth beneath the sea: New York, Wiley-Interscience, p. 411-436.

———, and Raitt, R. W., 1962, Topography and structure of the Peru-Chile Trench: Deep-Sea Res., v. 9, p. 423-443.

———, and Revelle, R., 1955, The trenches of the Pacific: Sci. Am., v. 193, p. 36-41.

Griggs, D., 1939, A theory of mountain-building: Am. Jour. Sci., v. 237, p. 611-650.

Grow, J. A., 1973, Crustal and upper mantle structure of the central Aleutian Arc: Geol. Soc. America Bull., v. 84, p. 2169-2192.

Gunn, R., 1947, Quantitative aspects of juxtaposed ocean deeps, mountain chains, and volcanic ranges: Geophysics, v. 12, p. 238-255.

Hayes, D. E., 1966, A geophysical investigation of the Peru-Chile Trench: Marine Geology, v. 4, p. 309-351.

———, and Ewing, M., 1971, Pacific boundary structure, *in* Maxwell, A. E., ed., The sea, v. 4, pt. 2, New concepts of sea floor evolution: New York, Wiley-Interscience, p. 29-72.

Hess, H. H., 1948, Major structural features of the western North Pacific, an interpretation of H.O. 5485, bathymetric chart, Korea to New Guinea: Geol. Soc. America Bull., v. 59, p. 417-445.

Hilde, T. W. C., and Raff, A. D., 1970, Evidence of a plunging ocean crust beneath the shoreward slope of the Japan Trench from seismic reflection and magnetic data (abstr.): EOS (Trans. Am. Geophys. Union), v. 51, p. 330.

Karig, D. E., 1973, Comparison of island arc-marginal basin complexes in the northwest and southwest Pacific, *in* Coleman, P. J., ed., The western Pacific: island arcs, marginal seas, geochemistry: Nedlands, Univ. Western Australia Press, p. 355-364.

Ludwig, W. J., Ewing, J. I., Ewing, M., Murauchi, S., Den, N., Asano, S., Hotta, H., Hayakawa, M., Asanuma, T., Ichikawa, K., and Noguchi, I., 1966, Sediments and structure of the Japan Trench: Jour. Geophys. Res., v. 71, p. 2121-2137.

Matthews, D. J., 1939, Tables of the velocity of sound in pure water and sea water for use in echo-sounding and sound-ranging (2nd ed.): London, Admiralty Hydrog. Dept. Publ. 282, 52 p.

Nichols, H., and Perry, R. B., 1966, Bathymetry of the Aleutian arc, Alaska: Environ. Sci. Serv. Admin. Monogr. 3.

Petelin, V. P., 1964, Hard rock in the deep water trenches of the southwestern Pacific Ocean, *in* Geology of Oceans and Seas, v. 16: Internatl. Geol. Congr., 22nd Sess. Repts. Soviet Geologists: p. 78-86 (in Russian).

Raitt, R. W., Fisher, R. L., and Mason, R. G., 1955, Tonga Trench: Geol. Soc. America Spec. Paper 62, p. 237-254.

Ross, D. A., and Shor, G. G., Jr., 1965, Reflection profiles across the Middle America Trench: Jour. Geophys. Res., v. 70, p. 5551-5571.

Scholl, D., von Huene, R., and Ridlon, J. B., 1968, Spreading of the ocean floor: undeformed sediments in the Peru-Chile Trench: Science, v. 159, p. 869-871.

———, Christensen, M. M., von Huene, R., and Marlow, M., 1970, Peru-Chile Trench sediments and seafloor spreading: Geol. Soc. America, Bull., v. 81, p. 1339-1340.

Shor, G. G., Jr., and Fisher, R. L., 1961, Middle America Trench: seismic refraction measurements: Geol. Soc. America Bull., v. 72, p. 721-730.

Sykes, L. R., Oliver, J., and Isacks, B., 1970, Earthquakes and tectonics, *in* Maxwell, A. E., ed., The sea, v. 4, pt. 1, New concepts of sea floor evolution: New York, Wiley-Interscience, p. 353-420.

Tayama, R., 1950, The submarine configuration off Shikoku: Bull. Hydrol. Office Japan, v. 7, p. 54-82 (in Japanese).

Udintsev, G. B., 1955, Topography of the Kuril-Kamchatka Trench: Tr. Inst. Okeanol. Aka. Nauk S.S.S.R., v. 12, p. 16-61 (in Russian).

———, 1960, Bottom relief of the western part of the Pacific Ocean: Oceanol. Res., v. 2, p. 5-32 (in Russian).

———, ed., in press, Geological-geophysical atlas of the Indian Ocean (IIOE): Moscow, Principal Administration, State Dept. Geodesy and Cartography (GUGK), pl. 41.

Vening Meinesz, F. A., 1954, Indonesian Archipelago: a geophysical study: Geol. Soc. America Bull., v. 65, p. 143-164.

Zeigler, J. M., Athearn, W. D., and Small, H., 1957, Profiles across the Peru-Chile Trench: Deep-Sea Res., v. 4, p. 238-249.

Part III

Transition from Continent to Ocean

Continental Margins, Freeboard and the Volumes of Continents and Oceans Through Time

Donald U. Wise

INTRODUCTION

Evidence is evaluated for freeboard of continents (relative elevation of continents with respect to sea level) as a function of time. Egyed's interpretation of steady continental emergence with time, based on changing areas of flooding shown on global paleogeographic atlases, seems unfounded on the grounds of inherent biases in the original maps, biases associated with changing time segments between successive maps, and by comparison with a freeboard versus time plot for North America compiled from Schuchert's more detailed atlas. Instead, a constant freeboard model or possibly one with long-term equilibria seems more appropriate. The North American plot is used as a basis for a quantitative estimate of the time distribution of deviations in freeboard. For over 80% of post-Precambrian time, freeboard has remained within ±60 m of a normal value 20 m above present sea level.

An equilibrium freeboard model of the earth is suggested with various feedback mechanisms continually maintaining this fine adjustment between volume of ocean basins and volume of ocean waters. From the model, a number of calculations and implications are drawn for continental and oceanic accretion, as well as for some rate relations in a global tectonic svstem. It is argued that approximately the present volumes and areas of continents and oceans have obtained over the last 2,500/my, producing neither net accretion nor retreat of the continental margins on a global scale over this time span. Instead, the margins should be considered as the focus of operation of a constant-volume system serving to weld escaping sediments back onto continental rafts of constant area and constant thickness with possible shifts in the equilibrium level modifying the details of the rewelding process.

PREVIOUS CONSIDERATIONS

Continental and oceanic evolution theories are numerous, commonly contradictory, and only rarely are they based on unequivocable facts. The implications of theoretical volumetric changes with time are profound for the long-term behavior of continental margins.

This paper is an attempt to reverse the logic process by using data of continental margins and freeboard (Kuenen's 1939, concept of relative elevation of continental platforms above sea level) to place some limits on the evolutionary theories. It is an examination of the tectonic implications of a simple fact: at present there is just about enough water to fill ocean basins.

In the static continents and ocean basins of the early twentieth century, the relationship would have reflected a most fortunate coincidence between the geometry of those basins and the amount of primordial volatiles captured by the earth in her fiery youth. This concept was critically discussed by Rubey (1951) and Brown (1952). Other theoretical degassing rates were proposed to form the present secondary atmosphere, hydrosphere, and excess volatiles of Rubey (1951). Simultaneously, remarkable growth rates were proposed for continents which varied with author and year from linear (Engel, 1963) to exponential (Hurley and Rand, 1969).

Some of the water displaced by continental growth could have been ingested back into the earth (Chase and Perry, 1972). The remainder was accommodated by an almost unbelievable expansion of the earth at rates of radius increase ranging from 0.5 mm/yr since the Cambrian (Egyed, 1956a, 1956b) to doubling since early Mesozoic, or 6 cm/yr (Carey, 1958). Egyed's rate increased the area of ocean basins so fast as to overcompensate for displaced waters and to produce a falling sea level with time. Finally, in the 1960s, the "new tectonics" began to play an important role, and the unlikelihood of catastrophic expansion was discussed by Francis Birch (1968).

Considering the possible diversity of behavior of almost every component of the system, the present nice adjustment of ocean waters to ocean basins seems very remarkable. It suggests that one must examine the present system for some highly effective but little recognized controls of global sea level through time. It is the thesis of this paper that such controls exist, that the continental margins are the focus of their operations, and that the long-term history of earth's continental margins should be interpreted as adjustments in a global tectonic equilibrium.

THE FREEBOARD CURVE

Regressions and transgressions of the sea representing freeboard variation have been recognized for a long time by a succession of geologists: Schuchert (1916), who presented a first plot of areas of North America emergent as a function of time; Stille

(1924), with his ideas of periodic transgressions and regressions; Grabau (1933), with a theory of pulsating sea levels; Kuenen (1939), who placed some quantitative estimates on the magnitude of eustatic changes; and Umbgrove (1939, 1947), who stamped his book's title "Pulse of the Earth," on these eustatic changes. All interpreted the record as one of essentially constant freeboard.

With the advent of two global paleogeographic atlases, by Strakhow (1948) and by Termier and Termier (1952), Egyed (1956a) produced a plot of areas of flooding shown on those maps versus time. The commonly reproduced Egyed freeboard curves, Figure 1, suggest a global emergence of the continents. Fitting the first few Paleozoic points to the Termier plot (Fig. 1) seems difficult, a problem which Holmes (1965, p. 969) resolved by assuming that required, but undiscovered, areas of Cambrian and Ordovician rocks were in the Antarctic and hence did not appear on the original atlases. These emergence curves are the basis of Egyed's hypothesis of an earth expanding at 0.5 mm/yr, increasing the area and volume of ocean basins at a rate adequate to cause the apparent secular drop in sea level.

Egyed's data of Figure 1 must bear some relationship to the late Precambrian, a time of general emergence and widespread planation of continents. In Figure 2 fictitious data points suggesting this older time of emergence have been added to Egyed's data points. The need for discovery of a few tens of millions of square kilometers of lost Cambrian sediments in the Antarctic seems somewhat diminished with the perspective of Figure 2. Egyed's biased line of Figure 1A could be replaced by many other lines, for example the biased lines superimposed on Figure 2 as average flooding for selected shorter time segments.

Unfortunately, measurement of secular change of freeboard by planimetry of paleogeographic maps is fraught with pitfalls and biases. (1) The record becomes increasingly blurred by erosion, burial, and metamorphism as a function of time. (2) Continental and upland deposits are more subject to erasure by erosion than deposits of marine basins. (3) Methods used do not take palinspastic considerations into account, an effect that is minor in cratonic areas but may be significant in orogenic belts. (4) Use of global atlases ensures inclusion of little-known regions and large segments of time so that shorelines become simpler and extrapolations grander as a function of age. As age increases, the atlas maker's bias toward a marine or a continental world is less and less hindered.

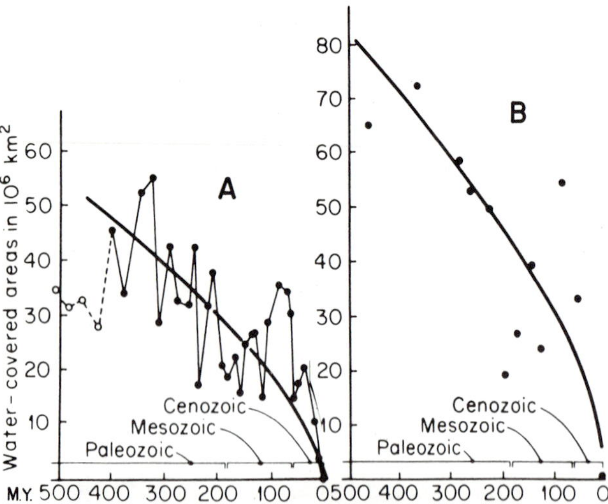

Fig. 1. Area flooded versus time after Egyed (1956a) for two global paleogeographic atlases. (A) Egyed plot based on Termier and Termier Atlas (1952). (B) Egyed plot based on Strakhow Atlas (1948).

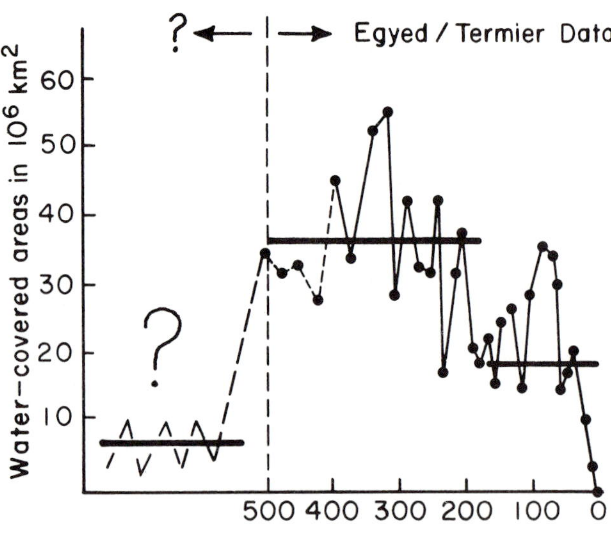

Fig. 2. Egyed data identical to Figure 1A viewed with the added perspective of late Precambrian continental emergence. The heavy lines are only a few of the many possible highly biased "average" lines that might be drawn through the data.

A more subtle pitfall is the length of time unit chosen for individual paleogeographic maps (Wise, 1972, 1973). In the extreme, one can imagine a map of North America with all areas containing any existing or inferred marine sediment of Cambrian to Recent age encircled as part of a flooded region. This paleogeographic map marked "Phanerozoic" would indicate 100% flooding of North America. With this phenomenon in mind, Egyed's curves for flooding can be reexamined. The Strakhow Atlas, with Phanerozoic time broken into 11 maps, shows a systematic flooding 30-40% greater at any given time than the Termier Atlas, with its 34 maps for the same overall time span (Fig. 1A and B). In addition, the slicing of time units on successive maps of both these global atlases becomes longer with increasing age (Table 1). It is obvious that a correction factor for increased apparent flooding with longer time segments would reduce the apparent Paleozoic flooding and decrease the slopes of the curves.

A method of reducing some of these inherent biases in Egyed's global treatment is the use of a single continent for which good control and finely sliced paleogeographic time segments are available. Schuchert's monumental paleogeographic work on North America resulted in an atlas of the continent (Schuchert, 1955) with 84 successive paleogeographic maps. Some details on some of the maps are certainly open to revision, but on the whole they are among the more complete coverages of this type for an extensive region of the world. The nearly uniform length of time segments on this atlas contrasts sharply with the changing length of time segments on the global atlases as indicated in Table 1.

Extent of flooding on each of the 84 maps of the Schuchert North American Atlas (1955) was measured to produce the curve of Figure 3 for the North American freeboard, plotted as a function of time using the revised (Harland et al., 1964) Holmes' time scale. The area measurements included the Gulf of St. Lawrence, Hudson Bay, and the Arctic Archipelago as having reasonably reliable geologic control but excluded all continental shelf data on the maps as being too speculative. Using this definition, North America is presently 8.9% flooded; variations have ranged from 2 to 39% flooding since the Precambrian, with an average value of 14.9% flooding. This curve has similarities to Schuchert's own curve (1916) of continental area versus time. However, his later refinements of the maps make the regressions of the Silurian, Mississippian, Triassic, and Cenozoic appear much more prominent (Fig. 3) than on his older curve.

The extent to which the North American freeboard, or eustatic tidal gauge, reflects global sea level is explored in Figure 3. On it, Egyed's plot of the Termier atlas is readjusted to the new time scale, and his areas are shown as a percentage of present continental area flooded. All points are plotted at the midpoints of their time span. Also included is Stille's (1944) qualitative and debatable curve estimating global flooding and regressions based on paleogeographic analyses. Various ages and cycles are included along the left margin. The curves, for the most part, are easily correlated one with another even to many of their details. The three great cycles of European geology have their orogenic separations marked by major continental emergence, a fact long recognized by Stille (1924), Bucher (1933), and Umbgrove (1947).

Many of Sloss's (1963) North American cratonic sequences, marked by regional emergences and unconformities, are detectable in the global plots to an extent at least as great as the traditional time boundaries between standard geologic periods. There are also differences between the North American and the global plots (the Sauk sequence seems either shifted or nonexistent on the global plots; the North American Cenozoic seems to have a greater average emergence than the global plots). On the whole, the oscillations of the North American freeboard plot seem to be a slightly more detailed version of a global pattern.

Table 1. Average number of years (in millions) between successive maps in several paleogeographic atlases.

	Schuchert (1955): North America	Strakhow (1948): Global	Termier and Termier (1952): Global
Cenozoic	7.5	30.0	10.0
Mesozoic	8.9	40.0	14.6
Paleozoic	6.2	60.0	21.2

Slope of the line of average flooding (Fig. 3) based on Schuchert's (1955) maps is only one-third that of the Termier-Egyed line (0.010 versus 0.032). An Egyed-style interpretation of this difference between the slopes of the global and the North American plots could include an expanding earth and falling average global sea level. However, within this Egyed system North America would have been sinking systematically with respect to other continents in order to slow its apparent rate of emergence by two-thirds. A simpler explanation is that detailed examination of one continent has eliminated two-thirds of the biases inherent in a "broad brush" global treatment. It is reasonable to expect that significant biases remain in a continent-wide treatment so that further detail might tend to reduce the remaining minimal slope of the line for North America.

I consider the above Egyed-type explanation of the North American curve too cumbersome, the late Precambrian emergence too obvious to neglect, and the biases inherent in Egyed's method of global atlas analyses too great to overthrow the conclusion of generations of geologists from Schuchert and Stille through Kuenen and Umbgrove to Hess that the geologic record is one of essentially constant freeboard through time.

REFINED FREEBOARD CURVE

In an attempt to reduce further the biases inherent in Egyed's curves, an area was selected bounded approximately by Big Bend National Park; Las Vegas, Nev.; Salt Lake City, Utah; Helena, Mont.; Dawson Creek, B.C.; Lake Athabaska, Alberta; Isle Royale National Park; the NE Adirondacks; Tuscaloosa, Ala.; Shreveport, La.; and Big Bend, Texas. This area of the craton, reasonably well known in Schuchert's day and reasonably undeformed, was planimetered on the Schuchert paleogeographic map series to produce the refined freeboard curve (shaded on Fig. 4).

With the use of transparent overlays, stacks of Schuchert maps were combined into larger time segments, duplicating the time spans used by Egyed for the Termier and Strakhow atlases. Planimetry of these combinations produced the indicated curves of Figure 4, plus a curve for more traditional period-by-period time slices. The systematically greater flooding with larger time segments is obvious.

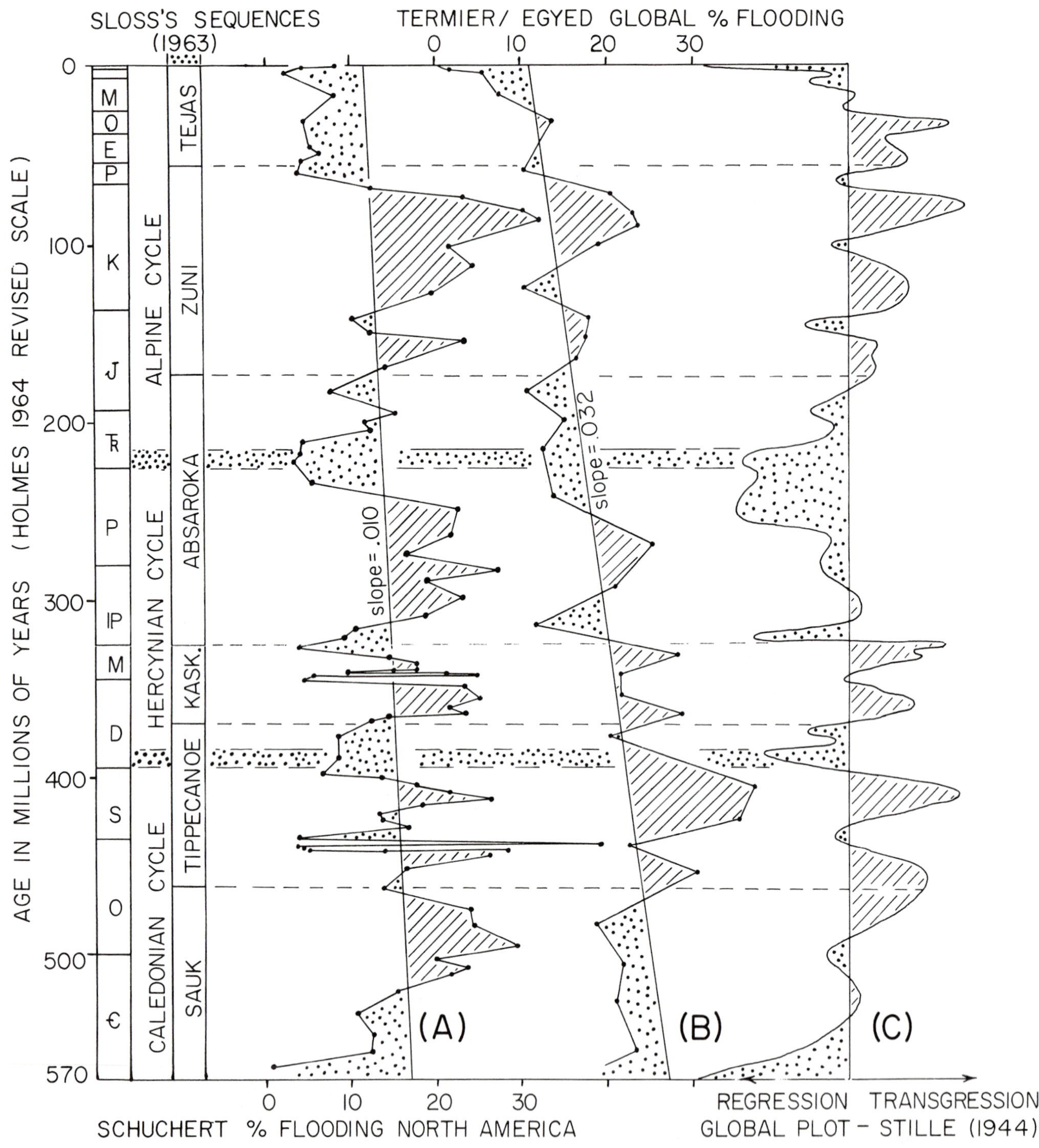

Fig. 3. Three versions of plots of freeboard versus time. (A) Schuchert-Wise plot for North America; (b) Termier-Egyed global plot with Egyed figures changed from actual area flooded to percent flooding of surficial area of continents; (C) Stille (1944) qualitative plot of transgressions-regressions with no scale implied. The lines of least-squares fit of the data are regarded as average sea level. As discussed in the text, the apparent slopes are interpreted as biases inherent in the data.

Additonal refinements are possible using other methods and data to derive a further improved freeboard-versus-time curve. Methods, such as those of Sloss (1972), using details of individual basins, and of Rona (1973), using steadily sinking continental margins, may hold promise. The present condition of this refined curve (Fig. 4) is adequate to place a few major limits on the system, but inadequate to probe many of the details. An important short-term goal in tectonics of the International Geodynamics Program might be the development of the best average trace of this curve of global sea level versus time from all continents and ocean basins. The curve is a tectonic glue, capable of holding together a wide variety of hypotheses and geologic observations. It is one of the more readily

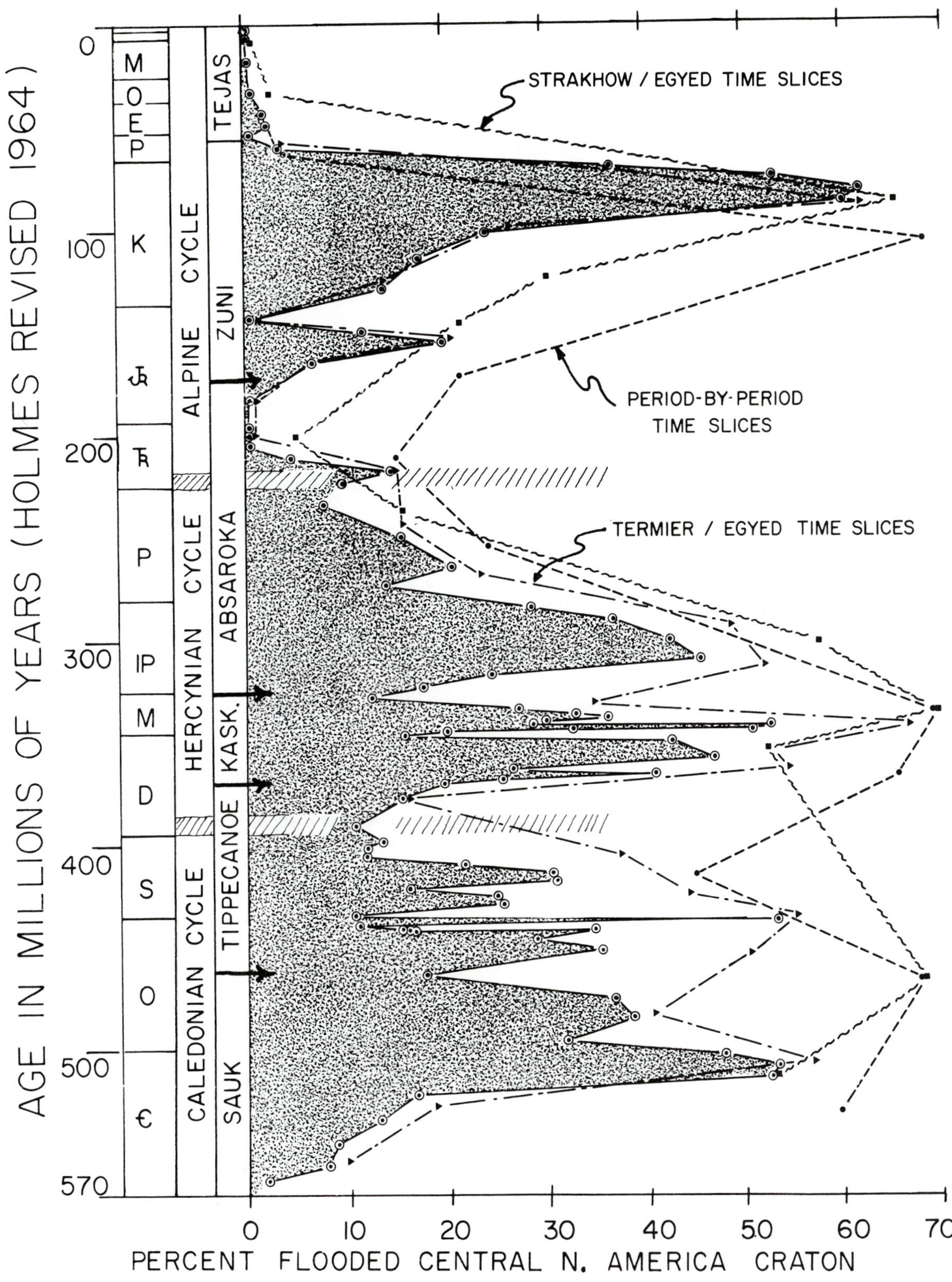

Fig. 4. Refined freeboard curve (shaded) for stable cratonic interior of United States and southern Canada. The area, defined in the text, was measured on Schuchert's map series in the same way as for Figure 3. The same maps were stacked for time spans similar to those used by Egyed for the Strakhow and the Termier atlases and for simple geologic period-by-period divisions. Total flooding areas were measured for these time slices and plotted as the unshaded curves indicated above. The effect of greater apparent flooding with cruder time slices is obvious.

obtainable facts linking the diverse elements of the global tectonic system.

RELATION OF SEA LEVEL TO PERCENT FLOODING

Kuenen (1939) first related the area of flooding shown on paleogeographic maps to hypsometric curves of continents to conclude that the magnitude of marine transgressions-regressions required at least 40 m of eustatic change. To apply similar methods to the Schuchert-derived curve of Figure 3 requires the construction of a hypsometric curve for North America keyed to the same flooding definitions as used for Figure 3. This plot (Fig. 5) is adjusted to have present sea level coincide with 8.9% flooding, a value defined in constructing Figure 3 as the present flooding of North America.

If the freeboard is constant, the slope of Figure 3A can be removed, and all measurements can be made as variations from the average flooding line, which intersects the present time at 11.8%. The 11.8% applied to Figure 5 suggests that present sea level is about 20 m below normal continental freeboard. This estimate might be modified by a decrease in area of about 5% by the continued isostatic rise of Hudson Bay, and decreased by about 10% for the potential 65 m of sea level rise from ice currently locked in the polar caps (Holmes, 1965). The net gain of 5% additional flooding would correspond to the 30 m sea level rise on Figure 5. Thus present freeboard with the shorter-term effects of glaciation removed would be very close to the average for the Phanerozoic.

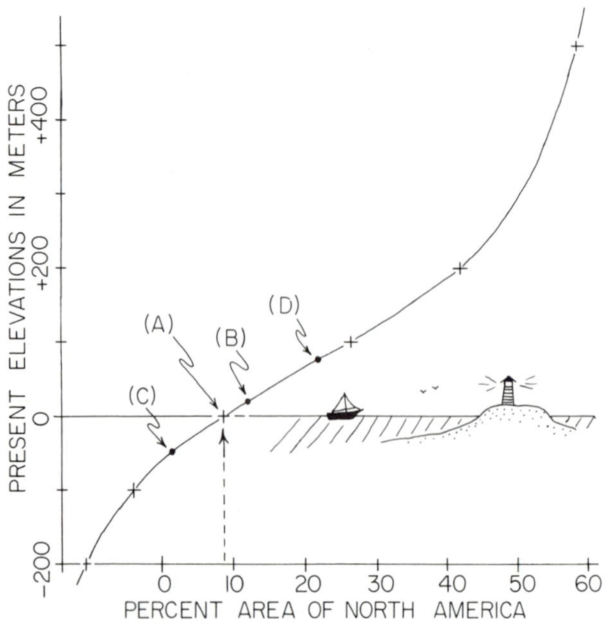

Fig. 5. Relation of area flooded to sea-level changes in present North America. The percentage areas correspond to the same total area as defined for the Schuchert paleogeographic maps, a definition under which North America is presently flooded by 8.9% (continental shelves are ignored). Accordingly, the horizontal scale is shifted so that 8.9% (A) corresponds to present sea level. Least-squares fit of the North America line at 11.8% flooding from Figure 3A is indicated by (B), with deviations of flooding of +10% from that line indicated by (C) and (D).

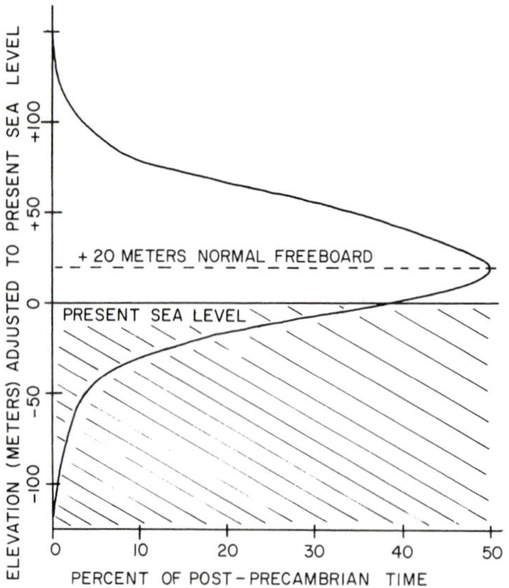

Fig. 6. Deviations in North American freeboard as a function of time.

Figure 5 suggests that the maximum 39% flooding of Figure 3 would correspond to a sea-level rise of 180 m. The apparent minimum value of 2% flooding, corresponding to a drop of 50 m, is a false value in that the paleogeographic definitions used for Figure 3 do not record extreme sea level drops in excess of 60 m.

The distribution of the deviations in flooding from the average freeboard line of Figure 3A can be measured for small increments of time. These flooding deviations can be converted to sea-level variation using Figure 5 and replotted in cumulative form as in Figure 6. The plot suggests that for 80% of Phanerozoic time, oscillations have remained within ± 60 m of a normal freeboard level about 20 m above present sea level. The values for percent flooding are derived from maps representing time segments averaging 7 my so that much larger variations are possible in the short term. Use of Figure 6 must be tempered by the inherent and debatable assumptions that the present hypsometric curve of North America is representative of the past, that the freeboard curve of North America is representative of the other continents, and that on the average constant freeboard has obtained over the Phanerozoic. These concepts are fundamental to the interpretations presented here.

Converted to volumes of water, the approximately 60 m general range in freeboard corresponds to about 1.6% of the volume of oceans. The extreme Ordovician flooding, with its 39% coverage and 180 m rise above present sea level, corresponds to a deviation of 4.2% of ocean volume. Thus the constant freeboard data suggest that for the last half-billion years, ocean waters have just about filled ocean basins, relative to continental masses, the maladjustment being within 2% and only rarely and briefly reaching 4%. Pitman and Hayes' (1973) suggestion of a 500 m deviation based on Cretaceous oceanic ridges is discussed later.

The North American freeboard curve (Fig. 3) also suggests that there may be a maximum time period of 60 to 70 my before the system is again brought back to adjustment. The three longest flooding cycles are 65, 73, and 52 my; the three longest complete emergence cycles are 65, 20, and 35 my, whereas the partially completed (?) Cenozoic cycle is 65 my; the incomplete lower Cambrian cycle is 570 my. The shape of most of these major deviations is relatively symmetric, suggesting that the system is not one of sudden perturbation and slow return to normal but one of slow perturbation and slow return.

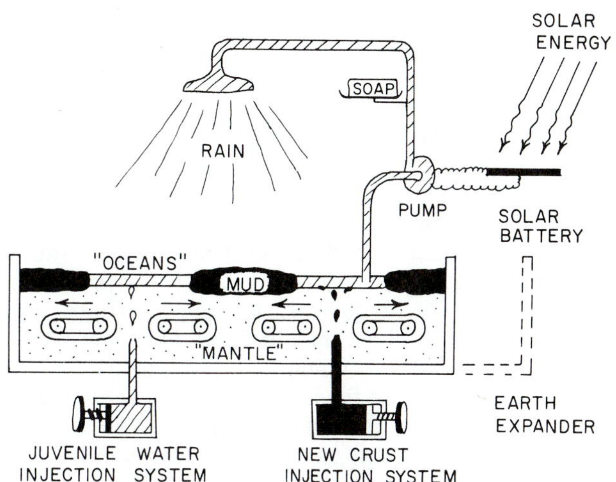

Fig. 7. Possible constant freeboard, global tectonic model, illustrating feedback mechanisms maintaining the volume of ocean basins equal to the volume of ocean water.

CONSTANT FREEBOARD MODEL

If one has a static view of the earth, a constant freeboard with time seems anomalous on a globe that might be evolving both surface water and crust with time, eroding continents at a geologically catastrophic rate (Judson and Ritter, 1964), and dumping debris into the oceans to displace additional water. If continents are not to submerge with time, a dynamic model is demanded with massive self-adjusting or negative feedback systems, continuously regulating the volume of ocean basins to coincide with the volume of ocean waters, an adjustment that the preceding section suggests has been maintained to within about 2% of volume over most of Phanerozoic time.

A possible constant-freeboard model capable of these self-adjustments is illustrated in Figure 7 (the debt to Harry Hess and a host of subsequent workers is obvious). It is designed fundamentally as an illustration of principles involved in constant freeboard feedback mechanisms, following more closely the model philosophy of Rube Goldberg than the model scaling of M. King Hubbert (1937). The earth's surface area is represented by the surface area of materials in a tank; a movable end permits "earth expansion." Mud "continents" are supported by a viscous paraffin "mantle" and separated by water-filled "ocean basins." A floating or isostatic link, dependent on the relative density contrasts of mud and water with respect to "mantle," controls the ratio of thickness of "continents" to depth of "oceans." Submerged conveyor belts shift crustal plates away from basins, causing continents to collect at collision points. A solar-powered "hydrologic cycle" is capable of eroding the continents to the limiting depth of sea level and spreading the debris into the ocean basins. Provision is made for injecting new water or crust through the mantle to the surface, whereas minor amounts of surface water and crust may be ingested back into mantle at the descending collision points of the plates. Speeds of operation of the various components of the system are obviously variable.

As an example of the possible operation of the model, we can use the most extreme (and probably least likely) of the possible changes, namely "global expansion," by movement of the tank wall. This increases the area and hence the volume of ocean basins, leaving the continents standing higher, the typical Egyed model. Increased continental elevation increases exposed area and stream gradients. This presumably increases regional rates of erosion and the resulting debris builds continents laterally at sea level, displacing ocean waters, and raising average sea levels. Rising sea levels presumably reduce rates and areas of erosion. The ultimate result will be a return to equilibrium, at constant freeboard, with thinner, broader continents and shallower, wider oceans having the same ratio of continental depth to oceanic depth as before (in this overly simplified isostatic system), so that the volume of ocean basins again equals the volume of ocean waters.

The reader is invited to play with the model mentally, making any of the possible changes to demonstrate for himself that any perturbation will result in short-term change in freeboard of the model but will trigger long-term feedbacks, thereby returning the system to normal. Some of the possible changes or limitations on the system are included in the work of Armstrong (1968), Birch (1968), Dietz and Holden (1966), Khain and Muratov (1969), Maxwell (1968), Menard (1964), Weertman (1963), and Pitman and Hayes (1973).

LONG-TERM EQUILIBRIUM FREEBOARD MODEL

Even the rigid concept of a constant freeboard must permit some deviations for limited time spans.

Deviations from the constant freeboard line in Figure 3A range up to 70 my. A less rigid and perhaps more realistic model might include even longer periods of deviation from an absolutely constant level. If the refined freeboard curve of Figure 4 could include the apparent long period of late Precambrian emergence, the Paleozoic might well stand out as an era of a few hundred million years of extensive average flooding with somewhat similar length time spans of general emergence preceding and following it. Figure 2, although drawn with tongue in cheek, has this idea of very long term variations in average freeboard implicit in it.

A more realistic freeboard model might involve a truly constant freeboard only when averaged over several of these deviation cycles lasting hundreds of millions of years. Within a cycle, long-term freeboard equilibria levels would be established for time spans on the order of 50 to 100 my. Shorter-term perturbations could drive the system away from the general equilibrium level. The most extreme instance of deviation from the long-term equilibrium is the Cretaceous flooding. That peak, on Figure 4, is interpreted following the model of Pitman and Hayes (1973), as the result of water displacement in a period of extremely high rates of oceanic ridge formation, far in excess of the control capability of the more sluggish feedback mechanisms.

The proposed model of long-term freeboard equilibrium cycles is clearly a simplification, but one that may lead into some useful future geologic models. For instance, one might develop the concept of levels of Paleozoic freeboard equilibria permitting the continents to be used as the major dumping place for carbonates carried into the oceanic system. The post-Permian shift of the equilibria into a new major cycle subjected to erosion the carbonates deposited at the earlier equilibrium levels, feeding additional carbonates into the oceanic system, simultaneously eliminating the former cratonic dumping areas. The result could be profound geochemical changes in the system, perhaps forcing the relatively abrupt geochemical shift in $CaCO_3$ content of shales or marked contrast of Mg/Ca ratio of limestones versus Na/Al ratios of shales of the Paleozoic versus the Mesozoic-Cenozoic (Garrels et al., 1972).

The slowly shifting equilibrium model also forces us to ask what kinds of changes could have produced such profound long-term shifts in global sea level pattern. Following the model of Figure 7, a plausible reason might be that processes of rebuilding continents were more effective after the Permian continental breakup than before, as a result perhaps of more highly active drift and plate tectonics. Whether this effectiveness was caused by actual differences in rates or patterns of mantle motion or was a reflection of a basic change away from supercontinent geometry, with fewer margins for active subduction, is mere speculation. If the post-Permian effectiveness of rebuilding of continents did differ, greater areal exposure and higher stream gradients would be required to keep pace. Accordingly, new freeboard equilibria with more emergent continents would be established for the new cycle.

To be placed on solid footing the multistage equilibrium model needs much better data for the freeboard curve, with detailed Fourier analysis of those data. The apparent pattern of very long time spans of one set of freeboard equilibrium levels, interrupted to begin a new long-term cycle with markedly different equilibrium levels, has a curious similarity in its timing to the tectonic resetting of age dates on a global scale, with periodicites of a few hundred million years.

EVOLUTIONARY IMPLICATIONS OF A CONSTANT FREEBOARD MODEL

A constant freeboard model provides a convenient basis for visualizing interdependence of some tectonically interesting variables discussed in the introduction: area of earth, A_e; area of continents, A_c; area of ocean basins, A_o; depth of continents, D_c; depth of oceans, D_o; volume of continents, V_c; volume of oceans, V_o; and the isostatic link, I_l, relating D_o to D_c. These eight variables are governed by four basic equations: (1) $A_e = A_c + A_o$, (2) $V_o = A_o \times D_o$, (3) $V_c = A_c \times D_c$, (4) $D_c = I_l \times D_o$. One of the variables, A_e, is ordinarily assumed constant, leaving seven variables, of which three are independent: V_o, V_c, and I_l.

The last variable, the isostatic link or I_l, is probably the most troublesome, requiring some geologic assumptions about the behavior of crust and of mantle layerings and densities. A simple model for I_l includes constant-density, single-layered continents in isostatic balance with an oceanic crust of constant thickness and density at present values through geologic time. If we use Hess's (1962, Fig. 2) crustal thicknesses and average densities with a constant 0.3 km of continent above sea level, then 9.1 km of continental crust is needed to counterbalance the oceanic crust and sediments with no water present. Beyond this thickness an increase in continental Moho depth must be balanced by increase in water depth, controlled by the ratio of density contrasts of water and continent with respect to mantle. Accordingly, a simple isostatic link using Hess's model would be:

$$D_c = 9.1 \text{ km} + 4.98\ D_o.$$

A more sophisticated link might include a two-layer continent for which additional assumptions of layer thickness, density, and method of variability would be required. Additional complexities of phase change limits on Moho, variable density mantle, and variable thickness or densities of multiple continental and oceanic layers are possible. Under any circumstances, some isostatic link of D_o to D_c must be devised until such time as past values of the other variables are known to a perfection such that I_l can be calculated from them.

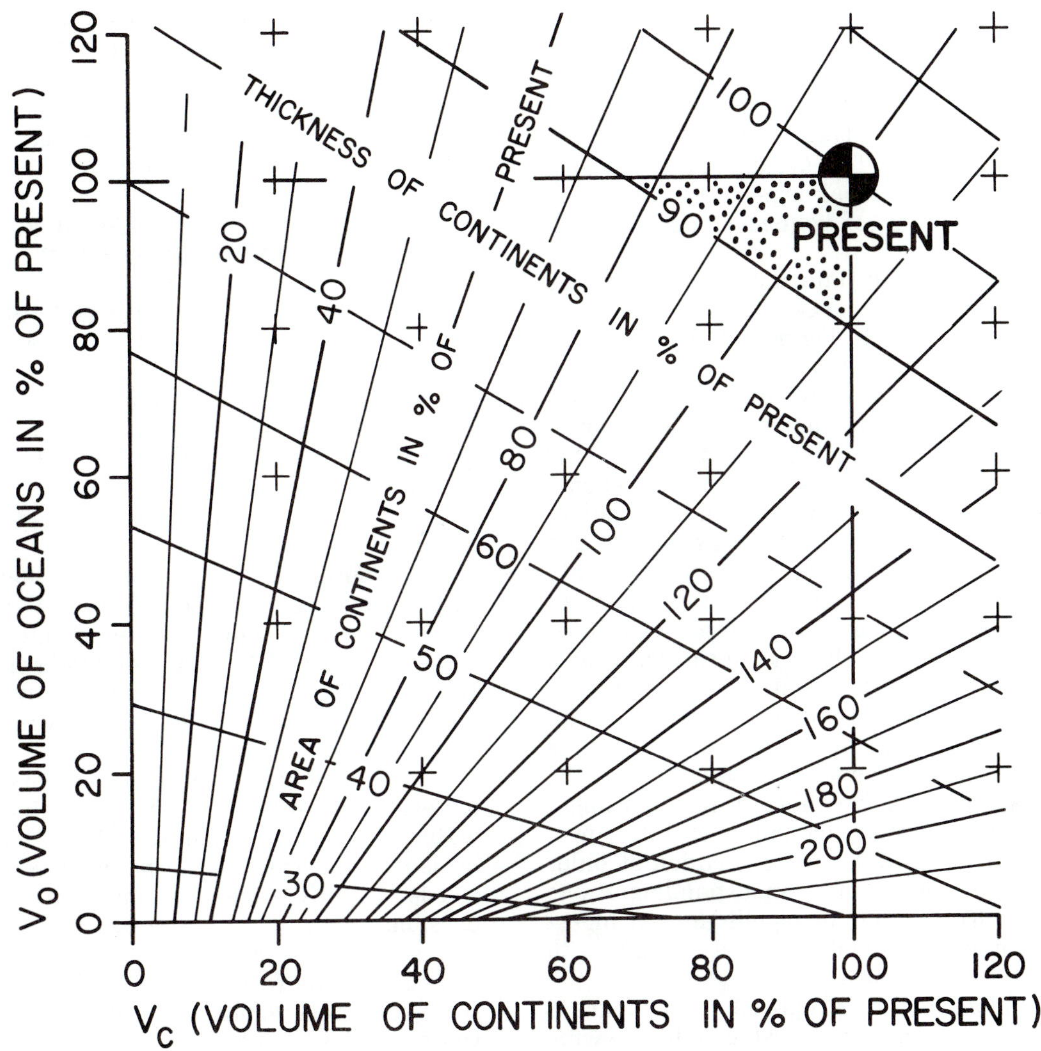

Fig. 8. Graph of solutions of the constant freeboard equations. Assumptions are constant area of earth, constant thickness of oceanic crust, and simple isostatic link given in the text. All values are percentages of present values, here defined as $A_C = 2.07 \times 10^9\,\mathrm{km}^2$, $D_O = 5$ km, $D_C = 34$ km, $V_C = 7.04 \times 10^{10}\,\mathrm{km}^3$, and $V_O = 1.52 \times 10^{10}\,\mathrm{km}^3$. The use of normal oceanic depth as 5 km (for isostatic reasons) results in a V_O 10% larger than the presently accepted values.

If simple I_l and constant A_e assumptions are made, the system is soluble for any past time at which two additional variables or certain ratios of variables are known. Solutions of the four basic equations are plotted in Figure 8 using the above assumptions, Hess's average depth of continents and oceans, and the division in area between continents and oceans at the 2-km isobath. All values are plotted as a percentage of present value.

If the reader believes that the volume of continents and of oceans has not decreased with geologic time, then all past values of the system should lie below and to the left of the "Present Values" point of Figure 8. The minimum D_C of 26% of present is defined by the requirement of isostatically balancing oceanic crust. Depending on the mode of formation of the earliest continents, this lower limit may or may not apply. There are many possible paths on Figure 8 leading from the primordial earth to the present 100% for all the variables. Insofar as the assumptions behind Figure 8 are valid, then values of areas, depths, and volumes are uniquely defined at any point along any path chosen, if two of the variables are known at that point.

An example of the way a constant freeboard model may be used to check various types of data paths and assumptions against one another is presented in Figure 9, using the system of assumptions and values inherent in Figure 8. Hurley's (1968) linear increase in V_C is plotted on Figure 9A, with time starting at 3.8 by, along with Condie and Potts's (1969) estimate of a range of 10 to 25 km crust (or 30-70% of present) at 2.6 by (based on chemistry of a Canadian Shield rock suite). Their 30% lower limit of D_C falls on a linear increase (Fig. 9), whereas the highest curve drawn through their upper limit of D_C is constrained by an upper limit of 100% V_O (as drawn). Points on these curves of D_C and V_C can be used with Figure 8 to determine the corresponding curves for V_O and A_C. The A_C curve for these upper limits is indicated (solid line A_C). If one wished to lower the V_O curve to permit a continuing evolution of oceans, the D_C curve could be lowered and the A_C curve raised, as shown by the dashed lines. The resulting 50% A_C for 2.6 by (dashed line A_C) is in rough accord with that of North America as implied by Muehlberger et al. (1967, Fig. 13A) as the minimum area at 2.6 by. This system of assumptions would seem to have internal consistency, including a reasonable evolution of volume of oceans with time.

Examples lacking in internal consistency are plotted in Figure 9B and 9C. Hurley's linear V_C curve is included in each. Figure 9B shows Hurley and Rand's (1969) estimate of A_C based on areas of Precambrian age provinces, whereas Figure 9C has Engel's (1963) estimate of linear growth in the area of North America, here extrapolated to a global representation. Each of these plots seems reasonable in itself until one includes V_O and D_C, calculated with the same assumptions as above and derived from Figure 8. Under this system the data would require an additional assumption of major decrease in volume of oceans in later geologic time. In Figure 9B the Moho would have to rise with time. Model 9C was built with the attractive feature of constant D_C or shields being created and maintained at essentially constant thickness but requires a rapidly decreasing volume of oceans toward the present. Ways of avoiding these difficulties might include: choice of more sophisticated isostatic links, bias in original areal data by obliteration of older crust by younger overprints, an expanding A_e to avoid the V_O discrepancy, or significant net ingestion of oceans or continents into the mantle with time.

Similar plots can be prepared from Figure 8 and checked for internal consistency for a wide range of assumptions or measurements of continental and oceanic growth as a function of time. However, the presence of consistency among several theories should not be mistaken for truth. The process is merely a way of finding mutually compatible sets of theories and detecting their hidden implications for the rest of the global system.

CONSTANT FREEBOARD RULES AND COROLLARIES

Either by use of Figure 8 or by intuition with the physical model of Figure 7, a number of basic tectonic rules appear valid and independent of any evolutionary rate models one happens to fancy.

1. The ratio of continental to oceanic area is a function of the volumetric ratio of continents to oceans. Accordingly, continental accretion cannot be considered independently of oceanic volumetric evolution.

2. Continents and oceans are thickening with time if any net increase in new surface material takes place, be it either water or continental crust or both (provided that A_e remains constant). Conversely, if they are not thickening with time, no net increase in continents or ocean waters is taking place.

3. If concurrent net increase in continental material is not included with any addition of juvenile water, the volume of continents will be maintained by decreasing area of continents with time. In other words, negative area of continental accretion is possible.

4. Further areal accretion of continents is possible only if the ratio of new V_C to new V_O exceeds some ratio determined by isostasy and the existing volume ratios of these materials at the surface. If new material is added at precisely this ratio, the area of continents is constant. For the simple model

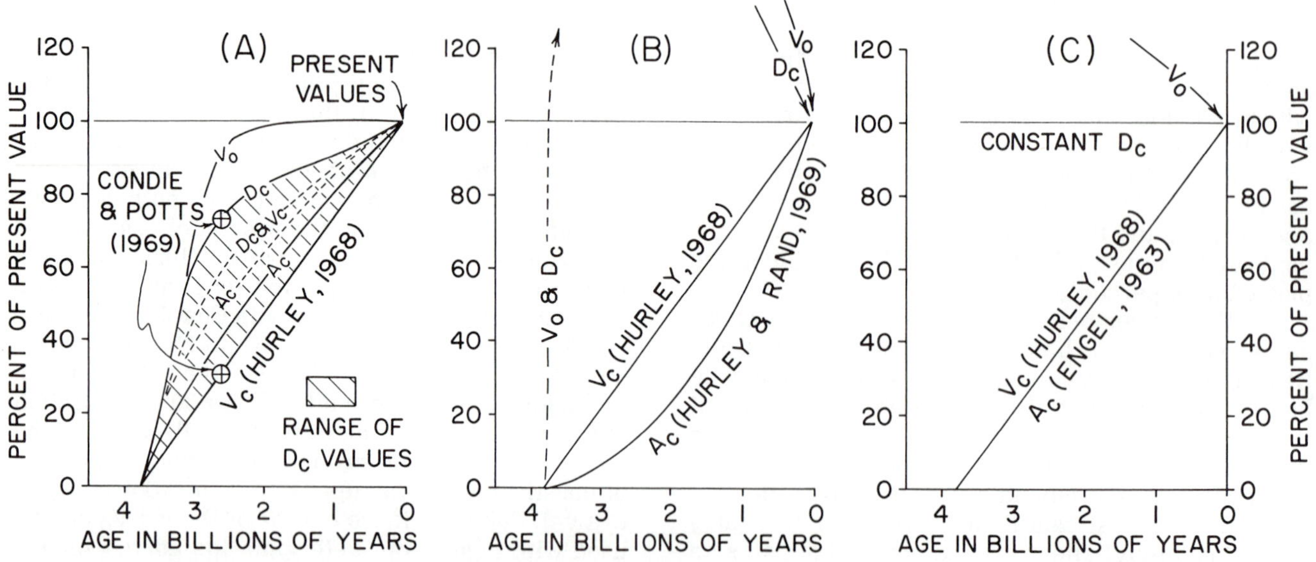

Fig. 9. Examples of internally consistent hypotheses (A) and mutually noncompatible hypotheses (B and C). The evolutionary curves suggested by various authors are indicated, and the additional curves are then derived from Figure 8. Note that B and C require precipitous decreases in volume of oceans toward the present time.

Table 2. Thickness of continents as a function of age of craton. Data extracted from Condie (1973).

Age (my)	0-225	225-600	600-1,200	1,200-2,500	+2,500
Number of seismic measurements	37	21	12	46	13
Average depth (km) to base of crust	38	40	39	39	38

of Figure 8, this ratio is 1 part new V_C to 2.87 parts new V_O in the present earth.

5. The model does not preclude a variable A_e with time. It merely says that on an expanding earth, areas of both continents and oceans will expand to maintain freeboard while depth ratios will be maintained according to isostasy. The change will be undetectable by ordinary paleogeographic flooding measurements. For this determination the fate of the average level of the sub-Cambrian unconformity might be of greater interest: if continents are to be progressively thinned, this unconformity will systematically be raised and eroded. The fact that the sub-Cambrian unconformity and some Phanerozoic sediments remain on most of the world's cratonic areas argues against progressive thinning of continents, and against significant Phanerozoic expansion of the earth.

In addition to these evolutionary-type implications of the constant freeboard model, another set of interdependences may be established based on rate phenomena.

1. Average rates of regional erosion (Judson and Ritter, 1964; Menard, 1961) over periods of a few tens of millions of years must equal the net rate at which continents are rebuilt, assuming negligible volume changes of crust and ocean on this time scale.

2. Long-term average elevation of continents will be determined by the gradients necessary to move material off a continent at the same long-term rate at which it is being rebuilt. For two continents enjoying the same rate of rebuilding, the more arid continent with its less efficient streams would stand higher in elevation to provide higher gradients for the streams.

3. Orogenic periods separating the great cycles of European geology seem to coincide with periods of maximum lowering of sea level on Figure 3, a relation recognized by many early workers. Gilluly (1949) raised many pertinent objections to Stille's (1944) numerous global orogenies affecting all continental geology. He argued instead for periodicity of orogeny when viewed for any single spot, but a constant orogenic rate when viewed for the entire globe. Nevertheless, the correlations among the various freeboard plots suggest that many aspects of ocean basin tectonics are recorded as unconformities or transgressions in local geologic records. In recent literature there have been many attempts to correlate changing rates of sea-floor spreading with a wide range of variations in geologic phenomena from sediment thickness on continental shelves (Rona, 1973) to rates of animal evolution (Valentine and Moores, 1972). The common link among such a variety of phenomena may be the freeboard changes generated by changing spreading rates rather than simply the spreading rates themselves.

4. Freeboard change associated with an extreme pulse of oceanic ridge development, as the one proposed by Pitman and Hayes (1973) for the extreme Cretaceous flooding, need not equal in volume the amount of new ridge produced. Even as the ridge is rising, the negative feedback processes in the system are attempting to restore the system to normal. The ultimate sea level reached becomes a function of the time constants of these feedback mechanisms as well as the time span of ridge development and decay. Pitman and Hayes's 10% rise in Cretaceous sea level is based on the assumption of active ridges operating in fixed-volume ocean basins. If the basins were simultaneously adjusting their volume, the actual value would be less than 10%, perhaps closer to the 4% suggested by the freeboard curves of Figures 3 and 4.

5. The long-term existence of continents above sea level is dependent on the operation of some system of this type. The presence of extensive continentally derived sediments in the Precambrian geologic record is evidence for past operation of this or a similar mechanism. Any suggestions of the absence of plate tectonics prior to 200 or 2,000 my should be tempered by the evidence of existence of ancient continents and the requirement of retaining their freeboard by one method or another.

PREFERRED EVOLUTIONARY MODEL

For me, the set of "facts" placing the most severe limitations on the system while standing the greatest chance of being substantially correct is Condie's (1973) compilation of 129 seismically determined thicknesses of cratonic areas in relation to the age of their last tectonic reworking. Condie's data, abbreviated as Table 2 of this paper, show that the present thickness of cratonic areas is independent of age. Of more debatable nature is Condie's use of a variety of geochemical indices on a large number of igneous rock suites of many ages to argue that present cratonic thicknesses are close to original thicknesses. His arguments are ingenious and, at the least, do not detract from the constant thickness conclusion derived from seismic and age data.

Another piece of evidence supporting constant thickness is Wynne-Edwards' (1969) quantitative study of areal extent of a variety of metasediments in a large area of the Grenville Province. He shows that sediments essentially at sea level on the pre-Grenville craton were run through the Grenville tectonism to return, on the average, to their original sea-level position. Isostasy would argue that the Grenville tectonic mill ground out cratons of the same thickness as did the preceding Hudsonian and Kenoran cratonic mills.

The data would suggest that cratonic crustal thicknesses for the last 2,500 my were at least 90% of present values. Constant continental thickness could be produced by the evolution rates of Figure 9C, which was calculated for this relation. Unfortunately, that model requires not only that the earth's oceans decrease in volume with time but that the decrease be precisely linked to the rate of evolution of new continental material from the mantle.

The most likely system would seem to utilize: (1) the above arguments that continents have been at least 90% of their present thickness for at least 2,500 my; (2) the assumptions of constant freeboard, constant area of earth, and the other assumptions inherent in Figure 8; (3) the assumption of no significant net ingestion of continents or ocean waters back into the earth in this time span. (In other words, volume of oceans and volume of continents in Figure 8 did not exceed 100 percent of present values.) Under these constraints, the global system had to remain within the stippled portion of Figure 8 for at least the last 2,500 my. The limitation on Figure 8 precludes changes in either area or volume of either continents or oceans by more than about 20% from present values during that time span. Recycling of oceanic or continental materials through the mantle as suggested by Armstrong (1968) is not precluded.

The argument implies that the earth arrived at essentially its present surficial volumes and areas relatively early in its history. Since then it has been maintaining the areas and volumes of its crustal domains in more or less a steady-state system. In this view, continental margins could be accreting or retreating locally, but their net effect would be to serve as the factories reworking dispersed materials back into continental rafts of constant thickness and total area.

CONCLUSIONS

Constant freeboard for extended periods is the best evidence that the continuing struggle between oceanic waters and continental crust for the total surface area of the globe has been a stalemate over much of geologic time. This is the result of the isostatic balance between typical oceanic and continental crusts with relative total areas being a function of respective volumes and volume ratios. This dynamic equilibrium is independent of shape or number of continents or ocean basins. Its rate of operation is controlled by the speed at which tectonic processes of the mantle drive the deeper half of the equilibrium toward wider ocean basins and thicker, smaller continents. Its surficial half pushes the system toward thin and widespread continents separated by shallow ocean basins, adjusting its rates of operation by changing average continental elevations or by changing the extent of eroding area above sea level so as to counterbalance any effect of the deeper half. An apparent long-term change in freeboard equilibrium at the Permian breakup of Gondwanaland is suggested as the system's response to a change in effectiveness of continental rebuilding.

The constant freeboard model coupled with an apparent constant thickness of continents in the last half of earth history can be used to argue that continental and oceanic volumes and areas have not changed significantly in this time span, during which the continental margins on global average have been neither accreting nor retreating.

The maintenance of a volume-area-isostasy equilibrium to within a few percent by volume is the unifying theme behind an endless number of seemingly random, seemingly purposeless geologic changes. The tortured geology of continents is a partial record of adjustments to this equilibrium or shifts of equilibrium position as a function of time. If ocean waters carried in their fabric a similar memory of past configurations, an equally complex history of ocean basin adjustment would appear. There is little in the first-order features of our planet's crustal and surficial geology that is not linked in some manner to this underlying theme of long times of constant freeboard. The process might well be called the master equilibrium of tectonics, with the continental margins acting as the anvils on which much of the equilibrium is forged.

ACKNOWLEDGMENTS

As a student the author was introduced to this problem by Harry H. Hess and John C. Maxwell. Much of the thought expressed here was done as a visiting scientist with W. H. Menard in 1961. This paper has benefited by many discussions with these men and with other friends and colleagues through the years. It includes insights from a large number of old and recent authors, only some of whom could be cited directly in this short summary.

BIBLIOGRAPHY

Armstrong, R. L., 1968, A model for the evolution of strontium and lead isotopes in a dynamic earth: Rev. Geophys., v. 6, p. 175-200.

Becker, R. H., 1973, Oceanic growth models: Science, v. 182, p. 601-602.

Birch, F., 1968, On the possibility of large changes in the earth's volume: Phys. Earth and Planetary Interiors, v. 1, no. 3, p. 141-147.

Brown, H., 1952, Rare gasses and the formulation of the earth's atmosphere, *in* Kuiper, G. P., ed., The atmospheres of the earth and planets, 2nd ed.: Chicago, Univ. Chicago Press, p. 258-266.

Bucher, W. H., 1933, The deformation of the earth's crust: Princeton, N.J.: Princeton Univ. Press, 518 p.

Carey, S. W., 1958, The tectonic approach to continental drift, *in* Carey, S. W., ed., Continental drift, a symposium: Univ. Tasmania, p. 177-354.

Chase, C. G., and Perry, E. C., 1972, The oceans: growth and oxygen isotope evolution: Science, v. 177, p. 992.

Condie, K. C., 1973, Archean magmatism and crustal thickening: Geol. Soc. America Bull., v. 84, p. 2981-2992.

———, and Potts, M. J., 1969, Calc-alkaline volcanism and the thickness of the early Precambrian crust in North America: Can. Jour. Earth Sci., v. 6, p. 1179-1184.

Dearnley, R., 1969, Crustal tectonic evidence for earth expansion, *in* Runcorn, K., ed., Application of modern physics to earth and planetary interiors: New York, Wiley-Interscience, p. 103-110.

Dietz, R. S., and Holden, J. C., 1966, Miogeoclines (miogeosynclines) in space and time: Jour. Geology, v. 74, p. 566-583.

Egyed, L., 1956a, Change of earth dimensions as determined from paleogeographical data: Geofisica Pura e Applicata, v. 33, p. 42-48.

———, 1956b, Determination of changes in the dimensions of the earth from paleogeographical data: Nature, v. 178, p. 534.

———, 1969, The slow expansion hypothesis, *in* Runcorn, K., ed., Application of modern physics to earth and planetary interiors: New York, Wiley-Intersicence, p. 65-75.

Engel, A. E. J., 1963, Geologic evolution of North America: Science, v. 140, no. 3563, p. 143-152.

Fanale, F. P., 1971, A case for catastrophic early degassing of the earth: Chem. Geology, v. 8, p. 79-105.

Garrels, R. M. Mackenzie, F. T., and Siever, R., 1972, Sedimentary cycling in relation to the history of continents and oceans: *in* Robertson, E. C., ed., The nature of the solid earth: New York, McGraw-Hill, p. 93-121.

Gilluly, J., 1949, Distribution of mountain building in geologic time: Geol. Soc. America Bull., v. 60, no. 4, p. 561-590.

Grabau, A., 1933, Oscillation or pulsation [with dis.]: 16th Internatl. Geol. Congr. Rept., v. 1, p. 533-539 [1936].

Harland, W. B., Smith, A. G., and Wilcock, I. B., eds., 1964, The Phanerozoic time scale, a symposium in honor of Arthur Holmes: Geol. Soc. London Quart. Jour., v. 1205, p. 260-262.

Hess, H. H., 1962, History of ocean basins, *in* Engel, A. E. J., James, H. L., and Leonard, B. F., eds., Petrologic studies: a volume in honor of A. F. Buddington: Geol. Soc. America, p. 599-620.

Holmes, A., 1965, Principles of physical geology, 2nd ed.: New York, Ronald Press, 1288 p.

Hubbert, M. K., 1937, Theory of scale models as applied to study of geologic structures: Geol. Soc. America Bull., v. 48, no. 10, p. 1459-1519.

Hurley, P. M., 1968, Absolute abundance and distribution of Rb, K and Sr in the earth: Geochim: Cosmochim. Acta, v. 32, p. 273-283.

———, and Rand, J. R., 1969, Pre-drift continental nuclei: Science, v. 164, no. 3885, p. 1229-1242.

Jordan, P., 1969, On the possibility of avoiding Ramsey's hypothesis in formulating a theory of earth expansion, *in* Runcorn, K., ed., Application of modern physics to earth and planetary interiors: New York, Wiley-Interscience, p. 55-63.

Judson, S., and Ritter, D., 1964, Rates of regional denudation in the United States: Jour. Geophys. Res., v. 69, p. 3395-3401.

Khain, V. E., and Muratov, M. V., 1969, Crustal movements and tectonic structure of continents, *in* Hart, P. J., ed., Earth's crust and upper mantle: Am. Geophys. Union Geophys. Monogr. 13, p. 523-538.

Kuenen, P. H., 1939, Quantitative estimations relating to eustatic movements: Geol. Mijnbouw, v. 18, no. 8, p. 194-201.

Maxwell, J. C., 1968, Continental drift and a dynamic earth: Am. Scientist, v. 56, no. 1, p. 35-51.

Menard, H. W., 1961, Some rates of regional erosion: Jour. Geology, v. 69, p. 154-161.

———, 1964, Marine geology of the Pacific: New York, McGray-Hill, 271 p.

———, and Smith, S., 1966, Hypsometry of ocean basin provinces: Jour. Geophys. Res., v. 71, no. 18, p. 4305-4325.

Muehlberger, W. R., Denison, R. E., and Lidiak, E. G., 1967, Basement rocks in continental interior of United States: Am. Assoc. Petr. Geol. Bull., v. 51, no. 12, p. 2351-2380.

Pitman, W. C., and Hayes, J. D., 1973, Upper Cretaceous spreading rates and the great transgression: Geol. Soc. America Abstr., v. 5, no. 7, p. 768.

Rona, P. A., 1973, Relations between rates of sediment accumulation on continental shelves, sea floor spreading, and eustacy inferred from the central North Atlantic: Geol. Soc. America Bull., v. 84, p. 2851-2872.

Rubey, W. W., 1951, Geologic history of sea water: Geol. Soc. America Bull., v. 62, no. 9, p. 1111-1147.

Schuchert, C., 1916, Correlation and chronology on the basis of paleogeography: Geol. Soc. America Bull., v. 27, p. 491-513.

———, 1955, Atlas of paleogeographic maps of North America: New York, Wiley, 177 p.

Sloss, L. L., 1963, Sequences in the cratonic interior of North America: Geol. Soc. America Bul., v. 74, p. 93-114.

———, 1972, Concurrent subsidence of widely separated cratonic basins: Geol. Soc. America Abstr., v. 4, no. 7, p. 668.

———, 1973, Tectonic and eustatic factors in late Precambrian-Phanerozoic global sea level changes: Geol. Soc. America Abstr., v. 5, no. 7, p. 813.

Stille, H., 1924, Grundfragen der vergleichenden tektonik: Berlin.

———, 1944, Geotektonische gliederung der erdgeschichte: Abh. Preuss. Akad. Wiss. Math. Nat. Klasse, v. 3, 80 p.

Strakhow, N. M., 1948, Outlines of historical geology: Moscow, U.S.S.R. Govt. Publ.

Termier, H., and Termier, G., 1952, Histoire geologique de la biosphere: Paris, Masson.

———, and Termier, G., 1969, Global paleogeography and earth expansions, *in* Runcorn, K., ed., Application of modern physics to earth and planetary interiors: New York, Wiley-Interscience, p. 87-101.

Umbgrove, J. H. F., 1939, On rhythms in the history of the earth: Geol. Mag., v. 76, p. 116-129.

———, 1947, The pulse of the earth: The Hague, M. Nijhoff, 357 p.

Valentine, J. W., and Moores, E. M., 1972, Global tectonics and fossil record: Jour. Geology, v. 80, p. 167-184.

Van Houten, F. B., 1969, Molasse facies: records of worldwide crustal stresses: Science, v. 166, p. 1506-1508.

Weertman, J., 1963, The thickness of continents: Jour. Geophys. Res. v. 68, no. 3, p. 929-932.

Wise, D. U., 1972, Freeboard of continents through time: Geol. Soc. America Mem. 132 (Hess vol.), p. 87-100.

———, 1973, Constant freeboard deviations and the master equilibrium of tectonics: Geol. Soc. America Abstr., v. 5, no. 7, p. 867.

Wynne-Edwards, H. R., 1969, Tectonic overprinting in the Grenville Province, S.W. Quebec: Geol. Assoc. Can. Spec. Paper 5, p. 163-182.

Standard Oceanic and Continental Structure

J. Lamar Worzel

INTRODUCTION

The floor of the ocean areas of the world has many major features, such as continental rises, oceanic basins, mid-ocean ridges, other submerged ridges, island ridges, deep-sea trenches, volcanic peaks, and seamounts, all with different crustal and mantle structures beneath them. What, then, represents a standard oceanic structure?

Hypsographic curves, which can be found, for instance, in Defant (1961) or Sverdrup et al. (1942), show one plateau in the sea water covering parts of the earth between the depths of 4.0 and 5.6 km. Sixty-five percent of the ocean falls in this sector of the curve, which is quite linear. The mean depth of these parts of the ocean is 4.8 km. The main ocean basins naturally fall within this part of the curves. Most of the other features mentioned above fall within the remaining 35% of the ocean-covered area.

Gravity and seismic studies show that features that are evident topographically usually extend several hundred kilometers laterally beyond their topographic limits. Consequently, we shall consider here the normal or standard ocean structure, that of ocean basins within the 4.0-5.6-km depth range, which are more than 200 km distant from continental topographic features. The continental areas of the world have many major features such as mountain peaks and ranges, elevated plateaus, rift valleys, basins, volcanic peaks, ridges and plateaus, and coastal plains. What then represents a standard continental structure?

Reference once more to a hypsographic curve (see Wise, this volume) shows that there is a plateau on the curve in the land areas near sea level in which the elevation compared to the world area shows a quite linear relationship from elevations of 0.3 km to sea depths of about 0.2 km. Forty-one percent of the land area (considered to extend to the 200-m water depth) falls within this plateau. Most of the features mentioned above fall within the other 59% of the land area. Consequently, we shall consider the "normal" or standard continental structure as that part of the continents within the 0.3-km-elevation to 0.2-km-ocean-depth range, which are more than 200 km distant from other topographic features.

PREVIOUS WORK

Before 1924 there had been surface geological studies of continental areas supplemented by small amounts of data at depth obtained from mines and a small amount of drilling information. Most subsurface information had come from inferences from drilled surface geology. Geophysics had contributed a very generalized view of continental structure from earthquake seismology and the concept of isostasy from gravity geodetic studies, principally of mountain ranges.

Then in 1924, Vening Meinesz (1932) made his epochal invention of a pendulum apparatus for measuring gravity onboard submarines. This quickly showed that the sea areas were in isostatic equilibrium and that there were large negative gravity anomalies associated with the deep trenches located near island arcs and continental margins. Although by this time the Airy type of isostasy was considered to most closely represent the structures delineated from earthquake seismic studies, no one made the inference, which was inherent, that the mantle beneath the oceans would be only about 15 km below sea level in order for the oceans to be in isostatic equilibrium with the 30-35-km-thick continent (for normal sea-level elevations).

In 1935, Maurice Ewing et al. (1937) introduced the use of seismic-refraction techniques for studies of the continental shelves. This can truly be said to be the birth of the study of continent margins. On land, earthquake seismology was being supplemented by explosion seismology, first from blasts associated with mining operations and later from explosives set off purely for seismic purposes.

A classical study of the topography of the continental shelf and slope along the east coast (Veatch and Smith, 1939) showed that the continental slope was greatly dissected by valleys that closely resembled stream valleys, some of which reached well back across the continental shelf.

World War II intervened and disrupted the scientific work along continental margins. It resulted, however, in many more scientists becoming interested in the world ocean and in the development of useful new techniques and in the funding of science. The influx of people, ideas, and funds led to a rapid expansion of earth science investigation beneath the sea.

A great many seismic-refraction profiles were made on the continental shelves resulting in

Table 1. Oceanic crustal structure summary

	Atlantic Ocean				*Indian Ocean*				*Pacific Ocean*			
Layer	Vel. (km/sec)	No. of obs.	Layer Thick(km)	No. of obs	Vel. (km/sec)	No. of obs.	Layer Thick(km)	No. of obs.	Vel. (km/sec)	No. of obs	Layer Thick(km)	No. of obs
Water	1.5		4.8		1.5		4.8		1.5		4.8	
Sediment	2.0		1.0		2.0		0.7		2.0		0.5	
Layer 2	4.93	21	2.28	21	5.13	14	1.26	14	5.23	29	1.42	28
Layer 3	6.66	65	4.74	35	6.80	18	4.44	16	7.79	35	5.07	22
Mantle	8.00	35			8.02	16			8.78	22		

structure sections to basement. Many of these are summarized by Drake et al. (1959).

One of the most significant investigations was the introduction of seismic refraction techniques into the deep ocean in 1949 (Ewing et al., 1950). This was the first seismic work to demonstrate that sediment thicknesses on the floor of the ocean basins were very thin and that basement velocities equivalent to the deepest part of the continental crust underlay them at shallow depth. Techniques were soon improved, and the shallow depth of the mantle beneath the oceans was soon demonstrated.

Worzel and Shurbet (1955b) summarized the new knowledge of oceans and continents and derived standard oceanic and continental crustal sections to aid in broad gravity interpretations. These were applied to show the very thin crustal section beneath ocean deeps, the thickness of the sedimentary trough in the Gulf Coast geosyncline, and the explanation of the shallow-water nature of the fossils in a well that had just been completed on Andros Island in the Bahamas.

Worzel and Shurbet (1955a) then used these standard columns to delineate the crustal transition zone between ocean and continent along the northeast coast of the United States. The rush was on, and in the intervening eighteen years the new geophysical tools were extensively used in most of the ocean-continent boundaries of the world. These are summarized by Worzel (1968) and Uchupi (1970), where references to most of the work may be found.

In 1973, Worzel and Watkins (1973) used the structure delineated for the Gulf Coast ocean-continent boundary to illustrate the evolution and development of the boundary since Cretaceous times.

OCEANIC STRUCTURE

Earthquake surface waves give good mean structures for oceanic paths but normally include a fair part of their path in the nonbasin portions of the ocean. For detailed crustal structure, seismic refraction data provide the best results. Good summaries of these data have been published by Ewing and Nafe, 1963; Raitt, 1963; Drake and Nafe, 1968; Ludwig et al., 1970; Woollard, 1968; Shor et al., 1970; and Laughton et al., 1970. References to original detailed publications can be obtained from these summaries.

Using our criterian for a standard ocean section, we have collected together and summarized the data for the Atlantic, Pacific, and Indian oceans in Table 1. In each case we have assumed that the standard ocean depth is 4.8 km. The velocities within the sediment layer are quite variable and in some cases can be divided into more than one layer. From sonobuoy refraction data, the mean velocity was chosen (Houtz et al., 1968). From reflection profiler sections we have determined the mean thickness of the sediment column visually. This seemed more rational than taking mean values from reflections since the basement beneath the sediments is often quite rough.

The mean values of the sound velocities for layer 2 for each ocean are surprisingly close, ranging from 4.93 to 5.23 km/sec. For all the "standard ocean" observations the mean is 5.12 km/sec, which will be the value that we adopt here.

The mean value for the thickness of layer 2 for all the "standard ocean" observations is 1.68 km. In the Atlantic Ocean this is 36% thicker, at 2.28 km; in the Indian Ocean, 27% thinner at 1.26 km; and in the Pacific Ocean, 16% thinner at 1.42 km. There does not appear to be any great significance to the variation in the thickness of this layer.

Layer 3 has a mean velocity of 6.70 km/sec for the "standard ocean" observations showing minor variations for each ocean of -0.6% for the Atlantic, +1.5% for the Indian Ocean, and +1.3% for the Pacific Ocean. The variations in layer thicknesses from the mean of 4.77 is also small at -0.6% for the Atlantic Ocean, +7.0% for the Indian Ocean, and +6.2% for the Pacific Ocean.

Table 2 shows this new "standard ocean structure" and compares it with that of Worzel and Shurbet (1955b), determined in 1954, updated by Worzel (1965). The main difference is, of course, the subdivision into layers 2 and 3, and the greater depth to the mantle. The latter is partially the result of the subdivision of the oceanic layer, but mostly to the much better data that are now available.

The densities for the layers in the 1955 work were chosen essentially by estimates guided by reference to the *Handbook of Physical Constants* (Birch et al., 1942). Those for the 1965 work and for this paper were derived from the seismic velocities using the Nafe-Drake velocity-density curve (Ludwig et al., 1970). As one might expect, since the data were sparse in 1955, there was a great deal of change in the ensuing decade, with a much more minor change in the succeeding decade.

Sutton et al. (1971) have reported a division of layer 3 into two separate layers for eight stations in the Pacific. They have suggested that this division of layers may have been overlooked in previous work because it appears mostly as second arrivals and appears for only a short distance as a first arrival. In normal refraction shooting it could easily be missed because of the shot spacing. They discovered it because they had nearly continuous data obtained using large airguns and sonobuoys. Only four of their stations fall within our definition of a standard oceanic structure. For these four stations Table 3 gives the appropriate mean values.

If these data were representative of the standard oceanic structure it would have the effect of increasing the depth to the mantle by about 1.25 km, of increasing the mean crustal density (layers 1, 2, and 3) by about 0.05 g/cm^3 and of increasing the mantle density by 0.17 g/cc. However, until it is confirmed that such data are representative of the normal oceanic structure throughout the world, we feel that it is better to use the section we have

Table 2. Standard ocean crustal structure.

Layer	Worzel and Shurbet, 1955: Depth (Km)	Worzel and Shurbet, 1955: DENSITY gm/cc	Worzel, 1965: Depth (Km)	Worzel, 1965: VELOCITY km/sec	Worzel, 1965: DENSITY gm/cc	This paper: Depth (Km)	This paper: VELOCITY km/sec	This paper: DENSITY gm/cc
Water	0–5.0	1.03	0–4.9	1.5	1.03	0–4.8	1.5	1.03
Sediment	5.0–6.0	2.30	4.9–5.6	2.50	2.30	4.8–5.6	2.0	1.90
			5.6–7.3	5.07	2.55	5.6–7.28	5.12	2.55
Oceanic	6.0–10.5	2.84	7.3–11.5	6.69	2.90	7.28–12.05	6.70	2.86
Mantle	10.5–	3.27	11.5–	8.13	3.40	12.05–	8.08	3.30

Table 3

Layer	Velocity (km/sec)	Thickness (km)	Density (g/cc)
3A	6.69	2.82	2.85
3B	7.37	3.18	3.06
4	8.28		3.47

derived above, since it is based on a much more considerable body of data.

Anisotropy of mantle velocity amounting to about 3.5% has been shown to occur near the California Coast by Raitt, et al. (1969). It could be argued that this anisotropy is related to the continental margin or to a "spreading axis" and might not be found in a normal oceanic section. If it does exist in a normal oceanic crust, it could possibly explain the variations in the mean velocities observed in the three oceans. If the anisotropy in a normal ocean section is no larger than that measured in their observations, it would have little effect on the purposes of the present investigation, although perhaps it would explain some of the variability of observations and perhaps some of the scatter of the velocity-density relationship.

CONTINENTAL STRUCTURE

The surface of the continents is a mélange of sedimentary, igneous, and metamorphic rocks, as is well known from surface geological surveys. The depth to which this mélange extends varies considerably from place to place. At depths of only a few kilometers, seismic velocities of 5.0-5.5 km/sec are usually encountered.

Both earthquake and explosion seismic-refraction studies have been interpreted to show the continental crust as a layered structure, commonly with two layers, but sometimes with as many as six or eight layers, and on some occasions as only one layer. In some areas data have been interpreted as showing two or more layers and later, other observers, or even sometimes the same observer, have interpreted it as one layer. Today, the prevalent view is that beneath the surface mélange, at only a few kilometers depth, a layer with a seismic velocity of 5.0-5.5 km/sec occurs. At about one-third of the way to the mantle this gives way to a layer with a velocity of 6.5-7.2 km/sec. At the mantle, velocities of the order of 8.0 km/sec are observed. There is disagreement about whether these changes of velocity are discontinuous or whether they are gradational. Some observers have argued that there are even layers of lower velocity within the crust beneath higher-velocity layers (e.g., Landisman et al., 1971).

In order to arrive at a standard continental

Table 4. Continental crustal structure summary.

Continent	V_c (km/sec)	No. of Obs.	D_m (km)	No. of Obs.	V_m (km/sec)	No. of Obs.
North America	6.40	48	32.98	52	8.10	48
Europe	6.34	12	30.55	12	8.19	12
Asia	6.34	13	39.3	13	8.10	13
Australia	6.04	2	27.8	2	8.17	2
Africa	6.3	6*	36.2	6*	8.05	6*
All continents	6.36	81	33.71	85	8.11	8

V_c mean velocity of crystalline crustal rocks.
D_m depth below sea level to mantle rocks.
V_m velocity of mantle rocks.
(* —These data were the results of earthquake studies over paths that substantially sampled six different regions.)

crust with these diverse data, we have derived the mean seismic velocity for the crystalline rock within the crust for those data in which a layered crust was provided. Good summaries of most of the data have been published by Steinhart and Woollard (1961), Woollard (1968), Closs (1969), Sollogub (1969), and Kosminskaya et al. (1969). References to particular studies may be found in these summaries. Additional data not included in these (in most cases because the work was completed after the summaries were made) were also used in this study as follows: Fuchs and Landisman (1966), Mereu and Hunter (1969), Gumper and Pomeroy (1970), Hales et al. (1970), and Dorman et al. (1972). Table 4 summarizes the data by continents for those seismic-refraction stations within normal continental crustal areas according to our definition above.

From Table 4 we have determined our normal continental structure. We have determined the densities from the Nafe-Drake velocity-density curve (Ludwig et al., 1970). Table 5 shows this normal continental structure and compares it with the earlier work of Worzel and Shurbet (1955b). It is obvious that much greater quantity of data now available does not require much change in the normal continental structure.

Table 5. Standard continental structure.

DEPTH Km	Worzel & Shurbet, 1955		This Work		
	DEPTH km	DENSITY gm/cc	DEPTH km	VELOCITY km/sec	DENSITY gm/cc
10-					
		2.84		6.36	2.84
20-					
30-					
	33.0		33.7		
		3.27		8.11	3.35
40-					

STANDARD OCEAN AND CONTINENTAL STRUCTURE

In the previous sections we have deduced standard structure sections for ocean and continents as shown in Table 6.

Some may wonder about the inclusion of the seawater in the oceanic crustal section; it is a crustal layer as much as any other seismic layer. It is a mineral with trace elements included, and it exhibits physical properties, such as sound velocity and density, which have major effects in seismic, gravity, and other measurements. It is considered here as a part of the crustal layering.

Since the ocean and continent are in close isostatic equilibrium, it is important to determine whether their standard structural sections should be in such equilibrium. If they are, the mass-per-unit-area for each column should be essentially the same. We can calculate them as shown in Table 7.

The total mass per unit area beneath our normal continental crust would be 95.71 x 10^5 g/cm^2, and beneath our normal ocean crust would be 95.83 x 10^5 g/cm^2—an agreement closer than one could hope for! The difference would amount to only about 5 mgal of gravity anomaly.

CONTINENTAL MARGIN HISTORICAL RECONSTRUCTION

Since the continental margins are substantially in isostatic equilibrium, and we have every reason to believe this was so in the past, we have a method to restore the margin to former geological times. Worzel and Watkins (1973) have done this for three

Table 6. Standard structure sections.

	Ocean				Continent			
	Depth (km)	Layer Thickness (km)	Velocity (km/sec)	Density (g/cm^3)	Depth (km)	Layer Thickness (km)	Velocity (km/sec)	Density (g/cm^3)
Sea level	0				0			
Crust								
Water		4.8	1.5	1.03				
	4.8							
Sediment		0.8	2.0	1.90				
	5.6					33.7	6.36	2.84
Layer 1		1.68	5.12	2.55				
	7.28							
Layer 2		4.77	6.70	2.86				
	12.05				33.7			
Mantle			8.08	3.30			8.11	3.35

sections across the continental margin south of Texas.

Figure 1 is a map of the Texas Gulf Coast showing the location of seismic data and of the section whose reconstruction we will discuss here. Figure 2 shows the reconstructions. The topmost section is the present structural section as delineated by the present data. The topmost layers have been identified by drilling to depths of up to 6 km near the left (north) side of the section. The author has extended these layers, in what is believed to be a rational manner, to the other areas where only seismic refraction data are available. Admittedly this is somewhat tenuous with the data that is presently available, and it must be considered at this stage as a first-order approximation. When better data are available, this method of reconstruction will produce more reliable results. We would like to plead for much additional structural data and, insofar as possible, data to identify geological ages, so that better results could be obtained.

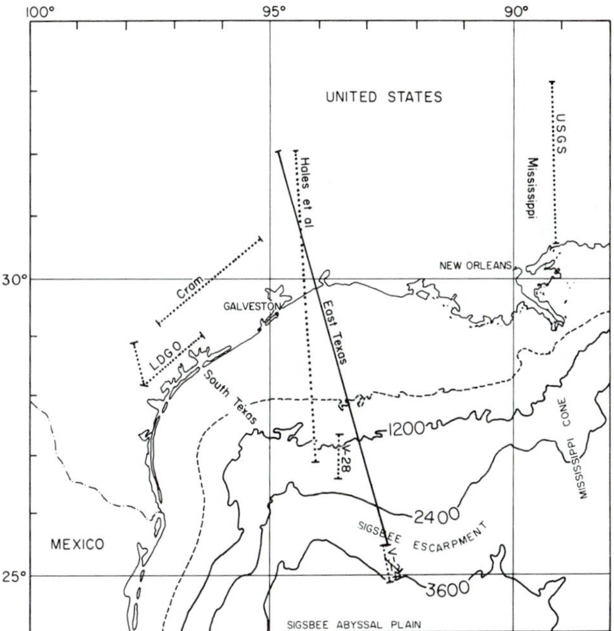

Fig. 1. Map of Gulf Coastal zone of Texas, Louisiana, and Mississippi, showing location of east Texas structural sections.

Assuming that the structure as drawn is accurate, and that sea level has not changed appreciably, we can successively strip off the Pleistocene layer, the layers deposited after the Oligocene, and the layers deposited after the Cretaceous. After each layer is stripped, the section is returned to presumed isostatic equilibrium and to the most probable continental margin of that time. Perhaps a better way to look at it would be to consider the pre-Tertiary section to be as represented and the other sections to represent the evolution of this coast with time, as sedimentation pro-

Table 7. Mass balance

Continent

$33.7 \text{ km} \times 2.84 \text{ g/cm}^3 = 95.71 \times 10^5 \text{ g/cm}^2$

Ocean

Water	$4.8 \text{ km} \times 1.03 \text{ g/cm}^3$	$= 4.94 \times 10^5 \text{ g/cm}^2$
Sediment	$0.8 \text{ km} \times 1.90 \text{ g/cm}^3$	$= 1.52 \times 10^5 \text{ g/cm}^2$
Layer 2	$1.68 \text{ km} \times 2.55 \text{ g/cm}^3$	$= 4.28 \times 10^5 \text{ g/cm}^2$
Layer 3	$4.77 \text{ km} \times 2.86 \text{ g/cm}^3$	$= 13.64 \times 10^5 \text{ g/cm}^2$
Mantle	$21.65 \text{ km} \times 3.30 \text{ g/cm}^3$	$= 71.45 \times 10^5 \text{ g/cm}^2$
Total	33.70	$95.83 \times 10^5 \text{ g/cm}^2$

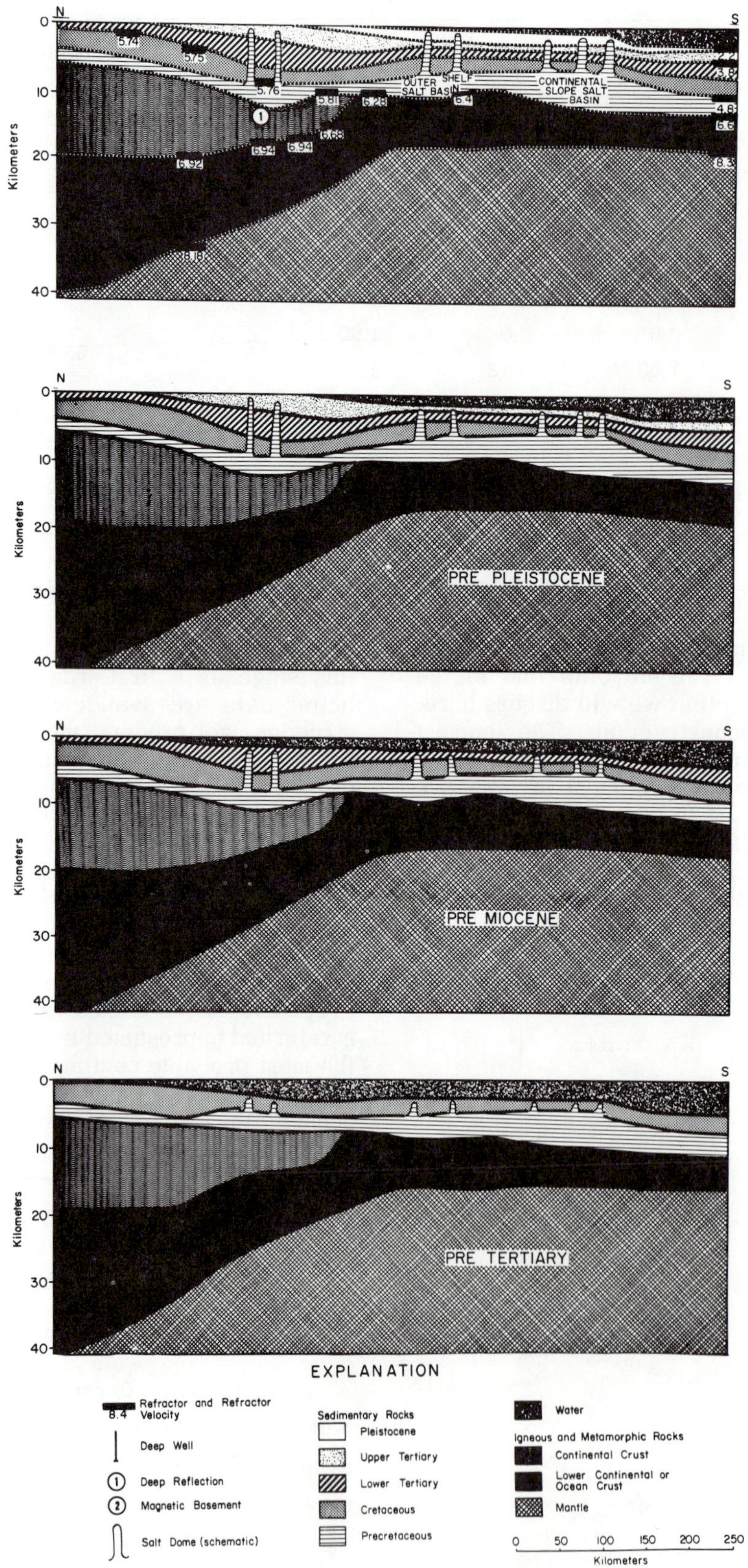

Fig. 2. East Texas historical reconstruction. Topmost section represents the present structural section.

gressed. The salt domes in these sections are diagrammatic, representing only regions where salt domes are known to exist. It has been necessary to show them as evolving with time in order that they be retained within the sedimentary framework. Perhaps this is more realistic than is immediately apparent.

It is interesting to see that the shoreline at the end of the Cretaceous would have had to be north of the end of the section (Fig. 2), so that the Cretaceous embayment would, in fact, be the normal continental margin rather than the usual explanation of a flooding of the continent. The Gulf Basin would have been about 6 km deep at that time, instead of the present 5 km. Where the present continental shelf is now located, would have been a deep basin of 3-4 km depth. It is easy to see that the main depocenter has moved seaward as the continental shelf has built outward through time.

ACKNOWLEDGMENTS

This is Contribution Number 39 of the Earth and Planetary Sciences Division of the University of Texas at Galveston.

BIBLIOGRAPHY

Birch, F., Schairer, J.F., and Cecil Spicer, H., eds., 1942, Handbook of physical constants: Geol. Soc. America Spec. Paper 36, 325 p.

Clöss, H., 1969, Explosion seismic studies in western Europe, *in* Hart, P. J., ed., The earth's crust and upper mantle, Am. Geophys. Union Monogr. 13.

Defant, A., 1961, Physical oceanography: Elmsford, N.Y., Pergamon Press, v. 1, 729 p.

Dorman, J., Worzel, J. L., Leyden, R., Crook, T. N., and Hatziemmanuel, M., 1972, Crustal section from seismic refraction measurements near Victoria, Texas: Geophysics, v. 37, p. 325-336.

Drake, C. L., and Nafe, J. E., 1968, The transition from ocean to continent from seismic refraction data, *in* Knopoff, Drake, and Hart, eds., The crust and upper mantle of the Pacific area, Am. Geophys. Union Geophys. Monogr. 12, p. 174-188.

———, Ewing, M., and Sutton, G. H., 1959, Continental margins and geosynclines: the east coast of North America north of Cape Hatteras, *in* Ahrens, L. H., Press, F., Runcorn, S. K., and Urey, H. C., eds., Physics and chemistry of the earth, v. 3: Elmsford, N.Y., Pergamon Press, p. 110-198.

Ewing, J., and Ewing, M., 1970, Seismic reflection, *in* Maxwell, A. E., ed., The sea, v. IV, pt. 1: New York, Wiley-Interscience, p. 1-52.

Ewing, J. I., and Nafe, J. E., 1963, The unconsolidated sediments, *in* Hill, M. N., ed., The sea, v. III: New York, Wiley-Interscience, p. 73-84.

Ewing, M., Crary, A. P., and Rutherford, H. M., 1937, Geophysical investigation in the emerged and submerged Atlantic Coastal Plain, pt. 1, Methods and results: Geol. Soc. America Bull., v. 48, p. 753-802.

Ewing, M., Worzel, J. L., Hersey, J. B., Press, F. and Hamilton, G. R., 1950, Seismic refraction measurements in the Atlantic Ocean basins (pt. 1): Seismol. Soc. America Bull., v. 4, p. 233-242.

Fuchs, K., and Landisman, M. 1966, Detailed crustal investigation along a north-south section through the central part of West Germany, *in* Steinhart, J. S., and Smith, T. J., eds., Am. Geophys. Union Monogr. 10.

Gumper, F., and Pomeroy, P. W., 1970, Seismic wave velocities and earth structure of the African continent: Seismol. Soc. America Bull., v. 60, p. 651-668.

Hales, A. L., Helsey, E. E., and Nation, J. B., 1970, Crustal structure study of the Gulf Coast of Texas: Am. Assoc. Petr. Geol. Bull., v. 54, p. 2040-2057.

Houtz, R., Ewing, J., and Le Pichon, X., 1968, Velocity of deep sea sediments from sonobuoy data: Jour. Geophys. Res., v. 73, p. 2615-2641.

Kosminskaya, I. P., Belyaevsky, N. A., and Volvovsky, I. S., 1969, Explosion seismology in the USSR, *in* Hart, P. J., ed., The earth's crust and upper mantle, Am. Geophys. Union Monogr. 13.

Landisman, M., Mueller, S., and Mitchell, B. J., 1971, Review of evidence for velocity inversions in the continental crust, *in* Heacock, J. G., ed., The structure and physical properties of the earth's crust: Am. Geophys. Union Monogr. 14.

Laughton, A. S., Matthews, D. H., and Fisher, R. L., 1970, The structure of the Indian Ocean, *in* Maxwell, A. E., ed., The sea, v. IV, pt. II: New York, Wiley-Interscience, p. 543-586.

Ludwig, W. J., Nafe, J. E., and Drake, C. L., 1970, Seismic refraction, *in* Maxwell, A. E., ed., The sea, v. IV, pt. 1, New York, Wiley-Interscience, p. 53-84.

Mereu, R. F., and Hunter, J. A., 1969, Crustal and upper mantle structure under the Canadian Shield from Project Early Rise data: Bull. Seismol. Soc. America Bull., v. 59, p. 147-165.

Raitt, R. W., 1963, The crustal rocks, *in* Hill, M. N., ed., The sea, v. III, New York, Wiley-Interscience, p. 85-109.

Raitt, R. W., Shor, G. G., Jr., Francis, T. J. G., and Morris, G. B., 1969, Anistropy of the Pacific upper mantle: Jour. Geophys. Res., v. 74, p. 3095-3109.

Shor, G. G., Jr., Menard, H. W., and Raitt, R. W., 1970, Structure of the Pacific Basin, *in* Maxwell, A. E., ed., The sea, v. 4, pt. 2: New York, Wiley-Interscience.

Sollogub, V. B., 1969, Seismic studies in southeastern Europe, *in* Hart, P. J., ed., The earth's crust and upper mantle: Am. Geophys. Union Monogr. 13.

Steinhart, J. S., and Woollard, G. P., 1961, Seismic evidence concerning continental structure. *in* Steinhart, J. S., and Meyer, R. P., eds., Explosion studies of continental structure: Washington, D.C., Carnegie Inst. Washington Publ. 622.

Sutton, G. H., Maynard, G. L., and Hussong, D. M., 1971, Physical properties of the oceanic crust, *in* Heacock, J. G., ed., The structure and physical properties of the earth's crust: Am. Geophys. Union Monogr. 14, p. 193-209.

Sverdrup, H., Johnson, M., and Fleming, R., 1942, The oceans: Englewood Cliffs, N.J., Prentice-Hall, p. 1045.

Uchupi, E., 1970, Atlantic continental shelf and slope of the United States—shallow structure: Geol. Survey Prof. Paper 529-1: Washington, D.C., Govt. Printing Office, 144 p.

Veatch, A. C., and Smith, P. A., 1939, Atlantic submarine valleys of the United States and the Congo Submarine Valley: Geol. Soc. America Spec. Paper 7, 98 p.

Vening Meinesz, F. A., 1932, Gravity expeditions at sea, 1923-1930: J. Waltman Jr., Delft, v. 1, 109 p.

Woollard, G. P., 1968, The interrelationship of the crust, the upper mantle and isostatic gravity anomalies in the United States, *in* Knopoff, Drake, and Hart, eds., The crust and upper mantle of the Pacific area: Am. Geophys. Monogr. 12, p. 312-341.

Worzel, J. L., 1965, Pendulum gravity measurements at sea, 1936-1959: New York, Wiley, 422 p.

———, 1968, Advances in marine geophysical research of continental margins: Can. Jour. Earth Sci., v. 5, p. 963-983.

———, and Shurbet, G.L., 1955a, Gravity anomalies at continental margins: Proc. Natl. Acad. Sci., v. 41, p. 458-469.

———, and Shurbet, G. L., 1955b, Gravity interpretations from standard oceanic and continental crustal sections, *in* Poldevaart, A., ed., Geol. Soc. America Spec. Paper 62, p. 87-100.

———, and Watkins, J. S., 1973, Evolution of the northern Gulf Coast deduced from geophysical data: Trans. Gulf Coast Assoc. Geol. Soc., v. 23, p. 84-91.

The Boundary Between Oceanic and Continental Crust in the Western North Atlantic

Philip D. Rabinowitz

INTRODUCTION

A poorly understood aspect of sea-floor spreading and continental drift involves the tectonic processes active during the earliest rifting history and separation of the continents. The identification of the boundary between oceanic and continental basement is required for any precise predrift reconstruction. No satisfactory criterion exists today for precisely determining continental reconstructions. Bullard et al. (1965) used selected isobaths for mathematically determining and evaluating the best geometric fits and others have used different criteria. Even though these fits appear visually satisfying at a large scale, many gaps and overlaps are present (Fig. 1).

Talwani and Eldholm (1973) have demonstrated that the boundary between oceanic and continental basement is not necessarily associated with constant bathymetric contours. They have shown that the margin off Norway includes a subsided block of continental materials which they can identify by seismic results together with the gravity and magnetic anomalies associated with its edges. They also describe the structures and geophysical signatures for the margin of South Africa on either side of the Agulhas Fracture Zone. They have noted the existence of magnetic quiet zones on these margins and cite evidence to suggest that the quiet zones are situated on continental basement. Furthermore, the gravity anomalies indicate that dense rocks are located beneath the shelf edge, marking a hinge line for the subsidence of the continental slope and rise. Numerous geological and geophysical lineaments are noted bordering the continental margin of the northwest Atlantic Ocean. These include:

1. A subsurface ridge defined by seismic compressional velocities and located near the seaward edge of the continental shelf (Drake et al., 1959).
2. A continuous free-air gravity high located near the shelf break (Emery et al., 1970; Rabinowitz, 1973).
3. A nearly continuous high-amplitude magnetic anomaly, called the east coast magnetic anomaly, which is located, in places, as far seaward as the continental rise and, in places, as far landward as the coastline (Taylor et al., 1968).
4. A magnetic quiet zone that is situated seaward of the east coast magnetic anomaly (Heirtzler and Hayes, 1967).

The data on the western North Atlantic margin show similarities with the data observed bordering Norway and South Africa, thus it will be concluded that the concepts of Talwani and Eldholm (1972, 1973) are applicable to the continental margin of the northwest Atlantic Ocean, and that this model may provide a working hypothesis for the study of the early rifting history of other passive continental margins.

MAGNETIC ANOMALIES

Perhaps the two most prominent characteristics of the magnetic field bordering the continental margin of eastern North America are the presence of a magnetic quiet (or smooth) zone and a nearly linear high-amplitude magnetic anomaly called the east coast magnetic "slope" anomaly (Fig. 2).

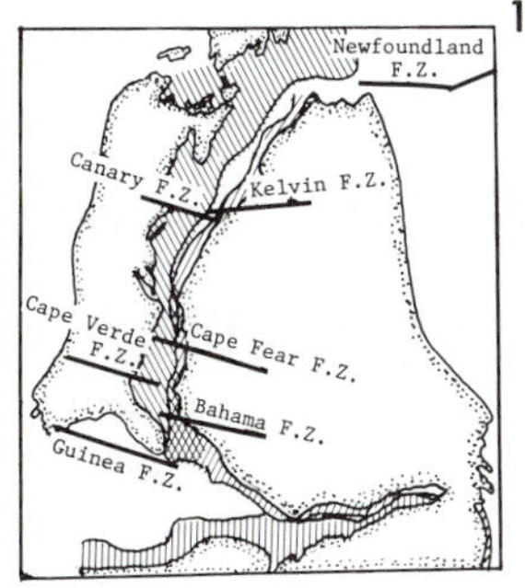

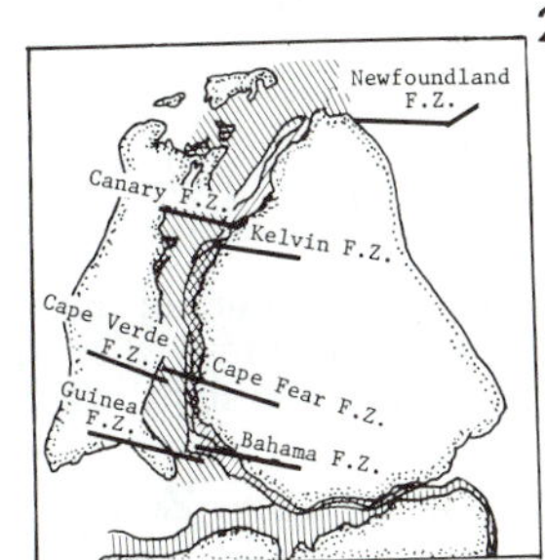

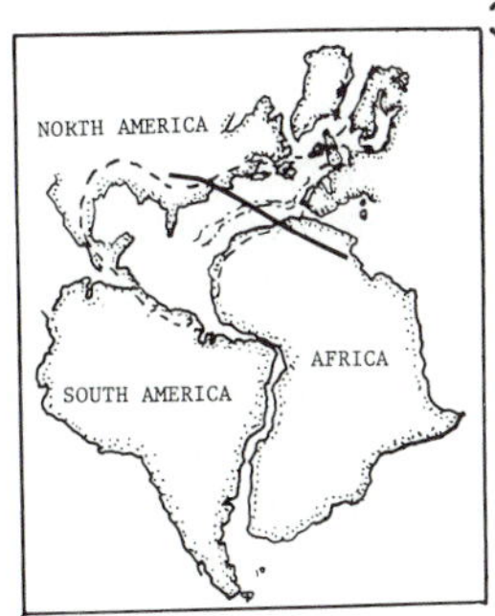

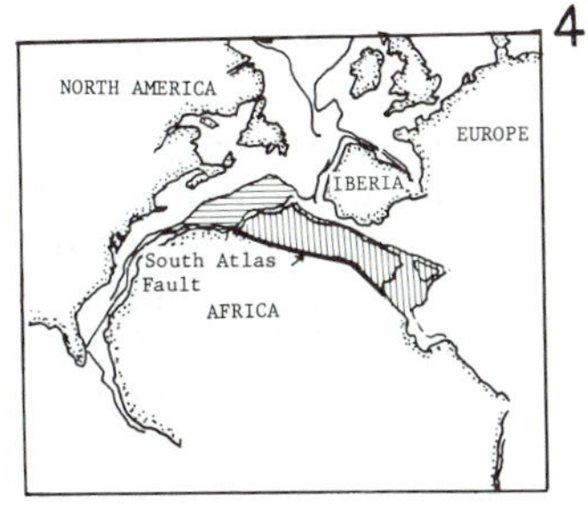

Fig. 1. Four reconstructions of the North Atlantic Ocean: (1) Bullard et al., (1965): a mathematical least-squares fit of the 500-fm isobath; (2) LePichon and Fox (1971): fracture ridges interpreted as defining the flow lines of early motion are predicted on either side of the ocean, with large mismatch is noted for the presumed Canary Island-Kelvin Seamount fracture zone; (3) Drake et al., (1968): matching the magnetic rough-smooth zone boundaries; (4) Dewey et al. (1973): similar to that of Bullard et al. (1965) south of the South Atlas Fault, with northern part of Morocco shifted to the east.

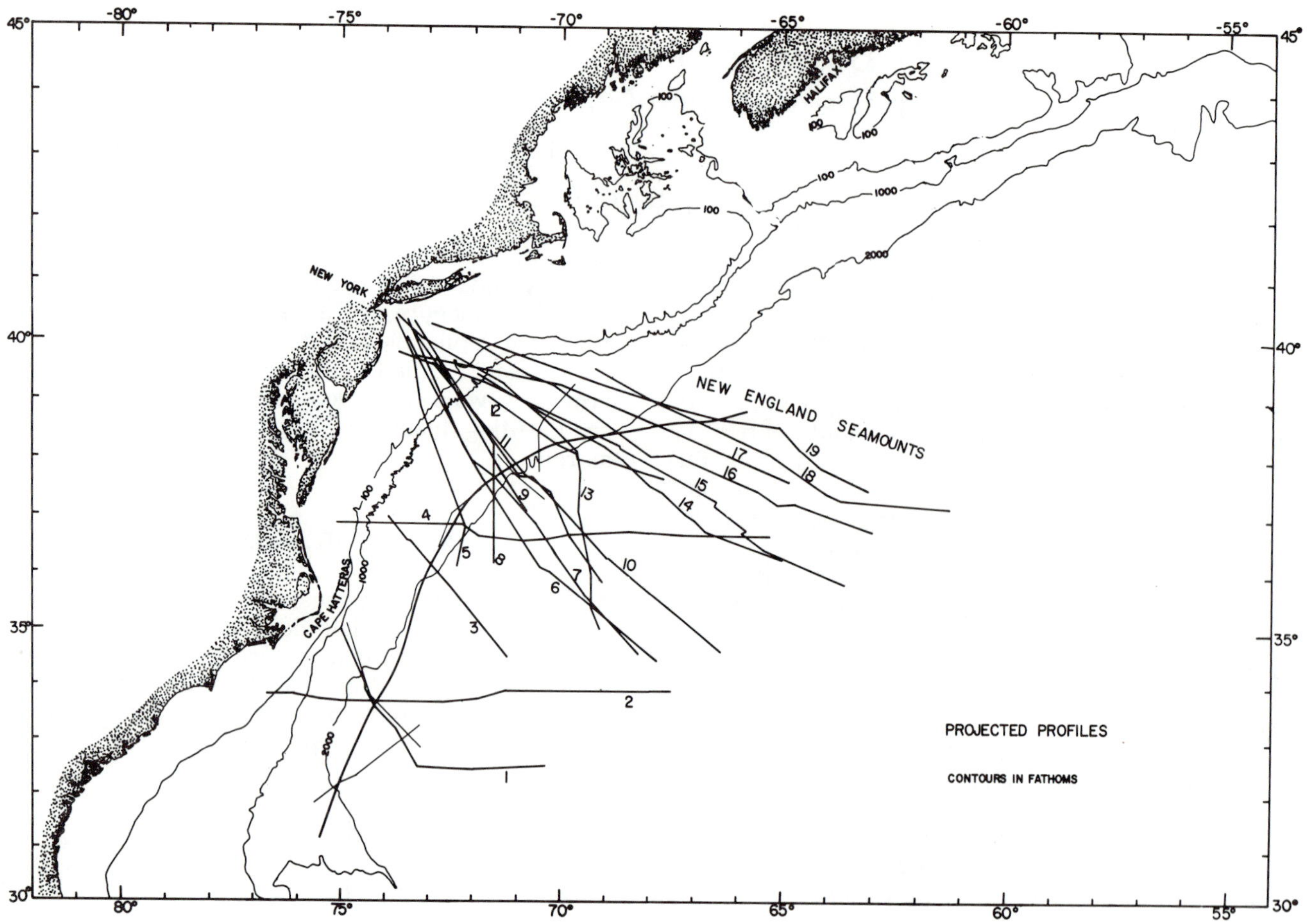

Fig. 2. Locations of ships' tracks where geophysical data are available and that are projected and shown in Figures 3-5. Some other tracks where data are available but which are not projected are shown also. The intersection of the bold curved line with the ships' tracks are the points along which the projections are aligned.

Magnetic Quiet Zone

King et al. (1961) first recognized a distinct quiet (or smooth) zone in the western North Atlantic Ocean bordered on the west by the east coast magnetic slope anomaly and on the east by the disturbed zone. Heirtzler and Hayes (1967) recognized a similar zone in the eastern North Atlantic Ocean. The smooth zones are characterized, in general, by anomaly amplitudes of ± 20 to 50 gammas in contrast to typical values of ± 100 to 300 gammas in the disturbed zone (Fig. 3). In the western North Atlantic Ocean, the smooth zone is observed at least from near the Grand Banks to the West Indies; in the eastern North Atlantic the smooth zone is observed at least from the Sierra Leone Rise to about Gibraltar (Heirtzler and Hayes, 1967).

At the eastern border of the smooth zone in the western North Atlantic a sequence of anomalies (Keathley sequence) have been identified by Heirtzler and Hayes (1967), Anderson et al. (1969), and Vogt et al. (1970a). This sequence has been identified bordering the quiet zone in the eastern North Atlantic by Vogt et al. (1969) and Rona et al. (1970). Recently, Larson and Pitman (1972) have correlated the Keathley sequence in the western North Atlantic with presumed Mesozoic lineations in the Pacific and established a working model of Mesozoic geomagnetic chronology.

The magnetic, gravity, and topographic data along the ship's tracks of Figure 2 have been projected normal to the coastline and are illustrated in Figures 3-5. The profiles are all aligned with respect to anomaly E. The continuity of anomaly E in the region between 32 and 34°N (south of profile 1, Fig. 3) and between about 34 and 36°N (between profiles 2 and 3, Fig. 2) is further substantiated by detailed magnetic surveys of Vogt et al. (1971) and Einwich (1972). This anomaly has an amplitude of generally less than 100 gammas compared to anomalies greater than 300 gammas in the disturbed zone farther seaward.

A number of important observations are noted from Figures 3-5:

1. In general, the magnetic field is much more subdued in the quiet zone landward of anomaly E than seaward of it (Fig. 3). Hence anomaly E separates the quiet zone into what will be termed a landward inner quiet zone and a seaward outer quiet zone.

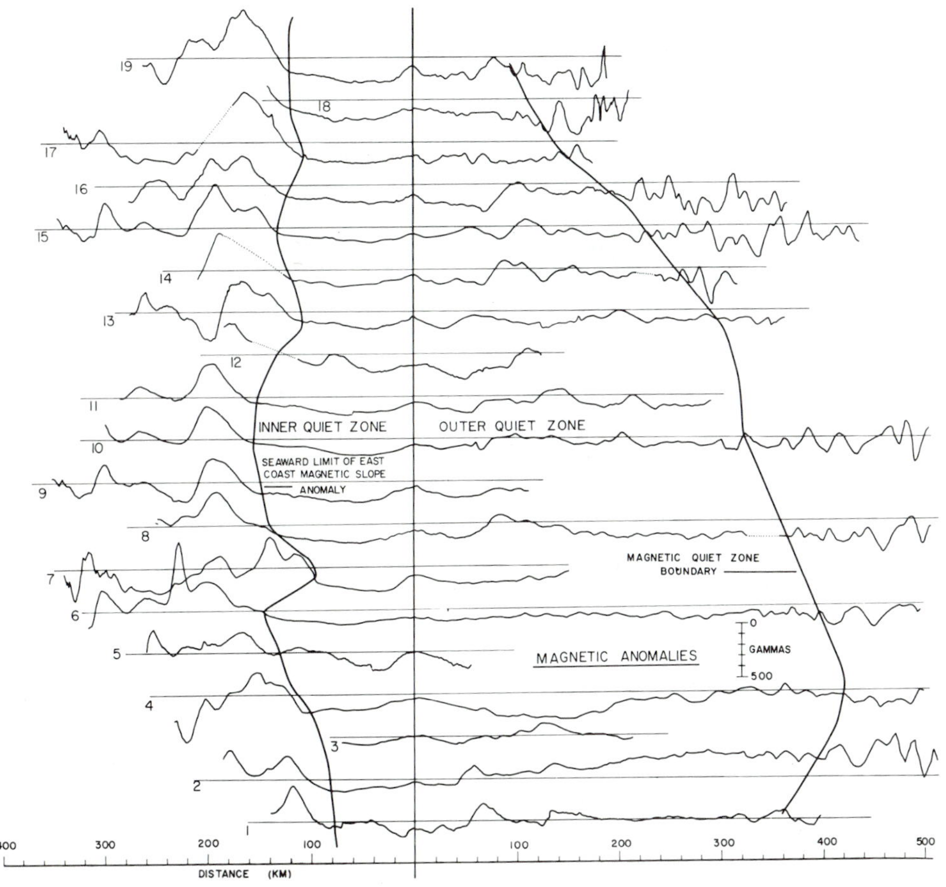

Fig. 3. Magnetic anomalies south of New England Seamounts chain for ship's tracks 1-19 (Fig. 2). The profiles are aligned with respect to a magnetic anomaly correlated within the quiet zone (anomaly E). Landward of anomaly E (inner quiet zone) the magnetic field is much more subdued than seaward of anomaly E (outer quiet zone).

2. Anomaly E is located seaward of the gravity low associated with the continental slope and over a broad (~150 km) low amplitude (~5-20 mgal) gravity high.

3. Anomaly E is observed very near the morphological change from the upper to lower continental rise and is located in water depths ranging from about 3 to 4.5 km.

4. There is a decrease in distance from anomaly E to the rough-smooth zone boundary from south to north.

Numerous interpretations have been proposed to explain the properties and formation of the quiet zone in the western North Atlantic Ocean and are outlined in Table 1.

Magnetic quiet zones are also observed bordering other rifted continental margins. On the Norwegian continental margin, Talwani and Eldholm (1972) have associated a magnetic quiet zone with a foundered continental basement. On the western margin of the southern Red Sea a magnetic quiet zone is observed extending to near the edge of the axial trough. Lowell and Genik (1972) have been able to trace a seismic reflector, associated with Precambrian rocks in Ethiopia, seaward across this magnetic quiet zone. This implies that the quiet zone is not situated on oceanic basement but on a subsided continental basement. A magnetic quiet zone is also observed seaward of the shelf break of the South African continental margin. Magnetic-edge-effect computations of Talwani and Eldholm (1973) on this margin are consistent with the concept that the magnetic quiet zone is situated on a subsided continental basement. On both the Norwegian and South African margins a steep gradient in the isostatic anomaly is observed at the boundary between oceanic and continental basement with an isostatic gravity low situated over the presumed foundered continental slope. On the western margin of the southern Red Sea a steep gradient in the Bouguer gravity anomaly is observed across the rough-smooth zone boundary.

The interpretation that the magnetic quiet zones bordering the rifted continental margins are situated on a subsided continental basement is appealing because it may offer an explanation for at least parts of all quiet zones bordering this type of margin. Most of the previous interpretations (Table 1) have application only to the North Atlantic Ocean. For example, it would be fortuitous if all margins that exhibit quiet zones rifted at times of nearly constant geomagnetic polarity. The opening

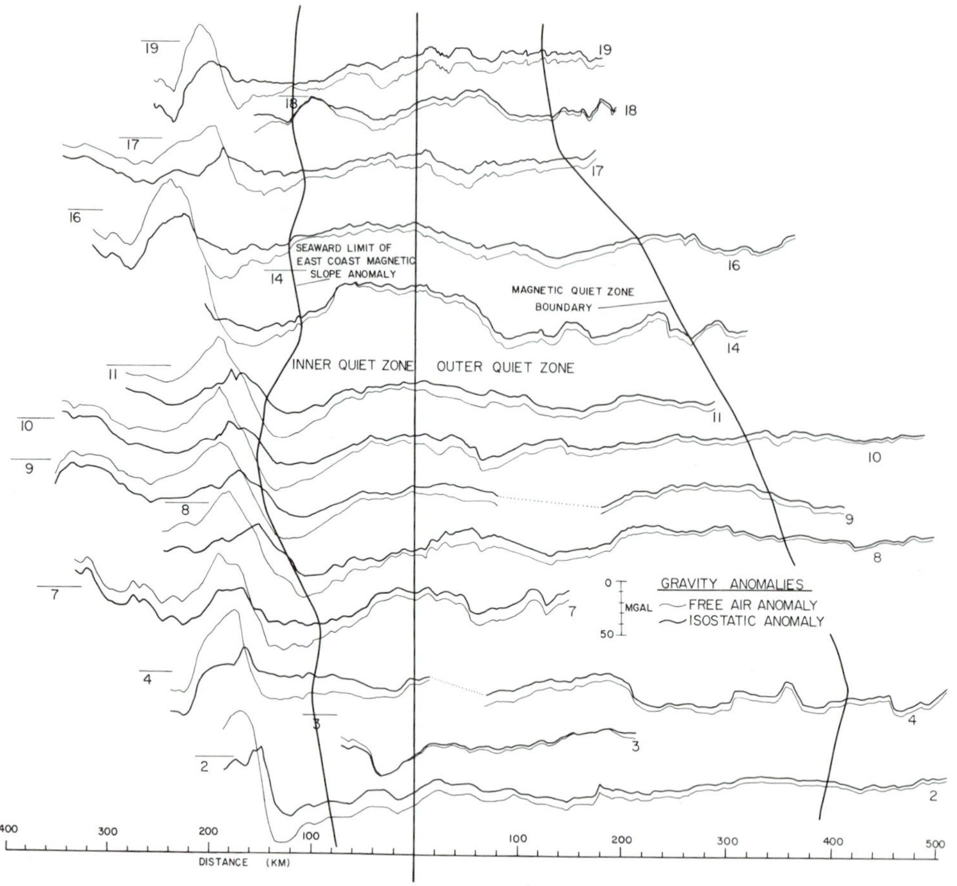

Fig. 4. Free-air and local isostatic gravity anomalies aligned with respect to magnetic anomaly E south of the New England Seamounts. The isostatic anomalies assume two-dimensionality and density contrasts of 1.57 and 0.47 g/cm^3 at the water-crust and crust-mantle interfaces, respectively, together with a depth of compensation of T = 30 km. On the ships' tracks 1, 5, 6, 12, 13, and 15 (Fig. 2), gravity was not recorded. Note that there is a relation between the axis of magnetic anomaly E and the seaward gravity high associated with the continental margin.

of the Norwegian Sea was about 60 my (Talwani and Eldholm, 1972), and the Red Sea may have opened up as late as 3.5 my (Lowell and Genik, 1972). At these times intervals of constant geomagnetic polarity are not known to exist. Also, the paleomagnetic data do not support the interpretation that all marginal quiet zones were formed as north-trending features at the magnetic equator. Thus, although the interpretations listed in Table 1 may explain a particular quiet zone, or parts of a particular quiet zone, none, with the exception of the interpretation of Talwani and Eldholm (1973), may offer an explanation for at least part of all quiet zones observed adjacent to the continental margins.

Emery et al. (1970) have noted that oceanic basement is observed on either side of the rough-smooth boundary in the western North Atlantic Ocean. It cannot be assumed, therefore, that subsided continental basement extends as far seaward as this boundary. We have noted, however, that anomaly E within the magnetic quiet zone separates the quiet zone into an inner and an outer quiet zone (Fig. 3). The inner quiet zone will be interpreted as being situated on a subsided continental basement in a fashion consistent with the interpretation of the quiet zones bordering the Norwegian, South African, and southern Red Sea margins. The outer quiet zone will be interpreted as being situated on an oceanic basement formed during a time period of nearly constant geomagnetic polarity (Graham or Newark interval).

Model magnetic-edge-effect calculations that assume a magnetized oceanic layer 2 with its landward edge terminating at the location of anomaly E are shown in Figure 6. The models are calculated assuming two-dimensionality, using the methods of Talwani and Heirtzler (1964) for a slab ½ km in thickness. The top of the slab is located at a depth of 7 km below sea level. The computations are made utilizing both the paleomagnetic pole positions for Upper Triassic (190 my) (Opdyke, 1961) and the Lower Jurassic (180 my) (Opdyke and Wensink, 1966). If the early stages of opening in the western North Atlantic took place between 155 my (the age established paleontologically for the rough-smooth zone boundary by Ewing et al., 1970) and 190-200 my ago (the age obtained for the intrusive rocks observed near the coastline of eastern United States

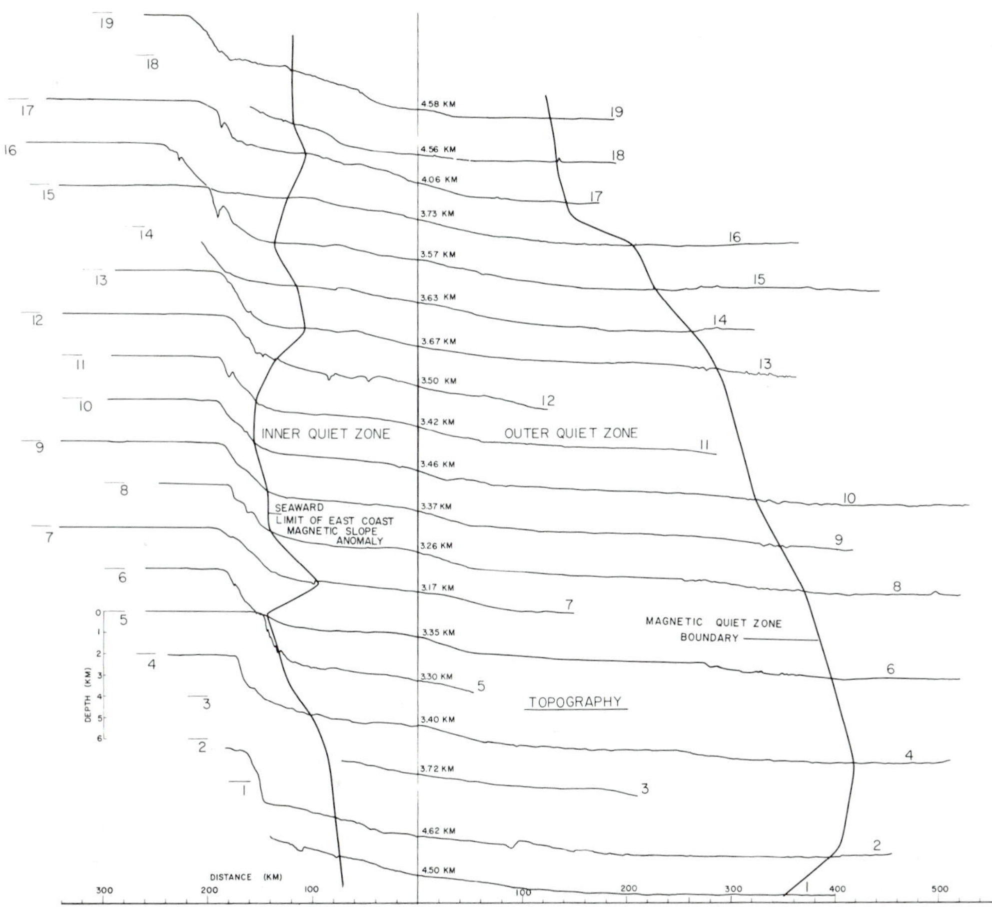

Fig. 5. Bottom topography aligned with respect to anomaly E. The water depth is noted at the axis of anomaly E. In most profiles anomaly E is located near the morphological change from upper to lower continental rise and is situated in water depths ranging from about 3.1 to 4.6 km.

by Erickson and Kulp, 1961), then the above paleomagnetic pole positions would be approximately correct. This is further substantiated by the recent work of Steiner and Helsley (1972) and Beck (1972), who indicate that the paleomagnetic pole position was reasonably stable during the Early Jurassic relative to North America. The computations assume reversed magnetization for the portion of the slab adjacent to the landward edge.

The results of the computations are consistent with the interpretation that anomaly E represents the boundary between continental and oceanic

Table 1. Previous conclusions with respect to the properties and formation of the magnetic quiet zone in the western North Atlantic.

Reference	Interpretation
Heirtzler and Hayes (1967)	Sea-floor spreading during late Carboniferous and Permian period of constant geomagnetic polarity (Kiaman interval).
Drake et al. (1968)	Quiet zone devoid of magnetic materials. A Paleozoic proto-Atlantic ocean existed in region of what is now a quiet zone.
Taylor et al. (1968)	Regional hydrothermal metamorphism, resulting in a reduction of the magnetic susceptibility of basalt, thereby creating a smooth magnetic field.
Vogt et al. (1970 b)	Sea-floor spreading generated by north-trending ridges at a low latitude near the magnetic equator.
Burek (1970); McElhinny and Burek (1971)	Sea-floor spreading during Late Triassic-Early Jurassic period of nearly constant geomagnetic polarity (Graham or Newark interval).
Pitman and Talwani (1972)	Rough-smooth quiet zone boundary is an isochron.
Luyendyk and Bunce (1973)	Slow sea-floor spreading produces smooth zone.

Table 2. Previous conclusions with respect to the origin of the east coast magnetic anomaly.

Reference	Interpretation
Keller et al. (1954)	Intrusive body, 48 km in width, that parallels the edge of the continent.
Drake et al. (1959)	Magnetic anomaly is associated with a basement ridge, defined by compressional wave velocities, located near the shelf break.
King et al. (1961)	Intrusive body near transition from continent to ocean. High susceptibilities necessary if magnetization is by induction only.
Drake et al. (1963)	Basement topography alone cannot account for anomaly. Intrabasement rocks are responsible for both the magnetic anomaly and shape of the basement.
Watkins and Geddes (1965)	Buried island arc is the source of the magnetic anomaly.
Keen (1969)	Anomaly arises as an "edge effect" resulting from a thick magnetic continental crust abutting against a thin magnetic oceanic crust and nonmagnetic oceanic mantle.
Taylor et al. (1968); Zietz (1971)	Magnetic source rocks felsic in composition and intrusive rather than extrusive.
Emery et al. (1970)	Sea-floor spreading models can account for east coast magnetic anomaly.
Luyendyk and Bunce (1973)	Concurs with Emery et al. (1970) that a sea-floor spreading model can account for the east coast magnetic anomaly. The anomaly represents a former mid-ocean crest which has jumped axes soon after its formation.

basement. Burek (1970) and McElhinny and Burek (1971) suggest that a long period of predominantly *normal* geomagnetic polarity existed from the Upper Triassic to the Upper Jurassic (Graham or Newark interval). However, important reversals do exist in this interval. Opdyke and McElhinney (1965) suggest that a reversal (Nuanetsi zone) occurs at the Triassic-Jurassic boundary. In order for anomaly E to represent a magnetic edge effect we must assume a reversed magnetization for the oceanic basement bordering the continental basement. This suggests that the Atlantic Ocean opened during a period of reversed magnetization, perhaps during the Nuanetsi zone.

East Coast Magnetic Anomaly

The east coast magnetic slope anomaly is a nearly linear anomaly that is observed along the continental margin of eastern United States and Canada (Taylor et al., 1968; Fig. 3). It is observed as far east as the longitude of Nova Scotia. Between the New England Seamounts and its eastern extension, the slope anomaly is located on the continental rise at depths, in places, exceeding 2,000 fm. South of the New England Seamounts, the slope anomaly is observed on either side of the shelf break and is located in shallower water depths than in the north.

East of New York the axis of the slope anomaly lies very near the shelf break. Farther south, at about 36°N, the magnetic slope anomaly splits in two parallel branches with the outer branch roughly following the contour at 800-1,000 fm. At 31°N the outer branch swings westward and terminates before reaching the coast, while the inner branch crosses the east coast near Brunswick, Georgia. Both the width and amplitude of the anomaly vary throughout its length. It has a minimum width of about 75 km off Cape May and a maximum of about 130 km off Savannah, Georgia. The amplitudes vary from about 400 to 700 gammas. In summary, the east coast "slope" anomaly does not necessarily follow the continental slope, but in places extends as far seaward as the continental rise and as far landward as the coastline.

Seaward of the magnetic slope anomaly is the magnetic smooth (or quiet) zone. Landward of the slope anomaly numerous magnetic anomalies are observed. There is some correlation of these landward magnetic anomalies with the Triassic basins observed near the coastline of eastern United States. Woollard (1943) suggests that the Triassic diabase sills in New Jersey are the cause of some anomalies under the coastal plain. Bonini and Woollard (1960) and MacCarthy (1936) reached similar conclusions for the coastal plain of South Carolina. Numerous interpretations have been proposed to explain the origin of the east coast magnetic anomaly, and these are outlined in Table 2.

Seismic-reflection profiles of Emery et al. (1970) have revealed the existence of a ridge complex situated near the continental rise north of the New England seamounts. South of the seamount chain, evidence for the ridge complex is vague. This ridge complex is *not* the same basement ridge observed by refraction studies of Drake et al. (1959), which is located near the shelf break. There is no evidence with respect to the age, origin, and composition of the ridge complex, although Emery et al. (1970) note that the length and location of the ridge complex suggest that it is not a salt diapiric structure but is related to the sea-floor spreading history of the western North Atlantic Ocean.

The location of the ridge complex in relation to the magnetic slope anomaly is shown in Figure 7. There is a systematic relationship of the ridge complex to the slope anomaly. This suggests that either ridge complex comprises, at least in part, the source rocks for the east coast magnetic anomaly or that a deeper source is responsible for both the ridge complex and the magnetic anomaly. South of the New England Seamount chain, the slope anomaly is located near the shelf break. The ridge complex is not observed in any of the profiler records obtained by either Lamont-Doherty or Woods Hole research vessels off the coast of New York, where the slope anomaly is on the continental shelf. With present techniques, the seismic multiples from the sea floor in shallow depths (~100 fm) obscure subbottom reflectors on the seismic profiler records. Thus, if the ridge complex is buried beneath the shelf sediments, we would not expect to observe it. Emery et al. (1970) show indication of the ridge complex in the regions off Cape Cod and Cape Hatteras. In these regions the presumed ridge complex is located in water depths of less than 1,000 fm (upper continental slope) and is associated with the magnetic slope anomaly.

Triassic intrusives into Precambrian and younger rocks exist near the present coastline of eastern United States and large-amplitude magnetic anomalies are associated with these intrusives (Woollard, 1943; Bonini and Woollard, 1960; MacCarthy, 1936). The correlation of the east coast magnetic anomaly with the ridge complex suggests

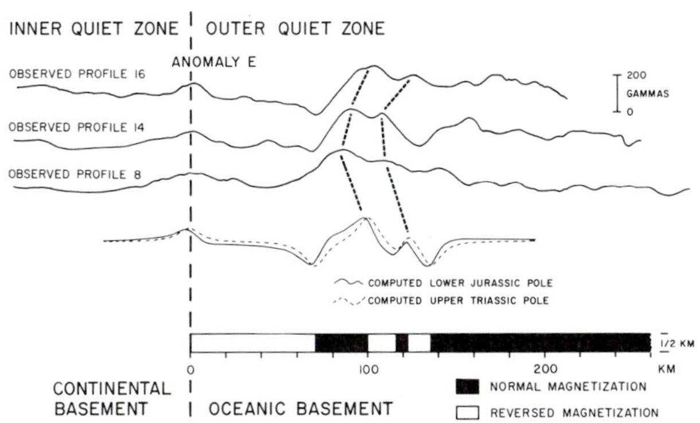

Fig. 6. Model of magnetic edge effect for a magnetic oceanic basement terminating against a nonmagnetic continental basement: assuming a magnetic source ½ km thick at a depth of 7 km, with a reversed magnetization of J = 0.008 emu. Profiles 8, 14, and 16 (Fig. 3) are shown with simulated profiles utilizing the North American Upper Triassic pole position (108°E, 63°N; Opdyke, 1961) and Lower Jurassic pole position (85.5°N, 124.5°E; Opdyke and Wensink, 1966), assuming a strike of N60°E. The simulated profiles utilizing either pole yield nearly symmetric positive magnetic anomalies which resemble the observed anomaly E. The double-peaked nature of the Blake Spur anomaly is simulated by assuming reversals within the outer quiet zone.

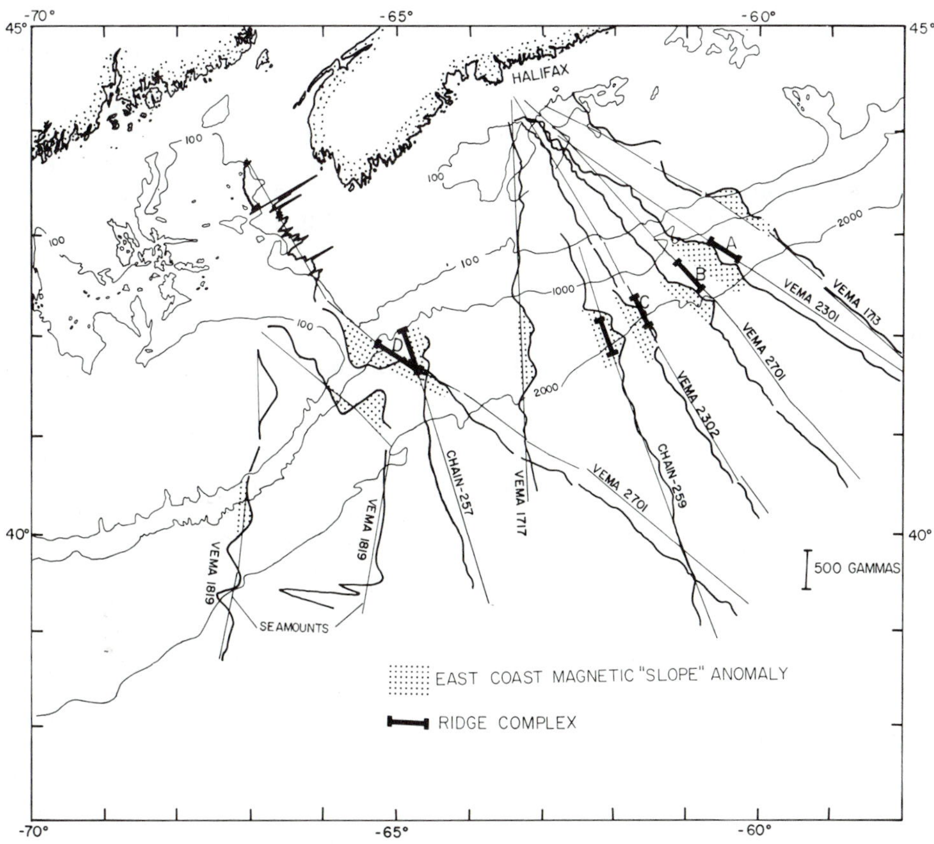

Fig. 7. East coast magnetic slope anomaly plotted along ship's tracks, with location of ridge complex noted. The CHAIN 257 and CHAIN 259 profiles are given in Emery et al. (1970). On VEMA cruises 1713, 1717, and 1819 the seismic profiler records were of poor quality in this region and were not useful for this study.

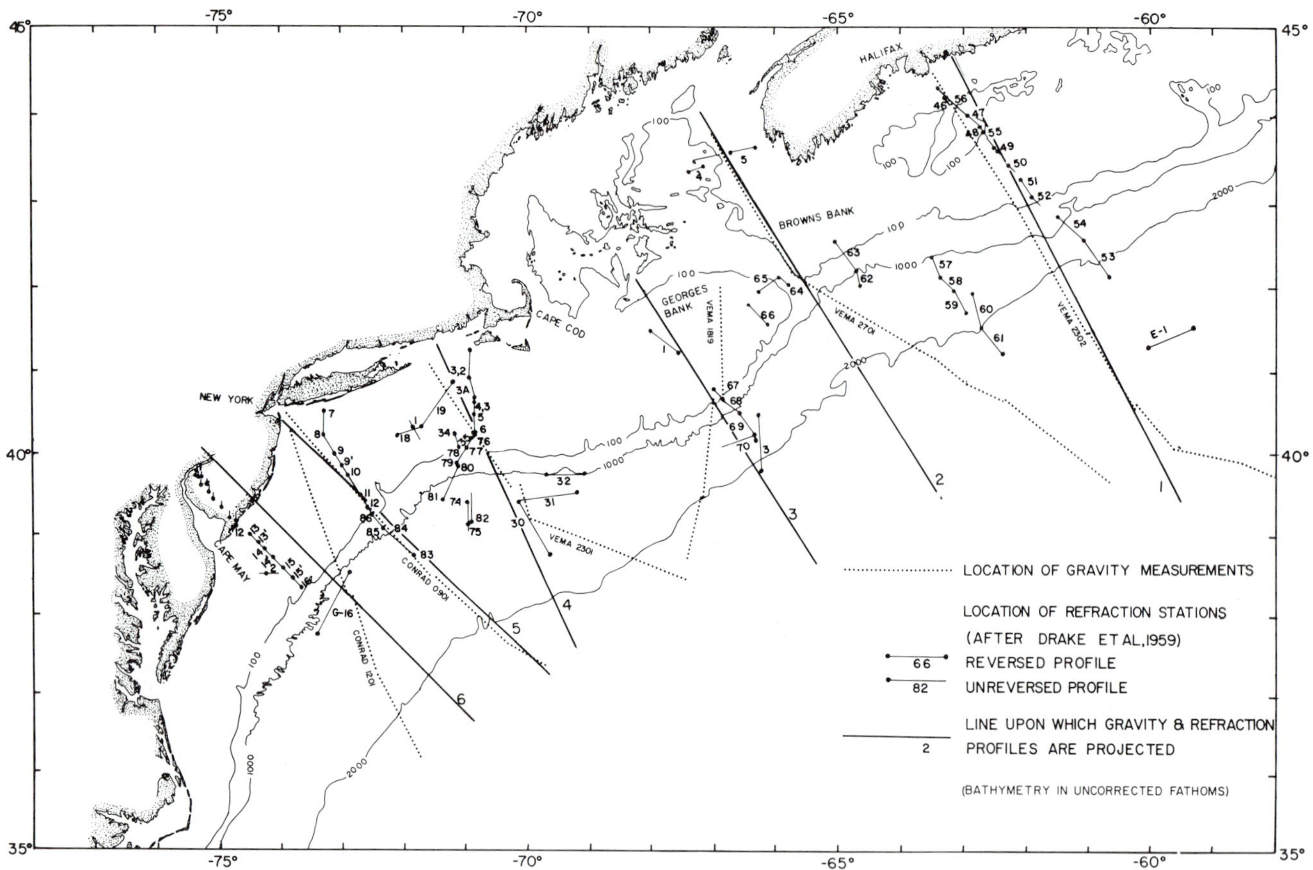

Fig. 8. Index map showing locations of gravity measurements and refraction stations on the east coast of North America.

perhaps that similar intrusives into continental basement may be located as far seaward as the continental rise.

The above interpretation of the slope anomaly is in great conflict with the interpretations that the anomaly is either a magnetic edge effect (Keen, 1969) or a sea-floor spreading anomaly (Emery et al., 1970; Luyendyk and Bunce, 1973). The first of these interpretations is plausible but improbable for three important reasons. First, if a magnetization contrast among oceanic crust, oceanic mantle, and continental crust exists in the fashion described by Keen (1969), we would expect to observe the slope anomaly along the entire east coast of North America. Second, as will be shown, the free-air gravity anomaly near the shelf break reflects for the most part an isostatic edge effect. We should therefore expect to observe a systematic spatial relationship between the peaks in the free-air anomaly associated with the shelf edge and the magnetic slope anomaly. This relationship is not present. Third, the magnetic slope anomaly in many regions branches into two parts, with the parts being somewhat similar in shape. An edge-effect model may explain one of the branches of the anomaly, but not both.

Luyendyk and Bunce (1973) indicate that a magnetic slope anomaly does not exist on the west African margin. However, isolated magnetic profiles near Cap Blanc and Cap Timiris on the African margin indicate a large-amplitude magnetic anomaly on the continental shelf bordering the magnetic quiet zone on the landward side (Rona et al., 1970). Near the coastline of Liberia, Behrendt and Wotorson (1970) have observed a linear high-amplitude magnetic anomaly associated with Triassic rocks intruded into Precambrian basement. Between Gran Canaria and the African coastline Roeser et al. (1970) have observed a magnetic anomaly extending for about 150 km, which shows a striking similarity in shape to the east coast magnetic slope anomaly. Seismic-refraction results of Roeser et al. (1970) indicate a continuation of a deep Upper Paleozoic sedimentary basin extending seaward from the Spanish Sahara. Their observed magnetic anomaly cuts obliquely across this basin. Seismic compressional velocities of 6.0 km/sec are observed at depths of about 10 km below sea level in the basin. The age of the oldest sediments in the basin (Upper Paleozoic), together with the deep seismic velocities observed (6 km/sec), suggest that the basement is continental and that the source rocks for the magnetic anomaly are situated within the continental basement. Thus the data on the African margin do not conflict with the concept that the magnetic slope anomaly arises from massive intrusions into continental basement, perhaps at the time when the continents first started to rift.

GRAVITY ANOMALIES AND SUBSURFACE RIDGES NEAR THE SHELF BREAK

Drake et al. (1959) have shown that two major sedimentary troughs, separated by a ridge, occur near the shelf break of eastern North America. The inner trough, located beneath the continental shelf, contains sediments up to 5 km or more in thickness; the outer trough under the continental rise and slope contains sediments, in places, greater than 6 km in thickness. A continuous free-air gravity high is also observed bordering the continental margin near the shelf break and is related to the subsurface ridge (Emery et al., 1970; Rabinowitz, 1973). The continuity of the free-air gravity high suggests that the ridge is continuous, as noted by Emery et al. (1970). Furthermore, local Airy isostatic computations indicate that the free-air gravity high does not solely reflect an isostatic edge effect (Emery et al., 1970; Rabinowitz, 1973).

Model structure sections were computed in the regions where gravity data are available near refraction profiles, in order to determine whether a *compensated* basement ridge can account for the isostatic anomaly. The locations of the seismic stations and gravity measurements are shown in Figure 8 and the computed models in Figure 9. Both the gravity and seismic data are projected onto a line roughly normal to the topographic contours. The local Airy isostatic anomaly assumes density contrasts of 1.57 g/cm^3 and 0.47 g/cm^3 at the water-crust and crust-mantle interfaces, respectively, together with a depth of compensation of 30 km.

A density of 2.30 g/cm^3 is chosen for the sediments in the calculations. In general, the velocity and hence the density, of the sediments increase with depth. However, Drake et al., (1959) have shown that for the sediment thicknesses of these sections, only very small differences ($\sim$a few mgal) in the gravitational attraction would be expected if we assume that the density is a function of depth rather than utilizing a constant density. The procedure used was to compensate the sediments at the crust-mantle interface in a fashion similar to compensating the topography in computing the local Airy isostatic anomaly. The gravitational effect of the sediments and its compensation is shown in Figure 9 with the local Airy isostatic anomaly and the isostatic anomaly with the sediments compensated. Although large accumulations of sediments are observed with large variations in basement relief, the effect of the sediments and its compensation is small (generally less than 10 mgal). When this effect is added to the local Airy isostatic anomaly, we are still left with a relative gravity high in the region of the basement ridge. The observations indicate that we still cannot explain totally the free-air gravity high associated with the shelf break by both the contribution of an isostatic edge effect resulting from the topography and its compensation and the basement ridge and its compensation. This suggests that intrabasement density highs are present.

It should be noted that basement in the above discussion is defined by a refractor with compressional wave velocities greater than 5.5 km/sec (Drake et al., 1959). If this refractor is not basement on the continental shelf but rather a high-velocity sedimentary layer, the above conclusions would not be altered substantially. The gravitational effect of the sediments and their compensation is small for the upper$\sim$5 km of crust (Fig. 9). Below this depth we would have to assume a smaller density contrast between the high-velocity sediments and basement, and hence yield an even smaller gravitational effect for the additional sediments and their compensation.

Seismic data together with gravity measurements have indicated basement ridges near the shelf edge of many rifted margins, e.g. the Gabon-Congo region of West Africa (Belmonte et al., 1965) and the Norwegian margin (Talwani and Eldholm, 1972). A feature common to many of the rifted margins thus far studied is that of isostatic gravity highs which generally appear as long, continuous trends near the shelf break. Off Norway, Talwani and Eldholm (1972) have associated these highs with high-density belts of Precambrian age. Off Angola, isostatic gravity highs near the shelf break trend onto a Precambrian basement high observed near the coastline (Rabinowitz, 1972). Isostatic gravity highs are also observed near the shelf break of Kenya (Rabinowitz, 1971). In this study gravity highs are shown to be associated with intrabasement density highs within a basement ridge near the shelf break.

Burk (1968) has suggested that the widespread existence of basement ridges on the rifted continental margins, and hence their associated gravity highs, indicates their common origin, possibly near the site of original rifting. Talwani and Eldholm (1972) have suggested that the gravity high associated with the shelf break off Norway results from dense rocks located at a "hinge" line for the subsidence of the continental slope and rise. The shelf break subsequently forms near this hinge line. Implicit in this interpretation is that the intrabasement density highs are located on a continental type of crust and are responsible for the location of the shelf break.

The subsidence of the region bordering the continental margin off eastern North America since the Late Jurassic and possibly earlier is well documented. Applin and Applin (1965), Swain (1947, 1952a, 1952b), Spangler and Peterson (1950), Dorf (1952), and Bartlett and Smith (1971) are among those who report deep drill holes on the coastal plain and continental shelf of eastern North America. The recovery of shallow-water microfauna and flora at depths, in places, in excess of 10,000 ft below sea level suggests a minimum subsidence of this amount. Heezen and Sheridan (1966) and Sheridan et al. (1969) report the recovery of Neocomian-Aptian shallow-water algal fragments from the precipitous Blake Escarpment at depths of about 5 km

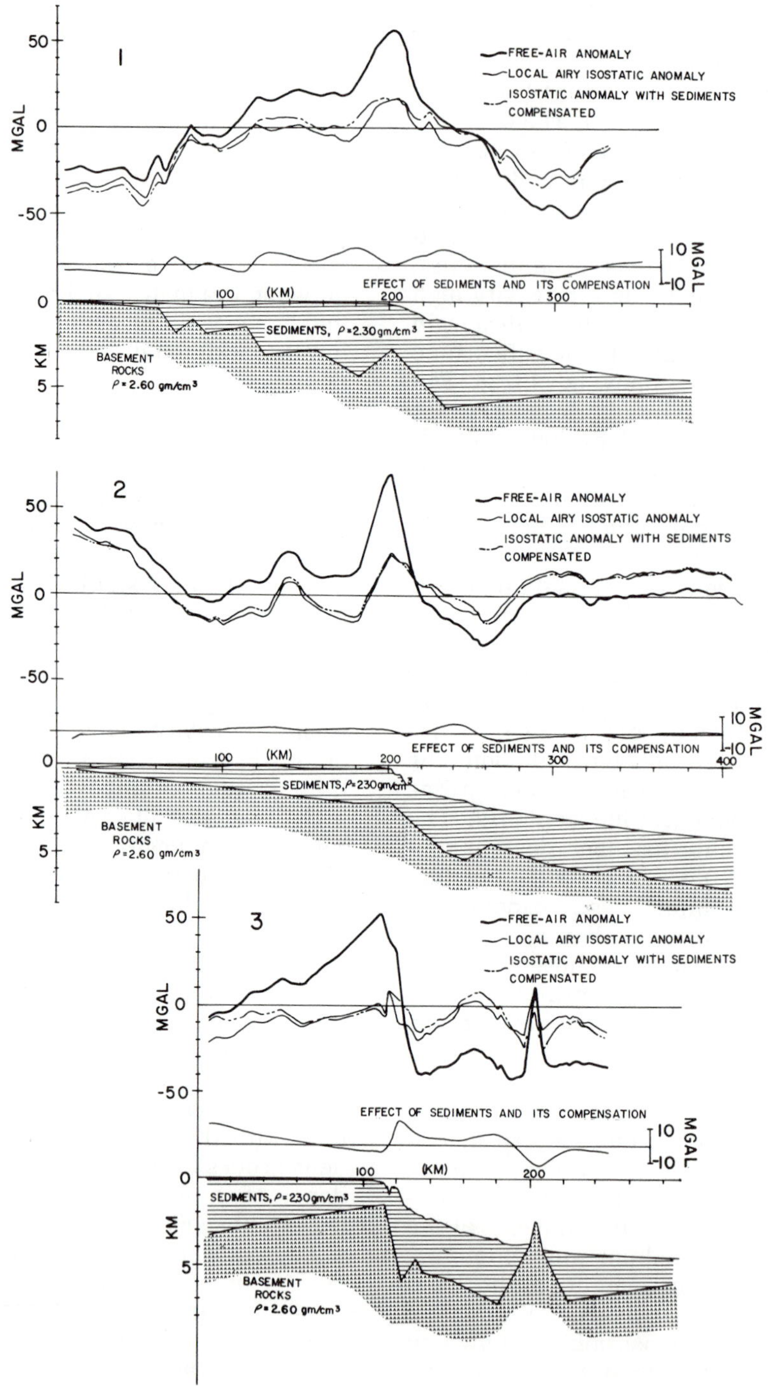

Fig. 9. Gravity computations over the subsurface ridge on the continental margin of the western North Atlantic Ocean (see Fig. 8 for locations). Each model assumes densities of 2.30 and 2.60 for the sediments and basement rocks, modified from Drake et al. (1959). The free-air anomaly is given together with a computed local Airy isostatic anomaly, compensating the topography

below sea level. Results of deep-sea drilling in the northwest Atlantic Ocean (Hollister et al., 1972) have revealed a continued subsidence of the continental rise from Late Jurassic to Late Cretaceous. From the Cenomanian to early Tertiary the sediments sampled indicate a period of stabilization followed by continental rise upbuilding from early Tertiary to the present (Lancelot et al., 1972).

Thus major portions of the continental margin of the east coast of North America have subsided considerably since at least Late Jurassic and possibly earlier. Furthermore, the gravity anomalies along this margin strongly indicate intrabasement density highs near the shelf edge. If these high-density belts predate the inception of rifting, as off Norway, then their location near the shelf edge

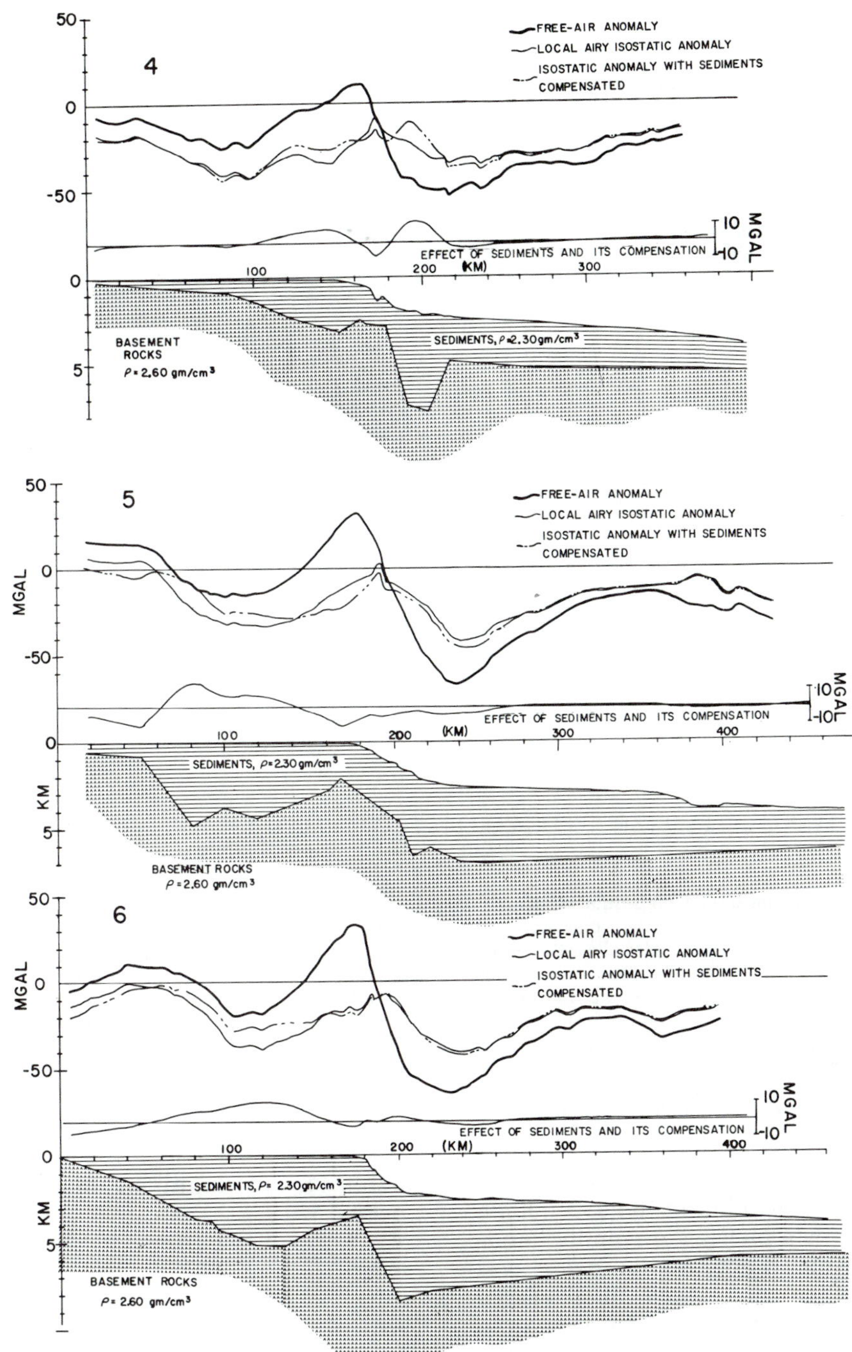

only, and an isostatic anomaly compensating both the topography and the sediments. Results indicate that intrabasement highs are present.

suggests that they are indeed important in the early development of the shelf edge.

The preceding discussion does not necessarily imply that the shelf edge has not been built out since its initial formation. Off many parts of the east coast the outbuilding has been relatively small. Seismic-reflection profiles of Uchupi and Emery (1967) indicate that the continental slope has been built out only about 5-10 km during the Tertiary for major portions of the margin north of Cape Hatteras. Between Miami and Cape Hatteras, Uchupi and Emery (1967) report as much as 35 km of progradation of the shelf edge. Emery et al. (1970) report that the continental slope off the entire eastern United States was built seaward an average of about 7 km during the Tertiary, even though the continental

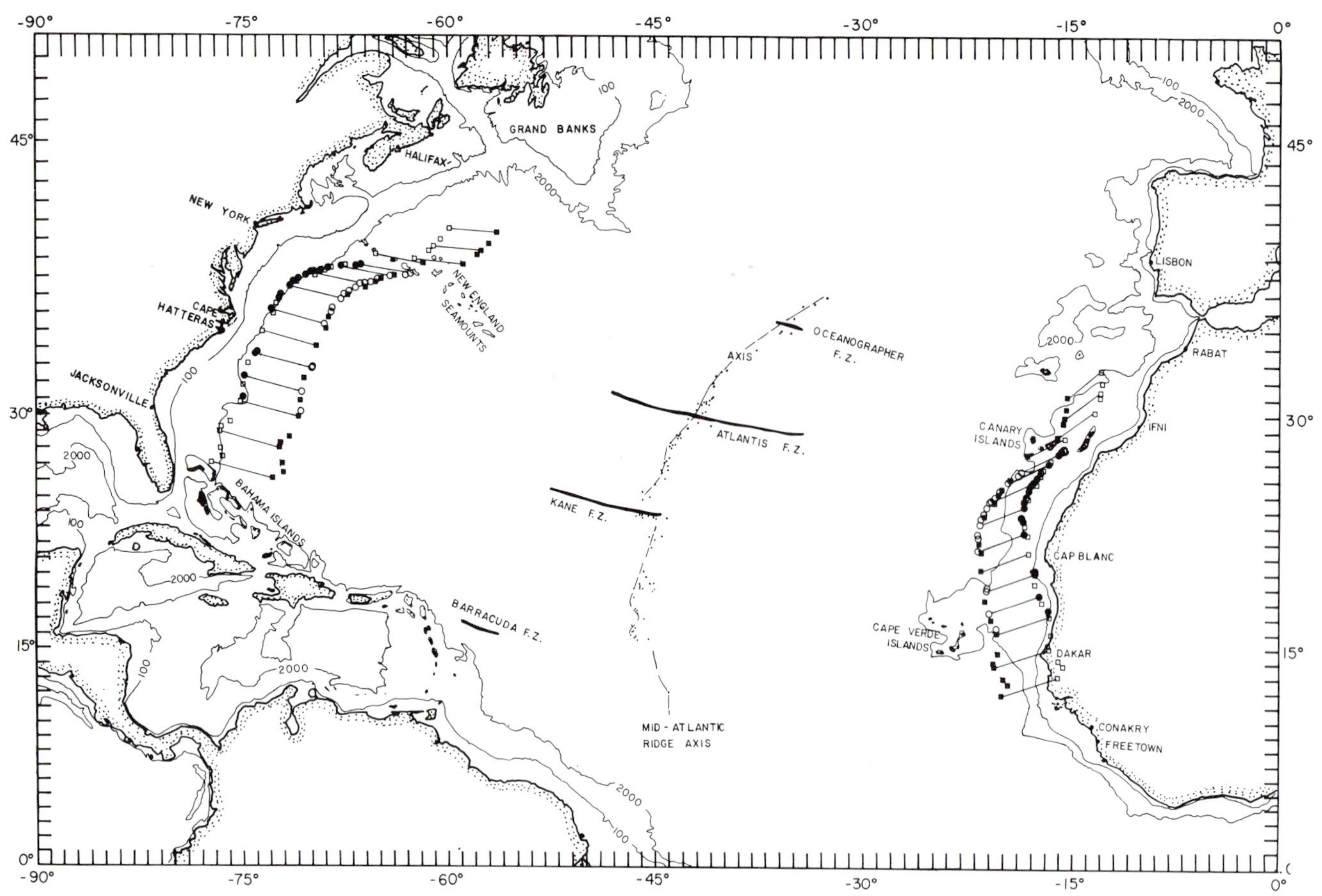

Fig. 10. Rotation of anomaly E onto the rough-quiet zone boundary and onto the conjugate positions on the African margin. On the North American margin the darkened circles and squares show the location of anomaly E and the rough-smooth zone boundary, respectively. The open circles show the fit of the rotation of anomaly E onto the rough-smooth boundary utilizing a pole of 64°N, 55°W, rotated 6.5°. The open squares show the location of the same rotation in the opposite direction, of the rough-smooth zone boundary onto anomaly E. In this fashion we should be able to predict the location of a presumed anomaly E in regions where more data are available for the location of the smooth-rough zone boundary than for the location of anomaly E. These boundaries are rotated to close the quiet zones (68° rotation about 63°N, 18.5°W; Pitman and Talwani, 1972) and are shown in their conjugate position on the African margin. The lines connecting the circles and squares are possible flow lines of early motion. The location of the ridge axis, earthquake epicenters (small dots along ridge axis), and fracture zones are afte Pitman and Talwani (1972).

slope was eroded by turbidity currents and large-scale sliding occurred. The relief of the subsurface ridge may have acted as a barrier preventing land-derived sediments from reaching the deep sea. When the landward troughs were filled, large amounts of land-derived sediments were then able to bypass the basement highs and deposit on the continental slope and rise. Emery et al. (1970) suggest that this occurred during the middle Eocene.

Dietz (1963) and Dietz and Holden (1966) suggest that modern eugeosynclinal continental rises may have grown through geologic time by continental accretion. They suggest a possible mechanism of isostatically induced subsidence resulting from thick continental rise sedimentary accumulations which cause a downwarp of the adjacent sialic continental shelf. The shelf edge subsequently progrades on this marginal flexure. This mechanism would apply after large accumulations of sediments are deposited on the continental rise and after an initial structural margin has been formed.

RECONSTRUCTION OF THE NORTH ATLANTIC OCEAN

Model magnetic-edge-effect calculations have been shown to be consistent with the concept that anomaly E is a magnetic signature representing the boundary between oceanic and continental basement. Stretching and thinning of the continental crust may have preceded the initial extrusion of oceanic crust. At the present time we have no means to take into account any presumed stretching of continental crust in the reconstruction. The reconstruction will therefore be only to the line of presumed initial rifting.

The magnetic lineation patterns in the disturbed zone of the North Atlantic Ocean have been formed at the Mid-Atlantic Ridge axis, are isochrons, and represent former plate margins. Therefore, if we fit together isochrons of the same age on either side of the ridge axis, we will observe the relative configuration of the continents and oceans at the time of the isochron. Utilizing trial-and-error techniques, a pole and a rotation about that pole was found that fits anomaly E onto the rough-smooth zone boundary (Fig. 10). The pole utilized for the rotation is located at 64°N, 55°W, with a required angle of rotation of 6.5°. The fact that such a good fit of anomaly E onto the rough-smooth zone boundary is obtained suggests that anomaly E is an isochron. South of about 30°N and also north of the New England Seamount chain, more data are available for the location of the rough-smooth zone boundary than for anomaly E. Therefore, the rough-smooth zone boundary was rotated about the same pole but in the opposite direction, to juxtapose onto anomaly E. In this fashion we should be able to predict the position of anomaly E in regions where the data are sparse. Such a rotation places the presumed location of anomaly E south of about 30°N, near the base of the Blake Escarpment (Fig. 10). This is a rather interesting position in view of the previously discussed interpretations pertaining to subsided continental margins. The location of the rotation north of the New England Seamounts is also shown in Figure 10.

Keen et al. (1973) indicate that a seismic reflector, interpreted as oceanic basement, extends as far landward as the ridge complex described earlier for the region north of the New England Seamounts. This is in conflict with the above *predicted* anomaly E, which is interpreted as representing the boundary between oceanic and continental basement. If oceanic basement extends farther landward than predicted for the region north of the New England Seamounts, and if anomaly E indeed is located at the boundary between oceanic and continental basement for the region south of the New England Seamounts, this would imply either that (1) the generation of oceanic basement by sea-floor spreading started perhaps sooner to the north, or (2) the rate of generation of this part of the North Atlantic during the time preceding the formation of the rough-smooth zone boundary was greater. In any case, both possibilities imply an initial three-plate configuration of the Early Jurassic North Atlantic.

The next step after the rotation of anomaly E onto the rough-smooth zone boundary is to rotate both these boundaries to their conjugate positions on the African side of the Atlantic Ocean. The pole of rotation found by Pitman and Talwani (1972) (68° rotation about 63°N, 18.5°W) was used in fitting the rough-smooth zone boundaries. The pole of rotation used to rotate anomaly E onto the rough-smooth zone boundary (6.5° rotation about 64°N, 55°W) was rotated about Pitman and Talwani's pole to determine the new pole position for the rotation of a presumed anomaly E onto the African margin. The new pole which determines the average direction of flow lines of Africa from the time of the rough-

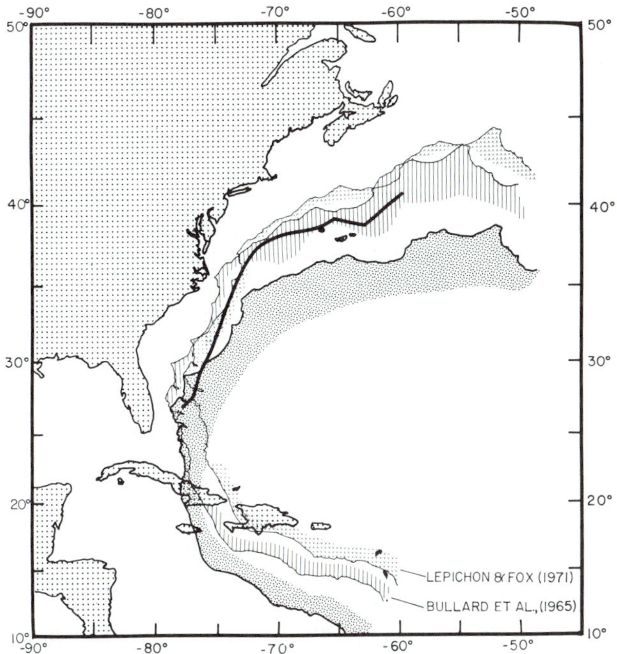

Fig. 11. Comparison of the reconstruction of the North Atlantic Ocean utilizing anomaly E as the ocean-continent boundary to reconstructions of Bullard et al. (1965) and LePichon and Fox (1971). North America is held fixed relative to Africa. The bold line represents the location of initial Mid-Atlantic Ridge axis for the reconstruction in this paper. Note that the relative position of Africa is farther south in the reconstruction that utilizes anomaly E than in the other two reconstructions.

smooth zone boundary to the presumed location of anomaly E on the African coast is calculated to be a rotation of 6.5° about a pole located at 49.7°N and 35.1°W. The positions of these rotations are shown in Figure 10.

Utilizing the above poles and rotation angles, and assuming that anomaly E represents the boundary between continental and oceanic basement, a reconstruction of the North Atlantic Ocean has been made to this boundary and is shown in Figure 11 together with two other reconstructions (Bullard et al., 1965; LePichon and Fox, 1971). The bold line in the reconstruction represents the region of the original separation of the continents. The total poles and rotations necessary to bring the continents to their original configuration for the three reconstructions are given in Table 3.

The important assumptions in making the reconstruction are:

1. Anomaly E is a magnetic edge effect and represents the boundary between oceanic and continental basement.

2. The stretching and deformation of the continental basement is not considered in the reconstruction. The continents are reconstructed utilizing assumption 1, and hence this reconstruction is only to the line of initial extrusion of oceanic crust.

3. Sea-floor spreading from the time of initial extrusion of oceanic crust to the time of the formation of the rough-smooth zone boundary has been

symmetric on either side of the ridge axis.

The reconstruction assuming that anomaly E represents the boundary between continental and oceanic basement places Africa in a position farther south than the reconstructions of Bullard et al. (1965) and LePichon and Fox (1971). This fit places certain features in such a fashion relative to the presumed edge of the continental basement as to support concepts of a subsiding continental crust. This will be discussed, as well as the apparent serious overlap of the ocean-continental boundary onto the Senegal Basin.

Blake Escarpment

It was noted earlier that shallow-water limestones of Early Cretaceous age have been dredged by Heezen and Sheridan (1966) and Sheridan et al. (1969) at water depths of about 5 km near the base of the Blake Escarpment. Thus the Blake Escarpment has undergone a minimum of about 5 km of subsidence since the Early Cretaceous. Seismic refraction velocities (Vp) obtained by Sheridan et al. (1966) indicate velocities typical of the oceanic layer 3 (6.65 km/sec) beneath the Blake-Bahama Basin seaward of the Blake Escarpment and a basement ridge beneath the seaward edge of the Blake Plateau with velocities typical of continental crystalline rocks (5.7-6.1 km/sec). In our reconstruction the boundary between oceanic and continental basement lies near the base of the Blake Escarpment. The geophysical and geological evidence thus provide favorable evidence in support of the above reconstruction and the concept of a subsiding continental margin consisting of a continental type of basement.

Senegal Basin

The conjugate position of the Blake Escarpment on the African coast lies within the Senegal Basin in our reconstruction. If our reconstruction is correct, then this implies that the Senegal margin must have prograded seaward about 200 km. Evidence from deep drilling within the Senegal Basin indicates that indeed the basin has undergone considerable subsidence since the Jurassic concommitant with considerable transgressive sedimentation.

Ayme (1965) noted that Precambrian basement consisting of slates, quartzites, and granites outcrop approximately 500 km east of the coastline. Seaward of the outcrop, but still at a distance greater than about 200 km from the coastline, the basement rocks are overlain by Paleozoic rocks consisting of Cambrian sandstones, Silurian shales, and sandstones and shales of Devonian age. Lying unconformably above the Paleozoic rocks are Mesozoic shales, sands, and limestones covered by thin (less than 600 m) Tertiary and Quaternary beds. Near the coastline, the pre-Mesozoic structure and lithology are *not* known; drilling has indicated that the Cretaceous in this region may be about 4.5 km in thickness compared to tens of meters farther east. Ayme (1965) notes that results of calculations performed on unpublished aeromagnetic profiles estimate the basement depth to be as great as 10 km near the coastline. Furthermore, seismic profiles show that the Cretaceous beds dip more steeply beneath the continental slope than the slope itself, indicating a prograding of the slope region. Castelain (1965) has noted that the basin sediments are essentially continental in the east and become thicker and more marine to the west.

Table 3. Parameters of the reconstruction of the North Atlantic Ocean.

	Position		
A. Rotations and poles to close up North Atlantic			
	Position		Degrees of Opening
Bullard et al., 1965	67.6°N, 14.0°W		74.8°
LePichon and Fox, 1971	66.0°N, 12.0°W		74.8°
This paper	62.2°N, 24.3°W		80.6°
B. Rotations and poles used for time of initial opening to time of rough-smooth zone boundary			
(1) With reference frame fixed to North America		Western Basin	Eastern Basin
LePichon and Fox, 1971	58.3°N, 21.8°W	8.80°	4.4°
Pitman and Talwani, 1972	56.8°N, 73.3°E	8.93°	(total for eastern and western basins)
This paper	64.0°N, 55.0°W	6.50°	6.5°
(2) With reference frame fixed to Africa			
LePichon and Fox, 1971	58.3°N, 1.0°W	8.8°	4.4°
Pitman and Talwani, 1972	72.7°N, 165.7°W	8.93°	(total for eastern and western basins)
This paper	49.7°N, 35.1°W	6.5°	6.5°
(3) Rotation and pole used to close up rough-smooth zone boundary on either side of ocean (after Pitman and Talwani, 1972)			
	63°N, 18.5°W		Rotation — 68°

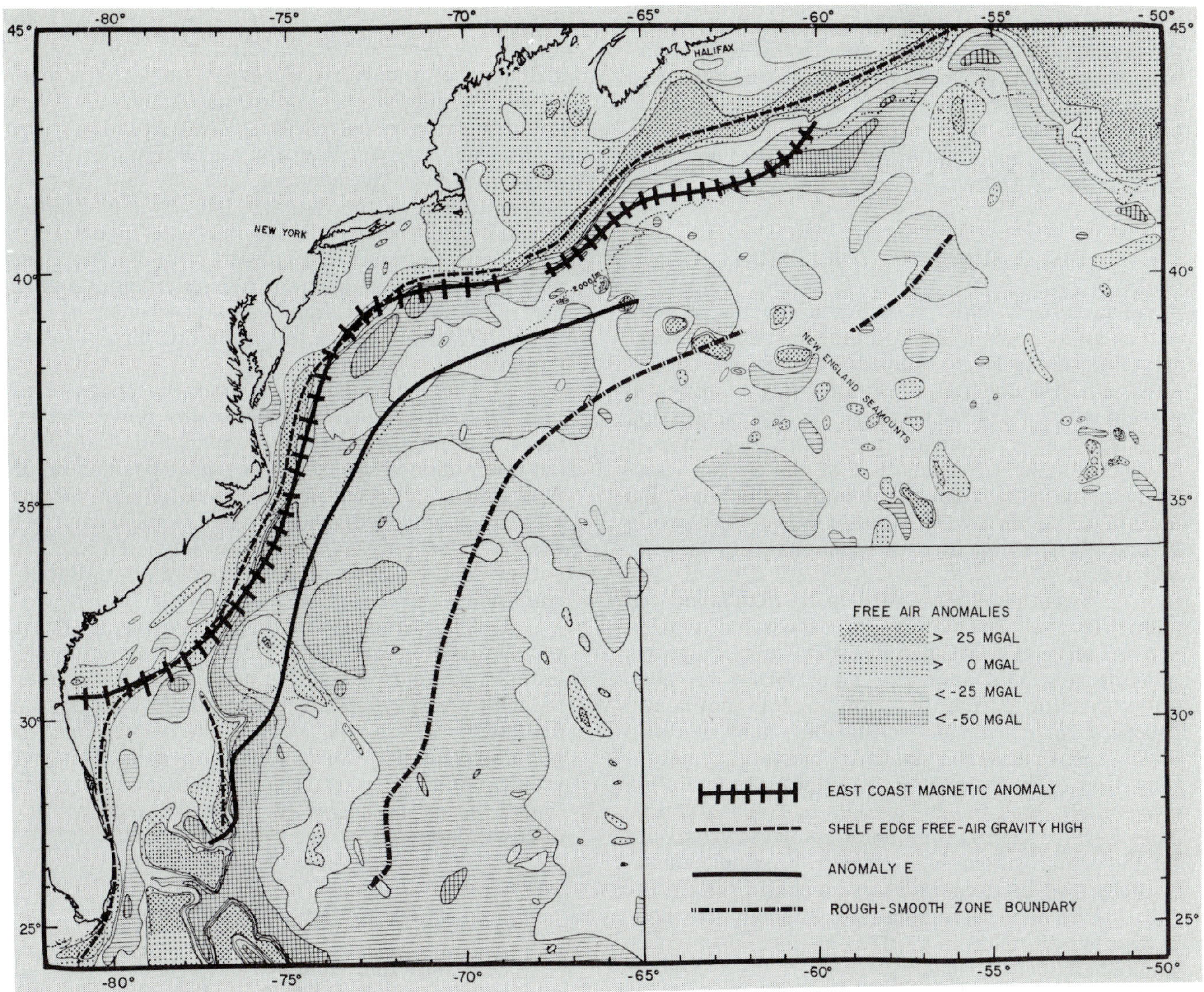

Fig. 12. Composite diagram showing the location of the geophysical lineaments along the continental margin of eastern North America (after Rabinowitz, 1973).

Ayme, (1965) states: "It would seem that the form of the Senegal Basin has been determined by regional basement features with vertical movements along hingelines. A major subsidence started possibly as early as the Jurassic. This was followed by a general transgression during Albian and Aptian times and the continuing subsidence and deposition of material carried from the east resulted in a thick monotonous and almost continuous sedimentation until the end of Cretaceous times."

McMaster and Lachance (1968) have noted that the deepest observable reflectors on seismic profiler records dip offshore more steeply than the present ocean bottom in the region south of Cap Blanc, indicating a prograding of the shelf edge. This is the region where anomaly E starts to curve landward. They note that the construction of the shelf results from uniform outbuilding and upbuilding, in contrast to the regions farther north.

These observations are consistent with the concept of a subsiding continental margin. If the reconstruction of the Atlantic Ocean is correct, then the western portion of the Senegal Basin is situated on an oceanic crust that has been filled with sediments and prograded seaward about 200 km.

The reconstruction utilizing anomaly E as the magnetic signature of the boundary between oceanic and continental basement differs somewhat from previously published reconstructions. Subsidence of large portions of the continental margin is inferred. The boundary between oceanic and continental basement is located near the base of the Blake Escarpment in the western North Atlantic Ocean and within the Senegal Basin of West Africa. Both these regions have undergone considerable subsidence, together with a considerable prograding of the shelf edge.

The utilization of the above pole and rotation for the reconstruction of the North Atlantic Ocean, and Bullard et al.'s (1965) pole for the reconstruction of the South Atlantic Ocean, does *not* resolve the serious overlap of South America onto Precambrian rocks of British Honduras, Nicaragua, and Mexico. This suggests that if the above model is correct, further work is required to obtain an improved pole position for the reconstruction of the South Atlantic Ocean.

DISCUSSION AND CONCLUSIONS

The interpretation presented for the gravity and magnetic anomalies, and the subsequent reconstruction of the North Atlantic Ocean, are consistent with the concept of a subsiding continental margin with part of the magnetic quiet zone situated on a subsided continental basement. A composite diagram showing the location of the various geophysical and geological lineaments located near the margin of the northwest Atlantic Ocean is shown in Figure 12. The conclusions summarized in this figure are:

1. A continuous free-air gravity high is located near the shelf break and is associated with a subsurface ridge. Local Airy isostatic calculations indicate that this anomaly is not totally an edge effect resulting from the thinning of continental crust as the margin is crossed but must relate to mass excess below the sea floor. Isostatic computations that take into account the thick accumulation of sediments and the subsurface ridge observed by seismic-refraction techniques cannot fully account for the gravity anomaly. Thus intrabasement density highs may be present near the shelf break.

2. The east coast magnetic anomaly has been shown to be related to a ridge complex observed on Lamont-Doherty seismic profiler records as well as on previously published seismic records of Emery et al., 1970, for the region north of the New England Seamounts. The continuity of the slope anomaly south of the New England Seamounts suggests that the ridge complex may be present in this region. Emery et al. (1970) show indications of this for the region south of Cape Hatteras.

A magnetic anomaly (anomaly E) has been identified within the magnetic quiet zone. This anomaly separates the quiet zone into an inner and outer quiet zone. The inner quiet zone, characterized by a very subdued magnetic field, is interpreted as being situated over a subsided continental basement. The outer quiet zone, characterized by smaller-amplitude anomalies than those observed in the rough zones, is interpreted as being located over oceanic basement formed during the Newark interval of predominantly normal geomagnetic polarity (Burek, 1970).

A pole and a rotation about that pole was found by trial-and-error techniques to rotate anomaly E onto the rough-smooth zone boundary. Both anomaly E and the rough-smooth zone boundary were rotated to their conjugate positions on the African side of the ocean utilizing the pole and the rotation determined by Pitman and Talwani (1972) to close up the rough-smooth zone boundaries, and a reconstruction of the North Atlantic Ocean has been made. A number of important observations are noted in this reconstruction. They include:

1. The derived flow lines of early motion are consistent with the concept that the New England Seamounts and the Canary Islands constitute a single fracture zone, the New England-Canary Fracture Zone (Pitman and Talwani, 1972). The three easternmost Canary Islands lie on continental basement. Three of the Canary Islands—Tenerife, Henno, and Gomera—align along the flow lines of early opening.

2. The reconstruction places the ocean-continent boundary near the base of the Blake Escarpment, and thus supports the concept of a subsiding continental basement. The conjugate position on the African margin is within the Senegal Basin. Geological and geophysical evidence in this region suggests considerable subsidence of the Senegal Basin, together with a major prograding of the continental shelf break.

3. The problem of the apparent asymmetry in the widths of the magnetic quiet zones on either side of the ocean is solved in the present reconstruction. Most of the asymmetry corresponds to a greater distance from the shelf edge to the ocean-continent boundary on the North American side. This may result, at least in part, from the seaward progradation of the shelf edge on the African margin south of about Dakar.

ACKNOWLEDGMENTS

The data and conclusions presented in this paper are a portion of a Ph.D. thesis submitted in the spring of 1973 to Columbia University. I wish to express my appreciation to Manik Talwani for his support and guidance, which were invaluable to the development of this research. J. Cochran, O. Eldholm, W. Ryan, and A. Watts critically read the manuscript and provided valuable comments. The data on which this paper is based were obtained over many years by Lamont ships, and their collection involved many scientists under the general direction of Maurice Ewing.

This work has been primarily supported by the National Science Foundation under grants GA-580, GA-10728, GA-17761, GA-27281, and GA-1434 and under Office of Naval Research contracts N00014-67-A-0108-0004 and NONR 266 (79). This article is Lamont-Doherty Geological Observatory Contribution 2097.

BIBLIOGRAPHY

Abdel-Monem, A., 1969, K-Ar geochronology of volcanism on the Canary Islands [Ph.D. thesis]: Columbia Univ., New York, 1969.

Anderson, C. N., Vogt, P. R., and Bracey, D. R., 1969, Magnetic anomaly trends between Bermuda and the Bahama-Antilles Arc (abstr.): EOS (Trans. Am. Geophys. Union), v. 50, p. 189.

Applin, P. L., and Applin, E. R., 1965, The Comanche series and associated rocks in the subsurface in central and south Florida: U.S. Geol. Survey Prof. Paper 447, 86 p.

Ayme, J. M., 1965, The Senegal Salt Basin, *in* Kennedy, W. Q., ed., Salt basins around Africa: London, Inst. Petroleum, p. 83-90.

Bartlett, G. A., and Smith, L., 1971, Mesozoic and Cenozoic history of the Grand Banks of Newfoundland: Can. Jour. Earth Sci., v. 8, p. 65-84.

Beck, M. E., Jr., 1972, Paleomagnetism of Upper Triassic diabase from Pennsylvania: further results: Jour. Geophys. Res., v. 77, p. 5673-5687.

Behrendt, J. C., and Wotorson, C. S., 1970, Aeromagnetic and gravity investigations on the coastal area and continental shelf of Liberia, West Africa, and their relation to continental drift: Geol. Soc. America Bull., v. 81, p. 3563-3574.

Belmonte, Y., Hirtz, P., and Wenger, R., 1965, The salt basin of the Gabon and Congo (Brazzaville), *in* Kennedy, W. Q., ed., Salt basins around Africa: London, Inst. Petroleum, p. 55-74.

Bonini, W. E., and Woollard, G. P., 1960, Subsurface geology of North Carolina-South Carolina coastal plain from seismic data: Am. Assoc. Petr. Geol. Bull., v. 44, no. 3, p. 298-315.

Bosshard, E., and Macfarelane, D. J., 1970, Crustal structure of the western Canary Islands from seismic refraction and gravity data: Jour. Geophys. Res., v. 75, p. 4901-4918.

Bullard, E. Everett, J. E., and Smith, A. G., 1965, The fit of the continents around the Atlantic, *in* A symposium on continental drift: Phil. Trans. Roy. Soc. London, v. 258, p. 41-51.

Burek, P. J., 1970, Magnetic reversals: their application to stratigraphic problems: Am. Assoc. Petr. Geol. Bull., v. 54, p. 1120-1139.

Burk, C. A., 1968, Buried ridges within continental margins: Trans. N.Y., Acad. Sci., v. 30, no. 3, p. 397-409.

Castelain, J. 1965, Aperçu stratigraphique et micropaleontologic: Mem. B.R.G.M., no. 32, p. 135-160.

Dash, B. P., and Bosshard, E., 1968, Crustal studies around the Canary Islands: 23rd Internatl. Geol. Congr. Prague, v. 1, p. 249-260.

Dewey, J. F., Pitman, W. C., III, Ryan, W. B. F., and Bonnin, J., 1973, Plate tectonics and the evolution of the Alpine system: Geol. Soc. America Bull., v. 84, p. 3137-3180.

Dietz, R., 1963, Collapsing continental rises: an actualistic concept of geosynclines and mountain building: Jour. Geology, v. 71, p. 314-333.

———, and Holden, J. C., 1966, Miogeoclines in space and time, Jour. Geology, v. 74, p. 566-583.

Dorf, E., 1952, Critical analysis of cretaceous stratigraphy and paleobotany of Atlantic Coastal Plain, Am. Assoc. Petr. Geol. Bull., v, 36, p. 2161-2184.

Drake, C. L., Ewing, M., and Sutton, G. H., 1959, Continental margins and geosynclines: the east coast of North America, north of Cape Hatteras, *in* Physics and geochemistry of the earth, v. 3: Elmsford, N.Y., Pergamon Press, p. 110-198.

———, Heirtzler, J., and Hirshman, J., 1963, Magnetic anomalies off eastern North America: Jour. Geophys. Res., v. 68, p. 5259-5275.

———, Ewing, J. I., and Stockard, H., 1968, The continental margin of the eastern United States: Can. Jour. Earth Sci., v. 5, p. 993-1010.

Einwich, A. M., 1972, A magnetic zone west of the "Smooth Zone," east coast of north America: Am. Geophys. Union Trans., v. 53, p. 365.

Emery, K. O., Uchupi, E., Phillips, J. D., Bowin, C. O., Bunce, E. T., and Knott, S. T., 1970, Continental rise off eastern North America: Am. Assoc. Petr. Geol. Bull., v. 54, no. 1, p. 44-108.

Erickson, G. P., and Kulp, J. L., 1961, Potassium-argon dates on basaltic rocks, Ann. N.Y. Acad. Sci., v. 91, p. 321-323.

Ewing, J., Hollister, C., Hathaway, J., Paulus, F., Lancelot, Y., Habib, D., Poag, C. W., Luterbacher, H. P., Worstell, P., and Wilcoxon, J. A., 1970, Deep Sea Drilling Project, Leg XI: Geotimes, v. 15, p. 14-16.

Heezen, B. C., 1968, The Atlantic Continental Margin, (Univ. Missouri at Rolla) Jour., no. 1, p. 5-25.

Heezen, B. C., and Sheridan, R. E., 1966, Lower Cretaceous rocks (Neocomian-Albian) dredged from Blake Escarpment: Science, v. 154, p. 1644-1647.

———, Hollister, C. D., and Ruddiman, W. F., 1966, Shaping of the continental rise by deep geostrophic contour currents: Science, v. 152, p. 502-508.

Heirtzler, J. R., and Hayes, D. E., 1967, Magnetic boundaries in the North Atlantic Ocean: Science, v. 157, p. 185-187.

Hollister, C. D., Ewing, J. I., Habib, D., Hathaway, J. C., Lancelot, Y., Luterbacker, H., Paulus, F. C., Poag, C. W., Wilcoxon, J. A., and Worstell, P., 1972, Initial reports of the Deep Sea Drilling Project, XI: Washington, D.C., U.S. Govt. Printing Office, 1077 p.

Hutchinson, R. W., and Engels, G. G., 1972, Tectonic evolution in the southern Red Sea and its possible significance to older rifted continental margins: Geol. Soc. America Bull., v. 83, p. 2989-3002.

Keen, M. J., 1969, Possible edge effect to explain magnetic anomalies off the eastern seaboard of the U.S.: Nature, v. 222, p. 72-74.

———, Keen, C. E., and Barrett, D. L., 1973, Changes in crustal properties near the continental margin: Am. Geophys. Union Trans., v. 54, no. 4, p. 332.

Keller, F., Jr., Meuschke, J. L., and Alldredge, L. R., 1954, Aeromagnetic Surveys in the Aleutian, Marshall and Bermuda islands: Am. Geophys. Union Trans., v. 35, no. 4, p. 558-572.

Khudoley, K. M., and Meyerhoff, A. A., 1971, Paleogeography and geologic history of Greater Antilles: Geol. Soc. America Mem. 129, 186 p.

King, E. R., Zietz, I., and Dempsey, W. J., 1961, The significance of a group of aeromagnetic profiles off the eastern coast of North America: U.S. Geol. Survey Prof. Paper 424-D, p. D299-D303.

Lancelot, Y., Hathaway, J. C., and Hollister, C. D., 1972, Lithology of sediments from the western North Atlantic, Leg XI, *in* Hollister, C. D., Ewing, J. I., et al., Initial reports of the Deep Sea Drilling Project, v. XI: Washington, D.C., U.S., Govt. Printing Office, p. 901-949.

Larson, R. L., and Pitman, W. C., III, 1972, World-wide correlation of Mesozoic magnetic anomalies, and its implications: Geol. Soc. America Bull., v. 83, p. 3645-3622.

LePichon, X., and Fox, P. J., 1971, Marginal offsets, fracture zones, and the early opening of the North Atlantic: Jour. Geophys. Res., v. 76, p. 6294-6308.

Lowell, J. D., and Genik, G. J., 1972, Sea-floor spreading

and structural evolution of southern Red Sea: Am. Assoc. Petr. Geol. Bull. v. 56, p. 247-259.

Luyendyk, B. P., and Bunce, E. T., 1973, Geophysical study of the northwest African margin off Morocco: Deep-Sea Res., v. 20, p. 537-549.

MacCarthy, G. R., 1936, Magnetic anomalies and geologic structures of the Carolina coastal plain: Jour. Geol. v. 44, no. 3, p. 396-406.

McElhinny, M. W., and Burek, P. J., 1971, Mesozoic, Palaeomagnetic stratigraphy: Nature, v. 232, p. 98-102.

McMaster, R. L., and Lachance, T. P., 1968, Seismic reflectivity studies on northwestern African continental shelf: Strait of Gibraltar to Mauritania: Am. Assoc. Petr. Geol. Bull., v. 52, p. 2387-2395.

Opdyke, N. D., 1961, The Paleomagnetism of the New Jersey Triassic: a field study of the inclination error in red sediments: Jour. Geophys. Res., v. 66, p. 1941-1949.

———, and McElhinny, M. W., 1965, The reversal at the Triassic-Jurassic boundary and its bearing on the correlation of Karroo igneous activity in southern Africa: Am. Geophys. Union Trans., v. 46, p. 65.

———, and Wensink, H., 1966, Paleomagnetism of rocks from the White Mountain plutonic-volcanic series in New Hampshire and Vermont: Jour. Geophys. Res., v. 71, p. 3045-3051.

Oversby, V. M., 1969, Lead isotope compositions in recent volcanic rocks from islands in the Atlantic Ocean, and from Troilite phase of meteorites [Ph.D. thesis]: Columbia Univ., New York.

Pitman, W. C., III, and Talwani, M., 1972, Sea-floor spreading in the North Atlantic: Geol. Soc. America Bull., v. 83, p. 619-646.

Rabinowitz, P. D., 1971, Gravity anomalies across the East African continental margin: Jour. Geophys. Res., v. 76, no. 29, p. 7107-7117.

———, 1972, Gravity anomalies on the continental margin of Angola, Africa: Jour. Geophys. Res., v. 77, p. 6327-6347.

———, 1973, The continental margin of the northwest Atlantic Ocean: a geophysical study [Ph.D. thesis]: Columbia Univ., New York, 181 p.

Roeser, H. A., Hinz, K., and Plaumann, S., 1970, Continental margin structure in the Canaries, *in* Delaney, F. M., ed., ICSU/SCOR Working Party 31 Symp., The Geology of the East Atlantic Continental Marine, v. 2: Inst. Geol. Sci., Africa Rept. 70/16, p. 28-36.

Rona, P. A., Brakl, J., and Heirtzler, J. R., 1970, Magnetic anomalies in the northeast Atlantic between the Canary and Cape Verde islands: Jour. Geophys. Res., v. 75, p. 7412-7420.

Rothe, P., and Schmincke, H. U., 1968, Contrasting origins of the eastern and western islands of the Canarian Archipelago: Nature, v. 218, p. 1152-1154.

Sheridan, R. E., Drake, C. L., Nafe, J. E., and Hennion, J., 1966, Seismic-refraction study of continental margin east of Florida: Am. Assoc. Petr. Geol. Bull., v. 50, p. 1972-1991.

Sheridan, R. E., Smith, J. D., and Gardiner, J., 1969, Rock dredges from Blake Escarpment near Great Abaco Canyon: Am. Assoc. Petr. Geol. Bull., v. 53, p. 2551-2558.

Sougy, J., 1962, West African fold belt: Geol. Soc. America Bull., v. 73, p. 871-876.

Spangler, W. B., and Peterson, J. J., 1950, Geology of Atlantic coastal plain, in New Jersey, Delaware, Maryland, and Virginia: Am. Assoc. Petr. Geol. Bull., v. 34, p. 1-99.

Steiner, M. B., and Helsey, C. E., 1972, Jurassic polar movement relative to North America: Jour. Geophys. Res., v. 77, p. 4981-4993.

Swain, F. M., 1947, Two recent wells in coastal plain of North Carolina: Am. Assoc. Petr. Geol. Bull., v. 31, p. 2054-2060.

———, 1952a, Ostracoda from wells in North Carolina: 1, Cenozoic Ostracoda: U.S. Geol. Survey Prof. Paper 234-A, p. 1-58.

———, 1952b, Ostracoda from wells in North Carolina: 2. Mesozoic Ostracoda: U.S. Geol. Survey Prof. Paper 234-B, p. 59-152.

Talwani, M., and Eldholm, O., 1972, Continental margin off Norway: a geophysical study: Geol. Soc. America Bull., v. 83, p. 3575-3606.

———, and Eldholm, O., 1973, The boundary between continental and oceanic basement at the margin of rifted continents: Nature, v. 241, p. 325-330.

———, and Heirtzler, J. R., 1964, Computation of magnetic anomalies caused by two dimensional structures of arbitrary shape: Computers Mineral Ind., p. 464-480.

Taylor, P. I., Zietz, I., and Dennis, L. S., 1968, Geologic implications of aeromagnetic data for the eastern continental margin of the United States: Geophysics, v. 33, p. 755-780.

Uchupi, E., and Emery, K. O., 1967, Structure of continental margin off Atlantic coast of United States: Am. Assoc. Petr. Geol. Bull., v. 51, p. 223-234.

———, Phillips, J. D., and Prada, K. E., 1970, Origin and structure of the New England Seamount Chain: Deep-Sea Res., v. 17, p. 483-494.

Vogt, P. R., Avery, O. E., Schneider, E. D., Anderson, C. N., and Bracey, D. R., 1969, Discontinuities in sea floor spreading: Tectonophysics, v. 8, p. 285-317.

———, Lorentzen, G. R., and Dennis, L. S., 1970a, An aeromagnetic survey of the Keathley magnetic anomaly sequence between 34°N and 40°N in the western North Atlantic (abstr.): Am. Geophys. Union Trans., v. 51, p. 274.

———, Anderson, C. N., Bracey, D. R., and Schneider, E. D., 1970b, North Atlantic magnetic smooth zones: Jour. Geophys. Res., v. 75, p. 3955-3967.

———, Anderson, C. N., and Bracey, D. R., 1971, Mesozoic magnetic anomalies, sea-floor spreading, and geomagnetic reversals in the southwestern North Atlantic: Jour. Geophys. Res., v. 76, p. 4796-4823.

Watkins, J. S., and Geddes, W. H., 1965, Magnetic anomaly and possible orogenic significance of geologic structure of the Atlantic shelf: Jour. Geophys. Res., v. 70, no. 6, p. 1357-1361.

Woollard, G. P., 1943, Geologic correlation of areal gravitational and magnetic studies in New Jersey and vicinity: Geol. Socl. America Bull., v. 54, p. 791-818.

Zietz, I., 1971, Eastern continental margin of the United States: pt. 1. A magnetic study, *in* Maxwell, A. E., ed., The sea, v. 4, pt. II: New York, Wiley, p. 293-310.

Seismicity at Continental Margins

Jack Oliver, Bryan L. Isacks, and Muawia Barazangi

INTRODUCTION

The global pattern of belts of high seismic activity is, in general, dissimilar to that of the continental margins. The seismic belts correspond to the plate boundaries of plate tectonics, for which the fundamental mechanical unit is the lithospheric plate and not the continents and ocean basins.

At many places where the plate boundaries and continental margins coincide, the seismicity of the margin is that of the particular type of plate boundary: convergent, divergent, or strike-slip. The effect of the margin on the process appears minor and manifested only in increased shallow activity behind arcs, if that.

The intraplate margins appear to be one type of boundary between more active and less active zones of seismicity of broad distribution, and do not serve as concentrators of seismic activity.

Most of the earth's seismicity is concentrated in a few narrow belts that encircle the earth and intersect so as to divide the earth's surface into a mosaic of large, relatively aseismic areas. These stable areas correspond to the plates of plate tectonics; the major seismic belts correspond to the recent plate boundaries.

The pattern of continental margins is also global in scope, but it differs geometrically and topologically from that of the seismic belts. The patterns of the continental margins are, in general, different from those of the seismic belts, and the margins do not intersect to form junctions of three belts as do the plate boundaries. Nevertheless, certain large segments of the seismic belts are coincident with continental margins, and hence a large part of the story of seismicity near continental margins concerns the highly active, and relatively well understood, major seismic belts.

Many segments of continental margins are not associated with highly active seismic belts, however, and the story of seismicity at these continental margins depends upon the limited information on the generally weak, infrequent, and relatively poorly understood earthquakes of plate interiors near the continental margins.

There is great contrast in the quantity, quality, and nature of the information on seismicity at these two types of margins. Our understanding of the relationship between seismicity and continental margins is best where seismic activity is high and the plate tectonics model in its present state of development can be readily applied. Hence this paper is divided into two parts according to the level of seismicity at or near the continental margin.

CONTINENTAL MARGINS AND THE BELTS OF MAJOR SEISMICITY

Most of the segments of the continental margins that coincide with major seismic belts, and hence plate boundaries, correspond to zones of plate convergence (Fig. 1). The belts of shallow seismic activity are broad compared to those at the spreading ridges. Some examples of localities where seismic belts coincide with continental margins are: large segments of the western margins of South America and Central America; parts of the Mediterranean; parts of Alaska, Kamchatka, and Japan; and, assuming the term "continental margin" is interpreted so as to apply here, many arcs or parts of arcs, such as the Aleutian, Kurile, Philippine, New Guinea, and East Indies arc.

In a very few places, such as the Red Sea and Gulf of Aden, the Gulf of California, and Iceland (if that island may be thought of as "continental" in the geological sense), belts of seismicity corresponding to zones of divergence of plates are associated with continental margins, largely because the plates have not yet spread apart very much. In these cases the nascent ocean is narrow and the mid-ocean spreading center is necessarily near both of the adjoining continental margins.

In the case of the Red Sea, the Gulf of Aden, and the Gulf of California, the zones of spreading, which include ridges and transform faults, in a gross sense parallel the continental margins. In fact, the case has frequently been made that the Red Sea is but a younger and smaller version of the Atlantic Ocean, each with its median spreading center and corresponding belt of seismicity that is approximately congruent with the shorelines. The seismicity of the regions does not deny this interpretation. Thus it is somewhat artificial from a tectonic viewpoint to associate the continental margins of the Red Sea with an active seismic belt, and to omit that association for the Atlantic margins, even though the Atlantic continental margins are far from the plate boundary at the mid-Atlantic ridge.

Iceland is a special case in which the spreading and the corresponding seismicity occur in a belt that crosses the island with a northerly trend intersecting the margins. The seismicity with an east-west trend along the northern margin of Iceland, and perhaps along the southern margin as well, may be associated with transform faults that are not spreading much, if any, at the present time.

At a few places, large, seismically active transform faults are near and subparallel to the

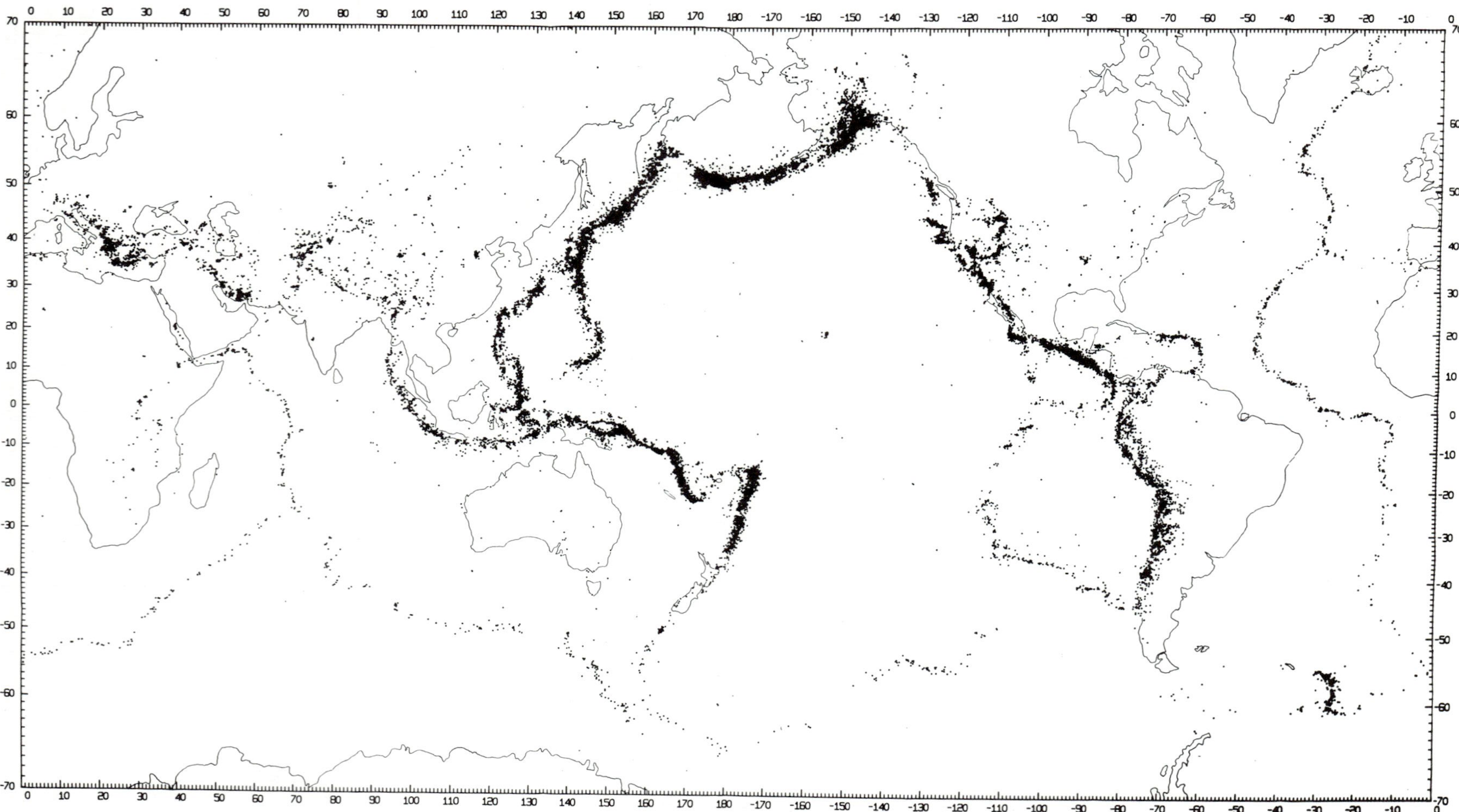

Fig. 1. Worldwide distribution of shallow epicenters (1961-1967; depth between 0 and 100 km) (after Barazangi and Dorman, 1969).

continental margins. Some obvious examples are the San Andreas fault in California, the faults along the western margin of Canada, the Philippine fault, and the Alpine fault of New Zealand. That these major fractures tend toward parallelism with the continental margin may be an observation of significance in the study of continental margins, but, if so, that significance is not yet well brought out.

The coincidence of some segments of seismic belts and continental margins is significant, but certainly the point of greatest importance concerning the relation between global seismicity and continental margins is that, in general, the major seismic belts do not follow the margins closely, nor are they related to them in an obviously simple manner. Recognition of the importance of this point was crucial in the development of plate tectonics.

In some places, such as in southern Chile, a continental margin-seismic belt relationship that prevails for thousands of kilometers to the north is effectively terminated at the intersection of that zone with an oceanic seismic belt (Fig. 1), even though there is no gross change in the margin there. In Southeast Asia, the northward extension of the western part of the East Indian seismic belt crosses the continental margin of Asia, giving little if any indication of the existence of that margin. At the western end of the Gulf of Aden, the seismic belt running into that zone from the Indian Ocean bifurcates and crosses the continental margin of Africa in its southern extension, again with essentially no indication in the geometrical pattern of the existence of the continental margin.

Thus the most significant point having to do with the patterns and relationships of major seismic belts and continental margins is that which stresses the difference between those patterns, and in so doing provides strong support for a fundamental tenet of plate tectonics, i.e., that the building blocks of global tectonics are not the continents and oceans but the large plates of lithosphere outlined by the seismic activity and other tectonic features.

Where the seismic belts and the continental margins coincide, something may be learned about the margins from detailed study of the seismicity and other seismic phenomena, particularly the focal mechanisms of the earthquakes. Consider a section through a convergent margin (Fig. 2). The zone of predominant shallow activity, i.e., the site of the greatest number of earthquakes and of the largest earthquakes, is associated with the thrust fault that corresponds to relative motion of the adjoining plates of lithosphere. Focal mechanism studies and related geological and geodetic observations verify this association. Huge shocks such as the Alaskan earthquake of 1964 and the Chilean earthquake of 1960 are associated with this zone. Additional moderate, shallow seismic activity occurs under the deep trenches where focal-mechanism studies and subbottom profiling show clear evidence of extension near the upper surface of the downgoing lithospheric plate in a horizontal direction normal to the trend of the seismic belt and the arc. This extention results from the curving of the plate as it enters the subduction zone.

On the landward side of the subduction zone shallow earthquakes often occur within a relatively broad zone. For example, a particularly active zone of crustal earthquakes is located on the eastern or landward side of the Andes in Peru and Ecuador (Figs. 1 and 2). Shallow earthquakes occur frequently in the near-coastal wedge of crust above the Benioff zone. This feature, a wedge of relatively low seismicity between the Benioff zone and an interior zone of shallow earthquake, may be characteristic of other regions of subduction as well. Among the most active shallow zones located landward of the subduction zone are those of the Ecuador-Peru-Chile system, Honshu (Japan), and the eastern end of the Aleutian-Alaskan system. In the first two of these for which data are available, focal-mechanism solutions indicate both strike-slip and thrust faulting, with maximum compressive stress oriented approximately perpendicular to the margin. It may also be significant that in these three regions of subduction the dip of the Benioff zone is quite small, varying from about 10-15° in Peru and Alaska to about 30° beneath Honshu.

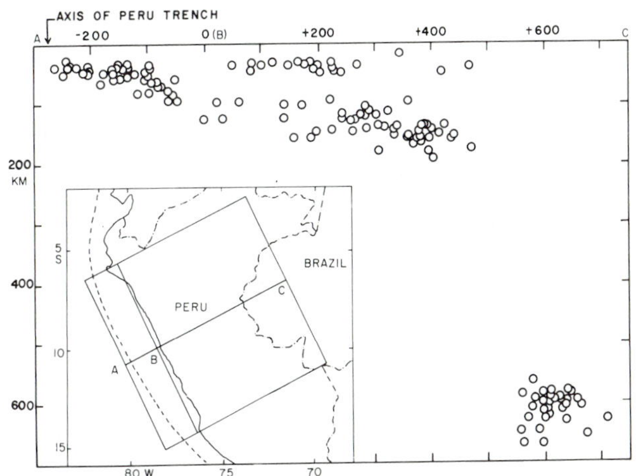

Fig. 2. Vertical section through Peru, including events within the area shown in the inset. The earthquakes are those with magnitudes greater than about 5 reported in the Preliminary Determination of Epicenters (1961-1967) (after Isacks and Molnar, 1971).

The occurrence of earthquakes at intermediate and great depths is generally taken as evidence that a subducting plate of lithosphere exists in those regions. There is a common misconception, however, that these shocks correspond to relative motion (thrusting) between the upper surface of the lithosphere and the adjoining mantle in a manner similar to that of the shallow shocks. This view, which appears intuitively reasonable, is not supported by the data, which indicate instead that the earthquakes occur within the descending lithospheric plates and are the result of a stress system in which either the axis of maximum or minimum stress is often oriented down the dip of the slab.

The global pattern of stresses deduced from study of the focal mechanism of such shocks is important in the study of plate dynamics. The pattern is consistent with generation of stresses due to the sinking of a lithospheric slab and hence extension in the intermediate depth range in some areas, and resistance to further sinking at still greater depths and hence down-dip compression for all deep shocks and for intermediate shocks where the lower end of the slab is sufficiently deep (Isacks and Molnar, 1971). For some arcs the pattern of deep seismicity and the transmission of seismic waves in the region indicate that the lower part of the slab is detached. The New Hebrides arc is the best example of this phenomenon.

Figure 1 shows that, grossly speaking, shallow seismic activity is greatest at the zones of plate convergence. More detailed studies show, however, that within and along such zones there are substantial spatial and temporal variations. In fact, the patterns of such variations are slowly becoming understood. The sites of a few major earthquakes have been predetermined on the basis that large gaps in recent activity of major seismic belts are likely locations for major events of the near future (Fedotov, 1965; Mogi, 1968; Kelleher et al., 1973).

This point may be carried a bit further. Evidence from seismicity, as well as certain geological evidence, suggests that the arcs are segmented. Successive large earthquakes in the same part of an arc have aftershock zones that terminate at the same position along the arc, and offsets in the chains of volcanoes and in certain surface features sometimes correspond to these same bounds. Hence it seems that the downgoing plate is perhaps fractured or divided along near-vertical surfaces oriented approximately normal to the arc. This hypothesis requires further testing, but it already has substantial support.

On the global and regional scales discussed above, it is clear that seismicity provides a great deal of information on the tectonics of seismically active margins. In fact, for well-studied areas such as Japan, Tonga, or South America, much more information is available than can possibly be discussed in a short review article. On a very detailed scale, however, results are limited, but there is clearly potential for further understanding. For example, thrusting in the sediments of the inner wall of the trench is rather well established now from subbottom profiling and geological studies (Ingle et al., 1973; Seely, Vail, and Walton, this volume). Whether any earthquakes are associated with these thrusts has not been established, although it might be possible if adequate observations and studies were made. If so, a tool would be available for learning more about this process.

CONTINENTAL MARGINS OUTSIDE THE MAJOR SEISMIC BELTS

As Figure 1 demonstrates, it is possible, within the space of a few years, to obtain a representative picture of the pattern of seismic activity in the major seismic belts. The same basic pattern is found independently of the time interval selected. Outside the major belts, however, the situation is quite different. Large shocks are so infrequent that only rarely have they occurred at or near the same location. Thus it is not clear that repeatable spatial patterns of seismicity have been determined for most parts of the plate interiors. Will the next large earthquake in the eastern United States occur near New Madrid, Missouri, or Charleston, South Carolina, where large shocks are known to have occurred in the past, or will it take place where there is nothing distinctive about the seismic pattern at present? This question cannot be definitely answered at the present time. The paucity of our knowledge of earthquakes in the plate interiors is evident.

This paucity of information is not solely the result of the low frequency of occurrence of shocks, although that is the principal factor. Observational capability based on seismographs is commonly poor in the intraplate areas, for the seismic hazard and hence interest in, and support for, seismology is lower there. Where the activity is low, and observations limited, a great deal of reliance must be placed on the historical, noninstrumental record. The quality of this record is strongly dependent upon population density and upon other factors, such as continuity of the written record, level of civilization, and forms of communication. Oceanic areas generally receive little or no coverage under such a system. Sparsely populated areas are almost as bad. Thus many uncertainties arise in dealing with seismicity of plate interiors and the uncertainties are generally much greater than in the case of the active belts.

To discuss the intraplate earthquakes, let us first examine Figure 1. The data of this figure are not of strictly uniform quality, for detection and location is better in some continental areas than others, and better, in general, on land than beneath the oceans. However, Figure 1 represents perhaps the most uniform set of data available on a global scale at present. When the major seismic belts at the plate margins are omitted from consideration, it is seen that the remaining activity is widespread, covering much of the world outside those belts. The pattern is not uniform in Figure 1, however, nor would it be even if a great deal more data were available. Substantial concentrations are seen on the Asian continent to the north and northeast of the Himalayan arc, in Africa in close association with the rift valleys in the north but in a more widely scattered fashion to the south, and in the United States. The apparent concentration in the United States must be due, at least in part, to more

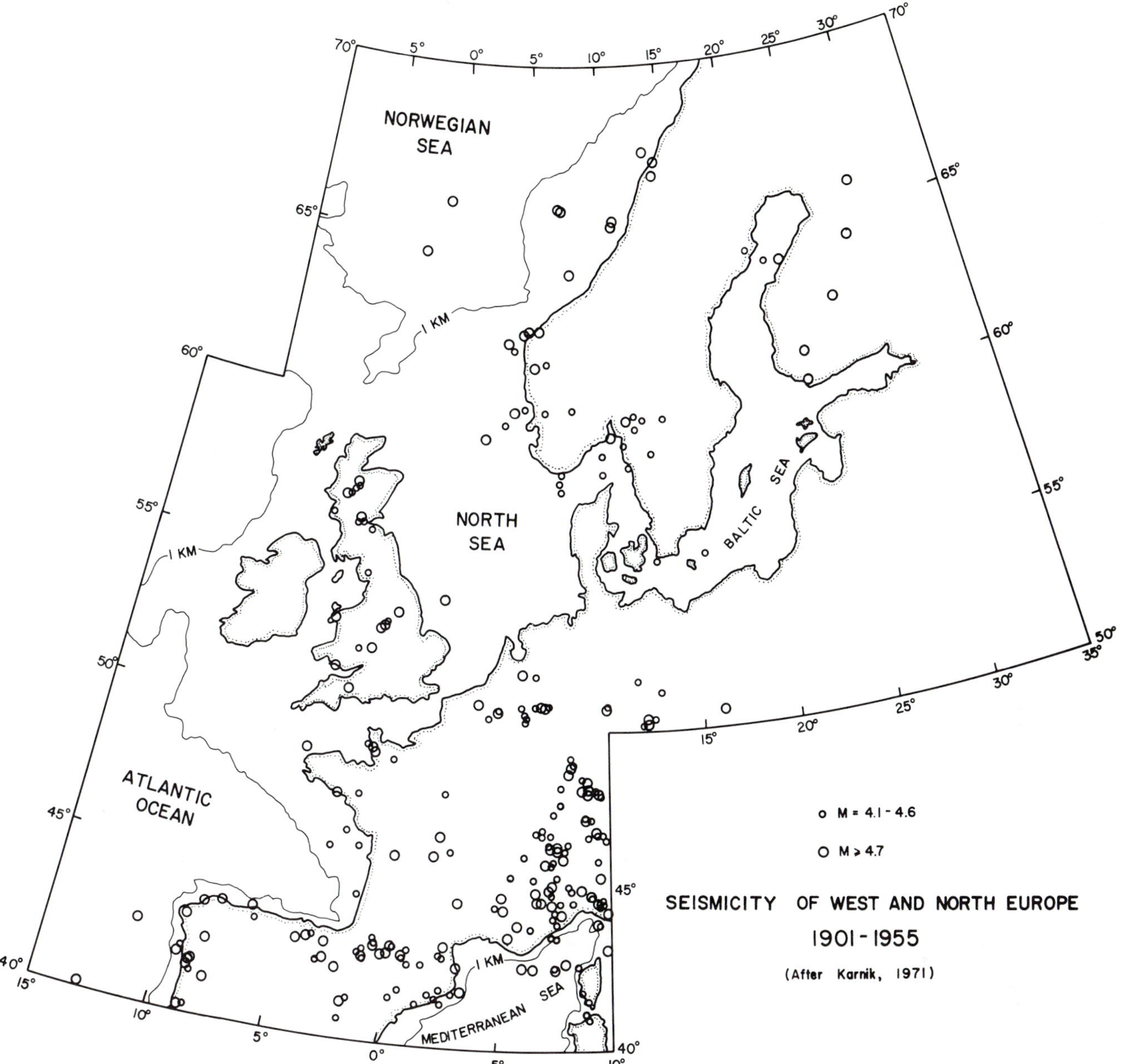

Fig. 3. Distribution of shallow epicenters in west and north Europe from 1901 to 1955. The earthquakes are those of magnitude 4.1 and larger. Note that the large number of events along the western Alps is probably some form of plate boundary (after Karnik, 1971).

thorough and complete reporting of events there and may not be so outstanding, as Figure 1 implies. Events are also scattered through Europe (Fig. 3), Australia (Fig. 4), Canada (Fig. 5), and other continental areas. A few events occur in the ocean basins away from the active plate margins.

The Asian zone behind the Himalayas is interesting, but it has little relation to a modern continental margin and hence will not be discussed further. In the case of the southern African zone, the earthquakes of Figure 1, with few exceptions, are confined to the land-covered areas. As these earthquakes are located largely from teleseismic data, there should be relatively little difference in detection capability from land to sea. Thus this set of data may be interpreted to indicate that the continental margin is some sort of boundary to the widely scattered seismic activity on land. However, there are other apparent bounds to such activity that are not continental margins, so the significance of this observation is not obvious from these data alone. In middle North America along the east coast, and perhaps, although surely to a lesser extent, in the case of Australia, there is also the suggestion that the continental margin is a boundary between a more active seismic zone and a less active one.

Figure 5 shows seismicity of the eastern United States and adjoining parts of Canada based not merely, as is Figure 1, on instrumentally located shocks of the past few years, but on all earthquakes of intensity 5 and larger known from the historical record. Thus this set of data has little overlap with that of Figure 1 and provides a more thorough

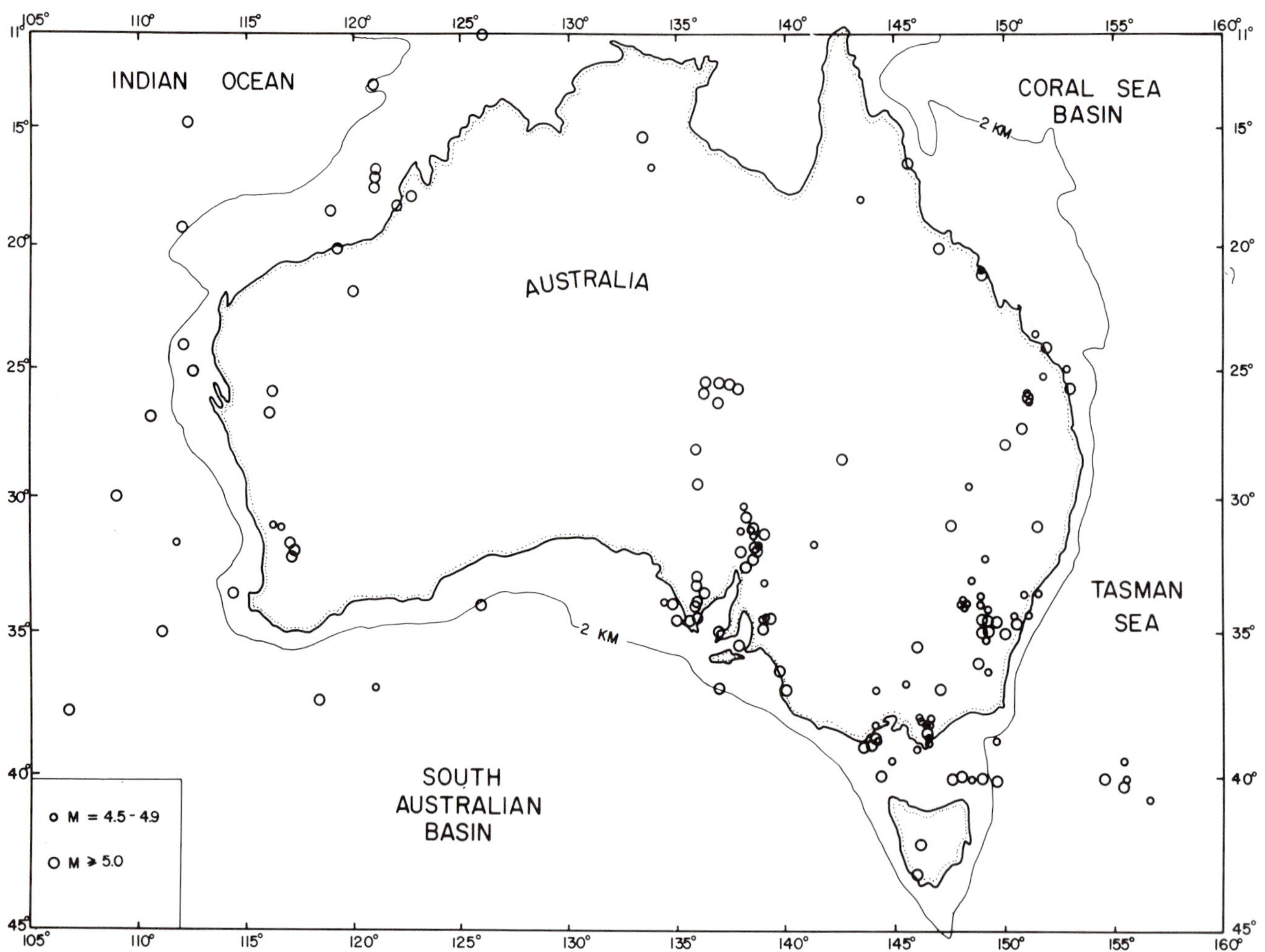

Fig. 4. Distribution of shallow epicenters in Australia from 1959 to 1966, including reliable epicenters prior to 1959. The events are those of magnitude 4.5 and larger (after Doyle et al., 1968).

sampling of seismic activity in the eastern United States over a long time interval.

Several trends are apparent in the data. There is a NE-trending belt of epicenters along the Appalachian mountains, and a possible, though disputed, subparallel trend from near New Madrid, Missouri, to the St. Lawrence Valley. Although these trends are also subparallel to the continental margin, at present there is little basis other than their spatial relationship for postulating a causal relation between the seismic trends and the margin.

There are also seismic trends oriented nearly normal to the continental margin. One running NW through Charleston, South Carolina, is evident in Figure 5 and, in fact, is more apparent in other sets of data not shown here. One trend runs through Ottawa to the southeast to the Boston area. Both of these trends terminate near the coast and certainly do not appear to continue beyond the margin. In fact, as with the data of Figure 1, the seismic activity indicated in Figure 5 is clearly less on the seaward side of the continental shelf than it is on the landward side. In Figure 5, no activity is indicated on the seaward side. But, to repeat an earlier point, there are areas within the continent, southern Florida for example, that also have no shocks indicated in Figure 5, so the margin cannot be thought of as the only type of boundary between active and inactive regions.

An outstanding feature of the margins of rifted continents is the absence of seismic activity along the boundary between continental and oceanic crusts (as defined recently, for example, by Talwani and Eldholm, 1973). This is true regardless of the age of the rifting episode. Figure 1 shows that no seismic activity exists along the margins of eastern South America, India, most of Africa, and Australia. The activity in the eastern United States begins about 200 km to the west of the continent-ocean boundary and most probably has no direct relationship to the margin, as discussed earlier. Also as noted earlier, the activity near the continental margins of the Red Sea, the Gulf of Aden, and the Gulf of California is associated with the spreading centers in these narrow bodies of water and not with the structural continental margin. There seems to be no obvious relationship between seismicity and continental margins of the

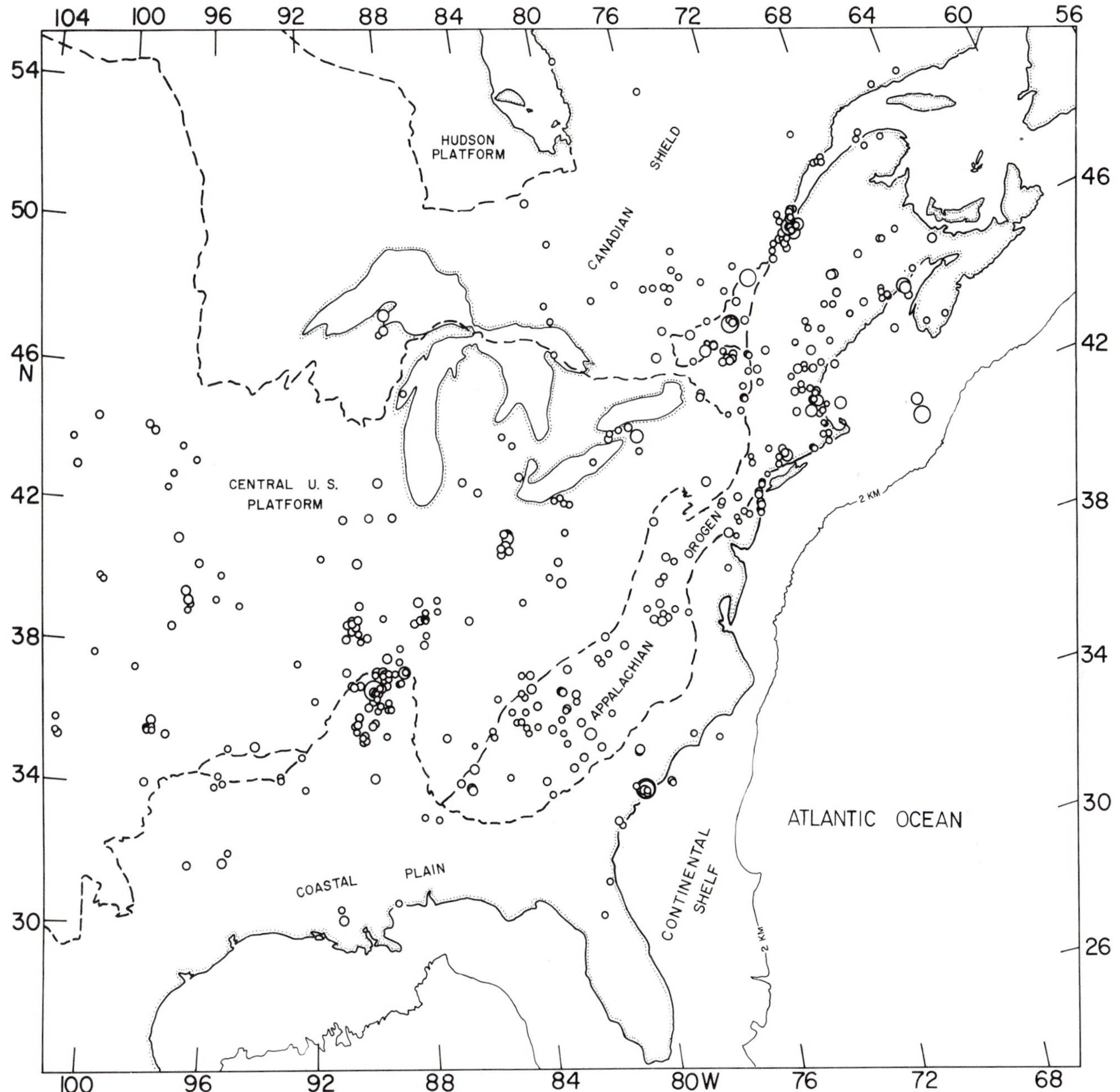

Fig. 5. Distribution of shallow epicenters in the eastern United States and Canada, including historical events of intensity of about 5 and larger (after Smith, 1962, 1966; and U.S. Coast and Geodetic Survey epicenter data).

rifting type.

The above observation is surprising for two reasons. First, the continental crust close to the margin was probably severely stretched and fractured during the rifting of the continent, and thus it is probably a major zone of weakness where seismic activity might be expected to occur. Second, major crustal thickness changes occur along the boundary between continental and oceanic crusts. These inhomogeneties would be the source of a considerable nonhydrostatic stress field that might cause renewed seismic activity (e.g., Bott and Dean, 1972; Artyushkov, 1973).

The depths and focal mechanisms of earthquakes in the relatively aseismic areas are quantities that, at least in principle, should provide information on the tectonics of these regions and, perhaps, related continental margins.

Commonly, depths of shocks in plate interiors are poorly known. For the eastern United States, for example, for many years it was held that the depth of earthquake foci were frequently subcrustal and of the order of 100 km. This conclusion was based on the unusually large felt area, which is commonly 15-25 times that of a shock of comparable magnitude in, say, the western United States. When better instrumentation and better observations became available, it became clear that the foci were in the crust and perhaps always in the upper crust, although this last point is not yet settled. The great size of the felt areas in this region is now attributed to highly efficient propagation of short-period surface waves. As it now appears that all intraplate continental shocks have their foci in the crust, the difference between the seismicity of continental areas and oceanic areas may be associated with the great change in the character and thickness of crustal rock at the continental margin, but a detailed explanation is not in hand.

The data on focal mechanisms of intraplate earthquakes are very sparse for two reasons: (1) intraplate shocks are infrequent and rarely large enough to provide good first-motion data at stations far from the source, and (2) seismograph stations are so widely spaced in aseismic regions that good close-in data are rarely available. Sbar and Sykes (1973) claim that the few focal mechanisms in the eastern United States plus some data of other types are consistent with E-W compressional stresses over much of the area, but the evidence is scanty. If correct, however, this hypothesis would be of interest in the study of continental margins.

Continental margins have a strong effect on the propagation of several types of seismic waves. Long surface waves of the Rayleigh and Love types are refracted, partially reflected, and perhaps partially converted to other types at the margin. Shorter surface waves, such as Lg waves, cannot cross the margin at all and must be absorbed or converted to other waves. Sn phases and T phases, propagating over oceanic paths are largely converted to other phases, primarily Lg, on striking the continental margin. P waves incident on a margin from the continental side are in some cases converted to T phases. Thus for many types of seismic waves the continental margins of the intraplate type are boundaries of much greater importance than they are with regard to seismicity.

ACKNOWLEDGMENTS

This work was supported by Air Force Office of Scientific Research grant AFOSR-73-2494, and by National Science Foundation grants GA-30473 and GA-34140. Cornell University, Department of Geological Sciences, Contribution No. 556.

BIBLIOGRAPHY

Artyushkov, E. V., 1973, Stresses in the lithosphere caused by crustal thickness inhomogeneties: Jour. Geophys. Res., v. 78, p. 7675.

Barazangi, M., and Dorman, J., 1969, World seismicity maps compiled from ESSA, Coast and Geodetic Survey, epicenter data, 1961-1967: Seismol. Soc. America Bull., v. 59, p. 369.

Bott, M. H. P., and Dean, D. S., 1972, Stress systems at young continental margins: Nature Physical Sci., v. 235, p. 23.

Doyle, H. A., Everingham, I. B., and Sutton, D. J., 1968, Seismicity of the Australian continent: Journ. Geol. Soc. Australia, v. 15, p. 295.

Fedotov, S. A., 1965, Regularities of the distribution of strong earthquakes of Kamchatka, the Kurile islands, and northeastern Japan: Inst. Fiz. Zemli Akad. Nauk SSSR Trans., no. 36, p. 66.

Ingle, J. C., et al., 1973, Western Pacific Floor. Leg 31 of the Deep Sea Drilling Project: Geotimes, October, p. 22.

Isacks, B., and Molnar, P., 1971, Distribution of stresses in the descending lithosphere from a global survey of focal mechanism solutions of mantle earthquakes: Rev. Geophys. and Space Phys., v. 9, p. 103.

Karnik, V., 1971, Seismicity of the European area: Amsterdam, D. Reidel Publishing Company, 218 pp.

Kelleher, J., Sykes, L., and Oliver, J., 1973, Possible criteria for predicting earthquake locations and their application to major plate boundaries of the Pacific and the Caribbean: Journ. Geophys. Res., v. 78, p. 2547.

Mogi, K., 1968, Migration of seismic activity: Earthquake Res. Inst. Bull., Tokyo University, v. 46, p. 53.

Sbar, M., and Sykes, L., 1973, Contemporary compressive stress and seismicity in eastern North America: An example of intra-plate tectonics: Geol. Soc. America Bull., v. 84, p. 1861.

Smith, W. E. T., 1962, Earthquakes of eastern Canada and adjacent areas, 1534-1927: Dominion Observatory Ottawa Pubs., v. 26, p. 271.

———, 1966, Earthquakes of eastern Canada and adjacent areas, 1928-1959: Dominion Observatory Ottawa Pubs., v. 32, p. 87.

Talwani, M., and Eldholm, O., 1973, Boundary between continental and oceanic crust at the margin of rifted continents: Nature, v. 241, p. 325.

Active Continental Margins and Island Arcs

T. Hatherton

INTRODUCTION

Certain regions of the earth are characterized by unusually high geophysical and geological activity, exemplified by deep earthquakes, volcanism, and mountain building. These regions form two fairly continuous belts, one around the Pacific Ocean (Fig. 1), the other extending through Asia to the Mediterranean Ocean. Although continuous, the belts are seen to consist of a number of elements, often arcuate in nature. These elements are sometimes marked by a series of islands (e.g., the Marianas and the Aleutians) and the term *island arc* was used to describe their physiographic and later, their geophysical features. The same geophysical phenomena are also found at the margins of continents, the principal examples being the west coast of South America and the south coast of Alaska, and the term *active continental margin* entered the nomenclature. However, much of the Alpide belt can hardly be characterized as island arc or continental margin. Since both island arcs and active continental margins are regions of intense geological activity, the term *orogenic belt* has also been used, which satisfactorily encompasses the Alpide elements. Because almost all the phenomena that contribute to the island arc or active continental margin are strongly asymmetric, Evison (1968) advocated the expression *asymmetric active regions* as the appropriate generic term. Some workers today may tend to prefer *subduction zone*, but this is an inferential rather than a descriptive definition. Active margins are also, in modern terms, "plate" margins or boundaries, but this phrase also has a wider meaning.

The present author prefers Evison's "asymmetric active region" as denoting the two prime factors, the intense geophysical activity and its asymmetrical distribution. However, the phrase has not entered common usage and indeed "island arc" remains the all-embracing term for such regions. In keeping with the title of this volume, the term "active margins" will be used throughout to denote the regions we are discussing, principally those of the circum-Pacific.

The history of our knowledge of active margins falls basically into three periods. Following many years of work, mainly by Japanese and Dutch scientists, Gutenberg and Richter (1945) were able to describe a "unilateral ordering of structures and attendant phenomena" which typified active margins. These were, going from oceanic (convex) to continental (concave) side of the arc:

1. An oceanic trench, trough, or foredeep.
2. Shallow earthquakes and negative gravity anomalies occurring in a narrow belt on the concave side of the submarine trough.
3. Positive gravity anomalies, earthquakes at depths near 60 km.
4. The principal structural arc of Late Cretaceous or of Tertiary age, with active or extinct volcanoes. Shocks at depth of the order of 100 km.
5. A second structural arc. Volcanism usually older and in a late stage. Shocks at depths of 200-300 km.
6. A belt of shocks at depths of 300-700 km.

The twenty years between 1945 and 1965 saw a large increase in the available data on gravity, earthquake morphology, and geochemistry. During this period Japanese petrologists and geochemists were beginning to recognize asymmetry in the volcanic chemistry and associated the volcanic products with earthquakes in the mantle, a relationship noted as early as 1935 by Wadati. The period was otherwise relatively arid in synthesis, with the exception of the work of Gunn (1947).

TRENCHES

Mountain ranges and deep trenches represent the greatest deviations from the general spheroidal shape of the earth. Since natural forces of erosion and sedimentation tend to oppose the formation of both these features, their very presence is the most telling indication of large-scale tectonic forces. The origins of mountains and trenches are far from clear even now.

A trench is one of the four major characteristics of an active margin. Depths are variable ranging from about 2 to 8 km. On the shoreward flank slopes range up to 10° or more but are smaller on the offshore flank, averaging about 5°. Although on a natural scale the slopes appear smooth, bench-like features do occur which can best be seen on a compressed scale.

Until recently few direct observations were available to supplement the bathymetric data and thus allow the development of a theory of the structure of trenches. The "tectogene" model of Kuenen (1936), which postulated downbuckling of light crustal material into the substratum under lateral compressive forces, was inspired mainly by the observed negative gravity anomalies. Seismic refraction investigations in the 1950s by Lamont, Scripps, and Woods Hole oceanographic workers provided gross information on seismic velocities and

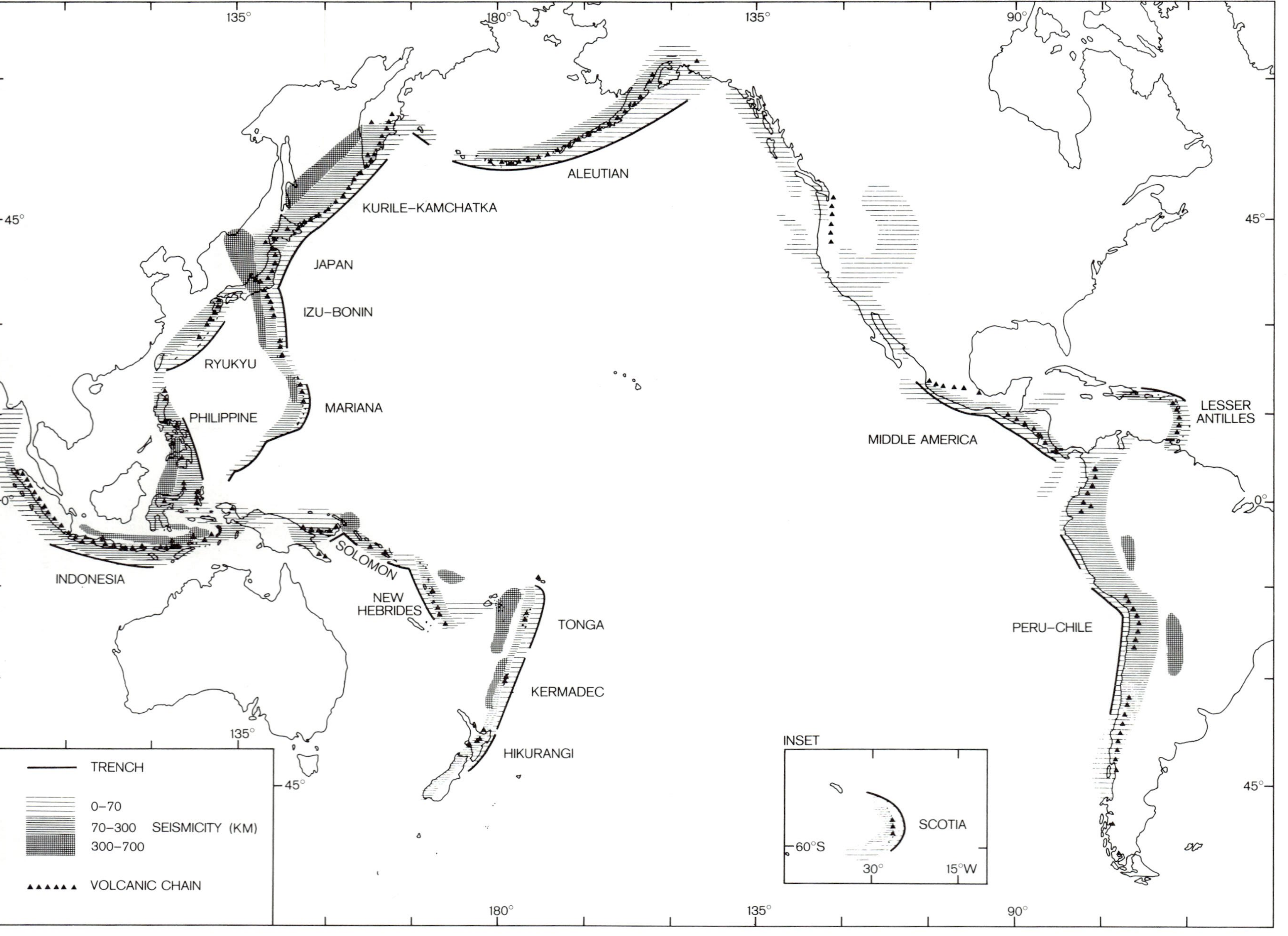

Fig. 1. Active continental margins and island arcs of the Pacific area.

thus, by inference, density, which enabled further models satisfying the gravity anomalies to be developed (see, for instance Worzel, 1965a). These models indicated crustal thinning beneath trenches and, together with the developing evidence for possible graben-style benches (Brodie and Hatherton, 1958), appeared to support an extensional mechanism of formation that superseded the earlier compressional hypothesis.

The development of plate tectonics reversed this process, with "a priori" acceptance of compression at the active margins to force down the new lithosphere created at the ocean ridges. The development of high-resolution reflection techniques allowed the internal structure of the trench sediments to be studied for the first time. The sediments in the bottoms of most trenches appear to be remarkably undeformed, and provide no evidence for a large thrust fault at the base of the continental slope, nor is there evidence that oceanic sediments have disappeared beneath the continents in later Tertiary to Recent time (von Huene and Kulm, 1974).

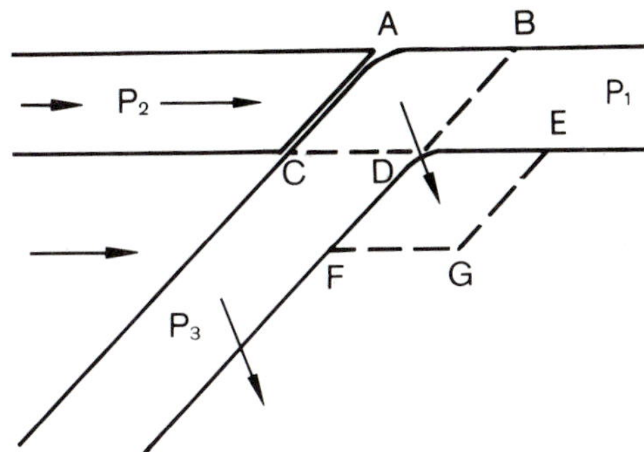

Fig. 2. Model for retrogressive motion producing tension at the trench and behind the active margin. (From Elsasser, 1971.)

Elsasser (1971) has objected to "the past intense fixation of geological reasoning on compressive stress patterns" and has attempted to develop a model which, while providing the subduction of oceanic lithosphere required by the mobilism of today, is consistent with the trench as a tensional feature. This model (Fig. 2) requires retrograde motion in which the rhomboid ABCD is carried down to DEFG. With such a motion, not only the lithosphere at P_2 but also the asthenosphere beneath it must be pulled oceanward. This model may also be consistent with the development of extensional features landward of the active margin (Karig, 1971).

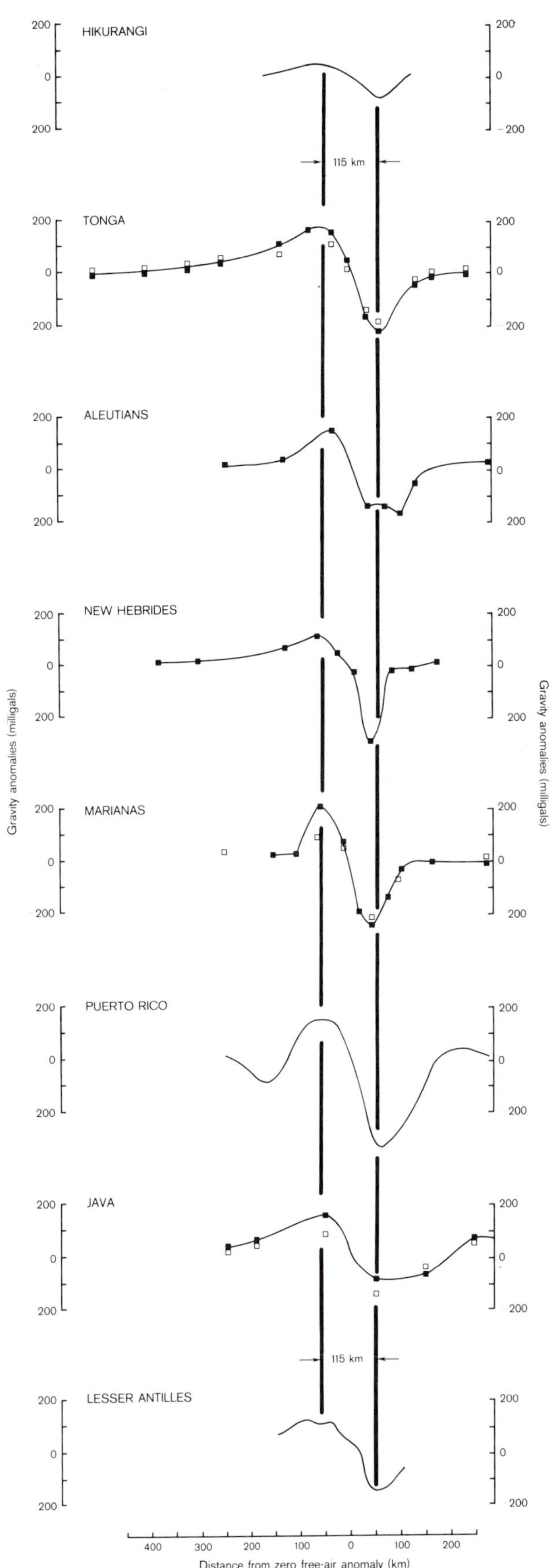

Fig. 3. Gravity anomalies across active margins and island arcs. Solid points represent free-air anomalies and open points isostatic anomalies. Observations on profiles 1, 6, and 8 are sufficiently dense for their distribution to be neglected. Data from Vening Meinesz (1948), Worzel (1965b), Reilly (1965), and Andrew et al. (1970).

GRAVITY ANOMALIES

The measurements of Vening Meinesz (1948) using pendulum apparatus in submerged submarines established the principal gravity features of the Indonesian arcs, which have been found to be common throughout the active margins (Fig. 3). Over, or on the inner side of the trench is a gravity minimum, the isostatic or free-air gravity anomalies reaching about -200 mgal. On the continental or island side of this negative anomaly is a positive anomaly, the amplitude of this gravity maximum, being rather less than the corresponding negative anomaly. Until relatively recently the cause of the negative anomaly dominated discussion of the gravity effects over active regions. Vening Meinesz considered it due to a "tectogene" or downbuckling and thickening of the "crust," from which orogeny later developed (Fig. 4b). Gunn (1947) discussed island arc structures in terms of a shear-fractured lithosphere which was assumed to be strong, elastic, and supported by an underlying deep magma. Overthrusting of the continental sector was held to account for the orogenic and geophysical observations.

Worzel (1965a) developed a pragmatic approach to gravity interpretation of island arcs, attributing no tectônic cause but considering the variations in depth of the crust-mantle interface as being the source of the anomalies (Fig. 4c). The Tonga-Kermadec "arc" superseded Indonesia as the region for exploration in the 1950s, and as a result of the coincidence of negative gravity anomaly and trench in that region, an acceptance of the identity of these two characteristics of island arcs grew; this, despite Vening Meinesz' work in the Java arc, where negative gravity anomalies occur on the "concave" side of the submarine trough. In New Zealand the belt of negative gravity anomalies passes through the center of the North Island where there is neither trench nor obvious tectogene. Hatherton (1969b) showed that rather than being primarily related to the trenches, the negative anomaly always occurred where the projection of the belt of deep and intermediate earthquakes (Fig. 7) intersected the crust.

Livshits (1965) was the first to suggest that the dipping slab or zone in which the intermediate and deep earthquakes occur has a slightly higher density (+0.06 gm/cm^3) than the surrounding mantle and in this he was followed independently by Hatherton (1969). Thus, the statement of Grow (1973), that the assignment of a higher density to the downgoing slab relative to the asthenosphere eliminates the "artificial thinning of the oceanic crust which was a paradox in all trench gravity models computed prior to the development of plate tectonics concepts," is only partly true, for Livshits's model was developed prior to plate tectonics.

The positive and negative anomalies form an anomaly pair that has a dipole character with the integrated anomalies approximating zero (Fig. 3). Hatherton (1970) then suggested that a possible

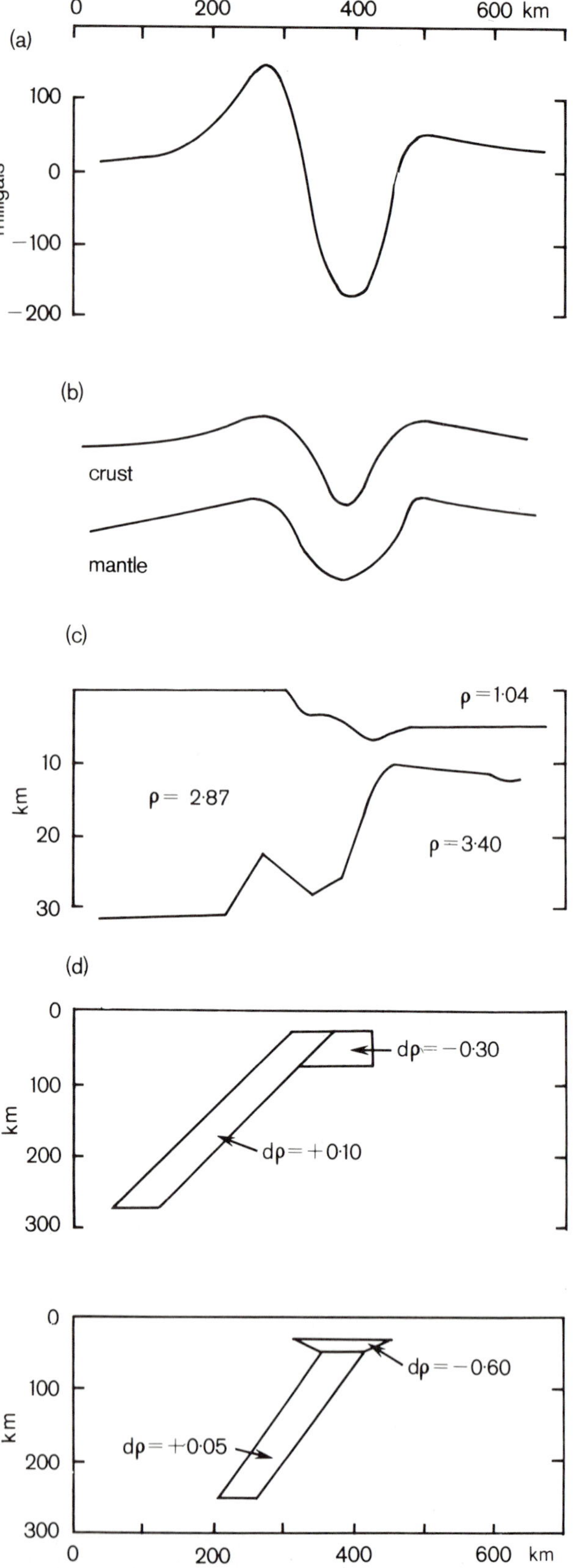

Fig. 4. Development of earth models to produce gravity anomalies at active margins and island arcs. (a) Typical gravity anomaly pair. (b) Tectogene model (Kuenen, 1936). (c) Model with variable crustal thickness only (Worzel, 1965a). (d) Model with mantle inhomogeneity (Hatherton, 1969b).

cause of the anomalies was differentiation within the upper mantle. However, majority opinion is that the two anomalies are independent in that one does not arise from the other, although both are due to whatever the geodynamic process causing active margins. The positive anomaly is ascribed to the higher density of a cool, downthrust (subducted) lithospheric slab. The negative anomaly is attributed to the accumulation and subsequent underthrusting of sediments at the trench, although there are problems with this interpretation and it is likely that the cause of the negative anomaly is, as yet, undiscovered.

A final interesting parameter of the gravity anomalies at active margins is that the distance between the peak of the positive anomaly and the trough of the negative anomaly tends to be constant at about 115 km (Hatherton, 1970; see also Fig. 3). Only one margin with a gravity anomaly pair departs significantly from this value, and that is in Fiordland, southern New Zealand, but the status of this margin is rather indeterminate. The constant value of 115 km for the separation of the gravity anomalies may play a significant role in determining the lateral dimensions of geosynclinal characteristics upon relaxation.

Even now there is one aspect of active region gravity anomalies that has not properly been analyzed. On the oceanic side the negative gravity anomaly recovers to a positive value usually less than half that of the continental-side positive anomaly, but still consistent in all island arcs and therefore significant. Vening Meinesz (1948) considered that these anomalies were due to compensating uplift of the mantle on the boundaries of the tectogene.

VOLCANISM

The great majority of volcanoes occurs at the active continental margins or island arcs in association with the other geophysical features. Gutenberg and Richter pointed out that the principal structural arc with active or extinct volcanoes lies above the seismicity of depth about 100 km. In fact, like most other features of the active margins or island arcs, the volcanism occurs asymmetrically. Moving from the convex (or oceanic) to the concave (or continental) side of the margin, the first volcanoes are met quite abruptly, occurring in a line along the arc which is termed the "volcanic front." The volume of recent eruptives is greatest at or immediately inland of this front and decreases rapidly in a direction away from the trench (Sugimura, 1968, Fig. 10).

It was realized early that there is a distinct contrast in petrological nature between intraoceanic and circum-oceanic provinces in the Pacific, and the line separating the two was named the "andesite line" (Marshall, 1912). The andesite suite includes rocks with a wide range of silica contents. At one extreme are the aluminous basalts containing about 50% silica and at the other are dacites with about 65% silica. Some authorities (Taylor, 1968; Dickinson, 1968) hold that the suite constitutes the greatest volume of eruptive products within the active margins. Others (e.g., Fitton, 1971) take the viewpoint that the eruptive rocks can be classified (from ocean toward continent) as tholeiitic, andesitic, or alkali basalt, and infer, since the volume of recent eruptives decreased rapidly away from the trench, that rocks having tholeiitic affinity are greatest in volume and the alkali basalts the least.

Irrespective of the terminology used, the chemistry of the eruptive rocks of active margins shows a characteristic asymmetry. After establishing that volcanic rocks became more alkaline to the concave side of the arc, Dutch geologists working in Indonesia, and Japanese petrologists, developed several numerical methods of characterizing variations using relationships involving SiO_2, $A1_2O_3$, and Na_2O, and K_2O. The two alkalis were sometimes added and on other occasions used in ratio. By the late 1950s Japanese petrologists were developing the idea that the composition of volcanic rocks in the active margins might be related to the focal depths of the intermediate and deep earthquakes beneath the volcanoes (see Fig. 5), an approach also used by Coats (1962).

The work of Taylor (1968) underlined the significance of potash in the calc-alkaline rocks of the active margin, the range of variation of K_2O being many times greater than any other oxide. Dickinson and Hatherton (1967) used the amount of K_2O (normalized to 60% SiO_2) as the most sensitive chemical parameter to investigate the suggested relationship of volcanic chemistry to the depth of the earthquakes beneath the volcano. The result was the K-h relationship shown in Figure 6.

Hatherton and Dickenson (1969) found this relationship to hold in most active regions, the Izu-Bonin arc being the only apparent area of discordance. The K-h relationship has also been inferred to be valid in the Mediterranean arcs (Ninkowich and Hayes, 1972), but the present author doubts the applicability of the principle in that region. Hatherton (1969) made two further points: first, the soda is invariant at 60% SiO_2 throughout the calc-alkaline volcanoes of the Pacific, and hence the more ancient K_2O/Na_2O ratio is an almost equivalent concept to the simple K_2O variation, but $K_2O + Na_2O$ is relatively insensitive to the potash variation; and second, the depth *h* of the earthquakes beneath the volcano can denote not only a point, but a distance, and that exchanges associated with path length, as well as original composition, can contribute to the chemical makeup of the calc-alkaline rocks.

Sugimura (1968) was independently developing a similar approach based on the "geotectonic position" of a volcano, defined as the relative depth of the active seismic plane with respect to a reference level at the depth of foci directly beneath the volcanic front. However, he attempted a much more compli-

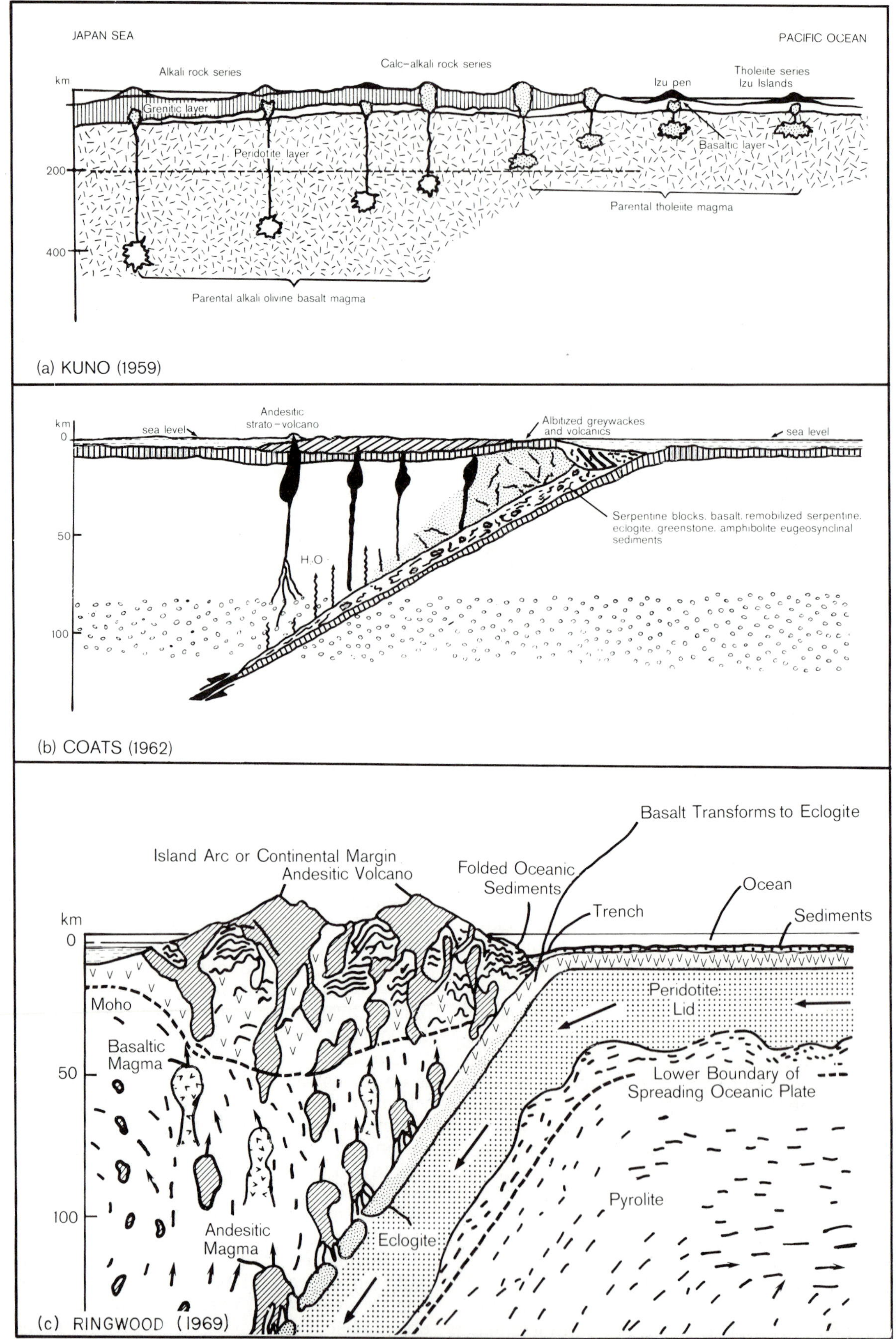

Fig. 5. Development of the concept of andesite produced by partial melting within the upper mantle. (After authors cited.)

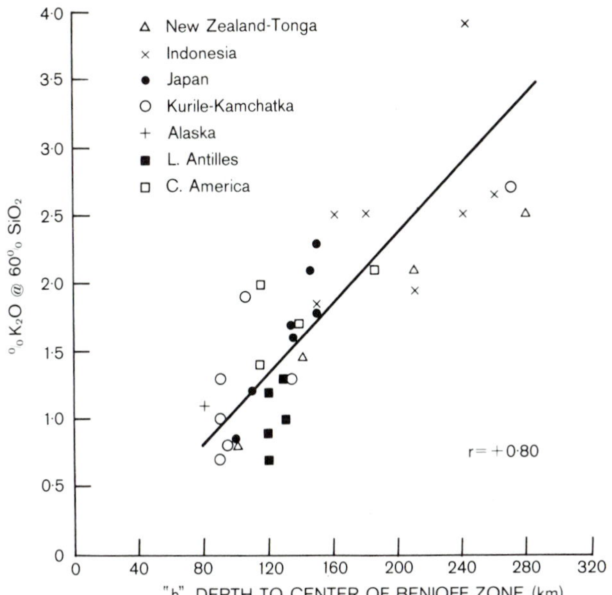

Fig. 6. Relationship of potash content of andesite to depth of earthquakes beneath volcanoes. (From Hatherton and Dickinson, 1969.)

cated correlation with volcanic chemistry, although it is significant that his best correlation of rock chemistry with geotectonic position occurs when K_2O is combined with the minimum number of other oxides.

Several mechanisms have been suggested for the origin of the island arc rock series in general. These are:

1. *Contamination.* This process envisages variable mixing of a basic parent magma with a contaminant, usually considered to be the sialic rock of the continental crust.

2. *Differentiation.* Fractional crystallization of a primary olivine basalt under appropriate oxygen partial pressure will produce a liquid similar to andesite with a crystal cumulate similar to alpine peridotites. Appropriate oxygen partial pressure is obtained (according to Osborn, 1969) by drawing water from wet geosynclinal sediments into which the primary magma, rising from the mantle, has intruded.

3. *Partial melting.* The results of high-pressure experimental work (Green and Ringwood, 1968) suggest that calc-alkaline magmas may be produced by the partial melting of oceanic crust (or its high-pressure equivalent) carried to great depth. Fitton (1971) has further suggested that the island arc tholeiitic series are produced by the breakdown of amphibole at shallow depths, and that between these two extreme types a continuum of transitional magma could be generated.

The contamination hypothesis has fallen into disfavor during recent years, mainly as a result of isotopic and trace element studies. Although differentiation still has its advocates, there is no doubt that, at present, partial melting is the most favored mechanism. The problems of heat and mass transfer involved in bringing the calc-alkaline magmas from depths in excess of 100 km to the surface have hardly been discussed. An interesting discussion of "gas transfer" as a mechansim of lava modification and development is given by Stanton (1967).

Ancient volcanic rocks can be used to infer the presence and attitudes of ancient Benioff zones (Hatherton, 1969a; Dickinson, 1970; Lipman et al., 1971) and assist in paleotectonic reconstruction.

HEAT FLOW

The volcanic front, as might be expected, marks a boundary in the heat-flow pattern of active margins. On the oceanic side the front values are about 0.7 cal/cm^2 sec, not much more than half the world-wide average. On the continental side of the front, heat-flow values are more erratic but appear to be about twice the world average. Within the Cenozoic volcanic zones, close to the volcanic front, the complex circulation patterns of groundwater under the influence of intense heat sources may produce highly variable values (Horai and Uyeda, 1963; Studt and Thompson, 1969). The heat-flow patterns induced by a downthrust slab of lithosphere have been investigated theoretically by Minear and Toksöz (1970) and Oxburgh and Turcotte (1970). Considering the limitations imposed by the rather arbitrary, but necessary, assumptions about the various heat-producing mechanisms, the overall agreement between observed and expected heat flow is good.

SEISMICITY

The compilation of Gutenberg and Richter (1945) and its subsequent versions outlined the general seismicity of active margins. In detail, their epicenter determinations suffered from the effects of inadequate, ill-distributed, and variable instrumentation. In the early 1960s the U.S. Coast and Geodetic Survey installed its World-Wide Standard Seismograph Network (WWSSN) and provided for the first time a standardized set of instruments which modern seismologists, particularly those of Lamont-Doherty Geological Observatory, have exploited with great effect.

Dominating the seismic geometry of the active margins are inclined seismogenic or "Benioff" zones of about 50 km width which dip toward the continental sides at angles between 15° and 90°, although this dip may vary along each zone. The seismicity is not distributed uniformly along the Benioff zone but usually reaches a secondary maximum at a depth that can vary between 200 and 500 km. In most margins the greatest seismicity, and the largest earthquakes occur within 40 km of the earth's surface.

The relationship between the inclined and shallow seismicity at an active margin is still not clear. In

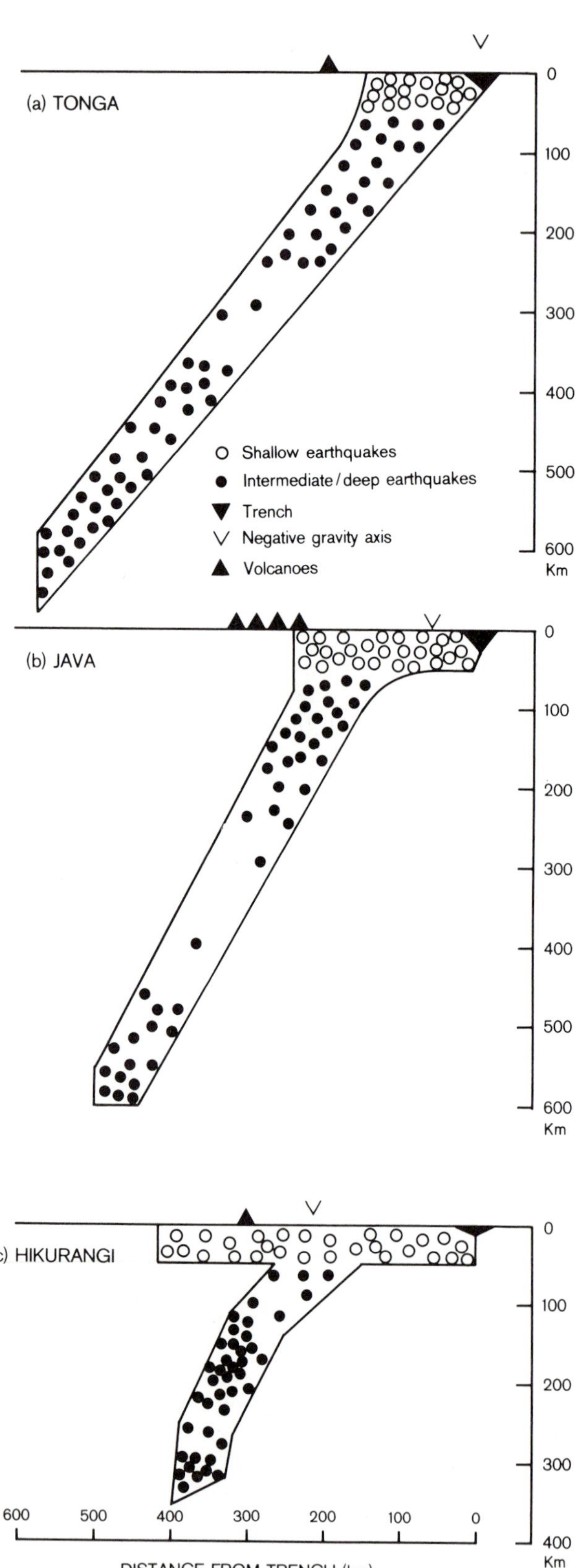

Fig. 7. Schematic representation of relative positions of trench, negative gravity anomaly, volcanism, and seismicity in three active margins.

some margins (e.g., Tonga, Kermadec), where the Benioff zone intersects the trench, the shallow seismicity is confined to an area about 200 km wide in the vicinity of the intersection of the Benioff zone and the earth's surface (Hatherton, 1971; Fig. 7a). In others, [e.g., Indonesia (Hatherton and Dickinson, 1969; Fitch, 1970) and Kamchatka (Fedotov, 1968)] the shallow seismicity occurs in an area between the trench and the Benioff zone, which intersects the shallow seismicity 150 km inland from the trench (Fig. 7b). In the Hikurangi margin (Hatherton, 1970; Gibowicz, 1974) the shallow seismicity is distributed virtually symmetrically about the upper part of the Benioff zone (Fig. 7c).

The reason for this variation is not clear, although underthrusting patterns involving horizontal translation of oceanic lithosphere below the continental lithosphere has been invoked (Sykes, 1972). The negative gravity anomalies everywhere coincide with the downturn of the Benioff zone and from the gradients of these anomalies limits can be placed on the depths of the low-density sources. These maxima tend to be about 30-50 km (i.e., less than lithospheric thickness), and if these anomalies are due to subducted sediments, as is commonly believed (although not by the present author), the sialic crust may have a mobile role independent of the thicker lithosphere.

In all cases, however, seismicity, whether shallow or deep, is confined to the continental side of the trench. Very few epicenters are placed on the oceanic side of the trench. Scholz (1969) has attributed the remarkably nonuniform geographical distribution of earthquakes not only to the presence of high differential motion but also to the necessary presence of relatively acidic rock.

Earthquake Mechanisms—Dipping Slab

Undoubtedly the most far-reaching effect of the WWSSN was in the impetus it gave to the study of mechanisms at the earthquake foci. Such studies have shown that the events occur in response to compressional or extensional stresses aligned parallel to the slab-like geometry of the inclined Benioff seismic zone. Isacks and Molnar (1971) produced a global summary of the distribution of down-dip stresses (Fig. 8). At intermediate depths extensional stresses parallel to the dip of the zone are predominant in zones characterized either by gaps in seismicity as a function of depth or by an absence of deep earthquakes. Compressional stresses parallel to the dip of the zone are prevalent everywhere the zone exists below about 300 km. The results indicate that subducted lithosphere sinks into the asthenosphere under its own weight, but encounters resistance to its downward motion below about 300 km.

Smith and Toksöz (1972) examined the stress distribution in the Benioff zones at active margins and demonstrated that gravitational sinking induced by thermal density anomalies can explain the directions of principal stress within a descending slab.

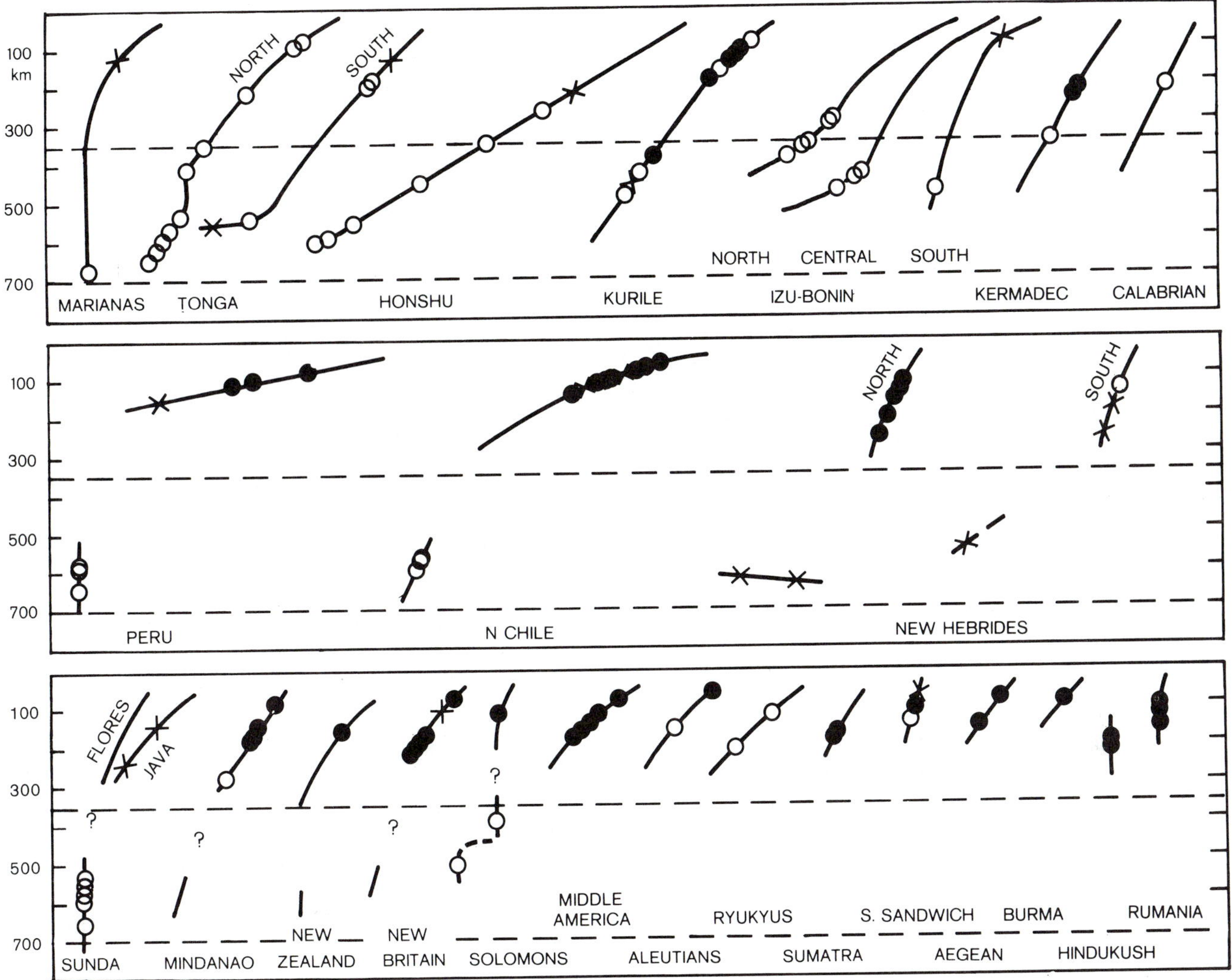

Fig. 8. Focal mechanisms in the Benioff zones. Filled circles represent downdip extension, open circles downdip compression. Crosses represent solutions that are not aligned along the zone. (From Isacks and Molnar, 1971.)

The rheologies of the slab and mantle are the dominant factors influencing the stress pattern, and significant convective effects in the mantle are not required. The computational results, according to Smith and Toksöz, "rely upon a variety of theoretical and experimental inferences that require further research," but Gibowicz (1974) shows that, for New Zealand, the variation of the seismic parameter b with depth shows clear inverse correlation wih the distribution of maximum stress as predicted by Smith and Toksöz and, further, that the maximum magnitude at different depths correlates distinctly with the distribution of the principal stress.

Earthquake Mechanisms—Shallow Events

Mechanisms of shallow events are by no means as well investigated as those in the dipping slab. Seismic events under trenches are usually consistent with normal faulting, which may be caused by tension in the upper surface of the oceanic lithosphere as it bends beneath the continental lithosphere (Stauder, 1968; Fitch, 1970; Kanamori, 1971). Earthquakes within the trench area are usually few, although heterogeneity of seismic wave transmission may tend to displace epicenters within this region to the continental side (Davies and McKenzie, 1969).

Inland of the trench the dominant mechanism of shallow earthquakes is generally of the thrust type, with the sense of motion on the shallower dipping of the two modal planes, consistent with underthrusting beneath the continental margin (e.g., Fitch, 1970). However, composite first-motion diagrams from microearthquake studies in New Zealand (W.J. Arabasz, personal communication) can only be interpreted satisfactorily by postulating a change in mechanism at a depth of about 20-25 km, with principally east-west compression above this level and east-west tension below. Detailed studies of shallow earthquakes in active margins are among the more important investigations still to be done in these areas.

SUBSEQUENT DEVELOPMENT OF ACTIVE MARGINS

The geophysical and geological manifestations at the active continental margins and island arcs are now ascribed to "subduction" of an oceanic lithosphere beneath the continental lithosphere. The concept of subduction is quite long-established. It appears, for instance, in Holmes (1944); the statement of Gutenberg and Richter (1945) that "The structural arcs may be interpreted as due to forces either pushing or drawing subcrustal material downwards toward the foredeeps, with compensating movements elsewhere" could not be altered even now. The diagram of Coats (Fig. 5b), showing underthrusting of the oceanic crust and overlying sediments beneath Alaska is almost identical with the many similar diagrams drawn since the "new global" or "plate tectonics" interpretations appeared.

What has really happened during the past decade is that the "corresponding movements elsewhere" have been identified. The lithosphere which is being subducted at the island arcs and active continental margins is interpreted to be created at the mid-ocean ridges. The unification brought about by the balancing of the two motions has resulted in a greatly increased number of workers studying the active margins. In the main, however, their interest has been kinematical and the major problems of active margins remain. Is the trench a tensional or compressional feature? Why are there limited earthquakes oceanward of the trench, yet shallow earthquakes occur up to 200 km on the continental side? How are andesites derived and transported? What is the origin of the negative gravity anomaly? And what is the consequence of "relaxation" of the state of physical perturbation exhibited by the island arc?

The concept of geosynclines originating in island arcs and active continental margins has also, like subduction, a long history. The impetus of plate tectonics has resulted in further attempts at geological unification by identifying the various geosynclinal components within the island arc system (e.g., Dickinson, 1971). Basically there are three elements in the geosyncline of the classical Appalachian type; the mildly deformed marginal or miogeosyncline, the grossly deformed eugeosyncline, and, separating them, a fairly continuous belt of ultramafic rocks. The geosynclines of New Zealand and California are very similar in structure and lateral dimensions to that of Appalachia. The problem is to find a common point so that the development of a geosyncline from an active margin may be established. Numerous attempts have been made to do this in recent years, but the present author is not satisifed that any conclusive matching has been made. In particular, little work has been done on the problem of relaxation of the hydrostatically disturbed system which an active margin represents and on the development of geological structures during the relaxation period.

BIBLIOGRAPHY

Andrew, E. M., Masson Smith, D., and Robson, G. R., 1970, Gravity anomalies in the Lesser Antillies: Inst. Geol. Sci. Geophys. Paper 5: London, H.M. Stationery Office.

Brodie, J. W., and Hatherton, T., 1958, The morphology of the Kermadec and Hikurangi trenches: Deep-Sea Res., v. 5, p. 18-28.

Coats, R. R., 1962, Magma type and crustal structure in the Aleutian arc, *in* Crust of the Pacific Basin: Geophys. Monogr. 6, p. 92-109.

Davies, D., and McKenzie, D. P., 1969, Seismic travel-time residuals and plates: Geophys. Jour., v. 18, p. 51-63.

Dickinson, W. R., 1968, Circum-Pacific andesite types: Jour. Geophys. Res., v. 73, p. 2261-2269.

———, 1970, Relations of andesites, granites and derivative sandstones to arc-trench tectonics: Rev. Geophys. Space Phys., v. 8, p. 813-860.

———, 1971, Plate tectonic models of geosynclines: Earth and Planetary Sci. Letters, v. 10, p. 165-174.

———, and Hatherton, T., 1967, Andesitic volcanism and seismicity around the Pacific: Science, v. 157, p. 801-803.

Elsasser, W. M., 1971, Sea-floor spreading as thermal convection: Jour. Geophys. Res., v. 76, p. 1101-1112.

Evison, F. F., 1968, Active regions of the southwest Pacific: Can. Jour. Earth Sci., v. 5, p. 1045-1049.

Fedotov, S. A., 1968, On the deep structure, properties of upper mantle and volcanism of Kuril-Kamchatka island arc according to seismic data, *in* The crust and upper mantle of the Pacific area: Geophys. Monogr. 12, p. 131-139.

Fitch, T. J., 1970, Earthquake mechanisms and island arc tectonics in the Indonesian-Philippine region: Seismol. Soc. America Bull., v. 60, p. 565-591.

Fitton, J. G., 1971, The generation of magmas in island arcs: Earth and Planetary Sci. Letters, v. 11, p. 63-67.

Gibowicz, S. J., 1974, Frequency-magnitude, depth and time relations for earthquakes in an island arc, North Island, New Zealand: Tectonophysics, (in press).

Green, T. H., and Ringwood, A. E., 1968, Genesis of the calc-alkaline igneous rock suite: Contr. Mineral and Petr., v. 18, p. 105-162.

Grow, J. A., 1973, Crustal and upper mantle structure of the central Aleutian arc: Geol. Soc. America Bull., v. 84, p. 2169-2192.

Gunn, R., 1947, Quantitative aspects of juxtaposed ocean deeps, mountain chains and volcanic ranges: Geophysics, v. 12, p. 238-255.

Gutenberg, B., and Richter, C. F., 1945, Seismicity of the earth: Geol. Soc. America Bull., v. 56, p. 603-668.

Hatherton, T., 1969a, The geophysical significance of calc-alkaline andesites in New Zealand: New Zealand Jour. Geol. Geophys., v. 12, p. 436-459.

———, 1969b, Gravity and seismicity of asymmetric active regions: Nature, v. 221, p. 353-355.

———, 1970, Gravity, seismicity and tectonics of the North Island, New Zealand: New Zealand Jour. Geol. Geophys., v. 13, p. 126-144.

———, 1971, Shallow earthquakes and rock composition: Nature, v. 229, p. 119-120.

———, and Dickinson, W. R., 1969, The relationship between andesitic volcanism and seismicity in Indonesia, the Lesser Antilles and other island arcs: Jour. Geophys. Res., v. 74, p. 5301-5310.

Holmes, A., 1944, Principles of physical geology: London, Nelson.

Horai, K., and Uyeda, S., 1963, Terrestrial heat flow in Japan: Nature, v. 199, p. 364-365.

Isacks, B., and Molnar, P., 1971, Distribution of stresses in the descending lithosphere from a global survey of focal-mechanism solutions of mantle earthquakes: Rev. Geophys. and Space Phys., v. 9, p. 103-174.

Kanamori, H., 1971, Seismological evidence for a lithospheric normal faulting—the Sanriku earthquake of 1933: Phys. Earth Planetary Int., v. 4, p. 289-300.

Karig, D. E., 1971, Structural history of the Mariana island arc system: Geol. Soc. America Bull., v. 82, p. 323-344.

Kuenen, P. H., 1936, The negative isostatic anomalies in the East Indies, with experiments: Leidsche Geol. Meded., v. 8, p. 169-214.

Kuno, H., 1959, Origin of Cenozoic petrographic provinces of Japan and surrounding areas: Bull. Volc., v. 20, p. 37-76.

Lipman, P. W., Prostka, H. J., and Christiansen, R. L., 1971, Evolving subduction zones in the western United States as interpreted from igneous rocks: Science, v. 174, p. 821-825.

Livshits, M. Kh., 1965, To the problem of the physical state of abyssal matter of the earth's crust and upper mantle, *in* The Kurile Zone of the Pacific Belt: Geol. i. Geofiz., v. 1, pp. 11-20.

Marshall, P., 1912, The structural boundary of the Pacific Basin: Rept. Australasian Assoc. Advan. Sci., v. 13, p. 90-99.

Minear, J. H., and Toksöz, M. N., 1970, Thermal regime of a downgoing slab and new global tectonics: Jour. Geophys. Res., v. 75, p. 1397-1419.

Ninkovich, D., and Hayes, J. D., 1972, Mediterranean island arcs and origin of high-potash volcanoes: Earth and Planetary Sci. Letters, v. 16, p. 331-345.

Osborn, E. F., 1969, The complementariness of orogenic andesite and alpine periodotite: Geochim. Cosmochim. Acta, v. 33, p. 307-324.

Oxburgh, E. R., and Turcotte, D. L., 1970, Thermal structure of island arcs: Geol. Soc. America Bull., v. 81, p. 1665-1688.

Reilly, W. I., 1965, Gravity map of New Zealand 1:4,000,000: Wellington, Dept. Sci. and Ind. Res.

Ringwood, A. E., 1969, Composition and evolution of the upper mantle: *in* Earth's Crust and Upper Mantle; Am. Geoph. Union Monograph 13, p. 1-17.

Scholz, C. H., 1969, Worldwide distribution of earthquakes: Nature, v. 221, p. 165.

Smith, A. T., and Toksöz, M. N., 1972, Stress distribution beneath island arcs: Geophys. Jour., v. 29, p. 289-318.

Stanton, R. L., 1967, A numerical approach to the andesite problem: Koninkl. Ned. Akad. Wetenschap. Proc., v. B70, p. 176-216.

Stauder, W., 1968, Tensional character of earthquake foci beneath the Aleutian Trench with relation to sea-floor spreading: Jour. Geophys. Res., v. 73, p. 7693-7702.

Studt, F. E., and Thompson, G. E. K., 1969, Geothermal heat flow in the North Island, New Zealand: New Zealand Geol. Geophys., v. 12, p. 673-683.

Sugimura, A., 1968, Spatial relations of basaltic magmas in island arcs, *in* Hess, H. H., and Poldervaart, A., eds., the Poldervaart treatise on rocks of basaltic composition: Interscience, N.Y., p. 537-572.

Sykes, L. R., 1972, Seismicity as a guide to global tectonics and earthquake prediction: Tectonophysics, v. 13, p. 393-414.

Taylor, S. R., 1968, Geochemistry of andesites, *in* Ahrens, L. H., ed., Origin and distribution of the elements: Elmsford, N.Y., Pergamon Press.

Vening Meinesz, F. A., 1948, Gravity expeditions at sea, 1923-1938, IV.: Netherlands Geodetic Comm. Publ.

von Huene, R. and Kuml, L. D., 1974, Tectonic summary of Leg 18 of Deep Sea Drilling Project; U.S. Govt. Printing Office, Washington, D.C., 975 p.

Wadati, K., 1935, On the activity of deep-focus earthquakes in the Japan Islands and neighbourhood: Geophys. Mag., v. 8, p. 305-326.

Worzel, J. L., 1965a, Deep structure of coastal margins and mid-oceanic ridges, Submarine Geol. and Geophys.: Proc. 17th Symp. Colston Res. Soc.: London, Butterworth.

———, 1965b, Pendulum gravity measurements at sea, 1936-1959: New York, Wiley.

The Segmented Nature of Some Continental Margins

M. J. Carr, R. E. Stoiber and C. L. Drake

INTRODUCTION

There is ample geological and geophysical evidence to show that certain continental margins and island arcs (converging plate margins) are broken by transverse features into segments a few hundred kilometers long. The segmentation varies between the different margins. Geological and geophysical data that define segments include lines of active volcanoes; clusters of volcanoes and cinder cones; changes in strike and offsets of trench axes; changes in strike and offsets of longitudinal geologic structures, such as normal faults or grabens parallel to the volcanic lines; transverse faults; the lateral margins of aftershock areas of great earthquakes; changes in strike and offsets of the deep seismic zone; focal mechanisms; P-wave travel-time anomalies; and concentrations of small and moderate-sized shallow earthquakes.

Recognition of tectonic subdivisions on the order of 100 km long may refine the plate tectonic model for converging plate margins and introduces new problems which can be resolved only by geological and geophysical studies on a detailed scale. In areas where large or great earthquakes can be expected, but have not occurred during the time of instrumental seismology, it may be possible to predict their areas of influence from several independent types of geological evidence.

HISTORICAL AND REGIONAL CONSIDERATIONS

Geologic interpretation of the island arcs and volcanic belts ringing the Pacific has greatly increased with the development of plate tectonics and the rapid qualitative and quantitative improvement of seismological data. The installation of the World Wide Standardized Seismograph Network (WWSSN) and detailed investigations of particular arcs, especially Tonga, Kamchatka-Kuriles, Japan, and the Aleutians, have enabled seismologists to define in detail the overall geologic and geophysical structure of these areas.

Maps of earthquake locations based on the WWSSN (Barazangi and Dorman, 1969) delineate plate margins. The inclined zones of mantle earthquakes were shown to mark the upper margin of a thick, anomalously rigid slab by Oliver and Isacks (1967). This lithospheric model for converging plate margins was refined and summarized in a later paper (Isacks, et al., 1968).

Several authors have examined the seismicity of specific island arcs in order to define the first-order geologic structures, the apparent stresses of the region, and their tectonic significance. Some examples for specific areas include Sykes (1966), Tonga and other areas; Molnar and Sykes (1969), middle America; Katsumata and Sykes (1969), the Phillipine Sea; Fitch (1970), Indonesia; and Santo (1969), South America. In general these papers are concerned with structures that strike parallel to the arc and that are on the order of a few thousand kilometers long. Some examples of these structures include trenches, inclined seismic zones, anomalous high-velocity slabs extending below the inclined earthquake zones, asymmetric gravity anomalies associated with the underthrust slabs, and active volcanic chains. These large parallel structures change significantly along the length of the arc. The changes have usually been portrayed as gradual and continuous, but there is steadily accumulating evidence that shows the existence of transverse structures at which the first-order parallel structures change abruptly and discontinuously. The recognition of such structures is a natural consequence of increasingly accurate geophysical data and geologists' desire to extend the new plate tectonic model to ever-smaller geologic features.

Very clear evidence for the existence of transverse discontinuities in island arc structures comes from studies of the focal areas of large, shallow earthquakes that occur along the upper 20-60 km of the inclined seismic zones. Fedotov (1965) recognized the spatially bounded character of great earthquake focal areas and emphasized their sharp lateral terminations. The focal area of the 1965 Rat Island earthquake in the Aleutian Islands, had a block-like pattern correlated with transverse geologic structures (Jordan et al., 1965). Mogi (1968, 1969b) studied great earthquakes throughout the circum-Pacific and related them to transverse geologic features.

The focal areas of great earthquakes divide island arcs into segments about 100-1,000 km long. Segments of similar scale were proposed by Carr et al. (1973) on the basis of abrupt changes in the strike and dip of the deep seismic zones beneath the Japanese islands. Volcanic lineaments 100-300 km long, first recognized by Sapper (1897), were used to segment the Central American arc (Stoiber and Carr, 1974). Stauder (1973) and Swift and Carr (1973) have subdivided the Chilean seismic zone using different seismological criteria. Swift and Carr based their subdivision solely on abrupt changes in the geometry of the deep seismic zone. Stauder found regional differences in seismic activity and intermediate depth earthquake focal mechanisms and proposed not only transverse discontinuities but also discontinuities between various depth ranges. These two interpretations differ in scale and in detail, as a result of the use of several types of data for making the subdivisions.

It is not yet possible to say which criteria are the most reliable for determining transverse structures because there has not been a systematic worldwide examination. As a first step we review here the various seismological, volcanological, and geological data that have been employed to interpret the presence of transverse structures in certain continental margins and island arcs.

Throughout the paper we use the term "deep seismic zone" in a loose sense to refer to any inclined zone of intermediate or deep earthquakes. We use the term "segment" to describe the tectonic subdivisions of seismically active continental margins; some authors have preferred the term "block." This term has a connotation of a closed area, which does not seem appropriate, since the landward sides of the segments are never clearly defined.

CONCEPTUAL MODELS OF TRANSVERSE STRUCTURES

Many of the transverse structures described in the literature have been attributed to tear faults or buckles, based on simple geometrical considerations. Frank (1968) and Stroback (1973) have pointed out that the surface area of an underthrust slab is conserved only if the dip of the slab is twice the radius of its surface trace. If the dip is too steep, the slab is subject to tensile stress in the strike direction and if the dip is too shallow the slab is subject to compressive stress in the strike direction. These stresses are apparently negligible at the surface and increase with depth. Models that describe tear faults or buckles (e.g., Isacks and Molnar, 1971) apply only to the underthrust slab. In areas where oceanic lithosphere underthrusts continental lithosphere, such as Middle America, South America, and southern Alaska, there is no approach to the ideal geometry of an arc, and these longitudinal stresses may be large.

The overlying plate also has transverse structures. These may be due to interaction with the underthrust lithosphere. One difficulty with this model is that some of the better evidence for segmentation—the boundaries of great earthquake focal areas and offsets of the trench axis—are rather shallow features, where the stresses that cause tear faulting and buckling are presumably smallest.

SEGMENTATION MODEL

Our present conceptual model for segmented converging plate margins is as follows: The zone of shallow earthquakes is divided into discrete areas by discontinuities in the underthrusting slab. Great earthquakes related to underthrusting result when a segment of the overlying plate rebounds after several decades of being elastically deformed by the underthrusting oceanic plate. The state of stress can be different in adjacent segments, depending on the time since the last great earthquake. At the segment boundaries deep crustal faults develop in the overlying plate. A segment of underthrust lithosphere separated from its neighbors by tear faults can descend into the mantle with a different strike and dip. The deep seismic zone associated with this segment would have a different strike or dip from that of adjacent segments. Similarly, a line of active volcanoes derived from melting along the slab would have a different strike, or if the dip changed, the melt zone could be at a different distance from the trench and the volcanic line would appear offset from adjacent lines.

The segmented pattern seems to affect crustal structures as far back as the volcanic chains 200 km or more from the trench. It seems probable that the deep crustal faults that segment the overlying plate extend as far away from the trench as the volcanic belt. At this distance the descending lithosphere is about 100 km deep and is considered unlikely to affect crustal structures, except through the rise of magma from the vicinity of the slab.

EVIDENCE FROM SEISMOLOGY

Most of the evidence cited to establish the presence of transverse structures in island arcs has been seismological. Carr et al. (1973) and Swift and Carr (1973) examined the morphology of the deep seismic zones of Japan and Chile, respectively. They used the better located earthquake foci from the Earthquake Data File of the National Oceanic and Atmospheric Administration (NOAA) to make epicentral maps and vertical sections. Changes in strike or dip of the deep seismic zone were interpreted as transverse faults segmenting the underthrust lithosphere. At present the lack of very accurate hypocentral locations and the large distances between adjacent locations do not allow determination of whether the strike and dip changes are the result of faults, as they have been interpreted, or folds. The important point is that discontinuities in the deep seismic zone have been recognized which correlate well with structural features on the surface. These interpretations of the Chilean and Japanese deep seismic zones, with the addition of the focal areas of recent great earthquakes, are reproduced here in Figures 1 and 2, respectively.

Discontinuities in the deep seismic zone are most clearly depicted by vertical sections which show the foci on either side of a discontinuity with different symbols. One difficulty is that projecting the foci from two segments with different strikes onto a plane perpendicular to one of the segments will cause the second segment to have an apparent dip. This causes widening and distortion of the pattern of foci from the second segment. Composite vertical sections for the north part of the Chilean seismic zone are given in Figure 4; for comparison vertical sections of the individual segments are given in Figure 3.

In making interpretations of the deep seismic zones a real attempt was made to not look at or be influenced by obvious discontinuities in the other tectonic features in the areas. This was a reasonable way to find out whether discontinuities in the deep seismic zone correlated to discontinuities in surficial structures. We now recognize several types of geophysical and geological data that indicate transverse structures, and it will be advisable in the future to consider all the data simultaneously. It is difficult to use NOAA earthquake locations to examine the morphology of deep seismic zones because of problems involved in locating earthquakes in island arc regions: lateral inhomogeneities in the seismic velocity structure of the upper mantle under island arcs can lead to large systematic errors in location, and large relative errors between events recorded by different sets of seismograph stations. Relocation schemes that take these lateral inhomogeneities into account will be necessary to define more precisely the presence of transverse structures in the deep seismic zone.

Stauder (1968, 1973) and Isacks and Molnar (1971) have deduced the presence of transverse structures in the descending lithosphere on the basis of focal mechanism studies. In some cases an intermediate depth earthquake has a focal mechanism, which is what would be expected of a tear fault, and it occurs in an area where there are abrupt changes in the strike of geologic structures. In other cases there is a discontinuous change in the orientation of intermediate depth focal mechanisms along the strike of the arc. If the spacing is close enough, it is possible to interpret an abrupt strike change between adjacent segments. One difficulty with both these methods is that few intermediate-depth earthquakes yield reliable focal mechanisms.

Another seismological method used to find transverse structures is the analysis of azimuthal variations in travel-time anomalies of P-waves from events located above a tear fault (Abe, 1972). P-waves that travel through low-velocity mantle in the tear between high-velocity segments of lithosphere will be slowed relative to those that pass through the inclined slabs of lithosphere. Abe used travel times from the longshot nuclear blast, which has a precisely known location and time of origin. The anomaly that was interpreted as due to passing through a tear was a second-order feature on the overall anomaly pattern. This method will be difficult to apply to earthquakes whose locations and origin times are less well known.

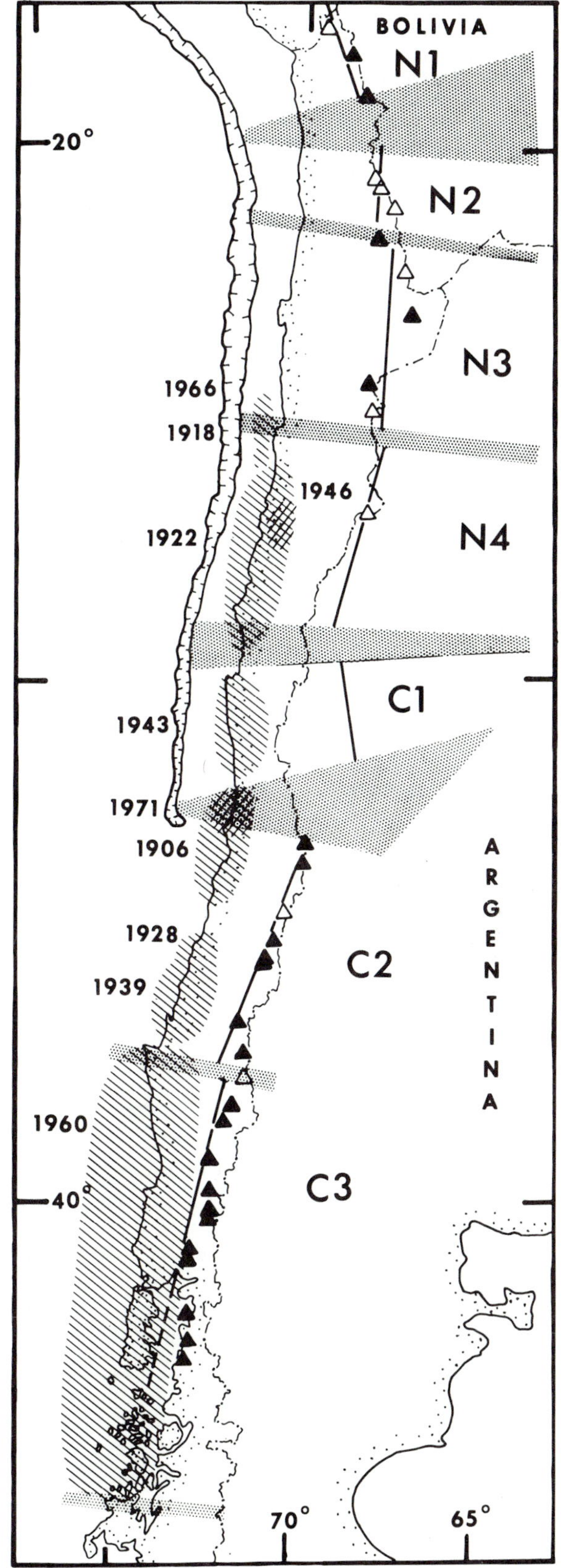

Fig. 1. Segments of the Chilean seismic zone. Solid triangles are volcanoes with historic eruptions; open triangles are volcanoes with solfatara activity (Casertano, 1963). Bathymetry is from Hayes (1966). The closed isobath is 5,500 m. The solid straight lines are the approximate trace of the 100-km depth of the inclined seismic zone. The stippled areas mark the boundaries between planar segments of the deep seismic zone. Lined areas are aftershock areas of great earthquakes from Kelleher (1972).

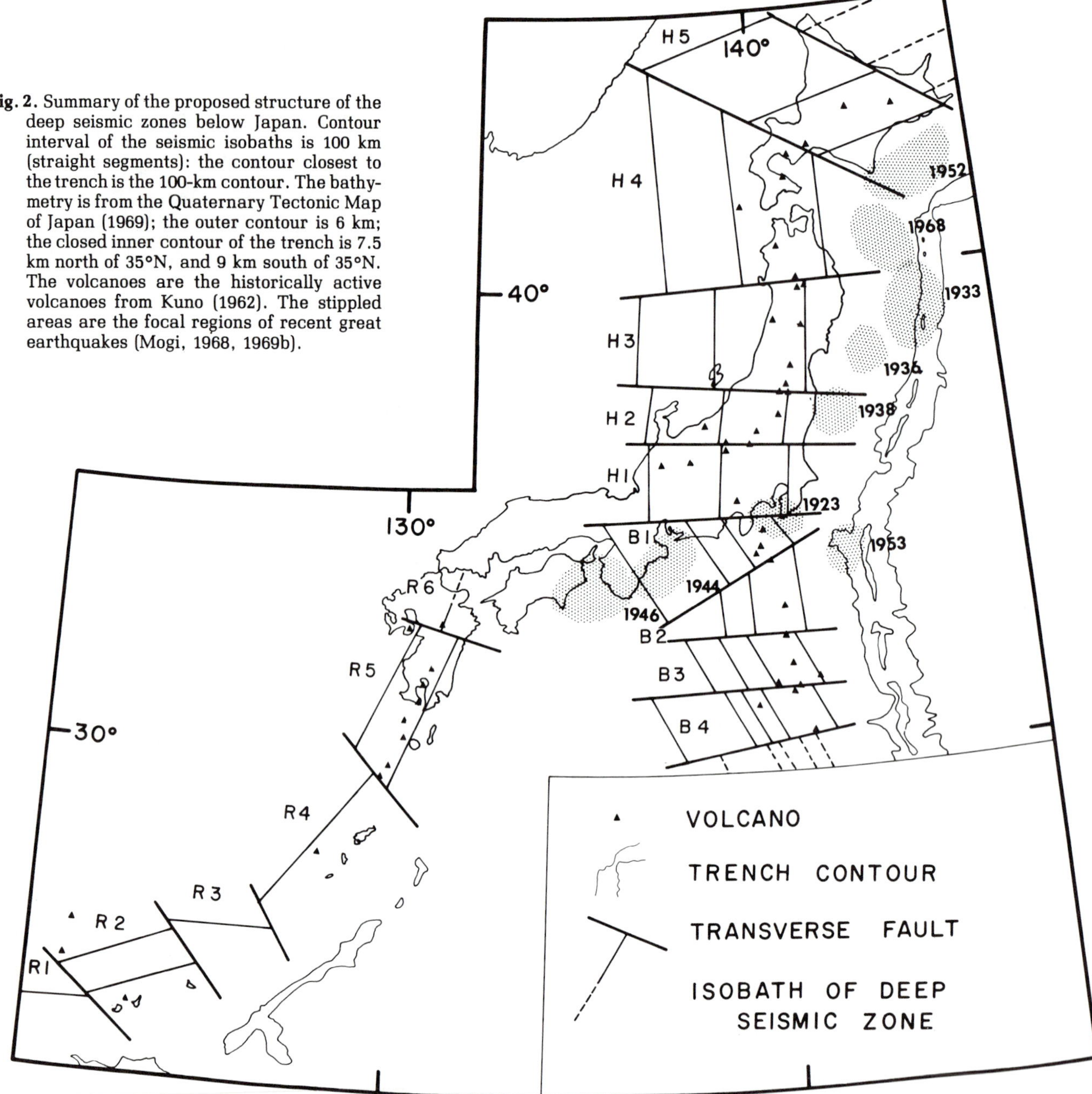

Fig. 2. Summary of the proposed structure of the deep seismic zones below Japan. Contour interval of the seismic isobaths is 100 km (straight segments): the contour closest to the trench is the 100-km contour. The bathymetry is from the Quaternary Tectonic Map of Japan (1969); the outer contour is 6 km; the closed inner contour of the trench is 7.5 km north of 35°N, and 9 km south of 35°N. The volcanoes are the historically active volcanoes from Kuno (1962). The stippled areas are the focal regions of recent great earthquakes (Mogi, 1968, 1969b).

Detailed examinations of the development of aftershock areas of great earthquakes provide strong evidence for the existence of transverse structures. Mogi (1969a) has demonstrated this in considerable detail. The clearest examples are from the 1957 Andreanof Island earthquakes and the 1965 Rat Island earthquakes in the Aleutian arc. Jordan et al. (1965) reported sharp eastern and western boundaries to the Rat Island earthquakes and suggested fault zones transverse to the arc to explain the abrupt termination. The eastern boundary abutted the western end of the 1957 Andreanof Island earthquakes in the vicinity of the Amchitka Pass (Brazee, 1965). Mogi (1969a) pointed out that the Aleutian arc between 165°E and 160°W is not a small circle, but is composed of three linear segments separated by transverse discontinuities at the Amchitka and Amkta deep-water passes. The focal area of the 1965 earthquakes filled the westernmost segment. The focal area of the 1957 earthquakes overspread the middle segment and part of the eastern segment, but the initial development of the aftershock area was discontinuous in the area of the Amkta Pass. Mogi proposed smaller-scale transverse structures subdividing the western segment into about 200-km-long units on the basis of the development of the aftershock area of the 1965 Rat Island earthquake and the fault system for the region deduced from submarine topography by Gates and Gibson (1956).

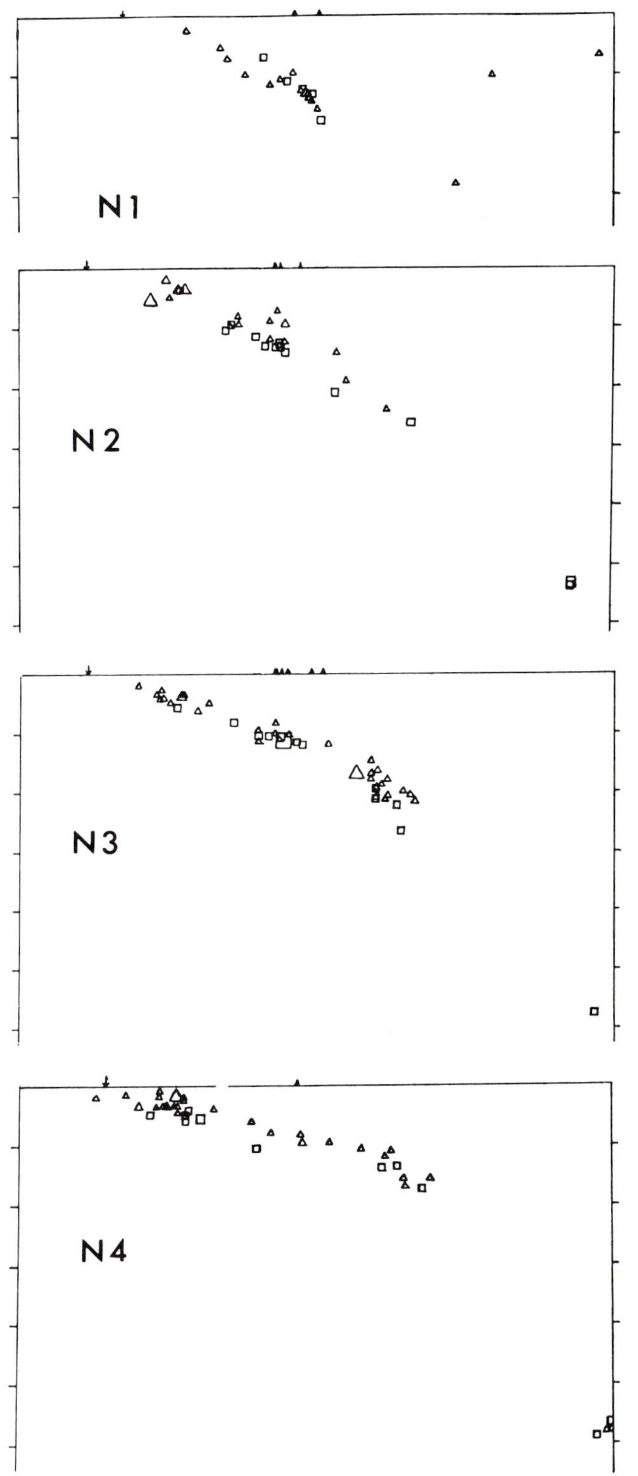

Fig. 3. Vertical sections of the segments of the seismic zone in northern Chile. Each section is perpendicular to the 100-km contour of the same segment in Figure 1. The approximate trench position is represented by down-pointed arrows. Active volcanoes are represented as solid triangles. Open triangles are earthquake foci. Open squares are earthquake foci whose depths were restrained by pP arrivals. The size of the symbols of earthquake foci is at least 10 km per side. Earthquakes of larger magnitude have larger symbols according to the magnitude versus diameter of after shock zone relationship of Utsu (1961).

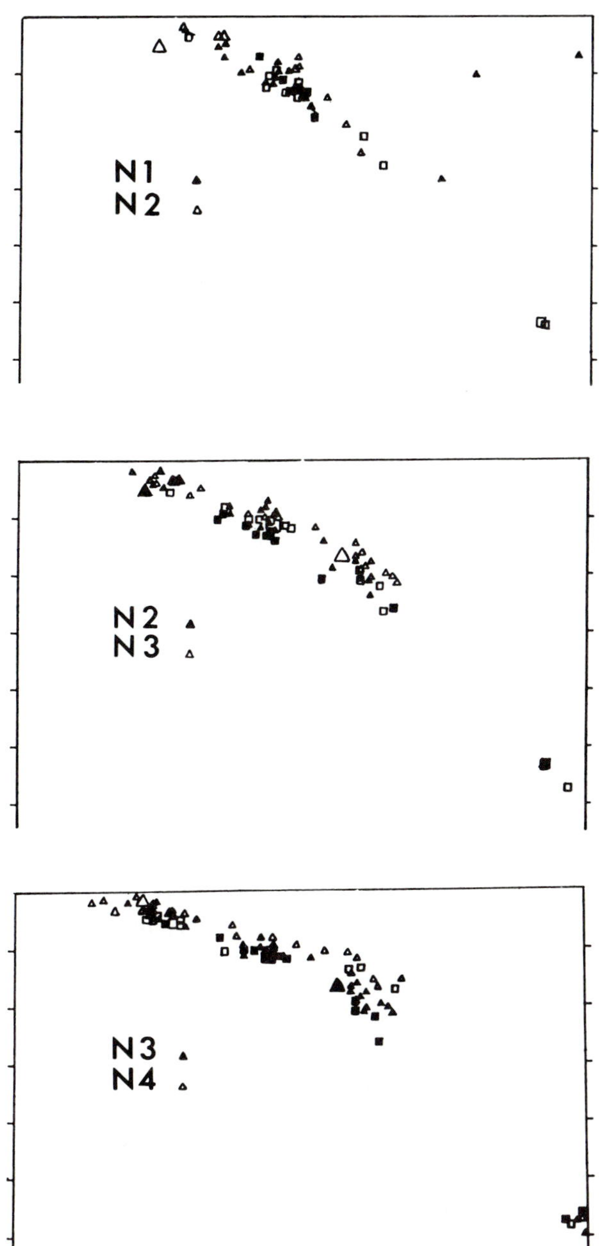

Fig. 4. Composite vertical sections in northern Chile. Foci from adjacent segments are plotted in the same projection. The section is perpendicular to the strike of the segment represented by solid symbols.

In Chile and Japan the surface projections of discontinuities in the deep seismic zone divide the shallow seismic zone into segments that frequently coincide closely with the focal areas of great earthquakes (Figs. 1 and 2). The focal regions of recent great earthquakes in Chile as summarized by Kelleher (1972) are superimposed on Figure 1. The earthquakes of magnitude greater than 8.2 (1960, 1906, 1922, 1928, 1939, 1943) tend to occur within individual segments. The 1906, 1928, and 1939 events all occur within segment C2, although Kelleher (1972) states that the lateral boundaries of the 1906 event are "determined from marginal evidence" and that the magnitude of the 1928 event may be overestimated.

The earthquakes with magnitude less than 8 (1966, 1918, 1946, 1971) tend to occur on or near one of the boundaries between the segments. The 1946 event is an exception but may not have been a major earthquake (Kelleher, 1972).

The focal area of recent great earthquakes in Japan as summarized by Mogi (1968, 1969b) is plotted in Figure 2. The projections of segments H5, H4, and H2 closely correspond to the focal regions of the 1952 Tokachi-oki and 1938 Fukushimaoki earthquakes, respectively. The surface projection of segment H3 is covered by the focal regions of the 1933 Sanriku-oki earthquake in the north and the 1934 Kinkazan-oki earthquake in the south. This lack of direct correspondence may be due to the fact that segment H3 is poorly defined because of the low density of deeper earthquakes. The area below the 1946 Nankaido and 1944 Tonankai earthquakes does not have a well-defined deep seismic zone. The 1926 Kanto and 1952 Boso-oki earthquakes may be related to transverse movements between the Philippine Sea and Eurasian plates and not be related to underthrusting (Kanamori, 1971; Ichikawa, 1971).

Stoiber and Carr (1974) proposed a method for determining transverse structures which is based on the distribution of shallow earthquakes in an area where there have been no recent great earthquakes. The Central American arc was broken into seven segments on volcanological evidence (Fig. 5).

A smoothed curve of energy release versus distance in Central America (Fig. 5) shows good correlation between peaks in the energy-release curve and the proposed transverse structures shown in Figure 5. The earthquake data are from NOAA for the period 1963-1971. Smoothing was done with a continuous normal-curve function with a standard deviation of 20 km. Since there were no large or great earthquakes during this period, the maxima in the energy-release curve reflect concentrations of small and moderate-sized earthquakes. We are currently examining the focal mechanism of the larger earthquakes in these concentrations to see if they are compatible with transverse faulting.

EVIDENCE FROM VOLCANOLOGY

The detailed distribution of active volcanoes has received rather little attention in recent tectonic studies. It is recognized that volcanoes in island arcs occur above the inclined deep seismic zones, where the depth to the zone is about 100-200 km. Sugimura (1961) and Kuno (1966) showed that there are chemical trends in recent lavas that can be correlated with depth to the seismic zone. A detailed study of the alignments of volcanic vents in Indonesia was carried out by Tjia (1967), who recognized several classes of volcanic lineaments.

The concept of 100- to 1,000-km subdivisions of island arcs is very compatible with the detailed distribution of active volcanoes. The Central American volcanic chain is particularly simple to interpret since all the active volcanoes are restricted to the "volcanic front" as defined by Sugimura (1960). Stoiber and Carr (1974) subdivided the Central American arc into seven segments, which coincided with the seven prominent volcanic lineaments (Fig. 6). Evidence from seismicity and Quaternary structures supported this interpretation. The volcanoes of

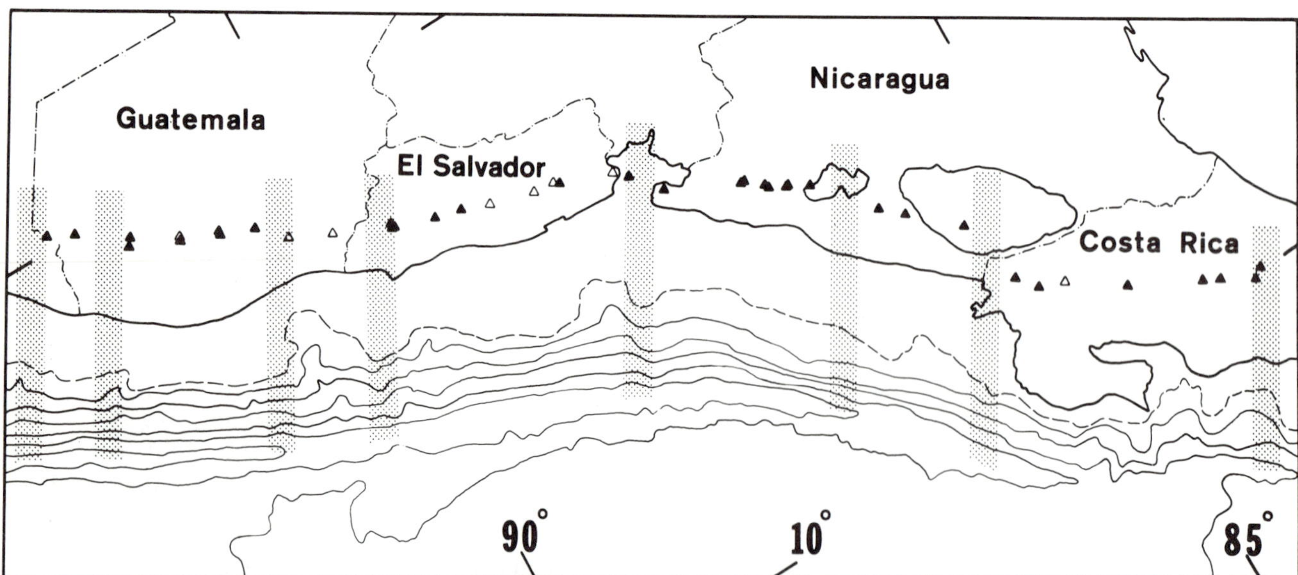

Fig. 5. Volcanic segments of Central America. Solid triangles are volcanoes with historic eruptions (from Mooser et al. 1958), with the addition of Arenal Volcano in Costa Rica. Open triangles are volcanoes with solfatara activity (from Mooser et al., 1958), with the addition of Moyuta Volcano in southeastern Guatemala and the deletion of Zuñil. Bathymetry is from Fisher (1961): the dashed contour is the 100-fathom contour; the next contour is the 500-fathom interval. Stippled areas represent boundaries between adjacent segments.

the Tonga-Kermedac chain have been divided into similarly scaled lineaments by Ewart and Bryan (1973).

The relation of other volcanic belts to the segmented structure of converging plate margins can take on different forms. In Mexico there are very few active volcanoes, considering the length of the volcanic belt. There are numerous prehistoric cinder cones that make up several clusters, each of which strikes transverse to the main volcanic axis. In Central America there are two transverse clusters of cinder cones, both of which occur at transverse breaks in the arc. The clusters of Mexican cinder cones probably mark transverse structures segmenting the Mexican arc (Fig. 7).

The diffuse volcanic activity in Mexico is in strong contrast to the simple lineaments in Central America. The available earthquake locations indicate a very shallowly dipping seismic zone extending to the volcanic front. Central America has a much more steeply dipping seismic zone, and this may explain the differences in volcanism. In Japan (Fig. 2) there are few volcanic lineaments, and the average spacing between active volcanoes is intermediate between that of Mexico and Central America. There is a tendency for active volcanoes to be clustered about the surface projections of the discontinuities in the deep seismic zone (Fig. 8).

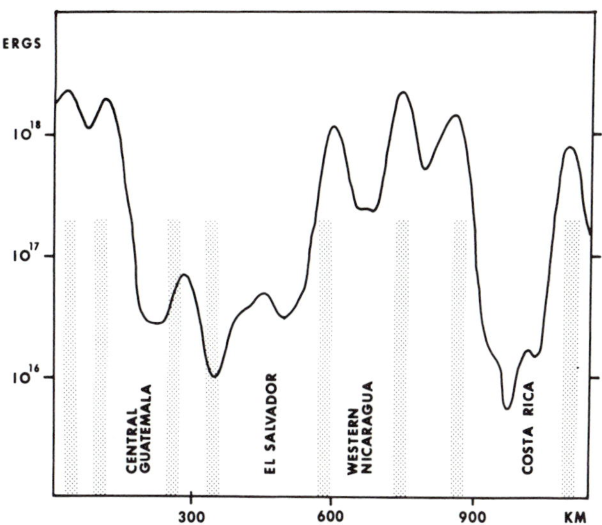

Fig. 6. Smoothed energy-release pattern for shallow (<70 km) earthquakes of Central America. Earthquake data from NOAA for 1963-1971. Smoothing was by a continuous normal-curve function, with a standard deviation of 20 km. This smoothing almost completely attenuates wave lengths smaller than 60 km, but retains 100-km wavelengths at about 50% of their original amplitude. Stippled bars represent the margins of the segments shown in Figure 5.

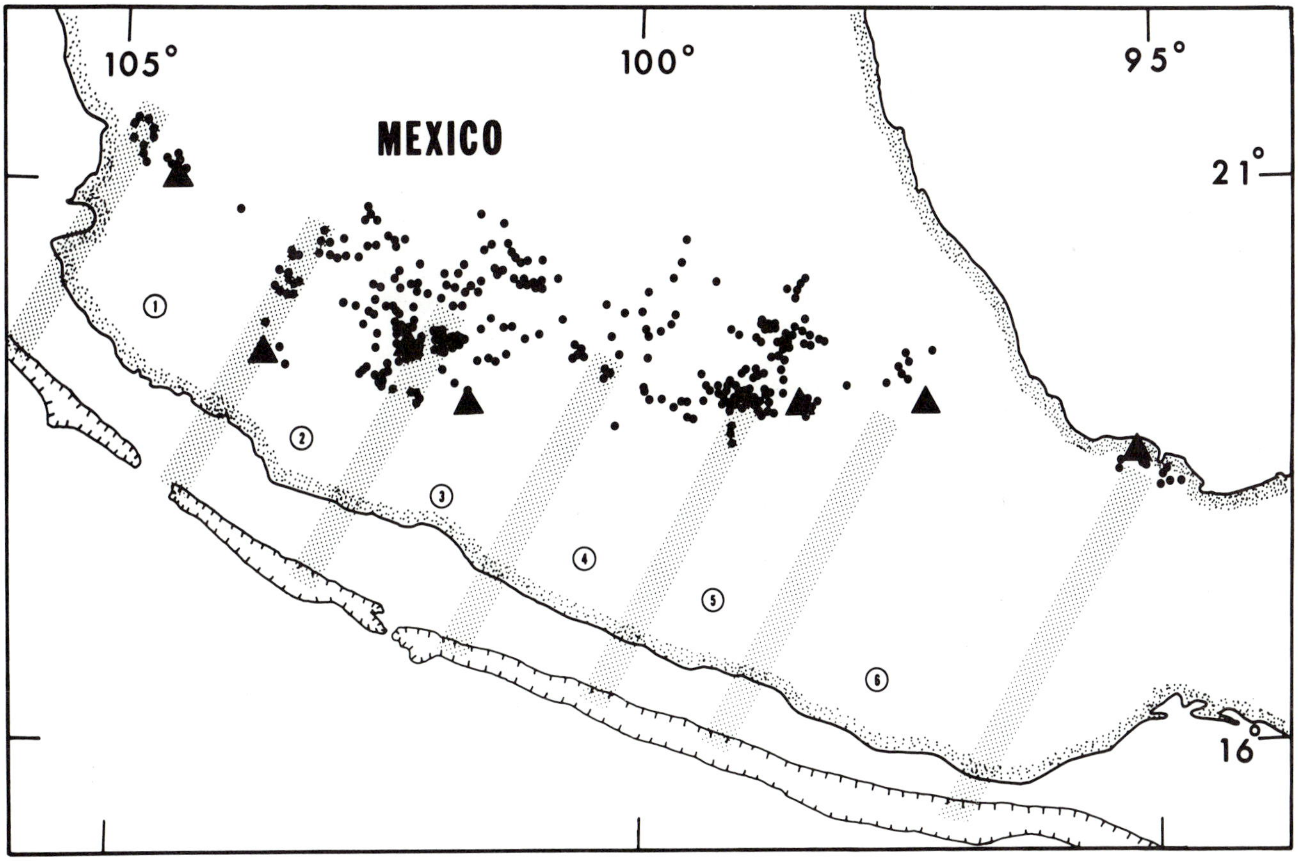

Fig. 7. Volcanic segments of Mexico. Solid triangles represent historically active volcanoes (Mooser et al., 1958). Small circles are cinder cones (from de Cserna, 1961). Stippled bars separate the proposed segments.

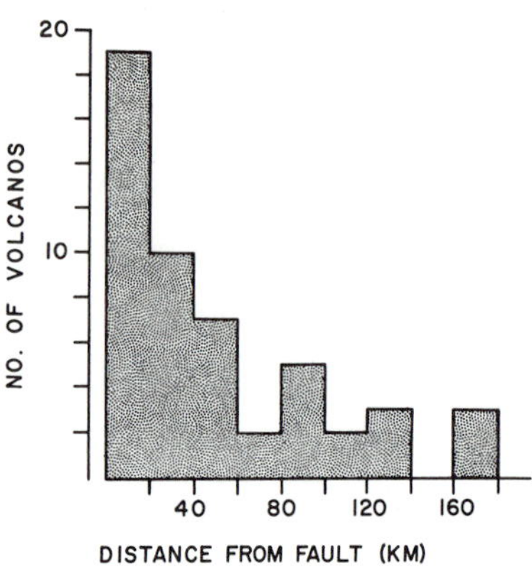

Fig. 8. Histogram relating the frequency of historically active volcanoes (Kuno, 1962) to their distance from the surface trace of the nearest proposed transverse structure.

Catastrophic volcanic eruptions may be related to the transverse structure of island arcs. Krakatau volcano is at a prominent structural break in the Indonesian arc. In Central America, Santa María Volcano (1902) and Cosigüina Volcano (1835) were the site of giant explosions of the Krakatauan type. Both are adjacent to transverse structures proposed by Stoiber and Carr (1974). A spectacular transverse volcanic alignment related to the Aleutian arc has been shown at Mt. Veniaminof in southern Alaska (Burk, 1965).

CORRELATION WITH MAPPED STRUCTURES

Transverse structures of active plate boundaries are frequently difficult to detect because most of the arc is below sea level. The parts above sea level are often covered with thick deposits of volcanic ash and debris. Nevertheless, there are several topographic and structural features that reflect subdivisions of converging plate boundaries.

In Chile (Fig. 1) transverse structures usually occur at abrupt bends in the coastline and separate linear stretches of coast that strike parallel to the deep seismic zone. The Japan Trench changes strike or is offset in several areas where discontinuities in the deep seismic zone project into it (Fig. 2). There are abrupt offsets between B2 and B3 and between B4 and B5. The discontinuity separating H1 and H2 narrowly misses projecting into another large offset of the trench axis. The trench axis changes strike near the projection of the discontinuities separating B3 and B4 and H2 and H3. In northern Japan (Fig. 1), the trench is relatively featureless from the area where the discontinuity separating H2 and H3 projects into it to the gross bend of the trench where the discontinuity separating H4 and H5 projects into it. Since the traces of the transverse discontinuities are drawn straight and since the seismic data were limited to depths of 85 km and greater, these coincidences are remarkable.

Perhaps the most striking comparison between surface structure and the transverse discontinuities occurs in the Ryukyu arc. The arc is geologically divisible into three structural units that correspond to segments R2, R4, and R5. In each of these segments, the strike of the deep seismic zone is parallel to the main topographic and tectonic elements represented by the linear island groups and lines of volcanoes, respectively. Konishi (1965) has proposed two major left-lateral transcurrent dislocations based on offsets of pre-Miocene tectonic belts. The northernmost of these is represented by the Tokara Channel, which lies above the proposed transverse discontinuity separating R4 and R5. The southernmost dislocation spans a broad area termed the Miyako depression. This corresponds almost perfectly with the area spanned by segment R3. The offsets of the 100-km isobaths for the top of the seismic zone do not show the same sense of offset as the pre-Miocene basement, but this may be attributed to the fact that the different segments have different dips. The main point is that major structural elements of the crust are mirrored in the structure of the deep seismic zone.

EXTENSIONS AND APPLICATIONS

The transverse structures subdividing island arcs into independent tectonic segments can significantly refine the plate tectonic model for these arcs. We can now suggest some extensions of plate tectonic methods to much-smaller-scale features.

In Central America, most prominent recent structures transverse to the arc are normal faults or alignments of volcanic vents striking north-south. Left-lateral strike-slip faults have occurred in the Managua area (Brown et al., 1973). These faults strike N30°-40°E, which is also the direction of plate convergence. The rate of convergence apparently increases to the southeast, so that at the transverse discontinuities there should be a slightly stronger stress on the southeasterly segment, which would give rise to a left-lateral shear striking N30°E. This model can explain the orientation of faults in the Managua area, which is in a transverse structure separating the east Nicaragua and west Nicaragua segments (Fig. 5). In Figure 9, which is oriented so that N30°E is vertical, the N30°-40°E strike-slip faults at Managua are represented by solid lines. North-south tensions, compatible with the left-lateral shear, are represented by the north-south alignment of volcanic vents.

The seismological, volcanological, and structural evidences of transverse structures in island arcs could be applied to the estimation of seismic risk and earthquake prediction. The lateral margins of focal areas of future great earthquakes can be predicted using the methods described above. Frequently, great earthquakes begin to rupture at one margin of their focal regions and propagate to the other margin. It may be advisable to monitor these margins extensively in order to record phenomena

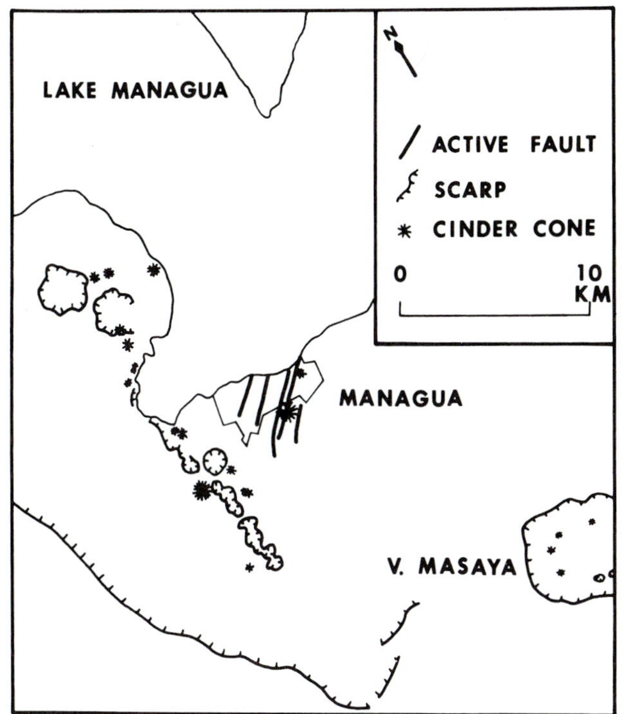

Fig. 9. Managua transverse structure. Data for this map from McBirney (1955) and Brown et al. (1973).

premonitory to great earthquakes. It should also be possible to postulate an upper limit on the magnitude of a future great earthquake based on the length of the segment in which it occurs.

Catastrophic volcanic eruptions of historic time have frequently occurred at long-dormant volcanoes on segment margins. The delineation of segment margins may allow predictions of which dormant volcano is more likely to erupt violently. The locations of active volcanoes appear to be strongly influenced by the segmented structural pattern of island arcs. Many Tertiary and earlier ore deposits are volcanogenic. It should prove helpful to look for segmented structures in Tertiary and earlier volcanic belts and determine if ore deposits occur in any particular position relative to segment boundaries.

The transverse structures segmenting island arcs have not yet been associated with oceanic fracture zones. There is no particular reason why they should be so associated, except for cases where oceanic fracture zones strike approximately parallel to the direction of plate convergence. With further study these special cases may have characteristics especially favorable to ore deposit formation.

ACKNOWLEDGMENTS

Financial support was supplied by National Science Foundation grant GA2-6211. Computer time was provided by the Kiewit Computation Center of Dartmouth College.

BIBLIOGRAPHY

Abe, K., 1972, Seismological evidence for a lithospheric tearing beneath the Aleutian arc: Earth and Planetary Sci. Letters, v. 14, p. 428-432.

Barazangi, M., and Dorman, J., 1969, World seismicity maps compiled from ESSA, coast and geodetic survey, epicenter data, 1961-1967: Seismol. Soc. America Bull., v. 59, p. 369-380.

Brazee, R. J., 1965, A study of *T* phases in the Aleutian earthquake series March and April, 1957: Earthquake Notes, v. 36, p. 9-14.

Brown, R. D., Ward, P. L., and Plafker, G., 1973, Geologic and seismologic aspects of the Managua, Nicaragua, earthquakes of Dec. 23, 1972: U.S. Geol. Survey Prof. Paper 838.

Burk, C. A., 1965, Geology of the Alaska Peninsula—island arc and continental margin: Geol. Soc. America Mem. 99, 3 pts., 250 p.

Carr, M. J., Stoiber, R. E., and Drake, C. L., 1973, Discontinuities in the deep seismic zones under the Japanese Arcs: Geol. Soc. America Bull., v. 84, p. 2917-2930.

Casertano, L., 1963, Catalogue of the active volcanoes and solfatara fields of the Chilean Continent, pt. XV, Chile: Rome, Internatl. Assoc. Volcanology, 55 p.

de Cserna, Z., 1961, Tectonic map of Mexico, scale 1:2,500,000: New York, Geol. Soc. America.

Ewart, A., and Bryan, W. B., 1973, The petrology and geochemistry of the Tongan Islands, *in* Coleman, P. J., ed., The western Pacific: island arcs, Marginal Seas, Geochemistry: Univ. Western Australia Press.

Fedotov, S. A., 1965, Regularities of the distribution of strong earthquakes in Kamchatka, the Kuril Islands, and northeastern Japan: Akad. Nauk SSSR Inst. Fiziki Zemli, no. 36 (203), p. 66-93.

Fisher, R. L., 1961, Middle America trench: topography and structure: Geol. Soc. America Bull., v. 72, p. 703-720.

Fitch, T. J., 1970, Earthquake mechanisms and island arc tectonics in the Indonesian-Philippine region: Seismol. Soc. America Bull., v. 60, p. 565-591.

Frank, F. C., 1968, Curvature of island arcs: Nature, v. 220, p. 363.

Gates, O., and Gibson, W., 1956, Interpretation of the configuration of the Aleutian ridge: Geol. Soc. America, v. 67, p. 127-146.

Hayes, D. E., 1966, A geophysical investigation of the Peru-Chile trench: Marine Geology, v. 4, p. 309-351.

Ichakawa, M., 1971, Re-analyses of mechanism of earthquakes which occurred in and near Japan, and statistical studies of the nodal planes obtained, 1926-1968: Geophys. Mag., v. 35, p. 207-274.

Isacks, B., and Molnar, P., 1971, Distribution of stresses in the descending lithosphere from a global survey of focal mechanism solutions of mantle earthquakes: Rev. Geophys. and Space Physics, v. 9, p. 103-174.

———, Oliver, J., and Sykes, L. R., 1968, Seismology and the new global tectonics: Jour. Geophys. Res., v. 73, p. 5855-5899.

Jordan, J. N., Lander, J. F., and Black, R. A., 1965, Aftershocks of the 4 February 1965 Rat Island earthquake: Science, June, p. 1323-1325.

Kanamori, H., 1971, Faulting of the Great Kanto earthquake of 1923 as revealed by seismological data: Bull. Earthquake Res. Inst. Tokyo Univ., v. 49, p. 13-18.

Katsumata, M., and Sykes, L. R., 1969, Seismicity and

tectonics of the western Pacific: Izu Mariana-Caroline and Ryukyu-Taiwan regions: Jour. Geophys. Res., v. 74, p. 5923-5948.

Kelleher, J. A., 1972, Rupture zones of large South American earthquakes and some predictions: Jour. Geophys. Res., v. 77, p. 2087-2103.

Konishi, K., 1965, Geotectonic framework of the Ryukyu Islands (Nansei-shoto): Jour. Geol. Soc. Japan, v. 71, p. 437-457.

Kuno, H., 1962, Catalogue of the active volcanoes of the world, pt. II: Japan, Taiwan and Marianas: Rome, Internatl. Assoc. Volcanology.

———, 1966, Lateral variations of basaltic magma across continental margins and island arcs: Can. Geol. Survey Paper 66-15, p. 317-335.

McBirney, A. R., 1955, The origin of the Nejapa pits near Managua, Nicaragua: Bull. Volcanol., v. 17, p. 145-154.

Mogi, K., 1968, Development of aftershock areas of great earthquakes: Bull. Earthquake Res. Inst. Tokyo Univ., v. 46, p. 175-203.

———, 1969a, Relationship between the occurrence of great earthquakes and tectonic structures: Bull. Earthquake Res. Inst. Tokyo Univ., v. 47, p. 429-451.

———, 1969b, Some features of recent seismic activity in and near Japan (2), activity before and after great earthquakes: Bull. Earthquake Res. Inst. Tokyo Univ., v.47, p. 395-417.

Molnar, P., and Sykes, L. R., 1969, Tectonics of the Caribbean and middle America regions from focal mechanisms and seismicity: Geol. Soc. America Bull., v. 80, p. 1639-1684.

Mooser, F., Meyer-Abich, H., and McBirney, A. R., 1958, Catalogue of active volcanoes of the world, pt. VI, Central America: Naples, Internatl. Assoc. Volcanology, 146 p.

Oliver, J., and Isacks, B., 1967, Deep earthquake zones, anomalous structure in the upper mantle, and the lithosphere: Jour. Geophys. Res., v. 72, p. 4259-4275.

Quaternary tectonic map of Japan, 1969, Research group for quaternary tectonic map: Tokyo, Natl. Res. Center Disaster Prevention, scale 1:2,000,000.

Santo, T., 1969, Characteristics of seismicity in South America: Bull. Earthquake Res. Inst. Tokyo Univ., v. 47, p. 635-672.

Sapper, K., 1897, Uber die raumliche Anordnung der mittelamerikanischen Vulkane, Z. Deut. Geol. Ges., v. 49, p. 672-682.

Stauder, W., 1968, Tensional character of earthquake foci beneath the Aleutian trench with relation to sea-floor spreading, Jour. Geophys. Res., v. 73, p. 7693-7701.

———, 1973, Mechanism and spatial distribution of Chilean earthquakes with relation to subduction of the ocean plate, Jour. Geophys. Res., v. 78, p. 5033-5061.

Stoiber, R. E., and Carr, M. J., 1974, (in press), Quaternary volcanic and tectonic segmentation of Central America, Bull. Volcanol.

Stroback, K., 1973, in press, Curvature of island arcs and plate tectonics: Z. Geophysik.

Sugimura, A., 1960, Zonal arrangement of some geophysical and petrological features in Japan and its environs: Tokyo Univ. Jour. Fac. Sci., sec. II, v. XII, pt. 2, p. 133-153.

———, 1961, Regional variations of the K_2O/Na_2O ratios of volcanic rocks in Japan and its environs: Jour. Geol. Soc. Japan, v. 67, p. 292.

Swift, S. A., and Carr, M. J., 1973, The segmented nature of the Chilean seismic zone (abstr.): Conference on geodynamics, Lima, Peru, August.

Sykes, L. R., 1966, The seismicity and deep structure of island arcs: Jour. Geophys. Res., v. 71, p. 2981-3006.

Tjia, H. D., 1967, Volcanic lineaments in the Indonesian island arcs: Bull. Volcanol., v. 31, p. 85-96.

Utsu, T., 1961, A statistical study on the occurrence of aftershocks: Geophys. Mag., v. 30, p. 521-605.

Part IV

Recent Sedimentation

Continental Shelf Sedimentation

Donald J. P. Swift

INTRODUCTION

In one of the first models of clastic sediments on continental shelves, Douglas Johnson (1919, p. 211) saw the shelf water column and the shelf floor as a system in dynamic equilibrium, in which the slope and grain size of the sedimentary substrate at each point controls, and is controlled by, the flux of wave energy into the bottom. The resulting surface is concave upward, steeper toward the shoreface. Grain size decreased with depth and distance seaward, and as a direct function of the diminishing input of wave energy into the sea floor. The model derived its sediment from coastal erosion rather than from river input. Despite its qualitative expression and limited applicability, the model was in advance of its time in its dynamical, systems analysis approach.

Shepard (1932) was the first to challenge it, noting that most shelves were veneered with a complex mosaic of sediment types rather than a simple seaward-fining sheet. He suggested that these patches were deposited during Pleistocene low stands of the sea rather than during recent time. Emery (1952, 1968) classified shelf sediments on a genetic basis, as *authigenic* (glauconite or phosphorite), *organic* (foraminifera, shells), *residual* (weathered from underlying rock), *relict* (remnant from a different earlier environment such as a now submerged beach or dune), and *detrital*, which includes material presently being supplied by rivers, coastal erosion, and aeolian or glacial activity. On most shelves a thin nearshore band of modern detrital sediment is supposed to give way seaward to a relict sand sheet veneering the shelf surface.

A third model for shelf sedimentation incorporates elements of both the Johnson and Emery models. It views the shelf surface as a dynamic system in a state of equilibrium with a set of process variables, but the rate of sea-level change is one of these variables and thus includes the effects of post-Pleistocene sea-level rise. The model may be referred to as a transgression-regression model, since it is generally expressed in these terms, or as a coastal model, since it focuses on the behavior of this dynamic zone (Grabau, 1913; Curray, 1964; Swift et al., 1972b).

The respective insights of these various models are by no means mutually exclusive. The process variables of the transgressive-regressive model are surely the dominant controls of the shelf sedimentary regime, and also the controls that shape the coastal boundary, a term used here to describe both the configuration of the inner shelf surface and the intracoastal zone of lagoons and estuaries.

The rate of sediment bypassing through the coastal boundary and the character of sediment bypassed are controlled by the equilibrium values attained by the process variables within the coastal boundary (Johnson model); hence this coastal equilibrium system itself modifies the sedimentary regime of the shelf and determines the petrographic types of Emery (1968) and stratigraphies of Curray (1964).

The shelf may be viewed as a dynamic system that is controlled by sediment flux through the coastal boundary, and also by the flux of energy through the water column into the substrate. This report will describe some characteristic patterns of shelf process and response. No text exists at this time which adequately deals with this material; however, Neuman and Pierson (1966) provide a lucid general treatment of physical oceanography. The mechanics of sediment transport have been ably summarized by Allen (1970a); see Bagnold (1963, 1966), Sternberg (1972), and Ludwick (1974) for advanced treatments.

COASTAL BOUNDARY

There are two basic categories of "valve" which regulate the passage of sediment from the continental sediment transport system into the domain of the continental shelf. The shoreface, during periods of coastal retreat, may erode and release sediment. This is an indirect process; the sediment released must first undergo storage as floodplain, lagoonal, or estuarine deposits, or be derived from an earlier cycle of sedimentation. River-mouth bypassing is more direct than shoreface bypassing, but sediment must still undergo storage. Sand is stored in the throat of the river mouth and fines are stored in marginal marshes and mud flats until the period of maximum river discharge, when the salt wedge moves to the shoal crest and stored sediment is bypassed to the shoreface of the shoal front. It may undergo a second period of storage on the shoreface and inner shelf until the period of maximum storm energy (Drake et al., 1972).

The mode of operation of these valves is dependent on basic parameters of coastal sedimentation: the absolute value of river discharge per unit length of coast, the ratio of salt water to fresh water discharge in river mouths, the wave climate, and finally the rate and sense of coastal translation as a function of sediment input, sea-level displacement, and energy input. The fluvial discharge per unit length of coast determines the spacing of river mouths and is the fundamental determinant of the relative roles of shoreface versus river-mouth bypassing. An intense tidal regime increases the efficiency of river-mouth bypassing, while an intense wave climate increases the efficiency of shoreface bypassing.

The rate and sense of coastal translation also strongly affects the relative roles of river mouth and shoreface bypassing. Rapid transgression results in disequilibrium estuaries which become sediment sinks, and shoreface and downdrift bypassing must dominate. The resulting sand sheet consists of a transient veneer of surf fallout on the upper shoreface and the residual sand sheet of the lower shoreface and adjacent shelf. These correspond to Emery's (1968) nearshore modern sand and shelf relict sand, respectively. Both are relict in the sense that they have been eroded from a local, pre-recent substrate, and both are modern, in the sense that they have been redeposited under the present hydraulic regime. They are, in fact, *palimpsest* sediments (Swift et al., 1971b), since they have petrographic attributes resulting from both the present and the earlier depositional environment. The term "relict" is best reserved for those specific textural attributes reflecting the earlier regime. Perhaps the most effective term for the provenance of these materials is *autochthonous*, and a shelf sedimentary regime characterized by rapid transgression and shoreface and downdrift bypassing will be henceforth described as a regime of *autochthonous shelf sedimentation*.

With a slower rate of translation, estuaries can equilibrate to their tidal prisms and river mouth, and shoreface bypassing becomes a significant source of sediment. More subtle, but equally important, is the effect of a slow transgression on the grain size of bypassed sediment. With a slower rate of shoreline translation, the intracoastal zone of estuaries and lagoons can aggrade nearly to mean sea level. The resulting surface of salt marshes (or in low latitudes, mangrove swamps) threaded by high-energy channels tends to serve as a low-pass or bandpass filter. Migrating channels preferentially select coarse materials for permanent burial in their axes. The surfaces of the tidal interfluves receive the finest materials for prolonged storage or permanent burial. However, fine sands and silts deposited as overbank levees tend to be reentrained by the migrating channels, hence have the highest probability of being bypassed to the shelf surface. This material is sufficiently fine to travel in suspension for long distances.

As the sense of coastal translation passes through stillstand to progradation, the shoreface becomes a sink rather than a mechanism for bypassing. Distributary mouths must further partition their prefiltered load, between sand sufficiently coarse to be captured by the littoral drift and buried on the shoreface and sand fine enough to escape in suspension in the ebb tidal jet and be entrained into the shelf dispersal system. The shoreface is now a total sediment trap, and bypassing is entirely through river mouths.

Shelves undergoing slow transgression or regression thus experience a contrasting regime of *allochthonous shelf sedimentation* characterized by significant river-mouth bypassing. In this regime there is a massive influx of river sediment filtered by passage through the coastal zone. Sheets of mobile fine sand and mud stretch from the coast toward the shelf edge. Shorefaces are broad and gentle and merge imperceptibly with a shallow inner shelf.

The allochthonous regimes, which build the broad constructional shelves, and the autochthonous regimes which periodically invade them, will be reviewed by means of a few representative examples whose dynamics are relatively well known.

CENTRAL ATLANTIC SHELF OF NORTH AMERICA: STORM-DOMINATED AUTOCHTHONOUS REGIME

Hydraulic Climate. The central Atlantic Shelf of North America is a storm-dominated shelf; midtide surface velocities are generally less than 20 cm/sec, except in the vicinity of tidal inlets and estuary mouths, (Redfield, 1956). As such, it experiences long periods of quiescence, mainly during the summer. At this time the shelf water mass is density-stratified and undergoes a slow, southerly, coast-parallel drift under the impetus of prevailing fair-weather winds and freshwater runoff (Harrison et al., 1967). The latter factor results in a nearshore elevation of the sea surface sufficient to induce a coast parallel, geostrophic flow of water, with an offshore surface component and a landward bottom component (Bumpus, 1973). This southward flow becomes entrained by the north-trending Gulf Stream at Cape Hatteras.

The central Atlantic Shelf is in the lee of the continent with respect to the prevailing planetary winds. Summer swells are far traveled and attenuated and are damped further at the shelf edge. Near-bottom wave surge on the shelf is occasionally able to ripple the bottom (McClennen, 1973) but becomes a significant agent of sediment transport only on the shoreface, in 15 m or less of water. Farther seaward, the resultant fair-weather velocity field of wave surge, wind drift, and thermohaline components is competent only to transport suspended fine sediments. However, summer concentrations of suspended sediment are low, usually less than 1 mg/liter except near estuary mouths and tidal inlets (Manheim et al., 1970), since the threshold velocities needed to suspend such materials are generally not exceeded.

By November the water column has cooled sufficiently to lose thermal stratification, and the increased frequency of storms has broken down

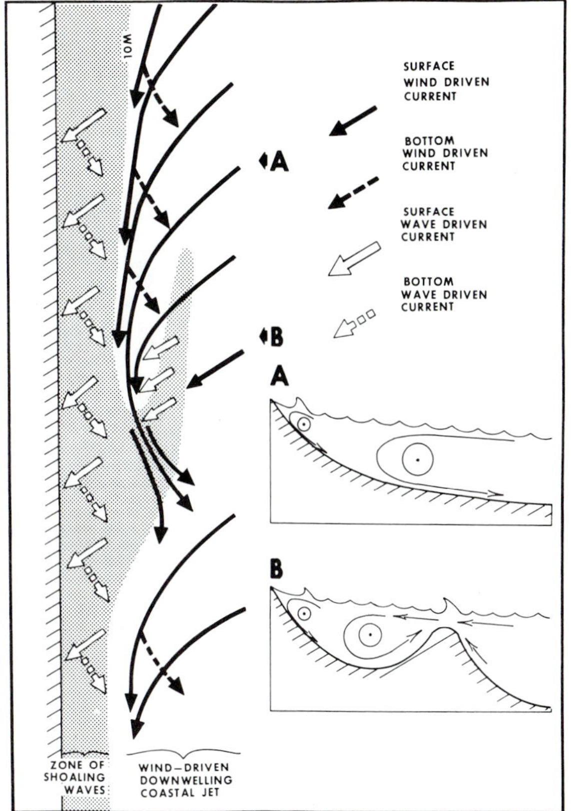

Fig. 1. Hypothetical model of the coastal boundary of the storm flow field during a period of onshore wind. Convergence of wind driven current with shoreline results in downwelling coastal jet.

salinity stratification. During such storms, wind stress on the sea surface drives Langmuir circulation in the mixed layer (Gordon and Gerard, in press). The stratified lower layer is eroded by this mechanism, so the mixed layer grows at its expense, until the stratified layer is entirely consumed and the rapidly flowing mixed layer is in contact with the sea floor.

The storm flow field of the central Atlantic Shelf is poorly understood. Storms tend to move northeastward up the shelf, parallel to the coast. As the storm approaches, winds rotate into the northeast and intensify as they do so, resulting in appreciable setup of water against the coast. Nearshore currents respond quickly to wind stress in this northward-migrating water bulge (Swift et al., 1973; B. Butman, M.I.T., unpublished data). The early work of Ekman (1905, in Neumann and Pierson, 1966, p. 203) suggests that under such circumstances streamlines of coastal flow should converge with each other as they converge with the coast, resulting in a coastal jet; the pressure head due to wind set up may be relieved by downwelling and obliquely seaward bottom flow as well as by downcoast flow (Fig. 1).

Current meter records indicate that storm flows are adequate to mobilize at least the inner shelf floor (Fig. 2). Values for sediment transport presented in Figure 2 are probably underestimated, since the lubricating effect of bottom wave surge was not taken into account. Wave surge is most intense during storms when unidirectional flow is also at a maximum. Wave surge generates steep sided ripples that increase the shear stress required to entrain sand (Bagnold, 1963), but field observations indicate that the sand of ripple fields is size-sorted with respect to the hydraulic microenvironment, with coarse sand in crests and finer sand in troughs, so that the whole surface is activated simultaneously as peak surge velocity is approached (Cook and Gorsline, 1972). Coupling between boundary surge and ripple is such that a burst of suspended sand is injected from the ripple crests into the boundary flow (Kennedy and Locher, 1972). In general, the effect of wave surge on the storm flow field is probably to depress the threshold values for sediment entrainment by the mean flow.

Origin of the Surficial Sand Sheet. The surficial sand sheet of the central and southern Atlantic Shelf was produced by the erosional retreat of the shoreface during the Holocene transgression, and its sediment textures and morphology faithfully re-

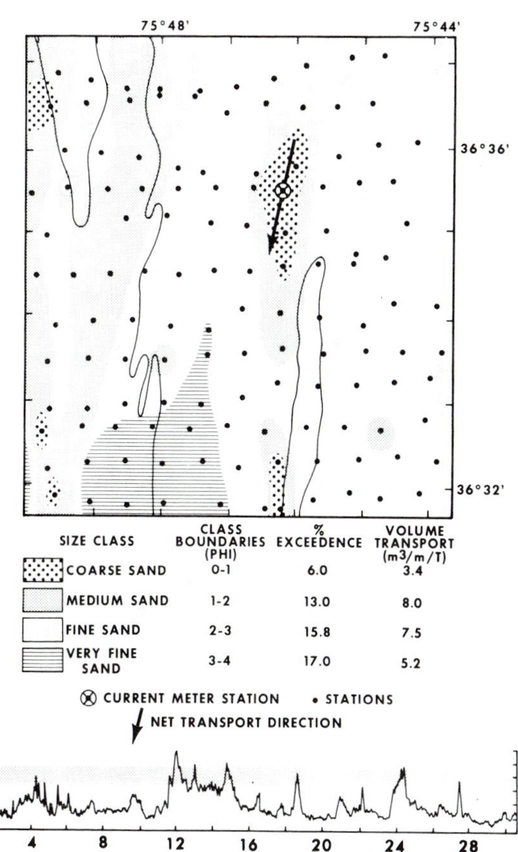

Fig. 2. Sediment transport during the month of November, 1972, in an inner shelf ridge field, False Cape Virginia. Estimates based on Shield's threshold criterion, a drag coefficient of 3 X 10^{-3} and Laursen's (1958) total load equation. Values expressed as m^3 of quartz per meter transverse to transport direction for time elapsed.

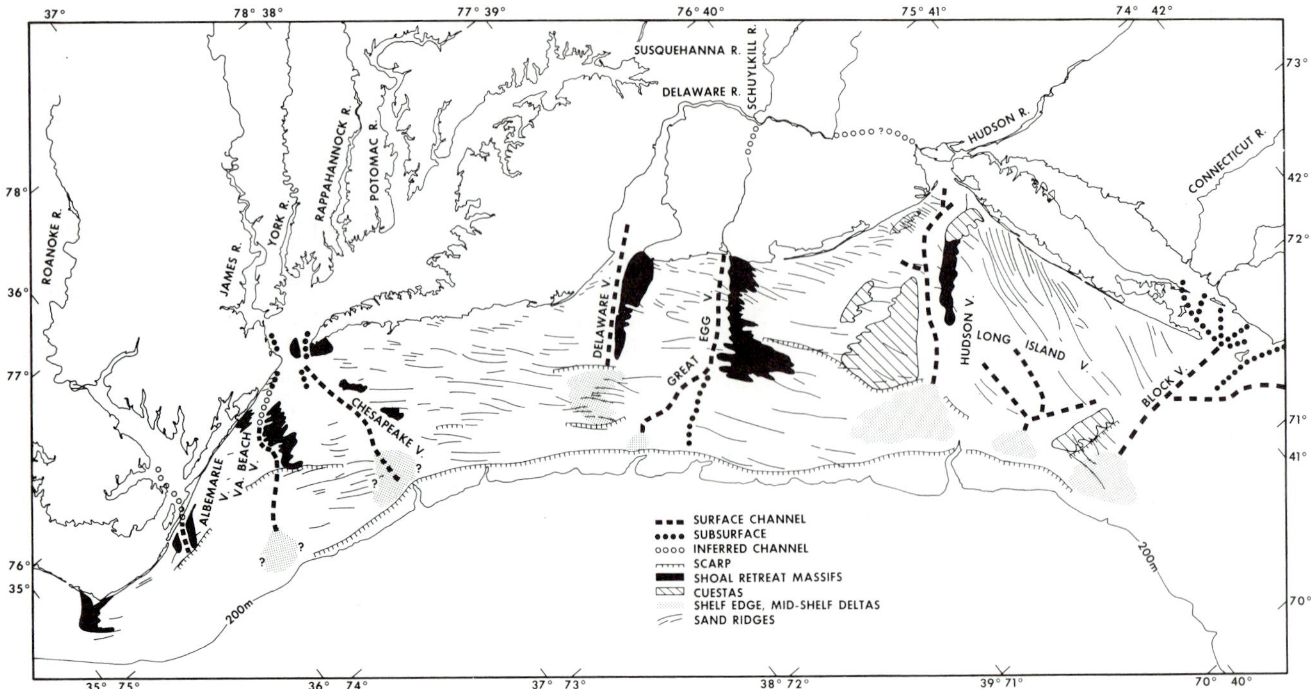

Fig. 3. Morphologic elements of the Middle Atlantic Bight. From Swift, in press.

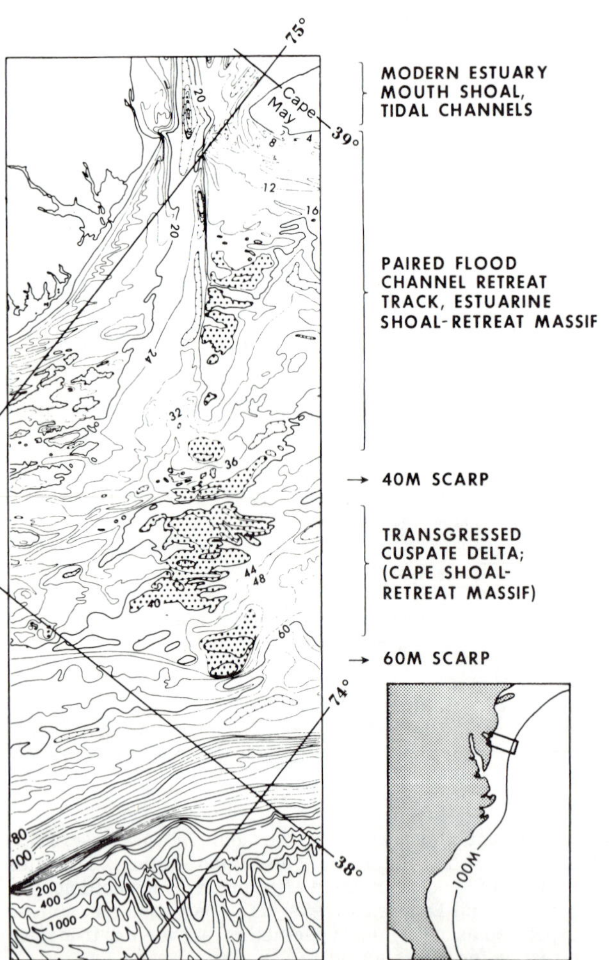

Fig. 4. Down-drift bypassing at the mouth of an erosional estuary (Delaware Bay, North American Atlantic Shelf). Southward littoral drift of New Jersey coastal compartment is injected into reversing tidal stream of mouth of Chesapeake Bay. The resulting shoal is stabilized as a system of interdigitating ebb and flood channels, north of the main couplet of mutually evasive ebb and flood channels. The shelf valley complex seaward of the bay mouth is the retreat path of the bay mouth sedimentary regime through Holocene time. Retreat of the main flood channel has excavated the Delaware Shelf Valley; retreat of the baymouth shoal has left a seaward trending shoal-retreat massif on the shelf valley's north flank; an example of down-drift bypassing. From Swift, 1973.

flect patterns of littoral sedimentation during this period. Zones of tidal scour at estuary mouths, and convergences in the littoral drift system and off cuspate forelands, have left records of their retreat as subdued, shelf transverse lows and highs, much as objects in a photograph leave streaks if the camera has moved while the picture is taken (Fig. 3). The sand sheet thus formed has continued to respond to the storm hydraulic regime, most notably by the overprinting of the relict nearshore topographic pattern by a ridge-and-swale topography that may in some respects be analogous with the fields of longitudinal dunes characteristic of subaerial sand seas (Wilson, 1972).

Relict Components of the Depositional Fabric. Uniformitarian principles may be readily applied to interpretation of relict morphologic, stratigraphic, and textural components in the depositional fabric of the surficial sand sheet. These components may be explained in terms of the modern littoral regime of the adjacent coast. The Middle Atlantic Bight is characterized by long, straight coast compartments, alternating with the mouths of master river systems that have been drowned to form large erosional estuaries. Relatively little sediment passes down these rivers from the temperate, glaciated hinterland, and the estuaries are able to efficiently trap it out and to trap out the littoral drift of the adjacent coastal compartments as estuary mouth shoals (Meade, 1969).

In the Middle Atlantic Bight, a large-scale depositional fabric consists of an alternation of shelf-transverse thickenings in the sand sheet (Swift and Sears, in press). These *shelf valley complexes* tend to consist of a partially filled river-cut valley paired with a shoal-retreat massif. The fill of the subaerial valley may have a narrow channel incised into it. Thinner sand blankets, the product of erosional shoreface retreat, occur on the plateau-like interfluves. Shelf valley complexes cannot always be traced to their littoral generating zone as a consequence of changes in the littoral sedimentation pattern attendant on the late Holocene reduction in the rate of sea-level rise (Milliman and Emery, 1968). The narrow Delaware Shelf Valley, however, can be traced directly into the flood-dominated channel that adjoins, in an échelon fashion, the ebb-dominated channel of the inner estuary mouth (Fig. 4). Such mutually evasive ebb-flood channel couplets are characteristic of estuary mouths (Ludwick, 1973) and the Delaware Shelf Valley is, in fact, the retreat path of the flood-dominated member of the pair. It only approximately follows the trend of the buried river-cut valley beneath it (R. Sheridan, personal communication). The north-rim high is similarly the retreat path of the north-side estuary mouth shoal that serves as a depositional focus for the New Jersey coastal compartment; this shoal-retreat massif constitutes a major zone of downdrift bypassing. Cores through the similar Albermarle Massif reveal a nuclear estuarine facies resting on a late Wisconsinan substrate and mantled by younger shelf sands (Fig. 5).

The intervening plateau-like massifs are veneered with 0-10 m of surficial sand. On long stretches of the coast, this material accumulates uniformly at the foot of the shoreface. Elsewhere,

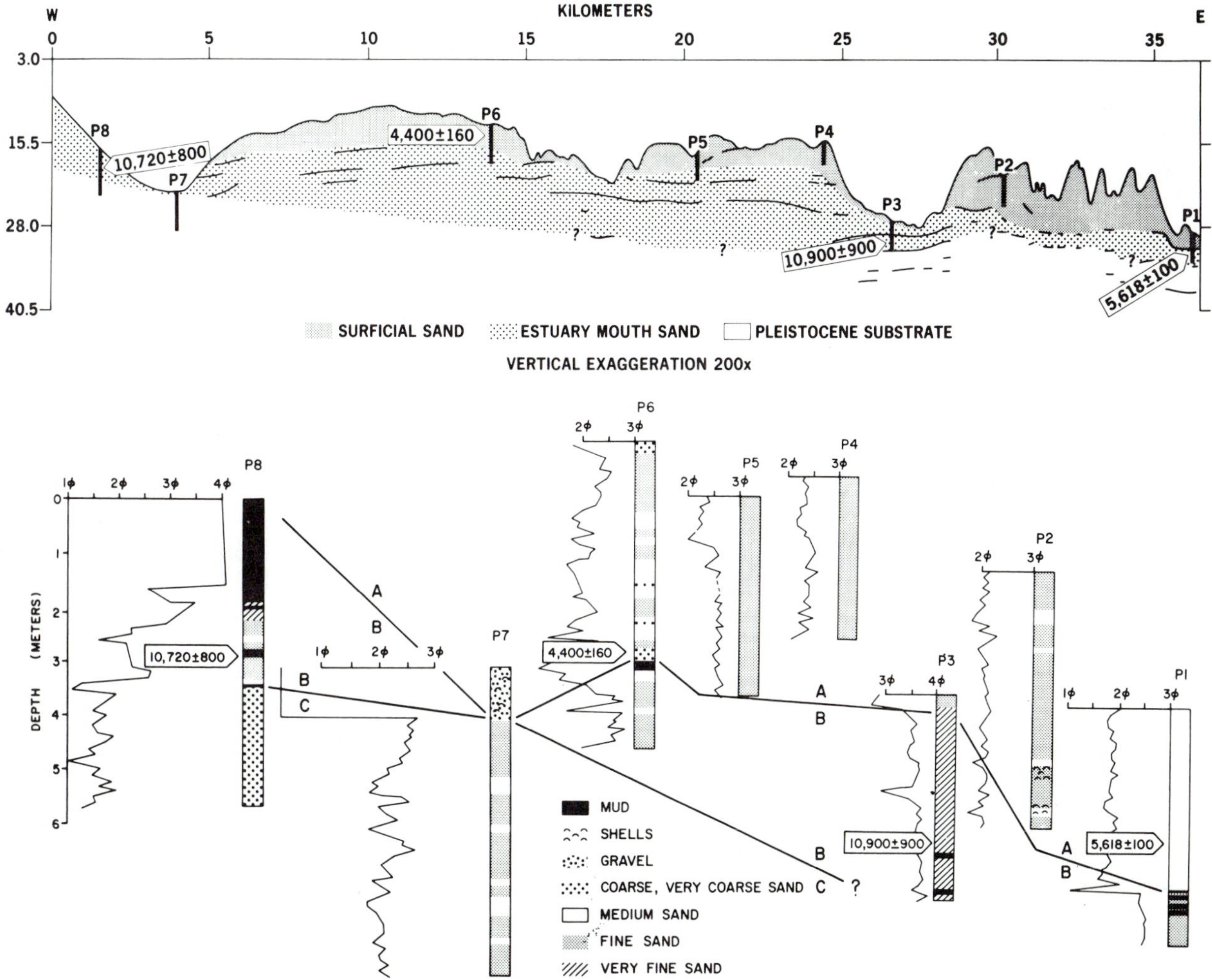

Fig. 5. Section through Platt Shoals, an estuarine shoal-retreat massif on the south flank of the Albermarle shelf valley complex. Note nucleus of estuary mouth deposits. Depositional environment was equivalent to down-current side of estuary mouth, Fig. 8. From Swift and Sears, in press.

rhythmic perturbations in the form of shoreface-connected sand ridges occur on the shoreface, in response to perturbations in the storm flow field of downwelling, southerly trending shoreface currents, and the nearly symmetrical surge of high storm waves (Figs. 1 and 6; Swift et al., 1972a; Duane et al., 1972). Each ridge is nourished by storm erosion of the up-current shoreface. Headward and axial erosion of the troughs during regional retreat of the shoreface results in periodic isolation of ridge segments on the deepening inner shelf floor, an important form of small-scale downdrift bypassing. The resulting stratum is a ridged sand sheet whose structures and textures reflect the course of the Holocene transgression.

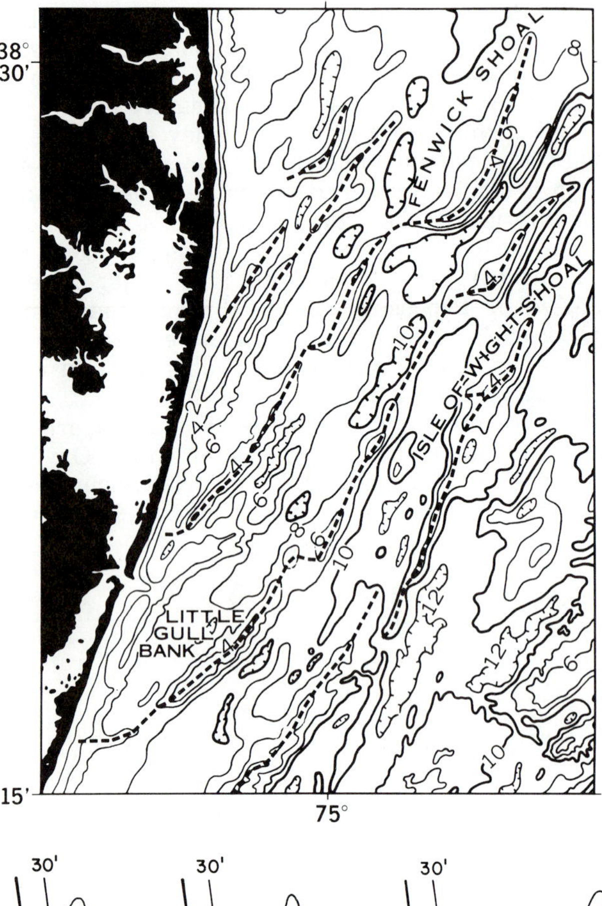

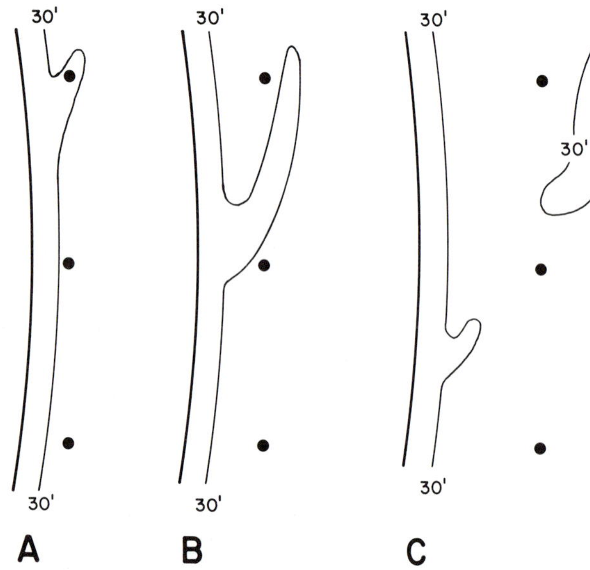

Fig. 6. Shoreface-connected ridge field of the Delmarva Coast. Ridges are nourished by storm current erosion of the shoreface. Ridges are migrating southeast (downcoast and offshore), while extending crest lines to maintain contact with the shoreface. As trough grows through headward and axial erosion, storm currents in trough become more intense, and eventually cut saddle. Perturbation of sea floor continued as new ridge downcoast, resulting in stepwise crest line (compare lower diagram with crestlines in map).

This pattern of shelf valley complex alternating with shoreface retreat blanket, characteristic of the Middle Atlantic Bight, is replaced by alternative patterns farther north and south (Swift and Sears, in press). To the north, the shelves have been glaciated and relief is greater. Here shelf highs are cuestas and similar erosional forms; river-cut valleys are only partially filled with estuarine deposits. Glacial basins in the Gulf of Maine and Scotian Shelf are presently accumulating mud.

To the south, off the Carolina salient, cape shoal-retreat massifs extend seaward from the littoral drift convergences off cuspate forelands. Off South Carolina, the massifs off small cuspate forelands merge as a cape shoal-retreat blanket. The closely spaced estuaries of the South Atlantic Bight may have similarly generated an estuarine shoal-retreat blanket. The smooth outer shelf from Cape Cod to Florida appears to be a zone of shelf-edge deltas.

Equilibrium Components of the Depositional Fabric. On the central Atlantic Shelf, then, the morphologic and stratigraphic framework of the shelf sand sheet is the consequence of erosional shoreface retreat and shoreface and downdrift bypassing. However, the sheet is clearly in a state of continuing response to the storm-dominated Holocene hydraulic regime. Inherited sand ridges continue to be maintained; bare Pleistocene substrate continues to be exposed in adjacent troughs. New ridges are constructed. Tide-built ridges on the retreating estuary mouth shoals may rotate from their initial coast-normal orientation to a more nearly coast-parallel orientation as the shoreline retreats, in response to weakening of the influence of estuary mouth tidal streams, and the increasing importance of coast-parallel storm currents. Where the older ridges are especially wide or deep, a new, smaller-scale ridge pattern may be imprinted obliquely across the old. Shelf-floor ridges shift landward or seaward and extend southward. A regional redistribution of sediment occurs, whereby fine and very fine sand is swept by storm flow out of ridge fields and out of the troughs incised into shoal-retreat massifs, and into zones of flow deceleration and expansion, in the shelf valleys beyond the massifs, and in downcoast reentrants (Fig. 7).

Sediment fractionation occurs on a smaller scale within the ridge topography (Stubblefield et al., in press). Crestal sands are uniformly medium

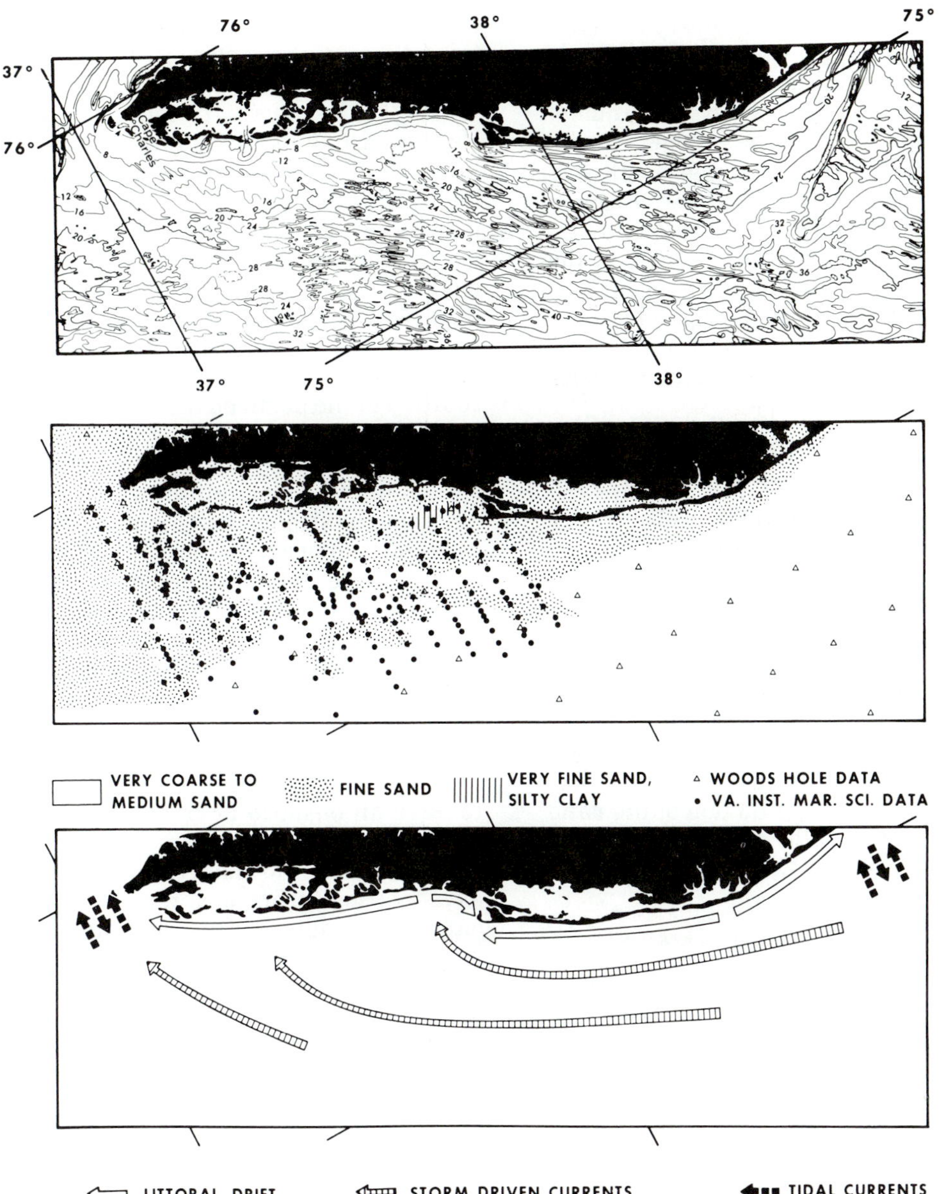

Fig. 7. Above: Bathymetry of the Delmarva Inner Shelf, from Uchupi, 1970. Center: Distribution of sediment, from Hathaway, 1971 and Nichols, 1972. Below: Inferred direction of sediment transport.

to fine grained; their relatively high percentages of finer interstitial sand suggest deposition by high-intensity flow (rheologic flow; Moss, 1972). Trough sands are highly differentiated. Fine and very fine sands in troughs may have settled from graded suspension. Coarse sands exposed in trough axes have the interstitial fine populations and very coarse laminae (traction clogs), indicating high-intensity flow (Moss, 1972). All three sand types occur on flanks, but fine to very fine sands are dominant.

In some troughs, sidescan sonar records reveal a dark axial band, interpreted as an elongate erosional window in the Holocene sand sheet, exposing a thin gravelly or shelly lag, resting on finer lagoonal deposits (McKinney, in press). Paired bands indicate that trough erosion has cut through a shelly bed in the pre-recent substrate. Sidescan records also reveal small-scale sand ribbon-like features in troughs, which are interpreted as responses to small-scale helical flow in the bottom boundary layer of the storm flow field.

The surface of the Middle Atlantic Bight of the North American Shelf appears to be in a state of equilibrium with the hydraulic regime, in terms of texture and morphology. Large-scale topographic features and textural patterns created during shoreface retreat have not been completely obliterated, however, so the equilibrium is imperfect. The equilibrium is a dynamic one in the sense that there is throughput of sediment across a surface of relatively stable morphology and texture.

In budgetary terms, this system is also in a state of near dynamic equilibrium. A finite amount of sediment is being introduced into the shelf surface by shoreface retreat and is flowing intermittently southward in response to storms (Swift et al., 1972b). The moving material must ultimately attain permanent storage on the shelf as current-adjusted deposits or be swept off the shelf edge. During Pleistocene low stands of the sea, the southward sand flux of the Middle Atlantic Bight was apparently tapped by the Hatteras canyon system, as the Hatteras abyssal plain is floored by material of this source (Horn et al., 1971). During the present high stand, however, the Hatteras shelf edge is capped by biogenic ooze deposited from the Gulf Stream, and the sand stream is instead aggrading the shelf north of Cape Hatteras.

SHELF AROUND THE BRITISH ISLES: A TIDE DOMINATED AUTOCHTHONOUS REGIME

Hydraulic Climate of Tidal Shelves. As the oceanic tide propagates onto the continental shelf, its maximum current velocity, U_{max}, is increased since the maximum orbital velocity of a shallow-water wave varies inversely with depth. Energy loss into the sea floor is rapid, however, and the wave will be rapidly damped. Thus the maximum tidal velocity on the shelf, U_{max}, is a function of the ratio of distance from shore to depth, X_H, as well as the amplitude of the tidal wave, C, and its period, T (Fleming and Revelle, 1939):

$$U_{max} = \frac{2\pi CX}{TH}$$

The tidal wave may propagate onto the shelf as a progressive wave. More commonly, sufficient energy is reflected from the shoreline so that the shelf wave is a standing oscillation that cooscillates with the oceanic tide. Current velocities are 90° out of phase with water level, so maximum velocities occur at midtide, minimum velocities at high and low tide. As a consequence of the Coriolis effect, shelf tides are rotary, with the flood (rising, landward-flowing) tide veering to the right in the Northern Hemisphere, and the ebb (falling, seaward-flowing) tide veering to the left. On embayed shelf sectors, bounded laterally as well as on the back by land masses, an amphidromic system results.

In the North Sea, the basic standing wave is modified in this fashion into several progressive edge waves that sweep counter-clockwise around the basin (Fig. 8). Another type of rotary tidal stream pattern may be set up by the effect of the rotation of the earth on a progressive tidal wave in a fairly narrow channel, where compensation is achieved by the development of transverse streams rather than by the setting up of subsidiary gradients. Such streams will rotate clockwise in the Northern Hemisphere, as in the English Channel.

The most intensive sediment streams are associated with the progressive edge waves that sweep around the margins of the amphidromic systems of marginal tidal seas. Amplitude of these edge waves increases toward shore, and so do their shallow-water distortions, resulting in net coast-parallel sediment transport. However, owing to the settling-lag phenomenon (Postma, 1967), transport will tend to have an onshore or offshore component (Stride, 1973).

In addition to inherent velocity and discharge asymmetries, transport inequalities in tidal seas may also be due to preferred patterns of wind-drift currents and storm surge. Storm surges moving as solitary edge waves pass along the western side of the North Sea, for instance, and markedly amplify the flood currents on that side (Ishiguro, 1966). Intensified wave surge associated with storms also greatly amplifies the transporting power of tidal currents (Johnson and Stride, 1969).

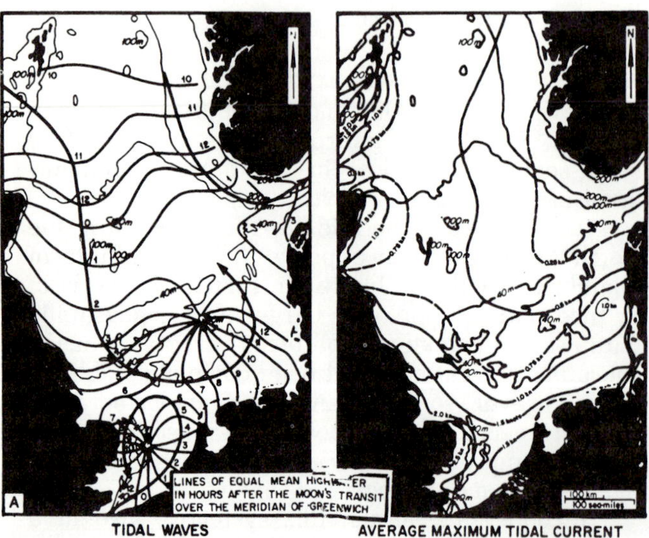

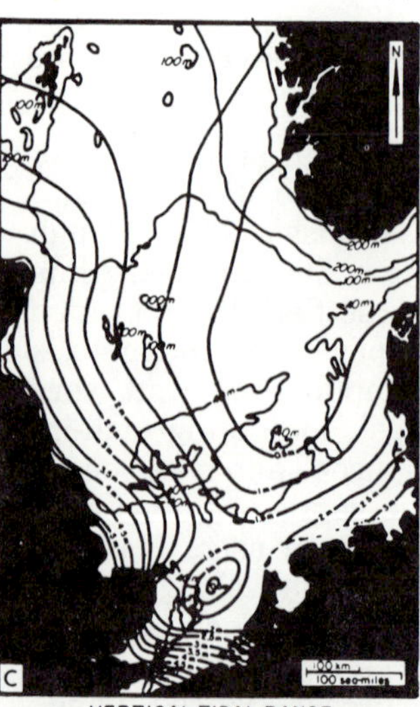

Fig. 8. Tidal regime of the North Sea. From Houbolt, 1968.

Sediment Transport Around the British Isles. The "tide-swept" shelf around the British Isles (Stride, 1963) is subjected to an autochthonous sedimentary regime. The Thames, Severn, and Humber debouch in estuaries; the Rhine "delta" consists at present of estuarine distributaries. Such dynamic bypassing as occurs in these estuaries is not sufficient to prevent the erosional retreat of the adjacent coasts, including the Dutch coast downdrift of the Rhine Delta (Van Straaten, 1965). The formation of the surficial sand sheet by tidal erosion of the shoreface and reconstitution of its materials under the tidal hydraulic regime are analyzed by Belderson and Stride (1966). Tide-induced erosional retreat of the Anglian coast is accomplished by the growth, migration, and detachment of shoreface-connected ridges similar in some respects to those of the Middle Atlantic Bight. However, they are maintained not primarily by storm flow but by residual tidal discharge that has a flood value on the inner flank and an ebb value on the outer flank (Fig. 9; Robinson, 1966). Storage and periodic detachment of these shoreface sand masses appears to have created a shoal retreat massif of tide-maintained sand ridges that extends for 200 km out into the North Sea (Fig. 10; Caston, 1972).

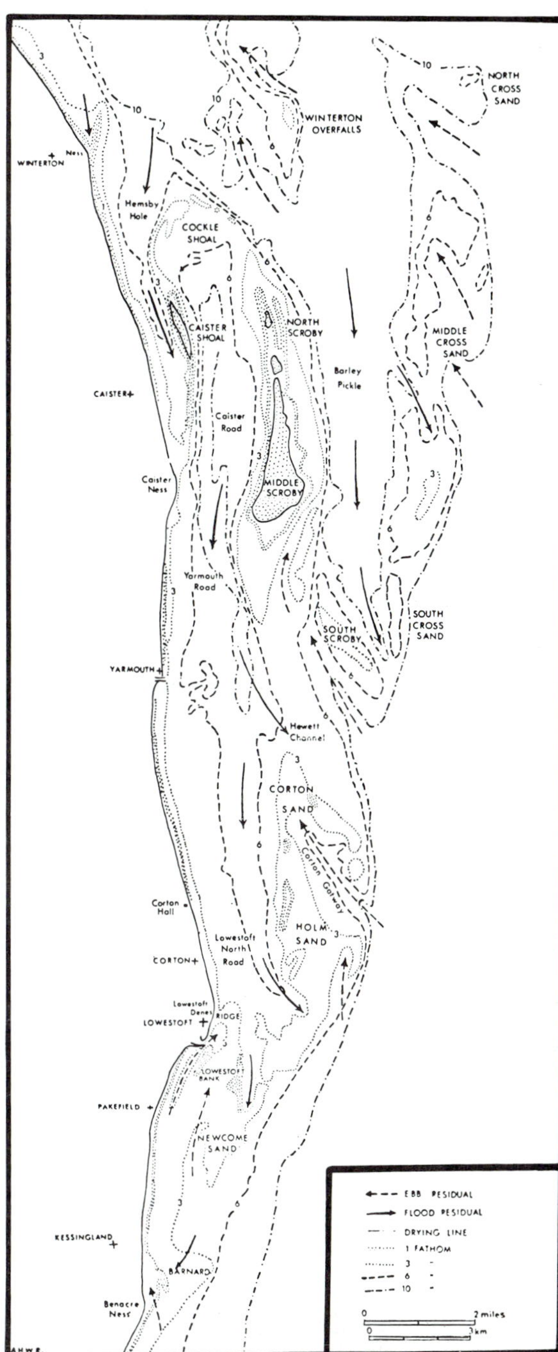

Fig. 9. Tide-maintained ridge topography on the inner Anglian Shelf. Shoreface-connected ridges separate ebb- and flood-dominated channels. Ridges tend to migrate southward with time, and to detach from retreating shoreface. Ridges are nourished at the expense of shoreface, hence constitute cases of down-drift bypassing. Offshore ridges are probably being nourished at expense of nearshore ridges; if so, sand is moving seaward more rapidly than are the ridge forms, and dynamic bypassing is occurring also. From Robinson, 1966.

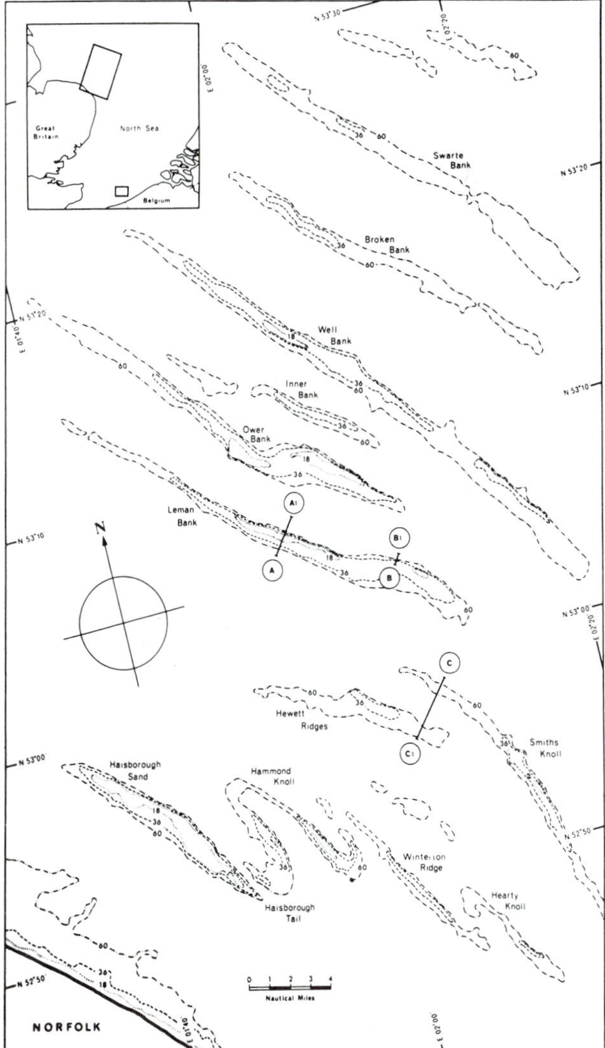

Fig. 10, Tidal ridge field of the Anglian Shelf. Ridges are confined to map area. These features appear to constitute a shoal-retreat massif, marking the retreat path of the nearshore tidal regime of the Anglian coast. From Caston, 1968.

In general, however, such stabilized morphologic traces of the retreat of nearshore sedimentation zones are less common on the British shelves, since the debris sheet generated by shoreface retreat has responded to the more intense, tide-dominated hydraulic climate with a much greater degree of mobility. The pattern of transport is surprisingly well organized, with sand streams diverging from beneath tide-induced "bedload partings" and flowing down the gradient of maximum tidal current velocities until either the shelf edge or a zone of "bed-load convergence" and sediment accumulation is reached (Stride, 1963; Kenyon and Stride, 1970; Belderson et al., 1970).

Each stream tends to consist of a sequence of more or less well-defined zones of characteristic

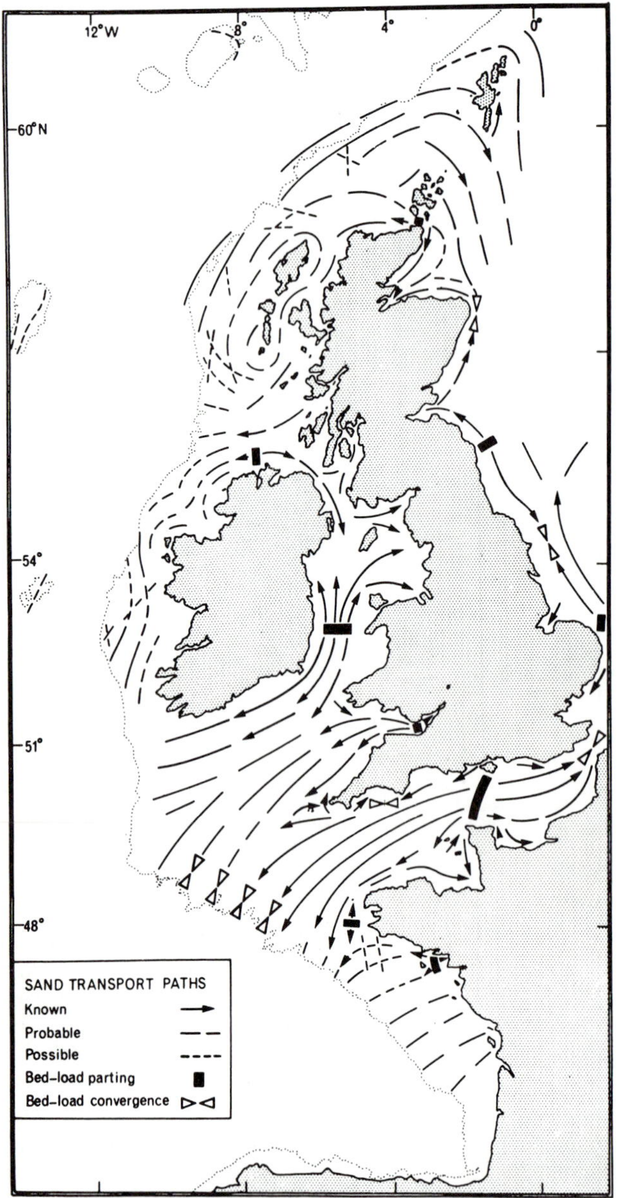

Fig. 11. Generalized sand transport paths around the British Isles and France, based on the velocity asymmetry of the tidal ellipse and the orientation and asymmetry of bedforms. From Kenyon and Stride, 1970.

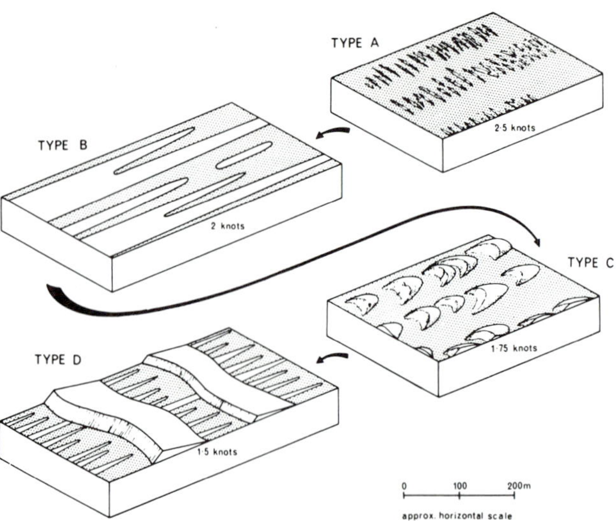

Fig. 12. Four main types of sand ribbons and the typical near-surface current veolcities at which they occur. From Kenyon, 1970.

bottom morphology and sediment texture (Fig. 11). Streams may begin in high-velocity zones [midtide surface velocities in excess of 3 knots (150 cm/sec)]. Here rocky floors are locally veneered with thin (centimeters thick) lag deposits of gravel and shell. Where slightly thicker, the gravel may display "longitudinal furrows" parallel to the tidal current, a bedform possibly related to sand ribbons (Stride et al., 1972).

Between approximately 2.5 and 3.0 knots (125-150 cm/sec) sand ribbons are the dominant bed form (Kenyon, 1970). These features are up to 15 km long and 200 m wide and usually less than 1 m deep. Their materials are in transit over a lag deposit of shell and gravel. Kenyon has distinguished four basic patterns which seem to correlate with maximum tidal current velocity and with the availability of sand (Fig. 12).

Farther down the velocity gradient, where midtide surface velocities range from 1 to 2 knots (50-100 cm/sec), sand waves are the dominant bed form. Where the gradient of decreasing tidal velocity is steep, or transport convergence occurs, this may be the sector of maximum deposition on the transport path. Over 20 m of sediment has accumulated at the shelf-edge convergence of the Celtic Sea, although it is not certain that this sediment pile is entirely a response to modern conditions.

The Hook of Holland sand wave field off the Dutch coast is one of the largest (15,000 km^2) and the best known (McCave, 1971). The sand body is anomalous in that it sits astride a bed-load parting; the sand patch as a whole may be a Pleistocene delta or other relict feature. Sand waves with megaripples on their backs grow to equilibrium heights of 7 m with wavelengths of 200-500 m in water deeper than 18 m; in shoaler water, wave surge inhibits or supresses them. Elongate tidal ellipses favor transverse sand wave formation, and

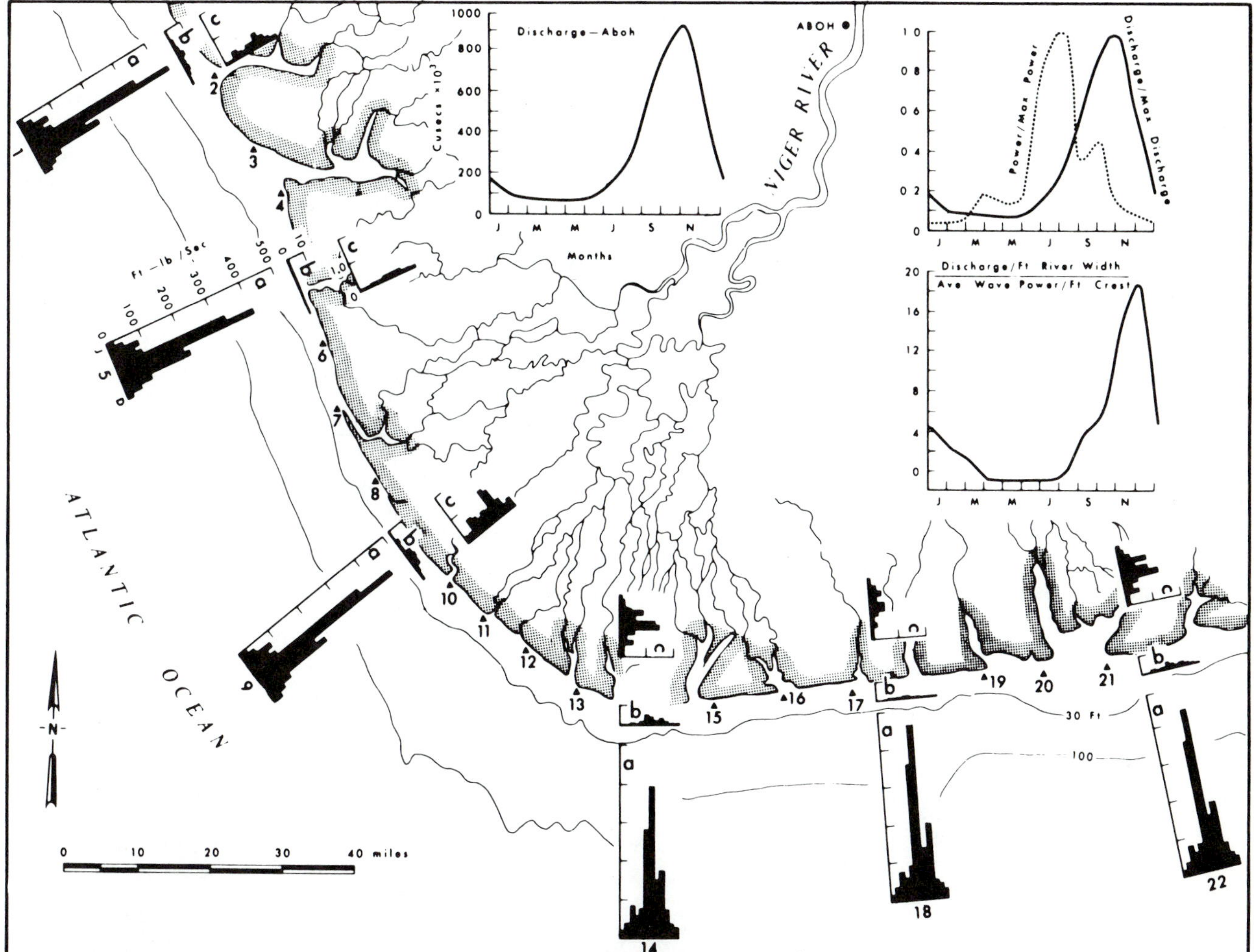

Fig. 13. Niger Delta and relation of river discharge to wave power over the yearly cycle. Histograms are wave power over the yearly cycle at the 30 foot contour (A) and nearshore (B). From Wright and Coleman, 1973.

the sand waves tend to be destroyed by midtide cross-flow when the ellipse is less symmetrical. Under the latter condition, linear sand ridges may be the preferred bedform, as midtide cross flow would tend to nourish rather than degrade them (Smith, 1969). The triangular sand wave field is limited by a lack of sand on the northwest, by shoaling of the bottom and increasing wave surge on the coast to the south, and by fining of sand to the point that suspensive transport is dominant to the north (McCave, 1971).

Farther down the velocity gradient, beyond the zones of obvious sand transport, there are sheets of fine sand and muddy fine sand and in local basins, mud. They lack bed forms other than ripples and appear to be the product of primarily suspensive transport (McCave, 1971) of material that has outrun the bed-load stream. These deposits may be as thick as 10 m (Belderson and others, 1966), but where they do not continue into mud, they break up into irregular, current-parallel or current-transverse patches of fine sand less a 2 m thick, resting on the gravelly substrate.

The complex pattern and mobile character of the shelf floor around the British Isles have led British workers to reject the relict model for the shelf sediments. They note that it correctly draws attention to the autochthonous origin of the sediment but that it fails to allow for its subsequent dynamic evaluation. They propose instead a dynamic classification:

1. Lower sea level and transgressive deposits, patchy in exposure but probably more or less continuous beneath later material; largely the equivalent of a blanket (basal) conglomerate.

2. Material moving as bed load (over the coarser basal deposits) mainly well-sorted sand and in places first-cycle calcareous sand.

3. Present sea-level deposits (category 2 sediment having come to permanent rest), consisting of large sheets to small patches, which range from gravel and shell gravel to sand and calcareous sands, muddy sands, and muds.

The implication is that of a shelf surface in a state of equilibrium with its tidal regime. The adjustment appears to be more effective than in the case of the North American Atlantic Shelf in that there is less preservation of nearshore depositional patterns. As a consequence of the intensity of the hydraulic climate, there is less on shelf storage (category 3) and more material in transit.

NIGER SHELF: STORM DOMINATED ALLOCHTHONOUS REGIME

General. A very different regime of shelf sedimentation, and probably one more representative of the allochthonous regimes that have built the broad constructional shelves of continental margins, is that of the Niger delta, as described by Allen (1964). The hydraulic climate of the Niger Shelf is probably most nearly analogous to that of the Central Atlantic Shelf, in that storm flow is more significant than tidal flow in driving shelf sedimentation. The shelf is dominated by the great arcuate Niger delta (Fig. 13), a concentric assemblage of terrestrial and transitional depositional environments that filter and modify the sediment load of the Niger River, before bypassing it to the Niger Shelf. Such a delta is by no means a prerequisite for allochthonous sedimentation, although the correlation between major river mouths and allochthonous sedimentation is probably higher in the present period of relatively rapid transgression than during the slow transgressions of the past.

Differential Bypassing in the Deltaic Environments. The Niger-Benue river system delivers about 0.9 X 10^6 m^3 of bed-load sediment and about 16 X 10^6 m^3 of suspended sediment (Allen, 1965) to its delta each year. During peak discharge from September to May, average flow velocities range from 50 to 135 cm/sec, and gravel as well as sand are in violent transport. During low stages, flow velocities decrease to 37 to 82 cm/sec, enough to transport sand and silt. In the higher part of the floodplain, the Niger is braided; in the remainder the Niger shows large meanders (Fig. 14). During high stages, levees are overtopped, crevasses develop, and bottom

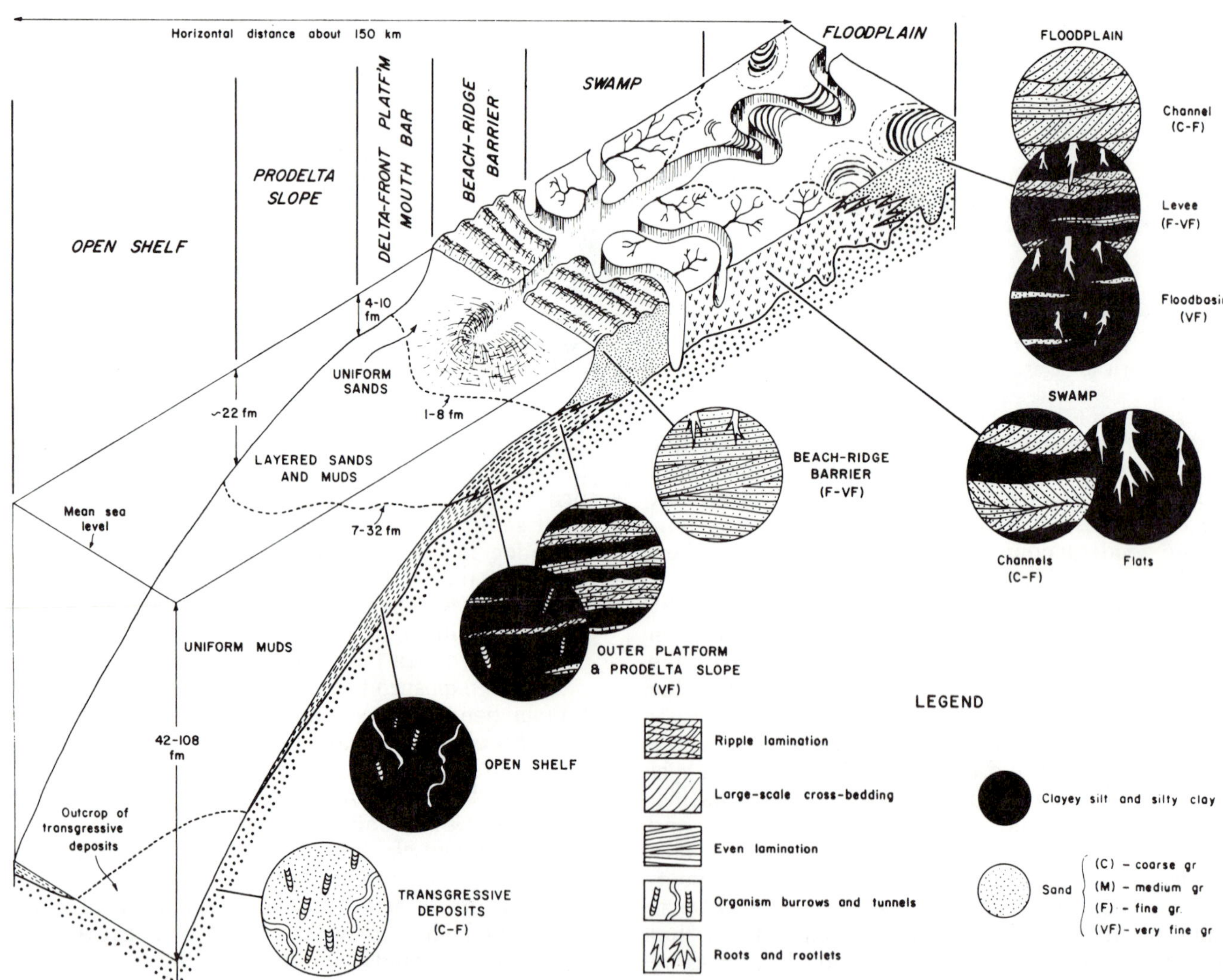

Fig. 14. Schematic illustration of the depositional environments and sedimentary facies of the Niger Delta and Niger Shelf. From Allen, 1970.

lands are flooded. Gravel and coarse sand are deposited as a substratum of braid bars and meander point bars, respectively, and are veneered with a top stratum of overbank clays. Silt undergoes temporary deposition in levees in the lower floodplain, but these tend to be undermined, so that their deposits reenter the transport system.

Thus the floodplain environment serves as a skewed bandpass filter, with preferential bypassing of the medium grades, entrapment of some fines over bank, and much coarse material deposited in channel axes. This process continues through the tidal swamp environment, where the entrapment of fines dominates. Reversing tidal flows generate velocities of 40-180 cm/sec in tidal creeks, enough to move sand and gravel. Entrapment of fines overbank in the mangrove swamps is enhanced by the phenomena of slack high water and the prolonged period of reduced velocity associated with it. Fines then deposited begin to compact, and require greater velocities to erode them than served to permit their deposition.

Major channels, which pass through the intertidal environment to the sea, must store their coarser sediment during low-water stages at the foot of the salt wedge, where the landward-inclined surface of zero net motion intersects the channel floor. During high-water stages, stored bottom sediment must be rhythmically flushed out of the estuary mouth by the tidal cycle. Sand coarser than the effective suspension threshold of 230 microns (Bagnold, 1966) will be deposited on the arcuate estuary mouth shoals (Oertel and Howard, 1972), where, after a prolonged period of residence in the sand circulation cells of the shoal, it leaks into the downcoast littoral drift system. Finer sand is entrained into suspension by large-scale top-to-bottom turbulence in the high-velocity estuary throat (Swift and Pirie, 1970, p. 75) and will be swept seaward with the ebb tidal jet, to rain out on the inner shelf (Todd, 1968), where it is accessible to distribution by the hydraulic regime.

The shoreface constitutes a second major zone of storage, since the periods of peak river discharge and peak wave power do not coincide on the Niger shelf (Fig. 13). Such poor coupling is a significant factor in coastal progradation, since the resulting sediment prism has a chance to consolidate prior to erosion.

Transport on Allochthonous Shelves. Drake et al. (1972) have presented a detailed case history of the seaward dispersal of such a nearshore prism of stored sediment (Fig. 15). In January and February 1969, southern California experienced two intense rainstorms, which resulted in a record flood discharge. The freshly eroded sediment was a distinctive red-brown in contrast to the drab hue of the reduced shelf sediments. The flood layer could therefore be repeatedly cored and isopached and its shifting center of mass traced seaward through time.

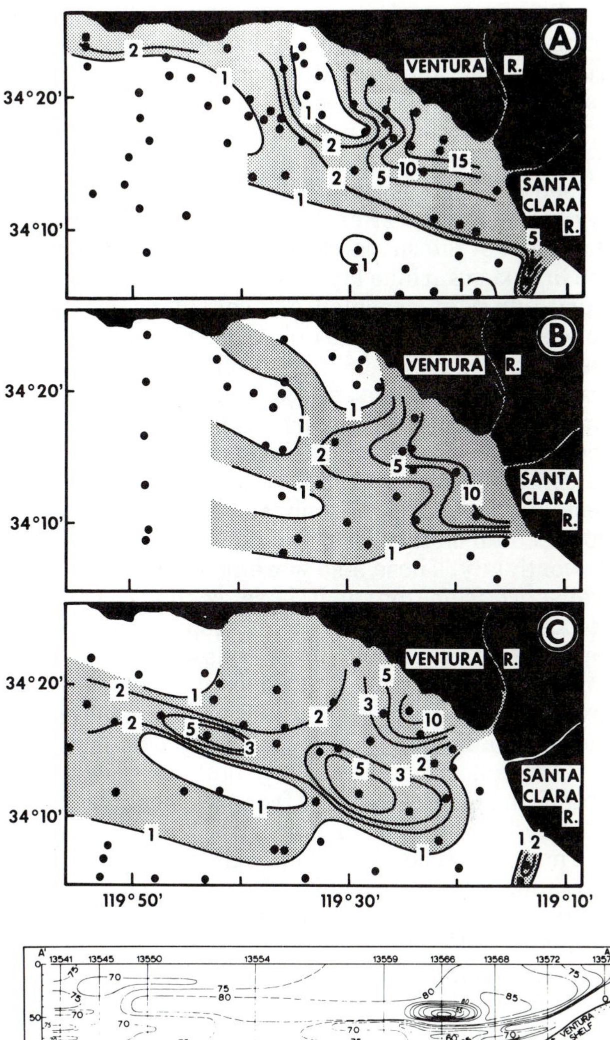

Fig. 15. Upper 3 diagrams: thickness of flood sediment (cm) on the Santa Barbara-Oxnard Shelf in March-April, 1969; May-August, 1969; and February-June, 1970, based on cores. Lowest diagram: East-west cross section showing vertical distribution of light attenuating substances over Santa Barbara-Oxnard Shelf. For clarity, the bottom 20 m of the water column is not contoured, but the percent transmission value at the bottom is noted. From Drake and Kolpack, 1972.

U.S. Geological Survey stream records show that 33-45 X 10^6 metric tons of suspended silt and clay and 12-20 X 10^6 metric tons of suspended sand were introduced by the Santa Clara and Ventura rivers. By the end of April 1969, more than 70% of this material was still on the shelf in the form of a submarine sand shoal extending 7 km seaward, and a westward thinning and fining blanket of fine sand, silt, and clay existed seaward of that (Fig. 15A).

By the end of the summer of 1969, the layer extended farther seaward, had thinned by 20%, and had developed a secondary lobe beneath the Anacapa current to the south (Fig. 15B). Eighteen months after the floods, the surface layer was still readily detectable. Considerable bioturbation, scour, and redistribution had occurred south of Ventura, but the deposit was more stable to the north (Fig. 15C).

A concurrent study of suspended sediment distribution in the water column revealed the pattern of sediment transport (Fig. 15D). Vertical transparency profiles, after 4 days of flooding, showed that most of the suspended matter was contained in the brackish surface layer, 10-20 m thick. Profiles in April and May revealed a layer 15 m thick, with concentrations in excess of 2 mg/liter and a total load of 10 to 20 X 10^4 metric tons. Since this load was equal to river discharge for the entire month of April, it must have represented lateral transport of sediment resuspended in the nearshore zone. Vertical profiles over the middle and outer shelf for the rest of the year were characterized by sharply bounded turbidity maxima, each marking a thermal discontinuity. These also were nourished by lateral transport from the nearshore sector, where the discontinuities impinged on the sloping bottom. The near-bottom nepheloid layer was the most turbid zone in the inner shelf. This nepheloid layer was invariably the coolest and was invariably isothermal, indicating that its turbidity was the result of turbulence generated by bottom-wave surge. Bottom turbidities ranged from 50 mg/liter during the flood to 4-6 mg/liter during the next winter, but were at no time dense enough to drive density currents. Thus transport of suspended sediment across shelves undergoing allochthonous sediment action would appear to be a matter of introduction by a river jet, deposition, resuspension, intervals of diffusion and advection by coastal currents in a near-bottom nephaloid layer, and further deposition and resuspension.

As a consequence of the intermittent nature of transport across the aggrading shelf surface, the fractionation of sediment by particle size, characteristic of the terrestrial environment, continues (Figs. 14 and 16). This process of *progressive sorting* is not a matter of the "fines outrunning the coarse"; it is a consequence of decreasing bottom-wave energy, and loss of the coarsest fraction each time there is resuspension is due to currents weaker than the preceding episode Swift et al., 1972c). The broad, gentle textural gradients of allochthonous shelves are characteristically coast-normal, although water mass advection is more nearly coast-parallel. The gradients reflect the greater role of bottom-wave energy gradients and sediment diffusion on fine-grained allochthonous shelves; grain-size gradients on coarse-grained autochthonous shelves are due primarily to sediment advection.

Like many shelves undergoing allochthonous sedimentation, the allochthonous deposits on the Niger Shelf appear to incompletely cover the older autochthonous deposits; "windows" of the latter show through (Fig. 14). This does not necessarily indicate a transient state, however; as suggested by McCave (1972), it may be a steady-state phenomenon. McCave notes that the localization of zones of mud deposition depends on the balance between the near-bottom "hydraulic activity" and the near-bottom suspended sediment concentration, rather than on either of the factors above. Bare outer shelves may be zones where fines are bypassed in steady-state fashion owing to increased tidal and wave energy on the shelf edge, breaking internal waves, or impinging oceanic currents.

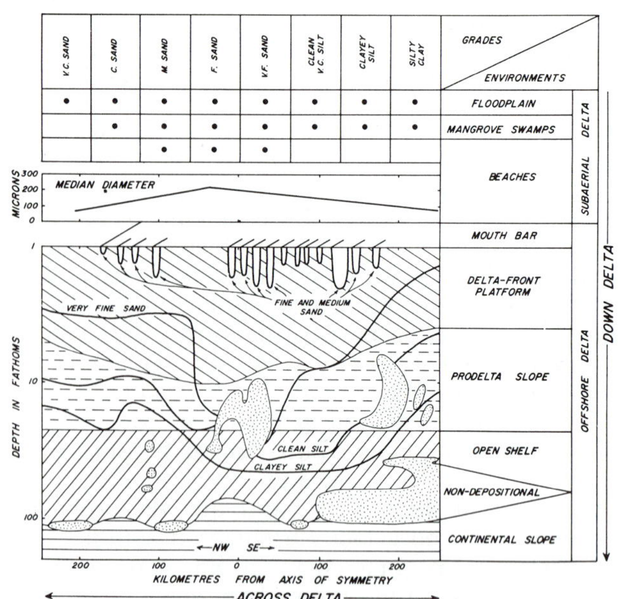

Fig. 16. Grain size in relation to sedimentary environments in Niger Delta area. In subaerial delta, all grades present are shown. In offshore part of delta, coarsest grade in near-surface layers is projected on to vertical plane perpendicular to axis of delta symmetry. From Allen, 1964.

SHELF SEDIMENTARY REGIMES AND CONTINENTAL SHELF CONSTRUCTION

The varying pattern of shelf sedimentation in time and space is compatible with the concept of plate tectonics. Inman and Nordstrom (1971) have classified coasts as *trailing-edge coasts* (facing a spreading center), *collision coasts* (bordering a subduction zone or transform fault), or *marginal coasts* (facing an island arc).

The full cycle of shelf development is seen on trailing-edge coasts. Neo-trailing-edge coasts (Inman and Nordstrom, 1971) of the Red Sea type are steep, as they are basically tectonic surfaces, as yet little modified by erosion and deposition, and they are high, presumably due to proximity to the thermally elevated spreading centers. The height of land is close to the shoreline; numerous steep small rivers deliver abundant coarse sediment. A tectonic margin regime prevails; because of the steep slope, gravity dispersal dominates. Shelves of collision coasts may never evolve beyond this stage, owing to consumption of sediments by subduction.

With continued spreading, the continental margin slowly subsides. The regional gradient of the continent reverses so that the drainage area serving the coast increases. Its sediment load increases and its drainage pattern becomes more integrated. As the gradient of the submarine slope decreases, gravity dispersal becomes less efficient; a sediment prism accumulates, whose upper surface is hydraulically maintained (graded) at a yet-lesser slope. Gravity dispersal is now confined largely to the continental slope. In zones where it has been particularly efficient, however, the original gradient is retained in the form of submarine canyons extending back into the prograding shelf (Rona, 1969). With further crustal subsidence, the allochthonous regime becomes dominant. Sedimentary grading of the continental margin surface extends into the subaerial zone, as fluvial and estuarine depositional environments. The grading process becomes largely self-maintaining. As long as the sediment load is adequate, the coastal environments prefilter it for maximum mobility and inject it onto the shelf by means of estuary jets, and shoreface and downdrift bypassing. All parts of the shelf surface are nourished and interact with the hydraulic regime so as to aggrade to the appropriate depth. Marginal coasts, if possessed of an independent mechanism of subsidence, may begin a cycle of shelf construction at this stage (Bally, in Isacks et al., 1973).

Trailing-edge shelves are most liable to allochthonous sedimentary regimes, but these are probably never uniformly developed over great distances. Differential subsidence results in continued integration of the stream net, and the resulting master streams tend to preferentially seek loci of maximum subsidence, where they build deltaic piles beneath surfaces undergoing allochthonous sedimentation. Intervening shelf sectors may be sufficiently starved of sediment to develop autochthonous carbonate or clastic regimes.

If the continent is of the African type (all trailing edges), areas of autochthonous sedimentation must expand on continental shelves at the expense of allochthonous sedimentation, as the eroding continent approaches base level and terrigenous sediment production declines. Unfortunately, the type example of Africa is a poor one, as its nascent rift system has resulted in continuing relief. The Cambrian continents are probably a better reference, as the basal Cambrian sandstones are good examples of the residual sands produced by autochthonous sedimentation (Swift et al., 1971a). Cambrian platforms were subjected to a mixed regime, however, as outer-shelf mud belts followed the landward retreat of the sands.

The earth appears to have been subject to cyclic eustatic sea-level fluctuations as a consequence of variations in the volume of ocean basins; continents have flooded as increased spreading rates have expanded the volume of mid-oceanic ridges at the expense of oceanic basin volumes. Slower spreading rates have reduced ridge volume and caused withdrawal of marginal seas (Rona, 1973). This cycle has proceeded at rates sufficiently slow (10-20 my) relative to the rates of other variables so that coastal boundaries during the transgressive and high-stand phases were permeable, resulting in sediment bypassing and the buildup of enormous marine sediment prisms. Regression has served mainly to introduce a subaerial erosional regime to the shelf surface, with gravity bypassing over the slope, and consequent aggradation of deep-water environments (Rona, 1973).

A second type of modulation has been the rapid eustatic fluctuations associated with ice ages. Here regressions have been the significant agents of marine deposition on continental shelves. The most recent deglaciation of the Quaternary ice age has transpired within the past 15 years (Milliman and Emery, 1968), resulting in a sea level rise so rapid that the intracoastal zone of estuaries and lagoons became an effective trap for the fluvial sediment input. Sediment was released to the sea floor only after stratigraphic storage, and only then in small volumes, and an autochthonous regime prevailed. During glacio-eustatic regressions, estuaries became deltas, and there was sufficient filtering and estuary jet injection of mobile fine sediment to significantly aggrade the shelf floor before passage of the retreating shoreline. Subaerial exposure apparently did not last long enough during the brief lowstands for extensive erosion of the regressive deposits. Emery and Milliman (1971) have noted that the Quaternary record within the Hudson shelf-edge delta consists of a series of acoustically opaque reflectors (mainly transgressive shoreline deposits) and acoustically transparent interbeds (finer high-stand and regressive deposits). Much of the neo-trailing-edge shelf of the Costa de Nayarit, Gulf of California, was formed in this way (Curray and Moore, 1964). The cyclothems of the Permo-Carboniferous glaciation are likewise dominated by regressive deposits (Fischer, 1961).

The cores of the great construction shelves of the trailing-edge coasts of the Atlantic rift ocean

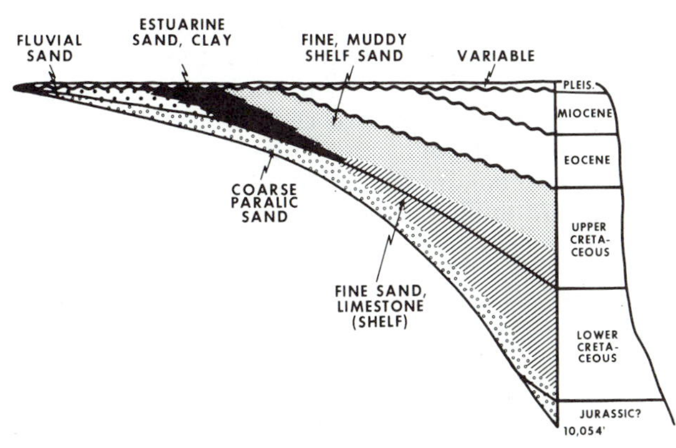

Fig. 17. Simplified schematic diagram of facies of the North American Atlantic Shelf at Cape Hatteras. Interpretation based on Swain (1952), Maher and Applin (1971), and Swift and Heron (1969).

were built during the Cretaceous when the slow postrifting subsidence coincided with a major transgression (Fig. 17). The Hatteras Light Well No. 1 bottoms at 10,054 ft in granitic rock (Maher and Applin, 1971). It is overlain by over 1,000 ft of coarse feldspathic sand and conglomerate, the product of a tectonic margin regime, and perhaps in part of gravity dispersal. This material is in turn overlain by up to 4,000 ft of interbedded fine sandstone and limestone of Early Cretaceous age. These beds are the offshore facies of the Cape Fear Formation (Swift and Heron, 1969), a coarse, pebbly sand of the inner coastal plain of fluvial, estuarine, and nearshore origin. The Early Cretaceous sedimentary regime was a mixed one, in which a coarse sediment input derived from a hinterland still possessing appreciable relief was not always efficiently bypassed to the outer shelf, and intervals of carbonate sedimentation occurred.

By Late Cretaceous time, the regional slope had decreased and a more extensive drainage net was delivering a finer sediment load. The littoral zone broke up into well-defined fluvial and estuarine depositional environments, and fine sand and mud were bypassed to all parts of the Cretaceous shelf in sufficient volume to overwhelm carbonate sedimentation. Regression veneered the resulting sediment pile with littoral deposits which were largely lost to erosion before the Eocene cycle of deposition.

The deposits of the Eocene and Miocene are thinner, and like the Lower Cretaceous have abundant interbedded limestones. Associated clastics are finer than those of the Lower Cretaceous, however, and the reversions to an autochthonous carbonate regime may reflect reduced sediment supply as a function of reduced relief in the hinterland rather than a lack of a mobile fine fraction in the sediment input. The coastal deposits of these later cycles have been largely lost to erosion.

Thus the vast sediment piles of the world's constructional (trailing edge) shelves appear to be the product of repeated depositional transgressions or "classic overlap," in which the coastal boundary filtered out the coarser fraction on an intensive sediment input, and the resulting mobile fine fraction accumulated beneath a surface undergoing allochthonous sedimentation. Episodes of autochthonous sedimentation are represented by carbonate strata, or if the climate was inappropriate, by thin, coarse, condensed horizons of lag sands which are difficult to recognize, especially in the subsurface.

ACKNOWLEDGMENTS

This paper is a contribution of the COMSED (Continental Margin Sedimentation) Program of NOAA'S Atlantic Oceanographic and Meteorologic Laboratories. Much of the information was assembled in the course of a study of shelf sedimentation conducted for NOAA's MESA (Marine Ecosystems Analysis Program). Data on the Virginia and Northern North Carolina Coast were collected on cruises E-21-71 and E-5-73 of Duke University Marine Laboratory's research vessel EASTWARD. The EASTWARD is supported by NSF grant GB-17545.

BIBLIOGRAPHY

Allen, J. R. L., 1965, Late Quaternary Niger delta, and adjacent areas: sedimentary environments and lithofacies: Am. Assoc. Petr. Geol. Bull., v. 49, p. 547-600.

———, 1970a, Physical processes of sedimentation: New York, American Elsevier, 248 p.

———, 1970b, Sediments of the modern Niger delta: a summary and review, *in* Morgan, J. P., ed., Deltaic sedimentation, modern and ancient: Tulsa, Okla., Soc. Econ. Paleontologists and Mineralogists Spec. Publ. 15, p. 138-151.

Bagnold, R. A., 1963, Mechanics of marine sedimentation, *in* Hill, M. N., ed., The sea: New York, Wiley-Interscience, p. 507-525.

———, 1966, An approach to the sediment transport problem from general physics: U.S. Geol. Survey Prof. Paper 422-1, 37 p.

Belderson, R. H., and Stride, A. H., 1966, Tidal current fashioning of a basal bed: Marine Geology, v. 4, p. 237-257.

———, Kenyon, N. H., and Stride, A. H., 1970, Holocene sediments on the continental shelf west of the British Isles: Inst. Geol. Sci. Rept. 70, p. 157-170.

Blackwelder, P. L., and Pilkey, O. H., 1972, Electron microscopy of quartz grain surface textures: the eastern U.S. continental margin: Jour. Sediment. Petrology, v. 42, p. 520-526.

Bruun, P., 1962, Sea level rise as a cause of shore erosion: Jour. Waterways Harbors Div. Am. Soc. Civil Engrs., v. 88, p. 117-130.

Bumpus, D. F., 1973, A description of circulation on the continental shelf of the east coast of the United States: Progr. Oceanography, v. 6, p. 117-157.

Caston, V. N. D., 1972, Linear sand banks in the southern North Sea: Sedimentology, v. 18, p. 63-78.

Cook, D. O., 1969, Occurrence and geologic work of rip currents in southern California: Jour. Sediment. Petrology, v. 39, p. 781-786.

———, and Gorsline, D. S., 1972, Field observations of sand transport by shoaling waves: Marine Geology, v. 13, p. 31-55.

Curray, J. R., 1964, Transgressions and regressions, *in* Milla, R. L., ed., Papers in marine geology: Shepard commemorative volume: New York, Macmillan, p. 175-203.

———, and Moore, D. G., 1964, Pleistocene deltaic progradation of the continental terrace, Costa de Nayarit, Mexico, *in* van Andel, J. J., ed., Marine geology of the Gulf of California: Am. Assoc. Petr. Geol. Mem. 3, p. 193-215.

———, Emmel, F. J., and Crampton, P. J. S., 1969, Lagunas costeras, un simposio, *in* Mem. Simp. Internatl. Lagunas Costeras, Nov. 28-30, 1967, Mexico, D.F.: UNAM-UNESCO, p. 63-100.

Drake, D. E., Kolpack, R. L., and Fischer, P. J., 1972, Sediment transport on the Santa Barbara-Oxnard shelf, Santa Barbara channel, California, *in* Swift, D. J. P., Duane, D. B., and Pilkey, O. H., eds. Shelf sediment transport: process and pattern: Stroudsburg, Pa., Dowden, Hutchinson & Ross, p. 301-332.

Duane, D. B., Field, M. E., Meisburger, E. P., Swift, D J. P., and Williams, S. J., 1972, Linear shoals on the Atlantic inner continental shelf, Florida to Long Island, *in* Swift, D. J. P., Duane, D. B., and Pilkey, O. H., eds., Shelf sediment transport: process and pattern: Stroudsburg, Pa., Dowden, Hutchinson & Ross, p. 447-499.

Emery, K. O., 1952, Continental shelf sediments of southern California: Geol. Soc. America Bull., v. 63, p. 1105-1108.

———, 1968, Relict sediments on continental shelves of the world: Am. Assoc. Petr. Geol. Bull., v. 52, p. 445-464.

———, and Milliman, J. D., 1971, Quaternary sediments of the Atlantic continental shelf of the United States: Quaternaria, v. 12, p. 3-18.

Faller, A. J., 1971, Oceanic turbulence and the Langmuir circulations: Ann. Rev. Ecology and Systematics, v. 2, p. 201-235.

Fischer, A. G., 1961, Stratigraphic record of transgressing seas in the light of sedimentation on the Atlantic coast of New Jersey, Am. Assoc. Petr. Geol. Bull., v. 45, p. 1656-1660.

Fleming, R. H., and Revelle, R., 1939, Physical processes in the oceans, *in* Trask, P. D., ed., Recent marine sediments: Tulsa, Okla., Am. Assoc. Petr. Geol., p. 48-141.

Gordon, A. L., and Gerard, R. D., in press, Wind drift surface currents and spread of contaminants in shelf waters: U.S. Coast Guard Res. Develop. Center, Groton, Conn., Rept. DOTCG 23339-A.

Grabau, A. W., 1913, Principles of stratigraphy. 1960 facsimile edition of 1924 revision. New York, Dover Books, 1185 p.

Harrison, W. Norcross, J. J., Pore, N. A., and Stanley, E. M., 1967, Shelf waters off the Chesapeake Bight: Environ. Sci. Services Admin. Prof. Paper 3, p. 1-82.

Hathaway, J. C., 1971, Data file, continental margin program, Atlantic coast of the United States, v. 2, Samples collection and analytical data: ref. 71-15, Woods Hole, Mass., U.S. Geol. Survey, Woods Hole Oceanogr. Inst., 446 p.

Horn, D. R., Ewing, M., Horn, B. M., and Delach, M. N., 1971, Turbidites of the Hatteras and Sohm abyssal plains, western North Atlantic: Marine Geology, v. 11, p. 287-323.

Houbolt, J. J. H. C., 1968, Recent sediments in the southern bight of the North Sea: Geol. Mijnbouw, v. 47, p. 245-273.

Hoyt, J. H., 1967, Barrier island formation: Geol. Soc. America Bull., v. 78, p. 1125-1136.

Inman, D. L., and Nordstrom, C. E., 1971, On the tectonic and morphologic classification of coasts: Jour. Geology, v. 74, p. 1-21.

Isacks, B., Mueller, I. T., Walcott, R. I., and Talwani, M., 1973, Vertical crustal motions and their causes: EOS (Trans. Am. Geophys. Union), v. 54, p. 1257-1260.

Ishiguro, S., 1966, Storm surges in the North Sea—an electronic model approach: Min. Agric., Fisheries and Food, Advisory Comm. on Oceanogr. Meteorolog. Res., London, Rept. 4, 57 p. (unpublished).

Johnson, D. 1919, [1938, 2nd ed.], Shore processes and shoreline development: New York, Wiley, 585 p.

Johnson, J. W., 1949, Scale effects in hydraulic models involving wave motion: Am. Geophys. Union Trans., v. 30, p. 517-527.

———, and Eagleson, P. S., 1966, *in* Ippen, A. J., ed., Estuary and coastline hydrodynamics: New York, McGraw-Hill, p. 404-492.

Johnson, M. A., and Stride, A. H., 1969, Geological significance of North Sea sand transport rates: Nature, v. 224, p. 1016-1017.

Kennedy, J. F., and Locher, F. A., 1972, Sediment suspension by water waves, *in* Meyer, R. E., Waves on beaches: New York, Academic Press, p. 249-296.

Kenyon, N. H., 1970, SAnd ribbons of European tidal seas: Marine Geology, v. 9, p. 25-39.

———, and Stride, A. H., 1970, The tide-swept continental shelf sediments between the Shetland Isles and France: Sedimentology, v. 14, p. 159-175.

Laursen, E. M., 1958, The total sediment load of streams: Proc. Am. Soc. Civil Engrs., v. 84 (HVI), p. 1530.

Leopold, L. B., Wolman, M. G., and Miller, J. P., 1964, Fluvial processes in geomorphology: San Francisco, W. H. Freeman, 522 p.

Longuet-Higgins, M. C. 1953, Mass transport in water waves: Phil. Trans. Roy. Soc. London, v. 245, p. 535-581.

Ludwick, J. C., 1973, Tidal currents and zig-zag shoals in a wide estuary entrance: Geol. Soc. America Bull. (in press).

———, 1974, Tidal currents, sediment transport, and sandbanks in Chesapeake Bay entrance, Virginia, *in* 2nd internatl. Estuarine Conf. Proc., Myrtle Beach, S.C., Oct. 15-18 (in press).

Maher, J. C., and Applin, E. R., 1971, Geologic framework and petroleum potential of the Atlantic coastal plain and continental shelf: U.S. Geol. Survey Prof. Paper 659, 98 p.

Manheim, F. T., Meade, R. H., and Bond, G. C., 1970, Suspended matter in surface waters of the Atlantic continental margin from Cape Cod to the Florida Keys: Science, v. 167, p. 371-376.

May, J. P., and Tanner, W. F., 1973, The littoral power gradient and shoreline changes, *in* Coates, D. R., ed., Publications in geomorphology: Binghamton, N.Y., State Univ. New York, 404 p.

McCave, I. N., 1971, Sand waves in the North Sea off the coast of Holland: Marine Geology, v. 10, p. 149-227.

———, 1972, Transport and escape of fine-grained sediment from shelf areas, *in* Swift, D. J. P., Duane, D. B., and Pilkey, O. H., eds., Shelf sediment transport: process and pattern: Stroudsburg, Pa., Dowden, Hutchinson & Ross, p. 225-248.

McClennen, C. F., 1973, New Jersey continental shelf near bottom current meter records and recent sediment activity: Jour. Sediment. Petrology, v. 43, p. 371-380.

McKinney, T. F., in press, Large-scale current lineations on the Great Egg Shoal Retreat Massif, New Jersey shelf investigation by side-scan sonar: Jour. Sediment. Petrology.

Meade, R. H., 1969, Landward transport of bottom sediments in estuaries of Atlantic coastal plain: Jour. Sediment. Petrology, v. 39, p. 229-234.

Menard, H. W., 1973, Epierogeny and plate tectonics: EOS (Trans. Am. Geophys. Union), v. 54, p. 1244-1255.

Milliman, J. D., and Emery, K. O., 1968, Sea levels during the past 35,000 years: Science, v. 162, p. 1121-1123.

Moody, D. W., 1964, Coastal morphology and processes in relation to the development of submarine sand ridges off Bethany Beach, Delaware [Ph.D. thesis], Johns Hopkins Univ., 167 p.Moss, A. J., 1972, Bed-load

Moss, A. J., 1972, Bed-load sediments: Sedimentology, v. 18, p. 159-220.

Neumann, G., and Pierson, W. J., Jr., 1966, Principles of physical oceanography: Englewood Cliffs, N.J., Prentice-Hall, 545 p.

Nichols, M.•M., 1972, Inner shelf sediments off Chesapeake Bay: I. General lithology and compositon: Gloucester Point, Va., Virginia Inst. Marine Sci. Spec. Sci. Rept. 64, 20 p.

Oertel, G. F., 1972, Sediment transport of estuary mouth shoals and the formation of swash platforms: Jour. Sediment. Petrology, v. 42, p. 858-863.

———, and Howard, J. P., 1972, Water circulation and sedimentation at estuary entrances on the Georgia coast, *in* Swift, D. J. P., Duane, D. B., and Pilkey, O. H., eds., Shelf sediment transport: process and pattern: Stroudsburg, Pa., Dowden, Hutchinson & Ross, p. 411-427.

Postma, H., 1967, Sediment transport and sedimentation in the marine environment, *in* Lauff, G. H., ed., Estuaries: Washington, D.C., Am. Assoc. Advan. Sci., p. 158-180.

Redfield, A. C., 1956, The influence of the continental shelf on the tides of the Atlantic coast of the United States: Jour. Marine Res., v. 17, p. 432-448.

Robinson, A. H. W., 1966, Residual currents in relation to shoreline evolution of the East Anglian Coast: Marine Geology, v. 4, p. 57-84.

Rona, P. A., 1969, Middle Atlantic continental slope of United States: deposition and erosion: Am. Assoc. Petr. Geol. Bull., v. 53, p. 1453-1465.

———, 1973, Relations between rates of sediment accumulation on continental shelves, sea floor spreading, and estuary inferred from the central North Atlantic: Geol. Soc. America Bull., v. 84, p. 2851-2872.

Schwartz, M. L., 1968, The scale of shore erosion: Jour. Geology, v. 76, p. 508-517.

Shepard, F. P., 1932, Sediments on continental shelves: Geol. Soc. America Bull., v. 43, p. 1017-1034.

———, 1973, Submarine geology: New York, Harper & Row, 517 p.

Smith, J. D., 1969, Geomorphology of a sand ridge: Jour. Geology, v. 77, p. 39-55.

Stahl, L., Koczan, J., and Swift, D., 1974, Anatomy of a shoreface-connected ridge system on the New Jersey shelf: implications for genesis of the shelf surficial sand sheet: Geology, v. 2, p. 117-120.

Stanley, D. J., 1969, Submarine channel deposits and their fossil analogs (fluxoturbidites), *in* Stanley, D. J., ed., The new concepts of continental margin sedimentation: Washington, D.C., Am. Geol. Inst., p. DJS-9-1-DJS-9-17.

Sternberg, R. W., 1972, Predicting initial motion and bedload transport of sediment particles in the shallow marine environment, *in* Swift, D. J. P., Duane, D. B., and Pilkey, O. H., eds., Shelf sediment transport: process and pattern: Stroudsburg, Pa., Dowden, Hutchinson & Ross, p. 61-82.

Stride, A. H., 1963, Current swept sea floors near the southern half of Great Britain: Geol. Soc. London Quart. Jour., v. 119, p. 175-199.

———, 1973, Interchange of sand between coast and shelf in European tidal seas (abstr.), *in* Abstracts symposium on Estuarine and shelf sedimentation, Bordeaux, France, July, 1972, p. 97.

———, Belderson, R. H., and Kenyon, N. H., 1972, Longitudinal furrows and depositional sand bodies of the English Channel: Mem. Bur. Recherches Geol. Minieres, no. 79, p. 233-244.

Stubblefield, W. L., Lavelle, W. J., McKinney, T. F., and Swift, D. J. P., in press. Sediment response to the hydraulic regime on the central New Jersey shelf: Jour. Sediment. Petrology.

Swain, F. M., Jr., 1951, Ostracoda from wells in North Carolina, pt. I, Cenozoic Ostracoda: U.S. Geol. Survey Prof. Paper 234-A, p. 1-58.

———, 1952, Ostracoda from wells in North Carolina, pt. II, Mesozoic Ostracoda: U.S. Geol Survey Prof. Paper 234-B, p. 54-93.

Swift, D. J. P., 1973, Delaware shelf valley: estuary retreat path, not drowned river valley: Geol. Soc. America Bull., v. 84, p. 2743-2748.

———, in press, Continental shelf sedimentation, *in* Fairbridge, R., ed., Encyclopedia of sedimentology: New York, Van Nostrand Reinhold.

———, and Heron, S. D., Jr., 1969, Stratigraphy of the Carolina Cretaceous: Southeastern Geology, v. 10, p. 201-245.

———, and Pirie, R. G., 1970, Fine-sediment dispersal in the Gulf of San Miguel, western Gulf of Panama: a reconnaissance: Jour. Marine Res., v. 28, p. 70-95.

———, and Sears, P., in press, Estuarine and littoral depositional patterns in the surficial sand sheet, central and southern Atlantic shelf of North America, *in* Allen, G. P., ed., Shelf and Estuarine sedimentation, a symposium: Talence, France, Univ. Bordeaux, Inst. Geol. Bassin d'Aquitaine.

———, Sanford, R. B., Dill, C. E., Jr., and Avignone, N. F., 1971b. Textural differentiation in the shoreface during erosional retreat of an unconsolidated coast, Cape Henry to Cape Hatteras, western North Atlantic shelf: Sedimentology, v. 16, p. 221-250.

———, Stanley, D. J., and Curray, J. R., 1971a, Relict sediments, a reconsideration: Jour. Geology, v. 79, p. 329-346.

———, Holliday, B. W., Avignone, N. F., and Schideler, G., 1972a, Anatomy of a shoreface ridge system, False Cape, Virginia: Marine Geology, v. 12, p. 59-84.

———, Kofoed, J. W., Saulsbury, F. P., and Sears, P., 1972b, Holocene evolution of the shelf surface, central and southern Atlantic coast of North America, *in* Swift, D. J. P., Duane, D. B., and Pilkey, O. H., eds., Shelf sediment transport: process and pattern: Stroudsburg, Pa., Dowden, Hutchinson & Ross, p. 499-574.

———, Ludwick, J. C., and Boehmer, R. W., 1972c, Shelf sediment transport, a probability model, *in* Swift, D. J. P., Duane, D. B., and Pilkey, O. H., eds., Shelf sediment transport: process and pattern: Stroudsburg, Pa., Dowden, Hutchinson & Ross, p. 145-224.

———, Schideler, G., Holliday, D. W., McHone, J., and Sears, P., 1973, Distribution and genesis of inner continental shelf sands, Cape Henry to Cape Hatteras: Coastal Eng. Res. Center Tech. Memo (in press).

———, Duane, D. B., and McKinney, T. F., 1974, Ridge and swale topography of the Middle Atlantic Bight, North America: secular response to the Holocene hydraulic regime: Marine Geology.

Todd, T. W., 1968, Dynamic diversion: influence of longshore current-tidal flow interaction on Chenier and Barrier Island plains: Jour. Sediment. Petrology, v. 38, p. 734-746.

Uchupi, E., 1970, Atlantic continental shelf and slope of the United States: shallow structure: U.S. Geol. Survey Prof. Paper 524-1, 44 p.

Van Straaten, L. M. J. U., 1965, Coastal barrier deposits

in south and north Holland—in particular in the area around Scheveningen and Ijmuden: Mededel. Geol. Sticht. n.s. 17, p. 41-75.

Wilson, I. G., Aeolian bedforms—their development and origins: Sedimentology, v. 19, p. 173-210.

Wright, L. D., and Coleman, J. M., 1972, River delta morphology: wave climate and the role of the subaqueous profile: Science, v. 176, p. 282-284.

———, and Coleman, J. M., 1973, Variations in morphology of major river deltas as functions of ocean wave and river discharge regimes: Am. Assoc. Petr. Geol. Bull., 57, p. 320-348.

Holocene Carbonate Sediments of Continental Shelves

R. N. Ginsburg and Noel P. James

INTRODUCTION

Carbonate sediments occur on most modern continental shelves. Because the various skeletal and nonskeletal grains are not moved far from their environments of formation, their distribution is a reliable tool on continental shelves for interpreting the history of Holocene deposition.

Eight examples of Holocene shelf carbonates are summarized, with maps of bathymetry, grain size, carbonate content, and predominant grain type. The examples are grouped into two intergrading categories: (1) open shelves—western North Atlantic, eastern Gulf of Mexico, Yucatan and Sahul; and (2) rimmed shelves—Great Barrier Reef, Belize, south Florida, and the Bahama Banks.

The thin accumulation of surface sediments on open shelves is largely relict and formed in shallow water earlier in the Holocene. The deposits of rimmed shelves, especially the shallower ones, are often thick, young (< 6,000 years old), and continuous, with contemporary deposits.

Open-shelf sediments most closely resemble the transgressive basal lag of fossil shelf sequences; the varied deposits of rimmed shelf lagoons are like the remainders of these repeated regressive or shoaling-upward cycles that occur throughout the geologic record.

Carbonate sediments and rocks occur in varying abundance on almost all modern continental shelves. Pure carbonates cover many shelves and platforms in tropical seas and even those shelves with major rivers have outer-shelf zones rich in carbonate. Although siliceous sediments predominate on shelves in temperate and polar seas, they also have substantial areas of carbonate-rich deposits.

Sand and granule-sized grains are the most frequent and widespread form of carbonate on modern shelves; coral algal reefs and algal hardgrounds are characteristic of the shelf margins in tropical seas; lime muds and mixtures of lime mud and sands are limited to lagoons rimmed by shelf-margin barriers.

Carbonate sands, reefs, and hardgrounds form in situ, and are not spread widely by waves and currents on the shelves; therefore, the nature and distribution of the various skeletal and nonskeletal components form a sensitive and reliable record of the environment and age of grain formation. Skeletal grains, whole and fragmented skeletons of benthic and planktonic organisms, are the most frequent and widespread carbonate element on modern shelves. Lees and Buller (1972) recognized two broad assemblages of skeletal grains related to latitude: the tropical chlorozoan assemblage, in which green algae and corals are the diagnostic types, and the foramol of temperate and polar seas, in which the principal contributors are molluscs and foraminifera, with echinoids, barnacles, bryozoans, and coralline algae often present.

Nonskeletal grains, ooids, peloids, and aggregates of these and other grains are restricted to the shelves and platforms of tropical seas. Extensive spreads of contemporary ooids and oolitically coated grains are forming only in agitated water less than 5 m deep; they are most prevalent on or near the margins of platforms or shelves (Ball, 1967). Contemporary, hard peloids, and rounded polygenetic grains of mud-sized carbonate are also restricted to shallow (usually <10 m) tropical seas; they may be cemented fecal pellets, the accretion of mud-sized particles, or the alteration of skeletal grains (micritization of Bathurst, 1966). Because the formation of ooids is limited to depths less than 5 m, these grains, so easily recognized, are one of the most valuable guides for reconstructing past conditions.

Extensive lime muds and muddy sands are accumulating now only in the protected lagoons of shelves and platforms with marginal rims. This restricted occurrence is source controlled: Holocene lime muds are derived chiefly from the disintegration of the fragile skeletons of green and red algae (Stockman et al., 1967; Lane, 1970; Patriquin, 1972), and possibly from direct precipitation (Cloud, 1962), both of which are limited to shallow tropical shelves. To illustrate the spectrum of variation of Holocene shelf carbonates eight examples will be reviewed.

The eight examples are grouped into two intergrading categories: open shelves and rimmed shelves. Open shelves are nearly flat, with extremely low-angle slopes from the shoreline to the break in slope that marks the shelf margin; rimmed shelves are those in which barriers along the shelf margin restrict circulation of the adjacent shelf lagoon. The most common barriers are coral-algal reefs, but shelf-margin buildups of carbonate sands as tidal bars and beach-dunes may form effective barrier rims.

OPEN SHELVES

Most shelves bordering continents in today's oceans are open, equivalent to Emery's (1968) category, "Deposition at Grade." The absence of an

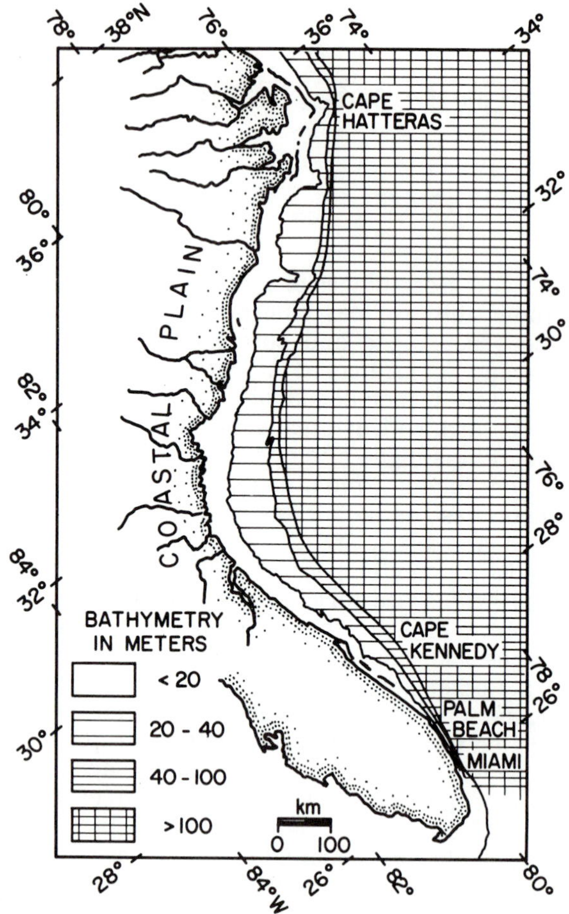

Fig. 1. Bathymetry of the western North Atlantic shelf from Cape Hatteras to Miami. (After Milliman et al., 1972.)

elevated rim allows tidal flushing and swell to extend across most or all of the inner shelf.

The four open shelves described and illustrated in the following sections fall into two groups: (1) the western North Atlantic and the eastern Gulf of Mexico shelves are characterized by mixtures of quartz and carbonate sands, with the carbonate content increasing from north to south, and by zonations of predominant carbonate grains that parallel the shoreline; and (2) the Yucatan and Sahul shelves have isolated banks or reefs along the shelf margin and inner shelf deposits that are entirely carbonate sands (Yucatan) or mixtures of terrigenous and carbonate sands and muds (Sahul).

Western North Atlantic Shelf

Most of the continental shelf south of Long Island, 40°N, is formed by the seaward extension of the Atlantic Coastal Plain, a wedge of almost flat-lying Tertiary and Quaternary sedimentary rocks (Fig. 1). This sector of the shelf has an extremely low eastward slope of about 0.4-1 m/km (Milliman et al., 1972). The shelf break is progressively shallower southward, ranging from 120 m off Cape Hatteras to 60 m off the southeastern United States, to less than 10 m near Miami. The shoreline is distinctly different north and south of Cape Lookout, with deeply embayed estuaries and rivers of high runoff and low suspended load to the north, and to the south, barrier islands with few rivers that have low runoff and high suspended load.

Because the shelf, extending continuously from 25°N to the Arctic Circle, spans the transition zone from temperate to tropical seas, the sediments show a southward increase in carbonate content unlike most of the other shelves discussed; north of the Cape the carbonate content is less than 5% (Milliman et al., 1972).

Surface Sediments. The surface sediments of the Atlantic shelf are relict and were deposited in shallow water earlier in the Holocene (Emery, 1968; Milliman et al., 1972). The only contemporary deposition is along the shoreline, and by bioturbation and the addition of tests of contemporary molluscs and foraminifera to the relict shelf sediments. From Cape Hatteras south to Miami the percentage of carbonate increases progressively (Fig. 2).

Milliman et al. (1972) mapped nine assemblages of carbonate sand grains on the southern shelf. A simplified version of their map (Fig. 3) shows that north of Jacksonville, molluscan debris predominates over the interior shelf, with zones of ooid-peloid and coralline algae near the margin. From Jacksonville to Miami, where the percentage of carbonate is higher, ooid-peloid sands are more extensive, barnacle fragments are a significant component, and the molluscan sand zone is narrower (Fig. 3). Off Onslow Bay ooid sand, now in depths of 25-40 m, has radiocarbon ages of 25,000-27,000 years BP (Milliman et al., 1968). Off Florida the age of the ooid sand near the shelf margin ranges from 9,000 to 14,000 years BP (Macintyre and Milliman, 1970).

From Cape Hatteras to Cape Kennedy the ridges of the shelf margin are interpreted as erosional remnants of Pleistocene limestone, capped by shallow-water calcarenite and coralline algal limestone. Two samples of the algal limestone had radiocarbon ages of about 12,000 and 27,000 years BP, respectively (Milliman et al., 1968). From Cape Kennedy to Palm Beach the ridges are relict dunes or beach ridges capped with prolific growth of living branched corals (*Oculina* sp.) (Macintyre and Milliman, 1970). From Palm Beach to Miami the single ridge is an "inactive" reef of hermatypic corals, octocorals, and sponges, with a narrow halo of carbonate sand rich in fragments of algae.

Although almost all the surface deposits of the Atlantic Shelf are relict, Milliman et al. (1968) showed that they are composite and include sediments and rocks formed at different times during the past 30,000 years. The oldest carbonate deposits are the ooid sands off Onslow Bay, and one specimen of algal limestone 25,000-27,000 years BP; both of these shallow-water deposits formed as sea level was falling during the last glacial period. The

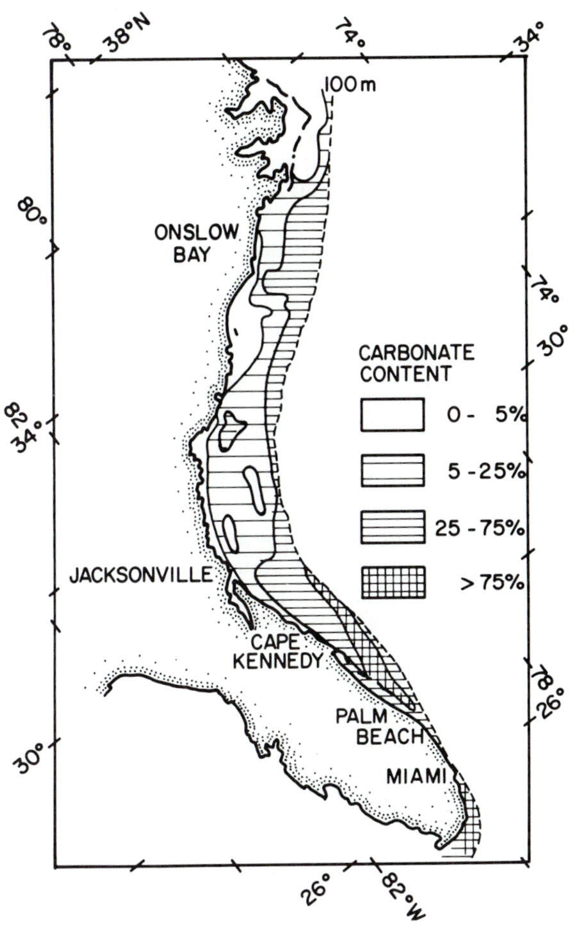

Fig. 2. Carbonate content of surface sediments of the western North Atlantic shelf from Cape Hatteras to Miami. (After Milliman et al., 1972.)

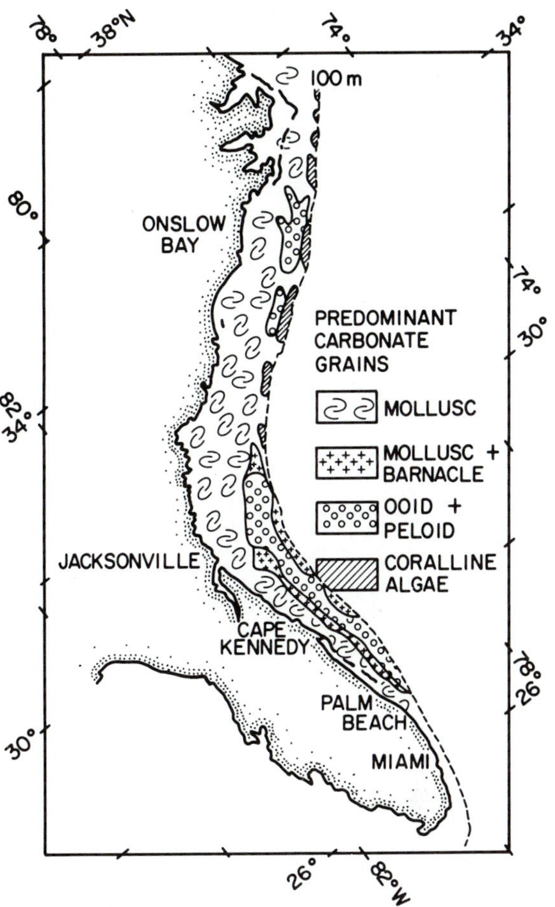

Fig. 3. Predominant carbonate grains in surface sediments of the western North Atlantic shelf from Cape Hatteras to Miami. (After Milliman et al., 1972.)

first shelf deposits of the rising Holocene sea are shallow-water carbonates near the shelf margin; ooid sands and algal limestone, and probably sands of coralline algae with radiocarbon ages of 9,000-15,000 years. Subsequently, when the rising sea flooded the inner shelf, molluscan debris began to accumulate there: at first from shallow-water species, but later as the water deepened from open-shelf species. As the rising sea approached its present position, reefs and branched corals began to flourish along the southern shelf margin, and planktonic foraminifera accumulated on the slope.

East Gulf of Mexico Shelf

The Eastern Gulf of Mexico is the 600-km-long, westward, submerged extension of low-lying peninsular Florida (Fig. 4). The shelf, from the shoreline to the 70-m shelf break, is about 130 km wide and encompasses an area of about 78,000 km^2. This shelf is characterized by its monotonous smooth slope and lack of relief, even at the margin, and by a zonation of predominant carbonate grains parallel to the shoreline and shelf margin.

Morphology. The Eastern Gulf Shelf is bounded landward by the flat, low-lying landmass of peninsular Florida, a terrain of karsted Tertiary and Pleistocene carbonates. Streams are short and drain only sandy limestones. The shelf is flat and smooth, extending from the shoreline to a depth of about 70 m, with a slope of only 0.4 m/km. The shelf margin at 70 m is marked only by an increase in slope to about 1.6 m/km. There are several north-trending ridges and domes between 60 and 100 m along the margin (Gould and Stewart, 1956).

Surface Sediments. Most of the shelf sediments are sand-sized, with little mud (Gould and Stewart, 1956; Chesser, personal communication, 1973) (Fig. 5). Irregular areas of shell coquina occur on the central part of the shelf, and lime muds only occur on the very outer shelf and in deeper water seaward (Fig. 5). The sediments grade from terrigenous sands in the nearshore zone to carbonates over the outer two-thirds.

The carbonate sand grains in these sediments occur in bands of varying width parallel to the shoreline and shelf break. Molluscs are the most

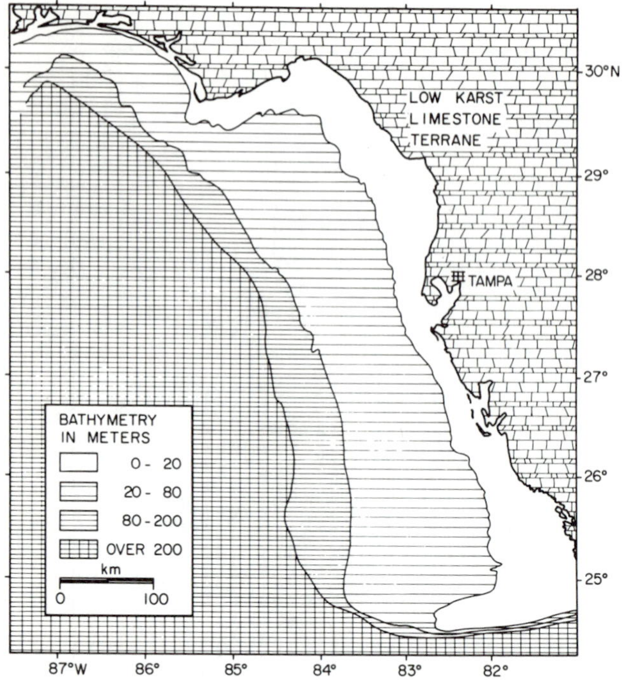

Fig. 4. Bathymetry of the eastern Gulf of Mexico shelf. (After Gould and Stewart, 1956.)

abundant grain type over the inner half of the shelf and are commonly mixed with terrigenous sands (Figs. 5 and 6). At about 60 m molluscan sands are replaced by sands in which branched coralline algae predominate (Fig. 6); the gradational contact comes near the change from mixed siliceous-carbonate sands to almost pure carbonate (Fig. 5). Between depths of 80 and 100 m the bottom is floored by deposits of ooid sand (Fig. 6); the seaward parts of these ooid sands contain both pelagic and benthic foraminifera, which in turn become predominant below a depth of 100 m (Gould and Stewart, 1956).

Local mounds or buildups of carbonate occur landward of the shelf margin; of the three shown on Figure 6 the most prominent is Florida Middle Ground, two irregular, north-trending ridges, rising 15 m from depths of about 40 m, covered by a growth of branched and massive corals, bryozoans, and *Halimeda*, and surrounded by molluscan sand and gravel (Back, 1972). The other two ridges shown on Figure 6 have a cover of coralline algae (Gould and Stewart, 1956). Scattered corals occur on the rock floor inshore along the northeastern part of the shelf (Fig. 6). A bar or reef of vermetid gastropods has formed nearshore on the southeastern part of the shelf (Shier, 1969) (Fig. 6).

Holocene Sedimentation. Holocene sediments over Tertiary and Quaternary bedrock are rarely more than 1 m thick, and most are considered to be relict (Gould and Stewart, 1956). The only radiocarbon age determination is from the crest of a ridge on the Florida Middle Ground, about 6,500 years BP (Back, 1972), indicating that it, too, is a relict feature. The ooid sands now in depths of 80 to 100 m are not contemporary, but formed in shallow water, ca. 5 m, early in the Holocene. The molluscan sands,

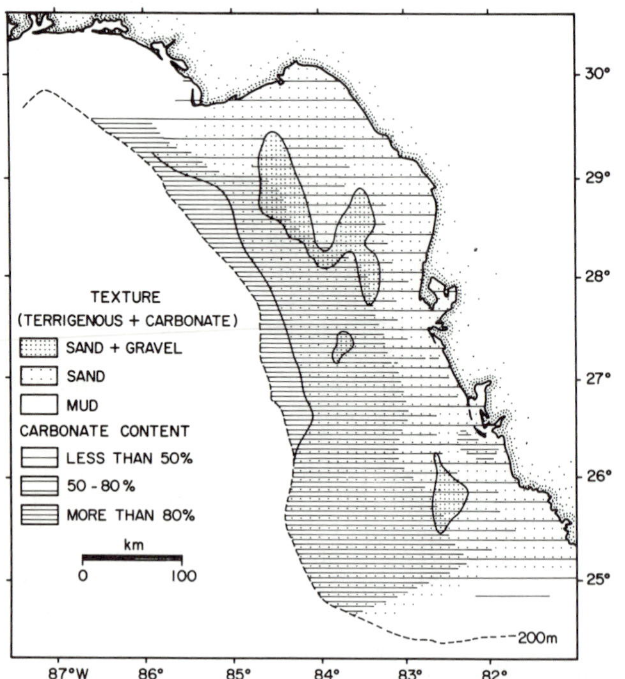

Fig. 5. Texture and carbonate content of surface sediments on the eastern Gulf of Mexico shelf. (After Gould and Stewart, 1956; Chesser, personal communication, 1973.)

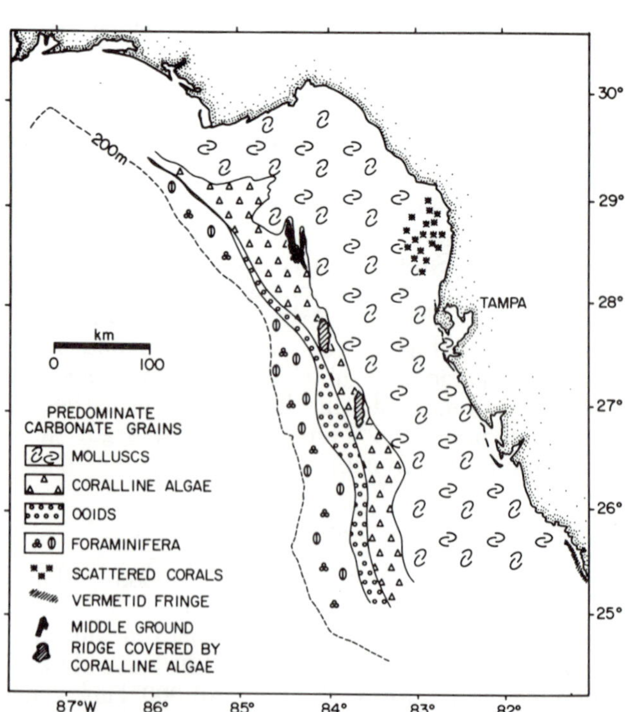

Fig. 6. Predominant carbonate grains in surface sediments and location of bathymetric highs on the eastern Gulf of Mexico shelf. (After Gould and Stewart, 1956; Ballard and Uchupi, 1970; Back, 1972.)

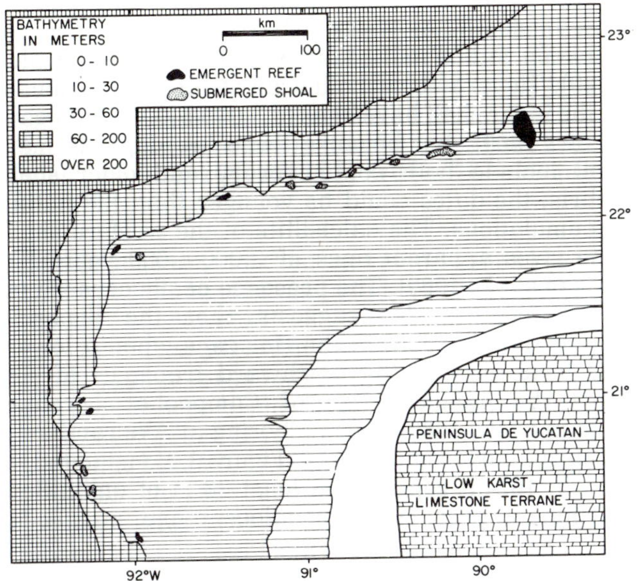

Fig. 7. Bathymetry of the Yucatan Shelf, Mexico. (After Logan et al., 1969.)

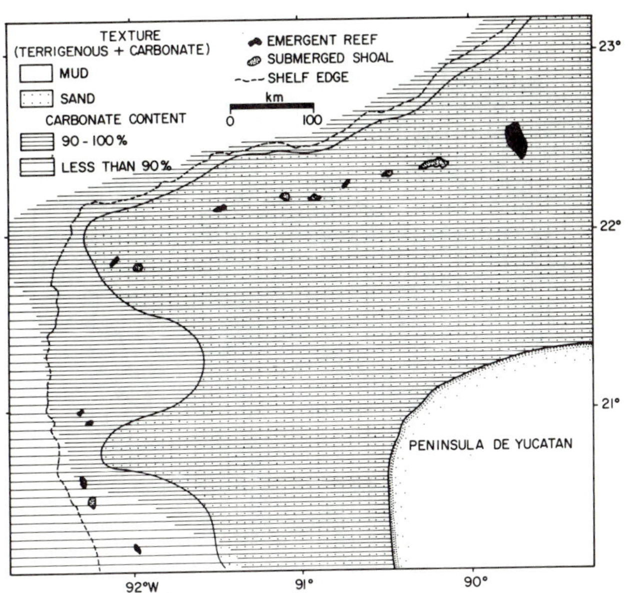

Fig. 8. Texture and carbonate content of surface sediments of the Yucatan Shelf, Mexico. (After Logan et al., 1969.)

like those of the western Atlantic shelf described above, are probably younger than the ooid sand and may include both nearshore and open-shelf species. The zone in which coralline algal grains predominate and the mounds and ridges veneered with coralline algae inside the shelf margin may also be somewhat younger than the ooid sands.

Yucatan Shelf

The Yucatan Shelf (Fig. 7) is the broad extension of the Yucatan Peninsula projecting northward into the Gulf of Mexico. The shelf, some 34,000 km^2 in area, is characterized by a low-angle, smooth slope, a series of isolated pinnacles or prominences along the shelf break, and a surface sediment veneer of pure carbonate sediments.

Morphology. Logan et al. (1969) divided the shelf into two bathymetric provinces (Fig. 7): (1) the inner shelf, from 0 to 60 m deep (130-190 km wide), with a series of reefs and submerged banks along the 60-m contour; and (2) an outer shelf from 60 to 210 m deep that widens rapidly from a few kilometers on the west to 30 km for most of the rest of the shelf (Fig. 7). The depth of the generally precipitous break in slope is variable; on the east it ranges from 82 to 210 m and along the north from 190 to 315 m. Logan et al. (1969) recognize three broad terraces at 35-41 m, 56-73 m, and 105-158 m, which are interpreted as eroded surfaces developed during the Holocene rise of sea level.

Surface Sediments. Over most of the shelf, Logan et al. (1969) found that changes in grain size and composition of surface sediments parallel changes in bathymetry (Figs. 8 and 9); the prominent exceptions are the two tongues of fine-grained sediment that extend onto the inner western shelf (Fig. 8).

Over the broad inner shelf the thin sediments overlying Quaternary limestones are skeletal sands and coquinas composed principally of molluscan debris (Figs. 8 and 9). Seaward of the discontinuous reef pinnacles and submerged banks along the 60-m contour, the lime sands are composed of ooids, peloids, and lithoclasts of the underlying limestone. From depths of about 90 m to the shelf break these nonskeletal sands are increasingly diluted with pelagic foraminifera. The anomalous tongues of

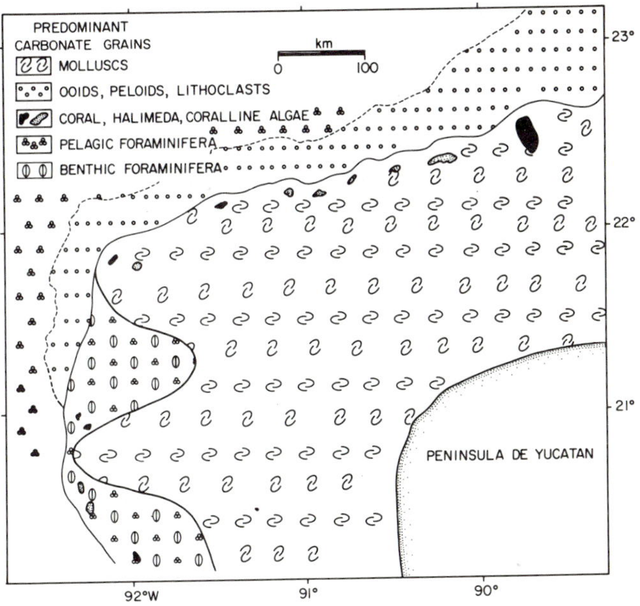

Fig. 9. Predominant carbonate grains in surface sediments of the Yucatan Shelf, Mexico. (After Logan et al., 1969.)

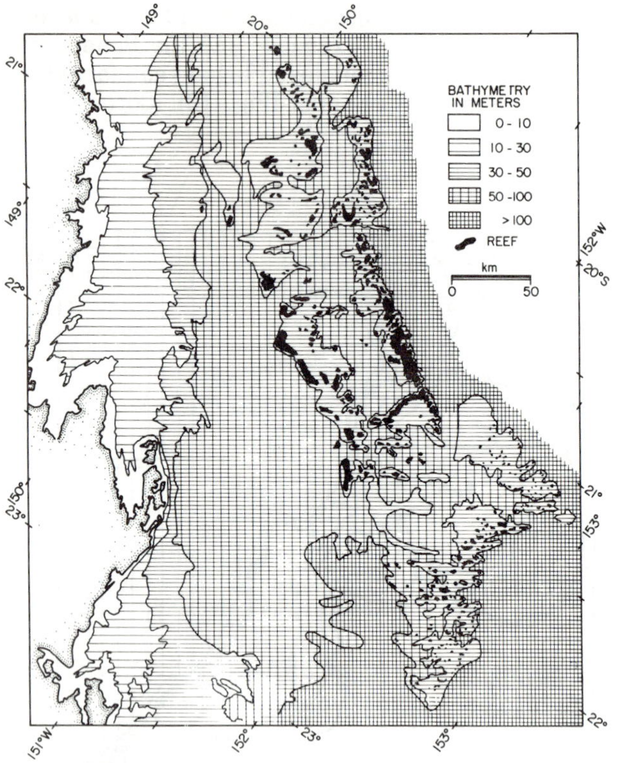

Fig. 15. Bathymetry of the southern part of the continental shelf off Queensland, eastern Australia. (After Maxwell, 1968.)

muddy sediment overlain by sandier sediment, often rich in carbonate on the shelf proper (van Andel and Veevers, 1967). This change in grain size and the increase in carbonate probably reflect the increased circulation produced by the flooding of the shelf as sea level rose. The local concentrations of quartz sand on the shelf parallel to the present shoreline are probably relict littoral zones of still-sands. Much of the relict carbonate, especially foraminifera and gastropod shells on the inner shelf, is glauconitized (van Andel and Veevers, 1967).

The origin of the banks is unknown; they may be eroded remnants of Tertiary coastal plain deposits or drowned reefs or banks (Fairbridge, 1950). The presence of relict coral debris around many banks (van Andel and Veevers, 1967) suggests a period of reef growth earlier in the Holocene. Contemporary reefs occur nearshore west of about 127°W and offshore west of 124°W (Teichert and Fairbridge, 1948).

Summary of Open Shelves

The relict surface sediments of open shelves are with one exception, the Sahul Shelf, sand-sized, and they show a consistent pattern of predominant carbonate grains: *inner shelf*, molluscs; *shelf edge*, ooids, peloids, and coralline algae; *slope*, planktonic foraminifera (Fig. 13). The deposits of these three zones are not strictly contemporary; instead, each began accumulating at a different stage in the Holocene flooding of the shelves.

The oldest Holocene deposits of ooids, peloids, lithoclasts, coralline algae, and reef-like banks formed along the shelf margins, in all probability at depths comparable to their contemporary counterparts, 3-10 m. However, the continued rise of seas was too rapid for most of these shelf-margin accumulations to keep pace; only locally on the Yucatan shelf was reef growth persistent. The flooding of the broad inner shelf transformed what was a coastal plain into a shallow sea, in which molluscs occupied the many varied habitats. As sea level continued to rise, nearshore molluscs were replaced by those of deeper water, making the inner shelf molluscan zone itself composite. The slope deposits rich in planktonic foraminifera probably did not begin to accumulate until sea level approached its present position. Contemporary bioturbation on the inner shelf is mixing modern skeletal debris with the relict sediments, so that on many shelves the upper few meters is truly palimpsest (Swift et al., 1971).

RIMMED SHELVES

Rimmed shelves are those in which a continuous or semicontinuous rim or barrier along the shelf margin restricts circulation and wave action on the adjacent shelf lagoon. In two of the examples presented below the rim is a barrier reef: Queensland Shelf or Great Barrier Reef, and the Belize (British Honduras) Shelf. The lagoon of the Great Barrier Reef and the southern part of the Belize Shelf is relatively deep and open, with numerous lagoon reefs; the northern part of the Belize Shelf is shallow and lime muds prevail. The South-Florida Shelf shows an interesting development of multiple barriers, an outer zone of discontinuous reefs and sand banks, and an inner continuous barrier of elevated islands of Pleistocene limestone. The rims of the Bahama Banks are also islands of Pleistocene limestone barring vast, shallow shelf lagoons.

Queensland Shelf—Great Barrier Reef

The shelf bordering the eastern side of the state of Queensland, eastern Australia, stretches from 10° to 24°S, a distance of 1,500 km. It is 24 to 290 km wide and encompasses an area of some 215,000 km^2. The shelf is characterized by the extensive development of shelf-edge reefs, a widespread zone of terrigenous sediment, and a relatively deep interior with open circulation.

Morphology. The shelf is narrowest and shallowest at its northern end (Fig. 14) and widens and deepens southward (Fig. 15). In the southern sector the landward and seaward parts of the shelf are separated by an embayment (Maxwell, 1968). The continental slope off the northern part of the shelf is abrupt and drops precipitously into the Queensland Trench; the slope off the southern shelf is gentle down to the Coral Sea platform. The landmass bordering the shelf comprises a coastal range flanked by a coastal plain which together fronts an inland, highly dissected plateau. The only three major river systems reach the coast between 20° and 25°S; the northern 1,000 km of the shelf is free of stream discharge.

Surface Sediments. In general, sediments throughout the shelf exhibit a uniform pattern; a nearshore zone of terrigenous sands and muds, and a central zone of mixed terrigenous and carbonate muddy sands and muds separating the two (Figs. 16 and 17).

Terrigenous sediments extend almost all the way across the northern part of the shelf near latitudes 14°, 14°30', and 15°30'S, and some inter-reef sediments on the shelf edge contain as much as 40% terrigenous material (Fig. 16). In the southern sector (Fig. 17), because of the width of the shelf, terrigenous sediments rarely extend more than halfway across, with most of the fine material deposited in the deeper axial depression (Maxwell, 1968).

The terrigenous sands and some gravels that floor the landward part of the inner shelf are considered relict (Maxwell, 1968). The fallout of terrigenous muds seems to be the only contemporary terrigenous sedimentation on the shelf. In areas of low runnoff and strong tidal flushing, such as the northern and southern sectors of the shelf, deposition of these muds is insignificant, with accumulation restricted to the middle of the shelf. However, in the central sector of the shelf, south of latitude 16°30'S (Fig. 16), where tidal action is reduced and stream discharge high, the muds are deposited nearshore and form a wedge of terrigenous muds prograding over relict coarse terrigenous sands (Maxwell and Swinchatt, 1970).

Carbonate grains on the southern sector of the shelf, the only portion described in detail to date, have a broad distribution of molluscan and bryozoan grains throughout, with concentrations of *Halimeda*, benthic foraminifers, coral, and coralline algae along the shelf margin (Fig. 18). In the nearshore terrigenous sands the most abundant carbonate particles are mollusc debris. Inter-reef sediments along the shelf edge are fine to coarse carbonate sands with local accumulations of gravel and mud, with mostly indigenous foraminifera, molluscs, *Halimeda*, and bryozoans (Fig. 18). Bryozoan debris is common around the reefs of the Pompey Complex; *Halimeda* is predominant southward in the Swains Complex (Fig. 18). The relatively large amount of bryozoan debris in sediments throughout the shelf, locally up to 30%, is far greater than any other reef-dominated shelf studied to date. Sediments on the reefs, both on the shelf edge and inner shelf, are characterized by carbonate sands and gravels on the reef flat, boulder ramparts on the reef crest, and fine sands and silts in the lagoons. Reef-associated sediments are mainly coral, coralline algae, *Halimeda*, and foraminifers (Maxwell, 1968; Maxwell and Swinchatt, 1970).

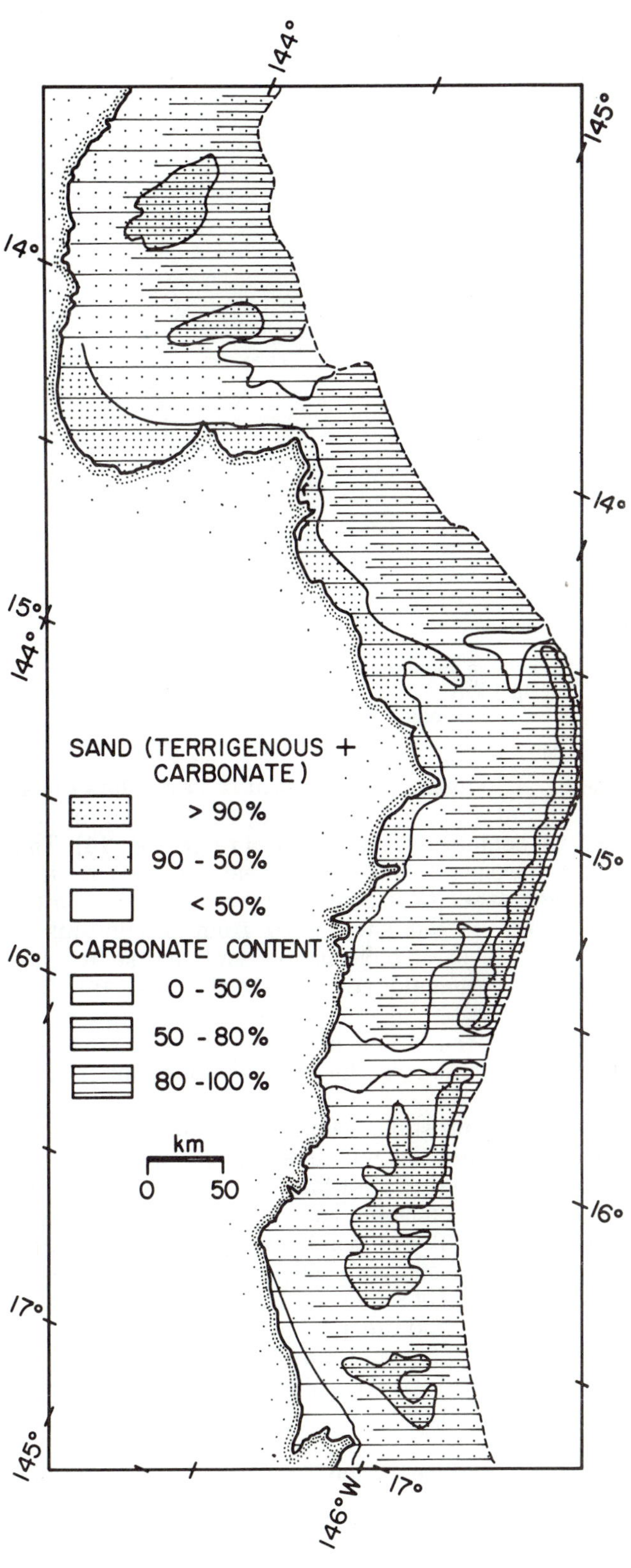

Fig. 16. Texture and carbonate content of the surface sediments on the relatively narrow northern part of the Queensland Shelf, eastern Australia. (After Maxwell, 1968.)

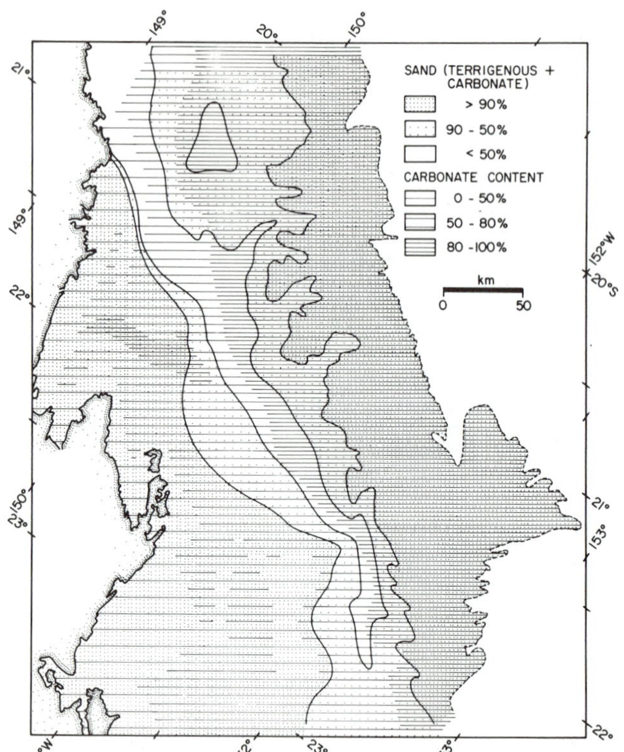

Fig. 17. Texture and carbonate content of surface sediments on the relatively wide southern part of the Queensland Shelf, eastern Australia. (After Maxwell, 1968.)

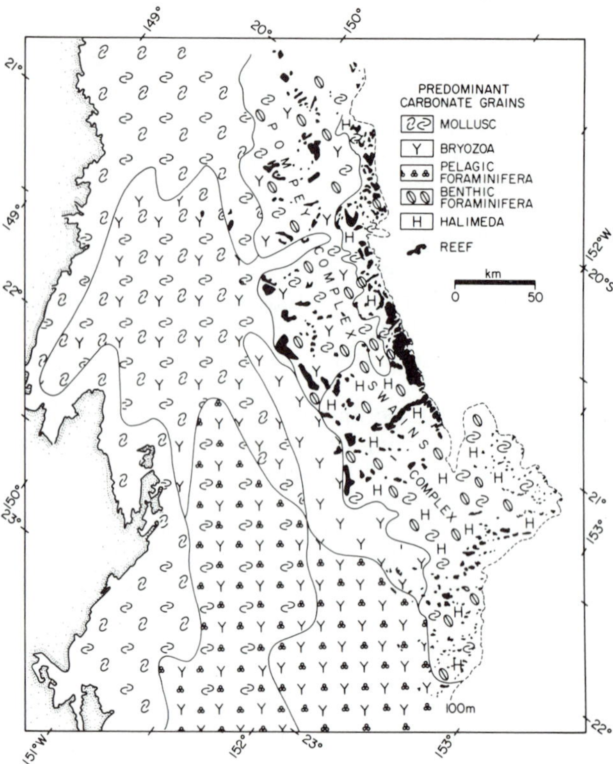

Fig. 18. Predominant carbonate grains in surface sediments on the southern part of the Queensland Shelf, eastern Australia. (After Maxwell, 1968; Maxwell and Swinchatt, 1970.)

Reefs. The northern part of the area has linear reefs that are less abundant southward, so the central sector has few marginal reefs (Fig. 14). This change from extensive linear reefs to fewer reefs in the central sector is coincident with a change in the character of the shelf edge from precipitous to gentle. The shelf edge in the southern sector has numerous elongate and ring reefs (Fig. 15). The greatest number of inner shelf reefs occurs in the lee of the barrier in the northern sector (Fig. 14) and to a lesser extent in the wide southern inner shelf (Fig. 15).

Most inner shelf reefs are variants on two basic forms, circular or equidimensional and linear to elongate, with the latter tending to curve at the ends and ultimately forming ring reefs or platform atolls (Maxwell, 1968). The windward parts of these reefs are a growth of branched corals and *Millepora*, with a reef crest veneered by coralline algae and the leeward and deeper portions dominated by growths of platey and massive forms. Behind the windward crescent of coral growth is a shallow sand and rubble flat.

Holocene Sedimentation. Because most of the terrigenous sediment on the shelf is sand and was clearly not deposited under present conditions, Maxwell (1968) concludes that these deposits are relict. Bathymetric anomalies at various depths, especially 64 m, are interpreted as strandlines, beaches, offshore bars, and dunes formed at still-stands during the Holocene sea-level rise (Maxwell and Swinchatt, 1970). The ancient courses of many rivers on the central and northern sectors of the shelf, which can be traced across the shelf, have no direct outlet to the sea, and the ends are surrounded by terrigenous sands, suggesting that sandy deltas likely extended over the shelf during low stands of sea level (Maxwell and Swinchatt, 1970). Rivers draining into the southern sector were apparently deflected southward and flowed seaward down the axial depression.

Of the three borings reported, only one gives the thickness of Holocene reef limestone, 18 m; in the other two the combined Holocene-Pleistocene reef is 121 and 154 m, respectively (Purdy, 1974a, Fig. 33). Maxwell (1968) suggests that much of the present relief was initiated during a still-stand at -64 m, with formation of various fluviatile and lacustrine features. Once sea level began to rise again after this pause, these depositional highs acted as loci for coral growth; thus the present distribution of many inner shelf reefs may reflect the topography formed during this still-stand of sea level.

Belize [British Honduras] Shelf

The Belize Shelf extends south from the Yucatan Peninsula a distance of nearly 250 km to the Gulf of Honduras (Fig. 19). The shelf is 20-40 km wide and bordered by a precipitous shelf margin. It covers an area of some 8,400 km^2.

Morphology. The northern part of the shelf is bordered by a low karsted surface of flat-lying Cenozoic and Cretaceous carbonates, with only a few small rivers draining onto the shelf. The inner shelf bathymetry, reflecting this surface, is flat, with a shallow depression rarely deeper than 8 m (Fig. 19). The southern part of the shelf is flanked by a mountainous terrain of uplifted Paleozoic sediments, metasediments, and acid intrusives, the Maya Mountains, surrounded by a coastal plain of terrigenous clastic sediments through which many rivers flow onto the shelf. Similarly, the inner shelf bathymetry reflects this topography, deepening southward from a depth of 20-25 m off Belize City near latitude 17°30'N to over 200 m in the Gulf of Honduras near latitude 16°S (Fig. 19). The entire margin of the shelf is a raised platform 3-10 km wide and rarely deeper than 3 m. The precipitous seaward margin of this platform is crowned by an almost continuous coral reef, and the leeward side of the platform is topped with many small low islands. The barrier platform narrows to less than 100 m near its southern end and terminates in a hook-shaped feature in the Gulf of Honduras (Stoddart, 1963, 1969).

Surface Sediments. The shallow northern part of the shelf, separated from the open ocean by the barrier reef, scattered islands, or the southern extension of the Yucatan Peninsula and constricted at its northern end into Chetumal Bay, is floored by pelleted lime muds containing abundant foraminiferal tests (Figs. 19 and 20). Because of the progressive restriction in circulation northward, the foraminifera grade from miliolids on the more open shelf lagoon to peneropolids to cryptocrystalline grains that are clearly altered peneropolids in Chetumal Bay (Purdy, 1974b). Terrigenous sediments are restricted to the nearshore zone (Fig. 20) (Wantland and Pusey, 1971; Purdy, 1974b).

The deeper southern lagoon is also floored by mud, but with an admixture of terrigenous clay. This clay, mostly montmorillonite, grades from more than 50% along the western shore to less than 20% adjacent to the barrier platform (Wantland and Pusey, 1971). The fine carbonate is derived from breakdown of larger carbonate skeletons (Matthews, 1965) and the fallout of calcareous nannoplankton (Scholle and Kling, 1972). Sand-sized and larger carbonate grains are mainly thin-walled molluscs, foraminifers, and near the southern end, *Halimeda*, with local concentrations of pteropods (Fig. 21). The nearshore zone is composed of terrigenous sands, mainly reworked Pleistocene quartzose sands, with local accumulations of fine muds in deltas; the carbonate fraction consists of reworked and indigenous mollusc debris (Fig. 21).

Isolated patch reefs and platform atolls that rise to the surface from the floor of the lagoon are surrounded by a halo of sand-size carbonate that grades into terrigenous-rich mud away from the reef (Wantland and Pusey, 1971; Purdy, 1974b).

Sediments on patch reefs and on the platform margin are almost all skeletal. The wide platform in the lee of the barrier is floored with *Halimeda*-rich sands (Fig. 21), except in *Thalassia* meadows, where the sediments contain an admixture of lime mud. In and on the reefs most sediments contain abundant coral and caralline algal debris (Wantland and Pusey, 1971; Purdy, 1974b).

Reefs. The barrier platform supports a semicontinuous reef whose shallow-water elements are growths of branched corals, *Millepora*, and calcareous algae. The growing spurs of the reef are surrounded by rubble of coral branches. This rubble accumulates behind the reef crest and is often cemented to form a pavement (James et al., 1973). The northern inner shelf contains no reefs, but the southern inner shelf contains a maze of linear reefs, platform atolls, and circular to ameboid "patch" reefs (Wantland and Pusey, 1971; Stoddart, 1962). Smaller reefs are unzoned, but larger reefs in open water are zoned with a windward crescent of coral growth protecting a leeward sand and rubble flat similar to the reefs at the shelf edge.

Holocene Sedimentation. Seismic measurements and drilling by Purdy (1974a) revealed that many of the lagoon reefs and the barrier platform are Holocene accretions over pre-Holocene highs. The amount of Holocene accretion on these high ranges from 8-20 m and averages 12 m (Purdy, 1974a).

Gradual flooding of the lagoon during the Holocene rise began at the southern end, and the sea did not begin to flood the northern shelf until it rose within 10 m of the present level. Throughout most of this time the southern lagoon may have been a large estuary, bounded on the seaward side by the elevated ridge of limestone that was the Pleistocene barrier, and receiving, for the most part, terrigenous mud from the mainland. Reefs developed and the effect of the barrier was lessened only when sea level rose above the tops of the karst highs some 20 m or so below present sea level. Only at this time would production of carbonate sediments increase dramatically. The total thickness of sediment in the southern lagoon is of the order of 3-5 m.

During almost all of sea level rise the northern shelf was a coastal plain floored by terrigenous clastics. Today the underlying terrigenous sediments are being mixed with contemporary lime muds by burrowing (Purdy, 1974b).

South Florida Shelf

The South Florida Shelf surrounds the southern end of peninsular Florida and connects the western North Atlantic shelf with the Eastern Gulf of Mexico shelf. Stretching in an arc that curves southward and westward from Miami for some 360 km, the shallow shelf is characterized by a string of low

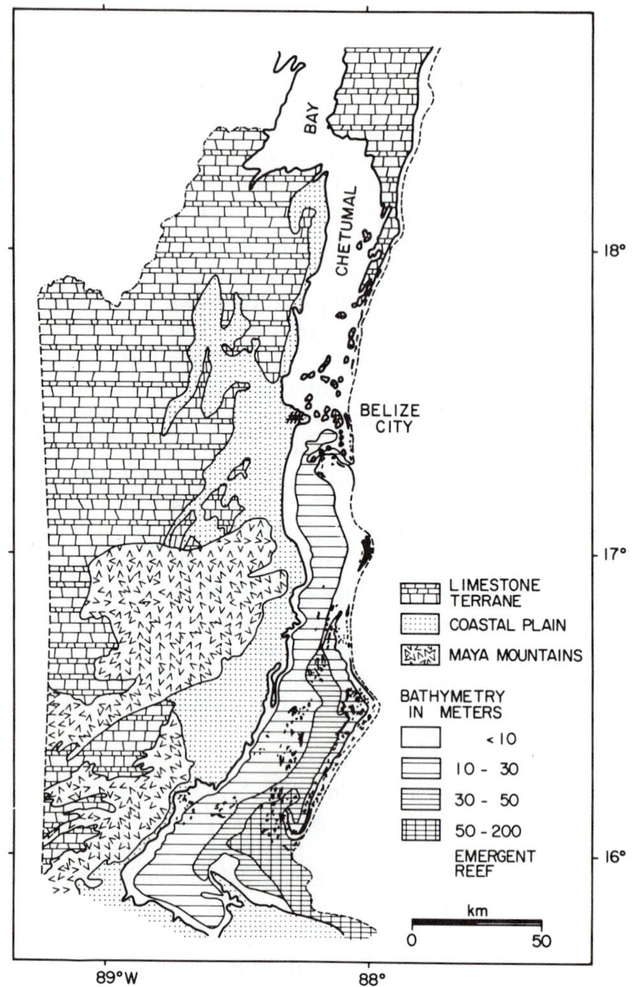

Fig. 19. Bathymetry of the continental shelf off Belize, British Honduras. (After Wantland and Pusey, 1971; Purdy, 1974b.)

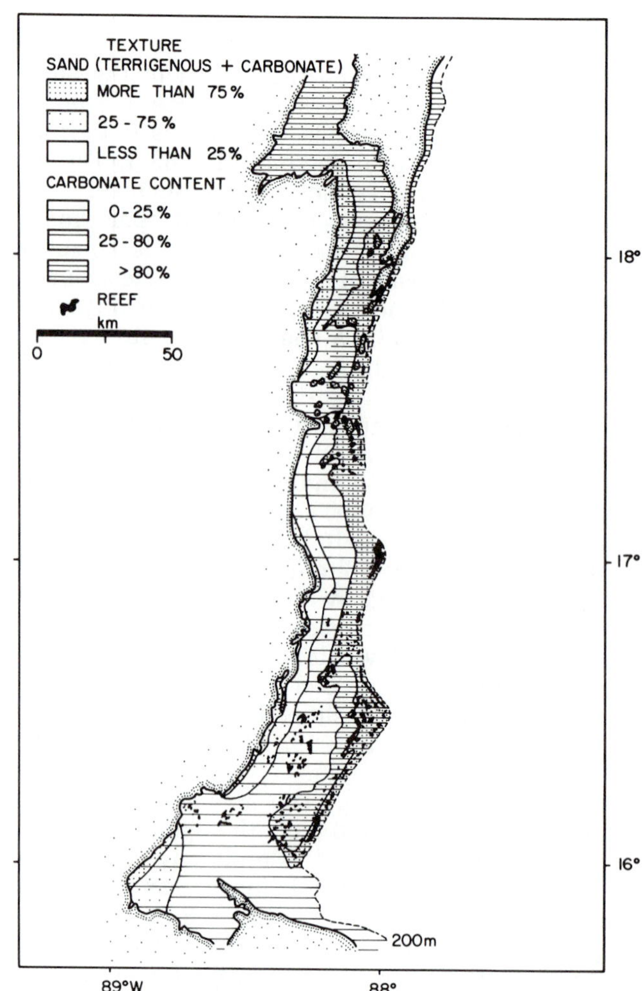

Fig. 20. Texture and carbonate content of surface sediments on the continental shelf off Belize, British Honduras. (After Wantland and Pusey, 1971; Purdy, 1974b.)

Pleistocene islands, the Florida Keys, lying in an arc parallel to, and some 10 km shoreward of, the shelf break. These islands separate the very shallow expanse of Florida Bay from the exposed reef tract (Fig. 22). Seaward of the reef tract the sea floor slopes gently down to the Pourtales Terrace, a flattish rocky surface, at depths of 200-300 m, which is bounded seaward by a precipitous slope into the axis of the Straits of Florida (Gomberg, 1973).

Morphology. The Reef Tract has a complex, small-scale topography of active coral reefs, sand shoals, grass-covered banks, and elongate depressions. Off the Upper Florida Keys, which are islands of Pleistocene reef limestone, there are two discontinuous but parallel ridges: an outer ridge of discontinuous shelf-margin reefs, hard banks, and coral rubble and skeletal sand; and an inner ridge, White Bank, of skeletal sand with scattered reefs along its seaward margin (Fig. 22).

Florida Bay is a broad, shallow area extending from the coastal swamps of the mainland to the Lower Keys, which are islands of Pleistocene oolitic limestone. The northern part of the Bay (Fig. 22), protected by a continuous island barrier, is a network of grass-covered mud banks less than ⅓ m deep; in the northeast corner of the bay these banks surround broad basins 2-3 m deep, in which the Pleistocene rock floor is bare or veneered with centimeters of shelly mud. In the southern part of Florida Bay the island barrier is discontinuous and allows tidal exchange between the Gulf of Mexico and the Atlantic; the resulting strong currents and wave action prevent accumulation of lime mud, so the limestone rock floor is bare or veneered with less than ⅓ m of muddy skeletal sand.

Surface Sediments. The broad pattern of surface sediments shown in Figures 23 and 24 has three contrasting deposits, each with distinctive grain size and faunas: (1) Interior Florida Bay, where restricted circulation allows lime mud to accumulate with indigenous molluscan and foraminiferal shells; (2) Outer Florida Bay, southwest of the mudbanks in Figure 22, where strong tidal currents and wave action from the open Gulf of Mexico winnow fine sediment and leave a thin lag, generally less than 30 cm thick, of contemporary skeletal debris and fragments of Pleistocene limestone; and (3) the reef tract (Fig. 22), where open circulation and tidal flushing promote the growth of lagoon and marginal reefs, lime sands, and muddy lime sands rich in algal debris. More detailed descriptions are given by Ginsburg (1956) and Swinchatt (1965).

Holocene Sedimentation. Because the Pleistocene surface lies at two successively shallower elevations, Holocene sedimentation occurred in two distinct stages. The first began when the sea rose to about -20 m and flooded the deeper parts of the reef tract. At this time reef growth probably began along the break in slope, and soon lime muds with shell debris accumulated in what was then shallow water, seaward of the present Keys. Reef growth along the margin continued and spread to the lagoon as sea level rose, and the lime muds were replaced first by muddy skeletal sands and finally by mud free sands (Enos, 1974). The second stage of Holocene deposition, the muds of Interior Florida Bay, did not begin until about 4000 years BP, when the rising sea flooded the rock floor. The first deposits were mangrove peats of the intertidal zone; they, in turn, are overlain by up to 4 m of shallow-water lime mud with foraminiferal and molluscan debris.

Bahama Banks

The Bahama Banks are a group of flat-topped shallow carbonate platforms surrounded by ocean depths; the array covers an area of 96,000 km^2 and extends from off northern Florida to near Puerto Rico (Fig. 25). The Bahama province is not considered a part of the Atlantic continental shelf and its basement is thought by some workers to be oceanic (Dietz et al., 1973).

Morphology. The size and configuration of individual banks is variable, but from west to east there is a trend of decreasing area and increasing relief (Fig. 25). This is probably the result of progressively increasing rates of subsidence toward the Atlantic Ocean basin. The marginal slopes of the Banks are steep, generally near-vertical in the upper few hundred meters, with overall slopes to 4,500 m of up to 40° (Emiliani, 1965). Steep margins that are open to ocean sea and swell, particularly east-facing margins, have islands that rise up to 50 m or more above sea level (Fig. 25). The islands are Pleistocene beach dunes of carbonate sand, eolianites, veneered with contemporary deposits: beaches, reefs, tidal sand bars.

The tops of the banks are shallow, flat, shelf lagoons; they are nowhere deeper than 15 m, and large areas are less than 6 m deep (Fig. 25). Where these shallow shelf lagoons are barred by continuous islands barriers, as west of Andros and

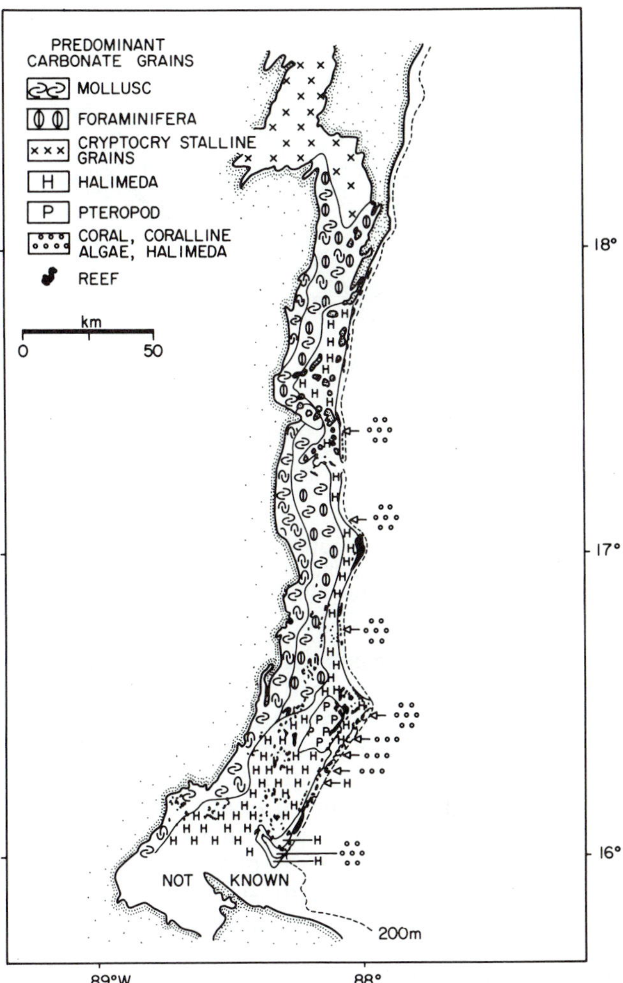

Fig. 21. Predominant carbonate grains in surface sediments on the continental shelf off Belize, British Honduras. (After Wantland and Pusey, 1971; Purdy, 1974b.)

Eleuthera islands (Fig. 26), the circulation is restricted, with attendant elevated salinity, mud-sized bottom sediment, and reduced populations of benthic organisms. Where marginal barriers are absent or discontinuous, the shallowness of the shelf lagoons amplifies wind-driven wave action and tidal currrents, and the lagoon is floored with mud-free, nonskeletal sands, ornamented by ripples and sand waves.

Surface Sediments. The surface sediments of Great Bahama Bank (Fig. 26), the most studied example, can be grouped into three major types: (1) reefs and skeletal sands rich in algal and coral grains; (2) nonskeletal sands of peloids, ooids, coated peloids (superficial ooids), and aggregates of these and skeletal grains into friable lumps, grapestone and irregular lithoclasts (Illing, 1954; Purdy, 1963; Traverse and Ginsburg, 1966; Taft et al., 1968); and (3) pelleted lime muds with less than 10% skeletal grains, principally molluscs and foraminifera (Cloud, 1962; Shinn et al., 1969). The skeletal sands are confined to a narrow marginal

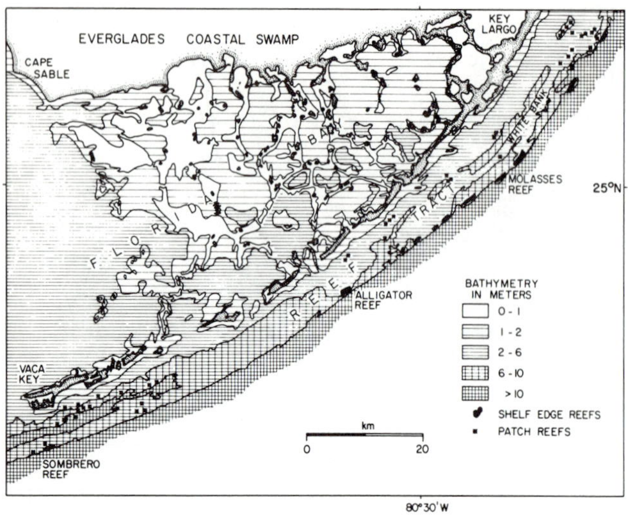

Fig. 22. Bathymetry of the south Florida shelf. South of the area shown here the island and reef rims are discontinuous and the shelf is open. (After Ginsburg, 1956.)

zone (Fig. 26); the reefs are well developed only on the east-facing, windward margins. The pelleted lime muds occur behind continuous island barriers such as Andros and Eleuthera islands (Fig. 26), and extend from depths of about 4 m to the extreme limit of storm flooding, 1 m or so above mean sea level. The nonskeletal sands over most of the broad interior shelf lagoon consist of peloids and coated peloids and aggregates of these grains, grapestone, friable lumps, and clasts. Ooid sands make up the extensive tidal bars on bank margins without island barriers (Fig. 26).

Holocene Sedimentation. Because the surface of the Pleistocene over most of the banks beyond the marginal islands is nearly flat and less than 10-15 m below present sea level, the banks were not flooded by the rising Holocene sea until about 6,000 years BP. The subsequent sequence of deposits is known only from three marginal sites (Ball, 1967; Buchanan, 1970); the initial sediments over the Pleistocene limestone floor are shallow-water deposits with appreciable lime mud, similar to the contemporary accumulations of the protected shelf lagoon. The overlying skeletal and nonskeletal sands reflect the more open circulation produced by the continued rise in sea level to its present position.

Summary, Rimmed Shelves

Shelf-margin barriers are restricted to tropical seas; some of the barriers are Pleistocene limestones: beach dunes (eolianites) of the Bahama Banks (Fig. 26) or the Pleistocene bank reefs of south Florida (Fig. 22). Others are Holocene reefs that overlie Pleistocene reefs: Belize (Fig. 19), the discontinuous outer reefs of South Florida (Fig. 22), and possibly the Great Barrier Reef (Figs. 14 and 15). Still others are banks of ooid sand: Great Bahama Banks (Fig. 26), or skeletal sand White Bank of the Florida Reef Tract (Fig. 22). All these barriers are produced by accelerated fixation and accumulation of carbonate, both skeletal and nonskeletal, on or near ocean-facing shelf margins, and all produce a variable but demonstrable effect on their lagoonal deposits.

The thickness of the Holocene part of barriers is known from limited borings in four areas: (1) Outer Florida Reefs, up to 16 m (Hoffmeister and Multer, 1968; Enos, 1974); (2) ooid sand bars of Great Bahama Bank, 8 m (Ball, 1967); (3) Belize Barrier Reef, 8-20 m, possibly 31 m (Purdy, 1974a); and (4) Great Barrier Reef, at least 18 m (Purdy, 1974a). These data suggest that Holocene barrier development may have been limited to areas where the pre-Holocene shelf margin was less than about 30 m below present sea level. However, the existence of a shallow antecedent platform is not

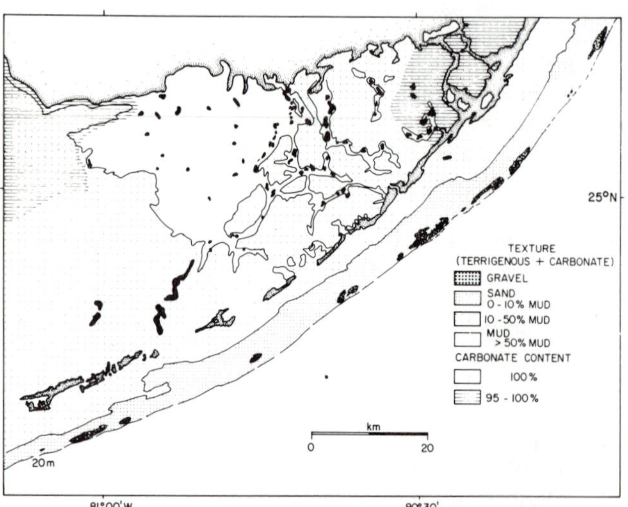

Fig. 23. Texture and carbonate content of surface sediments on the south Florida shelf. (After Ginsburg, 1956; Swinchatt, 1965.)

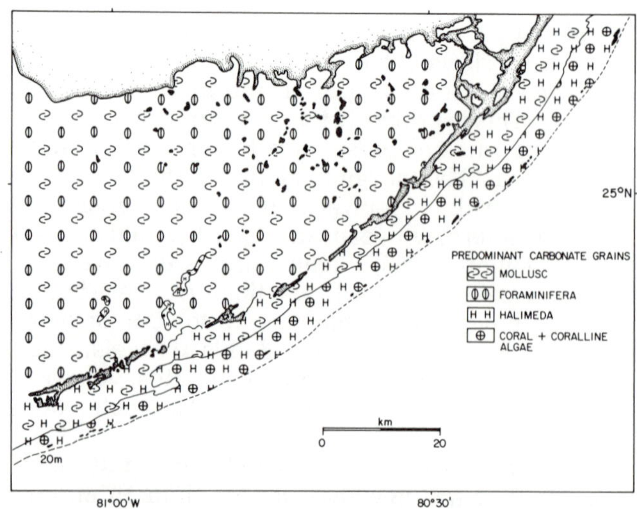

Fig. 24. Predominant carbonate grains in surface sediments on the south Florida shelf. (After Ginsburg, 1956.)

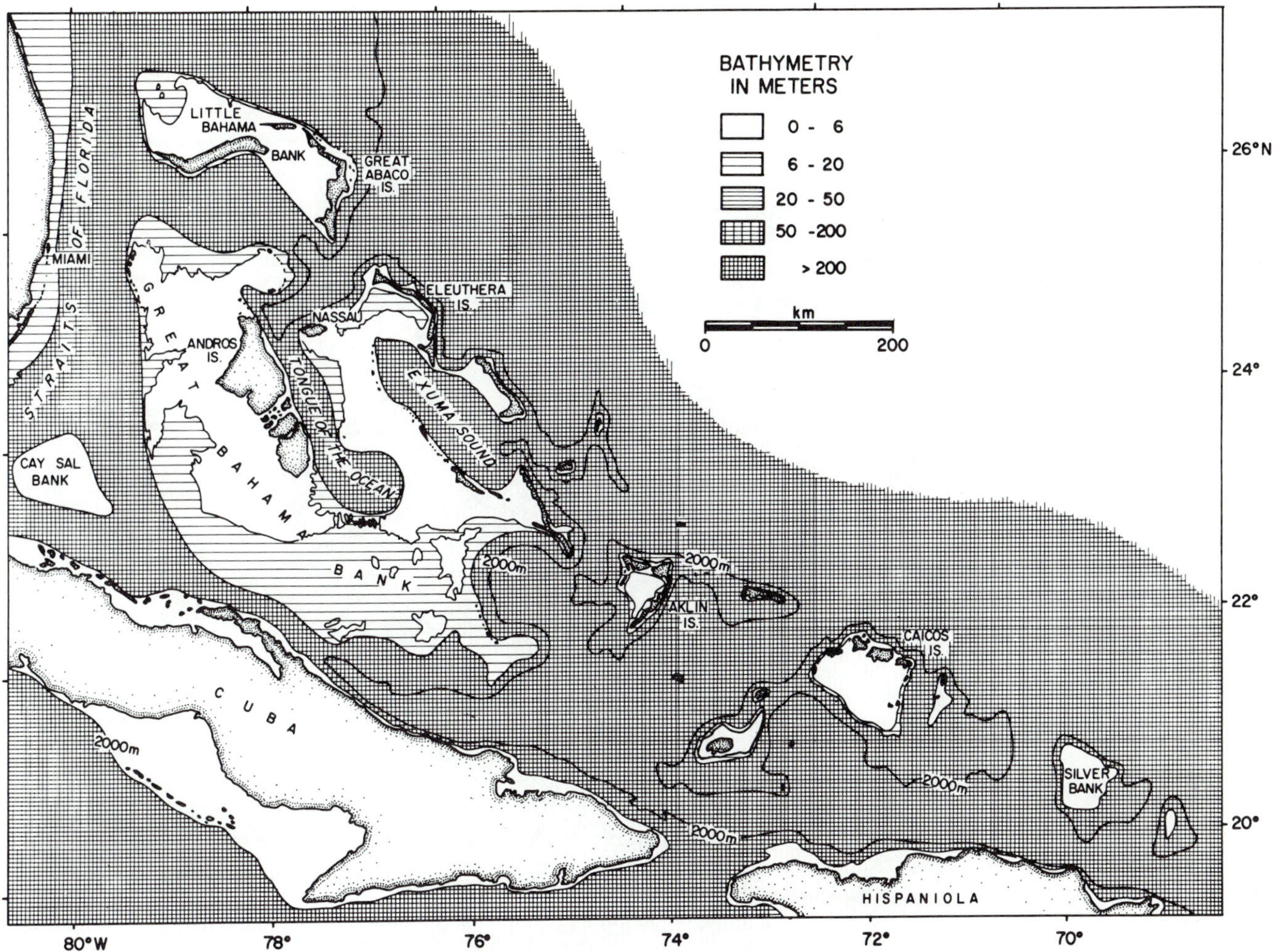

Fig. 25. Bathymetry of the Bahama Banks.

sufficient for the growth of a Holocene barrier, as shown by the inactive prominences of the Sahul Shelf (Fig. 10), now at depths of 20 m below sea level.

The characteristic deposit of lagoons shielded by a continuous barrier is lime mud with a restricted fauna of molluscs or foraminifera or both. These lagoonal lime muds are extensive where the pre-Holocene surface is less than 5 m below present sea level: west of Andros Island, Bahamas (Fig. 26); west of the Upper Florida Keys (Fig. 23); in the northern part of the Belize Lagoon (Fig. 20). In a deeper lagoon like that of southern Belize (Fig. 20), contemporary deposition of lime mud is limited to the outer zone near the barrier and around the numerous lagoon reefs; the rest of the lagoon is floored largely with terrigenous clays, relict deposits probably of early Holocene age, with admixed contemporary molluscs and foraminifera (Wantland and Pusey, 1971). The effect of the Great Barrier Reef on lagoon deposits has not been established, but it may be responsible for the abundance of bryozoans and molluscs and the accumulation of terrigenous mud in the axial deep (Fig. 17).

The effect of a discontinuous barrier is limited to a narrow zone behind the barrier. On the eastern side of Great Bahama Bank in the Exuma Islands (Fig. 26), tidal channels between the islands are sites of ooid sand formation as tongues that extend up to 10 km into the shelf lagoon. In the Florida Reef Tract, the wide breaks in the barrier permit the reef-building community of the marginal zone to extend 2 km shelfward (Fig. 24).

Barriers provide the protection necessary for growth of various types of coral-algal reefs in lagoons. Lagoon reefs are most abundant behind a continuous barrier in Florida (Fig. 22), the Bahamas (Fig. 26), and the Great Barrier Reef (Figs. 14 and 15), but not all continuous barriers have lagoon reefs, for example the central section of Belize (Fig. 19). Shelf-margin reefs are restricted to the windward, wave-swept margins of platforms like Great Bahama Bank, and are absent or poorly developed on the leeward sides (Newell and Rigby, 1957; Ginsburg, 1964). In south Florida and the Bahama Islands marginal reefs and their associated lagoonal reefs are most abundant seaward of continuous Pleistocene islands, because these islands shield the outer shelf from the waters of the inner lagoon that are unfavorable to coral growth.

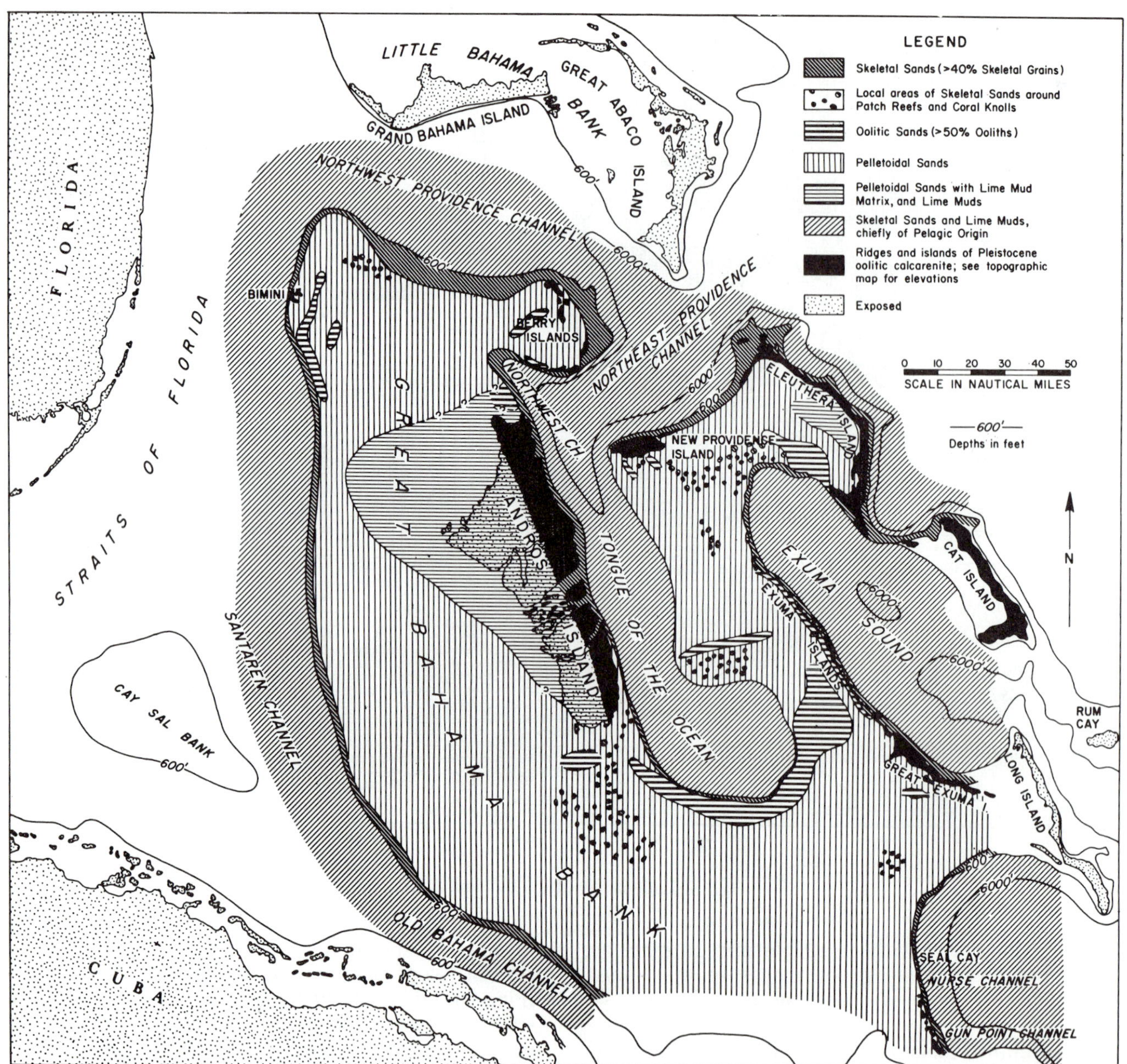

Fig. 26. Surface sediments of the Great Bahama Bank and surrounding deep sea floor. Contours in feet and scale in nautical miles. (From Traverse and Ginsburg, 1966.)

Where known, the Holocene sequence of sediments on shelves with marginal barriers is composite and consists of at least two stages of accumulation similar to the open shelves described above. The pattern of flooding and subsequent deposition, however, was determined by the pre-Holocene topography.

Where the pre-Holocene surface had pronounced relief, ca. 50 m, as in Belize, initial flooding was confined to the deeper southern lagoon (Figs. 19 and 20). Reef growth on the Pleistocene barrier platform and on top of karst remnants in the lagoon did not begin until they were flooded about 6,000 BP (Purdy, 1974a).

The shallow, flat-topped surfaces of the South Florida Shelf and Great Bahama Bank also were not flooded until about 6,000 years BP, and the initial deposits of the shelf lagoon were shallow-water lime muds and muddy sands with abundant mollusc debris (Buchanan, 1970; Enos, 1974). As the rising sea flooded the entire shelf or platform, reef growth flourished on the margin, while in the lagoon the earlier mud-rich sediments were replaced by skeletal sands, nonskeletal sands, or locally by lagoon reefs.

DISCUSSION

Comparison of Open and Rimmed Shelves. The sediments of open and rimmed shelves are distinctly different at the extremes; for example, the contrast between the parallel belts of thin lime sands on the Eastern Gulf of Mexico Shelf (Fig. 6) and the barrier-reef platform and mud-rich lagoon of Belize (Fig. 20). However, there is overlap: some examples, such as the South Florida Shelf (Fig. 22) and Great Bahama Bank (Fig. 26), have both open and rimmed segments; in others, such as the Sahul (Fig. 10) and Yucatan shelves (Fig. 7), the present-day open-shelf condition may be geologically transitory. In time the isolated, often inactive reefs and banks of the shelf margin might grow into an effective barrier.

Rimmed shelves are restricted to tropical seas because only in warm waters has the shelf-margin fixation of carbonate been able to keep pace with the rapid rise of Holocene sea level. Not all shelves or platforms in tropical seas have rims: for example, the Sahul and Yucatan shelves and parts of the Great Bahama Bank and the South Florida Shelf. One explanation is the absence of a suitable pre-Holocene elevation, the antecedent platform of earlier authors (see Purdy, 1974a, for a review). The absence of a pre-Holocene rim on the Yucatan Shelf might explain its present open-shelf condition. On the Sahul Shelf the failure of growth on the shelf-margin banks to keep pace with rising sea level may be the result of unfavorable hydrographic conditions.

Carbonates on Shelves in Temperate and Polar Seas. Chave (1967) first emphasized the abundance of carbonate on shelves in temperate and polar seas, and subsequent works have reinforced his conclusion (see also Lees and Buller, 1972). Coralline algae, as crusts and nodules (rhodolites), are abundant and widespread on shelves in high latitudes (Adey and Macintyre, 1973). Banks of the ahermatypic branched coral *Lophelia* up to 60 m high occur along much of the eastern North Atlantic shelf (Teichert, 1958). Most of these coral banks are found between 180 and 270 m; they extend to 70°N, and in addition to corals they contain worms, bryozoans, brachiopods, echinoids, and molluscs (Teichert, 1958). Most of the sediment on the wide southern shelf off Australia between 35° to 45°S is carbonate (Connolly and Von Der Borch, 1967). The principal contributor is bryozoa, 20-50%, of which an estimated two-thirds is Holocene.

Although temperatures below 15°C do not prevent the formation of extensive shelf carbonates, these low temperatures and covarying factors do restrict grain types, prevent the formation of lime mud in quantity, and possibly reduce the rate of fixation of carbonates.

Shelf Carbonates Limited by Terrigenous Mud. The input of terrigenous mud limits the occurrence of shelf carbonates because it is usually associated with large volumes of fresh water that reduce salinity, because muddy water reduces light penetration, because clay muds are an unfavorable substrate for most benthos, and because rapid deposition of terrigenous muds dilutes indigenous carbonate.

For these reasons shelves with major deltas have little or no carbonates. However, on the outer parts of these shelves, where mud deposition is reduced or absent and the water clears, there are carbonate-rich sediments. The shelf off the northeast coast of South America between the mouth of the Amazon River and Trinidad has terrigenous muds over its inner half, but on the outer half the carbonate content increases to 90% (Gibbs, 1973). Even in the northwestern Gulf of Mexico, off the Mississippi Delta, there are local banks of skeletal carbonate near the shelf margin (Curray, 1960).

Holocene and Fossil Shelf Deposits. The carpet of relict sediments that blankets many open shelves is the Holocene analogue of the basal transgressive lag of many ancient sequences. In both Holocene and fossil examples, these lags are thin; they include fossils and lithoclasts from the underlying unit, and have varied and abundant open marine faunas. The accumulations of Holocene carbonates in shallow water (<10 m) are analogues of the widespread and thick fossil shelf carbonates. The most common elements of these fossil sequences include shallow-lagoonal lime muds, banks of skeletal debris and lime mud, reefs, ooid and peloid sands, and tidal deposits. Similar deposits occur in the shelf lagoons of the Bahama Banks, South Florida, northern Belize, the southern part of the Persian Gulf, and on reefs and banks worldwide.

This broad-scale comparison shows that by combining successive stages of Holocene shelf deposition one can synthesize the familiar regressive or shoaling-upward sequence which is characteristic of the geologic history of carbonate deposition.

ACKNOWLEDGMENTS

Our research is supported by National Science Foundation grants GA-29302 and GA-25771, and by grants from the Marathon Oil Company and the Shell Oil Company. We thank S. A. Chesser for permission to use unpublished data on sediments of the eastern Gulf of Mexico, *Marine Geology* for permission to use Figure 26, and Richard Marra for drafting all the remaining illustrations.

BIBLIOGRAPHY

Adey, W. H. and Macintyre, I. G., 1973, Crustose coralline algae: a re-evaluation in the geological sciences: Geol. Soc. America Bull., v. 84, p. 883-904.

Back, R. M., 1972, Recent depositional environment of the Florida Middle Ground [M.S. thesis]: Florida State Univ., 224 p.

Ball, M. M., 1967, Carbonate sand bodies of Florida and the Bahamas: Jour. Sediment. Petrology, v. 37, p. 556-591.

Ballard, R. D., and Uchupi, E., 1970, Morphology and quaternary history of the continental shelf off the Gulf Coast of the United States: Bull. Marine Sci., v. 20, p. 547-560.

Bathurst, R. G. C., 1966, Boring algae, micrite envelopes, and lithification of molluscan biosparites: Geol. Jour., v. 5, p. 15-32.

Buchanan, H., 1970, Environmental stratigraphy of Holocene carbonate sediments near Frazers, Hog Cay, B.W.I. [Ph.D. thesis]: Columbia Univ., 229 p.

Chave, K. E., 1967, Recent carbonate sediments—an unconventional view: Jour. Geol. Educ., v. XV, p. 200-204.

Cloud, P. E., 1962, Environment of calcium carbonate deposition west of Andros Island, Bahamas: U.S. Geol. Survey Prof. Paper 350, p. 1-138.

Connolly, J. R., and Von Der Borch, C. C., 1967, Sedimentation and physiography of the sea floor south of Australia: Sediment. Geol., v. 1, p. 181-220.

Curray, J. R., 1960, Sediments and history of Holocene transgression, Continental Shelf, Northwest Gulf of Mexico, *in* Shepard, F. P., Phleger, F. B., and van Andel, T. H., eds., Recent sediments Northwest Gulf of Mexico: Am. Assoc. Petr. Geol., Tulsa, Oklah., p. 221-267.

Dietz, R. S., Holden, J. C., and Sprou, W. P., 1973, Geotectonic evolution and subsidence of Bahama Platform: Geol. Soc. America Bull., v. 81, p. 1915-1928.

Emery, K. O., 1968, Relict sediments on continental shelves of the world: Am. Assoc. Petr. Geol. Bull., v. 52, p. 445-465.

———, and Uchupi, E., 1972, Western North Atlantic Ocean, topography, rocks, structure, water, life, and sediments: Am. Assoc. Petr. Geol. Mem. 17, 504 p.

Emiliani, C., 1965, Precipituous continental slopes and considerations on the transitional crust: Science, v. 147, p. 145-148.

Enos, P., 1974, A carbonate shelf margin, the third dimension (abstr.): Geol. Soc. America, in press.

Fairbridge, R. W., 1950, Recent and Pleistocene coral reefs of Australia: Jour. Geology, v. 58, p. 330-401.

Gibbs, R. J., 1973, The bottom sediments of the Amazon shelf and tropical Atlantic Ocean: Marine Geology, v. 14, p. M39-M45.

Ginsburg, R. N., 1956, Environmental relationships of grain size and constituent particles in some South Florida carbonate sediments: Am. Assoc. Petr. Geol. Bull., v. 40, p. 2384-2427.

———, ed., 1964, South Florida carbonate sediments, guidebook, reprinted as Sedimenta II: Univ. Miami, 1972, 72 p.

Gomberg, D., 1973, Drowning of the Floridan Platform and formation of a condensed sedimentary sequence: Geol. Soc. America (abstr.): v. 5, p. 640.

Gould, H. R., and Stewart, R. H., 1956, Continental terrace sediments in the northeastern Gulf of Mexico, *in* Finding ancient shorelines: Soc. Econ. Paleontologists and Mineralogists Spec. Publ. 3, p. 2-19.

Hoffmeister, J. E., and Multer, G., 1968, Geology and origin of the Florida Keys: Geol. Soc. America Bull., v. 79, p. 1487-1502.

Illing, L. V., 1954, Bahaman calcareous sands: Am. Assoc. Petr. Geol. Bull., v. 38, p. 1-95.

James, N. P., Ginsburg, R. N., Marszalek, D. S., and Choquette, P. W., 1973, Early subsea cementation of shallow British Honduras reefs (abstr.): Am. Assoc. Petr. Geol. Bull., v. 57, p. 786.

Land, L. S., 1970, Carbonate mud: production by epibiont growth on *Thalassia testudinum*: Jour. Sediment. Petrology, v. 40, p. 1361-1363.

Lees, A., and Buller, A. T., 1972, Modern temperate-water and warm-water shelf carbonate sediments contrasted: Marine Geology, v. 13, p. M67-M73.

Logan, B. W., Harding, J. L., Ahr, W. M., Williams, J. D., and Snead, R. G., 1969, Carbonate sediments and reefs, Yucatan Shelf, Mexico: Am. Assoc. Petr. Geol. Mem. 11, p. 1-196.

Macintyre, I. G., and Milliman, J. D., 1970, Physiographic features on the outer shelf and upper slope, Atlantic Continental Margin, southeastern United States: Geol. Soc. America Bull., v. 81, p. 2577-2598.

Matthews, R. K., 1965, Genesis of lime mud in southern British Honduras: Jour. Sediment. Petrology, v. 36, p. 428-454.

Maxwell, W. G. H., 1968, Atlas of the Great Barrier Reef: New York, Elsevier, 258 p.

———, and Swinchatt, J. P., 1970, Great Barrier Reef: regional variation in a terrigenous-carbonate province: Geol. Soc. America Bull., v. 81, p. 691-724.

McMaster, R. L., Milliman, J. D., and Ashraf, A., 1971, Continental shelf and upper slope sediments off Portuguese Guinea; Guinea and Sierra Leone, West Africa: Jour. Sediment. Petrology, v. 42, p. 150-159.

Milliman, J. D., 1972, Atlantic continental shelf and slope of the United States—petrology of the sand fraction of sediments, northern New Jersey to southern Florida: U.S. Geol. Survey Prof. Paper 529-J, p. J1-J40.

———, Pilkey, O. H., and Blackwelder, B. W., 1968, Carbonate sediments on the continental shelf, Cape Hatteras to Cape Romain: Southeastern Geology, v. 9, p. 245-269.

———, Pilkey, O. H., and Ross, D. A., 1972, Sediments of the continental margin of the eastern United States: Geol. Soc. America Bull., v. 83, p. 1315-1334.

Newell, N. D., and Rigby, J. K., 1957, Geological studies on the Great Bahama Bank, *in* Leblanc, R. J., and Breding, J. C., eds., Regional aspects of carbonate deposition: Soc. Econ. Mineralogists and Paleontologists Spec. Publ. 5, p. 15-72.

Patriquin, D. G., 1972, Carbonate mud production by epibionts on *Thalassia*: an estimate based on leaf growth rate data: Jour. Sediment. Petrology, v. 42, p. 687-689.

Purdy, E. G., 1963, Recent calcium carbonate facies of the Great Bahama Bank, pt. I, Petrography and reaction groups: Jour. Geology, v. 71, p. 334-355; pt. II, Sedimentary facies: Jour. Geology, v. 71, p. 472-497.

———, 1974a, Reef configurations: Cause and effect, *in* LaPorte, L. F., Reefs in time and space: Soc. Econ. Paleontologists and Mineralogists Spc. Publ., v. 18, p. 9-77.

———, 1974b, Karst determined facies patterns in British Honduras: a Holocene carbonate sedimentation model: Am. Assoc. Petr. Geol. Bull. (in press)

Rao, M. S., 1964, Some aspects of continental shelf sediments off the east coast of India: Marine Geology, v. 1, p. 59-88.

Scholle, P. A., and Kling, S. A., 1972, Southern British Honduras: lagoonal coccolith ooze: Jour. Sediment. Petrology, v. 42, p. 195-205.

Shier, D. E., 1969, Vermetid reefs and coastal development in the Ten Thousand Islands, southwest Florida: Geol. Soc. America Bull., v. 80, p. 485-508.

Shinn, E. G., Lloyd, R. M., and Ginsburg, R. N., 1969,

Anatomy of a modern carbonate tidal flat, Andros Island, Bahamas: Jour. Sediment. Petrology, v. 39, p. 1202-1228.

Stockman, K. W., Ginsburg, R. N., and Shinn, E. A., 1967, The production of lime mud by algae in south Florida, Jour. Sediment. Petrology, v. 37, p. 633-648.

Stoddart, D. R., 1962, Three Caribbean atolls: Turneffe Islands, Lighthouse Reef, and Glover's Reef, British Honduras: Atoll Res. Bull., v. 87, 151 p.

———, 1963, Effects of Hurricane Hattie on the British Honduras Reefs and Cays, October 30-31, 1961: Atoll Res. Bull., v. 95, 142 p.

———, 1969, Post-hurricane changes on the British Honduras reefs and cays: resurvey of 1965: Atoll Res. Bull., v. 131, 34 p.

Swift, D. J. P., Stanley, D. J., and Curray, J. R., 1971, Relict sediments, a reconsideration: Jour. Geology, v. 79, p. 322-346.

Swinchatt, J. P., 1965, Significance of constituent composition, texture, and skeletal breakdown in some Recent carbonate sediments: Jour. Sediment. Petrology, v. 35, p. 71-90.

Taft, W. H., Arrington, F., Haimovitz, A., MacDonald, C., and Woolheater, C., 1968, Lithification of modern carbonate sediments at Yellow Bank, Bahamas: Bull. Marine Sci., v. 18, p. 762-828.

Teichert, C., 1958, Cold- and deep-water coral banks: Am. Assoc. Petr. Geol. Bull., v. 42, p. 1064-1082.

———, and Fairbridge, R. W., 1948, Some coral reefs of the Sahul shelf: Geograph. Rev., v. 38, p. 222-249.

Traverse, A., and Ginsburg, R. N., 1966, Palynology of the surface sediments of Great Bahama Bank, as related to water movement and sedimentation: Marine Geology, v. 4, p. 417-459.

van Andel, T. H., and Veevers, J. J., 1967, Morphology and sediments of the Timor Sea: Bur. Min. Resources, Geol. and Geophys. Bull., v. 83, 173 p.

Wantland, K. F., and Pusey, W. C., 1971, The southern shelf of British Honduras, guidebook: New Orleans Geol. Soc., 87 p.

Recent Sediments and Environment of Southern Brazil, Uruguay, Buenos Aires, and Rio Negro Continental Shelf

Carlos Maria Urien and Maurice Ewing

INTRODUCTION

The southern Brazil continental shelf is a narrow and gently sloping sediment-covered submerged coastal plain, with a width of 62 n.m. Southward it increases in width to 170 n.m. and steepens. The shelf break ranges from 35 fm in the north to 50 fm in the south. In the middle area the shelf-slope transition is steep, related to overlapping relict deltas.

The shelf topography is closely controlled by the morphostructural onshore units, mainly Precambrian and Cretaceous-Tertiary sedimentary basins, which extend as nuclei of the continental terrace.

Relict beach ridges, a barrier coast, and river-distributary channels are found, particularly in the Rio de la Plata and Rio Colorado-Negro areas.

Most of the shelf sediments are sandy (products of a succession of Holocene transgressive shorelines), forming a blanket more than 15 ft thick. Silty clay sediments are restricted to bays and estuaries or inner shelf channels close to the La Plata River. Shelf lutites are found only on the Rio Grande shelf break as relict of river drainage from the nearby Precambrian plateau. Several sea-level fluctuations which have modified sediment facies and bottom topography are believed to have occurred in the area before and after the Holocene transgressions, about 18,000 years bp.

On the shelf break, progradation layers were detected. These are related to the major rivers which drained there during low-sea-level stands, making estuaries and deltas.

One of the most remarkable features of the eastern South America continental margin is the gradual widening of the continental shelf south of 32°S (Fig. 1). The structural units that control the physiography and continental margin morphology are the Brazilian shield, the Buenos Aires hills, the Patagonian massif; and post-Jurassic negative units such as Pelotas basin and the Salado-Colorado basins (Urien, 1970). The uppermost expression of this basin group is present in the lowland and coastal plain complexes such as the Rio Grande-Rio de la Plata and the Bahia Blanca-Colorado coastal plains.

The coastal plain and continental shelf in this area have subsided slowly. During Late Tertiary times they were subject to alternate periods of transgression and regression, subsequently accumulating a considerable amount of sediment (Urien, 1972). About 18,000 years ago, a worldwide transgression from approximately the shelf border (-140 m) up to the present sea level took place. At that time the shoreline transgressed the coastal plain and was stabilized, after some fluctuations, approximately in its present position.

This article considers the events recorded in the morphologic features and sediments from the time of the worldwide transgression to the present, and is a synthesis of work started in 1960 and partially as reported by: Butler (1970); Delaney (1963, 1966); Fray and Ewing (1963); Martins and Urien (1967); Ottman and Urien (1966); Pierse et al. (1969); Remiro and Etchichury (1960); Schnack (1969); Urien (1967); Urien and Mouzo (1968); Urien and Ottman (1971); Urien and Zambrano (1972); Zambrano and Urien (1970).

GENERAL SETTING

The region is bordered by serveral morphostructural units, the most important of which is the Brazilian shield, which outcrops along the Brazilian and Uruguayan coast and, in some areas, extends offshore, forming the nucleus of the continental margin (Urien and Zambrano, 1972). South of Rio de la Plata, the Pampa plain is interrupted abruptly by a cliff, but south of Rio Colorado (Fig. 2) it has a transitional passage to the Patagonian plateau, which extends steeply from the cordillera toward the sea.

The Brazilian shield is interrupted by two basins, Pelotas and Salado. South of the Tandil-Ventana hills, the Colorado basin extends into the continental margin. The surface expressions of these sedimentary basins are partly seen in the Pleistocene-Holocene coastal plain as the Rio Grande do Sul, Rio de la Plata, and Bahia Blanca-Rio Colorado coastal plains (Figs. 1 and 2).

BOTTOM TOPOGRAPHY

Urien (1970) divided the continental shelf of southern Brazil, Uruguay, and Argentina into several zones, which have a close relationship to onshore morphostructural units. Three of these zones are observed in the region discussed in this paper: Rio Grande do Sul-Uruguay, Buenos Aires, and the upper part of Patagonia. Figure 3 shows the rapid change of gradient that occurs at the edge of the continental shelf at water depths between 80 and 130 m. The shelf is medium to wide, with a low rate of sediment supply (Curray, 1966). Most of the sediments and river mud are trapped in the

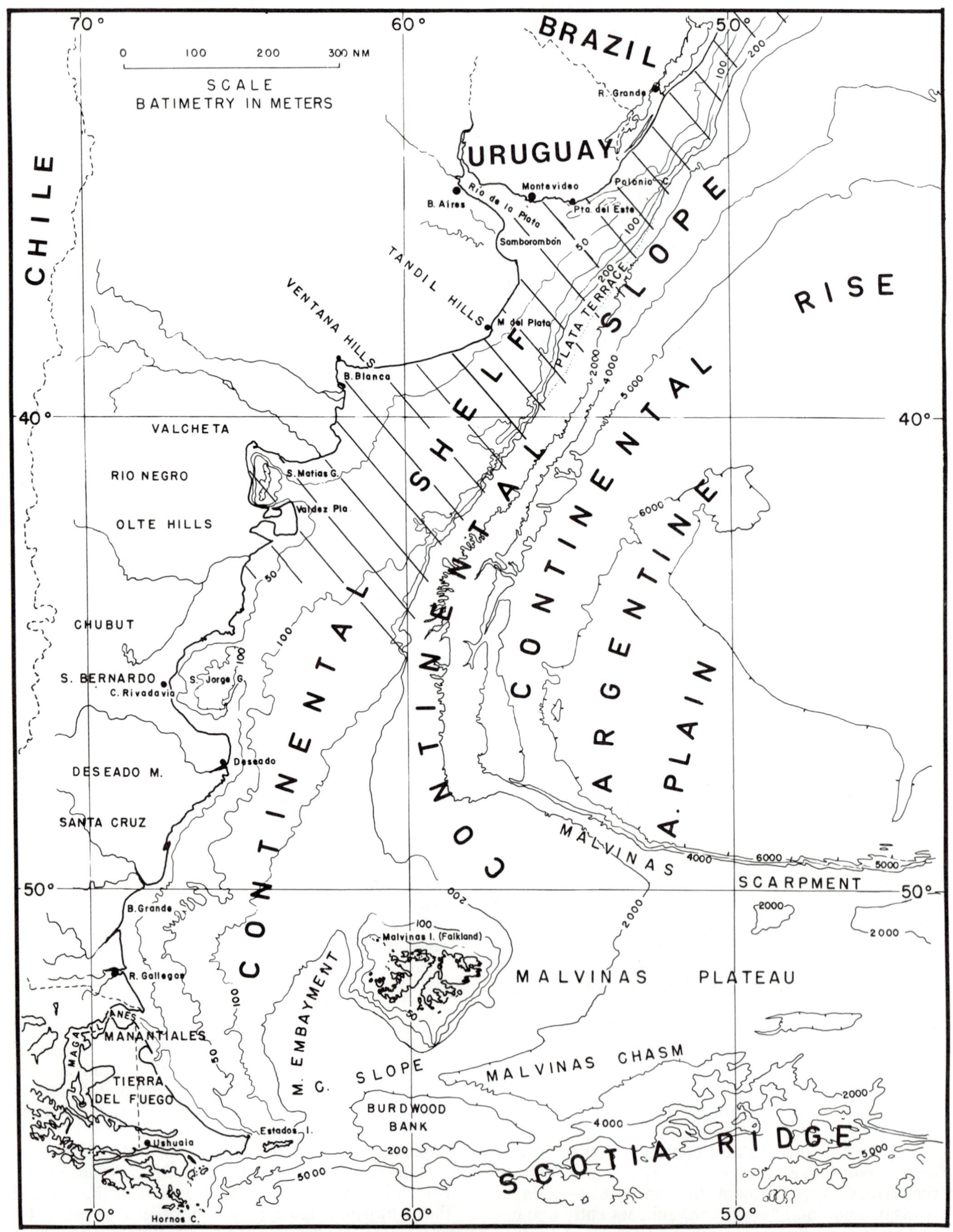

Fig. 1. Generalized topography and main physiographic units of the southeastern South American continental margin. Diagonal lines indicate the area studied (after Lonardi and Ewing, 1971).

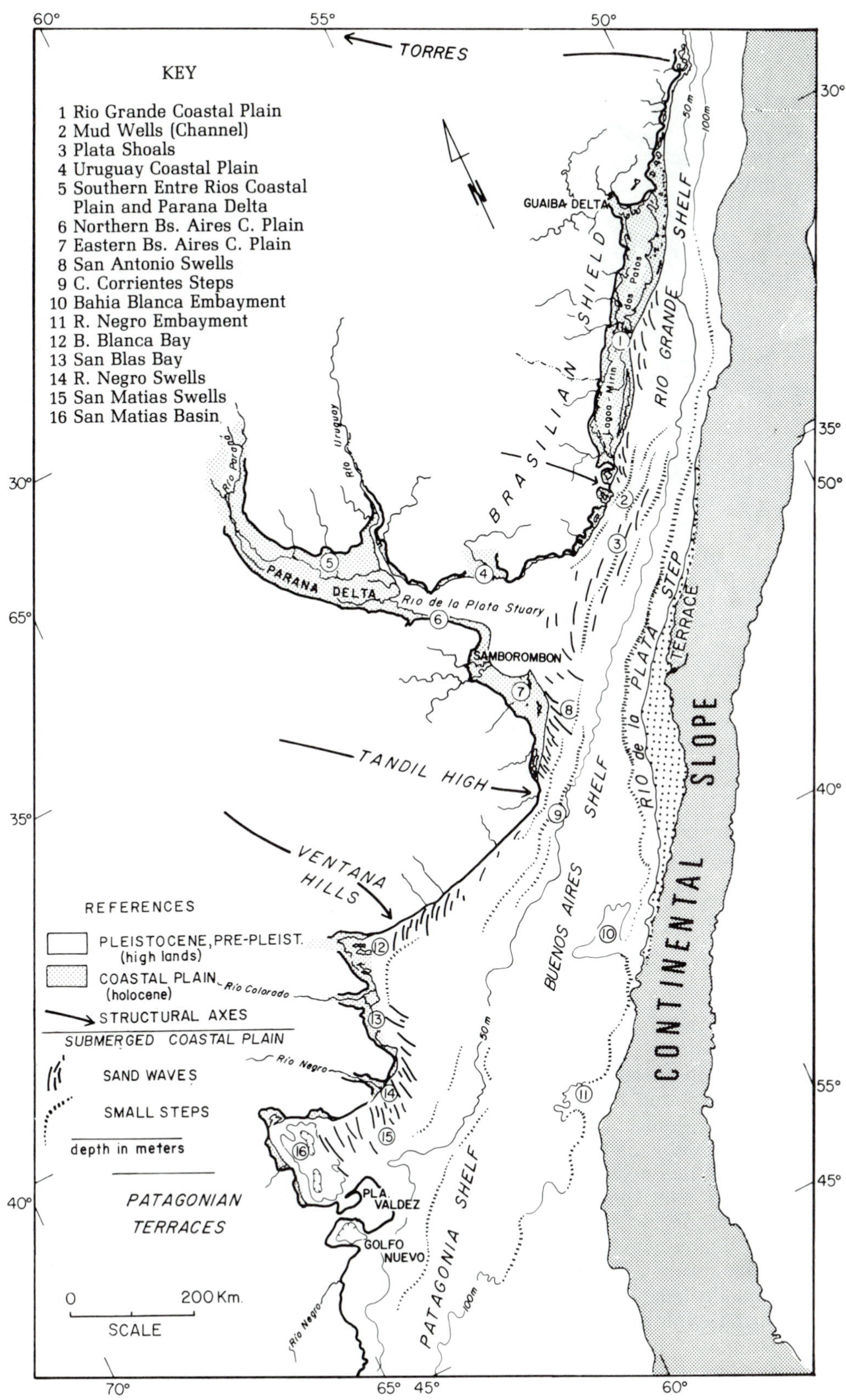

Fig. 2. Main physiographic and topographic features of the continental shelf and the salient morphostructural onshore units.

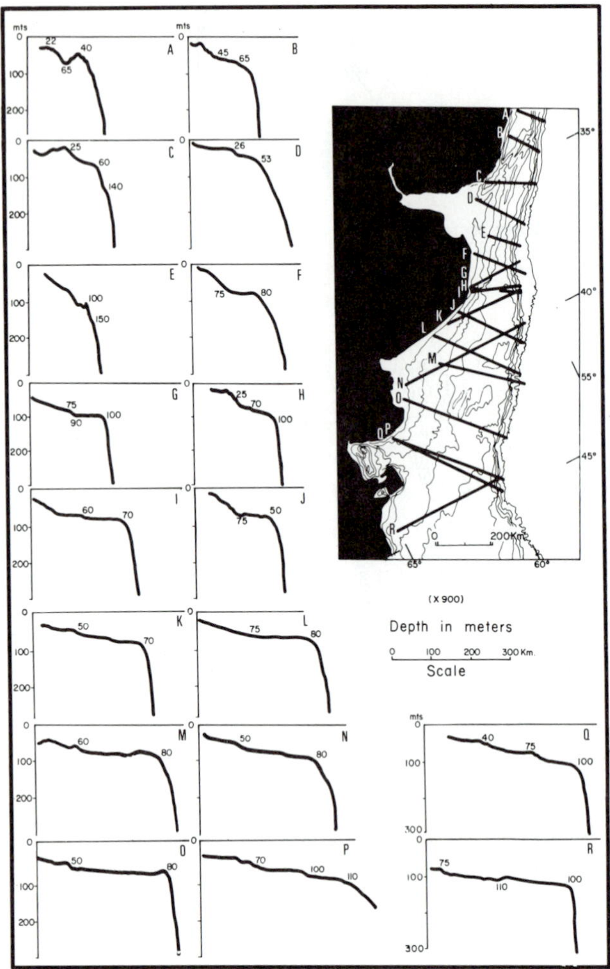

Fig. 3. Bottom profiles from PDR records, showing the shelf break. Vertical exaggeration x900. (From Urien & Zambrano, 1973, in "The Ocean Basins and their Margins", A.E.M. Nairn and F.G. Stehli, eds., p. 138, copyright by Plenum Press, N.Y.).

estuaries or partially bypass the shelf in suspension, and are deposited on the deeper continental slope.

The major depositional features are in the Rio de la Plata area, possibly related to an old delta system that had reached the shelf border, creating a series of remarkable steps between 60 and 620 m. The seismic reflection lines made by the *Zapiola-Vema/Zapiola* campaign in 1962 show a series of prograding layers in the modern shelf sediments. Lonardi and Ewing (1971) show similar features in the Rio de la Plata shelf border. These patterns suggest a very modern delta front, built up during the last sea-level lowering, which overlies several overlapping Quaternary and perhaps also late Tertiary deltas.

In meridional Brazil and Uruguay the width of the continental shelf is approximately 120 km (62 n.m.) and the depth of the shelf edge is 160 m. In general the surface is smooth; and only in the south are bottom furrows observed. In the Buenos Aires area, the width of the shelf increases to 200 and 300 km, with an average depth of 60 m. Channels, ridges, shoals, steps, and sandy waves are observed. The shelf break is at 100 m depth.

South of Rio Colorado (northern Patagonia) the width of the continental shelf reaches almost its maximum, 380 km (170 n.m.). The shelf gradient is 60 m in 200 km, with the 50-m, 100-m, and 200-m isobaths being approximately equidistant. South of 45°S, the 50-m and 100-m isobaths are deflected westward close to the shore. Here the shelf's dip increases sharply, perhaps as a seaward expression of the Patagonian terraces, but is smoother owing to wave action and sediment burial. Urien and Zambrano (1972) suggested a close relationship between the Precambrian to pre-Jurassic offshore basement's extension into the continental terrace and the extreme width of the continental shelf; they concluded that sedimentary accretion was not the only factor responsible for the great width of the shelf here.

The topography of the shelf border and upper slope is very irregular, with canyons, steps, terraces, and embayments, particularly between the depths of 100 and 450 m (Figs. 2 and 4).

Three principal topographic features are present on the continental shelf: ridges, steps, and channels (Fig. 2). These may be classified as depositional or erosional. The depositional features are ridges parallel to the shelf strike, marking sea-level stands. The most prominent of these are close to the Rio de la Plata mouth, forming a "barrier complex" which possibly closed the estuary in the recent past (Fig. 5). Between the complex "Plata Shoal" and the shore, a channel extends from the Rio de la Plata into the inner shelf (Butler, 1970; Urien and Mouzo, 1968; Urien, 1970). A *U.S. Coast and Geodetic Survey Oceanographer* reflection line at the latitude of Cabo Polonio (Fig. 6) shows a series of imbricated buried channels, which shift westward, perhaps as a result of sea-level rising.

South of the Plata Shoal, close to the shore, there are a series of sand waves or submarine dunes similar to those seen in desert areas, with heights of from 5 to 12 m (Fig. 7), some of which are oblique to the shore and more or less parallel to the tidal currents. These forms are also found in southern Buenos Aires, close to Bahia Blanca and San Blas, and are superimposed in front of the entrance to San Matias Gulf (Fig. 8), where they are 4 to 17 m high (Pierse et al., 1969). These waves may be active, but only in shallow areas with high tidal currents (100 to 206 cm/sec) which mobilize the sand bottom as in the San Matias and San Blas shallow areas.

The erosional features are small steps, 3-10 m high, most of them buried by sediments, following the topographic contours (Fig. 7). Some dredge hauls collected on these steps contained calcareous debris and beach rock fragments. One core (v16-158) taken off Bahia Blanca, in a water depth of 17 m, reached, at 2-8 m in penetration, "bedrock" formed of the same material outcropping on the shore. This

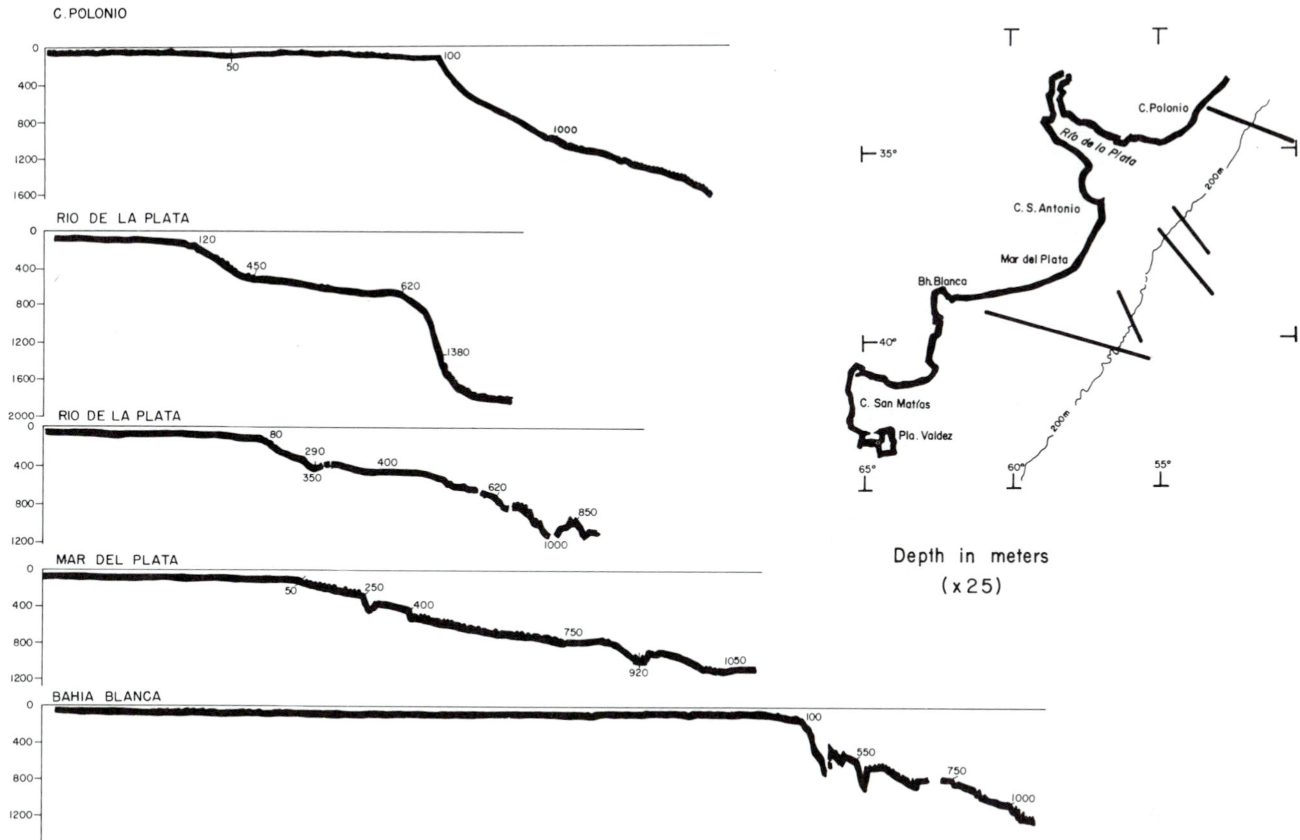

Fig. 4. PDR bottom profiles off Uruguay and Buenos Aires province (x25). The Rio de la Plata section shows the 400-600 terrace.

could be the extension of a wide, wave-cut abrasion platform in the northern part of Bahia Blanca bay. The most spectacular step occurs in the Rio de la Plata area at a water depth of 80 m. It is related to a delta front, probably built up during sea-level lowering.

At the boundary of the Buenos Aires-northern Patagonia outer continental shelf, the bathymetry shows a series of embayments, steps, and small terraces. These are also related to the deltaic apron built during a low sea-level stand. The shelf surface contours show inflections which may be buried channels (regressive estuaries) and which could be related to the eastward extension of the Rio Colorado and Rio Negro coastal plains during the maximum sea regression. Seismic reflection records on the shelf break show a considerable thickness of sediment structured in deltaic upbuilding.

A topographic low of more than 130 m depth is seen in the San Matias Gulf. It is a closed basin, separated from the neighboring inner shelf by a 37-m-deep sill whose seaward slope is covered by sand on which the sand wave systems appear (Pierse et al., 1969). In this area also, steeply wave-cut platforms were observed, in some cases partially buried by modern sediments.

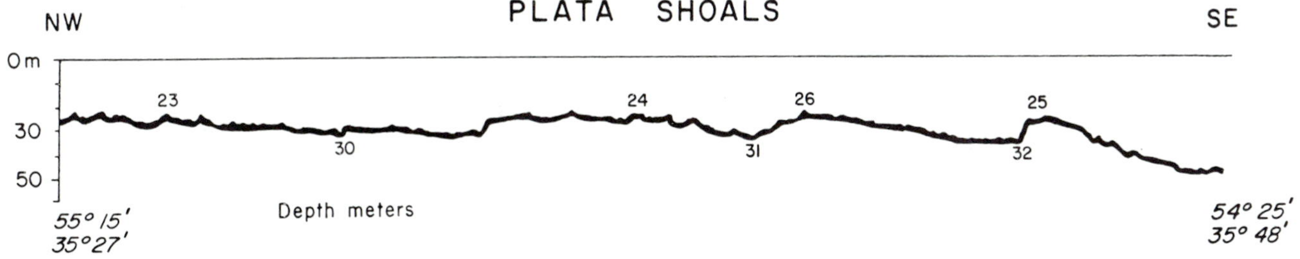

Fig. 5. Topographic section on the inner Rio de la Plata shelf which cut the "barrier island complex." the 5- to 7-m-high crests form scarps on the landward side, where dredge samples show the bottom to be shelly and rich in beach-rock debris.

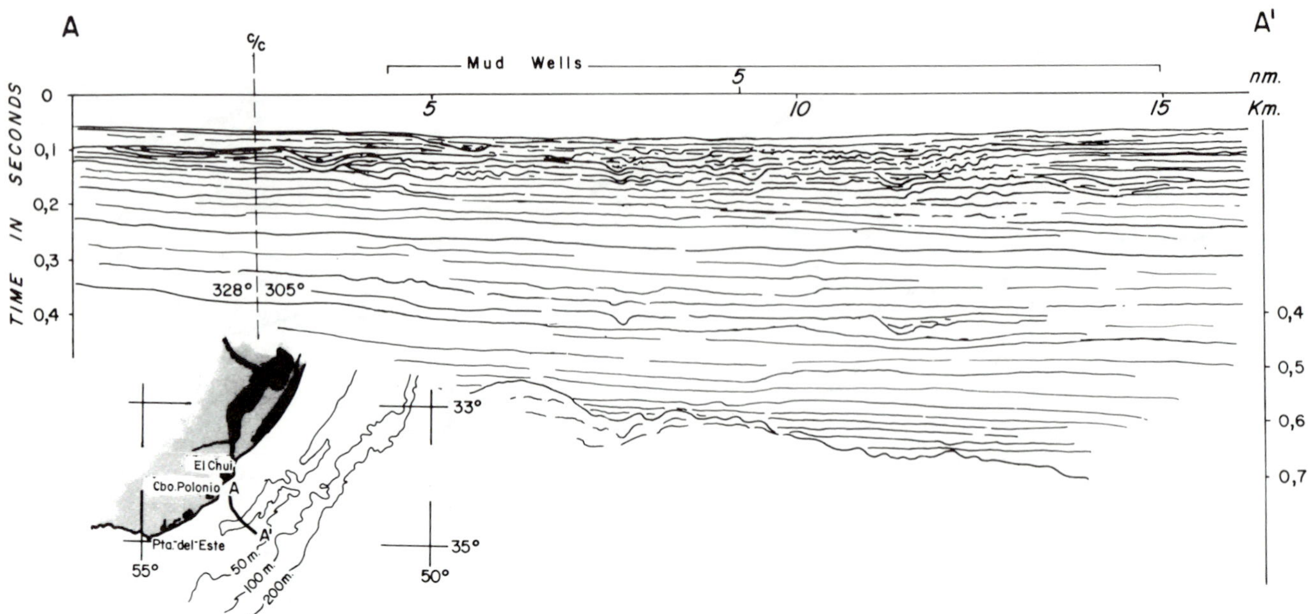

Fig. 6. Seismic reflection profile off Uruguay, crossing the "mud wells" area. This section was made by U.S.C.G.S. Oceanographer. The reflectors show a series of overlapping and buried channels. They indicate a rough bottom and a westward migration up to km 2.5. Reflection time is about ½s for 175-m water depth. Here the basement (possibly Precambrian), which appears at -150 m, is the offshore extension of the Cabo Polonio high, with a gentle eastward slope.

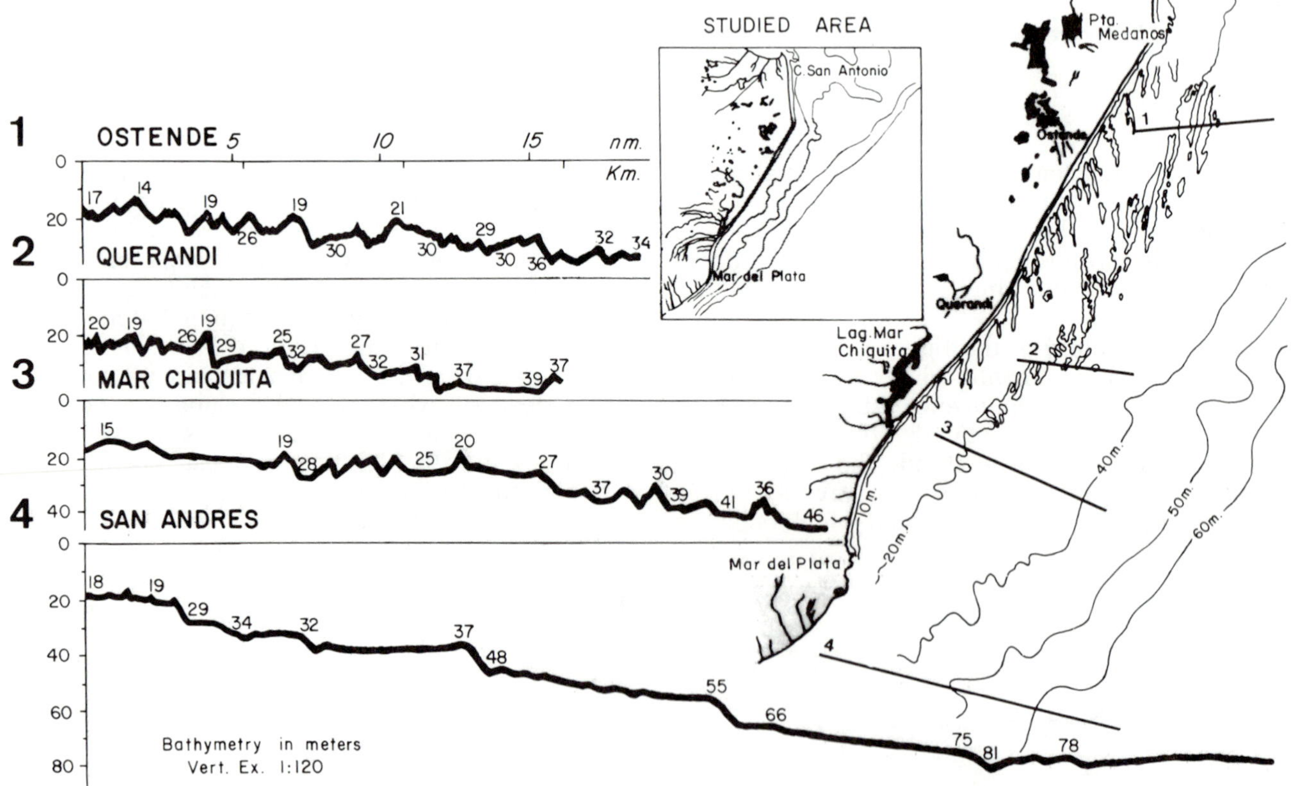

Fig. 7. Sections off Buenos Aires coastal plain. They partially show the San Antonio swells (sand waves). The San Andres section shows some of the Cabo Corrientes erosional steps, which are related to calcareous plates, like abrasion platforms. The profiles of the sand waves are irregular, probably as a result of erosion.

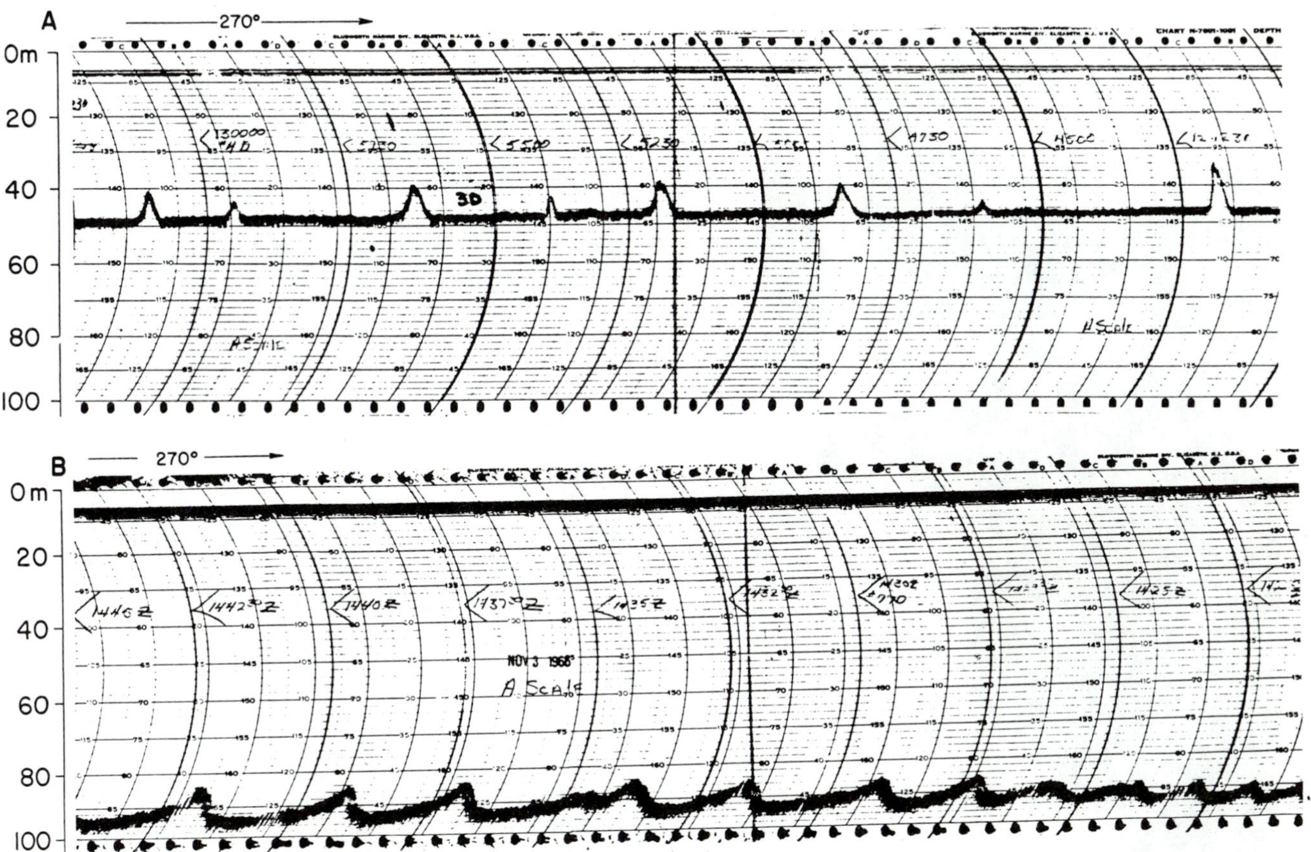

Fig. 8. Echograms showing some of the San Matias sand waves. These sections are turned 180°. (a) Symmetric and evenly spaced waves, becoming isolated waves. (b) Train of asymmetric waves, suggesting a westward bottom translation.

BOTTOM SEDIMENTS

Some 600 grab samples and 60 piston cores were collected in the area investigated, and a surface sediment map was compiled (Fig. 9). Size distribution was based on the triangle of sand, silt, and clay percentages. Figure 9 shows only the most representative textures. The material coarser than 2 mm (gravel and/or shells) is included in the texture map as a fourth component (shelly) of the Shepard triangle. Figure 10a shows the percentage of sand in the samples.

A brownish-gray, medium, well-sorted, quartzitic sand is predominant in the surficial bottom sediments. It is characterized by its homogeneous size, apparently due to the material being from a single sediment source and subsequently reworked by waves and currents along the successive landward migration of the shoreline.

The most important textural change in the Recent sediments occurs near the bays or estuaries, where the sediment becomes mainly silty (with mean diameters of 3 ø and 4 ø). Coarser bioclastic concentrations, containing shelly material and having mean diameters from 2ø up to greater than 1 o, appear in an elongated region trending northeast (Fig. 10b). Near the Rio Colorado-Negro area, as well as in the San Matias Gulf, the "shelly fraction" becomes coarser, rich in very well rounded igneous rocks, granules, and pebbles. These appear only in areas of high energy (waves and currents), clean of fine fractions or buried by sand. Presumably these spread gravels were the product of the sea's erosion of the Patagonian Cenozoic deposits and/or they were transported by rivers from the Patagonian terrace, which are rich in these sediments.

On the basis of the surficial bottom sediments it was possible to make five geographic divisions of the shelf from north to south as follows:

1. *Rio Grande do Sul area.* With greenish-gray silty clay sediments on the outer shelf, probably relicts of drainage during low sea level; and yellow sandy facies on the inner shelf and nearshore area (Martins and Urien, 1967).

2. *Uruguay and Rio de la Plata area.* On the inner Uruguayan shelf parallel to the coast, there appears a muddy area that covers a channel which extends from the Rio de la Plata. The greenish-tan clayey silt and sandy clay (with 20-50% sand) are products of the suspended solids in the estuary's runoff, most of which are deposited in this area. The nearshore area is rocky and sandy (pocket beaches). This yellowish sand is local and a product of the erosion of the Quaternary sand deposits that maintain the littoral sand budget. In the outer Rio de la Plata estuary a series of muddy to sandy mud arches are transformed seaward into a well-sorted medium to fine sand which is the extension of the

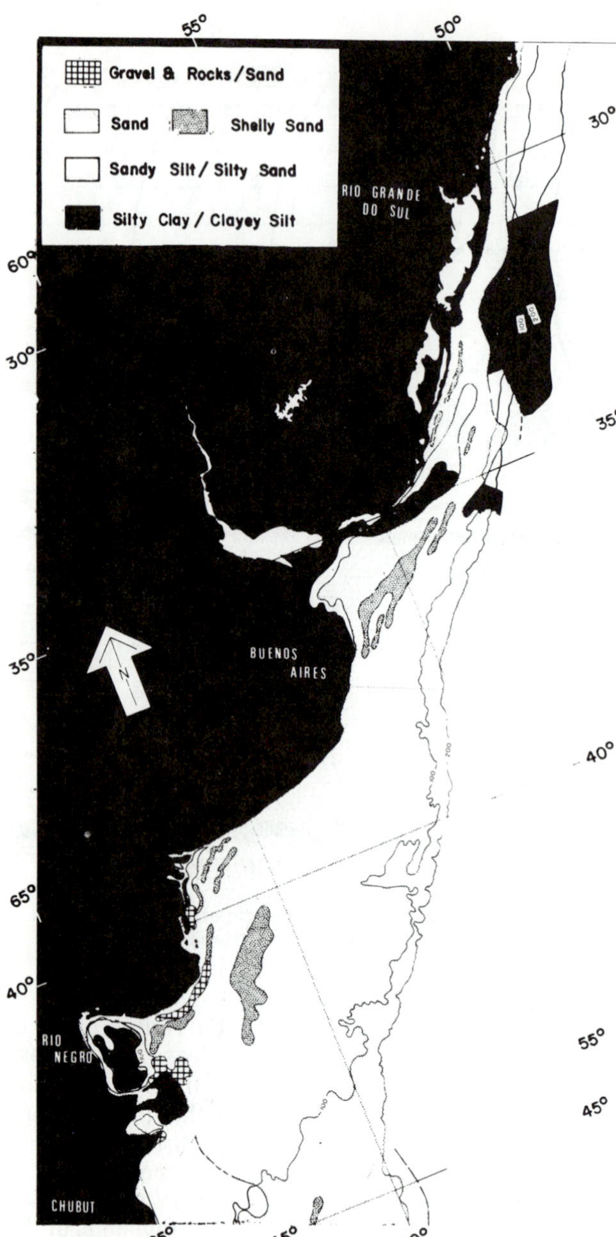

Fig. 9. Surficial bottom sediments. Generalized gross lithology based on sand-silt-clay percentages. Shelly sand and sandy gravel and/or pebbles are plotted independently of the Shepard triangle. This is the material coarser than 2 and 4 mm, respectively. The sand field is interpreted as a nearshore (relict or modern) sedimentary facies.

shelf sandy facies into the estuary (Ottman and Urien, 1966; Urien, 1967). This sand, designated "old sand," is the onlap facies, which contrasts with the silty fluvial deposits identified as offlap facies in the estuary evolution (Urien, 1972). The fluvial suspended solids are deposited within the Rio de la Plata estuary and on the inner Uruguayan shelf up to near Rio Grande do Sul. On the middle shelf there appears a concentration of shells, fragments, and calcareous debris (lime and cemented sandstone pebbles 10 mm in diameter) mixed with sand and spread in an elongated patch identified as the relict of an old shoreline, perhaps a barrier island in its destructive facies.

3. *Buenos Aires area.* Here, as on almost all the rest of the southern shelf, there is a blanket sand field which covers both inner and outer shelves. This sediment is a medium to fine well-sorted yellowish-gray to brownish-gray sand which in some sectors contains small shell fragments and/or calcareous debris. No muddy sediments were found here. Shelly textures are spread in nearshore elongated patches.

4. *Bahia Blanca-San Blas area.* The river's suspended solids are trapped within the bay and nearshore areas, making these bays highly lutitic with a nearshore olive silty sand belt. Both inner and outer Bahia Blanca, arcuate shelly bodies appear (like ridges), following the bathymetry. In the southern middle shelf there is an extended shelly blanket that is perhaps related to a relict barrier island similar to the one near Rio de la Plata.

5. *San Matias Gulf area.* This is a wide and deep embayment with a sandy and gravelly sill which separates the sandy shelf facies from the tannish-green clay that covers the main gulf depression. The gulf nearshore belt is silty sand and gravelly. Since no rivers supply sediments to this area, it is supposed that the mud must be coming in suspension from the Rio Negro and carried by tidal currents or, alternatively, it is partially a relict of a former drainage pattern within the gulf. Sieguel (personal communication) has dated Holocene shells beneath 2 m of homogeneous clay, proving the relatively young age of these sediments.

Thus the continental shelf in the area studied has predominantly a Recent sand cover of approximately 1-3 m thickness. The fine textures are related to the runoff (recent or ancient) of rivers on the continental shelf, the most important river being the Rio de la Plata, with a yearly runoff of 23,000 m3/sec (Rio Colorado, 180 m3/sec; Rio Negro, 950 m3/sec).

Since at the present time the fluvial deposits are concentrated in the main estuaries and bays or near the shore, most of the sandy shelf facies are free of fine sediments. Nevertheless, in the Rio de la Plata-Uruguay area, suspended sediments bypass the shelf and are probably deposited on the continental slope. In its passage, some material settled, covering the bottom, but very sparsely, with a fine dash of clay, as was observed in bottom photographs and samples.

SAND AND SHELLY CONSTITUENTS

The shelly areas are characterized by an abundance of bioclastic particles, principally mollusks, broken or well preserved, but with evidence of abrasion. They are mixed with sand in sandy areas of the shelf and are considered to be relicts of shallow-water areas (sublittoral) or shoreline deposits. Richards and Craig (1963) reported pelecypods, gastropods, barnacles, brachiopods, and fish bones

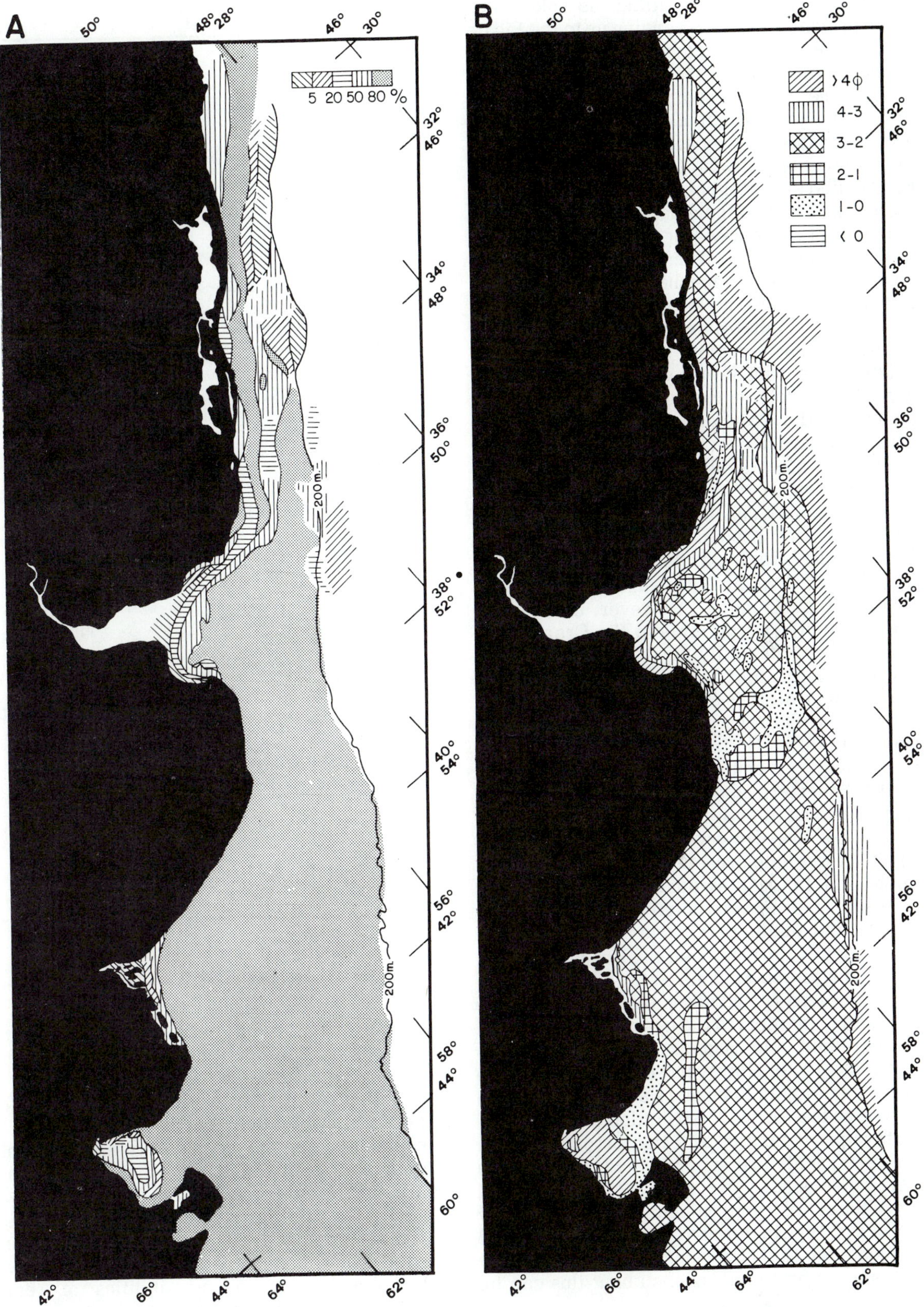

Fig. 10A. Sand percentage map of the whole sample. The sand is the chief component of the bottom sediments throughout the area, with the exception of the river mouths, bays, or estuaries, and the northern area, which is influenced by the Rio de la Plata or Rio Grande do Sul shelf facies. B. Mean diameters of the surficial bottom sediments. The areas in which mean diameters are higher than 1 ø are those in which the shelly content is high. These figures have a close relationship with the subbottom sediment properties.

in Lamont-Doherty cores, identified as cold-water (Magellan province) species, ranging up to the Rio de la Plata. Wave action spreads these particles and mixes neighboring sand with them, thus increasing the mean diameter of the sediments (up to 2 ø and 1 ø). Echinoid fragments (spines) and "sand dollars" are present, especially in areas with low or zero sedimentation rates (Urien, 1967).

Foraminiferal content of the sediment is moderate to low. The foraminiferal assemblage contains a large number of shallow-water species (Boltovskoy, personal communication) from littoral and sublittoral areas at the time of deposition (Fray and Ewing, 1963). Boltovskoy (1973) notes two kinds of foraminiferal assemblages, corresponding to coastal waters and to the outer neritic region. This difference is very clear in the cores from depths of 80 m or greater. In the topmost centimeter of the cores, the foraminifera are typical of the "Malvinas waters." Later the fauna became more like those in the nearshore area. This suggests a much shallower depth there at the time of deposition.

The terrigenous constituent of the sandy shelf sediments consists predominately of basic rock elements (plagioclase, hyperstene, augite, and rock fragments) described by Teruggi et al. (1959) as "Pampean-Patagonia suite." However, in the north of the area studied (Uruguay and Rio Grande) the minerals are predominantly derived from acid rock (quartz, epidote, garnet, and mica). Around 35°S latitude the mineralogy is mixed (Urien, 1967, 1972). In the Rio de la Plata estuary the two mineralogic zones present are (1) "northern shore" and (2) "southern river" assemblages (Urien, 1972).

In the "northern shore" assemblage, quartz, and epidote are as abundant as in the acid rocks of the Brazilian shore. However, Remiro and Etchechury (1960) report the presence of basic minerals in the beach sand. In Uruguay basic effusions crop out which would add other minerals (hypersthene, hornblende, and augite) to the acid mineralogic unit, which is evidently derived from the Uruguayan acid basement (gneiss and migmatites).

The "southern river" assemblage is rich in the Pampean suite, identical to the assemblage observed in the near-shelf and Buenos Aires shore areas (Teruggi et al., 1959). However, south of 35° latitude the quartz starts again to be more abundant than the plagioclase. In the vertical distribution a well-rounded quartz is predominant, and the ferromagnesium and lithic fragments are subordinated. Close to the Uruguay inner shelf, the sandy layers present different types of mineralogic suits, probably because of a change in the current patterns and sand provenance. But the mineralogic study is not yet finished, and no firm conclusions about this can be reached.

The distribution of these patterns is along the shore and is attributed to the longshore currents, where the sand is subject to transportation. This conclusion is for the sand only, the silt and clay being distributed independently. This is a reworking

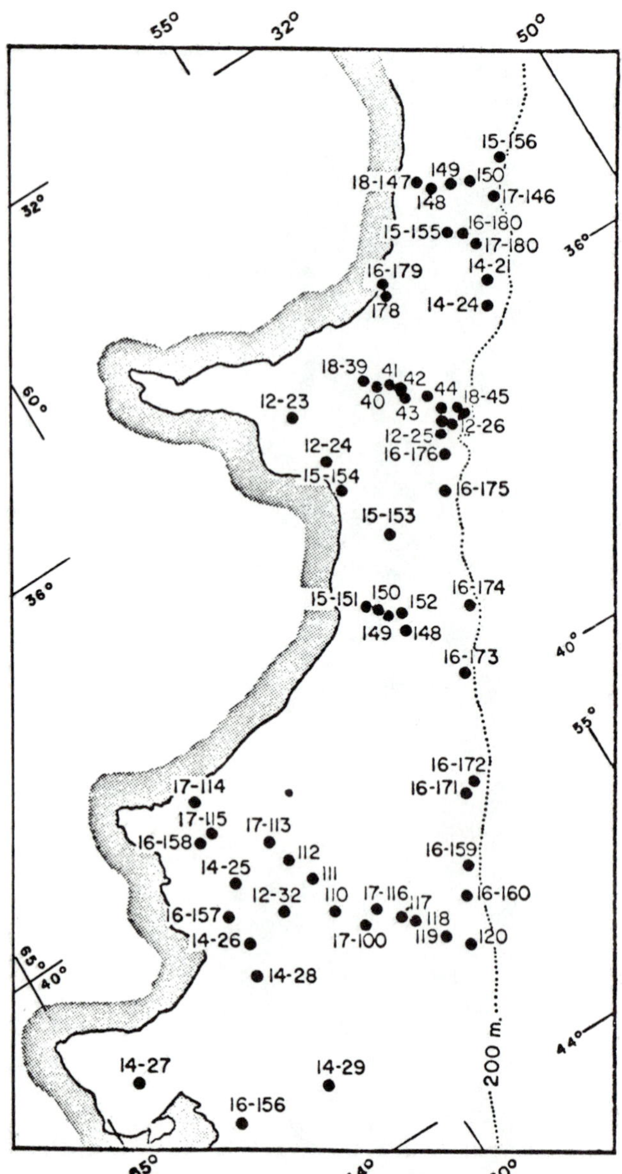

Fig. 11. Location of cores and bottom photographs.

process on the sand, because apparently at present there is no significant supply from the rivers along the present shoreline.

Neglecting the muddy matrix and following Folk's classification, north to Rio de la Plata the sediments can be considered as areas of orthoquartzite, south of it as subarkose. In the areas affected by muddy sediments (e.g., Uruguayan inner shelf), the sand turns into graywackes and subgraywackes, respectively. The mineral composition of sand would usually reflect the source provenance, but considering the possibilities of renewal, a significant portion of the minerals can be weathered and mixed in subaerial conditions, modifying the character of each mineralogic assemblage and creating a new combination suit.

VERTICAL SEDIMENTS CHARACTERISTICS

From 1957 to 1960 the Lamont-Doherty Geological Observatory collected numerous piston cores on the Argentine continental shelf. A description of them was reported by Fray and Ewing (1963). Later Schnack (1969) made a careful granulometric-mineralogic study of 17 cores near Rio de la Plata. In the present paper 60 cores are described (Fig. 11) each one studied in detail; their most remarkable aspects are synthesized in Table 1. Figure 2 shows the surface sediment facies that are related to the vertical distribution. Most of the bottom photographs show a moderately smooth bottom in the shallow areas, although some ripple marks appear (Table 2).

MINOR SEDIMENTARY STRUCTURES

Four main minor structures are identified: homogeneous (structureless deposits), mottled, conglomerated, and laminated. *Homogeneous* structures are found in many cores formed either by sand or mud. *Mottled* structures are also found in both sandy and muddy sediments, some of them due to burrowing organisms or plant roots. *Conglomerated* structures are characteristic of high biogeneous concentration, shell fragments, and calcareous pebbles. In areas with fluvial influence, the matrix is formed of clay. *Laminated* structures are formed by parallel laminations of sand and silty clay, alternating with thick individual layers of silty clay. The sand layers are from 1 to 3 mm thick. In some places shell hash alterations appear. Muddy beds sometimes show laminations, which result from the concentration of clay in very thin layers.

Figures 13 to 15 show some cores with some of the most distinctive minor sedimentary structures and sediment properties.

SEDIMENT FACIES

The continental shelf is subdivided according to the Holocene sedimentary facies which overlie the pre-Holocene regressive series. These facies are analyzed from the piston cores, assuming they penetrate the subbottom up to or through the basal Holocene discordance. The chance of sampling the late-Wisconsin regressive sediments is small because they are buried under the cover of transgressive deposits. Also, erosion subsequent to deposition could have removed part of the pre-Holocene regressive sequence.

1. *Facies predominantly sandy.* This is a homogenous brownish-gray to grayish-brown, fine to medium sand (2.5 ø). It is well sorted, with subrounded grains of quartz and plagioclase. Basic rock fragments, heavy minerals (ferromagnesium), and opaque and greenish lithic grains are concentrated in the finer sand sizes. This assemblage is common in cores obtained deeper than 100 m (V16-159 to 175, Fig. 11). Shell hash and fragments with sizes from 2 to 15 mm appear. Toward the core bottom they form layers which also include large pelecypod valves. Examples of such cores are those obtained between 20 and 100 m depth on the southern Buenos Aires shelf (V17-100 to 120, Fig. 13a, and V16-156/158).

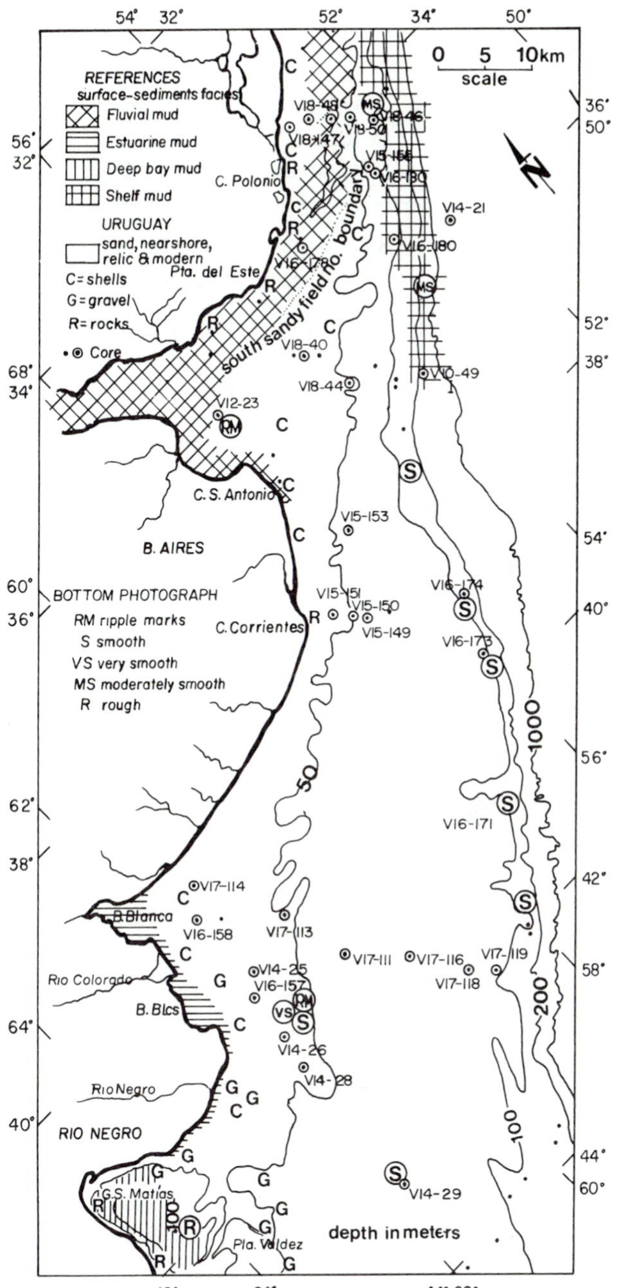

Fig. 12. Generalized bottom sedimentary facies map, based on the gross lithology. Bottom photographs are related to the bottom dynamics. Smooth: nondepositional areas. Ripple marks: bottom with relative mobilization. Rough: gravelly bottom under high currents.

Owing to the presence of some clay, the sand turns dark gray (dirty) with scattered shell flakes (clay ranges from 5 to 20%). This sand-mud texture generally appears to be structureless (homogeneous), and when the mud increases, it becomes

Table 1. Summarized characteristics of the depositional environments.

Environment	Sediments	Shells	Minor Structures	Laminae	Burrows	Remarks
Coastal Plain						
Tidal flats and marshes	Mud/shells and sand	Frequent to very frequent	Mottled and laminated	Common	Frequent	Fine sand laminae and organic clay
Estuaries	Mud/silty sand	Common	Homogeneous and laminated	Common	Common	Silty clay and organic clay with stratifications
Lagoons	Mud-sand	Common	Homogeneous and laminated	Frequent	Common	Sand Laminae in mud
Marine						
Holomarine						
Barrier face	Sandy shale/ shelly conglomerate	Very frequent	Shelly conglomerate Sand homogeneous	Very scarce	Scarce	Shelly gravel in irregular bedding and lenses of sand
Inner shelf	Sandy shale	Very frequent	Homogeneous to conglomerate	Very scarce	Common	Disturbed by burrows
Middle shelf	Sandy to muddy	Common	Homogeneous	Absent or scarce	Scarce	Disturbed by burrows
Fluviomarine						
Proximal	Mud and silty mud	Frequent	Homogeneous, mottled, some laminae	Scarce	Common	
Distal	Mud and sandy silt	Common to scarce	Mottled laminae	Common	Very scarce	Thin sand laminae interbedded with mud

Table 2. Bottom photographs.

Camera Station	South Latitude	West Longitude	Depth (m)	Frame (n)	Bottom Aspects: Sediments	Bottom Aspects: Texture	Description
V14-10	41°09'	61°36'	22-40	1	Sand	Smooth	Pits/mounds in smooth surface
V14-12	43°05'	61°42'	77	10	Sand	Smooth	Pits/mounds in smooth surface
V16-57	42°46'	63°11'	70	4	Gravelly	Rough	Pebbles on sandy, wavy bottom
V16-58	40°43'	61°41'	29-31	11	Shelly sand	Rhomboidal	Ripple marks in nearly rhomboidal shelly sand
V16-61	41°50'	57°53'	115	17	Sandy mud	Very smooth	Some mounds in very smooth lutaceous sandy bottom
V16-62	40°58'	57°12'	97	1	Sand	Very smooth	Smooth and flat with Echinoidea
V16-63	40°53'	57°10'	93	3	Sand	Smooth	Some flat ripples deformed by mounds
V16-64	39°43'	56°01'	99	17	Sand	Smooth	Some flat ripples deformed by mounds
V16-65	39°01'	55°44'	101	10	Sand	Smooth	Some flat ripples deformed by mounds
V16-66	37°36'	55°07'	115	17	Sand	Smooth	Some flat ripples deformed by mounds with some pits and Echinoidea
V18-5	36°06'	53°18'	192	12	Muddy	Medium smooth	Some pits in very smooth bottom
V18-65	34°18'	52°04'	85	9	Muddy sand	Medium smooth	Pits and mounds

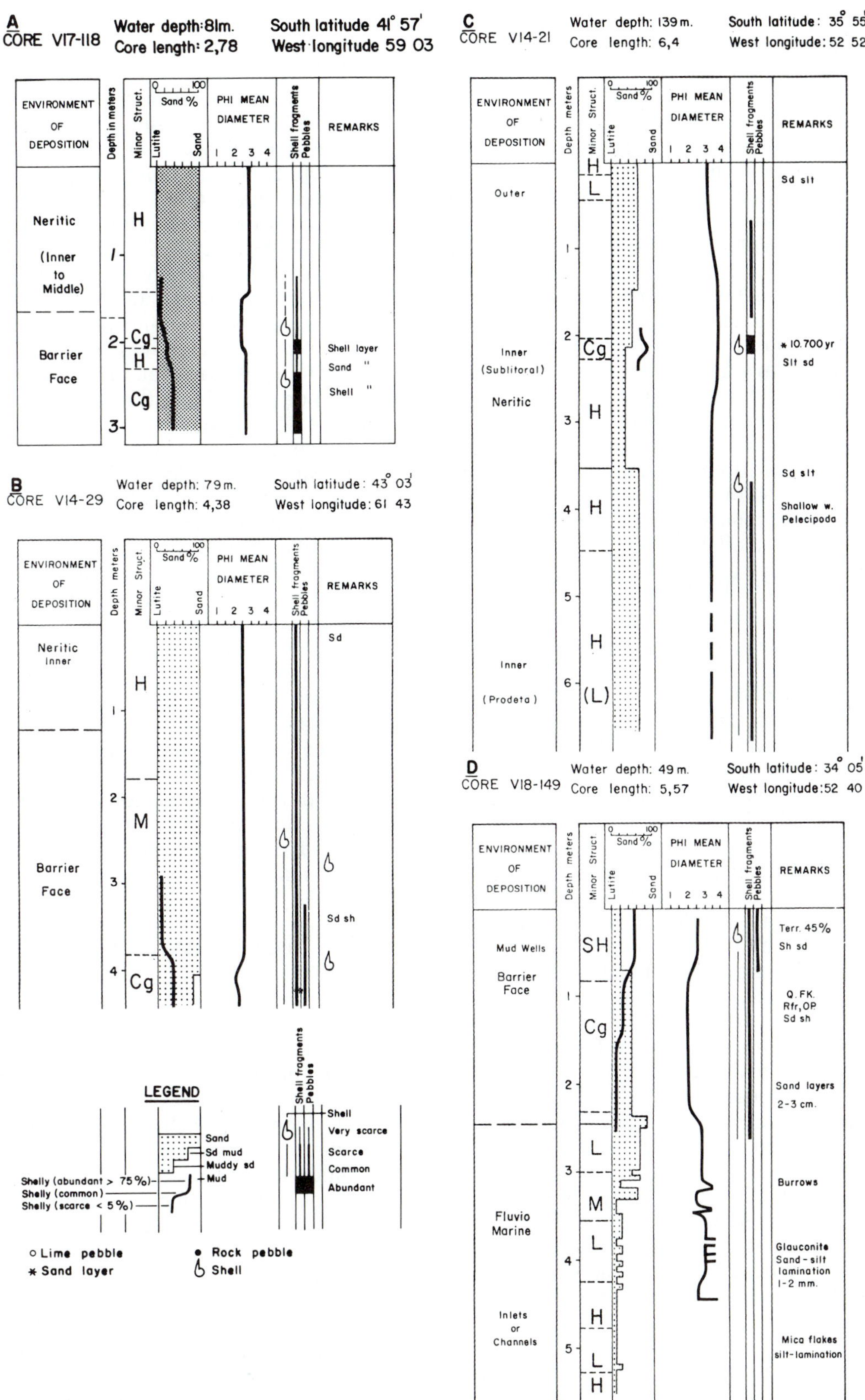

Fig. 13. A. Core V17-118, from the outer and inner shelf south of Bahia Blanca, shows the predominance of sand in the upper levels (onlap sand) and shell increasing toward the bottom, becoming conglomeratic, rich in large shell fragments (barrier face). B. Core V14-29. Fine to medium, gray, well-sorted quartzose onlapping sand, which turns conglomeratic toward the bottom; large shells are abundant. The conglomeratic layer suggests a nearshore depositional environment. C. Core V14-21. D. Core V18-149. Laminated, medium, dark-gray silty clay alternating with clear quartzose sand laminae with mica and shell fragments.

mottled. This facies forms because of the fluvial influence that "contaminated the onlapping shelf sands. The waves and bottom-current action structured and mixed the sediments, as in the cores obtained on the Rio de la Plata outer shelf (V16-176, 177, V18-39 to 49). South of Bahia Blanca, the grayish-brown sand becomes medium to coarse and poorly sorted, containing subangular grains of quartz (chalcedony and opal) and basic rock fragments. Deeper (after 1 to 2 m of sand), the sand started to be conglomeratic.

2. *Shelly and conglomeratic shell facies.* Subordinated to the sand facies, there appears in almost all the cores a shelly conglomerate layer. The sandy-shelly boundary is in general transitional, starting with shelly flakes, and reaching 70-80% of shell fragments. Large (50-mm) shell fragments appear, and cemented sandstone (beach rock) and calcareous pebbles of 10 mm diameter are also mixed with the shells (V17-115, V16-157, 158, and 180). Whole shells are not common, suggesting an area of high energy (beach or offshore bar), where bottom particles are highly abraded. Some layers of fragmented shells are well cemented with clay, acquiring a pale grayish-green color (V18-40, 43 to 48, 149, 150). Conglomeratic shell is another type of coarse clastic concentration; but, owing perhaps to the poor penetration of the core, it generally appears only in the core bottom. This facies is common on the shelf off Bahia Blanca and San Blas. The concentration is gradational, with an increase in the diameter of the andesitic pebbles from 3 to 5 mm up to 10 and 50 mm at the bottom of the core. In some cores (V14-29) the sandy fine conglomerate is more than 2 m thick (Fig. 13b). Shell fragments are present, and the sandy clastic material is around 30%. Near Bahia Blanca some cores (e.g., V16-158) are rich in pink clay pebbles and sandy matrix toward the bottom of the core. Some of the andesitic pebbles are encrusted by calcareous organisms; and whole shells, gastropods, and barnacles appear in the conglomeratic layer, indicating a shallow stable bottom with high biological activity. Camera station V16-57 at 70 m depth shows an exposure of this conglomeratic bottom, clean of sand cover. Perhaps here the strong bottom currents have mobilized the bottom sediments.

3. *Facies predominantly muddy.* Such facies present two characteristics. One is completely muddy and the other is interbedded and laminated mud with sand. A greenish-gray or blue clayey silt, completely homogeneous, with a few shell fragments, is the most typical sediment texture. Clam valves in living position were observed in a very homogeneous and plastic clay (Fig. 13C). This homogeneous muddy facies, in some places almost 10 m thick, is seen on the outer Rio de la Plata estuary main channel (Urien, 1967) and at mud wells (V16-178, 179).

In core V15-156, near the shelf break, this facies contains around 3.5 m of homogeneous to mottled muddy sand (3.4 ø) overlying muddy shell layers. The muddy facies are covered by a thin layer of sand and they contain interlayered homogeneous to mottled muddy sand with thin regular layers (25-50 mm) of clean yellow medium-sized sand (V16-176, 177, V14-21, and V18-48) (Fig. 13C).

This type of texture and the sand intercalation could signify the pre-delta facies, when the sea level was lower than 100 m. Lonardi and Ewing (1971) show a series of channels on the shelf border between 36°S and 37°5'S which could be identified as the outer delta gullies, where a pre-delta type of sedimentation is presumed.

In areas where sand is still abundant, muddy facies are interbedded with sand or vice versa. They appear as a pale-gray muddy silt with thin gray sand layers 1 or 2 mm thick; the mud is well laminated. A typical example is given in core V18-149 (Fig. 13d). In this case layered sand is interbedded between homogeneous to shelly mud in an area of considerable supply of fluvial sediments. In cores V15-149/150 layered mud is between mottled sand shelly sand. This facies differs from that described above and could correspond to a quiet environment with cyclic deposition of fluvial sediments (silty mud), as in the lagoons of the coastal plain subenvironment.

The cores collected on the continental shelf are far from being representative of a single environment of sedimentation. Vertical and horizontal changes are observed, which suggest the intervention of several events during the Holocene continental shelf evolution. Each of these sedimentary environments has a particular sedimentary assemblage, with minor sedimentary structures indicating variations in the hydraulic regime and conditions in the depositional process.

DEPOSITIONAL ENVIRONMENTS

As was explained before, it is possible to recognize a series of depositional environmental units which are the result of the Holocene transgression (and probably remnants of the young Pleistocene regression), but gaps in dating sediments from the shelf reduce the possibility of following the sea-level evolution step by step. The distinction made between the topographic features as either predominantly erosional or depositional, combined with the sedimentary properties, guided the mapping of the environmental units sketched in Fig. 14, which generalizes the extension of the relict topography and sediments according to the most conspicuous features built up during the shoreline migration.

These environments are classified as follows:

The *coastal plain* is formed near shore, where tidal flats, salt marshes, estuaries, and lagoons are recognized as subenvironments. This area, a temperate to dry coastal plain (slightly cooler than the present) (Groot et al., 1967), lies along almost all the margin studied and is now submerged because of the sea-level rise. In areas related to large rivers such as the Rio de la Plata (Buenos Aires), Guaiba

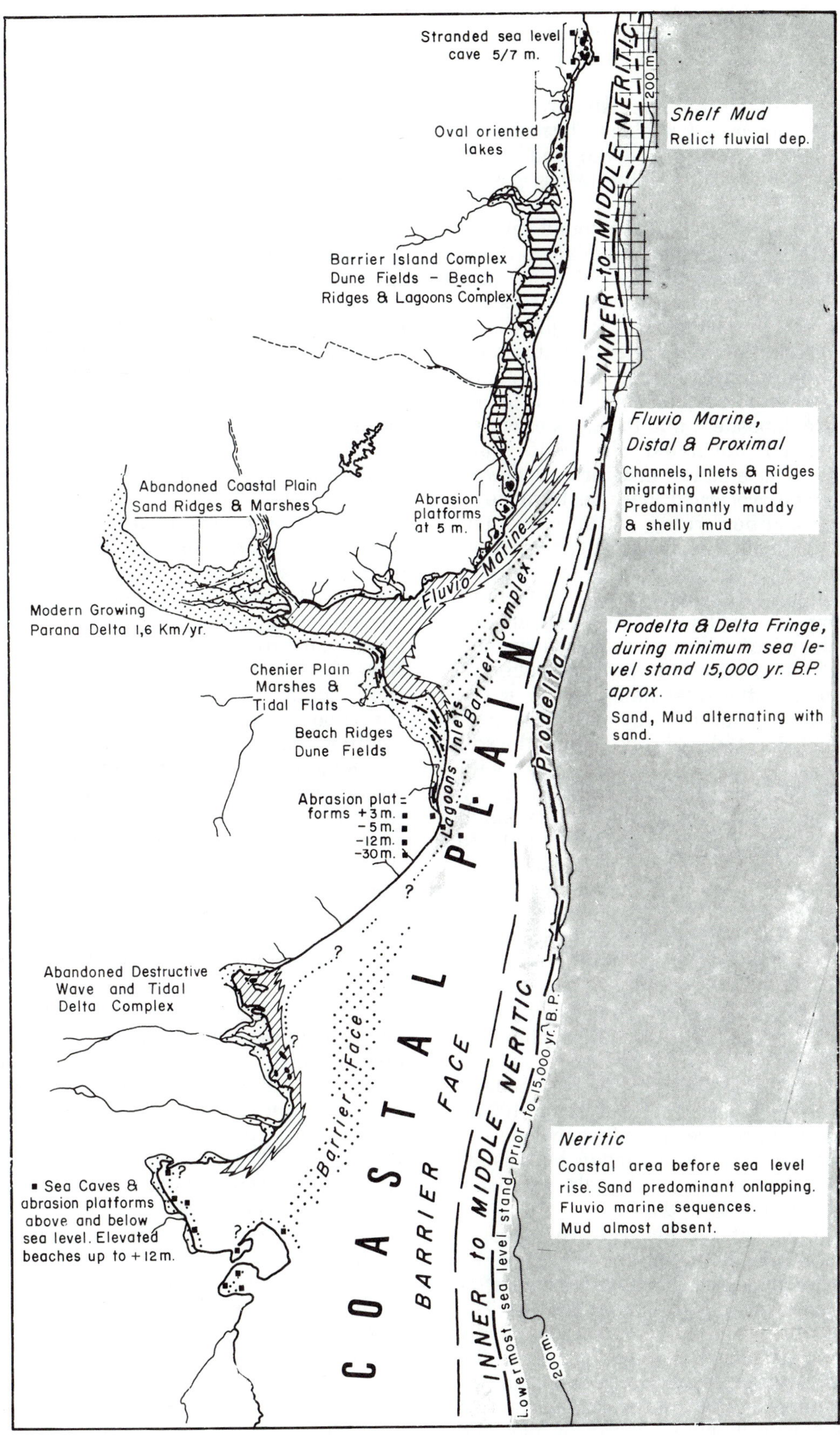

Fig. 14. Principal depositional environment units and the most remarkable features. Based on cores, topographic data, and seismic lines.

(Rio Grande do Sul), and Colorado-Negro (South Bahia Blanca), imposed relicts of estuaries and deltas remain on the shelf with morphology changed by the new hydraulic regime.

The substratum of this unit is indistinctly formed by pre-Young Pleistocene (or older) formations, which are buried over a large part of the shelf by the sediments of this environment or the transgressive sheet sand.

The *tidal flats and lagoons* are found on the landward side of the coastal plain, similar to those now found in San Blas, Bahia Blanca, and Samborombon bays. The tidal flats are connected with the open sea by channels through which sediments are discharged by strong tidal currents, maintaining a significant channel depth. Where currents are more active, basal poorly sorted shells and clay beds are deposited, and the slack of water produces occasional laminated clay with fine, well-sorted sand. These minor structures observed in the cores are well preserved, perhaps because of the rapid sedimentation rate, which prevented bottom-dwelling animals from disturbing the depositional laminations. In the tidal flats the rate of sedimentation appears lower, burrows are abundant, and complete shells in life position are frequent. The shell accumulations are often washed free of clay and mixed with indurate clay pebbles.

In this sequence the tidal flats sediments are covered by a thin suite of *barrier face* sands (V16-157, V18-149, Fig. 13d). Back tidal flats, the *salt marshes*, are characterized by sandy mud with occasional thin shell beds in accumulations that look like sand or shell ridges buried by mud. The lagoonal deposits look similar to those in tidal flats but are poorer in shells. Sandy layers interbedded with laminated mud are abundant, showing changes in the hydraulic process.

Some of the cores near Mar del Plata and Plata estuary (e.g., V15-148 to 151) show the succession of lagoonal or salt-marsh sediment facies overlain by barrier face sands, the product of landward shore movement onlapping the coastal environments in a typical retreating shore pattern.

According to the available cores, the coastal plain lay eastward up to the isobaths of 80 and 100 m. The 14C data show nearshore fauna (beach deposits) in deeper positions than is the case today, but seismic records suggest that the coastal plain extension could have been up to 140 m depth.

The *marine deposits*, in general are superimposed on those described above. It may be subdivided into *holomarine*, without fluvial influence; and *fluviomarine*, with fluviatile supply. The *holomarine* deposit is of two main types: highly bioclastic, and predominantly terrigenous. The highly bioclastic type is closely related to the *barrier face* subenvironment. This nearshore element is predominantly clastic (sandy), where the biogenic activity (due to the shallowness of the area) is prolific enough to provide predominance of organic over detritic deposition. These deposits are concentrated in longshore bars, ridges, or perhaps storm beach deposits, and are eventually winnowed by waves and currents in the nearshore area. Those accumulations often appear buried by sand. The contact may be well defined or transitional, as in cores V18-40 to 44. These cores were obtained near a shoal topography (Plata Shoals) identified as a relict barrier island complex (Figs. 2 and 5). In areas near the shore, the silt/clay alternations become common, possibly because of currents coming from inlets or channels. Most such areas are related to barrier spits and ridges, as are observed along the present-day shoreline (e.g., Punta Rasa sand spit, Urien, 1967; Rio Grande inlet and Bahia Blanca-San Blas) and in the shoal system of Mud Wells area.

In areas where fluvial pebbles are present, the shelly fraction is extremely gravelly, as was described for the southern Bahia Blanca area (Fig. 13a). The organic activity is also high. Consequently, shells and shell hashes are found mixed with the gravel or sandy gravel.

In the *inner and middle neritic* subenvironment, terrigenous sand is predominant, and the rate of deposition appears reduced. Waves and currents occasionally agitate and modify the bottom, as do the burrowing organisms. With the exception of some muddy mixtures, the sediments are homogeneous. Nevertheless, some storm waves and currents act sporadically on the bottom, activating it. Hence shelf-floor relict forms and sediments appear modified and sometimes show superimposed forms.

Relicts of inner neritic subenvironments are also found near the present-day outer shelf (border) as well as on the outer neritic (Fig. 14). Here its differentiation is based only on the silt content and the almost complete absence of biogenous remains. Near areas with active fluvial influence, structureless silty sand is common, but some alterations with sand appear, owing perhaps to high tidal current variations (gullies or channels?).

The *fluviomarine* deposit is a marine deposit with direct fluviatile supply. Here the rate of deposition was high enough to reduce biogenic development. In the *proximal fluviomarine* there appears a thick clay/sand lamination, similar to that in tidal channels, but muddier, owing perhaps to the salt water's precipitating the suspended solids (Ottman and Urien, 1966; Urien, 1967, 1972). Even when clay is free of sand, some laminations appear, resulting, perhaps, from current variation (V14-21, Fig. 13). The macroscopic fauna content decreases in an upward direction; and only in areas related to shoals (e.g., Plata shoals) is fluvial mud mixed with shell fragments (V18-147/149, Fig. 13d). This environment, predominantly muddy, is found on the Uruguayan shelf and extends far up to the middle shelf and shelf border, especially in front of Cabo Polonio, Off Rio Negro-Colorado, the cores do not show the same aspect as off Rio de la Plata. The modern sediments appear muddier and lie in a narrow fringe on the San Blas and Bahia Blanca bays.

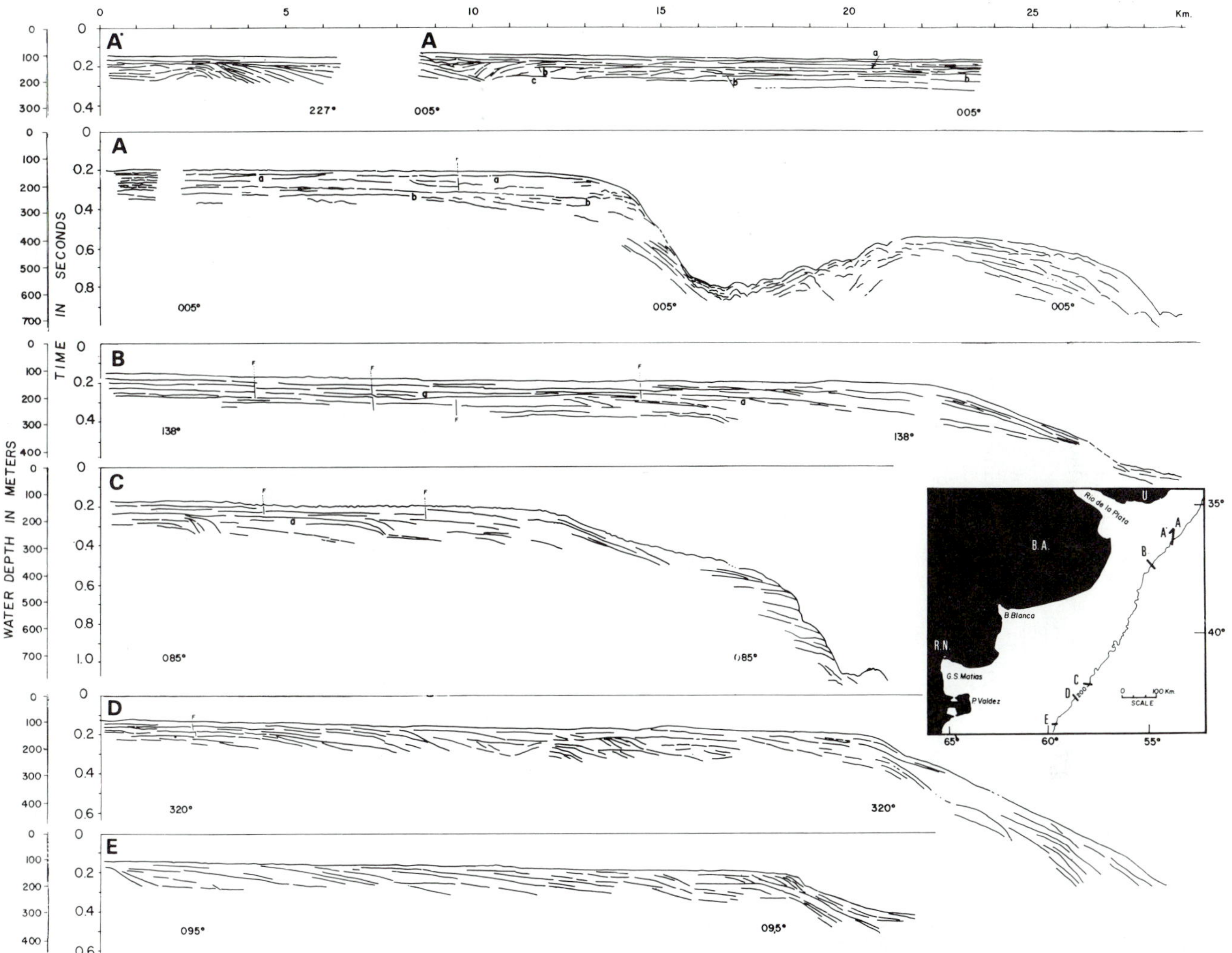

Fig. 15. Seismic reflection records across the shelf border area from off Rio de la Plata, Bahia Blanca, and Plata Valdez. Sparker lines made by the vessels ARA ZAPIOLA and VEMA in 1966. Section A'A, parallel to the shelf edge, shows a series of sediment-filled channels. Section A shows prograding layers, a delta-like front. At km 15-20, line A crosses a canyon and the Rio de la Plata terrace. The other sections also show deltaic prograding layers, some of them outcropping on the slope (Section C). From offshore exploratory well data, an early to middle Pleistocene age is inferred for these deltaic sequences.

The *fluviomarine distal* environment offlaps to the neritic because of the suspended load from rivers, such as Rio de la Plata. Here the transport is different. Whereas in the proximal area (fluvial), ebb currents are added to the fluvial discharge; in the distal, the marine semipermanent currents (tidal, wind) are the only agents transporting the suspended sediments.

These sediments are distributed widely in the area, but always with a north component. This is the case at present in the Bahia San Blas-Bahia Blanca areas and northeast off Rio de la Plata up to Rio Grande do Sul. At 35°S the cores appear muddy/homogeneous as a result of the influence of the Plata River in this area. Here minor textures of the cores and also the seismic reflection records (Fig. 6) show a landward shift of this depositional subenvironment which alternates with sand- and mud-covered barrier island (or shoal) forms.

It appears that the channel off Uruguay was extended toward the shelf border around 34°30'S, and it is assumed that the Rio de la Plata shelf valley had a greater influence in this sector, but evidently in the early Holocene or earlier.

In the Rio Grande do Sul shelf border, a *fluviomarine distal* subenvironment is identified, formed by dark-greenish silty clay that is, in part, rich in shell fragments. This environment is related to the Rio Grande do Sul rivers, especially the Guaiba, Camacúa, and Jaguarón rivers, which reached the shelf during the low-sea-level stand (Martins and Urien, 1967).

Two other subenvironments are recognized near the shelf border, but due to the scarcity of information, they are not completely outlined. They are *deltaic* and *bays*. The sediments consist of alternations of strongly disturbed sand lenses in homogeneous clay or totally silty clay/sand; this suggests intermediate environments. Both environments are found near the shelf border and are identified with particular bottom topographic features. In Figure 15 it is possible to see the typical

Table 3. Shell samples from the coastal plain.

	Lab. Det.	Sample	Locality	Level[a] (cm)	Water Depth (m)	Age (my)	Remarks
1	LDGO	Lss.	Plata shoals	Surface	- 22	1,500	Shale mixed with sand and limestone pebbles
2	1286 A. LDGO	Lss.	Plata shoals	Surface	- 40	6,000 ± 100	Shale mixed with sand and limestone pebbles
3	Gif 736 CNEA	D-5	Outer Rio Plata	Surface	- 10	3,250 ± 110	Shale mixed with clay (compact)
4	LDGO	V15-149	Mar del Plata inner shelf	60-80	- 66	15,000 ± 300	Shale interlayered with sand
5	LDGO	V14-21	Rio Plata shelf	210-220	-139	10,700[b]	
6	LDGO	V16-174	Rio Plata outer shelf	65-98	-115	11,000 ± 150	Shale interlayered with sand
7	LDGO	V17-109	Bahia Blanca middle shelf	130-160	- 73	9,800[b]	41°39'S, 59°53'W
8	LDGO	V17-109	Bahia Blanca middle shelf	670-690	- 73	9,900[b]	Provisory
9	LDGO	V17-112	Bahia Blanca middle shelf	100-125	- 58	17,250[b]	Shale interlayered with sand
10	LDGO	V17-112	Bahia Blanca middle shelf	330-320	- 58	35,000	Shale interlayered with sand
11	Smithsonian	Lss.	Punta Medanos		- 24	1,460	
12	Delaney, 1965		Rio Grande do Sul		+ 1.5	1,975 ± 150	Peat of Conceicao, Brazil
13	La Jolla 1399		Mar Chiquita	50	+ 5	3,400 ± 200	Abandoned beach ridge on backside of the lagoon near road
14	1016 LDGO		Mar Chiquita	30	+ 3	3,070 ± 150	Emerged beach ridge on the seaward lagoon side
15	Delaney, 1967		Montevideo	50	+ 2	2,450 ± 100	Elevated storm beach ridge in inner Montevideo bay (Arsenal)
16	Gif 737 CNEA		Samborombon	100	+ 3.5	3,770 ± 110	Samborombon, inner Chenier plain near coastal highway
17	Delaney, 1967		Punta del Indio	150	+ 4	4,460 ± 110	Chanier ridge, innermost Rio de la Plata coastal plain
18	Smithsonian		Cda. Arregui	100	+ 6	7,600 ± 600	Chanier ridge, innermost Rio de la Plata coastal plain
19	Delaney, 1967		Viscaino Islands	—	+ 1	5,980 ± 200	Beach ridge in island in the Uruguay River estuary
20	Van Der Molen		Parana delta	—	-24	8,640 ±	Peat deposited beneath the Parana River modern delta
21	Delaney, 1967		Soriano	—	+ 2	5,420 ± 110	Beach ridge on Uruguay coastal plain buried by sand
22	Groningen, Holland[c]		Los Talas	—	+ 2	4,600	Ridge on the Rio de la Plata coastal plain
23	Groningen, Holland[c]		Mar Chiquita	—	- 1.5	4,000 ± 200	Low-tide outcrop—shale and clay, very compact
24	Groningen, Holland[c]		Pipinas	—	+ 7	30,000 ± 1000	Coquina beneath Samborombon coastal plain cemented with limestone
25	Groningen, Holland[c]		La Plata	—	+ 5.5	3,530 ± 35	Back cliff, Rio de la Plata coastal plain

[a] Below the surface. [b] These ^{14}C dates are not commented upon in the present paper. [c] From Cortelezzi and Lerman, 1971.

prograding layers of an uppermost section of a recent deltaic sequence on the seismic records' overlay. Some upper horizontal reflectors and a rough bottom with wavy parallel topography indicate a covering onlapping transgressive barrier face sequence, which could be related to the sea-level rise and retreating shore face.

RADIOCARBON DETERMINATIONS

Table 3 presents a series of 14C dates of various shell samples obtained in the submerged and subaerial coastal plain. Although these data are not sufficient to make an adequate absolute sea-level-rise curve, they indicate that the sea level has definitely risen since 11,700 ± 500 years B.P. from approximately -139 m below the present sea level, and that between 9,000 and 2,000 yr. B.P., sea level oscillated until it stabilized about 2,000 yr. B.P.

These events are not necessarily synchronous with the time of glacial retreat, and probably here the sea's rise was delayed more than in other areas. But the facts from the available data are:

1. The continental shelf during the late Pleistocene was at least 150 m below the present sea level.

2. The sea level rose from approximately -140 m up to its present shoreline since 11,000 yr. B.P., which could be dated as the Holocene Transgression.

This is not only supported by the 14C data but also by the sedimentary facies and topographic features detected on the submerged coastal plain and the paleontologic record (Boltovskoy, 1973).

DISCUSSION AND CONCLUSION

From the evidence available on the continental shelf it is possible to analyze the recent sediments and the evolution of environments in the area studied since Upper Pleistocene time.

The principal factor in the evolution of the continental shelf was the sea-level rise and oscillations from the melting of glaciers during the last 18,000 years. More work is needed in this area; nevertheless, sedimentary, faunal, and topographic data, and some 14C dates, studied as a whole in the light of the environmental process and compared with present-day littoral models, make it possible to compose a more or less accurate picture of the history of the continental shelf during the last thousand years.

1. The continental shelf was under the action of postglacial relative sea-level fluctuations because of a universal glaciostatic process. The sea level was rising during the last 18,000 years, except for some periodic oscillations or epeirogenic movements, which we shall not attempt to evaluate here.

2. Most of the sediments were deposited within the estuaries and bays or near the mouths of rivers. The longshore currents dispersed these sediments along the shoreline.

3. Semipermanent currents, resulting from the tide or wind and the longshore drift due to wave convergence, were the main transporting agents. They have, apparently, the same pattern at the present time. However, the major ocean currents could have changed their convergence (cold and warm) areas either slightly northward or southward, consequently modifying some faunal habitats.

4. The rivers kept their basic patterns on the submerged coastal plain during the main transgression, perhaps as flooded regressive estuaries, and the littoral forms, either erosional or constructional, as the products of sea-level change.

Those events resulted in the lateral shoreline migration as a product of the balance between the relative sea level and the rate of sedimentation. The continental shelf sediments can be divided into relict sediments, reworked, and modern sediments. The former, which are not in equilibrium with the present environment, are identified with the transgressive Holocene sand sheet. Consequently, part of the shelf bottom is affected by constructional/erosional processes modifying the previous elements. The modern sediments are the product of the rivers' sediment supplies, and their rates of deposition are molded by the shoreline process.

Three main fluvial sources are observed: 1.) The *Rio Grande do Sul rivers*, which descending from the highlands (Precambrian hills and lava plateaus) have run across the coastal plain to the shelf border (Martins and Urien, 1967), prograding small deltas and estuaries until the beginning of the sea transgression, and subsequently leaving a belt of relict shelf mud. 2.) The *Rio de la Plata*, the most important system, because of its enormous basin (Urien, 1967; 1970, 1972), which has influenced its estuary area and the neighboring inner shelf on the Uruguayan middle shelf with an offlapping muddy carpet up to Rio Grande do Sul. In the late Pleistocene it reached the shelf border with the formation of a wide delta complex which was subsequently buried by retreating barrier face sediments because of the rising sea. 3). *The Colorado-Negro rivers* spread their load on the shore both southward into the San Matias Gulf and northward along the San Blas and Bahia Blanca bays, but with a very narrow muddy sand belt. When the load reached the shelf border, bays, estuaries, and deltas were built up. Perhaps this river system had a greater flow than at present and was related to the canyon of northern Patagonia (Lonardi and Ewing, 1971). It appears that the Patagonian rivers during the Holocene had higher load capacity, expelling a considerable sand stock onto the inner shelf, and being perhaps the main source of these sediments.

The case of the fine sediments is a bit different because they are partly flocculated and more extensively winnowed. The winnowing depends on the river's suspended load volume, which appears to be significant only in the Rio de la Plata.

The deep closed basin in the San Matias Gulf might be older than Pleistocene. The bottom of the basin is buried by a thick homogeneous clay that could bypass the sill from Rio Negro (Pierse et al., 1969) because of counterclockwise tidal currents into this gulf.

In the first phase the rivers and estuaries trapped the fluvial sediments. They were filled, and because the sea level stabilized, the fluvial sediments were spilled onto the shelf. The longshore current captured this material, spreading it along the shoreline, apparently with a very persistent northward drift component. Then with parallel retreat of the shore, an extensive onlap sandy blanket body was created, which barred the existing pre-Holocene and Holocene sediment facies. This onlap facies also invaded estuaries and covered deltas inland farther than the present shoreline which is known as the basal "old sands" (Urien, 1972). This is the case at the Rio de la Plata valley, where marine deposits were found near its head (Delanay, 1966; Urien, in preparation) and in the Rio Grande coastal plain (Delaney, 1963, 1965, 1966). When sea level stabilized, probably 2,000 years ago (Urien and Ottman, 1971), a progradation started because the deposition was balanced with the transgression. Consequently a seaward accretion commenced, building up modern deltas, offshore bars, etc.

In Rio Grande do Sul a seaward accretion of beach ridges and sand spits and dune fields built up a gigantic barrier island system more than 300 km long, forming lagoons, some more than 20 km wide (Delaney, 1966). For instance, the inner Rio de la Plata (head)-Parana River modern tidal delta is prograding into the estuary, and the fluvial load makes an offlap muddy facies that covers the "old sand" (Urien, 1970) even into the inner shelf. In the Colorado-Negro rivers the fluvial conditions are reduced (probably as a result of dry conditions in their upper basin). Here there is a high destructive dominant tide and wave delta system which has shifted southward from the Bahia Blanca embayment.

On the northeastern Buenos Aires coastal plain, beach ridges and chenier plains and lagoons are semicovered by a blanket of young fluvial sediments that are gradually filling the bays and lagoons.

These recent environments are the results of a persistent sea-level stand and are identical, with certain limitations, to those previously described environments of the submerged coastal plain which are the result of the events since the young Pleistocene sea-level retreat and subsequent Holocene transgression which created the retreating shoreline. Then the deltas, estuaries, and bays were flooded and partially buried with a quasi-autochthonous sand sheet which covered the coastal plain morphology. The modification of this new hydraulic regime superimposed on the shelf bottom a new relief by reactivation of the relict sediments in those areas shallow enough to be affected by waves and currents.

ACKNOWLEDGMENTS

We are indebted to many colleagues for assistance and useful suggestions: L.W. Butler, J. Delaney, H. Ewing, A.G. Lonardi, L. Martins, J. Pierse, H.G. Richards, F. Sieguel, and D. Thurber. We owe particular thanks to those at the Lamont-Doherty Core Laboratory, whose assistance was invaluable.

We thank officials of the Argentine Hydrographic Service for the use of their laboratories and for providing information essential to this study. We are grateful to Cabot Petroleum (Argentina), Inc., for help in the preparation of this paper. For the data provided by Columbia University's Lamont-Doherty Geological Observatory and for support of that part of the work carried out at the Observatory, we are indebted to the U.S. Navy, Office of Naval Research and to the National Science Foundation for their support under Contract N00014-67-A-0108-0004 and grants NSF-GA27281 and GA-29460, respectively.

Contribution Number 38, Earth and Planetary Sciences Division, University of Texas at Galveston.

BIBLIOGRAPHY

Boltovskoy, E., 1973, Estudio de los Testigos Submarinos del Atlantico Sudoccidental: Buenos Aires, Rev. Museo Cs. Nat. B. Rivadavia, v. 7, no. 4, 340 p.

Butler, L. W., 1970, Shallow structure of the continental margin, southern Brazil and Uruguay: Geol. Soc. America Bull., v. 81, no. 4, p. 1079-1096.

Curray, J., 1966, Continental terrace, *in* Fairbridge, R. W., ed., Encyclopedia of oceanography: New York, Van Nostrand Reinhold, p. 207-213.

Delaney, J., 1963, Outline of the geologic history of the coastal plain of Rio Grande do Sul; South American coastal studies: Coastal Studies Inst., Louisiana State Univ., 63 p.

———, 1965, Reef rock on the coastal platform of southern Brazil and Uruguay: Symp. on the ocean of western South Atlantic, Brazil, Sect. III, p. 9.

———, 1966, Geology and geomorphology of the coastal plain of Rio Grande do Sul, Brazil and Uruguay; South American coastal studies, 18B: Coastal Studies Inst., Louisiana State Univ., 58 p.

Fray, C., and Ewing, M., 1963, Pleistocene sedimentation and fauna of the Argentine shelf. I. Wisconsin sea level as indicated in Argentine continental shelf sediments: Acad. Nat. Sciences Phila., v. 115, p. 113-126.

Groot, J. J., Groot, C. R., Ewing, M., Burckle, L., and Conolly, J. R., 1967, Spores, pollen, diatoms and provenance of Argentine Basin sediments, *in* The quaternary history of the ocean basins: progress in oceanography: Elmsford, N.Y., Pergamon Press, v. 4, p. 179-216.

Harrington, H., 1956, Argentina, *in* Jenks, W. F., ed., Handbook of South American geology: Geol. Soc. America Mem. 65, p. 131-165.

Herman, H., 1948, Sounding taken during the *Discovery* investigations, 1932-1939: Discovery Report 25, p. 29-106.

Lonardi, A. G., and Ewing, M., 1971, Sediment transport

distribution in the Argentine basin. 4. Bathymetry of the continental margin, Argentine basin and other related provinces, canyons and sources of sediments, *in* Ahrens, L. H., et al., eds., Physics and chemistry of the earth: Elmsford, N.Y., Pergamon Press, v. 8, p. 79-121.

Martins, L., and Urien, C. M., 1967, Os sedimentos da plataforma continental do Rio Grande do Sul and Uruguay: Cong. Brasileiro de Geologia, Curitiba.

Ottman, F., and Urien, C. M., 1966, Sur quelques problèmes sédimentologiques dans le Río de la Plata: Rev. de Géogr. Physique et de Géol. Dynamique, v. fasc. 3, p. 209-224.

Pierse, J., Urien, C., and Sieguel, F., 1969, Topografia submarina del golfo de San Matias: Argentina, Journadas Geologicas Mendoza, abstract.

Remiro, J., and Etchichury, M., 1960, Muestras de fondo de la plataforma continental comprendida entre los paralelos 34° y 36°30' latitude South y meridianos 53°10' y 56°30' longitude Oeste: Bull. Mus. Cs. Nat. B., Rivadavia, t. 4.

Richards, H. G., and Craig, J. R., 1963, Pleistocene sedimentation and fauna of the Argentine shelf. II. Pleistocene mollusks from the continental shelf off Argentina: Acad. Nat. Sciences Phila., v. 115, no. 6, p. 113-152.

Schnack, E. J., 1969, Los sedimentos de la plataforma continental correspondientes al area externa del Rio de la Plata [thesis]: Argentina, Museo de la Plata.

Teruggi, M., Chaar, E., Remiro, J., and Limousin, T., 1959, Las arenas de la costa de la provincia de Buenos Aires entre Cabo San Antonio y Bahia Blanca: LEMIT, v. 12, p. 77, La Plata.

Urien, C. M., 1967, Los sedimentos modernos del Río de la Plata exterior: Bol. Servicio Hidrografia Naval, Buenos Aires, v. 4, no. 2, p. 113-213.

———, 1970, Les rivages et le plateau continental de Sud de Brésil, de l'Uruguay y de l'Argentine: Quaternaria, v. 12, p. 57-69.

———, 1972, Río de la Plata estuary environments, *in* Nelson, B. W., ed., Environmental framework of coastal plain estuaries: Geol. Soc. America Mem. 133, p. 213-234.

———, and Mouzo, F., 1968, Algunos aspectos morfológicos de la plataforma continental en las proximades del Río de la Plata: Bol. Servicio Hidrografia Naval, Buenos Aires, v. 4, no. 4, p. 8.

———, and Ottman, F., 1971, Histoire du Rio de la Plata au Quaternaire: Quaternaria, v. 14, p. 51-59.

———, and Zambrano, J. J., 1972, La estructura de la terraza continental de Brazil meridional, Uruguay y Argentina: Symposium on the results of upper mantle investigations with emphasis on Latin America, v. 2, 11 p.

Zambrano, J., and Urien, C. M., 1970, Geological outline of the basins in southern Argentina and their continuation off the Atlantic shore: J. Geophys. Res., v. 75, no. 8, p. 1363-1396.

Continental Slope Construction and Destruction, West Africa

Eugen Seibold and Karl Hinz

INTRODUCTION

According to the generally accepted concept, a continental margin of the Atlantic (passive) type develops by outbuilding and upbuilding of sediments. The sediment material is thought to be supplied mainly from land. A shallow-water terrace is always present between the coast and the continental slope. The shelf sediments may build up to a height of several tens of meters below sea level, determined by water movement (see Rona, 1973a, for numerous references). If more sediment material is supplied, it migrates onto the continental slope. Therefore, shallow-water sediments several kilometers in thickness, often encountered, require subsidence of the basement.

This generalized concept will be supplemented by several modifications, as shown on examples here from the west African continental margin. The slope and the upper rise there will be the center of our discussion as well as the processes that may cause the continental terrace to recede temporarily: erosion in closely spaced canyons, denudation through slumping, horizontal oceanic currents, and increased subsidence.

DESCRIPTION OF AREA

Morphology. It is well known that the continental margin off northwestern Africa is similar to that off eastern North America (Heezen et al., 1959). The most important differences are the considerably wider continental rise and the stronger volcanic activity off Africa. The widths of the continental shelves vary considerably, the shelf-breaks are always found at 100-110 m, and the inclinations of the slopes are between 1 and 3° (Table 1 and Fig. 1).

Regional Geology. The regional geology of coastal Africa is discussed by, among others: Choubert (1952, 1971), Furon (1960), and in Schiffers (1971), Sougy (1962), Haughton (1963), Aymé (1965), Kennedy (1965), Querol (1966), Spengler et al. (1966), Reyre (1966), Choubert (1968), Auxini (1969), Ratschiller (1970), Alia (1960), Rona (1970, 1973a), and Machens (1970). Summaries of the present knowledge on the west African continental margin are included in McMaster and Lachance (1968), Lowrie and Escowitz (1969), Rona (1970, 1971, 1973a), Summerhayes et al. (1971), Templeton (1971), Tooms et al. (1971), Beck (1972), Egloff (1972), and Luyendyk and Bunce (1973).

According to these references, the Requibat High separates the Aaioun (Tarfaya) Basin in the north from the Senegal Basin in the south. The Requibat High extends to the coast near Cape Barbas and Cape Blanc, where it, as well as the depth contours down to 3,000 m, bend sharply toward the southeast. The land portion of the Aaioun Basin is about 1,000 km long, up to 200 km wide, and covers an area of less than 200,000 km^2; for the Senegal Basin the respective values are 1,200 km length, 560 km width, 340,000 km^2 area. In landward directions these basins are surrounded by Paleozoic fold belts. However, there is no relationship between these Paleozoic basins and their facies, and the Meso-Cenozoic basins.

The marine history of the basins starts with a Jurassic transgression from the west after Triassic

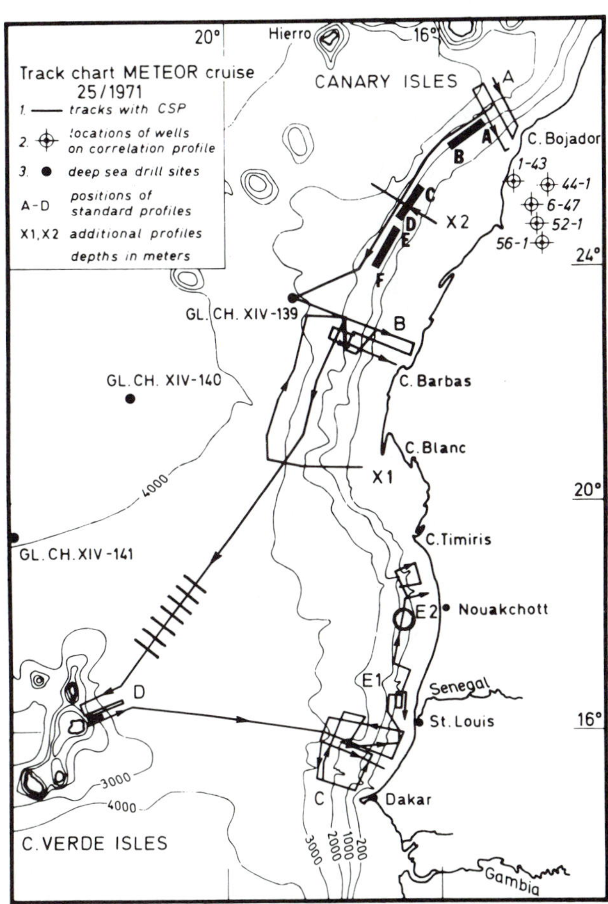

Fig. 1. Track chart, METEOR cruise 25/1971. Area of structural high is cross-hatched. Salt structure off Mauretania shown by a circle.

Table 1. Morphology.

Area	Width (km)	Gradient (%)	Shelf break (m)	Width (km)	Gradient		Average Lower Boundary (m)
					%	deg.	
Cape Bojador	40	2.5	110	70	37	2°10'	2,700
Cape Barbas	100	1.0	110	70	27	1°30'	2,000
Senegal	40	2.5	105	40	52	3°	2,200

precursors. Both basins underwent a similar development, with a maximum transgression during the Upper Cretaceous, a marked regression at the beginning of Tertiary times, and uplift and erosion during the Oligocene period. The general facies distribution is also similar. Near the present coast shales from deeper waters are found, followed in the landward direction by shallow-water sands, carbonates, beach and lagoonal sediments, and continental deposits. Sediment thicknesses are greater in the Senegal Basin than those in the Aaioun Basin, which thicken during the early half of their history, particularly from east to west. All this is related to growing differential subsidence.

Tensional features are present, such as faults and flexures, mostly parallel to the coast, but compressional deformation cannot definitely be excluded. Salt structures are known onshore and offshore in both basins.

Quaternary Sediments. Table 2 shows the average values for selected sediment characteristics off west Africa. The very high carbonate content here is essentially biogenic, i.e., of planktonic origin. When the remains of siliceous organisms are included, it turns out that up to three-fourths of the Quaternary sediment layers off the Spanish Sahara consist of plankton. Accordingly, the slope could be built upward and outward nearly without terrigenous supply. It must, however, be emphasized, that in these latitudes coastal upwelling increases biogenic production. Off the mouth of the Senegal River the carbonate content drops off to 10-20%, owing to increased terrigenous dilution, which in turn accelerates upbuilding and outbuilding.

Average rates of sedimentation during Holocene and late Pleistocene (Table 2) show the influence of climate, of the Senegal River supply, and the difference between upper and lower continental slope.

REFLECTION SEISMIC DATA

Cape Bojador. Figure 2 shows three vertically migrated depth profiles recorded offshore from Cape Bojador (area A in Fig. 1). The stratigraphic position of two distinct discontinuities appearing in the records was determined by extrapolating seaward Spanish Sahara well data (see also Fig. 3). It was assumed that the units in question extend horizontally and undisturbed across the coastal area (Hinz et al., 1974). Hence the older discontinuity, D_1, appears to be of Cenomanian age. Under the present continental rise at water depths of more than 2,500 m, we believe that D_1 represents the Lower Cretaceous continental slope. The Paleogene unconformity that occurs on land (Ratschiller, 1970) could not be identified on our records.

Discontinuity D_2 is of Oligocene to early Miocene age. It is characterized in the seismic records by an irregular trace and cuts deeply into pre-Miocene and even pre-Cenomanian sediments where the present continental slope grades into the continental rise.

Since broad areas of the Spanish Sahara were above sea-level in post-Oligocene times (Querol, 1966), the landward part of discontinuity D_2 was probably formed by erosion in shallow water or even subaerially during an Oligocene regression. Our profiles show further that submarine incisions first developed in the Oligocene/Miocene, that is after hiatus D_2. Very irregular and disturbed stratification characterizes the upper part of the Neogene sediments overlying discontinuity D_2 (Fig. 3). A structural high has been found beneath the present transition from slope to rise (Fig. 2). This feature seems to extend southward from Cape Bojador at least to a latitutde of 24°N (Hinz et al., 1974).

Wadi Craa. Discontinuities D_1 and D_2 can be recognized offshore from Wadi Craa (Fig. 2, line x_2 in Fig. 1); the same is true for the slope anticline (50-75 km offshore). Between 72 and 92 km offshore the evidence for the discontinuities is not so clear. The original seismic records of Figures 5 and 6 show that submarine incision first began after the development of D_2, the Oligocene hiatus.

It is apparent from Figure 6 that in post-Oligocene time canyons were filled and reexcavated and local slumping occurred (Fig. 5). Old, infilled canyons which could be over 10 km wide and 600 m deep have only been observed in the Wadi Craa area (between 24°N and 25°30'N). The broadening and westward extension of the 3,000-m depth contour in this area could have resulted from a major Neogene drainage system.

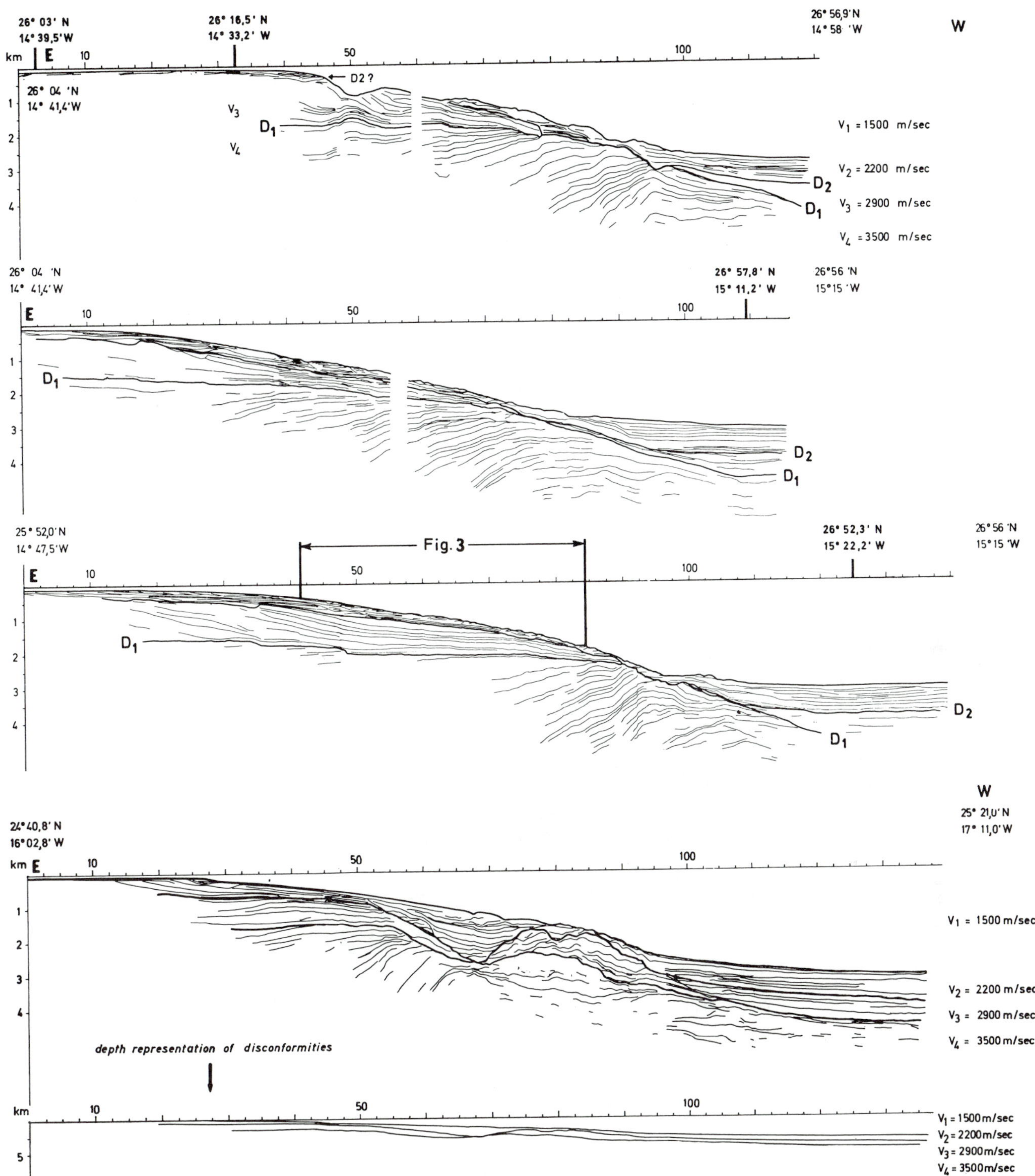

Fig. 2. Vertically migrated seismic reflection profiles of area A and profile x_2 (bottom). Velocities used for computation are indicated (see Fig. 1).

Cape Barbas. Several discontinuities are recognizable in the seismic records (Fig. 7) from the Cape Barbas area (area B in Fig. 1). Discontinuity D_1, which is assumed to be Cenomanian, cannot be unambiguously traced in the continental slope area. Also the slope anticline cannot be recognized here. Where D_1 can be observed beneath the slope, it is overlain by reflectors that dip slightly seaward. Discontinuity D_{1b}, which was found only in the upper slope area, may indicate that the late Cretaceous to early Tertiary prograding was apparently interrupted in the early Tertiary. In the Cape Barbas area, the presumably Oligocene to early Miocene D_2 discontinuity can be clearly traced in

Table 2. Sediment characteristics.

		(1) Sand/Silt/Clay$_b$ (wt%)						(2) Carbonates (wt%)		(3) Organic Carbon (wt%)		(4) Shear Strengh (g/cm^2)		(5) Sedimentation Rate (cm/1000y)	
		Water Depth (m)						Water Depth(m)		Water Depth(m)		Water Depth(m)		Water Depth (m)	
		800-2,000			>2,000			800-2,000	>2,000	800-2,000	>2,000	800-2,000	>2,000	800-2,000	>2,000
Portugal	Holocene	<1	25	74	<1	13	86	25	32	<1	<1	5-10	70-120	7-22	9-18
	Pleistocene	2	42	56	<1	22	78	-	-	-	-	-	-	15-40	12-30
Morocco	Holocene	16	12	72	.8-10	8-15	77-82	52	48	<1	<1	35-65	64-100	3-10	3-8
	Pleistocene	50	13	37	2-5	11-21	74-87	55	46	<1	<1	-	-	6-?	20-26
Cape Bojador	Holocene	14	18	58	12-24	18-27	54-61	-	55-70	-	0.4	-	60-120	-	4
	Pleistocene	-	-	-	4-10	50-60	30-40	-	45-55	-	1-2	-	-	-	9
Cape Barbas	Holocene	27-37	35-36	28-37	8-23	26-29	51-63	50-75	55-75	1	0.2-0.3	-	60-160	6	2-3
	Pleistocene	-	-	-	7-15	60-70	20-30	45-60	40-55	2-4	0.5-0.9	-	-	10-20	4-6
Senagal	Holocene	3-8	38-44	48-59	-	-	-	-	7-14	1-3	2.3	-	-	>20	3-15
	Pleistocene	-	-	-	-	-	-	- '	12-23	0.4-0.8	1-2	-	-	-	-

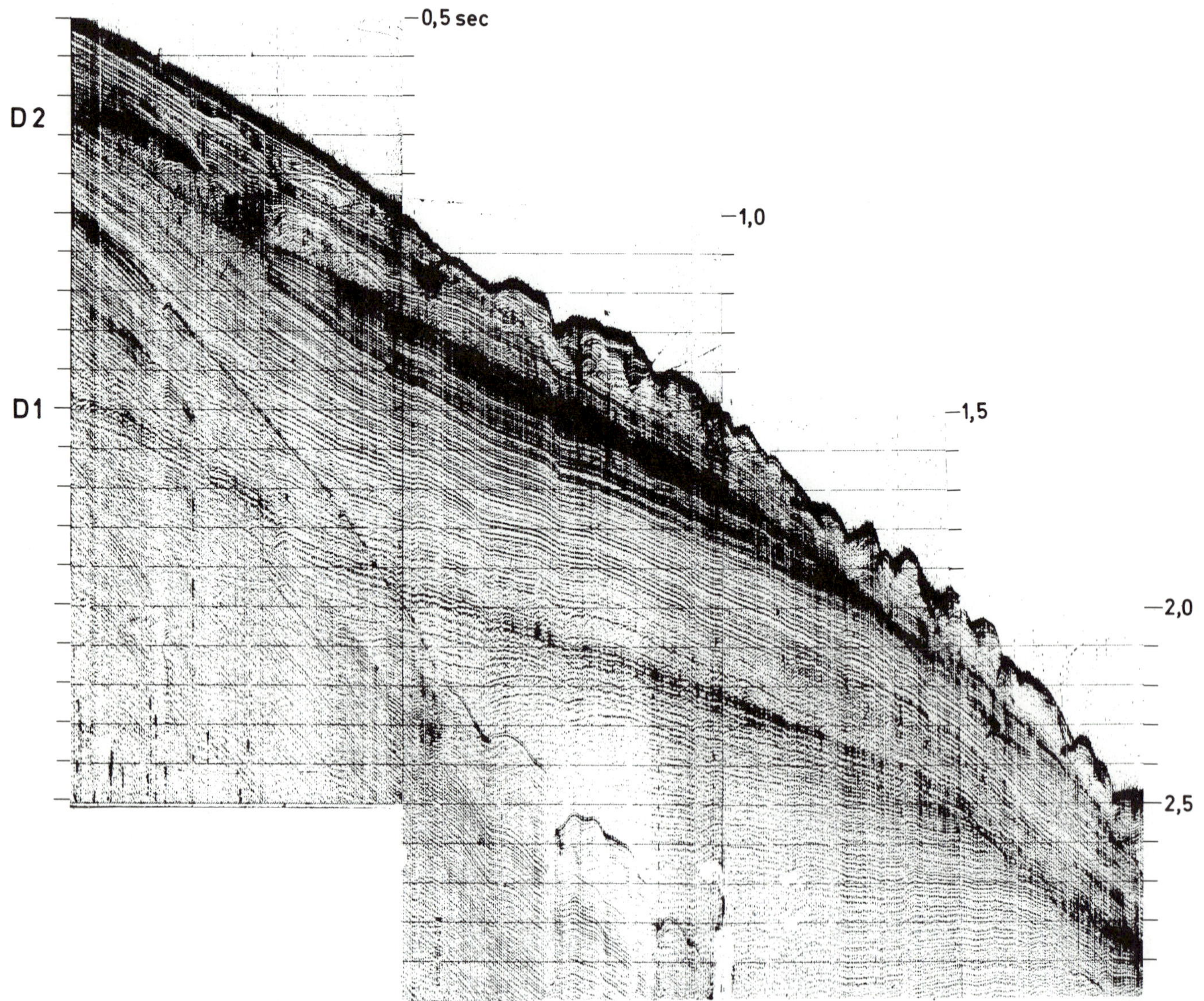

Fig. 3. Original record off Cape Bojador of section indicated in Figure 2; note discontinuities D_1 and D_2. Horizontal extent is about 43 km. (See also Fig. 7.)

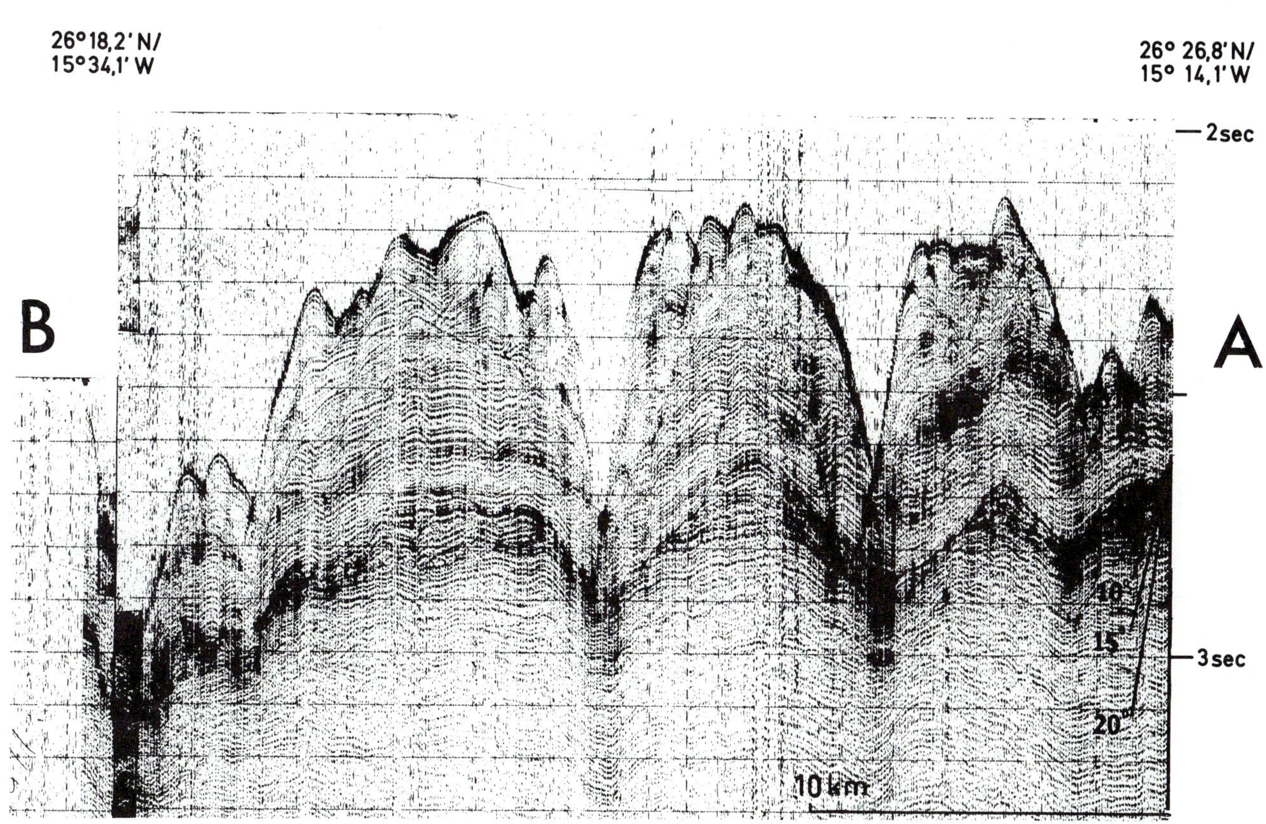

Fig. 4. Original record parallel to the continental slope off Cape Bojador. Section indicated in Figure 1.

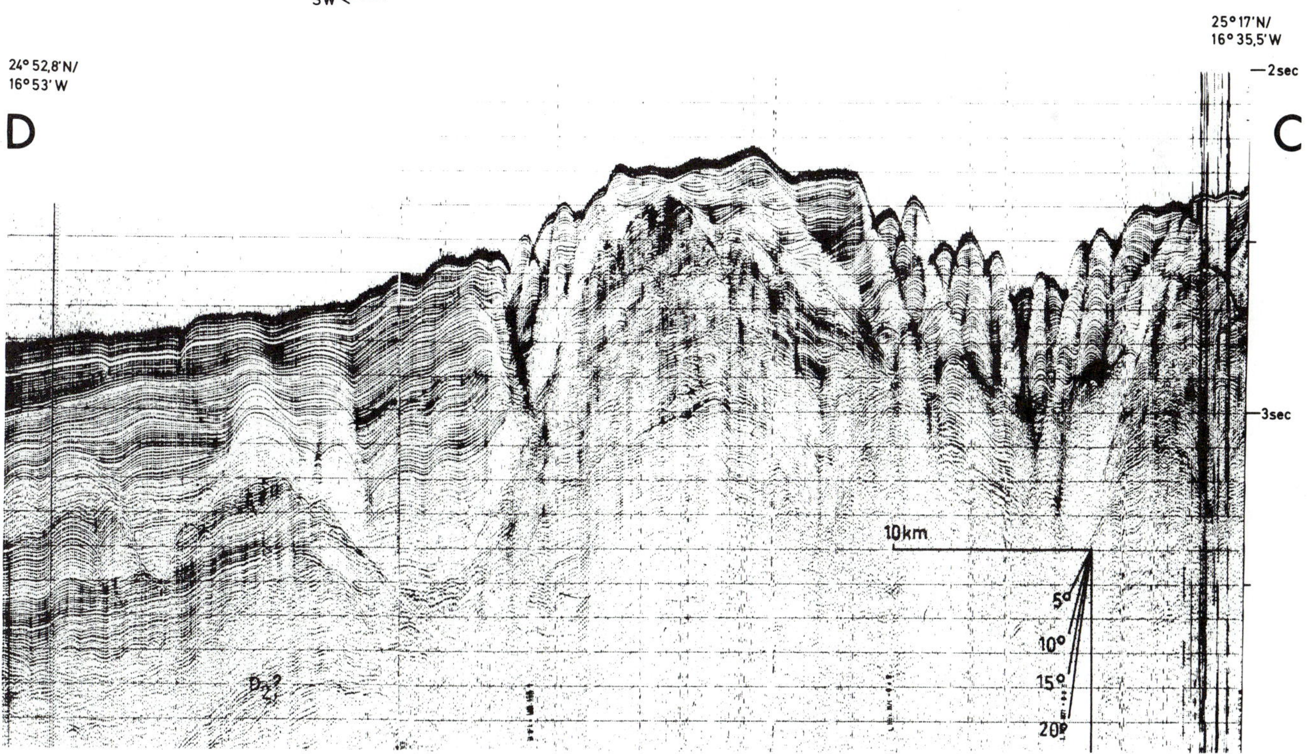

Fig. 5. Original record parallel to the continental slope off Wadi Craa. Section indicated in Figure 1.

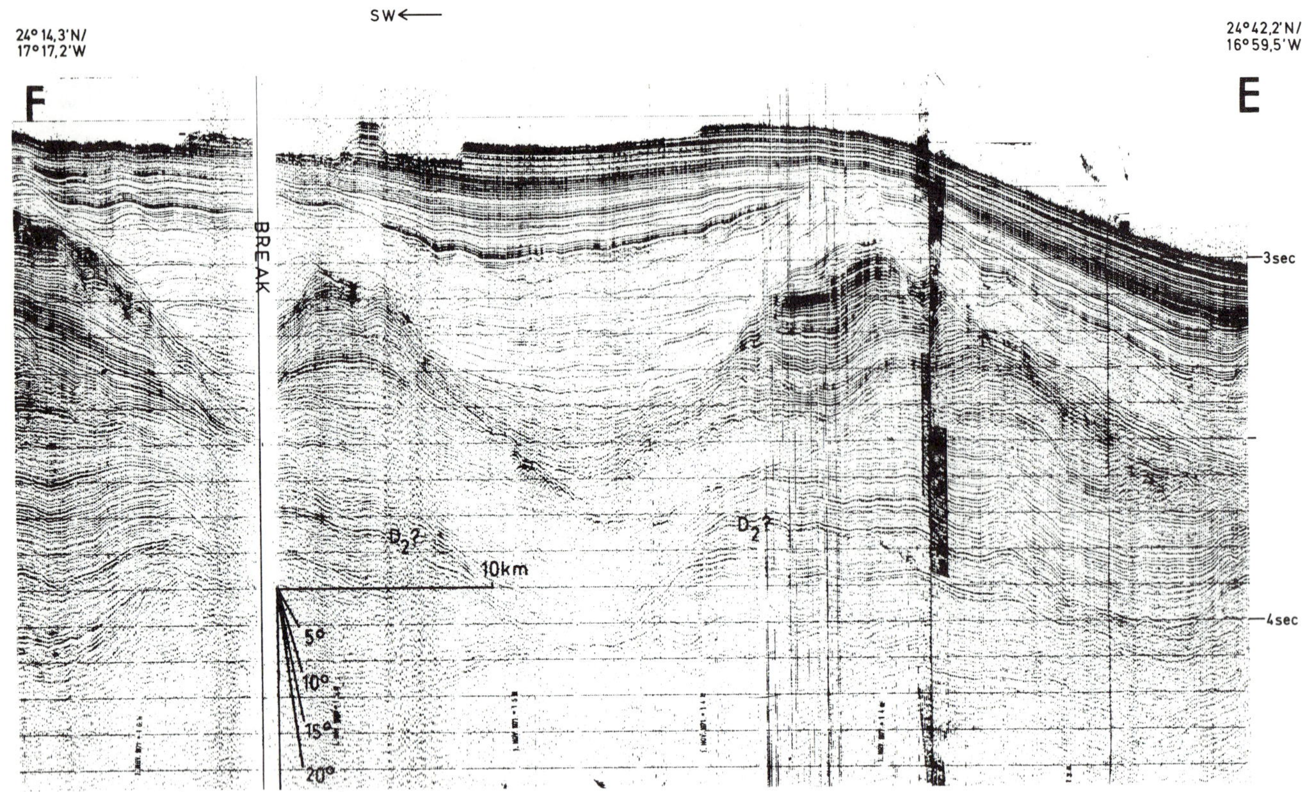

Fig. 6. Original record parallel to the continental slope off Wadi Craa. Section indicated in Figure 1. Note filling and reexcavation of large slope incisions and some features of young erosion.

the upper slope area, where it cuts into Upper Cretaceous to lower Tertiary sediments. At Deep Sea Drilling Project (DSDP) site 139, horizon D_2 is represented by a facies boundary between pelagic, chalky nanno-ooze, in the lower part of which terrestrial sediment may be intercalated, and lowermost Miocene with mature quartz sandstone. At DSDP site 140, Oligocene is missing (Berger and Rad, 1972).

Following the important Oligocene-Miocene erosion and regression phase, sedimentation was next interrupted by discontinuity D_3, possibly of Upper Miocene age. The sediments overlying discontinuity D_3 are characterized in the continental slope area off Cape Barbas by very disturbed and deformed stratification (Fig. 8). The deformed strata are overlain by undeformed ?Pliocene-Recent sediments.

Cape Blanc. Discontinuites D_1, D_2, and ?D_3 can also be recognized in profile x_1 (Figs. 1 and 8). The sudden drop of D_1 90 km seaward of today's shelf margin is probably the Cenomanian shelf edge. The overlying Oligocene-Miocene D_2 discontinuity, which cuts into the undisturbed Cretaceous to Lower Tertiary section, also dips sharply seaward here.

Dakar Area. The Cayar seamount (Fig. 10) and the Cayar channel (Fig. 11) are the dominant structural elements in the area offshore Dakar. The channel was formed after the double-reflection horizon D_2 (Fig. 12), the stratigraphic position of which is uncertain here. By correlation across the seismic profiles, it could be either Oligocene or Miocene. In the slope area, D_2 caps sediments that either dip less than the overlying beds or dip slightly to the east (Fig. 10, Profiles 3, 6, 9). The relief of D_2 is fairly regular. Only in the slope area at depths between 1,000 and 2,000 m does this discontinuity dip more steeply than approximately 5°. This could be the location of the Oligocene-Miocene continental slope. Since D_2 seems to be raised and pierced by the Cayar Seamount we assume that the volcanic seamount is younger. Oligocene-Miocene and Pleistocene volcanic effusions around Dakar are mentioned in Furon (1960) and Haughton (1963).

On the flanks of the Cayar Seamount, slump features are developed that may be counted as large-scale structures. A compilation of comparable examples may be found in Roberts (1972). The face of the slump scar (escarpment in Figs. 9, 10, 11, and 13) extends from the continental slope at a water depth of 1,000 m southward, dropping to water depths of 1,300 m. It then extends to the saddle between slope and seamount and surrounds its northern flank, dropping to water depths there of more than 2,600 m. Its entire length is more than 100 km. On the northwestern side of the seamount possibly two slump scars were observed (Figs. 9 and 10). The slumped sediment layer had a thickness of up to 300 m. Its sliding plane in its steepest upper part is inclined by 2°30' just below the crescent-shaped shear plane. The slumped sediments were transported as much as 1,700 m downslope and 55 km seaward and were deposited on the upper continental rise in a mass of acoustically diffuse material of 250 m thickness. No shear planes, folds, or other typical slumping features were observed. Its northern and western limits were not crossed by the *METEOR*.

The morphology of the entire large-scale slump structure appears fresh. In surface depressions no recent sediment infillings were observed on the airgun profiles (Fig. 14).

DISCUSSION

Outbuilding. If the rates of sedimentation from Table 2 and the slope inclinations from Table 1 are taken into account, some rates for outbuilding of the continental slope may be derived (Table 3). Some additional average outbuilding rates may be calculated from the sediment thickness above reflection horizon D_2 and its age of some 25 million years. The generally lower Neogene outbuilding rates illustrate the influence of erosional phases. Reference values are given by Emery et al. (1970, eastern U.S. margin, Tertiary = 10 cm/thousand years) and extreme values by Hospers (1971, Niger Delta, since Middle Eocene = about 500 cm/ty).

It is interesting to compare here the west African continental margin with the margin off Cape Hatteras and the Blake Plateau. The mostly complete blanket of Neogene sediments from the shelf edge to the continental rise (Figs. 2, 7, 10 and 11) is missing in the American examples although they are found off more humid regions. Differences in biogenic production and destructive processes may be responsible for this.

Still problematic is the extremely wide continental rise in the area of the Cape Verde Isles. Rona (1970) relates this to increased aeolian transport of sediments by the tradewind system—this would make the rise a sedimentary feature and would require an additional aeolian sediment accumulation of about 500-1,000 m. Sedimentological investigations and sedimentation rates in Table 2 off Spanish Sahara (Diester-Haass et al., 1973) and from a box core in 3,315-m water depth near 19°22'N, 19°55.8'W (*METEOR* 25/1971, core 12329, Diester-Haass et al., unpublished) suggest a maximal Holocene aeolian sedimentation rate of about 0.4 cm/ty. During the Pleistocene the maximum could have been about 0.6 cm/ty as a result of similar aridity and increased tradewind activity off Cape Blanc. This would mean about 85-170 my for the accumulation of the Cape Verde Rise there, even assuming this maximal value to be realistic during all the Tertiary, where more humid phases are known from the Sahara (Schwarzbach, 1953). Therefore, we believe that at least this part of the Cape Verde Rise is structurally determined.

We received additional information from the airgun profile between Cape Blanc and Cape Verde Isles (Fig. 1). The morphological rise can also be observed in the subsurface in an area hatched in Figure 1, where the uppermost 600 m of sediment shows pronounced parallel stratification. Below this and above a marked discontinuity, there are approximately 500 m of sediment to the south and more than 600 m of sediment to the north. The discontinuity is raised by 500 m at its highest point, which approximately corresponds to the morphological elevation of the Cape Verde Rise there above its surroundings. This situation suggests a very young uplift which may represent reactivation of an older structure. It is also noteworthy that this structure is situated on the connection between Cape Verde Islands and the Requibat Uplift. This, however, needs to be further substantiated through additional airgun profiles, especially perpendicular to the line reported in this paper.

Destruction by Slumping. *Small-Scale Structures.* Few slumping features have been observed in sediment cores taken by *METEOR* off West Africa, even though almost all cores were carefully investigated by radiographs. Some sediment characteristics may be seen from Table 2. Generally the Quaternary sediments are made up of clayey silts with some sand content. It should be pointed out again that the carbonate content indicates the essentially biogenic origin of the sand- and silt-sized fraction on the continental slope and rise. With respect to soil mechanical properties, this means increased shear strength due to rough and even spiney shapes and sharp edges of fragments. However, it seems also possible that occasional silt layers could be deposited which may be highly unstable.

According to criteria of standard soil mechanics, the sediments listed in Table 2 are stable on slopes up to 15-20°, even under a sediment cover of several meters (Kogler, personal communication). It should be taken into account that the shear strength data, which were all obtained on board ship immediately after retrieval from core cross sections of 15 by 15 cm and 30 by 30 cm, show considerable scatter, owing mainly to a varying degree of bioturbation. Comparable data on stability of continental margin sediments are given by, among others, Moore (1961) and Ross (1971). These findings exclude slumping on the continental slope on a scale of several meters up to tens of meters; slumping might be expected only on canyon walls, which receive locally, in addition, an increased sediment influx, i.e., show a higher rate of sedimentation. It is not very easy to identify from the airgun profiles undulating horizons with slumping features, particularly when an undulating morphology is present. Many features that so far have been interpreted as the results of slumping may be caused by infilled incisions that were cut obliquely by the airgun profile. The examples in Figures 3 and

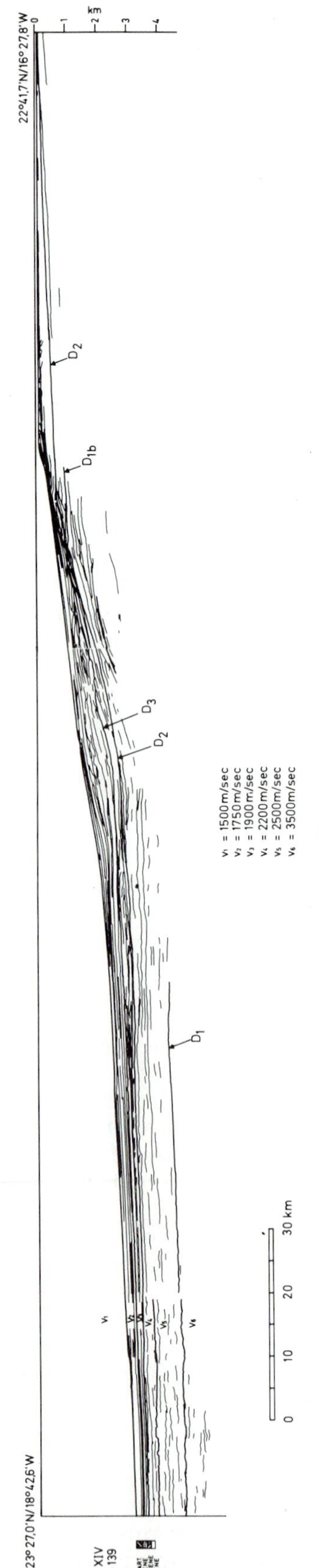

Fig. 7. Vertically migrated seismic reflection profile off Cape Barbas (see Fig. 1, area B, northern profile).

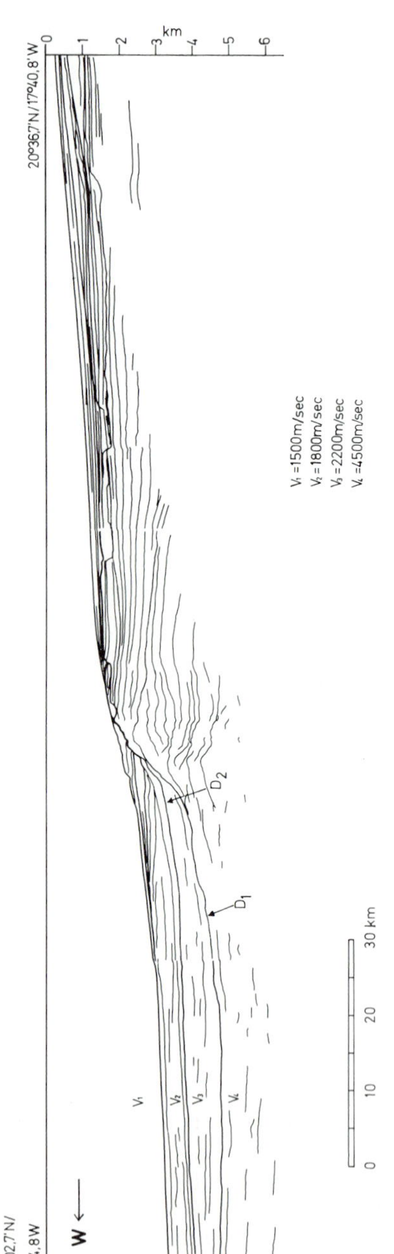

Fig. 8. Vertically migrated seismic reflection profile x_1 off Cape Blanc (see Fig. 1).

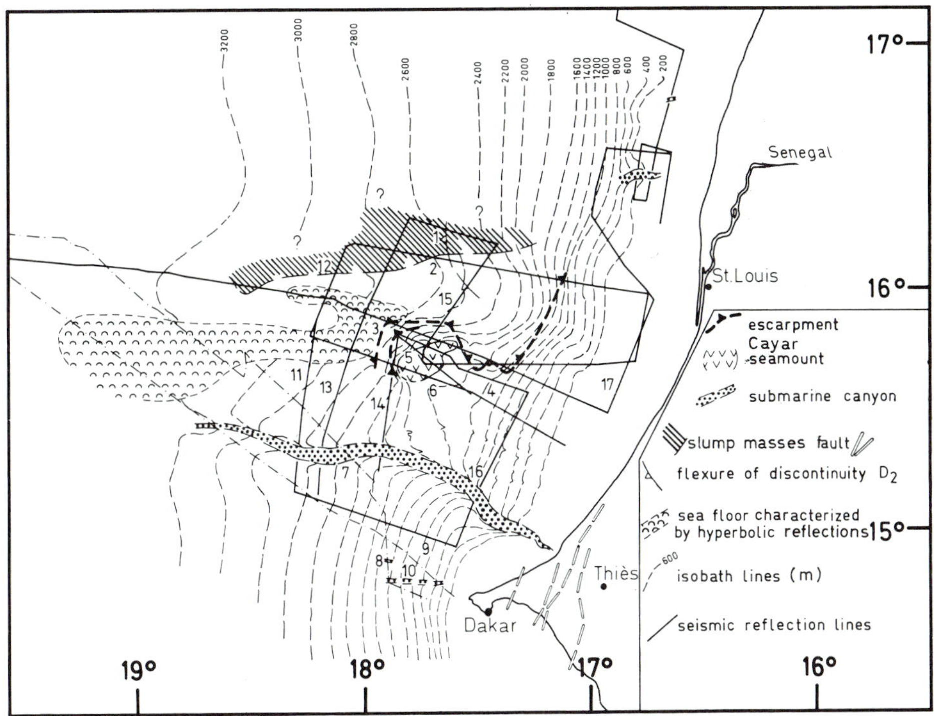

Fig. 9. Structural map off northern Senegal (area C, Fig. 1) compiled by H. U. Schluter. Profiles shown in Figures 10 and 11.

5 demonstrate that such features are quite common to the north of Cape Blanc. This is also the reason that the undulating sedimentary sequence in profile x_2 (Fig. 2) between D_1 and D_2 at 70-100 km offshore is difficult to interpret.

Large-Scale Structures. Earlier, a large slumping area off Senegal was mentioned. A sedimentological discontinuity might have facilitated slumping, possibly as the result of a varying sediment supply from the nearby ancient mouth of the Senegal River as well as sediment quality differences caused by climatic changes during the late Neogene. Volcanic ash might also have been responsible. It is unknown how and when the slumping was triggered.

So far we are unable to recognize in older sequences any indications for slumping of a similar magnitude. Perhaps characteristic slump scars—which have no opposite walls—and acoustically diffuse slumped sediment masses, might be less well preserved or might even be subsequently altered such that they will be difficult to recognize. No indications were found for a rotational slump as shown for the upper continental rise off Cape Hatteras in Rona (1969, Fig. 3).

Destruction by Incisions. *Erosion.* About 115 continental rise and slope incisions with a relief of more than 50 m were investigated by the *METEOR* narrow-beam echo sounder. About half of these are located north of Cape Blanc at about a 2,000-m water depth and half south of Cape Blanc in less than 1,000 m of water. The following results are from the evaluation of these data by Rust and Wienecke (1973). "Incision" was used as a neutral term, including submarine canyons from the shelf to the deep sea and slope valleys confined to the slope and rise. They may cut perpendicular or obliquely through the continental margin; a few might originate from slumping and thus may even parallel the continental margin. The average slope angle of the incisions was <10° for 63% of the examples, 10-20° for 30%, 20-30° for 6%, 30-40° for 1% (maximum value 37°). Many of these incisions are stepped. An example is shown in a profile across the Cayar Canyon (Fig. 12). Therefore, and by lines oblique to the incisions, higher slope angles are actually more frequent than is shown in the above-average values. The maximum floor-to-shoulder relief was 50-100 m (51%), 100-200 m (25%), 200-300 m (10%), >300 m (14%, maximum value 685 m). The width of shoulders <1 km was found in 17%, 1-2 km (35%), 2-3 km (21%), 3-5 km (13%), >5 km (14%). The maximum width was 13 km.

Thirty-eight incisions with a relief of more than 50 m were observed between the profiles of the investigation area A and x_2 over a distance of 550 km essentially parallel to the slope. With an average width of 2-3 km from shoulder to shoulder of each incision, this means that about 20% of the entire length of the continental slope is occupied by gaps. Off Cape Timiris at around 20°N about 40% is missing through such incisions (Fig. 4). If such a system of closely spaced incisions with their numerous branches is deepened and spreads landward and upward on the slope, the entire

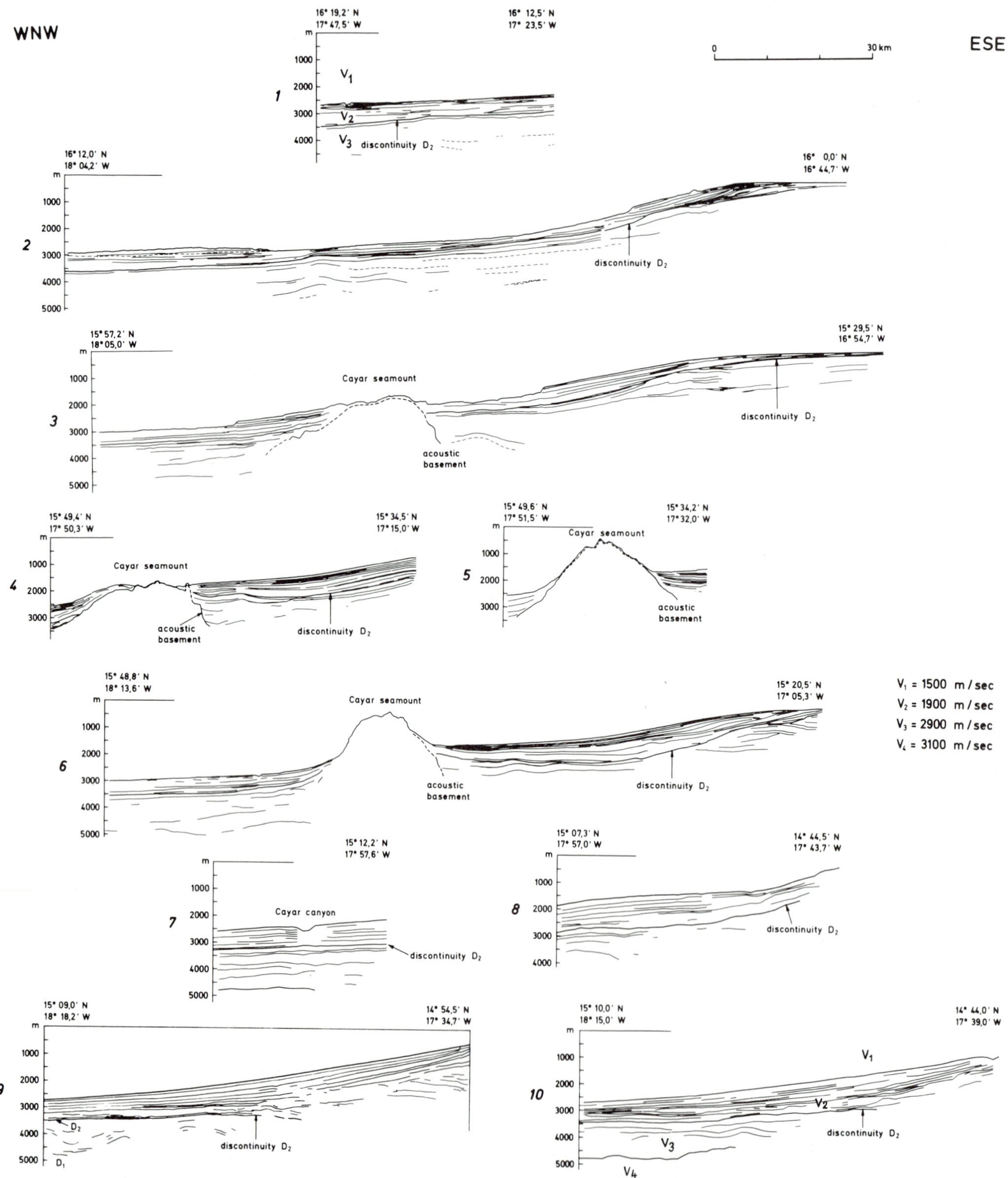

Fig. 10. Vertically migrated reflection profiles off northern Senegal, ESE-trending (see Fig. 9 for location). Vertical exaggeration 1:5.

continental margin recedes, just as the slopes of the Grand Canyon widen through deepening of the side canyons and their tributaries. This must be assumed, for example, in certain areas off Spanish Sahara (between profiles A and x_2) over a distance of more than 300 km. However, these incisions only cut into the lower continental slope. According to *METEOR'S* cruise track 19/1970, they extend upward to the 400-m depth contour only in four cases and reach the actual shelf break (*METEOR* cruise 13/1968) in only one case. Sometimes they cut through the D_2 horizon.

Ancient filled incisions are frequently found off Spanish Sahara. Examples are shown in Figures 5 and 6. They must have been developed after D_2 during the Neogene period and were then obviously much more effective in eroding the continental slope landward than the recent incisions are, because at comparable water depths shoulder-to-shoulder distances were longer (Fig. 6, up to 22 km) and thus slope angles are lower (4-6°). Figure 6 illustrates also that the ancient incisions were more closely spaced and thus could cause gaps of up to to 50% in the slope front. Rona (1971, Fig. 3) shows similar buried incisions farther to the south with shoulder-to-shoulder distances of around 30 km, a maximum relative relief of 700-1,000 m, and slope angles of 3-4°. Even the most spectacular open incision in the area investigated by *METEOR*, the Cayar Canyon, with a maximum shoulder-to-shoulder distance of 8 km (Dietz et al., 1968), does not reach the magnitude of the ancient incisions.

Sedimentation. The eroded and transported sediment from the canyons usually accumulates on the continental rise and on the adjacent abyssal plains. So far only a few indications of turbidites are known from the continental rise off West Africa. Horn et al. (1972) illustrated a few such layers in sediment cores of up to 12 m in length, of Pleistocene age. Even these are missing in the present cores from *METEOR'S* cruise 25/1971. However, on the abyssal plain they are reportedly frequent according to Horn et al. (1972). Berger and Rad (1972, p. 808)

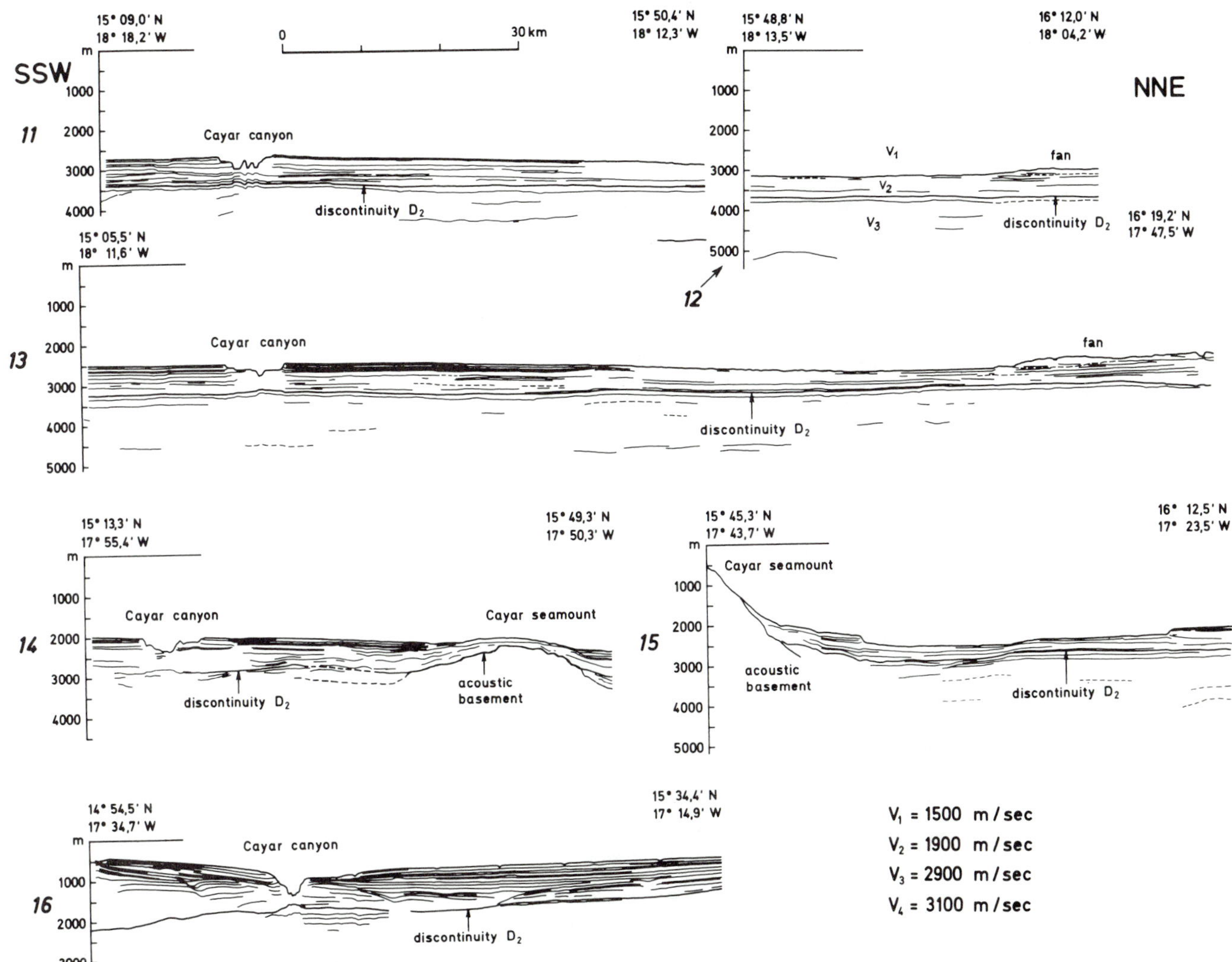

Fig. 11. Vertically migrated reflection profiles off northern Senegal, SSW-trending (see Fig. 9 for location). Vertical exaggeration 1:5.

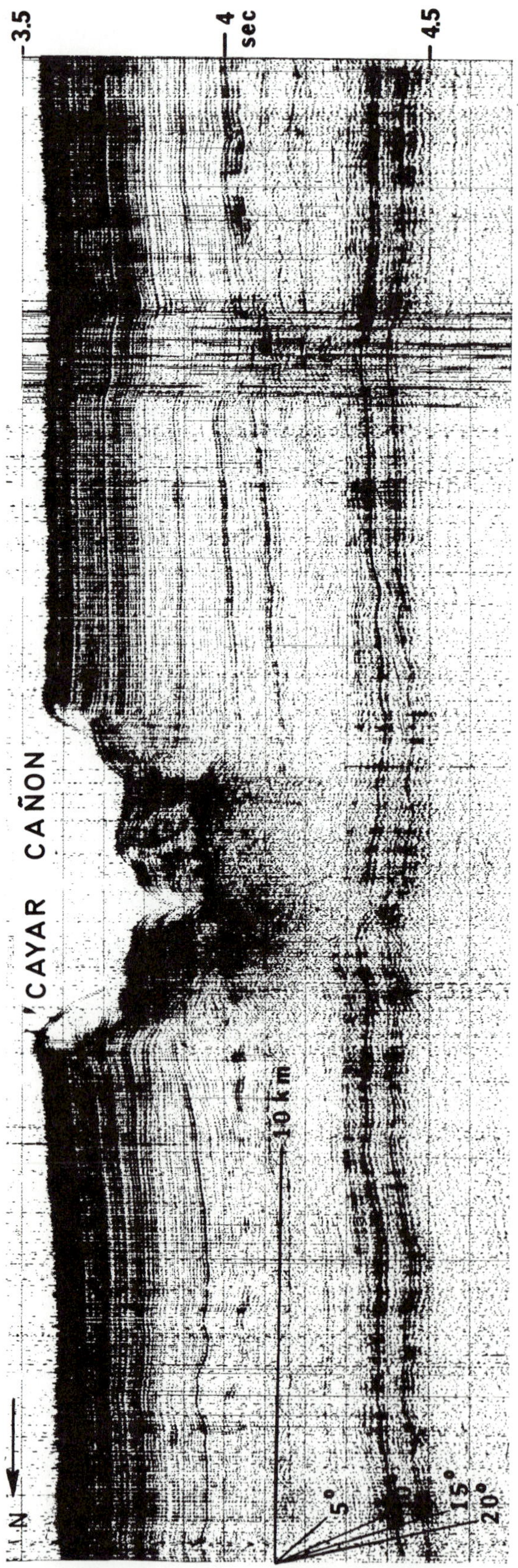

Fig. 12. Original record of the Cayar Canyon area (Fig. 11, profile 13). Note discontinuity D_2 (about 4.5 sec) characterized by double reflection horizons.

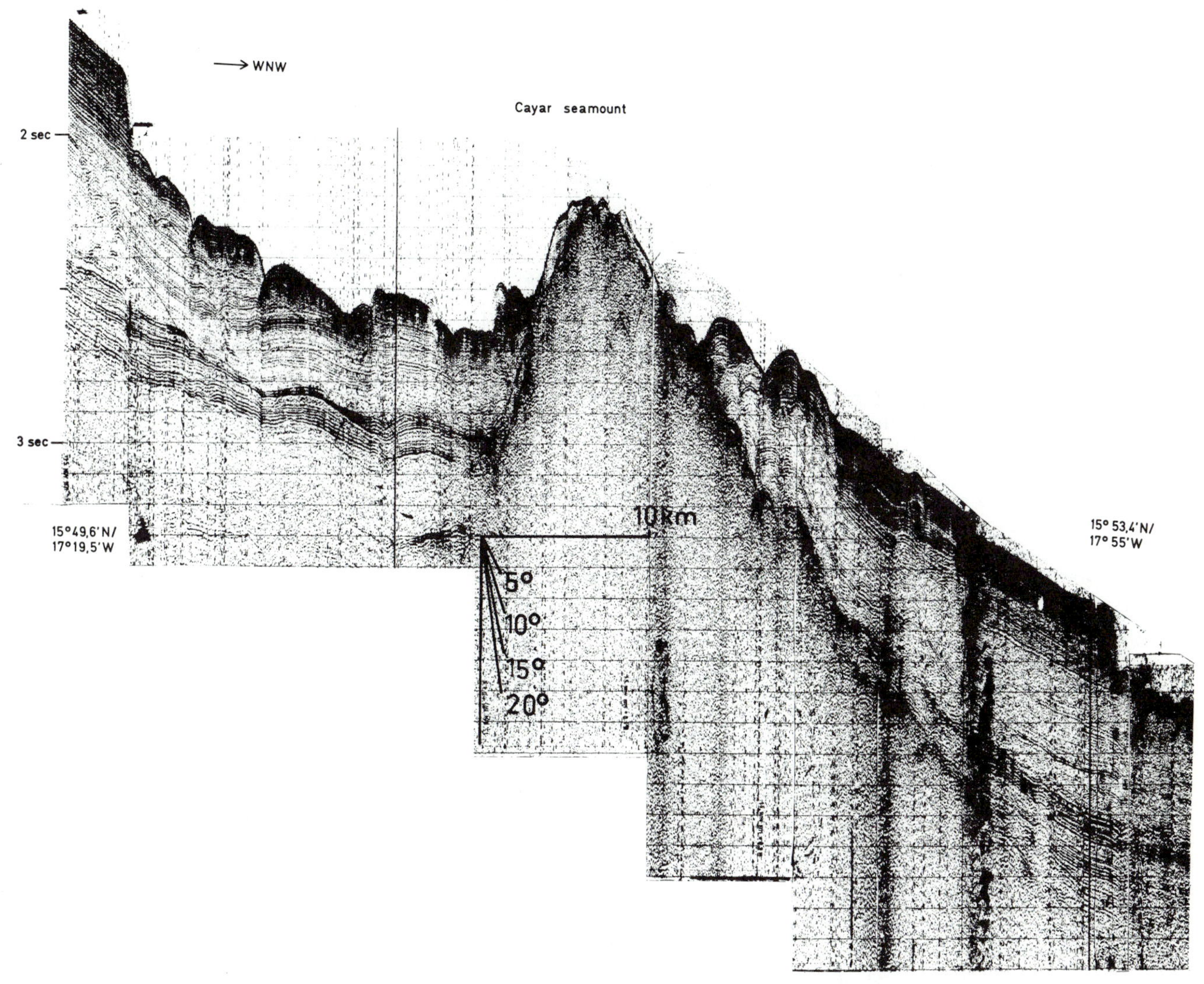

Fig. 13. Original record of Cayar Seamount area (Fig. 10, profile 3). Note escarpments caused by slumping.

report in DSDP site 140 an increased supply of terrigenous minerals in Lower Miocene strata. At site 139, which is located closer to the land, terrigenous minerals were reported from the lowermost Miocene and Lower Miocene sediments. The increased terrigenous influx may correspond to the event that cut the now-filled canyons.

This, however, would mean that after the event the supply of terrigenous material to the continental rise must have been drastically reduced and that the incisions were filled with sediments. This indicates that transport through the incisions outward practically ceased, and, according to data given above, large masses of sediments were retained on the continental slope. To the south of profile x_2 (Fig. 6) an uninterrupted, completely undisturbed sediment layer, approximately 200-300 m thick, covers the essentially filled incisions. It may occasionally contain erosional features, but in general this blanketing sediment layer also indicates reduced terrigenous flux seaward. This is also supported by findings of Berger and Rad (1972, p. 808) from DSDP site 138.

Three-fourths of the recent incisions with a relief of 50 m and more are V-shaped, one-fourth of them show a flat base which may indicate the beginning of sediment infilling, and in one case only was a completely filled young canyon observed, off a former mouth of the Senegal River (16°32'N, 16°39'W).

Spatial Distribution. Open incisions are more frequently found off former mouths of the Senegal River around 16°30'N, between 18 and 19°N to the south of Cape Timiris, which may be related to the former Inchiri Delta, and around 20°N between Cape Timiris and Cape Blanc. In these places a relationship to the fossil drainage system is also suspected, which had developed behind the present-day Banc D'Arguin.

The Cayar Canyon has developed unrelated to the Senegal River according to Dietz et al. (1968). It is situated, like many Californian examples, on the windward side of a cape relative to the preferred direction of sediment transport in shallow water. In a similar situation off Cape Barbas, however, no

Table 3. Continental slope outbuilding.

	Holocene		Late Pleistocene		Neogene	
					Present Shelf Edge	
	Sed. Rate (cm/1,000 y)	Outbuilding (cm/1,000 y)	Sed. Rate (cm/1,000 y)	Outbuilding (cm/1,000 y)	Thickness (m)	Sed. Rate (cm/1,000 y)
Cape Bojador	4 (>2,000-m water depth)	100	9 (>2,000-m water depth)	230	300	1.2
Wadi Craa	—	—	—	—	500	2.0
Cape Barbas	4 (average, 1,000-3,000-m water depth)	150	10 (average, 1,000-3,000-m water depth)	380	400	1.6
Cape Blanc	—	—	—	—	800	3.2
Senegal	15 (average, 1,000-3,000-m water depth)	>300	—	—	500	2.0

canyons have developed, probably owing to the wide shelf area there. The ancient and now-filled incisions we found so often between Wadi Craa and Cape Barbas (Fig. 1, between x_2 and B) could possibly be related to a fossil drainage system. In front of them the 3,000-m isobath extends far to the west. We cannot yet offer an explanation for the huge sediment accumulation necessary to fill these gaps. Perhaps increased biogenic production by upwelling phenomena has also contributed material.

The incisions between X_2 and area A (Fig. 1) were opened—or reopened?— during more humid phases of the Pleistocene, as has been proved for this area by Diester-Haass et al. (1973).

Destruction by Horizontal Currents. According to the above, the effect of the downslope currents on the morphology can be seen directly on the echo-sounding records and indirectly inferred from the airgun profiles. This becomes more difficult to establish for prevalent horizontal currents. So far, on the continental shelf off west Africa such indications have been observed by side-scan sonar investigations (Newton et al., 1973). There, sand presently is being transported by the Canary current down to water depths of 100 m. On the continental slope and rise, bottom photographs have not yet revealed any bottom-current activities except on seamounts and on their flanks (Lowrie et al., 1970; Johnson et al., 1971; *METEOR* cruise 25/1971). Pérès (1961) reports observations on board the bathyscape off Dakar of bottom currents of about 20 cm/sec at 1,450 m of water depth.

According to calculations from *METEOR*-I data, the geostrophic currents off west Africa reach maximum speeds of 3 cm/sec at 800 m of water depth, and 9 cm/sec at 2,000 m of water depth

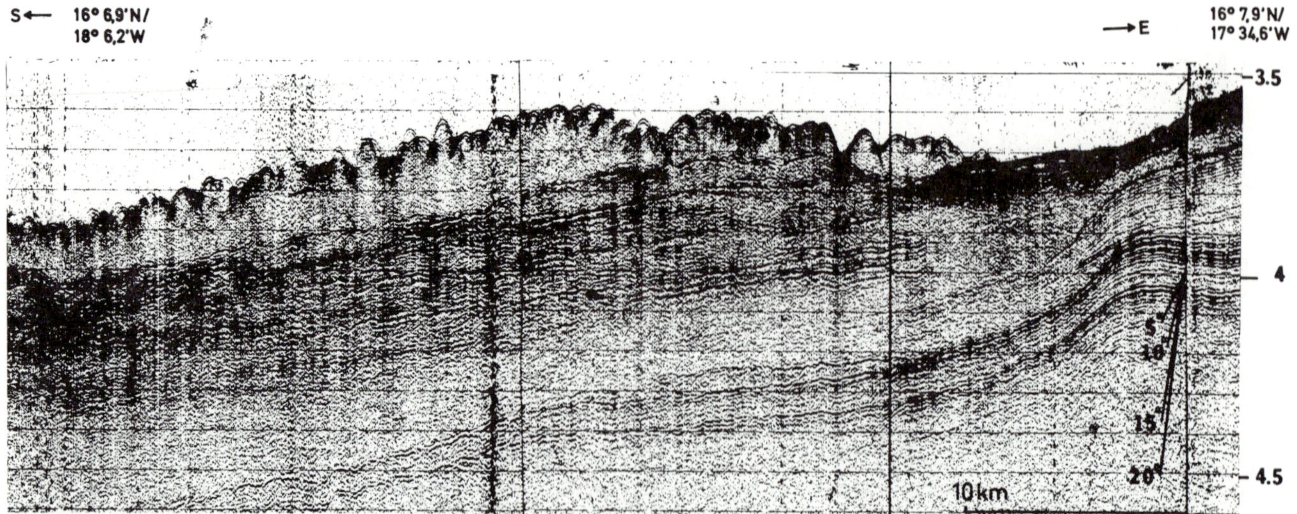

Fig. 14. Original record showing slump masses (Fig. 10, profile 2). Note flexure of D_2 (4/sec) in the right part.

Neogene							
Upper Slope (1,000-m Water Depth)			Lower Slope (2,000-m Water Depth)			Continental Rise	
Thickness (m)	Sed. Rate (cm/1,000 y)	Outbuilding Rate (cm/1,000 y)	Thickness (m)	Sed. Rate (cm/1,000 y)	Outbuilding Rate (cm/1,000 y)	Thickness (m)	Sed. Rate (cm/1,000 y)
300	1.2	30	200	0.8	20	800	3.2
>1,000	4.0	—	400-1,400	1.6-5.6	—	700	2.8
900	3.6	140	900	3.6	140	500	2.0
700	2.8	—	200	0.8	—	700	2.8
700-1,000	2.8-4.0	50-80	1,000	4.0	80	500-700	2.0-2.8

(Defant, 1941). These currents at the eastern boundary of the North Atlantic Ocean are small compared to those at the western boundary—this is also to be expected on theoretical grounds. For example, these currents were unable to fill a small 5-10-m subsidence area on top of a salt structure off Nouakchott in 700-m water depth found by *METEOR* (17°52'N, 16°43,3'W; circle in Fig. 1). How, then, are the periods of nondeposition and erosion to be explained which were found in the strata of profiles between Cape Bojador and Cape Blanc? Two possibilities will be discussed below.

Accentuated Oligocene-Miocene Subsidence. The lower Cretaceous continental slope was farther to the west, and today is marked by a relative steep dip (5-10°) of the discontinuity D_1, probably of Cenomanian age, as mentioned before. The distances between today's shelf edge and the Cenomanian continental slope are around 60-75 km off Cape Bojador, 40 km off Wadi Craa, 40 km off Cape Barbas, and 90 km off Cape Blanc.

The sediment thicknesses between discontinuities D_1 (about 100 my) and D_2 (about 25 my) in the outer part of the upper Cretaceous to Oligocene shelf indicate minimal sedimentation-subsidence rates of about 1.6 cm/ty off Cape Bojador, 1.3 cm/ty off Wadi Craa, 1.3 cm/ty off Cape Barbas, and 1.6 cm/ty off Cape Blanc.

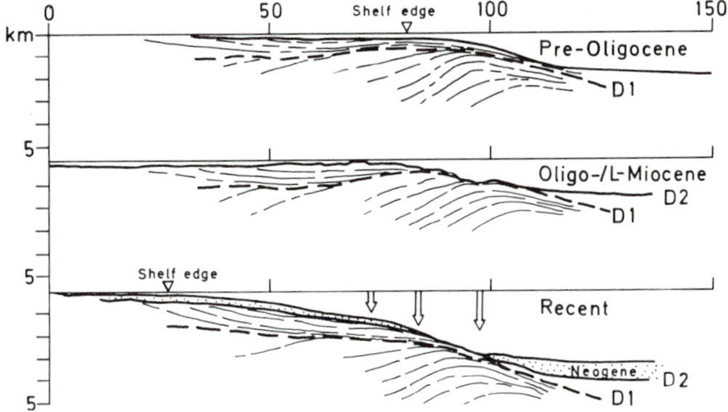

Fig. 15. Schematic evolution of the continental margin off Spanish Sahara (area A, Fig. 1).

The pre-Oligocene shelf edge was located near the lower Cretaceous one (Fig. 15). During the Oligocene regression the older strata were eroded subaerially or by horizontal currents under shallow water cover, particularly at the shelf break. It is a well-known phenomenon that strata thin toward the shelf edge. From erosion of this kind, an erosional horizon (D_2) has developed. At the same time and after the formation of D_2, the area around the former shelf edge began to subside drastically. Estimated amounts and average subsidence rates are 1,500 m (6 cm/ty) off Cape Bojador, the same for Wadi Craa and Cape Blanc, and 1,000-1,500 m (4-6 cm/ty) off Cape Barbas. This accentuated subsidence and possibly sea-level lowering (Rona, 1973b) caused the development of the incisions already mentioned, which later were filled. As shown earlier, areal erosion by these incisions certainly played an important role.

Many indications of tectonic-volcanic activities during this time are already known from this region and from the area to the north (Choubert and Marcais, 1956; Alia, 1960; Querol, 1966; Rothe, 1968; Rothe and Schmincke, 1968; Tooms et al. 1971; Robb, 1971; Abdel-Monem et al., 1972, volcanic activity, eastern Canary Isles). Also, filled-in canyons are missing in our records farther to the south, off Senegal, for example.

The rate of subsidence of the shelf edge for the post-Oligocene period would have been about 6 cm/ty—to give the average value for the last 25 million years. Matthews and Smith (1971) and Cressard (1972) report even higher rates for the Neogene subsidence of the Rockall area. The position of the present shelf edge can be explained by later progradation of sediments.

Oligocene-Miocene Oceanic Event. An alternative explanation for the Oligocene hiatus with only low rates of subsidence may be seen in a global event causing overall erosion. Erosional processes during Oligocene times in shallow and in deep waters were also observed in the Senegal Basin (Spengler et al., 1966), and off parts of East Africa (Kent et al., 1971; Beltrandi and Pyre, 1972; Simpson et al., 1972). Similar processes were important during the Oligocene in the Bay of Biscay (Davies and Laughton, 1972); Oligocene sediments are missing at DSDP site 140 (Berger and Rad, 1972). Two factors may be responsible for the drastic change in water movement at all depths: (1) the lowering of sea level caused by variations in the sea-floor spreading rate (Rona, 1973a) and/or by the beginning of glaciation with the buildup of an extensive ice shield on Antarctica about 20 my ago (Hayes et al., 1973); and (2) the development of the circum-Antarctic current system, the separation of Greenland from Europe, and changes of relief in the Panama region.

This would require considerable erosion at water depths of 1-3 km between Cape Bojador and Cape Blanc—not including any subsidence since then (1.8-3.2 km in profile A_1, 2.7-3.0 km in A_2, 2.2-3.0 km in A_3, about 1-2 km in profiles B, and about 1-3 km in profile x_1). Furthermore, similar events should have been responsible for the formation of the discontinuity—and sometimes angular unconformity, D_1 (Fig. 2).

We therefore prefer the first interpretation, until further attempts are made to clarify this global event and related problems such as the question of where the eroded sediments have accumulated during Oligocene times.

CONCLUSIONS

According to our reflection seismic results, it appears that a shelf-edge anticline, located 40-75 km to the west of the present-day shelf edge, and a zone of subsidence, located landward of that anticline, were responsible for a preferred accumulation of shelf sediments during Lower Cretaceous time between Cape Bojador and Cape Blanc. At the end of Lower Cretaceous a transgression (discontinuity D_1) followed a regression. Then until the Oligocene, the shelf and slope were built up and out during slow subsidence, except for a regression phase during the early Tertiary, indications for which were found, however, only off Cape Bojador.

At the beginning of the Oligocene, a strong regression and erosion period started on the shelf and slope. Large parts of the pre-Oligocene continental margin were removed (discontinuity D_2). This may be explained in two ways:

1. The present-day transitional zone between continental slope and rise was located 1.5 km higher and was thus subject to subaerial and/or shallow-water erosion. This was supplemented by large-scale erosion through submarine incisions. According to this interpretation during post-Oligocene time, the shelf edge further subsided at an increased rate through regional tectonics.

2. A global explanation could be the result of a worldwide lowering of the sea level and accompanying drastic changes in oceanic circulation during the Oligocene-Miocene.

The continental margin receded by subsidence more than 50 km, followed later by predominant upbuilding and outbuilding. Increased progradation of the continental slope was observed where sediment supply was particularly intense. This may have been through terrigenous influx by Neogene river systems or through increased biogenic production. Slope incisions were then filled in.

For the Quaternary, slope destruction was observed only around Cape Bojador in the form of narrowly spaced canyons and large-scale slumping to the north of the Cayar Seamount.

ACKNOWLEDGMENTS

Morphological, sedimentological, and geophysical data were collected on the R.V. *METEOR* (cruises 8/1967 (Closs et al., 1969), 13/1968, 19/1970, 25/1971). Cruise 25/1971 followed an agreement made during a symposium, "Geology of the East Atlantic Continental Margin," Cambridge, England, March 1970 (Seibold and Emery, 1970). First results were published in Seibold (1972), Diester-Haass et al. (1973), and Hinz et al. (1974).

We would like to express our thanks to the German Research Society (DFG); the German Hydrographic Institute (DHI); the German Geological Survey (BfB); the crew of the *METEOR* with Master E. W. Lemke; S. Garde, H. U. Schlüter, and G. Wissmann of BfB; to many members of the Geological Institute, Kiel University: L. Diester-Haass, M. Hartmann, F. C. Kögler, H. Lange, P. Müller, U. Pflaumann, H. J. Schrader, E. Suess (unpublished results), and I. Bornhöft (technical assistance); and to K. O. Emery, E. Uchupi of Woods Hole; and G. H. Keller and P. A. Rona of Miami.

BIBLIOGRAPHY

Abdel-Monem, A., Watkins, N. D., and Gast, P. W 1972; Potassium-argon ages, volcanic stratigraphy, and geomagnetic polarity history of the Canary Islands: Tenerife, La Palma, and Hierro: Am. Jour. Sci., v. 272, p. 805-825.

Alia, M., 1960, La tectonica del Sahara espanol: 21st Internatl. Geol. Congr. Copenhagen, p. 18, p. 193-202.

Auxini, A. E., 1969, Correlation estratigrafica de los sondeos perforados en el Sahara espanol: Bol. Geol. Minero, Madrid, v. 83, p. 235-251.

Aymé, J. M., 1965, The Senegal Salt Basin, *in* Salt basins around Africa: London Inst. Petr., p. 83-90.

Beck, R. H., 1972, The oceans, the new frontier in exploration: Austr. Pet. Expl. Soc. Jour., v. 12, p. 7-28.

Beltrandi, M. D., and Pyre, A., 1972, Geological evolution of South West Somalia, *in* ASGA-AAGS, Symposium of the sedimentary basins of Africa: Montreal, Canada.

Berger, W. H., and Rad, U. von, 1972, Cretaceous and Cenozoic sediments from the Atlantic Ocean, *in* Initial reports of the Deep Sea Drilling Project, v. 14;

Washington, D.C., U.S. Govt. Printing Office, p. 787-954.

Choubert, G., 1952, Histoire géologique du domaine de l'Anti-Atlas, *in* XIX Congr. Geol. Internatl. 3° Série, Maroc, no. 6, fasc. I, pt. 1, Rabat.

———, 1968, International tectonic map of Africa 1:5,000,000 Assoc. African Geol. Survey: Paris, UNESCO.

———, ed., 1971, Tectonique de l'Afrique: sciences de la terre, v. 6: Paris, UNESCO, 602 p. (West African coastal basins, p. 391-402.

———, and Marcais, J., 1956, Lexique stratigraphique du Maro: Geol. Survey Morocco Notes and Mem., v. 134, p. 165, Rabat.

Closs, H., Dietrich, G., Hempel, G., Schott, W., and Seibold, E., 1969, Atlantische Kuppenfahrten 1967 mit dem Forschungsschiff *METEOR* Reisebericht: *METEOR* Forsch. Ergeb. A, no. 5, p. 1-71.

Cressard, A. P., 1972, Etude géologique des régions sous-marines de Porcupine et Rockall (Atlantique Nord-Est): Univ. Rennes, C-251-81, thesis.

Davies, T. A., and Laughton, A. S., 1972, *in* Initial reports of the Deep Sea Drilling Project, 12: Washington, D.C., U.S. Govt. Printing Office, p. 905.

Defant, A. 1941, Die absolute Topographie des physikalischen Meeresniveaus und der Druckflächen, sowie die Wasserbewegungen im Atlantischen Ozean: Wiss. Erg. Deut. Atlantic Expedition *METEOR*, 1925-1927, v. 6, p. 2.

Delany, F. M., ed., 1970/1971, The geology of the east Atlantic continental margin. ICSU-SCOR Working Party 31 Symposium Cambridge 1970: Inst. Geol. Sci. Repts., London, no. 70/13, Gen. and Econ. Papers (1), no. 70/16, Africa (4).

Diester-Haass, L., Schrader, H. J., and Thiede, J., 1973, Sedimentological and paleoclimatological investigations of two pelagic ooze cores off Cape Barbas, North-West Africa: *METEOR* Forsch. Ergeb. C, no. 16, p. 19-66.

Dietz, R. S., Knebel, H. J., Somers, L. H., 1968, Cayar Submarine Canyon: Geol. Soc. America Bull., v. 79, p. 1821-1828.

Egloff, J., 1972, Morphology of ocean basin seaward of northwest Africa: Canary Islands to Monrovia, Liberia: Am. Assoc. Petr. Geol. Bull., v. 56, no. 4, p. 694-706.

Emery, K. O., Uchupi, E., Philips, J. D., Bowin, C. O., Bunce, E. T., and Knott, S. T., 1970, Continental rise off eastern north America: Am. Assoc. Petr. Geol. Bull., v. 54, p. 44-108.

Furon, R., 1960, Géologie de l'Afrique, 2nd Ed.: Paris, Payot, 400 p.

Haughton, S. H., 1963, The stratigraphic history of Africa south of the Sahara: New York, Hafner Press, 365 p.

Hayes, D. E., et al., 1973, Leg 28 Deep-sea drilling in the southern ocean: Geotimes, v. 18, no. 6, p. 19-24.

Heezen, B. C., Tharp, M., and Ewing, M., 1959, The floor of the oceans, pt. I, The North Atlantic: Geol. Soc. America Special Paper 65, 122 p.

Hinz, K., Seibold, E., Wissmann, G., 1974, Continental slope anticline and unconformities off West Africa: *METEOR* Forschung Ergebnise; vol. c-1, no. 17, p. 67-73.

Horn, D. R., Ewing, J. I., Ewing, M., 1972, Graded-bed sequences emplaced by turbidity currents north of 20°N in the Pacific, Atlantic and Mediterranean: Sedimentology, v. 18, p. 247-275.

Hospers, J., 1971, The geology of the Niger Delta area, *in* Delany, ed., Inst. Geol. Sci. Rept., London, no. 70-16, p. 121-142.

Johnson, G. L., Giresse, P., Dangeard, L., Jahn, W. H., 1971, Photographies de fonds bathyaux et abyssaux de l'océan Atlantique entre 30°N et l'Equateur. Mission du KANE 9: Bull. BRGM, sér. 2, sect. IV, 1, p. 59-95.

Kennedy, W. Q., 1965, The influence of basement structure on the evolution of the coastal (Mesozoic and Tertiary) basins of Africa, *in* Salt basins around Africa: London Inst. Petr.

Kent, P. E., Hunt, J. A., and Johnstone, D. W., 1971, Geophysics of coastal Tansania: Inst. Geol. Sci. London Geophys. Paper, v. 6.

Kudrass, H. R., 1973, Sedimentation am Kontinentalhang vor Portugal und Marokko im Spätpleistozän und Holozän: *METEOR* Forsch. Ergeb. C, v. 13, p. 1-63.

Lowrie, A., and Escowitz, E., 1969, KANE no. 9, Global ocean floor analysis and research data series: U.S. Naval Oceanograph. Office, v. 1.

———, Jahn, W., Egloff, J., 1970, Bottom current activity in the Cape Verde and Canaries Basin, eastern Atlantic: Trans. Am. Geophys. Union, v. 51, 336 p.

Luydendyk, B. P., and Bunce, E. T., 1973, Geophysical study of the northwest African margin off Morocco: Deep Sea Res., v. 20, p. 537-549.

Machens, E., 1970, Die Salinargürtel des afrikanischen Mesozoikums: Abhandl. Hess. Landesamtes Bodenforsch., v. 56, p. 97-111.

Matthews, D. H., Smith, S. G., 1971, The sinking of Rockall Plateau: Geophys. Jour., v. 23, p. 491-498.

McMaster, R. L., and Lachance, T. P., 1968, Seismic reflectivity studies on northwestern African continental shelf, strait of Gibraltar to Mauretania: Am. Assoc. Petr. Geol. Bull., v. 52, p. 2387-2396.

Moore, D. G., 1961, Submarine slumps: Jour. Sediment. Petr., v. 31, p. 343-357.

Newton, R. S., Seibold, E., Werner, F., 1973, Facies distribution patterns on the Spanish Sahara continental shelf mapped with side-scan sonar: *METEOR* Forsch. Ergeb. C, no. 15, p. 55-77.

Pérès, J. M., 1961, Océanographie biologique et biologie marine, I: Presses Universitaires Paris, 541 p.

Querol, R., 1966, Regional geology of the Spanish Sahara, *in* Reyre, ed., Sedimentary basins of the African coasts: Paris, Assoc. African Geol. Survey, p. 27-39.

Ratschiller, L. K., 1970, Lithostratigraphy of the northern Spanish Sahara: Mem. Mus. Tridentino Sci. Nat., Trento, v. 18, no. 1.

Reyre, D., ed., 1966, Sedimentary basins of the African coasts, pt. I, Atlantic coast: Paris, Assoc. African Geol. Survey, 304 p.

Robb, J. M., 1971, Structure of continental margin between Cape Rhir and Cape Sim, Morocco, northwest Africa: Bull. Am. Assoc. Petr. Geol., v. 55, no. 5, p. 643-650.

Roberts, D. G., 1972, Slumping on the eastern margin of Rockall Bank, north Atlantic Ocean: Marine Geology, v. 13, no. 4, p. 225-237.

Rona, P. A., 1969, Middle Atlantic continental slope of United States: deposition and erosion: Am. Assoc. Petr. Geol. Bull., v. 53, no. 7, p. 1453-1465.

———, 1970, Comparison of continental margins of eastern North America at Cape Hatteras and northwestern Africa at Cape Blanc: Am. Assoc. Petr. Geol. Bull, v. 54, p. 129-158.

———, 1971, Bathymetry off central northwest Africa: Deep Sea Res., v. 18, no. 3, p. 321-327.

———, 1973a, Relations between rates of sediment accumulation on continental shelves, sea-floor spreading, and eustacy inferred from the central North Atlantic: Geol. Soc. America Bull., v. 84, p. 2851-2872.

———, 1973b, Worldwide unconformities in marine sediments related to eustatic changes of sea level: Nature Phys. Sci., v. 244, p. 25-26.

Ross, D., 1971, Mass physical properties and slope stability of sediments of the northern Middle America Trench: Jour. Geophys. Res., v. 76, p. 704-712.

Rothe, P. 1968, Die Ostkanaren gehörten zum afrikanischen Kontinent: Umschau Wissenschaft Technik, nr. 4, p. 116-117.

———, and Schmincke, H. U., 1968, Contrasting origins of the eastern and western islands of the Canarian Archipelago: Nature, v. 218, no. 5147, p. 1152-1154.

Rust, U., and Wienecke, F., 1973, Bathymetrische und geomorphologische Bearbeitung von submarinen "Einschnitten" im Seegebiet vor Westafrika. Ein methodischer Versuch: Geograph. Inst. Univ. Munchen, Munchener Geograph. Abhandl., v. 9, p. 53-68.

Schiffers, H., ed., 1971, Die Sahara und ihre Randgebiete: Weltforum München, 664 p.

Schwarzbach, M., 1953, Das Alter der Wuste Sahara: Neues Jahrb. Geol. Palaentol. Monatsh., v. 4, p. 157-174.

Seibold, E., 1972, Cruise 25/1971 of R.V. *METEOR*: Continental margin of West Africa: general report and preliminary results: *METEOR* Forsch. Ergeb. C, no. 10, p. 17-38.

———, and Emery, K. O., 1970, Introduction: the geology of the east Atlantic continental margin, *in* Delany, ed., ICSU/SCOR Working Party 31 Symp.: Inst. Geol. Sci. Rept. 70/13, p. 1-2.

Simpson, E., et al., 1972, Leg 25 Deep Sea Drilling Project, western Indian Ocean: Geotimes, v. 17, no. 11, p. 21-24.

Sougy, Y., 1962, West African fold belt: Geol. Soc. America Bull., v. 73, p. 871-876.

Spengler, A., Castelain, J., Cauvin, J., Leroy, M., 1966, Le bassin secondaire tertiaire du Senegal, *in* Reyre, ed., Sedimentary basins of the African coasts: Paris, Assoc. African Geol. Survey, p. 99-114.

Summerhayes, C. P., Nutter, A. H., and Tooms, J. S., 1971, Geological structure and development of the continental margin of northwest Africa: Marine Geology, v. 11, p. 1-27.

Templeton, R. S. M., 1972, The geology of the continental margin between Dakar and Cape Palmas: *in* Delany, ed., Inst. Geol. Sci., London, no. 70/16, p. 43-60.

Thiede, J., 1973, Sedimentation rates of planktonic and benthic foraminifera in sediments from the Atlantic continental margin of Portugal and Morocco: *METEOR* Forsch. Ergeb. C, no. 16 (in press).

Tooms, J. S., Summerhayes, C. P., and McMaster, R. L., 1971, Marine geological studies on the northwest African margin: Rabat-Dakar, *in* Delany, ed., Inst. Geol. Sci., London, no. 70/16, p. 13-25.

Current-Controlled Topography on the Continental Margin Off the Eastern United States

Roger D. Flood and Charles D. Hollister

INTRODUCTION

Since the development of high-resolution precision echo sounders in the 1950s, marine geologists have been characterizing different types of sea floor on the character of the echo return. Short-ping (less than 5 msec) high-frequency (3.5-12 kHz) echograms have been used to provide much information on the nature of the microtopography and the sediment stratification and have been used as a basis for interpreting the erosional and depositonal processes acting in the deep sea (Heezen et al., 1959; Heezen et al., 1966; Hollister, 1967; Schneider et al., 1967; Heezen and Johnson, 1969; Hollister and Heezen, 1972).

The continental margin is one region of the sea floor where these techniques have proved especially useful in delimiting areas that appear to be quite active sedimentologically. The classification of the echo types from the margin is generally based on three criteria: (1) the coherence of the echo return, i.e., whether the echo is made up of one or more mushy or sharp returns; (2) the presence of topography below the limit of resolution of the echo sounder as evidenced by the presence of hyperbolic echoes; and (3) the wavelength, amplitude, and regularity of the larger-scale relief.

The echogram character of large areas of the western North Atlantic continental margin, especially the continental rise, can be mapped on the basis of an echogram classification similar to that first developed by Hollister (1967) (Figs. 1 and 2). Other parts of the ocean basins throught to be under the influence of current activity exhibit similar echo patterns, implying that to some extent the topographic forms developed may be related to the current activity.

Several characteristics of the microtopography, in particular, have often been cited as evidence of bottom-current activity. These characteristics are (1) hyperbolic echoes with wavelengths of up to several hundred meters and vertices approximately tangential to the sea floor, and (2) large abyssal mud waves (termed "giant ripples" by Ewing et al., 1971) with internal stratification often showing that these waves have migrated.

Many questions have arisen as to the actual nature of the topography that yields these types of echo traces:

1. What, in fact, are these features? What are their dimensions, degree of lineation, and orientation with respect to regional topography and to the currents observed?
2. Are there smaller features superimposed on the larger-scale features? If so, what are they and how are they related to the larger features?
3. Are these features, in fact, related at all to bottom-current activity, or are they a manifestation of other processes taking place on the continental margin?
4. If the topography is current-controlled and the features are some type of current bed form, are these bed forms currently active? If they are, what are the dynamics of their formation? If not, how recently have they been active and what can they tell us about the flow conditions at the time they were active?

In this study we shall attempt to address these questions through a newly organized program of sediment dynamics that is a combination of field investigations and laboratory studies.

FIELD TECHNIQUES

Previous investigations into the nature of the topographic forms present on the continental rise and on outer ridges have been severely hampered by the sampling techniques employed. These difficulties have been twofold. First, the features to be studied are larger than the field of view of a normal bottom photograph (several square meters), and the relationship between successive bottom photographs is usually not known. Second, these forms are also at or below the lower limit of detection by surface 3.5- and 12-kHz profilers. Often only a hyperbolic echo is recorded at the surface or, if the features are larger, side echoes tend to obscure the true form of the features. For the larger features (greater than several hundred meters) the navigational accuracy has not usually been good enough and the sounding lines have not been close enough to establish an unambiguous trend for these features. Hence features that are apparently similar have been mapped as being parallel to (Ballard, 1966), perpendicular to (Rona, 1969), and at 45° to (Heezen and Schneider, 1968) the regional contours. Although this variation could be real, it could also be a result of sampling technique. Since even the general form of the feature is not known, the sampling strategy has not been adequate, nor have the nature and structure of the bottom boundary layer as well as the sediments that make up these bed forms usually been studied.

Most of these difficulties have been overcome by the recent development of a deeply towed instru-

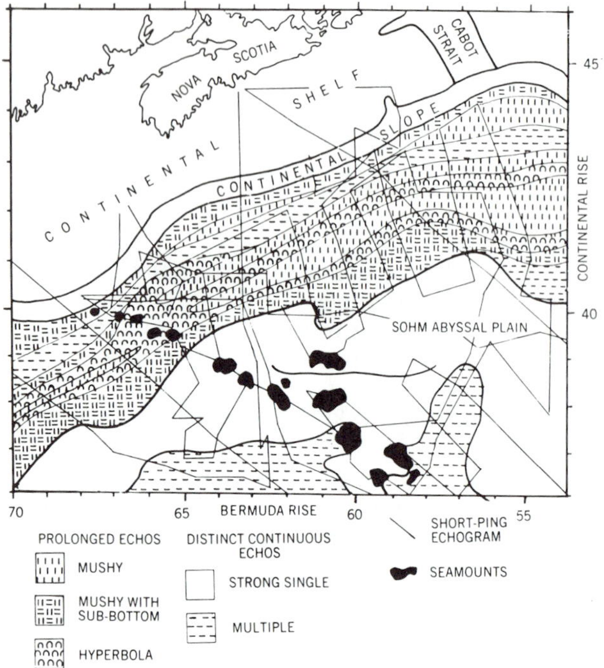

Fig. 1. Echogram character on the continental margin and Sohm Abyssal Plain off Nova Scotia. Zones of similar echo reflectivity generally follow bathymetric contours and can easily be correlated from profile to profile. The zones of hyperbolae and mushy echoes also run parallel to near-bottom isotherms, suggesting a relationship between echo character and water masses. (From Hollister and Heezen, 1972.)

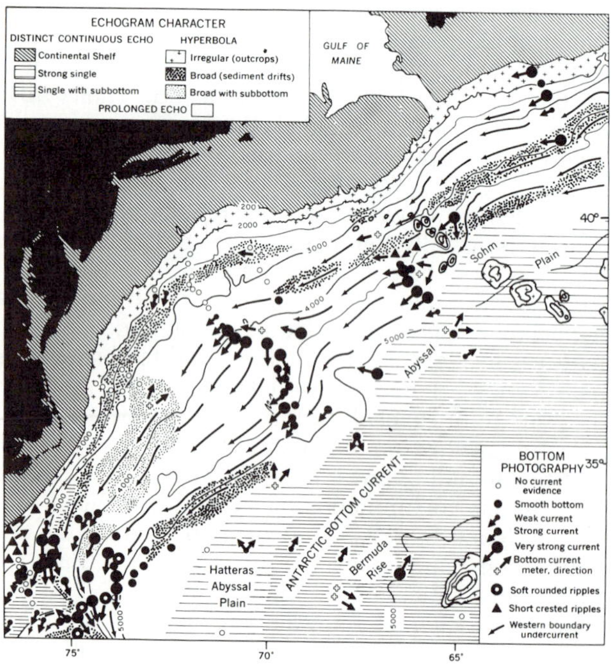

Fig. 2. Echogram character off the east coast of North America; can be related to the inferred and measured bottom currents of the Western Boundary Undercurrent (Schneider et al., 1967). Zones of prolonged and hyperbolic echoes on the continental rise parallel the regional contours. Above 3,500 m a tranquil, current-free bottom is usually seen, but below 3,500 m, swift contour-following currents appear to be transporting sediment toward the south (contours in meters).

ment package by the Marine Physics Laboratory of the Scripps Institute of Oceanography (Spiess and Mudie, 1970). This instrument package is towed at a height of 10-100 m above the sea floor, and with its side-scan sonar, narrow-beam echo-sounder, 4-kHz sub-bottom profiler, stereo camera, continuous-temperature recorder, and transponder navigation it can resolve sea-floor features such as bed forms with relatively small dimensions and can give information as to the temperature structure of the overlying water column.

It appears that an instrument of this type is almost required when studying features in the order of about 3 to several hundred meters. With good insight and adequate sampling procedures, good, though less complete, data can be collected by more conventional methods and a more subjective interpretation is usually involved.

In an effort to determine the character of the bottom morphology responsible for the hyperbolic echoes observed over the Blake-Bahama Outer Ridge and to search for answers to many of the questions listed above, two areas on the outer ridge were the objects of detailed investigations utilizing the MPL deep tow.

NEAR-BOTTOM INVESTIGATIONS OF MICROTOPOGRAPHY ON THE BLAKE-BAHAMA OUTER RIDGE

Clay and Rona (1964) noted the presence of hyperbolic echo returns in the Blake-Bahama Basin and related them to north-south corrugations which they thought could be sand waves. Later Bryan and Markl (1966) mapped the distribution of different echo types and determined, on the basis of the shape of the hyperbolic echoes, that these corrugations were generally parallel to the regional contours.

The Scripps deeply towed instrument package (Spiess and Mudie, 1970) was used to determine the nature of intermediate-scale bed forms responsible for the hyperbolic echoes on the Bahama Outer Ridge, the secondary ridge of the Blake-Bahama Outer Ridge complex, and to study the abrupt acoustic/sedimentologic contact between the Blake-Bahama Abyssal Plain and the Bahama Outer Ridge. Four giant piston cores (Hollister et al., 1973), 25 bottom-camera stations, 3 boomerang cores, and 7 current-meter records of up to 153 hours duration were obtained in the two areas studied on the Bahama Outer Ridge (Fig. 3). Area 1 is in a region of large (2-km wavelength) mud waves near the crest of the secondary ridge; Area 2 is at the contact between the Bahama Outer Ridge and the Blake-Bahama Abyssal Plain.

Previous studies of the Blake-Bahama Outer Ridge region have generally supported the concept that bottom currents have transported mud into the region and deposited it in the form of large lens-shaped ridges and swells (Heezen and Hollister,

1964; Heezen et al., 1966). Deep Sea Drilling results have confirmed that at least the upper kilometer of the Blake-Bahama Outer Ridge is composed of hemipelagic silty clay transported to the south from sources north of Cape Hatteras (Hollister et al., 1971). The entire suite of ridges and rises in the Blake-Bahama Outer Ridge area appears to have been formed since approximately the middle Tertiary, when the opening of the North Atlantic to sources of dense bottom water from the Norwegian Sea allowed the initiation of a vigorous bottom circulation which caused wholesale migration and construction of sediment drifts along the entire western side of the North Atlantic (Hollister and Heezen, 1972; Berggren and Hollister, 1974).

The preliminary results of our investigations have been reported (Flood et al., 1974; Hollister et al., 1974a) and are summarized below.

Small Furrows on Large Mud Waves [Area 1]. Transponder-navigated bathymetric profiles were collected in the area of large mud waves with the deeply towed instrument package. The large waves trend N10°E and have amplitudes of 20-60 m and wavelengths of about 2 km (Fig. 4). Available bathymetric data (Bryan and Markl, 1966; Markl et al., 1970) indicate that the large waves are not parallel

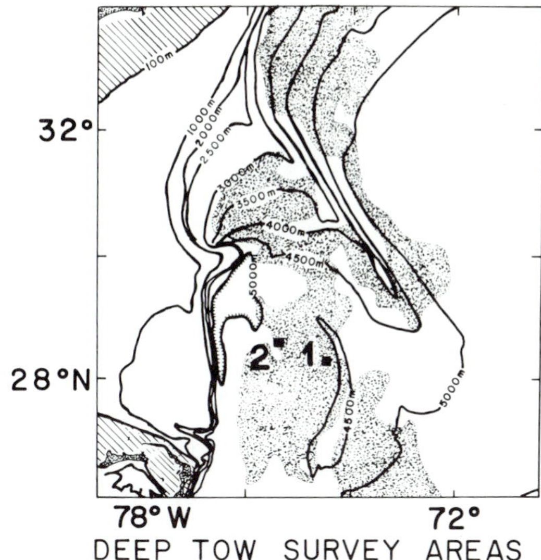

Fig. 3. Areas surveyed with the MPL deep tow on the Blake-Bahama Outer Ridge. Area 1 is in a region of large (2-km wavelength) mud waves near the crest of the secondary ridge. Area 2 is at the contact between the outer ridge and the Blake-Bahama abyssal plain. Distribution of fuzzy and hyperbolics from Bryan and Markl (1966) and echo-sounding records of the Woods Hole Oceanographic Institution.

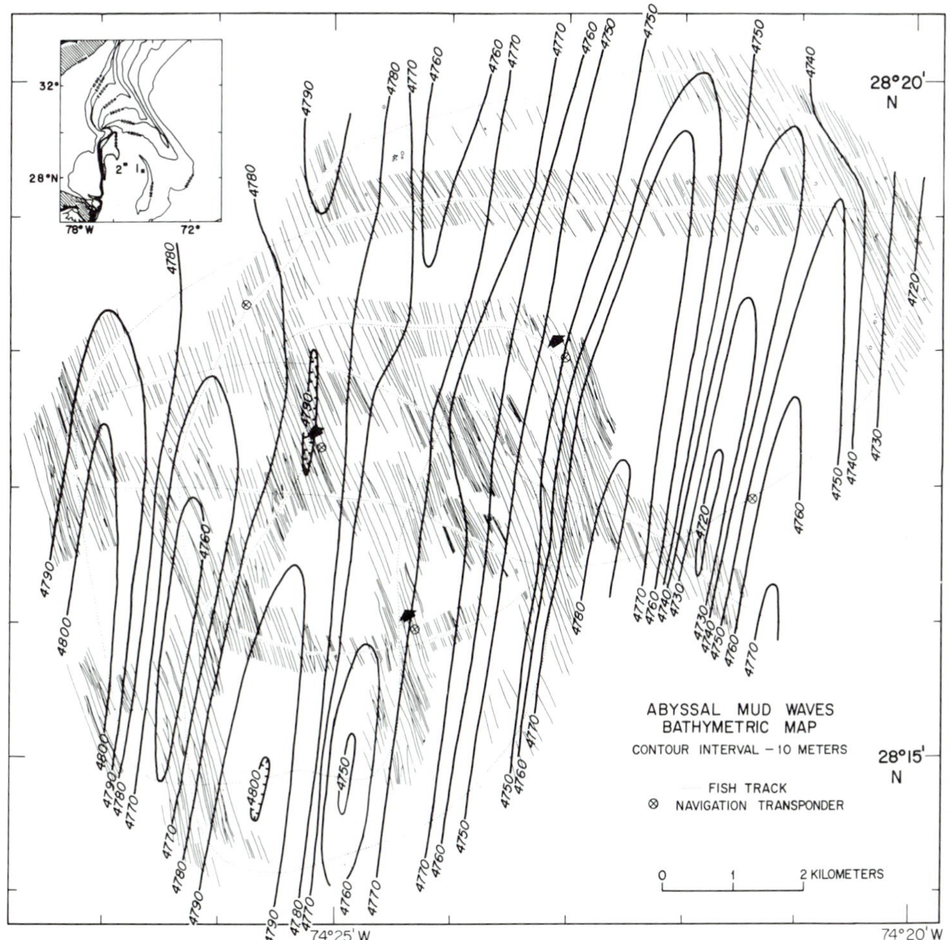

Fig. 4. Bathymetric map of abyssal mud waves area (Area 1) with furrows plotted from side-scan sonar records. Arrows indicate direction of maximum current velocities measured by current meters located at the navigation transponders.

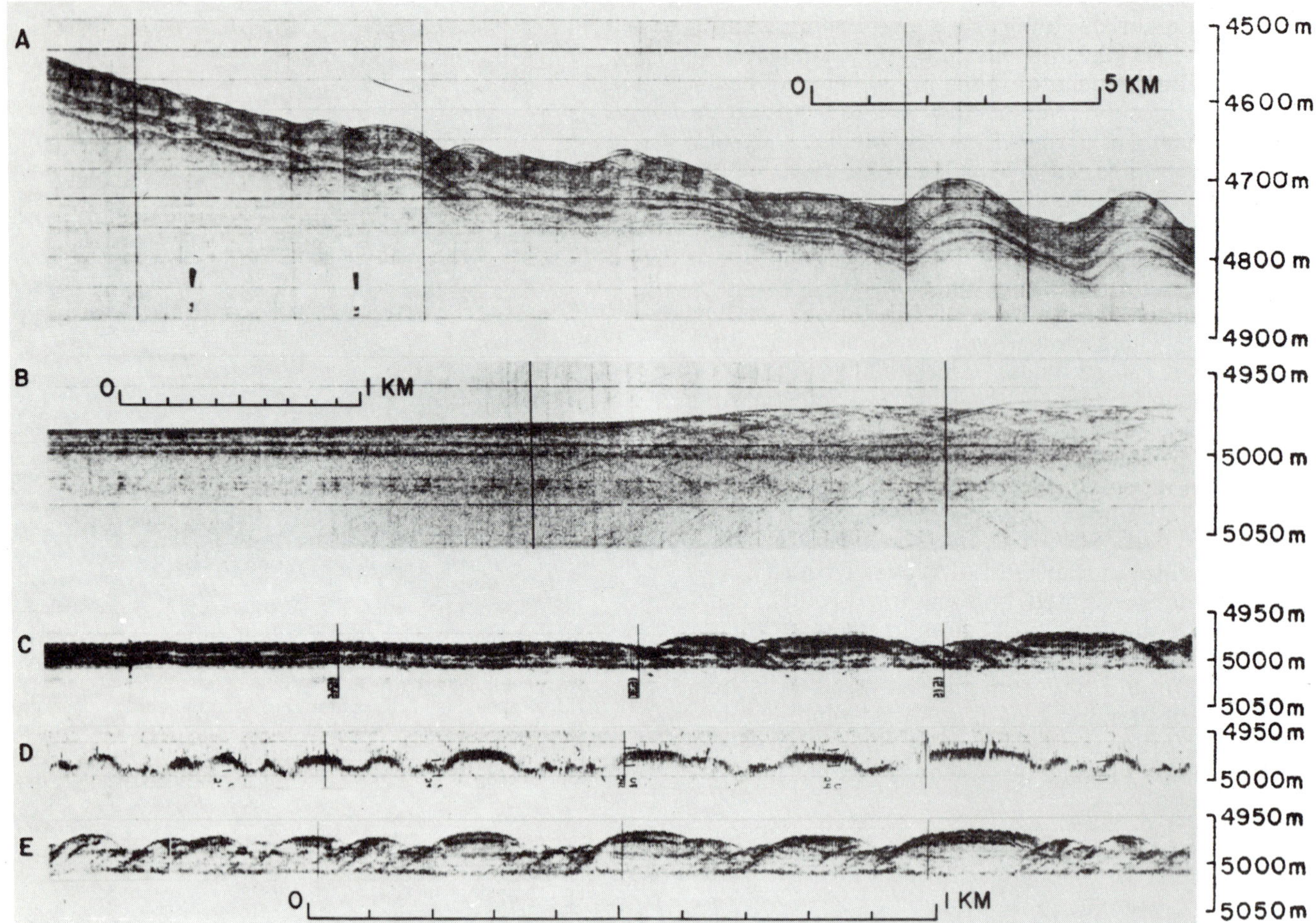

Fig. 5. A. Surface-ship 3.5-kHz record starting in Area 1 and continuing upslope toward the crest of the Bahama Outer Ridge. The wave farthest to the right corresponds to the wave at 74°21.8W, 28°16.9'N. B. Surface-ship 3.5-kHz record of the Bahama Outer Ridge-Abyssal Plain contact. C. Near-bottom 4-kHz record of the contact shown in B, demonstrating the abruptness of the contact. (Slight undulations in the record are caused by variations in fish elevation. The deeper reflectors are flat; see profile B.) D. Near-bottom narrow-beam echo-sounding (40-kHz) record of furrows responsible for hyperbolae seen on profile B in Area 2. E. Near-bottom 4-kHz record taken simultaneously with profile D. Note edge effects caused by wide beam angle.

to the regional contours, which trend N25°W in this area, but are at an angle of approximately 35° to this trend. High-resolution subbottom profiles indicate that the dunes have migrated to the east (Fig. 5A).

Superimposed on the large mud waves are long and remarkably straight furrows (Figs. 4, 6, 7, upper left and upper right; and 8). Measurements on a few stereo photographs indicate that the furrows are 1-4 m wide and 3/4-2 m deep (Fig. 8). The spacing of the furrows is 20-125 m. These furrows appear to be responsible for the very fine scale surface hyperbolic echoes seen on 3.5-kHz and 12-kHz records (Fig. 5A).

The furrows are steep-sided and have flat floors (Fig. 7, upper left; and Fig. 8) with an occasional median ridge present. The flat floors seem to correlate with a noticeably harder layer sampled by each of the cores taken in the furrowed area. This layer marks a striking change in lithology from a surface sediment of brownish calcareous hemipelagic mud to a deeper sediment of grayish slightly calcareous hemipelagic mud of Pleistocene age, containing scattered reworked Tertiary discoasters. The layer appears to be similar to iron-rich crusts observed by McGeary and Damuth (1973) and is thought to approximately mark the Pleistocene-Holocene boundary (McGeary and Damuth, 1973). It occurs at depths of 36-67 cm in our cores.

The furrows trend N25°W, parallel to the trend of the regional contours but at a 35° angle to the strike of the large waves. The furrows are parallel to the measured current directions and generally join in tuning-fork junctions which open into the measured current directions (Figs. 4 and 6). Some of the furrows can be traced for at least 5 km.

Current meters located 10 m off the bottom recorded maximum velocities of 9-10 cm/sec toward 325-335° for approximately three-fourths of the 102-108 hr of record. This is, in general, agreement with the currents calculated by Amos et al. (1971). Most of the furrows photographed seem to be presently active; their walls are sharp and sparsely tracked by bottom animals, and the associated ripples are well developed (Fig. 7, upper right and upper left). Occasional clumps of seaweed (possibly *Sargassum*) can be seen lying in the furrows.

Large Furrows Near the Contact Between Abyssal Plain and Outer Ridge [Area 2]. Surface-ship profiles across this important boundary show an abrupt transition between very pronounced hyperbolic echoes on the outer ridge and a strong reverberant coherent echo on the abyssal plain (Fig. 5B). The deep-tow profiles show that this contact is abrupt on a horizontal scale of several meters (Fig. 5C). The surface hyperbolic echoes are caused by a system of steep-sided erosional furrows (Figs. 5D, E and 9). These features, much larger (20 m deep by 50-150 m wide and spaced 50-200 m apart) than the furrows seen superimposed on the mud waves, are clearly erosional, since deeper reflectors, seen on both surface and deep-tow records, continue unbroken under these features (Fig. 5B, C, E) or outcrop on furrow walls (Fig. 7LR). Side-scan sonar records reveal branching or "tuning-fork" patterns similar to those of the smaller mud-wave furrows with junctions opening into the currents (Fig. 9).

Photographs of the sides of these furrows show outcropping more-resistant layers (Fig. 7, lower

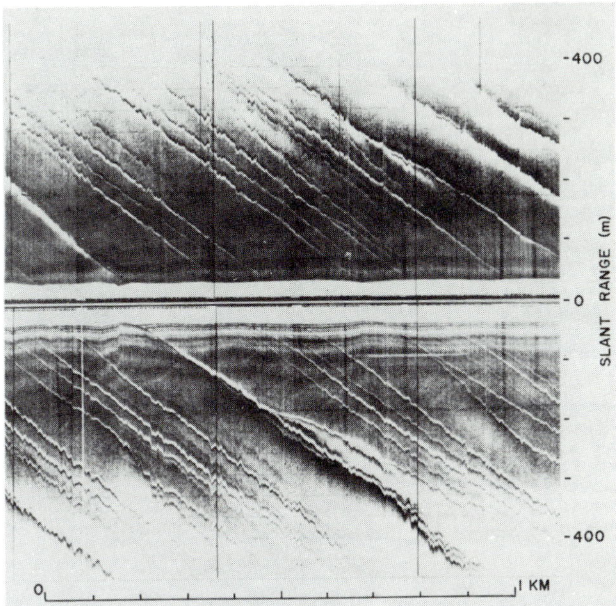

Fig. 6. Side-scan sonar record of the furrows in Area 1. Line down the center of the record represents the fish track. Measured currents are along the furrows from lower right to upper left. Note tuning-fork junctions opening into the current.

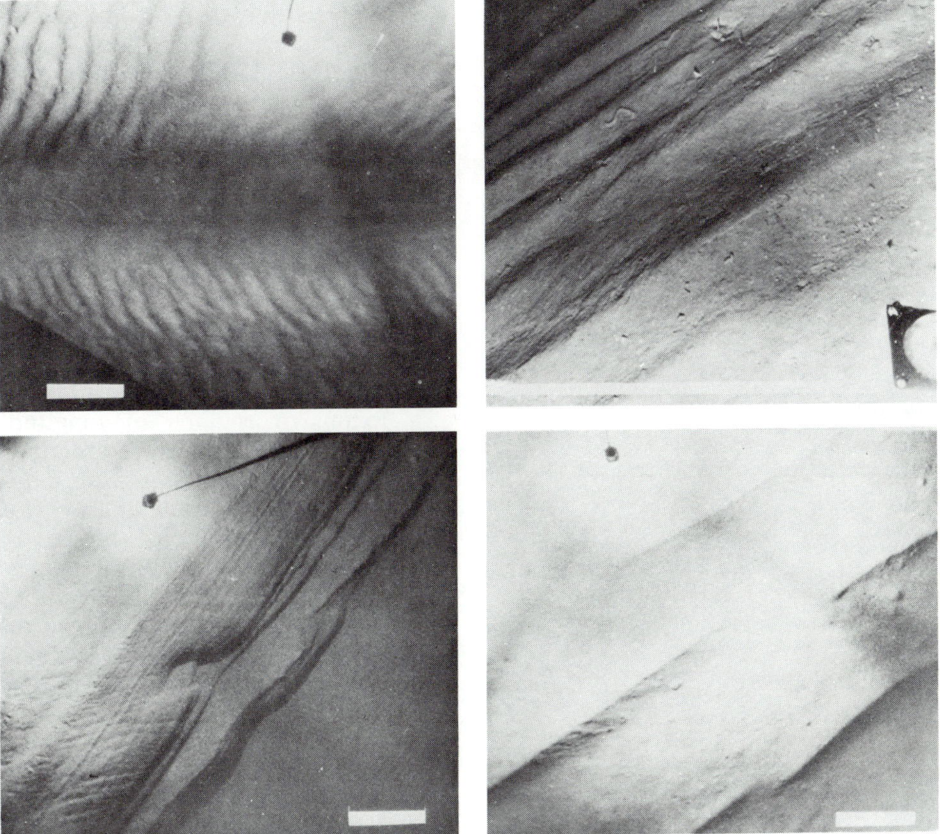

Fig. 7. Bottom photographs: UL, furrow in Area 1; measured currents are from right to left. UR, wall of a furrow in Area 1. LL, wall of a furrow in Area 2; note outcropping layers of a more-resistant sediment; the wall shown is 9 m high. LR, bed forms observed near the edge of the abyssal plain (area of coherent echoes). The scale bar is 1 m long.

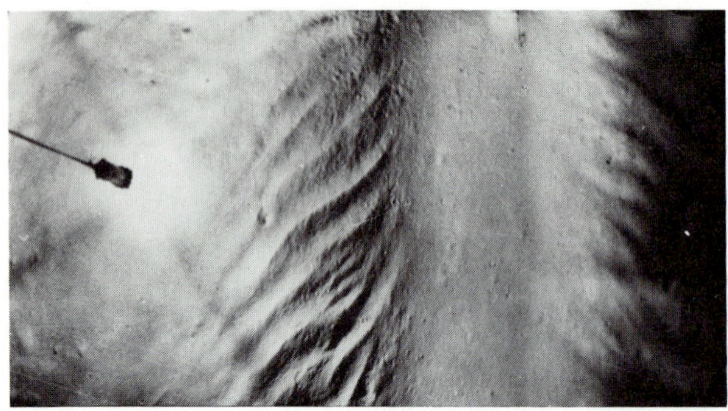

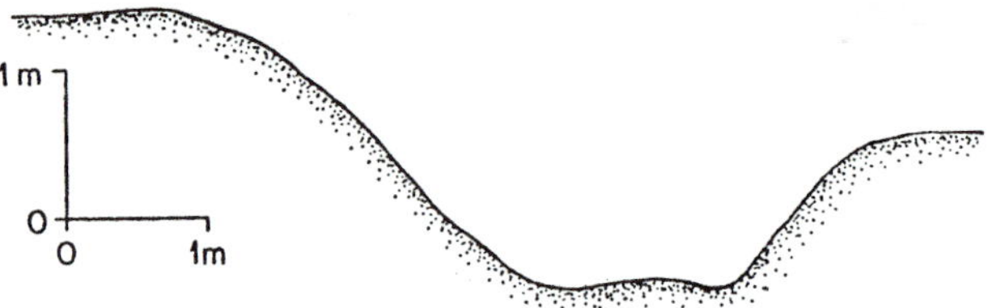

Fig. 8. Photograph and cross section of a furrow (Area 1). The cross section is based on depths measured from stereo photographs of the furrow. No vertical exaggeration.

right). Others show active erosion of material from beneath the more-resistant layers and downslope sliding of mud lumps. A few unidentifiable animals can be seen living on and in the precipitous side walls and may contribute to the process of erosion.

Current meters located 10 m above the bottom in the area of hyperbolic echoes recorded maximum current velocities of 8 cm/sec and average velocities of 5 cm/sec over a period of 153 hr, directed steadily toward the north. Maximum velocities of 4 cm/sec (average velocities of 2 cm/sec over a period of 144 hr) toward the north were recorded over the adjacent abyssal plain. These currents are also in rough agreement with the geostropic currents calculated by Amos et al. (1971).

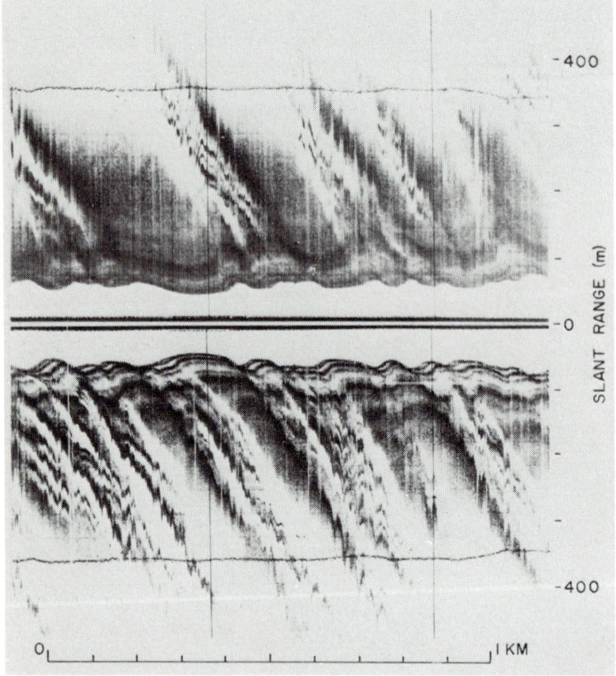

Fig. 9. Side-scan sonar record of the furrows in Area 2. Line down the center of the record represents the fish track. Measured current directions are along the furrows from lower right to upper left. Note tuning-fork junctions opening into the current.

Giant piston cores taken on the Bahama Abyssal Plain and in the area of hyperbolic echoes to the east show that the furrows are eroded into Pleistocene calcareous oozes and clays, similar to the sediment in Area 1. The underlying sediment is a Pliocene (Zone NN16) (D. Johnson, personal communication) highly calcareous ooze, with common fragmental carbonate debris and graded bedding. The abyssal plain sediments can be traced beneath the area of hyperbolic echoes to the east (Fig. 5B, C) and may determine the depth to which the furrows can erode.

On the abyssal plain, within several kilometers of the contact between the abyssal plain and outer ridge, smaller ridges with approximately 2-m spacings were photographed (Fig. 7, lower right) and recorded on side-scan sonar records. These ridges are developed in hemipelagic muds overlying more typical abyssal plain sediments.

These bed forms are identical to longitudinal ripples recorded in other areas of the ocean and are generally found to be parallel to the current directions. Here they are at an angle of 70° to the measured current directions. The significance of these features is not well understood.

In general, based on the data collected, especially the photography, none of the bed forms appear to have been active at the time our data were collected. However, the furrows and the smaller longitudinal ripples do seem to have been very recently active.

SIMILAR BED FORMS DEVELOPED IN OTHER ENVIRONMENTS

Extensive regions of longitudinal seif sand dunes exist in areas where steady winds blow over hot, dry, loose sand. These large dunes are thought to be the result of large helical-vortex circulations in the lower mixed or unstable layer of heated air (Folk, 1971; Hanna, 1969). Such large vortices have been observed in the atmosphere (Angell et al., 1968), and banded cloud patterns are thought to be indicative of a similar type of circulation (Kuettner, 1959). These aeolian longitudinal features are characterized by tuning-fork junctions opening into the prevailing wind direction (Folk, 1971).

Longitudinal bed forms with gross morphology similar to the abyssal furrows are found in shallow water in regions of strong currents. Furrows 5 m wide, 1 m deep, 4 km long, and 10-20 m apart with tuning-fork junctions that open into the stronger ebb tide have been reported from a shallow estuary (Dyer, 1970). Stride et al. (1972) report possibly similar features in gravel sediments on the floor of the English Channel.

Patches of bifurcating furrows have also been recorded by side-scan sonar at a depth of 5.8 km in the Samoan Passage (Lonsdale, 1973), where the Antarctic Bottom Current is the principal agent controlling the sediment dynamics (Hollister et al., 1974b). However, these features do not cause hyperbolic echoes on surface records, do not show up in bottom photographs, and are formed in a nannofossil ooze.

Experimental and natural features described by Dzulynski (1965) and Allen (1969) are parallel to water flow and have tuning-fork junctions opening upstream. The features produced by Allen (1969) show occasional well-defined median ridges where two furrows come together, and except for differences in scale (they are spaced 6 mm apart and are about 1 mm deep) these "longitudinal rectilinear grooves" are quite similar to the features we have described in the abyssal Atlantic.

Another comparison that seems relevant is to compare the abyssal furrows with the wind slicks formed on the surface of water bodies that have winds blowing across them. These surface slicks appear to be caused by "Langmuir circulations," large helical circulations in the upper layer of the water body. Photographs of the upwelling regions of the circulations taken by McLeish (1968) show some similarities to the abyssal furrows, especially the tuning-fork junctions opening into the direction of the wind.

ORIGIN OF ABYSSAL FURROWS

The similarity of the abyssal furrows to many bed forms described from other environments, and the rough way in which the abyssal furrows appear to imitate the forms of cloud streets and wind slicks, lead one to expect that the mechanism responsible for the forms of these features may also be responsible for the form of the abyssal furrows. The current view regarding most of the forms described above is that they are formed by secondary helical circulations within the fluid flow modifying the bed. The modified bed then tends to stabilize the secondary circulations, which then further modify the form of the bed. Helical circulations of this type have been predicted and observed in the atmosphere (Hanna, 1969), in laboratory experiments (Faller, 1963), and have long been postulated to exist in turbulent flows (Townsend, 1956). Recent numerical modeling of high-Reynolds-number turbulent planetary boundary-layer flows have indicated that this type of circulation may be present in turbulent boundary layers (Deardorff, 1972).

One important result that has come out of investigations made into the nature of the planetary boundary layer is the relationship between the depth of flow and the wavelength of the instabilities present, which should be equivalent to the spacing between the longitudinal bed forms. In all the cases reported, which are for neutrally stratified flows, the spacings reported are approximately two to four times the boundary-layer thickness. Thus, if this type of process is responsible for the bed forms observed on the Bahama Outer Ridge and if the furrows can be construed as being currently active, a similar relationship should be observed between the bottom boundary-layer thickness and the furrow spacing.

Salinity-temperature-density (STD) casts taken in regions of the Blake-Bahama Outer Ridge characterized by hyperbolic echoes and erosional furrows show that a layer of isothermal-isohaline water topped by relatively steep gradients of these parameters is usually developed near the bottom (Amos et al., 1971; unpublished deep-tow temperature measurements). This well-mixed boundary layer may be the result of mixing produced by bottom friction and provides a layer in which secondary circulations could develop, or which could be caused by secondary circulations.

Measurements made to date have shown that in Area 2, at the transition between the outer ridge and the abyssal plain, this boundary layer is approximately 100 m thick. The furrows developed in this region are spaced 50-200 m apart. In Area 1,the large mud-wave area, a bottom isothermal layer of varying thickness was encountered. Deep-tow temperature measurements show that such a layer was not more than 20 m thick at the time of our survey; however, an STD cast taken by Amos et al. (1971) shows a bottom isothermal layer approximately 100 m thick in a similar area. The furrows in this area are spaced 20-125 apart and the large mud waves are spaced approximately 2 km apart. Currents measured at 10 m above the bottom at the time of the deep-tow survey reached maximum velocities of 8-10 cm/sec. The tidal components of these currents are only 1-2 cm/sec, and the currents are quite uniform during the extended periods of maximum velocity.

The thickness of the boundary layer seems to be about the same as the spacings observed between the furrows. This indicates that a process of this sort could be responsible for the furrows and that they could be at least sporadically active. To determine this more precisely, however, many more measurements will have to be made of the thickness of the isothermal layer and the current velocities. The combination of a steady current with a well-mixed bottom boundary layer appears to provide almost ideal conditions for secondary circulations to develop.

DISTRIBUTION OF ABYSSAL FURROWS

The widespread occurrence of hyperbolic echoes in regions of suspected deep current activity leads one to suspect that the features and processes described above are also widespread. However, firm evidence for this statement is lacking. This is a result of the fact that the furrows are identifiable from the surface only as features that cause hyperbolic echoes. The trend of the linear feature responsible for the hyperbolic echo can be determined with some ambiguity from the surface (Bryan and Markl, 1966). All the trends as yet determined by this method are parallel to the known or inferred current directions. Photographs of the furrows or parts of furrows have most likely not been understood and thus have not been reported.

Our near-bottom investigations into the nature of the topography present on the continental rise have shown the existence of a current bed form, abyssal furrows, which has not been reported previously from the deep sea. The morphology of these furrows indicates that they have been formed by large-scale helical vortices developed in the bottom boundary layer of the ocean. These furrows, although they did not appear to be active at the time of our survey, do appear to have been active recently. As these furrows grow larger, sediments are eroded and made available for transport downstream.

BIBLIOGRAPHY

Allen, J. R. L., 1969, Erosional current marks of weakly cohesive mud beds: Jour. Sediment. Petrology, v. 39, p. 607-623.

Amos, A. F., Gordon, A. L., and Schneider, E. D., 1971, Water masses and circulation patterns in the region of the Blake-Bahama Outer Ridge: Deep-Sea Res., v. 18, p. 145-165.

Angell, J. K., Pack, D. H., and Dickson, C. R., 1968, A Lagrangian study of helical circulations in the planetary boundary layer: Jour. Atmospheric Sci., v. 25, p. 707-717.

Ballard, J. A., 1966, Structure of the lower continental rise hills of the western North Atlantic: Geophysics, v. 31, p. 506-523.

Berggren, W. A., and Hollister, C. D., (in press), Paleogeography and the history of circulation in the Atlantic Ocean, *in* Hay, W. W., ed., Studies in paleo-oceanography: Tulsa, Okla., Soc. Econ. Paleontologists and Mineralogists Spec. Publ. 20.

Bryan, G. M., and Markl, R. G., 1966, Microtopography of the Blake-Bahama Region: Lamont-Doherty Geological Observatory, Tech. Rept. 8, CO-8-66, 44 p.

Clay, C. S., and Rona, P. A., 1964, On the existence of bottom corrugations in the Blake-Bahama Basin: Jour. Geophys. Res., v. 69, p. 231-234.

Deardorff, J. W., 1972, Numerical investigations of neutral and unstable planetary boundary layers: Jour. Atmospheric Sci., v. 29, 91-115.

Dyer, K. E., 1970, Linear erosional furrows in Southampton water: Nature, v. 225, p. 56-58.

Dzulynski, S., 1965, New data on experimental production of sedimentary structures: Jour. Sediment. Petrology, v. 35, p. 196-212.

Ewing, M., Eittreim, S. L., Ewing, J. I., and LePichon, X., 1971, Sediment transport and distribution in the Argentine Basin: 3. Nepheloid layer and processes of sedimentation, *in* Physics and chemistry of the earth: Elmsford, N.Y, Pergamon Press, v. 8, p. 51-80.

Faller, A. J., 1963, An experimental study of the instability of the laminar Ekman boundary layer: Jour. Fluid Mech., v. 15, p. 560-576.

Flood, R. D., Hollister, C. D., Johnson, D. A., Southard, J. B., and Lonsdale, P. F., 1974, Hyperbolic echoes and erosional furrows on the Blake-Bahama Outer Ridge: Trans. Am. Geoph. Union, v. 55, p. 284.

Folk, R. L., 1971, Longitudinal dunes of the northwestern edge of the Simpson Desert, Northern Territory, Australia: 1. Geomorphology and grain size relationships: Sedimentology, v. 16, p. 5-54.

Hanna, S. A., 1969, The formation of longitudinal sand dunes by large helical eddies in the atmosphere. Jour. Appl. Meteorology, v. 8, p. 874-883.

Heezen, B. C., and Hollister, C. D., 1964, Deep-sea current evidence from abyssal sediments: Marine Geology, v. 2, p. 141-174.

———, and Johnson, G. L., 1969, Mediterranean undercurrent and microphysiography west of Gibraltar. Bull. Inst. Oceanog., v. 67.

———, and Schneider, E. D., 1968, The shaping and sediment stratification of the continental rise (abstr.): Washington, D. C. Marine Tech. Soc., Natl. Symp., Ocean Sci. and Eng. Atlantic Shelf Trans., Mar. 19-20, Philadelphia, p. 279-280.

———, Tharp, M., and Ewing, M., 1959, The floors of the oceans: 1. The North Atlantic: Geol. Soc. America Spec. Paper 65, 122 p.

———, Hollister, C. D., and Ruddiman, W. F., 1966, Shaping of the continental rise by geostrophic contour currents: Science, v. 151, p. 502-508.

Hollister, C. D., 1967, Sediment distribution and deep circulation in the western North Atlantic [Ph.D. thesis]: Columbia Univ., New York, 471 p.

———, and Heezen, B. C., 1972, Geological effects of ocean bottom currents, *in* Gordon, A. L., ed., Studies in physical oceanography: New York, Gordon & Breach, v. 2, p. 37-66.

———, Ewing, J. I., et al., 1971, Initial reports of the Deep Sea Drilling Project, v. XI: Washington, D.C., U.S. Govt. Printing Office, 1077 p.

———, Silva, A. J., and Driscoll, A., 1973, A giant piston-corer: Ocean Eng., v. 2, p. 159-168.

———, Flood, R. D., Johnson, D. A., Lonsdale, P. F., and Southard, J. B., 1974a, Abyssal furrows and hyperbolic echo traces on the Bahama Outer Ridge: Geology (in press).

———, Johnson, D. A., and Lonsdale, P. F., 1974b, Current-controlled abyssal sedimentation: Samoan Passage, equatorial west Pacific: Jour. Geology (in press).

Kuettner, J., 1959, The band structure of the atmosphere: Tellus, v. 11, p. 267-294.

Lonsdale, P. F., 1973, Erosional furrows across the abyssal Pacific floor: Am. Geophys. Union Trans., v. 54, p. 1110.

McGeary, D. F. R., and Damuth, J. E., 1973, Postglacial iron-rich crusts in hemipelagic deep-sea sediment: Geol. Soc. America Bull., v. 84, p. 1201-1212.

McLeish, W., 1968, On the mechanism of wind slick generation: Deep-Sea Res., v. 15, p. 461-489.

Markl, R. G., Bryan, G. M., and Ewing, J. I., 1970, Structure of the Blake-Bahama Outer Ridge: Jour. Geophys. Res., v. 75, p. 4539-4555.

Rona, P. A., 1969, Linear "lower continental rise hills" off Cape Hatteras: Jour. Sediment. Petrology, v. 39, p. 1132-1141.

Schneider, E. D., Fox, P. J., Hollister, C. D., Needham, D. H., and Heezen, B. C., 1967, Further evidence of currents in the western North Atlantic: Earth and Planetary Sci. Letters, v. 2, p. 351-359.

Spiess, F. N., and Mudie, J. D., 1970, Small scale topographic and magnetic features, *in* Maxwell, A. E., ed., The sea, v. 4, pt. 1, New York, Wiley, p. 205-250.

Stride, A. H., Belderson, R. H., and Kenyon, N. H., 1972, Longitudinal furrows and depositional sand bodies of the English Channel: Mem. B.R.G.M., v. 79, p. 233-240.

Townsend, A. A., 1956, The structure of turbulent shear flow: New York, Cambridge Univ. Press, 315 p.

Modern Trench Sediments

Roland von Huene

INTRODUCTION

Samples from the Aleutian Trench area and seismic records of the Aleutian and the Peru-Chile trenches are used to develop a sedimentary-facies distribution model in modern trenches. This simple model is useful to point out some features that might aid recognition of past trench environments. At the base of the continental slope, a distinctive trench-sediment sequence may consist of trench-fill turbidites, grading upward from a distal to a proximal facies. These turbidites rest on abyssal-plain sediments, similarly grading from pelagic (distal) to hemipelagic (proximal) deposits. The dominant trend is from an older mid-ocean environment to a younger continental slope environment. If modern trenches have the same characteristics as those of the past, this entire sequence will occur in a 200- to 1,500-m eugeosynclinal section.

Some orogenic belts contain eugeosynclinal rocks that may have been deposited in deep-ocean trenches. Recognition of trench environments by analogy with sedimentary sequences in modern trenches is difficult because the modern deposits are virtually unexplorable with commonly available sampling devices. A few piston cores or seismic-reflection records are inadequate to establish the vertical distribution of sediment types.

However, drilling in the eastern Aleutian Trench by the GLOMAR CHALLENGER during the Deep Sea Drilling Project provided cores of sufficient depth to span broad lithologic and paleontologic units. These units were correlated across the eastern Aleutian Trench area using good seismic records that ran through the drill sites. The data give better insight into the processes affecting sediments that filled the trench (von Huene and Kulm, 1973; Piper et al., 1973). Ideas from studies of CHALLENGER cores will be extended here to develop a possible general model of vertical sediment sequences in modern oceanic trenches.

Great sediment thickness in exposed orogenic belts is generally attributed to deposition in trenches. As a result of complex structure, such thicknesses are estimates. Of the trenches along the present Pacific continental margins and island arcs, however, trench-filling sediment as much as 1 km thick is found only in the Hikurangi Trench, eastern Aleutian Trench, and off southern Chile (Scholl and Marlow, 1974). (In areas where a basement trench has been completely filled, such as off Oregon and Washington, sediment patterns will change from those found in a long trough to those found on a broad fan on the continental rise.)

The Aleutian and Peru-Chile trenches are most analogous to ancient trenches and will be used here in the development of a model. In both of them, distribution of sediment is accomplished by lateral flow along the trench axis. In the eastern Aleutian Trench, axial transport is accomplished by flow and overbank spill in a current channel that runs along the landward wall (Piper et al., 1973), whereas in the Peru-Chile Trench, axial transport occurs in a current channel near the center of the trench fill.

ALEUTIAN TRENCH

Alaska's high mountain ranges are presently a great source of sediment, which is part of the explanation for the great thickness of sediment in the eastern Aleutian Trench. Another factor is glaciation, which is indicated by the fiord-indented coastline and the presently existing glaciers, as well as the large increases in late Quaternary rates of sedimentation on the Alaska Abyssal Plain (von Huene and Kulm, 1973). These sediments have partially filled the trench and form the flat trench floor, which is flanked landward by the steep (5°-40°) and rugged continental slope and seaward by the gently downbowed (>1°-3°) edge of the Alaska Abyssal Plain (Fig. 1).

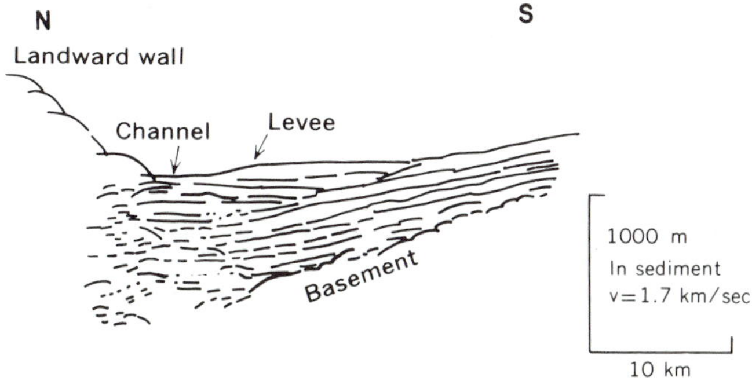

Fig. 1. Seismic-reflection record across the eastern Aleutian Trench near Kodiak Island, Alaska. (Modified from von Huene, 1972).

Seismic records show repetition of this morphology in the subsurface by a sequence of parallel downbowed reflections which are continuous with the Alaska Abyssal Plain, overlain by flat reflectors in the trench. The trench is a depressed section of the Alaska Abyssal Plain, which is presently the deepest feature and the locus of most-rapid sedimentation. From drill cores, the oldest trench fill is dated as 0.6 ± 0.1 my, whereas the existing orogenic system along this continental margin is at least 5 my old and probably became active 20 my ago (von Huene and Kulm, 1973).

Therefore, the present trench fill is only a part of the sediment that has been deposited in this trench since it first formed, and considerable earlier trench fill must have been incorporated into the continental slope. Just before incorporation, a distinct sequence of sedimentary facies may have formed, which should be recognizable if a trench such as this were only moderately deformed, uplifted, and exposed in an orogenic belt. This distinctive sequence would have resulted mainly from an axial channel in the trench that has no equivalent in the adjacent oceanic or continental environments.

Most seismic records across the eastern Aleutian Trench show a channel at the juncture of the trench floor and the landward wall that is 2.5 to possibly 6 km wide (Fig. 1). On its seaward side, a low levee has formed, such as those commonly found along deep-ocean turbidity-current channels (Hamilton, 1967). Variations in the width of the channel are attributed in some places to constriction or displacement by accumulations of sediment at the mouths of canyons and, in other places, to massive slumps (Piper et al., 1973). A profile in the channel along the trench axis (Fig. 2) shows irregularities in axial gradient that correspond to slumps and canyon mouths. Some slumps spread laterally across the trench floor as much as 15 km and must have hit the base of the slope with great momentum.

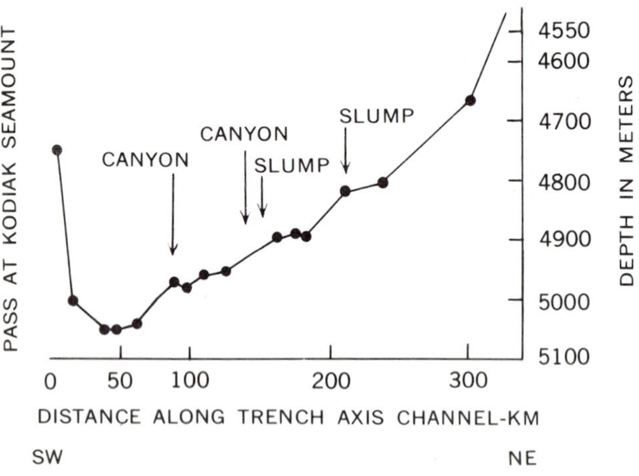

Fig. 2. Profile along the axial channel of the eastern Aleutian Trench. (From Piper et al., 1973).

Cores from Leg 18 of the Deep Sea Drilling Project were recovered from sites on the abyssal plain, the trench floor, and on the continental slope (Kulm, et al., 1973). Mud, containing varying amounts of sand and silt, is the dominant sediment type. Graded-sand turbidites are common only on the continental slope; graded siit and minor sand were recovered from the seaward channel levee of the trench floor, and mud is dominant in the upper part of the abyssal-plain section. Piper et al. (1973) have shown that if the recovered cores are representative of the slope, trench, and abyssal plain, at least 90% of the sediment filling the trench crosses the adjacent continental slope in turbidity currents, in slides, or by creep. The remainder is pelagic sediment, including ice-rafted and biogenous material (about 3%), and sediment transported across the abyssal plain from other highly glaciated parts of the Gulf of Alaska (about 7%). Rates of sedimentation in the trench average about 2,000 m/my, possibly becoming as much as 3,500 m/my during glacial periods, and as little as 200-300 m/my during interglacial periods.

Despite abundant sand in cores from the adjacent continental slope, cores seaward of the axial channel contain very little sand. Therefore, the sand must be confined to the axial channel, a conclusion supported by the apparent strength of currents. These appear to be turbidity currents strong enough to erode large slumps and to redistribute material from the slope, because only rarely can slumps and fans be identified below the sea floor in the seismic-reflection records (1 seismic record in a total of 17 examined). From these considerations, the sedimentary processes were inferred, and an idealized facies-distribution pattern for the Aleutian Trench was developed by Piper et al. (1973) (Fig. 3).

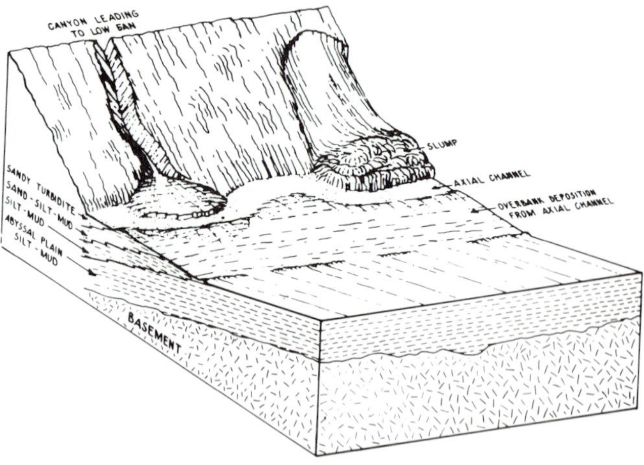

Fig. 3. Idealized facies-distribution diagram in the eastern Aleutian Trench. (From Piper et al., 1973).

Underlying the fill of the Aleutian Trench is a sequence of abyssal-plain sediments. Sedimentation rates in the present trench are about 10 times greater than those on the adjacent Alaska Abyssal Plain. The contact between abyssal-plain and trench-filling sediments is a facies change between relatively rapid and slow depositional environments. A reconstruction of stages of sedimentation across the contact is shown in Piper et al. (1973) (Fig. 2).

The sequence of sediments in the trench off Alaska is shown in Figure 4. Immediately above oceanic basement are fine clay-shale and chalk, which grade upward into mudstone broken by occasional silt and by rare accumulations of graded fine sand, from the lateral migration of a current channel. These graded beds become more frequent upward in the section. The upper part of the sequence consists of massive, glacially derived mud with occasional beds of diatom ooze or graded silt and sand. This increase of terrigenous material upward may reflect both the landward migration of the oceanic crust and increasing rates of late Neogene erosion on the continent. The Gulf of Alaska is rimmed on three sides by terrain that provides abundant terrigenous sediment, particularly during glacial periods; therefore, this section would probably have a more typical deep-oceanic character in other parts of the Pacific Ocean floor.

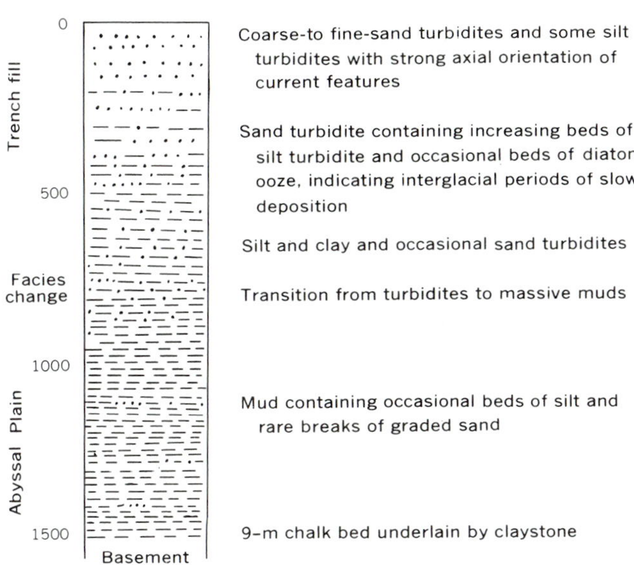

Fig. 4. Sediment sequence in the eastern Aleutian Trench. Depth of section given in meters.

The contact between abyssal-plain and trench-fill sequences is transitional: the types of sediment on either side are similar (Fig. 3); but even so, the contact separates beds deposited under different modes of deposition and with different rates of sedimentation.

The lower silt-mud turbidites in the trench-fill sequence grade upward into channel sands which probably dominate the upper one-fourth to one eighth of the trench section. These sands might be covered by the remains of slumps or other laterally introduced material that has not been completely reworked by the turbidity current. Trench fill probably has a dominantly axial paleocurrent pattern, in contrast to a more random pattern in the underlying abyssal section.

PERU-CHILE TRENCH

The southern Peru-Chile Trench, south of Valparaiso, Chile, is similar in configuration to the eastern Aleutian Trench, but seismic records across it are more widely spaced. This part of the trench receives sediment from Chile's largest rivers, and the amount of sediment supplied increases southward, where the fiord-indented coast was heavily eroded during Pleistocene glacial periods. The continental slope is steep, and the abyssal plain is gently bowed down against the continent, as in the Aleutian Trench (Fig. 5). A wedge of sediment filling the trench is responsible for its nearly flat floor. Along the southern trench axis, sediment is uniformly about 1.0 km thick, as shown in an axial profile (Fig. 6). However, the position of the axial channel and the thickness of the underlying abyssal sediment section are different from the Aleutian Trench.

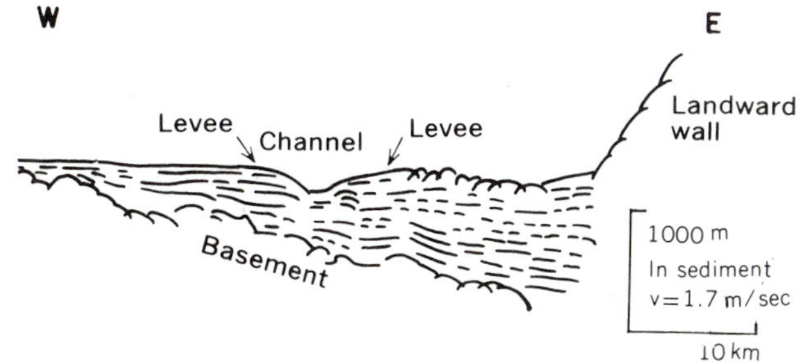

Fig. 5. Seismic-reflection record across the southern Peru-Chile Trench near latitude 38°S. Near the landward wall, a series of hyperbolic reflectors on the trench floor indicate rough topography on the landward levee.

Ten of 14 seismic records across the Chile Trench show a channel flanked by levees near the middle of the trench floor. The spacing of seismic records is insufficient to establish continuity of the channel or its pattern, but it is probably associated with the mechanism transporting great amounts of sediment from south to north. The smooth northward axial gradient, the uniform distribution of trench fill, and the suggestion of buried channels and levee deposits in some seismic records support this contention.

Sediments of the abyssal plain are 100-200 m thick immediately seaward of the trench (Fig. 5). Their topography-conforming configuration suggests a truly pelagic character, but in some areas adjacent to the trench they may have received larger amounts of hemipelagic material, as indicated by slightly greater thickness. Abyssal-plain sediments off Peru are pelagic and hemipelagic clays of Pleistocene age (Rosato, 1974). Only near the East Pacific Rise, where the sea floor is above the calcium carbonate compensation level, are carbonates and organic oozes mixed with the pelagic clays. Off Peru and Chile, oozes may only exist at the base of the abyssal-plain section, as most of the section has accumulated below the CCL.

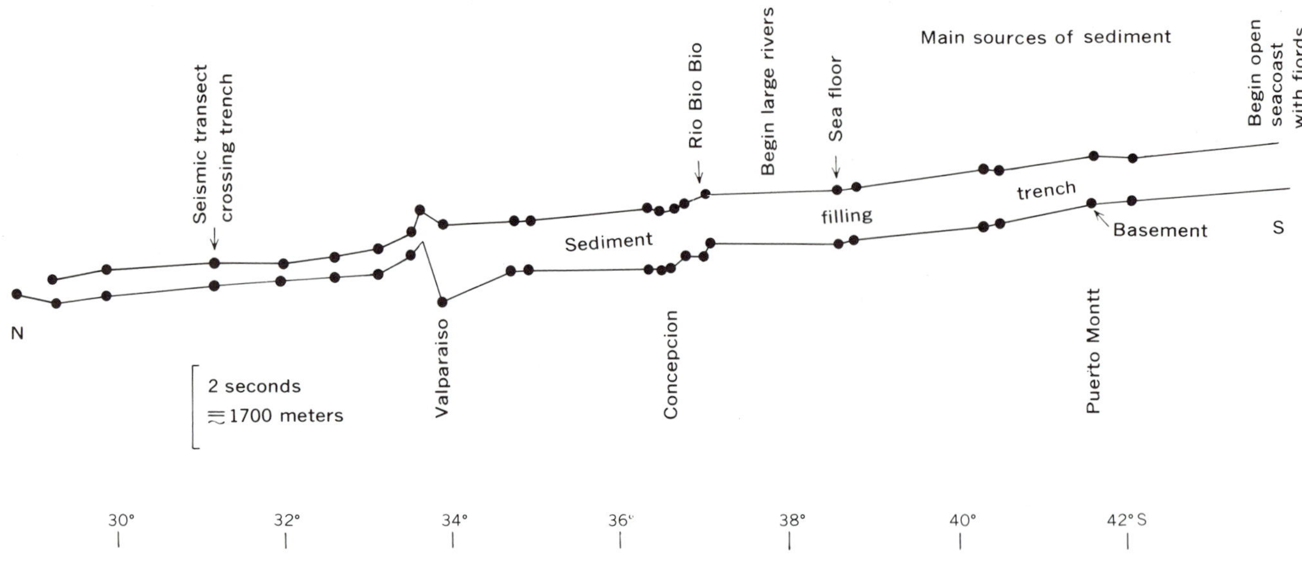

Fig. 6. Profile along the axis of the southern Peru-Chile Trench based on unpublished seismic records made by Scholl and von Huene.

The contact between hemipelagic and pelagic sediments of the abyssal plain and the trench-filling turbidites is not as clear as in records from the Aleutian Trench, and although a similar time-transgression is suggested, the contact appears more abrupt and variable. The contact may grade from pelagic through hemipelagic to turbidite, and much slower rates of abyssal-plain sedimentation (possibly 1/500 the rate of the trench fill) may occur. A distinct upward progression toward more terrigenous components was found in cores from the area off Peru (Rosato, 1974).

These observations and comparisons with the Aleutian Trench suggest that the facies pattern in the Peru-Chile Trench fill is also controlled by axial currents. The main difference from the Aleutian Trench is that the current channel is near the middle of the trench floor rather than against the landward wall. Therefore, slumps along the landward trench wall, and other accumulations around canyon mouths, remain uneroded, and only their distal parts are redistributed by the axial currents along the channel. The resulting vertical sequence of sediments will be similar to that in the Aleutian Trench, with the addition of an upper unit consisting of slides and other laterally introduced deposits that contain lateral and radiating current patterns (Fig. 4). The underlying abyssal section is also much thinner and consists of more pelagic sediments.

TRENCH-FILL THICKNESSES

It was noted earlier, that modern trenches usually contain less than 1,000 m of fill. Hayes and Ewing (1970) observed that modern trenches are generally devoid of fill. Scholl and Marlow (1974) have pointed out that even in modern filled trenches, such as the one off Oregon and Washington, the average thickness of sediment is only slightly more than 1,000 m. In their opinion, this thickness is abnormal in modern trenches, because it is probably caused by increased sedimentation during Pleistocene glaciation. They question whether such thicknesses were also found in Mesozoic and early Tertiary trenches.

The idea that present trench sediments are unusually thick is supported by the study of drill cores in the Aleutian Trench area. Sediments filling the trench are upper Pleistocene (von Huene and Kulm, 1973), and the rates of sedimentation have increased greatly during glacial periods. In fact, the thickest present trench fills around the Pacific are off mountainous, glaciated coasts where ice sheets may have extended across the continental shelf. Therefore, it appears that modern trench fills thicker than about 500 m are likely to have been affected by high-latitude coastal glaciation or by close proximity to large rivers, such as the Columbia River off Oregon and Washington.

The depth and configuration of modern trenches was shown by Hayes and Ewing (1970) to be surprisingly uniform in 35 selected profiles across Pacific trenches, in which the depth of a trench below the adjacent deep-sea floor ranges from 2.5 to 4.0 km. This uniformity suggested to them that all present Pacific trench configurations are constrained by a common mechanism. If modern trench-forming mechanisms result in sedimentary sections 1 km thick or less, could greater trenches have commonly formed in the past? Scholl and Marlow (1974) have suggested elongate basins on the continental slope as a possible alternative. If present world tectonism is characteristic of the past, the one identifying characteristic of trench deposits is their relatively limited thickness. This initial thinness and the ultimate great thickness of eugeosynclinal sequences has been discussed by Burk (1972), among others.

SUMMARY

In developing a facies-distribution model for trench fill, the axial-current channel provides a good basic mechanism. The likelihood of density currents in trenches is not only seen in the common occurrence of the channel, but it is also suggested by smooth axial gradients along trench floors that were formed by redistribution of fans and slumps. Sediment masses sometimes have great energy when they hit the foot of the continental slope, and these may turn into axial density currents. Local ponding in topographic lows, where density currents must cross a flat area and begin to flow uphill, is a possible exception.

A second process that affects the facies distribution in trenches is the steady seaward overlap of the trench fill. It has been assumed that the trench fill is continually incorporated into the slope or is covered by an outbuilding of the slope. In this case the channel and levee would be steadily displaced seaward with respect to previously buried channels. This would result in the general sequence of facies that may exist at the foot of the continental slope, as shown in Table 1, at a trench margin.

At the base of the trench fill are fine overbank deposits that grade upward into coarse channel deposits, which may be capped by slide deposits. The dominant trend is from fine distal to coarse proximal turbidites. The underlying abyssal-plain sediments follow a similar trend from pelagic (distal) to hemipelagic (proximal) deposits. The pelagic sediments may include oozes if they accumulated on oceanic ridges that were above the calcium carbonate compensation level. The more hemipelagic sediments, reflecting closer proximity to land, may even contain a rare turbidite. The contact between the pelagic and the turbiditic sequences can take various forms, from abrupt to gradational. It separates bodies of sediment that have accumulated under different modes and rates of sedimentation.

If modern trenches and tectonic processes are characteristic of those of the past, the vertical sequence of sedimentary facies in trenches may be recognizable in eugeosynclinal exposures on land. This sequence should be identifiable where about 1,000 m of coherent section can be examined, because fully compacted modern trench sequences (turbidites and pelagic) would be no more than 1,500 m thick. If the great thicknesses of eugeosynclinal sediments result from tectonic repetition of a basic trench section, a commonly repeated sequence of facies might help delineate fault zones (e.g., Burk, 1972).

BIBLIOGRAPHY

Burk, C. A., 1972, Uplifted eugeosynclines and continental margins: Geol. Soc. America Mem. 132, p. 75-85.

Hamilton, E. L., 1967, Marine geology of abyssal plains in the Gulf of Alaska: Jour. Geophys. Res., v. 72, no. 16, p. 4189-4213.

Hayes, D. E., and Ewing, M., 1970, Pacific boundary structure, *in* Maxwell, A. E., ed., The sea, v. 4, New concepts of sea floor evolution, pt. 2, Regional observations: New York, Wiley-Interscience, p. 29-72.

Kulm, L. D., von Huene, R., et al., 1973, Initial reports of the Deep Sea Drilling Project, v. XVIII: Washington, D.C., U.S. Govt. Printing Office, 1077 p.

Piper, D. J. W., von Huene, R., and Duncan, J. R., 1973, Late Quaternary sedimentation in the active eastern Aleutian Trench: Geology, v. 1, no. 1, p. 19-22.

Rosato, V., J., 1974, Peruvian deep-sea sediments: Evidence for continental accretion (M.S. thesis): Oregon State Univ.

Scholl, D. W., and Marlow, M. S., 1974, Sedimentary sequence in modern Pacific trenches and the deformed Pacific eugeosyncline, *in* Dott, R. H., Jr., Modern and ancient eugeosynclinal sedimentation: Soc. Econ. Paleontologists and Mineralogists Spec. Paper.

von Huene, R., 1972, Structure of the continental margin and tectonism at the eastern Aleutian Trench: Geol. Soc. America Bull., v. 83, no. 12, p. 3613-3626.

______, and Kulm, L. D., 1973, Tectonic summary of Leg 18, *in* Kulm, L. D., et al., Initial reports of the Deep Sea Drilling Project, v. XVIII: Washington, D.C., U.S. Govt. Printing Office, p. 961-976.

Deep-Sea Sedimentation

W. H. Berger

INTRODUCTION

In a volume on the geology of continental margins, a section on deep-sea sediments would seem in need of explanation. First, deep-sea sedimentation includes both the eupelagic and the hemipelagic facies domain, the latter being greatly influenced by continental margin effects. Second, all oceanic sedimentation is part of a global geochemical system, so processes in one realm have profound effects on sedimentation in the other. Third, the plate tectonics paradigm provides for long-term interaction between continental and oceanic crust at the continental margins. Thus a record can be preserved, however jumbled, of the workings of the biogeochemical fractionation apparatus that is the world ocean—be it obscure, as in the rocks and ore bodies of Andean mountains, or in ophiolite suites and ancient pelagic sediments of the Alpine chains.

Deep-sea deposits were first explored in a comprehensive fashion during the CHALLENGER Expedition 100 years ago, by Wyville Thomson, John Murray, and their associates, and the preliminary results of this expedition were first assimilated into contemporary geologic thought in the lectures of Thomas Huxley (see in Huxley, 1894). Many of the questions that are still with us were raised by these natural philosophers. The first part of this report is a general overview of Recent sediment patterns, sedimentation of red clay, of ferromanganese deposits, of calcareous oozes, and of siliceous oozes. The last part deals with pre-Recent sediments, which became accessible through coring during the last 30 years (Pleistocene record) and through deep-sea driling during the last 5 years (Tertiary and Cretaceous record).

Throughout I have attempted to focus on the basic questions that have been or are being asked in the study of deep-sea deposits. Inevitably, the range of questions determines the kinds of answers produced from the information at hand, and the answers themselves generate the new question; for instance, Walker (1973) has illustrated how entirely new questions led to the development of the turbidity current theory. However, some questions lead into dead-end streets, and some recently turned out to be formidable obstacles to research in marine geology (see Moore, 1972).

For a comprehensive introduction to the field of deep-sea sedimentation, I refer to the appropriate articles or chapters in Sears (1961), Hill (1963), Riley and Skirrow (1965), Maxwell (1970), Funnell and Riedel (1971), Lisitzin (1972), Shepard (1973), Goldberg (in press), Hay (in press), Riley and Chester (in press), and Ramsay (in press), as well as to the sediment syntheses in the Initial Reports of the Deep Sea Drilling Project (Beall and Fischer, 1969; and subsequent legs). For Mediterranean deposits, see Stanley (1973).

IN THE WAKE OF H.M.S. CHALLENGER: INVENTORY AND THE QUESTION OF ULTIMATE SOURCES

The unexcelled source book on deep-sea deposits by Murray and Renard (1891) at once defined and established the field of deep-ocean sedimentation. Their main task was to provide an inventory of deep-sea deposits, and their chief attempt at interpretation consisted of proposing ultimate sources for the sediment found. The deposits they described are classified in a mixed descriptive-genetic fashion (*Pelagic deposits*: red clay, radiolarian ooze, diatom ooze, Globigerina ooze, pteropod ooze; *terrigenous deposits*: blue mud, red mud, green mud, volcanic mud, coral mud). This classification is still in use today, modified to accommodate first Pleistocene and then pre-Pleistocene deposits, as well as the recognition of turbidites (Table 1).

On the basis of the CHALLENGER samples, as well as samples from later expeditions sent to Murray for study, he and his coworkers were able to provide the first reasonably reliable maps of surficial sediments on the sea floor. Subsequent investigations filled the gaps and provided more detail (Fig. 1). In brief, the calcareous facies outlines the oceanic rises and platforms, while the red clay facies is typical for the deep basins. Superimposed on this elevation-controlled dichotomy are the siliceous deposits, which are characteristic for areas of high fertility; that is, oceanic margins, equatorial belt, and polar front regions. In addition, relatively coarse terrigenous sediments from the continents invade the deep sea along the margins, in places to a considerable distance from the shelf (see Ericson et al., 1961; Heezen and Hollister, 1964).

Where the generalized distribution map shows sharp boundaries between well-defined facies, detailed surveys show complex patterns of overlapping mixtures of the various components of deep sea sediments listed in Table 2.

The bulk of the deep-sea deposits was found to consist of "materials of organic origin." Murray, who was responsible for the study of the biogenous sediments, described them in detail and outlined the

Table 1. Classification of Deep-Sea Sediments
Source: After Olausson, as modified by Berger and von Rad (1972).

I. (Eu-) pelagic deposits (oozes and clays)
<25% of fraction >5 μm is of terrigenic, volcanogenic, and/or neritic origin.
Median grain size 5<μm (excepting authigenic minerals and pelagic organisms).
  A. Pelagic clays. $CaCO_3$ and siliceous fossils <30%.
    (1) $CaCO_3$ 1-10%. (Slightly) calcareous clay.
    (2) $CaCO_3$ 10-30%. Very calcareous (or marl) clay.
    (3) Siliceous fossils 1-10%. (Slightly) siliceous clay.
    (4) Siliceous fossils 10-30%. Very siliceous clay.
  B. Oozes. $CaCO_3$ or siliceous fossils >30%.
    (1) $CaCO_3$ >30%. < ⅔ $CaCO_3$: marl ooze. > ⅔ $CaCO_3$: chalk ooze.
    (2) $CaCO_3$ <30%. >30% siliceous fossils: diatom or radiolarian ooze.

II. Hemipelagic deposits (muds)
>25% of fraction >5 μm is of terrigenic, volcanogenic, and/or neritic origin.
Median grain size >5 μm (excepting authigenic minerals and pelagic organisms).
  A. Calcareous muds. $CaCO_3$ >30%.
    (1) < ⅔ $CaCO_3$: marl mud. > ⅔ $CaCO_3$: chalk mud.
    (2) Skeletal $CaCO_3$ >30%: foram~, nanno~, coquina~.
  B. Terrigenous muds. $CaCO_3$ <30%. Quartz, feldspar, mica dominant. Prefixes: quartzose, arkosic, micaceous.
  C. Volcanogenic muds. $CaCO_3$ <30%. Ash, palagonite, etc., dominant.

III. Pelagic and/or hemipelagic deposits
    (1) Dolomite-sapropelite cycles.
    (2) Black (carbonaceous) clay and mud: sapropelites.
    (3) Silicified claystones and mudstones: chert.
    (4) Limestone.

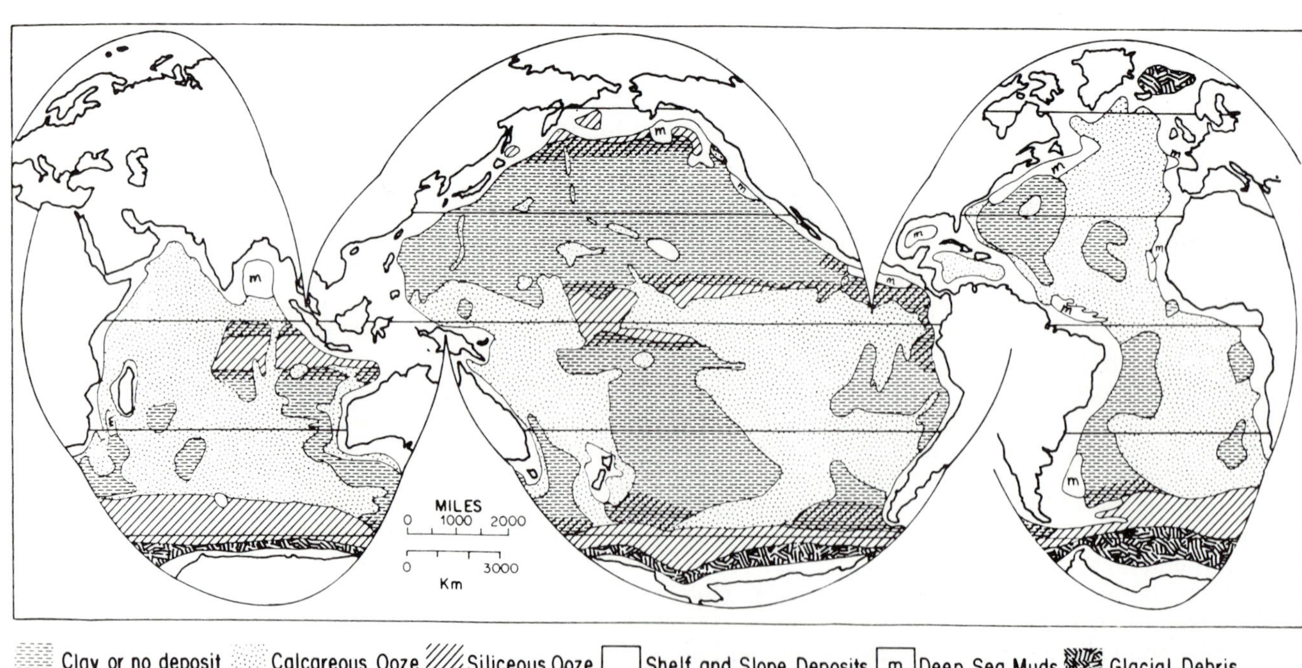

Fig. 1. Distribution of Recent sediments on the deep-ocean floor (modified from a large number of original sources).

Table 2. Main Constituents of Deep-Sea Deposits

Biogenous

Coccolithophorids (planktonic algae; calcite)
Calcareous foraminifera (planktonic and benthonic protozoans; calcite)
"Pteropods" (euthecosome and heteropod pelagic gastropods; aragonite)
Radiolaria (polycystine radiolarians; planktonic protozoans; occasionally phaeodarians; never acantharians; opaline silica)
Diatoms (siliceous algae; opaline silica; benthic forms redeposited from shallow water)
Silicoflagellates (siliceous algae; opaline silica)
Fish debris (teeth, bone fragments; calcium phosphate)
Agglutinated forams (benthic protozoans; inorganic particles with organic ligand)
Organic matter

Terrigenous

Quartz, feldspar, mica, clay minerals, heavy minerals, iron oxide, rock fragments, plant debris (carried by rivers, wind, or ice)

Authigenic

Oxides and hydroxides (ferromanganese concretions)
Silicates (zeolites, clay minerals; Na, Ca, K, Mg)
Heavy metal sulfides (fe, Zn, Cu, Ni, Co, Pb)
Sulfates (ba, Sr)
Carbonates (Ca, Mg, Mn, Fe)
Phosphates (fluorspar-apatite)
Phosphates (fluorspar-apatite)

Volcanogenic

Volcanic glass and palagonite (altered glass), pumice and other rock fragments ("ash"), transported by currents and wind (also certain chemical deposits, mainly iron oxides, on active mid-ocean ridges)

Cosmogenic

Meteoritic spherules

general framework of their distribution—depth, latitude, temperature, and fertility of surface waters, ocean currents, distance from land, amount of organic matter in sediments and associated CO_2 production and oxidation state, as well as activity of benthic organisms. The type of relationships that Murray sought to clarify are represented in Figure 2. In his vocabulary, the barren-versus-fertile contrast (Olausson, 1966) is the one between "mid-ocean" and "near-land," which was dubbed "oceanic" versus "neritic" by Haeckel. Bramlette's (1961) dichotomy stresses "highly oxidized, (eu)pelagic" versus "not highly oxidized, organic-rich, hemipelagic." These concepts are not identical, of course, but there is a large degree of congruency in their patterns, so that the diagram can be read accordingly.

As is evident from Figures 1 and 2, the major facies boundary in deep-sea deposits is the calcite compensation depth. The effects of depth on aragonite and calcite accumulation were recognized and carefully documented by Murray and Renard (1891). Thus, considering Globigerina ooze (= calcareous ooze) they state (p. 215): "This table shows generally a decrease in the quantity of carbonate of lime with increasing depth, but this fact would be still more strikingly exhibited had the samples been all from one region of the ocean, in which the surface conditions were the same, or had the samples from the shallower depths near continents and islands been eliminated, for in these there is always a large admixture of accidental matters derived from land."

The ultimate sources of the oozes may seem obvious today, but the topic was under much discussion at the time of the CHALLENGER Expedition. Being familiar with both the planktonic source and the sedimentary result, Murray easily established the principles of paleoclimatic interpretation of the fossil assemblages."... the predominating species in

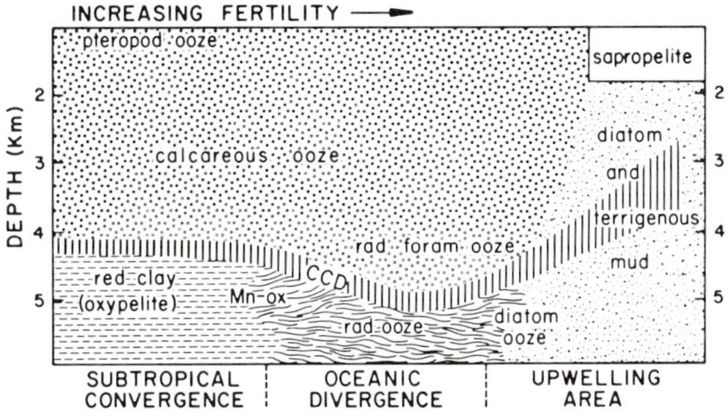

Fig. 2. Distribution of major facies in a depth-fertility frame (based on sediment patterns in the eastern central Pacific).

a deposit vary according to latitude, or more correctly according as the surface oceanic currents have a tropical or polar origin, along with other surface conditions of the locality." Refinements and application of this principle have become extremely important in the reconstruction of Pleistocene oceanic variations.

The ultimate source of red clay, according to Murray and Renard, is the in situ weathering of volcanic material. Wyville Thomson had earlier come to the opinion that the clay was the insoluble residue of foraminiferal shells, a view that was widely discussed at the time. Thomson's idea was easily refuted by dissolving shells from the plankton, but the answer to the question of the ultimate origin of the clay had to await systematic studies by X-ray analysis and other modern mineralogical methods.

Another important question that Murray tackled but failed to resolve is that of sedimentation rates. The first widely used rate estimates were those by Schott (1935), who identified the thickness of the postglacial sediment in the Atlantic by noting the presence of *G. menardii* and combined this thickness with the age of the Holocene. His results are still useful today (compare Lisitzin, 1972, Fig. 63). Rate estimates based on longer sequences show maximum values for terrigenous muds (several meters per thousand years), intermediate ones for calcareous ooze (several centimeters per thousand years), and very low ones for red clay (a few millimeters per thousand years) (see Table 3). Maps of rates, therefore, essentially duplicate facies maps (see Böstrom et al., 1973, for example).

Table 3. Rates of Accumulation of Recent and Sub-Recent Sediments

Facies	Area	mm/ty	Reference
Terrigenous mud	California Borderland	50-2,000	Bandy, 1960
	Ceara Abyssal Plain	200	Hayes et al., 1972
Calcareous ooze	North Atlantic (40-50°N)	35-60	McIntyre et al., 1972
	North Atlantic (5-20°N)	40-14	Schott, 1935
	Equatorial Atlantic	20-40	Schott, 1935; Ericson et al., 1956
	Caribbean	~28	Emiliani, 1966
	Equatorial Pacific	5-18	Hays et al., 1969 (last 1 my)
	Eastern Equatorial Pacific	~30	Blackman, 1966
	East Pacific Rise (0-20°S)	20-40	Blackman, 1966
	East Pacific Rise (~30°S)	3-10	Blackman, 1966
	East Pacific Rise (40-50°S)	10-60	Blackman, 1966
Siliceous ooze	Equatorial Pacific	2-5	DSDP (in Berger, 1973)
	Antarctic (Indian Ocean)	2-10	Lisitzin, 1972
	North and Equatorial Atlantic	2-7	Turekian, 1965
	South Atlantic	2-3	Maxwell et al., 1970
Red clay	Northern North Pacific (muddy)	10-15	Opdyke and Foster, 1970
	Central North Pacific	1-2	Opdyke and Foster, 1970
	Tropical North Pacific	0-1	DSDP (in Berger, 1973)

Table 4.

	N. Atl.	S. Atl.	N. Pac.	Indian	S. Pac.
Montmorillonite	16	26	35	41	53
Illite	55	47	40	33	26
Kaolinite	20	17	8	17	8
Chlorite	10	11	18	12	13

"continentality" "oceanicity"

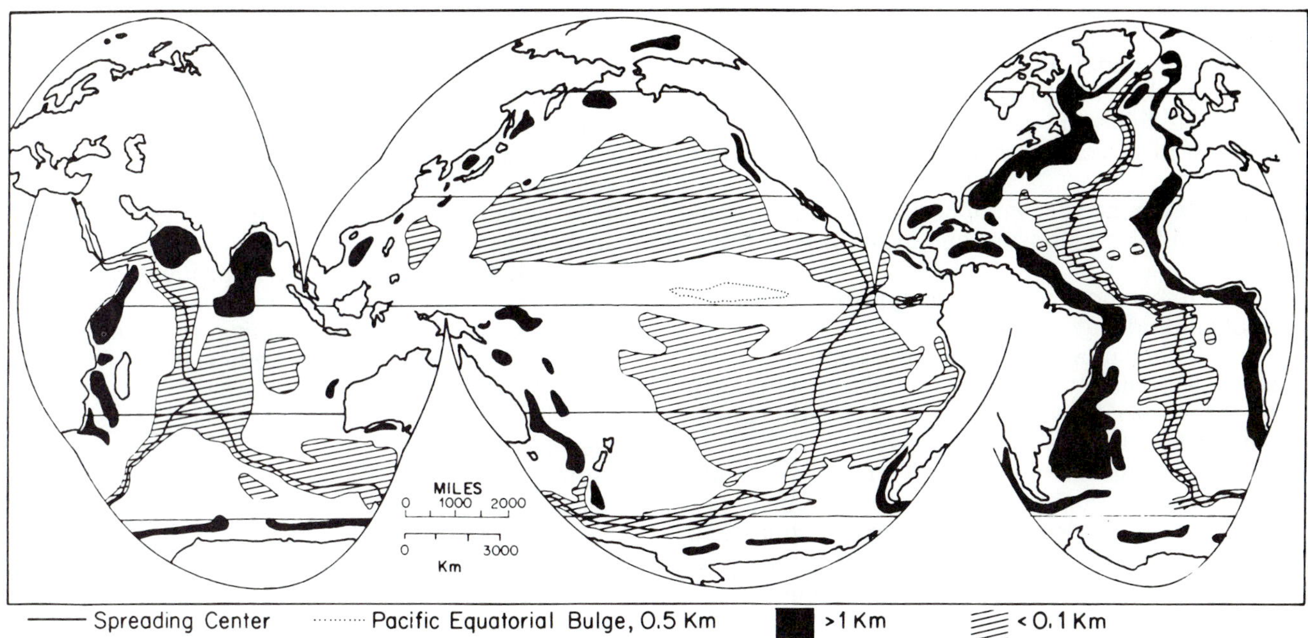

Fig. 3. Sediment thickness to "acoustic basement," assuming a sound velocity of 2 km/sec (compiled and simplified from many sources).

The next important step was the assessment of the total sediment thicknesses, by seismic methods. When it was found that the sedimentary columns for typical basins in the Atlantic average only about 500 m, and in the Pacific 300 m (Ewing and Landisman, 1961), it became clear that the weathering products from the continents are largely trapped around their margins (Fig. 3). However, the question about the age and dynamics of the ocean floor soon arose, which led Hess (1962) to his mantle convection model, which provided for a new sea floor "every 300 to 400 million years" and "accounts for the relatively thin veneer of sediments on the ocean floor" (Hess, 1962), p. 617. The large-scale lateral movements of the sea floor and the subsidence associated with its moving down the ridge flank makes for a peculiar geometry of deep-sea stratigraphy, to which we shall return when discussing results of the Deep Sea Drilling Project.

THE ORIGIN OF "RED CLAY" AND OF THE CLAY FRACTION IN GENERAL: A SHAKY CONSENSUS

Of all deposits, "Red Clay" is uniquely restricted to the deep-sea environment. The bulk of the components are extremely fine-grained, and the coarse silt and sand fractions consist of particles originating in the ocean: authigenic minerals, volcanogenic debris, ferromanganese concretions, and traces of biogenous particles such as fish teeth, arenaceous forams, and in cases spicules and radiolarians. John Murray came to believe "that the clayey matter in Marine Deposits far from land is principally derived from the decomposition of aluminous silicates and rocks spread over the ocean basins by subaerial and submarine eruptions," although A. F. Renard "is inclined to attribute a more important role to submarine eruptions than is admitted by Murray. Colloid clayey matter coming in suspension from the land may, it is admitted, play some part in the formation of this deposit" (Murray and Renard, 1891).

Murray's hypothesis, even in the absence of X-ray analysis, was more than a mere guess, since he had chemical analyses available, as well as physical properties of the clay, such as magnetic susceptibility, which suggested similarities of the clay with volcanogenic material. Also, the coarser material which could be identified microscopically (magnetite, manganese grains and fragments, feldspar, volcanic glass, augite, pumice, quartz, plagioclase, mica, zeolite, and other particles) could reasonably be assumed to give some indication of the nature of the clay-size fraction. On the basis of such coarser particles, Murray and Renard also identified eolian transport as a source of deep-sea clay, especially off the western coasts of Northern Africa and of Western Australia. Iron oxide-coated quartz grains, later termed "wüstenquartz" (see Radczewski, 1939), were one of the eolian indicators.

Essentially, Murray and Renard regarded Red Clay as a "deposit universally distributed in the ocean basins, but only appearing with its typical characters in the greatest depths far from continental land." They concluded that this clayey deposit is chiefly the result of the decomposition in situ of volcanic material, in particular basalt and andesite.

Analysis by X-ray diffraction began in the 1930s (R. Revelle in the Pacific, C. W. Correns in the Atlantic; see Mehmel, 1939; Biscaye, 1965) and has been greatly improved and systematically applied to deep-sea deposits, especially in the last 20 years (see Griffin and Goldberg, 1963; Biscaye, 1965; Griffin et al., 1968; Heath, 1969; Rateev et al., 1969). The following minerals were found to be important in the clay and fine-silt fractions of Red Clay:

1. Clay minerals: montmorillonite, illite, chlorite, kaolinite and gibbsite, and mixed-layer derivatives.

2. Lithogenics: plagioclase, pyroxene, quartz, and alkali feldspar.

3. Authigenics: clinoptilolite-heulandite, phillipsite, ferromanganese oxides, and hydroxides. The latter also provide much of the X-ray amorphous material present in Red Clay.

Concerning the nonclay minerals, there appears to be little doubt about their origin, since they can be identified also in the coarser fractions and compared to their parent rocks (e.g., basic or acidic volcanics) or peri-continental equivalents (e.g., "wüstenquartz"). The distribution of quartz, moreover, suggests eolian transport from desert belts (Rex and Goldberg, 1958; and eolian derivation of clay-size quartz in the North Pacific is further substantiated by evidence on size distribution and oxygen isotopes (Rex et al., 1969; Jackson et al., 1971).

The clay minerals warrant special attention, since they make up two-thirds of the clay fraction of nonbiogenous deep-sea sediments, and the clay-size fraction in turn provides some 90% of "pure" Red Clay. The main groups are montmorillonite, illite, chlorite, kaolinite, and their mixed-layered derivatives (see Berner, 1971, p. 165; Griffin et al., 1968). Montmorillonite is a weathering product of volcanic rocks. Illite, as originally defined, is not a specific mineral name but a general term for clay components belonging to the mica group and their derivatives. In a narrower sense, illite may be considered a fine-grained degraded muscovite. Chlorite is a common constituent of low-grade metamorphic rocks, which are widely exposed on glacially eroded shield areas and provide much of the material for glacial deposits. Kaolinite is a product of intense chemical weathering and represents an insoluble aluminosilicate residue remaining after cations are stripped from feldspars and other minerals by extensive leaching.

The clay minerals that are most abundant in deep-sea clay are montmorillonite and illite (Fig. 4; Table 4), and their distributions suggest that montmorillonite has a large oceanic component, at least in the Pacific, while illite is largely derived from the continents (Griffin et al., 1968). The continental origin of the illite receives strong support from the K-Ar ratios, indicating apparent "ages" of 200-400 million years (Hurley et al., 1963). The remaining two important clay minerals, kaolinite and chlorite, also are land-derived, kaolinite from chemical weathering in the tropics and chlorite from physical weathering in high latitudes.

From the evidence reviewed, we may conclude that the case for continental sources of fine-grained Red Clay components is very strong (see also Böstrom et al., 1973). The dominance of continental sources in deep-sea clay is a geologically recent phenomenon. Griffin and Goldberg (1963, p. 739) noted that "the older Tertiary sediments from the North Pacific are distinctly similar to the present-

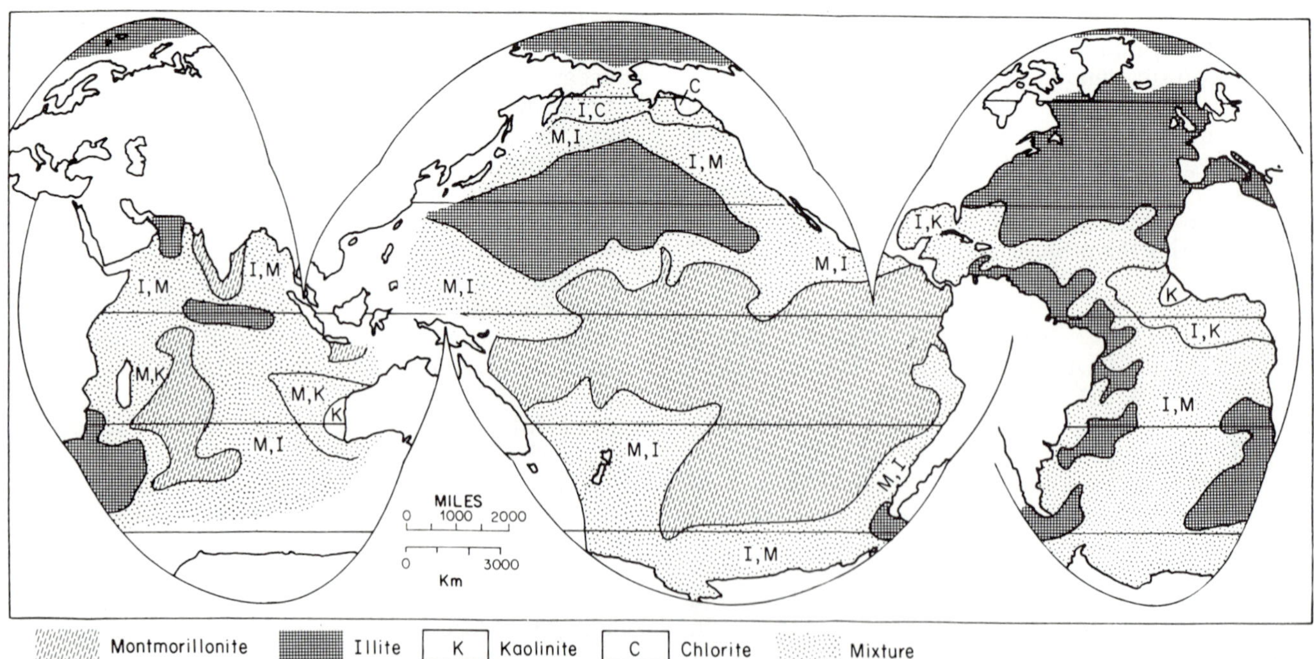

Fig. 4. Dominant clay minerals in the fraction less than 2μm. "Mixture" means that no one clay mineral exceeds 50% of the total (summarized and simplified from maps by Biscaye, 1965; Griffin et al., 1968; Goldberg and Griffin, 1970. Arctic Ocean: Carroll, 1970; Windom, 1969).

day South Pacific ones" and suggested that clays derived from pyroclastics dominated in both basins at that time. Heath (1969b), studying a large number of cores from the Central Pacific, came to the conclusion that tholeiitic debris largely was the source for the principal constituents of all middle Eocene to late Miocene deposits. About 10 million years ago, products from island-arc volcanism, indicated by abundant pyroxene and continent-derived material (quartz, illite, chlorite, kaolinite), begin to dominate the Red Clay mineral assemblage (Fig. 5; see also Jacobs and Hays, 1972). Thus the composition of Red Clay reflects global changes in the general tectonic and climatic setting. As sediments age, of course, a large part of this information is lost in diagenetic processes.

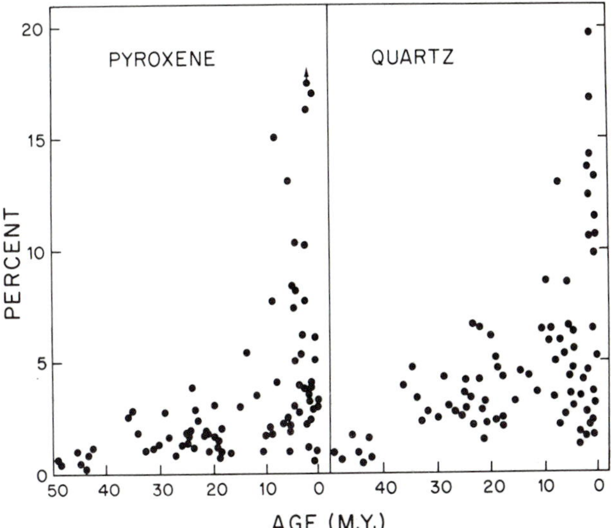

Fig. 5. The late Cenozoic change in mineral assemblages of fine-grained sediments (from Heath, 1969b).

"MANGANESE NODULES," METALLIFEROUS RIDGE SEDIMENTS, AND TRACE-ELEMENT DISTRIBUTIONS: PERENNIAL DISPUTATIONS

After careful description of the ferromanganese concretions and after reviewing the various opinions of the origin of ferromanganese, Murray rather forcefully advocated decomposition of volcanic rocks and minerals within the surficial sediment layer, with manganese and associated trace elements delivered as a by-product of formation of clay (Murray and Renard, 1891). To Renard, however, it appeared "that the greater part must have been derived from the manganese in solution in the seawater," presumably ultimately from continental weathering. Neither suggests which crucial observations could help decide the issue.

Together with von Gümbel's idea that the manganese comes from submarine springs (in modern parlance, "exhalations") and the suggestion by Murray and Irvine (1894) that manganese diffuses upward through the sediment to be precipitated near the interface, a wide spectrum of possible origins was available to choose from at the turn of the century, and more hypotheses were to follow—hypotheses that greatly influenced interpretation of ferromanganese concretions in ancient sediments on land (see Jenkyns, 1970).

Ferromanganese concretions are ubiquitous on the deep-sea floor (especially in areas with no or little sedimentation or erosional scour by bottom currents (Fig. 6). Micronodules range in size from microscopic specks to about 1 mm, nodules range between about 0.5 and 20 cm (Menard, 1964), and the size of slabs and crusts is limited mainly by the capacity of the dredge. Both nodules and slabs are laminated with clay-rich and metal-rich layers alternating. Generally, nodules form around a nucleus, consisting of volcanic debris, indurated clay, a shark's tooth or whale earbone, or any other solid object exposed on the sea floor. As a rule, the exposed surface of a concretion is rather smooth compared with the buried one. Nodules can coalesce to form aggregates or grow in an irregular fashion, producing "grapestones" or botryoidal masses. Radial, healed fractures are common within nodules.

The concretions are by no means unique for the deep sea, being common both in shallow marine areas (Fig. 6) and in periglacial lakes and bogs. A major part of the ferromanganese material is X-ray amorphous, using conventional analytical techniques (Mero, 1965, p. 152; Goodell et al., 1971; Bezrukov and Andrushchenko, 1973).

The Mn/Fe ratio is positively correlated with degree of oxidation and with depth when considering concretions from all environments (Manheim, 1965). In the deep sea, Mn/Fe ratios are slightly greater than unity on the average (Murray and Renard, 1891), but differ considerably between major oceanic regions. Encrustations on seamounts generally tend to have Mn/Fe ratios greater than unity (Cronan, 1972). The total manganese content varies, with an average of 14% for the Atlantic and 19% for the Pacific (Bender, 1972). Within ocean basins there are wide deviations from these averages. In the North Pacific, for example, Mn/Fe ratios range from greater than 10 to 0.7 or less, from off the equator to the northern pelagic province (Menard, 1964). High manganese content appears favored by both biogenous sediment supply and low rates of accumulation.

The enrichment of ferromanganese with cobalt, nickel, and copper (Murray and Renard, 1891), among other metals, has received much attention in connection with the "source problem" and of course also from a commercial point of view. Cobalt tends to be relatively high in nodules from elevated areas in the deep sea (Menard, 1964), especially in samples containing only MnO_2, while nickel and copper tend to be more concentrated in less highly oxidized ones rich in manganite (Barnes, 1967).

Recent studies of the growth rate of nodules, using various isotopic species (Bender et al., 1966; Somayajulu, 1967; Ku and Broecker, 1969; Bender et al., 1970), arrive at rates of several millimeters per million years. Another argument for extremely slow growth rates are K-Ar ages for volcanic nuclei (Barnes and Dymond, 1967), assuming that encrustation begins soon after formation of the nuclei and before alteration of the nucleus material. Despite all these efforts, the growth rate of nodules appears to be an open question, because it cannot be assumed that ratios between parent and daughter elements stay constant through time and because of evidence for geographic variations in incorporation rates and for differential precipitation of closely related isotopes (see Ku and Broecker, 1969; Patin and Tkachenko, 1972). Furthermore, the problem of diffusion of radio isotopes appears unresolved (Sackett, 1966; Arrhenius, 1967). The problem of how to keep slowly growing nodules from getting buried has been recognized for some time (see Goodell et al., 1971). Benthic activity and agitation by currents are among postulated processes that keep the nodules on top.

How fast does manganese accumulate *within* the sediment? Turekian and Wedepohl (1961; Table 2) give 6,700 ppm Mn for deep-sea clay, that is, 0.67%. For a sedimentation rate of 1-3 m per million years (Table 1.3), that is, about 300 mg/cm^2 10^3 yr, the accumulation rate becomes 1-3 mg/cm^2 10^3 yr, in agreement with the measurements by Bender et al. (1970). To match this rate, a typical nodule would have to grow at about 30-60 mm per million years. This is perhaps close enough to inferred rates to warrant the statement that nodule and sediment accumulation rates are very similar (Bender et al., 1966, 1970; Somayajulu et al., 1971).

A case for a "volcanic origin" of at least some ferromanganese can be found in studies of ridge crest sediments and basalts (Böstrom and Peterson, 1966, 1969; Bonatti and Joensuu, 1966; Hart, 1970; Corliss, 1971; Piper, 1973). Geographic distributions alone already suggest increased delivery of iron and manganese to sediments of the ridge crest (Skornyakova, 1965), whether through increased weathering of basalt or through precipitation from hydrothermal solutions, or both. Recent serious efforts to resolve the origin of metalliferous sediments from the East Pacific Rise (Dasch et al., 1971; Dymond et al., 1973) lead to the conclusion that the sediments "formed from hydrothermal solutions generated by the interaction of sea water with newly-formed basalt crust at mid-ocean ridges. The crystallization of solid phases took place at low temperatures and was strongly influenced by sea water, which was the source for some of the elements found in the sediment" (Dymond et al., 1973, p. 3355). However, their conclusions also rest mainly on the more-or-less familiar arguments associated with regional geography and mineralogic and chemical properties (e.g., X-ray amorphous phases, high iron content, and low Al) and not on REE patterns and isotopic composition. Their findings show that the rare element ratios are similar to those in seawater, oxygen isotope values of ferromanganese-rich phases correspond to those in manganese nodules formed at low temperatures in isotopic equilibrium with seawater, strontium isotopic ratios are those of seawater rather than basalt, sulfur isotopes suggest seawater derivation of associated barite, uranium data are difficult to interpret in the context, and so are lead isotopic ratios. The latter do suggest basaltic derivation of lead, whether derived from hydrothermal solutions or from weathering of volcanics

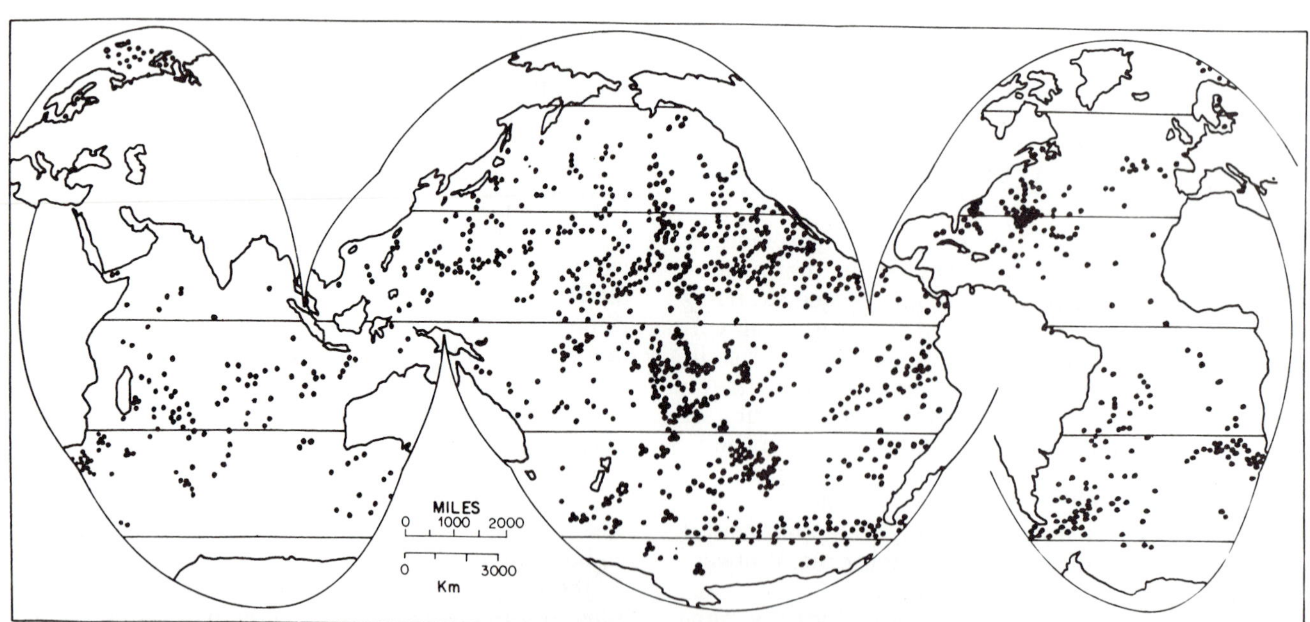

Fig. 6. Areas from which ferromanganese has been reported (after Horn et al., 1972, with additions from various other sources).

(see also Bender et al., 1971). Thus part of the metal enrichment—especially the very high iron content—would seem to derive from hydrothermal activity, while the rest is "scavenged" (Goldberg, 1954) by the freshly formed hydroxides. Unfortunately, manganese has only one isotope (55), precluding the type of analysis that can be done for lead and strontium, each of which appears to have its own geochemical pathways, which differ by unknown degrees from that of manganese.

Attempting to distinguish between the different types of origin, Arrhenius and Bonatti (1965) suggest a "rapid bulk precipitation process" for volcanogenic ferromanganese, and "slow precipitation of manganese oxide phases from dilute solution in sea water" for the nonvolcanic concretions. Their arguments concerning differential mobilization and precipitation of trace metals are greatly extended and widely applied by Price and Calvert (1970). They especially stress the influence of postdepositional migrations of manganese and associated elements (Murray and Irvine, 1894; Gripenberg, 1934; Ljunggren, 1953; Fomina, 1962; Hartmann, 1964; Lynn and Bonatti, 1965; Manheim, 1965; Skornyakova, 1965; Presley et al., 1967; Li et al., 1969) as a controlling factor of compositions. The ratio Mn/Fe, for example, is explained by the varying opportunity for fractionation within the sediment, with Mn being significantly more mobile, since Fe precipitates as oxide or a sulfide over a large range of conditions (Fig. 7). Such a fractionation of Fe from Mn has been nicely illustrated by Turner and Harris (1970) in sediments from the Kara Sea.

In summary, the regional variation in nodule composition (Mero, 1962, 1965; Menard, 1964) suggest to some workers regional differences in the type of volcanism producing the ferromanganese, and to other differences in the general sedimentation regime, chiefly differential mobilization during early diagenesis. The "diagenesis view" is able to marshall more observations, in part because sedimentary processes are easier to investigate than is active volcanism. However, as we have seen from the work of Dymond et al. (1973), different elements follow different geochemical pathways to an undetermined degree, so that no matter which opinions are accepted, the "ultimate source" of the manganese remains an open question, until the proper geochemical budgets are worked out. To argue origins from associations is a risky business, despite its common use in sedimentology.

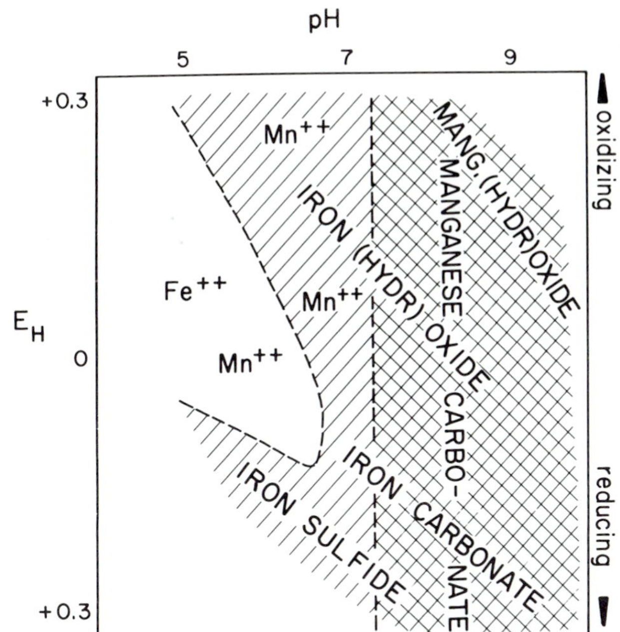

Fig. 7. Sketch of stability field of iron and manganese phases on the sea floor, illustrating the greater mobility of manganese (adapted from diagrams in Stumm and Morgan, 1970).

CALCAREOUS OOZE: THE SNOW-COVERED MOUNTAINS OF THE DEEP OCEAN

At the present time, the rivers that feed the oceans are essentially dilute solutions of calcium bicarbonate and silica. The fluxes are such that the amount of calcium in the oceans can be delivered within about 1 million years. To balance this input, the ocean precipitates calcium carbonate, releasing CO_2 in the process, which then becomes available for yet more weathering on land. The precipitation takes place near the surface, where ocean water is supersaturated with calcium carbonate. Virtually all crystallization takes place within shell-building organisms (coccolithophores, foraminifera, pteropods). Some of these shells find their way to the sea floor, where they are preserved on the more elevated parts and dissolved on the deeper ones, because the undersaturation of seawater increases with pressure and decreasing temperature. (For review of these processes from geological and chemical viewpoints, see Berger, 1970a, 1971; Berner, 1971; Bramlette, 1961; Broecker, 1971a; Chave and Schmalz, 1966; Edmond, 1972; Edmond and Gieskes, 1970; Gieskes, 1974; Harvey, 1957; Heath, 1969b; Heath and Culberson, 1970; Li et al., 1969; Lisitzin, 1971b, Lyakhin, 1968, Morse and Berner, 1972, Pytkowicz, 1968, Revelle and Fairbridge, 1957; Sillén, 1967; Turekian, 1965.)

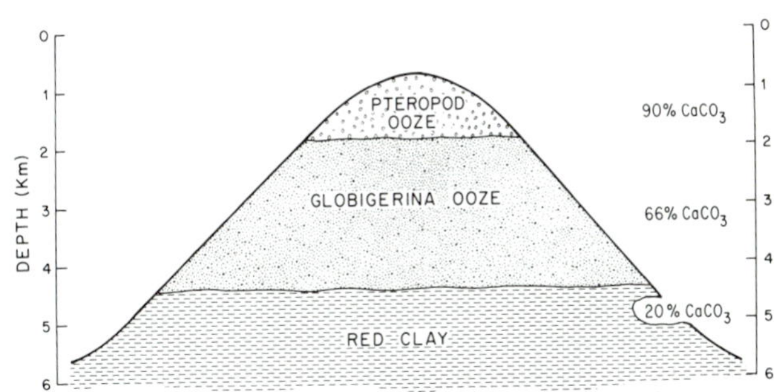

Fig. 8. Idealized bathymetric zonation of deep-sea deposits, produced by increasing dissolution of carbonate with depth, as envisaged by John Murray (see Murray and Hjort, 1912, p. 173).

Murray first recognized the general decrease of carbonate percentages with depth (Murray and Renard, 1891) and ascribed this distribution to dissolution effects (Fig. 8; see also Murray and Hjort, 1912, Fig. 142), although he also recognized the other two important factors controlling carbonate percentages: supply of shells, and dilution by detrital matter and clay. Ever since, the question of the relationship between carbonate percentages and depth has maintained high visibility in deep-sea sedimentology (Pia, 1933; Revelle, 1944; Revelle et al. 1955; Bramlette, 1961; Turekian, 1965; Smith et al., 1968; Heath and Culberson 1970; Berger, 1970; Peterson, 1966; Berger, 1967).

A variety of different dissolution rate profiles can reproduce the commonly encountered carbonate-depth profiles (Fig. 9). Thus the percent carbonate profiles based on core-top samples are poor indicators for dissolution-rate profiles, especially in view of other factors, such as differential dilution through redeposition of clay-size noncarbonate material. Such redeposition will tend to change carbonate percentages in a systematic manner with high values in winnowed deposits and low ones in "chaff"-enriched deposits. Differential dissolution of calcareous fossils, on the other hand, appears to offer rather more reliable evidence for the determination of dissolution profiles on the sea floor (see Yount and Berger, 1974).

Murray and Renard were well aware of the widespread dissolution of planktonic skeletons on the floor of the ocean and that the deep sea is undersaturated with both calcite and silica, a fact that was to be confirmed by various other means, including field and laboratory experiments as well as measurements and calculations (Wattenberg, 1935; Berner, 1965; Peterson, 1966; Pytkowicz and Fowler, 1967). In addition, Murray and Renard observed that dissolution acts selectively, first between major groups such as corals, mollusks, foraminifera, and calcareous algae (p. 277) and next between various species of foraminifera. The question of differential dissolution has received much elaboration since (Schott, 1935; Arrhenius, 1952; Berger, 1967; Ruddiman and Heezen, 1967; Berger, 1968 and subsequent studies, summarized in Berger, 1971). Analogous observations are found in diatoms (Murray and Ranard, 1891, p. 282; see also Schrader, 1972), in coccoliths (McIntyre and McIntyre, 1971; Berger, 1973c; Schneiderman, 1973; Roth and Berger, 1974), and in radiolarians (Moore, 1969; Johnson, 1974). Murray and Renard also discussed protection of the calcareous structures by organic envelopes, acceleration of dissolution by decomposition of organic matter, protection of skeletons by high accumulation rates, and the tendency for carbonate to accumulate in equatorial regions due to increased production of calcareous shells (1891, p. 277). These topics recur perennially when discussing carbonate distributions.

The calcite saturation level is the particular depth at any one place in the ocean, where seawater is neither supersaturated nor undersaturated with calcite; that is, rates of precipitation on the surface of calcite crystals brought in contact with seawater at this level are equal to rates of dissolution on this surface (Fig. 10). Operationally, this level is computed using thermodynamic equations with experimentally determined parameters to operate on measured properties of seawater. Thus the "saturation level," when mapped, describes the distribution of depths at which in situ seawater is in an idential state with respect to the carbonate system as defined within a particular framework.

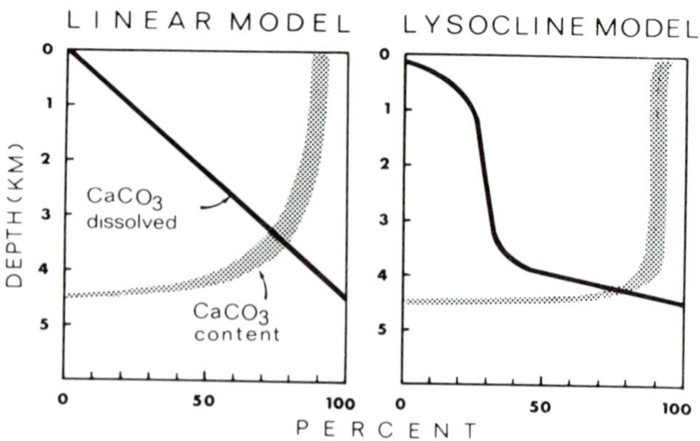

Fig. 9. Effects of hypothetical dissolution profiles on the carbonate content at depths. Losses of $CaCO_3$ were calculated according to $L = 1 - R_0/R$, with R_0 being the initial noncarbonate fraction, R the final one. Note that the depth profile of carbonate content is rather insensitive to the kind of model chosen, especially in consideration of other processes, such as winnowing and redeposition.

In concept, the calcite compensation depth (CCD) is the particular depth level at any one place in the ocean where the rate of supply of calcium carbonate to the sea floor is balanced by the rate of dissolution, so that there is no net accumulation of carbonate (Bramlette, 1961). In practice, the CCD is mapped as the level at which percent carbonate values drop toward zero. This method can lead to difficulties in areas with pre-Recent carbonate exposures (Broecker and Broecker, 1974).

The concept of the lysocline, implicit in the findings of Peterson (1966), Berger (1967), and Ruddiman and Heezen (1967), was introduced to denote a surprisingly well defined facies boundary zone between well-preserved and poorly preserved foraminiferal assemblages on the floor of the central Atlantic Ocean (Berger, 1968) and on that of the South Pacific (Berger, 1970b; Parker and Berger, 1971). In general, the lysocline is envisaged as the level at which there is a maximum change in the composition of calcareous fossil assemblages due to differential dissolution.

The term "lysocline," however, has also been used in contexts other than description of sedimentary facies. Thus, from the results of field experiments in the Pacific, showing a drastic increase of dissolution rates near 4,000 m, and from the cor-

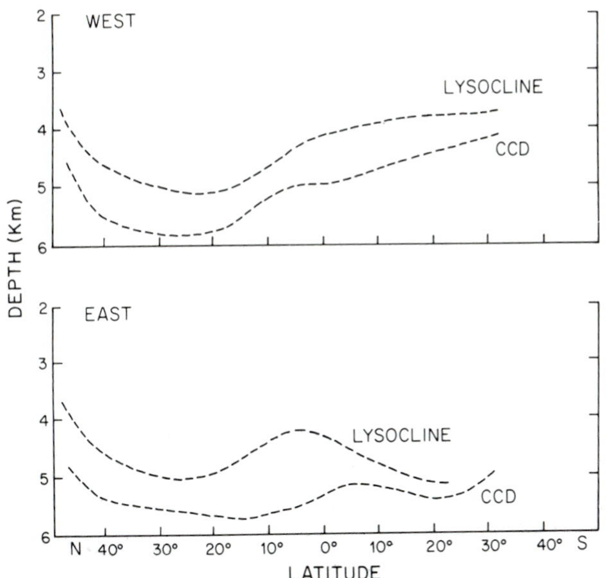

Fig. 10. Depth distributions of lysocline and carbonate compensation depth (CCD) in the Atlantic Ocean, west and east of the Mid-Atlantic Ridge (data from various sources).

relation of the lysocline with the top of the Antarctic Bottom Water in the Atlantic and apparently also in the Pacific, it appears very probable that over large areas the lysocline *on the sea floor* marks a level of increase in the propensity of the *surrounding water* to dissolve calcite (see also Edmond, 1971). This level in the water column has been called "hydrographic lysocline" to distinguish it from the sedimentologically defined "foraminiferal lysocline" or "coccolith lysocline."

The distinction between hydrographic and sedimentary lysocline is important both conceptually and operationally, because in fertile regions the sedimentary lysocline shallows and becomes diffuse. These effects of fertility in pericontinental areas are especially obvious in the rise of the CCD around the margins of the North Pacific (Fig. 11).

Paradoxically, high fertility along the Pacific equator leads to a depression of the CCD (Fig. 11), as recognized by Arrhenius (1952). The contrasting effects of fertility may be reconciled if it is admitted that fertility is positively correlated with calcite accumulation in some places and negatively in others. The striking difference in content of organic matter between coastal and deep-sea sediments (Kaplan and Rittenberg, 1963) offers a clue to this apparent contradiction. The negative correlation of fertility and calcite accumulation may be explained by the high supply of organic matter in the fertile pericontinental areas, which leads to highly increased benthic activity as well as development of relatively CO_2-rich interstitial waters, a mechanism first suggested by Murray. Thus calcite shells are attacked even at depths of a few hundred meters on continental slopes (Berger and Soutar, 1970; Kowsmann, 1973; Thiede, 1973), and dissolution is pronounced well above the open-ocean lysocline, especially in the areas beneath eastern boundary currents (Berger, 1970a; Parker and Berger, 1971). The positive correlation of equatorial high fertility on the one hand and equatorial calcite accumulation on the other indicates that the supply of calcareous shells outstrips the increased supply of organic matter in this region. Presumably, the reason is that the food web here is stable through time and utilization of food is efficient, leaving much less reactive organic matter for sedimentation, and consequently the ratio of calcitic shell to organic carbon produced is relatively high.

The question of where caicareous shells dissolve, in the water or on the ocean floor, has been a matter of much speculation. Both Kuenen (1950, p. 354) and Bramlette (1961, p. 356) postulated that most of the solution occurs on the bottom because of the long exposure times there. Box cores in several areas the eastern Pacific, between 10 and 15° north of the equator, at depths below the CCD, showed the presence of many delicate foraminifera in the surficial sediments mixed with heavily corroded resistant ones and their fragments, indicating that the suggestion of Kuenen and of Bramlette was correct. Net tows at abyssal depths also contained delicate forams, as well as the aragonitic shells of pteropods, which were not found in the sediment (Adelseck and Berger, 1974).

What factors ultimately control carbonate distribution? Fertility regulates both the supply of carbonate and the supply of organic matter, which largely determines the interstitial water "acidity" and the benthic activity, both inimical to calcite preservation. The Antarctic Bottom Water, either by reaching a "critical undersaturation," as implied by Morse and Berner (1972), or through rates of flow (Edmond, 1971), appears to aggressively dissolve calcareous shells wherever present. Fertility mainly depends on the upper circulation, whereas bottom water aggressiveness is strongly tied to abyssal circulation. Thus, in describing the distribution of calcareous deposits on the deep-sea floor, one gathers clues about the oceanic circulation, which ultimately controls these distributions. Redeposition processes introduce considerable "noise" into these general patterns. They largely depend on the interaction of local topography and bottom currents and appear to be responsible for much of the regional grain-size signature in deep-sea carbonates, which are also reflected in coccolithforam ratios (see Bramlette, 1958; Olausson, 1961; van Andel and Komar, 1969; Berger and von Rad, 1972; Moore et al., 1973).

The ultimate reason why the abyssal water dissolves calcite is that calcium carbonate is supplied to the bottom in excess of the amount that can be sedimented over the long run, an amount fixed by the influx from the continents (see Sillén, 1967). The shell supply to the ocean floor that exceeds this influx tends to deplete upper waters of calcium carbonate, to the extent that cooling, compression, and CO_2 uptake produce bottom water that is

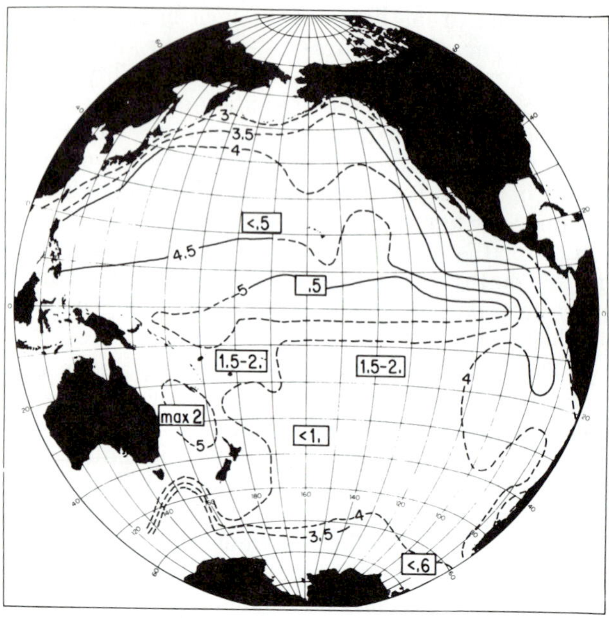

Fig. 11. Depth that separates core-top samples rich in carbonate from those with only a few percent of carbonate or none (CCD) level). In boxes: depths at which aragonitic shells disappear on the sea floor (ACD), in kilometers. The distance from ACD to CCD appears to be variable. (CCD from global map in Berger and Winterer, 1974; ACD estimated from pteropod data in various sources).

sufficiently undersaturated to redissolve this excess supply. In this fashion, a dynamic steady state is maintained (Revelle, 1944; Pytkowicz, 1968; Li et al., 1969). From this model it can be readily inferred that increased fertility leads to increased dissolution, and that the slowing of dissolution—by inhibitors (Chave and Suess, 1967; Morse and Berner, 1972) or by delivery of significant proportions of resistant shells—will increase the undersaturation of the ocean with respect to calcite. Similar arguments apply to silica (Bogoyavlenskiy, 1967) and, by analogy, to all other phases involved in biological cycling (see Arrhenius, 1963, p. 681; Broecker, 1971a).

SILICEOUS OOZE AND THE BIODYNAMICS OF OCEAN CHEMISTRY

The constituents of silicious deposits consist of the remains of diatoms, silicoflagellates and radiolarians, and sponge spicules, all of which are made of opal, a hydrated form of amorphous silicon dioxide. The first three types of organisms belong to the planktonic, the last to the benthic realm. Diatom oozes are typical for high latitudes, diatom muds for pericontinental regions, and radiolarian oozes for equatorial areas (Fig. 12). Both diatom and radiolarian oozes, of course, are mixtures, with one or the other form dominant. The deposits typically occur in areas of high fertility, that is, in regions of

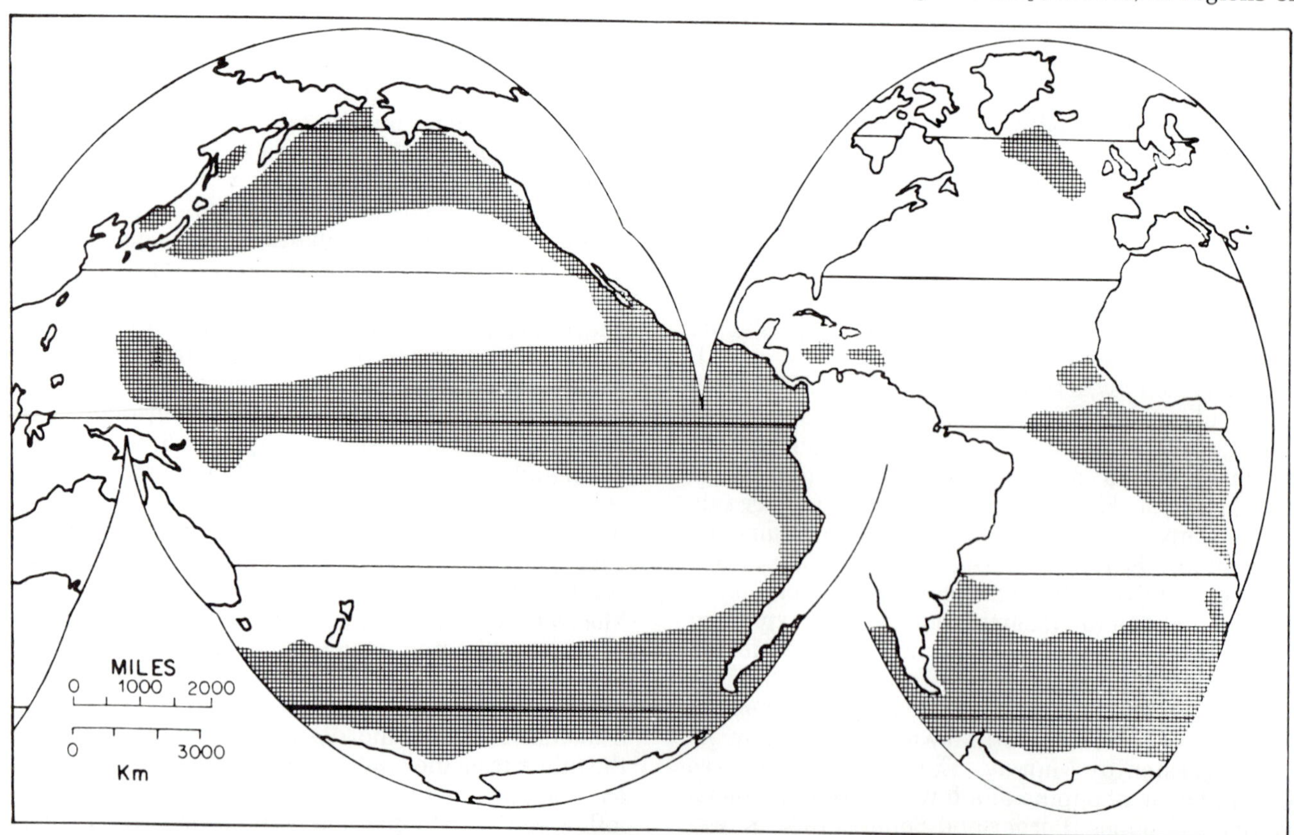

Fig. 12. Areas where siliceous fossils are abundant in deep-sea deposits (from numerous sources).

surface water divergence with relatively high phosphate values and, in the case of upwelling, cool waters (Pratje, 1951). This overall correspondence between fertility patterns and silica-rich deposits can be considerably modified by redeposition processes within individual regions. The silica frustules are light and easily transported, and the activity of benthic animals, which tends to resuspend fine sediment, is especially pronounced in fertile areas. Thus, aided by bottom currents, siliceous frustules tend to accumulate in depressions (Bramlette and Bradley, 1942; Bramlette, 1946; Moore et al., 1973; Johnson, 1974).

The production of siliceous shells attains its maximum in coastal regions, where productivities can exceed by a factor of 10 the values in the subtropical gyres. This phenomenon is chiefly due to a breakdown of the thermocline barrier to nutrient transport from deep to shallow sunlit waters and leads to the formation of a "silica ring" around the ocean basins. "Silica belts" are provided by the latitudinally arranged oceanic divergences, which are a result of atmospheric circulation (see Lisitzin, 1966, 1971a).

The degree of dilution of the siliceous material depends, of course, on the ratio between accumulation rates of nonsiliceous and siliceous particles. Since sedimentation rates of terrigenous material is highest around continents, where silica supply rates are also high, the dilution effect in coastal areas is less than one would perhaps expect and cannot mask the "silica rings" around the oceans. As concerns dilution by carbonate, one might expect that a high supply of calcareous shells would be accompanied by an equally high supply of siliceous shells (Cifelli and Sachs, 1966), so that concentrations of silica would simply be due to removal of calcite. This is not generally the case, however. Indeed, there is a distinct negative correlation between silica and calcite, which has been ascribed to opposing chemical requirements for preservation (Correns, 1939). We have already seen that increasing fertility from some point on, leads to decreasing preservation of calcite, but to increasing accumulation of silica. A similarly opposing trend is indicated for depth relationships, with silica corrosion being greatest in the highly undersaturated upper waters (Fig. 13). In addition, the tendency for silica to accumulate in the deep part of basins, and the tendency for carbonate to dissolve in such areas, tends to minimize the importance of carbonates as a dilutent.

A third factor believed by Riedel (1959) to control the abundance of siliceous fossils is the extent of dissolution. Maps showing both abundance and preservation patterns (Goll and Bjørklund, 1971) indicate that the preservation of siliceous shells is rather closely correlated with their relative abundance. A positive correlation between abundance and preservation may be ascribed to an increased supply of easily dissolved diatoms in fertile areas, which will "buffer" interstitial waters for the more robust skeletons, and to an otherwise favorable chemical environment in organic-rich sediments.

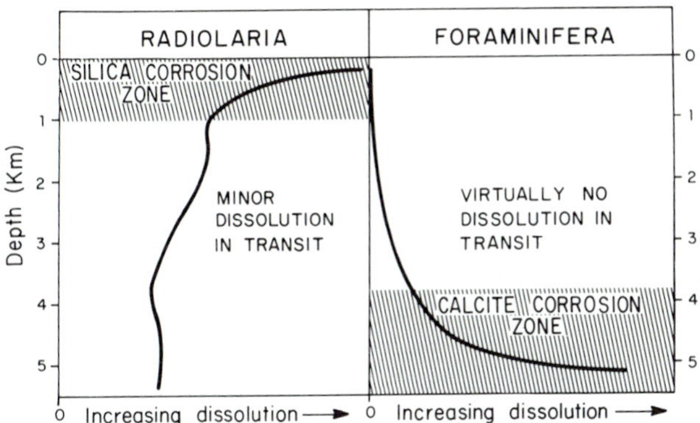

Fig. 13. Comparison of dissolution profiles of radiolarians and of foraminifera, based on field experiments. Most dissolution occurs whether in shallow waters (aided by predation) or on the sea floor (aided by reworking, CO_2 production in the case of carbonate, and by the long exposure times available).

Several other observations strongly indicate that dissolution plays an all-important role in restricting high abundances of siliceous fossils to the fertile regions. Over large areas of the ocean floor, radiolarians occur only in the uppermost few centimeters in Quaternary sediments (Riedel and Funnell, 1964, p. 361). It may be assumed that these sediments will ultimately be devoid of radiolarians (Riedel, 1971) and that their dissolution constitutes silica input to the overlying waters. Similarly, concentration gradients in interstitial waters and differences in concentration between these waters and the overlying seawater suggest that silica is diffusing out of the sediment (Fanning and Schink, 1969, p. 68), with siliceous fossils providing the source of the flux (see also Fanning and Pilson, 1971; Hurd, 1972; Bischoff and Sayles, 1972; Hurd, 1973; Heath and Dymond, 1973). Measurements in radiolarian clay sampled in box cores confirm that much of the difference is established within the uppermost layer of the sediment, so that gradients here are optimal for diffusion. This mixed layer at the interface still contains the most easily dissolved frustules and skeletons (see Schrader, 1972) and is in intensive exchange with the bottom water, so that it is expected to contribute the bulk of the silica reflux to the ocean.

Laboratory experiments show that fine-grained silicates react readily with seawater, releasing or sorbing silica, depending on conditions (Mackenzie and Garrels, 1965; Mackenzie et al., 1967; Siever, 1968; Siever and Woodford, 1973). Silica release by clays entering seawater and associated silica uptake by diatoms had been proposed by Murray and Irvine (1891), who were impressed by the dearth of this element in the photic zone.

Assuming a quasi-steady-state composition of seawater with respect to the major ions, the alkali metals delivered from the weathering of silicates with carbonic acid should recombine with silica in the ocean. This is deemed necessary not only to remove the alkali, but also to release the bicarbonate ions accompanying the alkali ions on their down-river voyage to the sea. Without such uptake of alkali and such release of bicarbonate, the ocean would presumably become a big "soda lake" (Garrels and Mackenzie, 1971).

Observations in estuaries purporting to show uptake of silica by detrital clays, although ingenious in principle (Bien et al., 1959; Liss and Spencer, 1970) are rather difficult to interpret (see Fanning and Pilson, 1973). Even is such uptake takes place, there is no reason to believe that any silica sorbed stays on the clay particles once they are exposed to undersaturated seawater at the site of deposition. On the other hand, the vigorous precipitation of silica by diatoms and silicoflagellates in estuarine settings is undisputed. Arguments based on budget calculations are unconvincing, whether they claim an important role of an abiological uptake (Burton and Liss, 1968, 1973) or deny such a role (Harris, 1966; Calvert, 1968). In both cases, the answer is contained in the assumptions about rates of supply to the ocean and of sedimentation on the sea floor, assumptions that are necessary for the subsequent calculations providing the answer sought.

The possibility and even the probability that *interstitial silica reacts within sediments* to form new phases (Wollast et al., 1968; Siever and Woodford, 1973; Heath, 1974; Calvert, 1974; Wollast, 1974) needs further study (see Heath and Dymond, 1973). A slow deterioration of opaline phases, and even silica-rich zeolite, with concomitant growth of stable silicate minerals (palygorskite, sepiolite) is indicated, for example, by the complicated diagenetic history recorded in zeolitic radiolarian casts in upper Cretaceous deep-sea deposits from the central Atlantic (Berger and von Rad, 1972, Plate 18). However, silica concentrations *within the waters of the open ocean* can easily be shown to be controlled by processes other than inorganic equilibrium reactions:

1. Profiles of dissolved silica with depth of water show the characteristic nutrient-type shape, which implies precipitation near the surface and dissolution at depth (see Redfield et al., 1963; Grill, 1970).

2. Plots of dissolved silica versus phosphate (Chow and Mantyla, 1965; Berger, 1970a) and versus barium and alkalinity (Bacon and Edmond, 1972) demonstrate coherence between chemically dissimilar systems that derive covariance from being involved in oceanic "biodynamics," that is, precipitation in skeletons and in organic matter and redissolution of such precipitates at depth, in the water or on the sea floor or both.

3. When the correlation between silica and phosphate in waters shallower than 1 km is extrapolated to greater depths, there remains an unaccounted-for "excess" of silica, which suggests dissolution of skeletons after the associated material has been oxidized. Combined with commonly accepted deep-water residence times (500-1,000 years), this "excess" indicates an input to the deep ocean which is greater than the river input by a factor of 10 (Berger, 1970a). Thus internal cycling of silica in the ocean dominates the system.

4. Profiles of dissolved silica with depth of water vary greatly between different ocean basins. Thermodynamic equilibrium with any one solid phase or phase combination does not exist, therefore. Instead these differences are in accord with the general abyssal circulation, high silica values being associated with "old," oxygen-depleted waters, and low silica values with waters that recently left the surface, where they were stripped of their silica. Thus the reason that silica concentrations in deep water are relatively low is not the uptake, if any, of dissolved silica by clay minerals, but the fact that the water remembers its condition at the surface and did not have time to saturate itself with respect to the actively dissolving opal.

In summary, elementary oceanographic considerations are sufficient to show that silica-clay interactions do not control silica concentration patterns in the ocean, and this is indeed the general consensus (Calvert, 1968, 1974; Heath, 1974; Wollast, 1974). As noted, such reactions may well ultimately control the concentration of alkali metals, especially if dissolved silica values rise drastically just below the sediment-water interface, so that diffusion contact with seawater can be maintained and alkali can be extracted from the seawater. The nature of such reactions is not understood at present. The presence of zeolites (and associated minerals) on the sea floor in areas of low sedimentation rates (Arrhenius, 1963, p. 698; Bonatti, 1963) is no proof that these silicates actually form at the sediment-water interface; they may be dissolving together with any associated Tertiary radiolarians. Such fossils are widespread over large areas of the Pacific (Riedel, 1971). Zeolite growth *within* the sediment is not disputed by these remarks, of course.

A biodynamic model of the ocean with regard to these questions may be sketched as follows (Fig. 14.).

The source of silica is commonly taken as the "river supply" (Calvert, 1968), although other subordinate ones are admitted (Heath, 1974). Stefansson (1966) has shown that volcanic eruptions can contribute considerable amounts of silica to surrounding ocean waters. Schutz and Turekian (1965) have speculated on the importance of Antarctica as a silica source, Burton and Liss (1968) on that of polar areas in general. The "river influx" of 4.3×10^{14} g/yr (Livingstone, 1963) corresponds to a balancing sedimentation rate of about 1 g/m^2yr. Doubling this value, we should obtain a very conservative upper limit of 2 g/m^2yr for the silica that can ultimately leave the ocean. Since siliceous organisms are oblivious to the influx but respond to the fertility of the ocean, they precipitate silica in excess of this influx, and seawater will therefore be undersaturated (Berger, 1970a; Broecker, 1971a).

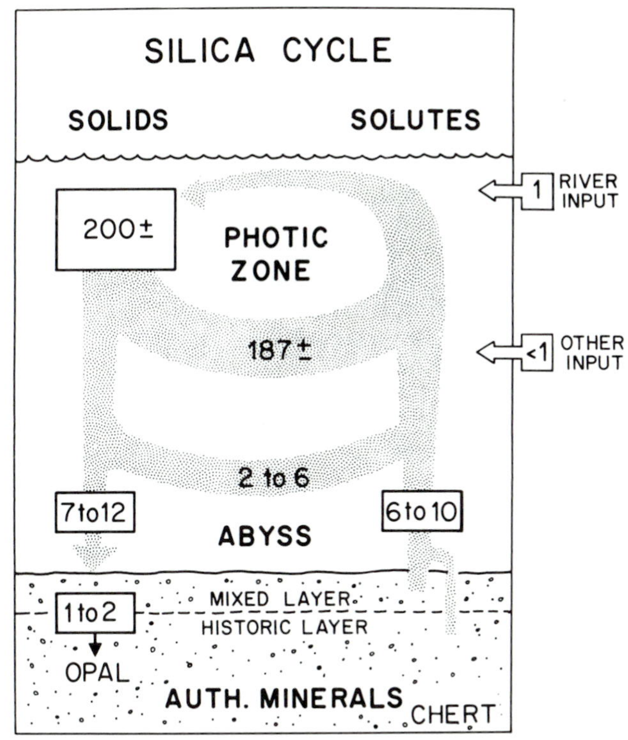

Fig. 14. Schematic representation of the oceanic silica cycle. "River input" is taken as 4.3×10^{14} g of SiO_2/yr (Livingstone, 1963), which corresponds to an average balancing deposition rate of 1 $g/m^2/yr$. The production of solids at 200 g/m^2 yr is estimated from data given by Lisitzin (1972) and is uncertain within a factor of 2. Any change in this estimate would alter the reflux value in upper waters and leave the other flux values unaltered. The flux from the sediment if thought to be largely from near the interface, in fertile regions.

The more resistant the shells are to dissolution, the greater will be the undersaturation toward opal for given deep-water residence times (Bogoyavlenskiy, 1967).

Thus the rate of dissolution of the "excess" precipitation over influx controls the concentration of dissolved silica, in complete analogy to the calcium carbonate system, with one important difference (Fig. 13): carbonate dissolution rates are positively correlated with depth, so deep waters are able to dissolve sediment even though upper waters are not undersaturated; silica dissolution rates show no such correlation, so deep waters can dissolve sediment only if upper waters are already greatly undersaturated. Also, for silica, as a nutrient mineral (Lewin, 1962), a feedback loop is established whereby the rate of dissolution of shells controls supply and the rate of supply controls dissolution. This general cycle has two components, one that runs parallel to phosphate, and one that runs independently, involving postoxidative dissolution of silica (Berger, 1970a). GEOSECS results on quasi-conservative behavior of dissolved silica in the deep ocean (Bacon and Edmond, 1972; Craig et al., 1972; Spencer, 1972) support the suggestion that the bulk of this postoxidative dissolution takes place at the sediment interface. As Heath (1974) has emphasized, the pericontinental areas are especially important as sinks and, by implication, as sources of dissolved silica from the sediments (see also Calvert, 1966). The buried opal may subsequently alter to chert, as documented by results from deep-sea drilling (Calvert, 1971; Heath and Moberly, 1971; von der Borch et al., 1971; von Rad and Rosch, 1972; Berger and von Rad, 1972; Heath, 1973; Lancelot, 1973) or may react to form silicates, as mentioned earlier.

The discovery of chert within deep-sea sediments has both fascinated geologists and frustrated them in efforts to drill and recover complete sections. A few words on these intriguing rock types are in order. The formation of deep-sea cherts (that is, silicified sediments cemented by cryptocrystalline and microcrystalline silica) appears to proceed from mobilization and reprecipitation of opal, generating a matrix of disordered cristobalite with unfilled opaline skeletons which eventually alters toward a quartzitic rock with mostly quartz-replaced and quartz-filled fossils as diagenesis progresses (Heath and Moberly, 1971; von Rad and Rosch, 1972), although recrystallization may proceed at various rates and somewhat divergent patterns, depending on the original sediment present (Lancelot, 1973). Thus at least some of the opal deposited on the sea floor will finally end up as microcrystalline quartz, with disordered cristobalite being an intermediate stage. Details of these diagenetic processes have been given a considerable amount of attention and study (see Calvert, 1974, for review). The complex of questions surrounding the "chert problem" requires investigation of a wide spectrum of evidence (Bramlette, 1946; Gartner, 1970; Calvert, 1971; Gibson and Towe, 1971; Heath and Moberly, 1971; Mattson and Pessagno, 1971; Berger and von Rad, 1972; Wise et al., 1972; Heath, 1973; Lancelot, 1973; Ramsay, 1973).

Among the questions of primary significance are the geologic setting in which cherts occur and questions of a more specialized chemical nature that need to be considered before the geologically important questions on original facies and on source can be answered with some confidence. There is every reason to believe that the answers to these questions will differ depending on the kinds of chert considered—pelagic chert ("clean," with planktonic remains and authigenic minerals only, or "dirty," with volcanic detritus, clay particles, or iron oxide) or hemipelagic chert (with much terrigenous silt and iron sulfide). Pelagic cherts tend to be nodular in carbonates and bedded in clay-rich sediments (Heath and Moberly, 1971; Heath, 1973; Lancelot, 1973), while hemipelagic cherts are bedded (Berger and von Rad, 1972), perhaps owing to rather pronounced differences in texture and chemistry of the original layers which reflect redeposition and sorting processes in the hemipelagic realm.

If siliceous fossils and/or silica-rich glass are to be available for later production of chert, important requirements besides a sufficiently high supply would seem to be a relatively silica-rich bottom water, a reasonably high burial rate and, perhaps, sufficiently high organic content and sufficiently low carbonate content to retard early mobilization and loss to the bottom water. If quick burial can be achieved without considerable dilution of the source particles, for example by high supply of opaline skeletons, whether indigenous or displaced, or by initial burial with allodapic carbonate from shallower depths which dissolves, the chances for later chert formation would appear much improved. During diagenesis, presumably, the opaline skeletons dissolve and volcanic material, if any, releases silica during devitrification. The silica-rich interstitial solutions migrate along bedding planes or fractures and vertically to areas of precipitation in nearby permeable lenses or layers (Heath and Moberly, 1971). The finding that interstitial concentrations of silica tend to be lower in reduced than in oxidized sediments (Heath and Dymond, 1973) suggests that precipitation in organic-rich layers is one mechanism by which a gradient can be created, allowing silica to migrate.

Large-scale changes in global geochemical conditions would seem necessary to account for stratigraphically fixed trends in chert abundances (Heath and Moberly, 1971; Lancelot, 1973). At times, such as in the mid-Eocene, flooded shelves may have tied up much carbonate but released the available silica to the deep ocean, while at present the deep sea swallows the carbonates (Kuenen, 1950), and the silica disappears largely in the rings around the continents (Heath, 1974). In addition, flooding of shelves prevents dilution of oceanic volcanogenic sediment with continental debris low in mobile silica. A link to global temperature variations (see Moore, 1972) in such a model would call for carbonate-rich, silica-poor deep-sea deposits during geocratic times, having cold climates and narrow shelves, with much mixing and upwelling at the shelf edges due to intensified seasonal winds, while silica-rich deep-sea deposits, providing for formation of cherts, would characterize times with warm equable climates and wide carbonate shelf seas with deep-water outflow.

EXPERIMENTS ON A PLANETARY SCALE: THE PLEISTOCENE RECORD

At the time the CHALLENGER samples were being studied, the "theory of the ice age," as proposed by Louis Agassiz and further developed by contemporary geologists, was a subject of intense discussion. The CHALLENGER samples were inadequate to test the idea of repeated large-scale climatic changes and associated changes in ocean currents. However, subsequent expeditions recovered short cores which clearly showed that deep-sea deposits did contain a record of changing depositional patterns, and Murray and Philippi (1908) and Philippi (1910) suggested shifts in oceanic circulation, both in surface waters and at depth, to account for these.

Schott (1935), of the METEOR Expedition, used variation in foraminiferal composition, in particular the presence and absence of *Globorotalia menardii*, to define a glacial-postglacial boundary [later extended (Ericson and Wollin, 1968) to cross-correlate cores from the tropical Atlantic]. The long cores gathered by the Piggot Expedition and the Swedish Deep Sea Expedition made it feasible to study a larger section of the record, and by the time initial results were published (Bradley et al., 1942; Phleger, 1948; Ovey, 1950; Arrhenius, 1952; Phleger et al., 1953; Emiliani, 1955) the main problems to be attacked in decades to follow were well defined.

The present-day surface circulation patterns are reflected in the sediments on the ocean floor, the critical boundaries of interest being between polar and subpolar waters and between temperate and tropical ones, mapped on the basis of a few dominant species of planktonic foraminifera (Fig. 15). Thus, where foraminifera are present, the corresponding boundaries for the last glacial can also be mapped, by finding their positions during the last cold peak. Although details remain to be worked out, it is quite clear that the boundaries, especially the polar fronts, were closer to the equator than now. The distribution of species of fossil groups other than foraminifera was, of course, similarly affected (Hays, 1965; Kanaya and Koizumi, 1966; Donahue, 1967; Nigrini, 1970; McIntyre, 1967), and the area of carbonate sedimentation was reduced, while that of ice-rafted debris was greatly extended (Bramlette and Bradley, 1942; Conolly and Ewing, 1965; McIntyre et al., 1972).

Faunal and floral temperature indices can be made to appear in the shape of actual temperatures, which makes them highly convenient for graphic representation. Appropriate conversions of compositional data to such temperature indices have been dubbed "paleoecologic equations" or "transfer equations" (Imbrie and Kipp, 1971). An assemblage is first indexed with a series of numbers representing its similarity ("loading") to various idealized standard assemblages for which the sampling space constitutes a mixture. This indexing sequence is then converted to a temperature estimate using regression equations derived from the distribution of present-day standard assemblages and associated surface-water temperatures.

Ultimately, the quality of all such estimates is limited by several fundamental problems: differential seasonal shell production, differential dissolution, adaptive responses of organisms to a change in conditions, and changing correlation between temperature and other ecologic factors controlling distributions, leading to changing correlations between the environment and the sedimentary record.

The change from a glacial-type ocean to the present interglacial one has been dated at between 12,000 years (Emiliani, 1955) and 11,000 years age

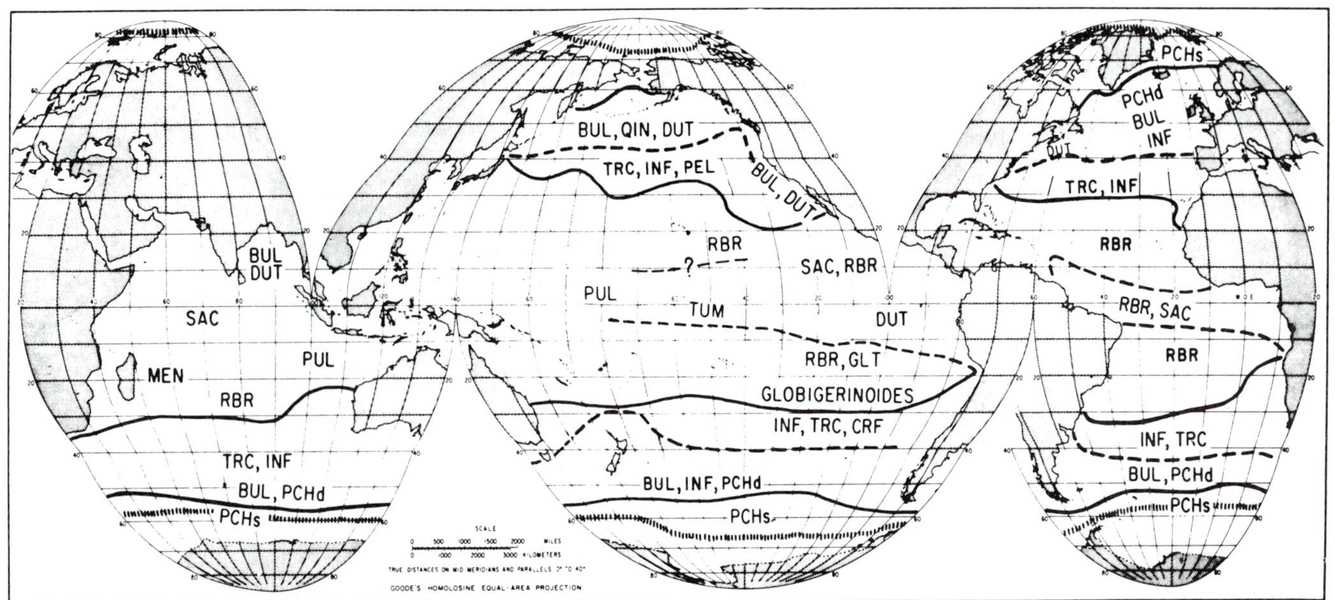

Fig. 15. Foraminiferal zonations in pelagic sediments. The major boundaries (tropical-extratropical, near 30°, and temperate-polar, near 55°) are climatic markers (Horse Latitudes and Polar Front) and follow isotherms (20° and 2-3°, winter temperature). The zones are: polar, subpolar, transition, subtropical, and tropical. Symbols: PCHs, *G. pachyderma*, sinistral; PCHd, *G. pachyderma*, dextral; QIN, *G. quinqueloba;* DUT, *G. dutertrei;* BUL, *G. bulloides;* PEL, *H. pelagica;* TRC, *G. truncatulinoides;* INF, *G. inflata;* RBR, *G. ruber;* SAC, *G. Sacculifer;* TUM, *G. tumida;* PUL, *P. obliquiloculata.* (Mainly from Imbrie and Kipp, 1971, and Hecht, 1973, for the Atlantic; Bradshaw, 1959, and Parker and Berger, 1971, for the Pacific; Be and Tolderlund, 1971, for the Indian Ocean.)

(Ericson et al., 1956; Broecker et al., 1958, 1960), mainly in tropical latitudes. This datum has been widely used to calculate sedimentation rates, following Schott (1935). At least beyond 40°N, however, the boundary probably is time transgressive (Fig. 16) and ranges in age from less than 8,000 years to more than 13,000 years, as shown by Ruddiman and McIntyre (1973). These authors also point out that if the oxygen isotope curve is controlled mainly by ice effects, it should be synchronous and should deviate systematically from the time-transgressive fossil record (see also Olausson, 1965; Imbrie and Kipp, 1971).

Arrhenius (1952) suggested that glacial strata are rich and interglacial strata poor in carbonate, and this correlation was confirmed by later work (Emiliani, 1955; Hays et al., 1969). As noted by Olausson (1965) and in subsequent studies (Olausson, 1967; Hays et al., 1969; Broecker, 1971b; Berger, 1973b), it appears that low carbonate (interglacial) stages correspond to increased dissolution and vice versa and that fertility variations play a secondary role, if any, in producing these carbonate cycles. Olausson (1965) compared his carbonate curves from the northern and central Atlantic with those of Arrhenius (1952) and concluded that high carbonate stages in one ocean correspond to low carbonate stages in the other, a correlation that has been abundantly confirmed. The glacial carbonate low in the Atlantic apparently is largely an effect of dilution by terrigenous material rather than of dissolution (Broecker, 1971b; Ruddiman, 1971).

The first workable absolute age scale was provided by Emiliani (1955), together with a generalized oxygen isotope curve for the tropical Atlantic. This scale was based on extrapolation of ^{14}C dates and an assumed locking-in to the Milankovitch curve (Fig. 17). Recently, Broecker and his associates (see Broecker and van Donk, 1970) have suggested that Emiliani's scale needs to be lengthened by about 25%. Broecker proposes an age of 127,000 years for the penultimate deglaciation; Emiliani (1972) prefers to retain an age of 100,000 years for this event, based on $^{231}Pa/^{230}Th$ dates (Rosholt et al., 1961, 1962; Rona and Emiliani, 1969). The information on the absolute ages of high sea levels from various authors given by Emiliani and Rona (1969, Fig. 1) agrees much better with the

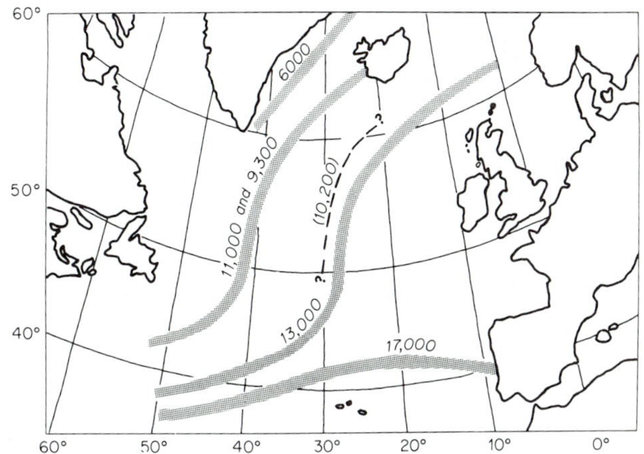

Fig. 16. Retreat of the polar front during deglaciation as recorded in deep-sea sediments, a prime example of climatic transgression (adapted from Ruddiman and McIntyre, 1973).

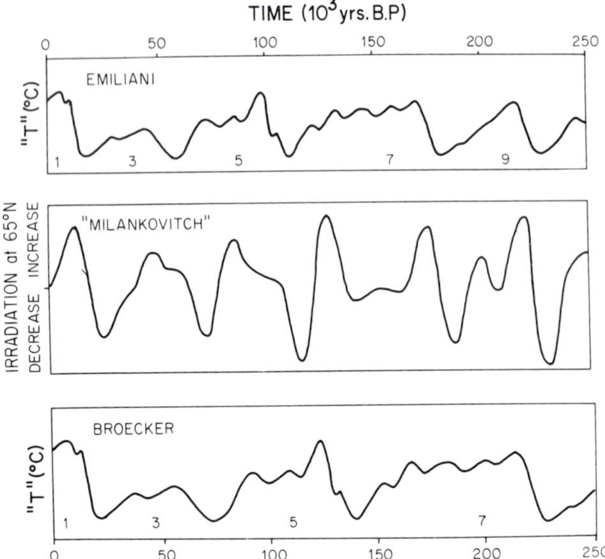

Fig. 17. Comparison of "generalized isotopic paleotemperature curve" of Emiliani (1966), with the Milankovitch irradiation curve and with the time scale of Broecker and van Donk (1970).

scale of Broecker and Ku (1969) than with their own. Apparently, either scale can be accommodated.

The paleomagnetic time scale in the deep sea has recently been reviewed (Opdyke, 1972; Watkins, 1972). In the context of this new kind of stratigraphy (see Hays et al., 1969), the Pleistocene is generally reckoned from the beginning of the Olduvai, 1.63-1.79 my (Berggren, 1972). Further refinement of Pleistocene stratigraphy is to be expected from determination of racemization of amino acids (Bada and Schroeder, 1972; Kvenvolden et al., 1973).

The problem of isotopic variation of seawater is at the base of the question about glacial-interglacial temperature amplitudes (Emiliani, 1971a). Recent consensus puts the range of variation in $\delta^{18}O/^{16}O$ near 1.2 (Olausson, 1965; Shackleton, 1967; Dansgaard and Tauber, 1969; van Donk and Mathieu, 1969; Duplessy et al., 1970); despite the objections of Emiliani over the years (see, however, Emiliani and Shackleton, 1974, for a more cautious attitude). Shackelton and Opdyke (1973), in an important contribution to this problem, follow the general consensus and are able to obtain internally consistent ressults for isotopic variations in seawater, planktonic foraminifera, and benthic foraminifera.

The oxygen isotope "paleotemperature" method is based on the face that calcite and seawater in thermodynamic equilibrium differ in their $^{18}O/^{16}O$ ratios, and that this difference decreases with increasing temperature. Shells precipitated in equilibrium with seawater are enriched in ^{18}O, but less so at increasing temperature. A detailed discussion of complications that tend to be ignored in paleotemperature studies has been given by Berger and Gardner (1974).

IN THE WAKE OF GLOMAR CHALLENGER: PLATE TECTONICS AND PLATE STRATIGRAPHY

Early stirrings of mobilist imagination which followed the recognition of crustal folding by horizontal contraction (A. von Humboldt and E. Suess; see Heim, 1878, pp. 190ff.), the concepts of continental drift (Taylor, 1910; Wegener in 1915 (1966); Runcorn, 1956), sea-floor spreading (Hess, 1962; Vine and Matthews, 1963; Wilson, 1965; Vine, 1966), and plate tectonics (McKenzie and Parker, 1967; Morgan, 1968; LePichon, 1968; Isacks et al., 1968) have revolutionized the earth sciences. Resistance to the mobilist view, based on cogent arguments as well as on Kelvin's fallacy (denial of a phenomenon because of lack of a known mechanism) has rapidly dwindled as a result of evidence collected by sea-going geologists, and plate tectonics has become indispensable as a coherent theoretical framework for geologic investigation.

In Hess's (1962) sea-floor model, which forms the core of this theory, oceanic crust is continually generated at the mid-ocean ridges, from whence it moves toward its sink in the island-arc system, accumulating sediment on the way. As a corollary, the sea floor is covered by sediments whose base is progressively younger toward the ridge crest and whose basinal facies succession shows abyssal Red Clay overlying the carbonates that originally accumulated on the ridge flank. The general acceptance of the concept of sea-floor spreading as a predictive tool was greatly facilitated by the results of the deep-sea drilling expeditions of GLOMAR CHALLENGER (Peterson et al., 1970; and subsequent DSDP reports). These results indeed showed the age progression with increasing distance from the ridge crest which was predicted by Hess (1962), and that the geometry of present facies boundaries is generally in accord with sea-floor subsidence and horizontal plate motions (see discussions by Hsü and Andrews, 1970, Leg 3; Benson et al., 1970, Leg 4; Fischer et al., 1970, Leg 6; Hays et al., 1972, Leg 9; Berger and von Rad, 1972, Leg 14; van Andel and Heath, 1973, Leg 16; and Heezen et al., 1973, Leg 20). The principles and operations appropriate to the treatment and interpretation of these conveyor belt facies relationships have been outlined (Berger, 1972; Winterer, 1971, 1973; Clague and Jarrard, 1973; Berger, 1973a; van Andel and Moore, 1974); They constitute the concept of "plate stratigraphy" (Berger and Winterer, 1974).

To interpret facies distributions with regard to depositional environment, we need to reconstruct depths and geographic positions as they appeared at the time of deposition. In relatively shallow marine waters, paleobathymetric analysis has traditionally relied on paleontologic and sedimentologic criteria (see Hallam, 1967). In the deep sea, rate of subsidence of sea floor on spreading ridge flanks can be used for direct paleodepth estimates. The rate of subsidence has been investigated by Menard (1969) and by Sclater et al. (1971). Sclater et al.

plotted sea-floor age against amount of subsidence (vertical distance to ridge crest) for various areas of the ocean. They found that the plots were very similar from one area to the other and concluded that the fit of age versus depth of (basaltic) sea floor is good enough in many cases to date a given area of actively spreading sea floor to within 2 million years, from a knowledge of bathymetry alone, for ocean floor younger than 40 million years. The great similarity among Pacific, Atlantic, and Indian ocean age-depth relationships suggests that the relationship is based on a fundamental law and was valid throughout Tertiary time and possibly earlier.

The hypothesis of age-depth constancy is based on this argument. The hypothesis states that at any point in geological time, the age of the sea floor determines how far it moved downward from the original ridge crest. This postulate provides a simple means of reconstructing the approximate depth of deposition of ancient deep-sea sediments, independent of spreading rates of the underlying basement, by a "backtracking" procedure. The method consists of positioning a given sediment sample in the age-depth diagram and moving it back through time parallel to the generalized subsidence curve (Fig. 18). A relatively minor adjustment is made for the weight of the sediment that is being unloaded during this backward voyage (see Berger and Winterer, 1974).

Combining the paleodepth method with the results of DSDP Leg 14 of northwestern Africa, it is possible to retrace the main trends in the evolution of major facies domains in the Central Atlantic (Fig. 19; see Berger and von Rad, 1972).

During the Pliocene and at present, pelagic and hemipelagic domains are of comparable width. The CCD stands relatively deep, but the ridge flank also reaches to great depths, so calcareous ooze and (oxidized) pelagic clay are represented to a similar extent. Continental slope and abyssal plain belong to the hemipelagic realm, where terrigenous influx of silt increases sedimentation rates and closer proximity to the continent increases fertility. This

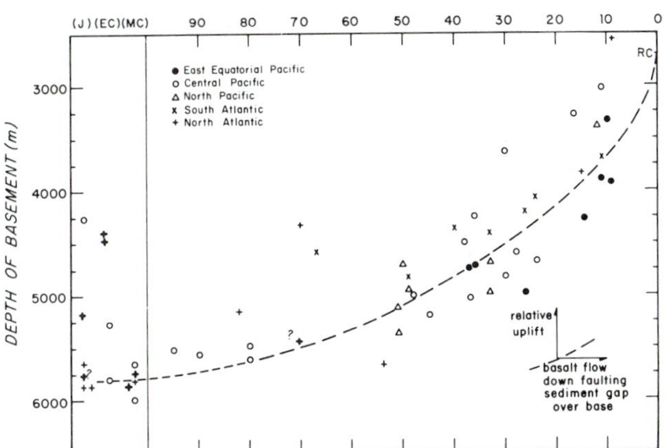

Fig. 18. Relationship between depth of basement and biostratigraphic age of overlying sediments in sites of the Deep Sea Drilling Project, compared with an age-depth curve based on present age-depth distributions as listed in Sclater et al. (1971) (from Berger, 1972).

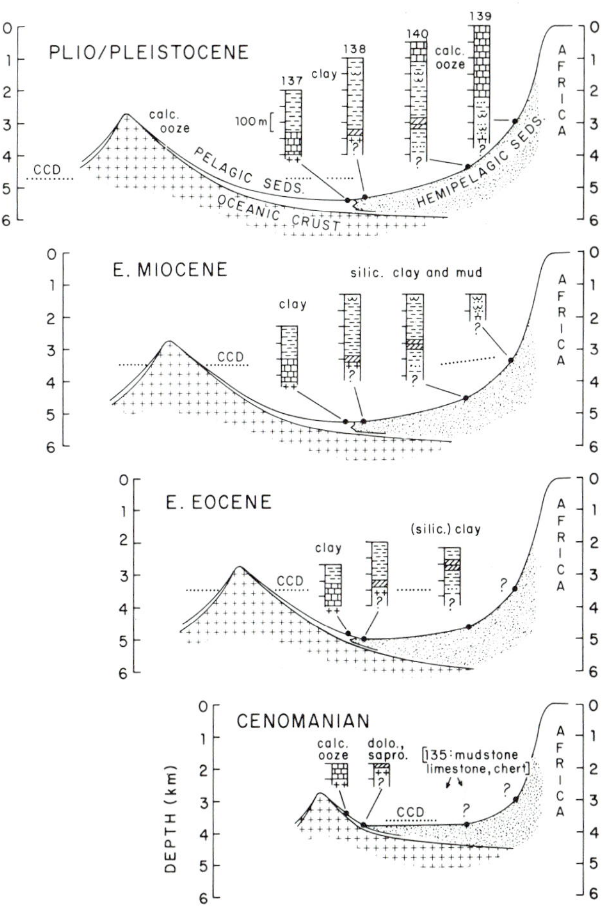

Fig. 19. Sedimentation in the eastern trough of the central Atlantic, off Africa, reconstructed from Leg 14 results, assuming sea-floor spreading and subsidence (from Berger and von Rad, 1972).

leads to preservation of some siliceous material and to reducing conditions within the sediment. The distribution of facies domains from the Pliocene on, with calcareous ooze, oxypelite, and hemipelagic clay containing but little admixture of siliceous remains, is typical for the present Atlantic facies regime in a "lagoon"-type deep-sea circulation, that is, a circulation with deep-water outflow.

During the Miocene, the ratio of pelagic to hemipelagic sediment domains is not very different from that in the Pliocene. The CCD is much shallower, however, distinctly rising toward the continent, as do other dissolution levels, which considerably decreases the proportion of carbonates during this time. Concomitantly with the generally greatly increased carbonate dissolution, the contribution from siliceous fossils becomes very significant, suggesting an "estuarine" deep-sea circulation pattern (deep-water inflow) according to the oceanic fractionation model (Berger, 1970a; Seibold, 1970). On the whole, Miocene sedimentation patterns are reminiscent of the present-day North Pacific facies regime, with shallow CCD, siliceous deposits, and considerable terrigenous influx, owing to orogenesis in the hinterland.

In the Eocene, the ridge flank domain is distinctly narrower than now, so hemipelagic sedimentation is more widespread than pelagic sedimentation. Otherwise, facies distributions appear quite comparable to those in the Miocene. The occurrence of calcareous oozes is restricted above a relatively shallow CCD. Sedimentation rates from terrigenous influx are smaller, however, and siliceous remains are less well preserved than in the Miocene, notwithstanding the formation of chert precursors elsewhere.

The Cenomanian sedimentary environment is characterized by hemipelagic sedimentation (black shale and layers of limestone and chert) in a trough that is significantly narrower and shallower than today, although for the most part it is still deeper than 3,000 m. At the time, a significant proportion of the world's sea floor may have been relatively young, thus increasing the average temperature and hence the average elevation of the oceanic crust (Sclater and Francheteau, 1970), with a consequent transgression of displaced water onto the continents (Menard, 1964, p. 240). The transgressions created shallow shelf seas, which should have trapped coarse terrigenous material while providing shallow-water carbonates for redeposition off the continent. The oxygen content of bottom waters apparently was quite low (occurrence of black pyritic shale). Bottom waters cannot be colder than the coldest oceanic surface waters. A generally warm climate (Schwarzbach, 1961; p. 137), therefore, would lead to warm bottom waters (Emiliani, 1961) with relatively low initial oxygen content. This initial oxyty could be further reduced if bottom waters were derived from haline convection rather than from a thermal one (Chamberlin, 1906). In conjunction with an increased residence time of abyssal waters, low oxygen values can then develop. This leads to sapropelitic deposition in the hemipelagic domain while still being able to provide brown pelagic clay in a narrow zone of low sedimentation rate, on the ridge flank below the CCD (site 137, cores 7-9). Redeposition processes apparently were active both on the ridge flank and on the continental slope, leading to calcareous layers within noncalcareous sediment in both pelagic and hemipelagic ("flysch") domain. Redeposition from the ridge flank could also have contributed to the dolomite-sapropel cycles at site 138 (core 6). Alternatively, the sapropelites developed in situ as the result of relatively high sedimentation rates in a low-oxygen environment, at the far end of the hemipelagic domain. In this case, much of the dolomite may have come from the continental slope or even from displaced shallow-water sediment. Alternatives to the redeposition hypothesis, such as hydrothermal acitivity, also can be envisaged.

In the equatorial Pacific, it is necessary to consider horizontal plate motions as well as vertical subsidence simultaneously, because both latitudinal positions and depth exert primary control on facies distributions, as first outlined by Arrhenius (1952).

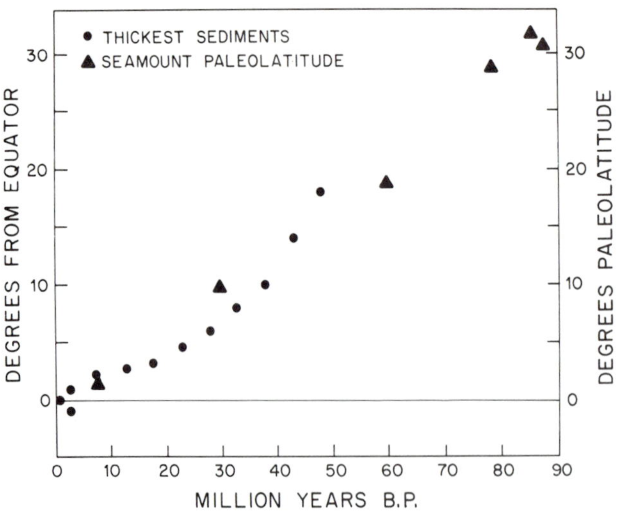

Fig. 20. Northward motion of the Pacific Plate, as evidenced in the shift of the thickest sediment stack, assumed to be equatorial, and in the difference between paleomagnetic latitudes of seamounts and their present latitudes (from van Andel and Moore, 1974).

The effects of horizontal plate motions on the distribution of Tertiary fossils were noted by Burckle et al. (1967) and Riedel (1967), among others. Results of deep-sea drilling have now permitted an answer to the questions raised by Arrhenius (1963, p. 714) and by Riedel (1963; p. 883) regarding changing productivity and dissolution patterns along the Equatorial Divergence and any changes in location of this divergence with regard to the present equator. Such a change in location leads to a predictable sequence of sedimentary facies, with horizontal distributions reflected in vertical stacking, in accord with Johannes Walther's well-known rule of the "correlation of the facies."

Vertical subsidence in the Central Pacific had been considered a possible reason for Tertiary carbonates underlying noncalcareous Quaternary sediments, especially carbonates of Oligocene age (Riedel and Funnell, 1964, p. 363; Heath, 1969a, p. 692). At least two other factors can theoretically be responsible for deeper-than-normal carbonate occurrences in the Tertiary: depression of the CCD in the past (as was recognized by these authors) and proximity to the equator at the time of deposition. To eliminate two of the variables, we first need to correct for the motion across latitudes, for example by assuming that maximum sedimentation rates are fixed on the equator (Winterer, 1973; van Andel and Heath, 1973; see Fig. 20). The rate of cross-latitude motion thus derived—about 1 degree in 4 million years for the area between 100 and 150°W longitude—allows us to assign a paleolatitude to each sample recovered by drilling, simply by noting the age and converting it to degrees traveled. The paleodepth is found by backtracking, as before, and the result is a paleolatitude-paleodepth frame, within which the facies distributions can now be plotted (Fig. 21). The highly generalized patterns, as figured, are only valid for post-Eocene time, and even

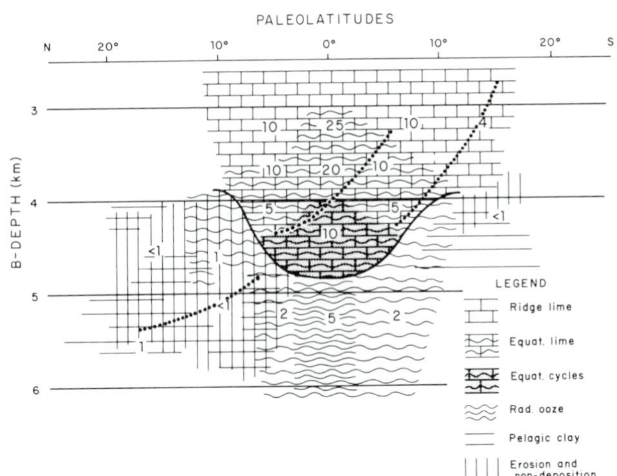

Fig. 21. Model of post-Eocene facies domains of the eastern central Pacific, west of the Rise, in a depth ("backtracking depth") versus latitude frame, highly generalized. Numbers are representative sedimentation rates in m/my. Equatorial lime-silica cycles appear to be well developed in Miocene and later. Heavity stippled lines are tracks of DSDP sites 68, 71, and 74, given as examples for the construction of the diagram. (Adapted from Berger, 1973a; based on DSDP results.)

within this time span there is considerable variation in distributions and sedimentation rates (Berger, 1973a; van Andel and Moore, 1974).

Rates during the earliest Tertiary tend to be essentially low in the central Pacific, and chert-rich deposits tend to be present from the mid-Eocene on back. Rates during the Late Cretaceous tend to be quite high. As Winterer emphasizes (in Winterer et al., 1973, p. 920), the explanation of these major features must probably be sought in global phenomena rather than regional subsidence or horizontal plate motions. The global nature of major trends is apparent from comparing Atlantic and Pacific carbonate deposition patterns with paleotemperature information from the oxygen isotope composition of foraminiferal shells, as suggested by Moore (1972; see Fig. 22).

Fluctuations of CCD levels in the extra-equatorial Pacific and in the Atlantic show parallelities with each other and with paleotemperature variations which are unlikely to be coincidental. Warm, equable climates appear unfavorable for carbonate deposition in the subtropical and tropical regions of the oceans, with the possible exception of the equatorial area proper. An attractive explanation for this relationship is that warm climates and widespread transgressions occur together and that transgressions lock up much carbonate on the shelves, making it unavailable for the deep sea. Such global interactions between shelf and deep ocean environments were envisaged by Kuenen (1950; see discussion in Berger and Winterer, 1974), and must be clarified to enable us to explain the trends of deep-sea deposition through the ages.

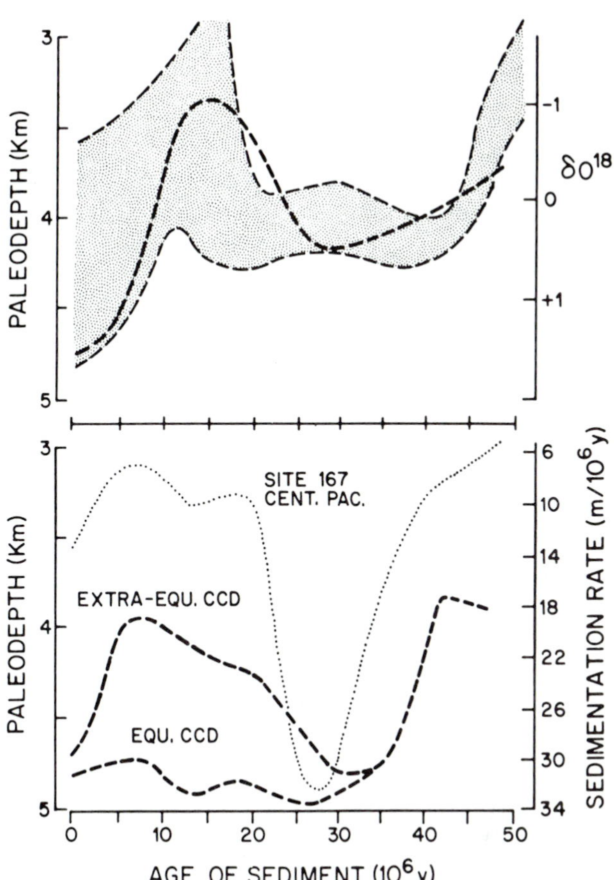

Fig. 22. Comparison of long-term sedimentation trends, as reflected in CCD variation (dashed lines) with oxygen isotopic composition. Sedimentation rates in DSDP site 167 reflect both general trends and the motion of the sea floor across the equator, between 25 and 30 my. (From numerous sources.)

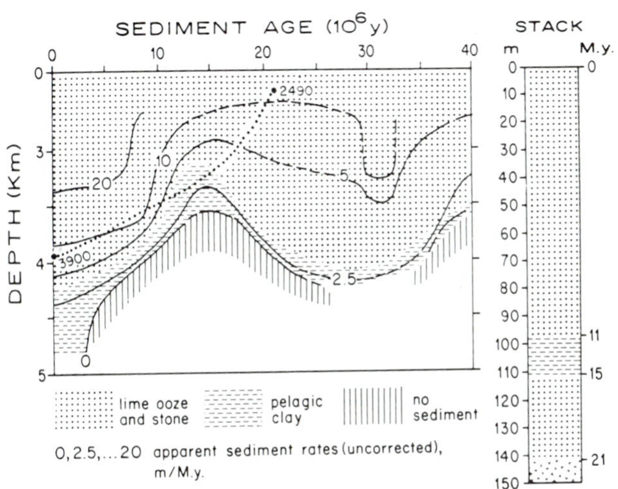

Fig. 23. Model of post-Eocene sedimentation patterns in the South Atlantic, based on backtracking of DSDP Leg 3 data, and illustration of "forward-tracking" (starting at 2,490 m, 21 my ago), producing the sediment stack at right.

An important step in achieving this goal is the construction of isopach maps of facies and age units. Once matrices of the type shown in Fig. 21 are established for a succession of reasonably brief time spans, they can be used to predict the stack of sediments by "forward-tracking." The simplest case, in which latitudinal changes are neglected, is illustrated for the South Atlantic matrix (Fig. 23). An arbitrary track, starting at a ridge crest elevation of 2,490 m, 21 my ago, is seen to result in the sediment column shown to the right ("stack"). To obtain this stack, both "normal" subsidence and "additional" subsidence from loading as the sea floor ages must be taken into account. This type of interpolation of drilling results will make it possible to construct reasonably accurate subsurface facies and age maps, which should greatly facilitate the interpretation of acoustic stratigraphy based on seismic profiling.

EPILOGUE

At the beginning of this report I defended talking about deep-sea sediments in a book on continental margins. One argument may now be added: The processes at the continental margins, as we have seen, have an entirely disproportionate influence on the chemistry of the ocean and hence on its pelagic deposits. Much of the chemical imprint of water masses appears to be generated at the oceanic margins, with their high fertility, high sedimentation rates, and intensive interaction between sediments and seawater. This imprint is largely preserved in the deep ocean as quasi-conservative properties, being altered mainly by mixing processes. The deposits generated by precipitation at the surface of the ocean (calcium carbonate, silica, organic matter) at the sea floor (ferromanganese), and in interstitial waters (zeolite, authigenic clay minerals) reflect ocean chemistry in their compositional and distributional patterns. In this sense, hydrogenous deep-sea sediments, to a large but undetermined extent, record chemical processes at the continental margins. For detrital deposits the link between deep-sea floor and continental margin is even more obvious, and the changing role of the shelves as traps and as sources of sediment, both hydrogenous and detrital, has been amply recognized. Thus, rather than considering this article as background for continental margin sedimentation, we might consider this book as providing part of the background knowledge for a fuller understanding of the enigmatic deposits of the abyssal ocean.

ACKNOWLEDGMENTS

My thanks go to the many colleagues who are my teachers in this vast field, in particular to Joris Gieskes, Ross Heath, Miriam Kastner, Frances Parker, Fred Phleger, Bill Riedel, Peter Roth, John Sclater, and E. L. Winterer. I also thank Sharon Weldy and Judy Lachmund Clinton, who drafted most of the illustrations, and Karen Berger, for help in the preparation of diagrams and the bibliography.

BIBLIOGRAPHY

Adelseck, C. G., and Berger, W. H., 1974, Planktonic foraminifera: dissolution during settling or on the sea floor? Cushman Found. Foram. Res. Spec. Publ. (in press).

Arrhenius, G., 1952, Sediment cores from the east Pacific: Rept. Swedish Deep Sea Exped. (1947-1948), v. 5, pts. 1-4, p. 1-228.

———, 1963, Pelagic sediments, *in* Hill, M. N., ed., The sea, v. 3: New York, Wiley-Interscience, p. 655-727.

———, 1967, Deep sea sedimentation, a critical review of U.S. work 1963-1967: Rept. 14th Gen. Assembly Internatl. Union Geodesy and Geophys. Scripps Inst. Oceanogr. Ref. 67-14.

———, and Bonatti, E., 1965, Neptunism and volcanism in the ocean, *in* Koczy, F., and Sears, M., eds., Progress in oceanography, v. 3: Elmsford, N.Y., Pergamon Press, p. 7-22.

Bacon, M. P., and Edmond, J. M., 1972, Barium at Geosecs III in the southwest Pacific: Earth and Planetary Sci. Letters, v. 16, no. 1, p. 66-74.

Bada, J. L., and Schroeder, R. A., 1972, Racemization of isoleucine in calcareous marine sediments: kinetics and mechanism: Earth and Planetary Sci. Letters, v. 15, p. 1-11.

Bandy, O. L., 1960, The geological significance of coiling ratios in the foraminifer *Globigerina pachyderma* (Ehrenberg): Jour. Paleontology, v. 34, no. 4, p. 671-681.

Barnes, S. S., 1967, Minor element composition of ferromanganese nodules: Science, v. 157, p. 63-65.

———, and Dymond, J. R., 1967, Rates of accumulation of ferromanganese nodules: Nature, v. 213, p. 1218-1219.

Bé, A. W. H., and Tolderlund, D. S., 1971, Distribution and ecology of living planktonic foraminifera in surface waters of the Atlantic and Indian oceans, *in* Funnell, B. M., and Riedel, W. R., eds., The micropalaeontology of oceans: London, Cambridge Univ. Press, p. 105-149.

Beall, A. O., and Fischer, A. G., 1969, Sedimentology: Initial reports of the Deep Sea Drilling Project, v. 1: Washington, D.C., U.S. Govt. Printing Office, p. 521-593.

Bender, M. L., 1972, Manganese nodules, *in* Fairbridge, R. W., ed., The encyclopedia of geochemistry and environmental sciences: New York, Van Nostrand Reinhold, p. 673-677.

———, Ku, T. L., and Broecker, W. S., 1966, Manganese nodules: their evolution: Science, v. 151, no. 3708, p. 325-328.

———, Ku, T. L., and Broecker, W. S., 1970, Accumulation rates of manganese in pelagic sediments and nodules: Earth and Planetary Sci. Letters, v. 8, p. 143-148.

———, Broecker, W., Gornitz, V., Middel, V., Kay, R., Scen, S. S., and Biscaye, P., 1971, Geochemistry of three cores from the East Pacific Rise: Earth and Planetary Sci. Letters, v. 12, p. 425-433.

Benson, W. E., Gerrard, R. D., and Hay, W. W., 1970, Summary and conclusions: Initial reports of the Deep Sea Drilling Project, v. 4: Washington, D.C., U.S. Govt. Printing Office, p. 659-673.

Berger, W. H., 1967, Foraminiferal ooze: solution at depths: Science, v. 156, p. 383-385.

———, 1968, Planktonic foraminifera: selective solution and paleoclimatic interpretation: Deep-Sea Res., v. 15, no. 1, p. 31-43.

———, 1970a, Biogeneous deep-sea sediments: fractionation by deep-sea circulation: Geol. Soc. America Bull., v. 81, p. 1385-1402.

———, 1970b, Planktonic foraminifera: selective solution and the lysocline: Marine Geology, v. 8, p. 111-138.

———, 1971, Sedimentation of planktonic foraminifera: Marine Geology, v. 11, p. 325-358.

———, 1972, Deep-sea carbonates: dissolution facies and age-depth constancy: Nature, v. 236, no. 5347, p. 392-395.

———, 1973a, Cenozoic sedimentation in the eastern tropical Pacific: Geol. Soc. America Bull., v. 84, p. 1941-1954.

———, 1973b, Deep-sea carbonates: Pleistocene dissolution cycles: Jour. Foram. Res., v. 3, no. 4, p. 187-195.

———, 1973c, Deep-sea carbonates: evidence for a coccolith lysocline: Deep-Sea Res., v. 20, p. 917-921.

———, and Gardner, J. V., 1974, On the determination of temperatures of the Pleistocene ocean: J. Foram. Res. (in press).

———, and Soutar, A., 1970, Preservation of plankton shells in an anaerobic basin off California: Geol. Soc. America Bull., v. 81, p. 275-282.

———, and von Rad, U., 1972, Cretaceous and Cenozoic sediments from the Atlantic Ocean: Initial reports of the Deep Sea Drilling Project, v. 14: Washington, D.C., U.S. Govt. Printing Office, p. 787-954.

———, and Winterer, E. L., 1974, Plate stratigraphy and the fluctuating carbonate line, *in* Hsü, K. J., and Jenkyns, H., eds., Pelagic sediments on land and under the sea: Spec. Publ. Internatl. Assoc. Sedimentologists (in press).

Berggren, W. A., 1972, A Cenozoic time scale—some implications for regional geology and paleobiogeography: Lethaia, v. 5, no. 1, p. 195-215.

Berner, R. A., 1965, Activity coefficients of bicarbonate, carbonate and calcium ions in sea water: Geochim. Cosmochim. Acta, v. 29, no. 8, p. 947-965.

———, 1971, Principles of chemical sedimentology: New York, McGraw-Hill, 240 p.

Bezrukov, P. L., and Andrushchenko, P. F., 1973, Iron-manganese concretions of the Indian Ocean: Internatl. Geol. Res., v. 15, no. 3, p. 342-356.

Bien, G. S., Contois, D. E., and Thomas, W. H., 1959, The removal of soluble silica from fresh water entering the sea, *in* Ireland, H. A., ed., Silica in sediments: Tulsa, Okla., Soc. Econ. Paleontologists and Mineralogists Spec. Publ. 7, p. 20-35.

Biscaye, P. E., 1965, Mineralogy and sedimentation of Recent deep-sea clay in the Atlantic Ocean and adjacent seas and oceans: Geol. Soc. America Bull., v. 76, p. 803-832.

Bischoff, J. L., and Sayles, F. L., 1972, Pore fluid and mineralogical studies of recent marine sediments: Bauer depression region of East Pacific Rise: Jour. Sediment. Petrology, v. 42, p. 711-724.

Blackman, A., 1966, Pleistocene stratigraphy of cores from the southeast Pacific Ocean [Ph.D. thesis]: Univ. California, San Diego, 200 p.

Bogoyavlenskiy, A. N., 1967, Distribution and migration of dissolved silica in oceans: Internatl. Geol. Rev., v. 9, no. 2, p. 133-153.

Bonatti, E., 1963, Zeolites in Pacific pelagic sediments: Trans. N.Y. Acad. Sci., ser. II, v. 25, no. 8, p. 938-948.

———, and Joensuu, O., 1966, Deep-sea iron deposits from the South Pacific: Science, v. 154, p. 643-645.

Boström, K., and Peterson, M. N. A., 1966, Precipitates from hydrothermal exhalations on the East Pacific Rise: Econ. Geology, v. 61, p. 1258-1265.

———, and Peterson, M. N. A., 1969, The origin of aluminum-poor ferromanganoan sediments in areas of high heat flow on the East Pacific Rise: Marine Geology, v. 7, p. 427-447.

———, Kraemer, T., and Gartner, S., 1973, Provenance and accumulation rates of opaline silica, Al, Ti, Fe, Mn, Cu, Ni, and Co in Pacific pelagic sediments: Chem. Geology, v. 11, p. 123-148.

Bradley, W. H., Piggot, C. S., Bramlette, M. N., Cushman, J. A., Henbest, L. G., Lohman, K. E., Tressler, W. L., Rehder, H. A., Clarke, A. H., Trask, P. D., Patrode, H. W., Stimson, J. L., Gay, J. R., Edgington, G., and Byers, H. G., 1942, Geology and biology of North Atlantic deep-sea cores between Newfoundland and Ireland: U.S. Geol. Survey Prof. Paper 196, p. 1-163.

Bradshaw, J. S., 1959, Ecology of living planktonic foraminifera in the North and Equatorial Pacific Ocean: Contrib. Cushman Found. Foram. Res., v. 10, no. 2, p. 25-64.

Bramlette, M. N., 1946, The Monterey Formation of California and the origin of its siliceous rocks: U.S. Geol. Survey Prof. Paper 212, p. 1-57.

Bramlette, M. N., 1958, Significance of coccolithophorids in calcium-carbonate deposition: Geol. Soc. America Bull., v. 69, p. 121-126.

———, 1961, Pelagic sediments, *in* Sears, M., ed., Oceanography: Am. Assoc. Adv. Sci. Publ. 67, p. 345-366.

———, and Bradley, W. H., 1942, Geology and biology of North Atlantic deep-sea cores, pt. 1, Lithology and geological interpertations: U.S. Geol. Survey Prof. Paper 196-A, p. 1-34.

Broecker, W. S., 1966, Absolute dating and the astronomical theory of glaciation: Science, v. 151, p. 299-304.

———, 1971a, A kinetic model for the chemical composition of sea water: Quaternary Res., v. 1, no. 2, p. 188-207.

———, 1971b, Calcite accumulation rates and glacial to interglacial changes in oceanic mixing, *in* Turekian, K. K., ed., The late Cenozoic glacial ages: New Haven, Conn., Yale Univ. Press, p. 239-265.

———, and Broecker, S., 1974, Carbonate dissolution on the flank of the East Pacific Rise: Soc. Econ. Geologists and Paleontologists Spec. Paper (in press).

———, and van Donk, J., 1970, Insolation changes, ice volumes, and the O^{18} record in deep-sea cores: Rev. Geophys. and Space Phys., v. 8, no. 1, p. 169-198.

———, Turekian, K. K., and Heeren, B. C., 1958, The relation of deep-sea sedimentation rates to variations in climate: Am. Jour. Sci., v. 256, p. 503-517.

———, Ewing, M., and Heeren, B. C., 1960, Evidence for an abrupt change in climate close to 11,000 years ago: Am. Jour. Sci., v. 258, p. 429-440.

Burckle, L. H., Ewing, J., Saito, T., and Leyden, R., 1967, Tertiary sediment from the East Pacific Rise: Science, v. 157, no. 3788, p. 537-540.

Burton, J. D., and Liss, P. L., 1968, Ocean budget on dissolved silicon: Nature, v. 220, no. 5170, p. 905-906.

———, and Liss, P. S., 1973, Processes of supply and removal of dissolved silicon in the oceans: Geochim. Cosmochim. Acta, v. 37, p. 1761-1773.

Calvert, S. E., 1966, Accumulation of diatomaceous silica in the sediments of the Gulf of California: Geol. Soc. America Bull., v. 77, no. 6, p. 569-596.

———, 1968, Silica balance in the ocean and diagenesis: Nature, v. 2219, p. 919-920.

———, 1971, Nature of silica phases in deep sea cherts of the North Atlantic: Nature Phys. Sci., v. 234, no. 50, p. 133-134.

———, 1974, Deposition and diagenesis of silica in marine sediments, *in* Hsü, K. J., and Kenkyns, H., eds., Pelagic sediments on land and under the sea: Spec. Publ. Internatl. Assoc. Sedimentologists, v. 1 (in press).

Carroll, D., 1970, Clay minerals in Arctic Ocean sea-floor sediments: Jour. Sediment. Petrology, v. 40, no. 3, p. 788-854.

Chemical Geology, v. 11, p. 109-116.

Chamberlin, T. C., 1906, On a possible reversal of deep-sea circulation and its influence on geologic climates: Jour. Geology, v. 14, p. 363-373.

Chave, K. E., and Schmalz, R. F., 1966, Carbonate-seawater interactions: Geochim. Cosmochim. Acta, v. 39, no. 10, p. 1037-1048.

———, and Suess, E., 1967, Suspended minerals in seawater: Trans. N.Y. Acad. Sci., ser. 11, v. 29, p. 991-1000.

Chow, T. J., and Mantyla, A. W., 1965, Inorganic nutrient anions in deep ocean waters: Nature, v. 206, no. 4982, p. 383-385.

Cifelli, R., and Sachs, K. N., 1966, Abundance relationships of planktonic foraminifera and radiolaria: Deep-Sea Res., v. 13, p. 751-753.

Clague, D. A., and Jarrard, R. D., 1973, Tertiary Pacific plate motion deduced from the Hawaiian-Emperor Chain: Geol. Soc. America Bull., v. 84, p. 1135-1154.

Conolly, J. R., and Ewing, M., 1965, Ice-rafted detritus as a climatic indicator in Antarctic deep-sea cores: Science, v. 150, p. 1822-1824.

Corliss, J. B., 1971, The origin of metal-bearing hydrothermal solutions: Jour. Geophys. Res., v. 76, p. 8128-8138.

Correns, C. W., 1939, Pelagic sediments of the North Atlantic Ocean, *in* Trask, P. D., ed., Recent marine sediments: Tulsa, Okla., Am. Assoc. Petr. Geol., p. 373-395.

Craig, H., Chung, Y., and Fiadeiro, M., 1972, A benthic front in the South Pacific: Earth and Planetary Sci. Letters, v. 16, no. 1, p. 50-65.

Croll, J., 1875, Climate and time: London, Daldy, Isbister & Co., 577 p.

Cronan, D. S., 1972, Regional geochemistry of ferromanganese nodules in the world ocean, *in* Horn, D. R., ed., Ferromanganese deposits on the ocean floor: Washington, D.C., Natl. Sci. Found., p. 19-30.

Dansgaard, W., and Tauber, H., 1969, Glacier oxygen-18 content and Pleistocene ocean temperatures: Science, v. 166, p. 499-502.

Dasch, E. J., 1969, Strontium isotopes in weathering profiles, deep-sea sediment, and sedimentary rocks: Geochim. Cosmochim. Acta, v. 33 (12), p. 1521-1552.

Dasch, E. J., Dymond, J. R., and Heath, G. R., 1971, Isotopic analysis of metalliferous sediment from the East Pacific Rise: Earth and Planetary Sci. Letters, v. 13, p. 175-180.

Donahue, J. G., 1967, Diatoms as indicators of Pleistocene climatic fluctuations in the Pacific sector of the southern ocean: Progr. Oceanogr., v. 4, p. 133-140.

Duplessy, J. C., Lalou, C., and Vinot, A. C., 1970, Differential isotopic fractionation in benthic foraminifera and paleotemperatures reassessed: Science, v. 168, p. 250-251.

Dymond, J., Corliss, J. B., Heath, G. R., Field, C. W., Dasch, E. J., and Veeh, H. H., 1973, Origin of metalliferous sediments from the Pacific Ocean: Geol. Soc. America Bull., v. 84, p. 3355-3372.

Edmond, J. M., 1971, An interpretation of the calcite spheres experiment (abstr.): EOS (Trans. Am. Geophys. Union), v. 52, p. 256.

———, 1972, The thermodynamic description of the CO_2 system in sea water: development and current status: Proc. Roy. Soc. Edinburgh, ser. B, v. 72, p. 371-380.

———, and Gieskes, J. M., 1970, On the calculation of the degree of saturation of sea water with respect to calcium carbonate under in situ conditions: Geochim. Cosmochim. Acta, v. 34, p. 1261-1291.

Emiliani, C., 1955, Pleistocene temperatures: Jour. Geology, v. 63, no. 6, p. 538-578.

———, 1961, The temperature decrease of surface seawater in high latitudes and of abyssal-hadal water in open oceanic basins during the past 75 million years: Deep-Sea Res., v. 8, p. 144-147.

———, 1966, Paleotemperature analysis of Caribbean cores P6304-8 and P6304-9 and a generalized temperature curve for the past 425,000 years: Jour. Geology, v. 74, no. 2, p. 109-126.

———, 1971, The amplitude of Pleistocene climatic cycles at low latitudes and the isotopic composition of glacial ice, *in* Turekian, K. K., ed., The late Cenozoic glacial ages: New Haven, Conn., Yale Univ. Press, p. 183-197.

———, 1972, Quaternary paleotemperatures and the duration of the high temperature intervals: Science, v. 178, p. 398-400.

———, and Rona, E., 1969, Caribbean cores P6304-8 and P6304-9: new analysis of absolute chronology. A reply: Science, v. 166, p. 1551-1552.

———, and Shackleton, N. J., 1974, The Brunhes Epoch: isotopic paleotemperatures and geochronology: Science, v. 183, p. 511-514.

Ericson, D. B., and Wollin, G., 1968, Pleistocene climates and chronology in deep-sea sediments: Science, v. 162, p. 1227-1234.

———, Broecker, W. S., Kulp, J. L., and Wollin, G., 1956, Late-Pleistocene climates and deep-sea sediments: Science, v. 124, no. 3218, p. 385-389.

———, Ewing, M., Wollin, G., and Heezen, B. C., 1961, Atlantic deep-sea sediment cores: Geol. Soc. America Bull., v. 72, no. 2, p. 193-286.

Ewing, M., and Landisman, M., 1961, Shape and structure of ocean basins, *in* Sears, M., ed., Oceanography: Am. Assoc. Adv. Sci. Publ. 67, p. 3-38.

Fanning, K. A., and Pilson, M. E. Q., 1971, Interstitial silica and pH in marine sediments: some effects of sampling procedures: Science, v. 173, p. 1228-1231.

———, and Pilson, M. E. Q., 1973, The lack of inorganic removal of dissolved silica during river-ocean mixing: Geochim. Cosmochim. Acta, v. 37, p. 2405-2415.

———, and Schink, D. R., 1969, Interaction of marine sediments with dissolved silica: Limnol. Oceanogr.,

v. 14, no. 1, p. 59-68.

Fischer, A. G., Heezen, B. C., Boyce, R. E., Bukry, D., Douglas, R. G., Garrison, R. E., Kling, S. A., Krasheninnikov, V., and Lisitzin, A. P., 1970, Geological history of the western north Pacific: Science, v. 168, p. 1210-1214.

Fomina, L. S., 1962, The oxidation-reduction processes in bottom sediments of the southwestern Pacific Ocean: Tr. Inst. Okeanol. Akad. Nauk S.S.S.R., v. 54, p. 158-169 (in Russian).

Funnell, B. M., and Riedel, W. R., eds., 1971, The micropaleontology of oceans: London, Cambridge Univ. Press, 828 p.

Garrels, R. M., and Mackenzie, F. T., 1971, Evolution of sedimentary rocks: New York, Norton, 397 p.

Gartner, S., Jr., 1970, Seafloor spreading, carbonate dissolution level, and the nature of horizon A: Science, v. 169, p. 1077-1079.

Gibson, T. G., and Towe, K. G., 1971, Eocene volcanism and the origin of horizon A: Science, v. 172, p. 152-154.

Gieskes, J. M., 1974, The alkalinity-total carbon dioxide system in seawater, *in* Goldberg, E. D., ed., The sea, v. 5: New York, Wiley-Interscience, (in press).

Goldberg, E. D., 1954, Marine geochemistry: 1. Chemical scavengers of the sea: Jour. Geology, v. 62, no. 3, p. 249-265.

______, 1974, The sea, v. 5: New York, Wiley-Interscience (in press).

______, and Griffin, J. J., 1970, The sediments of the northern Indian Ocean: Deep-Sea Res., v. 17, p. 513-537.

Goll, R. G., and Bjørklund, K. R., 1971, Radiolaria in surface sediments of the North Atlantic Ocean: Micropaleontology, v. 17, no. 4, p. 434-454.

Goodell, H. G., Meyland, M. A., and Grant, B., 1971, Ferromanganese deposits of the south Pacific Ocean, Drake Passage, and Scotia Sea: Antarctic Res. Ser., v. 15, p. 27-92.

Griffin, J. J., and Goldberg, E. D., 1963, Clay mineral distributions in the Pacific Ocean, *in* Hill, M. N., ed., The sea, v. 3: New York, Wiley-Interscience, p. 728-741.

______, Windom, H., and Goldberg, E. D., 1968, The distribution of clay minerals in the world ocean: Deep-Sea Res., v. 15, no. 4, p. 433-459.

Grill, E. V., 1970, A mathematical model for the marine dissolved silicate cycle: Deep-Sea Res., v. 17, p. 245-266.

Gripenberg, S., 1934, A study of the sediments of the north Baltic and adjoining seas: Havsforskningsinst. Skr., v. 96, p. 1-231.

Hallam, A., ed., 1967, Depth indicators in marine sedimentary environments: Marine Geology, v. 5, no. 5-6, p. 329-567.

Harris, R. C., 1966, Biological buffering of oceanic silica: Nature, v. 212, p. 275-276.

Hart, R., 1970, Chemical exchange between sea water and deep ocean basalts: Earth and Planetary Sci. Letters, v. 9, p. 269-279.

Hartmann, M., 1964, Zur Geochemie von Mangan und Eisen in der Ostsee: Meyniana, v. 14, p. 3-20.

Harvey, H. W., 1957, The chemistry and fertility of sea waters: London, Cambridge Univ. Press, 240 p.

Hay, W. W., ed., 1974, Geologic history of the oceans: Soc. Econ. Paleontologists and Mineralogists Spec. Publ. (in press).

Hayes, D. E., Pimm, A. C., Beckmann, J. P., Benson, W. E., Berger, W. H., Roth, P. H., Supko, P. R., and von Rad, U., 1972, Initial reports of the Deep Sea Drilling Project, v. 14: Washington, D.C., U.S. Govt. Printing Office, 975 p.

Hays, J. D., 1965, Radiolaria and late Tertiary and Quaternary history of Antarctic seas, *in* Biology of the Antarctic seas II: Am. Geophys. Union, Antarctic Res. Ser., v. 5, p. 125-184.

______, Saito, T., Opdyke, N. D., and Burckle, L. H., 1969, Pliocene-Pleistocene sediments of the equatorial Pacific: their paleomagnetic, biostratigraphic and climatic record: Geol. Soc. America Bull., v. 80, p. 1481-1514.

Heath, G. R., 1969a, Carbonate sedimentation in the abyssal Equatorial Pacific during the past 50 million years: Geol. Soc. America Bull., v. 80, p. 689-694.

______, 1969b, Mineralogy of Cenozoic deep-sea sediments from the Equatorial Pacific Ocean: Geol. Soc. America Bull., v. 80, p. 1997-2018.

______, 1973, Cherts from the eastern Pacific, Leg 16, *in* Initial reports of the Deep Sea Drilling Project, v. 16: Washington, D.C., U.S. Govt. Printing Office, p. 609-613.

______, 1974, Dissolved silica and deep sea sediments, *in* Hay, W. W., ed., Geologic history of the oceans: Soc. Econ. Paleontologists and Mineralogists Spec. Publ., (in press).

______, and Culberson, C., 1970, Calcite: degree of saturation, rate of dissolution, and the compensation depth in the deep oceans: Geol. Soc. America Bull., v. 81, no. 10, p. 3157-3160.

______, and Dymond, J., 1973, Interstitial silica in deep-sea sediments from the North Pacific: Geology, v. 1, p. 181-184.

______, and Moberly, R., Jr., 1971, Cherts form the western Pacific, Leg 7, *in* Initial reports Deep Sea Drilling Project, v. 7: Washington, D.C., U.S. Govt. Printing Office, p. 991-1007.

Hecht, A. D., 1973, A model for determining Pleistocene paleotemperatures from planktonic foraminiferal assemblages: Micropaleontology, v. 19, no. 1, p. 68-77.

Heezen, B. C., and Hollister, C., 1964, Turbidity currents and glaciation, *in* Nairn, A. E. M., ed., Problems in paleoclimatology: New York, Wiley-Interscience, p. 99-109, 11-112.

______, MacGregor, I. D., Foreman, H. P., Forristal, G., Hekel, H., Hesse, R., Hoskins, R. H., Jones, E. J. W., Kaneps, A., Krasheninnikov, V. A., Okada, H., and Ruff, M. H., 1973, Diachronous deposits: a kinematic interpretation of the post-Jurassic sedimentary sequence on the Pacific Plate: Nature, v. 241, no. 5384, p. 25-32.

Heim, A., 1878, Untersuchungen uber den Mechanismus der Gebirgsbildung, Bd. II, Schwabe, Basel, 246 p.

Hess, H. H., 1962, History of ocean basins, *in* Engel, A. E. J., et al., eds., Petrologic studies: a volume to honor A. F. Buddington: New York, Geol. Soc. America, p. 599-620.

Hill, M. N., ed., 1963, The sea, v. 3: New York, Wiley-Interscience, 693 p.

Horn, D. R., Horn, B. M., and Delach, M. N., 1972, Distribution of ferromanganese deposits in the world ocean, *in* Horn, D. R., ed., Ferromanganese deposits on the ocean floor: Harriman, N.Y., Arden House, p. 9-17.

Hsü, K. J., and Andrews, J. E., 1970, History of South Atlantic Basin, *in* Initial reports of the Deep Sea

Drilling Project, v. 3: Washington, D.C., U.S. Govt. Printing Office, p. 464-467.

Hurd, D. C., 1972, Factors affecting solution rate of biogenic opal in sea water: Earth and Planetary Sci. Letters, v. 15, no. 4, p. 411-417.

———, 1973, Interactions of biogenic opal, sediment, and seawater in the central equatorial Pacific: Geochim. Cosmochim. Acta, v. 37, p. 2257-2282.

Hurley, P. M., Heezen, B. C., Pinson, W. H., and Fairbairn, H. W., 1963, K-Ar age values in pelagic sediments of the North Atlantic: Geochim. Cosmochim. Acta, v. 27, p. 393-399.

Huxley, T. H., 1894, Discourses—biological and geological essays: New York, Appleton-Century-Crofts, 388 p.

Imbrie, J., and Kipp, N. G., 1971, A new micropaleontological method for quantitative paleoclimatology: application to a late Pleistocene Caribbean core, *in* Turekian, K. K., ed., The late Cenozoic glacial ages: New Haven, Conn., Yale Univ. Press, p. 71-181.

Isacks, B., Oliver, J., and Sykes, L. R., 1968, Seismology and the new global tectonics: Jour. Geophys. Res., v. 73, p. 5855-5899.

Jackson, M. L., Levelt, T. W. M., Syers, J. K., Rex, R. W., Clayton, R. N., Sherman, G. D., and Nehara, G., 1971, Geomorphological relationships of tropospherically derived quartz in the soils of the Hawaiian Islands: Proc. Soil Sci. Soc. America, v. 35, no. 4, p. 515-525.

Jacobs, M. B., and Hays, J. D., 1972, Paleo-climatic events indicated by mineralogical changes in deep-sea sediment: Jour. Sediment. Petrology, v. 42, no. 4, p. 889-898.

Jenkyns, H. C., 1970, Fossil manganese nodules from the west Sicilian Jurassic: Eclogae Geol. Helv., v. 63, no. 3, p. 741-774.

Johnson, T. C., 1974, The dissolution of siliceous microfossils in surface sediments of the eastern tropical Pacific: Deep-Sea Res., (in press).

Kanaya, T., and Koizumi, 1966, Interpretation of diatom thanatocoenoses from the North Pacific applied to a study of core V20-130: Studies of a deep-sea core V20-130, pt. IV: Sci. Rept. Tohoku Univ. Second Ser., v. 37, p. 89-130.

Kaplan, I. R., and Rittenberg, S. C., 1963, Basin sedimentation and diagenesis, *in* Hill, M. N., ed., The sea, v. 3, New York, Wiley-Interscience, p. 583-619.

Kowsmann, R., 1973, Coarse components in surface sediments of the Panama Basin, eastern equatorial Pacific: Jour. Geology, v. 81, p. 473-494.

Ku, T. L., and Broecker, W. S., 1969, Radiochemical studies on manganese nodules of deep-sea origin: Deep-Sea Res., v. 16, p. 625-637.

Kuenen, P. H., 1950, Marine geology: New York, Wiley, 568 p.

Kvenvolden, K. A., Peterson, E., Wehmiller, J., and Hare, P. E., 1973, Racemization of amino acids in marine sediments determined by gas chromatography: Geochim. Cosmochim. Acta, v. 37, p. 2215-2225.

Lancelot, Y., 1973, Chert and silica diagenesis in sediments from the central Pacific, *in* Initial Reports of the Deep Sea Drilling Project, v. 17: Washington, D.C., U.S. Govt. Printing Office, p. 377-405.

LePichon, X., 1968, Sea-floor spreading and continental drift: Jour. Geophys. Res., v. 73, p. 3661-3697.

Lewin, J. C., 1962, Silification, *in* Lewin, R. A., ed., Physiology and biochemistry of algae: New York, Academic Press, p. 445-455.

Li, Y-H., Bischoff, J., and Mathieu, G., 1969, The migration of manganese in the Arctic Basin sediment: Earth and Planetary Sci. Letters, v. 7, p. 265-270.

Li, Y. H., Takahashi, T., and Broecker, W. S., 1969, Degree of saturation of $CaCO_3$ in the oceans: Jour. Geophys. Res., v. 74, p. 5507-5525.

Lisitzin, A. P., 1966, Basic relationships in distribution of modern siliceous sediments and their connection with climatic zonation (in Russian), *in* Geockhimia kremnezema: Moscow, Nauka, p. 90-191; Engl. transl. (1967): Internatl. Geol. Res., v. 9, no. 5, p. 631-652; no. 6, p. 842-865; no. 7, p. 980-1004; no. 8, p.

———, 1971a, Distribution of siliceous microfossils in suspension and in bottom sediments, *in* Funnell, B. M., and Riedel, W. R., eds., The micropaleontology of oceans: London, Cambridge Univ. Press, p. 173-195.

———, 1971b, Distribution of carbonate microfossils in suspension and in bottom sediments, *in* Funnell, B. M., and Riedel, W. R., eds., The micropaleontology of oceans: London, Cambridge Univ. Press, p. 197-218.

———, 1972, Sedimentation in the world ocean: Soc. Econ. Paleontologists and Mineralogists Spec. Publ. 17, p. 1-218.

Liss, P. S., and Spencer, C. P., 1970, Abiological processes in the removal of silicate from seawater: Geochim. Cosmochim. Acta, v. 34, p. 1073-1088.

Livingstone, D. A., 1963, Chemical composition of rivers and lakes: U.S. Geol. Survey Prof. Paper 440-G, 64 p.

Ljunggren, P., 1953, Some data concerning the formation of manganiferous and ferriferous bog ores: Geol. Foren. Forh., v. 75, p. 277-297.

Lyakhin, Y. I., 1968, Saturation of the Pacific waters with calcium carbonate: Okeanologiia, Akad. Nauk SSSR, v. 8, no. 1, p. 58-68 (In Russian, with Engl. abstr.).

Lynn, D. C., and Bonatti, E., 1965, Mobility of manganese in the diagenesis of deep-sea sediments: Marine Geology, v. 3, p. 457-474.

Mackenzie, F. T., and Garrels, R. M., 1965, Silicates: reactivity with sea water: Science, v. 150, p. 57-58.

———, Garrels, R. M., Bricker, O. P., and Brickley, F., 1967, Silica in sea waters: control by silica minerals: Science, v. 155, no. 5768, p. 1404-1405.

Manheim, F. T., 1965, Manganese-iron accumulation in the shallow marine environment: Rhode Island, Narragansett Marine Lab., Symp. Marine geochemistry Occas. Publ. 3, p. 217-276.

Mattson, P. H., and Pessagno, E. A., Jr., 1971, Caribbean Eocene volcanism and the extent of horizon A.: Science, v. 174, p. 138-139.

Maxwell, A. E., ed., 1970, The sea, v. 4, pts. 1, 2: New York, Wiley-Interscience, p. 664, 791.

———, von Herzen, R. P., Andrews, J. E., Boyce, R. E., Milow, E. D., Hsu, K. J., Percival, S. F., and Saito, T., 1970, Initial reports of the Deep Sea Drilling Project, v. 3: Washington, D.C., U.S. Govt. Printing Office, 806 p.

McIntyre, A., 1967,Coccoliths as paleoclimatic indicators of Pleistocene glaciation: Science, v. 158, p. 1314-1317.

———, and McIntyre, R., 1971, Coccolith concentrations and differential solution in oceanic sediments, *in* Funnell, B. M., and Riedel, W. R., eds., The micropaleontology of oceans: London, Cambridge Univ. Press, p. 253-261.

———, Ruddiman, W. F., and Jantzen, R., 1972, Southward penetrations of the North Atlantic Polar Front:

faunal and floral evidence of large-scale surface water mass movement over the last 225,000 years: Deep-Sea Res., v. 19, p. 61-77.

McKenzie, D. P., and Parker, R. L., 1967, The North Pacific: an example of tectonics on a sphere: Nature, v. 216, p. 1276-1280.

Mehmel, M., 1939, Application of x-ray methods to the investigation of recent sediments, *in* Trask, P. D., ed., Recent marine sediments: Tulsa, Okla., Am. Assoc. Petr. Geol., p. 616-630.

Menard, H. W., 1964, Marine geology of the Pacific: New York, McGraw-Hill, 271 p.

———, 1969, Elevation and subsidence of oceanic crust: Earth and Planetary Sci. Letters, v. 6, p. 275-284.

Mero, J. L., 1962, Ocean-floor manganese nodules: Econ. Geology, v. 57, p. 747-767.

———, 1965, The mineral resources of the sea: Amsterdam, Elsevier, 312 p.

Moore, T. C., 1969, Radiolaria: change in skeletal weight and resistance to solution: Geol. Soc. America Bull., v. 80, p. 2103-2108.

———, 1972, Deep Sea Drilling Project: successes, failures, proposals: Geotimes, v. 17, p. 27-31.

Moore, T. C., Heath, G. R., and Kowsmann, O., 1973, Biogenic sediments of the Panama Basin: Jour. Geology, v. 81, no. 4, p. 458-472.

Morgan, W. J., 1968, Rises, trenches, great faults and crustal blocks: Jour. Geophys. Res., v. 73, p. 1959-1982.

Morse, J. W., and Berner, R. A., 1972, Dissolution kinetics of calcium carbonate in sea water: II. A kinetic origin for the lysocline: Am. Jour. Sci., v. 272, no. 9, p. 840-851.

Murray, J., and Hjort, J., 1912, The depths of the ocean: New York, Macmillan, 821 p.

———, and Irvine, R., 1891, On silica and the siliceous remains of organisms in modern seas: Proc. Roy. Soc. Edinburgh, v. XVIII, p. 229-250.

———, and Irvine, R., 1894, On the manganese oxide and manganese nodules in marine deposits: Trans. Roy. Soc. Edinburgh, v. 37, p. 721-742.

———, and Philippi, E., 1908, Die grandproben der Deutsches Tiefsee Expedition, 1898-1899: Wiss. Ergeb. Deut. Tiefsee-Exped. Valdivia, v. 10, p. 77-206.

———, and Renard, A. F., 1891, Report on deep-sea deposits based on the specimens collected during the voyage of H.M.S. Challenger in the years 1872 to 1876: Rept. Voyage Challenger: London, Longmans, 525 p. (reprinted (1965): New York, Johnson Reprint Co.).

Nigrini, C., 1970, Radiolarian assemblages in the North Pacific and their application to a study of Quaternary sediments in core V20-130: Geol. Soc. America Mem. 126, p. 139-183.

Olausson, E., 1961, Remarks on some Cenozoic core sequences from the central Pacific, with a discussion of the role of coccolithophorids and foraminifera in carbonate deposition. Gõteborg Kungl. Vetenskaps-och Vitterhets-Samhälles Handlingar: Sjätte Fõljden, ser. B., Bd. 8, no. 10, p. 1-35.

———, 1965, Evidence of climatic changes in North Atlantic deep-sea cores, with remarks on isotopic paleotemperature analysis: Progr. Oceanography, v. 3, p. 221-252.

———, 1966, Calcareous oozes, *in* Fairbridge, R. W., ed., The encyclopedia of oceanography: New York, Van Nostrand Reinhold, p. 76-78.

———, 1967, Climatological, geoeconomical and paleo-oceanographical aspects on carbonate deposition, *in* Sears, ed., Progress in oceanography, v. 4: Elmsford, N.Y., Pergamon Press, p. 245-265.

Opdyke, N. D., 1972, Paleomagnetism of deep-sea cores: Rev. Geophys. and Space Phys., v. 10, no. 1, p. 213-249.

———, and Foster, J. H., 1970, Paleomagnetism of cores from the North Pacific: Geol. Soc. America Mem. 126, p. 83, 119.

Ovey, D. C., 1950, On the interpretation of climatic variations as revealed by a study of samples from an equatorial Atlantic deep-sea core: Centenary Proc. Roy. Meteorolog. Soc., p. 211-215.

Parker, F. L., and Berger, W. H., 1971, Faunal and solution patterns of planktonic foraminifera in surface sediments of the South Pacific: Deep-Sea Res., v. 18, p. 73-107.

Patin, S. A., and Tkachenko, V. N., 1972, Biogeochemistry of thorium isotopes in the ocean: Geochemistry Internatl., v. 9, no. 4, p. 697-701.

Peterson, M. N. A., 1966, Calcite: rates of dissolution in a vertical profile in the central Pacific: Science, v. 154, p. 1542-1544.

Peterson, M. N. A., and Goldberg, Edward E., 1962, Feldspar distributions in South Pacific pelagic sediments: Jour. of Geophysical Res., v. 67: 9, p. 3477-3492.

———, Edgar, N. T., Cita, M., Gartner, S., Goll, R., Nigrini, C., and von der Borch, C., 1970, Initial reports of the Deep Sea Drilling Project, v. 2: Washington, D.C., U.S. Govt. Printing Office, 501 p.

Philippi, E., 1910, Die Gründproben der Deutschen Sudpolar-Expedition 1901-1903. Deutsch Sudpolar Expedition. II: Geographie Geologie, p. 415-616.

Phleger, F. B., 1948, Foraminifera of a submarine core form the Caribbean Sea: Medd. Oceanogr. Inst. i Goteborg, v. 16, p. 3-9.

———, Parker, F. L., and Peirson, J. F., 1953, North Atlantic foraminifera: Rept. Swedish Deep Sea Exped., v. 7, p. 1-122.

Pia, J., 1933, Die Rezenten Kalksteine: Leipzig, Akademischer Verlag, 420 p.

Piper, D. Z., 1973, Origin of metalliferous sediments from the East Pacific Rise: Earth and Planetary Sci. Letters, v. 19, p. 75-82.

Pratje, O., 1951, Die Kieselsaüreorganismen des Südatlantischen Oceans als Leitformen in den Bodenablagerungen: Deut. Hydrograph. Z., v. 4, p. 1-6.

Presley, B. J., Brooks, R. R., and Kaplan, I. R., 1967, Manganese and related elements in the interstitial water of marine sediments: Science, v. 158, p. 906-910.

Price, N. B., and Calvert, S. E., 1970, Compositional variation in Pacific Ocean ferromanganese nodules and its relationship to sediment accumulation rates: Marine Geology, v. 9, p. 145-171.

Pytkowicz, R. M., 1968, The carbon dioxide—carbonate system at high pressures in the oceans: Ann. Rev. Oceanogr. Marine Biology, v. 6, p. 83-135.

———, and Fowler, G. A., 1967, Solubility of foraminifera in sea water at high pressures: Geochem. Jour., v. 1, p. 169-182.

Radczewski, O. E., 1939, Eolion deposits in marine sediments, *in* Trask, P. D., ed., Recent marine sediments: Tulsa, Okla., Am. Assoc. Petr. Geol., p. 496-502.

Ramsay, A. T. S., 1973, A history of organic siliceous

sediments in oceans: Paleontology Spec. Paper 12, p. 199-234.

———, ed., 1974, Oceanic micropaleontology, (in press).

Rateev, M. A., Gorbunova, Z. N., Lisitzin, A. P., and Nosov, G. L., 1969, The distribution of clay minerals in the oceans: Sedimentology, v. 13, p. 21-43.

Redfield, A. C., Ketchum, B. H., and Richards, F. A., 1963, The influence of organisms on the composition of sea water, *in* Hill, M. N., ed., The sea, v. 2: New York, Wiley-Interscience, p. 26-77.

Revelle, R. R., 1944, Marine bottom samples collected in the Pacific Ocean by the CARNEGIE on its seventh cruise: Carnegie Inst. Wash. Publ. 556, p. 1-196.

———, and Fairbridge, R., 1957, Carbonates and carbon dioxide: Geol. Soc. America Mem. 67, no. 1, p. 239-296.

———, Bramlette, M. N., Arrhenius, G., and Goldberg, E. D., 1955, Pelagic sediments of the Pacific: Geol. Soc. America Spec. Paper 62, p. 221-236.

Rex, R. W., and Goldberg, E. D., 1958, Quartz contents of pelagic sediments of the Pacific Ocean: Tellus, v. X, no. 1, p. 153-159.

———, Syers, J. K., Jackson, M. L., and Clayton, R. N., 1969, Eolian origin of quartz in soils of Hawaiian Islands and in Pacific pelagic sediments: Science, v. 163, p. 277-279.

Riedel, W. R., 1959, Siliceous organic remains in pelagic sediments: Soc. Econ. Paleontologists and Mineralogists Spec. Publ. 7, p. 80-91.

———, 1963, The preserved record: paleontolgoy of pelagic sediments, *in* Hill, M. N., ed., The sea, v. 3: New York, Wiley-Interscience, p. 866-887.

———, 1967, Radiolaria evidence consistent with spreading of the Pacific floor: Science, v. 157, p. 540-542.

———, 1971, The occurrence of pre-Quaternary radiolaria in deep-sea sediments, *in* Funnell, B. M., and Riedel, W. R., eds., The micropaleontology of oceans: London, Cambridge Univ. Press, p. 567-594.

———, and Funnell, B. M., 1964, Tertiary sediment cores and microfossils from the Pacific Ocean Floor: Geol. Soc. London Quart. Jour., v. 120, p. 305-368.

Riley, J. P., and Chester, R., eds., 1974, Treatise on chemical oceanography, v. 3: New York, Academic Press, (in press).

———, and Skirrow, G., eds., 1965, Chemical oceanography, v. 2: New York, Academic Press, 508 p.

Rona, E., and Emiliani, C., 1969, Absolute dating of Caribbean cores P6304-8 and P6304-9: Science, v. 163, p. 66-68.

Rosholt, J. N., Emiliani, C., Geiss, J., Koczy, F. F., and Wangersky, P. J., 1961, Absolute dating of deep-sea cores by the Pa^{231}/Th^{230} method: Jour. Geology, v. 69, p.162-185.

Roth, P. H., and Berger, W. H., 1974, Comparison of dissolution patterns in coccoliths and foraminifera: Cushman Found. Foram. Res. Spec. Publ., (in press).

Ruddiman, W. F., 1971, Pleistocene sedimentation in the equatorial Atlantic: stratigraphy and faunal paleoclimatology: Geol. Soc. America Bull., v. 82, p. 283-302.

———, and Heezen, B. C., 1967, Differential solution of planktonic foraminifera: Deep-Sea Res., v. 14, p. 801-808.

———, and McIntyre, A., 1973, Time-transgressive deglacial retreat of polar waters from the North Atlantic: Quaternary Res., v. 3, p. 117-130.

Runcorn, S. K., 1956, Paleomagnetic comparisons between Europe and North America: Proc. Geol. Assoc. Can., v. 8, p. 77-85.

Sackett, W. M., 1966, Manganese nodules: thorium-230/protactinium-231 ratios: Science, v. 154, p. 646-647.

Schneiderman, N., 1973, Deposition of coccoliths in the compensation zone of the Atlantic Ocean, *in* Smith, L. A., and Hardenbol, J., eds., Proceeding of symposium on calcareous nannofossils: Houston, Texas, Soc. Econ. Paleontologists and Mineralogists, p. 140-151.

Schott, W., 1935, Die Foraminiferen in dem aquatorialen Teil des Atlantischen Ozeans: Deut. Atl. Exped. Meteor, 1925-1927, v. 3, no. 3, p. 43-134.

Schrader, H. J., 1972, Kieselsaure-Skelette in Sedimenten des iberomarokkanischen Kontinentalrandes und angrenzender Tiefsee-Ebenen: Meteor-Forsch. Ergebuisse, Reihe, C., no. 8, p. 10-36.

Schutz, D. F., and Turekian, K. K., 1965, The investigation of the geographical and vertical distribution of several trace elements in sea water using neutron activation analysis: Geochim. Cosmochim. Acta, v. 29, p. 259-313.

Schwarzbach, M., 1961, Das Klima der Vorzeit: Stuttgart, Ferdinand Enke, 275 p.

Sclater, J. G., and Francheteau, J., 1970, The implications of terrestrial heat flow observations on current tectonic and geochemical models of the crust and upper mantle of the earth: Geophys. Jour., v. 20, p. 509-542.

———, Anderson, R. N., and Bell, M. L., 1971, Elevation of ridges and evolution of the central eastern Pacific: Jour. Geophys. Res., v. 76, no. 32, p. 7888-7915.

Sears, M., ed., 1961, Oceanography: Washington, D.C., Am. Assoc. Adv. Sci. Publ. 67, 654 p.

Seibold, E., 1970, Nebenmeere im humiden und ariden Klimabereich: Geol. Rundschau, v. 60, no. 1, p. 73-105.

Shackleton, N., 1967, Oxygen isotope analyses and Pleistocene temperature reassessed: Nature, v. 215, no. 5096, p. 15-17.

———, and Opdyke, N. D., 1973, Oxygen isotope and paleomagnetic stratigraphy of equatorial Pacific core V28-238: oxygen isotope temperatures and ice volumes on a 10^5 year and 10^6 year scale: Quaternary Res., v. 3, p. 39-55.

Shepard, F. P., 1973, Submarine geology, 3rd ed.: New York, Harper & Row, 517 p.

Siever, R., 1968, Sedimentological consequences of a steady-state ocean-atmosphere: Sedimentology, v. 11, p. 5-29.

———, and Woodford, N., 1973, Sorption of silica by clay minerals: Geochim. Cosmochim. Acta, v. 37, p. 1851-1880.

Sillén, L. G., 1967, The ocean as a chemical system: Science, v. 156, no 3779, p. 1189-1197.

Skornyakova, I. S., 1965, Dispersed iron and manganese in Pacific Ocean sediments: Lithology and Mineral Resources, no. 5, p. 3-20; Engl. transl. (1967): Internatl. Geol. Rev., v. 7, no. 12, p. 2161-2174.

Smith, S. V., Dygas, J. A., and Chave, K. E., 1968, Distribution of calcium carbonate in pelagic sediments: Marine Geology, v. 6, p. 391-400.

Spencer, D. W., 1972, Geosecs II, the 1970 North Atlantic station: hydrographic features, oxygen and nutrients: Earth and Planetary Sci. Letters, v. 16, no. 1, p. 91-102.

Somayajulu, B. L. K., 1967, Beryllium-10 in a manganese nodule: Science, v. 156, p. 1219-1220.

———, Heath, G. R., Moore, T. C., Jr., and Cronan, D. S., 1971, Rates of accumulation of manganese nodules and associated sediment from the equatorial Pacific: Geochim. Cosmochim. Acta, v. 35, p. 621-624.

Stanley, D. J., ed., 1973, The Mediterranean Sea: a natural sedimentation laboratory: Stroudsburg, Pa., Dowden, Hutchinson & Ross, 765 p.

Stefańsson, U., 1966, Influence of the Surtsey Eruption on on the nutrient content of the surrounding seawater: Jour. Marine Res., v. 24, no. 2, p. 241-268.

Stumm, W., Morgan, J. J., 1970, Aquatic chemistry: an introduction emphasizing chemical equilibria in naural waters: New York, Wiley-Interscience, 583 p.

Taylor, F. B., 1910, Bearing of the Tertiary mountain belt on the origin of the earth's plan: Geol. Soc. America Bull., v. 21, p. 179-226.

Thiede, 1971, Planktonische Foraminiferen in Sedimenten vom iberomarokkanischen Kontinentalrand: Forsch.-Ergebnisse (C) 7, p. 15-102.

Thiede, J., 1973, Planktonic foraminifera in hemipelagic sediments: shell preservation off Portugal and Morocco: Geol. Soc. America Bull., v. 84, p. 2749-2754.

Turekian, K. K., 1965, Some aspects of the geochemistry of marine sediments, *in* Riley, J. P., and Skirrow, G., eds., Chemical oceanography, v. 2, New York, Academic Press, p. 81-126.

Turekian, K. K., and Wedepohl, K. H., 1961, Distribution of the elements in some major units of the earth's crust: Geol. Soc. America Bull., v. 72, p. 175-192.

Turner, R. R., and Harris, R. C., 1970, The distribution of non-detrital iron and manganese in two cores from the Kara Sea: Deep-Sea Res., v. 17, p. 633-636.

van Andel, T. H., and Heath, G. R., 1973, Geological results of Leg 16: the Central Equatorial Pacific west of the East Pacific Rise, *in* Initial reports of the Deep Sea Drilling Project, v. 16: Washington, D.C., U.S. Govt. Printing Office, p. 937-949.

———, and Komar, P. D., 1969, Ponded sediments of the Mid-Atlantic Ridge between 22° and 23° north latitude: Geol. Soc. America Bull., v. 80, p. 1163-1190.

———, and Moore, T. C., Jr., 1974, Cenozoic calcium carbonate distribution and calcite compensation depth in the Central Equatorial Pacific Ocean: Geology, v. 1, p. 87-92.

Vine, F. J., 1966, Spreading of the ocean floor: new evidence: Science, v. 154, no. 3755, p. 1405-1415.

———, and Matthews, D. H., 1963, Magnetic anomalies over ocean ridges: Nature, v. 199, p. 947-949.

von Rad, U., and Rosch, H., 1972, Mineralogy and origin of clay minerals, silica and authigenic silicates, in Leg 14 sediment, *in* Initial reports of the Deep Sea Drilling Project, v. 14: Washington, D.C., U.S. Govt. Printing Office, p. 727-751.

Walker, R. G., 1973, Mopping up the turbidite mess, *in* Ginsburg, R. N., ed., Evolving concepts in sedimentology: Baltimore, Johns Hopkins Press, p. 1-37.

Wallace, A. R., 1895, Island Lift: London, Macmillan and Co., 563 p.

Watkins, N. D., 1972, Review of the development of the geomagnetic polarity time scale and discussion of prospects for its finer definition: Geol. Soc. America Bull., v. 83, p. 551-574.

Wattenberg, H., 1935, Kalkauflösung und Wasserbewegung am Meeresboden: Ann. Hydr. Marit. Meteorology, v. 63, p. 387-391.

Wegener, A., 1966, The origin of continents and oceans; Engl., Tranl., Biram, J., (1929): New York, Dover.

Wilson, J. T., 1965, A new class of faults and their bearing upon continental drift: Nature, v. 207, p. 343-347.

Windom, H. L., 1969, Atmospheric dust records in permanent snowfields: implications to marine sedimentation: Geol. Soc. America Bull., v. 80, p. 761-782.

Winterer, E. L., 1971, History of the Pacific Ocean Basin (abstr.): Geol. Soc. America, v. 7, no. 3, p. 754.

———, 1973, Sedimentary facies and plate tectonics of Equatorial Pacific: Am. Assoc. Petr. Geol. Bull., v. 57, no. 2, p. 265-282.

———, Ewing, J. I., Douglas, R. G., Jarrard, R. D., Lancelot, Y., Moberly, R. M., Moore, T. C., Roth, P. H., and Schlanger, S. O., 1973, Initial reports of the Deep Sea Drilling Project, v. 17; Washington, D.D., U.S. Govt. Printing Office, 930 p.

Wise, S. W., Buie, B. F., and Weaver, F. M., 1972, Chemically precipitated sedimentary cristobalite and the origin of chert: Eclogae Geol. Helv., v. 65, no. 1, p. 157-163.

Wollast, R., 1974, The silica problem, *in* Goldberg, E. D., ed., The sea, v. 5: New York, Wiley-Interscience, (in press).

———, Mackenzie, F. T., and Bricker, O. P., 1968, Experimental precipitation and genesis of sepiolite at earth-surface conditions: Am. Mineralogist, v. 53, p. 1645-1662.

Yount, J., and Berger, W. H., 1974, Modelling of differential dissolution of planktonic foraminifera: Cushman Found. Foram. Res. Spec. Publ., (in press).

Acoustic Stratigraphy in the Deep Oceans

N. Terence Edgar

INTRODUCTION

Since the first successful continuous-reflection profiling at sea was reported by J. I. Ewing and Tirey (1961), our understanding of sediment distribution and processes in the oceans has advanced markedly. The combination of piston coring and reflection profiling provided a powerful combination in the study of marine geology; the profiler located outcrops of older sediments that could be cored, and the sediment recovered could be used to explain the nature of reflecting horizons (J. I. Ewing et al., 1966; M. Ewing et al., 1966). The locating and sampling of older sediments in this way was fortuitous, and the systematic study of acoustic horizons and intervals had to await the era of deep ocean drilling.

Seismic profiler records have provided the key to planning the drilling programs for the Deep Sea Drilling Project. In turn, the drilling results have increased the value of the profiler recordings, made them more meaningful, and provided insight for responsible interpretations on a regional basis. The emphasis of this section will be centered on the major reflecting horizons that dominate the acoustic stratigraphy of the North Atlantic and North Pacific oceans.

Acoustic stratigraphy of deep-sea sediments was first established by the recognition of an areally extensive subbottom reflector in the western North Atlantic Ocean called horizon A (J. I. Ewing et al., 1966). The deepest recorded reflector with a characteristically rough topography was called "acoustic basement." Near the continental margins this reflector, believed to be basalt, became relatively smooth, suggesting that it is composed, at least in part, of lithified sediments. As the reflection technique improved, other reflectors between A and basement, such as A* (J. I. Ewing and Hollister, 1972) and β (Windisch et al., 1968), were recognized. Similar reflectors in the Pacific Ocean and Caribbean Sea were called A′, B′ and A″, B″, respectively.

NORTH ATLANTIC OCEAN

The acoustic stratigraphy in the North Atlantic Ocean is best defined in the western part, the North American Basin. Figure 1 shows the major acoustic reflectors in the western Atlantic, a profiler record that shows these reflectors, and the corresponding lithologies for the profile as determined from DSDP Leg 11 drilling results.

Horizon A was a prime drilling objective in the early cruises of the D/V GLOMAR CHALLENGER, operating for the Deep Sea Drilling Project (JOIDES), and after completion of three cruises in the western North Atlantic (Leg 1: M. Ewing et al., 1969; Leg 2: Peterson et al., 1970; Leg 11: Hollister et al., 1972), the nature of the reflecting horizons became evident. Early drilling results from Legs 1 and 2 of DSDP indicated that horizon A corresponds to an early to middle Eocene chert interval; but Leg 11 data demonstrated that horizon A is a prominent reflector even where the drill encountered little or no chert. In this case the reflector may correspond to the acoustic impedance contrast associated with an early Tertiary hiatus beneath the cherts.

Chert in the deep ocean occurs in almost any lithology (radiolarian ooze, brown clay, or carbonates), but results of Leg 1 of DSDP indicate that parts of turbidites are preferentially silicified. Chert has been encountered in the eastern North Atlantic in Eocene (?) sediments, and on the Sierra Leone Rise and the Rio Grande Rise in the South Atlantic.

Horizon A can be traced beneath the lower continental rise as a fairly horizontal reflector. Considering that middle Miocene cores were recovered within 50 m of horizon A at sites 105 and 106, J. I. Ewing and Hollister (1972) point out that most of the lower continental rise was built after early Miocene time. Whether the rise westward of site 106 had been built prior to early Miocene time cannot be stated with certainty, but J. I. Ewing and Hollister (1972) note that one additional layer onlaps horizon A west of hole 105, possibly indicating a thickening of the late Eocene-early Miocene beds toward the slope.

Below horizon A another reflector called horizon A* (J. I. Ewing and Hollister, 1972) was identified from drilling results on Leg 11 of DSDP as the boundary between metal-rich clays of presumed early Tertiary age and underlying black clays of an euxinic interval. It is possible that these mineral-rich clays may be mapped by tracing horizons A and A* through the western North Atlantic Basin, but additional drilling will be necessary before such regional interpretations can be made with confidence.

The black euxinic clays of horizon A*-β interval underlie the volcanic clays at most sites drilled. In areas where the volcanic clays are missing, horizon A may correspond to the top of the black clays.

Horizon β corresponds to the basal contact of the black clays that overlie a limestone unit of Late

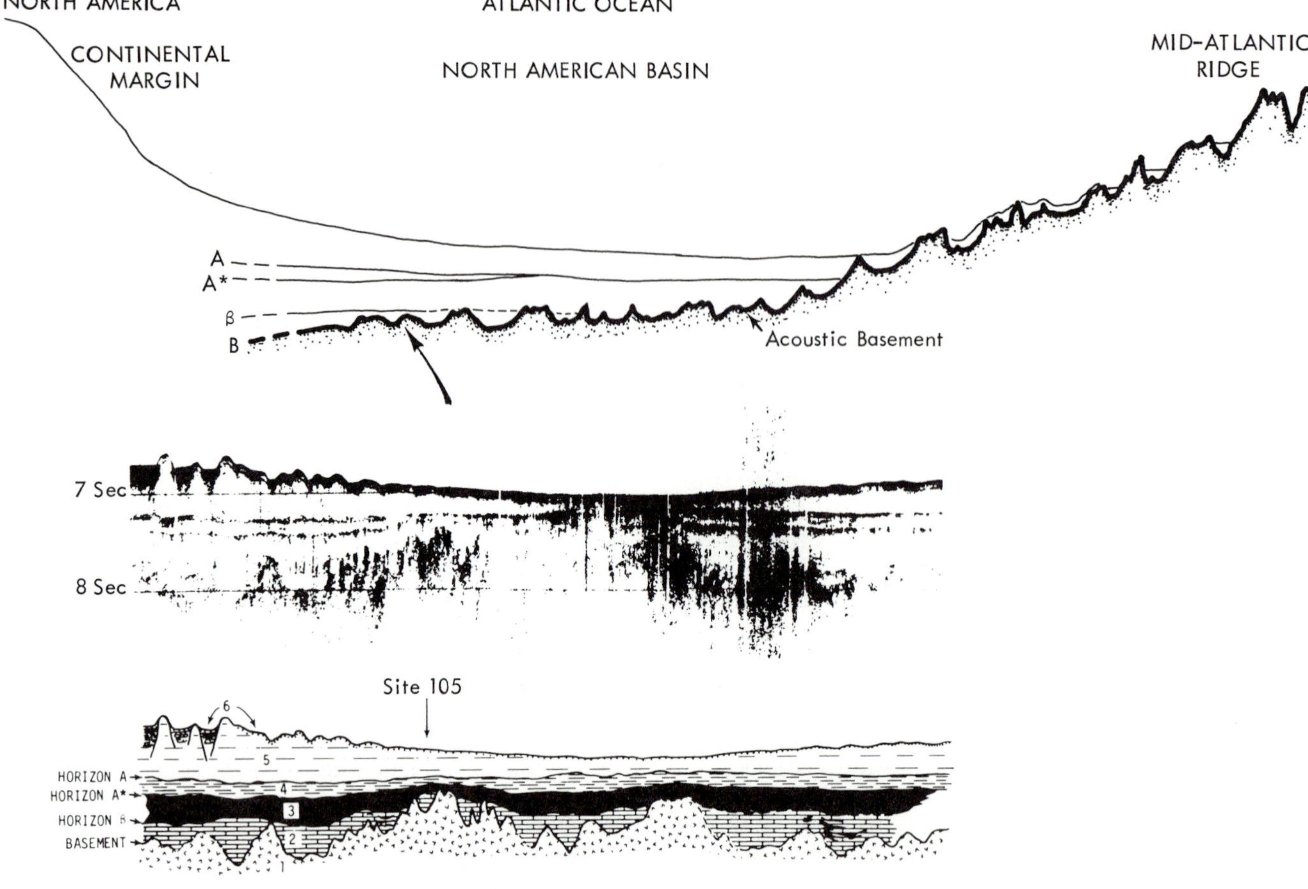

Fig. 1. Major acoustic reflectors in the western Atlantic Ocean (top); typical seismic profile of the region (center); and lithologies corresponding to the profile as determined from deep ocean drilling results (bottom).

Jurassic and Neocomian age (J. I. Ewing and Hollister, 1972). An earlier interpretation (M. Ewing et al., 1969) from Leg 1 of DSDP drilling suggested that horizon β is associated with a mid-Cenomanian unconformity, but subsequent site surveying and Leg 11 results identified it with the basal Jurassic-Cretaceous limestones.

The acoustically stratified horizon β to B unit conformably overlies horizon B in most, but clearly not all areas. The smoothness of horizon B suggested that it might consist of sedimentary rock rather than the basalt normally associated with the rough surface of acoustic basement. However, horizon B is a basalt that is not notably distinctive from other basalts recovered from acoustic basement which exhibit the normal rough surface.

PACIFIC OCEAN

A comprehensive analysis of seismic reflection horizons within the sediments of the Pacific was presented by J. I. Ewing et al. (1968). They recognized six units within the sedimentary column: an acoustically transparent layer, an upper opaque layer, a lower transparent layer, a lower opaque layer, acoustic basement, and turbidites. Based on drilling it is unlikely that there is any lithologic difference between the lower opaque layer and igneous acoustic basement. Figure 2 shows a profiler record made in the western Pacific showing the units referred to above.

The upper transparent layer is found throughout most of the Pacific Ocean and represents the mid- and late Cenozoic sedimentary cover. In the eastern Pacific Ocean, it overlies acoustic basement; the deeper and older sedimentary layers are found west of the Hawaiian Islands. The upper opaque layer appears to be almost ubiquitous in normal sedimentary basins of the western North Pacific Ocean, but the lower transparent layer is not. Where the lower transparent layer is missing, the acoustic basement is difficult to define beneath the upper opaque

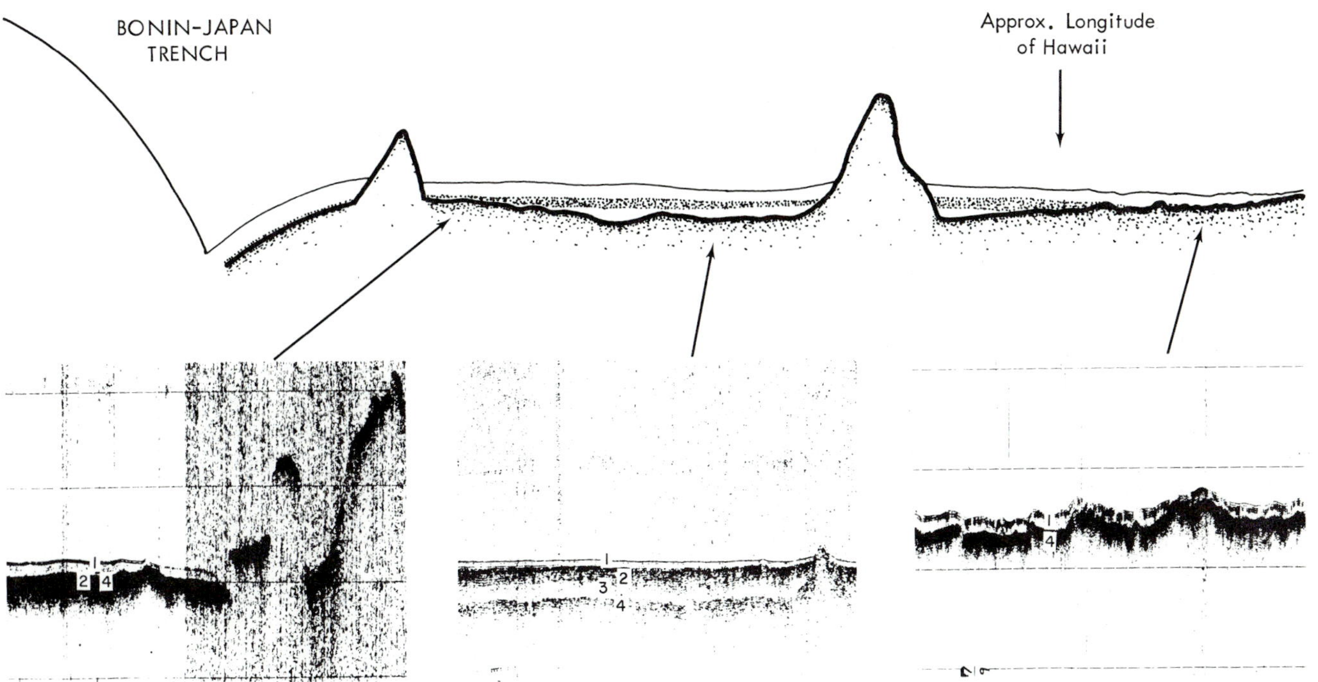

Fig. 2. Seismic profiler records illustrating sedimentary units identified in the western Pacific Ocean.

layer, and in some areas it is commonly impossible to distinguish between the two. The lower transparent layer is normally missing in the eastern part of the western Pacific Ocean west of Hawaii and in the extreme western part.

It was recognized after the completion of the first cruise of the D/V GLOMAR CHALLENGER into the western Pacific Ocean (Leg 6) that the nature of the upper surface of the opaque layer varied in lithology and age (Fisher et al., 1971). In the area just west of Hawaii the reflector was identified with middle Eocene chert in chalk in one case and mid-Tertiary volcanic ash in another case. In the western part of the basin the layer appears to correspond to soft Neogene or undated clays unconformably overlying Cretaceous cherty oozes, clays, and ash. Clearly, the concept of ocean-wide reflectors of seemingly uniform age and composition, which was reinforced by early Atlantic Ocean drilling, was subject to question.

The next cruise (Leg 7 DSDP, Winterer et al., 1971) reported the following identity for each of the layers. A radiolarian ooze corresponds to the upper acoustically transparent layer. The top of horizon B is correlated with basalt.

Scientists on Leg 17 (Winterer et al., 1973) drilled intensely the area southwest of Hawaii and recovered chert (middle Eocene) at approximately the level of the opaque layer at all sites successfully drilled. At a number of sites a major unconformity lay a few meters below the chert horizon, making it uncertain whether the top of the opaque layer corresponds to the chert or the unconformity. At only one site (167) could it be established that the reflector corresponded to the chert and not to the unconformity.

The lower transparent layer was clearly identified from seismic records only at site 169, where the nature of the acoustic "transparency" from the drilled data is obscure. The investigators suggested that it was caused by fewer chert layers at depth.

The D/V GLOMAR CHALLENGER returned to western Pacific Ocean on Leg 20 (Heezen et al., 1973) to drill a series of holes in the Pacific Basin just east of Japan and the Mariana-Bonin arc. Sparse coring makes identification and description of the acoustic intervals difficult, but the upper transparent layer is believed to extend to Late Cretaceous zeolitic clays. The top of the opaque zone is attributed to underlying Late Cretaceous cherts, chalks, limestones, and marls. The upper surface of the chalk-chert unit is time transgressive, ranging from Barremian in the north (east of Japan) to late Paleocene in the south (east of Guam).

SUMMARY

The nature of marine seismic reflectors in the Atlantic and Pacific oceans can vary over a relatively short distance, and they are time transgressive to varying degrees. Consequently, statements regarding the nature of even the major reflectors on an ocean-wide basis (time or lithology) are difficult to defend. However, there is a tendency for acoustic reflections to be associated with the early mid-Tertiary section, whether the result of silicification, other lithification processes, or unconformities.

BIBLIOGRAPHY

Ewing, J. I., Ewing, M., Aitken, T., and Ludwig, W., 1968, North Pacific sediment layers measured by seismic profiling, *in* The crust and upper mantle of the Pacific Area, Geophys. Monogr. 12: Washington, D.C., American Geophysical Union, p. 147-173.

———, and Hollister, C. D., 1972, Regional aspects of deep sea drilling in the western North Atlantic, *in* Hollister, C. D., et al., Initial reports of the Deep Sea Drilling Project, v. XI: Washington, D.C., U.S. Govt. Printing Office, p. 951-973.

———, and Tirey, G. B., 1961, Seismic profiler: Jour. Geophys. Res., v. 66, p. 2917-2927.

———, Worzel, J. L., Ewing, M., and Windisch, C. C., 1966, Ages of horizon A and oldest Atlantic sediments: Science, v. 154, p. 1125-1132.

Ewing, M., Saito, T., Ewing, J., and Burckle, L., 1966, Lower Cretaceous sediments from the northwest Pacific: Science, v. 152, p. 751-755.

———, Worzel, J. L., et al., 1969, Initial reports of the Deep Sea Drilling Project, v. I: Washington, D.C., U.S. Govt. Printing Office, 672 p.

Fischer, A. G., Heezen, B. C., et al., 1971, Initial reports of the Deep Sea Drilling Project, v. VI: Washington, D.C., U.S. Govt. Printing Office, 1330 p.

Heezen, B. C., MacGregor, Ian D., et al., 1973, Initial reports of the Deep Sea Drilling Project, v. XX, Washington, D.C., U.S. Govt. Printing Office, 1330 p.

Hollister, C. D., Ewing, J. I., et al., 1972, Initial reports of the Deep Sea Drilling Project, v. XI: Washington, D.C., U.S. Govt. Printing Office, 1078 p.

Peterson, M. N. A., Edgar, N. T., et al., 1970, Initial reports of the Deep Sea Drilling Project, v. II: Washington, D.C., U.S. Govt. Printing Office, 492 p.

Windisch, C. C., Leyden, R. J., Worzel, J. L., Saito, T., and Ewing, J. I., 1968, Investigation of horizon β: Science, v. 162, p. 1473-1479.

———, Ewing, J. I., et al., 1973, Initial reports of the Deep Sea Drilling Project, v. XVII: Washington, D.C., U.S. Govt. Printing Office, 930 p.

Winterer, E. L., Riedel, W. R., et al., 1971, Initial reports of the Deep Sea Drilling Project, v. VII: Washington, D.C., U.S. Govt. Printing Office, 1758 p.

Part V

Deformation at Continental Margins

Trench Slope Model

D. R. Seely, P. R. Vail, and G. G. Walton

INTRODUCTION

New data demonstrate oceanic underthrusting and enable construction of a model of trench inner slopes. Magnetic, earthquake, and deep-penetration reflection seismic data supplemented by data from JOIDES drill holes, dredge hauls, and piston cores leave little room for doubt that thrusts and folds are the primary structures on the landward slopes of trenches, and, in accord with plate tectonics theory, they are considered to be caused by underthrusting of the oceanic plate beneath continents or island arcs. The seismic data and magnetic anomalies document continuation of the oceanic plate beneath the landward slope of the trench. These seismic data also show structural separation between the oceanic plate and overlying sediments and reveal that it is caused by thrusting and/or folding. The resultant uplift is documented by paleobathymetric and compaction studies of sediments from the slopes of several trenches and by surface motion associated with recent earthquakes.

This relatively simple basic picture is a model of the structuring of trench inner slopes at a particular point in time and may be complexly modified as geologic conditions change with the passage of time. The model can be used to interpret the volumes of sediments in trenches, the rate of continental accretion, the development of fan structure, the depositional environments of sediments expected beneath trench inner slopes, and the phenomenon of diapirism in these sediments.

Accretion occurs by operation of the model through time. The trench slope builds outward while behind it the products of progressively older trench slopes first subside and then are uplifted while being incorporated into the continent by a plutonic belt.

PREVIOUS INTERPRETATIONS AND NEW DATA

New data make possible the testing of current diverse notions on the structuring of trench inner slopes and permit the development of a more detailed model of them, thus providing further insight into subduction processes. Until recently, most published seismic data in and adjacent to oceanic trenches have been of the shallow-penetration, high-resolution type. Deep-penetration seismic data of this zone are now becoming available, as are magnetic and earthquake data and widely spaced sampling from JOIDES drill holes, dredge hauls, and piston cores.

The trench inner slope is one of the major features of the oceanic, or fore-arc, side of volcanic arcs (Fig. 1). Other features were identified by Karig (1970) and elaborated by Dickinson (1971, 1973). They consist of a fore-arc basin or basin complex: an outer high (trench-slope break of Dickinson, 1973) constituting an outer arc or structural prominence usually at the edge of the shelf or terrace on the slope; and the trench outer, or seaward, slope. Locally the trench outer slope is separated from the abyssal plain by a low submarine ridge.

Previous interpretations of trench margin tectonics have debated variations on two contrasting concepts of the structuring of trench inner slopes—one in which the trench inner slope is underlain by down-dropped, normal fault blocks with no underthrusting of the oceanic plate, and a second in which the trench inner slope is underlain by compressional thrusts and folds and/or mélanges produced by underthrusting (e.g., Seyfert, 1969; Dickinson, 1971, 1973). Resolution of this controversy is critical to both the proponents and opponents of current theories of plate tectonics. This paper presents an interpretation of the newer data which was first presented at the Penrose Conference on Continental Margins, December 1972. It demonstrates underthrusting of the oceanic plate and provides the basis for construction of a model (Fig. 2) of the trench inner slope.

SEISMIC PROCESSING AND INTERPRETATION

A first glance at a time section of reflection seismic data across many trench inner slopes (e.g., the Mid-America trench, Fig. 3) often gives the impression that it consists of an undeciferable mass of diffractions (Fig. 4); thus, on shallow-penetration data, "acoustic basement" has commonly been placed near the sea floor. The diffractions, however,

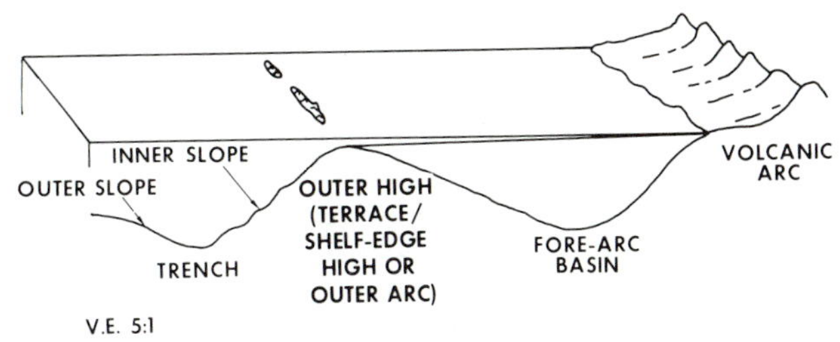

Fig. 1. Major features of the fore arc.

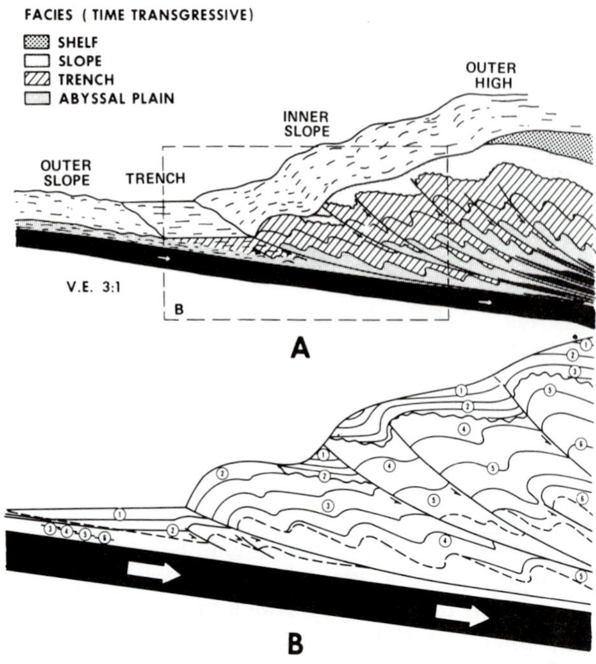

Fig. 2. Trench margin model: (A) facies patterns, (B) time stratigraphy. (Number sequence from youngest to oldest unit. Dashed line represents the boundary between trench turbidites and abyssal plain sediments; see Piper et al., 1973. Pinch-outs of trench turbidites at this boundary in thrust plates represent outer edges of older trenches.)

slopes because the reflection seismic technique generates more diffractions from faults and folds occurring at such depths than from the same features at shallower depths. Modern seismic processing techniques can reduce the exaggeration of this complexity and eliminate velocity distortions, such as the seaward tilting of reflections that is caused by the low-velocity water wedge above the slope.

Figure 4 is 12-second, 12-fold common depth point (CDP) data. To obtain the acoustic velocities necessary for generation of nonmigrated and migrated depth sections (Figs. 5, 6A, and 7A), calculations were made on every diffraction envelope. The determinations were not only useful for making the depth sections, but when broken down into interval velocity groupings, they indicated a thick sedimentary section beneath the slope, having interval velocities of 9,000 to 11,000 ft/sec (Fig. 6A).

The section across the Oregon slope (Figs. 8A and 8B) is 12-fold CDP data that were conventionally processed and converted to a nonmigrated depth section (Fig 8B) using velocities calculated from "gathers." "Gathers" were used here because diffractions are fewer and "gathers" are better than beneath the slope of the Mid-America trench (Fig. 4).

Because of the seismic processing, the strong reflector that *appears* to have a gentle slope toward the Mid-America trench axis on Figure 4 (see also Fig. 7B, where the reflector intersects the right-hand margin at about 8.3 sec) can be seen to have a *true* dip that is gently landward (Figs. 5 and 6A). The reflector appears to be a nearly straight-line continuation of the strong reflector that occurs at shallow depth beneath the ocean floor on the west side of the trench and extends more than 14 miles landward from the trench. This reveals the trench to be a bathymetric feature that is not formed by a major structural graben. Similar conditions occur in association with other trenches, such as the eastern Aleutian trench, as demonstrated by the continuation of ocean floor magnetic anomalies beneath the trench inner slope (Naugler and Wageman, 1973), the southeastern Puerto Rican trench as shown by reflection seismic data (Chase and Bunce, 1969; Marlow et al., 1972), and the Japan trench as indicated by refraction seismic data (Ludwig et al., 1966). They are also present in what we consider to be a filled trench off Oregon (Fig. 8; see also Kulm and Fowler, this volume).

The landward dip of the strong reflector beneath the slope of the Mid-America trench is not as great as that of the acoustic reflectors overlying it (Fig. 6A), so it is separated from them by structural disharmony. Similar disharmonies are interpreted to occur also within the overlying sequence of reflectors. It seems likely that the disharmonies represent thrust faults and are thus interpreted on Figure 6A.

The internal geometries of the proposed thrust plates are not clear, but the discontinuous reflectors within them almost certainly indicate that they contain many faults and/or folds. Similar faults and folds are present beneath the Oregon slope (Fig. 8) and appear to occur on the inner slope of the Java trench (Fig. 9). Owing to variations in the stratigraphy and underthrusting rates in other trenches, differing degrees of fault-versus-fold development can be expected beneath their inner slopes.

UPLIFT AS EVIDENCE OF THRUST INTERPRETATION

Further evidence that the trench inner slope is structured by thrust faults and compressional folds comes from rapidly accumulating core, dredge haul, and earthquake data which indicate that the trench inner slope and outer high are underlain by a thick section of uplifted sediments that overlie the uninvolved oceanic plate.

Deformed and dewatered sediments encountered at JOIDES site 181 near the foot of the inner slope of the Aleutian trench (169-369 m) have been interpreted as having been buried to depths of 1.0-1.5 km (Lee et al., 1973) before being uplifted and eroded prior to deposition of the overlying sediments. The latter sediments also give sedimentational evidence of uplift (von Huene and Kulm, 1973). More recently, at site 298 near the foot of the inner slope of the Nankai trough on the western margin of Japan, overcompacted sediments with distinct cleavage were found (JOIDES Leg 31 Staff, 1973).

Plafker (1969) documented sediments uplifted during the 1964 Alaska earthquake on and seaward

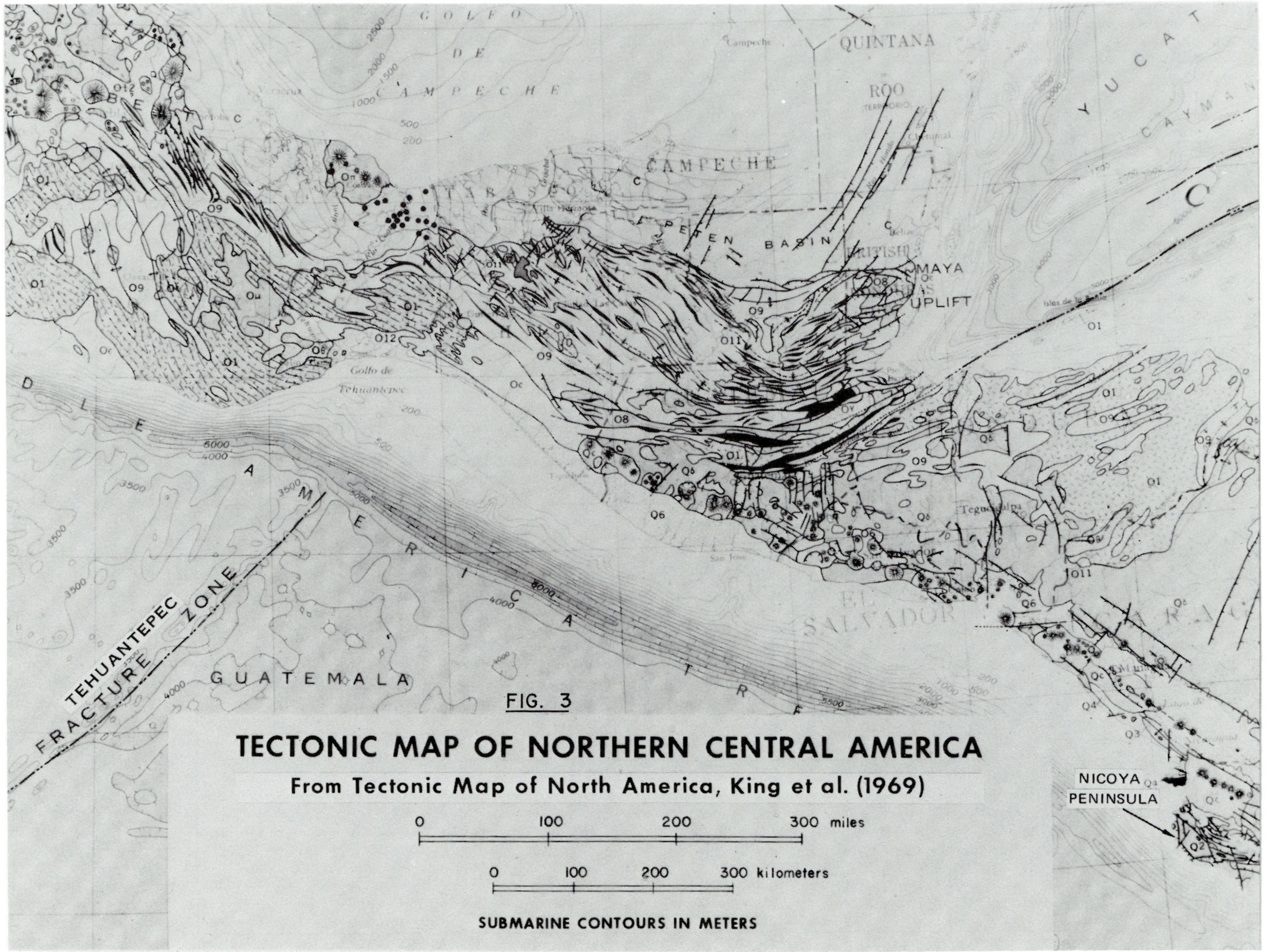
GOLFO DE CAMPECHE
QUINTANA ROO
CAMPECHE
PETEN BASIN
BRITISH
MAYA UPLIFT
YUCAT
CAYMAN
Golfo de Tehuantepec
MIDDLE AMERICA TRENCH
GUATEMALA
EL SALVADOR
TEHUANTEPEC FRACTURE ZONE
NICOYA PENINSULA
FIG. 3
TECTONIC MAP OF NORTHERN CENTRAL AMERICA
From Tectonic Map of North America, King et al. (1969)
0 100 200 300 miles
0 100 200 300 kilometers
SUBMARINE CONTOURS IN METERS

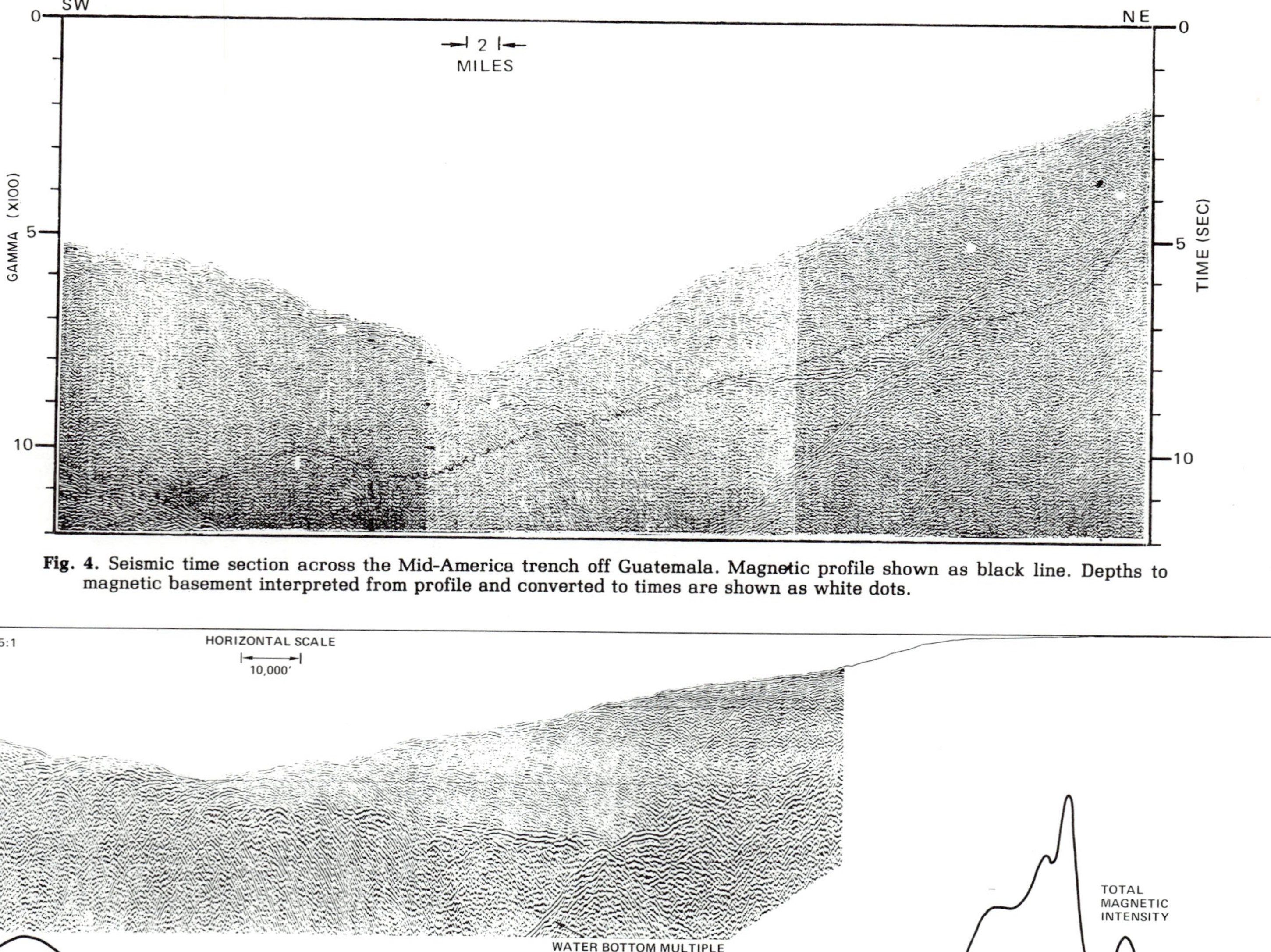

Fig. 4. Seismic time section across the Mid-America trench off Guatemala. Magnetic profile shown as black line. Depths to magnetic basement interpreted from profile and converted to times are shown as white dots.

Fig. 5. Seismic depth presentation of Figure 4 (nonmigrated). The magnetic profile has been extended landward, with regional gradient removed, to show evidence for shallow magnetic sources that occur locally near the Guatemalan shelf edge.

from the southeastern margin of Kodiak Island and the Kenai Peninsula. Similar belts of uplift and depression were also described by Plafker and Savage (1970) in an area adjacent to the Peru-Chile trench and by Fitch and Scholz (1971) adjacent to the Nankai trough. Uplifted deep-water sediments have also been discussed by Burk (1972).

Uplift of sediments at localities on the continental shelf and slope off Oregon is described by Kulm and Fowler (this volume). The rate of uplift is about 1000 m/my on the slope and 100 m/my on the shelf as determined from late Cenozoic sediments. Underthrusting of the Pacific plate, interpreted here to have caused the uplift of the slope sediments, is shown in Figure 8.

The island of Barbados exposes complexly deformed turbidite and pelagic sediments uplifted to form the Antillean outer high (referred to as the Barbados Ridge by Chase and Bunce,1969). The Mentawi Islands adjacent to the Java trench expose similar sediments and structures except that low-grade metamorphic and mafic igneous rocks are also present (van Bemmelen, 1949). Negative isostatic gravity anomalies characterize both of these occurrences (Bush and Bush, 1969; Vening Meinesz, 1954) and strongly suggest, along with magnetic and reflection seismic data from the Java trench (Beck, 1972), that they, like other outer highs are largely thickened piles of compressed sediments or metasediments (see Grow, 1973, for similar evidence on the Aleutian outer high).

The concept that "faulting and gravity folding during uplift" are responsible for the complex deformation of fore-arc outer highs (Mitchell and Reading, 1971) appears unreasonable to the present writers. If this concept was correct, such highs, once formed, should be characterized by extensional deformation with compressional deformation occurring downslope in opposite directions away from them toward both the trench and volcanic arc. However, folds and thrusts rather than normal faults characterize older sediments of outer highs, and the folds and thrusts can involve igneous rocks of the acoustic basement (e.g., Nicoya complex on the Nicoya Peninsula, Costa Rica, Fig. 3). Vergence does not change from one side of the highs to the other (e.g., on Figure 8 the folds shoreward from the outer high have westward vergence just as do the thrusts beneath the slope) as would be expected if gravitational sliding had occurred in opposite directions from them. It seems likely, therefore, that the outer highs are simply the most prominent expression of compressional deformation produced by underthrusting of the oceanic plate (Fig. 2).

THRUST AGE AND DIP RELATIONSHIPS

The thrust faults interpreted beneath the inner slope of the Mid-America trench appear to become progressively older and more steeply dipping in a landward direction, away from the trench (Fig. 6A). Those thrusts presently most active occur at the foot of the slope, where they have bathymetric expression. Similarly, location of the most active thrusting and/or folding near the foot of the slope off Oregon (Fig. 8) is strongly suggested by Kulm's and Fowler's observation (this volume) that the most rapid uplift occurs there.

Examples of older thrust faults, probably formed on a former slope of the Mid-America trench, occur on the Nicoya peninsula of Costa Rica (Fig. 3) in association with pillow basalts, basalt agglomerates, graywackes, chert, siliceous aphanitic limestones, and intrusions of diabase, gabbro, and peridotite of the Nicoya complex (Dengo, 1962). This assemblage of thrusts and rocks probably continues northwestward beneath the shelves of Nicaragua, El Salvador, and Guatemala. Similar, older thrusts and/or folds can be seen with various other rock associations adjacent to most trenches (e.g., the rim of the Gulf of Alaska, Barbados, Cuba, Andaman Islands, etc.)

The progressively steeper dip of the thrust faults may be attributed to the insertion of new, wedge-shaped slices, often containing asymmetric folds, beneath closely spaced, landward-dipping, concave-upward, thrust faults at the foot of the trench inner slope. As these slices underthrust older thrusts and sediments higher on the inner slope, the latter are tilted more and more landward (Fig. 10A) and produce fan structures. Because thrust surface configurations and the spacing between thrusts and/or asymmetric folds vary from place to place along trenches, the degree of development of fan structure can also be expected to vary.

Operation of the wedging mechanism may account for the steeply dipping to overturned Mesozoic strata on the Kenai Peninsula and Kodiak Island in Alaska. Some zones of steeply dipping strata in the internal parts of orogenic belts may also have had a similar origin, the dip and age relationships of orogenic overthrusts bearing the same relation to the basement of their foreland as the thrust plates of trench margins bear to the oceanic plate (Fig. 10B).

POSSIBLE CONSEQUENCES OF THRUSTING

Numerous possible stratigraphic and structural details can be related to thrust development on the trench inner slope (Fig. 2). The slope in most cases will probably be underlain in large part by trench deposits that are complexly folded and/or thrusted (Fig. 2A) and that increase in age in a landward direction (Fig. 2B). The intensity of the deformation can be extreme (e.g., see von Huene, 1972) and could well lead to the production at depth of mélanges as suggested previously by several authors (e.g., Page, 1972; Hsü, 1971; Dickinson, 1971, 1973).

Mélanges probably are more common than trench olistostromes in association with many modern trenches, particularly those adjacent to broad

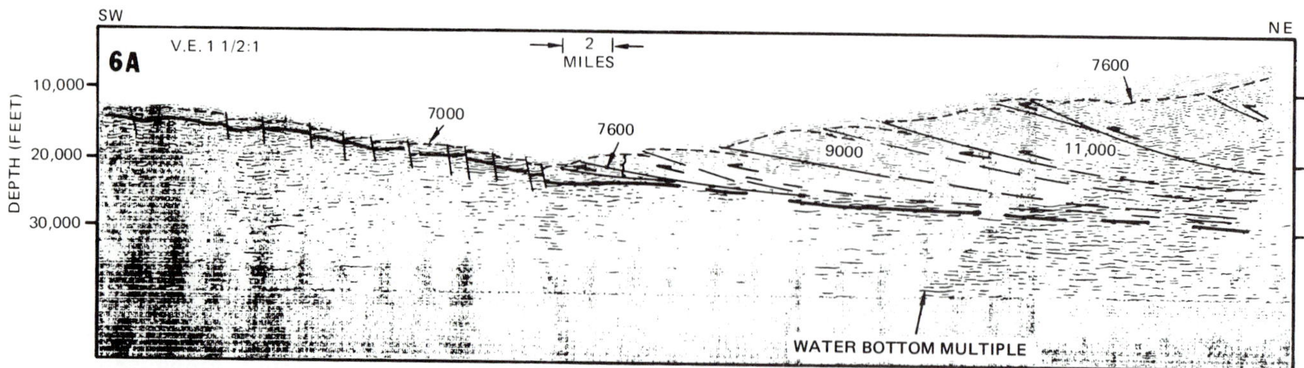

Fig. 6A. Interpretation of clipped amplitude, migrated depth section, showing interval velocities calculated from diffractions, normal faulting in the downbending oceanic plate on the trench outer slope (for a similar occurrence in the Japan trench, see Ludwig et al., 1966). Numbers are acoustic velocities in feet per second.

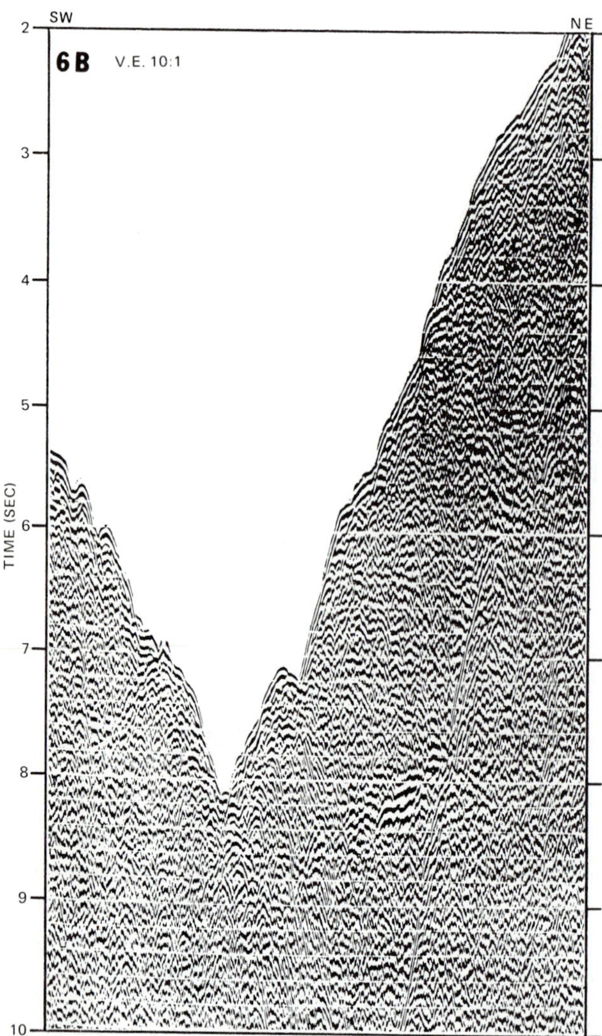

Fig. 6B. Effect of large vertical exaggeration on appearance of seismic section across Mid-America trench.

shelves underlain by accreted sediments. Sedimentary units consisting of chaotic zones containing large exotic blocks (having maximum dimensions exceeding a few tens of feet) are not apparent on published high-resolution seismic data from the Aleutian trench near JOIDES site 180 (Kulm et al., 1973a), from the Peru-Chile trench off central Peru (Kulm et al., 1973b), or from the filled trough off Oregon (Kulm and Fowler, this volume). Nor have they been found in the few sediment samples taken from these trenches (Piper et al., 1973; Kulm et al., 1973b; JOIDES Leg 31 Staff, 1973; Kulm and Fowler, this volume) or from the Mid-America trench southeast of Tehuantepec fracture zone (D.G. Moore, personal communication). To the northwest of the Tehuantepec fracture zone, however, very little accreted section is preserved beneath the narrow shelf, and a large crystalline boulder has been encountered near the base of the trench inner slope (D.G. Moore, personal communication).

Thrust faults can readily "sole-out" in the pelagic abyssal plain sediments on which the trench deposits accumulated (Fig. 2). These same pelagic sediments may form the cores of compressional diapiric folds above the detachment planes, as is perhaps the case for the folds reported by Shouldice (1971) off Vancouver Island.

Trench inner slopes, particularly in their lower reaches, may become areas of slumping which contribute to the trench turbidite sequence (Fig. 10A). Where folds grow or thrusts move and slumping does not remove the slope prominences thus formed, perched basins form upslope from them. Landward dips of sediments deposited in some perched basins are greater in older perched basin sediments than in younger ones (Figs. 2B and 8), possibly as a result of tilting caused by the wedging mechanism. Sediment accumulation is most likely on the upper parts of most trench inner slopes (Figs. 2, 6A, and 8), where thrust activity is diminishing or has ceased, and is least likely on the lower parts of trench inner slopes, where no more than a relatively thin veneer of slope sediments should generally be expected.

The volume of trench sediments appears to depend on the ratio between the rate of trench sedimentation and the rate of underthrusting (Dick-

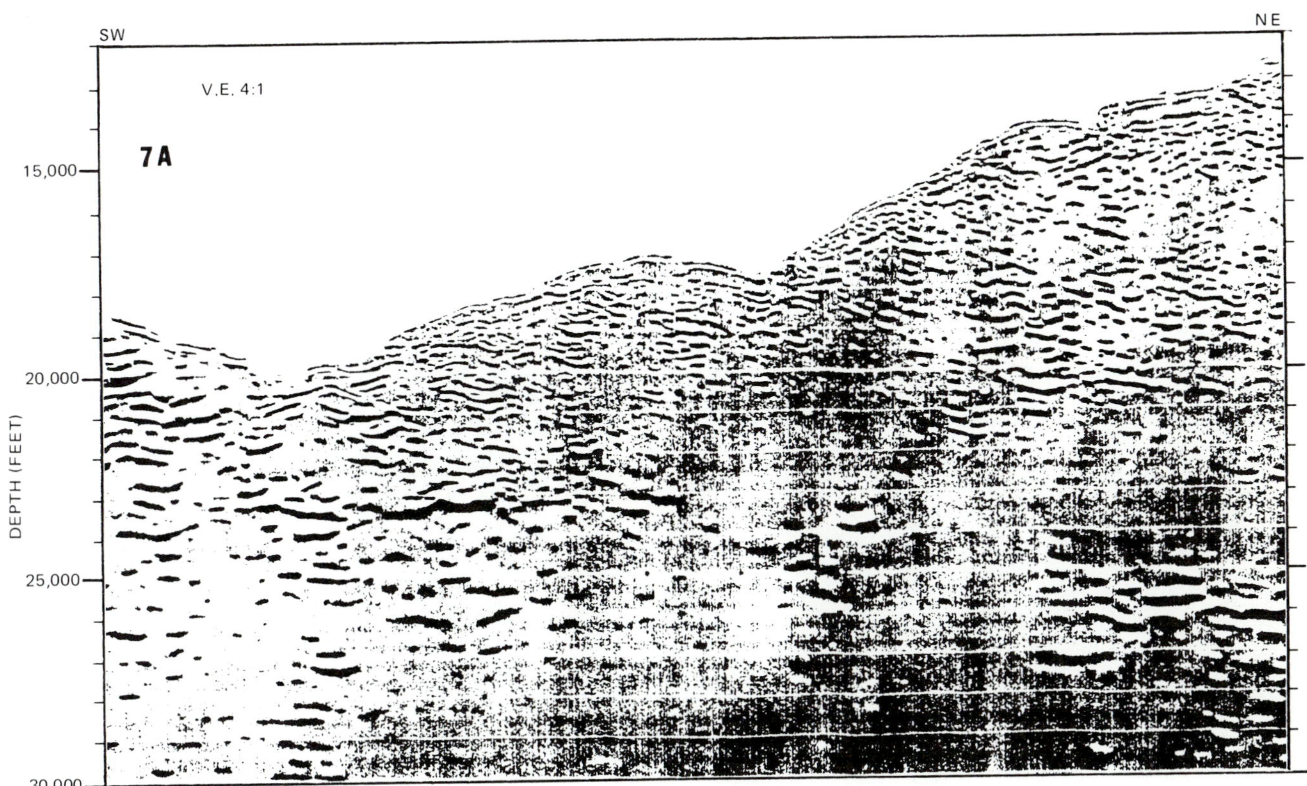

Fig. 7A. Detail of lower part of trench inner slope (a part of Fig. 4). Clipped amplitude, migrated depth display.

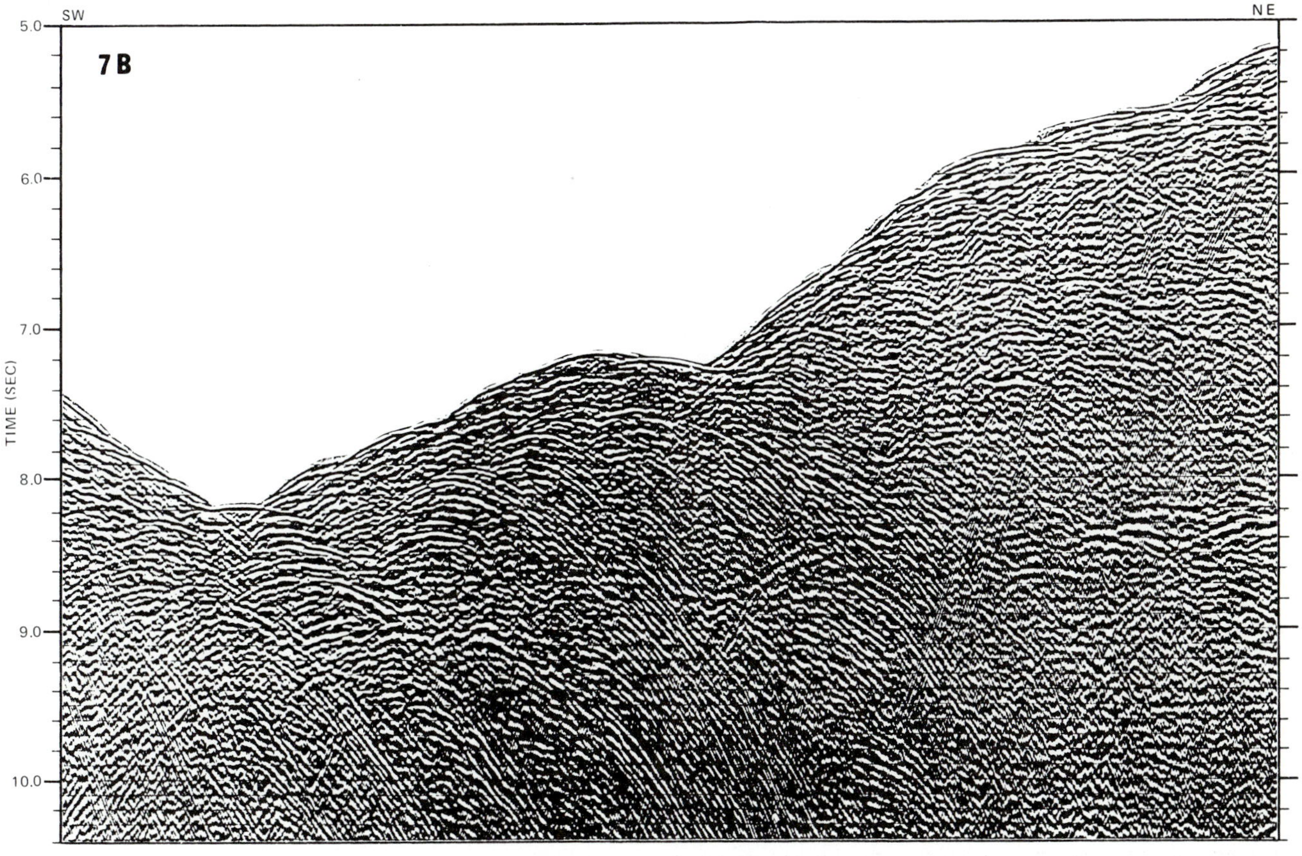

SW
NE
7B
5.0
6.0
7.0
8.0
9.0
10.0
TIME (SEC)

Fig. 7B. Time display of Figure 7A.

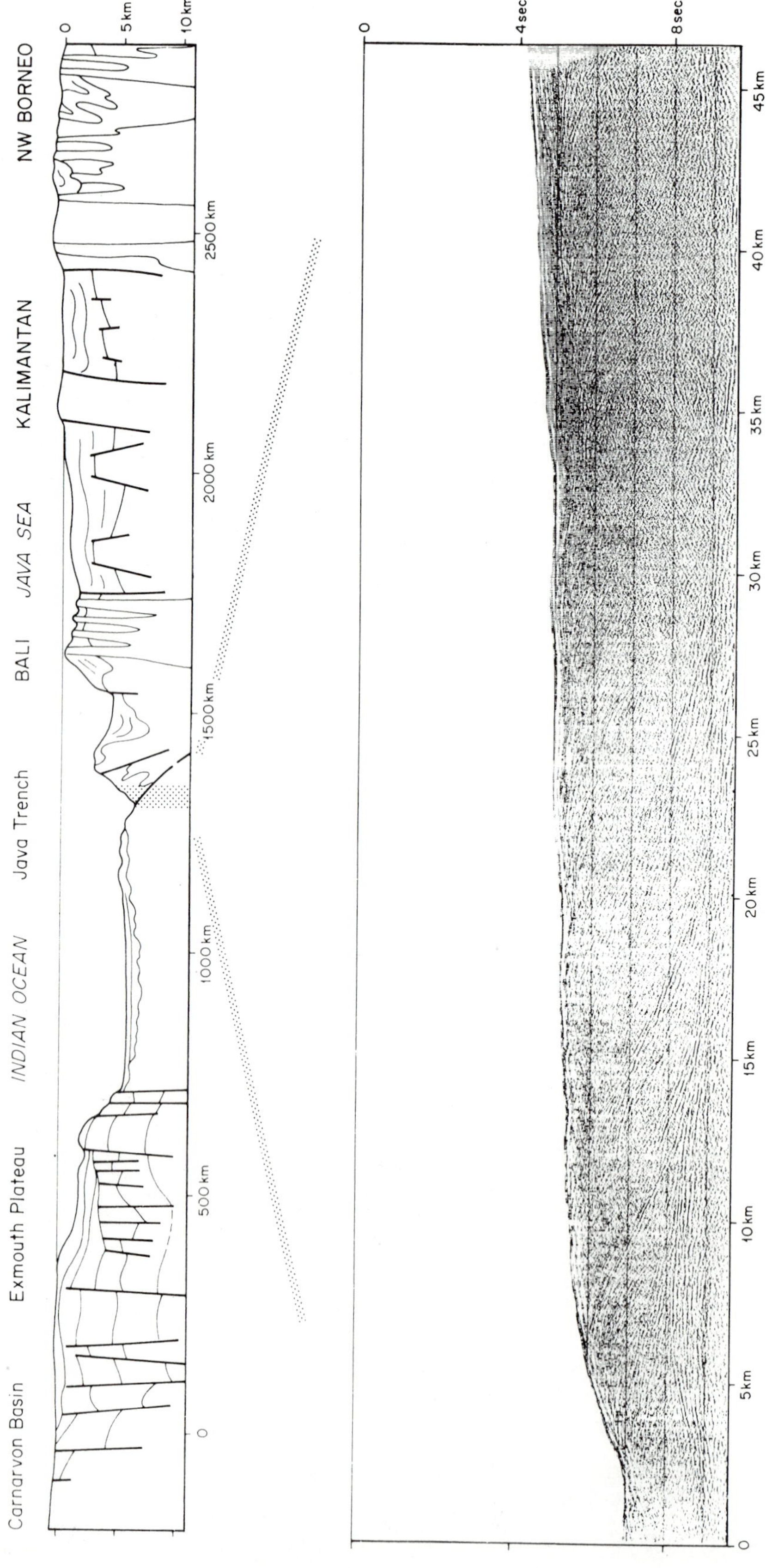

Fig. 9. Structure of the Java trench inner slope (Beck, 1972, courtesy of the author).

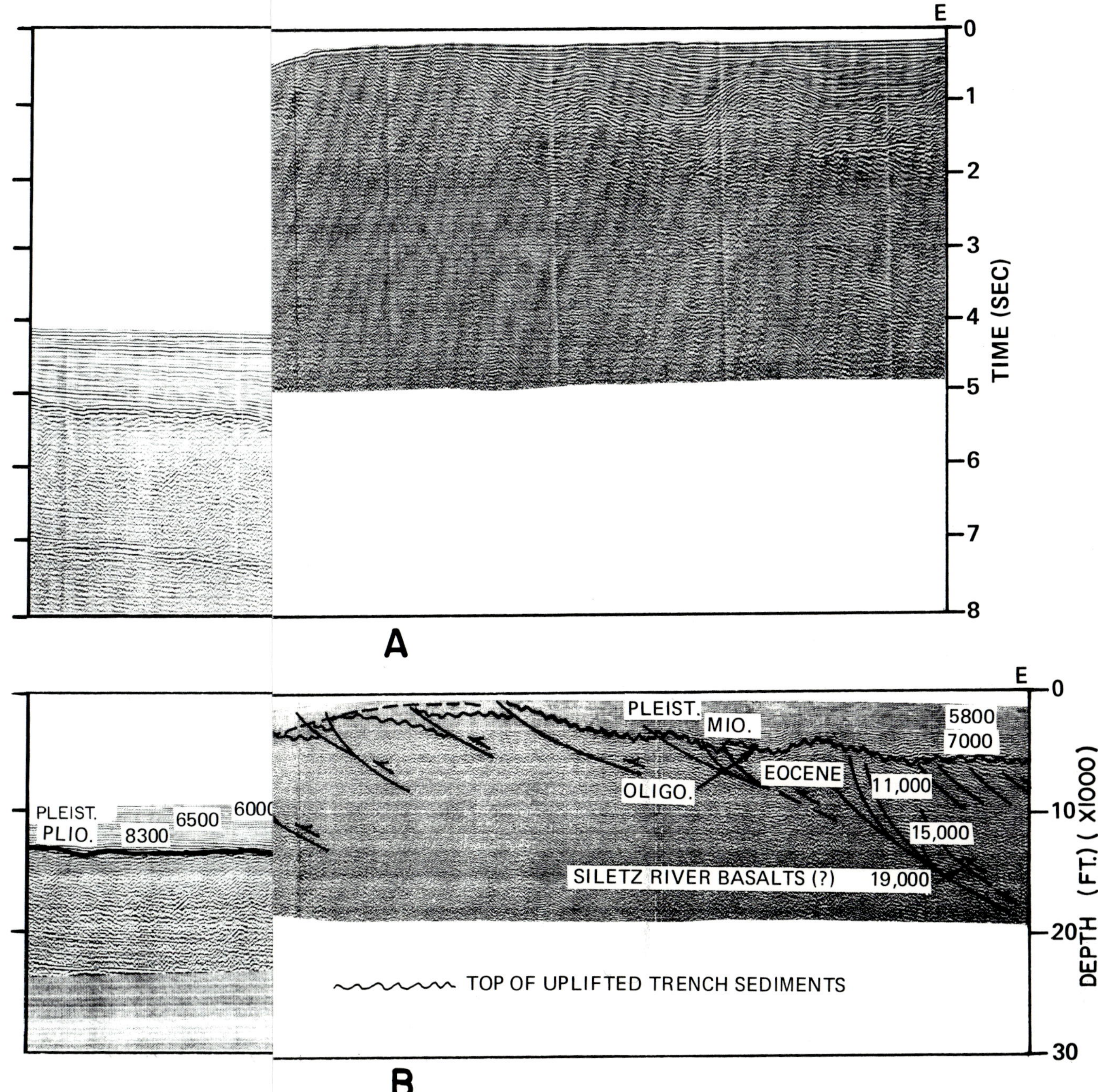

Fig. 8. Seismic time (A) and nor
are in feet per second. (Se

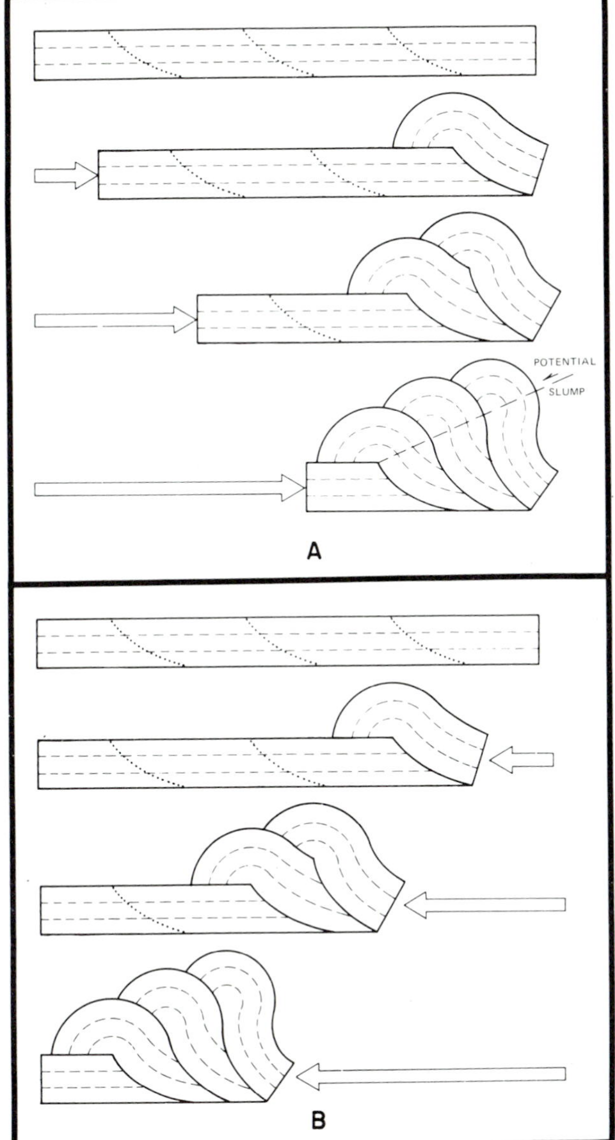

Fig. 10. Effect of wedging on thrust-fault dips and the development of fan structure. Figure 10A would be compared to wedging on a trench inner slope, and Figure 10B would be compared to wedging in an orogenic belt. Since motion is relative, these comparisons are interchangeable. The amount of rotation caused by wedging is related to the spacing between thrusts, the dips and configurations of the thrust surfaces, and the amounts of movement on those surfaces (up to a maximum, beyond which movement is not a factor; fault displacements on Figure 10 equal that maximum).

inson, 1973). Where this ratio is high, a well-developed sediment wedge accumulates, as it has in the Aleutian trench near JOIDES site 180. Where the ratio is low, little or no trench sediments are present, as illustrated by Figure 4.

Relatively rapid continental accretion should occur when the rates of both underthrusting and trench sedimentation are high. The trench inner slope then quickly builds seaward by the "sweeping up" of trench sediments through the thrust mechanism and by the outbuilding of shelf and upper slope sediments on top of the thrust "stack" (Fig. 11A). If, in the process of accretion, underthrusting is interrupted while a high depositional rate continued, thick sediments can accumulate in the trench and on the formerly active trench inner slope. Then, when underthrusting resumes, these sediments also become structured by thrusts and/or folds.

Accreted sediments constitute an additional load on the underthrusting oceanic plate. The load should tend to downbend that plate (Fig. 11B) and perhaps cause a seaward migration of the trench outer slope (such migration of ancient trenches has been noted by Burk, 1972). This effect, if it exists, seems most likely to be superposed on primary downbending of the oceanic plate caused by mantle processes that are thought to be responsible for subduction. It is, however, one way to explain how the trench persists in spite of seaward accretion of continents across its former sites.

It is also possible that the entire zone between the trench outer slope and uplifted plutonic roots of the volcanic arc subsides at depth as a result of seaward migration of the downbending oceanic plate (Fig. 11B). The subsidence would be interrupted, however, when the subduction rate is so slow that isostatic rebound occurs (possibly the current condition of the Oregon coast). During periods of active underthrusting, the uplift of sediments beneath trench inner slopes and outer highs would then be the net result of greater uplift due to detached folds and thrusts and lesser subsidence of the underthrust plate.

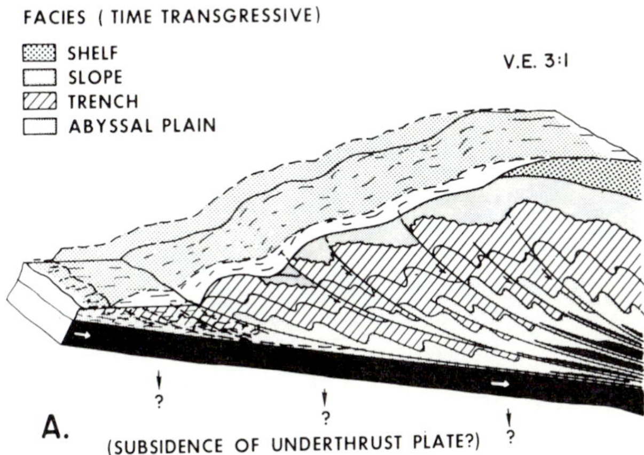

Fig. 11A. Model of trench inner slope at later evolutionary stage.

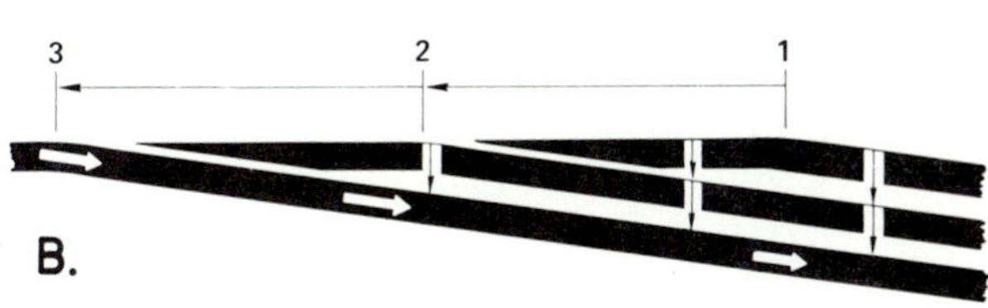

Fig. 11B. Subsidence caused by seaward migration of oceanic plate downbend: 1, 2, and 3 are successive stages of migration.

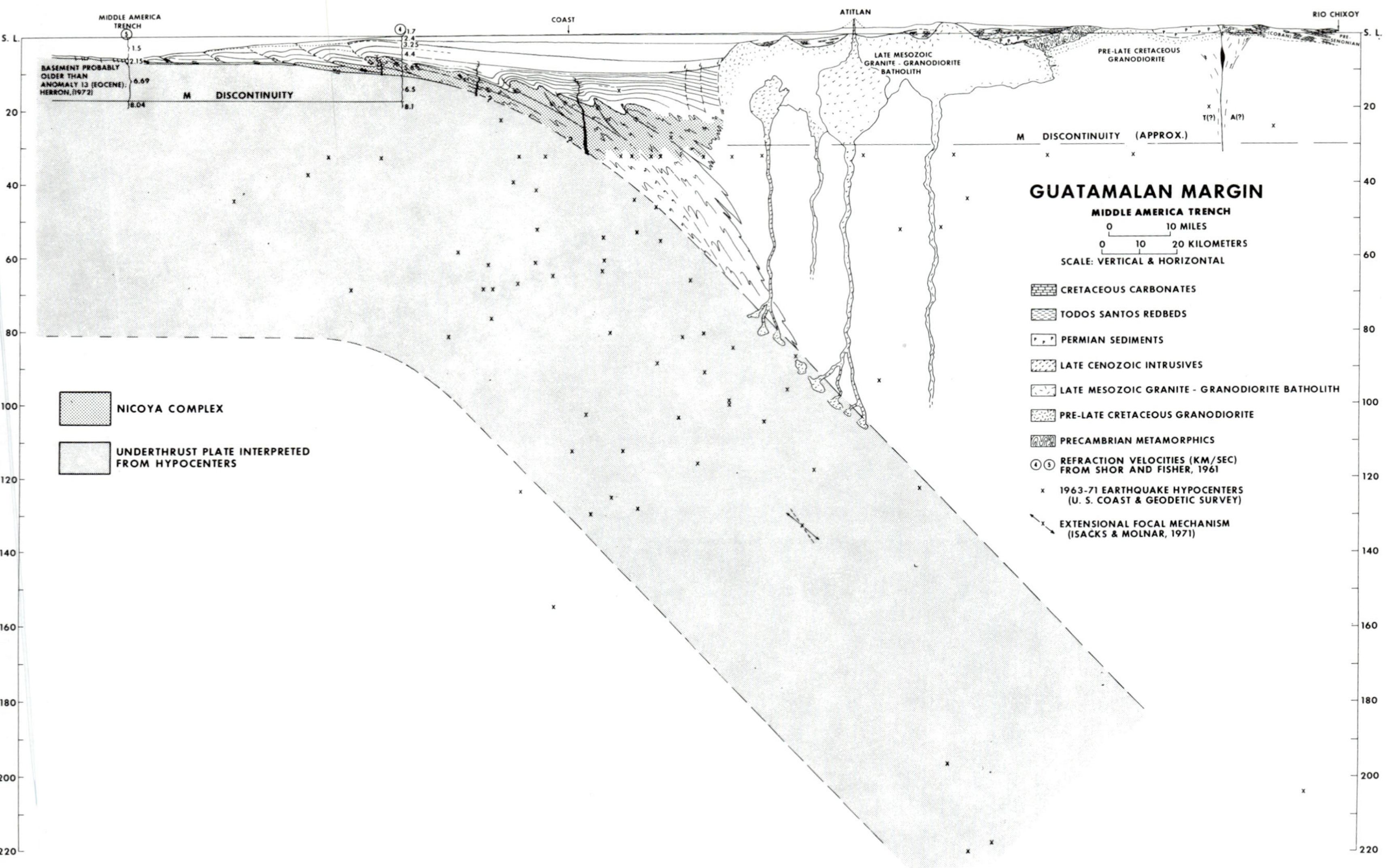

Fig. 12. Geologic cross section across the Guatemala margin of the Mid-America trench. This section is based on refraction-seismic and reflection-seismic, magnetic, earthquake, and surface geologic data.

TRENCH SLOPE MODEL AND THE ACCRETION PROCESS

Compressional structures formed on the inner slopes of ancestral trenches continue landward at depth beneath subsiding basins (Fig. 12). Where extensive accretion has occurred, these sediments and metasediments can compose the country rock into which the plutonic roots of volcanic arcs are intruded. The plutons are buoyant and cause uplift of the plutonic belt that destroys the inner borders of older subsiding basins while it forms new inner borders for younger subsiding basins. Uplift and erosion of this belt appear to occur during and soon after magmatism and can be so great that very little, if any, of the intruded trench sediments are preserved (Fig. 12).

As pointed out by several authors, the plutonism inferred to form the roots of present volcanic arcs appears to occur systematically relative to inferred depths to the underthrust oceanic plate in each arc (Dickinson, 1971; Nielson and Stoiber, 1973). Local intrusions, however, can also occur between the volcanic arc and trench (Fig. 12).

Grabens that form within the belt of uplifted plutons (e.g., Central America, Fig. 3) appear to be a result of contemporaneous volcanism. Cross trends that offset the magmatic and fore-arc areas are also common structural features of the uplifted belt (e.g. Stoiber and Carr, in press; Ranneft, 1972).

CONCLUSIONS

Compressional thrusting and folding beneath the trench inner slope is documented by (1) continuation of the oceanic plate beneath the slope, and (2) separation, structural thickening, and uplift of sediments overlying the plate. The compression cannot result from gravity sliding off the outer high because (1) the high is detached from autochthonous basement, (2) it is characterized by compressional rather than extensional deformation, and (3) vergence does not change across it. We conclude, therefore, that the compressional folding beneath trench inner slopes must be caused by underthrusting of the oceanic plate and that the outer high is but the most prominent structural expression of the underthrusting.

The underthrusting model of trench inner slopes explains the facies and time stratigraphy now being sampled on the slopes, the volumes of sediments in trenches, the rate of continental accretion, thrust-fault age relationships, and the development of fan structure. Persistence of the trench as a bathymetric feature in spite of accretion across its former sites is postulated to be due to loading effects.

Accretion occurs by outward building of the trench slope and by plutonic intrusion an uplift of the products of older trench slopes. The sediments of older basins of the fore arc are destroyed as a result of the uplift.

The emphasis in this paper has been on the structures developed by underthrusting of a relatively featureless sea floor beneath trench inner slopes during steady-state conditions. As time passes and conditions change, the processes active in this zone and their products adjust accordingly. The composite results are what we see today on trench margins. Studies of the evolutions of specific margins, the effects of collisions of sea-floor features, and further investigation of the concepts presented in this paper remain as subjects for future research.

BIBLIOGRAPHY

Beck, R. H., 1972, The oceans, the new frontier in exploration: Austral. Petr. Expl. Assoc., v. 12, pt. 2, p. 5-28.

Burk, C. A., 1972, Uplifted eugeosynclines at continental margins: Geol. Soc. America Mem. 132, p. 75-85.

Bush, S. A., and Bush, P. A., 1969, Isostatic gravity map of eastern Caribbean region: Trans. Gulf Coast Assoc. Geol. Socs., v. 19, p. 281-285.

Chase,, R. L., and Bunce, E. T., 1969, Underthrusting of the eastern margin of the Antilles by the floor of the western North Atlantic Ocean, and origin of the Barbados ridge: Jour. Geophys. Res., v. 74, p. 1413-1420.

Dengo, G., 1962, Tectonic-igneous sequence in Costa Rica, *in* Engel, A. E. J., James, H. L., and Leonard, B. F., eds., Petrologic studies: a volume in honor of A. F. Buddington: Geol. Soc. America, p. 133-161.

Dickinson, W. R., 1971, Clastic sedimentary sequences deposited in shelf, slope, and trough settings between magmatic arcs and associated trenches: Pacific Geology, v. 3, p. 15-30.

———, 1973, Widths of modern arc-trench gaps proportional to past duration of igneous activity in associated magmatic arcs: Jour. Geophys. Res., v. 78, p. 3376-3389.

Fitch, T. J., and Scholz, C. H., 1971, Mechanism of underthrusting in southwest Japan; a model of convergent plate interactions: Jour. Geophys. Res., v. 76, p. 7260-7292.

Grow, J. A., 1973, Crust and upper mantle of the central Aleutian arc: Geol. Soc. America Bull., v. 84, p. 2169-2192.

Herron, E. M., 1972, Sea-floor spreading and the Cenozoic history of the east-central Pacific: Geol. Soc. America Bull., v. 83, p. 1671-1692.

Hsü, K. J., 1971, Franciscan mélanges as a model for geosynclinal sedimentation and underthrusting tectonics: Jour Geophys. Res., v. 76, p. 1162-1170.

Isacks, B., and Molnar, P., 1971, Distribution of stresses in descending lithosphere from a global survey of focal-mechanism solutions of mantle earthquakes: Rev. Geophys. and Space Phys., v. 9, p. 103-174.

JOIDES Leg 31 Staff, 1973, Western Pacific floor: Geotimes, v. 18, no. 10, p. 22-25.

Karig, D. E., 1970, Ridges and basins of the Tonga-Kermadec island arc system: Jour. Geophys. Res., v. 75, p. 239-254.

King, P. B., et al., 1969, Tectonic map of North America: Washington, D.C., U.S. Geol. Survey.

Kulm, L. D., et al., 1973a, Initial reports of the Deep Sea Drilling Project, v. 18: Washington, D.C., U.S. Govt. Printing Office, p. 449-500.

———, et al., 1973b, Tholeiitic basalt ridge in the Peru trench: Geology, v. 1, p. 11-14.

Lee, H. J., Olsen, H. W., and von Huene, R., 1973, Physical properties of deformed sediments from site 181, *in* Kulm, L. D., et al., 1973, Initial reports of the Deep Sea Drilling Project, v. 18: Washington, D.C., U.S. Govt. Printing Office, p. 897-901.

Ludwig, W. J., et al., 1966, Sediments and structure of the Japan trench: Jour. Geophys. Res., v. 71, p. 2121-2137.

Marlow, M. S., Scholl, D. W., and Garrison, L. E., 1972, Evidence (?) of underthrusting from seismic reflection profiles across the Puerto Rico and Aleutian trenches (abstr.): Am. Geophys. Union Trans., v. 53, p. 1114.

Mitchell, A. H., and Reading, H. G., 1971, Evolution of island arcs: Jour. Geology, v. 79, p. 253-284.

Naugler, F. P., and Wageman, J. M., 1973, Gulf of Alaska: magnetic anomalies, fracture zones, and plate interaction: Geol. Soc. America Bull., v. 84, p. 1575-1584.

Nielson, D. R., and Stoiber, R. E., 1973, Relationship of potassium content in andesitic lavas and depth to the seismic zone: Jour. Geophys. Res. v. 78, p. 6887-6892.

Page, B. M., 1972, Oceanic crust and mantle fragment in subduction complex near San Luis Obispo, Calif.: Geol. Soc. America Bull., v. 83, p. 957-972.

Piper, D. J. W., von Huene, R., and Duncan, J. R., 1973, Late Quaternary sedimentation in the active eastern Aleutian trench: Geology, v. 1, p. 19-26.

Plafker, G., 1969, Tectonics of the March 27, 1964, Alaska earthquake: U.S. Geol. Survey Prof. Paper 543-I.

———, and Savage, J. C., 1970, Mechanism of the Chilean earthquakes of May 21 and 22, 1960: Geol. Soc. America Bull., v. 81, p. 1001-1030.

Ranneft, T. S. M., 1972, The effects of continental drift on the petroleum geology of western Indonesia: Austral. Petr. Expl. Assoc., v. 12, pt. 2, p. 55-61.

Seyfert, C. K., 1969, Undeformed sediments in oceanic trenches with sea floor spreading: Nature, v. 222, p. 70.

Shor, G. G., Jr., and Fisher, R. L., 1961, Middle America trench; seismic refraction studies: Geol Soc. America Bull., v. 72, p. 721-730.

Shouldice, D. H., 1971, Geology of the western Canadian continental shelf: Bull. Can. Petr. Geol., v. 19, p. 405-436.

Stoiber, R. E., and Carr, M. J., in press, Quaternary volcanic and tectonic segmentation of Central America: Bol. Volcanologique.

Tanner, W. F., 1973, Deep-sea trenches and the compression assumption: Am. Assoc. Petr. Geol. Bull., v. 57, p. 2195-2206.

van Bemmelen, R. A., 1949, Geology of Indonesia, v. 1A, General geology: The Hague, Govt. Printing Office, 732 p.

Vening Meinesz, F. A., 1954, Indonesian Archipelago: a geophysical study: Geol. Soc. America Bull., v. 65, p. 143-164.

von Huene, R., 1972, Structure of the continental margin and tectonism at the eastern Aleutian trench: Geol. Soc. America Bull., v. 83, p. 3613-3626.

———, and Kulm, L. D., 1973, Tectonic summary of Leg 18, *in* Kulm, L. D., et al., 1973, Initial reports of the Deep Sea Drilling Project, v. 18: Washington, D.C., U.S. Govt. Printing Office, p. 961-976.

———, et al., 1967, Geologic structures in the aftershock region of the 1964 Alaskan earthquake: Jour. Geophys. Res., v. 72, p. 3649-3660.

Oregon Continental Margin Structure and Stratigraphy: A Test of the Imbricate Thrust Model

L. D. Kulm and G. A. Fowler

INTRODUCTION

The Cenozoic structural and stratigraphic framework of the Oregon continental margin records several intervals of significant tectonism (uplift) with subsequent erosion and truncation of older structures. During late Cenozoic time this framework displayed many of the characteristics of a fore-arc structure defined by Karig (1970). It is also characterized by a substantial amount of continental accretion, interpreted to be the result of underthrusting of the oceanic plate. The compressional thrust model (Seely, et al., this volume; Burk, 1968) offers the best explanation for the late Cenozoic evolution of the Oregon margin.

Several lines of evidence indicate that the Pleistocene abyssal plain and fan deposits of the Cascadia Basin are being thrust beneath the earlier Cenozoic rocks that underlie the continental shelf. These abyssal deposits have been uplifted more than 1 km and incorporated into the lower and middle continental slope, where they are either exposed or covered by late Quaternary deposits. The stratigraphic position of these abyssal deposits in the continental slope and their age relationships strongly suggest imbricate thrusting of thick slices of sand turbidites typical of submarine fans, which alternate with silt turbidites characteristic of abyssal plains.

These underthrusting younger sediments are interpreted here as elevating the older deposits higher on the continental margin, the latter units attaining a progressively steeper dip in a landward direction. Consequently, Miocene to Pleistocene mudstone and sandstone of much shallower-water facies crop out on the outer continental shelf, forming the prominent submarine banks. These deposits have been uplifted largely in the late Miocene and early Pliocene. Contemporaneous *subsidence* of the inner continental shelf produced a broad, shallow syncline which filled with shallow-water Pliocene and Pleistocene sediments.

Imbricate thrusting of unconsolidated abyssal deposits is interpreted to produce rapid uplift of the lower continental slope at an average rate of 1,000 m/my. As continental accretion continues, the deposits are compacted and the rate of uplift in the older deposits on the outer shelf is reduced to only 100 m/my.

DATA COLLECTION

The structural and stratigraphic framework of the Oregon continental margin and the Cascadia Basin (Fig. 1) have been studied extensively since 1965 by the School of Oceanography, Oregon State University (e.g., Byrne et al., 1966; Fowler, 1966; Emilia et al., 1968; Dehlinger et al., 1968, 1971; Duncan, 1968; Mackay, 1969; Muehlberg, 1971; Fowler and Kulm, 1971; Kulm et al., 1973a, 1973b; Phipps, 1974). When the data presented from these previous studies are combined with the data described in this paper, we have sufficient information to test most models that offer a plausible explanation for the apparent evolution of the Oregon continental margin.

According to Atwater (1970), this area has been a subduction zone since at least early Cenozoic time. The early to middle Cenozoic record is obscured in most places by very late Cenozoic sediments. However, the tectonically most-active areas of the margin contain exposures with sufficient information to unravel a sequence of late Cenozoic events that have occurred in the uppermost part of the subduction zone.

Invaluable time-stratigraphic information allows resolution of tectonic events to within a few hundred thousand years from three Deep Sea Drilling Project sites (174, 175, 176; Fig. 1) positioned in the Cascadia Basin, the lower continental slope, and the outer continental shelf off Oregon. The imbricate thrust model presented by Seely, Vail, and Walton (this volume) at the Penrose Conference on Continental Margins in December 1972 provides an attractive solution to the structure observed in the fore-arc region that lies between the volcanic Olympia region and the filled trench.

The structure of the Oregon margin was determined with the aid of seismic reflection profiles (Fig. 2) taken with an air gun or sparker. Most of the rock samples were collected with a dart corer (Fig. 3); a few samples were obtained by dredging. The important stratigraphic units were located with the reflection profiling system, and buoys were dropped over appropriate structures to be sampled later. Navigation was by Loran A and radar.

MORPHOLOGY

The morphology of the Oregon continental terrace (shelf and slope) has been described by Byrne (1962, 1963a, 1963b), Maloney (1965), Carlson (1968), and Spigai (1971). The combined width of the terrace ranges from 75 km off Cape Blanco to 135 km off the Columbia River (Fig. 4). The shelf varies in width from about 17-74 km, has an inclination of 0°08'-0°43', and has a depth at its outer edge of

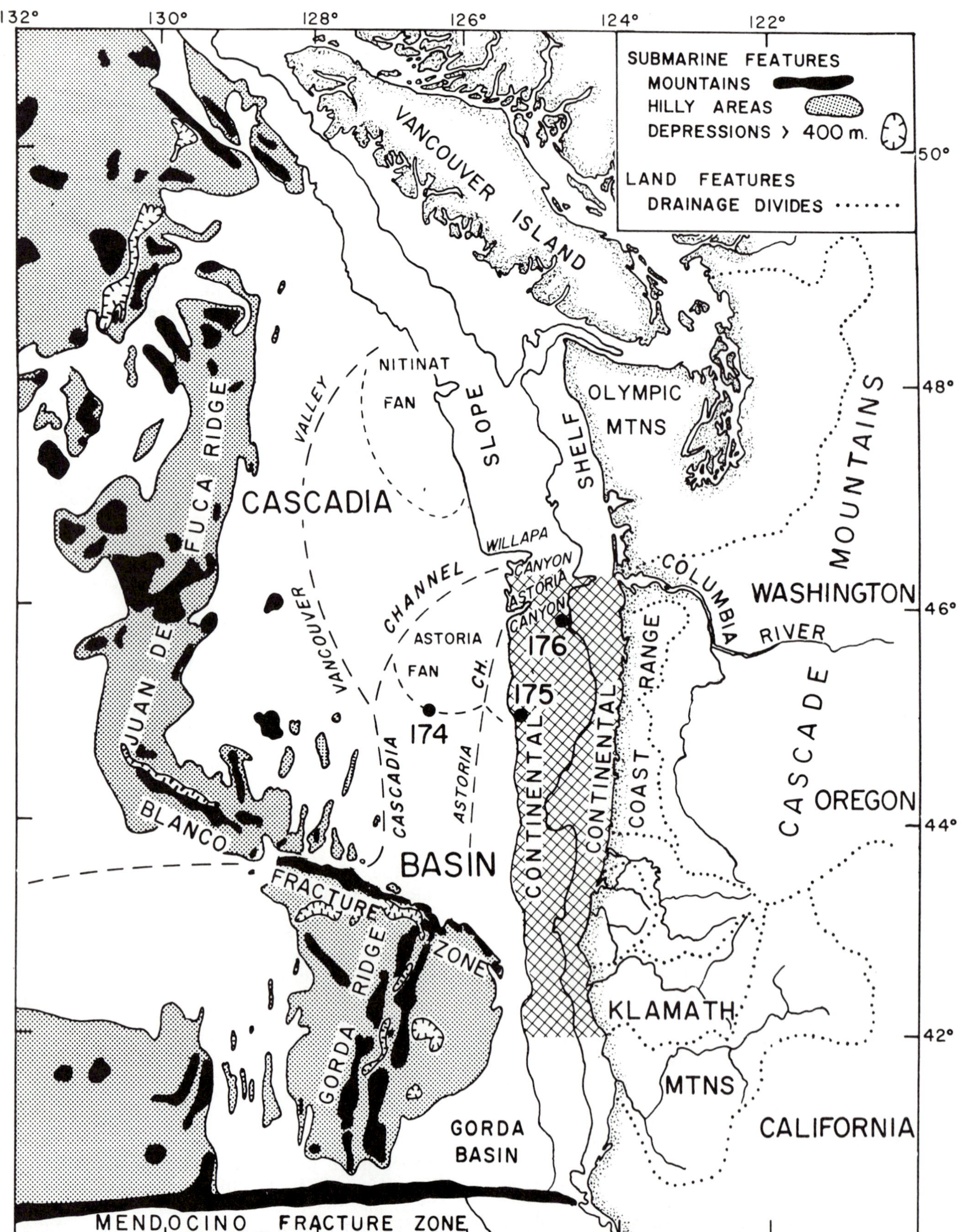

Fig. 1. Location map of the Oregon continental margin and surrounding features. Note location of DSDP sites 174, 175, and 176.

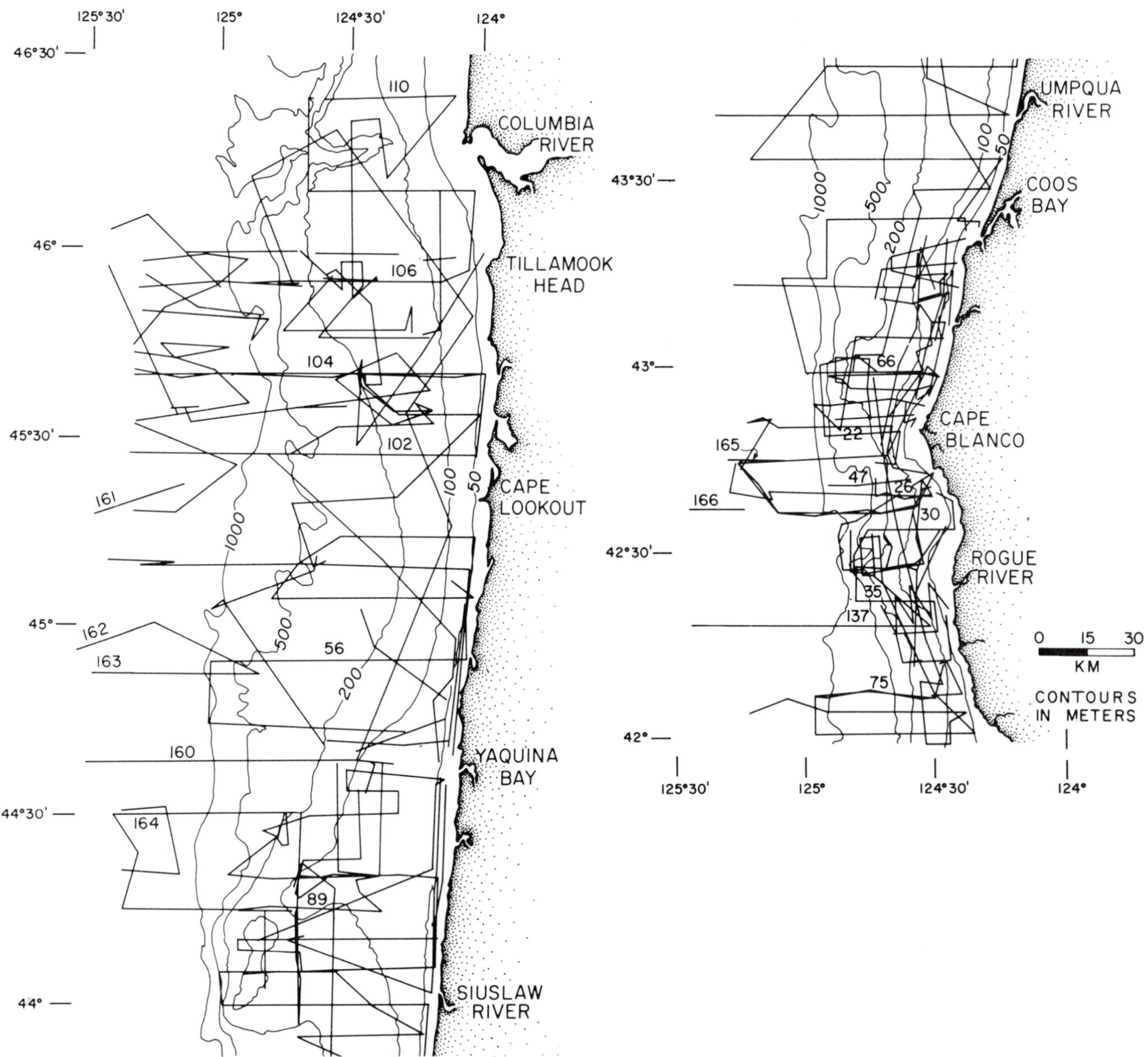

Fig. 2. Seismic reflection profile tracklines across the Oregon margin. (Illustrated profiles are numbered.)

145-183 m. These dimensions are narrower, steeper, and deeper than the average values given for the continental shelves of the world (Shepard, 1963). Prominent submarine banks (i.e., Nehalem, Stonewall, Heceta, Coquille) of exposed bedrock occur near the outer edge of the shelf and have as much as 75 m of relief (Figs. 3 and 4). Bedrock also crops out on the inner shelf, especially between Coos Bay and the Rogue River.

The upper continental slope (180-900 m) is characterized by benches and low relief hills (Fig. 4). The largest bench (Cascade Bench) occurs between Cascade Head and Tillamook Bay and is 400-600 m deep. Another prominent bench lies between 500 and 700 m and is located between Cape Sebastian and the California border (Spigai, 1971). It is a continuation of the Klamath Plateau found on the upper continental slope off northern California (Silver, 1971). Off central Oregon (Heceta Bank) the upper slope forms a steep escarpment. Astoria and Rogue River canyons and numerous small submarine valleys cross the outer shelf and upper slope. They are the most important avenues of sediment dispersal by turbidity currents originating on the continental terrace (Nelson, 1968; Carlson, 1968; Spigai, 1971).

Narrow benches, small hills, and valleys are typical of the lower continental slope off northern Oregon. A steep escarpment replaces these features off central Oregon. The most striking topographic features of the lower continental slope are the prominent N-NW-trending ridges and intervening troughs extending from off northern Oregon northward to the Washington margin. The ridges are 10-75 km long and have as much as 1,150 m of relief.

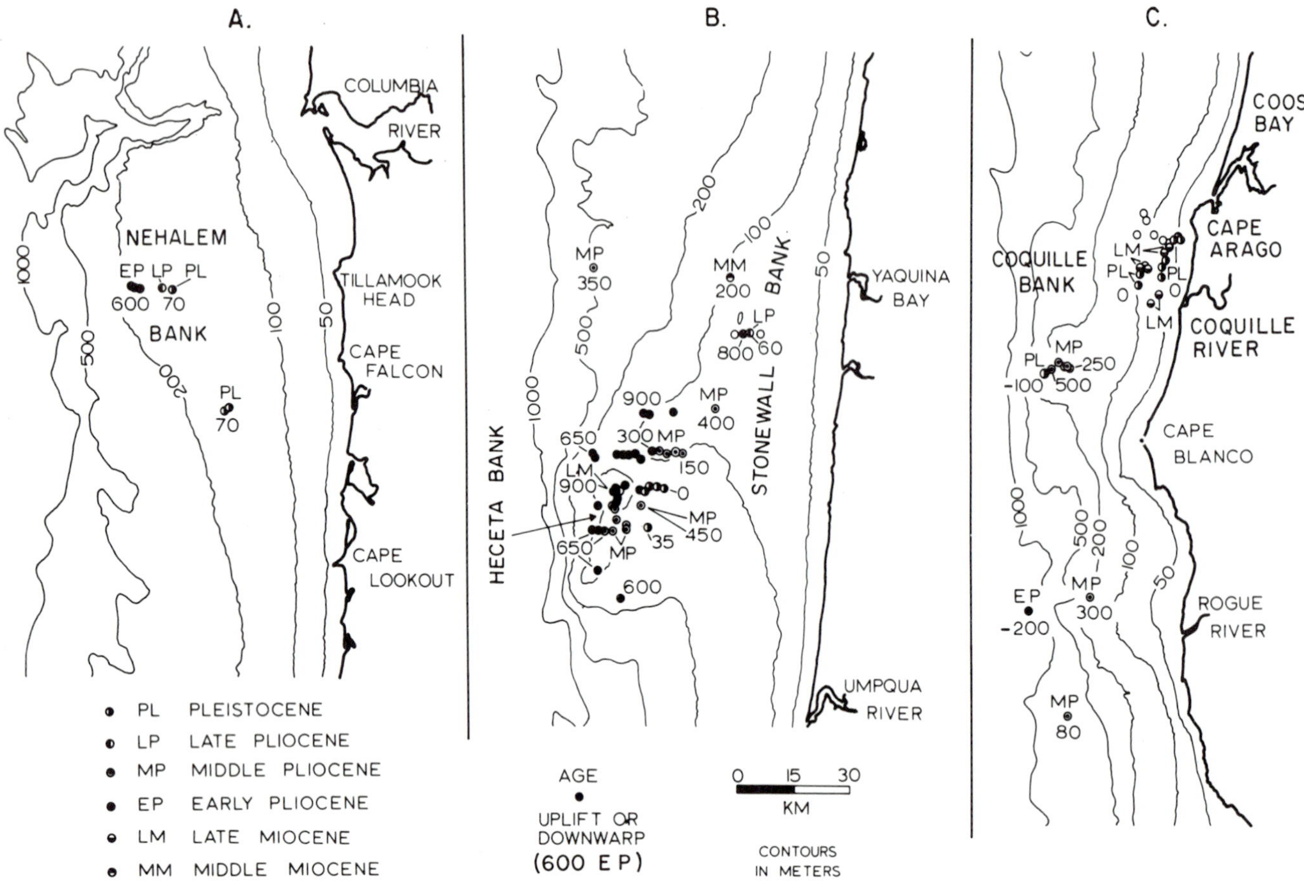

Fig. 3. Rock sample locations, ages, and relative amounts of tectonic movement in meters. Negative values indicate downwarp.

Astoria Canyon meanders through this topography to connect with the Astoria Fan (Figs. 3 and 4). The base of the continental slope is linear in plan view; it occurs at a depth of 1,800 m on Astoria Fan, but it extends to a depth of 3,100 m off southern Oregon. With the exception of the fan, the slope and abyssal plain meet at a rather large angle with no intervening continental rise.

The irregular width of the shelf, submarine banks, abundant benches, and linear ridges on the slope suggests strong structural control of the morphology and general configuration of the terrace.

CONTINENTAL SHELF STRUCTURE

General Framework

A generalized geologic map (Fig. 5) was constructed for the Oregon continental shelf with the aid of more than 8,000 km of seismic reflection profiles (Fig. 2) and more than 100 samples (Fig. 3). Regional correlation was aided by continuous tracing of the acoustic units laterally from control points with heavy reliance upon the ubiquitous angular unconformities for stratigraphic control.

Some of the most striking folded and faulted areas of the Oregon margin occur beneath submarine banks near the outer edge of the shelf (Figs. 6 and 7). The principal ones are Nehalem Bank (SP-106), Heceta-Stonewall Bank complex (SP-89), and Coquille Bank (SP-66). Pleistocene to pre and Coquille Bank (SP-66). Pleistocene to pre-late Miocene rocks are exposed on these banks (Fig. 5); the oldest known exposed sediment occurs in the vicinity of Heceta Bank on the upthrown side of faults.

Heceta and Coquille banks are characterized by positive free-air gravity anomalies, which suggests a slight mass excess beneath them. In contrast, the area around Nehalem Bank (about 46°N) shows a negative anomaly and implies a fairly thick, low-density sedimentary section (Dehlinger et al., 1968).

A series of interconnecting shallow synclines occurs between the shore and the outer banks off southern Washington and northern and central Oregon. The largest of these synclines is basinal and is centered off Cascade Head (Figs. 5 and 6). Along the seaward limb, sediments in the syncline have ponded behind and overlapped a series of subsurface folds (SP-56, 66). The most recent development of these synclines involves Pleistocene sediments that very nearly everywhere overlie older units unconformably. However, the synclines are generally best developed offshore from thick late Paleogene and early Neogene sedimentary sections on

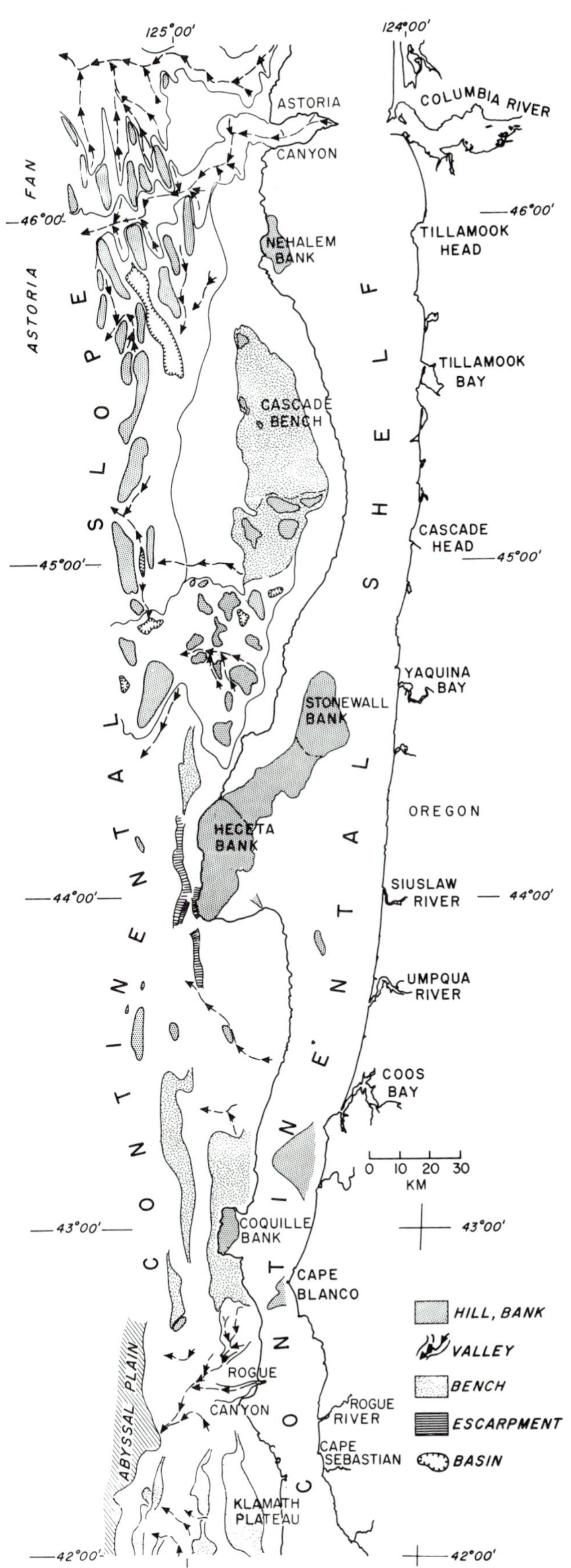

Fig. 4. Morphology of the Oregon continental terrace.

land that trend offshore.

Off Washington and north and central Oregon large negative free-air anomalies coincide with the inner shelf synclines (Fig. 8). From reflection profiles and gravity data, it seems clear that much thicker and older sedimentary basins underlie the shallow Pleistocene basins on the inner shelf such as off Cascade Head. Braislin et al. (1971, Fig. 2) show a known sedimentary section of 4 km and postulate a section in excess of 5 km thick over this area (Fig. 9B). An acoustic basement dips seaward beneath the inner part of the continental shelf along profile SP-106 (Fig. 6). It is marked by large positive magnetic anomalies (Emilia et al., 1968, Fig. 4) and is most likely the seaward extension of the Miocene volcanic rocks found onshore (Snavely et al., 1973). The possible importance of buried ridges beneath continental margins has been emphasized earlier (e.g., Burk, 1968).

Between Coos Bay and Cape Sebastian, the inner shelf is underlain by intensely folded and faulted sedimentary units (Fig. 5) (Mackay, 1969; Bales and Kulm, 1969). Erosion has exposed Mesozoic and younger strata. Most of the older structures have been truncated (e.g., Fig. 7, SP-35) and covered by a thin veneer of Pleistocene sediments. From the Rogue River to the California border, the shelf is underlain by a shallow basin consisting of undeformed sediments that unconformably overlie older units (Fig. 7, SP-75). This basin apparently trends perpendicular to the shoreline and is characterized by a large negative free-air gravity anomaly (Fig. 8).

Although the character of the outer shelf is in large part structurally controlled, it also has been shaped by considerable erosion and sediment progradation (Figs. 6 and 7). A Pleistocene sediment wedge unconformity overlies older truncated strata, which now occur at water depths as deep as 300 m (Fig. 7, SP-47). Subaerial erosion during the several Pleistocene lowerings of sea level has truncated the folded structures over much of the shelf (e.g., Figs. 6 and 7, SP-106, 89, 35). This suggests in turn that either the earlier sea-level lowerings were substantially lower than the negative -125 m indicated for the late Wisconsin regression (Curray, 1965), or there has been subsidence during the Pleistocene along parts of the outer shelf, as suggested by Kulm et al. (1973b) from deep-sea drilling.

Submarine Banks

Folds which developed in Pliocene siltstone and claystone that underlie Nehalem Bank are eroded and overlapped by younger sediments (Figs. 5 and 6, SP-106). Strong linear trends in bathymetry and faunal stratigraphy over Nehalem Bank indicate fault control. Deep-sea drilling on the seaward flank of Nehalem Bank (DSDP site 176) penetrated a sequence of Pleistocene greenish-gray clayey silts with abundant coarse debris and shallow-water megafossils, and bottomed in a Pliocene olive-gray fissile shale (Kulm et al., 1973b). The reflection

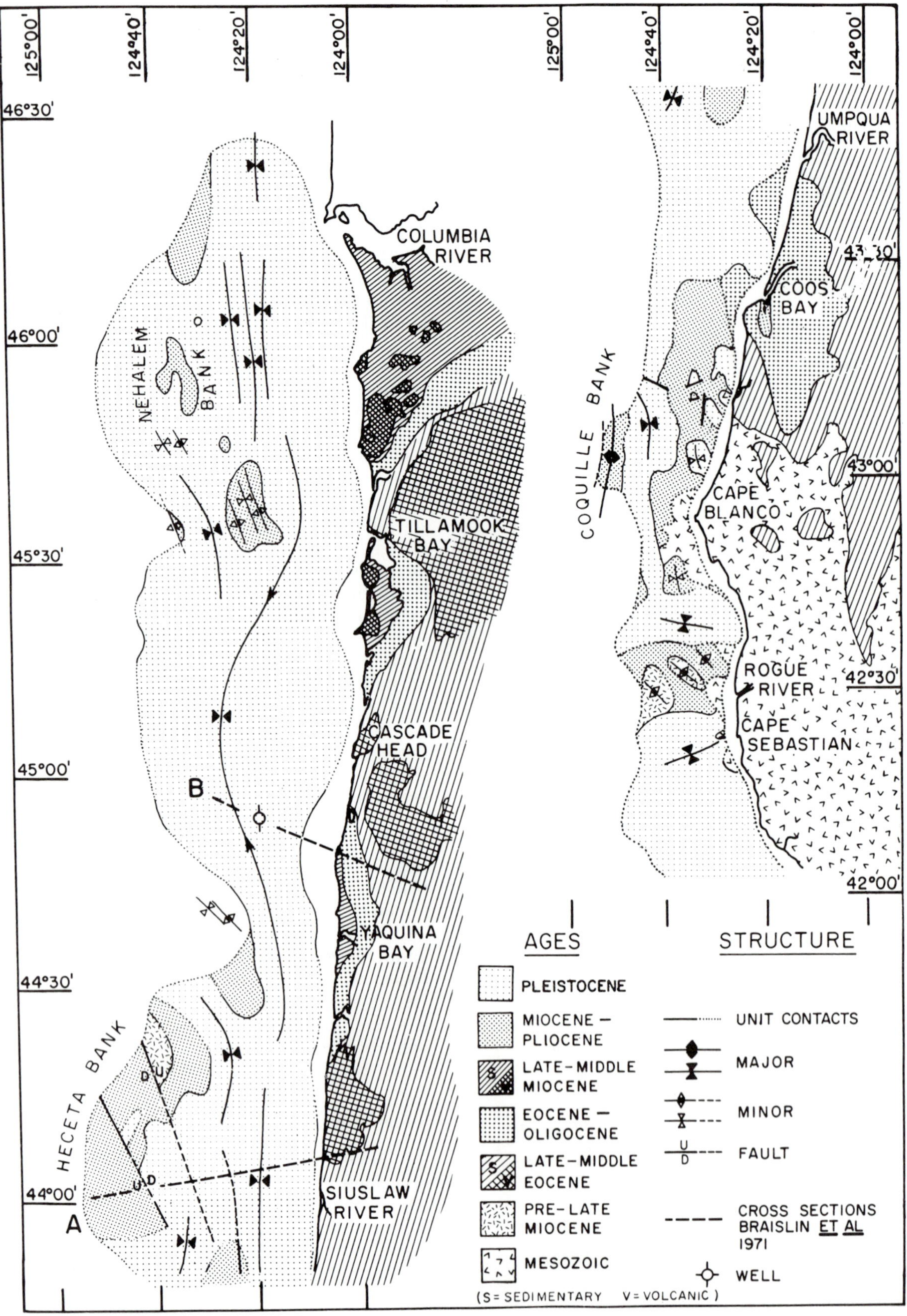

Fig. 5. Geologic map of the Oregon continental shelf and immediate onshore region.

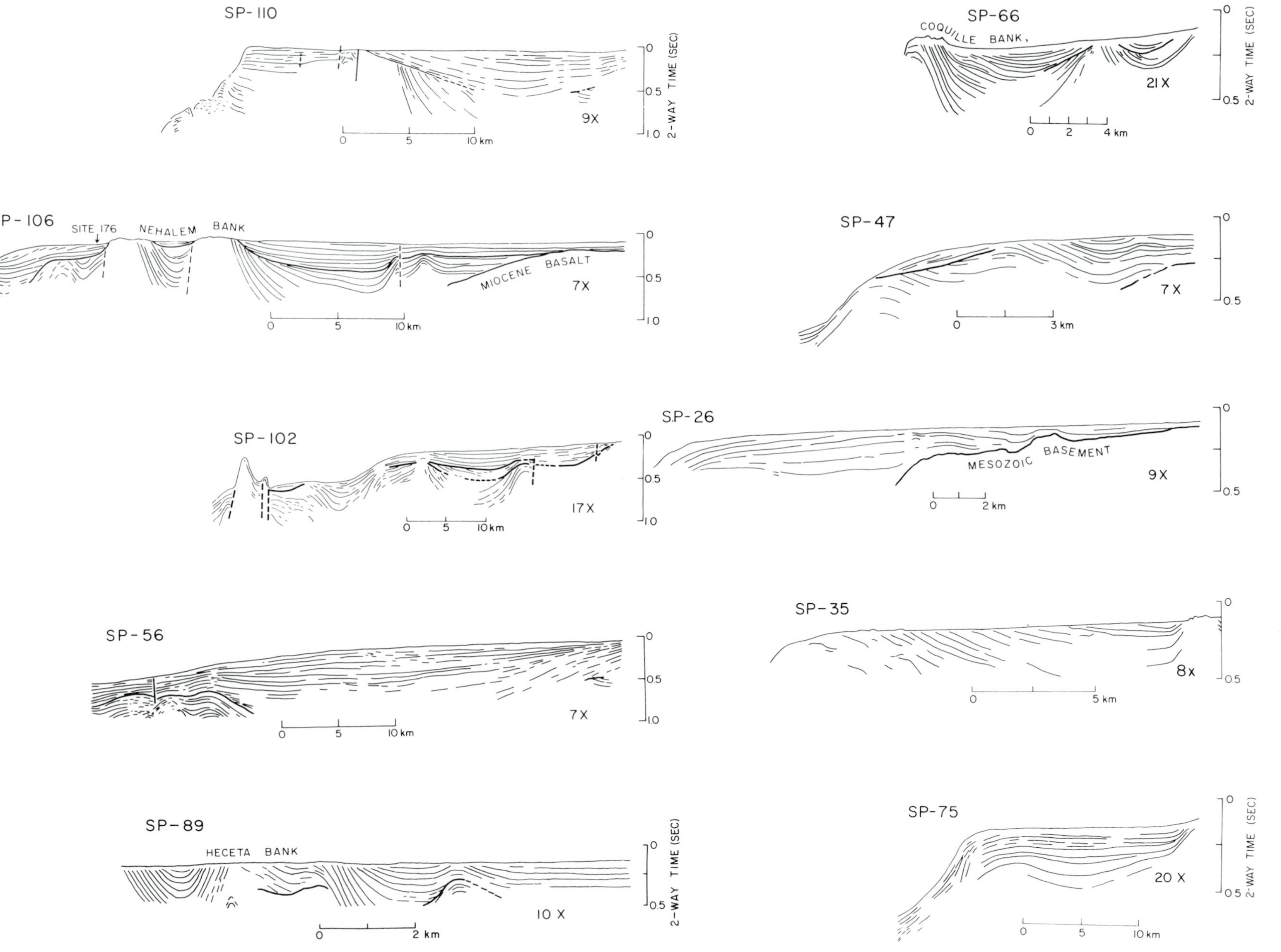

Fig. 6. Seismic reflection profiles of the northern Oregon shelf. (See Fig. 2 for location.)

Fig. 7. Seismic reflection profiles of the central and southern Oregon shelf. (See Fig. 2 for location.)

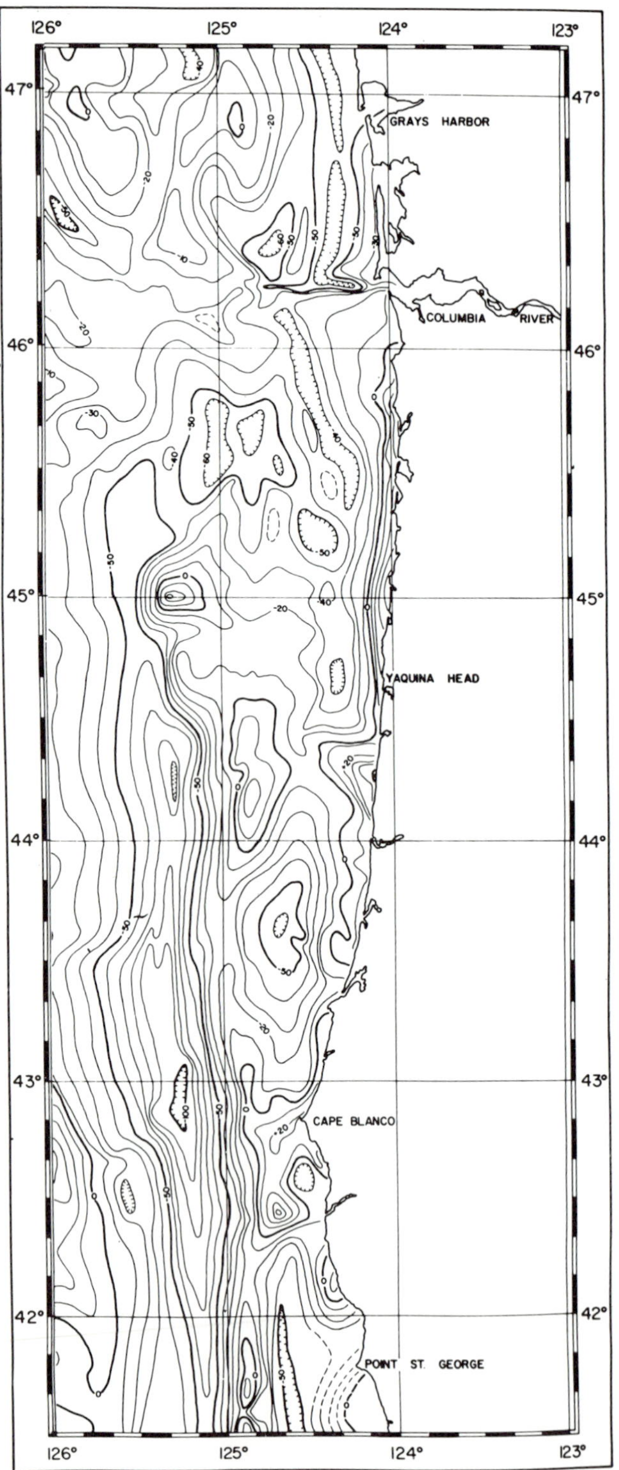

Fig. 8. Free-air gravity map (modified from Dehlinger et al., 1968, Fig. 2) and magnetic source depths (Emilia et al., 1968, Fig. 7b) for the Oregon shelf and slope. See Figure 4 for location of topographic features.

record (Fig. 6) and faunal data indicate an angular unconformity between these units. The Pliocene shale was deposited originally on the upper continental slope and has been uplifted at least 500 m. It was truncated during a former lowering of sea level with subsequent subsidence of more than 100 m (Kulm et al., 1973b).

The structure and stratigraphy of Heceta and Stonewall banks have been studied in some detail (Maloney, 1965; Fowler, 1966; Byrne et al., 1966; Fowler and Muehlberg, 1969; Snavely et al., 1968; Braislin et al., 1971; Muehlberg, 1971). Heceta Bank is structurally complex; it is difficult to trace acoustic units from one reflection profile to the next, and there are numerous areas of little or no seismic return, which probably indicates steeply dipping strata and fault zones. The older strata are more intensely deformed than are the younger strata. The topography along the western flank of Heceta Bank may be fault controlled.

The pre-late Miocene rocks indicated on the geologic map at the north end of Heceta Bank have not been sampled; the age is suggested by their stratigraphic position, noted in the reflection profiles. Diatom- and radiolarian-rich rocks of late Miocene age (Fowler et al., 1971) occur in the lowest parts of the stratigraphic unit that unconformably overlies pre-late Miocene rocks. The late Miocene rocks are similar to the siliceous mudstones and siltstones reported in the Pullen Formation of the Wildcat Group in northern California (Haller, 1967).

Coquille Bank is underlain by a N-S-trending, asymmetrical, doubly plunging anticline (Mackay, 1969). A thick section of Pliocene and Pleistocene siltstone and claystone forms its eastern limb (Figs. 3, 5, and 7). This thick section is absent on the western limb and a normal fault is inferred along the straight seaward edge of the bank. Pleistocene sandstone cored on the seaward side of the bank (Fig. 3) has a paleodepth shallower by 100 m than the present water depth, which suggests downfaulting of the sandstone.

Inner Continental Shelf

The most structurally complex area of the inner continental shelf lies between Coos Bay and the Rogue River, where the pre-Quaternary rocks either crop out or are covered by only a thin veneer of Quaternary sediment (Fig. 5). Additional rock sample and seismic reflection profile coverage have allowed us to date some of the acoustic units mapped by Mackay (1969, Plate I) for this area. Pleistocene sediments occur in the basin lows or in downfaulted blocks (Mackay, 1969). These Pleistocene sediments have been stripped off the structural highs or become thin over them. Late Miocene diatomaceous sediments were recovered from the inner shelf off the Coquille River and are correlated with the Pullen Formation of the Humboldt Basin, northern California (Fowler et al., 1971).

Mesozoic Basement

An irregular acoustic basement is shown in profiles made over the innermost part of the continental shelf between Cape Blanco and Sisters Rocks (Fig. 2, SP-22, 47, 26, and 30). The basement

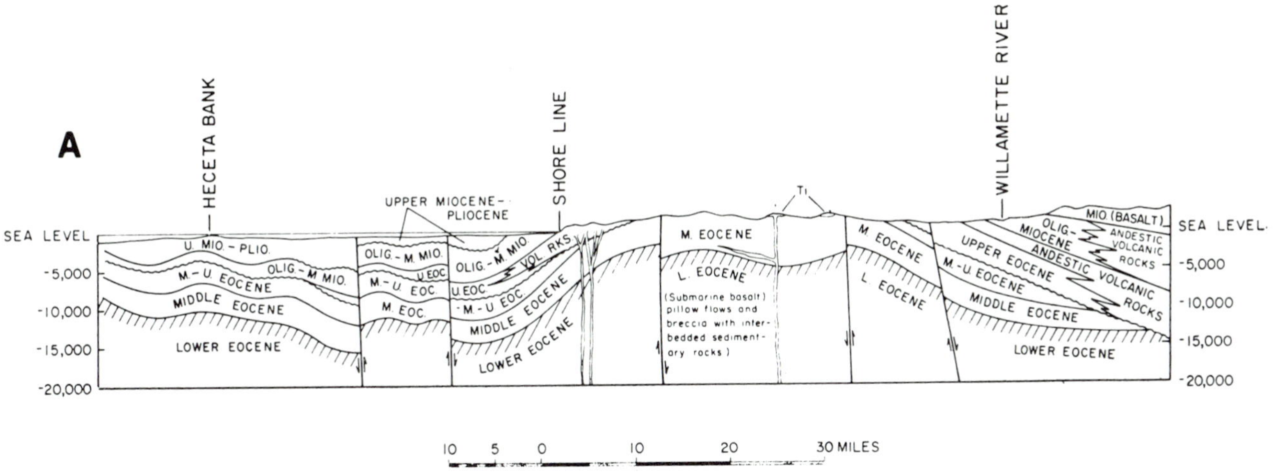

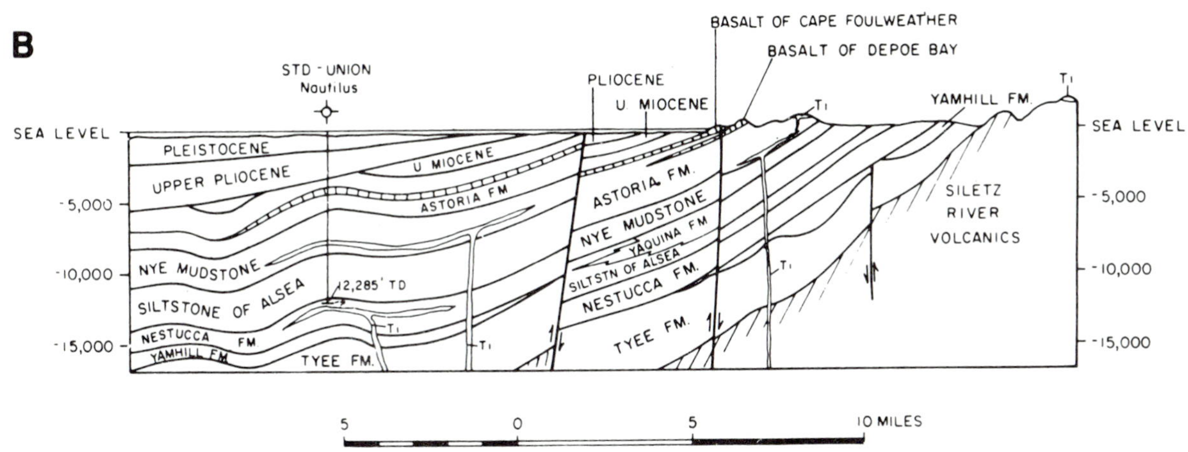

Fig. 9. A. Cross section from Willamette Valley to Heceta Bank (modified from Braislin et al., 1971, Fig. 3). See Figure 5 for location of section A. B. Cross section from central coast to continental shelf (modified from Braislin et al., 1971, Fig. 4). See Figure 5 for location of section B.

forms a platform which is shallowest near the shore and dips seaward away from Cape Blanco. In profile SP-26 (Fig. 7) the youngest deposits, and probably the acoustic basement, have been deformed. Probable upper Cretaceous sandstone and subordinate conglomerate occur in the sea stacks in the Blanco and Rogue River reefs (Hunter et al., 1970) (Fig. 5). These low-grade metamorphosed rocks may form the acoustic basement noted in the reflection records.

A 20-mgal positive free-air anomaly is associated with the Mesozoic rocks around Cape Blanco (Fig. 8). These rocks apparently have higher densities and are more resistant to erosion than the Tertiary sediments on the shelf, and define a narrow pre-Tertiary shelf or platform underlying the inner part of the shelf, which is adjacent to the Klamath Mountains. Uplift of Cape Blanco and vicinity (Janda, 1969) elevated the platform, and erosion during the Pleistocene stripped from it the Tertiary sediments.

SHELF UNCONFORMITIES

Most reflection profiles of the continental margin of this region display one or more unconformities (Figs. 6 and 7). They generally are marked by a degree of discordance, with the greatest amount noted in the vicinity of the submarine banks and near the present coastline.

At least two prominent angular unconformities, Pliocene-Pleistocene and middle to late Miocene, are widespread on the Oregon continental shelf (Fowler and Kulm, 1971; Kulm and Fowler, 1971). The youngest is present beneath most of the shelf and can be traced with fair confidence between adjacent reflection profiles. Strata above the unconformity in the vicinity of Nehalem Bank (Fig. 6, SP-106; Fig. 10A, SP-104), along the east flank of Heceta Bank (Fig. 7, SP-89), and in the area between Cape Arago and the Coquille River have been dated as Pleistocene. Sedimentary rocks beneath the unconformity have been dated as Pleistocene. Sedimentary rocks

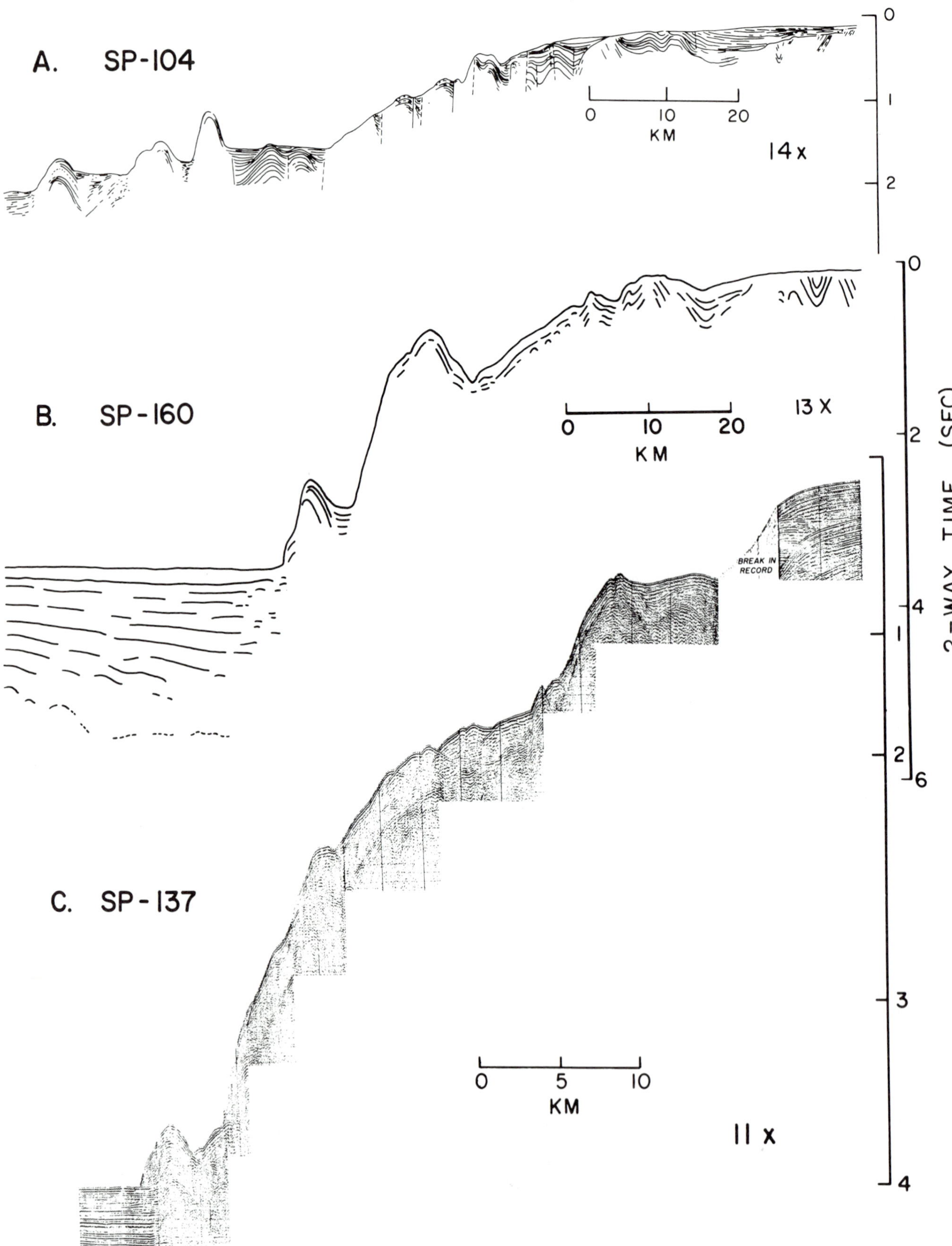

Fig. 10. Seismic reflection profiles from the northern (A), central (B), and southern (C) Oregon continental margin.

beneath the unconformity have been dated as late Miocene to late Pliocene. The biostratigraphy of DSDP site 176 shows that unconsolidated sediments of middle to late Pleistocene age (maximum age 1.3 my) overlie Pliocene shale (Kulm et al., 1973a) in the unconformity displayed on the seaward flank of Nehalem Bank (Fig. 6, SP-106). An unconformity of the same age probably occurs between the Pleistocene sediments and the middle Miocene basalt and on the east end of SP-106 off Tillamook Head and the Eocene-Oligocene rocks that underlie Pleistocene sediment at the landward end of SP-104 (Fig. 10A). Strata of Eocene-Oligocene age crop out along the coast between Nehalem Bay and Tillamook Bay (Fig. 5). The ubituitous nature of this unconformity and the wide range of ages of strata beneath it indicate that a tectonic event of considerable significance caused uplift and erosion on the Oregon shelf during Pliocene-Pleistocene time.

An older unconformity is seen in profiles SP-102 and 89 (Figs. 6 and 7) and it may also be present on SP-106. Sediments above the unconformity are as old as late Miocene, but we have no control on the age of the sediments below it. However, a significant middle to late Miocene unconformity is indicated in a well drilled on the central Oregon shelf (Fig. 9B). It may be as widespread as the Pliocene-Pleistocene unconformity over the Oregon shelf.

Perhaps the youngest unconformable surface is seen on the outer edge of the continental shelf off southern Oregon. A wedge-shaped deposit forms a discordant contact with the older underlying strata (Fig. 7, SP-47). These sediments are probably related to sea-level fluctuations during the Pleistocene (Mackay, 1969).

UPLIFT OF THE CONTINENTAL SHELF

Assemblages of benthic foraminifera from the sedimentary rocks collected on the Oregon continental slope and shelf generally are indicative of deeper environments of deposition than are found presently at the collection sites. Uplift of as much as 1 km has been reported for the central Oregon shelf and upper slope (Byrne et al., 1966; Fowler, 1966; Fowler and Muehlberg, 1969). The largest amounts of uplift (900-1,000 m) on the Oregon continental shelf involve the late Miocene and early Pliocene rocks in the vicinity of Heceta Bank (Figs. 3 and 11). Early to middle Pliocene strata on Nehalem and Coquille banks have been uplifted as much as 500-600 m. Pleistocene deposits on the submarine banks have an estimated uplift ranging from 0-100 m.

Downwarp of the outer shelf is indicated on the seaward flanks of Nehalem Bank (Kulm et al., 1973a) and Coquille Bank and on the upper continental slope off the Rogue River (Fig. 3). These negative movements range up to 200 m and in one case, Coquille Bank, are associated with faulting on the seaward flank of the bank. Figure 12 shows the calculated minimum rates of uplift for coastal terraces and the continental shelf and slope. These rates were determined by using the maximum age (using the time scale of Berggren, 1971) for the deposits and the amount of uplift indicated by the benthic foraminiferal assemblage or measured elevation change of coastal terraces above present sea level.

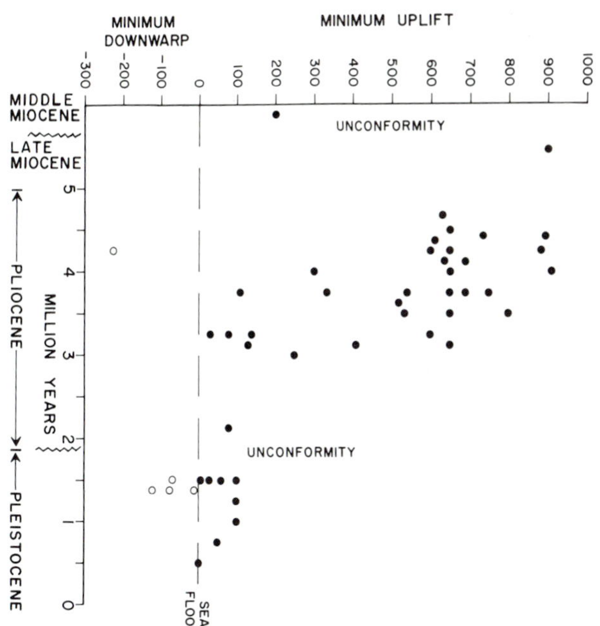

Fig. 11. Amount of relative tectonic movement on the Oregon continental margin during late Cenozoic time (in 100m/my). Lower continental slope samples (Table 3) are not included here; Pleistocene (dots), Pliocene (triangles), Miocene (stars). Time scale after Berggren (1971).

Average rates of uplift range from 100 to 1,000 m/my over the Oregon margin with the highest rates associated with coastal terraces at Cape Blanco and with the lower continental slope deposits off central Oregon (see section on Uplift of Abyssal Plain Deposits, especially Table 3). Although the largest amounts of uplift of shelf and upper continental slope rocks involve the late Miocene to early Pliocene strata, the rates are only 100 m/my and probably are not more than 200 m/my at a maximum. Rates of uplift over the entire shelf and upper slope are fairly constant which suggests a rather uniform rate of uplift during late Cenozoic time. On the other hand, the rate of uplift in the coastal region decreases to the north and south from Cape Blanco.

CONTINENTAL SLOPE STRUCTURE

Single channel seismic reflection records suggest the Oregon continental slope is underlain by folded and faulted structures that give rise to an irregular topography. Multiple channel (12-fold CDP) seismic records indicate that the N-trending ridges of the continental slope (Fig. 4) are imbricate

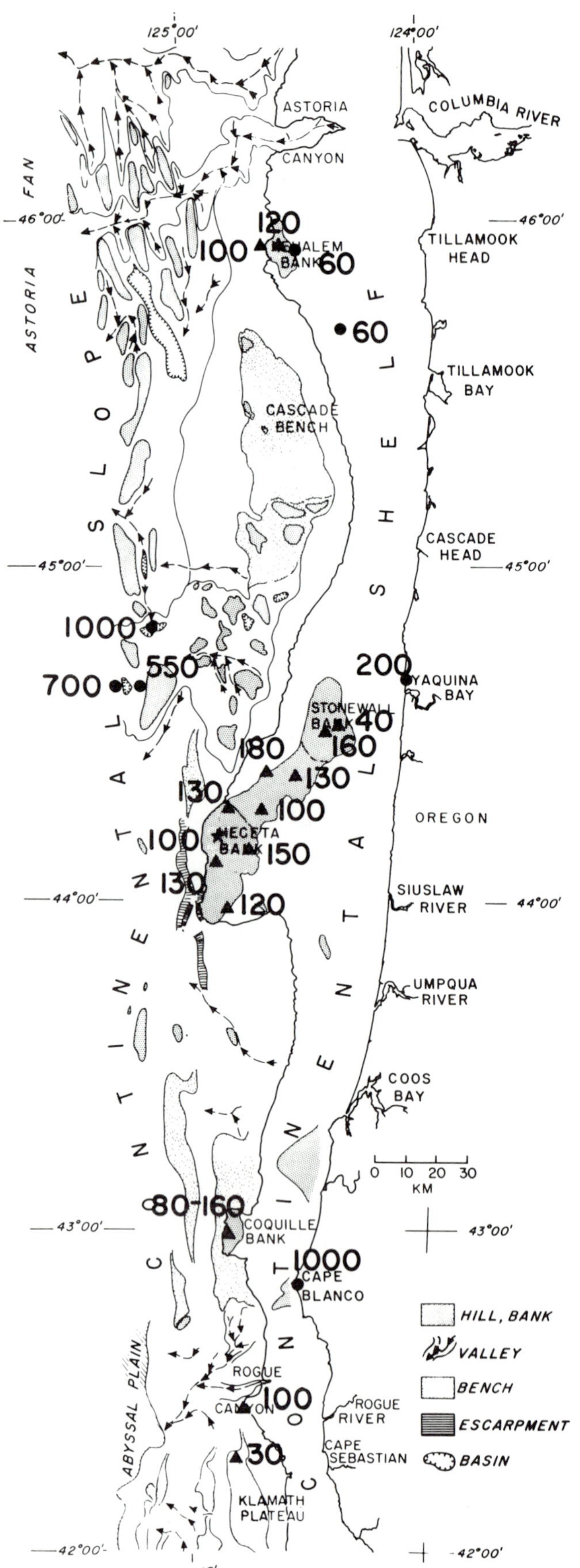

Fig. 12. Minimum rates of uplift on the Oregon continental margin and coastal terraces.

thrust sheets which probably contain faults and folds (Seely et al., Fig. 8, this volume). These structures provide the framework for sediment ponding in the intervening troughs or synclines with benches being created on the upper slope (Figs. 4, 6, and 10A).

Sediments on the lower continental slope off northern Oregon may be gently deformed into broad anticlines with substantial sediment fill between folds (Fig. 10A). The troughs are partially filled with sediment which is usually deformed by subsurface folding; some folds rise above the general level of the basins. On several profiles, there is an angular unconformity between the older underlying folds and the younger sediments above. The upper slope and outer shelf appear to be folded and faulted extensively due to the imbricate thrusts (Fig. 10A; Seely et al., Fig. 8).

Little structural information was obtained from single channel reflection profiles made across the slope off central Oregon due to the steep escarpments (Fig. 10B). However, the 12-fold CDP profile taken near the southern end of Heceta Bank at 44°S latitude (Fig. 4; Seely et al., Fig. 8) shows that it is created by a series of thrust faults which apparently steepen toward the shoreline. Between Cape Blanco and the California border, the slope is underlain by tight folds and faults (Fig. 10C) which are probably associated with the thrust sheets described above. Narrow structural benches result from sediment ponding behind these structures. The largest bench on the southern Oregon slope (Fig. 4), Klamath Plateau, probably formed in this manner.

CONTINENTAL SLOPE-ABYSSAL PLAIN LITHOLOGY AND STRUCTURE

Lithology

The stratigraphy of the eastern Cascadia Basin has been determined through deep-sea drilling at site 175 (Kulm et al., 1973; Fig. 14). The oldest sedimentary unit (A) above layer 2 is a late Miocene to early Pliocene calcareous-rich silty clay of pelagic and hemipelagic origin (Table 1). This unit grades upward into Pliocene and Pleistocene silt turbidites (units $C_{1,2}$); all these units have characteristics of an abyssal plain facies. Middle to late Pleistocene fine to medium sand turbidites with interbedded muds comprise unit E_1 of Astoria Fan. These sands unconformably overlie the abyssal plain facies, and form a wedge that thins southward and westward away from the apex of Astoria Fan.

The stratigraphy of the lower continental slope is more difficult to sample owing to the structural complexity of the region. DSDP site 175 was drilled in a narrow basin on the lower slope and lies about 87 km to the east of site 174 (Figs. 13 and 14). This section (Table 1, units C_3 and D) consists of early to middle Pleistocene silty clays, with interbedded silt turbidites, which grade upward into late Pleistocene silty clays of hemipelagic origin (Kulm et al., 1973b).

Fig. 13. Bathymetry off central Oregon. Note location of dredge stations and DSDP Site 175.

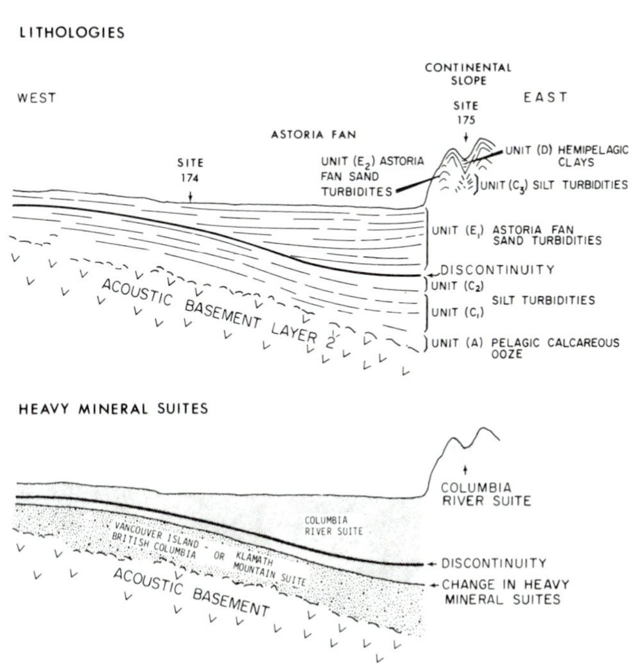

Fig. 14. Sedimentary deposits of the eastern Cascadia Basin and lower continental slope off Oregon (modified from von Huene and Kulm, 1973, Fig. 2).

Five dredge hauls were taken on the seaward scarps of the N-S-trending ridges that form the lower part of the continental slope (Fig. 15). Each dredge haul contained both Pleistocene sandstones and mudstones of varying character. Most sandstones are friable but some are cemented with calcium carbonate and probably represent portions of concretions and concretionary beds. Mica flakes and carbonaceous debris are frequently oriented in a subparallel fashion, producing a faint bedding. Flame structures, pull aparts, and mudstone clasts often are associated with general bedding characteristics. In at least one instance, there is clear evidence of graded bedding. A few samples of very fine grained silty sandstone and sandy siltstone are thinly laminated. Micromicaceous, carbonaceous mudstone occurs in the dredge samples as separate clasts and as interbeds in larger sandstone clasts. These sandstones have many typical turbidite characteristics.

Consolidation

Pleistocene turbidites of Astoria Fan (Fig. 14, unit E_1) were completely unconsolidated to a subsurface depth of 284 m and gave an interval velocity of 1.60 km/sec (Kulm et al., 1973b). Unit C_2 below unit E_1 contains compacted silty clays, but the basal very fine sand and silt layers of the turbidites were unconsolidated with no cement. An interval velocity of 1.9 km/sec was obtained for Pliocene units A, C_1 and C_2.

In contrast, the Pleistocene sandstones and mudstones dredged on the lower continental slope (Fig. 15) were fairly well consoldiated. Early to middle Pleistocene deposits at site 175 (Fig. 14, unit C_3) became noticeably consolidated at subsurface

Table 1. Stratigraphy of deposits cored at DSDP sites 174 and 175, and dredged from lower continental slope.

Unit	Location	Lithology	Age	Probable Source
E_2	Sample OC-0039 and -0045	Very fine to medium sandstone (turbidites) and mudstone	late Pleistocene (≤0.3 my)	Mixed source
E_1	Astoria Fan, Site 174	Fine to medium unconsolidated sand turbidites with interbedded muds	late Pleistocene (0.5 my-present)	Columbia River basin
D	Site 175	Hemipelagic silty clays and clay silts	late Pleistocene (0.6 my-present)	Columbia River basin
C_3	Site 175	Silt turbidites and interbedded muds	middle Pleistocene (1.2-0.6 my)	Columbia River basin
C_2	Site 174	Silt turbidites and interbedded muds	early Pleistocene to middle Pleistocene (1.85-0.6 my)	Columbia River basin
C_1	Site 174	Silt turbidites with interbedded muds grading downward into calcareous-rich muds	Pliocene (5.0-1.85 my)	North-South Sources[a]
B	Samples OC-0041, AD-93	Fine to medium sandstones (turbidites) and mudstones	late Pliocene to early Pleistocene (2.0-1.2 my)	North-South Sources
A	Site 174	Pelagic calcareous ooze grading upward into calcareous hemipelagic muds	late Miocene to early Pliocene (8-4 my)	—
Layer 2	Site 174	Probably tholeiitic basalt	late Miocene (10-8 my)	—

[a] Heavy mineral assemblages are derived from Klamath Mountains of southern Oregon and northern California or Vancouver Island-British Columbia area or from both source areas.

depths of 125-180 m, and their velocities increased rapidly from 1.6 to 1.9 km/sec within this short interval. The late Pleistocene unconsolidated hemipelagic sediments (unit D) recovered at site 175 have an interval velocity of 1.6 km/sec.

Ages of Lower Slope Deposits

The fauna of all lower slope samples were examined for age. The sandstones have no foraminifera or radiolaria, but the associated mudstones contain excellent benthic and planktonic assemblages (Table 2). Age assignments for foraminifera in these samples are based primarily upon the coiling direction of *Globigerina pachyderma* (Ehrenberg). Bandy (1964), Ingle (1967), Bandy and Ingle (1970), Ingle and Takayanagi (in press), and Ingle (1973) have determined the sequence of coiling changes displayed by *G. pachyderma* since the middle Miocene in southern and northern California. Sinistral populations predominate in the upper Miocene, upper middle to lower upper Pliocene, and glacial Pleistocene. Between these occurrences dextral forms are dominant. Age assignments for radiolaria are based largely on the work of Hays (1970) and Hays and Berggren (1971). The species composition and ages are given in Table 2.

Samples OC-0038 and OC-0039 contain 100% and 73% sinistral *G. pachyderma*, respectively. Because the radiolaria in sample OC-0038 give an age of ≤0.3 my, these mudstones are considered glacial Pleistocene in age. Sample OC-0041 contains 96% dextral *G. pachyderma* and a radiolarian assemblage with an age range of 0.9-2.0 my (late Pliocene or early Pleistocene). Work in progress on planktonic foraminiferal biostratigraphy of Juan de Fuca Ridge cores corroborates this date. Therefore, an upper limit of 1.2 my is indicated for sample OC-0041.

Sediment Sources

Several studies (e.g., Duncan and Kulm, 1970; Carson, 1971; Scheidegger et al., 1971, 1973) have outlined the heavy mineral provinces in the Cascadia Basin and on the Oregon continental margin. The

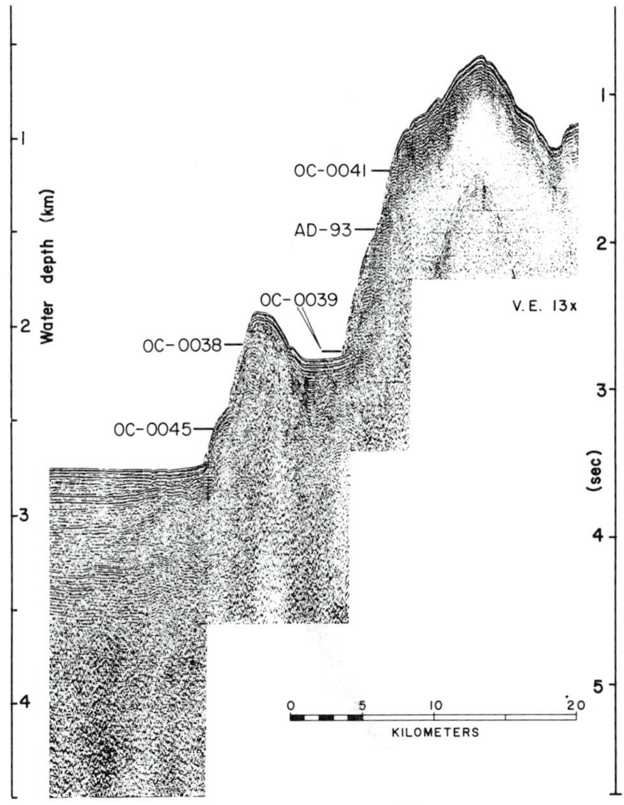

Fig. 15. Seismic reflection profile from continental slope off central Oregon. See Figure 13 for location (A-B). Dredge stations are indicated.

dominant sources are the Klamath Mountains of southern Oregon and northern California, the extensive Columbia River basin and the coastal drainage basins of Vancouver Island and British Columbia.

Scheidegger et al. (1973) show that the Pliocene silt turbidites of unit C_1 (site 174) were derived from either the Klamath Mountains to the south or Vancouver Island-British Columbia area to the north (Table 1, N-S sources; Fig. 14). In contrast, the sediment source changed about 2 my ago to one dominated by the Columbia River (unit C_2). Middle to late Pleistocene silt turbidites at site 175 (unit C_3) were also derived from the Columbia sources (Table 1).

The heavy mineral suites in the sandstones dredged from the lower continental slope are typical of immature sands found in the northeast Pacific and contain high percentages of metamorphic minerals such as kyanite, staurolite and blue-green hornblende. Some sandstones from station AD-93 have the same composition as sediments derived from the Klamath Mountain provenance or from the Vancouver-British Columbia area. Their gross composition is similar to the Pliocene sands at site 174. Other sandstones from AD-93 and those from OC-0045 are not strongly related to any of the source areas. They may be admixtures of sediments from several sources.

Structures

The nature of the abyssal plain-continental slope juncture off Oregon has been described by

Table 2. Microfaunal and microfloral zonations for lower continental slope lithologies.

Sample or Unit	Foraminifera[a]	Radiolaria[b]	Diatoms[c]	Age (my)
OC-0038	Sinistral *Globigerina pachyderma* (N22)	Post-*Stylatractus universus*, *Drappatractus acquilonius*	—	≤ 0.3
Unit D (0-100m), Site 175	Sinistral *Globigerina pachyderma* (N22)	*Artostrobium miralestense zone and* upper *Stylatractus universus* zone	NPD zone I and NPD zone II	<0.6
OC-0039	Sinistral *Globigerina* pachyderma (N22)	None	—	$\ll 1.2$
Unit C_3	Mainly sinistral *Globigerina pachyderma*	*Stylatractus universus* zone	NPD zone II and NPD zone III	0.6-1.2
OC-0041	Dextral *Globigerina pachyderma* (>1.2 my)	Eucyrtidium delimontensis, *Stylatractus universus*, *Eucyrtidium matuyama* (0.9-2.0 my)	—	1.2-2.0

[a] Foraminiferal zonations. Site 175, after Kulm et al. (1973b). Samples OC-0038, -0039, -0041 identified by G. A. Fowler, this study.
[b] Radiolarian zonations. Site 175, after Kulm et al. (1973). Samples OC-0038, -0039, -0041 identified by Ted C. Moore, School of Oceanography, Oregon State University.
[c] North Pacific diatom zones from Schrader (1973).

Table 3. Amount and rate of uplift on the lower continental slope off central Oregon.

Sample or Unit	Present Water Depth (m)	Paleodepth (m)		Maximum Age (my)	Rate of Uplift [a] (m/my)	
		Benthic Foraminifera	Present Abyss Plain Depth		Benthic Foraminifera	Abyssal-Plain Depth
OC-0038	2,100	> 2,300	2,750	0.3	666	2,166
OC-0039	2,150	> 2,300	2,750	1.2	125	500
OC-0041 (unit B)	1,200	> 2,300	2,750	2.0	550	775
Site 175 (unit C_3)	1,999	> 2,300	2,750	0.3	1,000	2,500

[a] Represents difference in present depth and paleodepth determined by fauna or present water depth of adjacent abyssal plain.

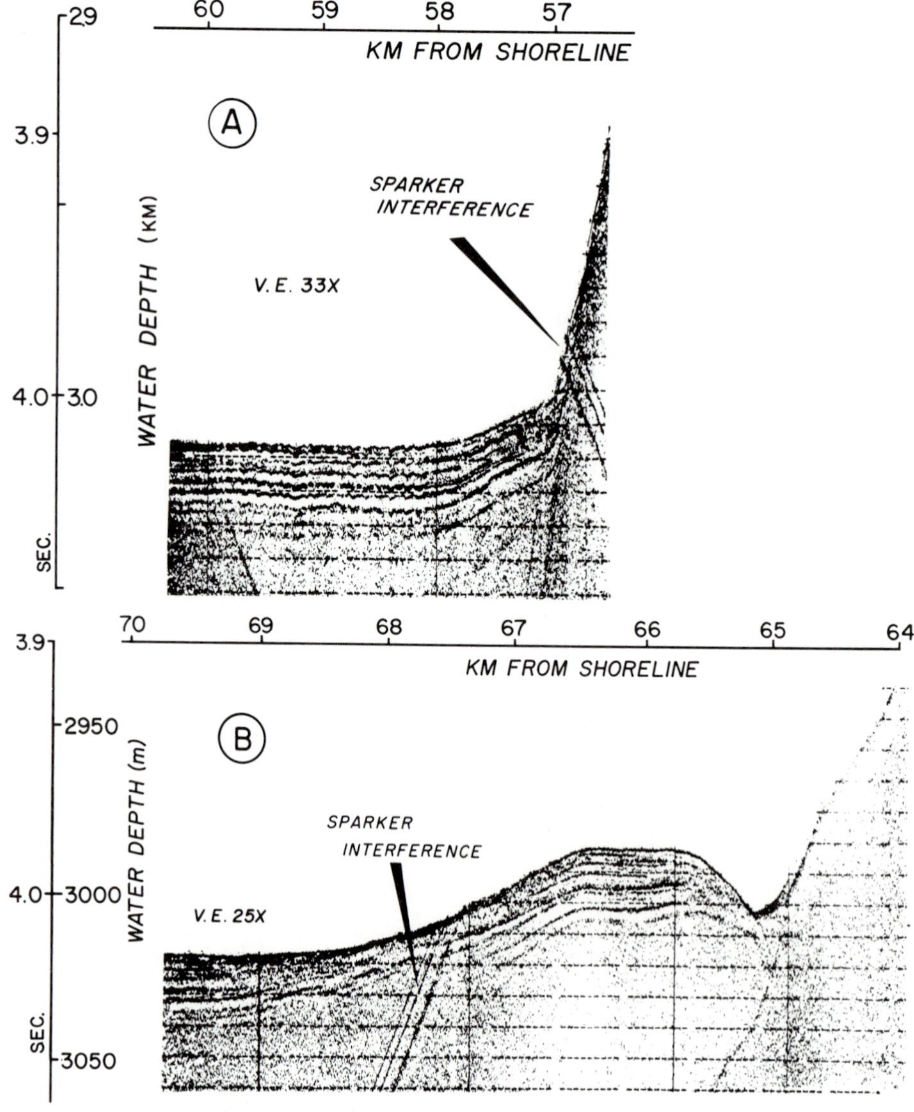

Fig. 16. High resolution 3.5 kHz records made at the base of the continental slope off southern Oregon (after Spigai, 1971, Fig. 18). A. Cascadia Basin sediments elevated at base of continental slope (SP-165). B. Small hill in basin sediments at base of continental slope (SP-166). See Figure 2 for profile (SP) locations.

Silver (1969, 1971, 1972), Spigai (1971), Kulm et al. (1973), and Seely et al. (this volume). The complex signals returned from the steeply dipping strata of the lower continental slope (eg. Fig. 15) make it difficult to determine the nature of the contact with the abyssal plain from single channel seismic records. On the other hand, the 12-fold CDP profile in this area (Seely et al., Fig. 8) clearly shows that the abyssal deposits are thrust beneath the lower slope with perhaps some folding of the abyssal plain deposits. Although the Oregon continental slope exhibits compressional thrusts and folds, the acoustic basement terminates a short distance landward of the continental slope-abyssal plain juncture. This may be an important factor in explaining the lack of seismicity observed for the Oregon-Washington region.

Although single channel seismic records provide only limited information on the structure of the continental slope, they are useful in detecting deformation of the adjacent abyssal plain deposits (Figs. 15, 16). In most profiles off southern and central Oregon, the sediments of Cascadia Basin seem to terminate abruptly against the lower slope with no apparent deformation at the juncture (Figs. 10C and 16C). Off northern Oregon Cascadia Basin sediments appear to be deformed near the base of the continental slope, and similar relationships occur near the Columbia River and farther north off Washington (Silver, 1972).

High resolution 3.5 kHz records at the base of the slope show sediments of Cascadia Basin to be bent upward against the slope (Fig. 16A) or folded into a small hill (Fig. 16B). This could represent slumping from the adjacent slope, but the coherent reflectors are of a uniform thickness suggesting no sediment wedging. Small hills probably reflect deformation with depth.

UPLIFT OF ABYSSAL PLAIN DEPOSITS

Benthic foraminifera in the lower slope mudstones (Fig. 15) are diagnostic of abyssal environments in excess of 2,300 m; the most noteworthy benthic species is *Melonis pompilioides* (Fichtel and Moll). We have not found this species in depths less than 2,300 m in modern sediments collected off Oregon. Bandy and Chierichi (1966, p. 267) give an upper limit for this species of 2,300 m, referring to studies in the Gulf of Mexico, the Gulf of California, off the coast of southern California, and in the Antarctic. Bandy and Echols (1964, p. 87), in an evaluation of Antarctic foraminiferal zonation, state that this species "...is especially prominent below 3000 m." Phleger (1964, Table 5, p. 387) lists its shallowest occurrence in the Gulf of California as 1,295 fm (2,830 m) and a maximum frequency of 17% of the total living foraminiferal fauna at 1,590 fm (2,930 m); Echols (1971) does not record it at depths less than 2,924 m in the Scotia Sea area.

Based upon the modern bathymetric range of *M. pompilioides* and its occurrence and abundance in the lower continental slope mudstones, samples 0041 and 0038 (Fig. 15) were clearly deposited deeper than 2,300 m and probably deeper than 3,000 m (Table 3). The water depths at the collection sites of OC-0038 and OC-0041 were 2,100 and 1,200 m, respectively. Thus, the mudstones and sandstones have been uplifted as much as 1,100 m during the late Pliocene and Pleistocene.

Using the 2,300-m upper limit for *M. pompilioides*, unit C_3 at site 175 records about 300 m of uplift during the late Pleistocene (Kulm et al., 1973b). The significant change in paleodepth occurred 0.3-0.45 my ago, according to diatom and radiolarian stratigraphy (Kulm et al., 1973b). Initial uplift of site 175 from the abyssal plain apparently occurred about 0.60 my ago because turbidite deposition ceased at that time (von Huene and Kulm, 1973).

A minimum rate of uplift of the lower continental slope can be calculated using the maximum age for the deposits (Table 2) and the amount of vertical displacement (Table 3). The rate ranges between 125 and 1,000 m/my using foraminifera, and between 500 and 2,500 m/my using the present water depth of the adjacent abyssal plain. The abyssal plain was probably deeper along the Oregon margin during the late Pleistocene since the Astoria Fan deposits accumulated during the glacial Pleistocene, partially filling the eastern part of the Cascadia Basin.

THE MODEL

The imbricate thrust-model for the Middle America Trench is based on multiple-channel seismic reflection work in both shallow and deep water (Fig. 17A; Seely, Vail, and Walton, this volume). Considerable computer processing was required to produce a valid structure section for the deep-water part of the reflection profile. Although the Oregon continental margin is far removed from the Middle America Trench, it has many of the gross characteristics of that region. We will use the geological data presented in the previous sections to test this imbricate thrust model for the Oregon margin:

1. A general uplift over the entire Oregon continental margin has occurred during late Cenozoic time. The rate of uplift is highest on the lower continental slope (1,000 m/my) and lowest on the outer edge of the continental shelf (100 m/my on the average). During any given time interval the amount of uplift is greatest for the oldest deposits.

2. Several major erosional unconformities (mid-late Eocene, mid-late Miocene, Plio-Pleistocene) occur in the continental margin section.

3. Late Pliocene to Pleistocene subsidence on the inner continental shelf created an elongate synclinal basin which subsequently was filled.

As discussed earlier, there is some evidence from seismic reflection records (Fig. 16) that the Cascadia Basin deposits are being deformed near the base of the continental slope. In a few cases, anticlines developed on the abyssal plain show apparent stratigraphic and structural continuity with the abyssal plain on the west limb and a probable fault on the east limb (Fig. 18B). If these structures are real and not artifacts created by diffraction from fault surfaces, the surface rupture of the thrust may occur higher on the slope.

Apparently Astoria Fan turbidites are exposed over the entire scarp of the lowermost ridge on the continental slope (Fig. 15). This suggests that the thrust slices may have a combined thickness of about 800 m in this area. The lower ridge was uplifted rapidly (Table 3, OC-0038) as the thick, younger turbidite sections were sliced off the descending oceanic plate. Von Huene and Kulm (1973, their Fig. 3) compared the cross-sectional areas of deformed sediments with the corresponding projections of the original undeformed areas and calculated that about 16 km of apparent compression has occurred seaward of site 175 (Fig. 13) since late Pleistocene time.

Seismic reflection profiles made over the ridge and trough topography of the lower slope off northern Oregon (Fig. 10A) display several unconformities. The discordance is variable and apparently depends upon variable rates of tectonism (uplift) and deposition. Astoria Fan turbidites are thickest in this region as a result of its position adjacent to the locus of deposition (Astoria Canyon). The lower slope is postulated to be accreting at a faster rate in this region because a thicker turbidite section accumulates over a shorter period of time and at a faster rate in the proximal than in the distal part of the fan.

EVOLUTION OF THE OREGON CONTINENTAL MARGIN

The several lines of evidence discussed above indicate that the Pleistocene deposits, abyssal plain, and fan of the Cascadia Basin are being thrust beneath the earlier Cenozoic rocks of the Oregon margin, uplifting the latter units to much shallower depths. According to our studies, late Pliocene and Pleistocene abyssal plain turbidites crop out on the lower and middle continental slope, whereas Miocene to Pleistocene mudstone and sandstone of much shallower water facies crop out on the upper continental slope and outer shelf. We have not sampled in the offshore Oregon area abyssal-plain deposits known to be older than late Pliocene.

Middle Eocene turbidites of the Tyee Formation exposed on land were deposited in a geosyncline at middle bathyal to deeper depths (1,500 to possibly more than 2,000 m), and they are similar to the late Cenozoic turbidites of the eastern Cascadia Basin (Kulm and Fowler, 1974). Terrigenous sediments dominated in the geosyncline of the western Oregon and Washington area as well as the Cascadia Basin throughout Cenozoic time. These deposits are characterized by mineralogically immature arkosic, lithic, feldspathic, or volcanic sands. Truly pelagic sediments are of minor importance, which indicates that the deeper-water environments were either relatively close to their continental sources or that turbidity currents transported large volumes of material into the adjacent ocean basin through a well-developed system of submarine canyons and connecting deep-sea channels. If a trench was present during the deposition of the Eocene turbidites, it was probably filled rapidly (i.e., similar to the present Cascadia Basin setting). The Tyee Formation is underlain by a thick sequence of middle Eocene pillow basalts and breccia that have the chemical composition of oceanic tholeiitic basalts (Snavely et al., 1968).

A major mid-late Eocene unconformity beneath the Oregon shelf (Fig. 9A) suggests that the area experienced an episode of tectonism which resulted in uplift and erosion. This may have been one of the major periods of imbricate thrusting by the oceanic plate. The thick Eocene turbidites would have provided ample material for continental accretion (note the importance of eugeosynclinal uplift at continental margins; e.g., Burk, 1972). Volcanism was renewed at this time within the depositional basin and in the ancestral Cascades to the east (Snavely et al., 1963). Volcanism continued in the Cascades into late Oligocene time, when the Coast Range was uplifted and thick gabbroic sills were emplaced. Marine deposition shifted westward and siltstone and mudstone were deposited on the then-existing Oregon margin during early Miocene time; thick nearshore deposits are also characteristic of the middle and late Miocene.

Another intense period of thrusting is interpreted to have occurred in mid-late Miocene time along a broad front, causing uplift of the marine deposits found on the outer continental margin. They were subsequently eroded to produce a major angular unconformity on the central Oregon shelf and in other parts of the Pacific Northwest. This unconformity is dated at about 10 my, which coincides with the well-documented worldwide change in plate motion at this time (e.g., Le Pichon, 1968).

The rate of subduction may have been increased substantially in this area at this time. It is postulated that a part of the early Eocene oceanic crust with overlying sedimentary deposits was scraped off the descending oceanic plate and uplifted higher in the vicinity of Heceta Bank (Fig. 9A) and possibly Coquille Bank than in the intervening areas of the continental margin. As noted previously, Heceta and Coquille banks are characterized by positive free-air gravity anomalies which suggest a fairly dense mass (probably Eocene basalt) below. It is important that magnetic source depths are also shallow in the vicinity of Heceta and Coquille banks (Emilia et al., 1968). To the north, the Eocene volcanics on land dip seaward and a thick sedimentary section is indicated by drilling records (Fig. 9B), seismic reflection records, and large negative free-air gravity anomalies.

A rather strong positive free-air anomaly also occurs northwest of Heceta Bank (45°N, 125°20'W) over the ridges on the lower continental slope (Fig. 8). Perhaps a thin slice of Miocene oceanic basalt (layer 2) has been scraped off the descending oceanic plate and subsequently uplifted with the overlying thick turbidite section to form one of the lower slope-ridges.

Pliocene deposits are widespread on the Oregon continental shelf and consist mainly of massive siltstone and claystone. They were originally deposited on the upper continental slope and have been uplifted to form the submarine banks; they probably form the subsurface folded structures that are covered with Pleistocene sediments (e.g., Cascade Bench, Figs. 4 and 6, SP-56). Subsidence of the inner shelf created a broad shallow syncline which filled with shallow-water Pliocene and Pleistocene sediments.

Late Pliocene to early Pleistocene abyssal fan deposits are interpreted as having been thrust beneath the lower continental slope as finer-grained deposits were accumulating in basins on the upper continental margin. These basins were probably created by the uplifting thrust sheets (Seely, Vail, and Walton, this volume) and were filled with hemipelagic sediments similar to unit D (Table 1) found at site 175 on the lower slope.

Through repeated thrust faulting of the wedge-shaped slices at the base of the continental slope, these late Pliocene to early Pleistocene turbidites may have been uplifted between 1 and 2 km to a present water depth of about 1,100 m on the upper continental slope. These deposits may be thrust beneath the early Eocene volcanics in the vicinity of Heceta Bank. As the underthrusting younger sediments raised the older deposits higher on the slope, the younger units attained a progressively steeper dip (Seely, et al., this volume).

The Cascadia Basin was never cut off from turbidite deposition for a significant period of time because turbidites occur throughout the Pliocene and Pleistocene section cored at site 174. If a trench was present during this time it was filled or there was access to the basin to the north in the vicinity of Vancouver Island. These abyssal-plain silt turbidites were presumably downwarped at the base of the continental slope and thrust beneath it from 1.2 to 0.5 my ago. Astoria Fan deposition commenced almost immediately and continued to the beginning of the Holocene; these deposits are more than 1 km thick off northern and central Oregon. As thrust faults presumably developed along the eastern edge of the fan, the wedge-shaped slices were uplifted to form the ridge topography found on the lower continental slope (Fig. 4). The structure discussed earlier suggests that the ridge-and-trough topography is relatively young.

Considerable continental accretion apparently has occurred off northern Oregon during late Cenozoic time (cf. J.C. Moore, this volume). Byrne et al. (1966) calculate that the maximum horizontal accretion of sedimentary rock to Heceta Bank area during the Pliocene has been from 5 to 50 km, with an average of about 16 km. Although accretion probably has occurred along the entire Oregon margin, it apparently has been accelerated along the northern Oregon slope, where the Astoria Fan deposits are thickest. The same rapid accretion is suggested for the Washington slope (Silver, 1972). Assuming a uniform rate of underthrusting in this area, relatively rapid continental accretion should occur where the trench sedimentation rates are highest (Seely et al., this volume). In summary, the Oregon continental margin has many of the structural and stratigraphic characteristics of a fore-arc structure defined by Karig (1970) and documented by Seely, et al. (this volume) for a segment of the Middle America Trench.

Despite the mounting geological evidence for very late Cenozoic thrusting beneath the Oregon continental margin, a well-defined Benioff Zone is absent beneath the northwest United States. Atwater (1970), Silver (1971, 1972), and McKenzie and Julian (1971) argue for compressional underthrusting in this region, whereas Dehlinger et al. (1971) suggest that the area is undergoing extension at the present time. Crosson (1972) suggests that the region may be in a state of transition between compression and extension. Because there are only a few scattered low-magnitude shallow-focus earthquakes beneath the margin and coastal region, the problem of modern underthrusting is still not resolved to the satisfaction of those studying this region, regardless of the various models suggested.

ACKNOWLEDGMENTS

Many staff members and students in the School of Oceanography were involved in various phases of this study. We are especially indebted to William E. Bales, Robert Buehrig, Carol Lantz, and Roger H. Neudeck for their assistance with the data collection at sea and the data reduction in the laboratory.

Numerous discussions with Peter R. Vail and Don R. Seely of ESSO Petroleum Corp. were instrumental in establishing geologic criteria for testing the imbricate thrust model on the Oregon margin. Informal discussions with Dick Couch on the crustal structure of the Oregon region were most helpful. We thank John V. Byrne for the use of Figure 4, which he compiled. David K. Rea and Ted C. Moore, Jr. reviewed the manuscript. T.C. Moore kindly supplied radiolarian age data for the lower slope rocks.

The Penrose Conference on Continental Margins provided new insight to the senior author regarding the evolution of the Oregon margin. We thank Creighton A. Burk and Charles L. Drake for convening the conference. This study was supported largely by the U.S. Geological Survey, Office of Marine Geology (contracts 14-08-0001-10766,-11941,-12187, and -12830).

BIBLIOGRAPHY

Atwater, T., 1970, Implications of plate tectonics for the Cenozoic tectonic evolution of western North America: Geol. Soc. America Bull., v. 81, p. 3513-3535.

Baldwin, E. M., 1964, Geology of Oregon: Ann Arbor, Mich., Edwards Brothers, Inc., 136 p.

Bales, W. E., and Kulm, L. D., 1969, Structure of the continental shelf off southern Oregon: Am. Assoc. Petr. Geol. Bull., v. 53, no. 2, p. 471.

Bandy, O. L., 1964, Cenozoic planktonic foraminiferal zonation: Micropaleo., v. 10, no. 1, p. 1-17.

______, and Chierichi, M. A., 1966, Depth-temperature evaluation of selected California and Mediterranean bathyal foraminifera: Marine Geology, v. 4, p. 259-271.

______, and Echols, R. J., 1964, Antarctic foraminiferal zonation, *in* Biology of the Antarctic seas: Am. Geophys. Union Antarctic Research Series, v. 1, p. 73-91.

______, and Ingle, J. C., Jr., 1970, Neogene planktonic events and radiometric scale, California, *in* Bandy, O. L., ed., Radiometric dating and paleontologic zonation: Geol. Soc. America Spec. Paper 124, p. 131.

Berggren, W. A., 1971, Tertiary boundaries and correlations, *in* Funnel, B., and Riedel, W. R., eds., Micropaleontology of the oceans: New York, Cambridge Univ. Press, 693 p.

Braislin, D. B., Hastings, D. D., and Snavely, P. D., Jr., 1971, Petroleum potential of western Oregon and Washington and adjacent continental margin: Am. Assoc. Petr. Geol. Mem. 15, v. 1, p. 229-238.

Burk, C. A. 1968, Buried ridges within continental margins: Trans. N.Y. Acad. Sci. Bull., Ser II, v. 30 (3), p. 397-409.

______, 1972, Uplifted eugeosynclines at continental margins: Geol. Soc. America, Mem. 132, p. 75-85.

Byrne, J. V., 1962, Geomorphology of the continental terrace off the central coast of Oregon: The Ore Bin, v. 24, p. 65-74.

______, 1963a, Geomorphology of the continental terrace off the northern coast of Oregon: The Ore Bin, v. 25, p. 201-209.

______, 1963b, Geomorphology of the Oregon continental terrace south of Coos Bay: The Ore Bin, v. 25, p. 149-157.

______, Fowler, G. A., and Maloney, N. M., 1966, Uplift of the continental margin and possible continental accretion off Oregon: Science, v. 154, no. 3757, p. 1654-1656.

Carlson, P. R., 1968, Marine geology of Astoria submarine canyon (Ph.D. thesis): Corvallis, Oregon State Univ., 259 leaves.

Carson, B., 1971, Stratigraphy and depositional history of Quaternary sediments in northern Cascadia Basin and Juan de Fuca Abyssal Plain, northeast Pacific Ocean (Ph. D. thesis). Seattle, Univ. Washington.

Crosson, R. S., 1972, Small earthquakes, structure, and tectonics of the Puget Sound region: Seismol. Soc. America Bull., v. 62, no. 5, p. 1133-1171.

Curray, J. R., 1965, Late Quaternary history, continental shelves of the United States, *in* Wright, H. E., and Frey, D. C., eds., The quaternary of the United States: Princeton, N.J., Princeton Univ., 723 p.

Dehlinger, P., Couch, R. W., and Gemperle, M., 1968, Continental and oceanic structure from the Oregon coast westward across the Juan de Fuca Ridge: Canada Jour. Earth Sci., v. 5, p. 1079-1090.

______, Couch, R. W., McManus, D. A., and Gemperle, M., 1971, Northeast Pacific structure, *in* Maxwell, A. E., ed., The sea: New York, John Wiley & Sons, Inc., v. 4, pt. 2, p. 133-189.

Duncan, J. R., 1968, Late Pleistocene and post-glacial sedimentation and stratigraphy of deep-sea environments off Oregon (Ph. D. thesis): Corvallis, Oregon State Univ.

______, and Kulm, L. D., 1970, Mineralogy, provenance, and dispersal history of late Quaternary deep-sea sands in Cascadia Basin and Blanco Fracture Zone off Oregon: Jour. Sed. Petr., v. 40, no. 3, p. 874-887.

Echols, R. J., 1971, Distribution of foraminifera in sediments of the Scotia Sea area, Antarctic waters, *in* Antarctic Oceanology. I: Am. Geophys. Union Antartic Res. Series, v. 15, p. 93-168.

Emilia, D. A., Berg, J. W., Jr., and Bales, W. E., 1968, Magnetic anomalies off the northwest coast of the United States: Geol. Soc. America Bull., v. 79, p. 1053-1061.

Fowler, G. A., 1966, Notes on late Tertiary foraminifera from off the central coast of Oregon: The Ore Bin, v. 28, no. 3, 53-60.

______, and Kulm, L. D., 1971, Late Cenozoic stratigraphy of the Oregon continental margin in relation to plate tectonics (abs.): Geol. Soc. America Ann. Meeting, v. 3, no. 7, p. 570-571.

______, and Muehlberg, G. E., 1969, Tertiary foraminiferal paleoecology and biostratigraphy of part of the Oregon continental margin (abs.): Am. Assoc. Petr. Geol. Bull., v. 53, p. 467.

______, Orr, W. N., and Kulm, L. D., 1971, An upper Miocene diatomaceous rock unit on the Oregon continental shelf: Jour. Geology, v. 79, no. 5, p. 603-608.

Griggs, G. B., and Kulm, L. D., 1970, Sedimentation in Cascadia deep-sea channel: Geol. Soc. America Bull., v. 81, p. 1361-1384.

Haller, C. R., 1967, Neogene foraminiferal faunas of Humboldt Basin, California (Ph. D. thesis): Berkeley, Univ. California, 204 leaves.

Hays, J. D., 1970, Stratigraphy and evolutionary trends of Radiolaria in North Pacific deep-sea sediments, *in* Hays, J. D., ed., Geological Investigations of the North Pacific: Geol. Soc. America Mem. 126, p. 185-218.

______, and Berggren, W. A., 1971, Quaternary boundaries and correlations, *in* Funnel, B. M., and Riedel, W. R., eds., The micropaleontology of Oceans: New York, Cambridge Univ. Press, p. 669-691.

Hunter, R.E., Clifton, H.E., and Phillips, R.L., 1970, Geology of the stacks and reefs off the southern Oregon coast. The Ore Bin, p. 185-201.

Ingle, J. C., Jr., 1967, Foraminiferal biofacies and the Miocene Pliocene boundary in southern California: Bull. Am. Paleontology, Miocene/Pliocene no. 236, p. 217.

______, 1973, Summary comments on Neogene biostratigraphy, physical stratigraphy, and paleo-oceanography in the marginal northeastern Pacific Ocean, Chapter 32 *in* Kulm, L. D., et al., Initial reports of the Deep Sea Drilling Project, vol. XVIII: Washington, D.C., U.S. Govt. Printing Office, p. 949-960.

______, and Takayanagi Y. (in press), Pliocene planktonic foraminifera from northern California: Science Repts., New Series, Geology, Tohoku, Univ. of Japan.

Janda, R. J. 1969. Age and correlation of marine terraces near Cape Blanco, Oregon (abs.): Program of the 65th Annual Meeting, Cordilleran Section, Geol. Soc. America, p. 29-30.

Karig, D. E., 1970, Ridges and basins on the Tonga-Kermadec Island arc system: Jour. Geophys. Res. v. 75, no. 2, p. 239-254.

Kulm, L. D., and Fowler, G. A., 1971, Shallow structural elements of the Oregon continental margin within a plate tectonic framework (abs.): 1971 Ann. Mtg. Geol. Soc. America, v. 3, p. 628.

______, and Fowler, G. A., 1974, Cenozoic sedimentary framework of the Gorda-Juan de Fuca Plate and adjacent continental margin—a review, in press.

______, Prince, R. A., and Snavely, P. D., Jr., 1973a, Site survey of the northern Oregon continental margin and Astoria Fan, *in* Kulm, L. D., et al., Initial reports of the Deep Sea Drilling Project, v. XVIII: Washington, D.C., U.S. Govt. Printing Office, p. 979-986.

______, von Huene, R., et al. 1973b, Initial reports of the Deep Sea Drilling Project, v. XVIII: Washington, D.C., U.S. Govt. Printing Office.

Le Pichon, X., 1968, Sea-floor spreading and continental drift: Jour. Geophys. Res., v. 73, no. 12, 3661-3697.

Mackay, A. J., 1969, Continuous seismic profiling investigations of the southern Oregon continental shelf between Coos Bay and Cape Blanco (M.S. thesis): Corvallis, Oregon State Univ., 118 leaves.

Maloney, N. J., 1965, Geology of the continental terrace off the central coast of Oregon (Ph.D. thesis): Corvallis, Oregon State Univ., 233 leaves.

McKenzie, D. P., and Julian, B., 1971, The Puget Sound, Washington earthquake and the mantle structure beneath the northwestern United States: Geol. Soc. America Bull., v. 82, p. 3519-3524.

Muehlberg, G. E., 1971, Structure and stratigraphy of Tertiary and Quaternary strata, Heceta Bank, central Oregon shelf (M.S. thesis): Corvallis, Oregon State Univ.

Nelson, C. H., 1968, Marine geology of Astoria deep-sea fan (Ph.D. thesis): Corvallis, Oregon State Univ.

Phipps, B., 1974, Sediments and tectonics of the Gorda-Juan de Fuca Plate (Ph.D. thesis): Corvallis, Oregon State Univ.

Phleger, F. B., 1964, Patterns of living benthonic foraminifera, Gulf of California: Am. Assoc. Petr. Geol. Memo, no. 3, p. 377-394.

Scheidegger, K. F., Kulm, L. D., and Runge, E. J., 1971, Sediment sources and dispersal patterns of Oregon continental shelf sands: Jour. Sed. Petr., v. 41, no. 4, p. 1112-1120.

______, Kulm, L. D., and Piper, D. J. W., 1973, Heavy mineralogy of unconsolidated sands in northeastern Pacific sediments: Leg 18, Deep Sea Drilling Project, *in* Kulm, L. D., et al., Initial reports of the Deep Sea Drilling Project, v. XVIII: Washington, D.C., U.S. Govt. Printing Office, p. 877-888.

Schrader, H. J. 1973, Cenozoic diatoms from the northeast Pacific, Leg 18, *in* Kulm, L. D., et al., 1973, Initial reports of the Deep Sea Drilling Project, v. XVIII: Washington, D.C., U.S. Govt. Printing Office, p. 673-797.

Shepard, F. P., 1963, Submarine geology: New York, Harper and Row, Publishers, 557 p.

Silver, E. A., 1969, Late Cenozoic underthrusting of the continental margin off northernmost California: Science, v. 166, p. 1265-1266.

______, 1971, Small plate tectonics in the northeastern Pacific: Geol. Soc. America Bull., v. 82, p. 3491-3496.

______, 1972, Pleistocene tectonic accretion of the continental slope off Washington: Marine Geology, v. 13, p. 239-249.

Snavely, P. D., Jr., and Wagner, H. C., 1963, Tertiary geologic history of western Oregon and Washington: Wash. Div. Mines Geol. Rept. Inv. 22, 25p.

______, MacLeod, N. S., and Wagner, H. C., 1968, Tholeiitic and alkalic basalts of the Eocene Siletz River volcanics, Oregon Coast Range: Am. Jour. Sci., v. 266, p. 454-481.

______, et al., 1973, Miocene tholeiitic basalts of coastal Oregon and Washington and their relation to coeval basalts of the Columbia Plateau: Geol. Soc. America Bull., v. 84, no. 2, p. 387-424.

Spigai, J. J., 1971, Marine geology of the continental margin off southern Oregon (Ph.D. thesis): Corvallis, Oregon State Univ.

von Huene, R., and Kulm, L. D., 1973, Tectonic summary of Leg 18, *in* Kulm, L. D., et al., Initial reports of the Deep Sea Drilling Project, v. XVIII: Washington, D.C., U.S. Govt. Printing Office, p. 961-976.

Eastern Atlantic Continental Margins: Various Structural and Morphological Types

Vincent Renard and Jean Mascle

INTRODUCTION

Eight seismic-reflection profiles across the Eastern Atlantic continental margin collected during several cruises of the Centre Océanologique de Bretagne are presented. They have been chosen to illustrate the fact that margins presumably related to continental breakup and later subsidence show large variations in their morphology and structure. This variety is due to the combined action of many factors, most of them presumably linked to the geometry and rate of the ocean opening. The most important of these are the relative motion of spreading oceanic crust and continental basement, fracture zones and their related marginal offsets and ridges (LePichon and Hayes, 1971), thermal evolution of the oceanic crust (Sleep, 1971), sediment supply and current action (Emery and Uchupi, 1972), eustatic sea-level changes (Rona, 1973), the presence of postopening closed basins of limited duration (Leyden et al., 1972; Pautot et al., 1973), the stress system developed across the margin (Bott and Dean, 1972), and sedimentary loading and consequent flexure (Walcott, 1972).

If the origin of continental margins is due to subsidence, no simple model can account for it and for the simultaneous crustal thinning. Consequently, terms such as "oceanization," "basification," "subcrustal erosion," and "cratonization," had to be created to compensate for the lack of understanding of the active mechanisms.

It is hoped that the data presented in this paper will help isolate some of the determining factors and lead to a classification based not only on morphology but also on its controlling factors.

BARENTS SEA MARGIN

Profile 1 crosses the continental margin at the transition from the Barents Sea to the Norwegian Sea. Its western extremity reaches the spreading Atka Ridge (Avery et al., 1968), where little sediment is present. The topographic valley seen on the western end of the seismic record lies at the intersection of the rift and a probable fracture zone. Toward the continent, the Atka Ridge is progressively buried under thick sedimentary units. Three main sedimentary series are observed: a deeper deformed unit filling basement depressions at 3 sec or more D.T.T. (double travel time) under the lower slope; an acoustically opaque intermediate unit with no visible reflectors and a slightly deformed top; and an upper well-stratified cover with maximum thickness under the upper third of the slope.

The maximum thickness of the sediments appears to shift with time toward the east, reflecting a shift of maximum subsidence possibly related to an eastward jump of the Atka Ridge. The oceanic ridge has acted as a dam for the sediments, and its proximity explains the absence of any abyssal plain and the gentleness of the continental slope.

According to proposed ocean-opening geometry (Avery et al., 1968; Pitman and Talwani, 1972; Phillips et al., 1972), early opening occurred around 60 my and was characterized by a lateral motion of the Greenland margin relative to the Barents Sea and Spitsbergen, which probably truncated Mesozoic and older series. The pre-opening fit suggests that the Barents Sea cone creates an important overlap due to the progradation of postopening Tertiary sediments, suggesting that part of the slope is not underlain by continental crust. Sediments were derived from the uplifted Barents Sea platform, which is almost bare of any Cenozoic sediments.

NORWEGIAN SEA MARGIN

Profile 2 is across the Norwegian continental margin down to the Norwegian Sea Basin. The main feature of this profile is the presence of a large marginal plateau, the Vøring Plateau, with an average depth of 1,500 m. Westward typical rough oceanic basement beneath about 0.8 sec sedimentary cover is observed. A scarp marks the transition to the Vøring Plateau, where basement rises upwarp the sedimentary cover. Small sedimentary basins lie in basement lows, and a large basement downfault limits the plateau eastward. An extensive sedimentary basin (2 sec or more) lies between this eastern limit and the Norwegian Shelf. In this basin layers are well stratified, with tight folds decreasing in amplitude upward. They do not affect the surface except in one zone, which has been correlated with possible piercing diapiric folds (Talwani and Eldholm, 1972). Our profile seems to indicate that these folds reflect structures of the underlying stratified series. These deeper folded layers plunge eastward under the continental slope, where they are buried under a thick unconformable series of prograded sediments.

From preopening reconstruction overlaps, the limit of the oceanic crust falls between the eastern limit of the Vøring Plateau (Talwani and Eldholm,

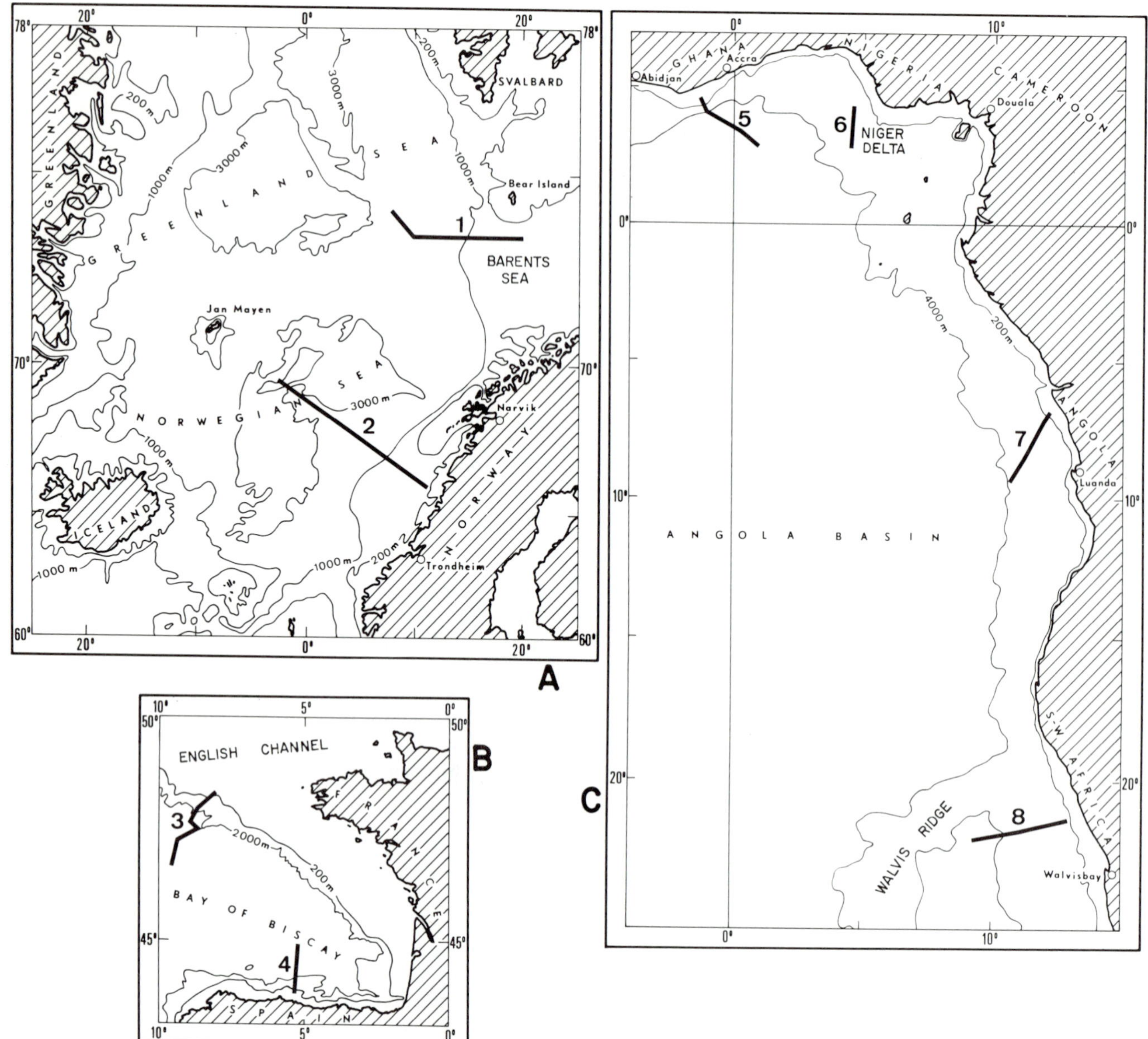

Fig. 1. Location of seismic-reflection profiles in the North Atlantic (A), Central Atlantic (B), and South Atlantic (C).

1973) and these deeper folds. These deeper layers do not extend under the Vøring Plateau, as suggested by Hinz (1972), indicating that the Vøring Plateau basement is not continental. Large, regular basement rises with upturned flanking sediments seen on the Vøring Plateau are not common and suggest that the original oceanic crust has undergone unusual modification since the shallower-than-normal crust must have thickened or become lighter in order to produce isostatic equilibrium.

At the time of the opening, the Mesozoic and Paleozoic basins of the Norwegian slope began to subside and thick series of prograding sediments accumulated during the Tertiary that were unconformable on the deeper folded sediments of probable Mesozoic age. Sedimentary lenses covered by smooth upper units seen on the outside scarp indicate that deformation or formation of the Vøring Plateau has taken place up to middle Tertiary. The southern extension of the Vøring Plateau appears to have been controlled by the Jan Mayen fracture zone. To the north, it narrows progressively and merges with the continental slope.

WESTERN ENGLISH CHANNEL

Profile 3 intersects the continental margin of the English Channel. It is characterized by two main topographic features: the Meriadzec Plateau and the Treveylyan Escarpment. The upper sedimentary cover is uniform except for some slumping. It can be followed from the shelf to the abyssal plain and represents post-Eocene deposits (Montadert et al., 1971). Erosion channels and canyons are numerous and cut across lower units which crop out along

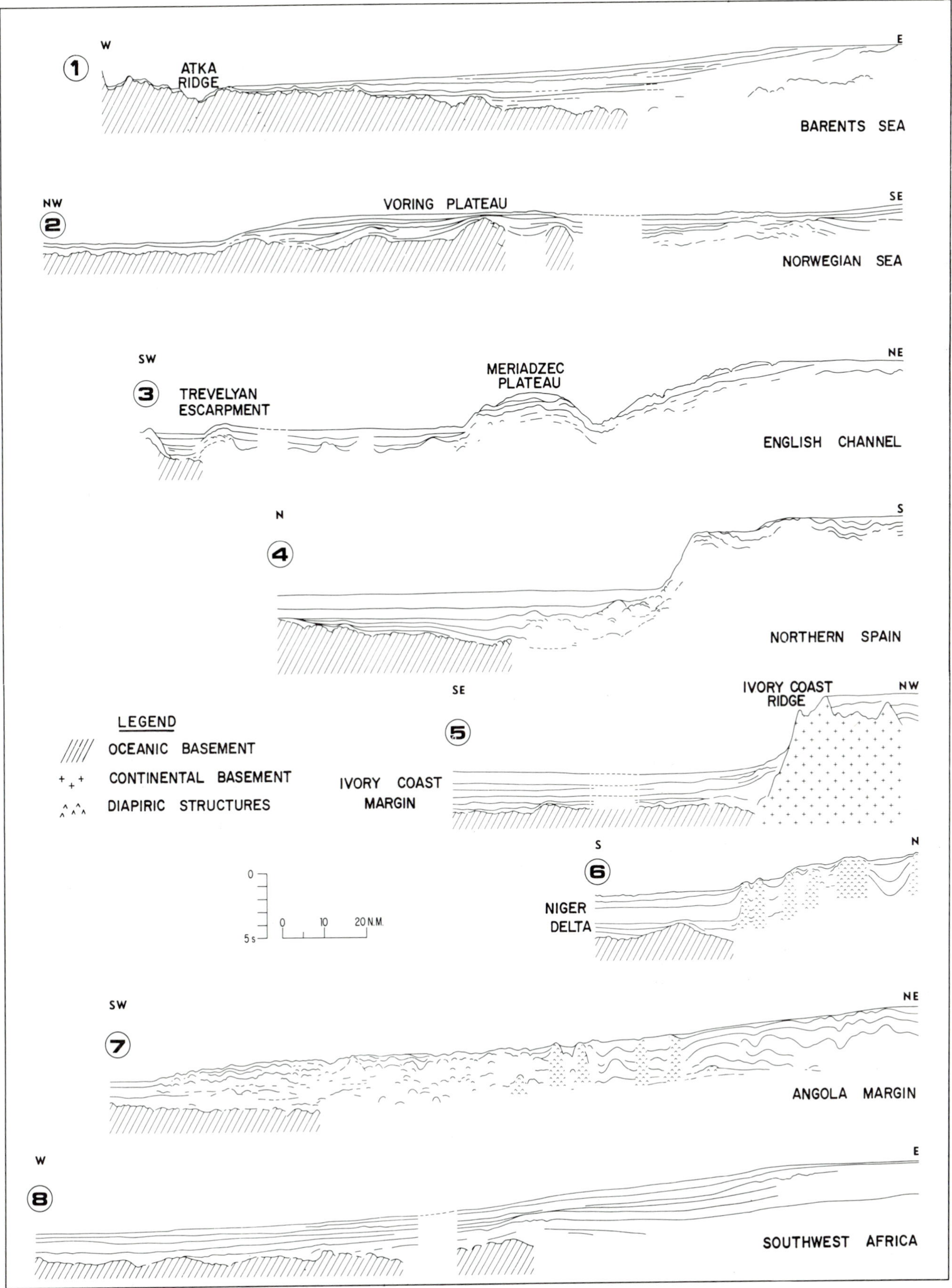

Fig. 2. Interpretation of seismic-reflection profiles of selected eastern Atlantic continental margins (locations in Fig. 1). Vertical exaggeration approximately X6.

steep slopes. Jurassic limestones have been dredged during a COB cruise. The lower sedimentary units are poorly stratified and have been tectonized.

This continental margin is an example of a collapsed margin formed of downfaulted structures (Cholet et al., 1968), related presumably to the Bay of Biscay opening. This opening was due to a counter-clockwise rotation of Spain (Jones and Ewing, 1969; LePichon et al., 1971). This rotation was probably induced by the spreading Atlantic crust and not by an active ridge system in the bay since no ridge, even fossil, has been found in the Bay of Biscay.

The collapsed origin of the margin with numerous horsts and grabens (Montadert et al., 1971) and the presence of a deep marginal basin south of the Treveylyan Escarpment are probably related to crustal extension created by the rotation of Spain. Continental structures extend at least to the Treveylyan Escarpment (Bacon and Gray, 1971).

NORTHERN SPAIN CONTINENTAL MARGIN

Profile 4 is perpendicular to the northern Spanish continental margin and crosses the fossil north Spanish marginal trough (Sibuet et al., 1971).

The continental shelf is divided in two regions: the shelf itself and a marginal plateau at about 1,000 m depth. Mesozoic units were folded in an east-west direction during the Eocene. Subsequent downfaulting created the marginal plateau and small basins filled with Tertiary units that were later affected by Mio-Pliocene horizontal dislocation. The continental slope is a major zone of downfaulting with outcrops of Cretaceous rocks reported west of the profile (Boillot et al., 1971). In the deep sea, the north Spanish marginal fossil trough is filled with three main units reaching at least 4 sec in thickness from Lower Cretaceous to Recent. The lower units dip to the south and wedge out on the northern end of the profile on a basement high, the South Biscay Ridge (Montadert et al., 1972). In the trough, the lower units are heavily disturbed and reflect compressional movements and possible diapiric activities (Sibuet et al., 1971).

The mechanism proposed for the creation of this margin is related to the opening of the Bay of Biscay by a rotation of the Iberian Peninsula around a pole close to Paris, followed by a phase of compression created by a northward movement of Spain, up to middle Tertiary, and subsequent infilling by younger Tertiary sediments (LePichon et al, 1971).

IVORY COAST MARGIN

Profile 5 is located between Ghana and Ivory Coast and crosses the Ivory Coast Ridge, which is the structural extension of the Romanche Fracture Zone (Fail et al., 1970; LePichon and Hayes, 1971; Francheteau and LePichon, 1972). On the continental shelf, sedimentary units consist of Tertiary overlying Mesozoic strata filling horsts and grabens within Paleozoic or older basement rocks (Cudjoe and Khan, 1972). The steep continental slope corresponds to the southern wall of the Ivory Coast Ridge. It is void of unconsolidated sediments and phillitic shales have been dredged along it during the COB WALDA cruise. In the abyssal plain, sedimentary units are flat and smooth and up to 2.5 sec thick. Several layers can be outlined over a typical oceanic basement (layer 2). The major part of the sedimentary sequence is Tertiary. Only the deepest poorly stratified horizons found in basement depressions are of Upper Cretaceous age, as inferred from proposed Atlantic opening chronology. These units progressively thin and terminate against the Ivory Coast Ridge.

The Ivory Coast Ridge is one of the best examples of a marginal ridge and associated coastal basin as proposed by Burk (1968), and LePichon and Hayes (1971). It has been present since the beginning of the South Atlantic opening and acted as a dam for continent-derived sediments.

NIGER DELTA

Profile 6 crosses part of the Niger Delta, from the foot of the delta to the upper continental slope. Two main regions are outlined. First, at the base of the delta there is a sedimentary sequence comparable to the one existing farther west in the abyssal plain except for thicker sedimentary units. This series covers a typical oceanic basement. The marked structural high corresponds to the extension of a fracture zone. At the base of the delta this sedimentary series reaches 4 sec. It is divided into an upper, well-stratified unit and a lower one ponded in basement depressions. Second, on the lower slope there are major structures with surface topography of up to 300 m, of diapiric origin (Mascle et al., 1973). They appear to be continuous seaward into a reflector close to the oceanic basement. From proposed Atlantic opening timing, this would point to a possible late Mesozoic age for the layer constituting the diapirs. On the upper slope of the delta, many other diapirs exist.

This deltaic margin has been built over oceanic crust, as evidenced by the important overlap in preopening reconstruction (Bullard et al, 1965). An early-opening evaporitic basin could have existed over oceanic crust. Subsequent discharge of the Niger River during the Cenozoic created the delta, and the sedimentary load induced diapiric activity.

ANGOLA MARGIN

Profile 7 intersects an important salt basin on the continental slope off Angola (Pautot et al., 1973; Leyden et al., 1972; see also Driver and Pardo, this volume). The slope is gentle and regular. Two zones with different underlying structures are delineated. The upper half of the slope is smooth and is underlain by long-wavelength anticlinal and synclinal folds with up to 4 sec of sediments. The lower half is more disrupted and its average slope is smaller. Diapirs are piercing and are reflected in the irregular bathymetry; fold wavelength is much shorter, and seismic horizons are discontinuous. Transition to the abyssal plain occurs over a narrow zone with complete disappearance of folding. A gravity high outlined at the base of the continental rise has been correlated with the ocean-continent boundary by Rabinowitz (1972).

This margin was presumably created by the subsidence of a basin formed during an early phase of South Atlantic opening and limited on the south by the Walvis Ridge. In this basin, salt deposits were formed during Aptian time (Brognon and Verrier, 1965; Baumgartner and van Andel, 1971).

On the shelf and upper slope salt was deposited on continental debris, filling block-faulted Precambrian basement (Beck, 1972). Seaward, salt could rest directly on oceanic basement, as suggested by the continuation of the oceanic layer under the continental rise. Such a possibility is in good agreement with the preopening reconstruction based on a 2,000-m isobath (Bullard et al., 1965), and also with the Red Sea analogy (Lowell and Genik, 1972). Salt thickness averages 2-3 km.

SOUTHWEST AFRICA MARGIN

Profile 8 crosses the southwest African continental margin south of Walvis Ridge. The margin forms a broad and regular structure with gentle slopes. Transition from the shelf to the slope and from the slope to the rise is very progressive, with no sharp gradient changes.

Under the upper slope lies a basin of more than 3.5 sec of well-stratified sedimentary layers, extending toward the continental shelf. Seaward, under the upper rise, deeper layers wedge out against a marked basement rise, while the upper layers cover it without marked thinning or deformation. Downslope, the sedimentary series is very regular and covers oceanic basement. Reflectors have uniform, well-stratified characteristics, except for the deepest layer, which fills the basement lows and corresponds probably to the limit of the Upper Cretaceous (LePichon, 1968).

This margin has been built out by the deposition of sediments mostly of continental origin. As noted by van Andel and Calvert (1971), erosion channels can be seen in the outer shelf but do not affect deeper layers. The main feature of this section is its great regularity and absence of any slumping or any significant tectonic movement. The basin formed against a basement high, as been reported on other profiles (Talwani and Eldholm, 1973; Uchupi and Emery, 1972). The absence of diapiric structures suggests that from the time of early opening, this margin was open to the deep ocean. Opening could be as early as Late Triassic.

DISCUSSION

Although acoustic basement cannot be followed continuously under continental margins without more sophisticated techniques than the one used here, seismic penetration is above the presently published average. These profiles outline surface and intermediate layers, which in turn often reflect deeper structures. This is especially true if the deepest layer outlined is preopening of the Atlantic, as seems to be the case in these profiles. Postopening series rest unconformably on this older series.

An attempt to differentiate the various eastern Atlantic margins on the basis of the opening mechanism is proposed. First, only when the opening is characterized by two blocks moving away from each other, normal to a spreading center, can simple two-dimensional models be considered. In such models, subsidence of the margin below sea level can possibly be linked to the cooling of the oceanic crust and/or to an increased sedimentary load. Decoupling between oceanic and continental crust presumably will not occur if the load is not excessive, and the continental margin should then exhibit downfaulted preopening sedimentary series covered by gentle postopening prograded sediments, as in profile 8. If for some time after the opening, closed basins exist with evaporitic deposits, these will lie over the downfaulted preopening formations, as in profile 6, and possibly on the oceanic crust, and later be covered by postopening sediments. Subsequent load will induce diapirism.

If the relative motion is parallel to the continental margin, instead of perpendicular, truncation along marginal offsets of the preopening marginal structures will occur, owing presumably to the cooling and subsiding of the oceanic crust. In profile 5 the margin is truncated, and Paleozoic rocks crop out on the slope. The margin structure is largely controlled by the extension of the Romanche Fracture Zone. The marginal ridge has formed a dam for a coastal basin.

Similar control of fracture zones is seen across the Niger Delta, however, the major supply of sediment from the Niger River has completely buried the original continental margin, as indicated by the large overlap in preopening reconstruction. A closed basin probably was formed during the early phase of opening, and evaporites were deposited over oceanic crust. This basin, or a succession of several small basins, extended southward up to the dam formed by the Walvis Ridge (LePichon and Hayes, 1971; Pautot et al., 1973). The salt layer of the Angolan diapir zone is Aptian. The relative shallowness of the diapir structures on the upper Niger Delta implies that if these diapirs are also Aptian, the original oceanic basement is not much deeper and therefore lies well above its theoretical isostatic depth. The original oceanic crust has therefore been thickened by some processes. Alternatively, the limit of the continental crust for the Atlantic may extend to greater depths. Another possibility is that these diapirs are more recent than those in the lower Niger Delta or Angolan continental margin.

On profile 1, the early opening phase was parallel to the margin of the Barents Shelf and was then followed by a normal spreading episode. It is therefore possible that the older sediments were truncated at the time of the opening, but were covered by a major transgressive sequence during the later opening phase. Existence of a marginal ridge related to the early opening should correspond possibly to the Nansen Fracture Zone. This marginal ridge however is hard to dissociate from the Spitsbergen-Northern Norway Caledonian basement. It could also have subsided during the second phase of opening during the cooling of the spreading Atka Ridge.

Profile 2, with its bordering marginal Vøring Plateau and Tertiary oceanic basement, appears to be atypical, unless there exists lateral evolution between a margin with an upper rise ridge (such as the Norwegian margin farther north) and a margin with a marginal plateau. The continuity of the outer rise ridge of the Norwegian margin and the Vøring Plateau Escarpment has been proposed by Talwani and Eldholm (1973). The evolution might correspond to a thickening of the sedimentary basin behind the upper margin ridge. The basin deepens and widens to the south toward the Jan Mayen Fracture Zone. Elastic failure may occur under excessive load with decoupling along the Vøring Plateau Escarpment and the Jan Mayen Fracture Zone. The southwest African margin, marked by an upper-rise ridge, as on profile 8, and the Vøring Plateau may have a similar evolution. The Bay of Biscay profiles are different from other profiles. On profile 3, continental crust extends at least to the base of the slope. The presence of horsts and grabens parallel to the margin, the absence of an obvious spreading center, the opening by rifting, all indicate a tensional origin. On the other hand, the North Spanish trough has a complex origin, since it is related to a reversal in sense of rotation, a rather unique situation.

In the profiles, the boundary between oceanic and continental crust can vary from the continental shelf (profiles 1 and 6) to the upper rise (profiles 8 and 7), at the limit of a marginal plateau (profiles 2 and 5), at the base of the risé or even more oceanward (profiles 3 and 4). Quite obviously no unique morphological feature corresponds to this limit. Furthermore, postopening processes can mask the original position of this limit as evidenced by large preopening overlaps.

The transition from oceanic crust to continental crust normally respects isostasy (Worzel, 1968). It is therfore progressive and implies continental crustal thinning, if this limit is chosen at great depth, or oceanic crust thickening if this limit is chosen up the slope. In any case, crustal structure under the slope involves deep-seated physicochemical changes (Drake and Kominskaya, 1969).

Obviously this problem of the ocean-continent transition is quite important. Long seismic-refraction lines and multichannel seismic profiles are needed to understand better the formation and subsidence of continental margins.

ACKNOWLEDGMENTS

This is Contribution 226 of the Département Scientifique, Centre Océanologique de Bretagne. (Editors' Note: Excellent reproductions of the original data were originally included with this summary. Unfortunately, because of their cumulative size they could not be included in the final version.

BIBLIOGRAPHY

Avery, D. E., Burton, G. D., and Heirtzler, J. R., 1968, An aeromagnetic survey of the Norwegian sea: Jour. Geophys. Res., v. 73, p. 4583-4600.

Bacon, M., and Gray, F., 1971, Evidence for crust in the deep ocean derived from continental crust: Nature, v. 229, p. 331-332.

Baumgartner, T. R., and van Andel, T. H., 1971, Diapirs of the continental margin of Angola, Africa: Geol. Soc. America Bull., v. 82, p. 793-802.

Beck, R. H., 1972, The oceans, the new frontier in exploration: A.P.E.A. Jour. v. 12, pt. 2.

Boillot, G., Dupeuble, P. A., Lamboy, M., D'Ozouville, L., and Sibuet, J. C., 1971, Structure et histoire géologique de la marge continentale au nord de l'Espagne (entre 4° et 9° W), *in* Histoire structurale du golfe de Gascogne: Paris, Ed. Technip, V.6.

Bott, M. H. P., and Dean, D. S., 1972, Stress systems at young continental margins: Nature, v. 235, p. 23-25.

Brognon, G., and Verrier, G., 1965, Tectonique et sédimentation dans le bassin du Cuanza (Angola): Bol. Serv. Geol. Angola, t. 1, p. 5-90.

Bullard, E. C., Everett, J. E., and Smith, A. G., 1965, The fit of the continents around the Atlantic: a symposium on continental drift: Phil. Trans. Roy. Soc. London, ser. A, v. 258, p. 41-51.

Burk, C. A., 1968, Buried ridges within continental margins: Trans. N.Y. Acad. Sci., ser. II, v. 30, p. 397-405.

Cholet, J., Damotte, B., Grau, G., Debyser, J., and Montadert, L., 1968, Recherches preliminaires sur la structure geologique de la marge continentale du golfe de Gascogne: Rev. Ins. Franc. Petrole, t. XXIII-9, p. 1029-1045.

Cudjoe, J. E., and Khan, M. H., 1972, A preliminary report on the geology of the continental shelf of Ghana: Fishery Res. Unit, TEMA, Ghana, Marine Fishery Res. Repts. 204.

Drake, C. L., and Kominskaya, I. P., 1969, The transition from continental to oceanic crust: Tectonophysics, v. 7, p. 363-384.

Eldholm, O., and Ewing, J., 1971, Marine geophysical survey in the southwestern Barents Sea: Jour. Geophys. Res., v. 76, p. 3832-3841.

Emery, K. O., and Uchupi, E., 1972, Western North Atlantic Ocean: topography, rocks, structure, water, life and sediments: Am. Assoc. Petr. Geol. Mem. 17.

Fail, J. P., Montadert, L, Delteil, J. R., Valery, P., Patriat, P., and Schlich, R., 1970, Prolongation des zones de fracture de l'océan Atlantique dans le golfe de Guinee: Earth and Planetary Sci. Letters, t. 7, p. 413-419.

Francheteau, J., and LePichon, X., 1972, Marginal fracture zones as structural framework of continental margins in South Atlantic Ocean: Am. Assoc. Petr. Geol. Bull., v. 56, p. 991-1007.

Gilluly, J., 1964, Atlantic sediments, erosion rates, and the evolution of the continental shelf: some speculations: Geol. Soc. America Bull., v. 75, p. 483-492.

Hinz, K., 1972, Der Krustenaufban des Norwegischen Kontinentalrandes (Voring Plateau) und der Norwegischen Tiefsee zwischen 66° und 68°N nach seismischen Untersuchungen: Meteor C, bd. 10, p. 1-16.

Jones, E. J. W., and Ewing, J. I., 1969, Age of the Bay of Biscay: evidence from seismic profiles and bottom samples: Science, v. 166, p. 102-105.

LePichon, X., 1968, Sea floor spreading and continental drift: Jour. Geophys. Res., v. 73, p. 3661-3697.

———, and Hayes, D. E., 1971, Marginal offsets, fracture zones and the early opening of the South Atlantic: Jour. Geophys. Res., v. 76, p. 6283-6293.

———, Bonnin, J., Francheteau, J., and Sibuet, J. C., 1971, Une hypothèse d'évolution tectonique du golfe de Gascogne, *in* Histoire structurale du golfe de Gascogne: Paris, Ed. Technip, VI.11.

Leyden, R., Bryan, G., and Ewing, M., 1972, Geophysical reconnaissance on African shelf: 2. Margin sediments from Gulf of Guinea to Walvis Ridge: Am. Assoc. Petr. Geol. Bull., v. 56, p. 682-693.

Lowell, J. D. 1972, Spitsbergen Tertiary orogenic belt and the Spitsbergen Fracture Zone: Geol. Soc. America Bull., v. 83, p. 3091-3102.

———, and Genik, G. J., 1972, Sea-floor spreading and structural evolution of southern Red Sea: Am. Assoc. Petr. Geol. Bull., v. 56, p. 682--693.

Mascle, J., Bornhold, B. D., and Renard, V., 1973, Diapiric structures off Niger delta: Am. Assoc. Petr. Geol. Bull., v. 57, p. 1672-1678.

Montadert, L., Damotte, B., Delteil, J. R., Valery, P., and Winnock, E., 1971, Structure géologique de la marge continentale septentrionale du golfe de Gascogne (Bretagne et entrée de la Manche) *in* Histoire structurale du golfe de Gascogne: Paris, Ed. Technip, III.2.

———, Damotte, B., Fail, J. P., Delteil, J. R., and Valery, P., 1972, Structure géologique de la marge continentale asturienne et cantabrique (Espagne du Nord), *in* Histoire structurale du golfe de Gascogne: Paris, Ed. Technip, V.7.

Pautot, G., Renard, V., Daniel, J., and Dupont, J., 1973, Morphology, limits, origin and age of salt layer along South Atlantic African margin: Am. Assoc. Petr. Geol. Bull., v. 57, p. 1658-1671.

Phillips, J. D., Fleming, H. S., and Feden, R., 1972, Aeromagnetic study of the Greenland and Norwegian sea (abstr.): Geol. Soc. America, v. 5, no. 7.

Pitman, W. C., III, and Talwani, M., 1972, Sea-floor spreading in the North Atlantic: Geol. Soc. America Bull., v. 83, p. 619-646.

Rabinowitz, P. D., 1972, Gravity anomalies on the continental margin of Angola, Africa: Jour. Geophys. Res., v. 77, p. 6327-6347.

Rona, P. A., 1973, Rlations between rates of sediment accumulation on continental shelves, sea-floor spreading, and eustasy inferred from the central North Atlantic: Geol. Soc. America Bull., v. 84, p. 2581-2872.

Sibuet, J. C., and LePichon, X., 1971, Structure gravimétrique du golfe de Gascogne et le fosse Nord-Espagnol, *in* Histoire structurale du golfe de Gascogne: Paris, Ed. Technip, VI.9.

———, Pautot, G., and LePichon, X., 1971, Interprétation structurale du golfe de Gascogne à partir des profils de sismique, *in* Histoire structurale du golfe de Gascogne: Paris, Ed. Technip, VI.10.

Sleep, N. H., 1971, Thermal effects of the formation of Atlantic continental margins by continental break up: Geophys. Jour., v. 24, p. 325-350.

Talwani, M., and Eldholm, O., 1973, Boundary between continental and oceanic crust at the margin of rifted continents: Nature, v. 241, p. 325-330.

Uchupi, E., and Emery, K. O., 1972, Seismic reflection, magnetic and gravity profiles of the eastern Atlantic continental margin and adjacent deep-sea floor: I. Cape Francis (South Africa) to Congo Canyon (Republic of Zaire): Woods Hole Inst. Oceanogr. Tech. Rept. (unpubl. ms.)

van Andel, T. H., and Calvert, S. E., 1971, Evolution of sediment wedge, Walvis shelf, Southwest Africa: Jour. Geology, v. 79, p. 585-602.

Walcott, R. I., 1972, Gravity, flexure, and the growth of sedimentary basins at a continental edge: Geol. Soc. America Bull., v. 83, pp. 1845-1848.

Worzel, J. L., 1968, Survey of continental margins, *in* Donoval, D. T., ed., Geology of shelf seas: Edinburgh, Oliver & Boyd, p. 117-152.

Seismic Traverse Across the Gabon Continental Margin

Edgar S. Driver and Georges Pardo

INTRODUCTION

A 150-mile geophysical traverse was recorded across the African Continental Margin by the R/V GULFREX in the spring of 1969. It trends southwest across offshore Gabon, pictured in Figure 1, southwest of Cap Lopez (Port Gentil).

A 24-channel seismic streamer 1 mile in length was employed with an array of four Aquapluse guns constituting the source. Vessel speed was 5 knots. Gravity was recorded by a La Coste Romberg stable-platform gravimeter and magnetics by a Varian proton-precession magnetometer. Satellite navigation was used.

In seismic processing, a 12-fold stack was followed by deconvolution. Elapsed time in half-hour increments in indicated along the top horizontal scale of the profile. The vertical scale is seconds of reflection time. Seismic events that are considered correlative are designated by the same symbol along the length of the section. Based on correlation with wells near the coast, the following stratigraphic identification is suggested.

STRATIGRAPHY

The fragmentary diamond horizon (Fig. 2) probably represents the Lower Cretaceous Gamba unconformity below an extensive salt formation. Below the Gamba unconformity lies the equivalent of the Lower Cretaceous Bucomazi Formation. Triangle reflections are Middle and Upper Cretaceous; one may correspond with the top of the Madiela Formation, as indicated. The star horizon is considered to be the base of the Neogene. A prograding sequence is evident in the Neogene beneath the outer shelf and upper slope. The youngest section, above the circle reflection, appears to be confined to the slope and rise. Some erosion of outer-shelf Neogene deposits may have preceded deposition of this youngest section.

Diapiric activity is the dominant structural mechanism deforming the Neogene and Middle and Upper Cretaceous sediments. Under the shelf, intrusion extends through the Middle and Upper Cretaceous. Proceeding down the slope the Middle and Upper Cretaceous formations are pierced, and the effects of intrusion move progressively higher in the section.

STRUCTURE

The structure at 69-2224 appears syngenetic with deposition throughout the Lower Neogene with attenuated uplift through the circle horizon, but terminating before the present. There is no sea-floor expression. Judging from the surface expressions, diapirs at 70-0112, 70-0236, 70-0412, 70-0590, 70-0700, and 70-0818 and the fault at the base of the slope appear to be currently active. A true-depth section along the seismic profile of Figure 2 is shown in Figure 2.

The northeast-dipping fault found intersecting the sea floor at 69-2324 appears to have been growing throughout the entire Neogene and recent periods. This is evidenced by increasing throw with depth. Fault growth is related to halokinetics.

The scarp at the base of the slope is a regional feature of a different origin and has been reported by numerous investigators. The profusion of small, presently active diapirs near the base of the slope suggests the front of a sedimentary prism sliding westward under the influence of gravity.

Figure 3 is a line drawing of the most important seismic events referred to the same depth scale along the entire section. The dips of the slope and underlying reflections are thereby presented in a more realistic fashion than on the seismic time sections. On the latter the apparent dips toward the abyssal plain are exaggerated by the wedge of slow-speed

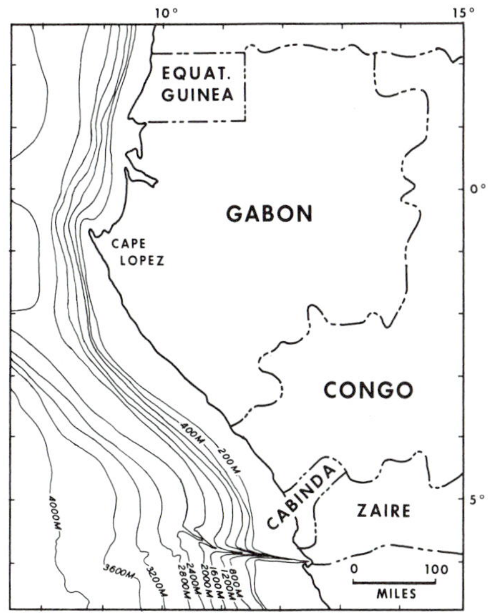

Fig. 1. Index map of offshore Gabon.

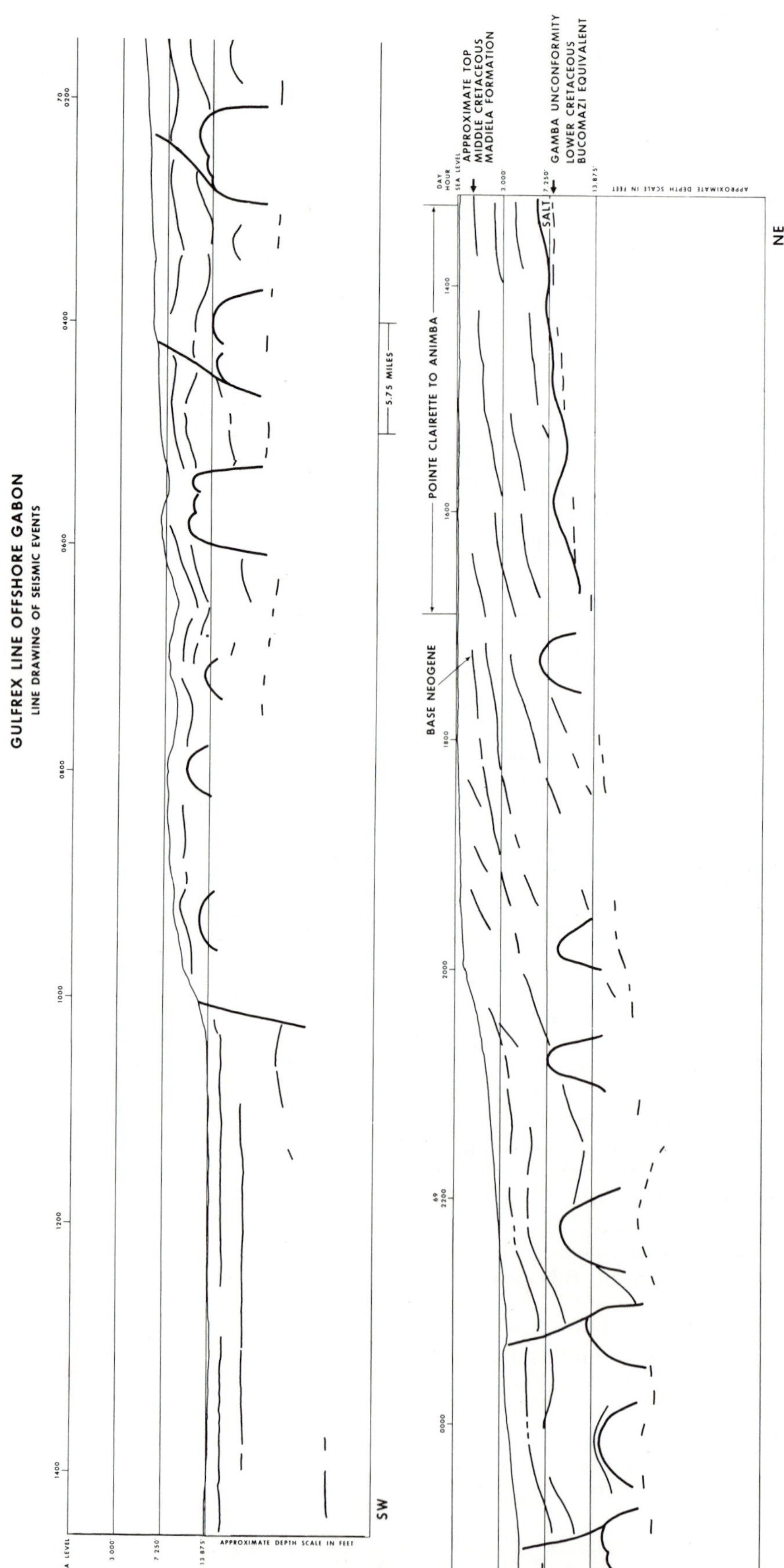

Fig. 3. Line drawing of seismic events, based on true-depth interpretation.

Continental Margin in the Northern Part of the Gulf of Guinea

Jean-Raymond Delteil, Pierre Valery, Lucien Montadert, Catherine Fondeur, Philippe Patriat, and Jean Mascle

INTRODUCTION

The continental margin of the West African Shield in the northern Gulf of Guinea may be divided into three characteristic areas:

1. The Liberia continental margin is linked to two structural trends. The first one corresponds to the escarpment of the Liberia marginal plateau, the second one to a deep structural trend that merges into the continental slope of the Ivory Coast to the west of Abidjan and is in alignment with it. This feature has been identified as the prolongation of the St. Paul's fracture zone.

2. The continental margin off the Ivory Coast and Ghana is characterized by the occurrence of a thick sedimentary basin which developed between the coastal fault system and the Romanche fracture zone. The seismic profiles reveal that this basin was weakly structured and contains sediments as old as the Late Cretaceous. The basin continues eastward on the marginal plateau of the Ivory Coast and Ghana. Here one can postulate the presence of older formations, probably Paleozoic in age. It is bounded to the south by a steep escarpment aligned in the Romanche fracture. One can point out the noteworthy similarity, from a tectonic standpoint, of the Liberia and Ivory Coast-Ghana marginal plateaus.

3. To the east the main feature of the Niger continental margin is, on the one hand, the development of diapiric phenomena that can be attributed to clay-shale diapirs, and, on the other, the existence of two fracture zones, the Chain fracture and a new one called the Charcot fracture zone.

These different regions of the margin thus correspond to crustal blocks separated by the oceanic equatorial fractures until the inner part of the gulf. In these structural blocks the sediment distribution shows that the geology has been controlled by the damming effect of the fracture zones, and consequently by the opening in Aptian time and the sea-floor spreading in the South Atlantic.

MARINE SURVEYS

The marine geology of the northern part of the Gulf of Guinea was relatively unknown prior to 1968, the date of the Guinea 1 cruise (CEPM), carried out on board the oceanographic ship JEAN CHARCOT of CNEXO, which led to a determination of the major Comité d'Études Pétrolieres Marines, composed of ELF-ERAP, Société Nationale des Pétroles d'Aquitaine, CFP-TOTAL, and the Institut Francais du Petrole.)

Special light was cast on the E-trending prolongations of the Romanche and Chain fracture zones, already studied at the ocean ridge (Heezen et al., 1964), and the role played by them in the individualization of the sedimentary basins of the continental margin off the Ivory Coast and Ghana (Fail et al., 1970; Arens et al., 1971).

The Walda (COB-CNEXO) and Benin (CEPM-structural features of this region. (CEPM is the IPGP) cruises of the N/O JEAN CHARCOT in 1971 in this region clarified these findings. The authors present a synthesis of these results here. Their interpretation has been checked against data provided by the Lamont-Doherty Geological Observatory (VEMA 27, CONRAD 13, VEMA 29) and the Woods Hole Oceanographic Institution (ATLANTIS II.75) (Fig. 1).

REGIONAL GEOLOGICAL FRAMEWORK AND BATHYMETRY

On land, north of the Gulf of Guinea, the African craton comprises two provinces bounded by an important NNE-trending fault (Burke and Dewey, 1970; Kennedy and Grant, 1969), passing through the Accra area, the origin of which has been discussed in terms of global tectonics (Hurley, 1972).

West of this fault, the Birrimian shield dates from 1.7 to 2.2 billion years; to the east, datings indicate an age of 500 to 650 my for the Dahomeyan shield (Clifford, 1968; Cahen and Smelling, 1966). The latter province seems to have been affected most by the Mesozoic and Cenozoic, responsible for the creation of extensive coastal and intracontinental sedimentary basins in West Africa (Radier, 1959; Furon, 1968; Kennedy, 1965; Karpoff, 1965; Reyre, 1966; Cratchley and Jones, 1965; Lemarechal and Vincent, 1971; Wright, 1968; Burke et al., 1970; Kennedy and Grant, 1971) and Brazil (Almeida, 1967).

The coastal basin of the Ivory Coast, individualized on the Birrimian craton, is only slightly developed on land, whereas its underwater prolongation, already suggested at the level of the continental shelf (Spengler and Delteil, 1966) is determined by the passage of the large equatorial oceanic fracture zones. The directions of these fractures (Saint Paul,

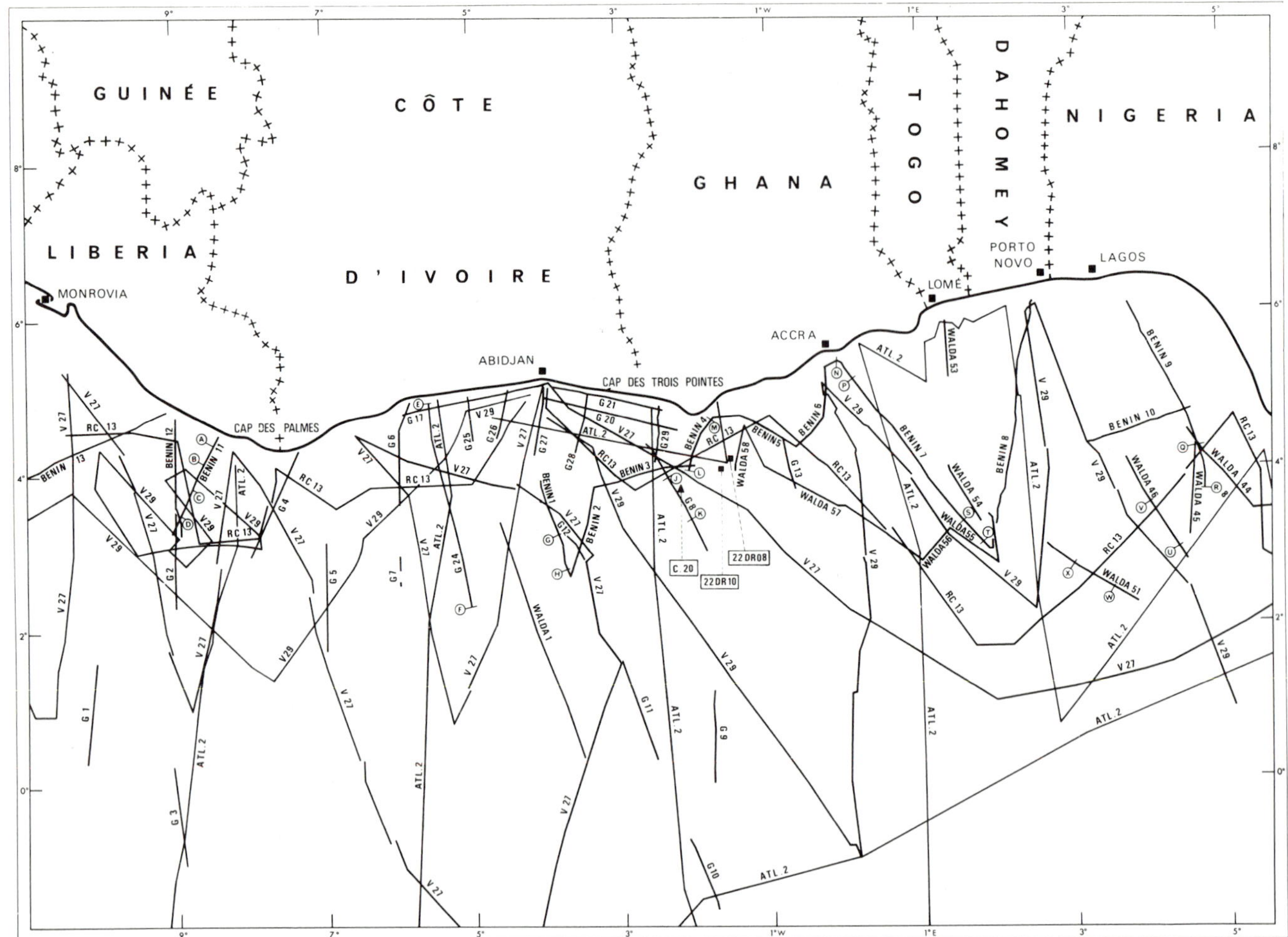

Fig. 1. Index map of cruise tracks. Letters show locations of profiles described in text.

Romanche, and Chain fracture zones) led various authors (Le Pichon, 1968; Le Pichon and Hayes, 1971; Francheteau and Le Pichon, 1972) to suggest evolutionary models for the opening of the South Atlantic, based on the initial reconstruction by Bullard et al. (1965).

The bathymetric map (Fig. 2) reveals the size of these fracture zones and permits the differentiation of the following morphological units:

1. The continental margin off Liberia, which may be divided into two sectors:

a. West of 8°30'W, the continental shelf has a mean width of 45 km, and a marginal plateau is observable to a depth of 3,000 m; it is bounded on the south by an escarpment.

b. East of 8°30'W, the continental shelf is far narrower (25 km); the continental slope exhibits a complex morphology and is furrowed by at least one substantial valley (Robbe et al., 1973).

2. The continental margin off the Ivory Coast and Ghana, which may also be divided into two sectors:

a. Off the Ivory Coast, the continental shelf is narrow and the upper continental slope steep and furrowed by several canyons (Martin, 1970; Dietz and Knebel, 1971). The passage to the abyssal plain occurs through a highly developed rise:

b. Off Ghana, the continental margin exhibits a highly individual morphology, distinguished by:

(1) A wide continental shelf (90 km to the southeast of the Cape of Three Points).

(2) A marginal plateau with a mean depth of 2,000 m, slightly inclined toward the southwest and continuing unbroken to the Ivory Coast Rise. On the south it is bounded by

(3) A WSW-trending escarpment, with a mean slope of 14°, reaching 27° locally, which continues between 3°W and 1°W.

3. The continental margin west of the Niger Delta. East of 1°W, the continental shelf fails to reveal any noteworthy prolongation except off the Niger Delta. The continental slope remains highly pronounced in its upper portion, while the striking morphological feature is the development of a very extensive rise connected to the Niger Delta.

4. The deep zone in the northern portion of the Gulf of Guinea. South of 2°N, this corresponds to the rough flanks of the mid-ocean ridge, which are cut by WSW-trending fracture zones. East of 5°W this unit is buried beneath an abyssal plain of limited extent.

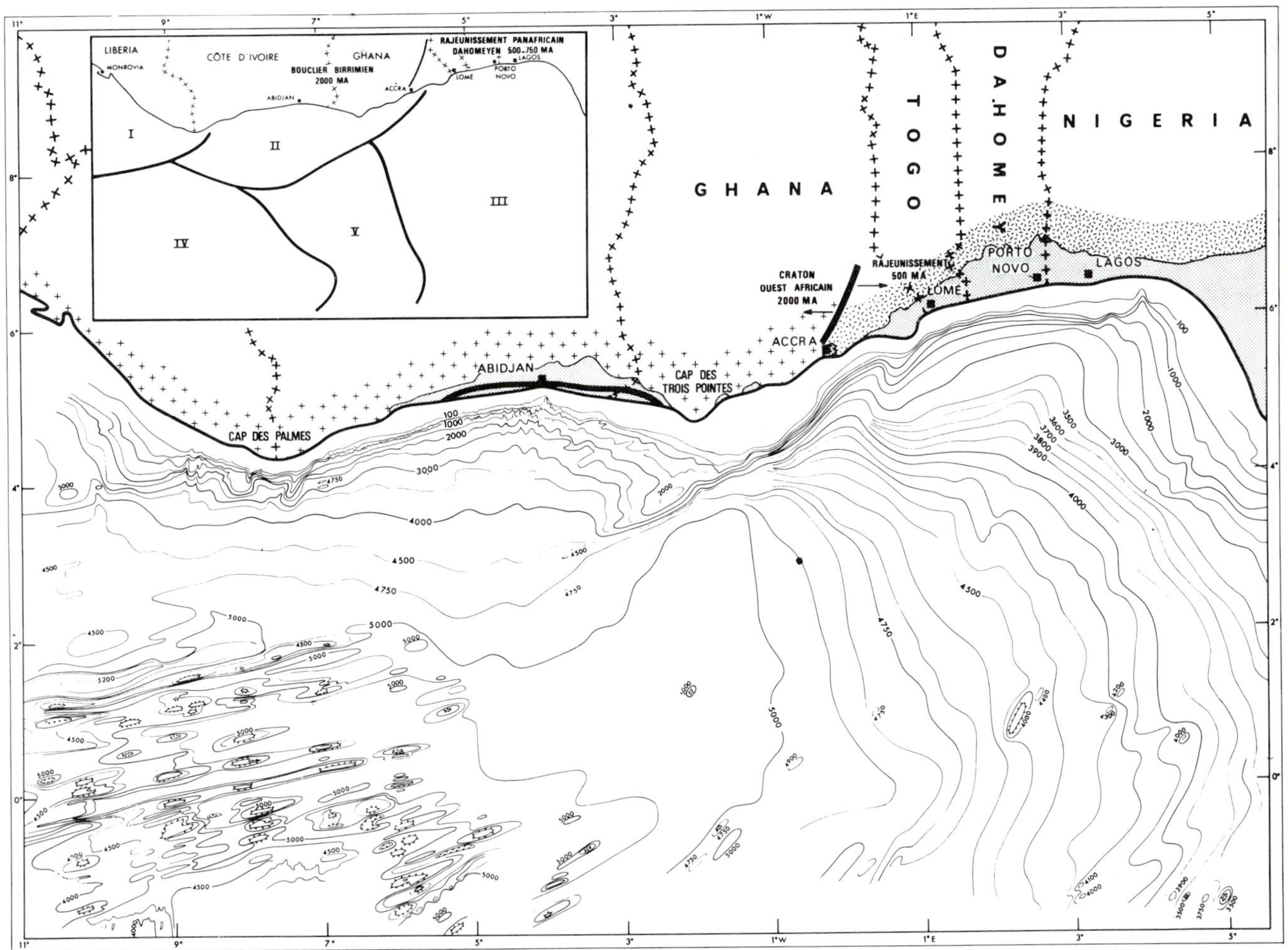

Fig. 2. Bathymetric map of northern Gulf of Guinea. Depth curves are in noncorrected meters.

STRUCTURE OF THE CONTINENTAL MARGIN SOUTH OF LIBERIA

The deep structure of the continental margin south of Liberia is caused essentially by two faults:

1. To the south, a deep fault that is not revealed by the topography and was previously described as the "rise fault" (Arens et al., 1971), coincides with a significant magnetic alignment (Fig. 3). Its throw may be estimated at 2 sDT (seconds, double time).

2. To the north, the fault corresponding to the escarpment of the "Liberian Ridge" is linked to an uplift of the basement and limits the Liberian marginal plateau. It does not exhibit a magnetic signature. East of 9°W this structure merges into the continental slope.

These structures are visible on the Benin 11 seismic profile, of which two sections are shown. Section AB to the north (Fig. 4) indicates the probable outcrop of basement on the continental shelf. The basement collapses in successive fault blocks, and supports a thick sedimentary basin on the continental slope. The stratigraphic series, which is 4 to 5 sDT thick, comprises three units:

1. An upper unit, I (mean thickness 2 sDT), which is acoustically transparent and unconformable, on

2. A central unit, II, distinguished by two strong low-frequency reflectors (mean thickness 0.6 sDT).

3. A lower unit, III, in which the reflectors are discontinuous and difficult to correlate. The structural framework of this unit suggests a sedimentary character, but the attribution of part of this unit to the substratum of the basin cannot be excluded.

To the south (Fig. 5) the oceanic substratum is only covered by the upper portion of unit I (1.3 sDT thick). West of 11°W the structure of the deep fault is different. It appears as a ridge of the oceanic basement bounded on both sides by a sedimentary series about 2 sDT thick. From the magnetic standpoint, it is distinguished by a characteristic signature, with a slope trending southward or northward, depending on the location of observations along this structural axis.

These different characteristics—fault prolongation and trending parallel to the Romanche and Chain fracture zones, geological structural framework associating basement variations in depth with sudden variations in sediment thickness, and magnetism—permit the identification of the two faults in the continental margin south of Liberia as

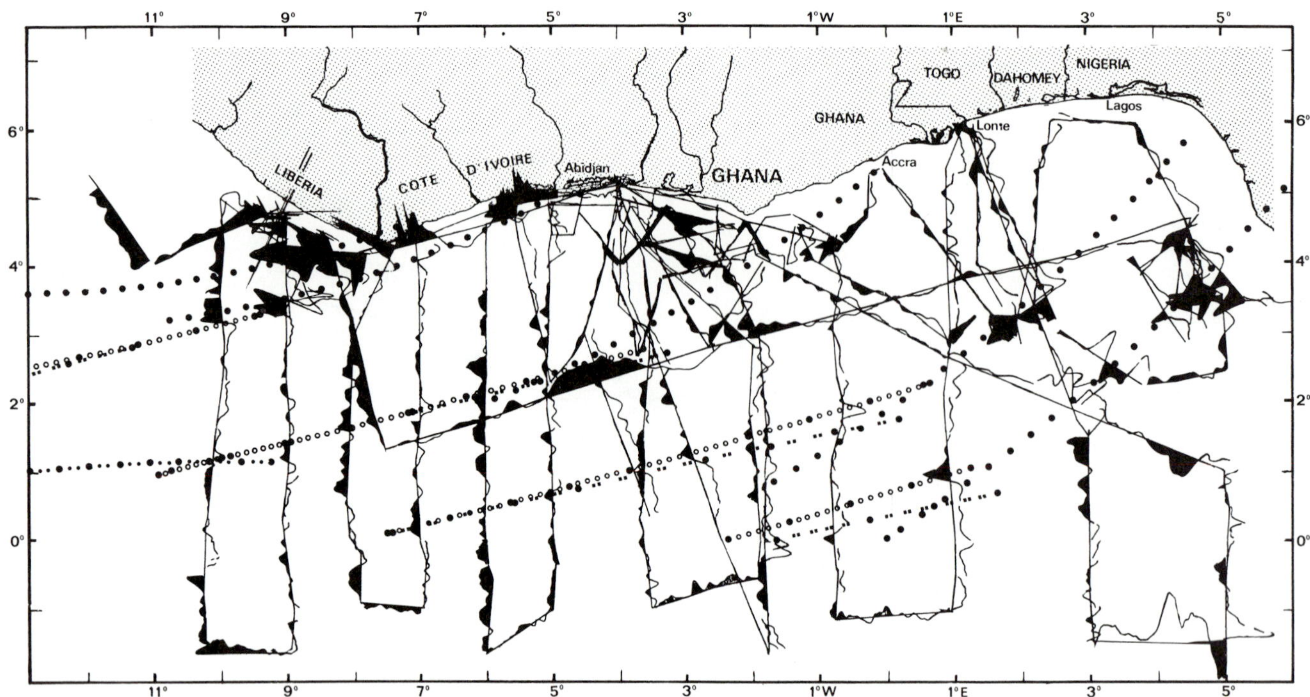

Fig. 3. Magnetic anomalies from Guinee (1968) and Walda and Benin (1971). 1 cm approxiamtes 400 gammas. Black areas are negative; dotted lines correspond to small circles plotted at 1° interval about different centers of rotation. Small black dots, 26°N i.e., 72°7N 170°E; black squares, 72°7N 139°⅓W; wide circles, 64°7N 38°1W; black circles only, 20°N 13°W.

fracture zones in the Mid-Atlantic Ridge. The deep fault with a pronounced characteristic magnetic signature is identified as the St. Paul fracture zone, rather than a fault restricting the craton, as was formerly suggested (Arens et al., 1971).

STRUCTURE OF THE CONTINENTAL MARGIN OFF THE IVORY COAST AND SOUTHWEST OF GHANA

The continental margin off the Ivory Coast and southwest of Ghana is linked to the existence of a very extensive sedimentary basin bounded by two remarkable structural features:

1. To the north the major fault of the Ivory Coast, corresponding to a system of normal faults (Spengler and Delteil, 1966), separating, with a highly substantial throw (>5000 m), the Birrimian basement and its thin cover of continental sediments, from the deep basin proper.

2. To the south, in the abyssal plain, the Romanche fracture zone (Arens et al., 1971), playing the role of a sedimentation dam and constituting the noteworthy tectonic and morphological element of the continental margin.

The Guinea 24 seismic profile (Fig. 6) illustrates the general prolongation of the deep basin of the Ivory Coast. This appears to be only slightly tectonized except on the continental slope. The sediment thickness is important (at least 4sDT) and, according to the authors, two seismic reflectors ascribed to the Cretaceous-Tertiary limit and to the Early Cretaceous-Late Cretaceous limit, in line with oil wells data in the Ivory Coast, have been traced as far as the Liberian basin described previously.

The most ancient series, dating from the Early Cretaceous, appears to lie on the oceanic crust in the southern part of the profile. This substratum is no longer visible below the continental slope, in which a sudden thickening of sediments is observable. This occurrence, in the deep structure of the basin, had initially been identified (Fail et al., 1970; Arens et al., 1971) as the southern limit of the African craton. The southern extremity of the profile reveals the closing of the basin by the Romanche barrier, beyond which the sedimentary cover is much thinner; the latter rests on an oceanic substratum which is younger than in the north, as implied by the offset of the Romanche fracture at the Mid-Atlantic Ridge (about 800 km).

To the east, the evolution of the basin of the Ivory Coast is illustrated by the Benin 1, Guinea 8, and Benin 4 profiles (Figs. 7-9).

The Benin 1 profile (Fig. 7) overlaps the Romanche fracture zone, which limits two clearly differentiated zones, as in profile 24 (Fig. 6). The thickness of the sedimentary cover in the northern zone is substantially greater than 3.5 sDT. Two reflectors may be observed there: the upper reflector is ascribable to the Cretaceous-Tertiary limit; the underlying reflector, in line with correlations with subsurface data, has been ascribed to the base of the Late Cretaceous or the upper limit of the early Cretaceous (Albian-Cenomanian). The oceanic substratum is relatively shallow in the southern zone, and the sedimentary series barely exceeds 1 sDT in thickness. The distinguishing features of this profile are the absence of structuration in the basin, as well as the presence of an oceanic substratum, which is still observable as far as 20 km to the north of the Romanche structure.

The Guinea 8 profile (Fig. 8) has revealed the existence of a suspended sedimentary basin bounded on the south by a substratum high, called the Ivory Coast-Ghana Ridge, and which corresponds to the eastern prolongation of the Ivory Coast basin. In this sector it comprises the following series: a disconformable Mio-Pliocene series; a Tertiary to late Cretaceous series, locally unconformable, on an early Cretaceous series of unknown thickness, fairly highly tectonized, and with perhaps more ancient series beneath it.

The uplift of the lower-unit reflectors on the Ivory Coast-Ghana Ridge has been confirmed by core samples, which have revealed a coarse detrital formation, dating from the Albian, below 7 m of recent sediments. The lower beds of the core provided a rich flora dating from the middle to late Albian.

These same levels also yielded some debris of foraminifera and microplankton, and several plant species were identified in the overlying beds suggesting deposition very close to a coastline (Benin: core 20, Fig. 1).

The Ivory Coast-Ghana Ridge, an escarpment with an average slope of 14° reaching 27° locally, connects the basin described above to the abyssal plain of the Gulf of Guinea. This escarpment corresponds to the boundary between the oceanic crust and the cratonic area (Fail et al., 1970; Arens et al., 1971), and constitutes the prolongation of the Romanche fracture zone. Especially noteworthy is the remarkable similarity between the Ivory Coast-Ghana Ridge and the Liberian Ridge, from the standpoint of origin and structure; it may also be compared with the continental margin of South Africa at the Agulhas fracture zone (Talwani and Eldholm, 1973).

The Benin 4 profile (Fig. 9) reveals that the structural framework of the central portion of the marginal basin is linked to the existence of a substratum uplift bounding two compartments: one west and south compartment, containing the series described above for the deep basin; together with a north and east compartment in which correlations indicate a considerable thinning of the post-Early Cretaceous series and the presence of a thick sedimentary series (noted J-Ci), which is ascribed to the early Mesozoic, but which probably contains Paleozoic formations. It should be recalled that the continental shelf of Ghana exhibits a Paleozoic and Mesozoic basin distinguished by a horst-and-graben architecture generally caused by NE- or NW-trending faults (Cudjoe and Khan, 1973). Observations carried out on the northern compartment of profile 4 appear to indicate a structural framework of this type, consequently suggesting geological continuity with the Ghanaian shelf from the horst mentioned above.

The deep structural framework of the Mesozoic and Tertiary basin off the Ivory Coast thus tends to become more complex toward the east, accompanied by the convergence of the two fundamental tectonic elements of the margin, the coastal fault in the north, and the Romanche fracture zone in the south.

At the present time it does not appear possible to locate the edge of the cratonic area in this basin. The available data only indicate a deep structural framework comprising faulted compartments affecting the pre-Late Cretaceous areas alone, and suggesting a continental structure in the western portion of the Ivory Coast-Ghana marginal plateau.

It may be taken for granted that part of the Paleozoic basin revealed by oil exploration drilling in the Ghanaian shelf, the outcroppings of which are well known on land (Crow, 1952), is prolonged in the eastern sector of the marginal plateau.

In effect, two types of samples were obtained from dredgings 22 DR 10 and 22 DR 08 carried out during the Walda cruise on the continental slope south of the Cape of Three Points (Fig. 1) at the eastern end of the Ivory Coast-Ghana Ridge:

1. Coarse-grained, feldspathic, micaceous sandstones and micaceous, ferruginous sandstones. These samples are comparable to the facies described by Crow in Takoradi sandstones (Devonian).
2. Schists and micaschists with abundant oriented-phylosilicates and feldspar.

The authors' interpretation thus shows good agreement with these results, which were obtained at a relatively short distance from profile Benin 4; these results also confirm the continental nature of at least one part of the Ivory Coast-Ghana Ridge.

CONTINENTAL MARGIN OFF THE NIGER DELTA

The Chain fracture zone has been identified as the southern extremity of the Benin 7 (Fig. 11) and Benin 8 profiles as well as the Walda 54 and 55 profiles. It corresponds to an uplift in the oceanic substratum separating two compartments distinguished by a sedimentary cover which is thicker in the north than in the south. However, the throw is low (0.5 sDT). On the other hand, a substantial magnetic anomaly is observable above the structure.

The seismic profiles located off the Niger Delta, particularly Walda 45, Walda 46 (Fig. 12), and Walda 51 (Fig. 13) reveal the existence of a highly characteristic structure at the level of the oceanic basement, about 150 km south of the Chain fracture zone.

As noted, this structure exhibits a significant variation in sediment thicknesses, with substantial thinning toward the south. This structural unit also coincides with a significant magnetic alignment clearly observable between 2°W and 5°E. These criteria and their relation to a structural axis parallel in direction to the Romanche and Chain fracture zones have led the authors to its definition as a new fracture zone, for which the name "Charcot" is proposed.

It appears that this fault was overlapped at the southern extremity of the Benin 9 profile (Fig. 14) by 3°N and 5°E, as the same magnetic anomaly is found here. However, in this profile, the substratum structure is obscured by the complex "diapiric" phenomena of the continental slope off the Niger Delta.

Its possible prolongation toward the continent was not studied. The existence of a gravimetric high (Hospers, 1965, 1971) in the region of the mouths of the Niger should be recalled. However, the existence of a relationship between these two structures cannot be positively affirmed.

From the standpoint of oceanic tectonics, it should be noted that the Charcot fracture appears to exceed the Chain fracture in importance by its influence on the sedimentation in the eastern region of the Gulf of Guinea. On the other hand, it cannot be positively identified in the direction of the Mid-Atlantic Ridge and at the central rift.

The main feature of the margin resides in the development of relatively intense diapiric tectonics visible on the Benin 9 profile (Fig. 14), which appears to persist at the foot of the continental shelf at a depth of about 3,500 m. The origin of this diapiric activity has been debated, and has been attributed by some authors to a salt diapir origin (Pautot et al., 1973; Mascle et al., 1973).

The authors feel that these phenomena should be attributed preferably to clay-shale diapirs. This hypothesis is in agreement with the ideas expressed by other authors (Burke, 1972; Beck, 1972), who connect the structures with the outcrop of the Akata Shale formation on the continental shelf of the Niger Delta. According to available data, these diapiric structures appear, moreover, to line up parallel to the slope, and may represent the underwater prolongation of the growth fault network of the Niger Delta (Merki, 1970). This structural organization is disturbed in the axial portion of the delta by the probably passage of the Charcot fracture zone.

To the west, i.e., on the downstream side of the Niger Rise, and south of the Romanche fracture zone, the sedimentary cover thins considerably and steadily. It has been estimated at over 5 sDT facing the delta and at 3.5 sDT off the Volta delta. Throughout this region, which overlaps the prolongation of the Chain and Charcot fracture zones, the sedimentary basin is weakly structured, except near these great oceanic trends. As for the major portion of the Ivory Coast-Ghana basin, the sediments have been deposited on a substratum of which the oceanic origin and the easterly extension beneath the delta are generally recognized (Hospers, 1965, 1971). The greatest sediment thicknesses have been observed north of the Chain fracture and of the Charcot fracture in the western portion of the Benin 10 profile.

GENERAL STRUCTURE OF THE GULF OF GUINEA

Substratum of the Basin

The deep structure of the Gulf of Guinea in the area under consideration is illustrated by the isochronous map at the substratum level in Figure 15. This map leads to the following observations:

1. The equatorial fracture zones extend to the internal portion of the Gulf of Guinea.

2. These fractures, whose positions have been clarified with respect to previous data, cause a division into parallel compartments. From north to south, the oceanic substratum exhibits a stepwise structure mounting toward the south. This agrees with the offsets observed at the Mid-Atlantic Ridge, implying that the substratum is more ancient on the northern than on the southern block for a given meridian, and that the subsidence has thus been more pronounced. This is probably accentuated by the damming effect of the fracture zones. In any given compartment, a deepening of the basement toward the continental margin is indeed observable. By comparison with the bathymetric map, it may be deduced that the sediment thickness is greatest north of the Romanche and off the Niger.

3. The orientation of the seismic profiles did not permit clarification of any noteworthy structural fault perpendicular to the fracture zones. The magnetic E-W line between the Romanche and the Chain trends failed to reveal any special anomaly of this type. However, despite the spacing of the profiles, the isochrone drawing seems to point to a new occurrence: the Chain and Charcot fracture zones appear to be subdivided into lengths that are slightly oblique to each other, causing modification of the general trend of the fracture. These gaps seem to disappear in the Chain and Charcot fractures at about 3°W and 2°W, respectively, i.e., in the region where a general change in direction of these fractures is observed.

The role of the Chain and Charcot fracture zones in the shaping of the continental margin off Nigeria cannot be understood because they are buried beneath the Niger continental rise. However, knowledge of the location of the extremity of the Romanche in the direction of the continent and the large offsets of the equatorial Mid-Atlantic Ridge should impose location of the extremities of these two fracture zones if the stresses of plate kinematics are applied to this region.

It should be recalled that any finite motion between two rigid bodies on a sphere can be described by a single rotation defined by one center and one angle, but that the real motion of these bodies are recorded by the fracture zones results from a succession of short-time-span rotations, if not an infinite number of quasi-instantaneous rotations.

A fracture zone thus represents a sequence of arcs of small circles about different centers. These centers of rotation can only be determined if corresponding small circles are known from at least two fracture zones. The determination of the corresponding small circles in the Gulf of Guinea is extremely difficult, owing to the large offsets of the ridge (the fracture parallelism fails to make any sense) and mainly because only one fracture zone (Romanche) is well known and because the ocean floor cannot be adequately dated on the basis of magnetic anomalies in their equatorial area.

To circumvent these difficulties, the rotation centers were determined from the Romanche fracture alone through the observation that its general direction remains constant over a long distance. Three series of possible poles were initially calculated for positions falling between 13°W and 10°W, 10°W and 5°W, and 5°W and 2°W, respectively [the most recent portion (Heezen et al., 1964), lying between 16°W and 13°W, is accurately defined by Morgan's pole: 56°N, 37°W]. These rotations were subsequently applied to neighboring fracture zones (topographic trends and magnetic anomalies) Figs. 3 and 15). In the portions corresponding to the "Romanche" between 13°W and 5°W, the Chain fracture zone is too ill defined or too close to permit a definitive choice between the possible poles. In the area lying between 5°W and 2°W, the best pole (20°N, 13°W) closely approaches that of Le Pichon and Hayes (1971), but this verification by the "Chain" and "Charcot" fractures is not significant. It should be observed that the arc of the small circle for this pole is only about 10°.

In the light of present knowledge, it is thus impossible to define accurately by this indirect method the terminations of the fracture zone by tracing their progress from the ridge. This ambiguity may be dispelled by a detailed study of the Ascension fracture zone, which is sufficiently distant from that of the Romanche, and also by a comparison with the unpublished center of rotation of Ladd et al. (1973), who suggested the paleogeographic reconstruction of the South Atlantic on the basis of magnetic anomalies.

On the other hand, the authors feel that the present Chain fracture is probably not the prolongation in space of that of the Chain observed beneath the Niger continental rise, implying that a substantial reorganization of the ridge about the line of epicenters by the creation of new offsets must occur at each important change in the direction of spreading. This change in the physiognomy of the ridge would tend to explain the "double" fracture of Saint Paul off Liberia and the lesser displacement (500 km instead of the current 800 km) on the Romanche, at the beginning of the opening. To understand fully the early opening phase, it would be necessary to possess more accurate knowledge of the Chain fracture zone between 1°W and 1°E, in which the direction changed significantly, together with knowledge of the prolongation of the equatorial fracture zones toward the American Continent and their relations with the North Brazilian Ridge.

Geological History of the Margin

The continental margin of the Gulf of Guinea was formed by a rift process, which was followed, after a substantial time span, by the expansion and renewal of the ocean floor. These events are considered to have occurred in the following manner:

The "weakness lines" of the African-South American craton during the Precambrian (McConnell, 1969) and at the beginning of the Paleozoic (Kennedy and Grant, 1969; Kennedy, 1965) led to the creation of the initial outline of the South Atlantic in the Gulf of Guinea (Reyre, 1966; Kennedy, 1965; Wright, 1968; Burke and Dewey, 1970; Burke et al., 1970; Kennedy and Grant, 1971; Reyment, 1969; Hurley, 1972).

The beginning of rift activity, corresponding to a subsidence zone (graben), is generally distinguished by the deposit of thick fluviolacustrine series with coarse detritic material at the edge of the depressed area, reflecting rapid subsidence (Wenger, 1973). These conditions definitely existed on African and Brazilian margin, from the end of the Jurassic, in basins located north of the Walvis Ridge, and from the Early Cretaceous, at least, in the northern region of the Gulf of Guinea. No current basis can be found for a more ancient dating of the beginning of the rift process: the magnetic anomalies fail to clarify this point, as it is certain that the opening began considerably earlier than the date indicated by the most ancient known anomaly, dating from 72 my (Ladd et al., in press).

In the northern region of the Gulf of Guinea, the initial rift was probably influenced, from the tectonic standpoint, by the expansion of the floor of the North Atlantic, as suggested by the existence of basic volcanic rocks on the continental shelf of Ghana (Cudjoe and Khan, 1972), dated 160-165 my (personal communication), and of those which generally crop out in Liberia, dated 180 my (Nyema Jones and Stewart, 1970); this does not necessarily imply the start of the expansion phase in the area of survey.

The filling of the grabens by essentially continental deposits stopped at different periods and under different conditions along the periphery of the Gulf of Guinea. Between the Walvis Ridge and the Niger, the initial arrival of the sea led to the development of a thick evaporitic basin dating from the Aptian and probably from the late Aptian (ELF, unpublished documents and Wenger, 1973). Farther to the north, in Nigeria and in the Benoué trough, the marine transgression dates from the middle Albian (Murat, 1970; Reyment, 1970, 1971), with the deposit of argillaceous and calcareous deposits. The first unconformable marine deposits in the Ivory Coast are argillaceous and also date from the middle and late Albian (ELF, unpublished documents).

These results suggest that the transgression of the sea occurred from the south of the Atlantic in the direction of the Gulf of Guinea, provided that one accepts that the junction with the North Atlantic and the epicontinental seas of Central Africa was not established until the Late Cretaceous (early Turonian: Reyment, 1969).

The problem raised in the study zone, one applying throughout the African margin of the Gulf of Guinea, is to know whether the marine transgression was associated with the graben-formation period, or whether it had already occurred as a result of the first drift movements. There is some basis for believing that the second explanation is the more likely one.

In the first place, core samples from the Ivory

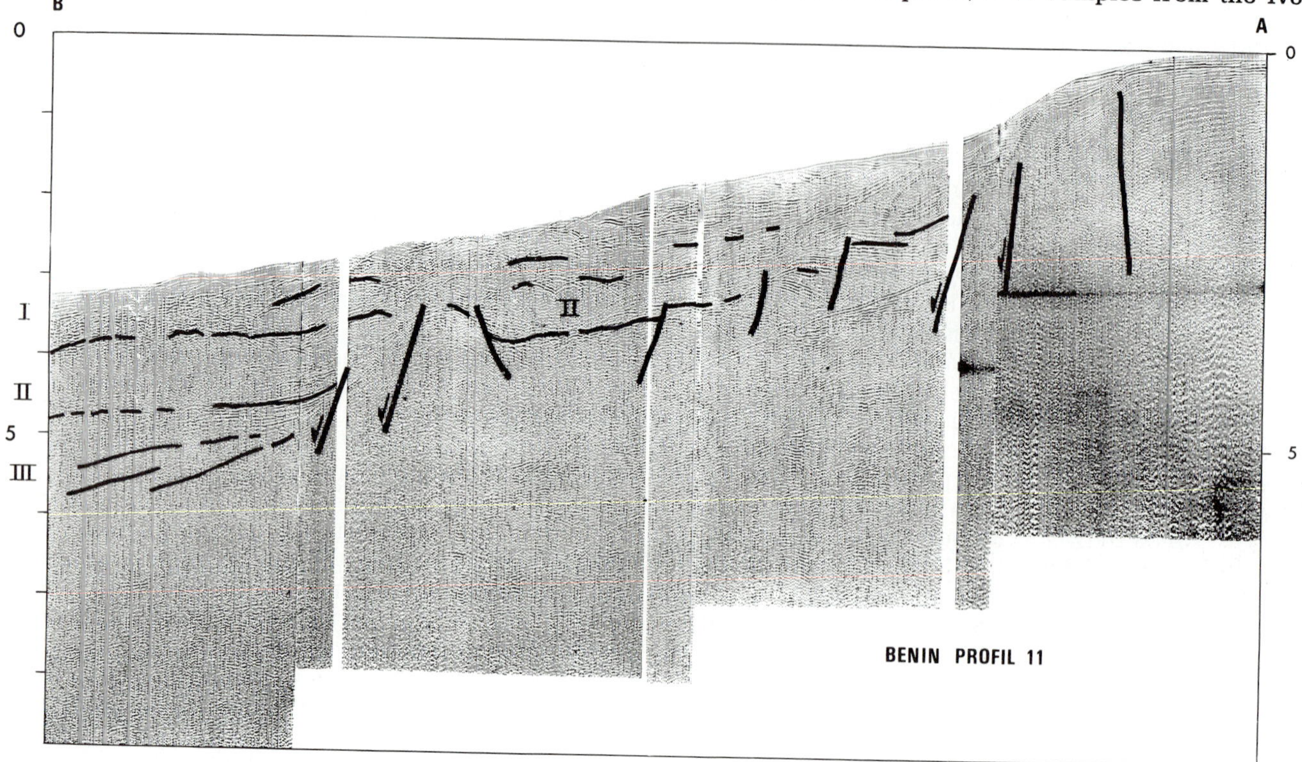

Fig. 4. Seismic reflection profile Benin 11 (section AB).

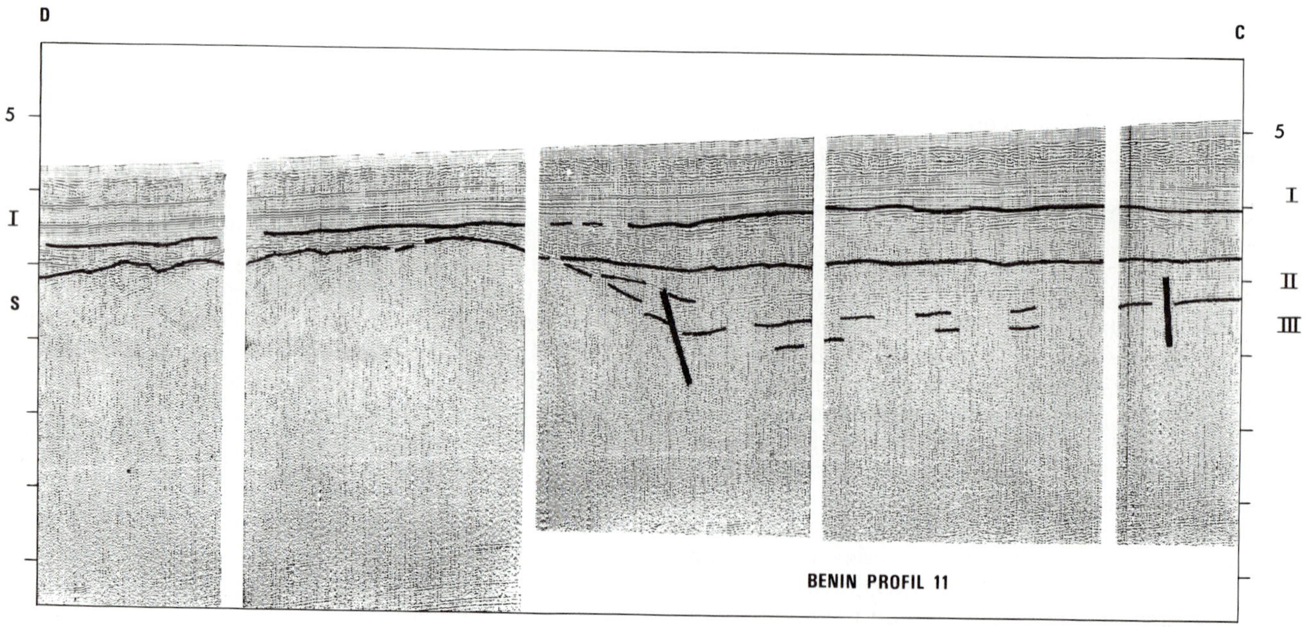

Fig. 5. Seismic reflection profile Benin 11 (section CD).

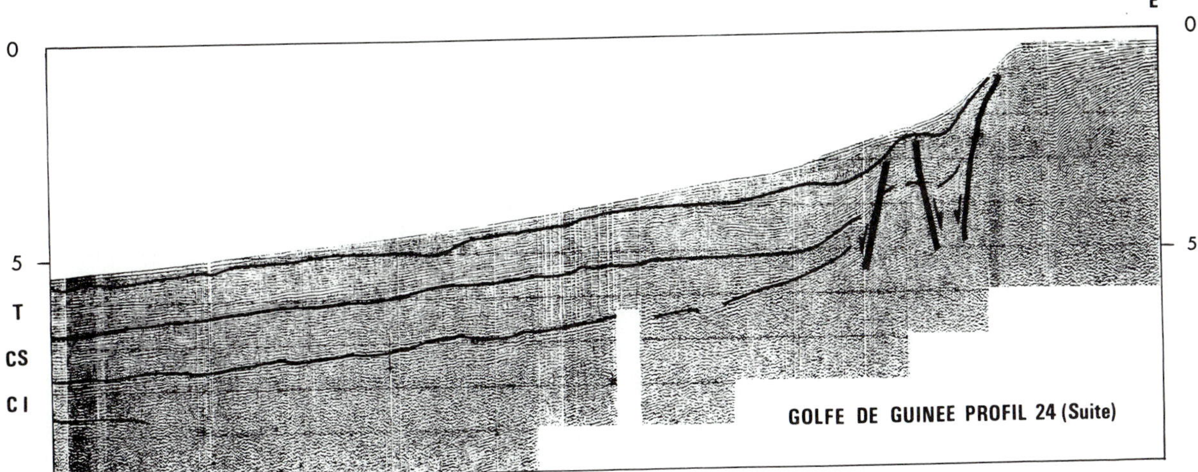

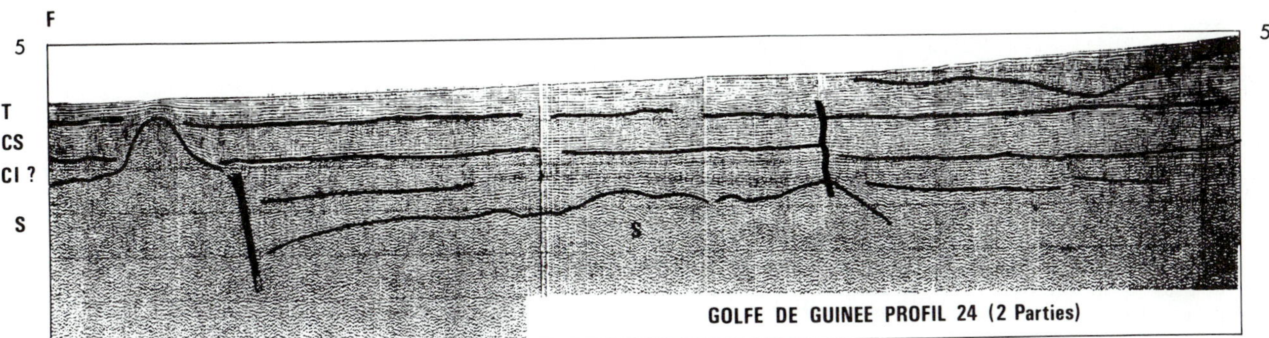

Fig. 6. Seismic reflection profile Guinee 24 (section EF).

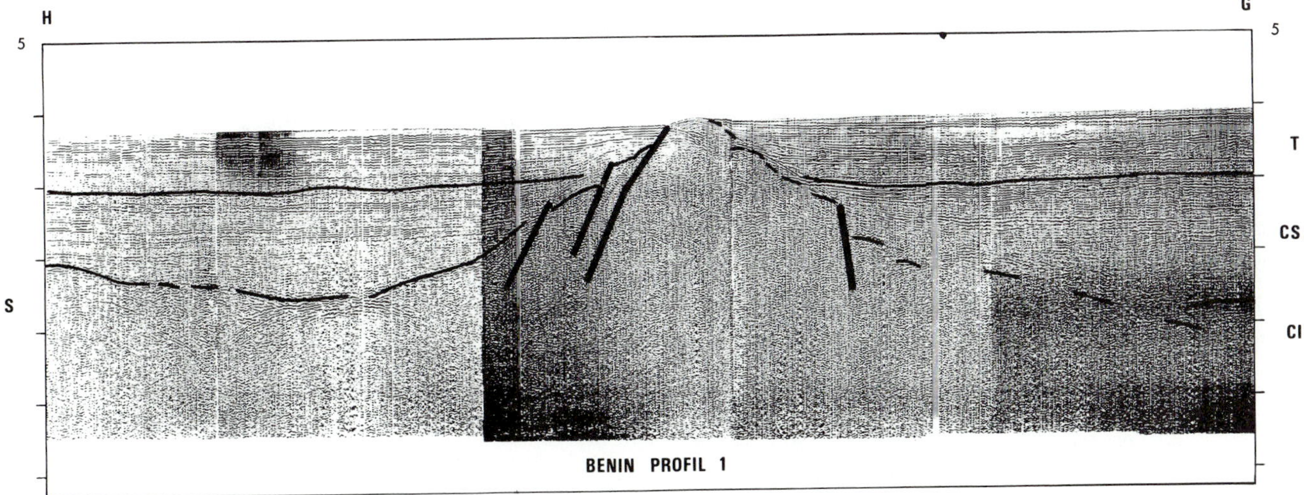

Fig. 7. Seismic reflection profile Benin 1 (section GM).

Coast-Ghana Ridge revealed the existence of a detrital marine formation dating from the middle to late Albian, indicating that this structure was already individualized as a high zone, since to the north and on the continental shelf of the Ivory Coast, the sedimentation of the Albian, post-unconformity, is essentially argillaceous. The tectonic relationship revealed between this structure and the Romanche fracture zone thus suggests that this oceanic fracture already existed at this time.

Second, the authors associate the radical change in sedimentation occurring during the Aptian

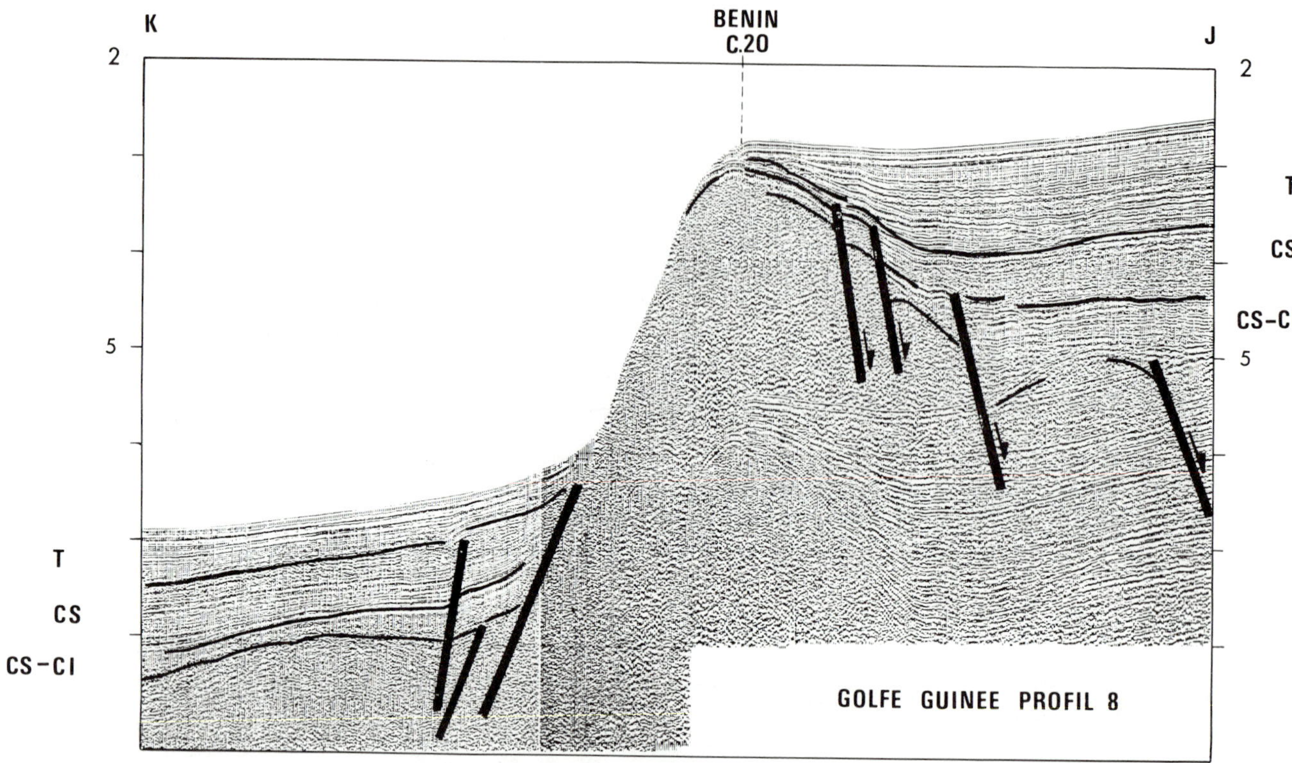

Fig. 8. Seismic reflection profile Guinee 8 (section JK).

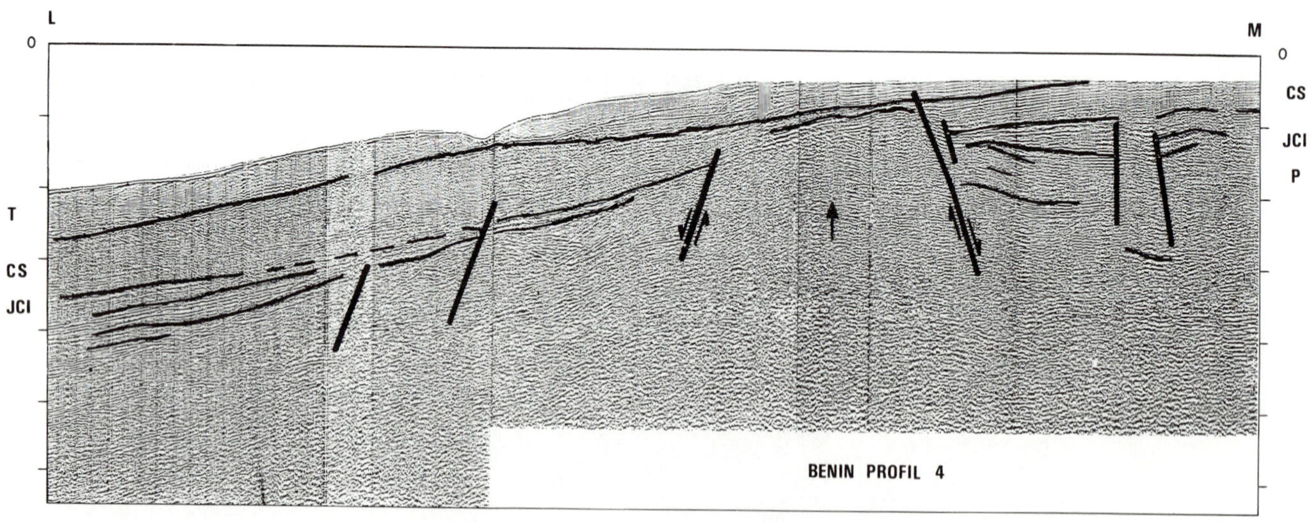

Fig. 9. Seismic reflection profile Benin 4 (section LM).

in the southern basins of the Gulf of Guinea (Belmonte et al., 1966; Wenger, 1973), with special tectonic conditions. In effect, the essentially vertical motions responsible for the deposit of the thick series of the early and middle Cocobeach in Gabon were followed by a new tectonic cycle leading to the establishment of sedimentation dominated by chemical processes. The authors attribute this new tectonic cycle to the opening of the ocean and the expansion of the ocean floor, which they consequently date as Aptian.

Finally, information recently obtained (ELF, unpublished documents) led the authors to believe that structures of plastic—and perhaps salt—origin could exist above an ocean substratum in certain sectors of the continental margin north of the Walvis Ridge, which would be in agreement with this hypothesis.

From the beginning of the opening, the future

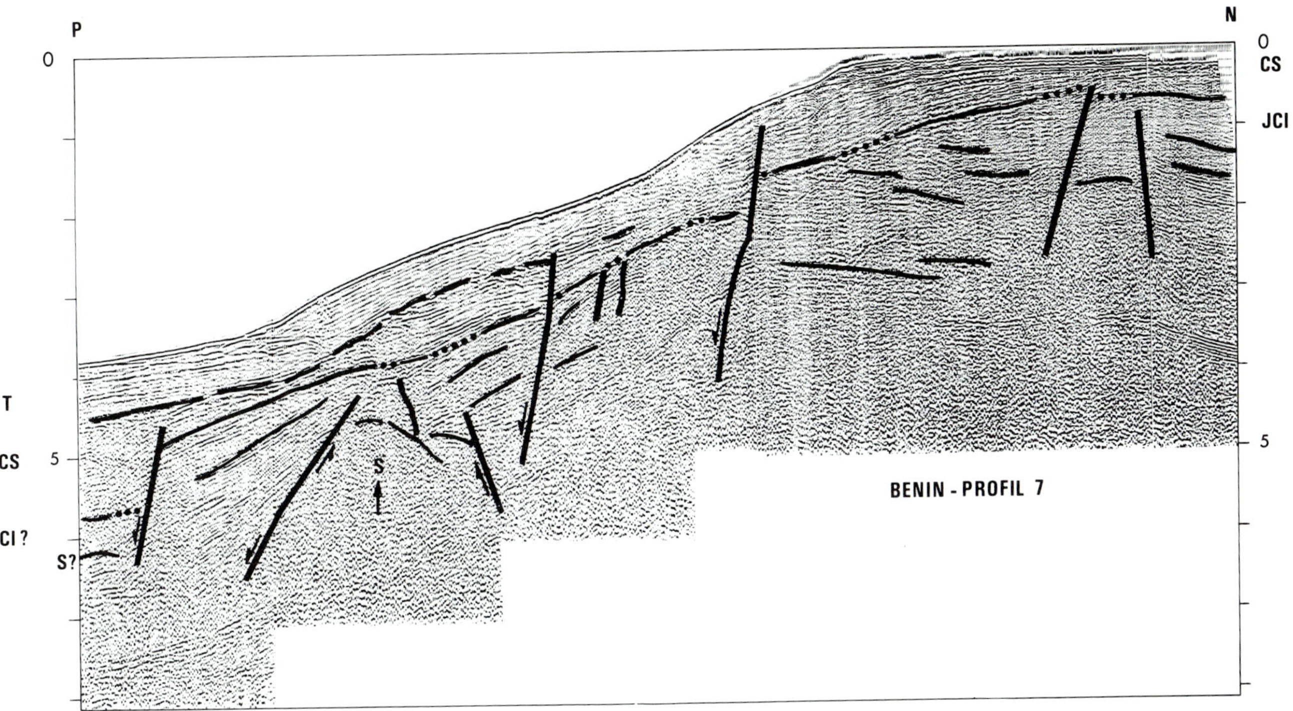

Fig. 10. Seismic reflection profile Benin 7 (section NP).

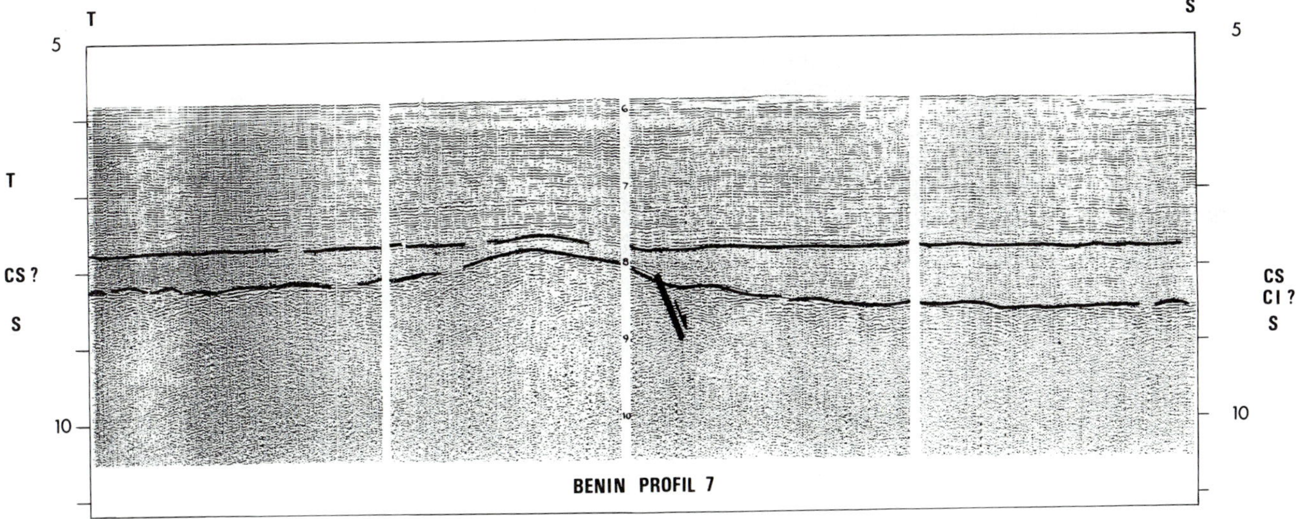

Fig. 11. Seismic reflection profile Benin 7 (section ST).

continental margin of the northern region of the gulf corresponds to an immense shear zone, probably characterized by rapid differentiation into downwarped and high blocks under the influence of the fracture zones.

The effect of these fracture zones on the tectonics of the edge of the craton and on the entrapment of sediments thus appears to be an essential factor in the construction of the margin.

From the Late Cretaceous and during the Tertiary, the different continental margins on the periphery of the Gulf of Guinea underwent a similar evolution. The current structural framework of these margins was not definitively established until the Miocene, following the epirogenic phase of the late Eocene-Oligocene, which affected the entire African framework.

Thus the continental margin of the northern part

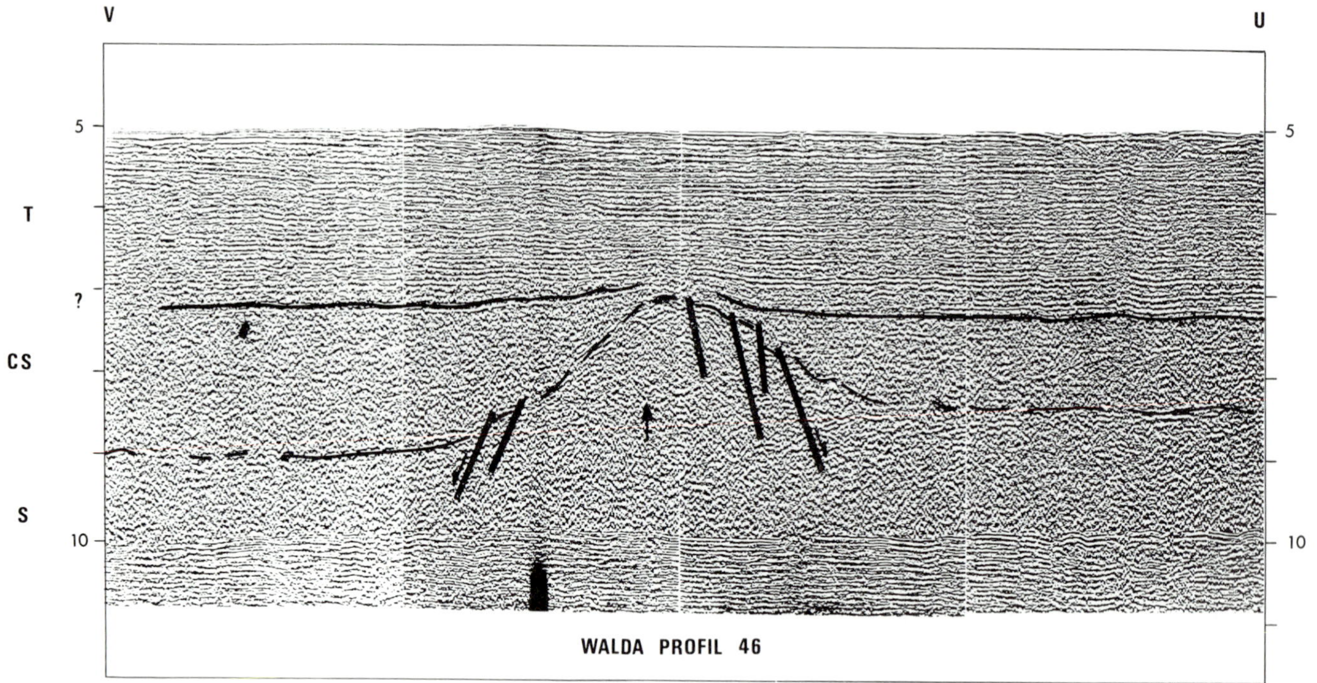

Fig. 12. Seismic reflection profile Walda 46 (section UV).

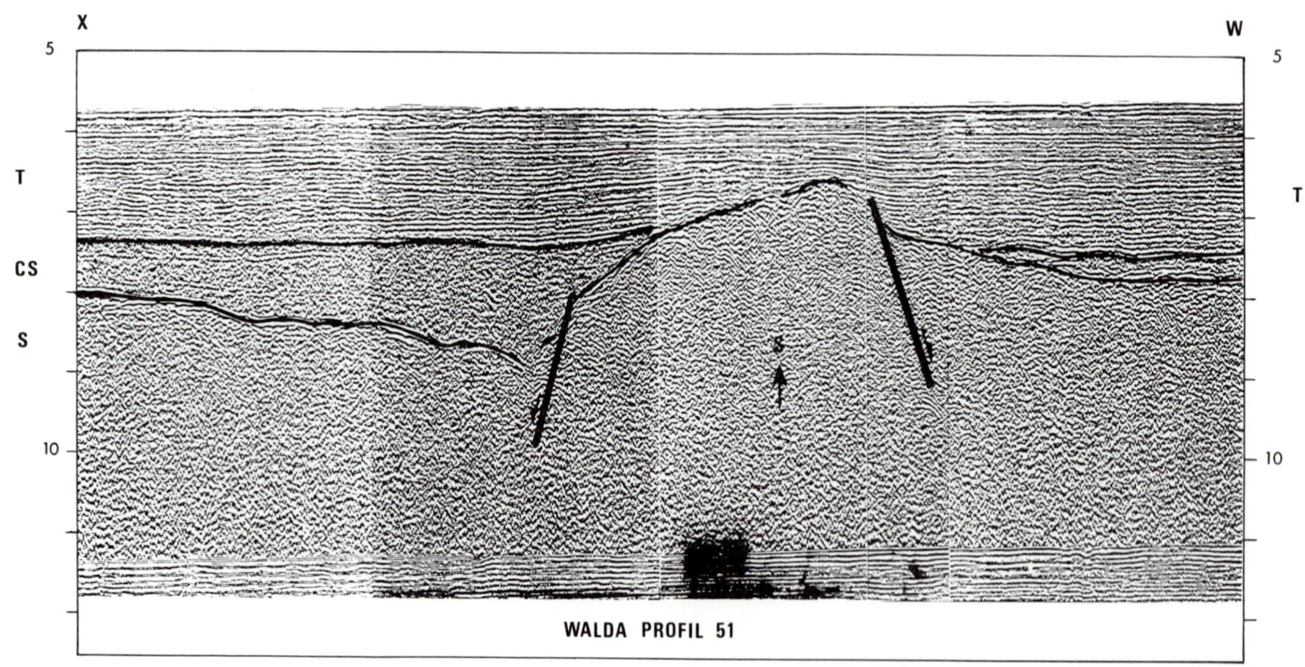

Fig. 13. Seismic reflection profile Walda 51 (section WX).

of the Gulf of Guinea corresponds to a stable margin on the edge of the craton, with a history associated with the opening of the South Atlantic. This margin is intersected diagonally by the large oceanic equatorial fracture zones, whose influence on the structure of the substratum and the sedimentation of this basin is observable as far as the Niger Delta region.

ACKNOWLEDGMENTS

This article is published with the kind agreement of the Comité d'Etudes Pétrolières Marines, the Société Nationale des Pétroles d'Aquitaine, CFP-TOTAL, and the Institut Francais du Petrole, who made several unpublished documents available to

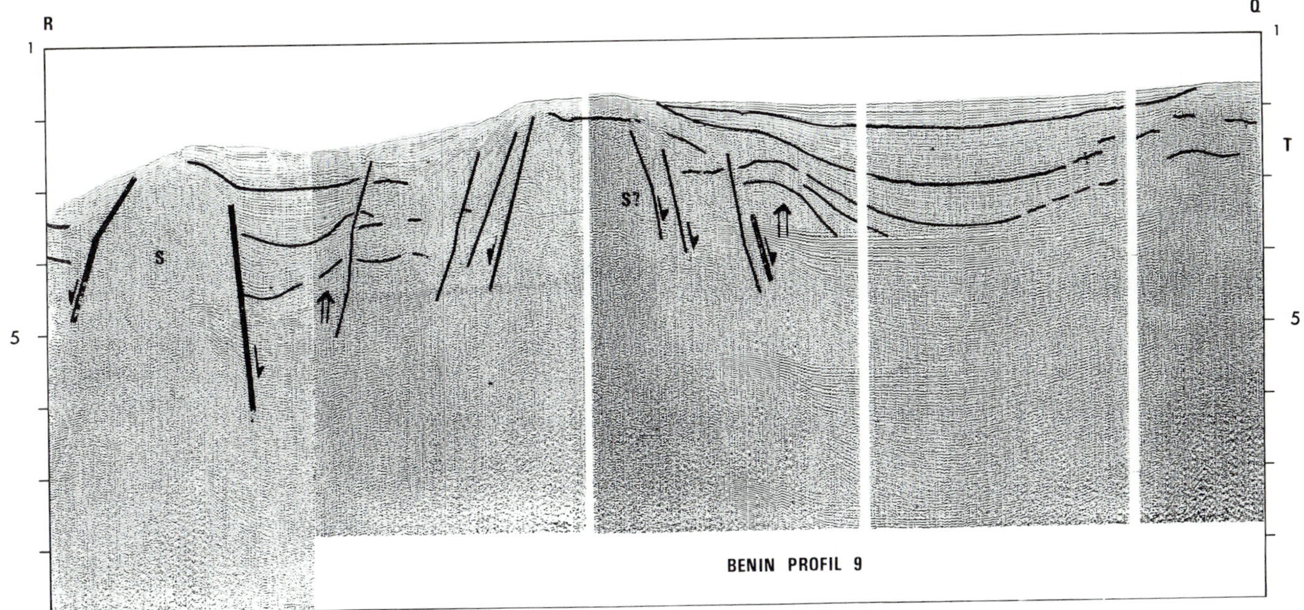

Fig. 14. Seismic reflection profile Benin 9 (section QR).

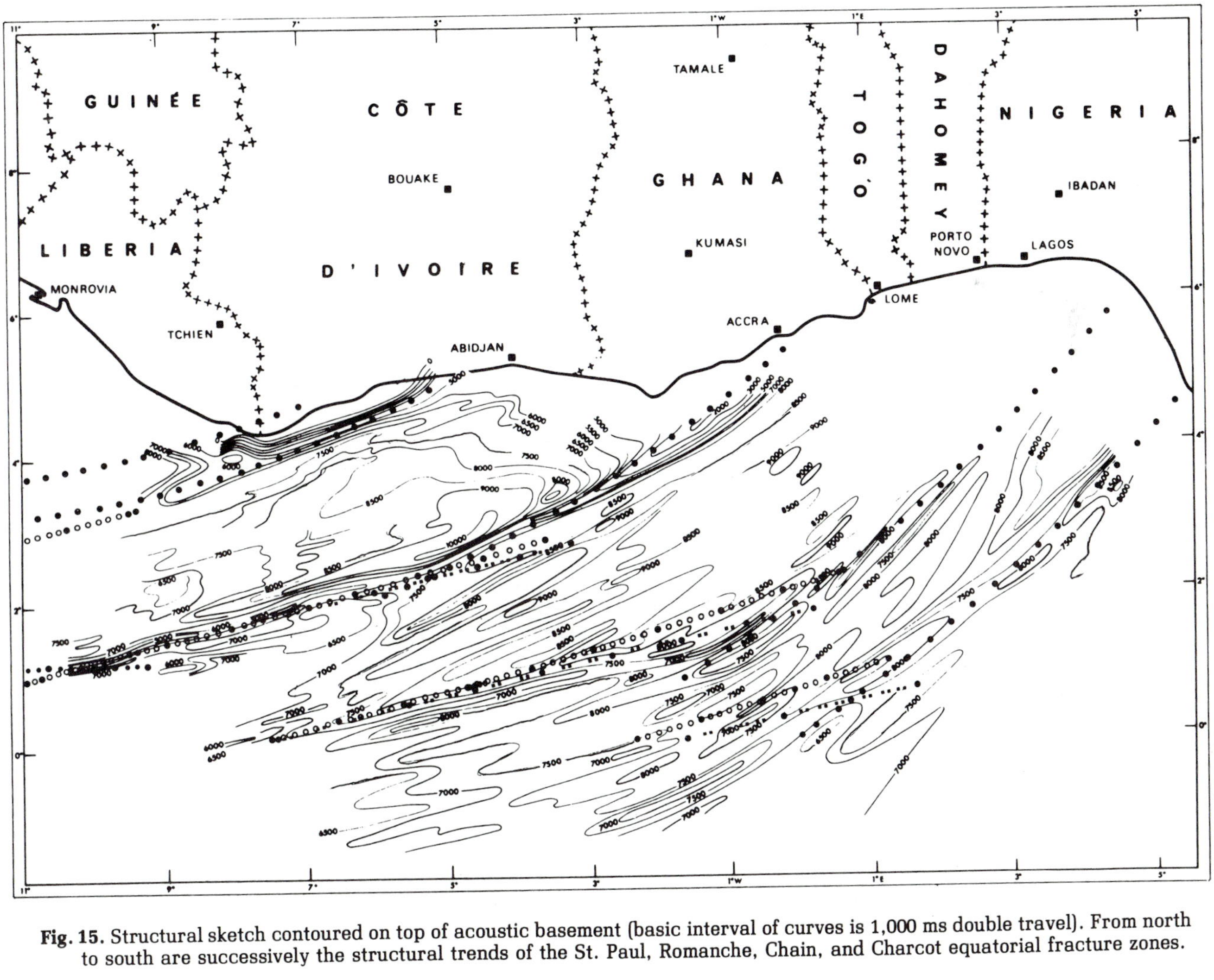

Fig. 15. Structural sketch contoured on top of acoustic basement (basic interval of curves is 1,000 ms double travel). From north to south are successively the structural trends of the St. Paul, Romanche, Chain, and Charcot equatorial fracture zones.

the authors. The authors wish to thank the Centre National d'Exploitation des Océans and the crew of the JEAN CHARCOT, the ship on which all the CEPM cruises in the Gulf of Guinea were carried out. They are greatly indebted to J. P. Fail, Donatien and Cassand, geophysicists of the Institut Francais du Petrole, who ran the seismic profiles. The authors also thank M. Talwani, J. Ewing, and R. Leyden of the Lamont-Doherty Geological Observatory, as well as K. O. Emery of the Woods Hole Oceanographic Institution, who supplemented the authors' seismic findings with those made during the cruises of the VEMA, CONRAD, and ATLANTIS II.

BIBLIOGRAPHY

Almeida, F. F. M. de, 1967, Origem e evolucao da plataforma brasileira Divisao de geologica e mineralogia, Bull. 241, pla. 35.

Arens, G., Delteil, J. R., Valery, P., Damotte, B., Montadert, L., and Patriat, 1971, The continental margin off the Ivory Coast and Ghana, *in* Delany, F. M., The geology of the East Atlantic continental margin: Great Britain, Inst. Geol. Sci. Rept. 70/16, p. 61-78.

Beck, R. M., 1972, The Oceans, the new frontier in exploration: APEA Jour., v. 12, pt. 2, p. 3.

Belmonte, Y., Hirtz, P., and Wenger, R., 1965. The salt basins of the Gabon and the Congo (Brazzaville): A tentative paleogeographic interpretation, *in* Salt basins around Africa: London Institute of Petroleum, p. 55-74.

Black, R., 1967, Sur l'ordonnance des chaines métamorphiques en Afrique occidentale: Chron. Mines Rech. Min. 364, p. 225-238.

Bullard, E. C., Everett, J. E., and Smith, A. G., 1965, The fit of the continents around the Atlantic: Roy. Soc. (London) Phil. Trans., Ser. A, v. 258, no. 1088, p. 41-51.

Burke, K. C., 1972, Longshore drift, submarine canyons and submarine fans in development of Niger delta: Am. Assoc. Petr. Geol., Oct.

———, Dewey, J. F., 1970, Orogeny in Africa, *in* African geology: Univ. Ibadan, p. 583-608.

———, Dessauvagie, T. F. J., and Whiteman, A. J., 1970, Geological history of the Benue Valley and adjacent areas, *in* African geology: Univ. Ibadan, p. 187-206.

Cahen, L., Snelling, N. J., 1966, The geochronology of equatorial Africa: Amsterdam, North-Holland.

Clifford, T. N., 1968, Radiometric dating and the pre-Silurian geology of Africa, *in* Hamilton, E. E., and Farquhar, R. M., eds., Radiometric dating for geologists: New York, Wiley-Interscience.

Cochran, J., 1973, Gravity and magnetic investigation in the Guiana basin, western equatorial Atlantic: Geol. Soc. America Bull., v. 84, p. 3249-3268.

Cratchley, C. R., and Jones, G. P., 1965, An interpretation of the geology and gravity anomalies of the Benne Valley, Nigeria: Overseas Geol. Surv. Geophys. Paper 1, 26 p.

Crow, A. T., 1952, The rocks of the Sekondi Series of the Gold Coast: Geol. Surv. Ghana Bull. 18, p. 12, 13, 28.

Cudjoe, J. E., and Khan, M. H., 1972, A preliminary report on the geology of the continental shelf of Ghana: Marine Fishery Res. Rept. 4, Apr. 1972, p. 22-31.

Dietz, R. S., and Knebel, H. J., 1971, Trou sans fond submarine canyon—Ivory Coast, Africa: Deep-Sea Res., v. 18, p. 441-447.

Fail, J. P., Montadert, L., Delteil, J. R., Valery, P., and Schlich, R., 1970, Prolongation des zones de fractures de l'ocean Atlantique dans le golfe de Guinee: Earth and Planetary Sci. Letters v. 7, no. 5, p. 413-419.

Francheteau, J., and Le Pichon, X., 1972, Marginal fracture zones as structural framework of continental margin in South Atlantic Ocean: Am. Assoc. Petr. Geol. Bull., v. 56, no. 6, p. 991-1007.

Furon, R., 1968, Geologie de l'Afrique: Paris, Payot Edit. 400 p.

Heezen, B. C., Bunce, E. T., Mersey, J. B., and Tharp, M., 1964, Chain and Romanche fracture zones: Deep-Sea Res. v. 11, p. 11-22.

Hospers, J., 1965, Gravity field and structure of the Niger delta (Nigeria) west Africa: Geol. Soc. America Bull., v. 76, p. 407-422.

———, 1971, The geology of the Niger delta area: *in* ICSU/SCOR Working Paper 31; Symposium, Inst. Geol. Sci. (London) Rept. 70/16, pt. 4, Africa, p. 125-142.

Hurley, P. M., 1972, Can subduction process of mountain building be extended to pan African and similar orogeny belts: Earth and Planetary Sci. Letters, v. 15, p. 305-314.

Karpoff, R., 1965, Les grandes époques de fracture et de bombement du Sahara central: Bull. Soc. Geol. France v. 7, p. 469-473.

Kennedy, W. Q., 1965, The influence of basement structures on the evolution of the Coastal (Mesozoic and Tertiary) basins, *in* Salt basins around Africa: London, Institute of Petroleum.

———, and Grant, N., 1969, The late Precambrian to early Paleozoic pan-Africa orogeny in Ghana, Togo, Dahomey and Nigeria: Geol. Soc. America Bull., v. 80, p. 45-56.

———, and Grant, N., 1971, South Atlantic, Benue Trough and Gulf of Guinea Cretaceous triple junction: Geol. Soc. America Bull., v. 82, p. 2295-2298.

Ladd, J. W., Dickson, G. O., and Pitman, W. C., III, in press, The age of South Atlantic (in press).

Lemarechal, A., and Vincent, P. M., 1971, Le fosse crétacé du Sud Adamaoua (Cameroun): Cahiers ORSTOM, Ser. Geol., v. 3, no. 1, p. 67-83.

Le Pichon, X., 1968, Sea floor spreading and continental drift: Jour. Geophys. Res., v. 73, p. 3661-3697.

———, and Hayes, D. E., 1971, Marginal offsets, fracture zones and the early opening of the South Atlantic: Jour. Geophys. Res., v. 76, p. 6283-6293.

McConnell, R. B., 1969, Fundamental fault zones in the Guiana and West African shields in relation to presumed axes of Atlantic spreading: Geol. Soc. America Bull., v. 80, no. 9, p. 1775-1782.

Martin, L., 1970, Premières investigations sur l'origine du trou sans fond, canyon sous marin de la Côte d'Ivoire: Paris, C.R. Hebd. Seances Acad. Sci. ser. D, v. 270, no. 1, p. 32-35.

Mascle, J. R., Bornhold, B. D., and Renard, V., 1973, Diapiric structures off the Niger delta: Am. Assoc. Petr. Geol. Bull., v. 57/9, no. 1973, p. 1672-1678.

Merki, P., 1970, Structural geology of the Cenozoic Niger delta, *in* African geology: Univ. Ibadan, p. 635-646.

Morgan, W. J., 1968, Rises trenches, great faults and crustal blocks: Jour. Geophys. Res., v. 73, p. 1959-1983.

Murat, R. C., 1970, Stratigraphy and paleogeography of the Cretaceous and lower Tertiary in southern

Nigeria, *in* African geology: Univ. Ibadan, p. 251-266.

Nyema Jones, A. E., and Stewart, W. E., 1970, General geology of Liberia, *in* African geology: Univ. Ibadan, p. 495-500.

Pautot, G., Renard, V., Daniel, J., and Dupont, J., 1973, Morphology, limits, origin and age of the salt layer along South Atlantic African margin: Am. Assoc. Petr. Geol. Bull., v. 57/9, p. 1658-1671.

Radier, M., 1959, Le bassin crétacé et tertiaire de Gao et le detroit Soudanais. Serv. Geol. Prosp. Miniere (A.O.F.). Bull. 26, v. 2, Dakar, Thèse.

Reyment, R. A., 1969, Ammonite biostratigraphy, continental drift and oscillatory transgression: Nature, v. 224, p. 137-140.

———, 1970, Cretaceous (Albian-Turonian) geology of the South Atlantic, *in* African geology: Univ. Ibadan, p. 505-512.

———, 1971, L'histoire de la mer transcontinentale saharienne pendant le Cenomamien-Turonien: Bull. Soc. Geol. France (7), v. 13, no. 5-6, p. 528-531.

Reyre, D., 1966, *in* Symposium sur les bassins sedimentaires du littoral africain: New Delhi, ASGA.

Robb, J. M., Schlee, J., and Behrendt, J. C., 1973, Bathymetry of the continental margin off Liberia, West Africa: Jour. Geophys. Res. U.S. Geol. Surv., v. 1, no. 5.

Rocci, G., 1965, Essai d'interprétation des mesures géochronologie. La structure de l'ouest africain: Colloque international de geochronologie absolve Nancy, May 3-8, 1965.

Schlee, J., Behrendt, J. C., and Robb, J. M., 1973, Shallow structure and stratigraphy of the Liberian continental margin.

Spengler, A. de, and Delteil, J. R., 1966, Le bassin secondaire-tertiaire de Cote d'Ivoire, *in* Reyre, D., ed., Bassins sedimentaires du littoral africain (Symposium lere partie: Littoral atlantique, New Delhi, 1964): Paris, IUGS-ASGA, p. 99-113.

Talwani, M., and Eldholm, O., 1973, Boundary between continental and oceanic crust at the margin of rifted continents: Nature, v. 241, Feb. 2.

Wenger, R. 1973, Le bassin sedimentaire gabonais et la derive des continents: Annales du XXVIIe Congres de Geologie du Bresil (sous press).

Wright, J. B., 1968, South Atlantic continental drift and the Benue Trough: Tectonophysics, v. 6, no. 4, p. 301-310.

Continental Margin of East Africa—A Region of Vertical Movements

P. E. Kent

INTRODUCTION

The concept that crustal plates are "structural units not being deformed" is at variance with evidence of large-scale vertical (epeirogenic) movements in the marginal areas of Atlantic-type coasts, of which east Africa provides a typical example.

Eastern Africa was deformed by widespread rifting in Karroo times (late Carboniferous to Early Jurassic), so thicknesses of 4,000 - 10,000 m of sediments accumulated in fault troughs or partially fault-bounded basins. Oceanic connection during these times is shown by marine transgressions in the Permian (major in northern Madagascar, minor in Tanzania) and by thick evaporite deposits in the Permo-Triassic (southern Tanzania and probably elsewhere).

Open-marine shelf conditions became established in the Lower Jurassic (Somalia, northeastern Kenya, northern Madagascar), Middle Jurassic (coastal Kenya, Tanzania, western Madagascar), and Upper Jurassic (Mozambique, probably also South Africa), and continued with few minor interruptions to the present day. Marine Mesozoic and Tertiary rocks, accumulated mainly in shallow water on a subsiding shelf, aggregating 10,000 - 12,000 m in coastal Kenya and Tanzania.

True oceanic conditions offshore are known from DSDP drilling to have been established at least as early as Lower Cretaceous (Somali Basin), significantly earlier than the oldest magnetic stripes of the sea-floor spreading process. There is evidence in one case (Madagascar Ridge) of subsidence of a shelf area to oceanic depth in post-Eocene times.

In the southerly coastal areas — Mozambique and Madagascar — sedimentation was interrupted by basalt extrusion, largely subaerial, varying from post-Jurassic (Neocomian-Aptian) to Middle and Late Cretaceous (Albian-Coniacian) and Miocene. These basalts are now reported to be continuous with the tholeiites of the ocean floor, dated as ranging from pre-Maestrichtian to post-Eocene by DSDP drilling. Sediments probably underlie the deep-water basalts in at least the westerly oceanic areas. Continuity of the basalt series from the land areas across the Mozambique Channel is further evidence in favor of the autochthoneity of Madagascar.

Knowledge of the sedimentary and structural history of the east African coast has made major progress over the last decade, with publication of many of the results of exploration for hydrocarbons in the continental margin of this area, and investigationof the adjoining ocean floor by Leg 25 of the Deep Sea Drilling Project. Exploration on the continental margin has included the drilling of many holes to depths of 3,000 - 4,500 m; on the ocean floor, basalt (tholeiite) has been reached in six places at 200-500 m depth; one oceanic penetration proved 1,200 m of sediment (base not reached). With the accompanying extensive seismic surveys of the coastal sedimentary basins and the ocean floor, the east African coast becomes one of the world's best-documented continental margins.

The description here covers events in historical order from the late Paleozoic, dealing with the area as a single geographical and geological unit.

KARROO BASINS

Deposition of the Karroo, extending from the late Carboniferous into the Jurassic, took place in eastern Africa under dominantly, but not entirely, continental conditions in a series of separate basins, largely fault controlled. The successions vary in detail from basin to basin, but show a broad similarity to the carefully studied South African sequence — an initial, mainly gray group (Ecca), locally coal-bearing, with basal glacial deposits in South Africa and, on a very minor scale, in Madagascar; a medium sandstone and shale group (Beaufort) and an upper coarse sandstone group (Stormberg), which includes extensive volcanic extrusions of approximately Liassic date in Mozambique and Malawi.

On land in Mozambique the Karroo is located in NE-NNW-trending rifts in the continental area. Inland, the series appears to be essentially continental, but limestones (?marine) are reported from the Lake Malawi area (Flores, 1973). A total thickness of 5.5 km has been estimated. In addition to the landward basins, there is seismic evidence of a pre-Cretaceous, probably Karroo, basin beneath the continental shelf east of Beira, with indications of an evaporite series which could be analogous to that of Mandawa, southern Tanzania (Fig. 2; personal communication; R. M. Sanford, Hunt International Pet. Co.).

In Madagascar, the Karroo series shows three major divisions. The lowest, the Sakoa Group, begins with black shales and very localized tillites and fluvioglacial deposits, followed by coal measures and red beds — totaling up to 1,400 m. A marine transgression followed, beginning with the 30-m Vohitolia Limestones, which contain a middle Permian brachiopod fauna, the first marine phase of the region.

The much-publicized glacial deposits of Madagascar have been over emphasized in the literature. They cover a minute area of "Gondwanaland," probably some 10 miles across, including the fluvioglacial gravels. It must be questioned whether this justifies extension of the Gondwanaland ice as a sheet extending continuously from South Africa over the intervening area.

The middle division, the Sakamena Group, in western Madagascar consists dominantly of buff and greenish shales and sandstones with rare conglomerates, thin green shales containing freshwater and marine fossils, followed by a further red-bed (argillite) series, totaling 2,000 m at outcrop, but significantly more at depth. The Sakamena Group crops out again near the northern tip of the island, where it contains a rich ammonite fauna, related to that of Australia, which presents an interesting paleogeographical problem.

The third division, the Isalo Group, consists characteristically of massive, continental sandstones (approximately 4 km thick), locally passing into a sandstone and red-clay series. This facies ranges up into the Lias and Middle Jurassic, varying with the local age of the first marine transgression.

According to Cliquet (1957) the Karroo sequence was deposited partly in a deep trough adjoining the eastern margin of the basin, elsewhere in a strongly and contemporaneously block-faulted area. The beds appear to thin westward, and the sinking of the basin adjoining the Mozambique Channel was largely associated with later, post-Karroo deposition. In Madagascar, as elsewhere in east Africa, we thus have contemporary subsidence of linear fault-controlled basins in Karroo times.

The former concept of a broad Karroo basin centered on, and extending across, the Mozambique Channel into Mozambique requires modification; it is at variance with thickness distribution in Madagascar and with the evidence from the Mozambique shelf that a deep sedimentary lens thins eastward at the continental shelf edge. There may have been continuous sedimentation across the Mozambique Channel, particularly in late Karroo (Isalo) times, but whether there are other independent Karroo fault troughs beneath the deep water remains undetermined.

In Tanzania and Kenya there is a broad similarity to the Madagascar sequence, with a basal coarse-grained, locally coal-bearing sequence below, a middle shale and sandstone group, and an upper coarsely clastic group. The marine episodes are only minimally represented — shales with *Estheria* in southern Kenya (Maji-ya-Chumvi beds), and a thin intercalation of shales with marine lamellibranchs at Kidodi in Tanzania. These reflect more nearly paralic or "estuarine" conditions; there is no open-water marine fauna.

A Karroo graben at Mandawa in southern Tanzania, separated from the present oceanic area by a broad fault block, is notable for a 3,000-m evaporite sequence (Kent, 1965). The top of this is an alternation of gypsum beds with Middle Jurassic marine shales; this overlies almost continuous halite, with rare Triassic fossils, resting on an alternation of evaporites and clastic rocks beneath.

There are indications of contemporary development of the graben, which subsequently must have shown only minor movements in Jurassic and Early Cretaceous times before becoming inoperative. Gravity surveys suggest that an analogous evaporite trough exists southwest of Dar-es-Salaam and that the horst-and-trough structural pattern continues offshore.

The importance of the Mandawa occurrence lies in a further demonstration of contemporary Karroo-age fault movements, and in providing an indication of access by Permo-Triassic marine waters as an evaporite source.

Seismic suggestions of comparable evaporite basins off Mozambique and off South Africa (Port Elizabeth) show that this marine link probably continued along the entire east African coast in early Mesozoic times.

KARROO - EARLY JURASSIC FAULTING

The faulting of the Karroo troughs continued into the Early Jurassic, for marginal boulder beds as well as coarse, pebbly sands are recorded on the basin edge at Ngerengere, west of Dar-es-Salaam (Aitken, 1961), and there is a striking case of transgression of Middle Jurassic reefal limestones onto crystalline basement rocks beyond the edge of the deep Karroo trough in northern Tanzania (Kent et al., 1971, p.11). In the latter case, aeromagnetic evidence indicates a pre-Middle Jurassic throw in excess of 6 km for the concealed fault, the largest dislocation known in the east African coastal region.

In southern Kenya (Mombasa), the later Callovian shallow-water shales and limestones show major slump structures, probably related to a faulted coast. Gradients on the continental margin were steep at least as late as Kimmeridgian (possibly on the shelf edge), as shown by large-scale contemporary sliding at Kilifi, 45 km to the north.

In Somalia, Beltrandi and Pyre (1973) show a north-south trough of pre-Jurassic (probably Karroo) age, which crosses the Horn of Africa in line with the Red Sea; this contains up to 3 km of sediments, including evaporites, and appears to have developed by progressive pre-Jurassic faulting. On the coast, the great thickness of Jurassic limestone found in deep test borings invites explanation as deposition controlled by rather late fault-basin subsidence.

Madagascar shows a comparable history of major Karroo basins bounded by faults, which are at least in part contemporary. Cliquet (1957) documented the progressive development of rifts throughout Karoo times, with movements measured in thousands

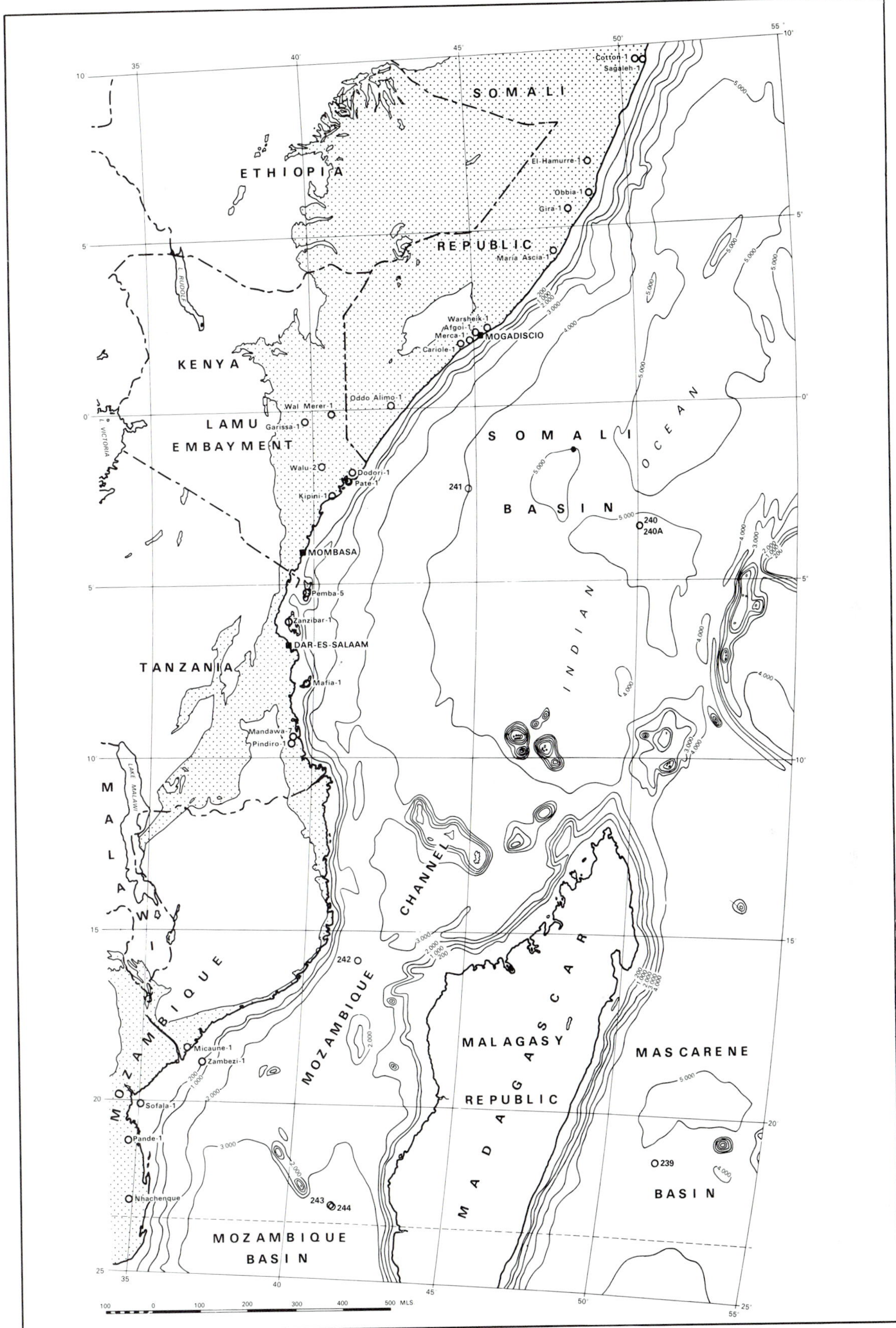

Fig. 1. Map showing coastal East Africa and the western Indian Ocean: boreholes shown by open circles.

of meters, which were largely completed by the Early Jurassic. In contrast, the Jurassic and Cretaceous were deposited on west- and north- facing open shelves, with an absence of indications of tectonic dislocation.

Thus in the East African coastal zone, major tectonic dislocations, large on a world scale, took place in Permian to Lower Jurassic, and effectively ended at that latter time. Later fault movements were much smaller and the main sea-floor spreading episode is not known to be associated with vertical movements of comparable magnitude.

MESOZOIC TRANSGRESSIONS AND REGRESSIONS

The first fully marine invasion in the Mesozoic took place in the Lower Jurassic. Beds of the lower Lias have been reported in deep borings in coastal Somalia; better known is a later Lias (Toarcian) fauna with the ammonite *Bouleiceras*, which is also known in Arabia, in northeastern Kenya and in the northern basin of Madagascar.

Middle Jurassic beds are more widespread, although their exact dating is rendered difficult by the scarcity of ammonites. Bajocian shales are present (probably locally developed) in southern Kenya (Spath, 1933; Arkell, 1956) and in both northern and western Madagascar. Limestones, often oolitic, with basal sands and occasional reef developments, which are broadly Bajocian/Bathonian, are present in Somalia (Baidoa Limestone), near Mombasa (Kambe Limestone), through much of the Tanzania coastal region (Amboni Limestone) and in Madagascar.

The Lower Jurassic is thus very localized; the sea spread farther in the Middle Jurassic, and it transgressed farther still onto basement in the early and later Upper Jurassic (Callovian), as at Tendaguru in southern Tanzania (Aitken, 1961). In Madagascar, the transgression was strongly oscillatory; there are reversions to red-beds (Karroo facies, not formally included in that group), as in the later Jurassic (Argovian).

The early Upper Jurassic Callovian is widely developed as a broad belt of shallow-water sandy deposits, through much of Kenya and Tanzania. The Oxfordian and Kimmeridgian sediments are largely open-water shales in Tanzania and coastal Kenya, with limestones dominant in northern (inland) Kenya and Somalia. These indicate much more extensive transgression, probably combined with planation of the topographic features produced by the Middle Jurassic fault episode, a transgression that may have extended far beyond the present coastal basin.

The situation changed again in the Lower Cretaceous. In Tanzania, estuarine conditions with an alternation of sandstones and marine shales (Neocomian-Aptian) spread across the Dar-es-Salaam embayment; red-beds (Makonde Beds, probably Aptian) were developed in inland southern Tanzania and northern Mozambique and also in the Aptian of Madagascar. This major regression, which corresponds broadly with the Wealden of northwestern Europe and the "continental intercalaire" of North Africa, was presumably controlled by world-wide eustatic movements, but it was also associated in east Africa with an end Jurassic - Early Cretaceous phase of faulting, which has been documented by Dixey (1956) in the southerly rift valleys.

In southern Mozambique the Early Cretaceous continental phase is represented by the Sena Formation, more than 2,500 m thick in the Zambesi graben (developed contemporaneously), with interleaved basalts in the lower part. This passes into a thinner marine facies (Albian-Aptian) in the north and south of the coastal belt. The marine episode continued in Mozambique, largely in a deeper water, argillaceous facies, and Albian shales overstep eroded Jurassic rocks in central Tanzania. Open-water shelf deposition continued in Mozambique, in Tanzania and Kenya through the Upper Cretaceous, in the two former areas into the Paleocene without a break. Sediments were mainly argillaceous. The maximum thicknesses of post-Aptian Cretaceous shales are quoted as around 2,000 m in coastal Kenya (Walters and Linton, 1973, p. 142), and more than 1,500 m in Tanzania (Kent et al., 1971) and in Mozambique (Flores, 1973).

JURASSIC - CRETACEOUS BASALT SERIES

The Karroo in South Africa ends with the Stormberg Basalts, some 1,500 m of basalts and rhyolites with intercalated sandstones. This series continues into Mozambique, but it has been found that the uppermost part of the series in the Lebombo Mountains is of Lower Cretaceous age, and thus links with the igneous intercalations found low in the sedimentary series of the coast, dated as Late Jurassic, Aptian and Turonian in different places (Flores, 1973). These flows continue beneath the sediments of the coastal plain and were encountered in Gulf Oil Company deep test boreholes at depths down to 3,000 m.

Offshore, tholeiitic basalts have been found in six of the eight DSDP borings in the western Indian Ocean, overlain by sediments varying in age from Early Cretaceous to Eocene. These rocks have been regarded as specifically ocean floor ("oceanic basement" of many authors, without consideration of possible or likely subbasement sediments), but the Royal-Dutch Shell Company has recently announced direct seismic continuity of these typical "oceanic" rocks with the Early Cretaceous basalts of Mozambique—and hence by implication with the inland continental basalt extrusions. The basalts beneath the coastal shelf are not much shallower than the oceanic tholeiites, and might have been regarded as ocean floor buried by sedimentary shelf outbuilding, but in any case it is clear that the shelf area is structurally part of the continent, with a deep subbasalt sedimentary sequence (probably Karroo).

The basalts are exposed again beyond the Mozambique Channel in Madagascar, where they are interbedded with marine sediments but were extruded subaerially. As in Mozambique, and as on the ocean floor, the date of the flows varies from place to place, from shortly post-Aptian and Coniacian to Santonian, with a final local Miocene episode. Central-type volcanoes developed in the Plio-Pleistocene on the high plateau. Offshore, the range of age of the oceanic basalt overlaps that of the landward flows, ranging from pre-Maestrichtian (DSDP site 239 in the Mascarene Basin) to post-Eocene (site 240 in the Somali Basin), with the Plio-Pleistocene volcanoes in the Mozambique Channel. There is no obvious difference in volcanic history—or hence, perhaps, in fundamental structure—between the Mozambique Channel and Madagascar.

Mesozoic basalts are not known on the east African coast north of Mozambique and Madagascar, but they continue a few hundred meters beneath the ocean floor northward across the Somalia basin.

PALEOGENE DEPOSITION

The marginal trough (miogeosyncline) of the African east coast continued to develop during the Tertiary. Thicknesses on land are variable and sometimes small, but beneath the shelf the post-Cretaceous sediments aggregate some 4,000 m. In Tanzania and Kenya, where this sequence is well documented, a large part is in shallow-water facies. All the Tertiary stages are present along the coast, but their development and distribution is affected by overlaps and disconformities, reflecting control of sedimentation by irregular subsidence of the shelf areas.

Paleocene and Focene follow the Cretaceous Maestrichtian without perceptible break in southern Tanzania, but in offshore Mozambique and in coastal Kenya (Lamu embayment), the end of the Cretaceous was marked by a regression and these beds are shallow-water type, transgressive, in overlapping/overstepping relationship to the Mesozoic. Where there was continuous deposition, the facies of the Paleocene is argillaceous and the fauna indicates outer shelf; elsewhere sands and sandy limestones are dominant. The maximum thickness of Paleocene (600 m) is found at Lindi (southern Tanzania) in the deeper-water area. In northern Madagascar, a transition from the Cretaceous is

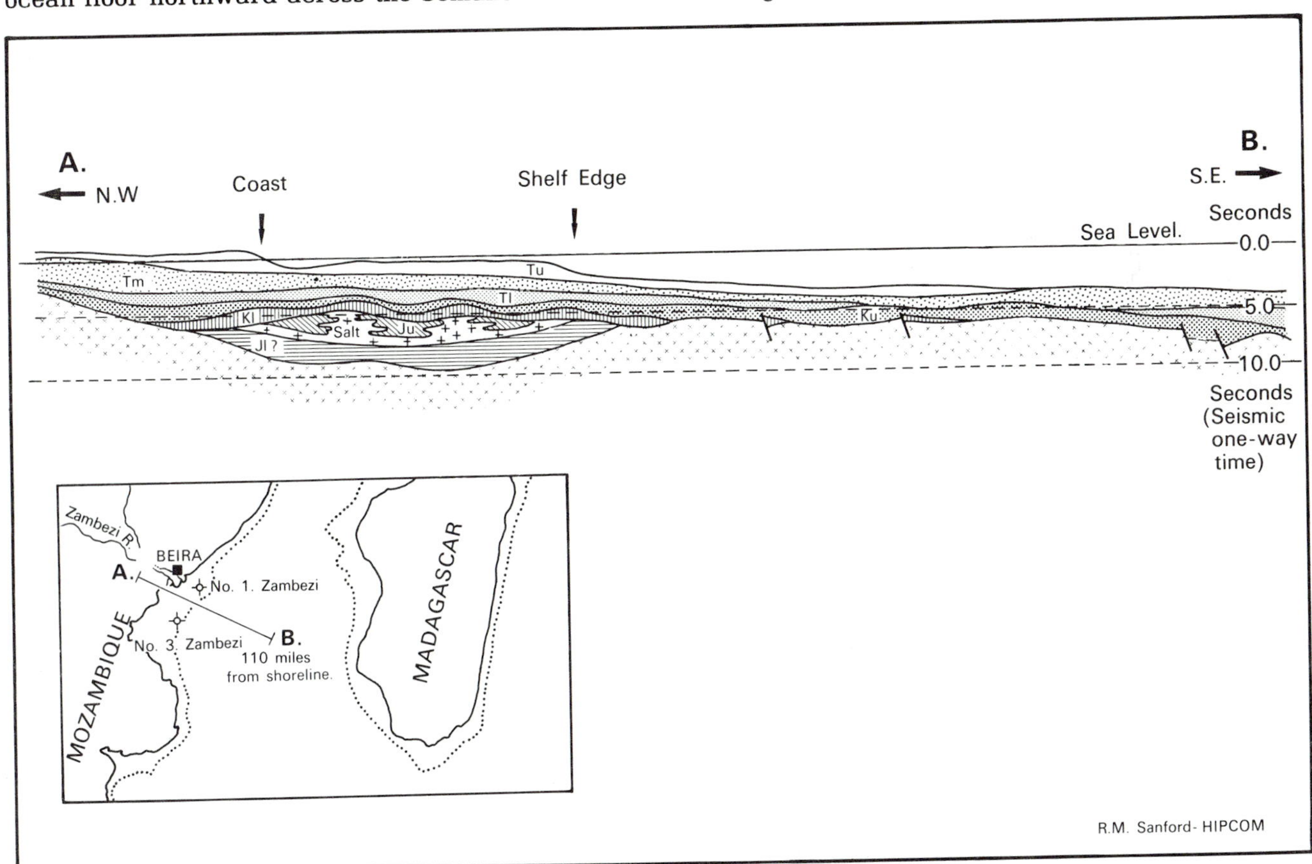

Fig. 2. Section from inland Mozambique into the Mozambique Channel (by R.M. Sanford); important boreholes noted.

reported locally, but elsewhere, lower Eocene is missing and middle Eocene is transgressive.

The Eocene is usually well represented and persistent on the mainland. In Mozambique, the sequence is relatively thin, with lower Eocene probably only present offshore; in Tanzania the Eocene expands at 1,000 m onshore (Kilwa), 900 m at Zanzibar, and 2,000 m at Pemba Island. On land, the sequence is mainly of shallow-water origin, an alternation of clays with *Nummulite* limestones (locally reefal). The facies is rather deeper water, reflecting neritic conditions at Zanzibar, but is largely deltaic in the thick sequence at Pemba (Kent et al., 1971). In the Lamu Embayment of Kenya, deposition was also "in a fluviolittoral, deltaic environment" (Walters and Linton, 1973), with an uncertain proportion of the later Eocene and Oligocene passing inland into a nonmarine, red-green, sandstone facies.

Somalia has a very limited area of Lower Tertiary rocks, but more than 2,700 m of early Tertiary terrigenous, mainly deltaic, clastics with a minor marine interval, are developed in a contemporary fault basin on the coast, possibly related to an ancestral Juba River (Beltrandi and Pyre, 1973).

In the deep-water oceanic areas, the DSDP borings have shown the Eocene to be widely present, and (with one exception) to be developed as deep-water clays and silts, largely with a nannoplankton flora.

Distribution of the Oligocene is discontinuous. Deposition corresponded with a regression and there appears to have been a limited supply of sediment, the latter probably related to the advanced state of peneplanation of the adjoining land areas. In Mozambique, there is a 670-m development (sands and marls, with a mainly pelagic fauna) on the north side of the Zambesi Delta, and a similar thickness offshore; upper, middle, and lower Oligocene are reported in the former area, but middle Oligocene only extends to the outer shelf. Oligocene is known only in one small area on Madagascar, which suffered the same regression as the mainland.

In Tanzania, only a thin selvedge of Oligocene (60-90 m) extends onto the present land; this is marine with local reefal development. It is generally thicker beneath the shelf, measuring 129 m at Mafia Island and 396 m at Pemba (in shallow-water marine facies). It may be absent at Zanzibar, or alternatively could be unrecognized in the shallow-water paralic succession at the base of the Miocene. The Kenya succession differs from those farther south, in that although marine Oligocene is limited to the outer part of the Lamu embayment, it forms the base of the transgressive "Miocene" limestones; marine incursion was evidently relatively early. Inland, the lateral equivalent may be present in the colored, post-middle Eocene continental wedge mentioned above (Walters and Linton, 1973).

It is remarkable that with quite large quantities of detrital sediment representing the Oligocene in parts of the continental shelf, the stage is unrepresented in most of the oceanic DSDP borings. Interruption of deposition by uplift of the ocean floor above sea level is excluded (there are no associated shallow-water deposits), and it can only be concluded that the quantities of detritus were insufficiently large to do more than fill the coastward sediment traps.

NEOGENE

During the Neogene, open-shelf conditions continued along the east African coast. The period, however, started with two major events: the dislocation of the inland "mid-Tertiary surface" (the name refers to the time of termination of its development), with an associated increase in sedimentary volume, and also with an extensive marine transgression which is partly world-wide but may also reflect coastal subsidence complimentary to inland elevation.

The mid-Tertiary ending of the main period of peneplanation was due to uplift, related to widespread faulting. Inland, early development of rift-valley movements was associated with (or followed by) the beginning of regional volcanic activity with extensive lava flows. Interbedded sediments in the volcanics there contain middle Miocene vertebrate faunas, so the dislocation of the older surface, including both the formation of the Kavirondo Trough and the uplift of the flanks of the Kenya rift, are dated as Lower Miocene (King et al., 1972).

In Mozambique, thick Oligocene-Miocene sediments (750 m or more) were deposited in two depocenters associated with the Zambesi graben. Late Oligocene was marine, but a local evaporite basin developed in the early Miocene. It was followed by widespread marine transgression, with Miocene overlapping inland onto Cretaceous rocks. On the Tanzania mainland relations were similar, with shallow-water limestones, sands, and clays predominating. Contemporary faulting is recognized at Lindi in the south, but the main sedimentary feature was the deposition of very thick deltaic beds centered on Zanzibar, where 2,500 m of Miocene beds were proved by deep boring. On the edge of this deltaic lens, limestone deposition was important, often of reefal type, and the general sedimentary assemblage was much like that of the present coast.

In Kenya and in Somalia, the main marine transgression began in the Oligocene (the basal part of the "Miocene" limestone unit of the coastward Lamu embayment contains *Nummulites*), but the widespread inland extension of marine waters took place in Lower and Middle Miocene. The dominant limestones of the coast give place to mudstones and sandstones inland—a demonstration of the sediment-trap mechanism, which is thought to account for the absence of late Eocene-Oligocene-early Miocene in the oceanic area.

Pliocene is not well known in the coastal area, partly because both the majority of exposures and most hydrocarbon exploration operations are limited to structural highs, where older beds outcrop. Development of late sedimentary basins between the positive structures was previously inferred in coastal Tanzania; it has now been extensively proved by later seismic work, which shows that post-Lower Miocene sediments in synclinal areas in and seaward of the Dar-es-Salaam embayment amount to some thousands of feet in thickness. These surveys also show a great deal of late Tertiary faulting contemporary with deposition. Elsewhere, Pliocene and Plio-Pleistocene deposits are poorly represented. They are very thin also in the ocean-floor borings of DSDP, possibly reflecting a generally arid climate inland, with thick deposition only in the neighborhood of the few large rivers (e.g., Zambesi and the Rufiji-Ruvu system). Off the Zambesi River, in particular, coarse sands and gravels were carried out into oceanic depths.

TECTONIC HISTORY OF THE EAST AFRICAN COASTAL REGION

Early concepts that compressional structures exist in the east African coastal region are no longer accepted. The structures are all related to vertical or tensional movements. Domal features known in Madagascar and southern Tanzania are recognized as due to injection at depth of igneous rock and salt flowage, respectively. Apart from these features, gentle folding is associated with fault movement and broad warping. Structural high areas are known from seismic work in Kenya and Tanzania to be cut by large numbers of strike faults, which in their final development postdate the Miocene beds. The faults are mostly, or entirely, steeply dipping, indicating control by vertical stresses rather than by major lateral extension.

The first dislocations were those of the Karroo—recognized in Madagascar as continuing from Permian times onward (Cliquet, 1957). In Tanzania the final movements of this phase are dated as Lower/Middle Jurassic; they were thus responsible for formation of at least one salt basin and a major (6,000 m) fault in the north (Kent et al., 1971). In Somalia, Beltrandi and Pyre (1973) have identified a north-south pre-Jurassic (probably Triassic) rift system containing continental sediments and evaporites. This system is cut obliquely by the later (Neogene) rift valley. The vertical movement is shown by stratigraphic overlaps to be almost entirely pre-Upper Jurassic, and to aggregate 3.0-4.5 km.

This phase of vertical movement, the largest known in the region, is thus identified throughout the east African coast, and is notable for the *absence* of rejuvenation of any significant scale in post-Jurassic times.

The Early Jurassic faulting led to dislocation of the early erosion surfaces inland; it was followed by a long period of quiescence until a later uplift, dated as end Jurrassic-Early Cretaceous (Dixey, 1956). This movement is recognizable in the coastal region from the sharp increase in coarse clastics, plus regressional features from Madagascar to Kenya, but only in the Zambesi rift is there certain indication of fault movements, although these might be deduced elsewhere from overlap of Albian onto Lias west of Dar-es-Salaam and abnormal thicknesses of Lower Cretaceous in inland Somalia.

A quiet period with peneplanation occupied most of the Cretaceous; it ended with uplift, which produced a sub-Tertiary (Paleocene) unconformity in Mozambique, Madagascar, and Kenya, but not in southern Tanzania. The structures in southern Kenya associated with this unconformity are strongly faulted. There may have been a minor uplift in the Eocene—known in Somalia and indicated elsewhere by development of shallow-water facies—but differential vertical movements in the coastal zone became strongly developed in the Oligocene - Lower Miocene. At this time, the Zambesi rift system suffered a further complex movement; the Lindi fault system of southern Tanzania was activated and movement of the coastal faults of Somalia led to thousands of meters of clastic deposition in a narrow trough.

Contemporary with these later Tertiary movements, the various positive features of the middle east African coast were progressively developed by relative uplift (mostly interrupted by slower subsidence), associated with closely space faulting. This led to the topographical highs, marked by Mafia, Zanzibar, and Pemba islands, with other features on the seaward mainland and also in deeper water.

Stratigraphic evidence shows that despite these differential movements, the dominant feature of the coast has been large-scale and continuous subsidence. In Mozambique, Tanzania, Kenya, and Somalia, proved and indicated Mesozoic and Tertiary thicknesses total 9-12 km and the sediments known are mostly of shallow-water facies. The continental edge is thus a miogeosyncline, filled by progradation associated with contemporary large-scale subsidence of the basement floor.

The inception of the Indian Ocean basin—as an ocean—is not yet dated. It was much earlier than the magnetic-stripe system, being certainly pre-middle Cretaceous and probably Jurassic. Vertical movements of the ocean floor do not appear to have matched those of the flanking miogeosyncline of the continental shelf, the history as indicated by DSDP borings indicating fairly constant depth conditions from Cretaceous onward, with minor depositional variation depending on external controls.

It is worth emphasis that the east African coastal miogeosyncline lies *within* a plate, for the structural edge of Africa is located near the Seychelles Islands, and its development was controlled by epeirogenic movements. The largest faults were normal, up to 10 km in throw, and were related to Karroo - early Jurassic deposition. Post-Cretaceous and Tertiary stresses were relieved by smaller faults, closer to the edge of the continental shelf, as the central axis of deposition shifted progessively seaward.

BIBLIOGRAPHY

Aitken, W. G., 1961, Geology and palaeontology of the Jurassic and Cretaceous of southern Tanganyika. Bull. Geol. Surv. Tanganyika 31.

Arkell, W. J., 1956, Jurassic geology of the world: London, Oliver & Boyd, 806 p.

Beltrandi, M. D., and Pyre, A., 1973, Geologic evolution of southwest Somalia, *in* Blant, G., ed., Sedimentary basins of the African coasts, pt. 2, South and east Coast: Paris, Association des Services Geologiques Africains, p. 151-178.

Cliquet, P. L., 1957, La Tectonique profonde du Sud Bassin de Morondawa Madagascar: Tananarive, C.C.T.A. Comites de Geologie Center—Est et Sud.

Dixey, F., 1956, The east African rift system: Colon. Geol. Miner. Res. Bull. Suppl. 1.

Flores, G., 1973, The Cretaceous and Tertiary sedimentary basins of Mozambique and Zululand, *in* Blant, G., ed., Sedimentary basins of the African coasts, pt. 2, South and east coast: Paris, Association des Services Geologiques Africains, p. 81-111.

Kent, P. E., 1965, An evaporite basin in southern Tanzania, *in* Salt basins around Africa: London, Institute of Petroleum, p. 41-45.

———, et al., 1971, The geology and geophysics of coastal Tanzania: Geophs. Paper Inst. Geol. Sci. 6.

King, B. C., Le Bas, M. J. and Sutherland, D.S., 1972, The history of the alkaline volcanoes and intrusive complexes of eastern Uganda and western Kenya: Jour. Geol. Soc. 128, p. 173-205.

Spath, L. F., 1933, Revision of the Jurassic cephalopod fauna of Kachh (Cutch): Mem. Geol. Survey India, Palaeont. Indica N.S. 9, Mem. 2, p. 659-945.

Walters, R. and Linton, R. E., 1973, The sedimentary basin of coastal Kenya, *in* Blant, G. ed., Sedimentary basins of the African coasts, pt. 2, South and east coast: Paris, Association des Services Geologiques Africains, p. 133-158.

Part VI

Geology of Selected Modern Margins: Atlantic Region

Continental Margins of Galicia-Portugal and Bay of Biscay

L. Montadert, E. Winnock, J.R. Deltiel, and G. Grau

INTRODUCTION

The continental margins in the Bay of Biscay and off Galicia-Portugal (Fig. 1) appear now to be stable margins, as are most Atlantic margins. But they have a singularity that greatly complicates their interpretation: the present structure is due not only to Atlantic history during the Mesozoic and Cenozoic but also to the Tethys (Mediterranean) history. They are the result of well-known distension movements on the margins formed by rifting as well as of compression movements responsible for building the Pyrenees.

Therefore, the following elements must be distinguished: the Armorican margin and its onshore prolongation in the Aquitaine Basin, both of which remained stable during their entire history; the North Spanish margin and the Pyrenean zone, which were active especially in the late Cretaceous and the early Cenozoic; and the Galicia-Portugal margin north of Nazaré, which remained stable during its entire history.

A great deal of geological and geophysical work has been done in this area. Most of it was brought together in 1970 at the Symposium on the Structural History of the Bay of Biscay and published in 1971. Important data were provided by Leg XII of the GLOMAR CHALLENGER (Laughton, et al., 1972). But the most important new data obtained since then come from geological reinterpretations made by oil companies in the Aquitaine Basin on the basis of new ideas on the Bay of Biscay. These data were the subject of a special meeting of the Geological Society of France, whose details were published in 1972.

CONTINENTAL MARGIN OFF GALICIA AND PORTUGAL, NORTH OF NAZARE

The bathymetric maps of this region, compiled by Berthois et al. (1964) and Black et al. (1964) reveal two units north of the Nazaré Canyon (Figs. 1 and 2): Offshore from the Portugal sedimentary basin, the continental shelf is 40-50 km wide. Farther north, all the way to Cape Finisterre, the continental shelf is about 30 km wide. In the west, it is prolonged for nearly 200 km by a marginal plateau at an intermediate depth, with seamounts such as the Galicia Banks and the Vigo and Porto seamounts. There is a considerable shift in latitude between this marginal plateau and the north Spanish continental margin.

The Portugal sedimentary basin has its northern end at about 40°30'N and is 4-5 km thick. It is made up of an evaporitic series from the Triassic and the Lias, a Jurassic series 3,000-4,000 m thick, and a Cretaceous and Tertiary series hardly more than 1,000-1,500 m thick. From the structural standpoint, it is disturbed by long diapiric structures trending north-south. It is affected also by a southwest-trending transverse-fault zone, probably transcurrent, located approximately in the prolongation of Nazaré Canyon. Offshore, south of the canyon, lie the Farilhoes and Berlingas islands, formed by outcrops of the Hercynian basement.

Black et al. (1964) were the first to study the Galicia Banks area by bathymetry, magnetism, seismic refraction, and dredging. They showed that the nonmagnetic submarine plateaus were probably collapsed blocks of continental origin. Stride et al. (1969) then made a sparker seismic-reflection profile from Cape Finisterre to the Galicia Bank. In 1969 a cruise carried out by IFP and SNPA provided new data on the continental margin between 40 and 43°N. The Flexotir seismic-reflection profiles were digitally recorded with a 12-trace streamer. They were then processed to obtain a CDP coverage. In addition, Lamboy and Dupeuble (1971), Boillot and Musellec (1972), and Boillot et al. (1972) studied the superficial structure of the Galicia-Portugal continental shelf using sparker seismic-reflection profiles as well as numerous cores and dredged samples. All these data define the major structural features of the continental margin in this area.

Margin Off the Portugal Sedimentary Basin (40°N)

Deep seismic-reflection profiles show the prolongation of the onshore Portugal basin onto the continental shelf with comparable thicknesses. Diapirs caused by Lias-Triassic salt are also visible. Some of them are piercing the bottom of the sea and cause the outcropping of a ring of older formations. These formations have been studied by Boillot et al. (1972), who revealed the presence of a probable Early Cretaceous with a Weald facies, a transgressive Late Cretaceous, and a mainly neritic Tertiary. On the whole, the Portugal basin can thus be seen to continue onto the continental shelf, with the deposits mainly remaining of the shelf type, although with more open marine facies.

Farther west, the break of the continental shelf corresponds to the edge of a Tertiary progradation slope. Beyond this, the horizons attributed to the Tertiary or Late Cretaceous become steadily deep-

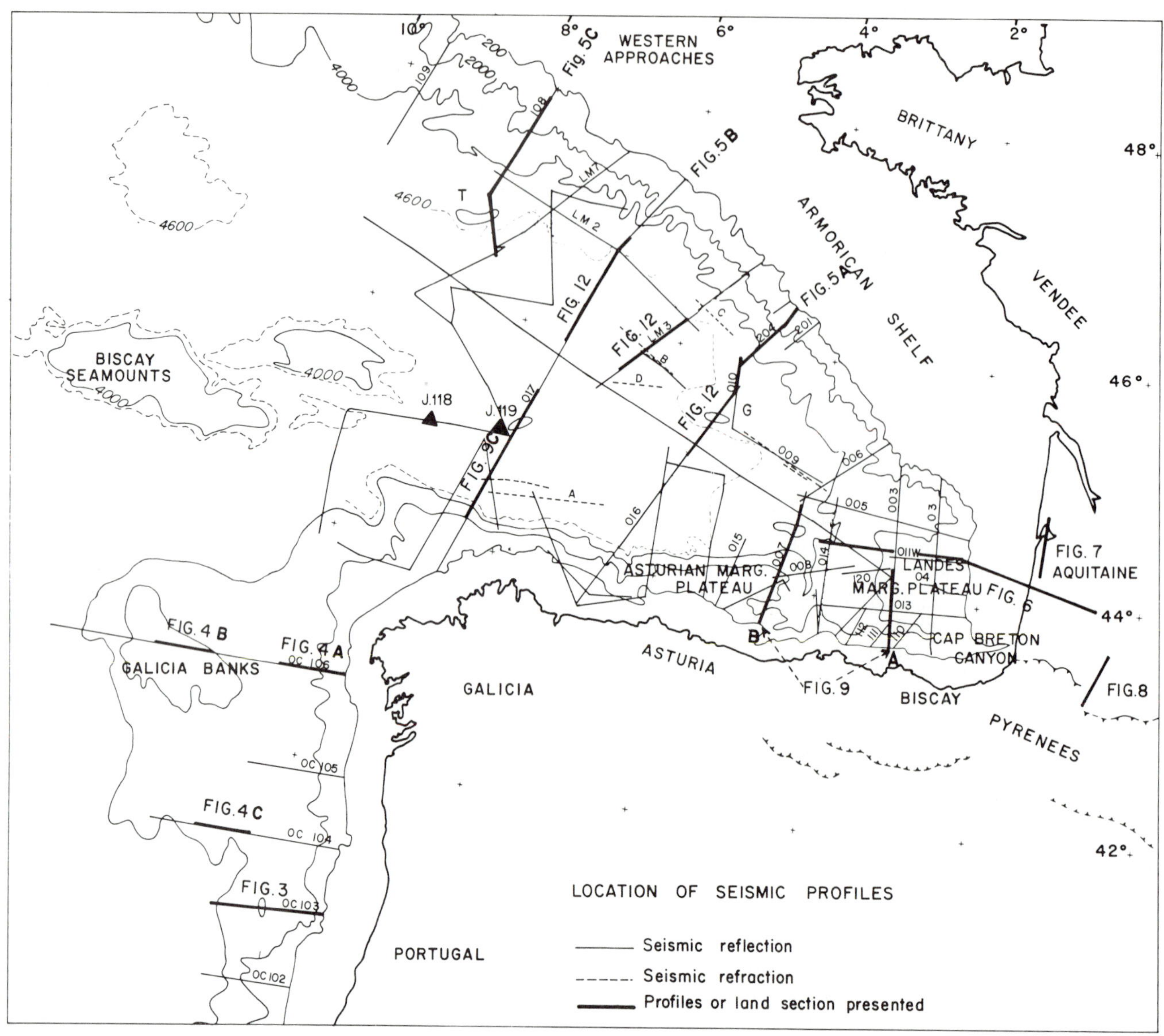

Fig. 1. Bathymetry and location of seismic profiles. DSDP sites shown by triangles.

er, to around 2,500-3,000 m. Here, a very large fault displaces the reflector attributed to the Late Cretaceous by nearly 3 sec DT (two-way time), while the sea-bottom quickly drops to a depth of 4,000 m. In the great depths a new series (formation 3) appears and becomes quite thick in the low points or grabens in the acoustic basement, where it reaches 1-1.5 sec. This series, which is found throughout the Galicia-Portugal margin (Fig. 2) is attributed mainly to a marine lower Cretaceous that has filled in the hollows in the substratum of the margin which collapsed during the distension movements linked to the opening up of the Atlantic at that time, as in the Bay of Biscay.

Continental Margin from the Northern Portugal Basin to Cape Finisterre

Farther north, at the latitude of the Porto Seamount, the pattern is different. On the continental shelf the sedimentary series are thinner than farther south and do not reveal any structures linked to diapirs. After the shelf break (profile OC 103, Fig. 3) the depths plunge quickly to 2,000-2,500 m in connection with a very large vertical fault zone. Between this escarpment and Porto Seamount, which is a basement horst and not a diapiric structure as assumed by Pautot et al. (1970), a very thick sedimentary basin of nearly 4 sec develops, i.e., at least 5,000-6,000 m. The underlying series is attributed to the Late Cretaceous-Tertiary (formations 1 and 2) with nearly 1.6 sec thickness; there is thus a thick fill of nearly 3 sec, attributed to the lower Cretaceous (formation 3).

The deep reflections may be Jurassic, and two quite visible piercing structures can perhaps be related to Lias-Triassic evaporites. On the Porto Seamount horst, where it is evident that the continental basement has been broken and collapsed by distension, only 0.4 sec of sediments can be found on

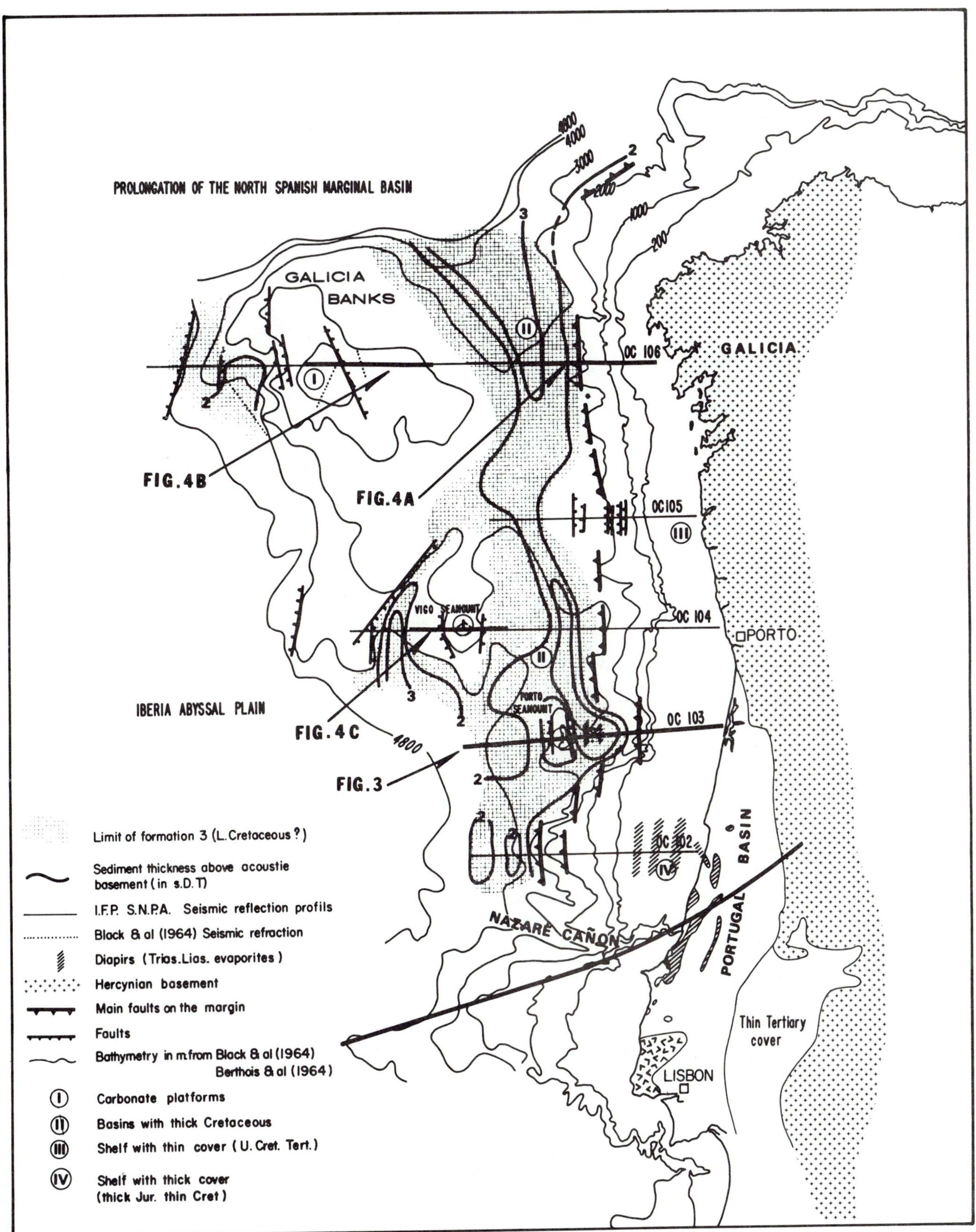

Fig. 2. Continental margin of Portugal-Galicia: bathymetry and structural sketch map.

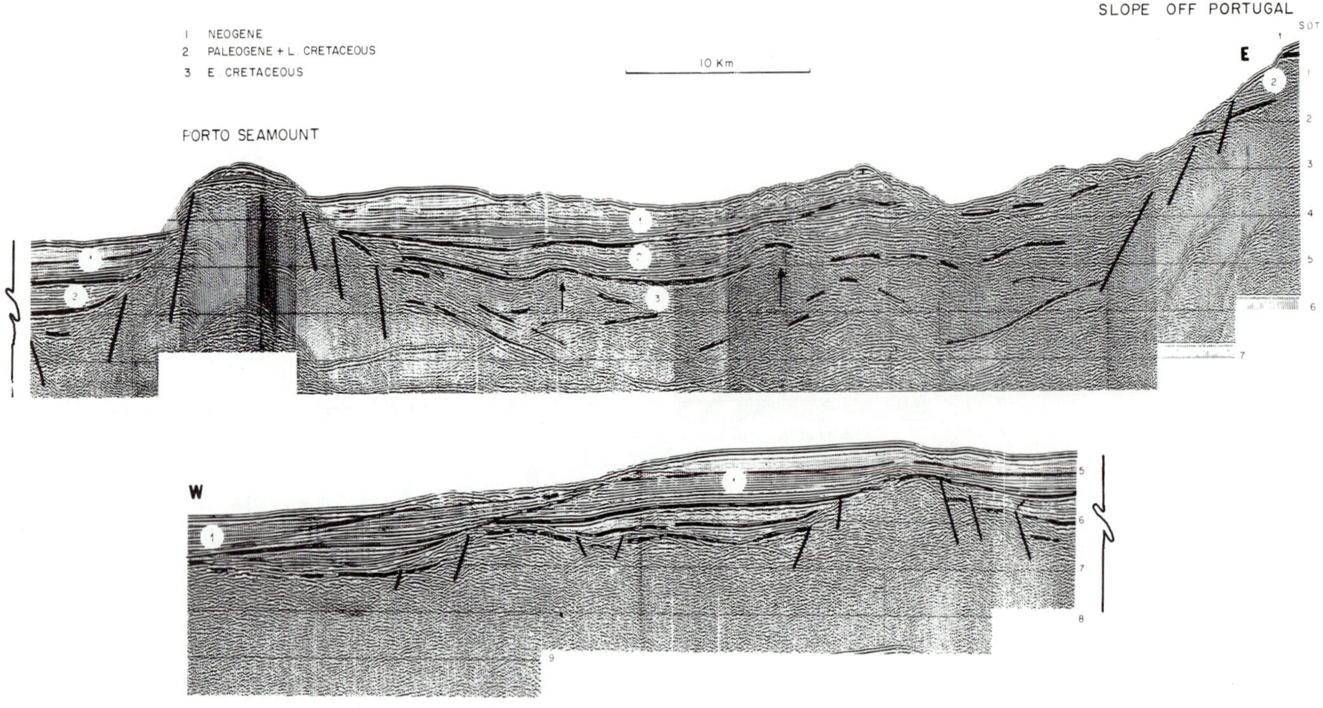

Fig. 3. Seismic profile on the Portugal margin, crossing the Porto Seamount. (see Figs. 1 and 2 for location.)

top of an acoustic substratum. Toward the west and with an appreciable vertical shift of the sea bottom, clearly indicating the dam effect caused by Porto Seamount, 1.2-2 sec of sediments are found on top of an acoustic substratum of an undetermined nature. The most noteworthy feature is the gradual westward thinning of the deep-lying series beneath the Tertiary-Late Cretaceous unit, which tends to lie directly on top of the acoustic basement. Such an arrangement could perhaps denote an oceanic-type substratum that becomes younger westward.

Farther north, from Oporto to Cape Finisterre, on the continental shelf and the slope, a constant structural pattern can be seen. The sedimentary cover on the shelf is thin and appears as a monocline dipping seaward. Two structural units are visible (Lamboy and Dupeuble, 1971; IFP-SNPA profiles OC 106, Fig. 4B): a Tertiary formation 200-300 m thick, lying unconformably on top of Late Cretaceous, dipping to the west and cropping out on the upper slope. Beyond, at a fair distance from the shelf break, a series of deeper large faults affect the deep-lying sedimentary series correlated farther south with the Lower Cretaceous. This causes the substratum to nearly crop out in the east.

Figure 2 shows that formation 3 follows the bathymetric depressions and skirts the plateaus. It is not present on Galicia Banks and Vigo and Porto seamounts. As a results, it outlines a complex elongate basin at the foot of the continental slope, joining the west Portugal region to the southern part of the deep basin that prolongs the north Spanish marginal trough. The superposition of the isopachs of the total sedimentary series and of the distribution area of formation 3 (Fig. 2) clearly shows that the areas of maximum thickness are linked to the development of formation 3.

Dredging on the horst that forms Vigo Seamount (Black et al., Fig. 4A, 1964) recovered Late Jurassic limestone pellets with *Clypeina* Jurassica algae on the lower rise of the escarpment. In the same zone there is Early Cretaceous (late Albian-Aptian according to Dupeuble in Pautot et al., 1970) with a neritic limestone facies. At the top of the horst, Black et al. (1964) took samples of middle Tertiary consolidated foraminiferal ooze. On the Galicia Banks, Black et al. (1964) recovered calcareous Late Cretaceous (Danian-Maestrichtian) porcellanous limestones, detrital limestones, and chalks. Tertiary limestone (middle Eocene) was also found. Reworked *Orbitolinae* in the Late Cretaceous indicate the probable existence of Early Cretaceous in the vicinity. The structure of the eastern flank of the Galicia Bank can be interpreted as a reef, possibly of Cretaceous or early Tertiary age (Fig. 4C).

These data suggest that horsts and grabens were formed during the Early Cretaceous distension of this margin. Thick Cretaceous detritus accumulated in the grabens, while sedimentation on the bathymetric highs was reduced and made up of neritic limestones. The limit of formation 3 must thus be considered rather as a facies variation (Fig. 2). The beds were lowered to their present depth by subsidence after the neritic Albian, as revealed by a more pelagic but still calcareous sedimentation in the Tertiary. In the case of Vigo Bank (Fig. 4A), subsidence can be estimated at 3,000 m since the Early Cretaceous, since neritic rocks were samples

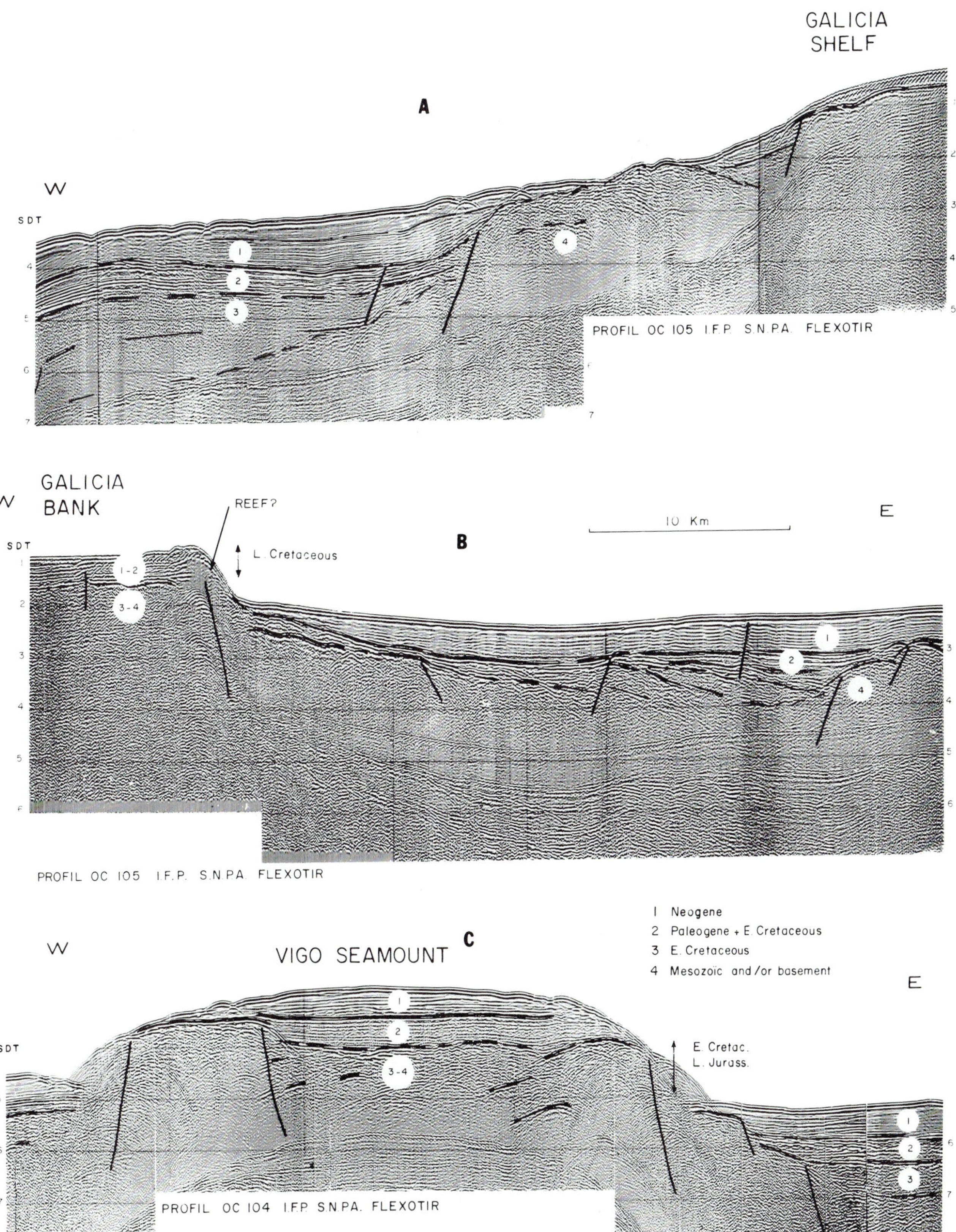

Fig. 4. Seismic profiles showing Vigo Seamount (A), the margin off Galicia (B), and the Galicia Banks (C). See Figures 1 and 2 for locations.

between 2,890 and 3,640 m. Likewise, in the Galicia Banks, the Jurassic quite probably exists, and in the basins it probably forms the acoustic basement.

To conclude, the Galicia-Portugal continental margin has all the features of an inactive margin, with numerous similarities to the Armorican margin in the Bay of Biscay. During the opening up of the North Atlantic, Late Jurassic (?)-Early Cretaceous tensional movements fragmented the Hercynian substratum and its Triassic-Jurassic overburden into horsts and grabens. While thick sedimentary series were being deposited in the grabens, the horsts received merely a reduced calcareous sedimentation. This arrangement is quite similar to the Newfoundland-Labrador margin (Ruffman and van Hinte, 1973; Grant, 1973) with the high Orphan Knoll and Flemish Cap showing reduced sedimentation and separated from the continental shelf by thick sedimentary basins. As on the Armorican margin in the Bay of Biscay, these distension movements created structures more or less parallel to the margin. The Portugal sedimentary basin on land and on the shelf, with the reduced Cretaceous, is thus not comparable to the Parentis and Adour basins, which are linked to the opening of the Bay of Biscay and contain thick Cretaceous sediments deposited along areas which were deep at that time.

The history prior to the Cretaceous is difficult to reconstruct. The continuation of Lias-Triassic evaporites beyond the shelf is probable but not certain, according to seismic data. The offshore transition to the oceanic basement is not clear. Oceanic basement may exist west of the Galicia Banks and at the ends of the southern profiles of Figure 1.

NORTHERN MARGIN OF THE BAY OF BISCAY AND ITS ONSHORE PROLONGATION IN THE AQUITAINE BASIN: A STABLE CONTINENTAL MARGIN

The northern continental margin of the Bay of Biscay, as it appears now, is limited to the Cap Ferret graben and the Landes marginal plateau (Fig. 1). This boundary must not be taken to be its actual end. However, studies by petroleum geologists have shown that this continental margin continues on land in the Aquitaine Basin and that the present shelf edge corresponds to the boundary of filling by prograding Tertiary sediments in the Bay of Biscay.

Armorican Continental Margin

Berthois et al. (1964, 1965) emphasized the morphological variety of the Bay of Biscay. In particular, they demonstrated the homogeneity of the Armorican continental margin morphology, between 47°30' and 45°N, to a well-defined geological unit linked to an inland Hercynian platform covered by reduced sediments and to a continental slope 40-70 km wide dissected by numerous canyons and having no typical continental rise (Fig. 1). The boundaries of this margin are clearly marked in the south by the Mesozoic and Cenozoic Parentis Basin and its western prolongation via the Cap Ferret graben (Fig. 13) and in the north by the Mesozoic and Cenozoic basin of the Western Approaches of the English Channel, expressed on the slope by an abrupt change in morphology with an advance of the margin being prolonged by the Trevelyan Escarpment.

Many studies of the superficial structure of the continental shelf have been made by sparker seismic-reflection surveys and core drilling (Robert, 1969; Vanney et al., 1971; Bouysse and Horn, 1972; Caralp et al., 1972; Lapierre, 1972). The deep structure has been studied by seismic refraction by the University of Montpellier (Martin et al., 1968) and Compagnie d'Exploitation Petroliere (1963) and has been discussed by Montadert et al. (1971b). After the surveys by Hersey and Whittard (1966), Curray et al. (1966), Jones (1968), and Stride et al. (1969), the geological structure of the continental slope and of the adjacent abyssal plain has been studied by higher-energy seismic-reflection profiles (Cholet et al., 1968; Damotte et al., 1968; Montadert et al., 1971a, 1971b), reinforced by coring and dredging.

The broad geological structure of the sedimentary overburden on the Armorican continental shelf appears to be quite simple. It forms a homocline that becomes steadily thicker offshore, affected by a few vertical faults and undulations trending southeast-northwest as prolongations of structures known on land between North Aquitaine and Brittany. The Hercynian basement crops out at different points, such as Rochebonne shoal. Coring shows that the Paleogene generally lies on top of the basement from Brittany to the Vendee region and is overlain in the west by Neogene.

An abrupt change appears in the northwest, beyond Brittany, in the Western Approaches Basin, where southwest-trending folds bring the Mesozoic to the surface (Vanney et al., 1971; Curray et al., 1966; Lapierre, 1972). These southwest-trending structures exhibit strong magnetic anomalies and may be linked with distension movements due to Atlantic rifting (Grau et al., 1973). This change in style occurs at the level of a Hercynian unconformity that extends westward onto the continental margin, where it separates the two above-mentioned morphological units.

Seismic reflection (Cholet et al., 1968; Damotte et al., 1971; Montadert et al., 1971a, 1971b) shows that this sedimentary overburden can be subdivided into three units near the slope (Fig. 5):

Unit I, of Late Tertiary (post-Eocene) age, 600-800 m thick at the shelf break, having the structure of prograding sedimentary talus. The shelf break corresponds to the edge of this talus and not to a structural boundary. This unit continues on the continental slope with slumping structures.

Unit II, of Early Tertiary-Late Cretaceous age, 800-1,000 m thick at the shelf break and becoming steadily thicker on the continental slope, as opposed to the previous unit. Its top is often eroded on the slope, probably in connection with the slumps that affected upper unit I.

Unit III, generally poorly visible on the continental shelf, but standing out clearly on the continental slope. It is unconformable beneath unit II and frequently fills grabens. It is probably mainly of Early Cretaceous age. Older Mesozoic formations probably are confused with the acoustic substratum.

Three seismic profiles reveal the structure of this continental margin and how it evolves from southeast to northwest toward the Western Approaches of the channel (Figs. 1 and 5). The margin has all the features of a stable area, with tension structures, horsts and grabens, tilted blocks, and evidence of considerable subsidence at the level of the continental slope and in the abyssal plain (Armorican marginal plain). Subsidence of about 3,000 m is demonstrated by coring and dredging of neritic Lower Cretaceous on the slope of the Armorican margin (Montadert et al., 1971a) and of the Western Approaches margin (Pastouret et al., 1974).

Formation 3 is interpretated as mainly Lower Cretaceous infilling grabens formed during the opening of the Bay of Biscay. Carbonate neritic deposits could be restricted to the upper part of the slope and to the top of the horsts. Thus the main part of the Amorican marginal basin could be Lower Cretaceous. Further faulting, probably during Upper Eocene, is related to an important phase of subsidence; it is responsible for the Trevelyan Escarpment. It is difficult to define accurately the continent-ocean boundary by seismic reflection. Faulted blocks or horsts prolong the continental slope beneath the abyssal plain; it is particularly evident under the Trevelyan Escarpment. Bacon et al. (1969), on the basis of gravimetric evidence, reach the same conclusion. Thus it is probable that off the Western Approaches of the channel the margin advances by 50-70 km when compared with the Armorican margin farther south. This could be related with an important transverse discontinuity due to a Hercynian structural feature. If so, it might mark the beginning of a transform fault. The opposite offset could be found between the Galicia Banks and the North Spanish margin.

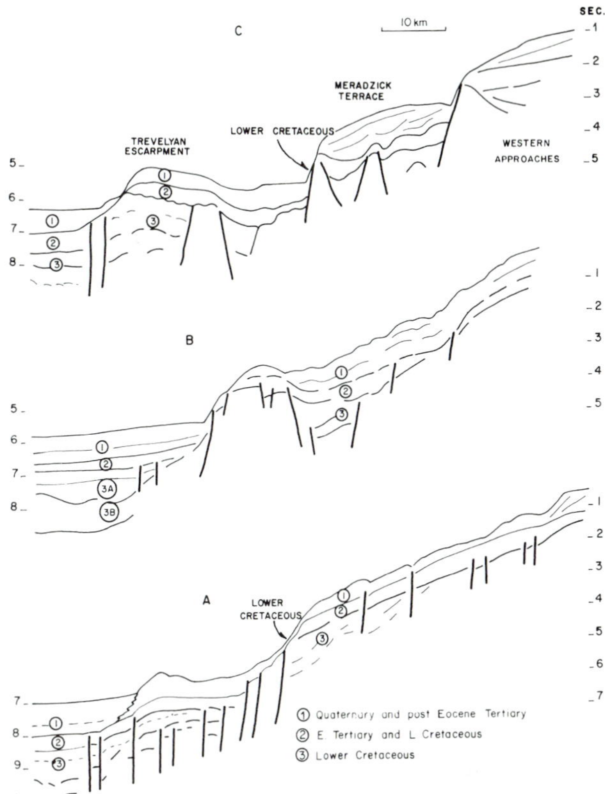

Fig. 5. Schematic profiles showing structure of the Armorican and Western Approaches margins (see Fig. 1 for location).

Southeast Corner of the Bay of Biscay—Prolongation of the Stable Armorican Margin in the Aquitaine Basin

Present Margin of Southeastern Bay of Biscay. The principal morphological and structural units have been defined in this areas by Cholet et al. (1968), Montadert et al. (1971a), and Valery et al. (1971) (Figs. 1 and 13).

In front of the north-trending Aquitaine shelf break, there is in the north the Cap Ferret depression, whose northern flank is the continuation of the Armorican margin; the Landes marginal plateau (between 1,000 and 2,000 m), which drops off westward in the Santander Canyon; and the Canyon of Cap Breton in the south at the foot of the Spanish continental shelf. It has been demonstrated that the limit of the Aquitaine continental shelf is not a structural boundary, but rather the present step of young prograding deposits (Fig. 6).

Cap Ferret depression forms a graben, acting as a direct continuation of the onshore Mesozoic Parentis Basin. It thus forms an indentation of the Bay of Biscay into the continental domain. This connection with the Parentis Basin throws considerable light on the chronology of the geological events responsible for forming the Bay of Biscay. Seismic surveys show that the deposits in this graben have a complex structure linked to slumps and channels, thus indicating that it is a transit zone for sediments on the way to the abyssal plain. The question arises of how far the oceanic crust extends into this indentation. No measurement of this exists. However, Limond et al., (personal communication) have shown that, farther west on profile 009, (Fig. 1), an entirely oceanic crust exists . We can also see that the North and South Gascony ridges end exactly in the axis of the graben. There is probably a cause-and-effect relation here.

The Landes marginal plateau corresponds to

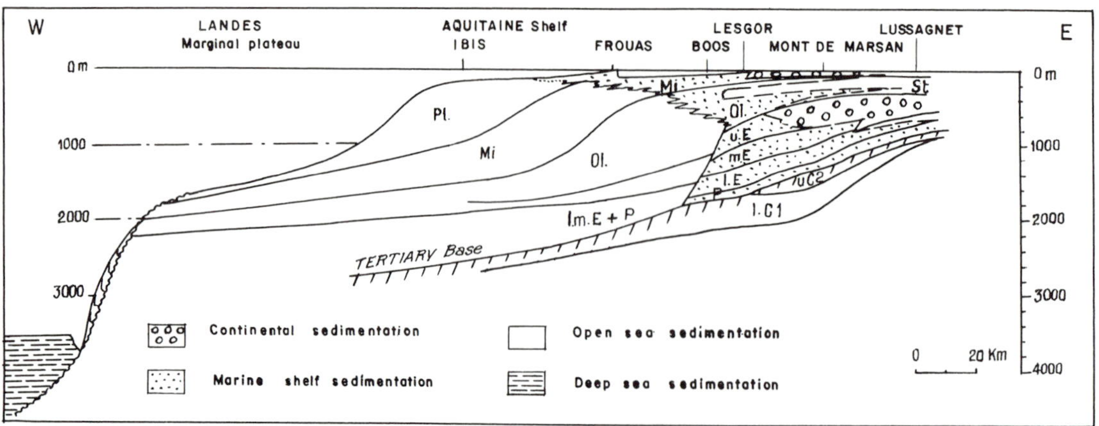

Fig. 6. Schematic cross section showing the progradation of Cenozoic and Recent deposits in Aquitaine (after Winnock et al., 1973).

the prolongation of a positive area (called Le Poteau) that is well known in Aquitaine and that separates the Parentis Basin from the Adour Basin (?). This high zone thus became an independent entity in the Late Malm-Early Cretaceous (Fig. 13). Underneath the relatively thin Cenozoic and Quaternary overburden (400-1,000 m) that corresponds to the bottom set of the present prograding slope, there is an early Cenozoic and a thin Late Cretaceous unconformably covering highly disturbed horizons with extensive diapiric activity linked to Triassic evaporites. A positive gravimetric anomaly exists in the northwestern plateau; it might be attributed to a rise in Paleozoic formations and/or the disappearance of evaporites (Valery et al., 1971). We suggest that it could be similar to the positive gravity anomaly existing on the Le Danois Bank and farther east. Seismic profiles show that this positive area during the Cretaceous stops in the west at the level of the Santander Canyon, where the early Tertiary to Cretaceous series again become thicker toward the abyssal plain.

There is a gradual thickening of the post-Eocene Tertiary toward the southwest (Fig. 9A) that may attain 2,500 m on the edge of the *Cap Breton Canyon*. The upper part of the Cenozoic crops out on the northern flank of the canyon, while the deep horizons continue regularly underneath. Beyond, continuity no longer exists, and the southern flank probably corresponds to a margin that was tectonized during and after the main Pyrenean movements, the Oligocene deposits being also folded by pre-Aquitanian movements. Levees stand out clearly on the edge of the canyon, which is probably a nondeposit or/and erosion zone for the transit of sediments toward the abyssal plain via the Santander Canyon. It is similar to a furrow located north of the Pyrenean front that gradually became filled by the westward advances of the Tertiary prograding deposits (Schoeffler, 1965; Kieken, 1973).

Former Margins in the Aquitaine Basin. It has been shown that the present limit of the Aquitaine continental shelf corresponds to the top of a prograding talus and not to a structural boundary. Kieken (1973) used data from wells to reconstruct the westward advance of the shorelines and of the continental shelf break during the Tertiary in Aquitaine. It is quite clear that, prior to the early Eocene, the shorelines still indicated the indentation of the Parentis and Adour basins. In the late Eocene they took on a northerly trend and were bounded in the south by the advancing Pyrenean front, which continued beneath the Cantabrian and Asturian continental shelves. There was thus an advance filling in by Cenozoic deposits of former margins, linked to the deep basins formed in the Early Cretaceous and making up the continuation of the present margins of the Bay of Biscay.

These former margins have been determined with remarkable clarity in the northern part of the Parentis Basin, which, as we have seen, is connected directly to the Bay of Biscay by the Cap Ferret graben. Winnock (1971) and Dardel and Rosset (1971) have shown that the northern boundary of the Cap Ferret-Parentis graben corresponds to a fracture that was active beginning in the Triassic, while in the south, thick evaporitic and volcanic series were accumulating. This is the northern edge of the initial rift that formed the Bay of Biscay (Montadert and Winnock, 1971).

The Parentis Basin became a separate entity in the late Malm with the beginning of subsidence, becoming deeper in the Early Cretaceous (especially the Albian). Fried and Morelot (1973) have studied the sedimentology and reconstructed the bathymetry during the history of the basin (Fig. 7). They have shown that the deposits are typical of margins with well-developed deposits on slopes whose depth is estimated to have been 2,000 m in the Albian (?). These onshore data can be used for the reconstruction of the geological history of the entire northern margin of the Bay of Biscay, which has remained stable since it was formed. The Cap Ferret-Parentis graben thus forms only an indentation, limited in the south by the Landes marginal plateau with very

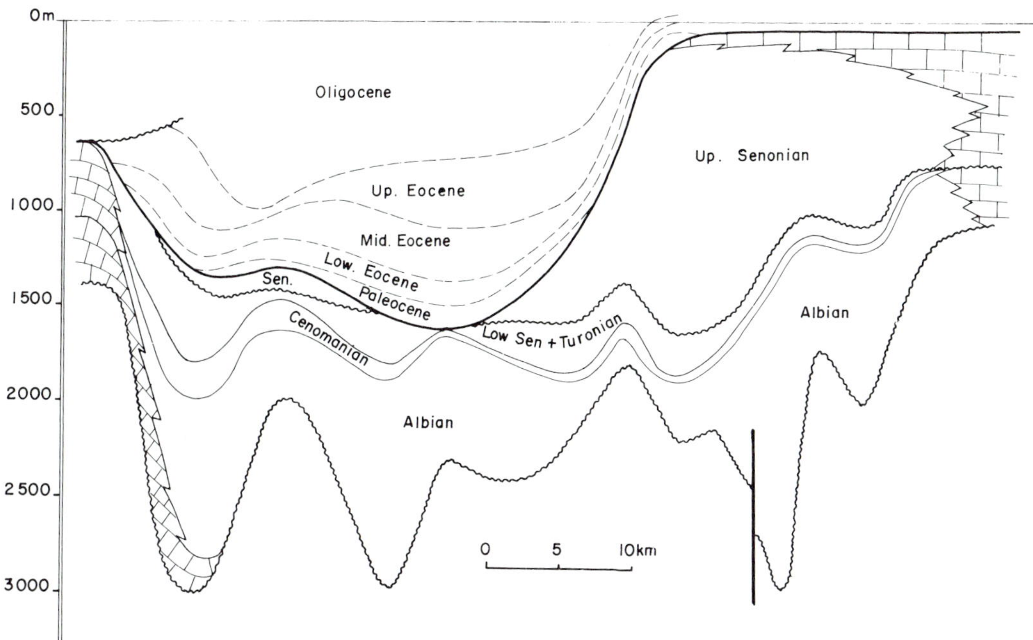

Fig. 7. Reconstruction of the Parentis Basin at the end of the Cretaceous (Senonian) on a transverse cross section (after Fried and Morelot, 1973).

important faults having a throw of at least 5,000 m of the Jurassic on either side (Dardel et al. 1971). In the Cretaceous the margin surrounded this positive area in the west, along the Santander Canyon, since the zone between the Landes plateau and the Asturian marginal plateau (Le Danois Bank) corresponded to a distension and collapsed zone in the Cretaceous.

The southern boundary is more complex because of the advance of the Pyrenean front, which affects the anterior structures in the Basque Country region (Fig. 13). Nonetheless, farther east, Winnock (1973), Winnock et al. (1973), and Feuillée et al. (1973) have shown that the Cretaceous South Aquitaine basins corresponded to deep basins at that time (Fig. 8): the continental margin of these deep basins remained stable only in the north, which was not affected by the Pyrenean orogenesis.

This fossil margin extends westward into the Atlantic where it continues under the Neogene basin of Cap Breton. Eastward it can be followed in the Comminges. To the north the platform is represented by a thin, neritic Upper Cretaceous (Cornes Mountain near Remy-les-Bains) transgressive upon Paleozoic at Monthoumet. Southward the basin is evident in the complex trough filled in by Cenomanian to Senonian flysch, overthrust by the North Pyrenean Front (Burgarach Range). Interpretation in this area is difficult because in addition to the tectonic complexity, there are numerous sedimentary complications, olistrolithes and olistostromes, owing to the proximity of the southern orogenic margin.

To conclude, new sedimentological interpretations in the Aquitaine Basin favor the hypothesis on the eastward prolongation of the Bay of Biscay that the deep marine areas reached the approaches of the Mediterranean during the middle Cretaceous. A deep and complex (several troughs) gulf occupied the area of the North Pyrenean Zone, and expanded in South Aquitaine, connecting westward to the abyssal plain via the Santander depression, also downfaulted at that time.

SOUTHERN MARGIN OF BISCAY AND PYRENEES: A FORMER ACTIVE MARGIN

North Spanish Continental Margin

The North Iberian continental margin extends westward to Galicia Banks, which are an extension of the Iberian block that collapsed during the Mesozoic and Cenozoic. There is a large offset between the northern edge of the Galicia Banks and the North Spanish margin beginning at Cape Ortegal Spur. Beyond Cape Ortegal the shallow structure of the continental shelf has been intensively studied (Boillot et al., 1971b, 1973b). These studies revealed that the Pyrenean system continues offshore onto the North Spanish continental shelf at least to Aviles Canyon, including a late Eocene, hence typically Pyrenean, compression phase. A second phase marked by vertical movements and probably also compression occurred in the late Oligocene. These authors consider that the slope beneath the continental shelf was built following this event, related to vertical faulting. The subsequent history is marked by the deposition of a Neogene progradation slope with further transcurrent and normal faults and the formation of canyons.

The outstanding feature of the North Spanish

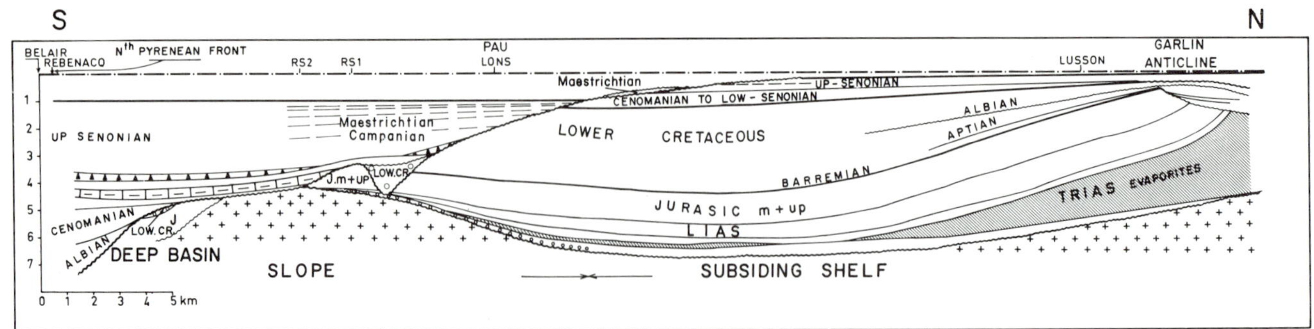

Fig. 8. Reconstruction of the South Aquitaine stable margin at the end of the Cretaceous on a transverse cross section (after Winnock et al., 1973).

continental slope is its steepness, as much as 20°. In addition, it has the shape of a regular curve extending northward over nearly 500 km, from 9°W to the meridian of Santander Canyon. In its western part the slope follows directly the continental shelf, but at Aviles Canyon there is an intermediate plateau, called the Asturian marginal plateau (Damotte et al., 1969). An uplifted zone exists on the edge of this plateau called Le Danois Bank; it is separated from the continental shelf by a depression occupied by the Lastres Canyon.

Damotte et al. (1969) and Montadert et al. (1971) have dicussed the structure of the margin in this area (Fig. 9). In particular, between the shelf and the Le Danois Bank, there is a thick Oligo-Miocene basin resting on older deep horizons. These horizons pinch out in the north on the Le Danois Bank, where various formations crop out: Late Cretaceous-Early Tertiary with neritic facies (Damotte et al., 1969), pelagic Early Cretaceous and Tithonian micrites with *Calpionellae* (Boillot et al., 1971a). These facies, just as those in the Cormoran I borehole on the Aquitaine continental shelf, show that in the Late Jurassic in the Bay of Biscay area there was an Atlantic Sea opened out westward (Durand-Delga, 1973). Recently, Capdevila et al. (1974) discovered the following on the slope at Cape Ortegal and the Le Danois Bank: nepheline syenites; lamprophyres; alkaline basalts unknown in northwestern Spain, but similar to the Late Cretaceous at Guipuzcoa and in the northern Pyrenees; and granulites and charnockites different from the ones in northwestern Spain, but similar to those in the northern Pyrenees. This area thus appears to be one of the most eroded ones in the Hercynian Range. Boillot et al. (1973b) deduce from this that the North Pyrenean Zone continues on to the southern part of the Bay of Biscay along Le Danois Bank to Ortegal Spur. At the level of the Asturian marginal plateau, the continental slope linking it to the abyssal plain is less steep than farther west. Horizons are visible underneath the Tertiary, and the overall structure is conventional, with a succession of collapsed blocks.

In the east, the Asturian marginal plateau is separated from the Landes marginal plateau by a deep depression occupied by the two large Torrelavega and Santander canyons (prolongation of Cap Breton Canyon) surrounding a promontory, the Santander Spur. The seismic profiles in this area (Montadert et al., 1971a, 1971c) have shown that beneath the post-Eocene, thick series extend toward the axis of the depression and toward the abyssal plain, as has already been mentioned west of the Landes marginal plateau. Farther east (Fig. 9A), the structural pattern is fairly similar to the one described at the Asturian continental shelf. The northern boundary of the Cantabrian continental shelf corresponds to the boundary of the north Pyrenean tectonized zone (pre-Oligocene in this area). Therefore, we interpret the limit of the continental shelf of Asturia, south of Le Danois Bank, not as a vertically faulted margin, but as the limit of the North Pyrenean Front.

Cap Breton Canyon is located in the axis of a Tertiary basin that thins northward onto the Landes marginal plateau. It thus seems logical to interpret the Santander depression as an ancient structure formed by distension during the opening up of the Bay of Biscay in the Cretaceous. A break occurred at this level between the high parts of the Le Danois Bank and the Landes marginal plateau. This depression was the transit zone for sediments being deposited in the abyssal plain at that time. Indeed, the Cretaceous North Pyrenean, Biscay, and perhaps Asturian deep basins (Fig. 13), with mainly Late Cretaceous to early Eocene flysch, converge in this region (Henry et al., 1971).

Active Margin South of Aquitaine

It has been said that the deep basins in Aqui-

Fig. 9. Seismic profiles through the North Spanish continental margin, showing (A) the Cantabrian shelf and slope corresponding to the North Pyrenean front, the Cap Breton Canyon and basin, and the Landes marginal plateau; (B) the Asturian shelf and first slope corresponding to the North Pyrenean front, the Asturian marginal plateau, and the slope and the marginal trough with the front of the North Spanish tectonized zone; and (C) the North Spanish marginal trough with the front of the tectonized zone, the South Gascony Ridge, the high area of Cantabria (Upper Eocene), and the central "young" area.

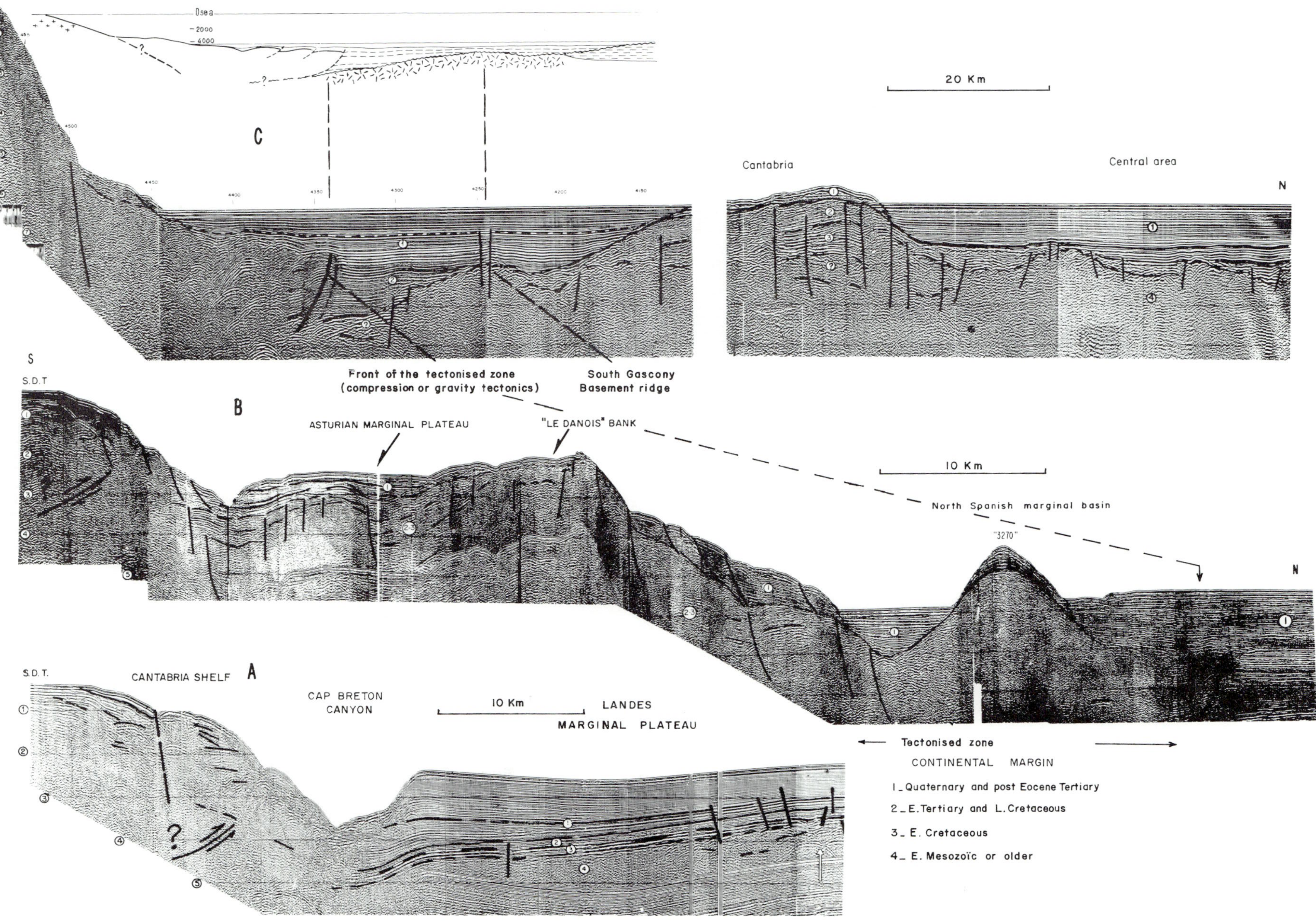
0sea
-2000
-4000
C
20 Km
Cantabria
Central area
N
S
S.D.T
Front of the tectonised zone
(compression or gravity tectonics)
South Gascony
Basement ridge
B
ASTURIAN MARGINAL PLATEAU
"LE DANOIS" BANK
10 Km
North Spanish marginal basin
"3270"
N
S.D.T.
CANTABRIA SHELF
A
CAP BRETON
CANYON
10 Km
LANDES
MARGINAL PLATEAU
Tectonised zone
CONTINENTAL MARGIN
1_Quaternary and post Eocene Tertiary
2_E.Tertiary and L.Cretaceous
3_ E. Cretaceous
4_ E. Mesozoïc or older

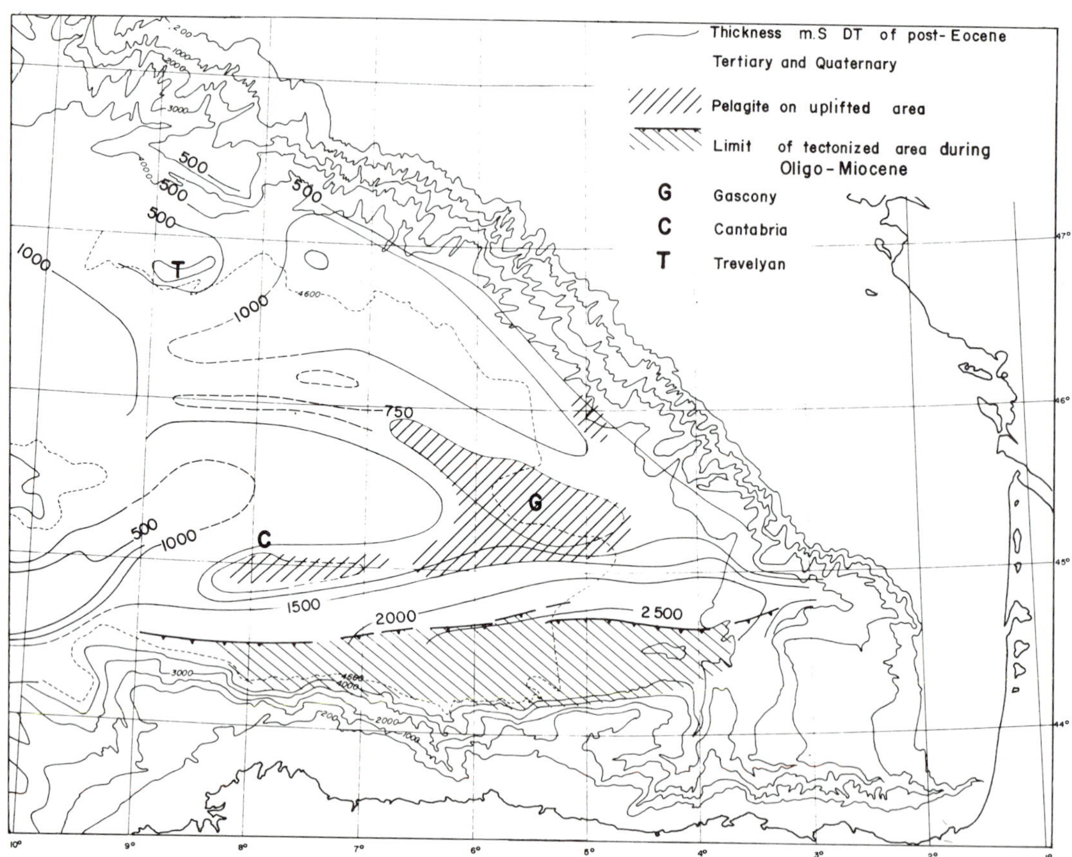

Fig. 10. Isopach map of formation 1 (post-Eocene) in the Biscay abyssal plain (contoured in seconds of two-way time).

taine were the result of a very active subsidence during the Lower Cretaceous. The existence of a continental slope to the north in Aquitaine, between a shelf with carbonate deposits and a deep basin, is evident since late Aptian. Since the Albian-Cenomanian a dissymmetry is also evident between a stable northern margin and an already active southern margin, marked by coarse detrital influx and olistolithic masses. This evidence of a southern tectonic margin continued to become more and more obvious. The southern part of the deep troughs overthrust the most northern troughs, which were not yet filled in, as shown by olistostromes of Campanian, Early Eocene, Late Eocene age which were deposited in a trough shifted more and more toward the north.

The southern margin of the deep Gulf of Aquitaine thus appears as an active margin, with the deepest and southest part of the troughs tectonized and uplifted by the North Pyrenean overthrusts. This overthrusting seems of different ages, but, in fact, is the result of a continuous activity probably since Albian-Cenomanian.

Structures in the North Spanish Marginal Basin

The question arises as to just how far north these Pyrenean movements can be recognized in the Bay of Biscay. At the foot of the North Spanish slope there is a thick marginal basin (Fig. 9), the southern part of which has been deeply disturbed. Montadert et al. (1971b) interpreted these structures (Fig. 9B, C) as resulting from the compression of the North Spanish Basin during the Pyrenean orogenesis by a relative movement toward the north of the Iberian block. Another reasonable explanation has been proposed by Winnock (1973), that these structures could be due only to gravity tectonics. This interpretation could explain, moreover, the regular curve of the continental slope as a sort of slip surface. Nevertheless, these structures could be only an epiphenomena of an active margin, representing deeper crustal shortening. Owing to the curvature of the Spanish slope, the linear front of the tectonized zone in the abyssal plain (Figs. 10A, B and 13) is limited westward at the Ortegal spur, and eastward at the northeastern corner of the Landes marginal plateau. Very important faults exist to the southern part of the Parentis Basin (throw up to 5,000 m).

Two hypotheses could thus be proposed for the North Spanish margin. A major overthrust exists at the foot of the Spanish slope; by a series of transcurrent faults it is related to the North Pyrenean front (Boillot et al., 1973b). Or, we suggest the

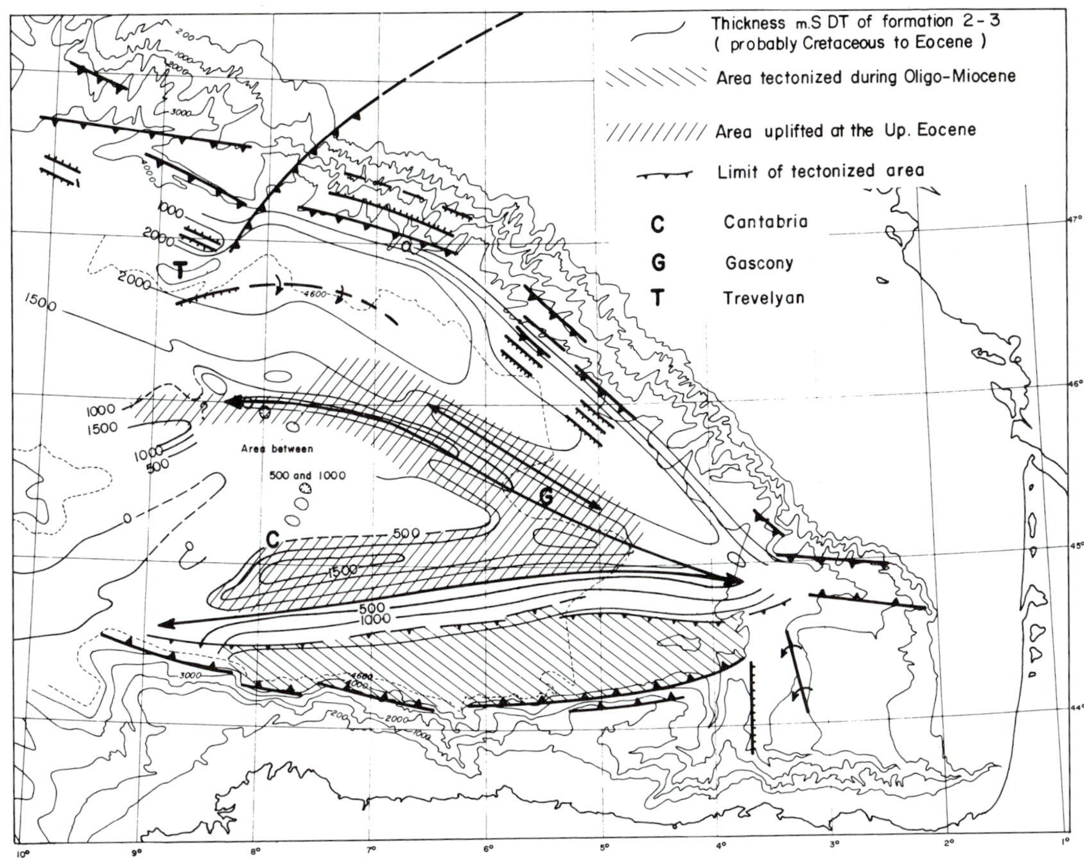

Fig. 11. Isopach map of formation 2 and 3 (early Cenozoic and Cretaceous) in the Biscay Abyssal Plain, showing the North and South Gascony basement ridges (contoured in seconds of two-way time).

existence of two overthrusts which relieve one another. The first one is the North Pyrenean overthrust, which can be followed to the west at the limit of Biscay, Cantabrian, and Asturian continental shelves; the second one could be a Biscay overthrust along the Spanish deep margin and extending perhaps to its end at the southern border of the Parentis Basin at the level of the great fault showed here by Dardel and Rosset (1971, Fig. 14C). Such an interpretation could give light on the southeast area of the Bay of Biscay, suggesting a relation between Le Danois Bank and the northwest part of the Landes marginal plateau, both showing strong positive gravity anomalies.

ABYSSAL PLAIN

The abyssal plain in the Bay of Biscay is an oceanic area that has been thoroughly explored: by magnetism (Matthews and Williams, 1968; Le Borgne and Mouel, 1970; Le Mouel and Le Borgne, 1971; Williams, 1971; Schouten et al., 1971); by gravimetry (Bacon et al., 1969; Sibuet et al., 1971; Coron and Guillaume, 1971); by seismic reflection (Jones and Ewing, 1969; Damotte et al., 1969; Montadert et al., 1971d; Sibuet et al., 1971; Frappa et al., 1970,; Malzac, 1970; Muraour et al., 1970, 1971; Grau et al., 1973); by seismic refraction (Bacon et al., 1969; Damotte et al., 1971; Limond et al., 1972). From the stratigraphic standpoint, if we exclude core drilling on the continental slope, data are limited. Information has mainly been obtained by core and dredging samples from the Cantabria Seamount Escarpment (Jones and Funnel, 1968; Montadert et al., 1971a, 1971d; Ewing et al., 1971) and especially from JOIDES Deep Sea Drilling (DSDP) sites 118 and 119 (Laughton et al., 1972). The main structural features and sedimentary units have been described from seismic reflection by Montadert et al. (1971a). New seismic profiles by IFP and ELF-ERAP have provided further information.

The age of formations 1, 2, and 3, as distinguished by seismic reflection, is confirmed by DSDP sites 118 and 119. The post-Eocene and Quaternary age of formation 1 is clearly established, and the boundary with the underlying formation corresponds to a tectonic episode with vertical faulting. Formation 2 includes Eocene and Paleocene in its upper part. Its basal part and the age of underlying formation 3 have not been determined directly. A conservative extrapolation of the high Paleocene sedimentation rate on Cantabria Seamount shows that at least the Late Cretaceous exists there. Formation 3, where it is clearly defined, is probably at least Early Cretaceous. Because of the late Eocene tectonic episode, when compiling isopach maps (Figs. 10 and 11) it is important to make a distinction between formation 1 and formations 2 and 3.

The important structural elements to be taken into consideration are the thick Armorican and North Spanish marginal sedimentary basins, the deep basement ridges called North and South Gascony ridges, and the central zone with a thinner sedimentary overburden surrounded by high zones formed during Eocene that are apparent in the topography, particularly by the Cantabria and Gascony seamounts (Figs. 10, 11, and 13).

The Armorican marginal basin, which is estimated to be between 5,000 and 7,000 m thick, hardly shows up on the isopachs (Fig. 10) of formation 1 (about 1,000 m thick), but stands out very clearly on the isopachs (Fig. 11) of formations 2 and 3. We find, in particular, the existence of a thick formation 3B (1 sec) at the bottom that seems to exist only in this basin. The Armorican marginal basin is thus an old one, probably with a well-represented Cretaceous. The boundaries of formation 3B more or less correspond to the isopach curve at 2 sec. Numerous intrusions affected formation 3A (Fig. 12A) prior to the deposition of formation 3B, which fills in the irregularities thus formed.

The basin is bounded in the southwest by a very large ridge in the acoustic basement, the North Gascony Ridge (Fig. 12A, B, C). This ridge is double in its southeastern part. Farther northwest its extension is less clear-cut. Formation 3B, which is the oldest, pinches out on this ridge. Refraction line B (Damotte et al., 1971; Limond et al., 1972) showed that the crust there was of an oceanic type (Fig. 1).

It is rather remarkable that the North Gascony Ridge is located at a high-magnetic-gradient zone between the area with numerous anomalies in the middle of the Bay of Biscay and the northwest-trending negative anomaly bounding it. Toward the southeast the ridge corresponds to a positive anomaly having the same trend. This means that there is a relation between magnetic anomalies and basement topography. The basin is prolonged toward the northwest along the margin of the Western Approaches of the English Channel (Fig. 13).

The North Spanish marginal basin has a quite different structure. The isopachs of formation 1 show that it was highly subsident after the Eocene since the formation 1 is locally 3,500 m thick. This can be explained by the proximity of detrital influx linked to the erosion of the Pyrenees Range. The boundaries of the Late Tertiary basin are: the high zones formed at the late Eocene in the north (Fig. 10), especially the Cantabria and Gascony seamounts; to the west it continues north of the Galicia Banks.

The pattern is entirely different for formations 2 and 3. The South Gascony Ridge (Montadert et al., 1971a, 1971d) follows the corner of Spain to the Cap Ferret-Parentis graben, where it connects with the North Gascony Ridge. This ridge forms the northern boundary of the deep horizons of the basin, which are thickening toward the margin. But the most remarkable fact is the existence of a tectonized zone affecting the entire southern part of the basin (Montadert et al., 1971a).

The Central part of the abyssal plain is for convenience, defined as the area located between the North and South Gascony ridges, which extend into the Cap Ferret-Parentis graben, but its structure is actually quite complex. The isopachs of formations 2 and 3 (Figs. 10 and 11) show that these formations are present with considerable thicknesses along an elongated edge from Cantabria to about 6°W, north of the South Gascony Ridge. In particular, they can be seen to pass over the North Gascony Ridge (Fig. 12B, C). From DSDP site 119, these formations probably include the Late Cretaceous (Fig. 1). It appears that, beyond this appreciably thick zone of formations 2 and 3, the oceanic substratum (DSDP site 118), rises abruptly and the sediments underlying formation 1 are not thick.

We think that this corresponds to a later stage of the formation of the oceanic floor in the Bay of Biscay. DSDP site 118 has shown the existence of Paleocene on basalts. It must be emphasized that the southern limit of this "young" area moves from an east-west to a southwest-northeast direction until the northwest corner of the Iberian block, and then again to an east-west direction. Its northern limit is at the North Gascony Ridge (Fig. 12A).

Consequently, it is possible that this area of the Bay of Biscay could correspond to a Late Cretaceous and even Early Tertiary episode of opening whose shape could be related with the shift of the margins near the Western Approaches in the north, and near Galicia Banks in the south.

Isopach maps (Figs. 10 and 11) allow us to define another geological event which affected the structure of the abyssal plain and margins. Vertical movements occurred at the late Eocene, as proved by DSDP sites 118 and 119. High areas were created, while adjacent zones foundered, and this event controlled post-Eocene sedimentation. Figure 10 shows that on the high areas, which are practically superimposed on the eastern part of the North and South Gascony basement ridges, there was reduced pelagic sedimentation (few hundred meters), while in the adjacent depressions thick turbidites were deposited. Cantabria and Gascony seamounts are the most evident high areas presently apparent in the topography. The development of the turbiditic sedimentation is particularly spectacular in the North Spanish marginal trough, where its thickness is at least 3,500 m. Laughton et al. (1972) demonstrated that the Paleocene and early Eocene pelagic sediments have been affected by basaltic intrusions, probably linked to this late Eocene event

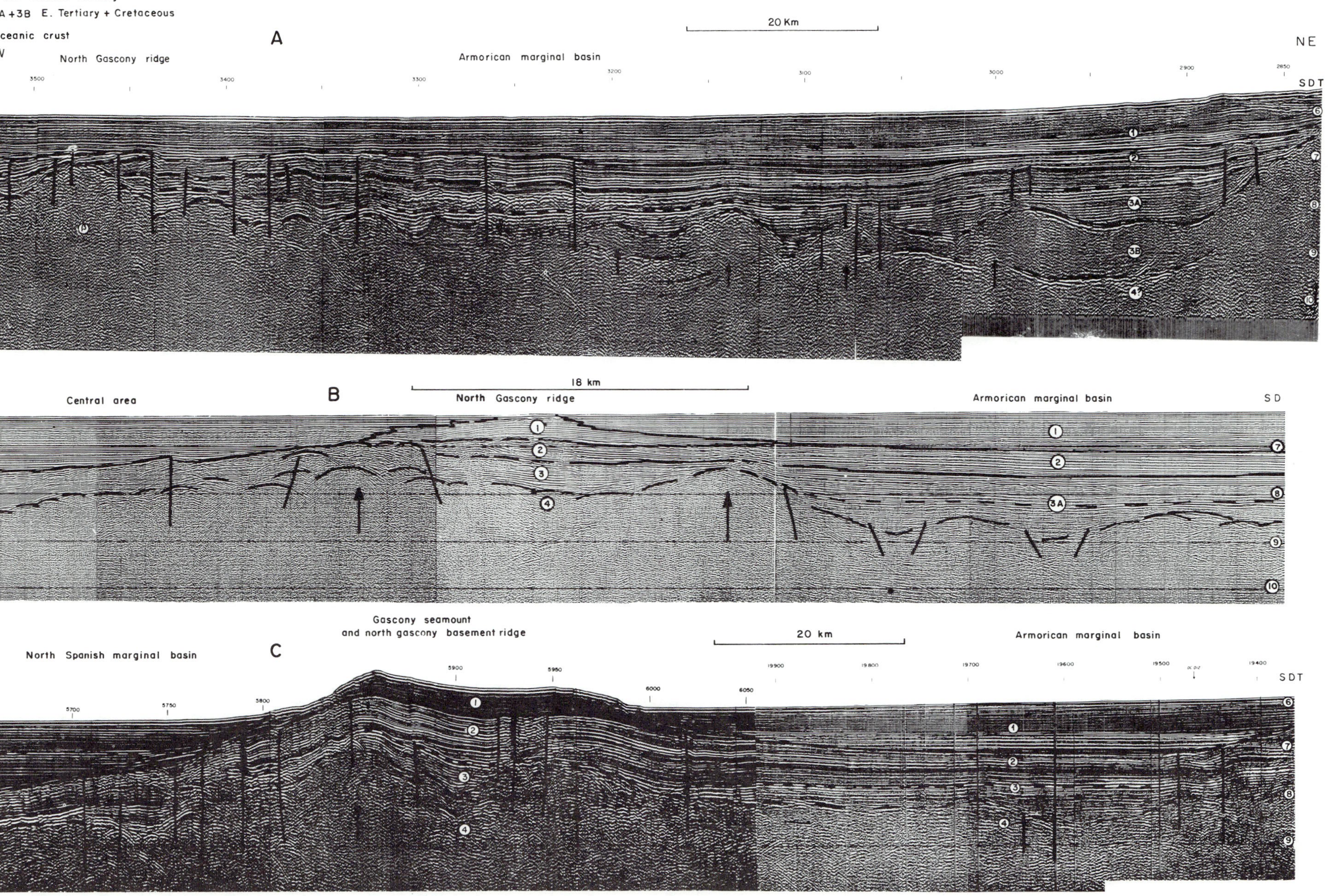

Fig. 12. Seismic profiles through the Armorican marginal basin and the North Gascony ridges (see Fig. 1 for locations).

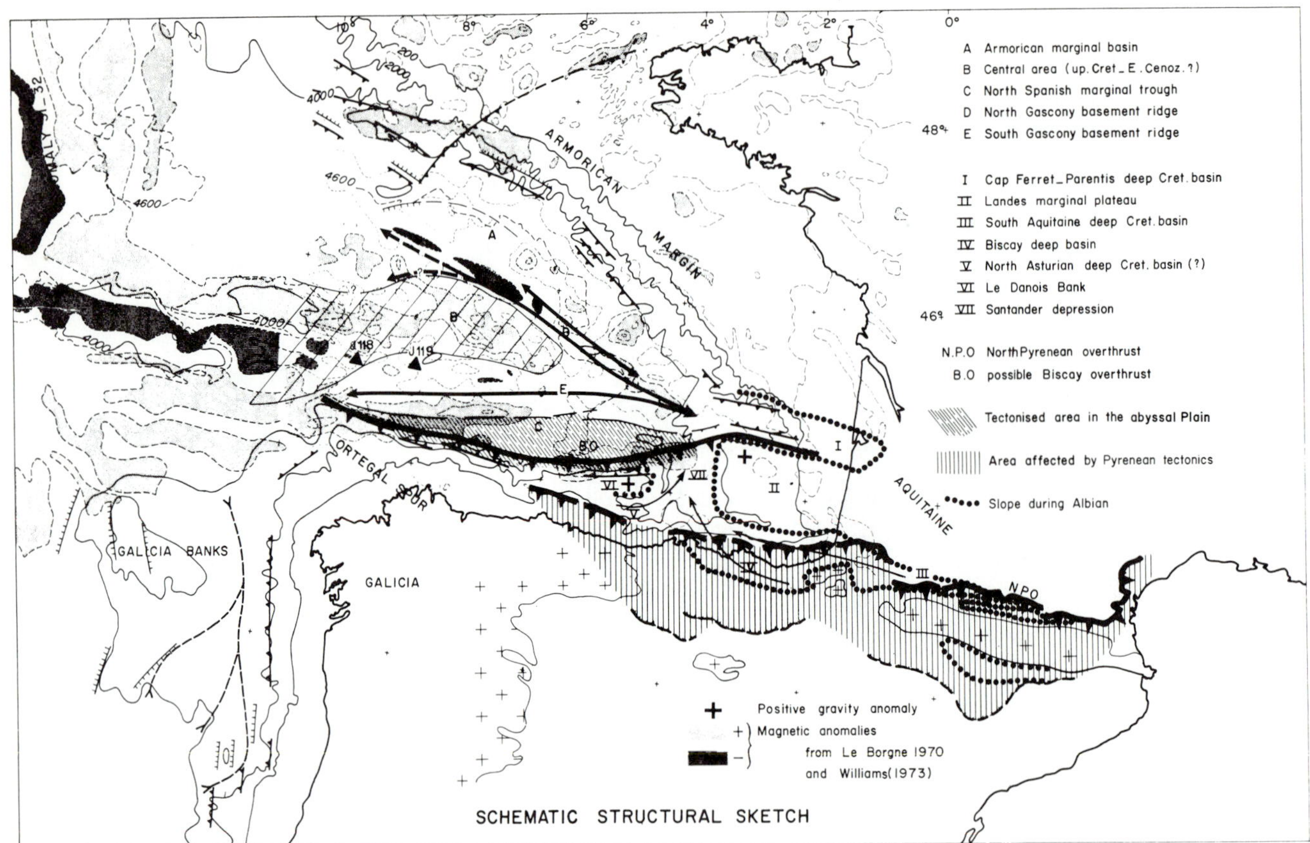

Fig. 13. Main structural features of the Bay of Biscay and their prolongation on land (magnetic data after LeBorgne and LeMouel, 1970; and Williams, 1973).

(DSDP 118). At the same time, large faulting is found at the margins (e.g., the fault bounding the Trevelyan Escarpment). Hence it is probable that, at that time, the abyssal plain was still under distension.

It seems now quite certain that Anomaly 31 exists in the Atlantic region adjacent to the Bay of Biscay (Williams and McKenzie, 1971; Pitman et al., 1971; Pitman and Talwani, 1972). Magnetic maps of the Bay of Biscay (Fig. 13) show some typically oceanic linear anomalies roughly located in a fan-shaped pattern. Recently, Williams (1973) showed that Anomaly 32 breaks off in front of the Bay of Biscay, and suggests that this area corresponds to a triple junction. The data given here favor this hypothesis. Whereas America, north of the Azores, was already moving away from Europe, a spreading center was still operating in the axis of the Bay of Biscay in the very Late Cretaceous, with volcanic events continuing up to the Eocene.

Seismic-reflection data are still insufficient, but do show that a link exists between the magnetic anomalies and the topography of the oceanic substratum. Hence it would appear dangerous to use the magnetic anomalies alone to deduce the mechanism of the formation of the Bay of Biscay without taking seismic data into consideration, which clearly delimit different oceanic areas with a different age. It should be pointed out that there is no evidence for the trend of the transform faults such as the ones assumed by LePichon et al. (1971). Data now available suggest a simpler pattern approaching the one proposed by Williams and MacKenzie (1971), and implying perhaps a correspondence between the southwest advance of the Armorican margin at the Western Approaches of the English Channel and the inturning of the Spanish margin north of the Galicia Banks.

CONCLUSION

The main events in the geological history of the margin around the Bay of Biscay can be summmarized as follows:

1. Rifting of the Hercynian platform occurred during the Triassic and early Lias with thick evaporitic and volcanic deposits. The northern boundary of the rift is well established in Aquitaine; the southern boundary must be located in northern Spain.

2. Following this, there was a quiescent period until the Malm.

3. During the Upper Malm, subsidence began in the Parentis Basin and grew more and more intense until its maximum during the Albian, both in this basin and in the South Aquitaine Basin. Since these basins are directly connected with the Bay of Biscay, it can be deduced that this one began to

open up at that time. The Galicia-Portugal continental margin was probably also shaped at that time. In the Biscay Abyssal Plain the floor of the marginal basins, if oceanic, was probably created at the same epoch.

4. Opening continued with creation of the central part of the Bay, probably at the end of the Upper Cretaceous, with volcanic activity until the Eocene. Strong subsidence with tilting of the continental platform occurred after the Albian (more than 3,000 m).

5. Probably since the beginning of the major opening phase, an active margin has existed at the northern boundary of the Iberian block (or subplate). This activity continued until the Oligocene and even perhaps later (Miocene), if the tectonic structures revealed in the North Spanish marginal trough are due to a compression.

6. During the Upper Eocene the abyssal plain deepened considerably, except for some high areas (e.g., Cantabria and Gascony seamounts) on which reduced pelagic sediments were deposited, in opposition to thick turbidites in the surrounding depressions, especially in the North Spanish marginal basin.

From a structural point of view, the Armorican margin and its prolongation in the Aquitaine, and also Galicia-Portugal margin, are quite similar to other stable margins described in the world—tilting of the platform, horsts and grabens on the slope, and marginal basins.. The North Iberian margin, now stable, was in fact an active one, probably from the Cenomanian until the Oligocene-Miocene. It is marked by overthrust structures and gravity tectonics. We suggest that there were two main overthrust fronts: the North Pyrenean one, which seems to exist from Provence to Asturia, and a Biscay one from the northwest corner of Spain possibly to the southern part of the Parentis Basin (west of Iberia, a possible prolongation is unknown).

The mechanism of formation of the Bay of Biscay has raised much discussion. Initially a counterclockwise rotation of Spain was proposed, with a center in the southeastern part of the Bay. This hypothesis (Carey, 1955) seemed difficult to accept because an opposite compressed area was necessary, whereas the Pyrenean orogenic movements were traditionally considered as late Eocene, and the Bay of Biscay was of Early Cretaceous age. LePichon et al. (1971), to overcome these objections, proposed a rotation of Iberia around a pole near Paris with a strike-slip motion of several hundred kilometers during the Mesozoic along the North Pyrenean fault, followed by a Cenozoic compression. This proposal has been developed (Choukroune et al., 1973) with the strike-slip motion during the Upper Cretaceous becoming a compression at the Pyrenees, accompanied by a displacement of the rotation pole.

Montadert and Winnock (1971) proposed that the distension area of the Bay of Biscay during the Lower Cretaceous has to be extended eastward in the Aquitaine Basin, perhaps all the way to Provence. These Cretaceous structures were considered to have been disturbed by later Pyrenean movements. An appreciable translation of Iberia in relation to Europe did not seem acceptable. Since then a reinterpretation of marine data, especially of land data from oil exploration, has contributed to improvements of this proposal, as follows.

There was no important translation of Iberia in relation to Europe; this is in agreement with the existence of an Ibero-Armorican Hercynian Arc. Moreover, a correspondence could be proposed between the salient of the margin near the western approaches and the reindentation of the North Spanish margin at the Galicia Banks. There are also the North and South Gascony basement ridges, which are continuous and converge to the east (Fig. 13). Cretaceous distension made itself felt very far to the east in the Aquitaine Basin and in a complex way, with the formation of deep furrows such as the Parentis, South Aquitaine, Biscay, and North Asturian basins, which converge (except the first one) toward the Santander depression, created at that time by a break between the Le Danois Bank and the Landes marginal plateau.

The opening of the Bay of Biscay during the Cretaceous was hampered almost from the beginning by the converging of Africa and Europe, causing compression in the whole Tethyan area. Thus the northern margin of the Iberian subplate became an active margin probably as early as the Cenomanian.

ACKNOWLEDGMENTS

The authors thank IFP, CFP, ELF-ERAP, SNPA, and Comité d'Etudes Marines, which kindly gave permission to publish this report. They are greatly indebted to Mrs. Fail, to Cassand, and Donatien, geophysicists at the Institut Français du Petrole, and the officers and crew of the FLORENCE, which made the surveys.

BIBLIOGRAPHY

Bacon, M., and Gray, F., 1971, Evidence for crust in the deep ocean derived from continental crust: Nature, v. 229, no. 5283, p. 331-332.

———, Gray, F., and Matthews, D. H., 1969, Crustal structure studies in the Bay of Biscay: Earth and Planetary Sci. Leetters, v. 6, no. 5, p. 377-385.

Berthois, L., ed., 1966, Cartes bathymétriques du talus du plateau continental: En onze feuilles éditées par Berthois L. avec le concours du CNRS Paris.

———, and Brenot, R., 1964, Bathymétrie du golfe de Gascogne et de la cote du Portugal. I. Commentaires sur le levé complémentaire des feuilles 9 et 10 des abords du plateau continental. II. Bathymétrie du talus du plateau continental à l'ouest de la péninsule Ibérique du cap Finisterre au cap Saint-Vincent: Cons. Internatl. Expl. Mer, P.V. 52e Réunion, Sept.-Oct. 1964, 77 p.

———, Brenot, R., and Ailloud, P., 1965, Essai d'interprétation morphologique et tectonique des leves bathymétriques exécutés dans la partie sud-est du golfe de Gascogne: Rev. Trav. Inst. Pêches Marit., t. 29, no. 3, p. 343-345.

Black, M., Hill, M. N., Laughton, A. S., and Matthews, D. H., 1964, Three non-magnetic seamounts off the Iberian coast: Geol. Soc. London, Quart. Jour., v. 120, p. 477-517.

Boillot, G., and Musellec, P., 1972, Géologie du plateau continental portugais au nord du cap Carvoeiro. Structure au nord et au sud du canyon de Nazaré: Compt. Rend., t. 274, May 15, ser. D, p. 2748-2751.

———, Dupeuble, P. A., Durand-Delga, M., and d'Ozouville, L., 1971a, Age minimal de l'Atlantique Nord d'apres la découverte de calcaire tithonique à Calpionelles dans la golfe de Gascogne: Compt. Rend., t. 273, no. 7, Aug. 18, ser. D, p. 671-674.

———, Dupeuble, P. A., Lamboy, M., d'Ozouville, L., Sibuet, J. C., 1971b, Structure géologique de la marge continentale asturienne et cantabrique (Espagne du Nord), in Histoire structurale du golfe de Gascogne, t. I-II: Paris, Ed. Technip, p. V.6-1-V6-52.

———, Berthou, P. Y., Dupeuble, P. A., and Musellec, P., 1972, Géologie du plateau continental portugais au nord du cap Carvoeiro. La série stratigraphique: Compt. Rend., t. 274, May 24, ser. D, p. 2852-2854.

———, Dupeuble, P. A., Hennequin-Marchand, I., Lamboy, M., and Leprêtre, J. P., 1973a, Carte géologique du plateau continental nord-espagnol entre le canyon de cap Breton et le canyon d'Aviles: Bull. Soc. Géol. France, 7e ser., t. 15, no. 3-4, p. 367-381.

———, Capdevila, R., Hennequin-Marchand, I., Lamboy, M., and Leprêtre, J. P., 1973b, La zone nord-pyrénéenne, ses prolongements sur la marge continentale nord-espagnole et sa signification structurale: Compt. Rend., t. 277, no. 24, ser. D, p. 2629-2632.

Bouysse, P., and Horn, R., 1972, La géologie du plateau continental autour du Massif Armoricain: Bull. BRGM (2 e ser.) sec. IV, no. 2, p. 3-17.

Capdevila, R., Lamboy, M., and Leprêtre, J. P., 1974, Découverte de granulites, de charnockites et de syénites néphéliniques dans la partie occidentale de la marge continentale nord-espagnole: Compt. Rend., t. 278, no. 1, ser. D, p. 17-20.

Caralp, M., Klingebiel, A., Latouche, C., Moyes, J., Prud'homme, R., et al., 1972, Bilan cartographique des éstudes effectuées sur le plateau continental Aquitain le 28 fevrier 1972: Bull. Inst. Géol. Bassin Aquitaine, no. spec., 25 p.

Carey, S. Warren, 1955, The orocline concept in geotectonics: Proc. Royal Soc. Tasmania, v. 89, p. 255-288.

Cholet, J., Damotte, B., Grau, G., Debyser, J., and Montadert, L., 1968, Recherches préliminaires sur la structure géologique de la marge continentale du golfe de Gascogne. Commentaires sur quelques profils de sismique réflexion "Flexotir": Rev. Inst. Franc. Pétrole, t. 23, no. 9, p. 1029-1045.

Choukroune, P., Le Pichon, X., Seguret, M., and Sibuet, J. C., 1973, Bay of Biscay and Pyrenees: Earth and Planetary Sci. Letters, v. 18, p. 109-118.

Coron, S., and Guillaume, A., 1971, Etude gravimétrique sur le golfe de Gascogne et les Pyrénées, in Histoire structurale du golfe de Gascogne, t. I-II: Paris, Ed. Technip, p. IV.9-1-IV.9-16.

Curray, J. R., Moore, D. G., Belderson, R. H., and Stride, A. H., 1966, Continental margin of western Europe: slope progradation and erosion: Science, v. 154, p. 265-266.

Damotte, B., Debyser, J., Montadert, L., and Delteil, J. R., 1969, Nouvelles données structurales sur le golfe de Gascogne obtenues par sismique réflexion "Flexotir": Rev. Inst. Franç. Pétrole, t. XXIV, no. 9, p. 1029-1060.

———, Grau, G., Gray, F., Limond, W. Q., and Patriat, Ph., 1971, Utilisation conjointe des méthodes sismique réflexion et refraction pour la détermination des vitesses, in Histoire structurale du golfe de Gascogne, t. I-II: Paris, Ed. Technip, p. VI.15-1-VI.15-26.

Dardel, R. A., and Rosset, R., 1971, Histoire géologique et structurale du bassin de Parentis et de son prolongement en mer, in Histoire structurale du golfe de Gascogne, t. I-II: Paris, Ed. Technip, p. IV.2-1-IV.2-28.

Durand-Delga, M., 1973, Les Calpionelles du golfe de Gascogne, témoins de l'ouverture de l'Atlantique Nord: Bull. Soc. Géol. France, 7e ser., t. XV, no. 1, p. 22-24.

Ewing, J., Burckle, L. H., Saito, T., and Poppe, H., 1971, Geophysical and geological studies of Cantabria Seamount and its environs, in Histoire structurale du golfe de Gascogne, t. I-II: Paris, Ed. Technip, p. VI.12-1-VI.12-14.

Feuillée, P., Villanova, M., and Winnock, E., 1973, La dynamique des fosses à "turbidites" et de leur contenu sédimentaire dans le système pyrénéen: Bull. Soc. Géol. France, 7è sér., t. XV, no. 1, p. 61-76.

Frappa, M., Klingebiel, A., Malzac, J., Martin, G., Muraour, P., and Vigneaux, M., 1970, Remarques sur la structure des montagnes sous-marines de Biscaye (golfe de Gascogne), à la suite d'une étude sismique par réflexion: Compt. Rend. Soc. Géol. France, fasc. 5, p. 149-150.

Fried, E., and Morelot, G., 1973, Evolution bathymétrique du bassin de Parentis au cours du Crétacé: Bull. Soc. Geol. France, 7e ser. t., XV, no. 1, p. 25-29.

Grant, A. C., 1973, Geological and geophysical results bearing upon the structural history of the Flemish Cap region: Earth Sci. Symp. on Offshore E. Can., Geol. Survey Can. Paper 71-23, p. 373-388.

Grau, G., Montadert, L., Delteil, R., and Winnock, E., 1973, Structure of the European continental margin between Portugal and Ireland, from seismic data, in Mueller, S., ed., The structure of the earth's crust: Tectonophysics, v. 20, no. 1-4, p. 319-339.

Henry, J., Lanusse, R., and Villanova, M., 1971, Evolution du domaine marin pyrénéen du Sénonien supérieur à l'Eocène inférieur, in Histoire structurale du golfe de Gascogne: Paris, Ed. Technip, t. IV-7, p. 1-8.

Hersey, J. B., and Whittard, W. F., 1966, The geology of the Western Approaches of the English Channel: V. The continental margin and shelf under the south Celtic Sea, in Continental margins and island arcs: Geol. Survey Can. Paper 66-15, p. 80-106.

Jones, E. J. W., 1968, Continuous reflection profiles from the European continental margin in the Bay of Biscay: Earth and Planetary Sci. Letters, v. 5, no. 2, p. 127-134.

———, and Ewing, J. L., 1969, Age of the Bay of Biscay: evidence from seismic profiles and bottom samples: Science, v. 166, no. 3901, p. 102-105.

———, and Funnel, B. M., 1968, Association of a seismic reflector and Upper Cretaceous sediments in the Bay of Biscay: Deep-Sea Res., v. 15, no. 6, p. 701-709.

Kieken, M., 1973, Evolution de l'Aquitaine au cours du Tertiaire: Bull. Soc. Géol. France, 7e ser., t. XV, no. 1, p. 40-50.

Lamboy, M., and Dupeuble, P. A., 1971, Constitution géologique du plateau continental espagnol entre la Corogne et Vigo: Compt. Rend., t. 273, Sept. 20, sér. D.

Lapierre, F., 1972, Etude structurale du plateau continental à l'Ouest de la Bretagne: Rev. Inst. Franç. Pétrole Ann. Combust. Liquides, t. 27, no. 1, p. 73-89.

Laughton, A. S., Berggren, W. A., et al., 1972, Initial reports of the Deep Sea Drilling Project, v. XII: Washington, D.C., U.S. Govt. Printing Office, Site 118 and 119.

Le Borgne, E., and Le Mouel, J., 1970, Cartographie aéromagnétique du golfe de Gascogne: Compt. Rend., t. 271, sér. D., no. 14, p. 1167-1170.

Le Mouel, J. L., and Le Borgne, E., 1971, La cartographie magnétique du golfe de Gascogne, *in* Histoire structurale du golfe de Gascogne, t. I-II: Paris, Ed. Technip, p. VI.3-1-VI.3-12.

LePichon, X., Bonnin, J., Francheteau, J., and Sibuet, J. C., 1971, Une hypothèse d'évolution tectonique du golfe du Gascogne, *in* Histoire structurale du golfe de Gascogne, t. I-II: Paris, Ed. Technip, p. VI.11-1-VI.11-44.

Limond, W. Q., Gray, F., Grau, G., and Patriat, P., 1972, Mantle reflections in the Bay of Biscay: Earth and Planetary Sci. Letters, v. 15, no. 4, p. 361-366.

Malzac, J., 1970, Apport d'une campagne de sismique reflexion Flexotir à l'étude géophysique du golfe de Gascogne: Bull. Inst. Géol. Bassin d'Aquitaine, no. 9, p. 265-267.

Martin, G., Muraour, P., and Ricolvi, M., 1968, Etude par sismique réfraction du plateau continental au large de Belle Ile: Trav. Lab. Géophys. Appl. l'Océanogr., Fac. Sci., Univ. Montpellier, fasc. 2, p. 1-5.
2, p. 1-5.

Mattauer, M., and Seguret, M., 1971, Les relations entre la chaine des Pyrénées et le golfe de Gascogne, *in* Histoire structurale du golfe de Gascogne, t. I-II: Paris, Ed. Technip, p. IV.4-1-IV.4-24.

Matthews, D. H., and Williams, C. A., 1968, Linear magnetic anomalies in the Bay of Biscay: a qualitative interpretation: Earth and Planetary Sci. Letters, v. 4, no. 4, p. 315-320.

Montadert, L., and Winnock, E., 1971, L'histoire structurale du golfe de Gascogne, *in* Histoire structurale du golfe de Gascogne, t. I-II: Paris, Ed. Technip, p. VI.16-1-VI.16-18.

______, Damotte, B., Debyser, J., Fail, J. P., Delteil, J. R., and Valéry, P., 1971a, Continental margin in the Bay of Biscay, *in* Delany, F. M., ed., The geology of the East Atlantic continental margin: Inst. Geol. Sci. Rept. 70/15, p. 43-74.

______, Damotte, B., Delteil, J. R., Valéry, P., and Winnock, E., 1971b, Structure géologique de la marge continentale septentrionale du golfe de Gascogne (Bretagne et Entrées de la Mancye), *in* Histoire structurale du golfe de Gascogne, t. I-II: Paris, Ed. Technip, p. III.2-1-III.2-22.

______, Damotte, B., Fail, J. P., Delteil, J. R., and Válery, P., 1971c, Structure géologique de la marge continentale asturienne et cantabrique (Espagne eu Nord), *in* Histoire structurale du golfe de Gascogne, t. I-II: Paris, Ed. Technip, p. V.7-1-V.7-16.

______, Damotte, B., Fail, J. P., Delteil, J. R., and Válery, P., 1971d, Structure géologique de la plaine abyssale du golfe de Gascogne, *in* Histoire structurale du golfe de Gascogne, t. I-II: Paris, Ed. Technip, p. VI.14-1-VI.14-42.

Muraour, P., Malzac, J., Frappa, M., and Martin, G., 1970, Contribution à l'étude géophysique du golfe de Gascogne: Compt. Rend., t. 270, no. 12, ser. D, p. 1552-1554.

______, Frappa, M., Malzac, J., and Vaillant, F. X., 1971, Quelques données structurales sur la zone ouest du golfe du Gascogne, *in* Histoire structurale du golfe de Gascogne, t. I-II: Paris, Ed. Technip, p. VI.6-1-VI.6-12.

Pastouret, L., Masse, J. P., Philip, J., and Auffret, G. A., 1974, Découverte de rudistes d'âge Crétacé inférieur sur la marge Armoricaine—implications paléogéographiques: Ile Réunion Ann. Sci. Terre, Pont-à-Mousson (Nancy, Apr. 22-26, Soc. Géol. France, Sp. V.

Pautot, G., Auzende, J. M., and LePichon, X., 1970, Continuous deep sea layer along North Atlantic margins related to early phase of rifting: Nature, v. 227, p. 351-354.

Pitman, W. C., III, and Talwani, M., 1972, Sea-floor spreading in the North Atlantic: Geol. Soc. America Bull., v. 83, Mar., p. 619-646.

______, Talwani, M., and Heirtzler, J. R., 1971, Age of the North Atlantic Ocean from magnetic anomalies: Earth and Planetary Sci. Letters, v. II, no. 3, p. 195-200.

Robert, J. P., 1969, Géologie du plateau continental français—données recueillies à l'occasion de la mise au point de l'Etinceleur et de son utilisation au profit de certains ports français et de l'Institut Français du Pétrole: Rev. Inst. Franç. Pétrole, t. XXIV, no. 4, p. 383-440.

Ruffman, A., and van Hinte, H., 1973, Orphan Knoll—A "chip" off the North American "Plate": Earth Sci. Symp. on Offshore E. Can., Geol. Survey Can. Paper 71-23, p. 407-449.

Schoeffler, J., 1965, Le "Gouf" de cap Breton, de l'Eocene inférieur à nos jours, *in* Submarine geology and geophysics, Colston Papers, v. XVII, p. 265-268.

Schouten, J. A., Collette, B. J., and Rutten, K. W., 1971, Magnetic anomaly symmetry in the Bay of Biscay, *in* Histoire structurale du golfe de Gascogne, t. I-II: Paris, Ed. Technip, p. VI.13-1-VI.13-10.

Sibuet, J. C., and LePichon, X., 1971a, Structure gravimétrique du golfe de Gascogne et le fossé marginal nord-espagnol, *in* Histoire structurale du golfe de Gascogne, t. I-II: Paris, Ed. Technip, p. VI.9-1-VI.9-18.

______, Pautot, G., and LePichon, X., 1971, Interprétation structurale du golfe de Gascogne à partir des profils de sismique, *in* Histoire structurale du golfe de Gascogne, t. I-II: Paris, Ed. Technip, p. VI.10-1-VI.10-32.

Stride, A. H., Curray, J. R., Moore, D. G., and Belderson, R. H., 1969, Marine geology of the Atlantic continental margin of Europe: Phil. Trans. Roy. Soc. London, ser. A, v. 264, no. 1148, p. 31-75.

Valéry, P., Delteil, J. R., Cottencon, A., Montadert, L., Damotte, R., and Fail, J. P., 1971, La marge continentale d'Aquitaine, *in* Histoire structurale du golfe de Gascogne, t. I-II, Paris, Ed. Technip, p. IV.8-1-IV.8-24.

Vanney, J. R., Scolari, G., Lapierre, F., Martin, G., and Dieucho, A., 1971, Carte géologique provisoire de la plate-forme continentale armoricaine (déc. 1970), *in*

Histoire structurale du golfe de Gascogne, t. I-II: Paris, Ed. Technip, p. III.1-1-III.1-20.

Williams, C. A., 1973, A fossil triple junction in the NE Atlantic west of Biscay: Nature, v. 244, no. 5411.

———, and McKenzie, D., 1971, The evolution of the north-east Atlantic: Nature, v. 232, no. 5307.

Winnock, E., 1971, Géologie succincte du bassin d'Aquitaine (contribution à l'histoire du golfe de Gascogne), *in* Histoire structurale du golfe de Gascogne, Tomes I-II: Paris, Ed. Technip, p. IV.1-1-IV.1-30.

———, 1973, Exposé succinct de l'évolution paléogéologique de l'Aquitaine: Bull. Soc. Géol. France, 7è sér., t. XV, no. 1, p. 5-12.

———, Fried, E., and Kieken, M., 1973, Les caractéristiques des sillons aquitains: Bull. Soc. Géol. France, 7è sér., t. XV, no. 1, p. 51-60.

Structural Development of the British Isles, the Continental Margin, and the Rockall Plateau

David G. Roberts

INTRODUCTION

The continental margin west of the British Isles is very complex. The shelf area consists of the English Channel, Celtic Sea, Bristol Channel, Irish Sea, and the Sea of Hebrides. West of Ireland, the Porcupine Bank is separated from the shelf by the Porcupine Seabight (Fig. 1). Farther to the west, the Faeroe Islands and Rockall Plateau microcontinents are separated from the margin by the Faeroe-Shetland Channel and Rockall Trough, divided at 60°N by the Wyville-Thomson Ridge. In addition, several seamounts occur in the Rockall Trough. The region is effectively aseismic, with the few minor earthquakes occurring in the Sea of the Hebrides; one earthquake has been observed on the Rockall Plateau (53°9.19'N, 19°7.13'W).

The Rockall Plateau is the only major microcontinent in the North Atlantic Ocean and therefore has great relevance to our understanding of continental margin processes. The evolution and ultimate isolation of a microcontinent can only be achieved in terms of plate tectonics by several phases of spreading about widely spaced axes. It is therefore more complex than that conceptually implied for margins formed by spreading about a single axis.

The later history of the young margin and oceanic part of the accreting plate is considered to be mainly by erosion, thermal subsidence, and epeirogenesis and is characterized by the development of a thick sedimentary wedge and regional warping. Although the structural framework of the margin is believed to be due to rifting, the pattern may be extensively influenced both by the age and tectonic fabric of the lithosphere. In addition, earlier faulting may have further influenced the final structure of the younger margins and the later history of the older ones. Other potentially important but much more speculative factors include rejuvenation of faults in response to hot creep or thermal subsidence and regional warping, epeirogenesis, and subsidence of both the oceanic and continental parts of the plate, possibly in response to spreading rate changes at the mid-ocean ridge axis.

To investigate fully the variable influence of these factors, it is necessary to establish the relationship between the structural framework of the continental lithosphere, the fabric imposed during development of the margin, the temporal relationships between stratigraphy of both margin and microcontinent, and the sea-floor spreading evolution of the area. The geological evolution of the British Isles, Rockall Plateau microcontinent, and continental margin is here reviewed in those terms.

Clearly, many important problems of the continental margin have been omitted from this review; they include the surficial sediments of the shelf, aspects of margin sedimentation, the geology of the North Sea, and many other details of geology that contribute to a complete perspective of margin development. Much of this work is included in the bibliography. An obvious depositional, structural, and stratigraphic continuum extends from the earliest formation of the continental margin to the present.

PRE-PERMIAN STRUCTURAL DEVELOPMENT

The main features of the pre-Permian structural framework of the British Isles are shown in Figure 2. The northeast-trending Caledonian mobile belt is the dominant feature and separates the northwestern foreland, composed of Precambrian gneisses overlain by late Precambrian and Cambro-Ordovician sediments, from a southeast foreland revealed by the isolated Precambrian inliers of central England. The fourth major structural division is the east-northeast-trending Hercynian orogenic belt of southern Wales, England, and Ireland (Dunning, 1966; Dalziel, 1969; Dewey, 1969; Dunning and Max, 1974).

The eastern boundary of the northwest foreland is the Moine thrust plane. On the foreland, the late Precambrian rests unconformably on the "Lewisian" basement and is overlapped by Cambro-Ordovician strata. The Lewisian gneisses are the remnants of the Laxfordian (1,650-2,200 my) and Scourian (2,200-2,500 my) orogenic belts (Sutton and Watson, 1951; Sabine and Watson, 1965). Laxfordian rocks sampled from the Rockall Bank prove the western continuation of the foreland and the continental nature of Rockall Plateau (Roberts et al., 1973). Grenvillian rocks from Rockall Bank and geophysical data suggest that the Grenville front crosses the Rockall Bank near 56°20'N, presumably to intersect the Caledonian front on the shelf west of the British Isles (Miller et al., 1973; Roberts and Jones, 1974). Grenvillian dates have been reported from western Scotland, but are questionable because of their proximity to the Caledonian front (Miller, 1965).

The Caledonian mobile belt can be divided into a northern orthotectonic zone largely separated

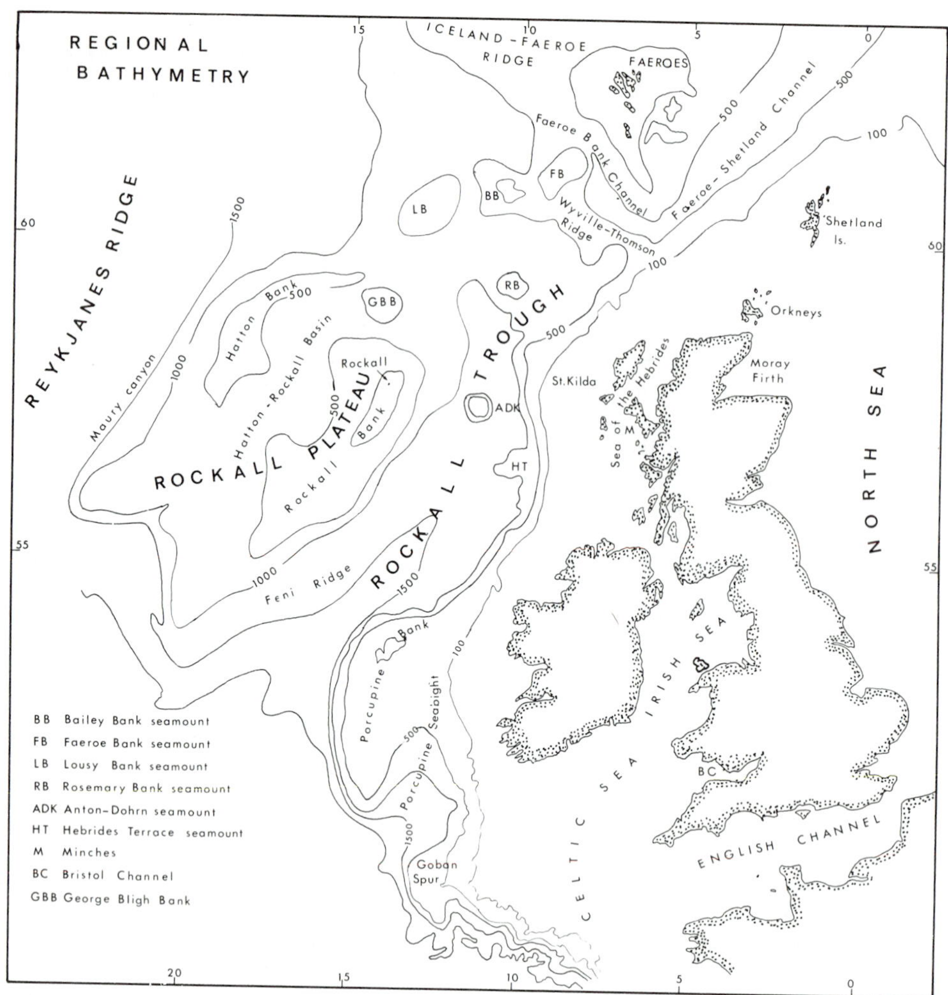

Fig. 1. Generalized bathymetry of the continental margin west of the British Isles.

from a subparallel paratectonic zone by the Highland Boundary Fault, although the orthotectonic zone also occurs to the south in Ireland (Dalziel, 1969; Dewey, 1969). The orthotectonic zone includes the Moine and Dalradian series and consists of a thick sedimentary sequence of late Precambrian to Lower Ordovician age characterized by large-scale recumbent folding and high pressure-temperature metamorphism; the age of the deformation is late Arenigian (Dewey, 1969). The paratectonic sequence was deposited as an irregular blanket in basins and troughs with thinning over the Irish Sea geanticline (Fig. 2); the deformation was Late Silurian-Early Devonian. Dewey (1969a) has presented a plate tectonic model for the evolution of the Caledonian-Appalachian system and considers the orthotectonic belt to arise from progressive deformation of a continental rise over a Benioff zone, where culmination of the orogeny resulted from a collision of the two continents represented by the northwestern and southeastern forelands. Major granitic intrusive activity took place in the early Devonian, producing, for example, the Leinster Granite of southeast Ireland. Upper Paleozoic faulting includes the development of northeast-trending sinistral strike-slip faults (Fig. 2) important in subsequent Mesozoic basin development. Among the best known are the Great Glen Fault (slip perhaps 160 km), the Minch Fault (slip 124 km), the Strathconnon Fault, and the Skerryvore-Camasunary .Fault (Kennedy, 1946; Dearnley, 1962; Winchester, 1973; Garson and Plant, 1973; McQuillin and Binns, 1973).

The offshore extension of these structures has been delineated by new geophysical data. Flinn (1961, 1969) postulated a northern extension of the Great Glen Fault to the edge of the Faeroe-Shetland Channel via the Walls Boundary Fault of Shetland. However, gravity highs on the shelf edge are not offset along this extension, and Avery et al. (1968) suggest that the fault splays eastward. A series of north-northeast-trending sedimentary basins and gravity highs mapped to the west of Orkney and Shetland are probably controlled by the Minch Fault and other parallel shears. The basins and highs can be traced to the shelf edge but do not seem to continue beneath the Faeroe-Shetland Channel (Bott and Watts, 1969; Watts, 1971; Flinn, 1973; Talwani and Eldholm, 1972). Evidence of offsets in the mar-

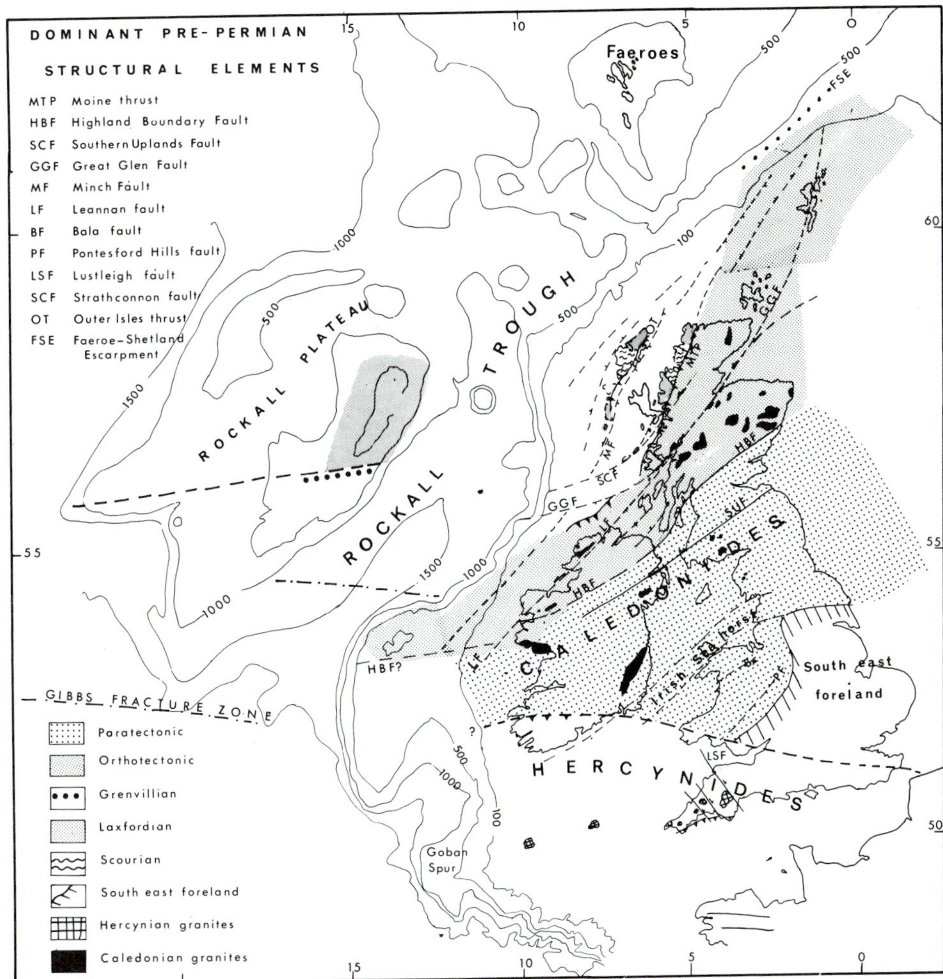

Fig. 2. Dominant pre-Permian structural elements of the British Isles and continental margin (after Dunning, 1966; Roberts, 1969; Dewey, 1969; Institute of Geological Sciences, 1972; Dunning and Max, 1974).

gin that might indicate structural control by these shears is masked beneath the thickly sedimented slope (Stride et al., 1969). Southwest of Mull, the Great Glen Fault splits into two branches; one continues west-southwest to the offset in the margin near 56°N and the other continues through northern Ireland as the Leannan Fault (Riddihough, 1968; Roberts, 1969b; Holgate, 1969; Pitcher, 1969; Bailey et al., 1974). The Highland Boundary Fault seems to continue to the west of Ireland but its offshore continuation is less certain. On Porcupine Bank, Vogt et al. (1971) show a west-southwest-trending magnetic lineament that lies a little north of the projected line of the fault, possibly because of sinistral strike-slip displacement by the Leannan Fault (Fig. 2).

The Hercynian mobile belt extends east-west through southern England and Wales but curves to the west-southwest in Ireland (Fig. 2) and is marked by thrusting. In southern England, the metamorphism was Acadian (360-390 my) and was followed at 250-300 my by intrusion of the Cornubian granitic batholith (Miller and Green, 1961). To the north of the Hercynian front, Upper Devonian and younger rocks were deposited on the Caledonian craton and folded along modified Caledonian trends, e.g., the north-south Pennine anticlinorium. Structures such as the Highland Boundary Fault were rejuvenated by Hercynian epeirogenic activity. In Ireland, the surficial geology consists of Dinantian limestone and many inliers of older rocks uplifted by the Hercynian movements.

In the Irish Sea, Carboniferous strata form a syncline striking east-northeast parallel to the Irish Sea geanticline (Eden et al., 1971). Carboniferous beds may also be preserved beneath the Minch (Eden et al., 1971). The offshore extension of the Hercynian front is poorly known. Cherkis et al. (1973a) have recently suggested a structural relationship with the Gibbs Fracture Zone, although there are no supporting magnetic and gravity data. Day and Williams (1970) adduce gravity data to show that the Cornish granites continue beyond the Scilly Isles and may comprise the Goban Spur midway down the slope. South of the English Channel, a considerable body of evidence shows the Hercynian (Variscan) arc extends to the shelf edge (Day et al., 1956; Hill and Vine, 1965; Le Mouel et al., 1971; Andreiff et al., 1972).

POST-CARBONIFEROUS STRUCTURAL AND STRATIGRAPHIC DEVELOPMENT

The post-Carboniferous history of the British margin is characterized by epeirogenesis, regional warping, and the development of thick sedimentary basins separated by areas of relative uplift (Fig. 3).

Permo-Triassic. Kent (1969) has pointed out that the earliest basin-faulting in the North Sea occurred in the Rotliegendes and was followed by Zechstein evaporite deposition. However, there is no evidence of actual extension on either local or continental scale, and the faulting may rather be related to post-Hercynian orogenic collapse accompanied by volcanism. In general, the basal Permian seems to infill irregularities in the Permian landscape (Wills, 1951; Rayner, 1967).

The paleogeography and stratigraphy of the Triassic has been reviewed and redefined by Audley-Charles (1970a, b). At the beginning of the period, most of the British Isles probably consisted of mountainous areas. The earliest beds were conglomerate deposited in narrow grabens that continued to subside and now underlie the Irish Sea—the Minch and Cheshire basins, for example. Detailed stratigraphic data are not available for the offshore areas, but the land data show a consistent overall pattern. The grabens were progressively infilled and then rejuvenated at intervals with local marine transgressions. Toward the top of the Triassic, extensive deposition of salt and saliferous mudstone took place, either from sea water or reworked evaporites. By the end of this period of salt deposition, faulting had almost ceased and the mudstones and evaporties of the uppermost Triassic were deposited over a much wider area. Toward the top of the Triassic, a return to more normal marine conditions is indicated by the Rhaetic transgression, which seems to have come from the southwest (Wills, 1951; Kent, 1969).

The pattern of onshore and offshore Triassic basin development has largely been controlled by Caledonian and Hercynian trends. In the Sea of the Hebrides, refraction data give a mean thickness for Permo-Triassic rocks of 1,000 m, although as much

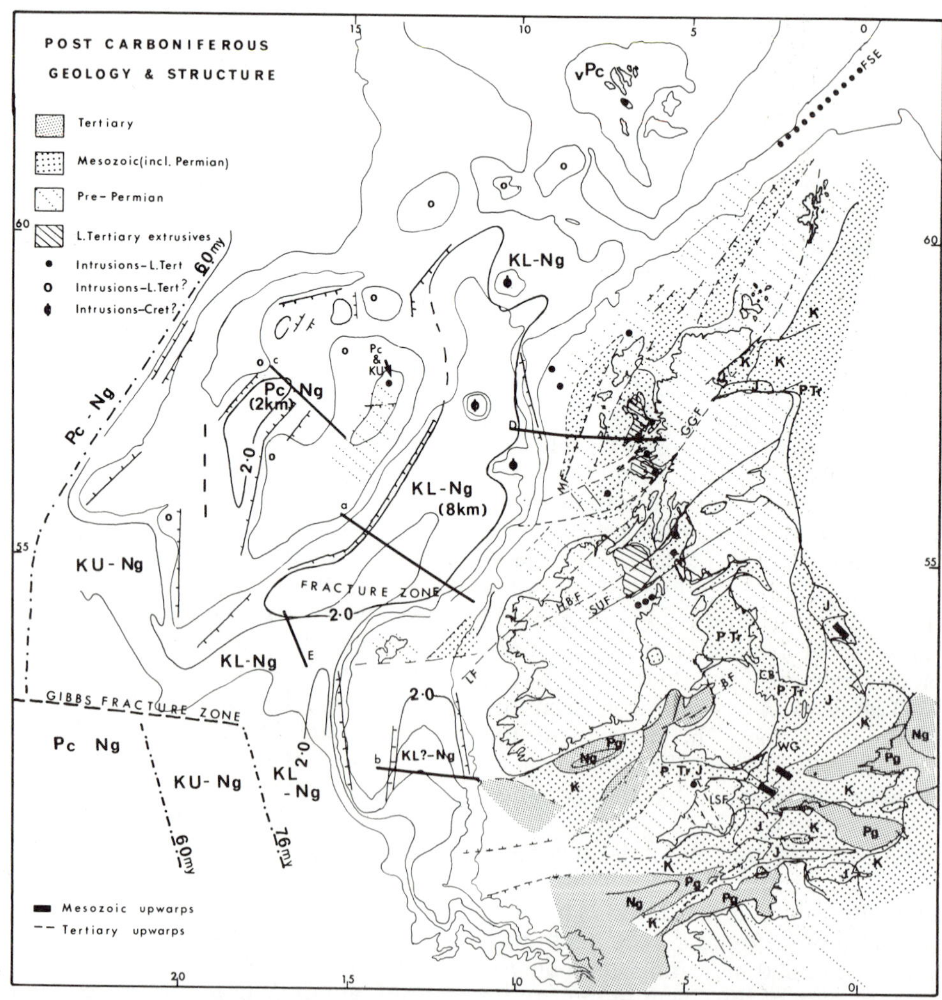

Fig. 3. Post-Permian structure and geology of the British Isles and continental margin. WG, Worcester Graben; CB, Cheshire Basin. Black lines show seismic sections illustrated in Figs. 6, 8, and 9. (Same symbols as in Fig. 2); 2.0-sec isopach is shown. Conventional geological symbols used in offshore areas (after Institute of Geological Sciences, 1972).

as 7,000 m may be present in Mull (Rast et al., 1968; Smythe et al., 1972). Sedimentation studies by Steel (1971) are consistent with deposition in a subsiding trough controlled either by the Minch Fault or the nearby Outer Isles thrust. Substantial thicknesses (6-7 km) of Permo-Triassic and other Mesozoic rocks may also be present in the north-northeast-trending sedimentary basins mapped by Bott and Watts (1970a), Watts (1971), Browitt (1972), and Flinn (1969), although these values may also include Devonian sediments downfaulted beneath them. Farther south, an estimated thickness of 1,800 m of sediments may have accumulated in the subsiding Antrim Basin, bounded by the northeast-trending Ericht-Laidon and Tyndrum-Glen Fyne faults (Eden et al., 1970; Bullerwell, 1967). In nearly all these basins, the Permo-Triassic thickens towards the marginal fault, indicating syndepositional subsidence.

In the Irish Sea and closer to the Hercynian mobile belt, there are different basin trends. For example, the Cheshire Basin-Worcester graben system trends north-south parallel to the axis of the Pennine anticlinorium (Fig. 3). South of the Hercynian front, the Permo-Triassic sediments of the Bristol and English Channel accumulated along east-west and east-northeast lines (Curry et al., 1970; Lloyd et al., 1973). The present south Irish Sea and Celtic Sea basins cut across both Hercynian and, to a lesser extent, Caledonian trends, possibly reflecting the influence of the Irish Sea geanticline and a weakness in the Precambrian basement (see Dunning and Max, 1974). Deep seismic profiles across the south Irish Sea suggest Triassic (?) salt diapirism (Bullerwell and McQuillin, 1968).

Jurassic. The Rhaetic and lower parts of the Jurassic were transgressive, although in the Toarcian there was an influx of clastics from the northwest (Arkell, 1956; Wills, 1951; Kent, 1969; Hallam, 1971). In post-Toarcian time, several stratigraphic and structural events that bear closely on the formation of the margin took place. In Middle Jurassic time, the main period of vertical movement occurred on the Great Glen Fault (Bacon and Chesher, in press) and was paralleled by rifting in the North Sea (Ziegler, 1974). In the Minch Sea, Jurassic deformation is evidenced by closures of isopachs drawn on the base of the Jurassic (R. McQuillin, personal communication). In the English Channel, Donovan (1972) has noted a progressive easterly tilt that began in the Middle Jurassic and continued into the post-Wealden/pre-Albian folding. Associated changes in sedimentation include the widespread development of nonmarine or estuarine Bajocian and Bathonian.

In northwest Scotland, estuarine deposition of the Middle Jurassic (Fig. 4A) was along northeast-southwest Caledonian lines (Hudson, 1964). In following Oxfordian and Callovian times, there was an extensive but shallow transgression. Faulting apparently continued through the Kimmeridgian (Bailey and Weir, 1932). Hallam (1971) has noted that, although the North American and European Jurassic marine invertebrate faunas argue against major oceanic separation, there is increasing provincialization through Jurassic time. There is a striking difference between the Middle Jurassic ammonite faunas of East Greenland and the British Isles, which implies a physical barrier.

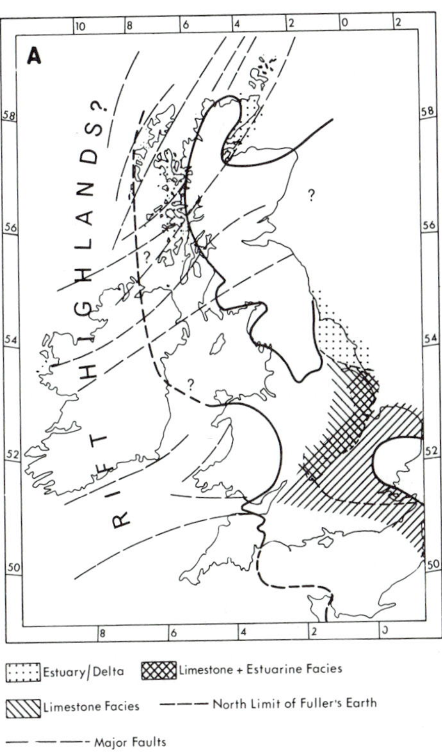

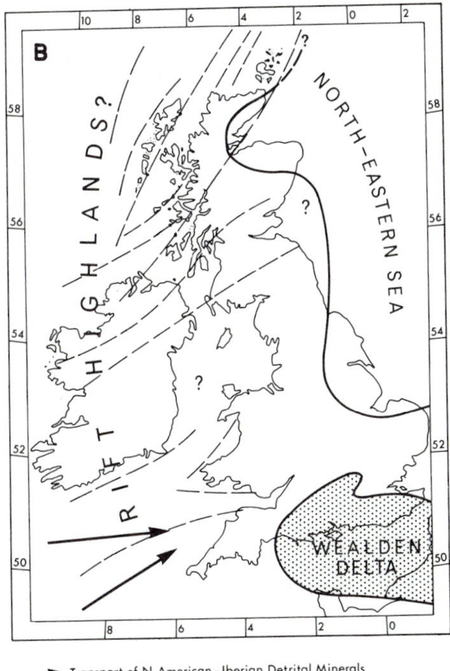

Fig. 4. A. Middle Jurassic paleogeography. B. Early Cretaceous paleogeography (modified from Wills, 1951).

Toward the end of the Jurassic there was an extensive period of faulting, perhaps marked by the rejuvenation of Hercynian northwest-trending faults of the Bristol Channel, Brittany, and Cornwall. The faulting continued through the Wealden in the western English Channel and may be related to the post-Wealden/pre-Albian (pre-Aptian?) folding (Curry et al., 1970; Donovan, 1972; Boillot et al., 1972). Evidence of associated volcanism is sparse but includes the 134 ± 2 my Wolf Rock Phonolite and a Bathonian bentonite tuff (Cowperthwaite et al., 1972; Hallam and Sellwood, 1968).

Cretaceous. In southern England and the English Channel, the end of the Jurassic and the early Cretaceous is marked by the nonmarine facies of the Wealden that is absent in Scotland and Ireland (Fig. 4B). Paleogeographic and mineralogic studies (Allen, 1969) suggest a westward provenance from a topographic high oriented roughly northwest-southeast and parallel to the present margin (Wills, 1951; Donovan, 1972; Larsonneur, 1972). In particular, detrital micas whose ages are consistent with a North American (or Iberian?) provenance extend through the basal Cretaceous and Weald Clay, reappearing in the Aptian to disappear finally by Lower Albian times (Allen, 1969). The Jurassic deposits were folded and tilted during Lower Cretaceous, followed by a period of erosion that may have removed much of the Jurassic and lowest Cretaceous from the middle channel, although more probably they were never deposited there.

The history of the rest of the Cretaceous is dominated by transgression from both the east and west, with overstep across lower members of the Mesozoic onto the Paleozoic (Fig. 5). Progressive westward retreat of the shoreline, which may be indicated by presumed Cenomanian marginal sandstones and dolomites north and west of the Scilly Isles, was followed by the establishment of the shelf carbonate facies of the chalk. Closer to the margin in northern Ireland, the Dalradian-North Antrim high and the country to the north and west remained emergent during the "Cenomanian" transgression, as evidenced by the presence of greensands. In Cenomanian-Upper Turonian time there was a major regression followed by a late Turonian-early Senonian transgression that submerged the Dalradian Ridge (Hancock, 1961). Evidence of a wider extent of the Senonian Sea is given by the outliers present in southern Ireland (Walsh, 1966).

It is worth noting that the general pattern of sedimentation and deformation in the Jurassic and Lower Cretaceous is entirely compatible with the initiation of the North Atlantic Ocean by rifting and spreading at this time (Hallam, 1971). Although the better-documented land and offshore areas lie distant from the present margin, the observed history may be a weaker reflection of the more intense movements that structured the margin. The body of evidence suggests that doming and rifting began in Middle Jurassic time (Fig. 4A), yielding a rift highland to the west of the British Isles later characterized by deltaic, lacustrine sedimentation on its distal eastern flanks (Wealden facies) and the influx of North American (Iberian?) detrital minerals. Phonolitic igneous activity is consistent with the uplift, suggesting the major break occurred around this time.

Persistence of the minerals through to Aptian time may be due to reworking during the Cretaceous transgression or to renewed erosion of the rift scarps at the continental edge. Subsequent subsidence and erosion of the margin is marked by the progressive westward transgression of the higher members of the Cretaceous. Note that the areas nearest to the new margin (such as northern Ireland) were not transgressed until as late as the Senonian, implying a strongly asymmetric drainage pattern consistent with later subsidence of the nearer (and presumably hotter) parts of the margin.

Until more data are available, it will be difficult to define precisely the onset of the primeval Atlantic. However, the Valanginian of East Greenland

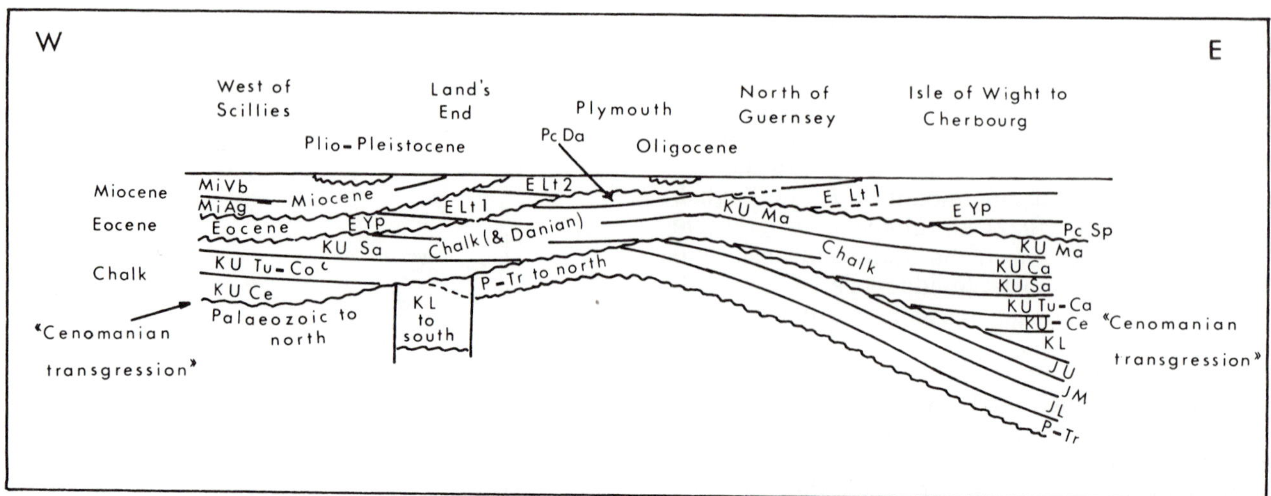

Fig. 5. Schematic section along the English Channel (after Curry et al., 1970).

contains Mediterranean faunas which are not found in the North Sea. This suggests the existence of a continuous barrier (rift mountains?) between Scotland and Norway, consistent with the absence of Lower Cretaceous in Scotland, indicating that an ocean west of the British Isles had formed by this time. Upper Cretaceous intrusive activity on Rockall Bank may be related to the opening of the Labrador Sea.

Tertiary. Much of the Tertiary geological record is missing, both onshore and offshore, but a consistent structural history can be gained from these rather limited data.

Off northwest Scotland and Ireland, thin Tertiary (Neogene?) marine sediments may unconformably overlie the Mesozoic sediments of the major basins, but their stratigraphy is as yet unknown and unproved. The Tertiary (Fig. 3) is represented by the well-known intrusive centers and thick basaltic lavas of these areas (Stewart, 1965; George, 1966, 1967). Between the latest Maestrichtian chalk and the earliest (65-66 my) lavas, there was a period of deformation, uplift, and widespread erosion (George, 1966, 1967; Evans et al., 1973). Along the Skerryvore-Camasunary Fault (Fig. 6), for example, the pre-basalt throw is 500 m, although some of this may be pre-Cretaceous. Throughout the accumulation of the lavas in Ireland, there was progressive regional deformation, shown both by faulting that does not affect the later analcite-natrolite and chabazite-heulandite zones, and by the overstep of the Oligocene Lough Neagh Clays onto the Mesozoic. Terminal volcanism occurred in the province at about 38 my. Post-Oligocene deformation is shown by the folding and 300 m of faulting along rejuvenated Caledonian lines of the Lough Neagh Clays. In northwestern Scotland (Fig. 6) post-Eocene faulting is shown by 330 m of movement in Skye and renewed movement of the Skerryvore-Camasunary Fault. Uplift is indicated by the exposure of dikes at high altitudes in both Ireland and Scotland.

A similar pattern of uplift, tilting, and erosion can be recognized in the offshore areas. In the English Channel (Fig. 5), Maestrichtian sedimentation continued into the Danian (Curry et al., 1970b; Donovan, 1972). A major unconformity marked by erosion and tilting separates the Danian and overlying Paleogene; in places Eocene rests on Santonian, and elsewhere Upper Paleocene (Sparnacian) overlies Maestrichtian. The same unconformity is present in the Celtic Sea, where Paleogene sediments rest on Upper Cretaceous, and in the Irish Sea, where the Paleogene rests on the Triassic (Wood and Woodland, 1968; Blundell et al., 1972). In Tremadoc Bay, Neogene sediments rest on the Paleogene and up to 600 m of Tertiary faulting (in part Neogene in age) has occurred (Wood and Woodland, 1968). Blundell et al. (1971) have observed up to 2 km of Paleogene subsidence in the south Irish Sea.

The faulting was primarily a renewal along existing lines, although reversal of basin tilt is indicated by preservation of Maestrichtian chalk in some areas and its absence in others. In the English Channel (Fig. 5) the main structural event was post-Eocene (more probably post-Lower Oligocene) and may be contemporaneous with the deformation in Ireland and Scotland. These movements were associated with renewed faulting along northwest lines in Oligocene time, shown by the accumulation of 1,300 m of Oligocene lagoonal-lacustrine beds against one such fault in the Bovey Tracey Basin (Fasham, 1971); these beds now lie 79 m above sea level, indicating subsequent uplift. In the western English Channel, the Eocene was downwarped by 500 m, although this movement had ceased by the Miocene unconformity (Curry et al., 1970, 1972).

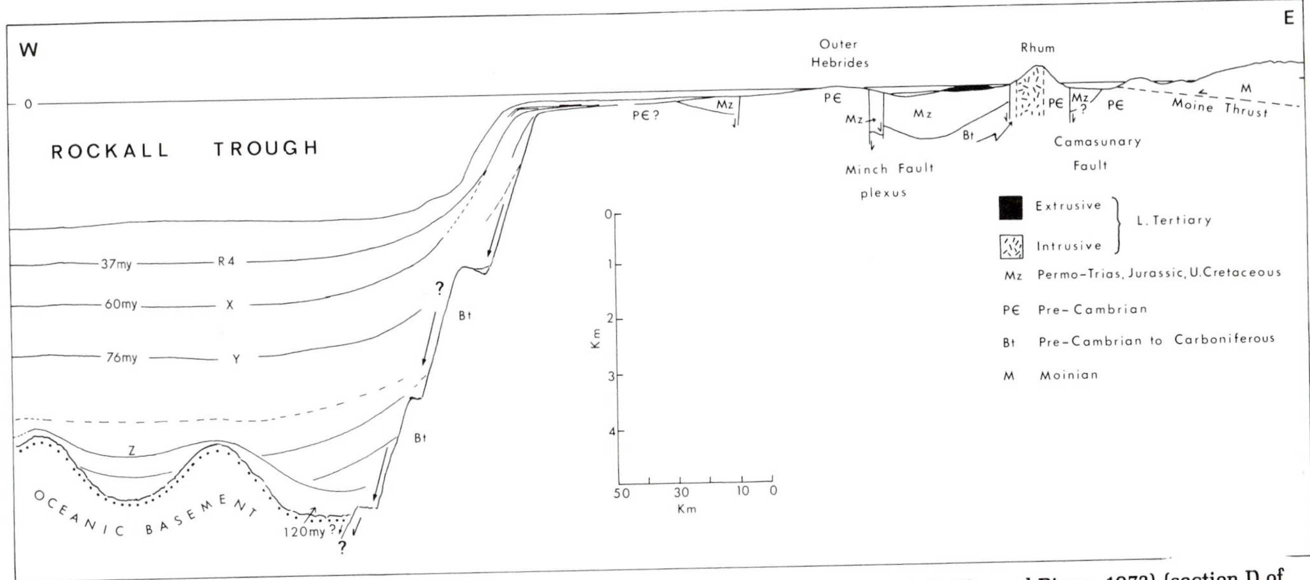

Fig. 6. Schematic section along 57°N across the margin west of Scotland (modified from McQuillin and Binns, 1973) (section D of Fig. 3).

SEA-FLOOR SPREADING INTERPRETATIONS

Syntheses of the plate tectonic evolution of the North Atlantic Ocean indicate that three distinct phases of spreading hived off and structurally isolated the Rockall Plateau microcontinent (Vogt et al., 1969, LePichon et al., 1972; Laughton, 1972). The syntheses are in broad agreement for the time since 76 my, based on successive closure to isochrons defined by oceanic magnetic anomalies. However, there is considerable doubt regarding the nature of the earliest phase partly because of the (then) largely unknown nature and structure of the Rockall Trough (see also Talwani and Eldholm, this volume). The oldest magnetic anomaly abutting the west margin of Rockall Plateau is 60 my and can be followed northeastward to the western edge of the Faeroes platform. Closely similar dates for igneous activity in East Greenland and Faeroes suggest opening began at 60 my, although older magnetic anomalies off East Greenland may indicate a slightly earlier date (Beckinsale et al., 1966; Tarling and Gale, 1968; Laughton, 1972).

Between 60 my and the present, there appear to have been two changes in spreading rate and direction. Prior to 45 my, spreading occurred on a northeast-southwest axis cut by few fracture zones, at a rate of about 1.0 cm/yr. Between 25 and 35 my spreading slowed to 0.6-0.8 cm/yr and reoriented to a north-south axis, cut by numerous east-west fracture zones, continuing in this mode until 22 my. Between 22 and 10 my, spreading accelerated to 0.97 cm/yr, reorienting to the present northeast-southwest axis cut by sparse fracture zones (Vogt et al., 1971). Identification of anomaly 20 close to the west end of the Iceland-Faeroes Ridge (Johnson and Tanner, 1971) suggests that the intital growth of Iceland may be related to the 45-my spreading discontinuity which also marks the termination of the 60-45-my spreading phase in the Labrador Sea.

The pre-60-my phase apparently began at about 76 my (anomaly 32), spreading the united Greenland-Rockall block away from North America to create the embryo Labrador Sea. Earlier distension may be indicated by northeast-trending dike swarms in West Greenland (Watt, 1969) and the contemporaneous igneous activity of the Cretaceous (ca. 80 my) microgabbros of Rockall Bank (Roberts et al., 1974b). The rectilinear southwest margin of Rockall Plateau probably represents an offset related to a fracture zone in the early Labrador Sea that was possibly localized by the Grenville front (Roberts et al., 1973). Reconstruction of the Atlantic to anomaly 32 completely closes the Labrador Sea, juxtaposing the Greenland-Rockall Plateau against North America (Fig. 7).

The absence of post-76-my magnetic anomalies in the Rockall Trough and Faeroe-Shetland Channel indicates that these areas, and indeed the whole margin west of the British Isles, are older than 76 my. A different pre-76-my spreading pattern is also indicated because the Gibbs Fracture Zone does not continue east beyond anomaly 32. However, there has been considerable difficulty in determining the evolution of the area, largely because of the unknown composition of the crust beneath the Rockall Trough. For example, Bullard et al. (1965), in closing the Rockall Trough, closed Hatton Bank and Rockall Bank together to minimize the overlap with Porcupine Bank.

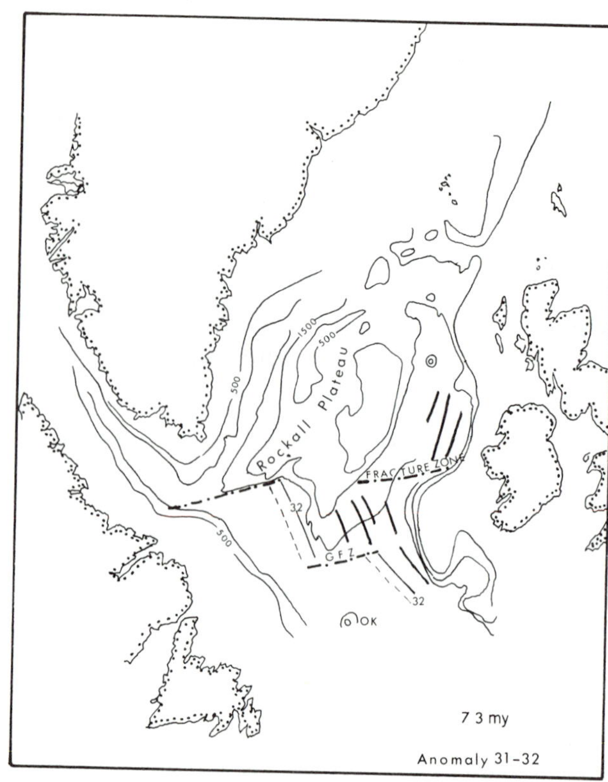

Fig. 7. Reconstruction of the Atlantic at 73 my. Heavy black lines indicate linear magnetic anomalies. Anomaly 32 is shown. (GFZ, Gibbs Fracture Zone; OK, Orphan Knoll).

Seismic-refraction studies indicate continental depths to the Moho on Rockall Plateau beneath the Hatton-Rockall Basin and Rockall Bank of 22 and 30 km, respectively (Scrutton and Roberts, 1970; Scrutton, 1972). Continental crust beneath the Faeroes (Casten, 1973) indicates a northward but narrowed extension of the continent beneath thick lavas. Porcupine Bank is underlain by 29 km of continental crust (R. B. Whitmarsh, personal communication). In the central Rockall Trough, refraction stations give oceanic velocities and a depth to the Moho of 14 km (Jones et al., 1970). Independently constructed gravity models give concordant Moho depths of 13 km both there and in the northern Rockall Trough (Roberts, 1971; Scrutton, 1972; Himsworth, 1973).

In the central part of Rockall Trough, linear, north-northeast-trending magnetic anomalies correlatable over 100 km afford further support for an oceanic origin, considering that their absence in the northern Rockall Trough may be due to Tertiary igneous activity (Fig. 7). The magnetic anomalies are truncated at 54°N by an east-west fracture zone (Fig. 3), also indicated by gravity and reflection profiles, extending the width of the trough to abut the northern foot of Porcupine Bank.

Gravity models across the north margin of Porcupine Bank do not show a trench invoked to reconcile the overlap of Porcupine Bank with Rockall Bank (Scrutton et al., 1971). South of this fracture zone, linear correlatable magnetic anomalies trending north to north-northwest characterize the pre-anomaly 32 oceanic crust. Seismic-reflection profiles across this area show a typically oceanic basement topography, lineated in a north-south direction, that merges close to the foot of Porcupine Bank with a tectonized area that may be the continent-ocean boundary. West of Porcupine Bank and south of the Gibbs Fracture Zone, the width of the pre-anomaly 32 oceanic crust is about 100 km (Fig. 7).

Beneath the western side of the Rockall Trough (Fig. 8a), a prominent subsurface escarpment with a relief of about 2.0 sec (ca. 2.5 km) extends the length of the trough and is offset by the fracture zone at 54°N. The scarp separates areas of contrasting geophysical character. To the west the basement is flat lying, has a refraction velocity of 4.34 km/sec, and higher amplitude and irregular magnetic anomalies. To the east the deeper basement is irregular, with a refraction velocity of 5.95 km/sec and is characterized by linear magnetic anomalies. Scrutton (1972) shows that the scarp marks an easterly thinning of the crust above Moho. The data strongly suggest that the scarp marks the continent-ocean boundary in the west of the Rockall Trough. On the east the continent-ocean boundary is not so clear but may be roughly defined by the 2.0-sec isopach (Fig. 3). In this area the width of presumed oceanic crust is approximately 200 km and is twice this width west of Porcupine Bank.

South of the fracture zone at 54°N, the oceanic crust seems to be about 270 km wide, although there is considerable uncertainty in the position of the continent-ocean boundary on both sides of the trough. The pre-anomaly 32 crust is roughly twice wide to the north of the Gibbs Fracture Zone than to the south. This may be because the 76-my spreading phase south of the Gibbs Fracture Zone began on the line of the pre-76-my ridge axis, resulting in the preservation of half the original width of crust to the west of Porcupine Bank.

In the Faeroe-Shetland Channel, gravity models (Bott and Watts, 1970; Watts, 1971) have given a substantially greater depth to the Moho of 20 km. The difference is surprising in view of the clear topographic continuity with the Rockall Trough and evidence of continental crust of the Faeroes and Rockall Plateau. More significantly, perhaps, the sediments beneath the east side of the channel are thicker and older than those overlying the Paleocene lavas of the Faeroes to the west (Stride et al., 1969); the channel is also narrower than the Rockall Trough. Further, the shelf gravity highs mapped by Bott and Watts do not extend beneath the channel. The difference may be due to burial of the older sediments (initially deposited on ocean crust) by the Paleocene lavas of the Faeroes as they built out into the channel. The presence of thick undetected prelava sediments would result in a spuriously deep Moho.

The body of geophysical evidence thus suggests a continuous oceanic crust extending northward from the Bay of Biscay into the Rockall Trough. The ocean presumably continues northward into the Norwegian Sea, where it may be found beneath the anomalously thick sequence of sediments underlying the wide Norwegian continental slope and the Voring Plateau (Talwani and Eldholm, this volume).

The important age and duration of spreading are difficult to determine because there are no pertinent drill and core data. However, four indirect, independent lines of evidence are in broad agreement. First, the reconstruction to 76 my places the Orphan Knoll on the western edge of the ocean (Fig. 7). Drilling on the knoll has revealed a geological history that closely mirrors that deduced for the British Isles. The Bajocian was deposited in a shallow-water coastal environment and contains anthracite fragments indicating a British provenance. The polynomorphs have both North American and British affinities, indicating continuity of the continents at that time. In post-Bajocian times, the area was uplifted and tilted to the southwest with a prolonged period of erosion before deposition of shallow-water Albian sediments. Sedimentation continued through to the middle Cenomanian. A prolonged hiatus followed before deposition of late Maestrichtian chalks in outer neritic or bathyal depths. The evidence of Albian marine conditions indicates formation of a seaway by that time. Second, the more complete geological history of the British Isles suggests a seaway, perhaps as early as the Valanginian. Third, we have mapped a reflector in the Rockall Trough that pinches out on crust dated at 76 my. The sediments beneath this reflector thicken and contain more reflectors toward the trough margins and are thus considered to be consistent with deposition in a spreading ocean. Estimated sedimentation rates suggest the formation of the ocean in early Cretaceous time. Fourth, the magnetic anomalies west of Porcupine Bank can be traced into the Bay of Biscay (Jones and Roberts, in preparation). Dewey et al. (1973) have reasoned that the Biscay anomalies must lie within the 148-110 my Keathley sequence.

The duration of spreading is difficult to assess. The interval between the Early Cretaceous spreading of the Rockall Trough and 76-my opening of the Labrador Sea was sufficiently long to permit fracturing of older annealed oceanic lithosphere along the new line of the Gibbs Fracture Zone. Removal of the heat source may be indicated by the Turonian transgression in northern Ireland, although this event may also be related to the formation of the Labrador Sea. The balance of the geological evidence suggests formation of the Rockall Trough, Faeroe-Shetland Channel, and margins of the Norwegian Sea by sea-floor spreading between the Valanginian and tentatively the Turonian.

STRATIGRAPHIC HISTORY OF THE ROCKALL PLATEAU AND TROUGH

The greatest sediment thicknesses underlie the Hatton-Rockall Basin, Rockall Trough, and Porcupine Seabight (Fig. 8). The Rockall Trough sediments have accumulated on oceanic crust, but the others have been deposited on continental crust.

The gross difference in thickness between the basins are a function of their ages, the thickest sediments being found beneath the Porcupine Seabight. However, the structural framework, pattern of subsequent deformation, and later sedimentation history all are closely similar (Fig. 8). For example, the basement within each basin is downfaulted, often associated with antithetic faulting, indicating early regional extension; unconformities are greatest at the basin margins, indicating progressive subsidence; a prominent reflector called R4 is common to all, and the stratigraphy of both overlying and underlying beds indicates downwarping in post-R4 strata, coupled with a major change in depositional regime.

In broad terms, the Rockall Plateau consists of a series of basement highs that partly enclose the Hatton-Rockall sedimentary basin. The basement

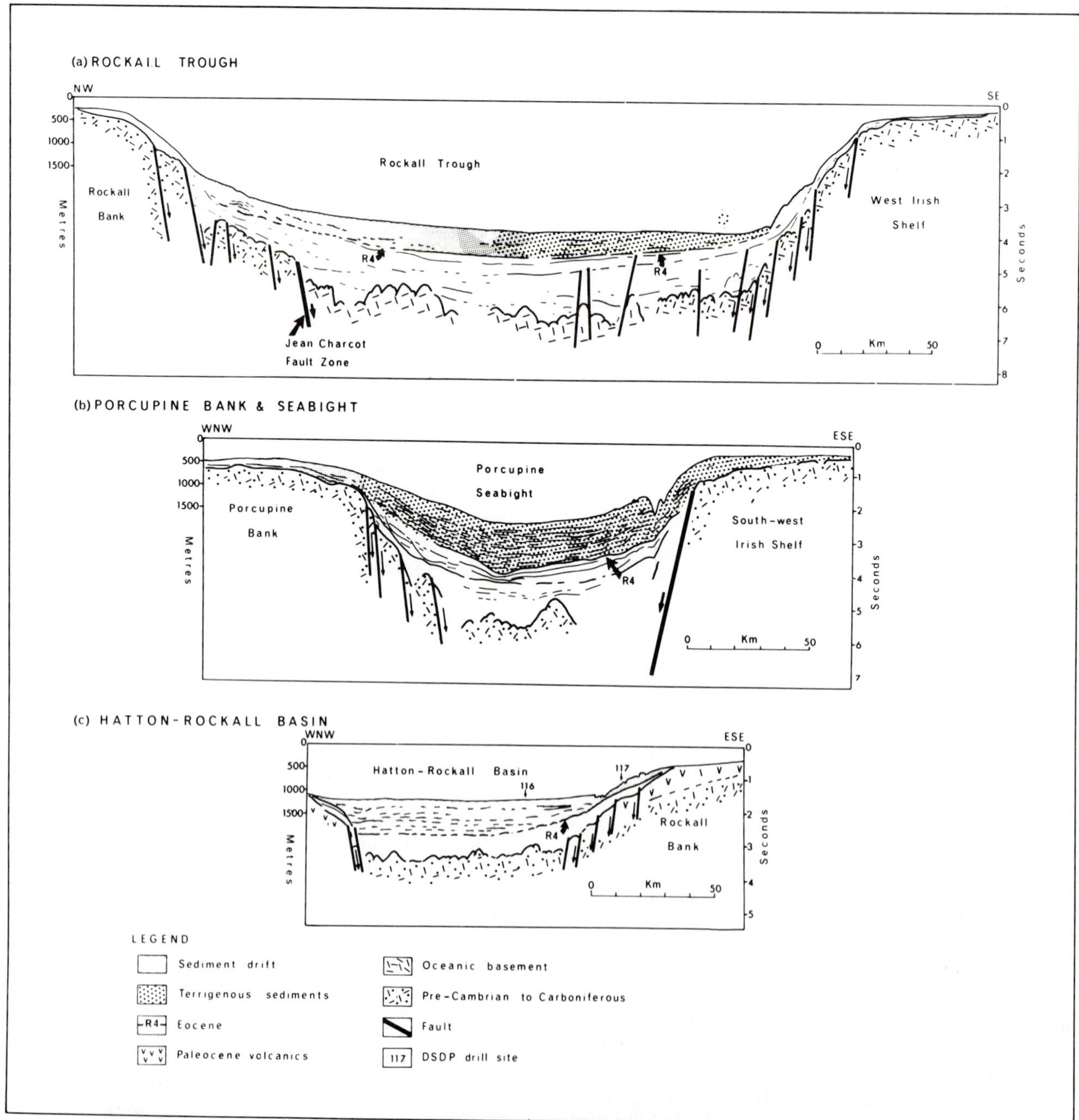

Fig. 8. Intercomparison of Rockall Trough, Hatton-Rockall Basin, and Porcupine Seabight (sections a, b, and c of Fig. 3).

structure and composition are largely unknown. On the Rockall Bank, Precambrian rocks occur as inliers in Cretaceous and/or Tertiary igneous rocks, but become more widespread to the south; Grenvillian ages suggest that the Grenville front may cross the Rockall Bank near 56°20'N.

Seismic profiles across the Hatton-Rockall Basin show the stratigraphy to consist of basement, a pre-R4 series, reflector R4, and a post-R4 series. The acoustic basement is structured into a broad graben with a relief of 2 km, downfaulted by as much as 1.5 sec at the margins. Isopachs on basement and seismic profiles show the controlling fault trends to be closely related to regional tectonic trends (Fig. 3). In the northern part of the basin, the north-northeast trend of the basin margin may be related to a transform fault inferred to the north of Hatton Bank. In the center, the basement faults trend northeast-southwest, parallel to the west margin of the plateau, and the magnetic anomalies of the adjacent post-60-my oceanic crust. South of 57°N, the basin margin fault trends north-south and the basin is formed into a half graben due to rotation and tilting about an axis parallel to the southwest margin of the plateau. Below the basin, basement is characteristically irregular, indicating faulting that bears an antithetic relationship to the basin margin faults.

The overlying pre-R4 series is acoustically transparent and in its lower parts infills the irregular acoustic basement. Faulting has frequently obscured stratigraphic relationships at the basin margins, but on some seismic sections there is evidence of progradation and internal overlap within the series.

Reflector R4 is an ubiquitous reflector within the basin but is cut out on the basin margins by overlapping. Within the basin itself, the post-R4 series and the pre-R4 series are conformable with R4. Toward the basin margins R4 is a major unconformity separating the pre-R4 and post-R4 series. Reflector R4 is a discrete reflector but is often masked by a thin (0.1-0.2 sec) sequence of overlying, very strong reflectors associated with numerous hyperbolas that comprise the base of the post-R4 series and pinch out against R4 on the basin margins. The overlying reflectors and R4 comprise the diffuse zone of energy called reflector 4 by Roberts et al. (1970) and Laughton et al. (1972). The rest of the post-R4 series thins progressively toward the basin margins, overstepping R4 to rest on Precambrian and lower Tertiary igneous rocks.

The post-R4 series is characterized by numerout nonsequences, unconformities, and cuspate reflectors indicative of differential deposition by bottom currents. Within the basin, R4 and the post-R4 series have been affected by only minor faults whose throw rarely exceed 0.1 sec (ca. 200 m), although at the margins larger throws along rejuvenated faults are present. In both cases the faulting only seems to affect the lowest part of the post-R4 series. R4 has been gently downwarped on lines mainly controlled by the basement structure, although there is evidence to suggest reversal of basin tilt.

The stratigraphy of the Hatton-Rockall Basin has been examined by JOIDES (DSDP) holes drilled at the basin margins (Fig. 8c). At DSDP site 116, the post-R4 sequence is largely conformable upon R4, which there marks the contact between Lower Oligocene cherts and underlying Upper Eocene lithified oozes; a minor unconformity or non-sequence may exist within the Oligocene cherts. At DSDP site 117, R4 is the contact between middle Oligocene cherts and lower Eocene clays. R4 is thus a major unconformity marking the overstep of the Oligocene cherts from Upper to Lower Eocene and ultimately onto pre-Upper Paleocene lavas; the overstep within the Oligocene cherts from lower Oligocene in the basin (at site 116) to middle Oligocene at the basin margins (at site 117) suggests that downwarping began at this time. Similar relationships are implied between the late and early Eocene.

One other unconformity is present; the post-Oligocene sequence thins from 600 m at DSDP site 116 to 50 m at site 117, indicating substantial internal overstep against the Oligocene cherts. These unconformities are greatest at the basin margin but decrease basinward to an apparently conformable sequence. The attitude of the beds and unconformities are compatible with deposition in an intermittently subsiding basin initially formed by extension in late Cretaceous?-Paleocene time. The absence of extensive faulting of R4 indicates that the downwarping was not accompanied by regional extension. Laughton et al. (1972), using paleontologic and stratigraphic data, have adduced a subsidence history for the Hatton-Rockall Basin. Their results show a late Paleocene-early Eocene phase of subsidence and an Oligocene phase that may have continued into the early Miocene.

To the west of the Rockall Plateau, the development of the pre-R4 and post-R4 series on post-60-my oceanic crust has been studied by Ruddiman (1972). His results show that R4 (his reflector R) has not been faulted, in contrast to the margins of the Hatton-Rockall Basin. R4 can be followed continuously until it wedges out on the continental basement of Hatton Bank and then appears deformed over the underlying transition from oceanic to continental crust. To the southwest, the pre-R4 series is thicker and includes an additional reflector, called X, that is absent in the Hatton-Rockall Basin and on post-60-my crust. The reflector lies between 76-my basement and the 37-my-old R4 reflector, and may therefore mark the 60-my initiation of spreading between Rockall Plateau and Greenland.

In the Rockall Trough the upper parts of the sequence are closely similar to those of the Rockall Plateau and areas to the west, even though the sequence as a whole is thicker and older (Figs. 6, 8, and 9). Arguments have already been presented in support of an Early Cretaceous acoustic basement age. The lowermost reflector, Z, drapes and infills the irregular oceanic basement, but toward the basin margins passes into prograded and tilted beds presumably representing the earliest clastics derived from the margin. Three reflectors or unconformities, called Y, X, and R4, occur above Z (Figs. 6 and 9). The unconformities are not associated with faulting except at the basin margins. Reflector Y pinches out on 76-my oceanic basement and marks a period of regional warping. Reflector X may be correlative with the reflector observed southwest of Rockall Plateau and, because of its stratigraphic position, may mark a period of warping at about 60 my. Reflector R4 also marks a major unconformity and change in depositional regime.

On the west side of the Rockall Trough, R4 is overlain by the transparent sediments of the Feni Ridge deposited under the influence of Norwegian Sea bottom water (Johnson and Schneider, 1969; Jones et al., 1970; Ellett and Roberts, 1973). The uniform undisturbed character of these sediments indicates that sedimentation was undisturbed by material derived from Rockall Plateau, implying that the Plateau was below sea level throughout this time (Fig. 9). In contrast, the post-R4 series deposited on the east side of Rockall Trough contains numerous closely spaced reflectors and forms two large fans. The influx of terrigenous material thus indicates regional uplift of the British Isles. On the outer shelf and upper slope, reflector R4 divides an older prograding sedimentary series from an upper series that exhibits both onlap and progradation.

Seismic profiles to the west of St. Kilda (Fig. 1) show that reflector R4 (previously identified as base Tertiary by Stride et al., 1969) pinches out close to the Paleocene St. Kilda igneous complex and is overlain by onlapping beds. Farther south, reflector R4 rests with marked unconformity on Mesozoic(?) sediments, abutting the extension of the Great Glen Fault (Stride et al., 1969). These relationships strongly suggest that R4 marks the post-Eocene deformation of Scotland on the outer shelf.

A similar pattern is present farther south in the Porcupine Seabight, where R4 indicates subsidence accompanied by relative uplift of Ireland shown by the terrigenous character of the post-R4 series (Stride et al., 1969; LePichon et al., 1970). It is interesting to note that Porcupine Bank, like Rockall Plateau, probably had tilted southward and subsided below sea level by Upper Eocene time, since the post-R4 series on the Bank is pelagic in character.

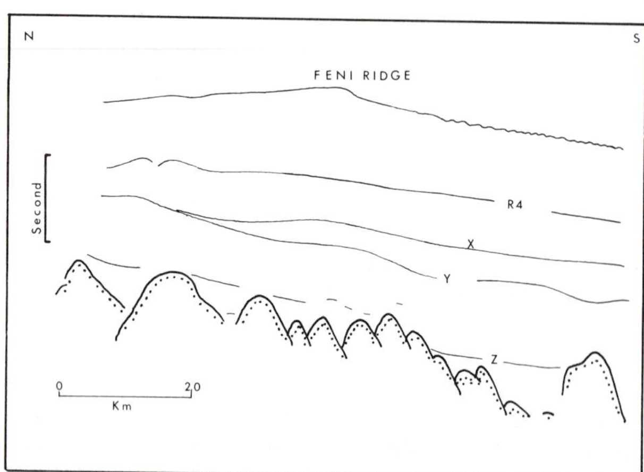

Fig. 9. Principal reflectors in the Rockall Trough (section E of Fig. 3).

GENERAL DISCUSSION

In the introduction, inhomogeneities in the continental lithosphere were considered to exert an important control on development of the continental margins. Such inhomogeneities are a consequence of both age and history of the lithosphere. For example, continental lithosphere resulting from Precambrian tectonics may be compositionally and structurally different from that resulting from the Hercynian and Caledonian orogenies. The continental lithosphere of Rockall Plateau and the British Isles was progressively modified (and generated) in the Keltidian, Scourian, Laxfordian, Grenvillian, Caledonian, and Hercynian orogenies. The rift margins cut across all these junctions (Fig. 10), suggesting that the age of the lithosphere does not here affect that pattern of margin development.

The same evidence also suggests that the boundaries of the orogenies also do not affect margin development, presumably because the complex low-angle thrusting and suturing inhibits subsequent, near-vertical normal faulting. However,

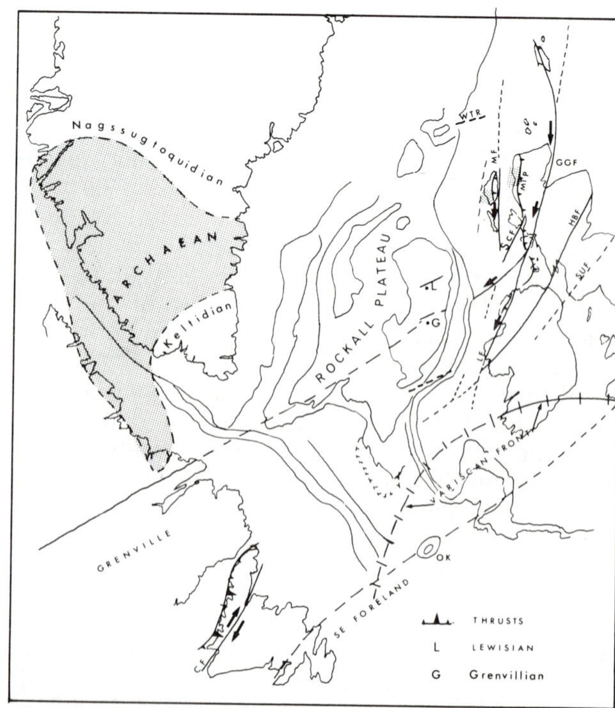

Fig. 10. Predrift configuration of the Rockall Plateau and British Isles and major structural divisions. Fit has been made using the 2.0-sec isopach in the Rockall Trough.

there is evidence to suggest that the Grenville front mav have done so.

Structures affecting the margin and controlling both initial and subsequent basin development include the Caledonian shears, the Highland Boundary Fault, and the Irish Sea geanticline. The Caledonian shears have horizontal slip of up to 160 km and must affect the whole lithosphere. These observations suggest that the main factor influencing the form of the margin is the presence or absence of a vertical fault fabric affecting the whole lithosphere. In the case of the west margin of Rockall Plateau, the stresses of spreading seem to have been more important, resulting in a linear rifted margin.

The influence of the fault fabric imposed during early rifting phases is more difficult to assess. Early basin formation in Permo-Triassic times was due to regional extension controlled by Hercynian and Caledonian trends. Subsequent rifting led to formation of oceanic lithosphere in Rockall Trough and also to crustal thinning beneath the Porcupine Seabight, Celtic Sea, and Hatton-Rockall Basin (Gray and Stacey, 1970; Blundell et al., 1971; Scrutton, 1972). Two basins, the Celtic Sea and English Channel, lie nearly perpendicular to the margin and are associated with offsets in the margin suggesting that the fabric of these basin may have influenced the form of the younger margin. Basins developed along Caledonian shears off northwest Scotland seem, however, to be truncated at the slope, although the Great Glen Fault may have exerted some influence at 56°N. In the Hatton-Rockall Basin, the north-south faulting may be related to the initial formation of the Labrador Sea.

The history of basins developed on both oceanic and continental lithosphere has been one of progressive and/or intermittent subsidence. Uplift and subsidence are contemporaneous over an extremely wide area and correlate with spreading rate changes of the mid-ocean ridge. For example, the post-Lower Oligocene subsidence of the Hatton-Rockall Basin is contemporaneous with downwarping of the shelf west of the British Isles, subsidence and faulting in the English Channel, Celtic Sea, and South Irish Sea, and also with uplift and faulting in Scotland, Ireland, and England and also correlates with the end of the 45-35 my spreading interval, marking a reorientation in spreading and decrease in spreading rate (Fig. 11). Further back in time, the Paleogene deformation of the English Channel and Celtic Sea correlates with warping of the 60-my reflector in the Rockall Trough and the Cretaceous-Paleogene deformation of Scotland and Ireland. These movements correspond to the initiation of spreading (also marked by a spreading discontinuity) between Greenland and Rockall.

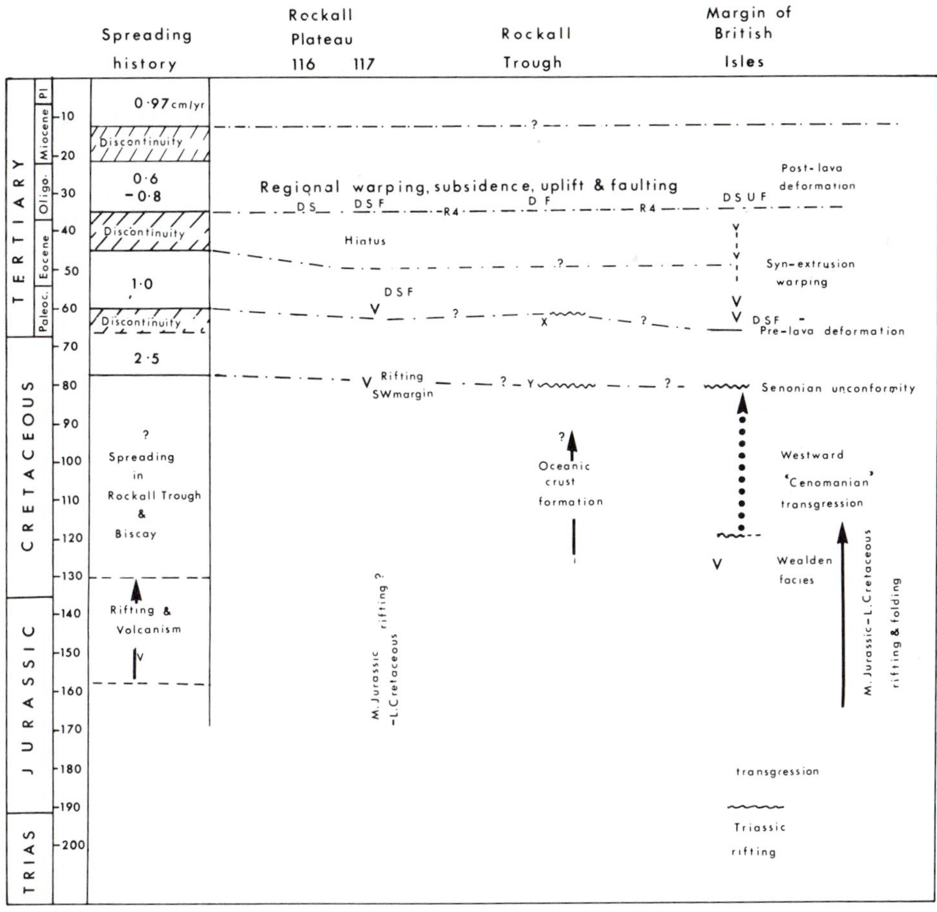

Fig. 11. Correlation among uplift, subsidence, and sea-floor spreading: D, downwarping; S, subsidence; U, uplift; F, faulting; V, volcanism. Sea-floor spreading data from Vogt et al. (1971).

In Scotland and Ireland, this deformation was the precursor of igneous activity lasting from 66 to 40 my and lying far from the accreting edge of the plate. The activity continued long after the formation of oceanic lithosphere and, perhaps more significantly its termination coincides with the 45-35 my spreading discontinuity.

The volcanism in Scotland and Ireland seems to have been localized at the intersection of Caledonian shears with Tertiary cross-faults and was associated with northeast-southwest distension evidenced by the dike swarms (Brooks, 1973). Attempts have been made to relate the volcanic activity in Iceland, but neither the ages nor areal distribution (Fig. 3) are consistent with such a model (Evans et al., 1973; Roberts, 1972). Clearly, the Scottish volcanism was localized by fractures extending through the lithosphere.

The most problematic feature in the framework of plate tectonics, is the uplift and subsidence of the margins far from the spreading axis contemporaneous with changes of spreading patterns and the absence of a contemporaneous intense deformation in the thinner oceanic crust. Two models have been developed to explain basin and continental margin subsidence. Sleep (1971) interprets that the continental margin subsides as it cools, with a time constant of about 50 X 10^6 yr, similar to that of the oceanic part of the plate. Walcott (1970, 1972) and Bott (1971, 1972) have suggested that differential gravitational loading across the margin causes flexure and faulting. The stress pattern is such that normal faulting is likely to develop parallel to the margin in the continental crust. Bott also suggests that the stress differences produce hot creep in the lower and middle continental crust, causing it to thin by flow toward the oceanic upper mantle. These mechanisms are not mutually exclusive, since differential loading will be concurrent with thermal subsidence of the margin.

In the case of the continental margin west of the British Isles, the progressive westward transgression of younger parts of the Cretaceous occurred over about 50 my from the presumed initiation of rifting and is thus consistent with Sleep's model. Subsequent uplift and subsidence has, however, occurred and, in the case of the Hatton-Rockall Basin, only 10 my and 23 my after the 60 my spreading phase. Hot creep is difficult to reconcile with a microcontinent (cf. Bott, 1972), since a sink is necessarily implied beneath the Rockall Trough to accommodate the thermal outflow, which must also occur in widely different directions beneath the west and southwest margins. Differential loading of the lithosphere rather nicely explains the minor deformation in the Rockall Trough and the absence of deformation on the flanks of the Reykjanes Ridge. However, it does not directly explain the regional stress imposed on the lithosphere to produce contemporaneous uplift and subsidence.

If it is assumed that, once the margin has subsided (presumably thermally), the stress due to differential loading is largely in equilibrium, further faulting and renewed warping can only occur by superimposing another stress on the lithosphere or by increasing the differential loading (for example, by increased sedimentation). Coincidence with spreading discontinuities suggests the former, and it is worth noting that localization of basin subsidence on old lithospheric fracture or on sites of crustal thinning implies that changes in the nature of the base of the lithosphere may be responsible, although the precise mechanism is unknown.

ACKNOWLEDGMENTS

I wish to thank A. S. Laughton and M. T. Jones for helpful discussions in the preparation of this review.

BIBLIOGRAPHY

Allen, P., 1969, Lower Cretaceous sourcelands and the North Atlantic: Nature London, v. 222, p. 657-658.

Al-Shaikh, Z. D., 1970, The geological structure of part of the Central Irish Sea: Geophys. J., 20, 208-233.

Andreiff, P., Bouysse, P., Horn, R., and Monciardini, C., 1972, Contribution a l'étude géologique des approches occidentales de la Manche, *in* Colloque sur la géologie de la Manche: Mem. B.R.G.M., France, t. 79, p. 31-38.

Arkell, W. J., 1956, Jurassic geology of the world: Edinburgh, Oliver & Boyd, 806 p.

Audley-Charles, M. G., 1970a, Stratigraphical correlation of the Triassic rocks of the British Isles: Geol. Soc. London Quart. Jour., v. 126, p. 19-48.

———, 1970b, Triassic palaeogeography of the British Isles: Geol. Soc. London Quart. Jour., v. 126, p. 49-90.

Avery, O. E., Burton, G. D., and Heirtzler, J. R., 1968, An aeromagnetic survey of the Norwegian Sea: Jour. Geophys. Res., v. 73, p. 4583-4600.

Bacon, M. and Gray, F., 1971, Evidence for crust in the deep ocean derived from continental crust: Nature London, v. 229, p. 331-332.

———, and Chesher, J., 1974, Geophysical surveys in the Moray Firth, *in* Geology of the North Sea and adjacent areas: Norsk Geol. Survey, (in press).

Bailey, E. B., and Weir, J., 1932, Submarine faulting in Kimmeridgian times: Trans. Roy. Soc. Edinburgh, v. 47, p. 431-467.

Bailey, R. J., Buckley, J. S., and Clarke, R. H., 1971, A model for the early evolution of the Irish Continental margin: Earth and Planetary Sci. Letters, v. 13, p. 79-84.

———, Grzywacz, J. M., and Buckley, J. S., 1974, Seismic reflection profiles of the continental margin bordering the Rockall Trough: Geol. Soc. London Quart. Jour., v. 130, p. 55-70.

Beckinsale, R. D., Brooks, C. K., and Rex, D. C., 1970, K-Ar ages for the Tertiary of East Greenland: Bull. Geol. Soc. Denmark, v. 20, p. 27-37.

Belderson, R. H., Kenyon, N. H. and Stride, A. H., 1970, Holocene sediments on the continental shelf west of the British Isles, *in* Geology of the East Atlantic Continental Margin: Inst. Geol. Sci. Rep. No. 70/14, p. 161-169.

Blundell, D. J., Davey, F. J., and Graves, L. J., 1971, Geophysical surveys over the S. Irish Sea and Nymphe Bank: Geol. Soc. London Quart. Jour., v. 127, p. 339-376.

Boillot, G., Horn, R., and Lefort, J. P., 1972, Evolution structurale de la Manche occidentale au Secondaire et au Tertiaire, *in* Colloque sur la géologie de la Manche: Mem. B.R.G.M., France, t. 79, p. 79-86.

Bott, M. H. P., 1971, Evolution of young continental margins and the formation of shelf basins: Tectonophysics, v. 1, p. 319-327.

______, 1972, Subsidence of the Rockall Plateau and continental shelf: Geophys. Jour., v. 27, p. 235-236.

______, and Watts, A. B., 1970a, Deep sedimentary basins proved in the Shetland-Hebridean continental shelf and margin: Nature London, v. 225, p. 265-268.

______, and Watts, A. B., 1970b, The Great Glen Fault in the Shetland area: Nature London, v. 227, p. 268-9.

______, and Watts, A. B., 1970c, Deep structure of the margin adjacent to the British Isles, *in* Geology of the East Atlantic continental margin: Inst. Geol. Sci. London Rept. 70/14.

______, Browitt, C. W. A. and Stacey, A. P., 1971, The deep structure of the Iceland-Faeroe Ridge: Mar. Geophys. Res., 1, p. 328-351.

Brooks, M., 1973, Some aspects of the Palaeogene evolution of western Britain in the context of an underlying mantle hot spot: Jour. Geol., v. 81, p. 81-88.

Browitt, C. W. A., 1972, Seismic refraction investigation of deep sedimentary basin in the shelf west of Shetlands: Nature Phys. Sci., v. 236, p. 161-163.

Bullard, E. C., Everett, J. E., and Smith, A. G., 1965, The fit of the continents around the Atlantic: Phil. Trans. Roy. Soc. London, ser. A., v. 258, p. 41-51.

Bullerwell, W., 1967, The gravity anomaly map of northern Ireland: Inst. Geol. Sci.

______, 1972, Geophysical studies relating to the Tertiary volcanic structure of the British Isles: Phil. Trans. Roy. Soc. London, ser. A., v. 271, p. 209-215.

Casten, U., 1973, The crust beneath the Faeroe Islands: Nature London, v. 241, p. 83-84.

Charlesworth, J. K., 1963, The historical geology of Ireland: Edinburgh, Oliver & Boyd, 565 p.

Cherkis, N. Z., Fleming, H. S., and Massingell, J. V., 1973a, Is the Gibbs Fracture Zone a westward continuation of the Hercynian front into North America?: Nature Phys. Sci., v. 245, p. 113-115.

______, Fleming, H. S., and Feden, R. H., 1973b, Morphology and structure of the Maury Channel, North East Atlantic Ocean: Geol. Soc. America Bull., v. 84, p. 1601-1606.

Chesher, J. A., Deegan, C. E., Ardus, D. A., Binns, P. E. and Fannin, N. G. T., 1972, IGS marine drilling with *M. V. Whitethorn* in Scottish waters 1970-71: Inst. Geol. Sci. Rep. no. 72/10.

Cole, A. J. G., 1897, On rock specimens dredged on the Rockall Bank: Trans. Roy. Ir. Acad., v. 31, p. 58-62.

Cowperthwaite, I. A., Fitch, F. J., Miller, J. A., Mitchell, J. G., and Robertson, R. H. S., 1972, Sedimentation, petrogenesis and radioisotopic age of the Cretaceous Fuller's Earth of Southern England: Clay Min., v. 9, p. 309-327.

Craig, G. Y., ed., 1965, The geology of Scotland: Edinburgh, London, Oliver and Boyd, 556 p.

Curry, D., Hamilton, D., and Smith, A. J., 1970a, Geological and shallow subsurface geophysical investigations in the western approaches to the English Channel: Inst. Geol. Sci. London Rept. 70/3, 12 p.

______, Hamilton, D., and Smith, A. J., 1970b, Geological evolution of the western English Channel and its relation to the nearby continental margin, *in* Geology of the East Atlantic continental margin: Inst. Geol. Sci. London Rept. 70/14, p. 133-142.

______, Hamilton, D., Smith, A. J., Murray, J. W., Channon, R. D., Williams, D. N., and Wright, C. A., 1972, Some research projects in submarine geology of the Bristol University group, *in* Colloque sur la geologie de la Manche: Mem. B.R.G.M., France, t. 79, p. 221-228.

Dalziel, I. W. D., 1969, Pre-Permian history of the British Isles—a summary, *in* Kay, M., ed., North Atlantic—geology and continental drift: Am. Assoc. Petr. Geol. Mem. 12, p. 5-31.

Day, A. A., Hill, M. N., Laughton, A. S., and Swallow, J. C., 1956, Seismic prospecting in the western approaches of the English Channel: Geol. Soc. London Quart. Jour., v. 112, p. 15-44.

Day, G., and Williams, C. A., 1970, Gravity compilation in the North East Atlantic and interpretation of gravity in the Celtic Sea: Earth and Planet. Sci. Letters, v. 8, p. 205-213.

Dearnley, R., 1962, An outline of the Lewisian complex of the outer Hebrides in relation to that of the Scottish mainland: Geol. Soc. London Quart. Jour., v. 118, p. 143-176.

Dewey, J. F., 1969a, Evolution of the Caledonian Appalachian orogen: Nature London, v. 222, p. 124-129.

______, 1969b, Structure and sequence in paratectonic British Caledonides, *in* Kay, M., ed., North Atlantic—geology and continental drift: Am. Assoc. Petr. Geol. Mem. 12, p. 309-335.

______, Pitman, W. C., Ryan, W. B. F., and Bonnin, J., 1973, Plate tectonics and the evolution of the Alpine system: Geol. Soc. America Bull., v. 84, p. 3137-3180.

Donovan, D. T., ed., 1968, Geology of Shelf Seas: Edinburgh, London, Oliver and Boyd, 160 p.

______, 1972, Geology of the central English Channel, *in* Colloque sur la géologie de la Manche: Mém. B.R.G.M., France, t. 79, p. 215-220.

Dunning, F. W., 1964, The British Isles, *in* Tectonics of Europe, eds. Bogdanoff, A. A., Mouratov, M. A., Shatsky, N. S.: Moskow, Nauk, p. 87-162.

______, 1966, Tectonic map of Great Britain and Northern Ireland: Inst. Geol. Sci., 1:158400.

______, and Max, L., 1974, The exposed and concealed Pre-Cambrian basement geology of the British Isles: Geol. Soc. London Quart. Jour., (in press).

Eden, R. A., Small, A. V. F., and McQuillin, R., 1970a, Preliminary report on marine geological and geophysical work off the east coast of Scotland 1966-1968: Inst. Geol. Sci. London Rept. 70/1.

______, Wright, J. B., and Bullerwell, W., 1970b, The solid geology of the east Atlantic continental margin: Inst. Geol. Sci. London Rept. 70/14, p. 114-128.

______, Ardus, D. A., Binns, P. E., McQuillin, R., and Wilson, J. B., 1971, Geological investigations with a

manned submersible off the west coast of Scotland 1969-1970: Inst. Geol. Sci. London Rept. 71/16.
Ellett, D. J., and Roberts, D. G., 1973, The overflow of Norwegian Sea Deep Water across the Wyville-Thomson Ridge: Deep-Sea Res., v. 20, p. 819-835.
Evans, A. L., Fitch, F. J., and Miller, J. A., 1973, Potassium-argon age determinations on some British Tertiary igneous rocks: Geol. Soc. London Quart. Jour., v. 129, p. 419-444.
Fasham, M. J. R., 1971, A gravity survey of the Bovey Tracey Basin, Devon: Geol. Mag., v. 108, p. 119-130.
Flinn, D. 1961, Continuation of the Great Glen Fault beyond the Moray Firth: Nature London, v. 191, p. 589-591.
———, 1969, A geological interpretation of the aeromagnetic maps of the continental shelf around Orkney and Shetland: Geol. Jour., v. 6, p. 279-292.
———, 1973, The topography of the sea floor around Orkney and Shetland in the northern North Sea: Geol. Soc. London Quart. Jour., v. 129, p. 39-59.
Funnell, B. M., 1964, Studies in North Atlantic Geology and Palaeontology: Upper Cretaceous: Geol. Mag., v. 101, p. 421-434.
Garson, M. S., and Plant, J., 1972, Possible dextral movement on the Great Glen and Minch Faults in Scotland: Nature Phys. Sci., v. 240, p. 31-35.
George, T. N., 1966, Geomorphic evolution in Hebridean Scotland: Scot. Jour. Geol., v. 2, p. 1-34.
———, 1967, Landform and structure in Ulster: Scot. Jour. Geol., v. 3, p. 413-448.
Gray, F., and Stacey, A. P., 1970, Gravity and magnetic interpretation of Porcupine Bank and Porcupine Seabight: Deep-Sea Res., v. 17, p. 467-475.
Hall, D. H. and Dagley, P., 1970, Regional magnetic anomalies, an analysis of the smoothed aeromagnetic map of Great Britain and Northern Ireland: Inst. Geol. Sci. Rep. no. 70/10.
———, and Smythe, D. K., 1973, Discussion of the relation of Palaeogene ridge and basin structures of Britain to the North Atlantic: Earth Planet. Sci. Lett., v. 19, p. 54-60.
Hallam, A., 1971, Mesozoic geology and the opening of the North Atlantic Ocean: Jour. Geol., v. 79, p. 129-157.
———, and Sellwood, B. W., 1968, Origin of Fuller's earth in the Mesozoic of Southern England: Nature London, v. 220, p. 1193, 1195.
Hancock, J., 1961, The Cretaceous system in Northern Ireland: Geol. Soc. London Quart. Jour., v. 117, p. 11-36.
Hersey, J. B., and Whittard, W. F., 1966, The geology of the western approaches to the English Channel, V. The continental margin and shelf of the South Celtic Sea: Continental margins and island arcs, Pap. no. 66-15, Geol. Surv. Can., p. 80-106.
Hill, M. N., 1952, Seismic refraction shooting in an area of the Eastern Atlantic: Phil. Trans. Roy. Soc., ser. A., v. 244, p. 561-596.
———, and Vine, F. J., 1965, A preliminary magnetic survey of the Western Approaches to the English Channel: Geol. Soc. London Quart. Jour., v. 121, p. 463-475.
Himsworth, E., 1973, The Wyville-Thomson Ridge (abstr.): Geol. Soc. London Quart. Jour., v. 129, p. 322-323.
Holgate, N., 1969, Palaeozone and Tertiary transcurrent movements along the Great Glen Fault: Scot. Jour. Geol., v. 5, p. 97-139.
Hudson, J. D., 1964, The petrology of the sandstones of the Great Estuarine Series and the Jurassic palaeogeography of Scotland: Proc. Geol. Assoc., v. 75, p. 499-527.
Institute of Geological Sciences, 1972, 1:2,500,000 map of the sub-Pleistocene geology of the British Isles and adjacent continental shelf, London.
Jones, E. J. W., Ewing, M., Ewing, J. I., and Eittreim, S., 1970, Influences of Norwegian Sea overflow water on sedimentation in the northern North Atlantic and Labrador Sea: Jour. Geophys. Res., v. 75, p. 1655-1680.
Johnson, G. L., and Schneider, E. D., 1969, Depositional ridges in the North Atlantic: Earth and Planetary Sci. Letters, v. 6, p. 416-422.
———, and Tanner, B., 1971, Geophysical observations on the Iceland-Faeroes Ridge: Jokull., v. 21, p. 45-52.
Kennedy, W. Q., 1946, The Great Glen Fault: Geol. Soc. London Quart. Jour., v. 102, p. 41-72.
Kent, P. E., 1969, The geological framework of petroleum exploration in Europe and North Africa and the implications of continental drift hypotheses, *in* Hepple, P., ed., The exploration for petroleum in Europe and North Africa: London, Inst. Petr., p. 3-17.
King, W. B. R., 1954, The geological history of the English Channel: Quart. Jour. Geol. Soc. London, v. 110, p. 77-102.
Lacroix, A., 1923, La constitution du Banc de Rockall: C. R. Hebd. Seanc. Acad. Sci. Paris, v. 177, p. 437-440. 437-440.
Larsonneur, C., 1972a, Données sur l'évolution palaéogéographique posthercynienne de la Manche, *in* Colloque sur la géologie de la Manche: Mem. B.R.G.M., Franch, t. 79, p. 203-214.
———, 1972b, Le modèle sédimentaire de la Baie de Seine à la Manche centrale dans son cadre géographique et historique, *in* Colloque sur la géologie de la Manche: Mem. B.R.G.M., France, t. 79, p. 241-256.
Laughton, A. S., 1971, South Labrador Sea and the evolution of the North Atlantic: Nature London, v. 232, p. 612-616.
———, 1972, The Southern Labrador Sea—a key to the Mesozoic and Early Tertiary evolution of the North Atlantic, *in* Laughton, A. S., et al., Initial reports of the Deep Sea Drilling Project, Leg XII: Washington, D.C., U.S. Govt. Printing Office, p. 1155-1179.
———, Berggren, W. A., et al., 1972, Initial reports of the Deep Sea Drilling Project, Leg XII: Washington, D.C., U.S. Govt. Printing Office, 1243 p.
Le Mouel, J. L., and Le Borgne, E., 1971, Magnetic map of the Bay of Biscay, *in* Debyser, E., et al., eds., Histoire structurale du Golfe de Gascogne, t. I-II: Ed. Technip., VI 11 p.
LePichon, X., Cressard, A., Mascle, J., Pautot, G., and Sichler, B., 1970, Structures sousmarines des bassins sedimentaires de Porcupine et de Rockall: Compt. Rend., t. 270, p. 2903-2906.
———, Hyndman, R., and Pautot, G., 1972, Geophysical study of the opening of the Labrador Sea: Jour. Geophys. Res., v. 76, p. 4724-4743.
Lloyd, A. J., Savage, R. J., Stride, A. H., and Donovan, D. T., 1973, The geology of the Bristol Channel Floor: Phil. Trans. Roy. Soc. London, ser. A., v. 274, p. 595-626.
McQuillin, R., and Binns, P. E., 1973, Geological structure in the Sea of the Hebrides: Nature Phys. Sci., v. 241, p. 2-4.
Miller, J. A., 1965, Geochronology and continental drift—

the North Atlantic: Phil. Trans. Roy. Sco. London, ser. A, v. 258, p. 180-191.
______, and Green, D. H., 1961, Age determination of rocks in the Lizard (Cornwall) area: Nature London, v. 192, p. 1175-1176.
______, Roberts, D. G., and Matthews, D. H., 1973, Rocks of Grenville age from Rockall Bank: Nature Phys. Sci., v. 246, p. 61.
Pitcher, W. S., 1969, North east trending faults of Scotland and Ireland and chronology of displacement, *in* North Atlantic—geology and continental drift: Am. Assoc. Petr. Geol. Mem. 12, p. 724-733.
Pitman, W. C. and Talwani, M. 1972, Sea floor spreading in the North Atlantic: Geol. Soc. America Bull., v. 83, p. 616-646.
Rast, N., Diggins, J. N., and Rast, D. E., 1968, Triassic rocks of the Isle of Mull; their sedimentation, facies, structure and relationship to the Great Glen Fault and the Mull Caldera: Proc. Geol. Soc. London, v. 1645, p. 299-304.
Rayner, D., 1967, Stratigraphy of the British Isles: New York, Cambridge University Press, 453 p.
Riddihough, R., Magnetic surveys off the north coast of Ireland: Proc. Roy. Irish Acad. v. 66, p. 27-41.
Roberts, D. G., 1969a, A new Tertiary volcanic centre on the Rockall Bank: Nature London, v. 223, p. 819-820.
______, 1969b, Recent geophysical investigations on the Rockall Plateau and adjacent areas: Proc. Geol. Soc. London, v. 1662, p. 87-93.
______, 1971, New geophysical evidence on the origins of the Rockall Plateau and Trough: Deep-Sea Res., v. 18, p. 353-359.
______, 1972, Slumping on the east margin of Rockall Bank: Mar. Geol., v. 13, p. 225-237.
______, 1973, The solid geology of the Rockall Plateau: Inst. Geol. Sci. London Rept., in press.
______, Bishop, D. G., Laughton, A. S., Ziolkowski, A., Scrutton, R. A., and Matthews, D. H., 1970, A newly discovered sedimentary basin on the Rockall Plateau: Nature London, v. 225, p. 170-172.
______, Ardus, D. A., and Dearnley, R., 1973, Pre-Cambrian rocks drilled from the Rockall Bank: Nature Phys. Sci., v. 244, p. 21-23.
______, and Jones, M. T., 1974, A bathymetric, magnetic and gravity survey of the Rockall Bank, H.M.S. Hecla, 1969: Adm. Mar. Sci. Publ. 19 (in press).
______, Flemming, N. C., Harrison, R. K., and Binns, P. E., 1974, Helen's Reef: a Cretaceous microgabbroic intrusion in the Rockall intrustive centre: Marine Geology, in press.
Ruddiman, W. B., 1972, Sediment distribution on the Reykjanes Ridge: seismic evidence: Geol. Soc. America Bull., v. 83, p. 2039-2062.
Sabine, P. A., and Watson, J. V., 1965, Introduction to isotopic age determination of rocks from the British Isles, 1955-1964: Geol. Soc. London Quart. Jour., v. 121, p. 477-523.
Scrutton, R. A., 1970, Results of a seismic refraction experiment on Rockall Bank: Nature London, v. 227, p. 826.
______, 1971, Gravity and magnetic interpretation of Rosemary Bank, Northeast Atlantic: Geophys. Jour. v. 24, p. 51-58.
______, 1972, The crustal structure of Rockall Plateau microcontinent: Geophys. Jour., v. 27, p. 259-275.
______, and Roberts, D. G., 1970, Structure of the Rockall Plateau and Trough, North East Atlantic, *in* Geology of the East Atlantic continental margin: Inst. Geol. Sci. London Rept. 70/14, p. 79-86.
______, Stacey, A. P., and Gray, F., 1971, Evidence for the mode of formation of Porcupine Seabight: Earth and Planetary Sci. Letters, v. 11, p. 140-146.
Sleep, N. H., 1971, Thermal effects of the formation of Atlantic continental margins by continental breakup: Geophys. Jour., v. 24, p. 325-350.
Smythe, D. K., Sowerbutts, W. T. C., Bacon, M., and McQuillin, R., 1972, Deep sedimentary basin beneath northern Skye and Little Minch: Nature London, v. 236, p. 87-89.
Stewart, F. H., 1965, Tertiary igneous activity, *in* Craig, G. Y., ed., Geology of Scotland: Edinburgh: Oliver & Boyd, p. 420-466.
Steel, R. J., 1971, New Red Sandstone movement on the Minch Fault: Nature Phys. Sci., v. 234, p. 158-159.
Stride, A. H., Curray, J. R., Moore, D. G., and Belderson, R. H., 1969, Marine geology of the Atlantic continental margin of Europe: Phil. Trans. Roy. Sco. London, ser. A, v. 264, p. 31-75.
Sutton, J., and Watson, J. V., 1951, The Pre-Torridonian history of the Loch Torridon and Scourie areas in the North West Highlands and its bearing on the chronological classification of the Lewisian: Geol. Soc. London Quart. Jour., v. 106, p. 241-307.
Talwani, M., and Eldholm, O., 1972, Continental margin off Norway: a geophysical study: Geol. Soc. America Bull., v. 83, p. 3575-3606.
Tarling, D. H., and Gale, N., 1968, Isotopic dating and palaeomagnetic polarity in the Faeroe Islands: Nature London, v. 218, p. 1043-1044.
Vogt, P. R., 1972, The Faeroe-Greenland-Iceland aseismic ridge and the Western Boundary undercurrent: Nature London, v. 239, p. 79-81.
______, Avery, O. E., Schneider, E. D., Anderson, C. N., and Bracey, D. R., 1969, Discontinuities in sea floor spreading: Tectonophysics, v. 8, p. 285.
______, Johnson, G. L., Holcombe, T. L., Gilg, J. G., and Avery, O. C., 1971, Episodes of sea floor spreading recorded by the North Atlantic basement: Tectonophysics, v. 12, p. 211-234.
Walcott, R. I., 1970, An isostatic origin for basement uplifts: Can. Jour. Earth Sci., v. 7, p. 931-937.
______, 1972, Crustal flexure and the growth of sedimentary basins at a continental edge: Geol. Soc. America Bull., v. 83, p. 1845-1848.
Walsh, P. T., 1966, Cretaceous outliers in south-west Ireland and their implications for Cretaceous palaeogeography: Geol. Soc. London Quart. Jour., v. 122, p. 63-84.
Watt, W. S., 1969, The coast parallel dyke swarm of south west Greenland in relation to the opening of the Labrador Sea: Can. Jour. Earth Sci., v. 6, p. 1320-1321.
Watts, A. B., 1971, Geophysical investigations on the continental shelf and slope north of Scotland: Scot. Jour. Geol., v. 7, p. 189-218.
Wills, L. J., 1951, Palaeogeographical atlas: Glasgow, Blackie, 64 p.
Winchester, J. A., 1973, Pattern of regional metamorphism suggests a sinistral displacement of 160 km along the Great Glen Fault: Nature Phys. Sci., v. 246, p. 81-84.
Worzel, J. L., 1965, Pendulum gravity measurements at sea, 1936-1959: London, John Wiley & Sons, 422 p.
______, 1968, Survey of continental margins, *in* Geology of Shelf Seas, Donovan, D. T., ed., London, Oliver and Boyd, p. 117-154.
Ziegler, P., 1974, Palaeographic development of the North Sea, *in* Geology of the North Sea and adjacent areas: Norsk Geol. Survey, (in press).

Margins of the Norwegian-Greenland Sea

Manik Talwani and Olav Eldholm

INTRODUCTION

The continental margins in the Norwegian-Greenland Sea consist of rifted and sheared segments that are related to the initial opening of the ocean. Parts of the rifted margins are characterized by basement highs underlying the lower continental slope. These highs are bounded landward by a steeply dipping basement surface—escarpments—which are interpreted to define the initial line of rifting separating oceanic crust from subsided continental crust. The North Sea Basin continues northward into the Norwegian Marginal Basin and the Barents Sea and contains thick sediments of Mesozoic and possibly late Paleozoic age. This basin also contained parts that now form the Greenland margin. We believe that prior to opening, deposition and subsidence kept pace with each other in an epicontinental marine environment. However, the continental slope has subsided further with the evolution of the Norwegian-Greenland Sea.

The margins have been complicated by the interpreted splitting off of the continental fragment comprising the Jan Mayen Ridge during a late phase of opening. They are also complicated by shearing and subsequent rifting of the Svalbard margin, as well as the shift of the Knipovich Ridge axis close to Svalbard.

GENERAL COMMENTS

In this study we review geological and geophysical data at the continental margins of the Norwegian-Greenland Sea and discuss their implications in inferring the history and development of these margins. The margins are of the passive type and apparently were originally created when Greenland separated from the Eurasian plate.

We have limited this review to the area between the Spitsbergen Fracture Zone in the north, and the Faroe-Iceland Ridge to the south (Fig. 1). Not all parts of these margins have been studied equally extensively. The area studied in most detail, by far, lies off Norway. The Greenland margin is much less accessible because a major part of the area is ice-covered even during the summer. The evolution of the apparently young Norwegian-Greenland Sea may be more easily traced back to the time of initial rifting than in many other deep-ocean regions.

These passive margins bordering the Norwegian-Greenland Sea are of sheared and rifted origins, and a complex combination of both processes.

EVOLUTION OF THE NORWEGIAN-GREENLAND SEA

The development of the Norwegian-Greenland Sea in terms of sea-floor spreading has been proposed by Johnson and Heezen (1967), Avery et al. (1968), and Vogt et al. (1970). The writers have considered recent and more extensive magnetic and other geophysical data to obtain a more precise history. The major structural elements related to this evolution are indicated in the bathymetric sketch (Fig. 1), and the various stages of evolution are summarized schematically in Figure 2.

Our study shows that opening of the Norwegian-Greenland Sea can be considered to have taken place in two successive phases. The initial opening took place along the Vøring Plateau and Faroe-Shetland escarpments, approximately 60 my to 40 my ago, with Greenland moving nearly northwest with respect to Europe. Greenland was also moving roughly north-northeast with respect to North America at this time with an opening of the Labrador Sea. The principal development of the Norwegian Sea margins occurred at this time. It is important to note that two different kinds of margin-forming processes were active—rifting and shearing. The marginal escarpments represent segments of rifted margins, whereas the oldest part of the Jan Mayen Fracture Zone, as well as the northeast boundary of the primitive Greenland-Lofoten Basin, represent sheared margins.

During the second phase of opening (~40 my to present) Greenland remained stationary with respect to North America and the direction of opening became approximately west-northwest. Greenland, instead of sliding along the Barents Sea and Svalbard as it had done earlier, split away, creating the Greenland Sea and further enlarging the Norwegian Sea. The opening of the Greenland Sea during this phase evidently took place along a line lying east of the Greenland Fracture Zone so that the fracture zone has old oceanic crust (~60-40 my) on one side and largely young oceanic crust (younger than ~40 my) on the other side. The Senja Fracture Zone, which forms the northeast boundary of the primitive Greenland-Lofoten Basin, has remained apparently unaltered through the second phase of opening and continues to have old oceanic crust (~60-40 my) on one side and continental crust on the other side.

The margins of the Greenland Sea are thus complicated, having changed from a shear zone during the initial opening of the Norwegian Sea to a rifted margin during the subsequent opening of the Greenland Sea.

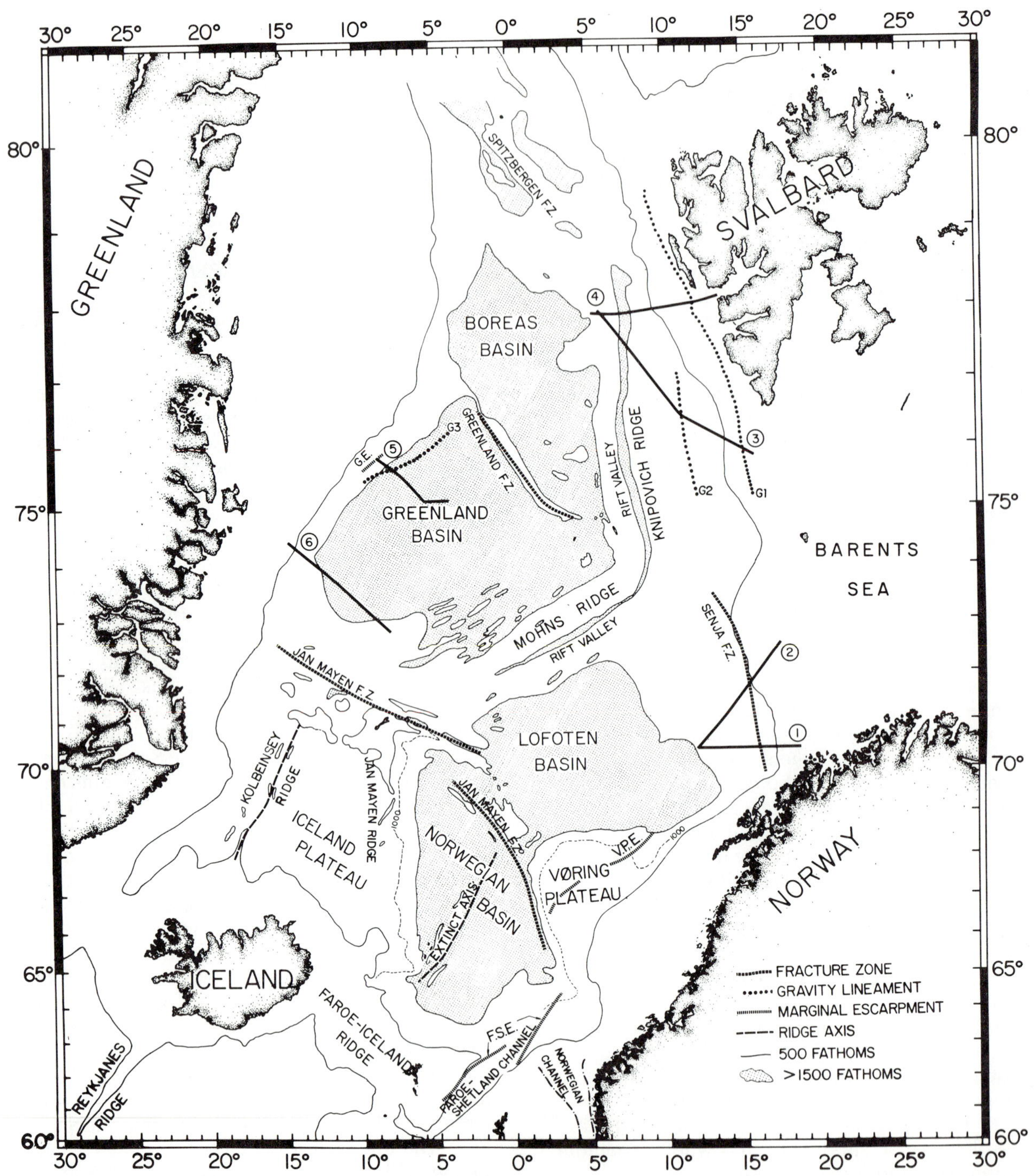

Fig. 1. Bathymetry and major structural features of the Norwegian-Greenland Sea. G1 and G3 indicate linear belts of large positive isostatic gravity anomalies; G2 represents an isostatic minimum. GE stands for a marginal escarpment. Circled numbers 1 through 6 represent segments of ships' track (Figs. 7-10). The 1,000-fm isobath outlines the Vøring Plateau and the eastern boundary of the Icelandic Plateau. V.P.E. and F.S.E. stand for Vøring Plateau Escarpment and Faroe-Shetland Escarpment (Talwani and Eldholm, 1972). The area south of the Greenland-Senja Fracture Zone is referred to as the Norwegian Sea; the area to the north is called the Greenland Sea.

A further modification of the Svalbard margin is caused by an eastward shift of the northern part of the ridge axis in the Greenland Sea. This eastward shift is suggested by the asymmetric position of the northern part of the Knipovich Ridge.

Even more important shifts of the position of the spreading axis have influenced the Greenland margin south of the Jan Mayen Ridge. One such shift apparently broke away a part of Greenland which now forms the Jan Mayen Ridge. There is some uncertainty as to the southern extent of this continental fragment. As a thin sliver it might extend almost as far south as the continental slope off northeast Iceland. Thus, south of the Jan Mayen Ridge, the margins off Greenland and Norway, as well as the eastern and western boundaries of the Jan Mayen

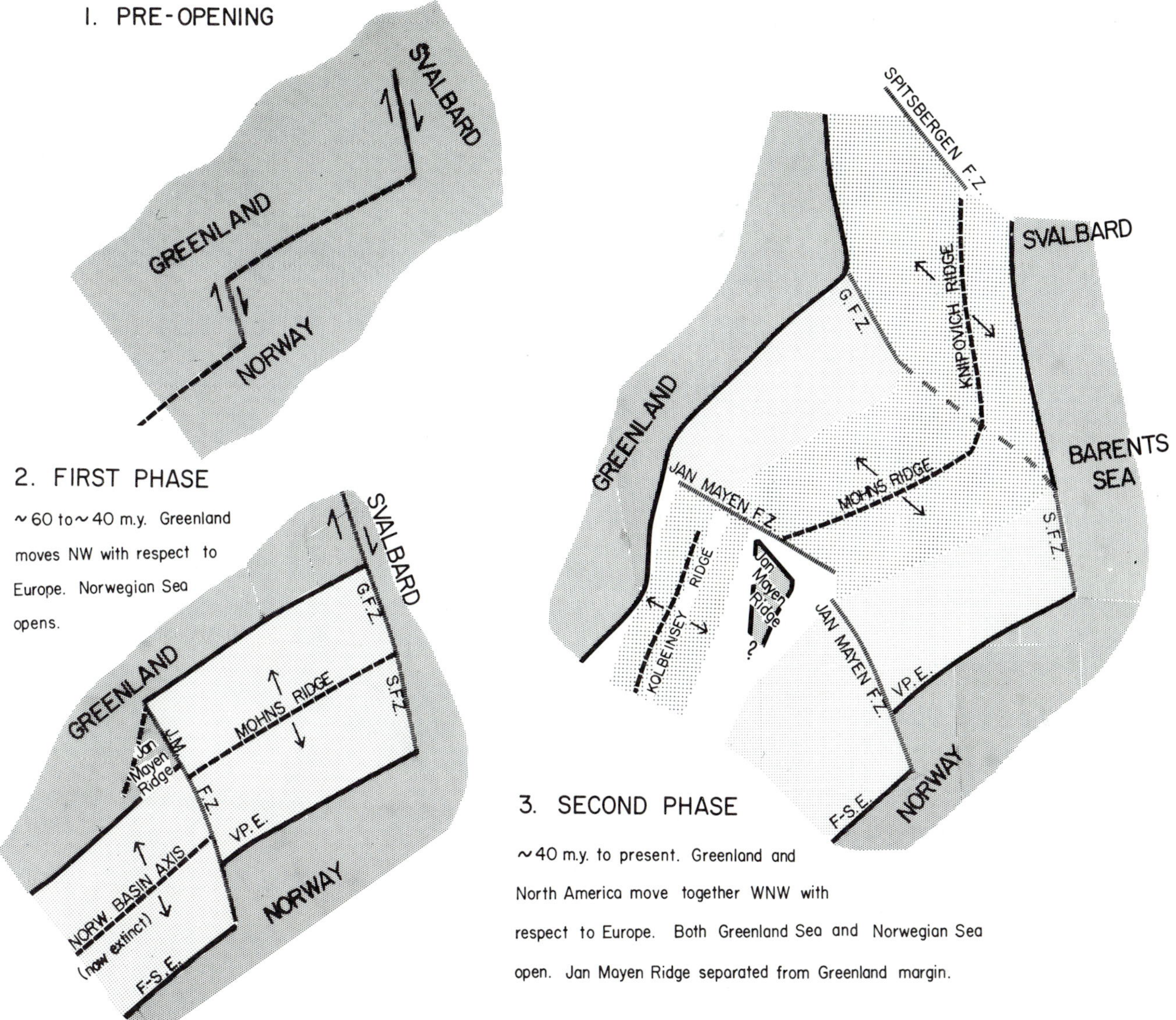

Fig. 2. Schematic disgrams of the evolution of the Norwegian-Greenland Sea region. For explanation of abbreviations, see Figure 1. The Norwegian Sea opened first along the Mohns Ridge, an axis which is still preserved, and another axis related to, but not identical with, the extinct axis in the Norwegian Basin shown in Figure 1. The continental margins of Norway and Greenland are of the rifted type. The margin off Svalbard is a complicated one, having undergone shearing during the first phase of opening and rifting during the second stage.

Ridge, are of the rifted type.

We next discuss in some detail the geophysical data on the margins that have been used to develop the concepts about the evolution of the Norwegian-Greenland Sea discussed above. Our choice of the order of presentation in this paper has not followed the order in which the deductions were made, but the order in which we felt a review would be most useful.

BATHYMETRY

The regional bathymetry in the Norwegian-Greenland Sea (Fig. 1) was constructed on the basis of Lamont-Doherty Geological Observatory tracks of R/V VEMA (cruises 23, 27, 28, 29, and 30), all of which employed satellite navigation. In some areas we have relied on other published information, largely from Johnson and Eckhoff (1966). This map shows only the first-order features of the regional bathymetry.

A few features related to the margins deserve special attention. The Greenland continental slope south of the Jan Mayen Fracture Zone has much less relief than its Norwegian counterpart, a fact that has been interpreted as indicating a significant age difference betwen the Norwegian Basin and the Icelandic Plateau. There is also a prominent marginal plateau, the Vøring Plateau, just northeast of the easternmost part of the Jan Mayen Fracture Zone.

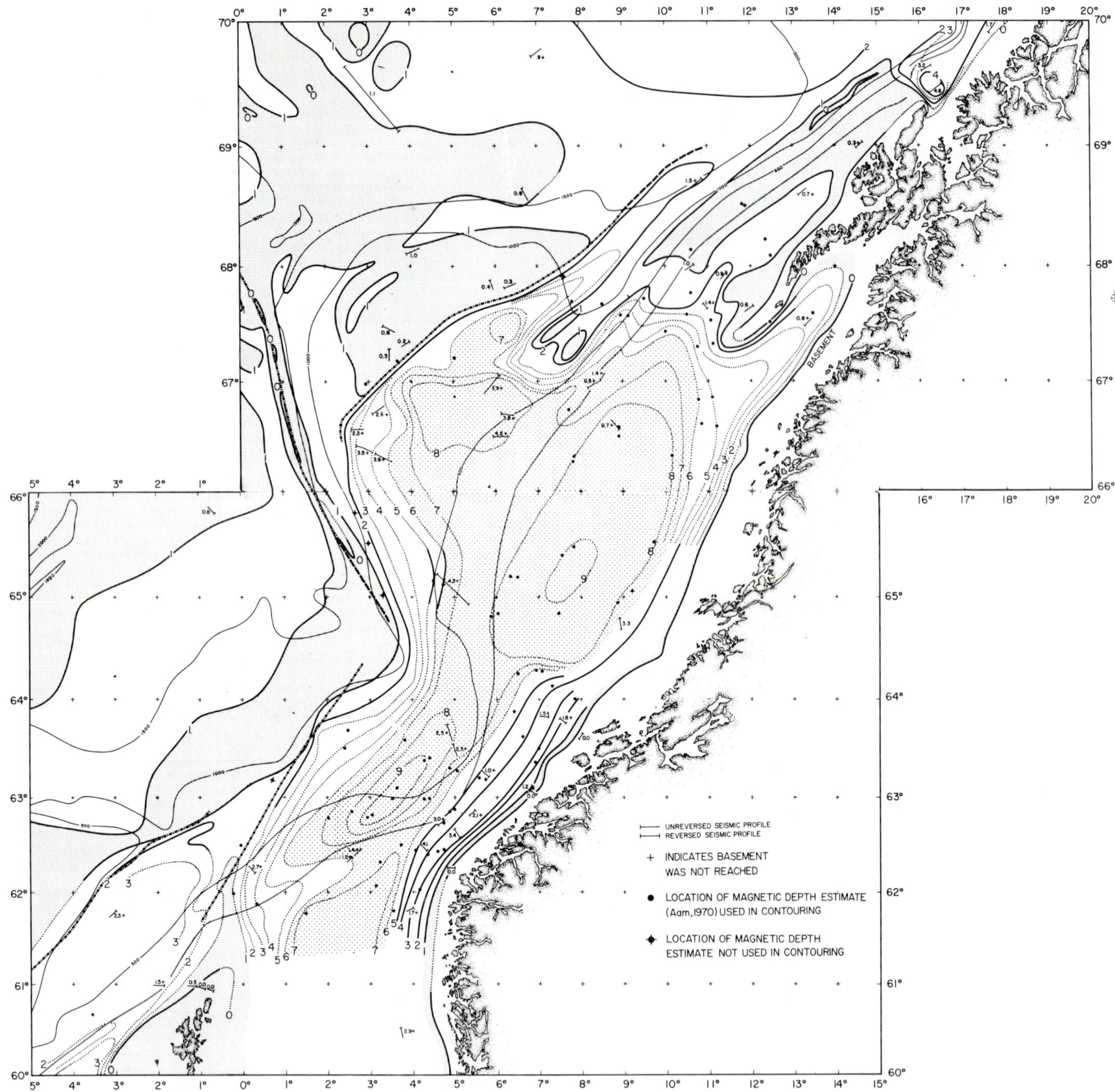

Fig. 3. Map showing total thickness of sediments on the continental margin off Norway (Talwani and Eldholm, 1972). Topographic contours are shown at intervals of 500 fm, sediment isopachs in km, and the location of the marginal escarpments indicated by patterned lines. Areas of exceptionally thick (>6 km) and thin (<1 km) sediments are shaded.

Farther north the gradient of the continental slope increases considerably with a very steep and narrow slope off the Lofoten-Vesteralen Islands. On the Greenland side, however, the continental slope in the segment between the Greenland Fracture Zone and the Jan Mayen Fracture Zone is steep in the south but gentler in the north, in the vicinity of the Greenland Fracture Zone.

West of Svalbard the Knipovich Ridge lies very close to the shelf. The continental slope extends directly onto the axial mountains without any basin between the ridge axis and the continental margin.

GEOPHYSICAL DATA

A geophysical study of the area between 60° and 70°N has been published by Talwani and Eldholm

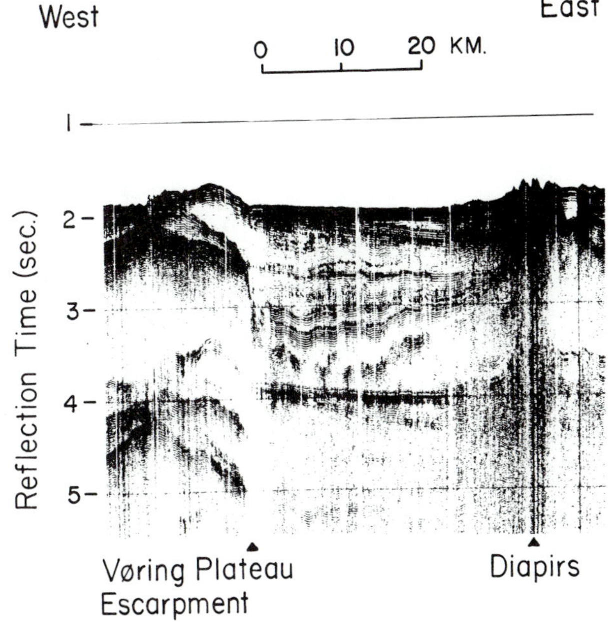

Fig. 4. Seismic profiler record across the Vøring Plateau Escarpment (Vema 30, 1973). The highly reflective surface to the left is the top of the basement high, and those on the right are believed to be sediment layers. A topographic high associated with a diapir field can be seen on the right.

(1972). The following studies are also included: aeromagnetic maps by the Norwegian Geological Survey for the continental shelf between 62 and 69°N (Aam, 1970); by the Institute of Geological Sciences (IGS) London for the Shetland region (Hall and Dagley, 1970); by the U.S. Naval Oceanographic Office (Project Magnet) for the Norwegian Sea (Avery et al., 1968); gravity values from Bakkelid (1959), Collette (1960), Anonymous (1970), Bott and Watts (1971), and Fleischer (1971); seismic-refraction measurements from Ewing and Ewing (1959), Kvale et al. (1966), Eldholm (1970), Eldholm and Nysaether (1969), and Sundvor (1971); and seismic-reflection data from Johnson et al. (1968), Eldholm and Nysaether (1969), Nysaether et al. (1969), and Johnson et al. (1971). Additional Lamont-Doherty data are being prepared by Talwani and Grønlie (in press) and Eldholm and Windisch (in press).

NORWEGIAN MARGIN

Figure 3 shows the total sediment thickness of the margin off Norway. The map was constructed with seismic data as the principal source of information; however, where basement could not be reached, depth estimates (Aam, 1970) determined from magnetic data, as well as gravity trends, were used as guidelines. The principal feature of the map is a large sedimentary basin underlying the margin. The basin appears to be a continuation of the North Sea Basin. The sediments decrease in thickness from the center of the basin toward the shoreline. On the seaward side the decrease in sediment thickness is more abrupt. The seaward boundary of the basin is characterized by prominent basement highs on the lower part of the continental slope. One of these basement highs underlies the outer part of the Vøring Plateau and another lies along the lower continental slope north of the Shetland Islands and along the eastern boundary of the Faroes. These two highs, called the Vøring Plateau Escarpment and the Faroe-Shetland Escarpment, are apparently offset by the Jan Mayen Fracture Zone and are bounded landward by steeply dipping basement surfaces.

The escarpments are important structural boundaries, which are reflected in changes in geophysical parameters across them. The seismic data indicate that the highly reflecting acoustic basement at the top of the escarpments can be followed westward into typical oceanic basement in the Lofoten and Norwegian basins, where both the structural fabric and the magnetic anomalies indicate a typical oceanic crust. On the landward side of the escarpments, on the other hand, we observe a sequence of horizontally or slightly landward-dipping sedimentary beds (Fig. 4). The contrast in structure across the Vøring Plateau Escarpment is also revealed by seismic refraction data shown in Figure 5. On the seaward side of the escarpments there is a thin section of unconsolidated and semiconsolidated material overlying a high-veloctiy basement refractor (5.2 km/sec). On the landward side several sedimentary layers, including a thick sequence of consolidated sediments, are observed. The deepest layer has an average velocity of 4.4 km/sec. The velocity structure therefore indicates that the escarpments mark a major structural boundary. The structural boundary at the northern part of the Faroe-Shetland Escarpment, as mapped by Talwani and Eldholm (1972), was less distinct; however, more recent seismic reflection crossings across this boundary have confirmed its existence.

The gravity and magnetic fields also reflect the presence of the escarpments. Although the gravity field is generally irregular, the majority of our crossings show a gravity gradient associated with the escarpments. Gravity values are low just landward of the escarpment and attain maximum positive values seaward. We call attention to the fact that we have calculated isostatic anomalies for all the sections shown in this paper; because passive margins are considered to be in isostatic equilibrium, these anomalies are good indicators of underlying mass excesses and deficiencies.

The magnetic anomaly field over the basement high is typical of the field associated with sea-floor spreading. However, in most profiles the character of the field changes markedly in the vicinity of the escarpments. In particular, at the Vøring Plateau a well-developed magnetic quiet zone exists on the landward side (Fig. 6).

The Vøring Plateau Escarpment is terminated at its southern end by the Jan Mayen Fracture Zone. At its northern end it does not extend as far north as the Senja Fracture Zone; the margin north of 68.5°N,

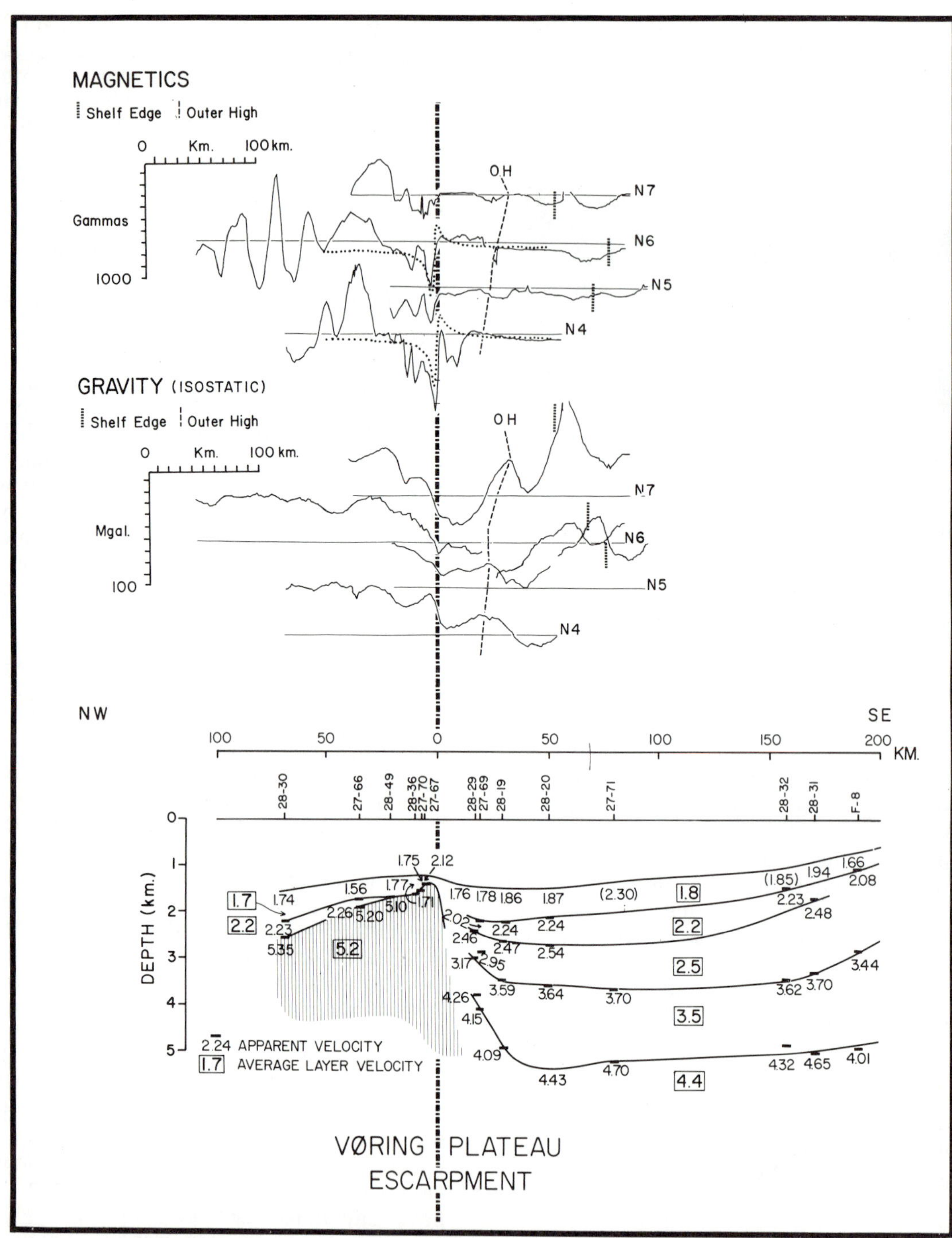

Fig. 5. Profiles of magnetic and isostatic gravity anomalies and a seismic structure section across the Vøring Plateau Escarpment (from Talwani and Eldholm, 1972). The escarpment is clearly associated with a gravity gradient and change in magnetic character and in seismic structure. The characteristic magnetic low at the escarpment was computed from the effect of the edge of a reversely magnetized, semiinfinite plate 0.5 km thick, suggesting formation during a period of reversed polarity—possibly between anomaly 24 and 25 (60-62 my). OH refers to a structural high which is sedimentary but may reflect a basement high.

where the Vøring Plateau Escarpment is not observed, is very narrow.

The seismic refraction data demonstrate that although the distribution of seismic velocities in the inner part of the Vøring Plateau is markedly different from that in the outer part, it is not significantly different from that on the slope and the continental shelf. The prominent North Sea sedimentary basin has no closure to the north and appears to continue into the Norwegian continental shelf and slope and the inner part of the Vøring Plateau (Fig. 3). Therefore, it seems reasonable to assume that age-velocity relationships in the North Sea Basin and in the Norwegian margin basin should be similar. In the North Sea Basin even thick Cenozoic sections have velocities below 2.5 km/sec (Hornabroek, 1967; Wyrobek, 1969). A considerable portion of the section landward of the escarpment has velocities higher than 2.5 km/sec (Fig. 5) indicating the presence of sediments of Mesozoic and perhaps even late Paleozoic age.

An anomalous area is found between 67.5 and 69°N, where the total sedimentary thickness is small (Fig. 3) and where Tertiary deposits exist only locally.

Eldholm and Nysaether (1969) and Nysaether et al. (1969) found that the Cenozoic section on the Norwegian shelf is composed of two depositionally different sequences separated by an erosional unconformity. The thickness of the upper sequence varies between 0 and 300 m, being thinnest in the coastal channels. Reflectors within the sequence are few and generally concordant with the sea floor except near the shelf edge, where foreset beds occur. The lower sequence consists of seaward-dipping reflectors striking approximately parallel to the coastline with maximum dips nearest the coast. On the outer shelf the sediments show evidence of progradation and outbuilding. The lower sequence has been interpreted as principally reflecting a differential uplift with subsequent erosion to account for the truncated dipping beds, whereas the top layer is believed to be deposited in the Quaternary. On the continental slope the bedding generally is uniform and is conformable to the surface of the sea floor toward the escarpments.

A characteristic feature of the free-air gravity field on the continental shelf is the existence of elongate belts of positive anomalies lying at or landward of the shelf edge. These belts are shown in Figure 6. A southern belt runs just landward of the shelf edge from west of Shetland to about 64°N. Farther north, another belt, which is offset some distance to the west, is found to consist of two maxima, one at the shelf edge and another passing through the Lofoten Islands (MS, ML, Fig. 6).

The magnetic anomalies on the shelf and upper slope also appear in belts (Fig. 6), and there is apparently a relationship between many of the magnetic and free-air gravity anomalies. Especially between 66 and 70°N this relationship is well developed, with the magnetic maxima systematically displaced to the east of the gravity maxima.

MARGIN OFF BARRENTS SHELF AND SVALBARD

The most prominent geophysical feature in this area is an elongate free-air gravity high between 70 and 73.5°N. This anomaly runs along the base of the slope south of 71°N but cuts across the bathymetric contours farther north. Two gravity profiles across this feature are shown in Figure 7. The magnitude of the positive isostatic anomalies approaches 80 mgal and would appear to indicate an important structural feature. We have only made a very limited exploration of this structural feature by shooting seismic refraction profiles on it, but were unable to penetrate through the sediments. Because of its conjugate position with respect to the Greenland Fracture Zone, we believe it is likely that this feature represents a fracture zone that is deeply buried under a thick cover of sediments. We note that other fracture zones in the Atlantic have been shown to possess a residual positive isostatic anomaly (Cochran, 1973), and have named this feature the Senja Fracture Zone.

The free-air map by Talwani and Grønlie (in press) also shows other prominent linear anomalies on this margin. G1 represents a prominent positive anomaly on the shelf off Svalbard (Fig. 1). G2 represents a minimum parallel to the continental slope. G1 and G2 both represent isostatic anomalies which are nearly coincident with free-air anomalies (Figs. 8 and 9). However, a free-air high coincident with the shelf edge (profile 4 in Fig. 9) is considerably reduced when the isostatic correction is applied.

The magnetic patterns are much less clear in this area than off Norway. The sea-floor spreading anomalies in the Lofoten Basin are subdued in the vicinity of the Senja Fracture Zone. East of the fracture zone, on the Barents Shelf, the magnetic field is generally very smooth. North of the fracture zone, the anomalies on the shelf and slope are of low amplitude; however, irregular anomalies of higher amplitude occur locally without a clearly defined pattern.

Seismic data for this part of the Barents margin have been analyzed by Sundvor (1971), Eldholm and Ewing (1971), and Kogan and Milashin (1970). Sundvor (1971) reported a layered sequence of sediments between 69 and 71°N with the beds cropping out just offshore and dipping gently seaward. An exception is the shelf outside Andøya, where faulting may have occurred. Talwani and Eldholm (1972) suggested that a disturbed area here might be associated with a landward extension of the Senja Fracture Zone.

The main part of the Barents Sea is underlain by layered sediments with velocity higher than 2.5 km/sec at the sea floor and increasing with depth. However, near the western shelf edge and on the slope there is a wedge of low-velocity sediments on top of beds with velocities similar to those of the higher velocity sediments to the east. Eldholm and Ewing (1971) interpreted this wedge as being formed by land-derived material deposited during the Tertiary, at which time the major part of the Barents Sea was probably emergent. The low-velocity sediments extend also into the shelves of Norway and Spitsbergen. The sediment distribution on the basins west of the margin indicate that the Barents Shelf was a major source of sedimentary deposits in the Cenozoic (Eldholm and Windisch, in press). North of the Senja Fracture Zone a large amount of sediments has been ponded between the Knipovich Ridge and the continent (Figs. 8 and 9).

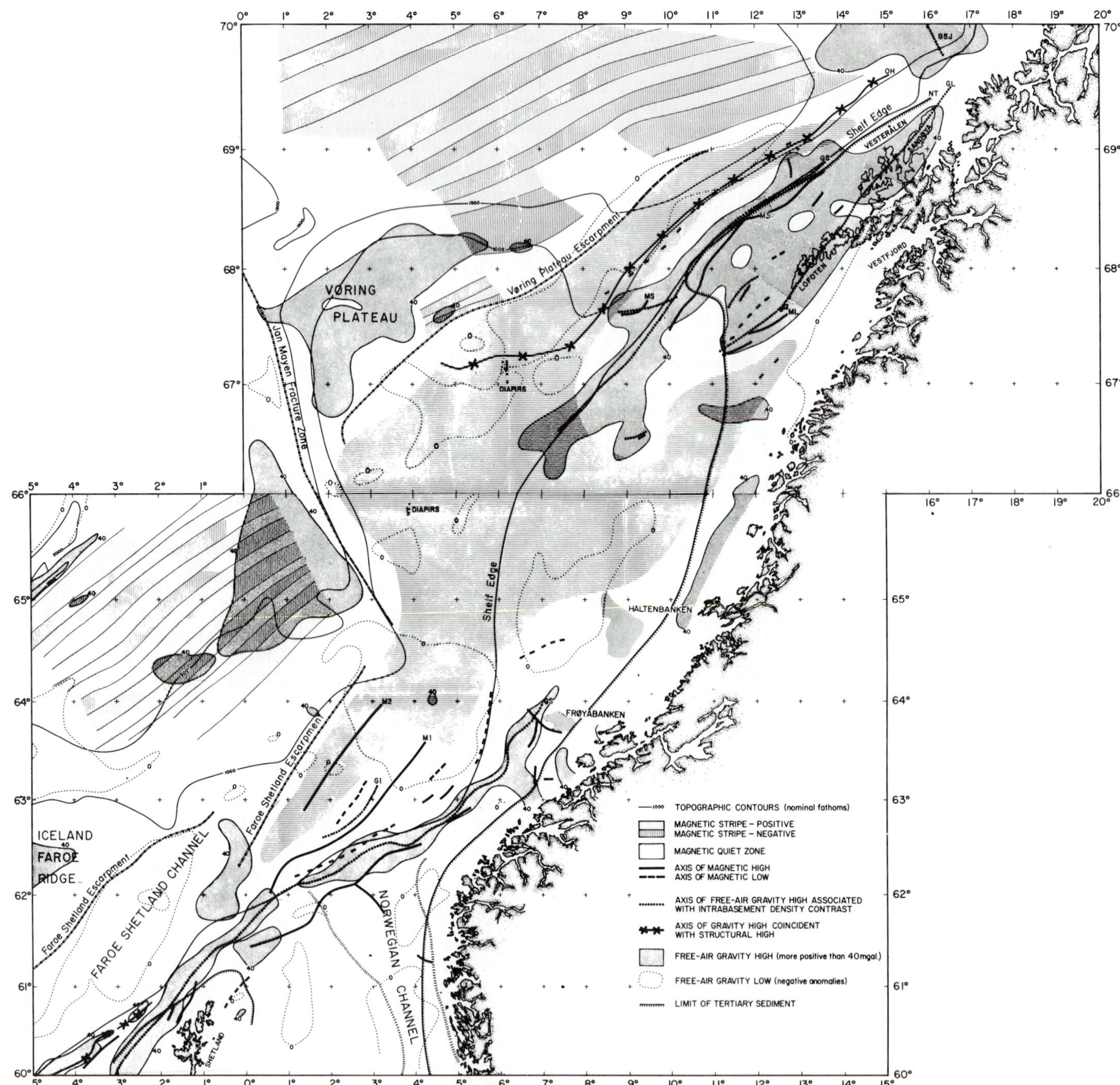

Fig. 6. Map showing prominent geophysical features of the Norwegian continental margin. The area seaward of the marginal escarpment has been produced by sea-floor spreading; the area landward is continental in character but has subsided. Near the shelf edge and on the shelf there are gravity and magnetic lineations which are caused primarily by intrabasement density and magnetization contrast, although a small (1 to 2 km) basement relief is probably also associated with these belts, which are Precambrian in age. The outer high and the gravity high closest to the shelf edge west of the Shetland Islands believed to be structural highs and therefore of different origin than the shelf highs (GS, MS). Note that gravity highs exist on the shelf edge seaward of the marginal escarpment. The area in between is characterized by gravity lows (Talwani and Eldholm, 1972).

GREENLAND MARGIN

Geophysical data collected on the ice island "Arlis" have been described by Ostenso (1968), and both Haines et al. (1970) and Ostenso and Wold (1971) have published aeromagnetic profiles crossing the margin. Although only a few of the Lamont-Doherty tracks cross the shelf edge, we have made some observations that are of interest.

Between the Jan Mayen and Greenland fracture zones the sedimentary structures on the continental rise and lower slope are generally conformable with the sea floor. However, between 75.3 and 76.4°N a basement high underlies the lower slope and is separated from the Greenland Fracture Zone by a sediment-filled trough. The high is covered by about

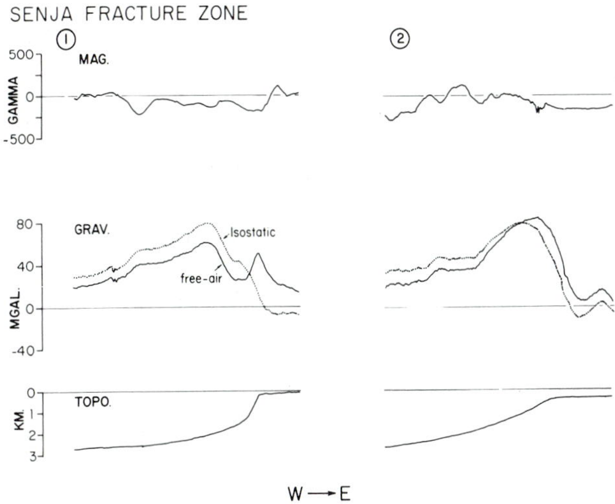

Fig. 7. Two geophysical profiles across the Senja Fracture Zone (see Fig. 1 for location). The profiles demonstrate that the prominent free-air high is not an edge effect and we interpret it as reflecting a major structural lineament, the Senja Fracture Zone.

0.6 km of sediment and its surface is smooth and highly reflective, similar to that of the Vøring Plateau basement high. In two crossings there are indications that this reflector dips steeply landward beneath a series of flat-lying beds on the landward side. We have no evidence of a basement high under the margin farther south. The basement apparently disappears underneath a thick cover of sediments near the base of the slope, although some of the crossings do indicate narrow basement peaks in this area. Figure 10 shows typical profiles for the northern and southern part of this margin. Seismic reflection data are not included; however, we have

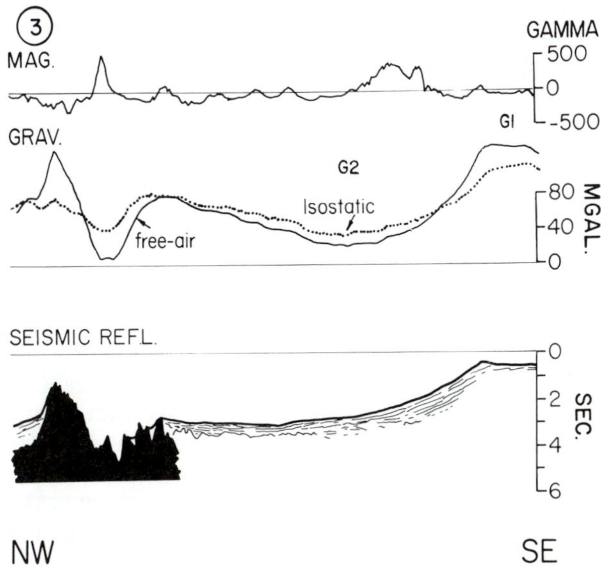

Fig. 8. Profile across the Svalbard margin (see Fig. 1 for location). Note that the sediments apparently are dammed by the axial mountains. The magnetic anomalies are small. Note also the positive gravity belt G1 and the gravity minimum G2.

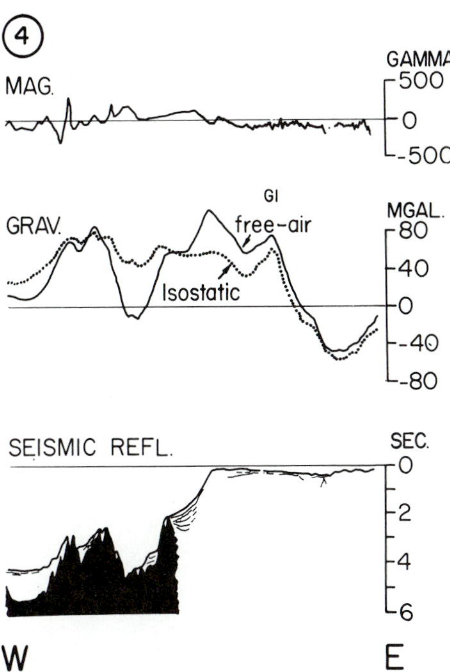

Fig. 9. Profile across the Svalbard margin at about 78°N (see Fig. 1 for location). The asymmetric position of the Knipovich Ridge in the northern part of the Greenland Sea places it very close to the margin.

shown the results of two sonobuoy stations. Station 112 is on top of the basement high, which has a velocity of 5.0 km/sec. Station 111 is on the continental shelf. If we consider these results in light of the section in Figure 5, the velocity structure at station 112 is typically oceanic and corresponds to the structure seaward of the Vøring Plateau Escarpment, whereas the velocity structure at station 111 corresponds to the landward side of the escarpments.

The free-air gravity field in the vicinity of the basement high (Fig. 10, profile 5) does not reflect the topographic relief of the lower slope. Computation of isostatic anomalies reveals a linear maximum (G3, Fig. 1) just seaward of the base of the slope. The high is bounded to the northwest by a steep gradient. On the other hand, profile 6 (Fig. 10), which lies south of the area of the basement high, reveals no prominent anomalies, and only a slight regional gradient is apparent in the isostatic gravity profile. The gravity map of Talwani and Grønlie (in press) reveals a linear free-air high near the shelf edge, but the isostatic calculations show that it is primarily an edge effect.

The Greenland margin between the Greenland Fracture Zone and the Jan Mayen Fracture Zone appears to be similar to the corresponding margin on the Norwegian side. The basement high on the Greenland side, established by a number of seismic refraction profile crossings, corresponds to the basement high on the Norwegian side. However, only two tracks define the marginal escarpment GE (Fig. 1) on the Greenland side. It might be possible to interpret the westernmost magnetic negative anoma-

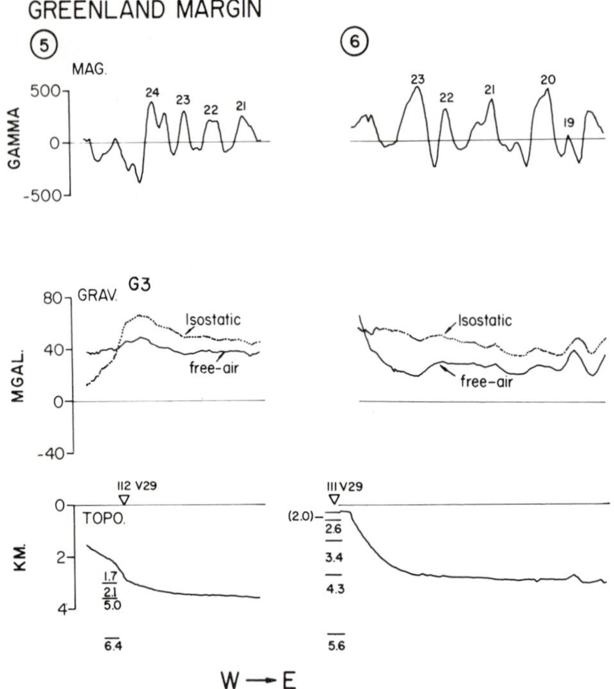

Fig. 10. Two geophysical profiles that typify the Greenland Margin between the Jan Mayen and Greenland fracture zones (see Fig. 1 for location). Seismic refraction information is projected onto the topographic profiles. There are indications of an escarpment (GE, Fig. 1) near the western end of profile 5, and the magnetic low is possibly an edge anomaly similar to the one at the Vøring Plateau escarpment (Fig. 5).

ly in profile 5 (Fig. 10) as an edge anomaly similar to the one at the Vøring Plateau Escarpment (Fig. 5). We also note that an aeromagnetic profile (Ostenso and Wold, 1971) indicates a change in character here with a smooth magnetic field on the landward side.

South of the Jan Mayen Fracture Zone there is a rapid increase in sediment thickness underneath the lower continental slope. The bedding is generally conformable with the sea floor on the slope and horizontal or dipping slightly seaward on the shelf. Sonobuoy stations on the shelf exhibit velocities similar to those in profile 111 (Fig. 10). A total sediment thickness of about 7 km is measured on the shelf at 69°N, overlying basement with velocity 5.2 km/sec. A common feature of the sonobuoy stations on the Greenland margin is that the low-velocity (less than 2.5 km/sec) sequence is thinner compared with the low-velocity sequency on the Norwegian margin.

The free-air gravity map shows a wide high (75-80 mgal) on the shelf. However, isostatic anomaly profiles across the margin are smooth with a low on the continental slope.

Magnetic anomaly 5 is well defined on either side of the Kolbeinsey Ridge. However, anomaly 5 on the western side does not follow the bathymetric trend of the margin (Johnson et al., 1972; Vogt et al., 1970), and older crust may exist between anomaly 5 and Greenland. Our profiles also show that the magnetic field changes character near the lower continental slope and that at least some landward portions of this margin may be classified as magnetically smooth.

RIFTED MARGIN OFF NORWAY

Talwani and Eldholm (1972) have described the importance of the Vøring Plateau and Faroe-Shetland escarpments. They noted, primarily on the basis of geophysical data, that these escarpments represent important structural boundaries. Sea-floor spreading types of magnetic anomalies exist seaward of these escarpments (anomaly 24 being the oldest identified anomaly), and typical oceanic basement, defined from both seismic refraction and reflection data, is found to continue west of the escarpments into the Norwegian Sea. All available evidence indicates that the sea floor west of the escarpments is oceanic in nature and has been generated by sea-floor spreading.

There is a magnetic quiet zone on the landward side of the escarpments (Fig. 6). The basement here is not only much deeper than the oceanic basement to the west but apparently also possesses different magnetic properties. The landward extent of the quiet zone is not characterized by any prominent structural boundary (Talwani and Eldholm, 1972).

The distribution of seismic velocities on the landward side of the escarpment is different from that reported in sediments typically deposited on oceanic crust. However, it is similar to the velocity distribution observed in the North Sea, Barents Sea, and the continental shelf off Norway. Sediments as old as Devonian exist in the North Sea, and the Upper Permian Zechstein evaporites are believed to exist between Scotland and Norway (Sorgenfrei, 1969) and are also found in Greenland (Haller, 1970). Hallam (1971) points out that a shallow seaway from the Arctic to Atlantic was well established by the Late Jurassic.

Studies of the sediments on Spitsbergen (Orvin, 1969), eastern Greenland (Haller, 1970), and the Barents Shelf (Freebold, 1951), together with the geophysical information, suggest that the margins are underlain by Mesozoic and possibly late Paleozoic sedimentary rocks. The Mesozoic ages are consistent with reports by Manum (1966), Os Vigran (1970), and Ørvig (1960) based on dredges and local outcrops on the continental shelf and upper slope. It is possible that the deepest sedimentary layer (4.4 km/sec) may represent evaporites and that the prominent diapiric structures found at the Vøring Plateau landward of the escarpment are associated with this layer. Piston cores taken above one of these diapir-like structures revealed recent sediment overlying sediments of Upper Eocene age (Bjørklund and Kellogg, 1972), but the age of the underlying diapiric material may, of course, be greater.

The seismic and magnetic evidence suggests that the marginal escarpments mark a major boundary both with respect to magnetic properties and basement age, favoring the idea that the area on the landward side is foundered continental crust. The fact that the boundary between magnetic and nonmagnetic (or weakly magnetic) crust is a sharp one is demonstrated by the agreement of the calculated edge anomaly with the observed negative anomaly over the escarpment (Fig. 5).

It is possible that while the marginal escarpments represent a major boundary, the area immediately to the east is not continental but represents oceanic crust formed during a much earlier phase of sea-floor spreading. Although we cannot discount this possibility, we have considered it less likely primarily for two reasons: first, it is difficult to obliterate the magnetization of a supposed oceanic crust, and second, there is a striking similarity of the velocity structure east of the escarpment to that of adjacent continental areas (see also Hinz, 1972).

Another important feature of this rifted margin is the presence of linear belts of gravity and magnetic anomalies lying in the vicinity of the shelf edge (Fig. 6). Grønlie and Ramberg (1970) interpreted the gravity anomalies in terms of basement relief, and Aam (1970) interpreted the magnetic anomalies as mainly being caused by intrabasement magnetization contrasts. Talwani and Eldholm (1972) discussed the available data and concluded that there was no evidence that these anomalies could be caused simply by varying depth to basement. The major gravity highs (GS and GL, Fig. 6) are believed to be related to intrabasement density contrasts, although a small amount of relief may be present, as seen in Figure 3.

Watts (1971) has shown that the gravity anomaly belt GS continues (Fig. 6) into the Lewisian basement of northwest Scotland, as does the magnetic belt (Flinn, 1969). In the Lofoten area, where the effect of the Caledonian orogeny has been minor, the anomaly may also be associated with Precambrian rocks. Although both the Lewisian basement and the Lofoten rocks have undergone a complex history of metamorphism, studies by Sutton and Watson (1969) and Heier and Compston (1969) indicate similar Precambrian dates for the metamorphic events in the two areas.

RIFTED MARGIN OFF GREENLAND

As mentioned earlier, geophysical data off Greenland are sparse. We note that profile 5 (Fig. 10) bears a resemblance to the profile across the Vøring Plateau Escarpment (Fig. 5). The isostatic gravity high lies on the seaward side of a basement high under which the seismic refraction data reveals oceanic structure. There is also some evidence that an escarpment similar to the Vøring Plateau Escarpment exists at the western end of the profile. Profile 6 (Fig. 10), which is farther south in an area where the oceanic magnetic anomalies exist quite close to the continental shelf, is similar to the Norwegian continental margin north of the Vøring Plateau Escarpment. Thus from the limited data available it seems reasonable to conclude that the Greenland margin is complementary to the Norwegian margin in the area lying between Greenland-Senja and the Jan Mayen Fracture Zone. On the Norwegian side the escarpment is developed in the southern portion of this segment, whereas on the Greenland side the indications for an escarpment exist in the northern part. On the other hand, sea-floor-spreading-type magnetic anomalies, which indicate oceanic floor, lie close to the shelf in the northern part on the Norwegian side but close to the shelf in the southern part on the Greenland side.

SHEARED MARGIN OFF THE SOUTHERN PART OF THE BARENTS SHELF

Figure 2 summarizes the interpreted origin of the Senja Fracture Zone as along the northern boundary of the primitive Greenland-Lofoten Basin. This figure suggests that a fracture zone that would be the counterpart of the Greenland Fracture Zone should be present along the continental slope off the Barents Shelf. Neither seismic reflection nor air-gun refraction measurements were able to penetrate to basement in this area of thick sedimentary cover, and hence we are unable to confirm the presence of this fracture. However, the gravity profile (Fig. 7) shows the presence of a very prominent isostatic anomaly high. Only two crossings are shown here, but several more were used to delineate this feature as shown in Figure 1, which is interpreted as a fracture zone that is the counterpart of the Greenland Fracture Zone.

MARGIN OFF SVALBARD

According to Figure 2, Greenland slid against Svalbard during the first phase of opening of the Norwegian Sea. With a small variation in the boundaries involved it is possible that the motion would have a compressive component, which would explain the evidence of compression during the early Tertiary along the western margin of Svalbard (Orvin, 1969). During the second phase of opening, Svalbard and Greenland split apart. The Knipovich Ridge, which at the present is the axis of sea-floor spreading, is asymmetrically close to Svalbard at its northern end, implying that the margin off Svalbard has been further modified by an eastward shift of the ridge axis. The above is not deduced from the measured geophysical parameters off Svalbard, but follows as a natural consequence if the proposed history of the evolution of the Norwegian Sea is correct.

This complicated history suggests that one would not expect the same crustal structure off Svalbard as off the relatively simple rifted margin off Norway. Profiles 3 and 4 (Figs. 8 and 9) show the presence of a positive gravity belt west of Svalbard. One can speculate that the gravity high originated in a manner similar to the Senja Fracture Zone during the early stage of opening of the Norwegian Sea. A likelier explanation, perhaps, is that it is similar to the shelf gravity high which extends from Scotland to the Lofoten-Vesteralen Islands. The gravity minimum G2 (Fig. 8) might possibly correspond to the minimum found east of the Vøring Plateau Escarpment. If that is true, G2 would roughly correspond to the line at which rifting was initiated in the Greenland Sea.

PRE-OPENING SEDIMENTARY BASIN

If one considers that the lines of initial rifting have been established, one can envisage the pattern of sediment deposition prior to the time rifting began. It appears that the rifting eventually began in nearly the center of a large basin that included the North Sea, the continental margin off Norway and Greenland, the Barents Shelf, and the Jan Mayen Ridge. Sediments started to accumulate in this basin in the late Paleozoic. Fluctuations of sea level with corresponding transgressions and regressions persisted until the end of the Cretaceous, forming an epicontinental sea in which deposition kept pace with the subsidence. (We believe that the water was shallow, because limited sampling has not turned up any deep-water deposits.) However, the variable thickness of pre-Tertiary sediments indicates that the amount of vertical movement was not uniform within this basin. It appears that the areas associated with high-density rocks have subsided the least (Talwani and Eldholm, 1972).

That a single large basin existed prior to the Tertiary is of considerable economic interest, since this basin included the oil and gas accumulations known in the North Sea. Certainly more detailed exploration on the Jan Mayen Ridge, off the coasts of Greenland and Norway, as well as on the Barents Shelf is important.

EPEIROGENIC MOVEMENTS

Vertical motions have played a very important role in the development of these continental margins. We have briefly mentioned above the vertical motions prior to rifting which had a major influence on distribution of the sediments. The vertical motions have also been of great importance subsequent to the time of opening, particularly a subsidence of the continental margins. The results on the Vøring Plateau suggest that sediments originally deposited in shallow water now lie at depths exceeding 1.5 km. The details of the exact timing and mechanism for the post-opening marginal subsidence are not known. One can speculate that as has been interpreted for other rifted margins, uparching occurred prior to rifting and the combined effects of erosion and cooling after the ridge axis moved away led to subsidence (Sleep, 1971).

If the material extruded at the time of rifting did not all come from directly below the rift, but came partially from the asthenosphere on either side, this would lead to subsidence and to a deficit of mass indicated by negative isostatic anomalies. Such negative anomalies exist on the Norwegian margin between the escarpments and the shelf edge, but are not large enough to indicate that all the subsidence has been caused by withdrawal of underlying material.

There appears also to be an important connection between the presence of linear gravity and magnetic anomaly belts coincident with the shelf edge and the extent of subsidence. These linear belts are associated with rocks of Precambrian age (Talwani and Eldholm, 1973). The circumstance of their location at the edge of the continental shelf of the Norwegian Sea, which opened much later (in the early Tertiary), suggests that these belts formed a hinge line with postopening marginal subsidence due to thermal and other effects confined to the seaward side. While overall subsidence and sedimentation took place over the entire margin, including the continental shelf and slope, the shelf break might be attributed to the larger differential subsidence occurring seaward of the hinge line. If these ideas are correct, rocks giving rise to the large gravity and magnetic anomalies at the edges of the continental shelf off Norway and elsewhere do not owe their existence to the circumstance of their location near the shelf edge; rather the location of the shelf edge is determined by the presence of these belts.

Hinz (1972) found velocities of 7.1 to 7.3 km/sec at a depth slightly greater than 10 km under the inner part of the Vøring Plateau. He attributes these velocities to an altered mantle under a ridge-type oceanic crust. However, we note that similar velocities lying between those associated with typical oceanic crust and oceanic mantle are often found in continental margin areas (Drake and Kosminskaya, 1969). Perhaps these velocities are related to the subsidence of the continental crust at the margin.

It seems plausible that the Cenozoic uplift of Fennoscandia and the east coast of Greenland is related to marginal subsidence, areas landward of the hinge lines having been elevated. The sedimentary structures on the Norwegian Shelf may be interpreted as indicating progressively increasing Cenozoic uplift landward; however, the rate of uplift appears to have decreased with time.

We have been concerned here with the vertical motions of the margin during the Tertiary, not with the well-known Fennoscandia uplift in postglacial times.

ACKNOWLEDGMENTS

This work was supported by grant GA-27281 from the National Science Foundation and by contract N00014-67-A-0108-0004 with the Office of Naval Research. We thank A.B. Watts and C. Windisch for critical reviews of the paper.

This report is Lamont-Doherty Geological Observatory Contribution 2098.

BIBLIOGRAPHY

Aam, K., 1970, Aeromagnetic investigations on the continental shelf of Norway, Stad-Lofoten (62-69°N): Norges. Geol. Undersokelse, no. 266, p. 49-61.

Anonymous, 1970, Anomalies de la pesanteur en mer de Nòrvege; résultats de mesures effectuées a bord du "Paul Goffény" 1965-1968: Cahier océanog., v. 22, p. 503-514.

Avery, O. E., Burton, G. D., and Heirtzler, J. R., 1968, An aeromagnetic survey of the Norwegian Sea: Jour. Geophys. Res., v. 73, p. 4583-4600.

Bakkelid, S., 1959, Gravity observations in a submarine along the Norwegian coast: Oslo, Norway Geograph. Survey, Geodetic Publ. 11, 28 p.

Bjørklund, K., and Kellogg, D., 1972, Five new Eocene radiolarian species from the Norwegian Sea: Micropaleontology, v. 18, p. 386-396.

Bott, M. H. P., and Watts, A. B., Deep structure of the continental margin adjacent to the British Isles, SCOR Symp., Cambridge: Inst. Geol. Sci. Rept. 70/14, p. 89-109.

Brooks, M., 1966, Regional gravity anomalies attributable to basic intrusions in orogenic belts: Geophys. Jour., v. 12, p. 29-31.

______, 1970, A gravity survey of coastal areas of West Finnmark, northern Norway: Geol. Soc. London Quart. Jour., v. 125, p. 171-192.

Cochran, J. R., 1973, Gravity and magnetic investigations in the Guiana Basin, western equatorial Atlantic: Geol. Soc. America Bull., v. 84, p. 3249-3268.

Collette, B. J., 1960, The gravity field of the North Sea, *in* Vening Meinesz, F. A., ed., Gravity expeditions, 1948-1958, v. 5, p. 47-96.

Drake, C. L., and Kosminskaya, I. P., 1969, The transition from continental to oceanic crust: Tectonophysics, v. 7, no. 5-6, p. 363-384.

Eldholm, O., 1970, Seismic refraction measurements on the Norwegian continental shelf between 62°N and 65°N: Norsk Geol. Tiddsskr., v. 50, p. 215-229.

______, and Ewing, J., 1971, Marine geophysical survey in the southwestern Barents Sea: Jour. Geophys. Res., v. 76, p. 3832-3841.

______, and Nysaether, E., 1969, Seismiske undersøkelser pa den norske kontinentalsokkel 1968: Tekn. Rapp. 2C, Jordskjelvstasjonen, Univ. i Bergen, 17 p.

______, and Windisch, C., (in press), The sediment distribution in the Norwegian-Greenland Sea: Geol. Soc. America Bull.

Ericson, D. B., Ewing, M., and Wollin, G., 1964, Sediment cores from the Arctic and the subarctic: Science, v. 144, p. 1183-1192.

Ewing, J., and Ewing, M., 1959, Seismic refraction measurements in the Atlantic ocean basins, in the Mediterranean Sea, on the Mid-Atlantic Ridge, and in the Norwegian Sea: Geol. Soc. America Bull., v. 70, p. 291-318.

Fleischer, U., 1971, Gravity surveys over the Reykjanes Ridge, and between Iceland and the Faroe Islands: Marine Geophys. Res., v. 1, p. 314-327.

Flinn, D., 1969, A geological interpretation of the aeromagnetic maps of the continental shelf around Orkney and Shetland: Geol. Jour. v. 6, p. 279-292.

Freebold, H., 1951, Geologie des Barentsschelfes: Abhandl. Deut. Akad. Wiss. Berlin., no. 5, p. 1-150.

Grønlie, G., and Ramberg, I. B., 1970, Gravity indications of deep sedimentary basins below the Norwegian continental shelf and the Vøring Plateau: Norsk Geol. Tidsskr., v. 50, p. 357-391.

Haines, G. V., Hannaford, W., and Serson, P. H., Magnetic anomaly maps of the Nordic countries and the Greenland and Norwegian seas: Dominion Observatory Ottawa Publs., v. 39, no. 5, p. 123-149.

Hall, D. H., and Dagley, P., 1970, Regional magnetic anomalies: an analysis of the smoothed aeromagnetic map of Great Britain and Northern Ireland: Inst. Geol. Sci. Ann. Rept. 70/10.

Hallam, A., 1971, Mesozoic geology and the opening of the North Atlantic: Jour. Geol., v. 79, p. 128-157.

Haller, J., 1970, Tectonic map of East Greenland: Medd. Grønland, v. 171, 286 p.

Harland, W. B., 1969, Contribution of Spitsbergen to understanding of tectonic evolution of North Atlantic region: Am. Assoc. Petr. Geol. Mem. 12, p. 817-851.

Heier, K. S., and Compston, W., 1969, Interpretation of Rb-Sr age patterns in high-grade metamorphic rocks, north Norway: Norsk Geol. Tidsskr, v. 49, p. 257-383.

Heybroek, P., Haanstra, U., and Erdmann, D. A., 1967, Observations of the geology of the North Sea area: 7th World Petr. Congr. Proc., v. 2, p. 905-916.

Hinz, K., 1972, The seismic crustal structure of the Norwegian continental margin in the Vøring Plateau, in the Norwegian deep sea and on the eastern flank of the Jan Mayen Ridge between 66 and 68°N; Internatl. Geol. Congr., Montreal, sec. 8, p. 28-37.

Holtedahl, H., 1955, On the Norwegian Continental Terrace, primarily outside Mør-Romsdal: its geomorphology and sediments: Univ. i Bergen, Arbok, Nat. Vit. Rekke, no. 14.

Holtedahl, O., 1960, On supposed marginal faults and the oblique uplift of the land mass in Cenozoic time, *in* Geology of Norway: Oslo, Norges Geol. Undersøkelser, no. 208, p. 351-357.

Hornabroek, J. T., 1967, Seismic interpretation problems in the North Sea with special reference to the discovery well: 48/6-1: 7th World Petr. Congr. Proc., v. 2, p. 837-856.

Ivanov, M. M. 1967, Some peculiarities of the magnetic field in the eastern Norwegian Sea: Oceanology, v. 7, p. 363-372.

Johnson, G. L., and Eckhoff, O. B., 1966, Bathymetry of the North Greenland Sea: Deep-Sea Res., v. 13, p. 1161-1173.

______, and Heezen, B. C., 1967, Morphology and evolution of the Norwegian-Greenland Sea: Deep-Sea Res., v. 14, p. 755-771.

______, Ballard, J. A., and Watson, J. A., 1968, Seismic studies on the Norwegian continental margin: Norsk Polarinst. Arb. 1966, p. 112-119.

______, Freitag, J. S., and Pew, J. A., 1971, Structure of the Norwegian basin: Norsk Polarinst. Arb. 1969, p. 7-16.

______, Southall, J. R., Young, P. W., and Vogt, P. R., 1972, The origin and structure of the Iceland Plateau and Kolbeinsey Ridge: Jour. Geophys. Res., v. 77, p. 5688-5696.

Kogan, L. I., and Milashin, A. P., 1970, Seismic investigations in the Greenland Sea: Oceanology, p. 358-360.

Kvale, A., Sellevoll, M. A., and Gammelsaeter, H., 1966, Seismic refraction measurements on the Norwegian continental shelf at 63°N, 06°30'E.: Univ. i Bergen, Jordskjelvstasjonen.

Litvin, V. M., 1965, Origin of the bottom configuration of the Norwegian Sea: Oceanology, v. 5, p. 90-96.

Manum, S., 1966, Deposits of probable Upper Cretaceous age offshore from Andøya, northern Norway: Norsk Geol. Tidsskr., v. 46, p. 246-247.

Nansen, F., 1904, The bathymetrical features of the North Polar Seas, with a discussion of the continental shelves and previous oscillations of the shoreline, Norwegian North Polar Expedition, 1893-1896: Sci. Rees., v. 4, p. 1-232.

Nysaether, E., Eldholm, O., and Sundvor, E., 1969, Seismiske undersøkelser av den norske kontinentalsokkel: Sklinnabanken-Andøya, Tekn. Rapp. 3, Jordskjelvstasjonen, Univ. i Bergen, 15 p.

Ørvig, T., 1960, The Jurassic and Cretaceous of Andøya in northern Norway, *in* Holtedahl, O., ed., Geology of Norway: Norges. Geol. Undersokelse, no. 208, p. 344-350.

Orvin, A. K., 1969, Outline of the geological history of Spitsbergen: Norges Svalbard-og Ishavs-undersøkelser, no. 78, 57 p.

Ostenso, N. A., 1968, Geophysical studies in the Greenland Sea: Geol. Soc. America Bull., v. 79, p. 107-132.

———, and Wold, R. J., 1971, Aeromagnetic survey of the Arctic Ocean: techniques and interpretations: Marine Geophys. Res., v. 1, p. 178-219.

Os Vigran, J., 1970, Fragments of a Middle Jurassic flora from northern Trøndelag, Norway: Norsk Geol. Tidsskr, v. 50, p. 193-214.

Romey, W. D., 1971, Basic igneous complex, margerite, and high-grade gneisses of Flakstadøy, Lofoten, northern Norway: I. Field relations and speculations on origins: Norsk Geol. Tidsskr., v. 5., p. 33-61.

Saito, T., Burckle, L. H., and Horn, D. R., 1967, Paleocene core from the Norwegian Basin: Nature, v. 216, p. 357-359.

Sleep, N. H., 1971, Thermal effects of the formation of Atlantic continental margin by continental break up: Geophys. Jour. v. 24, p. 325-350.

Sorgenfrei, T., 1969, Geological perspectives in the North Sea area: Geol. Soc. Denmark Bull., v. 19, p. 160-196.

Sundvor, E., 1971, Seismic refraction measurements on the Norwegian continental shelf between Andøya and Fugløbanken: Marine Geophys. Res., v. 1, p. 303-313.

Sutton, J. and Watson, J., 1969, Scourian-Laxfordian relationships in the Lewisian of northwest Scotland: Geol. Assoc. Can. Spec. Paper 5, p. 119-128.

Talwani, M., and Eldholm, O., 1971, The continental margin off Norway and the evolution of the Norwegian Sea (abstr.): Geol. Soc. America Ann. Meeting, v. 3, no. 7, p. 728.

———, and Eldholm, O., 1972, The continental margin off Norway: a geophysical study: Geol. Soc. America Bull., v. 83, p. 3575-3606.

———, and Eldholm, O., 1973, The boundary between continental and oceanic crust at the margin of rifted continents: Nature, v. 241, p. 325-330.

———, and Grønlie, G., (in press), A free-air gravity map of the Norwegian-Greenland Sea, Jour. Geoph. Res.

Vogt, P. R., Ostenso, N. A., and Johnson, G. L., 1970, Magnetic and bathymetric data bearing on sea floor spreading north of Iceland: Jour. Geophys. Res., v. 75, p. 903-920.

Watts, A. B., 1971, Geophysical investigations on the continental shelf and slope north of Scotland: Scot. Jour. Geol., v. 7, p. 189-218.

Wyrobek, S. M., 1969, General appraisal of velocities of the Permian Basin of northern Europe, including the North Sea: Jour. Inst. Petr. London, v. 55, p. 1-13.

Insular Margins of Iceland

Gudmundur Pálmason

INTRODUCTION

Iceland, with an area of about 100,000 km^2, straddling the Mid-Atlantic Ridge and the Greenland-Scotland rise, is surrounded by a shelf of shallow depth, mostly 100-300 m, with an area larger than that of Iceland itself (Fig 1). The shelf edge is in some places relatively steep, especially to the southeast and northeast. The bathymetry is more complicated where the crest of the Mid-Atlantic Ridge enters the shelf, to the southwest and to the north. In some parts of the shelf shallow, wide valleys are found with a relief of 100-150 m, many of them located in front of the main present-day rivers on the shore.

Recent geophysical surveys of the Icelandic insular shelf are described and some results presented, in particular from the southern part of the shelf. It is concluded that the shelf is built chiefly of basalts and that sediments from land have been deposited mainly seaward of the shelf edge. The margin has been modified in some places by later volcanism which, together with sediment deposition, has extended the shelf farther out.

THE INSULAR SHELF

Until recently, relatively little work has been done to investigate the nature of the insular margin. Einarsson (1963) discussed the possible origin of the shelf, mainly on the basis of available bathymetric data. He concluded that the outer part of the shelf would probably consist of sediments derived from land. The average total denudation of the present land area would amount to some 400 m. This would lead to isostatic readjustments, rise of the central part of the country, and sinking of the shelf area. The isostatic rise is consistent with the height of the highest base levels recognizable in the present topography of Iceland.

Geophysical surveys made on the Reykjanes Ridge (Talwani et al., 1971), the Iceland-Faeroes Ridge (Bott et al., 1971; Fleischer, 1971; Johnson and Tanner, 1971), and the Kolbeinsey Ridge (Meyer et al., 1972) have given information on certain parts of the shelf. In particular, the work of Johnson and Tanner (1971) shows that sediments up to 1,000 m thick are found off the edge of the southeastern insular shelf, but only very thin sediments are indicated on the shelf itself.

MODERN GEOPHYSICAL SURVEYS

The first comprehensive geophysical survey of the shelf was made in 1972-1973 as a collaborative effort between several institutions in Iceland and the United States. The survey covered an area some 100 km wide from the coast. Survey lines were spaced at 10-km intervals and oriented mainly perpendicular to the coastline. Several cross lines were also run to test the accuracy of the survey. Positioning was by Raydist distance measurements from two mobile stations on the shore. Gravity and bathymetric data were obtained for the whole area. Magnetic data were collected from the southwest and part of the northern shelf, and seismic reflection profiles were run along some of the survey lines. Some of the magnetic data from the western part of the shelf have been discussed by Thors and Kristjansson (1974).

GEOPHYSICAL INTERPRETATION

Most of the data from the above surveys have been only partly evaluated at the present time. The discussion of this paper will be mainly confined to the gravity data, collected in a cooperative effort between the National Energy Authority of Iceland (Orkustofnun) and the Defense Mapping Agency Topographic Center, Washington, D.C. Some seismic reflection data and borehole data will also be discussed.

A free-air anomaly map has been drawn in a preliminary form of the whole survey area. The southern part of this map is shown in Figure 2. The most conspicuous feature of the map is a relatively sharp gravity variation associated with the edge of the shelf, particularly off the southeastern and northeastern coasts. An elongated gravity maximum occurs landward of the shelf edge. The effect of the shelf edge is superimposed on a regional gravity field which probably reflects deeper compensating density variations, associated with the elevation of Iceland relative to the surrounding ocean floor. In looking at the free-air anomalies in the Iceland area it should be kept in mind that the regional gravity field is high over the whole ocean between Greenland and Europe (Talwani et al., 1971), possibly with a maximum in the area of Iceland.

The crestal zones of the Mid-Atlantic Ridge southwest and north of Iceland are clearly reflected

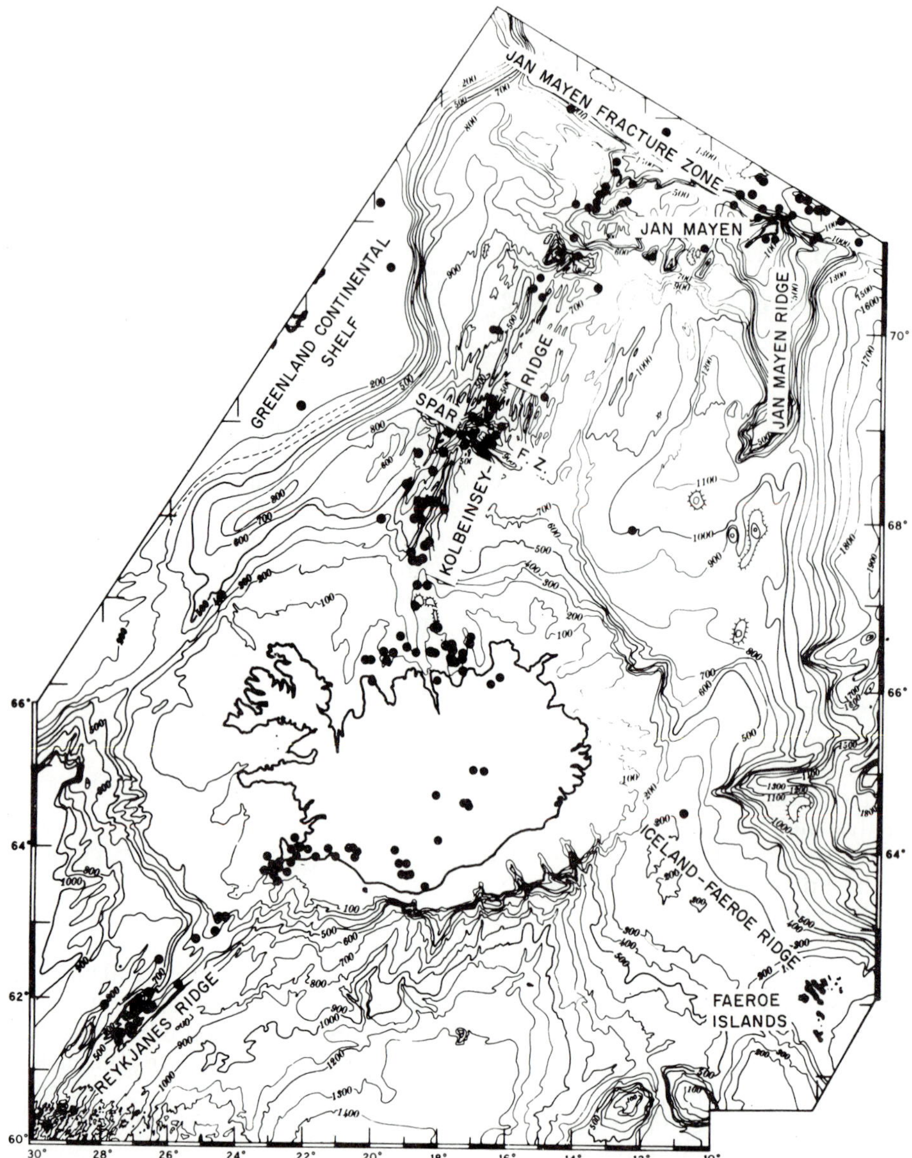

Fig. 1. Bathymetric map of the Iceland area, showing the insular shelf and the epicenter belt along the axis of the Mid-Atlantic Ridge. Bathymetry from Johnson et al. (1972). Depths are in nominal fathoms (1 fm/400 sec travel time). Modified from Pálmason and Saemundsson (1974).

in the free-air anomalies. The Reykjanes Ridge axis on the shelf southwest of Iceland is characterized by a gravity high of 60-70 mgal (Fig. 2), flanked by gravity lows on both sides. This could be an effect of deep compensation beneath the ridge axis, or it could be due to lighter material in the shelf adjacent to the ridge axis. A gravity field of this type can be produced by ridge models based on thermal expansion of the lithosphere (Lambeck, 1972).

North of Iceland, where the Kolbeinsey Ridge joins the Tjörnes Fracture Zone (Saemundsson, 1974) with the volcanic zone of northeastern Iceland, the ridge axis is also characterized by a gravity high (Fig. 3). The strongest gravity variations occur in the Tjornes Fracture Zone, and they seem generally to support the pattern of stepped faults as outlined by Saemundsson (1974). A particularly noteworthy feature in this area is a strong gravity minimum off Eyjafjordur in central northern Iceland. This minimum is elongated in a west-northwest direction and is probably related to the Húsavik faults, which form the southern boundary of the Tjornes Fracture Zone.

It is instructive to look closer at the insular margin south of Iceland between the Iceland-Faeroes Ridge and the Reykjanes Ridge. East of about 20°W longitude the shelf forms an oval shape, relatively steep-edged, mostly about 1,000 m high, but decreasing to some 200 m on the Iceland-Faeroes Ridge. The shelf is relatively flat, its main topographic features being several shallow valleys extending landward from the edge of the shelf. The width of the shelf increases eastward. To the west of 20°W the shelf broadens and the edge becomes smoother.

Several reflection profiles have been made across the edge of the insular shelf off southern Iceland. Johnson and Tanner (1971) measured a few

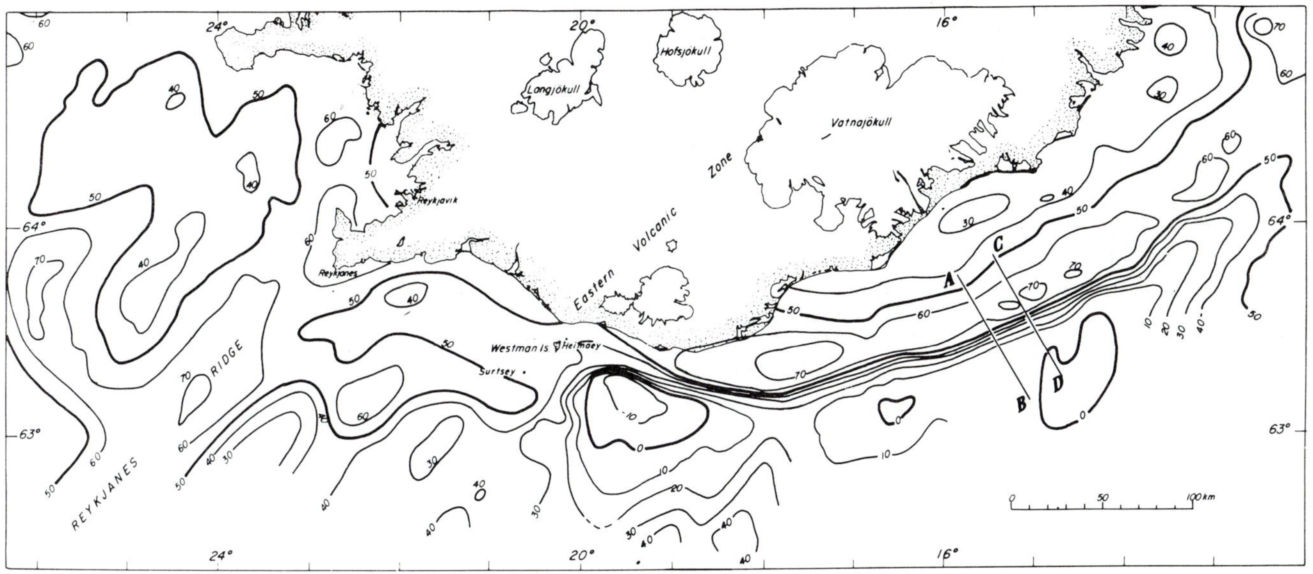

Fig. 2. Free-air gravity anomaly map of the southern insular shelf. A-B and C-D show the location of the profiles in Figure 4.

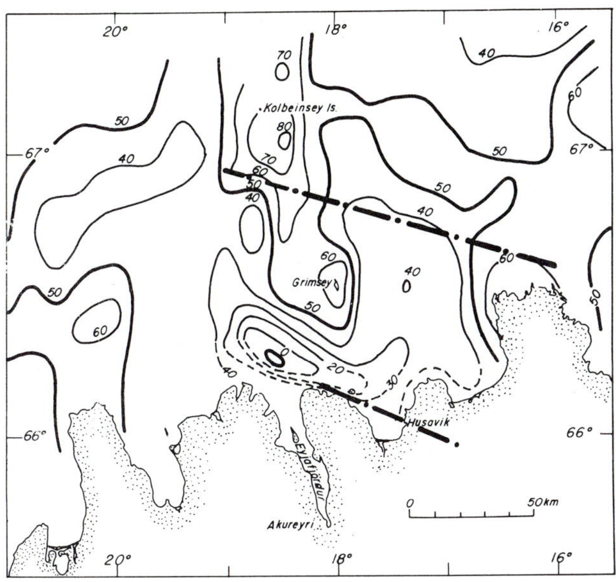

Fig. 3. Free-air gravity anomaly map of the central northern Icelandic insular shelf. The straight dash-dot lines show the boundaries of the Tjörnes Fracture Zone, as suggested by Saemundsson (1974).

east-west profiles crossing the eastern part of the shelf edge near the Iceland-Faeroes Ridge. Sediment thickness was small on the shelf, but up to 1,000 m of sediments was found just off the shelf. In 1973 two reflection profiles were run perpendicular to the shelf edge from the USSR research ship AKADEMIK KURCHATOV; the location of these profiles is shown in Figure 2 and bathymetric and free-air gravity anomaly profiles in Figure 4. The location of the two profiles was chosen to give information both along a shelf valley (western profile, A-B) and between valleys (eastern profile C-D). No major difference was found between the two profiles. Sediment thickness appears to be less than 100 m on the shelf.

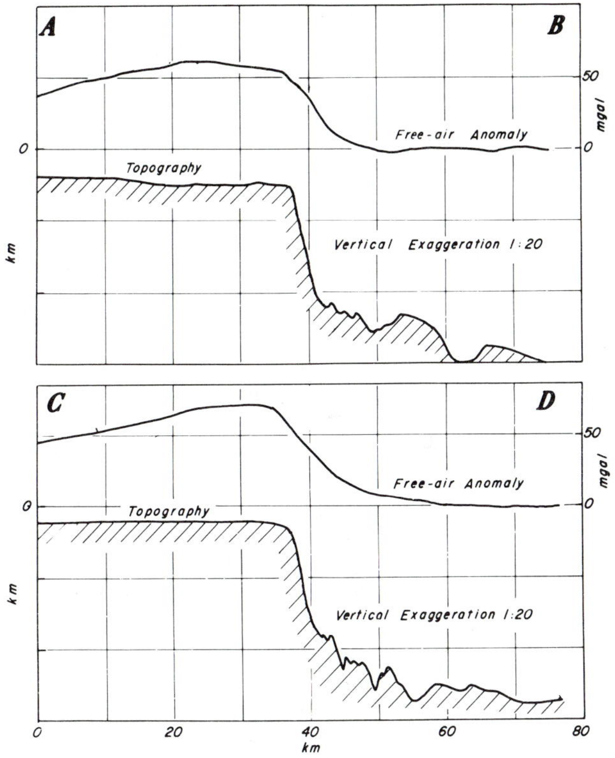

Fig. 4. Free-air gravity anomalies and topography along lines A-B and C-D in Figure 2.

The shelf edge has a slope of about 15° between about 200 and 900-1,000 m depth. At greater depths, the sediment bottom slopes about 1.5° near the shelf edge, decreasing outward from the shelf.

The gravity survey gives significant information concerning the nature of the shelf. The free-air anomaly profiles in Figure 4 are typical of similar profiles across the southeast Iceland shelf. From regional gravity surveys in Iceland (Einarsson, 1954;

and later unpublished work) it is known that the Bouguer anomaly has a minimum in central Iceland and that along the coast there is an outward gradient in the Bouguer values. This outward gradient is apparent over the more or less horizontal shelf in the profiles in Figure 4, and there to 0.7-0.9 mgal/km. The gravity effect of the shelf edge is superimposed on this regional trend. On the assumption that the gradient in the Bouguer values is due to deep compensation and does not vary appreciably over the shelf-edge area, it is possible to estimate the gravity variation due to the edge of the shelf. This is found to be about 70-90 mgal. By interpreting this in terms of a step model of thickness 1 km, it is found that the necessary density of the shelf material must be 2.7-3.1 g/cm^3. This is rather high, even for a basaltic lava pile, and shows that the shelf probably consists of basalt lavas without appreciable amounts of sediments. This result is suggested by the available seismic reflection data as well as by the relatively steep slope, about 15° of the shelf edge.

Turning to the part west of 20°W of the insular shelf south of Iceland, it is immediately clear that the physiography has a different character from that of the eastern part. The shelf widens suddenly and the edge slope is not as steep as farther east. The volcanic Westman Islands show that this is an area of active volcanism, the last eruptions being the Surtsey eruption in 1963-1967 and the Heimaey eruption in 1973. An important evidence of the nature of this part of the shelf comes from a 1,565-m deep hole, drilled in 1964 in Heimaey (Pálmason et al., 1965; Tómasson, 1967). The uppermost 180 m is comprised of volcanics, which form the base of the Westman Islands. This is followed by about 600 m composed mostly of sediments, containing foraminiferous shells, and thus deposited in water (Einarsson, 1968). The lowest 500 m or so in the hole is comprised of basalt lavas, considered to be of Tertiary age.

The structure of the shelf around the Westman Islands is thus different from that indicated by seismic reflection and gravity data for the eastern part of the shelf. Seismic-refraction measurements on the shore opposite the Westman Islands do not indicate major sediment thickness at shallow depth, comparable to that observed in the borehole (Pálmason, 1971). This makes it likely that the shelf edge as observed farther east continues westward between the Westman Islands and the southern coast of Iceland (Palmason et al, 1965). The widening of the western part of the shelf would then be a later event, caused by volcanism and sediment deposition.

Recent studies of the history of the volcanic zones in Iceland appear to throw some light on the sequence of events that led to the formation of the wider western part of the southern Iceland insular shelf. Stratigraphic and paleomagnetic studies (Saemundsson, 1974) and heat-flow studies (Pálmason, 1973) indicate that the eastern volcanic zone in Iceland is a relatively young feature, a few million years old. Paleomagnetic pole positions from Icelandic data also indicate an eastward shift of the spreading axis (Wilson and McElhinny, 1974). Most of the Icelandic lava pile should then, on the plate tectonics hypothesis, have been accreted in the western (Reykjanes-Langjökull) volcanic rift zone, which forms the landward continuation of the Reykjanes Ridge axial zone . The regular oval form of the eastern part of the southern shelf probably reflects the history of subaerial volcanism in the western zone, proceeding from a relatively small area near the present center of the Icelandic insular shelf and gradually extending southwest and northeast toward the edge of the present insular shelf. Shore erosion then gradually modified the coastline. When volcanic activity began in the present eastern zone, it extended somewhat the southern margin of the shelf edge at that time; this, together with sedimentary deposition from land, accounts for the present extra width of the shelf.

Our discussion has been confined mainly to the southern part of the Icelandic insular margin. The available data indicate that the southern margin of Iceland is built chiefly of basalts and that sediments from land have been deposited mainly in front of the shelf margin. Volcanic activity later modified the edge of the shelf. The shallow, more-or-less-constant-depth shelf would have been formed by shore erosion combined with isostatic readjustments.

It is still too early to discuss other parts of the Icelandic shelf in any detail. Further studies of the data collected in the 1972-1973 marine survey as well as data from other surveys will probably throw further light on the origin of the insular shelf. It appears clear at the present time that sedimentary thickness is small on large parts of the shelf itself but increases beyond the shelf edge. Recent unpublished data from the southern part of the western shelf, northwest of the Reykjanes Ridge axis, show that the outer part of the shelf there contains relatively thick sediments (Johnson and Pálmason, in preparation). This is in agreement with what one would expect in that area on the assumption that the original basaltic shelf material was formed by a process of ocean-floor spreading, symmetrically around the ridge axis in the direction of spreading. The greater width of the shelf northwest of the ridge axis, compared with the shelf southeast of the ridge, would then be due to greater sediment accumulation.

As an additional example of sedimentary thickness may be cited a reflection profile across the northwestern part of the shelf, measured in 1973 by the research vessel KURCHATOV. Very thin sediments are indicated to a distance of about 85 km from the coast. Beyond that the water depth increases and sedimentary thickness also. In the channel between the Iceland Shelf and the Greenland Continental Margin, where the water depth is about 640 m, some 500-600 m of sediments is indicated.

ACKNOWLEDGMENTS

The gravity data discussed in this paper were collected in a joint project by the Iceland National Energy Authority and the Defense Mapping Agency Topographic Center, Washington, D.C., within the framework of the comprehensive geophysical survey coordinated by the Icelandic Insular Shelf Committee. The author is also indebted to Gleb Udintsev, Chief Scientist on the R/V AKADEMIK KURCHATOV, for obtaining and making available seismic-reflection profiles across the insular margin.

M. Langseth, Jr., and A.B. Watts read the paper and made critical comments, for which the author is grateful. The paper was written during the author's stay at Lamont-Doherty Geological Observatory as a Vetlesen Visiting Professor. This is Lamont-Doherty Geological Observatory Contribution 2104.

BIBLIOGRAPHY

Bott, M. H. P., Browitt, C. W. A., and Stacey, A. P., 1971, The deep structure of the Iceland-Faeroe Ridge: Marine Geophys. Res., v. 1, p. 328-351.

Einarsson, T., 1968, Jardfraedi (geology): Mal og menning (Reykjavik), 335 pp.

Einarsson, T., 1954, A survey of gravity in Iceland: Soc. Sci. Islandica, rit 30, 22 pp.

______, 1963, On submarine geology around Iceland: Natturufraedingurinn, v. 32, p. 155-175. (In Icelandic with an English summary.)

Fleischer, U. 1971, Gravity surveys over the Reykjanes Ridge and between Iceland and the Faeroe Islands: Marine Geophys. Res., v. 1, p. 314-327.

Johnson, G. L., and Tanner, B., 1971, Geophysical observations on the Iceland-Faeroe Ridge: Jökull (Reykjavik), v. 21, p. 45-52.

______, Southall, J. R., Young, P. W., and Vogt, P. R., 1972, Origin and structure of the Iceland Plateau and Kolbeinsey Ridge: Jour. Geophys. Res., v. 77, p. 5688-5696.

Lambeck, K., 1972, Gravity anomalies over ocean ridges: Geophys. Jour., v. 30, p. 37-53.

Meyer, O., Vopper, D., Fleischer, U., Closs, H., and Gerke, K., 1972, Results of bathymetric, magnetic and gravimetric measurements between Iceland and 70° N.: Deut. Hydrograph. Z., v. 25, H. 5, p. 193-201.

Pálmason, G., 1971, Crustal structure of Iceland from explosion seismology: Sco. Sci. Islandica, rit 40, 187 pp.

______, 1973, Kinematics and heat flow in volcanic rift zone, with application to Iceland: Geophys. Jour., v. 33, p. 451-481.

______, and Saemundsson, K., 1974, Iceland in relation to Mid-Atlantic Ridge: Ann. Rev. Earth and Planetary Sci., v. 2,

______, Jonsson, J., Tomasson, J., and Jonsson, I., 1965, Deep drilling in the Westman Islands (In Icelandic): Reykjavik, Raforkumalastjori, 43 pp. (mimeographed).

Saemundsson, K., 1974, Evolution of the axial rifting zone in northern Iceland and the Tjornes fracture zone: Geol. Soc. America Bull., In press.

Talwani, M., Windisch, C. C., and Langseth, M.G., Jr., 1971, Reykjanes Ridge crest: a detailed geophysical study: Jour. Geophys. Res., v. 76, p. 473-517.

Thors, K., Kristjansson, L., 1974, Westward extension of the Snaefellsnes volcanic zone in Iceland. In press.

Tómasson, J., 1967, On the origin of sedimentary water beneath Westmann Islands: Jökull (Reykjavik), v. 17, p. 300-311.

Wilson, R. L., and McElhinny, M. W., 1974, Investigation of the large scale palaeomagnetic field over the past 25 million years: eastward shift of the Icelandic spreading ridge. In press.

The Continental Margins of Eastern Canada and Baffin Bay

C. E. Keen and M. J. Keen

INTRODUCTION

Canada's eastern continental margin extends over some 40 degrees of latitude from Georges Bank to Ellesmere Island. This account describes only a few of the features of this extensive continental margin, illustrative of the variations in geologic style. Full accounts have been written recently by a number of investigators (e.g., Barrett and C. E. Keen, 1973; Emery et al., 1970); consequently we have not put detailed references throughout this text.

The margin off Nova Scotia south of the Grand Banks is a rifted margin, according to the plate tectonics model, formed at a rather early stage in the history of the modern Atlantic Ocean. The margins of Labrador Sea and Baffin Bay also formed by rifting, but at a later time. The Nova Scotia margin and the Labrador Sea margin are separated from each by the Grand Banks, whose southern edge formed by transform faulting. Baffin Bay is separated from the Labrador Sea by a sill in the Davis Strait, which may have been a "hot-spot" in the early Tertiary. Baffin Bay terminates at its northern end in a series of fault-bounded, inter-island channels, much as the Red Sea terminates in the Gulf of Aquaba and the Gulf of Suez (Fig. 1).

MARGIN OFF NOVA SCOTIA

The Mesozoic and Cenozoic sediments of the Atlantic Coastal Plain extend beneath the shelf north of Cape Cod and, with Carboniferous and Permo-Triassic rocks off Cape Breton Island, form much of the shelf off Nova Scotia. The Coastal Plain sediments form a rather crude wedge which thickens seaward, but this wedge is broken into more-or-less isolated basins, and is intruded by diapiric structures. The sedimentary pile is thickest beneath the continental slope, where the thickness is known, from seismic refraction measurements, to be 8 km (Figs. 2 and 3). This sedimentary wedge is limited on its seaward side by a "ridge complex," seen clearly on seismic reflection records (Fig. 4). Shoreward of the ridge complex lies the sedimentary wedge, derived at least in part from terrestrial erosion. Seaward of the ridge complex is the typical sediment of the Sohm Abyssal Plain and the continental rise. Oceanic "basement" is overlain there by some 3-5 km of sediments which thin to the southeast, toward the Mid-Atlantic Ridge, and thicken to the north, toward the Grand Banks of the Laurentian fan.

The boundary marked by the ridge complex is also associated with the magnetic slope anomaly which can be traced with ease along much of the eastern seaboard of the United States and Nova Scotia (Fig. 4). The magnetic field east of the slope anomaly is devoid of large anomalies and is typical of the "quiet zone," where anomalies larger than 80 gammas are rare (Fig. 5).

The crust beneath the Sohm Abyssal Plain has a seismic structure typical of ocean basins. The velocities in the lower part of the crust here, and below the Mohorovicic discontinuity, are typical of layer 3 and of the mantle of ocean basins. The top of the oceanic crust—acoustic basement—becomes deeper to the west, toward the continent, and the depth to the Mohorovicic discontinuity increases concomitantly. Beneath the slope anomaly, however, the seismic results are no longer typical of ocean basins. No evidence has been found for "typical" layer 2; it must be thin or absent. Refraction data show that sediments appear to lie directly upon rocks in which the velocity is higher than is usual for layer 3 and lower than is usual for oceanic mantle (Fig. 3).

This area appears then to be the true transition zone between continent and ocean, some 70 km wide, as far as we know from the seismic lines that are available now. The results can be used as a guide in modeling the magnetic slope anomaly.

It seems likely that rocks comparable to the meta-sedimentary rocks of the Meguma Group of Nova Scotia underlie sediments west of the slope anomaly. Their magnetization is known. It is probable that basalts lie beneath sediments of the Sohm Abyssal Plain, and their magnetization can be estimated from results of sampling on the Mid-Atlantic Ridge. The seismic results in the boundary region suggest that this region is underlain by intrusive igneous rocks (with high velocities) which will have a much lower magnetization than in basalts typical of modern oceanic ridges. The juxtaposition of continental meta-sediments, intrusive igneous rocks, and oceanic basalts produces a double-peaked anomaly (Fig. 6).

The quiet zone beyond the slope anomaly appears to be quiet because few field reversals took place during the period in which the region formed. A multidata survey was made of the acoustic basement (the top of the buried oceanic crust) with a 15-km line spacing, supplemented with a more

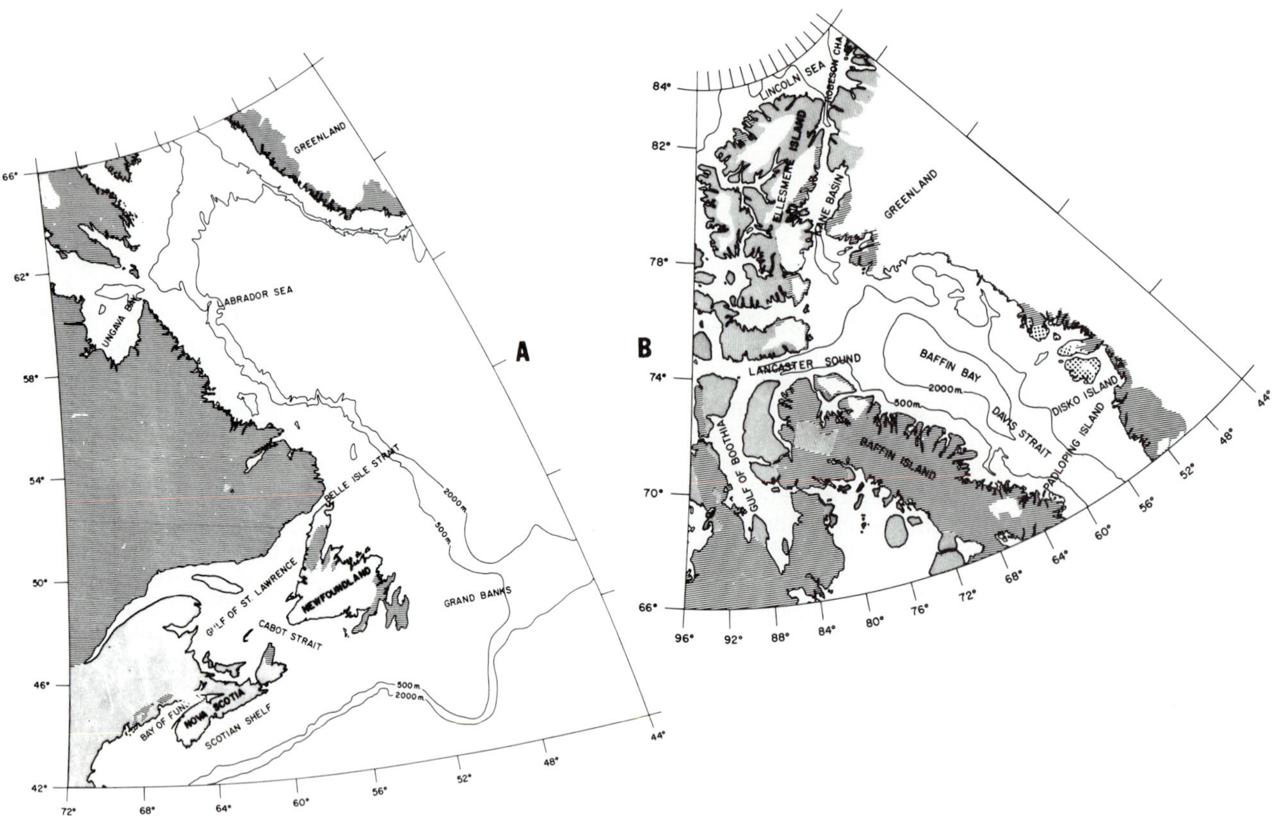

Fig. 1. Index maps to the continental margin of eastern Canada.

detailed magnetic survey of 7-km spacing; control was available to allow corrections for diurnal variation. This survey showed that the basement topography and the magnetic field anomalies are lineated, the principal trend of lineations being parallel to the boundary between the quiet zone and disturbed zone to the east (Fig. 5). The basement topography was used to model the magnetic anomaly field. It was found that if the basalts acquired their magnetization in a Triassic or Jurassic normal field, and have a rather high magnetization, the observed field can be accounted for, but that two reversals have to be incorporated (Fig. 7). These reversals (where magnetic "lows" cannot be accounted for by basement topography) also trend parallel to the boundary between quiet zone and disturbed zone.

It can be concluded therefore that this quiet zone is quiet because of formation at a time of few reversals. It is not necessary to postulate subdued anomalies because of loss of magnetization; the values of magnetization that have to be adopted to model the observations are higher, not lower than normal. The quiet-zone boundary, and the two reversals, may therefore be isochrons.

The nature of the ridge complex is still a question. It is known that there has been substantial subsidence of the margin off Nova Scotia, if only because shallow-water sediments are found in the lower parts of wells drilled offshore. The amount of subsidence is broadly consistent with theoretical models. Walcott (1972) showed that loading of a margin by sediment, accompanied by subsidence, would lead to tension at the margin. As a consequence, the ridge complex may be formed of diapir-like sedimentary bodies, which migrated upward from beneath the slope. This may have happened before the Late Cretaceous, since horizon A of the ocean basin terminates on the seaward side of the complex and appears to be undeformed in seismic profiles (Fig. 4).

SOUTHERN MARGIN OF THE GRAND BANKS

The Grand Banks of Newfoundland stick out like a sore thumb into the Atlantic Ocean. Their southern margin may be an extension of the system of faults that runs to the east from the northern part of mainland Nova Scotia, and the margin continues to the southeast as the Southeast Newfoundland Ridge. It bounds two parts of the Atlantic Ocean with very different histories, and it may have acted as a transform fault in a phase of spreading in the North Atlantic which started in the early Mesozoic. One modern analogue may be the Queen Charlotte Islands fault, bounding the margin off western Canada, along which there has been translational

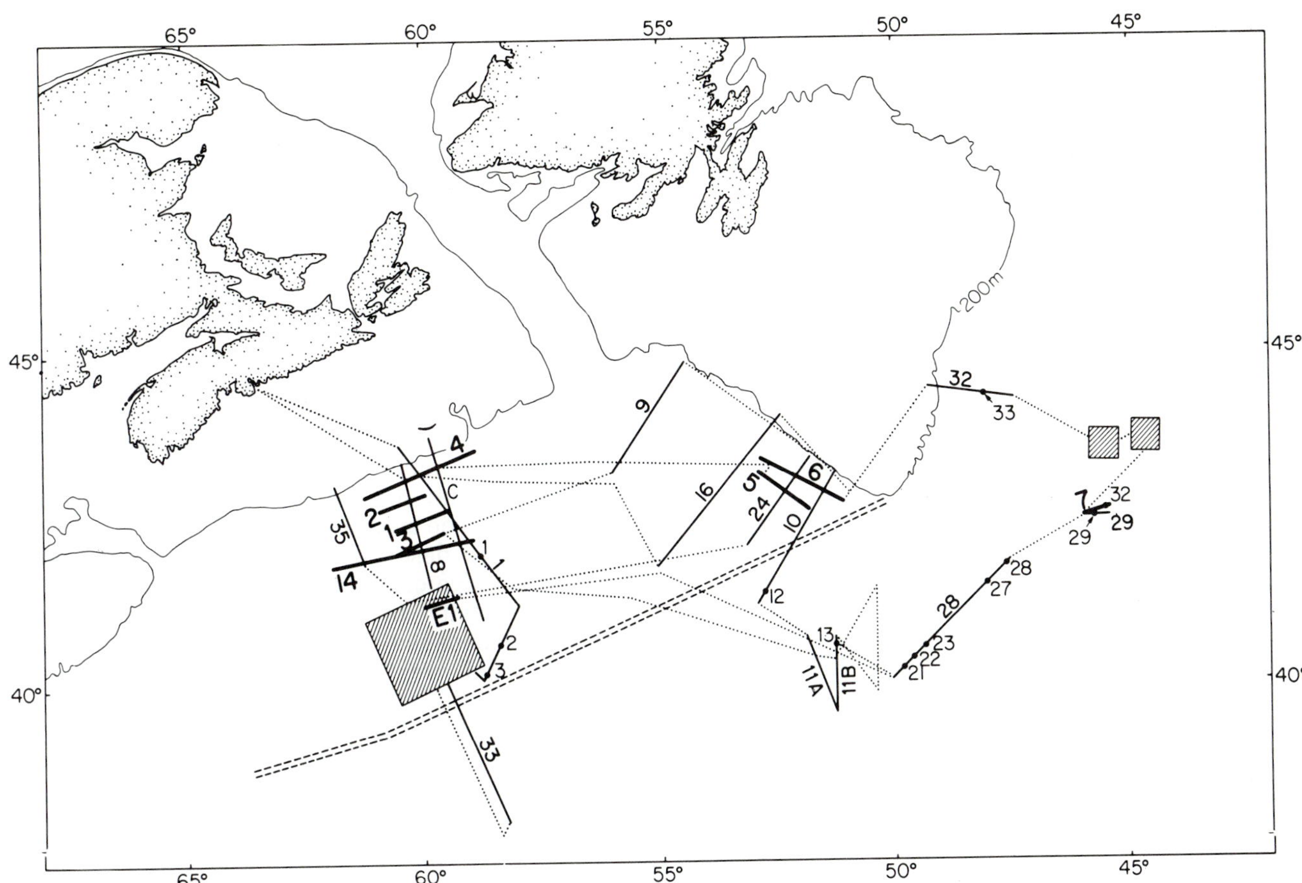

Fig. 2. Track chart showing location of geophysical data collected in the northwestern Atlantic in 1972 and 1973 from CSS HUDSON. The solid dots are sonobuoy stations; solid lines where seismic reflection, gravity, and magnetic data were collected; heaviest lines are reversed seismic refraction; shaded areas denote detailed surveys; dotted lines are tracks along which no seismic reflection data were obtained. The 200-m isobath is shown and the double dashed lines denote the quiet zone boundary.

motion. On that margin oceanic crust is found to be deformed in a characteristic manner, forming a number of horst- and graben-like features. It is possible that the southern margin of the Grand Banks possesses such characteristics, not evident now in the morphology because of sediment cover.

Reflection lines show a number of interesting features (Figs. 2, 8, and 9). The ridge complex of the south is not obvious on any crossings of this margin, although it could be present but buried more deeply. Buried seamounts and flat-topped blocks of layer 2, possibly faulted upward, characterize the oceanic basement and may have acted as sediment dams for material transported from the continental shelf and upper slope. These basement features are not observed beneath the rifted margin to the south. The sediments surrounding the seamounts are undisturbed, suggesting that these basement structures were formed earlier. Beneath the western part of this margin, oceanic basement is not observed (lines 9 and 16, Fig. 2), presumably because of the thick sediment deposits associated with transport through the Laurentian Chennel. However, faults cut the entire sedimentary section, particularly in the vicinity of line 9. They occur in the vicinity of the epicenter of the Grand Banks' earthquake of 1929.

A magnetic anomaly, similar to the slope anomaly discussed earlier, may be delineated, but with more difficulty. The character of the anomalies may be more complex because of the extremely irregular topography of oceanic basement which generates large magnetic anomalies and makes the slope anomaly less unique a feature by comparison with the Nova Scotian margin.

The width of the transition zone along the southern Grand Banks margin is not precisely known. However, oceanic basement can be traced to within 65 km of the shelf break beneath line 24. Further analysis of the refraction results may provide a more definitive estimate of the nature of this transition.

The results suggest that the geological characteristics of rifted (Nova Scotia) and faulted (southern Grand Banks) margins differ. The faulted margin appears to have features in common with the active Queen Charlotte transform fault along the margin of western Canada. The magnetic slope anomaly may be common to both the Nova Scotian and southern Grand Banks margins, although it is less clear along the latter.

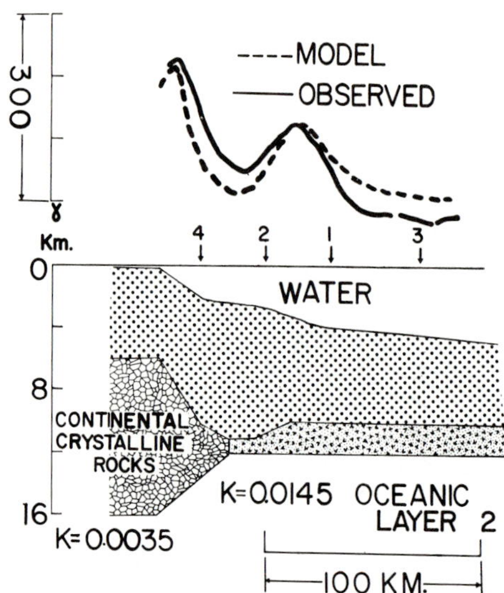

Fig. 6. Model used to calculate a magnetic anomaly across the Nova Scotia margin which is similar to the observed slope anomaly. An induced susceptibility of 0.0035 emu was used for the continental crystalline rocks. An effective susceptibility of 0.0145 emu was used for the remanent magnetization of oceanic layer 2. The strike and inclination of the field, corresponding to the Triassic pole, were 35° and 25°, respectively. The dotted pattern is nonmagnetic sediment. The positions of the four refraction lines are shown.

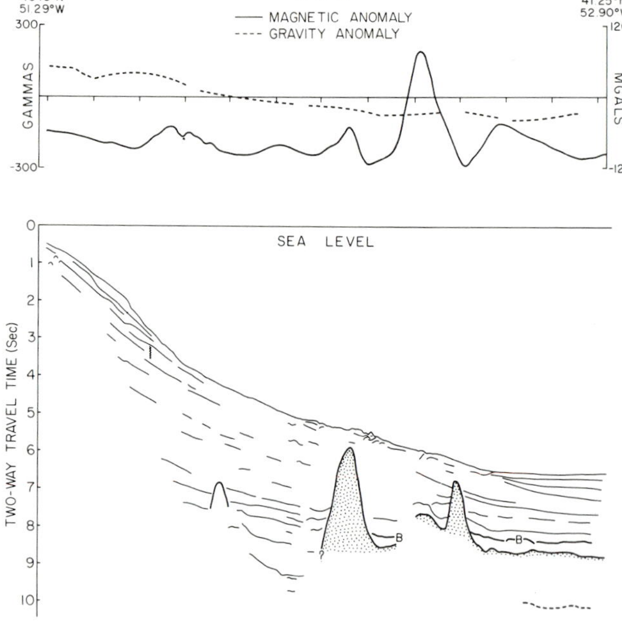

Fig. 8. Interpretive line drawing of seismic reflection results, gravity and magnetics, across the southern margin of the Grand Banks, (line 10 of Fig. 2). A strong reflector which may be horizon B is shown, as well as oceanic basement (shaded) and other sediment reflectors. The ticks on the horizontal axis represent a distance of 20 km.

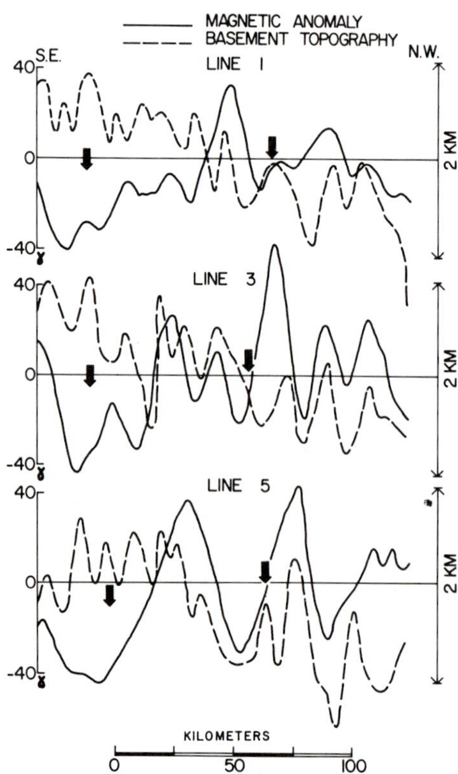

Fig. 7. The magnetic anomalies and corresponding basement relief along three survey lines, indicated in Figure 5. The arrows show negative correlation between basement topography and the magnetic anomalies.

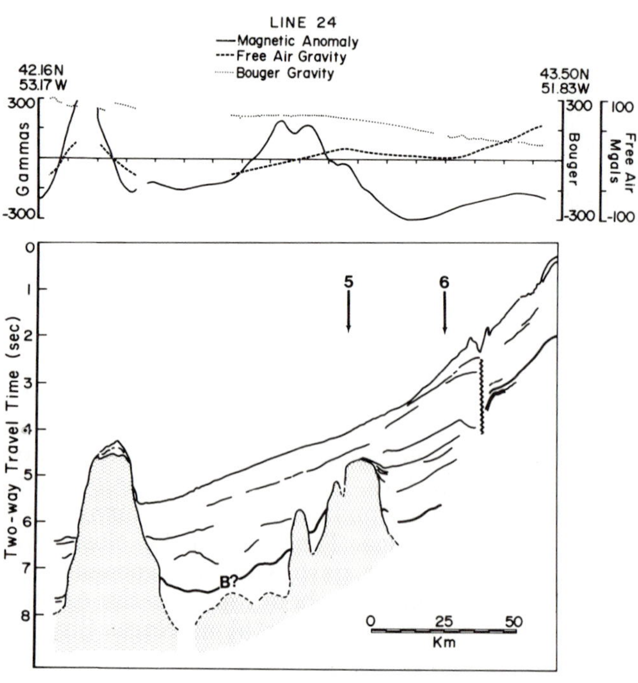

Fig. 9. Results across the southern margin of the Grand Banks (line 24) presented in the same way as in Figure 8. The positions of seismic refraction lines (not discussed here) crossing line 24 are also shown.

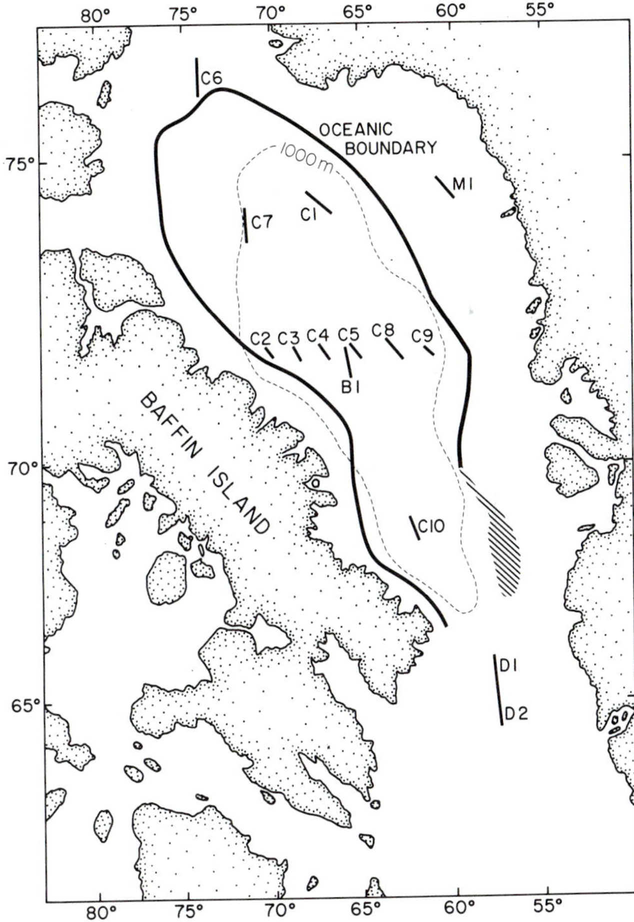

Fig. 10. Location of seismic refraction lines, two of the reflection lines, and the continent-ocean transition in Baffin Bay. The numbered lines are positions of reversed refraction profiles. The heavy black line is the position of the continent-ocean transition (see the text).

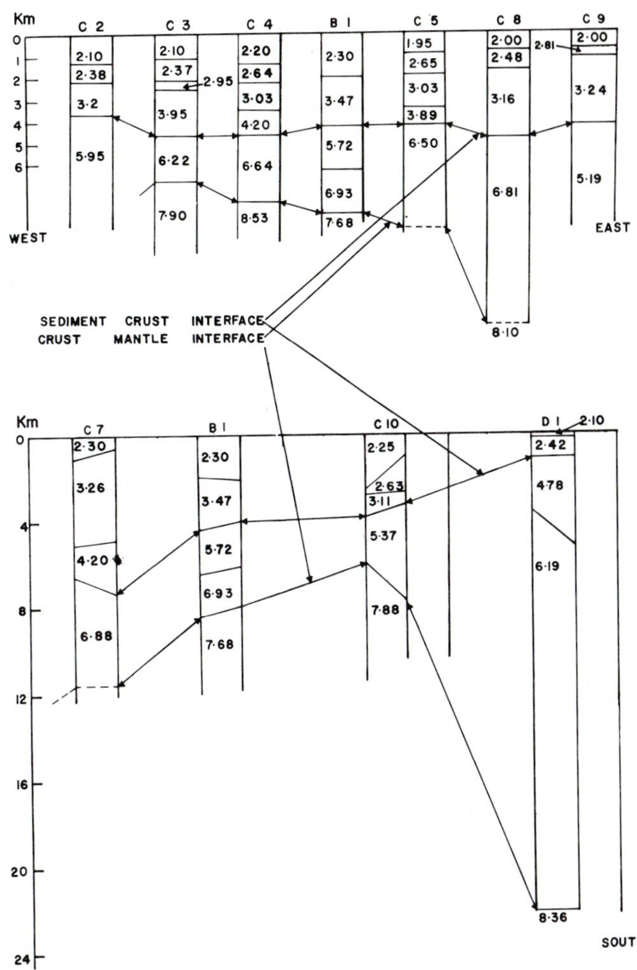

Fig. 11 Crustal sections in Baffin Bay from refraction lines (Fig. 10). Upper: west to east. Lower: north to south. Velocities are in km/sec. Dashed lines indicate minimum depth to the Mohorovicic discontinuity.

yields a picture consistent with seismic refraction observations: continental rocks are separated from magnetic oceanic crust by a nonmagnetic zone that may represent intrusive igneous rocks formed during initial rifting. A quiet magnetic zone beyond the slope anomaly develops where the oceanic crust formed in a period of few reversals, but the muted anomalies which are to be found are lineated as is the buried topography of the oceanic crust. The magnetic slope anomaly may not be so obvious where there is a fractured oceanic crust, or seamounts.

The southern margin of the Grand Banks does have the characteristics of a strike-slip fracture zone or transform fault. The southeast Newfoundland Rise is a natural continuation of the line of structural features extending from northern Nova Scotia: faulting east from Chedabucto Bay through the "Orpheus" anomaly to the Laurentian Channel; earthquake epicenters of the southeastern end of the Laurentian Channel; buried seamounts along the southern margin of the Grand Banks.

Baffin Bay is a small ocean basin formed in the Tertiary, perhaps in association with much older

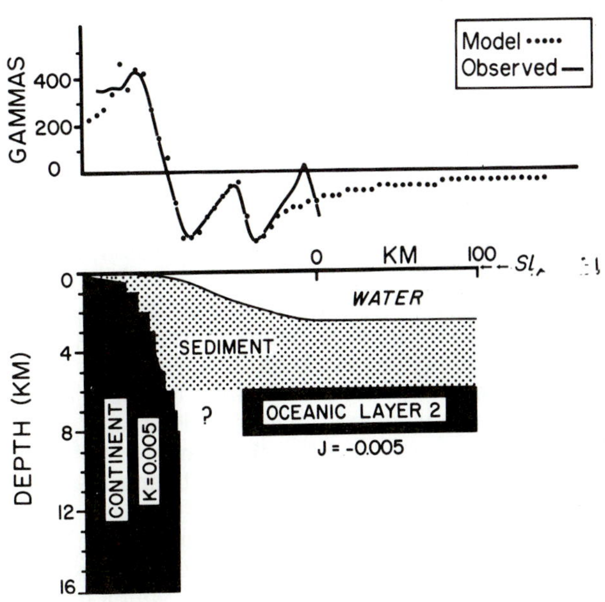

Fig. 12. Magnetic slope anomaly off eastern Baffin Island and interpretation.

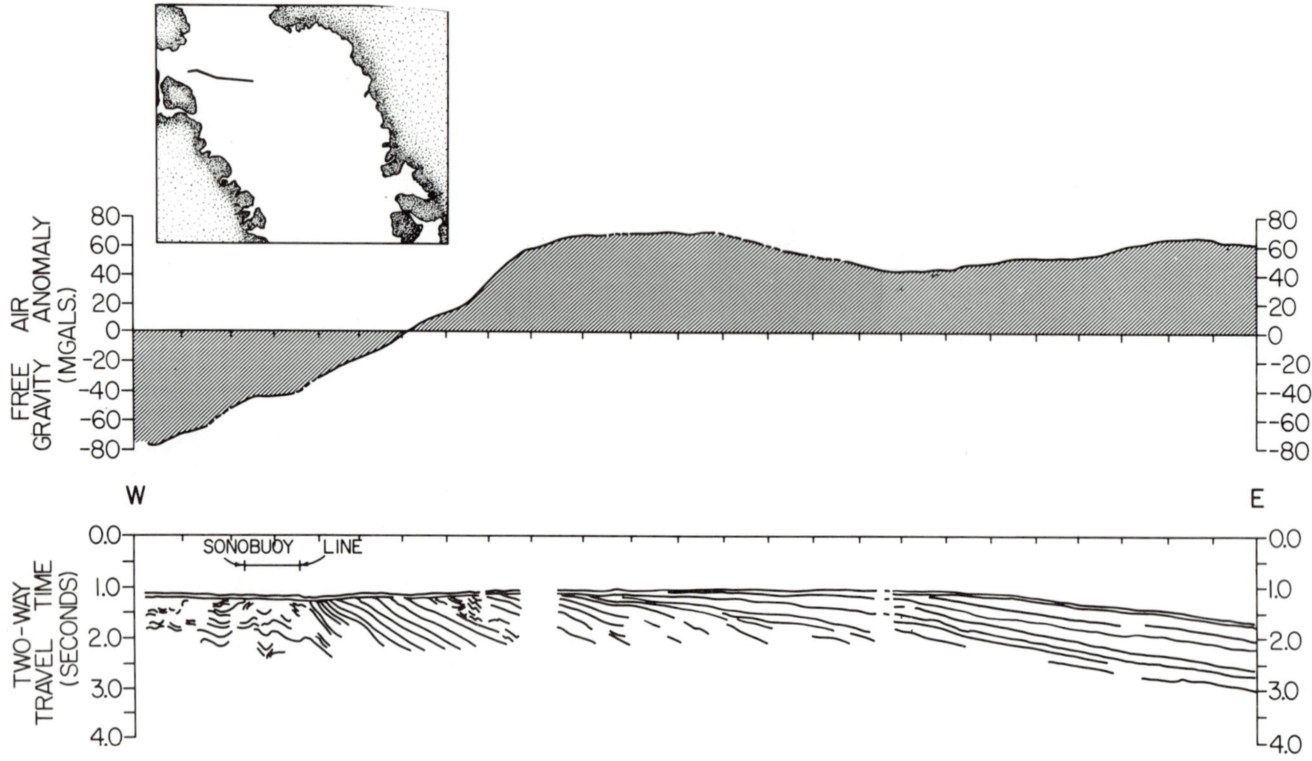

Fig. 13. Northern Baffin Bay: the free-air gravity anomaly and a line drawing of the seismic reflection record at the entrance to Lancaster Sound (index map). The location of a short sonobuoy refraction line is also indicated. The ticks along the horizontal scale are hour marks, so each division represents approximately 11 km.

zones of weakness within the crust. It is now partially filled with sediment, and there is little evidence that spreading is now an active process. Presumably in due time it will fill, and form a sedimentary basin within a single continental land mass of Canada and Greenland. It may be a rather modern analogue of comparable ancient intracontinental sedimentary basins.

There are a number of outstanding questions. (1) We have suggested that the Ridge Complex arose through migration of low-density sediments when tensional forces developed normal to the margin, these forces being caused by differential subsidence, as Walcott proposed, but the seismic evidence that the diapir-like bodies are sediment, not salt, is only sparse. (2) We have suggested, following Barret and Keen that the quiet-zone has arisen because of it and formed in a time of few reversals. But so far no comparison between an oceanic magnetic anomalies time-scale, and a reversals timescale from evidence of rocks on land has been made. (3) We have suggested, following Fenwick, Keen, Keen and Lambert, that the slope anomaly of eastern North America is an edge effect, and that high seismic velocities are associated with a zone in which basic rocks have been intruded, with low magnetization. It is difficult to see what experiment can be devised to test this hypothesis, which of itself does fit the observations known now.

BIBLIOGRAPHY

Barrett, D. L., 1966, Lancaster Sound shipborne magnetometer survey: Can. Jour Earth Sci., v. 3, p. 223-236.

______, and Keen, C. E., 1973, Lineations in the magnetic quiet zone of the northwest Atlantic: Jour. Geophys. Res. (in press).

Clarke, D. B., 1970, Tertiary basalts of Baffin Bay: Possible primary magma from the mantle: Contr. Mineral. and Petrol., v. 25, p. 203-224.

______, and Upton, B. J., 1971, Tertiary basalts of Baffin Island Field relations and tectonic setting: Can. Jour. Earth Sci., v. 8, p. 248-258.

Dainty, A. M., Keen, C. E., Keen, M. J., and Blanchard, J. E., 1966, Review of geophysical evidence on crust and upper mantle structure on the eastern seaboard of Canada, *in* Steinhart, and Smith, eds., The earth beneath the continents: Am. Geophys. Union Monogr. 10, p. 349-369.

Dawes, P. R., 1971, The north Greenland fold belt and environs: Bull. Geol. Soc. Denmark, v. 20, p. 197-239.

______, 1972, The north Greenland fold belt: a clue to the history of the Arctic Ocean basin and the Nares Strait lineament, *in* Tarling, D. H., and Runcorn, S. K., eds., Continental drift, sea-floor spreading and plate tectonics: New York, Academic Press (in press).

Emery, K. O., Uchupi, E., Phillips, J. D., Bowin, C. O., Bunce, E. T., and Knott, S. T., 1970, Continental rise off eastern North America: Am. Assoc. Petrol. Geol. Bull., v. 54, p. 44-108.

Fahrig, W. F., Irving, E., and Jackson, G. D., 1971, Paleomagnetism of the Franklin diabases: Can. Jour. Earth

Sci., v. 8, p. 455-467.

Fenwick, D. K. B., Keen, M. J. Keen, C. E., and Lambert, A., 1968, Geophysical studies of the continental margin northeast of Newfoundland: Can. Jour. Earth Sci., v. 5, p. 483-500

Fortier, Y. O., and Morley, L. M., 1959, Geological unity of the Arctic Islands: Trans. Roy. Soc. Can., ser. 3, v. 50, p. 3-12.

Henderson, G., 1971, The geological setting of the west Greenland basin in the Baffin Bay region: Geol. Surv. Can. Paper 71-23, p. 521-544.

Holtedahl, O., 1970, On the morphology of the west Greenland shelf with general remarks on the "marginal channel" problem: Marine Geology, v. 8, p. 155-172.

Hood, P. J., and Bower, M., 1971, Low-level aeromagnetic surveys of the continental shelves bordering Baffin Bay and the Labrador Sea: Geol. Surv. Can. Paper 71-23, p. 573-598.

______, Hyndman, R. D., and Pautot, G., 1971, Geophysical study of the opening of the Labrador Sea: Jour.

Keen, C. E. and Barrett, D. L., 1972, Seismic refraction studies in Baffin Bay: an example of a developing ocean basin: Geophys. Jour. Roy. Astron. Soc., v. 30, p. 253-271.

______, and Barrett, D. L., 1973, Structural characteristics of some sedimentary basins in northern Baffin Bay: Can. Jour. Earth Sci., v. 10, p. 1267-1279.

______, Barrett, D. L., Manchester, K. S., and Ross, D. I., 1972, Geophysical studies in Baffin Bay and some tectonic implications: Can. Jour. Earth Sci., v. 9 p. 239-256.

______, Keen, M. J., Barrett, D. L., and Heffler, D. E., 1973, Some aspects of the ocean-continent transition at the continental margin of eastern North America; Jour. Geophys. Res. (in press).

______, Keen, M. J., Ross, D. I., and Lack, M., 1973, Baffin Bay: small ocean basin formed by sea-floor spreading: Am. Assoc. Petrol. Geol. Bull. (in press).

Keen, M. J., 1969, Magnetic anomalies off the eastern seaboard of the United States: a possible edge effect: Nature, v. 222, p. 72-74.

______, and Keen, C. E., 1971, Subsidence and fracturing on the continental margin of eastern Canada: Geol. Surv. Can. Paper 71-23, p. 23-42.

______, Johnson, J., and Park, I., 1972, Geophysical and geological studies in eastern and northern Baffin Bay and Lancaster Sound: Can. Jour Earth Sci., v. 9, p. 689-708.

Kerr, J. W., 1967, A submerged continental remnant beneath the Labrador Sea: Earth and Planetary Sci. Letters, v. 2, p. 283-289.

King, L. H., 1972, Relation of plate tectonics to the geomorphic evolution of the Canadian Atlantic Provinces: Geol. Soc. America Bull., v. 83, p. 3083-3090.

Laughton, A. S., 1971, South Labrador Sea and the evolution of the North Atlantic: Nature, v. 232, p. 612-617.

Le Pichon, X., and Fox, P. J., 1971, Marginal offsets, fracture zones and the early opening of the North Atlantic: Jour. Geophys. Res., v. 76, p. 6294-6308.

______, Hyndman, R.D., and Pautot, G., 1971, Geophysical study of the opening of the Labrador Sea: Jour. Geophys. Res., v. 76, p. 4724-4743.

Løken, O. H., and Hodgson, D. A., 1971, On the submarine geomorphology along the east coast of Baffin Island: Can. Jour. Earth Sci., v. 8, p. 185-195.

Mayhew, M. A., Drake, C. L., and Nafe, J. E., 1970, Marine geophysical measurements on the continental margins of the Labrador Sea: Can. Jour. Earth Sci., v. 7, p. 199-214.

O'Nions, R. K., and Clarke, D. B., 1972, Comparative trace element geochemistry of Tertiary basalts from Baffin Bay: Earth and Planetary Sci. Letters, v. 15, p. 436-446.

Park, I., Clarke, D. B., Johnson J., and Keen, M. J., 1971, Seaward extension of the west Greenland Tertiary volcanic province: Earth and Planetary Sci. Letters, v. 10, p. 235-238.

Pelletier, B. R., 1966, Canadian Arctic Archipelago, pt. B, Bathymetry and geology, *in* Fairbridge, R. W., ed., The encyclopedia of oceanography: New York, Van Nostrand Reinhold, p. 160-168.

Pye, R. D., and Hyndman, R. D., 1972, Heat flow measurements in Baffin Bay and the Labrador Sea: Jour. Geophys. Res., v. 77, p. 938-944.

Renwick, G. K., 1973, Sea-floor spreading and the evolution of the continental margins of Atlantic Canada (M.Sc. thesis): Halifax, Nova Scotia, Dalhousie Univ.

Ross, D. I., 1973, Free air and simple Bouguer gravity maps of Baffin Bay: Marine Sciences Paper 11: Ottawa, Can., Information Canada.

______, and Henderson, G., 1973, New geophysical data on the continental shelf of central and northern west Greenland: Can. Jour. Earth Sci., v. 10, p. 485-497.

Taylor, P. T., Zietz, I., and Dennis, L. S., 1968, Geological implications of aeromagnetic data for the eastern continental margin of the United States: Geophysics, v. 33, p. 755-780.

Thorsteinsson, R., and Tozer, E. T., 1970, Geology of the Arctic Archipelago, *in* Douglas, R. J. W., ed., Geology and economic minerals of Canada: Geol. Surv. Can. Econ. Geol. Rept. 1, p. 547-590.

Trettin, H. P., 1972, The Innuitian Province, *in* Price, R. A., and Douglas, R. J. W., eds., Variations in tectonic styles in Canada: Geol. Assoc. Can. Spec. Paper 11, p. 83-180.

Vogt, P. R., and Ostenso, N. A., 1970, Magnetic and gravity profiles across the Alpha Cordillera and their relation to Arctic sea-floor spreading: Jour. Geophys. Res., v. 75, p. 4925-4938.

Walcott, R. I., 1972, Gravity, flexure and the growth of sedimentary basins at a continental edge: Geol. Soc. America Bull., v. 83, p. 1845-1848.

Whitham, K., Milne, W. G., and Smith, W. E. T., 1970, The new seismic zoning map for Canada: The Canadian Underwriter, June 15.

Atlantic Continental Margin of North America

Robert E. Sheridan

INTRODUCTION

The continental margin bordering eastern North America includes all the features common to Atlantic-type margins. Morphologically, the margin is characterized by a broad, flat continental shelf, a relatively gentle continental slope, and an even gentler-sloping continental rise. An exception to this rule is found east of Florida and the Bahama Islands.

Enough geological and geophysical data are available to conclude that these morphological features are formed wholly within the great sedimentary accumulations bordering the continent, but these features, especially the continental slope, are localized over major structural boundaries in the crust and upper mantle. The transition from oceanic crust to continental crust occurs in this margin zone.

Within the last five years, new geological and geophysical data have become available through petroleum exploration drilling and seismic studies. This information revealed that as much as 8-12 km of Mesozoic and Cenozoic sediments have accumulated in marginal, downfaulted basins. Lower Cretaceous and Jurassic carbonate and evaporite deposits, not known from the emerged Atlantic coastal plain, were found in the deeper basins to form high-velocity seismic layers, which masked the true basement structures on earlier seismic-refraction studies.

Interpretations can be made that extensive block-faulting has controlled the shape and location of the major marginal basins, with major wrench-fault motions producing tensional faults along the hingements. This pattern of faulting is compatible with the rifting of the North America and Africa-Europe plates in the Jurassic, perhaps beginning in the Triassic, and with the subsequent plate motions along oceanic fracture zones which formed major wrench faults transverse to the continental margin.

Coupled with this tensional faulting of the margin, lithospheric density changes in the Atlantic Ocean, near the North American margin, have caused drastic subsidence. Restoring the margin to its past shallow depth in the Jurassic would imply that the continental rise basement had also been relatively shallow, comparable in depth and velocity possibly to the present Red Sea. The sediments of the deeper part of the continental rise prism could well be of shallow-water facies. Such an environment could have been conducive to hydrocarbon accumulation.

REGIONAL DATA

The Atlantic continental margin of North America is one of the most extensively studied margins in the world. Beginning with the topographic studies of Veatch and Smith (1939), the sediment sampling of Stetson (1939) and Ericson et al. (1952), and including the early seismic refraction studies of Maurice Ewing and associates (M. Ewing et al., 1937, 1938, 1940), the basic ideas of thick, continental margin sediment accumulations were formed. In a classical paper on continental margins, Drake et al., (1959) mapped the basinal form of these marginal deposits off the northeast United States, and identified these thick deposits as modern geosynclines. The most recent seismic investigations (Mattick et al., 1973; Schultz and Grover, 1973), verify the general basinal form and locations mapped by Drake et al. (1959), but show the sediments to be probably much thicker. Acoustic masking of the full sequence was first interpreted east of Florida (Sheridan et al., 1966), where high-velocity Lower Cretaceous carbonates extended east and north of the Florida-Bahamas area.

Subsequent to Ewing's original studies, refraction data have been collected from Labrador to the Bahamas (Mayhew et al., 1970; Sheridan and Drake, 1968; Press and Beckmann, 1954; Officer and Ewing, 1954; Hersey et al., 1959; Sheridan et al., 1966; Ball et al., 1971). These data confirm the presence of thick Cretaceous and Cenozoic sediments along the margin, but the presence of older Jurassic and perhaps Triassic sediments was only a speculation: some of the higher velocity sediments, now thought to be of Cretaceous age, were assigned to Jurassic and Triassic.

Extensive dredging of the continental slope (Heezen and Sheridan, 1966; Sheridan et al., 1969b, 1970, 1971; Gibson, 1970; Marlowe, 1965) identified the Lower Cretaceous through Cenozoic sedimentary layers of the continental margin, while deep petroleum-exploration drilling proved the existence of a Jurassic section off Canada (Bartlett, 1969; McIver, 1973; Ayrton et al., 1973). Extrapolation of these geological data along strike now permits the correlation of most of the sedimentary seismic layers.

Recent seismic-reflection profiling has revealed the shallow structures of the continental margin (Ewing et al., 1963; J. I. Ewing et al., 1966; Uchupi and Emery, 1967; Emery et al., 1970; Uchupi et al., 1971; Garrison, 1970; Markl et al., 1970; King and McLean, 1970; Rona and Clay, 1967; and Grant,

1966, 1968). These data revealed features mostly of the Upper Cretaceous and Cenozoic, including prograding on the continental shelf and slope, reef damming at the shelf edge, unconformities, and truncated sedimentary layers on the continental slope.

Large slump blocks were identified on the slope (Heezen and Drake, 1964; Kelling and Stanley, 1970). Failure of metastable slope sediments (induced in part by earthquake tremors) has caused recurring slumping. These disturbed sediments contribute to large turbidity currents, which flow downslope to the continental rise and abyssal plains.

On the continental rise, great thicknesses of stratified sediments are identified in the seismic reflection profiles. Giant sand antidunes have been identified, revealing transport of the rise sediments by strong ocean-bottom currents (Fox et al., 1968). Deep-sea photographs proved the existence of these strong bottom currents, which Heezen et al. (1966) interpreted to be geostrophic contour currents, flowing southward. Subsequent sea-floor photography (Schneider et al., 1967), current calculations (Amos et al., 1971), and deep-sea drilling (Ewing and Hollister, 1972) have verified the presence of these contour currents. This mode of sediment distribution is thought to be the main factor in building the continental rise.

The deep crustal and mantle structure of the margin is known mostly from a few refraction lines (Dowling, 1968; Fenwick et al., 1968; Keen et al., 1973; G. N. Ewing et al., 1966; Ball et al., 1971) and from gravity measurements (Worzel and Shubert, 1955; Worzel, 1965). Mantle depths are consistently about 12-15 km under the outer rise, about 20 km under the slope, and about 30-35 km under the shelf. Worzel (1968) points out that this is characteristic of this continental margin. If the configuration of the upper mantle was used as a criterion, a question arises as to where in this broad 100-200 km transition would one place the edges of the continent.

Drake et al.(1963) reported a prominent positive magnetic anomaly, coincident with the edge of the shelf from Canada to the Florida-Bahamas area. Burk (1968) showed that buried basement ridges were common in continental margins throughout the world. Taylor et al. (1968), using newer data, mapped in detail the "east-coast" magnetic anomaly, which parallels the slope from Canada to Cape Hatteras, south of which it bifurcates with a prominent anomaly swinging into land in southern Georgia. Emery et al. (1970) show that this anomaly is localized above a buried basement ridge at depths of 6-7 km, with the magnetic properties of oceanic basalts which might well define the structural edge of the eastern North American Continent.

Beyond the east coast anomaly is a magnetic quiet zone, extending from the continental rise nearly to Bermuda, beyond which typical oceanic magnetic anomalies occur. Seismic-refraction data from the quiet zone show a more oceanic structure, and seismic reflection data reveal that acoustic basement crosses the eastern edge of the quiet zone without apparent structural interruption. These data suggest that the crust under the magnetic quiet zone was formed similarly to other oceanic crust.

Heirtzler and Hayes (1967) attributed the lack of magnetic anomalies in the quiet zone to the crust's formation during the Permian, when there were no magnetic reversals. However, recent DSDP drilling (Ewing and Hollister, 1972) indicate that this quiet zone can be of early Jurassic, perhaps Triassic age, Vogt et al. (1970) have suggested that the magnetic quiet-zone crust was forming in magnetically low latitudes, in a weak magnetic field, of a rapidly moving Triassic-Jurassic magnetic pole. Thus the basaltic basement would be only weakly magnetized and only a few of the more prominent reversals would be recorded. Vogt et al. (1970) and Brakl et al. (1968) show that there are linear correlatable magnetic anomalies of small amplitude within the quiet zone. Taylor et al. (1968) attribute the quiet magnetic character of the quiet zone to possible thermal metamorphism of the basalt, which has destroyed its remanent magnetism (RM). Recent measurements of the RM of these basalts appear to confirm this lower magnetization (Taylor et al., 1973).

Whatever the origin of the low magnetization of the basaltic crust of the quiet zone, most researchers agree that it was formed similarly to other oceanic crusts. Therefore, the edge of the eastern North American continent is thought to be under the slope, marked by oceanic basalts, forming the ridge of the east coast magnetic anomaly.

Off northeast Newfoundland, the slope has a linear, positive magnetic anomaly, somewhat similar to the east coast anomaly. Fenwick et al. (1968) attribute this anomaly to an edge effect of the thick continental slab of basic rocks against a nonmagnetic mantle. Again, the edge of the continent could be localized along this magnetic pattern.

GEOGRAPHICAL DISCUSSION

Ten cross sections of the Atlantic margin of North America provide the basis for this discussion (Fig. 1). Enough geophysical data, including seismic-reflection, seismic-refraction, gravity and magnetic data are available along these sections to permit a reasonable identification of the subsurface layers, especially the igneous and metamorphic crustal units. Enough geological data, including land drill holes, offshore petroleum exploration drilling, JOIDES Deep Sea Drilling Project holes, dredges, and piston cores, are available along these sections to provide reasonable correlations of the subsurface sedimentary units.

Some aspects of the cross sections are highly speculative and hypothetical. For instance, the exact position and throw of some of the proposed faults are actually inferred; the position of salt domes are generalized. However, recent petroleum-exploration

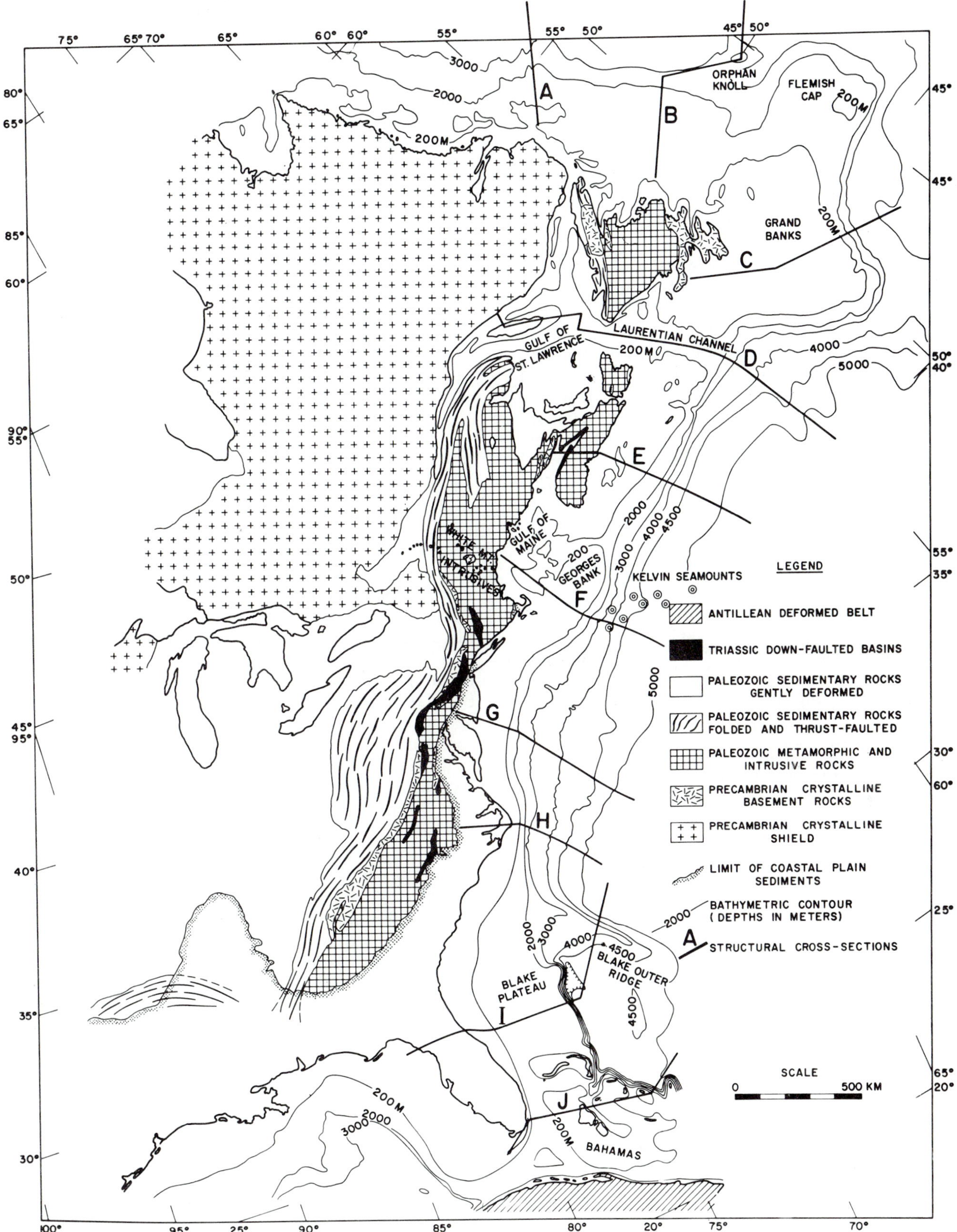

Fig. 1. Atlantic continental margin of North America, showing cross sections A-J. Generalized geologic and tectonic units are shown (after Sheridan and Drake, 1968).

reflection data show details of such diapirs (e.g., McMillan, 1973; Ayrton et al., 1973; Schultz and Grover, 1973). Such structures have thus been extrapolated into other areas with less control.

Labrador Shelf

Mayhew et al. (1970) found from seismic refraction data that the Precambrian basement sloping gently to the northeast, then dropping abruptly to depths of about 6 km beneath a great thickness of sedimentary rocks (Fig. 2). Depths to magnetic basement were found to be compatable with this interpretation.

Sedimentary layers of the area are correlated by seismic velocities to Cretaceous and Tertiary sediments drilled on the Grand Banks (Ayrton et al., 1973), and deeper sediments are compared to Cretaceous and Jurassic carbonates and sandstones, similar to those drilled in JOIDES (DSDP) hole 111 on nearby Orphan Knoll (Laughton et al., 1972), where Albian limestones are similar to the Albian lagoonal limestone facies from the Blake Plateau. Moreover, Jurassic conglomerates have now been reported from the coastal Labrador areas (McMillan, 1973), so that the presence of Jurassic sedimentary rocks in the deep basin offshore is very likely.

The presence of major normal faulting in the basement, midway across the shelf, is interpreted on the basis of the abrupt deepening of the basement in this region and the reversal of dip of the basement-interface seaward of the suggested fault. Such down-to-basin faulting is observed elsewhere along this margin (McMillan, 1973), so this interpretation is plausible. The longitudinal channel parallel to the Labrador coast (Fig. 2) has been suggested to be of possible fault origin (Holtedahl, 1958). However, seismic-reflection profiles (Grant, 1966) reveal that this is an erosional escarpment, probably scoured by Labrador shelf currents.

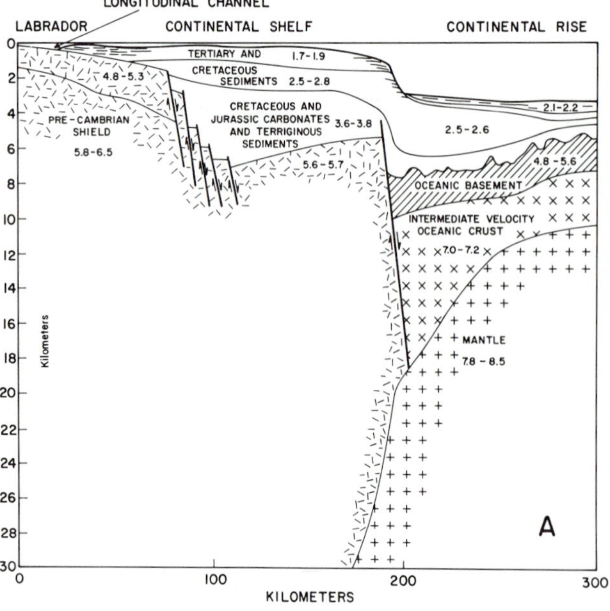

Fig. 2. Diagrammatic structural cross section across the Labrador continental margin; modified after Mayhew et al. (1970), Grant (1966), Fenwick et al., (1968) and McMillan (1973). Numbers refer to compressional wave velocities in km/sec.

The crust-mantle transition near the continental slope (Fig. 2) is based on several long refraction profiles made on the Newfoundland margin, southeast of this section (Fenwick et al., 1968). Normal mantle velocities and depths are found here, but the oceanic crust appears to have an intermediate velocity of 7.0-7.2 km/sec (Mayhew et al., 1970). Such intermediate velocities are observed elsewhere along this Atlantic Continental Margin.

Northeast Newfoundland Shelf

Acoustic basement is interpreted here to dip northeast and be downfaulted into a deep basin with about 8 km of Tertiary and Cretaceous sediments (Fig. 3).

The deeper 5.1-5.2 km/sec layer was originally interpreted to be meta-sedimentary rocks of possible Paleozoic age, but the drilling of Albian limestones on Orphan Knoll (Laughton et al., 1972) make it probable that these are high-velocity carbonates similar to those of the Blake Plateau (where there is also a similarity of facies).

The basement here is a two-layered structure with pre-Carboniferous crystalline rocks, including Devonian granites, above a layer of basic and ultrabasic rocks. Sheridan and Drake (1968) correlated these deeper velocities (7.2-7.4 km/sec) with ultrabasics exposed on Newfoundland. Dewey and Bird (1971) associated this layer with the ophiolites of the northern Appalachians, suggesting that it represents the compressed early Paleozoic oceanic crust (Wilson, 1966). The intermediate-velocity layer forms the root of the Appalachians under Newfoundland and the Gulf of St. Lawrence (Sheridan and Drake, 1968; G. N. Ewing, et al., 1966), and extends all the way to the continental slope northeast of Newfoundland.

The extension of middle Paleozoic crystalline rocks of the upper basement to the northeast is less certain. It is shown to extend to the slope as a thin layer, which could have been missed in refraction studies. The continuation of Devonian and later Paleozoic structures is possible from North America to England, and Rodgers (1970) feels there was such continuity.

The interpretations of the drilling results on Orphan Knoll by Laughton et al. (1972) suggest that this is a fragment of continental-shelf crust. The shallow-water, lagoonal, Albian facies of Orphan Knoll indicate a shelf environment for this time, after which the shelf subsided. Subsequently, Eocene to Holocene pelagic sediments draped over the knoll and filled the Labrador Sea in the area of drill hole 112.

The refraction and magnetic data presented by Fenwick et al. (1968) show normal oceanic crust and

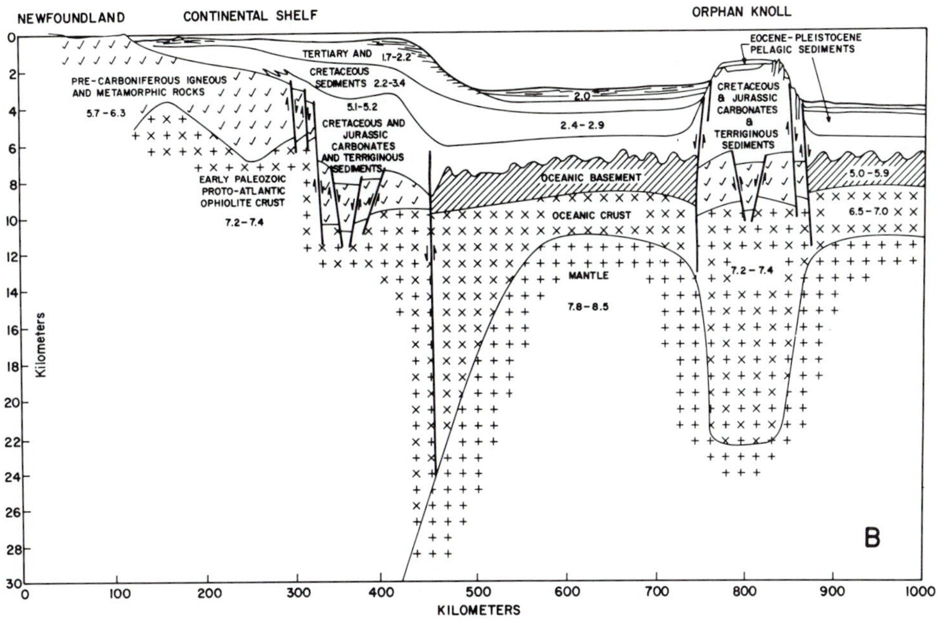

Fig. 3. Diagrammatic structural cross section across the northeast Newfoundland continental margin (modified after Sheridan and Drake, 1968; Laughton et al., 1972; Fenwick et al., 1968).

mantle up to the base of the Newfoundland slope, so that the knoll area is shown here to be apparently isolated (Fig. 3).

Grand Banks

The few seismic-refraction profiles on the Grand Banks (Press and Beckmann, 1954; Ewing and Ewing, 1959) are too widely spaced to reveal the true nature of the structure. Recently released geophysical and drilling results of petroleum exploration (Ayrton et al., 1973) reveal that the basement is downfaulted into several narrow, deep basins, reaching more than 6 km in depth (Fig. 4). Also, the consolidated sediments of Jurassic age are underlain by salt of possible Jurassic (?) age. The high-velocity salt apparently masked the true basement depth on some of the older refraction profiles.

The presence of salt on the Banks suggests its presence farther south along the Atlantic margin, and even farther north off Newfoundland and Labrador. The drilling of Albian lagoonal limestone on Orphan Knoll implies warm environments in this area even until Cretaceous (Laughton et al., 1972). The salt has now subsided to about 6 km in the basins under the Banks. If the salt continued oceanward at these depths, it might be expected to be found as domes under the continental rise. Schneider and Johnson (1970) have found possible domal salt structures on the rise off the Laurentian Channel.

Laurentian Channel

Data for this area (Fig. 5) are derived primarily from Sheridan and Drake (1968), Emery et al. (1970), Press and Beckmann (1954), G. N. Ewing et al. (1966), and Keen et al. (1973). This cross-section includes the Gulf of St. Lawrence, to reveal the basement structure of the Appalachian orogenic belt, and thus includes two continental margins, one ancient and one modern. The Cambrian-Silurian sediments of Anticosti Island must have dipped seaward off the Precambrian continental crust of the early Paleozoic proto-Atlantic Ocean (Wilson, 1966). In the Ordovician Toconian and the Devonian Acadian orogenies this oceanic crust was presuma-

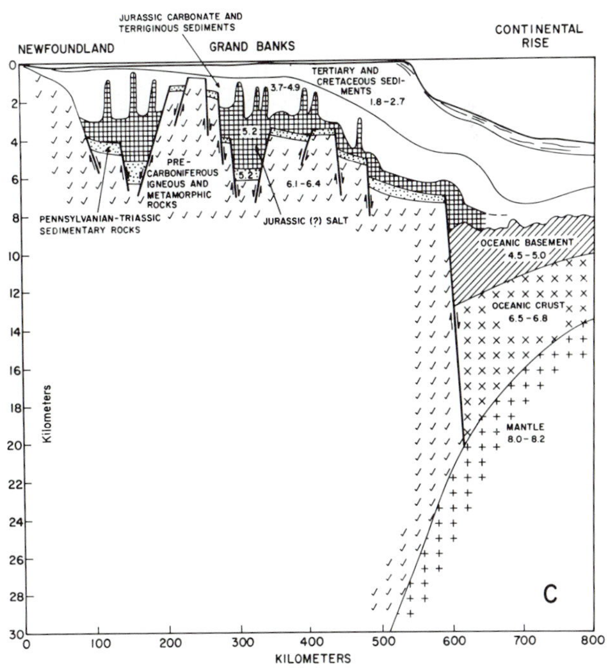

Fig. 4. Diagrammatic structural cross section across teh Grand Banks (modified after Press and Beckmann, 1954; Ewing and Ewing, 1959; Emery and Uchupi, 1972; Ayrton et al., 1973).

bly compressed into the intermediate velocity crustal layer forming the root of the northern Appalachians (Dewey and Bird, 1971). Subsequent to the formation of deep grabens in the Appalachian axial zone during the Carboniferous and Permian, the early Mesozoic rifting of the modern Atlantic margin presumably produced the downfaulting and accumulation of marginal geosynclines.

Slump scarps and rotated sedimentary blocks are evident in reflection profiles (Emery et al., 1970) close to the epicenter of the 1929 earthquake, which Heezen and Drake (1964) demonstrated to have broken submarine cables and generated large-scale turbidity currents.

Seismic-refraction data under the continental rise near this section (Press and Beckmann, 1954) reveal a dip of the oceanic crust and mantle toward the continent, to form a basin under the rise sediments to depths of 10 km. Shown on this section are two prominent reflectors which Emery et al. (1970) have mapped under the continental rise: "horizon A" and "horizon B" (J. I. Ewing et al., 1966). This deeper reflector ("B") has been drilled off the Bahamas (Ewing and Hollister, 1972) and was found to be basalt, but Emery and Uchupi (1972) still suggest that it might be sedimentary elsewhere along the margin.

Nova Scotian Shelf

Recent petroleum-exploration drillings to depths greater than 5 km on this shelf have produced the best knowledge of the offshore stratigraphy available for anywhere in the North American Atlantic continental margin (Emery and Uchupi, 1972; McIver, 1973). A thick section of Tertiary and Cretaceous sands and shales is found to overlie Jurassic carbonates, evaporites, and terrigenous deposits (Fig. 6). These deeper carbonates have high seismic velocities, which masked basement on older refraction profiles (Drake et al., 1959). Basement is shown on cross sections of Emery and Uchupi (1972) to be about 10 km deep, and may include rocks as old as Triassic.

Under the base of the slope and rise, Emery et al. (1970) find a buried-ridge structure with an associated magnetic anomaly. Magnetic properties of this ridge body suggest that it is made up of oceanic basalt and that it is at a depth of 6-7 km. Accordingly, some of the structures observed on reflection profiles shallower in the rise sediments are interpreted as deformation above this buried basement ridge. Whether such structures should be interpreted as diapiric salt ridges is uncertain. The presence of well-documented salt under the Scotian shelf and Grand Banks, however, makes this a possibility.

Deep-crustal refraction measurements on the rise (Keen et al., 1973) show a relatively normal distribution of mantle velocities and crustal thicknesses, with a gentle dip of the oceanic crust toward the continent.

Georges Bank

Although there is no exploratory drilling off the United States, as exists off Canada, the very thick sedimentary section found under Georges Bank (Fig. 7) by petroleum exploration surveys (Schultz and

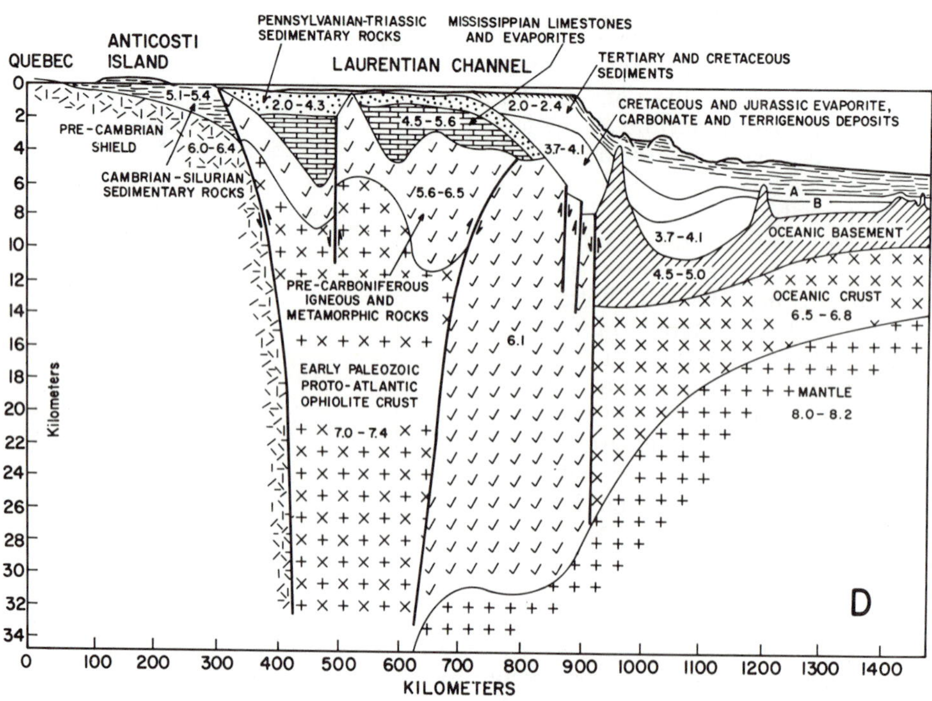

Fig. 5. Diagrammatic structural cross section across the Gulf of St. Lawrence and Laurentian Channel (interpreted from Sheridan and Drake, 1968; Press and Beckmann, 1954; Emery et al., 1970; G.N. Ewing et al., 1966; Emery and Uchupi, 1972).

Grover, 1973), and the higher-velocity sediments in the deeper part of the basin (Drake et al., 1959) are so similar to those of the Scotian Shelf that correlations can be projected. One difference has been suggested, however, by Schultz and Grover (1973), who feel that Triassic sedimentary deposits may exist in the bottom of the downfaulted basin. No direct evidence is available to prove such a speculation, but Triassic sediments in narrow basins within the basement are found in the Gulf of Maine (Ballard and Uchupi, 1972).

The Kelvin Seamounts, prominent in this area, are topped with sediments of Eocene age (Zeigler, 1955), suggesting a relatively young age with respect to the oceanic basement in this area. Perhaps they were intruded after the crust formed. The structure of this margin appears to be interrupted by some of the western seamounts (Emery et al., 1970).

New Jersey Shelf

Projection of the coastal-plain stratigraphy known from wells in this area (Fig. 8) indicates the presence of a thick Cretaceous and Tertiary sedimentary section under the continental margin (Kraft et al., 1971). Dredgings and corings of this age have been obtained from the continental slope (Gibson, 1970) and from DSDP holes off New Jersey (Ewing and Hollister, 1972).

The presence of high-velocity Jurassic limestones, and possible evaporites, can only be suggested from the knowledge of these deposits farther north on the Scotian shelf (although the bottom samples of the Cape May deep well have now been assigned a Jurassic age: Swain and Brown, 1972; Brown et al., 1972).

Nearly 12 km of sediments in the downfaulted Baltimore Canyon Trough (Mattick et al., 1973) have been reported. Previously discovered high-velocity refraction layers (Ewing et al., 1940; Drake et al., 1959) are related to strata within this sedimentary sequence. Relief in these high-velocity sediments is shown on the cross section (Fig. 8) to be due to a reef or carbonate platform. Such a reef, with very similar seismic velocities, has been well identified north of the Bahamas on the Blake Plateau (J. I. Ewing et al., 1966; Sheridan et al., 1966, 1970), and such a carbonate facies could easily have extended this far north in the Lower Cretaceous.

The presence of Triassic sedimentary rocks has also been suggested for the bottom of this basin (Mattick et al., 1973). Some red beds found in the bottom of coastal plain wells have been suggested to be Triassic, and the Triassic sedimentary basin of Connecticut appears to cross under Long Island (Grim et al., 1970) and to continue south under the New Jersey shelf.

On the continental rise, reflection data reveal a thick sequence of well-stratified Tertiary sediments above horizon A (Emery et al., 1970). Recent DSDP holes 105 and 106 in this area have correlated horizon A with Eocene chert, underlain by Cretaceous hemipelagic deposits, and finally by Jurassic limestones above basalt (Ewing and Hollister, 1972). This stratigraphy can be projected under the landward parts of the rise, where there are great thicknesses of sediments (Fig. 8).

It should be noted that the buried ridge in the oceanic basement is found on this section as it was on those off Nova Scotia (Emery et al., 1970; Emery and

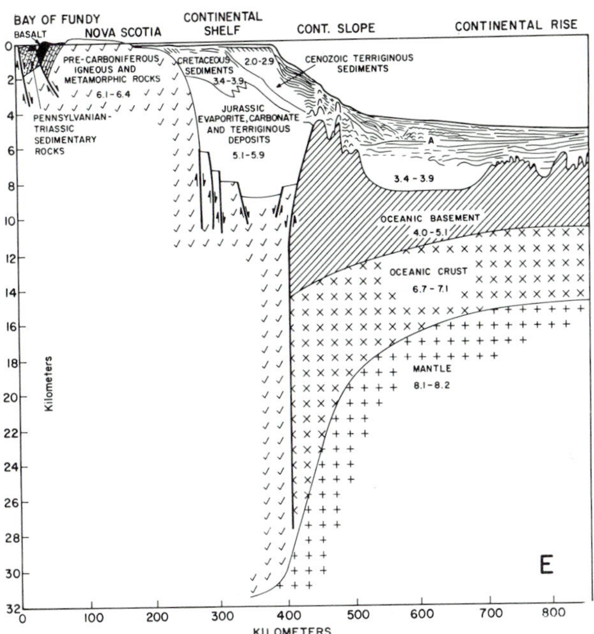

Fig. 6. Diagrammatic structural cross section of the Nova Scotian continental margin (data from Officer and Ewing, 1954; Drake et al., 1959; G.N. Ewing et al., 1966; Emery et al., 1970; Emery and Uchupi, 1972; McIver, 1973; Keen et al., 1973).

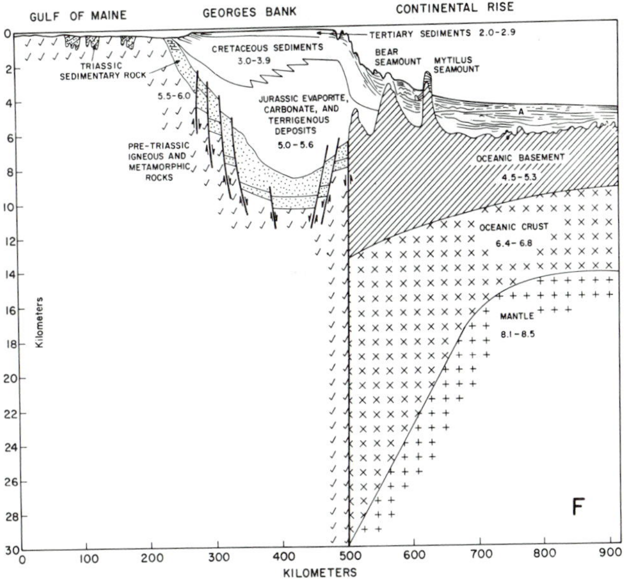

Fig. 7. Diagrammatic structural cross section across the Gulf of Maine and Georges Banks (modified after Ballard and Uchupi, 1972; Drake et al., 1959; Emery et al., 1970; Emery and Uchupi, 1972; Schultz and Grover, 1973).

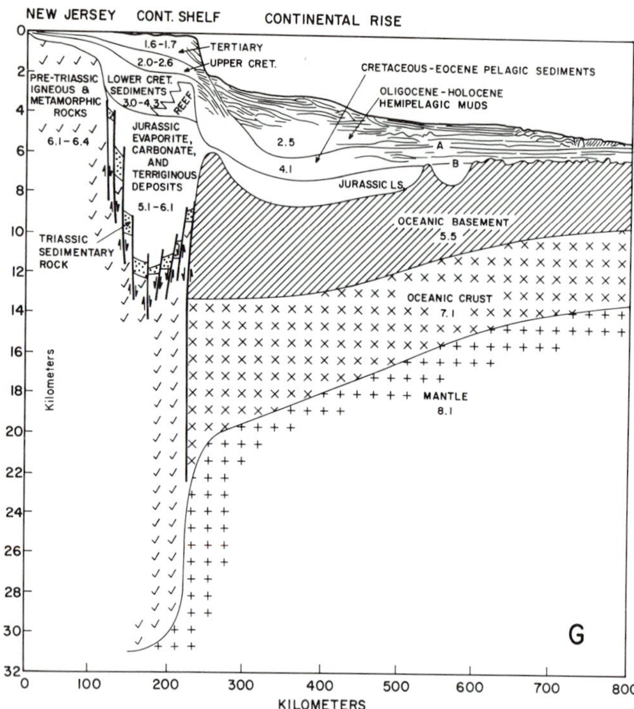

Fig. 8. Diagrammatic structural cross section of the New Jersey continental margin (modified after Drake et al., 1959; Ewing et al., 1938, 1940; Emery et al., 1970; Emery and Uchupi, 1972; Ewing and Hollister, 1972; Mattick et al., 1973).

Uchupi, 1972). This structure produces the east coast magnetic anomaly over the continental slope.

Cape Hatteras

The Cape Hatteras deep petroleum-exploration well provides knowledge of the deeper sedimentary section farthest seaward into the marginal geosyncline off the United States, practically at the shelf edge (Fig. 9). Brown et al. (1972) assign the bottom of this well to Jurassic; interpreted by Mattick et al. (1973) to exist as a thick section in the deeper part of the Baltimore Canyon Trough off Delaware and New Jersey (Fig. 8).

Between this well on Cape Hatteras and the continental slope, the suggested structure includes thickening Jurassic to Holocene sediments, truncated by erosion on the slope, and the presence of a possible Cretaceous and Jurassic reef (Emery et al. 1970; Emery and Uchupi, 1972). Although the presence of a reef is speculative, high velocities (3.7-5.0 km/sec) observed on the Hatteras slope (Houtz and Ewing, 1964) are compatible with reefal limestone.

Beyond the continental slope a thick, well-stratified sedimentary section is observed on reflection profiles (Emery et al., 1970; Rona and Clay, 1967). The Jurassic, Cretaceous through Eocene, and Oligocene to Holocene units can be correlated with the DSDP hole 105 (Ewing and Hollister, 1972).

Depth to the oceanic basement and crust is based on the seismic reflection data of Emery et al. (1970) and the seismic refraction data of Houtz and Ewing (1964). Once again a buried basement ridge, with a prominent magnetic anomaly, is suggested from the relief in the seismic reflector interpreted as basement.

Buried ridges within continental margins appear to be an important factor throughout a large part of the world (Burk, 1968).

Florida Blake Plateau

The narrow continental shelf off Florida drops off to the broad, flat, marginal Blake Plateau (Fig. 10), which is bordered on the east by the steep Blake Escarpment. Seaward is the Blake-Bahama Abyssal Plain, and beyond is the Blake Outer-Ridge.

Deep-sea drilling on the Florida shelf and Blake Plateau revealed that the Paleocene was at nearly equivalent depths in both areas (JOIDES, 1965), and consequently that the shallow-water limestones of the Florida shelf Tertiary give way to much thinner pelagic limestones on the Blake Plateau. Accordingly, the relief of the Blake Plateau was attributed to erosion and nondeposition, resulting from the Gulf Stream preventing sedimentation on the subsiding margin, while limestone accumulation continued on the Florida shelf (J. I. Ewing et al., 1966).

On the Plateau itself, reflection profiles (J. I. Ewing et al., 1966) revealed the shallow structure to be one of westward dipping beds, onlapping a rough reflector forming a reef-like feature along the eastern edge of the Plateau. Dredging recovered a lagoonal, Upper Cretaceous limestone from just below this reflector (Sheridan et al., 1970, 1971).

Deeper layers were interpreted to be high-velocity limestones bounded by a basement ridge which supported the reef along the east side of the Plateau (Sheridan et al., 1966). Dredging along the Blake Escarpment showed that at least the upper 5 km of these sediments included rocks of Lower Cretaceous (Neocomian) and younger ages (Heezen and Sheridan, 1966). These limestones are shallow-water, algal-bearing, reef talus. After some time in the Upper Cretaceous, the reef structure continued to subside and was mantled by a thin veneer of deeper-water-facies pelagic limestone.

Although sedimentary environmental characteristics apparently control the extent and detailed position of the relief, magnetic anomaly data (Taylor et al., 1968) and recent observations of minor faulting near the base of the Blake Escarpment (Sheridan et al., 1973), imply that this drastic relief may be localized over some deeper structural control.

On all the cross sections farther north, the actual physiographic relief of the continental margin is indeed confined within the sedimentary units and not, therefore, structurally originated. The position of the major relief is localized over the buried ridge forming the edge of the continent. This ridge may have influenced the position of initial shelf buildup by sediment entrapment or reef formation (Burk, 1968).

Reflection profiles reveal that for the Cenozoic, the Blake-Bahama Basin is essentially sediment-starved, while the Blake-Bahama Outer-Ridge has been built up by sediments transported by geostrophic currents (Heezen et al., 1966; Markl et al., 1970). Ewing and Hollister (1972), from the JOIDES holes 102, 103, and 104 on the Blake Outer-Ridge crest, have shown that reflector Y (Fig. 10) is not a stratigraphic boundary but rather is a diagenetic or gas-hydrate effect crossing nearly flat stratigraphic reflectors of late Miocene-Pliocene age. Reflector X was not penetrated, but extrapolation of sedimentation rates suggests its age to be early Miocene. With this information and the correlation of the Eocene, horizon A, which underlies the Blake Outer-Ridge without significant relief, Ewing and Hollister (1972) proposed a three-stage development of the Outer-Ridge crest.

First, the main body of the ridge formed by the A-X layer was deposited in Oligocene to early Miocene as a symmetrical pile. Later, the upper part of the ridge was built until late Miocene, when a recurved spit developed, signaling the time when the south-flowing contour currents finally swung around the ridge and north along its west slope. Finally, the crest was built in late Miocene to Holocene.

More recent piston coring in the Blake-Bahama Basin has recovered early late Miocene turbidites from a prominent reflector (Sheridan et al., 1974). West of the Blake Outer-Ridge, nearly 1,000 m of hemipelagic sediments overlie these late Miocene turbidite layers. These hemipelagic sediments were deposited by northerly flowing contour currents from late Miocene through Holocene. The late Miocene, therefore, marks an important change in the oceanic current regime.

Below the sediment accumulations is an unusually thick oceanic crust with intermediate seismic velocities, 7.2-7.4 km/sec (Hersey et al., 1959). In spite of the greater sediment load under the Blake Outer-Ridge, the oceanic crustal structure appears to have subsided to a greater depth in the Blake-Bahama Basin, implying that such subsidence was caused by subcrustal processes, and not sediment loading, although subsidence of the Blake Plateau may be depressing the adjacent Blake-Bahama Basin in spite of its lack of sediments.

The deeper crustal structure under the Blake Plateau (Fig. 10) is not directly observed by seismic refraction data, although Hersey et al. (1959) have reported mantle velocities on a single refraction profile north of this section. The existence of continental crust near the edge of the Blake Plateau is analogous to Orphan Knoll, off Newfoundland. Its presence would explain the buoyancy of the reef structure above this ridge simultaneous with subsidence to the west under the Blake Plateau, and would be compatible with the generally acceptable reconstructions of the North American and African continents for this area (Bullard et al., 1965; Dietz et al., 1970; Emery and Uchupi, 1972).

By projection of crustal models for the Bahamas

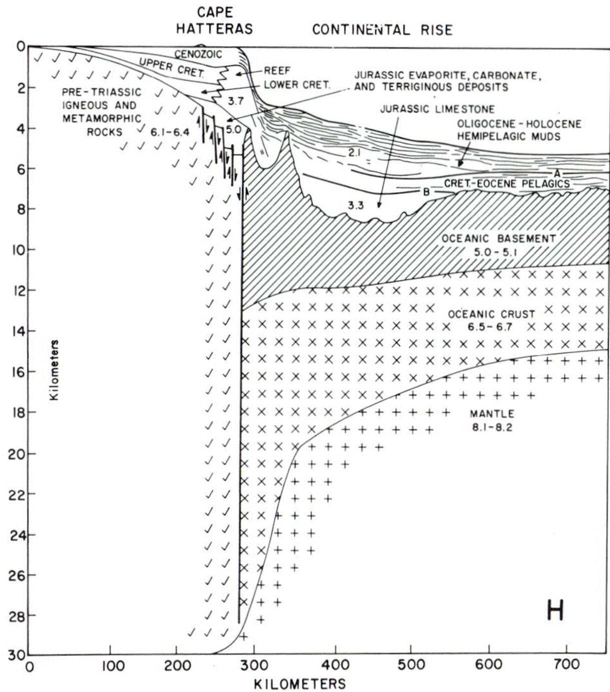

Fig. 9. Diagrammatic strucutural cross section of the Cape Hatteras continental margin (modified after Maher, 1965; Brown et al., 1972; Emery et al., 1970; Rona and Clay, 1967; Houtz and Ewing, 1964; Emery and Uchupi, 1972; Ewing and Hollister, 1972).

(Sheridan, 1972), the basinal area of the Blake Plateau is interpreted to be underlain by a thick oceanic crust of intermediate seismic velocity, 7.2-7.4 km/sec, and a mantle at about 23 km. The oceanic and denser nature of this crust would explain its subsidence, well documented since earliest Cretaceous (Sheridan, 1969) and probably occurring before that in the Jurassic. Such a crust would also be compatible with the Rayleigh wave-dispersion characteristics of this crust, which would be satisfied by such a model (Tarr, 1969; Sheridan, 1972). Tarr (1969) finds that the Blake Plateau crust gives dispersion curves similar to those of the continental rise areas.

Bahama Islands Area

The shallow sedimentary structures of the Bahama Platform were used to contruct Figure 11, based on Maher (1965), Sheridan et al. (1966), Ball et al. (1971), Uchupi et al. (1971), Paulus (1972), and Meyerhoff (1973). These data include a few deep oil wells in the Bahamas, where Lower Cretaceous and Upper Jurassic carbonates were penetrated at depths of 4.5 to 5.5 km, respectively; a few seismic-refraction profiles showing deep, high-velocity layers, 5.1-6.2 km/sec, associated with deep carbonates; the DSDP hole 98 in Northeast Providence Channel, which bottomed in Upper Cretaceous shallow-water limestone; and seismic reflection profiles in Exuma Sound, which reveal possible diapirs.

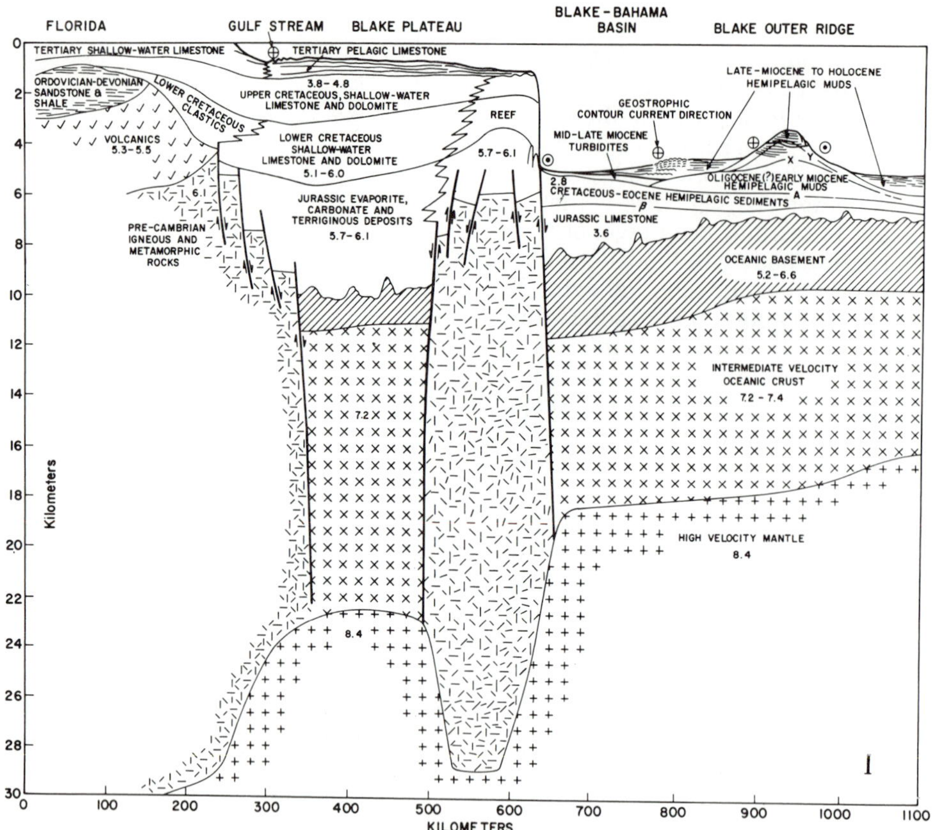

Fig. 10. Diagrammatic structural cross section across the Florida Platform, Blake Plateau, Blake-Bahama Basin and Blake-Outer-Ridge (modified after Sheridan et al., 1966; J.N. Ewing et al., 1966; JOIDES, 1965; Hersey et al., 1959; Heezen et al., 1966; Markl et al., 1970; Ewing and Hollister, 1972; Amos et al., 1971; Tarr, 1969; and Sheridan et al., 1973). Current direction symbols: cross denotes into section, dot denotes out of section.

Essentially, the upper part of the Bahama Platform is a great thickness of carbonates of Jurassic to Holocene age, accumulated by organic growth. In the Upper Cretaceous (certainly by the Tertiary) there were channels developed separating and isolating individual banks. Meanwhile, the reef and shallow-water limestone buildup continued on the banks, with continued subsidence.

Dietz et al. (1970) suggested that the channels are a natural consequence of the reef buildup. Other possibilities include the possible fault control of the general position and orientation of the channels (Sheridan, 1971). Geophysical anomalies, such as magnetics (Drake et al., 1963; Taylor et al., 1968) and gravity (Talwani, 1960; Uchupi et al., 1971) coincident with these channel walls implies basement control of position, and the linear nature of these anomalies suggests possible fault-block origins. Moody (1973) has indicated the position of some of these faults, and has suggested that they are major wrench faults. Indeed, the trend of these is N 50°-65°W, parallel to major Atlantic Ocean fracture zones, so that Uchupi et al. (1971) and Emery and Uchupi (1972) have implied this wrench-fault origin for these Bahama channels.

Continuation of the reef growth throughout the great subsidence since Jurassic could have maintained the channels in their present position. These faults also could have been active throughout Jurassic into the Upper Cretaceous (Sheridan et al., 1971) and thus offset the ocean floor in the latter part of the Upper Cretaceous to form the modern locus of reef growth and channel formation which began only then.

The bottom of the thick carbonate section is shown as a Jurassic salt deposit. As suggested by Dietz et al. (1970), these deposits are thought to be a continuation of the Louann salt of the Gulf of Mexico and the salt of Cuba (Meyerhoff and Hatten, 1968), of early Jurassic age. The presence of this salt is speculative, but doming of this salt may have occurred along fault-zone weaknesses to form the diapirs under Exuma South (Ball et al., 1968).

Beneath the sediments of the Bahamas, the basement is thought to lie at depths of around 10 km, based on the gradients of magnetic anomalies (Uchupi et al., 1971). The character of this basement is unknown from direct drilling and its origin is still problematic. Dietz et al. (1970) favor oceanic origin for the crust.

A 14-km crust of intermediate compressional wave velocity, 7.2 km/sec, below the high-velocity carbonates, 6.2 km/sec, and above a relatively high-velocity mantle, 8.4 km/sec, would give the appropriate Rayleigh wave dispersion characteristics observed for earthquake waves passing through

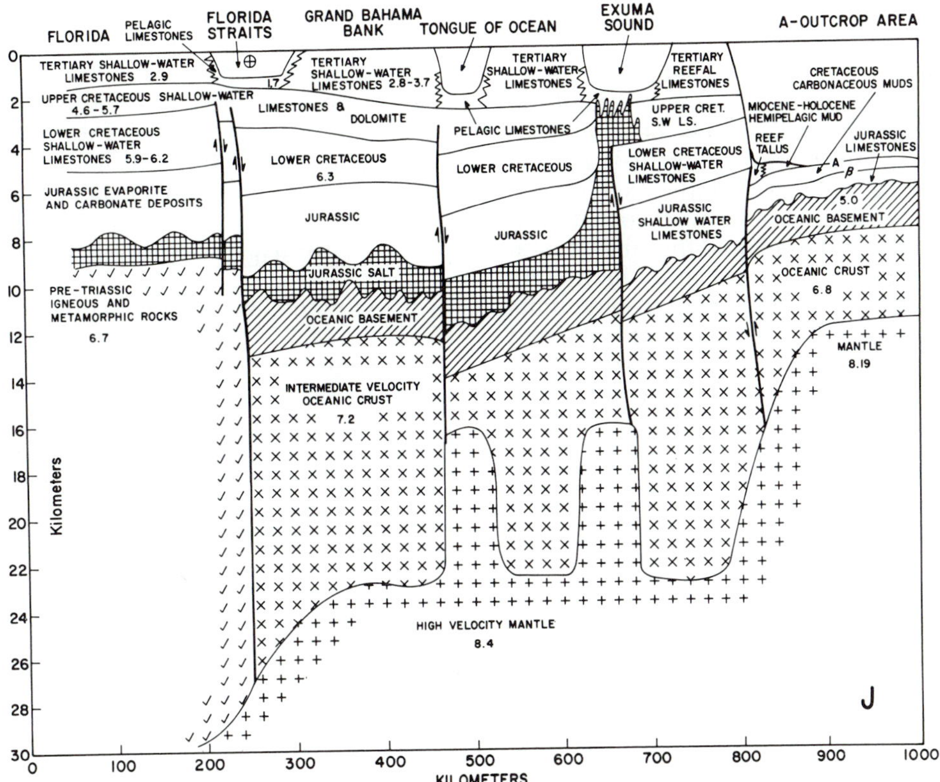

Fig. 11. Diagrammatic structural cross section across the Bahama Platform (based on Sheridan et al., 1966; Maher, 1965; Talwani, 1960; Ball et al., 1968, 1971; Uchupi et al., 1971; Sheridan, 1972; Ewing and Hollister, 1972; Dietz et al., 1970; Paulus, 1972; Meyerhoff, 1973).

the Bahamas (Sheridan, 1972). Such a structure would be more oceanic in character than continental. The gravity anomalies over the western Bahamas apparently favor a mantle depth deeper than 10 km (Talwani, 1960).

Although the origin and character of the basement and crust is problematic, given the fact that existing geophysical data could be satisfied by the presence of oceanic type crust, with intermediate velocities, 7.2 km/sec, it seems reasonable to accept this interpretation, since it also obviates the apparent overlaps of the Bahama salient and the African Precambrian shield in pre-drift reconstructions. Accordingly, the section (Fig. 11) shows a block-faulted oceanic crust under the Bahama Platform, upon which were deposited the thick, shallow-water evaporite and carbonate deposits of the Mesozoic and Cenozoic.

The Mesozoic and Cenozoic sediments are an order of magnitude thinner in the adjacent deep ocean than on the Bahama Platform, as evidenced in DSDP holes, 99, 100, and 101 (Ewing and Hollister, 1972), drilled in an area where horizon A crops out. Here, a few hundred meters of Cretaceous, carbonaceous, hemipelagic mud overlies a thin layer of Upper Jurassic pelagic limestones, horizon β.

STRUCTURAL FRAMEWORK

Cross sections A-J (Figs. 2-11) interpret the basement structure of the Atlantic continental margin to be of a block-faulted, rifted style, with the accumulation of great thicknesses of shallow-water sediments in subsided marginal basins. Such faulting would be compatible with the interpretation that North America and Africa, contiguous during the late Paleozoic in a reconstructed fit similar to that of Bullard et al. (1965), have rifted apart along lines near their continental slopes, to spread laterally with the formation of the Mesozoic and Cenozoic Atlantic Ocean (Pitman and Talwani, 1972).

On this basis, speculations can be made regarding timing of major events, similar to those proposed by Schneider (1969). Evolving in Triassic from an uplifted continental rift valley, similar perhaps to the present East African rifts, the Atlantic opened to be a structure similar to the Red Sea in the Jurassic. As the Atlantic Ocean spread apart with the new crust accreting at the Mid-Ocean Ridge, the margins subsided. Through the Jurassic and Cretaceous thermohaline circulation was restricted, and Jurassic salt may have been deposited in parts of the now-deeper ocean, under the continental rise, and carbonaceous sediments were left unoxidized on the Cretaceous ocean floor. Later in the Cenozoic, when the Norwegian Sea opened, circulation of swift Arctic bottom water transported great amounts of sediment to build the continental rise.

The evolution of the great geosynclinal accumulations of the Atlantic margin will be difficult to decipher until sufficient drilling has been accomplished. Brown et al. (1972) analyze the drilling data from North Carolina to Long Island, showing the evolutionary details which can be worked out with sufficient data. These authors found that the Coastal Plain basement in this area is broken by faults into distinguishable blocks. Different stratigraphic units thicken or thin in response to block movements, and the history of faulting can be deciphered for the Cretaceous through the Holocene. These blocks are interpreted to be related to a major right lateral wrench-fault system.

The basic thesis of Brown et al. (1972) can be applied to the entire Atlantic continental margin of North America (Fig. 12). The basement is interpreted as broken by faults into distinguishable blocks, having distinct geological histories. Major faults were identified using the magnetic data of Taylor et al. (1968), Sheridan and Drake (1968), Hood (1966), Fenwick et al. (1968), Mayhew et al. (1970), and Drake et al. (1963). Abrupt offsets in linear anomalies, linear gradients, and alignments of circular anomalies were identified as being produced by possible basement fractures. Also, published faults, such as those in the Gulf of Mexico and the S65°E Texas trends (Moody, 1973), the White Mountain volcanic trend (Ballard and Uchupi, 1972), the Cabot Fault, the Orpheous Fault structure (Loncarevic and Ewing, 1967), and the various oceanic fracture zones, such as the Gibbs, Newfoundland Ridge, Kelvin, Norfolk, Blake Spur (Emery and Uchupi, 1972), and Great Abaco (Sheridan et al., 1970), are located on the map. The generalized contours of the pre-Jurassic basement were made with the idea that the mapped faults controlled the major basins of the margin.

The Great Abaco and Blake Spur fracture zones appear to bound the Blake Plateau Basin, and the Norfolk and Kelvin fracture zones bound the Baltimore Canyon Trough. The possible connection of the Newfoundland Ridge fracture zone with the Orpheous fault may produce the structural discontinuity between the deeper parts of the Scotian Shelf Basin and those under the Grand Banks. Other fault relationships should also be apparent on Figure 12.

A pattern of right lateral wrench faulting along the transverse oceanic fracture zones and associated tensional rifting along the hinge zones of the marginal basins is consistent with the proposed overall clockwise rotation of North America throughout the Mesozoic and Cenozoic. The active right lateral shearing was evident for a long period in the Cretaceous through Eocene, but dormant at times from the Eocene to Holocene (Brown et al., 1972). This change might be related to the higher spreading rates interpreted for the Atlantic Mid-Ocean Ridge in the Cretaceous, and the subsequent decrease after the Eocene (Pitman and Talwani, 1972). If the North American plate is now undergoing E-W compression (Sbar and Sykes, 1973), the shearing along the major NW-trending oceanic fracture zones may have taken on left-lateral motion (Moody, 1973). From the evidence of the tensional faulting of the marginal basins in the Cretaceous, however, this EW compression could not have existed then.

SUBSIDENCE OF ATLANTIC MARGIN

As noted earlier, the sedimentary rocks recovered by drilling and dredging from beneath the continental shelf and Blake Plateau are of shallow-water facies to depths of more than 5 km. The presence of Jurassic (?) salt, seismically indicated at depths even greater than 6 km, suggests that shallow-water conditions persisted since the Jurassic for most areas that are today in less than 2,000 m of water. The basement under these areas has subsided in some places as much as 10-12 km (Fig. 12).

To accommodate this much subsidence, the crust must have undergone some redistribution in density, if isostatic equilibrium were to have prevailed while the basement went from near sea level to its present 8-12 km depth. Some of the lower-density crustal material must have been removed or replaced by denser material. Several possible mechanisms have been discussed to achieve such density changes, e.g., "basification" (Belousov, 1962), "subcrustal erosion" (Van Bemmelen, 1964), "deserpentinization" (Hess, 1955), and "eclogite-basalt phase changes" (Ringwood, 1966).

During the early rift stage of development, the granitic basement which formed the flanks of the rift, and which now underlies the shelf basins, is interpreted to have been uplifted into mountain ranges about 2 km high. Such uplifts may have been compensated at depth by anomalous low-density mantle with intermediate acoustic velocity, 7.2-7.4 km/sec. The granitic crust continuously became thinner by erosional stripping at the top, possibly removing as much as 10 km of crust (Olson, 1973).

An obvious problem here is how to account for the deposition of all the eroded material. For example, if this process had gone on around the present Red Sea flanks, there should be great volumes of clastics deposited there. To the contrary, much evaporite and limestone is present, but great thicknesses of clastics are not found (Lowell and Genik, 1972). Accordingly, a subcrustal crustal-thinning process might be better for this kind of tectonic mode.

If a continental crust originally 30 km thick was fractured pervasively by tensional rifts in its lower 20 km, and these fractures were intruded with ultrabasic magma from the mantle, then the sandwiched structure of ultrabasic (8.1 km/sec) and granitic rocks (6.2 km/sec) might yield an intermediate density and velocity (7.2-7.4 km/sec). The early Cambrian basic dike swarms of Newfoundland (Williams and Stevens, 1969) now make up 50-80% of

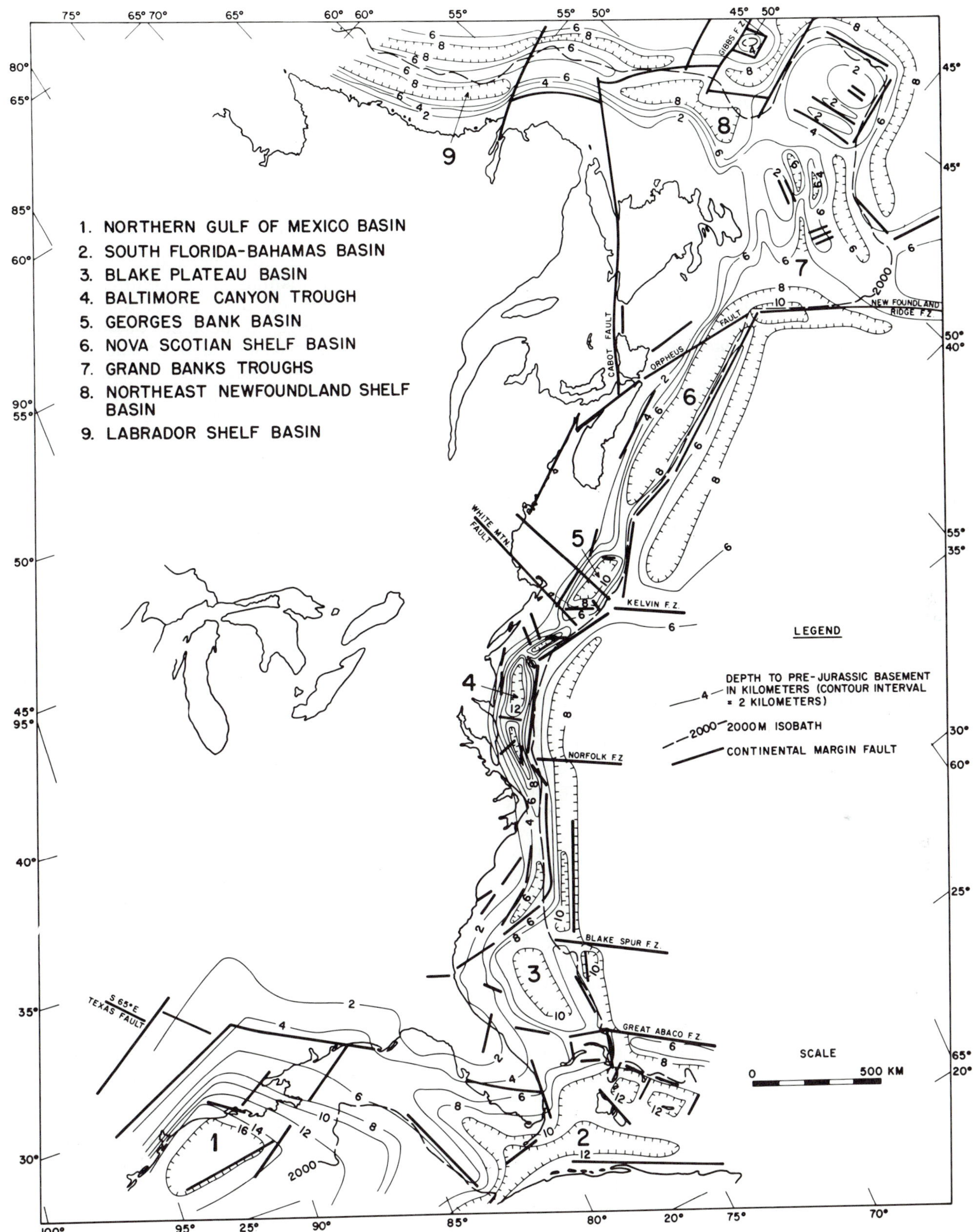

Fig. 12. Generalized basement map on the pre-Jurassic, based on Figures 2 to 11 and extrapolations. Major faults based on linear magnetic patterns of Taylor et al. (1968), Drake et al. (1963), Hood (1966), Sheridan and Drake (1968), Mayhew et al. (1970), and Fenwick et al. (1968), and other data from Moody (1973), Ballard and Uchupi (1972), Loncarevic and Ewing (1967), Sheridan et al. (1970), and Emery and Uchupi (1972). Also listed are the nine major sedimentary depocenters comprising the continental margin geosyncline.

the intruded rock, implying that a substantial density increase can occur. Refraction profiles near these intruded dikes have velocities of 7.2-7.4 km/sec for the deeper layers (Sheridan and Drake, 1968), suggesting a possible correlation. This pervasive intrusion of basaltic magma into granitic crust is similar to the process of basification described by Beloussov (1962), and thus this theory may have some application locally near continental margins.

An interesting aspect of the Atlantic Margin subsidence is that if the basement under the shelf and the Blake Plateau and Bahamas is restored to near sea level, the basement of the adjacent continental rise and Blake-Bahama Basin must also have been much shallower, perhaps within 1 km of sea level, unless there was large-scale vertical faulting. However, reflection horizons of Cretaceous and Tertiary age near the base of the continental slope (Emery et al., 1970) and the Blake-Bahama escarpment (Sheridan et al., 1970) suggest a passive onlap relationship and little or no differential shearing. Only along the base of the Bahama Escarpment, south of Great Abaco Canyon, is there some evidence of drag flexure in these sediments. Also, the Long Island well in the Bahamas indicates the top of Jurassic at depths below sea level nearly identical to the top of Jurassic limestones on the oceanic side of the scarp (Meyerhoff, 1973). These data appear to contradict large-scale differential throw between the shelf and rise and between the Blake Plateau-Bahamas and the Blake-Bahama Basin.

A pre-subsidence restoration of the basement surface, without major faulting, would imply perhaps a 1-km water depth for the continental rise and Blake-Bahama Basin during the Jurassic and a crustal structure similar to the present Red Sea. Such an environment, with limited circulation, would have been conducive to the accumulation of evaporites and highly carbonaceous sediments. This combination might have resulted in hydrocarbon accumulations of significant quantities, as suggested by Schneider (1969).

The preferred mechanism here for the subsidence of the continental rise (which may, in fact, be pulling down the attached Atlantic Margin) is the cooling of the 100-km-thick lithosphere and consequent eclogite-basalt phase changes (Wyllie, 1971). If only a part of the uper mantle lithosphere is gabbroic, this converts to eclogite as the lithosphere spreads far enough from the Mid-Ocean Ridge to cool. Only a 10% density change would contract the 100-km-thick lid on the asthenosphere to cause nearly 5 km of subsidence.

Earthquake-wave transmissions through the western Atlantic lithosphere show that the upper mantle has higher shear wave velocities (4.7 km/sec) for the area nearest the continental slope (Hart and Press, 1973). The lithosphere near the Mid-Ocean Ridge has lower velocities. These higher shear velocities are equivalent to the 8.4 km/sec compressional wave velocities found for the upper mantle under the Blake Outer Ridge (Hersey et al., 1959) (Fig. 10) and under the Bahamas (Sheridan, 1972) (Fig. 11). Such high velocities imply a denser upper mantle.

SUMMARY AND CONCLUSIONS

1. Major morphological features of the Atlantic border of North America, such as the slope, rise, and the Blake-Bahama escarpment, are formed totally within the thick sedimentary accumulations of the continental margin geosyncline.

2. The transition from continent to ocean occurs under the continental slope and is marked in places by a ridge of oceanic basement, which produces the east coast magnetic anomaly.

3. Jurassic to Holocene sediments reach thicknesses of 8 to 12 km in marginal downfaulted basins, aligned with the continental margin from the Bahamas to Labrador.

4. A structural model for the fault-block pattern of the North American Atlantic continental margin relates right-lateral shearing along major transverse fracture zones to tensional faulting along the hingements of the marginal basins. This pattern is compatible with the rotations of the North American margin as proposed in plate tectonics.

5. Drastic subsidence of the continental margin has occurred since the Jurassic, carrying shallow-water sediments to depths of 8-12 km under the continental shelf, Blake Plateau, and Bahamas. The continental rise has subsided in connection with the rest of the margin, and the deeper sediments of this province may be of shallower-water origin.

ACKNOWLEDGMENTS

The writing of this paper was supported by the University of Delaware Research Foundation Publication Fund. Credit must be given to all the many previous researchers, and apologies are given in advance for any results that have been unintentionally misinterpreted. The interpretations presented are the sole responsibility of the author. Special appreciation must be given to all the participants in the Penrose Conference on Continental Margins, whose discussions stimulated this reexamination of a well-studied subject. Also, much appreciation is felt for the recent AAPG East Coast Symposium, especially for the abundant petroleum-exploration data released there by the various industries. Without these data, much of this paper could not have been written.

BIBLIOGRAPHY

Amos, A. R., Gordon, A. L., and Schneider, E. D., 1971, Water masses and circulation patterns in the region of the Blake-Bahama Outer-Ridge: Deep Sea Res., v. 18, p. 145-165.

Ayrton, W. G., Birnie, D. E., Swift, J. H., Wellman, H. R., Harrison, D. B., Steuulak, J. G., Waylett, D. C., and Wilkinson, R. A. F., 1973, Grand Banks regional geology (abs.): Tech. Program, East Coast Offshore Symposium-Baffin Bay to the Bahamas, Eastern Sect. AAPG. Atlantic City, p. 5.

Ball, M. M., Gaudet, R. M., and Leist, G., 1968, Sparker reflection seismic measurements in Exuma South, Bahamas (abs.): EOS (Trans. Am. Geophys. Union), v. 49, p. 196.

______, Dash, B. P., Harrison, C. G. A., and Ahmed, K. O., 1971, Refraction seismic measurements in the northeastern Bahamas (abs.): EOS (Trans. Am. Geophys. Union), v. 52, p. 252.

Ballard, R. D., and Uchupi, E., 1972, Carboniferous and Triassic rifting: a preliminary outline of the Gulf of Maine's tectonic history: Geol. Soc. America Bull., v. 83, p. 2285-2302.

Bartlett, G. A., 1969, Cretaceous biostratigraphy of the Grand Banks of Newfoundland: Maritime Sediments, v. 5, p. 4-14.

Beloussov, V. V., 1962, Basic problems in geotectonics: New York, McGraw-Hill, 809 p.

Brakl, J., Clay, C. S., and Rona, P. A., 1968, Interpretation of a magnetic anomaly on the continental rise off Cape Hatteras: Jour. Geophys. Res., v. 73, p. 5313-5316.

Brown, P. M., Miller, J. A., and Swain, F. M., 1972, Structural and stratigraphic framework, and spatial distribution of permeability of the Atlantic Coastal Plain, North Carolina to New York: U.S. Geol. Survey Prof. Paper 796, 79 p.

Bullard, E. D., Everett, J. E., and Smith, A. G., 1965, The fit of the continents around the Atlantic, *in* Blackett, P. M. S., Bullard, E. C., and Runcorn, S. K., eds., A Symposium on continental drift: Royal Soc. London Phil. Trans., sec. A, v. 258, p. 41-51.

Burk, C. A., 1968, Buried ridges within continental margins: Trans. N.Y. Acad. Aci., ser. 2, v. 30, no. 3, p. 397-409.

Dewey, J. F., and Bird, J. M., 1971, Origin and emplacement of the ophiolite suite: Appalachian ophiolites in Newfoundland: Jour. Geophys. Res., v. 76, p. 3179-3206.

Dietz, R. S., Holden, J. C., and Sproll, W. P., 1970, Geotectonic evolution and subsidence of Bahama Platform: Geol. Soc. America Bull., v. 81, p. 1915-1928.

Dowling, J. J., 1968, The east coast onshore experiment, II. Seismic refraction measurements on the shelf between Cape Hatteras and Cape Fear: Seismol. Soc. America Bull., v. 58, p. 821-834.

Drake, C. L., and Girdler, R. W., 1964, A geophysical study of the Red Sea: Geophys. Jour. Roy. Astron. Soc., v. 8, p. 473-495.

______, Ewing, M., and Sutton, G. H., 1959, Continental Margins geosynclines; the east coast of North America north of Cape Hatteras, *in* Physics and chemistry of the earth, v. 3, Elmsford, N.Y., Pergamon Press, p. 110-198.

______, Heirtzler, J. R., and Hirshman, J., 1963, Magnetic anomalies off eastern North America: Jour. Geophys. Res. v. 68, p. 5259-5275.

Emery, K. O., and Uchupi, E., 1972, Western North Atlantic Ocean; Topography, rocks, structure, water, life, and sediments: Am. Assoc. Petr. Geol. Mem. 17, 532 p.

______, Uchupi, E., Phillips, J. D., Bowin, C. O., Bunce, E. T., and Knott, S. T., 1970, Continental rise of eastern North America: Am. Assoc. Petr. Geol. Bull., v. 54, p. 44-108.

Ericson, D. B., Ewing, M., and Heezen, B. C., 1952, Turbidity currents and sediments in North Atlantic: Am. Assoc. Petr. Geol. Bull., v. 36, p. 489-511.

Ewing, G. N., Dainty, A. M., Blanchard, J. E., and Keen, M. J., 1966, Seismic studies on the seaboard of Canada: The Appalachian system, I: Can. Jour. Earth Sci., v. 3, p. 89-109.

Ewing, J. I., and Ewing, M., 1959, Seismic-refraction measurements in the Atlantic Ocean basins, in the Mediterranean Sea, on the Mid-Atlantic Ridge, and in the Norwegian Sea: Geol. Soc. America Bull., v. 70, p. 291-318.

Ewing, J. I., and Hollister, C. D., 1972, Regional aspects of Deep Sea Drilling in the Western North Atlantic, *in* Initial reports of the Deep Sea Drilling Project, v. XI, Scientific Staff, eds., p. 951-976.

Ewing, J. I., LePichon, X., and Ewing, M., 1963, Upper stratification of Hudson apron region: Jour. Geophys. Res., v. 68, p. 6303-6316.

Ewing, J. I., Ewing, M., and Leyden, R., 1966, Seismic-profiler survey of Blake Plateau: Am. Assoc. Petr. Geol. Bull., v. 50, p. 1948-1971.

Ewing, M., Crary, A. P., and Rutherford, H. M., 1937, Geophysical investigations in the emerged and submerged Atlantic coastal plain: pt. I, Methods and results: Geol. Soc. America Bull., v. 48, p. 753-801.

Ewing, M., Woollard, G. P., and Vine, A. C., 1938, Geophysical investigations in emerged and submerged Atlantic coastal plain: pt. III: Barnegat Bay, N.J., section: Geol. Soc. America Bull., v. 50, p. 257-296.

Ewing, M., Woollard, G. P., and Vine, A. C., 1940, Geophysical investigations in the emerged and submerged Atlantic coastal plain: pt. IV: Cape May, N.J., section: Geol. Soc. America Bull., v. 51, p. 1821-1840.

Fenwick, D. K. B., Keen, M. J., Keen, C., and Lambert, A., 1968, Geophysical studies of the continental margin northeast of Newfoundland: Can. Jour. Earth Sci., v. 5, p. 483-500.

Fox, P. J., Heezen, B. C., and Harian, A. M., 1968, Abyssal antidunes: Nature, v. 220, p. 470-472.

Garrison, L. E., 1970, Development of continental shelf south of New England: Am. Assoc. Petr. Geol. Bull., v. 54, p. 109-124.

Gibson, 1970, Late Mesozoic-Cenozoic tectonic aspects of the Atlantic coastal margin: Geol. Soc. America Bull., v. 81, p. 1813-1822.

Grant, A. C., 1966, A continuous seismic profile on the continental shelf off NE. Labrador: Can. Jour. Earth Sci., v. 3, p. 725-730.

______, 1968, Seismic-profiler investigation of the continental margin northeast of Newfoundland: Can. Jour. Earth Sci., v. 5, p. 1187-1198.

Grim, M. S., Drake, C. L., and Heirtzler, J. R., 1970, Sub-bottom study of Long Island Sound: Geol. Soc. America Bull., v. 81, p. 649-666.

Hart, R. S., and Press, F., 1973, Sn velocities and the composition of the lithosphere in the regionalized Atlantic: Jour. Geophys. Res., v. 78, p. 407-411.

Hayes, D. E., Frakes, L. A., Barrett, D., Burns, D. A., Pei-Hsin Chen, Ford, A. B., Kaneps, A. G., Kemp. E. M., McCollum, D. W., Piper, D. J. W., Wall, R. E., and Webb, P. N., 1973, Leg 28: deep-sea drilling in the southern ocean: Geotimes, v. 18, no. 6, p. 19-24.

Heezen, B. C., and Drake, C. L., 1964, Grand Banks slump: Am. Assoc. Petr. Geol. Bull., v. 48, p. 221-225.
Heezen, B. C., and Sheridan, R. E., 1966, Lower Cretaceous rocks (Neocomian-Albian) dredged from Blake escarpment: Science, v. 154, p. 1644-1647.
Heezen, B. C., Hollister, C., and Ruddiman, W. F., 1966, Shaping of the continental rise by deep geostrophic contour currents: Science, v. 152, p. 502-508.
Heirtzler, J. R., and Hayes, D. E., 1967, Magnetic boundaries on the North Atlantic Ocean: Science, v. 157, p. 185-187.
Hersey, J. B., Bunce, E. T., Wyrick, R. F., and Dietz, F. T., 1959, Geophysical investigation of the continental margin between Cape May, Virginia, and Jacksonville, Florida: Geol. Soc. America Bull., v. 70, p. 437-466.
Hess, H. H., 1955, Serpentine, orogeny, and epeirogeny; *in* Crust of the earth: Geol. Soc. America Spec. Paper 62, p. 391-408.
———, 1962, History of the ocean basins, *in* Engle, A. E. J., James, H. L., and Leonard, B. P., eds., Petrologic studies: a volume in honor of A. F. Buddington: Geol. Soc. America, p. 599-620.
Holtedahl, H., 1958, Some remarks on geomorphology on continental shelves off Norway, Labrador, and southeast Alaska: Jour. Geology, v. 66, p. 461-471.
Hood, P., 1966, Magnetic measurements on the continental shelves of eastern Canada, *in* Poole, W. H., ed., Continental margins and island arcs: Geol. Survey Can. Paper 66-15, p. 19-32.
Houtz, R. E., and Ewing, J. I., 1964, Sedimentary velocities of the Western North Atlantic: Seismol. Soc. America Bull., v. 54, p. 867-895.
JOIDES, 1965, Ocean drilling on the continental margin: Science, v. 150, p. 709-716.
Keen, M. J., Keen, C. E., and Barrett, D. L., 1973, Changes in crustal properties near the continental margin (abs.): EOS (Trans. Am. Geophys. Union), v. 54, p. 332.
Kelling, G., and Stanley, D. J., 1970, Morphology and structure of the Wilmington and Baltimore submarine canyons, eastern United States: Jour. Geology, v. 78, p. 637-660.
King, L. H., and Maclean, B., 1970, Continuous seismic reflection study of the Orpheus gravity anomaly: Am. Assoc. Petr. Geol. Bull., v. 54, p. 2007-2031.
Kraft, J. C., Sheridan, R. E., and Maisano, M., 1971, Time-stratigraphic units and petroleum entrapment models in the Baltimore canyon basin of the Atlantic continental-margin-geosyncline: Am. Assoc. Petr. Geol. Bull., v. 55, p. 658-679.
Laughton, A. S., 1972, The southern Labrador Sea—a key to the Mesozoic and early Tertiary evolution of the North Atlantic, *in* Initial reports of the Deep Sea Drilling Project, v. XII, Scientific Staff, eds., p. 1155-1180.
———, Berggren, W. A., Benson, R., Davies, T. A., Franz, U., Musich, L., Peich-Nielsen, K., Ruffman, A., van Hinte, J. E., Whitmarsh, R. B., 1972, Site 111, *in* Initial reports of the Deep Sea Drilling Project, v. XII, Scientific staff, eds., p. 33-82.
Loncarevic, B. D., and Ewing, G. N., 1967, Geophysical study of the Orpheus gravity anomaly, *in* Proc. 7th World Petr. Congr., Mexico, v. 2, Origin of oil, geology and geophysics: Amsterdam, Elsevier, p. 827-835.
Lowell, J. C., and Genik, G. J., 1972, Sea floor spreading and structural evolution of the southern Red Sea: Am. Assoc. Petr. Geol. Bull., v. 56, p. 247-259.
McIver, N. L., 1973, Stratigraphy of the continental shelf, offshore Nova Scotia (abs.): Tech. Program, East Coast Offshore Symposium—Baffin Bay to the Bahamas, Eastern Sec. AAPG, Atlantic City, p. 7.
McMilliam, N. J., 1973, Geology of Labrador Shelf, Canada (abs): Tech. Program, East Coast Offshore Symposium—Baffin Bay to the Bahamas, Eastern Sec. AAPG, Atlantic City, p. 4.
Maher, J. C., 1965, Correlations of subsurface Mesozoic and Cenozoic rocks along the Atlantic coast: Am. Assoc. Petr. Geol. Cross Sec. Publ, 18 p.
Markl, R. G., Bryan, G. M., and Ewing, J. E., 1970, Structure of the Blake-Bahama Outer-Ridge: Jour. Geophys. Res., v. 75, p. 4539-4555.
Marlowe, J. I., 1965, Probable Tertiary sediments from a submarine canyon off Nova Scotia: Marine Geology, v. 3, p. 263-268.
Mattick, R. E., Weaver, N. L., Foote, R. Q., and Ruppel, B. D., 1973, A preliminary report on U.S. Geological Survey geophysical studies of the Atlantic outer continental shelf (abs.): Tech. Program, East Coast Offshore Symposium—Baffin Bay to the Bahamas, Eastern Sc. AAPG, ATlantic City, p. 9.
Mayhew, M. A., Drake, C. L., and Nafe, J. E., 1970, Marine geophysical measurements on the continental margins of the Labrador Sea: Can. Jour. Earth Sci., v. 7, p. 199-214.
Meyerhoff, A. A., 1973, Bahamas Salient of North America (abs.): Tech. Program, East Coast Offshore Symposium—Baffin Bay to the Bahamas, Eastern Sec. AAPG, Atlantic City, p. 10.
———, and Hatten, C. W., 1968, Diapiric structures in central Cuba, *in* Braunstein, J., and O'Brien, G. D., eds., Diapirism and diapirs, Am. Assoc. Petr. Geol. Mem. 8, p. 315-357.
Moody, J. D., 1973, Petroleum exploration aspects of wrench-fault tectonics: Am. Assoc. Petr. Geol. Bull., v. 57, p. 449-476.
Officer, C. B., and Ewing, M., 1954, Geophysical investigations of the emerged and submerged Atlantic coastal plain, pt. VIII, Continental shelf, continental slope and continental rise south of Nova Scotia: Geol. Soc. America Bull., v. 65, p. 653-670.
Olson, W. S., 1973, Structural history and oil potential of the offshore area from Cape Hatteras to the Bahamas (abs.): Tech. Program, East Coast Offshore Symposium—Baffin Bay to the Bahamas, Eastern Sec. AAPG, Atlantic City, p. 9.
Paulus, F. J., 1972, The geology of site 98 and the Bahama platform, *in* Initial reports of the Deep Sea Drilling Project, v. XI, Scientific Staff, eds., p. 877-900.
Pitman, W. C., III, and Talwani, M., 1972, Sea floor spreading in the North Atlantic: Geol. Sco. America Bull., v. 83, p. 619-646.
Press, F., and Beckmann, W., 1954, Geophysical investigations in the emerged and submerged Atlantic coastal plain, pt. VIII, Grand Banks and adjacent shelves: Geol. Soc. America Bull., v. 65, p. 299-314.
Ringwood, A. E., 1966, An experimental investigation of the gabbro-eclogite transformation and some geophysical implications: Tectonophysics, v. 3, p. 383-427.
Rodgers, J., 1970, The tectonics of the Appalachians, *in* De Setteo, L. U., ed., Regional geology series: New York, Wiley-Interscience, 271 p.
Rona, P. A., and Clay, C. S., 1967, Stratigraphy and structure along a continuous seismic reflection pro-

file from Cape Hateras, North Carolina to the Bermuda Rise: Jour. Geophys. Res., v. 72, p. 2107-2130.

Sbar, M. L., and Sykes, L. R., 1973, Contemporary compressive stress and seismicity in eastern North America: an example of intraplate tectonics: Geol. Soc. America Bull., v. 84, p. 1861-1882.

Schneider, E. D., 1969, The evolution of the continental margins and possible long term economic resources: Soc. Petroleum Engineers AIME, Offshore Technol. Conf., preprints, Houston, Tex., p. 257-264.

______, and Johnson, G. L., 1970, Deep ocean diapir occurrences: Am. Assoc. Petro. Geol. Bull., v. 54, p. 2151-2169.

______, Fox, P. J., Hollister, C. G., Needham, H. D., and Heezen, B. C., 1967, Further evidence for contour currents in the western North Atlantic: Earth and Planetary Sci. Letters, v. 2, p. 351-359.

Schultz, L. K., and Grover, R. L., 1973, Geology of the Georges Bank Basin (abs.): Tech. Program, East Coast Offshore Symposium—Baffin Bay to the Bahamas, Eastern Sec. AAPG, Atlantic City, p. 8.

Sheridan, R. E., 1969, Subsidence of continental margins: Tectonophysics, v. 7, p. 219-229.

______, 1971, Geotectonic evolution and subsidence of Bahama Platform: discussion: Geol. Soc. America Bull., v. 82, p. 807-810.

______, 1972, Crustal structure of the Bahama platform from Rayleigh wave dispersion: Jour. Geophys. Res., v. 77, p. 2139-2145.

______, and Drake, C. L., 1968, Seaward extension of the Canadian Appalachians: Can. Jour. Earth Sci., v. 5, p. 337-373.

______, Drake, C. L., Nafe, J. E., and Hennion, J., 1966, Seismic-refraction study of continental margin east of Florida: Am. Assoc. Petr. Geol. Bull., v. 50, v. 1972-1991.

______, Houtz, R. E., Drake, C. L., and Ewing, M., 1969a, Structure of continental margin off Sierra Leone, West Africa: Jour. Geophys. Res., v. 74, p. 2512-2530.

______, Smith, J. D., and Gardner, J., 1969b, Rock dredges from Blake escarpment near Great Abaco Canyon: Am. Assoc. Petr. Geol. Bull., v. 53, p. 2551-2558.

______, Elliott, G. K., and Oostdam, B. L., 1970, Seismic reflection profile across Blake escarpment near Great Abaco Canyon: Am. Assoc. Petr. Geol. Bull., v. 54, p. 2032-2039.

______, Berman, R. M., and Corman, D. B., 1971, Faulted limestone block dredged from Blake escarpment: Geol. Soc. America Bull., v. 82, p. 199-206.

______, Golovchenko, X., and Ewing, J. I., 1973, Late Miocene turbidite horizon in the Blake-Bahama Basin: Manuscript in prep.

Stetson, H. C., 1939, Summary of sedimentary conditions on the continental shelf off the east coast of the United States, *in* Trask, P. S., ed., Recent Maine sediments: Am. Assoc. Petro. Geol., p. 230-244.

Swain, F. M., and Brown, P. M., 1972, Some Lower Cretaceous, Jurassic (?), and Triassic Ostracoda from Atlantic coastal region for use in hydrogeologic studies: U.S. Geol. Survey Prof. Paper 795.

Talwani, M., 1960, Gravity anomalies in the Bahamas and their interpretation [Ph.D. thesis]: Geology Dept., Columbia Univ. 89 p.

Tarr, A. C., 1969, Rayleight wave dispersion in the North Atlantic Ocean, Caribbean Sea, and Gulf of Mexico: Jour. Geophys. Res., v. 74, p. 1591-1607.

Taylor, P. T., Zietz, I., and Dennis, L. S., 1968, Geologic implications of aeromagnetic data for the eastern continental margin of the United States: Geophysics, v. 33, p. 755-780.

______, Greenewalt, D., Watkins, N. D., and Ellwood, B. B., 1973, Magnetic properties analysis of basalt beneath the quiet zone (abs.): EOS (Trans. Am. Geophys. Union), v. 54, p. 255.

Uchupi, E., and Emery, K. O., 1967, Structure of continental margin off Atlantic coast of United States: Am. Assoc. Petr. Geol. Bull., v. 51, p. 223-234.

______, Milliman, J. D., Luyendyk, B. P., Bowin, C. O., and Emery, K. O., 1971, Structure and origin of southeastern Bahamas: Am. Assoc. Petr. Geol. Bull., v. 55, p. 687-704.

Van Bemmelen, R. W., 1964, The evolution of the Atlantic megaundation: Tectonophysics, v. 1, p. 385-430.

Veatch, A. C., and Smith, P. A., 1939, Atlantic submarine valleys of the United States and the Congo submarine valley: Geol. Soc. America Spec. Paper 7, 101 p.

Vogt, P. R., Schneider, E. D., and Johnson, G. L., 1969, The crust and upper mantle beneath the sea, *in* Hart, P. J., ed., The earth's crust and upper mantle: Am. Geophys. Union Monogr. 13, p. 556-617.

______, Anderson, C. N., Bracey, D. R., and Schneider, E. D., 1970, North Atlantic magnetic smooth zones: Jour. Geophys. Res., v. 75, p. 3955-3968.

Williams, H., and Stevens, R. K., 1969, Geology of Belle Isle northern extremity of the deformed Appalachian miogeosynclinal belt: Can. Jour. Earth Sci., v. 6, p. 1145-1157.

Wilson, J. T., 1966, Did the Atlantic close and reopen?: Nature, v. 211, p. 676-681.

Worzel, J. L., 1965, Pendulum gravity measurements at sea, 1936-1959: New York, Wiley, 422 p.

______, 1968, Survey of continental margins, *in* Donovan, D. T., ed., Geology of Shelf Seas: Edinburgh, Oliver & Boyd, p. 117-152.

______, and Shurbert, G. L., 1955, Gravity anomalies at continental margins: Proc. Natl. Acad. Sci., v. 41, p. 458-469.

Wyllie, P. J., 1971, The dynamic earth: New York, Wiley, 416 p.

Zeigler, J. M., 1955, Seamounts near the eastern coast of North America: Woods Hole Oceanog. Inst. Ref. 55-17, 16 p. (multilithed).

Geophysics of Atlantic North America

M. A. Mayhew

INTRODUCTION

On the basis of geophysical data, the East Coast margin can be divided into three structural provinces which are largely unrelated to the shelf-slope-rise morphological provinces. The eastern province extends from the abyssal plains to the magnetic slope anomaly. The central province centers on the outer shelf. Its seaward boundary is the magnetic slope anomaly, which in the northern area lies well out over the continental rise. Model calculations are consistent with the view that the anomaly marks the edge of oceanic crust. The western boundary is ill-defined but lies somewhere beneath the inner shelf. The central province is characterized by great mobility. Deeply subsided fault-block basins in this belt are filled with huge thicknesses of sediments, the basal sequences of which are at least as old as Late Triassic-Early Jurassic. Roughly the lower half of the section in these basins is of high seismic velocity and has previously been interpreted as basement. The western province, which included much of the Appalachian crystalline belt, is a zone of relatively minor vertical mobility and thin sediments. The subsidence of the margin can be seen as the result of a regional rise of the aesthenosphere prior to rifting, which partially followed the Appalachian lithospheric grain, and which was concentrated in the central province.

PREVIOUS STUDIES AND ANALYSES

In 1937 the results of the first offshore seismic refraction work, consisting of lines off Virginia and Massachusetts, were published (M. Ewing et al, 1937). A companion paper by Miller (1937) was "concerned with the geological interpretation of results obtained in a geophysical investigation of somewhat novel character." The novelty quickly became the basic tool for working out the structure of the continental margin, and only in recent years has it been surpassed in effectiveness by the reflection techniques used in petroleum exploration.

Ewing's study was followed by a long series of refraction studies of the margin between Cape Hatteras and the Gulf of Maine, which were ultimately synthesized and interpreted as a whole by Drake et al. (1959). Figure 2 shows three seismic cross sections from that study which are typical of the shallow velocity structure of the margin segment from Cape Hatteras to the Scotian Shelf. Under the shelf low-velocity, evidently sedimentary, layers occupy a series of linear basins as deep as about 5 km off New Jersey (section D). The floors of the basins are defined by an apparently continuous series of refractors with velocities primarily in the range 5.0-6.0 km/sec, which rises in a broad arch beneath the outer shelf. Numerous authors (e.g., Drake et al., 1959; Drake and Nafe, 1968; Sheridan et al., 1966; Sheridan and Drake, 1968; Emery et al., 1970; Austin and Howie, 1973; Hobson and Overton, 1973) have attempted to describe the configuration of this surface with structural contour maps on the basis of some velocity criterion. Until recently it has been considered that what was being delineated was the basement, because where it has been sampled in wells on the coastal plain it is the crystalline complex of the Appalachian piedmont.

Near the continental slope the "basement" surface forms a series of subsurface escarpments, plunging to depths as great as 12 km (Drake et al., 1968) beneath the sedimentary prism of the continental rise before shallowing again seaward. The surface commonly forms a secondary, more subdued arch under the upper continental rise.

Seismic velocities in the "basement" show a systematic decrease seaward to velocities as low as 5 km/sec within the shelf-edge arch (Fig. 3). This is well below the normal range of velocities in the upper continental crust, about 5.6-6.4 km/sec, and Drake et al. (1959) ascribed this to a transition from high-grade metamorphic to unmetamorphosed sedimentary rocks of Paleozoic age away from the Appalachian axis. While it is now considered that the "basement" beneath the outer shelf is a high-velocity sedimentary sequence of Mesozoic age, it is noteworthy that it has long been recognized that the rock forming the shelf-edge arch could hardly be identical with rocks of the Piedmont.

Seismic velocities within the "basement" are still lower, as low as 4.4 km/sec, under the continental rise. Their range (Fig. 3) is nearly identical with oceanic layer 2, known to be composed of basaltic volcanics, and also with a group of velocities recorded in the axial zone of the Red Sea, which Drake and Girdler (1964) identified with sedimentary rocks, evaporites, and pyroclastics.

Subbasement velocities beneath the continental rise are in the range of oceanic layer 3 and higher and are also like subbasement velocities in the Red Sea axial zone, and velocities in the normal range of upper continental crust have rarely been measured. Thus the consensus (e.g., see Dietz, 1966) has been that the basement of the continental rise is oceanic in character and that a rather abrupt boundary

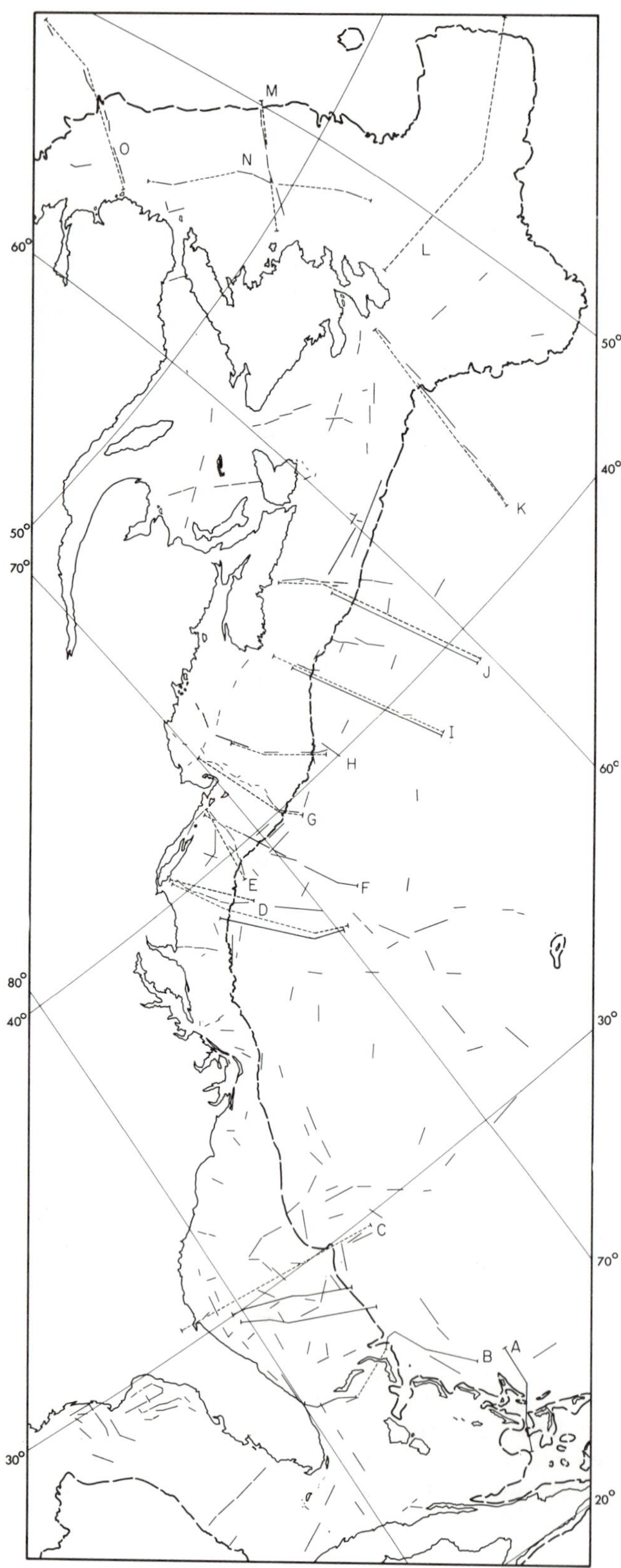

Fig. 1. Index map of Atlantic North America. Long dashed lines indicate profiles shown in Figs. 2, 4, 5, 6, and 7. Long solid lines are reflection traverses used to construct cross sections. Short solid lines are refraction profile locations compiled from many sources, included in the references.

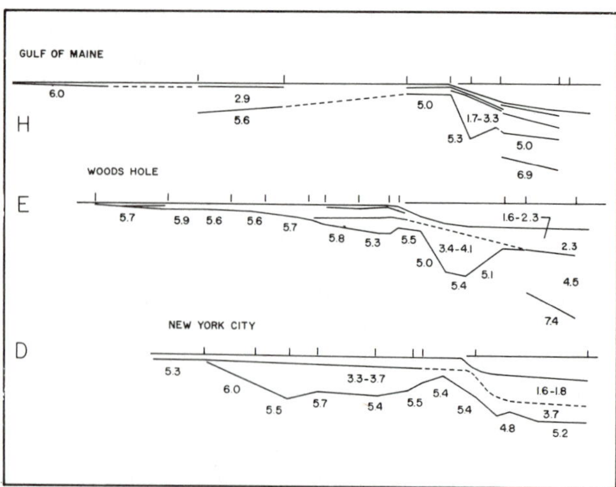

Fig. 2. Refraction from Drake et al. (1959) showing typical velocity structure of central margin segment. Numbers are velocities in km/sec. (Locations shown on Fig. 1.)

between continental and oceanic crust occurs in the vicinity of the continental slope, an idea that dates back to the isostatic arguments of Wegener (1924). There seems little doubt that the continental slope overlies an important structural boundary, but diverse kinds of data presently suggest that the transition from continental to oceanic crust is complex, and takes place over a very broad transition zone. This will be discussed further in a later section.

Drake et al. (1959) showed that in many respects the central segment of this continental margin is analogous to Kay's (1951) model of the early Paleozoic Appalachian geosyncline, with shelf basins representing the miogeosyncline, and the apparently much thicker continental rise sediments corresponding to the eugeosyncline. One of the important features of the eugeosyncline, the pre-orogenic volcanics not found in the miogeosyncline, might be analogous to what could be taken to be a thickened volcanic layer 2 beneath the continental rise. As the Appalachian geosyncline had come to be considered the "type" geosyncline, the east coast margin came to be thought of as the "type" continental margin and also as the prototype of modern geosynclines.

The model has been of inestimable value over the years, but at present our conceptions of continental margins are in a state of flux. Many studies of other parts of the Atlantic North American margin and of margins bordering other continents have revealed some important differences from the model, the significance of many of which are only now becoming clear with accelerated offshore exploration and drilling.

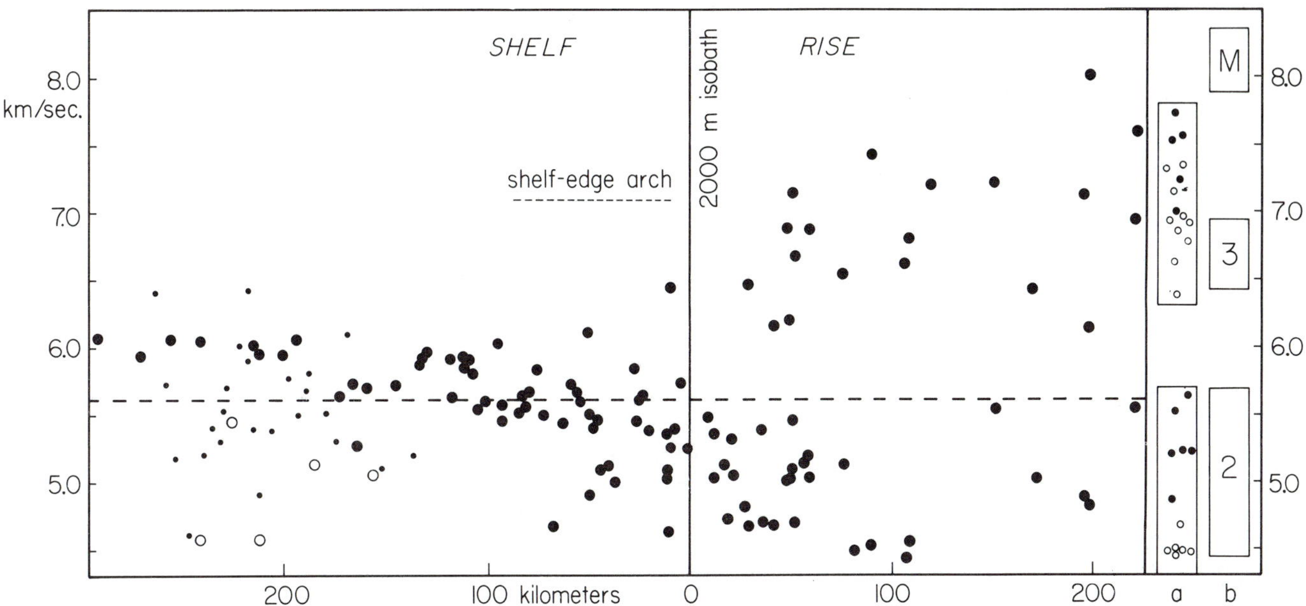

Fig. 3. Plot of velocities observed on refraction profiles of Fig. 1 versus distance from 2,000-m isobath for margin segment between Cape Hatteras and the Scotian Shelf. Small solid dots are values from short unreversed profiles on the coastal plain; open circles are from the Gulf of Maine, which may be over local sediment pockets. Column a on the right shows velocity measurements in the Red Sea axial trough as open circles (Drake and Girdler, 1964; Tramontini and Davies, 1969), and in the Labrador Basin as filled circles (Mayhew, 1969).Column b shows range of values in layers 2 and 3 and the mantle from the ocean basins as means ±1 S. D. (Raitt, 1963).

SOUTHERN AREA: BLAKE PLATEAU AND BAHAMAS PLATFORM

Seismic refraction and reflection coverage is extensive over the Blake Plateau (Hersey et al., 1959; J.I. Ewing et al., 1966; Sheridan et al., 1966; Emery and Zarudzki, 1967; Sheridan et al., 1971). It has been possible to a greater extent than for most parts of the margin to relate seismic layers observed in these surveys to rocks because a number of offshore holes have been drilled, and strata as old as Early Cretaceous has been sampled in outcrops along the Blake Escarpment (Figure 4). Some generalizations can be made from these studies which are relevant to interpretation of other areas of the margin.

First, the stratigraphy of the offshore is very complex; rapid facies transitions and unconformities are widespread, and both may act as refracting horizons. In this area the difficulties of using refraction for stratigraphic correlation and assignment of ages are well documented.

Second, high-velocity limestones are abundant in the deeper parts of the sedimentary section under the plateau. The velocities of these layers are principally in the range 5.0-6.0 km/sec, overlapping that usually assigned to basement rocks, and these strata commonly act to seismically mask the deeper rocks beneath them. In general, however, their velocities decrease somewhat, and they occur at shallower levels, toward the Blake Escarpment (Sheridan et al., 1966). This occurs in a manner similar to the "basement" of the outer shelf on sections to the north (compare Fig. 2). Where these strata crop out along the Blake Plateau they have proved to be algal reef carbonates of Early Cretaceous age (Heezen and Sheridan, 1966; Sheridan et al., 1969). Sheridan has argued that the escarpment was maintained by upbuilding of a great barrier reef complex through the Cretaceous while the foundation of the plateau and the Blake-Bahama Basin foundered from a shallow level with little relative vertical movement. The West Florida and Campeche escarpments of the Gulf of Mexico probably have a similar origin (Uchupi and Emery, 1968; Wilhelm and Ewing, 1972). Such a reef complex forms the acoustic basement beneath the Bahamas in section 1, Figure 4.

A very few profiles record a deeper layer under the plateau with velocities 5.7-6.6 km/sec. Whether this is more sediment or basement, it does indicate with certainty that the sediments are at least 8 km thick under the central part of the plateau, and that the reef complex is supported by some underlying structure.

Third, the foundation on which the large sediment accumulation of the plateau rests cannot be either typically oceanic or continental, but must be of a type unique to the marginal area, since otherwise large gravity anomalies would occur.

The offshore extent of the Eocambrian metamorphic basement rocks of Florida (Applin, 1951; Bass, 1969; Milton and Grasty, 1969) is unknown, but if they extend beneath the plateau, a drastically thinned continental crust is required to maintain isostatic equilibrium. Alternatively, Sheridan (1971) has suggested that remnant rift zone material

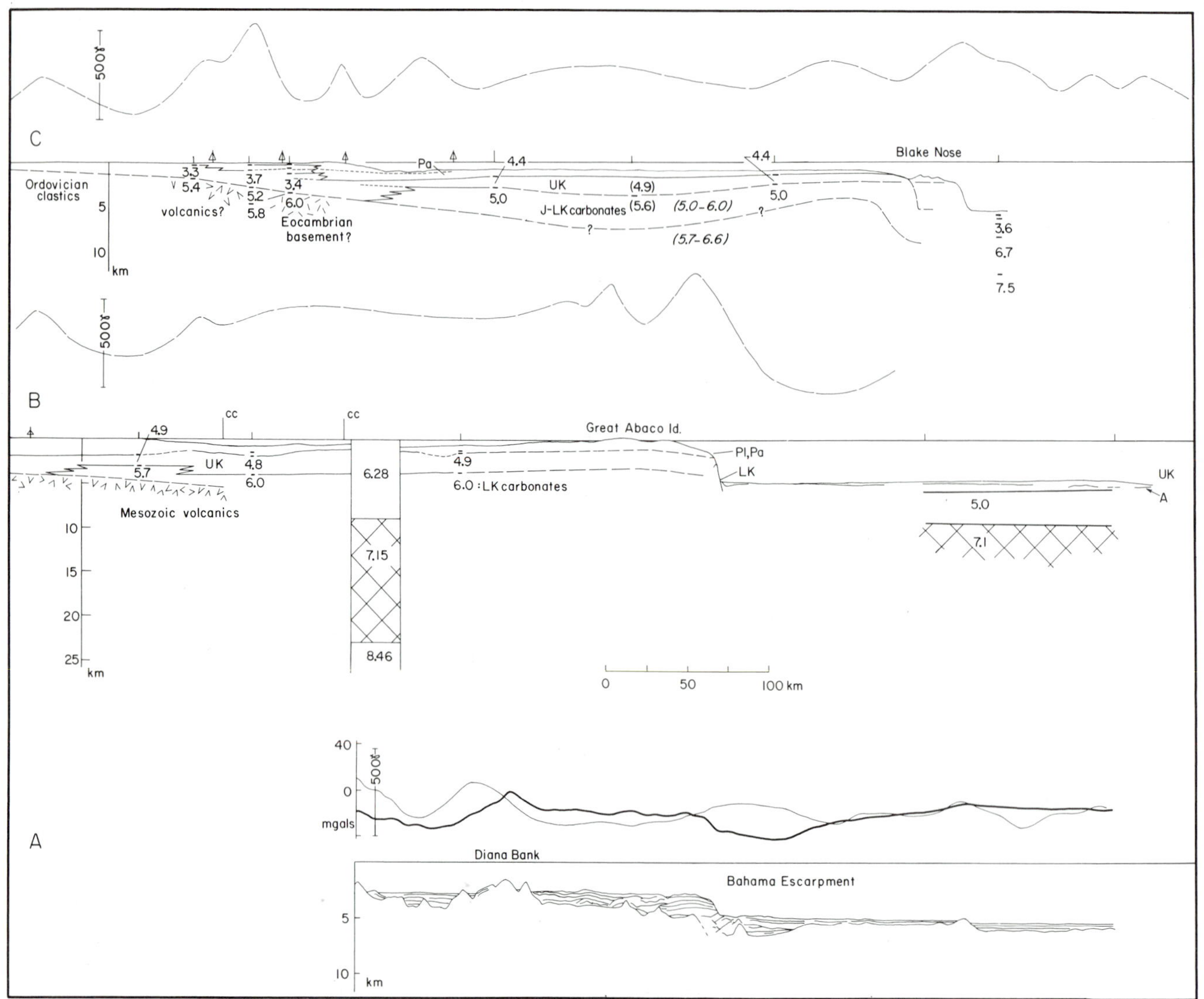

Fig. 4. Profiles of southern area. Section A across Bahama Escarpment based on reflection profile of Uchupi et al. (1971) assuming 2.0 km/sec sediment velocity (fig. 1). Free= air gravity shown as heavy line, magnetics as thin line. Sections B and C follow interpretations of Sheridan et al. (1966), based on their refraction data and from Hersey et al. (1959), Katz and Ewing (1956), and reflection data of J.I. Ewing et al. (1966), and Sheridan et al. (1969). Well data from J.I. Ewing et al. (1966), Applin (1951), and Emery and Zarudski (1967). Dredge data on Blake Escarpment from Sheridan et al. (1969); Pl; Pliocene; Pa: Paleocene; K; Cretaceous. Magnetic profiles along sections constructed from contours of Taylor et al. (1968). Refraction profile locations indicated by short vertical lines. Numbers are velocities in km/sec. Velocity model to fit Rayleigh wave dispersion data on section B from Sheridan (1971 a).

underlies the plateau and the adjacent basin, and that the two areas have subsided together from a shallow level by thermal contraction since the time of opening of the Atlantic. According to this interpretation, the basement of the Blake Plateau-northern Bahamas area is largely noncontinental.

Mayhew (in press) suggests that the reef complex originated over a linear fragment of continental crust rotated away from the Florida block during rifting, the intervening area being thinned and fragmented continental crust breached by volcanics, by analogy with Mohr's interpretation of the Afar block of the Red Sea. If this is correct, the analog of the Afar volcanics should form the basement deep beneath the central part of the plateau.

The Mesozoic volcanics sampled in a number of wells in central Florida are rhyolitic as well as basaltic, possibly owing to their presence within continental crust; measured ages are mainly Late Triassic to Late Jurassic (Milton and Grasty, 1969), but these are selective since all the volcanics were not suitable for determinations. A magnetic anomaly pattern associated with these volcanics appears to extend across the southern Blake Plateau to the northern Bahamas (Drake et al., 1963; Taylor et al., 1968; Bracey, 1968). Volcanic rocks of uncertain age in northern Florida and southern Georgia occur in the vicinity of an east-west belt of gravity highs and a zone of linear magnetic anomalies (see Fig. 9) which

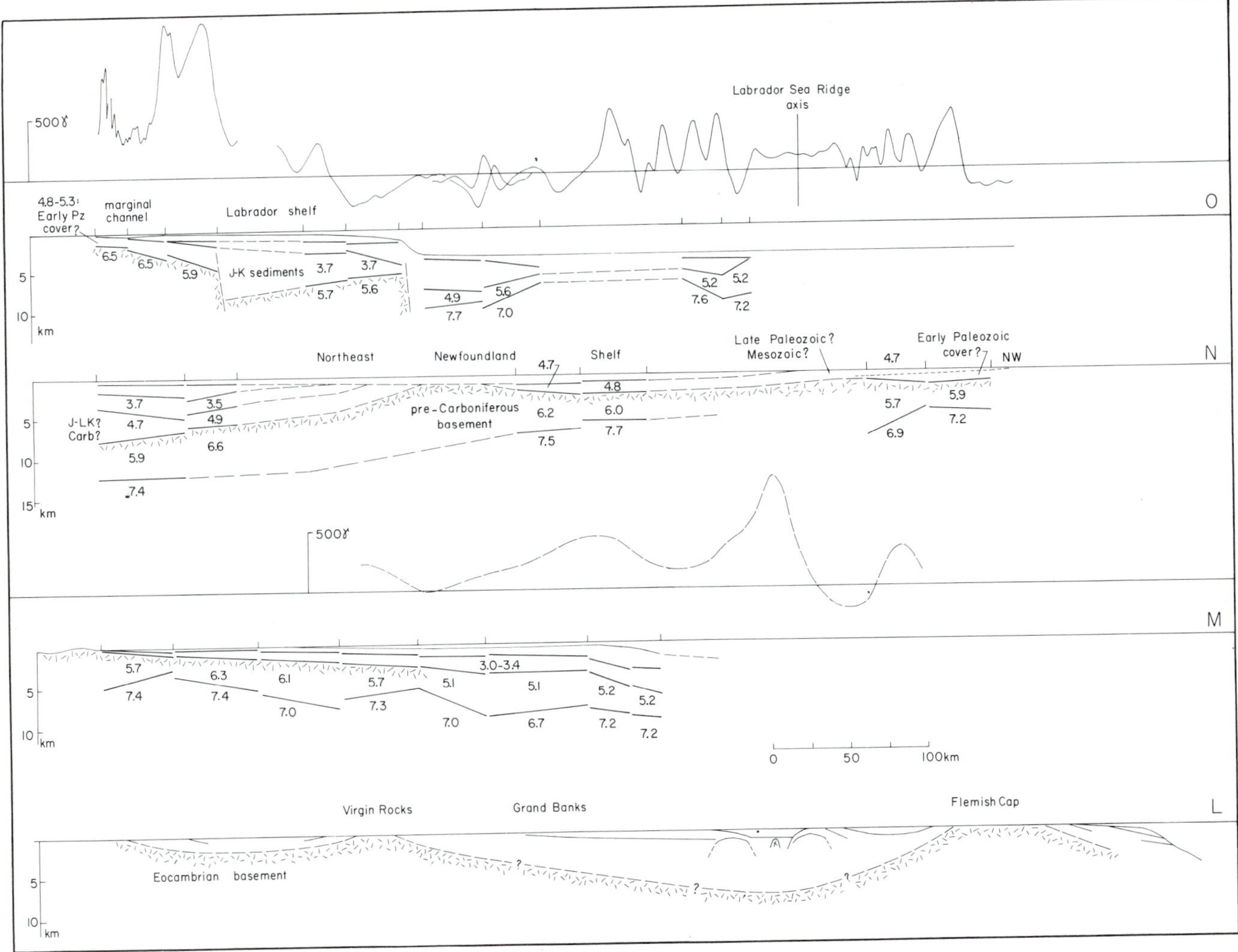

Fig. 5. Profiles of northern areas. Section L from Grant (1973); section M seismic data from Sheridan and Drake (1968); magnetic data constructed from contours of Fenwick et al. (1968); section N after Sheridan and Drake (1968) and Grant (1972); section O constructed from refraction and magnetic data from Mayhew (1969) and Mayhew et al. (1970).

extends out onto the northern Blake Plateau and may form a branch of the east coast magnetic anomaly (Taylor et al., 1968). Long and Lowell (1973) interpret this belt as a narrow pre-Cretaceous, probably Triassic, rift zone.

NORTHERN AREA: LABRADOR SHELF TO SOUTHWEST GRAND BANKS

Profile 0 of Figure 5 is a seismic section across the Labrador Shelf. Grant (1972) has argued that the 4.8-5.3 km/sec layer of the inner shelf corresponds, at least in part, to the early Paleozoic platform cover exposed along the shores of the Strait of Belle Isle. The underlying higher-velocity layer is probably the complex of Precambrian metamorphics which forms the bedrock all along the Labrador coast.

The probable continuation of this basement layer lies at a deep level under the outer shelf, and is overlain by sediments approaching 8 km in thickness, the oldest of which are at least as old as Jurassic (McMillan, 1973b; Bartlett, 1971). An important zone of basement faults probably lies between the inner and outer shelf areas: (Mayhew et al., 1970; Grant, 1972), and is undoubtedly related to the Late Triassic or Early Jurassic block faulting which controlled the early sedimentation patterns of the shelf (McMillan, 1973a). Magnetic data (Hood et al., 1967; Mayhew et al., 1970; Hood and Bower, 1973) also indicates a zone of rapid seaward deepening of basement beneath the Labrador Shelf. The basement of the outer shelf rises toward the Labrador Basin, and apparently ends in the vicinity of the continental slope.

The central part of the Labrador Basin has a linear, symmetrical magnetic pattern characteristic of ocean basins formed by sea-floor spreading (Godby et al., 1966; Mayhew, 1969). It is unclear whether this magnetic pattern extends all the way to the sides of the basin, but the most straightforward interpretation of the refraction results is that the transition from shelf to deep water corresponds to the seaward limit of continental crust. It is noteworthy that the velocity structure of the Labrador Basin is more like that of the Red Sea axial zone than typically oceanic, and is much like that of the central segment of the margin (Fig. 3).

Section N of Figure 5 presents an interesting contrast with Section O. The 5.7-6.3 km/sec layer of the inner shelf decreases rather abruptly to 5.1-5.2 km/sec under the outer shelf, a situation very much like that on lines across the shelf of the central segment of the margin (Fig. 2) and the Blake Plateau (Fig. 4), although there is only the slightest suggestion of a "basement" arch here. The magnetic pattern shows broad, smooth anomalies except near the coast (Fenwick et al., 1968; Hood and Bower, 1971; Grant, 1972). It is likely that at least part of the 5.1-6.3 km/sec layer is sedimentary, and that basement under the outer shelf is deeper, as under the Labrador Shelf. No comparable zone of downfaulting is evident in the underlying 6.7-7.4 km/sec layer, however.

Section M extends from the northern Grand Banks to the Labrador Shelf. The most interesting feature of this profile is the various interpretations that can be given to seismic layers of velocity 4.7-4.9 km/sec. At the southeast end of the section Grant (1972) has interpreted a layer of this velocity as sediments of Jurassic and/or Early Cretaceous age. The sedimentary section here, then, approaches 8 km in thickness, if the 5.9 and 6.6 km/sec layer is the basement. Further northwest, layers of comparable velocity have variously been interpreted as early Paleozoic metamorphics (Sheridan and Drake, 1968), Carboniferous and early Paleozoic sedimentary rocks (Grant, 1972), and Precambrian quartzite (Mayhew et al., 1970). All these interpretations may be correct for their respective areas (Mayhew, in press).

Section L, from Grant (1973), extends across the southern Grand Banks, and as Grant has noted, is higly schematic, although the essential features of the section are probably grossly correct. The most schematic part of the section is the basement configuration. Magnetic data indicates that the basement is as deep as 25,000 ft (~ 8 km) under parts of the eastern Grand Banks and Flemish Pass (Hood and Godby, 1965), but recent drilling has shown that the basement has been broken into a complex of block-fault basins and intervening highs (Ayrton et al., 1973), and does not have the simple downwarped form shown in section L. Grant (1973) has shown that Flemish Pass may also be fault-bounded. Basement of probable late Precambrian age is exposed at the sea floor on Virgin Rocks (Lilly, 1966) and on Flemish Cap, where granodiorite of comparable age has been drilled (Pelletier, 1971).

The sedimentary section of the southern Grand Banks (Armstrong et al., 1973; Brideaux and Williams, 1973) includes Mississippian and Devonian formations in some places, unconformably overlain by a complex Jurassic to Lower Cretaceous section which includes red beds underlying Lower Jurassic (?) salt. This salt forms complex domes and ridges in a number of areas (Ayerton et al., 1973). Numerous salt structures have been observed throughout the Grand Banks-Scotian Shelf area (Auzende et al., 1970; Grant, 1972; Keen et al., 1970; King and MacLean, 1970; Pautot et al., 1970; Watson and Johnson, 1970; Watts, 1972; Webb, 1973). The Jurassic-Lower Cretaceous section seems to correspond approximately to 4.9-5.2 km/sec layers measured by a few refraction profiles (Press and Beckmann, 1954) on Grand Banks (Grant, 1972). Limestones of comparable age have been dredged from the southwest flank of Flemish Cap (Grant, 1973).

The upper part of the section is Late Cretaceous and Cenozoic in age, and lies with angular unconformity on the Jurassic-Lower Cretaceous strata beneath. The stratigraphy of the Tors Cove and Grand Falls wells shows a number of cycles of emergence and submergence, but overall preserving a sedimentary wedge that thickens seaward (Bartlett and Smith, 1971). The 4.3-4.4 km/sec layer under the extreme southwestern Grand Banks shown in Figure 6, section K, may be the equivalent of the upper section, and the 5.3-km/sec layer the equivalent of the Jurassic-Lower Cretaceous section.

Flemish Pass is buttressed by nonmagnetic acoustic basement ridges in the subsurface which Grant (1973) has interpreted as buried reefs of probable Jurassic age. The ridges have associated gravity highs; the high associated with the western ridge extends along the northern edge of the Grand Banks, suggesting a northward extension of the buried reefs (Keen et al., 1970). The 5.1-5.2 km/sec layer on section N may include some of this material. Although the data are limited, the interpreted reef structures do not appear to extend to the Labrador Shelf (section O). The inference is that reef communities did not become established there in spite of the fact that tropical to subtropical marine conditions were present in that area as early as the Jurassic (McMillan, 1973b).

CENTRAL SEGMENT: LAURENTIAN CHANNEL TO THE CAROLINAS

The only part of this margin segment where exploratory offshore drilling has taken place is on the Scotian Shelf. Numerous holes have been drilled here, along with extensive seismic surveys. The stratigraphy revealed by this work has been outlined by McIver (1972, 1973). These reports have been of great value in suggesting the kind of stratigraphic complexities to be expected further south.

The oldest unit in this area is salt. As in the southern Grand Banks, the salt is of possible Early Jurassic age and occupies a series of subbasins in the crystalline basement. The salt is overlain by a Jurassic anhydrite-carbonate-shale sequence; this is, in turn, disconformably overlain by complex sequences of carbonate and clastic sediments ranging in age from Jurassic to Tertiary. These sediments

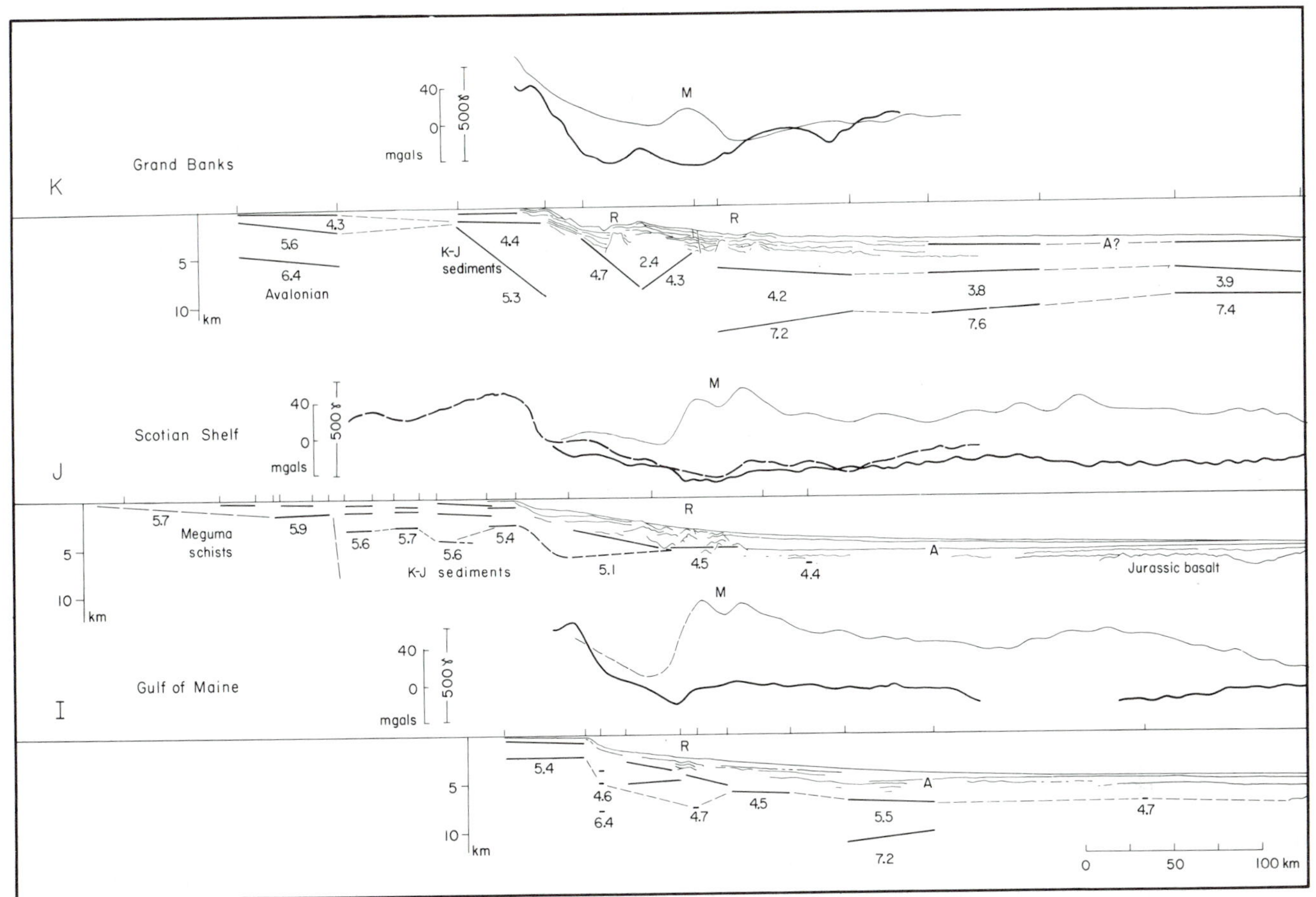

Fig. 6. Profiles of central segment, between southwest Grand Banks and Gulf of Maine. Selected reflectors from profiles of Emery et al. (1970) constructed using nearby refraction profiles (Fig. 1) projected to section lines. Free-air gravity data (heavy solid lines) and magnetics (light solid lines) from Emery et al. (1970). Heavy dashed line is gravity profile of Keen and Blanchard (1966). Long light dashes are constructed from magnetic contours of Taylor et al. (1968). Numbers are seismic velocities in km/sec. Ridge complex of Emery et al. (1970) indicated as R. Magnetic slope anomaly labled M. Refraction results of section K is a recomputation of the Grand Banks section of Bentley and Worzel (1956).

were deposited in a variety of depositional environments, and the facies relationships within them are complex, indicating an irregular history of subsidence. Similar inferences have been drawn from studies of well data in other parts of the margin (Armstrong et al., 1973; Keen and Keen, 1973; Kraft et al., 1972; Laughton et al., 1970; Sleep, 1971; Smith, 1973).

Whether the oldest sediments under the shelf farther south are of Early Jurassic age is presently speculative. The basal sediments recovered from wells onshore are as old as Late Jurassic (Maher, 1971). That deep, salt-filled subbasin areas are present seems more than likely, but this is also speculative. The presence of Mississippian in some of the Grand Banks wells suggests that late Paleozoic basins like those in the Gulf of St. Lawrence may not be confined to that area, and it has been suggested (Kraft et al., 1972) that late Paleozoic may make up the older part of the sedimentary section beneath areas of the shelf to the south, as well. This is still more speculative, but has not been disproved.

The age of the basal sediments beneath the outer shelf will not be known until drilling takes place, but recent exploration seismic surveys have shown far greater total sedimentary thicknesses than had previously been estimated. The section seaward of the Gulf of Maine has been estimated to have a thickness in excess of 8 km, that off New Jersey in excess of 10 km (Mattick et al., 1973). The latter figure is astonishing, for it represents fully one-third the thickness of normal continental crust. These two areas were previously identified by refraction as basins filled with low-velocity (<4 km/sec) sediments, with thicknesses around 3 and 5 km, respectively (Drake et al., 1959), which have been referred to as the Georges Bank and Baltimore Canyon troughs (Maher, 1971). The basins have apparently had somewhat independent subsidence histories extending into the Tertiary (Garrison, 1970).

It is noteworthy that the new estimates of sediment thicknesses for this part of the margin are comparable with those for the other basins in the northern and southern areas of the margin. These

Table 1. Thicknesses of sedimentary basins of the Atlantic seaboard: based on refraction data except for (m), magnetic and (g), gravity interpretations.

Location	Basin Name	Symbol (Fig. 11)	Minimum Thickness (km)	Reference
Labrador Shelf		LS	8	Mayhew et al., 1970
			15 (m)	Hood and Bower, 1971
Northern Grand Banks		NGB	7	Sheridan And Drake, 1968
Southern Grand Banks		SGB	7 (m)	Hood and Godby, 1965
Gulf of St. Lawrence	Magdalen Basin	MB	9 (g)	Watts, 1972
			9	Hobson and Overton, 1973
Prince Edward Island	Cumberland Basin	CB	9	Howie and Cumming, 1963
Scotian Shelf	Orpheus Basin	OB	6 (g)	Loncarevic and Ewing, 1967
Scotian Shelf	Emerald Bank Trough (Sable Island Trough)	EB	6 (m)	Hood, 1967
Gulf of Maine	Georges Bank Trough	GB	8	Mattick et al., 1973
Off New Jersey	Baltimore Canyon Trough	BC	10	Mattick et al., 1973
Blake Plateau	Blake Plateau Trough	BP	8	Sheridan et al., 1966

are summarized in Table 1.

Clearly, the 5.0-6.0 km/sec "basement" layer under the outer shelf in these areas is high-velocity sediment. The situation appears to be that suggested by Gibson (1970), who noted that a basal Upper Jurassic-Lower Cretaceous carbonate-anhydrite sequence found in wells around Cape Hatteras might be the feather edge of a much thicker unit farther offshore which had previously been mistaken for basement.

The shelf-edge "basement" arch is evidently within the sedimentary section of the outer shelf. The systematic seaward decrease in velocity to values around 5 km/sec in the arch, then, suggests a facies transition, and comparisons with the northern and southern areas of the margin indicate that the arch is a reef complex as suggested by Emery et al. (1970), presumably of Lower Cretaceous-Jurassic age. The "basement" surface of the arch may itself be a facies boundary rather than a structural surface (Dietz, 1966).

The fact that velocity measurements on the coastal plain and in the inner part of the shelf clustering around 6 km/sec are on crystalline rocks implies that somewhere in the middle part of the shelf there is a boundary within the surface previously considered the basement, separating these rocks from the high-velocity sediments under the outer shelf (Figs. 6 and 7). The position of this boundary is uncertain but seems most likely to occur where the low-velocity sediments start to thicken rapidly offshore.

The significance of this boundary cannot be overemphasized: it probably represents a zone of major crustal faulting, marking a transition from a relatively stable area of minor vertical movements and crustal changes to a zone in which major changes within the lithosphere have caused drastic thinning of the crust and the formation of mobile basins that have experienced major subsidence. It seems to be one of the two major boundaries of the continental margin, the other being that between continental and oceanic crust.

Although the "basement" ridge of Drake et al. (1959) is a sedimentary feature, there are indications that along the central margin segment a true basement ridge is present at greater depths, although its axis lies farther west (Mattick et al., 1973). Magnetic data suggest (Hood 1967) that a basement ridge underlies that arch defined by refraction off Halifax (Officer and Ewing, 1954), at roughly twice the depth, around 20,000 ft ($\sim$ 6 km). The arch rimming the Blake Plateau (section C) similarly appears to be underlain by structure (Sheridan et al., 1966). Basement rises seaward to a ridge under the edge of the Labrador Shelf (section O), although there is no sedimentary arch here. If high-velocity sediments were not present on the sections shown in Figures 2, 6, and 7, they might bear a greater resemblance to the Labrador Shelf section.

The features labeled "R" in the sections of Figure 6 are part of the "ridge complex" observed by seismic profiles beneath the continental rise north of the New England Seamounts (Emery et al., 1970). The ridge complex is quite distinct from the shelf edge "basement" arch. On some sections these features appear to be giant folds, whereas on others they appear to be piercement structures. In part they involve sediments above horizon A, which are therefore post-Eocene in age (Ewing and Hollister,

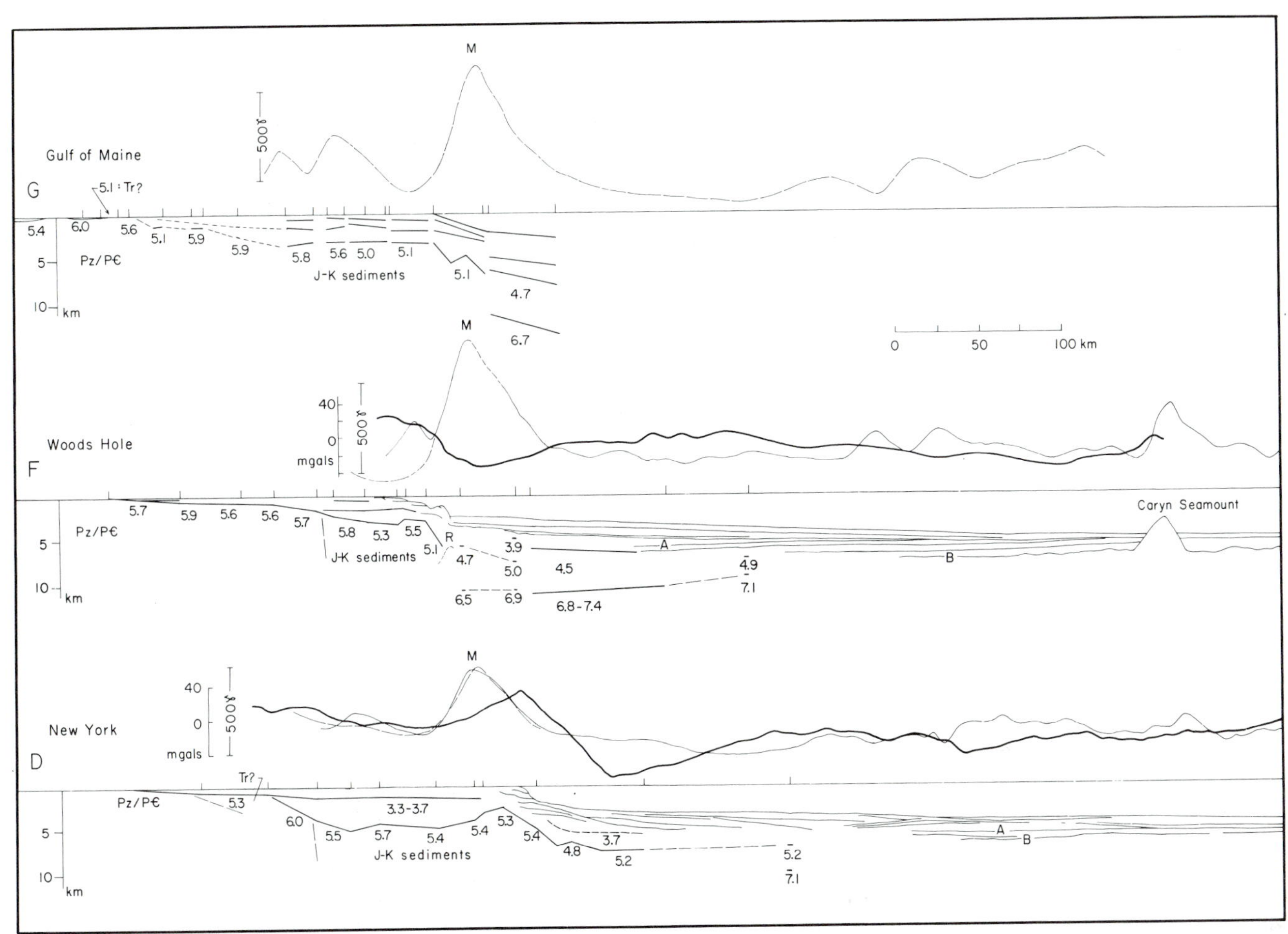

Fig. 7. Profiles of central segment, between the Gulf of Maine and Long Island (legend as in Fig. 5). Gravity and magnetics section E from Rabinowitz (1973).

1972). On profile records they resemble the ridges of the western Gulf of Mexico, which have been attributed to gravity sliding of sediments on decollment and/or to salt diapirism (Massingill et al., 1973). It is possible that they are, at least in part, salt ridges formed by flowage out from under the shelf. One may also speculate that they hypothesized reefs forming the shelf-edge "basement" arch originated over early salt structures as well as atop structures in the basement.

GRAVITY ANOMALIES

The gravity field outlined in Figure 8 is dominated by two parallel belts of positive anomalies, one along the edge of the shelf, the other within the Appalachians. The axis of the shelf-edge high coincides with the shelf break along the central margin segment (Rabinowitz, 1973). It is flanked by broad lows over the shelf and over the continental rise. Numerous crustal models have been constructed to fit the gravity data (e.g., Worzel and Shurbet, 1955; Worzel, 1968; Drake et al., 1959; Keen and Blanchard, 1966) in which the anomaly is mainly an edge effect due to a fairly abrupt change in crustal thickness under the continental slope and upper rise. In the simplest of such models, it is assumed that relief in a surface of constant density contrast between normal mantle and mean crustal densities is the means of isostatic compensation for changing water depth; this gives rise to a rapid transition from continental to oceanic crustal thickness.

Lacking reliable results from long refraction profiles over the outer margin, models based on gravity data can only be speculative, but two observations suggest that the transition from continental to oceanic crust is neither simple nor abrupt. First, the presence of very thick sediments under the shelf, having a mean density that is probably lower than mean crustal density, implies a thin continental crust and dense rocks at a relatively shallow depth. Second, Figure 3 indicates that normal mantle velocities have generally not been measured on short refraction profiles over the upper continental rise. Rather, the highest values form a continuous smear from the range of oceanic layer 3 up through the range 7.2-7.8 km/sec, which Drake and Nafe (1968) have suggested represents a transient petrology associated with vertical movements in tectonically

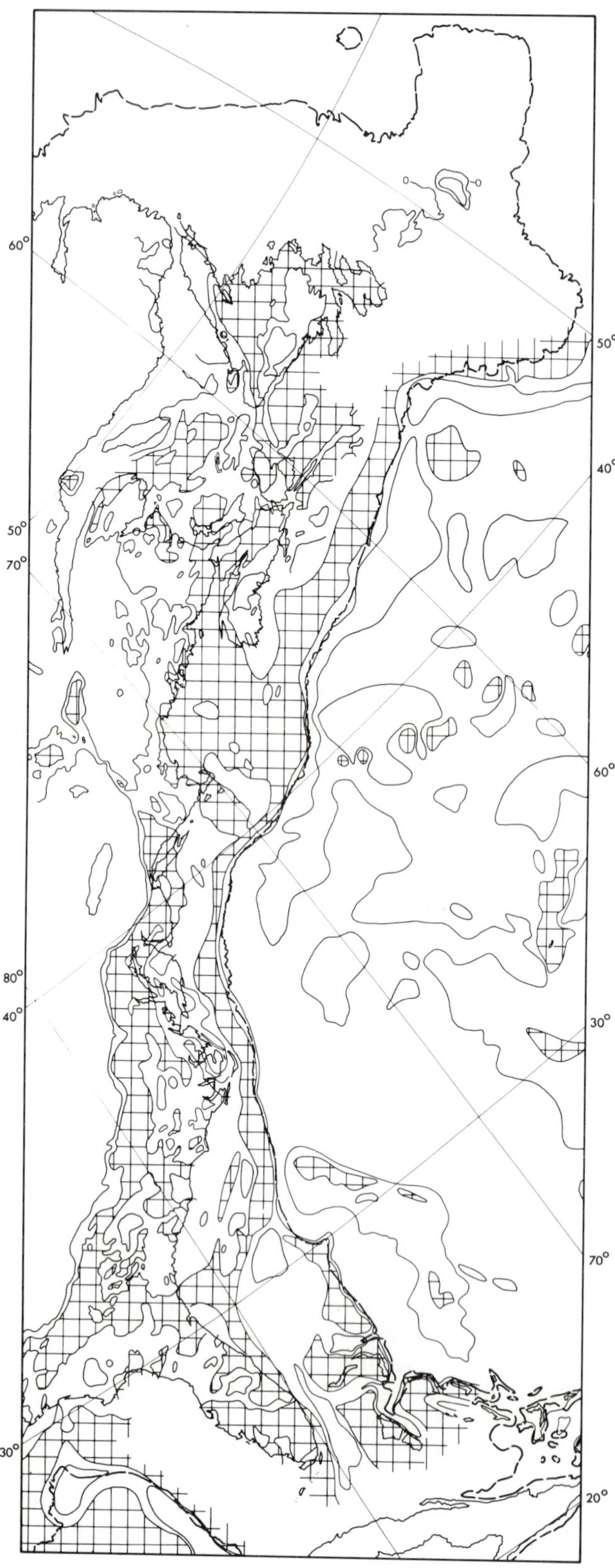

Fig. 8. Gravity anomalies from data of Woollard and Joesting (1964); Dehlinger and Jones (1965); Dept. Energy, Mines, and Resources (1969); Stephans et al. (1971); Keen et al. (1970); Uchupi et al. (1971); and Rabinowitz (1973). Zero- and -30 mgal Bouguer contours on land and 0- and -25-mgal free-air contours offshore are shown. Values greater than 0 mgal are crosshatched.

active regions. Sheridan (1969) has argued that this material represents "anomalous mantle" formed during Atlantic rifting, and that transformation to denser mineral assemblages in this material has led to the subsidence of the margin. Whatever the genesis of this material, it seems quite possible that it extends from beneath the continental rise far in under the shelf, and that lateral variation in thickness and density of these rocks is the principal mechanism of isostatic compensation.

The second major gravity feature is the Main Gravity High of the Appalachians. In the southern Appalachians, this high is centered on the Appalachian piedmont; the Triassic basins of the piedmont are distributed about its axis. A branch of the high extends from north of New York City parallel to the Triassic basin of western Connecticut and Massachusetts, follows the Berkshire-Green Mountain-Sutton Mountain axis and becomes the Gaspé Gravity High, and curves into the Gulf of St. Lawrence, where it joins a belt of highs through western Newfoundland. While this high follows the anticlinoria of the western part of the Appalachian crystalline belt along the way, its source is primarily with the lower crust and/or upper mantle (Diment, 1968).

Another belt of highs can be defined along the western Gulf of Maine (Kane et al., 1972), extending through the Triassic basin of the Bay of Fundy, along the Cobequid-Chedabucto fault zone, to the Orpheus Basin. Major movement along this fault zone occurred in probably Late Triassic time as a rejuvenation of older structures (King and MacLean, 1970). The Orpheus Basin is associated with a prominent local linear gravity low with flanking highs, which result from low-density salt fill in the basin (Loncarevic and Ewing, 1967). On the basis of drilling, the salt is of possible Early Jurassic age (McIver, 1972).

If one takes the zone of gravity highs along the western Gulf of Maine as the principal northern continuation of the Main Gravity High of the southern Appalachians, then the belt grossly parallels the continental margin between the latitudes of southern Georgia and northern Nova Scotia. A similar observation was made by Diment et al. (1972).

It is not a major jump to hypothesize that the Main Gravity High and the shelf-edge gravity high are closely related in time, as well as space, and that both are expressions of a regional rise of the aesthenosphere through the lithosphere and penetrating into the crust in the manner suggested by Girdler et al. (1969) for the East African rifts, and by Lowell and Genik (1972) for the Red Sea depression. Clearly, the process was to a large degree, although not perfectly, constrained by the Appalachian grain. Block faulting associated with the accompanying regional uplift involved partial rejuvenation of Paleozoic fault patterns.

The time of uplift presumably is specified by the age of the salt, clastic sediments, and volcanics that formed in the fault depressions. The age of the salt

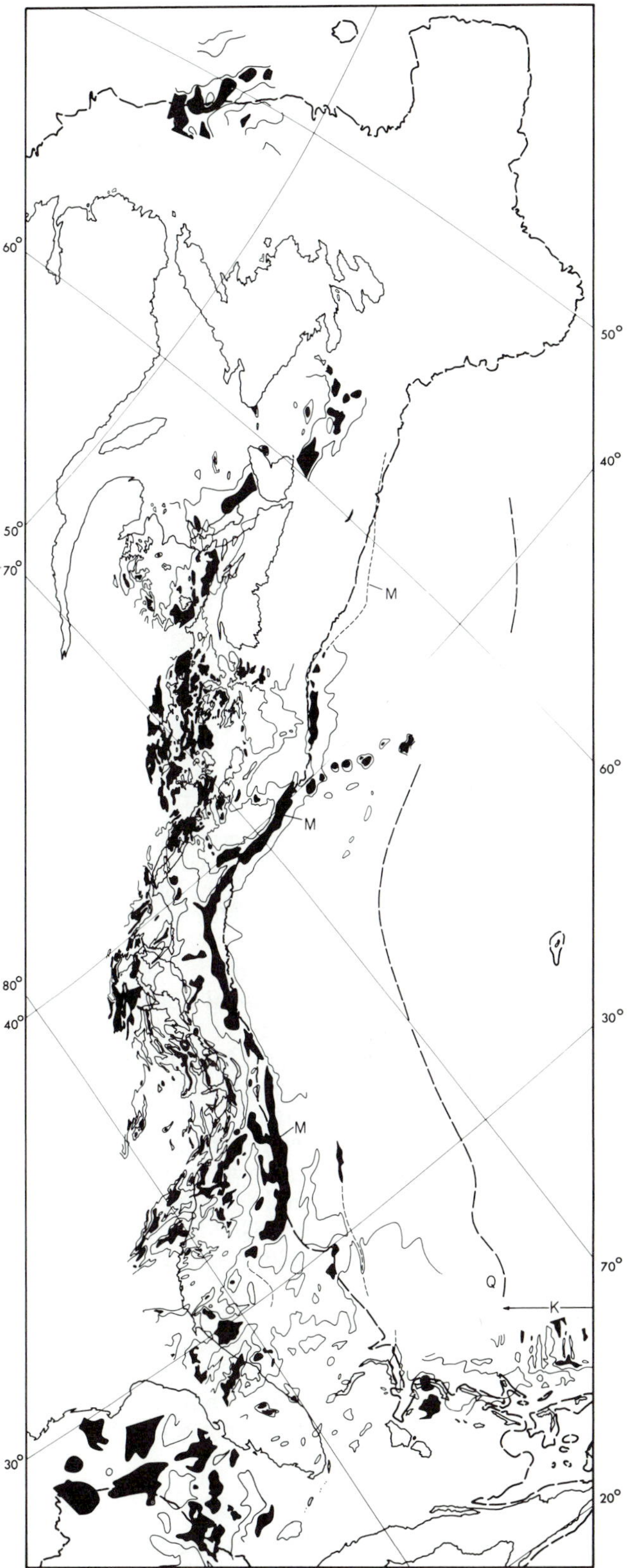

Fig. 9. Magnetic anomalies from contour maps of Taylor et al. (1968); Gough (1967); Bracey (1968); Dept. Energy, Mines, and Resources (1968); Sheridan And Drake (1968); and Fenwick et al. (1968). Zero- and +200-gamma contours shown; datums arbitrary. Anomalies greater than 200 gammas blackened. Continuation of slope anomaly from Rabinowitz (1973). Boundary of magnetic quiet zone from Pitman and Talwani (1972). Magnetic slope anomaly indicated by M, Keathley Sequence anomalies by K.

drilled on the Canadian Shelf is possibly Early Jurassic and, perhaps, Late Triassic. The grabens within the Appalachian piedmont are usually referred to as "Triassic," although paleontologic data from the sediments and dates on associated volcanics indicate that their formation extended well into the Early Jurassic (Cornet et al., 1973).

A regional uplift simultaneously affecting the whole area of the continental margin and much of the Appalachian region through a time interval bracketing the Triassic-Jurassic boundary is in accordance with the estimates of many authors arguing from diverse kinds of data about the time of breakup of the North Atlantic continents.

A similar history of Late Triassic-Early Jurassic block faulting affected the northwest African margin (Kanes et al., 1973). These movements created grabens that filled with arkosic sediments and salt. May (1973) has noted that basaltic intrusions and volcanics dating from the same time interval are present near the margins of all the continents around the North Atlantic. Thus the evidence points to Late Triassic rifting. However, it is quite possible that the earliest beginnings of the process lie well back in the Paleozoic (Kanes et al., 1973).

MAGNETIC SLOPE ANOMALY

One of the most important geophysical features of the margin is a continuous band of magnetic highs referred to as the East Coast Slope Anomaly. It is continuously present along the length of the central segment of the margin (Drake et al., 1963), except for a slight offset at the New England Seamounts. It is bounded on the seaward side by the magnetic "quiet zone" (King et al., 1961; Heirtzler and Hayes, 1967). The slope anomaly is most commonly a pair of broad, high-amplitude highs (Taylor et al., 1968; Emery et al., 1970), although the western high is not everywhere present, the eastern high being more continuous. This is suggested by the magnetic contours shown in Figure 9 off the Carolinas. It is also shown in section E of Figure 7 south of the Gulf of Maine where a marine survey (Emery et al., 1970) recorded a pair of highs, while the contours from aeromagnetic data of the East Coast Magnetic Survey (Taylor et al., 1968) indicate only a single broad high. The anomaly appears to be single off New Jersey (section D). To the north, the anomaly decreases in amplitude and appears to fade out near the Laurentian Channel. To the south, the anomaly hooks in toward southern Georgia at the northern end of the Blake Plateau, and changes character. The main part of the anomaly appears to curve back around the shoreward side of the Blake Plateau trough, and becomes indistinct. A possible thin branch of the anomaly runs onshore in southern Georgia in an area where volcanics of

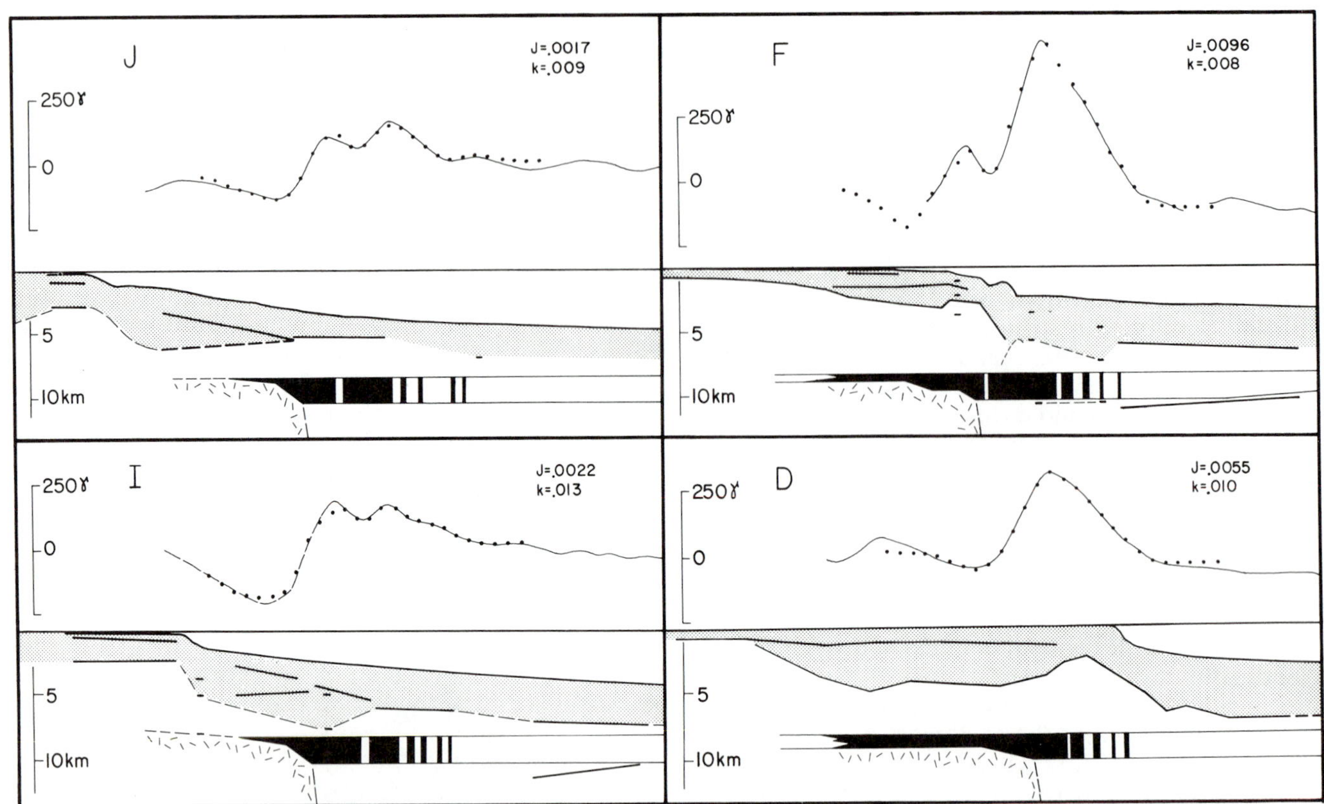

Fig. 10. Two-dimensional models of the magnetic slope anomaly for four profiles from Figures 6 and 7 computed using the method of Talwani and Heirtzler (1964). Remanent magnetization (J) and susceptibility (K) are indicated (emu/cm^3). Black blocks have normal remanent magnetization; white blocks are reversely magnetized. Magnetic profiles are projected perpendicular to magnetic trends from East Coast Magnetic Survey (Taylor et al., 1968). Low-velocity (4km/sec) sediments are stippled.

uncertain age have been drilled.

The slope anomaly in general does not overlie the continental slope. Off the Scotian Shelf it lies far out over the continental rise; off New Jersey it comes well in onto the shelf. Nor does it correlate with the shelf-edge gravity high (Figs. 6 and 7); the two anomalies must have different sources.

A variety of hypotheses have been advanced to account for the slope anomaly (Keller et al., 1954; King et al., 1961; Watkins and Geddes, 1965; Barrett et al., 1967; Taylor et al., 1968; Keen, 1969; Emery et al., 1970; Rabinowitz, 1973), which are summarized in part by Emery et al. (1970). In most of these the anomaly is some expression of the boundary between continental and oceanic crust. Certainly its amplitude and continuity indicate that it has some fundamental significance with respect to the structure of the margin.

MODEL OF THE MAGNETIC SLOPE ANOMALY

Figure 10 shows a new model of the sea-floor spreading type for the magnetic slope anomaly. It consists of positively and negatively magnetized blocks 2 km thick placed in the lower part of the low-velocity basement layer beneath the continental rise having seismic velocities equivalent to those of layer 2. Both remanent and induced magnetization were assumed. It was possible to make a good fit of essentially the same model to the four profiles shown by assuming different values of susceptibility and remanent magnetization, along with slight variations in the width of the model blocks.

Values of inclination, declination, and total field for the induced component were taken from the U.S. Navy Oceanographic Office charts for epoch 1965. Remanent magnetization directions were used which are consistent with the Early Jurassic pole determined by Opdyke and Wensink (1966) for the White Mountain igneous series. This is also consistent with the inclination value determined for basalt recovered from the bottom of JOIDES hole 155 at the seaward edge of the quiet zone (Taylor et al., 1973). Unsuccessful attempts were made at a fit assuming the Late Triassic and Early Jurassic poles for North America determined by Beck (1972) and by Steiner and Helsley (1972), which involve considerably lower inclinations.

Values of remanent magnetization assumed fall in the range of values measured for dredged oceanic basalts by Fox and Opdyke (1973), and are also consistent with values for the JOIDES 155 basalt. The assumed susceptibility values, however, are an order of magnitude higher than values determined in the above studies. An explanation for this may be that the apparent induced component is actually a substantial near-axial viscous remnant component.

Such an effect has been recently observed on oceanic basalt samples (Lowrie et al., 1973; Kitazawa et al., 1973) and on continental sedimentary rocks (Steiner and Helsley, 1972).

The basic model consists of two wide, normally magnetized blocks on the west with a narrow negative block between, passing through a transition zone of mixed polarity to a wide negative block under the landward part of the quiet zone. The model is quite similar to that of Emery et al. (1970), except that it is more detailed and lies within layer 2, whereas those authors placed their model within layer 3. In the present model, it is assumed that no significant contribution to the anomaly arises from layer 3. Low-amplitude, correlatable anomalies are present within the quiet zone (Emery et al., 1970), which may be due either to brief reversals or to excursions of the field (Larson and Pitman, 1972), but no attempt was made to model them.

It is possible that the western normal blocks were magnetized during the same polarity interval in which the basalts of the "Triassic" basins onshore were magnetized; their polarity is uniformly normal (Beck, 1972). This would date at least the outer portion of these blocks as Early Jurassic. The outer part of the magnetic quiet zone bordering the Keathley Sequence (Vogt et al., 1971) probably corresponds to spreading during a long interval of normal polarity during the Late Jurassic (Larson and Pitman, 1972). Therefore, the model implies that the inner part of the quiet zone was formed during a fairly long interval of reversed polarity around the middle of the Jurassic. The compilations of Irving (1964) and McElhinny and Burek (1971) indicate, however, that this was a time of mixed but mainly normal polarity, while Creer's (1971) reversal column indicates roughly equal epoch lengths for this period. These observations represent a conflict with the model which cannot be satisfactorily resolved at present.

In order to fit the landward side of the model to the observed profiles, it was assumed that the western magnetic block thins from the bottom up. Other types of models might fit as well, for example, one involving a lateral change in mean susceptibility, as might be expected for volcanics interfingering with sediments. In any case, there is a strong indication that volcanics underlie large parts of the shelf. This is supported by magnetic anomaly patterns suggestive of volcanic centers in a number of areas (Drake et al., 1963; Hood, 1967; Taylor et al., 1968).

If the model is correct, a straightforward interpretation is that the volcanics thin where they lap up onto the continental block, in which case the slope anomaly specifies the continent edge. Since the positon of the slope anomaly varies with respect to the continental slope, it follows that the continental slope is not directly related to the boundary between oceanic and continental crust.

COMMENT ON CONTINENTAL MARGINS AS GEOSYNCLINES

According to Kay's (1951) classification, the essential characteristic that distinguishes a eugeosyncline from a miogeosyncline is the presence of volcanics in the former and their absence in the latter. Additionally, eugeosynclines are belts of greater magnitude and rates of subsidence; i.e., they are more mobile, and as a result they are most commonly filled with thick sequences of immature clastic sediments. Sediment lithology is not a definitive criterion, however, for the stratigraphy of eugeosynclines may be quite varied depending on factors such as sediment source, climatic and depositional environments, relative sea level, and peculiarities of the tectonic history of the belts.

The analogy of Drake et al. (1959) between the miogeosyncline-eugeosyncline and shelf-rise couples was based primarily on a favorable comparison between apparent sediment thicknesses. Although evidence concerning the distribution of volcanics was lacking, the possibility of their presence low in the continental-rise section was discussed. An essential feature of Drake's model is that the boundary between miogeosyncline and eugeosyncline lies in the vicinity of the continental slope.

For two reasons, a modification of this model may be necessary. First, the total sediment thicknesses of the outer shelf are now known to be far greater than previously believed; they are of continental rise and eugeosynclinal proportions. A fairly abrupt transition must lie between these accumulations and the much thinner sequences of the inner shelf and coastal plain. Second, as suggested in the previous section, volcanics may be extensive deep beneath large parts of the outer shelf.

Clearly, the outer shelf, a zone of highly mobile basin areas that have experienced great subsidence and have received great thicknesses of sediments and probably volcanics, has all the important attributes of a eugeosyncline. Accordingly, it can be argued that the boundary between eugeosyncline and miogeosyncline lies not at the continental slope, but somewhere beneath the inner shelf. It can further be argued that this boundary marks an important transition zone in the whole of the lithosphere from a stable cratonic area to a zone of great vertical mobility. Seen this way, the thin coastal plain sequence (the miogeosyncline) is a relatively insignificant indicator of where this transition takes place. Inasmuch as the outer shelf zone experienced the greatest mobility during its history at an accreting plate margin, it seems likely that the same will be true if at some future time the margin becomes a consuming plate margin.

The most important implication of this model is that it suggests a mechanism for the formation of ensialic eugeosynclines. Such a model may account for the persistent evidence of sialic basement in eugeosynclinal belts.

CONCLUSIONS

A number of the principal features of the east coast margin are summarized in Figure 11. Figure 12 shows an interpretive cross section based on information discussed in this paper, along with a similar section for the Red Sea based on data and interpretations of Drake and Girdler (1964), Hutchinson and Engels (1972), and Lowell and Genik (1972).

A fairly coherent picture of the east coast margin emerges from the information presently available, and applies to the whole of the margin discussed in this paper. Block faulting occurred in Late Triassic-Early Jurassic time, possibly as a result of a regional rise of the aesthenosphere, a process that may have begun much earlier. This produced an initial uplift and crustal thinning, which was concentrated in a zone beneath the present outer shelf and to a relatively minor extent in the periferal area of the inner shelf and Appalachian piedmont. The process may have proceeded to a greater degree around the Blake Plateau, leading to the splitting off of a sialic sliver from the Florida platform.

The fault basins filled with clastic sediments and, at least in the northern area, with salt. Volcanics were erupted during this time. They were probably concentrated beneath the outer part of the outer shelf, adjacent to the locus of continental breaching, but may be scattered through much of the lower part of the sedimentary section of the outer shelf.

Irregular subsidence of the margin took place as it moved away from the young mid-Atlantic ridge, and is continuing (Grant, 1970). The total subsidence is of massive proportions beneath the outer shelf.

Barrier reefs formed subsequently in some areas, probably over basement highs or salt ridges, which were not necessarily situated at the outer edge of the continent. In at least the southern area, these reefs continued to develop through the Late Cretaceous. The reefs probably determined the position of the continental slope. Most of the continental rise sediments were deposited in Miocene and later times (Ewing and Hollister, 1972). This may have been the result of increased bottom-water circulation at this time (Ewing and Hollister, 1972), resulting in increased deep contour current deposition (Heezen et al., 1966).

If it were not for the fact that in the Blake-Bahama basin area sediments were stranded on the Blake Outer Ridge, the Blake Escarpment would probably be draped with continental rise sediments.

The Red Sea section of Figure 12 follows the view that continental crust extends to the margin of the axial trough. The section is similar in a number of respects to the east coast margin section. The

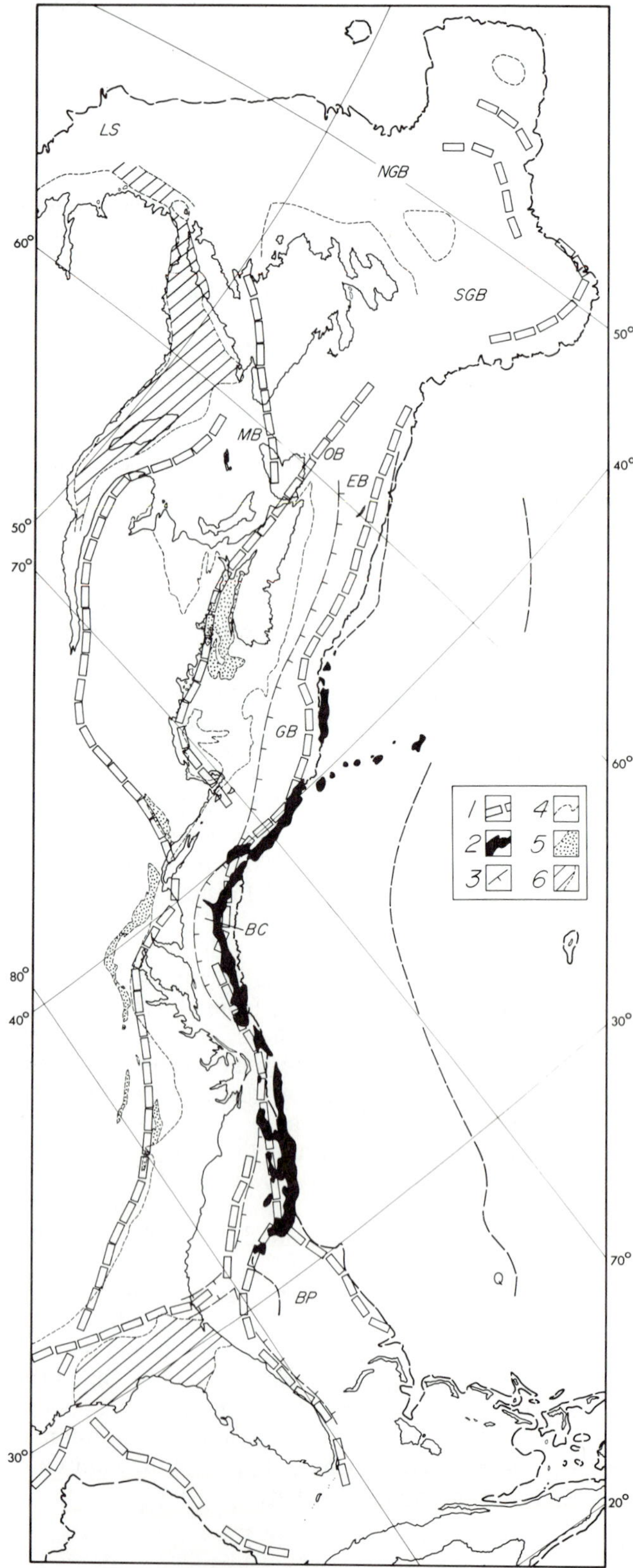

Fig. 11. Summary map of principal features of the east coast margin. Symbols are as follows: 1, gravity highs; 2, magnetic highs; 3, boundary between zones A and B (position is largely speculative); 4, limit of crystalline basement exposure (beneath Quaternary cover offshore), based on U.S. Geological Survey (1965), Uchupi (1966) and Grant (1972); 5, Triassic basins from U.S. Geological Survey (1965) and Uchupi (1966); 6, distribution of early Paleozoic platform cover, mostly in the subsurface, modified from Applin (1951), Sheridan and Drake (1968), and Grant (1972). Position of magnetic quiet zone boundary (Q) from Pitman and Talwani (1972).

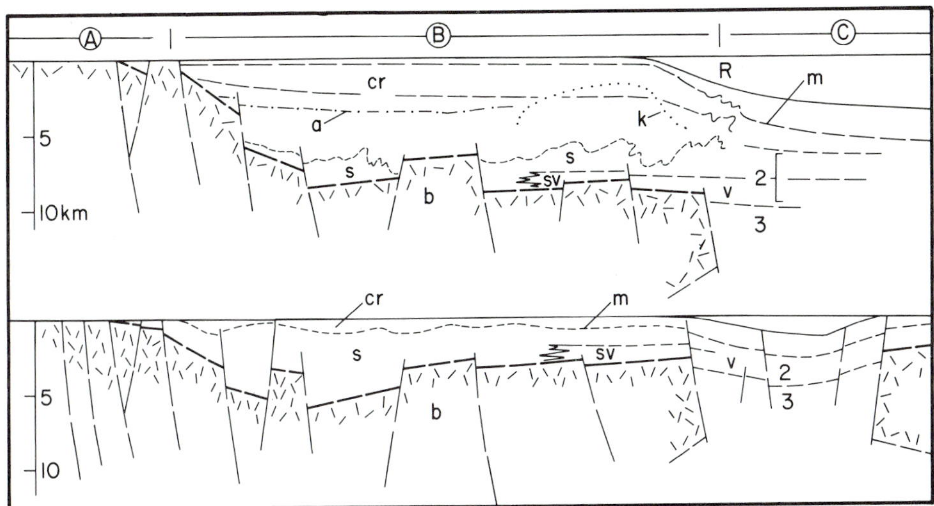

Fig. 12. Schematic and hypothetical cross sections of east coast margin (top) and southern Red Sea (bottom). Symbols as follows: cr; clastics and carbonates; m; Miocene horizon; k; top of inferred barrier reef complex; a; top of high-velocity (5.5 km/sec) sediments; s; salt; v; volcanics (probably includes much sediment); sv; mixed salt and volcanics; R; "ridge complex"; b; basement; 2 and 3; correlatives of oceanic layers 2 and 3 for the continental margin. Zones A, B, and C (discussed in the text) are indicated.

principal differences are (1) sediment thickness, which is ascribable to the different age of these margins; and (2) sediment fill, which is related to environmental differences. It has been argued that the outer shelf of the east coast margin has the important criteria to be classified as a eugeosyncline. By this reasoning, the Red Sea Basin is a eugeosyncline whose principal sediment fill is salt.

ACKNOWLEDGMENTS

This work was partially supported by the Arts and Sciences Research Fund at New York University. Computer time at the Academic Computer Center, New York University, and the Computing Services Division, University of Wisconsin-Milwaukee is gratefully acknowledged. Thanks to Mary Polzin for typing the manuscript. The help and encouragement of C.L. Drake and C.A. Burk is especially appreciated.

BIBLIOGRAPHY

Antoine, J. W., and Ewing, J., 1963, Seismic refraction measurements on the margins of the Gulf of Mexico: Jour. Geophys. Res., v. 68, no.7, p. 1975-1996.

Antoine, J. W., and Harding, J. L., 1965, Structure beneath the continental shelf, northeast Gulf of Mexico: Am. Assoc. Petr. Geol. Bull., v. 49, p. 157-171.

Antoine, J. W., and Henry, V. J., Jr., 1965, Seismic refraction study of shallow part of continental shelf off Georgia coast: Am. Assoc. Petr. Geol. Bull., v. 49, p. 601-609.

Applin, P. L., 1951, Preliminary report on buried pre-Mesozoic rocks in Florida and adjacent states: U. S. Geol. Survey Circ. 91, 28 p.

Armstrong, W. E., Creath, W. B., Kidson, E. J., Sanderson, G. A., Upshaw, C. F., Craig, J., Gradstein, F. M., Van Hint, J., Jenkins, W. A. N., Pocock, S. A. J., Staplin, F. L., and Sulek, J. A., 1973, Biostratigraphic framework of the Grand Banks: Am. Assoc. Petr. Geol. East Coast Offshore Symp., Tech. Progr., 5.

Austin, G. H., and Howie, R. D., 1973, Regional geology of offshore eastern Canada: Geol. Surv. Can. Paper 71-23, p. 73-108.

Auzende, J. M., Oliver, J. L., and Bonnin, J., 1970, La marge du Grand Banc et la fracture de Terre-Neuve: Compt. Rend. v. 271, p. 1063-1066.

Ayrton, W. G., Birnie, D. E., Swift, J. H., and Wellman, H. R., 1973, Grand Banks regional geology: Am. Assoc. Petr. Geol. East Coast Offshore Symp., Tech. Progr., 5.

Barrett, D. L., Ewing, G. N., Keen, M., Loncarevic, B. D., and Manchester, K. S., 1967, The continental margin of eastern Canada: Nova Scotia to Nares Strait (abstr.): IUGG General Assembly Progr.

Bartlett, G. A., 1973, The Canadian Atlantic continental margin: paleogeography, paleoclimatology and seafloor spreading: Geol. Survey Can. Paper 71-23, p. 43-72.

———, and Smith, L., 1971, Mesozoic and Cenozoic history of the Grand Banks of Newfoundland: Can. Jour. Earth Sci., v. 8, p. 65-84.

Bass, M. N., 1969, Petrography and ages of crystalline basement rocks of Florida—some extrapolations: Am. Assoc. Petr. Geol. Mem. 11, p. 283-310.

Beck, M. E., Jr., 1972, Paleomagnetism of Upper Triassic diabase from southeastern Pennsylvania: further results: Jour. Geophys. Res., v. 77, p. 5673-5687.

Bentley, C. R., and Worzel, J. L., 1956, Geophysical investigations in the emerged and submerged Atlantic Coastal Plain, pt. X: continental slope and continental rise south of Grand Banks: Geol. Soc. America Bull., v. 67, p. 1-18.

Blanchard, J. E., Keen, M. J., McAllister, R. E., and Tsong, C. F., 1965, Geophysical observations on the sediments and basement structure underlying Sable Island, Nova Scotia: Am. Assoc. Petr. Geol. Bull., v. 49, p. 959-965.

Bonini, W. E., and Woollard, G. P., 1960, Subsurface

geology of North Carolina-South Carolina coastal plain from seismic data: Am. Assoc. Petr. Geol. Bull. v. 44, p. 298-315.

Bracey, D. R., 1968: Structural implications of magnetic anomalies north of the Bahama-Antilles Islands: Geophysics, v. 33, no. 6, p. 950-961.

Brideaux, W. W., and Williams, G. L., 1973, Palynological analysis of Mesozoic-Cenozoic sediments of the Grand Banks of Newfoundland: Am. Assoc. Petr. Geol. East Coast Offshore Symp. Tech. Progr., 5.

Bryant, W. R., Antoine, J., Ewing, M., and Jones, B., 1968, Structure of Mexican continental shelf and slope, Gulf of Mexico: Am. Assoc. Petr. Geol. Bull., v. 52, p. 1204-1228.

Bunce, E. T., and Falquist, D. A., 1962, Geophysical investigations of the Puerto Rico Trench and outer ridge: Jour. Geophys. Res., v. 67, p. 3955-3972.

Burk, C. A., 1968, Buried ridges within continental margins: N.Y. Acad. Sci. Trans., ser. 2, v. 30, p. 397-409.

Cornet, B., Traverse, A., and McDonald, N. G., 1973, Fossil spores, pollen, and fishes from Connecticut indicate Early Jurassic age for part of the Newark Group: Science, v. 182, p. 1243-1247.

Creer, K. M., 1971, Mesozoic paleomagnetic reversal column: Nature, v. 233, p. 545-546.

Dehlinger, P., and Jones, B. R., 1965, Free-air gravity anomaly map of the Gulf of Mexico and its tectonic implication (1963 edition): Geophysics, v. 30, p. 102-110.

Dept. Energy, Mines, and Resources of Canada, 1968, Magnetic anomaly map of Canada (1255A).

———, 1969, Bouguer gravity map of Canada (GMS-69-1).

Dietz, R. S., 1966, Passive continents, spreading sea floors, and collapsing continental rises: Am. Jour. Sci., v. 264, p. 177-193.

———, Holden, J. C., and Sproll, W. P., 1970, Geotectonic evolution and subsidence of Bahama Platform: Geol. Soc. America Bull., v. 81, p. 1915-1928.

Diment, W. H., 1968, Gravity anomalies in northwestern New England, *in* E-An Zen, White, W. S., Hadley, J. B., and Thompson, J. B., Jr., eds., Studies of Appalachian geology: northern and maritime: New York, Wiley-Interscience, p. 399-414.

———, Urban, T. C., and Revetta, F. A., 1972, Some geophysical anomalies in the eastern United States, *in* Robertson, E. C., ed., The nature of the solid earth: New York, McGraw-Hill, p. 544-574.

Drake, C. L., 1969, Continental margins, *in* The earth's crust and upper mantle, Geophys. Monogr. 13: Washington, D.C., Am. Geophys. Union, p. 549-556.

———, and Girdler, R. W., 1964, A geophysical study of the Red Sea: Geophys. Jour., v. 8, p. 473-495.

———, and Nafe, J. E., 1968, The transition from ocean to continent from seismic refraction data, *in* Knopoff, L., Drake, C. L., and Hart, P. J., eds., The crust and mantle of the Pacific area, Geophys. Monogr. 12: Washington, D. C., Am. Geophys. Union, p. 174-186.

———, Ewing, M., and Sutton, G. H., 1959, Continental margins and geosynclines—the east coast of North America north of Cape Hatteras, *in* Physics and chemistry of the earth: Elmsford, N.Y., Pergamon Press, v. 3, p. 110-198.

———, Heirtzler, J., and Hirshman, J., 1963, Magnetic anomalies off eastern North America: Jour. Geophys. Res., v. 68, p. 5259-5275.

———, Ewing, J. I., and Stockard, H., 1968, The continental margin of the eastern United States: Can. Jour. Earth Sci., v. 5, p. 993-1010.

Emery, K. O., and Zarudzki, E. F. K., 1967, Seismic reflection profiles along the drill holes on the continental margin off Florida: U. S. Geol. Survey Prof. Paper 581-A, 8 p.

———, Uchupi, E., Phillips, J. D., Bowin, C. O., Bunce, E. T., and Knott, S. T., 1970, Continental rise off eastern North America: Am. Assoc. Petr. Geol. Bull., v. 54, p. 44-108.

Ewing, G. N., and Hobson, G. O., 1966, Marine seismic refraction investigation over the Orpheus gravity anomaly off the east coast of Nova Scotia: Geol. Survey Can. Paper 66-38.

———, Dainty, A. M., Blanchard, J. E., and Keen, M. J., 1966, Seismic studies of the eastern seaboard of Canada: the Appalachian system, pt. I: Can Jour. Earth Sci., v. 3, p. 89-109.

Ewing, J. I., and Ewing, M., 1959, Seismic refraction measurements in the Atlantic Ocean basins, in the Mediterranean Sea, on the Mid-Atlantic Ridge, and in Norwegian Sea: Gol. Soc. America Bull., v. 70, p. 291-317.

———, and Hollister, C. D., 1972, Regional aspects of deep drilling in the western North Atlantic, *in* Initial reports of the Deep Sea Drilling Project, v. XI, Washington, D.C., U.S. Govt. Printing Office, p. 951-976.

———, Ewing, M., and Leyden, R., 1966, Seismic-profiler survey of the Blake Plateau: Am. Assoc. Petr. Geol. Bull., v. 50, p. 1948-1971.

Ewing, M., Crary, A. P., and Rutherford, H. M., 1937, Geophysical investigations in the emerged and submerged Atlantic coastal plain, pt. I: methods and results: Geol. Soc. America Bull., v. 48, p. 753-802.

Fenwick, D. K. B., Keen, M. J., Keen, C., and Lambert, A., 1968, Geophysical studies of the continental margin northeast of Newfoundland: Can. Jour. Earth Sci., v. 5, p. 483-500.

Fox, P. J., and Opdyke, N. D., 1973, Geology of the oceanic crust: magnetic properties of oceanic rocks: Jour. Geophys. Res., v. 78, p. 5139-5154.

Garrison, L. E., 1970, Development of continental shelf south of New England: Am. Assoc. Petr. Geol. Bull., v. 54, no. 1, p. 109-124.

Gibson, T. G., 1970, Late Mesozoic-Cenozoic tectonic aspects of the Atlantic coastal margin: Geol. Soc. America Bull., v. 81, p. 1813-1822.

Girdler, R. W., Fairhead, J. D., Seale, R. C., and Sowerbutts, W. T. C., 1969, Evolution of rifting in Africa: Nature, v. 224, p. 1178-1182.

Godby, E. A., Baker, R. C., Bower, M. E., and Hood, P. J., 1966: Aeromagnetic reconnaissance of the Labrador Sea: Jour. Geophys. Res., v. 71, p. 511-517.

Gough, D. I., 1967, Magnetic anomalies and crustal structure in eastern Gulf of Mexico: Am. Assoc. Petr. Geol. Bull., v. 51, p. 200-211.

Grant, A. C., 1966, A continuous seismic profile on the continental shelf off northeast Labrador: Can. Jour. Earth Sci., v. 3, p. 725-730.

———, 1968, Seismic profiler investigation of the continental margin northeast of Newfoundland: Can. Jour. Earth Sci., v. 5, p. 1187-1198.

———, 1970, Recent crustal movements on the Labrador shelf: Can. Jour. Earth Sci., v. 7, p. 571-575.

———, 1972, The continental margin off Labrador and eastern Newfoundland—morphology and geology: Can. Jour. Earth Sci., v. 9., p. 1394-1430.

———, 1973, Geological and geophysical results bearing

upon the structural history of the Flemish Cap region: Geol. Survey Can. Paper 71-23, p. 373-387.

Heezen, B. C., and Sheridan, R. E., 1966, Lower Cretaceous rocks (Neocomian-Albian) dredged from Blake Escarpment: Science, v. 154, no. 3757, p. 1644-1647.

______, Hollister, C., and Ruddiman, W. F., 1966, Shaping of the continental rise by deep geostrophic contour currents: Science, v. 152, p. 502-508.

Heirtzler, J. R., and Hayes, D. E., 1967, Magnetic boundaries in the North Atlantic Ocean: Science, v. 157, p. 185-187.

Hersey, J. B., Bunce, E. T., Wyrick, R. F., and Dietz, F. T., 1959, Geophysical investigation of the continental margin between Cape Henry, Virginia, and Jacksonville, Florida: Geol. Soc. America Bull., v. 70, p. 437-466.

Hobson, G. D., and Overton, A., 1973, Sedimentary seismic surveys, Gulf of St. Lawrence: Geol. Survey Can. Paper 71-23, p. 325-336.

Hood, P. J., 1967, Geophysical surveys of the continental shelf south of Nova Scotia: Maritime Sediments, v. 3., no. 1, p. 6-11.

______, and Bower, M. E., 1971, Low-level aeromagnetic surveys in the Labrador Sea: Geol. Survey Can. Paper 71-1, pt. B, p. 37-39.

______, and Godby, E. A., 1965, Magnetic profile across the Grand Banks and Flemish Cap off Newfoundland: Can. Jour. Earth Sci., v. 2, p. 85-92.

______, Sawatsky, P., and Bower, M. E., 1967: Progress report on low-level aeromagnetic profiles over the Labrador Sea, Baffin Bay, and across the North Atlantic Ocean: Geol. Survey Can. Paper 66-58, 11 p.

Hoskins, H., 1967, Seismic reflection observations on the Atlantic continental shelf, slope, and rise southeast of New England: Jour. Geology, v. 75, no. 5, p. 595-611.

Houtz, R. E., and Ewing, J., 1963, Detailed sedimentary velocities from seismic refraction profiles in the western North Atlantic: Jour. Geophys. Res., v. 68, p. 5233-5258.

______, and Ewing, J., 1964, Sedimentary velocities of the western North Atlantic margin: Seismol. Soc. America Bull., v. 54, p. 867-895.

Howie, R. D., 1970, Oil and gas exploration—Atlantic coast of Canada: Am. Assoc. Petr. Geol. Bull., v. 54, no. 11, p. 1989-2006.

______, and Cumming, L. M., 1963, Basement features of the Canadian Appalachians: Geol Soc. Can. Bull., v. 89.

Hutchinson, R. W., and Engels, G. G., 1972, Tectonic evolution in the southern Red Sea and its possible significance to older rifted continental margins: Geol. Soc. America Bull., v. 83, p. 2989-3002.

Irving, E., 1964, Paleomagnetism and its application to geological and geophysical problems: New York, Wiley-Interscience, 339 p.

JOIDES, 1965, Ocean drilling on the continental margin: Science, v. 150, p. 709-716.

Kane, M. F., Simmons, G., Diment, W. H., Fitzpatrick, M. M., Joyner, W. B., and Bromery, R. W., 1972, Bouguer gravity and generalized geologic map of New England and adjoining areas: U.S. Geol. Survey Geophys. Inv. Map. 6P-839.

Kanes, W. H., Saadi, M., Ehrlich, E., and Alem, A., 1973, Moroccan crustal response to continental drift: Science, v. 180, p. 950-952.

Katz, S., and Ewing, M., 1956: Seismic refraction measurements in the Atlantic Ocean, pt. VII: Atlantic Ocean Basin, west of Bermuda: Geol. Soc. America Bull., v. 67, p. 475-510.

Kay, M., 1951, North American geosynclines: Geol. Soc. America Mem. 48.

Keen, C., and Loncarevic, B. D., 1966, Crustal studies on the eastern seaboard of Canada: studies on the continental margin: Can. Jour. Earth Sci., v. 3, p. 65-76.

Keen, M. J., 1969, Possible edge effect to explain magnetic anomalies off eastern seaboard of the U.S.: Nature, v. 222, p. 72-74.

______, and Blanchard, J. E., 1966, The continental margin of eastern Canada: Geol. Survey Can. Paper 66-15, p. 9-18.

______, and Keen, C. E., 1973, Subsidence and fracturing on the continental margin of eastern Canada: Geol. Survey Can. Paper 71-23, p. 23-42.

______, Loncarevic, B. D., and Ewing, G. N., 1970, Continental margin of eastern Canada: Georges Bank to Kane Basin, in Maxwell, A. E., The sea, v. 4, pt. 2: New York, Wiley-Interscience, p. 251-292.

Keller, F., Meuschke, J. L., and Alldredge, L. R., 1954, Aeromagnetic surveys in the Aleutian, Marshall, and Bermuda islands: Am. Geophys. Union Trans., v. 35, p. 558-572.

King, E. R., Zietz, I., and Dempsey, W. J., 1961, The significance of a group of aeromagnetic profiles off the eastern coast of North America: U.S. Geol. Survey Prof. Paper 424-D, p. D299-D303.

King, L. H., 1972, Relation of plate tectonics to the geomorphic evolution of the Canadian Atlantic Provinces: Geol. Soc. America Bull., v. 83, p. 3083-3090.

______, and MacLean, B., 1970, Continuous seismic-reflection study of the Orpheus gravity anomaly: Am. Assoc. Petr. Geol. Bull., v. 54, p. 2007-2031.

Kitazawa, K. Ryall, P. J. C., and Ade-Hall, J. M., 1973, Widespread viscous magnetization in oceanic igneous rocks (abstr.): EOS (Trans. Am. Geophys. Union), v. 54, p. 256.

Kraft, J. L., Sheridan, R. E., and Morino, M., 1972, Time-stratigraphic units and petroleum entrapment models in Baltimore Canyon Basin of Atlantic continental margin geosynclines: Am. Assoc. Petr. Geol. Bull., v. 55, p. 658-679.

Larson, R. L., and Pitman, W. C., III, 1972, World-wide correlation of Mesozoic magnetic anomalies and its implications: Geol. Soc. America Bull., v. 83, p. 3645-3662.

Laughton, A. S., Berggren, W. A., Benson, R., Davies, T. A., Franz, V., Musich, L., Perch-Nielsen, K., Ruffman, A. S., Van Hinte, J. E., and Whitmarsh, R. B., 1970, Deep Sea Drilling Project Leg 12: Geotimes, v. 15, p. 10-14.

Lilly, H. D., 1966, Submarine surveys on the Great Bank and in the Gulf of St. Lawrence: Maritime Sediments, v.2, p. 12-14.

Loncarevic, B. D., and Ewing, G. N., 1967, Geophysical study of the Orpheus gravity anomaly: 7th World Petr. Congr. Proc., p. 828-835.

Long, L. T., and Lowell, R. P., 1973, Thermal model for some continental margin sedimentary basins and uplift zones: Geology, v. 1, p. 87-88.

Lowell, J. D., and Genik, G. J., 1972, Sea-floor spreading and structural evolution of the southern Red Sea: Am. Assoc. Petr. Geol. Bull., v. 56, p. 247-259.

Lowrie, W., Opdyke, N. D., and Løvlie, R., 1973, Magnetic properties of DSDP basalts from the Pacific Ocean (abstr.): EOS (Trans. Am. Geophys. Union), v. 54, p. 256.

Magnusson, D. H., 1973, The Sable Island deep test of the Scotian Shelf: Geol. Survey Can. Paper 71-23, p. 253-266.

Maher, J. C., 1971, Geologic framework and petroleum potential of the Atlantic coastal plain and continental shelf: U. S. Geol. Survey Prof. Paper 659.

Massingill, J. V., Bergantino, R. N., Fleming, H. S., and Feden, R. H., 1973, Geology and genesis of the Mexican Ridges: western Gulf of Mexico: Jour. Geophys. Res., v. 78, p. 2498-2507.

Mattick, R. E., Weaver, N. L., Foote, R. Q., and Ruppel, B. D., 1973, A preliminary report on the U.S. geological survey geophysical studies of the Atlantic outer continental shelf: Am. Assoc. Petr. Geol. East Coast Offshore Symp., Tech. Progr., 9.

May, P. R., 1973, Pattern of Triassic-Jurassic diabase dikes around the North Atlantic in the context of predrift position of the continents: Geol. Soc. America Bull., v. 82, p. 1285-1292.

Mayhew, M. A., 1969, Marine geophysical measurements in the Labrador Sea: relation to Precambrian geology and sea floor spreading [Ph.D. thesis]: Columbia Univ., New York.

———, (in press), "Basement" to the east coast continental margin: Am. Assoc. Petr. Geol. Bull.

———, Drake, C. L., and Nafe, J. E., 1970, Marine geophysical measurements on the continental margins of the Labrador Sea: Can. Jour. Earth Sci., v. 7, p. 199-214.

McElhinny, M. W., and Burek, P. J., 1971, Mesozoic palaeomagnetic stratigraphy: Nature, v. 232, p. 98-102.

McGrath, P. H., Hood, P. J., and Cameron, G. W., 1973, Magnetic surveys of the Gulf of St. Lawrence and the Scotian Shelf: Geol. Survey Can. Paper 71-23, p. 339-358.

McIver, N. L., 1972, Cenozoic and Mesozoic stratigraphy of the Nova Scotia Shelf: Can. Jour. Earth Sci., v. 9, p. 54-70.

———, 1973, Stratigraphy of the continental shelf, offshore Nova Scotia: Am. Assoc. Petr. Geol. East Coast Offshore Symp., Tech. Progr., 7.

McMillan, N. J., 1973a, Geology of Labrador Shelf, Canada (abstr.): Am. Assoc. Petr. Geol. East Coast Offshore Symp., Tech. Progr., 4.

———, 1973b, Surficial geology of the Labrador and Baffin Island shelves: Geol. Survey Can. Paper 71-23, p. 451-469.

Miller, B. L., Geophysical investigations in the emerged submerged Atlantic Coastal Plain, pt. II: Geological significance of the geophysical data: Geol. Soc. America Bull., v. 48, p. 803-812.

Milton, C., and Grasty, R., 1969, "Basement" rocks of Florida and Georgia: Am. Assoc. Petr. Geol. Bull., v. 53, p. 2483-2493.

———, and Hurst, V. J., 1965, Subsurface "basement" rocks of Georgia: Georgia Geol. Survey Bull., v. 76, 56 p.

Minard, J. P., Perry, W. J., Weed, E. A., Robbins, E. I., Mixon, R. B., and Rhodehamel, E. C., 1973, Preliminary geological report on the U.S. Northern Atlantic continental margin: Am. Assoc. Petr. Geol. East Coast Offshore Symp., Tech. Progr., 8.

Mohr, P. A., 1970, The Afar triple-junction and sea-floor spreading: Jour. Geophys. Res., v. 75, p. 7340-7352.

Officer, C. B., and Ewing, M., 1954, Geophysical investigations in the emerged and submerged Atlantic Coastal Plain, pt. 7: Continental shelf, continental slope and continental rise south of Nova Scotia: Geol. Soc. America Bull., v. 65, p. 653-670.

———, Ewing, M., and Wuenschel, P. C., 1952, Seismic refraction measurements in the Atlantic Ocean, pt. IV: Bermuda, Bermuda Rise and Nares Basin: Geol. Soc. America Bull., v. 63, p. 777-808.

Opdyke, N. D., and Wensink, H., 1966, Paleomagnetism of rocks from the White Mountain plutonic-volcanic series in New Hampshire: Jour. Geophys. Res. v. 71, p. 3045-3051.

Pautot, G., Auzende, J. M., and Le Pichon, X., 1970, Continuous deep sea salt layer along North Atlantic margins related to early phase of rifting: Nature, v. 227, p. 351-354.

Pelletier, B. R., 1971, A granodioritic drill core from the Flemish Cap, eastern Canadian continental margin: Can. Jour. Earth Sci., v. 8, p. 1499-1503.

Pitman, W. C., III, and Talwani, M., 1972: Sea-floor spreading in the North Atlantic: Geol. Soc. America Bull., v. 83, p. 619-646.

Press, F., and Beckmann, W. C., 1954, Geophysical investigations in the emerged and submerged Atlantic Coastal Plain, pt. VII: Grand Banks and adjacent shelves: Geol. Soc. America Bull., v. 65, p. 299-313.

Rabinowitz, P. D., 1973, The continental margin of the northwest Atlantic Ocean: a geophysical study [Ph.D. thesis]: Columbia Univ., New York.

Raitt, R. W., 1963, The crustal rocks *in* Hill, M. N., ed., The sea, v. 3: New York, Wiley-Interscience, p. 85-102.

Schultz, L. K., and Grover, R. L., 1973, Geology of the Georges Bank Basin: Am. Assoc, Petr. Geol. East Coast Offshore Symp., Tech. Progr., 8.

Sheridan, R. E., 1969, Subsidence of continental margins: Tectonophysics, v. 7, p. 219-229.

———, 1971a, Crustal structure of the Bahama Platform from Rayleigh wave dispersion: Jour. Geophys. Res., v. 77, p. 2139-2145.

———, 1971b, Geotectonic evolution and subsidence of Bahama Platform: discussion: Geol. Soc. America Bull., v. 82, p. 807-810.

———, and Drake, C. L., 1968, Seaward extension of the Canadian Appalachians: Can. Jour. Earth Sci., v. 5, p. 337-373.

———, Drake, C. L., Nafe, J. E., and Hennion, J., 1966, Seismic-refraction study of continental margin east of Florida: Am. Assoc. Petr. Geol. Bull., v. 50, p. 1972-1991.

———, Smith, J. D., and Gardner, J., 1969, Rock dredges from Blake Escarpment near Great Abaco Canyon: Am. Assoc. Petr. Geol. Bull., v. 53, no. 11, p. 2551-2558.

———, Elliott, G. K., and Oostdam, B. L., 1971, Seismic-reflection profile across Blake Escarpment near Great Abaco Canyon: Am. Assoc. Petr. Geol. Bull., v. 54, no. 11, p. 2032-2039.

Sleep, N. H., 1971, Thermal effects of the formation of Atlantic continental margins by continental breakup: Geophys. Jour., v. 24, p. 325-350.

Smith, L., 1973, Late Mesozoic and Cenozoic of the Sable Island Bank and Grand Banks: Geol. Survey Can. Paper 71-23, p. 267-284.

Spencer, M., 1967, Bahamas deep test: Am. Assoc. Petr. Geol. Bull., v. 51, p. 263-268.

Steiner, M. B., and Helsley, C. E., 1972, Jurassic polar movement relative to North America: Jour. Geophys. Res., v. 77, p. 4981-4993.

Stephens, L. E., Goodacre, A. K., and Cooper, R. V., 1971, Results of underwater gravity surveys over the Nova Scotia continental shelf, with Map No. 123—Halifax-Burgeo: Ottawa, Can., Dept. Energy, Mines, and Resources.

Tagg, A. R., and Uchupi, E., 1966, Distribution and geologic structure of Triassic rocks in the Bay of Fundy and the northeastern part of the Gulf of Maine: U.S. Geol. Survey Prof. Paper 550-B, p. B95-B98.

Talwani, M., 1959, Gravity anomalies in the Bahamas and their interpretation [Ph.D. thesis]: Columbia Univ., New York.

______, and Heirtzler, J. R., 1964, Computation of magnetic anomalies caused by two-dimensional structures of arbitrary shape, *in* Parks, G., ed., Computers in mineral industries: Stanford Univ. Press, p. 464-480.

Taylor, P. T., Zietz, I., and Dennis, L. S., 1968, Geologic implications of aeromagnetic data for the eastern continental margin of the United States: Geophysics, v. 33, no. 5, p. 755-780.

______, Greenewalt, D., Watkins, N. D., and Ellwood, B. B., 1973, Magnetic properties analysis of basalt beneath the quiet zone (abstr.): EOS (Trans. Am. Geophys. Union), v. 54, p. 255.

Tramontini, C., and Davies, D., 1969, A seismic refraction survey in the Red Sea: Geophys. Jour., v. 17, p. 225-241.

Uchupi, E., 1966, Structural framwork of the Gulf of Maine: Jour. Geophys. Res., v. 71, p. 3013-3028.

______, 1969, Marine geology of the continental margin off Nova Scotia, Canada: N.Y. Acad. Sci. Trans., ser. 2, v. 31, p. 56-65.

______, and Emery, K. O., 1968, Structure of continental margin off Gulf Coast of United States: Am. Assoc. Petr. Geol. Bull., v. 52, p. 1162-1193.

______, Milliman, J. D., Luyendyk, B. P., Bowin, C. O., and Emery, K. O., 1971, Structure and origin of the southeastern Bahamas: Am. Assoc. Petr. Geol. Bull., v. 55, p. 687-704.

U.S. Geological Survey, 1965, Geologic map of North America.

Vogt, P. R., Anderson, C. N., and Bracey, D. R., 1971, Mesozoic magnetic anomalies, sea-floor spreading, and geomagnetic reversals in the southwestern North Atlantic: Jour. Geophys. Res., v. 76, p. 4796-4823.

Watkins, J. S., and Geddes, W. H., 1965, Magnetic anomaly and possible orogenic significance of geological structure of the Atlantic Shelf: Jour. Geophys. Res., v. 70, p. 1357-1361.

Watson, J. A., and Johnson, G. L., 1970, Seismic studies in the region adjacent to the Grand Banks of Newfoundland: Can. Jour. Earth Sci., v. 7, p. 306-316.

Watts, A. B., 1972, Geophysical investigations east of the Magdalen Islands, southern Gulf of St. Lawrence: Can. Jour. Earth Sci., v. 9, p. 1504-1528.

Webb, G. W., 1973, Salt structures east of Nova Scotia: Geol. Survey Can. Paper 71-23, p. 197-218.

Wegener, A., 1924, The origin of continents and oceans: New York, Dutton.

Wilhelm, O., and Ewing, M., 1972, Geology and history of the Gulf of Mexico: Geol. Soc. America Bull., v. 83, p. 575-600.

Willmore, P. L., and Scheidegger, A. E., 1956, Seismic observations in the Gulf of St. Lawrence: Trans. Roy. Soc. Can., v. 50, ser. III, p. 21-38.

Woollard, G. P., and Joesting, H. P., 1964, Bouguer gravity anomaly map of the U.S. (exclusive of Alaska and Hawaii): Am. Geophys. Union and U.S. Geol. Survey, 1:2,500,000.

Worzel, J. L., 1968, Advances in marine geophysical research of continental margins: Can. Jour. Earth Sci., v. 5, p. 963-983.

______, and Shurbet, G. L., 1955, Gravity measurements at continental margins: Natl. Acad. Sci. Proc., v. 41, p. 458-469.

Bahamas Salient of North America

A. A. Meyerhoff and C. W. Hatten

INTRODUCTION

Radiometric dates from the sialic, continental, Florida basement demonstrate a minimum range in age for that basement of early Paleozoic to late Proterozoic. Geophysical data—seismic, gravimetric, and magnetic—suggest that this continental crust extends unbroken from Florida to the eastern Bahamas and northern Cuba. The total area involved, 300,000 km^2 (116,000 mi^2), overlaps Africa in "pre-drift" reconstructions. This overlap extends more than 1,500 km (960 mi) in an east-west direction. Because of this, and the probable great age of the continental crust beneath the Bahamas, such "pre-drift" reconstructions are unacceptable, and alternative explanations ranging from the Nafe-Drake drift model to a fixist model must be reconsidered.

The sialic crust of the Bahamas has played an important role in the geologic development of the Cuba-Bahamas-Florida region: (1) a pre-Jurassic basement ridge in central Cuba—extending to the longitude of the eastern Bahamas—played a major role in localizing the salt deposits of the Early and Middle Jurassic Punta Alegre Formation; (2) the Early to Middle Jurassic San Cayetano Group of western, south-central and possibly southeastern Cuba was derived from a sialic terrane—presumably in Florida, Cuba, and Yucatan; (3) the southern and eastern margins of this continental salient, in Cuba and north of the Dominican Republic, determined the location of the western half of the Late Jurassic-middle Eocene Greater Antilles orthogeosyncline; the eugeosyncline borders the oceanic-continental crust boundary, whereas the miogeosyncline and associated basins overlie the continental crust beneath northern Cuba and the southern Bahamas.

Since the beginning of Early to Middle Jurassic time, Cuba, the Bahamas, and south Florida have subsided 10-11 km. Extensive carbonate and evaporite deposition characterized northern Cuba, all of the Bahamas, and southern Florida for approximately 200 my. Although the gross lithologies would suggest the persistence of uniform depositional conditions with few facies changes, lateral and vertical facies changes in the carbonate-evaporite sequence are extensive. Thrusting from the Caribbean Sea during Late Jurassic (or earlier) through middle Eocene time produced large, gentle structures on the site of the present Bahamas which behaved as a foreland platform during the development of the Greater Antilles orthogeosyncline.

GEOGRAPHY

The Bahamas consist of about 300,000 km^2 (116,000 mi^2) of limestone islands, shallow-water carbonate banks, and deep-water channels which indent and separate various banks (Fig. 1). The land area is only 11,406 km^2 (4,404 mi^2). The shallow-water banks generally are about 10 m (33 ft) deep, and inside the 200-m bathymetric contour (656 ft) there are about 124,716 km^2 (48,135 mi^2) of territory.

The Bahama carbonate platform is unusual not only for its size, but also for the nearly 6-km-thickness (3.75 mi) of carbonate and evaporite sedimentary rocks which are known to underlie the platform—easily one of the thickest carbonate sequences in the world. An additional 5 km (3 mi) of section probably underlies the section that has been penetrated (Furrazola et al., 1964). Tectonically the area has remained remarkably stable with only gentle warping movements recorded in the attitudes of the carbonate-evaporite rocks.

PREVIOUS HYPOTHESES OF BAHAMAS ORIGIN

Many origins have been postulated for the Bahamas, and the literature on this topic is becoming very extensive:

1. The Bahamas are a great delta whose materials were "apparently thrown down by the waters of the Gulf Stream on their receiving a check from the Atlantic, as they emerge in full strength from the Gulf of Mexico..." (Nelson, 1853, p. 202).

2. The Bahamas are isolated fragments of the Gulf of Mexico-Florida plate, composed of nearly horizontal beds of Tertiary coral rock (Hilgard, 1871, 1881; Gabb, 1873; Suess, 1888, p. 160-161; 1908, p. 701).

3. The Bahamas are underlain by a foldbelt, on which the carbonate platform section is a veneer. Woodring (1928) suggested that the folds are Cretaceous, related to Cuban orogenesis. Hess (1933, 1960), basing his conclusions on ideas first suggested by Guppy (1917), also proposed that the Bahamas overlie an ancient foldbelt, and he ascribed the trellis-like pattern of some deep-water channels to a stream pattern originally determined by the strike of the supposed folds. E. R. King (1959, p. 2853-2854) suggested that the Bahamas are a salient of North America and that the northwest southeast trend of the islands and the northwest trends of the magnetic anomalies of the Bahamas

manian and downward into the Neocomian and Portlandian. A Neocomian-Albian age is well established for the Blake Plateau region (Heezen and Sheridan, 1966) and for northern Cuba (Meyerhoff and Hatten, 1968). The presence of an Early Cretaceous reef proves nothing about the type of crust beneath the Bahamas. However, this reef is associated with continental crust in Cuba. The elevation of the reef with respect to sea level is surprisingly constant except in the Gulf Coast geosyncline subsurface, and also where it comes to the surface in eastern Mexico and northern Cuba, which were deformed severely during the "Laramide" orogeny. There are no large changes in the reef level except in the geosynclinal belts.

The growth of such a reef through a distance of more than 3,000 km is indicative of geologic continuity, uniform geologic history, and uniform subsidence of the foundation on which the reef grew. The great diversity of the Cretaceous faunas of the eastern Atlantic margin of the Bahamas-Cuba-North America region shows that the Atlantic was a broad ocean by Neocomian time, since the diversity would not have been possible in a narrow, restricted seaway (Curt Teichert, written communication, March 28, 1973).

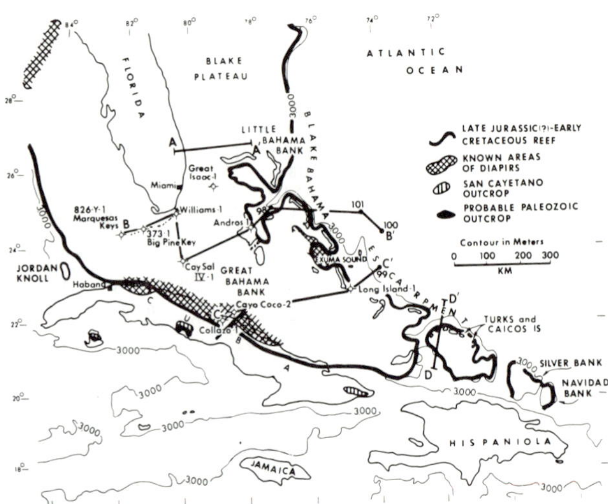

Fig. 5. Known and inferred Neocomian-Cenomanian reef trends, known areas of diapirs of Punta Alegre and Louann salts, Jurassic San Cayetano Group outcrops in Cuba, and areas of probable Paleozoic outcrops in Cuba.

Louann Salt-Punta Alegre Formation. The southeastward extent of the Louann Salt from the western Florida Panhandle is unknown, but it is a fact that an age-correlative evaporite unit, the Punta Alegre Formation, is widespread in northern Cuba (Meyerhoff and Hatten, 1968) (Fig. 5). The recent discovery by Ball et al. (1971, p. 252; see Lidz, 1973, p. 845) of possible salt diapirs in Exuma Sound suggests the possibility that the salt is continuous from northern Cuba to the northeastern Bahamas, and possibly from the Florida Panhandle to northern Cuba beneath the western Florida shelf. The abundant surface expressions for salt diapirism in the northern Gulf of Mexico are absent beneath most of the carbonate bank area, presumably because of the great competence of the carbonate cover.

A barrier between the Atlantic Ocean and the Bahamas has not been found, i.e., a barrier behind which salt deposition would have been possible. Drake et al. (1963, p. 5272) commented on this problem and offered a possible solution: ". . .The magnetic anomalies suggest a relationship to the termination of the Appalachian system and an extension of the Ouachita system that would favor the concept of a crust-thickening mechanism. Extension of the Ouachita trend to Navidad Bank, north of Hispaniola, would provide a foundation for the entire chain of islands and banks. Another consequence of this extension of the Ouachita system would arise if the system was above sea level during late Paleozoic-early Mesozoic time. Ewing et al. (1962) demonstrated the presence of large domes in ...part of the Gulf of Mexico ... Salt deposition requires arid conditions and either limited or no access to the open ocean. At the present time the passages between the Gulf of Mexico and the Atlantic Ocean are the Yucatan Channel on the south and the Straits of Florida on the east. The postulated Ouachita extension would have closed the Straits of Florida and, together with the land mass, which might have been vary narrow . . .could have effectively sealed off the Gulf of Mexico from the open ocean."

In Cuba, south of the area of deposition of the Punta Alegre Formation, a basement high reaches the surface and evidently had pronounced relief during Late Jurassic through early Eocene time. There is no evidence to extend the salt south of this basement high. It is almost certain that a barrier was present across the present Yucatan Channel, where Pyle et al. (1973) found that the Cuban fold system continues to offshore British Honduras. We assume, without proof, that the basement high of Cuba, or a high like it, continued around the outer rim of the continental block underlying the Bahamas during Early to Middle Jurassic time.

The alternative—that Africa's alleged position against North America provided a barrier for salt deposition—encounters formidible difficulties:

1. The volume of salt underlying the northern Gulf of Mexico area alone would require the continuous introduction of seawater for many millions of years, and a large ocean would have to be close by to provide the salt water.

2. The Louann Salt occupies the northern part of the Gulf of Mexico, and does not occupy the deepest part of the Gulf of Mexico (Bryant et al., 1968; Amery, 1969; Antoine and Bryant, 1969; Ewing et al., 1969, Fig. 2, p. 12). Thus, there is no apparent explanation for the existence of different and largely separated Jurassic salt basins, such as the Isthmian Saline basin, the basins of the northern Gulf, and the North Cuba salt basin.

3. The Bahamas overlap of Africa poses a major problem, inasmuch as the salt diapirs of Ex-

uma Sound reported by Ball et al. (1971) and Lidz (1973) overlap the African continent in the Bullard et al. (1965) reconstruction of the continents.

4. There appear to be no equivalent salt deposits in Africa from Southwest Africa to Portuguese Guinea. This is the part of Africa where equivalent Jurassic deposits ought to be found if, indeed, the continents did break up during Triassic or Jurassic time. The offshore Senegal salt domes (Ayme, 1965) are too far north to have been opposite the Gulf of Mexico (Fig. 2). It is true that marine (reworked) Triassic is present in Cameroun (Müller and Mosher, 1971), but in Rio Muni, just south of Cameroun, marine Cretaceous directly overlies marine Silurian in deep wells (Teichert, in press); and in offshore Ghana, marine to paralic Portlandian-Neocomian directly overlies a diabase sill that is intrusive into marine Devonian (Meyerhoff and Meyerhoff, 1972, in press). Thus marine and lagoonal Jurassic deposits equivalent to the Punta Alegre-Louann are unknown in those parts of Africa where they should be found.

San Cayetano Group. The San Cayetano Group includes the San Cayetano, Azucar, and Jagua formations (Gutierrez-Domech, 1968), overlain by the Vinales Formation. These units are typically developed in Pinar del Rio Province, western Cuba, but may extend across at least the southern half of the island.

The mineralogy of the thick Early to Middle Jurassic San Cayetano Group (shale, siltstone, sandstone) indicates a provenance from a mature sialic crust. The chemistry of the San Cayetano Group appears to match that of possible provenance terranes in central Florida (Khudoley and Meyerhoff, 1971, p. 44-49), Yucatan, and the Isle of Pines (Rigassi, 1963, p. 341), although chemical data are not abundant. The possibility that local uplifts within the San Cayetano area of deposition (Fig. 5) provided some of the debris for the San Cayetano cannot be eliminated.

The San Cayetano Group has been involved in a number of large-scale nappe-like thrust sheets (Hatten, 1957, 1967; Rigassi, 1963), and the depositional features that characterize the San Cayetano in each of the thrust sheets commonly are distinctly different. Large-scale displacement from south to north has caused an apparent reversal in the pattern of deposition (i.e., the northernmost thrust sheets are actually from the southernmost area of original deposition).

Primary sedimentary features in the San Cayetano include cross-bedding, truncation, and cut-and-fill in the more coarse-grained, thick- to medium-bedded sandstone deposits now exposed in the present northern thrust sheets. There is a general increase in coarser-grained sediments in the westernmost exposures of the San Cayetano. This indicates that a western source, in addition to a southern source, contributed material (composed predominantly of quartz sand) to the unit. The statements by Khudoley and Meyerhoff (1971, p. 44-49) that the main source was the present region of Florida are not supported by observations in San Cayetano outcrops of western Cuba and therefore may not be valid.

In the central or middle thrust sheets, more rhythmic, thin-bedded sandstone alternating with phyllitic shale predominates. Some minor cross-bedding is present in the sandstone beds of the central thrust sheets, but slump features are very common and well developed. These include load casts and intraformational contortion from subaqueous slumping. In the thrust sheets now exposed along the south and southeastern area of San Cayetano exposure, the sediments are predominantly finer grained siltstone, thin-bedded phyllitic shale, and some calcareous mudstone. It therefore appears that in western Cuba one is dealing with a disrupted cross section of a sedimentary sequence whose depositional environment ranged from deltaic to floodplain on the south and west to prodelta slope and deep water on the north and east.

Therefore, if the San Cayetano provenance actually was the Isle of Pines or Yucatan, as field evidence indicates, the Dietz et al. (1970) explanation for the source of the San Cayetano sediments in Africa and in North America requires major modification.

Basement in Yucatan, Florida, and Cuba. The Yucatan basement consists of Paleozoic metamorphic rocks which, in Pemex's Yucatan No. 1 well (Fig. 1), gave radiometric ages ranging from 290 to 420 my (Dengo, 1969, p. 312-313; Viniegra, 1971, p. 482). In Florida, basement rocks range in age from 143 to 530 my (Jurassic to Cambrian: Bass, 1969, p. 304-307; Milton and Grasty, 1969, p. 2485-2487; Milton, 1972, p. 5-9). In Cuba the oldest dated rock is 180 my (Late Triassic to Early Jurassic: Meyerhoff et al., 1969, p. 2498; Khudoley and Meyerhoff, 1971, p. 27), but rocks collected more recently by geologists of the Cuban Academy of Sciences appear to be late and middle Paleozoic. Studies of these rocks still are in progress and reliable radiometric dates will not be forthcoming for a few years. Recent collecting on the Isle of Pines resulted in the discovery of nautiloids "of Silurian-Devonian aspect" (V. Housa, letter to A.A.M., July 27, 1973).

In all three basement areas, the rocks are a mixture of typically sialic (continental crust) igneous and metamorphic types. For this reason, as well as for reasons enumerated in preceding paragraphs, it is difficult to conceive of the Bahamas being underlain by oceanic crust when all the data appear to indicate that the area is underlain by continental crust. This conclusion applies mainly to the area of the Bahamas, but may apply to the Turks and Caicos as well.

Clay Content of the Florida Cretaceous. Weaver and Stevenson (1971) showed that the montmorillonite content of Late Cretaceous carbonates and illite content of Early Cretaceous in Florida increases from north to south. They concluded (p. 3460) that: "The clay mineral facies in the Cretaceous of southern Florida are those that would be expected if Cuba was the major source of the detritus." The authors also concluded that the illite clays originated from a mica-schist terrane, such as that exposed in southern Cuba, and that the younger montmorillonite came from a volcanic terrane, similar to that exposed in Cuba. These conclusions, if valid, imply a geologic unity between Florida and Cuba.

Significance of Greater Antilles Orthogeosyncline. This Portlandian-middle Eocene orthogeosyncline began to form on pre-Jurassic continental crust and on the San Cayetano Group in Cuba during Portlandian time (Khudoley and Meyerhoff, 1971). The orthogeosyncline extended eastward to the Lesser Antilles island of La Desirade where Late Jurassic (139-143 my) and older rocks are known (for age dates and different interpretations, see Fink, 1970a, p. 326-327; 1970b; 1971, p. 261; 1972, p. 275; Meyerhoff and Meyerhoff, 1972). West of Cuba, the Greater Antilles orthogeosyncline extends to offshore central British Honduras (Belize) (Pyle et al., 1973). Thus the total length of the Greater Antilles orthogeosyncline, once thought to be about 2,300 km from western Cuba to St. Croix, now is known to extend more than 3,000 km from offshore Belize to La Desirade (Fig. 6).

The orthogeosyncline has a definite relation to the age and origin of the Bahamas. The Yucatan Peninsula and the Bahamas, as far east as the Turks and Caicos islands, comprise the foreland of the Greater Antilles orthogeosyncline. The Greater Antilles were thrust from south to north against this foreland—at least once in pre-Early Jurassic time; once during middle Cretaceous time (Subhercynian orogeny of Hatten et al., 1958; Furrazola-Bermudez, 1963, 1964, 1968; Khudoley, 1967; Meyerhoff and Hatten, 1968; and Khudoley and Meyerhoff, 1971); and several times from the Campanian through the middle Eocene. The foreland nature of the Bahamas was recognized early by Schuchert (1935) and by H. A. Meyerhoff (1946, 1954).

Although we cannot deny the possibility that a near-sea-level foreland underlain by oceanic crust can exist, we know of no other example and doubt that the Bahama foreland is an exception (a possible exception could be a part of the West Florida shelf: Krivoy and Pyle, 1972). It is a fact that the Yucatan part of the foreland is underlain by continental crust. Yucatan is continuous with the Bahamas through Cuba (Bryant et al., 1969; Baie, 1970; Pyle et al., 1973). Therefore, this is another reason for believing that the Bahamas are underlain by continental crust. Moreover, the fact that this foreland area was present when the orthogeosyncline formed is indicative of some antiquity. The fact that the San Cayetano Group appears to extend from Oriente Province, eastern Cuba, to the offshore of Belize is evidence of even greater antiquity. The mature nature of the San Cayetano sediments also shows that the source was an older continental terrane.

Conclusions. The entire Bahamas, including the Turks and Caicos Islands, are considered here to be underlain by continental crust. Thus the problem of the Bullard et al. (1965) fit remains unsolved, unless one adopts a fixist view of ocean basins and continents, or accepts a hypothesis like those proposed by Wilson (1966), in which the Atlantic Ocean closes and reopens, or by Nafe and Drake (1969, p. 83), in which the presence of a Paleozoic proto-Atlantic

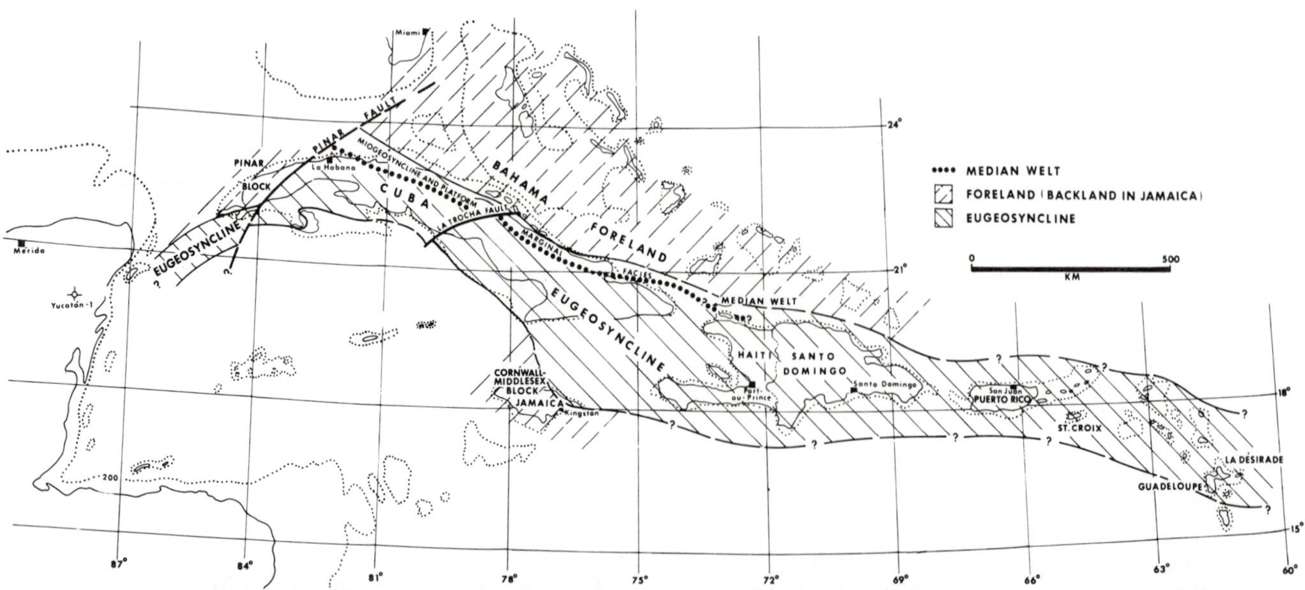

Fig. 6. Greater Antilles orthogeosyncline and Bahama foreland. Miogeosyncline and platform marginal facies separates eugeosyncline on south from foreland on north. Based on H. A. Meyerhoff (1954), Khudoley and Meyerhoff (1971), Meyerhoff and Meyerhoff (1972), Pyle et al. (1973), and Fink (1972).

Ocean is required. The Wilson hypothesis encounters formidable problems related to a complete closing of the Atlantic (Meyerhoff and Meyerhoff, 1972); the Nafe and Drake hypothesis overcomes some, but not all, of these problems. The Dietz et al. (1970) hypothesis—or the Uchupi et al. (1971) modification—may be, as Sheridan wrote (1971, p. 807), "a liberation from older ideas," but both these hypotheses suffer in that many geologic and geophysical facts are not carefully considered. The ultimate answer will be provided by the drill, and we expect that the basement will turn out to be Paleozoic, like that beneath Cuba, Florida, and Yucatan.

STRATIGRAPHY

Only four deep wells have been drilled in the Bahamas. General stratigraphic information can be deduced from reflection-seismic records. The stratigraphic section of the four Bahamas wells is related very closely both to that of northern Cuba and southern Florida. A brief summary of the Bahamas wells and seismic characteristics follows.

Bahamas Wells and Seismic Data

Andros-1. This, the first well in the Bahamas (Fig. 1), was drilled in 1946-1947 to a total depth of 4,448 m (14, 585 ft). The well reached a total depth in Albian (Early Cretaceous) backreef carbonates. No evaporites were penetrated. Lithologic details were given by Spencer (1967) and Goodell and Garman (1969).

Cay Sal IV-1. This well (Fig. 1) was drilled in 1958-1959 by the Standard Oil Company of California and the Gulf Oil Corporation. At total depth 5,766 m (18,906 ft) the well was in shallow-water carbonates and anhydrites of early Neocomian or late Portlandian age. Generalized descriptions of all or parts of the well were published by Furrazola-Bermudez (1964, 1968), Holser and Kaplan (1966), Khudoley (1967), Goodell and Garman (1969), and Khudoley and Meyerhoff (1971).

Long Island-1. Mobil, Standard of California, and Gulf drilled this well (Fig. 1) in the southeastern Bahamas during 1970. It reached total depth in reef and backreef carbonates of early Neocomian or late Portlandian age. Total depth is 5,355 m (17,557 ft). The lithologic section is very similar to that in the Andros well. No anhydrite, gypsum, or salt was penetrated. Top of the Cretaceous is at 1,220 m (4,000 ft). The oldest fauna is Albian and contains *Nummoloculina heimi* at 4,057 m (13,300 ft). The section below 4,057 m to total depth 5,355 m is barren of fossils, but consists of dolomite which closely resembles the Neocomian-Portlandian Perros Formation of Cuba. Moreover, the barren unit is

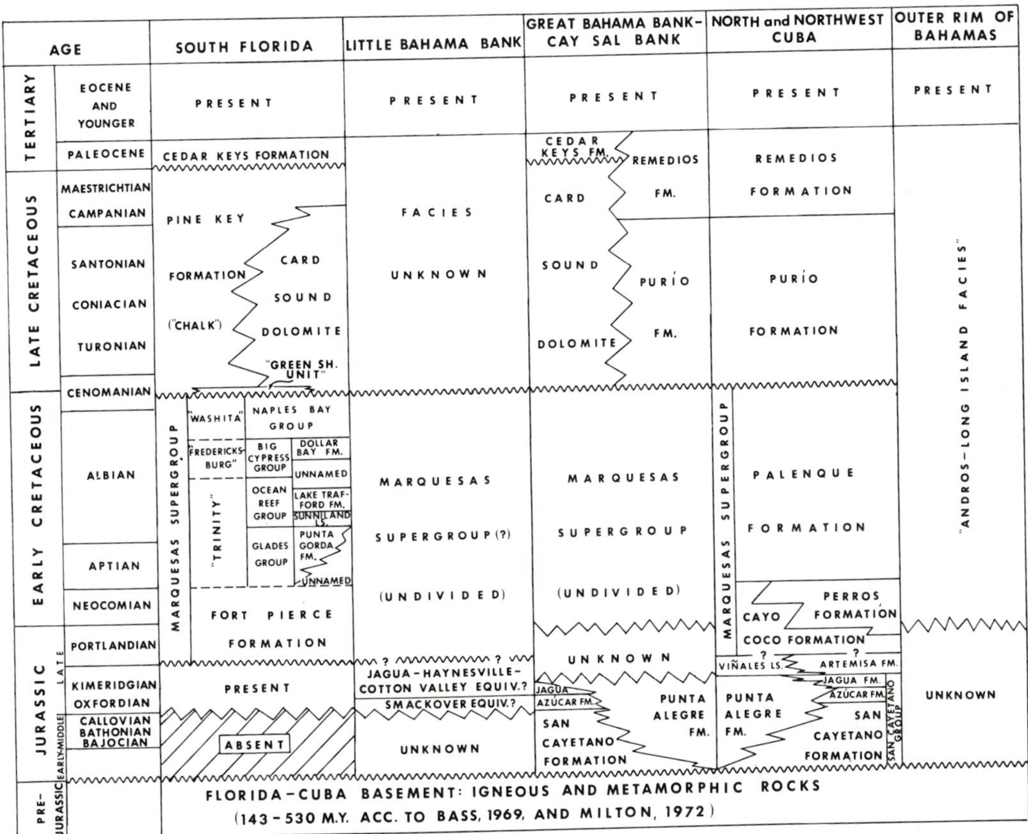

Fig. 7. Stratigraphic chart for northern Cuba, Bahamas, and southern Florida.

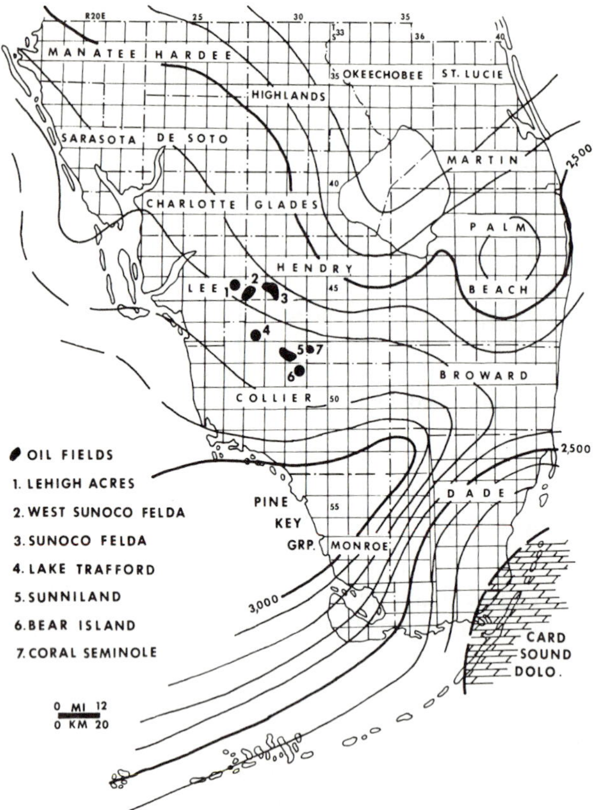

Fig. 10. Isopach map, Upper Cretaceous, southern Florida, from Winston (1971b), essentially the Pine Key Group ("chalk"); note the Card Sound Dolomite in the southeastern part of area. Card Sound extends eastward and southward to Andros-1 and Cay Sal IV-1 (see Figs. 1 and 8).

and (2) we have not studied the samples from this well. In contrast, the correlation with the Cay Sal IV-1 well (2,521-2,815 m) seems to be good, based on sample studies by us and several colleagues.

The fauna collected from the Palenque includes *Bolivinopsis* sp., *Cuneolina* spp., *Nummoloculina heimi*, *Dictyoconus walnutensis*, and *Orbitolina* cf. *O. concava*. The presence of these fossils suggests a late Aptian through Cenomanian age for the type Palenque.

During Late Jurassic (Portlandian) time, a reef barrier surrounded the present Bahamas, swung through northern Cuba, and turned northward in the vicinity of the West Florida Escarpment (Fig. 5). Lagoonal conditions were widespread. During Early Cretaceous time, the area occupied by evaporitic lagoons diminished gradually in size. Lagoonal conditions no longer characterized northern Cuba by middle Aptian time and the Cay Sal area during late Albian time. The most widespread lagoons during late Albian-Cenomanian times were present in southern Florida and the northern Bahamas.

As lagoonal, evaporitic conditions became more restricted, open-water bank limestone was deposited over an increasingly wide area. These open-water bank limestone conditions are represented by the Perros and Palenque formations, which became more widespread at the expense of the Marquesas Supergroup.

Late Cretaceous-Paleocene

As a result of the middle Cretaceous ("Subhercynian") orogeny in Cuba, the Bahamas area began to break up into a series of isolated banks. Many of the modern reentrants into the banks were already in existence, as proved by the presence near the mouth of the Tongue of the Ocean (JOIDES site 98, Figs. 1 and 9) of deep-water Cenomanian carbonate mudstone. Deep-water carbonates also were deposited within the Cayo Coco bank area of northern Cuba (Bryant et al., 1969; Khudoley and Meyerhoff, 1971). The "proto-Florida Straits" formed during the early part of Late Cretaceous time (Bryant et al., 1969), as proved by the widespread occurrence of Late Cretaceous deep-water "chalk" in southern Florida. This "chalk" is largely a deep neritic to upper bathyal deposit, as can be determined in many localities, particularly Jordan Knoll in the western part of the Florida Straits (Bryant et al., 1969; see Fig. 1).

However, Meyerhoff (*in* Bryant et al., 1969) appears to have been wrong in stating that the modern Straits of Florida originated during Late Cretaceous time; Chen's (1965) view that modern straits originated during Eocene time is better founded. We base this conclusion on the careful lithofacies mapping of the Late Cretaceous by Winston (1971b, p. 21, his Fig. 8; Fig. 10 of this paper). Winston noted that the comparatively deep-water Late Cretaceous "chalk" grades laterally in southeasternmost Florida into a shallow-water dolomite facies, the Card Sound Dolomite, which is on the Florida side of the Florida Straits, and which also is present in the Cay Sal and Andros wells. These facts *suggest* that southern Florida, northwest of the type locality of the Card Sound Dolomite, was the site of the proto-Straits of Florida, whose present position was not fixed until Eocene or Paleocene time (Chen, 1965; JOIDES, 1965). The possibility that the present Straits formed during Late Cretaceous time certainly is not eliminated by Winston's (1971b)

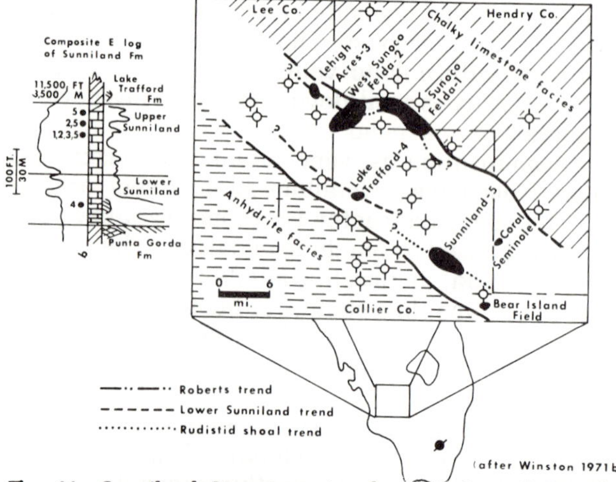

Fig. 11. Sunniland Limestone trends, type log of Sunniland (composite), and oil fields in Sunniland trend of southwestern Florida. From Winston (1971b), with modifications. Location is shown on Fig. 1.

data, but the data that he presented do suggest a very late Cretaceous or post-Late Cretaceous date of their origin.

PETROLEUM POTENTIAL

All early Tertiary and Late Jurassic-Cretaceous strata in this region, except those directly overlying basement and the anhydrite-halite beds, are limestone or dolomite. Marl and shale are scarce. Therefore, the principal source materials for hydrocarbons in the region would have had to be deposited with the carbonates. Study of two areas—one in Mexico and one in Florida—suggests that basinal ("deep-water"), lithographic to sublithographic, pelagic-microfossil-bearing limestones comprise one type of limestone in which organic materials may be buried and from which oil and gas may be generated.

The first example is the El Abra Limestone (Golden Lane) of Mexico, where Early to middle Cretaceous reef limestones grade westward down the El Abra reef flank into interbedded dolomitized reef talus and deep-water limestone (Tamabra Limestone) and, farther west, into the deep-water Tamaulipas Limestone (Barnetche and Illing, 1956; Viniegra and Castillo, 1970). East of the Golden Lane reef, in the middle of the Golden Lane atoll, hydrocarbon accumulations are small and difficult to find—possibly because of a lack of structure, absence of porosity, and permeability, or a lack of source materials. The second example is the Sunniland and associated trends of south Florida (Fig. 11), where the Sunniland limestone mounds grade westward into thin-bedded, dark-gray, basinal limestone (which is overlain by the thick anhydrite that Winston, 1971b, termed the "anhydrite facies"; see Fig. 11); and eastward into chalky, shallow-water, evaporite-bearing, lagoonal carbonate rocks.

In both examples—the Golden Lane and Sunniland areas—a deep-water, dark-gray to black, petroliferous limestone is present on one side of and grades laterally into the reservoir beds. Although proof is lacking, the inference is that the basinal limestones contained the original organic materials that were transferred into petroleum. Hedberg (1964, p. 1791) came to an identical conclusion regarding the Cretaceous La Luna Formation of Venezuela—a formation very similar in origin to the basinal limestones of eastern Mexico and south Florida.

The basinal, deep-water limestones south of the Late Jurassic and Cretaceous reefs or bank edges of northern Cuba long have been assumed to be the sources of the oil seeps and small oil fields of Cuba (Meyerhoff and Hatten, 1968). However, this assumption is far from proved. Several geologists (e.g., Hedberg, 1964) have suggested that petroleum can be generated in other types of carbonate muds, regardless of the depth of deposition.

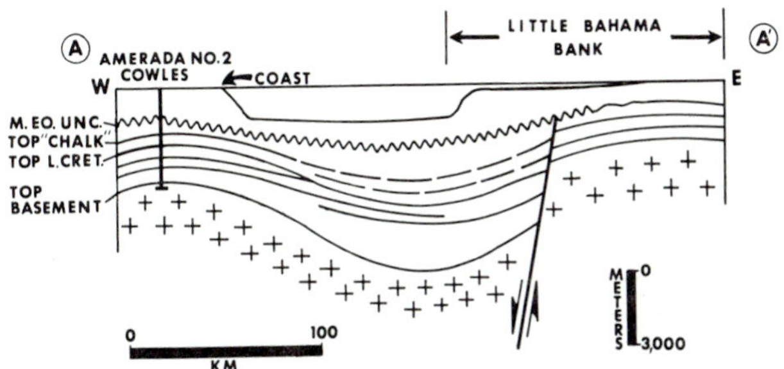

Fig. 12. West-east cross section A-A' (Fig. 1), Florida to Little Bahama Bank. Fault on right-hand side of cross section may be continuation of Pinar fault (see Fig. 4). Cross section is based on reflection-seismic information.

DISCUSSION

Little Bahama Bank. The seismic data are of high quality from Little Bahama Bank, in contrast to the poor-quality data from other areas of the southern Florida-Bahamas-northern Cuba area, and it is probable that a terrigenous clastic sequence of rocks, possibly an extension of the strata known to be present in the Southeast Georgia embayment, underlies the Little Bahama Bank area. Part of the reason for the better reflections may be that the older part of the section (Jurassic, Paleozoic) rises closer to the surface from south to north. A generalized cross section is shown on Figure 12.

Southern Florida-Great Bahama Bank-Inagua-Northern Cuba. This area encompasses most of the Bahamas platform and the edge of the deformed northern margin of the Greater Antilles orthogeosyncline. The main Aptian-Albian (Neocomian-Cenomanian in some areas) "Golden Lane type" reef appears to be exposed around the margins. JOIDES hole 98 (Hollister et al., 1972) was drilled in the same general area as core 167-51 (Ericson et al., 1952; see Figs. 1 and 8) northwest of New Providence Island, in Northeast Providence Channel, at the mouth of the Tongue of the Ocean.

The oldest sedimentary rock in site 98 is early Campanian-late Santonian (Hollister et al., 1972, p. 15). This agrees with the age given by Ericson et al., 1952, p. 505) for the shallow core A167-51. An Eocene deep-water sample was collected close to site 98 (Stehman, 1970). A study of core A167-51 by A. Loeblich, M. A. Furrer, and P. Brönnimann from 1957 through 1959 yielded important additional information: strata of late Cenomanian age also are present in A167-51. The facies of the Cenomanian material is very deep water, probably bathyal. This discovery has two corollaries: (1) the deep-water channels of at least part of the Great Bahama Bank already were in existence by Cenomanian time; and (2) the Hollister et al. (1972) remarks on pages 15-16 of their paper are very appropriate: ". . .The comparison of the depth of [the oldest] horizon with its stratigraphic equivalent in the Andros Island well . . .suggests that pelagic rates of deposition do not apply to sediments laid down at Site 98 prior to late Cretaceous time. If they were applied to the lower Cretaceous deposits, an apparently abnormal structural condition would result, wherein deep-water beds beneath the Bahama Channels would be structurally higher than the correlative shallow-water bank deposits in the Andros Island well." Examination of Figure 8 should help to resolve this problem; it is not nearly as serious as Hollister et al. implied.

Cay Sal Tonavidad Bank. Air-magnetometer data show only a large magnetic "high" along the north side of the Turks and Caicos islands (Meyerhoff and Treitel, 1959). However, local, sharp anomalies are absent along the trend of this "high." In contrast, ground magnetometer information shows three sharp negative anomalies 1 km wide—one on West Caicos Island, two on Turks Bank. The anomalies are about 100 gammas in magnitude. They are not salt-generated, inasmuch as the gravity anomalies are positive.

Four possible causes are: (1) the anomaly sources are anhydrite, which has negative magnetic susceptibility; (2) the sources are reverse-polarized rocks whose magnetization was induced by intense compression; (3) the anomalies are produced by intrusives with reversed magnetic polarity; and (4) iron buried a few feet below the surface produces the anomalies. This last possibility is not eliminated, inasmuch as military bases have been built close to the sites of the anomalies. Possibility (1) is eliminated because anhydrite would produce a negative anomaly of about only -4 gammas. Possibility (2) is eliminated because seismic data show that compression in this area is nil. Possibility (4) is not eliminated and requires further study.

The third possibility is the most likely (Fig. 13). The sharp negative anomalies are believed to be the result of a strongly paramagnetic rock source, probably mafic or ultramafic, with reverse polarization. A strong magnetic susceptibility is required to depress the recorded values to -100 gammas. The

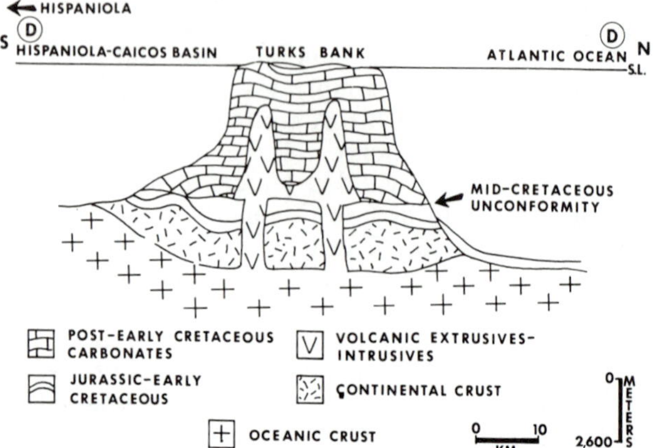

Fig. 13. South-north cross section D-D' (Fig. 1), Hispaniola to Turks-Caicos islands. Cross section is based on magnetometer and reflection-seismic information.

probability that limestone is the host rock for the postulated intrusives eliminates the concept that acidic intrusives produce the anomalies. The magnitude of the anomalies suggests that the source is 800 m (2,600 ft) or less in depth.

Figure 13 is a possible interpretation of the geology of the Turks-Caicos. At least 1,800 m (6,000 ft) of post-Eocene rocks is present. These rocks are intruded by shallow mafic to ultramafic bodies. The crust beneath the sedimentary column is sialic but thin. This is in keeping with the Uchupi et al. (1971) results. The Cauto fault (Meyerhoff, 1966) extends from Cuba between Great Inagua and the Turks-Caicos, an interpretation supported by Uchupi et al.'s seismic data (Fig. 4).

The Turks-Caicos basin south of the Turks and Caicos islands is underlain by crust transitional between crust and crust (Ewing and Heezen, 1955; Meyerhoff and Treitel, 1959), but the Turks and Caicos islands are fundamentally continental (despite Uchupi et al.'s opinions based on their *lack* of geophysical data over the Turks and Caicos islands and the Turks-Caicos basin). Farther east, seismic data obtained by the various companies show that, south of Silver and Navidad banks (Fig. 1), the crust is intermediate to continental and probably there is a continental crust connection between Hispaniola and the Silver-Navidad banks area (Meyerhoff and Treitel, 1959).

Figure 5 shows a different reef interpretation for this area than for the area of the Turks and Caicos islands. Unpublished 1967 seismic data show the presence, in the Silver-Navidad area, of 1,800-2,100 m (6,000-7,000 ft) of north-dipping Tertiary and younger sediments above several thousand meters of Cretaceous and probably older sedimentary rocks. The pre-middle Eocene section forms a gently arched east-striking anticlinal trend which is overlain unconformably by the middle Eocene and younger strata.

It is our opinion that the Cretaceous of the Navidad and Silver banks area may be of terrestrial origin—at least in part, and in this sense (even though the provenance terrane was volcanic) is somewhat comparable with the Little Bahama Bank. The principal difference is that terrigenous sedimentary rocks beneath the Little Bahama Bank are likely to be mature sedimentary rocks, whereas those beneath the Silver-Navidad banks area are less likely to be mature sediments.

ACKNOWLEDGMENTS

We thank Y. Bonillas, Jules Braunstein, George V. Cohee, L. W. Funkhouser, H. A. Meyerhoff, J. D. Moody, Lyman D. Toulmin, Jr., and G. O. Winston for critical review and useful suggestions. W. R. Oglesby, R. O. Vernon, and G. O. Winston provided valuable information. We are grateful to the American Association of Petroleum Geologists for granting permission for reprinting parts of the original manuscript from its June 1974 *Bulletin* (Meyerhoff and Hatten, 1974).

BIBLIOGRAPHY

Amery, G. B., 1969, Structure of Sigsbee Scarp, Gulf of Mexico: Am. Assoc. Petr. Geol. Bull., v. 53, no. 12, p. 2480-2482.

Antoine, J. W., and Bryant, W. R., 1969, Distribution of salt and salt structures in Gulf of Mexico: Am. Assoc. Petr. Geol. Bull., v. 53, no. 12, p. 2543-2550.

______, Bryant, W. R., and Jones, B., 1967, Structural features of continental shelf, slope, and scarp, northeastern Gulf of Mexico: Am. Asso. Petr. Geol. Bull., v. 51, no. 2, p. 257-262.

Applin, P. L., 1960, Significance of changes in thickness of the Sunniland Limestone, Collier County, Fla.: U.S. Geol. Survey Prof. Paper 400-B, p. 209-211.

______, and Applin, E. R., 1944, Regional subsurface stratigraphy and structure of Florida and southern Georgia: Am. Assoc. Petr. Geol. Bull., v. 28, no. 12, p. 1673-1753.

______, and Applin, E. R., 1947, Regional subsurface stratigraphy, structure, and correlation of middle and early Upper Cretaceous rocks in Alabama, Georgia and north Florida: U.S. Geol. Survey Oil and Gas Inventory, Prelim. Chart 26.

______, and Applin, E. R., 1965, The Comanche Series and associated rocks in the subsurface in central and south Florida: U.S. Geol. Survey Prof. Paper 447, 84 p.

Ayme, J.-M., 1965, The Senegal salt basin, in Salt Basins Around Africa, London, Inst. of Petroleum, p. 83-90.

Babcock, C., 1969, Geology of the Upper Cretaceous clastic section, northern peninsular Florida: Florida Div. Geol., Inform. Circ. 60, 44 p.

Baie, L. F., 1970, Possible structural link between Yucatan and Cuba: Am. Assoc. Petr. Geol. Bull., v. 54, no. 11, p. 2204-2207.

Ball, M. M., 1967, Tectonic control of the configuration of the Bahama Banks: Trans. Gulf Coast Assoc. Geol. Soc., v. 17, p. 165-267.

______, Gaudet, R. M., and Leist, G., 1968, Sparker reflection seismic measurements in Exuma Sound, Bahamas (abstr.): Am. Geophys. Union Trans., v. 49, no. 1, p. 196-197.

______, Dash, B. P., Harrison, C. G. A., and Ahmed, K. O., 1971, Refraction seismic measurements in the northeastern Bahamas (abstr.): EOS (Trans. Am. Geophys. Union), v. 52, no. 4, p. 252.

Banks, J. E., 1959, Limestone conglomerates (recent and Cretaceous) in southern Florida: Am. Assoc. Petr. Geol. Bull., v. 43, no. 9, p. 2237-2243.

______, 1960, Petroleum in Comanche (Cretaceous) section, Bend area, Florida: Am. Assoc. Petr. Geol. Bull., v. 44, no. 11, p. 1737-1748.

Barnetche, A., and Illing, L. V., 1956, The Tamabra of the Poza Rica oil-field, Veracruz, Mexico: 20th Internatl. Geol. Congr., Mexico, 38 p.

Bass, M. N., 1969, Petrography and ages of crystalline basement rocks of Florida—some extrapolations: Am. Assoc. Petr. Geol. Mem. 11, p. 283-310.

Bracey, D. R., 1968, Structural implications of magnetic anomalies north of the Bahama-Antilles islands: Geophysics, v. 33, no. 6, p. 950-961.

Braunstein, J., 1958, Habitat of oil in eastern Gulf Coast, *in* Weeks, L. G., ed., Habitat of oil: Am. Assoc. Petr. Geol., p. 511-522.

Bridge, J., and Berdan, J. M., 1952, Preliminary correlation of the Paleozoic rocks from test wells in Florida and adjacent parts of Georgia and Alabama: Florida Geol. Survey Guidebook, Assoc. Am. State Geologists, 44th Ann. Mtg. Field Trip, Apr. 18-19, p. 29-38.

Browning, W. F., and Welch, S. W., 1970, Developments in southeastern states in 1969: Am. Assoc. Petr. Geol. Bull., v. 54, no. 6, p. 1030-1035.

______, and Welch, S. W., 1972, Developments in southeastern states in 1971: Am. Assoc. Petr. Geol. Bull., v. 56, no. 7, p. 1303-1309.

Bryant, W. R., Antoine, J., Ewing, M., and Jones, B., 1968, Structure of Mexican continental shelf and slope, Gulf of Mexico: Am. Assoc. Petr. Geol. Bull., v. 52, no. 7, p. 1204-1228.

______, Meyerhoff, A. A., Brown, N. K., Jr., Furrer, M. A., Pyle, T. E., and Antoine, J. W., 1969, Escarpments, reef trends, and diapiric structures, eastern Gulf of Mexico: Am. Assoc. Petr. Geol. Bull., v. 53, no. 12, p. 2506-2542.

Bullard, E., Everett, J. E., and Smith, A. G., 1965, The fit of the continents around the Atlantic, *in* Blackett, P. M. S., Bullard, E., and Runcorn, S. K., eds., A symposium on continental drift: Phil. Trans. Roy. Soc. London, ser. A, v. 258, no. 1088, p. 41-51.

Burk, C. A., 1968, Buried ridges within continental margins: Trans. N.Y. Acad. Sci., ser. 2, v. 30, no. 3, p. 397-409.

Carroll, D., 1963, Petrography of some sandstones and shales of Paleozoic age from borings in Florida: U.S. Geol. Survey Prof. Paper 454-A, p. 1-15.

Chen, C. S., 1965, The regional lithostratigraphic analysis of Paleocene and Eocene rocks of Florida: Florida Geol. Survey Bull. 45, 105 p.

Cole, W. S., 1944, Stratigraphic and paleontologic studies of wells in Florida: Florida Geol. Survey Bull. 26, 168 p.

Daly, R. A., Manger, G. E., and Clark, S. P., Jr., 1966, Density of rocks: Geol. Soc. America Mem. 97, p. 19-26.

DeGolyer, E., 1918, The geology of Cuba petroleum deposits: Am. Assoc. Petr. Geol. Bull., v. 2, p. 133-167.

Dengo, G., 1969, Problems of tectonic relations between Central America and the Caribbean: Trans. Gulf Coast Assoc. Geol. Soc., v. 19, p. 311-32

Dietz, R. S., and Sproll, W. P., 1970, Overlaps and underlaps in the North America to Africa continental drift fit, *in* Delany, F. M., ed., The geology of the east Atlantic continental margin, v. 1, General and economic papers: London, Nat. Environ. Res. Council, Inst. Geol. Sci., p. 147-151.

______, and Holden, J. C., 1970, Reconstruction of Pangea: Jour. Geophys. Res., v. 75, p. 4939-4956.

______, and Sproll, W. P., 1970, Geotectonic evolution and subsidence of Bahama platform: Geol. Soc. America Bull., v. 81, no. 7, p. 1915-1927.

Drake, C. L., Heirtzler, J., and Hirshman, J., 1963, Magnetic anomalies off eastern North America: Jour. Geophys. Res., v. 69, no. 18, p. 5259-5275.

Echevarria, G., and Veliev, M., 1967, La perforacion de los pozos profundos "Frances 5" y "Fragoso 1": La Habana, Min. Ind., Rev. Tecnol., v. 5, no. 1, p. 49-54.

Ericson, D. B., Ewing, W. M., and Heezen, B. C., 1952, Turbidity currents and sediments in North Atlantic: Am. Assoc. Petr. Geol. Bull., v. 36, no. 3, p. 489-511.

Ewing, J. I., Worzel, J. L., and Ewing, M., 1962, Sediments and oceanic structural history of the Gulf of Mexico: Jour. Geophys. Res., v. 67, no. 6, p. 2509-2527.

Ewing, M., and Heezen, B. C., 1955, Puerto Rico Trench topographic and geophysical data, *in* Poldervaart, A., ed., Crust of the earth: Geol. Soc. America Spec. Paper 62, p. 255-267.

———, Worzel, J. L., Beall, A. O., Berggren, W. A., Bukry, D., Burk, C. A., Fischer, A. G., and Pessagno, E. A., Jr., 1969, Initial reports of the Deep Sea Drilling Project, v. I: Washington, D.C., U.S. Govt. Printing Office, 672 p.

Field, R. M., et al., 1931, Geology of the Bahamas: Geol. Soc. America Bull., v. 42, no. 3, p. 759-784.

Fink, L. K., Jr., 1970a, Evidence for the antiquity of the Lesser Antilles island arc (abstr.): EOS, (Trans. Am. Geophys. Union), v. 51, no. 4, p. 326-327.

———, 1970b, Field guide to the island of La Desirade with notes on the regional history and development of the Lesser Antilles island arc: Am. Geol. Inst., Internatl. Field Inst. 1970 Guidebook, 17 p.

———, 1971, Evidence in the eastern Caribbean for mid-Cenozoic cessation of sea-floor spreading (abstr.): EOS (Trans. Am. Geophys. Union), v. 52, no. 4, p. 261.

———, 1972, Bathymetric and geologic studies of the Guadeloupe region, Lesser Antilles island arc: Marine Geology, v. 12, no. 4, p. 266-288.

Furrazola-Bermudez, G., Judoley, C. M., and Solsona, J. B., 1963, Generalidades sobre la geologia de Cuba: La Habana, Min. Ind., Rev. Tecnol., v. 2, no. 10, p. 3-22.

———, Judoley, C. M., Mijailovskaya, M. S., Miroliubov, Y. S., and Solsona, J. B., 1964, Geologia de Cuba: La Habana, Min. Ind., 239 p.

———, Judoley, C. M., Mijailovskaya, M. S., Miroliubov, Y. S., and Solsona, J. B., 1968, Geology of Cuba: Washington, D.C., U.S. Dept. Commerce, Joint Publ. Res. Serv., Transl. on Cuba, no. 311, 207 p.

Gabb, W. M., 1873, Topography and geology of Santo Domingo: Trans. Am. Phil. Soc., v. 15, p. 49-259.

Goodell, H. G., and Garman, R. K., 1969, Carbonate geochemistry of Superior deep test well, Andros Island, Bahamas: Am. Assoc. Petr. Geol. Bull., v. 53, no. 3, p. 513-536.

Gordon, W. A., 1973, Marine life and ocean surface currents in the Cretaceous: Jour. Geology, v. 81, no. 3, p. 269-284.

Gough, D. I., 1967, Magnetic anomalies and crustal structure in eastern Gulf of Mexico: Am. Assoc. Petr. Geol., v. 51, no. 2, p. 200-211.

Guppy, H. B., 1917, Plants, seeds, and currents in the West Indies and Azores: London, Williams, and Norgate, 531 p.

Gutierrez-Domech, M. R., 1968, Breve resena sobre el periodo Jurasico en las Provincia de Pinar del Rio, Cuba: La Habana, Inst. Nacl. Recursos Hidraulicos, Publ. Esp. 5, p. 3-23.

Halbouty, M. T., Meyerhoff, A. A., King, R. E., Dott, R. H., Sr., Klemme, H. D., and Shabad, T., 1970, World's giant oil and gas fields, geologic factors affecting their formation, and basin classification: Am. Assoc. Petr. Geol. Mem. 14, p. 502-555.

Hatten, C. W., 1957, Geologic report on Sierra de los Organos: La Habana, Min. Ind., unpubl. rept., 140 p.

———, 1967, Principal feature of Cuban geology: discussion: Am. Assoc. Petr. Geol. Bull., v. 51, no. 5, p. 780-789.

———, and Meyerhoff, A. A., 1965, Pre-Portlandian rocks of western and central Cuba (abstr.): Geol. Soc. America Spec. Paper 82, p. 301.

———, Schooler, O. E., Giedt, N., and Meyerhoff, A. A., 1958, Geology of central Cuba, eastern Las Villas and western Camaguey Provinces: La Habana, Min. Ind., unpubl. rept., 250 p.

Hedberg, H. D., 1964, Geologic aspects of origin of petroleum: Am. Assoc. Petr. Geol. Bull., v. 48, no. 11, p. 1755-1803.

Heezen, B. C., and Sheridan, R. E., 1966, Lower Cretaceous rocks (Neocomian-Albian) dredged from Blake Escarpment: Science, v. 154, no. 3757, p. 1644-1647.

———, Tharp, M., and Ewing, M., 1959, The floors of the oceans: I. The North Atlantic: Geol. Soc. America Spec. Paper 65, 122 p.

Herrera, N. M., 1961, Contribucion a la estratigrafia de la Provincia del Pinar del Rio: La Habana, Soc. Cubana Ing. Rev., v. 61, no. 1-2, p. 2-24.

Hess, H. H., 1933, Interpretation of geological and geophysical observations, *in* U.S. Hydrographic Office-Navy-Princeton Gravity Expedition to the West Indies in 1932: Washington, D.C., U.S. Hydrog. Office, p. 27-54.

———, 1960, Caribbean research project: progress report: Geol. Soc. America Bull., v. 71, no. 3, p. 235-240.

Hilgard, E. W., 1871, On the geological history of the Gulf of Mexico: Am. Jour. Sci., ser. 3, v. 2, p. 391-404.

———, 1881, The later Tertiary of the Gulf of Mexico: Am. Jour. Sci., ser. 3, v. 22, p. 58-65.

Hill, R. T., 1898, Cuba and Puerto Rico, with other islands of the West Indies: New York, Century, 429 p.

Hollister, C. D., Ewing, J. I., Habib, D., Hathaway, J. C., Lancelot, Y., Luterbacher, H., Paulus, F. J., Poag, C. W., Wilcoxson, J. A., and Worstell, P., 1972, Initial reports of the Deep Sea Drilling Project, v. IX: Washington, D.C., U.S. Govt. Printing Office, 1077 p.

Holser, W. T., and Kaplan, I. R., 1966, Isotope geochemistry of sedimentary sulfates: Chem. Geology, v. 1, no. 2, p. 93-135.

Howell, E. E., 1904, Model of the Bay of North America including the Gulf of Mexico and the Caribbean Sea: Washington, D.C., U.S. Coast and Geodetic Survey, relief map (scale not specified).

Imlay, R. W., 1942, Late Jurassic fossils from Cuba and their economic significance: Geol. Soc. America Bull., v. 53, no. 10, p. 1417-1477.

———, 1944, Correlation of the Cretaceous formations of the Greater Antilles, Central America, and Mexico: Geol. Soc. America Bull., v. 55, no. 8, p. 1005-1045.

Ipatenko, S., and Sashina, N., 1971, Sobre el levantamiento gravimetrico en Cuba, *in* Serie de levantamientos gravimetricos en Cuba: La Habana, Min. Mineria, Combust. Met., p. 3-14.

JOIDES, 1965, Ocean drilling on the continental margin: Science, v. 150, no. 3697, p. 709-716.

Judoley, C. M., and Furrazola-Bermudez, G., 1965, Estratigrafía del Jurásico Superior de Cuba: La Habana, Min. Ind., Inst. Cubano Rec. Mineral. Publ. Esp. no. 3, 32 p.

———, and Furrazola-Bermudez, G., 1968, Estratigrafía y fauna del Jurásico de Cuba: La Habana, Acad. Cienc. Cuba, 126 p.

Khudoley, K. M., 1967, Principal features of Cuban

geology: Am. Assoc. Petr. Geol. Bull., v. 51, no. 5, p. 668-677.

______, and Meyerhoff, A. A., 1971, Paleogeography and geological history of Greater Antilles: Geol. Soc. America Mem. 129, 199 p.

______, and Meyerhoff, A. A., 1974, Middle Cretaceous nappe structures in Puerto Rican ophilites and their relation to the tectonic history of the Greater Antilles: Geol. Soc. America Bull., v. 85, (in press).

King, E. R., 1959, Regional magnetic map of Florida: Am. Assoc. Petr. Geol. Bull., v. 43, no. 12, p. 2844-2854.

Knebel, G. M., 1968, AAPG research committee: Am. Assoc. Petr. Geol. Bull., v. 52, no. 3, p. 536-537.

Knipper, A. L., and Puig, M., 1967, Protrusiones de las serpentenitas en el noroeste de Oriente: La Habana, Acad. Cienc. Cuba, Rev. Geol., Ano 1, no. 1, p. 122-137.

Kozary, M. T., 1968, Ultramafic rocks in thrust zones of northwestern Oriente Province, Cuba: Am. Assoc. Petr. Geol. Bull., v. 52, no. 12, p. 2298-2317.

Krivoy, H. L., and Pyle, T. E., 1972, Anomalous crust beneath west Florida shelf: Am. Assoc. Petr. Geol. Bull., v. 56, no. 1, p. 107-113.

LePichon, X., and Fox, P. J., 1971, Marginal zones, fracture zones, and the early opening of the North Atlantic: Jour. Geophys. Res., v. 76, no. 26, p. 6294-6308.

Lidz, B., 1973, Biostratigraphy of Neogene cores from Exuma Sound diapirs, Bahama Islands: Am. Assoc. Petr. Geol. Bull., v. 57, no. 5, p. 841-857.

Lewis, J. W., 1932, Geology of Cuba: Am. Assoc. Petr. Geol. Bull., v. 16, no. 6, p. 533-553.

Lynts, G. W., 1970, Conceptual model of the Bahamian platform for the last 135 million years: Nature, v. 225, no. 5239, p. 1226-1228.

______, and Stehman, C. F., 1969, Deep-sea Eocene in Northeast Providence Channel, origin of Bahamas and sea-floor spreading (abstr.): Geol. Soc. America, v. 1, pt. 7, p. 281-282.

Maher, J. C., 1965, Correlation of subsurface Mesozoic and Cenozoic rocks along the Atlantic coast: Am. Assoc. Petr. Geol., Cross Sect. Publ., 3, 18 p.

______, and Applin, E. R., 1971, Geologic framework and petroleum potential of the Atlantic coastal plain and continental shelf: U.S. Geol. Survey Prof. Paper 659, 98 p.

Mattson, P. H., 1973, Middle Cretaceous nappe structures in Puerto Rican ophiolites and their relation to the tectonic history of the Greater Antilles: Geol. Soc. America Bull., v. 84, no. 1, p. 21-37.

Meyerhoff, A. A., 1966, Bartlett fault system—age and offset, *in* Robinson, E., ed., Trans. 3rd Caribbean Geol. Conf., Kingston: Jamaica Geol. Survey Publ., 95, p. 1-9.

______, 1967, Future hydrocarbon provinces of Gulf of Mexico-Caribbean region: Trans. Gulf Coast Assoc. Geol. Soc., v. 17, p. 217-260.

______, 1970, Continental drift: implications of plaeomagnetic studies, meteorology, physical oceanography, and climatology: Jour. Geology, v. 78, no. 1, p. 1-51.

______, and Hatten, C. W., 1968, Diapiric structures in central Cuba: Am. Assoc. Petr. Geol. Mem. 8, p. 315-357.

______, and Hatten, C. W., 1974, Bahamas salient of North America: tectonic framework, stratigraphy, and petroleum potential: Am. Assoc. Petr. Geol. Bull., v. 58, no. 6.

______, and Meyerhoff, H. A., 1972, "The new global tectonics": major inconsistencies: Am. Assoc. Petr. Geol. Bull., v. 56, no. 2, p. 269-336.

______, and Meyerhoff, H. A., 1974, "The new global tectonics": confusion worse confounded: Am. Assoc. Petr. Geol. Mem. 21, (in press).

______, and Treitel, S., 1959, Geological preliminary evaluation of the Turks and Caicos Islands: La Habana, Min. Ind., unpub. rept., 10 p.

______, Khudoley, K. M., and Hatten, C. W., 1969, Geologic significance of radiometric dates from Cuba: Am. Assoc. Petr. Geol. Bull., v. 53, no. 12, p. 2494-2500.

Meyerhoff, H. A., 1946, Tectonic features of the Greater Antilles (abstr.): Am. Assoc. Petr. Geol. Bull., v. 30, no. 5, p. 744.

______, 1954, Antillean tectonics: Trans. N.Y. Acad. Sci., ser. 2, v. 16, no. 3, p. 149-155.

Milton, C., 1972, Igneous and metamorphic basement rocks of Florida: Florida Bur. Geology Bull. 55, 125 p.

______, and Grasty, R., 1969, "Basement" rocks of Florida and Georgia: Am. Assoc. Petr. Geol. Bull., v. 53, no. 12, p. 2483-2493.

Müller, K. J., and Mosher, L. C., 1971, Post-Triassic conodonts: Geol. Soc. America Mem. 127, p. 467-470.

Nafe, J. E., and Drake, C. L., 1969, Floor of the North Atlantic—summary of geophysical data, *in* Kay, M., ed., North Atlantic—geology and continental drift: Am. Assoc. Petr. Geol. Mem. 12, p. 59-215.

Nelson, R. J., 1853, On the geology of the Bahamas and on coral formations generally: Geol. Soc. London Quart. Jour., v. 9, p. 200-215.

Newell, N. D., 1955, Bahamian platforms, *in* Poldevaart, A., ed., The crust of the earth: Geol. Soc. America Spec. Paper 62, p. 303-315.

Oglesby, W. R., 1965, Folio of South Florida basin: a preliminary study: Florida Geol. Survey, map ser. 19, 3 p. (10 maps and cross sections).

Paine, W. R., and Meyerhoff, A. A., 1970, Gulf of Mexico basin: interactions among tectonics, sedimentation, and hydrocarbon accumulation: Trans. Gulf Coast Assoc. Geol. Soc., v. 20, p. 5-44.

Palmer, R. H., 1945, Outline of the geology of Cuba: Jour. Geology, v. 53, no. 1, p. 1-34.

Pardo, G., 1966, Stratigraphy and structure of central Cuba (abstr.): New Orleans Geol. Soc. Log, v. 6, no. 12, p. 1,3.

Parker, T. J., and McDowell, A. N., 1955, Model studies of salt-dome tectonics: Am. Assoc. Petr. Geol. Bull., v. 39, no. 12, p. 2384-2470.

Powell, L. C., and Culligan, L. B., 1955, Developments in southeastern states in 1954: Am. Assoc. Petr. Geol. Bull., v. 39, no. 6, p. 1004-1014.

Powers, R. W., 1962, Arabian Upper Jurassic carbonate reservoir rocks: Am. Assoc. Petr. Geol. Mem. 1, p. 122-192.

Pratt, R. M., and Heezen, B. C., 1964, Topography of the Blake Plateau: Deep-Sea Res., v. 11, p. 721-728.

Pressler, E. D., 1947, Geology and occurrence of oil in Florida: Am. Assoc. Petr. Geol. Bull., v. 31, no. 10, p. 1851-1862.

Pyle, T. E., Antoine, J. W., Fahlquist, D. A., and Bryant, W. R., 1969, Magnetic anomalies in straits of Florida: Am. Assoc. Petr. Geol. Bull., v. 53, no. 12, p. 2502-2505.

______, Meyerhoff, A. A., Fahlquist, D. A., Antoine, J. W., McCrevey, J. A., and Jones, P. C., 1973, Metamorphic rocks from northwestern Caribbean Sea: Earth and Planetary Sci. Letters, v. 18, p. 339-344.

Rainwater, E. H., 1971, Possible future petroleum potential of peninsular Florida and adjacent continental shelves, *in* Cram, I. H., Sr., ed., Future petroleum provinces of the United States—their geology and potential, v. 2: Am. Assoc. Petr. Geol. Mem. 15, p. 1311-1341.

Reyre, D., 1966, Particularités géologiques des bassins cotiers de l'ouest Africain, *in* Reyre, D., ed., Sedimentary basins of the African coasts, pt. 1, Atlantic coast: Paris, Assoc. Serv. Geol. Africains, p. 253-301.

Rigassi, D., 1961, Quelques vues nouvelles sur la géologie cubaine: Chron. Mines et Recherche Minière, 29 ann., no. 302, p. 3-7.

———, 1963, Sur la géologie de la Sierra de los Organos, Cuba: Arch. Sci., Soc. Phys. et d'Histo. Nat. Genève, t. 16, fasc. 2, p. 339-350.

Schuchert, C., 1935, Historical geology of the Antillean-Caribbean region or the lands bordering the Gulf of Mexico and the Caribbean Sea: New York, Wiley, 811 p.

Scott, K. R., Hayes, W. E., and Fietz, R. P., 1961, Geology of the Eagle Mills Formation: Trans. Gulf Coast Assoc. Geol. Soc., v. 11, p. 1-14.

Seiglie, G. A., 1961, Contribución al estudio de las microfacies de Pinar del Río: La Habana, Soc. Cubana Ing. Rev., v. 61, nos. 3-4, p. 87-109.

Sheridan, R. E., 1969, Subsidence of continental margins: Tectonophysics, v. 7, no. 3, p. 219-229.

———, 1971, Geotectonic evolution and subsidence of Bahama platform: Geol. Soc. America Bull., v. 82, no. 3, p. 807-809.

———, Drake, C. L., Nafe, J. E., and Hennion, J., 1966, Seismic-refraction study of continental margin east of Florida: Am. Assoc. Petr. Geol. Bull., v. 50, no. 9, p. 1972-1991.

———, Smith, J. D., and Gardner, J., 1969, Rock dredges from Blake Escarpment near Great Abaco Canyon: Am. Assoc. Petr. Geol. Bull., v. 53, no. 12, p. 2551-2558.

Shreveport Geological Society, 1968, Stratigraphy and selected gas-field studies of north Louisiana: Am. Assoc. Petr. Geol. Mem. 9, v. 1, p. 1009-1175.

Soloviev, O. N., Skidan, S. A., Skidan, I. K., Pankratov, A. P., and Judoley, C. M., 1964a, Comentarios sobre el mapa gravimétrico de la Isla de Cuba: La Habana, Min. Ind. Rev. Tecnol. v. 2, no. 2, p. 8-19.

———, Skidan, S. A., Pankratov, A. P., and Skidan, I. K., 1964b, Comentarios sobre el mapa magnetométrico de Cuba: La Habana, Min. Ind. Rev. Tecnol., v. 2, no. 4, p. 5-23.

Spencer, M., 1967, Bahamas deep test: Am. Assoc. Petr. Geol. Bull., v. 51, no. 2, p. 263-268.

Stehman, C. F., 1970, Eocene deep-water sediment from the Northeast Providence Channel, Bahamas: Maritime Sediments, v. 6, no. 2, p. 65-67.

Suess, E., 1888, Das Antlitz der Erde, Zweiter Bank: Vienna, F. Tempsky, 704 p.

———, 1908, Das Antlitz der Erde, Erster Band: Bienna, F. Tempsky, 779 p.

Talwani, M., 1960, Gravity anomalies and their interpretation [Ph.D. thesis]: Columbia Univ., New York, 89 p.

———, Worzel, J. L, and Ewing, W. M., 1960, Gravity anomalies and structure of the Bahamas: Trans. 2d Caribbean Geol. Conf., Mayaguez, P.R., p. 156-161.

Teichert, C., 1974, Marine sedimentary environments and their faunas in the Gondwana area: Am. Assoc. Petr. Geol. Mem. 21, in press.

Tijomirov, I. N., 1967, Formaciones magmáticas de Cuba y algunas particularidades de su metalogenia: La Habana, Min. Ind. Rev. Tecnol., v. 5, no. 4, p. 13-22.

Uchupi, E., Milliman, J. D., Luyendyk, B. P., Bowin, C. O., and Emery, K. O., 1971, Structure and origin of southeastern Bahamas: Am. Assoc. Petr. Geol. Bull., v. 55, no. 5, p. 687-704.

Viniegra, F., 1971, Age and evolution of salt basins of southeastern Mexico: Am. Assoc. Petr. Geol. Bull., v. 55, no. 3, p. 478-494.

———, and Castillo-Tejero, C., 1970, Golden Lane fields, Veracruz, Mexico: Am. Assoc. Petr. Geol. Mem. 14, p. 309-325.

Wassall, H., 1956, The relationship of oil and serpentine in Cuba: 20th Internatl. Geol. Congr., Mexico, sect. 3, p. 67-77.

Weaver, C. E., and Stevenson, R. G., Jr., 1971, Clay minerals in the Cretaceous rocks of Florida: Geol. Soc. America Bull., v. 82, no. 12, p. 3457-3460.

Welch, S. W., 1969, Developments in southeastern states in 1968: Am. Assoc. Petr. Geol. Bull., v. 53, no. 6, p. 1280-1284.

———, and Browning, W. F., 1971, Developments in southeastern states in 1970: Am. Assoc. Petr. Geol. Bull., v. 55, no. 7, p. 1074-1079.

———, and Browning, W. F., 1973, Developments in southeastern states in 1972: Am. Assoc. Petr. Geol. Bull., v. 57, no. 8, p. 1542-1547.

Wilson, J. T., 1966, Did the Atlantic close and then reopen?: Nature, v. 211, no. 5050, p. 676-681.

Winston, G. O., 1971a, The Dollar Bay Formation of Lower Cretaceous (Fredericksburg) age in south Florida: Florida Bur. Geology, Spec. Publ., 15, 99 p.

———, 1971b, Regional structure, stratigraphy, and oil possibilities of the south Florida basin: Trans. Gulf Coast Assoc. Geol. Soc., v. 21, p. 15-29.

Woodring, W. P., 1928, Tectonic features of the Caribbean region: Proc. 3d Pan-Pacific Sci. Congr., Tokyo, v. 1, p. 401-431.

Worzel, J. L., 1965, Deep structure of coastal margins and mid-oceanic ridges, *in* Whittard, W. F., and Bradshaw, R., eds., Submarine geology and geophysics—Proc. 17th Colston Soc. Symp., Bristol, Eng.: London, Butterworths, p. 335-359.

———, and Shurbet, G. L., 1955, Gravity interpretations from standard oceanic and continental crustal sections: Geol. Soc. America Spec. Paper 62, p. 87-100.

Geology of the Brazilian Continental Margin

C. W. M. Campos, F. C. Ponte, and K. Miura

INTRODUCTION

The Brazilian continental margin has been divided into three distinct physiographic provinces: the Amazon-Maranhão continental shelf, including the Amazon Cone; the north-northeastern coast continental shelf, covering the Barreirinhas, Ceará, Potiguar, Recife-João Pessoa, Sergipe-Alagoas, South Bahia, and Jequitinhonha basins; east-southeastern coast continental shelf, in which lie the coastal and offshore basins of Cumuruxatiba, Espírito Santo, Campos, Santos, and Pelotas.

The Brazilian stratigraphic section has eight major units: the Precambrian Brazilian Shield, composed of folded metamorphic rocks and igneous intrusives; and early Precambrian to Cambro-Ordovician metasedimentary platform cover, and six lithostratigraphic sequences: two Paleozoic, two Mesozoic, and two Cenozoic in age, bounded by interregional unconformities.

After a period of tectonic stability, prevailing since early Paleozoic, the general tectonic configuration of the Brazilian continental margin was established in the Early Cretaceous, by the "Wealdian Reactivation," a severe taphrogenic tectonism accompanied by intense basaltic magmatism, 110-140 my old. Another peak of igneous activity, associated with the tectonic evolution of the continental margin, occurred in Late Cretaceous to Early Tertiary, 50-80 my ago.

Precambrian basement lineations imposed strong tectonic control in the structural pattern of gravity-faulted step blocks, horsts, and grabens which compose the structural framework of the entire Brazilian continental margin. The pre-Campanian Cretaceous reverse faults and compressional folds, in the Equatorial continental margin, are the only known anomalous features superimposed to this taphrogenic structural style.

The main stages of tectonic evolution of the Brazilian coastal and offshore basins are well documented in the stratigraphic column, basically composed of three intervals: a *lower continental interval*, correspondent to the Neocomian-Barremian continental rift stage; an *intermediate evaporitic interval*, resulting from restricted environments prevailing during the Aptian intracontinental gulf stage; and an *upper marine interval*, correspondent to the coastal basin stage. The evaporitic interval is absent in the basin of the equatorial coast, as well as in the Pelotas basin, south of the Rio Grande Rise.

This paper deals with the general geology of the Brazilian continental margin. PETROBRÁS exploration activities in the search for petroleum has provided most of the data that led to the present geologic synthesis.

Briefly, the Brazilian continental margin comprises the Equatorial continental margin, consisting of the Amazon Cone and Barreirinhas, Ceará, and Potiguar basins, and the Eastern continental margin, consisting of the Recife-João Pessoa, Sergipe-Alagoas, South Bahia, Camamu, Almade, Jequitinhonha, Cumuruxatiba, Espírito Santo, Campos, Santos, and Pelotas basins.

The systematic study of the continental shelf started in 1957, with seismic-reflection reconnaissance work and with a gravimetric survey in 1963. The second phase of exploration of the continental shelf started in 1968 with an air-magnetometric survey covering almost all the offshore basins, and detailed seismic surveys. The first offshore well was drilled in 1968 (Campos, 1970).

Asmus and Ponte (in press) made a general synthesis of the geology of the Brazilian continental margin. Asmus and Porto (1972) developed the classification of the Brazilian sedimentary basins and its evolution, following the plate tectonics concepts. Ponte et al. (1971) studied the paleogeology and the evolution of this continental margin.

From the regional point of view, in the equatorial continental margin, Rezende and Ferradaes (1971) studied the development of the Amazon Cone, and Miura and Barbosa (1972) studied the geology of Maranhão to Rio Grande do Norte offshore area. In the eastern continental margin, Ojeda and Bisol (1971) studied the regional extension of the offshore Sergipe-Alagoas Basin; Ferradaes and Souze (1972) studied the continental shelf of the Camamu and Almada basins; Gomes et al. (1973) studied the Cumuruxatiba Basin; Asmus et al. (1971) and Bacoccoli and Morales (1973) developed the study of the geology of Espírito Santo Basin; Asmus (1969) and Bacoccoli and Saito (1973) studied the Campos Basin; Miranda (1970) and Ojeda (1973) studied the geology of offshore Santos and Pelotas basins; and Baccar (1970) analyzed the development of the São Paulo Plateau.

PHYSIOGRAPHY

The Brazilian continental margin, with an extension of about 6,900 km, reaches from Cape Orange in the north to Chui Creek in the south (Fig. 1). It represents a total area of 2,430,000 km^2, 710,000 km^2 of continental shelf and 1,720,000 km^2

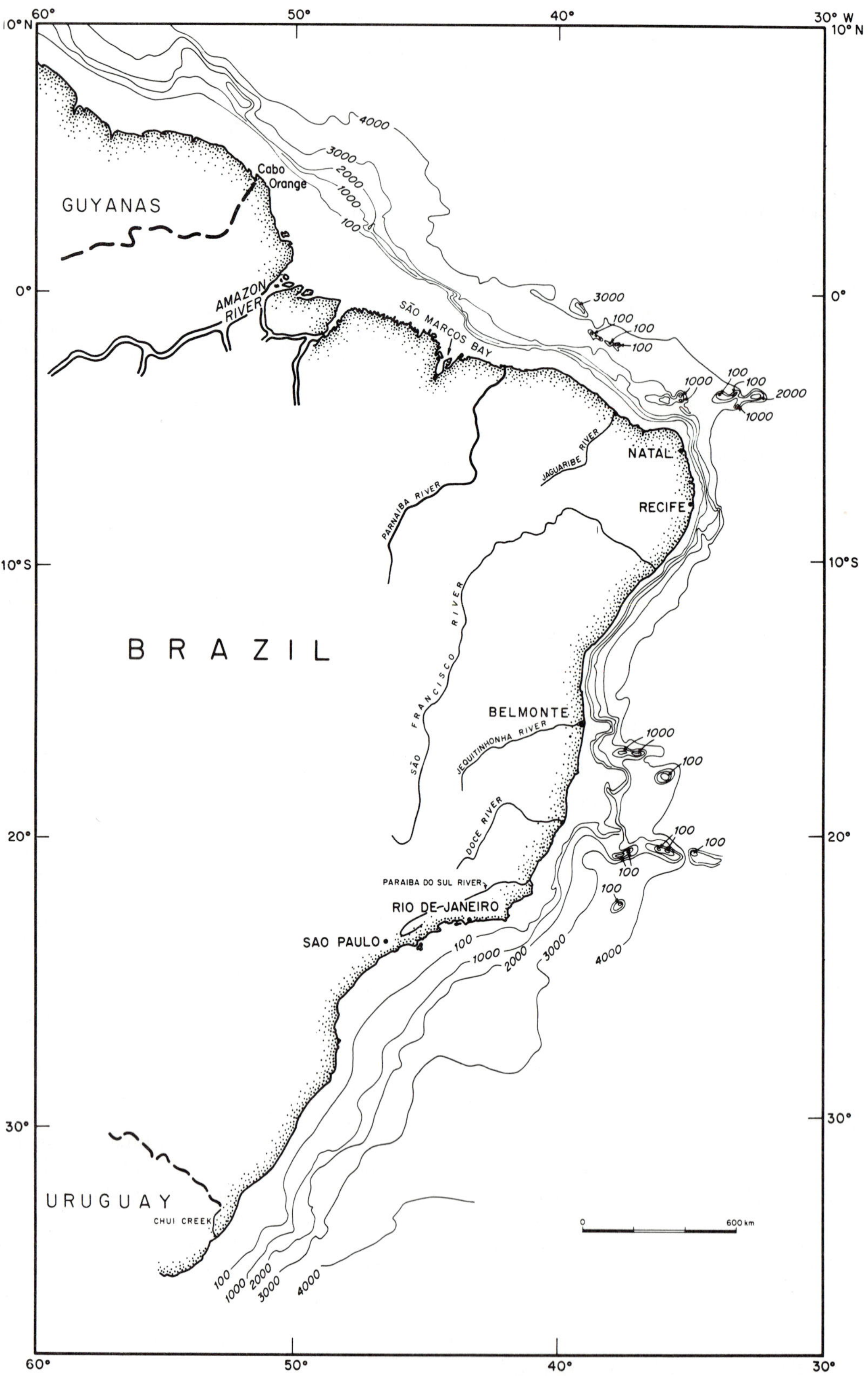

Fig. 1. Bathymetric map, with contour intervals in meters.

of continental slope and rise (Zembruscki et al., 1972).

Barreto and Milliman (1969) subdivided the Brazilian continental margin into three distinct provinces:

1. The Amazon-Maranhão continental shelf, which extends 820 km from Cape Orange to S. Marcos Bay, and is 80-320 km wide. The shelf break depth is about 80 m. The topographic surface is smooth and gentle, forming broad terraces. At the mouth of the Amazon River the slope is less than 1°.

2. The north-northeastern coast continental shelf, a narrow shelf 980 km long, from S. Marcos Bay to Natal, and an irregularly shaped shelf from Natal to Belmonte, 1,100 km in length. The shelf width varies from 11 to 100 km and the shelf break depth from 55 to 100 m. The topographic surface presents small magnitude features made up of sandstone reefs (Mabesoone, 1964), calcareous algae, and other biogenic accumulations. The Ceará and Rio Grande do Norte continental slope has prominent morphological features due to volcanic activity along the Fernando de Noronha ridge, a probable extension of the Chain fracture zone. On the continental slope eastward from Recife is the Pernambuco Plateau (Fig. 2).

3. The east-southeastern coast continental shelf reaches over 2,000 km, from Belmonte to Chui Creek, the width varying from 40 to 240 km. It can be divided into three subprovinces:

a. The Royal Charlotte Bank, a shallow platform of 7,200 km^2, 90-105 km wide, with an irregular surface and shelf slope averaging 0° 22'.

b. The Espirito Santo Shelf, 54,600 km^2, 240 km wide with an irregular outer shelf forming the Caravelas and Vitoria submarine salients. The shelf slope averages 0°08'. Small seamounts in the inner shelf originated from basic Cretaceous and Tertiary volcanism. The Vitoria, Hotspur, and Jaseur banks form an east-west alignment from the continental slope to the oceanic basins.

c. Rio de Janeiro-Chui Creek Shelf, having a width of 50-205 km and a shelf break depth generally from 55 to 100 m, but up to 165 m in the southern part. The topographic surface is irregular in the northern part, becoming smooth southward. The shelf slope varies from 0°27' to 0°59', and the continental slope is generally smooth and discrete. Offshore from Sao Paulo is the Sao Paulo Plateau, with an area of 200,000 km^2 (Baccar, 1970).

Sediment Distribution in the Continental Shelf

From available data, Bareto and Milliman (1969) constructed a general map of recent sediment distribution, grouping it into three major types (Fig. 3):

1.a. Terrigenous mud of silt-clay size, predominating in the nearshore environments, in the inner shelf area in front of low coastal plains, and in estuarine zones and in the inner shelf area sheltered from marine and tidal currents.

2.b. Carbonate-rich sediments, predominating where the drainage system presents insufficient energy to transport clastic sediments and where there are abundant organisms.

3.c. Terrigenous sand and gravel, predominating where there is more river density, the coarser sediments being related to more rapid deposition.

The Brazilian continental margin presents seven well-characterized Holocene deltas. They are marine deltas, the Amazon River delta being dominated by tides, and the Parnaiba River, the Jaguaribe River, the S. Francisco River, the Jequitinhonha River, Doce River, and the Paraiba do Sul River deltas being dominated by waves (Bacoccoli, 1971).

STRATIGRAPHY

The Brazilian stratigraphic section may be divided into eight major units: the Brazilian shield, the metasedimentary platform cover, and six stratigraphic sequences. These sequences, defined by Ghignone and Northfleet (1971; 1972), according to Sloss's concept (1963), are "stratigraphic units of higher rank than group, megagroup or supergroup, traceable over major areas of a continent, and bounded by unconformities of interregional scope." The two lower sequences, Paleozoic in age, are better developed on intracratonic basins. They are formed by blanket-like strata that exhibit extensive geographical distribution and great lateral lithologic continuity. The younger sequences, two Mesozoic and two Cenozoic, are the most important in the coastal basins. Their origin is considered to be closely related to the tectonic evolution of the South Atlantic continental margin. In these sequences the bounding unconformities, although very distinct onshore, are in some cases obscure or even absent basinward in offshore areas. The brief description of each of these eight units is complemented by Table 1, which indicates the formations and groups occurring in the main Brazilian basins.

The Brazilian Shield is composed of folded metamorphic Precambrian rocks and igneous intrusives, which cover most of the Brazilian territory. Almeida (1971), based on recent information on geotectonics, stratigraphy, and radiometric dating, proposed the following geochronological division of the Precambrian of South America:

Lower Precambrian: more than 2,600 million years

Middle Precambrian: 1,800-2,600 million years

Upper Precambrian: 570-1,800 million years

The upper Precambrian is subdivided in three geotectonic cycles: Espinhaco, Uruacuano, and Brazilian, from oldest to youngest. This last cycle extended beyond the limit of Precambrian—to the Cambro-Ordovician periods.

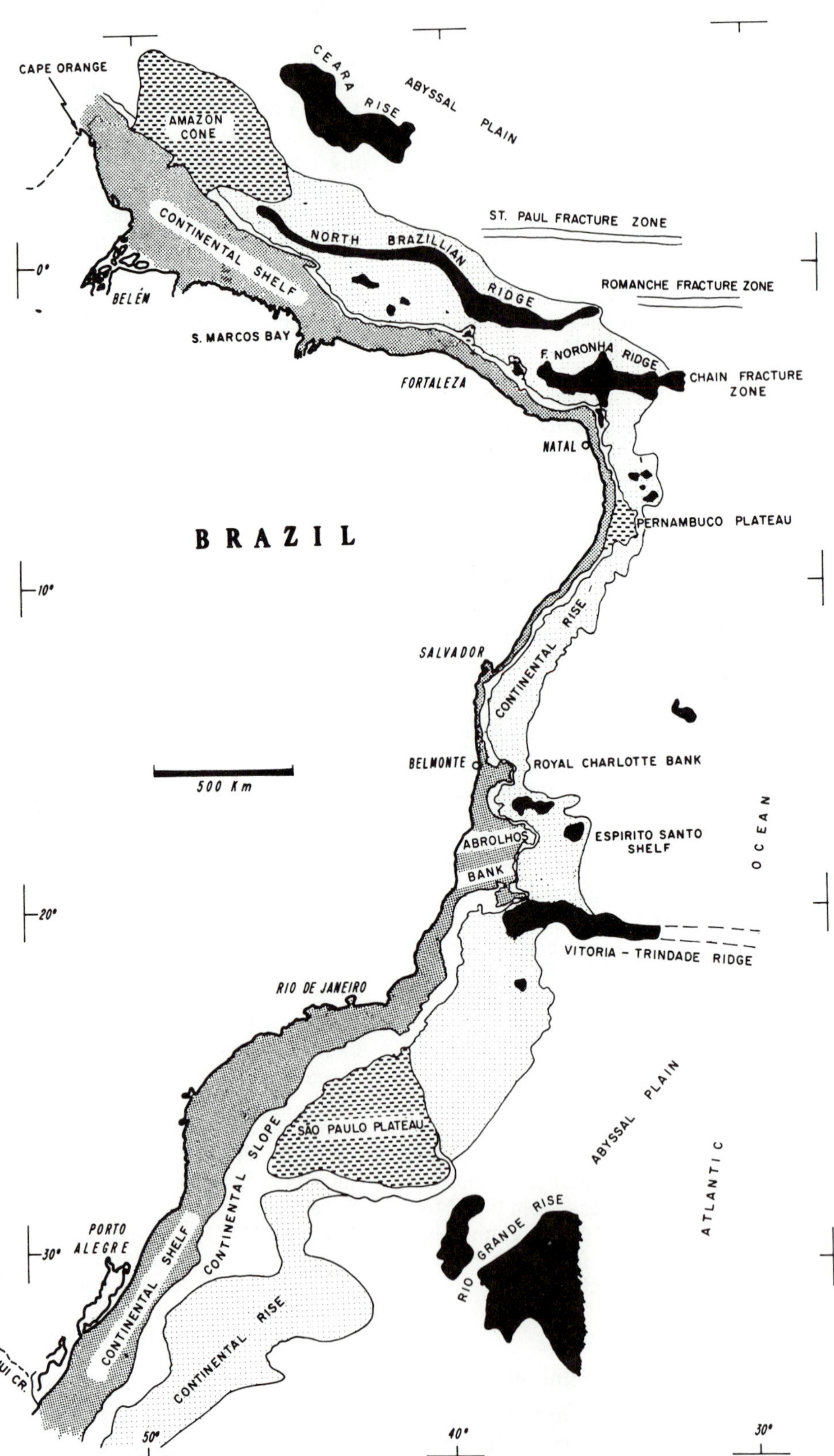

Fig. 2. Physiographic map showing the main provinces and prominent features (after Asmus, 1973).

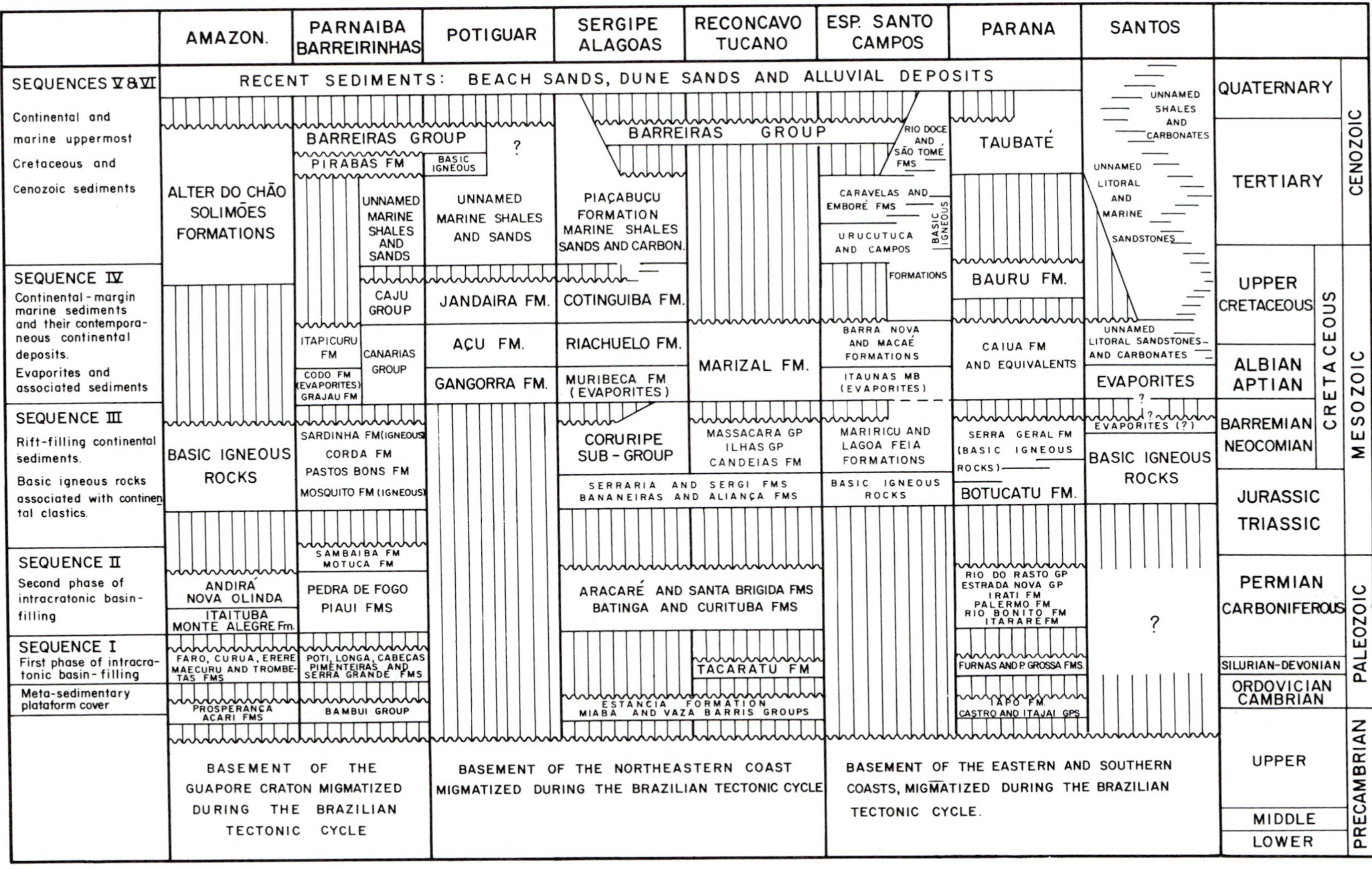

	AMAZON.	PARNAIBA BARREIRINHAS	POTIGUAR	SERGIPE ALAGOAS	RECONCAVO TUCANO	ESP. SANTO CAMPOS	PARANA	SANTOS	
SEQUENCES V & VI Continental and marine uppermost Cretaceous and Cenozoic sediments	RECENT SEDIMENTS: BEACH SANDS, DUNE SANDS AND ALLUVIAL DEPOSITS							UNNAMED SHALES AND CARBONATES	QUATERNARY (CENOZOIC)
	ALTER DO CHÃO SOLIMÕES FORMATIONS	BARREIRAS GROUP PIRABAS FM UNNAMED MARINE SHALES AND SANDS	BARREIRAS GROUP BASIC IGNEOUS ? UNNAMED MARINE SHALES AND SANDS	BARREIRAS GROUP PIAÇABUÇU FORMATION MARINE SHALES SANDS AND CARBON.	BARREIRAS GROUP	RIO DOCE AND SÃO TOMÉ FMS CARAVELAS AND EMBORÉ FMS URUCUTUCA AND CAMPOS FORMATIONS BASIC IGNEOUS	TAUBATÉ	UNNAMED LITORAL AND MARINE SANDSTONES	TERTIARY (CENOZOIC)
SEQUENCE IV Continental-margin marine sediments and their contemporaneous continental deposits. Evaporites and associated sediments		CAJU GROUP ITAPICURU FM CANARIAS GROUP CODO FM (EVAPORITES) GRAJAU FM	JANDAIRA FM. AÇU FM. GANGORRA FM.	COTINGUIBA FM. RIACHUELO FM. MURIBECA FM (EVAPORITES)	MARIZAL FM.	BARRA NOVA AND MACAÉ FORMATIONS ITAUNAS MB (EVAPORITES)	BAURU FM. CAIUA FM AND EQUIVALENTS	UNNAMED LITORAL SANDSTONES AND CARBONATES EVAPORITES	UPPER CRETACEOUS ALBIAN APTIAN (CRETACEOUS, MESOZOIC)
SEQUENCE III Rift-filling continental sediments. Basic igneous rocks associated with continental clastics.	BASIC IGNEOUS ROCKS	SARDINHA FM (IGNEOUS) CORDA FM PASTOS BONS FM MOSQUITO FM (IGNEOUS)		CORURIPE SUB-GROUP SERRARIA AND SERGI FMS BANANEIRAS AND ALIANÇA FMS	MASSACARA GP ILHAS GP CANDEIAS FM	MARIRICU AND LAGOA FEIA FORMATIONS BASIC IGNEOUS ROCKS	SERRA GERAL FM (BASIC IGNEOUS ROCKS) BOTUCATU FM.	? EVAPORITES (?) BASIC IGNEOUS ROCKS	BARREMIAN NEOCOMIAN (CRETACEOUS) JURASSIC TRIASSIC (MESOZOIC)
SEQUENCE II Second phase of intracratonic basin-filling	ANDIRÁ NOVA OLINDA ITAITUBA MONTE ALEGRE Fm.	SAMBAIBA FM MOTUCA FM PEDRA DE FOGO PIAUI FMS		ARACARÉ AND SANTA BRIGIDA FMS BATINGA AND CURITUBA FMS			RIO DO RASTO GP ESTRADA NOVA GP IRATI FM PALERMO FM RIO BONITO FM ITARARÉ FM	?	PERMIAN CARBONIFEROUS (PALEOZOIC)
SEQUENCE I First phase of intracratonic basin-filling	FARO, CURUA, ERERE MAECURU AND TROMBETAS FMS	POTI, LONGA, CABEÇAS PIMENTEIRAS AND SERRA GRANDE FMS			TACARATU FM		FURNAS AND P. GROSSA FMS		SILURIAN-DEVONIAN (PALEOZOIC)
Meta-sedimentary plataform cover	PROSPERANÇA ACARI FMS	BAMBUI GROUP		ESTANCIA FORMATION MIABA AND VAZA BARRIS GROUPS			IAPÓ FM CASTRO AND ITAJAI GPS		ORDOVICIAN CAMBRIAN (PALEOZOIC)
	BASEMENT OF THE GUAPORE CRATON MIGMATIZED DURING THE BRAZILIAN TECTONIC CYCLE		BASEMENT OF THE NORTHEASTERN COAST MIGMATIZED DURING THE BRAZILIAN TECTONIC CYCLE			BASEMENT OF THE EASTERN AND SOUTHERN COASTS, MIGMATIZED DURING THE BRAZILIAN TECTONIC CYCLE.			UPPER MIDDLE LOWER (PRECAMBRIAN)

Table 1. Stratigraphic correlation chart of the main Brazilian sedimentary basins.

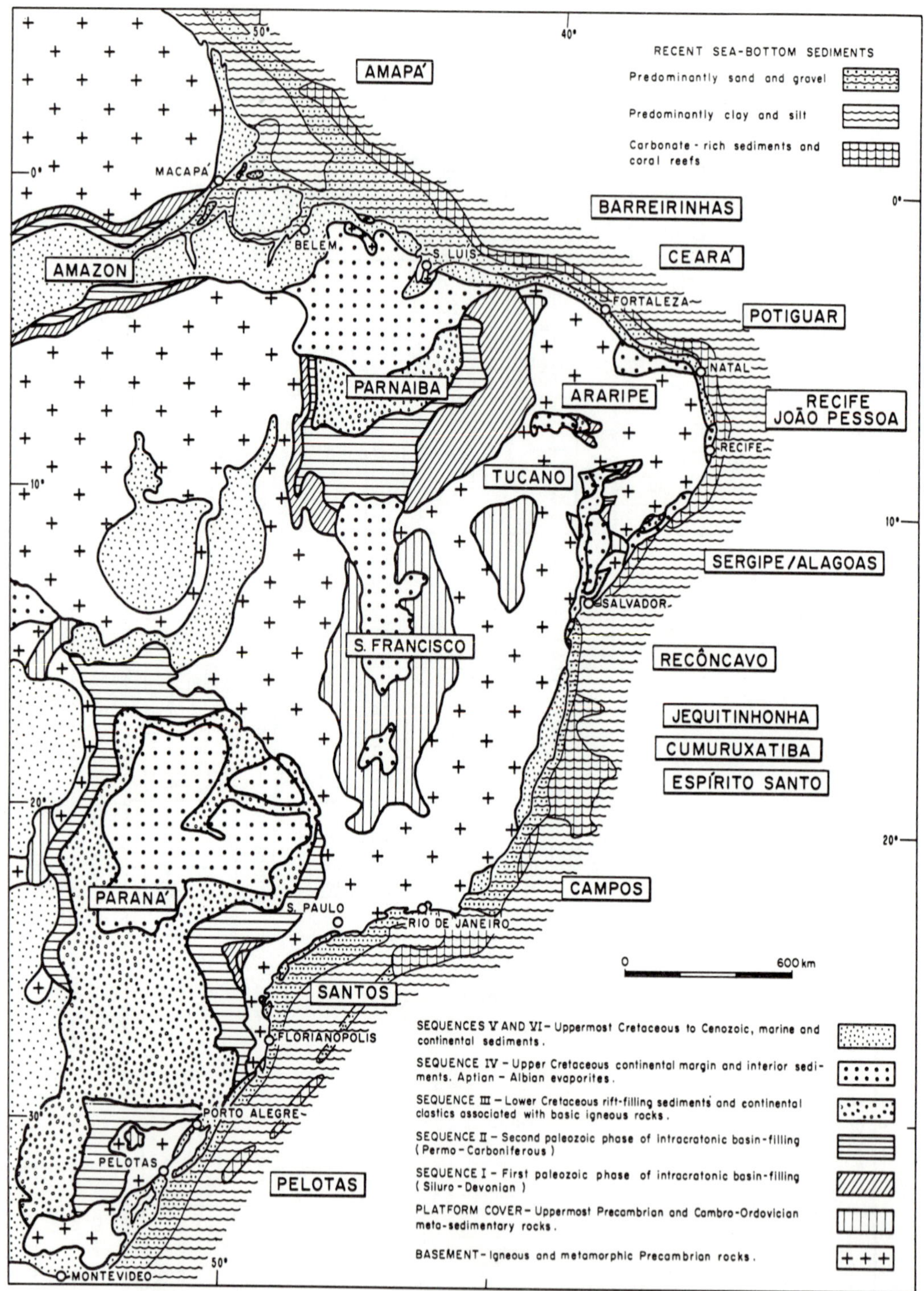

Fig. 3. General outline of the Brazilian geology, simplified from Ghignone and Northfleet (1972), with indications of the coastal basins, and the recent sea-bottom sediment distribution (after Barreto & Milliman, 1969).

Under the designation Metasedimentary Platform Cover are included two related stratigraphic units: the early Precambrian to Late Cambrian metasedimentary rocks (meta-arkoses, slates, and meta-limestones), partly deformed during the last episodes of the Brazilian Orogeny, and the Cambro-Ordovician (?) molassic complexes, associated in minor extent with acid igneous rocks, such a rhyolite and andesite. The limit between the Metasedimentary Platform Cover and the Brazilian Shield is an interregional angular unconformity older than 650 my found at the base of the Bambui Group and correlatable units. For practical reasons Ghignone and Northfleet (1972) in their geologic map of Brazil took this unconformity as the mapable limit between the Precambrian and Paleozoic, in spite of its being older than the limit internationally accepted. The upper part of this unit is considered to be the sedimentary record of the transitional stage betwen the Brazilian orogeny and the following stage of cratonic stability.

Sequence I ranges in age from Silurian to Devonian. The strata are not disturbed by the structural deformations that affected the older rocks underneath. Thick blankets of coarse sandstones and basal conglomerates deposited over the Ordovician (?) erosional surface form an interregional angular unconformity. The rocks of sequence I, deposited under stable cratonic conditions, represent the first stage of sedimentation, filling the Amazon depression and the large Paleozoic sineclises of Paraná and Parnaiba depressed on ancient Brazilian foldbelts. Outcrops of basal conglomerates of sequence I, found on the borders of the Tucano and Araripe basins, are considered as remnants of these earlier continuous Paleozoic basins.

Sequence II includes upper Carboniferous to Upper Permian or Permo-Triassic detritic and immature sediments. In the Parana basin, the basal unconformity of sequence II is on the glacially eroded surface of sequence I. In the Parnaiba basin this unconformity, although existent, is less conspicuous. The strata of sequence II were deposited in the slowly subsiding intracratonic basins of Paraná, Parnaiba, and Amazon. The superposition of sandstones, evaporites, carbonates, and black shales in this sequence reflects the alternation of climatic conditions, from arid to humid and arid again, prevailing during its sedimentation.

The strata of sequence III, uppermost Jurassic to lowermost Cretaceous (Neocomian to Barremian) in age, are bounded at the base and at the top by two major interregional unconformities. The lower erosional surface is specially evident in northeastern Brazil, where a long period of uplift promoted the deep erosion of the entire sedimentary cover, exposing the Precambrian basement over large areas. In this region sequence III marks the beginning of sedimentation into a depression elongated in the north-south direction, oriented by the tectonic lineations of the Precambrian basement. Continuous blankets of red beds (shales and sandstones) and coarse sandstones with striking lateral lithologic continuity were deposited from South Bahia to Araripe, including the areas presently occupied by the Almada, Recôncavo, Tucano, Sergipe-Alagoas, and Araripe basins. The upper part of sequence III in these basins is composed by typical rift-filling lacustrine and fluvial sediments: thick wedges of conglomerates exhibiting rapid lateral facies change to dirty sandstones and shales.

In the Paraná, Parnaiba, and Amazon basins, sequence III consists of continental clastics, commonly eolian sandstones, associated with thick igneous layers of basaltic extrusives and diabase dikes and sills, 110-148 my old. Basaltic rocks, about the same age, are also found in the Santos and Espírito Santo basins, associated with syntectonic sediments. This widespread volcanism is contemporaneous with the Wealdian Reactivation (Almeida, 1967), the taphrogenic tectonism that affected the entire Brazilian continental margin. The volcanism, however, did not affect the northeastern basins: Reconcavo, Tucano, Sergipe-Alagoas, Recife-João Pessoa, and Potiguar.

Sequence IV, Aptian to Santonian in age, is composed of continental and marine sediments. In the interior basins flat-top "tabuleiros" or "chapadas" are formed by horizontal blankets of coarse continental sandstone with great geographical extent locally underlain by thin-bedded limestones and evaporites (gypsum and anhydrite). In the coastal basins the pericratonic sediments of sequence IV are the first record of marine sedimentation in the present Atlantic continental margin of Brazil. From Santos to Sergipe-Alagoas, the lower part of this sequence bears a thick section of evaporites, including massive salt layers, commonly underlain by basal conglomerates and sandstones of transitional environment. Upward the evaporites grade to littoral sandstones and marine shales and limestones. In some places, basinward, the pre-Aptian unconformity seems to be absent. In Alagoas, for example, the Neocomian-Barremian lacustrine shales grade upward to euxinic shales of brackish-water origin, interbedded with salt rock, and then to a typical marine section. Aptian evaporites are not known in the Equatorial coastal basins, in spite of being present in the intracratonic Parnaiba basin. They also seem to be absent in the southernmost Pelota basin.

Sequences V and VI are especially well developed along the coastal margins and offshore basins. They include all the Uppermost Cretaceous and Cenozoic sediments in Brazil, composed of continental coarse clastics and marine shales associated with sandstone and limestones. The lower boundary of sequence V is the pre-Campanian unconformity, a remarkable feature on most of the Brazilian coastal basins, which becomes obscure or even absent farther offshore. Onshore the continental Barreiras Group and equivalent units lie on a pre-Miocene interregional unconformity, and are unconformably overlain by Quaternary sand beaches, dunes, and alluvial sediments. Offshore, the marine sedimentation is mostly continuous in the whole Cenozoic. Basic igneous rocks occur associated with sediments of sequence V in the Espirito Santo and Potiguar basins.

STRUCTURAL CONFIGURATION

The general tectonic configuration of the Brazilian continental margin was established in the Early Cretaceous (Neocomian-Barremian), during the Wealdian Reactivation (Almeida, 1967), an intense taphrogenic tectonism which affected the present coastal regions after a long period of tectonic stability that prevailed since the Early Paleozoic.

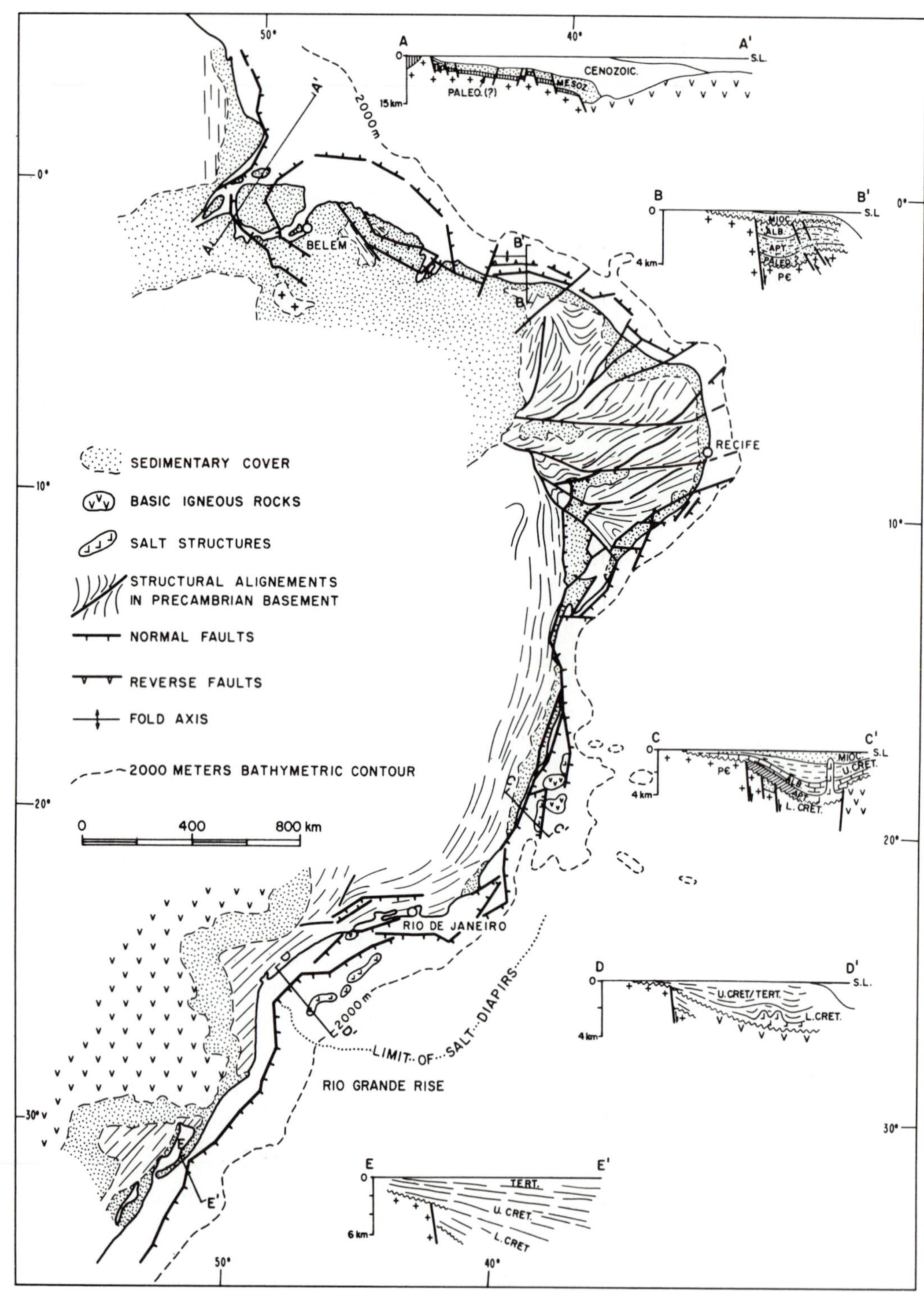

Fig. 4. Structural framework of the Brazilian Shield and continental shelf. Cross-sections show the general profiles of some typical coastal basins.

Precambrian lines of crustal weakness, reactivated by tensional forces developed a gravity-faulted structural pattern building the floor of rifts and coastal basins depressed in the basement. The strong structural control of Precambrian lineation, reflected in the basin floors, is also evident in the present configuration of the eastern shoreline, from the southern boundary up to Alagoas, which is essentially parallel to the basement alignments. In contrast, the northeastern and equatorial coastlines cut those alignments transversally. Despite this fact, the equatorial coastal basins exhibit in their structural framework strong basement lineation control (Fig. 4).

The Paleozoic strata preserved on the floor of many of the Brazilian coastal basins (Barreirinhas,

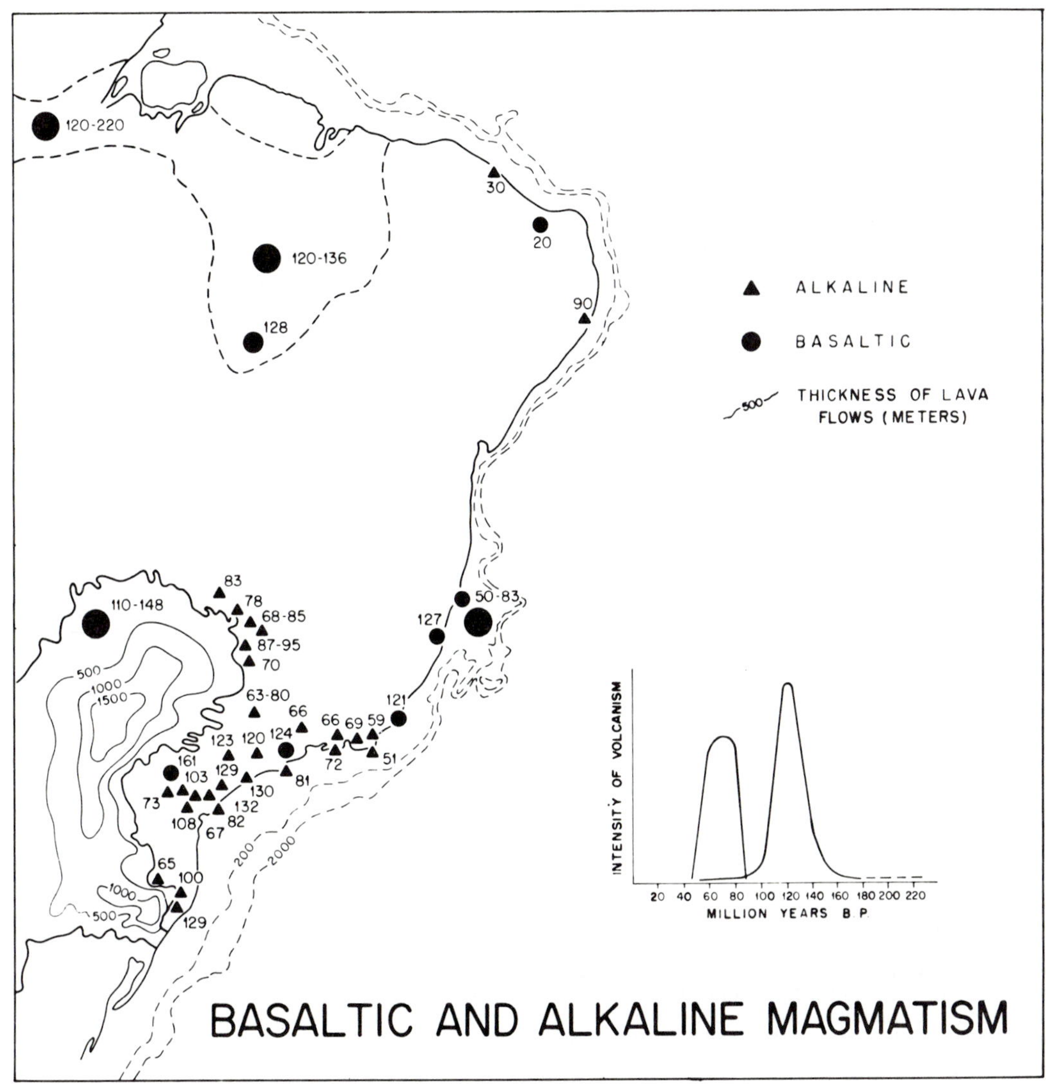

Fig. 5. Geographic and age distribution of basaltic and alkaline magmatism in Brazil. Two prominent peaks of volcanic activity are observed: Early Cretaceous (110-140 my) and Late Cretaceous to Early Tertiary (50-80 my) (after Asmus, 1973).

Sergipe-Alagoas, Recôncavo-Tucano) are relicts of older cratonic or epicontinental basins, partly destroyed by the Wealdian Reactivation. The main stages of the tectonic evolution of these taphrogenic basins are well documented in the basin-filling sediments, which compose three major intervals: a *lower continental interval*, corresponding to the Neocomian-Barremian continental rift stage; a *middle evaporitic interval*, resulting from arid climate and restricted environments prevailing in the Aptian intracontinental-gulf stage and interior lakes; and an *upper marine interval*, corresponding to the coastal basin stage. The middle evaporitic interval is absent in the equatorial coastal basins, indicating a somewhat different history of structural evolution. In the Reconcavo and Tucano, which remain as rift basins, only the lower interval and the lower-most part of the middle interval are present.

Many of the fault systems that developed step blocks and a horst-and-graben pattern during the Early Cretaceous taphrogeny were later rejuvenated by the progressive subsidence and seaward tilting of the continental margin. Presently the main fault systems may have a vertical displacement of 3-5 km.

Widespread magmatism accompanied the tectonic evolution of the Brazilian continental margin. Radiometric age determinations indicate two main peaks of igneous activity: an Early Cretaceous, 110-140 my old, and a Late Cretaceous to Early Tertiary, 50-80 my old (Fig. 5).

The Early Cretaceous magmatism seems to have been synchronous with the Wealdian Reactivation. It is more important in the intracratonic Amazon, Parnaiba, and Paraná basins. Basaltic lava flows in Paraná basin reach thicknesses greater than 1,500 m, attesting to the tremendous intensity of volcanism in this region. The formation of the Rio Grande Rise is supposed to be associated with this volcanism. Great thickness of Early Cretaceous

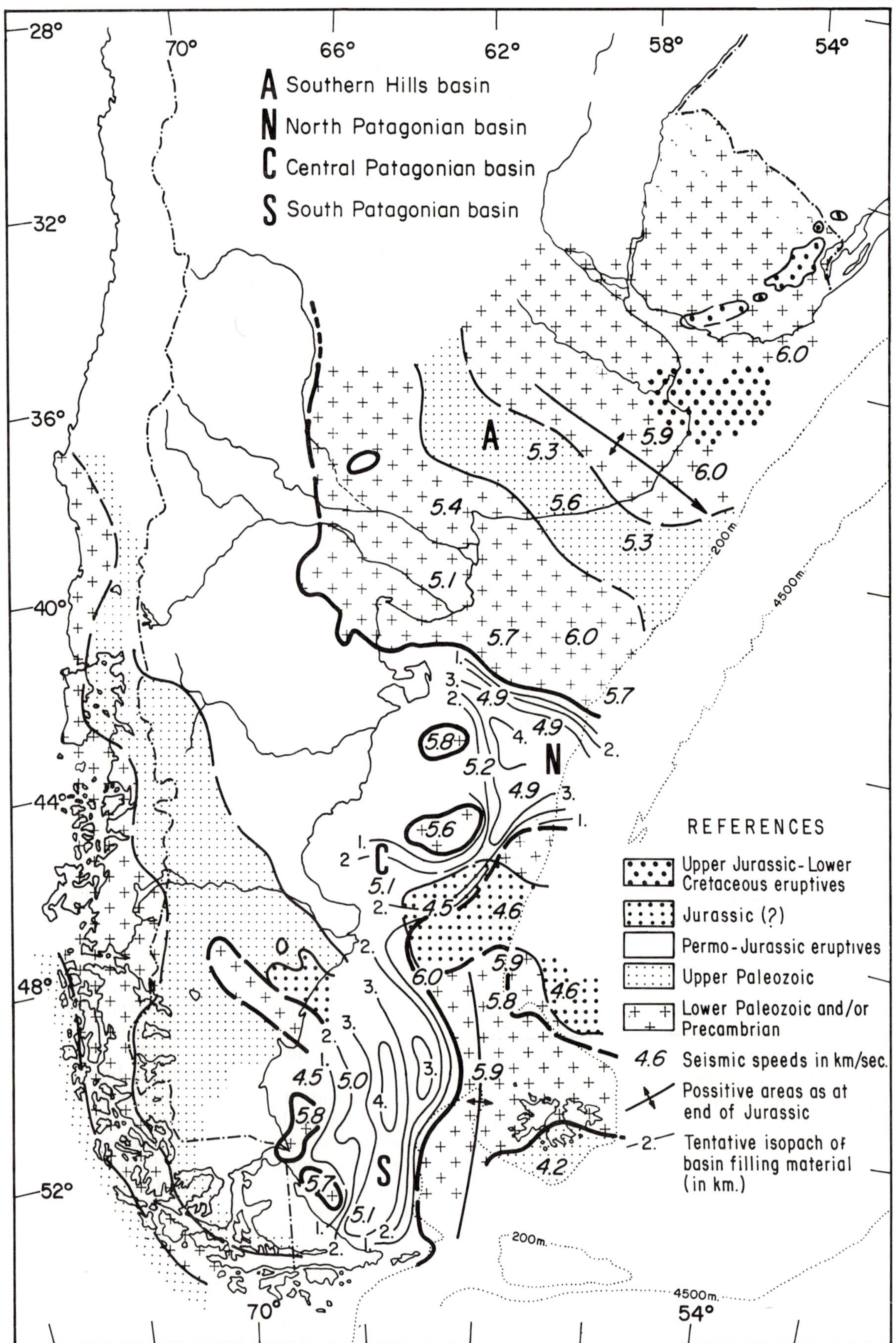

Fig. 1. Tentative paleogeologic map of the pre-Cretaceous units showing the areal extent of metamorphic basement (Precambrian north of 38°S, and partially lower and probably middle Paleozoic south). Isopachs show tentative thickness of Paleozoic and lower Mesozoic sedimentary extrusives and pyroclastic material.

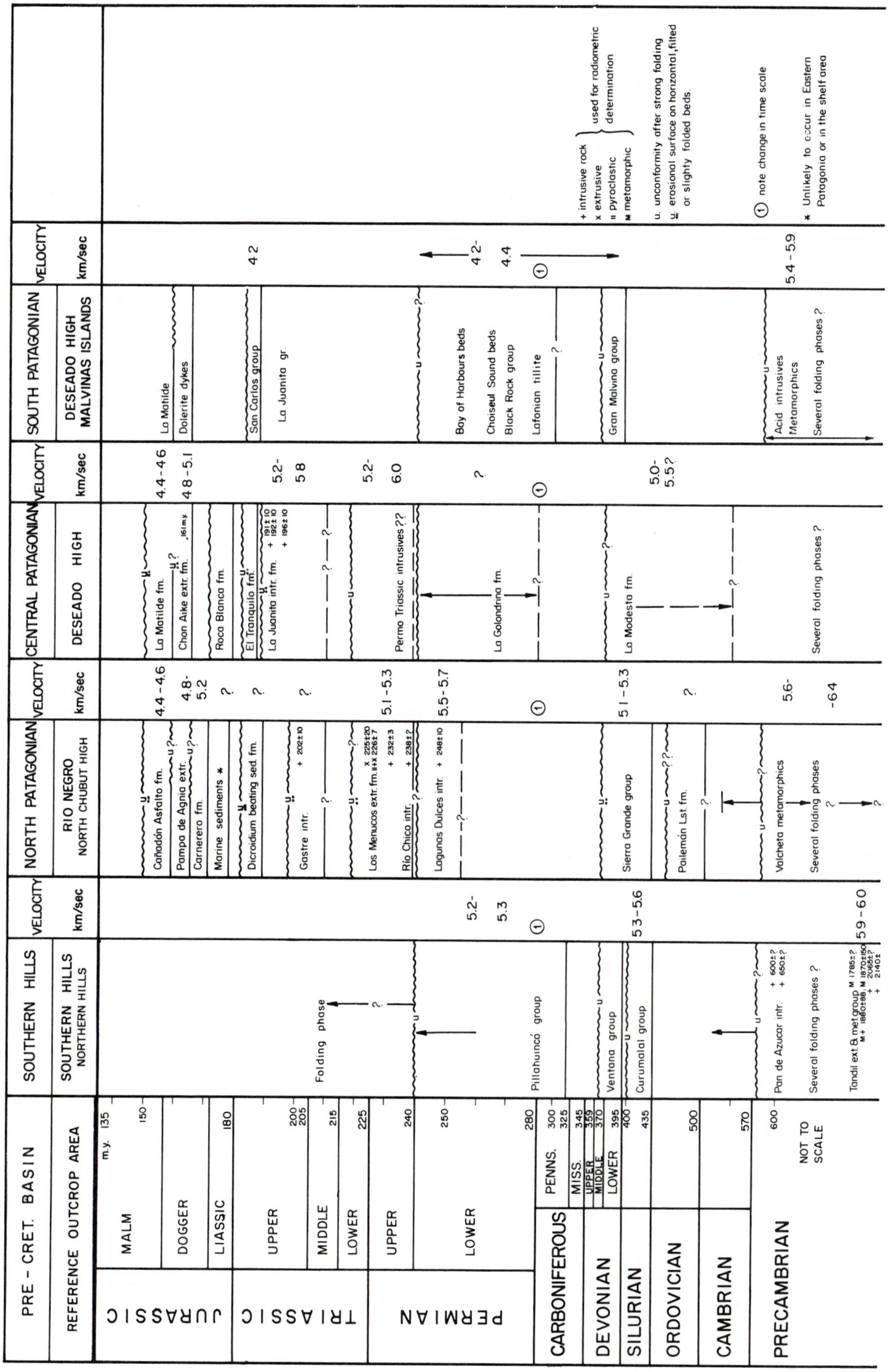

PRE-CRET. BASIN			SOUTHERN HILLS	VELOCITY	NORTH PATAGONIAN	VELOCITY	CENTRAL PATAGONIAN	VELOCITY	SOUTH PATAGONIAN	VELOCITY
REFERENCE OUTCROP AREA		m.y.	SOUTHERN HILLS NORTHERN HILLS	km/sec	RIO NEGRO NORTH CHUBUT HIGH	km/sec	DESEADO HIGH	km/sec	DESEADO HIGH MALVINAS ISLANDS	km/sec
JURASSIC	MALM	135, 150			Cañadón Asfalto fm.	4.4-4.6	La Matilde fm.	4.4-4.6	La Matilde	
	DOGGER				Pampa de Agnia extr.; Carnerero fm.	4.8-5.2	Chon Aike extr. fm. .161 m.y.	4.8-5.1	Dolerite dykes	
	LIASSIC	180			Marine sediments *	?	Roca Blanca fm.			
TRIASSIC	UPPER	200, 205			Dicroidium bearing sed. fm.	?	El Tranquilo fm.		San Carlos group	4.2
					Gastre intr. + 202±10	?	La Juanita intr. fm. + 191±10 + 192±10 + 196±10	5.2-5.8	La Juanita gr.	
	MIDDLE	215	Folding phase							
	LOWER	225	?							
PERMIAN	UPPER	240			Los Menucos extr. fm. x 225±20, ıı+x 226±7, + 232±3; Río Chico intr. + 238±?	5.1-5.3	Permo Triassic intrusives ??	5.2-6.0		
	LOWER	250, 280		5.2-5.3	Lagunas Dulces intr. + 248±10	5.5-5.7	La Golondrina fm.	?	Bay of Harbours beds; Choiseul Sound beds; Black Rock group	4.2-4.4
CARBONIFEROUS	PENNS.	300, 325	Pillahuincó group	(1)		(1)		(1)	Lafonian tillite	(1)
	MISS.	345								
DEVONIAN	UPPER / MIDDLE / LOWER	359, 370, 395	Ventana group		Sierra Grande group	5.1-5.3	La Modesta fm.		Gran Malvina group	
SILURIAN		400, 435	Curumalal group	5.3-5.6						
ORDOVICIAN		500			Pailemán Lst. fm.	?		5.0-5.5?		
CAMBRIAN		570								
PRECAMBRIAN	NOT TO SCALE	600	Pan de Azucar intr. + 600±? + 650±?; Several folding phases ?; Tandil ext. & met. group M 1785±?, M+ 1880±88, M 1870±150, + 2065±?, + 2140±	5.9-6.0	Valcheta metamorphics; Several folding phases ?	5.6-6.4	Several folding phases ?		Acid intrusives; Metamorphics; Several folding phases ?	5.4-5.9

+ intrusive rock, x extrusive, ıı pyroclastic, M metamorphic } used for radiometric determination

u. unconformity after strong folding

u̲. erosional surface on horizontal, tilted or slighty folded beds.

(1) note change in time scale

* Unlikely to occur in Eastern Patagonia or in the shelf area

Fig. 2. Comparative stratigraphic columns for each basin according to nearest outcrop and well data. The P-wave velocities assigned to each main rock unit are based on well data and correlations with onshore refraction tests.

The youngest pre-middle Cretaceous unit in the area consists of basaltic rocks, which were found overlying the metamorphics in a well drilled near Samborombom Bay (35°41'S, 57°19'30"W). These basalts could be foretold by seismic velocities (5.0-5.2 km/sec) and regional geological studies and may be correlated with the early Cretaceous extrusives of widespread occurrence in the Parana Basin (northeast Argentina, northwestern Uruguay, southwestern Brazil, and eastern Paraguay). The subsurface distribution of these basalts (Fig. 1) has been sketched on the basis of seismic data.

On the Argentine shelf these units have been fractured, so that the present regional features consist of cratonic blocks with displacement along regional fractures, frequently with northwest trends, readily observed in seismic records. In the areas where the blocks subsided, the Cretaceous basins were formed. It is also possible that some fault zones were developed in the location of previous tectonic lines. In other areas of the Argentine shelf, the existence of pre-Cretaceous faults can be inferred from some basin edges sketched in Figure 3. There is little doubt that the boundaries of the North Patagonian and South Patagonian basins are structural, fault edges, already existing before the movements giving rise to the Cretaceous basins took place. However, there is still practically no evidence for the existence of such earlier faults in this northern geological province, where it is doubtful that Mesozoic basins older than Cretaceous ever occurred.

The area lying between 38° and 43°S, which corresponds mainly to the offshore prolongation of the Colorado Basin (Urien and Zambrano, 1972). Here, Paleozoic folded sedimentary sequences—Ventana system—occur in the subsurface, as revealed by seismic data. These sediments are exposed in the Southern Hills (Harrington, 1970), where they rest upon an igneous basement. The folded sediments have undergone slight metamorphism, but this has been enough to compact the material and give it P-wave velocities between 5.0 and 5.5 km/sec. Therefore, the seismic differences between the Paleozoic sequences and the earlier basement have been blurred. However, the prolongation of these sediments into the continental shelf can be inferred from the structural trends onshore and from the presence of a folded unit in reflection-seismic sections. No less than seven wells drilled offshore in the Colorado basin (Fig. 2) reached the late Paleozoic basement, thus furnishing the only direct proof of the presence of these sediments. It is likely, then, that the folded sediments trending southeast reach the continental slope, which would then truncate them. Thus the remnant of a Paleozoic basin can be identified in the continental terrace. It is interesting to remark that inland, northwest of the Southern Hills, the prolongation of this basin in the subsurface is not known (Harrington, 1969), owing to lack of pertinent subsurface data, but a connection with the Permo-Carboniferous sediments of the Parana Basin cannot be ruled out, and is suggested by refraction test results.

The region to the south of 44°S shows a greater variety of seismic layers than the previous ones, since it is possible to distinguish on the refraction lines a unit with velocities between 5.5 and 6 km/sec which is considered to represent various sequences of various ages, such as the metamorphic rocks disclosed by several wells drilled in the San Jorge basin area (Urien and Zambrano, 1972); possibly also the predominantly acidic intrusives, which range in age from Permian and Middle Triassic (Halpern et al., 1971; Stipanicic et al., 1971). As more radiometric determinations are made, it becomes increasingly evident that these late Paleozoic early Mesozoic igneous rocks make up the greater part of the intrusive bodies present in central and southern Patagonia (Figs. 1 and 2). These intrusions probably helped to build up the nearly cratonic areas known as North Patagonia and Deseado massifs. This igneous activity produced the intrustion of thousands of cubic kilometers of magmatic material. Thereby vast extensions in Patagonia became quasi-cratonic. This is the most likely origin of the North Patagonia and Deseado massifs, which formerly had been considered as remnants of a Precambrian shield. However, materials of this age have not been found in the Deseado area, their presence in North Patagonia is doubtful, and even these doubtful occurrences occur in areas far more restricted than previously assumed (Stipanicic, 1972; Urien and Zambrano, 1972).

A 4.8-5.1 km/sec layer is locally overlain by a 4.2-4.6 km/sec interval. It is considered that the first layer corresponds to:

1. A sequence with acid and mesosilicic extrusives and pyroclastics, with subordinate continental sediments in between. This is the Chon Aike Formation (Urien and Zambrano, 1972) of Middle Jurassic age (Fig. 2). 2. In Northern Patagonia, it is possible that the Siluro-Devonian iron-bearing sediments cropping out in the Sierra Grande (west San Matias Gulf, at about 41°30'S) are represented in this seismic layer. In the North Patagonia area, both Triassic and Jurassic acid extrusives with continental clastics and pyroclastics also occur, and may well be represented in this 4.8-5.1 km/sec seismic interval.

Middle Jurassic sediments and pyroclastics (La Matilde Formation) and time equivalents, whose most likely expression offshore from the Deseado River area is the layer with seismic velocities between 4.2 and 4.6 km/sec. This layer rests upon the previously considered 4.8-5.1 km/sec seismic interval, as shown in Figures 1, 2, and 4.

To the south of the Malvinas Archipelago, near the slope area, a 4.2-km/sec layer can be correlated with the late Paleozoic-Triassic sediments cropping out in the islands. This relatively low speed indicates the slight to moderate degree of tectonic deformation in these sediments.

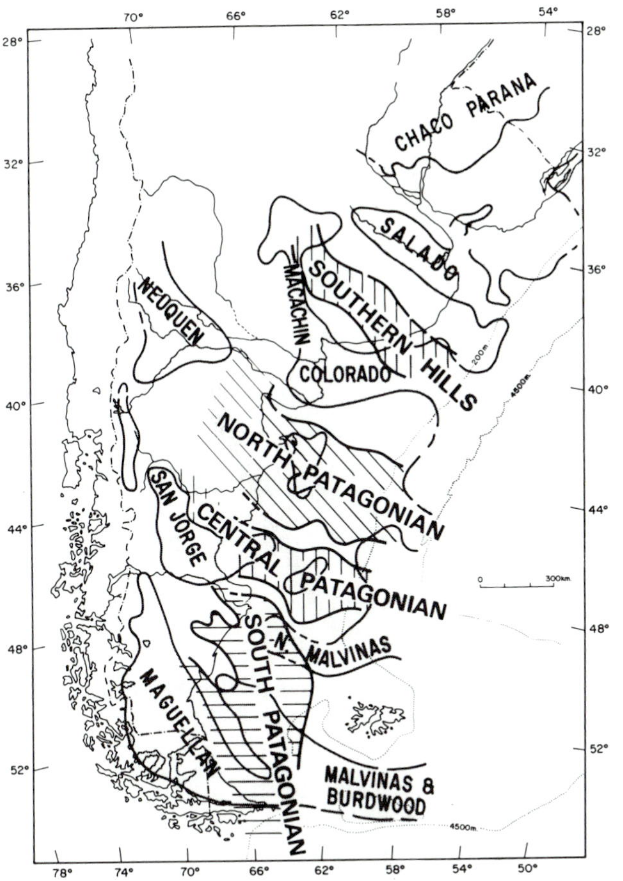

Fig. 3. Map showing the comparative situation of Cretaceous (shaded) and pre-Cretaceous basins (crosshatched).

AREAL DISTRIBUTION OF PRE-CRETACEOUS BASINS

Figure 1 shows the thicknesses of the 4.4-4.6 and 4.8-5.1 km/sec seismic layers and the most likely extension of the Southern Hills folded sequences, as inferred from regional studies. Wherever possible, tentative isopachs have been traced. These lower velocity layers obviously do not make up a continuous sheet, but are interrupted by subcrops with higher velocities, thus forming paleohighs and arches. This is the case of the Dungeness high (east of Magellan Strait), which not only interrupts the Jurassic layers, but also separates the Magallanes and Malvinas Cretaceous basins. It is interesting to point out that these structural units maintain a positive character since no later than late Paleozoic time. These isopachs disclosed the presence of several negative areas, the pre-Cretaceous basins whose fill is lying upon a Triassic, Paleozoic, or Precambrian basement.

These basins, from north to south, are (Figs. 1 and 3) (1) Southern Hills aulacogen (Harrington, 1970, 1972); (2) North Patagonian Basin; (3) Central Patagonian Basin; (4) South Patagonian Basin, which makes up the offshore prolongation of the "Carboniferous Jurassic composite basin" in southern Patagonia, described by Ugarte (1966).

The stratigraphical columns to be expected in the offshore are outlined in Figure 2. A net difference between the Southern Hills "aulacogen" and the remaining three basins is shown in the age and composition of the sedimentary columns. The tectonic styles are also widely different.

STRUCTURAL FEATURES

In spite of the wide information gaps it is possible to reconstruct the general structural framework of the area, with the following conclusions:

1. A Precambrian cratonic basement related to the Brazilian shield does not seem to occur to the south of 38°S. However, in the continental terrace there is a ridge of basement rock (Urien and Zambrano, 1972; see also Burk, 1968). As a structural unit, this ridge is the result of Late Jurassic-Early Cretaceous fracturing and subsequent rising of the resultant blocks. In the basement ridge, the presence of different seismic layers is observed (Fig. 4); of special interest an interval with velocities between 5.5 and 6.4 km/sec. Based on the interpretation outlined previously (Figs. 1 and 4), these high velocities suggest the likelihood of Precambrian rocks in the Patagonian offshore area; this interpretation does not exclude the possibility of younger intrusives. Then, if there is a Patagonian shield, its existence should be sought in this ridge of basement rock and not in the Deseado nor in the North Patagonian massifs.

2. Strong folding phases seem to be restricted to the Southern Hills area (Harrington, 1972). Harrington considers that the sedimentary filling was folded in a single tectonic phase which took place after late Permian and before Miocene times. Radiometric data point to a thermal event that took place in Late Jurassic time, which might be coeval with the folding phase. The available regional data, either geological or geophysical, give no consistent indication of the presence of any subduction zone related to this folded area, nor any other hint of significant crustal shortening. Thus, with the present state of knowledge, it can be assumed that this tectonic event may have been the result of a relatively short-lived compressional episode between two large blocks within a continental mass or at a continental edge. Another remarkable feature of these hills is the absence of field evidences for through-going faults. However, in photographs taken from spacecraft, a marked east-west joint pattern appears (Amos and Urien, 1968).

3. To the south of this folded belt, the Permo-Jurassic igneous activity gave rise to a seemingly cratonic-type area. But it did not remain stable too long, because it was fractured probably in Late Jurassic time, that is, at about the time the aforementioned thermal event took place.

4. Igneous material covers Paleozoic sediments, whose tectonic style is not easy to observe.

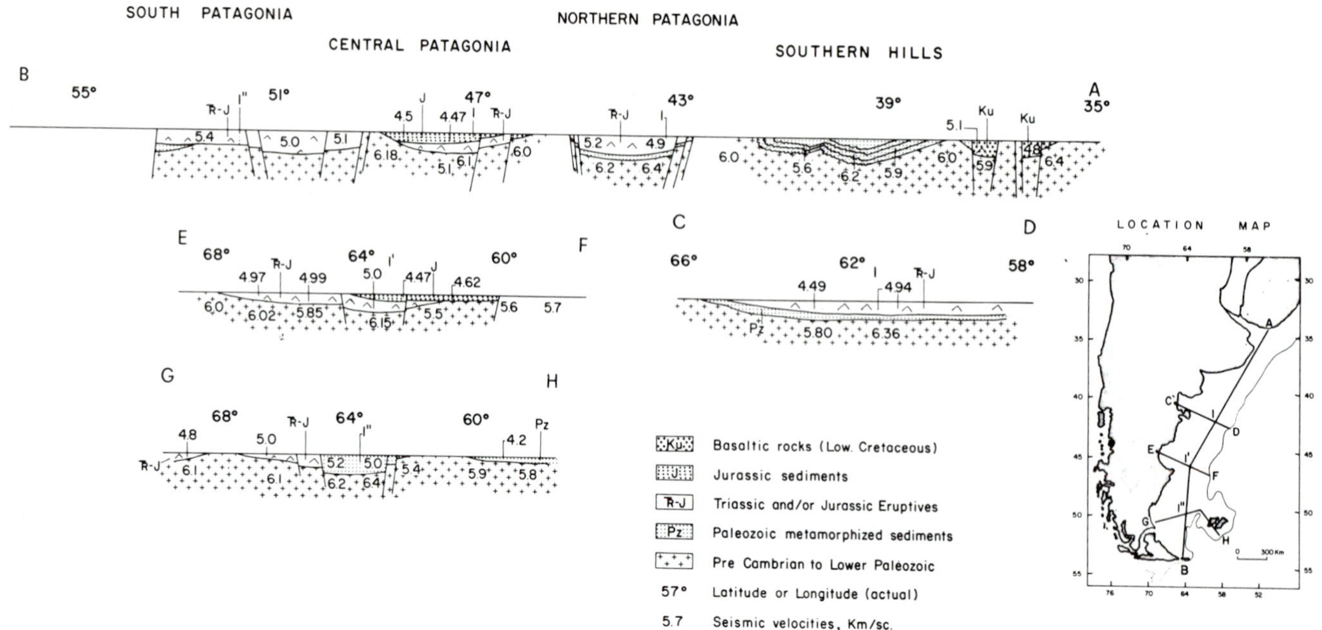

Fig. 4. Cross sections with base of Cretaceous as datum. The faults are the most adequate interpretation for the abrupt change of thickness observed in the seismic layers, interpreted as pre-Cretaceous.

However, in the Malvinas Islands, the outcropping Paleozoic sediments have only been moderately deformed, and the same is the case with the scattered exposures in the extra-Andean (eastern and central) Patagonia. Therefore, it is possible that in the continental shelf a strong folding phase should be restricted areally to the prolongation of the Southern Hills region, from early Paleozoic times onward.

5. The Triassic-Jurassic eruptives have been fractured, as suggested by sharp thickness changes observed in the 4.8-5.2 km/sec layer. This can be interpreted as a structural basin cage and suggests an important faulting phase which took place after the extrusion of the prophyries and the deposition of the related pyroclastics. However, the overlying 4.0-4.6 km/sec layer (Fig. 2) apparently has not been affected by these faults, which would then give a younger age boundary to the tectonic episode, suggesting a Late Jurassic age for the faulting phase. The relations between this and the movements that gave birth to the Cretaceous basins need further study before any more statements can be made.

6. The features, such as the Dungeness high, situated near the northeastern end of the Magellan Strait (Fig. 1) are remarkable, since they appear to have persisted since middle Mesozoic or even earlier. This may also be the case of the already mentioned basement ridge present between 42 and 50°S (Urien and Zambrano, 1972), although this ridge was fractured in Cretaceous and later times. Further study and data are also needed to resolve this problem.

7. Near the southern edge of the Argentine shelf, the 4.8-5.2 km/sec layer extends into a tectonically mobile area, close to the prolongation of the Andean belt. Here, its correlation with geological units becomes doubtful, since the presence of a thick, partly folded Cretaceous-Cenozoic cover obscures deeper data; this seismic layer may represent younger units than elsewhere in Patagonia. Therefore, the isopachs (Fig. 1) probably do not represent the outlines of a pre-Cretaceous basin, but include younger units in part.

8. The Malvinas platform represents a relatively rigid block, apparently separated by tectonic lines from the remainder of the Argentine shelf.

9. The igneous activity which either began, or was greatly increased, in Permian times is related to tectonic episodes (Fig. 2), and continued to the present day, intermittently related to the phases that gave rise to the Andean Cordillera.

DISCUSSION AND ADDITIONAL REMARKS

In the Patagonian region (excluding the Andean area) Paleozoic sediments occur in three main areas (Fig. 1). The eastern area contains the previously considered folded sequences exposed in Buenos Aires Northern and Southern Hills, and the slightly deformed Siluro-Devonian sediments cropping out at Sierra Grande, eastern Rio Negro, and few tens of kilometers west of the San Matias Gulf (about 41°30S). These sequences are separated by an area 400-500 km wide extending over the southern part of Buenos Aires province, covered by a Cretaceous and Cenozoic sedimentary blanket, and where no proved Paleozoic has been reported from the exploration wells. If any connection between the two sedimentary areas did exist, it has very likely been destroyed by erosion (Fig. 1).

The western area, where probable Devonian (Lancha Formation) and proved Permo-Carboniferous (Tepuel Group) crop out, has a considerable thickness (up to 5,200 m in Sierra Tepuel), apparently with a regional trend to the southeast, thus extending into the Deseado River area and the adjoining offshore region (Fig. 1). Although this sequence is folded, it does not show a development of secondary cleavage or metamorphism comparable to the Southern Hills folded sediments.

The southern area has nearly undeformed Devonian and younger Paleozoic sequences in the Malvinas Archipelago. These southern sediments are separated from any other occurrence of Paleozoic sediments. Lesta and Ferello (1972) suggest a connection with the Deseado River outcrops in the western area, through the continental shelf. Locally great thicknesses in the 4.8-5.1 km/sec seismic layer (Figs. 1 and 2) of the South Patagonian Basin may favor this interpretation. However, the possibility of important lateral or rotational displacements of the Malvinas area should not be overlooked, so for the time being it seems to be convenient to consider these sediments separately from the western occurrences.

This distribution of Paleozoic sequences shows the existence of a central high, with a regional north-northwest trend, which extended in pre-Cretaceous times over the central part of the present-day emerged Patagonia and continental terrace, as far as the Malvinas Islands area (Figs. 1 and 2). In this old high, some scattered small outcrops of low-metamorphic rocks have been reported (La Modesta and La Enriqueta in the Deseado River area; Valcheta and near the Limay River, in the North Patagonian region) which are ascribed to the early Paleozoic in recent papers (Stipanicic and Methol, 1972).

Rocks with higher metamorphism have been encountered in some exploration wells in the northern marginal part of the Cretaceous San Jorge Basin (Urien and Zambrano, 1972), and probably are Precambrian in age. It should be remarked that in case this age could be proved, these rocks would then be the only occurrences of Precambrian rocks in the emerged portion of Patagonia.

This structural high remained a mobile zone during middle and later Paleozoic times, but reacted to the tectonic stresses by fracturing, rather than by folding. We have already mentioned the igneous activity, mainly in Permo-Triassic times, that gave the area a quasi-cratonic character. However, it never became a stable neocraton, since it was fractured in Cretaceous and Cenozoic times, which caused the extrusion of important amounts of lava flows, predominantly basic (especially in Neogene), which cover wide extensions of land in extra Andean Patagonia. Part of this high was deformed by folding along the San Bernardo folded belt area (Urien and Zambrano, 1972), during several tectonic phases in Cretaceous, Paleogene, and Neogenic times.

A reconstruction of the original shape and areal relationships of the depositional pre-Cretaceous basins is highly uncertain with the present state of knowledge. A myriad of data must be obtained before attempting any reliable, consistent scheme. This applies especially to the possibility of finding evidence of pre-Cretaceous, long-distance drifting of blocks and wrench-type faulting in the area. At any rate, the classical reconstructions of the Gondwana continent, based mainly on geographical affinities, do not fit in many aspects in the area discussed.

SUMMARY

Geophysical and well data from the Argentine shelf, and new geological information from the onshore area, have made it possible to determine the presence of pre-Cretaceous basins.

The Southern Hills Basin, filled with about 4,000 m of Paleozoic sediments, strongly folded and slightly metamorphosed, rests upon a metamorphic and igneous Precambrian basement.

The North Patagonian Basin, filled with Siluro-Devonian sediments, Permo-Triassic eruptives, with continental intercalations, upon which rests a sequence of continental sediments, pyroclastics, and acid-to-mesosilicic extrusions. The basement probably consists of intruded Precambrian to early Paleozoic metamorphics.

The Central Patagonian Basin is less known. Its fill probably consists of younger Triassic and Jurassic continental sediments lying upon a basement of Permo-Triassic intrusives and perhaps early Paleozoic or Precambrian metamorphics.

The South Patagonian Basin is even less known. Its sedimentary fill may be Carboniferous-Triassic, resting upon an older intrusive and/or metamorphic basement, probably Paleozoic or, less likely, Precambrian.

Conclusions from the present study are: (1) The situation of the pre-Cretaceous basins is unrelated to the Cretaceous and younger ones; (2) the only marine sediments older than Cretaceous found in the area are Paleozoic in age; (3) pre-Cretaceous folding phases are restricted to the Southern Hills Basin areas; (4) no evidence for a Precambrian craton related to the Brazilian shield have been found south of 38°S; (5) the basement ridge in the eastern part of the shelf, near the present continental slope, is the only place where a Precambrian Patagonian craton may exist; (6) the area was tectonically mobile, but reacted to tectonic forces, at least since middle Paleozoic times, by fracturing rather than by folding, except in the Southern Hills area; (7) a reconstruction of the Gondwana continent based solely on geographical affinities does not fit, in most aspects, for the Patagonian area.

Northwest Pacific Trench Margins

Seiya Uyeda

INTRODUCTION

The northwest Pacific Ocean margins are rimmed by two major series of trench-arc systems (Fig. 1). From the north, one system is composed of the Kamchatka-Kuril, Japan, Izu-Bonin (Ogasawara), Mariana, Yap, and Palau trenches. The other system, defining the northwestern margin of the Philippine Sea, is formed by the Nankai Trough and Ryukyu Trench. The latter system continues southward via the Taiwan-Luzon and Mindanao systems. This report will review some of the information about these systems. It would be appropriate, from the tectonic point of view, to call the above two systems the East Japan system and the West Japan system, respectively (Sugimura and Uyeda, 1973). That the trenches are the expression of compressive tectonic forces, which probably are due to converging mantle convection currents, has long been maintained (e.g. Vening Meinesz, 1947). Recent synthesis of island arcs has also been em- 1947). Recent synthesis of island arcs has also Recent synthesis of island arcs has also emphasized the same view (Sugimura and Uyeda, 1973). Combining such a aview of trench-arc systems, that the sea floor is spreading from mid-oceanic ridges (Hess, 1962; Dietz, 1961), global tectonics has become a self-contained concept now called plate tectonics (McKenzie and Parker, 1967; Morgan, 1968; LePichon, 1968). In this concept the trench is the place where an oceanic plate underthrusts or subducts under another plate. However, some observations of the existence of undisturbed sediments in trenches, and the balance of supply of terrigeneous sediments with the existing amount of trench sediments, have pointed up some problems with plate convergence (Scholl et al., 1968; Scholl et al., 1970).

On the other hand, a contrary view that trenches are extensional features has also been held by a number of investigators. This view has been somewhat supported by such observations as normal faults and tensional earthquakes in the trench areas. In this review we wish to examine which of these conflicting ideas may be supported by the information available for the northwest Pacific trench margins.

TOPOGRAPHIC FEATURES

The general topographic features are remarkably similar in all the trench margins in the northwest Pacific (Fig. 2). The principal features of these trench margins will briefly be reviewed for each trench, mainly after Iwabuchi (1970), Mogi (1972), and Iwabuchi and Mogi (1973).

East Japan System. The Kamchatka-Kuril Trench extends from the eastern coast of the central Kamchatka Peninsula to Cape Erimo in Hokkaido, where it meets the northern end of the Japan Trench. At its northeastern end, the Kamchatka-Kuril Trench meets the Aleutian Trench. At this junction the Emperor Seamount Chain also joins from the south. Along this 2,200-km-long trench lies the volcanically active Kuril Arc. Here the average distance between the volcanic front and the trench axis is about 240 km (about 200 km in the Kamchatka area and about 250 km in the Hokkaido area). The greatest water depth reported is 10,542 m by the VITIAZ in the area south of Urup Island. However, a later detailed survey by the TAKUYO of the Japan Maritime Safety Agency found the greatest depth to be 9,550 m in the same area (Iwabuchi, 1968). Along the Pacific Ocean side, there is a prominent marginal swell called the Hokkaido or Zenkevich Rise. It has a width of 200-400 km and a height of several hundred meters.

In the southwestern Kuril Arc, the Shiretoko Peninsula of Hokkaido, the Kunashiri Islands, the Etorofu Islands, and the Urup Islands, all of which are volcanic, form en echelon the Inner or the Great Kuril Chain. In front of this chain lies the nonvolcanic Outer or Lesser Kuril Chain, which consists of Namuro Peninsula, Habomai Islands, and its submarine extension (Vitiaz Rise). These two chains make the southern Kuril a typical double arc (Fig. 1). The central Kuril Arc is a single arc, but the north Kuril Arc has an outer submarine rise similar to the Vitiaz Rise in the south. In the northernmost part, the submarine topography between eastern Kamchatka and the trench is complicated. Extensions of three capes and bays of the peninsula appear to intersect the trench obliquely. The Okhotsk Sea side of the south and central Kuril Arc has a steep slope to the 3,300-m-deep Kuril Basin. The water depth of the Okhotsk Sea outside the Kuril Basin is less than 2,000 m.

On the trench side of the central Kuril Arc, there are deep sea terraces on the continental slope at depths from 2,500 to 3,000 m, some having a width of 75 km and length of 130 km. The trench slope appears to start at the depth of 3,000 m in this part, while it seems to start at about 500 m depth where there are no terraces. Topographic profiles

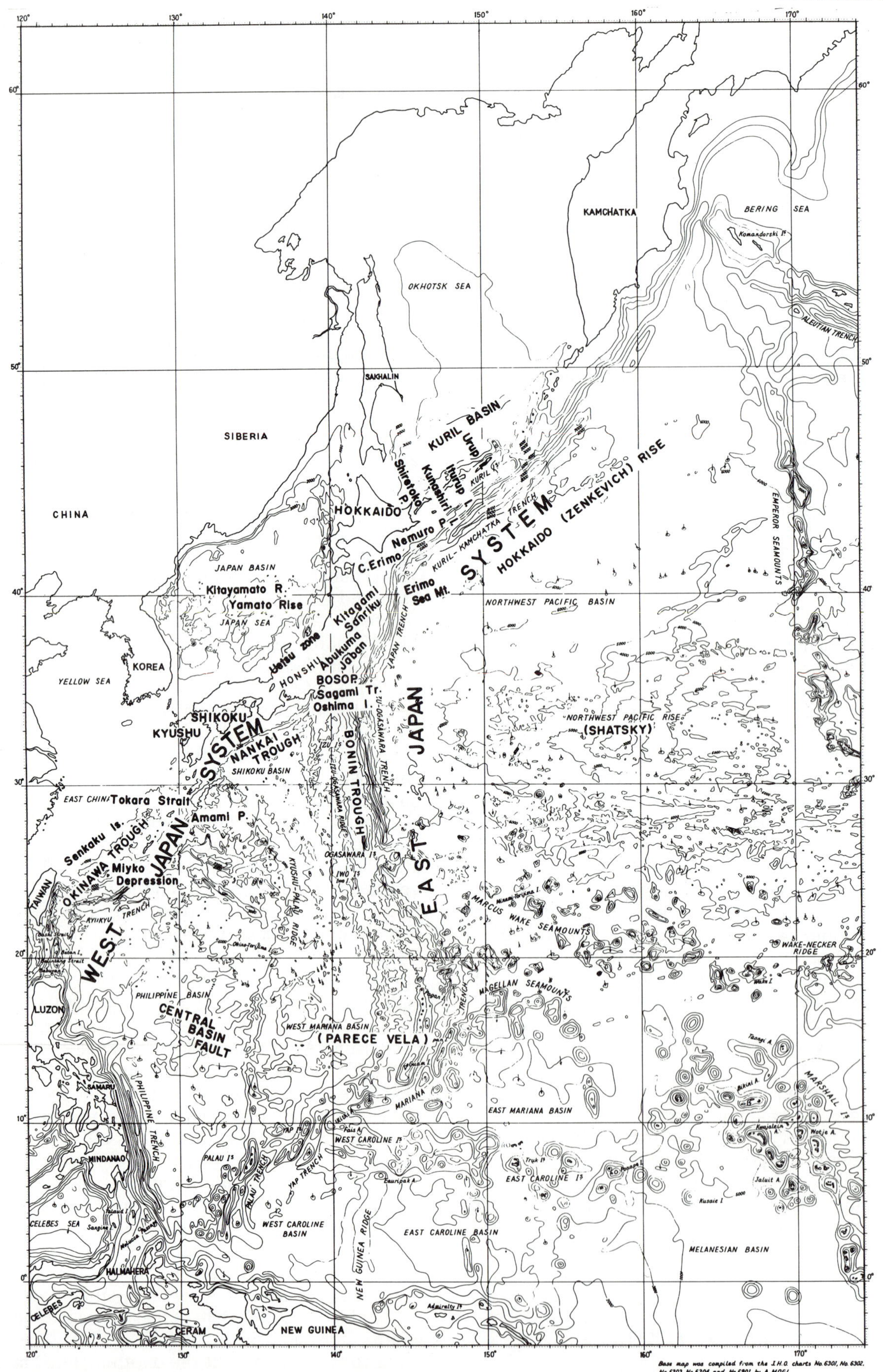

Fig. 1. Index map of the northwestern Pacific, compiled from Japan Hydrographic Office charts 6301, 6302, 5303, 6304, and 6901 by A. Mogi.

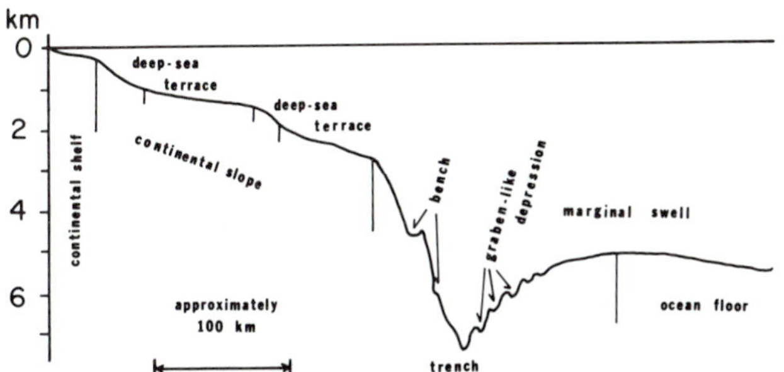

Fig. 2. Typical cross section of a trench margin (Iwabuchi, 1970).

of the trench are roughly in the shape of an asymmetric V, the landward wall being steeper. The trench wall has a mean inclination of 7° (upper part 5-6°, lower part 15-20°). The trench wall has benches that may be taken as a combined result of gentle and steep slopes. In the south Kuril Trench at depths of 7,000-7,400 m an 18-km-wide bench is found. Its outer rim has an elevation of 200-400 m. The trench floor in some portions is flat and as wide as 20 km.

The Japan Trench trends in an almost north-south direction with a slight convexity toward the Pacific Ocean, from off Cape Erimo to off Boso Peninsula. Its length is about 900 km and width is about 100 km, the greatest depth being 8,412 m. At its south end, the Japan Trench meets the Izu-Bonin (Ogasawara) Trench and the Sagami Trough protrudes into the Sagami Bay. This trough probably is the boundary between the Japanese and the Philippine Sea plates (Sugimura, 1972; Kanamori, 1972). The great Kanto earthquake in 1923 is believed to be due to a major fault at this boundary (Kanamori, 1971, 1972). The Japan Trench is associated with the northeast Japan Arc, the northeastern half of Honshu Island, and has a nonvolcanic Kitakami-Abukuma plateau to the seaward and a volcanic inner arc often called the Green Tuff Zone to the landward (Minato et al., 1965).

If sea level were raised by a few hundred meters (Miyashiro, 1967), northeastern Honshu would look like a typical double arc. Here the trench-volcanic front gap is 170-320 km. Continentward of the arc is the Japan Sea, containing the 3,000-m-deep Japan Basin and Yamato and Kita-yamato Banks (Rises). The Japan Sea has an oceanic crust and high heat flow, as will be shown later.

There are many deep-sea terraces on the continental slope facing the Japan Trench. Those at a depth of 1,000-1,800 m off the Sanriku area, 1,600-1,900 m and 2,500-3,000 m off the Joban (Zyoban) area are notable, having a width over 50 km and length over 100 km (Fig. 4). The size of terraces becomes smaller southward. The continental slope becomes a trench wall at a depth of about 2,000-3,000 m by an abrupt steepening. Topographic profiles over the Japan Trench off the Sanriku District are shown by Iwabuchi (1968). In addition to the deep-sea terraces mentioned above, benches and rifts on the trench wall can be seen that are most conspicuous on the landward wall in the northern profiles. At a depth of 4,700-4,800 m, benches with a width of some 20 km exist in the north. To the south, the width of benches appears to become smaller, but the depth of depression increases.

On the oceanward wall, several narrow but long depressions are found; in the case of Sanriku District, those at the depth 6,800 m are notable. The nature of these features is probably important in relation to the tectonic evolution of trench-arc systems. Here the trench cross section is generally V-shaped, and no wide, flat bottom seems to exist. The well-known profiles of the Japan Trench (Ludwig et al., 1966) clearly show that the landward slope has thick sediments and that the oceanward wall sediments are cut by many postdepositional faults, predominantly normal, showing that the deepening of the trench is recent. Outside the trench, a marginal swell exists as a continuation of the Zenkevich Rise in the north (Fig. 5).

South of the Japan Trench, a chain of trenches continues as the Izu-Bonin (Ogasawara) Trench, which extends some 850 km southward (Fig. 1). The Izu-Bonin Trench is remarkably straight in shape. The deepest water of 10,347 m is said to be in the Ramapo Deep (Udintsev, 1959), but more reliable surveys indicated that the depth does not exceed 9,695 m (Nasu et al., 1960). Deepest water in this trench now known is 9,810 m at 29°05'N, 142°53.5'E northeast of the Bonin Islands. South of about 26°N is the Mariana Trench, which has a prominent convexity toward the east. The trench system continues to the Yap and Palau trenches southward and meets the southern end of the Philippine Trench.

The greatest depth of the Mariana Trench, 11,034 m at 11°20.9'N, 142°11.5'E (Hanson et al., 1959), appears to be the world's record so far reported. West of the Izu-Bonin Trench lies the Izu-Bonin Arc and other linear elevations. On the axis of the arc, there is a zone of active volcanoes extending from Oshima to Iwo Jima. South of 30°N, a nonvolcanic Bonin Ridge becomes evident to the east of the volcanic inner arc, forming a double arc (Fig. 1). These features are well described by Iwabuchi, 1969. The outer nonvolcanic arc may be the emerged part of the "mid-slope basement high" of Karig (1971b).

Farther south, the Mariana Arc develops with its volcanic ridge and nonvolcanic frontal arc. Transition from insular slope to trench wall is at a depth of 2,000-3,000 m in the Izu-Shichita area and 4,000-5,000 m in the south. The trench wall has benches (terraces) with walls several hundred meters high. Behind the arc lie the Mariana Trough and the West Mariana Ridge (the third arc). The Mariana Trough between the two arcs may be of great significance for the development of interarc basin (Karig, 1971a, 1971b).

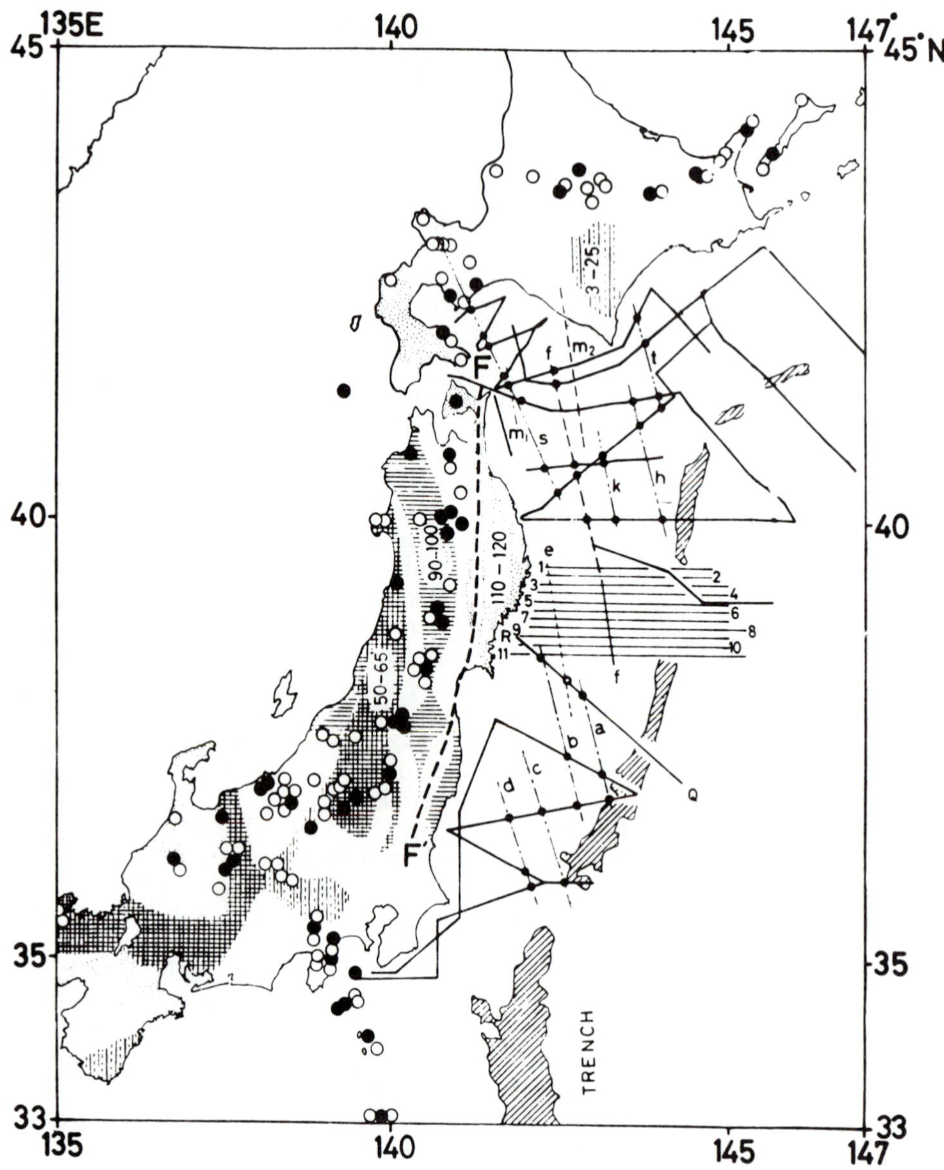

Fig. 3. Geological features of Japan and in the sea east of Japan: numbers are isotopic in my of granitic rocks on land (Kawano et al., 1966; Shibata, 1968). F-F', Miocene volcanic front; closed circles, active volcanoes; open circles, other Quaternary volcanoes. North-northwest-trending lines indicate the strike of offshore anticlines (Murauchi et al., 1973). Marine seismic traverses also shown.

West Japan System. Along the southern coast of southwest Japan, there is a trough in the Philippine Sea, the Nankai Trough, or Southwest Japan Trench (Hoshino, 1963), but it is relatively shallow for a trench (maximum depth = 5,736 m at 31°45.5'N, 133°28.5'E). Probably related to this is the fact that southwest Japan (Honshu) is not a typically active island arc in that it lacks a deep seismic zone or volcanic front (Sugimura and Uyeda, 1973). There are two different views on the nature of the trough: one considers it to be a sediment-filled ancient trench because of the thick sediment cover on the landward slope (Den et al., 1968; Murauchi et al. 1968) and the other postulates it to be a juvenile trench from reflection studies (Hilde et al., 1969; Ludwig et al., 1973) and seismic and geodetic studies (Kaneko, 1966; Fitch and Scholz, 1971; Kanamori, 1972). According to these studies, young sediments are being folded and an embryo Benioff Zone is being formed to a depth of about 100 km, and the southern coasts of Shikoku and Kii Peninsula have been uplifting during the Quaternary period, all indicating the recent start of the Philippine Sea plate subduction. Since geological structures, notably the existence of Mesozoic paired metamorphic belts (Miyashiro, 1961), indicate past active subduction (Matsuda and Uyeda, 1970; Uyeda and Miyashiro, 1974), an old trench could have existed while the abovementioned evidence suggests subduction activity probably rejuvenated in recent times (1-2 my) and perhaps both views are correct (see also Kimura, this volume).

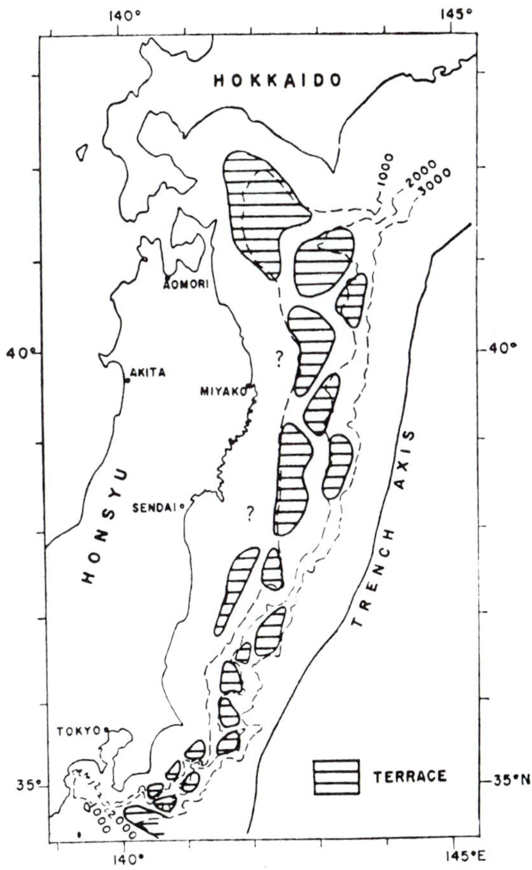

Fig. 4. Distribution of deep-sea terraces along the Japan Trench (Iwabuchi, 1968).

The southwestern end of the Nankai Trough meets the Ryukyu Trench southeast of Kyushu, where the Kyushu-Palau aseismic ridge also joins. Existence of such uplifts as the Amami Plateau makes the trench topography of the northern part of the Ryukyu Trench obscure. The Ryukyu Arc runs parallel with the trench, showing typical island arc features. Although the deep seismic zone reaches only to about 300 km depth. The Ryukyu Arc consists of several zones (Konishi, 1965). From the continent side, they are the East China Sea Shelf and its eastern rim the Senkaku Islands, which contain Miocene to Quaternary volcanics. Then the Okinawa Trough, having a water depth of 2,270 m, lies behing the Ryukyu volcanic arc. The Ryukyu Trough may be an embryo-marginal basin, having very high heat flow (see later). The nonvolcanic Ryukyu outer arc lies oceanward, showing signs of uplift in the Quaternary period. Outside the nonvolcanic arc, features such as continental slope, trench, and the ocean basin are found in a more or less similar fashion as other systems. The trench axis-volcanic front gap here is 180-230 km. One notable feature (Konishi, 1965) is the possible existence of transverse left-lateral faults that offset the arc system. They are suspected at the Miyako depression and Tokara Strait (Fig. 1).

Deep-Sea Terraces. As depicted above, one characteristic aspect of continental slopes facing trenches is the occurrence of deep-sea terraces. Various hypotheses have been put forth to explain the nature of these terraces. In some of these hypotheses deep-sea terraces are regarded as the drowned plains of erosional or depositional surfaces formed at ancient sea level: Hoshino (1962), for instance, ventured to hypothesize that the world-wide sea level in late Miocene time was some 2,000 m lower than at present, while Iijima and Kagami (1961), in their often-cited paper, postulated that the continental slope off Sanriku coast subsided since the latest Miocene or early Pliocene. Iijima and Kagami's argument was based on the observation that rock samples dredged on the outer rim of the deep-sea terrace about 150 km off the Sanriku coast were wind- and wave-eroded sedimentary rocks, containing late Miocene or early Pliocene

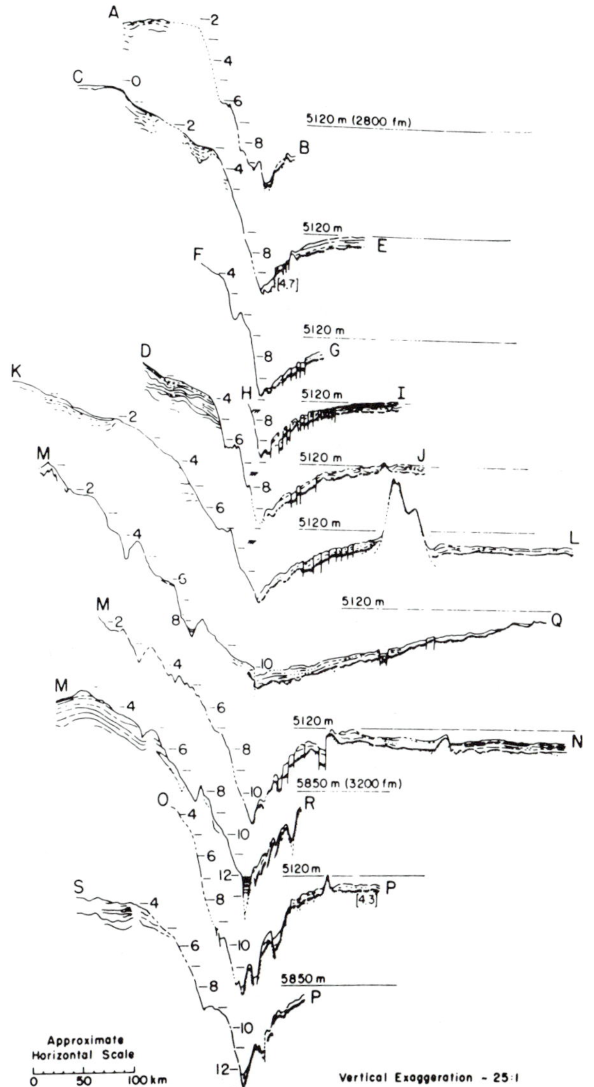

Fig. 5. Seismic-reflection profiles across the Japan Trench. From above to below, the profiles cover the latitudes from about 41°N to 31°N. The vertical scale represents two-way reflection time in seconds (Ludwig et al., 1966).

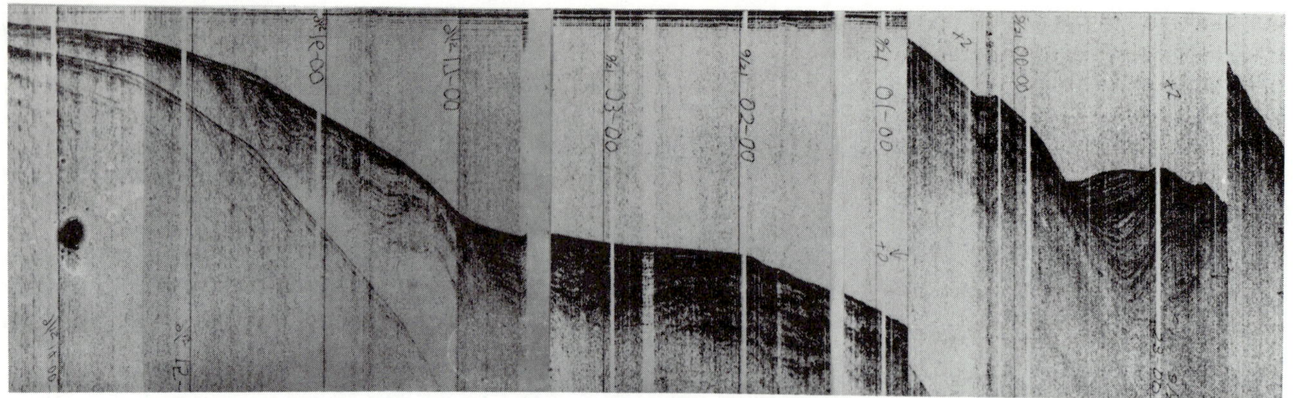

Fig. 6. Seismic-reflection profile of the landward slope of the Japan Trench off the Sanriku District. Ship speed 10 knots; 3-sec sweep; vertical exaggeration X 24 (Sato, 1973).

diatoms. The rocks had holes bored by shallow-water organisms. They inferred that the area is covered by the strata of this rock, i.e., the Sanriku Formation. Dredged samples also contained many gravels that were considered to cover the Sanriku Formation. These gravels were supposed to have been supplied from the then-subareal terrace to the west. Since the present water depth of the dredge site is about 2,350 m, it was suggested that the site has subsided over 2,000 m during the last 10^6-10^7 years, probably associated with the late Cenozoic deepening of the Japan Trench itself.

In recent years (since 1967), the Hydrographic Department of the Japan Maritime Safety Agency has been conducting a series of multidisciplinary detailed surveys, "Project Basin Map of the Sea," over the continental margins around the Japanese Islands. The R/V MEIYO and R/V SYOYO have been engaged in this program, which includes topographic, geological (sampling and profiling), magnetic, and gravity measurements carried out at 2-mile intervals. Following Sato (1973), some considerations will be made of the deep-sea terraces using this new information, and that by Ludwig et al., 1966, Yoshii et al., 1973, and others.

First, seismic-reflection profiles obtained on continental slopes do not support the idea that deep-sea terraces are drowned erosional plains or depositional surfaces near sea level. If the deep-sea terraces were formed at sea level in the past, the landward slope of the terraces must have been above sea level at that time and there must be extensive unconformities. But, in fact, the continental slopes are covered by the continuation of sediments forming the terraces (Fig. 6). Sediments are usually folded, but their surface is parallel to the bedding plane, indicating that folding took place underwater.

Second, deep-sea terrace sediments probably are not pelagic but terrigenous turbidites dammed by buried highs, as observed in the cases of Tosa Terrace (Hilde et al., 1969; Ludwig et al., 1973; Yoshii et al., 1973), Tokachi-oki and off Sanriku Terrace (Den et al., 1971; Murauchi et al., 1968b; 1973), off Boso area (Murauchi and Asanuma, 1970), Mariana area (Karig, 1971), and Ryukyu area (S. Murauchi, personal communication). Therefore, if the postulate of Iijima and Kagami (1961), that the ancient shoreline was about 150 km east of the present Sanriku coast, is to be reconciled with the new profiler data, structural evolution of the continental slope off Sanriku coast may be modeled as shown in Figure 7. Since the deep-sea terrace could not be an erosional surface, the source of Iijima-Kagami's gravels must have been the vicinity of the site of dredging. Thus Sato (1973) infers that there was land off the Sanriku coast which trapped the terrigenous sediment now forming the deep-sea terrace. Karig (1971) called the buried high "mid-

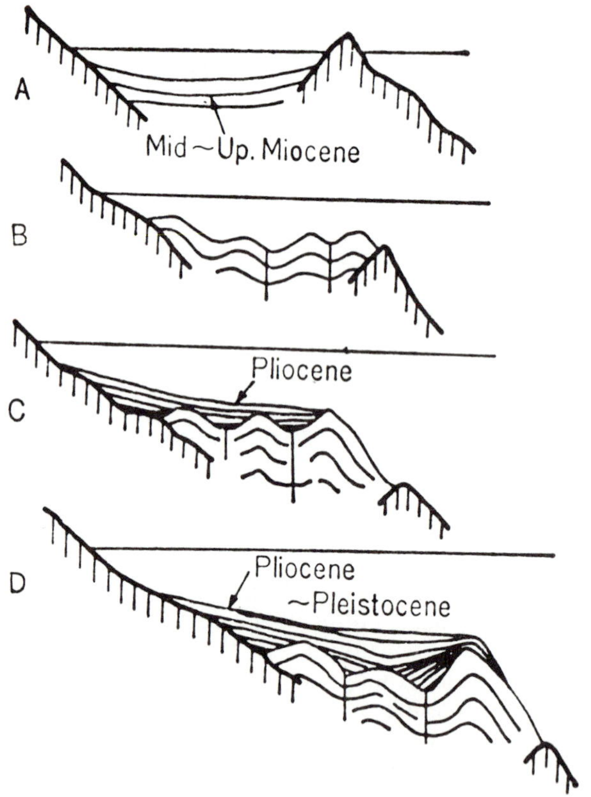

Fig. 7. Model of the evolution of the deep-sea terrace off the Sanriku-Joban districts, Japan (Sato, 1973).

slope basement high," and Dickinson (1973) termed the general feature "trench slope break." Subsidence of Sato's hypothetical offshore land may well be related to the deepening of trenches, which in turn may be due to the dragging of the subducting oceanic plate. Both Murauchi and Asanuma (1970) and Sugimura (1972) propose that a considerable amount of continental crust has been carried down into the mantle with the subducting Pacific plate during Cenozoic time. Sato's eastern land may be a part of the sinking landward plate. Karig (1971b), however, suggests on some indirect evidence that the mid-slope basement high has essentially uplifted throughout the late Cenozoic (see also Seely et al., this volume).

Perhaps the most important problem concerning the deep-sea terraces at this stage may be the mechanism of production of the mid-slope basement high that resulted in the evolution of the terraces. Is it a sinking feature of a rising one, and why is it where it is? Occurrence of deep-sea terraces is not restricted to the continental slopes facing trenches. They are found, for instance, on the continental slopes on the Japan Sea side of the Japanesse Islands (Iwabuchi and Mogi, 1973), although they are definitely less well developed than those on the Pacific side (Hoshino, 1970). Common features are that the surface of thick dammed sediments, filling swales, forms the terrace surface, but the way the swales or dams were formed may be different for the Pacific and Japan Sea margins.

Although various suggestions were made for the possible mechanism of producing buried ridges which act as dams in continental margins (e.g., Burk, 1968), the dynamics of the "mid-slope basement high" is not clear at present; it is not clear whether it is formed by surficial mechanisms such as large-scale rotational submarine land sliding (Burk's B mechanism; A. Sugimura and K. Nakamura, personal communication) or by a more deep-rooted one such as block movements of the deeper crust (Burk's A?; Murauchi, 1971). Den and Yoshii (1970) suggest that the mid-slope basement highs (their outer ridges) are the hardened pelagic sediment piles scraped off the intermittently subducting oceanic plate (Fig. 8). This is a case not explicitly mentioned in Burk (1968). In such a case, subsidence of a former land is not needed, but generally the outer ridges have uplifted.

The tectonic importance of deep-sea terraces as possible sites of present-day geosynclines (see later) and of the mid-slope basement high as a hitherto unestablished integral component of arc-trench systems should not be overlooked.

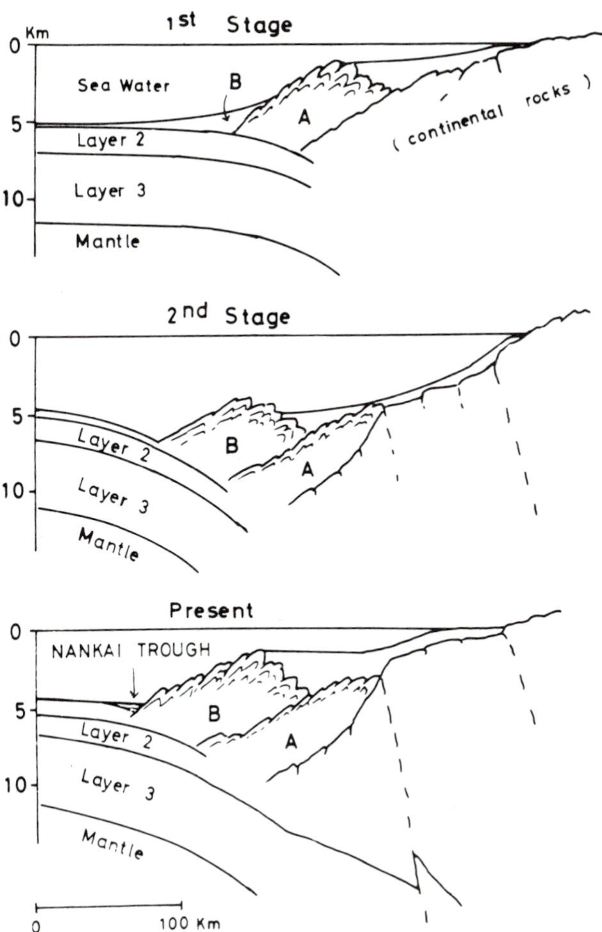

Fig. 8. Model of the evolution of the Tosa Terrace (Den and Yoshii, 1970).

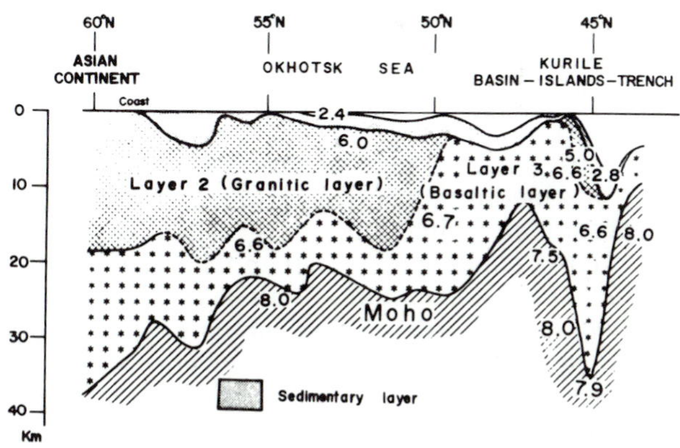

Fig. 9. Crustal structure of the Kuril Arc (Kosminskaya et al., 1963).

SOME GEOPHYSICAL FEATURES

Crustal and Upper Mantle Structures. Seismic refraction and reflection studies in the area of the northwest Pacific trench margins have been carried out mainly by Soviet, U.S., and Japanese scientists (Kosminskaya et al., 1963; Ludwig et al., 1966; Murauchi et al., 1968a; Hotta, 1970; Den et al., 1968; Yoshii et al., 1973; Ludwig et al., 1973). International cooperative programs have been quite successful. Some representative crustal structure profiles are shown in Figure 9-11. The northwest Pacific Basin itself has a typical oceanic crust, and the trench margins are characterized by the oce-

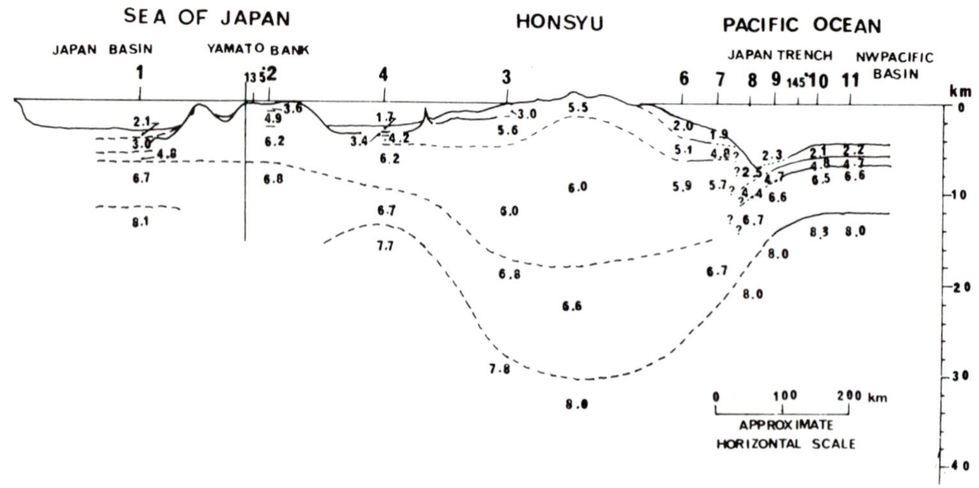

Fig. 10. Crustal structure of northeast Japan (Murauchi and Yasui, 1968).

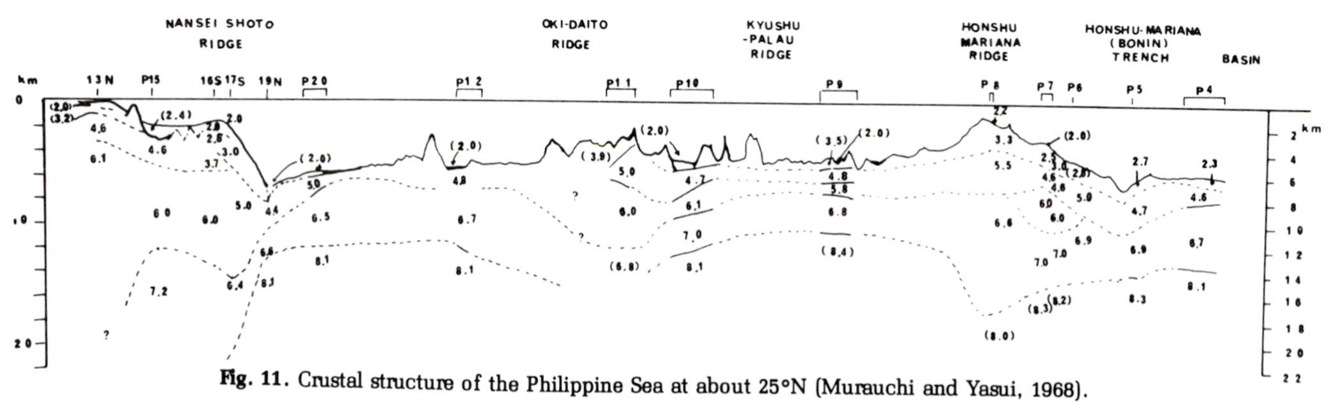

Fig. 11. Crustal structure of the Philippine Sea at about 25°N (Murauchi and Yasui, 1968).

anic crust bending downward approaching the trench. Surface sedimentary layers, as disclosed by reflection measurements, show numerous faults, predominantly normal (e.g., Fig. 5).

The mechanisms of earthquakes occurring oceanward of trenches suggest that extensional forces are acting here. This tension, however, is interpreted as being due to the bending of the lithosphere (Isacks et al., 1968) and does not conflict with the idea of subduction. Another main feature of trenches is the thickening of crust under the trench-arc gap or the continental slope where deep-sea terraces are developed. A thick sediment layer is seen at this place for each trench; in particular, the Tosa Terrace facing the Nankai Trough has very thick sediments (Den et al., 1968). In the case of the Kuril Trench (Fig. 9), layers with P velocity of 5.0 km/sec and 6.3 km/sec, typical of continental layers, are found. Similar features are observed in other trench-arc systems. The continental crust clearly extends to the oceanward rim of the continental slope, where it contacts with the oceanic crust.

Pn velocity under the Kuril Arc is given as 7.3-7.5 km/sec, which is considerably lower than the normal value for the uppermost mantle. It is well known that the Pn velocity beneath the Japanese arc is also low (e.g., Asada and Asano, 1972). This low Pn may be a characteristic of the mantle under island arc volcanic zones, as well as under mid-oceanic ridges (LePichon et al., 1965). The crust of island arcs have a continental "granitic" layer and a "basaltic" layer which is thicker than in the nearby ocean basin. This seems to apply to such uplifts as the Kyush-Palau Ridge in the Philippine Sea (Fig. 11), and Shatsky Rise, but not the Emperor Seamount Chain (Den et al., 1969). These differences of crustal structure should be related to the differences in the nature of these features; i.e., the Kyushu-Palau Ridge may be the landward half of a former island arc or a remnant arc (Karig, 1971a, 1971b, 1972; Uyeda and Ban-Avraham, 1972); Shatsky Rise may be an extinct island arc (Hilde, 1973) and the Emperor Seamount Chain may be the trace of a hotspot (Morgan, 1971).

The marginal seas have oceanic crust in their deep basins. Typical examples are the Kuril Basin; Japan and Yamato basins; Shikoku, Parece Vela, and Philippine basins. From this fact and others, such as topography, gravity, and high heat flow in these basins (see later), the extensional origin of marginal basins behind island arcs has been postu-

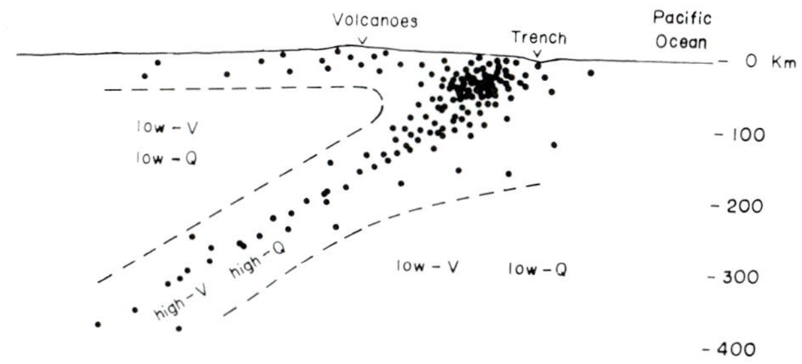

Fig. 12. Upper mantle model under the northeast Japan Arc (Utsu, 1971).

lated by a number of investigators (Murauchi, 1966; Rodolfo, 1969; Hasebe et al., 1970; Matsuda and Uyeda, 1970; Packham and Falvey, 1971; Karig, 1971a, 1971b; Sleep and Toksöz, 1971; Uyeda and Ben-Avraham, 1972; Uyeda and Miyashiro, 1974). The Okinawa Trough, which is considered to be an embryo marginal sea from its high heat flow (Yasui et al., 1970; Watanabe et al., 1970) does not show evidence of oceanic crust (Fig. 11). It would be very interesting to know the crustal structure of such an interarc basin as the Mariana Trough, which is believed to be a recent extensional opening (Karig, 1971b).

The upper mantle structure of the region concerned has been studied intensively. Based on the long-noted anomalous geographical distribution of earthquake intensities and the amplitude attenuation of seismic waves, as well as the regional travel-time variations, Utsu (1971) proposed a model of the upper mantle under the northeast Japan Arc (Fig. 12), which is quite similar to that proposed by Oliver and Isacks (1967) for the Tonga Arc. Such a model is compatible with models of the upper mantle derived from heat flow and electrical conductivity anomaly (see below) and supports the plate tectonic concept of subduction of oceanic plate. Distribution of the upper mantle seismic wave velocity similar to the above and the existence of regional magma chambers were also found under the Kuril and Kamchatka region (Fedotov, 1968, 1973).

Kanamori and Abe (1968) and Abe and Kanamori (1970a, 1970b) obtained the velocity structure of the upper mantle in front of and behind active arcs through the group-velocity analysis of long-period Rayleigh and Love waves: the general results are that velocities under marginal seas are much less than those under oceanward basins: for instance, S-wave velocity at 30-80 km depth under marginal seas, such as the Japan Sea and Philippine Sea, is about 0.3 km/sec smaller than the standard value. It may be presumed that the plate is anomalously thin in these seas.

Seismicity. The northwest Pacific is one of the most active seismic zones of the world (Fig. 13). Benioff zones are clearly defined for all the arcs, except Southwest Japan; there is also an intermediate-depth zone, where seismicity is definitely weak for each Benioff zone. Vertical profiles of relocated hypocenters for the Kuril, Northeast Japan, Izu-Bonin, and Ryukyu arcs have been shown by Miyamura (1972). Ishida (1970) showed that the Benioff zone is probably as thin as 20 km and remarkably straight under the Northeast Japan Arc.

Seismicity of the ocean floor has been investigated in more detail by the use of ocean bottom seismographs in the last few years (Nagumo et al., 1970a, 1970b). Preliminary results in the area off Sanriku indicate seismicity of microearthquakes is highest under the inner trench slope (Nagumo et al., 1974). Further results, however, are still to be analyzed.

Focal mechanisms of earthquakes have long been studied in Japan by many authors, notably Honda (1962) and his colleagues. Recent studies have revealed facts important to tectonics (Isacks et al., 1968; Isacks and Molnar, 1969; Katsumata and Sykes; Ichikawa, 1971). That the shallow shocks occurring oceanward of the trench axis are extensional or normal fault in type may be interpreted, in terms of the new global tectonics, as being due to bending of oceanic plate. One of the principal stress axes is usually parallel with the Benioff zone and, often, it is tensional at shallower depth and compressional at greater depth (Isacks and Molnar, 1969). This has been well discussed by Isacks et al., 1968.

More recently, Kanamori (1970, 1971a, 1971d) made exhaustive investigations of the great earthquakes (m 8) in the northwestern Pacific margin. As has been noted by Mogi (1968), Fedotov (1970), and Sykes (1971), great earthquakes appear to occur in such a way that the whole seismic belt of the north and northwest Pacific margin is covered by their focal areas evenly in a period of the order of 10^2 years. If we take the area of aftershocks as the focal area, focal areas do not overlap (Fig. 14).

Through analysis of long-period surface waves (T 100 sec), which best describe the nature of the fault (dimension, L, and displacement, u) of great earthquakes, showed that the L and u of great earthquakes are a few hundred kilometers and several meters. If the recurrence period of these great earthquakes is of the order of 10^2 years and the displacement for a single event 10^2 cm, the relative motion of the Pacific and continental plate would be of the order of 10^0 cm/yr. Kanomori also showed that the Sanriku earthquake in 1933 was a normal-fault type, whereas all the other great earthquakes studied were underthrust type, and the the normal-fault event may be interpreted as being due to the total break of the oceanic lithosphere.

Gravity, Heat Flow, and Geomagnetism

The Japan Trench is one of the classical areas where submarine pendulum measurement of gravity was made decades ago (Matuyama, 1936). Since the advent of surface-ship gravity meters, data in the general area of the northwestern Pacific and mar-

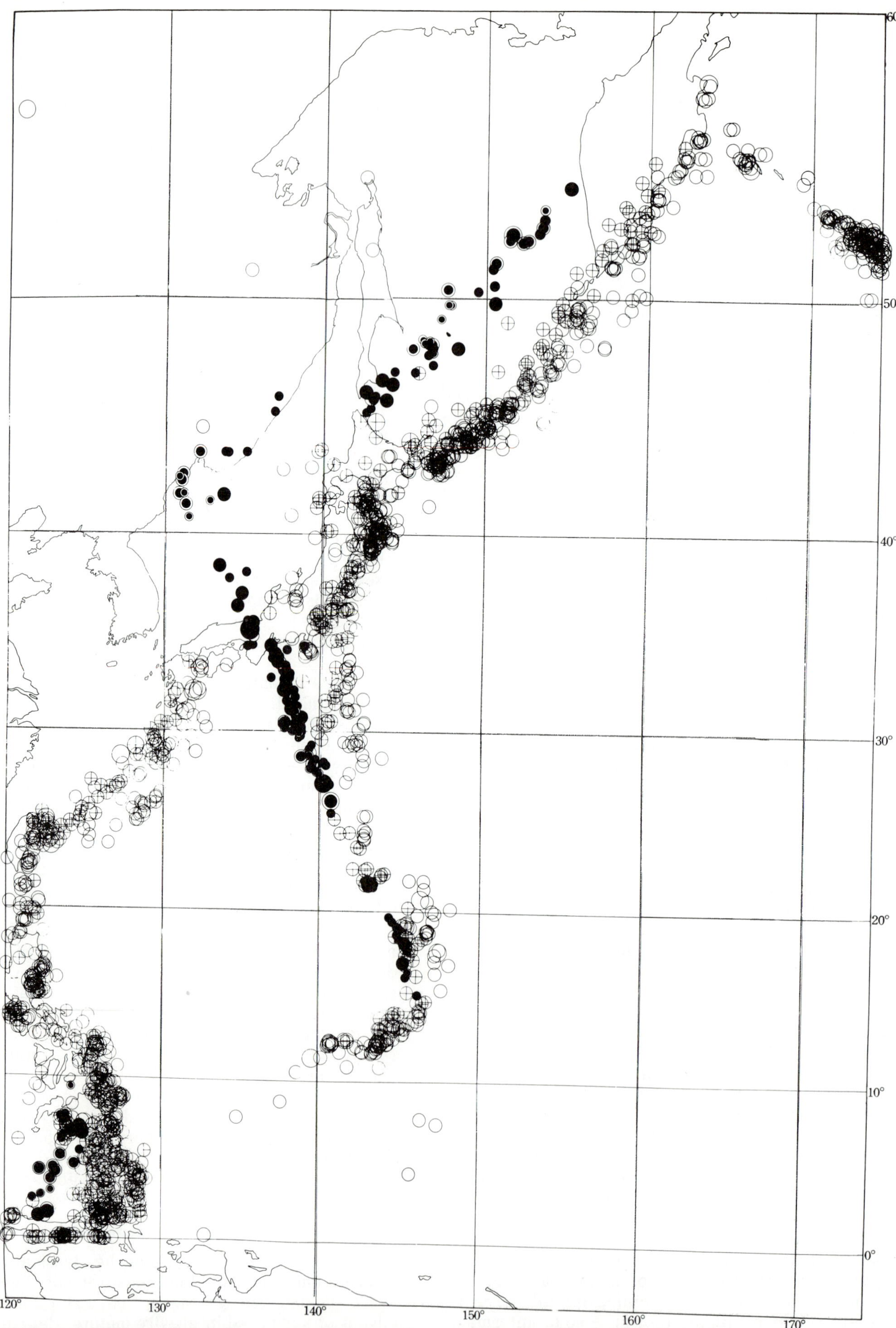

Fig. 13. Seismicity map of the northwest Pacific margin. Epicenters of shallow and intermediate earthquakes with m 5 and epicenters of transient and deep earthquakes with m 4.5 for 1965-1970 are plotted after the data given by the U.S. National Ocean Survey (Miyamura, 1972).

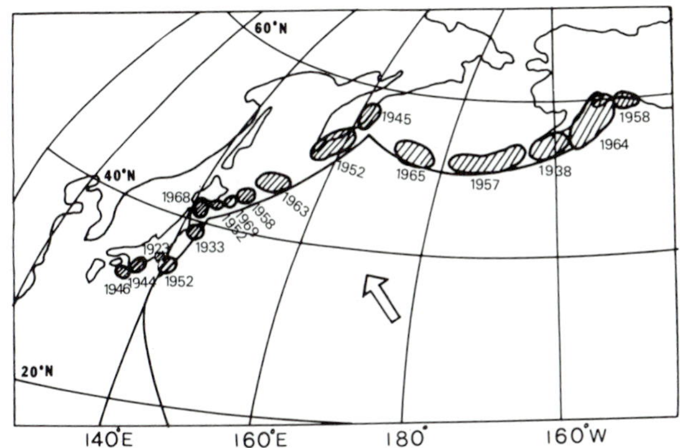

Fig. 14. Distribution of focal regions, as defined by after-shock areas of great earthquakes in the northwest Pacific margin (Mogi, 1968).

ginal seas have been accumulated, mainly by U.S., Soviet, and Japanese investigators (Worzel, 1965; Tomoda et al., 1970; Pavlov et al., 1971). The island arc and the trench appear to form a pair of positive and negative anomaly belts of almost equal intensity and spatial spread (Fig. 15); i.e., they form a mass dipole. The minimum axis of the gravity anomaly is slightly but definitely displaced landward of the topographic trench axis. One interesting observation is that gravity profiles taken off southwest Japan showed two minima, one over the terrace and the other over the Nankai Trough. While the latter is attributed largely to the topography, the former should be related to the thick light sediment (Ludwig et al., 1973). In marginal seas bordered by an island arc and a continent, the free-air anomalies are positive (10-20 mgal on the average for the Japan Sea). Apparently isostasy is not achieved in the marginal seas.

Another interesting aspect of gravity is the positive anomaly belt associated with the outer swell, such as the Hokkaido Rise. Watts and Talwani (1973) showed that this positive anomaly, reaching 50-60 mgal (outer gravity high), can be explained by a stress system associated with the convergence of lithosphere at trenches. They showed further that the Southern Bonin and Mariana Outer gravity highs can be explained in the absence of horizontal compressive stress, which is required for the explanation of the outer gravity highs of all other trenches. This observation may be related with the nearly vertical Benioff zone and scarce shallow earthquakes of the Southern Bonin and Mariana zones.

Heat-flow measurements in the northwest Pacific margin have also been conducted intensively, mainly by Japanese, U.S., Soviet, Korean, and Chinese (Taiwan) scientists (Uyeda and Horai, 1964; Uyeda and Vacquier, 1968; Yasui et al., 1968, 1970; Watanabe et al., 1970; Mizutani et al., 1970). Here again international cooperative work has been most successful. Published data around Japan as of 1972 have been summarized in Uyeda (1972). A more recent account of heat flow in the region has been made by Watanabe (1973). Data are dense enough in the Japan Sea but are still insufficient in almost all the other areas, in particular the East China Sea and most of the Asian Continent lack any data.

Notable features of heat flow are (1) uniformly subnormal heat flow in the northwest Pacific Basin, including the Shatsky Rise and Emperor Seamount ridge; (2) minimum heat flow almost coinciding with the trench axis; and (3) zone of high heat flow in the marginal seas, such as the Japan Sea, Okhotsk Sea, Okinawa Trough, and the eastern Philippine Sea (Parece Vala Basin).

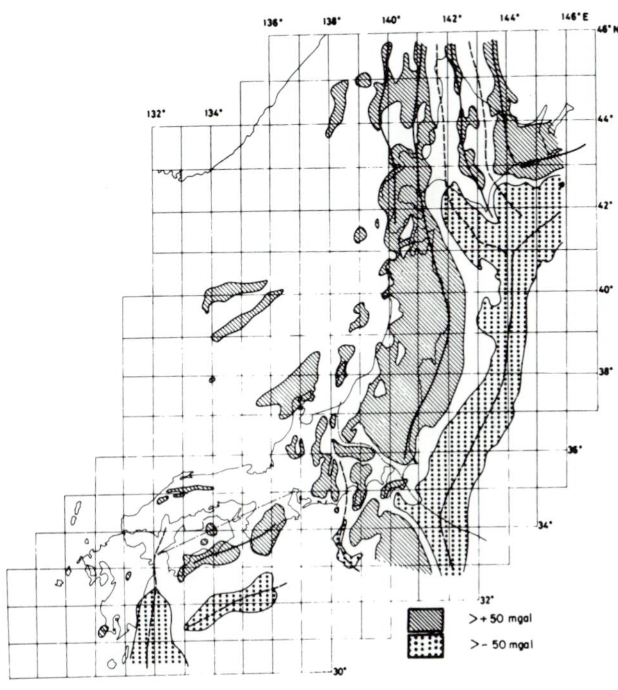

Fig. 15. Free-air gravity anomaly map in and around Japan (Tomoda, 1973).

Some complications to the above general observations are found in the Philippine Sea. Although heat flow is higher toward the east in the sea and lower to the west, agreeing with the general tendencies, limited areas of anomalous heat flow are also found, typical of which is the pronounced high-heat-flow area in the Nankai Trough area (Fig. 16). Obviously, subnormal and low heat flows in the northwest Pacific and the trench areas are compatible with the ideas of subduction of the Pacific plate. However, the high heat flow in the inner side of the arc, together with the existence of volcanic activity, is not easily explained. Various hypotheses have been put forth on the important problem of the thermal regime of subducting margin (Turcotte and Oxburgh, 1969; Minear and Toksöz, 1970; Hasebe et al., 1970). All these hypotheses assume heat generation on the subducting plate by a frictional mechanism.

Hasebe et al. (1970) suggested that if the observed high heat flow is to be explained by such a

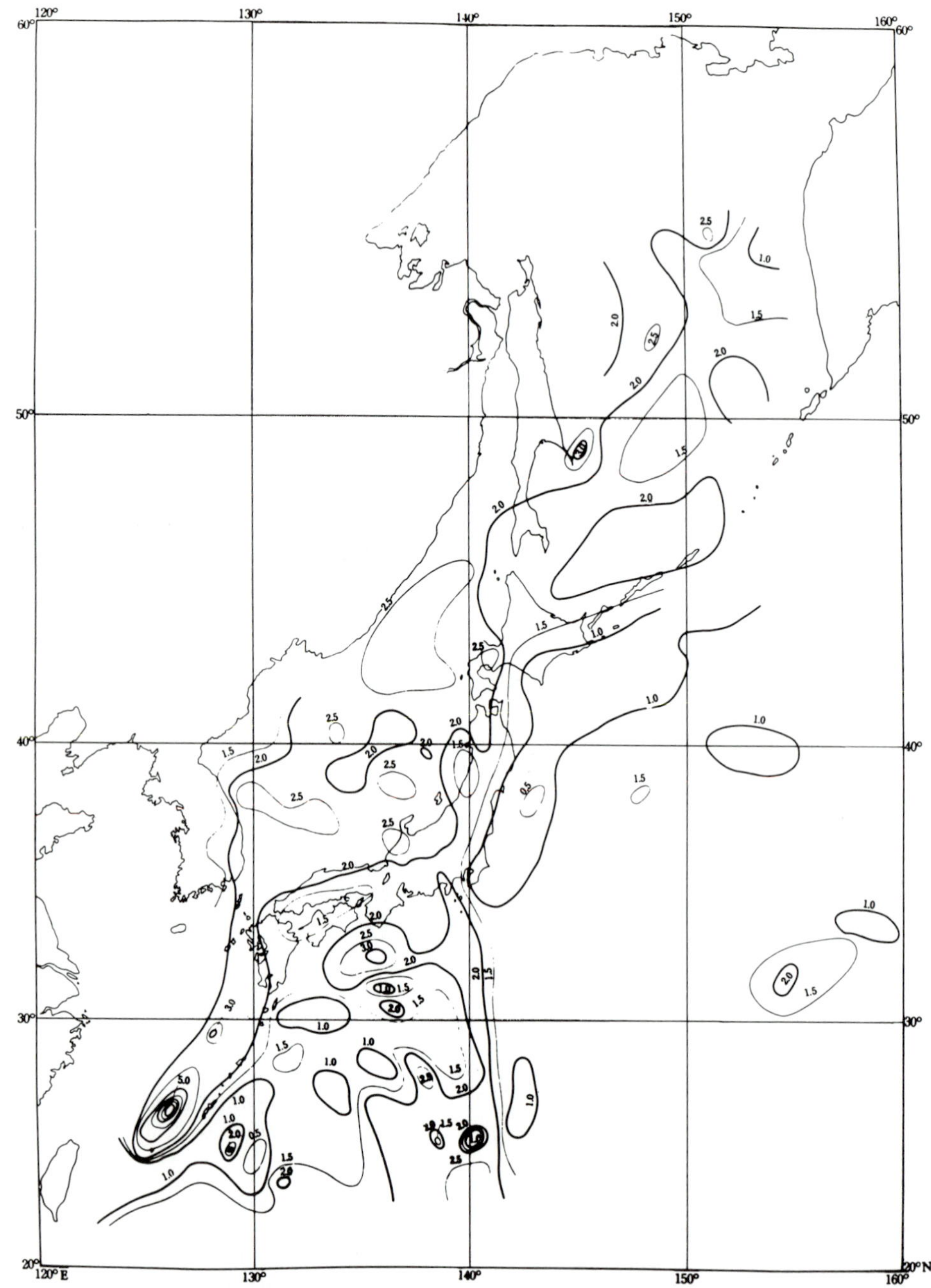

Fig. 16. Smoothed contours of heat flow in 10^{-6} cal/cm^2 sec (Uyeda, 1972).

mechanism, the rate of magma rise has to be so high that the island arc as well as the subduction zone must be pushed oceanward and an extensional marginal sea has to be developed behind the arc. The possibility seems to be concordant with the hypotheses of Karig (1971) and others for the opening of interarc basins. An alternative mechanism for the opening of the marginal sea calls for a hydrodynamic flow induced in the upper mantle above the subducting plate by the motion of the subducting plate (Sleep and Toksöz, 1971). Andrews and Sleep (1974) have made numerical calculations to show that this model can account for the thermal aspects of island arcs and marginal seas as well.

The thermal state in the upper mantle under continental margins has been studied also by means of measuring rapid magnetic variations. Variations such as bay disturbances and daily variations are affected by the electromagnetic induction of the earth's interior layers, which depends on the electrical conductivity distribution. From the observation of magnetic time variations, therefore, information on the electrical conductivity of the earth can be obtained. For the solid parts of the earth, information of electrical conductivity yields some information on the temperature. In continental margin areas, however, the separation of effects of sea water and an electrically high conducting layer in

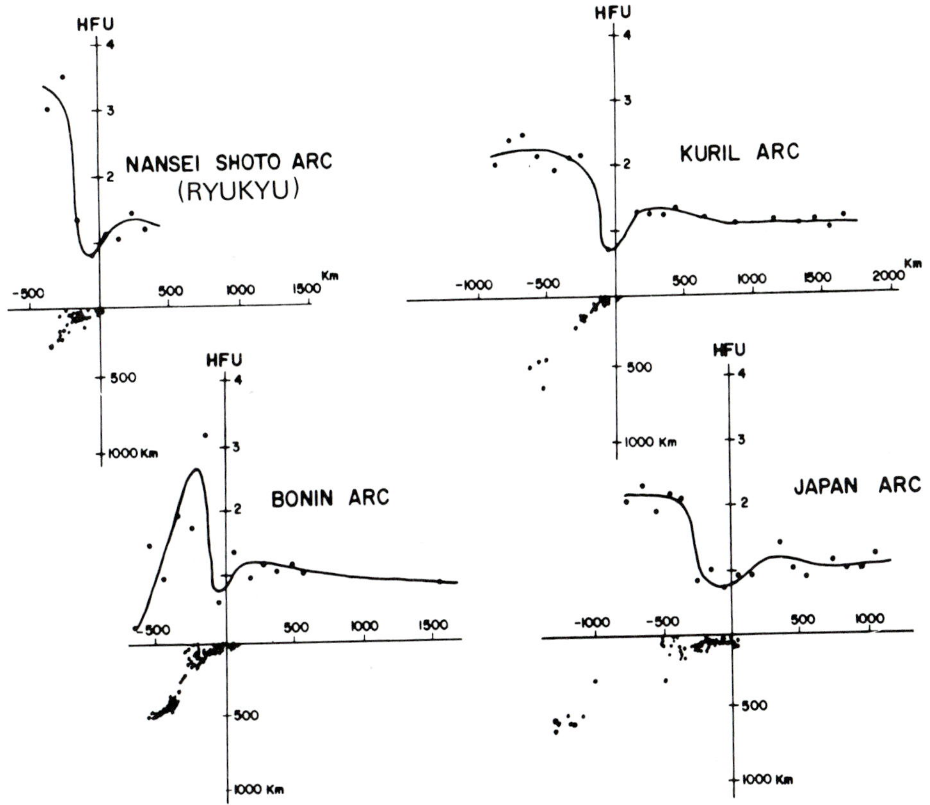

Fig. 17. Heat-flow profiles and deep seismic zones across various trenches in the northwestern Pacific (Yasui et al., 1970).

the upper mantle is a difficult program.

Uyeda and Rikitake (1970) put forth qualitative models to interpret the observations made at continental margins of various types in terms of the earth's temperature distribution. The temperature distribution that fits the magnetic observations is compatible with that derived from the heat-flow data: Figure 18 shows the depth of the 1200°C isotherm obtained mainly from the magnetic data. So far, however, magnetic observations have been possible only on land. In order to obtain more meaningful information, it is highly desirable that the same types of measurements are carried out in the oceanic areas. Ocean-bottom magnetometers are not yet in use in the western Pacific.

Magnetic anomaly patterns in the northwest Pacific area are extremely complex. Magnetic lineations of oceanic type off the northeast Japan-Kuril Arc (Uyeda et al., 1967) were not datable with the known Cenozoic time-scale patterns established in younger oceans until Larson and his colleagues (Larson and Chase, 1972; Larson and Pitman, 1972) showed that they are probably Mesozoic in age, produced by a spreading center that migrated underneath the northeast Japan-Kuril margin. They also showed that the northwest Pacific Basin migrated northward by about 40° in latitude since Cretaceous time, confirming earlier results obtained from paleomagnetism of seamounts of the area (Uyeda and Richards, 1966; Vacquier and Uyeda, 1967). Paleontological study of one of the seamounts

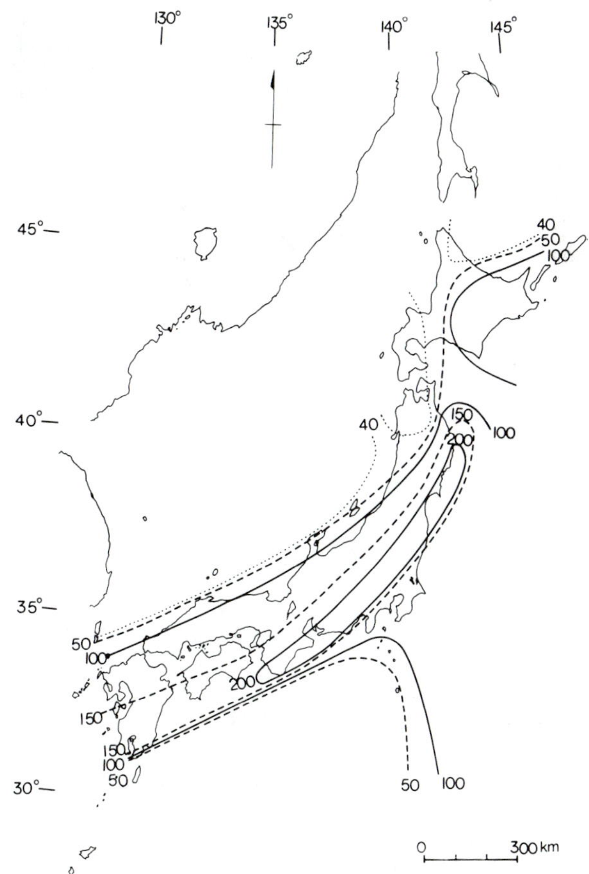

Fig. 18. Contours of depth of 1200°C isotherm, deduced mainly from magnetic variation measurements (Rikitake, 1969).

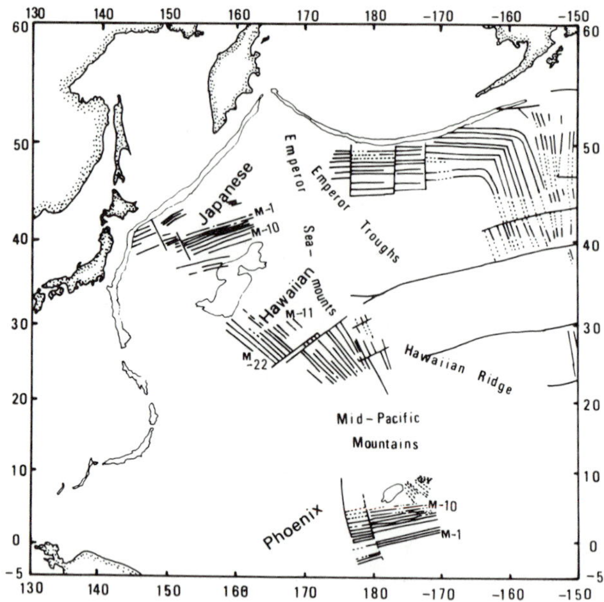

Fig. 19. Magnetic lineations in the Pacific (Larson and Chase, 1972).

(Seamount Sisoev or Erimo) supported the northward movement (Tsuchi and Kuroda, 1973).

The Japanese lineations were correlated with the Hawaiian and Phoenix sets of lineations (Fig. 19). More recently, Hilde (1973) has worked out further details of the anomalies in the area and discussed the evolution of the northwest Pacific Basin. His compilations are yet to be published, but it may be stated here that the Pacific Basin west of the Emperor Seamount Chain and northwest of the Shatsky Rise is characterized by the Mesozoic lineations M-22 to M-1 (and Hild's CL), with numerous fracture zones.

Magnetic anomalies in the marginal seas are even more complex. Soloviev and Gainanov (1963) showed anomalies in the Okhotsk Sea (Fig. 20), and Isezaki and Uyeda (1974) showed linear magnetic anomalies in the Japan Sea (Fig. 21). These linear anomalies are much smaller than those in the Pacific, both in intensity and dimensions. They are subparallel to the trend of Honshu Island, but not quite so with the lineations in the Pacific just on the other side of the arc. No apparent center of symmetry was noticed, although numerical elaboration indicated the existence of a spreading center in the Japan Basin (Isezaki, 1973a).

Much magnetic survey work is being carried out in the Philippine Sea, the genesis of which is still highly controversial. Kobayashi (1973) showed the existence of linear anomalies in the Shikoku Basin, but the details appear to require further surveys. In the western Philippine Basin, Ben-Avraham et al.

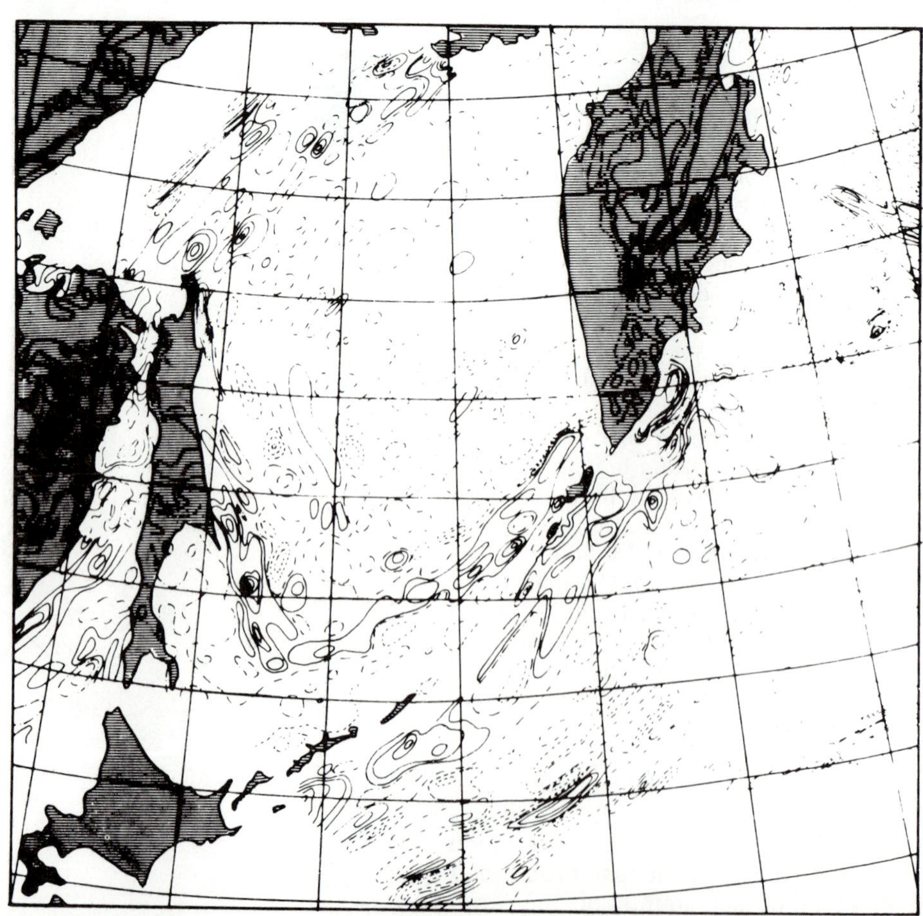

Fig. 20. Geomagnetic total force anomalies of the Okhotsk Sea; aeromagnetic survey results shown as solid lines. (Soloviev and Gainanov, 1963).

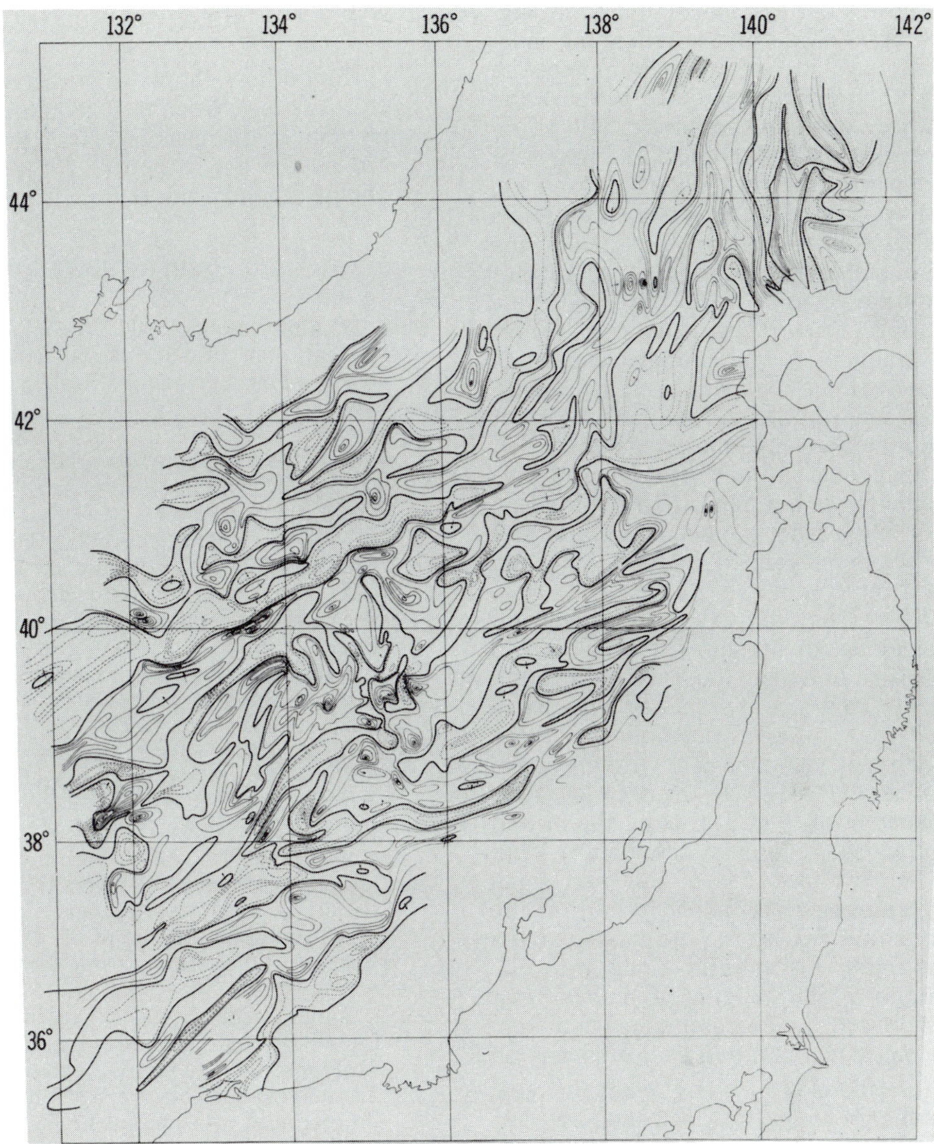

Fig. 21. Geomagnetic total force anomalies of the Japan Sea: contour interval, 50 gammas (Isezaki and Uyeda, 1974).

(1972) showed that the magnetic anomaly profiles transverse to the central Basin Fault are highly symmetric, indicating that the fault may be an extinct spreading center.

One interesting aspect of magnetic lineations is the way the lineations fade away across trenches. Some lineations are known to continue landward of the Japan and Kuril trench axes (e.g. Uyeda et al., 1967; Murauchi et al., 1973; Isezaki, 1973b). Rather clear examples were given by Hilde and Raff (1970) for northeast Japan. Along the same tracks seismic reflection records were also taken. These pieces of evidence appear to indicate that magnetic layer of the Pacific ocean plate does extend beneath northeast Japan.

Subduction. Slip vectors of earthquakes in the northwest Pacific margins were shown by McKenzie and Parker (1967) to be almost parallel. This observation was one of the major contributions on which plate tectonics is founded, because it says that the Pacific plate is thrusting under Asian and North American plate as a rigid body. LePichon (1968) computed, from the information on the spreading rates at mid-oceanic ridges, the rates of underthrusting or subduction across the trenches, obtaining 8-8.5 cm/yr for the Kuril Trench and 9 cm/yr for the Japan and Mariana trenches. These rates should represent average values for the Cenozoic period. On the other hand, Brune (1968) computed the subduction rates directly from the present earthquake data in the area, obtaining 3.8 cm/yr for the Aleutian Trench and 16 cm/yr for Japan Trench. LePichon, for simplicity, had to assume that all the relative motion between the Pacific and Asian plates is taken up by the Mariana Trench, ignoring the Ryukyu Trench.

Katsumata and Sykes (1969) examined the seismicity of the Philippine Sea and surrounding margins and showed that, in fact, the Philippine Sea plate is underthrusting northwestward beneath the southwest Japan arc systems. Kanamori (1971), on the other hand, clarified that the slip vectors of great earthquakes occurring in the northern margin of the Philippine Sea are also parallel to MacKenzie and Parker's vectors. Present-day subduction at the Izu-Bonin-Mariana arcs seems to deserve further seismic investigations.

As for the Japan-Kuril margins, some 6,000 km of the Pacific Ocean plate, including a spreading center, is believed to have underthrust since Middle Jurassic time (Larson and Chase, 1972). The budget of sediment related to the long-continued subduction on this part of trenches has not been worked out: How much sediment was supplied from the land by erosion and how much from the oceanic plate, and has the sediment been consumed or accreted to the landward wall of the trench?

Murauchi (1971) noted, that if the basement highs observed along their ship's tracks in the trench-arc gap area between the northeast Japan and the Japan Trench are correlated, they are aligned generally north-northwest oblique to the trench (Fig. 3), which is in agreement with the strike of the late Mesozoic (110-120 my in age) low-temperature metamorphic belt in northeast Japan and its possible offshore extension (Sugimura and Uyeda, 1973, Fig. 75). Because of this and some other evidence, Murauchi (1971) proposed that the Mesozoic continental crust of the arc has subsided and disappeared into the mantle along with the subducting Pacific plate. Sugimura and Uyeda (1973, Fig. 76) inferred the same phenomenon from the geologic information on land. If real, this may be related with Sato's ideas (Fig. 7) on the development of the thick sediment traps in the trench-arc gap area. This is the area where Matsuda et al., (1967) suspected that a present-day geosyncline (Off-Zyoban-Sanriku geosyncline) is in its growth stage. Kimura (see this volume) has suggested that the locus of geosynclinal accumulation in Japan has continually moved seaward through time.

Sugimura and Uyeda (1973, p. 107) consider that the sedimentary zones in the trench-arc gaps in other systems may be of the same character and formed as an integral member of the island arc system. At present, however, so little is known about the trench-arc gap area, especially the sedimentary structures of the terraces and nature of rocks of the basement highs. Any geophysical measurement would help to reveal the secret. For instance, magnetic anomalies over these highs are too small for them to be igneous in origin (Ludwig et al., 1973). Only through detailed investigations of these, would it be possible to understand the real tectonic significance of the features described so far.

ACKNOWLEDGMENTS

The authors thank Takahiro Sato, Yoshio Iwabuchi, A. Mogi, Kazuaki Nakamura, and Sadanori Murauchi for discussions. The manuscript was prepared by the assistance of N. Mizushima and K. Noguchi, to whom the author is grateful.

BIBLIOGRAPHY

Abe, K. and Kanamori, H., 1970a, Mantle structure beneath the Japan Sea, as revealed by surface waves: Bull. Earthquake Res. Inst., v. 48, p. 1011-1021.

———, and Kanamori, H., 1970b, Upper mantle structure of the Philippine Sea, *in* Hoshino, M., and Acki, H., eds., Island arc and ocean: Tokyo, Tokai Univ. Press., p. 85-91.

Asada, T., and Asano, S., 1972, Crustal structures of Honshu, Japan, *in* Miyamura, S., and Uyeda, S., eds., The crust and upper mantle of the Japaneses area, pt. 1, Geophysics: Japanese Comm. for Upper Mantle Project, p. 45-56.

Ben-Avraham, Z., Segawa, J., and Bowin, C., 1972, An extinct spreading center in the Philippine Sea: Nature, v. 240, p. 453-455.

Brune, J. M., 1968, Seismic moment, seismicity and rate of slip—along major fault zones: Jour. Geophys. Res., v. 73, p. 777-784.

Burk, C. A., 1968, Buried ridges within continental margins: Trans. N.Y. Acad. Sci., p. 2-30, 397-409.

Den, N., 1972, Crustal structures in the western Pacific Ocean, *in* Miyamura, S., and Uyeda, S., eds., The crust and upper mantle of the Japanese area, pt. 1, Geophysics: Japanese Comm. for Upper Mantle Project, p. 57-68.

———, and Yoshii, T., 1970, Crustal structure of Tosa Deep-Sea Terrace and Nankai Trough, pt. 2, Interpretations, *in* Hoshino, M., and Aoki, H., eds., Island arc and ocean: Tokyo, Tokai Univ. Press, p. 95-100, (in Japanese with Eng. abstr.).

———, Murauchi, S., and Hotta, H., 1968, A seismic refraction exploration of Tosa Deep-Sea Terrace off Shikoku: Jour. Phys. Earth, v. 16, p. 7-10.

———, Ludwig, S., Murauchi, S., Ewing, M., Hotta, H., Edgar, N. T., Yoshii, T., Asanuma, T., Hagiwara, K., Sato, T., and Ando, S., 1969, Seismic refraction measurements in the Northwest Pacific basin: Jour. Geophys. Res., v. 74, p. 1421-1434.

———, Hotta, H., Asano, S., Yoshii, T., Sakajiri, N., Ichinose, Y., Motoyama, M., Kakiichi, K., Beresnev, A. F., and Sagaievitch, A. A., 1971, Seismic refraction and reflection measurements around Hokkaido, pt. 1, Structure of the continental slope off Tokachi: Jour. Phys. Earth, v. 19, p. 329-345.

Dickinson, W. R., 1973, Width of modern arc-trench gaps proportional to past duration of igneous activity in associated magmatic arcs: Jour. Geophys. Res., v. 78, p. 3376-3389.

Dietz, R. S., 1961, Continent and ocean basin evolution by spreading of the sea floor: Nature, v. 190, p. 854-857.

Fedotov, S. A., 1968, On deep structure, properties of the upper mantle and volcanism of the Kuril-Kamchatka island arc according to seismic data, *in* Knopoff, L,

Drake, C. L., and Hart, P. J., eds., The crust and upper mantle of the Pacific Area, Geophys. Monogr. 12: Washington, D.C., Am. Geophys. Union, p. 131-9.

______, 1973, Deep structure under the volcanic belt of Kamchatka, *in* Coleman, P. J., ed., The western Pacific: island arcs, marginal sea, geochemistry: Univ. Western Australia Press, p. 247-254.

______, et al., 1970, Investigations on earthquake prediction in Kamchatka: Tectonophysics, v. 9, p. 249-258.

Fitch, T. J., and Scholz, C. H., 1971, Mechanism of underthrusting in southwest Japan: a model of convergent plate interactions: Jour. Geophys. Res., v. 76, p. 7260-7292.

Hanson, P. P., Zenkevich, N. L., Sergeev, U. V., and Udintsev, G. B., 1959, Maximum depth of the Pacific Ocean: Priroda, no. 6, p. 84-88 (in Russian).

Hasebe, K., Fujii, N., and Uyeda, S., 1970, Thermal processes under island arcs: Tectonophysics, v. 10, p. 335-355.

Heezen, B. C., 1967, Suboceanic trenches, in Runcorn, S. K., ed., International dictionary of geophysics: Elmsford, N.Y., Pergamon Press, p. 1475-1480.

Hess, H. H., 1962, History of ocean basins, *in* Petrologic studies, Buddington volume: Geol. Soc. America, p. 599-620.

Hilde, T. W. C., 1973, Mesozoic sea floor spreading in the North Pacific [thesis]: Univ. Tokyo.

______, and Raff, A. D., 1970, Evidence of oceanic crust beneath the shoreward slope of the Japan Trench from magnetic and seismic reflection data (abstr.): 51st Am. Geophys. Union Meeting.

______, Wageman, J. M., and Hammond, W. T., 1969, The structure of Tosa Terrace and Nankai Trough off southeastern Japan: Deep-Sea Res., v. 16, p. 67-75.

Honda, H., 1962, Earthquake mechanism and seismic waves: Jour. Phys. Earth, v. 10, p. 1-197.

Hoshino, M., 1962, Pacific Ocean: Tokyo, Chigaku Dantai Kenkyukai (Coop. Res. Group Earth Sci.), 136 p. (in Japanese).

______, 1963, Southwest Japan Trench: Jour. Marine Geol., v. 1, p. 10-15 (in Japanese with Engl. abstr.).

______, 1970, Continental slopes, *in* Deep Sea Geology: Tokyo, Tokai Univ. Press, p. 1-43 (in Japanese).

Hotta, H., 1970, A crustal section across the Izu-Ogasawara arc and trench: Jour. Phys. Earth, v. 18, p. 125-141.

von Huene, R., 1972, Structure of continental margin and tectonism at the Eastern Aleutian Trench: Geol. Soc. America Bull., v. 83, p. 3613-3626.

Ichikawa, M., 1971, Re-analysis of mechanism of earthquakes which occurred in and near Japan and statistical studies on the nodal plane solutions obtained, 1926-1968: Geophys. Mag., v. 35, p. 207-274.

Iijima, A., and Kagami, H., 1961, Cainozoic tectonic development of the continental slope, northeast of Japan: Jour. Geol. Soc. Japan, v. 67, p. 561-577 (in Japanese).

Isacks, B., and Molnar, P., 1969, Mantle earthquake mechanisms and sinking of the lithosphere: Nature, v. 223, p. 1121-1124.

______, Oliver, J., and Sykes, L. R., 1968, Seismology and the new global tectonics: Jour. Geophys. Res., v. 73, p. 5855-5899.

Isezaki, N., 1973a, Geomagnetic anomalies and tectonics around the Japanese Islands: Oceanogr. Mag., v. 24, p. 107-158.

______, 1973b, Geomagnetic anomalies in the western Pacific and Japan Sea: paper read at IAGA General Assembly, Kyoto.

______, and Uyeda, S., 1974, Geomagnetic anomaly of the Japan Sea: Jour. Marine Geophys. Res. (in press).

Ishida, M., 1970, Seismicity and travel-time anomaly in and around Japan: Bull. Earthquake Res. Inst., v. 48, p. 1032-1051.

Iwabuchi, Y., 1968, Topography of trenches east of the Japanese Islands: Jour. Geol. Soc. Japan, v. 74, p. 37-46 (in Japanese).

______, 1969, Trench topography of the northwest Pacific: Umi (The sea), v. 7, no. 3, p. 15-23 (in Japanese).

______, 1970, Trench, *in* Deep-sea geology: Tokyo, Tokai Univ. Press, p. 145-220 (in Japanese).

______, and Mogi, A., 1973, Summarization of submarine geology in each zone of Japanese upper mantle project, *in* Gorai, M., and Igi, S., eds., The crust and upper mantle of the Japanese area: Japanese Comm. for Upper Mantle Project.

Kanamori, H., 1970, The Alaska earthquake of 1964, Radiation of long-period surface waves and source mechanism: Jour. Geophys. Res., v. 75, p. 5029-5040.

______, 1971a, Faulting of the Great Kanto earthquake of 1923 as revealed by seismological data: Bull. Earthquake Res. Inst. Tokyo Univ. v. 49, p. 13-18.

______, 1971b, Focal mechanism of Tokachi-Oki earthquake of May 16, 1968: Tectonophysics, v. 12, p. 1-13.

______, 1971c, Great earthquakes at island arcs and the lithosphere: Tectonophysics, v. 12, p. 187-198.

______, 1971d, Seismological evidence for a lithospheric normal faulting—the Sanriku earthquake of 1933: Phys. Earth Planetary Interiors, v. 4, p. 289-300.

______, 1972, Tectonic implications of the 1944 Tonankai and 1946 Nankaido earthquakes: Phys. Earth Planetary Interiors, v. 5, p. 129-139.

______, and Abe, K., 1968, Deep structure of island arcs as revealed by surface waves: Bull. Earthquake Res. Inst., v. 46, p. 1001-1025.

Kaneko, S., 1966, Rising promontories associated with a subsiding coast and sea-floor in south-western Japan: Trans. Roy. Soc. New Zealand, v. 4, p. 211-218.

Karig, D. E., 1971a, Origin and development of marginal basins in the western Pacific: Jour. Geophys. Res., v. 76, p. 2542-2560.

______, 1971b, Structural history of the Mariana Island arc system: Geol. Soc. America Bull., v. 82, p. 323-344.

______, 1972, Remnant arcs: Geol. Soc. America Bull., v. 83, p. 1057-1068.

Katsumata, M., and Sykes, L. R., 1969, Seismicity and tectonics of the western Pacific: Izu-Mariana-Caroline and Ryukyu-Taiwan regions: Jour. Geophys. Res., v. 74, p. 5923-5948.

Kawano, Y., et al, 1966, K-A ages of the granites in Japan: read before Japan Geol. Soc., Kanazawa, Oct.

Kobayashi, K., 1973, Magnetic limations in the Philippine Sea: paper read at the IAGA General Assembly, Kyoto.

Konishi, K., 1965, Geotectonic framework of the Ryukyu Islands (Nansei-Shoto): Jour. Geol. Soc. Japan, v. 71, p. 437.

Kosminskaya, I. P., Zverev, S. M., Vverv, P. S., Tulina, Yu. V., and Krashina, R. M., 1963, Basic feature of the crustal structure of the Sea of Okhotsk and the

Kuril-Kamchatka zone of the Pacific Ocean from deep seismic sounding data: Bull. Acad. Sci. S.S.S.R, Geophys. Ser., v. 1, p. 20-41; Engl. transl., p. 11-27.

Larson, R. L., and Chase, C. G., 1972, Late Mesozoic evolution of the western Pacific: Geol. Soc. America Bull., v. 83, p. 3645-3662.

———, and Pitman, W. C., III, 1972, World-wide correlation of Mesozoic magnetic anomalies and its implications: Geol. Soc. America Bull., v. 83, p. 3627-3644.

LePichon, X., 1968, Sea-floor spreading and continental drift: Jour. Geophys. Res., v. 73, p. 3661-3697.

———, Houtz, R. E., Drake, C. L., and Nafe, J. E., 1965, Crustal structure of the Mid-ocean ridges: 1. Seismic refraction measurements: Jour. Geophys. Res., v. 70, p. 319-340.

Ludwig, W. J., Ewing, J. I., Ewing, M., Murauchi, S., Den, N., Asano, S., Hotta, H., Hayakawa, M., Ichikawa, K., and Noguchi, I., 1966, Sediments and structure of the Japan Trench: Jour. Geophys. Res., v. 71, p. 2121-2137.

———, Den, N., and Murauchi, S., 1973, Seismic reflection measurements of southwest Japan margin: Jour. Geophys. Res., v. 78, p. 2508-2516.

Matsuda, T., and Uyeda, S., 1970, On the Pacific-type orogeny and its model—Extension of the paired belts concept and possible origin of marginal seas: Tectonophysics, v. 11, p. 5-27.

———, Nakamura, K., and Sugimura, A., 1967, Late Cenozoic orogeny in Japan: Tectonophysics, v. 4, p. 349-366.

Matuyama, M., 1936, Distribution of gravity over the Nippon Trench and related areas: Proc. Imp. Acad. Tokyo, v. 12, p. 93-95.

McKenzie, D., and Parker, R. L., 1967, The north Pacific: an example of tectonics on a sphere: Nature, v. 216, p. 1276.

Minato, M., Gorai, M., and Hunabashi, M., 1965, The geologic development of Japanese Islands: Tokyo, Tsukiji Shoka.

Minear, J. W., and Toksoz, H. N., 1970, Thermal regime of a downgoing slab and new global tectonics: Jour. Geophys. Res., v. 75, p. 1397-1419.

Miyamura, S., 1972, Natural earthquake, *in* Miyamura, S., and Uyeda, S., eds., Crust and upper mantle of the Japanese area, pt. 1, Geophysics: Japanese Comm. for Upper Mantle Project, p. 15-44.

Miyashiro, A., 1961, Evolution of metamorphic belts: H. Petrol., v. 2, p. 277-311.

———, 1967, Orogeny, regional metamorphism and magmatism in the Japanese Islands: Medd. Dansk Geol. Foren., v. 17, p. 390-446.

Mizutani, H., Baba, K., Kobayshi, N., Chang, C. C., Lee, C. H., and Kang, Y. S., 1970, Heat flow in Korea: Tectonophysics, v. 10, p. 183-204.

Mogi, A., 1972, Bathymetry of the Kuroshio region, *in* Kuroshio—its physical aspects: Tokyo, Univ. Tokyo Press, p. 53-79.

Mogi, K., 1968, Sequential occurrences of recent great earthquakes: Jour. Phys. Earth, v. 16, p. 30-36.

Morgan, W. J., 1968, Rises, trenches, great faults and crustal blocks: Jour. Geophys. Res., v. 73, p. 1959-1982.

———, 1971, Convection plumes in the lower mantle: Nature, v. 230, p. 42-43.

Murauchi, S., 1966, On the origin of the Japan Sea: read at Earthquake Res. Inst. Tokyo Univ. Monthly Meeting.

———, 1971, The renewal of island arcs and the tectonics of marginal seas, *in* Asano, S., and Udintsev, G. B., eds., Island arc and marginal sea: Tokyo, Tokai Univ. Press, p. 39-56 (in Japanese with Engl. abstr.).

———, and Asanuma, T., 1970, Studies on seismic profiler measurements off Boso-Jyoban district, northeast Japan: Mem. Natl. Sci. Museum, v. 13, no. 2, p. 337-356 (in Japanese).

———, and Yasui, M., 1968, Geophysical investigations in the seas around Japan: Kagaku, v. 38, no. 4 (in Japanese).

———, Den, N., Asano, S. Hotta, H., Yoshii, T., Asanuma, T., Hagiwara, K., Ichikawa, K., Sato, T., Ludwig, W. J., Ewing, J. I., Edgar, N. T., and Houtz, R. E., 1968a, Crustal structure of the Philippine Sea: Jour. Geophys. Res., v. 73, p. 3143-3171.

———, Asanuma, T., and Hotta, H., 1968b, Studies on sediments off Sanriku area by seismic profiler: Natl. Sci. Museum, no. 1, p. 37-40 (in Japanese).

———, Asanuma, T., and Ishii, H., 1973, Geophysical studies on the sea round Hokkaide: Mem. Natl. Sci. Museum, no. 6, p. 163-182.

Nagumo, S., et al., 1970a, Ocean bottom seismographic observation at the off-side of Japan Trench near the Erimo Seamount: Bull. Earthquake Res. Inst., v. 48, p. 769-792.

———, et al., 1970b, Ocean bottom seismographic observation off-Sanriku: Bull. Earthquake Res. Inst., v. 48, p. 793-809.

———, Kasahara, J., and Koresawa, S., 1974, Structure of microearthquake activity around Japan Trench, off Sanriku, obtained by ocean-bottom seismographic network observation (preprint).

Nasu, N., Iijima, A., and Kagami, H., 1960, Geological results in the Japanese Deep Sea Expedition in 1959: Oceanogr. Mag., v. 11, p. 201-214.

Oliver, J., and Isacks, B., 1967, Deep earthquake zones, anomalous structure in the upper mantle, and the lithosphere: Jour. Geophys. Res., v. 72, p. 4259-4275.

Packham, G. H., and Falvey, D. A., 1971, An hypothesis for the formation of marginal seas in the western Pacific: Tectonophysics, v. 11, p. 79-109.

Pavlov, U. S., Sychev, P. M., Tuyesou, I. K., Gainanov, A. G., and Stroev, P. A., 1971, Gravity anomaly fields in far east seas and Northwest Pacific, *in* The crust of island arcs and Far East seas: Moscow, Akad. Nauk.

Rikitake, T., 1969, The undulation of an electrically conductivity layer beneath the islands of Japan: Tectonophysics, v. 7, p. 257-264.

Rodolfo, K. S., 1969, Bathymetry and marine geology of the Andaman Basin and tectonic implications for southeast Asia: Geol. Soc. America Bull., v. 80, p. 1203-1230.

Sato, T., 1973, Several considerations on the deep sea plains: Marine Sci. Monthly, v. 5, no. 10, p. 55-59 (in Japanese with Engl. abstr.).

Scholl, D. W., von Huene, R., and Ridlon, J. B., 1968, Spreading of the ocean floor: underformed sediments in the Peru-Chile Trench: Science, v. 159, p. 869-871.

———, Christensen, M. N., von Huene, R., and Marlow, M. W., 1970, Peru-Chile Trench sediments and seafloor spreading: Geol. Soc. America Bull., v. 81, p. 1339-1360.

Segawa, J., 1968, Measurement of gravity at sea around Japan: off southwestern part of Japan and east China Sea: Bull. Geod. Soc. Japan, v. 13, p. 53-65.

Shibata, K., 1968, K-Ar age determinations on granitic and metamorphic rocks in Japan, Geol. Survey Japan, Rept. 227.

Sleep, N., and Toksoz, M. N., 1971, Evolution of marginal basins: Nature, v. 233, p. 548-550.

Soloviev, O. N., and Gainanov, A. G., 1963, Geological structure in the zone of transition from the Asiatic continent to the Pacific Ocean in the region of Kuril-Kamchatka Island arc: Soviet Geology, v. 3, p. 113-123.

Sugimura, A., 1972, Plate boundaries around Japan: Kagaku, v. 42, p. 192-202 (in Japanese).

———, and Uyeda, S., 1973, Island arcs: Japan and its environs: New York, Elsevier.

Sykes, L. R., 1971, Aftershock zones of great earthquakes, seismicity gaps, and earthquake prediction for Alaska and the Aleutians: Jour. Geophys. Res., v. 70, p. 8021-8041.

Tomoda, Y., 1973, Gravity anomalies in the Pacific Ocean, *in* Coleman, P. J., ed., The western Pacific: island arcs, marginal seas, geochemistry: Univ. Western Australia Press, p. 5-20.

———, Segawa, J., and Tokuhiro, A., 1970, Free-air gravity anomaly around Japan: Proc. Japan Acad., v. 46, p. 1006-1010.

Tsuchi, R., and Kuroda, N., 1973, Erimo (Sysoev) seamount and its relation to the tectonic history of the Pacific Ocean Basin, *in* Coleman, P. J., ed., The western Pacific: island arcs, marginal seas, geochemistry: Univ. Western Australia Press, p. 57-64.

Turcotte, D. L, and Oxburgh, E. R., 1969, A fluid theory for the deep structure of dip-slip fault zones: Phys. Earth Planetary Interiors, v. 1, p. 381-386.

Udintsev, G. B., 1955, Topography of the Kuril-Kamchatka Trench: Tr. Inst. Okeanol., v. 12, p. 16-61 (in Russian).

———, 1959, Relief of abyssal trenches in the Pacific Ocean (abstr.): Internatl. Oceanogr. Congr. Preprints, Am. Assoc. Adv. Sci., Washington, D.C.

Utsu, T., 1967, Anomalies in seismic wave velocity and attenuation associated with a deep earthquake zone (I): Jour. Fac. Sci. Hokkaido Univ., ser. VII, v. 2, p. 359-374.

———, 1971, Seismological evidence for anomalous structure of island arcs with special reference to the Japanese region: Rev. Geophys. Space Phys., v. 9, p. 839-890.

Uyeda, S., 1972, Heat Flow, *in* Miyamura, S., and Uyeda, S., eds., Crust and upper mantle of the Japanese area, pt. i, Geophysics: Japanese Comm. for Upper Mantle Project, p. 97-105.

———, and Ben-Avraham, Z., 1972, Origin and development of the Philippine Sea: Nature Phys. Sci., v. 240, p. 176-178.

———, and Horai, K., 1964, Terrestrial heat flow in Japan: Jour. Geophys. Res., v. 69, p. 2121-2141.

———, and Miyashiro, A., 1974, Plate tectonics and Japanese Islands, Geol. Soc. America Bull. (in press).

———, and Richards, M., 1966, Magnetization of four Pacific seamounts near the Japanese Islands: Bull. Earthquake Res. Inst., v. 44, p. 179-213.

———, and Rikitake, T., 1970, Electrical conductivity anomaly and terrestrial heat flow: Jour. Geomag. Geoelectr., v. 22, p. 75-90.

———, and Vacquier, V., 1968, Geothermal and geomagnetic data in and around the Island Arc of Japan, *in* Knopoff, L, et al., eds., The crust and upper mantle of the Pacific Area, Geophys. Monogr. 12: Washington, D.C., Am. Gephys. Union, p. 349-366.

———, Vacquier, V., Yasui, M., Sclater, J., Sato, T., Lawson, J., Watanabe, T., Dixon, F., Silver, E., Fukao, Y., Sudo, K., Nishikawa, M., and Tanaka, T., 1967, Results of geomagnetic survey during the cruise of R/V ARGO in western Pacific, 1966, and the compilation of magnetic charts of the same area: Bull. Earthquake Res. Inst., v. 45, p. 799-814.

Vacquier, V., and Uyeda, S., 1967, Paleomagnetism of nine seamounts in the western Pacific and of three volcanoes in Japan: Bull. Earthquake Res. Inst., v. 45, p. 815-848.

Vening Meinesz, F. A., 1947, Major, tectonic phenomena and the hypothesis of convection currents in the earth: Geol. Soc. London Quart. Jour., v. 103, p. 191-207.

Watanabe, T., 1973, Heat flow in the western margin of the Pacific [thesis]: Univ. Tokyo.

———, Epp, D., Uyeda, S., and Langseth, M., 1970, Heat flow in the Philippine Sea: Tectonophysics, v. 10, p. 205-224.

Watts, A. B., and Talwani, M., 1973, Gravity anomalies seaward of deep-sea trenches and their tectonic implications (preprint).

Worzel, J. L., 1965, Pendulum gravity measurements at sea, 1936-1959: New York, Wiley.

Yasui, M., Kishii, T., Watanabe, T., and Uyeda, S., 1968, Heat flow in the Sea of Japan, *in* Knopoff, L., et al., eds., The crust and upper mantle of the Pacific area, Geophys. Monogr. 12: Washington, D.C., Am. Geophys. Union, p. 3-16.

———, Epp, D., Nagasaka, K., and Kishii, T., 1970, Terrestrial heat flow in the seas around the Nansei Shoto (Ryukyu Islands): Tectonophysics, v. 10, p. 225-234.

Yoshii, T., Ludwig, W. J., Den, N., Murauchi, S., Ewing, M., Hotta, H., Buhl, P., Asanuma, T., and Sakajiri, N., 1973, Structure of southwest Japan Margin off Shikoku: Jour. Geophys. Res., v. 78, p. 2517-2525.

shallowest, about 4,700 m, and from its eastern end the thalweg of the trench slops westward for a distance of about 2,300 km, to its deepest sector, located in the vicinity of the southernmost swing (178°W-178°E) of the trench's arcuate path. From here, the floor of the trench climbs gently to the west, but, near its western end, the thalweg ascends more rapidly to intersect the floor of the Kuril-Kamchatka Trench at a depth near 5,500 m (Fig. 4). The trench's longitudinal profile (which has been a key factor influencing the distribution of its axial deposits), the stratigraphic relationships between major subbottom sedimentary sequences, and the longitudinal configuration of its deeply buried oceanic basement, are illustrated diagrammatically in Figure 4.

Originally surmised by von Huene and Shor (1969) and von Huene (1972), and subsequently established at DSDP sites 178 and 180 (Fig. 1; Kulm et al., 1973b), the sedimentary section underlying the floor of the eastern Aleutian Trench can be divided into a lower hemiterrigenous sequence and an overlying terrigenous one (the turbidite wedge; Fig. 4). Along much of the eastern trench segment the seaward-thinning wedge is 30-40 km wide, and, adjacent to the trench's inner wall, approximately 1,000 m thick. Unlike the Washington-Oregon Trench, the turbidite deposits of the Aleutian wedge do not extend outward to (or beyond) the seaward limits of the structural trench; hence a flat-floored geomorphic trench has been preserved above the axis of the structural one (Figs. 2 and 4). Beneath this flat floor, the hemiterrigenous sequence is also about 1 km thick; here the sequence rests conformably on basaltic crust and slopes gently landward (1-3°). South of the trench the hemiterrigenous sequence can be traced updip to the Alaskan Abyssal Plain, which occupies a large area of the Gulf of Alaska (Figs. 1 and 4; Hurley, 1960; Hamilton, 1967, 1973; Ness and Kulm, 1973). The structural framework of the eastern Aleutian Trench, and the sequence of sedimentary deposits partly filling it, are noticeably similar to those of the Washington-Oregon Trench.

At DSDP site 180, located near the southern edge of the trench's flat floor (Fig. 4), the turbidite wedge is approximately 460 m thick and consists chiefly of mud and silt deposits. Because deep-sea channels occur along the inner or northern side of the trench floor, and the wedge is thickest here,

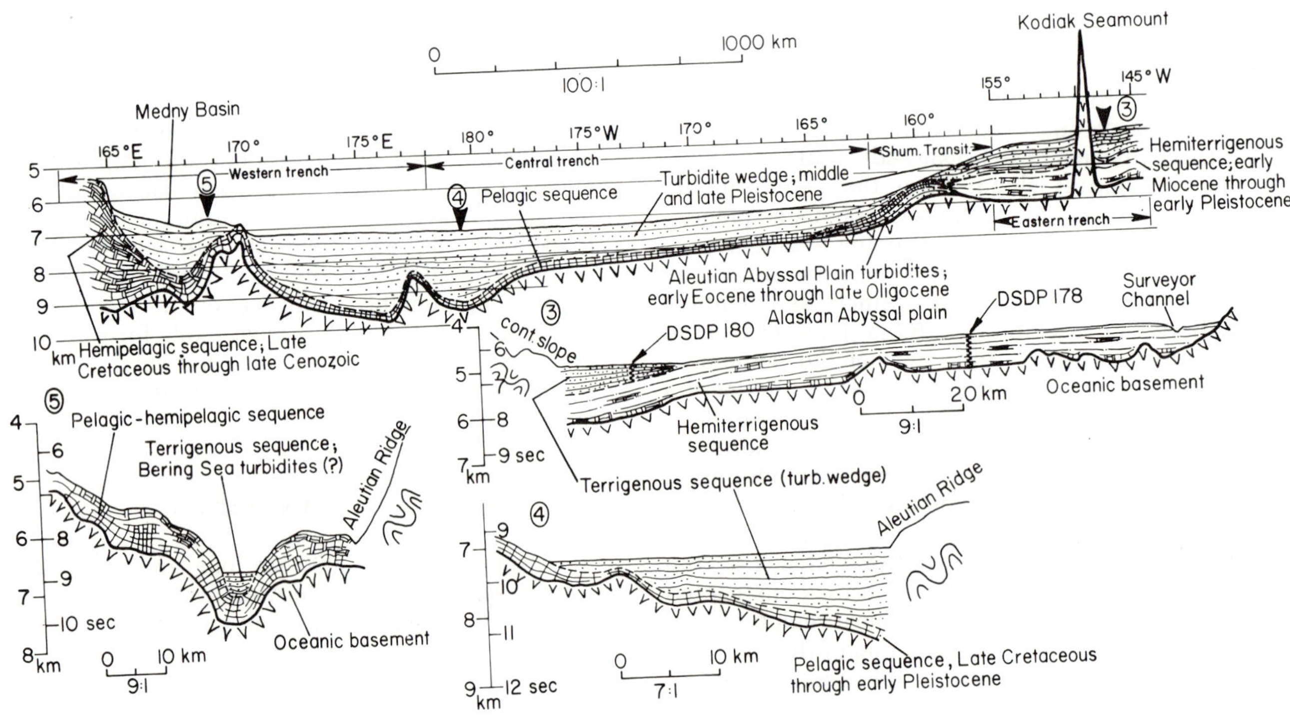

Fig. 4. Generalized drawings of reflection profiles 3, 4, and 5, which traverse the eastern, central, and western segments of the Aleutian Trench, respectively (Fig. 1); and a diagrammatic representation of the trench's longitudinal cross section (upper drawing). (Profile 3 is adapted from von Huene and Kulm, 1973; profile 4 is adapted from Holmes et al., 1972; profile 5 is adapted from Buffington, 1973.) The longitudinal section is approximately along the trench's thalweg, which typically lies adjacent to the base of the continental or insular slope and above the thickest part of the turbidite wedge. Sound-velocity-corrected thicknesses were used in constructing the cross section. However, reflection profiles 3, 4, and 5 are uncorrected drawings; their depth scales are for water only (1,500 m/sec).

turbidites overlying the trench's structural axis are presumably richer in beds of sand and coarse silt (Piper et al., 1973; see also von Huene, this volume). Of special importance is the observation by von Huene and Kulm (1973) and Piper et al. (1973) that the 1-km-thick terrigenous wedge is a lithofacies of the upper 120-130 m of silty and clayey terrigenous and hemiterrigenous beds underlying the adjacent Alaskan Abyssal Plain (Fig. 4). These beds, which are rich in ice-rafted debris, accumulated during the past 0.7 my, during the corresponding time-equivalent (but much thicker) facies of the turbidite wedge formed during middle and late Pleistocene time (von Huene and Kulm, 1973). The average sedimentation rate for the thickest part of the wedge is 1,780 m/my; during glacial epochs the rate increased to more than 3,000 m/my (von Huene and Kulm, 1973; see also von Huene, this volume). These high rates were in part maintained by the reworking of large slump masses that cascaded into the trench from its landward wall (von Huene, 1972; Piper et al., 1973). The terrigenous detritus of the wedge was chiefly derived from adjacent and extensively glaciated southern Alaska; only a small fraction, about 7% reached the trench from the east and south via distal transportation across the Alaskan Abyssal Plain (Piper et al., 1973; Ness and Kulm, 1973).

The hemiterrigenous sequence underlying the terrigenous section of the turbidite wedge was penetrated and sampled at DSDP site 178, located about 70 km south of the trench axis along the northern perimeter of the Alaskan Abyssal Plain (Figs. 1 and 4). Because the thickness of the hemiterrigenous sequence at site 178 (about 800 m) is equivalent to that underlying the trench, it is presumed that the stratigraphic relations found at site 178 are applicable to the axial area of the trench. The sequence's basal strata (50-60 m) are pelagic beds of claystone and chalk of early Miocene and possibly older age. Overlying these pelagic units are 400-500 m of muddy sediment interbedded with silt and fine-sand turbidite and diatomaceous layers, a succession of beds of middle Miocene through early Pliocene age. Above this section are middle Pliocene through early Pleistocene beds, probably about 150 m thick, composed of terrigenous and hemiterrigenous mud rich in ice-rafted debris; these strata are probably not turbidites (Horn et al., 1969, 1970; Gershanovich, 1968). Younger beds belong to the thick overlying terrigenous sequence of the turbidite wedge (Fig. 4). Because the lower hemiterrigenous sequence does not thicken toward the base of the continental slope, beneath the trench floor this turbidite-dominated sequence cannot represent an axial trench accumulation. Clearly, as von Huene and Kulm (1973) state, the pelagic-rich but chiefly terrigenous beds of this sequence formed beneath part of the Alaskan Abyssal Plain (see also Hamilton, 1973; Ness and Kulm, 1973).

Separating the eastern and central segments of the Aleutian Trench is a narrow, relatively steeply sloping (westward) trench segment (155°-162°W), the Shumagin transition of von Huene and Shor (1969) (Figs. 1 and 4). Reflection profiles reveal a poorly formed, or locally absent, turbidite wedge (Ewing et al., 1968; Hayes and Ewing, 1970; unpublished records, author's file). Sedimentary units overlying the axial zone of the structural trench are landward-dipping turbidites of Eocene and Oligocene age conformably overlain by an equal thickness of Neogene pelagic beds (Fig. 4, 160°W). Together about 88 m thick, these terrigenous and pelagic sequences can be traced seaward up the southern flank of the trench to DSDP site 183, located 80 km south of the trench along the northern perimeter of the Aleutian Abyssal Plan (Creager et al., 1973; Scholl and Creager, 1973). As the studies of Hamilton (1967), Grim and Naugler (1969), Mammerickx (1970), Jones (1971), and especially Hamilton (1973) imply, the older sedimentary sequences underlying the floor of the Shumagin transition are not axial trench deposits but tectonically downwarped pelagic and terrigenous sequences of the Aleutian Abyssal Plain (Fig. 4). Local absence of the turbidite wedge, a characteristic feature of the Aleutian Trench, can be ascribed to the greater westward slope of the transition area. Although the slope is gentle, 0.1°, it is evidently sufficient to promote sediment bypassing rather than deposition by westward-flowing axial turbidity currents.

As the reflection records of Ewing et al. (1968), Hayes and Ewing (1970), Holmes et al. (1973), Marlow et al. (1973a, 1973b), Grow (1973), Buffington (1972), and Scholl and Marlow (1974; in press a) document, the flat floor of the central segment of the Aleutian Trench and nearly all of the eastern segment is underlain by a thick section of late Cenozoic turbidites (Fig. 4). Between about 162°W and 179°E, a distance of about 2,000 km, this terrigenous sequence forms a wedge-shaped body that thins seaward against the trench's southern flank. The turbidite wedge is thickest, typically 1,000-2,000 m, at the base of the southern margin of the Aleutian Ridge, i.e., above the trench's structural axis (Fig. 4). The flat-lying deposits of the wedge rest on an older pelagic sequence (Horn et al., 1970) that dip landward parallel to the surface of the underlying oceanic crust (Hayes and Ewing, 1970). Greatly compacted by the superimposed wedge, the pelagic sequence is typically 200-400 m thick and comprises beds of latest Cretaceous through late Cenozoic age (Fig. 4).

The trench's lengthwise or longitudinal cross section (Fig. 4) shows that the turbidite wedge is absent near 170°E. In this area the smooth flat floor of the trench is disrupted and partly constricted by a deeply buried basement knoll. The crest of the knoll is covered by a 500-m-thick blanket of pelagic beds. However, west of the knoll, which may have served as a dam for axial-flowing turbidity currents, a thick (500-1,500 m) section of turbidites underlies Medny Basin (Fig. 4), the flat-floored trench sector lying south of the Komandorsky Islands (Figs. 1). As Buffington (1973) points out, the Medny Basin sector of the Aleutian Trench is not underlain by a landward-dipping basement, but by a graben-shaped trough in the oceanic crust (Fig. 4, profile 5). The axis of the structural trench therefore lies about 20 km seaward of the base of the Aleutian Ridge. As Figure 2 illustrates, the structural axis (i.e., the loci of points of greatest depth to basement) of north Pacific trenches is more typically located directly below the base of their flanking continental and island-arc margins, which the landward-dipping oceanic crust plunges beneath. Because of its unusual basement configuration, the transverse shape of the Medny Basin turbidite section is trough- rather than wedge-shaped.

Beginning beneath Medny Basin, the thickness of the lower pelagic sequence appears to increase markedly to the west (Fig. 4). At the western end of the trench (164°E), below the rapidly ascending trench floor, the sequence thickens to about 3,000 m. Drilling at nearby DSDP site 192 (Fig. 1; Creager et al., 1973), surface cores (Conolly and Ewing, 1970), and the regional correlations of Buffington (1973) establish that this greatly thickened sequence is a hemipelagic one that chiefly involves Neogene beds. Beneath the trench the basal beds of the sequence are presumably pelagic strata of chalk and claystone of late Cretaceous through Eocene and possibly Oligocene age. The overlying beds, which comprise about 80% of the section, are terrigenous mudstone and diatomaceous mudstone of middle Miocene through early late Miocene age, diatom ooze of late Miocene and Pliocene age, and younger hemiterrigenous beds of detrital silt and clay, rich in diatomaceous, volcanic ash, and ice-rafted debris (Creager, et al., 1973). Where the slope of the trench's thalweg is relatively steep, late Cenozoic turbidites are missing above the hemipelagic sequence. The hemipelagic sequence itself is part of a thick, widespread blanket of similar deposits mantling the northwestern corner of the Pacific basin (Ewing et al., 1968; Buffington, 1973; Scholl and Creager, 1973).

Although the thick, turbidite deposits of the central and western segments of the Aleutian Trench have not been drilled, it is very likely of Pleistocene age and probably began to form within the last million years (Marlow et al., 1973b; Scholl and Marlow, 1974; in press a). Because the Aleutian Terrace, a deep (3,500-5,000 m),50-km-wide basin-shaped platform, disrupts the trenchward slope of the Aleutian Ridge (Grow, 1973; Scholl and Creager, 1973, p. 906), the bulk of the terrigenous debris filling the central and much of the western trench segments was presumably derived from Alaska. The implication is that a large quantity (about 17,000 km^3; Scholl and Marlow, 1974) of the erosional detritus originally contributed to the trench's eastern segment from immediately adjacent southern Alaska, and from southeastern Alaska via Surveyor Channel (Fig. 1; Piper et al., 1973; Ness and Kulm, 1973; Hamilton, 1973), was ultimately conveyed to the central segment by westward-flowing turbidity currents. These currents, which swept down the Shumagin transition, continued westward along the smoothly graded trench floor for a distance of abut 1,400 km to its deepest sector. West of here (178°W-178°E), as Figure 4 reveals, the trench floor shallows. The shoaling is slight, amounting to only about 80 m in 800 km, and could be a product of Holocene tectonics; hence it is reasonable to suspect that westward-flowing turbidity currents may have contributed significant amounts of Alaskan detritus as far west as about 170°E.

There is little chance that the terrigenous beds of the turbidite sequence underlying Medny Basin could have been derived from Alaska. Possibly the Komandorsky region of the Aleutian Ridge supplied these deposits, a sector not skirted by a well-formed Aleutian Terrace. More likely, they were contributed by turbidity currents originating in the Bering Sea. The Kamchatka Basin region of the Bering Sea is open to the Pacific via Kamchatka Gap, the 4,500-m-deep pass (Lisitsyn, 1966) separating the Aleutian Ridge and the Kamchatka Peninsula (Fig. 1). Reflection profiles (Ludwig et al., 1971; Horn et al., 1972; Creager et al., 1973; Scholl et al., 1974) and drilling at DSDP site 191 (Scholl and Creager, 1973) establish that during the middle of late Pleistocene the floor of Kamchatka Basin was flooded with sandy and silty turbidites derived from northern Kamchatka and eastern Siberia. Some of this detritus may have exited the basin via Kamchatka Gap, swinging eastward and following the eastward-sloping trough of the Aleutian Trench to Medny Basin. The subtrench basement knoll at 170°E could have been bypassed by these currents. Accordingly, Bering Sea sediment may have been transported eastward as far as the trench's central segment.

When roughly corrected for compaction, the thickness of the pelagic sequence underlying the turbidites of the central and western segments of the Aleutian Trench is not noticeably different from thicknesses immediately seaward of its axis. This may mean that the subtrench pelagic beds, which range in age from late Cretaceous through early Pleistocene, may not constitute an *in situ* or axially deposited sequence. As discussed elsewhere below, this interpretation may not be justified because, except locally, pelagic and even hemipelagic sequences may not thicken axially. However, it is likely that at least the upper part of the extraordinarily thick hemipelagic sequence underlying the western end of the trench is in part related to depositional conditions controlled by the topography of a preexisting trench. The important terrigenous component of this sequence was derived from nearby Kamchatka and, via the Kamchatka current (Lisitsyn, 1966; Conolly and Ewing, 1970; Reid, 1973), from eastern Siberia. These adjacent areas began to contribute large quantities of pelitic detritus to the northwestern Pacific in the middle Miocene, possibly in the early Miocene (Buffington, 1973; Scholl and Creager, 1973).

KURIL-KAMCHATKA TRENCH

Approximately 2,000 km in length, the gently eastward-convex Kuril-Kamchatka Trench fringes the northwestern margin of the Pacific (55.5°-42°N) and lies about 180 km seaward of Kamchatka, the Kuril Islands, and Hokkaido (Udinstev, 1954; 1955; Chase et al., 1970). The Kuril-Kamchatka Trench is the deepest of the north Pacific trenches; all its floor lies below 6 km and much (about 40%) below 8 km. Vitiaz Deep (44.3°N; Fig. 1), the trench's maximum depression, descends to 10,540 m (Menard, 1964), a depth comparable to the greatest known beneath the world's oceans (11,022 m; Mariana Trench). The great depth of the Kuril-Kamchatka Trench, and its U to V shaped transverse profile, signify that it lacks a thick sedimentary fill, specifically a turbidity wedge that is so typical of the Aleutian Trench (Fig. 2; Scholl and Marlow, 1974).

At the northern, relatively shallow end of the Kuril-Kamchatka Trench, its floor is probably underlain by several thousand meters of hemipelagic debris. Southward, seaward of the northeastern coast of Kamchatka, the axial region of the trench is underlain by a hemipelagic sequence as much as 2 km thick; however, south of about 53°N the thickness decreases to about 500 m (Fig. 5; Buffington, 1973). The sequence can be traced eastward to the summit of nearby Meiji Guyot; here the sequence, approximately 1,000 m of biogenic ooze and terrigenous mud, has been penetrated and sampled at DSDP site 192 (Fig. 1; Creager et al., 1973). Drawing of this information, and regional correlations or stratigraphic units identified on seismic reflection profiles, Buffington (1973) established that the lower part of the trench's hemipelagic sequence is of late Cretaceous age (Maestrichtian). The remainder and the bulk of the sequence accumulated after the early Miocene.

Opposite the southern tip of Kamchatka (49°N), E. L. Hamilton (written communication, 1972) reports a thin turbidite sequence evidently overlying a thicker hemipelagic or pelagic one. However, farther south, the reflection records of Reynolds (1970) show only a relatively thin (300-400 m) mantle of pelagic (possible hemipelagic) beds in the Kuril Island segment of the trench. Seaward of Hokkaido, short segments of the southernmost reaches of the trench have a thin turbidite sequence (Fig. 5; Reynolds, 1970). Interestingly, as Figure 5 illustrates, in some sectors the turbidites are ponded in a trough located about 15 km seaward of the base of the insular slope. The inner wall of the trough is a

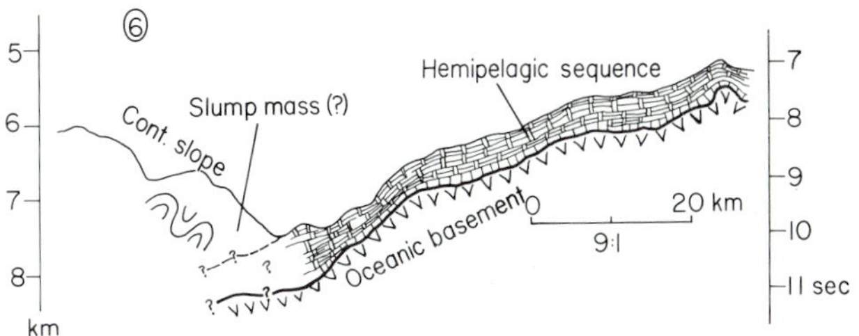

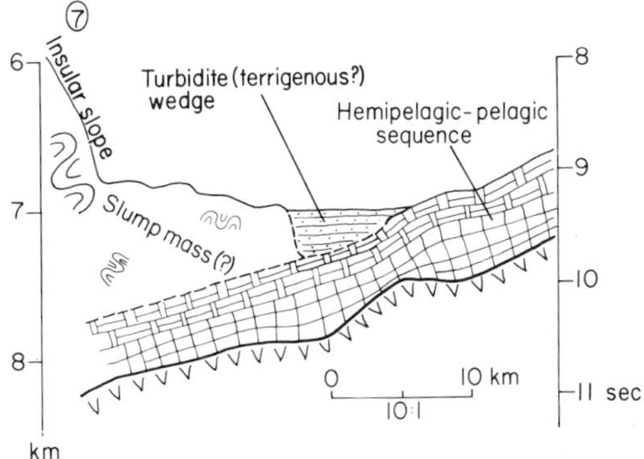

Fig. 5. Generalized drawings of acoustic reflection records across Kuril-Kamchatka Trench (Fig. 1.). (Profile 6 is adapted from Buffington, 1973; profile 7 is adapted from Reynolds, 1970.)

rampart of deformed strata apparently resting on beds of the pelagic or hemipelagic sequence. The rampart is either slumped masses of insular slope deposits, or possibly offscrapings related to subduction of Pacific lithosphere (Buffington, 1973).

The all-but-missing turbidite wedge from the axis of the Kuril-Kamchatka Trench has important implications, especially its absence seaward of the extensively glaciated mountains of Kamchatka. Evidently, as suggested by Buffington (1973) and Scholl and Marlow (1974), coarse, clastic detritus shed eastward from Kamchatka was mostly trapped in the large midslope basins (the "Kamchatka Terrace") that disrupts the seaward descent of the continental margin (Fig. 1; Chase et al., 1970). Also, as earlier speculated, if turbidity currents exited from the Bering Sea via Kamchatka Gap, they evidently did not flow southward into the northern head of the trench.

GEOLOGICAL IMPLICATIONS OF TRENCH SEQUENCES

Geometrically distinctive (as recorded on reflection profiles), axial deposits of north Pacific trenches are virtually restricted to those of the turbidite wedge, a terrigenous sequence of middle through late Pleistocene age (Fig. 2). An exception may be the thickened mass of hemipelagic beds underlying the western reaches of the Aleutian Trench (Fig. 4). As previously emphasized, the axial thickness of older (pre-Pleistocene) pelagic, hemipelagic, or hemiterrigenous sequences is not appreciably different from that typical of the trench's seaward flank. It may be supposed, therefore, that north Pacific trenches are either exceptionally young geologic features, or that they are long-established structures that through tectonic processes continually lose axially distinctive depositional sequences.

Two factors favor the latter interpretation for the missing axial wedges that would correspond to the landward-dipping turbidite-dominated hemiterrigenous sequences of the Washington-Oregon and eastern Aleutian trenches. First, regional tectonic considerations (based on extrapolated plate motions, Pitman and Hayes, 1968; Atwater, 1970, Grow and Atwater, 1970; Hayes and Pitman, 1970; Silver, 1971a; Marlow et al., 1973b; Scholl and Creager, 1973; Scholl et al., 1974) imply that most segments of north Pacific trenches have been in existence since at least the middle Miocene. Second, evidence has been recognized for tectonically offscraped ocean-floor deposits along the inner wall of these trenches. For example, it seems certain from the investigations of Silver (1969, 1971c, 1972), Barr (1972), Kulm et al., (1973a), and von Huene and Kulm (1973) that the lower hemiterrigenous sequence of the Washington-Oregon Trench has been deformed and elevated to form the base of the continental slope. Some of the uplifted beds include those of the overlying axial wedge of Pleistocene turbidites (see especially Silver, 1972, Fig. 3; Carson, 1973, Fig. 6). Evidently, older axial trench deposits are absent because they have been tectonically welded into the base of the continental slope by the underthrusting action of the subadjacent oceanic lithosphere (von Huene and Kulm, 1973, Fig. 3).

Unfortunately, evidence for offscraping (or subduction) of trench sequences is equivocal for the Aleutian Trench. Although, as von Huene and Kulm (1973) note, a reasonable argument can be made that part of the turbidite wedge has been accreted tectonically to the inner wall of the eastern Aleutian Trench, a circumstance possibly also true for its central segment (Holmes, et al., 1972; Grow, 1973), information gathered and interpreted by Marlow et al. (1973a, 1973b), Hamilton (1973), and Scholl and Marlow (in press a) indicate considerable doubt that any significant offscraping has occurred along and west of the Shumigan transition (Figs. 1 and 4).

Relative to the tectonic implications of the missing axial hemiterrigenous deposits, a mitigating factor is the likelihood that the existing turbidite wedge is a special deposit of glacial age (Scholl et al., 1968, 1970). Scholl and Marlow (1974, in press a) have proposed that the youthfulness (1.0-0.5 my) of the turbidite wedge is attributable to the initiation, in the early middle Pleistocene, of extensive continental glaciation (Kent et al., 1971), which greatly lowered sea level and enabled rivers burdened with the products of glacial denudation to debouch at the summit of the continental slope. This resulted in the rapid filling of the Washington-Oregon and Aleutian trenches with detritus conveyed into them by turbidity currents and slower downslope slumping and gliding of larger sediment bodies (Piper et al., 1973). Conceivably, in the absence of glacially lowered sea levels and greatly increased continental erosion rates, thick wedges might not have formed prior to the middle Pleistocene, or thin ones may have formed only locally. The tacit assumption by many geologists that axial turbidite sequences formed in Tertiary trenches must therefore be questioned. If, in general, the wedges did not form, then tectonic deductions concerning their absence in modern trenches may be partly misleading. This same cautionary note applies to the recognition of ancient trench sequences, which are presumed by many geologists to be exposed in belts of deformed eugeosynclinal rocks (Burk, 1972; Scholl and Marlow, 1972, 1974, in press a, in press b).

An equally significant note should be added here concerning the efficacy of midslope basins (e.g., the Aleutian and Kamchatka terraces) to trap terrigenous detritus shed toward adjacent trenches. Menard (1964, p. 113), Horn et al. (1969, 1972), and Scholl and Marlow (1972, 1974, in press b) stress the geologic difficulties involved in filling trenches with coarse, terrigenous detritus if intervening upslope basins are present.

Midslope basins are largely absent opposite the Washington-Oregon Trench and, in part because of this, it has been completely filled with turbidites of glacial age. The eastern Aleutian Trench, also lacking an adjacent midslope bench, would have been similarly filled had not the sediment-dispersing turbidity currents been shunted to the west (via the Shumigan transition) into the deeper central and possibly the western segments of the trench. The smoothly graded, westward-inclined axial floor of the central Aleutian Trench implies that very little glacial-age detritus was derived from the adjacent Aleutian Ridge, which borders the western 2,400 km of the trench. The ridge is flanked by an exceptionally large midslope terrace, the Aleutian Terrace, which trapped the bulk of the erosional debris shed southward toward the trench (Scholl and Marlow, 1974). It is further likely that turbidites partly filling the western reaches of the trench, i.e., Medny Basin (Fig. 4), represent the outflow of turbidity currents from the Bering Sea via Kamchatka Gap (Fig. 1). In fact, the trench's western and central segments would probably lack a thick section of glacial-age turbidites had not turbidity currents been able to enter the Aleutian Trench laterally.

Except in local areas, the Kuril-Kamchatka Trench lacks a terrigenous sequence of glacial-age turbidites. As noted by Buffington (1973) and Scholl and Marlow (1974), this circumstance probably reflects the trapping of erosional detritus by the Kamchatka Terrace, actually a series of large midslope basins disrupting the trenchward descent of the continental slope. Because coarse terrigenous debris did not reach the floor of the trench, it lacks a geometrically distinctive axial deposit.

Unlike turbidite-dominated sections (e.g., many terrigenous and hemiterrigenous sequences), there is no particular reason why pelagic or perhaps even hemipelagic deposits should thicken over the trench's axial region. This interpretation is based on the realization that whereas the rate of infilling pelitic detritus may increase as the axis is approached from its seaward side, the rate of deposition of biogenic ooze will correspondingly decrease, owing to solution of calcareous and siliceous matter settling through the greater water column. Currents may also be strengthened along the trench's axial region. The lack of an axially distinctive sedimentary sequence in the Kuril-Kamchatka Trench cannot therefore be construed as evidence that such deposits have been either offscraped or subducted, the substance of the argument offered that this must have taken place along the inner wall of the Washington-Oregon and possibly the eastern Aleutian trenches. Even reflection profiles of the inner wall of the Kuril-Kamchatka Trench provide only inconclusive evidence that these tectonic processes have occurred here. As the interpreted drawings in Figure 5 illustrate, the deformed strata forming the inner wall may simply be slumps of insular or continental slope deposits rather than offscraped pelagic or hemipelagic beds (Buffington, 1973). Perhaps, as von Huene (1972) laments for the eastern Aleutian Trench, evidence for deformation has been obscured by the slumps.

It is apparent that the study of the sedimentary sequences in modern Pacific trenches has in only a few areas provided easily interpreted evidence useful in deciphering the formation and tectonic histories of these structural features. However, and perhaps more importantly, these investigations have resulted in an improved understanding of the origin of their infilling sedimentary sequences and indirectly, thereby, identified those factors that controlled sedimentation in ancient trenches. Many geologists suspect that ancient trenches were an important, if not the major, geomorphic element of Mesozoic (and older) eugeosynclines. Obviously, the far-reaching concept that eugeosynclinal rock assemblages are related to the offscraping and partial subduction of the sedimentary sequences of these trenches can only be tested by the continued investigations of modern trenches. These vital studies must include the collection of better-quality geophysical data than have been published heretofore, and, most importantly, the drilling of several stratigraphic test holes through the entire sedimentary section underlying trenches and their inner walls.

BIBLIOGRAPHY

Atwater, T., 1970, Implications of plate tectonics for the Cenozoic evolution of western north America: Geol. Soc. America Bull., v. 81, p. 3511-3536.

Barr, S. M., 1972, Geology of the northern end of the Juan de Fuca Ridge and adjacent continental slope [dissert.]: Univ. of British Columbia, Dept. of Geology, 286 p.

Buffington, E. C., 1973, Aleutian-Kamchatka Trench convergence: an investigation of lithospheric plate interaction in the light of modern geotectonic theory [dissert.]: Univ. of Southern California, Dept. Geol. Sci., 364 p.

Burk, C. A., 1972, Uplifted eugeosynclines and continental margins: Geol. Soc. America Mem. 132, p. 75-85.

Carson, B., 1973, Acoustic stratigraphy, structure, and history of Quaternary deposition in Cascadia Basin: Deep-Sea Res., v. 20, p. 387-396.

Chase, R. L., and Tiffin, D. L., 1972, Queen Charlotte Fault-Zone, British Columbia: Intern. Geol. Congr., 24th Session, Montreal, sec. 8, Marine geology and geophysics, p. 17-27.

Chase, T. E., Menard, H. W., and Mammerickx, J., 1970, Bathymetry of the north Pacific: Scripps Inst. Oceanography and Inst. Marine Resources, Tech. Rept. Ser. TR-7, 10 sheets.

Conolly, J. R., and Ewing, M., 1970, Ice-rafted detritus in northwest Pacific deep-sea sediments, *in* Hays, J. D., ed., Geological investigations of the North Pacific: Geol. Soc. America Mem. 126, p. 219-231.

Creager, J. S., Scholl, D. W., et al, 1973, Initial reports of the Deep Sea Drilling Project, v. 19: Washington, D.C., U.S. Govt. Printing Office, 913 p.

Ewing, J., Ewing, M., Aitken, T., and Ludwig, W. J., 1968, North Pacific sediment layers measured by seismic profiling, *in* Knopoff, L., Drake, C. L., and Hart, P. J., eds., The crust and upper mantle of the Pacific Area: Geophys. Mem. 12, Washington, D.C., Am. Geophys. Union.

Gershanovich, D. W., 1968, New data on geomorphology and recent sediments of the Bering Sea and the Gulf of Alaska: Marine Geology, v. 6, p. 281-296.

Grim, P. J., and Naugler, F. P., 1969, Fossil deep-sea channel on the Aleutian Abyssal Plain: Science, v. 163, p. 383-386.

Grow, J., 1973, Crustal and upper mantle structure of the central Aleutian Arc: Geol. Soc. America Bull., v. 84, p. 2169-2192.

——, and Atwater, T., 1970, Mid-Tertiary tectonic transition in the Aleutian Arc: Geol. Soc. America Bull., v. 81, p. 3715-3722.

Hamilton, E. L., 1967, Marine geology of abyssal plains in the Gulf of Alaska: Jour. Geophys. Res., v. 72, p. 4189-4213.

——, 1973, Marine geology of Aleutian Abyssal Plain: Marine Geology, v. 14, p. 295-325.

Hayes, D. E., and Ewing, M., 1970, Pacific boundary structure, *in* Maxwell, A. E., ed., The sea, v.4, The earth beneath the sea—concepts, pt. 2, Regional observations: New York, Wiley-Interscience, p. 29-72.

Hayes, D. E., and Pitman, W. C., III, 1970, Magnetic lineations in the north Pacific, *in* Hays, J. D., ed., Geological investigations of the North Pacific: Geol. Soc. America Mem. 126, p. 291-314.

Holmes, M. L., von Huene, R. E., and McManus, D. A., 1972, Seismic reflection evidence supporting underthrusting beneath the Aleutian Arc near Amchitka Island: Jour. Geophys. Res. v. 77, p. 959-964.

Horn, D. R., Delach, M. N., and Horn, B. M., 1969, Distribution of volcanic ash layers and turbidites in the north Pacific: Geol. Soc. America Bull., v. 80, p. 1715-1724.

Horn, D. R., Horn, B. M., and Delach, M. N., 1970, Sedimentary provinces of the north Pacific, *in* Hays, J. D., ed., Geological investigations of the North Pacific: Geol. Soc. America Mem. 126, p. 1-21.

Horn, D. R., Ewing, J. I., and Ewing, M., 1972, Graded-bed sequences emplaced by turbidity currents north of 20°N in the Pacific, Atlantic and Mediterranean: Sedimentology, v. 18, p. 247-275.

Hurley, R. J., 1960, The geomorphology of abyssal plains in the northeast Pacific Ocean: Scripps Inst. Oceanography, SIO Ref. 60-7, 105 p. (unpubl. ms).

Jones, E. J. W., Ewing, J., and Truchan, M., 1971, Aleutian Plain sediments and lithospheric plate motion: Jour. Geophys. Res., v. 76, p. 8121-8127.

Kent, D., Opdyke, N. D., and Ewing, M., 1971, Climate change in the north Pacific using ice-rafted detritus as a climatic indicator: Geol. Soc. America Bull., v. 82, p. 2741-2754.

Kulm, L. D., von Huene, R., et al., 1973b, Initial reports of the Deep Sea Drilling Project, v. 18: Washington, D.C., U.S. Govt. Printing Office, 1077 p.

Lisitsyn, A. P., 1966, Recent sedimentation in the Bering Sea: Instit. Oceanol., S.S.S.R., Acad. Sci. (in Russian). English version: Israel Program Sci. Transl., Jerusalem; National Sci. Foundation, Washington, D.C., 1969, 614 p.

Ludwig, W. J., Houtz, R. E., and Ewing, M., 1971, Sediment distribution in the Bering Sea: Bowers Ridge, Shirshov Ridge, and enclosed basins: Jour. Geophys. Res., v. 76, p. 6367-6375.

Mammerickx, J., 1970, Morphology of the Aleutian Abyssal Plain: Geol. Soc. America Bull., v. 81, p. 3457-3464.

Marlow, M. S., Scholl, D. W., and Buffington, E. C., 1973a, Discussion of paper by M. L. Holmes, R. von Huene, and D. A. McManus, "Seismic reflection evidence supporting underthrusting beneath the Aleutian Arc near Amchitka Island": Jour. Geophys. Res., v. 78, p. 3517-3522.

Marlow, M. S., Scholl, D. W., Buffington, E. C., and Alpha, T. R., 1973b, Tectonic history of the western Aleutian Arc: Geol. Soc. America Bull., v. 84, p. 1555-1574.

McManus, D. A., Holmes, M. L., Carson, B., and Barr, S. M., 1972, Late Quaternary tectonics, northern end of Juan de Fuca Ridge (northeast Pacific): Marine Geology, v. 12, p. 141-164.

Menard, H. W., 1964, Marine geology of the pacific: New York, McGraw-Hill, 217 p.

Ness, G. E., and Kulm, L. D., 1973, Origin and development of the Surveyor Deep-Sea Channel: Geol. Soc. America Bull., v. 84, p. 3339-3354.

Nichols, H., and Perry, R. B., 1966, Bathymetry of the Aleutian Arc, Alaska; scale 1:400,000, 6 maps: U.S. Coast and Geodetic Survey, Monogr. 3.

Piper, D. J. W., von Huene, R., and Duncan, J. R., 1973, Late Quaternary sedimentation in the active eastern Aleutian Trench: Geology, v. 1, p. 19-22.

Pitman, W. C., III, and Hayes, D. E., 1968, Sea-floor spreading in the Gulf of Alaska: Jour. Geophys. Res., v. 73, p. 6571-6580.

Reid, J. L., Jr., 1973, Northwest Pacific ocean waters in winter: The John Hopkins Ocean. Studies, no. 5, Baltimore, Md., John Hopkins Univ. Press, 96 p.

Reynolds, L., 1970, Oceanographic cruise report, northwest Pacific, July-August, 1970: Washington, D.C., U.S. Navy Oceanographic Office, OCR 931003, 6 p.

Scholl, D. W., and Marlow, M. S., 1972, Ancient trench deposits and global tectonics: a different interpretation [abs.]: Geol. Soc. America, v. 4, no. 3, p. 232-233.

Scholl, D. W., and Marlow, M. S., 1974, Sedimentary sequence in modern Pacific trenches and the deformed Pacific eugeosyncline, *in* Dott, R. H., Jr., and Shaver, R. H., eds., Modern and ancient geosynclinal sedimentation: Tulsa, Okla., Soc. Econ. Paleontologists and Mineralogists Spec. Paper 19.

———, and Marlow, M. S., in press a, Global tectonics and the sediments of modern and ancient trenches; some different interpretations, *in* Kahle, C. F., and Meyerhoff, A. A., eds., Sea-floor spreading, other interpretations: Tulsa, Okla., Am. Assoc. Petr. Geol. Mem.

———, and Marlow, M. S., in press b, Discussion of paper by V. Matthews III and D. Wachs, Mixed depositional environments in the Franciscan geosynclinal assemblage: Jour. Sed. Petrology, v. . p.

Scholl, D. W., von Huene, R., and Ridlon, J. B., 1968, Spreading of the ocean floor—undeformed sediments in the Peru-Chile Trench: Science, v. 159, p. 869-871.

Scholl, D. W., Christensen, M. N., von Huene, R., and Marlow, M. S., 1970, Peru-Chile Trench sediments and sea-floor spreading: Geol. Soc. America Bull., v. 81, p. 1339-1360.

Scholl, D. W., Buffington, E. C., and Marlow, M. S., 1974, Plate tectonics and the structural evolution of the Aleutian-Bering Sea region, *in* Forbes, R. B., ed., The geophysics and geology of the Bering Sea region: Geol. Soc. America Spec. Paper 151, p.

Silver, E. A., 1969, Late Cenozoic underthrusting of the continental margin off northernmost California: Science, v. 166, p. 1265-1266.

———, 1971a, Small plate tectonics in the northeastern Pacific: Geol. Soc. America Bull., v. 82, p. 3491-3496.

———, 1971b, Tectonics of the Mendocino triple junction: Geol. Soc. America Bull., v. 82, p. 2965-2978.

———, 1971c, Transitional tectonics and Late Cenozoic structure of the continental margin off northernmost California: Geol. Soc. America Bull., v. 82, p. 1-22.

———, 1972, Pleistocene tectonic accretion of the continental slope off Washington: Marine Geology, v. 13, p. 239-249.

Tiffin, D. L., Cameron, B. E. B., and Murray, J. W., 1972, Tectonics and depositional history of the continental margin off Vancouver Island, British Columbia: Can. Jour. Earth Sci., v. 9, p. 280-296.

Udinstev, G. B., 1954, Novye dannye o relefe dna Kurilo-Kamchatskoi Vpadiny: S.S.S.R., Dokl. Akad. Nauk, v. 94, p. 315-318. (New data on the topography of the Kuril-Kamchatka Trench; translation available from Defence Research Board, Canada, T 217 R, 1956, p. 5-8)

———, 1955, Relef Kurilo-Kamchatskoi Vpadiny: Trudy Inst. Okean., Akad. Nauk, S.S.S.R., v. 12, p. 16-61.

von Huene, R., 1972, Structure of the continental margin and tectonism at the eastern Aleutian Trench: Geol. Soc. America Bull., v. 83, p. 3613-3626.

———, and Kulm, L. D., 1973, Tectonic Summary of Leg 18, *in* Kulm, L. D., von Huene, R., et al., Initial reports of the Deep Sea Drilling Project, v. 18: Washington, D.C., U.S. Govt. Printing Office, p. 961-976.

———, and Shor, G. G., Jr., 1969, The structure and tectonic history of the eastern Aleutian Trench: Geol. Soc. America Bull., v. 80, p. 1889-1902.

Geology of the Lau Basin, a Marginal Sea Behind the Tonga Arc

James W. Hawkins, Jr.

INTRODUCTION

The Lau Basin is a small marginal sea located in the southwestern Pacific Ocean. It is bounded on the west by the Lau Ridge, an extinct andesitic island arc, and on the east by the still-active andesitic-dacitic volcanic arc of the Tonga Ridge. The basin is situated above the zone of deep seismicity ($\sim$250 km), associated with the presumed zone of subduction along the Tonga Trench. The Tonga Trench-Arc system is believed to mark a zone of consumption of the Pacific Plate and therefore may be considered to be a zone of crustal shortening. The Lau Basin, however, exhibits many lines of evidence which indicate that it is an area of crustal dilation related to the generation of oceanic lithosphere by processes similar to those which form oceanic lithosphere at mid-ocean ridges.

The basin appears to have generally high (up to 6.75 HFU), but varied heat-flow values, as do other marginal seas, such as the seas of Okhotsk and Japan and the Philippine Sea. There are two depth levels of seismicity in the basin—the deep ($\sim$250 km) zone, presumably related to the descending Pacific Plate, and many shallow events ($<$ 50 km), some of which are aligned along the axis of a 300-km-long ridge. This ridge apparently acts as a transform fault in the dilation of the basin. The basin is anomalously shallow, less than 3.4 km depth, with the average depth being about 2.25 km. Several ridges and seamounts rise to 1 km or less of water depth, and at least one seamount reaches the surface (Niuafo'ou Island). Sediment cover in the basin is very thin, generally less than 0.1 sec acoustic penetration, with a few narrow ponds that reach 0.1-0.2 sec. Thick sedimentary wedges up to 0.7 sec thick are found along the edges of the basin parallel to the Lau and Tonga ridges. This volcanic-derived, silica-rich, clastic material appears to be ponded by ridges, seamounts, and scarps; some of the scarps are probably fault-controlled.

Rocks dredged from ridges, seamounts, and scarps in the central part of the basin show chemical and petrologic similarities to tholeiitic basalts formed at mid-ocean ridges. The average composition of least-altered samples is (in weight-percent): SiO_2, 48.8%; TiO_2, 1.2%; K_2O, 0.18%; P_2O_5, 0.08%; H_2O+, 0.30%; Fe(III)/Fe(II), 0.26; and CaO/Al_2O_3, 0.77. Trace elements averages are (ppm): Ni, 160; Cr, 390; Sr, 100; Ba, 1; Rb, 0.15. K/Rb, in a least-altered and unfractionated sample, is 860; Ba/Sr is 0.1; Ba/Rb is 8. These data are essentially identical to typical values for mid-oceanic ridge basalt, and indicate derivation either by partial melting of mantle material at depths on the order of 50 km or by separation of liquid from residual crystals at that depth.

The bathymetry and magnetic anomaly patterns in the basin have been mapped with an extensive ship survey and provide a test of tectonic models. Both the magnetic and bathymetric evidence support plate divergence as an explanation for basin generation in some areas, but not for the whole basin. In general the basin may be characterized as a zone of multiple intrusion of basaltic material which, in some parts, is localized along the boundaries of rigid plates. The wide areas of newly intruded basalt mapped in the basin clearly requires that there has been an increase in the volume of lithosphere of the basin and helps to account for the high heat flow, lack of sediment cover, and ridge-seamount topography.

Unless the Lau Basin and the other marginal seas are features unique to the Cenozoic, there must be remnants of ancient marginal seas preserved in the record of orogenic belts. The lithology of the Lau Basin shows many similarities to the rock types recognized as ophiolite suites. It seems likely that some, if not most, ophiolite suites were formed in marginal basins behind volcanic island arcs and trenches, and thus are significant in the interpretation of ancient continental margins—inasmuch as they may represent a region of transition between ancient oceanic and continental crust.

REGIONAL SETTING AND PREVIOUS INVESTIGATIONS

One of the more striking geologic characteristics of the western Pacific Ocean basin is the festoon-like arrangement of the island arc-trench systems. The arcs are separated from the Asian mainland, or from other arcs, by marginal seas. Except for the Scotia and Antilles arcs of the Atlantic, these arcs with their marginal seas appear to be a feature distinctive of the Pacific Basin. Alfred Wegener (1929) called attention to this configuration and postulated that "the island arcs, and particularly the eastern Asiatic ones, are marginal chains which were detached from the continental masses when the latter drifted westwards and remained fast in the old sea floor, which was solidified to great depths. Between the arcs and the continental margin, later, still-liquid areas of sea floor were exposed as windows."

The increased knowledge of the geology and geophysics of the ocean basins has helped substanti-

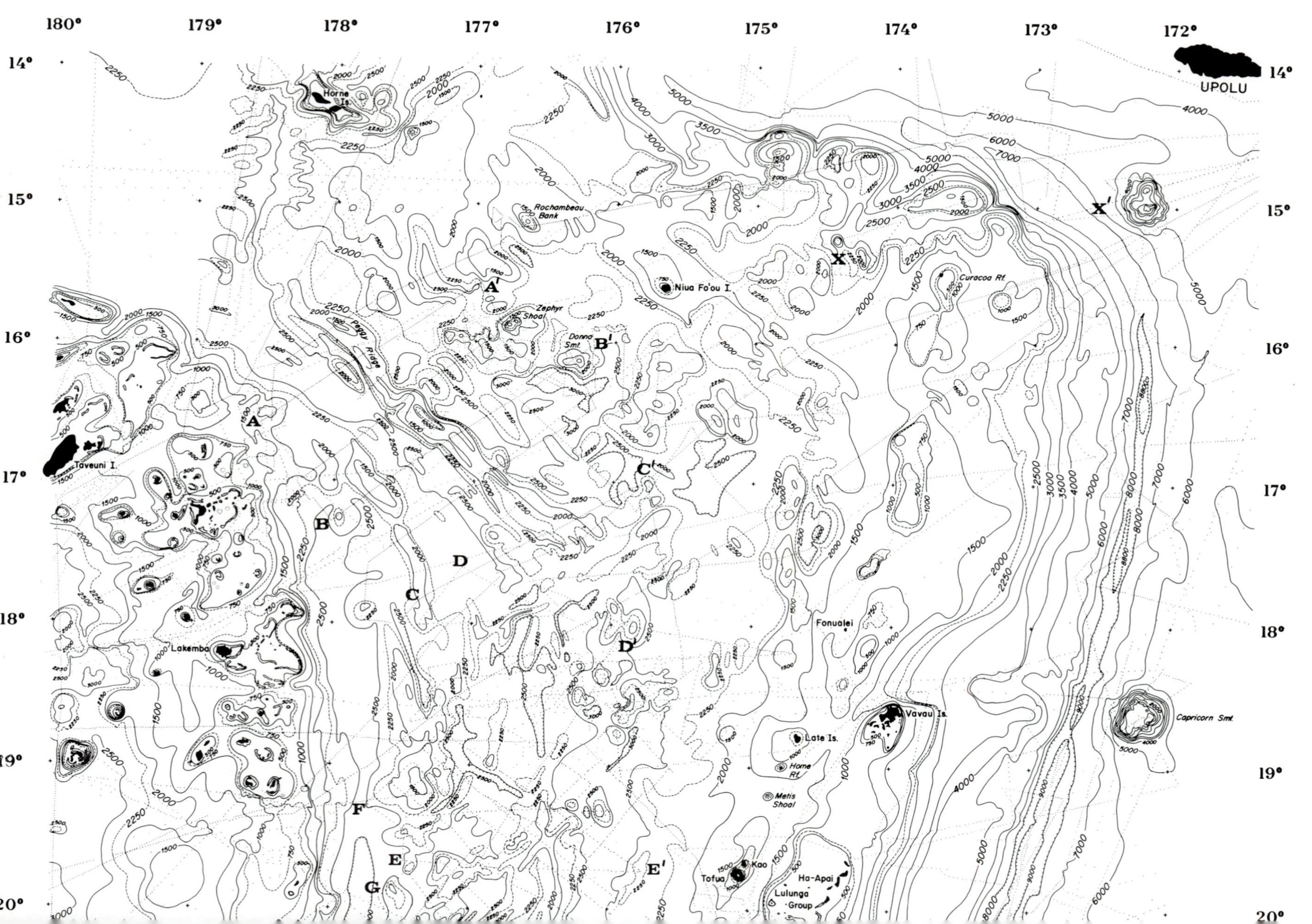
180°
179°
178°
177°
176°
175°
174°
173°
172°
14°
15°
16°
17°
18°
19°
20°
UPOLU
Horne Is.
Rochambeau Bank
Zephyr Shoal
Donna Smt.
Peggy Ridge
Niua Fo'ou I.
Curacoa Rf.
Taveuni I.
Lakemba
Fonualei
Vavau Is.
Capricorn Smt.
Late Is.
Home Rf.
Metis Shoal
Kao
Tofua
Ha-Apai
Lulunga Group
A
A'
B
B'
C
C'
D
D'
E
E'
F
G
X
X'

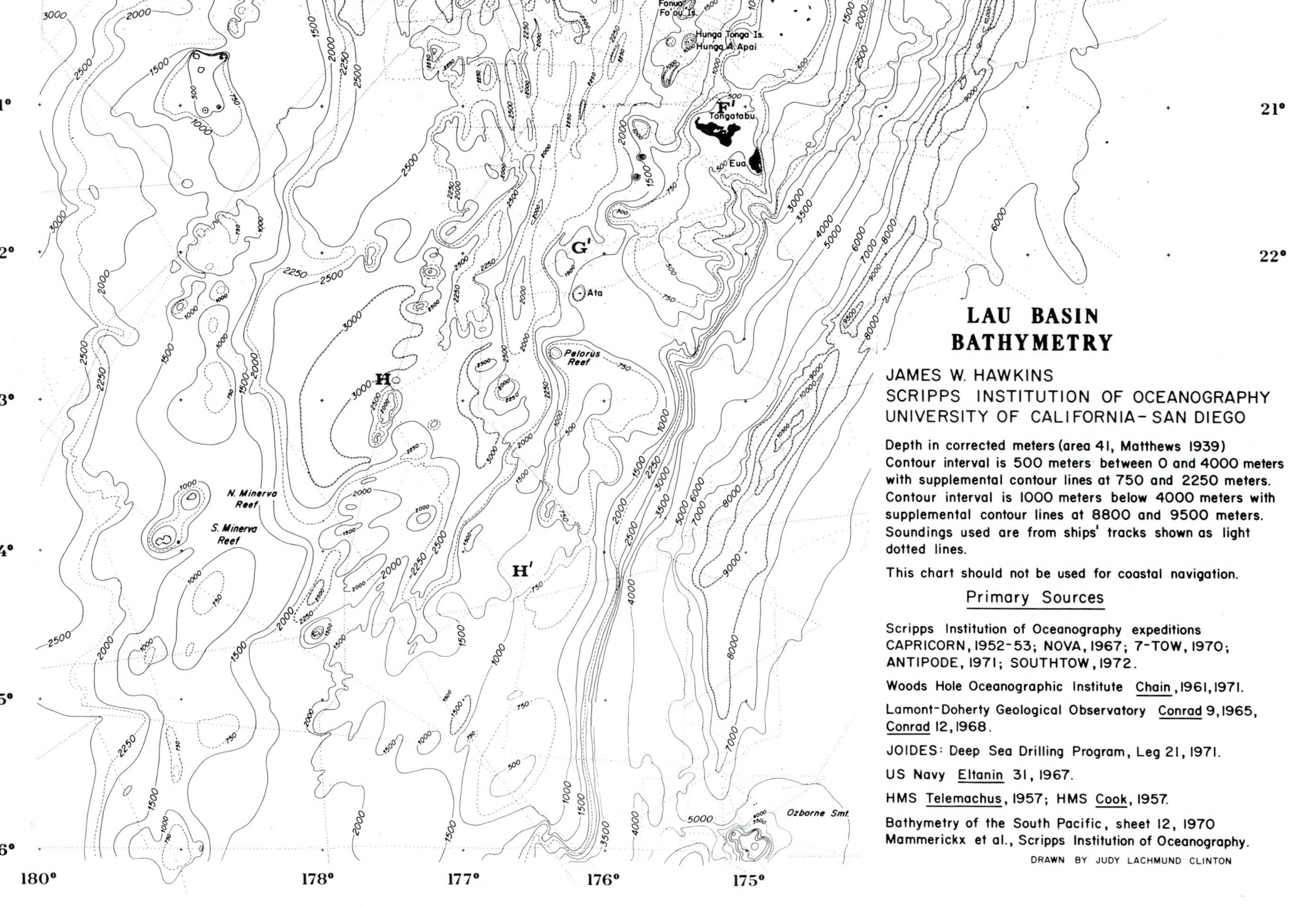

Fig. 1. Bathymetric chart of Lau Basin. This chart is an updated version of part of "Bathymetry of the South Pacific," Sheet 12, Mammerickx et al., 1971. Revised depth contours in the area between the Tonga and Lau Ridges are based primarily on soundings taken along the profiles shown in dotted lines. Depths are in meters corrected for area 41 (Matthews, 1939).

ate the validity of Wegener's postulate, although it appears from the literature that the idea has been rediscovered in recent years rather than having been tested. Several authors have discussed the nature and possible origin of marginal seas; for example, Karig (1970) postulates "that the frontal arc and trench position migrate, generally away from the continents. . .creating new oceanic crust, without mid-ocean ridges. . . ." Packham and Falvey (1971) propose that "creation of oceanic crust took place along rifts, associated with olivine-basalt extrusion, along or behind the andesite arc, producing a style of asymmetrical seafloor spreading." Those whose work bears directly on the origin of the Lau Basin include Kibblewhite and Denham (1967), Karig (1970), Packham and Falvey (1971), Hawkins et al. (1970), Sclater et al. (1972), and Hawkins (1973).

Packham and Falvey (1971) summarized the characteristics of the marginal seas of the western Pacific and they noted in particular that these are regions of oceanic crust on the continental side of the trench-arc systems. Menard (1967) called attention to the high heat flow in the western Pacific marginal seas and noted that in general the areas of high heat flow in oceanic basins were closely correlated with vulcanism at the mid-ocean ridges. The seas of Japan and Okhotsk (Uyeda and Vacquier, 1968) and the Philippine Sea (Watanabe et al., 1970) have been extensively studied and have anomalously high heat flow when compared with deep-ocean basins. Packham and Falvey (1971) suggested that the high heat flow of marginal seas indicated that they were regions of high-energy tectonic processes and also observed that these high-heat-flow areas were not found on the landward sides of arc-trench systems which lacked marginal seas. Thus there seems to be an association of arc-trench systems in the ocean basin with marginal seas having an oceanic crust and the high-heat-flow values commonly found otherwise on mid-ocean ridges.

The high heat flow of the western Pacific was evaluated by McKenzie and Sclater (1968), who concluded that either the observed values were due to conduction from mantle sources at 3000°C over a time span of 300 my, or were due to magma injection at the average rate of 0.5 km3/yr. Murauchi and Den (1966) previously had invoked a hypothetical rift system to permit the extension of the Sea of Japan in an easterly direction and to allow magma to well up to form the hot oceanic crust. Oxburgh and Turcotte (1971) modeled the thermal structure of the crust and upper mantle in island arc areas and estimated the volume of magma required to explain the high-heat-flow values observed in the Sea of Japan. The magmatic intrusion required to maintain a 4-HFU (heat-flow-units) average over a 500-km-wide belt would be sufficient to maintain a 5 cm/yr extension rate. The geometric-bathymetric arguments and the interpretation of heat-flow data have been the main lines of evidence to suggest that at least some marginal basins owe their origin to dilational processes (Karig, 1970; Sugimura and Uyeda, 1973).

If sea-floor spreading actually takes place due to the separation of rigid lithosphere plates, one should find symmetric magnetic anomalies similar to those formed at mid-ocean ridges. Van der Linden (1969) suggested that the magnetic patterns from some east-west profiles across the Dampier Ridge supported his interpretation that the ridge was a dormant or fossil spreading center which was the source of new sea floor to form the Tasman Sea. He admitted that the patterns could not be matched with known anomalies and did not present any synthetic patterns for a possible test of this interpretation. To date, the only magnetic anomaly patterns that appear to support sea-floor spreading in back-arc basins are for the Scotia Sea, Woodlark Basin and for the Lau Basin (this report). If new sea floor is indeed being produced by the injection of basaltic magmas into the oceanic crust of marginal seas, a critical test of the concept would be to recover samples of fresh oceanic tholeiites. To date the only rocks from marginal seas are the samples from the Woodlark Basin (Luyendyk et al. 1973) and the Lau Basin; these are petrologically similar to tholeiites from mid-ocean ridges (Hawkins et al., 1970; Hawkins and Nishimori, 1971; Sclater et al., 1972; Hawkins, 1973; Luyendyk, et al., 1973).

The main purpose of this paper is to present a summary of the geochemical, petrologic, and geophysical data for the Lau Basin and to suggest some possible models to explain its origin.

The Lau Basin lies between the "andesitic" arcs of the Tonga Ridge and the Lau Ridge; both areas are on the west (or concave) side of the Tonga Trench. It covers the region between 15°-25°S and 174°-178°W. It is important to note its position relative to the Tonga Trench (Fig. 1, and profile X-X' Fig. 2), as perhaps all marginal seas owe their origin to stresses set up in the mantle by the subduction of an oceanic plate. The Tonga Trench also is important in that it has offered a cross section of the upper 10 km of lithosphere on which the Tonga Arc and Lau Basin are built. Rocks dredged from the trench wall by R.L. Fisher comprise basalt, gabbro, and ultramafic rocks (Fisher and Engel, 1969; Hawkins et al., 1972). There is no evidence of sialic material or metamorphosed "eugeosynclinal" sediments under the Tonga Arc. The Lau Basin cannot be a foundered continental plate.

The Lau Basin may be typical of many small marginal seas; it clearly has the main characteristics of most of these seas, but also has a very thin sedimentary cover and thus permits detailed sampling of the underlying igneous rocks. Its relatively small size has made it possible to make a fairly detailed survey of the basin and to prepare large-scale maps of the magnetics, bathymetry, and sediment distribution. Because of its location adjacent to a zone of active tectonism and vulcanism, the Lau Basin is an important component of the geologic system, which may be the inital stages of evolution of an orogenic belt. If the Tonga Trench-Arc system becomes involved in an intense folding and plutonic event, the Lau Basin presumably will be incorporated into the resulting orogenic belt. Older orogenic belts may contain remnants of similar ancient marginal basins.

BATHYMETRY

The average depth of ocean basins ranges from about 4 to 6 km. At active ridge crests, the depth is about about 1-2 km and at the margins of ocean basins, close to trenches, the depth is typically about 5.5 km. The average depth of the Lau Basin is only about 2.3 km. The depth range is from about 3.3 km to about 0.8 km, although the island of Niuafo'ou is essentially a seamount which rises 260 m above sea level. Average relief in the Lau Basin is about 1 km, but maximum relief of about 2.5 km is reached near some of the shoals and on ridge segments or seamounts adjacent to deeps. Thus the Lau Basin has anomalously high-standing sea floor, and large areas of the basin are more shallow than at the crest areas of many typical mid-ocean ridges. Although far removed from continents, the basin floor is anomalous when compared with typical oceanic areas.

Lau Basin bathymetry is dominated by several linear ridge-like features and by numerous seamounts. The seamounts are clearly the result of volcanism localized above a point source of magma. Most of the ridges are also volcanic and were formed by eruptions along linear magma sources. Tectonic activity has played an important role in the development of the basin, and at least some of the ridges may owe their present form to faulting. Figure 2 shows a series of profiles aligned generally in a E-W direction across the basin. The profiles were selected both to show a regional overview of the basin and to be representative of typical basin morphology.

Two major linear trends are apparent in the bathymetry—northwest and north; a northeast trend is also seen in some parts of the basin. The northwesterly trending Peggy Ridge extends for about 300 km. It has about 1.3 km relief and is the dominant bathymetric element between 15°S and 17°30'S. A lower and less continuous broad ridge runs parallel to it on the southwest side. Several large seamounts, including Donna Seamount, Zephyr Shoal, and Niuafo'ou lie northeast of the Peggy Ridge and their location and morphology suggest that they may be volcanoes built on disrupted fragments of originally continuous ridge systems. One of these ridge systems probably trended northeast and the other west-northwest. The steep scarps and high relief in the vicinity of Zephyr Shoal and on the south and west sides of Donna Seamount are probable fault scarps. Rocks dredged from one of these scarps show evidence of cataclasis and recrystallization.

Between 19°30'S and 24°S a discontinuous central ridge appears to be broken by several lower divides which may be related to an en échelon fault system. Bathymetric data are relatively poor in this region and the details of basin morphology are not as well known as to the north. However, because of the limited distance between the Lau and Tonga ridges at this latitude, the morphology cannot deviate too widely from that shown on the chart.

The area between 17°30'S and 19°30'S has quite complex morphology and also includes the deepest soundings in the Lau Basin ($\sim$3.4 km). A north-south trend is prominent, but northeasterly trending ridges are also present. Near 19°S and 177°W a northwesterly trending ridge causes a rectilinear system of ridges and deeps. The profiles show the shallow depths, the isolated seamounts, and the irregular topography of ridges. If these are analogous to mid-oceanic ridges, the irregular topography suggests slow spreading rates.

SEDIMENT FILLING

The Lau Basin has a relatively thick wedge of sediment ranging from 0.4 to 0.7 sec of two-way travel time, or 400-700 m (assuming a 2.0 km/sec velocity), along the flanks of the Lau and Tonga ridges. The seismic-reflection profiles (Fig. 2, profiles F-F', G-G', and H-H') show that this filling is located in narrow ponds bounded by steep scarps. The thickness of fill drops off quite markedly in successive ponds toward the center of the basin until in the central part of the basin much of the floor is essentially bare rock. Although there are many thin and narrow sediment ponds in the basin, it is safe to characterize the basin as having a very thin veneer of sediment overlying a strong acoustic reflector.

The geometry of the basin and the appearance of the air-gun seismic profiles suggest that the sediment ponds are in fault-controlled depressions. It is also possible that seamounts and ridges of volcanic origin have acted as dams to trap the sediment derived from the two ridges. A generalized sediment-thickness map based on air-gun profiles is shown in Figure 3. Generally, the bare-rock areas coincide with ridges and seamounts, but there are broad areas with very thin sediment cover in the deeper parts of the basin.

Data for the sediments of the Lau Basin are limited to 12 gravity core samples plus sediment-catcher samples from 20 rock-dredge sites. All these samples are from the central part of the basin, where existing sediment cover is very thin. Deep Sea Drilling Program site 203 (Leg 21) was located on the thick sedimentary wedge adjacent to the Lau Ridge at 22°09.22'S, 177°32.77'W. This hole penetrated 409 m of ash and volcanic sands. In the central part of the basin, the surface sediments are predominantly brown, volcanic-derived silts and muds with 10-20% calcareous material, composed of foraminifera and coccoliths. Some diatoms and radiolaria, as well as sponge spicules, are present also. Shards of silicic volcanic glass are quite abundant in many of the samples; the glass has a refractive index of 1.498-1.502, indicating a probable silica content of about 73%. The dominant clastic minerals are plagioclase and pyroxene. Montmorillonite and clusters of

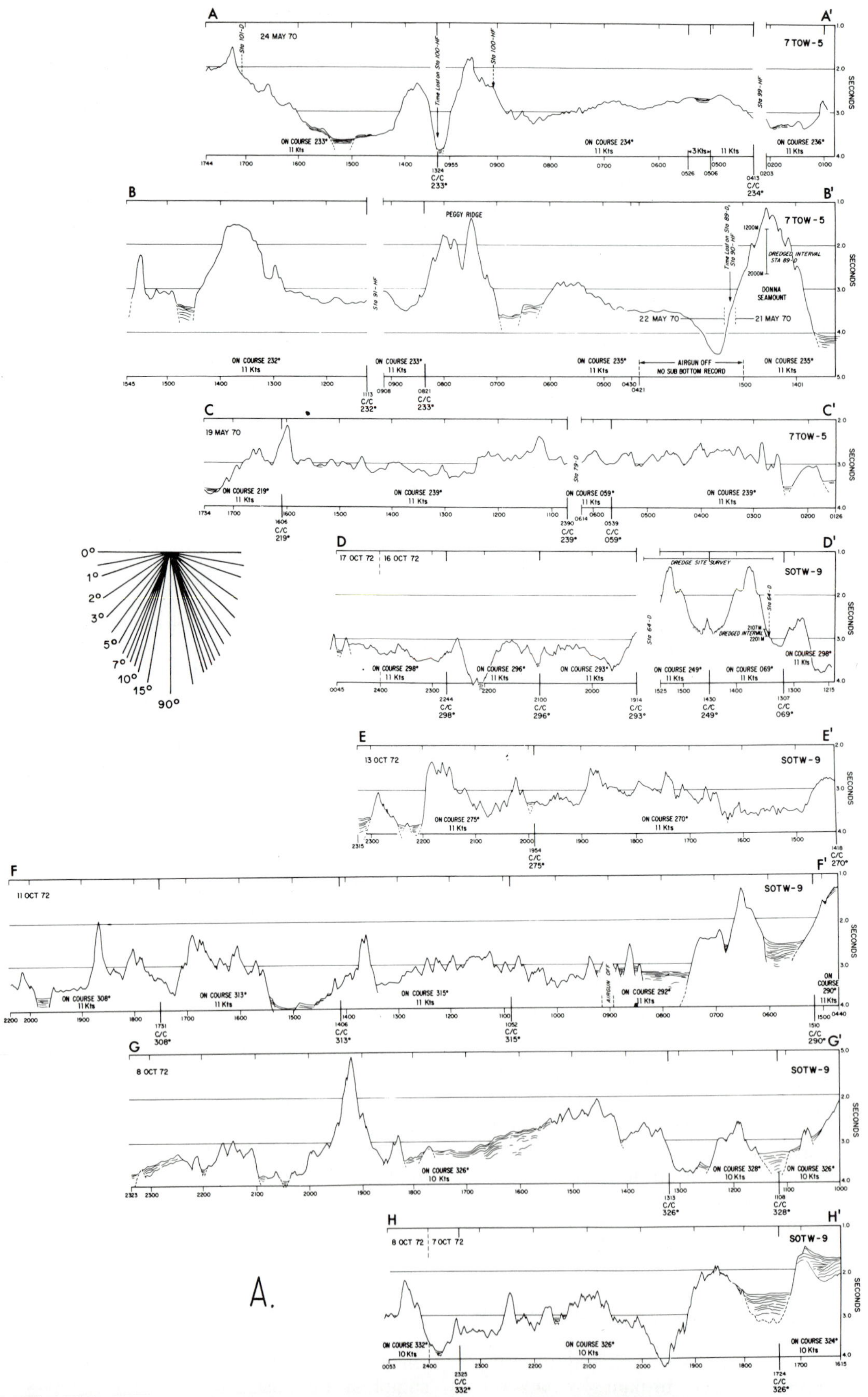

Fig. 2a. Selected bathymetric and sediment-thickness profiles (Fig. 1). Depths are shown as seconds of two-way travel time and correspond to depths of approximately 750 m/sec in water and 1 km/sec in sediments (assuming 2 km/sec velocity for the sediments). Vertical exaggeration is 20x at 11 knots.

zeolites are products of local bedrock weathering and sediment diagenesis, respectively. Data for the sediment composition and sample location are given in Table 1.

The intermediate to silicic composition volcanic arcs of the Tonga and Lau groups probably were the source areas for the clastic components of the basin filling. The thick sedimentary wedges on either side of the basin clearly were derived from the volcanic arcs and the thin cover of the central basin must represent material that bypassed the sediment traps along the basin margins. Drift pumice, undoubtedly from the Tonga arc, is widespread throughout the basin and the silicic glass shards, as well as much of the plagioclase and pyroxene, probably was derived from mechanical breakdown of the pumice. Aeolian transport of some of these components must have been important during eruptions. Mineralogic and geochemical data for weathered Lau Basin basalts and the sediments (Griffin et al., 1972; Berthine, 1974) indicate that some of the sedimentary filling of central basin ponds was locally derived and probably has slumped off the ridges.

All the surface sediments in the basin appear to have a fairly large component of pumice fragments. On a carbonate-free basis the sediments probably would have about 60-65% SiO_2, low Ca, Fe and Mg, and relatively high Na and K. If the Lau Basin is representative of present and historic marginal basins, the sedimentary fillings of these basins may be distinctive in that they have slightly higher SiO_2 content than "average abyssal sediments," which have about 54% SiO_2. If lithified and eventually incorporated in a fold mountain belt, the Lau Basin sediment filling might form a silicic shale.

GEOPHYSICAL DATA

The Lau Basin lies in a region of intense seismic activity (Baranzangi and Dorman, 1969); it is situated above a well-defined, inclined seismic zone, having seismic events down to 700 km, which extends westward from the Tonga Trench at about 173-175°W to Fiji at about 178°E. The seismic zone is about 50 km thick and is inclined at about 45°. The upper boundary of this zone is at about 250 km depth under the Lau Basin, but numerous shallow seismic events (< 50 km) have also been recorded in this region. Many of these shallow earthquakes are concentrated along a zone following the trace of the Peggy Ridge (Baranzangi and Isacks, 1971). A second, well-defined group of shallow earthquakes follows a trend more-or-less east-west along the north end of the basin. This northern group of events does not lie in the area of greatest depth, but is at the top of the scarp bounding the westward extension of the depression of the Tonga Trench. Many of these earthquakes appear to line up along submarine canyons which are incised into the steep slopes of the Trench wall and the north end of the basin (Fig. 4). There are no first-motion studies for most of these earthquakes, although a left lateral strike-slip interpretation was made for one shallow event of the north set and a dilational event was reported for an earthquake in the Lau Basin. No first-motion studies have been reported for the Peggy Ridge earthquakes.

The seismicity in this area seems to be related to at least two types of lithosphere deformation. The dominant style of deformation appears to be due to relative lithosphere shortening, presumably as the Pacific Plate is consumed in the Tonga Trench. The deep seismic events are considered to be related to the subduction of the Pacific Plate, moving on a trend of N87°W relative to the India Plate, as it is consumed at about 9.0 cm/yr (Sclater et al., 1972).

However, these deep seismic events probably are not directly related to the shallow events noted earlier and no seismicity has been reported for depths between 50 and 250 km under the basin. The second style of deformation appears to be dilational; the lithosphere of the Lau Basin appears to be extending in a northwesterly-southeasterly direction with the Peggy Ridge operating as a transform fault (Sclater et al., 1972). Although the North end of the Lau Basin is bounded by a postulated trench-trench transform fault, the actual nature of this boundary must be far more complicated than a simple zone of lateral slip. Air-gun reflection profiles across this northern boundary made on SIO expedition ANTIPODE, Leg 16, showed a series of sediment-covered step-like features, suggesting vertical movement of blocks. Thus this northern boundary is an interface between relative east-west strike-slip motion, as the Pacific Plate passes by on a path oblique to the bathymetry of the trench; it is further complicated by a northwestern component of movement of the Lau Basin as the basin is dilated; and it is also experiencing vertical displacement as well as horizontal movement to the north, as shown by the down-faulted step-like features on the trench wall.

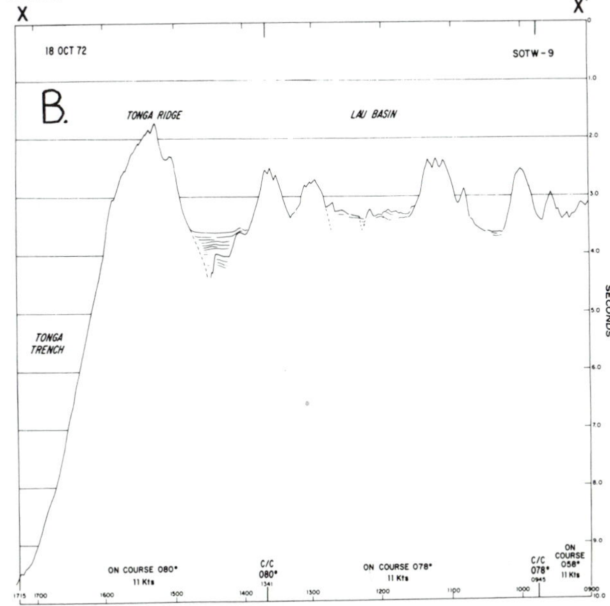

Fig. 2b. Detailed profile in Lau Basin.

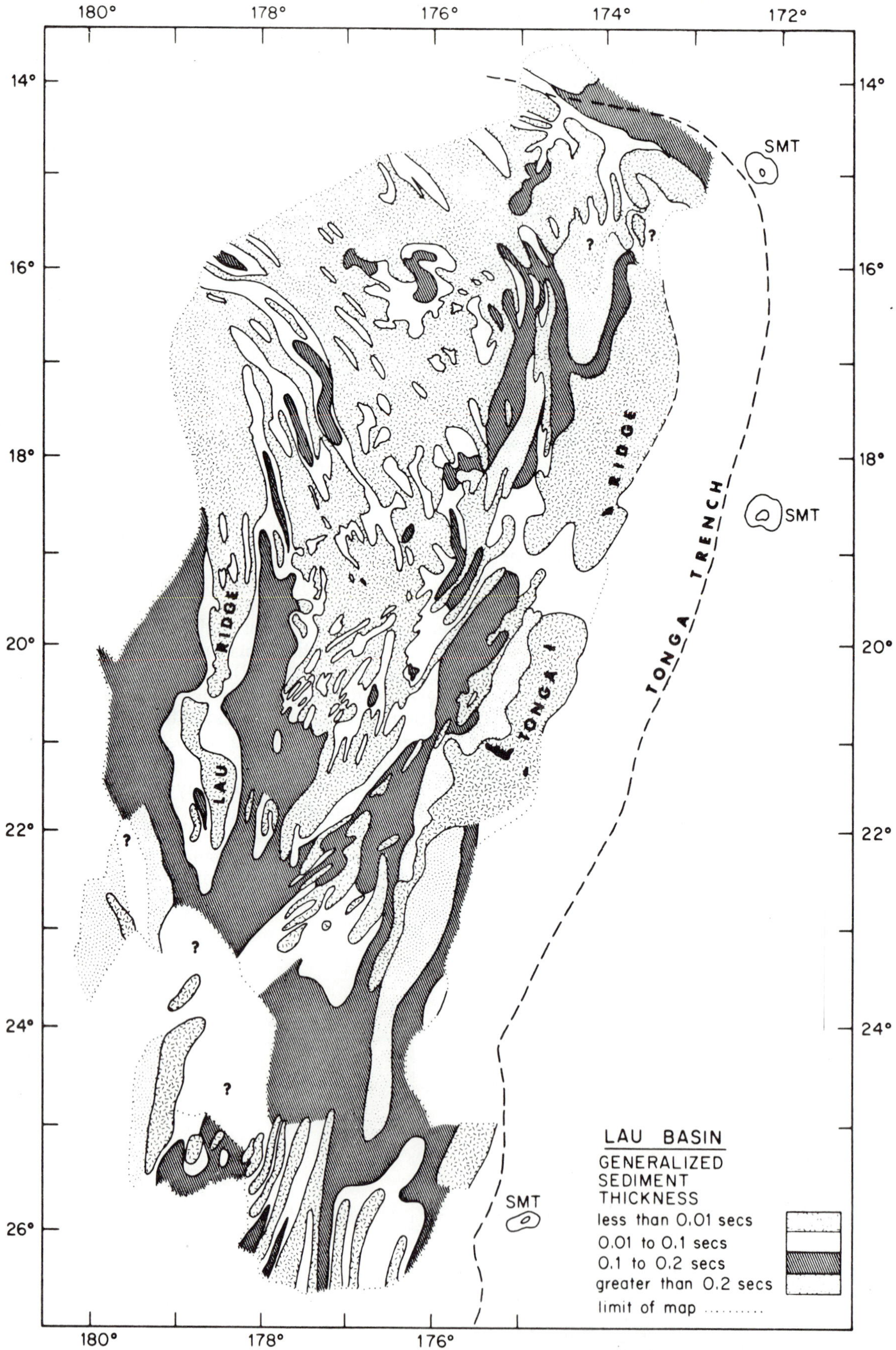

Fig. 3. Generalized sediment-thickness map for the Lau Basin based on air-gun profiles along ship tracks (Fig. 1). Thicknesses are shown in seconds of penetration. The geometry of the sediment basins has been interpreted on the assumption that they follow the bathymetry. Sediment thicknesses on the Lau and Tonga ridges are schematic. Data are limited in these areas but locally there are thick sediments.

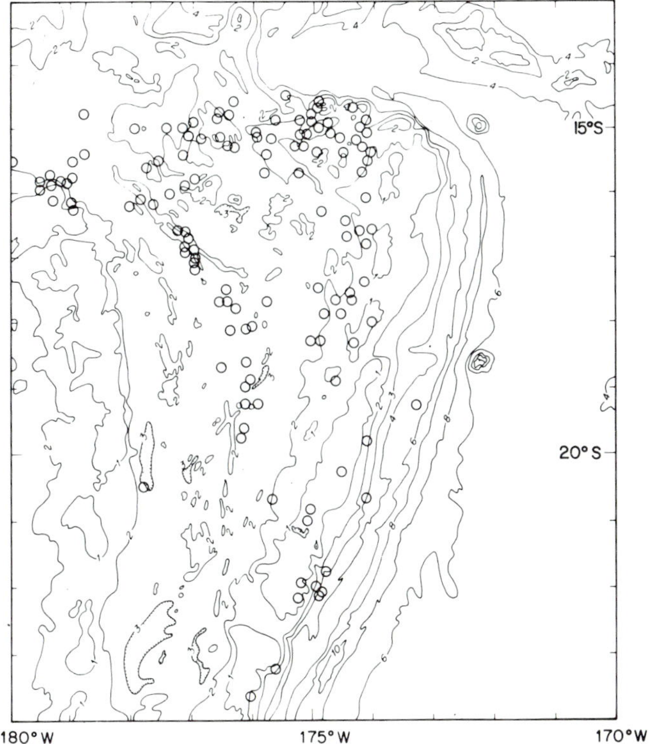

Fig. 4. Generalized chart of the Lau Basin showing shallow-focus seismic events. Circles are epicenters based on relocated events listed by Sykes et al., 1969. Depths are in corrected km. Contour interval is 1 km between the Tonga and Lau Basin ridges and 2 km in other areas. See Figure 1 for names of geographic features.

Heat-flow data for the basin have been summarized in an earlier report (Sclater et al., 1972), and data collected since then have substantiated the original interpretation. The basin has irregular heat flow but in general it is high; the arithmetic mean is about 1.96 HFU and 12 of the 20 measurements are higher than 1 HFU. The range in values is 0.3-6.7. Some of the low values may have been due to slump areas; the high values seemed to be readily explained in terms of proximity to young volcanic features. For example, a value of 3.2 was obtained 75 km southeast of Niuafo'ou, a volcanic island that last erupted in 1946, and 6.75 HFU was measured at the base of Rochambeau Bank which is capped with fresh basalt vitrophyre. The high heat flow in the basin implies that the relatively high elevation of the Lau Basin floor is due to upwelling of hot material. The basin may be situated above a relatively hot part of the mantle from which mantle material rises and partly melts. Barazangi and Isacks (1971) presented a model cross section of the lithosphere beneath the Lau Basin which showed a zone of low Q in a region between the Peggy Ridge and the northeastern edge of the basin. Of the possible explanations which they present for this, the presence of a high-temperature (partly melted?) mass seems to fit best with the bathymetry, tectonic setting, and the basaltic nature of the sea floor.

Over 13,000 km of magnetic profiles have been collected in the Lau Basin along tracks selected to test models of the geometry of the basin and its ridges in terms of plate-tectonic concepts. While it is clear from the morphology of the basin that there are ridges, deeps, scarps, and seamounts that tend to be aligned in directions that fit possible plate models, the magnetic profiles show that any plate geometry must be complex. There seems to be quite a good fit between observed and computer-generated anomaly patterns for the area between 17°30'S and 20°S when the profiles are projected to a trend of 125° (spreading about an axis oriented 035°). This spreading direction is very close to the topographic trend and probably is a very close fit in terms of the limits imposed by ship-track spacing. Neither the bathymetry nor the orientation of tracks permits good control on the orientation of possible spreading centers in the area between 15°30'S and 17°30'S. Disrupted ridges, formerly trending northeast and west-northwest, are suggested by the bathymetry, but they cannot be tested by our magnetic profiles. Details of the magnetic patterns and their origin will be the topic of a separate report (L. Lawver, SIO, in progress, 1973), but some representative magnetic profiles are shown in Figure 5.

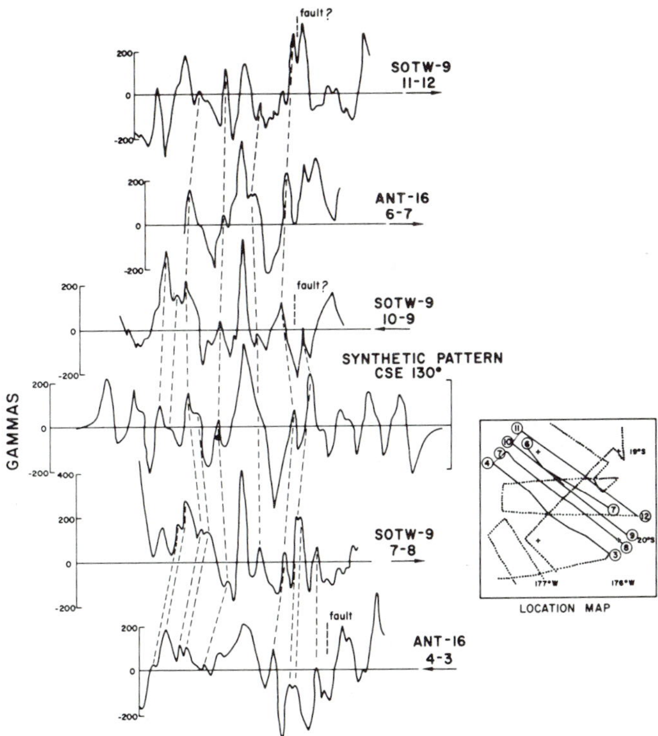

Fig. 5. Observed and model magnetic profiles. Location shown in the inset and tracks correspond to ship tracks shown on Figure 1. Arrows indicate the direction of the ship's heading. The model was generated by assuming a half-spreading rate of 1 cm/yr for a ridge oriented 030° at 19°30's. The two peaks on the right of the fault in ANT-16, 4-3 appear to be the central anomaly and first two positive peaks on the left side of the synthetic model. The fault is interpreted as a transform striking approximately NW.

The conclusion, based on analysis of the available data, is that symmetric anomalies have been developed by northwest-southeast spreading in the area between 18°S and 24°S. A probable triple junction is implied by a swing in the pattern near 18°30'S, 176°W. Between 15°30'S and 17°30'S, the present style of deformation appears to result in the formation of a number of point sources of vulcanism

Table 1. Lau Basin Dredge Stations, Volcanic and Plutonic Rocks

Station	Latitude	Longitude	Depth Interval (m)	Description[a]	
7 TOW-74	18°45.6′ S	175°57.2′ W	2413-2384	R	OTL
7 TOW-79	17°21.0′ S	176°25.7′ W	2290-2054	R	OTL
7 TOW-86	16°55.4′ S	176°49.5′ W	1990-1664	R	OTL
7 TOW-89	16°12.2′ S	176°15.4′ W	1352-1195	S	OTH
7 TOW-95	16°23.6′ S	177°28.3′ W	1322- 999	R	OTL
7 TOW-98	15°50.2′ S	176°42.3′ W	1390-1244	S	D*
7 TOW-101	16°31.8′ S	178°25.8′ W	1600-1446	R	OTL
7 TOW-103	15°55.5′ S	178°01.9′ W	1400-1150	R	QTH
7 TOW-106	15°09.9′ S	176°37.7′ W	1400-1250	S	OTH
ANT-223	20°23.8′ S	176°59.6′ W	2667-2235	R	OTL
ANT-225	19°15.7′ S	175°49.9′ W	2537-2236	R-S	QTL
ANT-229	17°38.0′ S	175°51.0′ W	2556-2220	R	OT
ANT-233	15°53.5′ S	177°05.7′ W	2013-1558	F	G
SOTW-61	19°26.7′ S	176°24.9′ W	2612-2300	R	OTH
SOTW-64	18°02.7′ S	176°01.2′ W	2201-2107	R	OTL
		Drift Pumice and Volcanic-clastic Sediments			
7 TOW-81	17°58.9′ S	177°41.9′ W	2316-1649	R	Pumice
ANT-221	21°23.6′ S	176°44.2′ W	1734-1567	R	Volcaniclastics
ANT-222	20°45.6′ S	177°10.8′ W	2360-2160	R	Pumice
ANT-224	19°07.6′ S	177°33.8′ W	2050-1882	R	Mn-crusted siltstone
ANT-228	18°10.7′ S	175°51.0′ W	1511-1400	R	Pumice and tuff

Lau Basin Gravity Cores

Core No.	Latitude	Longitude	Core Length (cm)	Water Depth	Composition
7 TOW-72G	20°24′ S	176°46′ W	157	2730	B*
7 TOW-84G	17°36.3′ S	177°45.4′ W	150	2406	B*
7 TOW-88G	16°47.5′ S	176°39.5′ W	110	2698	B*
7 TOW-94G	16°29′ S	177°33′ W	67	2328	B*
7 TOW-105G	15°15.5′ S	176°49.5′ W	66	2163	B*
ANT-226G	18°35′ S	176°45′ W	162	2472	B
ANT-230G	17°33.4′ S	175°10.9′ W	——	2353	P
ANT-231G	17°09.5′ S	175°53.8′ W	130	2238	B
ANT-237G	16°18′ S	174°48′ W	——	2181	B
ANT-238G	14°52′ S	174°53′ W	26	2172	B
SOTW-62G	18°20.0′ S	177°17.5′ W	160	2626	B
SOTW-65G	16°01.2′ S	175°19.8′ W	77	2193	B

[a]Explanation for rock descriptions: R = Ridge; S = Seamount; F = Probable fault scarp; OT = Olivine Tholeiite; QT = Quartz Tholeiite; H = High Alumina; L = Low Alumina; G = Gabbro and Greenstone; D = Dacite; * = Dredge haul included back reef limestone.

Key for Composition:

P = Pumice fragments

B = Brown calcareous mud with volcanic glass fragments and clay minerals. Sponge spicules, radiolarians and diatoms common. Rare quartz and abundant plagioclase and pyroxene. Zeolites are common. Calcareous component comprises coccoliths and foraminifera.

B* = Composition as above, less than 2μ consists of montmorillonite and opal, greater than 2μ fraction consists of calcite, glass, plagioclase, clinopyroxene, cristobalite-tridymite, montmorillonite and analcime or clinoptilolite (Griffin et al., 1972).

which give rise to seamounts and short ridges. These point sources are superimposed on what seem to be remnants of an older, more or less orthogonal, ridge-fracture zone system having northeast and west-northwest trends. For the last million years or so, basin dilation has operated with the Peggy Ridge acting as a transform fault. A well-defined northeast-trending bathymetric ridge associated with this spreading has not yet formed, but this is not surprising considering the slow spreading rates, which presumably imply slow rates of volcanism.

PETROLOGY

The floor of the Lau Basin has been sampled at 20 localities from dredge sites representing seamounts, ridges, and steep scarps; scarps probably are fault controlled. On the basis of these samples, the exposed rocks of the basin are almost exclusively basalt or gabbro, and it seems almost certain that the acoustic basement beneath the thin sediment cover throughout the basin must be basaltic as well. Zephyr Shoal (15°50'S, 176°42'W) is an exception to this uniform distribution of basalt; it is (capped by?) dacite vitrophyre. Siliceous pumice, probably rhyodacite in composition, is widely distributed throughout the basin, but it is of drift origin, probably having come from the volcanoes of the Tonga Arc.

Silicic scoria dredged from the flank of the Tonga Ridge and volcanically derived siltstone-sandstone, from a small scarp on the flanks of the Lau

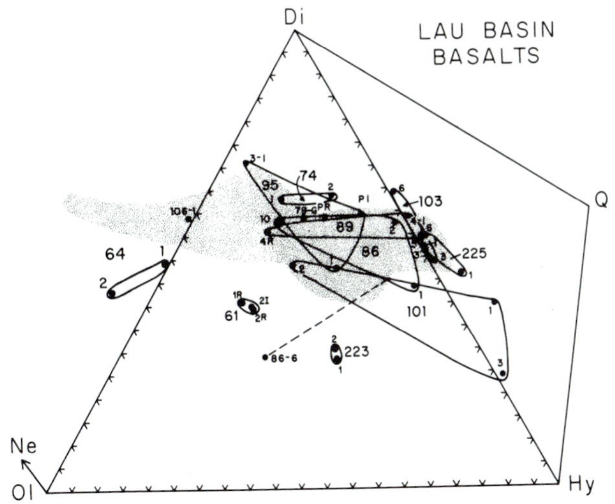

Fig. 6. Composition of Lau Basin basalts plotted in the normative system Ne-Ol-Di-Hy-Q. Samples from individual dredge hauls are encircled and individual samples are shown as points. The shaded area represents the range of analyzed material from other oceanic ridge areas (Kay et al., 1970). The role of crystal fractionation in these samples is seen in the Ol-enriched sample (86-6), which has about 15% olivine phenocrysts.

Ridge, are exceptions to the generalization of basalt basin floor, but these samples are clearly parts of the volcanic arcs bordering the basin. A summary of dredge locations and rock types is given in Table 1.

The basaltic material in the Lau Basin is anomalous if one subscribes to the old idea of an

Table 2. Average Rock Analyses

	A	B	C	D	E	F	G	H	I
SiO_2	48.8	50.01	49.21	48.8	49.55	51.1	49.1	63.26	61.97
TiO_2	1.2	1.37	1.39	0.6	1.49	1.3	2.0	0.53	0.94
Al_2O_3	16.4	16.18	15.81	16.8	15.72	17.2	13.8	14.56	16.00
Fe_2O_3	2.0	2.32	2.21	2.7	—	4.1	3.9	3.09	3.22
FeO	6.9	7.07	7.19	4.5	7.56	5.3	7.7	2.55	3.57
MnO	0.2	—	0.16	0.1	0.16	0.2	0.2	0.09	0.09
MgO	8.6	7.71	8.53	8.4	9.27	4.6	6.1	3.42	2.43
CaO	12.6	11.33	11.14	12.9	11.16	10.0	9.4	5.16	3.24
Na_2O	2.4	2.79	2.71	2.2	2.70	2.9	3.3	4.03	5.55
K_2O	.18	0.22	0.26	.20	0.08	.87	0.4	0.98	0.75
H_2O^+	.30	0.87	—	1.9	2.22	1.2	2.4	1.40	1.28
H_2O^-	.27	—	—	0.5	2.22	.9	0.4	0.15	
P_2O_5	.08	0.13	0.15	.06	0.03	.2	0.2	0.05	0.22
CO_2	.14	—	—	.11	—	.1	0.8	0.26	<0.1
S	.04	—	—	.04	—	trace	—	trace	—
Total	100.11	100.00	98.76	99.81	99.94	99.97	99.7	99.53	99.26

A. average of 11 unaltered basalt samples, Lau Basin
B. average oceanic tholeiite, Engel et al., 1963
C. average of 33 basalts from greater than 1 km depth on Mid-Atlantic Ridge, Melson et al., 1968
D. average of 5 feldspathic gabbros, Lau Basin
E. average gabbro from Mid-Atlantic Ridge, Miyashiro et al., 1970
F. average of 5 altered basalts, Lau Basin
G. average of 6 greenstones from Franciscan Formation, Bailey et al., 1964
H. average of 2 dacite vitrophyres, Zephyr Shoal, Lau Basin
I. Diorite from Mid-Atlantic Ridge, Aumento, 1969

"andesite line" as an actual indication of the boundary between oceanic and continental crust. The Tonga Arc is the trace of the andesite line in this region, but the Lau Basin lies to the west of it and is sandwiched between the Tonga Arc and the Lau Ridge—another andesitic arc. Since basaltic eruptions may form in both oceanic and continental settings, it is important to compare the chemistry of Lau Basin basalt (LBB) with other types, such as ocean ridge basalt (ORB) and flood basalts of continents (CFB). Petrologic details of LBB and its assumed origin have been presented elsewhere (Hawkins et al., 1970, 1972; Sclater et al., 1972; Hawkins, 1973, 1974); only the major points will be summarized here.

The petrology and geochemistry of basalt from ocean ridges has been well studied. Ocean ridge basalts form a distinct class of rocks in terms of elements such as TiO_2, K_2O, P_2O_5, Sr, Ba, Rb, and Ni; element ratios such as CaO/Al_2O_3, Fe/Fe + Mg, K/Rb, and K/Ba; and in mineralogy. Strontium isotope ratios and rare earth element data give further support for this distinction, and in all respects the ridge basalt may be considered representative of essentially primitive partial melts of mantle material. The remarkable similarities of ORB samples, and the distinctive chemical characteristics which set them apart from CFB and alkalic baslats, have helped delineate P-T and chemical limits on the likely source regions for basalts leaking out at ocean-ridge crests. These data may be used to compare LBB with other basalt types to help define the type of magma generation and evolution in the Lau Basin.

Chemical data for the LBB indicate that all samples are subalkaline or tholeiitic basalts. Four classes of basalt were recognized on the basis of degree of alteration and normative composition (Hawkins, 1973, 1974). Three of the classes represent different degrees of oxidation, hydration, and selective removal of elements during alteration; the fourth class may show the effects of both strong fractional crystallization and alteration. The least-altered samples fit criteria for fresh submarine basalt (Shido et al., 1971) and, since they contain only microphenocrysts, they are considered to be representative of liquid compositions. These samples offer an opportunity to recognize possible physical and chemical characteristics of the parent material.

The data for least-altered samples of LBB show very close similarity to unaltered ORB samples. The average composition, based on 11 least-altered samples, is given in Table 2. The average oceanic tholeiite (Engel et al., 1963) and an average for Mid-Atlantic Ridge samples (Melson et al., 1968) are given for comparison. Also given in Table 2 is an average for the LBB-altered basalts and for greenstones from an ophiolite complex (Franciscan formation, Bailey et al., 1964). The only significant difference are the lower Al_2O_3 and higher total iron of the Franciscan.

In addition to the major element similarities seen in Table 2, LBB resemble ORB in trace elements,

Table 3. Average Values, Trace Elements (ppm)

	A	B
Cr	390	310
Co	100	100
Ni	170	160
Cu	100	100
Rb	0.15	2.9
Sr	100	100
Cs	0.002	0.04
Ba	1	30

A. Averages for 6 least altered basalts (data for transition metals by atomic absorption)

B. Averages for 6 least altered basalts plus 4 basalts with slight alteration. Rb, Cs and Ba data were done by isotope dilution by S. Hart. These data are for one least altered sample and three slightly altered samples and may not be representative of all of the samples.

especially Cr, Ni, Sr, Ba, Rb, and in the ratios K/Rb and K/Ba (Table 3). Sr-isotope data (Hart, 1971) show LBB to be slightly more radiogenic ($^{87}Sr/^{86}Sr$ = 0.7033-0.7051) than typical ORB values (0.7024). The enrichment could be due to contamination by crustal Sr or to an origin in an ^{87}Sr enriched source area. Because of the very low Sr abundances in the LBB samples (< 100 ppm), only a small amount of crustal Sr contamination could markedly effect the isotope ratios.

The analytic data for LBB samples are summarized in Fig. 6 in which the normative compositions of the samples are plotted on the Ne-Ol-Di-Hy-Q diagram. The shaded area includes the normative data for ORB reported by Kay et al. (1970). This diagram also points out the variation in chemistry within individual dredge hauls and between dredge hauls in the Lau Basin.

These differences are discussed in more detail in a separate paper (Hawkins, 1974), but, in summary, they are ascribed to minor differences in the ratio of clinopyroxene and anorthite in the mantle source material. Low-pressure fractional crystallization of liquidus phases (olivine and Ca-plagioclase) have been an additional control on the differences in chemistry.

SPECULATIONS ON THE ORIGIN OF THE LAU BASIN

The tectonic processes responsible for the formation of the Lau Basin, and perhaps for other marginal seas, such as the Philippine Sea and the seas of Japan and Okhotsk, may be related to the stress field set up in the interface between the descending slab and the low-velocity zone. Upwelling of hot mantle material behind the arcs results in partial melting and the production of basaltic magmas which are intruded to form new oceanic crust in the marginal seas. The source of the upwelling mantle material may be in the aesthenosphere rather than in the mantle of the lithosphere beneath the sea or in the lithosphere slab subducted in the inclined seismic zone. The main reasons to support this inference are: (1) the source material and the melting process appears to be similar to that which forms mid-ocean ridges; (2) there must be a very large magma source in order to continually generate oceanic tholeiites to floor the basins; and (3) the melts derived from the subducted lithosphere, or the mantle adjacent to it, appear to be andesitic or dacitic rather than basaltic, (i.e., they form the volcanic island arcs: Bryan et al., 1972; Ewart et al., 1973). Thus the probable source material for generating the marginal basin sea floor is likely to be in the low-velocity layer (LVL). The problem remains as to how and why it should melt and come to the surface. Sleep and Toksöz (1971) presented a model to explain the dilational forces in marginal basins as the result of a convective cell due to counterflow set up in the LVL. The hydrodynamic drag of the descending slab of rigid oceanic lithosphere would set up these cells in the low-viscosity material of the LVL.

This counterflow in the LVL is shown schematically in the cross sections of Figure 7. It is postulated that the drag exerted by the counterflow, plus the upward convective movement of hot mantle material, causes the separation and intrusion in the marginal basin. Figure 7 suggests two possible explanations for the geologic setting of the Lau Basin. The original configuration is shown in A of models I and II with the Lau Ridge as an active island arc above a descending slab of oceanic lithosphere. In late Miocene time the arc probably looked like this (Gill and Gorton, 1973). Model I-B shows the initial stages of the opening of the Lau Basin—perhaps in late Miocene or early Pliocene time. The original volcanic arc has been split as a result of the counterflow, the Lau Ridge is an abandoned volcanic arc, it has been replaced by the nascent Tonga Arc and the Lau Basin has started to open between the arcs. The position of the seismic zone has migrated relatively to the east; its former position is shown by a dashed line. This model is essentially that suggested by Karig (1970).

Model II-B is a slight modification of IB in that it does not require splitting of the old volcanic arc (Lau Ridge) but postulates that the inclined seismic zone in II-A is replaced by a second seismic zone a few tens or a hundred kilometers to the east. Counterflow above the seismic zone is again responsible for the opening of the basin between the arcs. In either case the position of the older (Lau Ridge) inclined seismic zone is abandoned and the active arc migrates relatively to the east. Both models present mechanical problems; Model I requires a splitting of the arc down the middle with the active half moving eastward. There is evidence of normal faulting on the east side of the Lau Ridge (Karig, 1970), but it is not clear what its age is, nor is it apparent from the morphology that the volcanic arc itself was separated into active and inactive segments. The best evidence to support this is the presence of upper Eocene foraminiferal limestone which overlies "rhyolitic flows and pyroclastics" of Eua Island in the Tonga Arc (Hoffmeister, 1932). The geology of Eua is unique in the Tonga Arc, and since it is not yet clear whether Eua represents a seamount, an andesitic island, or lifted sea floor, its significance in interpreting the history of the arcs is uncertain. Data presented by Ewart and Bryan (1972) support the idea that it is formed, at least in part, of uplifted sea-floor mafic rocks and thus is "ophiolitic."

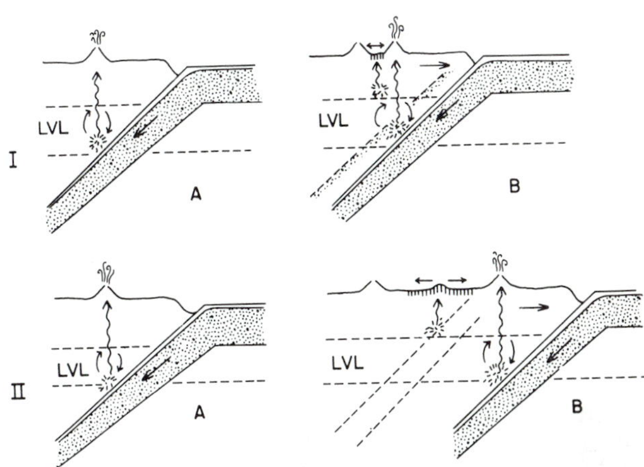

Fig. 7. Hypothetical models for explaining the origin of the Lau Basin and other marginal seas. In each case, Figure A represents an older single volcanic arc situated above an inclined seismic zone. Figure II-B represents the present Lau Basin-Tonga Trench tectonic setting. Figure I-B would illustrate an early stage in the opening of the basin due to the splitting of the older volcanic arc (Lau Ridge), as suggested by Karig (1970). The present depth of magma generation, which forms Lau Basin crust, is shown as coinciding with the upper edge of the abandoned slab of lithosphere in Figure II-B. This coincidence is the result of the geometry shown in the sketch and probably has nothing to do with the petrogenetic processes in the basin.

The writer favors Model II as an equally reasonable model, although there is a problem in explaining why the seismic zone should become reoriented by stepping out to a new location. Possible factors involved in this reorientation include a change in the angle of inclination of the Lau Basin seismic zone which made it mechanically unstable, a change in rate or direction of lithosphere subduction, or inability of the mantle to assimilate the incoming slab of lithosphere. The geometry shown in II-B implies that the Tonga Arc is encroaching on the location of the Tonga Trench. This should cause the trench to migrate east and either flatten the inclination of the seismic zone or more likely to cause the whole zone to migrate easterly while maintaining its present inclination (P. Molnar, oral communication, 1973). It is speculative, but this may be an important factor in explaining the apparent migration of island arcs and may actually be the reason for the termination of the Lau Ridge subduction zone.

MARGINAL BASINS AND THE ORIGIN OF OPHIOLITES

The ophiolite suite comprises rock types that are generally accepted as being typical of the lithologic units that form the oceanic crust and upper mantle. For example, pillow basalts, mafic volcanic breccia, diabase dikes, gabbro, greenstone, serpentinite, and various types of massive and layered cumulate ultramafic and anorthositic rocks have been recovered by dredging steep scarps on oceanic ridges and fracture zones. These rock types, plus radiolarites or bedded cherts, are the diagnostic components of the ophiolite sequences.

Ophiolites commonly show great disruption due to tectonism, but those which are least disturbed (e.g., Dewey and Bird, 1971) show a stratigraphic sequence like the layering of the sea floor, as implied by geophysical data and the recovery of carefully located dredge hauls. However, there are two major discrepancies between the geologic characteristics mapped in classic ophiolite sequences and our interpretation of the nature of the deep-ocean floor:

1. The rhythmically layered ribbon chert—"green shale" units, which are a relatively thin but distinctive part of most ophiolites, have not been sampled by dredging in the deep ocean, although DSDP data suggest that they may be a rare component of the sea floor.

2. The total preserved thickness of layers 2 and 3 in many ophiolite sequences (pillow basalts, diabase, and gabbro) rarely exceeds 1 or 2 km (e.g., Aubouin, 1965). This is in striking contrast to the aggregate thickness of seafloor layers 2 and 3 which is typically at least 4 km and possibly up to 7 km. The interpretation of the thickness of layers 2 and 3 in ophiolite sequences is based on the assumption that the petrologic transition from the mafic volcanic and plutonic assemblages to the serpentinized or unaltered massive and cumulate textured ultramafic rocks (a "petrologic Moho") actually is equivalent to the type geophysical Moho, as detected under present ocean basins. It is obvious that the thickness discrepancy that arises may be due to the assumption that the ultramafic part of ophiolite sequences is mantle material and not part of the layer 3.

Many descriptions of ophiolites lack definite data about the thicknesses of the components, probably because interpretation of layer thickness is made difficult by the intense faulting typical of these exposed occurrences. Most measured sections emphasize the relatively thin mafic layers (Auboin, 1965). The Vourinous ophiolite sequence of Greece has a maximum mafic volcanic rock thickness of only 3 km (Moores, 1969). The Great Valley, California, ophiolite sequence has only about 900 m of mafic plutonic and volcanic rocks (Bailey et al., 1970). A summary of 10 Franciscan and Great Valley ophiolite occurrences (Bailey et al., 1970) shows none with more than about 1.7 km of mafic rocks overlying ultramafic material. The Lush's Bight (ophiolite) Group of Newfoundland has at least 1,500 m of mafic rocks, although the true thickness is obscured by faulting (Strong, 1973). Smitheringale (1972) discusses the possibility that the reported 4.6 km of pillow basalts is probably too high because of unrecognized folding or faulting. The aggregate thickness of gabbro, basalt, and spilite of the Point Sal, California, ophiolite is only 300 m (Hopson and Frano, 1973, and oral communication). This anomalously thin "crust" has led Hopson to suggest that we see a "petrologic Moho" in these ophiolites rather than the geophysical Moho.

It appears that the thickness of oceanic crustal components of many ophiolites is varied but commonly is thin in comparison with the thickness of oceanic crust of the present-day ocean basins. Parker and Oldenburg (1973) have presented a model for lithosphere evolution and suggest an equation ($Z = 9.4t^{1/2}$ km, where Z = lithosphere thickness and t = age in my) that gives plate thickness with age. If the equation is applicable it is easy to see how thin oceanic lithosphere would be expected at distances of only a few hundred kilometers from ridge crests. In a marginal basin 300 km wide, spreading at a half-rate of 2 cm/yr, the maximum lithosphere thickness would be about 25 km. It would be unusual to expect a "normal" 6-km-thick crust on such thin lithosphere, and a 2- or 3-km crust might be more reasonable in a basin of this size. The point is that young marginal basins should be characterized by thin crust if thermal gradients are steep throughout the basin.

The structural position of marginal basins with respect to island arcs and subduction zones (Fig. 7) places them in a favorable position for incorporation in an orogenic belt. If marginal basins existed in pre-Cenozoic convergence zones, we should possibly expect to find their remnants preserved in orogenic belts; Dewey and Bird (1971) suggest this explanation for some of the Newfoundland ophiolites. Either subduction or imbrication of the marginal basin lithosphere, at either side of the basin presumably would be sufficient to incorporate it in the folded rocks of the resulting orogenic belt.

The silica-rich clastic material from the island arc forms a thin blanket on the Lau Basin floor. It is speculative to say that this could form ribbon cherts by diagenesis, but the sediments could certainly be the protolith of siliceous shales. Experimental studies on chert genesis by Keene and Kastner (1974) suggest that with increasing time these siliceous sediments could undergo diagenetic alteration, leading to formation of silica-rich layers and to bedded cherts.

There is no reason to suspect that all ophiolites were formed in marginal oceanic basins, but the model helps to explain the anomalously thin "oceanic crust" of a number of well-known occurrences, such as the Franciscan and Mediterranean examples. The Papuan Ultramafic Belt (Davies, 1971) is an example of an ophiolite assemblage with 8-10 km of basaltic and gabbroic crust. This could represent deep sea floor about 60 my old; at a half-spreading rate of 1 cm/my, the sea would have been at least 1,200 km wide. The 4- to 5-km-thick Newfoundland ophiolites are of oceanic thickness, but structural relations to older metamorphic rocks and their internal characteristics led Dewey and Bird (1971) to suggest that at least some of them formed in marginal basins. Thus there is evidence, both field and theoretical, to support the contention that marginal basins are an important geologic setting for ophiolite generation. The petrologic data for deep-sea-floor, marginal-basin, and ophiolite-suite mafic materials do not seem to permit making a distinction between these rocks. A very critical factor in attempting such a comparison is thatophiolite chemistry apparently has been modified by alteration and varying intensities of metamorphism. Field relations and layer thickness may prove to be the most useful criteria for recognizing relict crust from marginal basins.

SUMMARY

The Lau Basin is underlain by young sea floor formed of tholeiitic basalt. Chemically and petrologically it resembles the basalts formed at mid-ocean ridges. It is inferred that the same petrologic processes and the same type of source area which generates new sea floor at oceanic spreading centers have operated in the Lau Basin. The basin has a very thin sedimentary cover except for wedges of volcanic-derived sands and silts (to 0.7 sec thick) adjacent to the Tonga and Lau ridges. Heat-flow data show a wide range in values, but generally they are high, and in this respect the Lau Basin resembles many other marginal seas of the western Pacific. Although situated above the inclined seismic zone related to the presumed subduction of the Pacific Plate, the basin appears to have been formed by dilation through separation of the Lau and Tonga ridges. This separation probably has been operating since at least Pliocene time but probably no longer than since mid-Miocene. The high heat flow seems to be explained by the emplacement of large volumes of basaltic magma which generate the sea floor. The source of the magma probably is in the low-velocity layer (LVL) above the inclined seismic zone.Counterflow in the LVL may exert a drag on the lithosphere and be important in initiating and maintaining the stresses that cause basin generation.

The sea floor of the Lau Basin comprises tholeiitic pillow basalts, dikes, gabbros, and greenstones. It is overlain by thin layers of silica-rich volcanic-clastic sediments. Presumably ultramafic residue are deeper in the crust or upper mantle. The rock materials of the Lau Basin resemble components of ophiolite suites. If eventually the Lau Basin were to be caught up in an orogenic belt, it possibly would resemble typical deformed and sheared ophiolites.

ACKNOWLEDGMENTS

The research was supported by NSF grants GA-16120, GA-30315, and GA-33227, and ship operations for SIO expeditions 7-TOW, ANTIPODE, and SOUTHTOW were supported by NSF block funding. The author especially wishes to acknowledge the assistance of Captains T. Hansen, N. Ferris, and J. Bonham and the ships crews of R/V WASHINGTON and R/V MELVILLE. James Natland, Richard Nishimore, Lawrence Lawver, John Keene, and Rodey Batiza all assisted in data collection and analysis; George Hohnhaus and Fred Dixon gave invaluable help aboard ship in the sampling operations. The Lau Basin chart was drafted by Judy Clinton. The paper has been helped by numerous discussions with J. Sclater, R.L. Fisher, J. Mammerickx, D. Karig, E. Winterer, and P. Molnar. I wish to thank the Government of the Kingdom of Tonga for providing port clearances and encouraging the studies.

BIBLIOGRAPHY

Aubouin, J., 1965, Geosynclines: New York, American Elsevier, 335 p.

Aumento, F., 1969, Diorites from the Mid-Atlantic Ridge at 45°N: Science, v. 169, p. 1112-1113.

Bailey, E., Blake, C., and Jones, D., 1970, Onland Mesozoic oceanic crust in California coast ranges, *in* U.S. Geol. Survey Prof. Paper 700-c, p. c70-c81.

Bailey, E. H., Irwin, W. P., and Jones, D. L., 1964, Franciscan and related rocks and their significance in the geology of western California: Calif. Div. Mines Geol. Bull., v. 183, 177 p.

Barazangi, M., and Dorman, J., 1969, World Seismicity map of ESSA Coast and Geodetic Survey epicenter data for 1961-1967: Seismol. Soc. America Bull., v. 59, p. 369-380.

Barazangi, M., and Isacks, B., 1971, Lateral variations of seismic-wave attenuation in the upper mantle above the inclined earthquake zone of the Tonga Island arc: deep anomaly in the upper mantle: Jour. Geophys. Res., v. 76, p. 8493-8516.

Berthine, K., 1974, Submarine weathering of tholeiitic basalts and the origin of metalliferous sediments: Geochim. Cosmochim. Acta (in press).

Bryan, W. B., Stice, G. D., and Eward, A., 1972, Geology, petrography, and geochemistry of the volcanic islands of Tonga: Jour. Geophys. Res., v. 77, p. 1566-1585.

Davies, H., 1971, Peridotite-gabbro-basalt complex in eastern Papua: an overthrust plate of oceanic mantle and crust: Commonwealth Austral., Bur. Mineral Resources, Geol. and Geophys. Bull., v. 128, 48 p.

Dewey, J., and Bird, J., 1971, Origin and emplacement of the ophiolite suite: Appalachian ophiolites in Newfoundland: Jour Geophys. Res., v. 76, p. 3179-3206.

Engel, A. E., Engel, C. G., and Havens, R. G., 1963, Chemical characteristics of oceanic basalts and the upper mantle: Geol. Soc. America, v. 76, p. 719-734.

Ewart, A. and Bryan, W.B., 1972, Petrography and geochemistry of igneous rocks from Eua, Tongan Islands: Geol. Soc. America Bull., v. 83, p. 3281-3298.

Ewart, A., Bryan, W., and Gill, J., 1973, Mineralogy and geochemistry of the younger volcanic islands of Tonga, S.W. Pacific: Jour. Petr., v. 14, p. 429-466.

Fisher, R. L., and Engel, C. G., 1969, Ultramafic and basaltic rocks dredged from the nearshore flank of the Tonga Trench: Geol. Soc. America Bull., v. 80, p. 1373-1378.

Gill, J., and Gorton, M., 1973, A proposed geological and geochemical history of eastern Melanasia, *in* Coleman, P. J., ed., The western Pacific: Univ. Western Australia Press, p. 543-566.

Griffin, J. J., Koide, M., Hohndorf, A., Hawkins, J., and Goldberg, E., 1972, Sediments of the Lau Basin—rapidly accumulating volcanic deposits: Deep-Sea Res., v. 19, p. 139-148.

Hart, S. R., 1971, Dredge basalts: some geochemical aspects: Am. Geophys. Union Trans., v. 52, p. 376.

Hawkins, J. W., 1973, Petrology of marginal basins and their possible significance in orogenic belts: the Lau Basin as an example (abstr.): Geol. Soc. America, v. 5, p. 51.

———, 1974, Petrology and geochemistry of the basaltic seafloor of the Lau Basin: ms.

———, and Nishimori, R. K., 1971, Oceanic ridge-type tholeiitic rocks from the Lau Basin: their petrology and significance (abstr.): Geol. Soc. America, v. 3, p. 594.

———, Sclater, J., and Hohnhaus, G., 1970, Petrologic and geophysical characteristics of the Lau Basin Ridge: a spreading center behind the Tonga Arc (abstr.): Geol. Soc. America Ann. Meeting, v. 2, no. 7, p. 571.

———, Fisher, R. L., and Engel, C. G., 1972, Ultramafic and mafic rock suites exposed on the deep flanks of the Tonga Trench (abstr.): Geol. Soc. America, v. 4, p. 167-168.

Hoffmeister, J. E., 1932, Geology of Eua Island, Tonga: Bernice P. Bishop Museum Bull., v.96, p. 1-93.

Hopson, C., and Frano, C., 1973, Jurassic oceanic crustal sequence of Point Sal, Calif.: EOS Trans. Am. Geophys. Union (abstr.), v. 54, p. 1220.

Karig, D. E., 1970, Ridges and basins of the Tonga-Kermadec island arc system: Jour. Geophys. Res., v. 75, p. 239-254.

Kay, R., Hubbard, N., and Gast, P., 1970, Chemical characteristics and origin of oceanic ridge volcanic rocks: Jour. Geophys. Res., v. 75, p. 239-254.

Keene, J. B., and Kastner, M., 1974, Diagenesis of clays and the formation of deep-sea chert: Science (in press).

Kibblewhite, A. C., and Denham, R. N., 1967, The bathymetry and total magnetic field of the South Kermadec ridge seamounts: New Zealand Jour. Sci., v. 10, p. 52-67.

Luyendyk, B.P., MacDonald, K.C., and Bryan, W.B., 1973, Rifting history of the Woodlark Basin in the Southwest Pacific: Geol. Soc. America Bull., v. 84, p. 1125-1134.

Mammerickx, J., Chase, T. E., Smith, S. M., and Taylor, I. L., 1971, Bathymetry of the south Pacific: Scripps Institution of Oceanography, Sheet 12.

Matthews, D. J., 1939, Tables of the velocity of sound in pure water and sea water for use in echo-sounding and sound-ranging: Hydrographic Dept., Admiralty, London.

McKenzie, D., and Sclater, J., 1968, Heat flow inside the island arcs of the northwestern Pacific: Jour. Geophys. Res., v. 73, p. 3173-3179.

Melson, W. G., Thompson, G., and van Andel, T. H., 1968, Volcanism and metamorphism in the Mid-Atlantic Ridge, 22°N latitude: Jour. Geophys. Res., v. 73, p. 5925-5941.

Menard, H. W., 1967, Transitional types of crust under small ocean basins: Jour. Geophys. Res., v. 72, p. 3061-3073.

Miyashiro, A., 1970, Crystallization and differentiation in abyssal tholeiites and gabbros from mid-oceanic ridges: Earth and Planetary Sci. Letters, v. 7, p. 361-365.

Moores, E., 1969, Petrology and structure of the Voarinos ophiolitic complex of northern Greece: Geol. Soc. America Special Paper 118, 74 p.

Murauchi. S.. and Den, N., 1966, Origin of the Japan Sea: Paper presented at Earthquake Res. Inst., Univ. Tokyo, Monthly Colloquium.

Oxburgh, E. R., and Turcotte, D. L., 1971, Origin of paired metamorphic belts and crustal dilation in island arc regions: Jour. Geophys. Res., v. 76, p. 1315-1327.

Packham, G. H., and Falvey, D. A., 1971, An hypothesis for the formation of marginal seas of the western Pacific: Tectonophysics, v. 11, p. 79-109.

Parker, R., and Oldenburg, D., 1973, Thermal model of ocean ridges: Nature, 242, p. 137-139.

Sclater, J., Hawkins, J., Mammerickx, J., and Chase, C., 1972, Crustal extension between the Tonga and Lau ridges: petrologic and geophysical evidence: Geol. Soc. America, v. 83, p. 505-518.

Shido, F. Miyashiro, A., and Ewing, M., 1971, Crystallization of abyssal tholeiites: Contrib. Mineral. Petr., v. 31, p. 251-266.

Sleep, N., and Toksoz, M. N., 1971, Evolution of marginal basins: Nature, v. 233, p. 548-550.

Smitheringale, W. G., 1972, Low-potash Lush's Bight tholeiites: ancient oceanic crust in Newfoundland: Can. Jour. Earth Sci., v. 9, p. 574-588.

Strong, D. F., 1973, Lush's Bight and Roberts Arm groups of Newfoundland: possible juxtaposed oceanic island-arc volcanic suites: Geol. Soc. America Bul., v. 84, p. 3917-3928.

Sugimura, A., and Uyeda, S., 1973, Island arcs, Japan and its environs: develoments in geotectonics, v. 3, New York, Elsevier, 247 p.

Sykes, L. R., Isacks, B. L. and Oliver, J., 1969, Spatial distribution of deep and shallow earthquakes of small magnitude in the Fiji-Tonga region: Seismol. Soc. America Bull., v. 59, p. 1093-1113.

Uyeda, S., and Vacquier, V., 1968, Geothermal and geomagnetic data in and around the island arc of Japan, *in* Knopoff, L., Drake, C. L., and Hart, P. J., eds., The crust and upper mantle of the Pacific area: Washington, D. C., Am. Geophys. Union, AGU Monogr. 12, p. 349-366.

Van der Linden, W., 1969, Extinct mid-ocean ridges in the Tasman Sea and the western Pacific: Earth and Planetary Sci. Letters, v. 6, p. 483-490.

Watanabe, T., Epp, D., Uyeda, S., Langseth, M., and Yusui, M., 1970, Heat flow in the Phillipine Sea: Tectonophysics, v. 10, p. 205-224.

Wegener, A., 1929, The origin of continents and oceans (translated by J. Biram, 1966), New York, Dover Publications, 246 p.

Continental Margins Near New Caledonia

J. Dubois, C. Ravenne, A. Aubertin, J. Louis, R. Guillaume, J. Launay and L. Montadert

INTRODUCTION

A survey of the New Caledonian margins reveals several regions between the proper oceanic domain in the central Pacific and the continental domain forming the Australian continent. Earth physics and, in particular, seismology can be used to differentiate the following three units of the uppermost mantle and the crust: (1) the submerged continental zone (made up of subsea basins and ridges) to which New Caledonia and the Loyalty Islands belong; (2) the interarc expansion basins in which an oceanic lithosphere takes form (Lau Basin, North Fiji Plateau); and (3) the island arcs (New Hebrides, Tonga-Kermadec). The relationships between these units are what determine the tectonics of these margins and enable their geological history to be better understood.

The main facts showed off by the seismic-reflection profiles are the following.

Structural Point of View

It is important that the existence of a Fairway rise can be interpreted as an outer arc related to New Caledonia-Loyalty island arc. The New Caledonian area, where the structural directions (northwest-southeast) are different from the Australian (north-south), is explained as a fossil island arc that was active at least during Oligocene and probably since Early Cretaceous.

Referring to the present patterns about island

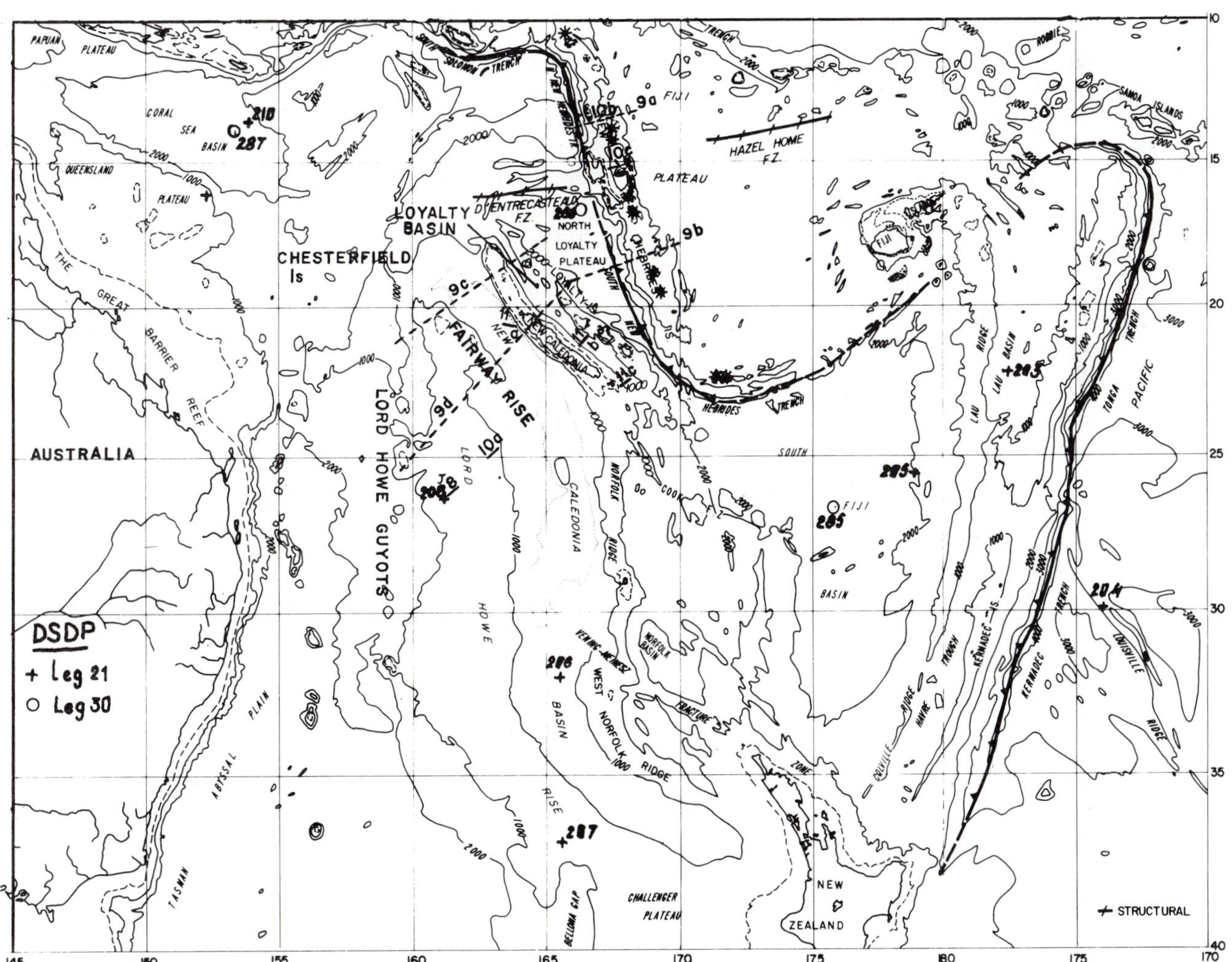

Fig. 1. The margins near New Caledonia. Bathymetry, site location of DSDP, JOIDES (Legs 21 and 30), and location of the cross sections and of some sections of the seismic profiles.

arcs, we put forward the following equivalence: (1) New Caledonia, nonvolcanic outer arc; (2) Loyalty Basin, inter deep basin; (3) Loyalty islands, volcanic inner arc; and (4) North Loyalty Plateau, interarc or marginal basin. The trench could have been located west of New Caledonia, in the deep basin.

Throughout the history of the area, vertical and horizontal movements occurred; the vertical ones are emphasized by the series of emergences and submergences and the great uplifting of the erosion terraces in the islands. The erosion disconformity, observed on Norfolk Ridge, connected with the Oligocene disconformity on Lord Howe Rise, appears to be synchronous with the Miocene peneplanation in New Caledonia. The horizontal movements are closely tied to the history of the remnant island arcs, as for New Caledonia, and active verse faults through the New Caledonia-Loyalty system.

Sedimentary Point of View

The sedimentary basins are conformable to the bathymetric basins. Thicknesses of sediments greater than 2.4 sec (two-way time, D.T.T.) occur in the Lord Howe Basin, New Caledonia Basin, and Loyalty Basin.

The erosion of the relief of New Caledonia seems responsible for the sedimentary filling of the New Caledonia Basin (where only the Neogene is important) and partly of the Loyalty Basin (which also received some volcano-clastic sediments from the Loyalty Chain). A strong subsidence zone runs along New Caledonia, and the thickness of sediments is far greater than 3 sec D.T.T.

We notice a thin thickness of sediments on the Fairway Rise and North Loyalty Plateau. There is a difference between the constitution of the Fairway Rise and Lord Howe Rise. The substratum of the Lord Howe Rise is sedimentary, for there are continuous coherent reflections at depth.

In the New Hebrides island arc, we observed a great thickness of sediments (more than 2 sec D.T.T.), especially in the broken-down basins in the north and the south. This is not compatible with the existence of an active interarc basin at this place. Difference between arc-trench gap in the New Caledonia and New Hebrides system can possibly be explained by difference in age.

DEEP STRUCTURE

New Caledonia and the regions around it are located in a zone halfway between the oceanic Pacific domain in the east and the continental Australian domain in the west. This continental margin area roughly corresponds to the zone west of the andesite line plotted by MacDonald et al. (1953). The sea floors surrounding New Caledonia have various structures (basins, trenches, rises, seamounts, etc.) (Fig. 1). An initial approach to a survey of these margins consists in considering the lateral heterogeneities that can be observed by earth physical/techniques, in defining the principal structural units in the region.

Bathymetry shows that this region is made up of a succession of submarine basins and ranges generally trending southeast, interrupted north of 16°S by the d'Entrecasteaux Fracture Zone. A deep trench runs along the New Hebrides Island arc, which is known for its great seismic and volcanic activity. It stops after curving eastward toward the Fiji Islands. Another arc farther eastward, the Tonga-Kermadec arc, bounds the area of the purely oceanic domain. Between these two arcs, monotonous plateaus (2,500 m), the North Fiji Plateau and the Lau Basin, stretch west and east of the Fiji Islands. The South Fiji Basin is deeper (4,500 m), as is the Tasman Sea in the southwestern part of the region, belonging to the oceanic domain.

The shape of the geoid (Guier and Newton, 1965) as determined from satellite tracking shows an appreciable swelling (+50 m) having its center in New Caledonia. The axis of this swelling runs southeast and generally covers the entire area being examined. The +40-mgal contour line follows the trend of the New Hebrides arc. We can thus expect to find an abnormal uppermost mantle in this region.

Magnetic anomalies such as those found during cruises in this area show that the deep-lying structural features of the crust closely follow those of the bathymetry, i.e., trending with the axes of the southeast anomalies (van der Linden, 1968; Lapouille and Henry, 1971). However, analysis of the seismicity (Fig. 2) and of the volume of body wave and surface-wave propagation are the most useful to distinguish the main units making up this region on the scale of the uppermost mantle and the crust. A separation is thus made between the zone directly surrounding New Caledonia, the interarc-basin region (North Fiji Plateau, Lau Basin) and the seismic arcs (New Hebrides and Tonga-Kermadec).

Area Surrounding New Caledonia and the Loyalty Islands

An analysis of Rayleigh-wave dispersion (Dubois, 1968, 1969, 1971) (Fig. 3) shows that the crustal structure in this area has a continental nature, even though it is underneath a deep-water layer (up to 3,500 m). A comparison between the observations made at Noumea, New Caledonia, and the theoretical dispersion curves of Rayleigh waves (Saito and Takeuchi, 1966) suggests that the crust here has a mean thickness of 20 km, that it becomes thicker underneath the ranges to attain 25 km (Norfolk Ridge and Lord Howe Rise), and then becomes thinner underneath the basins (New Caledonia Basin, Loyalty Basin, South Fiji Basin: Macquarie-Nouméa, New Zealand-Nouméa, and Kermadec-Nouméa ray paths). On the New Guinea-Noumea ray paths in the northwestern part, the crust appears to be the thickest, 25 km at the most.

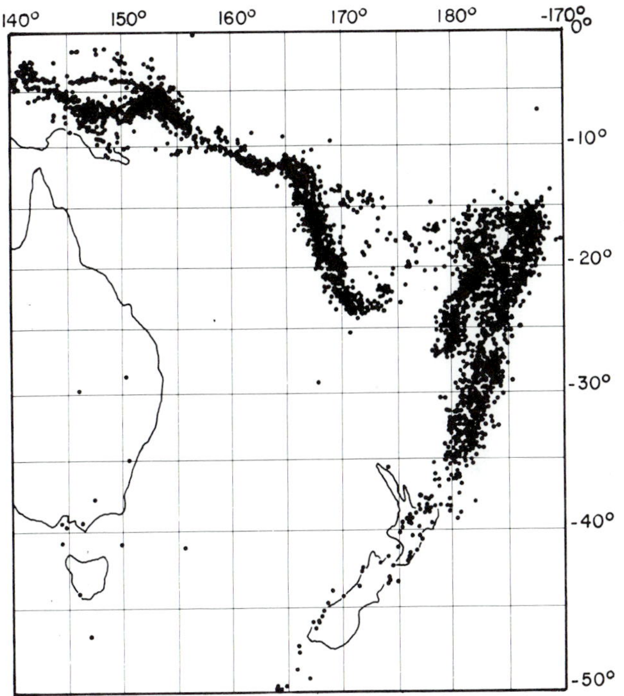

Fig. 2. Seismicity map from ESSA, Coast and Geodetic Survey. Epicenter data, 1961-1967, southwestern Pacific area.

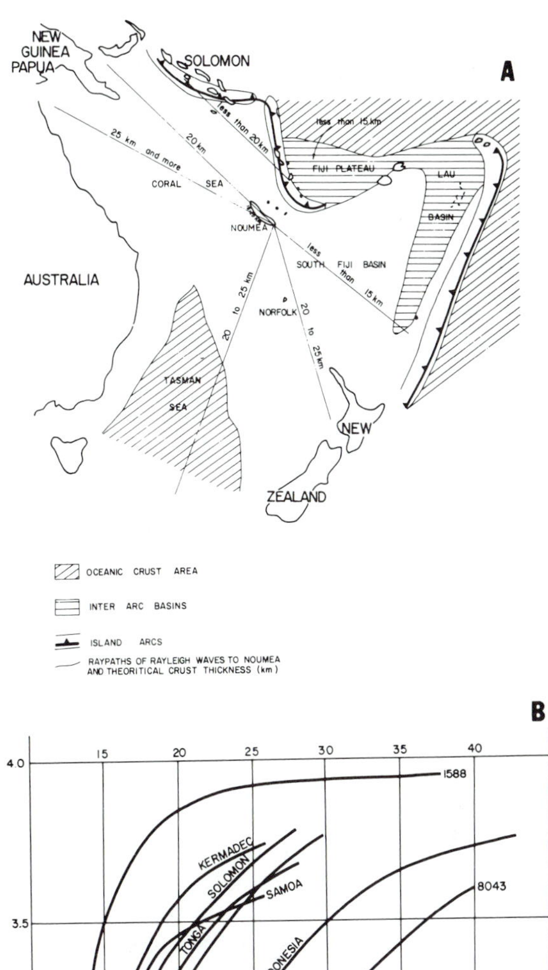

Fig. 3. New Caledonia-Fiji-Loyalty seismicity. (a) The four structural units: oceanic, continental, expansion basins, and four island arcs. Rayleigh-wave ray paths in the continental area and theoretical crust thickness (km). (b) Dispersion curves of Rayleigh waves to Noumea from different earthquakes belts. Comparison with oceanic 1588 and continental 8043 theoretical models.

The reason for this is that it crosses the Queensland Plateau, which is the submarine prolongation of Australia.

The Rayleigh-wave phase velocities between Noumea and Koumac, along the western coast of New Caledonia, suggest a crustal thickness of 18-20 km, whereas underneath the central range the residual time variations in the P waves, along the azimuth, reveals a thickening of the crust to 35 km (Dubois, 1969; Dubois et al., 1973). In addition, the relatively slow propagation velocities of the long Rayleigh wavelengths also show that, in this entire zone, the layer with the slowest S-wave velocity in the uppermost mantle is at a depth of about 60 km (Saito and Takeuchi, 1966; Dubois, 1969).

The sedimentary basins show negative smooth magnetic anomalies, while the subcontinental elevated structures produce long positive anomalies with smaller ones superimposed (Lapouille and Henry, 1973). Refraction and sonobuoy profiles complete the patterns given by seismology. In the Coral Sea, Shor (1967) finds a crust thickness of 15 km underneath 4 km of water. Layer 3, called the oceanic layer, reaches 11 km thickness there, and Raitt (1956) feels that there might be a transformation process whereby the oceanic crust turns into a continental crust.

The refraction profiles by Shor et al. (1971) in the Lord Howe Rise clearly reveal the thickening underneath the rise (Fig. 4), as brought out by seismology (thickness between 16.5 and 28 km), while Woodward and Hunt (1970) use gravimetry to determine a thickness of 26 km and qualify this crust as quasi-continental. Likewise, Layer 3 (P-wave velocity of 6.8 km/sec) is present underneath the rise, and so the problem is to determine whether the Lord Howe Rise is a fragment of the continental crust or a volcanic ridge built up on an oceanic foundation (see also Bentz, this volume). Underneath the Norfolk Ridge, the refraction profiles by Shor et al. (1967) show a thickness of 21 km, and Layer 3 is still present.

In the New Caledonia Basin, the uppermost mantle is at a depth between 10 and 16 km according to refraction profiling (Shor, 1971), whereas Solomon and Biehler (1969) use gravimetry to locate

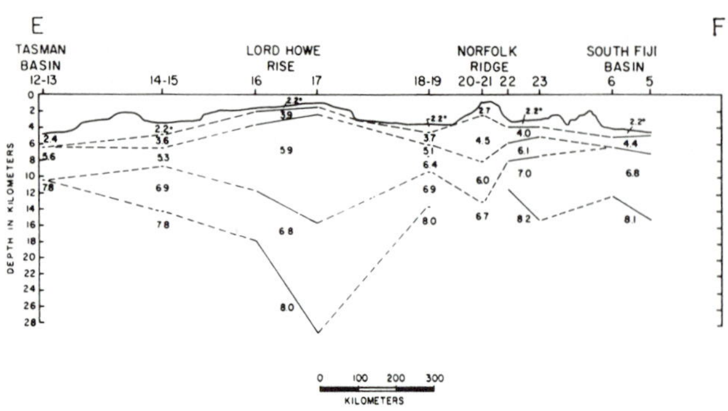

Fig. 4. Structure section across the western part of the Melanesian border (from Shor et al., 1971).

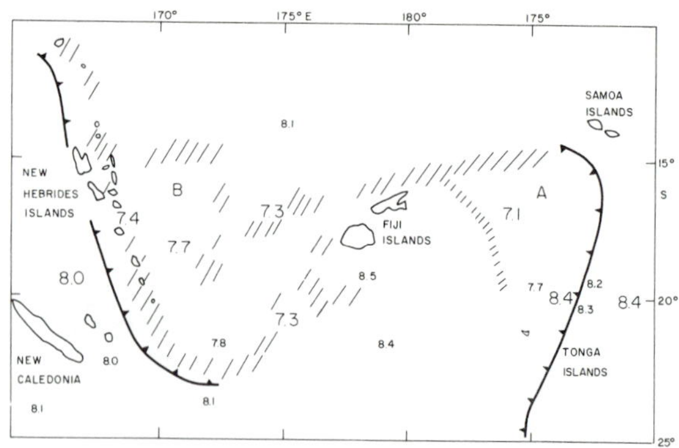

Fig. 5. Fiji Plateau and Lau Basin; P-wave velocities in the uppermost mantle, showing locations of shallow earthquakes (dashed lines). Data from Shor et al. (1971) (small numbers), Aggarwal et al. (1972), Dubois et al. (1973) (large numbers).

it at 17-18 km, and Woodward and Hunt (1970) find it between 9 and 16 km by gravimetry. In the Loyalty Basin crustal thickness appears to vary between 15 and 20 km (Shor, 1967). The South Fiji Basin has a more oceanic nature, but profiles by Raitt (1956) and Shor et al. (1971) show that the "oceanic" and sedimentary layers are thicker than usual there (6-12 km).

North Fiji Plateau

The structure of this basin is quite different from that of the regions described above; it is actively seismic, and the superficial and normal foci are distributed according to broad zones of deformation (Sykes et al., 1969; Chase, 1971; Dubois et al., 1973). Propagation of P and S waves in the uppermost mantle reveal the actual seismic plateau for which these velocities underneath the Moho are 7.6 and 4.3 km/sec (with attenuation of the S). The boundaries of the plateau are defined by a seismic belt made up of a line east of the New Hebrides, the Hunter Fracture Zone, a line west and north of Fiji, and the Hazel Home Fracture Zone, all of which are characterized by their low P-wave velocity (7.2 km/sec) and the very high attenuation of the S (Dubois et al., 1973; Barazangi et al., 1973) (Fig. 5). Likewise, the dispersion of Rayleigh waves from these earthquakes to the Port Vila seismological station clearly reveals these differences according to the azimuth considered. However, an analysis of P- and S-wave propagations does not show the existence of a thin crust underneath the Plateau. For Rayleigh wave propagation, the low S velocities in the mantle make up for the influence of the thin crust, and the dispersion curves are of a type halfway between oceanic and continental (Dubois, 1969). A refraction profile (Shor et al., 1971) and a sonobouy survey (Sutton et al., 1969) show a crust thickness of 5.5 km underneath a water layer of 2.5 km.

On the basis of these seismological data and considering observations on the high heat flow (Sclater and Menard, 1967; Sclater et al., 1972; MacDonald et al., 1973) on gravity measurements indicative of low-density uppermost mantle (Solomon and Biehler, 1969), as well as on the shallow-water depth (2.5 km) and the thinness of the sediments, we can deduce (Dubois et al., 1973) that the Fiji Plateau is not an oceanic lithosphere belonging to the Pacific Ocean, but a recently generated crust. The focal mechanisms of earthquakes on the Hunter Fracture Zone, with left-lateral strike-slip motion along the fracture (Johnson and Molnar, 1972), fit in with this pattern of opening up. The Lau Basin has the same features (Sykes et al., 1969; Karig, 1970; Barazangi and Isacks, 1971; Sclater et al., 1972a, 1972b; Mitronovas and Isacks, 1971).

Island Arcs

Between the two domains described above, the New Hebrides Island arc is situated. With regard to its seismicity, this arc is characterized by two singularities: (1) the eastward inclination of the seismic plane, i.e., toward the middle of the Pacific, as opposed to all the other circum-Pacific arcs (except the Solomon Islands arc); and (2) the seismic gap between the depths of 350 and 600 km (Fig. 6).

Structurally, the oceanic trench along the arc in the west is discontinuous. It breaks off at the latitude of Espiritu Santo at the level of the Hazel Fracture Zone and the d'Entrecasteaux Fracture Zone, both trending east-west. This break does not exist along the seismic plane, which is quite continuous from north to south. The major structural line appears as an active volcanic alignment parallel to the trench and perpendicular to the 200-km isodepth line of the seismic Benioff Zone (Fig. 6). The northern part of the volcanic line (north of Epi Island) crosses the basins bounded in the east and west by former volcanic islands. In the south the volcanic line is bounded in the east by en échelon troughs, of which the largest is the Coriolis Trough (Puech and Reichenfeld, 1969). The velocity of the P waves is high (7 km/sec) in a crust about 20 km thick, but it is low in the uppermost mantle (7.4 km/sec), i.e., in the part of the asthenosphere located between the

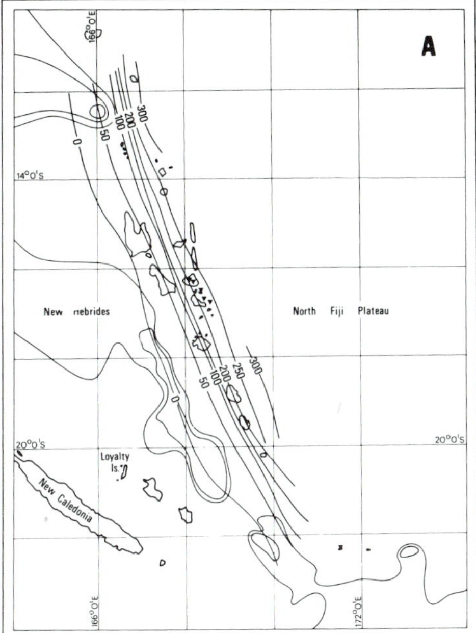

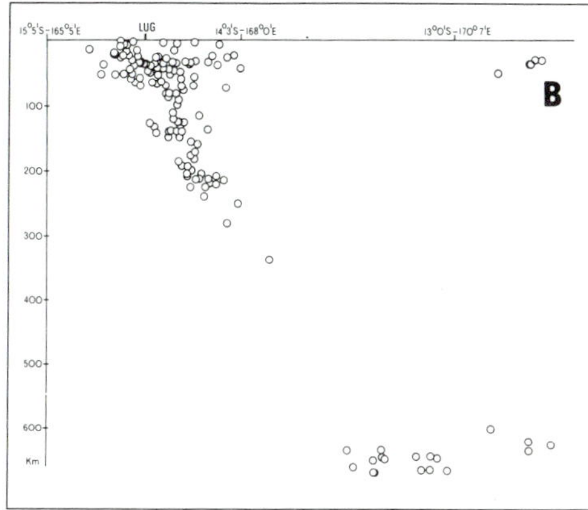

Fig. 6. New Hebrides island arc. (a) Map of isodepth lines that define Benioff Zone. (b) Typical cross section of earthquake hypocenters (Luganville, Espiritu Santo profile) (from Dubois, 1971).

Benioff plane and the crust. Attenuation of the S waves is great there (Barazangi et al., 1973).

The possible geometry and structure of an underthrusting slab were determined in detail from an analysis of the P and S propagations by the ray-tracing method between the deep foci underneath the Fiji Plateau and the seismological stations in the New Hebrides (Pascal et al., 1973). The break in seismicity at 350 km is interpreted (Barazangi et al., 1973) by a break in the oceanic lithospheric slab at this depth. Below that, the asthenosphere (low velocity of S waves, great attenuation) appears to extend to 600 km, and everything seems to indicate that the deep earthquakes in the Fiji Plateau are associated with a piece of the lithosphere that has become detached from the underthrusting oceanic slab.

Gravimetric measurements made on land (Malahoff and Woolard, 1969) and offshore (Luyendyk et al., 1973) show that the positive Bouguer anomaly in this area increases from 100 mgal in the west (Espiritu Santo, Malekula) to more than 200 mgal in the islands in the east. This is in good agreement with the presence of an upper mantle "corner" in the Benioff zone. The offshore profiles show this increase in the Bouguer anomaly from 100 mgal on the east coast of Vate (isodepth line 100 km) to 250 mgal 170 km offshore to the northeast.

The Tonga-Kermadec island arc has been the subject of a great many seismological and structural studies (Sykes, 1966; Oliver and Isacks, 1967; Isacks et al., 1967, 1969; Sykes et al., 1968, 1969; Mino et al., 1968; Barazangi et al., 1972; Mitronovas and Isacks, 1971; Isacks and Molnar, 1971; Barazangi and Isacks, 1971; Aggarwal et al., 1972; Karig, 1970). Its seismicity in its northern part extends continuously to 700 km depth, and an analysis of the wave propagation by Oliver and Isacks (1967) in this area is the basis for the fundamental hypothesis of the underthrusting of the oceanic lithosphere under the island arcs. Mention should be made of the symmetrical pattern, compared with the Fiji Islands, of both the New Hebrides and Tonga-Kermadec arcs with opposite-facing lithospheric consumption zones and the existence, in their concavity, of an expansion basin where the oceanic lithosphere is formed. Other important features are the basic difference in these two underthrusting systems with an oceanic lithosphere under a subcontinental lithosphere for the Tonga-Kermadec arc, and a continental lithosphere under a recently generated oceanic lithosphere for the Hebrides arc.

An analysis of the marine fossil levels in New Caledonia and the Loyalty Islands led us (Dubois et al., 1973) to try to interpret the uplifting of the atolls that form the Loyalty Islands in the lithospheric bulge of Australo-Tasmania, prior to its subduction at the New Hebrides. Indeed, the entire marginal area of New Caledonia and the Loyalty Islands appears to have been affected by this downthrust of the continental plate under the Fiji Plateau. In addition to topographic arguments (uplifted atolls) there are geophysical arguments (Bouguer anomaly corresponding to a bulge of the upper mantle and a magnetic anomaly) along with arguments of a dynamic nature (such as dating the different atolls at different stages of the uplifting according to their position in relation to the axis of the bulge).

Conclusions

These margins in the vicinity of New Caledonia have been classified in three main structural units on the basis of seismological data as well as of bathymetry, magnetism, gravimetry, heat flow, etc. These three units are as follows:

1. The oceanic region, with a continental-type crust, that surrounds New Caledonia and extends, via a succession of basins and ranges, from Australia to the New Hebrides and Tonga-Kermadec island arcs (except for the northern part of the Tasman Sea).

2. The North Fiji Plateau and the Lau Basin, in the concave parts of the island arcs, are highly singular areas at a shallow depth, with an oceanic crust and an abnormal uppermost mantle, that are interpreted as having been formed from the recently generated lithosphere (the Woodlark Plateau in the north appears comparable).

3. The island arcs make up the boundaries of these different zones and form two opposite-facing lithospheric consumption zones.

STRUCTURAL HISTORY AND SEDIMENTARY RECORD

New data on sediment distribution in the southwestern Pacific were obtained by seismic reflection during the Austradec cruises. These cruises were carried out by IFP, CFP, SNPA, and ELF-ERAP as part of a project sponsored by the Comité d'Etudes Pétrolières Marines in collaboration with ORSTOM and using CNEXO's vessels. The profiles were obtained with the IFP Flexichoc implosion source, a numerical recording laboratory and a 12-traces streamer for performing CDP shooting of 1,200 per 100. The profiles used in this report cover the area between the alignment of the Lord Howe Guyots and the Chesterfield Islands in the west, and the north Fiji Plateau in the east.

Historical Elements Provided by Onshore Geology

New Caledonia. The oldest formations identified are ante-Permian (Avias and Tonord, 1973). In the Permian, the sediments are made up of a volcanic series with a few calcareous interbeds and sandy deposits. The Triassic is marked by a marine transgression. The first graywackes appear and will go on to make up most of the Jurassic deposits. The absence of the Bathonian-Bajocian suggests, as in New Zealand, that the island emerged. The sedimentation changes abruptly in the Cretaceous, with argillaceous and sandy deposits. Also in the Cretaceous, there is a formation of emerged ridges with the presence of coal deposits and a thick conglomeratic and red arkosic formation encompassing the future Paleogene "sillon" (Routhier, 1953), in which the Cretaceous and Tertiary formations are well developed, folded but nonmetamorphic. The Eocene is present in three lithologic types: the external flysch facies (Tissot and Noesmoen, 1958) against the west coast; the internal facies (Tissot and Noesmoen, 1958), with sediments and fossils that are markers of deep sedimentation (phtanites, globigerinids) east of the external facies; and the basalt and dolerite volcanic outflows in two places—one in the "sillon" (Routhier, 1953) between the external facies and the middle range, and the other east of the island.

During the Eocene tectonic movements began. They went on to reach their climax in the Oligocene with the emplacment of peridotites. This compressional phase continued until the beginning of the Miocene. The Neogene and Quaternary are marked by vertical movements (transgressions and uplifts).

Loyalty Island Ridge. This range is volcanic and is topped off by reefs. The last major outflows date from the late Miocene (oral report by the ORSTOM Geology Section in Nouméa). This does not exclude an earlier date for this range, as shown by Chevalier's (1968) dating of a marine basalt indicating a late Oligocene age.

New Hebrides. These mountains are divided into three ranges (Dugas, 1971). The central range is the most recent (Plio-Quaternary) and is formed of still-active or recently active volcanoes. The eastern (Pentecost, Aurora) and western (Espiritu Santo, Malekoula, Torres) ranges are older. Andesite from Torres Island has been dated in the late Eocene (Warden, 1968), showing that the arc was already active there at that time. Pre-Miocene red argillites (Mitchell, 1970) are found west of Malekula. Their makeup reveals a deep-water environment located beneath the carbonate compensation depth. These two lateral ranges are essentially made up of volcano-sedimentary deposits with a few calcareous interbeds, linked to the existence of reefs in the early Miocene. The activity of the arc probably started later on the emerged folds in this part of the New Hebrides, around the Oligo-Miocene.

Bathymetry is of great interest for the area around New Caledonia because we find that the boundaries of the sedimentary and bathymetric basins are the same and that the isobath curves reflect structural features (Fig. 1).

Western New Caledonia Region

Bathymetry is of great interest for the area around New Caledonia because we find that the boundaries of the sedimentary and bathymetric basins are the same and that the isobath curves reflect structural features (Fig. 1). Two zones can be distinguished according to the trend of the structural elements between 19°S and 24°S (Fig. 7):

1. A zone in which the trends are north-south, called the "Australian" zone because this trend is the same as that of the Australian continental slope as well as of the structures between this slope and the area studied. It includes the alignment of the Lord Howe Guyots, Lord Howe Basin, and Lord Howe Rise.

2. A New Caledonian zone, in which the trend is northwest-southeast and which includes Fairway Rise (previously unknown structural element), New Caledonia Basin, New Caledonia, Loyalty Basin, and the Loyalty Island Range.

Alignment of the Lord Howe Guyots. These guyots pierce into the Lord Howe Basin and represent an aspect of the volcanism that affected the southwest Pacific. Their age is unknown, but is probably at least middle Tertiary (Conolly, 1969). Several

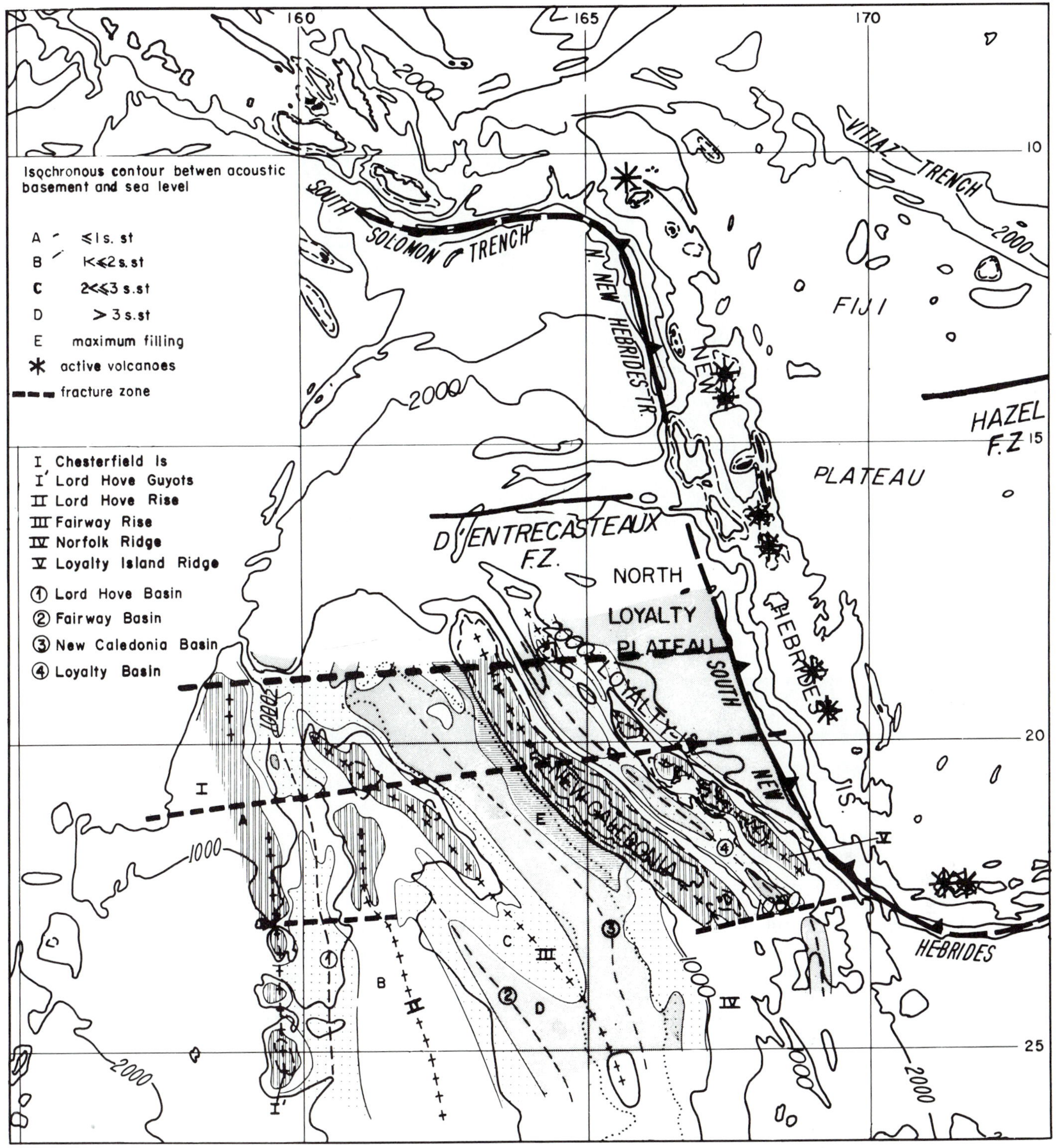

Fig. 7. Structural map of greater New Caledonia region.

authors (Vogt and Conolly, 1971; van der Linden, 1969) have tried to explain the genesis of this alignment by the Wilson (1965)-Morgan (1972) theory (northward movement of the Australian plate on top of a fixed hot spot).

Lord Howe and Fairway Rises. The Fairway Rise extends at least from Fairway Reef to 25°S, and our more recent investigations apparently link it to West Norfolk Ridge (Figs. 1, 8, and 10A). The comparable morphological evolution of these two folds is the reason why they are studied together. From south to north they can be seen to shrink and grow progressively more shallow, until they emerge, because of the reefs which succeeded in forming on them. Lord Howe Rise disappears at 21°S against

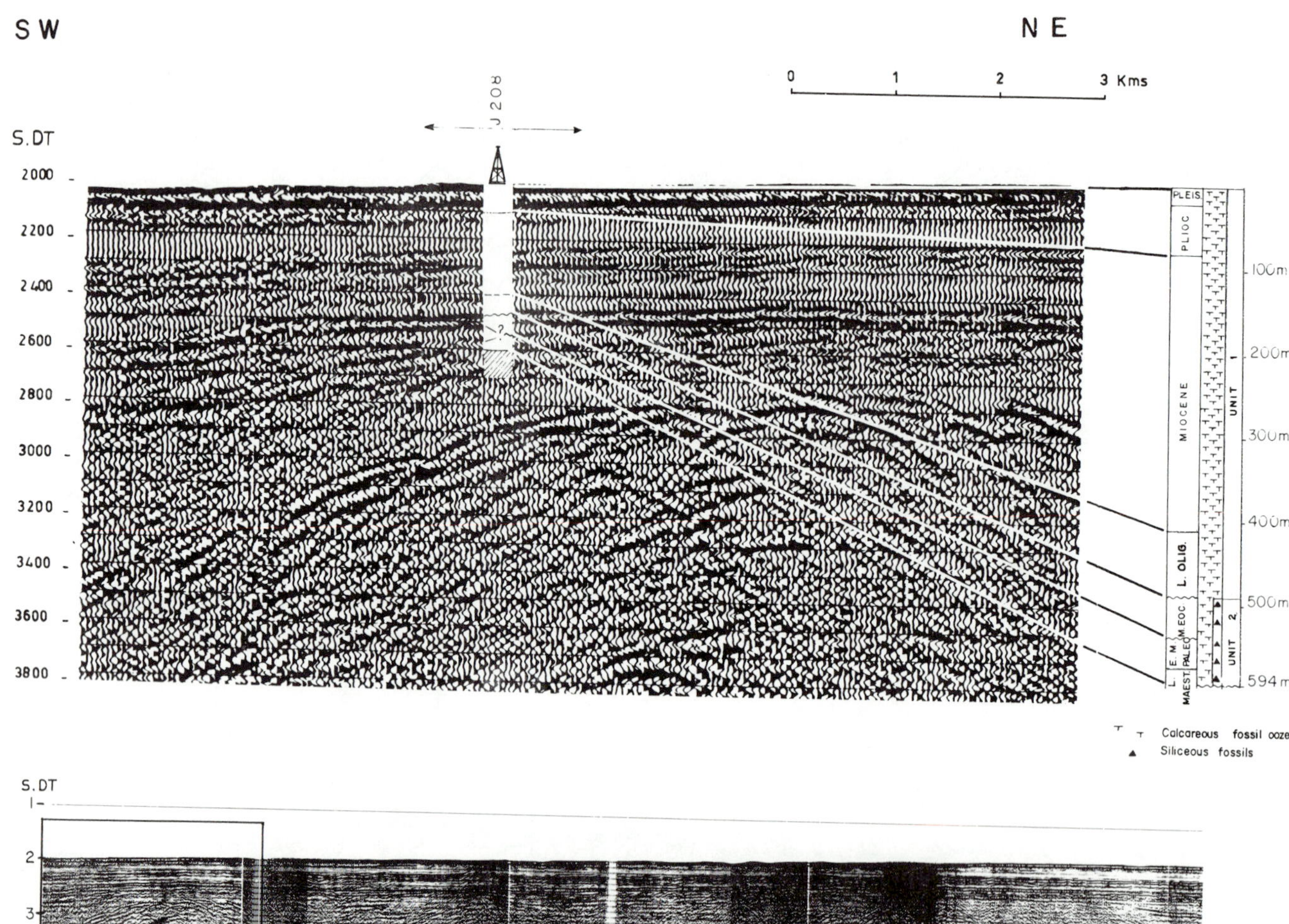

Fig. 8. Location of DSDP site 208 on the profile AUS-101; the relationship between the disconformities observed in the drill hole and in the reflectors of the seismic profiles is apparent.

the Fairway Rise, which extends northward to 20°S. Numerous volcanic outcrops can be seen at the end of Lord Howe Rise just before it disappears. The acoustic basements of these rises are different. The one for Lord Howe Rise is apparently made up of a folded sedimentary series, thus implying that this zone is quite old. Site 208 (DSDP, 1972) stopped in the Late Cretaceous at least 150 m above these folded formations that form a sedimentary substratum (Fig. 10), whereas the basement of Fairway Rise may possibly be a nonsedimentary substratum whose age and nature is unknown.

Seismic velocity obtained on Lord Howe Rise near site 208 by sonobuoy above the strong deepest reflector is 3 km/sec (Andrews et al., 1973), which is in good agreement with a substratum of sedimentary nature. Their sedimentary covering is not very thick but may be as much as 3 km in the small basins cutting into the rises (Fig. 8). Different tectonic episodes are revealed by vertical movements of the rises (unconformities observed in the sedimentary series) and by volcanism.

Basins. The fact that the boundaries of the bathymetric and sedimentary basins are the same is the result of fault movements on the boundaries of the sedimentary basin, causing a slope break in the bathymetry. There are four sedimentary series at the most, separated by major unconformities (Fig. 11A). Lord Howe and Fairway basins, which show that the lower series are well developed, are thus older than New Caledonia Basin, in which only the upper series (Series I) is well developed, if not the only one there. The New Caledonia Basin seems to have been filled in by debris from the erosion of the island, as shown by the maximum sediment thickness observed in the subsident trough along the west coast of New Caledonia. The deposits there are probably of the turbidite type, whereas west of Fairway Rise the data from JOIDES site 208 indicate the presence of pelagic deposits (chalk with nannofossils: van der Lingen et al., 1973), which apparently are widespread, as shown by the uniform seismic character of this series observed on the profiles in this area.

Age of the Sediments. JOIDES site 208 (Fig. 8) revealed two sedimentary gaps (DSDP, 1972; Webb, 1973). The upper is in the Oligocene, which appears to be worldwide and here extends to the late Eocene, and coincides with the major tectonic movements that affected New Caledonia. It corresponds to the bottom of "Series I." "Series II" is probably middle Eocene, while the second gap covers the middle and late Paleocene and the early Eocene. "Series III" is made up of Mesozoic; the DSDP hole stopped at the Late Cretaceous. There are older and relatively undeformed sediments, as well as orderly horizons, underneath the deepest mapable unconformity.

Norfolk Ridge

Several narrow and relatively calm sedimentary basins are visible (Fig. 1), but the successive tectonic phases and the volcanic episodes either deform the sediments or mask them and prevent any orderly seismic reflections. The numerous deformations affecting the sedimentary basins show that these latter are not entirely post-tectonic. They may have begun earlier in the Cretaceous, but apart from several horizons the visible levels are more probably post-Eocene (Fig. 11C).

A major subhorizontal erosional unconformity, covered with slightly undulated sediments, is attached to the Oligocene unconformity. This unconformity, at 24°S, is already at a depth of more than 1,000 m, thus implying that major vertical readjustments took place in the ridge. These vertical movements of the ridge are known on land since fragments of the Miocene peneplain (Davis, 1925) appear at an altitude of 1,000 m in the middle of New Caledonia. It is possible that the erosion unconformity seen in Norfolk Ridge may be connected with the Miocene peneplain since other fragments of the peneplain are found at an altitude of no more than 300 m on the edges of the island; hence the unconformity descends toward the edges.

The Oligocene age of the marine unconformity on Lord Howe Rise does not act as a boundary for the entire area of Lord Howe and Fairway rises. This unconformity disappears in New Caledonia Basin. On Norfolk Ridge it corresponds to the same event, and the angular unconformity between "Series I and II" is horizontal and continuous, implying that the unconformity there corresponds to an erosion phase later than the major Oligocene tectonic episode and that an identical phenomenon occurred at the same time on the Lord Howe Rise. It is possible that beneath this unconformity there may be all the formations known in New Caledonia, perhaps even with marine Cretaceous. Hence Neogene tectonics apparently operated by blocks, causing a rise of New Caledonia and a collapse of its edges. Profiles north of New Caledonia, in the "Grand Passage," reveal the existence of a central basin framed by two lateral ranges.

Loyalty Island Region

This region is bounded in the west by New Caledonia, in the north by the d'Entrecasteaux fracture zone, and in the east by the beginning of the New Hebrides trench (Fig. 9C, D). From west to east there are (Fig. 1) the Loyalty Basin, Loyalty Island Ridge, and North Loyalty Plateau.

Loyalty Basin. The sediments attain maximum thickness between New Caledonia and the Loyalty Islands. Rises of the basement divide the basin into several parts in the south. In the north, opposite Cook Reef, the shallow sedimentary fill reveals various secondary structures that trend north-northwest. From southwest to northeast we find the following (Figs. 11B, 11C): a graben disturbing the slope down to New Caledonia toward Loyalty Basin; a horst dividing the basin; a rise of the Loyalty Island Ridge via faults; and a large flexure fault that runs along the entire Loyalty Island Range.

The very thick sedimentary fill attains 6 km between the emerged land areas, thus showing that the fill comes partly from erosion of New Caledonia and partly from the Loyalty Island volcanoes. The upper series is particularly well represented. How-

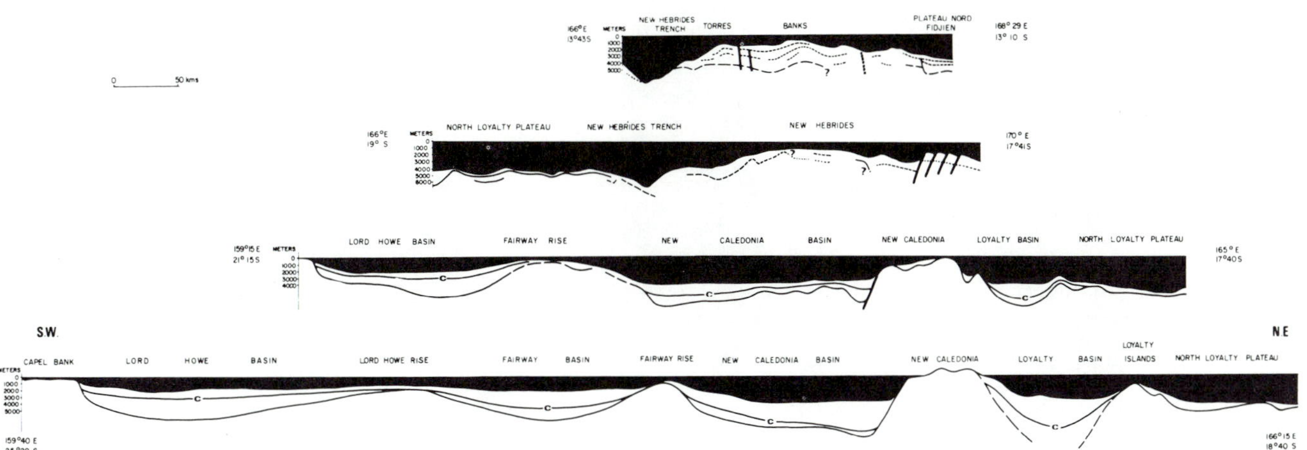

Fig. 9. Cross sections across New Hebrides Island arc and New Caledonia.

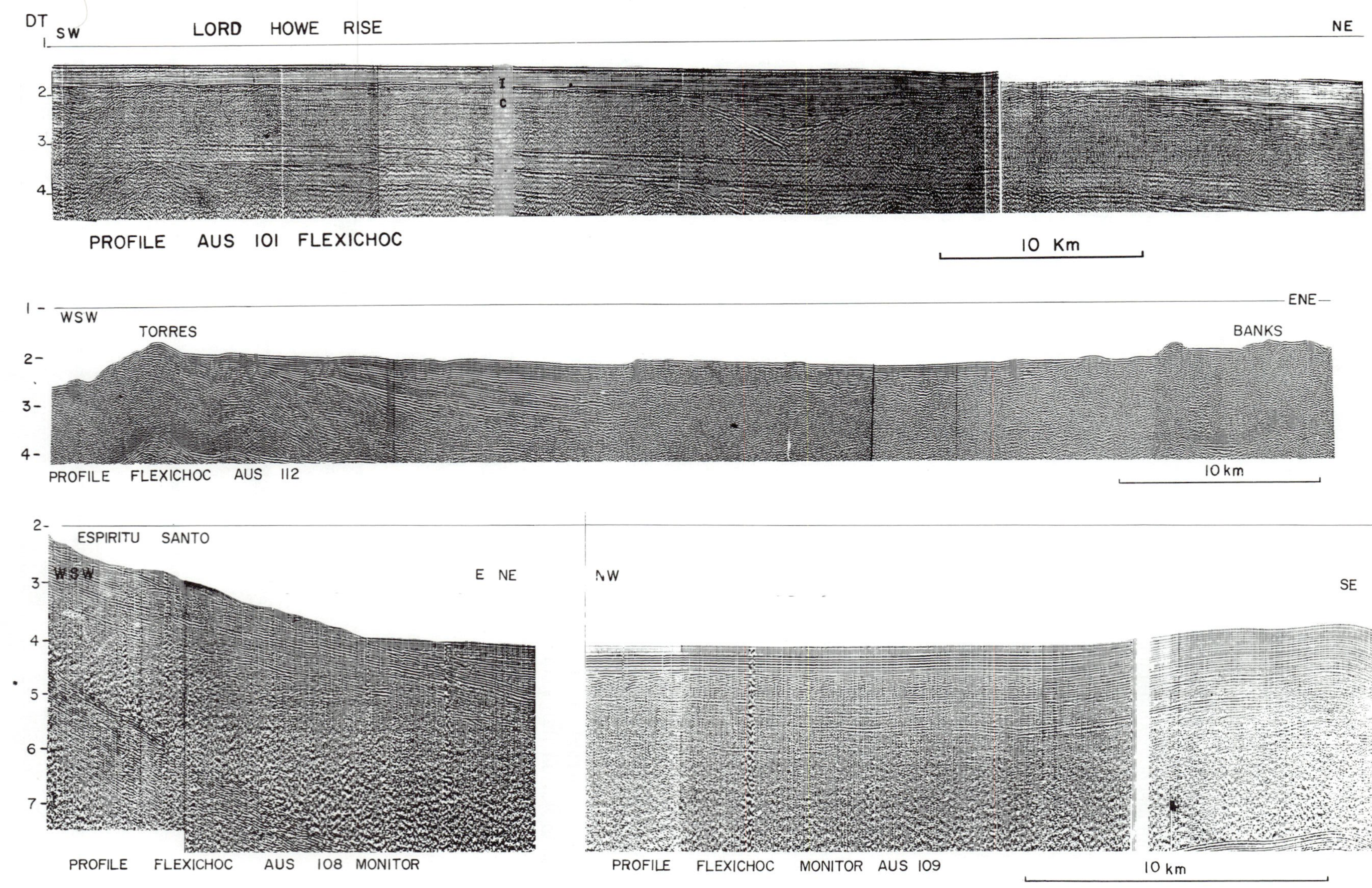

Fig. 10. Seismic-reflection profiles of (a) western New Caledonia area: Lord Howe Rise; (b) New Hebrides area, and between southern Torres and Banks islands; (c) between Espiritu Santo and Aurora.

ever, underneath the eastward prolongation of the erosion unconformity outlined on Norfolk Ridge, which quickly disappears, a thick series can still be seen, followed by a new and deeper unconformity.

Loyalty Island Ridge. Profiles in the submerged parts of this range show that it is divided by a central graben. The sedimentary covering is very thin here, except in the graben. In addition to volcanic elements, it probably contains calcareous elements coming from reefs.

North Loyalty Plateau. This plateau joins the Loyalty Island Range at the New Hebrides trench via a succession of faulted blocks having an opposite inclination from the general slope (Fig. 9B). The very thin sedimentary covering (maximum of 2 km) on this plateau appears to be recent. JOIDES site 286 on Leg 30 (DSDP, 1973) reached an intrusive gabbro inside a basalt, overlain by volcano-sedimentary deposits (450 km) of middle to late Eocene age, followed by chalks and muds with late Eocene to Oligocene nannofossils (110 m), in turn overlain by Oligocene red clays and then cinerites. The nature of the sediments shows that this area was already deep in the Oligocene (underneath the carbonate compensation zone). This plateau has an oceanic appearance because the substratum under the shallow sediments is highly diffracting and has a high velocity (5.5 km/sec).

Interpretation of the New Caledonian Zone

In the Southwest Pacific, the active island arcs highlight the convergence zone of the Australian and Pacific plates. The distinction between the north-south "Australian" structural directions and the northwest-southeast "New Caledonian" ones helps to bring out the individuality of the New Caledonian area. The island arc nature of this zone is supported by the thrust structure and overthrusting of the peridotites in New Caledonia (Guillon, 1973), the Loyalty Island volcanic arc, and the geographic arrangement of the different structural elements. Referring to the present concepts on island arcs (Karig, 1971, 1972; Uyeda, 1972), and even though the arc has probably been active previously, the following succession can be reconstructed at the end of the Paleogene (late Eocene, Oligocene).

New Caledonia apparently made up the external, nonvolcanic arc (supposed by Geze in 1963). The Loyalty Basin was the interdeep basin (more or less developed in the arc-trench gap; Dickinson, 1973); the Loyalty Island Ridge was the volcanic inner arc (Geze, 1963), and the North Loyalty Island Plateau was the interarc or marginal basin. This pattern assumes the existence of a subduction zone at that time, at the latitude of New Caledonia. The emplacement of peridotites complicates this pattern (obduction). The trench was probably located between the western edge of New Caledonia and the Loyalty Basin, and more certainly at the site of the subsequent trench along the west coast of the island. The polarity of the structural features implies that the Australian plate plunged underneath the Pacific plate at that time. Fairway Rise might be a vestige of the swell or outer arch that often seems to precede the trenches. The major disturbances roughly trending east-west (Fig. 7) that cut across and offset the structures also appear to be linked to the island arc history. The vertical movements later than this phase are probably readjustments. It should be mentioned that a system of ridges existed at the beginning of the Cretaceous at the site where New Caledonia is located.

Seismic data have shown that the substratum of Lord Howe Rise is made up of ancient sedimentary formations: the disappearance of the Lord Howe Rise in the north, in front of Fairway Rise, which cuts across it, is a result of the more recent island arc history of New Caledonia. Distribution of the Neogene sediments around New Caledonia is controlled by the structures formed when this island arc was active. Hence the erosion of New Caledonia is responsible for the fill in New Caledonia Basin and for part of the fill in Loyalty Basin, with the other part being provided by the volcano-clastic deposits from the Loyalty Islands. The "old series" appear clearly only in the "Australian" zone (Lord Howe Basin, Lord Howe Rise), which was not touched by this tectonic episode. The lower part of these series is older than the Late Maestrichtian. The erosional discordance revealed on Lord Howe Rise and Norfolk Ridge, separating the "upper series (I)" from the lower ones, and dated in the Oligocene by DSDP site 208, should apparently be attached to the New Caledonia Miocene peneplain.

New Hebrides. The seismic-reflection profiles (see Figs. 1, 9a, 9b, 10b, and 10c) reveal that the sediments on the arc itself are quite thick (more than 2 sec D.D.T.), especially in the collapsed basins between the two lateral ranges. In connection with what is known on land, these are thought to be volcano-sedimentary formations with interstratified calcareous series coming from the reefs on top of the rises. The layers observed are subhorizontal, but blocks with a different seismic nature succeed one another laterally (Fig. 10b). As at Malekula (Mitchell, 1970), major faults apparently bring different layers in contact with one another. These faults were again active recently as shown by offsets visible in the bathymetry. Few reflections are visible on the western flank of the arc, except in the small basin located at mid-slope (see Bentz, this volume).

One of the problems raised by the plate tectonics theory is that of interarc basins, as defined by Karig (1970). Studies of the Tonga, Kermadec, and New Hebrides arcs has led Dubois et al. (1972) to consider the grabens in the southern New Hebrides (studied during the Coriolis cruises in 1968; Puech and Reichenfeld, 1969) as en échelon grabens that can be assimilated with interarc basins. The Kimbla and Coriolis cruises (1971), especially the

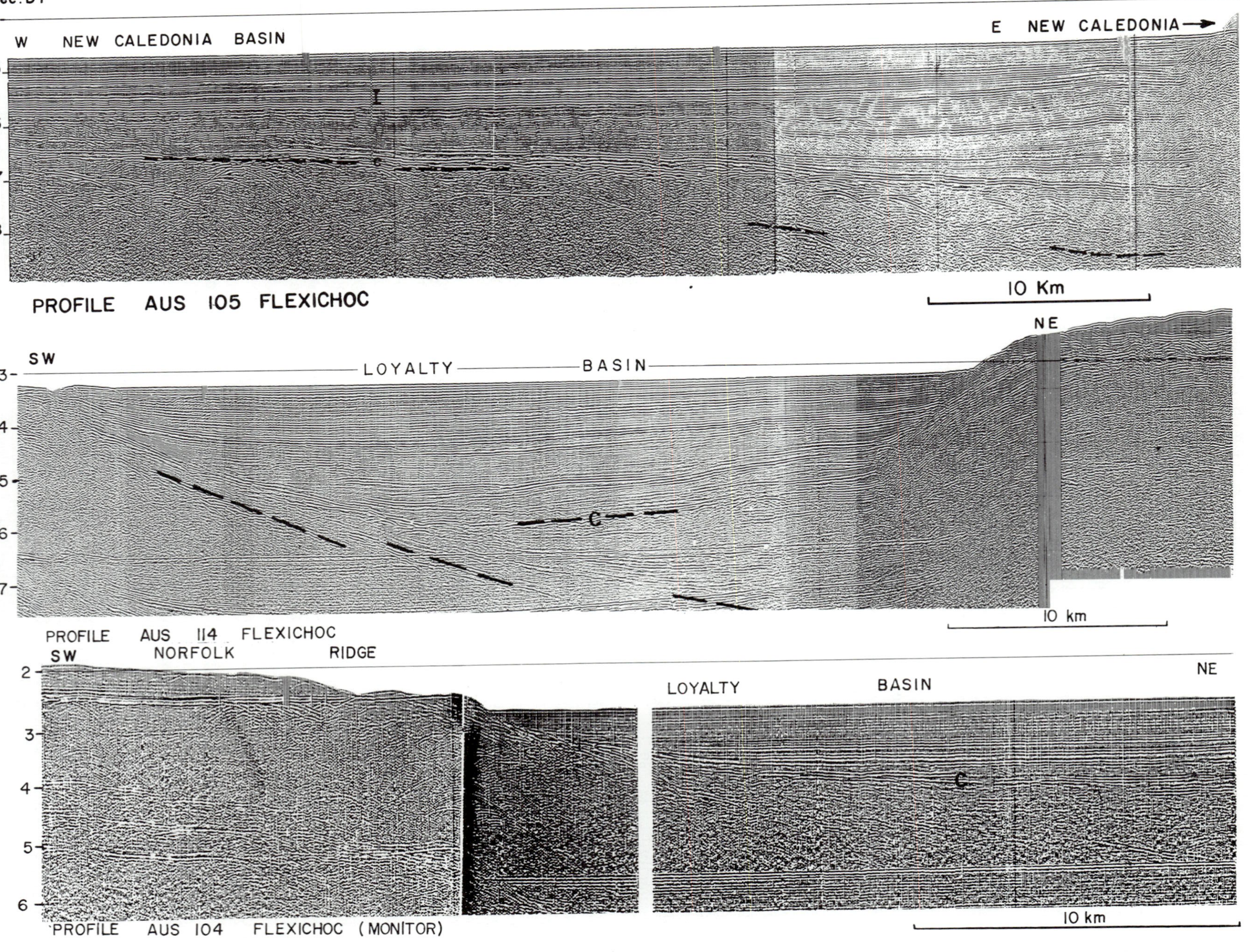

Fig. 11. Seismic-reflection profiles of (a) western New Caledonia area (axis of maximum filling); (b) Loyalty Islands area, and Loyalty Basin in its central part; (c) Norfolk Ridge extending to southern New Caledonia.

first Austradec cruise, enabled two sorts of collapse basins to be distinguished in this region: the narrow grabens in the south are certainly quite recent and are poor in sediments; the much wider basins north and south of Aoba contain thick, old sediments. We feel that these collapse basins, of which both kinds could be caused by a slight opening up, are not comparable to the Tonga and Kermadec interarc basins, but that they possibly represent recent secondary disturbances or grabens as often described inside volcanic arcs. Indeed, we have already seen that an analysis of seismic-wave propagations and attenuations appears to show that the opening up, in the concave part of the New Hebrides, should be sought for in the North Fiji Plateau, which probably played the same role as the Lau interarc basin for Tonga-Kermadec.

CONCLUSION

The deep structure and distribution of sediments in the New Caledonia-New Hebrides region can be explained within the context of fossil or active island arcs. But on the contrary, the Lord Howe Rise belongs to an older domain, and its origin is still unexplained.

Following this interpretation, the same tectonic characteristics of an arc-trench system seem to exist in both the New Caledonia and New Hebrides systems, but with a different development of the arc-trench, i.e., the distance between outer nonvolcanic arc and inner volcanic arc. In the New Hebrides the gap is very small and these two tectonic components are close together, with no interdeep basin well developed; here the outer nonvolcanic arc is similar to the mid-slope basement high described by Karig (1971). On the contrary, in the New Caledonia system the arc-trench gap is very large, with a very well developed interdeep basin (ref. APEA). Following Dickinson (1973), this could be related to the different age of these two island-arc systems. Indeed, the New Hebrides island arc seems to be of early Miocene age, while the New Caledonia island arc could be as old as Late Jurassic-Early Cretaceous. Using Dickinson's relation between age and arc-trench gap, we obtain for the beginning of New Caledonia island-arc activity an age of 125 my, which is in good agreement with land geology.

ACKNOWLEDGMENTS

We thank ORSTOM, IFP, CFP, ELF/ERAP, and SNPA for permission to publish this paper. We are indebted to the officers and crew of the N.O. CORIOLIS and N.O. NOROIT from CNEXO for their efficiency during Austradec cruises.

This paper results from a cooperative scientific program for the geological and geophysical study of the Southwest Pacific called Austradec, grouping the Oceanographic Center of Noumea of the Office de la Recherche Scientifique et Technique Outre Mer and Institut Francais du Petrole, Compagnie Francaise des Petroles, ELF/ERAP, and Societe Nationale des Petroles d'Aquitaine sponsored by the Comite d'Etudes Petrolieres Marines.

The following persons participated in the cruises or in the study of the data: D. Bosquet, J. Dubois, F. Dugas, A. Lapouille, B. Larue, J. Launay, J. Recy, C. Reichenfeld, J. Daniel, and J. Dupont from ORSTOM; R. Cassand, L. Donatien, J. P. Fail, L. Montadert, and J. Ravenne from IFP; J. Louis, from ELF/ERAP; F. Aubertin and R. Dhinnin from CFP; and J. Guillaume from SNPA.

BIBLIOGRAPHY

Aggarwal, Y. P., Barazangi, M., and Isacks, B., 1972, P and S travel times in the Tonga Fiji region: a zone of low velocity in the uppermost mantle behind the Tonga island arc: Jour. Geophy. Res., v. 77, p. 6427-6434.

Andrews, J. E., et al., 1973, Deep Sea Drilling Project, Leg 21; Tasman Sea-Coral Sea, *in* Oceanography of the South Pacific 1972: Comp. R. Frazer, New Zealand Natl. Commission for UNESCO, Wellington.

Avias, J., 1953, Contribution à l'étude stratigraphique et paléontologique des formations antécrétacees de la Nouvelle-Calédonie Centrals: Sci. Terre, t. 1, no. 2, p. 1-276.

Barazangi, M., and Isacks, B. L., 1971, Lateral variations of seismic-wave attenuation in the upper mantle above the inclined earthquake zone of Tonga Island arc: deep anomaly in the upper mantle: Jour. Geophys. Res., v. 76, no. 35, p. 8493-8516.

———, Isacks, B., and Oliver, J., 1972, Propagation of seismic waves through and beneath the lithosphere that descends under the Tonga island arc: Jour. Geophys. Res., v. 77, no. 3, p. 952-958.

———, Isacks, B. L., Oliver, J., Dubois, J., and Pascal, G., 1973, Descent of lithosphere beneath New-Hebrides, Tonga-Fiji and New Zealand: evidence of detached slabs: Nature, v. 242, no. 5393, p. 98-101.

Chalvron, M. De, Dubois, J., Puech, J. L., and Reichenfeld, C., 1966a, Croisière bathymétrique à bord du Coriolis: Nouméa, New Caledonia, ORSTOM, June (multigr.).

Chase, C. G., 1971, Tectonic history of the Fiji Plateau: Geol. Soc. America Bull., v. 82, p. 3087.

Chevalier, J. P., 1968, Expédition française sur les récifs coralliens de la Nouvelle-Calédonie: Paris, Ed. Fondation Singer Polignac, t. 3.

Conolly, J. R., 1969, Western Tasman sea floor: New Zealand Jour. Geol. Geoph., v. 12, p. 310.

Davis, W. M., 1925, Les côtes et les récifs coralliens de la Nouvelle-Calédonie: Ann. Géogr. t. 34, p. 244-269, 332-359, 424-441, 521-558.

Deep Sea Drilling Project, Leg 21, 1972 (Burns, R. E., and Andrews, J. E., Chief scientists): Geotimes, v. 17, no. 5.

———, Leg 30, 1973 (Andrews, J. E., and Packham, G., Chief scientists): Geotimes, v. 18, no. 9.

Dickinson, W. R., 1973, Widths of modern arc-trench gaps proportional to past duration of Igneous. Activity in associated magmatic arcs: Jour. Geophys. Res., v. 78, no. 17, p. 3376-3389.

Dubois, J., 1965, Sur la vitesse de propagation des ondes P le long de l'arc séismique des Nouvelles Hébrides: Compt. Rend., t. 200, p. 2275-2277.

———, 1968, Etude de la dispersion des ondes de Rayleigh dans la région du Sud Ouest Pacifique: Ann. Geophys., t. 24, no. 1.

———, 1969, Contribution à l'étude structurale du Sud Ouest Pacifique d'après les ondes séismiques observees en Nouvelle Caledonie et aux Nouvelles Hébrides: Ann. Geophys., t. 25, no. 4, p. 923-972.

———, 1971, Propagation of P waves and Rayleigh waves in Melanesia: structural implications: Jour. Geophys. Res., v. 76, no. 29, p. 7217-7240.

———, Launay, J., and Revey, J., 1972, Les mouvements verticaux en Nouvelle-Calédonie et aux iles Loyaute et l'interprétation de certains d'entre eux dans l'optique de la tectonique des plaques: Cahiers ORSTOM, t. 5, no. 1.

Dugas, F., 1971, Apercu structural et bibliographie geologique de l'ard insulaire des Nouvelles Hebrides: Noumea, New Calidonia, ORSTOM, 43 p. (multigr.).

Geze, B., 1963, Observations tectoniques dans le Pacifique (Hawaii, Tahita, Nouvelles-Hébrides): Bull. Soc. Géol. France, t. 7, p. 154-164.

Guier, W., and Newton, R. R., 1965, The earth's gravity field as deduced from the Doppler tracking of 5 satellites: Jour. Geophys. Res., v. 70, p. 4613-4625.

Guillon, J. H., 1973, Les massifs péridotitiques de Nouvelle Calédonie. Modèle d'appareil ultrabasique stratiforms [thèse doct. Sci.]: Natl. Fac. Sci. Paris, Univ. Paris, 125 p.

Isacks, B., and Molnar, P., 1971, Distribution of stresses in the descending lithosphere from a global survey of focal mechanism solutions of mantle earthquakes: Rev. Geophys. Space Phys., v. 9, no. 1, p. 103-174.

———, Sykes, L. R., and Oliver, J., 1967, Spatial and temporal clustering of deep and shallow earthquakes in the Fiji-Tonga-Kermadec region: Seismol. Soc. America Bull., v. 57, no. 5, p. 935-958.

———, Sykes, L. R., and Oliver, J., 1969, Focal mechanisms of deep and shallow earth-quakes in the Tonga-Kermadec region and tectonics of island arcs: Geol. Soc. America Bull., v. 60, p. 1443-1470.

Johnson, T., and Molnar, P., 1972, Focal mechanisms and plate tectonics of the southwest Pacific: Jour. Geophys. Res., v. 77, p. 5000-5032.

Karig, D. E., 1970, Ridges and trenches of the Tonga-Kermadec island arc system: Jour. Geophys. Res., v. 75, p. 239-254.

———, 1971, Origin and development of marginal basins in the western Pacific: Jour. Geophys. Res., v. 76, no. 11, p. 2542-2561.

———, 1972, Remnant arcs: Geol. Soc. America Bull., v. 83, no. 4, p. 1057-1068, 6 fit.

Luyendyk, B. P., Bryan, W. B., Jezek, P. A., 1973, Shallow structure of the New Hebrides island arc: Woods Hole Oceanogr. Inst., Contrib. no. 3096, in press.

MacDonald, K. C., Luyendyk, B. P., and Von Huren, R. P., 1973, Heat flow and plate boundaries in Melanesia: Jour. Geophys. Res., v. 78, no. 14, p. 2537-2546.

Malahoff, A., and Woollard, G. P., 1969, The New Hebrides islands gravity network, pt. I: Hawaii Inst. Geophys., Univ. Hawaii (Final rept.), 26 p.

Mino, K., Onoguchi, T., and Mikumo, T., 1968, Focal mechanism of earthquakes on island arcs in the Southwest Pacific region: Bull. Dist. Prev. Res. Inst., Kyoto Univ., v. 18, p. 78-96.

Mitchell, A. H. G., 1970, Facies of an early Miocene volcanic arc, Malekula Island, New Hebrides: Sedimentology, v. 14, no. 3-4, p. 201-243.

Mitronovas, W., and Isacks, B. L., 1971, Seismic velocity anomalies in the upper mantle beneath the Tonga-Kermadec island arc: Jour. Geophys. Res., v. 76, no. 29, p. 7154-7180.

Morgan, W. J., 1972, Deep Mantle convection plumes and plate motions: Am. Assoc. Petr. Geol. Bull., v. 56, no. 2, p. 203-213.

Oliver, J., and Isacks, B., 1967, Deep earthquake zones, anomalous structures in the upper mantle, and lithosphere: Jour. Geophys. Res., v. 72, no. 16, p. 4259-4275.

Pascal, G., Dubois, J., Barazangi, M., Isacks, B. L., and Oliver, J., 1973, Seismic velocity anomalies beneath the New Hebrides island arc: evidence for a detached slab in the upper mantle: Jour. Geophys. Res., v. 78, no. 29, p. 6998-7004.

Raitt, R. W., 1956, Seismic-refractions studies of the Pacific ocean Basin: Geol. Soc. America Bull., v. 67, p. 1623-1640.

———, Fisher, R. L., and Mason, R. G., 1955, Tonga trench: Geol. Soc. America Spec. Paper 62, p. 237-254.

Routhier, P., 1953, Etude géologique du versant occidental de la Nouvelle-Calédonie entre le col de Boghen et la pointe d'Arama: Mém. Soc. Géol. France (nouv. sér.), t. 32, no. 67 fasc. 1-3, feuilles 1-34, p. 1-271.

Saito, M., and Takenchi, H., 1966, Surface waves across the Pacific: Seismol. Soc. America Bull., v. 56, no. 5, p. 1067-1091.

Sclater, J. G., and Menard, H. W., 1967, Topography and heat flow of the Fiji Plateau: Nature, v. 216, p. 991.

———, Hawkins, J. W., Mammerickx, J., and Chase, C. G., 1972a, Crustal extension between the Tonga and Lau ridges: petrologie and geophysical evidence: Geol. Soc. America Bull., v. 83, p. 505-518.

———, Ritter, U. G., and Dixon, F. S., 1972b, Heat flow in the Southwestern Pacific: Jour. Geophys. Res., v. 77, no. 29, p. 5697-5704.

Shor, G. G., 1967, Seismic refraction profiles in Coral Sea Basin: Science, v. 158, p. 911-913.

———, Kirk, H. K., and Maynard, G. L., 1971, Crustal structure of the Melanesian area: Jour. Geophys. Res., v. 76, no. 11, p. 2562-2586.

Solomon, S., and Biehler, S., 1969, Crustal structure from gravity anomalies in the Southwest Pacific: Jour. Geophys. Res., v. 74, no. 27, p. 6696-6701.

Sutton, G. H., Maynard, G. L., and Hussong, D. M., 1973, Widespread occurrence of a high velocity basal layer in the Pacific crust found with repetitive sources and sonobuoys: Honolulu, Hawaii Inst. Geophys., Univ. Hawaii, (in press).

Sykes, L. R., 1966, The seismicity and deep structure of island arc: Jour. Geophys. Res., v. 71, no. 12, p. 2981-3006.

———, Oliver, J., and Isacks, B., 1968, Earthquakes and tectonics, *in* Maxwell, A. E., et al., eds., The sea, v. 4, pt. I: New York, Wiley-Interscience.

———, Isacks, B. L., and Oliver, J., 1969, Spatial distribution of deep and shallow earthquakes of small magnitudes in the Fiji Tonga region: Seismol. Soc. America Bull., v. 59, no. 3, p. 1053-1113.

Tissot, B., and Noesmoen, A., 1958, Les bassins de Nouméa et de Bourail: Rev. Inst. Franc. Petrole, t. 13, no. 5, p. 739-760.

Uyeda, S., 1972, Dérive des continents et tectonique des plaques: Recherche, t. 3, no. 25, p. 649-664.

Van der Linden, W. J. M., 1968, Western Tasman Sea geomagnetic anomalies 1:2,000,000: New Zealand Oceanogr. Inst. Chart Misc. Ser. 19.

______, 1969, Extinct Mid. Ocean ridges in the Tasman Sea and in the western Pacific: Earth and Planetary Sci. Letters, v. 6, no. 6, p. 483-490.

Van der Lingen, G. J., et al., 1973, Lithostratigraphy of eight drill sites in the south-west Pacific: preliminary results of Leg 21 of the Deep Sea Drilling Project, *in* Oceanography of the South Pacific 1972: Comp. R. Fraser, New Zealand Natl. Commission for UNESCO, Wellington.

Vogt, P. R., and Conolly, J. R., 1971, Tasmantid Guyots: the age of the Tasman Basin and motion between the Australia plate and the mantle: Geol. Soc. America Bull., v. 82, no. 9, p. 2577-2584.

Warden, A. J., 1968, Age determination Malekula, Torres, *in* Annual report for 1966: New Hebrides Geol. Survey, p. 54.

Webb, P. N., 1973, Preliminary comments on Maastrichtian Paleocene Foraminifera from Lord Howe Rise Tasman Sea, *in* Oceanography of the South Pacific 1972: Comp. R. Fraser., New Zealand Natl. Commission for UNESCO, Wellington.

Wilson, J. R., 1965, Evidence from ocean islands suggesting movement in the earth: Symp. on Continental Drift, Phil. Trans. Roy. Soc. London, v. 258, p. 145-165.

Woodward, D. J., and Hunt, T. M., 1970, Crustal structure across the Tasman Sea: New Zealand Jour. Geol. Geophys., v. 14, no. 1, p. 39-45.

Marine Geology of the Southern Lord Howe Rise, Southwest Pacific

Felix P. Bentz

INTRODUCTION

In April 1972 Mobil Oil Corporation's research vessel M/V FRED H. MOORE traversed several JOIDES Leg 21 drill sites on the southern Lord Howe Rise. The combination of these drilling results and Mobil's high-quality seismic records provides important insights into the geology of this area. Neogene and Paleogene sedimentary thicknesses observed on our seismic records and derived from sonobuoy refraction measurements show a much greater degree of diversity than could be deduced from JOIDES data alone. Cenozoic thicknesses of about 3,000 m recorded in the New Caledonia Basin are most likely due to large-scale slumping from the flanks of the Lord Howe Rise and the Norfolk Ridge. Rapid sedimentary thickness changes from 0 to 1,300 m on top of the Lord Howe Rise are largely caused by a strong Neogene unconformity, beveling the highly block faulted Paleogene and older section. New seismic evidence strongly suggests the occurrence of continuous, explosive-type deep-water vulcanism throughout the Cenozoic. This region represents a margin between oceanic and "intermediate" crust (which in Late Cretaceous was at sea level).

With the October 1973 publication of Volume XXI of the Deep Sea Drilling Project (JOIDES) substantial contributions have been made toward a better understanding of the geology of the Lord Howe Rise and the New Caledonia Basin. Shortly after the actual drilling of Leg 21 had been carried out between November 1971 and January 1972, Mobil Oil Corporation's research vessel M/V FRED H. MOORE was conducting marine reconnaissance work in the Tasman Sea. Knowing the locations of the JOIDES drill sites the survey track of our vessel was adjusted to pass close to sites 206, 207, and 208. We believe the quality of the reflection seismic profiles recorded with one of our many systems employed on board the FRED H. MOORE permits further analysis and interpretation of JOIDES results, furnishing an even clearer picture of the stratigraphy, structure, and extensive vulcanism in the area of site 207 on the southern Lord Howe Rise, an area of "intermediate" crustal properties.

GEOPHYSICAL SYSTEMS AND DATA

The standard geophysical systems employed in Mobil's worldwide marine reconnaissance survey were reflection and refraction seismic, gravity, and magnetics. Using as a sound source four Mobil-designed 1,560-in^3 air guns, discharged in unison, our reflection seismic signals were recorded with a 2,600-ft-long SEI six-trace streamer, our refraction seismic signals with Magnavox sonobuoys. In addition, however, a high-frequency/high-resolution system (HI/FES) was employed using a low-noise streamer, towed close to the surface, and consisting of a 200-ft leader and a 50-ft live section containing 20 hydrophones at 1-in. spacing. The shipboard output of this system consisted of an analog paper printout of a Varian 1900 data acquisition and display system.

HI/FES data were not recorded on tape, and therefore the scale and other parameters chosen by the shipboard scientist at the time of recording can no longer be modified. From our far-more-extensive network of seismic lines in the Tasman Sea, we have selected for this review only a number of those line segments, where our HI/FES data were printed at a vertical scale of 1 s (two-way travel time) per 10 in., since they show by far the greatest stratigraphic detail, particularly in the area of relatively thin sediment cover around site 207.

On the normal seismic profiles recorded at a more natural vertical scale of 8 s/10 in., only gross structure and little or no stratigraphic detail can be recognized in this area. Even on those HI/FES records where the vertical scale was changed to 2 s/10 in., much of the geologic detail is lost; only line 113, recorded at this particular scale, is being included in our presentation of interpreted cross sections.

The horizontal scale used on our shipboard printouts was 1 h of ship's travel per 4½ in. of paper record. At an average speed of about 7 knots, this amounts to about 8 miles or 13 km of data per hour, resulting in a vertical exaggeration of approximately 30 times.

The location of the seismic line segments and the position of sonobuoy refraction data discussed in this review are shown on Figure 1.

STRATIGRAPHY

Our stratigraphic control is based entirely on the JOIDES Leg 21 drilling results, particularly the section cored at site 207. The stratigraphic column described from this site is shown in Figure 2, superimposed on the HI/FES recording of Mobil's line 108. A preliminary description of the stratigraphy of site 207 was published in the May

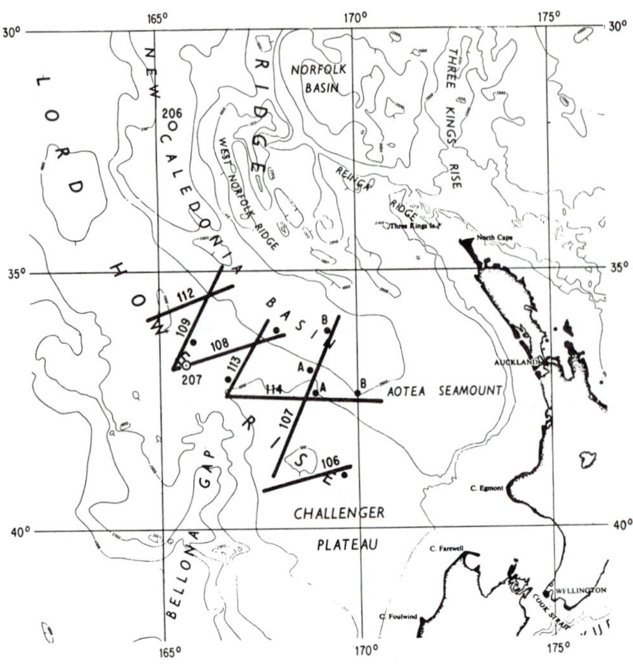

Fig. 1. Location of Mobil's Marine Reconnaissance seismic line segments, sonobuoy refraction surveys (dots), and JOIDES sites (circles).

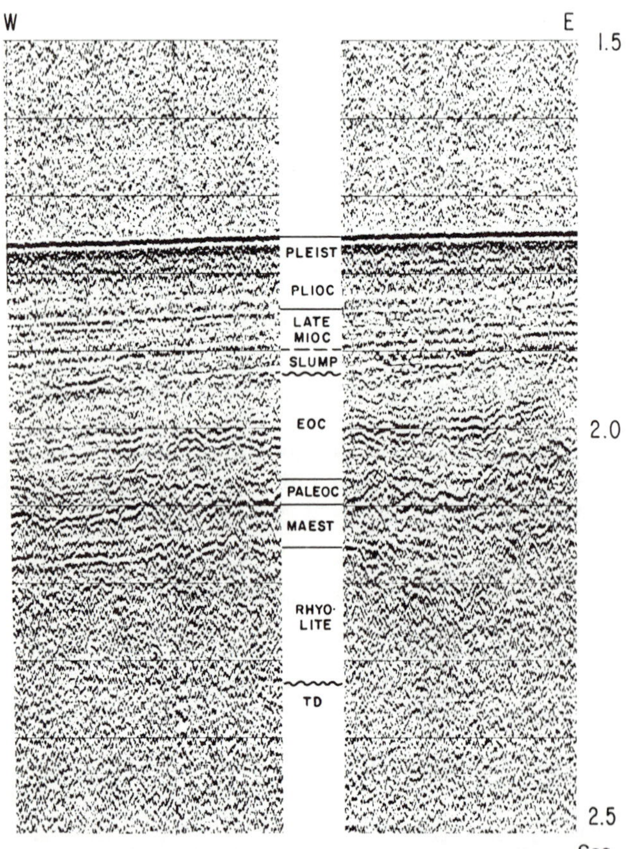

Fig. 2. Stratigraphy of JOIDES drill hole 207, superimposed at the drill site on a high-frequency seismic (HI/FES) record of line 108.

1972 issue of Geotimes by R. E. Burns et al.; a much more detailed and slightly revised version by D. Burns et al. is now available in the Leg 21 JOIDES Volume (1973). Our own remarks will be restricted to a description of the way our high-quality seismic data might supplement the published information.

Figure 2 demonstrates that most of the stratigraphic subdivisions established on the basis of paleontology are also readily identifiable on our HI/FES seismic record covering site 207. However, rapid changes in sedimentary thickness due to differential subsidence, angular unconformities, and recurrent vulcanism place severe limitations on any attempt at regional correlation of small units. Only the following stratigraphic units can be correlated with some degree of certainty over longer distances:

1. The relatively undisturbed Neogene section above the regional unconformity.
2. The highly faulted Paleogene and Late Cretaceous below the regional unconformity.
3. The acoustic basement.

Neogene Section

At JOIDES site 207 the Neogene consists of a 119-m sequence of late Miocene through Pleistocene calcareous fossil oozes. There is no distinguishable change of seismic character within the Plio/Pleistocene sequence; but at site 207 the top of the Miocene seems to be marked by the beginning of a more-banded-appearing series of seismic reflectors, although no lithologic differences between these three stratigraphic units have been reported.

The most remarkable identifying feature of the Neogene section is its lack of pronounced deformation, particularly when compared with the underlying, highly disturbed Paleogene and older rocks. We shall discuss the lateral changes in thickness and lithology later.

Neogene/Paleogene Unconformity

An unconformity of regional extent and significance occurs between the Neogene and Paleogene. It was encountered to a greater or lesser degree in all JOIDES drill sites from the New Caledonia Basin westward. Edwards (1973) discusses in detail the stratigraphic gaps encountered in JOIDES cores. The greatest gap was encountered at site 207, extending from the top of middle Eocene into the base of late Miocene. At other sites the stratigraphic gap normally covers a shorter interval, mostly extending from within the late Eocene into middle or late Oligocene. Our HI/FES seismic lines (Figs. 3 and 4) illustrate that this great variability in the amount of missing sequence is caused by highly local changes in pre-unconformity topography and post-unconformity differential subsidence.

Using as an example our HI/FES line 108 (Fig. 3, 5, and 6), which crosses site 207, and following it about 20 km to the west of the drill site, we observe

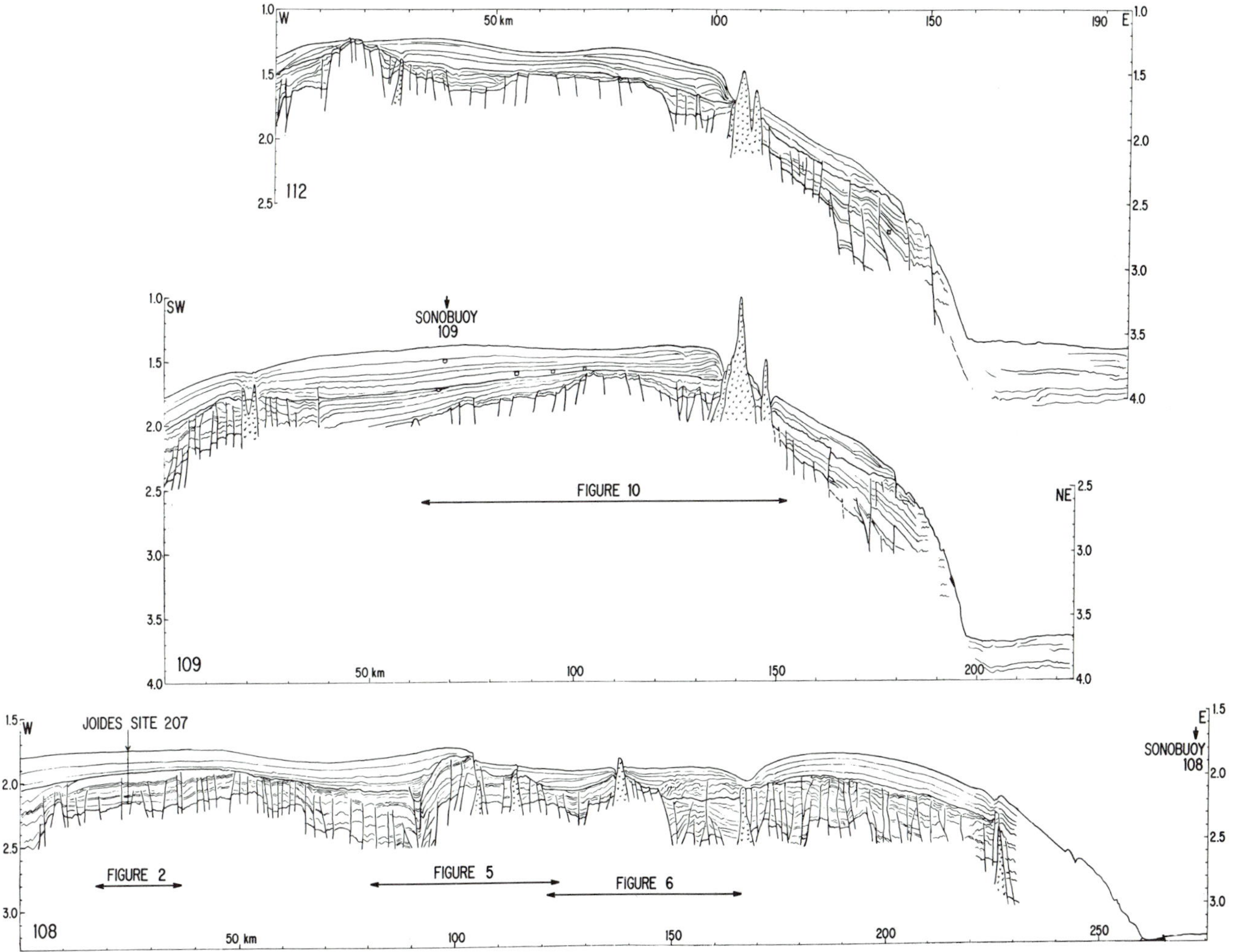

Fig. 3. Interpreted HI/FES lines 112, 109, and 108. Refractors indicated by circles, magnetic anomalies in sedimentary sequence by small squares.

a younger Eocene section preserved below the unconformity, above the middle Eocene of site 207. At the same place we also notice above the unconformity additional sediments occurring below the late Miocene sequence recovered at site 207.

The changes along this unconformity are even more pronounced as we move about 25 km east of site 207. There the unconformity has cut across an old structural high area, removing sediments down to or near the top of the Maestrichtian. Yet scarcely 30 km farther to the east we are in an area of consistent downwarping which has not only resulted in thicker sequences but also in the preservation of sediments below and some addition of sediments above the unconformity. While the time gap at this point may reflect a relatively short interval from late Eocene to early Oligocene, at another point 20 km farther to the east the gap may again extend from mid-Eocene to Pleistocene or younger, caused by a high area that is apparently still active.

Thus over a distance of only 100 km we have seen a great variety of unconformable relationships, and similar observations can be made on all other cross sections. But, while the amount of section missing is controlled by strictly local structural events, the regional significance of the unconformity is not diminished. It clearly demarcates a distinct change in structural regimes, from an era of accentuated block faulting to a relatively stable system of mostly gentle downwarps.

Late Eocene/Mid-Miocene Reverse Sequence

At site 207, a 16-m-thick displaced sequence, consisting of late Eocene underlain by mid-Miocene, was recovered from the level of the main unconformity. Edwards (1973), in analyzing this anomalous sediment slice, favors actual post-unconformity slumping, instead of possible distortion and movements within the core barrel during drilling. The foregoing review of the Neogene/Paleogene unconformity along line 108 indicates that late Eocene as well as mid-Miocene sediments could have been present nearby, although they were not drilled in the normal sequence of site 207. The

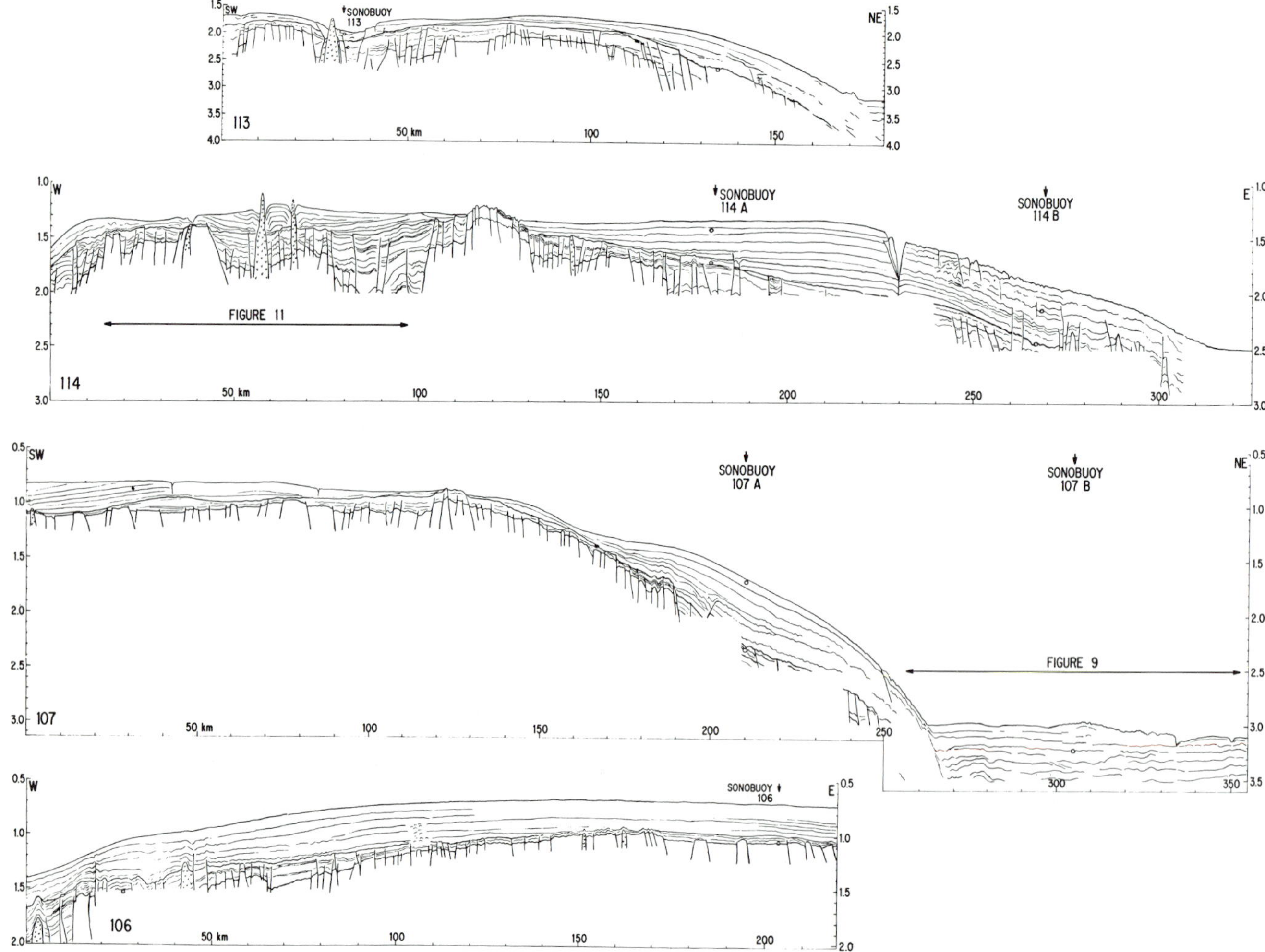

Fig. 4. Interpreted HI/FES lines 113, 114, 107, and 106. Vertical exaggeration 30 times, except for line 113, where it is 15 times.

frequent occurrence of slump features throughout the section will be discussed later, and supports a slump origin of this anomalous feature.

Paleogene and Older Section

At site 207 the Paleogene and older section below the regional unconformity consists of 167 m of Paleocene and early and middle Eocene calcareous fossil oozes, as well as a 48-m section of Maestrichtian silty claystone.

Eocene. At site 207 the middle Eocene is characterized by a band of three closely spaced reflectors, probably caused by chert layers. This characteristic event can be observed on a number of our seismic lines, occasionally increasing to a band of four or more reflectors. Although this event is useful in places to identify the presence of mid-Eocene sediments with reasonable assurance, it is not consistent enough to use as a marker for regional correlation. The remainder of the middle and lower Eocene sequence does not have any distinguished seismic characteristics.

Paleocene/Eocene Unconformity. This minor unconformity has been described from sites 206 and 208 and less certainly so from site 207. Our HI/FES record at site 207 does also suggest a slight unconformity at the base of the early Eocene, with the underlying Paleocene and older rocks being somewhat more deformed. A number of other unconformities within the Paleocene can be observed on our cross sections. Since most of them are probably related to volcanic activity, they will be discussed in that context.

Maestrichtian. The calcareous fossil oozes representing the stratigraphic sequence extending from Paleocene through Pleistocene were deposited in a deep-water environment far from land. The terrigenous silty claystones of the Maestrichtian, on the other hand, indicate that at the close of Cretaceous the Lord Howe Rise was situated in a shallower water environment and close to a land mass, probably representing an ancient continental margin.

Acoustic Basement

At site 207 the acoustic basement reflector marks the top of an Upper Cretaceous rhyolite sequence (Andrews, 1973). Van der Lingen (1973) provides a detailed description of these rhyolite units and concludes that at least the lapilli tuffs of the upper unit probably were extruded subaerially or in very shallow water, again suggesting possible ancient continental crust.

Whether the horizon identified as acoustic basement on our HI/FES records consists of this same lithologic unit, throughout the area, is questionable. Considering the localized distribution of subsequent volcanic events, and the probability of an irregular topography at the time of the Late Cretaceous extrusive activity we are inclined to believe that the Lord Howe Rise basement consists of a variety of rock units of continental crust composition. Local variations in the seismic character and velocity of the acoustic basement tend to support this view.

VELOCITIES AND SEDIMENT THICKNESS

At JOIDES sites 206, 207, and 208 the total thicknesses of the drilled sedimentary columns vary only slightly, from 357 m at site 207 and 590 m at site 208, both on top of the Lord Howe Rise, to the 734 m recovered at site 206, presumably located in the deepest portion of the New Caledonia Basin. Our own reflection and refraction seismic data indicate much greater variations in sedimentary thickness, particularly between the highest points of the Lord Howe Rise and the deepest parts of the New Caledonia Basin. Our tabulation of possible thicknesses (Fig. 8) shows a spread of sedimentary accumulations from 260 m on the southern Lord Howe Rise to at least 3,230 m in the New Caledonia Basin, and possibly more. Sediment thickness of 2-3 km in the New Caledonia Basin have been calculated from geophysical measurements by Woodward and Hunt (1971).

Acoustic Velocities

Average velocities for certain intervals recorded in this area by JOIDES and Mobil (Fig. 7) are quite comparable (see Andrews, 1973). In the Neogene section JOIDES measured velocities in the range 1,500-1,900 m/s, Mobil's averaged 1,650-2,040 m/s. The Paleogene-Cretaceous velocities range from 1,661 to 2,375 m/s for JOIDES and 2,150-3,380 m/s for Mobil. The higher values in Mobil's measurements were from the deeper parts of the New Caledonia Basin, and most likely are due to greater compaction caused by thick overburden. JOIDES velocities for the two rhyolite units drilled at site 207 average 3,348 m/s for the lower unit and, oddly enough, 4,921 m/s for the upper unit. We have interpreted only those intervals where we recorded velocities of 4,800-4,980 m/s as good evidence for rhyolite flows. Whether any of our other measurements in the 3,000-4,000 m/s range could be compared with the JOIDES rhyolite measurement of 3,348 m/s for the lower rhyolite unit is open to speculation.

In addition to these rhyolite velocities usually recorded on top of acoustic basement, we have measured velocities in each location in the range 5,000-6,000 m/s, which probably indicate metamorphic or igneous basement.

Sediment Thickness

Based on a correlation of these velocity intervals and on the interpretation of our reflection seismic data we arrive at the possible thickness values shown in Figure 8. Neogene thicknesses on top of the Lord Howe Rise drilled by JOIDES and ranging from 119 m at site 207 and 434 m at site 208 compare readily with our refraction-calculated thicknesses of 260-400 m. Our Neogene thicknesses of 440 and 530 m calculated at the flank of the Lord Howe Rise compare with the JOIDES site 206 value of 475 m. However, our measurements in the New Caledonia Basin range from 820 to 1,120 m thickness for Neogene sediments.

The Paleogene-Cretaceous sedimentary thicknesses of 157-259 m drilled by JOIDES are considerably thinner than anything indicated by our refraction data. Our measurements range from 950 to 1,020 m on top of the Lord Howe Rise to as much as 2,410 m in the New Caledonia Basin and could be slightly greater, since the refractor that we correlate with the Neogene-Paleogene unconformity probably corresponds in many places to the top of the Eocene chert beds well within the Paleogene sequence. However, in reviewing our HI/FES interpretations we can indeed find numerous places on the Lord Howe Rise where Paleogene thicknesses comparable to JOIDES drilling results can be found. The reason for the thin sequence drilled by JOIDES in the New Caledonia Basin is that site 206 was purposely located on the flank of a hill, interpreted by us as a volcanic feature.

Thickness Changes

Our HI/FES records show clearly the various causes responsible for the erratic thickening or thinning of sedimentary sequences in this area. We can distinguish four principal causes for these thickness changes.

1. The Neogene-Paleogene unconformity, causing pronounced stratigraphic gaps in one locality and scarscely noticeable in another, has already been discussed at length. A glance at the interpreted HI/FES sections (Figs. 3 and 4) should suffice to illustrate the highly variable thicknesses of either Neogene or Paleogene sediments at any given locality. There are even places where one or the other unit, or both, are completely absent. Thus any regional implications drawn from a few drill

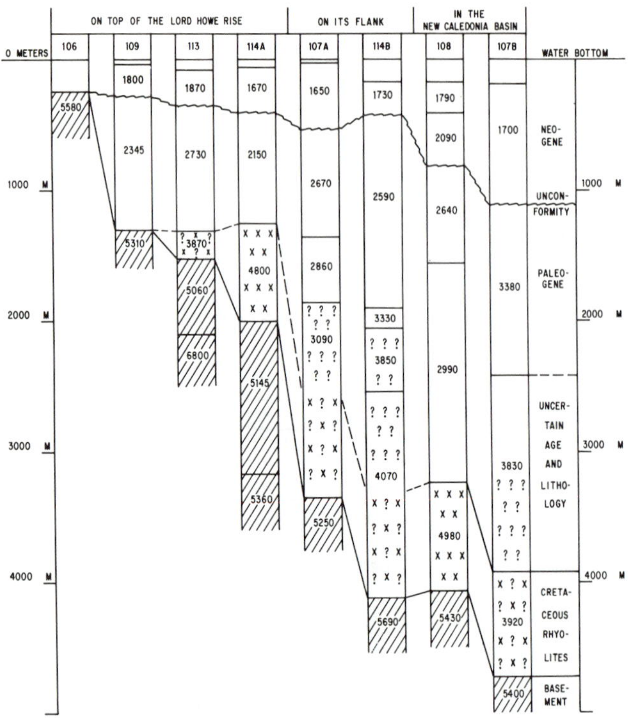

Fig. 7. Velocities in meters per second and possible correlations of intervals, based on sonobuoy refraction data.

	ON TOP OF THE LORD HOWE RISE				ON ITS FLANK		IN THE NEW CALEDONIA BASIN	
	106	109	113	114A	107A	114B	108	107B
NEOGENE	260	280	350	400	530	440	820	1120
PALEOGENE	0	1020	950	840	1340	1610	2410	1290
UNCERTAIN	0	0	0	0	? 730	1250	0	1500
RHYOLITE	0	0	220 ?	760	? 740	? 800	830	800
TOTAL	260	1300	1520	2000	3340	4100	4060	4710

Fig. 8. Possible thicknesses in meters, interpreted from refraction data.

time of proposed opening of the Tasman Sea (Hayes and Ringis, 1973) and could have been extruded subaerially or in a shallow-water environment, as during the subsequent deposition of the Maestrichtian claystones.

With the onset of Paleocene time a rapid subsidence into a deep-water environment took place throughout the area, which continued through Eocene time and then remained more or less static to this day. Thus the bulk of the ubiquitous and continuous vulcanism on the southern Lord Howe Rise must have occurred in relatively deep water.

Some of the more striking examples of such submarine volcanoes are shown as reproductions of our actual HI/FES records in Figures 6, 10, and 11. The vertical exaggeration of 30 times tends to accentuate greatly the apparent steepness of the volcaniç peaks; the peak on line 109 (Fig. 10), protrudes only 500 m above the sea floor from a 5.5-km-wide base. These examples show that there must have been rather recent explosive eruptions, since the blow-out craters have not yet been covered over or smoothed out by submarine currents. They also suggest that similiar eruptions have occurred throughout Neogene time, resulting in a circular deposit of volcanic debris. Indeed, even below the Neogene-Paleogene unconformity these vents have probably been actively pouring out rhyolitic flows and other extrusives, as suggested by some of the seismically transparent intervals and by irregular dipping beds.

Similar volcanic phenomona have been studied extensively on North Island of New Zealand (Brown et al., 1968). Eruptions of basaltic pillow lavas, breccias, and tuffs, and intrusions of dolerite and teschenite, appear to have continued spasmodically from the middle Cretaceous to middle Eocene time in the Northland area. During the whole of the Neogene, mainly andesitic and dacitic volcanic activity occurred throughout northern North Island, followed at times by eruptions of rhyolites and some basaltic vulcanism.

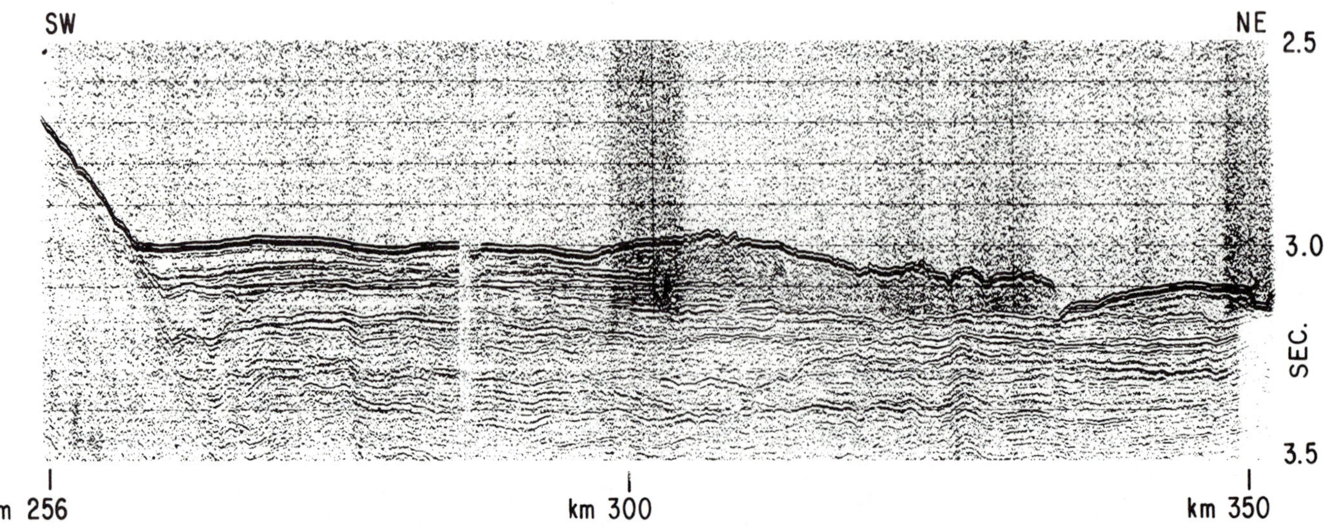

Fig. 9. Detail of line 107 showing sediment slumps on the floor of the New Caledonia Basin.

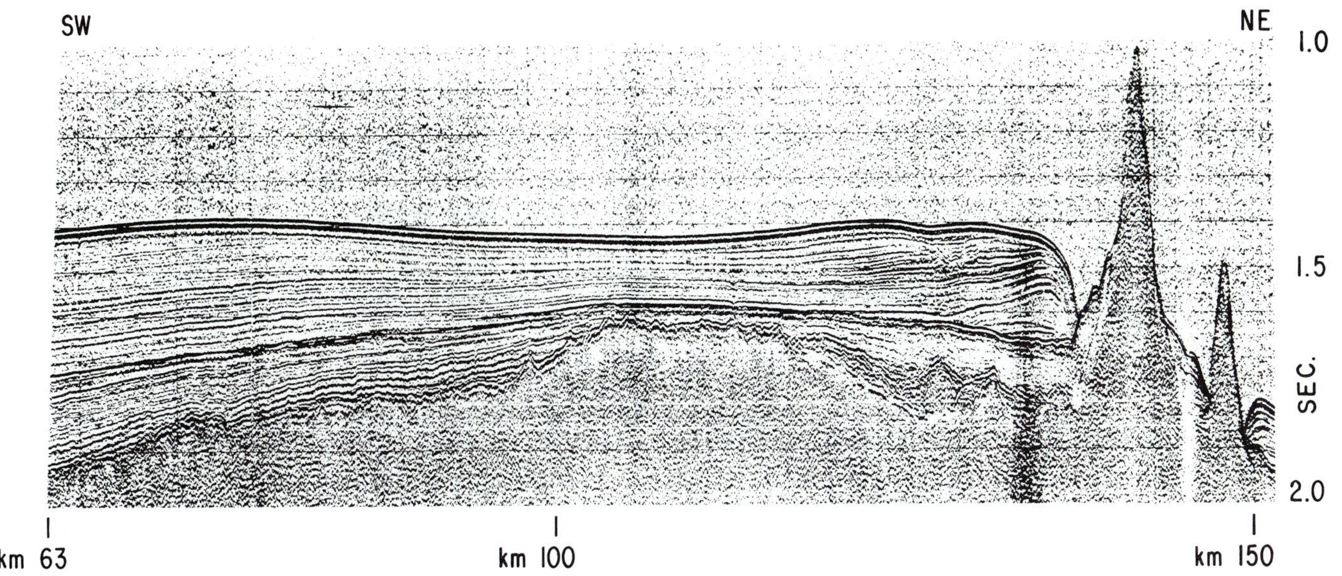

Fig. 10. Detail of line 109, showing submarine vulcanism.

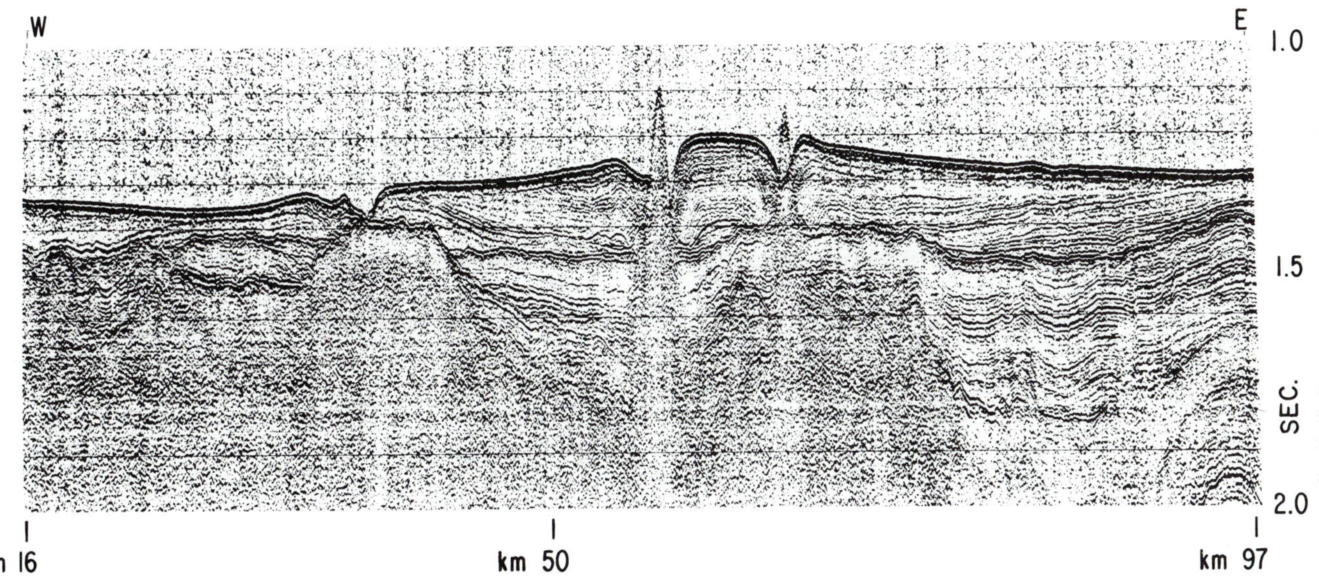

Fig. 11. Detail of line 114, showing submarine vulcanism.

Of the great variety of volcanic landscapes resulting from this vulcanism, the tuff rings of the Auckland volcanic field most closely resemble the features visible on our cross sections. As described by Cotton (1969), "the tuff-making ash has come to rent with regular bedding, at low angles, parallel to the outer slopes and in some cases also towards the center of the ring." The only thing missing in these tuff rings of New Zealand are the central cones or Mt. Pelee-type plugs visible on our seismic records.

The critical problem is whether this volcanic morphology, known from subaerial eruptions, can be adopted for our subaqueous volcanism. The wide spread even bedding within the volcanic-cone morphology and the low velocities (Fig. 7) seen on our seismic records strongly suggest an explosive nature of eruption. It is likely that only the larger and heavier extrusive products were deposited in the immediate vicinity of the volcanic vent, and that most of the ash and lighter pumice was carried away by currents. This may be one of the reasons why only relatively small amounts of volcanic material were recovered in the JOIDES cores in this area so pockmarked with active volcanoes.

The great number of these volcanic peaks crossed by our random lines leads one to suspect that we may be actually dealing with dikes rather than circular volcanoes. The double peaks of lines 109 and 112, occurring very close to where these two lines crossed, do indeed suggest a lateral extent. On the other hand, the evidence on line 114 (see actual HI/FES record, Fig. 11) indicates that the eastern one of the twin peaks is located just slightly north of the seismic line, since the water-bottom reflection is visible across the central volcanic cone.

Cotton (1969) states that alignment of closely spaced small vents suggests the presence of a fissure at shallow depth. The presence of numerous major and minor faults on the Lord Howe Rise would provide ample avenues of escape for a magmatic source.

GRAVITY AND MAGNETICS

A few additional clues to this volcanic activity can be found on the Bouguer gravity and total magnetic intensity profiles shown in Figure 12. While the majority of the high magnetic anomalies coincide with elevations in the acoustic basement, there are a number of places where such anomalies fall within the Neogene and Paleogene sedimentary section, suggesting the presence of interbedded volcanic materials. Some of the anomalies that fall within the depth range of our HI/FES records are indicated on our interpreted cross sections. Two rather small negative gravity anomalies appear to coincide with the volcanic twin peaks on lines 109 and 112, indicating the lesser density of the eruptive volcanic materials in comparison with the surrounding hard-rock basement.

Most of the Bouguer gravity curves are rather uneventful, portraying largely the general crustal composition and thickness. Values around 100 mgal characterize the semicontinental crust of the Lord Howe Rise, gradually increasing to 200 mgal in the central New Caledonia Basin. Woodward and Hunt (1971) give a thickness of 26 km for the crust under the Lord Howe Rise and call it "quasi-continental." The 9-km-thick crust under the New Caledonia Basin is called "oceanic" by them.

GEOLOGIC SUMMARY

The results of the JOIDES drilling Leg 21 and our observations based on seismic HI/FES and refraction data enable us to outline the geologic history of the southern Lord Howe Rise briefly as follows:

1. In the late Upper Cretaceous, rhyolitic extrusions occurred subaerially or in very shallow water. The overlying Maestrichtian silty claystones recovered at JOIDES site 207 indicate that at the close of Cretaceous the Lord Howe Rise was still at moderate water depth and close enough to a land mass to receive terrigenous sediments. These Late Cretaceous developments correspond closely to the time (80 my) proposed by Hayes and Ringis (1973) for the opening of the Tasman Sea.

2. At the beginning of Cenozoic time a rather sudden foundering of the Lord Howe Rise occurred, evidenced by the deposition of Paleocene deep-water fossil oozes. This sinking process apparently continued throughout Paleogene time until it reached an average depth of around 1,500 m below sea level on the crest of the Rise. In the process of this downward movement, the Lord Howe Rise minicontinent was shattered by numerous faults and differential subsidence of individual horsts and grabens, accompanied by acidic volcanic eruptions.

3. It is also likely that the New Caledonia Basin was created through crustal thinning during

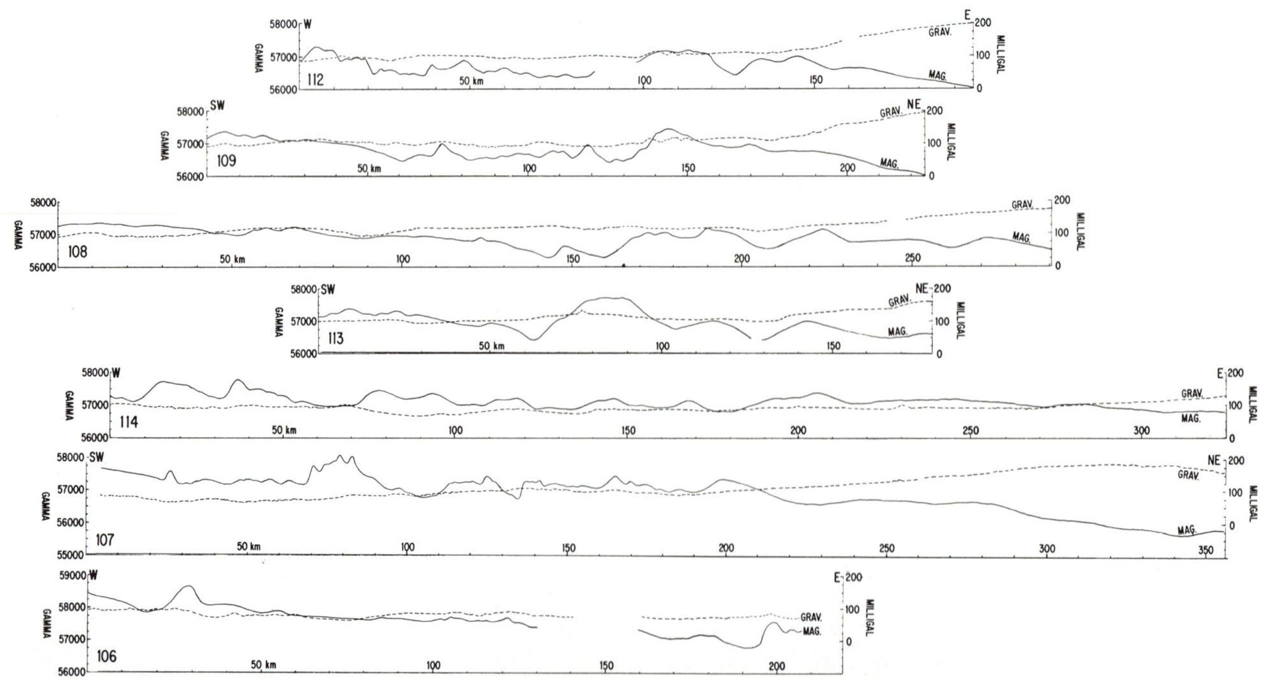

Fig. 12. Bouguer gravity and total magnetic intensity profiles.

Paleogene, separating the Lord Howe Rise and the Norfolk Ridge. As a result of isostatic forces, the relatively thinner crust of the New Caledonia Basin subsided at a faster rate than the adjacent ridges composed of thicker continental crust. This caused continued slumping of the highly water saturated sediments deposited on the flanks of the Lord Howe Rise and Norfolk Ridge, resulting in an anomalously thick deep-water sediment sequence in the New Caledonia Basin.

4. Sometime toward the latter part of the Paleogene a rather drastic change occurred. The foundering process of the Lord Howe Rise stopped and with it most of the faulting activity. During this stagnant period, centered around late Oligocene to early Miocene, strong submarine currents probably swept over the ridge, beveling the highly irregular Paleogene topography and creating a conspicuous regional unconformity.

5. However, not all movements came to a halt at the onset of Neogene time. The New Caledonia Basin kept on sinking and on top of the Lord Howe Rise gentle downwarping continued locally; coinciding in places with the sites of earlier Paleogene grabens. Some of the major faults remained active; many of them possibly serving as feeder vents for continued volcanic eruptions. The pattern of differential subsidence, accompanied by explosive volcanic eruptions, both of which contributed to massive sediment slumping off the flanks of the Lord Howe Rise, appear to have continued throughout the Neogene into the present.

The Southern Lord Howe Rise can no longer be regarded as the structurally stable and volcanically dormant area it once was assumed to be. Particularly, the continued submarine volcanic activity may add a new dimension to the geologic interpretation of this fascinating area. Interpretations of geological history, based on new geophysical techniques and deep-sea coring, provide new dimensions to the understanding of this area of possible ancient continental margin that may be of value in analyzing objectively the history of similar areas throughout the world.

ACKNOWLEDGMENTS

The writer gratefully acknowledges permission to publish this information by Mobil Oil Corporation International Division.

BIBLIOGRAPHY

Andrews, J. E., 1973, Correlation of seismic reflectors, *in* Initial reports of the Deep Sea Drilling Project, v. XXI: Washington, D. C., U.S. Govt. Printing Office.

Brown, D. A., Campbell, K. S. W., and Crook, K. A. W., 1968, The geological evolution of Australia and New Zealand: Elmsford, N.Y., Pergamon Press.

Burns, D., Watters, W. A., and Webb, P. N., 1973, Site 207, *in* Initial reports of the Deep Sea Drilling Project, v. XXI: Washington, D.C., U.S. Govt. Printing Office.

______, and Webb, P. N., 1973, Site 206 and Site 208, *in* Initial reports of the Deep Sea Drilling Project, v. XXI: Washington, D.C., U.S. Govt. Printing Office.

Burns, R. E., and Andrews, J. E., 1973, Regional aspects of deep-sea drilling in the Southwest Pacific, *in* Initial reports of the Deep Sea Drilling Project, v. XXI: Washington, D.C., U.S. Govt. Printing Office.

______, et al., 1972, *Glomar Challenger* down under: Geotimes, May 1972.

Cotton, D. A., 1969, Volcanoes as landscape forms: New York, Hafner Press.

Edwards, A. R., 1973, Southwest Pacific regional unconformities encountered during Leg 21, *in* Initial reports of the Deep Sea Drilling Project, v. XXI: Washington, D.C., U.S. Govt. Printing Office.

Hayes, D. E., and Ringis, T., 1973, Seafloor-spreading in the Tasman Sea: Nature, v. 243, June 22.

Houtz, R., Ewing, J., Ewing, M., and Lonardi, A. G., 1967, Seismic reflection profiles of the New Zealand Plateau: Jour. Geophys. Res., v. 72, no. 18.

Van der Lingen, G. J., 1973, The Lord Howe Rise rhyolites, *in* Initial reports of the Deep Sea Drilling Project, v. XXI: Washington, D.C., U.S. Govt. Printing Office.

Woodward, D. J., and Hunt, T. M., 1971, Crustal Structures across the Tasman Sea: New Zealand Jour. Geol. Geophys., v. 14, no. 1.

Margins of the Southwest Pacific

H.-R. Katz

INTRODUCTION

The region here considered lies between Antarctica in the south and the Tonga Islands in the north, and extends roughly along the 180th meridian (Fig. 1). The continental margins referred to border the Pacific basin proper; marginal oceanic areas like the Tasman and South Fiji basins are excluded.

These margins are not synonymous with plate boundaries as defined by the plate tectonic theory. They may coincide, however, as along the Tonga-Kermadec ridge-trench system, but in the New Zealand region they are entirely within the Pacific plate. Here the Pacific-Indian plate boundary is assumed to pass through the middle of a continental block (LePichon, 1968), while south of New Zealand this boundary separates oceanic crust on both sides (Fig. 2).

In this region much informative knowledge has accumulated in recent years, in particular from holes drilled by the Deep Sea Drilling Project (DSDP), from predrilling site surveys, and from extensive offshore surveys undertaken by oil companies. Above all, information from a regional marine reconnaissance by Mobil Oil Corporation has been made freely available for the purpose of this study and is widely incorporated in it. Since many of the data here presented are too new to have all their implications fully taken into account, this study should be considered a preliminary progress report.

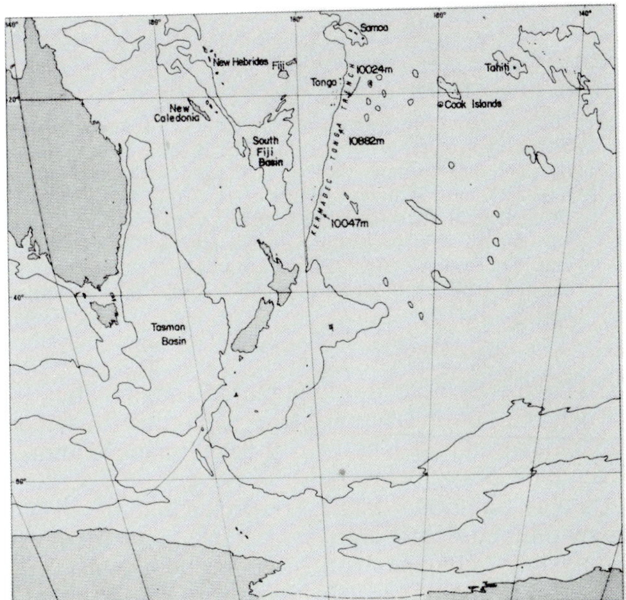

Fig. 1. Southwest Pacific region, showing land areas and 4,000-m isobath.

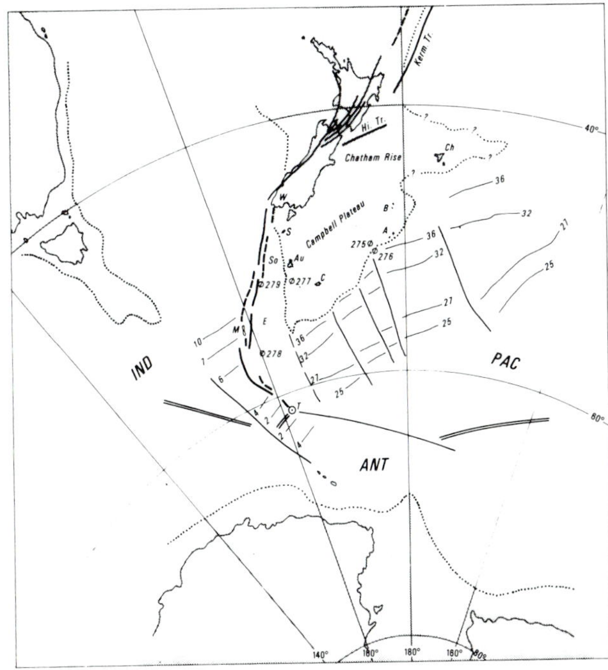

Fig. 2. Area south of New Zealand (after Christoffel and Falconer, 1972; Falconer, 1972; Hayes and Talwani, 1972). T, triple junction between Indian (IND), Antarctic (ANT) and Pacific (PAC) plates. Heavy continuous lines, oceanic trenches. Heavy dashed lines, ridges. Next-thinner lines, fracture zones and faults. Double lines, oceanic spreading ridges. Thin numbered lines, oceanic magnetic anomalies. Shaded area, underlain by continental crust. M, Macquarie Island; E, Emerald Basin; So, Solander Trough; W, Waiau Basin; S, Snares Islands; Au, Auckland Islands; C, Campbell Island; A, Antipodes Islands; B, Bounty Islands; Ch, Chatham Islands; Hi.Tr., Hikurangi Trench (renamed "Hikurangi Trough" in this paper). Kerm. Tr., Kermadec Trench. 275-279, DSDP Leg 29 sites (Geotimes, 1973).

CAMPBELL PLATEAU

A vast area of over 800,000 km^2, which lies at a water depth between 500 and 2,000 m, is part of New Zealand by virtue of both structure and geologic history. The crustal thickness, however, is considerably reduced—in the same way as happens in other depressed but quasi-continental plateaus and rises in the southwest Pacific (Fig. 3), which may be due to stretching associated with continental breakup (Bott, 1971) and/or a process of oceanization. Compositionally, the sialic character of the Campbell Plateau is beyond doubt.

The basement exposed on and around the various islands on the Campbell Plateau consists of

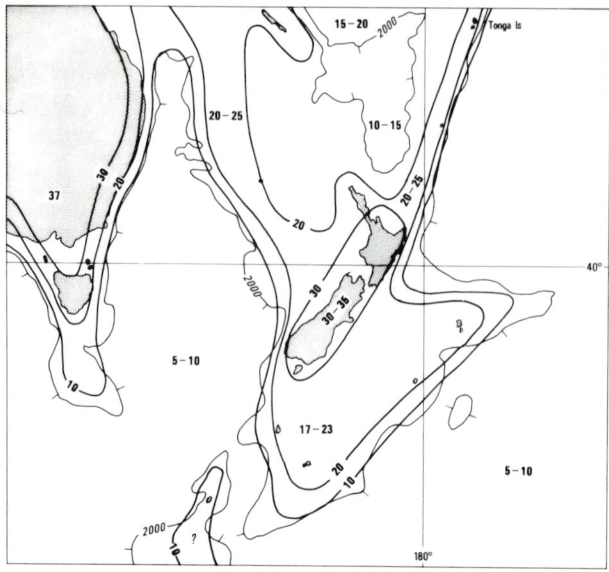

Fig. 3. Crustal thickness (in kilometers) in the southwest Pacific region (after Brodie, 1964).

Paleozoic and Mesozoic schist, granite, and gabbro (Summerhayes, 1969). An extensive peneplain was established in Cretaceous time, as shown by seismic data (Houtz et al., 1967) and is overlain by a marine Maestrichtian to Lower Tertiary sedimentary sequence up to 1 km thick. Some basal sediments may be terrestrial, as on Campbell Island (Oliver et al., 1950), but a fully marine, oceanic environment was soon established over probably the whole area. The Late Cretaceous transgression reached into the eastern part of the South Island, New Zealand (Katz, 1968). The region was gently deformed in Oligocene time, and volcanic centers developed on several broad rises during the late Tertiary (Summerhayes, 1969). The olivine basalt of the Antipodes Islands is possibly Quaternary, as suggested by the good preservation of volcanic cones. As far as is known, the volcanism is alkaline in most areas, and appears closely related to the late Cenozoic volcanism along the east coast of the South Island. Characteristic of this major volcanic province is its location in an isostatically and tectonically stable and mainly aseismic region (Eiby, 1958).

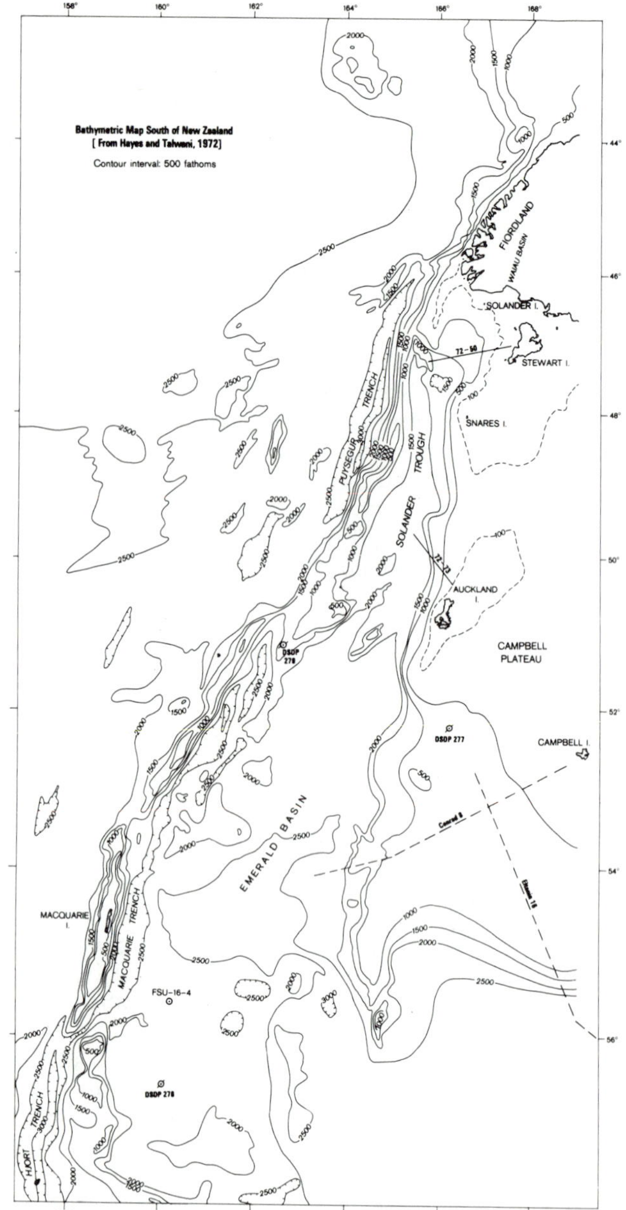

Fig. 4. Bathymetric map south of New Zealand (Emerald Basin-Solander Trough) (after Hayes and Talwani, 1972). Contour interval, 500 fathoms. FSU-16-4 location of Miocene core at top of 600 m of sediment (Houtz et al., 1971). DSDP 277-279, DSDP Leg 29 sites (Geotimes, 1973). Conrad 8 and Eltanin 16, Lamont seismic sections across edge of Campbell Plateau (Houtz et al., 1967, Figs. 12 and 14). 72-50 and 72-73, seismic sections Solander Trough (Mobil Oil Corporation; Fig. 5).

Western Margin

A steep slope defines the western edge of the Campbell Plateau (Figs. 2 and 4), along which it drops to the floor of the Solander Trough-Emerald Basin, 3,000-4,000 m deep. This slope is controlled mainly by down-to-basin normal faults (Fig. 5a, b; Summerhayes, 1969; Houtz et al., 1971; Hayes and Talwani, 1972). Near the edge of the plateau, i.e., on southern Stewart Island, the Snares and Auckland islands, the crystalline basement consists of gneissic granites the last heating of which, if not emplacement, occurred during the early Late Cretaceous (Aronson, 1968; unpublished data of Mobil Oil Corp.). Structurally, the plateau edge both in the west and south is marked by a shallow basement high covered only locally by thin sediments (Houtz et al., 1967; Katz, in press b).

On Campbell Island these sediments overlie peneplained schist and consist of 15 m of quartz sandstone, conglomerate, and carbonaceous mudstone of probably Late Cretaceous to Paleocene age, which pass upward into Eocene to Oligocene pelagic, foraminiferal limestone and chalk with chert

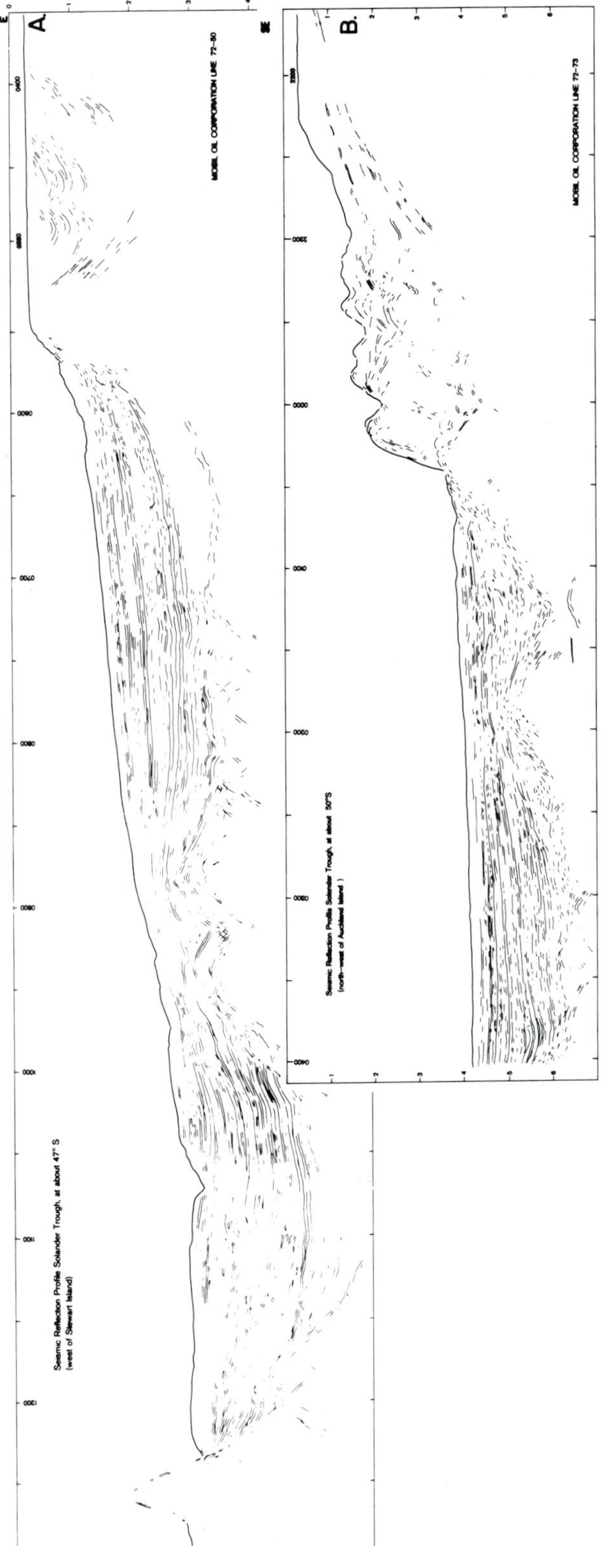

Fig. 5. Tracings of seismic-reflection profiles, Solander Trough (see Fig. 4 for location). Vertical scale, seconds of two-way reflection time. Horizontal scale, hours of ship time.

nodules (Oliver et al., 1950). A near-identical sequence ranging in age from Paleocene to middle-Late Oligocene was penetrated for 462 m in DSDP site 277 south of Auckland Island (Fig. 4); it is unconformably overlain by thin Plio-Pleistocene foraminiferal ooze (Geotimes, 1973). This Late Oligocene limestone is the youngest rock in the area which has been tectonically deformed; it was slightly folded, faulted and uplifted, and eroded, but submerged again in the Pliocene, when explosive volcanicity on Campbell Island unconformably covered it. Middle to Late Miocene tuffaceous sandstone in the Auckland Islands (Fleming, 1968) may belong to the same volcanic cycle.

The Solander Trough to the west is filled with turbidite-type sediments (Houtz et al., 1971), which northwest of Auckland Island are 2.5-3 m thick (Fig. 5b). Younger volcanic complexes occupy large areas near its head, where the Solander Islands, which are composed chiefly of hornblende andesite, are thought to be late Pleistocene (Harrington and Wood, 1958). Northward the Solander Trough partly continues into the South Island, where the Tertiary Waiau Basin (Katz, 1968) is structurally related to it. Across the southern part of the Solander Trough magnetic profiles (Hatherton, 1967) are increasingly complex and can partly be correlated with sea-bottom topography. A shallow, narrow trench down the axis of the trough, and an alignment of three seamounts, of which one at least has magnetic characteristics that favor a basic volcanic origin (Summerhayes, 1969), may imply rifting along the trough. Pillow basalt interpreted as oceanic crust was found in the Emerald Basin farther south (DSDP site 278; Geotimes, 1973); it is overlain by a continuous sedimentary sequence, 428 m thick, consisting of nanno-chalk, nanno and siliceous oozes, and radiolaria-diatom ooze ranging from the Middle Oligocene up to the late Pleistocene. About 600 m of sediments are reported from location FSU-16-4 (Fig. 4), which are Miocene at the top (Houtz et al., 1971).

The Macquarie Ridge Complex (Hayes and Talwani, 1972) to the west of the Solander Trough-Emerald Basin is considered entirely oceanic. Varne and Rubenach (1972) suggest that on Macquarie Island a faulted section through different levels of oceanic crust is exposed. At the top a layer of unmetamorphosed extrusive rocks overlies greenschist facies extrusive rocks; some of the freshest volcanic rocks are associated with sediments that appear to be Middle or Early Miocene. Oceanic crust of the same age, i.e., basalt interpreted as lava flow overlain by middle Early Miocene sediments (20 my old), was drilled in DSDP site 279 (Geotimes, 1973) farther north on the ridge (Fig. 4). A large number of *in situ* dredge samples from the Macquarie Ridge (Summerhayes, 1969) are predominantly volcanic or plutonic, basic igneous rocks, including olivine basalts, basalts, dolerites, gabbros, peridotites, basic agglomerates, and undifferentiated basic volcanics; the general modal mineralogy indicates a broadly tholeiitic character for many of these rocks.

Both the oceanic Macquarie Ridge Complex and the Emerald Basin-Solander Trough downwarp (which at least in the Emerald Basin is also underlain by oceanic crust) continue northward into distinct tectonic elements that are part of the New Zealand continental block: the Macquarie Ridge complex, thought to represent essentially a zone of predominant transcurrent motion (Hayes and Talwani, 1972), continues into the Alpine fault (Suggate, 1963), while the Solander Trough continues into the Waiau and adjacent Tertiary basins of Southland and Fiordland (New Zealand Geological Survey, 1972). Thus the continental margin around the southwest corner of New Zealand must cut obliquely across these major tectonic features, but its nature and location are unknown. Obviously, complex interactions can be expected between the Macquarie Ridge-Alpine Fault shear zone, which has been interpreted as a Recent plate boundary (Christoffel, 1971), and the continental margin.

The Solander Trough-Emerald Basin structure would appear to represent a marginal rift, with sea-floor spreading along the trough-basin axis similar to the Red Sea. The short sequence of magnetic anomalies recorded by Hayes and Talwani (1972) between Macquarie Ridge and Solander Trough, although of questionable correlation and relationship, may support this interpretation. Also, shallow earthquake seismicity in the Solander Trough (Summerhayes, 1969) indicates crustal acitivity of the type known along spreading ridges. If New Zealand once was joined with Australia, it could be claimed that the thick sediments in the Solander Trough filled a marginal rift originally lying between the two continents.

The downwarp of this trough probably was initiated during the Eocene or earlier. Subsidence in the Waiau Basin at the northern extremity began in the Late Eocene (Katz, 1968), while in the Middle Oligocene the Emerald Basin to the south would have sunk to over 2,600 m below sea level, as is suggested by the generation at that time of new oceanic crust (Sclater et al., 1971). The facies of Oligocene sediments in the Emerald Basin (DSDP site 278, Geotimes, 1973), much thinner than the corresponding sequence on the Campbell Plateau (DSDP site 277), also suggests deeper water than on the plateau, while sedimentation in the basin was continuous from the Middle Oligocene onward but was interrupted on the plateau in the Late Oligocene. On tectonic grounds it is reasonable to assume, too, that downwarping occurred at or before the time of last deformation on the adjacent plateau. It is concluded that probably over its entire length of 1,100 km the Solander Trough-Emerald Basin began subsiding in the early Tertiary, which resulted in final rifting and the generation of new oceanic crust in the southern part, during the Oligocene.

Southeastern Margin

In the southwest Pacific basin linear magnetic anomalies extend to the steep slope of the Campbell Plateau (Fig. 2); Christoffel and Falconer (1972) were able here to expand their time scale to anomaly 36, deducing that New Zealand separated from Antarctica just prior to 81 my ago. This implies a discontinuity of 50 my between the middle Oligocene sea floor in the Emerald Basin, and the Late Cretaceous floor below the Campbell Plateau immediately to the east (Fig. 2).

Here the basement is flat right to the base of the continental rise (Houtz and Markl, 1972); there is no distinct steepening of dip as shown below the continental rises of Antarctica and Australia, where a type of rift valley is filled with thick sediments; nor is there massive accumulation of an extensive sedimentary wedge that obscures the continent-ocean boundary. Indeed, the southeastern margin of the Campbell Plateau is remarkably unobscured and undisturbed. A seismic cross section at 176°30' east (Houtz and Markl, 1972, Fig. 10B) suggests erosion of sediments by a current of considerable force. DSDP site 276 (Geotimes, 1973) located on the abyssal plain near the continental rise (Fig. 2), where the magnetic anomalies indicate a crustal age of 80-85 my, confirmed such erosion to have cut down into the Paleogene, which is covered by a surficial layer of sand-gravel formed by winnowing action. Fragments of plutonic and metamorphic rocks in the Paleogene sediments suggest proximity of the site to the Campbell Plateau since at least the Oligocene.

DSDP site 275 (Geotimes, 1973), on the plateau about 50 km northwest of site 276, found Maestrichtian marine sediments below a thin Pleistocene veneer of foraminiferal ooze and manganese nodule pavement. This and the fact that Tertiary sediments, which seismic data indicate exist farther north and northwest, appear to be removed by erosion at this site near the plateau edge indicate that a western boundary current (Warren, 1970) of considerable width and reaching across the edge of the Campbell Plateau to relatively shallow depth is active at the present day.

Rifting with normal faults at and near the continental margin seems to be present in the basement of the Campbell Plateau, but it does not affect the overlying Late Cretaceous sediments. All evidence thus points to a mode of separation from Antarctica—if that is how this continental margin originally was formed—quite different from the Australian-Antarctic mode of separation. Houtz and Markl (1972) suggest that the Campbell Plateau and the western Ross Sea (and West Antarctica) shelves were sheared apart as a result of a large component of translation in the initial stages of separation.

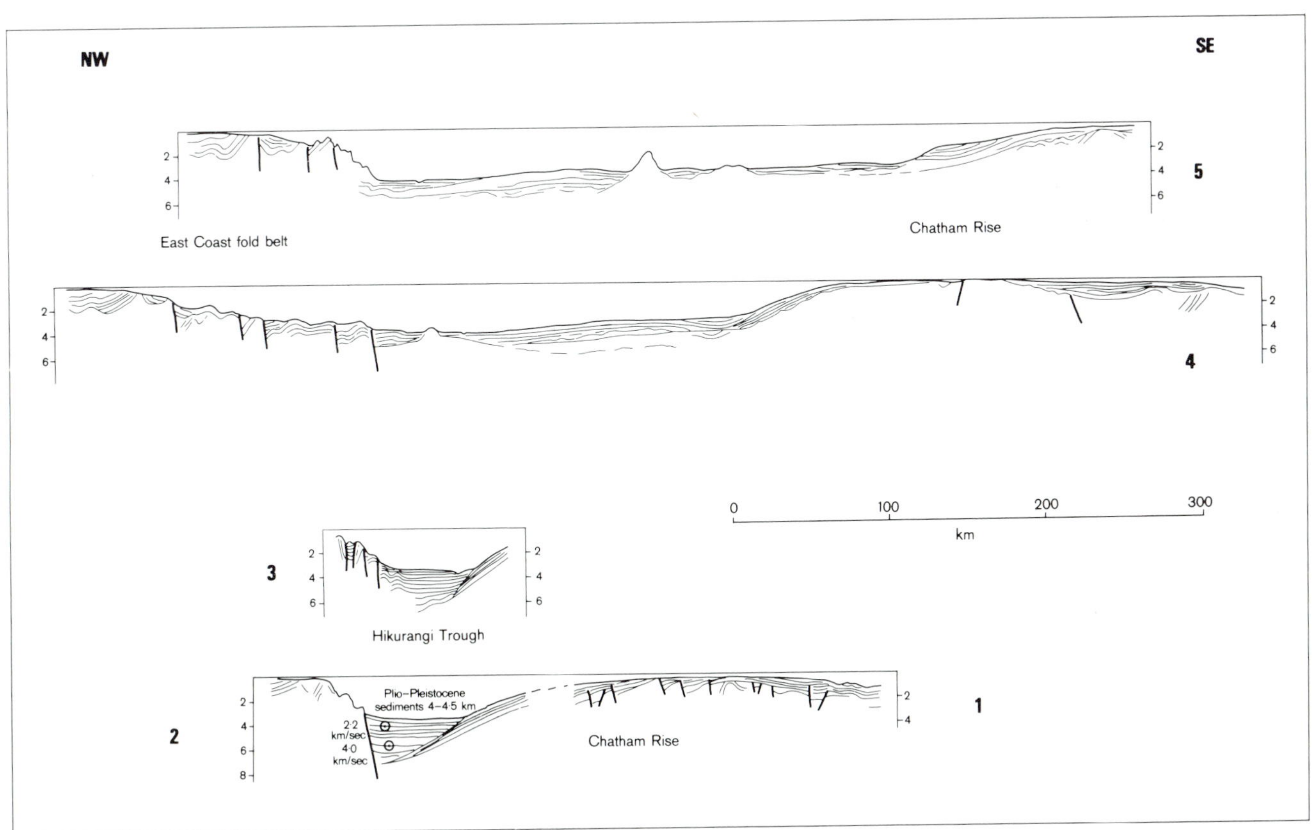

Fig. 6. Profiles from the East Coast foldbelt, North Island, New Zealand, to the Chatham Rise (see Fig. 7 for location). Vertical scale, seconds of two-way reflection time.

CHATHAM RISE AND HIKURANGI TRENCH

Separated from the Campbell Plateau by the broad depression of the Bounty Trough (Fig. 2), the Chatham Rise extends in a straight west-east direction for about 1,000 km. Its geologic history is similar to the Campbell Plateau, except that it has been a separate arched feature since the Late Cretaceous (Katz, in press b). The basement of Paleozoic to early Mesozoic quartzo-feldspathic schist and graywacke and mid-Cretaceous sediments, exposed on Chatham and Pitt islands (Hay et al., 1970), sporadically outcropping on the rise farther west (Fleming and Reed, 1951; Cullen, 1965), and well identified on marine seismic sections (O'Halloran, 1971; Austin et al., 1973), has been folded and block-faulted, and peneplained probably in Late Cretaceous time. The Tertiary is relatively thin, although reactivation of mainly longitudinal tensional faults has resulted in local increased thickness of Eocene and/or Miocene sequences. The entire area is dominated by horst and graben tectonics; antithetic step faulting accompanies the structure on its northern flank, where it bends down to the Hikurangi Trench (Fig. 6, profile 1; profile locations, Fig. 7).

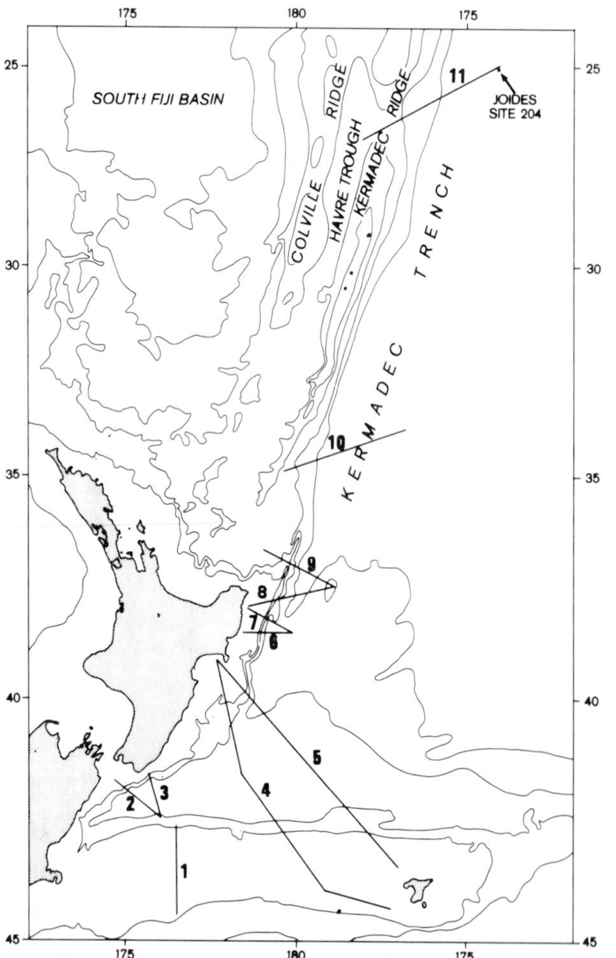

Fig. 7. Location of profiles in Figs. 6 and 9. Bathymetric contour interval, 1,000 m.

A significant proportion of the Tertiary exposed on the Chatham Islands is made up of volcanic rocks of Eocene and Miocene-Pliocene age. Offshore, seismic and magnetic evidence (Austin et al., 1973) suggests widespread occurrence of volcanic necks of limited extent. According to Watters (in Hay et al., 1970) the older volcanic group comprises abundant olivine-basalt flows and intervening ash beds and includes submarine and terrestrial lavas. More silica-rich rocks—nearly all of which are alkaline and probably represent differentiates of the basaltic magma—are found as rare phonolites in the upper part of the sequence, or as fairly numerous dikes of trachyte. The younger group occurs widely as flows and plugs, and several pipes of explosive origin; the rocks comprise mainly limburgites (both as lavas and agglomerates), with minor monchiquite and nephelinite.

The Hikurangi Trench (Brodie and Hatherton, 1958) has invariably been considered to be related to the Kermadec-Tonga Trench farther north, which defines the boundary between a quasicontinental island arc and oceanic crust. As the development of the downthrust lithosphere model along active continental margins (Oliver and Isacks, 1967; Isacks et al., 1968) was based to a large extent on observations in the Tonga and Kermadec regions, the similar pattern of geophysical features in the North Island, New Zealand (Hamilton and Gale, 1968), has led to the assumption of an asymmetric active region typical of island arcs (Hatherton, 1969), the eastern margin of which would be defined by the Hikurangi Trench.

The "Hikurangi active margin" (Hatherton, 1970b, 1970c) has been found, however, to differ in important aspects from the model proposed by Isacks et al. (1968). Although the projected Benioff zone, as is invariably the case in other active regions (Hatherton, 1969), intersects the surface of the earth close to the axis of the negative gravity anomaly, this intersection lies in the middle of the North Island continental crust and 200 km to the west of the Hikurangi Trench (Hamilton and Gale, 1968). If the negative gravity anomalies of the North Island are an extension of the negative gravity anomalies of the Kermadec Trench, the latter must swing to the southwest before meeting the East Cape (Fig. 8); they certainly do not continue into the Hikurangi Trench, which is covered by positive anomalies (Hatherton, 1970c). Together with this apparent change in trend, there is also a gap in seismicity between the North Island and the Kermadec Ridge (Hamilton and Gale, 1968).

Whereas shallow earthquakes in the Tonga-Kermadec region are confined to the continental side of the Benioff zone intersection with the earth's surface (Sykes, 1966; Mitronovas et al., 1969; Hatherton, 1971), crustal seismicity in the North Island, New Zealand, is symmetrical about this axis (Hatherton, 1970b). If the cause of shallow earthquakes above the Benioff zone is sought in terms of overthrusting of lithospheric plates, such symmetry is certainly unexplained. Moreover, no evidence of underthrusting of the eastern half of the continental crust, along the Benioff Zone in New Zealand, is found from geological information. The change in direction of the negative gravity anomaly axis in Cook Strait (Fig. 8), apparently accompanied by the main belt of shallow earthquakes (Hatherton, 1970a), and its trend at that point, which is at right angles to the Hikurangi Trench, is incompatible with an interpretation of the latter as the morphological expression of a downgoing lithospheric slab. There is therefore strong doubt that the Hikurangi Trench represents the locus of an active continental margin to the east of the North Island.

From seismic sections across the Hikurangi Trench (Fig. 7) it is apparent that in the Cook Strait-Wairarapa region (Fig. 6, profiles 2, 3) it is filled with 4-4.5 km of turbidite-type, virtually flat-lying sediments which probably have been brought into the trench from the northern Canterbury shelf and the Kaikoura-East Coast foldbelt (Kaikoura and Cook Strait canyons, etc., Fig. 8). They abut against the northwestern wall of the trench, which appears to be faulted, and unconformably overlap the Miocene-Eocene sequence of the Chatham Rise to the southeast.

The trench sediments, probably of Plio-Pleistocene age, gradually thin to the northeast, while in the same direction the trench becomes wider and flatter. In a northwest section from Hawke Bay to the Chatham Islands (Fig. 6, profiles 4,5) it has disappeared completely, both structurally and morphologically. There remains only a flat shallow depression between the abyssal plain to the east and the steep and abrupt irregular slope to the west; this depression is filled with a sedimentary sequence of limited thickness that is horizontal and unconformably underlain by a slightly deformed series. The latter continues into the slope of the Chatham Rise, thus revealing a probable (Lower) Miocene age. Underneath the northwestern slope both series become involved in the structural deformation of the East Coast foldbelt, where they cannot be identified further, as the strong topographic irregularities cause significant dispersal of the seismic signal.

In the same cross section (profiles 4 and 5, Fig. 6) the sediment cover is pierced in a few places by what seem to be volcanic necks, probably related to similar occurrences found along the northern slope of the Chatham Rise and around the Chatham Islands (Austin et al., 1973). Such piercement structures were lined up by Houtz et al. (1967) to form a 300-km-long "outer ridge," the existence of which now appears extremely doubtful. The detailed bathymetry available (van der Linden, 1968) does not suggest such a ridge to be real.

Structural relationships (Fig. 8) and tectonic history suggest that the Hikurangi Trench is part of the Kaikoura-East Coast late Cenozoic foldbelt. Elongate basins of similar structural position were repeatedly formed in this foldbelt during the Late Miocene and Pliocene and were filled with partly thick turbidite sequences (Kingma, 1958, 1960; Katz, 1968); comparable rate and amount of subsidence are found, for instance, in the late Cenozoic basin between Wairarapa and Hawke Bay (Katz, 1974), where depth of deposition (Vella, 1963; Beu, 1967, 1969) compares with present water depth in the Hikurangi Trench. Thinning and locally complete wedging out of Miocene or older Tertiary series underneath the trench (profiles 2 and 3, Fig. 6), and the presence of Mesozoic basement of similar type—of undoubtedly continental crust—on both sides and probably also underneath it, indicate that the Hikurangi Trench indeed is a tectonic element closely related to the East Coast foldbelt, of entirely continental tectonic affinities.

It is concluded that the Hikurangi Trench is not a continental margin. Detailed mapping (Fig. 8) also shows that it is not connected with the Kermadec Trench, but constitutes a depositional basin unrelated to it, of locally great subsidence and thick sediment fill. It is proposed therefore to abandon the designation "trench" and name the feature "Hikurangi Trough."

KERMADEC-TONGA REGION

The island arc-trench system north of New Zealand has become a classic for the concept of the "new global tectonics," which is based on a large amount of geophysical data obtained here by various workers (e.g., Raitt et al., 1955; Talwani et al., 1961; Sykes, 1966; Oliver and Isacks, 1967; Isacks et al., 1968, 1969; Mitronovas et al., 1969; Sykes et al., 1969; Karig, 1970). The crustal structure is well known from gravity data and seismic velocities, while the inclined seismic zone, which is extremely thin and dipping west underneath the island arc, is one of the best defined and can be traced continuously from the surface to 650 km deep. Very little is known, however, of the geologic-tectonic structure and history of the uppermost part of the lithosphere, and its relationships with adjacent regions.

Head of the Kermadec Trench near East Cape, New Zealand

One of the most critical, though least understood areas lies to the east of East Cape (Fig. 8). In this area, i.e., south of 38°S, the Kermadec Trench looses its identity, merging south and upward into the rather featureless abyssal plain at 3,500 m. Piercement structures similar to those found farther south and probably of igneous-volcanic origin are the only disturbances in this area, and at 38°30'S occur right at the head of the Kermadec Trench. Southward to about 40°S (Mahia High) they seem to line up, according to seismic and bathymetry, in a way so as to trap sediment between them and the slope of the East Coast foldbelt; up to 3 km of

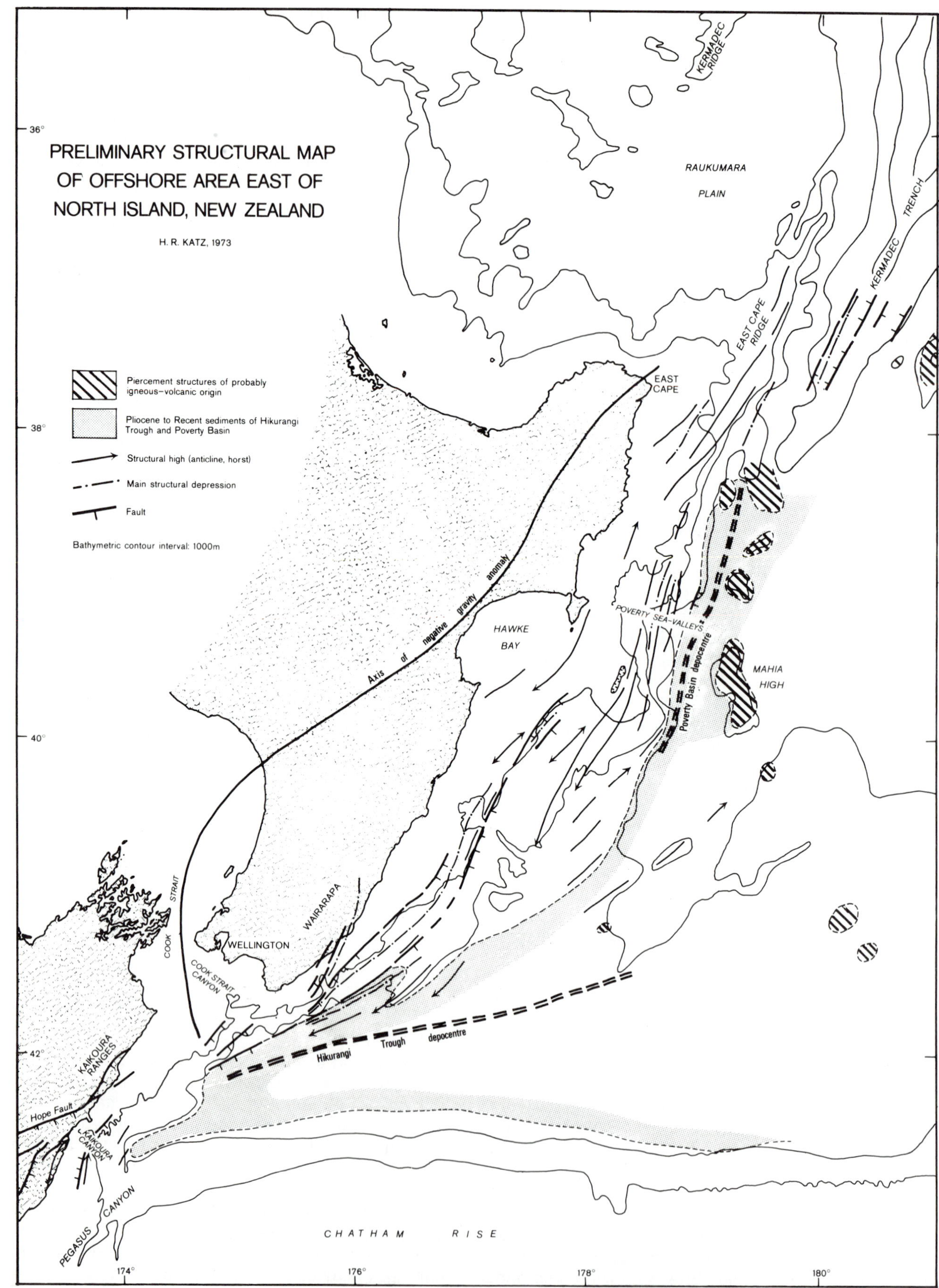

Fig. 8. Structural map of offshore area east of North Island, New Zealand. Contour interval, 1,000 m. Axis of negative gravity anomaly after Hatherton (1970a).

tectonically undisturbed, ponded sediments have accumulated in front of the Poverty Sea-Valleys (van der Linden, 1968), which descend from the shelf north of Mahia Peninsula. They thin and partly seem to be faulted against Mahia High in the east, and are deformed also at the margin of the East Coast foldbelt in the west. From seismic lines across the slope and shelf their age is considered to be Pliocene to Recent. This "Poverty Basin" (Fig. 8) is not, however, continuous with or related to the Hikurangi Trough nor the Kermadec Trench but is a separate area of major deposition, probably of turbidites, adjacent to and below the East Coast foldbelt.

The slope, which in this area has markedly steepened and also changed its direction, is formed by a tight bundle of fault-bounded, horst-type anticlines. As a rule they have acoustic basement in their core consisting of sediments of older Tertiary to Late Cretaceous age (Katz, in press a). The East Cape Ridge, which here emerges from this foldbelt immediately below the shelf, is seismically characterized as basement, yet it could thus be a long and narrow, upfaulted horst of Lower Tertiary and Upper Cretaceous rocks. From 37° northward, however, where a large and deeply subsided sedimentary basin (Raukumara Plain, Fig. 8) lies to the west of the East Cape Ridge, seismic velocities below its deepest reflectors, where they onlap onto the ridge, are 4.5 and 6.15 km/sec (Fig. 9, profile 9). The ridge therefore is believed to be true basement, at least in its northern part, and probably igneous.

In its southernmost portion the Kermadec Trench is an asymmetric feature that is distinctly V-notched (Fig. 9, profiles 8 and 9). It is here entrenched in the 3,500-m-deep abyssal plain which extends between the North Island and the Chatham Islands and is flanked to the east by basement highs of probably igneous origin, elevated above that plain. The steeper western slope of the Kermadec Trench is devoid of sediments, but the more gently dipping slope to the east is covered by a series 600-900 m thick, apparently provenant from the east, that can generally be subdivided into an upper transparent layer underlain by a turbidite-type sequence. Slight folding is seen, particularly in the lower series on the upper slope (which may or may not be tectonic), while block faulting affects all the sediments, extending right to the bottom of the trench; here the sediments are synclinally bent, apparently by drag along normal faults on both sides, and abut sharply against the western slope. This marked difference of sediment distribution, thickness, and provenence between the Kermadec Trench and the Poverty Basin and Hikurangi Trough is further evidence in support of the argument that the latter two are unrelated to the first.

In this area between 38 and 37°S, for the first time when going north, the characteristic elements and structural situation of the Kermadec-Tonga ridge-trench system, and as such of an active continental margin, become apparent. In the west the sedimentary basin of the Raukumara Plain (which is in direct continuation with the basins of the East Coast foldbelt) is nearly certainly continental, whereas the Kermadec Trench to the east marks the oceanic domain; associated with this structure from here to as far north as Tonga are a west-dipping Benioff zone and negative gravity anomaly (Talwani et al., 1961; Mitronovas et al., 1969). In this setting the East Cape Ridge and its continuation in the Kermadec-Tonga terrace seem to compare with similar ridges along other continental margins (Burk, 1968); significant and characteristic, too, are the much greater subsidence in younger geological history on the continentward (western) side of the East Cape Ridge, and the accumulation there of a thick sedimentary series, whereas on the oceanic side to the east, sediments are thin or nonexistent.

South of here the continental margin ridge-trench couple disappears as a whole, and somehow merges with the complex structure of the East Coast foldbelt, which is continental throughout and which has thick sediment accumulation, particularly in its young basins to the east. In this area, too, the negative gravity anomaly and Benioff zone, still dipping west, have shifted far to the west of the broad continental-type foldbelt; thus the locus, structure, and tectonic relationships of the continental margin east of it, south of 38°S, remain unknown.

Geological Structure of the Continental Margin Along the Kermadec-Tonga Ridge

From its head at 38°S, the Kermadec Trench extends north-northeast in a straight line for about 1,700 km. A shallow area and a slight offset separate it from the Tonga Trench, which continues in nearly the same direction for another 1,000 km as far north as 16°S, where it sharply bends to the west.

In profiles 10 and 11 (Fig. 9) the central part of the Kermadec Trench and the southernmost part of the Tonga Trench expose a typical graben-type structure. Here the abyssal plain of the southwest Pacific to the east of the trench is below 5,000 m and is underlain by a comparatively flat-lying sediment cover in which seismic penetration generally is about 0.4 sec. The sediments seem to extend into the block-faulted graben of the trench, but the seismic coverage is not detailed enough to distinguish them in the deepest parts of the trench. On the abyssal plain to the east, however, seismic sections (Fig. 9, profile 11; Fig. 10) indicate a low-angle unconformity near the site, where DSDP site 204 (24°57.27'S, 174°06.69'W; Geotimes, 1972; Burns et al., 1973) drilled 103 m of abyssal brown clay with intercalated ash, dated from Recent to Oligocene (but including reworked Late Cretaceous nannofossils and *Radiolaria*), underlain along a sharp contact by volcanogenic sandstone and conglomerate of possible early Tertiary to Cretaceous age. Basement was not reached, nor is it distinguished on the seismic

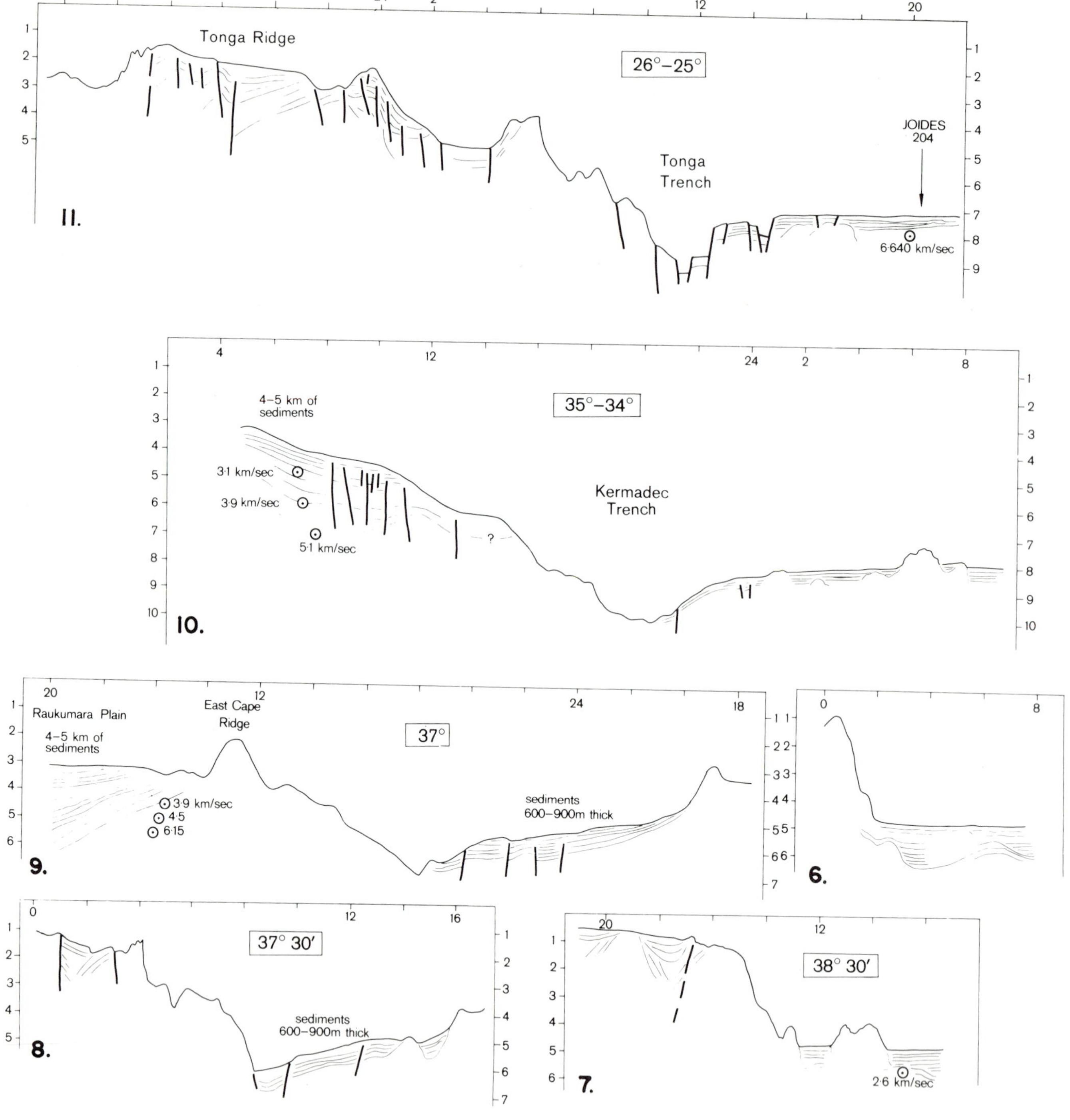

Fig. 9. Profiles east of East Cape, North Island, New Zealand, and across the Kermadec-Tonga ridge and trench (see Fig. 7 for location). Vertical scale, seconds of two-way reflection time; horizontal scale, hours of ship time.

sections, but a sonobuoy velocity of 6.640 km/sec was obtained close to this drill site about 0.53 sec below sea bottom (Fig. 10), indicating that layer 3 underlies these sediments at shallow depth.

No sediments are recognized west of the trench up to and generally including an extensive terrace at a depth of about 4,000-5,000 m (Eade, 1971). Ultramafics (primarily peridotite and some dunite, as well as basalt) have been dredged from the western wall at a depth of 9,150-9,400 m (Fisher and Engel, 1969, Fisher, this volume); it has been argued that they may be part of an ophiolite sequence, of which the pre-late Eocene basalt and gabbro on 'Eua Island may represent the uppermost exposed part (Ewart and Bryan, 1972). Since ophiolite sequences currently are closely linked with ocean-floor igneous rocks, these possible relationships, which are fully discussed by Ewart and Bryan (cf.

also Bryan et al., 1972), may be highly significant; it is implied that the Tonga Ridge, being part of a circum-Pacific orogenic zone, apparently has developed out of oceanic crust. Ewart and Bryan (1972) have stressed that, while it cannot be proved that the basement beneath the Tonga has not been affected by even older tectonic and volcanic activity, the 'Eua igneous suite represents a very early stage of island arc evolution, the rocks being decidedly less fractionated (or more "primitive") than the Recent Tongan basaltic andesite-dacite suite erupted from volcanoes that are lined up along the western side of the ridge, i.e., about 20 km west of the axis of the Tonga Trench (Fig. 11).

The Tonga-Kermadec Ridge is formed by a layered and tectonically deformed series of considerable thickness; this has been recognized on all the seismic traverses between the North Island, New Zealand, and Vava'u, which is the northernmost of the main Tongan islands. Underneath the Raukumara Plain in the south (36°S; profile 9, Fig. 9), where no basement is seen on seismic records, it is at least 4.5 km thick in the center of an extensive flat basin that is only slightly deformed. Deformation becomes gradually stronger northward along the Kermadec Ridge, being characterized by fault-block tectonics combined with lesser folding (germanotype "Bruchfaltung"); the same situation prevails through the Tonga Ridge (profile 11, Fig. 9; Fig. 11), and is clearly also seen in the exposed part of it on 'Eua Island. Seismic sections suggest the local presence of igneous rocks, both intrusive and extrusive, and a large proportion of the sediments is probably volcanogenic. Their total thickness near Tongatapu Island is over 4-5 km (Fig. 11).

On 'Eua the oldest sediments are Eocene limestone (Hoffmeister, 1932; Stearns, 1971), but it is important to note that an unconformity separates them from the overlying Miocene tuffaceous sandstones, with Oligocene mainly missing due to nondeposition and/or erosion. The dark red-brown weathering Miocene is also deformed, as can be seen in good sections on the western slope of the high ridge along the island and is separated by another unconformity from the upper limestones of probable late Pliocene to Quaternary age. Thick Plio-Pleistocene limestones (200 m+) on Vava'u Island are faulted and tilted to the south and are unconformably overlain by the youngest, raised terrace-forming coral limestones. From this situation as recognized in Tonga by seismic and surface geology combined (Fig. 11), and from possible correlation of sections in the southern Kermadec Ridge with known geology in the North Island, New Zealand, it is concluded that the greater part of the sedimentary series of the Tonga-Kermadec Ridge probably is Middle to Upper Tertiary but includes Lower Tertiary in its lowermost portion.

To the west the Tonga Ridge is bordered by a major fault zone which has been mapped for more than 200 km; here the deformed series of the ridge complex are downfaulted by 1,500-2,000 m (Fig. 11). They appear to extend underneath the Tofua Trough and the Lau Basin, although extensively faulted and pierced by younger igneous-volcanic masses. Well known are the volcanoes emerging above sea level and lined up along the western side of the Tonga Ridge (Bryan et al., 1972), but many more are seen on seismic sections also much farther to the west.

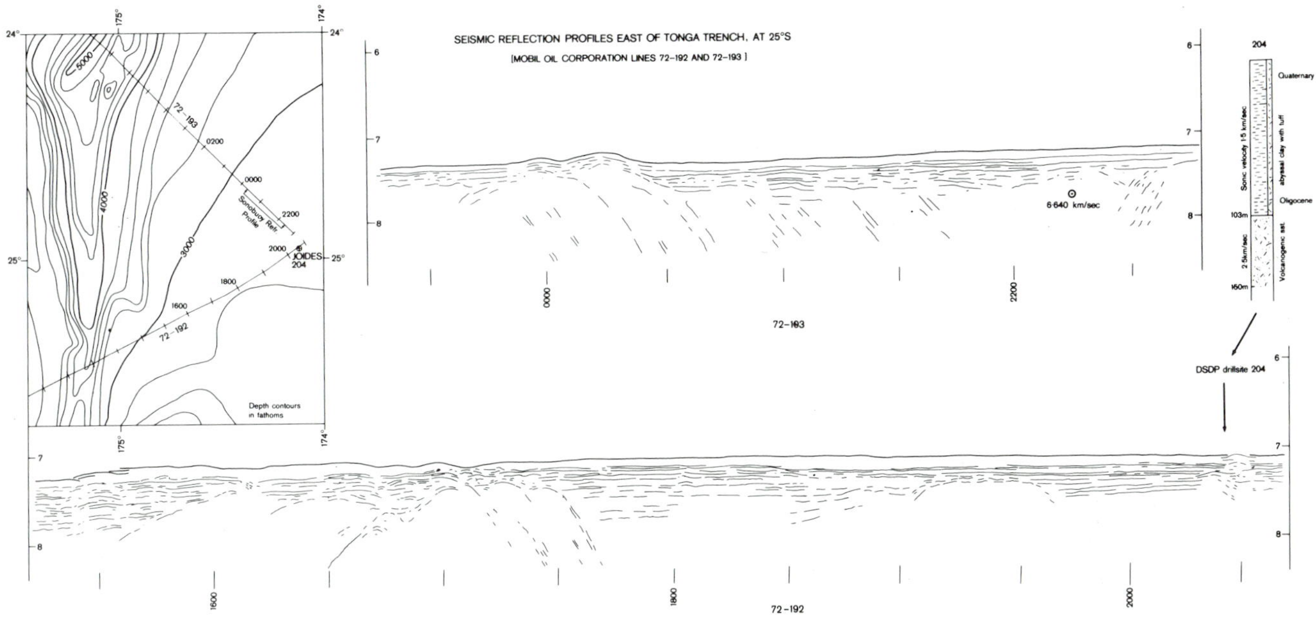

Fig. 10. Tracings of seismic-reflection profiles east of Tonga Trench, at 25°S. Vertical scale, seconds of two-way reflection time; horizontal scale, hours of ship time. Well log of DSDP site 204 after Burns et al., 1973.

Although the Lau Basin is thought to be underlain by new oceanic crust (Karig, 1970; Sclater et al., 1972), good seismic evidence now available suggests a much more complex tectonic history that probably dates back into the early Tertiary. While younger and locally confined basins such as the Tofua Trough and many others scattered throughout the Lau Basin are filled with sediments that in most cases are undeformed, the structure of a deformed basement complex which includes layered sequences is highly suggestive, in part at least, of an evolution similar and related to the regions covered by the Lau (Ladd and Hoffmeister, 1945) and Tonga ridges. DSDP site 203 (22°09.22'S, 177°32.77'W; Burns et al., 1973) drilled 409 m into one of these younger basins, where it bottomed in the Middle Pliocene about halfway through the entire sequence that conformably fills the central part of the particular basin. This suggests an age not younger than Miocene for the unconformity underlying it, and probably early Tertiary for the basement complex.

While no gravity profiles across the Lau Basin have been published, the results of a single seismic-refraction station in its southern extension, the Havre Trough close to New Zealand (Karig, 1970), has been interpreted as showing an oceanic crustal structure (Sclater et al., 1972; Hawkins, this volume). However, it is clearly different from the structure of the adjacent deep ocean basins of the southwest Pacific and the South Fiji Basin. The presence of an intermediate layer several kilometers thick under the Colville and Kermadec ridges (5.1 and 5.4 km/sec, respectively), also under the Havre Trough (4.4 km/sec), corresponding to the 5.1-km/sec layer of comparable thickness underneath the Tonga Ridge (Raitt et al., 1955), probably indicates a similar crustal structure, and therefore a comparable tectonic history across the entire region. Thus evidence from seismic wave velocities does not seem to provide a basis for considering the crust between the two ridges as different and oceanic; nor does its somewhat reduced overall thickness provide a valid argument for such a distinction.

It is concluded that probably in early Tertiary time a quasi-continental crustal regime was established over a much wider region in this part of the southwest Pacific, extending far west from the Tonga-Kermadec Ridge and embracing the Lau Basin, Lau Ridge, and Fiji. The Tonga-Kermadec Ridge is interpreted as a marginal orogenic belt of this continent, the eastern boundary of which was located along the Tonga-Kermadec Trench in early Tertiary if not Cretaceous time. Throughout its history of repeated up and down movements, and faulting and folding accompanied by lasting episodes of volcanism, the Tonga Ridge has basically remained high, whereas the Lau Basin region to the west suffered major subsidence in the late Cenozoic, similar to the central graben of the North Island, New Zealand. Subsidence and rifting across an area as large as the Lau Basin probably has resulted in local upwelling of mantle material, thus initiating a process that may tend to destroy older continental crust and cause it to revert to an oceanic regime (see also Hawkins, this volume). The occurrence of high but variable heat flow, confused magnetic anomaly patterns, and the presence of interspersed young and sharp ridges such as the Peggy Ridge from which fresh tholeiitic basalt has been dredged (Sclater et al., 1972) seem to testify to a Recent tectonic evolution of this kind, peculiar in the Lau Basin. The concept of its formation, during the past 10 my, by the single process of intrusion of oceanic crust as a result of lateral spreading and separation of the two ridges, which in the north would amount to 400 m (Karig, 1970; Sclater et al., 1972), probably is an oversimplification not easy to reconcile with the complex pre-Miocene basement structure of the Lau Basin, with its close geologic-tectonic as well as geophysical affinities with both the Tonga and Lau ridges.

As a corollary the Tonga Ridge is not, in a tectonic sense, an oceanic island arc, but seems to represent the uplifted margin of a more extensive continental block which has undergone a varied and complex history. The fact that it is not arcuate in shape but straight across 22° latitude or nearly 2,700 km in length, uncommon for island arcs throughout the world but similar to other young Pacific margins of even old continental blocks like southern South America (which is also controlled by a marginal orogenic belt), may not be fortuitous but related to this particular situation.

CONCLUSIONS

From this brief analysis of continental margins in the southwest Pacific, between 56 and 16°S, it is evident that there are various segments very different in age, structure, and tectonic regime as well as in sedimentation patterns.

South of New Zealand the continental margins are of two types. Although both are basically rifted margins, the subantarctic slope to the southeast of the Campbell Plateau is older, of Late Cretaceous age, and appears to have remained inactive ever since. It probably formed by early separation from Antarctica, possibly with a strong component of lateral shear. Oceanic crust, which on the basis of linear magnetic anomalies is 80-85 my old (Christoffel and Falconer, 1972), is juxtaposed to the sialic crust of the plateau, the basement of which consists of late Paleozoic to early Mesozoic schist on Campbell Island (Oliver et al., 1950), and of early Mesozoic granite on Bounty Island (Wasserburg et al., 1963). Late Cenozoic alkaline volcanicity and lack of seismicity characterize the Campbell Plateau as a tectonically and isostatically stable region. Its crustal thickness as obtained from surface-wave dispersion patterns (Adams, 1962) is only 17-23 km. This may be due to stretching in the course of continental fragmentation, and its low elevation would then be

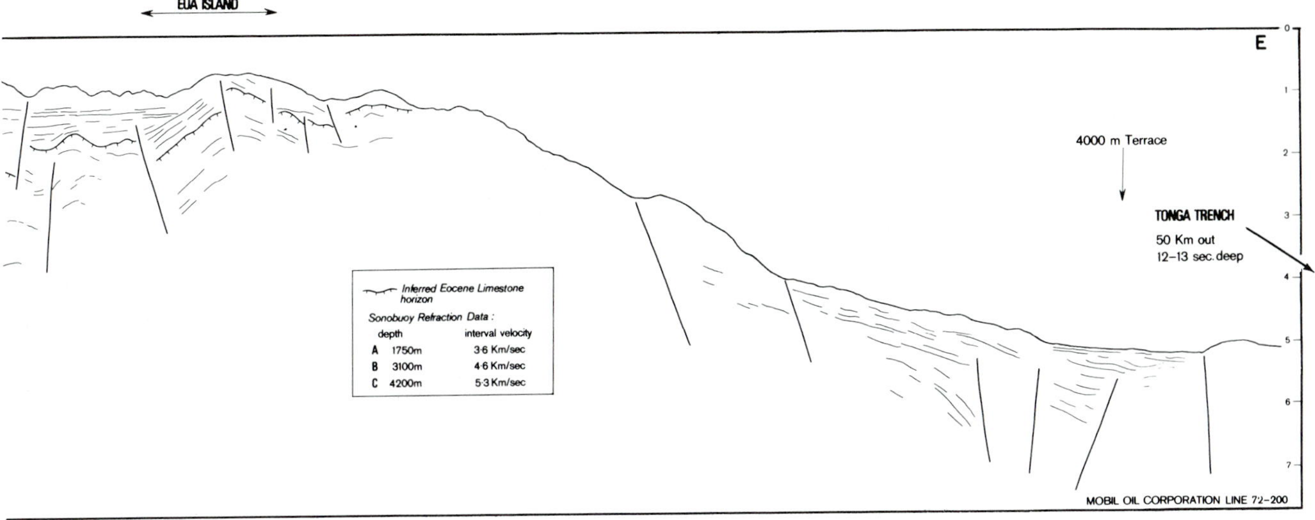

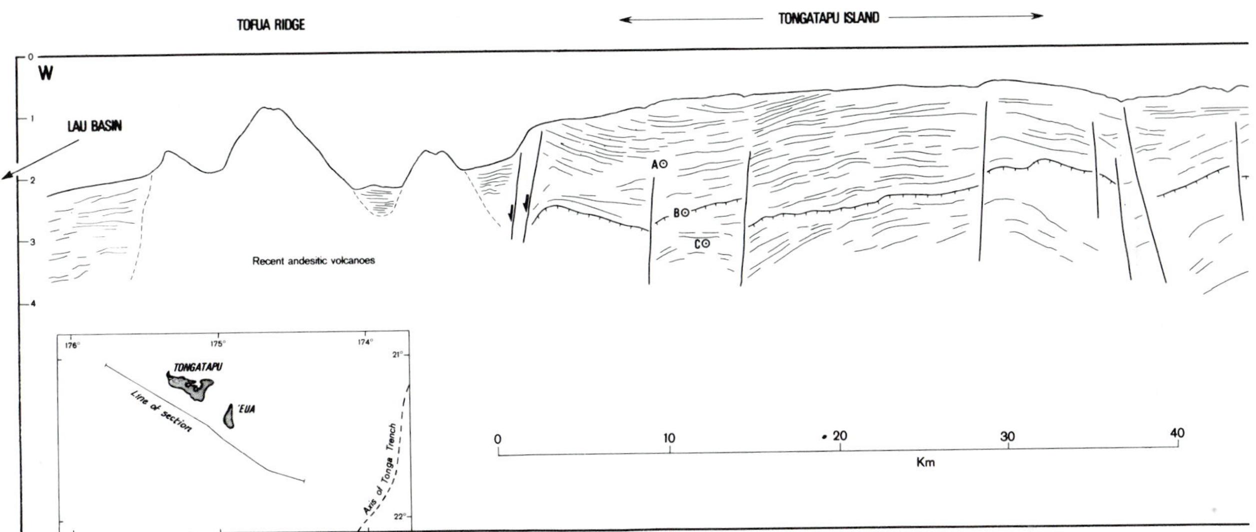

Fig. 11. Interpretive seismic cross section of Tonga Ridge between 21 and 22°S. Vertical scale, seconds of two-way reflection time.

normal if isostatic equilibrium is to be maintained (Bott, 1971).

The western margin of the Campbell Plateau is formed by downwarp and normal faulting, probably initiated in the Eocene. This marginal rift of the Emerald Basin-Solander Trough possibly is related to the separation of New Zealand and Australia. Oceanic crust in the Emerald Basin is middle Oligocene, but it is early to middle Miocene in parts of the complex shear zone of the Macquarie Ridge west of it, which has been interpreted as a younger, ocean-ocean plate boundary with essentially transcurrent motion along it (Hayes and Talwani, 1972). Both the Macquarie shear zone, which trends into the Alpine Fault, and the Emerald Basin-Solander Trough marginal rift, which northward branches into the Tertiary Waiau Basin, are distinct tectonic features that traverse from an oceanic crustal environment into the New Zealand continent. This suggests complex relationships and interactions between oceanic and continental crust around the southwest corner of New Zealand; the complicated tectonic processes in general are highlighted by the fact that young oceanic crust on the Macquarie Ridge is at the same, or higher, elevation as old continental crust, including Late Cretaceous plutonics of the Campbell Plateau only a short distance farther east.

No thick sedimentary wedge exists on the eastern margin of the Campbell Plateau, as occurs along other inactive margins. This is apparently for two reasons: (1) great distance (700-800 km) from any sizable source area, hence a limited supply of land-derived materials; (2) a strong western boundary current impeding sedimentation. At the base of the continental slope, erosion has cut down to a Paleogene sequence, while scouring action above the plateau edge may actually have removed the entire Tertiary section, leaving a thin veneer of Pleistocene foraminiferal ooze with manganese nodule pavement immediately above Maestrichtian silty diatom-radiolarian ooze (DSDP sites 275 and 276; Geotimes, 1973).

Sharply contrasting is the sedimentary pattern along the western margin. While the plateau near its edge generally is a basement high with little or no sediment cover, it is overlain elsewhere by Late Cretaceous to Early Tertiary sediments, which, after an initial period of terrestrial deposition, indicate remarkably uniform and fully oceanic conditions over a period of 30 my (Middle Paleocene to middle-Late Oligocene, DSDP site 277; Geotimes, 1973). Below the steep escarpment, the Solander Trough is filled with turbidite-type sediments 2.5-3 km thick, which probably range from Eocene to Mio-Pliocene. Later tectonic activity along this rift is indicated by postdepositional deformation of these sediments, by further downfaulting below the plateau edge, volcanicity, and shallow earthquake seismicity; current action has caused considerable erosion of sediments down the center of the trough (Fig. 5a). Terrigeneous material, however, has not reached as far as the Emerald Basin farther south, where a fully oceanic sequence overlying pillow basalt ranges from Middle Oligocene to Late Pleistocene. The irregular and rough topography of the floor of the Emerald Basin (Summerhayes, 1969) suggests tectonically rather unstable conditions here also, which may be consistent with the lack of sediments younger than Miocene at core sample location FSU-16-4 (Fig. 4; Houtz et al., 1971).

The tectonic evlution of the *Chatham Rise* northeast of the Campbell Plateau is similar and undoubtedly closely related to the latter. Its structural relationship with New Zealand, however, is still enigmatic, while the position and structure of the continental margin around this particular feature, which projects east far out into the southwest Pacific basin, is entirely a matter of speculation.

The *Hikurangi Trench* east of the North Island, New Zealand, was regarded as a continuation of the Kermadec Trench, and was thought to define the eastern boundary of an asymmetric active region typical of island arcs (Hatherton, 1969, 1970a, 1970c). However, seismic surveys have revealed a close association, both structurally and in sedimentation patterns, with the East Coast foldbelt, while the tectonic framework and history is intimately related to the evolution of adjacent areas which are entirely continental. Also, the Hikurangi Trench does not continue toward and into the Kermadec Trench, but is a local basin of thick Plio-Pleistocene deposition whose axis trends east-northeast, in which direction the basin opens up and merges with the essentially flat abyssal plain (Fig. 6, profiles 4 and 5; Fig. 8). Since the area called Hikurangi Trench is not associated with a Benioff zone either—the North Island axes of seismicity and negative gravity anomaly apparently being unrelated to it—it is concluded that there is no continental margin here. The particular sedimentary basin is renamed "Hikurangi Trough." Thus the continental margin east of the northern half of New Zealand remains undefined.

North of New Zealand a well-defined ridge-trench system, virtually straight for 2,700 km, marks the boundary between the deep ocean of the southwest Pacific and the continental crust of the Kermadec-Tonga Ridge (Talwani et al., 1961). The latter appears to be a marginal orogenic belt of a crustal block that extends much farther to the west, including the regions of the Lau Basin, Lau Ridge, and Fiji, all of which seem to have developed, probably out of a fully oceanic crustal environment, from the Early Tertiary onward. The continental margin along the Tonga-Kermadec Trench can thus be considered an orogenic-type margin. Unlike other orogenic margins of the Pacific (Japan, Chile) it has, however, a much shorter and less complex history.

If the pre-Late Eocene basalt and gabbro on 'Eua Island represent part of an old ocean floor igneous suite (Ewart and Bryan, 1972), the Tonga-Kermadec Trench is the site of a major suture within such original oceanic crust; this suture has been separating two different tectonic environments from at least the Eocene onward, as shown by the greatly different thickness and facies of Lower Tertiary to Recent sediments to the east and west of it. While in the east the 103 m of abyssal clay with volcanic ash ranging from Oligocene to Recent (Burns et al., 1973) indicates a deep ocean environment through all this time similar to the one at present, in the west there was increased crustal mobility, resulting in pronounced differential vertical movements, volcanicity, and the deposition of a markedly thicker sedimentary sequence (5 km) at comparatively shallow depth (60-200 m and possibly down to bathyal, as based on different faunal groups in the late Eocene of 'Eua; Ladd, in Cole, 1970). In the course of this process, a continental-type crust was established west of the trench. Folding and faulting combined with further uplift from the Mio-Pliocene onward, the downfaulting of the Tofua Trough-Lau Basin in the west (new marginal deep?), Recent andesitic-dacitic volcanism (Bryan et al., 1972), and strong shallow to deep seismicity characterize the area as a typical orogenic belt still fully active. On the oceanic side of it, the Tonga-Kermadec Trench has been downfaulted to a deep graben structure in comparatively recent time (Fig. 9, profiles 8-11); with respect to the orogenic belt it may represent a backdeep of dominantly taphrogenic breakdown, such as occurs behind marginal orogenic belts in the southeast Pacific (Katz, 1971, 1972, 1973).

Little if any sedimentation seems to occur in the trench or its slope on the continentward side, except locally on the broad terrace halfway up the slope (Fig. 11). This general lack of sedimentation along the continental margin probably is due, as it is along the subantarctic slope of the Campbell Plateau, to limited supply from the mainly submerged ridge, in particular to the presence of a pronounced north-flowing boundary current. Temperature-salinity profiles at 28°S (northern Kermadec Trench) indicate an extreme intensification of this current within a narrow zone along the flank of the Tonga-Kermadec Ridge (Warren, 1970).

The structure of the Tonga-Kermadec Ridge, which appears to be mainly controlled by horst and graben tectonics, is indicative of tensile stress across the continental margin. This seems difficult to reconcile with the commonly held view that Pacific-type orogenic margins are areas of collision and consumption of major crustal elements. There is certainly no evidence of compression, as has been found also along the margin of Chile, where the tectonic history throughout the Cenozoic indicates a state of extension and breakdown (Katz, 1971); this is consistent with a thinning of the crust underneath the Chile Trench (Worzel, 1965). Although gravitational forces may, in connection with subduction in the Benioff zone, exert a downward pull and therefore extension in the tectonosphere (Isacks and Molnar, 1969), there are no visible features from seismic or otherwise, to suggest that due to the (postulated) large differential movement of compression between the two plates along the Tonga-Kermadec Trench, "the Pacific tectosphere has decoupled itself from the adjacent block and sinks on its own, creating the surface tensional features" (LePichon, 1968, p. 3694).

Downward movement along the continental margin, however, cannot be denied, both from the geological structure and the large negative gravity anomaly observed (Talwani et al., 1961). Since such downward movements commonly occur behind orogenic belts, the trenches along orogenic margins may indeed be comparable to the backdeep of an orogen. If so, they are of local significance and closely related to the tectonic evolution of the orogenic margin itself, rather than being the result of large-scale collision of plates driven from far away. Thus there is no need to advocate an ocean-floor spreading and conveyor-belt theory for their formation. Indeed, and as shown by Elsasser (1968), "the forces that generate and maintain the trenches originate locally rather than by transmission from very far away."

SUMMARY

The margin of the southwest Pacific basin can be subdivided into a number of segments very different in age, structure, and tectonic regime, as well as sedimentation.

The margins of the Campbell Plateau south of New Zealand were formed by rifting. To the west a rift valley that originates within continental crust in the south of New Zealand deepens and widens southward; subsidence began in the early Tertiary, probably Eocene, resulting in final rifting and the generation of new oceanic crust in the southern part during the Oligocene. While the edge of the continental margin is a basement high of Paleozoic-Mesozoic metamorphic and intrusive rocks with little or no sediment cover, the downfaulted marginal rift is filled with thick turbidites probably brought in mainly from the north and transported down the rift. Postdepositional deformation of these sediments, further subsidence by normal faulting, volcanicity, and shallow earthquake seismicity suggest continuing tectonic activity along this margin. The Subantarctic Slope to the southeast apparently results from a Late Cretaceous continental breakup, and has remained inactive ever since. Great distance from source area is a limiting factor for deposition along this margin, while a strong western boundary current actuallv maintains erosion.

The continental margins around Chatham Rise and east of North Island, New Zealand, remain undefined. The Hikurangi Trench does not compare with oceanic trenches and is unrelated to the Kermadec Trench, but is a structural feature entirely within the framework of continental tectonics. Its name is therefore changed to "Hikurangi Trough."

The Kermadec-Tonga ridge-trench system north of New Zealand represents an orogenic belt. Lower Tertiary to Recent thin abyssal clays east of the trench compare with 5-km-thick shelf to bathyal sediments, including two unconformities, deposited within about the same period to the west. In pre-late Eocene times the site of the trench was already an important suture within original oceanic crust, to the west of which a crustal block of greater mobility was apparently converted into continental-type crust. The structure of the Tonga Ridge is controlled by horst-and-graben tectonics; continuing fault-block movements, andesitic-dacitic volcanism, and shallow to deep seismicity in a west-dipping Benioff zone associated with a negative gravity anomaly suggest that this orogenic belt is still active. Also, the trench to the east, which is a downfaulted graben structure, is affected by very recent faulting. The tectonics on the whole indicate tensile stress across this continental margin, which is difficult to reconcile with the concept of collision and consumption of major structural elements. Sedimentation in the trench and along its western slope is negligible to nonexistent, probably because of restricted source area but mainly resulting from the presence of a pronounced north-flowing boundary current.

ACKNOWLEDGMENTS

To a large measure this study is based on recent results from a regional marine geophysical survey by Mobil Oil Corporation. Permission to use these data is gratefully acknowledged. The author is also indebted to Mobil Oil Corporation for the opportunity to take part in this survey between New

Zealand, Tonga, and Fiji. Valuable assistance was received from the Government of the Kingdom of Tonga during field work on the islands of 'Eua and Vava'u.

Wendy White, New Zealand Geological Survey, and the Cartographic Section of the Department of Scientific and Industrial Research, Wellington, drew the figures.

BIBLIOGRAPHY

Adams, R. D., 1962, Thickness of the earth's crust beneath the Campbell Plateau: New Zealand Jour. Geol. Geophys., v. 5, no. 1, p. 74-85.

Aronson, J. L., 1968, Regional geochronology of New Zealand: Geochim. Cosmochim. Acta, v. 32, p. 669-697.

Austin, P. M., Sprigg, R. C., and Braithwaite, J. C., 1973, Structure and petroleum potential of eastern Chatham Rise, New Zealand: Am. Assoc. Petr. Geol. Bull., v. 57, no. 3, p. 477-497.

Beu, A. G., 1967, Deep-water Pliocene Mollusca from Palliser Bay, New Zealand: Trans. Roy. Soc. New Zealand Geol. Ser., v. 5, no. 3, p. 89-122.

______, 1969, Additional Pliocene bathyal Mollusca from South Wairarapa, New Zealand: New Zealand Jour. Geol. Geophys., v. 12, no. 2/3, p. 484-496.

Bott, M. H. P., 1971, Evolution of young continental margins and formation of shelf basins: Tectonophysics, v. 11, no. 5, p. 319-327.

Brodie, J. W., 1964, Bathymetry of the New Zealand region: New Zealand Dept. Sci. Ind. Res. Bull., v. 161.

______, and Hatherton, T., 1958, The morphology of Kermadec and Hikurangi trenches: Deep-Sea Res., v. 5, no. 1, p. 18-28.

Bryan, W. B., Stice, G. D., and Ewart, A., 1972, Geology, petrography and geochemistry of the volcanic islands of Tonga: Jour. Geophys. Res., v. 77, no. 8, p. 1566-1585.

Burk, C. A., 1968, Buried ridges within continental margins: Trans. N.Y. Acad. Sci., ser. II, v. 30, no. 3, p. 397-409.

Burns, R. E., Andrews, J. E., Churkin, M., Davies, T. A., Dumitrica, P., Edwards, A. R., Galehouse, J. S., Kennett, J. P., Packham, G. H., and van der Lingen, G. J., 1973, Initial reports of the Deep Sea Drilling Project, v. XXI: Washington, D.C., U.S. Govt. Printing Office.

Cole, W. S., 1970, Larger Foraminifera of late Eocene age from Eua, Tonga: U.S. Geol. Survey, Prof. Paper 640-B, 15 p.

Christoffel, D. A., 1971, Motion of the New Zealand Alpine fault deduced from the pattern of sea floor spreading: *in* Recent crustal movements: Collins, B. W., and Fraser, R., eds., Roy. Soc. New Zealand, p. 25-30.

______, and Falconer, R. K., 1972, Marine magnetic measurements in the southwest Pacific Ocean and the identification of new tectonic features: *in* Hayes, D. E., ed., Antarctic oceanology: II. The Australian-New Zealand sector: Antarctic Res. Ser., v. 19, p. 197-209.

Cullen D. J., 1965, Autochthonous rocks from the Chatham Rise: New Zealand Jour. Geol. Geophys., v. 8, no. 3, p. 465-474.

Eade, J. V., 1971, Tonga bathymetry: New Zealand and Oceanog. Inst. Chart, Ocean. Ser., 1: 1,000,000.

Eiby, G. A., 1958, The structure of New Zealand from seismic evidence: Geol. Rundschau, v. 47, p. 647-662.

Elsasser, W. M., 1968, Submarine trenches and deformation: Science, v. 160, no. 3831, p. 1024.

Ewart, A. and Bryan, W. B., 1972, Petrography and geochemistry of the igneous rocks from Eua, Tongan Islands: Geol. Soc. America Bull., v. 83, no. 11, p. 3281-3298.

Falconer, R. K., 1972, The Indian-Antarctic-Pacific triple junction: Earth and Planetary Sci. Letters, v. 17, p. 151-158.

Fisher, R. L., and Engel, C. G., 1969, Ultramafic and basaltic rocks dredged from the nearshore flank of the Tonga Trench: Geol. Soc. America Bull., v. 80, no. 7, p. 1373-1378.

Fleming, C. A., 1968, Tertiary fossils from the Auckland Islands: Trans. Roy. Soc. New Zealand Geol. Ser., v. 5, no. 11, p. 245-252.

______, 1969, The Mesozoic of New Zealand: chapters in the history of the circum-Pacific mobile belt: Geol. Soc. London Quart. Jour., v. 125, p. 125-170.

______, and Reed, J. J., 1951, Mernoo Bank, E. of Canterbury, New Zealand: New Zealand Jour. Sci. Tech. sect. B., v. 32, no. 6, p. 17-30.

Geotimes, 1972, GLOMAR CHALLENGER Down Under: v. 17, no. 5, p. 14-16.

______, 1973, Deep-sea drilling in the roaring 40s: v. 18, no. 7, p. 14-17.

Hamilton, R. M., and Gale, A. W., 1968, Seismicity and structure of North Island, New Zealand: Jour. Geophys. Res., v. 73, no. 12, p. 3859-3876.

Harrington, H. J., and Wood, B. L., 1958, Andesitic volcanism at the Solander Islands: New Zealand Jour. Geol. Geophys., v. 1, no. 3, p. 419-431.

Hatherton, T., 1967, Total magnetic force measurements over the North Macquarie Ridge and Solander Trough: New Zealand Jour. Geol. Geophys., v. 10, no. 5, p. 1204-1211.

______, 1969, Gravity and seismicity of asymmetric active regions: Nature, v. 221, p. 353-355.

______, 1970a, Gravity seismicity, and tectonics of the North Island, New Zealand: New Zealand Jour. Geol. Geophys., v. 13, no. 1 [3rd Spec. Pacific Issue], p. 126-144.

______, 1970b, Symmetry of crustal earthquakes above Benioff zones: Nature, v. 225, p. 844-845.

______, 1970c, Upper mantle inhomogeneity beneath New Zealand: surface manifestations: Jour. Geophys. Res., v. 75, no. 2, p. 269-284.

______, 1971, Shallow earthquakes and rock composition: Nature Phys. Sci., v. 229, no. 4, p. 119-120.

Hay, R. F., Mutch, A. R., and Watters, W. A., 1970, Geology of the Chatham Islands: New Zealand Geol. Survey Bull., n.s. 83.

Hayes, D. E., and Ringis, J., 1973, Seafloor spreading in the Tasman Sea: Nature, v. 243, no. 5407, p. 454-458.

______, and Talwani, M., 1972, Geophysical investigation of the Macquarie Ridge Complex: *in* Hayes, D. E., ed., Antarctic oceanology. II: The Australian-New Zealand sector: Antarctic Res. Ser., v. 19, p. 211-234.

Hoffmeister, J. E., 1932, Geology of Eua, Tonga: Bernice P. Bishop Museum Bull. Honolulu, v. 96, 93 p.

Houtz, R., Ewing, J., Ewing, M., and Lonardi, A. G., 1967, Seismic reflection profiles of the New Zealand Plateau: Jour. Geophys. Res., v. 73, no. 18, p. 4713-4729.

______, Ewing, J., and Embly, R., 1971, Profiler data from the Macquarie Ridge area: Reid, J. L., ed., *in* Antarctic oceanology I: Am. Geophys. Union, Antarctic Res. Ser., v. 15, p. 239-245.

Houtz, R. E., and Markl, R. G., 1972, Seismic profiler data between Antarctica and Australia: *in* Hayes, D. E., ed., Antarctic oceanology: II. The Australian-New Zealand Sector: Antarctic Res. Ser., v. 19, p. 147-164.

Isacks, B., and Molnar, P., 1969, Mantle earthquake mechanism and the sinking of the lithosphere: Nature, v. 223, p. 1121-1124.

______, Oliver, J., and Sykes, L. R., 1968, Seismology and the new global tectonics: Jour. Geophys. Res., v. 73, no. 18, p. 5855-5899.

______, Sykes, L. R., and Oliver, J., 1969, Focal mechanisms of deep and shallow earthquakes in the Tonga-Kermadec region and the tectonics of island arcs: Geol. Soc. America Bull., v. 80, p. 1443-1470.

Karig, D. E., 1970, Ridges and basins of the Tonga-Kermadec island arc system: Jour. Geophys. Res., v. 75, no. 2, p. 239-254.

Katz, H. R., 1968, Potential oil formations in New Zealand, and their stratigraphic position as related to basin evolution: New Zealand Jour. Geol. Geophys., v. 11, no. 5, p. 1077-1133.

______, 1971, Continental margin in Chile—is tectonic style compressional or extensional?: Am. Assoc. Petr. Geol. Bull., v. 55, no. 10, p. 1753-1758.

______, 1972, Plate tectonics and orogenic belts in the Southeast Pacific: Nature, v. 237, no. 5354, p. 331-332.

______, 1973, Contrasts in tectonic evolution of orogenic belts in the South-east Pacific: Jour. Roy. Soc. New Zealand, v. 3, no. 3, p. 333-362.

______, 1974, Recent exploration for oil and gas: *in* Williams, G., ed., Economic geology of New Zealand (2nd ed.), IV, Chap. 24: 8th Commonwealth Mining and Met. Congr.

______, in press a, Ariel Bank off Gisborne—an offshore late Cenozoic structure and the problem of acoustic basement on the East Coast, North Island, New Zealand: Austral. Petr. Expl. Assoc. Jour., v. 14.

______, in prep., Hikurangi Trough—a tectonic analysis: New Zealand Jour. Geol. Geophys.

Kingma, J. T., 1958, The Tongaporutuan sedimentation in central Hawke's Bay: New Zealand Jour. Geol. Geophys., v. 1, no. 1, p. 1-30.

______, 1960, Outline of the Cretaceous-Tertiary Sedimentation in the eastern basin of New Zealand: New Zealand Jour. Geol. Geophys., v. 3, no. 2, p. 222-234.

Ladd, H. S., and Hoffmeister, J. E., 1945, Geology of Lau, Fiji: Bernice P. Bishop Museum Bull. Honolulu, v. 181, 399 p.

Lawrence, P., 1967, New Zealand region, bathymetry: New Zealand Oceanog. Inst. Chart, 1:6,000,000 Misc. Ser. 15.

LePichon, X., 1968, Sea-floor spreading and continental drift: Jour. Geophys. Res., v. 73, no. 12, p. 3661-3697.

Mitronovas, W., Isacks, B., and Seeber, L., 1969, Earthquake locations and seismic wave propagation in the upper 250 km of the Tonga Island Arc: Seismol. Soc. America Bull., v. 59, no. 3, p. 1115-1135.

New Zealand Geological Survey, 1972, Geological map of New Zealand: New Zealand Dept. Sci. Ind. Res., 1:1,000,000.

O'Halloran, N. G., 1971, Interpretation report, marine seismic reflection survey within license area 878, Chatham Rise, New Zealand: BP Petr. Develop. Ltd., Tech. Note 354, Open File New Zealand Geol. Survey, Petr. Rept. 301A.

Oliver, J., and Isacks, B., 1967, Deep earthquake zones, anomalous structures in the upper mantle, and the lithosphere: Jour. Geophys. Res., v. 72, no. 16, p. 4259-4275.

Oliver, R. L., Finlay, H. J., and Fleming, C. A., 1950, The geology of Campbell Island: New Zealand Sept. Sci. Ind. Res., Cape Exped. Ser. Bull., v. 3.

Raitt, R. W., Fisher, R. L., and Mason, R. G., 1955, Tonga Trench: Geol. Soc. America Spec. Paper 62, p. 237-254.

Sclater, J. G., Anderson, R. N., and Bell, M. L., 1971, Elevation of ridges and evolution of the central eastern Pacific: Jour. Geophys. Res., v. 76, no. 32, p. 7888-7915.

______, Hawkins, J. W., Mammerickx, J., and Chase, C. G., 1972, crustal extension between the Tonga and Lau ridges: petrologic and geophysical evidence: Geol. Soc. America Bull., v. 83, no. 2, p. 505-518.

Stearns, H. T., 1971, Geologic setting of an Eocene fossil deposit on Eua Island, Tonga: Geol. Soc. America Bull., v. 82, no. 9, p. 2541-2552.

Suggate, R. P., 1963, The Alpine fault: Trans. Roy. Soc. New Zealand Geol. Ser., v. 2, no. 7, p. 105-129.

Summerhayes, C. P., 1969, Marine geology of the New Zealand subantarctic sea floor: New Zealand Dept. Sci. Ind. Res. Bull., v. 190.

Sykes, L. R., 1966, The seismicity and deep structure of island arcs: Jour. Geophys. Res., v. 71, p. 2981-3006.

______, Isacks, B., and Oliver, J., 1969, Spatial distribution of deep and shallow earthquakes of small magnitudes in the Fiji-Tonga region: Seismol. Soc. America Bull., v. 59, no. 2, p. 1093-1113.

Talwani, M., Worzel, J. L., and Ewing, M., 1961, Gravity anomalies and crustal section across the Tonga Trench: Jour. Geophys. Res., v. 66, no. 4, p. 1265-1278.

Varne, R., and Rubenach, M. J., 1972, Geology of Macquarie Island and its relationship to oceanic crust, *in* Hayes, D. E., ed., Antarctic oceanology: II. The Australian-New Zealand sector: Antarctic Res. Ser., v. 19, p. 251-266.

Van der Linden, W. J. M., 1968, Cook bathymetry: New Zealand Oceanog. Inst. Chart. Ocean. Ser. 1: 1,000,000.

Vella, P., 1963, Foraminifera from Upper Miocene turbidites, Wairarapa, New Zealand: New Zealand Jour. Geol. Geophys., v. 6, no. 5, p. 775-793.

Warren, B. A., 1970, General circulation of the South Pacific: *in* Wooster, W. S., ed., Scientific exploration of the South Pacific: Washington, D.C. Natl. Acad. Sci.

Wasserburg, G. J., Craig, H., Menard, H. W., Engel, A. E., and Engel, C. G., 1963, Age and composition of a Bounty Islands granite, and age of a Seychelles Island granite: Jour. Geology, v. 71, p. 785-789.

Worzel, J. L., 1965, Structure of continental margins and development of ocean trenches: Geol. Survey Can. Paper 66-15, p. 357-375.

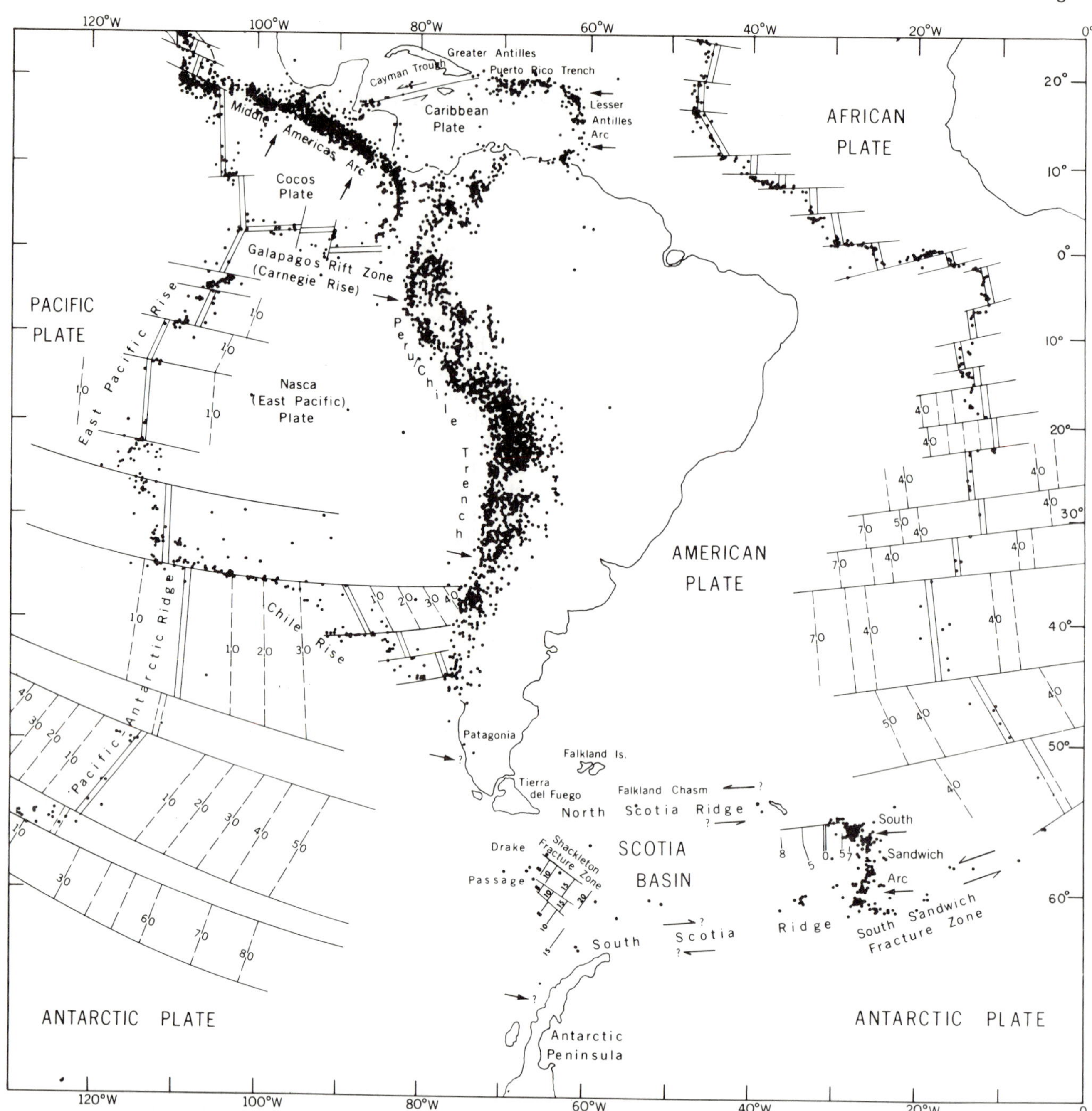

Fig. 1. Geotectonic setting of the Scotia Sea. Numbers give the apparent magnetic age of the sea floor in millions of years (from Dalziel and Elliot, 1973, The Ocean Basins and Their Margins, Vol. 1, A.E.M. Nairn and F.H. Stehli, eds, copyright by Plenum Press, N.Y.)

ded graywackes, shales, and massive fusulinid limestones (Cecioni, 1955; Dalziel, 1972b; Dalziel and Elliot, 1973). The graywackes and shales of the well-known Trinity Peninsula "Series" ("series" as used here has no time-stratigraphic significance), which forms the spine of the Antarctic Peninsula, are certainly pre-Middle or Upper Jurassic as they are unconformably overlain by silicic volcanics of that age at the tip of the Peninsula (Dalziel, 1972a). Lithic and structural comparison indicates that sedimentary sequences that can broadly be correlated with the Madre de Dios and Trinity Peninsula rocks crop out in the South Shetland Islands and the South Orkneys Group (Fig. 2). Previous reports that the Sandebugten sedimentary sequence of South Georgia Island at the east end of the North Scotia Ridge is correlative with the Trinity Peninsula "Series" (Adie, 1964; Skidmore, 1972) can now be discounted. The contact between the Sandebugten rocks and the structurally overlying Lower Cretaceous Cumberland Bay sedimentary sequence is a thrust plane, and not an unconformity as previously supposed (Dalziel et al., 1973a. in press b).

The tectonic significance of the Late Paleozoic

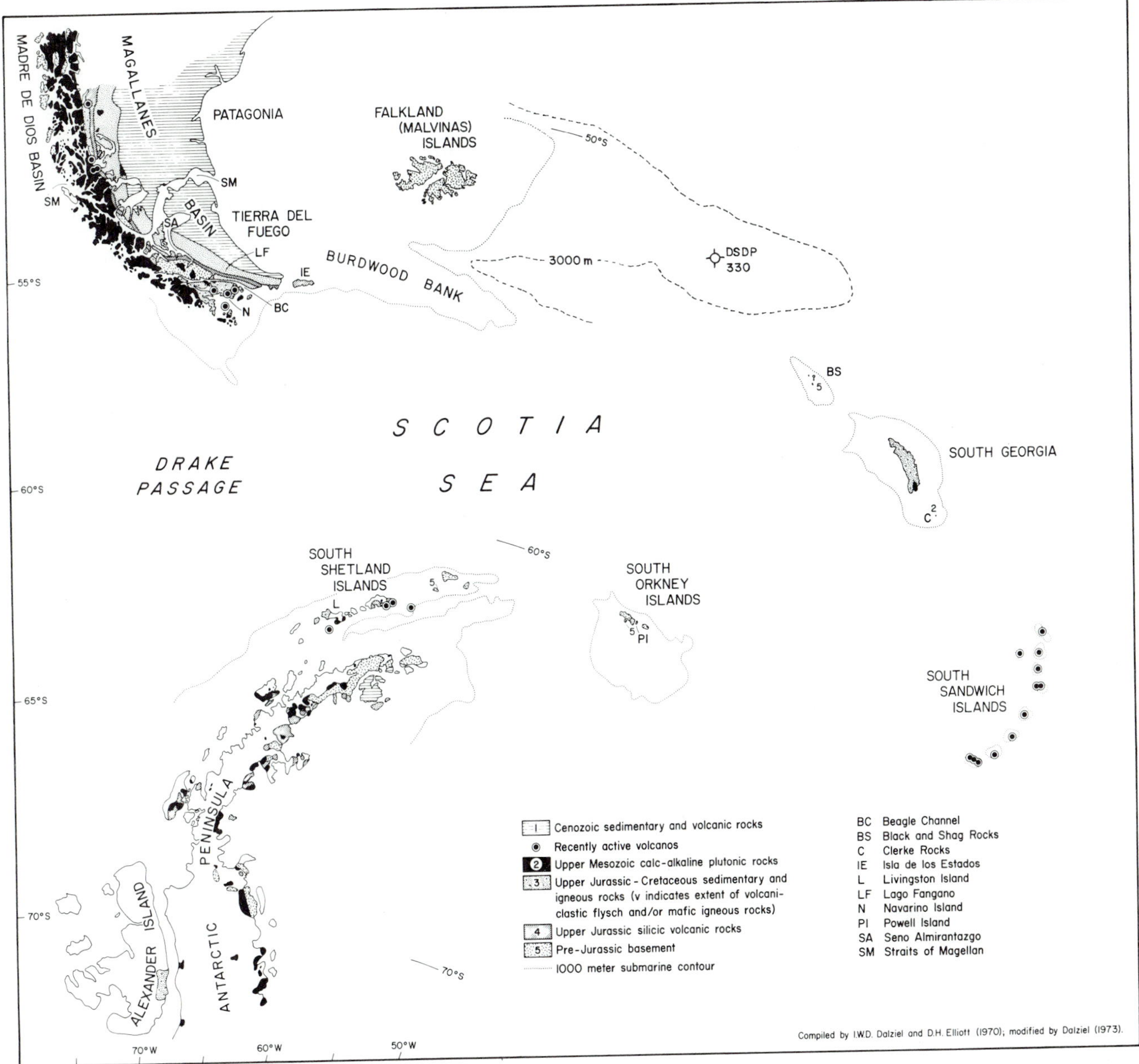

Fig. 2. Geologic map of the margins of the Scotia Sea (modified from Dalziel and Elliot, 1973, on the basis of more recent work).

sedimentary sequences in the Scotia Arc region is as yet not fully understood. They clearly form part of the Early Mesozoic Gondwanide foldbelt (Du Toit, 1937), which includes the Cape foldbelt in South Africa, the Sierra de la Ventana in Argentina, and the Ellsworth and Pensacola Mountains in Antarctica. The rocks are all rich in quartzose detritus and deficient in volcanic rock fragments. Many of the sequences contain pebbly mudstones that could be of glacial origin, and the fold geometry of the Trinity Peninsula "Series" and the Miers Bluff Formation of the South Shetland Islands indicates a close relationship to the Pacific continental margin of West Antarctica at the time of deformation in the Early Mesozoic (Dalziel, 1972b). Exactly how these different occurrences of Upper Paleozoic sedimentary sequences on three separate continents are related is a major problem, but one whose resolution is likely to be an important clue to our understanding of the earlier history of the continental margins of the Scotia Sea area.

Elsewhere in the region the Middle-Upper Jurassic silicic volcanics unconformably overlie metamorphic rocks whose relationship to the Late Paleozoic sedimentary sequences is unclear. Structural relationships on Powell Island in the South Orkneys Group (Fig. 2) suggest that at least some of the metamorphics predate the Late Paleozoic sedimentary rocks. However, recent mapping in the Patagonian Andes by Maarten J. de Wit and Keith F. Palmer has indicated that some of the basement there may be the metamorphosed equivalent of the Upper Paleozoic sedimentary sequences. Underlying the Mesozoic-Cenozoic Magallanes Basin to the east of the Andean Cordillera there is a more gneissose basement that, based on radiometric evidence, could be as old as Early Paleozoic (Halpern, 1973).

Hence little can be said at this time of the pre-Middle Jurassic history of the continental margins of the Scotia Sea region. Upper Paleozoic and probably older sedimentary and metasedimentary rocks and some gneisses form the basement throughout the region. They were all completely deformed and to a variable extent metamorphosed prior to the extrusion of vast quantities of silicic volcanics in the Middle-Late Jurassic. Deposition of the Upper Paleozoic sedimentary sequences probably took place along the Pacific continental margin of Gondwanaland (Dalziel and Elliot, 1973).

LATE JURASSIC [EXTENSION AND SILICIC VOLCANISM ASSOCIATED WITH CONTINENTAL FRAGMENTATION]

Many details of the timing of the fragmentation of Gondwanaland are still obscure. It has, however, long been recognized that there are vast quantities of Late Jurassic-Early Cretaceous extrusive and intrusive igneous rocks in the southern continents that are associated with extensional tectonics. These include:

1. The Karoo, Lebombo, and associated volcanics in southern Africa.
2. The Ferrar dolerites and Dufek layered intrusive in Antarctica.
3. The Serra Geral tholeiites of northeastern South America.
4. The silicic volcanics of southern South American and the Antarctic Peninsula.

It seems clear from magnetic anomalies and results of the Deep Sea Drilling Project (DSDP) that the South Atlantic began to form in the latest Jurassic or earliest Cretaceous (Larson and Ladd, 1973). Throughout the Scotia Arc region the metamorphic and sedimentary basement is unconformably overlain by silicic volcanics dated as Middle-Upper Jurassic (Dalziel, 1972b, Figs. 6 and 7).

From recent field, laboratory, and literature research, it has become clear that the rocks consist of homogeneous tuffs with subordinate lavas, are mainly of rhyolite to quartz-latite composition, and, at least on Isla de los Estados at the southern extremity of the Andes (Fig. 2), contain hardly any mafic minerals (Dalziel et al., in press a). Moreover, the volcanics extend almost to the Atlantic continental margin of South America and are clearly associated with deep graben whose fault boundaries began to operate in the Late Jurassic and controlled sedimentation well into the Cenozoic (Palmer and Dalziel, in press). The fact that these silicic volcanics extend all the way to the Atlantic continental margin even at the latitude of Buenos Aires (over 1,000 km from the Pacific) seems to indicate that they are not all related to subduction of Pacific crust. However, the situation is complicated by the spatial and temporal association of the following:

1. Separation of South America, Africa, and Antarctica.
2. Subduction along the Pacific margin of South America and the Antarctic Peninsula, as evidenced by ages on the granodioritic batholith (see the next section).
3. Initiation of a Japan Sea-like marginal basin along the Pacific margin behind the volcanic chain that overlay the batholith (see the next section).

Isotopic data notwithstanding, global geologic relations indicate that generation of vast quantities of silicic volcanic products, as in the central Andes, the Taupo volcanic zone of North Island, New Zealand, the Lebombo monocline in southern Africa, and southern South America and West Antarctica, requires partial melting of continental crust. The heat can be located in a subduction zone. The generation of this type of volcanic product above a subduction zone is clearly demonstrated by the Cenozoic history of the Central Andes. At the southern end of the Lau Basin-Havre trough marginal basin spreading center behind the Tonga-Kermadec Trench, it seems certain that melting of the continental crust of North Island, New Zealand, has resulted in the generation of the highly silic Taupo volcanic zone. There appears to be no example outside southern South America and West Antarctica in the Late Jurassic where vast volumes of silicic volcanics were generated by continental rifting without the presence of extensive contemporaneous mafic volcanics, as is in the case of the rocks along the Lebombo monocline in Southern Africa. However, the distance (>1,000 km) between the subduction zone that must have developed along the western margin of South America by the Late Jurassic (see the next section) and some of the Upper Jurassic silicic volcanics along the eastern continental margin make it seem unlikely that the generation of these volcanics was directly related to subduction. It seems likely that the close proximity of an active divergent plate margin in the South Atlantic to a subduction zone along the western margin of South America could have resulted in this apparently unique occurrence of silic volcanics.

LATE JURASSIC-EARLY CRETACEOUS [PLATE CONVERGENCE ALONG THE PACIFIC MARGINS]

A granodioritic batholith over 60-km wide and dated as late Jurassic-late Cretaceous extends along the Pacific margin of southern South America. Plutons of the same general age and of identical composition occupy the spine of the Antarctic Peninsula and the South Shetland Islands. The batholithic rocks were clearly emplaced in the South American and West Antarctic continental crust, as xenoliths, screens, and pendants of Paleozoic sedimentary and metasedimentary rocks are common (Dalziel et al., 1973c, in press c). The presence immediately adjacent to the batholith in South America and South Georgia of extensive Lower Cretaceous volcaniclastic flysch with andesitic detritus and local interbedded andesitic flows and breccias indicates that the batholith represents the root of a Late Mesozoic calc-alkaline volcanic chain comparable to that of the Central Andes of today (Dalziel et al., 1973c, in press c). This means that subduction of Pacific crust beneath southern South America and West Antarctica was initiated at least by the Late Jurassic, the oldest published date on the Patagonian batholith being 150 my (Halpern, 1973).

The Lower Cretaceous volcaniclastic flysch in South America crops out in a belt along the eastern margin of the batholith (Fig. 2). It has recently been established that the flysch rests conformably upon a series of mafic lenses up to 120 km long and 40 km wide that parallel the Cordillera (Dalziel et al., 1973c, in press c). These lenses are funnel-shaped in cross section and represent the upper portions of ophiolite suites. A lower unit of layered gabbro at sea level, with steeply dipping igneous contacts into the Paleozoic metasedimentary basement, passes upward into a sheeted dyke complex parallel to the Cordillera. This in turn passes into an extrusive unit of pillow lavas. The extrusive unit overlaps onto the continental crust and is overlain by chert and jasper layers and the volcaniclastic flysch.

The belt of volcaniclastic flysch and mafic lenses separates the batholith from the main outcrops of Upper Jurassic silicic volcanics and the underlying Paleozoic basement (Fig. 2). The mafic rocks can be interpreted as the floor of a marginal basin that opened up behind the calc-alkaline volcanic chain that was developing along the South American continental margin (Dalziel et al., 1973c, in press c; see Fig. 3). This is regarded as being a situation comparable to the Japan Arc and Japan Sea of today, at least in the sense that the late Mesozoic volcanic chain was clearly established on South American continental crust. It is important to emphasize that no evidence of a comparable Mesozoic marginal basin exists in the Antarctic Peninsula. However, until we know something about the nature and age of the floor of the Weddell Sea, the possibility exists that it might represent a marginal basin that opened up behind the Late Mesozoic volcanic chain along the Antarctic Peninsula.

The precise time relations between inception of the volcanic chain along the western margin of South America and develoment of the marginal basin there must await further field work. The gabbros and dykes of the mafic complexes intrude not only Paleozoic basement but also the unconformably overlying silicic volcanics. Throughout southern South America, as discussed in the last section, there are upper Jurassic volcanics of the same highly silicic composition unconformably overlying the Paleozoic basement. Moreover, the sediments overlying the mafic complexes contain early Cretaceous fossils. However, the highly silicic volcanics along the Pacific margins could well represent the early products of the calc-alkaline volcanic chain (as in the Cenozoic of the Central Andes), or the products of partial melting of continental crust when the marginal basin started to open (as in the Taupo volcanic zone), rather than the widespread Upper Jurassic unit. At the present time, therefore, it is safer to say only that the basin formed in a single major extensional phase around the Jurassic-Cretaceous boundary and early in the history of the volcanic chain. As pointed out in the previous section, the proximity in time and space of a divergent plate boundary in the South Atlantic and a convergent plate boundary along the Pacific margin hampers the identification of processes uniquely related to one or the other. A developing marginal basin between the two major plate boundaries makes the distinctions even more difficult.

MID-LATE CRETACEOUS [ANDEAN OROGENY; DEFORMATION AND UPLIFT]

The youngest preserved volcaniclastic sediments of the marginal basin are the Upper Aptian graywackes and shales of South Georgia Island at the eastern end of the North Scotia Ridge (Trendall,

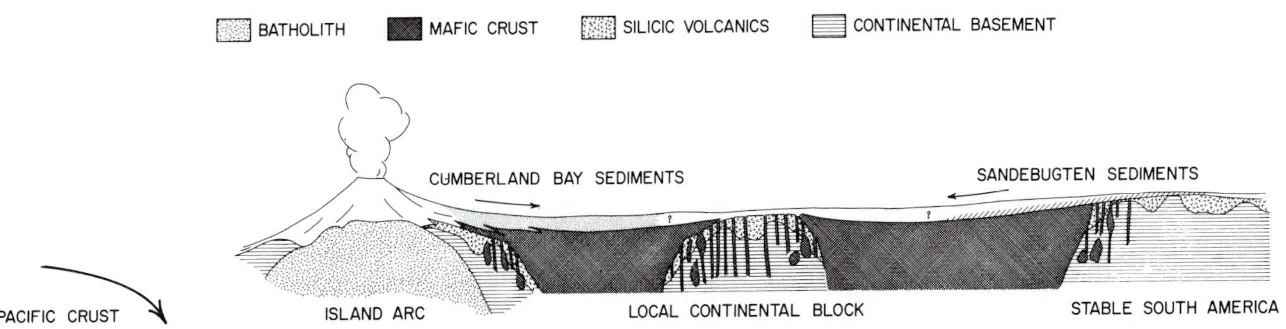

Fig. 3. Restored section across the Pacific continental margin of southernmost South America in the Early Cretaceous (after Dalziel et al., in press b).

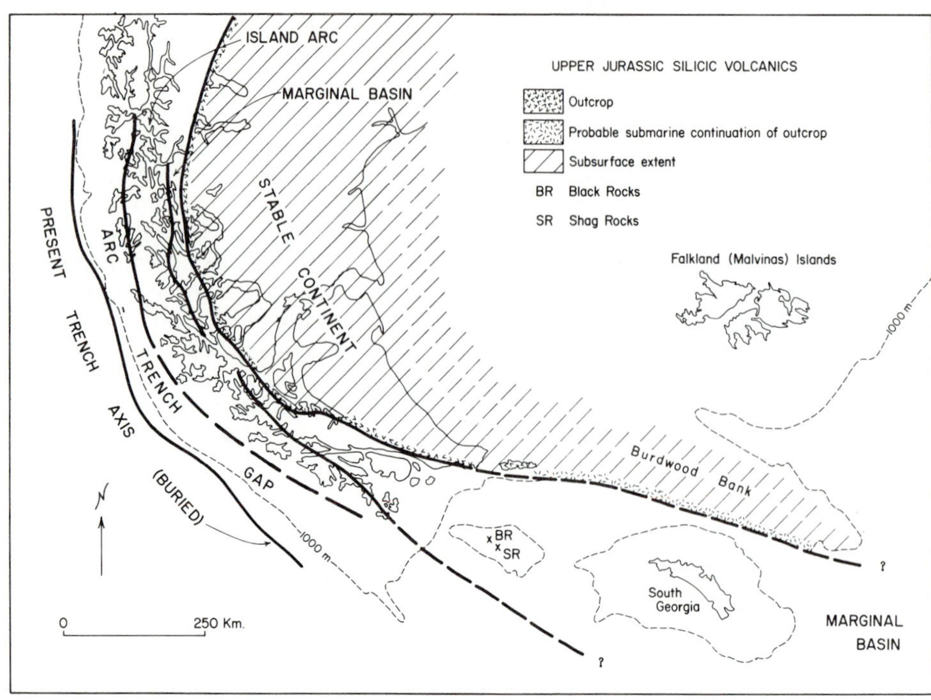

Fig. 4. Map showing the location of early Cretaceous geotectonic units in the southernmost Andes (after Dalziel et al., in press b).

1953, 1959). South Georgia was apparently separated from the Pacific side of the southernmost Andean Cordillera by a strike-slip fault of considerable displacement activated in the latest Cretaceous or later (see the next section). A reconstruction is shown in Figures 3 and 4. Subsequently, the whole Pacific margin of the continent underwent considerable deformation and uplift. The youngest rocks involved in this event, the main deformational phase of the Andean orogenesis in the southern part of South America, are Albian black shales that were deposited on the gently subsiding stable shelf of South America east of the marginal basin. The deformed sedimentary rocks of the basin fill are cut by post-tectonic granitic plutons 80-90 my old (Halpern, 1973). Hence the deformation occurred during the mid-Late Cretaceous (late Albian-Turonian).

Classically, deformation in the Andean Cordillera is thought to be characterized by open folding, block faulting, and uplift. This concept does, in fact, apply to the South Scotia Ridge and the Antarctandes as well as to the Central Andes, but the rocks of the southernmost Andes have undergone much more intense deformation that resulted in the formation of penetrative fabrics, even in the Paleozoic basement. South of the Scotia Sea such fabrics are totally absent from the Mesozoic cover as well as the Paleozoic basement (Dalziel and Cortes, 1972) except for the southernmost part of the Peninsula, where the Mesozoic sedimentary rocks are highly cleaved once again (Williams et al., 1972).

Deformational structures in the arc rocks of southern South America (batholith and Paleozoic rocks, and overlying (?)Upper Jurassic silicic volcanics into which it was emplaced), include shear zones, a steeply dipping slaty cleavage striking parallel to the Cordillera, and widespread foliations in the plutonic rocks. The foliations are likely to be at least in part the result of regional stress systems operating during batholithic emplacemtn. The rocks of the basin floor are very inhomogeneously deformed. To a considerable exent the mafic lenses lack any sign of a penetrative fabric. Deformation seems to have been concentrated along lithic contacts, and in places localized shear zones show "polyphase" deformation. The most striking fabric again is a steeply dipping slaty cleavage mainly striking parallel to the Cordillera.

The sedimentary basin fill shows a more homogeneous structural style and geometry than the basin floor. Where the Andes trend north-south, the sedimentary rocks have a steeply dipping slaty cleavage that is axial-planar to folds of varying tightness with new horizontal hinge lines parallel to the Cordillera. Where the present outcrop area of marginal basin rocks widens, in the east-trending part of the Cordillera (Fig. 2), the folds become more asymmetric, with a strong vergence directed away toward the continent. On South Georgia Island, which presumably was once situated immediately east of Cape Horn (Dalziel et al., 1973a, in press b), the same vergence persists. A prominent thrust carries the Cumberland Bay andesitic volcaniclastic flysch NNE (present coordinates) away from the Scotia Sea over the more silicic Sandebugten volcaniclastic sequence (Fig. 5). The Sandebugten dominantly has opposed vergence.

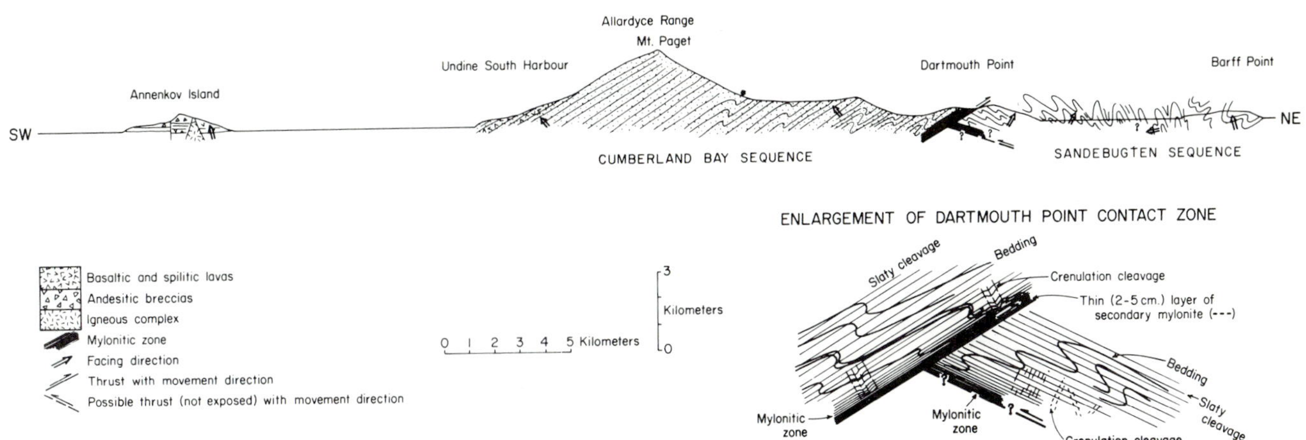

Fig. 5. Section across the island of South Georgia (Fig. 2) (after Dalziel et al., in press b).

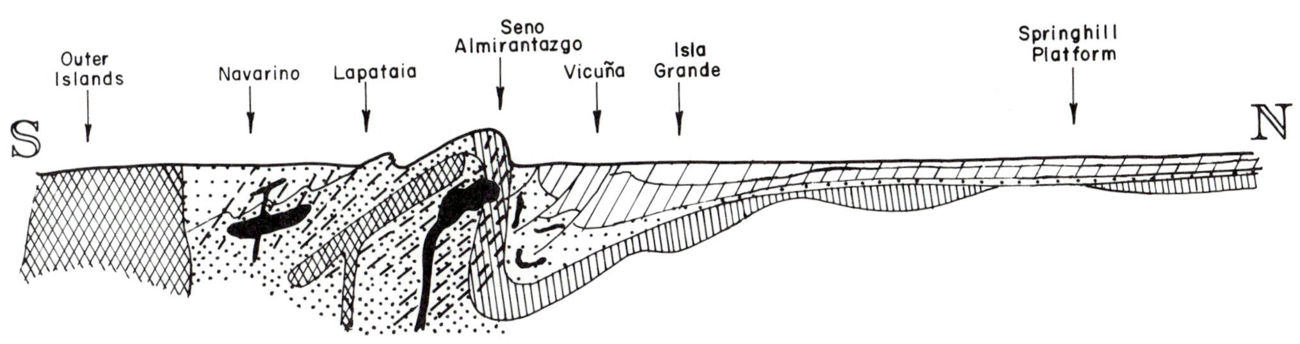

Fig. 6. Section across Cordillera Darwin in southern Tierra del Fuego (after Palmer and Dalziel, 1973).

The rocks on the continental side of the marginal basin in the Early Cretaceous, the Paleozoic basement, and its cover of Upper Jurassic silicic volcanics and Lower Cretaceous shales were also very highly deformed before the intrusion of the 80-90-my-old plutons. Much work remains to be done on the detailed structural geometry. In the south the High Cordillera that flanks the marginal basin rocks consists of large asymmetric folds verging toward the continent with a penetrative axial planar Andean cleavage that can be traced from the Mesozoic cover down into the reactivated Paleozoic basement in the anticlinal cores (Palmer and Dalziel, 1973; see Fig. 6). The deformation dies away very rapidly on the continental side of the Cordillera.

It has already been pointed out that the structures referred to above had to be formed between the mid-Albian (youngest sediments involved) and the Coniacian (minimum age of post-tectonic plutons). There is independent geologic evidence that the Andean Cordillera experienced its initial tectonic uplift during this interval: the Cenomanian Punta Barrossa Formation on the continental side of the High Cordillera contains the first nonvolcanic flysch; the overlying Cerro Terro Formation contains huge slump folds, indicating that an easterly dipping paleoslope from the rising Cordillera to the contemporaneously downwarping Precordilleran trough (Scott, 1966), and thick wildflysch-type conglomerates containing cordilleran rocks (Paleozoic basement, batholith, mafic igneous rocks, and silicic volcanics), were deposited in this trough during the Campanian (Fig. 7).

The present exposure level of the calc-alkaline volcanic chain, in particular that of the marginal basin floor, indicates substantial vertical uplift since the Early Cretaceous. The floor of the marginal basins of the Western Pacific are 2-3 km below sea level. The thickness of the Lower Cretaceous volcaniclastic flysch filling of the southern Andean marginal basin is uncertain because of the subsequent intense strain, but it is unlikely to have been less than 3 km. Hence uplift in excess of 5 km is likely to have occurred since the Early Cretaceous, and the presence of the mafic igneous clasts in the Campanian conglomerates of the Precordillera trough makes it likely that much of this uplift occurred by the Late Cretaceous.

Consideration of the general character and geometry of structures in the marginal basin rocks,

the South Sandwich fracture zone. There is little seismicity along the South Sandwich Ridge and the western margin of West Antarctica. Active volcanoes of a mildly alkaline cast along the deep Bransfield Trough that separates the South Shetland Islands from the Antarctic Peninsula are associated with some seismic activity and seem to indicate that the trough is an extensional feature (Griffiths and Barker, 1972).

Geophysical studies (D. H. Griffiths and P. F. Barker at the University of Birmingham, England, in particular) have shown that the western part of the Scotia Sea and the Drake Passage are underlain by a NE-trending spreading center active from the mid-Cenozoic to the present day. Indications are that the present rate is very low indeed, less than 1 cm/yr (P. F. Barker, personal communication, 1973), but faster rates in the Tertiary were apparently associated with subduction in the trench flanking the South Shetland Islands, where there was calc-alkaline volcanicity at that time (Dalziel and Elliot, 1973). It is tempting to think of the Bransfield Trough as a narrow back-arc spreading center. Extension is apparently continuing there now, even though subduction beneath the South Shetland Islands, as indicated by seismicity, has ceased.

Barker's work has shown that the eastern part of the Scotia Sea is underlain by a N-S-oriented spreading center that is still active and has operated for some 8 my (Barker, 1972). The oldest dated rock from the South Sandwich Islands is 4 my (Baker, 1968); hence Barker regards this spreading center immediately west of the volcanic arc as a marginal basin or back-arc spreading center comparable to those of the western Pacific. The eastern Scotia Sea spreading center is, however, unusual in that it has a well-defined magnetic lineation pattern that is apparently missing in the western Pacific marginal basins.

Hence it is only the central part of the Scotia Sea floor between the South Georgia Platform and the South Orkney Islands Platform whose nature remains virtually unknown. Magnetic anomalies are indistinct, and from the sediment cover it appears to be older than the rest of the Scotia Sea floor. Barker and Griffiths (1972) believe that there may be large foundered continental blocks along the north-south broad ridge between South Georgia and the South Orkneys that has been compared with the Aves Swell in the Caribbean.

It has long been considered that the Andean Cordillera was once continuous with the West Antarctic Cordillera (Barrow, 1831). In the 1950s Hawkes (1962) and Matthews (1959) both showed reconstructions with a straight "land bridge" between the two continents that was eventually fragmented to form the continental fragments on the North and South Scotia ridges. Halpern (1968) and Dott (1972) have also discussed the similarity between the geology north and south of the Scotia Sea and the reasons for considering an ancestral connection between the two continents.

A more detailed reconstruction based on new field work and a detailed literature survey was recently proposed (Dalziel and Elliot, 1971, 1973). This also showed the Andean Cordillera continuing as a rectilinear feature into the West Antarctic Cordillera until at least the Late Cretaceous. Katz (1972), and this volume has objected to this reconstruction and to the term "Antarctandes" for the West Antarctic Cordillera because of the differences in tectonic style north and south of the Drake Passage pointed out by Dalziel and Cortes (1972). However, as discussed earlier, major changes in tectonic style take place along the Andean Cordillera within South America itself, and would not therefore seem to preclude original continuity of the Andean and West Antarctic Cordilleras.

Barker and Griffiths (1972) have taken a different approach, based on their analysis of the geophysical data in the Scotia Sea and Drake Passage. They have reconstructed the original connection between the two continents by closing up the Cenozoic spreading centers. Undoubtedly this indicates an easterly directed cusp-shaped configuration, although, as pointed out by Dalziel and Elliot (1971, 1973), this is unsatisfactory as an *original* configuration because it leaves the South Georgia Platform, clearly part of the Pacific margin of the Andean Cordillera (Fig. 4), on the Atlantic side.

An analysis of British Antarctic Survey paleomagnetic data (Blundell, 1962) by Hamilton (1966) had indicated that oroclinal bending had occurred in the Antarctic Peninsula since the emplacement of the batholith, presumably in the late Mesozoic. Taken with new data from the batholith in South America (Dalziel et al., 1973b), that supports a post-Late Cretaceous origin for the Patagonian orocline, this seemed to indicate that the correct original configuration of the connection between the two continents was an essentially straight cordillera. This did not preclude the possibility of the later existence of the cusp-shaped continental mass indicated by Barker and Griffiths' analysis of the marine geophysical data. However, reanalysis of Blundell's data from the Antarctic Peninsula and excellent new data from a pluton at least 45 my old from the South Shetland Islands contradicted Hamilton's (1966) interpretation and indicated that the shape of the Antarctic Peninsula has not changed significantly since the emplacement of the batholith (Dalziel et al., 1973b). Many of the specimens were obtained from what is now considered to have been an island arc separated from continental South America in the early Cretaceous by a marginal basin of unknown width. It is therefore possible that the data reflect the shape of the arc of islands and not the shape of the southern part of the South American continent.

The evolution of the Scotia Arc is now known to have been very complex. Until more work is done on the early Cretaceous marginal basin in South America and the floor of the Scotia Sea, between South Georgia and the South Orkney Islands in particular, it would be rash to speculate on an evolutionary sequence of events. Having said that, however, one speculation that is hard to resist is the possibility that the original development of a westerly dipping subduction zone in the Scotia Arc may have been related to the closure of the early Cretaceous marginal basin in southern South America and the major strike-slip faulting along the North Scotia Ridge that resulted in the separation of the South Georgia Platform.

PROBLEMS AND POSSIBILITIES

The problems concerning the evolution of the continental margins of the Scotia Sea are of much broader concern. The configuration of the continental masses during the late Mesozoic and Cenozoic is important for the migration of mammals between the southern continents, especially for the question of how the marsupials reached Australia. The continental configuration also reflects directly on the develoment of the circumpolar current, with its attendant effects on world climate.

Finally, it is becoming clear that the margins of the Scotia Sea form a unique laboratory for the study and eventual understanding of any tectonic processes.

THE FAULKLAND PLATEAU

Note Added In Proof

As mentioned in the text, *D/V GLOMAR CHALLENGER* sailed into the Scotia Sea region on Legs 35 and 36 of the Deep Sea Drilling Project from February through May, 1974, while this volume was in preparation. Unfortunately, the bad weather for which the region is notorious limited the scientific results of the cruises, and prevented any sites in the Scotia Sea itself from being drilled. The results will be published shortly in the Initial Reports of the Deep Sea Drilling Project (Craddock, et al., in preparation; Barker, et al., in preparation). One important result of Leg 36 was the recovery of continental crust (gneiss and granite) of probable Precambrian age at site 330 (Fig. 2) on the Faulkland (Malvinas) Plateau, over 800 km east of the Faulkland Islands (Barker, et al., in press).

Although it was suspected that the relatively shallow Faulkland Plateau had a continental basement, the discovery provided an interpretation of the reconstruction of this part of Gondwanaland. The basement rocks are overlain by middle Jurassic, Upper Jurassic and Cretaceous sediments. The Faulkland Fracture Zone lines up with the margin of southeast Africa prior to continental drift, the continental crust of the Faulkland Plateau having presumably extended as far to the northeast as the present location of the city of Durban.

Peter F. Barker of the University of Birmingham, England and the author were Co-Chief Scientists for Leg 36. Other scientists in the shipboard party were: David H. Elliot (The Ohio State University), Christopher C. von der Borch (The Flinders University, South Australia), Robert W. Thompson (Humboldt State College, California), George Plafker (U.S. Geological Survey, Menlo Park, California), R.C. Tjalsma (Woods Hole Oceanographic Institution, Massachusetts), Sherwood W. Wise, Jr., Menno G. Dinkelman and Andrew M. Bombos, Jr. (Florida State University), Alberto Lonardi (Buenos Aires, Argentina), and John Tarney (University of Birmingham, England).

ACKNOWLEDGMENTS

The work described in this review was very largely supported by the National Science Foundation's Office of Polar Programs (OPP) under grants GV 19543, GA 12301, and GA 4185. Much of our work in southern South America would have been impossible without the logistic support and cooperation of the Chilean Empresa Nacional del Petroleo (ENAP). Additional financial support came from the Foundation's International Decade of Oceanographic Exploration (IDOE) program under Grant GX 34410. I am particularly grateful to Robert L. Dale, Robert Elder, Philip M. Smith, and Mort D. Turner of OPP; to Eduardo Gonzalez P. of ENAP; and to Edward Davin of IDOE for their help and understanding.

Scientists who have taken part in the work described include David P. Price, Robert H. Dott, Jr., Keith F. Palmer, Maarten J. de Wit, Ronald L. Bruhn, and Robert D. Winn. In particular, I should like to record my gratitude to Prof. R. H. Dott of the University of Wisconsin, Madison, who took part in our field season on South Georgia and without whose encouragement and previous experience in the Scotia Arc region I would never have started the project. The extent to which all these friends have been involved in the scientific output is reflected in the number of coauthored papers listed in the bibliography.

Finally, it is necessary to record my appreciation of the efforts of various masters and crews of R/V *Hero*, USCGC *Edisto*, USCGC *Glacier*, RRS *Bransfield* and the cutters *Ivan* and *Viking* in giving us an opportunity to do detailed geology in inaccessible places.

Lamont-Doherty Geological Observatory Contribution No. 2102.

BIBLIOGRAPHY

Adie, R. J., 1964, Geological history, *in* Priestly, R., Adie, R. J., and Robin, G. de Q., eds., Antarctic Research: London, Butterworth, p. 118-162.

Baker, P. E., 1968, Comparative volcenology and petrology of the Atlantic island arcs: Bull. Volc., v. 32, p. 187-206.

Barker, P. F., 1972, A spreading centre in the East Scotia Sea: Earth and Planetary Sci. Letters, v. 15, p.

123-132.

———, and Griffiths, D. H., 1972, The evolution of the Scotia Ridge and Scotia Sea: Phil. Trans. Roy. Soc. London, v. 271, p. 151-183.

———, Dalziel, I. W., D., et al., in preparation, Initial Reports of the Deep Sea Drilling Project, v. XXXVI.

———, et al., in press, Preliminary results of Deep Sea Drilling Project, Leg 36: Southernmost Atlantic Ocean. Geol. Soc. Amer., Abstracts with Programs, 1974 Annual Meeting, Miami Beach, Fla.

Barrow, J., 1831, Introductory note (to an article on the Islands of Deception): Roy. Geograph. Soc., v. 1, p. 62.

Blundell, D. J., 1962, Paleomagnetic investigations in the Falkland Islands Dependencies: British Antarctic Survey Sci. Reprt. 39, 24 p.

Burk, C. A., 1965, Geology of the Alaska Peninsula—Island Arc and Continental Margin: Geol. Soc. America Mem. 99 (3 parts), 250 p.

Butterlin, J., 1972, Comparaison des caracteres structuraux de la Cordillere Caraibe (Venezuela) et de la Cordillere Magellanienne (Chili-Argentine), *in* Petzall, C., ed., Trans. VI Caribbean Geol. Conf., Caracas, p. 265-273.

Cecioni, G., 1955, Prime notizie sopra l'esistenza del Paleozoico Superiore nell'Archipelago Patagonico tra i paralleli 50° e 52° S: Atti della Soc. Toscana di Scienze Naturali Mem., v. 62, sec. A, p. 201-224.

Cobbing, E. J., and Pitcher, W. S., 1972, Plate tectonics and the Peruvian Andes. Nature Phys. Sci., v. 240, p. 51-53.

Craddock, C., Hollister, C., et al., in preparation. Initial Reports of the Deep Sea Drilling Project, v. XXXV.

Dalziel, I. W. D., 1972a, Large scale folding in the Scotia Arc, *in* Adie, R. J., ed., Antarctic geology and geophysics: Oslo, Universitetsforlaget, p. 47-55.

———, 1972b, The tectonic framework of the Southern Antilles (Scotia Arc): its possible bearing on the evolution of the Antilles, *in* Petzall, C., ed., Trans. VI Caribbean Geol. Conf. Caracas, p. 300-301.

———, and Cortés, R., 1972, The tectonic style of the southernmost Andes and the Antarctandes: Proc. 24th Intern. Geol. Congr., Montreal, Can., sec. 3, p. 316-327.

———, and Elliot, D. H., 1971, The evolution of the Scotia Arc: Nature, v. 233, p. 246-252.

———, and Elliot, D. H., 1973, The Scotia Arc and Antarctic margin, *in* Stehli, F. H., and Nairn, A. E. M., eds., The ocean basins and their margins: I. The South Atlantic: New York, Plenum Publishing Corporation, p. 171-245.

———, Dott, R. H., Jr., Winn, R. D., Jr., and Bruhn, R. L., 1973a, Tectonic relations of South Georgia Island to the southernmost Andes (abs.), Geol. Soc. America Abstracts with Programs, p. 590.

———, Kligfield, R., Lowrie, W., and Opdyke, N. D., 1973b, Palaeomagnetic data from the southernmost Andes and the Antarctandes, *in* Tarling, D. H., and Runcorn, S. K., eds., Implications of continental drift to the earth sciences: New York, Academic Press, p. 87-101.

———, de Wit, M. J., and Palmer, K. F., 1973c, A fossil marginal basin in the southern Andes (abs.): Geol. Soc. America, p. 589.

———, Caminos, R., Palmer, K. F., Nullo, F., and Casanova, R., in press a, Further observations on the geology of Isla de los Estados, Argentine Tierra del Fuego.

———, Dott, R. H., Jr., Winn, R. D., Jr., and Bruhn, R. L., in press b, South Georgia Island, North Scotia Ridge: Preliminary results of recent field work and their significance.

———, de Wit, M. J., and Palmer, K. F., in press c, A fossil marginal basin in the southern Andes.

Deuser, W. G., 1970, Hypothesis of the formation of the Scotia and Caribbean seas: Tectonophysics, v. 20, p. 391-402.

Dott, R. H., Jr., 1972, The antiquity of the Scotia Arc (abs): Am. Geophys. Union Trans., v. 53, p. 178-179.

Du Toit, A. C., 1937, Our wandering continents: Edinburgh, Oliver & Boyd, 366 p.

Gansser, A., 1973, Facts and theories on the Andes: Geol. Soc. London Jour., v. 129, p. 93-131.

Griffiths, D. H., and Barker, P. F., 1972, Review of marine geophysical investigations in the Scotia Sea, *in* Adie, R. J., ed., Antarctic geology and geophysics, Oslo, Universitetsforlaget, p. 3-11.

Halpern, M., 1968, Ages of Antarctic and Argentine rocks bearing on continental drift: Earth and Planetary Sci. Letters, v. 5, p. 159-167.

———, 1973, Regional geochronology of Chile south of 50° latitude: Geol. Soc. America Bull., v. 84, p. 2407-2422.

———, and Rex, D. C., 1972, Time of folding of the Yahgan formation and age of the Tekenika beds, southern Chile, South America: Geol. Soc. America Bull., v. 83, p. 1881-1886.

Hamilton, W., 1966, Formation of the Scotia and Caribbean arcs: Geol. Survey Can. Paper 66-15, p. 178-187.

Hawkes, D. D., 1962, The structure of the Scotia Arc: Geol. Mag., v. 99, p. 85-91.

Katz, H. R., 1961, Algunas notas acerca de la intrusion granitica en las Cordillera del Paine, Provincia de Magallanes: Santiago de Chile, Minerales, v. 16, no. 74, p. 1-15.

———, 1972, Plate tectonics and orogenic belts in the South-east Pacific: Nature, v. 237, p. 331-332.

Larson, R. L., and Ladd, J. W., 1973, Evidence for the opening of the South Atlantic in the Early Cretaceous: Nature, v. 246, p. 209-212.

———, and Pitman, W. C., III, 1972, World-wide correlation of Mesozoic magnetic anomalies, and its implications: Geol. Soc. America Bull., v. 83, p. 3645-3662.

Matthews, D. H., 1959, Aspects of the geology of the Scotia Arc: Geol. Mag., v. 96, p. 425-441.

North, F. K., 1965, The curvature of the Antilles: Geol. en Mijnbouw, v. 44, p. 73-86.

Palmer, K. F., and Dalziel, I. W. D., 1973, Structural studies in the Scotia Arc: Cordillera Darwin, Tierra del Fuego: Antarctic Jour. U. S., v. 8, p. 11-14.

Rex, D. C., and Baker, P. E., 1973, Age and petrology of the Cornwallis Island granodiorite: British Antarctic Survey Bull. 32, p. 55-61.

Scott, K. M., 1966, Sedimentology and dispersal pattern of a Cretaceous flysch sequence, Patagonian Andes, southern Chile. Am. Assoc. Petr. Geol. Bull., v. 50, p. 72-107.

Skidmore, M. J., 1972, The geology of South Georgia: III. Prince Olav Harbour and Stromness Bay areas: British Antarctic Survey Sci. Rept. 73, 50 p.

Trendall, A. F., 1953, The geology of South Georgia I: Falkland Islands Dependencies Sci. Rept. 7, 26 p.

———, 1959, The geology of South Georgia II: Falkland Islands Dependencies Sci. Rept. 19, 148 p.

Williams, P. L., Schmidt, D. L., Plummer, C. C., and Brown, L. E., 1972, Geology of the Lassiter Coast area, Antarctic Peninsula: preliminary report, *in* Adie, R. J., eds., Antarctic geology and geophysics, Oslo, Universitetsforlaget, p. 143-148.

Wilson, J. T., 1966, Are the structures of the Caribbean and Scotia Arc regions analogous to ice rafting?: Earth and Planetary Sci. Letters, v. 1, p. 335-338.

Continental Margin of Western South America

Dennis E. Hayes

INTRODUCTION

The continental margin of the west coast of South America marks the seismically active, tectonic boundary between the converging Nasca and South American plates. The Peru-Chile Trench is the dominant morphologic element of the margin and persists as a structural feature from the coast of western Colombia to Tierra del Fuego, at the southern tip of South America. Variations in the trench morphology, sediment distribution, volcanism and orogenesis in the Andes, and the seismicity are interpeted to be the likely consequence of variations in the rate and direction of plate convergence and the effectiveness of geologic boundary conditions within the continental lithosphere. Some outstanding problems are cited also.

SUMMARY OF INVESTIGATIONS

This brief review was undertaken to provide a broad framework in which this margin might be considered in future studies and to provide geographic continuity to the collective studies of this volume. It is not intended as an exhaustive summary of existing knowledge of the continental margin of western South America and makes no pretense at incorporating any major new ideas or insights.

The continental margin of western South America is relatively poorly investigated. The presence of the Peru-Chile Trench system has been known for some time (Zeigler et al., 1957; Fisher and Raitt, 1962). The large free-air gravity anomaly minimum associated with the trench was originally described by Wuenschel (1952) using submarine pendulum-gravity measurements. The presence of the seismic Benioff zones was established long ago through investigations of the distribution of earthquake hypocenters along the continental margin (Benioff, 1954). The region was the subject of a geophysical-geological investigation by Hayes (1966), prior to the general acceptance of the concepts of plate tectonics. The evolution of this margin was later the focus of additional investigations by Scholl et al. (1968, 1970), and is most pertinent to the plate tectonic studies of the deep ocean basin by Herron and Heirtzler (1967), Herron and Hayes (1969), and Herron (1971, 1972a). Considerable work has been done in defining the nature of the deep crustal structure and seismic zones beneath the Andean margin by Aldrich et al. (1958), Fisher and Raitt (1962), Sykes and Hayes (1971), Ocola et al. (1971), James (1971), and Stauder (1973).

Most of this margin is bounded on the west by the Nasca plate, the subject of a National Science Foundation (IDOE) project involving joint investigations by the Hawaii Institute of Geophysics, Pacific Oceanographic Laboratory (NOAA), and Oregon State University. At the present time these investigations are still in progress and relatively few results have been published, except as agency reports. This project encompasses studies of the entire Nasca plate (~15 million km^2), and even though the continental margin portion constitutes only about one-fourth of the perimeter of the Nasca plate, it is unfortunate that the results of this work, the only new and concentrated field studies, cannot be included here.

GENERAL OBSERVATIONS

The continental margin of the west coast of South America can be characterized as one of the most seismically and tectonically active boundaries separating a major continent and a deep ocean basin (Fig. 1). Physiographically the margin is spectacular in its relief and is unique in that a deep-sea trench can be traced (as a structural feature) from its northern limit in Colombia some 8,000 km to the south to Tierra del Fuego (Fig. 2). A few hundred kilometers to the east, the Andean mountain range climbs to elevations of about 7 km and, in concert with the trench system, defines some of the largest regional topographic gradients found anywhere in the world.

The continental margin is unusually narrow and fits into category III, a continental margin based on the standard physiographic province classification of Heezen et al. (1959). Because of the presence of the Peru-Chile Trench, which persists as a major topographic deep and the dominant morphologic element along most of the extent of the margin, this "continental margin" is unique. Although the trench system exists with many common features through the entire length of the margin, there are also many important differences in the morphology and in the geophysical and geological properties of the trench and associated continental margin.

The tectonic expression of the continental margin is considered to be largely the consequence of the collision of the South American plate with the

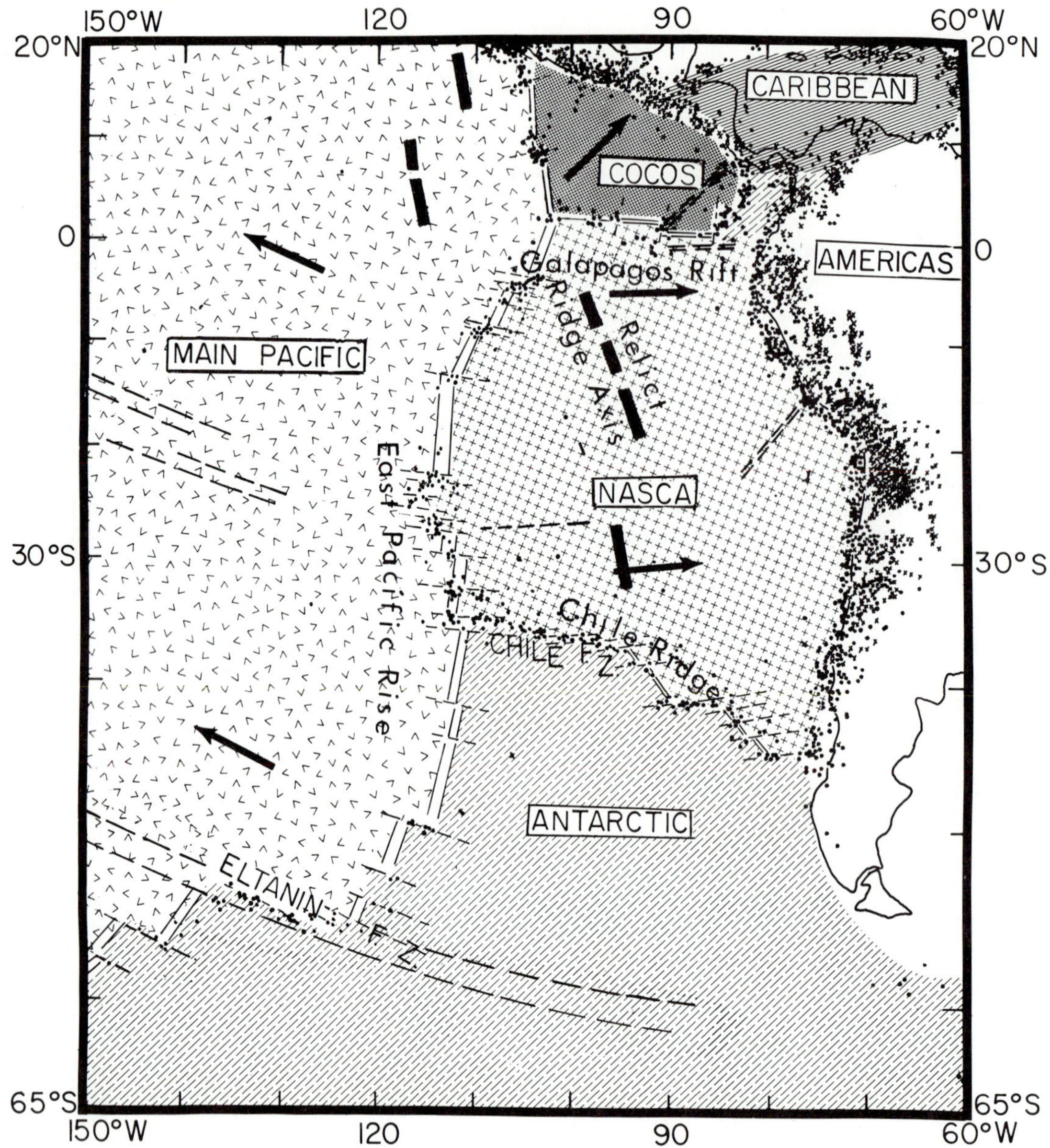

Fig. 1. Active and ancient plate boundaries and plate direction as defined by earthquake epicenters and marine geophysical data (modified after Herron, 1972a).

Nasca plate to the west, and with smaller plates both to the north of the Carnegie Ridge and to the south of the Chile Ridge (Fig. 1). Variations in tectonism reflect variations in the nature of the relative motion between these plates as a function of space and time. It is now well established (e.g., Herron, 1972a) that the pattern of sea-floor spreading, as recognized within the Pacific basin to the west of the margin, has varied considerably throughout the Cenozoic and that these changes have given rise to different modes of tectonism and have influenced the timing of orogeny at various places along this margin. The continental margin as we see it today is thus considered to be the integrated product of all geological processes, but with the most obvious effects resulting from plate collisions throughout the Cenozoic.

SUBPROVINCES AND SPECIAL PROBLEMS

The continental margin can conveniently be subdivided into four broad subprovinces which are recognized primarily on the basis of differences in those properties, such as trench morphology, seismicity, volcanism, gravity anomalies, sediments, and continental geology (see Figs 1, 3, and 4).

The deep-sea sediments provide a surficial blanket on the underlying ocean crust and in some instances thereby complicate our recognition of subtle structural and morphologic properties of the basement rocks along the length of the trench system (Fig. 3). Variations in sediment within the trench and on the continental slope are interpreted to be largely due to differences in the postorogenic geology of the continent and in particular to the

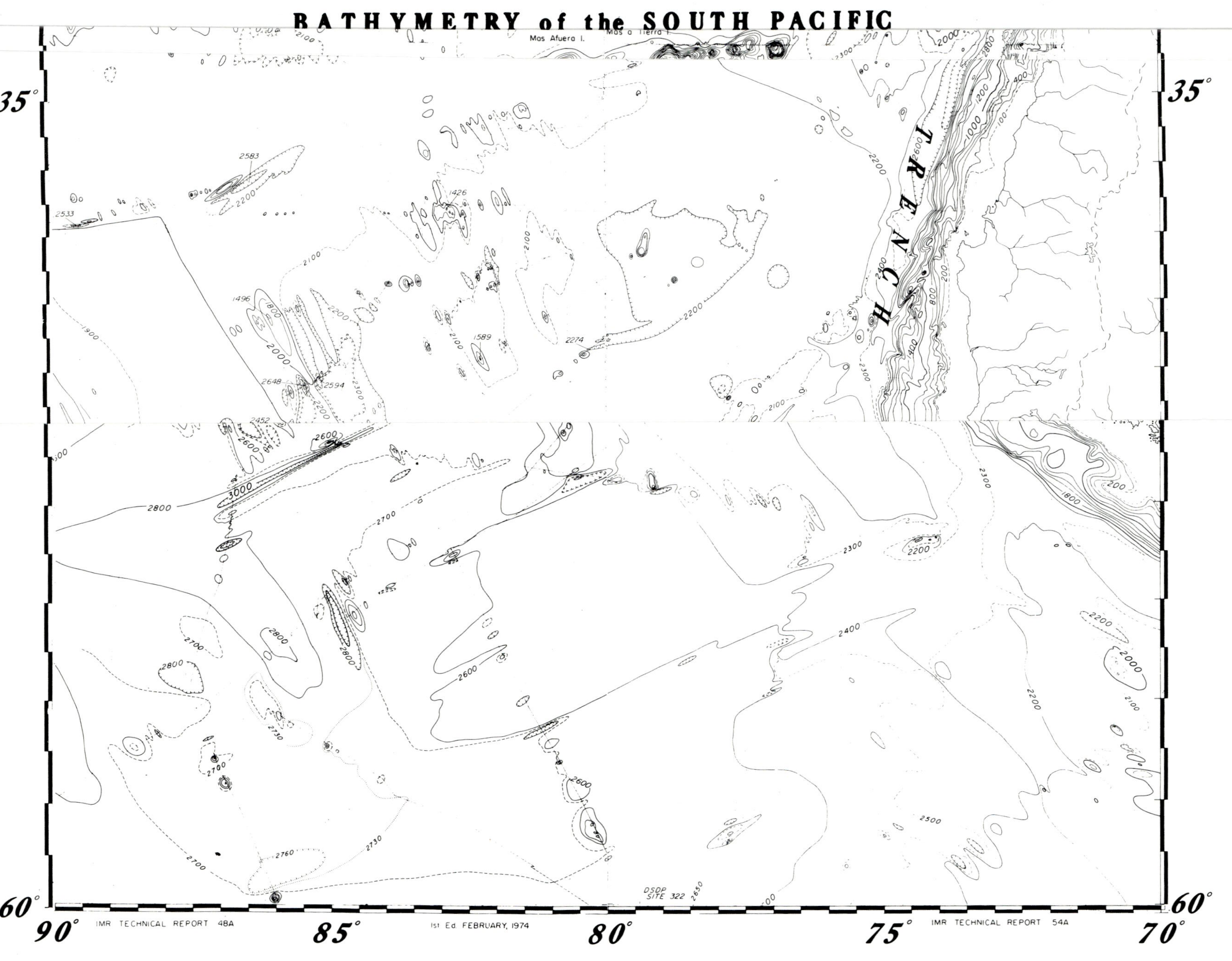

Fig. 2. Morphology of the Peru-Chile trench, continental margin of the west coast of South America and the adjacent deep Pacific basin. Including sheets 21a, 21b, and portions of sheets 15 and 20 from Mammerickx, et al., (1974). Contours are in nominal fathoms; mercator projection.

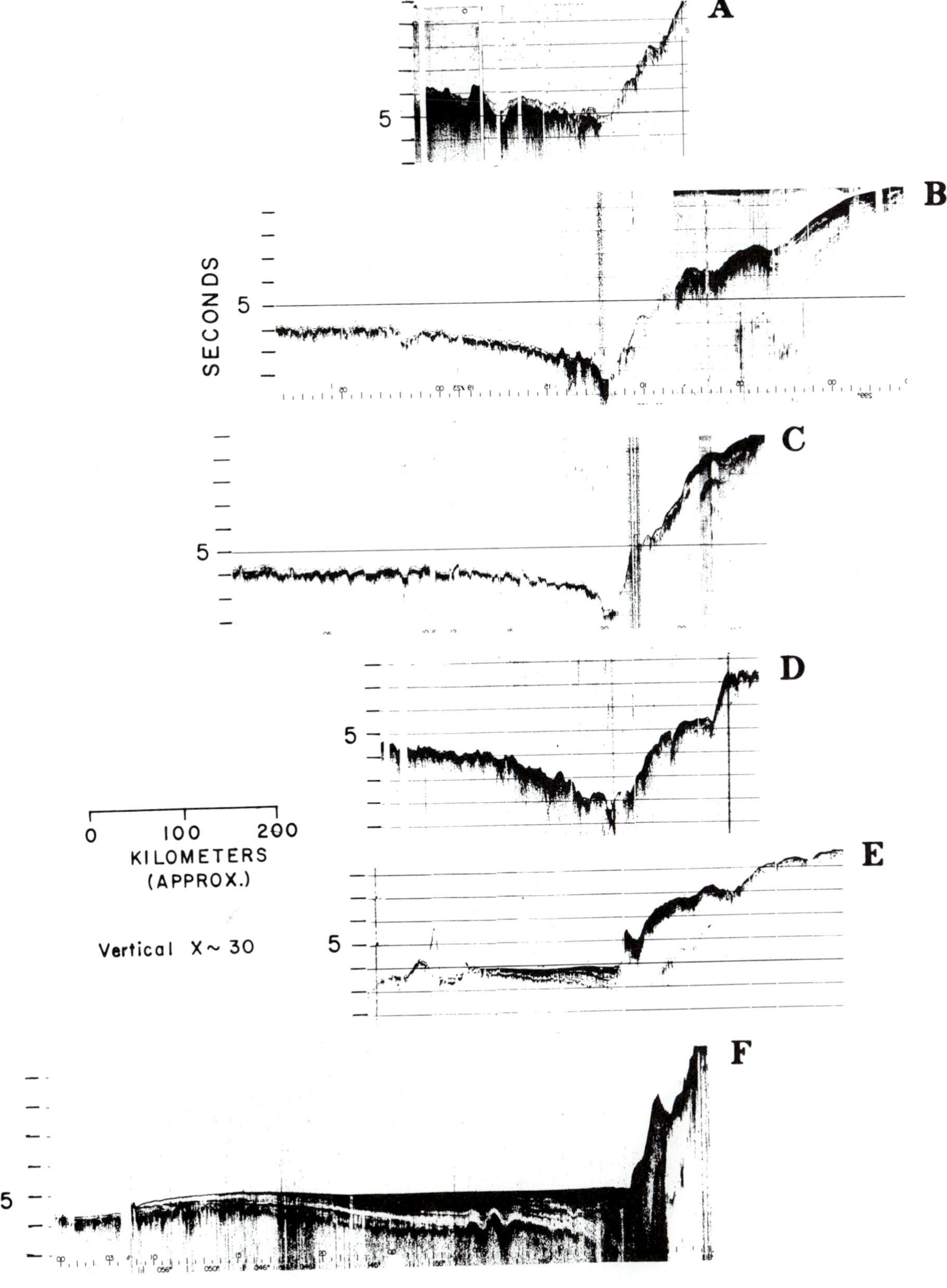

Fig. 3. Seismic-reflection profiles taken from Hayes and Ewing (1970), Houtz et al. (1973), and unpublished data from the Lamont-Doherty marine seismic department; 1 sec of two-way reflection time is 400 fm, or 750 m (in water). (See Figure 4 for locations.)

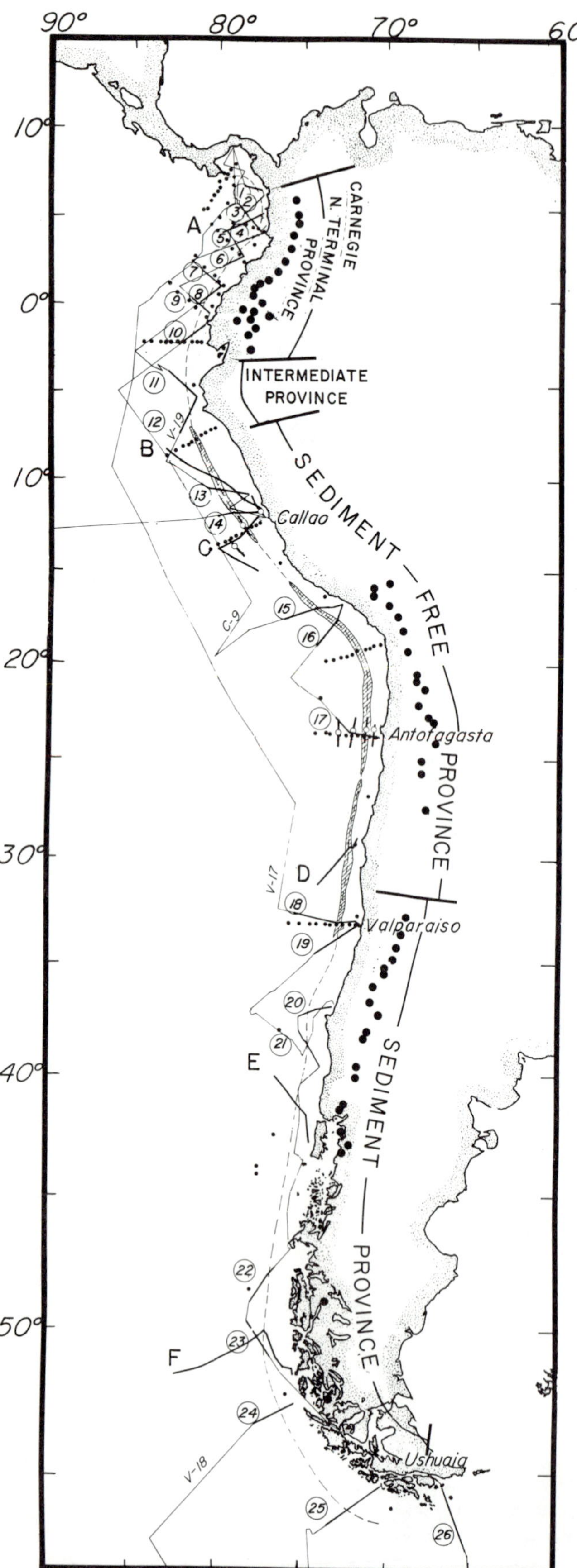

Fig. 4. Index map showing location of profiles in Figure 3 and 5, subprovinces of the Peru-Chile Trench. The locations of active volcanoes are shown as black dots.

average geologic "climate" of the area; i.e., those factors, such as elevation, rock type, amount of rainfall, and vegetation, which control the rate of denudation of the continent.

The present-day seismicity is by no means uniform throughout the area; there are marked north-south variations in the presence of deep earthquakes (see Stauder, 1973) which are confined to rather narrow latitudinal zones. Similarly, there is a conspicuous absence of intermediate-focus earthquakes between about 300 and 550 km along the entire margin. Even the shallow seismic earthquake activity which appears to be generally uniformly distributed, on closer examination outlines small, relatively aseismic regions such as the triangular area centered near 26°S, 67°W and the entire margin south of about 46°S, where the Chile Ridge intersects the continent (Fig. 1). Furthermore, detailed examination of hypocenter data shows that the spatial distribution of earthquakes is quite different within various latitudinal belts (Sykes and Hayes, 1971; Stauder, 1973). The nature, in fact the presence or absence of a well-defined Benioff zone, is systematically different along different portions of the margin (cf. Carr et al., this volume).

As mentioned, the major part of the margin is bounded by the Nasca plate from about the equator to approximately 46°S. Since this entire portion of the margin presently represents the interaction of only two plates (the Nasca and South American plates), one would intuitively suspect that the tectonic manifestation of this plate collision would, to a first approximation, be similar. Even considering that the relative geometry along the collision boundaries changes by about 30°, the apparent interaction of the plates as recorded both by the present and Cenozoic volcanism (Noble et al., 1974), and the distribution of earthquakes, is dramatically different. This contrast suggests that the variations in the gross continental geology of the margin serve to play an important role, perhaps even in determining boundary conditions, in the final expression of the tectonic interaction at this boundary. It is also probable that at least some of the seismic and volcanic activity records not only the present interaction of the Nasca and South American plates, but their interaction in response to plate motions that existed prior to 10 my, and were substantially different from present conditions (Herron, 1972a; Noble et al., 1974; Stauder, 1973).

According to Allen (1965) the Atacama fault, which subparallels the central Chilean margin, is one of the major strike-slip (wrench) faults in the world and has been compared to the San Andreas, the Alpine, and the Philippine faults. The San Andreas and Alpine fault systems have been analyzed as transform fault segments, connecting other major plate boundaries (e.g., Atwater, 1970; Hayes and Talwani, 1972). In view of the rather straightforward evidence for crustal subduction along the margin of northern Chile, there seems no obvious way to force the Atacama fault system to fit into a similar transform-fault classification, and further investigation perhaps leading to an alternative explanation is badly needed.

The physiography and geology of the Andes themselves are quite different from north to south, and consist of at least two major branches in Colombia, Peru, Bolivia, and northern Chile, but narrows down into a narrower linear chain in central Chile that extends southward and plunges gradually into the southernmost portions of Patagonia. Many detailed studies have been published on this subject, and the reader is referred to International Upper Mantle Project Scientific Report 37-1 (1970) and the extensive references contained therein. The active volcanism associated with the Andes is also largely variable, with the main activity concentrated in southern (~32-43°S) and northern Chile and in southern Peru, Ecuador, and Colombia (Fig. 4). No attempt is made here to analyze this distribution, but readers are referred to Casertano (1963), Hantke and Parodi (1966), and Noble et al. (1974).

Through recent seismic-refraction and earthquake surface-wave analysis, it has been shown that rocks with seismic-wave velocity characteristics of continental crust (6.0-6.6 km/sec, granitic-type rocks) extend to depths in excess of 70 km under parts of the Andean Range and constitute the thickest sections of continental crust found anywhere (James, 1971; Ocola et al., 1971).

There are major gravity anomalies associated with the trench system, as with most typical island arc and trench systems of the world. The free-air gravity minimum is greatest west of the central Chilean regions, where it reaches values of about -260 mgal (Hayes, 1966). The gravity signature is thought to reflect an anomalous mass distribution resulting from several processes. One process is associated with the development of the trench topography; another process is associated with the accumulation of relatively low-density sediments on the landward flank of the trench system (e.g., Hayes and Ewing, 1970), and a third process is the plunging of the lithosphere beneath the trench system (Watts and Talwani, 1974; Grow and Bowin, in press). As all these factors vary along the length of the continental margin, the gravity anomaly is correspondingly variable. However, the very presence of a major free-air gravity minimum was used in a qualitative way to predict the existence of a structural trench system which extends all the way from Panama to the Drake Passage (Hayes, 1966).

Perhaps equally important is the presence of secondary gravity minimum, which has been shown by Hayes (1966) to be the likely consequence of thick accumulations of low-density sediment in deep basins, whose axes oscillate back and forth with respect to the present coastline. These gravity anomalies may serve to map the axis of maximum sedimentation of the present and ancient Bolivar geosyncline or its counterpart to the south. This secondary gravity minimum is confined to the region

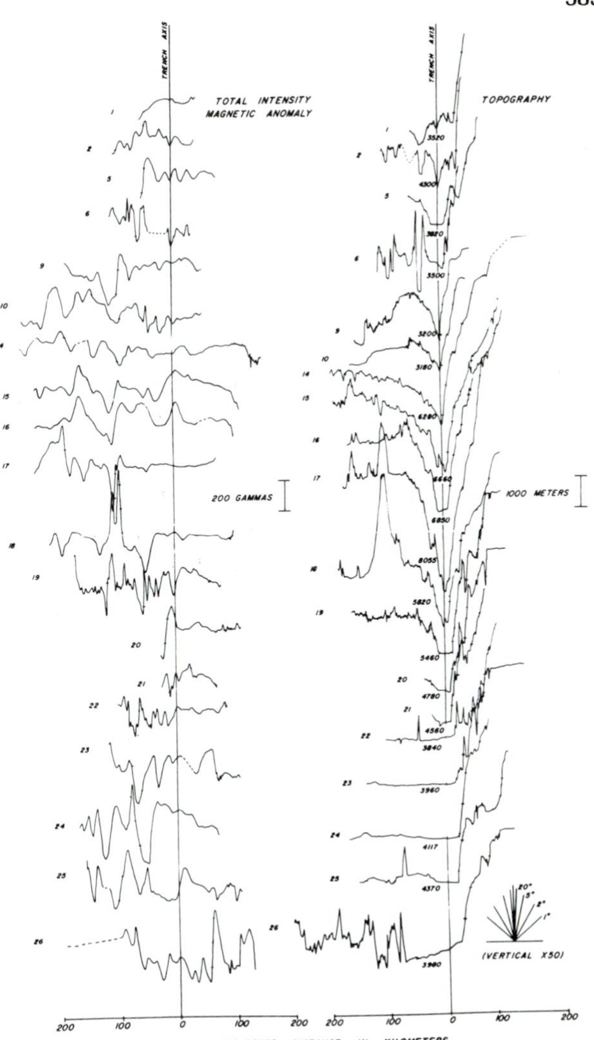

Fig. 5. Projected topographic and magnetic anomaly profiles across the Peru-Chile Trench (after Hayes, 1966).

north of Chile along the coasts of Peru, Ecuador, and Colombia and enters the continental margin near 8°N in the Choco Province of Colombia. The presence of terraces, or secondary sedimentary basins, in association with other trench-arc continental margin systems is quite common. The Aleutian terrace of the Aleutian trench (see Hayes and Ewing, 1970) provides an excellent example, as do the trench-arc gap of the Sumatra-Java-Andaman arc system and to a lesser extent the terraces of the Philippine Trench and the Tonga-Kermadec Trench. Karig (1974) has recently evaluated the process of "tectonic erosion" in attempting to account for these and other properties of inner walls of the trenches by proposing that rifting occurs in these areas.

Magnetic anomalies in the deep ocean basin are typically of the sea-floor spreading variety and date the crust as approximately 40-50 my old adjacent to the trench between the Carnegie Ridge near 0°, and the Chile Ridge near 45°S (Herron, 1972a). As shown in Figure 5, the amplitudes of these sea-floor spreading lineations become rapidly attenuated at or near the topographic trench axis

(Hayes, 1966; Hayes and Ewing, 1970). It has been shown for other areas that, in general, this attenuation in amplitude cannot be explained solely by the increased depth from the detector to source that might be associated with the plunging oceanic crust beneath the trench system. Hence some additional process is acting near the trench which effectively serves to destroy at least a part of the natural remanent magnetization of the oceanic basalts.

As mentioned, the sediments within the trench are highly variable from north to south along the trench system (Fig. 3), and the trench has been classified into four provinces: a sedimentary-free province off the central parts of Chile and Peru between 8°S and 32°S; a sedimentary province which begins about 33°S and extends to the south; an intermediate province lying between 4°S and 8°S; and a Carnegie Ridge-northern terminal province, lying between 4°S and 5°N (Fig. 4).

The sedimentary-free province lies adjacent to the Atacama Desert, where there is virtually no detritus supplied from the continent to the trench system. Similarly, the entire ocean basin area to the west of the trench typically has a very thin sediment cover, so few pelagic sediments are being brought into the trench system from the east. There has been no significant sediment accumulation recorded within the main sedimentary-free province of the trench. South of Valparaiso, the trench begins to be filled with sediments and by about latitude 45°S (the sediment province), the trench is not only filled but overflowing with sediment (profile E, Fig. 3). Here there are two contributions of sediment, one source being terrigenous materials deposited in the trench as the result of downslope transport of material from the continent, and this is intermixed with pelagic sediments, the consequence of higher near-surface primary productivity and the subsequent higher pelagic sedimentation. These later sediments are brought into the trench system in the area north of the Chile Ridge by settling of planktonic tests. To the south of the Chile Ridge, the only pelagic sediment contribution comes from the settling of tests through the water column because at present, no relative motion is indicated between the deep ocean basin and the continental margin, and hence there is no sediment contribution.

In the regions south of Valparaiso large accumulations of essentially flat-lying terrigenous sediments were initially used as evidence that there was no major tectonic compression between the ocean basin and the continental margin (Scholl et al., 1968, 1970). This question was also addressed by Hayes and Ewing (1970), and they concluded that the undisturbed or only slightly tilted terrigenous sediment horizons, the absence of coherent seismic reflections beneath the lower continental slope, and the dramatic contrast in the apparent morphology of the lower continental slope as compared with other trench systems were indications that a sharp front of deformation existed within these semiconsolidated sediments.

They also concluded that the sediments were highly deformed into a mélange sequence just slightly shoreward of the structural axis, and that a part of these disturbed sediments was, in effect, plastered onto the lower continental slope. Furthermore, this melange inhibited the recording of coherent seismic-reflecting horizons and contributed to the overall gravity signature and displacement of the gravity minimum from the trench axis. This explanation still seems plausible and eliminates the previous paradox of the presence of active compressive forces along the plate boundary, yet the absence of a clear record of deformed sediments. More recent seismic-reflection profiling, particularly obtained along profiles at low incidence to the continental margin, (see Seely et al., this volume), shows that the top of the oceanic crust can be traced well beneath the axis of the structural trench and plunges beneath the continental margin (profile E, Fig. 3). These records confirm the occurrence of subduction of oceanic lithosphere at some time within the recent geologic past.

There are numerous features that are oriented transverse to the continental margin (see Carr, et al., this volume). The most prominent of these include seismic ridges such as the Carnegie and Chile ridges and also the aseismic Nasca Ridge. Furthermore, the Chile Ridge system and the East Pacific Rise are dissected by major transform faults, and the relict fracture zone traces can be extrapolated toward their intersection with the western South American margin (see Herron, 1972a, and Fig. 2).

The aseismic Nasca Ridge and the seismic Carnegie-Galapagos ridge system are two major features whose effects in modulating the expression of the continental margin are dramatically different. The Nasca Ridge may be a relic hot-spot trail which has no direct relationship to a spreading ridge system throughout its entire history, and is perhaps analogous to the Walvis or the Rio Grande ridge systems in the Atlantic. No magnetic anomalies have been identified that parallel this system. The Nasca Ridge appears to be relatively old, as its intersection with the continental margin of the Peru-Chile Trench does not manifest itself in any obvious way in the morphology of the trench or tectonics of the continental margin. In contrast, the Carnegie Ridge-Galapagos system marks an east-west center of sea-floor spreading about a ridge axis that defines one branch of the triple junction located near 102°W, 2°S (Herron and Heirtzler, 1967; Raff, 1968; Herron, 1972a). This particular branch of the spreading mid-ocean ridge system has only been active for about the last 10 my. As a result of this recent spreading activity, great quantities of heat have been brought to the surface, elevating the ridge system and in effect modulating the morphology of the trench system where the ridge intersects the continental margin near 0° latitude.

The actual intersection of the Galapagos system with the continental margin is somewhat ambigu-

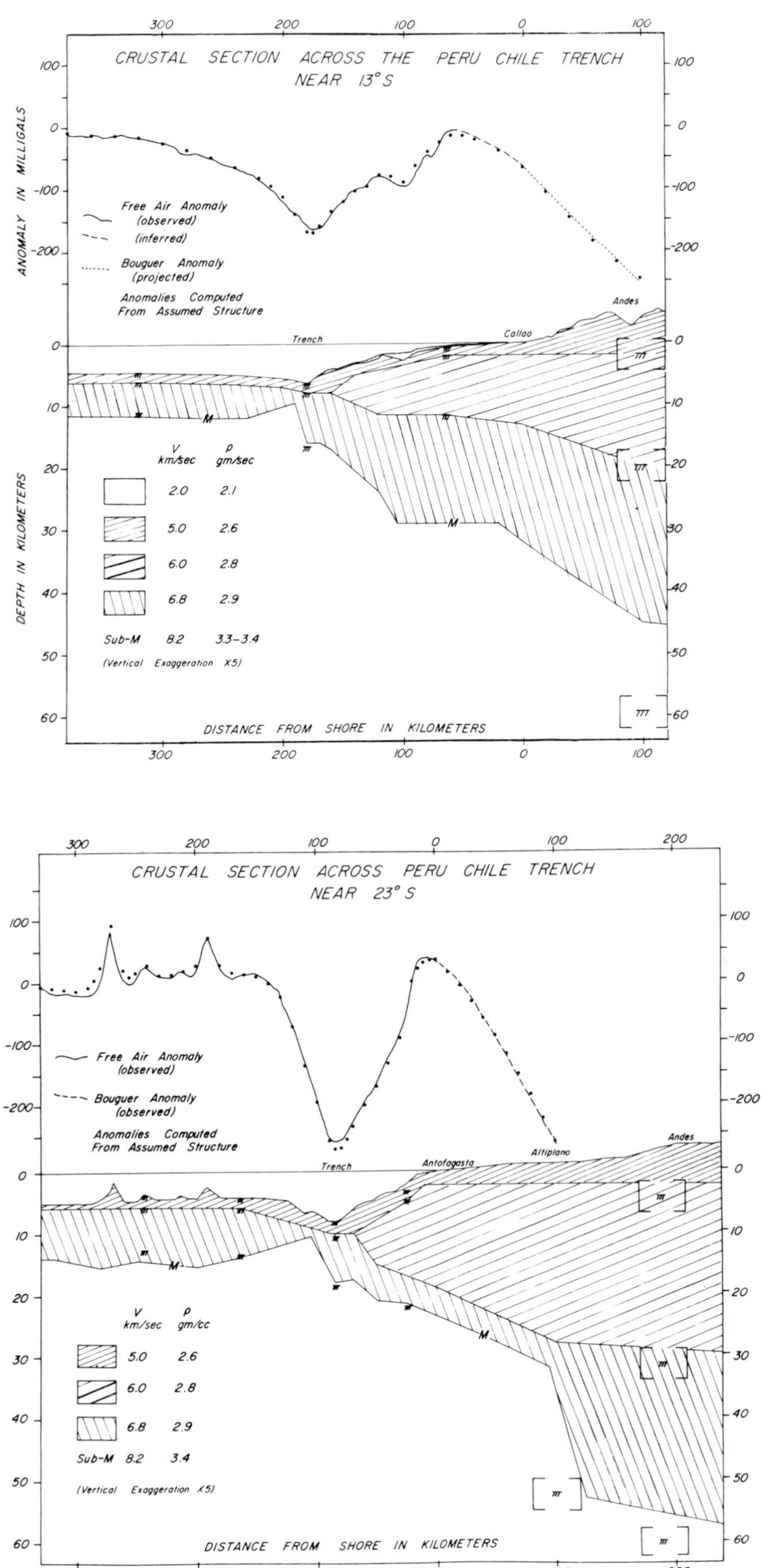

Fig. 6. Gross crustal structure sections from seismic refraction and gravity data (after Hayes, 1966).

ously defined. The major plate bounding the Galapagos system to the north is the Cocos plate which is considered to being subducted now along the Mid-America Trench. Motion of a sub-plate lying to its east and within the Panama basin and east of the Cocos Ridge is not well understood. This small region must represent the complicated interaction at several minor plate boundaries (e.g., Sass et al., 1974).

The crustal parameters that can be defined by the combination of a few widely spaced refraction measurements and surface-wave studies beneath the Andes can be used together with gravity measurements to determine the gross crustal structure and transition from oceanic to continental crust across the trench system. The crustal structure is in part shown by Fisher and Raitt (1962) and was further interpreted by Hayes (1966), utilizing continuous gravity measurements. The crustal structure (along two sections) as inferred by Hayes (Fig. 6) was constrained to be compatible with the seismic-refraction measurements presented by Fisher and Raitt. The details, particularly of the crust and mantle interface, were inferred by modeling crustal boundaries to account for the observed variations in gravity.

A major conclusion was that either there was a pronounced crustal thinning just seaward of the axis of the trench, as illustrated in these figures, or that high-density subcrustal material (mantle) was present in this region to account for the extremely high gravity gradients observed. A recent paper by Grow and Bowin (in press) has reinvestigated this problem. They present a model which allows for density changes within the subducting oceanic lithosphere as a consequence of phase changes that result from the pressure-temperature variations associated with subduction. They have attempted to show that the apparent thinning is an artifact of assuming a constant density across this section. They conclude that crustal thinning is not necessary if the lateral variations in density are allowed.

At the present time, there are no seismic-refraction data located strategically enough to clearly resolve these two alternative interpretations. The results of Grow and Bowin (in press) indicate that if their effects are properly evaluated, the need for crustal thinning is, for the most part, eliminated. Their model does, however, show residual gravity anomalies which would imply that some crustal thinning or high-density mantle material is still required in order to mutually satisfy the seismic-refraction and gravity observations. An alternative is that their model could be slightly modified, thereby removing the need to postulate any crustal thinning beneath the outer flank of the trench.

Unpublished seismic-refraction measurements using sonobuoys near profile C of Figure 4 were conducted as part of the Nasca plate/IDOE project (Nazca Plate Project Progress Report). These data provide some indication that a crustal thinning may in fact be present along this particular traverse. The thinning is less severe than that proposed by Hayes (1966).

Since the surface morphology of the basin outside the trench typically consists of an outer ridge, as shown elsewhere (e.g., the Aleutian trenches; Watts and Talwani, 1974), it is likely that some up-buckling of the crustal material within the lithosphere is accompanied by an up-bowing at the crust-mantle interface. This up-bowing may appear as an effective crustal thinning, and it is possible that both this effect and the effect of lateral density changes proposed by Grow and Bowin (in press) most plausibly explain both the seismic-refraction measurements and the continuous gravity measurements.

Regardless of the details in the crustal transition, it is clear that there is a gradual change from crustal thicknesses typical of oceanic basins, about 5-6 km, to continental crustal thicknesses of the order of 30-40 km near the coast but here extending down to as much as 70 km beneath the axis of the high Andes Mountains. Any variations in this gradual transition are relatively minor. It also seems apparent from the gravity measurements that the nature of the transition from oceanic to continental type crust is, to the first order, not dramatically different along the entire continental margin of western South America.

FUTURE STUDIES

Major new efforts need to be devoted to further study of this continental margin. New seismic techniques employing multichannel recording and sophisticated data processing are needed in order to investigate the nature of the sediments and crustal rocks underlying the continental slope and narrow shelf of western South America. Furthermore, the presence of commercial copper mines throughout north-central Chile are important to investigate in terms of the possible metallogenic relationship of ore minerals to processes at such a continental margin, such as subduction, subsequent volcanism, and concentration of metals (e.g. Sillitoe, 1972).

The presence of large aseismic zones suggests that either these zones represent a unique aspect of the continental geology, or underthrust lithosphere, which is not favorable for the gradual and systematic release of the earthquake energy, or that strain is being built up and that these aseismic zones are the most likely locales for additional destructive earthquakes.

Major strike-slip faults—such as the Atacama in Chile—must be considered as an important result of any major stress system in regions of such continental margins.

According to the hypothesis of Herron and Hayes (1969), the general plate configuaration in the region south of the Chile Ridge suggests that the deep ocean basin between the Chile Ridge and the East Pacific Rise is part of the Antarctic plate.

There is no indication of major seismicity or an active plate boundary between the "South American plate" and the "Antarctic plate" south of the Chile Ridge (Fig. 1). There is an effective aseismic gap along the boundary of these plates, except to the southeast along the Scotia arc (see Dalziel, this volume).

For some reason there is no substantial seismicity or other evidence of current plate interaction in the corridor to the north (45-55°S). The relative motion between these two plates, apparently essentially the same along the Scotia Ridge (Forsyth and Chapman, 1974) as it is in this corridor, presents a difficult problem and may indicate some very recent and undetected change in relative motion between the Antarctic and South American plates. There has been recent sea-floor spreading activity within the Scotia Sea region as described by Barker (1970), and this is the subject of current investigations by the Deep Sea Drilling Project. The logical southward extension of this margin includes the Scotia Arc and Antarctic Peninsula, and the tectonic evolution of these regions is discussed in a comprehensive study by Dalziel and Elliot (1973).

At the present time the Chilean section of the circum-Pacific margin is one of the least studied and least understood. It is also one of the most interesting because of the unique aspect of a major collision boundary well removed from a spreading ridge system and the presence of a deep trench system adjacent to a major continental margin with large mountain chains and extreme crustal thicknesses (see Katz, this volume).

ACKNOWLEDGMENTS

The work was supported by the Office of Naval Research under contract N00014-67-A-0004, and the National Science Foundation through grants GA-27281 and GD-40896. I thank J. Mammerickx Winterer for the prepublication release of the map shown in Figure 2. J. Ewing kindly provided unpublished seismic-reflection data incorporated into Figure 3, and N. Opdyke and E. Herron reviewed the manuscript and provided valuable comments. Several key references are incorporated in the selected bibliography which were not cited in the text.

This is Lamont-Doherty Geological Observatory Contribution 2100.

BIBLIOGRAPHY

Aldrich, L. T., Tatel, H. E., Tuve, M. A., and Wetherill, G. W., 1958, The earth's crust: Carnegie Inst. Wash. Yearbook, v. 57, p. 104-111.

Allen, C. R., 1965, Transcurrent faults in continental areas: Phil. Trans. Roy. Soc. London, v. 258, p. 82-89.

Atwater, T., 1970, Implications of plate tectonics for the Cenozoic tectonic evolution of western North America: Geol. Soc. America Bull., v. 81, p. 3513-3536.

Barker, P. F., 1970, Plate tectonics in the Scotia Sea region: Nature, v. 228, p. 1293-1296.

Benioff, H., 1954, Orogenesis and deep crustal structure—additional evidence from seismology: Geol. Soc. America Bull., v. 65, p. 385-400.

Casertano, L., 1963, Catalogue of the active volcanoes of the world including Solfatara Fields, pt. XV: Rome, Internatl. Assoc. Volcanology, 55 p.

Dalziel, I. W. D., and Elliot, D. H., 1973, The Scotia arc and Antarctic margin, *in* Nairn, A. E. M., and Stehli, F. G., eds., The ocean basins and margins, v. 1: New York, Plenum, p. 171-246.

Fisher, R. L., and Raitt, R. W., 1962, Topography and structure of the Peru-Chile Trench: Deep-Sea Research, v. 9, p. 423-443.

Forsyth, D., and Chapman, E., 1974, Focal Mechanisms in the South Atlantic and Scotia Sea (abstr): Am. Geophys. Union Trans., v. 55, no. 4, p. 446.

Grow, J. A., and Bowin, C. O., 1974, Evidence for high density mantle beneath the Chile Trench due to the descending lithosphere: Jour. Geophys. Res., in press.

Hantke, G., and Parodi, I. A., 1966, Catalogue of the active volcanoes of the world including Solfatara Fields, pt. XIX: Rome, Internatl. Assoc. Volcanology, 73 p.

Hayes, D. E., 1966, A geophysical investigation of the Peru-Chile Trench: Marine Geology, v. 4, p. 309-351.

______, and Ewing, M., 1970, Pacific boundary structure, *in* Maxwell, A., ed., The sea, v. 4: New York, Wiley, p. 29-72.

______, and Talwani, M., 1972, A geophysical study of Macquarie-Ridge-Complex, *in* Hayes, D. E., ed., Antarctic oceanology: II. The Australian-New Zealand sector: Washington, D.C., Am. Geophys. Union, Antarctic Res. Ser., v. 19, p. 211-234.

______, Talwani, M., Houtz, R., Pitman, W. C. III, and Meijer, R. R. II, 1972, Preliminary report of volume 22, U.S.N.S. ELTANIN, *in* Ewing, M., ed., Lamont-Doherty survey of the World Ocean: Tech. Rept. CU-1-72, 232 p.

Heezen, B. C., Tharp, M., and Ewing, M., 1959, The floor of the oceans: I. The North Atlantic: Geol. Soc. America Spec. Paper 65, 122 p.

Herron, E. M., 1971, Crustal plates and sea floor spreading in the southeastern Pacific, *in* Reid, J. L., ed., Antarctic oceanology I: Washington, D.C., Am. Geophys Union, Antarctic Res. Ser., v. 15, p. 229-237.

______, 1972a, Sea-floor spreading and the Cenozoic history of the east-central Pacific: Geol. Soc. America Bull., v. 83, p. 1671-1692.

______, 1972b, Two small crustal plates in the South Pacific near Easter Island: Nature Phys. Sci., v. 240, p. 35-37.

______, and Hayes, D. E., 1969, A geophysical study of Chile Ridge: Earth and Planetary Sci. Letters, v. 6, p. 77-83.

______, and Heirtzler, J. R., 1967, Sea-floor spreading near the Galapagos: Science, v. 158, p. 775-780.

Houtz, R., Ewing, M., Hayes, D., and Naini, B., 1973, Sediment isopachs in the Indian and Pacific ocean sectors (195°E to 70°W): New York, Am. Geogr. Soc., Antarctic Map Folio Ser., Folio 17, p. 9-12.

International Upper Mantle Project Scientific Report 1970, Planning seminar on the Andean geophysical program and geologic and geophysic related problems: Conf. on Solid Earth Problems, Buenos Aires, Argentina, Oct. 26-31, Rept. 37-I, v. 1, 232 p.

James, D. E., 1971, Andean crustal and upper mantle structure: Jour. Geophys. Res., v. 76, p. 3246-3271.

Julivert, M., 1973, Les traits structuraux et l'evolution des Andes Colombiennes: Rev. Geogr. Phys. et Geol. Dynamique, v. XV, p. 143-156.

Karig, D., 1974, Tectonic erosion at trenches: Earth and Planetary Sci. Letters, v. 21, p. 209-212.

Katz, H. R., 1971, Continental margin in Chile—is tectonic style compressional or extensional?: Am. Assoc. Petr. Geol. Bull., v. 55, p. 1753-1758.

Kulm, L. D., Resig, J. M., Moore, T. C., Jr., and Rosato, V. J., 1974, Transfer of Nazca Ridge pelagic sediments to the Peru continental margin: Geol. Soc. America Bull., v. 85, p. 769-780.

Lister, C. R. B., 1971, Tectonic movement in the Chile Trench: Science, v. 173, p. 719-722.

Mammerickx, J., Smith, S. M., Taylor, I. L., and Chase, T. E., 1974, Bathymetry of the South Pacific: Scripps Inst. Oceanography, IMR Tech. Rept. 48A, 52A, 53A, 54A.

Myers, J. S., 1974, Cretaceous stratigraphy and structure, western Andes of Peru between latitudes 10°-10°30': Am. Assoc. Petr. Geol. Bull., v. 58, p. 474-487.

Nazca Plate Project Progress Report, 1973, Appendix A: Joint Proj. Hawaii Inst. Phys. and Oregon State Univ., Program Internatl. Decade Ocean Expl., 128 p.

Noble, D. C., McKee, E. H., Farrar, E., and Petersen, U., 1974, Episodic Cenozoic volcanism and tectonism in the Andes of Peru: Earth and Planetary Sci. Letters, v. 21, p. 213-220.

Ocola, L. C., Meyer, R. P., and Aldrich, L. T., 1971, Gross crustal structure under Peru-Bolivia altiplano: Earthquake Notes, v. XLII, no. 3-4, p. 338-48.

Raff, A. D., 1968, Sea-floor spreading: another rift: Jour. Geophys. Res., v. 73, p. 3699-3706.

Sass, J. H., Munroe, R. J., and Moses, T. H., Jr., 1974, Heat flow from eastern Panama and northwestern Colombia: Earth and Planetary Sci. Letters, v. 21, p. 134-142.

Scholl, D. W., von Huene, R., and Ridlon, J. B., 1968, Spreading of the ocean floor: undeformed sediments in the Peru-Chile Trench: Science, v. 159, p. 869-871.

———, Christensen, M. N., von Huene, R., and Marlow, M. S., 1970, Peru-Chile Trench sediments and sea-floor spreading: Geol. Soc. America Bull., v. 81, p. 1339-1360.

Sillitoe, R. H., 1972, A Plate Tectonic Model for the Origin of Porphyry Copper Deposits, Economic Geology, v. 67, pp. 184-197.

Stauder, W., 1973, Mechanism and spatial distribution of Chilean earthquakes with relation to subduction of the oceanic plate: Jour. Geophys. Res., v. 78, p. 5033-5061.

Sykes, L. R., and Hayes, D., 1971, Seismicity and tectonics of South America and adjacent oceanic areas (abstr.): Geol. Soc. America, v. 3, p. 206.

Vergara, M., and Munizaga, F., 1974, Age and evolution of the Upper Cenozoic andesitic volcanism in central-south Chile: Geol. Soc. America Bull., v. 85, p. 603-606.

Watts, A., and Talwani, M., 1974, Gravity anomalies seaward of deep sea trenches: Geophys. Jour., v. 36, p. 57-90.

Wuenschel, P. C., 1952, Gravity measurements and their interpretation in South America between latitudes 15° and 33° south [thesis]: Columbia Univ., New York, 191 p.

Zeigler, J. M., Athearn, N. D., and Small, H., 1957, Profiles across the Peru-Chile Trench: Deep-Sea Res., v. 4, p. 238-249.

Geology of a Part of the Pacific Margin of Chile

Carlos Mordojovich K.

INTRODUCTION

An exploration program of the continental shelf of the Chilean Pacific Coast has been carried on since 1970 by Empresa Nacional del Petróleo of Chile (ENAP). This program included bathymetry, aeromagnetic and seismic surveys, and exploration drilling. The area investigated is located between 35 and 40°S, a distance of about 600 km.

The stratigraphic sequence of the Arauco inshore well, Tubul 1, and of the six recent offshore wells are summarized. A regional map showing the coastal zone geology, the shelf morphology, the location of the offshore exploration wells, and three offshore seismic sections illustrate the text.

A sedimentary basin located between the metamorphic rocks of the coastal range and the shelf edge is described in a very general fashion, showing that more than 3,000 m of clastic sediments ranging from upper Cretaceous to Recent have accumulated in a rather irregular depression along the shelf.

Most of the coastal zone is a high range formed by granitic and old metamorphic rocks. Nevertheless, several small areas are present where sedimentary rocks, ranging from Pliocene to upper Cretaceous, are found. From Valparaiso to Isla Chiloé we find:

1. *Algarrobo:* outcrops of Upper Cretaceous shales and sandstones are found unconformably on top of granitic rocks along the coast (33½°S).

2. *Navidad:* several hundred meters of Miocene and Pliocene shales and sands are found in a tectonic depression, lying on top of a granitic basement (34°S).

3. *Chanco Bay:* a graben-type depression in metamorphic slates, filled by up to 700 m of Upper Cretaceous and Eocene marine sandstones and shales (35½°S).

4. *Arauco Peninsula:* part of the shelf is uplifted by faulting, where a thick series of continental and marine Pliocene to Upper Cretaceous sediments crops out. Exploration drilling has shown about 2,200 m of sedimentary rocks, lying on top of metamorphic slates. This is the main coal mining district of Chile.

5. *Toltén Delta:* at the mouth of the Tolten River some Pliocene marine sediments have been found. Geophysical surveys show that about 1,000 m of Tertiary sediments might exist above the metamorphic rocks.

6. *Chiloé Island:* on the southwest coast, facing the Pacific Ocean, some cliffs are made up of Pliocene to Miocene sedimentary rocks; most of the island has a metamorphic core.

All these outcrops, while discontinuous, pointed toward the existence of a more-or-less continuous sedimentary basin in the continental shelf, but no positive evidence was obtained until geophysical surveys were conducted by several research vessels. This is an unknown part of the complex continental margins facing the Pacific Ocean. As soon as the presence of a thick sedimentary sequence of folded and faulted rocks was evident, ENAP became interested in further studies of the continental shelf.

In close collaboration with the United Nations Development Programme a geophysical exploration program was defined in 1969, and started in 1970 with an aeromagnetic survey, followed in 1971 by a detailed seismographic survey that covered the shelf from Constitución (35°S) to Valdivia (40°S) and from the coast out to the 200-m isobath. The aeromagnetic survey and interpretation was assigned by contract to Aeroservice Corp. and the seismographic survey was assigned to United Geophysical Co. Following the geophysical survey, a contract was signed with Santa Fe Drilling Co. for the rental of the semisubmersible drilling barge "Blue Water II," which accomplished the drilling of six exploration wells in the Chilean Pacific shelf, from February to November 1972. The six wells reached the metamorphic basement complex and provided valuable geological information.

AEROMAGNETIC SURVEY

A semidetailed aeromagnetic survey was flown over the shelf from 35 to 40°S with east-west lines every 5 km and some tielines north-south. Some reconnaissance lines were also flown north and south of this area. Its purpose was to investigate the depth to magnetic basement in the shelf, as a first step for a more detailed petroleum exploration program. A careful sampling of sedimentary, metamorphic, and granitic rocks cropping out along the coastal zone was conducted, and magnetic susceptibility was measured.

The sampling showed surprising results. The granitic and metamorphic rocks had very low magnetic susceptibilities: 14 granite samples with an average of 24 K X 10^{-6} and 35 samples of mica slate with an average of 40 K X 10^{-6} units, while the sediments showed rather high susceptibilities. Tertiary and Cretaceous rocks showed 100-200 K X 10^{-6}

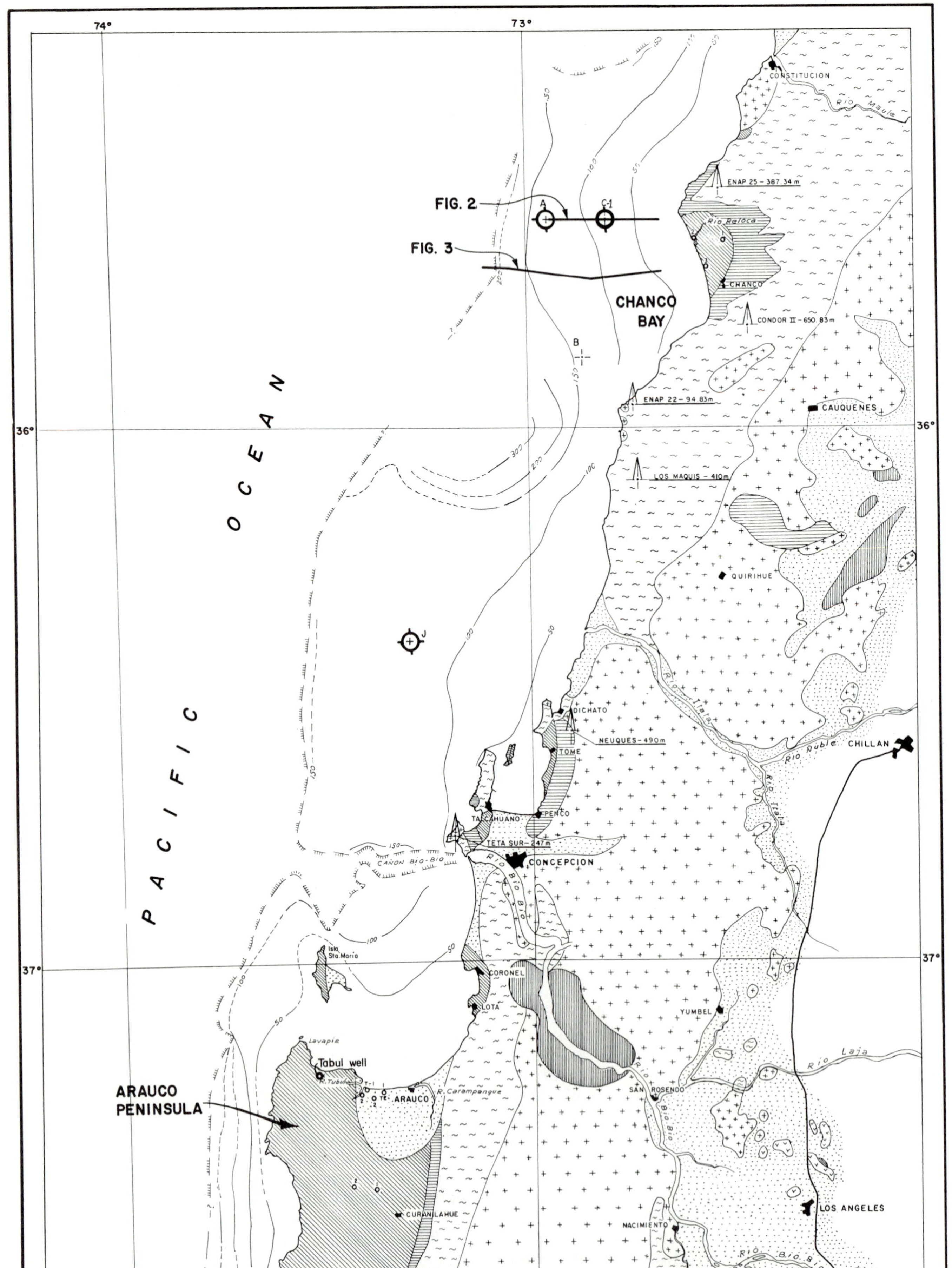

Fig. 1a. Geological map of coastal Chile between 35°S and 37°35'S.

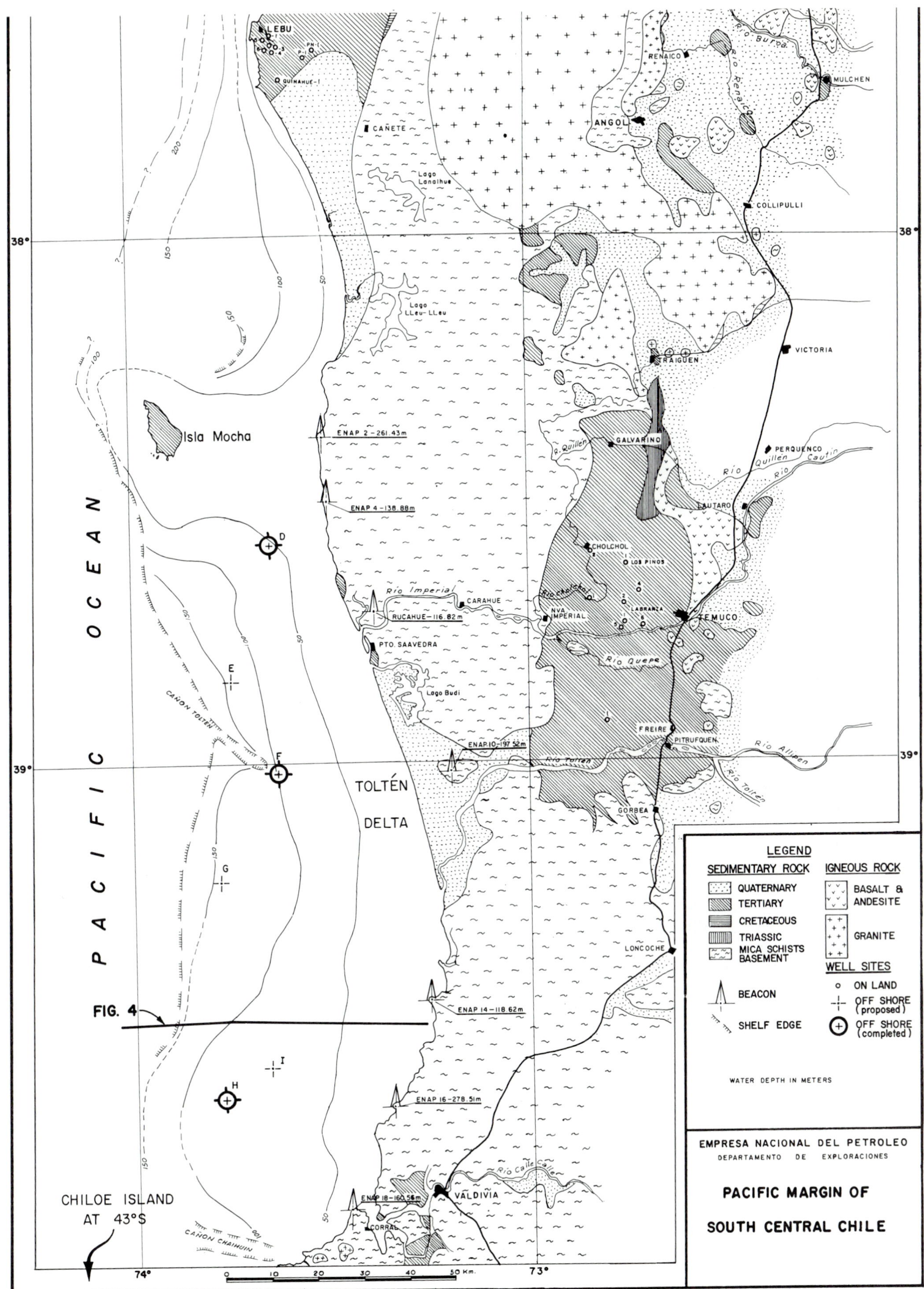

Fig. 1b. Geolgical map of coastal Chile between 37°35'S and 40°S.

units, and Pleistocene to Recent deposits, made up of volcanic sands carried by the rivers from the Andes to the coast, showed values well over 1,000 K X 10^{-6} units. These abnormal magnetic conditions caused a difficult interpretation problem in relation to total thickness of sediments, but they gave some interesting information about the distribution pattern of recent sediments on the shelf.

High-frequency anomalies are present, aligned west-northwest from the mouths of the most important rivers (e.g., Rio Itata and Bio Bio). Nevertheless, filtering techniques allowed computations of sedimentary thicknesses of up to 3,000 m in certain sectors of the shelf. Also some fault patterns were postulated, giving the general picture of a graben-like sedimentary basin parallel to the coast, complicated by some east-west faults.

MORPHOLOGY OF THE SHELF

The seismic survey produced the first detailed bathymetric map of the Chilean shelf, in addition to the very interesting deeper data (Fig. 1). Between 35°15'S and 40°S, the shelf has an area of 22,638 km^2. The outer border (shelf edge) appears at 40-55 km from the coast, except in front of the Arauco Peninsula, where it is not wider than 12 km. If we consider that this is an uplifted part of the shelf, we still have about 40-50 km from the shelf edge to the metamorphic rocks cropping out to the east of the Arauco Peninsula. About 26% of the shelf has less than 50 m of water, 27% has between 50 and 100 m, and 47% has more than 100 m of water.

Four repetitive geomorphologic features are present on this shelf. Arauco Bay, a wide depression open toward the north (37°10'S), repeats itself under the sea in Constitucion (35°S), north of Itata River (36°10'S) and north of Mocha Island (38° 15'S). The real meaning of this peculiar morphology is not well understood, but it is certainly related to deep tectonic phenomena. It is also apparent that each sector between these depressions is about 110 km long and has had different stratigraphic and tectonic development.

A deep and well-known canyon is present at the mouth of Bio Bio River, the largest and most important river of this region (near 37°S). A very interesting canyon starts 30 km offshore at 39°S, in front of Tolten River, but no evidence of this canyon is seen closer to shore; it only exists below the 120-m isobath. It probably developed during glacial times, when much stronger erosion and low-sea-level conditions prevailed. Another canyon, just south of Valdivia, was detected (near 40°S), but its origin does not seem related to the Rio Valdivia, rather to a downfaulted block seen in the coastal range south of Valdivia (Chaihuin Canyon).

Between Mocha Island and Toltén River (38°-30'S to 39°15'S) a distinct fracture zone is seen parallel to the coast and 7-10 km from it. Bottom relief dips smoothly to the west at a rate of 3-4 m/km, but the fracture zone shows one or more very sharp cliffs, down to the east, with a relief of 2-7 m. It has all the appearance of very recent faulting. The bottom sediments here are nonconsolidated sand and mud. It must be pointed out that during the Valdivia earthquake of May 1961, the coastal zone went down about 1.5 m. Since then, ocean-going boats can dock at the Valdivia River port.

SEISMIC SURVEY AND EXPLORATION DRILLING

From April to July 1971, 117 lines with 4,574 km of continuous reflection profiles between 35 and 40°S were digitally recorded. Subsequent processing was performed by Geocom, and interpretation was provided by the United Geophysical Interpretation Center, and later by ENAP and United Nations geophysicists. Line locations were determined by Satellite and Doppler-Sonar Navigation System. Gas exploders were used as energy sources, and the data were recorded digitally, with multifold coverage. A few experimental refraction lines were also shot. The quality of the sections is variable, depending mostly on local geological conditions. Although some rough weather was encountered, it did not seem to be fundamental in relation to data quality.

Chanco Area

A strong reflection from metamorphic basement could be mapped (35°15'S to 35°50'S). The area is strongly fractured by high-angle and large-throw faults, usually trending northeast. The sedimentary section is thicker at the center of the shelf, thinning both east and west. This basin plunges toward the north, where more than 3,000 m were measured, while in the southern part no more than 1,000 m are present. A basement high flanks the shelf border. To the west of this high, very poor reflections were obtained, but it is our opinion that thick sediments are also present here, obscured by very complex tectonics and highly indurated sediments.

On land, the Chanco area appears as a downfaulted block, limited by two main northeast-trending faults. Up to 700 m of Eocene to Upper Cretaceous sediments have been mapped and logged by exploration drilling. The sediments consist of about 400 m of Eocene shales with a rather loose basal sand of medium- to fine-grained quartz and glauconite. It is followed by about 200-300 m of soft sandy shale with Upper Cretaceous ammonites. At the base of the section a very thick, clean, coarse-grained quartz sand is present, usually about 80-100 m thick. The Cretaceous rocks rest unconformably on the metamorphic slates. North and south of the bay, some granite stocks which intruded the slates have been mapped. On the east flank of the coastal range the granite intrusive is present as a continuous batholith. Exploration wells A and C-1 were drilled on the shelf (Fig. 1) close to the seismic line of Figure 2.

Well A, 29 km offshore, found some 100 m of Recent to Pliocene sediments, mainly tuffaceous, light-colored, silty clays; then about 1,350 m of Miocene marine clays, with a few limestone beds; followed by about 240 m of shales with some fine-grained, glauconite sands. Mica schist was found at the base of this sequence, with no signs of a basal conglomerate or sand. The age of the lower sedimentary section is presumable Eocene or Upper Cretaceous. It is probable that the lowest section has been cut out by faulting.

Well C-1 was drilled to the east of Well A (Fig. 2). It found probable Pliocene sands and sandy shales down to 355 m and then about 230 m of Miocene shales, followed by shales and siltstones of probable Eocene age. Cretaceous sediments, mainly fine, calcareous sands, were found at 1,075 m, and a coarse-grained, clean, quartz sand about 60 m thick was found at the base of the section, on top of the metamorphic slates. A heavy flow of salt water with dissolved gas came from this sand. This sand is identical to the basal Cretaceous sand found in the onshore outcrops and wells.

Rio Itata-Rio Bio-Bio Area

Here the shelf is wider, reaching a maximum of 60 km. Metamorphic basement is present at the coast. East-west sections show a rapid deepening of the basement to over 3,000 m at the center of the shelf, and then it seems to come up toward the shelf edge. South and west relationships are obscured by very poor reflection quality.

The basin has an upper section of nonfolded sediment with a maximum thickness of about 1,100 m, underlain by folded and faulted rocks of great thickness. On the east side of the shelf, basement reflections show a general west dip with step faults down to the east. This horizon is obscured on the west flank and probably marked by the complex structure of the lower sedimentary section.

Well J was drilled on the west flank of the basin (Fig. 1). This well found Recent to Pliocene sediments down to 360 m, made up of soft, tuffaceous, and fossiliferous clays, underlain by about 1,450 m of Miocene marine shales, slightly glauconitic. Between 1,807 and 2,460 m of glauconitic fine-grained sands and sandy shales of possible Eocene age were found. Below the glauconitic sediment, gray silty shales with frequent *Inoceramus* prisms were found (Upper Cretaceous). Some limestone beds are present in the lower part of the Cretaceous shales. At 3,192 m a fine and tight quartz sand 40 m thick was drilled, followed by 20 m of coarse-grained, hard, well-cemented sand, resting unconformably on micaceous slates. This lower section had interesting shows of natural gas.

Arauco Basin and Arauco Peninsula

The geology of this area (37 to 38°S) is well known as the main coal basin of Chile, which has been in production for more than a century. In the eastern foothills of the coastal range, some sporadic outcrops of upper Cretaceous, usually shaly sands and conglomerates, rest unconformably on the metamorphic basement, but they are usually overlapped by lower Eocene coal-bearing rocks. A maximum of about 2,200 m of sediments has been found by exploration drilling in the Arauco Peninsula, with a sequence that extends from Pliocene to Upper Cretaceous.

The stratigraphy of the Arauco area is rather uniform, but quite different from the rocks found to the north. Apparently Araúco has been a site of continuous deposition, but it has been filled in part with continental sediments during Eocene time. Miocene and Pliocene are present north and south in the Arauco Peninsula, but are absent or very thin elsewhere (Fig. 1). Oligocene seems to be absent, or represented only by thin glauconitic sands below the Miocene marine shales.

The Eocene and Cretaceous rocks are highly indurated and compact (density of shales is 2.4 -2.5 g/cm^3 and acoustic velocities of 3.5-4.0 km/sec). No compressional folding is present, but strong normal faulting, usually north-trending, is very frequent. An erosional unconformity has been reported at the base of the Miocene beds, and Pliocene sediments are lying flat on tilted and faulted Miocene or older rocks.

The highly indurated Eocene and Upper Cretaceous rocks of Arauco contrast sharply with rocks of the same age in Chanco and the rocks found in the offshore "J" well. This might well be due to a very thick sedimentary load that was partly removed by uplift and erosion before Miocene in the Arauco Peninsula.

Well Tubul 1, 10 km west of the town Arauco (Fig. 1) encountered the following section: 200 m of light-gray, tuffaceous, Pliocene marine clay with some coarse, well-rounded quartz sand near the base, unconformably overlying 470 m of light-gray shale, with few fine-grained light color sand beds of Miocene age. These are underlain successively by 182 m of glauconitic shale, 97 m of fine glauconitic sandstone and sandy shale of Eocene age, and 210 m of continental sands with frequent volcanic ash beds also of Eocene age. (This Eocene formation is coal-bearing in the Lebu district, 50 km southwest, and at Trihueco mine, on the east flank of this basin.) This is underlain by 421 m of hard marine Eocene shale, and then by 347 m of hard, greenish sandstones and shale, mostly continental, with some shallow-marine and coal beds all of Eocene age.

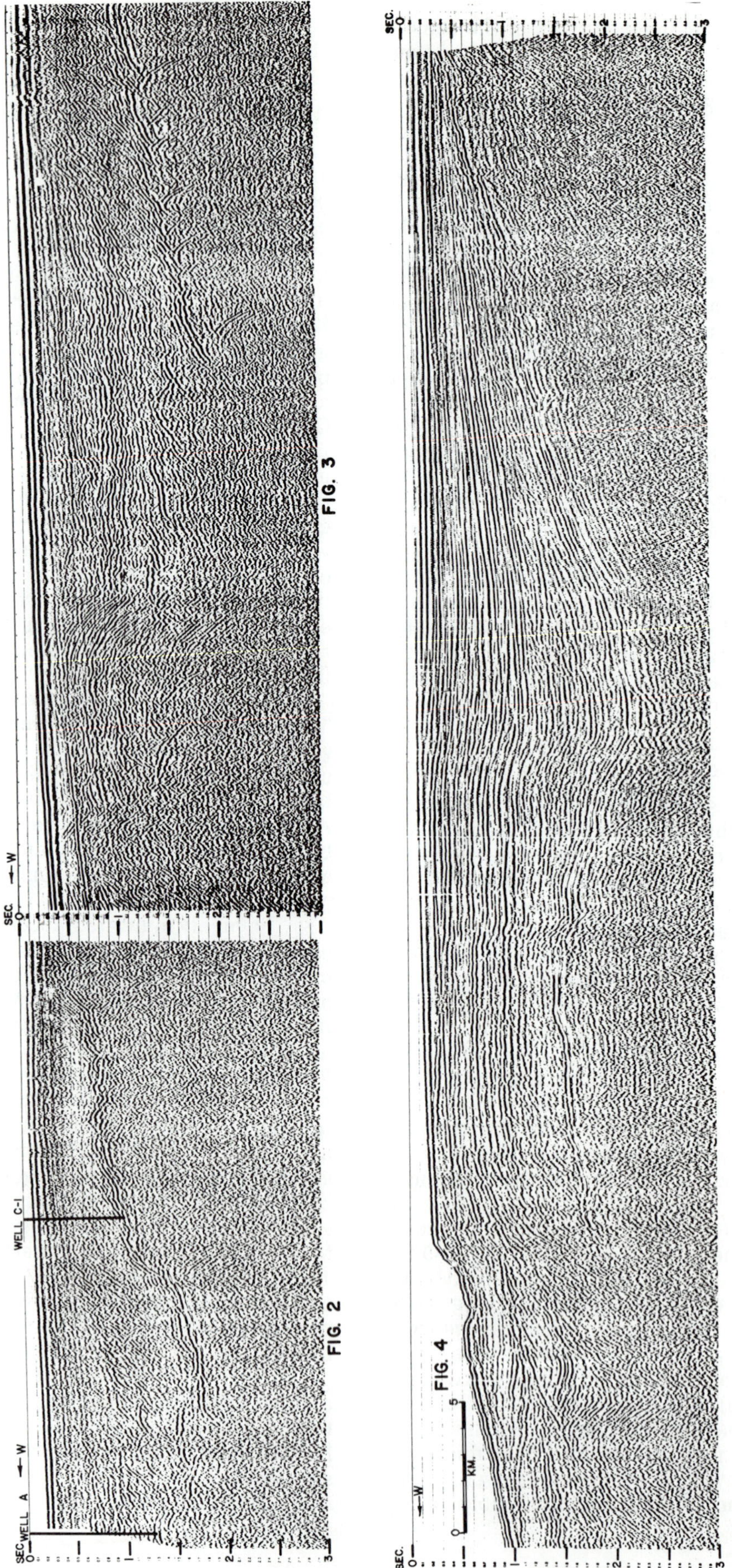

Fig. 2-4. Seismic profiles on Chilean continental margin. Upper two profiles in northern part of region, and lower profile in southern part (location shown on **Figure 1**).

This is the main coal-producing formation of the Arauco Basin, and coal is presently being produced on the east flank of the basin. The underlying Cretaceous strata consists of 268 m of micaceous shale with some brown limestone beds; the lower 73 m are quite sandy. A coarse quartz sandstone (12 m thick) is the base of the sedimentary section, resting unconformably on metamorphic slate. This lower sandy section has some gas production in the Lebu area.

Isla Mocha to Valdivia

South of Arauco Peninsula the metamorphic rocks of the coastal range reach the Pacific Coast. Before the recent offshore drilling, very little was known of the stratigraphy of this zone, with the exception of Isla Mocha. Here, more than 1,000 m of sediments are exposed along the coast. The rocks dip about 20° west or northwest and consist mainly of Miocene tuffaceous sandstones and shales. The west coast of the island is highly faulted, and some hard, carbonaceous shale outcrops have been tentatively assigned an Eocene age. Several gas seepages have been found all along the coast of the island. The core of the island consists of flat-lying Pliocene beds, similar to the Pliocene in the Arauco area. Undeformed Pliocene beds are also present over the metamorphic basement in a few places along the coast in this mainland.

The seismic survey shows a north-south elongated and oval sedimentary basin in the shelf between Mocha and Valdivia. It is a very shallow basin between Mocha and the mainland, plunging gradually southward to a maximum depth of about 2,000-2,200 m in the center, emerging toward Valdivia. Sections also show a rapid deepening toward the west, and then shallowing again at the west edge of the shelf (Fig. 4).

The reflection character of the seismograms of this area shows two sedimentary sections: an upper one, unfolded, thickening gradually toward the center of the basin, where it reaches about 1,200 m thick, and a lower, slightly folded series up to 1,000 m thick. A very evident angularity exists between these two sedimentary series.

Of the three exploration wells drilled in this sedimentary basin, *Well "F"* was the most interesting. It showed that the upper, unfolded Pliocene series, consisting mainly of shales and glauconitic sands, was about 700 m thick. At 600 m a basaltic layer about 5 m thick, probably a breccia, was found. Below the Pliocene, marine Miocene shales were found. At the base of this series, a 23-m-thick sand, made up of rounded quartz grains, was found. A drill-stem test showed potential gas production at a rate of 220,000 m^3/day.

The age of the 160 m of sediments found below the porous sand in Well "F" is unknown; it might represent the Eocene. No evidence of Cretaceous sediments has been found in this subbasin.

Between the base of this sand (1,500 m) and the basement, (1,660 m) tight, fine-grained sandstones and sandy shales were found. A 10-m basal conglomerate of quartz and slate fragments was present over the basement micaceous slate.

CONCLUSIONS

There is a more-or-less continuous sedimentary basin on the south-central Chilean Pacific Shelf. It seems to be divided by tectonic movements into separated subbasins, approximately 110 km long.

Sediments of volcanic origin are abundant in the Pliocene and Quaternary and are sparsely present in the rest of the sedimentary column. No direct evidence of in situ volcanic rocks has been found. Limestones are practically absent; the main part of the sediments are shales, glauconitic shales, and quartz sandstones. Continental facies are known only in the Arauco subbasin, formed during Eocene time. Upper Cretaceous and younger sediments are present from Arauco northward. Eocene sediments are known as far south as Isla Mocha, south of which only Miocene and younger sediments are known.

The structure is very complex from Chanco to Arauco, but apparently much simpler in the Mocha-Valdivia subbasin. The basement dips from the coast line to a maximum depositional depth at the center of the shelf, and then seems to rise westward toward the shelf edge.

Metamorphic and acoustic basement is easily recognized in the Valdivia subbasin, where Eocene and Cretaceous rocks are absent and the structure is simple. Maximum thickness of unconsolidated sediments is about 2,200 m. North of Mocha Island, Miocene sediments overlap with angular unconformity older Eocene and Cretaceous sediments, which are highly indurated and faulted, and the acoustic basement cannot be mapped with certainty. Nevertheless, sporadic reflections and evidence from drilling shows that the thickness of the sedimentary section is well over 3,000 m.

ACKNOWLEDGMENTS

A very important contribution to our knowledge of the Chilean continental shelf was the survey conducted by the U.S. R/V CHARLES H. DAVIS in 1967. Through the particularly fine cooperation of D. W. Scholl and R. von Huene, ENAP geologists and geophysicists were able to collect information during and after this survey.

Published by permission of Empresa Nacional del Petróleo. (Chile)

BIBLIOGRAPHY

Bello, A., ed., 1968, El Terciario de Chile; Zona Central: Santiago, Sociedad Geológica de Chile.

Di Biasi F., F., and Lillo, R., F., 1968, Geología regional, geoquímica de drenaje minería de la Provincia de Valdivia: Santiago, Instituto de Investigaciones de Recursos Naturales.

Brüggen, J., 1950, Fundamentos de la geología de Chile: Santiago, Instituto Geográfico Militar.

Gonzáles-Bonorino, F., 1962, Nuevos datos de edad absoluta del basamento cristalino de la cordillera de la costa, Chile central: Fac. Ciencias Fís. y Mat. Univ. Chile.

Mapa Geológico de Chile, 1968: Instituto de Investigaciones Geológicas.

Munoz Cristi, J., 1950, Geografía Económica de Chile: Santiago, Corporación de Fomento de la Producción, Chap. 3.

———, 1969, Estado actual del conocimiento de la geología de la provincia de Arauco: Anales Fac. Cienc. Fís. y Mat. Univ. Chile, t. 3, p. 30-63.

Scholl, D. W., and von Huene, R., 1969, Geologic implications of Cenozoic subsidence and fragmentation of continental margins (abstr.): Am. Assoc. Petr. Geol. Bull., v. 53, p. 740.

———, Christensen, M. N., von Huene, R., and Marlow, M. S., 1970, Peru-Chile trench sediments and sea-floor spreading: Geol. Soc. America Bull., v. 81, p. 1339-1360.

Tavera, J., 1968, Contribución al estudio de la estratigrafía y paleontología del Terciario de Arauco: Anales Primer Congr. Pan Americano Minas y Geología, v. 2, p. 580-632.

———, and Veyl, C., 1958, Reconocimiento geológico de la Isla Mocha: Anales Fac. Cienc. Fís. y Mat. Chile, v. 14-15.

Continental Margin of Middle America

George G. Shor, Jr.

INTRODUCTION

The continental margin of "Middle America" is one of the classic areas of "Pacific Margin" structure. Defined by its dominant feature, the Middle America Trench (named by Heacock and Worzel, 1955), it extends from the coast of the Mexican state of Jalisco to the southern coast of Costa Rica (Fig. 1). Structurally, it forms the boundary between the Americas Plate and the small Cocos Plate. It possesses most of the elements postulated by Gutenberg and Richter (1954) as being typical of a Pacific continental margin: high mountains, a volcanic belt, a deep oceanic trench, and earthquake foci of varying depths, ranging from shallow shocks directly beneath the trench to intermediate-depth shocks on the landward side, forming a classic "Benioff zone" (Benioff, 1949, 1954, 1962). Features lacking that were thought to be typical of "Pacific margins" (now known as subduction zones) are an island arc and truly deep-focus earthquakes.

Geophysical exploration of the continental margin of Middle America was begun earlier than that of most of the other trench-bounded margins, primarily because of the easy accessibility of the area. In 1926 Vening Meinesz (1948) measured the gravity field and found the typical negative anomalies to be widely associated with trenches. During the late 1940s to early 1960s the margin was the subject of additional gravity investigations (Worzel and Ewing, 1952), detailed topographic surveys (Fisher, 1957; Shor and Fisher, 1961), heat-flow surveys (Bullard et al., 1956), and seismic refraction (Shor and Fisher, 1961) and seismic reflection (Fisher, 1953; Ross and Shor, 1965). Most of these studies were summarized by Fisher and Hess (1963). Perhaps because of these early investigations, the area has been somewhat neglected in recent years; current geophysical investigations of crustal structure have been concentrated mostly on the western and southern boundaries of the Cocos Plate and along the continental margins of the Gulf of California to the north and the Nasca Plate to the south.

BATHYMETRY

The topography of the area is relatively uniform. The continental shelf is narrow in the north and widens at the Gulf of Tehuantepec, where the coastline swings to the northeast; the outer edge of the shelf and the trench axis are almost straight. The continental slope is extremely steep, particularly along the northwestern half, with occasional benches part way down the slope. The trench itself is narrow and has only small accumulations of recent sediment in the bottom, except at the Tres Marias Basin at the northwestern end. The outer "wall" of the trench is topographically rough in most areas. There is ponding of sediment in the bottom of the trench, particularly in the northern half, but the sediment apparently has not filled all depressions to form a continuous downward slope of the axis from the ends toward the deepest locality (off Guatemala). The apparent topography of the trench axis is shown in Figure 2, along with the sub-bottom structure.

Although the trench has been described as arcuate or as sigmoid, the deviation from a straight line on bearing 112° is not great, except at the northern end, where the trench turns north to form the Tres Marias Basin, and at the southern end, where it turns south against the Cocos Ridge. It may continue as a buried feature southeast to Cape Mala, Panama (Ross and Shor, 1965). Seaward of the trench the topography and structure are approximately orthogonal to the trench itself, with a low outer ridge paralleling the trench along much of its length, and two aseismic ridges (the Cocos Ridge and Tehuantepec Ridge) trending perpendicular to the trench axis.

GEOPHYSICAL DATA

Magnetic anomalies 700 km west of the trench have been reported by Herron (1972) with a trend of 140°, about 30° off of a line parallel to the trench; one seismic anisotropy measurement on the Cocos Plate at 09°N 92°W (within Herron's anomalies) has been reported by Shor et al. (in press) with the high-velocity vector 053°, at right angles to the magnetic anomalies; another at 14°N 100°W has the high-velocity vector oriented at 029°, nearly at right angles to the trench. Farther westward the topography and the magnetic anomalies change direction, becoming nearly north-south (parallel to the East Pacific Rise), indicating a past change in spreading direction.

The 1954 seismic refraction studies by Shor and Fisher (1961) southeast of the Gulf of Tehuantepec provided a cross section (Fig. 2), showing a slight thickening of the crust beneath the trench and a nearly horizontal Mohorovicic discontinuity from

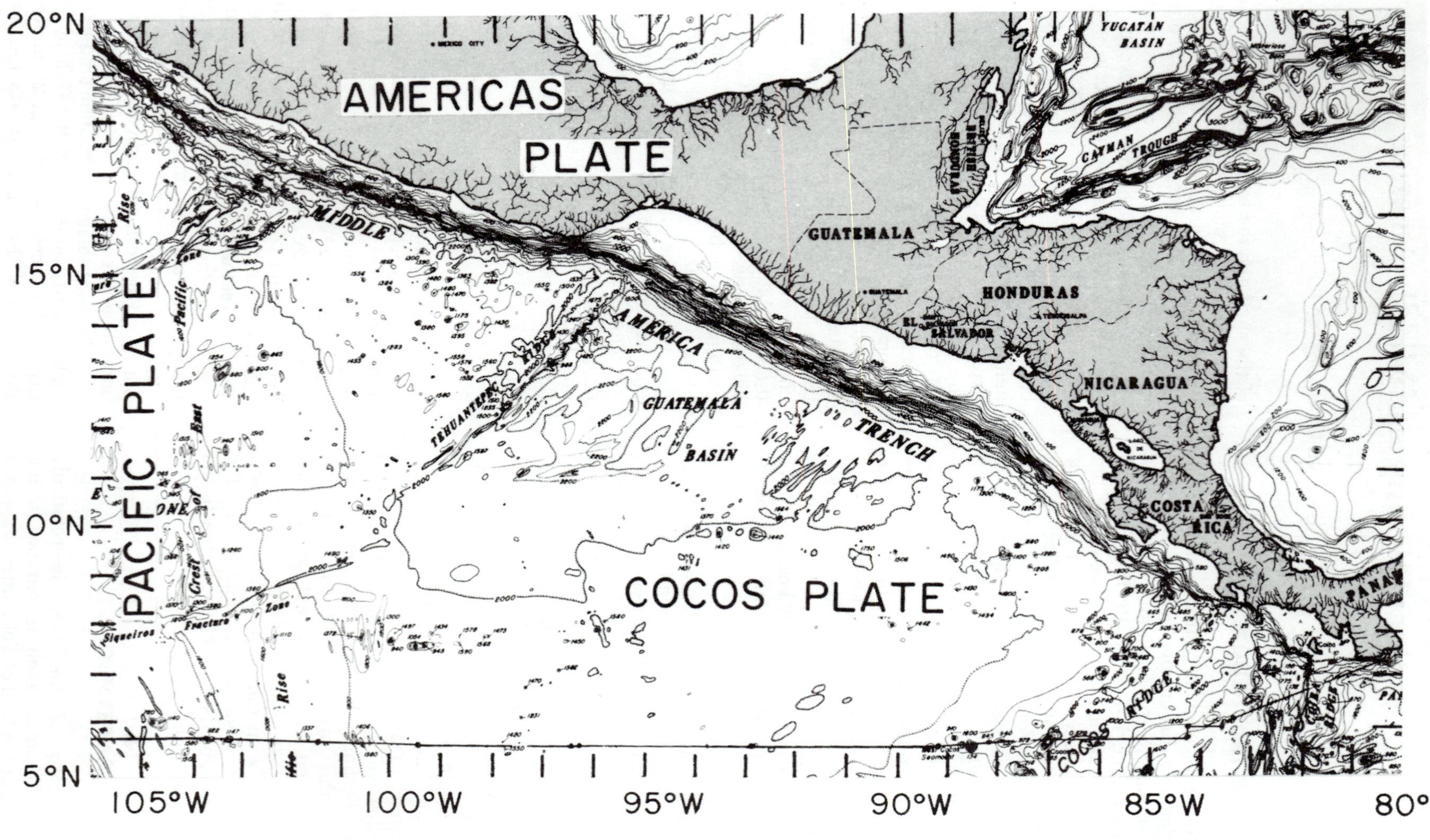

Fig. 1. Topography in the vicinity of Middle America based primarily on data from Fisher (1961).

the trench to the outer shelf. This was a surprise at the time, since it did not conform to the geometry called for by the tectogene hypothesis (Kuenen, 1936); it conforms far better to the present interpretations of the geomety of underthrusting plates in a subduction zone. Shor and Fisher (1961) considered the material under the Tehuantepec shelf, with velocity near 4 km/sec, to be possible basement. It could as well be older sedimentary deposits, and may represent material from the sea floor accreted to the continental mass of North America (see Seely, Vail, and Walton, this volume).

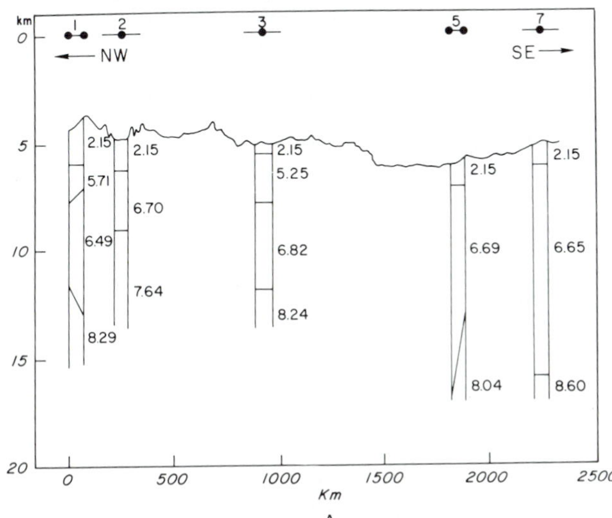

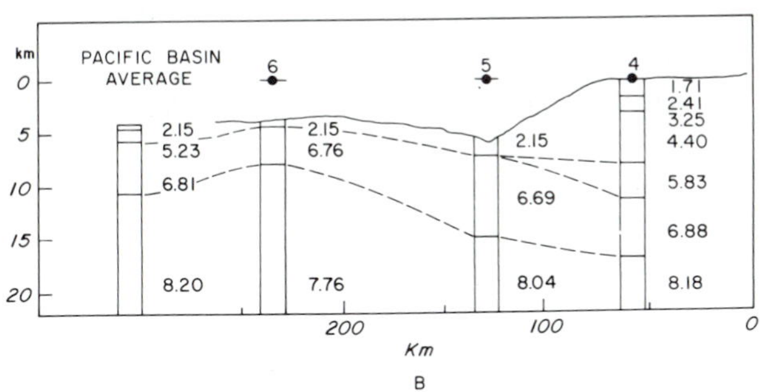

Fig. 2. Structural sections along and across the Middle America Trench modified from refraction solutions in Shor and Fischer (1961). Upper figure is along trench axis. Lower figure is across the trench south of Champerico, Guatemala.

Numerous reflection profiles (mostly unpublished) have been made across the trench and continental shelf of Middle America. The most detailed of these surveys (by Ross and Shor, 1965) showed little evidence of sedimentary material under the continental slope but thick sediments in basins on the shelf itself. In several cases there was evidence for a structural high along the shelf edge that defined the seaward boundary of a shelf basin filled with younger sediments (see Fig. 3 for one example); similar features have been found in many areas elsewhere on the continental shelves of the world. In cases where such a structure was not seen in the profiles by Ross and Shor, there is still a

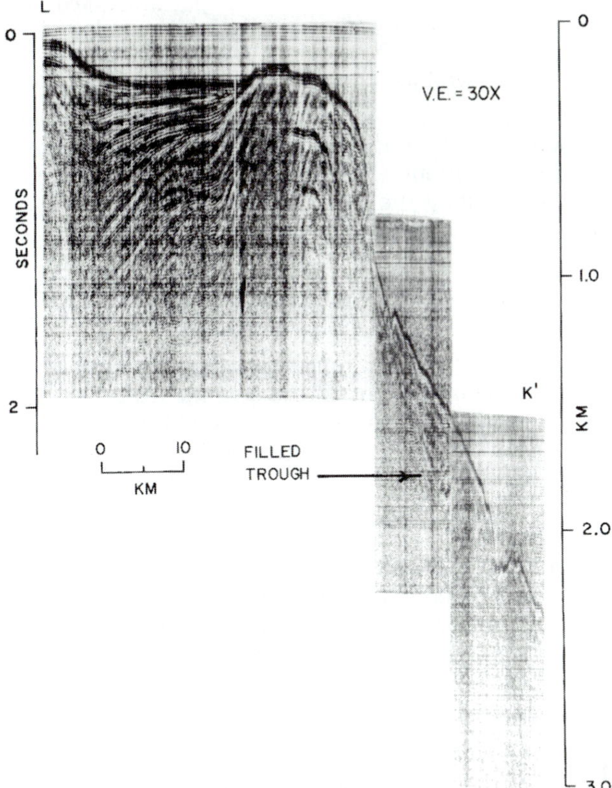

Fig. 3. Arcer profile across Middle America continental margin at Gulf of Tehuantepec. Note the thick sequence of shelf sediments which form a syncline in the middle of the shelf. (Section L of Ross and Shor, 1971, Jour. Geoph. Res., v. 70, Fig. 16, p. 553, copyright by American Geophysical Union).

possibility that the feature is there but was missed because of the low acoustic penetration and lack of detail possible with the equipment used at that time.

A record taken off the coast of Costa Rica in 1972 (Shor, unpublished), using an airgun system with a fast sweep and an AGC-equipped recording system, showed dip reversals immediately inshore of the continental slope. These indications of closure around a shelf-edge high would have been missed with the type of recording used in the earlier surveys. These records and other more recent records of the continental shelf near the Gulf of Tehuantepec (from CATO Expedition; T. Chase, personal communication) show strong evidence for folding and distortion of a thick sequence of sediments under the continental shelf, giving credence to the belief that this part of the shelf, at least, is generally underlain by a thick layer of sedimentary material accreted as the oceanic crust with its sedimentary overburden is carried down beneath the continental plate.

More sophisticated records using multitrace recording and extensive data processing have been reported (by Seely, Vail, and Walton, this volume) from a section of the margin southeast of Tehuantepec; these show a deep reflector under the continental slope and shelf, in about the position of some areas of igneous rock outcrops; there are also greater amounts of sediment fill in the trench bottom here.

The southeastern half of the Middle America continental margin has nearly all the features expected of a subduction zone; the northwestern half lacks evidence for an accumulation of marine sediments along the inner side of the trench.

ACKNOWLEDGMENTS

Much of the work reported here was carried out under various contracts and grants from the U.S. Navy Bureau of Ships, Office of Naval Research, and the National Science Foundation to the Scripps Institution of Oceanography, University of California at San Diego.

I thank Robert L. Fisher for his extensive constructive comments.

This is a contribution from the Scripps Institution of Oceanography, new series; reproductions for any purpose by the U.S. government or the Scripps Institution is permissible despite any copyright notation appearing hereon.

BIBLIOGRAPHY

Benioff, Hugo, 1949, Seismic evidence for the fault origins of oceanic deeps: Geol. Soc. America Bull., v. 60, p. 1837-1856.

———, 1954, Orogenesis and deep crustal structure — additional evidence from seismology: Geol. Soc. America Bull., v. 65, p. 385-400.

———, 1962, Movements on major transcurrent faults, *in* continental drift, v. 3, International geophysics series: New York, Academic Press.

Bullard, E. C., Maxwell, A. E., and Revelle, R. R., 1956, Heat flow through the deep sea floor *in* Advances in geophysics, v. 3: New York, Academic Press.

Fisher, R. L., 1953, Sedimentary fill in two Mexican foredeeps (abs.): Geol. Soc. America Bull., v. 64, p. 1422-1423.

———, 1957, Geomorphic and seismic-refraction studies of the Middle America Trench, 1952-1956 (Ph.D. thesis): Los Angeles, Univ. of California.

———, and Hess, H. H., 1963, *in* Hill, M. N., ed., The sea, v. 3: New York, Wiley—Interscience.

Gutenberg, B., And Richter, C. F., 1954, Seismicity of the earth: Princeton, N.J., Princeton Univ. Press.

Heacock, J. G., Jr., and Worzel, J. L., 1955, Submarine topography west of Mexico and Central America: Geol. Soc. America Bull., v. 66, p. 773-776.

Herron, E. M., 1972, Sea-floor spreading and the Cenozoic history of the east-central Pacific: Geol. Soc. America Bull., v. 83, p. 1671-1692.

Kuenen, P. H., 1936, The negative isostatic anomalies in the East Indies (with experiments): Leidsche Geol. Neded., v. 8, p. 169-214.

Ross, D. A., and Shor, G. G., Jr., 1965, Reflection profiles across the Middle America Trench: Jour. Geophys. Res., v. 70, p. 5551-5572.

Shor, G. G., Jr., and Fisher, R. L., 1961, Middle America Trench: Geol. Soc. America Bull., v. 72, p. 721-730.

———, Raitt, R. W., Henry, M., Bentley, L. R., and Sutton, G. H., in press, Anisotropy and crustal structures of the Cocos Plate: Proc. Symposium on Geodynamics, July 20-24, 1973.

Vening Meinesz, F. A., 1948, Gravity expeditions at sea, 1923-1938: Waltman, Delft, Pub. Neth. Geod. Comm., v. 4.

Worzel, J. L., and Ewing, M., 1952, Gravity measurements at sea, 1948 and 1949: Trans. Am. Geophys. Union, v. 33, p. 453-460.

Part VIII

Geology of Selected Modern Margins: Indian Ocean Region

Western Continental Margin of Australia

J. J. Veevers

INTRODUCTION

Seaward of a wide (300 km) shelf in the north and a narrow (50-100 km) shelf in the south, the 3,000-km-long western margin of Australia is dominated by marginal plateaus. The northern Scott and Exmouth Plateaus are terraces that lie between an upper low-gradient part of the slope and a lower steeper part, and the southern Wallaby Plateaus and the Naturaliste Plateau are separated from the slope by a trough 3-4 km deep. The entire margin has a near-surface structure of an inshore or coastal graben, filled mainly with about 10 km of Permian and Mesozoic nonmarine and shallow marine detrital sediments and minor alkaline volcanics, flanked oceanward by a rise that is highest structurally near the outer edge of the slope or plateau. Deep-sea drilling dates the sea floor off the margin as Late Jurassic to Early Cretaceous.

The margin developed in three phases: (1) rifting: in the Permian to late Jurassic or Early Cretaceous, a broad intracontinental uplift parallel to what later became the margin was rifted into a complex of grabens; (2) rupture (juvenile ocean): in the Late Jurassic to Early Cretaceous, continental crust ruptured along the rift zone, newly generated sea floor started spreading from this rupture, and the newly formed margin subsided and became blanketed by marine shale and in places was covered by tholeiite; (3) maturity: by the Santonian (Late Cretaceous), the Indian Ocean had spread sufficiently to develop a permanent circulation, and this caused the onset of progradation of coarser carbonates on the shelf and upper slope and of deposition of pelagic carbonates on the marginal plateaus, which had subsided more rapidly than the shelf during the Early Cretaceous; this phase has persisted to the present day.

The continuity from the shelf to an adjacent abyssal plain of an Early Cretaceous shallow marine shale implies that at this site the original margin subsided through 5.5 km across the entire slope and not by displacement along a few major faults, and that since the Early Cretaceous the mean density of the edge of margin above the level of isostatic compensation has increased uniformly across a transition zone.

DESCRIPTION OF REGION

Scientific interest in the western margin of Australia climaxed in 1972 with (1) the drilling by the Deep Sea Drilling Project (DSDP) of two sites on the Naturaliste Plateau and of four sites in abyssal areas adjacent to the margin, and (2) the completion by the Bureau of Mineral Resources, Geology and Geophysics (BMR) of a marine geophysical survey of the margin out to a depth of 4 km. The results of the BMR survey are not yet published, but the drilling results and their preliminary interpretation appeared in 1973 and 1974 (Luyendyk et al., 1973, 1974; Heirtzler et al., 1973; Veevers et al., 1973, 1974a; Hayes et al., 1973, 1974). This and other recently published work, such as that on seismic profiles of the Exmouth and Scott Plateaus by Veevers et al. (1974b), on the physiography of the margin north of latitude 29°S by Falvey and Veevers (1974), and on regional geological studies by Johnstone et al. (1973), Martison et al. (1973), and Veevers and Evans (1973, 1974), provide sufficient material for a preliminary account of the evolution of the continental margin.

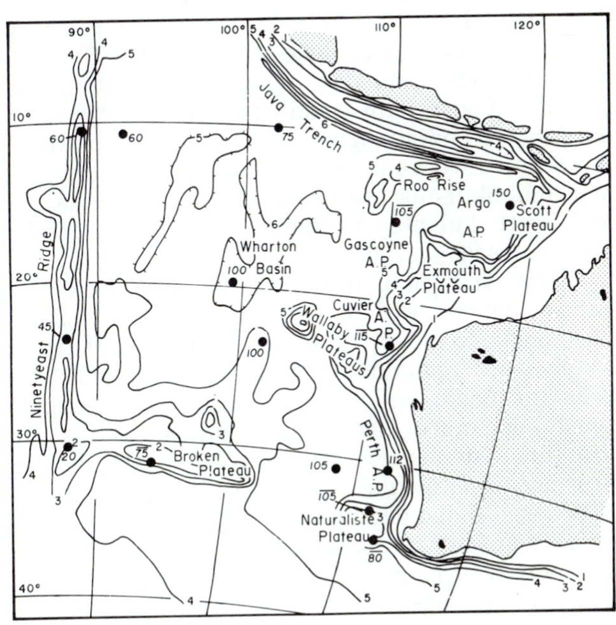

Fig. 1. Regional setting of the western margin of Australia on the eastern edge of the Wharton Basin. Bathymetry (in km) after Udintsev (1973), by courtesy of V.F. Kaneav, Institute of Oceanology, Moscow. Land higher than 1 km (shown for Australia only) in solid black. Drilling sites of DSDP Legs 22 (von der Borch et al., 1972; Sclater et al., 1973), 26 (Luyendyk et al., 1974), 27 (Veevers et al., 1974a), and 28 (Hayes et al., 1974) shown by solid circles, with ages (in my) of sediment overlying basalt. Ages with a bar above (e.g., $\overline{105}$) indicate the age of the deepest sediment penetrated in holes that did not reach basalt basement.

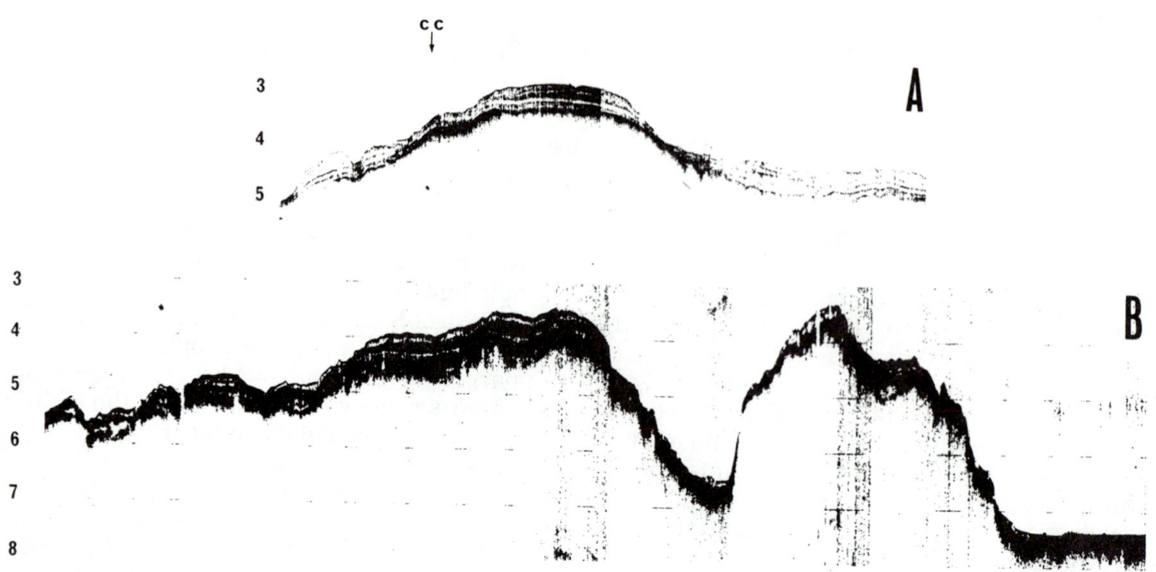

Fig. 2. Seismic (airgun) profiles taken by the *GLOMAR CHALLENGER*. Location shown on Fig. 5. Vertical exaggeration of sea-floor profile about 25:1. Two-way reflection time in seconds. *A*, innermost Wallaby Plateaus (from Veevers and Heirtzler, 1974b), showing probably calcareous sediments draped over a faulted basement. *B*, Roo Rise (from Veevers, 1973b), showing probably calcareous sediments draped over basement.

General

The western margin of Australia (Fig. 1) extends through 23 degrees of latitude, from south of Timor (12°S) to the Naturaliste Plateau (35°S). The landward boundary can be approximated by two straight lines: a south-trending segment (22°S) and a northeast-trending segment to the north. In contrast, the oceanward boundary has a complex outline of marginal plateaus (Naturaliste, Wallaby, Exmouth, and Scott) indented by the abyssal plains (Perth, Cuvier, Gascoyne, and Argo).

The western margin of Australia marks the eastern boundary of the roughly polygonal Wharton Basin, whose other sides are the Broken Plateau, Ninetyeast Ridge, and the Java Trench. Parts of the Wharton Basin and its margins have an obscure crustal structure. According to Kanaev (1973), the Wallaby Plateaus (Fig. 2A) and the Broken Plateau are "microcontinents" and the Roo Rise (Fig. 2B) and Ninetyeast Ridge are "block uplifts." Adding to the difficulty of interpretation of the Wharton Basin is the paucity of magnetic anomalies decipherable as being due to sea-floor spreading. Deep-sea drilling (Fig. 1) shows that the oldest known oceanic sediments in the region are Late Jurassic (150 my) to Early Cretaceous (112 my), along the eastern edge of the Wharton Basin at its boundary with the Australian margin, and the youngest sediments are Paleocene (60 my) along its western boundary near the Ninetyeast Ridge. Owing to the obscure but presumably complex history of the Wharton Basin, the comparatively well known western margin of Australia assumes a key role in elucidating the history of this region of the Indian Ocean. Petroleum exploration of the onshore and offshore basins has provided most of the knowledge of the margin, and because of a government subsidy on exploration up to 1974, this information became public six months after completion of the operation.

Whereas the shelf and upper slope are fairly well known by drilling and seismic profiling, other geophysical studies, such as gravity and magnetics, were neglected until the 1972 BMR survey. This account therefore leans heavily on the side of geology, and although it is thus unbalanced, it is fairly comprehensive, thanks to the recent deep-sea drilling and the exploration of the shelf. An account of geophysical work is given in the Appendix.

Physiography

With the exception of original work south of latitude 29°S and west of longitude 108°E, this account follows the physiographic analysis of Falvey and Veevers (1974).

In the simplest terms, the margin is divisible physiographically into shelf, slope, and marginal plateau. The shelf comprises a narrow shelf proper, with its outer edge, either a notch or change in gradient, at depths between 54 and 160 m, and a broad shelf-edge zone, which extends out to depths of 100-700 m. Near the Scott Plateau [Figs. 3B (1-4) and 4A] the shelf edge has a mean depth of 535 m, near the Exmouth Plateau [Figs. 3B (4-7) and 4C] it is shallower at 142 m, and south of North West Cape it deepens to depths of 150-300 m. The shelf is correspondingly widest (300 km) in the north, narrowest off North West Cape (15 km), and of

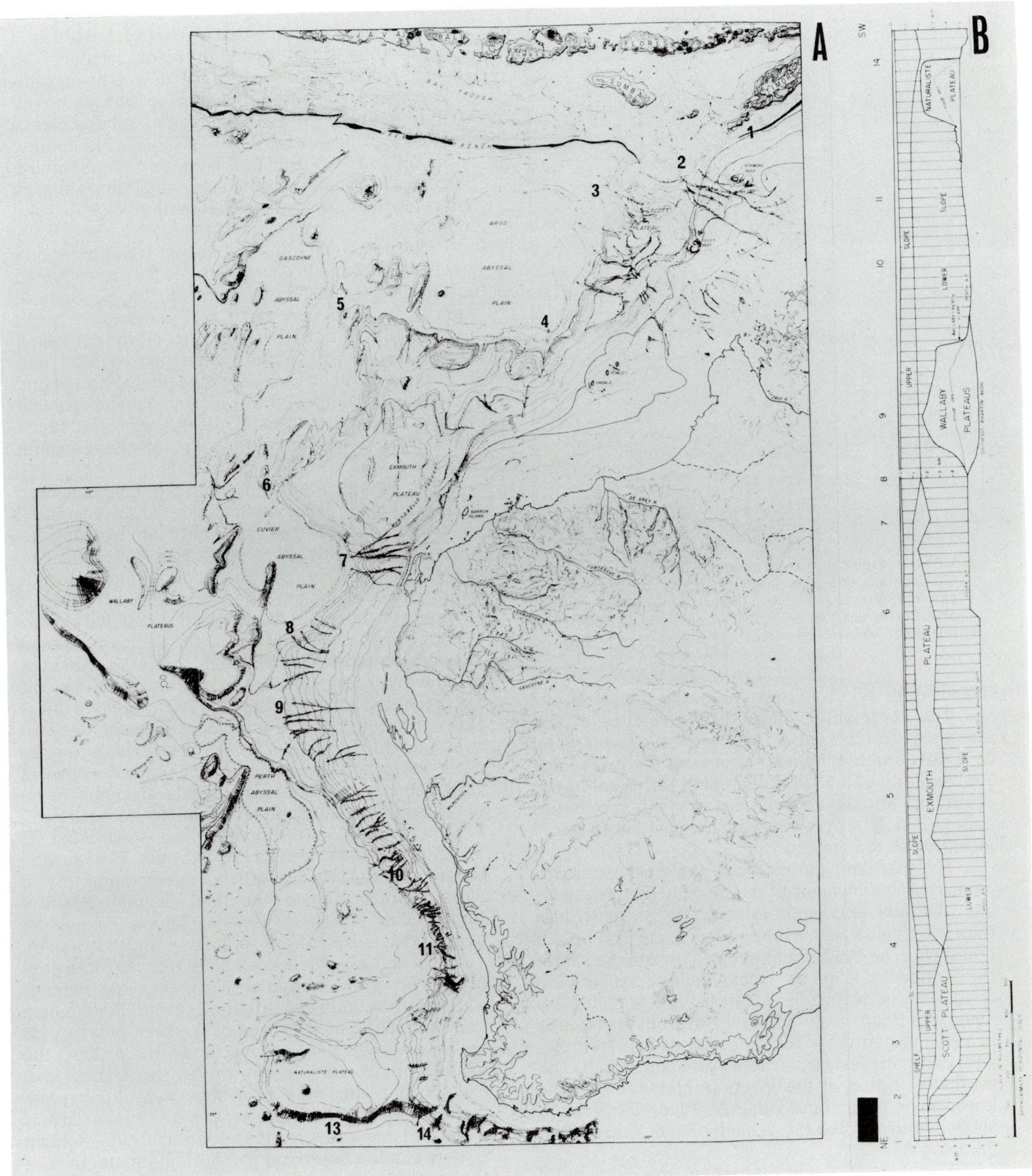

Fig. 3. Physiography of the western margin of Australia and adjacent ocean. *A*, physiographic map, with bathymetric contour interval of 500 m (solid line) and of 100 m (dashed line) inshore. Relief on land shown by 100-, 200-, 500-, and 1,000-m contours. Lines of relict drainage on land shown by dashed lines. After Falvey and Veevers (1974), except for physiography of area west of longitude 108°E and south of latitude 29°S. *B*, side view of margin. Vertical exaggeration X 45. Numbers 1-11 and 14 refer to locations shown on *A*. Vertical lines indicate continental slope.

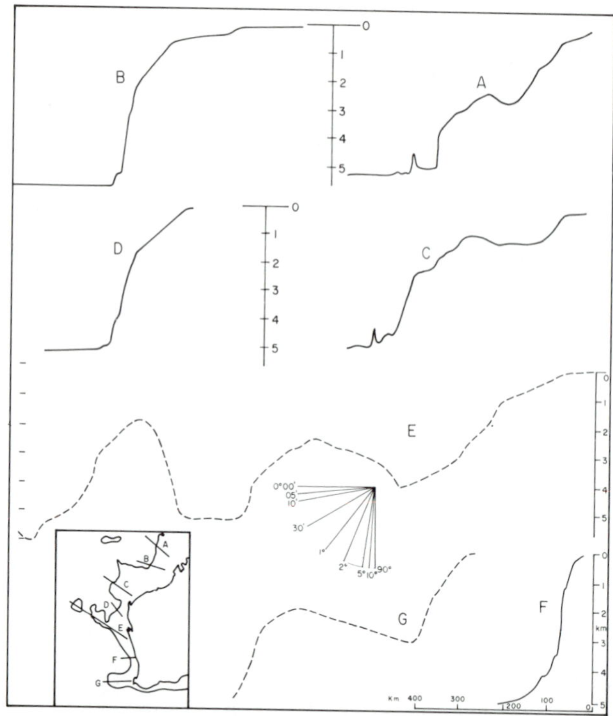

Fig. 4. Physiographic profiles of margin. Vertical exaggeration X 60. Location shown in inset. E and G constructed from physiographic map (Fig. 3A); the rest are single profiles, all after Falvey and Veevers (1974) except F, by courtesy of the Bureau of Mineral Resources.

intermediate width (about 100 km) to the south.

The continental slope also is composite: the upper slope has a gradient of 1° to 2°, the lower slope 2° to 20°, and the boundary between the two varies from a depth of 3.5 km in the north to 1-2 km in the south (Fig. 3B). As we shall see later, this systematic change in depth correlates with the postulated age of inception of the margin.

The marginal plateaus are the most extensive physiographic element of the margin. The Naturaliste, Wallaby, and Scott plateaus rise to slightly less than 2 km below sea level and the Exmouth Plateau to 0.9 km. The Wallaby Plateaus are detached from the continental slope by a trough whose axis is 4 km deep, and the Naturaliste Plateau by a trough 3 km deep. The Scott, Exmouth, and Naturaliste plateaus are single, and the Wallaby Plateaus are multiple, with two main parts separated by a 5-km trough. The southwest flanks of the Wallaby Plateaus and the lower slope next to the Perth Abyssal Plain form a single feature called the Wallaby-Perth Scarp. (Fig. 5).

The abyssal plains that indent the continental margin lie at a common depth of 5.7 km except the Cuvier Abyssal Plain, at 5.0 km. The boundary between continental and oceanic crust has an orthogonal pattern of trends, exemplified by the Wallaby-Perth Scarp, southeast and northeast boundaries of the Cuvier Abyssal Plain, and western boundary of the Exmouth Plateau.

Geology

The geology of the western part of Australia is dominated by three Archean blocks, called the Yilgarn, Pilbara, and Kimberley blocks. The marginal Perth and Carnarvon basins lie alongside the western boundary of the Yilgarn and Pilbara blocks and are the sites of thick (10 km) Paleozoic (mainly Permian) and Mesozoic sedimentation. The Canning Basin lies between the Kimberley and Pilbara blocks and is a broad downwarp filled with thick Paleozoic sediments.

The northeast- and northwest-trending structural grains of the Precambrian blocks are reflected in the pattern of lineaments of the continental margin, in particular by the physiographic lineaments of the Wallaby-Perth Scarp and the southeast and northeast flanks of the Cuvier Abyssal Plain, and by the structural lineaments of the offshore folds and faults, in particular those off the northern Perth Basin, the Rankin Trend (Fig. 6C), and the faults parallel to it on the outer part of the Exmouth Plateau.

The structural style of the margin, summarized in a series of six cross sections (Fig. 6), is dominated by longitudinal ridges and rifts or troughs, as described already for southwest Australia by Veevers et al. (1971, Fig. 1b). In detail the pattern is made up of a basin filled with thick Mesozoic sediment on the inner or landward side of the margin and thin Cretaceous and younger sediment over a rise of faulted older sediments on the outer or oceanward side, as brought out well in the northernmost section (Fig. 6A). Here, in the Browse Basin, 6 km of Jurassic and younger sediments rest over faulted Triassic and older sediments on the inner side, and pass oceanward into a wedge of Cretaceous and Tertiary sediments draped over an outer rise, with separate culminations beneath Scott Reef and the Scott Plateau. The same kind of structure is found across the Exmouth Plateau and the adjacent shelf (Fig. 6C); as seen clearer in the more northern section, the inner Dampier Basin of faulted Permian to Jurassic sediments draped by Cretaceous and younger sediments and the Rankin Trend of upfaulted Triassic and thin cover rocks pass oceanward into a wedge of Cretaceous and younger sediments that thin beneath the Exmouth Plateau to a vanishing point at the crest of the rise, only 2.5 km below sea level, near the top of the lower slope; the more southern section shows the same structure, except that the pre-Cretaceous east-dipping sequence beneath the outermost part of the rise is displaced down toward the ocean by faults. Section B, between the Scott and Exmouth plateaus, follows the general pattern of an inner basin of Jurassic and younger sediments draped over faulted Triassic and older sediment and an intermediate ridge (at Bedout-1), but beyond an oceanward-thickening sequence of Cretaceous and younger sediments, homologous to the inner part of the Exmouth Plateau, the outer ridge (or main part of the marginal plateau) is lacking.

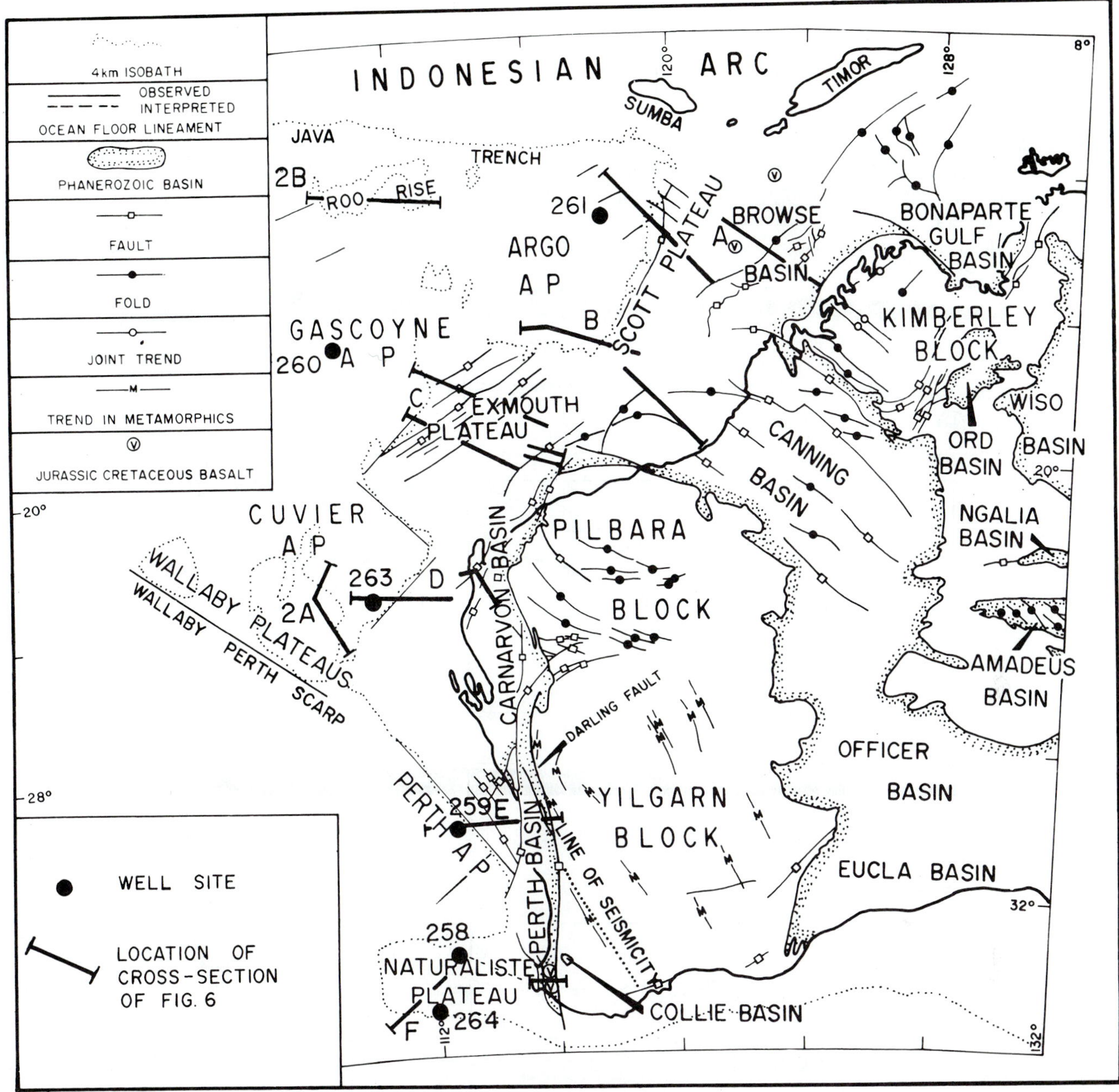

Fig. 5. Tectonic map, after Veevers and Johnstone (1974), with location of the cross sections of Figs. 2 and 6.

South of the Exmouth Plateau (Fig. 6D), the inner basin is represented by a graben (the Carnarvon Basin) filled with 8 km of faulted Permian to Jurassic sediments draped by Cretaceous and younger sediments, and the outer ridge by the Pendock-1 area, with Early Cretaceous sediments resting direct on early Carboniferous sediments. With its inner basin of faulted Permian to Jurassic sediments divided internally by a basement ridge at Dongara and draped by Cretaceous and younger sediments, and an outer ridge with a presumably thinner sedimentary section, the section (Fig. 6E) across the northern Perth Basin maintains the pattern. The section across the Naturaliste Plateau (Fig. 6F) is not as well known as the others. On the landward side, the narrow southern Perth Basin is confined between the Precambrian Yilgarn Block on the east and the Naturaliste Block on the west, and contains 8 km of faulted nonmarine Permian to Jurassic sediments draped by Early Cretaceous basalt and marine sediments. Despite the drilling of two sites on the Naturaliste Plateau, its geology remains obscure. A superstructure of Cretaceous sediment and presumably early Cretaceous basalt breccia is known from drilling, but what lies beneath these rocks, whether a Precambrian block, an older Mesozoic and Permian sedimentary basin, or a pile of oceanic volcanic rocks, is unknown. The same possibilities

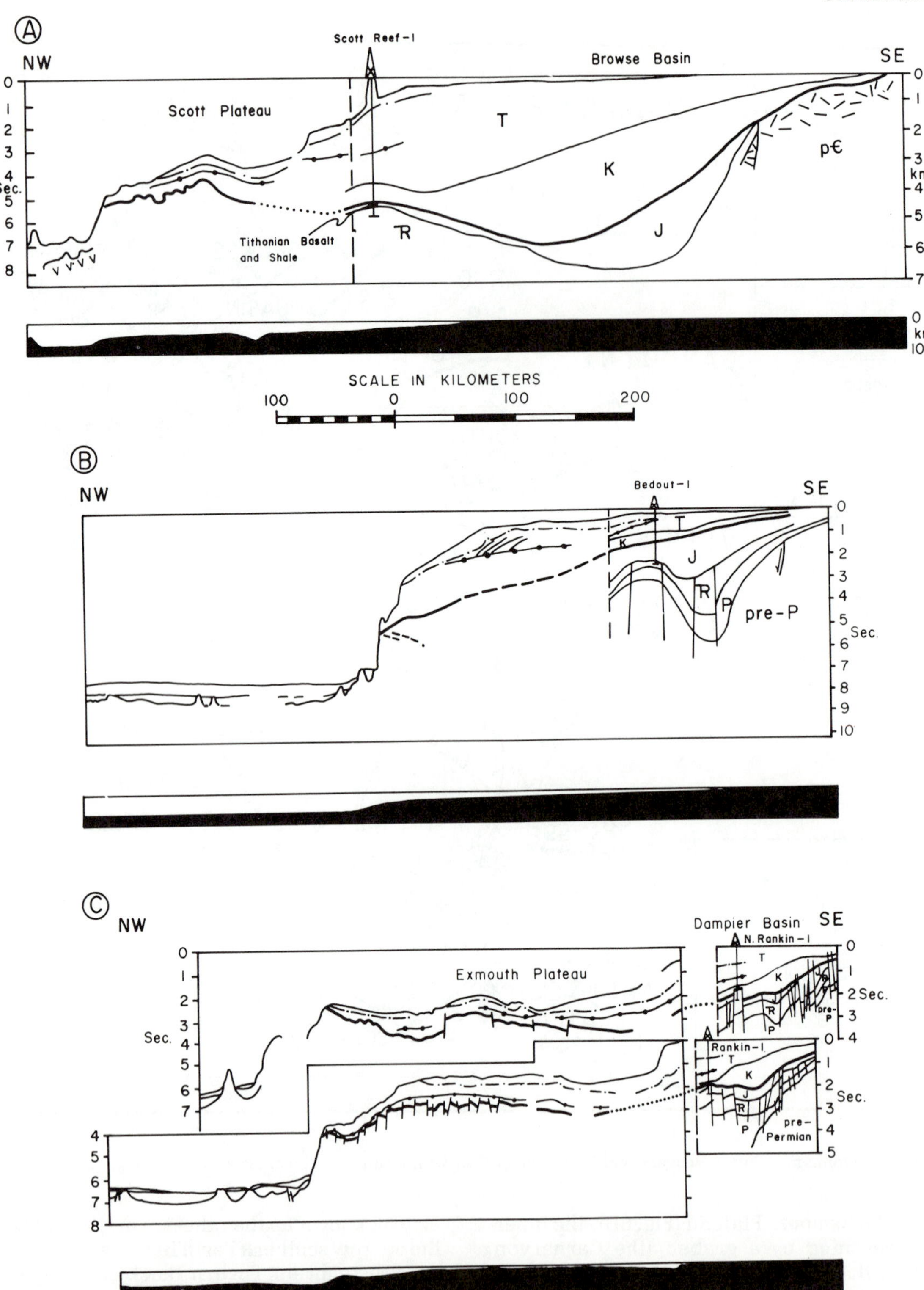

Fig. 6. Cross sections and two—way time sections (locations on Fig. 5). Vertical exaggeration X 25; 2.5 VE of physiographical silhouette, each 10 km high. *Sources: A*, southeast of dashed line: cross section after Halse and Hayes (1971) and B.O.C. of Australia (1971); northwest of dashed line: time section after Veevers et al. (1974b). The heavy line beneath the Scott Plateau indicates reflector 4 (R_4) of Veevers et al. (1974b), which is interpreted as an unconformity separating Aptian and younger sediments from older rocks below. Other reflectors are R_3 (full line with dots), interpreted as basal Oligocene, and R_2 (dots and dashes), at the Early/Middle Miocene boundary. Basalt (V's) and overlying Late Jurassic and younger sediment of Argo Abyssal Plain dated by Veevers et al. (1974a). T, Tertiary; J, Jurassic; K, Cretaceous; R, Triassic; pC, Precambrian. *B*, southeast of broken line: Veevers and Johnstone (1974); northeast of broken line: Veevers et al. (1974b). Reflectors as in *A*. Note the prograding structure of the layers above R_3. Letter symbols as in *A*; pre-P, pre-Permian. *C*, southeast side: time sections after Kaye et al. (1972); northeast side: time sections after Veevers et al. (1974b). Reflectors beneath Exmouth Plateau and letter symbols as in *A* and *B*. *D*, east side: cross section after Veevers and Johnstone (1974); west side: time sections after Veevers (1974), by courtesy of the Bureau of Mineral Resources. Letter symbols as in *A-C*, with, additionally:

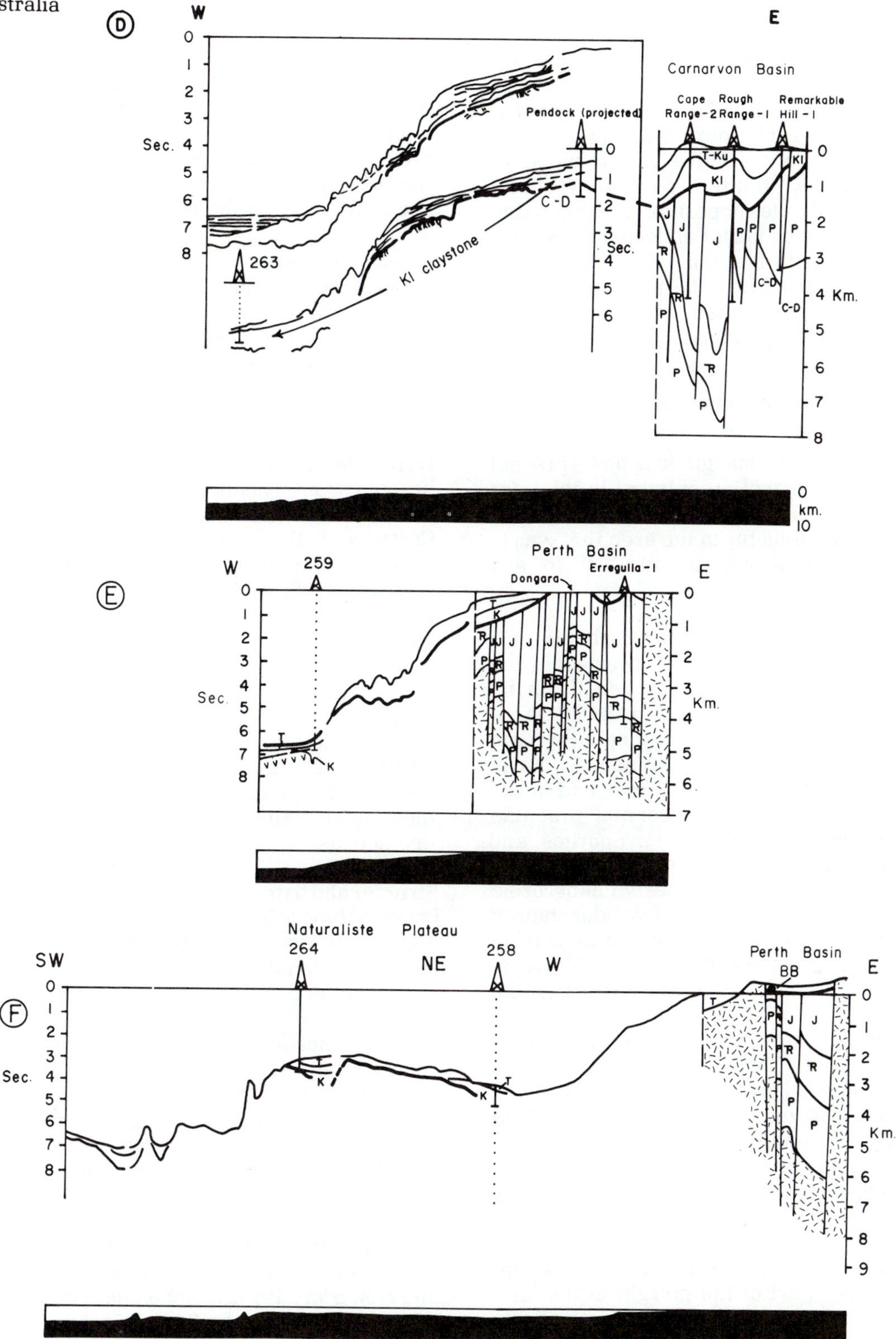

Ku, Late Cretaceous; K1, Early Cretaceous; C-D, Carboniferous and Devonian. Pendock-1 Well (Geary, 1970) penetrated a shallow marine sequence of Santonian to Recent carbonates, Aptian to Cenomanian shale (the Winning Group), and early Carboniferous and older carbonates. At DSDP site 263, drilling terminated near the base of claystone equivalent in age and facies to the Winning Group. Veevers and Johnstone (1974) showed that, in the absence of a significant gravity anomaly, the apparent continuity across the continental slope of the pre-Aptian basement (shown by acoustic basement beneath site 263) and its continuous increase in depth imply an increase since the Early Cretaceous across a transition zone of the mean density of the crust and mantle at DSDP site 263 above the level of isostatic compensation. *E*, east side: cross section after Jones and Pearson (1972); west side: time section after Veevers (1974), by courtesy of Bureau of Mineral Resources, with well data (site 259) from Veevers et al. (1974a). Letter symbols as in *A-C*. Precambrian shown by pattern. *F*, east side: cross section after Jones and Pearson (1972); west side: time section (southwest of site 258) from Burckle et al. (1967), drilling information from Luyendyk et al. (1973, 1974) and Hayes et al. (1974). BB, Bunbury Basalt.

apply to the Wallaby Plateaus and the Roo Rise, except that even their superstructure is unknown. The seismic profiles (Fig. 2) and the general bathymetry provide the only sources for speculation. The seismic character of reflective layers draped on basement (apparently faulted on the Wallaby Plateau) probably indicates deposition of carbonates above the carbonate compensation level and resembles that of the Exmouth Plateau.

The uniform structural pattern along the western margin, as described above, does not necessarily imply continuity of individual structures.

Phanerozoic, Sedimentary, Igneous, and Structural History

I argue below that the margin took on its present configuration at the edge of a continent in the later part of the Mesozoic. Reference here to the margin before this time relates simply to the area that was to become the continental margin and not to any contemporary geography.

Sedimentary History. Up to late Carboniferous what was to become the margin was overlain in the north Perth Basin and south Carnarvon Basin by probable Ordovician to Early Silurian fluvial sandstone, in the Carnarvon Basin additionally by Late Silurian limestone and Late Devonian to Early Carboniferous carbonate and sandstone, and in the Canning Basin by Ordovician carbonate, probable Silurian and Devonian red beds and evaporites, and Late Devonian to Early Carboniferous carbonates and shales. The sequences of this early Phanerozoic phase of deposition show the trend from dominantly nonmarine facies in the south to dominantly marine facies in the north (Veevers et al., 1971; Veevers, 1971), and this trend is maintained during the succeeding phase.

Starting in the Late Carboniferous, the entire margin was the site of a common set of sedimentary, volcanic, and structural events that persisted almost to the end of the Mesozoic (Fig. 7), details of which are given by Veevers and Evans (1974). In summary, nonmarine glacigene deposits in the Late Carboniferous and earliest Permian were succeeded by alternating shallow marine and nonmarine deposits to the end of the Triassic, and in the Early Jurassic by nonmarine deposits. A marine transgression swept across the northern part of the margin in the Late Jurassic, and across the southern part in the Early Cretaceous (Aptian). At the extreme south of the margin, in the south Perth Basin, the entire stratigraphic column is nonmarine until the Aptian transgression (Veevers, 1971). Marine deposition continued from the Aptian to the Cenomanian, and, after a hiatus in the Coniacian and locally in the Turonian also, returned in the Santonian. The Santonian marked a radical change in the source of the sediments: from almost wholly detrital before the Santonian to almost wholly calcareous during the Santonian and later (Veevers and Johnstone, 1974), presumably due to the initiation of strong thermohaline circulation in the youthful Indian Ocean in the Santonian. Except for a regional hiatus in the latest Maestrichtian or earliest Paleocene, carbonate deposition has persisted from the Santonian to the present day, and comprises calcarenite on the shelf and ooze on the marginal plateaus.

Igneous History. The only known outcrops of igneous rocks on the margin are the Early Cretaceous Bunbury Basalt (tholeiite) of the south Perth Basin, and the Miocene Fitzroy Lamproites. All other igneous rocks are known from drilling. Beneath the adjacent deep ocean floor, tholeiitic basalt at DSDP site 259 (Fig. 5) is taken to be earliest Aptian (112 my), and at site 261 latest Oxfordian (150 my). At DSDP site 264, on the Naturaliste Plateau, the basalt breccia or interbedded basalt and tuff is pre-Santonian, presumably the same age as the Early Cretaceous Bunbury Basalt. Late Jurassic basalt occurs at the north end of the margin at Scott and Ashmore reefs, and Late Triassic (196 my) dolerite in the onshore Canning Basin. Rhyolite in the Dampier Basin (Fig. 6C) is latest Permian, and a carbonatite association of phonolite, lamprophyre, and trachyte in the offshore Carnarvon Basin is Permian. To be noted in this history are the narrow range of dates of igneous activity, from Permian to Early Cretaceous and Miocene, and the petrological associations of paralkaline igneous rocks in the Permian, and of tholeiitic basalts in the Late Jurassic and Early Cretaceous.

Structural History. Rapid subsidence of fault-bounded blocks in the Carnarvon and Canning basins and gentle downwarping elsewhere accompanied the late Carboniferous to earliest Permian phase of glacigene deposition. Rifting in the Perth Basin started in the Late Permian and persisted into the Early Cretaceous, whereas in the Dampier Basin it ceased in the Late Jurassic and in the Bedout and Browse basins in the Early Jurassic.

Sea-floor spreading from northeast-trending axes, indicated by magnetic anomalies south of the Java Trench (Falvey, 1972) and north of Naturaliste Plateau (Veevers and Heirtzler, 1974b), started off the northern part of the margin in the Late Jurassic or 160 my ago, and off the southern part in the Hauterivian (123 my) to Barremian (116 my) (Veevers and Heirtzler, 1974a). Accompanying and following the inception of sea-floor spreading, the newly formed continental margin subsided. As shown by a decrease in interval velocity in the Aptian and younger sediments from the Exmouth Plateau (ooze) to the adjacent shelf (calcarenite), the Exmouth Plateau subsided rapidly in the Aptian, Albian, and Late Cretaceous, and thereafter has slowed (Veevers et al., 1974b).

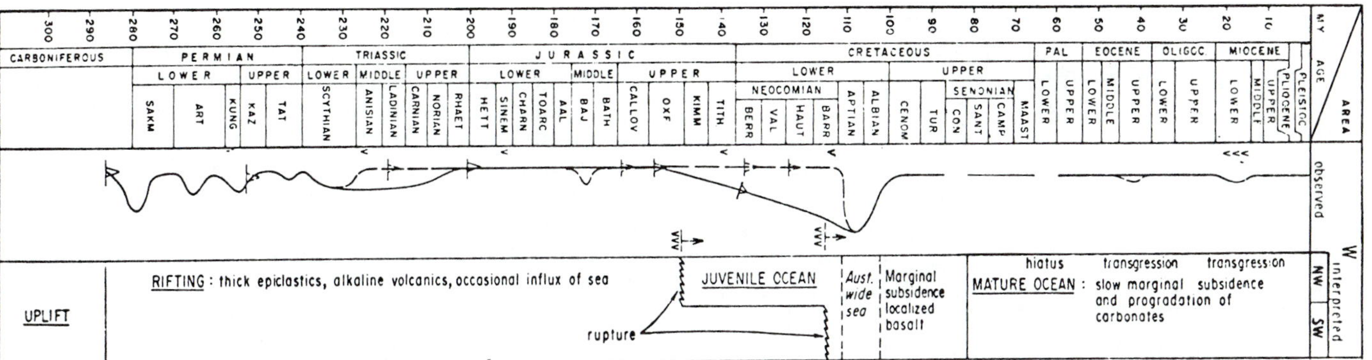

Fig. 7. Summarized observations and interpretations of principal geological events along the western margin (after Veevers and Evans, 1974). Time scale from Harland et al. (1964, 1971). Sedimentary events: nonmarine on left, marine on right. Solid line and symbols refers to northwest, dashed line to southwest. Volcanics: V's. Rifting: A. Inception of sea-floor spreading: VVV.

HISTORICAL SYNTHESIS

As shown in the right-hand column of Figure 7, the events since the Late Carboniferous are interpreted as reflecting a sequence of intracontinental rifting, rupture, and sea-floor spreading to form a continental margin facing a juvenile and then a mature ocean. The cessation of rifting, the outpouring of basalt, the onset of a marine transgression, and the generation by spreading of the adjacent sea floor coincide and indicate the time of rupture: Late Jurassic along the northwest, and Early Cretaceous (Aptian) along the southwest. Following rupture, the older northwest margin has subsided deeper than the southwest margin.

The long Phanerozoic history of marine conditions in the northwest led Veevers et al. (1971) to conclude that the northwest margin is old, and that the change in the Late Carboniferous to Early Cretaceous from marine facies in the north to nonmarine in the south followed by uniform marine deposition from the Aptian shows that the southern margin came into being in the Early Cretaceous. The ridge from which the new sea floor was generated in the north probably lay at or near a previous margin (see Fig. 1 of Veevers and Heirtzler, 1974a), and any continental slivers broken off the continent in this way may be represented by the oceanic ridges and plateaus north and northwest of the Exmouth Plateau.

Formation of Rifted Margins

Observations of the western margin support what may be called the classical view of the formation of rifted margins, as summarized by Schneider (1969) and Wonfor (1972), and shown diagrammatically in Figure 8.

1. A phase lasting 190 my from the late Carboniferous (300 my) to the Late Jurassic (150 my) or Early Cretaceous (110 my), of intracontinental rifting of a broad arch, with a fluvial (in the south) to marine (north) fill, and minor alkaline volcanics.
2. Rupture of the arch in the Late Jurassic (north) or Early Cretaceous (south) with local effusion of basalt, followed by marginal subsidence and marine transgression.
3. Since the Santonian, progradation of carbonates along a slowly subsiding margin.

Figure 8 (A and B) shows the arch in continental crust. The dotted half represents a postulated subsequently dispersed continental fragment (?India), and this part of the diagram applies to that part of the margin south of the Exmouth Plateau only. If the postulated continental neighbor also contained a deep graben on its flank, then the arch would have resembled that part of the present African Rift System that bifurcates into western and eastern rifts on either side of Lake Victoria. North of the Exmouth Plateau, I speculate that the Late Jurassic margin formed near the site of the previous Paleozoic margin that faced Tethys. Most of the west side of the arch consisted of oceanic crust, and any slivers of continental crust on this side would have been split off Australia as Baja California is being split off Mexico at the present day.

ACKNOWLEDGMENTS

T. Stewart of the R.A.N. Hydrographic Office provided sounding sheets of the area south of latitude 29°S and west of longitude 108°E, and S. Robins prepared the physiographic maps of these areas. V. F. Kanaev, Institute of Oceanology, Moscow, kindly supplied an advance copy of a bathymetric map of the eastern Indian Ocean, and the Bureau of Mineral Resources, Canberra, gave me access to selected seismic profiles before publication. This work was supported by the Australian Research Grants Committee.

J. G. Jones and C. M. Powell critically reviewed the manuscript. L. A. Frakes let me see details of DSDP site 264 before publication.

APPENDIX: GEOPHYSICS

McElhinny (1973, Fig. 123 and p. 230) describes the Phanerozoic apparent polar-wander path for

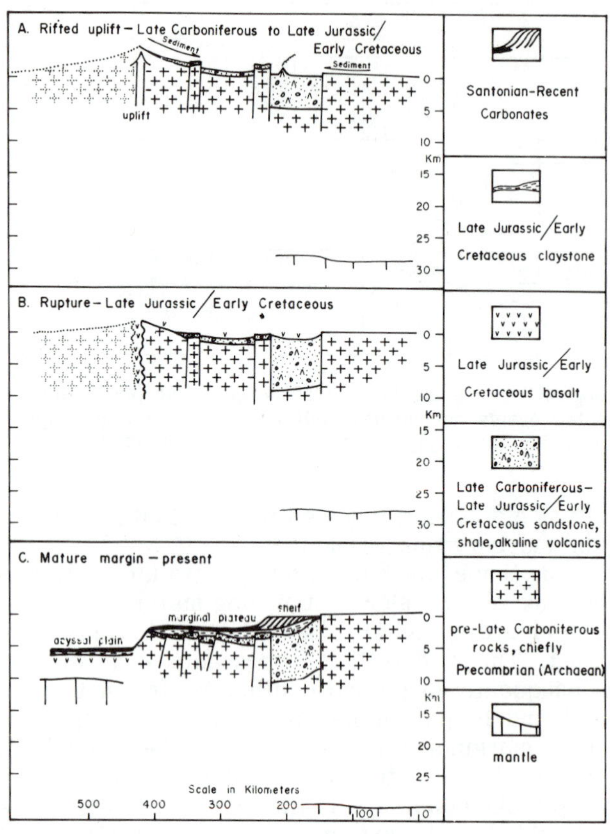

Fig. 8. Diagrammatic evolution of the western Australian continental margin.

Australia and states that "from a position just south of Australia during the Mesozoic, the pole moved progressively towards the present south pole during the Tertiary."

A well-marked zone of seismicity cuts across the southwest tip of Australia (Fig. 5; Cleary and Simpson, 1971; Fitch et al., 1973) parallel to the Wallaby-Perth Scarp and to the Archean structural grain. The intersection of this trend and the Darling Fault (Fig. 5) determines (1) the sole occurrence of Permian sediments on the Yilgarn Block at Collie, on line with the Wallaby-Perth Scarp, and (2) the unique occurrence in the Perth Basin of Early Cretaceous basalt at and south of Bunbury. Another, but less clearly defined, zone of seismicity crosses the offshore Canning Basin.

Hawkins et al. (1965) investigated the offshore Perth Basin by gravity, magnetic, and seismic refraction methods. The deep seismic refraction measurements in the Indian Ocean by Francis and Raitt (1967) and heat-flow measurements by Langseth and Taylor (1967) do not extend to the continental margin. Burckle et al. (1967) pioneered study of the Naturaliste Plateau, and Beck (1972) published a seismic profile of the Exmouth Plateau. Many accounts of petroleum exploration of the margin are based largely on seismic surveys, and detailed studies of seismic profiles are given by Veevers (1973a, 1973b), Veevers and Heirtzler (1974b), and Veevers et al. (1974b). Veevers and Heirtzler (1974b) also present some magnetic profiles. But for a comprehensive geophysical survey of the margin, including gravity and magnetic as well as seismic reflection studies, we must await the publication of the 1972 BMR survey. Even so, the chief gap in knowledge remains the crustal structure of the marginal and oceanic plateaus, and this should be determined by deep seismic-refraction measurements.

BIBLIOGRAPHY

Balme, B. E., 1969, The Triassic System in Western Australia: Austral. Petr. Expl. Assoc. Jour., v. 9, no. 2, p. 67-78.

Beck, R. H., 1972, The oceans, the new frontier in exploration: Austral. Petr. Expl. Assoc. Jour., v. 12, no. 2, p. 5-28.

B.O.C. of Australia Ltd., 1971, Scott Reef No. 1 Completion Report: Bur. Mineral Resources Austral. Petr. Search Subsidy Acts Publ. (in press).

Boutakoff, N., 1963, Geology of the off-shore areas of north-western Australia: Austral. Petr. Expl. Assoc. Jour., v. 3, p. 10-18.

Brown, D. A., Campbell, K. S. W., and Crook, K. A. W., 1968, The geological evolution of Australia and New Zealand: Elmsford, N.Y., Pergamon Press.

Burckle, L. H., Saito, T., and Ewing, M., 1967, A Cretaceous (Turonian) core from the Naturaliste Plateau, southeast Indian Ocean: Deep-Sea Res., v. 14, p. 421-426.

Carrigy, M. A., and Fairbridge, R. W., 1954, Recent sedimentation, physiography, and structure of the continental shelves of Western Australia: Jour. Roy. Soc. Western Austral., v. 38, p. 65-95.

Challinor, A., 1970, The geology of the offshore Canning Basin, Western Australia: Austral. Petr. Expl. Assoc., Jour., v. 10, no. 2, p. 78-90.

Cleary, J. R., and Simpson, D. W., 1971, Seismotectonics of the Australian continent: Nature, v. 230, p. 239-241.

Condon, M. A., 1965-1968: Geology of the Carnarvon Basin, Western Australia: Bur. Mineral Resources Austral. Bull., v. 77.

Dietz, R. S., and Holden, J. C., 1971, Pre-Mesozoic oceanic crust in the eastern Indian Ocean (Wharton Basin)?: Nature, v. 229, p. 309-312.

Ewing, M., Eittreim, S., Truchan, M., and Ewing, J. I., 1969, Sediment distribution in the Indian Ocean: Deep-Sea Res., v. 16, p. 231-248.

Fairbridge, R. W., 1953, The Sahul Shelf, northern Australia: its structure and geological relationships: Jour. Roy. Soc., Western Austral., v. 37, p. 1-33.

———, 1955, Some bathymetric and geotectonic features of the eastern part of the Indian Ocean: Deep-Sea Res., v. 2, p. 161-171.

Falvey, D. A., 1972, Sea-floor spreading in the Wharton Basin (northeast Indian Ocean) and the breakup of eastern Gondwanaland: Austral. Petr. Expl. Assoc. Jour., v. 12, no. 2, p. 86-88.

———, and Veevers, J. J., 1974, Physiography of the Exmouth and Scott plateaus, Western Australia, and adjacent northeast Wharton Basin (in prep.).

Fitch, T. J., Worthington, M. H., and Everingham, I. B., 1973, Mechanisms of Australian earthquakes and contemporary stress in the Indian Ocean Plate: Earth and Planetary Sci. Letters, v. 18, p. 345-356.

Francis, T. J. G., and Raitt, R. W., 1967, Seismic refraction measurements in the southern Indian Ocean: Jour. Geoph. Res., v. 72, no. 12, p. 3015-3041.

Geary, J. K., 1970, Offshore exploration of the southern Carnarvon Basin: Austral. Petr. Expl. Assoc. Jour., v. 10, no. 2, p. 9-15.

Halse, J. W., and Hayes, J. D., 1971, The geological and structural framework of the offshore Kimberley Block (Browse Basin) area, Western Australia: Austral. Petr. Expl. Assoc. Jour., v. 11, no. 2, p. 64-70.

Harland, W. B., Smith, A. G., and Wilcock, B., eds., 1964, The Phanerozoic time-scale: Geol. Soc. London Quart. Jour., 120 S.

———, Smith, A. G., and Wilcock, B., eds., 1971, Supplement to the Phanerozoic time-scale: Geol. Soc. London.

Hawkins, L. V., Hennion, J. F., Nafe, J. E., and Thyer, R. R., 1965, Geophysical investigations in the area of the Perth Basin, Western Australia: Geophysics, v. 30, no. 6, p. 1026-1052.

Hayes, D. E., Frakes, L. A., Barrett, P., Burns, D. A., Pei-Hsin Chen, Ford, A. B., Kaneps, A. G., Kemp, E. M., McCollum, D. W., Piper, D. J. W., Well, R. E., and Webb, P. N., 1973, Leg 28: deep-sea drilling in the southern ocean: Geotimes, June, p. 19-24.

———, Frakes, L. A., Barrett, P., Burns, D. A., Pei-Hsin Chen, Ford, A. B., Kaneps, A. G., Kemp, E. M. McCollum, D. W., Peper, D. J. W., Well, R. E., and Webb, P. N., 1974, Initial reports of the Deep Sea Drilling Project, vol. 28: Washington, D.C., U.S. Govt. Printing Office.

Heezen, B. C., and Tharp, M., 1966, Physiography of the Indian Ocean: Phil. Trans. Roy. Soc. London, ser. A, v. 259, p. 137-149.

Heirtzler, J. R., Veevers, J. J., Bolli, H. M., Carter, A. N., Cook, P. J., Krasheninnikov, V. A., McKnight, B. K., Proto-Decima, F., Renz, G. W., Robinson, P. T., Rocker, K., and Thayer, P. A., 1973, Age of the eastern Indian Ocean floor: Science, v. 180, no. 4089, p. 952-954.

Johnstone, M. H., Lowry, D. C., and Quilty, P. G., 1973, The geology of southwestern Australia—A review: Jour. Roy. Soc. Western Austral., v. 56, no. 1,2, p. 5-15.

Jones, D. K., and Pearson, G. R., 1972, The tectonic elements of the Perth Basin: Austral. Petr. Explo. Assoc. Jour., v. 12, no. 1, p. 17-22.

Jones, H. A., 1971, Late Cenozoic sedimentary forms on the northwest Australian continental shelf: Marine Geology, v. 10, M20-26.

Kanaev, V. F., 1973, Fig. 2, *in* Pushcharovskiy, Y. M., Principles of tectonic classification of the oceans: Geotectonics, v. 6, p. 336-340.

Kaye, P. Edmond, G. M., and Challinor, A., 1972, The Rankin Trend, Northwest Shelf, Western Australia: Austral. Petr. Expl. Assoc. Jour., v. 12, no. 1, p. 3-8.

Langseth, M. G., and Taylor, P. T., 1967, Recent heat flow measurements in the Indian Ocean: Jour. Geoph. Res., v. 72, no. 24, p. 6249-6260.

Laughton, A. S., Matthews, D. H., and Fisher, R. L., 1971, The structure of the Indian Ocean, *in* Maxwell, A. E., ed., The sea, v. 2: New York, Wiley, p. 543-586.

Le Pichon, X., and Heirtzler, J. R., 1968, Magnetic anomalies in the Indian Ocean and sea-floor spreading: Jour. Geoph. Res., v. 73, no. 6, p. 2101-2117.

Luyendyk, B. P., Davies, T. A., Rodolfo, K. S., Kempe, D. R. C., McKelvey, B. C., Leidy, R. D., Horvath, G. J. Hyndman, R. D., Theirstein, H. R., Boltovskoy, E., and Doyle, P., 1973, Across the southern Indian Ocean aboard GLOMAR CHALLENGER: Geotimes, Mar., p. 16-19.

———, Davies, T. A., Rodolfo, K. S., Kempe, D. R. C., McKelvey, B. C., Leidy, R. D., Horvath, G. J., Hyndman, R. D., Theirstein, H. R., Boltovskoy, E., Doyle, P., 1974, Initial reports of the Deep Sea Drilling Project, v. 26: Washington, D.C., U.S. Govt. Printing Office.

Martison, N. W., McDonald, D. R., and Kaye, P., 1973, Exploration on continental shelf off northwest Australia: Am. Assoc. Petr. Geol., v. 57, no. 6, p. 972-989.

McElhinny, M. W., 1973, Palaeomagnetism and plate tectonics: New York, Cambridge Univ. Press.

McKenzie, D., and Sclater, J. G., 1971, The evolution of the Indian Ocean since the Late Cretaceous: Geophys. Jour. Roy. Astron. Soc., v. 25, p. 437-528.

McWhae, J. R. H., Playford, P. E., Lindner, A. W., Glenister, B. F., and Balme, B. E., 1958, The stratigraphy of Western Australia: Jour. Geol. Soc. Austral., v. 4, no. 2.

Mollan, R. G., Craig, R. W., and Lofting, M. J. W., 1969, Geological framework of the continental shelf off northwest Australia: Austral. Petr. Expl. Assoc. Jour., v. 9, no. 2, p. 49-59.

———, Craig, R. W., and Lofting, M. J. W., 1970, Geologic framework of the continental shelf off northwest Australia: Am. Assoc. Petr. Geol. Bull., v. 54, no. 4, p. 583-600.

Schneider, E. D., 1969, Models for rifted and compressional continental margins (Abs.): Geol. Soc. America, v. 7, p. 291-292.

Sclater, J. G., von der Borch, C. C., Gartner, S., Hekinian, R., Johnson, D. A., McGowan, B., Pimm, A. C., Thompson, R. W., and Veevers, J. J., 1973, Initial reports of the Deep Sea Drilling Project, v. 22: Washington, D.C., U.S. Govt. Printing Office.

Teichert, C., and Fairbridge, R. W., 1948, Some coral reefs of the Sahul Shelf: Geograph. Rev., v. 38, no. 2, p. 222-249.

Udintsev, G. B., ed., 1973, International Indian Ocean expedition atlas (in press).

Veevers, J. J., 1967a, Cartier Furrow, a major structure along the continental margin of north-western Australia: Nature, v. 215, p. 265-267.

———, 1967b, The Phanerozoic geological history of northwest Australia: Jour. Geol. Soc. Austral., v. 14, no. 2, p. 253-272.

———, 1971, Phanerozoic history of Western Australia related to continental drift: Jour. Geol. Soc. Austral., v. 18, no. 2, p. 87-96.

———, 1973a, Stratigraphy and structure of the continental margin between North West Cape and Seringapatam Reef, northwest Australia: Bur. Mineral Resources Austral. Bull., v. 139, p. 79-102.

———, 1973b, Seismic profiles made underway on Leg 22, *in* Sclater et al. (1973).

———, 1974, Regional site surveys—259, 262, 263, *in* Veevers et al., 1974a.

———, and Evans, P. R., 1973, Sedimentary and magmatic events in Australia and the mechanism of world-wide Cretaceous transgressions: Nature Phys. Sci., v. 245, p. 33-36.

———, and Evans, P. R., 1974, Late Palaeozoic and Mesozoic history of Australia, 3rd Int. Symp. Gondwana Series: Canberra, A.N.U. Press (in press).

———, and Johnstone, M. H., 1974, Comparative stratigraphy and structure of the Western Australian margin and the adjacent deep ocean floor, *in* Veevers et al., 1974a.

———, and Wells, A. T., 1961, The geology of the Canning Basin, Western Australia: Bur. Mineral Resources Austral. Bull., v. 60.

———, and Heirtzler, J. R., 1974a, Tectonic and paleogeographic synthesis of Leg 27, *in* Veevers et al. (1974a).

———, and Heirtzler, J. R., 1974b, Bathymetry, seismic profiles, and magnetic anomaly profiles, *in* Veevers et al. (1974a).

———, Jones, J. G., and Talent, J. A., 1971, Indo-Australian stratigraphy and the configuration and dispersal of Gondwanaland: Nature, v. 229, no. 5284, p. 383-388.

———, Heirtzler, J. R., Bolli, H. M., Carter, A. N., Cook, P. J., Krasheninnikov, V. A., McKnight, B. K., Proto-Decima, F., Renz, G. W., Robinson, P. T., Rocker, K., and Thayer, P. A., 1973, Deep Sea Drilling Project, Leg 27, in the eastern Indian Ocean: Geotimes, Apr., p. 16-17.

———, Heirtzler, J. R., Bolli, H. M., Carter, A. N. Cook, P. J., Krasheninnikov, V. A., McKnight, B. K., Proto-Decima, F., Renz, G. W., Robinson, P. T., Rocker, K., and Thayer, P. A., 1974a, Initial reports of the Deep Sea Drilling Project, v. 27: Washington, D.C., U.S. Govt. Printing Office.

———, Falvey, D. A., Hawkins, L. V., and Ludwig, W. J., 1974b, Seismic reflection measurements of the northwest Australian margin and adjacent deeps: Am. Assoc. Petr. Geol. Bull. (in press).

Vening Meinesz, F. A., 1948, Gravity expeditions at sea, 1923-1938: Publ. Neth. Geod. Comm., v. 4.

von der Borch, C. C., Şclater, J. G., Gartner, S., Hekinian, R., Johnson, D. A., McGowan, B., Pimm, A. C., Thompson, R. W., and Veevers, J. J., 1972, Deep Sea Drilling Project, Leg 22: Geotimes, June, p. 15-17.

Wonfor, J. S., 1972, New global tectonics: its exploration application to the eastern section of the Australia Plate: Austral. Petr. Expl. Assoc. Jour., v. 12, no. 2, p. 34-45.

Sedimentary and Tectonic Processes in the Bengal Deep-Sea Fan and Geosyncline

Joseph R. Curray and David G. Moore

INTRODUCTION

Extensive geophysical studies have shown that the modern Bengal Deep-Sea Fan is the uppermost 4 km of the geosynclinal pile of sediments filling the Bay of Bengal, northeast Indian Ocean. The fan postdates the first collision of India and Asia and uplift of the ancestral Himalayas at the end of the Paleocene. Underlying the fan are continental rise sediments up to 12 km thick, which extend into the Bengal and Assam valleys, deposited off the margin of India following its separation from Antarctica and Australia in the Cretaceous. Deposits of the modern fan are structurally complex, particularly in the proximal part, where overlapping and interleaving natural levees and channel deposits make up the bulk of the section. Modern and buried channels in the proximal fan are tens of kilometers wide and hundreds of meters deep and are bounded by extensive natural levees, which terminate abruptly about 450 km downslope from the canyon mouth. Channel size and levee development are significantly less beyond this point. Turbidity currents which top the large channels of the proximal fan spread largely as sheet flow. Thus, in contrast to the proximal fan, most of the central and distal fan is sheet-flow-derived and shows lateral continuity in section, with isolated channels and channel deposits. Deformation is occurring simultaneously with deposition, as the fan and geosyncline pass obliquely northeast into the subduction zone of the Sunda Arc and Indoburman Ranges.

REGIONAL SETTING

The Bay of Bengal, northeastern Indian Ocean, is bordered by Sri Lanka, India, Bangladesh, Burma, the Andaman and Nicobar Islands, the Andaman Sea, and Sumatra (Fig. 1). It is presently floored by the Bengal Deep-Sea Fan, the largest deep-sea fan in the world (Curray and Moore, 1971), which extends from approximately 20°N to 7°S, a total length of approximately 3,000 km, with a width of 1,000 km. Sediments of the Bengal Fan have been derived predominantly from drainage of the Ganges and Brahmaputra rivers, confluent in Bangladesh. The Ganges River drains much of the south slope of the Himalayas, while the Brahmaputra drains much of the north slope of the Himalayas. The confluent rivers deliver an enormous supply of sediment, some of which enters the submarine canyon "Swatch of No Ground," to be distributed by turbidity currents through a system of turbidity current channels or fan valleys that extend the length of the Bengal Fan. The fan is divided by the Ninetyeast Ridge into two lobes: the main Bengal Fan and the eastern lobe, which we have named the Nicobar Fan (Curray and Moore, 1971). The eastern margin of these fans is the Andaman-Sunda Trench. Off Sumatra, the arc boundary is formed by the outlying Mentawai Islands and to the north by the Nicobar and Andaman sedimentary ridge and the Indoburman Ranges of Burma. These extend to the eastern Himalayan syntaxis towering above the Assam Valley, where the compressive plate edge extends to the west into the high Himalayan Mountains.

Underlying the Bengal Fan and its northern subaerial extension, the Bengal Basin, is one of the thickest sediment sections in the world, exceeding 16 km of sediment beneath the continental shelf of Bangladesh (Moore et al., 1974). In this report we shall discuss the sedimentary and tectonic processes that have acted during 130 my of history of this sediment wedge. We shall do this in three parts. First, our reconstruction of the geological history of the Bay of Bengal and Bengal geosyncline is presented. Second, we describe the facies distribution and discuss the sedimentary processes of the modern Bengal Fan. Third, the present tectonics and the orogenic fate of this geosynclinal wedge of sediments are discussed.

Our plate tectonics model for the geological history of this region has been developed to fit best the large amount of geological and geophysical data we and our colleagues have collected during Scripps' (SIO) expeditions to the region in 1968, 1971, and 1973. Almost 50,000 km of ship's track, with continuous seismic-reflection profiling, magnetics, and 3.5-kHz echo sounding have been completed (Fig. 2). Continuous gravity profiling was also carried out on the last cruise. Sampling included both piston cores of the sediments and dredge hauls of outcropping rock. We have made over 100 seismic stations; these include airgun, wide-angle reflection and refraction for determination of velocity of the sedimentary section, and over 60 explosion seismic-refraction stations for deeper crustal structure. We and our colleagues have studied sedimentology of the cores, and we have made our own interpretations of the descriptions in the literature of the geology of the Himalayan and Indoburman Mountains, the Andaman and Nicobar Islands, and the Mentawai Islands off Sumatra.

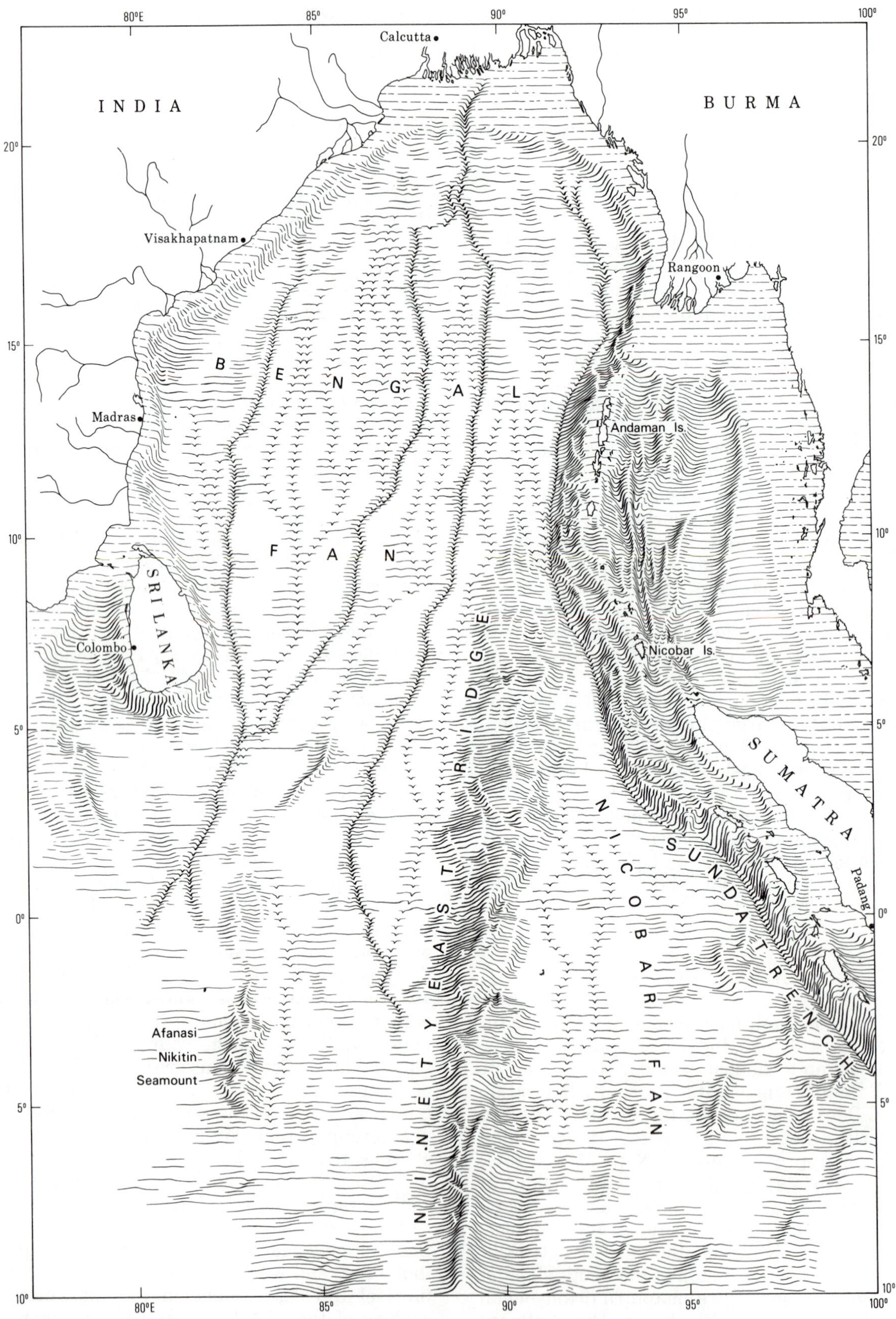

Fig. 1. Present topography and channel system of the Bengal Fan. Channel immediately west of Ninetyeast Ridge is being abandoned today.

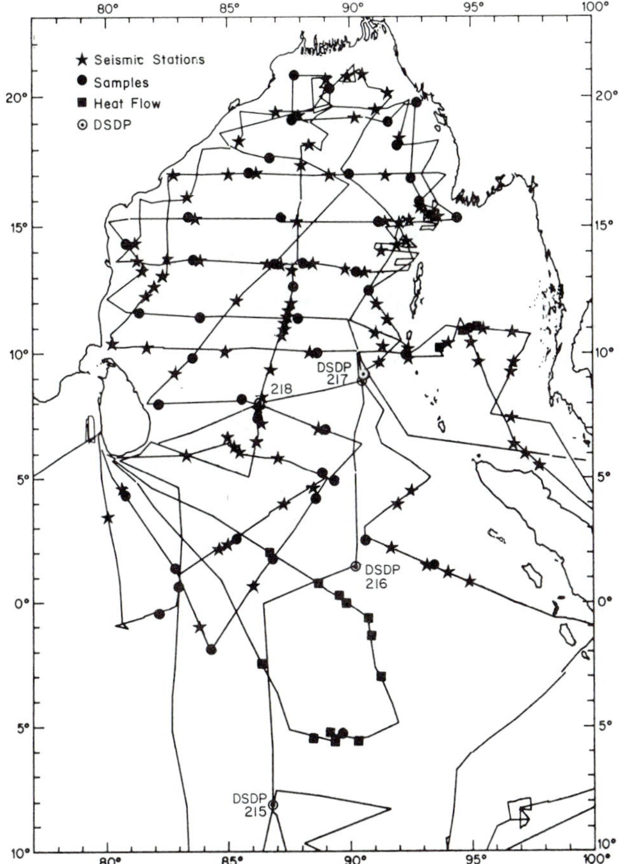

Fig. 2. Geophysical control and Deep Sea Drilling sites in the Bengal Fan region.

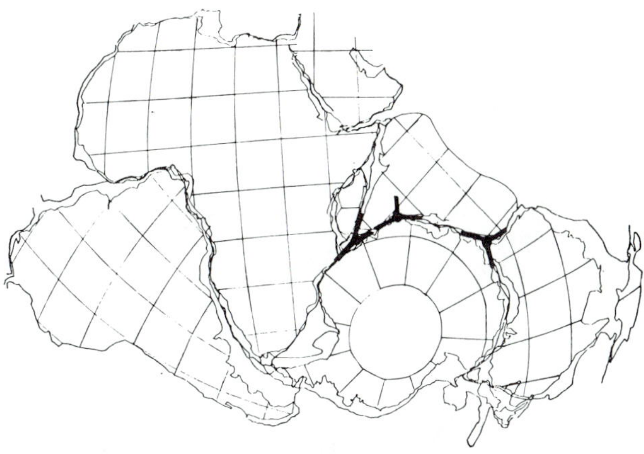

Fig. 3. Gondwanaland reconstruction modified from Smith and Hallam (1970) by increasing the area of Indian continental crust believed to have underthrust Asia. Three or more hot spots, shown as rrr tripled junctions in heavy lines, may have lain between India, Antarctica, and Australia.

GEOLOGICAL HISTORY

Our interpretation of the geological history starts with the arrangement of continents in Gondwanaland (Fig. 3). We will ignore for the present purposes the fit between western and eastern Gondwanaland, and discuss only the relationship among India, Australia, and Antarctica. We have modified the Smith and Hallam (1970) reconstruction by enlarging Gondwanic India to the area that we believe has underthrust southern Asia to form the Tibetan Plateau and Tsaidam Basin to the northeast. We have subtracted approximately half of the area of Tasidam Basin and the Tibetan Plateau to allow for shortening of the southern margin of Asia which existed prior to the collision of India and southern Asia. The resulting configuration of greater Gondwanic India produces an excellent fit in the Smith and Hallam reconstruction, closing the gap that has existed in most reconstructions between India and Australia. This closure also appears to satisfy some geological requirements of Australia (Veevers et al., 1971).

Our reconstruction also includes what we believe were the positions of three important hot spots, which joined arms to form the rifts between Australia, India, and Antarctica. We believe that the hot spot from the bight of India formed the Ninetyeast Ridge and lies today under St. Paul and Amsterdam islands. The hot spot from the corner between India, Antarctica, and Australia was responsible for forming Broken Ridge and Kerguelen Plateau after separation of India from Antarctica and Australia (Fig. 4). One arm of the rift junction at the bight of India failed and produced an aulacogen, which is occupied by the Godavari River valley of India (Fig. 4). A third hot spot is postulated at the ill-defined corner between India, Africa, and Antarctica. The exact position of Sri Lanka (Ceylon) is probably erroneous in this reconstruction, as are the position of Madagascar and the relationship between eastern and western Gondwanaland.

The initial breakup of this part of Gondwanaland we believe occurred in the early Cretaceous, with separation of India from combined Australia and Antarctica in a direction approximately perpendicular to the northeast-trending continental margin of India. New sea floor was created in what is now the Bay of Bengal, as well as sea floor that we believe has disappeared beneath the Sunda Arc subduction zone. After separation, sediments were shed off all sides of India to form the Indus Flysch, Naga Hills Flysch, Indoburman Flysch, Andaman Flysch, and part of the section that now underlies the present Bay of Bengal (Fig. 4). The northern 15° of the Ninetyeast Ridge consists of en echelon topography resulting from the offset pattern of this initial split, similar to the geometry of rifting of the present Gulf of California. Deposition of sediments was probably rapid, accompanying a subsidence of the sea floor adjacent to the southeastern continental margin of India.

In Late Cretaceous time, approximately 75 my ago, direction of sea-floor spreading changed to become approximately north-south, parallel to the present Ninetyeast Ridge. This northerly movement of the Indian plate persisted until about 35 my ago and was taken up on the Chagos-Laccadive transform on the west and on the Ninetyeast Ridge

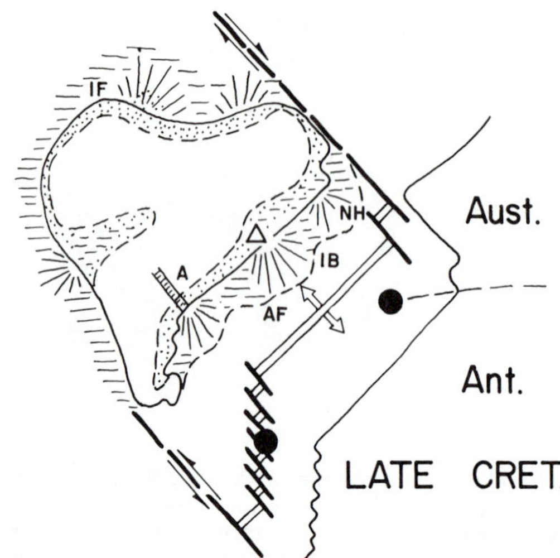

Fig. 4. Late Cretaceous paleogeography. Locations of the possible hot spots are shown as circles. Spreading directions indicated by arrows from spreading ridge, shown with double lines, and transform faults, shown with single heavy lines. Present delta of Ganges-Brahmaputra shown by open triangle. NH, Naga Hills Flysch; IB, Indoburman Flysch (Brunnschweiler, 1966); AF, Andaman Flysch (Karunakaran et al., 1964); IF, Indus Flysch (Gansser, 1964); A, Godavari River aulacogen.

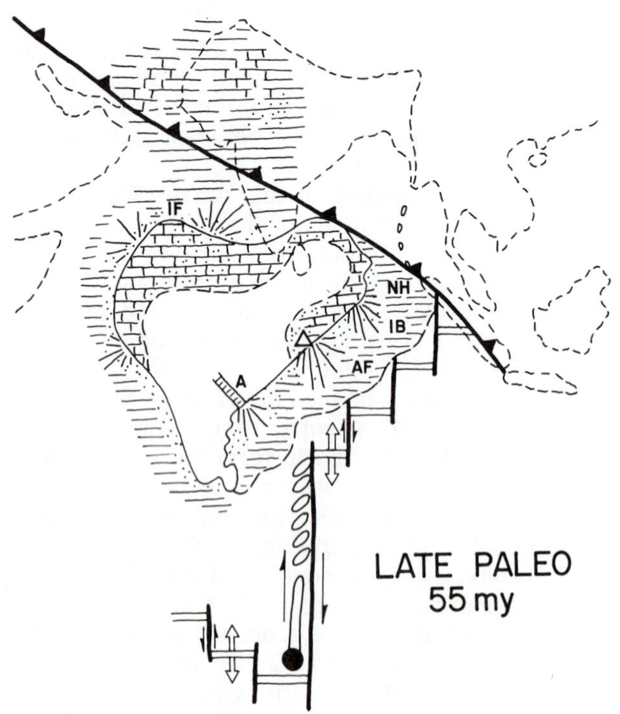

Fig. 5. Late Paleocene paleogeography, time of first collision of India and the subduction zone at the southern margin of Asia. This subduction zone may have been either an Andean-type continental margin or an island arc separated from Asia by a marginal basin. Prior to collision, relative motion had been north-south, parallel to the transform on the east side of Ninetyeast Ridge. (Abbreviations as in Fig. 4.)

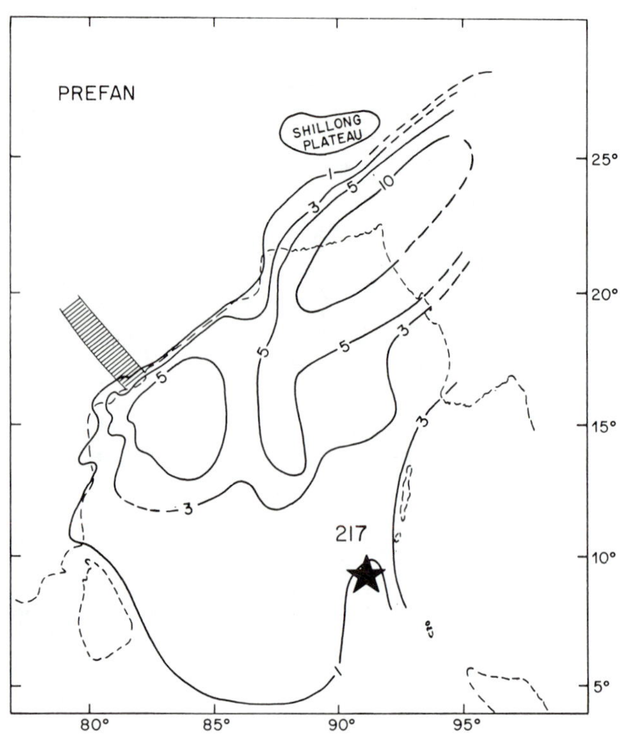

Fig. 6. Isopachs of sediments (in kilometers) deposited before collision at 55 my. Assam Valley (Fig. 15) is restored 250 km left lateral along Dauki fault (Evans, 1964).

transform at the eastern plate edge (McKenzie and Sclater, 1971). The transform fault formed on the east side of what became the Ninetyeast Ridge, flanking the hot spot that originally lay in the bight of India. The relatively northerly motion of the Indian plate over this hot spot resulted in the massive intrusion of magma into the now dormant and offset spreading center. This built an island chain which has since subsided to form the relatively shallow northern Ninetyeast Ridge. The Ninetyeast Ridge has been shown in Deep Sea Drilling Project (DSDP) to have once been shallow-water to subaerial and to be older; that is, Campanian age—75 my at the northern end and becoming progressively younger to the south (von der Borch et al., 1974).

Our projections suggest that the initial contact or collision between India and southern Asia occurred in late Paleocene time, approximately 55 my ago (Fig. 5). This was a time of change in relative plate motion in the entire Indian Ocean, with slowing of the rate of sea-floor spreading (McKenzie and Sclater, 1971; Sclater and Fisher, 1974). It coincides with the time of an extensive regional unconformity in the sedimentary section underlying the present Bay of Bengal. The unconformity has been traced in our seismic records, both as an unconformable surface in seismic reflection profiles and as a pronounced velocity contrast in our seismic-refraction measurements.

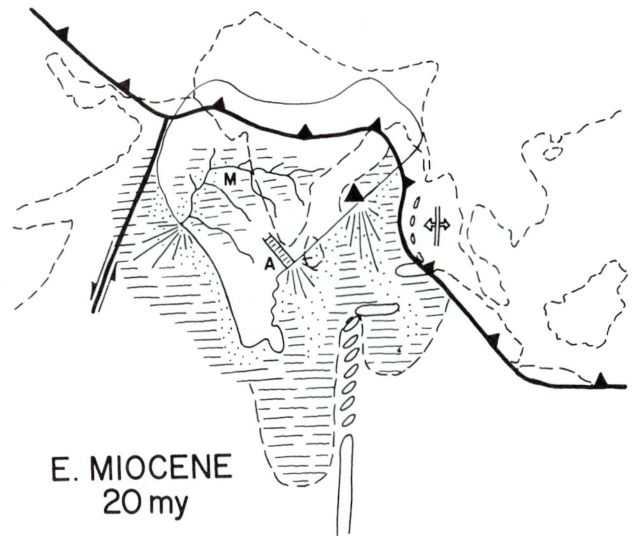

Fig. 7. Early Miocene paleogeography with India "plowing" into Asia or into island arc — marginal sea complex of southern Asia. M, Murrees (Gansser, 1964) derived from continental India. Tectonic extension has probably started in Andaman Sea (see Fig. 15).

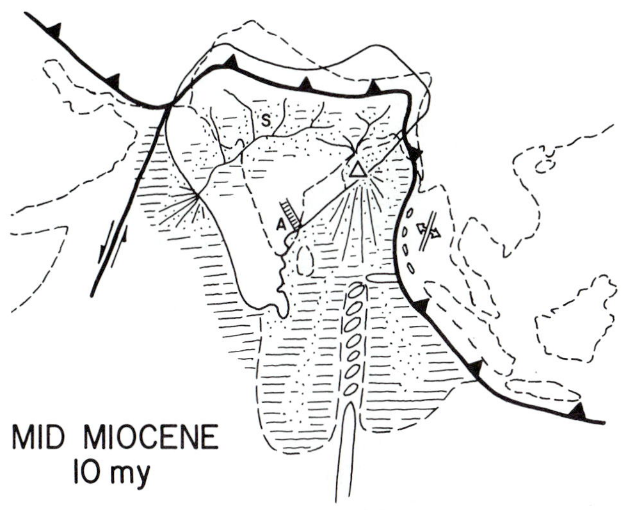

Fig. 8. Middle to late Miocene paleogeography. S, Siwaliks (Gansser, 1964) derived from rising Himalayas.

We have extrapolated the age of the lower regional unconformity from drilling of an admittedly anomalously thin part of the sediment section at DSDP site 217, located on the northern end of the Ninetyeast Ridge (Fig. 6; Moore et al., 1974, Fig. 5). From our interpretation of the geology of the Andaman-Nicobar islands (Karunakaran et al., 1964), of the Indoburman Ranges of Burma (Brunnschweiler, 1966), and from published results of drilling and seismic work in the Bengal Basin and Assam Valley (Sengupta, 1966; Evans, 1964), we conclude that this unconformity spans the time from late Paleocene to middle Focene time in the drilling at DSDP site 217. Sediments that were deposited prior to the time of collision and before development of the unconformity in late Paleocene time are isopached on the basis of velocity and structure to indicate the volumetric distribution of this great sediment mass (Fig. 6). The isopachs show that the sediments comprise a continental rise-like wedge lying parallel to the elongate northeast-southwest continental margin of India, with thicknesses in excess of 10 km extending from the present bay into the Bengal Basin, Assam Valley, and Naga Hills. A second depocenter lies off the ancient aulacogen of the Godavari River valley (the "Guntur" of Burke and Dewey, 1973).

Plate motion resumed or accelerated in a slightly different direction in early Oligocene time, approximately 35 my ago (McKenzie and Sclater, 1971; Sclater and Fisher, 1974), with India converging toward Asia in a more northeasterly direction. Motion along the transform fault on the east side of the Ninetyeast Ridge had stopped by this time, and sediments of the Bengal Fan now extended east of the Ninetyeast Ridge to form the Nicobar Fan. India was "plowing" itself into the southern margin of Asia or into a bordering island arc and marginal sea. During Oligocene and early Miocene time (Fig. 7), however, sediments now found in the sub-Himalayan Mountains consisted of the Murrees, derived predominantly from continental India in the south. The Himalayan Mountains had apparently not yet been uplifted to sufficient height to provide much detritus. In contrast, the early sections of the modern Bengal Fan are probably derived from these ancestral Himalayan Mountains, and also during this time the Andaman and Nicobar Island Ridge first began to form, presumably by the offscraping of fan sediment over the Sunda Arc subduction zone.

By middle to late Miocene time (Fig. 8), the modern Himalayan Mountains were forming. Sediments on continental India deposited at that time constitute the Siwalik series, derived from the rising youthful mountains. Sediments were actively shed off both sides of India into the present Indus Fan to the west and the modern Bengal Fan to the east, where both lobes of the fan, the Bengal and Nicobar, were active. By Pliocene time, turbidites of the Nicobar Fan were carried as far as the southern end of Sumatra and have been sampled in DSDP site 211, showing identical mineralogy to sands in the Bengal Fan (Thompson, 1974). During all this time, from late Cretaceous to the present, pelagic sediments were deposited over the subsiding Ninetyeast Ridge and intermittently over inactive portions of the distal parts of the fan.

At the present time (Fig. 9), India is still moving northeasterly into Asia. Seismic activity continues along the Sunda Arc up to the eastern Himalayan syntaxis, across the Himalayan Mountains, under the Tibetan Plateau and southern China, and around to Iran. The eastern lobe, or Nicobar Fan, was cut off from its supply of sediments in

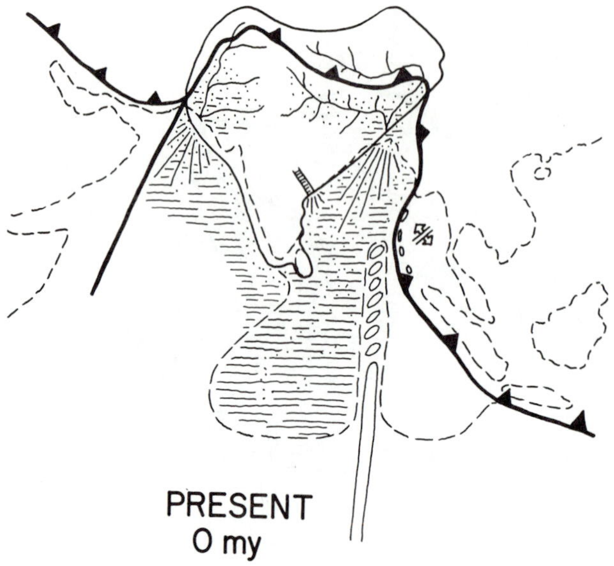

Fig. 9. Present geography and depositional patterns with continuing convergence. Nicobar Fan was cut off from supply in Quaternary by convergence of Ninetyeast Ridge with Sunda-Andaman Trench. Spreading in Andaman Sea is now oriented northwest-southeast.

middle Pleistocene time, as the northern end of the Ninetyeast Ridge converged on the Sunda-Andaman Trench. All sediment that now passes between the northern end of the Ninetyeast Ridge and Andaman-Nicobar Ridge pours down the axis of the Sunda Trench off Sumatra and Java, and the still open channels and interchannel areas of the Nicobar Fan receive only pelagic sediments.

Isopachs of sediment thickness of the modern Bengal Fan were prepared largely on the basis of seismic-refraction and reflection data (Fig. 10). The isopachs still show some excess thickness parallel to the continental margin of eastern India, but also now show a main trend of deposition extending down the central part of the fan between the northern part of the Ninetyeast Ridge and Sri Lanka. Greatest thicknesses of the modern fan exist at the north and extend into the now-filled Bengal Basin and Assam Valley. Our extrapolation (Moore et al., 1974) of stratigraphic ages from DSDP sites 217 and 218 northward to published information on drilling from the Bengal Basin (Sengupta, 1966) suggests that initiation of the modern fan part of the section started in middle Eocene time.

SEDIMENTARY PROCESSES AND FACIES DISTRIBUTION

At present there is a single large turbidity current channel issuing from the submarine canyon, "Swatch of No Ground," at the head of the Bay of Bengal, but many beheaded and abandoned channels of major dimensions are also mapped (Fig. 1). Sediments of the confluent Ganges and Brahmaputra rivers today are deposited predominantly in the huge subaerial delta of these rivers, but during the lowered sea levels of Pleistocene time they probably poured largely into the existing submarine canyon, and into similar older, abandoned and now-filled canyons, to flow down the turbidity current distributary channels. Channel dimensions are therefore now adjusted to conditions of Pleistocene lowered sea level, and turbidity current activity is today considerably reduced because the head of the present submarine canyon is now somewhat removed from depocenters off the delta. As a rule, channel development of the Bengal Fan does not show bifurcation and braiding. The main active channel today has great continuity without bifurcation as it passes to the southwest of Ceylon, nearly 3,000 km from the canyon.

A major central channel is now cut off near its head, and parts of the fan previously supplied by this channel are now receiving predominantly pelagic sediments. Active channel abandonment and migration occur only in the proximal part of the fan (about the northern one quarter). We must stress here that scale and dimensions are important. The proximal part of this fan exceeds 500 km in length, larger than almost any other complete deep-sea fan in the world. The entire La Jolla submarine fan, certainly one of the best studied fans in the world, would fit into the "Swatch of No Ground" submarine canyon cutting the continental shelf.

Within the proximal fan area, there has been a continuing process of channel-jumping, with new, large natural levee complexes commonly being built on the flanks of previously active channels (Fig. 11). Virtually the entire pattern and history of turbidite

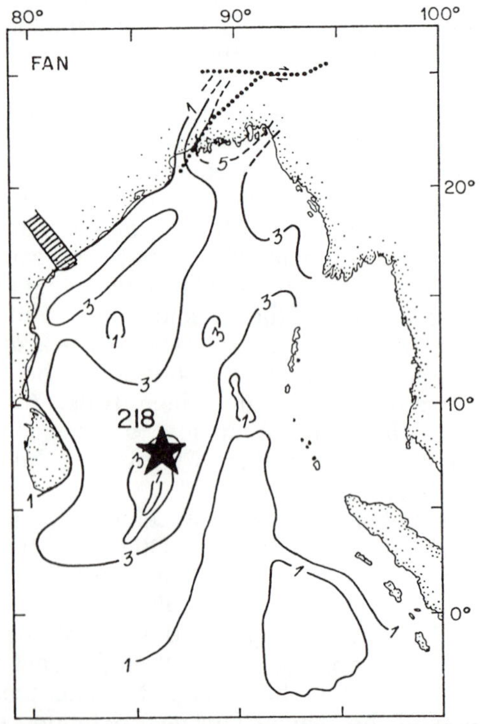

Fig. 10. Isopachs (in kilometers) of Bengal Fan and Bengal Basin sediments above the Paleocene-Eocene unconformity.

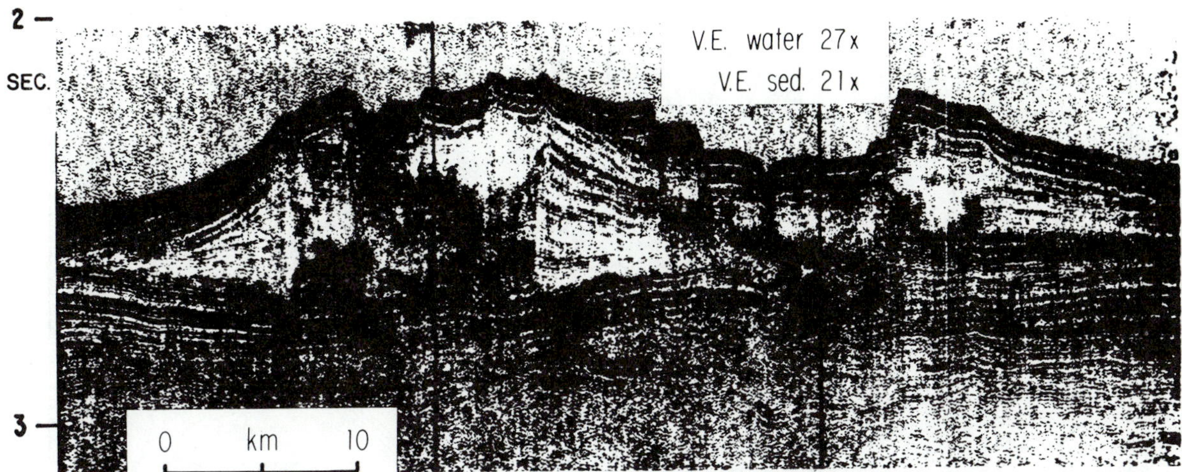

Fig. 11. Reflection profile across channel-levee complex in proximal portion of Bengal Fan at 19°N. The Ganges-Brahmaputra bed load is very high. Some of this sand must be deposited in the channel-levee complex.

deposition on the fan is controlled by the channel changes within the huge natural levee-channel complexes of this proximal fan. It is known that the bedload of the confluent Ganges and Brahmaputra rivers is very high. We presume, although we cannot prove, that much of the sand of this bedload is deposited in these levee-channel complexes, with probably clean sands forming on the bottom of some of the channels and with muddy sands forming the flanks or levee deposits.

A composite cross section through the proximal portion of the fan (Fig. 12) shows interleaved channel-levee sediment wedges, interspersed with sheet-flow or interchannel fan sediments. Our piston cores, which are at the most 6-7 m in length, penetrate only the Holocene and uppermost portion of the Pleistocene and could hardly be represented on a diagram of this scale. Nevertheless, we do sample sandy sediments adjacent to the channels, and we believe further that extensive parts of these channel-levee complexes are sandy, perhaps ranging from muddy sands to very clean channel sands. If this interpretation is correct, a sand-shale ratio for this part of the fan could approach 1:1.

The levee-channel complex sediments, although largely confined to the proximal or northern part of the fan, are also associated with the total length of these channels as isolated linear deposits within the vast, monotonous, horizontally layered sheet-flow strata. The sheet-flow parts of the fan comprise the interchannel areas and are believed to be muddy sediments deposited from broad-fronted turbidity flows originating where major, primary turbidity currents debouch from the canyon mouth to overtop the large levees and spread over the fan. Other sheet-flow currents must also form where the high levees and broad deep valleys terminate abruptly on the proximal fan about 450 km from the canyon (Fig. 13).

Smaller sheet flows may originate on the lower parts of the fan wherever channel overtopping occurs on a broad front. Channels passing through

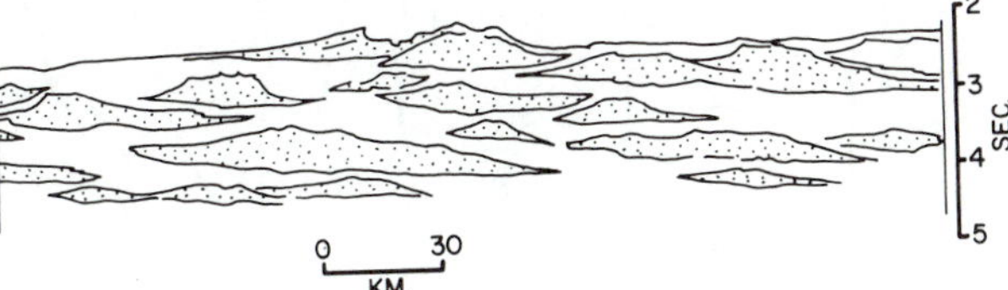

Fig. 12. Line drawing of east-west section of proximal portion of fan at 19°N, showing interleaved channel-levee complexes (stippled) and fan sheet-flow turbidites.

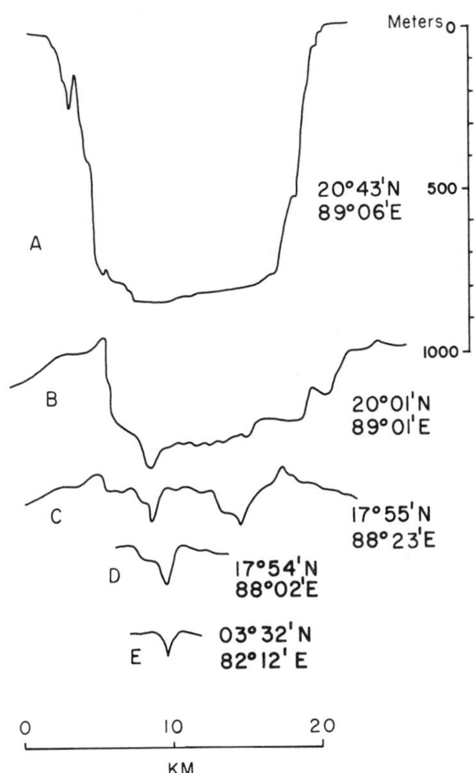

Fig. 13. Series of bathymetric cross sections of canyon (A) and connecting channel on the proximal (B and C), central, and distal (D and E) parts of the fan. Note that the cross-sectional area, and hence capacity of the channel, decreases abruptly as the large elevated natural levees terminate on the proximal fan.

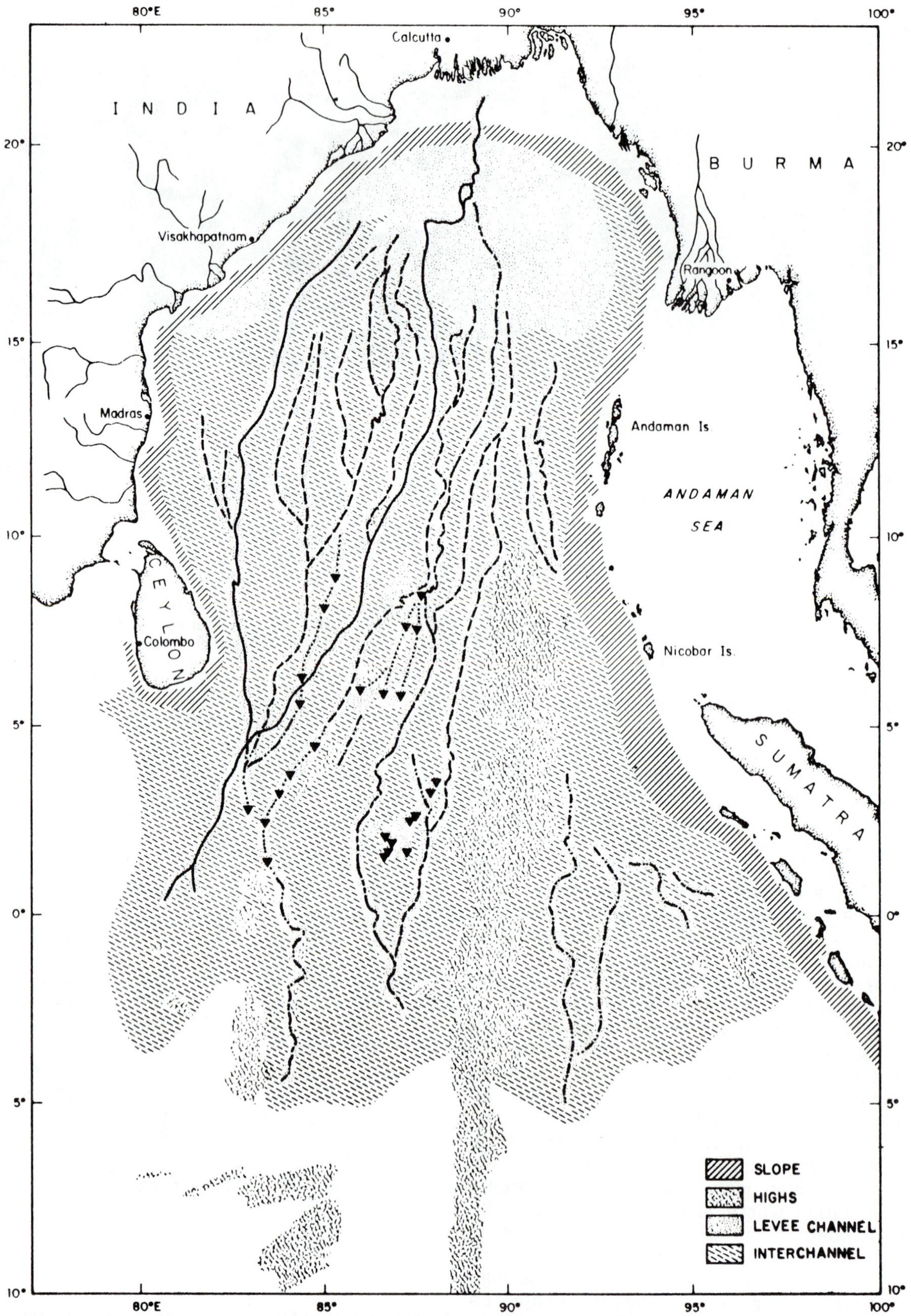

Fig. 14. Generalized distribution of facies in the Bengal Fan (post-Paleocene) and associated areas. Levee-channel deposits and interchannel sheet-flow deposits are fan turbidites. Pelagic sediments are deposited on highs. Slope sediments are derived from adjacent terrigenous sources and pelagic source. Sediments forming marginal slope from Burma to beyond Sumatra may be either fan sediments scraped off over Sunda-Andaman Arc subduction zone or a veneer of in-place slope and slope basin deposits. Small triangles show crossings of buried channels.

the medial and distal portions of the fan are smaller, more widely spaced, and show many of the characteristics of river valleys, with relatively small natural levees, terraces, and inner channels showing changes in flow regime, just as rivers do. Still, however, they are several kilometers in width and up to 100 m in depth. The extensive medial and distal horizontally stratified sediments which comprise the bulk of these parts of the fan are predominantly turbidites. Distal parts of the fan, however, and portions of the fan that are periodically cut off from a direct supply by the turbidity current channels do include larger proportions of pelagic sediments.

The areal distribution of the fan facies is summarized in Figure 14. Proximal facies comprise predominantly levee-channel complexes interleaved with interchannel deposits and with each other. Isolated channels and abandoned channel systems, with their relatively small flanking levees, lie within the vast interchannel sheet flow or muddy fan sediments, making up the bulk of the medial and distal facies. Pelagic sediments occur mainly on topographic highs at the Ninetyeast Ridge and on isolated hills, and the continental shelf and slope sediments ring the bay on the west, north, and northeast. Present deposition on the Nicobar Fan is largely pelagic, as it is also on parts of the southern or distal part of the Bengal Fan, now cut off from active distribution by turbidity currents.

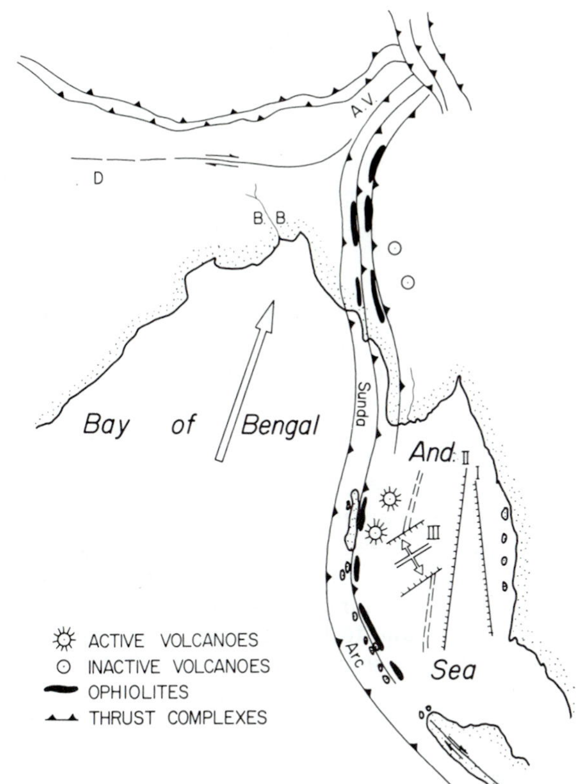

Fig. 15. Diagrammatic sketch of present tectonics of the head of the Bay of Bengal, Bengal Basin (B.B.), Assam Valley (A.V.), and the Andaman Sea. Subduction occurs along the Sunda-Andaman Arc, which passes into the Indoburman Ranges to the eastern Himalayan syntaxis. Large arrow shows direction of plate convergence. Extension in the Andaman Sea started in about Oligocene to Miocene time (I), with direction of extension progressively rotating clockwise (II, III). Present rifting in the Andaman Sea, northwest-southeast (III), gives a component of subduction even along the northern Andaman Islands. Strike-slip faulting parallel to thrust fronts is shown by the Semangko fault (S) in Sumatra and the Dauki fault (D) north of the Bengal Basin.

PRESENT TECTONICS AND OROGENIC FATE

The present tectonics of the Bay of Bengal region have important implications for the future orogenic fate of the continental rise and fan sediments underlying the bay. Present relative plate motion between the Indian plate and the Asian plate is apparently in a northeast-southwest direction at a rate of convergence of between 5 and 6 cm per year. The plate edge lies along the Sunda Arc, which is fronted by the Java Trench off Java and Sumatra and the filled trench off the Andaman-Nicobar islands (Fig. 15). The plate edge passes into the Indoburman Ranges and to the eastern Himalayan syntaxis, where it swings to the westward, along the suture of the Himalayan Mountains. The plate edge is marked by seismicity. The seismicity, however, becomes very diffuse in the region of the Andaman-Nicobar Ridge and the Andaman Sea, Burma, and southern Asia, possibly as a result both of extension in the Andaman Sea and continent collision and underthrusting in southern Asia.

The Mentawai Islands off Sumatra, the Andaman-Nicobar Ridge, and the Indoburman Ranges are composed primarily of sedimentary rocks of ages and lithologies suggestive of sediments and sedimentary rocks of the Bengal geosyncline lying to the west (Karunakaran et al., 1964; Brunnschweiler, 1966; H. Wories, personal communication). We interpret these as Bengal geosyncline and fan sediments which have been scraped off above the subduction zone. Active volcanism occurs in the Andaman Sea (Rodolfo, 1969), and inactive volcanoes of apparent andesitic composition lie in the central valley of Burma. Interspersed through the sediments fronting the volcanic arc are ophiolite bodies (probable Cretaceous sea floor), presumably representing early stages of the opening of the ocean between India, Australia, and Antarctica.

Our detailed seismic-reflection profiles of the Andaman-Nicobar Ridge suggest that it is composed almost entirely of folded Bengal Fan sediments, arranged grossly in eastward-dipping thrust sheets or nappes. These folds, in turn, form sediment dams or basins, which pond sediments passing downslope from higher parts of the ridge (Fig. 16). Similar folding, increasing in intensity to thrust faulting to the east is shown in subaerially exposed parts of the Chittagong, Arakan, and Naga hills of the Indoburman Ranges.

Tectonics of this region are even more complicated when examined in the light of major plate motions. Relative motion between the Indian and the Asian plates would suggest only transcurrent motion along the northern part of the Sunda Arc. Structure of the sediments, however, suggests compression, and a first-motion study of an earthquake beneath the southern tip of Burma (Molnar et al., 1974) suggests southeasterly subduction of the Indian plate beneath the arc. We believe that this compression occurs because the Andaman Sea is presently, and has been throughout much of Tertiary time, an extensional basin. The present extension is in a northwest-southeast direction, perpendicular to a complex central Andaman rift valley. Prior to the present era of spreading, we believe that the spreading or extension directions (Fig. 17) were east-west, with the first period of spreading representing rifting at the edge of the continental shelf off the Malay Peninsula. The second episode of spreading occurred in a west-northwest direction away from the edge of the Mergui Terrace in the Andaman Sea, and finally, the last has rotated around to the present northwest-southeast spreading direction.

The presently active rift valley lies within the older, broad basin floor. Beneath the flat-lying sediments that were deposited prior to the opening of this rift lies an older section with upturned broken edges and butt slopes facing the active open rift. Underlying this are still older sediments. We do not yet know the ages of these units, but they constitute proof of a kind of episodic spreading, even during this latest period of northwest-southeast extension of the Andaman Sea.

In conclusion, we present a diagrammatic cross section from the eastern Bengal Fan through the Andaman-Nicobar Ridge, the active volcanic arc, and the spreading axis of the Andaman Sea to the Malay Peninsula (Fig. 18). Sediments of the Bengal Fan, ranging in age from Early Cretaceous to Holocene, are now passing on the subducting plate into the Sunda-Andaman Arc. Portions of these sediments and the underlying ocean floor are subducted, and other parts are scraped off and form the predominantly sedimentary arc of the Andaman and Nicobar Islands, Mentawai Islands, and Indoburman Ranges. Ophiolites of Cretaceous age represent fragments of ocean floor formed during the earliest episode of opening between India, Antarctica, and Australia. The sediments are roughly arranged in eastward-dipping nappes or thrust plates, with deformed folded sediments arranged internally within these nappes.

The topography has created sedimentary basins for ponding of arc-trench gap and trench slope sediments. During the latter period of Bengal Fan

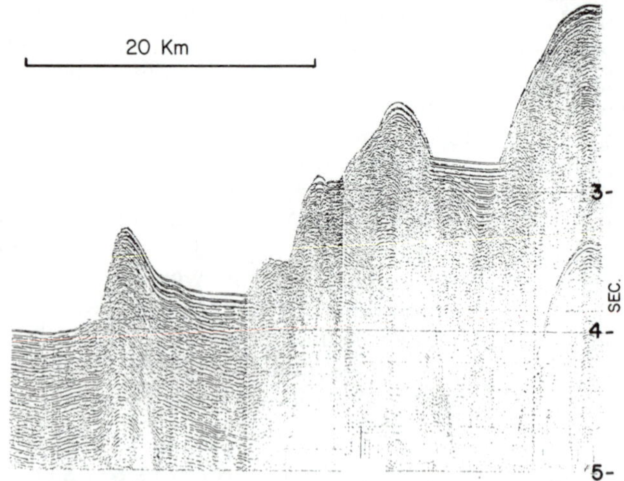

Fig. 16. Early deformation of fan sediments converging with Sunda-Andaman Arc subduction zone off the Andaman Islands. Slope basins are formed by folded fan sediments above the subduction zone. Basins are tilted landward by continuing tectonic addition of fan material to slope base by folding and thrusting.

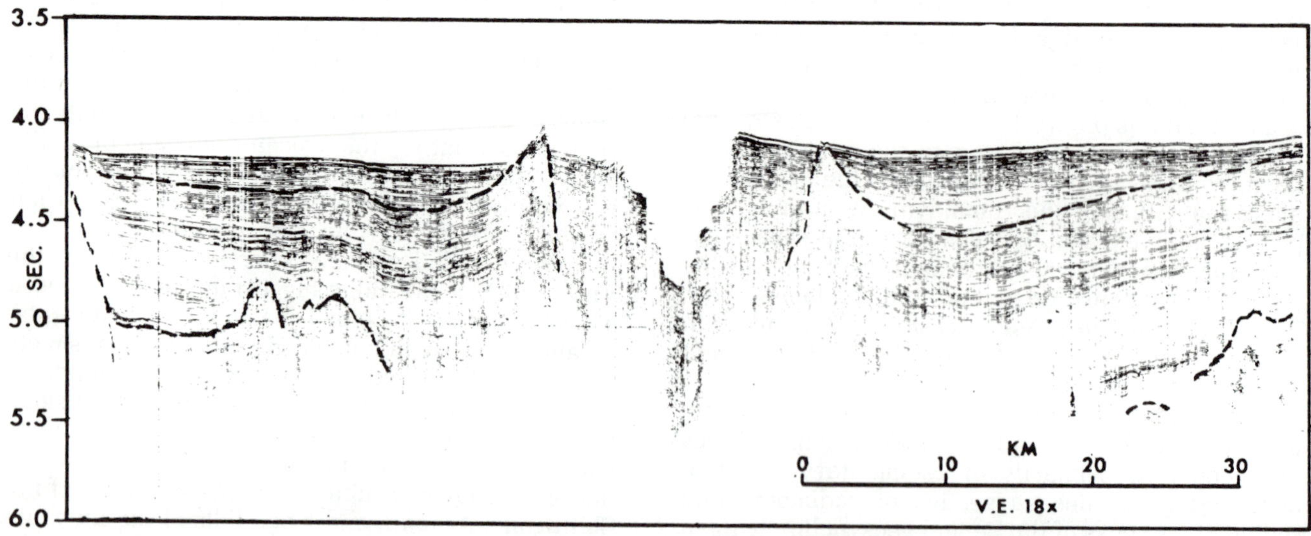

Fig. 17. Reflection profile across central Andaman Valley, showing episodic extension and the present rift valley, which has high heat flow. Three episodes of rifting can be seen.

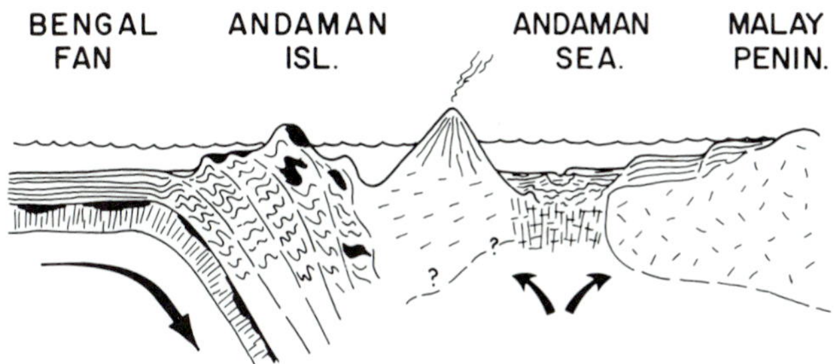

Fig. 18. Diagrammatic present-day cross section from the Bengal Fan to the Malay Peninsula.

deposition and the formation of the Andaman-Nicobar Islands, extension has occurred in the Andaman Sea with successive rifting and down-dropping of the continental shelf, Mergui Terrace, and central Andaman Basin. Today, the sedimentary ridge of the Mentawai Islands off Sumatra, the Nicobar and Andaman Islands, and the Indoburman Ranges form a mountain range comparable in size, dimensions, and volume to the Appalachian Mountains. To the west of the Sunda-Andaman Arc, an even greater volume of sediments lies beneath the present Bay of Bengal.

ACKNOWLEDGMENTS

This work was undertaken with financial support from the Office of Naval Research. We gratefully acknowledge the assistance of our many colleagues with whom we have conducted field operations, interpretation and analysis of data. We especially mention R. W. Raitt, F. J. Emmel, P. J. Crampton, and M. Henry. We have previously presented many of these thoughts in public oral presentations, including Houston, September 1972; La Jolla, October 1972; Sydney, January 1973; La Jolla, March 1973; Madison, November 1972; and San Antonio, April 1974.

BIBLIOGRAPHY

Brunnschweiler, R. O., 1966, On the geology of the Indoburman Ranges: Jour. Geol. Soc. Australia, v. 13, p. 137-194.

Burke, K., and Dewey, J. F., 1973, Plume-generated triple junctions: key indicators in applying plate tectonics to old rocks: Jour. Geology, v. 81, p. 406-433.

Curray, J. R., and Moore, D. G., 1971, Growth of the Bengal Deep-Sea Fan and denudation in the Himalayas: Geol. Soc. America Bull., v. 82, p. 563-572.

Evans, P., 1964, The tectonic framework of Assam. Jour. Geol. Soc. India, v. 5, p. 80-96.

Gansser, A., 1964, Geology of the Himalayas: New York, Wiley-Interscience, 289 p.

Karunakaran, C., Ray, K. K., and Saha, S. S., 1964, Sedimentary environment of the formation of Andaman Flysch, Andaman Islands, India: Rept. 22nd Internatl. Geol. Congr., India, pt. XV, p. 226-232.

McKenzie, D. P., and Sclater, J. C., 1971, The evolution of the Indian Ocean since the Late Cretaceous: Geophys. Jour., v. 25, p. 437-528.

Moore, D. G., Curray, J. R., Raitt, R. W., and Emmel, F. J., 1974, Stratigraphic-seismic section correlations and implications to Bengal Fan history, *in* von der Borch, C. C., Sclater, J. G., et al., Initial reports of the Deep Sea Drilling Project, v. XXII: Washington, D.C., U.S. Govt. Printing Office, p. 403-412.

Rodolfo, K. S., 1969, Bathymetry and marine geology of the Andaman Basin, and tectonic implications for Southeast Asia: Geol. Soc. America Bull., v. 80, p. 1203-1230.

Sclater, J. G., and Fisher, R. L., 1974, Evolution of the east central Indian Ocean, with emphasis on the tectonic setting of the Ninetyeast Ridge: Geol. Soc. America Bull., v. 85, p. 683-702.

Sengupta, S., 1966, Geological and geophysical studies in western part of Bengal Basin, India: Am. Assoc. Petr. Geol. Bull., v. 50, p. 1001-1017.

Smith, A. G., and Hallam, A., 1970, The fit of the southern continents: Nature, v. 225, p. 139-144.

Thompson, R. W., 1974, Mineralogy of sands from the Bengal and Nicobar Fans, sites 218 and 211, eastern Indian Ocean, *in* von der Borch, C. C., Sclater, J. G., et. al., Initial reports of the Deep Sea Drilling Project, v. 22: Washington, D.C., U.S. Govt. Printing Office, p. 711-714.

Veevers, J. J., Jones, J. G., and Talant, J. A., 1971, Indo-Australian stratigraphy and the configuration and dispersal of Gondwanaland: Nature, v. 229, p. 383-388.

von der Borch, C. C., Sclater, J. G., et al., 1974, Initial reports of the Deep Sea Drilling Project, v. 22: Washington, D.C., U.S. Govt. Printing Office, 890 p.

Continental Margins of India

H. Closs, Hari Narain, and S. C. Garde

INTRODUCTION

The first magnetic observations at sea included the west coast of India, during the voyage of João de Castro, 1538-1541 (Hellman, 1897). He sailed around the Cape of Good Hope, up the west coast of India and into the Red Sea, making 43 determinations of declination. After this cruise the Indian Ocean did not attract much attention except for its partial coverage by the nonmagnetic vessels GALILEE (1905-1908), CARNEGIE (1909-1929), and the Russian ship ZARYA (1956-1961). Systematic investigations since then have been conducted by various agencies under the International Indian Ocean Expedition (IIOE) during 1961-1965, and by a U.S. Coast and Geodetic Survey global expedition on OCEANOGRAPHER in 1967. Investigations till the end of the IIOE program have been reviewed by Narain et al. (1968) and deal mainly with geophysical investigations on the continental shelf and slope.

This report is concerned mainly with bringing the review of Narain et al. up to date and presenting new or relevant results of marine or nearshore geology and geophysics.

WESTERN MARGIN OF INDIA

Physiography. The continental shelf near Karachi widens from about 100-160 km southwest of the Gulf of Cutch (Pepper and Everhart, 1963). The shelf north of Bombay broadens to a maximum width of about 300 km. Three topographic profiles from the western coast near Bombay, Mangalore, and Cochin (Narain et al., 1968) show the bottom topography (Fig. 1). Near Bombay, the continental shelf is about 220 km wide, narrowing to about 60 km near 10°N. Near Cape Comorin the shelf is again about 100 km wide. From there it bends northeastward, narrowing to about 25 km.

Recent echo-sounding by the National Institute of Oceanography (NIO) has shown that the inner shelf, to a depth of about 60 m, is smooth and featureless (Varadachari, personal communication). Seaward to a depth of 200 m the bottom becomes irregular by the presence of pinnacles (1-8 m in height) and adjoining troughs that have originated through constructional and erosional processes operating upon algal and oolitic carbonates of Holocene age (about 9,000-11,000 yr; Nair, 1971b). The shelf break along the western shelf is generally found at 130 m.

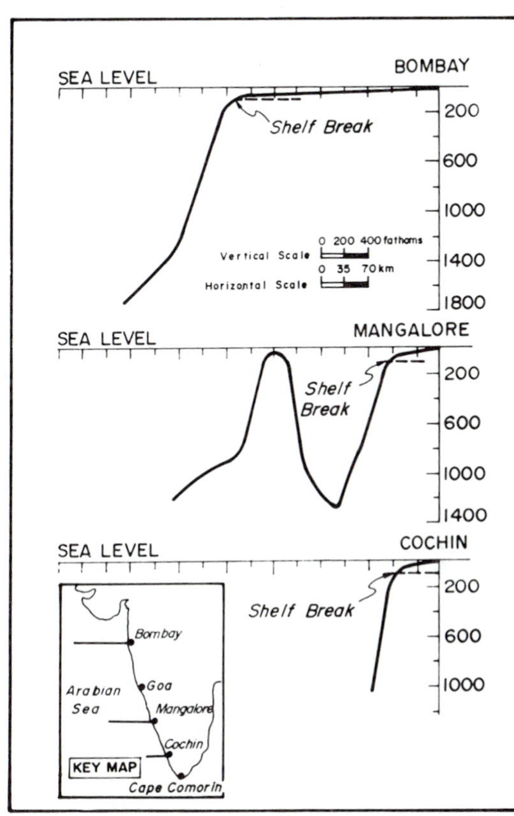

Fig. 1. Topographic profiles on the continental margin of the west coast of India (after Narain et al., 1968).

Sea-Level Fluctuations. Former low stands of sea level are indicated by the presence of submerged terraces at depths of 110 m (Shepard, 1963), 92 m, 85 m, 75 m, and 55 m; the most prominent is that at 92 m (Nair, 1971b). These low stands, according to radiocarbon dating, are of Holocene age. The outer reefs of the Laccadive Islands also show the existence of submerged terraces at depths of about 7, 25, and 40 m (Siddiquie, 1973). Numerous ^{14}C measurements were made with the material collected from METEOR II during the I.I.O.E., but it was not possible to establish a time-sea level relationship. The maximum eustatic sea-level changes reached about 150 m (von Stackelberg, 1972).

There is clear indication of a transgression at the end of the Pleistocene (Mattiat et al., 1973). The end of the youngest cold period, from microbiological evidence in combination with ^{14}C data, can be fixed on the basis of the METEOR cores as 10,500-13,000 yr which is about 2,000 years earlier than in Europe or North America (Zobel, 1973).

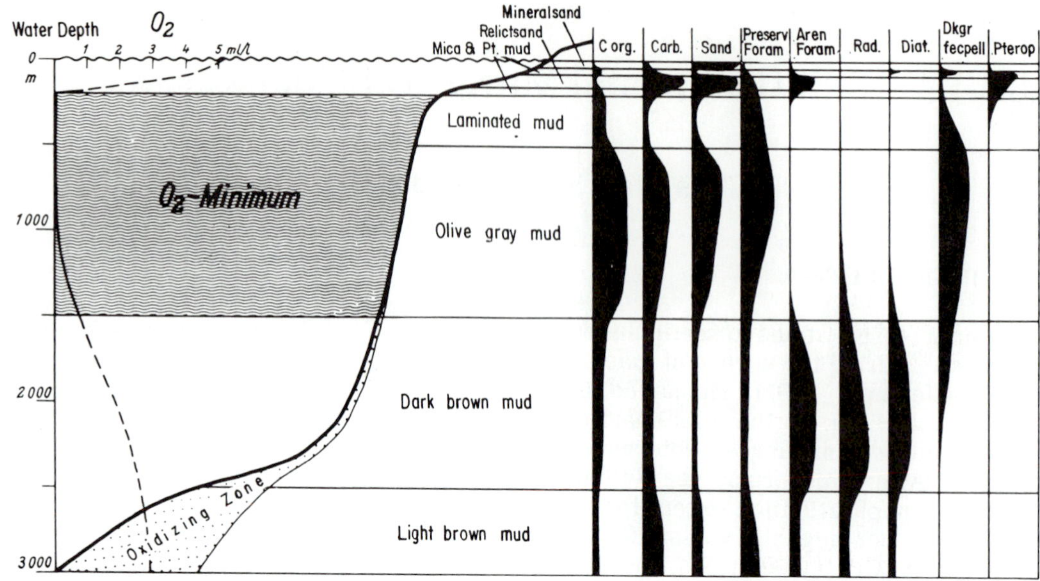

Fig. 2. Schematic section across the Indian-Pakistan continental margin. Oxygen contents in the seawater and qualitative contents of selected components of the surface sediments (after von Stackelberg, 1972).

Geology. Tertiary and Cretaceous sediments have been encountered in the Indian Ocean by the Deep Sea Drilling Project (DSDP), whereas marine Jurassic and Triassic sediments of the old Tethys are known onshore in India. A marine transgression of central India during Permian time has been reported (Gee, 1928; Ghosh, 1954; Ahmad, 1970). Similarities of faunas with those of western Australia have been reported for the Permian (Thomas, 1954; Dickins and Thomas, 1959).

The Indian Peninsula is a relatively stable platform composed of Archean and Precambrian igneous and metamorphic gneisses and schists. Approximately 518,000 km^2 of the Peninsula are covered by the extensive Deccan trap series, primarily basaltic lava flows of Late Cretaceous or early Eocene age. The seaward extension of these immense volcanic masses is as yet unknown. Faulting, broad uplifting, and downwarping have created sediment-filled basins and shelves on the northern shield area of the Peninsula. The Cutch and Cambay basins, Kathiawar Peninsula, and the Shelf off Bombay are regional depositional areas.

The continental shelf off Bombay has low relief, with the shelf break occurring at about 90 m. This shelf is most probably underlain by Tertiary sediments, since it is just south of the Cambay Basin and offshore seismic work suggests that Tertiary formations in the Cambay Basin extend south onto the shelf off Bombay (Sengupta, 1967). In the Bombay area the surface rocks consist of lavas that are more acidic than the Deccan trap basalts. The 1,800-m-thick Deccan trap series abruptly terminates to the west, indicative of a downfault along the coast of Bombay. However, recent seismic investigations by the Oil and Natural Gas Commission of India do not substantiate these indications. The western extension of this downfaulted trap series lies beneath the shelf, according to Krishnan (1953) and Pepper and Everhart (1963). To the north, Kathiawar Province (around 22°N, 71°E) is covered by the Deccan traps, which on the coastal margin are overlain by Tertiary sediments extending probably into the adjoining shelf. Poddar (1964) supposes the coastline of Kathiawar to be marked by a fault system. The broad shelf off the Gulf of Cambay forms part of the Cambay Basin. A number of deep wells drilled on land here have revealed a maximum thickness of about 2,500 m of sediments overlying the Deccan traps (Mathur and Kohli, 1963).

South of Bombay, normal basalts of the Deccan traps are found overlying the metamorphic and igneous rocks of the Archean shield of peninsular India. Along the southern part of the west coast Pleistocene and Recent sediments are found in narrow belts.

Sediments of the Ocean Floor. The Indus River transports yearly 440 million tons of suspended matter into the Arabian Sea and, in addition, the monsoons transport 80 million tons of dust (Goldberg and Griffin, 1970). Investigations on the shelf by Rao (1968), Nair and Pylee (1968), and Murty et al. (1969) can be summarized as follows:

1. The nearshore region, demarcated by the 15-m isobath, is composed of quartz and heavy mineral sands.

2. From 15 to 60 m the shelf is floored by recent terrigenous silts and clays which are low in calcium carbonate (0-20%) and high in organic matter (up to 5%).

3. The outer shelf (60-200 m) is composed of relict carbonate sand and carbonate rocks in the form of algal and oolite limestones of Holocene age.

The results of the METEOR cruise have been presented by Beiersdorf (1972), Cepek (1973), Mattiat et al. (1973), von Stackelberg (1972), and Zobel (1973). A schematic vertical section across the Indian-Pakistan continental margin and the O_2 content in the seawater is given in Figure 2. According to von Stackelberg, the sediment on the outer shelf is mostly coarse, recent deposits are generally absent, and the fine fraction of the sediments has been separated and deposited on the uppermost part of the continental slope, where recent sedimentation reaches its maximum. The coarse material of the outer shelf is rich in oolite, which formed when the sea had its lowest level at the end of the Pleistocene. The influence of the Indus River is clearly reflected in the quartz distribution on the sea floor (Fig. 3).

In the deep sea, sedimentation rates are 1 cm/10^3. Near the coast it averages greater than 35 cm/10^3 yr, with a maximum of 1.5 m/10^3 yr (von Stackelberg, 1972; Zobel, 1973).

National Institute of Oceanography of India (NIO) has recently carried out geochemical investigations of the margin sediments (Varadachari, personal communication). The general findings are:

1. The phosphate content is as follows: Kerala coast, 52-430 μg/g; Mangalore coast, 40-430 μg/g; Bombay coast, 190-540μg/g.

2. The fine-grained sediments in the inner shelf have higher concentrations of iron, manganese, nickel, cobalt, and copper than the sediments in the outer shelf and slope.

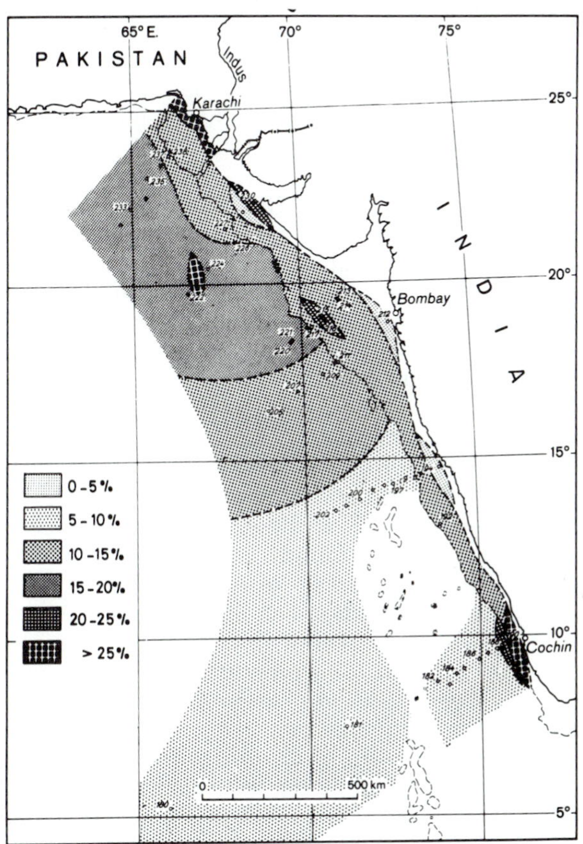

Fig. 3. Quartz distribution on the sea floor of the western margin of the Indian Peninsula (after Mattiat et al., 1973).

The organic content decreases with depth from shelf to slope. Although the average values, 7.25-11.12%, on the upper continental slope of the Bombay coast are far above the world average of 2.5%, according to Marchig (1972) the maximum content of organic substance is reached in the upper part of the continental slope, where the recent sediments have maximum organic content and the oxygen content of the sea water is remarkably low. The content in organic carbon is generally <1% in the abyssal plains. In the Laccadive Trough it is 1.5%. Off Cochin, on the slope at about 2,000 m depth, it averages >3%, and at 1,000 m depth on the same profile, from 3 to >6%. In the Indus Delta a correlation of the grain size of the sediment and the content of organic matter has been found, the organic content increasing with smaller grain size.

Water Currents and Influences on Sedimentation. Many features of the sediments are influenced by currents parallel to the coast, e.g., a tongue of detritic material reaching far to the south on the continental slope and a strong montmorillonite mixed layer influence in the nearshore mud. Seasonal upwelling causes high plankton production, corresponding sedimentation, O_2 consumption, and preservation of organic material in the sediments (von Stackelberg, 1972).

Deep Sea Drilling. DSDP site 219 between Laccadive and Maldive islands revealed that 2,000 m of mostly late Paleocene to early Eocene subsidence occurred at this locality. There is no evidence of subsequent uplift. The sediments appear to have been derived from altered volcanic rocks and may be related to one of the phases of Deccan trap volcanic activity.

In the western part of the Indian Ocean (Zobel, personal communication) there are regional unconformities similar to that observed in the south Pacific in the Oligocene, on DSDP leg 25 (Kennett et al., 1972). It can be supposed that from the Miocene onward the paleo-environment in the Arabian Sea was not too much different from the situation at present time.

Geophysical Investigations. In 1965, 17 seismic refraction profiles were shot by the German research vessel METEOR II on the Gulf of Oman, the Murray Ridge, the continental shelf area between the Indus Delta and the Gulf of Cambay, including parts of the Indo-Arabian abyssal plain (Closs et al., 1969). Location of these groups of profiles is marked in Figure 4 as A, B, C. Profile A reveals four consistent and well-defined layers across the entire traverse. Profile B (Fig. 5) shows a much more complex structure. Profile C (Fig. 6) shows a group of refraction profiles that have been recorded in a two-ship operation involving METEOR II and the Indian reserach vessel I.N.S. KISTNA. On the continental shelf, basement with a velocity 6+ km/sec lies about 5,000 m deep; below the continental slope it plunges to a depth of about 8,000 m.

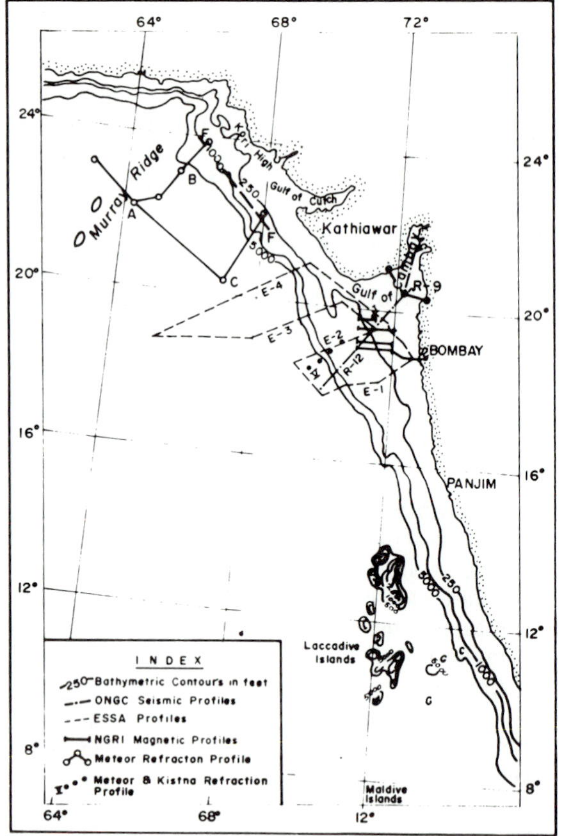

Fig. 4. Bathymetric map of the western margin of India, showing the location of geophysical profiles.

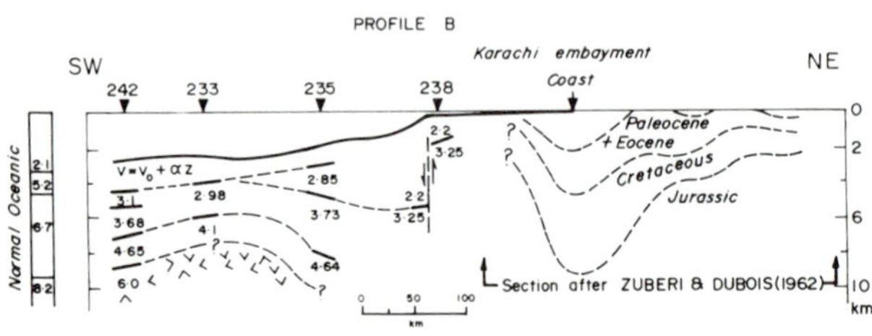

Fig. 5. Crustal cross section along profile B (Fig. 4); velocities in km/sec (after Closs and Hinz, 1967).

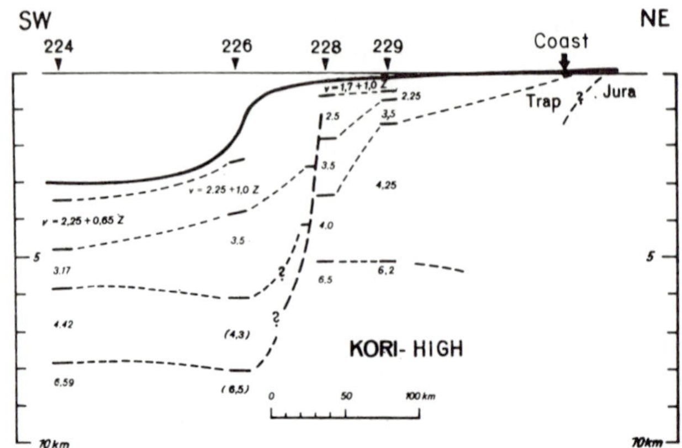

Fig. 6. Crustal cross section along profile C (Fig. 4); velocities in km/sec (after Closs and Hinz, 1967).

The distance between the two short seismic profiles B and C is about 300 km. Nevertheless, there seem to exist some similarities: the deepest refractor has a velocity of about 6 km/sec. Under the edge of the shelf there is an abrupt downthrow of the order of 3-4 km and an 8-km-thick series of sediments under the continental slope thins to 5-6 km toward the deep sea (Kori Trough, west of Kori High). There is some probability that a fault system parallels the coast from about 50 km west of the shoreline of Kathiawar up to the area of Karachi.

Outside of the Gulf of Cambay a group of five refraction profiles were shot during the Indo-German cooperative program. The edge of the shelf is similar to that described above, with a sedimentary thickness of the order of 7-8 km. The layers with velocities less than 4 km/sec have been correlated with Tertiary and Quaternary sediments, increasing in thickness from practically zero nearshore to about 3,000 m in a distance of about 150 km. Under the deeper part of the shelf the maximum thickness of Cenozoic sediments may reach nearly 6,000 m. Downslope, at a water depth of 2,000 m and a distance 230 km from the coast, the Cenozoic sediments are 4,000 m thick. The layer with velocity greater than 4 km/sec can be associated with Cretaceous sediments and, inshore, to a combination of sediment and trap layers. Jurassic or older sediments may also form a part of this layer.

The basement layer, with velocities>6 km/sec near the coast, is presumed to be metamorphics. On the seaward section, under a cover of 5 km of sediments and more, it can represent ocean crustal material. If so, and assuming that the data from these first seismic experiments are reliable, a margin some 120 km outside the present shoreline may have existed.

The Oil and Natural Gas Commission of India conducted thorough geophysical investigations from 1963 to 1965 in the Gulf of Cambay and the Arabian Sea (Sengupta, 1967). Reflection profiles R-9 and R-12 (Fig. 4) revealed that the Cambay Basin extends into the Gulf of Cambay and the adjoining Arabian Sea with a maximum possible thickness of Tertiary sediments of about 5 km. Profile R-9 indicated a depression in a northeastly direction. Profile R-12 (Fig. 7) revealed a faulted uplift northwest of Bombay, called the "Bombay High."

No evidence of a western margin fault close to the shore has been found. However, a prominent fault parallel to and west of the coast about 50 km from 18°15'N, extending southward, has been indicated. Besides the present shelf edge, another shelf has been identified. This paleo-shelf is always nearer to the shore than the present-day shelf, meaning that the shelf has been prograding outward.

Scripps Institute of Oceanography carried out seismic refraction shooting in the northwest Indian Ocean (Francis and Shor, 1966), which indicated that the Maldive Ridge has a layer, probably volcanic, about 4-5 km thick throughout its entire length. The Mohorovicic discontinuity is deepest at

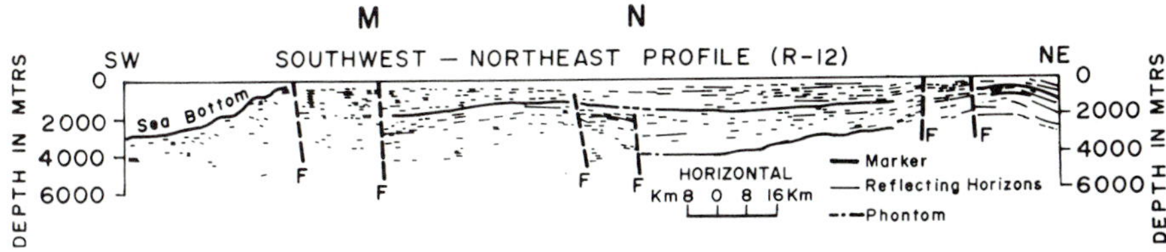

Fig. 7. Seismic reflection section along profile R-12 (Fig. 4) in the Gulf of Cambay (after Sengupta, 1967).

its southern end, and this has led to the idea that the Maldive Ridge must have developed northward, starting from the Chagos island group. Narain et al. (1968) suggested from these observations that the Chagos-Maldive Ridge continues north under the sea floor along the western coast of India up to Karachi, and represents an arcuate volcanic ridge feature.

As part of the 1967 global expedition of the U.S. research ship OCEANOGRAPHER, approximately 1,700 km of reconnaissance seismic reflection and magnetic traverses were made between 17 and 21°N over the shelf and slope off Bombay (Harbison and Bassinger, 1970). These investigations were undertaken to explain the anomalously wide shield off the Bombay coast.

Near the landward edge of the continental shelf, the seismic reflection profiles show strata dipping west from Bombay. The configuration of the continental slope is influenced by anticlinal structures, suggesting that the western Indian shelf is a sediment-filled structural basin. Within the basin, a buried north-trending anticlinal high is present about 75 km west of Bombay. The structural basin and subsurface anticlinal trends off Bombay may continue northward into the Cambay Basin, where similar features are present. Recently, Harbison and Bassinger (1973) reexamined these data and suggested that the anticlinal trends at the continental slope represent a basement ridge trending north-northwest and believe it to be extending into the Chagos-Laccadive Ridge. No magnetic lineations are observed in their study area.

Magnetic observations off the Bombay coast were made in 1969 by the National Geophysical Research Institute (NGRI) between 19-20°N and 70-72°E (Garde et al., 1970). An isogam map was prepared (Fig. 8) using the data from this survey and those obtained on board OCEANOGRAPHER in 1967. The depth to the sources of the anomalies was computed as 5 km. Two magnetic profiles off the Goa coast, along 17°30'N and 17°45'N between 71°31'E and 72°30'E, were undertaken during the summer of 1971 on board I.N.S. DARSHAK by NGRI (Garde and Joshi, 1971). A geomagnetic field reversal was found and discussed (estimated age 65 my).

Roeser (in press) gave an interpretation of the flight lines of the world magnetic survey over the Arabian Sea (Fig. 9) using the irregular fluctuations of the total intensity of the earth's magnetic field. Roeser's results support the idea (Narain et al., 1968) that the Kori High, Bombay High, and Lacca-

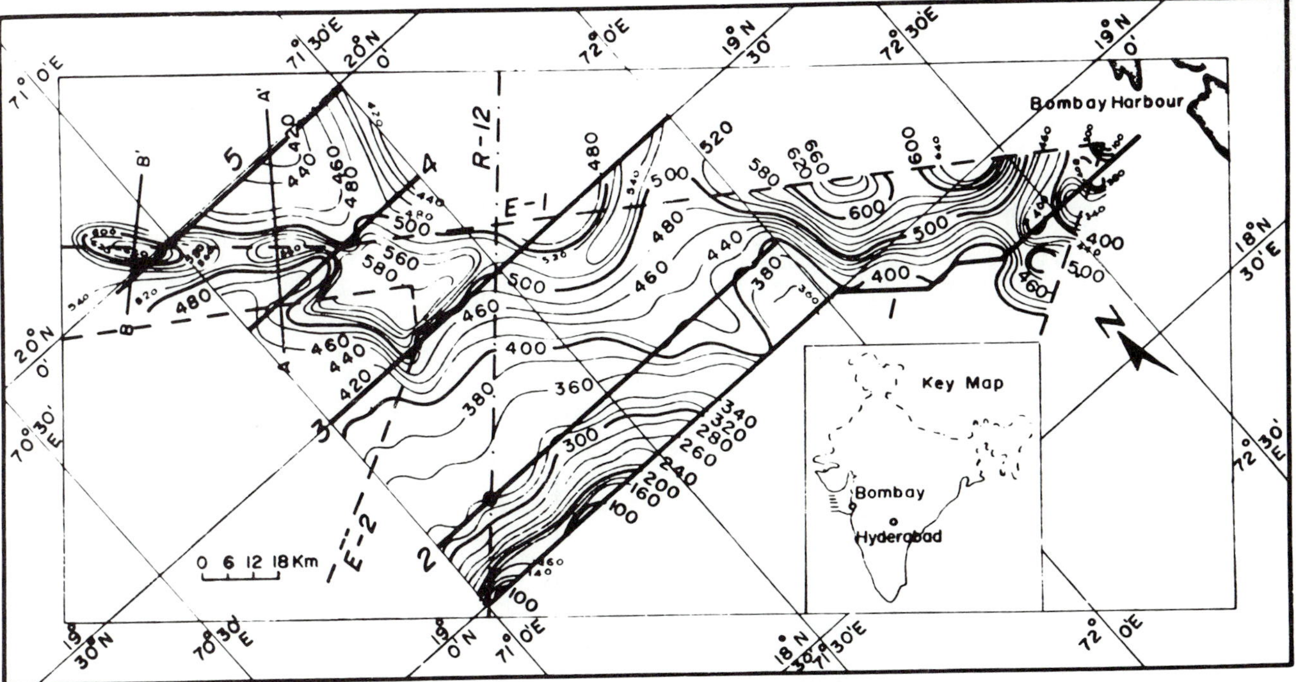

Fig. 8. Isogam map off the Bombay coast, India (after Garde et al., 1970).

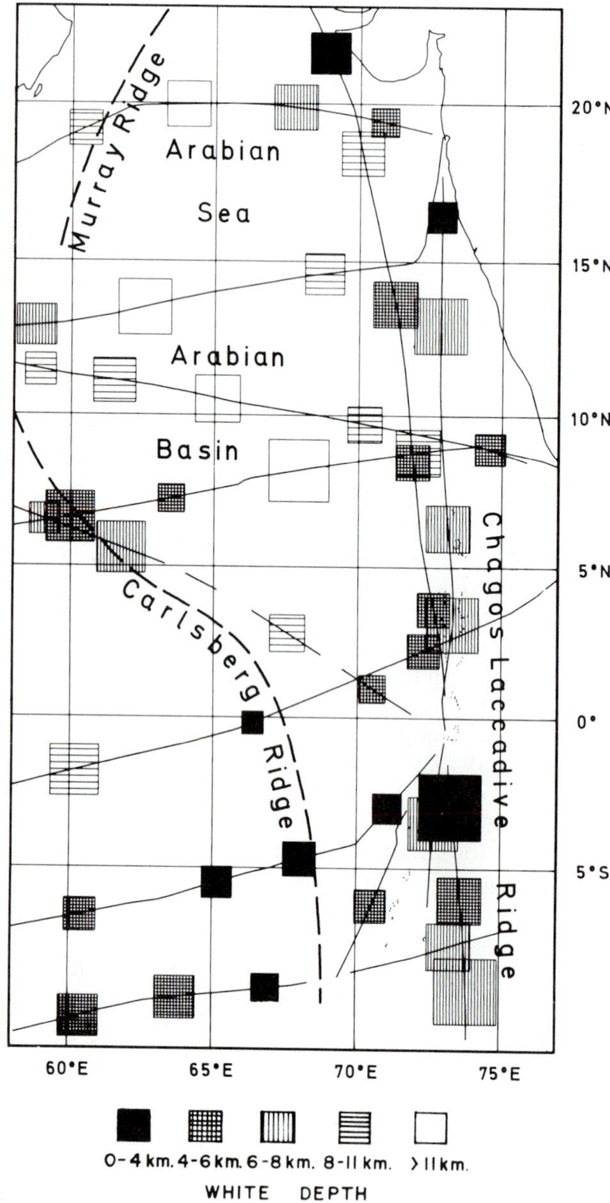

Fig. 9. Interpretation of the world magnetic survey over the Arabian Sea (after Roeser, 1973). The size of the squares are proportional to the average magnetic amplitudes in the "white depths" (roughly the center of the magnetic bodies).

dive Ridge can more or less be considered as one important lineament-type structure.

Recently, Kahle and Talwani (1973) analyzed the free-air gravity map of the Indian Ocean and indicated a ridge-like structure south of Cape Comorin ("Comorin Ridge"). The Comorin Ridge is shown to continue northwestward as the edge of the western Indian shelf. Gravity observations off the Bombay coast were made with an Askania shipborne gravity meter on board H.M.S. OWEN during IIOE 1961-1962 and 1962-1963. The Bouguer anomaly map prepared on the basis of these data reveals a positive gravity anomaly (about 70 mgal) in the area of Bombay (Takin, 1966). It is interpreted to be due to a mass of high-density rocks (probably olivine gabbro) emplaced in the crust just below the Deccan traps in the Bombay area.

EASTERN MARGIN OF INDIA

Physiography. In Figure 10 three topographic profiles have been drawn from the bathymetric data collected by the Indian Naval Hydrographic Office, starting from the coast near Calcutta, Visakhapatnam and Madras (Narain et al., 1968). The width of the shelf has its minimum value of 2.5 km about 150 km south of Madras. At the Ganges River, near Calcutta, the width increases to a maximum of 210 km near the delta.

Canyons. The Ganges submarine canyon, also called the Swatch-of-No-Ground, has been surveyed in detail by the Pakistan Navy (Hayter, 1960). The GALATHEA, during the Danish deep-sea expedition of 1950-1952, made five crossings of the Swatch-of-No-Ground, which showed the canyon to extend at least to 19°N, some 150 km south of the shelf break (Kilerich, 1958). A crossing of a leveed channel at 5°N by the Swedish vessel ALBATROSS in 1948 led Dietz (1953) to suggest that this channel was formed by turbidity currents, possibly originating at the

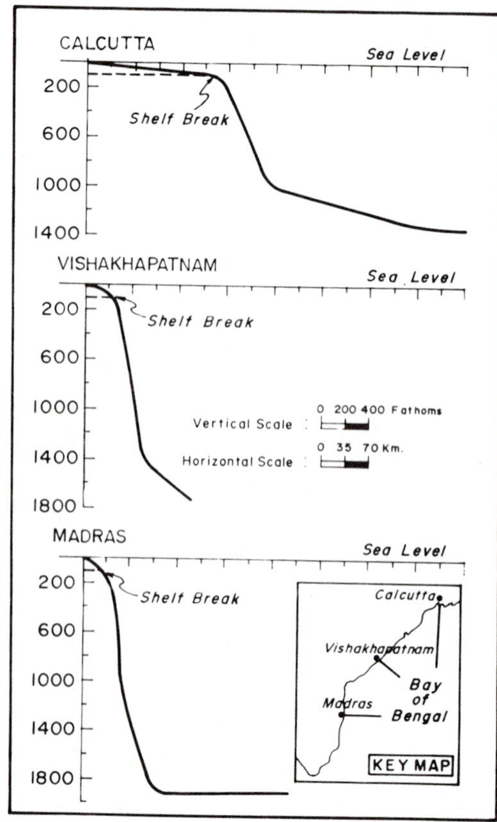

Fig. 10. Topographic profiles on the continental margin of the east coast of India (after Narain et al., 1968).

head of the Bay of Bengal. A core sample taken during the PIONEER cruise, in the center of the Swatch-of-No-Ground at a depth of 513 m, had fine silt in its surface layer underlain by fine to medium sand. A dredge haul on the steep northwest wall of the canyon contains still blue-green clay. The explanation for this formation of this trough given by Stewart et al. (1965) attests to the action of turbidity currents.

The results of depth soundings (Stewart et al., 1965) showed that the Swatch-of-No-Ground is a long narrow trough incised in the continental shelf at the head of the Bay of Bengal. It differs from other submarine canyons in that it has no major tributaries and is generally U-shaped in cross section rather than having the V-shape of typical canyons. In addition, the trough does not extend directly down the shelf normal to the slope, but crosses the slope at an angle of 45°, trending northeast. The floor of the trough is as much as 800 m below the surrounding shelf surface and averages 15 km in width. Farther south from the head, the canyon becomes narrower and the relief less, until broad natural levees are found along both sides. At the end the leveed channel or valley is found similar to those of broad fans seaward of many submarine canyons.

Three submarine canyons 5-8 km apart at about 18°N were discovered on the continental slope off the Andhra coast during the biological cruises of the U.S. research vessel ANTON BRUNN (La Fond, 1964). They are typically V-shaped, and their narrow bottom indicates that they are not being filled with sediments. It is not known whether they extend up to the break in slope or if they cut across the shelf and intersect the longshore transport of nearshore sediments. Farther south off the Coromandel coast, three similar canyons have been reported (Varadachari et al., 1967).

Geology. Near Cape Comorin there are largely Archean rocks with a thin cover of Tertiary sediments, which are also found along the northwest coast of Ceylon. Farther north the shelf consists of Cretaceous and Tertiary rocks overlain by Pleistocene and Recent sediments. In the coastal areas of southern Madras, an important middle Cretaceous transgression is reported by Krishnan (1968). Continuing northward, the coast is formed mostly by Pleistocene and Recent sediments; this continues up to the Ganges Delta.

The most southern drillhole in the Ganges Delta reached the Pliocene-Miocene transition about 4,000 m below sea level. Other drillholes onshore in the Bengal Basin found as basement the Tertiary traps, but in some places Lower Cretaceous may occur below the trap (Biswas, 1962). The assumption of a maximum total thickness of sediments of about 6,000 m in this area seems to be realistic. They may overlie a basement of lava flows ranging in age from Jurassic to Cretaceous (Narain et al., 1968). According to Ghosh (1959) the continental shelf off West Bengal may contain at depth marine Cretaceous and Tertiary sediments extending southward from the Bengal Basin.

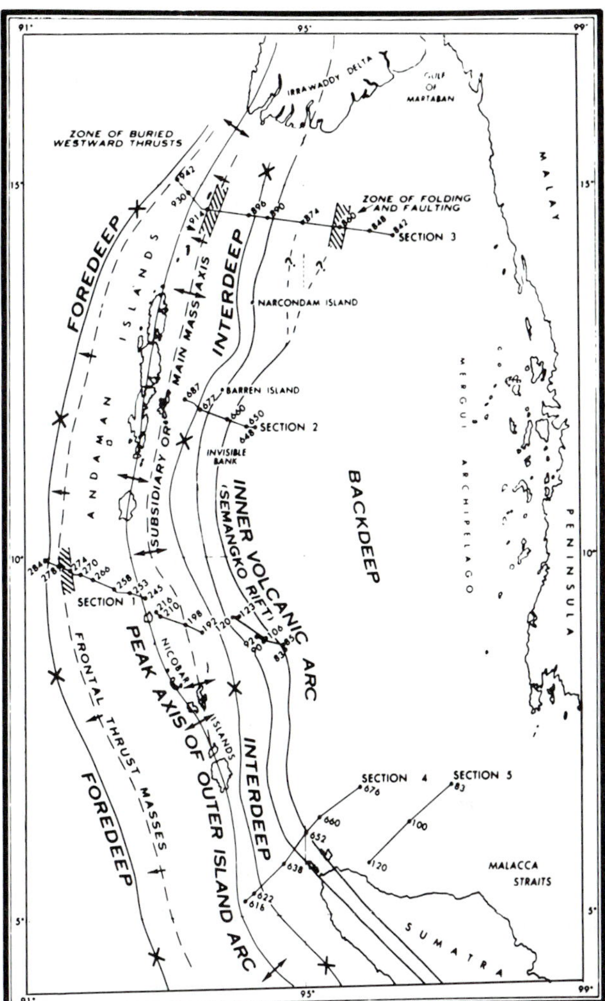

Fig. 11. Schematic presentation of the principal features of the Andaman Arc (after Peter et al., 1966).

Between Madras and the mouth of the Ganges, Subba Rao (1964) has investigated the sediments of the shelf. He could find a zonation of sediments, in some respects similar to the west coast. From about 60 m depth on to the shelf edge, relict sediments are a characteristic feature. The low sea level in Pleistocene time and a resulting wave action is thought to be responsible for the characteristics of those sediments.

ANDAMAN-NICOBAR ISLAND ARC

A detailed bathymetric picture of the Andaman-Nicobar island arc was obtained by Peter et al. (1966) during the cruise of PIONEER during the IIOE. Figure 11 shows the principal bathymetric features in Andaman Sea area. The dominant structural features are related to the Indonesian island arc, which consists of a primary double arc, an inner volcanic trend, and an outer sedimentary arc.

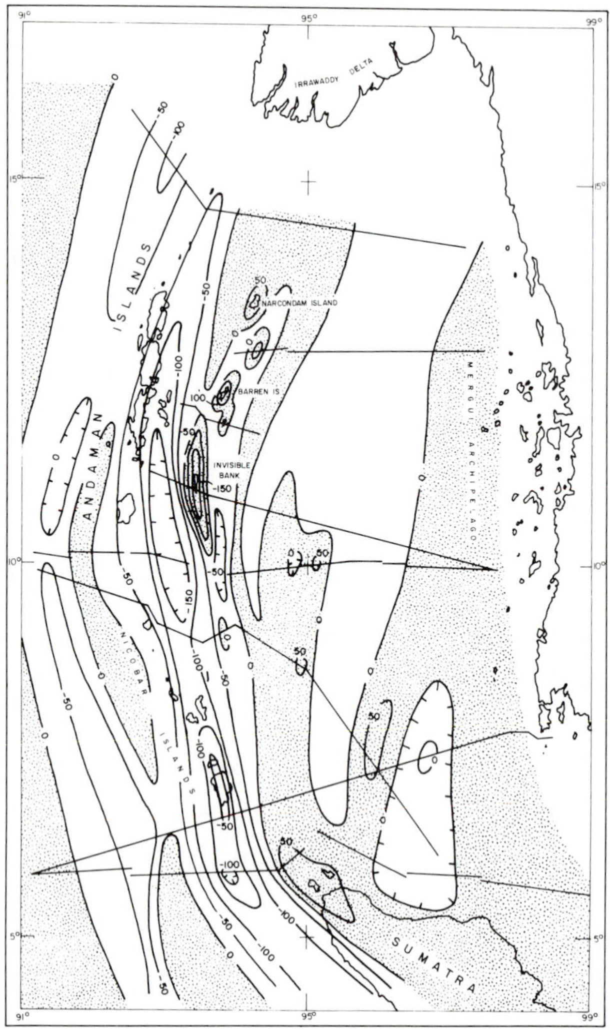

Fig. 12. Free-air gravity anomalies and major geologic trends in the Andaman Sea (after Peter et al., 1966).

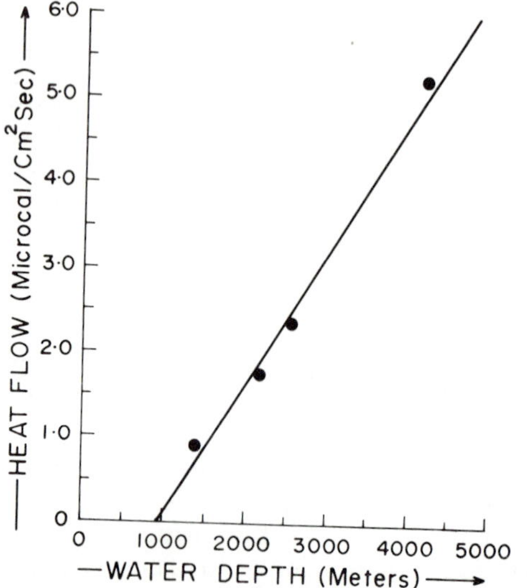

Fig. 13. Plot of heat flow versus water depth, showing the linear relationship between them.

The inner volcanic trend consists of a belt originally folded during Cretaceous, which developed during late Tertiary and Quaternary into a volcanic arc system still marked by active volcanism.

Geophysical Investigations. Russian research vessels carried out seismic work in the Bay of Bengal and Andaman Sea to determine the thickness of the sedimentary layers (Neprochnov, 1964). The results showed the thickness in the Bay of Bengal to be 2.5-3.0 km, decreasing to 1.5 km southward. The average seismic velocity in the sedimentary layer was 2.1 km/sec. The thickness of sediments in the Andaman Sea is approximately 1.5 km.

A reconnaissance marine geophysical survey by the U.S. research ship PIONEER (Fig. 12) revealed the following features (Peter et al., 1966):

1. Parallel belts of positive and negative gravity anomalies extend over the inner igneous arc and the island platforms, respectively. The axis of the negative anomaly belt passes through the Nicobar Islands and slightly to the eastern side of the Andaman Islands.

2. The crustal thickness deduced from Bouguer gravity studies shows it to be of continental dimensions beneath the arc.

3. Airy-Heiskanen isostatic anomalies indicate that the islands are not in isostatic equilibrium.

4. The magnetic anomalies over the igneous belt are relatively large and negative.

Heat flow was measured at four stations in the Andaman Sea during the PIONEER cruise (Burns, 1964). There is good correlation between water depth and heat flow (Fig. 13). The highest heat flow is associated with the deepest portion of the basin. There is no apparent effect reflecting the volcanism at Barren Islands (12°16'N, 93°50'E). The measurements, which are parallel to the Indonesian primary arc system, show great variability, contrary to the relationship of heat flow and island arcs suggested by Uyeda et al. (1962) and Yasui et al. (1963).

The analysis of shallow earthquakes recorded at Port Blair yielded a Pg velocity of 5.3 km/sec and Pn velocity of 7.8 km/sec, with Pn time intercept of 3.4 sec. The value of Pg is rather low for granite, but it may be quite appropriate for the volcanic trap rocks that may form the basement in the Andaman-Nicobar region. The crustal thickness obtained by Narain et al. (1968) is of the order of 20 km.

DEEP SEA DRILLING IN THE BAY OF BENGAL

The GLOMAR CHALLENGER drilled eight sites in the eastern Indian Ocean (Von der Borch et al., 1972). The central and deepest part of the Wharton Basin is probably not older than Cretaceous. Ninetyeast Ridge was part of the old Indian plate in the Late Cretaceous. Sites on the ridge show that this plate moved rapidly north during this epoch. Sites on either side of the ridge show a unified history and little or no relative motion since early Tertiary. Basal sediments on Ninetyeast Ridge become older to the north and sediments at all three sites were deposited close to sea level. Early in its history the ridge had reached sea level, and it is inferred that it has subsided as it migrated northward.

In the Bengal fan the sedimentary section consists of silts, sandy silts, and clayey silts, interbedded with nanno-ooze layers. These range in age from Quaternary to middle Miocene at the bottom of the hole. On the basis of grain sizes of the terrigenous sediments, four distinct pulses of turbidity current activity are identified. The youngest pulse is overlain by several meters of clayey pelagic ooze, suggesting a decline of turbidity current activity in the area. (See also Curray and Moore, this volume.)

Over the main block of the Peninsula only in small areas can sediments younger than Paleozoic be found from which geologic events of the past can be deduced. Therefore, a huge amount of research in the offshore areas has yet to be undertaken, since many consequences of the younger history of India are buried under the waters of the surrounding Indian Ocean.

ACKNOWLEDGMENTS

We are grateful to S. N. Sengupta, Chief of Exploration Geophysics, O.N.G.C.; D. N. Awasthi, Joint Director, Institute of Petroleum Exploration, O.N.G.C.; V. V. R. Varadachari, Acting Director, National Institute of Oceanography; and H. N. Siddiquie, Head, Geological Oceanography Division, N.I.O., with whom we had stimulating discussions resulting in many new ideas and suggestions. We acknowledge the help of S. M. Naqvi, for suggesting references on marine transgression in the Permian; C. P. Pandey and R. K. Drolia for assisting in correcting the manuscript; J. Andres, Wintershall A. G., Kassel for providing documentation; and B. Zobel, Geological Survey, Hannover, for stimulating discussions.

Dr. C. T. Klootwijk from the Vening Meinesz Lab. Utrecht, had prepared a special chapter dealing with the drift of the Indian continent as it can be derived from paleomagnetic data, but for editorial reasons it was not possible to print it here. Nevertheless we thank him for his remarkable effort and stimulating cooperation.

BIBLIOGRAPHY

Ahmad, F., 1970, Marine transgression in the Gondwana System of Peninsular India—a reappraisal: Proc. 2nd Gondwana Symp. South Africa, p. 179-183.

Beiersdorf, H., 1972, Heavy mineral analysis of sediments from West Pakistan and the adjacent continental shelf: Meteor Res. Results, C, no. 9, p. 74-83.

Biswas, B., 1963, Results of explorations for petroleum in the western part of the Bangal Basin, *in* Symposium on the development of petroleum resources: U.N. Econ. Comm. for Asia and the Far East, 2nd Session Proc., v. 1, pt. 2, p. 241.

Burns, R. E., 1964, Sea bottom heat flow measurements in the Andaman Sea: Jour. Geophys. Res., v. 69, p. 4918-4919.

Cepek, P., 1973, The special *Pontosphaera indooceanica* n.sp. and its importance for the stratigraphy of the youngest sediments of the Indian Ocean: Meteor Res. Results, C, no. 12, p. 1-3.

Chaterji, G. C., Karunakaran, C., and Siddiquie, H. N., 1968, Exploration for Minerals on the continental Margin of India—an appraisal of the existing data: Symp. Indian Ocean, Natl. Inst. Sci. India, New Delhi, Bull. Natl. Inst. Sci., India, v. 38, p. 552-562.

Closs, H., and Hinz, K., 1967, Refraction seismic measurements in the northern Arabian Sea: Paper presented at Symp. upper mantle project, Hyderabad, India, Jan. 4-8.

———, Bungenstock, H., and Hinz, K., 1969, Results of seismic refraction measurements in the northern Arabian Sea, a contribution to the International Indian Ocean expedition: Meteor Res. Results, C, no. 2, p. 1-28.

Dickins, J. M., and Thomas, G. A., 1959, The marine fauna of the lyons group and the Carrandinbby formation of the Carnarvon basin, Western Australia, Australia Bur. Mineral Resources Geol. Geophy. Rept. 38, p. 65-96.

Dietz, R. S., 1953, Possible deep sea turbidity current channels in the Indian Ocean: Geol. Soc. America Bull., v. 64, p. 375-378.

Drake, C. L., Ewing, M., and Sutton, G. H., 1959, Continental margins and geosynclines: the east coast of North America north of Cape Hatteras, *in* Ahrens, L. H., Press, F., Rankama, K., Runcorn, S. K., eds., Physics and chemistry of the earth, v. 3: Elmsford, N.Y., Pergamon Press, p. 110-198.

Dutta, A. B., 1965, *Fenestella* sp. from the Talchir of Daltonganj coal field, Bihar, Quart. Jour. Geol. Mining Met. Soc. India, v. 37, p. 133-134.

Emery, K. O., 1968, Relict sediments on continental shelves of world: Am. Assoc. Petr. Geol., v. 52, no. 3, Mar., p. S.445.

Ermenko, N. A., and Gagelganj, A. A., 1966, New data on the tectonic frame work of the Indian Peninsula: Bull. ONGC (India), v. 3, p. 1-8.

Francis, T. J. G., and Shor, G. G., Jr., 1966, Seismic refraction measurements in the north-west Indian Ocean: Jour. Geophys. Res., v. 71, p. 427-449.

Garde, S. C., and Joshi, M. S., 1971, Tectonic implications of the Kori-Laccadive-Chagos monoematath: Marine Geology, v. 11, p. M58-M62.

———, Rastogi, B. K., and Gupta, C. P., 1970, Magnetic anomalies and the sub-shelf geologic structure off the Bombay coast (India): Marine Geology, v. 9, p. 355-363.

Gee, E. R., 1928, The geology of the Umaria coal field: Recent Geol. Survey India, v. 60, p. 399-410.

Ghosh, A. M. N., 1959, Possible oil bearing regions of India: 5th World Petr. Congr., New York, Proc., sect. 1, p. 1023-1035.

Ghosh, S., 1954, Discovery of a new locality of marine Gondwana formations near Manendragarh in M.P.: Sci. and Cult., 19(12), p. 620.

Goldberg, E. D., and Griffin, D. L., 1970, The sediments of the Northern Indian Ocean: Deep-Sea Res., v. 12, no. 3, p. 513-537.

Harbison, R. N., and Bassinger, B. G., 1970, Seismic reflection and magnetic study off Bombay, India: Geophysics, v. 35, p. 603-612.

———, and Bassinger, B. G., 1973, Marine geophysical study off Western India: Jour. Geophys. Res., v. 78, p. 432-440.

Hayter, P. J. D., 1960, The Ganges and Indus submarine canyons: Deep-Sea Res., v. 6, p. 184-186.

Hellman, G., 1897, Die Anjange der magnetic schen Beobachtungen: Z. Ces. Erak., v. 32, p. 112-136.

Henderson, R. G., and Zietz, I., 1948, Analysis of total magnetic intensity anomalies produced by point and line sources: Geophysics, v. 13, p. 428-436.

Heye, D., 1970, Dating of deep-sea cores of the Indian Ocean with the Jo/Th- and the Pa/Jo-method: Meteor Res. Results, C, no. 3, p. 15-22.

———, 1970, Correlation of sedimentary cores from the Indian Ocean on the basis of their magnetization: Meteor Res. Results, C, no. 3, p. 23-27.

Kahle, H. C., and Talwani, M., 1973, Gravimetric Indian Ocean Geoid: Z. Geophys., Bd. 39, p. 167-187.

Kalinin, N. A., 1965, Problems of oil and gas geology in India: Tokio Econ. Comm. Asia Far East Symp. Develop. Petr. Resources 3rd Session, Tokyo, p. 1-16.

Keen, C. E., and Loncarvic, B. D., 1966, Crustal structure on the eastern sea-board of Canada: studies on the continental margins: Can. Jour. Earth Sci., v. 3, p. 65-76.

Kennett, J. P., 1972, Biostratigraphic, climatic paleomagnetic record of the Miocene to early Pleistocene sediments of New Zealand, (abstr.): Internatl. Geol. Congr. Congr. Geol. Internatl. Resume, no. 24, p. 538.

———, Burns, R. E., Andrews, J. E., Churkin, jun. M., Davies, T. A., Dumitrica, P., Edwards, A. R., Galehouse, J. S., Packham, G. H., and Lingen, G. J. van der, 1972, Australian-Antarctic continental drift, Palaeocirculation changes and Oligocene deep-sea erosion: Nature Phys. Sci., v. 239, no. 91, p. 51-55.

Kilerich, A., 1958, The Ganges submarine canyon: Andhra Univ. Mem. Oceanogr. Ser., v. 2, no. 62, p. 29-32.

Kohli, G., and V. Raghavendra, Rao, 1965, Status of offshore exploration in India: 3rd ECAFE Symp., Tokio.

Krishnan, M. S., 1953, The structural and tectonic history of India: Calcutta: Geol. Survey India Mem. 81, p. 109.

———, 1968a, The evolution of the coasts of India: Symp. Indian Ocean, Natl. Inst. Sci. India, New Delhi, Bull. Natl. Inst. Sci. India, v. 38, p. 398-404.

———, 1968b, Geology of India and Burma, 4th ed.: Madras, India, Higginbothams, Ltd., 604 p.

La Fond, E. C., 1964, Andhra, Mahadevan, and Krishna submarine canyons and other features of the continental slope off the east coast of India: Jour. Indian Geophys. Union, v. 1, no. 1, p. 25-32.

Marchig, V., 1972, Contributions to the geochemistry of recent sediments from the Indian Ocean: Meteor Res. Results, C, no. 11, p. 1-10.

Mathur, L. P., and Kohli, G., 1963, Exploration and development for oil in India: 6th World Petr. Congr., Frankfurt, Proc., sect. 1, p. 633-658.

———, Rao, K. L. N., and Chouble, A. N., 1966, Tectonic framework of Cambay basin, India: Paper presented to 7th World Petr. Congr., Mexico.

Mattiat, B., Peters, J., and Eckhardt, F. J., 1973, Results of petrographical analysis on sediments from the Indian-Pakistran continental margin (Arabian Sea): Meteor Res. Results, C, no. 10, p. 1-50.

McKenzie, D. P., and Sclatter, J., 1971, The evolution of Indian Ocean since the Late Cretaceous: Geophys. Jour., v. 25, p. 437-528.

Misra, J. S., Srivastava, B. P., and Jain, S. K., 1961, Discovery of marine Permo-Carboniferous in western Rajasthan: Current Sci. India, v. 30, no. 7, p. 262-263.

Murty, P. S. N., Reddy, C. V. G., and Varadachari, V. V. R., 1969, Distribution or organic matter in the marine sediments off the west coast of India: Proc. Natl. Inst. Sci. India, v. 35 B, p. 167.

Nair, R. R., 1971a, Beachrock and associated carbonate sediments on the fifty fathom flat, a submarine terrace on the outer continental shelf off Bombay: Proc. Indian Acad. Sci., sect. B, p. 148-154.

———, 1971b, Outer shelf carbonate pinnacles and troughs on the western continental shelf: I.G.U. seminar on scientific technological and legal aspects of the Indian continental shelf, Panjim, Nov. 11-13.

———, and Pylee, A., 1968, Size distribution and carbonate content of the sediments of the western shelf of India: Bull. Natl. Inst. India, v. 38, no. 1, p. 411-420.

Narain, H., Kaila, K. L., and Verma, R. K., 1968, Continental margins of India: Can. Jour. Earth Sci., v. 5, p. 1051-1065.

Neprochnov, Y. P., 1964, Structure and thickness of the sedimentary layer of the Arabian sea, Bay of Bengal and Andaman Sea: Trend. Inst. Okeanol., v. 64, p. 214-226.

Pepper, J. F., and Everhart, G. N., 1963, The Indian Ocean: the geology of its bordering lands and the configuration of its floor: U.S. Geol. Survey Misc. Geol. Inv. Map 380, p. 1-33.

Peter, G., Weeks, L. A., and Burns, R. E., 1966, A reconnaissance geophysical survey in the Andaman sea and across the Andaman-Nicobar island arc: Jour. Geophys. Res., v. 71, p. 495-509.

Poddar, M. C., 1964, Mesozoics of Western India—their geology and oil possibilities: 26th Internatl. Geol. Congr. I, Proc. sect. 1: Geology Petrology, p. 126-143.

Rahman, H., 1963, Geology of petroleum in Pakistan: 6th World Petr. Congr., Frankfurt, sect. 1, paper 31-PD 3.

Rao, G. P., 1968, Sediments of the near shore region off the Neendakara-Kayankulan coast and the Ashtamudi and Valla estuaries, Kerala, India: Bull. Natl. Inst. India, v. 38, no. 1, p. 513-551.

Rao, T. C. S., 1967, Seismic refraction measurements in the gulfs of Cambay and Cutch: Proc. Symp. Upper Mantle Project, Hyderabad, India, p. 342-347.

Roeser, H. A., in press, A representation of the character of the spatial fluctuations of the earth's magnetic field measured along widely spaced long profiles: Geol. Jahrb.

Sengupta, S. N., 1967, The structure of the Gulf of Cambay: Proc. Symp. Upper Mantle Project, Hyderabad. India, p. 334-341.

Shepard, F. P., 1963, Submarine canyons, *in* Maxwell, A. E., ed., The sea, v. 3, New York, Wiley-Interscience, p. 480-504.

———, Inman, D., and Goldberg, E., 1963, Submarine geology, 2nd ed.: New York, Harper & Row.

Siddiquie, H. N., 1973, Submerged terraces in the Laccadive islands: (in press).

———, Srivastava, P. C., Mallik, T. K., Ray, I., and Vankatesh, K. V., 1972, Bottom sediment map of the seas bordering India (1:2,000,000) 1972: Presented at 24th Internatl. Geol. Congr., Montreal.

Stackelberg, von U., 1972, Facies of sediments of the Indian-Pakistan continental margin (Arabian Sea): Meteor Res. Results, C, no. 9, p. 1-73.

Stewart, H. B., Jr., Dietz, R. S., and Shepard, F. P., 1965, Submarine valleys off the Ganges delta: IIOE, U.S.C.G.S. ship PIONEER 1964: cruise narrative and scientific results, v. 1, p. 77-79.

Subba Rao, M., 1964, Aspects of continental shelf sediments of the east coast of India: Marine Geology, v. 1, p. 59-87.

Sykes, L. R., 1970, Seismicity of the Indian Ocean and a possible nascent island arc between Ceylon and Australia: Jour. Geophys. Res., v. 75, p. 5041.

Takin, M., 1966, An interpretation of the positive gravity anomaly over Bombay on the west coast of India: Geophys. Jour., v. 11, p. 527-537.

Thomas, G. A., 1954, Correlation and faunal affinities of the marine Permian of western Australia: Proc. Pan. Indian Ocean. Sci. Congr., Perth Meeting, Geol. sect. 5-6.

Ulrich, J., 1968, Echo-sounding sections of the course of R. V. METEOR in the Arabian Sea during the international Indian Ocean expedition: Meteor Res. Results, C, no. 1, p. 1-12.

Uyeda, S., Horai, K., Yasui, M., and Akamatsu, H., 1962, Heat flow measurements over the Japan trench: Jour. Geophys. Res., v. 67, p. 1186-1188.

Valdiya, K. S., 1973, The tectonic framework of India: a review and interpretation of recent structural and tectonic studies: Geophys. Res. Bull., v. 11, no. 2, (in press).

Van Bemmelen, R. W., 1949, The geology of Indonesia: The Hague, Govt. Printing Office.

Varadachari, V. V. R., Nair, R. R., and Murty, P. S. N., 1967, Submarine canyons off the Coromandal coast: Proc. Symp. Indian Ocean: Bull. Natl. Inst. Sci. India, no. 38, p. 457-462.

Vonder Borch, C. C., Sclater, J. G., Stefan, G., Jr., Hekinian, R., Johnson, D. A., McGownan, N., Pimm, A. C., Thompson, R. W., and Veevers, J. J., 1972, Deep Sea Drilling Project, leg 22: Geotimes, v. 17, no. 6, p. 15-17.

Von Herzen, R. P., and Uyeda, S., 1963, Heat flow through the eastern Pacific Ocean floor: Jour. Geophys. Res., v. 68, p. 4219-4250.

Yasui, M., Horai, K., Uyeda, S., and Akamatau, H., 1963, Heat flow measurements in the western Pacific during the JEDS-5 and other cruises in 1962 abroad M/s RYOFU-MARU: Oceanogr. Mag. Japan Met. Agency, v. 14, p. 147-156.

Zobel, B., 1973, Biostratigraphic investigation of sediments from the Indian and Pakistan continental margin (Arabian Sea): Meteor Res. Results, C, no. 12, p. 9-73.

Atlantic and Indian Ocean Margins of Southern Africa

W. G. Siesser, R. A. Scrutton, and E. S. W. Simpson

INTRODUCTION

This review summarizes published data from the continental margins of South Africa and South West Africa between the Angolan border (17° 15'S—Kunene River) on the west coast, and the Mozambique border (26°51'S—Punta do Ouro) on the east coast (Fig. 1).

Apart from the study of isolated samples, marine geological work in southern Africa commenced in 1959 with the founding of the C.S.I.R. OCeanographic Research Unit at the University of Cape Town. Also, since 1961 numerous samples from the Atlantic Margin of South and South West Africa were collected and investigated by the Institutes of Oceanology and of Fisheries and Oceanography of the U.S.S.R. Commercial diamond exploration also began offshore in 1961, initially on the shallow shelf north of the Orange River mouth.

Detailed sea-floor mapping started in 1967 with the establishment of the Marine Geology Unit at the University of Cape Town. Since then this unit has been engaged in two main projects: (1) a bathymetric, bottom-sampling, and shallow seismic-reflection survey of the continental shelf and upper slope, and (2) an investigation of the distribution, composition, and origin of potentially economic deposits, with special emphasis on marine phosphorite and glauconite. A reconnaissance survey of the shelf and upper slope (down to 1,500 m) has now been completed between the Kunene River and Punta do Ouro, but almost all the data published thus far concern the margin between Walvis Bay and Cape Recife (Fig. 1).

Geophysical work on the margins began in 1960 with isolated gravity and magnetic recordings. Deep crustal structure studies were initiated in 1962 by the Bernard Price Institute of Geophysical Research at the University of the Witwatersrand as part of the South African contribution to the International Indian Ocean Expedition. In 1965 extensive shallow crustal investigations were begun by the Agulhas Bank Project (now designated the Marine Geophysics Unit) at the University of Cape Town. Throughout the 1960s and the early 1970s American groups, notably Lamont-Doherty Geological Observatory and Woods Hole Oceanographic Institution, collected geophysical data of all types on a reconnaissance basis.

The margin is naturally divided at Cape Agulhas into two morphological, sedimentological, and structural provinces; the following discussion of these aspects will therefore treat the Atlantic Ocean Margin (Kunene River to Cape Agulhas) first, then the Indian Ocean Margin (Cape Agulhas to Punta do Ouro).

BATHYMETRY

Detailed bathymetric maps of the continental shelf and upper slope between Walvis Bay and Cape Recife have been produced by Dingle et al. (1971), Dingle (1970, 1973a, 1973c), van Andel and Calvert (1971), Rogers and Bremner (1973), and Birch and Rogers (1973). The following paragraphs are mostly summaries of their interpretations (Fig. 1).

Atlantic Ocean Margin

Between Walvis Bay and Cape Agulhas the continental shelf is wide and deep—certainly one of the deepest in the world. From Walvis Bay to the vicinity of Lüderitz the shelf maintains a width of about 130 km, narrowing to about 55 km north of the Orange River, then widening to over 220 km southwest of the Orange. The shelf narrows again to about 37 km west of Cape Town before widening to about 150 km off Cape Agulhas.

The shelf break is well defined at a depth of about 400 m from Walvis Bay to Meob Bay. There it loses its distinctiveness, becoming a gently convex zone lying between 200 and 450 m as far south as Lüderitz, where it again becomes sharply defined at about 500 m. Double shelf breaks are common off the west coast. The inner break is at 130-160 m west of Walvis Bay, at 200 m just south of the Orange Banks, and at 380 m south of Childs Bank, whereas the outer break lies at 400-500 m. A normal, well-developed shelf break is again found farther south, where it shallows to 200 m southwest of Cape Town. The break has an average depth of about 270 m between Cape Town and Cape Agulhas.

Several large topographic features occur on the Atlantic margin. The Lüderitz Bank (L.B., Fig. 1) is a mid-shelf elevated area (depth of 205 m) some 37 km wide and 92 km long. Farther south, two shallow areas occur at the very edge of the shelf. One of these is an elongate shallow area with several distinct banks named the Orange Banks (O.B., Fig. 1, depth of 160 m). The other, Childs Bank (C.B., Fig. 1), is a flat-topped plateau (depth of 200 m) with a gentle landward edge and a steep, slump-generated seaward face. Tripp Seamount occurs on the upper slope southwest of the Orange Banks. It is relatively small (6 km across the top) and appears to have a

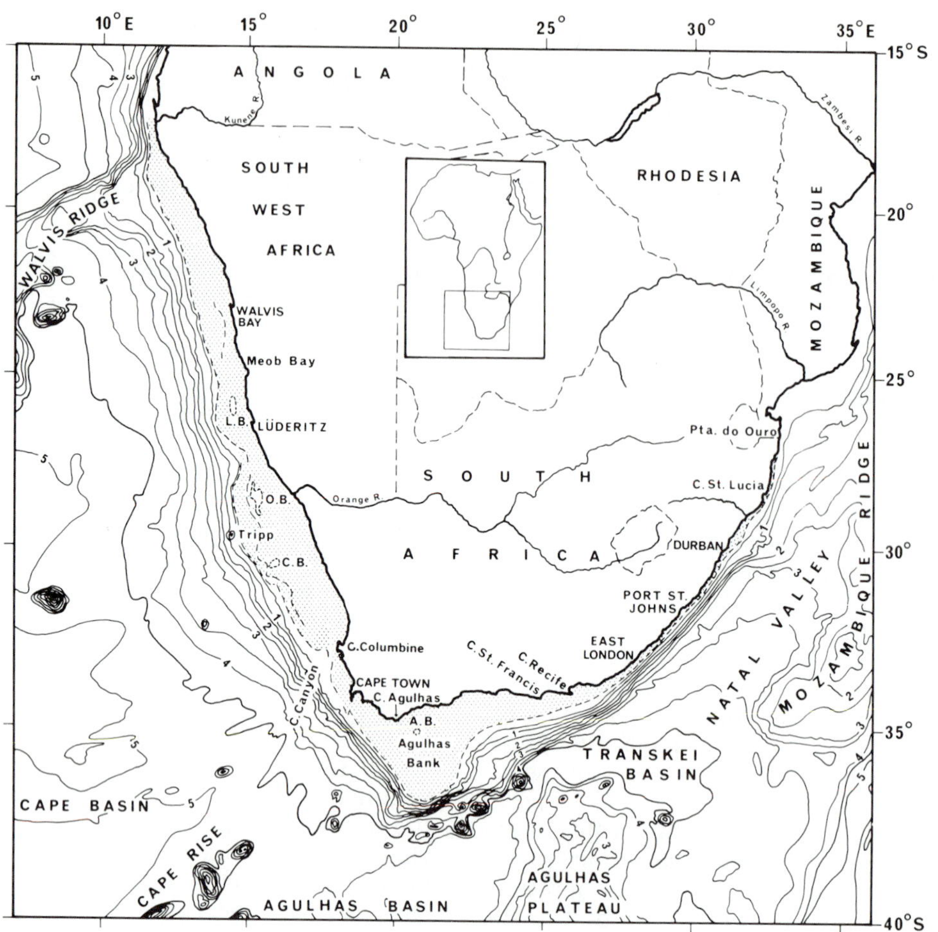

Fig. 1. Location and bathymetry map, modified from Simpson (1970, 1971). Isobaths are in kilometers (the 4.5-km isobath is arbitrarily designated as the base of the continental rise). The shelf break is shown by the dashed line. Topographic features shown on the shelf are: Luderitz Bank (L.B.), Orange Banks (O.B.), Childs Bank (C.B.), Alphard Banks (A.B.).

wave-cut ledge at about 200 m and a sediment-filled crater-like depression at the top (-170 m).

The Cape Canyon (Simpson and Forder, 1968; Dingle, 1971b) cuts diagonally across the shelf and slope from Cape Columbine to the vicinity of Cape Town. Dingle reported that this canyon has a maximum relief of about 800 m and can be traced to a depth of 3,600 m; its steep walls are more-or-less free of Quaternary sediments. Although the present morphology is probably the result of submarine erosion during the Pleistocene (presumably by turbidity currents channeled through the depression), the canyon has a complicated history of erosion and infilling dating from Paleogene times.

Indian Ocean Margin

On the south coast, the shelf widens from Cape Agulhas to a maximum of 240 km at about 21°E, then narrows markedly to 37 km off Cape Recife, 22 km off East London, and 15 km off Port St. Johns, a width it maintains to just north of Durban, where it widens to almost 55 km. From Cape St. Lucia to Punta do Ouro the shelf is very narrow, ranging between 4 and 9 km. Detailed bathymetric maps of the margin between Cape Recife and Punta do Ouro have not yet been published.

The shelf break is much shallower and better defined off the south coast than off the west coast. Between Cape Agulhas and Cape Recife the break is sharp and varies from 120 to 180 m in depth, averaging about 140 m. East of Cape Recife the break shallows to less than 50 m in places.

The only prominent topographic features known on the southern shelf are the Alphard Banks (a group of volcanic pinnacles: Fig. 1) and a series of west-trending ridges in the vicinity of Cape St. Francis. The latter features have a maximum relief of 26 m and are probably submerged beach ridges.

Outer Margin Bathymetry and Morphological Zones

The bathymetry of the rest of the slope and rise is known (Simpson, 1970, 1971), although in less detail. A marked distinction exists between the Atlantic and Indian ocean margins; a gentle slope and a well-developed rise exist on the former, but a steep slope and only a poorly developed rise on the latter. A major feature of the outer Atlantic margin is the northeast-trending Walvis Ridge (Fig. 1),

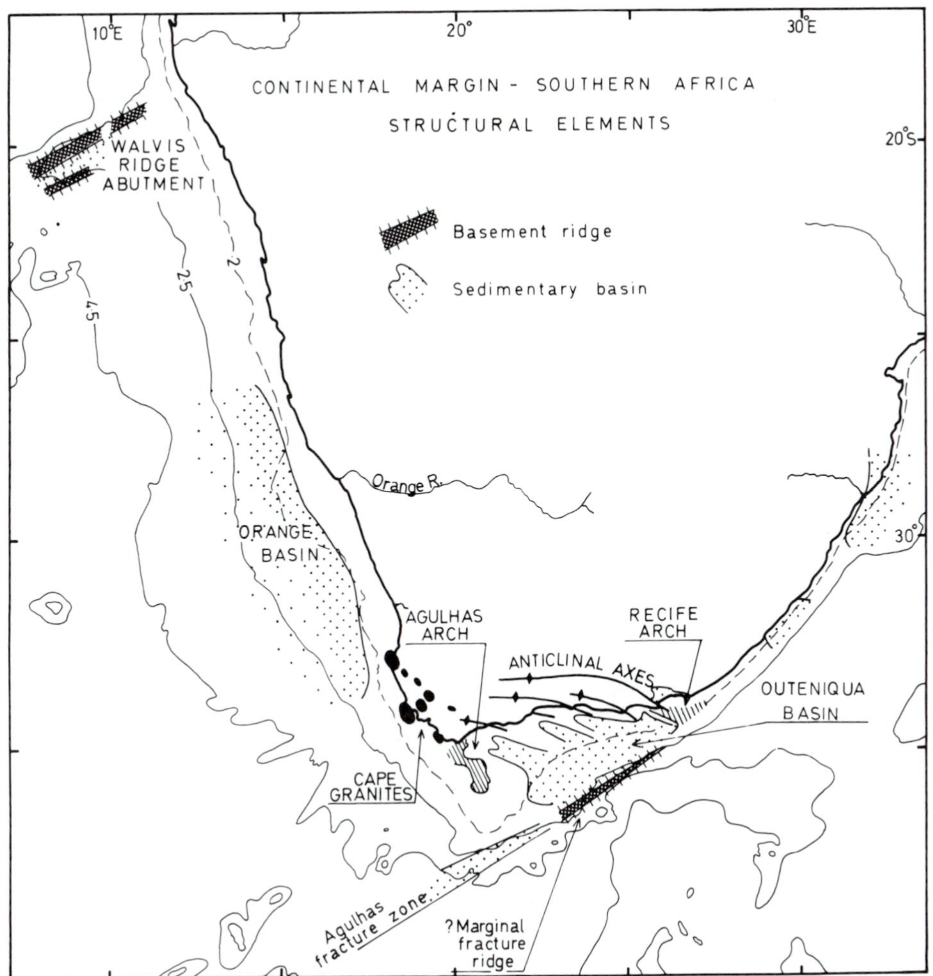

Fig. 2. Structural elements of the southern African continental margin, modified from Scrutton et al. (in press). All features are referred to in the text. Anticlinal structures are located in the Cape foldbelt. The 0.2-, 2.5-, and 4.5-km isobaths are shown.

which abuts the continent at 20°S, dividing subdued topography of the south from more rugged topography to the north. In the Indian Ocean, the 2,000-m-high Agulhas Plateau occurs beyond the slope at 26°E; from its northern tip a chain of seamounts extends southwestward to the Cape Rise (Fig. 1).

In general, the shelf and upper slope off both coasts can be divided into three morphological zones from the shoreline seaward: a narrow nearshore platform, a gently sloping (1-2°) shelf, and a steeper (2-15°) slope. The nearshore platform is rocky and is bounded on its seaward edge by a steep cliff which varies from 5 to 25 m in height off thewest coast to 50-70 m off the south coast. The inner shelf has a comparatively thick sediment cover and its relief is thus generally subdued. Outer parts of the shelf are rough and rocky, owing to a patchy, thin sediment cover. The slope is generally smooth because of its thick mantle of sediment, but several prominent notches do occur, which result from hard rock outcrops.

GEOPHYSICAL RESULTS AND CRUSTAL STRUCTURE

Atlantic Ocean Margin

The Walvis Ridge Abutment (Fig. 2) is the point at which the easternmost section of the northeast-trending Walvis Ridge joins the continental margin. Simpson (1971) has commented on the remarkable change in margin morphology from north to south across this junction. To the north, gravity data collected over the steep continental slope indicate a mass excess below the sea floor, which is interpreted as a basement ridge of varying height damming back at least 1.5 km of sediment (Rabinowitz, 1972). Over the abutment itself, and as far south as Walvis Bay (Fig. 1), gravity data have been compiled and contoured by Goslin et al. (in press), but not yet interpreted. However, their contour map and the depth-to-basement contours of Barnaby (in preparation) suggest that a rifted margin structure, where the gravity edge effect overlies the slope,

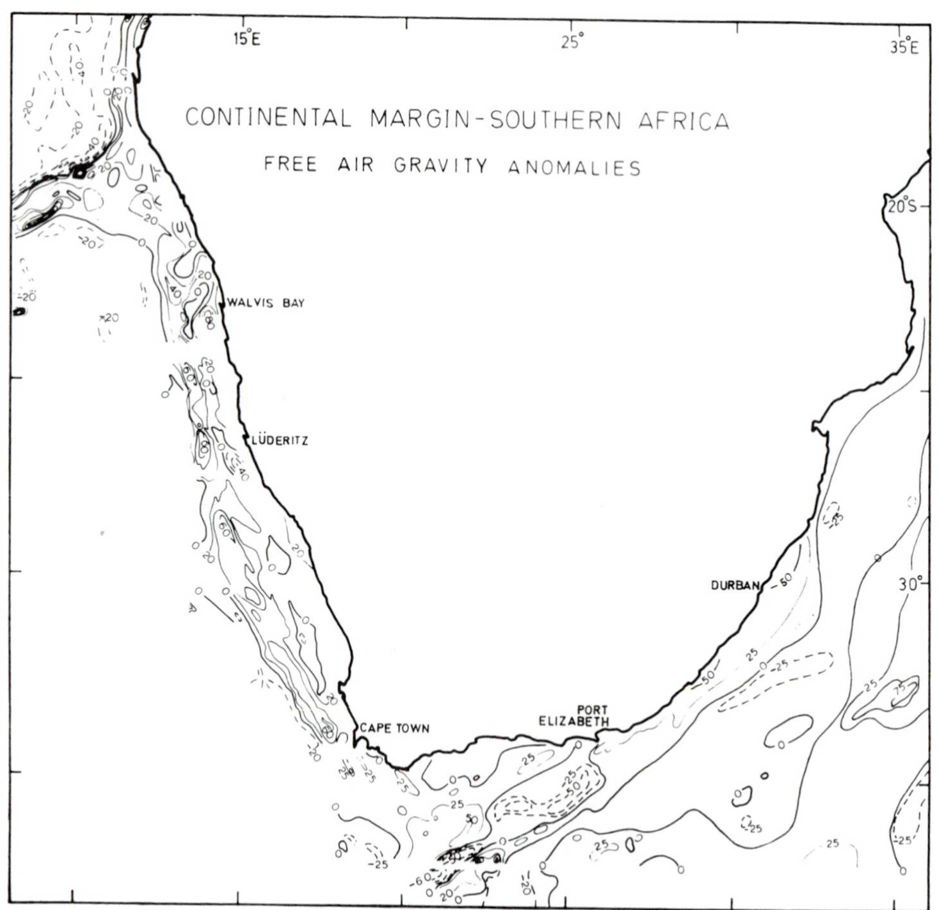

Fig. 3. Available free-air gravity anomaly data for the continental margin of southern Africa; compiled from Rabinowitz (1972) north of 17°S in the Atlantic Ocean; Goslin et al. (in press) between 17°S and 24°S; Simpson and du Plessis (1968) between 24°S and 34°S in the Atlantic; Graham and Hales (1965), the area south of 37°; and Talwani and Kahle (in press), the remaining areas east of Cape Town and in the Indian Ocean. Contours are at 20-mgal intervals except those redrawn from Talwani and Eldholm (1973), which are at 25-mgal intervals. Positive anomaly contours, solid; negative anomaly contours, dashed.

does not develop to the south until 21½°S (Fig. 3). North of this latitude they swing southeast-northwest to skirt the southwestern edge of a basement plateau that exists beneath the abutment area.

This plateau shows large-scale roughness and is overlain by approximately 2 km of sediment. Toward the coast the sediment thickens before wedging out against rapidly rising continental basement. Trending west-southwest from the northern parts of the plateau, the Walvis Ridge consists of two steep-sided ridges of probable volcanic material (Hekinian, 1972), enclosing a sedimentary basin about 2 km in depth (Fig. 2). The southern ridge begins at the western edge of the plateau at 10°E, but the northern one appears to be an extension of another ridge beneath the northern scarp of the abutment (Barnaby, in preparation).

There is a consensus that the Walvis Ridge has existed as a barrier to deep oceanic circulation since the time the South Atlantic opened at the beginning of the Cretaceous period (Francheteau and LePichon, 1972; Leyden et al., 1972; Scrutton and Dingle, in preparation). If this is so, the damming effect of the ridge and abutment basement plateau on northward-moving sediment (Goslin et al., in press) may well have been a factor contributing to the heavy sedimentation, and resulting subdued topography, over the continental margin to the south. The structure of the sediment wedge (at least 3 km thick) in the vicinity of Walvis Bay has been described in some detail by van Andel and Calvert (1971). They discovered that the erosion features being created today on the shelf and upper slope (down to 600 m below sea level) are repeated in the unconformities between depositional units to a depth of 1 or 2 km below the sea floor. To explain this cyclic erosion and deposition during the Cenozoic, they appealed to varying rates of margin subsidence, fluctuating sediment supply, and changes in the strength of the eroding bottom currents.

Gravity and magnetic data from the region between Walvis Bay and Lüderitz (Figs. 3 and 4) reveal some intense anomalies (amplitudes up to 60 mgal and 2,000 gammas, respectively: Simpson and du Plessis, 1968; Uchupi and Emery, 1973; Goslin et al., in press; Barnaby, in preparation) which are indicative of rapid variation in basement topo-

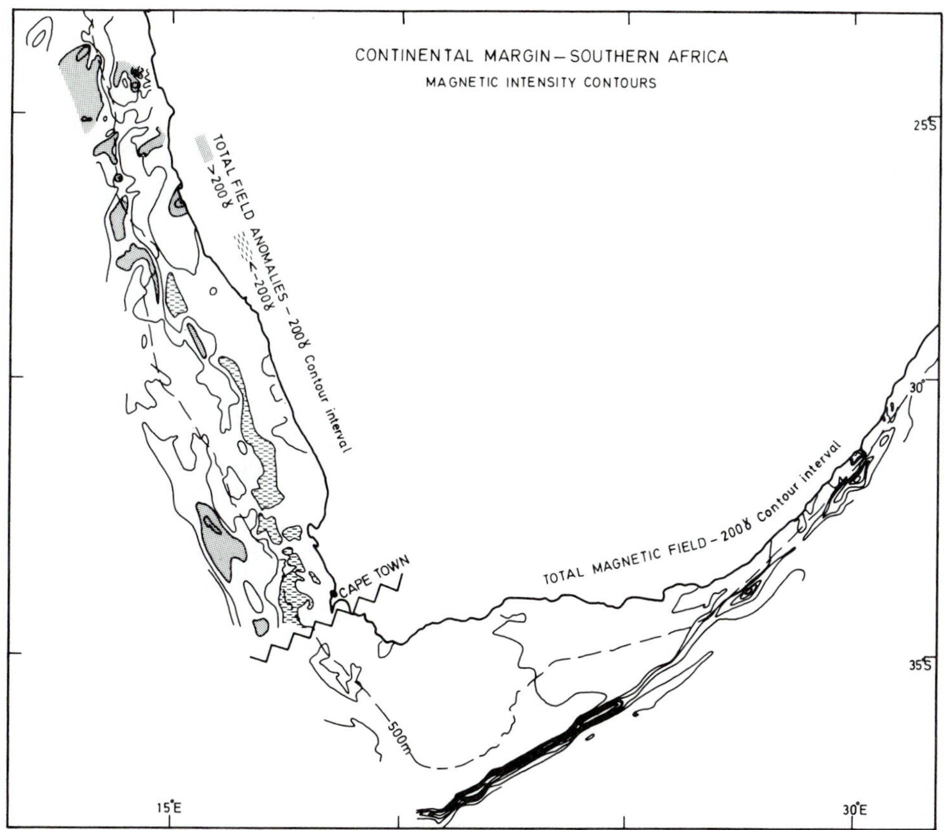

Fig. 4. Available magnetic intensity contour data for the continental margin of southern Africa; total field anomaly contours redrawn at 200-gamma intervals from Simpson and du Plessis (1968) north of Cape Town, and from du Plessis and Simpson (in preparation) south and east of Cape Town. The 0.5-km isobath is shown.

graphy or composition or both. The nature of the magnetics and a change in structure of the sediment wedge led du Plessis et al. (1972) to postulate the presence of a basement rise in this area. Although the data to the south of Walvis Bay are not as good as they are to the north, it appears that the general trend of gravity and magnetic anomalies over the shelf changes from northeast in the north to north-northwest near Lüderitz. This change in direction matches the change in strike of the late Precambrian Damara orogenic belt onshore, and a correlation between the two is probable.

South of Lüderitz the crustal structure of the margin is better understood. Between Lüderitz and Cape Columbine the coastline forms a broad embayment and the continental margin widens (Fig. 2). Beneath the shelf and slope lies the Orange Basin (Dingle, 1973a), which is really a thickening to about 5 km of the sediment wedge (rather than a true basin), whose structure is known from continuous seismic profiling (du Plessis et al., 1972; Uchupi and Emery, 1973; Dingle, 1973a), seismic-refraction measurements (Bryan and Simpson, 1971), and gravity results (Scrutton, in press). Within the area of the Basin, the gravity interpretation gives an indication of the changes in deep crustal structure across the margin. A mass excess exists beneath the outer parts of the shelf and has been attributed to a density increase in the upper crust, whereas the crust as a whole thins gradually from 30 km at the coastline to 14 km beneath the continental rise. The margin is clearly of the rifted type.

At 32½°S the crustal structure changes dramatically. The coastline moves seaward (Figs. 1 and 2) in response to the presence of the northwest-trending Cape Granites-Agulhas Arch basement high (Scrutton and Dingle, in preparation). The late Precambrian Cape Granites are intruded into metamorphic country rocks, causing a negative gravity anomaly (Smit et al., 1962), whereas the Agulhas Arch, an antiform of Cape Granite and Cape Supergroup rocks, is surrounded by low-density post-Paleozoic sediments, producing a positive anomaly (Graham and Hales, 1965; Talwani and Kahle, in press) (Figs. 2 and 3). The southwestern flank of the basement high carries a narrow, relatively thin (about 3-4 km) sediment wedge (Leyden et al., 1971).

Over the continental slope, Talwani and Eldholm (1973) have traced a positive magnetic anomaly that they consider to be the first of the sea-floor spreading sequence and, therefore, an indicator of the onset of oceanic crust. When traced northwestward to 31°S, this anomaly does, indeed, lie adjacent to the continent-ocean boundary as deduced from gravity data (Scrutton, in press). Thus the edge of the continent may well have been located in this region. If so, the sequence of linear or en echelon magnetic anomalies observed over the continental shelf at 31°S (Simpson and du Plessis, 1968) has its origin in continental- or transitional-type crust.

There are structural features of the Atlantic continental margin that are present throughout its length. The most notable are the present-day outbuilding of the sediment wedge at the expense of upbuilding (du Plessis et al., 1972), and the apparent lack of large basement ridges paralleling the margin which could act as sediment dams (Scrutton and Dingle, in preparation).

Indian Ocean Margin

At the tip of the Agulhas Bank the continental slope turns through an angle of 90° to face the Indian Ocean (Fig. 1), abruptly truncating the southeast-trending Agulhas Arch. Between 20½°E and 22½°E the truncated basement produces a very steep slope (up to 15°), which is covered by only a thin veneer of sediment (Scrutton and du Plessis, 1973). To the southwest this steep slope merges into a bathymetric spur which has a small basement nucleus (du Plessis, unpublished data). The profiles of Talwani and Eldholm (1973) show that there is little or no isostatic gravity anomaly associated with the spur, but the magnetic field is not as quiet as that over the adjacent Agulhas Arch, suggesting some compositional difference in basement between the two features.

Along the base of the steep continental slope, there is a narrow trough containing at least 2 km of sediment (Scrutton et al., in press); free-air gravity values over this trough are strongly negative (down to -80 mgal) because of the large continental edge effect at this point and the mass deficiency of the sediment (Graham and Hales, 1965). A seamount chain lying upon oceanic crust (du Plessis and Simpson, in preparation) borders the trough to the southeast (Fig. 1) and, together with the trough, constitutes the continental rise along this part of the margin. The trough marks the position of the Agulhas Fracture Zone (Emery, 1972; Talwani and Eldholm, 1973), which controls the trend and nature of the entire Indian Ocean margin (Francheteau and LePichon, 1972; Scrutton, 1973).

Along longitude 20°49'E a large-scale seismic-refraction experiment has been carried out over the continental shelf. The sedimentary structure to a depth of 10 km has been determined by Spence (1970) and the deep structure by Hales and Nation (1972). The profile runs obliquely over the northeast flank of the Agulhas Arch and reveals the Lower Paleozoic and Precambrian basement downfaulted to the north and northeast to produce a synform, which has its northern edge near the coastline. Mesozoic and Cenozoic deposits infill this synform and extend and thicken eastward (Leyden et al., 1971; Dingle, 1973c) to form the Outeniqua Basin beneath the continental shelf (Fig. 2). Hales and Nation (1972) discovered continental crust of about 32 km in thickness beneath the Agulhas Arch. They attempted to extrapolate their structure seaward using Graham and Hales (1965) gravity data but were unable to take into consideration the large variations in shallow structure that have recently been discovered. The extrapolated structure is, therefore, unlikely to be correct.

The margin east of 22½°E, as far as Cape Recife, possesses a completely different structure. Beneath the continental shelf and upper slope lies the Mesozoic-Cenozoic Outeniqua Basin previously mentioned. Seismic investigations of the basin have been reviewed by Dingle (1973c, 1973d) and show that the basin fill reaches a maximum of 6.2 km southwest of Cape Recife. This basin is dammed by a 350-km-long, nonmagnetic basement ridge (Fig. 2). Scrutton and du Plessis (1973) have suggested that this ridge is a marginal fracture ridge (LePichon and Hayes, 1971) formed at the time the Agulhas Fracture Zone was active as a transcurrent fault between the African and South American continental blocks. This marginal fracture ridge appears to give way southwestward to the narrow sediment-filled trough at the foot of the truncated Agulhas Arch, as would be expected if both ridge and trough were formed within the Agulhas Fracture Zone.

In this region the continental rise is either poorly developed or absent. Southeast of the margin and separated from it by oceanic crust (Ludwig et al., 1968) lies the Agulhas Plateau (Fig. 1), an area at about 3,000 m depth between the oceanic Agulhas and Transkei basins. The northern parts of the plateau are rugged, owing to block faulting in the basement and only thin sediment cover, but the southern parts have a thicker overburden and are, therefore, of low relief (Nicolaysen, 1973). Until recently, the plateau had been regarded as a continental fragment (Graham and Hales, 1965; Laughton et al., 1973), but now there is some seismic refraction evidence that it is underlain by thickened oceanic crust (Hales et al., 1970).

The final sector of the South African continental margin to be discussed here extends from Cape Recife to Durban. Another change in margin structure (Fig. 3) takes place at Cape Recife, indicated by the change in free-air gravity anomalies (Talwani and Kahle, in press). In this sector the margin is extremely narrow and clearly delineated by the Agulhas Fracture Zone, which runs along the base of the steep slope (Fig. 2). On the shelf, the structure is not well understood, but seismic refraction results reveal the gross features (Ludwig et al., 1968). It appears that the sea floor is underlain by a few hundred meters of Cenozoic deposits and that immediately beneath these is a block-faulted basement (Precambrian, Cape and Karroo ages).

At 32°S, however, a depression occurs in the basement (Fig. 2), containing about 1 km of presumed Mesozoic rocks (Ludwig et al., 1968), which may be represented onshore by a tiny outlier of Neocomian beach deposits. Seaward, the depression could be open ended, as suggested by Ludwig et al. (1968) and Dingle and Scrutton (in preparation), since the proposed marginal fracture ridge mentioned above appears not to continue past Cape Recife. In the vicinity and to the east of Durban another, but very much larger, sediment-filled depression occurs (Ludwig et al., 1968), coinciding with another change in margin structure (Fig. 2). The Agulhas Fracture Zone stops at this point and the steep slope gives way to a wide, possibly downfaulted, terrace seaward of Cape St. Lucia and southern Mozambique (Fig. 1), which is underlain by the larger depression described by Dingle and Scrutton (in preparation). There are, as yet, few published geophysical results from this region.

Adjacent to the continental margin in this region lies the Natal Valley. Nicolaysen (1973) has recently reported the results of seismic-refraction studies which show that beneath the valley the depth to the base of the crust increases toward the northeast from oceanic values south of Durban to more than 20 km east of Durban, at 30°S. This lends support to the hypothesis that a segment of rifted continental margin strikes southeast from Durban, between the end of the Agulhas Fracture Zone and the Mozambique Ridge (Scrutton, 1973), separating oceanic crust in the Natal Valley from continental or transitional crust beneath the terrace to the north.

The entire Indian Ocean margin between Durban and the tip of the Agulhas Bank is characterized by a large positive magnetic anomaly overlying the poorly developed continental rise (Fig. 4). A wealth of magnetic data collected by the Marine Geophysics Unit clearly defines its shape and extent. It is interpreted (du Plessis and Simpson, in preparation) as being caused by the abrupt truncation of the magnetic oceanic basement of the Agulhas and Transkei basins against nonmagnetic or continental basement to the northwest. There seems little doubt that faulting along the line of the Agulhas Fracture Zone is responsible for this abrupt truncation.

POST-PALEOZOIC BEDROCK STRATIGRAPHY AND HISTORY

Van Andel and Calvert (1971) and du Plessis et al. (1972) have reported on some geologic aspects of the continental margin north of Lüderitz (discussed in the last section), but at present only the margin between Lüderitz and Cape Recife has been mapped geologically (Dingle, 1970, 1971b, 1973a, 1973c; Dingle et al., 1971). Simpson (1971) has reviewed the general geology of the southwest African margin, Dingle (1973d), Simpson and Dingle (1973) have presented information on the offshore sedimentary basins on the Indian Ocean part of the margin, and Dingle and Scrutton (in preparation) have discussed the relationship of sedimentary basins to the breakup of Gondwanaland. The petrology and origin of offshore limestones have been described by Siesser (1970, 1971b, 1972d, 1972e), dolostones by Siesser (1972c), volcanic rocks by Dingle and Gentle (1972), and phosphorites by Parker (1971) and Parker and

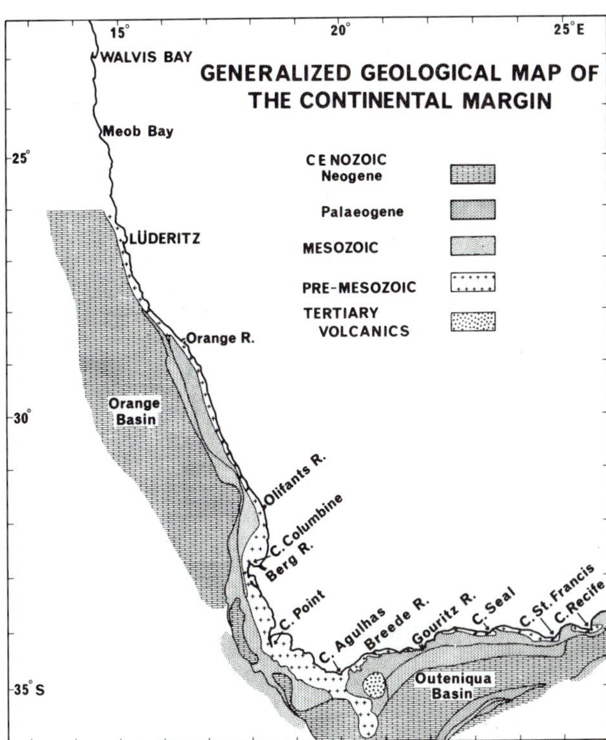

Fig. 5. Generalized bedrock geology of the shelf and upper slope between Luderitz and Cape Recife; after Dingle (1970, 1971b, 1973a, 1973c).

Siesser (1972). Mesozoic paleogeography of the southern Cape has been discussed by Dingle (1973b). The following summary of post-Paleozoic geology draws most heavily on data and interpretations from Dingle's publications.

Atlantic Ocean Margin

During Cretaceous time the continental margin of the west coast experienced rapid upbuilding and outbuilding onto a subsiding, tension-rifted basement. In post-Cretaceous time outbuilding became progressively more important. The Orange Basin (Figs. 2 and 5) represents a prograding wedge of Mesozoic sediment built out over the margin of the continent onto oceanic basement by discharge from the Orange, Olifants, and Berg rivers. The Orange River was, and has remained, the major sediment source for this segment of the margin. Mesozoic sediments have a maximum thickness of 3.8 km in this basin; their lithology is poorly known. The rate of sedimentation has decreased progressively since the Cretaceous, Cenozoic sediments being only slightly over 1 km thick. Paleogene sediments (up to 800 m thick) are mostly glauconitic clays and quartzose limestones, whereas Neogene sediments (up to 300 m) are glauconitic and phosphatic shelly limestones and sandstones.

Cretaceous sedimentation on the west coast began after the rift between the African and South American plates, and ended with a major regression, folding, and erosion. Marine sedimentation began in Aptian times around Angola; the margin to the south lay nearer the equator of spreading, and marine incursions there were in all likelihood somewhat earlier, probably Neocomian. Tertiary sedimentation commenced with an Eocene transgression, followed by a major regression and erosion at the end of the Eocene, which probably cut the proto-Cape Canyon. Subsequent Miocene and Pliocene transgressions laid down mostly thin, shallow-water facies on the shelf, the bulk of the sediments being deposited on the continental slope. Vigorous erosion during a major pre-Quaternary regression removed most of the Tertiary rocks, which are now found only as isolated outliers.

Indian Ocean Margin

Off the south coast, the Agulhas Bank consists of a thick sediment wedge built upon block-faulted, foundered, continental basement. The Outeniqua Basin (Figs. 2 and 5) lies between two basement highs, the Agulhas and Recife arches, and behind the marginal fracture ridge (Fig. 2). The Agulhas Arch has little topographic expression and is only patchily covered by unconsolidated sediments; the Recife Arch also has little topographic expression, being covered by several hundred meters of sediment. The latter arch strikes west-northwest and is probably made up of Cape Supergroup rocks. The Agulhas Basin is subdivided by at least two smaller west-northwest-trending basement ridges (Cape St. Blaize and Cape Seal Arches), which are anticlinal extensions of the Cape Fold Belt (Fig. 2). Most Mesozoic sediments in the Outeniqua Basin were trapped behind basement strauctures, but the very narrow continental margin farther east allowed Mesozoic (and later Cenozoic) sediments to be carried beyond the margin and into the Natal Valley and Transkei Basin. Consequently, these repositories have a greater than average thickness of sediment for oceanic regions (Green and Hales, 1966).

Mesozoic sediments in the Outeniqua Basin are up to 5.7 km thick. The oldest known sediments related to this basin are Upper Jurassic clays found onshore (Dingle and Klinger, 1971, 1972). The oldest Mesozoic rocks dredged offshore thus far are Lower Cretaceous clays and pyritic sandstones, but these are rarely exposed. Upper Cretaceous rocks (clays and glauconitic quartzose limestones) are more common, probably forming a continuous cover over the major part of the Agulhas Bank beneath the Tertiary beds.

As in the Orange Basin, Cenozoic sediments have a maximum thickness of about 1 km. Paleogene units are up to 500 m thick and are mostly calcareous clays, glauconitic-quartzose limestone, and glauconitic silt- and sandstones. Neogene sediments are thinner—up to 400 m—and are mostly foraminiferal-bryozoan limestones. In both the Orange and Outeniqua basins, Cenozoic sedimentation was concentrated on the outer shelf and slope, and much of it was transported by turbidites into adjacent deep-water basins.

The sedimentary history of the Agulhas Bank probably commenced with continental sedimentation following after the formation of the Cape Fold Belt during the Middle Triassic. Marine sediments were laid down in late Jurassic/early Cretaceous times; Dingle and Scrutton (in preparation) believe marine sediments may have been deposited slightly earlier (at least by late Jurassic), owing to epicontinental transgression. Sedimentation was temporarily interrupted in the middle Cretaceous, but extensive marine deposits were laid down during the Late Cretaceous. A major unconformity resulting from regression marks the close of the Cretaceous. This regression was accompanied by folding, faulting, erosion, and, on the northeastern side of the Agulhas Arch, by intrusion of basalt and trachyte plugs (the Alphard Banks). Aegerine trachyte from these plugs has been dated at 58 + 2.4 my (Paleocene) by the K-Ar method.

Cenozoic sedimentation commenced with a late Paleocene-early Eocene transgression. The Agulhas and Recife basement arches were still positive features at this time, as shown by facies changes and thinning of sediments toward them. The Oligocene must have been a time of major regression around southern Africa. No rocks of this age have been reported either onshore or offshore, and seismic records show a well-developed unconformity all around the continental margin which probably represents this epoch and possibly the early Miocene

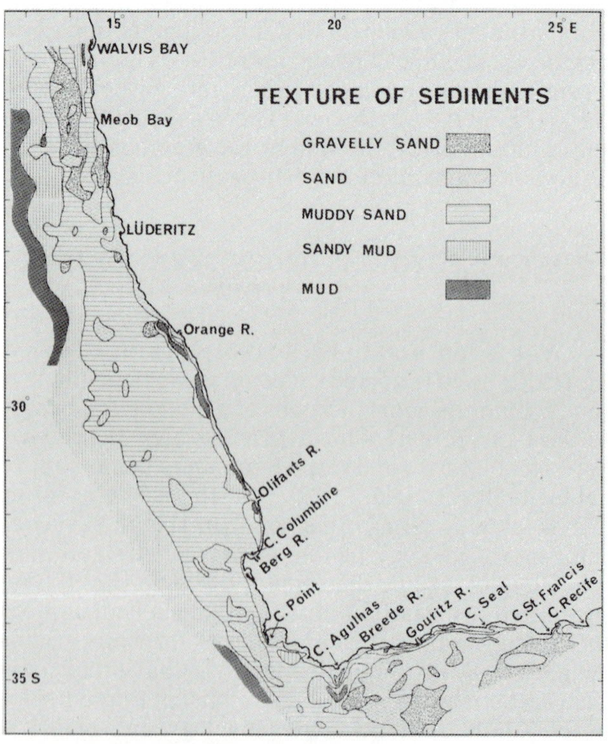

Fig. 6. Generalized sediment texture map of the shelf and upper slope between Walvis Bay and Cape Recife; after Birch (1973), Bremner (1973), Rogers (1973), and Birch and Rogers (1973).

as well. Many of the lower Paleogene rocks were eroded and destroyed during this emergence and during the subsequent Miocene transgression when seas returned again to cover the margins of southern Africa. This was followed by regression at the end of the Miocene and a further transgression during the Pliocene.

The Tertiary-Quaternary boundary cannot always be determined from seismic records, particularly where both units dip at very low angles. However, in many areas there is an angular unconformity at the end of the Pliocene. Two thin Pleistocene limestone facies are exposed on the Agulhas Bank: a sublittoral limestone consisting mostly of broken shell fragments, phosphorite, and glauconite grains, and (on the inner shelf) a lithified calcareous dune deposit, similar to the coastal limestones exposed onshore.

Mesozoic and Tertiary successions on both coasts have been subdivided by correlating unconformities, which are assumed to be contemporaneous. This has allowed division into a number of post-Paleozoic time-equivalent lithologic sequences. These units have recently been given group, formation, and member names by Dingle (1973c) to replace his earlier suffix designations (Dingle, 1971b, 1973a).

SEDIMENTS

General mapping of surficial sediments on various parts of the continental margin has been done by a number of workers (Senin, 1968; Nekritz and Busch, 1969; Emelyanov and Senin, 1969; Bowie et al., 1970; Rogers, 1971, 1973; O'Shea, 1971; Birch, 1973; Bremner, 1973; Gershanovich et al., 1972; Birch and Rogers, 1973). Specialized studies have been made of sediment texture by Fuller (1961, 1962) and Fuller and Lamming (1967), of the petrology and classification of glauconite by Lloyd and Fuller (1965) and Birch (1971), and of the occurrence of phosphate in sediments by Summerhayes et al. (1973) and Summerhayes (in press). The petrology and origin of phosphorite "nodules" on the continental margin have been discussed by Baturin (1971), Baturin et al. (1972), Senin (1970), Parker (1971), and Parker and Siesser (1972). Relict algal nodules were reported by Siesser (1972f).

The carbonate fraction of these sediments has been studied with respect to abundance and distribution (Siesser, 1972a), mineralogy and diagenesis (Siesser, 1971a, 1972b), and geochemistry (Siesser, 1973a). Information on the west coast diamondiferous sands and gravels has been published by Borchers et al. (1969), Murray (1969), and Murray et al. (1970). Calvert and Price (1970, 1971) have investigated the organic and metal-rich muds off South West Africa, and Rogers et al. (1972) summarized some economic aspects of sea-floor deposits. Studies relating to Pleistocene paleogeography have been made by Dingle and Rogers (1972a, 1972b) and Pleistocene paleoclimatology and piston core stratigraphy by Siesser (1973b).

Atlantic Ocean Margin

Sediments on the west coast margin form texturally distinct belts parelleling the coast (Fig. 6). The nearshore rocky platform between Walvis Bay and Cape Agulhas is patchily covered by sand and mud. A lens of mud or sandy mud up to 10 m thick is often found at the base of the small cliff at the seaward edge of this platform. Seaward of the nearshore zone, sand and muddy sand form a broad belt covering the middle and outer shelf; mud and sandy mud mantle the upper slope.

The inner shelf sands and muds are Quaternary deposits chiefly composed of quartz (Fig. 7). Sands on parts of the middle and outer shelf are largely Tertiary residual and relict deposits composed predominantly of quartz and shell fragments with locally abundant glauconite and phosphorite grains. Quaternary planktonic foraminifera and coccoliths are the dominant components of the sediments on most of the outer shelf and all the upper slope.

Residual deposits on the shelf are believed to have been derived by erosion during Tertiary sea-level fluctuations. They are now being covered on their seaward edge by deposition of modern oceanic sediment and on their inshore edge by modern river and aeolian sediments. The Berg and Olifants rivers

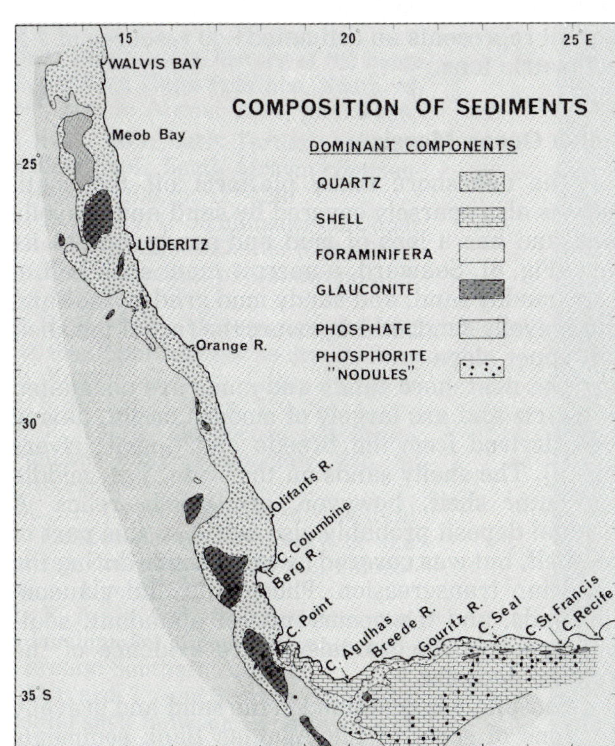

Fig. 7. Generalized sediment composition map (sand fraction) of the shelf and upper slope between Walvis Bay and Cape Recife; after Birch (1973), Bremner (1973), Rogers (1973). Areas where phosphorite "nodules" are most abundant are also shown; after Parker and Siesser (1972).

———, 1973b, Post-Palaeozoic stratigraphy of the eastern Agulhas Bank, South African continental margin: Marine Geology, v. 15, p. 1-23.

———, 1973d, Regional distribution and thickness of post-Palaeozoic sediments onthe continental margin of southern Africa: Geol. Mag., v. 110, p. 97-102.

———, 1973b, Mesozoic palaeogeography of the southern Cape, South Africa: Palaeogeog. Palaeoclimatol. Palaeoecol. v. 13, p. 203-213.

du Plessis, A., Scrutton, R. A., Barnaby, A. M., and Simpson, E. S. W., 1972, Shallow structure of the continental margin of southwestern Africa: Marine Geology, v. 13, p. 77-89.

———, and Simpson, E. S. W., in preparation, Total magnetic field anomalies over the continental margin of South Africa—Cape Agulhas to Durban.

Emelyanov, E. M., and Senin, M., 1969, Composition of the sediments of the South-West Africa shelf: Litolog. i Poleznye Tskopaemye, v. 2, p. 10-25.

Emery, K. O., 1972, EACM program of IDOE (GX-28193): some results of 1972 cruise of R/V Atlantis II: Woods Hole Ocean. Inst. Ref. 72-54, 11 p.

———, Milliman, J. D., and Uchupi, E., 1973, Physical properties and suspended matter of surface waters in the southeastern Atlantic Ocean: Jour. Sediment. Petrology, v. 43, p. 822-837.

Ewing, M., Eittreim, S., Truchan, M., and Ewing, J. I., 1969, Sediment distribution in the Indian Ocean: Deep-Sea Res., v. 16, p. 231-248.

Francheteau, J., and LePichon, X., 1972, Marginal fracture zones as structural framework of continental margins in the South Atlantic Ocean: Am. Assoc. Petr. Geol. Bull., v. 56, p. 991-1007.

Fuller, A. O., 1961, Size distribution characteristics of shallow marine sands from the Cape of Good Hope, South Africa: Jour. Sediment. Petrology, v. 31, p. 256-261.

———, 1962, Systematic fractionation of sand in the shallow marine and beach environment off the South African coast: Jour. Sediment. Petrology, v. 32, p. 602-606.

———, 1971, South Atlantic fracture zones and lines of old weakness in southern Africa: Nature, v. 231, p. 84-85.

———, 1972, Possible fracture zones and rifts in southern Africa: Geol. Soc. America Mem. 132, p. 159-172.

———, and Lamming, P. J., 1967, The hydraulic equivalence of quartz and zircon in coastal deposits from the south western districts of the Cape Province, and its application as an environmental indicator: S. African Jour. Sci., v. 63, p. 521-526.

Gershanovich, I. K., Avilov, I. P., and Zarikhin, I. P., 1972, Bottom sediments of continental margins in the South Atlantic, *in* Bonitation of the World Ocean: Moscow, Pischevaya Promyshlevnost, p. 166-190.

Goslin, J., Mascle, J., Sibuet, J. C., and Hoskins, H., in press, Geophysical study of the easternmost Walvis Ridge (South Atlantic), pt. I, Morphology and shallow structure: Am. Assoc. Petr. Geol. Bull.

Graham, K. W. T., and Hales, A. L., 1965, Surface ship gravity measurements in the Agulhas Bank area south of South Africa: Jour. Geophys. Res., v. 70, p. 4005-4011.

Green, A. G., 1972, Sea-floor spreading in the Mozambique Channel: Nature Phys. Sci., v. 236, p. 19-21.

Green, R. W. E., and Hales, A. L., 1966, Seismic refraction measurements in the southwestern Indian Ocean: Jour. Geophys. Res., v. 71, p. 1637-1647.

Hales, A. L., and Nation, J. B.,1972, A crustal structure profile on the Agulhas Bank: Seismol. Soc. America Bull., v. 62, p. 1029-1051.

———, Barrett, D., and Spence, D. L., 1970, The Indian Ocean seismic programme: a reivew, *in* Oceanography in South Africa: CSIR Symp., Durban.

Hekinian, R., 1972, Volcanics from the Walvis Ridge: Nature Phys. Sci., v. 239, p. 91-93.

Houtz, R., Ewing, J., and LePichon, X., 1968, Velocity of deep-sea sediment from sonobuoy data: Jour. Geophys. Res., v. 73, p. 2615-2641.

Hoyt, J. H., Oostdam, B. L., and Smith, D. D., 1969, Offshore sediments and valleys of the Orange River (South and South West Africa): Marine Geology, v. 7, p. 69-84.

Larson, R. L., and Pitman, W. C., 1972, World-wide correlation of Mesozoic magnetic anomalies, and its implications: Geol. Soc. America Bull., v. 83, p. 3645-3662.

Laughton, A. S., Sclater, J. G., and McKenzie, D. P., 1973, The structure and evolution of the Indian Ocean, *in* Tarling, D. H., and Runcorn, S. K., eds., Implications of continental drift to the earth sciences, v. I, NATO symposium, Newcastle: New York, Academic Press, p. 203-212.

LePichon, X., and Hayes, D. E., 1971, Marginal offsets, fracture zones and the early opening of the South Atlantic: Jour. Geophys. Res., v. 76, p. 6283-6293.

Leyden, R., Ewing, M., and Simpson, E. S. W., 1971, Geophysical reconnaissance on the African shelf: Cape Town to East London: Am. Assoc. Petr. Geol. Bull., v. 55, p. 651-657.

———, Bryan, G., and Ewing, M., 1972, Geophysical reconnaissance on African shelf: 2. Margin sediments from Gulf of Guinea to Walvis Ridge: Bull. Am. Assoc. Petr. Geol. Bull., v. 56, p. 682-693.

Lloyd, A. T., and Fuller, A. O., 1965, Glauconite from shallow marine sediments off the South African coast: S. African Sci., v. 61, p. 444-448.

Ludwig, W. J., Nafe, J. E., Simpson, E. S. W., and Sacks, S., 1968, Seismic refraction measurements on the southeast African continental margin: Jour. Geophys. Res., v. 73, p. 3707-3719.

Mascle, J., and Phillips, J. D., 1972, Magnetic smooth zones in the South Atlantic: Nature, v. 40, p. 80-84.

Murray, L. G., 1969, Exploration and sampling methods employed in the offshore diamond industry: IX Commonwealth Mining and Met. Congr., Mining and Petr. Geol., v. 14.

———, Joynt, R. H., O'Shea, D. O'C., Foster, R. W., and Kleinjan, L., 1970, The geological environment of some diamond deposits off the coast of South West Africa: Inst. Geol. Sci. London Rept. 70/13, p. 119-141.

Nekritz, R., and Busch, D. J., 1969, Distribution of bottom sediments off the coasts of Republic of South Africa and South West Africa: U.S. Navy Ocean. Office, Informal Rept. 69-9.

Nicolaysen, L. O., 1973, Progress in marine geology and geophysics in South Africa during the past three years (abstr.): Div. Sea Fisheries, Cape Town, S. African Nat. Ocean. Symp. p. 8-9.

Oguti, T., 1964, Geomagnetic anomaly around the continental shelf margin southern offshore of Africa: Jour. Geomag. Geoelec., v. 46, p. 65.

O'Shea, D. O'C., 1971, An outline of the inshore submarine geology of southern South West Africa and Namaqualand [M.Sc. thesis]: Univ. Cape Town.

Parker, R. J., 1971, The petrography and major element geochemistry of phosphorite nodule deposits on the Agulhas Bank, South Africa: Dept. Geol. Univ. Cape Town, SANCOR Marine Geol. Prog. Bull. 2, 94 p.

———, and Siesser, W. G., 1972, Petrology and origin of some phosphorites from the South African continental margin: Jour. Sediment. Petrology, v. 42, p. 4334-440.

Rabinowitz, P., 1972, Gravity anomalies on the continental margin of Angola, Africa: Jour. Geophys. Res., v. 77, p. 6327-6347.

Rogers, J., 1971, Sedimentology of Quaternary deposits on the Agulhas Bank: Dept. Geol. Univ. Cape Town, SANCOR Marine Geol. Prog. Bull. 1, 117 p.

———, 1973, Texture, composition, and depositional history of unconsolidated sediments from the Orange-Lüderitz shelf, and their relationship with Namib desert sands: Dept. Geol., Univ. Cape Town, Joint GSO/UCT Marine Geol. Tech. Rept. 5, p. 67-88.

———, and Bremner, J. M., 1973, Bathymetry of the Lüderitz-Walvis continental margin: Dept. Geol. Univ. Cape Town, Joint GSO/UCT Marine Geol. Tech. Rept. 5, p. 7-9.

———, Summerhayes, C. P., Dingle, R. V., Birch, G. F., Bremner, J. M., and Simpson, E. S. W., 1972, Distribution of minerals on the seabed around South Africa and problems in their exploration and eventual exploitation: ECOR Symp. on Ocean's Challenge to S. African Eng., Stellenbosch: C.S.I.R. 571.

Scrutton, R. A., 1973, Structure and evolution of the sea floor south of South Africa: Earth and Planetary Sci. Letters, v. 19, p. 250-256.

———, in press, Gravity results from the continental margin of South-western Africa: Marine Geophys. Res.

———, and du Plessis, A., 1973, Possible marginal fracture ridge south of South Africa: Nature Phys. Sci., v. 242, p. 180-182.

———, and Dingle, R. V., in preparation, Basement control over sedimentation on the continental margin west of Southern Africa.

———, du Plessis, A., Barnaby, A. M., and Simpson, E. S. W., in press, Contrasting structures and origins of the western and southeastern continental margins of Southern Africa, *in* Campbell, K. S. W., ed., 3rd Internatl. Gondwana Symp., Canberra: Austral. Nat. Univ. Press.

Senin, Yu. M., 1968, Specific features of sedimentation on the shelf in South-Western Africa: Litolog. i Poleznye Iskopaemye, v. 4, p. 108-111.

———, 1970, Phosphorus in bottom sediments of the South West African shelf: Litolog. i Poleznye Iskopaemye, v. 1, p. 11-26.

Siesser, W. G., 1970, Carbonate components and mineralogy of the South African coastal limestones and limestones of the Agulhas Bank: Trans. Geol. Soc. S. Africa, v. 73, p. 49-63.

———, 1971a, Mineralogy and diagenesis of some South African coastal and marine carbonates: Marine Geology, v. 10, p. 15-38.

———, 1971b, Petrology of some South African coastal and offshore carbonate rocks and sediments: Dept. Geol. Univ. Cape Town, SANCOR Marine Geol. Prog. Bull. 3, 232 p.

———, 1972a, Abundance and distribution of carbonate constitutents in some South African coastal and offshore sediments: Trans. Roy. Soc. S. Africa, v. 40, p. 261-278.

———, 1972b, Carbonate mineralogy of Bryozoans and other selected South African organisms: S. African Jour. Sci., v. 68, p. 71-74.

———, 1972c, Dolostone from the South African continental slope: Jour. Sediment. Petrology, v. 42, p. 694-699.

———, 1972d, Limestone lithofacies from the South African continental margin: Sediment. Geology, v. 8, p. 83-112.

———, 1972e, Petrology of the South African Coastal limestones: Trans. Geol. Soc. S. Africa, v. 75, p. 177-185.

———, 1972f, Relict algal nodules (rhodolites) from the South African continental shelf: Jour. Geology, v. 80, p. 611-616.

———, 1973a, Ca/Mg and Sr/Ca ratios of some South African coastal and offshore carbonate sediments: Am. Assoc. Petr. Geol. Bull., v. 57, p. 930-932.

———, 1973b, Stratigraphic and palaeoclimatic analysis of continental slope sediment cores (abstr.): Div. Sea Fisheries, Cape Town, S. African Nat. Ocean Symp., p. 30-31.

———, and Rogers, J., 1971, An investigation of the suitability of four methods used in routine carbonate analysis of marine sediments: Deep-Sea Res., v. 18, p. 135-139.

Simpson, E. S. W., 1966, Die Geologie van die Vastelandsplat: Tegnikon, v. 15, p. 168-176.

———, 1968, Marine geology: progress and problems: Proc. Geol. Soc. S. Africa, v. 71, p. 97-111.

———, 1970, Southeast Atlantic and southwest Indian Oceans: Dept. Geol. Univ. Cape Town, Chart 12A Bathymetry, 1:10,000,000 at 33°.

———, 1971, The geology of the south-west African continental margin: a review: Inst. Geol. Sci. London Rept. 70/16, p. 153-170.

———, and Dingle, R. V., 1973, Offshore sedimentary basins on the south-eastern continental margin of South Africa, *in* Blant, G., ed., Sedimentary basins of the African coasts: Paris, Assoc. African Geol. Survey, p. 63-68.

———, and du Plessis, A., 1968, Bathymetric, magnetic and gravity data from the continental margin of south-western Africa: Can. Jour. Earth Sci., v. 5, p. 1119-1123.

———, and Forder, E., 1968, The Cape submarine canyon: Fish. Bull. S. African, v. 5, p. 35-37.

———, and Heydorn, A. E. F., 1965, Vema seamount: Nature, v. 207, p. 249-251.

———, du Plessis, A., and Forder, E., 1970, Bathymetric and magnetic traverse measurements in False Bay and west of the Cape Peninsula: Trans. Roy. Soc. S. Africa, v. 39, p. 113-116.

Smit, P. J., Hales, A. L., and Gough, D. I., 1962, The gravity survey of the republic of South Africa: Geol. Survey Publ., Pretoria, 486 p.

Spence, D. L., 1970, A seismic refraction study of sedimentary structure on the Agulhas Bank, south of Cape Infanta [M.Sc. thesis]: Univ. Witwatersrand.

Summerhayes, C. P., in press, Distribution and origin of phosphate in sediments from the Agulhas Bank, South Africa: Trans. Geol. Soc. S. Africa.

———, Birch, G. F., Rogers, J., and Dingle, R. V., 1973, Phosphate in sediments off southwestern Africa: Nature, v. 243, p. 509-511.

Talwani, M., 1962, Gravity measurements on H.M.S. "Acheron" in South Atlantic and Indian Oceans: Geol. Soc. America Bull., v. 73, p. 1171-1182.

———, and Eldholm, O., 1973, Boundary between continental and oceanic crust at the margin of rifted continents: Nature, v. 241, p. 325-330.

———, and Kahle, H. G., in press, Free air gravity charts of the Indian Ocean, *in* Udintsev, G., ed., The atlas of geology and geophysics of the international Indian Ocean expedition.

Uchupi, E., and Emery, K. O., 1973, Seismic reflection, magnetic and gravity profiles of the eastern Atlantic continental margin and adjacent deep-sea floor: I. Cape Francis (South Africa) to Congo Canyon (Republic of Zaire): Woods Hole Ocean. Inst. Ref. 72-95, 9 p.

van Andel, T. H., and Calvert, S. E., 1971, Evolution of sediment wedge, Walvis shelf, Southwest Africa: Jour. Geology, v. 79, p. 585-602.

Continental Margin of Antarctica: Pacific-Indian Sectors

Robert E. Houtz

INTRODUCTION

Structure sections based on seismic profiler and sonobuoy data from the continental margin of Antarctica are presented and briefly discussed. Sections from the Bellingshausen Sea-Antarctic Peninsula region of West Antarctica show an undisturbed, nearly horizontal basement surface (5-6 km below sea level) that extends up to the foot of the continental slope. No evidence of subduction accompanying the development of the early Mesozoic foldbelt of the Antarctic Peninsula has been recorded.

In marked contrast, the continental margin of East Antarctica (a stable shield) is characterized by a marginal basement depression that is at least 8-10 km below sea level. This "pan-Antarctic rift" seems to extend from Victoria Land westward just past the Kerguelen Plateau. This implies that there is no structural connection between the Kerguelen Plateau and the mainland, and is further supported by the lack of correspondence between the Cenozoic volcanics of the plateau and the shield rocks of the mainland.

REGIONAL DATA

Sonobuoy records have been obtained from the Antarctic margin since 1969 when the Lamont-Doherty group first installed a sonobuoy system aboard the U.S.N.S. ELTANIN. During the four years of our Antarctic sonobuoy operations, more than 700 sound velocity and layer thicknesses were computed. These solutions are used with profiler data to convert subbottom reflection profiles into thickness profiles. The sonobuoy technique is most useful along stable margins that have thick accumulations of sediment, such as the Antarctic margin, where the vertical profiler system usually does not record basement reflections through the thickest sediments. Sound-velocity data are critically needed along continental margins where the high sound speeds within the deeper layers produce significant deviations from the apparent structures observed in reflection profiles.

Six continental margin profiler sections, supplemented by sonobuoy data, are located in Figure 1. Two of the sections are located off West Antarctica, a relatively young province of Paleozoic and younger sediments that were folded most severely during the early Mesozoic. Three of the sections are located off East Antarctica, a stable Precambrian shield that is distinctly different from West Antarctica. One section is located in the Ross Sea, a region that is tectonically disturbed and occurs in the boundary area between East and West Antarctica. It is therefore appropriate to compare our structure sections from the continental margins of the three major tectonic elements of Antarctica. Previous work includes several profiler/sonobuoy crossings of the Ross Sea margin (Houtz and Davey, 1973) and one crossing of the Wilkes Land margin (Houtz and Markl, 1972).

REDUCTION OF SONOBUOY DATA

The sonobuoy solutions shown in the sections are based on variable-angle reflection data wherever possible. The slopes and intercepts of the deeper refraction lines have been used to supplement the variable-angle data. Occasionally reflection events are observed at depth but are too incoherent to provide solutions. In such cases the refractions from the top of the layer are used to estimate mean velocity, and the thickness is computed on the basis of the observed interval time (reflection time within the layer at vertical incidence).

If no velocity data are available from certain parts of the section, reasonable values are assumed, and are indicated by parentheses in the figures. Thicknesses computed from refraction data can be identified by the occurrence of a velocity below the final interface. Hachures beneath the deepest interface indicate that it is interpreted as basement. The interpretation is made on the basis of the velocity if there are refraction data from the interface, or on the characteristic appearance of the variable-angle reflection data from the typically rough basement surface. Section C (Fig. 3) is taken from Houtz and Davey (in press) and employs a velocity function (integrated to compute depth from reflection time) to compute depths of the reflection interfaces; hence there can be small disagreements between basement depth obtained from the individual sonobuoy station data and the generalized velocity function.

DISCUSSION OF RESULTS

West Antarctic Margin

Section A (Fig. 2) was obtained from the Pacific margin of the Antarctic Peninsula. The section

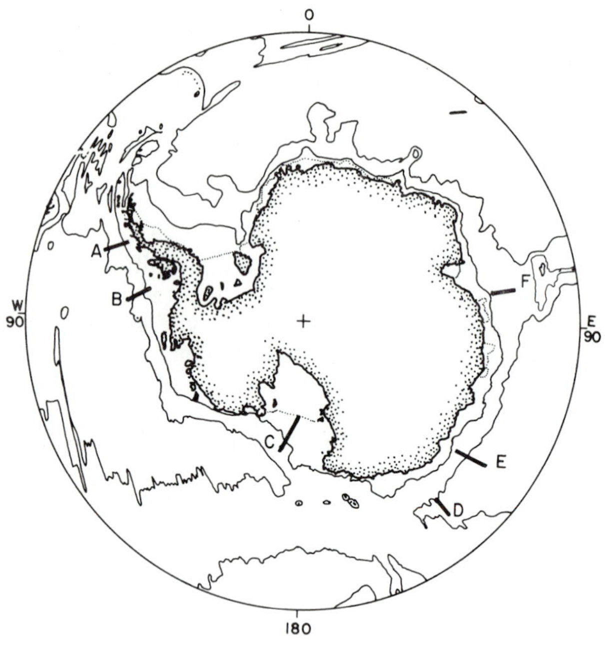

Fig. 1. Generalized bathymetry of Antarctic region (after Heezen et al., 1972), showing locations of structure sections.

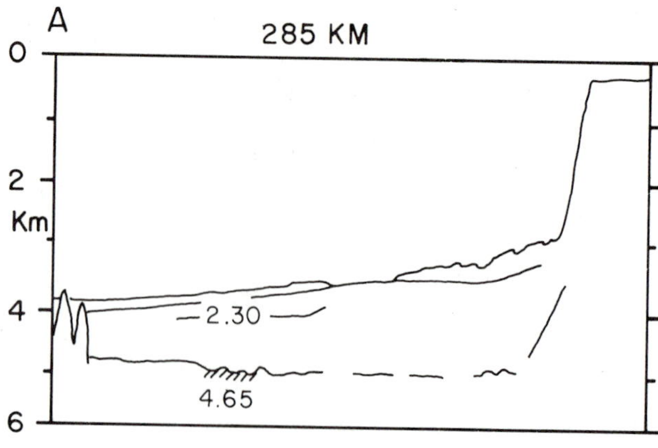

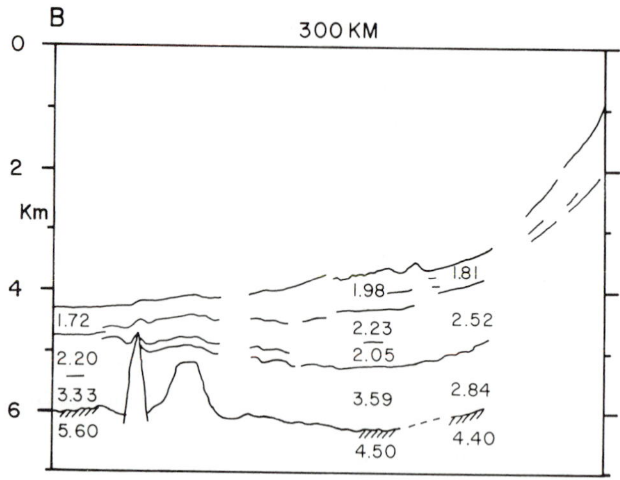

Fig. 2. Structure sections from Palmer Peninsula margin (upper) and Bellingshausen Sea (lower).

reveals a basement surface that is essentially horizontal right up to the foot of the continental shelf. Although the sediment cover is much thicker, section B from the Bellingshausen margin also reveals a fairly horizontal basement surface. However, a basement prominence near the left edge of the section breaks the regularity of the basement. The interval velocity in the layer marked 2.84 in the right-hand sonobuoy station is poorly determined (the standard deviation of the velocity is 0.5 km/sec). Hence the basement surface would be even flatter if the more reliable values of 3.3 to 3.6 km/sec from the adjacent stations were applied here. A third profiler crossing of the continental rise between sections A and B confirms the unusually flat character of the basement surface at the continental margin.

It seems likely that (in spite of the undisturbed appearance of the profiler data) crustal destruction occurred in this region when the Antarctic Peninsula foldbelt was developed during the early Mesozoic (Dalziel and Elliott, in press). Magnetic anomalies in the Bellingshausen Sea are interpreted as becoming younger toward land (Ellen Herron, personal communication); if confirmed, this observation also strongly implies crustal destruction. Apparently no trench-arc structure was formed at this time, or else it is now covered by sediments that have built out from the peninsula. Although precipitation rates here (hence sedimentation) are much higher than elsewhere in Antarctica (Dalrymple, 1966), there seems to be insufficient provenance from the peninsula to obscure an ancient trench.

Ross Sea Sector

Houtz and Davey (in press) have studied the structure of the Ross Sea shelf and its seaward margin in some detail. Section C (Fig. 3) is from their work and shows a possible fault midway up the continental rise. The 4.5-4.6 km/sec basement appears to have been displaced 2-3 km vertically in the plane of the section. To the west of this section the basement morphology is more complex, with major tensional features both parallel and perpendicular to the coastline. Since the Ross Sea is in the boundary region between East and West Antarctica, and is bordered in the west by the Transantarctic Mountains, the complexity is predictable. Hayes and Ringis (in press) have speculated that the Ross Sea is a sphenochasm resulting from differential movements along a possible suture between East and West Antarctica. The basement morphology remains comparatively complex for about 600 km west of section C (Houtz and Davey, in press).

Eastern Antarctic Margin

In marked contrast to the sections discussed above, section D (Fig. 3) shows a marginal depression of the basement that deepens toward the continent to at least 10 km below sea level. This depth compares with maximum depths of 5 km in section A and 6 km in B. Section E (Fig. 4) is located to the west of D, off the central Wilkes Land coast. Here the total depth to igneous basement may not have been measured because the deepest interface below the 4.20 km/sec layer appears to be too coherent for a typical basement reflection. However, the basement is fairly well identified in the records of the stations to the north.

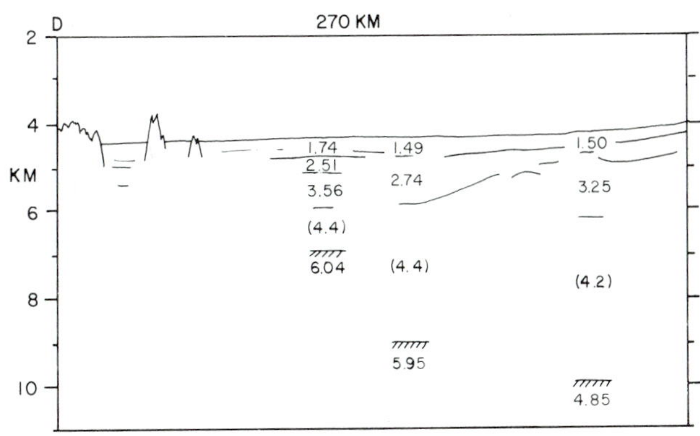

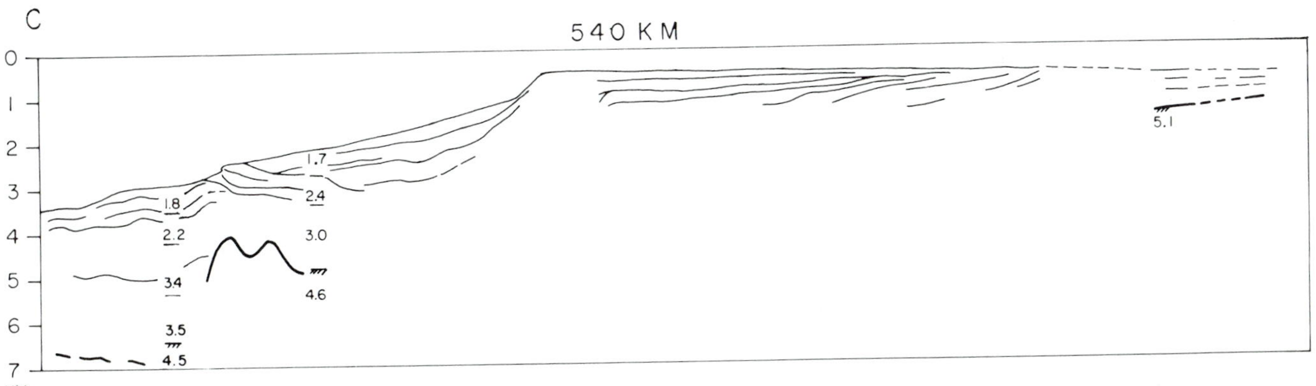

Fig. 3. Structure sections of Ross Sea (upper; from Houtz and Davey, 1973) and the Wilkes Land margin (lower).

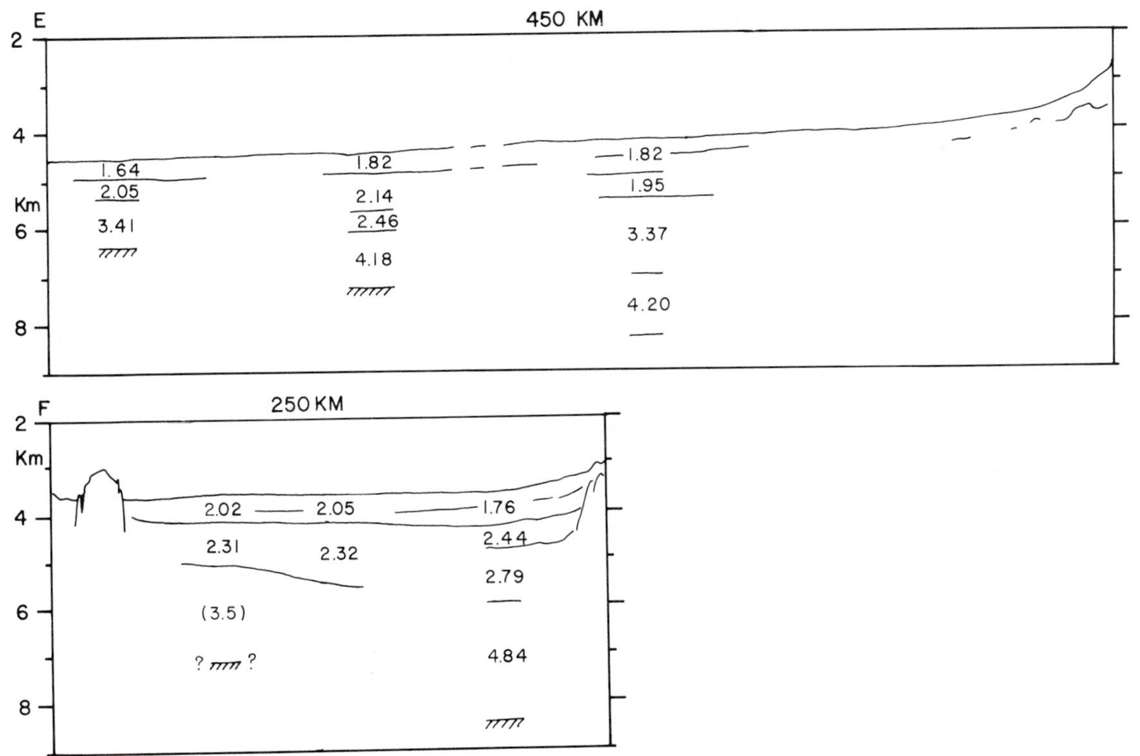

Fig. 4. Structure sections of Wilkes Land margin (upper) and south of the Kerguelen plateau (lower).

The basement depression in section F (Fig. 4) occurs in the region between the Kerguelen Plateau and the Antarctic mainland. Note that the deepest reflector in the northern part of the section is poorly observed. The existence of the depression between the plateau and the mainland makes it possible to speculate that the marginal depression is a continuous feature from sections D to F. If so, the Kerguelen Plateau has no structural connection with Antarctica. This proposed configuration is supported by the complete lack of correspondence between the Precambrian shield rocks of Antarctica and the Cenozoic volcanics that crop out on Heard and Kerguelen Islands on the plateau (Stephenson, 1970; Nougier, 1970).

The well-documented rift or graben structure of the Otway, Gippsland, and Bass "basins" (von der Borch et al., 1970; Weeks and Hopkins, 1967; Sprigg, 1967) and the seismic sections south of Adelaide and also farther west (Houtz and Markl, 1972) seem to represent a fairly continuous rift structure along the southern Australian margin. The data of this paper strongly suggest that the Australian marginal rift is mirrored along the Antarctic coast. The existence of the rift and the lack of any volcanic activity since the Precambrian show that significant subduction has not occurred since the rift was formed. This rift (presumably without later subduction) may extend westward beyond the Kerguelen Plateau, as suggested by the data in section F. If the Indian subcontinent detached from Antarctica just west of Kerguelen 225 my ago, as suggested by Dietz and Holden (1970), the sea floor here could be the oldest yet dated in the deep sea.

ACKNOWLEDGMENTS

The seismic program aboard the *U.S.N.S. Eltanin*, which provided the basis for this report, and the subsequent data reduction and analysis were supported by the Office of Polar Programs. The sonobuoys were provided by the Office of Naval Research Contract No. N00014-67-A-0108-0004. The useful suggestions of Dennis Hayes and Jeff Weissel, the reviewers of this paper, are gratefully acknowledged.

BIBLIOGRAPHY

Dalrymple, P., 1966, A physical climatology of the Antarctic Plateau: Antarctic Res. Ser., v. 19, p. 195-231.

Dalziel, I., and Elliot, D., in press, *in* Stehli, F., and Nairn, A., eds., The ocean basins and continental margins: 1. The South Atlantic: New York, Plenum.

Dietz, R., and Holden, J., 1970, Reconstruction of Pangea: breakup and dispersion of continents, Permian to present: Jour. Geophys. Res., v. 75, no. 26, p. 4939-4956.

Hayes, D., and Ringis, J., in press, Sea floor spreading in a marginal basin: the Tasman Sea: Nature.

Heezen, B., Tharp, M., and Bentley, C., 1972, Morphology of the earth in the Antarctic and Subantarctic, Antarctic Map Folio Series 16: New York, Am. Geograph. Soc.

Houtz, R., and Davey, F., in press, Seismic profiler and sonobuoy measurements in Ross Sea, Antarctica: Jour. Geophys. Res.

______, and Markl, R., 1972, Seismic profiler data between Antarctica and Australia, Hayes, D.,ed., *in* Antarctic oceanology II: The Australian-New Zealand sector: Antarctic Res. Ser., v. 19, p. 147-164. Washington, D.C., Am. Geophys. Union.

Nougier, J., 1970, Geochronology of the volcanic activity in Archipel De Kerguelen (abs.), *in* Adie, R., ed., Antarctic geology and geophysics, SCAR/IUGS Symposium, Oslo, Norway.

Sprigg, R., 1967, A short geological history of Australia: Australian Petr. Explor. Assoc. Jour., p. 59-82.

Stephenson, R., 1970, Geochemistry of some Heard Island igneous rocks (abs.), *in* Adie, R., ed., Antarctic geology and solid earth geophysics, SCAR/IUGS Symposium. Oslo, Norway.

von der Borch, C., Conolly, J., and Dietz, R., 1970, Sedimentation and structure of the continental margin in the vicinity of the Otway basin, southern Australia: Marine Geology, v. 8, p. 59-83.

Weeks, L., and Hopkins, B., 1967, Geology and exploration of three Bass Strait basins, Australia: Am. Assoc. Petr. Geol. Bull., v. 51, no. 5, p. 742-760.

Part IX

Geology of Selected Small Ocean Basins

Structure of the Bering Sea Basins

William J. Ludwig

INTRODUCTION

This report summarizes knowledge of the structural framework of the deep Bering Sea basins and intervening ridges from the published results of geophysical measurements and JOIDES deep-sea drilling through 1973. It includes a free-air gravity map of the south-central Bering Sea and schematic structure sections of the various physiographic units, based on gravity and seismic data.

Since the last review of geophysical studies of the Aleutian-Bering Sea region by Stone (1968) and by Gaynanov, et al (1968), a considerable amount of geological and geophysical data have been collected that bears on the structure and composition of the deep-water Bering Sea basins and the ridges that separate them. Scholl, et al (1974) have presented an up-to-date account of knowledge of the Aleutian-Bering Sea region within a framework of the expected consequences of plate tectonics. In this article the important geologic and geophysical properties are assembled to provide a factual basis for developing further lines of inquiry into the origin and development of the sea and its possible economic potential. A bibliography of papers that deal with the Bering Sea shelf is included along with those cited in the text.

GENERAL DISCUSSION AND BATHYMETRY

The deep-water Bering Sea is separated by Bowers ridge and Shirshov ridge into the Bering Sea (or Aleutian) basin, Bowers basin, and the Kamchatka basin (Fig. 1). Maps of the general bathymetry of the sea are given by Scholl, et al (1968, 1974), Chase et al (1971), and in the Bathymetric Atlas of the north-central Pacific Ocean (1971). Descriptions of the various physiographic features are given by Udintsev, et al (1959) and Gershanovich (1968), among others.

The Kamchatka basin, Bering Sea basin, and Bowers basin are floored by abyssal plains whose gradients indicate that they belong to a single abyssal plain that slopes inward toward a center. Ludwig, et al (1971a) noted that the abyssal floor of the deep-water Bering Sea (exclusive of the ridges) has the approximate form of a half-saucer-shaped depression with Bowers basin occupying the lower-central and deepest part. The floors of the Kamchatka and Bering Sea basins have water depths of 3,700-3,840 m; Bowers basin generally has depths about 70 m deeper. Some bathymetric maps (e.g., Scholl, et al., 1968) show a subtle depression in the sea floor that lies adjacent to the convex side of Bowers ridge.

Bowers ridge is a tightly arcuate submarine elevation of the Bering Sea floor that projects counterclockwise from the Aleutian Islands ridge at approximately 180° longitude, almost (back) to the Aleutian ridge in the vicinity of 170°E longitude. Possible analogs of Bowers ridge are the Beata ridge of the Caribbean Sea and the Yamato ridge of the Japan Sea. Bathymetric maps of the central Aleutians-Bowers Ridge area have been presented by Ludwig, et al (1971a, 1971b) and Marlow, et al (1973), based on the detailed maps of Nichols, et al (1964) and Nichols and Perry (1966). Marlow, et al include large-scale physiographic views (diagrams) of Bowers Ridge racing northward and southward. Generally, these maps show that the ridge is asymmetrical and has three crestal surfaces (pedestals) which decrease in depth northward. The crestal surfaces are topped by banks or hilly plateaus, indicating that they may have undergone subaerial erosion followed by susidence. West of 176°E the ridge becomes abruptly deeper, narrower, steep-sided, and symmetrical in cross section. Severe bending of the ridge during its formative history is indicated by transverse topographic depressions resembling rifts along the convex side (extension) and transverse depressions resembling folds along the concave side (compression).

Shirshov submarine ridge extends southward with gradually increasing water depth from the continental margin of Siberia (Cape Olyutorskiy) to the vicinity of the Aleutian ridge, where it becomes abruptly deeper and curves eastward toward Bowers ridge. The extremities of Bowers ridge and Shirshov ridge are not connected, but instead closely overlap below the floor of the channel that separates the emerged portions (L-DGO, unpublished data). Like Bowers ridge, Shirshov ridge maintains a fairly uniform width and surface trend along the major portion of its length and then abruptly tapers, slightly reverses the surface curvature, and becomes more steep-sided.

SEISMICITY

The Bering Sea basins and ridges are seismically inactive; almost all earthquake activity is confined to the Aleutian ridge (Barazangi and Dorman, 1969; Kienle, 1971). A number of shallow earthquake epicenters have been plotted in Bowers basin, but

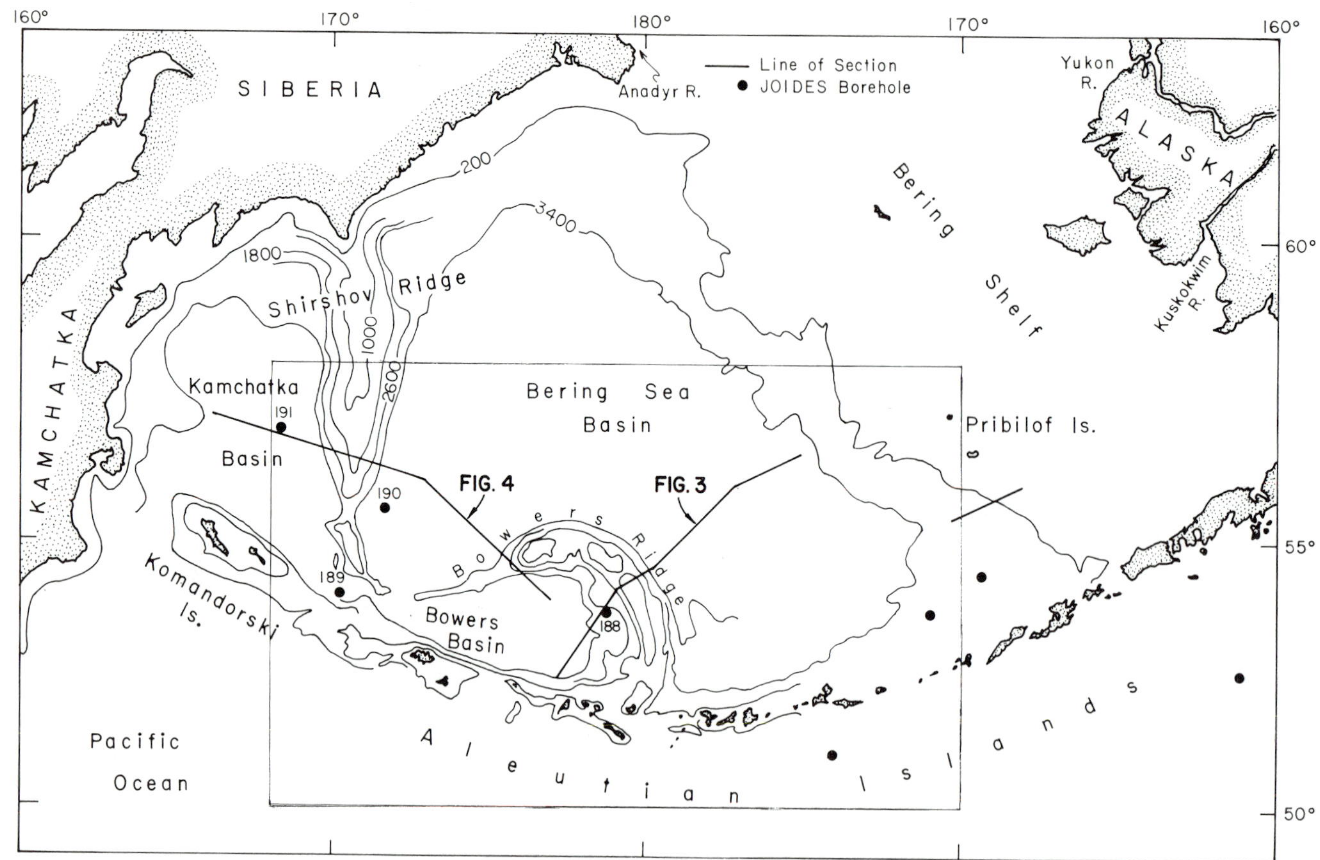

Fig. 1. Index map of the Bering Sea. Generalized bathymetry in meters. Detail of smaller area shown in Figure 2.

these may be mislocated north of the true epicenter (beneath the Aleutian ridge). East of Bowers ridge, intermediate-to-deep focus earthquakes are located in a Benioff zone that dips beneath the Aleutian ridge and extends a short distance into the southern Bering Sea basin to a depth of about 250 km; to the west the earthquakes do not extend below 100 km (Kienle 1971; Grow, 1973).

HEAT FLOW

Values of heat flow in the Bering Sea basin increase from about 1.1-1.2 HFU (heat flow units) in the east to about 1.5-1.6 HFU in the west (including Bowers basin), the average heat flow being near the average heat flow of 1.3 HFU of all ocean basins (Foster, 1962; Langseth and Von Herzen, 1971). The Kamchatka basin, however, has a higher than normal average heat flow of about 3.0 HFU.

GRAVITY

Variations in the gravity field strength of the deep-water Bering Sea have been studied and described by Kienle (1971). A contour map of free-air gravity anomalies (Fig. 2) and selected profiles of the gravity (Figs. 3 and 4) show positive free-air anomalies greater than 200 mgal over Bowers ridge that correspond with the topography, negative anomalies greater than 100 mgal bordering the entire length of the ridge on its convex side, steep gradients between the gravity high and low, and near-zero anomalies in the Bering Sea basin. The amplitude of the gravity low is independent of the adjacent ridge elevation. Negative anomalies up to 30 mgal characterize the eastern part of Bowers basin, whereas the western part has positive values to 30 mgal.

Shirshov ridge has smaller amplitude positive anomalies (up to 90 mgal) associated with it. There is a small gravity low with values slightly more than -25 mgal paralleling Shirshov ridge on either side that Kienle (1971) attributes to edge effects of an isostatically compensated ridge. The Kamchatka basin has positive gravity anomalies of 10-30 mgal.

MAGNETICS

Profiles of the total intensity magnetic anomaly observed at sea level (Hayes and Heirtzler, 1968; Kienle, 1971; Figs. 3 and 4) show the short-wave-length and large-amplitude anomalies (up to 1500 gammas) that exist over the crestal surfaces of Bowers ridge and Shirshov ridge, the magnetic low paralleling the convex side of Bowers ridge that coincides with the gravity low there, a short-wave-

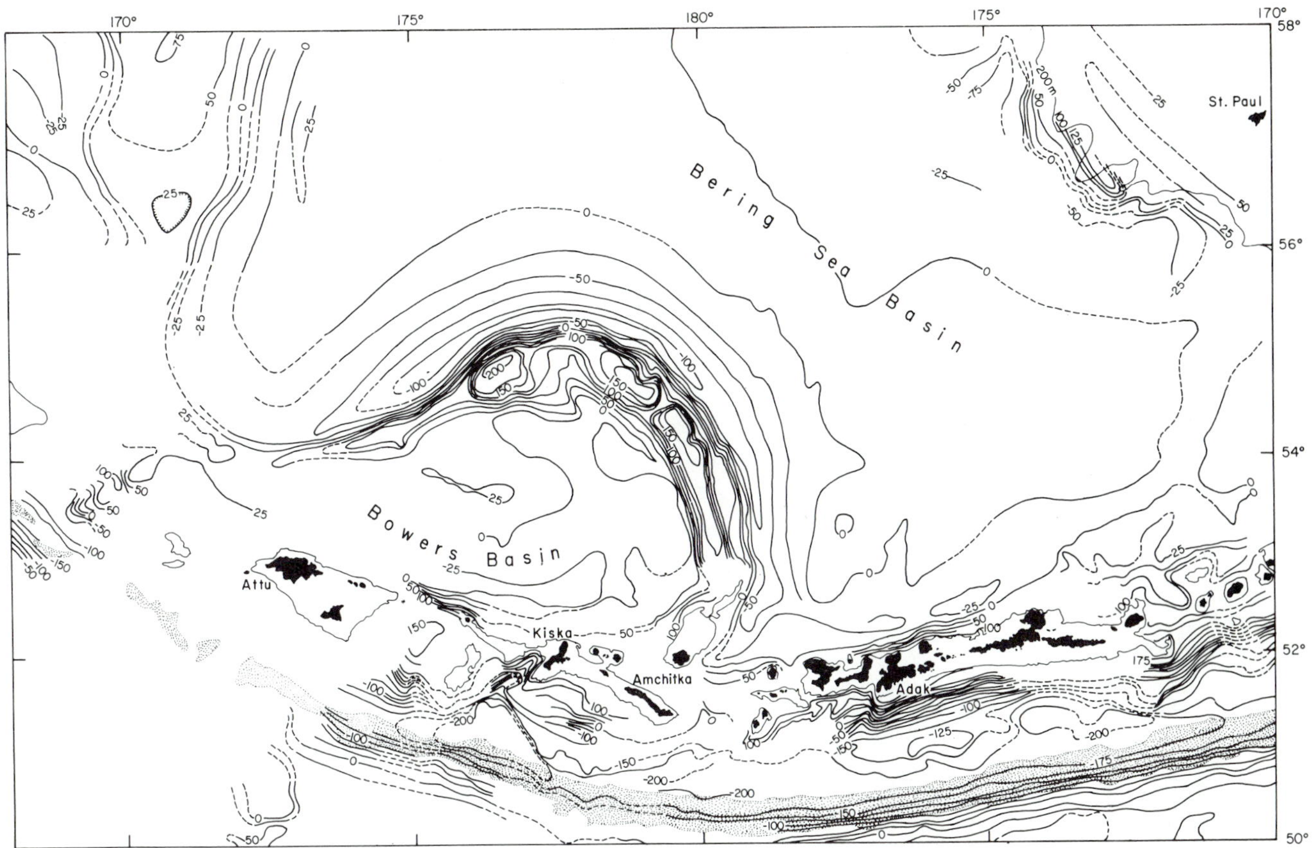

Fig. 2. Inset map of Figure 1: Free-air gravity map of the south-central Bering Sea, adapted from Kienle (1971). Contour interval is 25 mgal. Stippled areas represent water depths greater than 6,400 m in the Aleutian trench.

length magnetic high that borders the low, and the relatively long-wavelength and small-amplitude anomalies of about 100-300 gammas that are characteristic of the basins.

CRUSTAL STRUCTURE

Figures 3 and 4 show the seismically determined structure of the deep-water Bering Sea along lines of traverse extending southwestward from the Bering shelf to Bowers basin, and westward to the Kamchatka basin (locations shown in Fig. 1). Beneath Bowers ridge the lower crustal layer and crust-mantle interface were not determined seismically but were deduced from gravity data by "balancing" the crustal section of the ridge against the section of the Bering Sea basin (see Kienle, 1971, for details). In the interpretation of the gravity data all density inhomogeneities were assumed to be uniform.

Bowers ridge is a thickened, raised welt of 5.8-6.2 km/sec crustal material capped by sediments and (or) volcanics of veloctiy 3.3-4.3 km/sec. There is a lower crustal layer between the 5.8-6.2 km/sec layer and the upper mantle whose velocity has not been determined, but is assumed to be about 7.0 km/sec on the basis of measurements made in the basins on either side of the ridge. Seismic-refraction measurements of the Aleutian ridge revealed a 6-km-thick layer of velocity 5.5 km/sec overlying a 6.6 km/sec layer between it and the upper mantle (Shor, 1964). In all probability the basic velocity structure of Bowers ridge and the Aleutian ridge are identical. The close structural affinity of these two ridges is further demonstrated by the existence of a sediment-filled trench bordering Bowers ridge on its convex (eastern) side. Like the Aleutian ridge, Bowers ridge appears to be an island arc-trench system having typical island arc gravity and magnetic anomalies. Unlike the Aleutian ridge, Bowers ridge is seismically inactive and its foredeep has been completely filled by sediments.

There are no seismic refraction data available on the deep structure of Shirshov ridge. Both Shirshov and Bowers ridges are characterized by high-amplitude, short-wavelength magnetic anomalies, perhaps indicating some lithologic similarity between them. A contrasting aspect of these two ridges, denoting gross differences in structure and (or) differences in the degree of isostatic equilibrium, is their respective gravity signatures. The values of positive free-air gravity anomalies over Bowers ridge (and the Aleutian ridge) are much higher than those over Shirshov ridge for the same elevation. Furthermore, there is no gravimetrically expressed sediment-filled trench bordering Shirshov ridge, as is the situation along the convex side of Bowers ridge. The

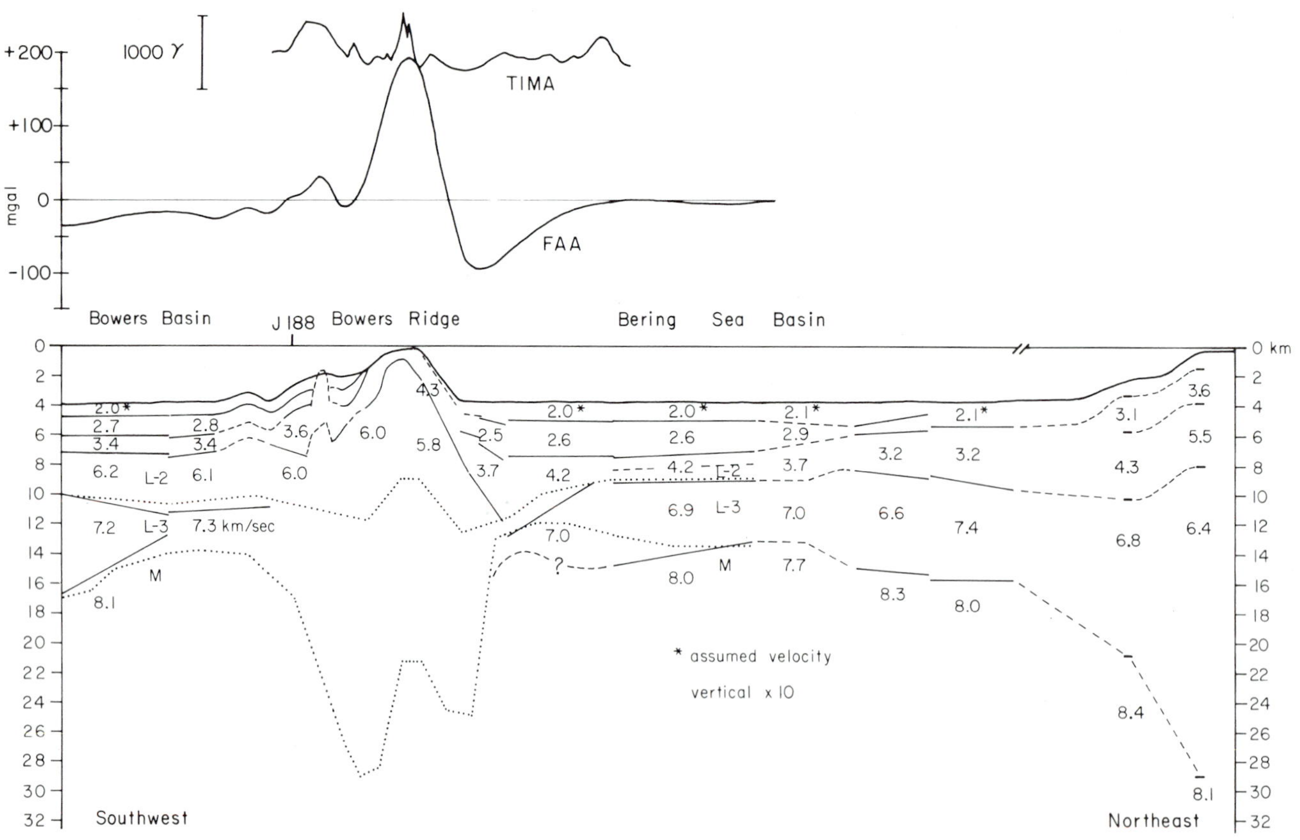

Fig. 3. Seismic structure section southwest-northeast between Bowers basin and the Bering shelf. Profiles of the total intensity magnetic anomaly (TIMA) and free-air gravity anomaly (FAA) are also shown. The dotten lines represent gravimetrically determined layer boundaries. Extrapolated boundaries are indicated by dashed lines. The gravity anomaly profile computed for the structure agrees with the observed profile. Seismic data from Shor (1964) and Ludwig et al. (1971a, 1971b); magnetic and gravity data from Kienle (1971). Note that the line of section crosses Bowers basin near the junction of Bowers ridge and the Aleutian ridge (Fig. 1). Therefore, the velocity structure shown for the lower crustal layers of Bower basin may reflect that of the adjacent ridge system.

fact that Shirshov ridge connects with the continental margin of Siberia (off Cape Olyutorskiy) in line with the Koryak Mountains indicates that it may be a promontory of the mountain system.

The Bering Sea basin has normal oceanic crust over which has been deposited 3-4 km of sediment and volcanics (?) having velocities generally less than 3.9 km/sec, but which range to 4.2 km/sec. Oceanic basement (or seismic layer 2) is 1-2 km thick, has a rough upper surface, and has velocities of 4.7-5.4 km/sec in the eastern part of the basin and velocities of 5.5-6.2 km/sec in the western part. The velocities and thicknesses of layers beneath Bowers basin are similar to those in the (western) Bering Sea basin, except that the velocities measured in layers 2 and 3 along the line of section represent the high end of the range of values. This "difference," noted by Ludwig et al. (1971b), is more apparent than real because the measurements were made close to the juncture of Bowers ridge and the Aleutian ridge. A more typical section for Bowers basin might be found in the central portion. Velocities near 5.8 km/sec represent the oceanic basement of the Japan Sea as well (Ludwig et al., 1974) and generally are characteristic of oceanic basement in the North Pacific basin (Houtz et al., 1970).

The Kamchatka basin has considerably less sediment (1.0-1.5 km) than Bowers basin and the Bering Sea basin, although the water depths in all three basins are about the same; i.e., the surface of the oceanic basement beneath the Kamchatka basin is shallower than elsewhere. This may seem to suggest correspondence between basement elevation and heat flow (shallow basement=high heat flow). In the Japan Sea, however, the values of heat flow increase with *increasing* sediment thickness and depth to basement (Ludwig, et al., 1974). Velocity data for the Kamchatka basin are sparse. From profiler-sonobuoy measurements Ludwig, et al. (1971a, 1971b) showed the velocity in the acoustic basement (layer 2?) to be 6.64 km/sec at one locality. Reexamination of the record (buoy 398) reveals that the velocity reported is erroneous; the correct value is 5.64 km/sec.

Removal of the sediment of Bowers trench would reveal a narrow depression with basement depths comparable to those in the Aleutian trench and other trenches of the Pacific Ocean. In this respect it is

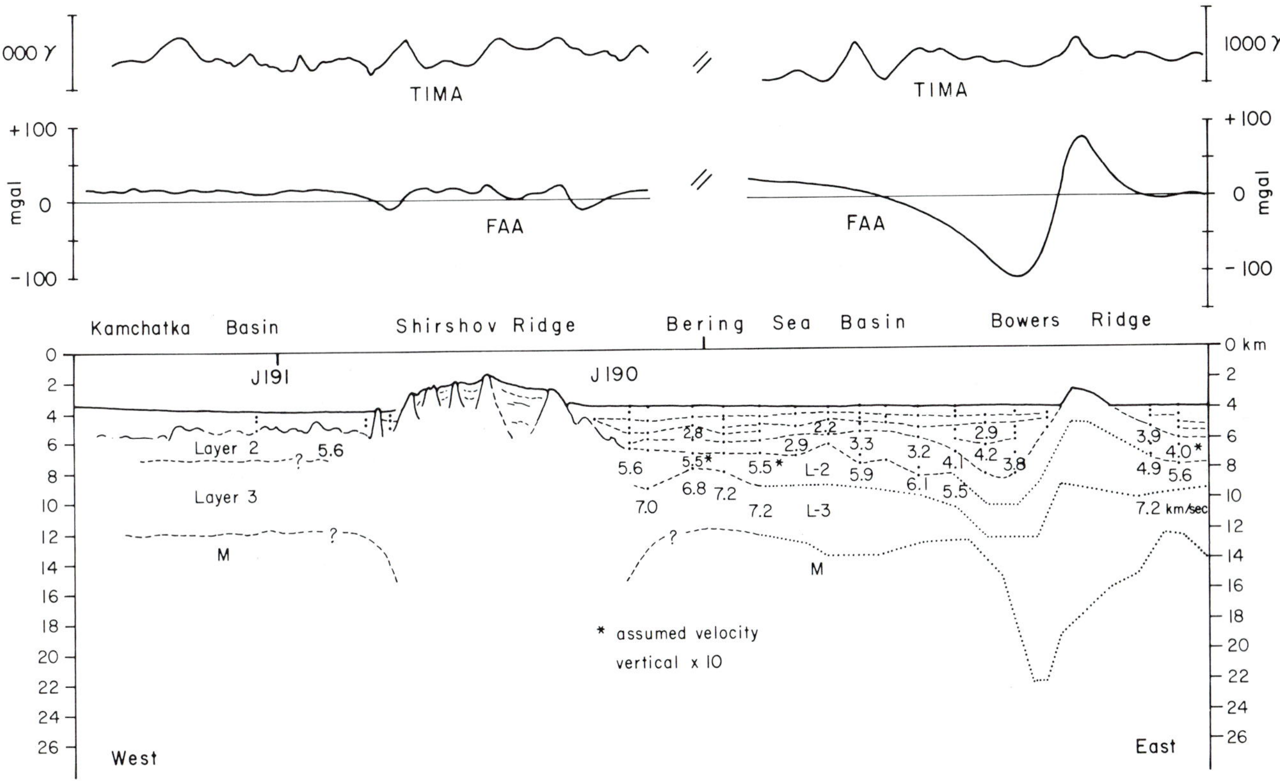

Fig. 4. Seismic structure section generally west-east between the Kamchatka basin and Bowers basin. Explanation same as for Figure 2. Seismic data from Ludwig et al. (1971a, 1971b). Magnetic and gravity data from Kienle (1971).

interesting to note that, as in almost all schematic structure models of deep-sea trenches computed from gravity data on the assumption of uniform mantle density, Kienle's (1971) model requires a marked thinning of oceanic layer 3 beneath Bowers "trench" in order to balance the gravity section. However, as was mentioned by Kienle, the model does not require thinning of layer 3 if there is a local increase in mantle density beneath the "trench." Grow (1973) eliminates the gravimetrically thinned layer 3 beneath the Aleutian trench (and all other trenches) by assigning a higher density to a supposed slab of descending lithosphere beneath the trench and adjacent island arc.

Seismic refraction measurements of the Ryukyu (or Nansei Shoto) trench indicate a marked thinning of oceanic layer 3 beneath the structural axis of the trench (Murauchi et al., 1968; Ludwig et al., 1973). The important point here is that the thinning occurs beneath the structural axis, which may or may not correspond to the topographic axis of the trench. In the middle part of the Ryukyu trench the structural and topographic axes are coincident, whereas in the northern part the structural axis of the trench lies landward of the topographic axis. In both localities, however, the structural axis of the trench has a gravity minimum centered over it. The close agreement between the results of refraction and gravity measurements of crustal structure (e.g., Fig. 3 and 4) and the strong possibility that the upper mantle has uniform density (indicated by uniform mantle velocity) suggests that a large part of gravity anomalies are indeed caused by density inhomogeneities within the crust.

Therefore, the gravimetrically thinned layer 3 of deep-sea trenches may be a real effect. This thinning or stretching of oceanic layer 3 beneath Bowers "trench" and others connotes formation of deep-sea trenches by extensional forces acting in a near-horizontal plane (see Tanner, 1973). The seaward (east) wall of the Japan trench has a complex pattern of normal and antithetic faults, indicating clearly that extension and settling of the strata have taken place (Ludwig, et al, 1966).

SEISMIC STRATIGRAPHY AND JOIDES [DSDP] DRILLING

Seismic reflection profiling (Ewing et al., 1965; Ludwig et al., 1971a; Fornari, et al., 1973) reveals that the Bering Sea basin and Bowers basin have highly to moderately stratified sediments over a weakly stratified sequence on a generally rough basement (layer 2). The upper sedimentary unit has a fairly uniform thickness of about 1 km and consists of flat-lying beds that indicate relative stability of the basins during the entire period of deposition. The base of the upper unit, designated P by Ewing, et al (1965), corresponds closely to the top of the layer of velocity near 2.7 km/sec (Figs. 3 and 4). The lower sedimentary unit has variable thickness and also shows no signs of internal deformation in the seismic records. Similar sedimentary layering exists in the Kamchatka basin but, as was mentioned earlier, the sediments are much thinner there because the basement is shallower than elsewhere in the deep-water Bering Sea.

Thick accumulations of moderately to weakly stratified sediments of low velocity drape the middle and lower western slopes of Bowers ridge and appear to extend below a younger highly stratified sequence in the adjacent Bowers basin. The unconformity thus developed marks a change in sedimentation. By contrast, there are much lesser amounts of these sediments on the crestal surfaces and eastern slope of Bowers ridge, indicating some mode of concentration and deposition of sediments that favors the western side. West of 178°E the sediments draped onto the flanks of Bowers ridge decrease in thickness with corresponsive changes in the topography of the ridge and become more nearly equal in thickness on either side. The configuration of the sediments of Shirshov ridge is, in general, similar to that of Bowers ridge. Sediments that appear to be continuous with those in the adjacent basins are draped onto the flanks of Shirshov ridge with the greater amounts located on the eastern side.

JOIDES (DSDP) drilling in the North Pacific Ocean during 1971 provided information on the Miocene to Holocene history of the deep-water Bering Sea (Fullam, et al., 1973; Scholl and Creager, 1973). At site 188 located on the lower western slope of Bowers ridge (Fig. 1), the section drilled and cored consists of Pleistocene to upper Miocene unlithified, terrigenous-rich, diatom ooze overlying mudstone or claystone lacking siliceous microfossils. The mudstones cored represent the top of the 2.7 km/sec layer, or reflector P. No evidence of turbidite deposition was found in the section cored, indicating that the sediments draping Bowers ridge are wholly pelagic and (or) hemipelagic in nature, and most likely postdate the elevation of the ridge. A similar lithologic transition from hemipelagic sediments to mudstone was found at other sites in the southeastern part of the Bering Sea basin, indicating a change in oceanographic conditions during the middle or late Miocene that allowed massive productions of siliceous microorganisms and subsequent deposition of vast quantities of their remains. This expansion of biogenic productivity has apparently continued up to the present (Lisitsyn, 1959, 1966).

Drilling at sites 189, 190, and 191 in the basins (Fig. 1) penetrated Pleistocene to Pliocene silty or clayey diatomaceous layers, of which many are graded units (turbidites). Therefore, it is likely that the leveling of the Bering Sea abyssal plain resulted from the combined deposition of pelagics and turbidites. Borehole 191 in the Kamchatka basin bottomed in extrusive basalt of middle Oligocene age. The mean value of the compressional wave velocities measured in the basalt is 5.6 km/sec (Christensen, 1973), which is within the range of velocities determined for seismic layer 2 from profiler-sonobuoy measurements.

The nature and age of the deeper lithified sediments and basement rocks of the Bering Sea basin and Bowers basin are not yet known. The acoustic basement of the Bering shelf has velocities of 3.1-3.6 km/sec (Shor, 1964; Scholl, et al., 1968) and, where dredged by Hopkins et al (1969) in the Pribilof canyon, it consists of lithified turbidites of lower Cretaceous age.

CONCLUDING REMARKS

The total evidence indicates that the deep-water Bering Sea basin is a remnant of the North Pacific Ocean basin that was separated from it by the elevation of the Aleutian ridge in late Cretaceous or early Tertiary time (see Scholl, et al., 1974). The contemporaneous or subsequent growth of Bowers ridge and Shirshov ridge partitioned the Bering Sea basin into three smaller basins which eventually received massive amounts of pelagic and turbidite material. Indications are that the Kamchatka basin had a somewhat different history from that of Bowers basin and the Bering Sea basin because the level of its basement surface is appreciably higher. The question of the relationship of the basins and intervening ridges and the question of why the ridges end where they do are important problems to be solved by future drilling.

The effects of large-scale tectonic processes such as behind-the-arc sea-floor spreading and underthrusting of the lithosphere are not evident in the deep-water Bering Sea. The level-bedded character of the sediments below the abyssal floor implies undisturbed sedimentation from at least upper Miocene time to the present. In terms of plate tectonics, Bowers "trench" (and others) represents extension along the upper part of a lithospheric plate caused by flexing of the plate as it bends and descends beneath the island arc (Bowers ridge). Although it seems likely that Bowers "trench" (and others) is an extensional feature, it is rather doubtful if the strain caused by a 2°-5° flexing of a plate to form the trench could produce the observed thinning or stretching of oceanic layer 3 beneath it.

ACKNOWLEDGMENTS

Preparation of this summary report was aided by grant GA-27281 from the Oceanography Section of the National Science Foundation. This paper is Lamont-Doherty Geological Observatory Contribution 2096.

BIBLIOGRAPHY

Barazangi, M., and Dorman, J., 1969, World seismicity maps compiled from ESSA, Coast and Geodetic Survey, Epicenter data: Seismol. Soc. America Bull., v. 59, no. 1, p. 369-380.

Bathymetric atlas of the northcentral Pacific Ocean, 1971, H.O. Publ. 1302-S: Washington, D.C., U.S. Naval Oceanographic Office.

Chase, T. E., Menard, H. W., and Mammerickx, J., 1971, Topography (map) of the north Pacific: Scripps Inst. Oceanogr. and Inst. Marine Res. Tech. Rept. Ser. TR-7.

Christensen, N. I., 1973, Compressional and shear wave velocities and elastic moduli of basalts, DSDP, Leg 19, *in* Creager, J. S., Scholl, D. W., et al., eds., Initial reports of the Deep Sea Drilling Project, v. 19: Washington, D.C., U.S. Govt. Printing Office, p. 657-659.

Creager, J. S., and McManus, D. A., 1967, Geology of the floor of Bering and Chukshi seas, *in* Hopkins, D. M., ed., The Bering land bridge: Stanford, Calif., Stanford Univ. Press, 495 p.

Ewing, M., Ludwig, W. J., and Ewing, J., 1965, Oceanic structural history of the Bering Sea: Jour. Geophys. Res., v. 70, no. 18, p. 4593-4600.

Fornari, D. J., Inliucci, R. J., and Shor, G. G., Jr., 1973, Preliminary site surveys in the Bering Sea for the DSDP, Leg 19, *in* Creager, J. S., Scholl, D. W., et al., eds., Initial reports of the Deep Sea Drilling Project, v. 19: Washington, D.C., U.S. Govt. Printing Office, p. 569-613.

Foster, T. D., 1962, Heat flow measurements in the northeastern Pacific and Bering Sea: Jour. Geophys. Res., v. 67, no. 7, p. 2991-2993.

Fullam, T. J., Supko, P. R., Boyce, R. E., and Steward, R. W., 1973, Some aspects of late Cenozoic sedimentation in the Bering Sea and north Pacific Ocean, *in* Creager, J. S., Scholl, D. W., et al., eds., Initial reports of the Deep Sea Drilling Project, v. 19, Washington, D.C., U.S. Govt. Printing Office, p. 887-896.

Gaynanov, A. G., Kosminskaya, I. P., and Stroyev, P. A., 1968, Geophysical studies of the deep structure of the Bering Sea: Izv. Akad. Nauk S.S.S.R., no 8, p. 3-11; Eng. trans. (1968): Washington, D.C., Am. Geophys. Union, p. 461-465.

Gershanovich, D. E., 1968, New data on geomorphology and recent sediments of the Bering Sea and Gulf of Alaska: Marine Geology, v. 6, no. 4, p. 281-296.

Grim, M. S., and McManus, D. A., 1970, A shallow seismic profiling survey of the northern Bering Sea: Marine Geology, v. 8, p. 293-320.

Grow, J. A., 1973, Crustal and upper mantle structure of the central Aleutian arc: Geol. Soc. America Bull., v. 84, no. 7, p. 2169-2192.

Hatton, C. W., 1971, Petroleum potential of Bristol Bay Basin, Alaska, *in* Cram, I. H., ed., Future petroleum provinces: Am. Assoc. Petr. Geol. Mem. 15, p. 105-108.

Hayes, D. E., and Heirtzler, J. R., 1968, Magnetic anomalies and their relation to the Aleutian ridge, Jour. Geophys. Res., v. 73, no. 14, p. 4637-4646.

Helmberger, D. V., 1968, The crust-mantle transition in the Bering Sea: Seismol. Soc. America Bull., v. 58, no. 1, p. 179-211.

Hoare, J. M., 1961, Geology and tectonic setting of lower Kuskokwim-Bristol Bay region, Alaska: Am. Assoc. Petr. Geol. Bull., v. 45, p. 595-611.

Hopkins, D. M., 1959, Cenozoic history of the Bering land bridge: Science, v. 129, p. 1519-1928.

______, 1967, The Cenozoic history of the Beringia: a synthesis, *in* Hopkins, D. M., ed., The Bering land bridge: Stanford, Calif., Stanford Univ. Press, p. 451-484.

Hopkins, D. M., Scholl, D. W., Addicott, W. O., Pierce, R. L., Smith, P. B., Wolfe, J. A., Gershanovich, D., Kotenev, B., Lohman, K. E., Lipps, J. H., and Obradovich, 1969, Cretaceous, Tertiary and early Pleistocene rocks from the continental margin in the Bering Sea: Geol. Soc. America Bull., v. 80, p. 1471-1480.

Houtz, R., Ewing, J., and Buhl, P., 1970, Seismic data from sonobuoy stations in the northern and equatorial Pacific, Jour. Geophys. Res., v. 75, no. 26, p. 5093-5111.

Kienle, J., 1971, Gravity and magnetics measurements over Bowers ridge and Shirshov ridge, Bering Sea: Jour. Geophys. Res., v. 76, no. 29, p. 7138-7153.

Kummer, J. T., and Creager, J. S., 1971, Marine geology and the Cenozoic history of the Gulf of Anadyr, Marine Geology, v. 10, p. 257-280.

Langseth, M. G., and Von Herzen, R. P., 1971, Heat flow through the floor of the world oceans, *in* The sea, v. 4, pt. 1: New York, Wiley-Interscience, p. 299-352.

Lisitsyn, A. P., 1959, Bottom sediments of the Bering Sea, *in* Bezrukov, P. L., ed., Geographical description of the Bering Sea: Akad. Nauk SSSR Inst. Okeanol. Trudy, v. 29; Engl. transl. (1964): Washington, D.C., U.S. Dept. Commerce, p. 65-188.

Lisitsyn, A. P., 1966, Recent sedimentation in the Bering Sea: Moscow, Izd. Nauka; Engl. trans. (1969): Jerusalem, IPST Press, 614 p.

Ludwig, W. J., Ewing, J. I., Ewing, M., Murauchi, S., Den, N., Asano, S., Hotta, H., Hayakawa, M., Asanuma, T., Ichikawa, K., and Noguchi, I., 1966, Sediments and structure of the Japan Trench: Jour. Geophys. Res., v. 71, no. 8, p. 2121-2137.

Ludwig, W. J., Houtz, R. E., and Ewing, M., 1971a, Sediment distribution in the Bering Sea: Bowers ridge, Shirshov ridge, and enclosed basins: Jour. Geophys. Res., v. 76, no. 26, p. 6367-6375.

Ludwig, W. J., Murauchi, S., Den, N., Ewing, M., Hotta, H., Houtz, R. E., Yoshii, T., Asanuma, T., Hagiwara, K., Sato, T., and Ando, S., 1971b, Structure of Bowers ridge, Bering Sea: Jour. Geophys. Res., v. 76, no. 26, p. 6350-6366.

Ludwig, W. J., Murauchi, S., Den, N., Buhl, P., Hotta, H., Ewing, M., Asanuma, T., Yoshii, T., and Sakajiri, N., 1973, Structure of East China Sea—West Philippine Sea margin off southern Kyushu, Japan: Jour. Geophys. Res., v. 78, no. 14, p. 2526-2536.

Ludwig, W. J., Murauchi, S., and Houtz, R. E., 1974, Sediments and structure of the Japan Sea: Geol. Soc. America Bull.

Marlow, M. S., Scholl, D. W., Buffington, E. C., Alpha, T. R., 1973, Tectonic history of the central Aleutian Arc: Geol. Soc. America Bull., v. 5, p. 1555-1575.

Moore, D. G., 1964, Acoustic reflection reconnaissance of continental shelves: eastern Bering and Chukchi Seas, *in* Miller, R. L., ed., Papers in marine geology: New York, Macmillan, p. 319-362.

Murauchi, S., Den, N., Asano, S., Hotta, H., Yoshii, T., Asanuma, T., Hagiwara, K., Ichikawa, K., Sato, T., Ludwig, W. J., Ewing, J. I., Edgar, N. T., and Houtz, R. E., 1968, Crustal structure of the Philippine Sea: Jour. Geophys. Res., v. 73, no. 10, p. 3143-3171.

Nichols, H., and Perry, R. P., 1966, Bathymetry of the

Aleutian Arc, Alaska, scale 1:4,000,000, Coast and Geodetic Survey Monogr. 3: U.S. Dept. Commerce, ESSA, 6 maps.

———, Kofoed, J. W., 1964, Bathymetry of Bowers Bank, Bering Sea: Surveying and Mapping, v. 24, p. 443-448.

Pratt, R. M., Rutstein, M. S., Walton, F. W., and Buschur, J. A., 1972, Extension of Alaskan structural trends beneath Bristol Bay, Bering shelf, Alaska: Jour. Geophys. Res., v. 77, no. 26, p. 4994-4999.

Scholl, D. W., and Creager, J. S., 1973, Geologic synthesis of Leg 19 (DSDP) results: far north Pacific, and Aleutian ridge, and Bering Sea, *in* Creager, J. S., Scholl, D. W., et al., eds., Initial reports of the Deep Sea Drilling Project, v. 19: Washington, D.C., U.S. Govt. Printing Office, p. 897-913.

Scholl, D. W., and Hopkins, D. M., 1969, Newly discovered Cenozoic basin Bering Sea shelf, Alaska: Am. Assoc. Petr. Geol. Bull., v. 53, no. 10, p. 2067-2078.

Scholl, D. W., and Marlow, M. S., 1970, Diapirlike structures in southeastern Bering Sea: Am. Assoc. Petr. Geol. Bull., v. 54, p. 1644-1650.

Scholl, D. W., Buffington, E. C., and Hopkins, D. M., 1966, Exposure of basement rock on the continental slope of the Bering Sea: Science, v. 153, no. 3739, p. 992-994.

Scholl, D. W., Buffington, E. C., and Hopkins, D. M., 1968, Geologic history of the continental margin of North America in the Bering Sea: Marine Geology, v. 6, p. 297-330.

Scholl, D. W., Buffington, E. C., Hopkins, D. M., and Alpha, T. R., 1970, The structure and origin of the large submarine canyons of the Bering Sea: Marine Geology, v. 8, p. 187-210.

Scholl, D. W., Buffington, E. C., and Marlow, M. S., 1974, Plate tectonics and the structural evolution of the Aleutian-Bering Sea region, *in* Forbes, R. B., ed., The geophysics and geology of the Bering Sea region: Geol. Soc. America Mem. 151.

Shor, G. G., 1964, Structure of the Bering Sea and the Aleutian ridge: Marine Geology, v. 1, p. 213-219.

Stone, D. B., 1968, Geophysics in the Bering Sea and surrounding areas: a review: Tectonophysics, v. 6, no. 6, p. 433-460.

Tanner, W. F., 1973, Deep-sea trenches and the compressional assumption: Am. Assoc. Petr. Geol. Bull., v. 57, p. 2195-2206.

Udintsev, G. B., Boichenko, I. G., and Kanaev, V. R., 1959, Bottom relief of the Bering Sea, *in* Bezrukov, P. L., ed., Geographical description of the Bering Sea: Akad. Nauk S.S.S.R. Inst. Okeanol. Trudy, v. 29; Engl. transl. (1964): Washington, D.C., U.S. Dept. Commerce, p. 14-64.

The Black Sea

David A. Ross

INTRODUCTION

The Black Sea is an oval basin situated between the folded alpine belts of the Pontic Mountains to the south and the Caucasus and Crimea ranges to the north. It is connected via the Bosporus, having a sill of about 50-m depth, to the Mediterranean Sea. The Black Sea basin has an area of 432,000 km^2 and a volume of 534,000 km^3.

The Black Sea can be divided into four major physiographic features: shelf, basin slope, basin apron, and abyssal plain. A large depositional feature, the Danube Fan, extends across the basin slope and apron, dividing the abyssal plain into two unequal parts. Three distinct sediment units prevail in the Black Sea, and these are clearly related to recent environmental changes occurring over the last 25,000 years. Prior to 9,000 years and extending back to about 22,000 years B.P. the Black Sea was a fresh or brackish-water lake. Anoxic H_2S conditions started about 7,300 years ago, and the most recent inflow of Mediterranean water into the Black Sea was initiated about 9,000 years ago.

Structurally, the Black Sea is recently an area of subsidence, except the western margin of the basin. Magnetic data indicate that coastal structures such as those associated with the Caucasus and Pontic mountains extend partly into the Black Sea. The deep crustal structure of the basin is intermediate to that of continental and oceanic regions, with 8-12 km of sediment overlying as much as 18 km of material having seismic velocities similar to that of basalt.

There are generally two hypotheses for the origin of the Black Sea crustal structure: it is either a relict ocean crust, or it has been newly formed in place. The latter idea has two variants: either *in situ* conversion of low-density, thick continental crust into a denser, thinner, more oceanic-like crust; or the replacement of continental crust by upwelling mantle material along fractures or extensions. Present data lean toward the second hypothesis but do not strongly favor either variant. Further answers must await additional seismic studies and deep-sea drilling.

PREVIOUS INVESTIGATIONS

The Black Sea has been studied by Russian scientists for almost 100 years. Important early works concerning its general shape include those of Andrusov (1890), Arkhangel'skiy (1928, 1930), Arkhangel'skiy and Strakhov (1932), Goncharov (1958), and the modern detailed echo-sounding surveys of Goncharov and Neprochnov (1967). Important sediment studies have been made by Arkhangel'skiy and Strakhov (1938), Strakhov (1954, 1961), Barkovskaya (1961a, 1961b), and Nevesskiy (1967). Recent geophysical studies include those of Goncharov et al. (1966), Malovitskiy et al. (1969a, 1969b), Neprochnov et al. (1964), Melikov et al. (1969), and Malovitskiy et al. (1969 a). In 1969, scientists aboard the research vessel ATLANTIS II of the Woods Hole Oceanographic Institution studied the bathymetry, water chemistry, sediments, and general structure of the Black Sea. Much of the data presented in this paper is from a recent book (Degens and Ross, 1974), based on the results of that expedition, and also including several summary papers by Russian, Bulgarian, and Rumanian scientists (see also Emery and Hunt, 1974).

GENERAL SETTING

Hydrography

The Black Sea is an anoxic basin with the present interface between anaerobic and aerobic water at about 200 m depth. During the Würm, the Black Sea was a fresh or brackish-water lake. Present environmental conditions were initiated about 9,000 years ago when inflow of saline Mediterranean water into the then relatively fresh Black Sea caused it to develop a density stratification. This stratification, combined with organic material falling to the bottom and being oxidized, eventually led to the depletion of the oxygen in the bottom waters. Deuser (1974) calculated that anoxic conditions began in the deepest part of the basin about 7,300 years ago and since then the anaerobic-aerobic interface has been slowly rising to its present position. Presently, a layer of relatively low salinity (about 18 o/oo) surface water (generally less than 300 m thick) overlies more-saline (about 22 0/00) deep water. The thickness and salinity of the upper water is generally related to distance from large rivers and the season.

Bathymetry

A recent bathymetric chart (Fig. 1) shows that the Black Sea basin can be divided into four major physiographic provinces (Fig. 2): shelf, basin slope, basin apron, and abyssal plain. The shelf is generally delineated by the 100-m isobath, and varies in width from more than 190 km west of the Crimea Peninsula

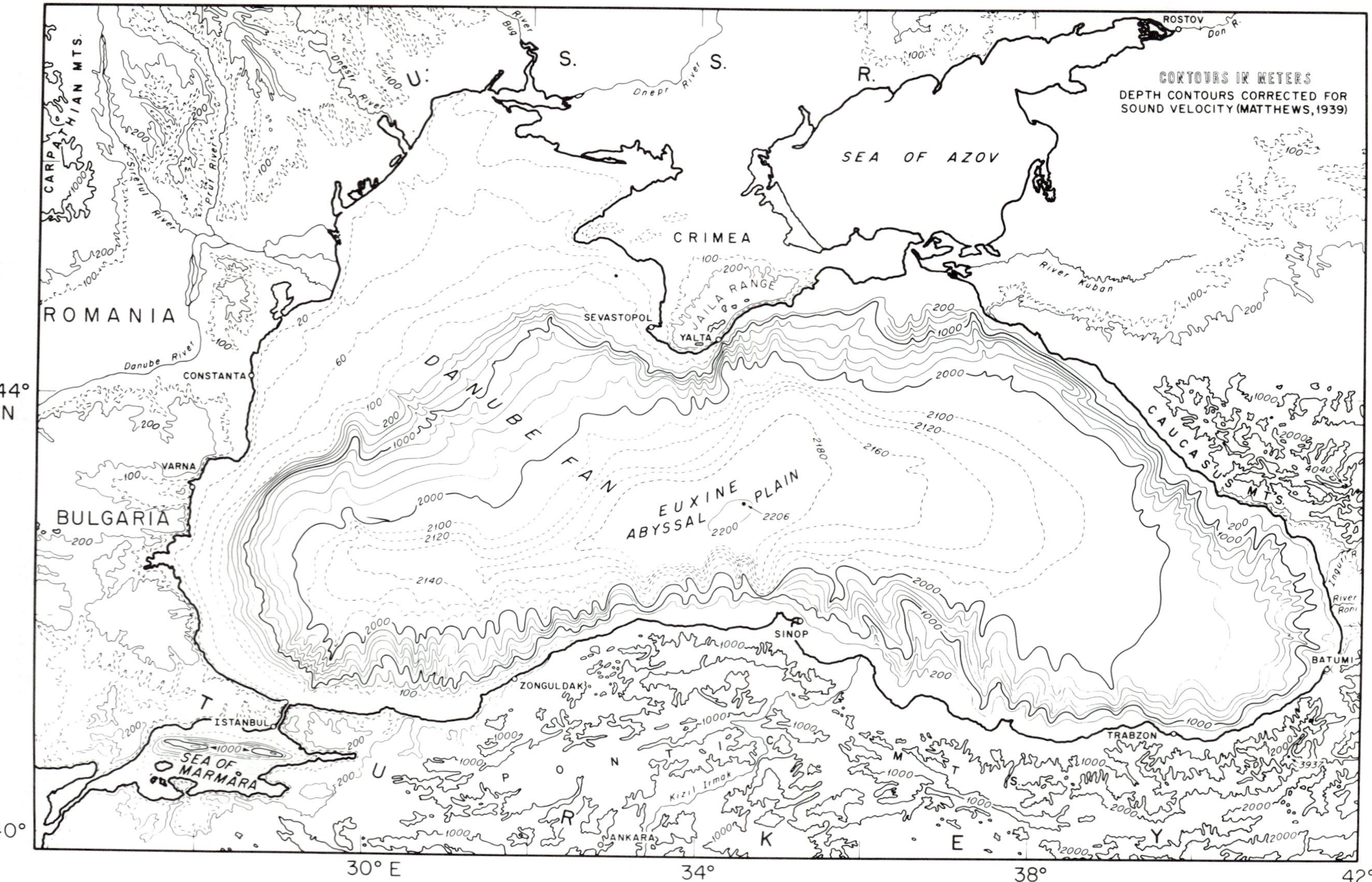

Fig. 1. Bathymetric chart of the Black Sea (modified from Ross et al, 1974). Note the change in contour interval at 200 m and 2,000 m. Data from the 1969 ATLANTIS II cruise, a Russian chart supplied by Pavel Kuprin of the University of Moscow, and U.S. plotting sheets 108N and 3408N. Land topography is from the Morskoi Atlas, Tom 1, Navigatsionne-Geographicheski I Izdanie Morskogogeneralnogo Shtaba. The map was contoured by Elazar Uchupi of the Woods Hole Oceanographic Institution.

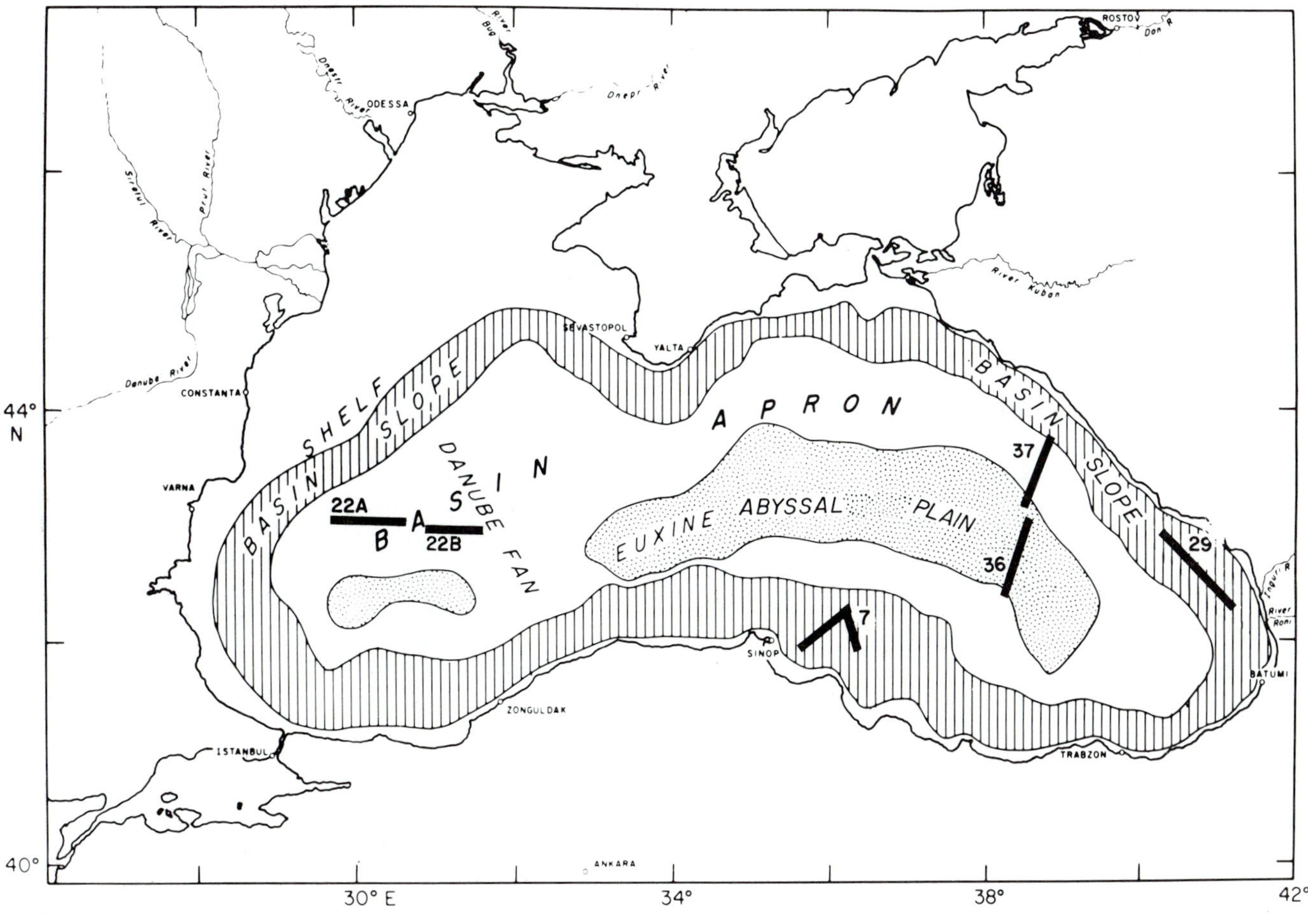

Fig. 2. Main physiographic features of the Black Sea (modified from Ross et al, 1974).

to 20 km or so along the Turkish coast and parts of the Russian coast. Off the Danube, the wide shelf is due to outbuilding by sediments carried by the numerous rivers (Danube, Dnestr, Bug, and Dnepr) that enter the area. There are also suggestions of old river channels extending across the shelf and onto the basin slope. The basin slope is generally of two types, either steep and dissected by submarine canyons such as off most of the Turkish coast and part of the Russian coast (Fig. 3), or relatively smooth as found off Rumania and Bulgaria. Basin slopes have gradients generally greater than 1:40. The basin apron, seaward of the basin slope, has a gradient varying from 1:40 to 1:1000, a gradient similar to that of continental rises. The Danube Fan, an apron deposited by the rivers that built the broad shelf off the Danube area (Fig. 4), extends across the Black Sea basin slope and apron dividing the abyssal plain in the center of the basin into two unequal parts. Presently, no major deposition seems to be occurring on the Danube Fan.

The Euxine Abyssal Plain in the center of the Black Sea basin has a gradient of less than 1:1000 and a maximum depth of 2,206 m (corrected for sound velocity). Recent sediment studies (Ross et al., 1970) and seismic profiles show that the eastern part of the basin had greater incidence of turbidity current deposits than the western part (Fig. 4). Earlier workers (Andrusov, 1890) have suggested that the abyssal plain was divided into two distinct depressions; recent mapping, however, has not shown this to be the case (Fig. 1).

Sediments

Recent sediment studies of the Black Sea include a detailed mineralogical and textural analysis by Müller and Stoffers (1974), a summary of recent Russian work by Shimkus and Trimonis (1974), and a stratigraphic study by Ross and Degens (1974). The Shimkus and Trimonis work, based on over 700 shallow-water and 300 deep-water stations, clearly shows the importance of organic productivity and river detritus in the sediment budget of the Black Sea (Fig. 5). Stratigraphic studies based on piston cores collected during the 1969 ATLANTIS II expedition show that three main sediment units were deposited over the last 25,000 years (Ross et al., 1970; Degens and Ross, 1972). These units can be correlated (Table 1) with the sediments described by Arkhangel'skiy and Strakhov (1938) from deep water, and with the shallow-water sediment units of Neveskiy (1967). Helpful in such correlation is the work of Nevesskaya (1974), who used shallow-water mollusk

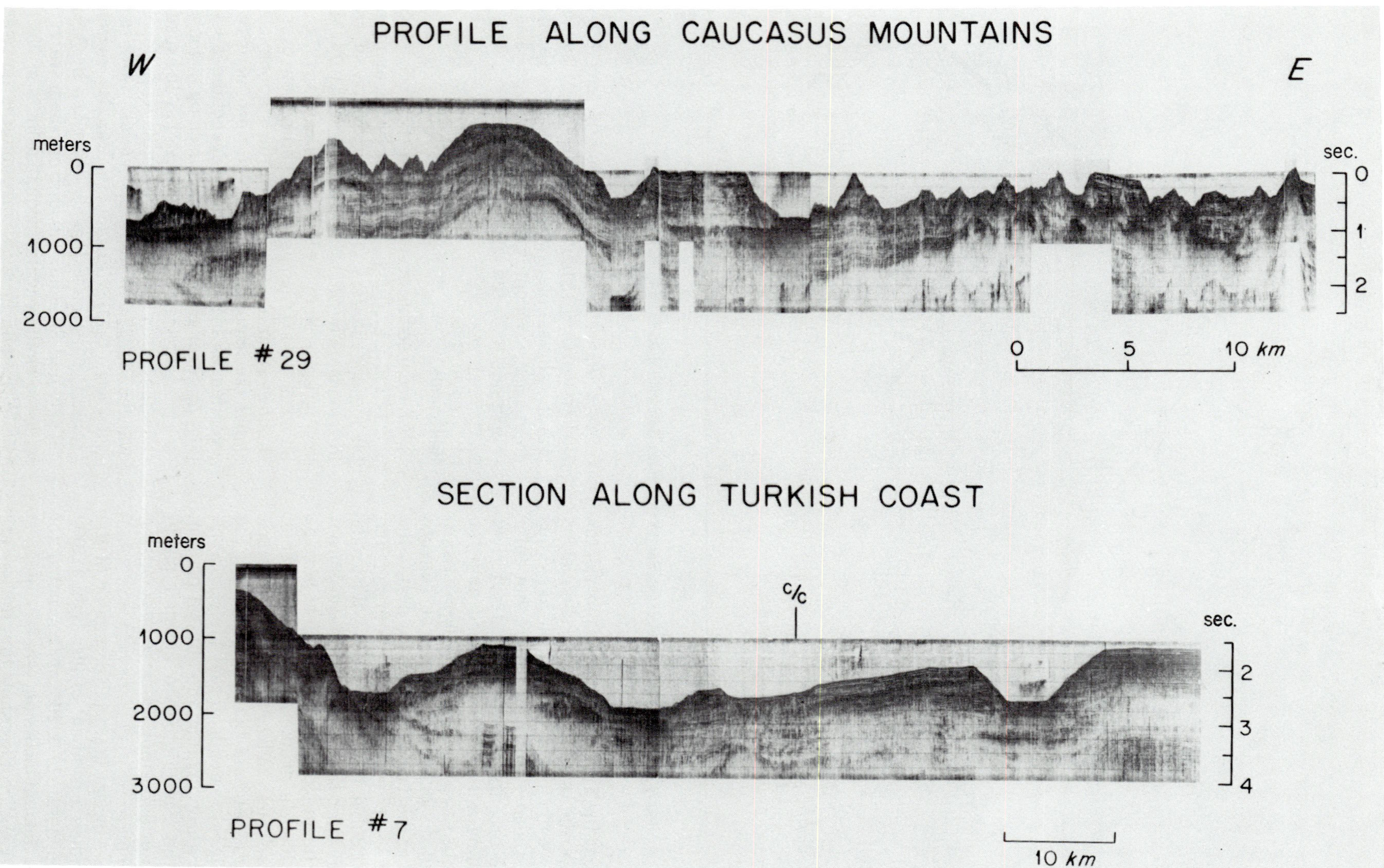

Fig. 3. Continuous seismic reflection profiles made across the basin slope area of the Black Sea. See Figure 2 for location of profiles.

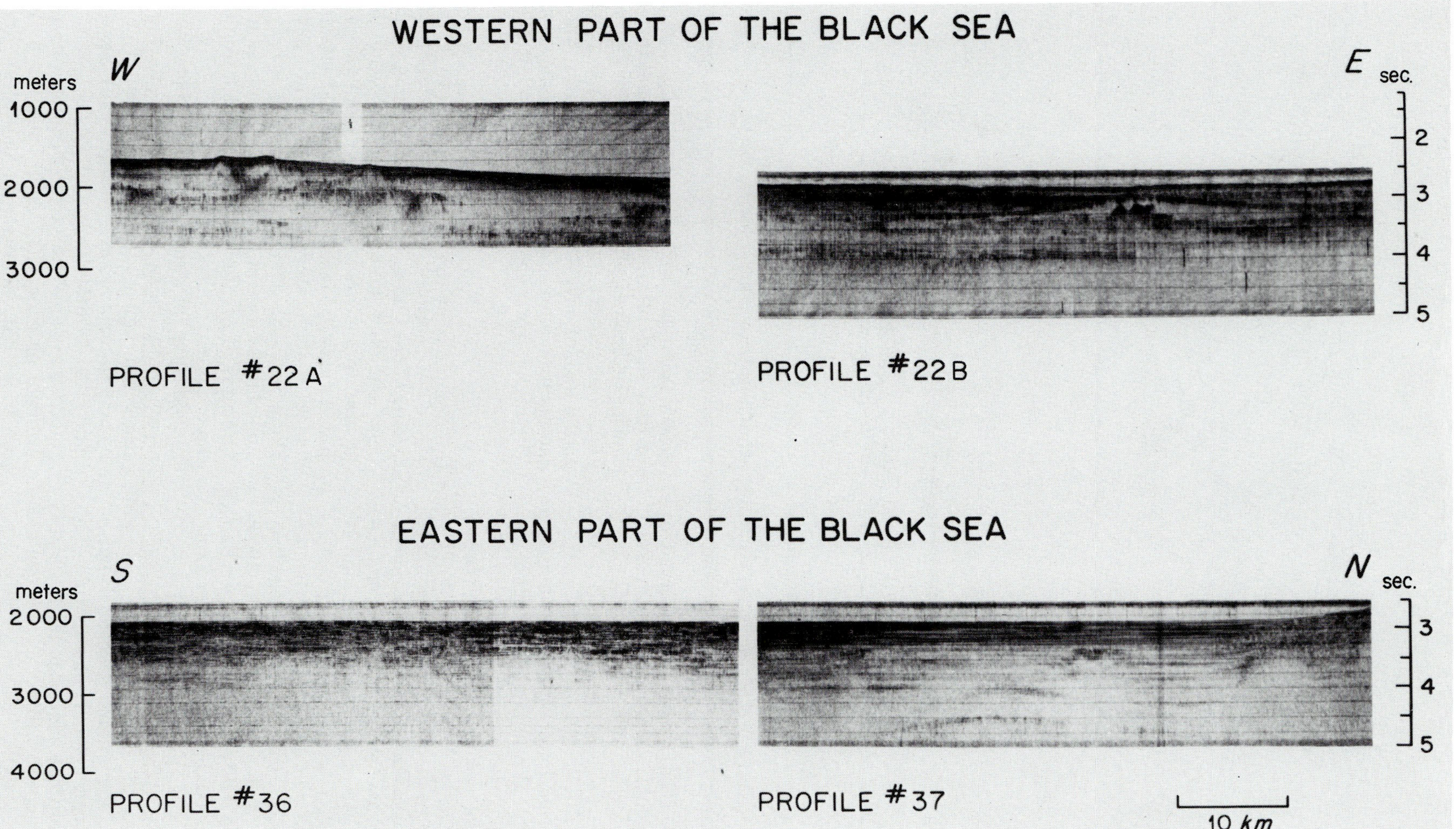

Fig. 4. Continuous seismic reflection profiles made across the basin apron and abyssal plain area of the Black Sea. See Figure 2 for location of profiles.

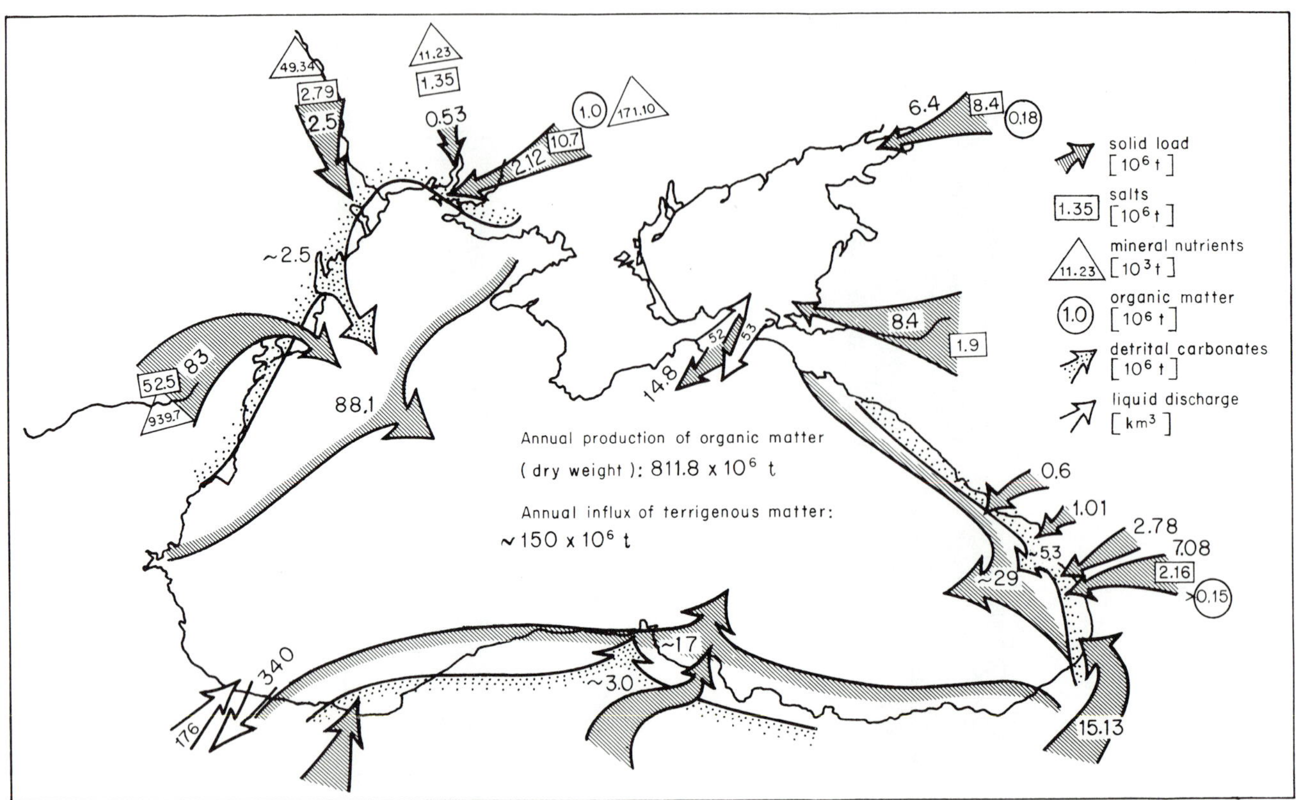

Fig. 5. Annual supply of sedimentary material (modified from Shimkus and Trimonis, 1974).

assemblages present in deep-water turbidites as a method of correlation. These three units, referred to as units 1, 2, and 3 (Fig. 6), are distinguished as follows:

Unit 1. This uppermost unit, generally about 30 cm thick, is composed of alternating white and black layers (Fig. 6a); there are between 50 and 100 layers per cm. The unit has a high calcium carbonate content, as much as 80%. White layers are composed almost entirely of the coccoliths of *Emiliania huxleyi* (Fig. 6b), whereas the darker layers appear to have somewhat higher contents of organic material. The radiocarbon age at the base of this unit is about 3,000 years B.P.

It was thought previously that this carbonate material had an inorganic origin, but recent electron microscope studies (Müller and Blashchke, 1969; Bukry et al., 1970; and Bukry, 1974) have shown that the carbonate is biogenic, being composed of *Emiliania huxleyi*.

Unit 2. The middle unit is about 40 cm thick and is a dark brown and somewhat jelly-like organic-rich sediment, similar to sapropels found in the Mediterranean Sea. Organic matter within this unit can be as high as 50%.

Electron microscope studies of the organic material show that it consists of protein crystals and numerous membranes of different shapes (Degens et al., 1970; Degens, 1971; Fig. 6c and d). It is surprising to find such crystals and membranes present in marine sediments, since they are rarely preserved. However, preservation in the Black Sea may be due to the lack of oxygen in the pore water, and heavy metals may have stabilized the organic structure. The radiocarbon age at the base of Unit 2 is about 7,000 years B.P.

Unit 3. This unit was not completely penetrated by any of our cores. It is an alternating sequence of light and dark lutite. Occasionally sand and silt layers, which are sometimes graded, are found within the unit. The graded sands and silts are more typical of the eastern than the western part of the Black Sea. The dark layers are apparently due to metastable forms of iron sulfide, such as mackinawite and greigite (Berner, 1974), which oxidize when the cores are opened.

Because of the absence of bottom-dwelling organisms in the Black Sea, sediments tend to be well preserved. Indeed, the three sedimentary units can be correlated over most of the Black Sea (Fig. 7). Individual layers, some as thin as 1 mm, can be traced from one end of the basin to the other, a distance of over 1,000 km. Thus there is strong evidence that conditions at any given time were essentially uniform over the Black Sea during the last 25,000 years. Since Unit 1 (Fig. 7) was detected in all

Table 1.
Stratigraphic Correlation of Black Sea Sediments.

Arkhangel'skiy & Strakhov (1938)	Neveskiy (1967)	Ross, et al (1970)	
		Sediment units	Years B.P.
Resent deposit	Dzhemetinian	Unit 1: Coccolith ooze	2000
Old Black Sea beds	Kalamitian		3000
	Vityazevian	Unit 2: sapropel or organic-rich	5000
			6000
	Bugazian		7000
Neoeuxinian sediments	Neoeuxinian	Unit 3: banded lutite	10,000
	Karkinitian ? ? ? ? ? ? Tarkhankutian		11,000
			25,000

the cores and we know the age of the base of the unit (about 3,000 years B.P.), and assuming that the age of the organic matter at the sediment-water interface is essentially zero (Östlund, 1974), the sedimentation rate of this unit can be estimated during the last 3,000 years (Fig. 8).

High rates occur along the Turkish coast, apparently due to the large supply of sediment from the rivers, in part due to high relief of the source area that extends to the coast, and the presence of a narrow shelf without any major estuaries that could trap sediments, but with numerous canyons that can transport sediments to the deeper parts of the basin. Sedimentation rates are also high in the eastern part of the basin, probably due to the higher frequency of turbidite deposition in this area. Lower sedimentation rates are common in the western part of the Black Sea, probably due to the lower-relief, broad coastal plain, and because the major rivers in this area are presently depositing their sediments within their estuaries rather than in the deeper part of the basin. However, in the past, sedimentation rates were probably higher in the western part of the basin, as indicated by the broad shelf and gentle slope area and the Danube Fan.

RECENT GEOLOGICAL HISTORY

Using sea-level curves and the stratigraphy established from recent sediment cores from the Black Sea that have been radiocarbon-dated, it is possible to reconstruct the recent geological history of the region (Fig. 9). Apparently, about 25,000 years ago (the age of the deepest layer penetrated), the Black Sea was in the process of changing from a marine basin to a freshwater or brackish lake. Sea level was probably at least about 30-40 m lower than today, thus isolating the Black Sea at the Bosporus from the Mediterranean. At this time, the Black Sea was also aerobic at depth. This lake phase continued for about 12,000-13,000 years.

Starting about 9,000 years B.P. and continuing to about 7,000 years B.P. there was a gradual shift from a freshwater environment to a marine environment and from a well-aerated to a stagnant condition. This was due to some Mediterranean sea water occasionally entering the Black Sea via the Bosporus as a result of rising sea level (Deuser, 1972). These occasional inflows of saline water formed deep bottom layers of water. Eventually, reducing conditions became established which resulted in better preservation of organic detritus.

About 7,000 years ago the H_2S zone became well established and started to increase in thickness. Deposition of sedimentary Unit 2 started at this time. The deposition rates decreased as most of the rivers dropped their sediments within their estuaries, due to rising sea level. It was not until about 3,000 years B.P. that environmental conditions presently found in the Black Sea became established and the deposition of sedimentary Unit 1 was initiated.

STRUCTURE

Gravity, magnetic, and seismic reflection data show that the Black Sea has been an area of recent subsidence that has been coincident with faulting along the southern, eastern, and northern parts of the basin. The western margin seems to have been tectonically relatively quiet during the subsidence. A large Bouguer anomaly is typical of much of the Black Sea, and free-air gravity anomalies are usually negative. Local patterns in the gravity field suggest a separate structure for the eastern and western parts of the basin (Bowin, in press; Ross et al., 1974b). The residual magnetic field parallels the surrounding Caucasus and Pontic mountains, suggesting that some of these structures extend into the Black Sea itself (Fig. 10). Present-day earthquake activity is generally shallow and restricted to the margins of the sea. A distinct seismic zone occurs along the northern part of the Anatolian Fault that extends parallel to the southern margin of the Black Sea. Focal depths there are usually less than 30 km (Canitez and Toksöz, 1970).

Neprochnov et al. (1970) have summarized the Soviet deep seismic refraction data and show that

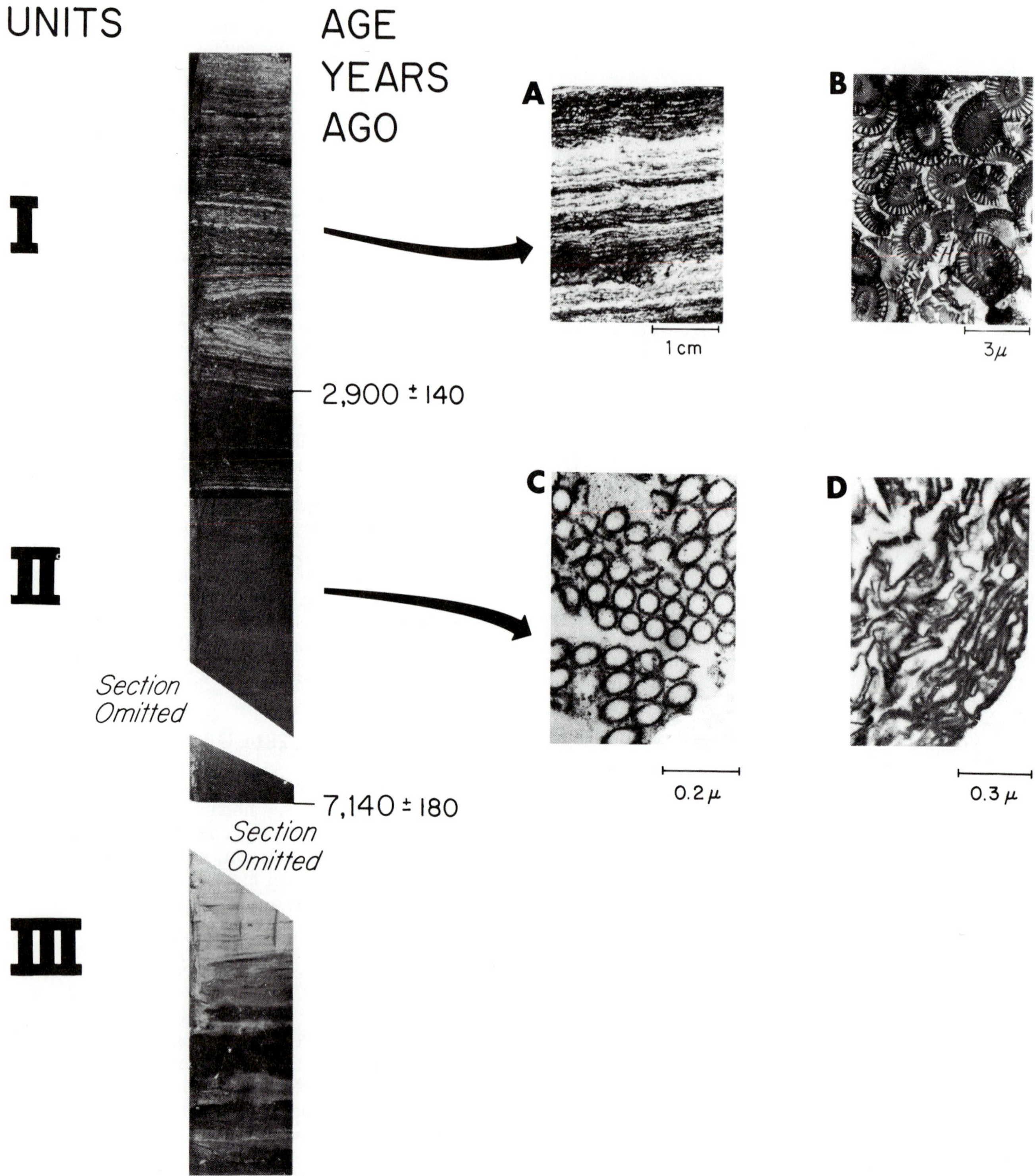

Fig. 6. Portions of a core collected from the Black Sea. The three distinct units are indicated by numbers (see text for a more detailed discussion). Ages are based on carbon 14 dates. Insert A shows the alternating light and dark microlaminated layers; both layers are mainly composed of the coccolith *Emiliania huxleyi*, which is shown enlarged in insert B (from Degens and Ross, 1972). Insert C shows some large tubular membranes having a diameter of about 700-800 A. Insert D shows branched tubular membranes aggregated together and bound by a limiting membrane (from Degens et al. 1970).

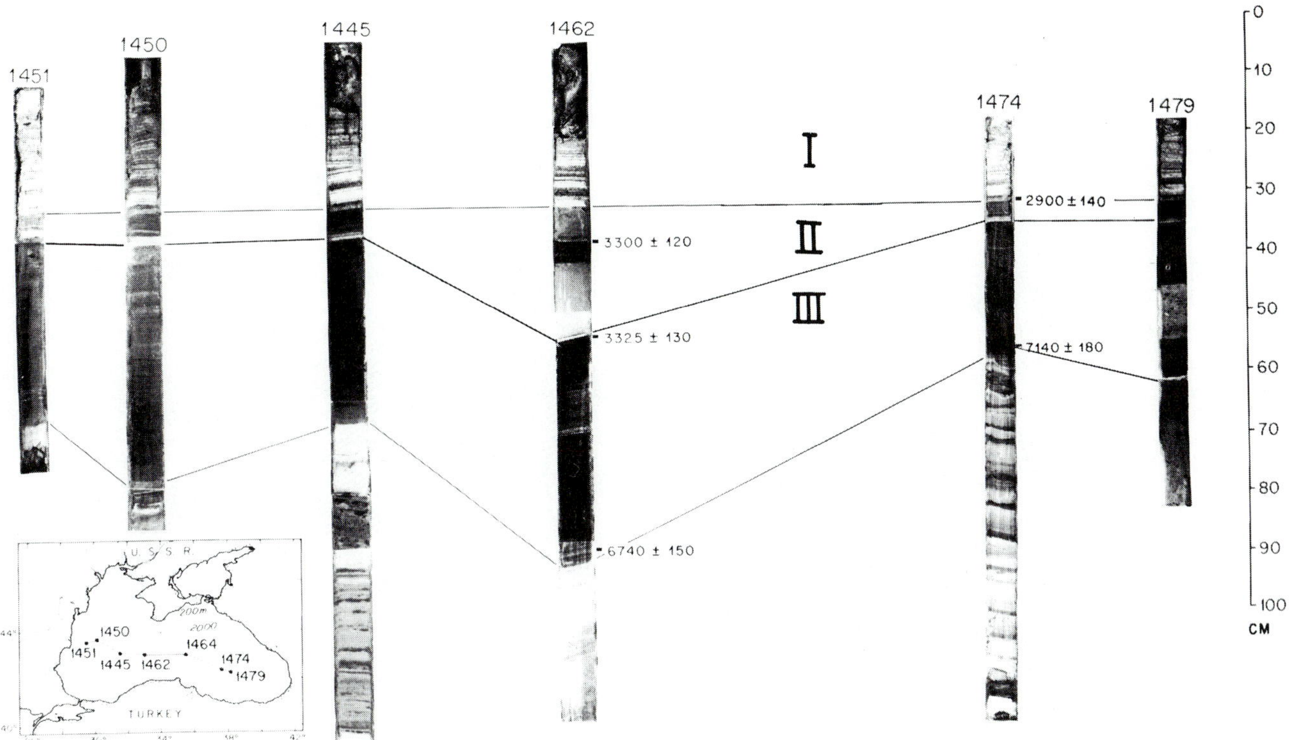

Fig. 7. Sediment profile of some cores collected from the Black Sea. Note abrupt changes at about 3,000 and 7,000 years B.P. Three distinct sediment units (discussed in the text) are indicated. Individual sequences can be correlated across the basin (adapted from Ross et al, 1970).

although the crust beneath the central Black Sea is "somewhat oceanic" in structure, it has a total thickness of 18-20 km (Fig. 11). Sediments make up a large part of this thickness (Neprochnova, 1970) and can reach a maximum of 8-12 km. Crustal thicknesses tend to increase toward the margin of the Black Sea, in part because of the appearance of material having compressional velocities of about 6.0-6.4 km/sec, suggestive of granitic rocks (Neprochnov et al., 1967; Garkalenko, 1970).

Continuous seismic profiles generally show deeply channelled faults that are especially common along the southern and eastern parts of the Black Sea (Fig. 3). Seaward of these areas, the transition from basin slope to apron or abyssal plain is usually abrupt and is indicated by small faults, slumps, or even diapiric structures (Ross et al., 1974b). Within the abyssal plain, reflectors can be traced from one edge of the basin to the other (Fig. 4). A deep anticlinal structure was observed south of the Crimea which has also been detected from Russian deep-seismic sounding and may be related to the Caucasus orogeny.

Heat-flow measurements in the Black Sea (Erickson and Simmons, 1974) yield values having an average of 0.92 $\pm$0.23 μcal/cm2/sec (16 observations). However, this value is only about 50% of the geophysically relevant heat flow, because a significant fraction of the geothermal flux is absorbed by the rapidly accumulating sediments. In addition, the lower thermal conductivity of the sediments has a blanketing effect and has further reduced flux by as much as 15% Erickson and Simmons (1974) conclude that the mean heat flux for the Black Sea after correcting for these two effects is about 2.2 μcal/cm2/sec.

ORIGIN OF THE BLACK SEA

Estimates of the age of the Black Sea range from as early as Precambrian (Milanovskiy, 1967) to as recent as early Quaternary (Nalivkin, 1960). However, most workers favor a middle to late Mesozoic age (Brinkmann, 1974). Neprochnov et al. (1967) have noted similarities in the crustal structures of the Mediterranean, Caspian, and Black seas, and that the Black Sea is intermediate in crustal thickness and thickness of sedimentary rocks compared to the other two basins. These authors suggest a sequence of sinking and sediment accumulation, beginning (in the Caspian Sea at least) perhaps as early as the Paleozoic.

The crustal structure of the Black Sea is intermediate to that of continents and ocean (Fig. 11). The origin of this type of structure in this and other marginal seas is not well understood, although some evolutionary mechanisms have been proposed (Menard, 1967). There are data to suggest that the present Black Sea area was in the past a topographic high from which Paleozoic sediments were derived (Brinkmann, 1974). The general east-west alignment

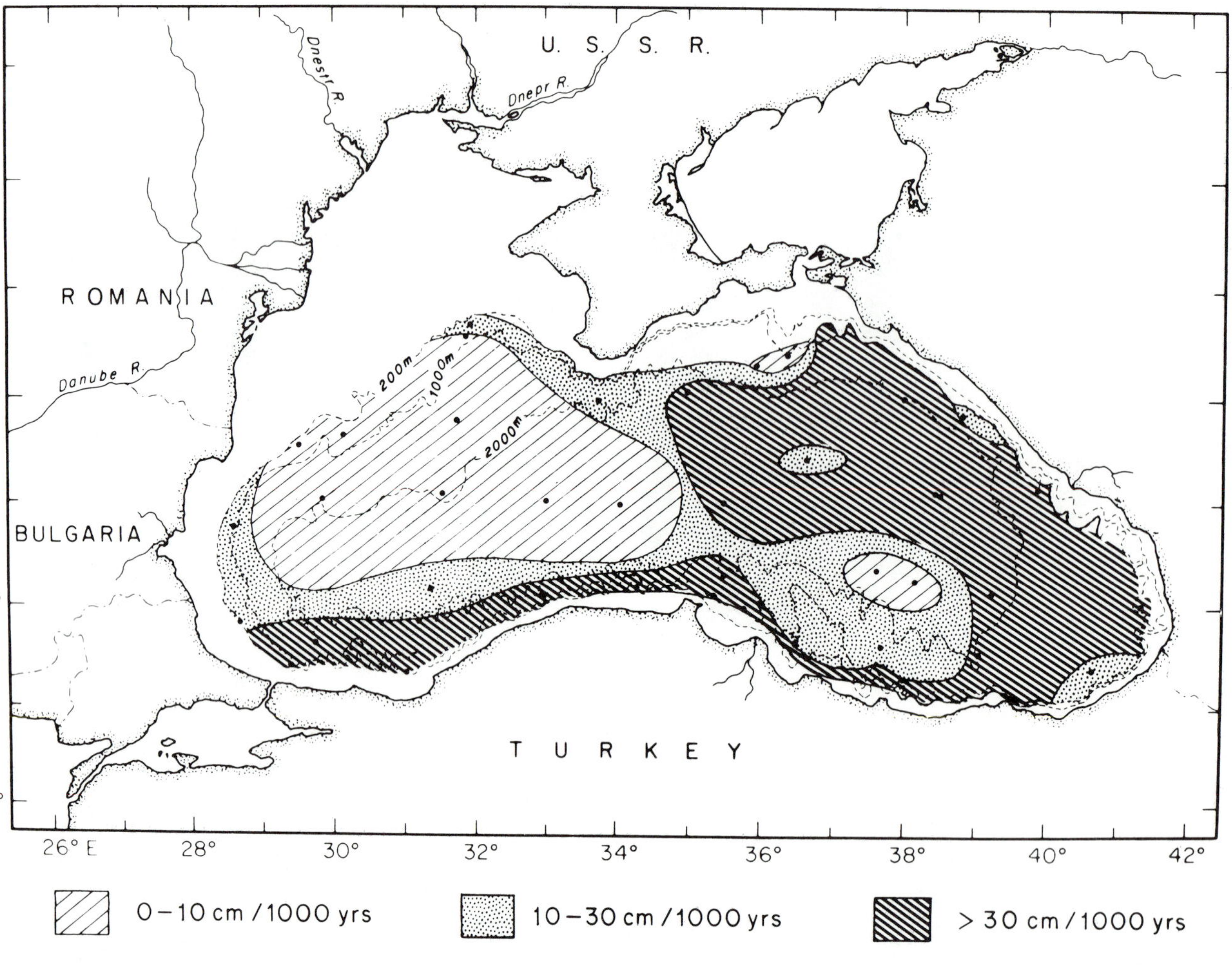

Fig. 8. Sedimentation rates in the Black Sea over the past 3,000 years. Dots indicate sample locations (from Ross et al, 1970).

of the surrounding Caucasus and Pontic mountains indicates that during the Mesozoic and Cenozoic the Black Sea area was a zone of north-south compression. More recent development of the Black Sea is mainly by subsidence of coastal and platform segments, including some parts of the western Caucasus (Belavadze et al., 1966; Tzagareli, 1974). Residual magnetic anomalies (Fig. 10) in the northern part of the Black Sea are parallel to the structural trends of the Caucasus (Ross et al., 1974b), suggesting an extension of these structures into the Black Sea. The presence of "granitic"-type material (as suggested by its seismic velocity) along the marginal part of the basin may indicate intense erosion of a prior continental crust followed by subsidence and sedimentation (Rezanov and Chamo, 1969).

However, it has been noted that the "granitic" layer is nonmagnetic and could also consist of metamorphosed sedimentary rocks (Borisov, 1967; Sollogub, 1968; Neprochnov et al., 1974). The "granite-free" area within the central part of the Black Sea has a seismic velocity of 6.6-7.0 km/sec and is thought to be gabbroic in nature (Neprochnov et al., 1974). The origin of this "granite-free" area could be of two main possibilities: either a relict

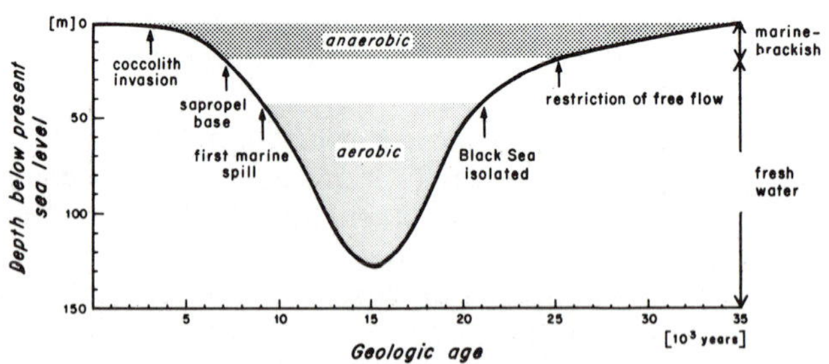

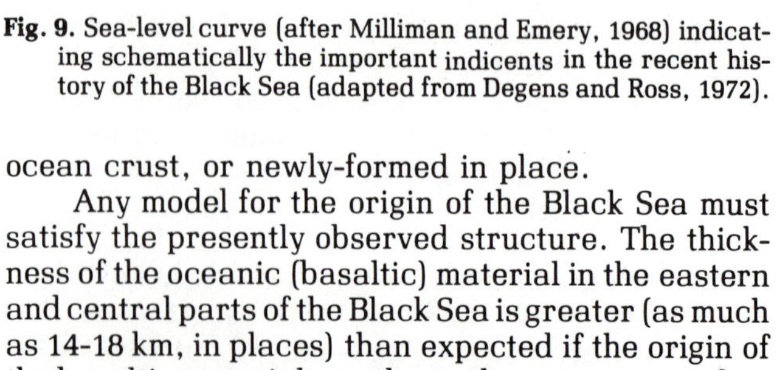

Fig. 9. Sea-level curve (after Milliman and Emery, 1968) indicating schematically the important indicents in the recent history of the Black Sea (adapted from Degens and Ross, 1972).

ocean crust, or newly-formed in place.

Any model for the origin of the Black Sea must satisfy the presently observed structure. The thickness of the oceanic (basaltic) material in the eastern and central parts of the Black Sea is greater (as much as 14-18 km, in places) than expected if the origin of the basaltic material was due to the entrapment of an ocean crust, as suggested by Milanovskiy (1967).

Hypotheses that consider the transformation of a continental crust into a more intermediate or

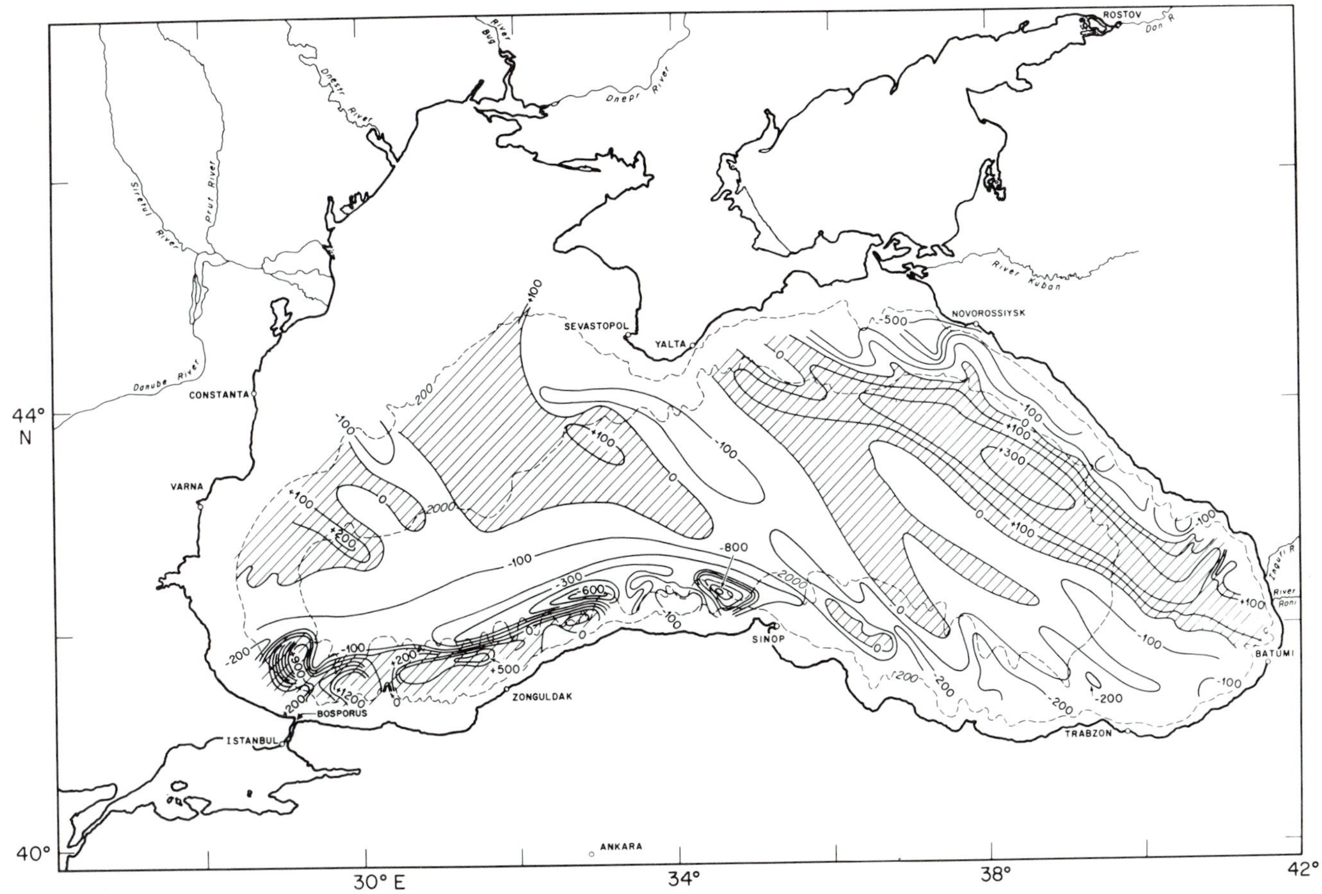

Fig. 10. Black Sea magnetic anomalies based on measurements made aboard ATLANTIS II (Ross et all, 1974b). Anomalies calculated by removal of a reference field (Cain et al., 1968); contour interval is 100 gammas. Positive anomaly areas are shaded.

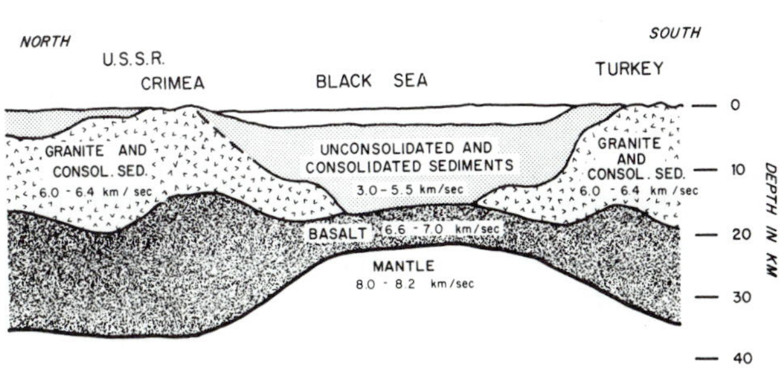

Fig. 11. Crustal north-south section across the Black Sea (adapted from Subbotin et al, 1968; and Neprochnov et al, 1967, 1970).

oceanic type of crust can be of two main types. Either *in situ* conversion of low-density, thick continental crust into a denser, thinner, more oceanic crust; or replacement of the continental crust by upwelling mantle material along cracks or extensions. The evidence does not appear definitively to favor either of these suggestions. For example, the continuity of geologic structures around the Black Sea and magnetic trends paralleling these structures and extending into the basin argue against an extension origin. *In situ* "oceanization" presents several difficult chemical and mechanical problems. The presence of a landward-thickening ring of "granitic" or metamorphosed sediment presents additional complications.

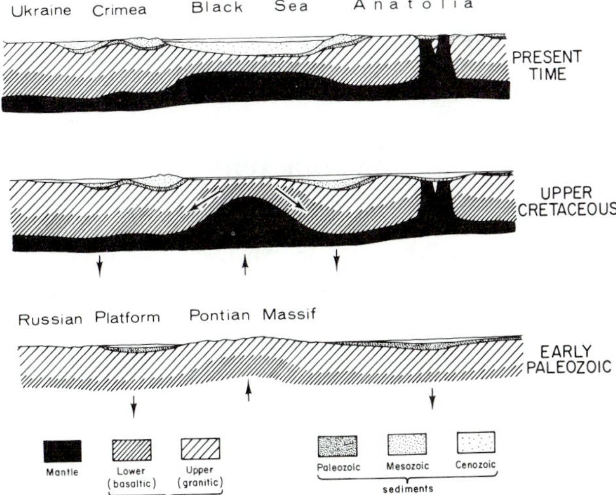

Fig. 12. Schematic sections across the Black Sea showing its geological evolution according to mechanisms proposed by Brinkmann (1974).

Recent hypotheses include that of Erickson and Simmons (1974), who suggest that the Black Sea began with the initiation or renewal of convergence of the Eurasian and African plates. Subaerial erosion would have removed the upper parts of the crust. Once subduction slowed or ceased, temperature decreases in the mantle and reversible phase and density changes caused the lithosphere and crust to sink and be covered by sediment. Brinkmann (1974) visualizes a long period of uplift and erosion, followed by crustal tension resulting in rising volcanic material (Fig. 12). Increased tension results in a suboceanic thickness of the crust and sinking and formation of a "soft" mantle (Vp = 7.8 km/sec). Decreasing temperatures within the crust and mantle result in increased sinking of the basin, and as volcanic activity ceases the seismic velocity at the Mohorovicic discontinuity increases to about 8 km/sec. As this sinking occurs the basin is filling up with sediment.

All the above is speculation, however, and better understanding of the origin and evolution of the Black Sea awaits further seismic studies and results of deep-sea drilling.

ACKNOWLEDGMENTS

I wish to thank the National Science Foundation (Grant GA-25234), who sponsored the 1969 ATLANTIS II Cruise and the original research program in the Black Sea. Support for this paper came from the Office of Naval Research (Contract N00014-66-CO241 NR 083-004). I would also like to thank the officers and crew of R/V ATLANTIS II and the numerous scientists and technicians who participated in the cruise and later research on the data collected. Elazar Uchupi and Werner Deuser kindly reviewed this paper; Francy Forrestel typed the manuscript. This paper is Woods Hole Oceanographic Institution Contribution 3251.

BIBLIOGRAPHY

Andrusov, N. I., 1890, O neobkhodimosti glubokovodnykh issledovanii Chernogo morya (Need for deep-water investigations in the Black Sea): Izv. Russk. Geogr. Obshch., v. 26, p. 171-185.

Arkhangel'skiy, A. D., 1928, Karta i razrezy osadkov dna Chernogo morya (Map and cross sections of Black Sea sediments): Moskov. Obshch. Ispytateley Prirody Byull. Otdel. Geol., v. 6, no. 1, p. 77-108.

———, 1930, Opolzanie osadkov na dne Chernogo morya i geologicheskoe z actrenie etogo yavleniya (Slumping sediments on bottom of Black Sea and geological significance of this phenomenon): Moskov, Obshch. Ispytateley Prirody Byull. Otdel. Geol., v. 8, no. 1-2, p. 32-79.

———, and Strakhov, N. M., 1932, Geologischeskoe istoria Chernogo morya (Geological history of Black Sea): Moskov. Obshch. Ispytateley Prirody Byull. Otdel. Geol., v. 10, no. 1, p. 3-104.

———, and Strakhov, N. M., 1938, Geologicheskoe stroyeniye i istoria razvitiya Chernogo morya (Geological structure and history of the evolution of the Black Sea): Izv. Akad. Nauk S.S.S.R., v. 10, p. 3-104.

Barkovskaya, M. G., 1961a, Zakonomernosti raspredeleniya donnykh osadkov na shel'fe sovetskikh beregov Chernogo morya (Regularities in the distribution of bottom sediments on the shelf of the Soviet shores of the Black Sea): Akad. Nauk S.S.S.R. Inst. Okeanol. Trudy, v. 53, p. 123-148.

———, 1961b, Zakonomernosti respredeleniya terrigenogo materiala v priurezovoy polose sovetskogo poberzh'ya Chernogo morya (Regularities in the distribution of terrigenous material in the littoral zone of the Soviet shore of the Black Sea): Akad. Nauk S.S.S.R. Inst. Okeanol. Trudy, v. 53, p. 64-94.

Belavadze, B. K., Tvaltradze, Y. K., Shengelaya, Y. Sh., Sikharulidze, D. I., and Kartrelishvili, K. M., 1966, Geophysical investigations of the earth's crust and upper mantle of the Caucasus: Geotektonika, v. 3, p. 30-40.

Berner, R. A., 1974, Iron sulfides in Pleistocene deep Black Sea sediments and their paleo-oceanographic significance, *in* Degens, E. T., and Ross, D. A., eds., The Black Sea—geology, chemistry, and biology: Am. Assoc. Petr. Geol. Mem. 20., p. 524-531.

Borisov, A. A., 1967, Glubinnaya struktura territorii S.S.S.R. po geofizicheskim (Deep structure of territory of U.S.S.R. based on geophysical data): Moskow, Izd. Nedra.

Bowin, C. O., (in press), Gravity anomalies of the Black Sea: Jour. Geophys. Res.

Brinkmann, R., 1974, Geologic relations between Black Sea and Anatolia, *in* Degens, E. T., and Ross, D. A., eds., The Black Sea—geology, chemistry, and biology: Am. Assoc. Petro. Geol. Mem. 20, p. 63-76.

Bukry, D., 1974, Coccoliths as paleosalinity indicators—evidence from Black Sea, *in* Degens, E. T., and Ross, D. A., eds., The Black Sea—geology, chemistry, and biology: Am. Assoc. Petr. Geol. Mem. 20, 353-363.

———, Kling, S. A., Horn, M. K., and Manheim, F. T., 1970, Geological significance of coccoliths in fine-grained postglacial carbonate bands of Black Sea sediments: Nature, v. 226, p. 156-158.

Cain, J. C., Hendricks, S., Daniels, W. E., and Jensen, D. C., 1968, Computation of the main geomagnetic field from spherical harmonic expressions: NASA Data Center. Data Users Note NSSDC 68-11, 46 p.

Canitez, N., and Toksöz, M. Nafi, 1970, Source parameters of earthquakes and regional tectonics of the eastern Mediterranean (abstr.): Am. Geophys. Union Trans., v. 51, p. 420.

Degens, E. T., 1971, Sedimentological history of the Black Sea over the last 25,000 years, *in* Campbell, A. S., ed., Geology and history of Turkey: Tripoli, Petro. Expl. Soc. Libya, p. 407-429.

———, and Ross, D. A., 1972, Chronology of the Black Sea over the last 25,000 years: Chem. Geology, v. 20, p. 1-16.

———, and Ross, D. A., eds., 1974, The Black Sea—geology, chemistry, and biology: Am. Assoc. Petr. Geol. Mem. 20, 633 p.

———, Watson, S. W., and Remsen, C. C., 1970, Fossil membranes and cell wall fragments from a 7000-year-old Black Sea sediment: Science, v. 168, p. 1207-1208.

Deuser, W. G., 1972, Late-Pleistocene and Holocene

history of the Black Sea as indicated by stable-isotope studies: Jour. Geophys. Res., v. 77, no. 6, p. 1071-1077.

———, 1974, Evolution of anoxic conditions in Black Sea during the Holocene, *in* Degens, E. T., and Ross, D. A., eds., The Black Sea—geology, chemistry, and biology: Am. Assoc. Petr. Geol. Mem. 20, p. 133-136.

Emery, K. O., and Hunt, J. M., 1974, Summary of Black Sea investigations, *in* Degens, E. T., and Ross, D. A., eds., The Black Sea—geology, chemistry, and biology: Am. Assoc. Petro. Geol. Mem. 20, p. 575-590.

Erickson, A., and Simmons, G., 1974, Environmental and geophysical interpretation of heat-flow measurements in Black Sea, *in* Degens, E. T., and Ross, D. A., eds., The Black Sea—geology, chemistry, and biology: Am. Assoc. Petr. Geol. Mem. 20, p. 50-62.

Garkalenko, I. A., 1970, The deep-seated crustal structure in the western part of the Black Sea and adjacent areas: seismic reflection measurement: Tectonophysic, v. 10, p. 539-547.

Goncharov, V. P., 1958, Novyye dannyye o rel'yefe dna Chernogo morya (New data on topography of bottom of Black Sea): Akad. Nauk S.S.S.R. Dok., v. 121, no. 5, p. 830-833.

Goncharov, V. P., and Neprochnov, Yu. P., 1967, Geomorphology of the bottom and tectonic problems in the Black Sea, *in* Runcorn, S. K., ed., International dictionary of geophysics: Elmsford, N.Y., Pergamon Press, 2 v., 1728 p.

———, Neprochnova, A. F., and Neprochnov, Yu. P., 1966, Geomorfologiya dna i glubinnoye stroyeniye Chernogo morya vpadiny (Geomorphology of the floor and deep structure of the Black Sea basin), *in* Glubinnoye stroyeniye Kavkaza (Deep structure of the Caucasus): Moscow, Izd. Nauka.

Malovitskiy, Ya., P., Uglov, B. D., and Osipov, G. V., 1969a, Geomagnitnoye pole Chernomorskoy vpadiny (Geomagnetic field of the Black Sea depression): Akad. Nauk. Ukr. S.S.R. Geofiziki, Geofiz. Sbornik., v. 32, p. 28-38.

Malovitskiy, Ya. P., Neprochnov, Yu. P., Garkelenko, I. A., Starskinov, E.A., Milashina, K.G., Komornaiam, Ya., Ryunov, L. N., Kloiopov, B. V., and Sedov, V. V., 1969b, Stroenie zemnoi kory v zapadnoi chasti Chernogo morya (Structure of the earth's crust in the western part of the Black Sea): Akad. Nauk S.S.S.R. Dokl., v. 186, no. 4, p. 905-907.

Matthews, D. J., 1939, Tables of the velocity of sound in pure water and sea water for use in echo sound and sound ranging: Hydrog. Dept. Ad. H. D. 383, Charts 62 a-b.

Melikhov, V. P., Mirlin, E. G., Uglov, B. D., and Shreider, A. A., 1969, Otsenka raspredeleniya magnitovozumushchayuikh gel v kore glubokovodnoi kotloviny Chernogo morya s pomoshch transformatsii v nizhnea poluprostranstvo (Estimate of the distribution of magnetic perturbing bodies in the crust of the deep-water basin of the Black Sea with the aid of transformations in the lower half-space), *in* Morskaya Geol. Geofizika, v. 2.

Menard, H. W., 1967, Transitional types of crust under small ocean basins: Jour. Geophys. Res., v. 72, no. 12, p. 3061-3073.

Milanovskiy, Ye. Ye., 1967, Problema proiskhozhdeniya Chernomorskoy vpadiny i yeye mesto v strukture al'piyskogo poyasa: Moskov. Unv. Vestnik. Ser. Geol. 4, v. 22, no. 1, p. 27-43; Engl. transl. (1967) Problem of origin of Black Sea depression and its position in structure of the Alpine belt: Internatl. Geol. Rev., v. 8, no. 1, p. 36-43.

Milliman, J. D., and Emery, K. O., 1968, Sea levels during the past 35,000 years: Science, v. 162, no. 3858, p. 1121.

Müller, G., and Blaschke, R., 1969, Zur Entstehung des Tiefsee-Kalkschlammes im Schwarzen Meer: Naturwissenschaften, v. 56, p. 561-562.

Müller, G., and Stoffers, P., 1974, Mineralogy and petrology of Black Sea basin sediments, *in* Degens, E. T., and Ross, D. A., eds., The Black Sea—geology, chemistry, and biology: Am. Assoc. Petr. Geol. Mem. 20, p. 200-248

Muratov, M. V., and Neprochnov, Yu. P., 1967, Stroyeniye dna Chernomorskoy kotloviny i yeye proiskhozhdeniye (Structure of Black Sea basin and its origin): Moskov. Obshch. Ispytateley Prirody Byull. Otdel. Geol., v. 42, no. 5, p. 40-59.

Nalivkin, D. V., 1960, The geology of the USSR (Engl. transl.): Elmsford, N.Y., Pergamon Press, 170 p.

Naprochnov, Yu. P., 1962, Rezul'taty glubinnogo seysmicheskogo zondiravaniya na Chernom more (On results of deep seismic sounding in the Black Sea), *in* Glubinnoye seysmicheskoye zondirovaniye zemnoy kory v S.S.S.R. (Deep seismic sounding of the earth's crust in the USSR): Leningrad, Gostekhizdat, 271 p.

———, Neprochnova, A. F., Zverev, S. M., Mironova, V. I., Bokuni, R. A., and Chekunov, A. V., 1964, Novye dannye o stroenii zemnoi kory Chernomorskoi vpadiny k yugu ot Kryma (New data on the structure of the Black Sea crust south of the Crimea): Akad. Nauk. S.S.S.R. Dok., v, 156, no. 3, p. 561-564.

———, Naprochnova, A. F., Zverev, S. M., and Moronova, V. I., 1967, Deep seismic sounding of the earth's crust in the central part of the Black Sea depression, *in* Zverev, S. M., ed., Problems in deep seismic sounding; Leningrad, Gostekhizdat, 271 p.

———, Kosminskaya, I. P., and Malovitskiy, Ya. P., 1970, Structure of the crust and upper mantle of the Black and Caspian seas: Tectonophysics, v. 10, p. 517-538.

———, Neprochnova, A. F., and Mirlin, Ye. G., 1974, Deep structure of Black Sea Basin, *in* Degens, E. T., and Ross, D. A., eds., The Black Sea—geology, chemistry, and biology: Am. Assoc. Petr. Geol. Mem. 20, p. 35-49.

Neprochnova, A. F., 1970, Izucheniye stroyeniya i skorostnoy kharakteristkiy osadochnogo sloya Chernomorskoy vpadiny po dannym GSZ (Study of the structure and velocity characteristics of the sedimentary layer of the Black Sea basin based on DSS data): Akad. Nauk S.S.S.R. Inst. Okeanol. Trudy, v. 87.

Neveskiy, E. N., 1967, Protsessy osadkoobrazovaniya v pribrezhnoi zone morya (Processes of sediment formation in the near-shore zone of the sea): Moscow, Izd. Nauka, 255 p. (in Russian).

Nevesskaya, L. A., 1974, Molluscan shell in deep water sediments of the Black Sea, *in* Degens, E. T., and Ross, D. A., eds., The Black Sea—geology, chemistry and biology: Am. Assoc. Petr. Geol. Mem. 20, p. 349-352.

Nevesskiy, Ye. N., 1967, Protsessy osadkoobrazovaniya v pribrezhnoi zone morya (Sediment deposition in nearshore region): Moscow, Izd. Nauk, 255 p.

Östlund, H. G., 1974, Expedition ODYSSEUS 65: radiocarbon age of Black Sea deep water, *in* Degens, E. T., and

Ross, D. A., eds. The Black Sea—geology, chemistry and biology: Am. Assoc. Petr. Geol., Mem. 20.

Rezanov, I. A., and Chamo, S. S., 1969, O prichinakh otsutstviya "granitnogo" sloya vo vpadinakh tipa Yuzhno-Kaspiyskoy i Chernomorskoy: Akad. Nauk S.S.S.R. Izv. Ser. Geol., no. 2, p. 3-11; Engl. transl. (1969) Reasons for absence of a "granite" layer in basins of the south Caspian and Black Sea type: Can. Jour. Earth Sci., v. 6, no. 4, pt. 1, p. 671-678.

Ross, D. A., and Degens, E. T., 1974, Recent sediments of Black Sea, *in* The Black Sea—geology, chemistry, and biology; Am. Assos. Petr. Geol. Mem. 20, p. 183-199.

———, Degens, E. T., and MacIlvaine, J. C., 1970, Black Sea: recent sedimentary history: Science, v. 170, no. 3954, p. 163-165.

———, Uchupi, E., Prada, K. E., and MacIlvaine, J. C., 1974a, Bathymetry and microtopography of the Black Sea, *in* Degens, E. T., and Ross, D. A., eds., The Black Sea—geology, chemistry, and biology; Am. Assoc. Petr. Geol. Mem. 20, p. 1-10.

———, Uchupi, E., and Bowin, C. O., 1974b, Shallow structure of the Black Sea, *in* Degens, E. T., and Ross, D. A., eds., The Black Sea—geology, chemistry, and biology: Am. Assoc. Petr. Geol. Mem. 20, p. 11-34.

Shimkus, K. M., and Trimonis, E. S., 1974, Modern sedimentation in the Black Sea, *in* Degens, E. T., and Ross, D. A., eds., The Black Sea—geology, chemistry, and biology: am. Assoc. Petr. Geol. Mem. 20.

Sollogub, V. B., 1968, Opirode seysmicheskikh granits zemlor kory (On the nature of seismic boundaries within the earth's crust): Akad. Nauk Ukr. S.S.R. Inst. Geofiziki, Geofiz. Sbornik, v. 25.

Strakhov, N. M., 1954, Osadkoobrazovaniye v Chernom more (Sediment formation in the Black Sea), *in* Belyankin, D. S., ed., Obrazovaniye osadkov v sovremennykh vodoemakh (Formation of sediments in contemporary basins): Moscow, Akad. Nauk S.S.S.R. Izd., p. 81-136.

———, 1961, O znachenii serovodovodnogo zarazheniya naddonnoy vodoy basseyna dlya autigennogo mineral-oobrazovaniya v ego osadkakh-na primere Chernogo morya (Significance of hydrogen sulfide contamination of water overlying the bottom in formation of authigenic materials in sediments in the Black Sea), *in* Strakhov, N. M., ed., Sovremennye osadki morey i okeanov (Recent sediments in seas and oceans): Moscow, Akad. Nauk S.S.S.R. Izd., p. 521-548.

Subbotin, S. I., Sollogub, V. B., Prosen, D., Dragasevic, T., Mituch, E., and Posgay, K., 1968, Junction of deep structures of the Carpatho-Balkan region with those of the Black and Adriatic seas: Can. Jour. Earth Sci., v. 5, p. 1027-1035.

Tzagareli, A. L., 1974, Geology of the western Caucasus, *in* Degens, E. T., and Ross, D. A., eds., The Black Sea—geology, chemistry and biology: Am. Assoc. Petr. Geol. Mem. 20, p. 77-89.

Continental Margins of the Gulf of Mexico

John W. Antoine, Ray G. Martin, Jr., T. G. Pyle, and William R. Bryant

INTRODUCTION

The Gulf of Mexico is a small ocean basin whose continental margins are structurally complex and in some cases rather unique. The origin of the Gulf Basin and the subsequent construction of the continental margins are somewhat in contention. The prominent theories contain one of four basic ideas that the Gulf represents: (1) a foundered and oceanized continental mass; (2) a downwarp related to a thermally controlled phase change in the crust and mantle; (3) a gigantic tensioned rift formed in relation to Mesozoic opening of the Atlantic Ocean; and (4) a Paleozoic or older ocean basin.

The structure of the continental margins of the Gulf of Mexico are the results of tectonic activity related to salt movement, reef growth, current activity, and the massive uppouring of sediments along its northern boundaries. The continental margins of the Gulf are divided into two distinct physiographic and sedimentological provinces, separated physically by two submarine canyons. The DeSoto Canyon in the northeast and the Campeche Canyon in the southwest. These two canyons are the dividing line between the limestone platforms of the West Florida and Yucatan platforms and the clastic embayments of the northern and western Gulf of Mexico.

CARBONATE PLATFORMS

West Florida Platform

The subsurface geology of the west Florida platform is dominated by a structural basin that has subsided while accumulating more than 15,000 ft (4.6 km) of shallow-water, primarily carbonate-evaporite sediments since Late Jurassic-Early Cretaceous time. The South Florida Basin, best defined beneath peninsular Florida by oil well tests, can be traced westward beneath the continental shelf by correlation with seismic-refraction and seismic-reflection profiles (Fig. 1; Antoine and Ewing, 1963; Oglesby, 1965; Pyle and Antoine, 1973).

The seismic data indicate that the western structural rim of the South Florida Basin during Early Cretaceous time was at the Florida Escarpment. Reflection profiles combined with core and dredge samples indicate that an algal reef flourished along this rim during late Aptian-Albian time. Although this reef was not necessarily one continuous barrier, it restricted circulation sufficiently to lead to the deposition of evaporites in the basin. The reef was either buried or deeply submerged during Late Cretaceous time and transgressive "chalks" were deposited over the Lower Cretaceous cyclic carbonate-evaporite sequence. Conditions favorable for reef growth were reestablished during Paleocene time, and evaporites were again deposited across the peninsula. South Florida was separated from areas of clastic deposition by the Suwannee Strait, whose western end may have coincided with the head of DeSoto Canyon. After Paleocene time, reef growth did not keep pace with subsidence, and sediment sampled along the escarpment indicates that bathyal conditions were established here before middle Miocene time (Bryant et al., 1969). It is concluded that total subsidence of the Lower Cretaceous reef and of the western edge of the Florida platform has been more than 6,000 ft (1,830 m).

Greater subsidence and faulting of the reef front along the strike has occurred in the southern Straits of Florida. Continuity of the Lower Cretaceous reef trend on the north-central coast of Cuba suggests that the southern margin of the straits has been downfaulted, or possibly that the Cuban part of the trend has been thrust above the Florida part. In either case, it is concluded that the southern wall of the Straits of Florida is one tectonic margin of a Florida-Bahamas trough. In response to the great subsidence and faulting of the southern margin, reef fronts of the Florida-Bahamas region and of the southern straits in particular probably migrated northward during Cretaceous, Tertiary, and Quaternary time. The Pourtales and Mitchell escarpments on the northern slope of the straits are proposed to be expressions of these reef fronts.

Draping of southward-transported sediments over reef buttresses in the straits and downslope movement of unconsolidated sediments at their bases have produced examples of gravity tectonics in the form of slump blocks and apparent normal faults that may curve toward bedding planes at depth (see Fig. 2). Although depositional sedimentary ridges occur along the floor of the northern Straits of Florida, erosion has shaped many of the features of the western end of the southern straits, especially the surface of the deep-sea fan noted by Hurley (1964).

Major tectonic events in the region are reflected in the magnetic anomaly patterns and in the structure of the Florida Escarpment. Offshore parts of the Florida and Yucatan platforms exhibit magnetic anomalies suggestive of volcanic sources esti-

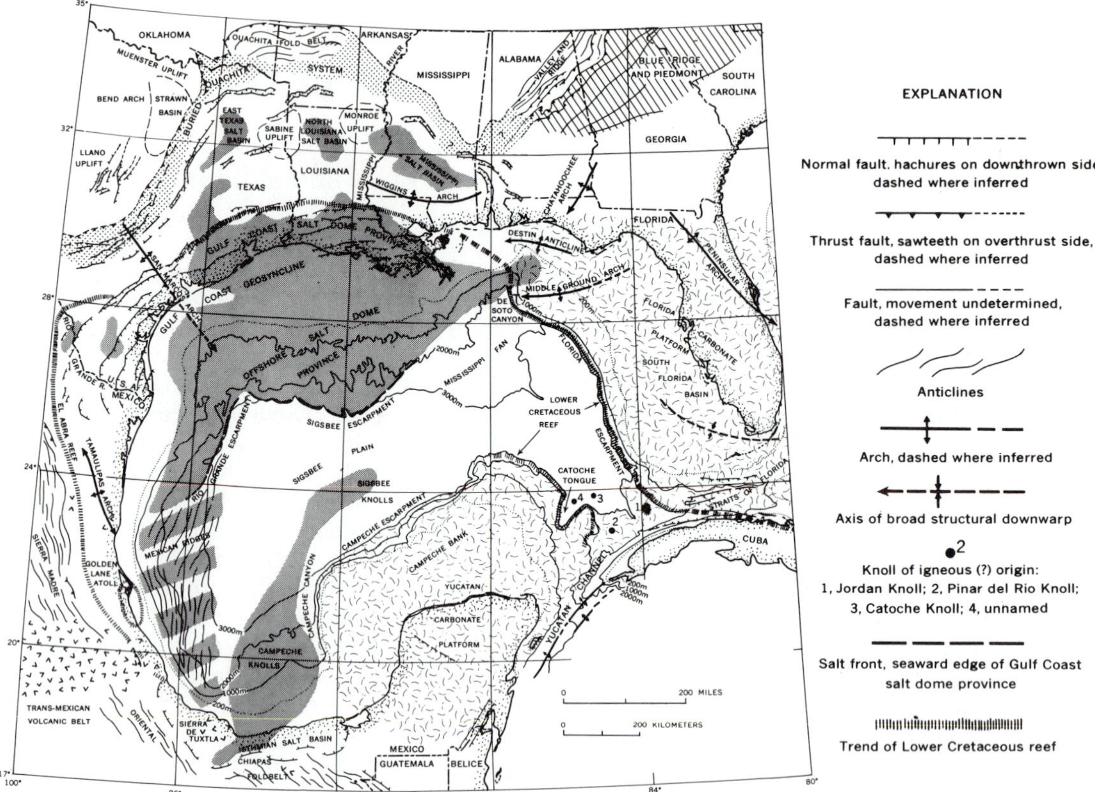

Fig. 1. Tectonic sketch map showing major structural trends and physiographic features in the Gulf of Mexico region. Distribution of salt structures shown by shaded areas.

mated to be buried at depths of approximately 20,000-36,000 ft (6-11 km) below sea level (Gough, 1967; Pyle and Antoine, 1973, p. 109). Fragmentary evidence suggests that the source of some of the magnetic anomalies may be the 6.8 km/sec crustal layer observed on a few marine seismic-refraction profiles (Antoine and Ewing, 1963; Sheridan et al., 1966; Pyle and Antoine, 1973). The age of these presumed volcanic rocks is unknown.

Beneath the Florida peninsula, wells have penetrated unaltered basaltic rocks of Mesozoic age (Milton and Grasty, 1969) at much shallower depths (11,000-13,000 ft; 3.4-4 km). Older igneous rocks penetrated by wells in east-central Florida are generally altered or metamorphosed (Bass, 1969; Milton and Grasty, 1969). In Yucatan, late Cretaceous andesite has been reported in two wells and a preliminary early to middle Paleozoic radiometric date has been obtained from metamorphosed acidic volcanic rocks in another well (Dengo, 1968; Bass and Zartman, 1969).

The majority of rocks sampled by deep drilling are of types unlikely to produce significant magnetic expression at the surface. This tends to support Gough's (1967) suggestion that the anomalies of the eastern Gulf of Mexico are related to Appalachian structural trends and originate deeper in the Paleozoic or Precambrian basement. However, the basalts more recently described by Bass and by Milton and Grasty from wells in central Florida are probably shallow enough and sufficiently magnetic to be the source of surface magnetic expression. Therefore, it is concluded that the eastern Gulf of Mexico anomalies may be, at least in part, due to Mesozoic volcanism.

The presence of pre-Lower Cretaceous igneous rocks in the western approaches of the Straits of Florida is suggested by magnetic and gravity profiles over Jordan and Pinar del Rio knolls. These same two knolls are adjacent to a "disturbed belt" delineated by the buried southwestern edge of the Florida platform, the projection of the northeast Campeche Escarpment trend and the northern margin of Cuba. The time of formation of this disturbed belt is unknown, but tentative correlation of reflection profiles with the stratigraphy of a nearby drilling site suggests that it was post-Early Cretaceous. Depths to what may be the same reflector on either side of the disturbed belt are greater on the east than on the west. This may indicate a westward extension of the vertical offset previously noted in the Lower Cretaceous reef trend. Bryant et al. (1969) reported that the Jordan Knoll area deepened steadily from Cenomanian through Santonian times and that since Santonian (Late Cretaceous), bathyal depths prevailed. This leads to the conclusion that the post-Early Cretaceous deformation that produced the disturbed belt and faulting of the southern margin of the Straits of Florida occurred during the early phases of the late Cretaceous-middle Eocene "Laramide" orogeny in Cuba.

Northward along the Florida Escarpment this activity appears to have lead to large-scale erosion or downfaulting of the platform margin. The irregularity of the scarp south of 27°N, the existence of the Florida canyon system, evidence of slump blocks in the seismic records, and the general absence of a Lower Cretaceous marginal reef south of 27°N where the stratigraphy of south Florida suggests that it once existed, all support this concept but do not pinpoint the mechanism responsible. In addition, recent studies by others indicate that erosion and faulting of carbonate platform margins in this region has been widespread.

Wilhelm and Ewing (1972) stated that "remarkable rockslides" from the Campeche Escarpment into the deep Gulf of Mexico basin were observed on several of their reflection profiles. The age of these rockslides could only be determined as pre-Pleistocene. The same authors speculated that terraces on the margins of the Yucatan and Florida platforms are due to erosion by submarine currents. For Florida at least, this seems an unlikely explanation of the origin of the terraces because of the highly indurated nature of the reef rocks that would have had to be eroded.

Ewing et al. (1966) suggested that the Blake "nose" at the northern edge of the reef-capped Blake Escarpment may be "...a very large slumped block." The time of slumping was not given, but the authors did suggest a correlation of layers which indicates that "reflector 1" is offset by the slumping. This reflector was described as "...a zone of reflecting surfaces between middle and basal Eocene" (p. 1951) and as "...a horizon at near the top of the Paleocene" (p. 1960). Either of these correlations suggests a synchroneity of the slumping with the later part of the "Laramide" orogeny in Cuba.

Sheridan et al. (1971) dredged Upper Cretaceous limestone blocks from the Blake-Bahama escarpment in the vicinity of Great Abaco Canyon. They described these blocks as having well developed slickensides and shear planes indicative of faulting. Although the occurrence of these limestones was attributed to transform faulting by Sheridan et al., the coincidence of these samples with Great Abaco Canyon might be explained by downfaulting of the platform margin in this conspicuous gap along the reef front.

On the basis of drilling results in the Northeast Providence Channel adjacent to Great Bahamas Bank, Paulus (1972) suggests that "...slumping could have been an active erosional process in the Late Cretaceous as it was in the Tertiary...." The deep-water drilling recovered fragments of shallow-water "perireef" limestone that may have undergone subaerial diagenesis (Purdy, in Paulus, 1972, p. 884-885).

Results from DSDP Leg X of the Deep Sea Drilling Project (Worzel et al., 1970) reveal a hiatus between Cenomanian and Eocene strata which may possibly reflect "Laramide" orogenic events in Cuba. Drilling of site 97 in the western approach to the Straits of Florida confirmed results from Jordan Knoll (Bryant et al., 1969) indicating that the area was deep in Cenomanian time. Shallow-water limestone rubble of late Albian-early Cenomanian age within the Cenomanian pelagic sediments suggests significant erosion of a nearby platform margin, perhaps subaerially. Drilling on the Campeche Escarpment recovered shallow-water carbonates indicative of a low-energy, interior platform environment. The lack of reef or other higher-energy bank edge deposits along this scarp might be the result of faulting or massive slumping, i.e., the "remarkable rockslides" of Wilhelm and Ewing (1972).

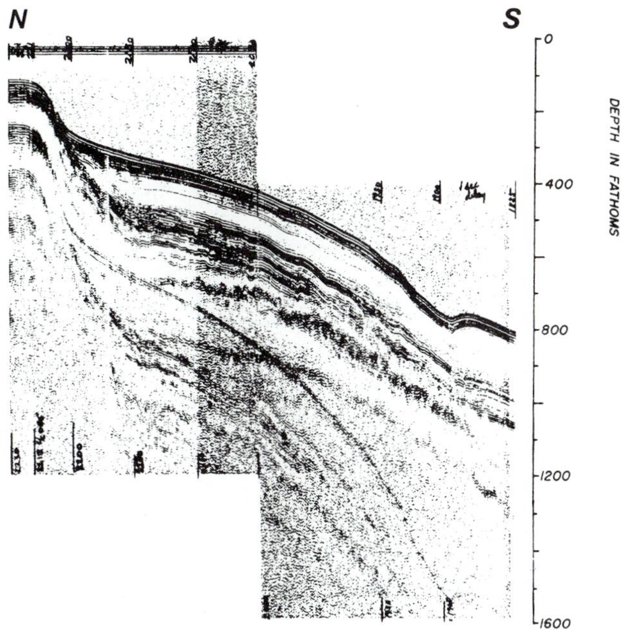

Fig. 2. Texas A&M University profile F-178 across the northern slope of the Straits of Florida along long 82°20'W (from Pyle and Antoine, 1973).

Thus a convergence of evidence from various sources supports the reasonableness of one of the main conclusions of this study—that post-Early Cretaceous erosion or faulting has strongly modified the topography and structure of the western margin of the Florida platform.

The evidence for subaerial diagenesis of Mesozoic bank limestones found in DSDP sites in the Bahamas and Straits of Florida, as well as the improbability of massive submarine erosion of lithified reef carbonates, suggests that if erosion was involved it occurred after uplift and subaerial exposure of the shallow-water deposits of the outer continental shelf. If further study supports the limited evidence cited above, for a late Mesozoic or younger age of the volcanic anomaly belt, the volcanism and uplift may be related. The approximate coincidence of the portion of the Florida Escarpment where the reef is generally absent, with the limits of the volcanic anomaly belt, suggests at least an indirect relationship.

If faulting is partially or completely responsible

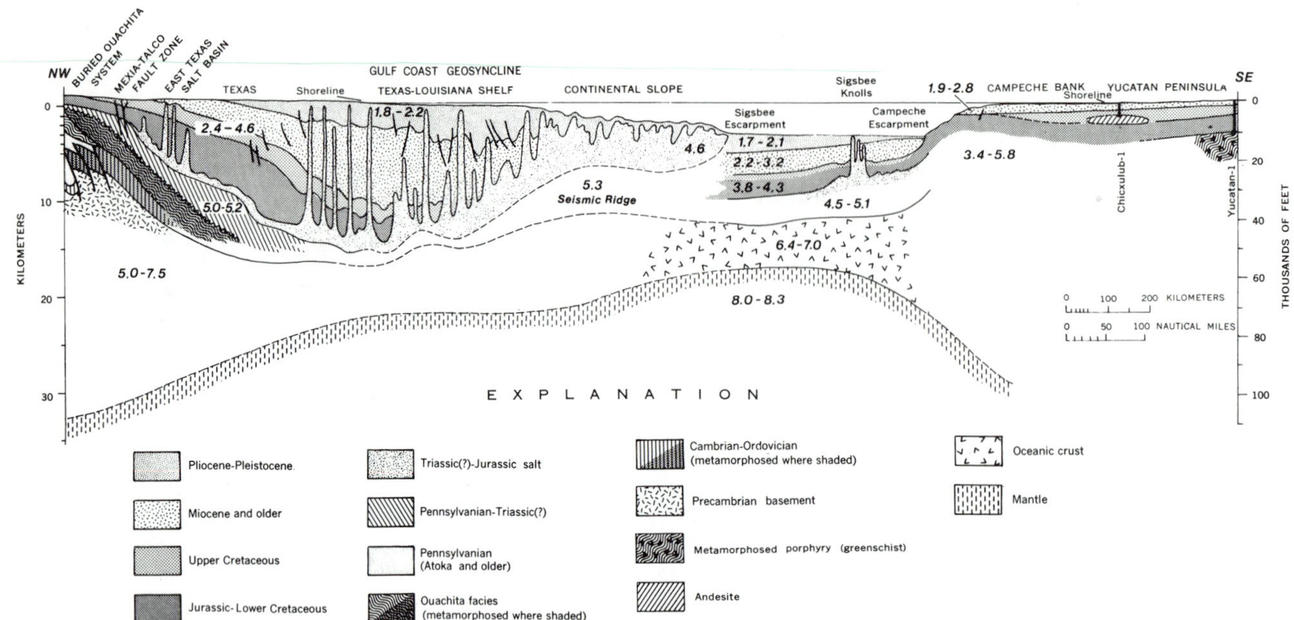

Fig. 3. Cross section of Gulf of Mexico from eastern Texas to central Yucatan showing generalized stratigraphy and crustal structure. Velocities in km/sec. (Adapted from numerous sources.)

for removal of the Lower Cretaceous marginal reef, it probably took the form of large slumps and rockslides. Downfaulting of solid masses of the platform margin seems less likely than slumping of oversteepened slopes produced by the uplift and subsequent accumulation of erosional debris. A gradation can exist from erosion to slumping to faulting, depending upon scale. Therefore, a clear recognition of the process or processes responsible for modification of the structure of the Florida Escarpment is not possible with our present, generalized knowledge of the region.

Yucatan Platform

The Yucatan platform is a broad region of peninsular lowland and shallow shelf, fronted in the Gulf Basin by steep slopes and underlain by a thick section of shallow-water carbonate and evaporite deposits. The geologic framework of its offshore region, Banco de Campeche, has been inferred from a limited number of wells on the Yucatan Peninsula and outer bank and from seismic-refraction profiles on the shelf (Ewing et al., 1960; Antoine and Ewing, 1963; Paine and Meyerhoff, 1970; Worzel et al., 1973). Dredge hauls and seismic-reflection profiles across the escarpments that border the platform have revealed evidence of early Cretaceous reefing along the northeast face (Fig. 2) and a history of faulting, slumping, and erosion on the northwest face (Bryant et al., 1969; Wilhelm and Ewing, 1972; Garrison and Martin, 1973). The late Mesozoic-Cenozoic history of the platform appears to be analogous to that of the Florida platform.

Scattered well data across the Yucatan Peninsula indicate that the platform is composed of a thick section of Cretaceous and Tertiary carbonate and evaporite beds (mainly anhydrite which accumulated over carbonates), evaporites, and redbeds of Jurassic age (Paine and Meyerhoff, 1970; Wilhelm and Ewing, 1972). Mesozoic and Cenozoic units thicken northwestward across the platform from structurally high basement composed of metamorphosed silicic volcanic rocks of early-middle Paleozoic age (Fig. 3; Paine and Meyerhoff, 1970; Dengo, 1968; Bass and Zartman, 1969). In northwestern Yucatan wells have penetrated an extensive andesite sill or flow overlain by a sequence of Upper Cretaceous carbonate and clastic rocks and underlain by limestone and interbedded anhydrite of early to middle Cretaceous age (Bass and Zartman, 1969; Paine and Meyerhoff, 1970; Viniegra and Lopez-Ramos, in press).

Marine geophysical investigations over Banco de Campeche indicate that the submerged part of the platform is composed of velocity layers that suggest an offshore continuity with Cretaceous and Tertiary units onshore and a Mesozoic-Cenozoic depositional history basically similar to the Florida platform (Antoine and Ewing, 1963). The nature of the deep-seated structure of the bank is not well known. Bouguer anomalies over the peninsula and Banco de Campeche (Krivoy et al., in press) suggest continental crustal thickness over much of the northern and western part of the region, but considerably thinner crust over the easternmost margin (Martin and Case, in press). Large positive magnetic anomalies over the bank may indicate the presence of volcanic masses in the foundation of the platform (Miller and Ewing, 1956; Gough, 1967), or structural relief on deeply buried magnetic basement rocks (Heirtzler et al., 1966; Martin and Case, in press).

Seismic-reflection profiles across the northeastern face of Campeche Escarpment have revealed evidence of a reef of early Cretaceous age extending from south of Catoche Tongue northwestward to near 25°N—87°W (Ewing and Ewing, 1966; Uchupi and Emery, 1968; Bryant et al., 1969; Garrison and Martin, 1973). Dredge samples from the escarpment in the vicinity of Catoche Tongue contained forereef or forebank fauna of Albian or late Aptian age (Bryant et al., 1969). The distribution of evaporite beds within the sedimentary section on Yucatan strongly implies that a barrier rimmed the platform in early to middle Cretaceous time, but evidence of an extension of the eastern reef trend to the western sectors of the bank is lacking. The northwest face of the platform has been strongly scarred by erosion, slumping, and possibly by faulting. Canyons incised to depths of more than 1 km and massive blocks slumped into the abyssal plain characterize the escarpment from west of 91°W to the Campeche Canyon (Wilhelm and Ewing, 1972; Garrison and Martin, 1973), and may explain the apparent terminus of the Campeche reef trend at the northeast corner of the bank.

DSDP sites 86, 94, and 95 along the top of the Campeche Escarpment have revealed Lower Cretaceous (Albian) shallow-water sediments at depths of 1,500-1,800 m in unconformable contact with deep-water deposits of Late Cretaceous to Paleocene age (Worzel et al., 1973). Faunal assemblages in the overlying Tertiary to Quaternary deposits indicate little change in the depositional environment of the outer Campeche Bank since the Maestrichtian (Worzel et al., 1973). These data, together with the rather consistent occurrence of the Lower Cretaceous reef top at 1,800-2,000 m water depth on seismic-reflection profiles across the eastern scarp, suggest a relatively abrupt and uniform foundering of the platform at the close of Early Cretaceous time followed by a post-Laramide history of gradual subsidence.

TRANSITION ZONES

Northeastern Gulf of Mexico

The boundary between the Florida carbonate platform and the terrigenous clastic continental margin of the northern Gulf of Mexico approximates the northern extension of the Florida Escarpment and associated Lower Cretaceous reef trend (Fig. 1). There are actually two sedimentary transition zones present between these geologically distinct regions, the first represented by the abrupt change from Mesozoic carbonate rocks to Cenozoic clastic rocks, east to west, across the escarpment. This is the effect of the prograding sediments of the clastic embankment, filling the old ocean basin of the Gulf of Mexico and abutting the edge of the adjacent carbonate platform (Martin, 1972).

The other transition is a more subtle, west to east, clastic to carbonate change that affects the Tertiary sediments. The offshore part of this zone is roughly outlined by the shoreline on the north, the northern extension of the Florida Escarpment on the southwest, and the Middle Ground arch on the southeast (Fig. 1). Within this area there is a steady increase in the seismic velocities characteristic of the Tertiary sediments, mainly clastic, from the western edge of the platform toward the east, where the entire Tertiary section is carbonate (Antoine and Harding, 1963a).

The major basement features of this zone are the Destin anticline and the Middle Ground arch (Fig. 1). The arch is the structural feature that separates the clastic zone of the northern Gulf from the carbonate platform and is known from magnetic measurements (Heirtzler et al., 1966), seismic-refraction studies (Antoine and Harding, 1965), and reflection profiling (Antoine, 1972; Martin, 1972). It has been suggested by Antoine (1972) that this is the structure that limited the eastern distribution of Mesozoic salt in the northern Gulf Coast and in this sense it represents the eastern terminus of the Gulf Coast geosyncline.

The evidence that salt is present beneath the carbonate platform from the Middle Ground arch toward the west is from seismic-reflection measurements (Harbison, 1968), which indicate the presence of numerous piercement salt domes at the head of DeSoto Canyon (Fig. 1). A possible explanation for the restricted local distribution of these domes is that although buried salt is present west of the Middle Ground arch, the thick sequence of carbonate rocks overlying the salt limits vertical migration except in the area of DeSoto Canyon, where erosion and nondeposition have restricted the carbonate thicknesses. Seismic-reflection data indicate that erosional processes have been in operation in portions of DeSoto Canyon during parts of all geologic ages between Late Cretaceous and the present.

It is suggested that the early erosion, and/or nondeposition, is related to the formation of the Florida and Yucatan straits during the Laramide orogeny in Cuba and the subsequent initiation of the early Gulf Stream. The path of this early stream would have been through the Suwannee Strait (Hull, 1962), over Florida and Georgia, and across the southern extension of the Cape Fear Arch into the Atlantic (Ewing et al., 1966). The latest erosional or nondepositional cycles are apparently related to the great gyre (eastern loop current) that is presently active (Nowlin, 1971).

Southwestern Gulf of Mexico

The Campeche Canyon and Escarpment and their southern extensions mark the boundary between the carbonates of the Yucatan Platform and the clastics toward the west. The main difference between this clastic-carbonate transition zone and the one in the northeastern gulf is that this lineament represents a distinct boundary for both Tertiary and older sediments; i.e., there is no wide zone of gradual transition included along the western edge of the bank. This is particularly clear on the submerged part of the platform (Banco de Campeche), where recent detrital carbonate rocks blanket the western platform margin.

In three major respects these northeastern and southwestern transition zones are similar: (1) the boundary is the edge of a large carbonate platform adjacent to a partially filled terrigenous clastic basin, (2) the adjacent clastic zone is structurally dominated by salt tectonics, and (3) there is evidence that the salt of the clastic zone extends beneath the carbonate platform (Contreras and Castillón, 1968).

CLASTIC EMBANKMENT

The continental margin of the northern and western Gulf of Mexico is composed of a huge prism of clastic sediments deposited in off-lapping wedges seaward from Mesozoic carbonate banks during Cenozoic time. This terrigenous province of the Gulf Basin includes the coastal plains, shelves, and slopes around the perimeter of the western Gulf from DeSoto Canyon in the north to Campeche Canyon in the south (Fig. 1). The principal basins of deposition in the province are the Gulf Coast geosyncline in the north, the Tampico and Veracruz embayments in the west, and the Isthmian Embayment in southern Mexico.

Large volumes of salt and perhaps low-density, overpressured shale beneath the margins and abyssal plain have played an important role in the structural and morphological development of the basin (Fig. 1). In the northern Gulf, salt masses and shale ridges pierce and uplift the sedimentary prism from DeSoto Canyon to northern Mexico and terminate abruptly at the foot of the continental slope along the Sigsbee and Rio Grande escarpments. Diapiric structures extend from the coastal plain of Tabasco Province in southern Mexico northward to the Sigsbee Knolls in the abyssal plain of the central Gulf Basin and form the steep, hummocky topography of the continental slope in the Golfo de Campeche. These diapiric provinces are separated from one another along the western Gulf margin by a system of almost symmetrically folded sediments expressed on the sea floor by ridge-and-valley topography, which reflects the mobility of either salt or shale masses at depth.

Northern Gulf of Mexico

The terrigenous margin of the northern Gulf of Mexico extends from DeSoto Canyon westward into northeastern Mexico and from more than 300 km inland in the central Gulf Coast to the foot of the continental slope. The primary structural element within the region is the Gulf Coast geosyncline, whose structure and history have been described by Bornhauser (1958), Williamson (1959), Murray (1961), Hardin (1962), Rainwater (1967), Lehner (1969), and many others. The axis of the geosyncline is just south of the present shoreline and extends westward from the northeastern Gulf to northern Mexico (Fig. 1). The landward limit of the geosyncline is considered to be the updip edge of basal Tertiary deposits in the inner coastal plain, and its southern limit probably is near the Sigsbee Escarpment, where a geanticlinal ridge of high-velocity material separates low-velocity sediments of the geosyncline from those in the Sigsbee Plain (Fig. 3; Martin and Case, in press; Ewing et al., 1960; Lehner, 1969).

The Gulf Coast geosyncline is principally a Cenozoic basin where a great thickness of clastic sediments has accumulated since late in the Cretaceous (Moody, 1967). The stratigraphic sequence in the geosyncline (Fig. 3) consists of thick transgressive and regressive sections of Tertiary-Quaternary clastic sediments deposited in off-lapping wedges over mainly carbonate beds of Cretaceous age. Dark-colored paralic sediments, redbeds, and evaporites range in age from Late Pennsylvanian to Jurassic and lie beneath the Cretaceous section (Lehner, 1969; Vernon, 1971). Along the inner margin of the coastal plain these units overlie truncated rocks of the Ouachita foldbelt that rims the Gulf Basin from Alabama to the Rio Grande River (Flawn et al., 1961).

In early Mesozoic time, the northern rim of the Gulf Basin was characterized by shallow evaporite basins, carbonate shelves, and a slow rate of subsidence. After the Laramide orogeny, increased subsidence in the geosyncline and the rapid influx of clastic materials from the north and northwest have steadily prograded the shelf edge as much as 400 km into the Gulf basin (Lehner, 1969; Woodbury et al., 1973). In general, the beds dip and thicken southward and are greatly disrupted by diapiric structures and by flexures and contemporaneous faults of regional extent. Regional Gulf Coast flexures may represent zones of increased basinward dip and thickening of sediments (Colle et al., 1952) and each is generally located basinward of the preceding older one, indicating a progressive southerly migration of the geosyncline through time (Antoine and Pyle, 1970). An eastward migration of sediment depocenters within the geosyncline from south Texas to Louisiana occurred throughout the Tertiary in response to the changing influence of river systems delivering sediments to the basin (Hardin, 1962).

The structural complexity of the northern Gulf margin has resulted from the mobility of salt, at least partly of Middle to Late Jurassic age (Kirkland and Gerhard, 1971), which has pierced and uplifted the younger sedimentary section from DeSoto Canyon to northeastern Mexico and from the inner coastal plain to the Sigsbee Escarpment (Fig. 1). In the southwestern region of the geosyncline under the coastal plain and shelf, where salt diapirism is not dominant, large uplifts and complex fault systems in the Tertiary section have resulted from the mobility of linear shale masses characterized by low bulk density and high fluid pressure (Bruce, 1973).

Salt domes beneath the Texas-Louisiana shelf and slope may be grouped into morphological belts (Woodbury et al., 1973; Garrison and Martin, 1973; Martin, 1973). Small, isolated salt spines that rise several tens of thousands of meters above their bases dot the inner shelf and adjacent coastal plain. Many large isolated salt stocks interconnected by intricate networks of growth faults characterize the middle shelf and lower Mississippi River delta region. Broad, semicontinuous diapiric uplifts associated with plastic shale masses dominate the outer shelf in the central part of the region. Virtually the entire continental slope from the Mississippi Fan to northeastern Mexico is underlain by massive salt structures that interconnect at relatively shallow subbottom depths. These structures tend to be in the form of broad stocks and swells whose diameters range from 10 to 15 km and whose crests rise only a few thousand feet above their bases.

Salt structures on the upper slope off east Texas and Louisiana are small plug-like masses that project above the semicontinuous uplifts similar to those noted by Woodbury et al. (1973) under the outer shelf. Salt structures on the middle slope region appear as very broad, flat-topped, steep-flanked massifs that often have coalesced to surround deep topographic depressions containing thick sections of bedded sediment. In marked contrast, salt masses under the lower slope occur as large pillow-like swells with broad sedimentary basins perched in shallow depressions on the salt mass (Lehner, 1969; Garrison and Martin, 1973; Martin, 1973).

Off southernmost Texas and northeastern Mexico diapiric structures result from the mobility of undercompacted shale (Bruce, 1973). Broad anticlinal masses of salt trend northeasterly beneath the adjacent slope. The size and relief of these structures decrease abruptly toward the northwest corner of the Gulf margin, where an extension of the San Marcos arch may have affected the distribution of evaporite basins and subsequent diapiric growth (Garrison and Berryhill, 1970). Along the Rio Grande Escarpment, salt structures plunge steeply beneath the Sigsbee Plain, where folded sediments as young as Pleistocene indicate continued growth of the mobile salt mass.

Salt structures of the Texas-Louisiana slope province terminate abruptly at the steep salt front under the Sigsbee Escarpment from near 27°N on the east to the southward bend in the continental margin on the west (Fig. 1). Seismic-reflection profiles across the Sigsbee Scarp have shown it to be underlain by a wall of salt along most of its length (Ewing and Antoine, 1966; Uchupi and Emery, 1968; Lehner, 1969; Wilhelm and Ewing, 1972; Garrison and Martin, 1973). In the salient where the scarp trend lies south of 26°N, de Jong (1968) and Amery (1969) have reported evidence of salt extrusion onto sedimentary deposits that appear to correlate with relatively young beds of the continental rise. The vertical or near-vertical attitude of the salt wall seen elsewhere along the Sigsbee Scarp in reflection profiles may be misleading. In the area of observed overflow the salt mass is only several hundred meters thick (Amery, 1969). It may be that the salt mass elsewhere along the scarp also overflows abyssal beds, but that the distance of overflow is small and that the thickness of the salt is great enough to mask acoustically any evidence of underlying beds (Martin, 1973).

Ewing and Antoine (1966) and Wilhelm and Ewing (1972) consider that the northern Gulf margin evolved under the control of a salt-flow process, whereby the accumulation of sediment updip has created a horizontal pressure gradient which, in turn, has caused the underlying salt to migrate seaward. In a regional sense, the front of the mobile salt layer is thought to form a ridge or a system of salt structures behind which subsequent sedimentary deposits accumulate and induce further growth of the system (Fig. 4).

Lehner (1969) has discussed a similar mechanism for the growth of individual domes and systems of diapirs on the slope. The groupings of salt structures according to relief, size, and shape described by Woodbury et al. (1973) and Martin (1973) for the Texas-Louisiana coastal plain, shelf, and slope define morphological belts of decreasing diapiric maturity from the Gulf coastal plain to the foot of the continental slope. The Sigsbee Escarpment south of Texas and Louisiana and the Rio Grande Escarpment to the west are the least mature features in the cycle of salt tectonism in the northern gulf and

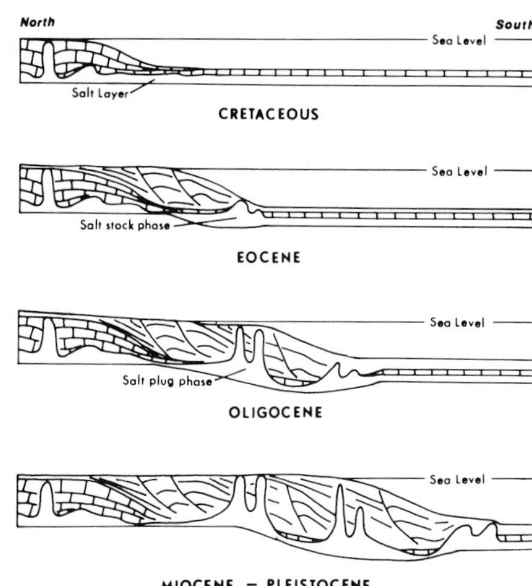

Fig. 4. Schematic diagram showing hypothetical evolution of Gulf Coast geosyncline as related to lateral and vertical displacement of salt under progressive loading (from Wilhelm and Ewing, 1972; reprinted by permission of Geological Society of America).

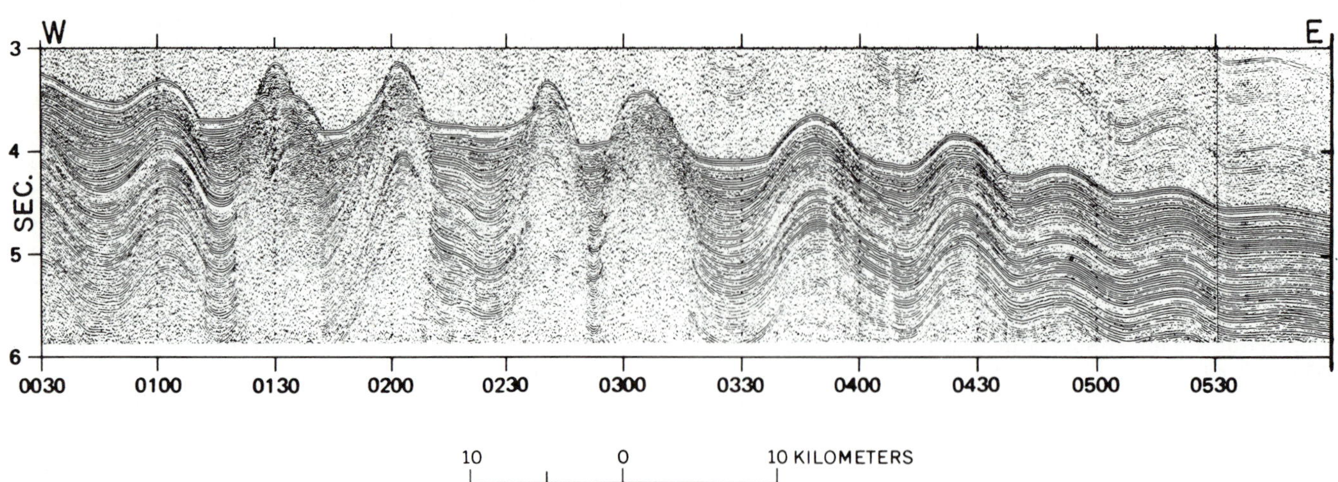

Fig. 5. Seismic reflection profile across central region of the Mexican Ridges near 23°N (from Garrison and Martin, 1973).

represent the fronts of an advancing salt "wave" that is presumably responding to the load of sediments that have accumulated in the Gulf Coast geosyncline. The presence of large volumes of Upper Triassic (?) to Upper Jurassic (?) salt that subsequently imparted mobility to the geosyncline, the large volumes of sediment supplied to the Gulf Coast region after the Laramide orogeny, and subsidence of the geosyncline in pace with deposition have been the principal factors controlling the evolution of the northern Gulf of Mexico margin.

WESTERN GULF OF MEXICO

The narrow continental margin of the western Gulf of Mexico is characterized by a series of gentle folds that generally parallel the shoreline. These folds form a ridge system that extends between 24 and 19°N. Figure 1 illustrates the strike of the individual ridge segments based on available topographic and seismic surveys. The ridges are bordered on the west by the Tamaulipas carbonate platform, which is of the same age as the platforms of Florida and Campeche. The El Abra reef trend forms the western border of the carbonate platform and is of Early Cretaceous age, as are the reefs that border the Florida and Campeche banks. These reefs of the eastern and western carbonate regions are apparently connected by reef trends across the northern Gulf coastal plain, as shown in Figure 1. The southern extent of the Tamaulipas Platform is cut by the east-trending trans-Mexican volcanic belt, which ends near the shoreline or turns toward the south (Fig. 1). In any case, it does not impinge on the southern extent of the ridge system. This entire margin complex of ridges, platform, and reef is flanked on the west by the Laramide Sierra Madre Oriental mountain belt of mainland Mexico.

A striking feature of the folds forming the ridge system is the persistence of conformable bedding across the features (Fig. 5). Garrison and Martin (1973) have shown at least 3 km of bedded sediment to be involved in the folding. The folds are observed as topographic features on the outer continental slope but are buried by sediment on the shelf and inner slope. The ridges have acted as a barrier to gulfward migrating sediments from continental Mexico, i.e., sediments collect between the exposed ridges nearest shore until the valley is filled and sediments spill toward the east into the next depression. Relief across the individual ridges is generally 400-500 m.

Toward the north the most easterly folds trend toward and abut the Rio Grande Escarpment, while the shallow water structures change from simple linear folds to individual diapiric features. This change is reasoned to be related to an increasing sedimentary overburden toward the north, which has activated secondary vertical movement of the salt.

A change in trend of the ridge system occurs near 22°N, marked by an 1,800-m scarp (Bryant et al., 1968) and an area of low-amplitude folds (Garrison and Martin, 1973). It has been suggested by Bryant et al. (1968) that this may reflect the offshore extension of a transcurrent fault (Zaetaros Fracture Zone) which is reported to extend from Tampico to Baja California (Murray, 1961).

Bryant et al. (1969) suggested that the origins for the development of the ridges were one of four possibilities: (1) sliding of sediment or rocks on a décollement surface, including gravity sliding; (2) folding associated with compressional tectonic stresses; (3) vertical movement of shale or salt masses related to static loading; and (4) folds controlled by faulting. They favored the idea that the ridges are related to an underlying layer of salt whose initial deformation under sediment loading created a series of salt anticlines and a subsequent increase in sediment overburden and upward migration of secondary salt structures evolved into anticlinal ridges. The evolution of salt structures into anticlinal ridges gains support from the work of

Selig and Wermund (1966), who demonstrated that salt ridges can develop from faulting and may also form as a result of inhomogeneous loading on a depositional slope.

Garrison and Martin (1973) suggested from the foregoing hypothesis that the ridges are the result of salt tectonics is plausible, but suggested that the evidence for a salt layer was sketchy. However, Moore and Del Castillo (1974) found that simple Bouguer gravity analyses, when considered at the wavelength of the folds, indicated that relatively low-density material lies near the sea floor on the ridges. Using a longer wavelength and therefore deeper analysis, their data also showed that the low-density material could result from a chance average from interlayered material of several densities, but it was also compatible with a thick layer of salt under the continental rise. From their analyses Moore and Del Castillo (1974) agree with the concept that the Mexican Ridge System is salt anticlines.

Garrison and Martin (1973) favor the gravity sliding concept for the ridge formation. They show that at least 3 km of bedded sediments are involved in the folding, with no indication of a salt basement to that depth, and that on many ridges the form of the fold is repeated internally by each individual reflector, with no indication that the fold is dying out upward as would be the case were the beds folded over the crest of a rising diapir or draped over a buried anticline.

Regardless of the origin of the Mexican Ridge System, the East Mexico continental slope represents one of the more unique continental slopes of the world.

GOLFO DE CAMPECHE

The physiographic and structural styles of the continental margin in the Golfo de Campeche (Fig. 1) are characterized by diapiric structures, which are in frequency and dimensions similar to those of the Texas-Louisiana continental slope. Diapirs that underlie the rugged topography of the slope are essentially continuous southward beneath thick shelf sediments into the Isthmian basin of southern Mexico (Fig. 1; Ballard and Feden, 1970), where salt has been drilled in similar structures (Murray, 1961). Northward, at the foot of the slope, the Campeche Knolls province narrows and trends toward the Sigsbee Knolls in the abyssal plain, but domal structures are not known to be continuous between the two areas (Garrison and Martin, 1973). The Campeche Knolls are separated on the east from the Yucatan Platform by the Campeche Canyon (Creager, 1958). To the west the knolls are bordered by a narrow trough of undisturbed abyssal sediments which separates the province from the Mexican Ridges.

From the occurrence of salt domes and bedded evaporites onshore in southern Mexico (Viniegra, 1971) and the recovery of caprock from one of the Sigsbee Knolls (Burk et al., 1969) it is generally assumed that the Campeche Knolls are cored with salt and that the Sigsbee-Campeche-Isthmian diapirs define a contiguous salt basin. Gravity models by Moore and Del Castillo (1974) support the assumption that salt underlies the Golfo de Campeche slope, and studies by Kirkland and Gerhard (1971) and Viniegra (1971) suggest that the salt is of Middle to Late Jurassic age.

The origin of the Sigsbee-Campeche-Isthmian salt is very difficult to explain in terms of either old or new oceanic basins. The salt is apparently continuous from onshore to the abyssal plain, over a general elevation change of some 4 km, its age is essentially identical at both ends, and it is confined to a relatively narrow belt which is peripheral to the western and northwestern edges of the Yucatan Platform. The apparent physical and temporal continuity of the basin from the deep Gulf of Mexico to onshore argues against deep-water salt deposition (Ewing et al., 1962; Schmaltz, 1969) and suggests that the salt was deposited in a shallow marginal basin, possibly restricted from open waters by a structural rim (Worzel et al., 1970) and later subsided differentially to its present levels. Antoine and Bryant (1969) have suggested that the salt basin extended south and east from the Sigsbee-Campeche Knolls province under a large part of the Yucatan Platform and that diapirism in the Sigsbee Knolls region was initiated by lateral flow away from the bank as the thickness of carbonate deposits increased and the salt attained mobility under the resulting overburden pressures.

BIBLIOGRAPHY

Amery, G. B., 1969, Structure of Sigsbee scarp, Gulf of Mexico: Am. Assoc. Petr. Geol. Bull., v. 53, p. 2480-2482.

Antoine, J. W., 1972, Structure of the Gulf of Mexico: *in* Rezak, R., and Henry, V. I., eds., Contributions on the geological and geophysical oceanography of the Gulf of Mexico: Houstin, Gulf Publishing Co.

———, and Bryant, W. R., 1969, Distribution of salt and salt structures in Gulf of Mexico: Am. Assoc. Petr. Geol. Bull., v. 53, no. 12, p. 2543-2550.

———, and Ewing, J. I., 1963a, Seismic refraction measurements on the margins of the Gulf of Mexico: Jour. Geophys. Res., v. 68, no. 7, p. 1975-1996.

———, and Harding, J. L., 1963b, Structure of the continental shelf, northeastern Gulf of Mexico: Texas A & M Univ. Tech. Rept. 63-13T, 18 p.

———, and Pyle, T. E., 1970, Crustal studies in the Gulf of Mexico: Tectonophysics, v. 10, no. 5-6, p. 477-494.

Ballard, J. A., and Feden, R. H., 1970, Diapiric structures on the Campeche shelf and slope, western Gulf of Mexico: Geol. Soc. America Bull., v. 81, no. 2, p. 505-512.

Bass, M. N., 1969, Petrology and ages of crystalline basement rocks of Florida—some extrapolations, *in* Other papers on Florida and British Honduras: Am. Assoc. Petr. Geol. Mem. 11, p. 283-310.

———, and Zartman, R. E., 1969, The basement of Yucatan Peninsula (abstr.): EOS (Trans. Am. Geophys. Union), v. 50, no. 7, p. 1204-1228.

———, and Zartman, R. E., 1969, The basement of Yucatan Peninsula (abstr.): EOS (Trans. Am. Geophys Union), v. 50, no. 4, p. 313.

Bornhauser, M., 1958, Gulf Coast tectonics: Am. Assoc. Petr. Geol. Bull., v. 42, no. 2, p. 339-370.

Bruce, C. H., 1973, Pressure shale and related sediment deformation: mechanism for development of regional contemporaneous faults: Am. Assoc. Petr. Geol. Bull., v. 57, p. 878-886.

Bryant, W. R., Antoine, J. W., Ewing, M., and Jones, B. R., 1968, Structure of Mexican continental shelf and slope, Gulf of Mexico: Am. Assoc. Petr. Geol. Bull., v. 52, no. 7, p. 1204-1228.

———, Meyerhoff, A. A., Brown, N. K., Jr., Furrer, M. A., Pyle, T. E., and Antoine, J. W., 1969, Escarpments, reef trends and diapiric structures, eastern Gulf of Mexico: Am. Assoc. Petr. Geol. Bull., v. 53, no. 12, p. 2506-2542.

Burk, C. A., Ewing, M., Worzel, J. L., Beall, A. O., Jr., Berggren, W. A., Bukry, D., Fischer, A. G., and Pessagno, E. A., Jr., 1969, Deep-sea drilling into the Challenger Knoll, central Gulf of Mexico: Am. Assoc. Petr. Geol. Bull., v. 53, no. 7, p. 1338-1347.

Colle, J., Cook, W. F., Penham, R. L., Ferguson, H. C., McGuirt, J. H., Reedy, F., and Weaver, P., 1952, Sedimentary volumes in Gulf Coastal plain of United States and Mexico, pt. IV, Volume of Mesozoic and Cenozoic sediments in Western Gulf Coastal Plain of United States: Geol. Soc. America Bull., v. 63, no. 12, p. 1193-1199.

Contreras, V. H., and Castillón, B., 1968, Morphology and origin of salt domes of Isthmus of Tehuantepec, *in* Braunstein, J., and O'Brien, G. D., eds., Diapirism and diapirs: Am. Assoc. Petr. Geol. Mem. 8, p. 244-260.

Creager, J. S., 1958, A canyon-like feature in the Bay of Campeche: Deep-Sea Res., v. 5, no. 2, p. 169-172.

deJong, A., 1968, Stratigraphy of the Sigsbee scarp from a reflection survey (abstr.): Soc. Expl. Geol., Fort Worth Meeting Program, p. 51.

Dengo, G., 1968, Estructura geologica, historia tectonica y morfologia de America Central: Mexico, Centro Regional de Ayuda Techica (AID), 50 p.

Ewing, J. I., Antoine, J. W., and Ewing, M., 1960, Geophysical measurements in the western Caribbean Sea and in the Gulf of Mexico: Jour. Geophys. Res., v. 65, no. 12, p. 4087-4126.

———, Worzel, J. L., and Ewing, M., 1962, Sediments and oceanic structural history of the Gulf of Mexico: Jour. Geophys. Res., v. 67, no. 6, p. 2509-2527.

———, Ewing, E., and Leyden, R., 1966, Seismic profiler survey of Blake Plateau: Am. Assoc. Petr. Geol. Bull., v. 50, p. 479-504.

Ewing, M., and Antoine, J. W., 1966, New seismic data concerning sediments and diapiric structures in Sigsbee Deep and continental slope, Gulf of Mexico: Am. Assoc. Petr. Geol. Bull., v. 50, no. 3, pt. 1, p. 479-504.

———, and Ewing, J. I., 1966, Geology of the Gulf of Mexico, *in* Exploiting the ocean: Marine tech. Soc., 2nd Ann. Conf. and Exhibit, 1966, Trans. Suppl.: Washingtin, D. C., Marine Tech. Soc., p. 145-164.

Flawn, P. T., Goldstein, A., Jr., King, P. B., and Weaver, C. E., 1961, The Ouachita system: Texas Univ. Publ. 6120, 401 p.

Garrison, L. E., and Berryhill, H. L., Jr., 1970, Possible seaward extension of the San Marcos Arch (abstr.): Geol. Soc. America, v. 2, no. 4, p. 285-286.

———, and Martin, R. G., 1973, Geologic structures in the Gulf of Mexico basin: U.S. Geol. Survey Prof. Paper 773, 85 p.

Gough, D. I., 1967, Magnetic anomalies and crustal structure in eastern Gulf of Mexico: Am. Assoc. Petr. Geol. Bull., v. 51, no. 2, p. 200-211.

Harbison, R. N., 1967, DeSoto Canyon reveals salt trends: Oil and Gas Jour., v. 65, no. 8, p. 124-128.

———, 1968, Geology of the DeSoto Canyon: Jour. Geophys. Res., v. 73, no. 16, p. 5175-5185.

Hardin, G. C., Jr., 1962, Notes on Cenozoic sedimentation in the Gulf Coast geosyncline, U.S.A., *in* Geology of the Gulf Coast and central Texas and guidebook of excursions: Geol. Soc. America, 1962 Ann. Meeting: Houston, Tex., Houston Geol. Soc., p. 1-15.

Heirtzler, J. R., Burckle, L. H., and Peter, G., 1966, Magnetic anomalies in the Gulf of Mexico: Jour. Geophys. Res., v. 71, p. 519-526.

Hull, J. P. D., Jr., 1962, Cretaceous Suwannee Strait, Georgia, and Florida: Am. Assoc. Petr. Geol. Bull., v. 46, p. 118-122.

Hurley, R. J., 1964, Bathymetry of the Straits of Florida and the Bahama Islands, pt. 3, Southern Straits of Florida: Bull. Marine Sci. Gulf and Caribbean, v. 14, no. 3, p. 373-380.

Kirkland, D. W., and Gerhard, J. E., 1971, Jurassic salt, central Gulf of Mexico, and its temporal relation to circum-Gulf evaporites: Am. Assoc. Petr. Geol. Bull., v. 55, no. 5, p. 680-686.

Krivoy, H. L., Pyle, T. E., and Eppert, H. C., Jr., (in press), Bouguer gravity anomaly map of Gulf of Mexico: U.S. Geol. Survey Prof. Paper.

Lehner, P., 1969, Salt tectonics and Pleistocene stratigraphy on continental slope of northern Gulf of Mexico: Am. Assoc. Petr. Geol. Bull., v. 53, no. 12, p. 2431-2479.

Martin, R. G., Jr., 1973, Structural features of the continental margin, northeastern Gulf of Mexico: U.S. Geol. Survey Prof. Paper 800-B, p. B1-B8.

———, and Case, J. E., in press, Geophysical studies in the Gulf of Mexico, *in* Stehli, F., and Nairn, A., eds., Ocean Basins and Margins, v. III, The Gulf of Mexico and Caribbean Sea: New York, Plenum.

Miller, E. T., and Ewing, M., 1956, Geomagnetic measurements in the Gulf of Mexico and in the vicinity of Caryn Peak (Atlantic Ocean): Geophysics, v. 21, no. 2, p. 406-432.

Milton, C., and Grasty, R., 1969, Basement rocks of Florida and Georgia: Am. Assoc. Petr. Geol. Bull., v. 53, p. 2483-2493.

Moody, C. L., 1967, Gulf of Mexico Distributive Province: Am. Assoc. Petr. Geol. Bull., v. 51, p. 179-199.

Moore, G. W., and Del Castillo, L., 1974, Tectonic evolution of the southern Gulf of Mexico: Geol. Soc. America Bull., v. 85, no. 4, p. 607-618.

Murray, G. E., 1961, Geology of the Atlantic and Gulf Coastal Province of North America: New York, Harper & Row, 692 p.

Nowlin, W. D., 1971, Water masses and general circulation of the Gulf of Mexico: Oceanology, v. 6, no. 2, p. 28-33.

Oglesby, W. R., 1965, Folio of south Florida basin—a preliminary study: Florida Geol. Survey Map Ser. 19, 3 p.

Paine, W. R., and Meyerhoff, A. A., 1970, Gulf of Mexico basin interactions among tectonics, sedimentation, and hydrocarbon accumulation: Trans. Gulf Coast Assoc. Geol. Soc., v. 20, p. 5-44.

Paulus, F. J., 1972, The geology of site 98 and the Bahama Platform, *in* Hollister, C. D., et al., eds., Initial reports of the Deep Sea Drilling Project, v. XI: Washington, D.C., U.S. Govt. Printing Office, p. 877-897.

Pyle, T. E., and Antoine, J. W., 1973, Structure of the West Florida Platform, Gulf of Mexico: Texas A & M Tech. Rept. 73-7-T, 168 p.

______, Bryant, W. R., and Antoine, J. W., 1968, Geophysical studies of the south Florida continental margin and western Straits of Florida (abstr.): Gulf Coast Assoc. Geol. Soc., v. 18, p. 50.

______, Antoine, J. W., Fahlquist, D. A., and Bryant, W. R., 1969, Magnetic anomalies in Straits of Florida: Am. Assoc. Petr. Geol. Bull., v. 53, no. 12, p. 2501-2505.

Rainwater, E. H., 1967, Resumé of Jurassic to Recent sedimentation history of the Gulf of Mexico basin: Trans. Gulf Coast Assoc. Geol. Soc., v. 17, p. 179-210.

Schmalz, R. F., 1969, Deep-water evaporite deposition—a genetic model: Am. Assoc. Petr. Geol. Bull., v. 53, no. 4, p. 798-823.

Selig, F., and Wermund, E. G., 1966, Families of salt domes in the Gulf Coastal province: Geophysics, v. 31, no. 4, p. 726-740.

Sheridan, R. E., Drake, C. L., Nafe, J. E., and Hennion, J., 1966, Seismic refraction study of continental margin of Florida: Am. Assoc. Petr. Geol. Bull., v. 50, p. 1972-1991.

______, Berman, R. M., and Corman, D. B., 1971, Faulted limestone block dredged from Blake Escarpment: Geol. Soc. America Bull., v. 82, p. 199-203.

Uchupi, E., and Emery, K. O., 1968, Structure of continental margin off Gulf Coast of United States: Am. Assoc. Petr. Geol. Bull., v. 52, no. 7, p. 1162-1193.

Vernon, R. C., 1971, Possible future petroleum potential of pre-Jurassic, western Gulf Basin, *in* Cram, I. H., ed., Future petroleum provinces of the United States—their geology and potential, v. 2: Am. Assoc. Petr. Geol. Mem. 15, p. 954-979.

Viniegra, O. F., 1971, Age and evolution of salt basins of southeastern Mexico: Am. Assoc. Petr. Geol. Bull., v. 55, no. 3, p. 478-494.

______, and Lopez-Ramos, J., (in press), Geology of Yucatan Peninsula, *in* Stehli, F., and Nairn, A., eds., Ocean basins and margins, v. III, The Gulf of Mexico and Caribbean Sea: New York, Plenum.

Wilhelm, O. and Ewing,, M., 1972, Geology and history of the Gulf of Mexico: Geol. Soc. America Bull., v. 83, no. 3, p. 575-600.

Williamson, J. D. M., 1959, Gulf Coast Cenozoic history: Trans. Gulf Coast Assoc. Geol. Soc., v. 9, p. 15-29.

Woodbury, H. O., Murray, I. B., Jr., Pickford, P. J., and Akers, W. H., 1973, Pliocene and Pleistocene depocenters, outer continental shelf Louisiana and Texas: Am. Assoc. Petr. Geol. Bull., v. 54, p. 2428-2439.

Worzel, J. L., et al., 1973, Deep Sea Drilling Project, Leg 10, preliminary report: Geotimes, v. 15, no. 6, p. 11-13.

Worzel, J. L., and Bryant, W. R., et al., 1973, Initial Report of the Deep Sea Drilling Project, v. S: Washington, D.C., U.S. Govt. Printing Office, 748p.

______, Leyden, R., and Ewing, M., 1968, Newly discovered diapirs in Gulf of Mexico: Am. Assoc. Petr. Geol. Bull., v. 52, no. 7, p. 1194-1203.

Geology of the Mediterranean Sea Basins

B. Biju-Duval, J. Letouzey, L. Montadert, P. Courrier, J.F. Mugniot, and J. Sancho

INTRODUCTION

The geological knowledge of the Mediterranean area has been advanced recently as the result of reinterpretation and synthesis of the considerable amount of land data and extensive geophysical surveys using deep-seismic-reflection methods. The present Mediterranean basin shows a certain unity related to the recent geological history. Several Mediterranean basins formed before the late Miocene, whose age, structure, and genesis are different:

1. Cenozoic basins (Western, Tyrrhenian, Aegean, North Cyprus basins) located in areas that were tectonized during the Mesozoic. These basins are superimposed on, or secants to, these Alpine folded belts. The genesis of some of them could be explained in an island-arc-system context, complicated by the existence of small rigid blocks between Europe and Africa.

2. Mesozoic-Cenozoic basins (eastern Mediterranean south of Sicily-Crete-Cyprus and Adriatic basins), in areas that have been slightly affected by Alpine folding. They could be the northern prolongation of the African crust.

Several features are emphasized: 1. The generality of the Pliocene-Quaternary foundering; 2. The importance of the recent deltas; 3. The widespread distribution of Upper Miocene evaporites and the location of salt basins; 4. The importance of large gravity nappes of different ages (since the Late Cretaceous) offshore: Cyprus Arc, Mediterranean Ridge, Messina Cone, West Gibraltar Arc, and their onshore prolongations.

REGIONAL FRAMEWORK

The Mediterranean covers nearly 2,000,000 km2 between the continents of Europe and Africa, connected in the west with the Atlantic via the Strait of Gibraltar and, in the east, with the Black Sea via the Dardanelles and the Bosphorus. A maximum depth of 5,093 m has been found by Hersey (1965) in the Hellenic Trough, the mean depth being approximately 1,500 m, but great differences exist, depending on the region. The sea is divided into several subbasins that are more or less separated from one another by thresholds, peninsulas, or islands (Fig. 1).

Geologically speaking, some of the Mediterranean basins continue onshore in the form of sedimentary basins, whose history and structure are now becoming well known. Others have their own individuality, and most information about them comes from marine geophysical data. Our understanding of the deep Mediterranean basins has continued to evolve rapidly since the latest synthesis articles (Ryan et al., 1971; Ryan et al., 1973).

Figure 1 shows that there are roughly two categories of basins, according to their location inside the Alpine belt:

1. Those contained in or secant to the belt, i.e., the Western Basin, the Tyrrhenian Basin, the Aegean Basin, the Antalya Basin, and the Adana Basin, all of which could be compared with other basins inside the same belt (Pannonian Basin, Central Turkish basins, Black Sea, South Caspian Sea).

2. Basins on the margin of this orogenesis, located between the stable African platform and the belt itself: i.e., the Ionian Basin, the Levantine Basin, and the Adriatic Basin.

This subdivision is rough, considering that the present physiography of these basins is the end result of a complex geological history (sedimentary and structural) and, to a great extent the result of quite recent evolution.

It is considered here that the Alpine arc is the result of the African and European plates drawing together, causing the disappearance of the ocean Tethys, which originally separated them. Reconstituting the movements of these plates, using data from the Atlantic Ocean, can be done starting from the beginning of the Mesozoic. Such reconstructions assume not only that the two plates were moving toward each other, but also that they were subject to lateral movements linked to the differential opening of the Atlantic on both sides of the Azores-Gibraltar fracture zone (Pittman and Talwani, 1972; Dewey et al., 1973). Geological data on the Alpine belt show that important compression phases occurred at intervals between the late Jurassic and the Quaternary (e.g., Aubouin et al, 1970a, 1970b; Blumenthal, 1963; Caire, 1970; Dercourt, 1970; Durand-Delga, 1960; Glangeaud, 1956; Klemme, 1958; Laubscher, 1971; Smith, 1971). The present Mediterranean basins, with the exception of part of the eastern Mediterranean, are recent ones when compared with this long and complex history (e.g., Bourcart, 1963; Glangeaud et al., 1967; Aubouin, 1971). Hence they can be studied without first reconstructing all the stages in the Mesozoic history.

In the late Cretaceous, north of the present eastern Mediterranean, compression movements led to the creation of Triassic, Jurassic, or Cretaceous

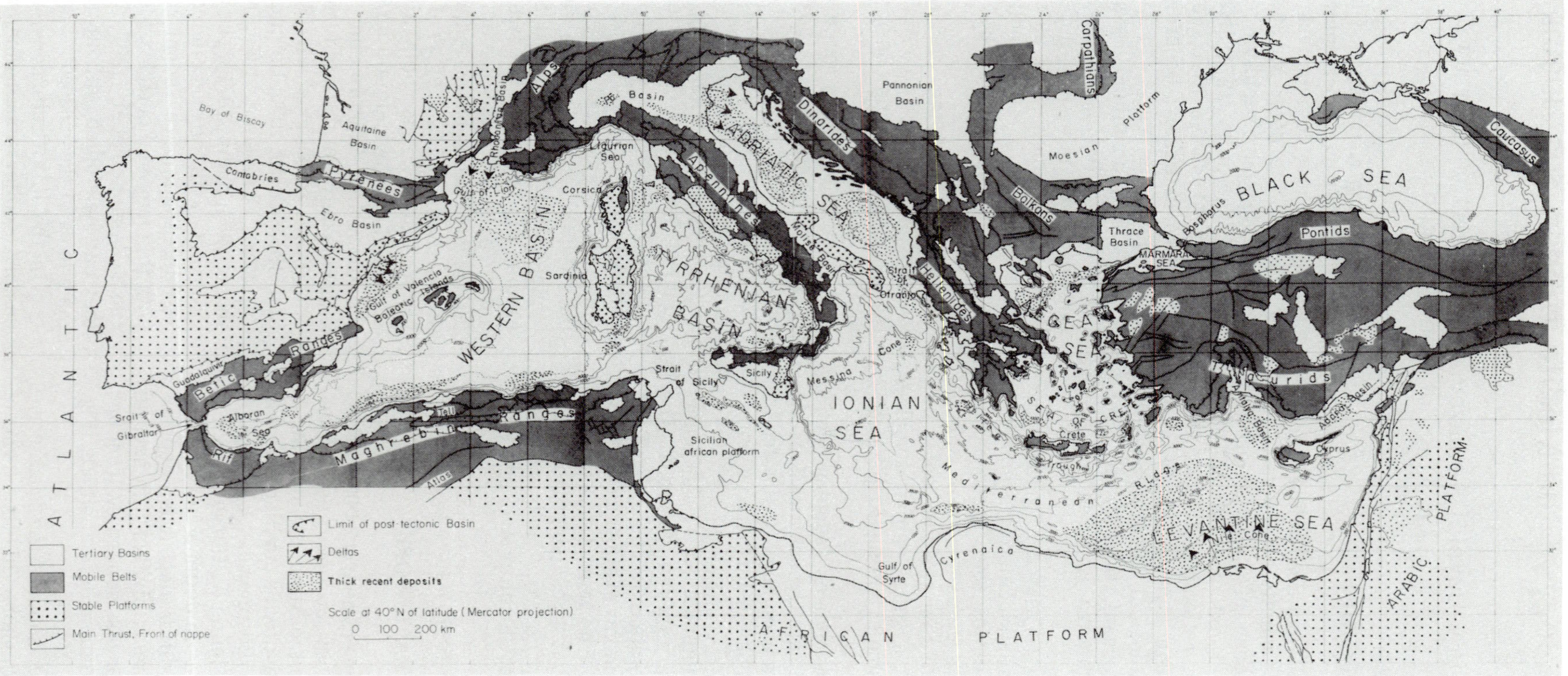

Fig. 1. Mediterranean area, showing major geological features.

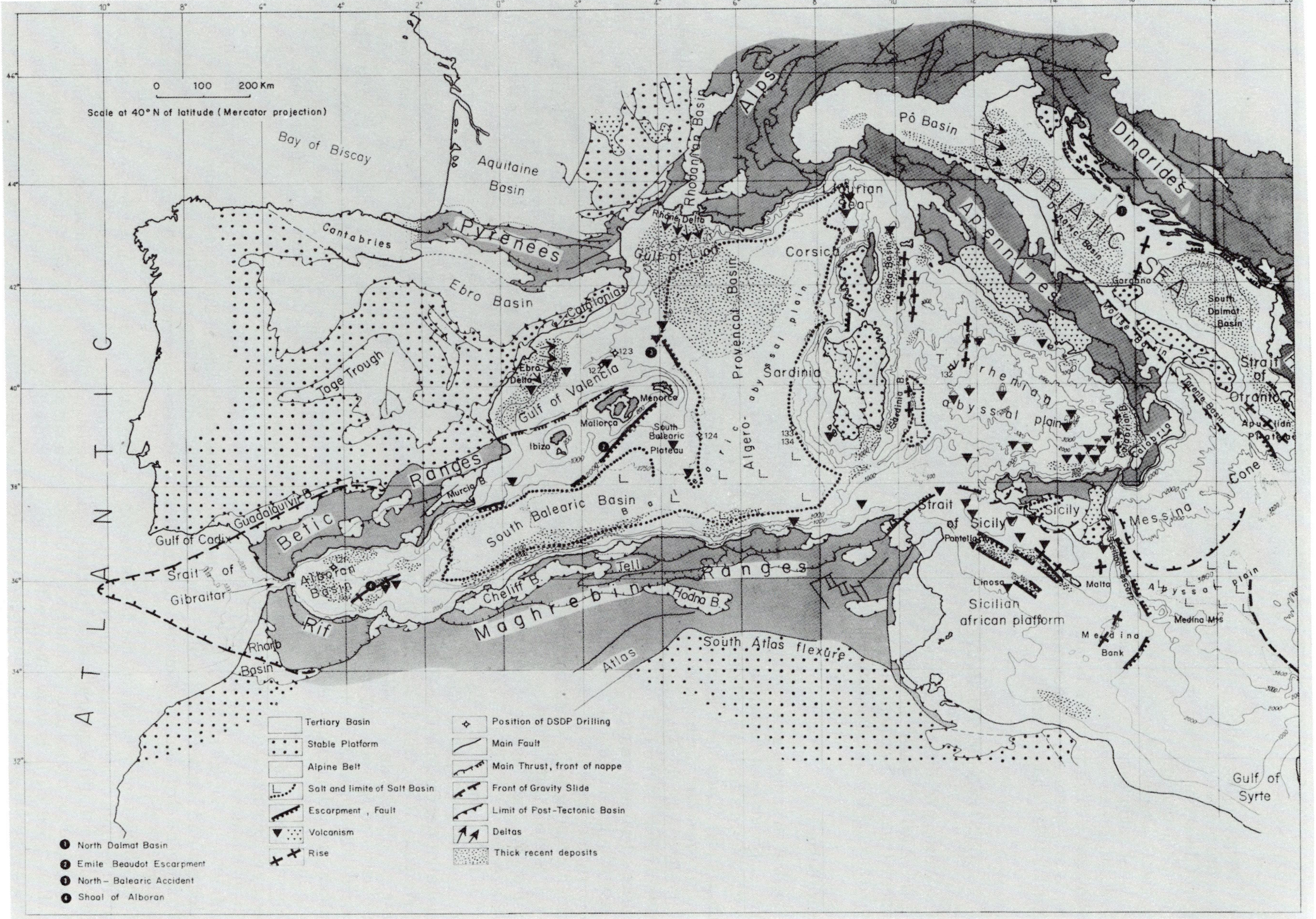

Fig. 2. Structural map of the western Mediterranean region.

ophiolitic massifs or melanges along a line running from Oman to Cyprus (Ricou's "croissand ophiolitique," 1971; Stonely, this volume), as evidence of the former Tethys oceanic realm. The ophiolites are known throughout the entire belt all the way to the Gibraltar arc but in a less clear-cut structural setting because of the superposition of later tectonic events (e.g., Grandjacquet et al., 1972; Lemoine, 1972). During the Cenozoic, throughout the entire Alpine range, local compression or distension movements followed after one another. They governed the formation and evolution of the basins. The tectonic activity is still going on, as shown by earthquakes (McKenzie, 1972) and volcanism.

WESTERN MEDITERRANEAN

The Western Mediterranean includes two major units: the Western Basin (Balearic Basin and its dependents) and the Tyrrhenian Basin. These units are separated by the Corsica-Sardinia block (Fig. 2).

Western Basin

The first structural and morphological sketch of the Western Basin was made about ten years ago by Bourcart (1963) on the basis of bathymetric data and bottom samples. Then the use of seismic reflection helped advance our understanding of this basin. After the first work by Menard (1965) and Hersey (1965) that revealed the first salt domes, along with the work by the Geodynamics Center in Villefranche-sur-Mer (Glangeaud et al., 1967), the Monaco Oceanographic Institute (Alinat et al, 1966), and American oceanographic institutions (Ryan, 1969), further research has been actively pursued, in particular by the Institut Francais du Pétrole (Montadert et al., 1970), French oil companies (C.F.P., Elf-Re, and S.N.P.A.) (Alla et al., 1972; Delteil et al., 1972b), the Centre National pour L'Exploitation des Océans (CNEXO) (Le Pichon et al., 1971; Auzende et al., 1971), and the Osservatorio Geofisico Sperimentale of Trieste (Morelli et al., 1969; Finetti and Morelli, 1972b). All these institutions used deep seismic reflection to obtain information about the sedimentary series and deep structures in the basins.

These seismic-reflection data were complemented by gravimetric data (Allan and Morelli, 1971) and magnetic data (Vogt et al., 1971; Allan and Morelli, 1971; Le Borgne et al., 1971). Since the work by Fahlquist and Hersey (1969) and Muraour et al. (1966), seismic-refraction experiments were carried on by Hinz (1972b).

On the basis of structural criteria, the basin can be subdivided into several distinct geographic units. Figure 2 shows that the central part of the basin is made up of a large abyssal plain located at a depth of more than 2,500 m. It is very even, with a thick sedimentary series. The two subbasins that can be defined are the Algero-Provencal Basin and the South Balearic Basin. On the edges there are either broad marginal areas such as the Gulf of Valencia and the Alboran Sea or else more-or-less steep margins such as the Ligurian and the Provencal margins.

Sedimentary Series. On the basis of the first deep-seismic-reflection surveys done in the northern part of the basin (Montadert et al., 1970), the sedimentary series found throughout the abyssal plain and extending over part of the margins and into some onshore basins can be divided as follows (Fig. 3).

Upper Series A. This series corresponds to the Plio-Quaternary. Interval velocities range from 1.7 to 2.8 km/sec. This series was penetrated by boreholes in Leg 13 of the Deep Sea Drilling Project, DSDP (Ryan et al., 1973), which encountered Pliocene sediments of pericontinental or hemipelagic mud, made up of fine terrigenous clastics and biological elements (mainly nannoplankton and some foraminifers) with turbidite and contourite features. This series is deeply affected in the abyssal plain by halokinetic phenomena. The extent of this marine series is limited on land to various localities in the vicinity of the present coastline.

Evaporitic Series. This series is divided into two terms, B and C (Montadert et al., 1970). The top of the series (term B) was drilled during Leg 13 of the DSDP, and marls were found to alternate with evaporites made up of anhydrite, gypsum, dolomite, and perhaps halite. The mean velocity for this group is about 3.5 km/sec, and the thickness can attain 600 m. At the bottom, a layer (term C) which is at the origin of the halokinetic phenomena (Figs. 3 and 7) is supposed to be mainly salt (halite or potash beds). Calculated velocities are about 4.5 km/sec, and thicknesses are extremely variable and may be greater than 1,000 m.

This series is considered to be of Messinian age (Upper Miocene) according to DSDP data and by analogy with neighboring onland basins where evaporites from this age are known to exist (Cheliff Basin in North Africa, Murcia and Sorbas basins in the Betic Ranges, etc.). However, data from DSDP boreholes concern only the upper part of the evaporites of term B.

Infrasalt Series D. In the abyssal plain, the age and lithology of this series are not known because no borehole has penetrated this deep. Some idea of its thickness (as much as 4,000 m) can be obtained from velocity analysis (Mauffret et al., 1973). However, on the margins in the northern part of the basin, sediments that are probably partly equivalent are known. They were penetrated by petroleum boreholes in the Gulf of Valencia and the Gulf of Lion, and show gray detritic, marly facies of Miocene age (Aquitanian to Tortonian). In the onshore basins the Miocene post-tectonic series are well known on all

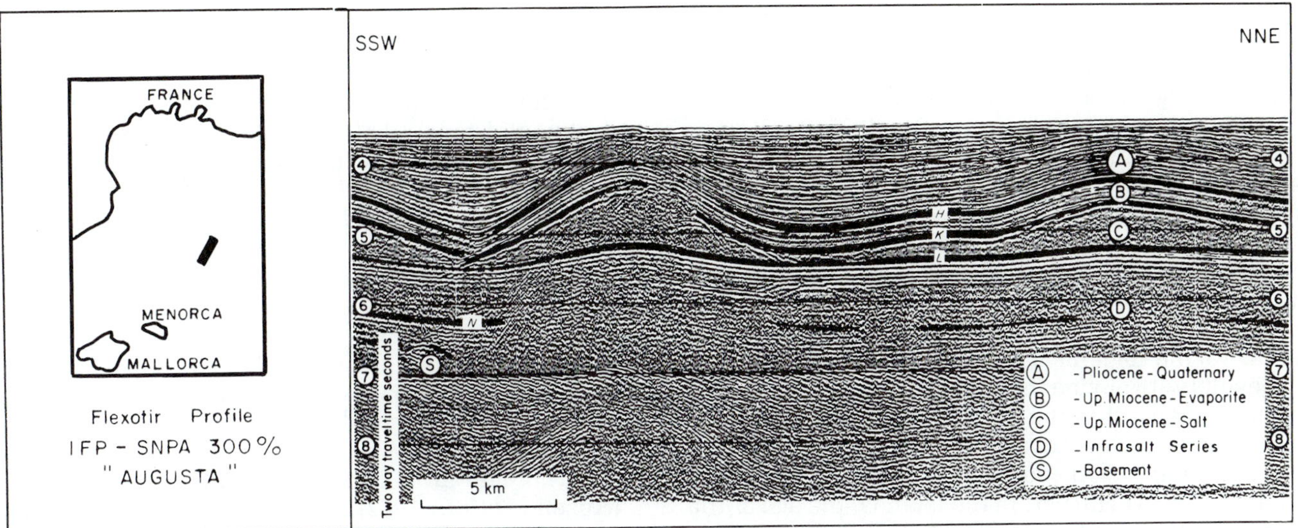

Fig. 3. Typical sedimentary section in the north of the Algero-Provencal Basin with salt anticlines.

the edges. They vary greatly in age with the region and show great facies variations.

The nature of the acoustic substratum observed on the profiles could be different in the margins and in the abyssal plain. It could be oceanic in the abyssal plain and volcanic, metamorphic or sedimentary in the margins (Cenozoic series, or older and more or less tectonized).

Regional Description and Marginal Structure. The extent of the Algero-Provencal and South Balearic abyssal plains generally corresponds to the extent of the salt layer and continuous infrasalt layers. Thus the limit of the abyssal plain is a feature at least as old as the Upper Miocene. On the edges of the salt basin, the infrasalt series mainly pinch out or are partly eroded. This erosion may have begun at the time the salt was formed on the margins. These series exist underneath evaporitic and Pliocene sediments, which are generally transgressive with relation to the salt and gradually cover over an erosion surface. In the Plio-Quaternary, sedimentation was controlled in the northern part of the basin by the Ebro and Rhone deltas. Halokinesis occurred in the abyssal plain. Canyons were being formed and synsedimentary slides and faults were affecting the margins.

Gulf of Lion. Recent sedimentation of the Gulf of Lion is related to the activity of the Rhone Delta. About 1,000 m of Plio-Quaternary are known in petroleum boreholes (Burollet and Dufaure, 1972), and progradation structures can be seen. Its thickness attains 1,500 m at the foot of the slope in the abyssal plain. Along the slope, deep Plio-Quaternary canyons (Fig. 4) trending mostly southeast can be seen. There are also extensive Plio-Quaternary slidings linked to the presence of the evaporitic base and to the slope.

Underneath these superficial series, the salt layer (term C) can be seen to pinch out. The evaporites (term B) go beyond the salt and cover over the "Pontian" unconformity. Beyond they also pinch out, and there Pliocene or Quaternary layers rest directly on the infrasalt series more or less eroded. The infrasalt series in the abyssal plain extend onto a rise near the edge of the shelf which represents the Pyrenees-Provence junction. On and to the north of this axis, petroleum boreholes have revealed a complete Miocene series (Burollet and Dufaure, 1972). Oligocene is presumed in some troughs.

Gulf of Valencia. The Gulf of Valencia occupies a special position between the prolongation of the Betic Range in the Balearic Islands and northeast Spain. The Ebro Delta is important in recent sedimentation. The Plio-Quaternary thicknesses attain 1,200 m and present a typical prograding shelf structure. A great many deep canyons are found on the Spanish continental rise, with some extending into the deepest part (Mauffret and Sancho, 1970). In the axis of the basin, DSDP boreholes 122 and 123, located on a canyon and on a height, show reduced thicknesses of recent sediments (160-268 m) (Ryan et al., 1973).

The salt series (term C) in the Algero-Provencal abyssal plain undergoes a reduction in thickness toward the gulf to the north of Minorca. This North Balearic boundary (identified by Mauffret et al., 1973) corresponds to a strong northwest magnetic anomaly and was active as a fault during deposition of the salt. West of the North Balearic boundary, the thinned salt extends to 3°E. Here, the gentle slope probably explains the absence of slidings in the superficial series. But the reflectors that characterize evaporites (term B) recognized in the DSDP boreholes can be followed into the western part of the gulf.

The infraevaporitic series also extends into this area, usually thinner than in the abyssal plain. Onshore, in Spain, the transgressive Tortonian deposits are known to cover the Mesozoic or older substratum. Onshore and offshore, south of Valencia the series is completed by Middle and Lower Miocene that are increasingly influenced by tectonics as they approach the south (Betic Range). Nevertheless, offshore, the age and nature of the ancient sedimentary series are not known.

Volcanic features occupy an important position in the gulf. Some of them are known on the surface in the Columbretes Islands (Plio-Quaternary). Others are visible on the seismic and bathymetric profiles and are mainly of upper Miocene age (Fig. 5) but may be older: DSDP borehole 123 attained volcanic ash dated 21 my±2 on the side of one of these volcanoes. Samples show that the nature of the volcanism ranges from trachyandesites to basalts with olivine (Bellaiche et al., in press). It is quite probable that the Iberian continental substratum extends widely throughout the entire area. The North Balearic slope corresponds to the front of the Betic Range overthrust. In the Islands, the age of the last tectonic event is Middle Miocene, but before there were also earlier events during Upper Eocene and Aquitanian (Bourgois et al., 1970). The prolongation of the Betic Range east of Menorca raises various unsolved problems.

Catalan and Provencal Margins. The Catalan margin, between the Gulf of Valencia and the Gulf of Lion, and the Provencal margin are very steep (about 20°). These recent abrupt slopes are also marked by deep canyons and outcrops of the substratum. They correspond to a former margin because they mark the sudden disappearance of the evaporitic and salt series as well as of the infrasalt series in the abyssal plain (Delteil et al., 1972a). Figure 6 shows an example of this steep margin with a Plio-Quaternary sedimentary ridge at the foot of the slope.

Ligurian Sea and Corsica-Sardinian Margin. The Ligurian Sea corresponds to the closing off of the abyssal plain in the Algero-Provencal Basin by topographic steps linked to faults (Rehault, 1968). The salt and evaporites that extend for a great distance in the gulf are clearly offset by these faults (some of which have a throw of more than 1,000 m). Some small hanging basins exist, including ones that may be considered as prolongations of Italian coastal basins.

Bottom samples have shown that some magnetic anomalies in the Ligurian Sea correspond to volcanism (dacites, andesites, basalt with olivine; Bellaiche et al., in press). A remarkable singularity is the great northward extension of the Corsican basement, which continues out into the abyssal plain underneath a thin sedimentary cover. This is not revealed by bathymetry.

In the south, along the Corsica-Sardinia block, the infrasalt series and the salt layer pinch out rather abruptly along an escarpment located far west of the islands (Fig. 2). Hanging basins on this margin may include ancient sedimentary series (Lower Miocene from the Bonifacio and Sassari coastal basins) that are more or less eroded underneath recent or evaporitic sediments. It should be borne in mine that the Paleozoic basement of Sardinia was found at the bottom of DSDP borehole 134 (Ryan et al., 1973).

The west and northwest extension of a shallow continental-type substratum, in particular between Corsica and Provence, introduces new constraints in reconstructing a possible rotation of the Corsica-Sardinian block. The possible displacement of this block cannot be as great as is generally thought in most existing schemes compiled on the basis of bathymetry and magnetism (e.g., Carey, 1958).

North Algerian Margin. The Sardino-Tunisian strait marks the eastward closing off of the deep basin in the abyssal plain, with the gradual disappearance of salt and evaporites. The North Algerian margin is characterized by its steep slope (Auzende et al., 1972), which cuts across "internal zones" of the folded Maghrebian Range, extending from the Gibraltar arc to Calabria. The age of the main structures in this southward-overthrust range runs from the late Eocene to the Burdigalian and even to the Tortonian in the southern parts (e.g., Durand-Delga, 1972; Kieken, 1962; Ogniben, 1970).

The substratum (volcanic, granitic, or tectonized sedimentary series) drops in a short distance, and at the foot of the slope there are great thicknesses of recent sediments and pinching out of the infrasalt and salt series. Series of this type, including the development of evaporites in the Messinian, are known on land, especially in the Cheliff Basin, where the post-tectonic Mio-Pliocene can attain a thickness of more than 5,000 m.

South Minorcan [Balearic] Marginal Plateau. The South Minorcan Marginal Plateau forms a north-trending promontory stretching far southward to the vicinity of the Algerian margin (Mauffret et al., 1972). Part of this zone, magnetically quiet, probably corresponds to the offshore extension of the South Balearic series (tectonized basement and reduced, post-tectonic overburden). The few magnetic anomalies found correspond to volcanoes that are visible in the bathymetry. On the edge of the southern margin of this plateau, a structure cutting across the sedimentary series corresponds to a strong magnetic anomaly.

Algero-Provencal Basin. The Algero-Provencal plain is circumscribed inside the margins noted above. The thick sedimentary series has been described earlier (as much as 7,000 m thick). Structurally speaking, this basin can be subdivided into several zones: 1. A zone with thick salt and wide domes (Fig. 3); 2. A zone with a thin salt layer; and 3. Highly disturbed zones (Fig. 7).

In the disturbed zones, halokinesis assumes varying forms, such as narrow domes and injections along faults. These faults are linked to an intra-Pliocene faulting event. The Algero-Provencal Basin has a certain degree of unity and is separated from the South Balearic Basin by a narrow abyssal plain,

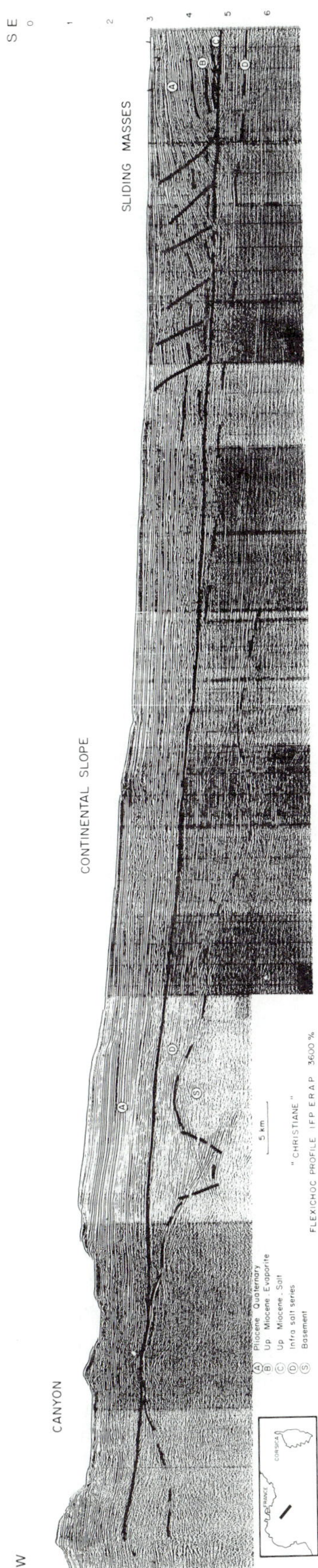

Fig. 4. Seismic profile in the Gulf of Lion, showing canyon and sliding masses along the continental slope.

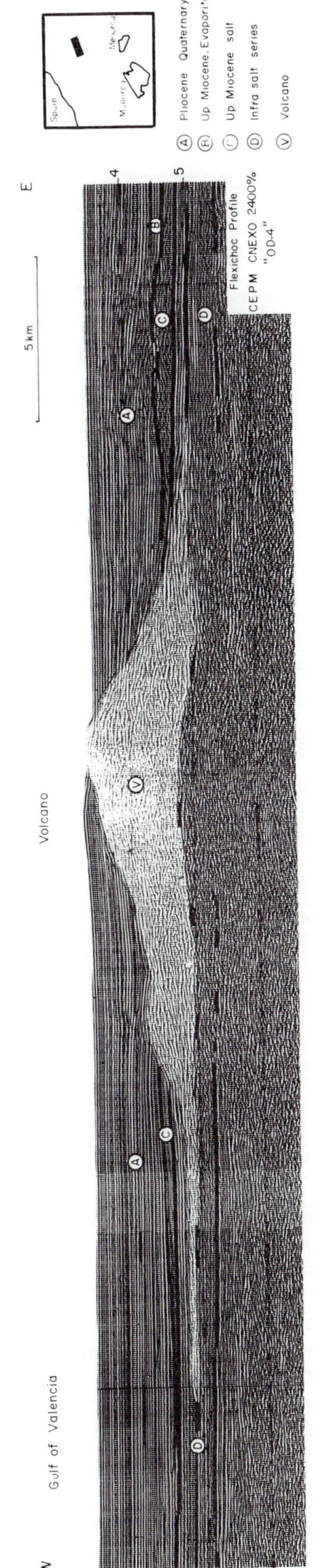

Fig. 5. Volcano on the eastern edge of Valencia Gulf, showing the interbedded volcanic rocks in the Upper Miocene layers on the flank of the volcano.

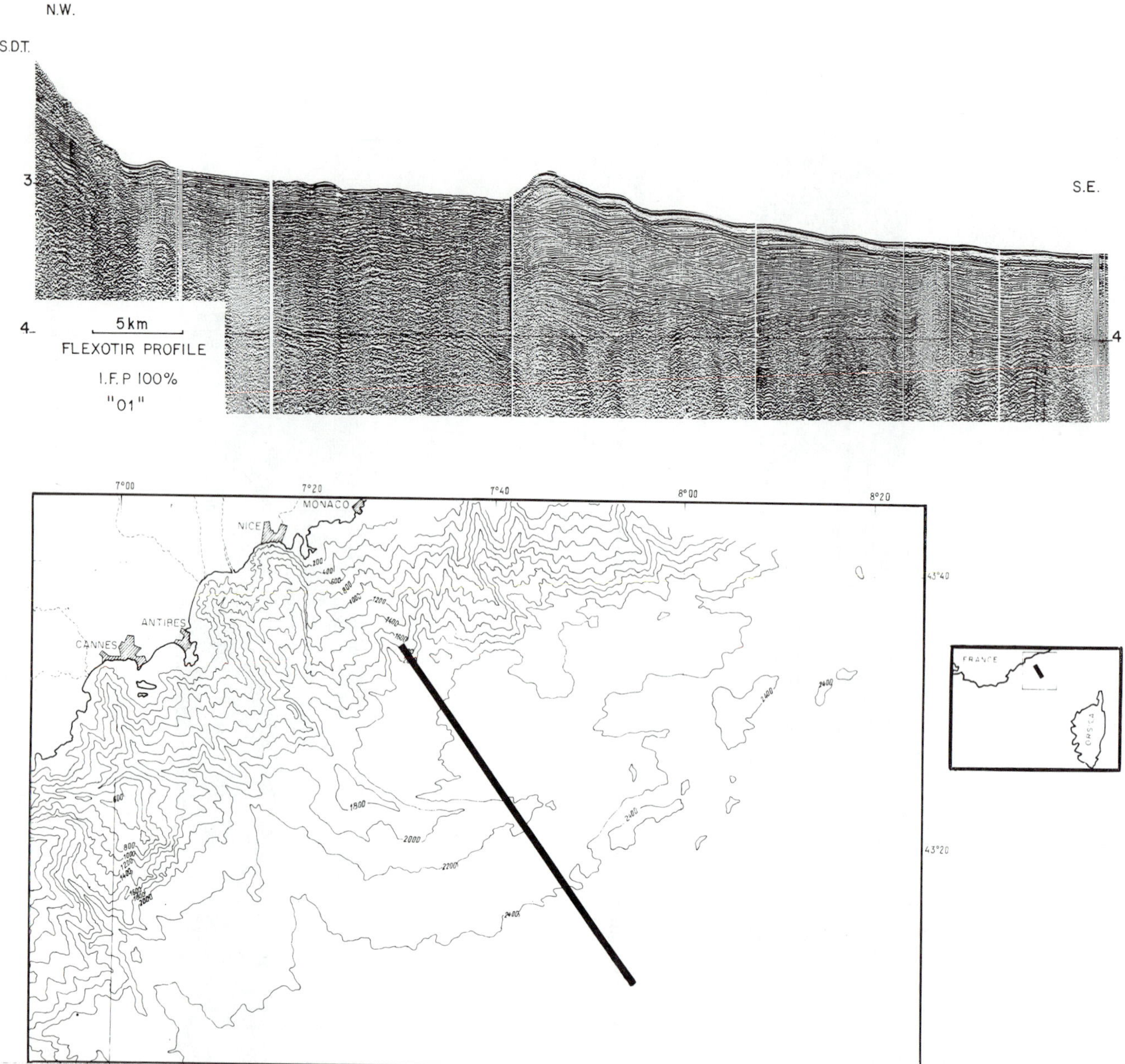

Fig. 6. Steep margin of Provence with canyons and local development of a sedimentary ridge in Plio-Quaternary sediments.

south of the Minorcan plateau, which contains a reduced sedimentary series.

In an attempt to retrace the structural history of the Algero-Provencal Basin, we note prominent events:

1. During the Plio-Quaternary, there was extensive halokinesis, often linked to faults.

2. There was a foundering and subsidence along the present margins, whose formation dates back to upper Miocene when the salt was deposited. Some faults may shift the salt layer by more than 1,000 m, thus clearly demonstrating the importance of the Messinian and post-Messinian vertical movements.

3. Below the salt, regular sedimentary filling took place to a thickness of several thousand meters (at least in the northern part) in a zone well circumscribed in the abyssal plain, corresponding to the salt basin. The age of this filling in is uncertain. On the northern margin (Gulf of Lion) the early Miocene (Aquitanian) has been found and the Oligocene is presumed to be present. It is possible that the entire Miocene series is represented in the abyssal plain, but often basement heights along the margins prevent any correlation from being made between the abyssal plain and the marginal basins. In addition, because of the great disturbances caused by the salt tectonics, hiding possible pinch-outs in the lower sedimentary series, there is the problem of the continuity of the sedimentary

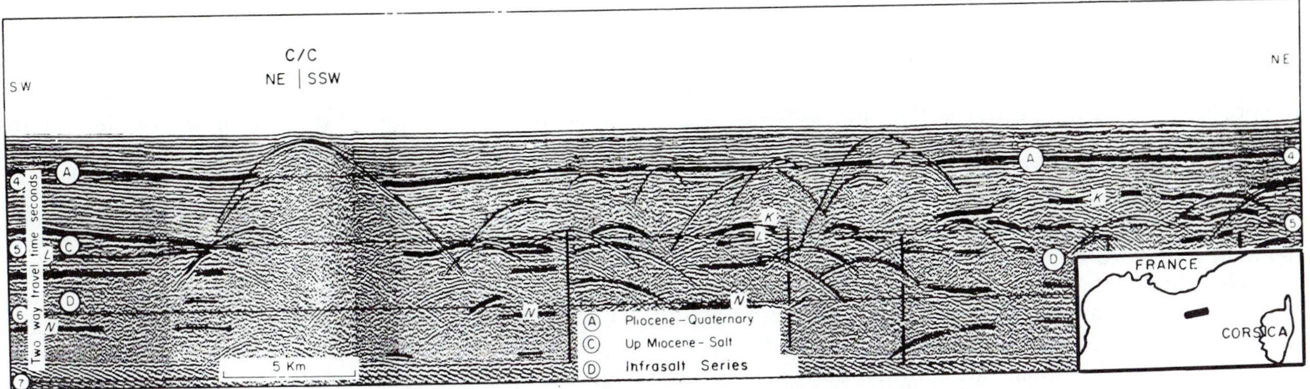

Fig. 7. Piercing salt domes in the Algero-Provencal Basin.

series in the southern part of the Algero-Provencal Basin in an area where large-scale tectonic events are known to have influenced the sedimentation at several intervals during the Paleogene and Miocene (Gibraltar-Calabria zone).

Therefore, the evolution of the Algero-Provencal Basin since the upper Miocene is well known, but before this we are mostly reduced to making hypotheses in recreating this basin. We know that (Fig. 8):

1. Data concerning the Atlantic indicate that the closeness of Africa and Europe has caused various compressions and shearings since the Jurassic and that an Iberian plate may have been displaced in connection with the opening of the Bay of Biscay since the late Jurassic-early Cretaceous.

2. Some paleomagnetic data apparently show a rotation of the Iberian and Corsica-Sardinia block in relation to a stable Europe since the Permian (Zijderveld and Van Der Voo in press). Most authors assume that there was a rotation or a translation of the Corsica-Sardinia block during the Oligocene (Bayer et al., 1973). Refraction and surface-wave studies (Rahlquist and Hersey, 1969; Payo, 1967) postulate the existence of a crust of an "intermediate" or oceanic type in the deep part of the basin. Magnetic maps show a certain degree of unity in the abyssal plain with relatively slight anomalies, sometimes lineated (Le Mouel and Le Borgne, 1970; Le Borgne et al., 1971).

3. Geological data show that during the Cretaceous a disappearance of an oceanic realm (Tethys) occurred between Corsica-Sardinia, probably linked to Iberia at that time, and the Adriatic block. This disappearance is highlighted by the presence of ophiolite blocks of supposed oceanic origin inside Cretaceous overthrusts (Grandjacquet et al., 1972). The exact reconstruction of the Cretaceous phases is still extremely difficult. Then, the Alpine range, s. 1., is the result of different folding movements and superposed phases, with the main ones occurring in this area in the late Eocene, early Miocene (Aquitanian-Burdigalian) and middle Miocene (Tortonian).

Different mechanisms have been suggested to explain the formation of the Western Basin, (Bemmelen, 1969; Ritsema, 1970; Le Pichon et al., 1971), and various proposals have been made, mainly based on bathymetric and magnetic data, for a possible rotation of the Corsica-Sardinia block. These include the formation of a collapsed zone with a "thinning out" of the continental crust, with almost no rotation; the formation of a symmetrical rift of an oceanic type, with the opening and rotation of the Corsica-Sardinia block, the angle varying with the supposed position of the rotation pole (Carey, 1958, etc.); and the formation of a basin behind an arc with distension and thinning of a continental crust and formation of an oceanic crust. In this hypothesis, displacement may have been less than the present width of this basin.

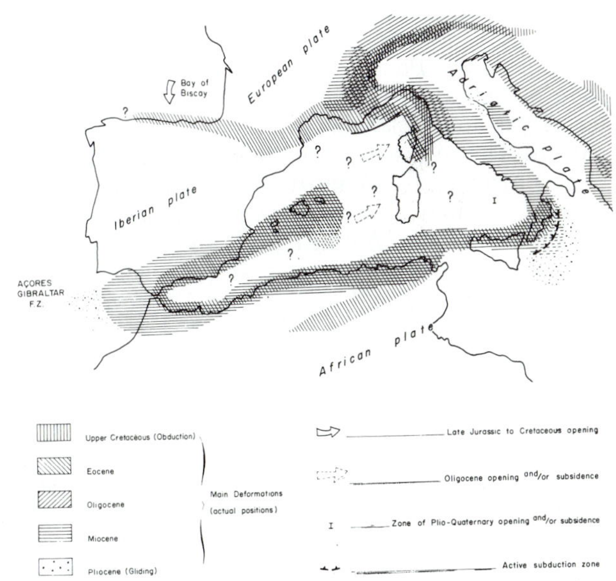

Fig. 8. Schematic distribution of main deformations in the western Mediterranean.

Different ages have been proposed for the occurrence of these phenomena. Geological constraints lead us to suppose that, if there was any movement of the Corsica-Sardinia block, it must have taken place in the Oligocene, and the kinematics proposed until now are not satisfying, considering the extent of the margins on the basis of recent seismic data and magnetism. The rotation pole must have been far and the extent of the true displacement is hard to specify, as is shown by the proposed scheme in Figure 8.

How the Algero-Provencal Basin evolved during the Lower Miocene and its relation to the tectonized zones are still enigmatic. After the Eocene, post-Aquitanian and then post-Langhian, folding is known on land, but it is not known how they end in this basin east of the Balearic Islands.

For this problem of recreating the history of the Algero-Provencal Basin, consideration should be given to the events that occurred in the Alboran region and the South Balearic zone as well as in the folded ranges surrounding them, and in particular to the influx of "Numidian" flysch (Oligocene to Burdigalian ages) such as are found from Gibraltar to Sicily and for which many authors have been looking for an offshore origin.

South Balearic and South Betic Margin. This margin is extremely steep along the Emile Baudot escarpment. It still corresponds to the limit of salt extension. Toward the islands and continental Spain, small hanging basins exist and doubtless contain sediments known on land in some basins (Murcia, Baleares) that have a tectonized substratum (Betic Range) that continue offshore. These basins include an argillaceous marine Pliocene, an evaporitic or calcareous Messinian, and a chalky or detrital Tortonian.

South Balearic Basin. The series here is thinner than in the Algero-Provencal Basin. The Plio-Quaternary is not thick except at the foot of the Algerian margin. The salt is often disturbed, and a great many domes are visible. Wherever the infrasalt series is visible it is quiet, and the substratum can sometimes be seen between Algeria and Spain (Finetti and Morelli, 1972b). Toward the east, the abyssal plain is related to the abyssal plain of the Algero-Provencal Basin via a faulted, shallow area on which all the series appear reduced (Fig. 9); the rises in the substratum correspond to magnetic anomalies. The basin ends in the west, toward the Alboran Sea, with the pinching out of the salt series (Fig. 2).

Auzende et al (1972) described a down-warping of the sedimentary layers at the limit of the Algerian margin. They explained it by the existence of a trench being formed by compression. Nevertheless, such features are common around Mediterranean and seem to be related often only to the thickening of the recent deposits near the shore.

Alboran Basins. The Alboran Sea corresponds to a complex structure where the following can be observed in a very schematic manner. First can be seen Plio-Quaternary series that is locally thick (Stanley, 1973), with internal unconformities and progradation at the margins, in particular in the northwestern part of the basin. Underneath the generally unconformable Plio-Quaternary, the evaporitic series has not been found, but locally in the western part of the basin the presence of strong reflectors suggests such evaporites. DSDP borehole 121 attained Messinian marls (new redetermination by G. Bizon, personal report). This infra-Pliocene series contains at its base a thick series (several thousand meters), filling more-or-less closed depressions of an indeterminate age, separated by rises in the acoustic substratum. The Alboran border corresponds to the dominant northeastern structural trend. It separates a thick western basin and an eastern basin that represents an extension of the South Balearic Basin in the direction of the Rif Mountains. These basins are cut by crosswise trends. The region contains strong magnetic anomalies. The volcanic nature (calc-alkaline rhyolites, ranging to basalts) of these anomalies has been revealed locally by dredgings (Giermann et al., 1968), in the Alboran and Habibas islands, and on the continental edges, where this volcanism has been dated to be of late Miocene age. However, it is still rather difficult to determine what proportion is really volcanic in the acoustic substratum as a whole.

To retrace the history of the present South Balearic and Alboran basins, the following facts should be taken into consideration:

1. The highly special situation of these basins inside the folded Betic-Rif and Maghrebian arc. From the Mesozoic to the Miocene, this part of the Western Basin made up part of a larger unit, extending at least to Calabria. Paleogeographical evolution all along the folded range located in the south shows the close analogies of the sedimentary and structural units from Liassic limestones ("Dorsale calcaire") to early Miocene flysch (Numidian). The South Balearic Basin with its Alboran dependence thus appears as a recent feature of the Mediterranean Basin.

2. The Strait of Gibraltar is a recent morphological feature that cuts across the structures of the Gibraltar arc, whose similarity on both shorelines has recently been established (Didon et al., 1973). The differential movements between Iberia and Africa since the Jurassic probably did not occur at the present location of the Gibraltar strait, but such evidence should be looked for farther north or south. Likewise, communication between the Mediterranean and the Atlantic in the Messinian could have existed only farther north (Guadalquivir) or south (external Rif). It can also be seen that the gravity nappes and olistostromes found on land in the outside part of the arc at the level of these furrows have also

been found in the Atlantic, and so act as proof of the continuity of the arc (Beck, 1972; Bonnin et al., 1973).

3. The acoustic substratum probably partly includes, in addition to its volcanism, sedimentary or metamorphic series belonging to the folded arc in which the main compression phases occurred between the late Eocene and the Tortonian. Just like the Algero-Provencal Basin, it can include an oceanic part.

Numerous hypotheses have been proposed to explain the formation of the Alboran and South Balearic basins. They include the following:

1. Opening by rotation during the Lower Miocene, with transform faults corresponding to the northeasterly tectonic trends of the Alboran area (Le Pichon et al., 1972).

2. Opening as a marginal basin behind an active subduction zone during the Lower Miocene (Auzende et al., 1972; Olivet et al., 1973). In this scheme the same transform faults are used. Glangeaud (1971) proposed a development by foundering on an area with thin crust behind a subduction zone (here southward); thus it corresponds also to a marginal basin.

3. Migration of an "Alboran plate," squeezing an oceanic area located between Spain and Africa and creating the Gibraltar arc during the Cenozoic (Andrieux et al., 1971).

The following is apparent: The Betic-Rif-Maghrebin ranges have been created along an active margin at least since the Eocene. Older history is not yet well defined. Behind this active margin, a marginal basin could have existed since the Upper Eocene. The succeeding closeness of Africa and Iberia-Europe affected this previous hypothetic marginal basin during the Miocene and created the Gibraltar arc as it appears now. Thus the Alboran Sea and South Balearic Basin can be considered as marginal basins related to this last event.

Tyrrhenian Basin

Exploration by Italian institutions (Selli and Fabbri, 1971; Morelli, 1970; Finetti et al., 1970) has enabled definition of the major geological features of the Tyrrhenian Sea, as shown in Figure 2.

Abyssal Plain. Its morphology is opposed to the very flat one of the Western Basin. Here its bottom is studded with seamounts that are mostly along a north-south trend; some have yielded very old samples: Paleozoic and Triassic (Heezen et al., 1971; Selli and Fabbri, 1971). A great many magnetic anomalies have been found, mostly corresponding to volcanism (rhyolitic to basaltic) that occurred on the surface at a recent time, less than 5 my ago.

The sedimentary series visible in the plain are mainly shallow and are found beneath an overburden of recent and not very thick sediment (190 m in DSDP site 132), Messinian evaporites have been found in some seismic profiles. Locally, a salt layer has developed. As a result, a small basin with thick salt exists in the southwestern part and is closed by the Sardino-Tunisian Strait (Finetti and Morelli, 1972b).

Peri-Tyrrhenian Margins and Basins. The abyssal plain is closed off along the ridges or mountains parallel to the present coastlines. For example, on the western margin, there is a high north-south axis running from the island of Elba to South of Sardinia. To the east and south the rise is less even and the margin often bounded by submarine volcanoes or volcanic islands. Peripheral basins, which are often quite thick, have been found on the various margins (Fig. 10).

In the sedimentary series observed on the seismic profile, a distinction can be made between superficial series of Quaternary and Pliocene age (on the margins, there are internal unconformities, one of which may correspond to the middle Pliocene unconformity known on land above blue marls of Trubi) and a bottom series that is more or less thick and may or may not include evaporites (this series is quite thick in the Corsican Basin).

In the middle of the basins, the series may be continuous. On the margins, pinch-outs can be seen. This pattern is similar to the unconformity between the Pliocene and the Miocene in the island of Pianosa. The substratum of these basins may be at a greater depth than that of the abyssal plain. It is difficult to be sure of the age of the lower series in the absence of accurate data, and reference must be made to the geological history of the edges. With this in mind, the following points should be taken into consideration:

The compressive phases observed in the southern part of the Alps, in the Apennines, in Corsica, and perhaps in Sardinia are dated from the Cretaceous to the Paleogene (e.g., Abbate et al., 1970; Barbier et al., 1963; Grandjacquet and Glangeaud, 1962; Haccard and Lemoine, 1970). They may correspond the partial or complete closing off of an oceanic area between the Adriatic block and the Corsica-Sardinian block, perhaps joined to the Iberian block at that time.

The South of the Tyrrhenian Sea was marked by a compression in the late Eocene (Dubois, 1970), followed by various compressions (emplacement of nappes) during the Miocene (e.g., Wezel, 1970; Caire, 1970; Mascle, 1973). This is still going on in the Ionian Sea.

The consequences in the Tyrrhenian area of a possible Oligocene opening of the Algero-Provencal Basin are not clear. They should be sought after farther east, because the movement has been absorbed in the Apennines and in the Dinarides, where compressions are known from the period (Fig. 8). In Sardinia and east of Corsica, on the contrary, troughs exist that are perhaps related to formation of the Western Basin.

The Tyrrhenian collapse was mainly post-Early

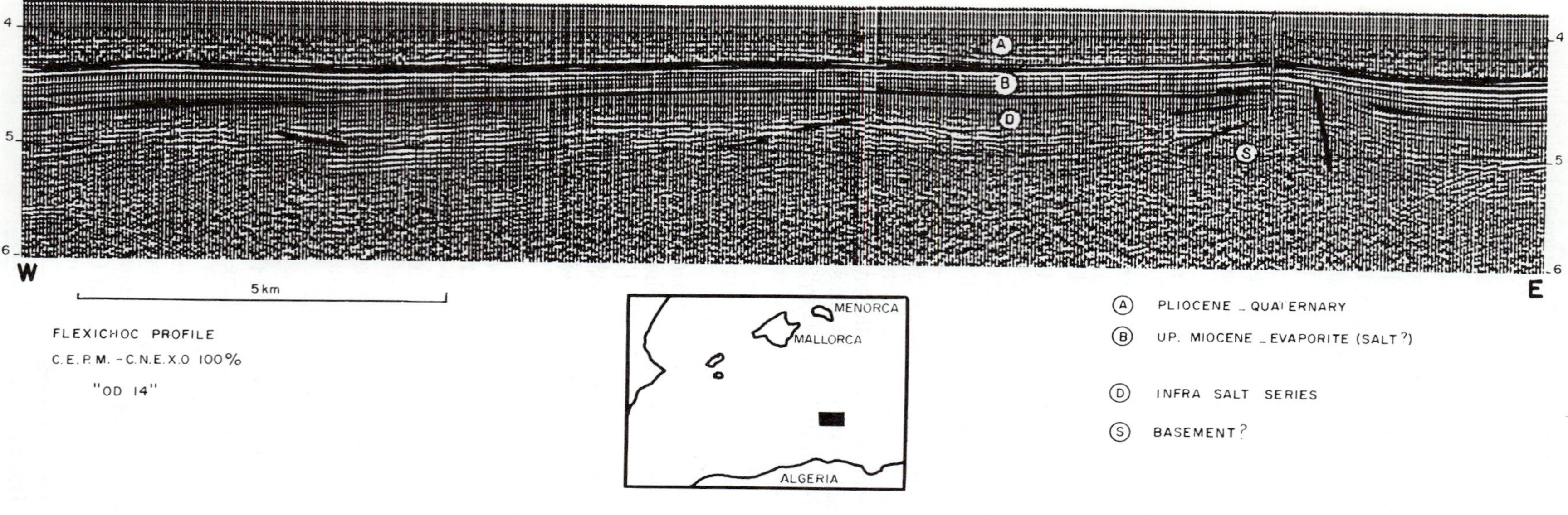

Fig. 9. Abyssal plain between South Menorca Plateau and Algeria, showing reduced sedimentary sequences.

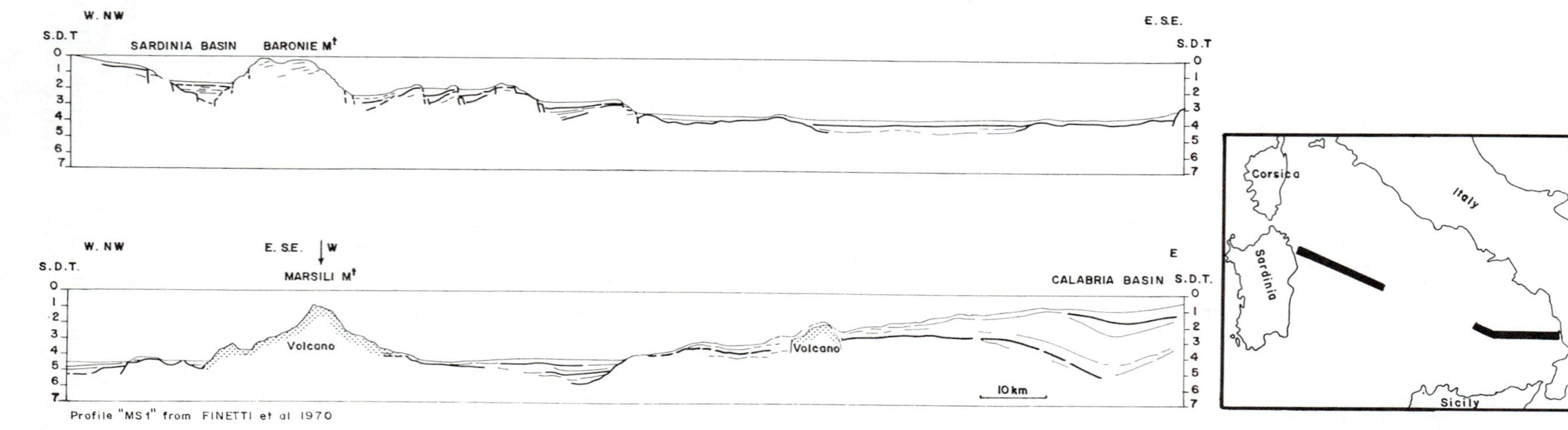

Fig. 10. Schematic sections across the margins of the Tyrrhenian Sea.

Pliocene (Selli and Fabbri, 1971; Morelli, 1970). At that time an important volcanic event began, which is still going on. A difference can be seen between this basin and the Algero-Provencal Basin, which underwent subsidence beginning in the Miocene and continued to collapse during the Plio-Quaternary.

With regard to the present deep structure of the Tyrrhenian Basin and its development, the interpretation based on geophysics (gravimetry, refraction) showing the presence of dense material at shallow depth (Collombi et al., 1973) can be summarized as follows: either the recent oceanic crust was formed from a continental crust by surface erosion or assimilation and basification; or it was formed by distension and the intrusion of a magma through a preexisting continental crust, followed by subsidence according to a mechanism identical to the one proposed for the formation of the marginal basins (Karig, 1971).

Seismicity in this area at present is very active and reveals the existence of intermediate and deep foci roughly trending along a meridian direction representing a Benioff plane dipping from the Ionian Sea under Calabria. The Calabrian arc can be considered at present as probably representing an island arc, and the Tyrrhenian Sea probably corresponds to a back-arc basin with its related volcanism (Ustica, Lipari, etc.), (Caputo et al., 1970 Ritsema, 1970; Bousquet, 1972).

Thus it appears that the structure of the Tyrrhenian Sea and its borders corresponds to a complex evolution since the Cretaceous. Different compressionnal events occurred at different places. The last one is the most evident; it corresponds to the existence of an active margin in Calabria, the Tyrrhenian Sea being the back-arc basin. The peripheral basins (especially the Calabrian basins) may be interpreted as arc-trench gap basins. Nevertheless, the significance of the Corsica-Sardinian basins is still doubtful.

EASTERN MEDITERRANEAN

The Eastern Mediterranean can be divided into several areas, in which subbasins can be defined (Fig. 11). The first physiographic sketches of the eastern Mediterranean were made by Emery et al (1966) and Giermann (1969); then, geophysical studies (refraction, gravimetry, seismic reflection) by Russian and American scientists (e.g., Emelyanov et al., 1964; Wong et al., 1971; Ryan, 1969; Ryan et al., 1971) gave the first structural results. In a second stage, the Ionian Sea was explored by the cruises of the *Meteor* (refraction, seismic reflection; Closs and Hinz, 1972; Hinz, 1970, 1972a) and by the Osservatorio Geofisico Sperimentale of Trieste (deep seismic, Finetti and Morelli, 1972b). The Levantine Sea was studied by the University of Cambridge, using gravimetry, magnetism, refraction, and seismic reflection (Lort, 1972; Woodside and Bowin, 1970); by Vogt et al (1971; aeromagnetism); and by Israeli institutions (Neev et al., 1966). More recently, deep-seismic studies have been carried out, in particular by the Institut Francais du Pétrole and French oil companies (C.F.P., Elf-Re, and S.N.P.A.; Sancho et al., 1973), the Centre National pour l'Exploitation des Océans (CNEXO), and the Osservatorio Geofisico Sperimentale of Trieste.

Adriatic Basin

Figure 11 shows that the present basin can be divided into two parts.

Northern Part [Po Basin, Dinaric Basin]. Underneath relatively shallow water, a thick sedimentary basin stands out. Its structure is not symmetrical. In the east it gently touches the stable Istrian Platform with progressive pinch-outs (Vercellino, 1970). Its structure is not so well known on the Dinaric margin, on the edge of which a narrow, recent basin may exist (North Dalmat Basin), which should be compared with the South Adriatic Basin. In the west, the overthrusted Apennine structures affect the entire series all the way to the Pliocene. In the south, at depths greater than 200 m, the basin is almost closed off by a structural high (prolongation of Mount Gargano), oblique in relation to the Dinaric and Apennine structures. Southward it continues again on land via the Molise Basin.

Very thick Pliocene and Quaternary formations are found, especially in the north (Po Basin, where more than 7,000 m of recent deposits are known) and along the Italian coast, where Pliocene subsidence was also very active (E.N.I., 1962-1971). There is a progradation of recent sediments from the Po and along the margins. Gravity tectonics are found at the southern edge of the Po Plain, extending to the entire outer margin of the offshore Apennines, just as on land. It has considerable influence in the early and middle Pliocene and may even be continuing at present. We will see that it extends into the Ionian Sea via the Messina Cone in the direction of Sicily. Evaporitic layers from the top of the Miocene are also included in the gravity tectonics.

Offshore, underneath the Cenozoic overburden, a folded Mesozoic series extends westward. It is calm in the east, where Cretaceous, Jurassic, and Triassic (including evaporitic layers) are known. These sediments are also found on land on the outer margin of the Dinarides, where they are overthrust in the opposite westward direction.

South Adriatic or South Dalmat Basin. This deep-water basin also has asymmetrical structure. In the west, it gently touches against the Apulian platform and plateau; and in the east, it is bounded by reverse faults from the front of the Dinarides (Fig. 12). The following features are mainly observed:

1. Very thick Plio-Quaternary sediments (sometimes more than 2,500 m). This series gradually pinches out toward the west, with evidence of progradation in the vicinity of the coastline (Fig. 12).

2. The thick underlying series, Cenozoic and complete Mesozoic, that becomes condensed as it approaches the Apulian platform. This platform is made up of a Mesozoic and Paleogene shelf carbonate series (d'Argenio et al., 1971). Offshore it extends toward the southeast via the Apulian plateau. This promontory forms a wide arch that comes against a recent bathymetric feature bounding the islands of Zante and Cephalonia (Fig. 12). It is bounded in the west by a faulted zone and a trench.

3. The front of the Dinaric folded zone can be interpreted as reverse faults. This possibly indicates a recent compression, but the main compressional movements occurred in the late Eocene-Oligocene (Aubouin et al., 1970b). However, farther south, in Epirus, we know that the movement continued during the Miocene, Pliocene (Triassic overthrust on Pliocene conglomerates), and even more recently (I.F.P., 1966; Mercier et al., 1972).

Seismological data reveal an earthquake belt around the Adriatic, although not clearly closed southward (Lort, 1972; McKenzie, 1972). Nevertheless it appears that during recent geological history, the Apenninic margin was much more active than the Dinaric one. It is the prolongation of the Calabrian margin; this is marked by large gravity slidings, thick marginal troughs, and young volcanism. The Adriatic is surrounded by overthrust belts with ophiolites tectonically emplaced during different phases of the Jurassic and the Cretaceous and remobilized during the Cenozoic (e.g., Abbate et al., 1970; Aubouin et al., 1970a). If ophiolites have an oceanic origin, the existence of a surrounding ocean (Tethys) during the Mesozoic is indicated. The Adriatic area, which is an old continental area, was thus probably much larger than is apparent today.

Ionian Basin

Messina Cone or Calabrian Margin. The Messina Cone is located on the outside margin of Calabria (Fig. 11). It is quite probably the geological extension of structures known on land in Sicily and in southern Italy. This opinion is confirmed by seismic data, which show:

1. The existence of small hanging basins, comparable to those in Calabria, sometimes containing a Pliocene or Quaternary series.

2. The development of wide allochthonous masses found in the north in the Taranto Basin (Finetti and Morelli, 1972a). These masses extend all the way to the abyssal plain (Sancho et al., 1973; Fig. 13).

It can be suspected that the entire Messina Cone, like the external Gibraltar Arc (Beck, 1972), corresponds to a slip structure. It affects at least the late Miocene (evaporites) at its front and it was probably always active (Letouzey et al., 1974). It possibly represents the offshore extension of gravity nappes known on land in the Molise Basin in continental Italy or the Caltanisetta Basin in Sicily, whose age is as young as middle Pliocene. Behind the slide front, the faults, the numerous diffractions, and the absence of reflection coherence reveal a highly tectonized zone in which large sedimentary masses are in disequilibrium. This vast structure can be related to the existence of a Benioff plane beneath Calabria. The absence of the present trough should be pointed out.

Abyssal Plain and Gulf of Sirte. Figure 12 shows a thick sedimentary series under the gravity nappe of the Messina Cone. This covers the entire abyssal plain. Several thousand meters of sediments can be seen, and they are generally thought to correspond to a complete Cenozoic series and also to the Mesozoic (Finetti and Morelli, 1972b). At the level of the abyssal plain underneath a thin Plio-Quaternary layer, Messinian evaporites are present and there is even salt, without any halokinetic phenomenon.

The southern part of the abyssal plain is bounded by faults and structures (Medina Seamounts) where the existence of strong magnetic anomalies trending east-west are found and show that volcanism is present. The thick Cenozoic sedimentary series move steadily toward the Gulf of Sirte. Onshore, in Libya, this series is known to turn into a more carbonate facies (e.g., Burollet and Magnier, 1970; Conant and Goudarzi, 1967).

In the east, the abyssal plain is bounded by the Mediterranean Ridge (Fig. 11). Beyond the Cyrenian Plateau, where strong magnetic anomalies are found, there are thick sediments in a narrow trough between the Cyrenaica and the Ridge, probably equivalent to the Cenozoic and Mesozoic of the Ionian abyssal plain.

Sicilo-Tunisian Margin

In the west, the Ionian abyssal plain is abruptly closed off on the Sicilian escarpment, which borders the Sicilo-African Plateau (or the Pelagian Platform), where the water depth is rarely more than 500 m except in the west-northwest grabens of Pantellaria and Malta (Finetti and Morelli, 1972b; Morelli, 1972). This plateau corresponds to the extension of the African platform, where relatively calm clastic or carbonate Cenozoic or Mesozoic sediments are found with several internal pinch-outs (Burollet, 1967). North of this plateau there are recent slidings (southwestern Sicily), equivalent of Apenninic-Messina Cone allochthonous masses. Some zones are intensely fractured along the main northwesterly trends of the grabens in the Sicilo-Tunisian Strait.

These grabens correspond to recent faulting, affecting the Pelagian Platforms (Zarudzki, 1972). They are accompanied by extensive volcanism from

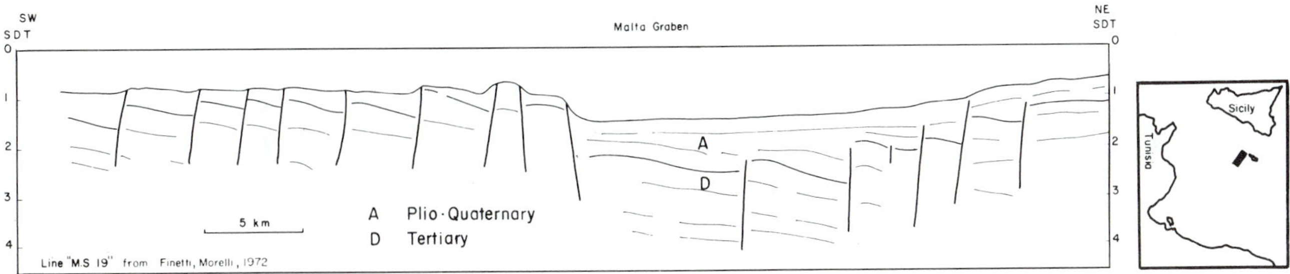

Fig. 14. Schematic section across the Malta Graben.

the late Miocene to the Quaternary (Pantellaria, Linosa). The troughs are filled by recent sediment (Fig. 14) that may be more than 1,500 m thick. It is difficult to follow them all the way to the Campidano trough, near Sardinia, because they are interrupted by the Tunisia-Sicily folded belt. Likewise, in the east they end before the top of the Sicilian escarpment (Medina plateau). Creation of this faulted system is linked to a recent distension, compared with the age of the Alpine range in north Africa and Sicily, and is transversal in relation to this range. Most of these faults are probably late Miocene to present (Finetti and Morelli, 1972a; Morelli, 1972). A system of grabens having the same age and trend exists in Tunisia (Burollet, 1967).

North African and Levantine Margin

Cyrenaican Margin. Figure 15 shows anticlinal structures with a faulted northern flank that are known on land in Cyrenaica (Conant et al., 1967), extending offshore with small recent hanging basins. These structures were developed continuously since the middle Cretaceous. Underneath the recent overburden, the entire Tertiary series must exist in a more-or-less condensed form (with erosions in the Paleocene and Oligocene), as well as the Cretaceous series that outcrops on land in the anticlines. It is generally assumed that the Mesozoic series suddenly thickens toward the sea, as is the case on the Tripolitan margin (Klitzsch, 1970). This is possibly evidence of a former continental margin.

The present continental margin is marked offshore by the existence of a trough in which very thick, subhorizontal sediments have been found. This trough corresponds to the westward extension of the plain of Herodotus and probably communicates with the Messina abyssal plain. At least 5,000 m of sediments is visible (Sancho et al., 1973). The recent escarpments prolonging the Cyrenaican margin toward the Nile deltaic cone are probably of a similar type. In the north, this narrow plain is bounded by the Mediterranean Ridge.

Nile Deltaic Cone. The Nile Delta extends for a long way offshore in the form of an asymmetric cone. Recent sediments are relatively reduced in the east (Damietta Fan) and pinch out in the north on a former anticlinal structure, the Erathosthenes plateau, well indicated in the bathymetry, corresponding to a strong magnetic anomaly (Fig. 16). Toward the west the Rosetta Fan corresponds to a considerable thickening of Plio-Quaternary sediments. This series is more than 2,000 m thick and extends widely in the Herodotus Plain and gradually thins on the Mediterranean Ridge, where they have been drilled by DSDP borehole 130. Historically, there is evidence that the north Nile Delta was active at least since Oligocene, but it migrated during Miocene to the northeast.

North of the Rosetta Cone, to the Mediterranean Ridge, there are large folds in this superficial series. Such a phenomenon is often described in deltas, and here it could be related to the presence of the Messinian salt layer.

Levantine Margin and Abyssal Plain. The margin of the Levantine coast is steep, marked by recent faulting (Neev et al., 1973 a) with canyons. On land, there are Cenozoic series, mainly Neogene, covering the Mesozoic and opening out seaward (Neev, 1966, 1973b). The location of this margin is controlled by two main structural trends:

1. The Aqaba-Dead Sea-Kara Su Graben line, which is at least as old as Oligocene; it is marked by an important sinestral displacement, especially during Plio-Pleistocene, and by strong volcanic activity (Dubertret, 1967, 1969; Freund et al., 1968; Quennell, 1958);
2. The Negev folded zone, which trends northeast, then east, until the Gulf of Suez, north of the Red Sea. The age of folding is Oligocene, as in the Palmyrids of Syria. The northern prolongation of Suez Gulf faulted zone is unknown. It is important that during lower and middle Miocene, the Mediterranean Sea and Red Sea communicated.

At the foot of the continental margin, a very calm basin opens out and contains a thick sedimentary series in which a thin Plio-Quaternary series can be distinguished (about 500 m), along with a late Miocene evaporitic series (1,000-1,500 m), and calm and very thick lower layers that are probably largely Miocene, but whose bottom may represent older strata. The important point in the Levantine Basin is

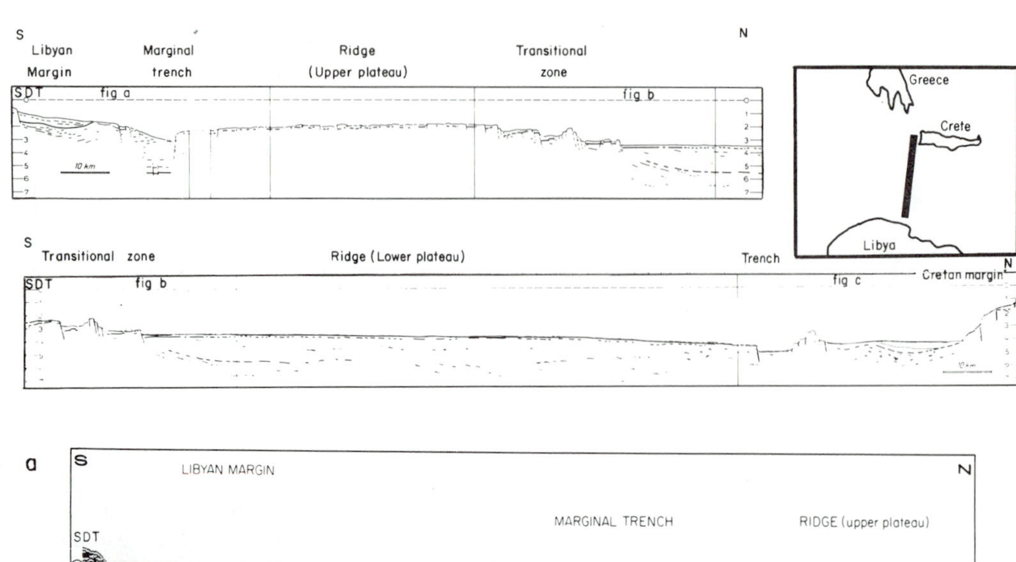

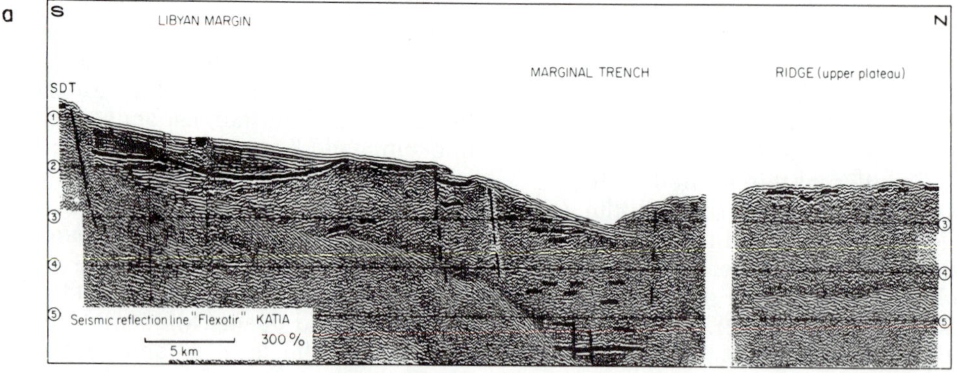

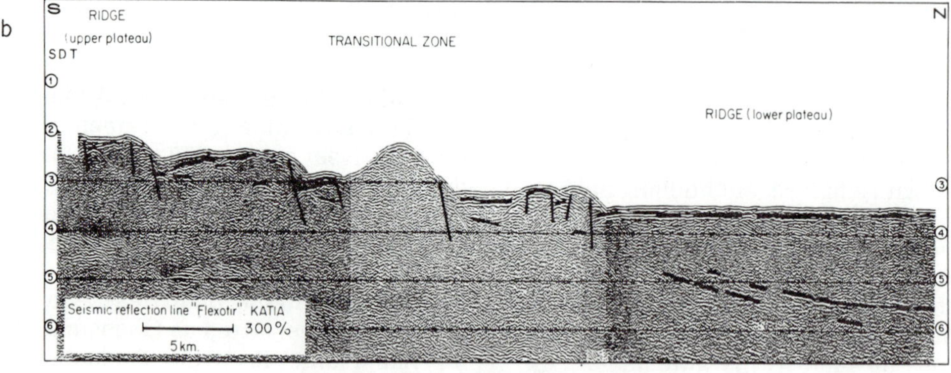

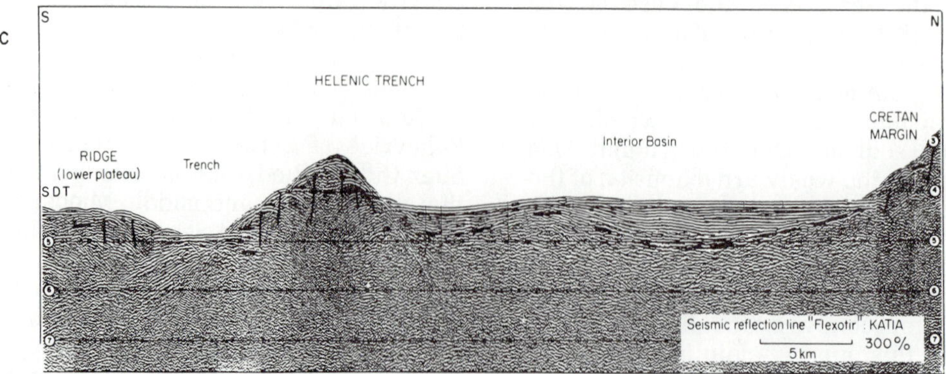

Fig. 15. Schematic sections between Cyrenaica and Crete across the Mediterranean Ridge, and some examples of seismic profiles: (a) Libyan margin and front of Mediterranean Ridge, (b) Mediterranean Ridge, and (c) Hellenic trench and Cretan margin.

the existence of a thick Tertiary or older sedimentary series that has been confirmed by the few refraction surveys made in this area. Two possible explanations can now be considered; i.e., either these sediments overlie the remainder of a former oceanic crust (Mesozoic Tethys), or else they correspond to a prolongation of the African platform as well as the Arabian platform. The latter possibility would mean that in the Mesozoic the Tethys Sea was located farther north, beyond the present position of the Cyprus ophiolites, which have been carried southward. There is also the problem of the origin of the Erathosthenes plateau, on which Upper Miocene and Plio-Quaternary strata pinch out.

MEDITERRANEAN RIDGE, TRENCHES, AEGEAN ARC, AND AEGEAN SEA

Mediterranean Ridge. The Mediterranean Ridge is a submarine topographic feature that forms a swelling with a large curve radius on the outside of the Aegean Arc. It can be followed from south of the Apulian Plateau all the way to just south of the Florence Rise (Fig. 11). On seismic profiles it is characterized by strong diffractions caused by the irregular topography of the surface and of the bottom of the Pliocene (e.g., Giermann, 1969; Hieke, 1972, Weigel, 1972; Closs and Hinz, 1972). At the southern limit of the ridge, beginning at the Ionian abyssal plain, there is a bathymetric rise accompanied by an irregular topography that is quite similar to the one observed on the front of the Messina Cone. Surface deformations affect the sedimentary series, at least to the bottom of the evaporites. Underneath the surface diffractions, it is not possible to continue the deep and calm reflectors of the thick sedimentary series of the abyssal plain. This type of structure is quite widely found east of the Ionian Sea and north of Cyrenaica (Fig. 15A). In this area, the Plio-Quaternary sediments are not very thick, generally 0-100 m.

The surface topography contains relief differences of about 100 m, mainly trending in the same direction as the Mediterranean Ridge. Toward the northwest the ridge joins with the allochthonous masses of the Calabrian margin (Messina Cone), and the quite comparable seismic nature is such that they cannot be told apart. At the other end, in the Levantine Sea, the thick sedimentary series and the numerous deformations in this series do not enable the exact extension of the ridge to be determined. In this eastern part, the thickening of the recent sediments southward is due to sediment influxes from the Nile, while to the north, the series thins out toward the trench. Surface topography is influenced by this thickening, and seismic profiles reveal some folded structures in these recent sediments, whose origins may be partly due to the presence of an underlying Messinian salt layer affected by collapses and dissolution.

Generally speaking, seismic reflections on the ridge reveal very few coherent reflections underneath the diffracting surface of the bottom of the Pliocene. DSDP boreholes 125, 126, and 129 revealed different Miocene levels, ranging from Messinian evaporites to Langhian, underneath the Plio-Quaternary pelagic deposits (marls, clays, silts, and sandstones, with interbedded volcanic ashes), (Ryan et al., 1973). The presence of salt, very widespread in the eastern Mediterranean, is possible also in some basins to the north of the ridge. South of the Cretan Arc, basins may contain more than 4,000 m of sediments (Sancho et al., 1973; Fig. 15).

Hellenic Trough. In the present morphology, the Hellenic trough corresponds to a succession of several depressions. These depressions extend from south of the Apulian Plateau to a point south of Crete, and are filled with recent sediments that may locally be thicker than 1,500 m; the bottoms of the depressions are generally flat. South of Crete and Rhodes (Fig. 11), a series of recent seamounts are separated from one another by depressions that form basins containing recent sedimentary fill. It is probable that this area of troughs and seamounts is evidence of the Paleogene-folded Cretan arc, which is bisected by recent tectonics (Aubouin and Dercourt, 1965). The Strabo trough is the outside boundary of this system; south of Rhodes it curves toward a point south of the Florence Rise.

At the present time this arc is active, as shown by earthquake distribution (Papazachos and Comninakis, 1971; Galanopoulos, 1969; McKenzie, 1972), and isostatic gravity anomalies (Woodside and Bowin, 1970).

Aegean Sea. North of the islands of the Cretan Arc, depressions are seen and are filled with recent and highly faulted sediments in the eastern part. Beyond these depressions, there is the volcanic Aegean Arc, which dates from the Burdigalian to the present (andesites and basalts), (Borsi et al., 1972), with recent ashes (Santorian) that have been widely dispersed throughout the entire eastern Mediterranean.

The Aegean Sea is a geologically complex area located within the Alpine system, a key point between the Hellenides and the Turkish ranges (Pontids, Central Ranges, and Taurids). As in the Tertiary basins in the neighboring land areas or islands (Meso-Hellenic Trough, Thrace Basin), some interior basins developed with marine sedimentation from the middle Eocene to Oligocene (after the late Cretaceous-early Eocene tectonic events), (Brunn and Desprairies, 1965; Mercier, 1966). It is difficult to determine accurately the present contours of these basins, and to draw a paleogeographic image of this period. Subsequently, during the Miocene, the Aegean area, like central Turkey, underwent extensive dislocations and collapses that broke it up into various small basins (Ikaria, Lesbos, Northern Crete) or troughs (Skiros, Northern Aegean), (Maley and Johnson, 1971). Here, a marine sedimentation developed with clastic influxes, considerable thickness differences, and probable facies variations, and also including the late Miocene evaporitic event (400 m of evaporites near Thassos). The Cyclades crystalline massif forms an ancient nucleus.

North of the Aegean Sea, the North Aegean Trough represents a recent distension zone (McKenzie, 1972). It is considered by some to be a prolongation of the North Anatolian Fault; nevertheless, the prolongation of this fault could possibly be south of Imbros-Lemnos Islands.

Ancient Oceanic Crust. The tectonic setting of ophiolites in the Alpine range dates back to the Jurassic and Cretaceous. They are interpreted as remnants of an old Tethyian oceanic crust that was initially localized in an area within the northern part of the eastern Mediterranean. The structures visible in the Cretan Arc were formed in the Oligocene, subsequently accompanied by metamorphism, indicating a subduction zone between the Aegean and African areas. In the Miocene it is possible that part of the superficial sedimentary masses that were in disequilibrium may have slid toward the outside of the present range: a sedimentary rim. The Mediterranean Ridge could have been formed by gravity sliding (Mulder, 1973), according to a mechanism comparable to the one occurring in the Miocene in Turkey (Lycian nappes) or the more recent one in the Messina and Gibraltar cones. Such a mechanism is opposed to the one proposed by others (Ryan et al., 1971; Rabinowitz and Ryan, 1970). They supposed that the ridge was a thickening formed by a rim of recent tectonized sediments pushed to the front of the trench.

In our hypothesis, such an arc, formed of imbricate sedimentary piling up, must be located northward at the level of the Cretan Arc. The allochthonous masses might be found underneath the present Mediterranean Ridge, thus partially explaining the absence of coherent deep reflections; the southern front of the ridge could correspond to a recent activity of this sliding. Since the end of the Miocene, faulting has been active in the whole area and seismological data led McKenzie (1972) to define here a boundary of a present plate. This interpretation in no way rules out the presence or absence of an oceanic crust underneath the entire present eastern Mediterranean.

All the post-Oligocene structural elements described could be interpreted within the context of an active insular arc: 1. The Mediterranean Ridge apparently corresponds to a thickening of a mainly Miocene gravity nappe, but it is also probably a large recent swelling of tectonic origin, i.e. an outer arc, related to the ones found frequently forward of a subduction zone; 2. The Hellenic trenches correspond to the present subduction zone; 3. The islands and seamounts in the Cretan Arc probably represent an external nonvolcanic arc; 4. The trough north of Crete possibly corresponds to the arc-trench gap in front of the volcanic arc, made up of the volcanoes south of the Cyclades; 5. The Aegean Sea in the north might well be a back-arc basin. The structural elements related to the older history of this area are difficult to define accurately in the same way.

Cyprus Arc and Associated Basins

Late Cretaceous Cyprus Arc. Geophysical and geological data can be used to define a Cyprus Arc stretching from south of the Anaximander Mountains to the Levant Coast (Fig. 11) and representing the offshore extension in the eastern Mediterranean of the "croissant ophiolitique péri-arabe" (Ricou, 1971), which constitutes the outside edge of the arc that was folded in the late Maestrichtian. It is made up of gravity nappes that include vast ophiolitic masses stretching from Oman to the allochthonous masses at Kizil Dag-Baasit, Turkey, and along the Cyprus Arc. In Cyprus, the presence of allochthonous masses (Mamonia nappes) has been fully demonstrated (Lapierre, 1972). In this island, the Troodos Massif has been interpreted as evidence of a Cretaceous oceanic crust, related either to the activity of a midoceanic ridge or to an oceanic basin behind an insular arc (e.g., Vine et al., 1973; Miyashiro, 1973). Different studies have demonstrated the asymmetry of the massif (Lapierre, in press), and an obduction is suggested as an explanation of its position. Geological similarities and geophysical data enable the Cyprus structures to be linked with certainty to those on the Levant Coast. Therefore, extensive allochthony of the outcrops on Cyprus is a distinct possibility (Biju-Duval et al., in press). Offshore, the front of this arc is marked by escarpments separating broad plateaus that are more or less parallel to the southward overthrust structures, known onshore, but it passes clearly south of Cyprus.

In the west, the "Florence Rise" stands out clearly in the bathymetry and follows this general structure in the direction of the Anaximander Mountains (Fig. 17). Beyond, the Antalya nappes and Lycian nappes (Brunn et al., 1971), containing ophiolites, could represent the western extension of the Late Cretaceous Arc. On this rise, there is a reduction in the Tertiary sedimentary series of the Antalya Basin, and even the disappearance of some parts of the sequence. The rise forms a wide asymmetric arch that may correspond to the southward-overthrusted structures known in Cyprus. Strong magnetic anomalies (Woodside and Bowin, 1970; Vogt and Higgs, 1969) are localized along this arc and may be interpreted as the probable presence of rocks of the ophiolitic suite overthrusted at the end of the Cretaceous. Southwest of the Cyprus Arc, seismic profiling has revealed a

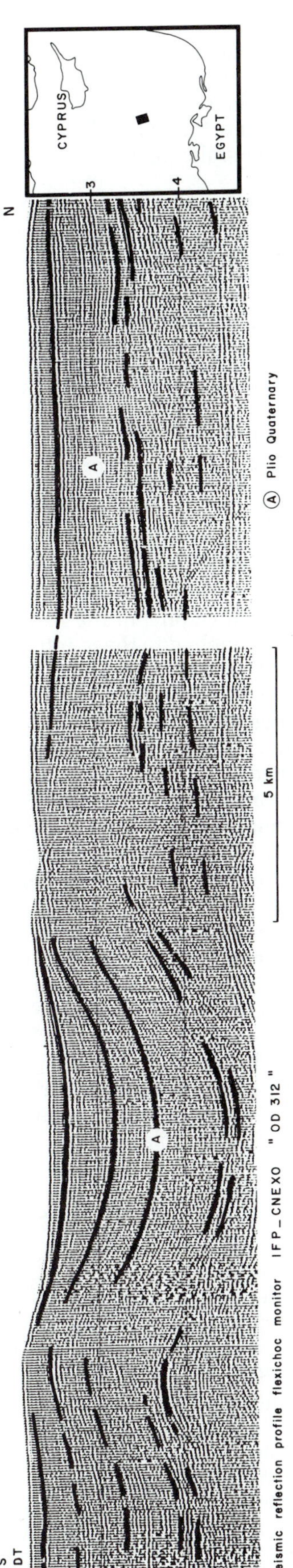

Fig. 16. Large structures (due to salt?) from the Nile Cone, and thinning of the sedimentary series near the Erathosthenes Plateau.

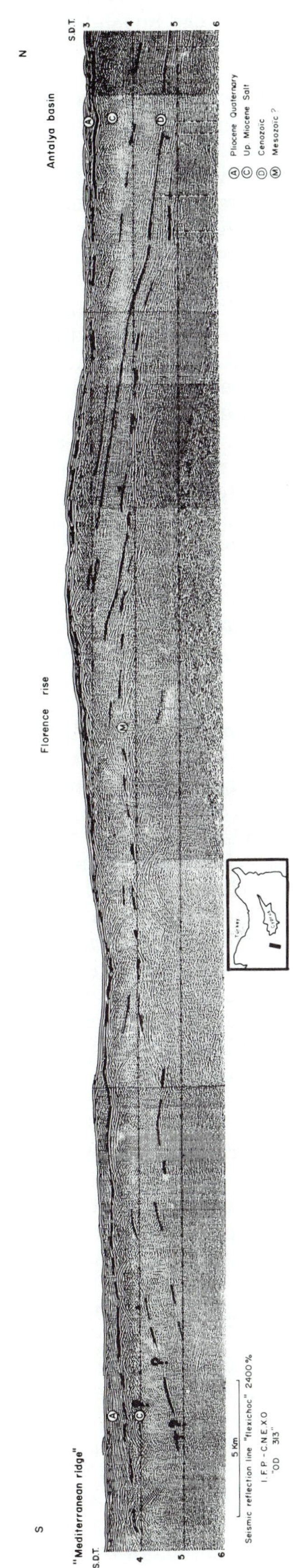

Fig. 17. Pinching out and thinning of the sedimentary series of Antalya Basin on the Florence Rise.

highly disturbed area that is possibly the front of a recent allochthonous mass beyond which the calm sedimentary series from the Levantine Sea extend.

Basins Located Between the Cyprus Arc and the Neogene Arc. *Antalya Basin.* This basin covers a reduced area on land, but extends over a wide one in deep water. In the west, recent north-south tectonics makes it impossible to establish its relations with the Antalya nappes. According to the data gathered on land, this basin may include a complete Miocene and Pliocene series, highly structured in some places. Two subbasins can be defined, which may be related with the extension of the Neogene Kyrenia Arc. Underneath the Miocene, Oligocene series may exist, along with more or less complete Eocene, Mesozoic and, perhaps, even Paleozoic strata, constituting an autochthonous substratum (Brunn et al., 1971; Magne and Poisson, 1974; Bizon et al., in press). These series can include allochthonous masses and can be affected in the northern part, corresponding to the Antalya-Mamonia nappes, by more recent tectonics (Eocene or Neogene overthrusts). It must be noted that the Upper Miocene gypsum known on land extends into the deep marine basin, where a very thick salt layer developed.

North Levantine Basins. These basins (Mesaorea, Iskenderun, deep Antakya) are located between Cyprus and the Levant Coast. They are more-or-less separated because of recent faulting. They include a recent series of varying thickness, a late Miocene evaporitic series possibly including salt, and a thick infrasalt series which is assumed to have Eocene and even Late Maestrichtian deposits at the base, as on the land edges. The substratum can be represented by allochthonous masses under which the Mesozoic, and even the Paleozoic, of the Arabic Platform may be present (cf. the Amanos or Kessab inliers that are found on land along these basins on the Levant Coast).

In conclusion, the ophiolites in the Cyprus Arc, like those known in the Taurus Mountains and probably in the substratum of the basins, could be evidence of an ancient oceanic crust formed from the Triassic to the Cretaceous, obducted at the end of the Cretaceous, whose "racines" must be sought as far as Turkey. The extensive allochthony of the outer part of the arc is such that the northward extension of the African block cannot be determined precisely.

Neogene Arc. Toward the north, these basins are bounded by a Pliocene overthrusting zone running from the south to the eastern Taurus Mountains (Bitlis overthrust, Elazig Nappe; Rigo de Righi and Cortesini, 1964), up to the Misis Mountains, and continuing offshore until the Kyrenia Range (North of Cyprus), and then extending toward the northwest in the eastern part of the Gulf of Antalya until onshore perhaps the small Gebiz Mountains. Toward the south in the Neogene Arc we find exotic sedimentary masses in flysch of various ages, thus providing evidence of an active margin between the Eocene and the Miocene. Recent vertical faulting affects this complex area.

Cilicia-Adana Basin. The offshore Cilicia Basin is located north of the Neogene Arc and continues with the onshore Adana Basin. The sedimentary series in the latter is well known (Schmidt, 1961). It contains great thicknesses of Pliocene and Quaternary, a development of the Messinian with evaporites changing very quickly to thick salt (nearly 1,000 m, traversed by boreholes), and a calm and relatively unstructured Miocene (several thousand meters thick). These features stand out on seismic profiles. Great accumulations of recent sediments are linked to the deltas of rivers coming from the Taurus Mountains and leaving accumulations on the present margin of the basin. It should be noted that in the Messinian there was quite probably some communication with the basins located farther south. The eastern part of the Antalya Basin should probably be attached to these internal basins, being situated north of the Neogene arc.

To the north of these basins, in both onshore and offshore development, the folded Taurus Range is present (e.g., Blumenthal, 1963; Ozgul and Arpat, 1972), then a zone with Neogene basins having a thick continental infilling, as well as a recent highly developed basaltic volcanism (Jung and Keller, 1972; Westerveld, 1957). This general pattern is similar to the succession of typical structural elements of an island arc, with the Turkish continental basins representing the equivalent of the Aegean Sea basins. However, we should bear in mind that here, again, there is the superposition of structures having various ages, from the Late Cretaceous to the Pliocene.

It must be noted that during Eocene to Upper Miocene time, there was a large communication between the Mediterranean area and the northern part of the Arabian platform.

CONCLUSIONS

Our geological understanding of the Mediterranean Sea has evolved considerably in recent years, especially because of numerous deep-seismic-reflection studies and from data of the GLOMAR CHALLENGER (DSDP, Leg 13), which have enabled both offshore and onshore data to be better integrated.

The recent structures and distribution of Plio-Quaternary sediments are controlled to a considerable extent by influxes from the Nile, Po, Rhone and Ebro deltas. The Plio-Quaternary tectonics are prominent and locally manifested either by distension or compression movements. In the western Mediterranean, this consists mainly of a collapse of the basin, accompanied by the uplifting of some margins. Extensive volcanism occurs in particular in the Tyrrhenian Sea and in the Sicilo-Tunisian Strait; there are various causes for this, depending on the area. In the eastern Mediterranean, sliding phenomena, comparable to those in the Gibraltar Arc, and compression movements occur offshore along the Calabrian, Aegean and Cyprus arcs, or onshore, such as in the Apennine or Kyrenian-Misis mountains.

The widespread nature of evaporite development in the late Miocene has been established both onshore and offshore. Distribution covers quite varied structural realms in relation to Alpine tectonic elements: either on the inside (i.e., Adana, Tyrrhenian, Transylvania), or on the outside (eastern Mediterranean, North Western Basin, outside boundary of the Carpathian).

Many interpretations have been proposed regarding the formation of these evaporitic basins. Are such evaporitic layers: deep-water deposits (Schmalz, 1969); shallow-water deposits (Ogniben, 1957); or precipitations by desiccation of a deep Mediterranean basin (e.g., Ryan et al., 1973; Hsu, 1972)?

In these hypotheses, there is always a reference to communication with the Atlantic (constant or partial inflow from a hypothetical strait of Gibraltar). But is is noticed that communications could have existed more or less with the Red Sea during the middle Miocene, easily with the Euphrate Basin and Iran by the North Levantine area (Antakya Basin, Maras Basin, etc.) all during Miocene time, and briefly with the Para-Tethys by the Aegean Sea.

On the other hand, it appears that the communication with the Atlantic was not the present Strait of Gibraltar but a North Betic Strait and/or a South Rif Strait. Moreover, the salt layers invariably developed in a basinal position: the topography, related to more-or-less old structures, controlled the extension of evaporites. This basinal position of salt does not allow definition of the absolute depth of the basin at that time. It is proved that very important down-faulting and subsidence occurred during the Pliocene and Quaternary. The extension of evaporites, especially of salt, generally prefigures the present Mediterranean Basin (Figs. 2 and 11). Halokinetic phenomena occur in some basins, especially in the Western Mediterranean Basin, the Antalya Basin, and probably in the Heordotus Plain.

Reconstruction of how the preevaporite deposition evolved, however, is much less clearly established, because correlations between what is known onshore and the seismic profiles is still difficult. Great differences appear between the western and eastern Mediterranean. In most of the Western Basin and the Tyrrhenian Basin, the visible sediments are "post-tectonic" in relation to the folded belts existing all around, and their age may range from the Oligocene to the Miocene. On the other hand, in the eastern Mediterranean the sedimentary series may include older units (Paleogene and Mesozoic), some of which represent the extension of African and Adriatic deposits.

Hence the age and genesis of the different Mediterranean basins appear to be different, even though the most recent geological events give them a certain degree of unity. Understanding their genesis implies an overall reconstruction of the history of the Mediterranean and Alpine realm. Data taken from the neighboring Atlantic and Indian oceans can be used to consider the African and European movements during the Mesozoïc and Cenozoic. Between these two major plates, the movements and translations (Dewey et al., 1973) have affected a realm that should be complex and probably includes a certain number of subplates.

This type of reconstruction, although somewhat risky, should be based on synthesis of land geological data and on additional offhsore data, i.e., an oceanic crust formed by extension, an oceanic or intermediate crust formed in a back-arc basin, or a collapsed and eventually oceanized continental crust. It is possible, in particular with the seismicity (McKenzie, 1972) and offshore seismic profiling, to define the present boundaries of the different crustal blocks, their recent movements, and the different kinds of recent basins; but difficulties quickly arise when considering older history.

ACKNOWLEDGMENTS

This paper is published by permission of IFP, CNEXO, CFP, ELF-ERAP, and SNPA. We are grateful to J. P. Fail, R. Donatien, and J. Cassand, geophysicists from IFP, and the officers and crew of the M.S. FLORENCE, which made the seismic survey. We thank G. Alla, V. Apostolescu, G. and J. J. Bizon, and A. Mauffret for their collaboration in this study, and J. Lort and R. Gonnard for useful comments on the manuscript.

BIBLIOGRAPHY

Abbate, E., Bortolotti, V., Passerini, P., and Sagri, M., 1970, The geosyncline concept and the northern Apennines: Sediment. Geology, v. 4, no. 3-4, p. 625-636.

Alinat, J., Giermann, G., and Leenhardt, O., 1966, Reconnaissance sismique des accidents de terrain en Mer Ligure: Compt. Rend., t. 262, ser. B, no. 19, p. 1311-1314.

Alla, G., Byramjee, R., and Didier, J., 1972, Structure géologique de la marge continentale du Golfe du Lion (abstr.): 23 ème Congr. CIESM, Athènes.

Allan, T. D., and Morelli, C., 1971, A geophysical study of

the Mediterranean Sea: Bull. Geofis. Teor. Appl., v. 13, no. 50, p. 99-141.

Andrieux, J., Fontbote, J. M., and Mattauer, M., 1971, Sur un modèle explicatif de l'arc de Gibraltar: Earth and Planetary Sci. Letters, v. 12, no. 2, p. 191-198.

Aubouin, J., 1971, Réflexion sur la tectonique de faille plio-quaternaire: Geol. Rundschau, v. 60, no. 3, p. 833-848.

———, and Dercourt, J., 1965, Sur la géologie de l'Egée: regard sur la Crète (conclusions générales): Bull. Soc. Géol. France, sér. 7, t. 7, no. 5, p. 787-821.

———, Blanchet, R., Cadet, J. P., Celet, P., Charvet, J., Chorowicz, J., Cousin, M., and Rampnoux, J. P., 1970b, Essai sur la géologie des Dinarides: Bull. Soc. Géol. France, sér. 7, t. 12, no. 6, p. 1060-1095.

———, Bonneau, M., Celet P., Charvet, J., et al., 1970a, Contribution à la géologie des Hellénides: le Gavrovo, le Pinde et la zone ophiolitique subpélagonienne: Ann. Soc. Géol. Nord, t. 90, no. 4, p. 277-306.

Auzende, J. M., Bonnin, J., Olivet, J. L., Pautot, G., and Mauffret, A., 1971, Upper Miocene salt layer in the western Mediterranean basin: Nature Phys. Sci., v. 230, no. 12, p. 82-84.

Auzende, J. M., Olivet, J. L., and Bonnin, J., 1972, Une structure compressive au Nord de l'Algérie: Deep-Sea Res., v. 19, no. 2, p. 149-155.

Barbier, R., Bloch, J. P., Debelmas, J., Ellenberger, F., Fabre, J., Feys, R., et al., 1963, Problèmes paléogéographiques et structuraux dans les zones internes des Alpes occidentales entre Savoie et Méditerranée, Livre à la mémoire du Prof. P. Fallot: Mém. Soc. Géol. France, t. 2, p. 331-378.

Bayer, R., Le Mouel, J. L., and Le Pichon, X., 1973, Magnetic anomaly pattern in the western Mediterranean: Earth and Planetary Sci. Letters, v. 19, no. 2, p. 168-176.

Beck, R. H., 1972, The oceans, the new frontier in exploration: A.P.E.A. Journal, v. 12, p. 2, p. 1-21.

Bellaiche, C., Gennesseaux, M., Mauffret, A., and Rehault, J. P., 1974, Prélèvements systématiques et caractéristiques des réflecteurs acoustiques: nouvelle étape dans la compréhension de la géologie de la Méditerranee occidentale: Marine Geology (in press).

Bemmelen, R. W., van, 1969, Origin of the western Mediterranean sea: Verhandel. Ned. Geol. Mijnbouwk. Genoot., v. XXVI, p. 13-52.

———, 1972, Driving forces of Mediterranean orogeny: Geol. Mijnbouw, v. 51, no. 5, p. 548-573.

Biju-Duval, B., Lapierre, H., and Letouzey, J., 1974, Le massif ophiolitique du Troodos (Chypre), est-il allochtone comme son prolongement en Turquie (Nappe de Kevan, Kizil Dag): 2eme Réunion Ann. Sci. Terre, Nancy (to be published).

Bizon, G., Biju-Duval, B., Letouzey, J., Oztumer, E., and Poisson, A., 1974, Nouvelles précisions stratigraphiques concernant les bassins tertiaires du Sud de la Turquie (Antlya, Mut, Adana): Rev. Inst. France Pétrole (to be published).

Blumenthal, M. M., 1963, Le système structural du Taurus sud-anatolien, Livre à la mémoire du Prof. P. Fallot: Mém. Soc. Géol. France, t. 2, p. 611-662.

Bonnin, J., Auzende, J. M., and Olivet, J. L., 1973, L'extrêmité orientale de la zone Acores-Gibraltar: structure et Evolution (abstr.): Réunion Ann. Sci. Terre, Paris, p. 91.

Borsi, S., Ferrara, G., Innocenti, F., and Mazzuoli, R., 1972, Petrology and geochronology or recent volcanics of eastern Aegean Sea (west Anatolia and Lesvos Island): Z. Deut. Geol. Ges., Bd. 123, Tl. 2, p. 521-522.

Bourcart, J., 1963, La Méditerranée et la révolution du Pliocène, Livre à la mémoire du Prof. P. Fallot: Mém. Soc. Géol. France, t. 1, p. 103-116.

Bourgois, J., Bourrouilh, R., Chauve, P., et al., 1970, Données nouvelles sur la géologie des Cordillères bétiques: Ann. Soc. Géol. Nord, t. 90, no. 4, p. 347-393.

Bousquet, J. C., 1972, La tectonique récente de l'Apennin Calabro-Lunanien dans son cadre géologique et géophysique [thèse]: Montpellier, 172 p.

Brunn, J. H., and Desprairies, A., 1965, Esude sédimentologique préliminaire de formations à caractères flysch et molasse (Flysch du Pinde et molasse du sillon méso-hellénique): Rev. Géograph. Phys. Geol. Dyn., sér. 2, v. 7, fasc. 4, p. 339-354.

Brunn, J. H., Dumont, J. F., Graciansky, P. C. de, Gutnic, M., Juteau, T., Marcoux, J., Monod, O., and Poisson, A., 1971, Outline of the geology of the western Taurids, *in* Campbell, A. S., ed., Geology and history of Turkey: Tripoli, p. 225-255.

Burollet, P. F., 1967, General geology of Tunisia and Tertiary geology of Tunisia, *in* Guidebook to the geology and history of Tunisia: Amsterdam, Holland-Breumelhof, p. 51-58, 215-225.

———, and Dufaure, P., 1972, The Neogene series drilled by the Mistral no. 1 well in the Gulf of Lion, *in* Stanley, D. J., ed., The Méditerranean Sea: a natural sedimentation laboratory, p. 91-98.

———, and Magnier, P., 1970, Mésozoique et Tertiaire en Libye: Compt. Rend. Soc. Géol. France, fasc. 5, p. 141-142.

Caire, A., 1970, Tectonique de la Méditerranée centrale: Ann. Soc. Géol. Nord, v. 90, no. 4, p. 307-346.

Caputo, M., Panza, G. F., and Postpischl, D., 1970, Deep structure of the Mediterranean basin: Jour. Geophys. Res., v. 75, no. 26, p. 4919-4923.

Carey, S. W., 1958, A tectonic approach to continental drift, *in* Carey, S. W., ed., Continental drift: Hobart, Tasmania, P. 177-355.

Closs, H., and Hinz, K., 1972, Seismiche und Bathymetrische Ergebnisse von Mediterranean Rucken und Hellenischen Graben, Communication orale., Aegean Symp. Hannover.

Colombi, B., Giese, P., Luongo, G., Morelli, C., Riuscetti, M., Scarascia, S., Schutte, K. G., Strowald, J., and Visintini, G. de., 1973, Preliminary report on the seismic refraction profile Gargano-Salerno-Palermo-Pantelleria (1971): Boll. Geofis. Teor. Appl., v. 15, no. 59, p. 225-254.

Conant, L. C., and Goudarzi, G. H., 1967, Stratigraphic and tectonic framework of Libya: Am. Assoc. Petr. Geol. Bull., v. 51, no. 5, p. 719-730.

D'Argenio, B., Radoicic, R., and Sgrosso, I., 1971, A paleogeographic section through the italo-dinaric external zones during Jurassic and Cretaceous times: Nafta Zagreb, v. 22, no. 4-5, p. 195-207.

Delteil, J. R., Durand, J., Semichon, P., Montadert, L., Fondeur, C., and Mauffret, A., 1972a, Structure géologique de la marge continentale catalane (abstr.): 23ème Congr. CIESM, Athenes.

Delteil, J. R., Durand, J., Semichon, P., Montadert, L., Letouzey, J., and Fail, J. P., 1972b, Structure de la marge continentale à l'Ouest du Golfe de Bonifacio (abstr.): 23ème Congr. CIESM, Athènes

Dercourt, J., 1970, L'expansion océanique actuelle et fossile: ses implications géotectoniques: Bull. Soc. Géol. France, ser. 7, t. 12, no. 2, p. 261-317.

Dewey, J. F., Pitman, W. C., III, Ryan, W. B. F., and Bonnin, J., 1973, Plate tectonics and the evolution of the Alpine system: Geol. Soc. America Bull., v. 84, no. 10, p. 3137-3180.

Didon, J., Durand-Delga, M., and Kornprobst, J., 1973, Homologies géologiques entre les deux rives du détroit de Gibraltar: Bull. Soc. Géol. France, Ser. 7, t. 15, no. 2, p. 77-105.

Dubertret, L., 1967, Remarques sur le fossé de la Mer Morte et ses prolongements au Nord jusqu'au Taurus: Rev. Géograph. Phys. Géol. Dyn., sér. 2, v. 9, fasc. 1, p. 3-16.

———, 1969, Le Liban et la dérive des continents: Hannon, Rev. Libanaise Géograph., v. 4, p. 53-61.

Dubois, R., 2970, Phases de serrage, nappes de socle et métamorphisme alpin à la jonction Calabre-Apennin: la suture calabro-appeninique: Rev. Géograph. Phys. Géol. Dyn., sér. 2, v. 12, fasc. 3, p. 221-254.

Durand-Delga, M., 1960, Mise au point sur la structure du Nord-Est de la Berbérie: Publ. Serv. Géol. Algérie (N.S.), no. 39, p. 89-131.

———, 1972, La courbure de Gibraltar, extrémité occidentale des chaînes alpines, unit l'Europe et l'Afrique: Eclogae Geol. Helv., v. 65, no. 2, p. 267-278.

Emelyanov, Y. M., Mikhailov, O. V., Moskalenko, V. N., and Shimkus, K. M., 1964, Basic features of the tectonic structure of the Mediterranean floor: 22ème Congr. Geol. Internatl. New Delhi, Compte rendu des géologues soviétiques, p. 97-111.

Emery, K. O., Heezen, B. C., and Allan, T. D., 1966, Bathymetry of the eastern Mediterranean Sea: Deep-Sea Res., v. 13, no. 2, p. 173-192.

Ente Nazionale Idrocarburi, 1962-1971, Colombo, Carlo, ed., Enciclopedia del Petrolio e del gaz naturale, v. 1-8.

Faculty of Science, University of Libya, 1971, Symp. on the Geology of Libya, Gray, C., ed., 522 p.

Fahlquist, D. A., and Hersey, J. B., 1969, Seismic refraction measurements in the western Mediterranean Sea: Bull. Inst. Océanog., v. 67, no. 1386, 52 p.

Finetti, I., and Morelli, C., 1972a, Regional reflection seismic exploration of the Strait of Sicily, *in* Oceanography of the strait of Sicily: Proc. Saclant Conf., no. 7, p. 208-223.

———, and Morelli, C., 1972 b, Wide scale digital seismic exploration of the Mediterranean Sea: Boll. Geofis. Teor. Appl., v. 14, no. 56, p. 291-342.

———, Morelli, C., and Zarudzki, E., 1970, Reflection seismic study of the Tyrrhenian Sea: Boll. Geofis. Teor. Appl., v. 12, no. 48, p. 311-346.

Freund, R., Zak, I., and Garfunkel, Z., 1968, Age and rate of the sinistral movement along the Dead Sea rift: Nature, v. 220, no. 5164, p. 253-255.

Galanopoulos, A. G., 1969, The seismotectonic regime in Greece: Izv. Acad. Sci. S.S.S.R., Physics of the solid earth, no. 7, p. 455-460 (trad. Am. Geophys. Union).

Giermann, G., 1969, The eastern Mediterranean ridge: Rapp. Comm. Internatl. Mer Medit., v. 19, p. 605-607.

———, Pfannenstiel, M., and Wimmenauer, W., 1968, Relations entre morphologie, tectonique et volcanisme en Mer d'Alboran (Méditerranée occidentale): Resultats préliminaires de la campagne du Jean Charcot: Compt. Rend. Soc. Géol. France, fasc. 4, p. 116-118.

Glangeaud, L., 1956, Correlation chronologique des phénomènes géodynamiques dans les Alpes, l'Apennin et l'Atlas Nord-Africain: Bull. Soc. Géol. France, sér. 6, v. 6, p. 867-891.

———, 1971, Evolution géodynamique de la mer d'Alboran et de ses bordures: la phase messino-plio-quaternaire (résumé): Compt. Rend. Soc. Geol. France, fasc. 8, p. 431-433.

———, Alinat, J., Agarate, C., Leenhardt, O., and Pautot, G., 1967, Les phénomènes ponto plio-quaternaires dans la Méditerranée occidentale, d'après les données de Géomède 1: Compt. Rend., t. 264, sér. D, no. 2, p. 208-211.

Grandjacquet, C., and Glangeaud, L., 1962, Structures mégamétriques et évolution de la Mer Tyrrhénienne et des zones pérityrrhéniennes: Bull. Soc. Géol. France, sér. 7, t. 4, no. 5, p. 760-773.

———, Haccard, D., and Lorenz, C., 1972, Essai de tableau synthétique des principaux évènements affectant les domaines alpin et appenin à partir du Trias: Compt. Rend. Soc. Géol. France, fasc. 4, p. 158-163.

Haccard, D., and Lemoine, M., 1970, Sur la stratigraphie et ales analogies des formations sédimentaires associees aux ophiolites dans la zone piémontaise des Alpes ligures (zones de Sestri-Voltaggio et de Montenotte) et des Alpes cottiennes (zone de Gondran, Queyras, Haute-Ubaye): Compt. Rend. Soc. Géol. France, fasc. 6, p. 209-211.

Haccard, D., Lorenz, C., Grandjacquet, C., 1972, Essai sur l'évolution tectogénétique de liaison Alper-Apennins (de la Ligurie à la Calabra): Mém. Soc. Géol. Ital., v. 11, p. 309-341.

Heezen, B. C., Gray, C., Segre, A. G., Zarudski, E. F. K., 1971, Evidence of foundered continental crust beneath the central Tyrrhenian Sea: Nature, v. 229, no. 5283, p. 327-329.

Hersey, J. B., 1965, Sedimentary basins of the Mediterranean Sea, *in* Whittard, W. F., and Bradshaw, R., eds., Submarine geology and geophysics: Colston Papers, v. 17, p. 75-92.

Hieke, W., 1972, Erste Ergebnisse von stratigraphisch-sedimentologischen und morphologishtektonischen Untersuchungen auf dem Mediterranean Rucken (Ionisches Meer) Z. Deut. Geol. Ges., Bd. 123, Tl. 2, p. 567-570.

Hinz, K., 1970, A low velocity layer in the upper crust of the Ionian Sea (abstr.): 22eme Congr. CIESM, Roma.

———, 1972a, Refraktionsseismische Untersuchungen im ostlichen Mittlemeer: das unternehmen Erdmantel Deutsch.: Weisbaden, Franz Steiner Verlag, p. 174-175.

———, 1972b, Results of seismic refraction investigations (Project Anna) in the western Mediterranean Sea, south and north of the Island of Mallorca, *in* Leenhardt, O., et al., eds., Results of the Anna cruise, Bull. Centre Rech. Pau—SNPA, v. 6, no. 2, p. 405-426.

Hsu, K. H., 1972, Origin of Saline Geants: a critical review after the discovery of the Mediterranean evaporite: Earth Sci. Rev. v. 8, p. 371-396.

Institut Français du Pétrole et Institut de Géologie de Grèce, 1966, Etude géologique de l'Epire, Grèce nord-occidentale: Technip, 306 p.

Jung, D., Keller, J., 1972, Young volcanic rocks of the Konya-Kayseri region (central Anatolia): Z. Deut. Z. Deut. Geol. Ges., Bd. 123, t. 2, p. 503-512.

Karig, D. E., 1971, Origin and development of marginal basins in the Western Pacific. Jour. Geophys. Res., v. 76, no. 11, p. 2542-2561.

———, 1972, Remnant arcs: Geol. Soc. America Bull., v. 83, no. 4, p. 1057-1068.

Kieken, M., 1962, Les traits essentiels de la géologie algérienne, Livre a la mémoire du Prof. P. Fallot:

Mém. Soc. Géol. France, t. 1, p. 545-614.

Klemme, H. D., 1958, Regional geology of circum-Mediterranean region: Am. Assoc. Petr. Geol. Bull., v. 42, no. 3, p. 477-512.

Klitzsch, E., 1970, Die Strukturgeschichte der Zentralsahara: neue Erkenntnisse zum Dau und zur Paläogeographic eines Tafellandes: Geol. Rundschau, Bd. 59, 2, p. 459-527.

Lapierre, H., 1972, Les formations sédimentaires et éruptives des nappes de Mamonia et leurs relations avec le massif du Troodos (Chypre) [thèse d'Etat]: Nancy, 420 p.

———, 1974, Troodos allochthonous sedimentary and volcanic cover: the Mamonia nappes (Cyprus); a probable Late Cretaceous subduction zone: Earth and Planetary Sci. Letters (to be published).

Laubscher, H. P., 1971, The large-scale kinematic of the Western Alps and the Northen Apennines and its palinspastic implications: Am. Jour. Sci. v. 271, p. 193-226.

Le Borgne, E., Le Mouel, J. L., and Le Pichon, X., 1971, Aeromagnetic survey of south western Europe: Earth and Planetary Sci. Letters, v. 12, no. 3, p. 287-299.

Lemoine, M., 1972, Eugeosynclinal domains of the Alps and the problem of past oceanic areas. Internatl. Geol. Congr., 34th Session, Montreal, sect. 3, p. 476-485.

Le Mouel, J., and Le Borgne, E., 1970, Les anomalies magnétiques du Sud-Est de la France et de la Méditerranée occidentale: Compt. Rend., t. 271, ser. D, no. 16, p. 1348-1350.

Le Pichon, X., Pautot, G., Auzende, J. M., and Olivet, J. L., 1971, La Méditeranée occidentale depuis l'Oligocène. Schéma d'évolution: Earth and Planetary Sci. Letters, v. 13, no. 1, p. 145-152.

———, Pautot, G., and Weill, J. P., 1972, Opening of the Alboran Sea: Nature Phys. Sci., v. 236, no. 67, p. 83-85.

Letouzey, J., Biju-Duval, B., Courrier, P., and Montadert, L., 1974, Nappes de glissement actuelles au front de l'arc calabrais en Mer Ionienne (d'après la sismique réflection): 2ème Réunion Ann. Terre, Nancy, (in press).

Lort, J. M., 1972, The crustal structure of the eastern Mediterranean [these]: Univ. Cambridge, 117 p.

McKenzie, D., 1972, Active tectonics of the Mediterranean region: Geophys. Jour. v. 30, no. 2, p. 109-185.

Magne, J., and Poisson, A., 1974, Présence de niveaux oligocènes dans les formations sommitales du massif des Bey Daglary, près de Korkuteli et de Bucak (autochtone du Taurus lycien, Turquie): Compt. Rend., t. 278, ser. D, no. 2, p. 205-208.

Maley, T. S., and Johnson, G. L., 1971, Morphology and structure of the Aegean Sea: Deep-Sea Res., v. 18, no. 1, p. 109-122.

Mascle, G. H., 1973, Etude géologique des Monts Sicani (Sicile) [these doct.]: Univ. Paris VI, 691 p.

Mauffret, A., and Sancho, J., 1970, Etude de la marge continentale au nord de Majorque (Baléares, Espagne): Rev. Inst. Franc. Pétrole, v. 25, no. 6, p. 714-730.

Mauffret, A., Auzende, J. M., Olivet, J. L., and Pautot, G., 1972, Le bloc continental Baléare (Espagne)—Extension et évolution: Marine Geology, v. 12, no. 4, p. 289-300.

Mauffret, A., Fail, J. P., Montadert, L., Sancho, J., and Winnock, E., 1973, Northwestern Mediterranean sedimentary basin from seismic reflection profile: Am. Assoc. Petr. Geol. Bull., v. 57, no. 11, p. 2245-2262.

Menard, H. W., 1965, The world-wide oceanic rise-ridge system: Phil. Trans. Roy. Soc. London, ser. A, no. 1088, p. 109-122.

Mercier, J., 1966, Paléogéographie, orogenèse, métamorphisme et magmatisme des zones internes des Hellénides en Macédoine (Grece): vue d'ensemble: Bull. Soc. Géol. France, 7ème sér. t. 8, no. 7, p. 1020-1049.

———, Bousquet, B., Delibasis, N., Drakopoulos, I., Keraudren, B., Lemeille, F., and Sorel, D., 1972, Déformations en compression dans le Quaternaire des rivages ioniens (Céphalonie, Grèce): données néotectoniques et séismiques: Compt. Rend., ser. D, t. 275, no. 21, p. 2307-2310.

Miyashiro, A., 1973, The Troodos ophiolitic complex was probably formed in an island arc: Earth and Planetary Sci. Letters, v. 19, no. 2, p. 218-224.

Montadert, L., Sancho, J., Fail, J. P., Debyser, J., and Winnock, E., 1970, De l'âge tertiaire de la série salifère responsable des structures diapiriques en Méditerranée occidentale (Nord-Est des Baléares): Compt. Rend., sér. D, t. 271, no. 10, p. 812-815.

Morelli, C., 1970, Physiography, gravity and magnetism of the Tyrrhenian Sea: Boll. Geofis. Teor. Appl., v. 12, no. 48, p. 275-308.

———, 1972, Bathymetry, gravity and magnetism in the Strait of Sicily, *in* Oceanography of the Strait of Sicily: Proc. Saclant. Conf., no. 7, p. 193-208.

———, Carrozzo, M. T., Ceccherini, P., et al., 1969, Regional geophysical study of the Adriatic Sea (with an appendix by G. Giogetti, and F. Mosetti: General morphology of the Adriatic Sea): Boll. Geofis. Teor. Appl., v. 11, no. 41-42, p. 3-56.

Moskalenko, V. N., 1966, New data on the structure of the sedimentary strata and basement in the levant sea: Oceanology, 1966, v. 6, no. 6, p. 828-836.

Mulder, C., 1973, Tectonic framework and distribution of Miocene chemical sediments with emphasis on the eastern Mediterranean: Congr. Utrecht.

Muraour, P., Marchand, J. P., Ducrot, J., and Ceccaldi, X., 1966, Remarques sur la structure profonde du précontinent de la région de Calvi (Corse) à la suite d'une étude de sismique réfraction: Compt. Rend., ser. D, no. 262, no. 1, p. 17-19.

Neev, D., Edgerton, H. E., Almagor, G., and Bakler, N., 1966, Preliminary results of some continuous seismic profiles in the Mediterranean shelf of Israel: Israel Jour. Earth Sci. v. 15, p. 170-178.

Neev, D., Almagor, G., Arad, A., Benavraham, Z., Ginzburg, A., and Hall, J. K., 1973a, The geology of the southeastern Mediterranean Sea: Internatl. Rept. Geol. Survey Israel, 32 p.

Neev, D., Bakler, N., Moshkovitz, S., Kaufman, A., Magaritz, M., and Gofna, R., 1973b, Recent faulting along the Mediterranean coast of Israel: Nature, v. 245, no. 5423, p. 254-256.

Ogniben, L., 1957, Petrographia della serie solsifera-siciliana e considerazion geotoche relative: Mem. Descrit. Carta Geol. Ital., v. 33, 275 p.

———, 1970, Paleotectonic history of Sicily, *in* Geology and history of Sicily: Petr. Expl. Soc. Libya, 12th Ann. Field Conf., Tripoli, p. 133-144.

Olivet, J. L., Auzende, J. M., and Bonnin, J., 1973, Structure et évolution tectonique du bassin d'Alboran: Bull. Soc. Géol. France, 7ème sér., t. 15, no. 2, p. 108-112.

Ozgul, N., and Arpat, E., 1972, Structural units of the Taurus orogenic belt and their continuation in neighbouring regions (abstr.): 23eme Congr. CIESM, Athenes.

Papazachos, B. C., and Comninakis, P. E., 1971, Geophysical and tectonic features of the Aegean Arc: J. Geophys. Res., v. 76, no. 35, p. 8517-8533.

Payo, G., 1967, Crustal structure of the Mediterranean Sea by surface waves, pt. 1., Group velocity: Seismol. Soc. America Bull., v. 57, no. 2, p. 151-172.

Pitman, W. C., III, Talwani, M., 1972, Sea-floor spreading in the North Atlantic: Geol. Soc. America Bull., v. 83, no. 3, p. 619-646.

Quennell, A. M., 1958, The structural and geomorphic evolution of the Dead Sea Rift: Quart. Jour. Geol. Soc. London, v. 114, p. 1-24.

Rabinowitz, P. D., and Ryan, W. B. F., 1970, Gravity anomalies and crustal shortening in the eastern Mediterranean: Tectonophysics, v. 10, no. 5-6, p. 585-608.

Rehault, J. P., 1968, Contribution à l'étude de la marge continentale au large d'Imperia et de la plaine abyssale [these 3eme cycle]: Paris, 115 p.

Ricou, L. E., 1971, Le croissant ophiolitique péri-arabe: une ceinture de nappes mises en place au Crétacé supérieur: Rev. Geograph. Phys. Géol. Dyn., 2ème sér., v. 13, fasc. 4, p. 327-350.

Rige de Righi, M., and Cortesini, A., 1964, Gravity tectonics in foothills structure belt of southeast Turkey: Am. Assoc. Petr. Geol. Bull., v. 48, no. 12, p. 1911-1937.

Ritsema, A. R., 1970, On the origin of the western Mediterranean Sea basins: Tectonophysics, v. 10, no. 5-6, p. 609-623.

Ryan, W. B. F., 1969, The floor of the Mediterranean Sea, pt. 1, Structure and evolution of the sedimentary basins, pt. II, The stratigraphy of the eastern Mediterranean [Ph.D. thesis]: Columbia Univ., New York, 421 p.

———, Stanley, D. J., Hersey, J. B., Fahlquist, D. A., and Allan, T. D., 1971, The tectonics and geology of the Mediterranean Sea, *in* Maxwell, A. E., ed., The Sea, p. 387-492.

———, Hsu, K. J., et al., 1973, Initial reports of the Deep Sea Drilling Project, v. XIII: Washington, D.C., U.S. Govt. Printing Office, 1447 p.

Sancho, J., Letouzey, J., Biju-Duval, B., Courrier, P., Montadert, L., and Winnock, E., 1973, New data on the structure of the eastern Mediterranean Basin from seismic reflection: Earth and Planetary Sci. Letters, v. 18, no. 2, p. 189-204.

Schmalz, R. F., 1969, Deep-water evaporite deposition: a genetic model: Am. Assoc. Petr. Geol. Bull., v. 53, no. 4, p. 798-823.

Schmidt, G. C., 1961, Stratigraphic nomenclature of the Adana region petroleum district VII: Petr. Admin. Publ. Bull Ankara, no. 6, p. 47-62.

Schuiling, R. D., 1972, Oceanization-geothermal models: Geol. Mijnbouw, v. 51, p. 546-547.

Selli, R., and Fabbri, A., 1971, Tyrrhenian: a Pliocene deep sea: Atti. Accad. Nazl. Lincei, v. 50, sér. 8, fasc. 5, p. 579-592.

Smith, A. G., 1971, Alpine deformation and the oceanic areas of the Tethys, Mediterranean, and Atlantic: Geol. Soc. America Bull., v. 82, no. 8, p. 2039-2070.

Stanley, D. J., 1973, The Mediterranean Sea—a Natural sedimentation laboratory: Stroudsburg, Pa., Dowden, Hutchinson & Ross, 765 p.

Vercellino, J., 1970, Here's what's known about the geology of the Italian Adriatic: Oil and Gas Int., v. 10, no. 11, p. 70-78.

Vine, F. J., Poster, C. K., and Gass, I. G., 1973, Aeromagnetic survey of the Troodos igneous massif, Cyprus: Nature Phys. Sci., v. 244, no. 133, p. 34-38.

Vogt, P. R., and Higgs, R. H., 1969, An aeromagnetic survey of the eastern Mediterranean Sea and its interpretation: Earth and Planetary Sci. Letters, v. 5, no. 7, p. 439-448.

———, Higgs, R. H., and Johnson, G. L., 1971, Hypotheses on the origin of the Mediterranean basin: magnetic data: Jour. Geophys. Res., v. 76, no. 14, p. 3207-3228.

Weigel, W., 1972, Preliminary results of refractional seismic measurements in the eastern Ionian Sea: Z. Deut. Geol. Ges., Bd. 123, t. 2, p. 571.

Westerveld, J., 1957, Phases of Neogene and Quaternary volcanism in Asia Minor: Congr. Geol. Internatl., 22ème Session, Mexico, sect. 1, Volcanologie du Cenozoique, t. 1, p. 103-119.

Wezel, C., 1970, Interpretazione dinamica della "eugeo-sinclinale meso-mediterranean:" Rivista Mineraria siciliana, v. 21, no. 124-126, p. 187-198.

Wong, H. K., Zarudzki, E. F. K., Phillips, J. D., and Giermann, G., 1971, Some geophysical profiles in the eastern Mediterranean: Geol. Soc. America Bull., v. 82, no. 1, p. 91-100.

Woodside, J., and Bowin, C., 1970, Gravity anomalies and inferred crustal structure in the eastern Mediterranean Sea: Geol. Soc. America Bull., v. 81, no. 4, p. 1107-1122.

Zarudzki, E. F. K., 1972, The Strait of Sicily—a Geophysical study: Rev. Geograph. Phys. Geol. Dyn., ser. 2, v. 14, fasc. 1, p. 11-27.

Zijderveld, J. D. A., and Van Der Voo, R., (in press) Palaeomagnetism of the Mediterranean area: Palaeomagnetic Rev.

Structure of the Western Mediterranean Basin

Jean-Marie Auzende and Jean-Louis Olivet

INTRODUCTION

The purpose of this paper is to describe margins of some Western Mediterranean basin zones that have played a particular tectonic role during the Oligo-Miocene phases of creation of the basin or that have undergone recent tectonic activity. In the following description, five zones will be considered: Ligurian Sea, Gulf of Valencia, Alboran Sea, Algerian-Balearic Basin, and Sardinia-Tunisia Straits. For each zone an acoustic basement isochron map and one or several seismic profiles will be presented. The acoustic basement might include crystalline continental rocks, Paleozoic sediments or consolidated Mesozoic or Cenozoic layers, as well as volcanic basement.

The western Mediterranean basin is bounded or intersected by several orogenic belts. On the northeastern side, the Ligurian Sea is a gap between the French-Italian alpine chain and its Corsican continuation. At the northwestern side, the Gulf of Lyon interrupts the eastern extension of the Pyrenees Mountains. In the central part of the basin, the Balearic Islands, a prolongation of the Betic Cordillera, separate the Gulf of Valencia from the north African basin. The southern part of the basin is bordered by the north African alpine chain. Such a complex structural framework enables us to propose a single model for the western Mediterranean margins. The Provencal Basin has been least affected by postcreation tectonism and therefore would be most "typical." Profiles in this basin also illustrate the relationship between margin basement and the sedimentary cover as observed all around the western Mediterranean basin (Auzende et al., 1971). Three sedimentary units can be distinguished: an upper Plio-Quaternary unit extending toward the shelf; a middle evaporitic and a salt layer of Messinian age (the continuity of this layer with those of similar age found in coastal basins is hard to ascertain from seismic-reflection profiles); and a lower unit of pre-Messinian sediments absent or very thin on the margins but well developed (1-2 sec) in the abyssal plain.

According to several authors, the western Mediterranean basin originated at the beginning of the Pliocene on a foundation of collapsed alpine structures. Our data show the existence of an extensive Messinian evaporitic layer as well as older nontectonized sedimentary layers, in a disagreement with this hypothesis (Auzende et al., 1971; Olivet et al., 1973). Other authors proposed that the basin was formed at the time of a major Triassic-Lias phase of distension. The deepest part of the basin is oceanic, as suggested by magnetic anomalies (Le Borgne et al. 1971) and seismic refraction (Fahlquist and Hersey, 1969). The magnetic basement is 8-10 km deep (Le Borgne et al., 1971), while seismic-reflection profiles indicate acoustic basement at about 6-7 sec. D.T.T. (double travel time) below sea level; a Trias-Lias oceanic basement would most probably lie at greater depth (Sclater et al., 1971), and the relatively thin sedimentary cover also suggest a younger age for the observed basement. A Middle to Upper Oligocene age and a Lower to Middle Miocene age have been proposed for the creation of the Provencal Basin and the Algero-Balearic Basin, respectively (LePichon et al., 1971; Auzende et al., 1973).

All seismic profiles are presented together at the conclusion of this text. Location of profiles is shown on Figures 1-5.

LIGURIAN SEA

The continental border of the Ligurian Sea is geologically complex. To the north it is bordered by the Maures and Esterel massifs, the Provencal folds, and the Alps-Apennine link, and to the south by the Corsican crystalline massif. Isochrons on acoustic basement from reflection profiles are given in Figure 1.

The northwestern or "Provencal" margin is abrupt and downfaulted in blocks parallel to the coastline. The southern, or "Corsican" margin is quite different. Its average slope is rather gentle with a general north-south orientation. Several basement highs of volcanic origin (Rehault, personal communication) crop out along the slope. They seem to be related to a northwest-southeast transverse fault system outlined by the trend of the Provencal margin canyons (Fig. 2). The Cap Mele fracture zone, with its vertical offset of more than 1,000 m, is the most important of these transverse faults (Fig. 2, profile 1). Northeast of this structure, the margin trend shifts toward the north. A decoupling seems to appear across this fracture zone between a southern part with an oceanic basement and a northern part characterized by a continental crust less affected by the Oligo-Miocene distension (LePichon et al., 1971). Dredging in the northern part reveals various continental material (Corsican granites, as well as sedimentary rocks, "*flyschs a Helmintoides*."

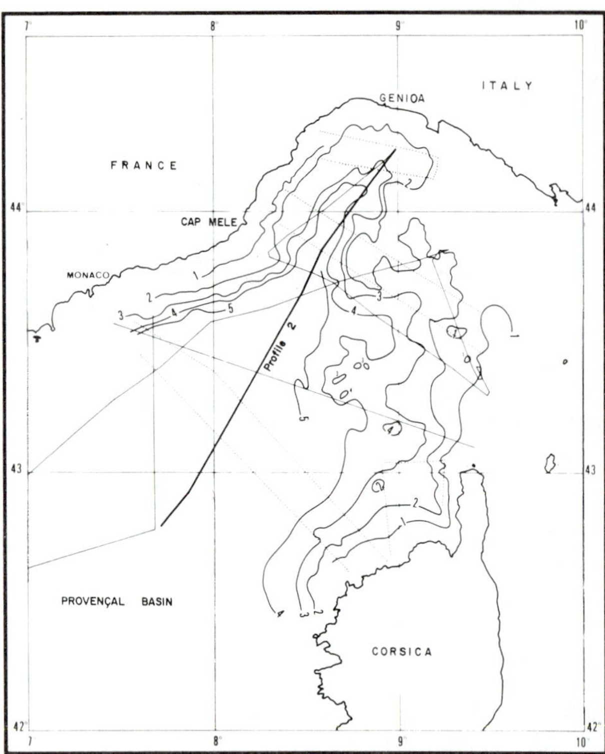

Fig. 1. Isochrons to acoustic basement of the Lugurian Sea (in seconds of two-way time). Continuous line, Flexotir profile; dotted line, airgun profile.

The flexotir profile along the axis of the Ligurian sea illustrates the transition between the elevated northern zone and the southern zone of the Ligurian sea (profile 2, Figs. 1 and 2). The difference in level reaches about 1 sec. The continuity of the evaporitic layer can be followed from the Ligurian Abyssal Plain to the northern end of the Gulf of Genoa. Several parallel profiles allow it to be traced to within 6 miles of the coastline.

GULF OF VALENCIA

The Gulf of Valencia is limited to the south by the Balearic Islands, which constitute the eastern prolongation of the Betic Cordillera (Bourrouilh, 1970) with the possible exception of Minorca. To the north, the continental border is formed by the Iberian chain perpendicular to the northeast-southwest axis of the Gulf of Valencia and by coastal Neogene basins.

The acoustic basement isochron map (Fig. 3) shows three main zones:

1. West of 2°30 an axial depression about 30 km wide, with a basement shallower than the Provencal Basin, trends west-southwest. A strong positive magnetic anomaly corresponds to a 25-km-long basement high (Vogt et al., 1971; Auzende et al., 1972). Eastward, this depression widens and is shifted toward the north by a lateral offset extending between the Ibiza and Majorca margins.

2. East of 3°E an eastern zone opens toward the Provencal Basin. The Balearic continental margin orientation changes from east-west at the latitude of Minorca to northwest-southeast toward the east. A series of basement highs cuts across the southern part of this zone, corresponding to an important northwest-southeast positive magnetic anomaly (Le Borgne et al., 1971) which is probably associated with the Balearic fracture zone (B, Fig. 2) (Auzende et al., 1973).

3. An intermediate zone between 40° and 41°N, where the basement forms two large elevated features, a western one cone-shaped with steep-slopes and the other one in the continuation of the Minorca Plateau, is characterized by a smoother slope.

Seismic-refraction results favor a continental origin for the basement of the Gulf of Valencia (Hinz, 1972). However, the basement highs of zone C are characterized by strong magnetic anomalies which point to important volcanic activity. The transition to the Provencal Basin oceanic basement is poorly defined but could be related to the northwest-southeast volcanic axis.

The sedimentary cover of the Gulf of Valencia is characterized by a thin salt layer which consists of evaporites that were cored during Deep Sea Drilling Project (DSDP) Leg XIII (site 122, Pautot et al., 1972). This layer (1 sec higher than in the Provencal Basin) lies over more than 0.5 sec of pre-Messinian sediments.

Profile 3, located between Minorca and the Spanish margin (Figs. 2 and 3), illustrates the morphology of the eastern Gulf of Valencia margins and their relationship to the sedimentary cover. In the central part of the profile the basement high is the northwestern extremity of the Balearic fracture zone volcanic trend (B, Fig. 2).

ALBORAN SEA

The Alboran Sea is a narrow basin (about 200 km wide) closed to the west by the Gibraltar Arch and open to the east toward the Algerian-Balearic Basin. Its northern boundary is the Betic Cordillera, the southeastern one is the Rif. The Atlas foreland limits it to the south.

The acoustic basement isochron map (Fig. 4) shows that the Alboran Basin is divided in two parts by a double line of basement highs, known as the Alboran Ridges (C, D, Fig. 2) (Olivet et al., 1972; Olivet et al., 1973). The western basin is 1,000 m shallower than the Algerian-Balearic Basin. The basement consists of a succession of ridges and lows. The most important of the latter lies under the Spanish continental margin between 3 and 5°W and is limited eastward by the Northern Alboran Ridge (C on Fig. 2), characterized by a northeast-southwest trend (Giermann et al., 1968; Olivet et al., 1972). Other depressions lie parallel to the Rif margin and are limited by the narrow northeast-

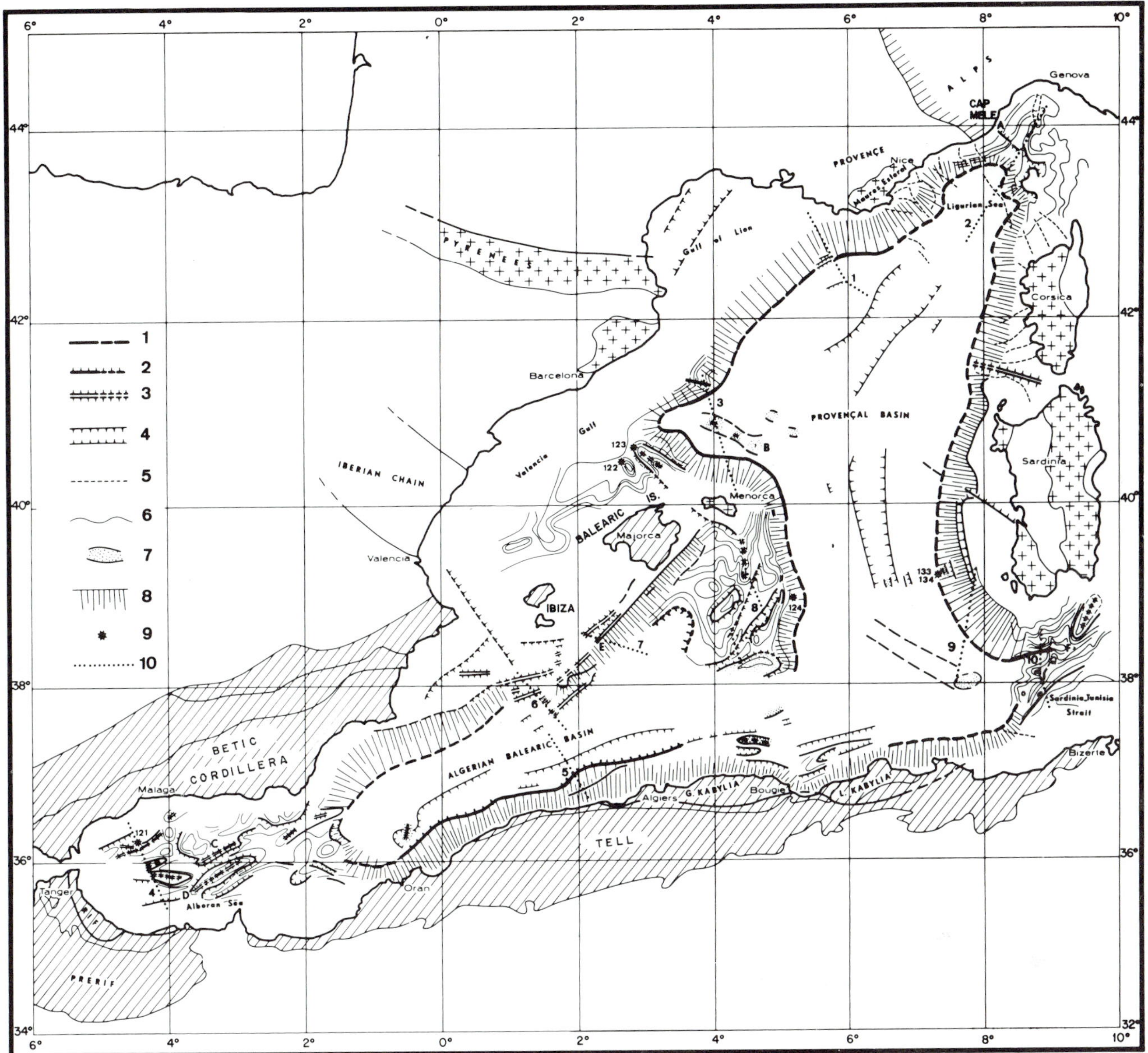

Fig. 2. Structural sketch map of western Mediterranean basin: 1, assumed limit between continental and oceanic basement; 2, major faults; 3, crest line; 4, in the Provencal basin: deeper axis of the basin; 5, canyons; 6, schematic isochrons to the basement (see other figures for more details); 7, basement highs in the oceanic part of the western Mediterranean basin or at the boundary between continental and oceanic crust; 8, slopes; 9, location of DSDP Leg XII sites; 10, location of profiles shown at end of text.
Important features include A, Cap Mele fracture zone; B, North Balearic fracture zone; C, Northern Alboran Ridge; D, Southern Alboran Ridge; E, Emile Baudot fracture zone; F, North Tunisian fracture zone.

trending southern Alboran Ridge (D, Fig. 2). These ridges are known to be for a great part volcanic (Olivet et al., 1972).

The eastern basin is a narrow east-west depression limited to the north by the extension of the Alboran Ridge and to the south by a large plateau extending north of the Algeria-Morocco margin with depths ranging from 200 to 500 m. Outcrops of continental rocks are numerous in this area. This basin is deeper than the western one and dips gently eastward to 1°W, where a sharp 500-m break marks the transition with the Algerian-Balearic Basin.

Profile 4 (Figs. 2 and 4) across the western basin in a north-northeast direction shows the difference between the Spanish and Moroccan margins: thick sedimentary accumulation covers the gentle northern slope. During DSDP Leg XIII, site 121 sampled a complete Plio-Quaternary section resting over a Messinian erosion level. This thickness of eroded sediments at the drill site has been estimated to range between 500 and 800 m (Olivet et al., 1973). This erosional phase occurred at the time of deposition of salt and evaporites in the eastern Alboran and Algerian-Balearic basins. Under this discontinuity at least 1-1.5 sec of undisturbed sedi-

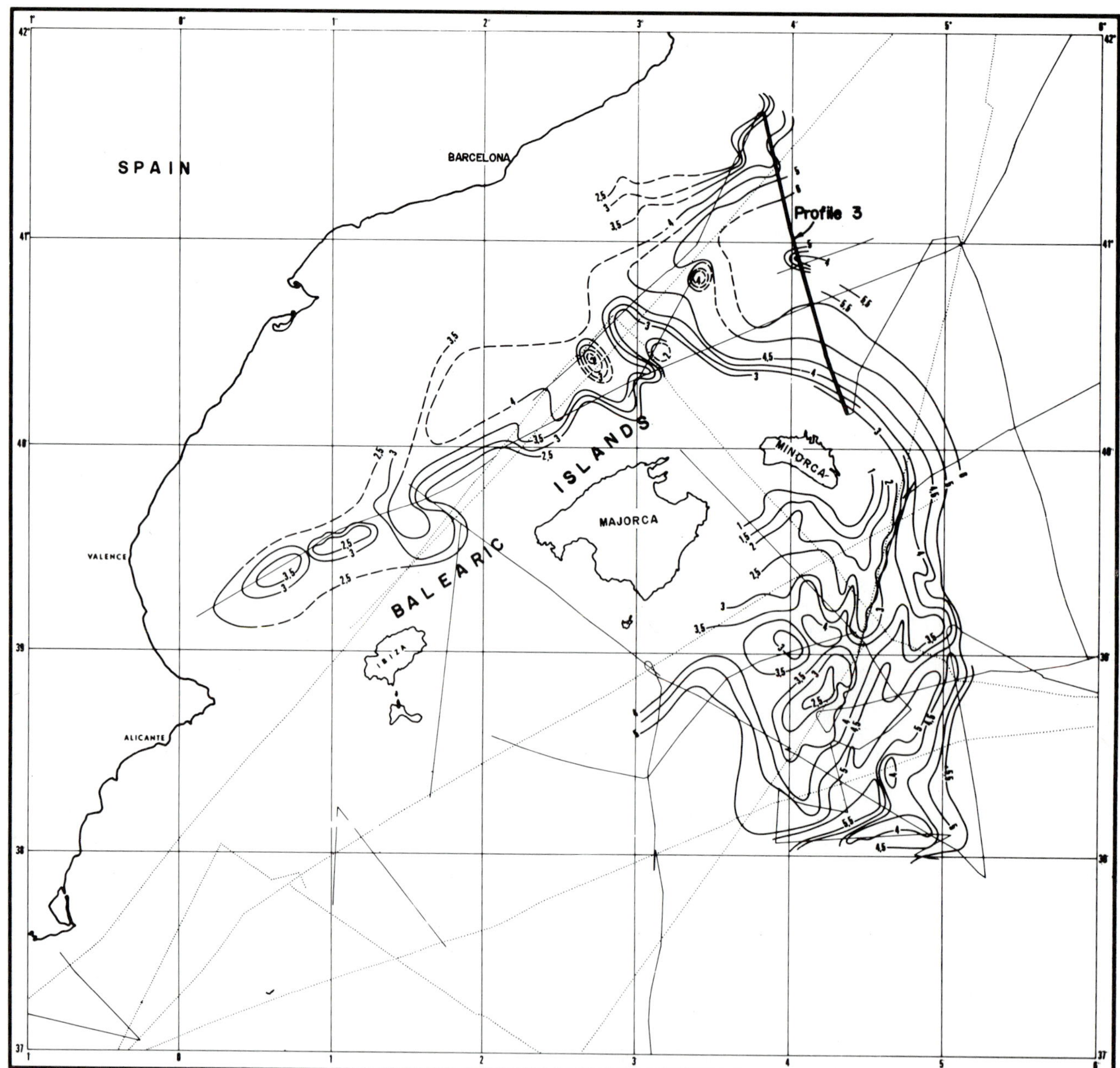

Fig. 3. Isochrons to acoustic basement around the Balearic Islands (in seconds of two-way time). Continuous line, Flexotir profile, dotted line, airgun profile.

ments are observed on the profile 4. The deepening of the Messinian erosion surface toward the south is associated with a thickening of the Plio-Quaternary cover. This deepening is observed everywhere along the north African margins and has been related to a recent compressional phase (Auzende et al., 1972).

ALGERIAN-BALEARIC BASIN

The Algerian-Balearic Basin constitutes the southern part of the western Mediterranean between the Alboran Sea and 6°E. Its northern limit is the Balearic Chain, its southern limit is the North African Alpine Chain. The asymmetry of the two continental margins can be seen on the acoustic basement isochron map (Fig. 5). The Algerian margin is narrow and displays structures parallel to the coastline. In the western zone, offshore from Oran, the slope consists of steps limited by normal faults. The general trend is northeast-southwest. East of 1°30'E the margin widens and trends in an east-west direction. Profile 5, northwest of Algiers, shows a large basement block at the base of the slope. Farther east, in the central part of the basin north of Great Kabylia and Gulf of Bougie, there are basement highs displaying strong positive magnetic anomalies (Le Borgne et al., 1972).

The Balearic continental margin widens from about 180 km south of Ibiza to more than 200 km

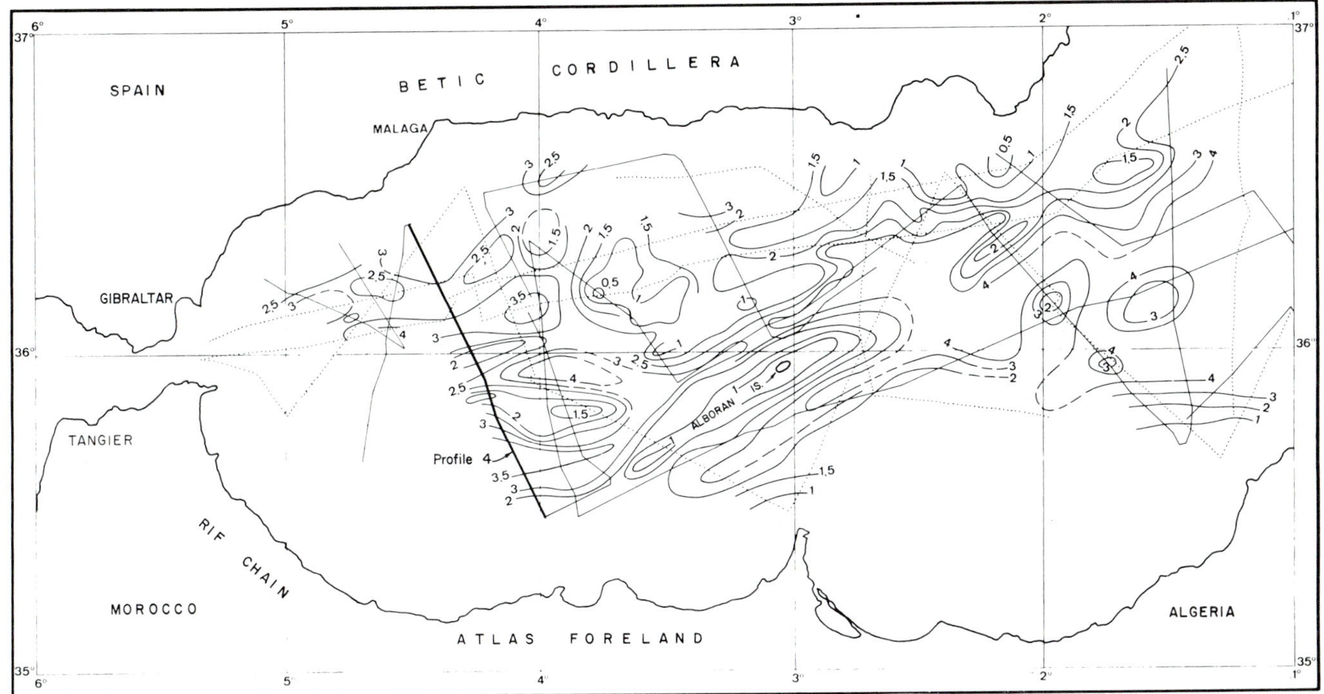

Fig. 4. Isochrons to acoustic basement of the Alboran Sea (in seconds of two-way time). Continuous line, Flexotir profile; dotted line, airgun profile.

south of Minorca. Three main morphological zones can be distinguished:

1. To the southwest of Ibiza a zone of alternating east-west basement ridges and lows (profile 6), is thought to be the seaward prolongation of the Betics (Mauffret et al., in press).

2. An intermediate zone corresponding to the Emile Baudot fracture zone (E, Fig. 2) limits the above basement ridges to the east. It consists of northeast-trending structures: to the northeast, steep slopes; in the center, a double-ridge system; and to the southwest, a series of circular and elongated highs (profile 7).

3. To the east, the "South Balearic Promontory" is probably a dislocated continental basement plateau with horst and graben structures along a dominant northeast-southwest trend. It extends southward to 38°N. Profile 8 crosses one of the longest of these northeast-trending grabens. An east-northeast-trending basement ridge marked by a strong positive gravity anomaly limits this plateau to the south.

The sedimentary cover in the axial part of the Algerian-Balearic Basin displays the same three layers as those found in the Provencal Basin. The top of the Messinian salt layer lies at the same level in both basins, but the basement is 0.5 sec higher in the Algerian-Balearic Basin than in the Provencal Basin. The pre-Messinian sediments are therefore thinner. This fact favors a younger age for the Algerian-Provencal Basin.

In the axial part of the basin (north of the Gulf of Bougie), the sedimentary sequence is very thin, owing to the presence of basement highs between the Gulf of Bougie and the south of Minorca.

As shown on Profiles 5, 6, and 7 (and particularly on profile 6 across the northeast-trending graben of the south Balearic margin), the evaporitic layer extends far up on the margins.

Along the Algerian margin, as well as in the Alboran Sea, the Messinian and Plio-Quaternary layers dip toward the "north Algerian trough" (Auzende et al., 1972; location on Fig. 2). In some places the trough is characterized by large salt-dome structures, as on profile 5.

SARDINIA-TUNISIA STRAIT

The Sardinia-Tunisia Straits form a transition area between the Western Mediterranean Sea and the Tyrrhenian Sea. It is located between southern Sardinia, structurally constituted by an Oligocene graben and two Paleozoic horsts, and northern Tunisia, essentially covered by the Numidian nappe, which is related to the Lower Miocene Alpine orogenic event (Biely et al., 1971).

Three main structural units can be distinguished (Auzende et al., in press): the Tunisian continental margin, the Sardinian continental margin, and the North Tunisia fracture zone (F, Fig. 2). Elsewhere in the strait the basement appears to be of continental nature, except on the North Tunisian fracture zone, which is of volcanic origin.

The Tunisian continental margin consists of a large plateau (500-1,000 m deep, Fig. 6), connecting northern Tunisia to Sicily, with tectonic trends the same as on land, and a relatively smooth slope

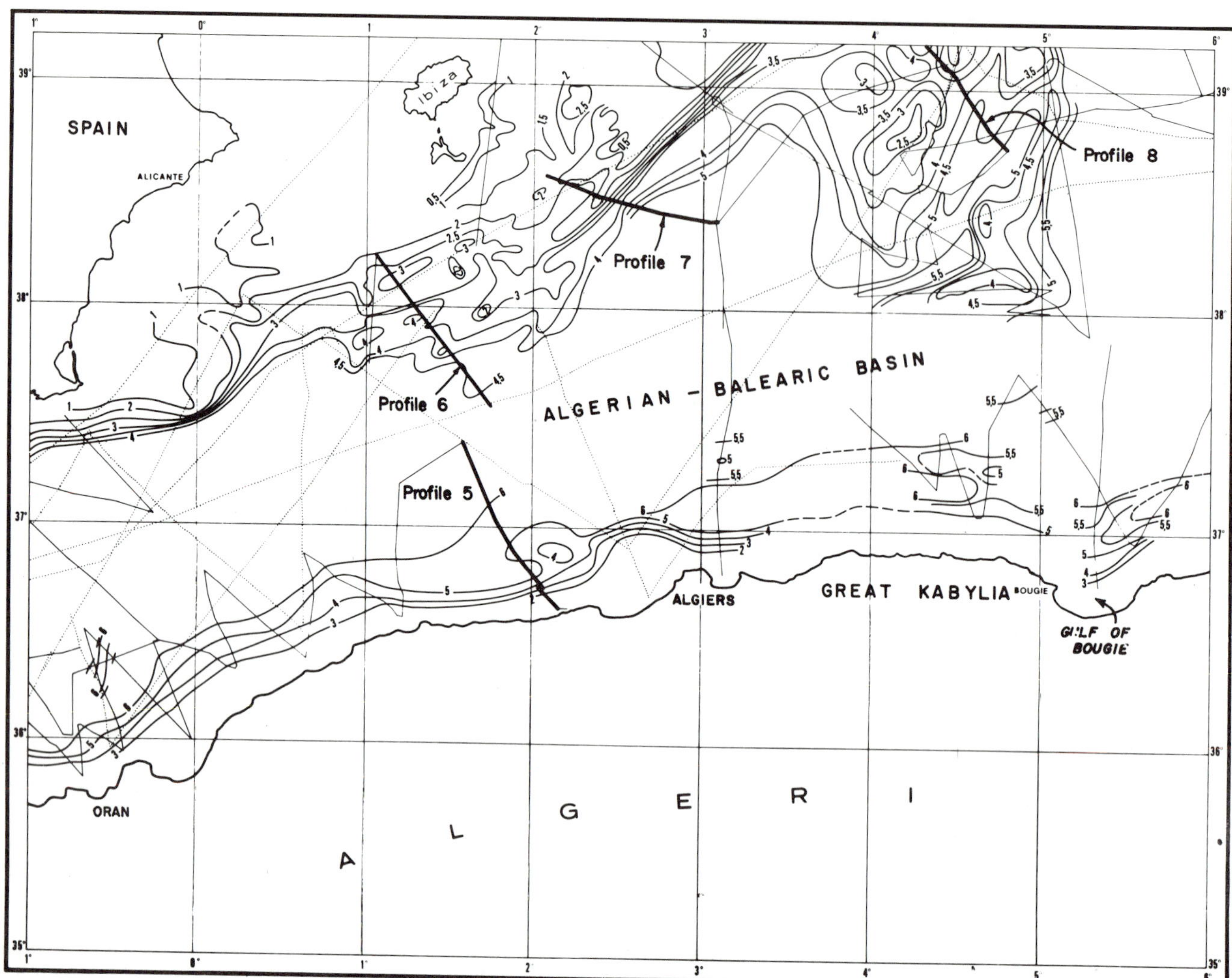

Fig. 5. Isochrons of the Algerian-Balearic Basin (in seconds of two-way time). Continuous line, Flexotir profile; dotted line, airgun profile.

trending west-southwest and structurally broken in its central part, corresponding to the Sardinia and Tunisia basement junction.

The northeast-southwest North Tunisian fracture zone consists of a line of volcanic highs approximately 300 km long and 40 km wide. Southeast of Sardinia, an elongated volcanic ridge constitutes the most obvious relief in the system (Fig. 6).

The southwestern continental margin consists of a marginal plateau (700 m deep on average) and two basement steps, which constitute the continental slope (profile 9). During DSDP Leg XIII, fragments of Paleozoic basement were sampled about 100 km from shore (site 134). The data from the aeromagnetic survey of Le Borgne et al. (1971) show a possible limit between an oceanic and a continental crust and an important magnetic north-south positive anomaly on the slope. To the south, the dominant west-southwest structural directions interrupt against the North Tunisian fracture zone.

Profile 10 crosses the Sardinia-Tunisia Straits and shows the transition between the Sardinian and the Tunisian basement, a volcanic seamount probably related to the North Tunisian fracture zone, a sedimentary basin characterized by the three previously described layers, the salt being present only under the evaporite facies.

CONCLUSIONS

From our data we can point out several facts concerning the evolution of the Western Mediterranean continental margins.

1. An Oligo-Miocene distension formed the western Mediterranean Sea (LePichon et al., 1971). The distension has been more important in the central part of the Provencal and Algerian-Balearic Basin, where oceanic crust was created. In the Gulf of Valencia and the Ligurian Sea, these processes were apparently less accentuated, and most of the basement is of continental origin. Volcanic activity is the only evidence of the distension.

2. An average depth difference of 1 sec exists between the basement in the Ligurian Sea, in the Gulf of Valencia, and in the Alboran Sea on one side and the Provencal and the Algerian basins on the other side.

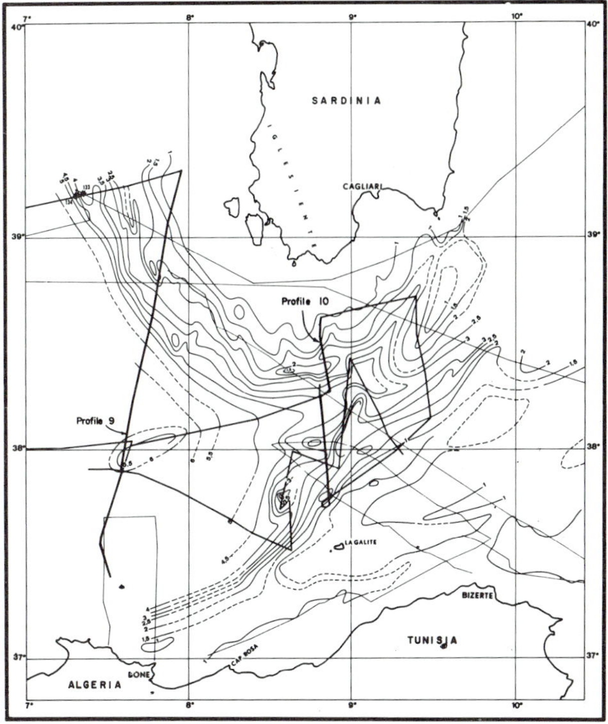

Fig. 6. Isochrons of the Tunisia-Sardinia straits (in seconds of two-way time). Heavy line, Flexotir profile; thin line, airgun profile. Locations shown on Leg XIII, sites 133 and 134.

3. Recent tectonic activity is mainly located on the north African margin. Seismicity data (McKenzie, 1970), as well as the geology (Guiraud, 1967; Devaux, 1969), show present-day tectonic activity in the north African area. In the Alboran Sea and in the Algerian-Balearic Basin, Messinian and Plio-Quaternary strata dip toward the south, the North Algerian trough (Fig. 2). A recent compression phase related to a shortening between Africa and Eurasia has been proposed to explain the existence of this trough (Auzende et al., 1972).

4. The Sardinia-Tunisia Straits forms the contact zone between two different welded continental basements. This welding is related to the Corsica-Sardinia rifting and the consequent creation of the Provencal Basin (LePichon et al., 1971).

ACKNOWLEDGMENTS

We thank N. Guillo, D. Carre, and S. Monti for their help in working out this paper. G. Auffret, L. Pastouret, and J. C. Sibuet have read critically the manuscript. The help of G. Auffret, N. Guillo, J. Mascle, and V. Renard was much appreciated for the translation.

This is Contribution 225 of the Departement Scientifique, Centra Océanologique de Bretagne.

BIBLIOGRAPHY

Alla, G., Dessolin, D., and Leenhardt, O., 1971, Données du sondage sismique continu concernant la sédimentation plio-quaternaire en Méditerranée occidentale: Symp. Méditerranée, Heidelberg (in press).

Allan, T. D., and Morelli, C., 1971, A geophysical study of the Mediterranean: Bol. Geof. Teor. Appl. V. 13, p. 99-142.

Alvarez, W., 1972, Rotation of the Corsica-Sardinia microplate: Nature Phys. Sci., v. 235, p. 103-105.

Andrieux, J., 1970, Structure du Rif central [thèse]: Univ. Montpellier, 284 p.

———, Fontbote, J. M., and Mattauer, M., 1971, Sur un modèle explicatif de l'arc de Gibraltar: Earth and Planetary Sci. Letters, v. 12, p. 191-199.

Argand, E., 1924, La tectonique de l'Asie: 13th Internatl. Geol. Congr., Brussels, p. 171-372.

Arthaud, F., 1970, Etude tectonique et microtectonique comparée de deux domaines hercyniens: les nappes de la Montagne Noire (France) et l'anti-clinorium de l'Inglesiente (Sardaigne) [thèse]: Univ. Montpellier.

Auzende, J. M., 1969, Etude par sismique réflexion de la bordure continentale algéro-tunisienne entre Bougie et Bizerte [thèse 3ème cycle]: Univ. Paris, 117 p.

———, 1971, La marge continentale tunisienne: résultats d'une étude par sismique réflexion: sa place dans le cadre tectonique de Méditerranee occidentale: Marine Géophys. Res., t. 1, p. 162-177.

———, and Pautot, G., 1970, La marge continentale algérienne et al phénomène de subsidence: exemple du golfe de Bougie: Compt. Rend., t. 271, p. 1945-1948.

———, Bonnin, J., Olivet, J. L., Pautot, G., and Mauffret, A., 1971, Upper Miocene salt layer in the western Mediterranean Basin: Nature Phys. Sci., v. 238, p. 82-84.

———, Olivet, J. L., and Pautot, G., 1972a, Balearic islands: southern prolongation, *in* Initial Reports of the Deep Sea Drilling Project, v. XIII: Washington, D.C., U.S. Govt. Printing Office, 48-53, p. 1441-1447.

———, Olivet, J. L., Mauffret, A., and Pautot, G., 1972b, La dépression nord-baleare (Espagne): Compt. Rend., ser. D., t. 274, p. 2291-2294.

———, Bonnin, J., and Olivet, J. L., 1973, The origin of the western Mediterranean basin: Jour. Geol. Soc. London, v. 129, p. 607-620.

———, Olivet, J. L., and Bonnin, J., (in press), Le détroit sardano-tunisien et al zone de fracture nord-tunisienne: Tectonophysics.

Bayer, R., Le Mouel, J. L., and LePichon, X., 1973, Magnetic anomaly pattern in the western Mediterranean: Earth and Planetary Sci. Letters, v. 19, p. 168-176.

Bellon, H., and Brousse, R., 1971, L'âge oligo-miocène du volcanisme ligure: Compt. Rend., t. 272, p. 3109-3111.

Biely, A., Lajmi, T., and Rouvier, H., 1971, Les unités allochtones du pays de Bizerte (Tunisie septentrionale): Compt. Rend., t. 273, p. 2052-2055.

Boccaletti, M., and Guazzone, G., 1970, La migrazione terziara dei bacini toscani e la rotazoine dell'Appenino Settentrionale in una "zona du torzione" per deriva continentale: Mem. Soc. Geol. Italiana, v. IX, p. 177-195.

Bourcart, J., 1960, Carte topographique du fond de la Méditerranée occidentale: Bull. Inst. Océanog. Monaco, t. 1163, 20 p.

———, 1962, La Méditerranée et al révolution du Pliocene, *in* Livre à la mémoire du Pr. Fallot: Mém. Soc. Géol. France, sér. H, t. 1, p. 103-116.

Bourrouilh, R., 1970, *in* Bourgeois et al., Données nouvelles sur la géologie des cordillères bétiques: Ann. Soc. Geol. Nord, v. XC, no. 4, 347-393.

Caire, A., 1965, Morphotectonique de l'autochtone présaharien et de l'allochtone tellien: Rev. Géogr. Phys. Géol. Dyn., t. 2, no. 7, p. 267-175.

———, 1970, Tectonique de la Méditerranée centrale: Ann. Soc. Géol. Nord, t. XL, no. 4, 307-346.

Caputo, M., Panza, G. F., and Postpischl, D., 1970, Deep structure of the Mediterranean Basin: Jour. Geophys. Res., v. 75, p. 4919-4923.

Carey, W. S., 1958, The orocline concept in geotectonics: Papers Proc. Roy. Soc. Tasmania, v. 89, p. 255-288.

Castany, G., 1959, La géologie profonde du territoire Tunisie-Sicile, *in* Topogr. Géol. Profondeurs Océan., Colloque Centre Nat. Rech. Sci., t. 83, p. 165-183.

Colom, G., and Escandell, B., 1962, Evolution du géosynclinal baléare, *in* Livre à la mémoire du Pr. Fallot: Mém. Soc. Géol. France, sér. H., t. 1, p. 125-136.

Devaux, J., 1969, Recherche de l'organisation des contraintes dans les tréfonds de l'Algérie du Nord: Publ. Serv. Géol. Algérie Bull. (n. sér.), no. 39, p. 46-69.

Dubordieu, G., 1962, Dynamique Wégénérienne de l'Afrique du Nord, *in* Livre à la mémoire du Pr. Fallot: Mém. Soc. Géol. France, sér. H., t. 1, p. 627-644.

Durand-Delga, M., 1967, Structure and geology of the north-east Atlas, *in* Guidebook to the geology and history of Tunisia: Petr. Expl. Soc. Libya, Tripoli, p. 59-83.

———, 1969, Mise au point sur la structure du NE de la Berbérie: Publ. Serv. Géol. Algérie, t. 39, p. 89-131.

Fahlquist, D. A., and Hersey, J. B., 1969, Seismic refraction measurements in the western Mediterranean Sea: Bull. Inst. Océanogr. Monaco, v. 67, 52 p.

Giermann, G., Pfannenstiel, M., and Wimmenauer, W., 1968, Relations entre morphologie, tectonique et volcanisme en mer d'Alboran (Méditerrranée occidentale). Résultats préliminaires de la campagne Jean Charcot (1967): Compt. Rend. Somm. Soc. France, t. 4, p. 116-118.

Glangeaud, L., 1932, Etude géologique de la région littorale de la province d'Alger: Bull. Serv. Carte Géol. Algérie, t. 8, 627 p.

———, 1952, Les éruptions tertiaires nord-africaines et leurs relations avec la tectonique méditerranéenne: Compt. Rend. 19th Congr. Geol. Internatl. Alger, t. 17, p. 71-101.

———, 1962, Paléogéographie dynamique de la Méditerranée et de ses bordures. Les rôles des phases Ponto-Plio-Quaternaires, *in* Océanographie Géologique et Géophysique de la Méditerranée Occidentale: Collect. Natl. C.N.R.S., p. 125-165.

———, Bobier, C., and Bellaiche, G., 1967, Evolution néotectonique de la mer d'Alboran et ses conséquences paléogéographiques: Compt. Rend., t. 165, p 1672-1675.

Guiraud, R., 1967, La transversale de Colbert, accident majeur de la région du Hodna (Algérie du Nord): Compt. Rend., t. 264, p. 1245-1248.

Hasebe, K., Fujii, N., and Uyeda, S., 1970, Thermal processes under island arcs: Tectonophysics, v. 10, p. 335-355.

Hinz, K., 1972, Results of seismic refraction investigations (project Anna) in the western Mediterranean Sea south and north of the Island Mallorca: Bull. Centre Rech. Pau, S.N.P.A., v. 6, p. 405-526.

Karig, D. E., 1971a, Origin and development of marginal basins in the western Pacific: Jour. Geophys. Res., v. 76, p. 2542-2561.

———, 1971b, Structural history of the Mariana Island arc system: Geol. Soc. America Bull., v. 82, p. 323-344.

Kieken, M., 1962, Les traits essentiels de la géologie algérienne, *in* Livre à la mémoire du Pr. Fallot, Mém. Soc. Géol. France, sér. H., t. 1, p. 545-614.

Le Borgne, E., Le Mouel, J., and LePichon, X., 1971, Aeromagnetic survey of southwestern Europe: Earth and Planetary Sci. Letters, v. 12, p. 287-299.

———, Bayer, R., and Le Mouel, J. L., 1972, La cartographie magnétique de la partie sud du bassin algéro-provencal: Compt. Rend., t. 274, p. 1291-1294.

Leclaire, L., 1968, Contribution à l'étude géomorphologique de la marge continentale algérienne: Cah. Océanogr., t. XX, no. 6, p. 451-521.

Le Mouel, J., and Le Borgne, E., 1970, Les anomalies magnétiques du sud-est de la France et de la Méditerranée occidentale: Compt. Rend., sér. D., t. 271, p. 1348-1350.

LePichon, X., Pautot, G., Auzende, J. M., and Olivet, J. L., 1971, La Méditerranée occidentale depuis l'Oligocène: schéma d'évolution: Earth and Planetary Sci. Letters, v. 13, p. 145-152.

McKenzic, D. P., 1970, Plate tectonics of the Mediterranean region: Nature, v. 226, p. 239-243.

Mattauer, M., 1963, Le style tectonique des chaînes tellienne et rifaine: Geol. Rundschau, bd. 53, p. 296-813.

Mauffret, A., Auzende, J. M., Olivet, J. L., and Pautot, G., 1972, Le bloc continental baléare (Espagne). Extension et évolution: Marine Geology, t. 12, p. 289-300.

———, Olivet, J. L., Auzende, J. M., and Lajat, D., (in press), La marge sud-baléare et la zone de fracture de l'Emile Baudot: Rapp. P.V. réunion CIESM, Athenes, Nov. 3-11, 1972.

Montadert, L., Sancho, S., Fail, J. M., Debyser, J., and Winnock, E., 1970, De l'âge tertiaire de la série salifère responsable des strauctures diapiriques en Méditerranée occidentale: Compt. Rend., t. 271, p. 812-815.

Olivet, J. L., Pautot, G., and Auzende, J. M., 1972, Alboran sea: structural framework, *in* Initial reports of the Deep Sea Drilling Project, v. 13: Washington, D.C., U.S. Govt. Printing Office, 48-1, p. 1417-1430.

———, Auzende, J. M., and Bonnin, J., 1973, Structure et évolution tectonique du bassin d'Alboran: Bull. Soc. Geol. France, 7e ser., t. XV, no. 2, p. 108-112.

Pautot, G., Auzende, J. M., Olivet, J. L., and Mauffret, A., 1972, Valencia Basin, *in* Initial reports of the Deep Sea Drilling Project, v. 13: Washington, D.C., U.S. Govt. Printing Office, 48-52, p. 1430-1441.

Rehault, J. P., 1968, Contributions à l'étude de la marge continentale au large d'Imperia et de la plaine abyssale ligure [thèse 3ème cycle]: Univ. Paris.

Ryan, W. B. F., 1969, The floor of the Mediterranean Sea [Ph.D. thesis]: Columbia Univ., 196 p.

———, Hsü, K. J., Nesteroff, W. O., Pautot, G., Wezel, F. C., Lort, J. M., Cita, M. B., Mayne, W., Stradner, H., and Dumitrica, P., 1971, Deep Sea Drilling Project, Leg 13: Geotimes, v. 15, no. 10, p. 12-15.

Sclater, J. G., Anderson, R. N., and Bell, M. L., 1971, Evolution of ridges and evolution of the central eastern Pacific: Jour. Geophys. Res., v. 76, no. 32, p. 7888-7916.

Smith, A. G., 1971, Alpine deformation and the oceanic areas of the Tethys, Mediterranean and Atlantic: Geol. Soc. America Bull., v. 82, p. 2039-2070.

Stanley, D. J., and Mutti, E., 1968, Sedimentological evidence for an emerged land mass in the Ligurian sea during the Palaeogene: Nature, v. 218, p. 32-36.

Vardabasso, S., 1963, Die Ausseralpine Taphrogenese in Kaledonisch Variszich konsolidierden Sardischen Vorlande: Geol. Rundschau, bd. 53, p. 613-630.

Vogt, P. R., Higgs, R. H., and Johnson, E. L., 1971, Hypotheses on the origin of the Mediterranean Basin: magnetic data: Jour. Geophys. Res., v. 76, p. 3207-3228.

Westphal, M., 1967, Etude paléomagnétique de formations volcaniques primaires de Corse. Rapports avec la tectonique du domaine Ligurien [thèse 3ème cycle]: Univ. Strasbourg.

Wezel, F. C., and Ryan, W. B. F., 1971, Flysch, margini continentali e zolle litospheriche: Boll. Soc. Italiana, v. 90, p. 249-270.

Major Basins Along the Continental Margin of Northern South America

J. E. Case

INTRODUCTION

Seismic-reflection and gravity surveys indicate a major folded sedimentary basin, the South Caribbean basin, immediately south of the Venezuelan and Colombian topographic basins. Toward the east, deformed sedimentary rocks compose Curacao Ridge and Los Roques Basin, north of the horst-like blocks that make up the Netherlands and Venezuelan Antilles. Toward the west, the South Caribbean basin can be traced north of the Guajira Peninsula and the Santa Marta uplift, and then across the Magdalena Delta. The basin evidently extends onshore to form the Sinu-Atlantico basin of Colombia. Many subsidiary basins splay southward from the South Caribbean basin, and many of them may be in graben-like depressions between basement blocks. Some of the known or inferred sedimentary basins trend northwest: (1) between La Orchila and La Blanquilla, (2) between Aruba and Curacao, (3) between Aruba and the Paraguana Peninsula, and (4) between Los Monjes and the Guajira Peninsula. The Baja Guajira basin, in contrast, trends nearly east, and the Lower Magdalena basin trends nearly north or northeast. Some of these splay basins serve to connect the South Caribbean basin with the Bonaire, Falcon, and Maracaibo basins.

Most of the basin fill is of Cenozoic age. Sediments of the northernmost basin are probably deformed pelagic and turbidite material that was deposited in deep water along the south Caribbean margin.

Deformation of the South Caribbean basin indicates relative compressive strain between the Caribbean and South American plates, especially during the later Cenozoic; such compressive components may have been achieved in any one of three ways: (1) by stress redistributions across a broad zone of right-lateral shear between the Caribbean and South American plates; (2) by underthrusting of the Caribbean plate beneath the South American plate; or (3) by gravitative downslope movement of material. Splay and interior basins to the south may be extensional features. Clockwise rotation of a Bonaire block may account for compressional deformation along the Venezuelan borderland.

GEOLOGICAL AND GEOPHYSICAL DATA

This report briefly describes some major structural features of the continental margin of northern South America, summarizing more extensive offshore data presented by Edgar et al. (1971b), Krause (1971), Galavis and Louder (1970), Roemer et al. (1973), Silver et al. (1972, in press), Emery and Uchupi (1972), and Shepard (1973). Geologic Structure of the Tortuga-Margarita region has been described by Peter (1972) and by Ball et al. (1971) and will not be reviewed here.

Immediately south of the Caribbean abyssal plain, a highly deformed thick sequence of sediments forms the northern continental slope or borderland of South America (Fig. 1). Seismic-refraction and gravity surveys indicate that as much as 5 km of low-velocity material underlies Los Roques Basin and 14 km (probably a mixture of sediments and low-velocity igneous or metamorphic rocks) underlies Curacao Ridge (Edgar et al., 1971b; Worzel, 1965). From east to west this deformed sequence extends as a broad arc from north of the Venezuelan Antilles, along the Venezuelan and Colombian borderland, past the Magdalena Delta, and extends onshore as the Sinu-Atlantico basin of northwestern Colombia.

This fold belt probably includes rocks of Late Cretaceous and Cenozoic age and is a nearly continuous feature, as shown by representative seismic-reflection profiles (Figs. 2-4) and negative free-air anomalies (Talwani, 1966; Bowin, in press). This deformed complex will be informally termed the S. Caribbean basin (Venezuelan-Colombian Trough of Galavis and Louder, 1970), and, as discussed later, is partly composed of folded trench deposits.

Beneath the abyssal plains of the southernmost part of the Caribbean Sea, the sedimentary sequence has been involved in moderate faulting and folding (Edgar et al., 1971b; Silver et al., in press; Krause 1971; Holcombe and Matthews, 1973; Edgar et al., 1973b), but the deformation is much less extensive than that along the Continental Slope. In deep drill holes of the Deep Sea Drilling Project (DSDP), the sedimentary sequence ranges from Turonian-Late Coniacian to Recent in age, although several depositional hiatuses or erosional intervals have been recognized (Edgar et al., 1971a; Edgar et al., 1973b). Most of the sediments are deep-water deposits. The lower part of the section consists mostly of pelagic and hemipelagic sediments, but turbidites are widespread in the upper part of the sequence in the Colombian Basin, in Aruba Gap, and along the extreme southern part of the Venezuelan Basin. Sediments are only 170 m thick in DSDP site 150 in the Venezuelan Basin (Edgar et al., 1971a), but thicken in all directions away from the site.

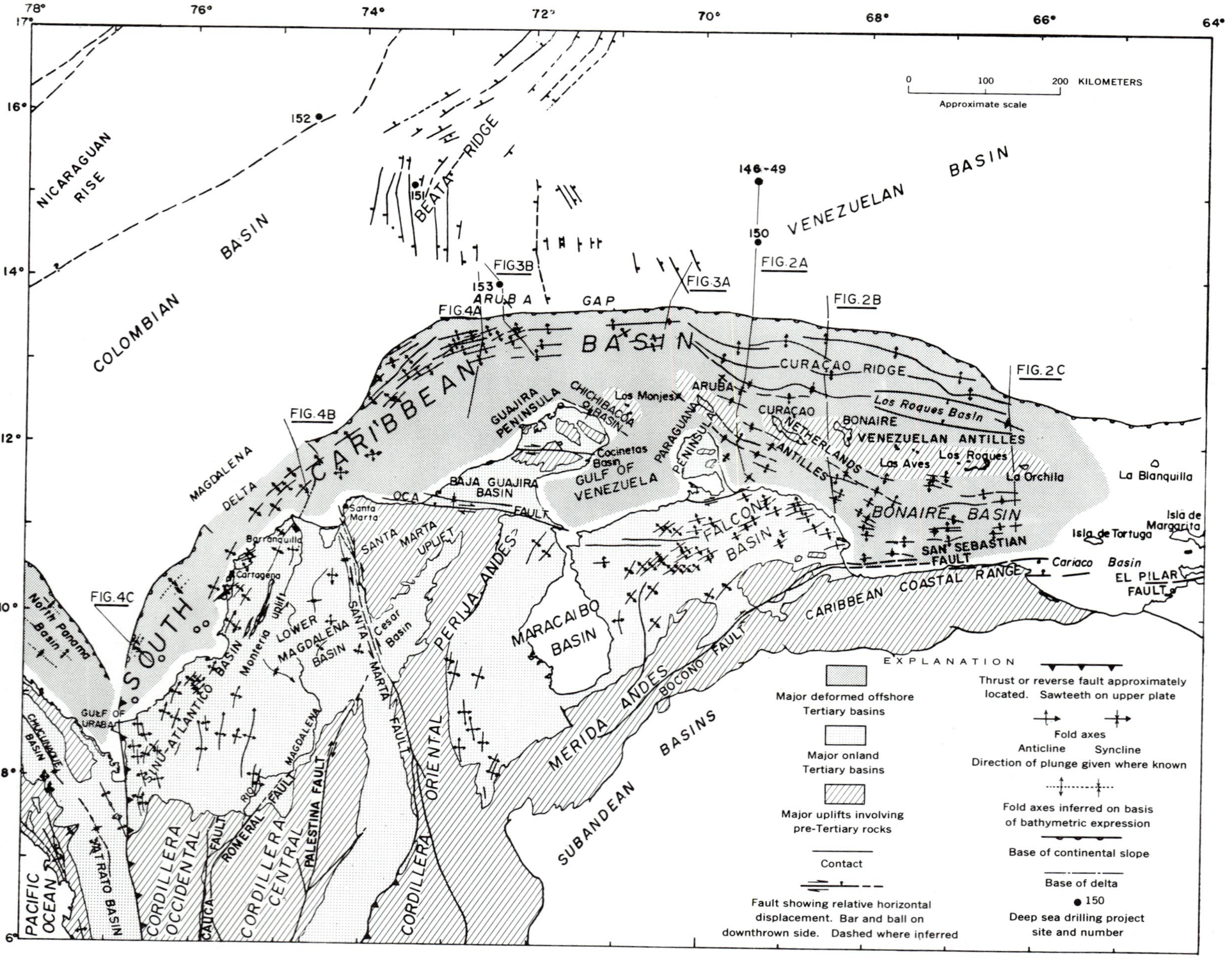

Fig. 1. Index and simplified tectonic map of the continental margin of part of northern South America. Principal sources of data: northern Colombia, Irving (1971); Venezuela, King (1969); Beata Ridge, Roemer et al. (1973); eastern Panama, Panama, Direccion General de Recursos Minerales (1972); fold trends off Colombia, Krause (1971), Shepard (1973), Emery and Uchupi (1972), Roemer et al. (1973); fold trends in vicinity of Bonaire Basin and Curacao Ridge from data of Silver et al. (1972; in press) and Scientific Staff (1973).

Northwest of La Orchila (Fig. 1; Fig. 2c), sediments beneath the southern topographic Venezuelan Basin attain thicknesses of about 4 km (4 sec of two-way travel time) or more. On Figure 2c seismic horizon "A" (approximately of middle Eocene age; Edgar et al., 1971a) or horizon "B" (approximately of Coniacian to Campanian Age) dips southward and is not observed on reflection profiles below the toe of the slope. In the lower part of the profile, sediments have the reflective characteristics (discontinuous, incoherent returns) of pelagic sediments with some interlayered beds (coherent reflectors), probably turbidites. Most of the upper part of the section consists of turbidites. The uppermost horizontal turbidites of the abyssal plain are continuous with the folded beds of Curacao Ridge. Farther west (Fig. 2b), near Bonaire, the sedimentary sequence is thinner, about 2 km ±; horizon "B" clearly passes beneath deformed sediments of the outer Curacao Ridge on adjacent seismic profiles (Silver et al., 1972). The geometrical relations observed on this and adjacent seismic profiles can be interpreted to mean that the base of the slope is a zone of underthrusting.

Still farther west, the sediments above horizon "B" at the southern margin of the Venezuelan Basin are only about 1.5 km± thick (Fig. 2a), and many small faults cut the section below the turbidites on adjacent profiles (Silver et al., 1972). Horizon "B" again can be traced beneath the toe of the outer slope (see original profiles by Silver et al., 1972). On Figures 2a and 2b, the Curacao Ridge is shown to be a complex double feature with a deeper, highly deformed anticline to the north and a broader, less-deformed anticline making up the main shallow ridge. At its eastern end (Fig. 2 c) Los Roques Basin is a graben-like feature, but westward (Figs. 2a and 2b) it is a structural basin near Aruba. Negative free-air and isostatic anomalies indicate that Curacao Ridge and Los Roques Basin are underlain by a great thickness of low-density material (Scientific Staff 1973). Still farther south (Figs. 2b and 2c), locally faulted scarps form the northern flank of the horst-like platform of the Netherlands and Venezuelan Antilles.

Geology of the island platform has been recently

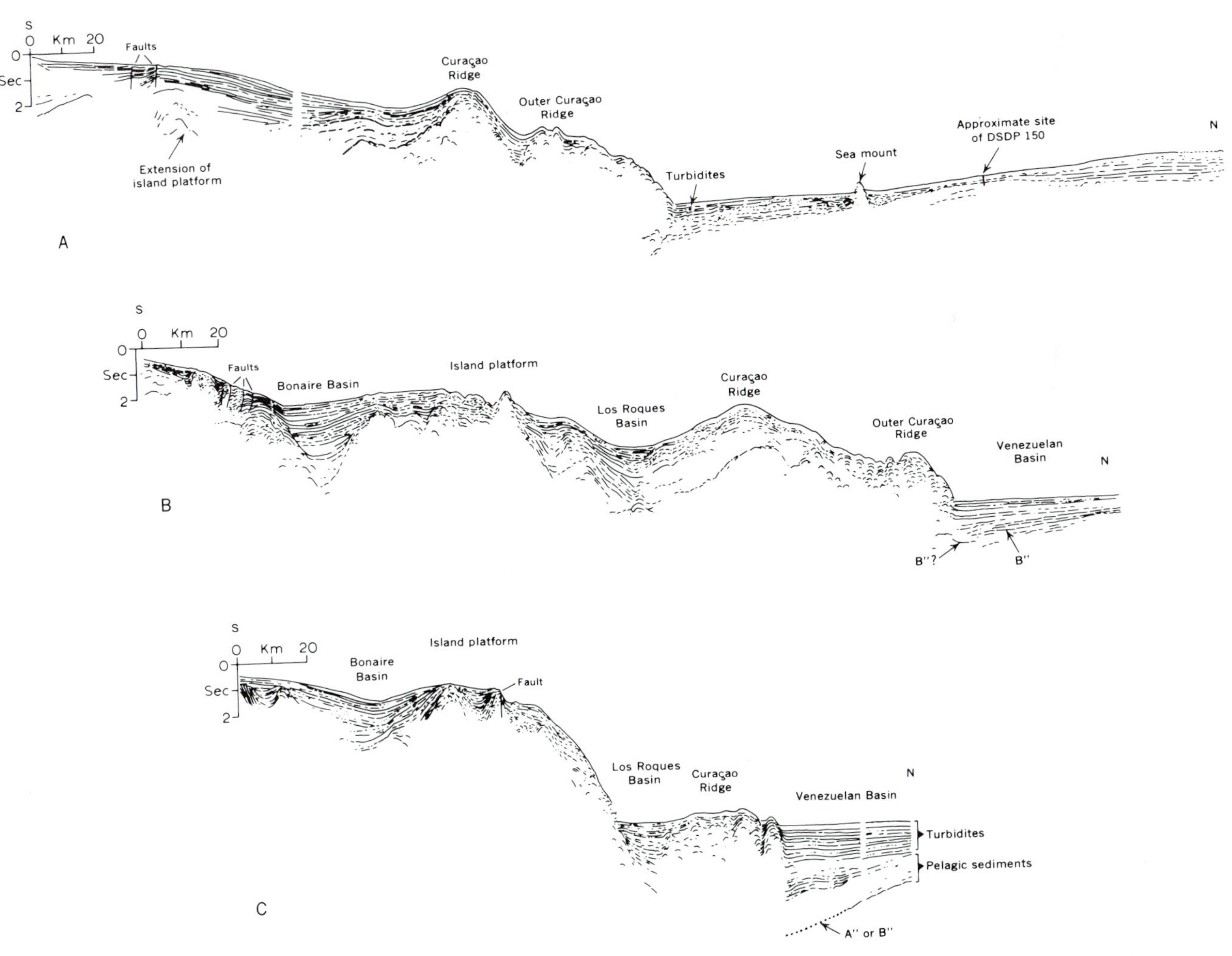

Fig. 2. Seismic-reflection profiles across the Venezuelan borderland (modified from Silver et al., in press). Profile lines shown on Figure 1.

reviewed by Beets (1972) and by Schubert and Moticska (1972). Basement rocks of the Netherlands Antilles include basalt and diabase and, on Aruba, metamorphosed mafic volcanic rocks. These rocks have been intruded by dioritic to granodioritic bodies of Late Cretaceous age (Beets, 1972). Associated sedimentary rocks interlayered with or overlying the igneous rocks are of Late Cretaceous age (Turonian-Danian). On the smaller islands of the platform, a variety of metamorphic, ultramafic, and mafic rocks have been intruded by granitic rocks (Schubert and Moticska, 1972). Bonaire Basin, a structural and topographic basin south of the island platform, contains more than 2 km (±2 sec) of sediments in some places. As much as 4,500 m of sediments [Upper Cretaceous (?)-Tertiary] in the basin have been folded and faulted in many places (Galavis and Louder, 1970; Silver et al., 1972; Silver et al., in press). South of the Bonaire Basin, metamorphosed Mesozoic rocks form the Caribbean Coastal Ranges of Venezuela.

West of the Netherlands Antilles, the northern

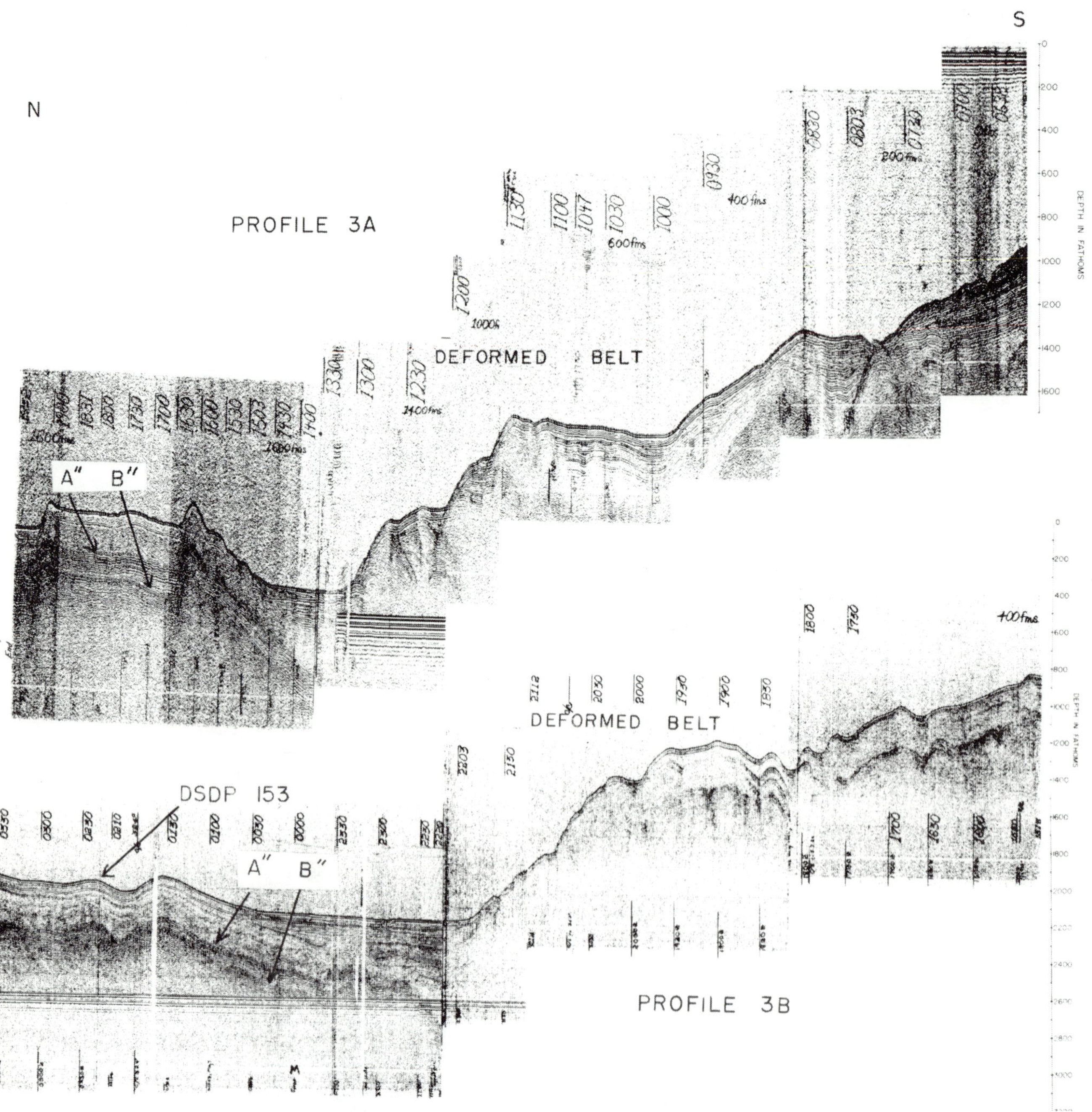

Fig. 3. Seismic-reflection profiles near Aruba Gap (from Roemer et al., 1973). In DSDP Site 153 (lower profile), hard silicified limestone of Eocene age was found corresponding to horizon A'' at a depth of 570 m below the sea floor. At 760 m, the drilling reached amygdaloidal basalt; the level corresponds to horizon B'' (Roemer et al., 1973). Length of upper profile about 145 km. Length of lower profile about 135 km. (Published with permission of the authors and Texas A & M Univ.)

deformed belt can be traced north of the Golfo de Venezuela (Fig. 3). Figure 3a shows a series of faulted blocks and gently folded sediments in the outer slope, and a major northwest-trending shear zone may pass through this region. Ponded turbidites occur at the extreme southern margin of the Venezuelan Basin in Aruba Gap, and horizons A'' and B'' can be seen dipping southward toward the toe of the slope (Roemer et al., 1973). On Figure 3b the outer slope again is folded, and A'' and B'' dive southward toward the slope off the Guajira Peninsula. This deformed belt is continuous with that identified by Krause (1971) off the Santa Marta-Guajira area (Fig. 4a) and by Shepard (1973) in the Magdalena Delta area. Krause identified turbidites in the upper part of the section in the southern Colombian Basin, the outer fold belt, and prograded sediments that dip seaward from the Guajira Peninsula. Although complicated by diapiric intrusions and slump folds, linear faults and folds appear in young sediments of the Magdalena Delta (Shepard et al., 1968; Shepard, 1973; Krause, 1971).

Southwest of the Magdalena Delta, the topographic trends, well data, and a reflection profile (Fig. 4c, Uchupi, in press) indicate that the deformed belt extends onshore, where it strikes about N30°E, and is continuous with the folded Sinu-Atlantico basin (Duque Caro, 1972) between Cartagena and the Golfo de Uraba. Offshore drill holes between Cartagena and the Gulf of Uraba have penetrated more than 4 km of Tertiary (?) sedimentary rocks.

Onland, the Sinu-Atlantico basin (Fig. 1) contains as much as 10 km of Tertiary sediments (Campbell, 1968; Irving, 1971; Duque Caro, 1972). Superimposed folds are clearly present (Irving, 1971), involving beds as young as Miocene in age: one set trends north and the other trends northeast, striking directly toward the offshore folds described by Krause (1971) and Shepard (1973). Duque Caro (1972) has presented a concise summary of the lithofacies, paleobathymetry, and paleoecology of sediments in the Sinu-Atlantico basin. He inferred

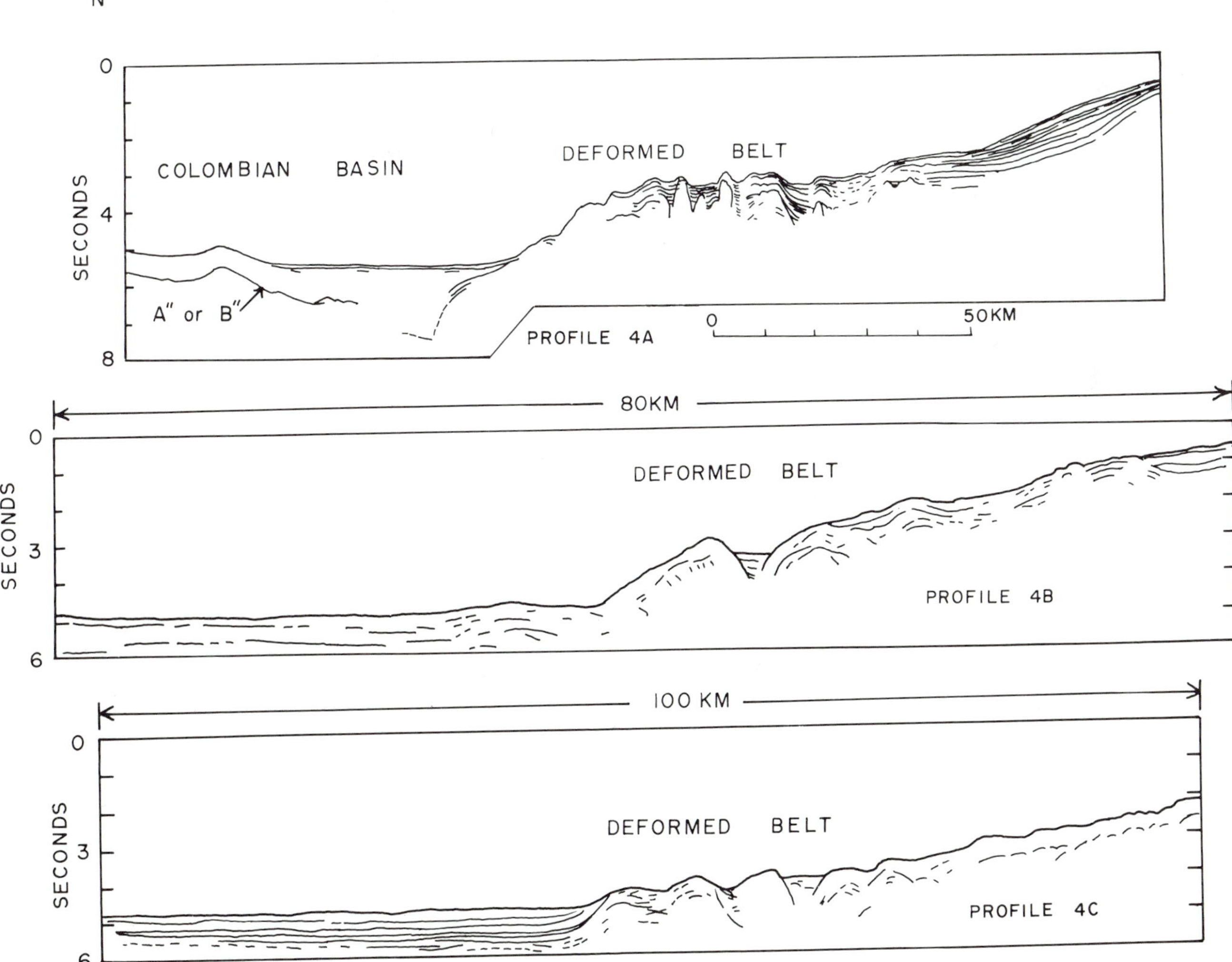

Fig. 4. Seismic-reflection profiles off Colombian borderland. Upper profile from Krause (1971), lower profiles from Bowin (in press) and Uchupi (in press). Note that vertical exaggeration and horizontal scales vary on the three profiles. Published with permission of the authors, the Geological Society of America, and Plenum Publishing Corporation.

that a substantial segment of the sedimentary section, especially Upper Cretaceous and Oligocene to middle Miocene beds, was deposited in deep water. Middle and Upper Eocene deposits accumulated in shallower water.

The southwestern end of the Sinu-Atlantico basin is evidently terminated by an eastward-dipping thrust, according to Irving (1971). [A northwest-trending basin lies off the north coast of Panama (Bowin and Folinsbee, 1970; Bowin, in press; Uchupi, in press). A paramount question is whether this basin is continuous with the South Caribbean basin or whether it formed completely independently, but discussion of the tectonic problem is beyond the scope of this report.]

TECTONIC IMPLICATIONS

In summary, an enormous deformed belt, involving at least 4 km and possibly as much as 10 km of Tertiary sediments, extends 1,500 km from the Golfo de Uraba almost to Aves Ridge. It may represent a deformed sediment-filled trench, and the deformation of the sediments has continued into recent time, as evidenced by folded sediments of the Magdalena Delta on the west and young folded turbidites (Fig. 2c) on the east. Geometrical underthrusting of the Caribbean stratigraphic units beneath the marginal deformed belt is indicated on many reflection profiles (Hopkins, 1973). On a larger scale, this style of deformation may be reflected in the Santa Marta uplift and Guajira Peninsula. F.K. North (written communication, 1973) has likened the Santa Marta uplift to the prow of a ship; it has right-lateral displacement on the north side and left-lateral displacement on the west side. The strike-slip displacements pass into a major thrust fault beneath the uplift. Relative displacement of the Santa Marta was toward the northwest. This may account for the minimum structural relief of about 12 km of Upper Cretaceous horizons between the Aruba Gap and the crest of the Santa Marta uplift. This concept is reinforced by the evidence of strongly positive gravity anomalies over the Santa Marta and Guajira uplifts (Case and MacDonald, 1973).

The major offshore deformed belt may have originated in several ways (Fig. 5):

1. Gravitative downslope sliding could produce chaotic folds of the outer slope and the apparent underthrusting of horizon B'' beneath the fold belt.

2. Tectonic compression related to southerly underflow of the Caribbean plate could produce some of the observed relations. If the fold axes are normal to the maximum compressive stress, then apparent regional stress axes around the arcuate foldbelt change orientation with respect to fixed geographic coordinates, perhaps related to changing stress orientations with time.

3. The fold belt could be partly related to broad regional right-lateral shear between an eastward-moving Caribbean plate and a westward-moving South American plate. Where apparent underthrusting is observed, this would imply oblique subduction rather than normal subduction (Krause, 1971).

In fact, all three of the general processes could have been operative at various times.

A key question, here, regards the antiquity of this deformational pattern. Has it continued since Late Cretaceous, middle Eocene, or middle Miocene—all times of widespread deformation, nondeposition, or erosion—or is it a relatively recent geologic phenomenon? Reflection-seismic evidence of southward sediment thickening indicates that a trench-like feature has existed near the site of the fold belt since at least horizon B'' time (± Coniacian-Campanian; Figs. 2-4) Hopkins (1973) has postulated pre-Coniacian age of the feature near Aruba Gap.

The evidence for deposition of a thick wedge of sediments coincident with a subsequent site of extensive folding argues against production of the fold belt by only simple gravitative sliding. The broad anticlinal form of Curacao Ridge likewise could not have been readily attained by simple gravitative sliding on shallow detachment surfaces. It seems that gravitative sliding has played an important role in the deformational process, particularly in the vicinity of the Magdalena Delta, but it probably should be regarded as a subsidiary effect to regional tectonism. The broadly arcuate pattern of the deformed South Caribbean basin is parallel to structural trends of the whole Andean System of northern South America and hence may have formed under the influence of the same regional stress systems. A critical problem is the extent to which older structural trends have influenced the orientation of younger trends.

SUBSIDIARY BASINS

Many Tertiary basins of varying size splay southward, eastward, and southeastward from the South Caribbean basin (Fig. 1). Sedimentary deposits in these basins are of relatively shallow-water to terrestrial facies (see summaries by Campbell, 1968; Vasquez and Dickey, 1972; Galavis and Louder, 1970; Zambrano et al., 1971). To the west, the Lower Magdalena basin is separated from the Sinu-Atlantico basin by a northeast-trending structural highland (Monteria uplift, southeast of Cartagena), but the Lower Magdalena basin may connect with the South Caribbean basin north of Barranquilla and Santa Marta (Fig. 1). The Baja Guajira basin, containing at least 3 km of Tertiary sediments (Irving, 1971), extends eastward from the South Caribbean basin into the Gulf of Venezuela. The Chichibacoa basin (Thomas, 1972), containing at least 4 km of Tertiary sediments, extends southeast between the Alta Guajira Peninsula and Los Monjes Islands. Other basins lie between the Paraguana Peninsula and Aruba (minimum sediment thickness is 1,500 m,

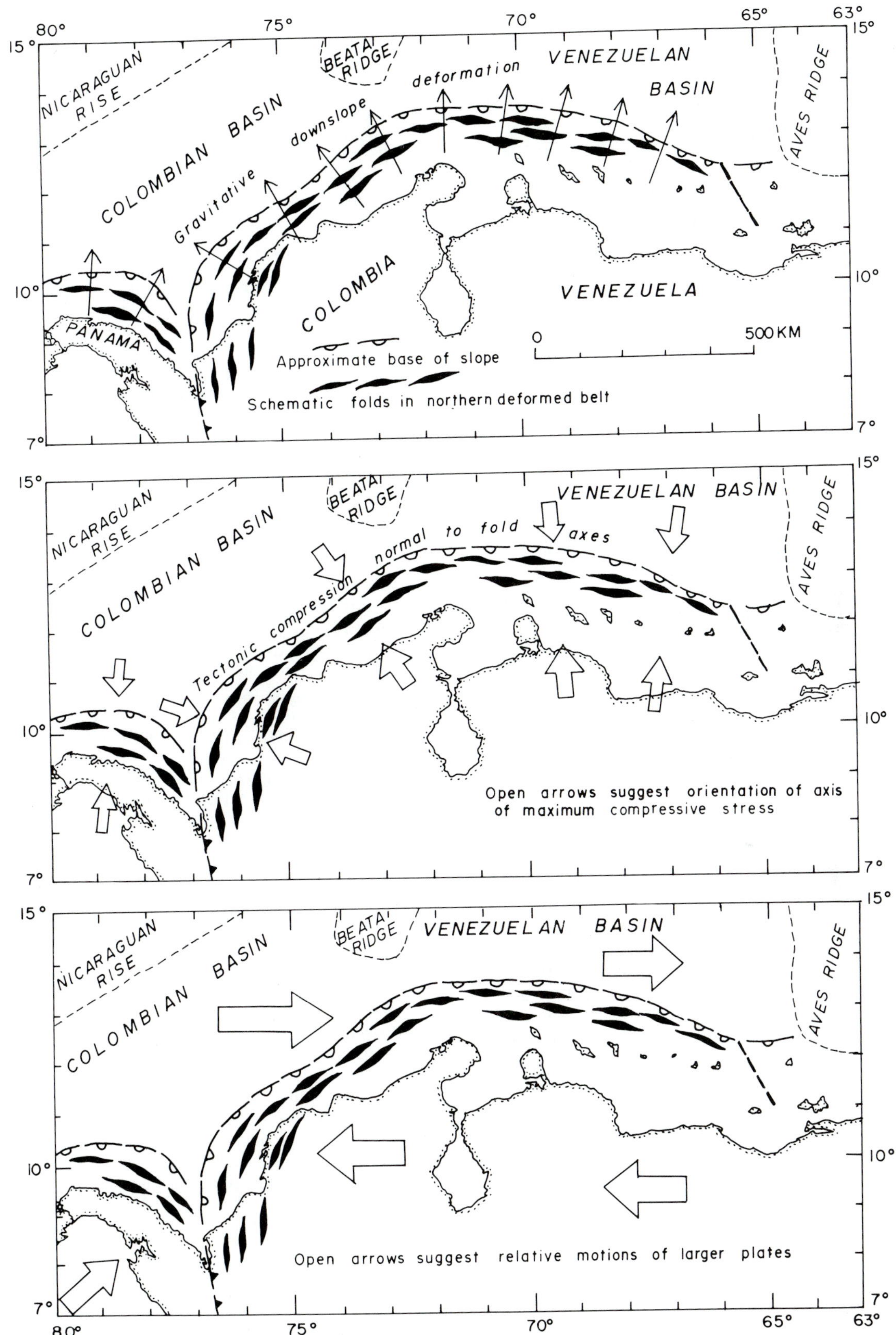

Fig. 5. Highly schematic diagrams illustrating possible stress distributions causing deformed belt of South Caribbean basin. Upper diagram, gravitative downslope movement away from South American continent. Middle diagram, tectonic compression related to underflow of Caribbean plate beneath nuclear South America. Lower diagram, deformation and oblique subduction in a broad zone between an east-moving Caribbean plate and a west-moving South American plate.

Feo-Codecido, 1971), and between Aruba and Curacao (minimum sediment thickness, 2 km) connecting the outer fold belt with the Bonaire basin.

In a very general way, many of the splay basins are "paired" with prominent basement uplifts (Fig. 1). For example, the horst-like ridge of the Venezuelan and Netherlands Antilles, the structural highs of the Paraguana and Alta Guajira Peninsulas, and the Santa Marta uplift occur north or northwest of the Bonaire Basin, Falcon basin, Baja Guajira basin, and Cesar basin, respectively. Silver et al. (in press) suggested that these Tertiary basins are partly of extensional origin. They appear to have formed south or southeast of basement blocks that moved north or northwest. However, major blocks have been uplifted south of these basins as well: the Caribbean coastal range, the Cordillera de Merida, and the Serrania de Perija along the Venezuelan-Colombian border are all sites of major post-Cretaceous uplift.

TECTONIC SUMMARY

Vertical displacements of this type have occurred all along the northern continental margin during the Cenozoic, especially during the later Cenozoic. Thus we are faced with the tectonic dilemma of explaining the compressive deformation of the outer fold belt and combined extensional-compressional deformation in the southerly basins. This strain pattern is not consistent with simple gravity tectonics, nor is it consistent with simple north-south compression (and subduction) between a Caribbean and a South American plate. It is, however, mechanically feasible to obtain such dislocations in a broad zone between an eastward-moving Caribbean plate and a westward-moving South American plate, especially if the motion vectors are not exactly parallel. Similarly, superposition of compressive features on extensional features may well reflect major changes in the stress regimes with time—the southerly basins may have formed under extension during the earlier Cenozoic and may have been subjected to compression during the later Cenozoic.

Still a fourth deformational model has been proposed for the Venezuelan borderland by Silver et al. (in press): They have suggested that clockwise motion of a small crustal block—the Bonaire block—about a pole near Caracas may account for the structures of Curacao Ridge and vicinity. The rotation of the Bonaire block probably results from dextral movements between the much larger Caribbean and South American plates.

ACKNOWLEDGMENTS

Helpful review of this manuscript was provided by E. A. Silver and F. P. Shepard. C. O. Bowin, Elazar Uchupi, and Lamar Roemer provided unpublished data for use in compilation of several illustrations. Published with permission of the Director of the U.S. Geological Survey.

BIBLIOGRAPHY

Ball, M. M., Harrison, C. G. A., Supko, P. R., Bock, W., and Maloney, N. J., 1971, Marine geophysical measurements on the southern boundary of the Caribbean Sea, *in* Donnelly, T. W., ed., Caribbean geophysical, tectonic, and petrologic studies: Geol. Soc. America Mem. 130, p. 1-33.

Beets, D. J., 1972, Lithology and stratigraphy of the Cretaceous and Danian Succession of Curacao: Uitgaven "Natuurwetenschappelijke studiekring voor Suriname en de Nederlandse Antillen": Utrecht, no. 70, 165 p.

Bowin, C. O., (in press) The Caribbean gravity field and plate tectonics: Geol. Soc. America Spec. Paper.

———, and Folinsbee, A., 1970, Gravity anomalies north of Panama and Columbia (abstr.): EOS (Trans. Am. Geophys. Union), v. 51, no. 4, p. 317.

Campbell, C. J., 1968, The Santa Marta wrench fault of Colombia and its regional setting: Trans. Caribbean Geol. Conf., 4th, Trinidad, 1965, p. 247-260.

Case, J. E., and MacDonald, W. D., 1973, Regional gravity anomalies and crustal structure in northern Colombia: Geol. Soc. America Bull., v. 84, no. 9, p. 2905-2916.

Donnelly, T. W., Rogers, J. J. W., Kay, R., and Melson, W., 1971, Basalt and dolerite from the central Caribbean; preliminary results from Deep Sea Drilling Project, Leg 15 (abstr.): Geol. Soc. America, v. 3, no. 7, p. 548.

Duque Caro, H., 1972, Relaciones entre la bioestratigrafia y la cronoestratigrafia en el llamado geosinclinal de Bolivar: Colombia Ser. Geol. Nacl., Bol. Geol., v. 19, no. 3, p. 25-68.

Edgar, N. T., Saunders, J. B., et al., 1971a, Deep Sea Drilling Project, Leg 15: Geotimes, v. 16, no. 4, p. 12-16.

Edgar, N. T., Ewing, J. I., and Hennion, J., 1971b, Seismic refraction and reflection in Caribbean Sea: Am. Assoc. Petr. Geol. Bul., v. 55, no. 6, p. 833-870.

Edgar, N. T., Saunders, J. B., et al., 1973a, Initial reports of the Deep Sea Drilling Project, v. 15: Washington, D.C., U.S. Govt. Printing Office, 1137 p.

Edgar, N. T., Holcombe, T. Ewing, J., and Johnson, W., 1973b, Sedimentary hiatuses in the Venezuelan Basin, *in* Edgar, N. T., Sanuders, J. B., et al., 1973, Initial reports of the Deep Sea Drilling Project, v. 15: Washington, D.C., U.S. Govt. Printing Office, p. 1051-1062.

Emery, K. O., and Uchupi, E., 1972, Caribe's oil potential is boundless: Oil and Gas Jour., v. 70, no. 50, p. 156-162.

Feo-Codecido, G., 1971, Geologia y recursos naturales de la Peninsula de Paraguana, *in* Symposium on investigations and resources of the Caribbean Sea and adjacent regions: Paris, UNESCO, p. 231-240.

Galavis, J. A., and Louder, L. W., 1970, Preliminary studies on geomorphology, geology and geophysics on the continental shelf and slope of northern South America: 8th World Petr. Congr., Caracas, preprint, 26 p.

Holcombe, T. L., and Matthews, J. E., 1973, Structural fabric of the Venezuela Basin, Caribbean Sea (abstr.): Geol. Soc. America, v. 5, no. 7, p. 671-672.

Hopkins, H. R., 1973, Geology of the Aruba Gap abyssal plain near DSDP site 153, *in* Edgar, N. T., Saunders, J. B., et al., 1973, Initial reports of the Deep Sea Drilling Project, v. 15: Washington, D.C., U.S. Govt. Printing Office, p. 1039-1050.

Irving, E. M., 1971, La evolucion estructural de los Andes mas septentrionales de Colombia: Colombia Serv.

Geol. Nacl., Bol. Geol., v. 19, no. 2, 89 p.

King, P. B., compiler, 1969, Tectonic map of North America: Washington, D.C., U.S. Geol. Survey, scale 1:5,000,000.

Krause, D. C., 1971, Bathymetry, geomagnetism, and tectonics of the Caribbean Sea north of Colombia, *in* Donnelly, T. W., ed., Caribbean geophysical, tectonic, and petrologic studies: Geol. Soc. America Mem. 130, p. 35-54.

Panama, Direccion General de Recursos Minerales, 1972, Mapa geologico preliminar, Republica de Panama: Panama Direccion General Recursos Minerales, open-file map, scale 1:1,000,000.

Peter, G., 1972, Geologic structure offshore, north-central Venezuela: Trans. Caribbean Geol. Conf., 6th, Margarita, Venezuela, 1971, p. 283-294.

Roemer, L. B., Bryant, W. R., and Fahlquist, D. A., 1973, Geology and geophysics of the Beata Ridge—Caribbean: Texas A & M Univ., Dept. Oceanog. Tech. Rept. 73-14-T, 92 p.

Schubert, C., and Moticska, P., 1972, Geological reconnaissance of the Venezuelan islands in the Caribbean Sea between Los Roques and Los Testigos: Trans. Caribbean Geol. Conf., 6th, Margarita, Venezuela, 1971, p. 81-82.

Scientific Staff, 1973, Leg 4, 1971, cruise UNITEDGEO I, International Decade of Ocean Exploration: Regional gravity anomalies, Venezuela continental borderland: U.S. Geol. Survey open-file report, 24 p.

Shepard, F. P., 1973, Sea floor off Magdalena delta and Santa Marta area, Columbia: Geol. Soc. America Bull., v. 84, no. 6, p. 1955-1972.

______, Dill, R. F., and Heezen, B. C., 1968, Diapiric intrusions in foreset slope sediments off Magdalena delta, Colombia: Am. Assoc. Petr. Geol. Bull., v. 52, no. 11, pt. 1, p. 2197-2207.

Silver, E. A., et al., 1972, Acoustic reflection profiles—Venezuela continental borderland: U.S. Dept. Com., Natl. Tech. Inform. Service, PB2-07597, 38 p.

Silver, E. A., Case, J. E., and MacGillavry, H. J., (in press), Geophysical study of the Venezuelan Borderland: Geol. Soc. America Bull.

Talwani, M., 1966, Gravity anomaly belts in the Caribbean (abstr.): *in* Poole, W. H., ed., Continental margins and island arcs: Canada Geol. Survey Paper 66-15, p. 177.

Thomas, D. T., 1972, Tertiary geology and paleontology (phyllum Mollusca) of the Guajire Peninsula, northwestern Venezuela and northeastern Colombia [Ph.D. diss.]: State Univ. New York, Binghamton, N.Y., 150 p.

Uchupi, E., (in press), Physiography of the Gulf of Mexico and Caribbean Sea, *in* Stehli, F. G., and Nairn, A. E. M., eds., Ocean basins and margins, v. III, Caribbean and Gulf of Mexico: New York, Plenum.

Vasquez, E. E., and Dickey, P. A., 1972, Major faulting in northwestern Venezuela and its relation to global tectonics: Trans. Caribbean Geol. Conf., 6th, Margarita, Venezuela, 1971, p. 191-202.

Worzel, J. L., 1965, Pendulum gravity measurements at sea, 1936-1959: New York, Wiley, 421 p.

Zambrano, E., Vasquen, E., Duval, B., Latreille, M., and Coffinieres, B., 1971, Sintesis paleogeografica y petrolera del occidente de Venezuela: Venezuela Ministerio de Minas e Hidrocarburos, Boletin de Geologia, Publicacion Especial, no. 5, t. I, p. 483-546.

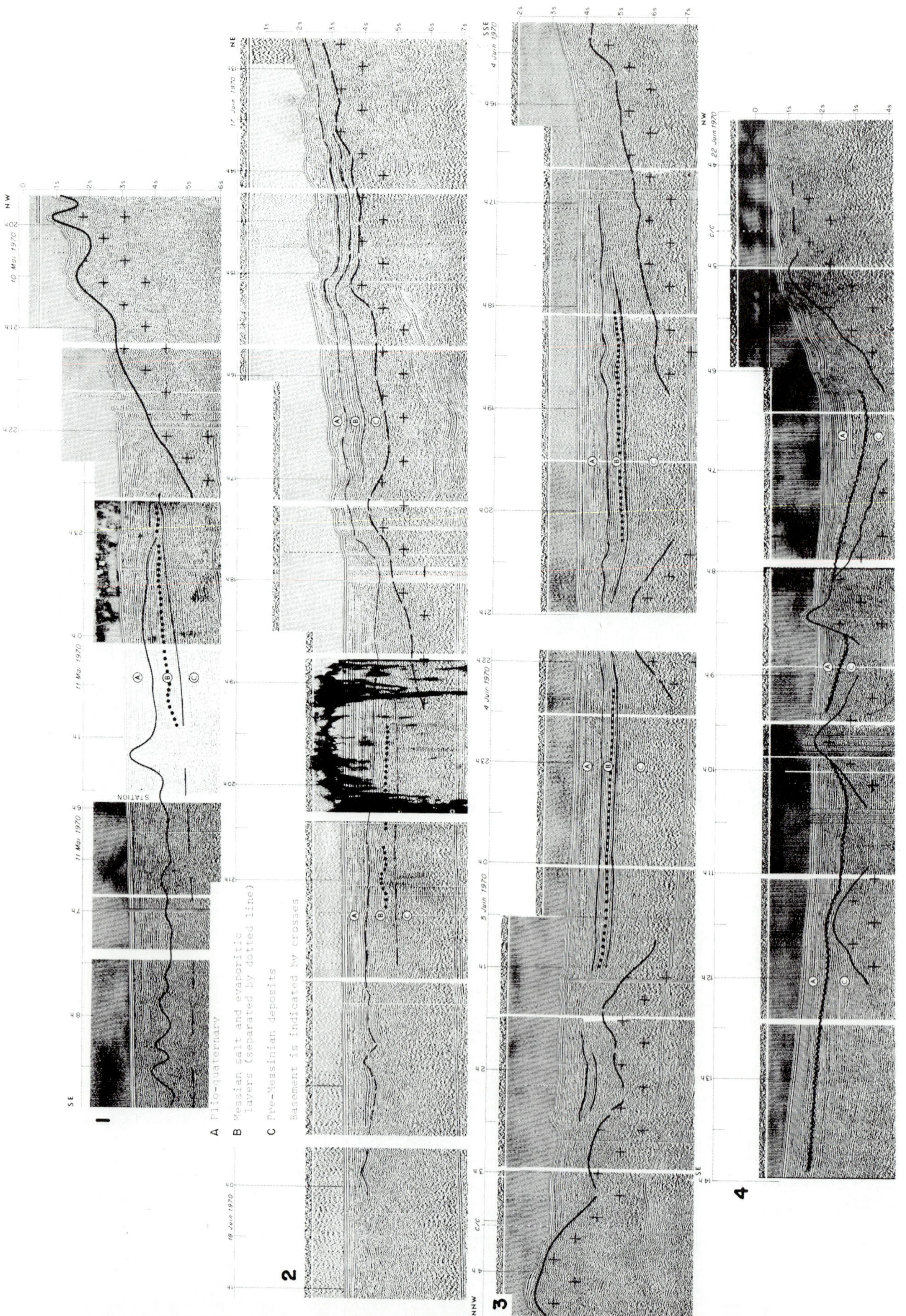

A Plio-quaternary
B Messian salt and evaporitic layers (separated by dotted line)
C Pre-Messinian deposits
Basement is indicated by crosses
1
2
3
4

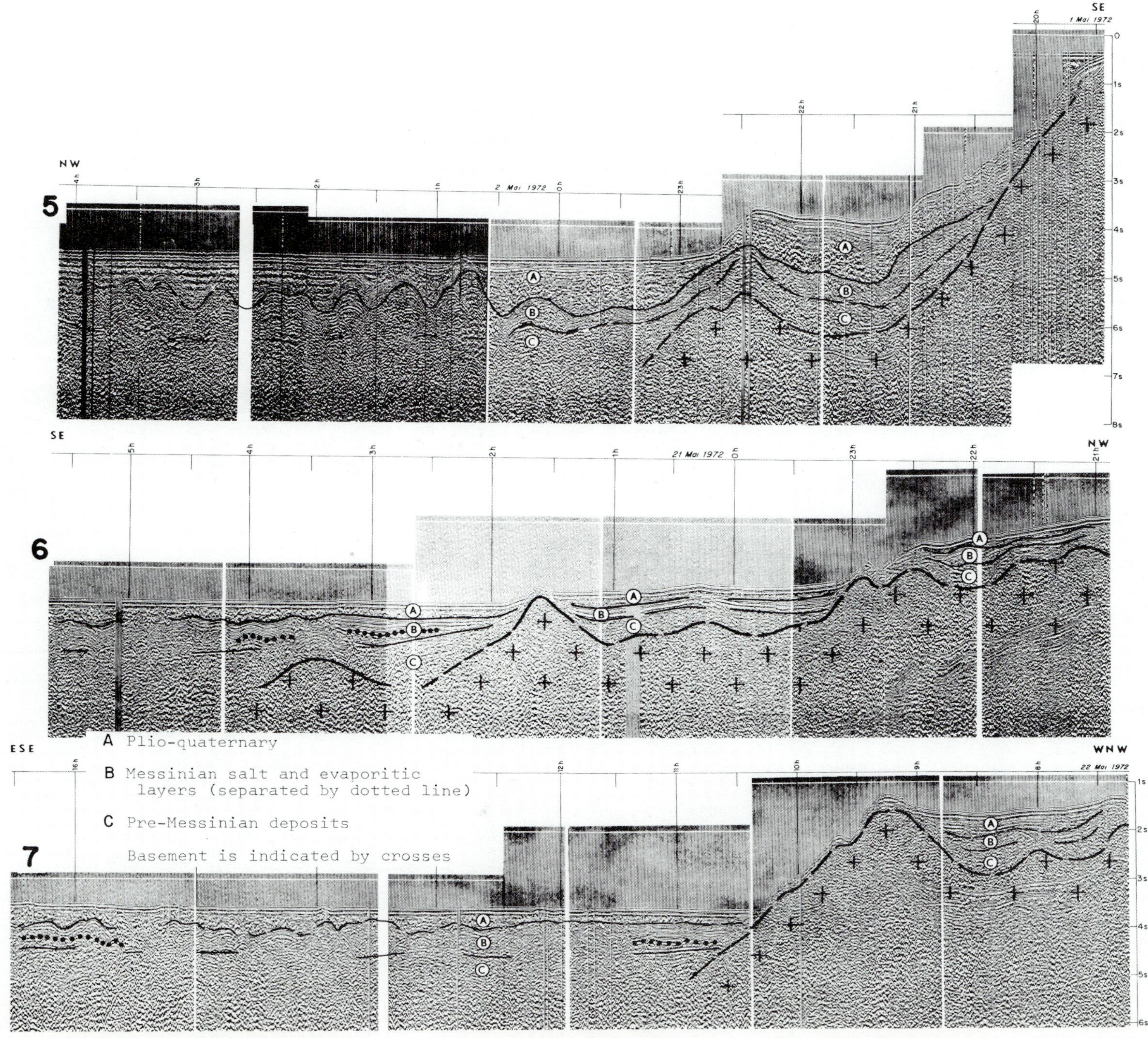
5
NW
SE
2 Mai 1972
1 Mai 1972
6
SE
NW
21 Mai 1972
7
ESE
WNW
22 Mai 1972
A Plio-quaternary
B Messinian salt and evaporitic
layers (separated by dotted line)
C Pre-Messinian deposits
Basement is indicated by crosses

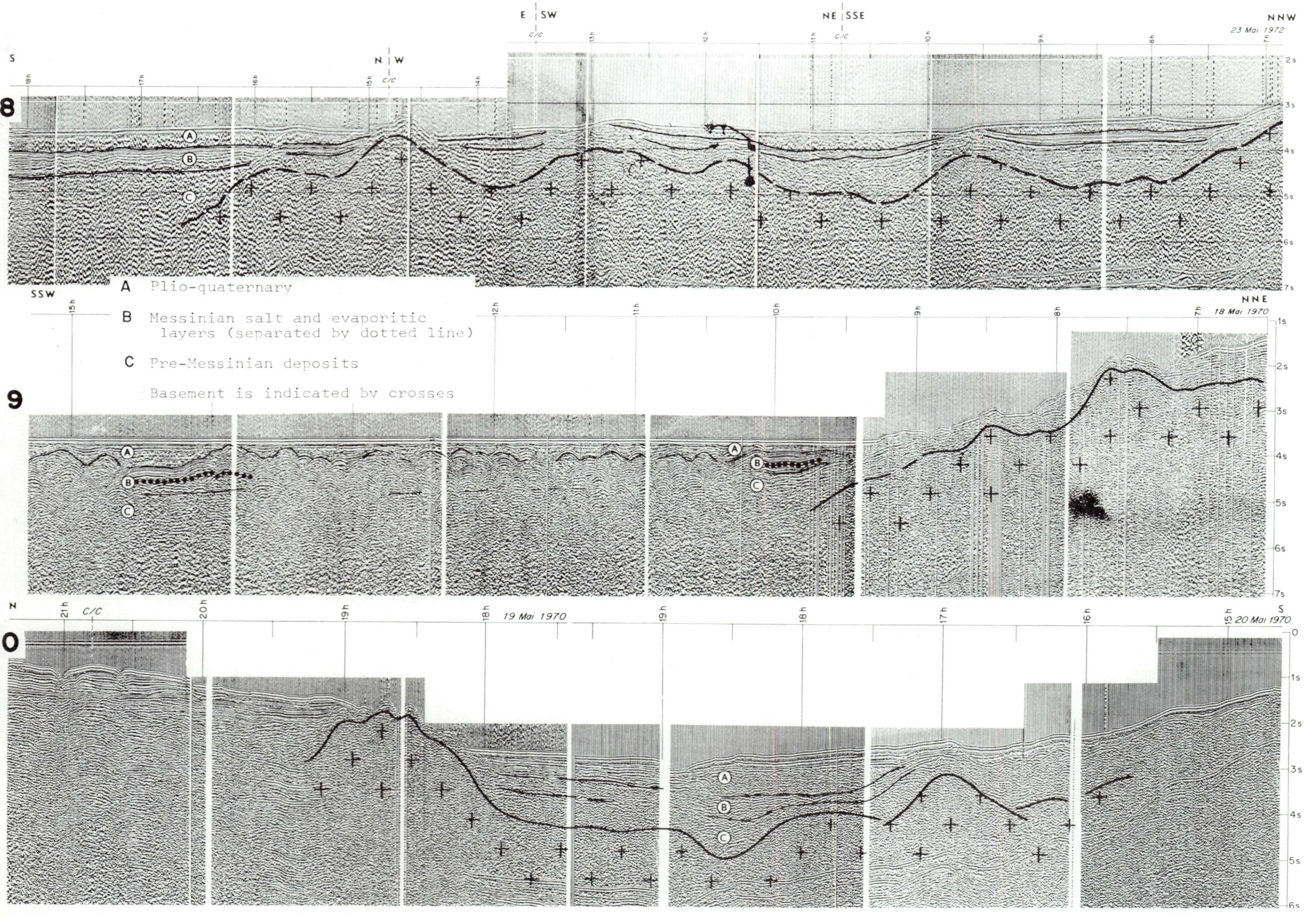
8
S
N | W
E | SW
NE | SSE
NNW
23 Mai 1972
9
SSW
NNE
18 Mai 1970
10
N
S
19 Mai 1970
20 Mai 1970
A Plio-quaternary
B Messinian salt and evaporitic layers (separated by dotted line)
C Pre-Messinian deposits
Basement is indicated by crosses

Geologic Background of the Red Sea

R. G. Coleman

INTRODUCTION

New concepts related to the mobility of the earth's crust have focused attention on the Red Sea area. The geographic outline of western Arabia and northeast Africa shows the apparent jigsaw fit that appears possible in closing the Red Sea. Indeed, the modern paradigm of plate tectonics would have the Arabian Peninsula drifting away from the African continent. The early formation of the Red Sea is somewhat obscure and is not clearly a result of sea-floor spreading. This is an attempt to summarize much of the recent geologic and geophysical data to form a basis for further speculation on the actual formation of this most-interesting small ocean basin.

PHYSIOGRAPHY

The Red Sea occupies an elongate depression over 2,000 km long. In the northern part, shorelines are only 180 km apart; southward, the gap widens to 360 km. At the south end, the islands of the flat-lying, coral-topped Farasan and Dahlak Archipelagoes rise slightly more than 60 m above sea level and are part of a shallow shelf extending nearly to the center of the Red Sea (MacFadyen, 1930). The southern part of the Red Sea narrows (28 km) to the straits of Bab el-Mandeb, where it connects with the Gulf of Aden. North of the straits, a series of volcanic islands occupy the medial part of the Red Sea (MacFadyen, 1932). The most recent bathymetric chart of the Red Sea (Laughton, 1970) reveals a very distinct *main trough* extending northward from the Zebayir Islands to the southern tip of the Sinai Peninsula. The main trough is cut by an *axial trough*, varying from 30 km in width near 20°N and narrowing southward, where it may be 5-14 km wide (Fig. 1). Selected topographic cross sections reveal steep-sided walls in the axial trough and a very irregular bottom topography. A line connecting the deepest parts of the axial trough reveals sharp changes in strike-direction, with apparent offsets. High-temperature brine pools and hydrothermal sediments are present within the axial trough.

The bottom topography is shallow and flat on the Red Sea shelf. South of Jidda, Saudi Arabia, the shelf merges imperceptibly with the coastal plain; north of Jidda it is narrower and interrupted by rather sharp topographic breaks. The coastal plains broaden south of Jidda and widen to 50 km in Yemen and Saudi Arabia. Volcanic flows cover parts of the plain in Arabia and Ethiopia. North of 22°N, the coastal plain tends to be narrower, with terraces well above the present shoreline. The general aspect is one of shoreline emergence to the north and of submergence to the south.

Landward from the littoral zone and coastal plain, the topography varies considerably and is best described in segments. A magnificent erosional escarpment extends unbroken southward from the latitude of Jidda, Saudi Arabia, to Yemen, forming the crest of the tilted Arabian block. The crest follows an extremely irregular erosional pattern that is subparallel to the coastline, extending inland as far as 120 km in some places and less than 50 km in others. In Yemen, altitudes of over 3,700 m have been measured, and in Saudia Arabia the scarp crest averages more than 2,000 m (U.S. Geological Survey, 1963). No major normal faults parallel to the Red Sea or to the escarpment have been recognized between the coastal plain and the crest. In Ethiopia, a similar scarp bounds the coastal plain on the west, reaching altitudes of over 3,000 m. Geologic evidence here shows a close relationship between faulting and the scarp (Mohr, 1962). At Massawa (15°N) the scarp turns southward and leaves the Red Sea trend, connecting with the East African rift system. South of Massawa, broadening of the East African rift system has produced the magnificent Afar Depression. Extensive volcanism and block faulting in the Afar Depression have produced an incredible mixture of interior basins and volcanic landscape.

The coastal plains in Sudan and southern Egypt are narrow and not bordered by scarps. Instead, the lower slopes merge seaward as interdigitating fans and pediments broken by outliers of older rocks (Whiteman, 1971). A similar situation exists along the coastal plain on the east side of the Red Sea at these latitudes. The northern end of the Red Sea bifurcates into the Gulf of Suez and the Gulf of Aqaba, the former controlled primarily by normal faulting, the latter by wrench faulting. The narrow coastal plains here climb directly into fault-bounded blocks of basement rock, less than 5-10 km from the shoreline.

GEOLOGY

Stratigraphic Record

Paleogeographic reconstructions (Swartz and Arden, 1960; Said, 1962; Mohr, 1962; Brown, 1972) for the Mesozoic show Jurassic seas spreading northward to the latitude of Massawa and Jizan, and the shoreline trending directly across the Red Sea at

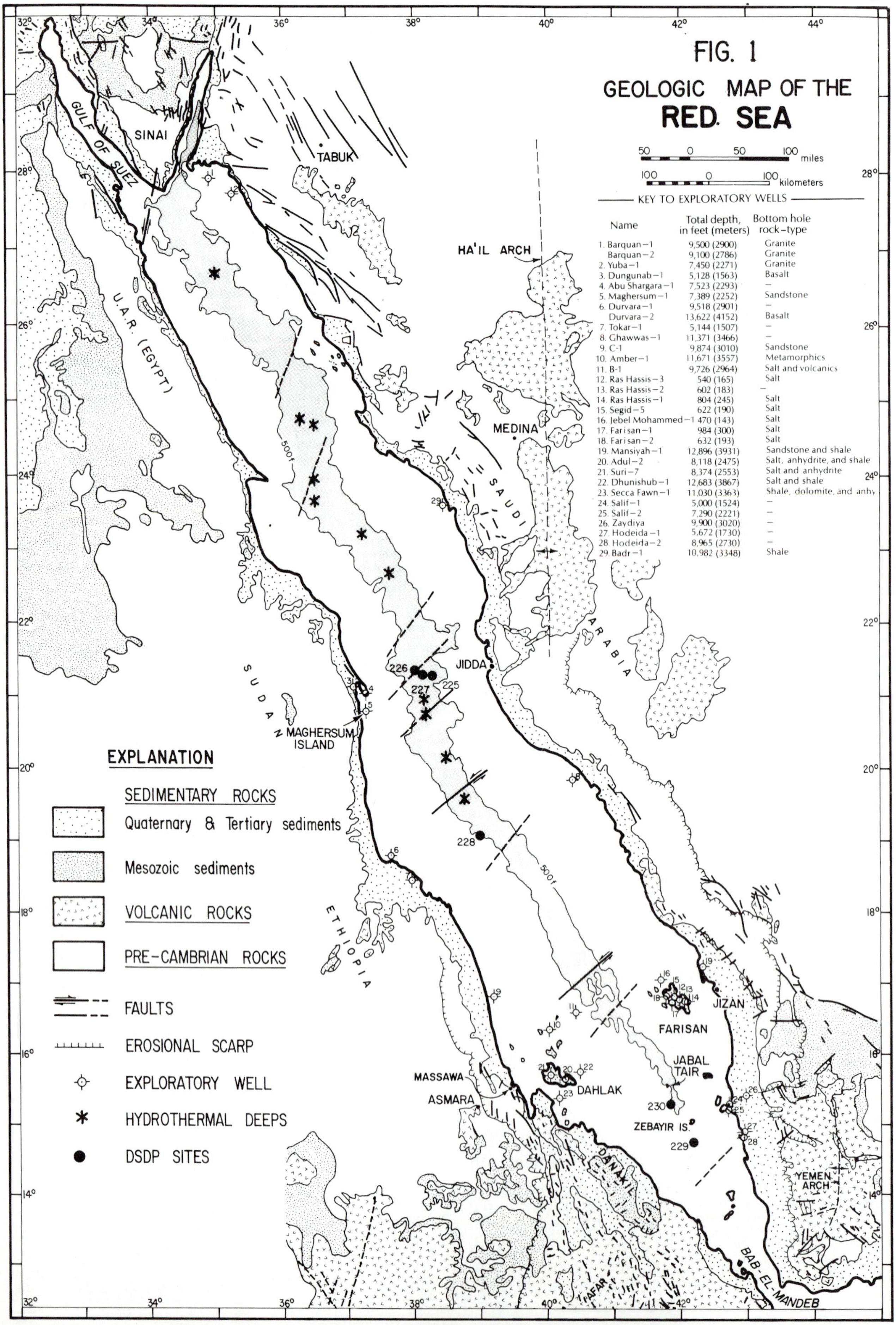

FIG. 1

GEOLOGIC MAP OF THE RED SEA

50 0 50 100 miles

100 0 100 kilometers

KEY TO EXPLORATORY WELLS

Name	Total depth, in feet (meters)	Bottom hole rock-type
1. Barquan–1	9,500 (2900)	Granite
Barquan–2	9,100 (2786)	Granite
2. Yuba–1	7,450 (2271)	Granite
3. Dungunab–1	5,128 (1563)	Basalt
4. Abu Shargara–1	7,523 (2293)	–
5. Maghersum–1	7,389 (2252)	Sandstone
6. Durvara–1	9,518 (2901)	–
Durvara–2	13,622 (4152)	Basalt
7. Tokar–1	5,144 (1507)	–
8. Ghawwas–1	11,371 (3466)	–
9. C-1	9,874 (3010)	Sandstone
10. Amber–1	11,671 (3557)	Metamorphics
11. B-1	9,726 (2964)	Salt and volcanics
12. Ras Hassis–3	540 (165)	Salt
13. Ras Hassis–2	602 (183)	–
14. Ras Hassis–1	804 (245)	Salt
15. Segid–5	622 (190)	Salt
16. Jebel Mohammed–1	470 (143)	Salt
17. Farisan–1	984 (300)	Salt
18. Farisan–2	632 (193)	Salt
19. Mansiyah–1	12,896 (3931)	Sandstone and shale
20. Adul–2	8,118 (2475)	Salt, anhydrite, and shale
21. Suri–7	8,374 (2553)	Salt and anhydrite
22. Dhunishub–1	12,683 (3867)	Salt and shale
23. Secca Fawn–1	11,030 (3363)	Shale, dolomite, and anhy
24. Salif–1	5,000 (1524)	–
25. Salif–2	7,290 (2221)	–
26. Zaydiya	9,900 (3020)	–
27. Hodeida–1	5,672 (1730)	–
28. Hodeida–2	8,965 (2730)	–
29. Badr–1	10,982 (3348)	Shale

EXPLANATION

SEDIMENTARY ROCKS

Quaternary & Tertiary sediments

Mesozoic sediments

VOLCANIC ROCKS

PRE-CAMBRIAN ROCKS

FAULTS

EROSIONAL SCARP

EXPLORATORY WELL

HYDROTHERMAL DEEPS

DSDP SITES

GULF OF SUEZ
SINAI
TABUK
HA'IL ARCH
U.A.R. (EGYPT)
MEDINA
SAUDI
ARABIA
JIDDA
SUDAN
MAGHERSUM ISLAND
ETHIOPIA
FARISAN
JIZAN
JABAL TAIR
MASSAWA
ASMARA
DAHLAK
ZEBAYIR IS.
DANAKIL
AFAR
YEMEN ARCH
BAB-EL-MANDEB
500f

17°N. To the northwest the Jurassic shoreline crosses the Sinai Peninsula, trending east-west at about 29°N. By the end of the Cretaceous, the Tethyan Sea reached southward to approximately 21°N (Whiteman, 1968, 1971; Karpoff, 1957; Carella and Scarpa, 1962). In the south, regressive Cretaceous marine facies are present in Ethiopia and southern Yemen, extending northward only to 13°N. The marine sedimentary record for the Mesozoic provides no strong evidence that the Red Sea depression existed during this time. Indirect evidence of tectonic activity during Mesozoic times in and around the Red Sea is generally lacking. However, interlayering of basaltic rocks with Cretaceous sedimentary rocks signals the beginning of emplacement of the thick Trap Rock sequences in Yemen (Geukens, 1966).

The early Tertiary sedimentary record in the Red Sea area is incomplete and has been the subject of debate. It is generally agreed that during Eocene time, transgressive marine deposits developed in the northern Sinai and Egypt and that Eocene rocks may occur as far south as Maghersum Island, Sudan (Said, 1962; Whiteman, 1968; Sestini, 1965; Carella and Scarpa, 1962). Shorelines of early and middle Eocene age are shown to extend across the southern end of the Red Sea, and by the end of the Eocene, the horn of Africa had completely emerged (Mohr, 1962). Oligocene sediments have not been identified positively along most of the Red Sea coastal plains; however, deposition of the Red Series in the Afar Depression is considered to have begun in the late Oligocene (Bannert et al., 1970; Barberi et al., 1972). Oligocene sediments were found in exploratory drilling for iron ore in Wadi Fatima, east of Jidda (Shanti, 1966). Oligocene and Miocene marine sediments are found on the south coast of the Arabian Peninsula (Beydoun, 1964).

Deep wells along the western margins of the Red Sea have penetrated shale containing Oligocene fossils and interlayered basalt that gives Oligocene radiometric ages (Whiteman, 1968). Further deep drilling will resolve the question of whether Oligocene sediments underlie the Red Sea depression (Lowell and Genik, 1972; Hutchinson and Engels, 1970). Separate, great uplifts on either side of the Red Sea during Oligocene time were accompanied by outpourings of alkaline basalts that now form a thick trap series as high plateaus behind the Arabian and Ethiopian scarps. The paucity of Oligocene marine sediments in and around the Red Sea depression has prompted some geologists to suggest that during the Oligocene the Red Sea trend occupied the crest of a large Afro-Arabian dome or domes (Gass, 1970; Swartz and Arden, 1960).

Miocene sediments have been described along the coastal plains of the Red Sea from the Gulf of Suez to 14°N (Said, 1962; Heybroek, 1965; Brown, 1970). Numerous deep-exploratory hydrocarbon wells have penetrated great thicknesses of Miocene clastics (2-3 km) and evaporites (3-4 km) without reaching basement rocks (Ahmen, 1972; Frazier, 1970). This new geologic information, combined with that obtained on Leg 23B, Deep Sea Drilling Project, demonstrates that the Red Sea depression received marginal clastic sediments, interfingering with evaporites, during Miocene time. The Miocene marine invasion is considered to have come from the Mediterranean rather than from the Indian Ocean (Montanaro, 1941). During the Miocene, volcanic activity continued in the Afar Depression and in western Saudi Arabia. As the Red Sea depression continued to subside, uplift continued in the Arabian-Yemen block on the east and the Ethiopian segment on the west.

At the beginning of the Pliocene, marine oozes and marginal clastics began to be deposited on the Miocene evaporite sequence, signaling a drastic change in the sedimentary record. The Pliocene marine invasion is considered to have come from the Indian Ocean rather than the Mediterranean, which was separated from the Red Sea during the early Pliocene by uplift of the Isthmus of Suez (Swartz and Arden, 1960). The important lack of Miocene salt deposits in the axial trough of the Red Sea has been confirmed by drilling on Leg 23B, Deep Sea Drilling Project. In the axial trough, a thin veneer of sediments overlies basalt that is considered to have been extruded into this axial trough (Gass, 1970).

Seismic-reflection profiles made across the shelf of the Red Sea confirm the presence of a strong reflector at depths up to 500 m, which is now considered to represent the top of the evaporite section (Phillips and Ross, 1970; Ross et al., 1972). Assuming that the salt once extended across the axial trough, the seismic profiles reveal a separation of 48-74 km across the deep main trough. Rates of deposition of Pliocene and Pleistocene sediments in the southern part of the Red Sea were distinctly higher than those observed at the latitude of Jidda, indicating continued uplift and erosion along the southern margins of the Red Sea (Ross et al., 1972). Within the Afar Depression, a thick section of evaporites (~ 975 m) overlies the Red Series (Holwerda and Hutchinson, 1968). These evaporites are considered to be Pleistocene in age and are probably evidence of intermittent connections with the Red Sea through the Zula Gulf (Hutchinson and Engels, 1970).

All along the present shorelines are recent coral-reef limestone deposits, some of which are as much as 300 m thick. Along the east margin near Jizan, these deposits are rapidly buried by the clastic debris brought into the littoral zone from the high erosional scarps.

Structure

The divergent trends of the Red Sea depression and adjacent Precambrian basement clearly indicate that the Red Sea depression is not related to these early structures (Picard, 1970; Schürmann, 1966; Brown and Jackson, 1960). Recent structural maps (Brown, 1972; Choubert, 1968) further amplify the discordance and show that Paleozoic to Paleogene sediments and related structures cannot be part of the young Red Sea evolution.

Uplift in Ethiopia and Arabia took place on an immense scale immediately after the late Eocene regression (Mohr, 1962). The Hail Arch, an early Paleozoic north-trending headland in the Precambian basement of the Arabian Shield, was deformed during pre-Permian times, and during late Cretaceous was folded parallel to its north-south axis (Greenwood, 1972). Fractures that developed as part of the uplift provided feeder fissures for the Oligocene Trap Series in Ethiopia and Arabia. A monoclinal flexure developed along the eastern margin of the Red Sea as the Yeman and Hail Arches continued to rise. East of Jizan, along the hinge lines of the monoclinal warp, zones of stretched and fractured rock were invaded by tholeiitic magma, forming locally differentiated layered gabbros, granophyres, and diabase dike swarms, ranging in age from 20 to 25 my (Coleman and Brown, 1971).

Uplift and arching in Ethiopia coincided with the intrusion of granite and rhyolite (22-25 my), marginal to the Afar Depression (Barberi et al., 1972). In Ethiopia, continued uplift during the Oligocene and Miocene produced a complex pattern of major faults that gave rise to the down-faulted Afar Depression, the Danakil Horst, and the highlands west of the Depression (Tazieff et al., 1972). Deposition of the Red Series, confined to the Afar Depression, began in the late Oligocene and marked the warping of the depression (Bannert et al., 1970; Barberi et al., 1972). The normal faults bounding the depression are still active (Gouin, 1970). During Oligocene and Miocene time, uplift was sustained in the Yemen and Hail arches with continued monoclinal flexuring and gentle downwarping of the coastal plain. The thick Miocene clastic sections discovered in deep wells along the coastal plains were deposited unconformably on older, downwarped units and demonstrate rapid erosion of the Yemen, Hail, and Ethiopian arches (Gillmann, 1968). Steep dips of Jurassic and early Miocene sedimentary rocks which toward the Red Sea axis along the monocline indicate downwarping of the Red Sea depression (Coleman and Brown, 1971; Gass and Gibson, 1969).

Major normal faults have not been found along the eastern margin of the Red Sea, even though they are shown on numerous generalized maps (e.g., Mohr, 1962, p. 160; Picard, 1970; Holmes, 1965). Whiteman's comments (1968, p. 235) on a presumably fault-bounded Red Sea are appropriate: "It is not possible here to give all the details relating to the nature of the bounding escarpments but analysis reveals that many of the faults shown on general maps do not exist on the ground and have, in fact, been put in on the assumption that most of the escarpments are of fault origin. It is indeed curious that almost every map of the Red Sea Depression showing faults shows a different pattern." Important to this argument is the recognition that the contacts between the Mesozoic and Tertiary sediments and the Precambrian basement are unconformities, and that the sediments dip steeply toward the axis of the depression.

Perhaps some geologists hold to the view that the Red Sea is bounded by faults because the Suez and Aqaba grabens and the Afar Depression are all well-documented, fault-bounded structures. The matching shorelines across the northern Red Sea, along with the subparallel southern erosional scarps in the Arabian Peninsula and Ethiopia, give quasi-tectonic evidence that permits this interpretation. Minor normal faults are associated with the monoclinal flexures along the Red Sea, and continued exploratory drilling may reveal major normal faults concealed by the thick Miocene evaporite-clastic deposits (Lowell and Genick, 1972). However, the Oligocene and Miocene development of the Red Sea depression deduced from the stratigraphic and structural record indicates that the northwest-trending depression developed as a trough between the Arabian and African swells, rather than as a down-faulted block (Brown and Coleman, 1972).

Mobility of the Miocene evaporites is indicated by the numerous salt domes developed in the marginal parts of the Red Sea south of 18°N (Brown, 1972). Salt domes may have begun to develop during the Pliocene and may be forming currently. Present-day subsurface salt flowage has been recorded at depths greater than 3 km (Frazier, 1970). At bottom-hole temperatures above 200°C, and under the pressure of thick overburdens, salt acts very nearly like "soft butter." Salt could easily flow into the axial trough, and the irregular topography at 16°40'N suggests that this may well have happened. Strong deformation of the upper surface of the evaporites, as deduced from seismic profiles, clearly indicates salt flowage (Phillips and Ross, 1970).

Formation of the axial trough at the beginning of the Pliocene appears to have been related to a major rift, a marked departure from the Oligocene and Miocene development of the Red Sea depression. The physiography of the trough, a steep-walled depression with a floor of basalt bears out the possible similarity to other active mid-ocean spreading centers. Concentration of seismic epicenters within the axial trough, and the alignment of active volcanoes along the trough axis, also indicate present-day rifting. Solutions of first-motion movements on recent epicenters in the axial trough indicate movement along northeast-trending faults (Fairhead and Girdler, 1970; McKenzie et al., 1970). These first-motion solutions, combined with geologic studies, have shown that recent left-lateral shear along the Dead Sea rift would fit with the general northeast movement of Arabia away from Africa. In summary, synthesis of the present structural knowledge on the evolution of the Red Sea indicates at least two stages of development: (1) pre-Miocene downwarping and crustal thinning, and (2) post-Miocene rifting.

Volcanology

Distribution of Tertiary volcanic rocks provides further clues as to the origin of the Red Sea. One of the largest areas of alkaline basalts in the world is centered over Ethiopia, Yemen, and western Saudi Arabia. These are "plateau basalts." Eruption of these lavas accompanied the development of the Ethiopian and Arabian swells, probably starting in early Eocene in Yemen and culminating in the Oligocene in both Ethiopia and Arabia (Baker et al., 1972). The Ethiopian Trap Series near Asmara has been dated as 36 my, and the Sirat plateau basalts of Arabia have been bracketed between 25 and 29 my (Brown, 1970). These flood basalts are predominately alkaline olivine basalt, with some interbedded silicic lavas, agglomerates, and pyroclastic rocks (Baker et al., 1972). In the Ethiopian plateau, the Ethiopian Trap Series reaches a thickness of 3,500 m; in Yemen, 1,000 m. The Yemen Trap Series extends northward into Saudi Arabia, where these rocks are called the Sirat Plateau Basalts, and thin to about 400 m.

No evidence has been discovered that shows whether the Ethiopian and Arabian Peninsula plateau basalts were coextensive across the southern part of the Red Sea; however, drilling could resolve this question. The Yemen Trap Series forms an anticlinal swell, with the western margin dipping under the coastal plain (Brown, 1970). On the Danakil Horst, pre-Miocene basalts cover Mesozoic deposits and appear to dip toward the Red Sea axis, but these basalts may not be equivalent to the Eocene and Oligocene trap series.

In both Ethiopia and Arabia, the basal flows have been extruded upon an extensive laterite that is considered to be Jurassic (Portlandian) to Eocene in age (Mohr, 1962). The chemical and petrographic continuity of the Yemen and Ethiopian Trap Series has been demonstrated by Mohr (1971), and recent work by Coleman and Brown (1971) on the Sirat plateau basalts extends the area of this great effusion of alkali olivine basalts. Northward, on both sides of the Red Sea depression, there are no extensive areas of trap rock of this age. Tromp (1950) reports minor, but widespread, Oligocene basalt flows within the marine sediments of eastern Egypt; however, no petrographic details are available.

In Miocene time (~22 my), differentiated layered gabbros and tholeiitic dike swarms invaded the hinge line of monoclinal downwarps along the south Arabian coast (Coleman and Brown, 1971). East of Jizan, a canoe-shaped, layered gabbro approximately 300 m thick is associated with diabase dikes cutting Jurassic sandstones. Dike swarms several hundred meters across trend parallel to the Red Sea in this area, and linear magnetic anomalies associated with sporadic dike outcrops have been traced as far north as Jidda (Brown, 1972). Within the Afar Depression, mainly along its margins, small granitic intrusive and silicic volcanic rocks (22-25 my) are considered to be related to the downwarping and marginal faulting that created the Afar Depression (Barberi et al., 1972).

As the Miocene evaporite and clastic deposits accumulated in the Red Sea depression, great volumes of alkali olivine lavas erupted along the crest of the Hail Arch, north of Jidda. These flows generally are less than 50 m thick; nonetheless, they spread over vast areas, some even extending westward over the coastal plain. The bulk of these flows lies approximately 150-200 km inland from the Red Sea, and the north-south alignment of vents on the flows parallels the Hail Arch (Greenwood, 1972).

Contemporaneous with these inland flows are other alkali olivine basalt extrusions along the same north-south trend, situated between Amman in Jordan and Turayf in northern Saudi Arabia, more than 250 km from the Red Sea. Exploratory deep drilling in the Miocene evaporite and clastic section of the Red Sea shelf has penetrated interlayered basalts in the southern Red Sea (Lowell and Genik, 1972), and interlayered Miocene basalts are reported in the Red Series of the Afar Depression (Bannert et al., 1970), indicating the possibility of discontinuous rifting in the Red Sea depression during Miocene times.

As the axial rift in the Red Sea began to form, sometime early in the Pliocene (~5 my), subalkaline oceanic tholeiites began to fill the axial trough, and numerous submarine volcanoes contributed pyroclastic rocks and lava (Chase, 1969). While this was happening, many small alkaline olivine basalt eruptive centers developed along the hinge line of the flexure zone of the eastern margin of the Red Sea south of 19°N. Mantle xenoliths common to these basaltic eruptions indicate a source much different from that of the basalts filling the axial trough about 170 km to the west. Additional widespread alkaline olivine basalts were erupted along the crest of the Yemen and Hail Arches from Yemen northward to Tabūk in Saudi Arabia. Mantle xenoliths have also been reported from these rocks. This volcanic activity has continued until the present, with historical eruptions reported near Medina, Saudi Arabia, and in Yemen.

Within the Afar Depression, there was a great outpouring of floor basalts—the Stratoid Series (Barberi et al., 1972). This volcanism seems to have extended from the late Miocene to Quaternary, and is represented by transitional basalts and peralkaline silicic rocks (Barberi et al., 1972). Continuing up to the present, volcanism in the axial trough of the Afar Depression produces what are considered to be "transitional basalts," with affinities toward subalkaline oceanic basalts and including some silicic, welded ash-flow tuffs and lavas. Marginal vents from which silicic rocks have erupted remain active today. In the Red Sea depression, within the straits of Bab el-Mandeb and northward to Jabal Tair, Holocene volcanic islands have developed along the axial rift. Rock types progress from oceanic tholeiite at Jabal Tair southward to alkali basalt, trachybasalts, and trachyandesites within the Zukur-Hanish Islands (Gass et al., 1965). Basaltic and peralkaline eruptive centers extend eastward along the south coast of Aden. Individual, small submarine peaks within the axial trough are also considered to be extinct volcanoes.

GEOPHYSICS

Many geophysicists have worked in the Red Sea area, but mostly in conjunction with oceanographic expeditions rather than land surveys. Political problems have prevented a complete magnetic or gravity survey of the whole region. Most geophysical interpretations are therefore made from local surveys and provide only partial answers.

Interpretations of magnetic surveys (Girdler, 1970a; Kabbani, 1970; Drake and Girdler, 1964; Allan, 1970) within the Red Sea seem to provide a consistent picture. Strong magnetic anomalies associated with the axial trough have been correlated with the interpreted, world-wide geomagnetic polarity history and are considered to represent a time span of approximately 4 my (Vine, 1966). Displacement of the magnetic anomalies of the axial trough between 19°N to 23°N suggests to some workers the presence of active transform faults (Allan, 1970). Away from the axial trough, magnetic expressions are diffuse and fail to provide clear correlations with geomagnetic polarity history. These axial-trough anomalies do not extend even to the northernmost part of the Red Sea or southward into the straits of Bab el-Mandeb.

Seismic events recorded in this area since 1953 reveal a consistent but incomplete pattern related to possible rifting. Most of the epicenters are along the axial trough, but surprisingly few are in the northern Red Sea (Fairhead and Girdler, 1970). Most of the events appear to be shallow (<100 km), and possibly related to the development of the axial trough. Epicenters within the Afar Depression are mostly along the margins of the depression, rather than in the axial part (Gouin, 1970). Fault-plane solutions in the Red Sea axial trough indicate a strike-slip motion, directed northeast-southwest, perhaps indicating that the axial trough is offset by oblique faults. A strong recent quake (March 1969) in the northern Red Sea, at the opening of the Gulf of Suez, provided additional information on tectonic movements in this area (Ben-Menahem and Aboodi, 1971). The tectonic solution shows normal sinistral motion, with dip-slip and strike-slip components, and a trend of motion 10°NE at a depth of 10 km.

The Red Sea area appears to be in isostatic equilibrium, according to Girdler (1969). Positive Bouguer anomalies, some as high as 140 mgal, have been detected in the central part of the axial trough. However, negative gravity anomalies in the Gulf of Aqaba may possibly result from active left-lateral shear (Allan, 1970). The Afar Depression shows Bouguer gravity values that generally are negative, and apparently are related to relief (Gouin, 1970). Positive gravity anomalies are present over volcanic centers and along the coast northeast and parallel to the Danakil Horst, but they are not nearly so strong as those observed over the axial trough of the Red Sea. Thick accumulations of dense oceanic crust appear therefore to be restricted to the Red Sea axial trough.

Interpretations of seismic velocities measured within the Red Sea are numerous, but controversy exists regarding the deep structures (Knott et al., 1966; Davies and Tramontini, 1970; Girdler, 1969). In the axial trough, at depths of less than 5 km below the surface, velocities are from 6.6 to 7.0 km/sec, and it is generally agreed that the material at these depths is oceanic crust. Other seismic traverses along the Red Sea shelf show a sedimentary cover to be 2-5 km thick (3.5-4.5 km/sec), consisting of a mixture of evaporites and clastic and volcanic material. The controversy centers on the nature of the materials below the sedimentary cover. Davies and Tramontini (1970), surveying a 1° area on the east shelf, between 22° and 23°N, measured velocities of 6.6 km/sec at average depths of 4.6 km and concluded that this material was also oceanic crust similar to that in the axial trough; however, they may possibly have transected one of the many greenstone belts within the Precambrian, which could also account for such high velocities. Others have measured velocities of from 5.84 to 6.97 km/sec (average 6.1 km/sec) around the margins of the Red Sea, and interpreted them as representing Precambrian basement (Girdler, 1969; Knott et al., 1966). Gravity measurements appear to support the latter interpretation, not the former.

Continuous seismic profiles reveal a strong reflector at depths from 0 to 500 m in the marginal parts of the Red Sea but not in the axial trough (Phillips and Ross, 1970; Knott et al., 1966). This reflector is now interpreted as the top of the evaporite section, commonly correlated with the Miocene-Pliocene boundary.

Very high values of heat flow have been measured in the axial trough of the Red Sea. Plotted as a function of closeness to the axis, heat flow increases from $\sim$3.0 mcal//cm^{-2}/sec^{-1} at the margins to $\sim$79 mcal/cm^{-2}/sec^{-1} in the axial trough (Girdler, 1970b). Formation of new basaltic crust along the axial trough presumably relates to this apparent high heat flow. Heat flow in the entire Red Sea area however, appears to be high, with values of 2.7 to 3.0 mcal/cm^{-2}/sec^{-1} along the margins, and general geothermal gradients of 40°C/km, and probably higher. Hot springs are common along the east margin hinge line of the Red Sea and in the Afar Depression.

SUMMARY

Formation of the Red Sea and its surrounding features is the result of many interrelated events. The fact that Red Sea structures can be fitted into a world-wide plate-tectonic system has obscured its earlier geologic development. Its time of origin is poorly known, but all evidence points toward the early Tertiary.

Late Eocene and early Oligocene marked the origin of the Ethiopian and Arabian swells, with the proto-Red Sea depression separating them. Vast outpourings of alkali olivine basalts formed thick Oligocene plateaus along the crests of these swells. Monoclinal flexures along the eastern margin related to the downwarping of the Red Sea produced a fractured hinge line that filled with differentiating tholeiitic gabbros and diabase dike swarms. These early Miocene dike swarms (~22 my) might suggest "rifting" in the Red Sea depression at this time, but the geologic situation clearly shows that they developed in the fractured hinge line of a monoclinal warp.

Therefore, there is no compelling evidence of steady-state "rifting" prior to the Pliocene. Invasion of the Mediterranean seas southward into a closed depression provided conditions under which a thick (2-5 km) evaporite-clastic series could form. The continued development of the evaporite section during the Miocene is further evidence against large-scale opening of the Red Sea at this time, even though continued subsidence is inferred with possible minor extension.

Interfingering with the evaporites along the margins of the sea are thick (2-3 km) clastic sequences, produced by erosion of the Arabian-Ethiopian swells. Continued uplift in the Miocene produced block faulting in the Afar Depression in Ethiopia, while the east margin of the Red Sea was marked by the continued depression of the marginal sediments. Erosional retreat of the edges of the Arabian and Ethiopian swells produced the present scarps. Northward, the Miocene grabens of the gulfs of Suez and Aqaba extended southward into the Red Sea depression.

Localized Miocene basalt lavas within the Red Sea depression, interlayered with the evaporites and clastic deposits, indicate localized magma generation rather than Miocene sea-floor spreading. Seismic velocities indicate that much of the material underlying the thick Miocene evaporite-clastic sequence could be Precambrian crust rather than oceanic crust generated at the axial trough during Miocene times.

Sometime during the early Pliocene the Red Sea became part of the world rift system. Apparently, Arabia broke away from Africa, opening the straits of Bab el-Mandeb, allowing the Indian Ocean to enter the Red Sea depression from the south. The unconformity between the evaporite section and the overlying marine oozes, widespread throughout the Red Sea, marks the beginning of drift. Axial rifting in the main trough gave rise to new oceanic crust, consisting of subalkaline basalts, gabbros, and diabases. The strong positive Bouguer gravity anomalies and seismic velocities of 6.6-7.0 km/sec within the present axial trough show that (apparently from the Pliocene to the present) the trough is related to volcanic activity along an active spreading center. Contemporaneous with rifting, continued uplift along the Yemen-Arabian swell seems to have given rise to outpourings of alkali olivine basalt.

Although volcanic events within the Afar Depression coincide with the development of the Red Sea, the volcanic rocks include large amounts of silicic eruptives and differentiated basalts. Vertical faults in the Afar Depression remain active, but rifting and the development of oceanic crust similar to the Red Sea axial trough have not been demonstrated. The negative Bouguer gravity anomalies, combined with the spectrum of differentiated lavas, seem to suggest that the Afar Depression is the northern termination of the East African continental rift system rather than a continuation of any oceanic rift.

Present-day seismic activity is concentrated in the axial trough of the Red Sea, and fault-plane solutions suggest strike-slip movement along faults in a northeasterly direction for the Arabian plate. Within the Afar Depression, the earthquake epicenters are concentrated along the marginal block faults, but insufficient fault-plane solutions prevent statements regarding motions.

The combined geophysical and geological data demonstrate that the Red Sea axial trough is a rift in which new oceanic crust is forming. This rifting appears to have begun sometime in the Pliocene. Other features of the Red Sea originated early in the Tertiary and cannot be tied directly to the world rift system as it is visualized today.

Several outstanding problems remain in this fascinating area. When did the actual Red Sea depression form, after the swells developed or contemporaneous with them? Are the marginal parts of the Red Sea underlain by continental Precambrian crust or by oceanic crust developed during the Miocene? Can the volcanic products be related to the megatectonics of the area? When did the separation of Arabia from Africa begin?

Although in this summary some of these questions appear to be answered, the geologic and geophysical evidence is too sparse to be convincing. Future efforts of research in this area are important to the development of natural resources and to the documentation of a newly formed and active rift in the earth's crust.

ACKNOWLEDGMENTS

Work on which some of the paper is based was carried on while the author was a member of a U.S.G.S. project in Saudi Arabia, which operates under an agreement funded by the Saudi Arabian government. Permission to publish this data is approved by the Deputy Minister, Directorate General for Mineral Resources, Saudi Arabia. D. A. Ross of the Woods Hole Oceanographic Institute provided new information on the Red Sea, and his encouragement in this work is appreciated. A.S. Laughton allowed the use of his Red Sea bathymetric chart as the base for the geologic map, and R. Whitmarsh reviewed an early draft; both are from the National

Institute of Oceanography, Great Britain. G.F. Brown, U.S. Geological Survey, provided the author with considerable information on the geology of this area, and his help in the preparation of the map was invaluable. The use of information from the Mobil Oil Corp. has been helpful in the geologic synthesis of the area. We are indebted to Edward Blanton of Sun Oil Co. for providing information on the exploratory wells Badr No. 1 and Ghawwas No. 1. R.W. Girdler, University of Newcastle-upon-Tyne, provided the author with a continuous geophysical perspective during the preparation of this paper.

BIBLIOGRAPHY

Ahmed, S. S., 1972, Geology and petroleum prospects in eastern Red Sea: Am. Assoc. Petr. Geol. Bull., v. 56, no. 4, p. 707-719.

Allan, T.D., 1970, Magnetic and gravity fields over the Red Sea: Phil. Trans. Roy. Soc. London, ser. A, v. 267, p. 153-180.

Baker, B. H., Mohr, P. A., Williams, L. A. J., 1972, Geology of the Eastern Rift System of Africa: Geol. Soc. America Spec. Paper 136, 67 p.

Bannert, D., Brinckmann, J., Käding, K., Knetsch, G., Kürsten, M., and Mayrhofer, H., 1970, Sur geologie der Danakil Senke: Geol. Rundschau, v. 59, p. 409-443.

Barberi, F., Borsi, S., Ferrara, G., Marinelli, G., Santacroce, R., Tazieff, H., and Varet, J., 1972, Evolution of the Danakil Depression (Afar, Ethiopia) in light of radiometric age determinations: Jour. Geology, v. 80, p. 720.

Ben-Menahem, A., and Aboodi, E., 1971, Tectonic patterns in the northern Red Sea region: Jour. Geophys. Res., v. 76, no. 11, p. 2674-2689.

Beydoun, Z.R., 1964, The stratigraphy and structure of the eastern Aden Protectorate: Brit. Overseas Geol. and Mineral Resources Suppl. Ser. Bull. 5, 107 p.

Brown, G. F., 1970, Eastern Margin of the Red Sea and coastal structures in Saudi Arabia: Phil. Trans. Roy. Soc. London, ser. A, v. 267, p. 75-87.

———, 1972, Tectonic map of the Arabian Peninsula: Kingdom Saudi Arabia, Min. Petr. and Mineral Resources, Map AP-2.

———, and Coleman, R. G., 1972, The tectonic framework of the Arabian Peninsula: 24th Internatl. Geol. Congr., Montreal, Proc., sect. 3., p. 300-305.

———, and Jackson, R. O., 1960, The Arabian Shield: 21st Internatl. Geol. Congr., Copenhagen, Proc., sect. 9, p. 69-77.

Carella, R., and Scarpa, N., 1962, Geological results of exploration in Sudan by A.G.I.P. Mineraria Ltd.: 4th Arab Petr. Congr., Beirut.

Chase, R. L., 1969, Basalt from the axial trough of the Red Sea, *in* Degens, E. T., and Ross, D. A., eds., Hot brines and recent heavy metal deposits in the Red Sea: New York, Springer-Verlag, p. 122-128.

Choubert, G., 1968, International tectonic map of Africa, 1:5,000,000: Paris, UNESCO.

Coleman, R. G., and Brown, G. F., 1971, Volcanism in southwest Saudi Arabia (abstr.): Geol. Soc. America, v. 3, no. 7, p. 529.

Davies, D., and Tramontini, C., 1970, The deep structure of the Red Sea: Phil. Trans. Roy. Soc. London, ser. A, v. 267, p. 181-189.

Drake, C. L., and Girdler, R. W., 1964, A geophysical study of the Red Sea: Geophys. Jour., v. 8, no. 5, p. 473-495.

Fairhead, J. D., and Girdler, R. W., 1970, The seismicity of the Red Sea, Gulf of Aden and Afar triangle: Phil. Trans. Roy. Soc. London, ser. A, v. 267, p. 49-74.

Frazier, S. B., 1970, Adjacent structures of Ethiopia: that portion of the Red Sea coast including Dahlak Kebir Island and the Gulf of Zula: Phil. Trans. Roy. Soc. London, ser. A, v. 267, p. 131-141.

Gass, I. G., 1970, The evolution of volcanism in the junction area of the Red Sea, Gulf of Aden and Ethiopian rifts: Phil. Trans. Roy. Soc. London, ser. A, v. 267, p. 369-381.

———, and Gibson, I. L., 1969, Structural evolution of the rift zones in the Middle East: Nature, v. 221, p. 926-930.

———, Mallick, I. J., and Cox, K. G., 1965, Royal Society volcanological expedition to the south Arabian Federation and the Red Sea: Nature, v. 205, no. 4975, p. 952-955.

Geukens, F., 1966, Geology of the Arabian Peninsula, Yemen: U.S. Geol. Survey Prof. Paper 560-B, 23p.

Gillmann, M., 1968, Preliminary results of a geological and geophysical reconnaissance of the Gizan coastal plain in Saudi Arabia: AIME Symp. Khahran, Saudi Arabia, p. 198-208.

Girdler, R. W., 1969, The Red Sea—a geophysical background, *in* Degens, E. T., and Ross, D. A., eds., Hot brines and recent heavy metal deposits in the Red Sea: New York, Springer-Verlag, p. 38-58.

———, 1970a, An aeromagnetic survey of the junction of the Red Sea, Gulf of Aden and Ethiopian rifts—a preliminary report: Phil. Trans. Roy. Soc. London, ser. A, v. 267, p. 359-368.

———, 1970b, A review of Red Sea heat flow: Phil. Trans. Roy. Soc. London, ser. A, v. 267, p. 191-203.

Gouin, P., 1970, Seismic and gravity data from Afar in relation to surrounding areas: Phil. Trans. Roy. Soc. London, ser. A, v. 267, p. 339-358.

Greenwood, W. R., 1972, The Hail Arch: a key to the Arabian Shield during evolution of the Red Sea Rift (abstr.): Geol. Soc. America, v. 4, no. 7, p. 520.

Heybroek, R., 1965, The Red Sea evaporite basin, *in* Salt basins around Africa: Inst. Petr. London, p. 17-40.

Holmes, A., 1965, Principles of physical geology: London, Nelson (new ed.), 1288 p.

Holwerda, J. G., and Hutchinson, R. W., 1968, Potash-bearing evaporites in the Danakil area, Ethiopia: Econ. Geology, v. 63, p. 124-150.

Hutchinson, R. W., and Engels, G. G., 1970, Tectonic significance of regional geology and evaporite lithofacies in northeastern Ethiopia: Phil. Trans. Roy. Soc. London, ser. A, v. 267, p. 313-329.

Kabbani, F. K., 1970, Geophysical and structural aspects of the central Red Sea rift valley: Phil. Trans. Roy. Soc. London, ser. A, v. 267, p. 89-97.

Karpoff, R., 1957, Sur l'existence du Maestrichtien au nord de Djeddah (Arabie Sédouite): Compt. Rend., v. 245, no. 2, p. 1322-1324.

Knott, S. T., Bunce, E. T., and Chase, R. L., 1966, Red Sea seismic reflection studies, *in* The world rift system: Can. Geol. Survey Paper 66-14, p. 33-61.

Laughton, A. S., 1970, A new bathymetric chart of the Red Sea: Phil. Trans. Roy. Soc. London, ser. A, v. 267, p. 21-22.

Lowell, J. D., and Genik, G. J., 1972, Sea-floor spreading and structural evolution of southern Red Sea: Am. Assoc. Petr. Geol. Bull., v. 56, p. 247-259.

MacFadyen, W. A., 1930, The geology of the Farsan Islands, Gizan, and Kamaran Island, Red Sea, pt. I, General geology: Geol. Mag., v. 67, p. 310-315.

______, 1932, On the volcanic Zebayir Islands, Red Sea: Geol. Mag., v. 69, p. 63-66.

McConnel, R. B., 1967, The East African rift system: Nature, v. 215, p. 578-581.

McKenzie, D. P., Davies, D., and Molnar, P., 1970, Plate tectonics of the Red Sea and East Africa: Nature, v. 226, p. 243-248.

Mohr, P. A., 1962, The geology of Ethiopia: Univ. College Addis Ababa Press.

______, 1971, Ethiopian rift and plateaus: some volcanic petrochemical differences: Jour. Geophys. Res., v. 76, p. 1967-1984.

Montanaro, G. E., 1941, Foraminiferi, posizione stratigraphica e facies di un calcare a "operculina" dei Colli di Ebud (Sahel Eritreo): Palaeontogr. Ital., v. 40 (n.s. 10), p. 67-75.

Phillips, J. D., and Ross, D. A., 1970, Continuous seismic reflexion profiles in the Red Sea: Phil. Trans. Roy. Soc. London, ser. A, v. 267, p. 143-152.

Picard, L., 1970, On Afro-Arabian graben tectonics: Geol. Rundschau, v. 59, p. 337-381.

Ross, D. A., Whitemarsh, R. B., Ali, S., Boudeaux, J. E., Coleman, R. G., Fleischer, R. L., Girdler, R., Manheim, F., Matter, A., Nigrini, C., Stoffers, P., and Supko, P., 1972, Deep sea drilling project in the Red Sea: Geotimes, v. 17, no. 7, p. 24-26.

Said, R., 1962, The geology of Egypt: Amsterdam, Elsevier, 377 p.

Schürmann, H. M. E., 1966, The Precambrian along the Gulf of Suez and the northern part of the Red Sea: Leiden, Netherlands, E. J. Brill, 404 p.

Sestini, J., 1965, Cenozoic stratigraphy and depositional history, Red Sea coast, Sudan: Am. Assoc. Petr. Geol. Bull., v. 49, p. 1453-1472.

Shanti, M. S., 1966, Oölitic iron ore deposits in Wadi Fatima between Jeddah and Mecca, Saudi Arabia: Jeddah, Saudi Arabia, Saudi Arabia Mineral Resources Bull., v. 2.

Swartz, D. H., and Arden, D. D., Jr., 1960, Geologic history of Red Sea area: Am. Assoc. Petr. Geol. Bull., v. 44, p. 1621-1637.

Tazieff, H., Varet, J., Barberi, F., and Giglia, G., 1972, Tectonic significance of the Afar (or Danakil) depression, Nature, v. 235, p. 144-147.

Tromp, S. W., 1950, The age and origin of the Red Sea graben: Geol. Mag., v. 37, p. 385-392.

U.S. Geological Survey, 1963, Topographic map—Arabian Peninsula 1:2,000,000: U.S. Geol. Survey Misc. Geol. Inv. Map I-270, B-2.

Vine, F. J., 1966, Spreading of the ocean floor: new evidence: Science, v. 154, p. 1405-1415.

Whiteman, A. J., 1968, Formation of the Red Sea depression: Geol. Mag., v. 105, p. 231-246.

______, 1970, The existence of transform faults in the Red Sea depression: Phil. Trans. Roy. Soc. London, ser. A, v. 267, p. 407-408.

______, 1971, The geology of the Sudan Republic: New York, Oxford Univ. Press, 290 p.

Arctic Ocean Margins

Ned A. Ostenso

INTRODUCTION

There are few geological and geophysical data from the Arctic continental margins. Accordingly, the morphology of the Arctic Ocean Basin and extrapolation of known terrestrial structure becomes very important. The approach of this review will be from the most speculative to the more substantive insights available from the known terrestrial tectonic framework, and finally, to see what conclusions are supported by the meager available observations.

Prime sources of information summary are a number of geologic and tectonic atlases that have been published over the past decade. These include "New tectonic map of the Arctic" (Atlasov et al., 1964), "Tectonic map of the Arctic and Subarctic" (Atlasov et al., 1964, 1967), "Tectonic map of the polar regions of the earth" (Egiazarov et al., 1969); "Geomorphic map of the Arctic Ocean" (Dibner et al., 1965); "Geological map of the Arctic" (Link et al., 1960); "Map of tectonic elements of the north polar region" (Shell Oil Co., 1959); and a "New Canadian bathymetric chart of the western Arctic Ocean north of 72°" (de Leeuw, 1967).

GENERAL FEATURES OF THE ARCTIC OCEAN BASIN

The existence of a deep ocean basin was first discovered by Fridtjof Nansen in 1904 with wire-soundings from the FRAM. Little further insight was gained about the nature of this basin until the past two decades, when data became available through various submarine traverses, airborne and airlifted surveys, and drifting ice stations (Webster, 1954; Panov, 1955a, 1955b; Somov, 1955; Gakkel, 1957; Dietz and Shumway, 1961; Hunkins, 1961; Heezen and Ewing, 1961; Ostenso, 1962; Atlasov et al., 1964; Dibner et al., 1965; Kutschale, 1966; Hall, 1973). We now know that the Arctic Basin is divided into two principal subbasins by the trans-Arctic Ocean Lomonosov Ridge, which extends 2,000 km and attains a maximum known elevation to within 954 m of the ocean surface. The unequal subdivisions of the Arctic Basin formed by the Lomonosov Ridge are called the Amerasian Basin (the larger) and the Eurasian Basin (the smaller).

The Amerasian Basin is, in turn, transected by the Alpha Cordillera, which divides this basin into the Canada Basin and the much smaller Marakov Basin. The Alpha Cordillera is a broad fractured ridge of rugged topography which reaches to within 1,800 m of the ocean's surface and is believed to be a dormant axis of sea-floor spreading (Vogt and Ostenso, 1970; Vogt et al., 1970; Ostenso and Wold, 1971; Hall, 1973). However, Pitman and Talwani (1972) do not accept spreading along the Alpha Cordillera, and Herron et al (1973) postulate that it is a remnant of an early subduction zone (Fig. 1).

The Eurasian Basin is transected by the seismically active Nansen Cordillera, which is an extension of the Mid-Atlantic Ridge into the Arctic Basin through the Lena Trough between Spitsbergen and Greenland (Fig. 2) (Heezen and Ewing, 1961; Sykes, 1965; Johnson and Heezen, 1967; Vogt et al., 1970; Ostenso and Wold, 1971; Rassokho et al., 1967; Karasik, 1968; Demenitskaya and Karasik, 1969). Distinctive features of the Arctic Ocean segment of the Mid-Ocean Ridge are its low elevation, subdued magnetic signature, and apparent lack of fracture zone offsets. The Nansen Cordillera further subdivides the Eurasian basin into the Fram and Nansen basins.

The aseismic Lomonosov Ridge was discovered by a Russian expedition in 1948 but not reported until 1954 (Webster). Early Soviet reports (Panov, 1955b; Saks et al., 1955; Gakkel, 1961) postulated the ridge to be a Mesozoic or Tertiary structure. Later papers ascribed the feature to Caledonian folding (Gakkel and Dibner, 1967; Egiazarov et al., 1969; Meyerhoff, 1970). However, it is now generally believed that the Lomonosov Ridge is a former slice of the Eurasian continental margin that has been translated northward as a consequence of opening of the Eurasian Basin along the locus of the Nansen Cordillera (Wilson, 1965; Harland, 1966; Johnson and Heezen, 1967; Demenitskaya and Karasik, 1969; Vogt and Ostenso, 1970; Ostenso and Wold, 1971).

In summary, some features of the Arctic Ocean Basin can be ascribed to sea-floor spreading from the Tertiary to the present, although details of its morphology differ among investigators (Demenitskaya and Karasik, 1966; Karasik, 1968; Demenitskaya and Hunkins, 1970; Vogt and Ostenso, 1970; Vogt et al., 1970; Ostenso and Wold, 1971; Pitman and Talwani, 1972; Ostenso and Wold, 1973; Ostenso, 1973; Herron et al., 1973; Hall, 1973). There is additional evidence that at least a part of the southern Canadian Basin dates back to the early

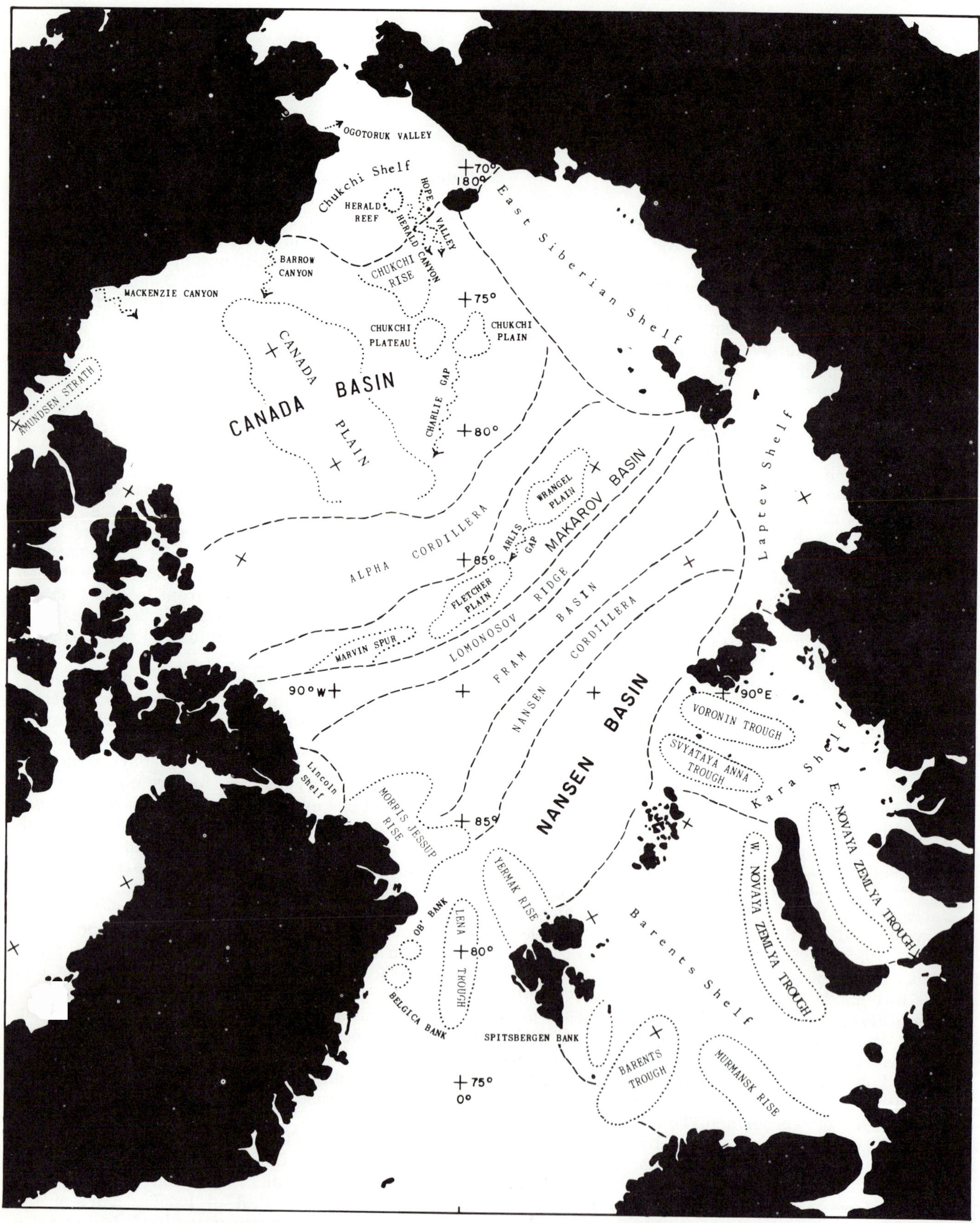

Fig. 1. Major features of the floor of the Arctic Ocean and adjacent seas (after Beal et al., 1966).

Fig. 2. Schematic presentation of the major physiographic features in deep-water areas of the Arctic (after Vogt et al., 1970). Triangles denote earthquake epicenters. Profile segment A is shown in Figure 4.

Paleozoic or older (Churkin, 1969, 1970, 1973; Ostenso, 1973; Ostenso and Wold, 1973; Hall, 1973; Hamilton, 1970; Harland, 1973).

The occurrence of Late Cretaceous sediments on the Alpha Cordillera indicates that the Arctic Ocean existed at least 70 my ago (Clark, 1974). This is the only direct evidence for a minimum age of the basin. Clark also has evidence that the Cretaceous Arctic Ocean was warmer than today and not ice covered. At that time, volcanism occurred extensively along the Alpha Cordillera. The Arctic Ocean is known to have been continuously ice covered from Pliocene time to the present (Clark, 1971), with variations in thickness in response to minor climatic fluctuations. Clark (1974) hypothesized that initial formation of the ice pack occurred between Eocene and Pliocene time.

From a continuing coring program on ice island T-3 facts are available about sediments in the Canada Basin and Alpha Cordillera that have extrapolative relevance to the adjacent continental margins (Steuerwald et al., 1968; Clark, 1971; Hunkins et al., 1971). Sediments accumulated on the Alpha Cordillera at an average rate of 1.0-1.5 mm/1,000 years (ty), whereas the adjacent Chukchi Plain received between 2.2 and 2.6 mm/ty. Sedimentation rates in the deeper Canada Basin are higher yet, ranging from 8 mm/ty to 39 cm/ty. These are the largest accumulation rates yet found anywhere in the Arctic Basin and the sediments contain considerable thicknesses of turbidity deposits. Clark (1973) observed that the sediment regime in all parts of the Arctic Ocean is unique because of the abundance of glacially rafted debris. Many of Clark's cores contained more than 30% ice-rafted sediments. Another 10% of the core material may be attributed to atmospheric dust that has fallen on the ice pack and subsequently been released through melting.

ALASKA-CANADA-GREENLAND MARGIN

The Alaska-Canada-Greenland continental margin is anomalous in the Arctic because it is only of normal width (ranging between 50 and 150 km wide, except north of Greenland, where it attains a width of 350 km). Further, the shelf edge is deeper than usual, from 300 to 500 m. This 100 to 300-m excess depth is generally attributed to remanent isostatic depression resulting from Pleistocene ice loading. The general tectonic framework for northern Alaska has been developed and/or summarized by Payne (1951, 1955), Lathram (1973), Churkin (1970), Brosge and Dutro (1973), Detterman (1973), Tailleur (1969), and Churkin (1969, 1972, 1973). The geology and structure of northern Canada bordering the Arctic Ocean has been described by Thorsteinsson and Tozer (1960, 1962), Tozer (1960), Trettin (1967, 1969, 1973), Drummond (1973), and Bourne (1973). The geology and structure of northern Greenland is summarized in Koch (1920, 1925), Dawes and Soper (1973), and Dawes (1973). Also see Drummond (this volume) for a broad geological history of arctic North America.

Contrary to numerous theories intended to place Alaska within a speculative scenario of plate tectonics, geologic evidence argues that it is not formed by rifted continental segments. Rather, Alaska is a continental mass formed by the accretion of the foldbelts consequent to Precambrian and Paleozoic interactions between the Canadian and Siberian shields and the ancient Arctic and Pacific basins. The northern coast of Alaska is structurally dominated by an early Paleozoic and Precambrian arctic (Innuitian) foldbelt which extends eastward across the Canadian Arctic Archipelago (Fig. 3). Along the northern flank of the Archipelago, Paleozoic marine formations thicken northward and westward, becoming miogeosynclinal in character. These formations make up the Franklin geosyncline, which underwent major orogenic disturbances between the late Devonian and middle Pennsylvanian, producing first a north-trending foldbelt along the northern extension of the Boothia uplift and, second, westerly-trending foldbelts that parallel the geosynclinal axis. The northern limb of the Franklin geosyncline west of Banks Island has unconformably superimposed on it a very thick sequence of shallow-water clastic deposits ranging in age from Permo-Carboniferous to early Tertiary which form the Sverdrup basin. Deformation of these rocks in late Cretaceous and Tertiary times resulted in a great variety of north-trending structures (see also Drummond, 1973, and this volume).

In northern Greenland the Precambrian crystalline shield is bounded on the north by the North Greenland foldbelt, in which mainly lower Paleozoic rocks are exposed. As this belt is part of the Innuitian system, it shows the effects of both late Precambrian and Paleozoic tectonism. Sedimentation on the edge of the continental shield began in the late Precambrian and continued into the early Paleozoic, with interruptions by periods of uplift. In the extreme north, the North Greenland geosyncline became a zone of fairly rapid subsidence and the site of thick lower Paleozoic sediment accumulation. These sediments were folded during the Late Silurian to Late Devonian to form a linear belt paralleling the north coast of Greenland. The exposed part of this foldbelt has an asymmetric tectonic pattern with northern overturning facing toward the Arctic Ocean. The foldbelt shows a long and complex structural and metamorphic history which intensifies from south to north. Unmetamorphosed and undeformed platform strata in the south grade northward through partially and simply folded, slightly metamorphosed strata to complexly folded schistose rocks along the Arctic coast. Magnetic data (Ostenso, 1962, 1963; King et al., 1966) suggest that the folded geosynclinal sediments continue northward under the Arctic Ocean, possibly to the edge of the continental shelf. Subbottom seismic profiling from the drift of ice island ARLIS-II shows structures indicative of a northward continuation of

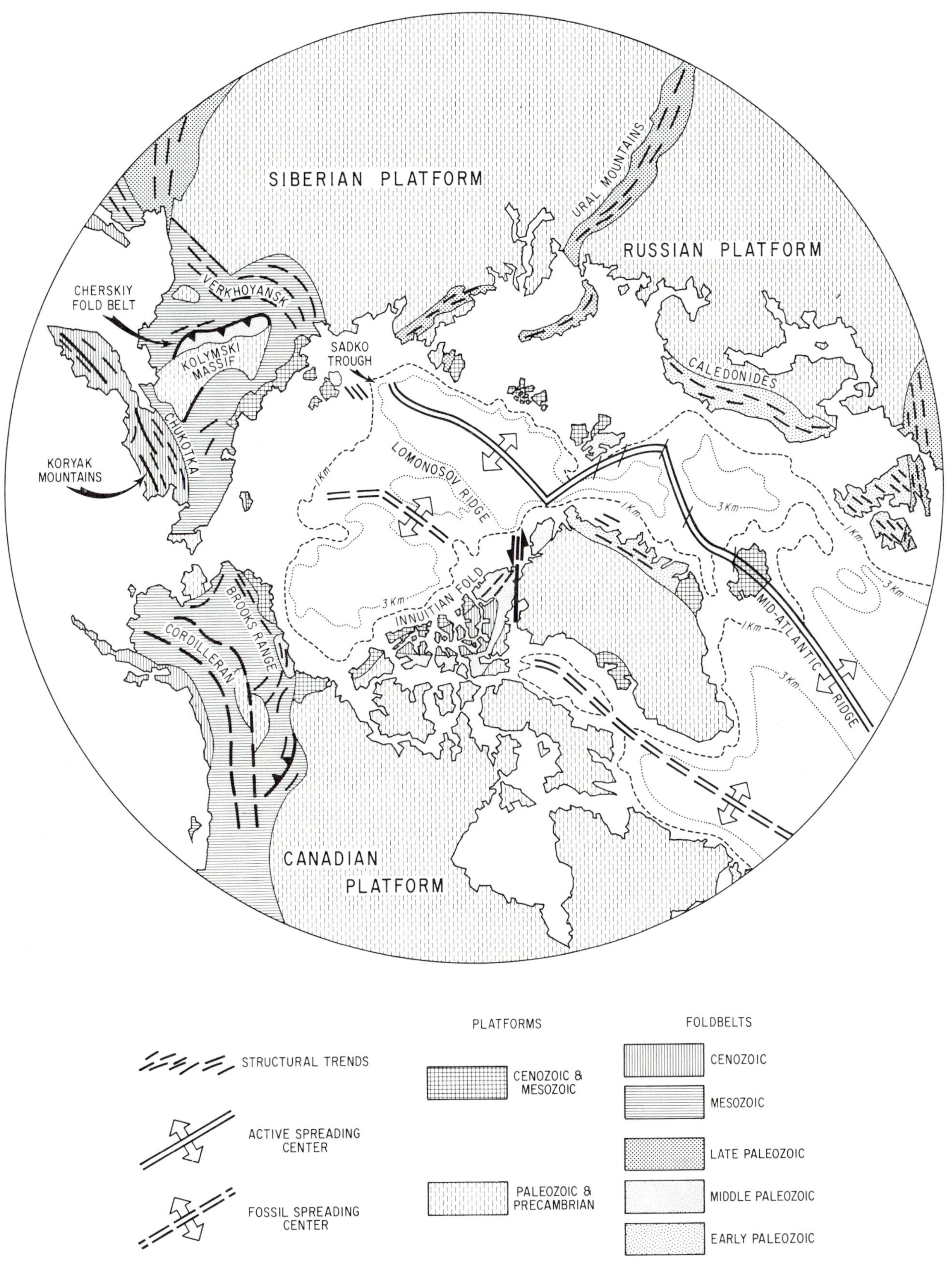

Fig. 3. Major tectonic features of the Arctic (modified from Churkin, 1972).

the East Greenland (Caledonian) geosyncline across the continental shelf (Ostenso, 1968a).

Gravity surveys along the Arctic coast of North America (Wold et al., 1970; Barnes 1970; Ostenso and Wold, 1971; Sobczak and Weber, 1973) show a series of large elliptically shaped positive anomalies extending eastward from Point Barrow to at least as far as the middle of the Archipelago. The anomaly amplitudes exceed 100 mgal and have axial flank gradients as high as 2.5 mgal/km. They parallel the continental shelf edge being offset slightly oceanward. These observed anomalies can be matched by a model that includes a basement ridge paralleling the 200-m isobath in addition to crustal thinning along the continental margin. Such a basement ridge appears to be a common feature of continental margins (Worzel, 1968; Burk, 1968; Emery, 1968) and may be a consequence of continental separation. However, the ridge bordering the Canada Basin is unique in being nonmagnetic.

Wold et al. (1970) noted that, compared to continental margins elsewhere, the transition from continental to oceanic crust occurred in an unusually narrow zone centered on the 200-m isobath rather than the usual broader transition centered on the 2,000-m isobath. They postulated that an uncommonly thick accumulation of sediments may have displaced this transition zone landward. Such an accumulation could result from either a great age for the Canada Basin or a fast rate of sedimentation (or some degree of combination of both). In support of rapid sedimentation as a possible explanation, Wold et al. (1970) estimated that 30 million tons of solid material was transported annually into the Beaufort Sea, which is nearly one-third of the sediment load carried by the Mississippi River. Further evidence is given by Beal (1968), who observed numerous apparent slump structures on submarine bathymetric profiles.

Sobczak and Stephans (1970) infer from gravity data that the Sverdrup Basin extends seaward as a 7-km-thick wedge of low-density clastic sediments. Their data also show elliptical gravity highs lying offshore.

Riddihough et al. (1973) observed sublinear magnetic anomalies northwest of Ellesmere Island which corresponded with those of the Alpha Cordillera, notwithstanding some possible transverse dislocation between the two sets. Trettin (1969) suggested that the north-south trends of northern Ellesmere could be related to the Lomonosov Ridge. However, aeromagnetic data show the area northwest of Ellesmere Island to be tectonically complex, and there is as yet no clear relationship of the Lomonosov Ridge to the continental margin.

A major dislocation has been proposed between Greenland and Ellesmere Island by Wegener (1924), DuToit (1937), and Wilson (1963), to mention but a few. This feature has been variously referred to as the Wegener fault and the Nares Strait lineament. Thus far, there has been no success in tracing this proposed major feature across the continental margin north of Greenland. The aeromagnetic data of Riddihough et al. (1973) shows a continuity of sublinear trends across the strait that would substantiate Kerr's (1967) suggestion that, although Nares Strait was a major rift, there has been little lateral offset along it and that what motion has occurred can be divided into extensional and compressional, pivoting about a point near 80°N (also see Dawes, 1973).

In summary, the North American Arctic continental margin is bordered by one of the earth's major (Innuitian) foldbelts. From Point Barrow, Alaska, eastward to about 90°W the margin appears to have been stable since the beginning of Cenozoic time and is marked by (1) an abnormally narrow transition zone between continental and oceanic crusts, (2) a basement ridge lying just oceanward of the continental shelf break that produces a large gravity anomaly but no magnetic anomaly, and (3) rapid sediment accumulation. Between 90°W and the Lena Trench (0° longitude) there are few direct data about the continental margin, but sparse geophysical data and tectonic inferences imply structural complexity (Fig. 4).

BARENTS-KARA MARGIN

A natural subdivision of the circum-arctic continental margin is the Barents-Kara Sea segment (Fig. 1). This segment is bounded on the west by the Lena Trough, which forms the only break in the Arctic continental margin. The Lena Trough is an extensional feature (Laktionov, 1959; Heezen and Ewing, 1961; Ostenso, 1962). The actual transition of the mid-ocean ridge from the Greenland Sea into the Arctic Ocean is not a simple linear feature but a complicated and not thorougly documented series of offset ridge sediments forming an en échelon pattern (Johnson and Heezen, 1967; Vogt et al., 1970; Ostenso, 1973; Johnson and Vogt, 1973). East of the Vøronin trough (Fig. 1) natural shelf segment boundary is the Taymyr Penninsula-Severnaya Zemlya island group, which together cut across the continental shelf at 100°E. The Barents-Kara shelf contains numerous islands and archipelagos, which include Spitsbergen, Bear Island, Novaya Zemlya, and Franz Josef Land. Many of the Islands in the Arctic marginal seas are called "Land," a vestige from the early belief that the central Arctic was a land mass and that new island discoveries were, in fact, segments of a greater coastline.

The Barents-Kara continental shelf reaches a maximum width of 1,700 km at the same longitude as the entrance to the White Sea (42°E). Waters of this shelf sea communicate freely with the Arctic Ocean to the north. Barents Sea circulation is dominated by the eastward- and northward-flowing Murman-Novaya Zemlya current, which has its origin in the Gulf Stream. Hydrology in the Kara Sea is dominated by the inflow of fresh, silt-laden water from the Ob and Yenisey rivers (Milligan, 1969). The

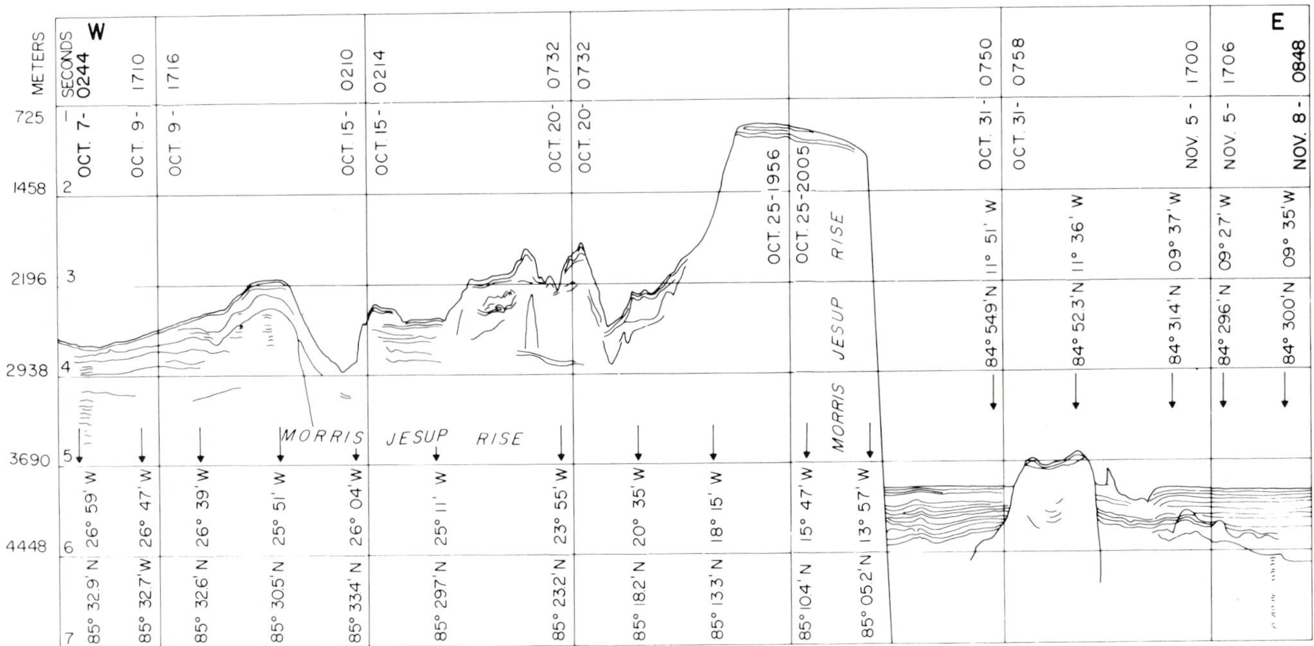

Fig. 4. Seismic subbottom profile across the Morris Mesup Rise north of Greenland. Location is shown in Figure 3 as line segment A.

shelf depth varies between 100 and 200 m, with local topographic features, such as the East and West Novaya Zemlya trenches reaching depths as great as 500 m. These topographic features appear to have a variety of origins, including structural, ice influence, erosional, and depositional. (Vinogradova, 1957; Vogt and Ostenso, 1973).

The terrestrial geology of northwestern Siberia bordering the Barents-Kara seas and their islands has been summarized by such investigators as Atlasov and Sokolov, (1961), Rabkin and Ravich (1961), Markov and Tkachenko (1961), Sachs and Strelkov (1967), Harland (1961, 1973), Atlasov et al. (1964, 1967), Hamilton (1970), Vinogradov et al. (1973), Semenovich et al. (1973), and Sachs et al. (1973).

Structurally, the northern part of the Barents shelf belongs to the crustal block of the Barents-Kara epi-Caledonian platform (Fig. 3). The Barents region covers the central and northern parts of the shelf and includes Franz Josef Land plus several islands of the Spitsbergen Archipelago. The folded basement is composed mainly of pre-Baykalian complexes with Caledonian structures along the southern and southeastern margins. The lower part of the sedimentary mantle is composed of Silurian to Permian carbonates 2.5-3 km or more thick. The upper part is a terrigenous and terrigenous-volcanic sequences of Triassic, Jurassic, Cretaceous, and Cenozoic age with total thicknesses up to 2-2.5 km.

The southern part of the Barents shelf is a northward extension of the Pechora block of the Russian platform. Its folded basement is composed of Baykalian age rocks with the overlying sedimentary mantle composed of carbonate and terrigenous-carbonate Paleozoic, Mesozoic, and Cenozoic deposits having a total thickness of more than 6 km.

Thus the Barents-Kara shelf has a heterogeneous basement consisting of folded structures of different ages, the youngest being Paleozoic. Subsequently, the region was subjected to repeated terrestrial and marine sedimentation (Klenova, 1960, 1961). The Kara shelf is incised by the St. Anna and Vøronin troughs, which appear to have been formed by major faulting (Gakkel and Dibner, 1967).

Geophysical data (Vogt and Ostenso, 1973) show the crustal structure of the Barents shelf to be continental, having a thickness of over 30 km. On the continental slope the shelf structures are buried under an accumulated mantle of sediments. Only terrigenous deposits—sands, coarse aleurites (silts), and fine aleurite muds—occur on the shelf. Finer material are carried away by strong permanent and tidal currents and wave action. The contribution of ice-rafted material is negligible, owing to the rapid melting of ice under the influence of northward-flowing warm Atlantic water. The principal source of terrigenous sediments are northward-flowing rivers that are only seasonally active.

LAPTEV-EAST SIBERIAN-CHUKCHI MARGIN

The next natural subdivision of the Arctic continental margin is the section extending eastward from the Taymyr Peninsula-Severnaya Zemlya salient in central Siberia to Point Barrow, Alaska (Fig. 1). This section of shelf ranges in width from 700 to 800 km, except for the Chukchi shelf, which extends for 1,200 km at the Bering Strait, and contains several islands, most notable of which are the New Siberian Islands and Wrangel Island.

The shelf has gentle topographic relief, some of which may be attributed to earlier subaerial drainage patterns (Creager, 1963; Creager and McManus, 1965; McManus and Creager, 1963; Hopkins, 1967). The most notable feature of this shelf segment is the Chukchi Rise, believed to be a continental outlier (Hunkins, 1966; Ostenso, 1973) that is separated from the shelf by a saddle (Hall, 1973). A positive magnetic anomaly, which borders the western and northern margins of the Rise, is interpreted (Ostenso and Wold, 1971) to reflect a buried basement ridge similar to that along the continental margin of eastern North America. The Chukchi shelf is incised by two submarine valleys. The largest, Barrow valley, lies just off Point Barrow in the northwest coast of Alaska. The smaller Hope valley extends between Wrangel Island and Herald reef. The Laptev shelf is incised by the Sadko trough (Fig. 3), which is postulated to be a manifestation of tensional spreading along the Nansen Cordillera ocean ridge system (Demenitskaya and Karasik, 1969; Churkin, 1972).

The geologic framework of eastern Siberia and western Alaska adjacent to the Labtev-East Siberian-Chukchi margin has been described by Atlasov and Sokolov (1961), Rabkin and Ravich, (1961), Markov and Tkachenko (1961), Sachs and Strelkov (1967), Atlasov et al. (1964, 1967), Semenovich et al. (1973), Sachs et al. (1973), Hamilton (1970), Lathram (1973), Brosge and Dutro (1973), Detterman (1973), Churkin (1969), Hopkins (1959), Partunov and Samoilovich (1973), Demenitskaya et al. (1973), and Tkachenko et al. (1973).

The Laptev block of the Siberian platform underlies most of the Laptev shelf, with the exception of the southern and eastern margins, which include extensions of the Verkhoyansk-Kolyma and New Siberian-Chukotsk Mesozoic fold systems, respectively. The basement of the Laptev block is composed of folded lower Proterozoic rocks. This foundation is covered by upper Proterozoic and lower to middle Paleozoic carbonates totaling in thickness at least 4-5 km, and terrigenous Upper Cretaceous and Cenozoic sediments 0.5-1 km in total thickness. Two to three kilometers of terrigenous Permian and Triassic sediments lie along the southern margin of the block, in addition to at least 3 km of Jurassic and Lower Cretaceous strata. The Verkhoyansk-Kolyma and New Siberian-Chukotsk fold systems are almost completely covered by unconsolidated sediments. Exposures are found only on the islands. The shelf sediments are terrigenous Upper Cretaceous and Cenozoic and range in thickness from 0.5 km on structural uplifts to 8 km in the troughs and grabens.

Structurally, the western part of the East Siberian shelf and the southern part of the Chukchi Sea belong to the New Siberian Mesozoic fold systems. The remaining parts of the shelves are extensions of the Hyperborean platform. Geophysical evidence (Ostenso, 1968b; Grantz, 1970; Bassinger, 1968) supports continuation of the Brooks Range structure across the Chukchi shelf to the Chukchi-Anadyr foldbelt. The shelf was subject to moderate Tertiary folding, which produced a dry-land interconnection between North America and Asia until about a million years ago, when erosion finally reduced the low mountain barrier to a near peneplain. Subsequently, the shelf has been modified by numerous changes in sea level and deposition of a thin, often patchy veneer of sediments.

BIBLIOGRAPHY

Atlasov, I. P., 1964, Tektonicheskaya karta Arktiki i Subarktiki: Leningrad. Sci. Res. Inst. Geology Arctic.

———, and Sokolov, V. N., 1961, Main features of the tectonic development of the central Soviet Arctic, *in* Roasch, G. O., ed., Geology of the Arctic: Univ. Toronto Press, p. 5-17.

———, Yegiazarov, B. K., Dibner, V. D., Romanovich, B. S., Zimkin, A. V., Vakar, V. A., Demenitskaya, R. M., Levin, D. V., Karasik, A. M., Gakkel, Y. Y., and Litvin, L. M., 1964, Tectonic Map of the Arctic and Subarctic: Internatl. Geol. Congr. Leningrad, Res. Inst. Geol. Arctic., Rept. XXII.

———, Egiazarov, B. K., Dibner, V. D., Romanovich, B. S., Zimkin, A. V., Vakar, V. A., Demenitskaya, R. M., Gakkel, Y. Y., and Litven, V. M., 1967, The tectonic map of the Arctic and Subarctic, *in* The tectonic maps of continents: Moscow, Akad. Nauk, p. 154-165.

———, Dibner, V. D., Egiazaroe, B. K., Romanovich, B. C., Bakar, V. A., Gakkel, Y. Y., Demenitskaya, R. M., Zimkin, A. V., Karasik, A. M., Levin, D. V., and Litvin, B. M., 1970, Obyasnitelnaya zapiska k tektonicheskoy karte Arktiki i Subarktiki: Moscow, Ministry of Geology of the S.S.S.R.

Barnes, D. F., 1970, Gravity and other regional geophysical data from northern Alaska, *in* Adkison, W. L., and Brosge, M. M., eds., Proc. Geol. Seminar on the North Slope of Alaska: Am. Assoc. Petr. Geol., Pacific Sect., Los Angeles, p. 11-120.

Bassinger, B. G., 1968, A marine magnetic study in the northeast Chukchi Sea: Jour. Geophys. Res., v. 73, p. 683-687.

Beal, M. A., 1968, Bathymetry and structure of the Arctic Ocean [Ph.D. thesis]: Oregon State Univ., Corvallis, 204 p.

———, Edvalsen, F., Hunkins, K., Molloy, A., and Ostenso, N., 1966, The floor of the Arctic Ocean: geographic names: Arctic, v. 19, p. 215-219.

Belov, N. A., and Dibner, V. D., 1968, Rezultaty gedogogeomorfologicheskikh issledsvaniy Arkticheskogo Basseyna: Problemy Arktiki i Antarktiki, p. 94-111.

Bourne, S. A., and Pallister, A. E., 1973, Offshore areas of Canadian arctic islands—geology based on geophysical data, *in* Pitcher, M. G., ed., Arctic geology: Am. Assoc. Petr. Geol. Mem. 19, p. 48-56.

Brosge, W. P., and Dutro, J. T., Jr., 1973, Paleozoic rocks of northern and central Alaska, *in* Pitcher, M. G., ed., Arctic Geology, Am. Assoc. Petr. Geol. Mem. 19, p. 361-375.

Burk, C. A., 1968, Buried ridges within continental margins: Trans. N.Y. Acad. Sci., ser. 2, v. 30, p. 397-409.

Carey, S. W., 1955, Orocline concept in geotectonics: Tasmania Univ. Dept. Geol. Publ. 28, Royal Soc. Tasmania Papers and Proc., v. 59, p. 255-288.

Churkin, M., Jr., 1969, Paleozoic tectonic history of the Arctic Basin north of Alaska, Science, v. 165, no. 3893, p. 549-555.

Churkin, M., Jr., 1970, Fold belts of Alaska and Siberia and drift between North America and Asia, *in* Adkinson, W. L., and Brosge, M. M., eds., Proc. Geol. Seminar on the North Slope of Alaska: Am. Assoc. Petr. Geol., Pacific Sect., Los Angeles, p. G1-G12.

______, 1972, Western boundary of the North American continental plate in Asia: Geol. Soc. America Bull., v. 83, no. 4, p. 1027-1036.

______, 1973, Geologic concepts of Arctic Ocean Basin, *in* Pitcher, M. G., Arctic geology: Am. Assoc. Petr. Geol. Mem. 19, p. 485-499.

Clark, D. L., 1971, Arctic Ocean ice cover and its late Cenozoic history: Geol. Soc. America Bull., v. 82, p. 3313-3324.

______, 1973, Arctic Ocean studies progress: Oil and gas Jour. Oct. 22, p. 104-106.

______, 1974, Late Mesozoic and early Cenozoic sediment cores from the Arctic Ocean: Geology, v. 2, p. 41-44.

Crary, A. P., and Goldstein, N., 1957, Geophysical studies in the Arctic Ocean: Deep-Sea Res., v. 4, p. 185-201.

Creager, J. S., 1963, Sedimentation in a high energy, embayed, continental shelf environment: Jour. Sediment. Petrology, v. 33, p. 815-830.

______, and McManus, D. A., 1965, Pleistocene drainage patterns on the floor of the Chukchi Sea: Marine Geology, v. 3, p. 279-290.

Dawes, P. R., 1973, The north Greenland fold belt: a clue to the history of the Arctic Ocean Basin and the Nares Strait lineament, *in* Implications of continental drift to the earth sciences, v. 2: London, Academic Press, p. 925-948.

______, and Soper, N. J., 1973, Pre-Quaternary History of North Greenland, *in* Pitcher, M. G., ed., Arctic geology: Am. Assoc. Petr. Geol. Mem. 19, p. 117-134.

de Leeuw, M. S., 1969, New Canadian bathymetric chart of the western Arctic Ocean north of 72°: Deep-Sea Res., v. 14, p. 489-504.

Demenitskaya, R. M., and Hunkins, K. L., 1970, Shape and structure of the Arctic Ocean, *in* Maxwell, A. E., ed., The sea, v. 4, pt. 2: New York, Wiley-Interscience, p. 223-249.

______, and Karasik, A. M., 1969, The active rift system of the Arctic Ocean: Tectonophysics, v. 8, p. 345-351.

______, and Karasik, A. M., 1966, Magnetic data that confirms the Nansen-Amundsen Basin is of normal oceanic type, *in* Poole, W. H., ed., Continental margins and island arcs: Geol. Survey Can. Paper 66-15, p. 191-6.

______, Karasik, A. M., Kiselev, Y. G., Litvinenko, I. N., and Ushakov, S. A., 1968, The transition zone between the Eurasian continent and the Arctic Ocean: Can. Jour. Earth Sci., v. 5, p. 1125-1129.

______, Gaponenko, G. I., Keselev, Y. G., and Ivanov, S. S., 1973, Features of sedimentary layers beneath the Arctic Ocean, *in* Pitcher, M. G., ed., Arctic geology: Am. Assoc. Petr. Geol. Mem. 19, p. 332-335.

Detterman, R. L., 1973, Mesozoic sequences in arctic Alaska, *in* Pitcher, M. G., ed., Arctic geology: Am. Assoc. Petr. Geol. Mem. 19, p. 376-387.

Dibner, V. D., Gakkel, Y. Y., Litoin, V. M., Martynov, V. T., and Shurgayeva, V. T., 1965, Geomorphic map of the Arctic Ocean: Trudy Nauchn. Issled., Res. Inst. Arctic Geol., Leningrad.

Dietz, R. S., and Shumway, G., 1961, Arctic Basin geomorphology: Geol. Soc. America Bull., v. 72, p. 1319-1330.

Drummond, K. J., 1973, Canadian Arctic Islands, *in* McCrossan, R. G., ed., The future petroleum provinces of Canada: Can. Soc. Petr. Geol. Mem. 1, p. 443-472.

DuToit, A. L., 1937, Our wandering continents: a hypothesis of continental drifting: Edinburgh, Oliver & Boyd, 366 p.

Egiazarov, B. K., Atlasov, I. P., Ravich, M. G., Demenitskaya, R. M., Dibner, V. D., Karaski, A. M., Kulakov, Y. N., Puminov, A. P., Romanovich, B. S., and Tkachenko, B. V., 1969, Tectonic map of the polar regions of the earth: Res. Inst. Geol. Arctic, Leningrad, 4 sheets.

______, Atlasov, I. P., Ravich, M. G., Girkov, G. E., Demenitskaya, R. M., Znachko-Yavorsky, G. Z., Puminov, A. P., and Romanovich, B. S., 1971, Tectonic map of the polar regions of the earth: Res. Inst. Geol. Arctic, Leningrad.

______, Atlasov, I. P., Ravich, M. G., Girkuvov, G. E., Demenitskaya, R. M., Znachiko-Yavorsky, G. Z., Karasik, A. K., Kulakov, Y. N., Pumivov, A. P., and Romanovich, B. S., 1973, Tectonic map of the earth's polar regions and some aspects of comparative analysis, *in* Pitcher, M. G., ed., Arctic geology: Am. Assoc. Petr. Geol. Mem. 19, p. 317-322.

Emery, K. O., 1968, Shallow structure of continental shelves and slopes: Southeastern geology, v. 9, p. 173-194.

Gakkel, Y. Y., 1957, Nauka i osvoyeniye Arktiki: Leningrad, Morskoi Transport Press.

______, 1961, Modern presentation of the Lomonosov Ridge: Material po Arktike i Antarktike, Leningrad.

______, and Dibner, V. D., 1967, Bottom of the Arctic Ocean, *in* Runcorn, S. K., International dictionary of geophysics,, v. 1: Elmsford, N.Y., Pergamon Press, p. 152-65.

Grantz, A., 1970, Geology of the Chukchi Sea as determined by acoustic and magnetic profiling, *in* Adkinson, W. L., and Brosge, M. M., eds., Proc. Geol. Seminar on the North Slope of Alaska: Am. Assoc. Petr. Geol., Pacific Sect., Los Angeles, p. F1-F28.

Hall, J. K., 1973, Gcophysical evidence for ancient seafloor spreading from the Alpha Cordillera and Mendeleyev Ridge, *in* Pitcher, M. G., ed., Arctic geology: Am. Assoc. Petr. Geol. Mem. 19, p. 542-561.

Hamilton, W., 1970, The Uralides and the motion of the Russian and Siberian platforms: Geol. Soc. America Bull., v. 81, p. 2553-2576.

Harland, W. B., 1961, An outline structural history of Spitsbergen, *in* Roasch, G. O., ed., Geology of the Arctic: Univ. Toronto Press, p. 68-132.

______, 1966, A hypothesis of continental drift tested against the history of Greenland and Spitsbergen: Cambridge Res., v. 2, p. 18-22.

______, 1973, Tectonic evolution of the Barents shelf and related plates, *in* Pitcher, M. G., ed., Arctic geology: Am. Assoc. Petr. Geol. Mem. 19, p. 599-608.

Heezen, B. C., and Ewing, M., 1961, the Mid-Oceanic Ridge and its extension through the Arctic Basin, *in* Roasch, G. O., ed., Geology of the Arctic, v. 1: Univ. Toronto Press, p. 622-42.

Herron, E. M., Dewey, J. F., and Pitman, W. C., III, 1973, Permian to present day evolution of the Arctic (abstr): Geol. Soc. America Ann. Meeting.

Hopkins, D. M., 1959, Cenozoic history of the Bering land bridge: Science, v. 129, p. 1519-1528.

———, 1967, The Cenozoic history of Beringia, a synthesis, *in* Hopkins, D. M., ed., Bering land bridge: Stanford, Stanford Univ. Press, p. 451-484.

Hunkins, K., 1961, Seismic studies of the Arctic Ocean floor, *in* Roasch, G. O., ed., Geology of the Arctic, v. 1: Univ. Toronto Press, p. 645-65.

———, 1966, The Arctic continental shelf north of Alaska, *in* Continental Margins and Island Arcs; Poole, W. H., ed., Geol. Survey Canada Paper 66-15, p. 197-205.

———, Herron, T., Kutschale, H., and Peter, G., 1961, Geophysical studies of the Chukchi Cap, Arctic Ocean: Jour. Geophys. Res., v. 67, p. 234-247.

———, Be, A. W. H., Opdyke, N. D., and Mathieu, G., 1971, The late Cenezoic history of the Arctic Ocean, *in* Turekian, K. K., ed., The late Cenezoic glacial ages: New Haven, Yale Univ. Press, p. 215-237.

Johnson, G. L., and Heezen, B. C., 1967, The Arctic mid-oceanic ridge: Nature, v. 215, p. 725-5.

———, and Vogt, P. R., 1973, Marine geology of the Atlantic Ocean north of the Arctic Circle, *in* Pitcher, M. G., ed., Arctic geology, Am. Assoc. Petr. Geol. Mem. 19, p. 161-170.

Karasik, A. M., 1968, Magnetic anomalies of the Gakkel Ridge and origin of the Eurasian subbasin of the Arctic Ocean: Geophys. Methods Prospecting in Arctic, v. 5, p. 8-19.

Kerr, J. W., 1967, A submerged continental remnant beneath the Labrador Sea: Earth and Planetary Sci. Letters, v. 2, p. 283-289.

King, E. R., Zietz, I., and Alldredge, L. R., 1966, Magnetic data on the structure of the central Arctic region: Geol. Soc. America Bull., v. 77, p. 619-646.

Klenova, M. V., 1960, The Barents Sea geology: Moscow, Akad. Nauk, 367 p.

———, 1961, Recent sedimentation in the Barents Sea, *in* Recent sediments in the seas and oceans: Moscow, Akad. Nauk, p. 419-436.

Koch, L., 1920, Stratigraphy of northwest Greenland: Medd. Dansk Geol. Foren., v. 5, 78 p.

———, 1925, The geology of North Greenland: Am. Jour. Sci., ser. 5, v. 9, p. 271-285.

Kutschale, H., 1966, Arctic Ocean geophysical studies: the southern half of the Siberia Basin: Geophys, v. 31, p. 683-710.

Laktionov, A. F., 1959, Bottom topography of the Greenland Sea in the region of Nansen's Sill: Priroda, v. 10, p. 95-97: DRB transl., Hope, E. (1959), no. T 333R.

Lathram, E. H., 1973, Tectonic framework of northern and central Alaska, *in* Pitcher, M. G., ed., Arctic geology: Am. Assoc. Petr. Geol. Mem. 19, p. 351-360.

Link, T. A., Downing, J. A., Roasch, G. O., Byrne, A. W., Wilson, D. W. R., and Reece, A., 1960, Geological map of the Arctic: Alberta Soc. Petr. Geol., Calgary.

Markov, F. G., and Tkachenko, B. V., 1961, The Paleozoic of the Soviet Arctic, *in* Roasch, G. O., ed., Geology of the Arctic: Univ. Toronto Press, p. 31-47.

McManus, D. A., and Creager, J. S., 1963, Physical and sedimentary environment of a large spit-like shoal: Jour. Geology, v. 71, p. 498-512.

Meyerhoff, A. A., 1970, Continental drift: II. High-latitude evaporate deposits and geologic history of Arctic and North Atlantic oceans: Jour. Geology, v. 78, p. 406-44.

———, 1973, Origin of Arctic and North Atlantic oceans, *in* Pitcher, M. G., ed., Arctic geology: Am. Assoc. Petr. Geol. Mem. 19, p. 562-582.

Milligan, D. B., 1969, Oceanographic survey results of the Kara sea—summer and fall 1965: U.S. Naval Oceanogr. Office Tech. Rept. 217, 64 p.

Ostenso, N. A., 1962, Geophysical investigations in the Arctic Ocean Basin: Univ. Wisconsin Polar Res. Center Res. Rept. 62-4, 124 p.

———, 1963, Aeromagnetic survey of the Arctic Ocean Basin: Proc. 13th Alaska Sci. Conf., Alaska Div. Am. Assoc. Advan. Sci., Jueneau, p. 115-148.

———, 1968a, Geophysical studies in the Greenland Sea: Geol. Soc. America Bull., v. 79, p. 107-132.

———, 1968b, A gravity survey of the Chukchi Sea region, and its bearing on westward extension of structures in northern Alaska: Geol. Soc. America Bull., v. 79, p 241-254.

———, 1973, Sea-floor spreading and the origin of the Arctic Ocean Basin, *in* Implications of continental drift to the earth sciences, v. 1: London, Academic Press, p. 1965-1973.

———, and Wold, R. J., 1971, Aeromagnetic survey of the Arctic Ocean: techniques and interpretations: Marine Geophys. Res., v. 1, p. 178-219.

———, and Wold, R. J., 1973, Aeromagnetic evidence for origin of Arctic Ocean Basin, *in* Pitcher, M. G., ed., Arctic geology: Am. Assoc. Petr. Geol. Mem. 19, p. 506-365.

Overton, A., 1970, Seismic refraction surveys, western Queen Elizabeth Islands and polar continental margin: Can. Jour. Earth Sci., v. 7, no. 2, p. 346-365.

Panov, D. C., 1955a, Neotectonic movements in the Arctic region: Dokl. Akad. Nauk., v. 104, p. 462-465; DRB transl., Hope, E. (1955), no. T204R.

Patrunov, D. K., and Samoilovich, Y. G., 1973, Middle Paleozoic reefs of Siberian north: potential oil and gas reservoirs, *in* Pitcher, M. G., ed., Arctic geology: Am. Assoc. Petr. Geol., Mem. 19, p. 275-279.

Payne, T. G., ed., 1951, Geology of the Arctic Slope of Alaska, oil and gas investigations: U.S. Geol. Survey Map OM 126, 3 sheets.

———, 1955, Mesozoic and Cenezoic tectonic elements of Alaska: U.S. Geol. Survey Misc. Geol. Inv. Map, p. 1-84.

Pitman, W. C., III, and Talwani, M., 1972, Sea-floor spreading in the North Atlantic: Geol. Soc. America Bull., v. 83, p. 619-646.

Rabkin, M. I., and Ravich, M. G., 1961, The Precambrian of the Soviet Arctic, *in* Roasch, G. O., ed., Geology of the Arctic: Univ. Toronto Press, p. 18-30.

Rassokho, A. I., Senchura, L. I., Demenitskaya, R. M., Karasik, A. M., Kiselev, Y. G., and Timashenko, N. K., 1967, The Mid-Arctic Ridge as a unit of the Arctic Ocean mountain system: Dokl. Akad. Nauk. S.S.S.R., v. 172, p. 659-662.

Rickwood, F. K., 1970, The Prudhoe Bez field, *in* Adkinson, W. L., and Brosge, M. M., eds., Proc. Geol. Seminar on the North Slope of Alaska: Am. Assoc. Petr. Geol., Pacific Sect., Los Angeles, LI-LII.

Riddihough, R. P., Haines, G. V., and Hannaford, W., 1973, Regional magnetic anomalies of the Canadian Arctic: Can. Jour. Earth Sci., v. 10, p. 1957-1963.

Sachs, V. M., Basov, V. A., Dagis, A. A., Dagis, A. S., Ivanova, E. F., Meledina, S. V., Mesezhnikov, M. S.,

Nalnyayeva, T. I., Zakharov, V. A., and Shulgina, N. I., 1973, Paleozoo-geographic of Boreal-Realm seas in the Jurassic and Neocomian, *in* Pitcher, M. G., ed., Arctic geology: Am. Assoc. Petr. Geol. Mem. 19, p. 219-229.

———, and Strelkov, S. A., 1967, Mesozoic and Cenozoic of the Soviet Arctic, *in* Roasch, G. O., ed., Geology of the Arctic: Univ. Toronto Press, p. 48-67.

Saks, V. N., Belov, N. A., and Lapina, N. N., 1955, Present concepts of the geology of the central Arctic: Priroda, v. 44, p. 13-22.

Semenovich, V. N., Gramberg, I. S., and Nesterov, I. I., 1973, Oil and gas possibilities in the Soviet Arctic, *in* Pitcher, M. G., ed., Arctic geology: Am. Assoc. Petr. Geol. Mem. 19, p. 194-203.

Shatsky, N. S., 1935, O tektonike Arktiki, *in* Geologiya i polyeznye is Kopaemye Severa S.S.S.R., Glavsevmorputi, v. 1, Geologiya.

Shaver, R., and Hunkins, K., 1964, Arctic Ocean geophysical studies: Chukchi Cap and Chukchi abyssal plain: Deep-Sea Res., v. 11, p. 905-916.

Shell Oil Co., 1959, Map of tectonic elements of the North Polar region: Shell Oil Co.

Sobczak, L. W., and Stephans, L. E., 1970, Gravity field over northeastern Ellesmere Island, Greenland and Lincoln Sea with map: Can. Dept. Energy, Mines and Resources, Ottawa.

———, and Weber, J. R., 1973, Crustal structure of Queen Elizabeth Islands and polar continental margin, Canada, *in* Pitcher, M. G., ed., Arctic geology: Am. Assoc. Petr. Geol. Mem. 19, p. 517-25.

———, Krasilshchikov, A. A., Livshitz, Y. Y., and Semevsky, D. V., 1973, Structural history of Spitsbergen and adjoining shelves, *in* Pitcher, M. G., ed., Arctic geology: Am. Assoc. Petr. Gol. Mem. 19, p. 269-274.

Somov, M. M., 1955, Observational data of the scientific-research drifting station of 1050-51: Morskoi Transport Press; Engl. transl., Hope, E., Am. Met. Soc., Boston.

Steuerwald, B. A., Clark, D. L., and Andrew, J. A., 1968, Magnetic stratigraphy and faunal patterns in Arctic Ocean sediments: Earth and Planetary Sci. Letters, v. 5, p. 79-85.

Sykes, L. R., 1965, The seismicity of the Arctic: Seismol. Soc. America Bull., v. 55, p. 519-36.

Tailleur, I. L., 1969, Speculations on North Slope geology: Oil and Gas Jour., v. 67, no. 38, p. 215-220.

———, 1973, Probable rift origin of Canada Basin, Arctic Ocean, *in* Pitcher, M. G., ed., Arctic geology: Am. Assoc. Petr. Geol. Mem. 19, p. 526-535.

———, and Brosge, W. P., 1970, Tectonic history of northern Alaska, *in* Adkinson, W. L., and Brosge, M. M., eds., Proc. Geol. Seminar on the North Slope of Alaska: Am. Assoc. Petr. Geol., Pacific Sect., Los Angeles, p. E1-E20.

Thorsteinsson, R., and Tozer, E. T., 1960, Summary account of structural history of the Canadian Arctic Archipelago since Precambrian time: Can. Geol. Survey Paper 60-7, 25 p.

———, and Tozer, E. T., 1962, Banks, Victoria, and Stefansson islands, Arctic Archipelago: Can. Geol. Survey Mem. 330, 85 p.

Tkachenko, B. V., Egiazarov, B. K., Atlasov, I. P., Lazurkin, V. M., Markov, F. G., Polkin, Y. I., Ravich, M. G., Romanovich, B. S., and Sokolov, V. N., 1973, Main geologic structures of the Arctic, *in* Pitcher, M. G., ed., Arctic geology: Am. Assoc. Petr. Geol. Mem. 19, p. 336-347.

Tozer, E. T., 1960, Summary account of Mesozoic and Tertiary stratigraphy, Canadian Arctic Archipelago: Can. Geol. Survey Paper 60-5, 24 p.

Trettin, H. P., 1967, Devonian of the Franklin eugeosyncline, *in* Oswald, D. H., ed., Internatl. Symp. on the Devonian System: Alberta Soc. Petr. Geol., Calgary, v. 1, p. 693-701.

———, 1969, Paleozoic-Tertiary foldbelt in northernmost Ellesmere Island aligned with the Lomonosov Ridge: Geol. Soc. America Bull., v. 80, p. 143-148.

———, 1973, Early Paleozoic evolution of northern parts of Canadian Arctic Archipelago, *in* Pitcher, M. G., ed., Arctic geology: Am. Assoc. Petr. Geol. Mem. 19, p. 57-75.

Vinogradov, V. A., Gramberg, I. S., Pogrebitsky, Y. E., Rabkin, M. I., Ravich, M. G., Sokolov, V. N., and Sorokov, D. S., 1973, Main features of geologic structure and history of north-central Siberia, *in* Pitcher, M. G., ed., Arctic geology: Am. Assoc. Petr. Geol. Mem. 19, p. 181-188.

Vinogradova, P. S., 1957, New data on the bottom relief of the Barents Sea: Tr. Polyarnogo Nauchn. Issled. Inst. Ocean., v. 10, p. 244-259.

Vogt, P. R., and Ostenso, N. A., 1970, Magnetic and gravity profiles across the Alpha Cordillera and their relations to Arctic sea-floor spreading: Jour. Geophys. Res., v. 75, p. 4925-37.

———, and Ostenso, N. A., 1973, Reconnaissance geophysical studies in Barents and Kara Seas—summary, *in* Pitcher, M. G., ed., Arctic geology: Am. Assoc. Petr. Geol. Mem. 19, p. 588-598.

———, Ostenso, N. A., and Johnson, G. L., 1970, Magnetic and bathymetric data bearing on sea-floor spreading north of Iceland: Jour. Geophys. Res., v. 75, p. 903-920.

Webster, C. T., 1954, The Soviet expedition to the central Arctic, 1954: Arctic, v. 7, p. 59-80.

Wegener, A., 1924, The origin of the continents and oceans: London, Methuen, 212 p.

Wilson, J. T., 1963, Continental Drift: Sci. Am., v. 208, p. 86-100.

———, 1965, Evidence from ocean islands suggesting movements in the earth, *in* A symposium on continental drift: Phil. Trans. Roy. Soc. London, v. 1088, p. 145-67.

Wold, R. J., and Ostenso, N. A., 1971, Gravity and bathymetry survey of the Arctic and its geodetite implications: Jour. Geophys. Res., v. 76, p. 6253-6264.

———, Woodzik, T. L., and Ostenso, N. A., 1970, Structure of the Beaufort Sea continental margin: Geophysics, v. 35, p. 849-861.

Worzel, J. L., 1968, Advances in marine geophysical research of continental margins: Can. Jour. Earth Sci., v. 5, p. 963-983.

Part X

Ancient Continental Margins

The Most Ancient Continental Margins

A. M. Goodwin

INTRODUCTION

Each continent of the world contains a Precambrian core or continental platform that includes one or more early Precambrian or Archean cratons. A pre-drift reconstruction of continents indicates that (1) the Precambrian continental crust formed a single asymmetric (one-sided) mass upon the earth's sphere, and (2) Archean crust was confined to a narrow, irregular cresent, concave to the east, within it.

Detailed studies of local Archean volcanic belts lead to the conclusion that miniaturized, "soft-plate" tectonics, involving interaction of numerous small, thin, comparatively supple lithospheric plates, operated in development of Archean crust. On a larger scale, numerous elliptical to quasi-circular basins in Archean crust, each approximately 600 miles in diameter, are attributed to volcano-tectonic development resembling in mode of origin extensional basins of the western Pacific, such as the Sea of Japan.

Analysis of the Canadian Shield suggests that it grew from a number of small, primitive, sialic units (protocontinents), through development of larger cratons, to a single continental shield; and that growth of the shield occurred about a long-lived crustal anomaly situated at the geographic heart of the shield in Hudson Bay, the site of repeated vertical fluctuations of the crust. Other shields of the world may have had similar histories.

The crescentic pattern of Archean crust within the global Precambrian reconstruction referred to above is attributed in origin to major terrestrial meteorite impacting in the period 4,200-3,800 my, coincident with development of lunar maria. Deep-seated, long-lived, thermal convection patterns including mantle plumes, initiated by such terrestrial impacting, promoted mantle-crust differentiation, resulting in surficial concentration of sialic end products at the sites of the growing shields. The original impact sites are interpreted to lie in the center of shields adjoining Archean terrain, in some cases marked by topographic depressions such as the Hudson, Chad, and Congo basins.

PRECAMBRIAN SHIELDS

The purpose of this paper is to present a brief review of ancient terrestrial crust, predating the emergence of large modern continents, with still larger ocean basins and widespread intervening continental margins. What was the nature of this ancient crust and what tectonic processes operated in its development? Answers to these fundamental questions must be sought mainly in Precambrian shields of the world and particularly in those very ancient nuclei, the Archean cratons.

Each continent of the world comprises a Precambrian core with Phanerozoic additions (Fig. 1), either in the forms of cover rocks, (e.g. Hudson Platform) or marginal accretions, situated either at continental-oceanic interfaces (e.g., Cordilleras) or between such converging continental plates as the Alps. The present distribution of continents with Precambrian cores is manifestly a product of a long, varied history of fragmentation and dispersal of evolving continental crust. Fragmentation occurred repeatedly in response to global tectonic processes.

To understand these early Precambrian, crustal-spawning processes, we must reconstruct the Precambrian fragments and interpret the resulting global patterns. This process of reconstruction becomes increasingly difficult with increasing geologic age, and is most difficult for Archean crust.

The exposed Precambrian core of a continent is conventionally called a shield. The buried extension—the crypto-shield—may exceed it in area. It is the total Precambrian crust—exposed and buried—that forms the Precambrian core, or stable continental platform. In addition, some Precambrian crust representing the extreme margins of the platforms has been incorporated in Phanerozoic mountain-building processes (e.g., interior Precambrian belts in the Rocky Mountains, Urals, and Himalayas) to form integral parts of these younger peripheral mobile belts.

The Precambrian core, relative to the accretion, may by central and predominant (Africa), or offset and more nearly equal (e.g. Americas, Eurasia). Most continents include a fragment only of an originally larger Precambrian mass. For example, in the several continents resulting from the fragmentation of Gondwanaland, the major Phanerozoic accretions, each situated on the leading edge of a Precambrian fragment, face respectively west (South America), south (Antarctica), east (Australia), and north (Africa and India), conforming to the particular direction of plate motion.

In the Northern Hemisphere the conventional Precambrian shields from west to east are: Canadian Greenland, Baltic, Ukrainian, Angara or Siberian (Aldan and Anabar, Chinese or Korean, Guianan, west African, Arabian, and Indian. In the Southern Hemisphere they are: Brazilian, African (central

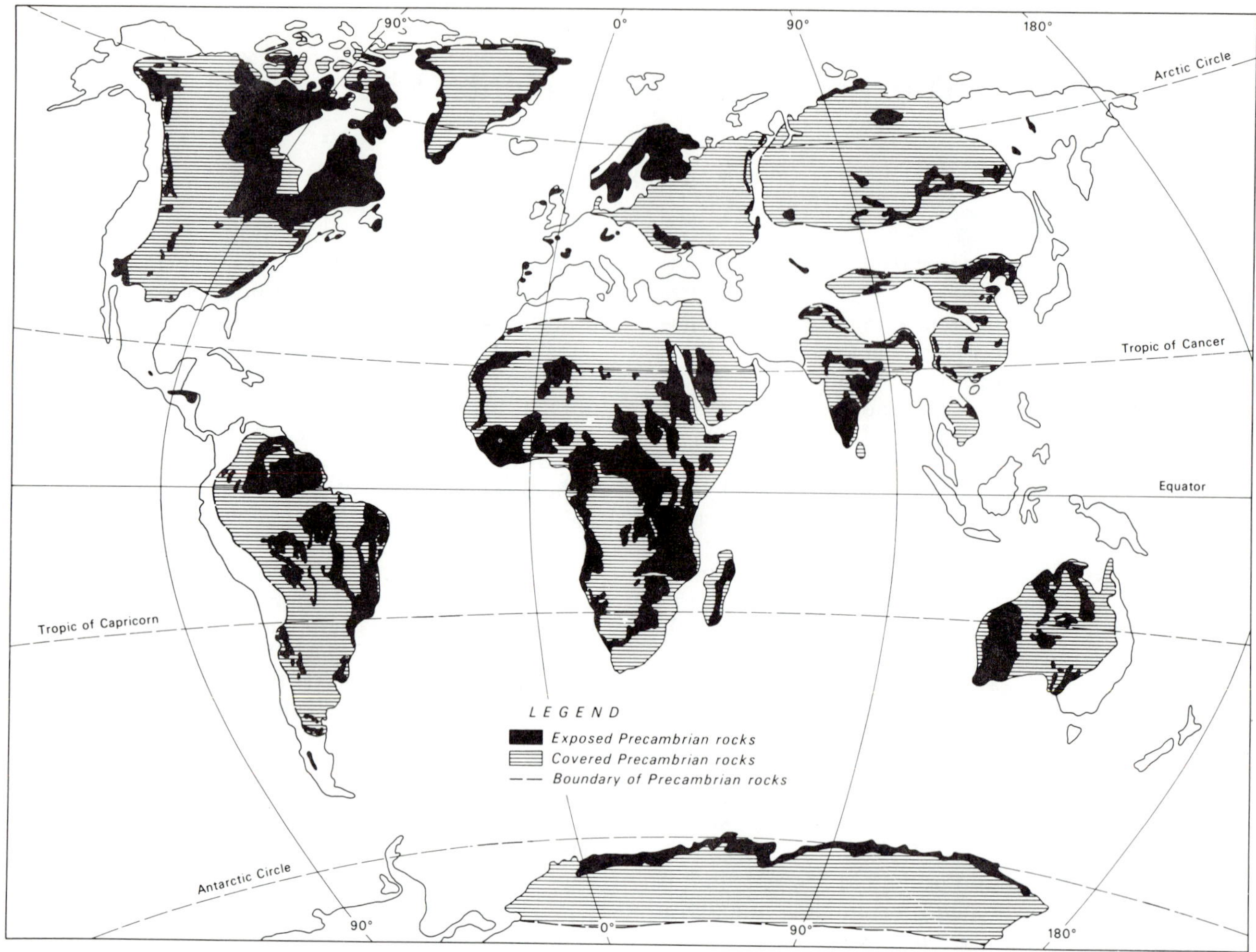

Fig. 1. Precambrian crust of the world.

and southern), Australian, and Antarctic. In addition, micro-shields (e.g., Madagascar), representing island slivers, are present.

The pre-drift reconstruction of Precambrian crust as interpreted for the end of Precambrian time is shown on the Dietz-Holden (1970) fit of Pangaea, with modern continental margins shown for ease of reference (Fig. 2). The apparent continuity of Precambrian crust within this reconstruction is the outstanding feature. There is growing evidence that the northern and southern supercontinents of Laurasia and Gondwanaland were, in fact, united about the "gaping gore" of the future Tethys, as illustrated, rather than separated into two distinct land masses. "New" or Phanerozoic crust, as illustrated in Figure 2, constitutes (1) the circum-Pacific and Mediterranean-Himalayan fold belts (Alpine) of Cenozoic age, situated at the leading edge of modern continents; (2) certain restricted older Phanerozoic fold belts (e.g., Caledonian and Hercynian), mainly in Laurasia; and (3) extensive basin and platform covers, including those at continental margins (e.g., Grand Banks).

ARCHEAN CRATONS

Most, if not all, Precambrian shields include Archean crust (> 2,5000-2,7000 my) arranged in a number of ancient nuclei or cratons. Additional Archean crust is present either as smaller undoubted Archean outliers (e.g., Rankin Inlet-Ennadai belt in Churchill Province of the Canadian Shield), or as widely distributed, overprinted crust of uncertain Precambrian age. With due allowance for this type of metamorphic overprinting, including more extreme processes of crustal digestion (e.g., anatexis), the total quantity of Archean crust undoubtedly exceeds by a substantial amount that presently recognized. However, the recognized and undisputed Archean cratons, by their very state of excellent preservation, provide most insight into the nature of Archean crust and Archean tectonic processes.

Most Precambrian shields contain at least one Archean craton (Fig. 2). In the northern continents, Archean cratons occur in the Canadian (Superior and Slave provinces), southern Greenland, eastern Baltic, Ukranian and Aldan shields; and in the southern continents in the Guianan, Brazilian, African (western, central and southern), Indian,

Australian, and, possibly, Antarctic shields. Archean crust is is also present in the island of Madagascar.

The distribution of Archean cratons within pre-drift global reconstruction is shown in Figure 2, together with the larger postulated total Archean presence. The latter, which forms a narrow irregular crescent of ancient crust, concave to the east, is notably asymmetrical in distribution upon the earth's sphere.

The best preserved, exposed, accessible, and studied Archean cratons, lie in the Canadian, African, and Australian shields. Of these, the Canadian Shield provides the example used herein. For present purposes, exclusive attention will be directed to Archean volcanic and sedimentary rocks as indicators of Archean tectonic environments and processes, despite the general predominance of granite rocks in the cratons. Two parameters will be briefly reviewed: (1) evidence within an Archean volcanic belt indicating some form of "plate" motion and (2) the presence in Archean crust of numerous large basins of apparent volcano-tectonic derivation. Finally, an interpretation is offered relating the primitive crescentic Archean presence to cataclysmic impact of large cosmic bodies during pre-Archean time (>4,200-3,900 my). This obviously relates to the most ancient of continental margins.

ARCHEAN VOLCANIC BELTS OF THE CANADIAN SHIELD

Volcanic rocks are principal components of widespread Archean greenstone belts present in

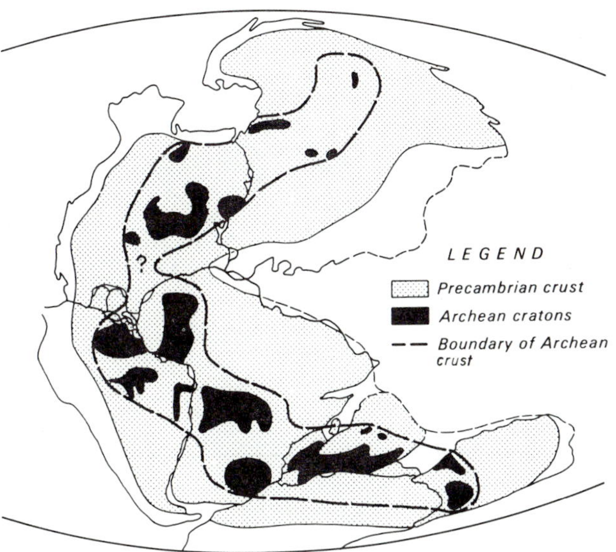

Fig. 2. Reconstruction of Precambrian continental crust with modern coastlines included for convenience. The apparent continuity of Precambrian crust in the reconstruction is the outstanding feature. Archean cratons (2,500 my) are approximately located, forming a crescent, concave to the east.

most Archean cratons. In the Canadian Shield Archean these have received wide attention since they are disproportionately rich in valuable mineral deposits. The classical field studies of Lawson (1885), Cooke et al. (1931), Gunning and Ambrose (1940), Wilson (1914, 1941), and others have established their gross similarities to moden volcanic rocks. Accelerated field and laboratory studies (e.g., Goodwin 1973b) have shed much light on Archean volcanic processes. One important result has been to establish their basic similarity to processes operating in the development of Cenozoic orogenic belts, particularly those at continental margins and island arcs.

The Canadian Shield is used here as a suitable example, because of the size and variety of Archean volcanic belts, the long history of examination, and the pioneer patterns established therein which have served as world-wide models for many other scientists.

Distribution

Archean volcanic belts are widely distributed across the Canadian Shield (Fig. 3). The units range from isolated ribbons a few kilometers long to large irregular belts up to 700 km long. The units are characteristically east-trending in the southern shield (Superior Province), east to northeast trending in the central part (Churchill Province), and north trending in the northwestern part (Slave Province). The largest continuous belt, the Abitibi in southeastern Superior Province, is 650 km long by 200 km broad. Other belts of substantial dimensions are present in Superior Province. The belts in Churchill Province are widely scattered, although comparatively small, with the exception of the Rankin Inlet-Ennadai belt west of Hudson Bay; in Slave Province, the belts are comparatively thin and discontinuous.

Lithology and Stratigraphy

Archean volcanic belts typically contain a wide variety of supracrustal and plutonic rocks, with volcanic components predominating. Volcanic rocks typically comprise lavas and fragmentals of the basalt-andesite-dacite-rhyolite association, commonly arranged in mafic to felsic cycles or sequences (Goodwin, 1961, 1962, 1968). The volcanic rocks are intercalated, especially in upper stratigraphic parts, with volcanogenic sediments, mainly graywacke, mudstone, tuff, conglomerate, and "iron formation." Granitoid plutons are common. Mafic to ultramafic intrusions in the form of discrete sills, dikes, and small irregular plutons are widespread, although in small proportions.

A variety of volcanic rocks is present in the belts. Flows and pyroclastics of tholeiitic to calc-alkaline chemical affinity predominate. Peridotitic and other compositionally primitive volcanics are present locally in the lower parts, as are such highly

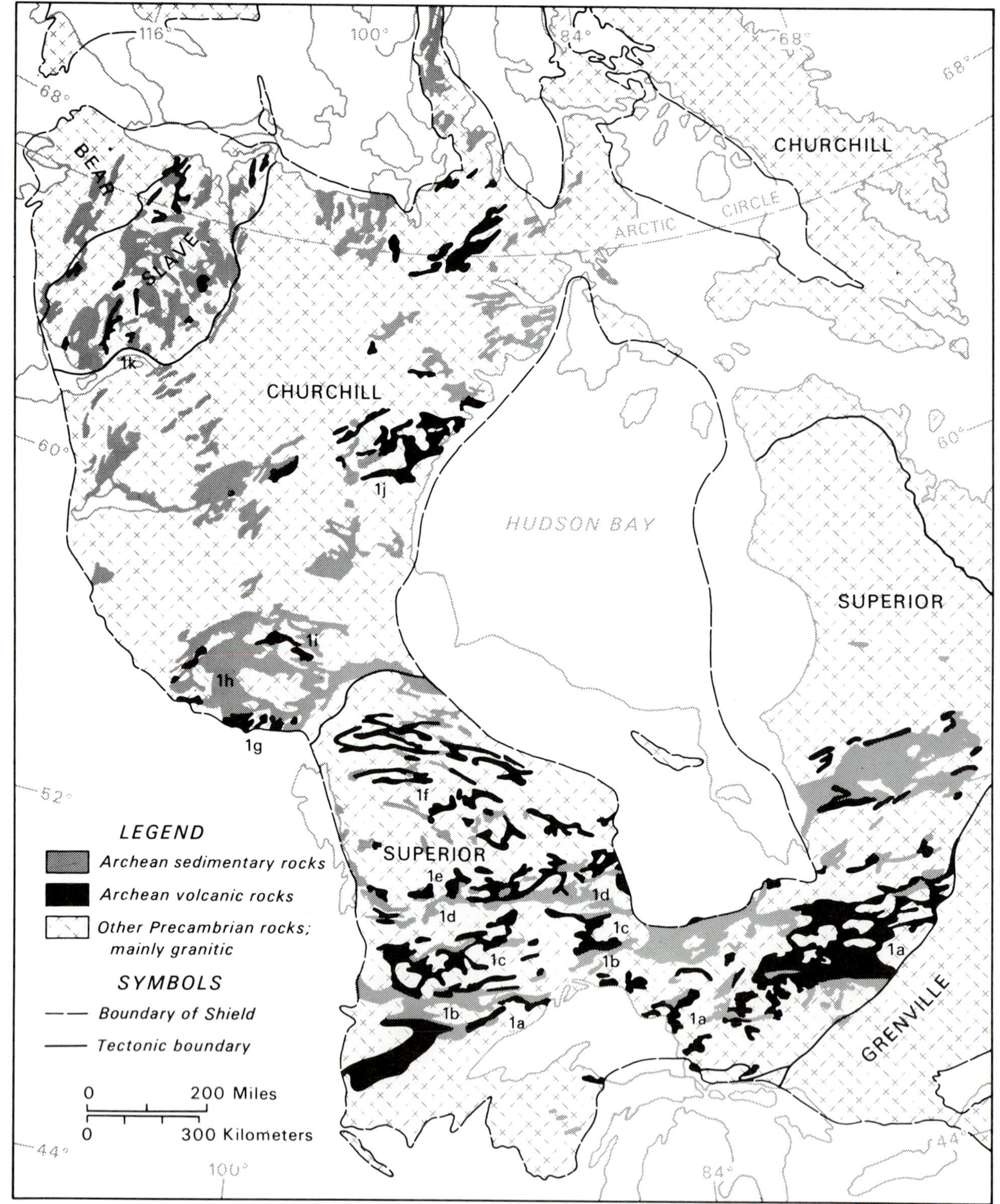

Fig. 3. Distribution of Archean volcanic and sedimentary rocks of the Canadian Shield. The principal tectonic provinces are named.

differentiated and undersaturated volcanic rocks as trachyte and leucitite in the upper parts. Basaltic lava flows predominate in the lower stratigraphic parts of typical volcanic cycles. Andesite flows and fragmentals are intercalated with basalt and, in many assemblages, increase proportionately upward. Felsic rocks, mainly pyroclastics, but locally massive and fragmental flows, are generally present in the upper stratigraphic parts. This general mafic-to-felsic progression forms the fundamental feature of the Canadian model of Archean volcanism, to which other models and modifications have been arranged. Many Archean assemblages display a single generalized mafic-to-felsic cycle; some have, in addition, a thick mafic capping; still others contain two or more superimposed mafic-to-felsic cycles. The total stratigraphic thickness of an Archean volcanic pile is generally in the order of 12,000 m but locally attains 20,000 m.

Although the mafic-to-felsic compositional trend is the common first-order stratigraphic feature of Archean piles, not all assemblages display the complete compositional trend. The mafic, basalt-rich lower part of the pile is present in nearly all belts. The upper, more felsic parts may be absent, because of nondeposition or later removal by erosion.

The larger Archean belts, such as the Abitibi, contain a number of such mafic-to-felsic volcanic assemblages, each representing a composite volcanic pile. Thus a belt in general comprises a complex of coalescing volcanic piles, with intercalated sediments, now deformed, metamorphosed, and intruded by younger, mainly granitic, plutons. Of particular economic interest are base metal sulfide, gold-silver, and iron deposits present in the upper stratigraphic felsic parts of the volcanic piles, as well as nickel, asbestos, and rare chrome deposits in the stratigraphically lower, mafic-to-ultramafic parts.

Abundance of Classes

The abundance of volcanic classes present in three well-studied Archean volcanic belts of the southern Canadian Shield are shown in percentages in Table 1. The proportions of Archean volcanic classes, in summary are basalt to andesite to felsics = 57:30:13. The general proportions of dacite to rhyolite = 3:1.

The corresponding proportions in modern island arcs of tholeiitic association are as follows: basalt to andesite to "dacite" = 50:35:15 (Jakes and White, 1971, Table 1, p. 225). Although Archean assemblages contain more basalt and less andesite than modern arcs of tholeiitic association, their general similarities are remarkable and suggest similar histories of origin and tectonic development.

Chemical Compositions by Belts

The average composition of three representative volcanic belts in Superior Province and the calculated average of Superior Province are illustrated in Table 2. The three Archean volcanic belts are remarkably similar to each other in average chemical composition. They correspond closely to the computed average composition of a "developed island arc" (Jakes and White, 1971, p. 227, Table 3) and even more closely to that of the more primitive or tholeiitic (young) island arc (e.g., South Sandwich Islands, Central Kurile Islands), which are the oceanward part of a fully developed island arc. The main distinguishing characteristics are significantly higher FeO, Ni, V, Cr, and CO_2 and somewhat lower Na_2O and CaO in Archean rocks.

Thus Archean volcanic belts present close analogies to modern island arcs in terms of (1) prevailing mafic-to-felsic stratigraphic successions; (2) similar chemical compositions and abundances of volcanic classes; (3) widespread calc-alkaline volcanism, including much highly explosive felsic eruptives; (4) predominantly subaqueous accumulation of volcanic rocks and sedimentary associates; and (5) parallel arc-like arrangement of successively younger linear volcanic belts.

Table 1. Archean volcanics of Canada

Volcanic Belt	Basalt	Andesite	Dacite	Rhyolite
Birch-Uchi	57.8	29.2	13.0	
Lake of the Woods-Wabigoon	55.7	26.4	13.3	4.6
Timmins-Kirkland Lake-Noranda	58.9	32.4	7.2	1.7
Average Superior Province	57.3	29.4	10.2	3.1

ABITIBI VOLCANIC BELT

A characteristic feature of a Cenozoic island arc, such as the islands of Japan, is a systematic lateral variation across the arc and continental margin of several distinctive igneous suites; this is marked particularly by increasing alkalinity of volcanic components from ocean side (younger) to continent side(older) of the arc. Lateral variation of this type is commonly attributed to increasing depth to source of magma, a linear function of vertical distance to the continental-inclined Benioff zone, and horizontal distance from the trench (Dickinson and Hatherton, 1967). According to the classical studies of Kuno (1966), three main volcanic series—tholeiitic, high alumina, and alkaline—are arranged in that order across an arc, from ocean to continent side. This constitutes a measurable chemical polarity.

The southwestern part of the Abitibi belt in the Superior Province of the Canadian Shield has been studied in stratigraphic and geochemical detail (Goodwin, in press a). The Abitibi belt contains all the distinctive igneous suites of modern island arcs. Furthermore, in at least one south-trending section across this belt, the three main volcanic series are distributed according to the classical chemical polarity, with increasing alkalinity from north to south thereby suggesting a corresponding distribution of "oceanic" and "continental" crusts as briefly described below.

Timmins-Kirkland Lake-Noranda Region

The most carefully studied Archean volcanic terrain underlies the southwest quarter of the Abitibi

Table 2. Average chemical composition of Archean volcanic belts in the Superior Province of the Canadian Shield.

Major Elements (percent)

Volcanic Belt	N	SiO_2	Al_2O_3	Fe_2O_3	FeO	MgO	CaO	Na_2O	K_2O	TiO_2	MnO	CO_2	H_2O	P_2O_5
Birch-Uchi	116	54.3	14.5	2.62	7.06	4.93	6.84	2.70	0.67	0.83	0.16	4.56	—	0.26
Lake of the Woods	420	54.3	14.9	2.37	7.3	4.81	7.25	2.78	0.75	0.87	0.17	2.06	1.98	
Timmins-Noranda	1,080	53.7	15.1	2.21	6.93	4.96	6.92	3.27	0.74	1.07	0.18	1.17	2.93	0.16
Average		54.1	14.8	2.40	7.10	4.90	7.00	2.92	0.72	0.92	0.17	1.61	2.46	0.21

Minor Elements (ppm)

Volcanic Belt	N	Sr	Ba	Cr	Zr	V	Ni	Cu	Co	Zn	Pb	Ga	Sn	Ag
Birch-Uchi	116	278	188	307	252	285	112	88	32	121	6	32	1.99	0.15
Lake of the Woods	420	210	195	161	140	271	131	76	32	95	5	24	1.19	0.11
Timmins-Noranda	1,080	150	200	170	110	290	140	80	70	83	6	21	1.68	0.14
Average		212	194	213	167	282	128	81	45	100	6	26	1.62	0.13

Belt (Fig. 4). It is underlain by a complex assemblage of Archean volcanic and sedimentary rocks, granitic intrusions, and younger Precambrian cover rocks (Cobalt Group).

In Figure 4 the volcanic rocks are subdivided into four classes—basalt, andesite, dacite (plus rhyolite), and trachyte. Trachytes are confined to the Timiskaming Group at Kirkland Lake in the south.

The volcanic assemblage was studied in detail along numerous close-spaced stratigraphic sections; 1,086 individual volcanic units, mainly massive lava flows, but including fragmentals, were sampled. Each sample was chemically analyzed for major and minor elements and the volcanic class determined using the Barth-Niggli normative classification of Irving and Baragar (1971).

These volcanic rocks divide naturally into a number of volcanic piles, each with a thick, widespread mafic platform and a comparatively restricted felsic edifice, the latter representing vent and near-vent accumulations. Three major volcanic piles are present: (1) the Deloro-Tisdale pile near Timmins in the west, (2) the Skead-Timiskaming complex near Kirkland Lake in the south, and (3) the Blake River pile in the center-east. The latter extends for 150 km from Nighthawk Lake, near Timmins on the west, to Noranda on the east; it features a large concentration of rhyolite and andesite flows at Noranda which together form the upper stratigraphic part of the pile underlain to the west by progressively more mafic lavas.

Blake River volcanic rocks are deformed about an east-trending and plunging, upright, isoclinal syncline which in the east divides, to wrap around the Noranda antiformal dome. As a result of this fold pattern, Blake River volcanic rocks generally "young" ("face") both southward from the northern, lower stratigraphic contact of the pile as far as the main east-trending syncline as well as eastward along the trace of this east-plunging axis. To the north of the Blake River volcanic rocks, and separated therefrom by a pronounced shear zone of undetermined fault movement (Porcupine-Destor "break"), are peridotite-bearing volcanic flows, including those of Munro Twp. To the south of Blake River volcanic rocks, at Kirkland Lake-Larder Lake, are unconformably overlying trachytic and phonolitic volcanic rocks of the Timiskaming Group. The chemical polarity referred to above involves volcanic rocks of, successively from north to south, Munro Twp., Blake River Group, and Skead and Timiskaming groups.

More specifically, volcanic rocks in the central part of the Timmins-Kirkland Lake Noranda region display a systematic variation in chemical composition along a 30-mile-long section extending from Munro Twp. on the north to Kirkland Lake on the south (Fig. 4). The three distinctive volcanic series—(1) tholeiite, (2) high-alumina calc-alkaline, and (3) alkaline—are distributed as follows: (1a) Mg-rich tholeiite (Munro Twp.); (1b) Fe-rich tholeiite (Garrison Twp.), (2) high-Al calc-alkaline (Magusi), and (3) alkaline [Timiskaming trachyte (3b) and Catherine basanites (3a)]. In stratigraphic terms, (1a) represents primitive tholeiites lying north of the Porcupine-Destor "break," as in Munro Twp.; (1b) and (2), respectively, form the lower tholeiitic and intermediate to upper calc-alkaline parts of the Blake River Group; and (3) represents alkalic rocks of the Timiskaming and Skead (Catherine Basalts) groups to the south. These four representations are briefly reviewed below.

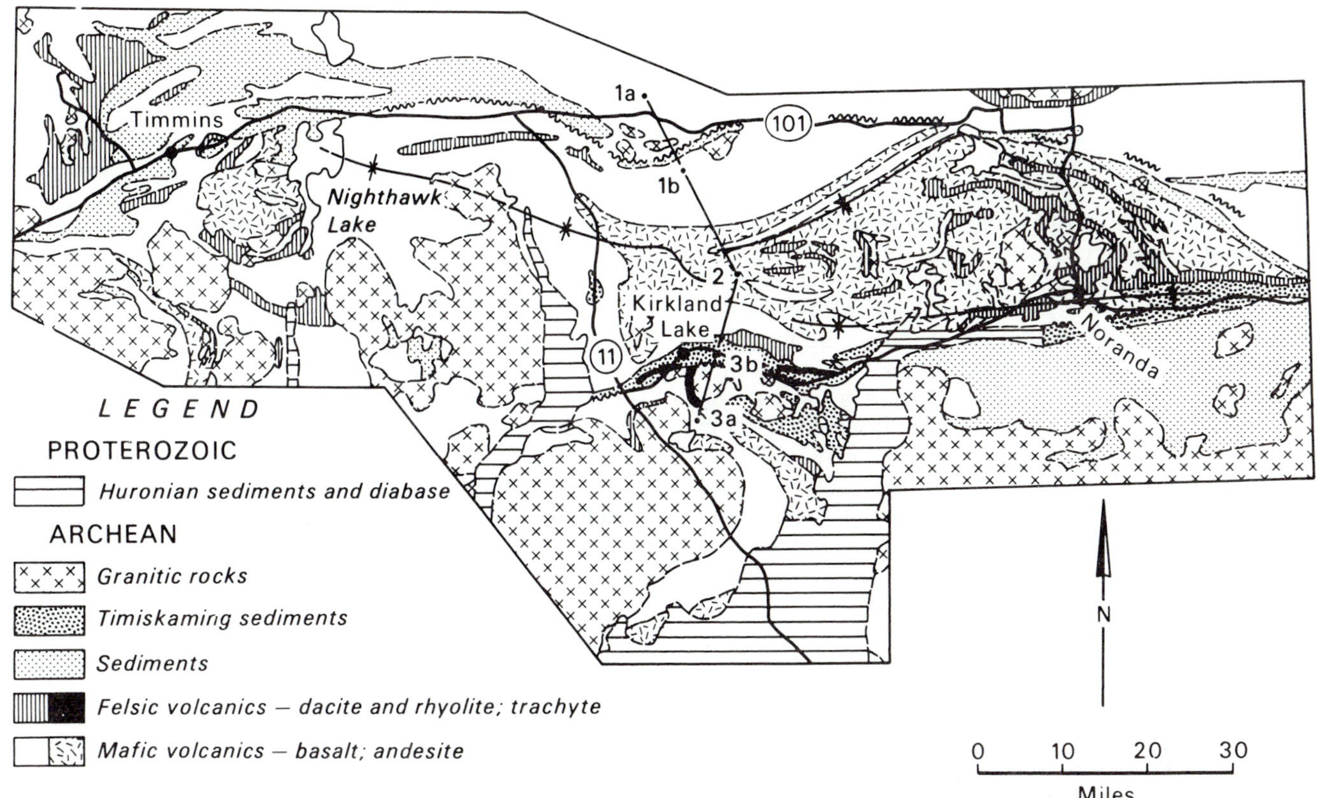

Fig. 4. Geology of the Timmins-Kirkland Lake-Noranda region, Ontario and Quebec, Canada.

1a. Munro Twp. Ultramafic lava flows exhibiting quench ("spinifex") texture indicative of rapid cooling are unusually fresh and well exposed in Munro Twp. (Fig. 4) (Pyke et al., 1973). Numerous superimposed ultramafic units occur ranging from 1/2 to 17 m thick with a strike length varying from a few meters up to nearly 200 m. The ultramafic units are part of a sequence of pillowed basalts and andesites, and felsic pyroclastics. According to Naldrett (1972), the concordance of the ultramafic units, their asymmetry, the quench textures, and the nature of the surfaces interpreted as flow tops are taken as evidence that the units are individual flows. If so, the liquid portion of the ultramafic lava was extruded subaqueously at temperatures on the order of 1350-1400°C. Their association with pillowed basalt, together with the absence of significant clastic sediments and cherty oxide iron formation, has been interpreted as evidence of a comparatively deep-water environment of accumulation (Goodwin, 1967). This type of assemblage may represent Archean ocean-floor material, the Archean analogue of modern oceanic crust. Variation diagrams and mean compositions of the normal lava flows and of peridotites representative of this series have been previously published (Goodwin, 1973b, p. 1044, Table 1).

1b. Garrison Twp. A 7-km sequence of superimposed mafic flows form the stratigraphically lower part of the Blake River Group along the north limb of the regional syncline (Fig. 4). An MgO-Fe_2O_3 variation diagram (Goodwin, 1973b, Fig. 3, p. 1046) and mean chemical composition (Goodwin, 1973b, Table 1, column 3, p. 1044) illustrate a moderately high iron-enrichment trend, which is characteristic of Fe-rich tholeiites.

2. Magusi River. A thick succession of high-Al mafic flows is present in the upper stratigraphic parts of the Blake River Group north of Kirkland Lake (Fig. 4). The high-Al character has been well illustrated by alkali-SiO_2 and alkali-Al_2O_3 variation diagrams (Goodwin, 1973b, Figs. 4, 5a, 5b). The calc-alkaline character (no iron enrichment or Cascade trend) of the series is illustrated in a MgO-Fe_2O_3 variation diagram (Goodwin, 1973b, Fig. 6).

3. Timiskaming and Skead Area. The Skead Group south of Kirkland Lake (Fig. 4) includes a substantial thickness of alkalic mafic volcanics, including basanite and alkali andesite (Ridler, 1969, p. 18-23), as illustrated by the alkali-SiO_2 variation diagram and table of mean chemical compositions (Goodwin, 1973b, Fig. 7 and Table 1, col. 5).

Stratigraphically overlying the Skead and Blake River volcanic rocks is the Timiskaming Group, which includes substantial thicknesses of trachyte, phonolite, and trachy basalt (Ridler, 1969, p. 35). An alkali-SiO_2 variation diagram (Goodwin, 1973b, Fig. 8) illustrates a significant concentration of mafic to intermediate volcanic rocks of strong alkalic affinity. The mean chemical composition supports this (Goodwin, 1973b, Table 1, col. 6).

Tectonic Interpretation

In summary, Archean volcanic rocks of the Timmins-Kirkland Lake-Noranda region contain a variety of igneous types, including the major volcanic series characteristic of modern island arcs, namely, tholeiite, high-Al, and alkaline. Within this particular region the alkali content (K_2O) of the volcanic rocks increases from north (Munro Twp.) to south (Kirkland Lake). This north-south variation may be analogous to Kuno's (1966) classical lateral variation of magma types across a modern volcanic island arc. If so, the ocean side of the Archean section would have been to the north, Archean oceanic crust being represented by the primitive, Mg-rich tholeiites and peridotite lavas of Munro Twp. Other similarly primitive, Mg-rich volcanic assemblages have been identified elsewhere in the Abitibi belt (Naldrett, 1973), as have other highly developed calc-alkaline volcanic assemblages. In brief, a patch pattern of the main igneous series is present in the Abitibi belt. It is suggested that the Archean crust featured interaction of many small, thin comparatively supple, lithospheric plates in a process termed "miniaturized soft-plate tectonics." This prevailing crustal state promoted development of numerous, large tectonic basins with characteristic distribution of Archean iron formation as described below.

ARCHEAN IRON FORMATION AND TECTONIC BASINS

Iron formations of Archean age are widely distributed in supracrustal assemblages of the Canadian Shield. They are abundant in Superior Province, substantial in parts of Churchill Province, but limited in Slave Province. The pattern of distribution and mode of origin provide valuable insight into the nature and tectonic style of early Precambrian crust.

As recently summarized (Goodwin, 1973a), iron formations, by their composition, mineralogy, and oxidation state, are particularly sensitive indicators of depositional environment. The depositional environments have been studied in detail and it has been demonstrated that pH and Eh of the aqueous medium are major factors in controlling the kind and amount of chemical deposition. James (1954) has applied these concepts and proposed that oxide-facies iron formation usually indicates a relatively high Eh; sulfide facies, a strongly negative Eh; and carbonate and silicate facies are intermediate.

Briefly stated, oxide-facies iron formation is interpreted as a comparatively shallow-water deposit. Conditions favorable for the deposition of siderite would be found in deeper water. Pyrite or pyrrhotite beds of the sulfide facies apparently formed, commonly in the deeper parts of sedimentary basins, where strongly reducing conditions permitted preservation of carbon, and where H_2S was released.

Distribution patterns of Archean iron formation by facies have been analyzed in terms of Archean paleoslopes and basins. These patterns, in conjunction with other paleoenvironmental indicators, serve to identify a number of primitive depositional basins in Archean crust of the Canadian Shield.

Elements of Archean Basins: Michipicoten Area

The Michipicoten assemblage contains assorted mafic and felsic volcanic rocks, clastic sedimentary rocks, and banded iron formation (Fig. 5). The supracrustal rocks contain widespread mafic effusion intermittent felsic pyroclastics, accompanied by tectonic subsidence, clastic sedimentation, and extensive exhalative activities, leading to chemical precipitation of iron formation (Goodwin, 1973a). The resulting volcano-tectonic basins are distinctive features of the Archean crust. Accordingly, iron formation facies must be key elements in the recognition of basins in Archean crust.

The Michipicoten iron formation is arranged from west to east in transitional oxide, carbonate, and sulfide facies (Fig. 5, inset). This pattern corresponds to the common oxide-carbonate-sulfide zonal model defined by Borchert (1960) and James (1966, p. W15). Although the chemical components of Michipicoten iron formation are considered to be volcanic derivatives, the environmental parameters exerted by basin configuration were sufficient to have produced a general shore-to-depth facies pattern that conforms to a world-wide mode.

This typical Archean basin margin features a triple lithofacies association of (1) oxide-carbonate-sulfide facies transition; (2) proximal to distal clastic transition; and (3) volcanic piles, featuring thick, comparatively coarse-grained, felsic-volcanic concentrations. The general patterns of basin configuration established in the Michipicoten area can now be used in search for other tectonic basins in the Canadian Shield (Goodwin, 1973a).

Archean Iron Formation of the Canadian Shield

Archean iron formation is particularly abundant in the Superior Province, with two principal distribution characteristics: (1) iron formation is almost exclusively concentrated in volcanic-rich segments, and (2) within a particular volcanic-rich segment there is a preferential association of iron formation and felsic volcanic centers. These relationships, common to the shield at large, provide strong supporting evidence for a direct genetic link between Archean iron formation and volcanism.

A preliminary distribution of Archean iron formation by facies (Goodwin, 1973a, Figs. 4-9) reveals that they are not distributed indiscriminately but occur in regional patterns which serve to outline deformed tectonic basins within Archean crust.

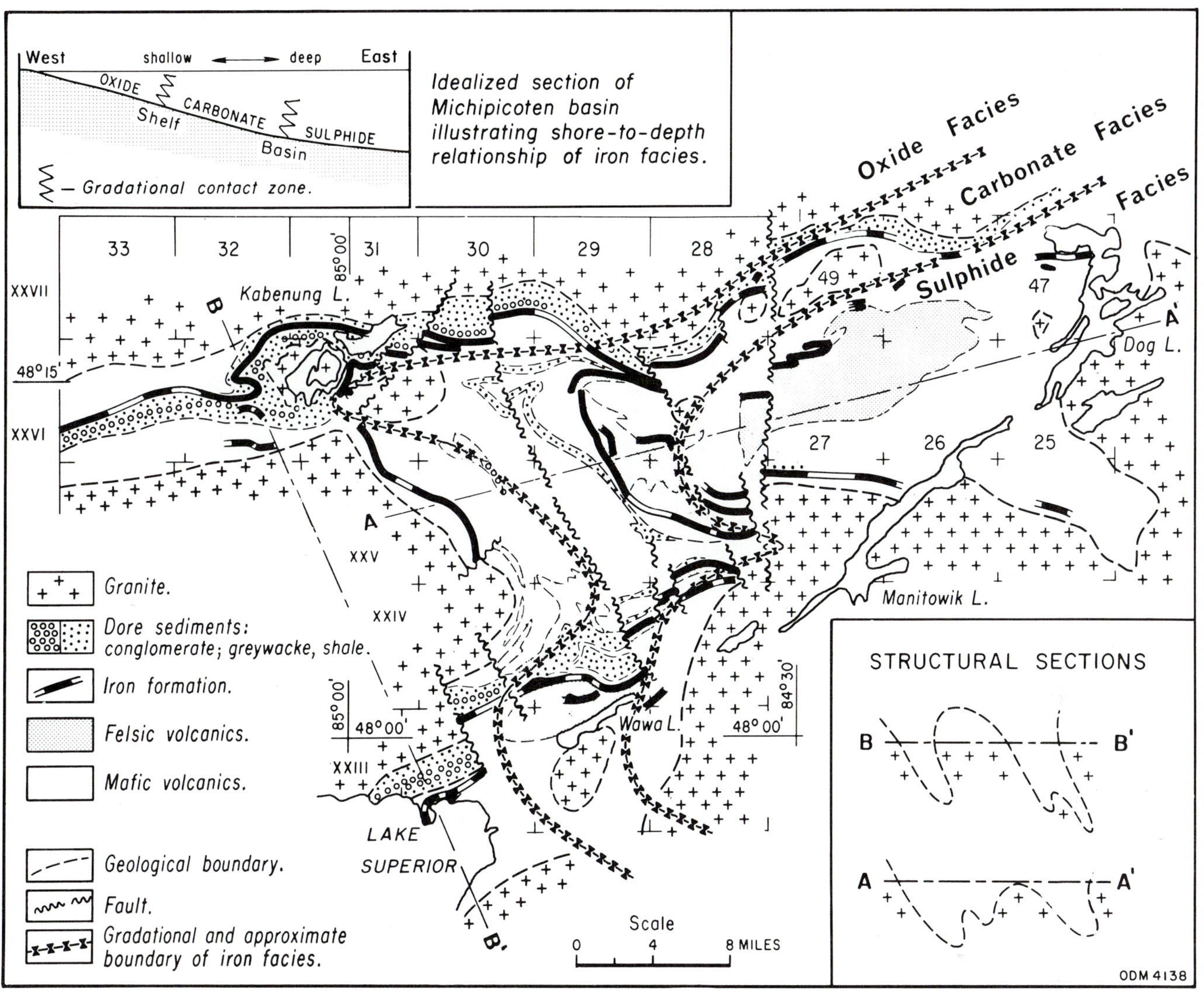

Fig. 5. Geology of the Michipicoten area, Ontario, Canada, showing the distribution of oxide, carbonate, and sulfide facies of iron formation, relative to felsic volcanic concentrations and proximal (conglomerate) and distal (shale) sediments, together with an idealized section of shore-to-depth relationships of the iron facies.

Tectonic Basins

On the basis of the principles established in the Michipicoten area and briefly reviewed above for recognition of Archean basin margins, 10 Archean basins have been provisionally identified in Archean crust of the Canadian Shield (Fig. 6). Seven basins (nos. 1-7) are present in Superior Province, two (nos. 8 and 9) in Churchill Province, and one (no. 10) in Slave Province.

Lithofacies Association. The 10 basins each exhibit the characteristic lithofacies association of an Archean basin: a triple association of oxide-carbonate-sulfide transition, "arc"-type calc-alkaline volcanic piles with felsic concentrations, and coarse-grained proximal clastics. The interior parts, as preserved, feature predominant sulfide-facies iron formation, finer-grained distal clastics, and uniform tholeiitic volcanic rocks.

Oxide-facies iron formation features both hematitic (jaspilite) and magnetitic bedded cherts; these two subfacies appear to be preferentially related, the hematitic subfacies marking a higher original oxidation state (and apparently shallower water environment) compared with the magnetitic subfacies. The latter is the more common. Locally jaspilite clasts in conglomerate beds are the only evidence of preexisting shallow-water hematitic iron formation.

The principal concentrations of arc-type calc-alkaline volcanics with felsic concentrations are similarly distributed at or near the margins of the basins, commonly coincident with the oxide-sulfide facies transitions. Their distribution is suggestive of rim volcanoes at the margins of subsiding basins.

Closely associated with the iron-facies transitions and arc-type volcanic concentrations are thick beds of coarse conglomerate, containing volcanic clasts but locally including concentrations of grani-

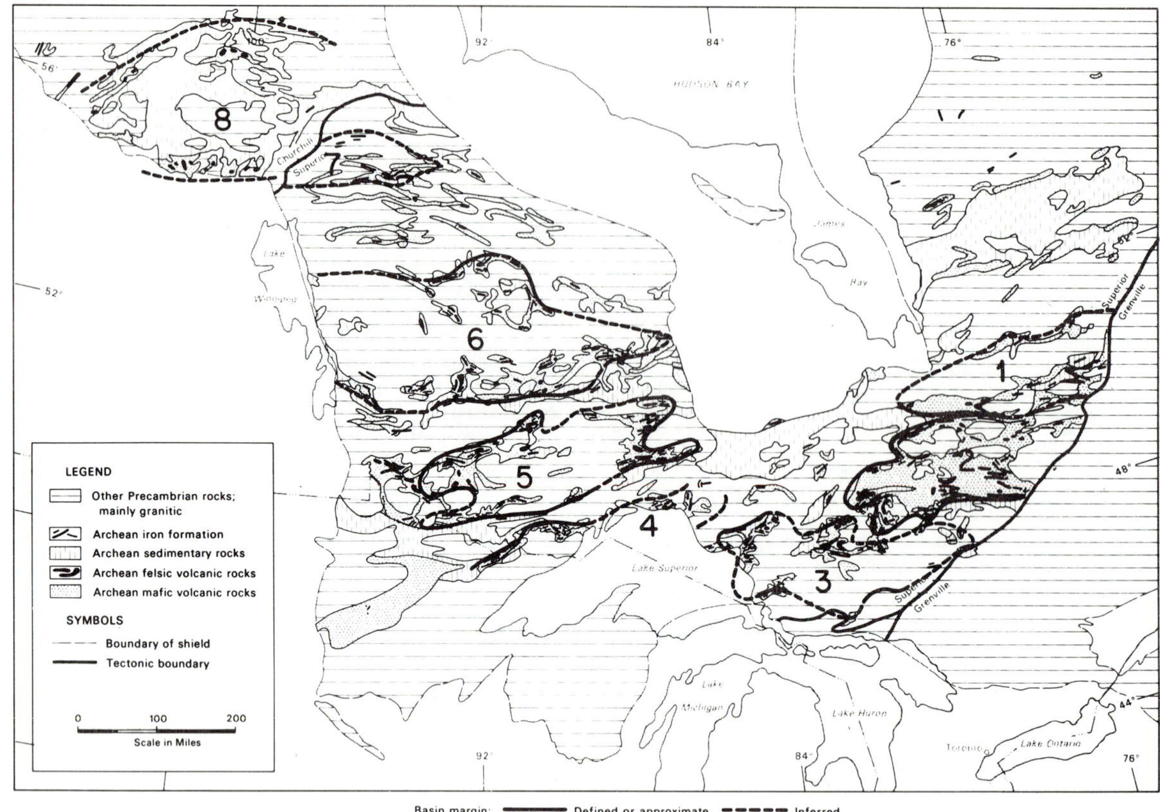

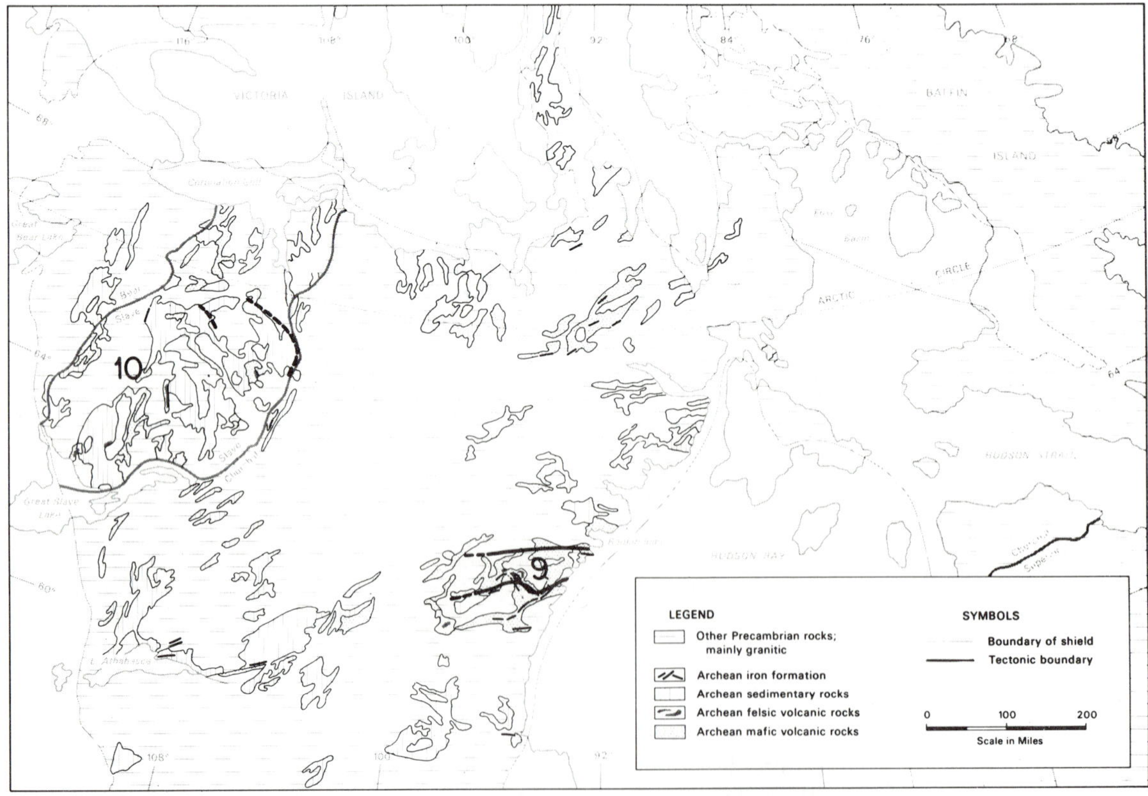

Fig. 6. Archean Basins in the Canadian Shield: Location of ten volcano-tectonic basins in the southern (a) and northern (b) parts of the Canadian Shield. The margins of the basins are generally marked by oxide-facies iron-formation, felsic volcanics, and local conglomerates.

toid clasts of undetermined provenance. Typically these coarse, proximal clastics grade basinward to more distal graywacke-mudstone assemblages. Volcanic exhalation, particularly active in proximity to felsic centers, contributed large quantities of dissolved chemical components to the basins, which were selectively precipitated in the form of oxide, carbonate, and sulfide facies.

Dimensions. All 10 postulated Archean basins are crudely elliptical, by a factor of 3 or 4:1. The present shape is thought to reflect considerable horizontal foreshortening. If this interpretation is correct, the original basins were larger and more nearly circular.

Horizontal compression in all directions is indicated by east-trending isoclinal folds within the "greenstone" segments, with moderately to severely folded hinge lines. Double buckling is the most reasonable mechanism to explain these vertical to subvertical folds.

The present major axes of the basins are from 300 to 400 miles long. The original major axis may have been 450 to 600 miles long. The original major axis may have been 450 to 600 miles long, assuming 50% foreshortening. The present minor axes, 100 to 150 miles long, may in turn have been substantially longer, even approaching 450 to 600 miles long, allowing a possible 4-to-1 foreshortening in a northerly direction.

The general picture that emerges is of a number of closely spaced elliptical to quasi-circular tectonic basins, each 450 to 600 miles in diameter, distributed in what is generally interpreted to have been comparatively thin, tectonically mobile Archean crust.

Interpretation

Two theories of origin are considered for the Archean basins: (1) meteorite "impact scar" theory and (2) volcano-tectonic ("hot spot") theory. The presence in Archean crust of numerous, closely spaced, possibly originally circular depressions requires careful examination of possible impact scarring. Archean basins are of possible appropriate original size and shape to contend as mega-impact scars comparable in order of magnitude to the lunar Maria, such as Mare Imbrium. Also, the presence within Archean volcanic belts of "spinifex-textured," peridotite lava flows, and ostensibly high heat flow, has been interpreted by Green (1972, p. 264) as evidence of a meteorite impact origin for Archean greenstone belts.

However, the "impact scar" theory is considered here as unlikely, if not untenable, for two reasons: (1) U-Pb isotope dating of zircons from rocks in Archean basins number 1 to 6 (Fig. 6) support the interpretation (Goodwin, 1968) that the three major volcanic-rich belts of Superior Province are arranged in a simple, systematic age progression, with the oldest belt situated on the north and the youngest belt situated on the south. Thus the Abitibi-Wawa belt (including basins 1-4) is indicated to be approximately 2,750 my old; the Wabigoon belt (basin 5), 2,800 my old; and the Uchi belt (basin 6), 2,950 my old (Krogh and Davis, 1972). Such an apparently orderly age progression in development of the basins (Fig. 6) is not in accord with indiscriminate meteorite scarring. (2) Ultramafic lava is interpreted on the basis of experimental data as being able to flow subaqueously possibly at a temperature of about 1350°C (Naldrett, 1972, p. 149), a not-unreasonable surface temperature in Archean time. If so, this would remove the only indicated evidence of extreme extraterrestrial energy state ("energy overkill") in development of Archean basins.

The alternative and preferred theory of origin relates Archean basins to a more conventional volcano-tectonic derivation. According to this theory each basin formed in response to normal, terrestrial, volcano-tectonic activities. Thus the basins may have formed in a similar manner to modern inter-arc or back-arc basins of the western Pacific (Karig, 1970; Sugimura and Uyeda, 1973). Accordingly, each basin would represent accumulation of oceanic-type volcanics in the interior and of arc-type volcanics at the margins of the basins.

Archean plates may have behaved in a manner similar to modern plates; in contact with sialic crust at continental margins, Archean oceanic plates were underthrust with consequent development of arc-type volcanics. Volcanic exhalations and volcanic clastics were deposited mainly within the basin as chemical (iron formation) and clastic sediments. Thermal plumes may have been the controlling feature in development of Archean basins: an isolated plume produced an isolated basin; a linear alignment of close-packed plumes produced a linear array of coalescing basins to form a typical "greenstone" belt.

TECTOGENESIS

Evidence has been presented above favoring some type of plate motion, albeit local and constrained, in development of Archean crust. Supporting evidence includes the presence (1) in the Abitibi volcanic belt of the diagnostic igneous series of modern island arcs arranged in appropriate order from north to south, and (2) in Archean crust of the Canadian Shield of a number of elliptical to quasi-circular basins, each approximately 600 miles long, which are of probable volcano-tectonic development resembling modern origins of such basins (e.g., the Sea of Japan). The direction of these studies, then, is toward using modern island arcs as the model for interpreting major aspects of Archean crust. In this regard, full allowance must be made for the fact that the age difference of 2,500 my between these Archean rocks and Cenozoic island arcs exceeds half the total age of the earth.

The Canadian Shield appears to have grown from a number of small primitive crustal units

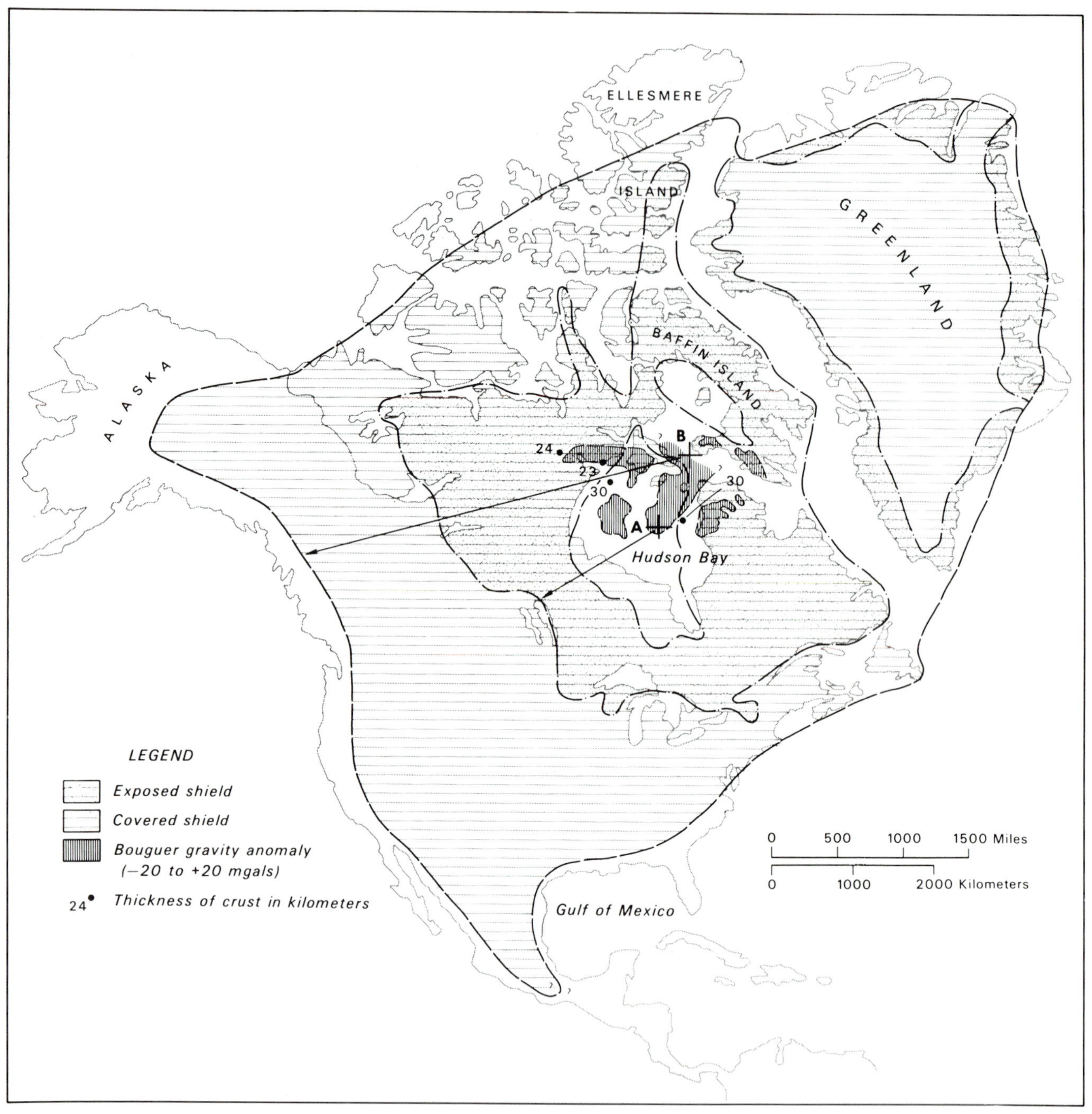

Fig. 7. Hudson Bay and the Canadian Shield: the Laurentian (Canadian-Greenland) Supershield in pre-drift reconstruction, showing the central location of Hudson Bay. Point A is at the geographic center of the exposed Canadian Shield, point B at the center of the reconstructed supershield. Hudson Bay is underlain in part by unusually thin (23-24 km) crust with high-gravity expression.

(protocontinents) by processes of unit enlargement and aggregation, through the development of larger cratons, to form a single continental shield, and growth of this shield occurred about a long-lived crustal anomaly situated at the geographic center of the shield in Hudson Bay, the site of repeated vertical fluctuations of the crust (Fig. 7). Hudson Bay may represent a paleoplume, a key element in growth of the shield.

Such shield growth featured (1) early development of comparatively small unstable protocontinental nuclei about Hudson Bay in Archean time (ca. 4,000-2,500 my); (2) their enlargement and aggregation to form a number of metastable-to-stable cratons, similarly grouped around Hudson Bay in Aphebian (i.e., early Proterozoic) time (ca. 2,5000-1,700 my); (3) attainment of a continuous shield platform, featuring a broad central uplift or plateau over Hudson Bay in Helikian-Hadrynian times (ca. 1,700-600 my); and, finally, (4) collapse of the Hudson Bay region to its present negative crustal status.

On the global scale, pre-Paleozoic crust has been interpreted in terms of six supershields. A major implication is that global "plate tectonics" of Cenozoic type, involving widespread fragmentation and global dispersal of the continental crust, did not operate in Precambrian time, especially in the Archean, to destroy the fundamental symmetry of growing shields.

The pattern of Archean crust illustrated in Figure 2, especially its crescentic, asymmetric pattern (one-sided) is difficult to explain. This particular pattern in this most ancient crust, predating the oldest presently recorded geological age of 3,750 my, represents a fundamental global process.

Recourse may be had to the moon, especially the nature and distribution of the large meteorite impact scars from which the moon itself accreted in the time interval 4,200-3,800 my. Such impact scars are asymmetrically concentrated on the near side of the moon in a crude horseshoe pattern about the highland projection bounded by maria Nectaris, Tranquillitatus, Serenitatis, Imbrium, and Nubium.

Earth may also have experienced even larger major meteorite impacting during the same interval. Precambrian shields may have developed by way of deep mantle plumes formed in direct response to major asymmetric meteorite impacts in the interval 4,200-4,000 my. Such long-lived thermal convection patterns, once initiated, profoundly influenced mantle-crust differentiation.

The sites of major pregeologic impacting, located in the shields, lack direct casual evidence, owing to their very long ensuing crustal histories. Using Hudson Bay of the Canadian Shield as an example, the main stages of thermal response to impact may have been as follows: (1) initiation of a deep mantle plume, or cluster of related plumes, beneath each impact site with gradual development of a cluster of small sialic cratons; (2) progressive aggregation of such sialic seeds to form, consecutively, large shield-type cratons and, eventually, a plateau-like sialic carapace above the plume, including substantial granulite in deeper crustal levels, with consequent peripheral erosion and sedimentation on a continental scale; and, finally (3) with demise of lateral migration of the shield-bearing lithospheric plate off the still-active plume, collapse and fragmentation of the sialic remnant to form a topographic depression or basin, such as the present Hudson Bay. Paradoxically, original impact depressions are now, as a result of these growth stages, at the center of high-standing shields or continental platforms.

Other shield-contained basins or topographic elements that warrant examination in this light include the Siberian and Russian platforms in Laurasia and the Tadoueni, Chad, Congo, Botswana, and possibly Karroo basins in Gondwanaland.

BIBLIOGRAPHY

Borchert, H., 1960, Genesis of marine sedimentary iron ores: Inst. Mining Met. Bull., v. 640 p. 261-279.

Cooke, H. C., James, W. F., and Maudsley, J. B., 1931, Geology and ore deposits of the Rouyn-Harricanaw region, Quebec: Geol. Survey Can. Mem. 166.

Dickinson, W. R., and Hatherton, T., 1967, Andesite volcanism and seismicity around the Pacific: Science, v. 157, p. 801-803.

Dietz, R. S., and Holden, J. C., 1970, Reconstruction of Pangaea: break-up and dispersion of continents, Permian to present: Jour. Geophys. Res., v. 75, p. 4939-4956.

Goodwin, A. M., 1961, Genetic aspects of Michipicoten iron formation: Can. Inst. Mining Met. Trans., v. LXIV, p. 3236.

———, 1962, Structure, stratigraphy and origin of iron formations, Michipicoten area, Algoma District, Ontario, Canada: Geol. Soc. America Bull., v. 73, p. 561-586.

———, 1967, Volcanic studies in the Timmins-Kirkland Lake-Noranda region of Ontario and Quebec: Geol. Survey, Can., Paper 68-1, pt. A, p. 135-137.

———, 1968, Evolution of the Canadian Shield: Presidential address: Geol. Assoc. Can. Proc., v. 19, p. 1-14.

———, 1973a, Archean iron-formations and tectonic basins of the Canadian Shield: Econ. Geology, v. 68, p. 915-933.

———, 1973b, Plate tectonics and evolution of Precambrian crust, *in* Tarling, D. H., and Runcorn, S. K., eds., Implications of continental drift to the earth sciences: Academic Press, p. 1047-1069.

———, (in press a), Precambrian belts, plumes and shield development: Am. Jour. Sci.

———, (in press b), Volcanic relations in the Timmins-Kirkland Lake-Noranda region, Ontario and Quebec: Geol. Survey Can.

Green, D. H., 1972, Archean greenstone belts may include terrestrial equivalents of lunar maria?: Earth and Planetary Sci. Letters, v. 15, p. 263-270.

Gunning, H. C., and Ambrose, J. W., 1940, Malartic area, Quebec: Geol. Survey Can. Mem. 222.

Hutchinson, R. W., Ridler, R. H., and Suffel, G. G., 1971, Metallogenic relationships in the Abiti belt, Canada: a model for Archean metallogeny: Can. Mining Met. Bull., v.64, p. 48-57.

Irving, N. T., and Baragar, W. R. A., 1971, A guide to the chemical classification of the common volcanic rocks: Can. Jour. Earth Sci., v. 8, p. 523-548.

Jakes, P., and White, A. J. R., 1971, Composition of island arcs and continental growth: Earth and Planetary Sci. Letters, v. 12, p. 224-230.

James, H. L., 1954, Sedimentary facies of iron-formation: Econ. Geology, v. 49, p. 235-293.

———, 1966, Chemistry of the iron-rich sedimentary rocks, *in* Fleischer, M., Data of geochemistry, 6th ed.: U.S. Geol Survey Prof. Paper 440-W, p. W1-W60.

Karig, D. E., 1970, Ridges and basins of the Tonga-Kermadic Island arc system: Jour. Geophys. Res., v. 75, p. 239-254.

Krogh, T. E., and Davis, G. L., 1972, Zircon U-Pb ages of Archean metavolcanic rocks in the Canadian Shield: Carnegie Inst. Yearbook 70, Geophys. Lab., preprint.

Kuno, H., 1966, Lateral variation of basalt magma across continental margins and island arcs, *in* Continental margins and island arcs: Geol. Survey Can. Paper 66-15, p. 327-335.

Lawson, A. C., 1885, Report on the geology of the Lake of the Woods region with special references to the Keewatin (Hunonian?) belt of the Archean rocks: Geol. Survey Can. Ann. Rept. (new series), v. 1, Rept. CC.

Naldrett, A. J., 1972, Archean ultramafic rocks: Earth Physics Branch, Ottawa, v. 42, no: 3, p. 141-152.

———, 1973, Nickel sulfide deposits—their classification and genesis with special emphasis on deposits of volcanic association: Bull. Can. Inst. Mining Met., p. 183-201.

Pyke, D. R., Naldrett, A. J., and Eckstrand, O. D., 1973, Archean ultramafic flows in Munro Township, Ontario: Geol. Soc. America Bull., v. 84, p. 955-978.

Ridler, R. H., 1969, The relationship of mineralization to volcanic stratigraphy in the Kirkland Lake area, Ontario [Ph.D. thesis]: Univ. Wisconsin.

Sugimura, A., and Uyeda, S., 1973, Island arcs: Japan and its environs: Develop. Geotectonics, v. 3, 247 p.

Wilson, M. E., 1914, Kewagama Lake map-area, Quebec: Geol. Survey Can. Mem. 39.

———, 1941, Noranda district, Quebec: Geol. Survey Can. Mem. 229.

The Ancient Continental Margin of Eastern North America

Harold Williams and R. K. Stevens

INTRODUCTION

Continental Margins and the Siting of Orogens

It has long been recognized in North America that orogens represent deformed ancient continental margins, although the basic concepts have changed greatly from the early views of Hall, Dana, and Schuchert to later ones of Kay, Drake, and Dietz. This North American view stemmed from the fact that on this continent all the Phanerozoic orogens are at or near the present edge of the continent. The doctrine that geosynclines are open toward the ocean (one-sided) and that continents grow by the outward addition of younger orogenic belts was popular in North America during the 1950s and early 1960s, and it was enhanced by the interpretation of Paleozoic volcanic belts as island-arc complexes (Kay, 1951). The doctrine was further supported by early oceanographic studies at the present Atlantic margin, e.g., Drake et al. (1959), and by the models of Dietz (1963) and Dietz and Holden (1966) based upon actual comparisons between ancient and modern continental margins.

These concepts were not popular outside North America, for in many other parts of the world orogens appear to be intracratonic, e.g., the Alps and the Urals. Even in North America there were growing suspicions in some geological circles that the Appalachian Orogen, North America's type Mountain System, is in fact two-sided, and that this idea refuted Paleozoic accretion (Williams, 1964). How, then, can orogens form at continental margins if they are bounded by older basement and found inside some continental blocks?

Of course, there can be no reconciliation of these concepts without accepting some form of continental drift. Wilson (1966) was most instrumental in solving this dilemma when he utilized the theory of continental drift to postulate a proto-Atlantic Ocean that separated the opposite sides and contrasting early Paleozoic faunal domains of the Appalachian-Caledonian geosyncline. Wilson thus suggested continental drift in the remote geological past to reconcile the ideas of continental accretion with the present siting of some orogens within continental blocks. The emergence of plate tectonics fully endorsed Wilson's views, so that geosynclines are interpreted now as sedimentary troughs and volcanic islands along rifted continental margins, and those geosynclines that presently occur as foldbelts within continental blocks represent sutures where oceans closed and where continents collided. This model not only explains the siting of orogens at continental margins but also provides a mechanism for mountain building through subduction and continental collision.

The conclusion that orogens are a continental-margin phenomenon can also be deduced by comparing some special rock assemblages found in them, such as exceptionally coarse lime-breccias, flysch sequences, calc-alkaline volcanic assemblages, and the ophiolite rock suite, with similar rocks at present continental margins or accreting oceanic ridges. Furthermore, the structural styles developed at young active margins have counterparts in ancient orogens. An understanding of present continental margins is therefore a key to the understanding and delineation of ancient margins. Conversely, an understanding of ancient continental margins, their depositional and structural evolution, and the position and nature of the transition between continental and oceanic crust aids in understanding recent margins. Exactly how, then, is the ancient continental margin of eastern North America identified? How far is it possible to delineate the edge of the continental crust, to restore sedimentary facies during constructional phases of the continental margin, and to interpret its structural development during subsequent destructional stages?

Probably the most characteristic overall feature of any ancient continental margin is that it represents a zone where things change. The change may be the presence of crystalline basement rocks on one side of the zone and their absence on the other, or there may be marked facies changes and a lack of stratigraphic continuity between contrasting rock groups, or the zone may represent a marked change in structural style or metamorphic facies. Compounding the difficulties in locating and interpreting ancient continental margins are local problems of structural and metamorphic style and erosional level, for in some cases where the sedimentary record is obliterated or removed, juxtaposed margins may be represented only by a narrow mylonite zone that separates contrasting basement complexes.

Definition of the Ancient Continental Margin of Eastern North America

All orogens must necessarily embrace at least one continental margin if, as stated, they originate at continental margins. Some orogens that involve continental collision and repeated opening and closing of ocean basins ("Harry Hibbs effect" after the well-

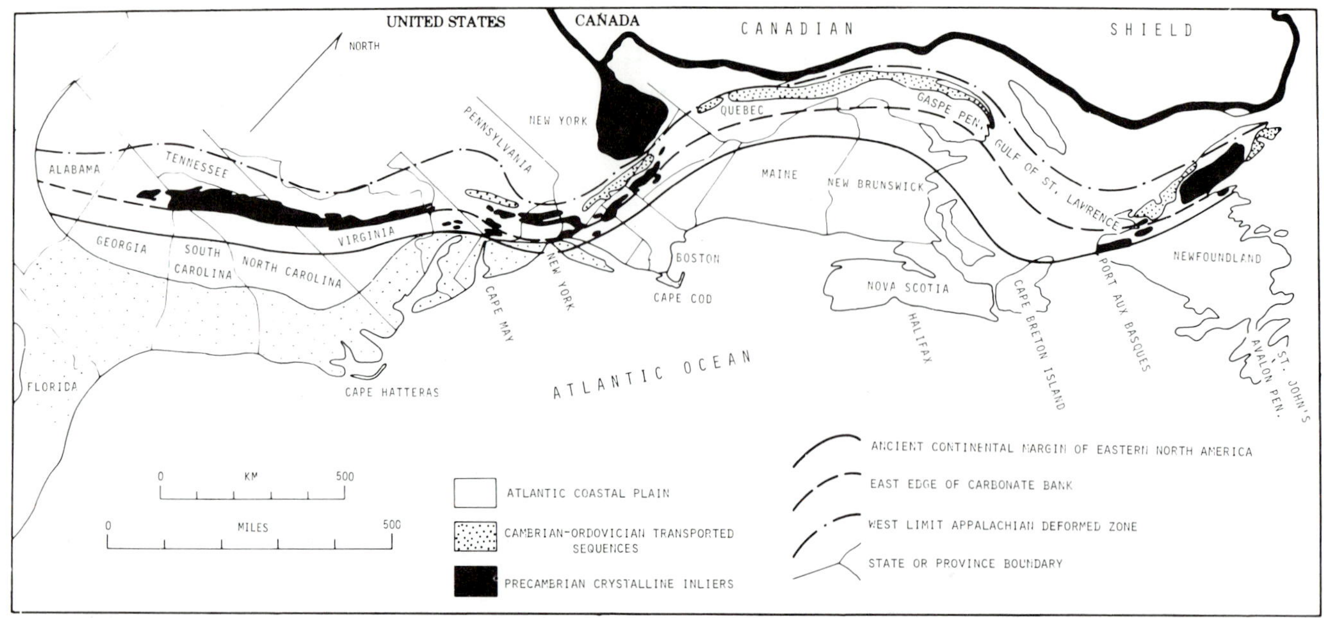

Fig. 1. Salient features of the ancient continental margin of eastern North America.

known Newfoundland accordionist) may embrace two or more continental margins, all structurally telescoped and in different stages of destruction. It is not enough to say that the early Paleozoic continental margin of eastern North America is represented by the Appalachian Orogen. This is but a half-truth, for the Appalachian System contains stratigraphic-tectonic zones in its eastern parts that had no connection with North America during much of their development. Recent views on the development of the Appalachians in the north indicate that at least two margins are represented on opposite sides of the system in Newfoundland, each underlain by a Precambrian crystalline basement and with the opposing margins separated by much younger (Early Ordovician) oceanic crust and mantle. Possibly a third continental margin is represented in Nova Scotia, where the southeasterly-derived Meguma Group forms a zone in a still-more-easterly disposed part of the system.

The ancient continental margin of eastern North America is defined as that belt of deformed rocks along the western side of the Appalachian System that was geologically an integral part of North America in the early Paleozoic. The eastern limit of this zone coincides (or nearly so) with several significant geologic and geophysical changes as follows: (1) it is the eastern limit of Precambrian basement inliers, which in the north are followed eastward by a volcanic terrane that is at least locally built upon the ophiolite suite of rock units; (2) it is a zone of thick clastic sedimentary rocks of late Precambrian to Early Cambrian age with associated mafic dikes and extrusions; (3) it is close to the eastern limit of exposure of the early Paleozoic carbonate sequence that is so well developed along the western border of the Appalachian System; (4) it is a zone of faunal contrasts; (5) it is a zone of faulting, contrasting structural style and contrasting metamorphic facies; (6) the zone is broadly coincident with a regional gravity gradient from positive Bouguer anomalies on the ocean side to negative values on the continent side; and (7) it is marked by contrasting seismic-refraction profiles that reflect deep structural contrasts. The eastern edge of the carbonate bank and the east limit of continental crust are depicted in Figure 1.

The northeastern extension of the same ancient margin is found along the northwest flank of the Caledonides in Ireland and Scotland and along the eastern seaboard of Greenland, and it once continued beyond Spitzbergen. Its southern extension must trend westward along the Ouachita System and continue southwestward to southern Mexico, where it is truncated by the present Pacific ocean.

The ancient margin of eastern North America is sharply truncated by the present Atlantic margin in northeast Newfoundland, but throughout the 2,000-mile length of the Appalachian System and westward through the Ouachitas, the ancient and modern margins of eastern North America are essentially parallel. The same parallelism of ancient and modern continental margins is evident in the Caledonides of east Greenland and Scandinavia. Furthermore, the early Paleozoic margin of eastern North America parallels the older Grenville Structural Province. The pattern suggests that the opening of oceans and the establishment of successively younger continental margins is at least partly controlled by zones of weakness within existing orogens.

Purpose and Scope

The purpose of this paper is to summarize the most salient geologic features of the western part of the Appalachian System and to interpret their significance and origin in terms of the development of an ancient continental margin. Many of these features have been summarized and interpreted in this way already, especially by Rodgers (1968, 1972), Stevens (1970), Bird and Dewey (1970), Ross and Ingham (1970), Strong and Williams (1972), Hatcher (1972), Burke and Dewey (1973), Williams et al. (1972, 1974), and several others. The present paper is intended as a general overall appraisal. It is based mainly upon the authors' field experience in the Canadian Appalachians and a review of the literature, but it includes as well first-hand data collected over the past 10 field seasons in western Newfoundland. Attempts are made to extend some of the conclusions drawn in the north to southern parts of the Appalachian System of which the authors have no direct knowledge. These extrapolations are offered only as debatable suggestions, in the hope that they will stimulate further work and direct attention toward remaining problems.

CONSTRUCTIONAL FEATURES OF THE ANCIENT CONTINENTAL MARGIN

Some features of the ancient continental margin of eastern North America are recognized all the way along the western side of the Appalachian System, some others are represented only in a few widely separated areas, and still others are confined to certain unique areas. In general, those tectonic, sedimentologic, or igneous features that are associated with continental rifting and the construction of the continental margin are most widely recognized. These are (1) the east limit of crystalline Precambrian inliers that represent easternmost exposures of North American basement; (2) a thick prism of clastic sediments that intervenes between the crystalline basement and an overlying, Cambrian and later, mainly carbonate succession; (3) mafic extrusions and dike intrusions that are related in space and time to the clastic sequences; and (4) an east-thickening carbonate sequence that overlies the clastic sequences and disappears eastward where rocks of equivalent age are lime-breccias, and shales with interbedded lime-breccia units.

Eastern Limit of Crystalline Precambrian Inliers

Precambrian crystalline rocks characterized by isotopic dates of about 800-1,000 my appear as basement inliers along the western margin of the Appalachian Orogen. The crystalline rocks clearly formed as part of the Grenville Structural Province of the nearby Canadian Shield, and in the Appalachians the largest massifs are exposed in uplifted horsts or anticlines; e.g., the Long Range and Indian Head Range of Newfoundland, the Green Mountains of Vermont, the Berkshire, Housatonic, New Milford, and Hudson Highlands massifs from Massachusetts to eastern Pennsylvania, and the Blue Ridge from Maryland through Virginia to North Carolina (Fig. 1). Most of these Precambrian massifs are faulted along their western boundaries and some have moved westward along east-dipping thrusts, e.g., the Blue Ridge, Berkshire, and Hudson massifs. Grenvillian rocks also occupy the cores of gneiss domes east of the anticlines or transported massifs where they are deformed and retrograded so that in places the basement is difficult to distinguish from the cover sequences. Examples are known in the Burlington Peninsula of Newfoundland, in at least some of the gneiss domes in western New England, in the Baltimore gneiss domes in Maryland, and probably in domes still farther southwest.

These inliers mark the eastern limit of identifiable Grenvillian basement. In Newfoundland, the Burlington domes are bordered to the east by ophiolites of the Baie Verte Lineament and the Betts Cove Complex of Notre Dame Bay. In this northern area, then, the eastern edge of continental basement can be delineated within relatively narrow limits.

The presence of transported ophiolites in Quebec and the occurrence of ophiolite detritus in the easterly-derived Normanskill of New York and Martinsburg of Pennsylvania indicate that oceanic crust and mantle lay to the east of these southern areas as well.

Farther south in the Appalachian Piedmont, the eastern edge of the ancient North American continent cannot be delineated with any certainty. However, this former continental margin probably lies to the west of the Carolina Slate Belt, because the late Precambrian rocks and a local occurrence of fossiliferous Cambrian rocks in that belt correlate best with the Avalon Zone that lay on the eastern side of the proto-Atlantic in the Northern Appalachians (Williams et al., 1972, 1974). If there is no suture between the Blue Ridge and Carolina Slate Belt, the opening of the system in the north is but a local phenomenon (Chidester and Cady, 1972). Yet the western margin of the Appalachian System is similar throughout, which implies that, if it evolved as a continental margin bordered by an ocean in the north, it was probably bordered by an ocean in the south as well.

Clastic Sequences at the Ancient Continental Margin

Thick clastic sequences that postdate the metamorphism and intrusion of the billion-year-old Grenville rocks overlie the crystalline inliers along the western margin of the Appalachians with profound unconformity. The oldest of these clastic cover rocks increase in age from Upper Cambrian in the west to late Precambrian in the east, for in the east, upper parts of the thick clastic successions are locally dated as Lower Cambrian. The clastic rocks now exhibit a pronounced metamorphic gradient across strike from little deformed and metamorphosed in the west to multideformed rocks of greenschist or amphibolite metamorphic facies in the east.

The rocks are of variable thickness both across and along strike. The thickest successions occur in the easternmost and now more metamorphosed parts of the sequences, e.g., Fleur de Lys Supergroup (30,000 ft) in Newfoundland, Rosaire and Caldwell Groups of Quebec, Mendon Group (3,000 ft) and Pinnacle Formation of Vermont, Glenarm Series of Maryland, Lynchburg Formation (10,000 ft) of Virginia, and Ocoee Group (25,000 ft) in the Great Smoky Mountains of east Tennessee and northern Georgia. Other parts of the clastic sequences are found in allochthonous terranes of western Newfoundland (Maiden Point and Summerside Formations), Quebec (Sillery Group), and New York (Nassau Formation).

The clastic rocks are mostly quartzites in the upper parts of the sequences or in westerly exposures where the sequences are thin, e.g., Bradore (500 ft) and Bateau Formations of Newfoundland, Potsdam Formation of New York, Cheshire Quartzite (1,000 ft) of Vermont, and Chilhowee Group of Virginia. These are well sorted with well-rounded grains and thin, even beds. Crossbeds are common and the sediments represent shallow-water deposits in areas of generally low relief.

Where thick, the clastic sequences are monotonous and poorly sorted with conglomerate, graywacke, siltstone, shale, feldspathic sandstone, and arkose predominating. Graded beds are common, but directional current features are poorly known. Blue quartz like that in the crystalline Grenvillian inliers is abundant, and granite and metamorphic rock fragments are recorded in coarser beds in most places. Many of these rocks are interpreted as typical turbidites, deposited in deep water.

The increasing maturity of the late Precambrian sequences from bottom to top and from east to west implies rugged terrane with sediment accumulation in deep water at the onset of deposition followed by gradual erosion and infilling of the depositional troughs, concluding with the deposition of Lower Cambrian mature beach sands.

The surface onto which the Cambrian seas transgressed cannot everywhere be described as a peneplain, for there is considerable local relief (Ambrose, 1964). Since an early Paleozoic marine transgression is a worldwide phenomenon and most likely the result of a prolonged eustatic rise in sea level (Russell, 1968; Pitman and Hayes, 1973), it is difficult to separate the local tectonic and regional eustatic influences on sedimentation during this early development of the ancient continental margin.

A lack of directional features in the lower clastic units precludes accurate paleo-basin analyses, but where such studies have been made in the higher quartzitic parts of the sequences, currents are from the west, locally with additional components along the trend of the depositional troughs, e.g., Chilhowee Group of Virginia and Tennessee (Brown, 1970), Weverton Formation in Maryland (Whitaker, 1955).

The distribution, thickness variation, provenance, and directional features all indicate that the clastic sedimentary sequences were deposited in a continuous, although irregular trough that can be interpreted as a rifted continental margin. The sediments were derived from the Grenvillian basement and transported eastward toward the ocean, probably with local redistribution by bottom contour currents and longshore currents. The rough coincidence of the regional belt of clastic rocks with easternmost exposures of Precambrian basement (Fig. 1) strongly suggests this model, and locally where the clastics are followed eastward by well-defined ophiolite sequences, a continental margin is demanded. Comparisons between the sedimentary records at the modern and restored ancient margins of eastern North America (Fig. 2) further support this interpretation.

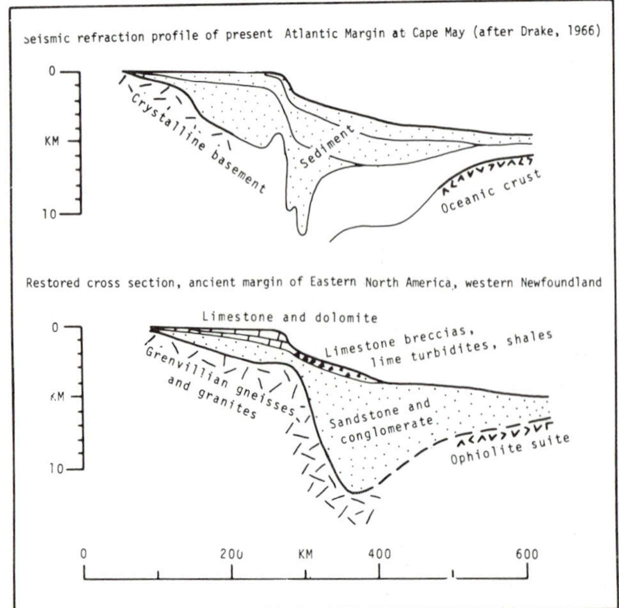

Fig. 2. Comparison between the sedimentary records at the modern and restored ancient margins of eastern North America.

Mafic Volcanism and Dike Intrusion at the Ancient Continental Margin

Volcanic rocks and mafic dikes are intimately associated with late Precambrian to Early Cambrian clastic sequences along the western border of the Appalachians. In the extreme northeast, mafic flows of the Lighthouse Cove Formation (500 ft) are fed by a ramifying network of mafic dikes that cut the Grenvillian basement in Newfoundland. A bimodal volcanic assemblage occurs in the Blue Ridge of Virginia, where the much thicker Catoctin Formation (2,000-12,000 ft) overlies Precambrian basement or is interlayered with clastic sedimentary rocks. Probable correlatives of these volcanics are the mafic pyroclastics and flows of the transported Maiden Point Formation in Newfoundland, the Tibbit Hill volcanics of Quebec, the volcanic rocks of the Bull Formation in the transported Taconic sequence of New York, and the Mount Rogers volcanics of southwest Virginia.

The volcanic rocks are chiefly mafic flows in Newfoundland with well-developed columnar structures and local amygdules. The Catoctin Formation includes rhyolite flows and ignimbrites. All these rocks are altered in the greenschist facies in the southern examples. A lack of pillow structures suggests that most are terrestrial, particularly in western exposures. Others are marine, as they include pillow lavas interlayered with deep-water graywackes, e.g., Maiden Point Formation.

Locally in Newfoundland, the mafic flows are separated from Grenvillian basement in eastern exposures by a thick quartzite-conglomerate unit (Bateau Formation). The Bateau is cut by vertical mafic dikes that fed subhorizontal flows, and variable attitudes of the quartzite-conglomerate beds compared with consistently vertical attitudes of mafic dikes indicate tilting of the sediments in fault blocks either before or during mafic dike intrusion (Bostock, 1973).

The mafic volcanic rocks of the Lighthouse Cove and Catoctin Formations are tholeiitic basalts of similar petrochemical characteristics, and all are chemically similar to coeval mafic dikes. Mafic volcanic rocks of the Maiden Point Formation in northern Newfoundland, although lithologically distinct, are chemically similar to the Lighthouse Cove Formation (Smyth, 1973).

Strong and Williams (1972) interpreted the Lighthouse Cove Formation, and by analogy the Catoctin Formation, as plateau volcanics formed in an environment of continental rifting, distension, and separation. This interpretation is based mainly on the tholeiitic nature of the volcanic rocks, which resemble tholeiites so widely distributed at continental margins along which substantial crustal separation has taken place; e.g., Karroo basalts of East Africa, Deccan traps of India, Ferrar dolerites of Antarctica, Serra Geral lavas of Brazil, and Tertiary basalts of west Greenland. In contrast, alkaline volcanic rocks characterize rifted zones within continental blocks that have not undergone extensive separation, e.g., east African rifts, Benue trough, and the Scottish Midland Valley.

This interpretation of the volcanic rocks fits well with that of the clastic sedimentary rocks with which the volcanics are interlayered. The age of the volcanics further indicates the time of continental rifting.

Zircons from the Catoctin Formation and related rocks yield an age of 820 my (Rankin et al., 1969). The Newfoundland examples have yielded a wide range of potassium/argon ages, but recent determinations of 805 my for dikes that are possibly related to the Lighthouse Cove Formation (Pringle et al., 1971) suggest correlation with the Catoctin Formation.

These ages are embarrassingly old for volcanic rocks that form basal parts of continuous stratigraphic sections that include fossiliferous Cambrian beds. Furthermore, they imply the existence of a continental margin shortly following the completion of the latest orogenic events in the Grenville Structural Province. They therefore present the problem of establishing a continental terrace wedge in the Early Cambrian, 200 million years after the first indications of rifting.

Recent $^{40}Ar/^{39}Ar$ age determinations of 600 my for the Lighthouse Cove Formation, and the recognition of excess argon in all specimens studied, imply that in Newfoundland the rifting was initiated probably in latest Precambrian time (Peter Reynolds and Vidas Stukas, personal communication, 1974). Similarly, $^{40}Ar/^{39}Ar$ studies of basement gneisses in the Central Blue Ridge indicate a long cooling history before the deposition of unconformable cover sequences, so that a more favorable age for the Catoctin Formation is 650-700 my (R. D. Dallmeyer, personal communication, 1974).

A supporting age for the time of initial rifting and continental separation is provided by carbonatites and lamprophyres in widely separated areas along the St. Lawrence River and Labrador coast that are isotopically dated at 565 my (Doig, 1970). These North American examples have been correlated with alkaline intrusions of similar age in coastal Greenland and Scandinavia, and all have been interpreted as the result of rifting during initiation of the proto-Atlantic (Doig, 1970). The St. Lawrence occurrences mark the failed arm of a triple rift junction (Burke and Dewey, 1973), and the persistent alkaline igneous activity in this zone (Currie, 1970) may relate to a subsequent abortive attempt toward separation during the opening of the present Atlantic Ocean.

Geophysical Changes Across the Ancient Continental Margin

The deformed and metamorphosed parts of the Appalachians are underlain in most places by a thick, dense, presumably mafic crust, whereas the western, less-deformed zone is underlain by continental crust comparable to that of the nearby Canadian Shield (Ewing et al., 1966; Dainty et al., 1966; Sheridan and Drake, 1968). The change between the two types of crust roughly corresponds with a pronounced gravity gradient (King, 1964) on which local gravity highs are superposed (Innes and Argun-Weston, 1967). The gravity highs are interpreted as mafic or ultramafic intrusions within the continental crust (Diment, 1953; Fitzpatrick, 1953). These hypothetical intrusions are presumably deep seated and associated with mafic volcanism controlled by initial rifting. The presence of mafic intrusions and ramifying dike swarms at or near the ancient continental edge might explain the difficulty in the seismic delineation of the boundary between continental and oceanic crust at present continental margins.

Carbonate Bank at the Ancient Continental Margin

A thick sequence of carbonate rocks that ranges in age from Early Cambrian to Middle Ordovician occurs in most places along the western side of the

Appalachians from Newfoundland to Alabama. Its lithology, thickness, and relationships have been well documented by Rodgers (1968), and only its main features are summarized here. The rocks are best exposed in the Valley and Ridge Province of the south, in the Hudson and Champlain valleys in central parts of the system, and in western Newfoundland in the north. Their absence or nonexposure in the Quebec City-Gaspe segment of the system may be partly because of erosion, although the carbonate sequence may be hidden by overlying transported rocks south of the St. Lawrence River. The sequence thins rapidly to the west of the deformed Appalachian belt and thickens eastward from 3,000 to 10,000 ft across its exposures in the deformed zone.

The carbonate sequence rests conformably above Cambrian clastic rocks, and its Middle Ordovician top grades upward and eastward into black graptolitic shales overlain in turn by westward transgressive clastic wedges. In most places an erosional disconformity is recorded at the Lower Ordovician-Middle Ordovician contact, but the Ordovician-Cambrian boundary is not marked by a lithological or structural break.

Algal mounds, desiccation cracks, and local erosional disconformities, combined with crossbedding in interlayered mature sands, all indicate that the carbonate rocks accumulated in shallow water. Cloud and Barnes (1948) and Rodgers (1968) suggested that the carbonates formed on a bank not unlike the present Great Bahama Bank. Cambrian shallow-water carbonates pass westward into shales, sandstones, and conglomerates, and within the bank there seems to have been a central zone of dolomite flanked by limestones at the inner and outer bank margins (Palmer, 1971). The thickness of the carbonates across the western Appalachians indicates that the continental margin subsided much more rapidly in the east, although shallow-water conditions were also maintained there.

Rodgers (1968) delineated the eastern edge of the carbonate bank and interpreted it as an abrupt declivity, like the margin of the present Bahama Bank. This explains the marked facies change and the lack of interlayering between eastern exposures of the carbonate sequence and clastic sediments or volcanic rocks farther east in the Appalachians. Possibly the bank edge outcrops as an abrupt lithological boundary in northwestern Vermont and in Lancaster County, Pennsylvania (Rodgers, 1968), but in most places the eastern exposures of the carbonates are in a zone of metamorphism and thrusting so that original relationships are obscured.

Transported sequences in the western Appalachians that originated east of the bank but that now structurally overlie the carbonates exhibit Middle Cambrain to Lower Ordovician shales with limestone-breccia units that most likely built up offshore just beyond the bank edge. The spectacularly coarse limestone-breccias of the Cow Head Group in western Newfoundland form part of a thin (1,000 ft) condensed sequence that spans the same interval as the carbonate sequence. Thinner, finer, and shalier units of the same age within the transported Humber Arm Supergroup (Cooks Brook Formation) are interpreted as sediments formed in deeper water and away from the bank edge. Similar lime-breccias and shales are known in transported sequences in Quebec, in the Taconic sequence of Vermont and eastern New York, and in the Hamburg klippe of Pennsylvania.

Rodgers (1968) pointed out that the eastern edge of the carbonate bank closely follows the locus of Precambrian inliers along the western margin of the Appalachians, so the edge of the carbonate bank may itself have been localized by the original eastern extent of Grenvillian basement. Rodgers also suggested that the drop-off from bank to deep water approximates the ancient edge of the North American continent. However, the bank is slightly west of the easternmost exposures of Grenvillian basement in gneiss domes, and the clastic sequences previously described extend well east of the bank edge, where they are still underlain in most places by continental crust. We therefore delineate the continental edge as somewhat east of the bank edge (Fig. 1).

Most graphic attempts to reconstruct a continental fit before the opening of the present Atlantic Ocean assume that the continental margins coincide with the drop-off from shallow to deep water, or the 3,000-ft bathymetric contour. There is no assurance, however, that the present continental margins exactly match this morphological feature, and comparison with the ancient margin of eastern North America suggests that continental crust extends seaward beyond the continental slope. A lack of coincidence between the continental slope and the limit of continental crust like that in the ancient example may explain apparent gaps and overlaps in present continental reconstructions, e.g., Bullard et al. (1965).

Faunal Changes at the Ancient Continental Margin

The most prominent faunal change at the Lower Paleozoic continental margin is the change from shelly faunas of the carbonate bank to offshore graptolitic and related faunas of the shale, turbidite, and volcanic sequences. During Cambrian time, two faunal realms were concentric around North America, the cratonic realm and the extracratonic-intermediate realm (Lochman-Balk and Wilson, 1958). These are correlated with a restricted shelf sea and a shelf margin-open ocean environment, respectively (Palmer, 1971).

More subtle changes involve provincialism across the bank itself. The Whiterock genera of brachiopods and trilobites, referred to the Toquima-Table Head Faunal Realm, are allied environmentally to the continental shelf or slope so that they were interpreted to mark a transitional bank-edge facies (Ross and Ingham, 1970). The Table Head Formation of Newfoundland, however, overlies Lower Ordovician shallow-water carbonates with erosional disconformity across the former bank, and conglomerates in its upper part are overlain by graptolite-bearing shales and graywackes. This formation, therefore, more likely represents a facies that migrated westward across a subsiding bank during early stages of Taconic orogeny. The Middle Ordovician Toquima-Table Head Faunal Realm in Newfoundland seems to represent the tectonically-disturbed migrating bank edge rather than the transitional offshore stable bank edge that existed from Cambrian to Early Ordovician. In fact, the stable bank-edge facies is difficult to locate, although it appears to have been the main source for fossiliferous blocks in bank-foot breccias such as the Cow Head (Stevens, 1970). Possibly it is represented in Lancaster County, Pennsylvania, and in northwestern Vermont.

Epstein et al. (1972) suggested that limestone slide blocks in the Hamburg klippe of Pennsylvania, which contain a North Atlantic province conodont fauna, originated at a northwest-facing bank that was far removed from the ancient margin of eastern North America. However, the Cow Head Group and the Table Head Formation of Newfoundland both contain a distinctive North Atlantic conodont assemblage (L. Fahraeus, personal communication 1973), which, like the Whiterock genera, seems to be an environmentally controlled continental-margin assemblage. This removes the basis for suggesting that contrasting faunas in Pennsylvania imply distant transport.

DESTRUCTION OF THE ANCIENT CONTINENTAL MARGIN: TACONIC OROGENY

The events that led to the destruction of the continental margin are those that are attributed to Taconic orogeny (Rodgers, 1971). The earliest movements are recorded in the east and at the edge of the continental margin. Regional metamorphism that accompanied ophiolite obduction predated the emplacement of allochthonous sequences upon the carbonate bank. Certain of the allochthonous masses have early west-facing recumbent folds that relate to the displacement of the sequences. Flysch wedges progressed across the carbonate bank as forerunners of the allochthons and, locally, autochthonous rocks in upper parts of the stratigraphic sequences display recumbent folds that resulted from kneading beneath the structurally overriding slices. Following klippe emplacement the rocks were further telescoped by westward thrusting accompanied by penetrative deformation, and this event locally involved the crystalline Precambrian basement.

Clastic Wedges

The first intimation of Taconic orogeny and destruction of the continental margin is the pre-Middle Ordovician unconformity and the appearance of black shales and then siltstones and graywackes above the carbonates. The pre-Middle Ordovician unconformity reflects uplift and warping of the shelf that resulted in a regional break in sedimentation along the entire western side of the Appalachian System. In Newfoundland, the carbonate bank probably rose as much as 300 ft locally with the development of a karst topography (Collins and Smith, 1972). In New England, the uplift was irregular and apparently reflects the formation of a horst-and-graben topography. Residual iron deposits locally occur at the top of the carbonate succession there, and erosion has cut down to the Precambrian basement in some horsts. In nearby grabens continuous sections are preserved (Zen, 1968). Following fragmentation, the bank subsided rapidly so that deeper water facies migrated landward in response to its collapse. The amount of subsidence is difficult to estimate, but where transported rocks above the carbonate bank in Newfoundland are unconformably overlain by Middle Ordovician shallow-water carbonates, the subsidence may have been as much as the total structural thickness of the allochthon. This implies that the former bank sank to oceanic depths.

Contemporaneous with the breakup and collapse of the bank, flysch wedges were built out from tectonic lands in the east. These are first recorded in the Lower Ordovician parts of transported sequences in western Newfoundland and transgress westward across the Cow Head breccias. Farther west they are of early Middle Ordovician age where they overlie the carbonate bank sequence. Elsewhere in the western Appalachians there are significant time differences in the first appearance of flysch. In Quebec the Cloridorme Formation is mainly of late Middle Ordovician age, and in New York and Pennsylvania the Normanskill and Martinsburg clastics are of comparable age (Trenton).

The flysch wedges vary greatly in thickness and may exceed 10,000 ft in Pennsylvania. Thicknesses decrease westward, where the graywackes grade out into shalier beds. Paleocurrent analyses show that the sediments were derived from sources that lay generally to the east, although longitudinal transport directions are common (McBride, 1962; Stevens, 1970; Enos, 1969). The provenance seems to be much the same for all the flysch wedges: sedimentary rocks of the continental margin, Precambrian basement, mafic volcanic rocks, and possibly Paleozoic intrusive rocks. Chromite grains and serpentine fragments are locally common, suggesting that ophiolites had already been obducted onto the continental margin and had been uplifted above sea level. A few occurrences of shallow-water limestone clasts even in the earliest flysch suggest that the tectonic lands that supplied detritus were ringed by local carbonate banks. In Newfoundland, the flysch becomes coarser upward, probably reflecting the gradual encroachment of the Humber Arm allochthon. Similarly, in New York, progressively larger pieces of the Taconic sequence slid into the sedi-

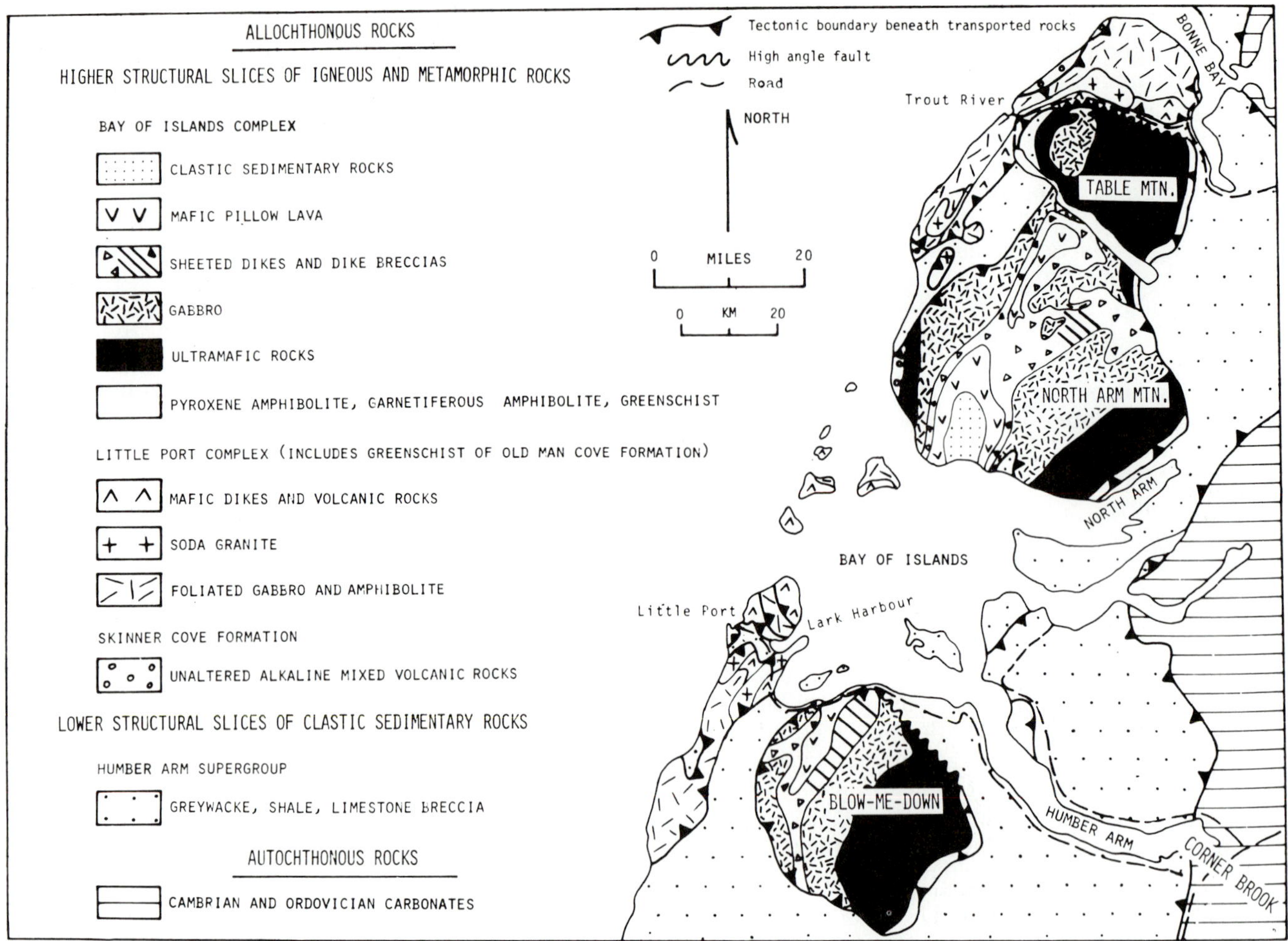

Fig. 3. Geology of part of the Humber Arm Allochthon, western Newfoundland.

mentary basin during Trenton time, thus heralding the impending emplacement of the Taconic allochthon (Zen, 1968).

Emplacement of Allochthons

Transported Cambrian-Ordovician clastic sequences that comprise the lowest structural slices above the autochthonous flysch wedges and carbonate bank represent large undigested masses embedded in the flysch wedges, e.g., parts of the Hare Bay and Humber Arm allochthons (Fig. 3) of western Newfoundland, parts of the Taconic allochthon of Vermont and east New York, the Hamburg klippe of Pennsylvania, and a large area of transported rocks between Quebec City and Gaspé Peninsula that is probably more extensive than all the others combined (Fig. 1). The transported clastic sequences, such as the Humber Arm sequence of western Newfoundland and the Taconic sequence of New York, have been interpreted as continental margin deposits; lower clastic units that contain detritus from the crystalline basement were derived from the west, overlying condensed shale and lime-breccia units reflect the development of the carbonate bank, and upper clastic units signal uplift and instability toward the east that eventually resulted in the transport of the sequences.

Ophiolite suites that are interpreted as oceanic crust and mantle overlie the transported clastic sequences. The clearest examples are found in Hare Bay and Bay of Islands in western Newfoundland and the Thetford Mines serpentinite belt of Quebec. Other examples may be represented at Mount Albert of Gaspé Peninsula and possibly the Baltimore Gabbro in Maryland (Crowley, 1969; Ina Alterman, personal communication, 1972). It is not yet clear whether or not these transported ophiolites originated in a series of small marginal ocean basins (Dewey and Bird, 1971; Kennedy, 1973) or else represent the crust of a major ocean comparable in size to the present Atlantic (Church and Stevens, 1971). The continuity of the margin and similarities throughout its length seem to favor a uniform major ocean rather than a chain of small ocean basins with expectable regional irregularities.

In the Bay of Islands area of western Newfoundland the ophiolite slice is locally underlain by foliated gabbros, amphibolites, mafic dikes, and volcanic rocks (Little Port slice assemblage, Williams, 1973) that in turn are underlain by a distinctive unaltered alkaline volcanic suite (Skinner Cove slice assemblage, Strong, in press) (Fig. 4). Toward the north at Hare Bay, mafic pillow lavas (Cape Onion Slice), undated polymictic conglomerates and sandy limestones (Grandois slice), and a mega-mélange, which has huge recycled volcanic and rodingitized amphibolite and gabbro blocks now in a black shale matrix (Milan Arm Mélange), all intervene between lower structural slices of clastic sediments and the overlying ophiolite slice. These assemblages and their structural stacking orders serve to indicate the variety of rocks that must have lain at or near the continental margin before Ordovician transport.

Among the lower structural slices of clastic sedimentary rocks the lowest slices contain the youngest parts of the stratigraphic sections. In New York and Vermont, the two lowest slices are underlain by wildflysch-type conglomerate whose matrix is marine black shale and whose fragments are dominantly rocks of the Taconic sequence. This suggests that the lower slices were emplaced in a marine environment, probably as a moving archipelago, and that the final emplacement was a near-surface gravity-sliding phenomenon. Fossils in the matrix of the wildflysch-type conglomerate date the emplacement of these slices to the *Orthograptus truncatus* var. *intermedius* zone (Zen, 1967). In Newfoundland, the lower structural slices are unconformably overlain by the Middle Ordovician neoautochthonous Long Point Formation (Rodgers, 1965), which grades downward into autochthonous flysch at the western leading edge of the allochthon. It is older than upper parts of the Taconic sequence in New York, indicating that the Newfoundland clastic slices were emplaced earlier than their New York analogues.

The youngest transported rocks in the Taconic allochthon of New York (Pawlet Formation) are equivalent to the youngest rocks of the underlying autochthonous flysch. In the allochthon, the Pawlet rests unconformably on rocks of various ages. This unconformity is interpreted to reflect the beginning of the movements that led to gravity sliding, and conceivably the Pawlet Formation was deposited partly on a moving substrate and partly after the allochthon had come to rest (Zen, 1968). Soft-rock deformation accompanied the emplacement of these lowest structural slices.

Recumbent folding and penetrative deformation characterize only the upper parts of the autochthonous sequences, where they were locally overridden by more rigid higher structural slices, e.g., Canada Bay, Newfoundland.

All the transported rock groups of western Newfoundland, regardless of lithology or structural position in the stacking order, are underlain by thin zones of shaly mélange with sedimentary, volcanic, and plutonic exotic blocks. The Newfoundland mélanges are in most respects similar to the wildflysch-type conglomerates of New York, and they are interpreted as formed during the later stages of transport when the structural slices (including the ophiolite) moved across a shaly sedimentary terrane by gravity sliding. The mélanges are thought to represent the combined effects of surficial mass wastage (Brückner, 1966) and tectonic mixing (Stevens, 1965) at the soles of the advancing slices. The local occurrence of large serpentinite, gabbro, diorite, and volcanic blocks within even the lower mélanges that separate sedimentary rock slices clearly attests to the proximity of the ophiolite slice during formation of the mélanges. This implies that in Newfoundland the lowermost mélange zones are the surfaces of latest movement and that the transported rocks were emplaced along them as an already assembled allochthon. The ophiolite slice, now the structurally highest, is interpreted as the first to have moved.

New fossil discoveries in western Newfoundland indicate that the volcanic rocks in at least two of the higher structural slices (Skinner Cover and Cape Onion) are Lower Ordovician, and therefore equivalent to parts of the transported clastic sequences in lower structural slices (Williams, 1971; A. Berger, personal communication, 1973). Determination of the order of structural stacking combined with facies considerations indicate that the structurally highest slices are the farthest traveled; i.e., the on-land oceanic crust-mantle rocks in the highest slice lay most easterly at the time of their formation. If the occurrence of ultramafic debris in the flysch within sedimentary slices of the Humber Arm allochthon can be taken as evidence that the Bay of Islands ophiolite slice was moving during flysch deposition, then the assembly and transport of the slices progressed from east to west, and it was a relatively slow process that extended over about five graptolite zones. The geology of part of the Humber Arm

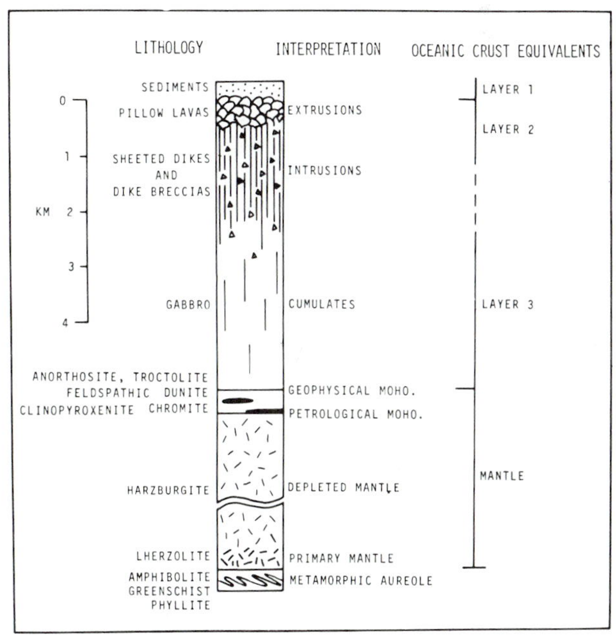

Fig. 4. Interpretation of the Bay of Islands Complex as oceanic crust and mantle.

allochthon is summarized in Figure 3, and a restored section of the Bay of Islands ophiolite complex and its interpretation as oceanic crust and mantle is depicted in Figure 4.

Displacement of Allochthons

In western Newfoundland the ophiolites are underlain by thin (1,000 ft) metamorphic aureoles that are interpreted as the result of initial uprooting and earliest transport. The aureoles parallel the stratigraphic base of the ophiolite complexes and the rocks exhibit decreasing metamorphic grade and decreasing intensity in subhorizontal schistosities downward from the stratigraphic base of the ultramafic rocks. Williams and Smyth (1973) suggested that the aureole rocks represent volcanic rocks and graywackes like those of the Maiden Point Formation and, as that formation originally lay at the edge of the continent, the aureoles must result from the obduction of hot ophiolite across the continental margin. The contacts between the aureoles and ultramafic rocks, which represent the level of earliest transport of the ophiolites, are now steep zones in the Bay of Islands area that are truncated by the present subhorizontal structural bases of the ophiolite slices. The absence of metamorphism and penetrative fabrics (like those in the aureoles) in sedimentary rocks of lower structural slices supports the interpretation that the aureoles formed during the earliest and most intense deformation that occurred farthest east, and that this event predated gravity sliding. This conclusion is also supported by the occurrence of schistose aureole blocks in shaly mélange beneath some of the ophiolite slices. Initial obduction can be regarded as the subduction of the continental margin beneath oceanic lithosphere.

The direction of sheeted dikes within the ophiolite complexes should roughly parallel the direction of the oceanic ridge at which the ophiolites originated. This direction is expectedly northeast, as the Appalachian System and ancient continental margin trend in that general direction. The best examples of sheeted dikes in the Bay of Islands Complex (Williams and Malpas, 1972), however, trend northwest and suggest that either the ophiolites were rotated during transport or else they represent oceanic crust that was striped normal to the direction of the continental margin. Studies are presently in progress to ascertain the direction of ophiolite transport as indicated by the facing direction of early recumbent folds in the aureole rocks with respect to both the direction of sheeted dikes in the same structural slices and the direction of the ancient continental margin.

In most respects the tectonic setting of the transported ophiolites in western Newfoundland is similar to that of the Semail ophiolite nappe in the Oman Mountains of the Persian Gulf (Glennie et al., 1973). In both areas, the ophiolite sequences form the highest structural slices and they were transported soon after their formation, during the Ordovician in the northern Appalachians and during the Cretaceous in the Oman.

Regional metamorphism and polyphase deformation, which affected eastern parts of the late Precambrian clastic sequences that were marginal to the continent, e.g., Fleur de Lys Supergroup in Newfoundland, are either related to earliest ophiolite obduction or else to the closing of a small ocean basin at the continental margin (Kennedy, 1973). The transported Maiden Point Formation, which is at least partly equivalent to the Fleur de Lys Supergroup, displays west-facing recumbent folds that predate its Middle Ordovician emplacement.

Postemplacement Deformation

Thrusting and the local development of penetrative cleavage are later events that affected the already emplaced allochthons. At Becraft Mountain in New York, Silurian rocks unconformably overlie deformed Taconic sequence rocks that were uncleaved at the time of their arrival, as inferred from the fact that shale blocks in wildflysch-type conglomerates were uncleaved at arrival. Similarly, along strike in western Newfoundland, autochthonous Cambrian and Ordovician rocks were deformed penetratively before the deposition of Silurian beds at White Bay (Lock, 1969). Locally at Canada Bay in Newfoundland, prograde greenschist metamorphism was developed in autochthonous rocks (Sugarloaf Schists) and thrust faulting and cataclasis affected basement rocks that occur in thrust slices among the Paleozoic successions (Smyth, 1973).

The close of Taconic orogeny marked an end to the major developmental stage of the ancient continental margin, for the margin was now transformed from a rifted zone of active deposition to a relatively stable deformed zone. Parts of the deformed margin in the Northern Appalachians were affected only slightly by later orogenic events, although Acadian (Devonian) deformation is recognized in most northern localities. The proto-Atlantic was virtually closed in Newfoundland during the Middle Ordovician, as Late Ordovician and Silurian rocks there record a change from deep-water marine to terrestrial conditions (Williams, 1967). Farther south in New Brunswick and New England, Silurian land masses and shelf areas bordered the northeast-trending Fredericton Trough (McKerrow and Ziegler, 1971), which possibly represented a much-contracted proto-Atlantic Ocean.

Acadian orogeny represents shortening and lateral compression that probably resulted from continued closing and tightening of the already contracted proto-Atlantic Ocean. Where intense along the western margin of the Appalachians, it may be the result of the restoration of gravitational stability after the obduction of dense ophiolite sheets in a way similar to that demonstrated by Ramberg (1967) in centrifuge experiments.

Paleomagnetic evidence suggests that the proto-

Atlantic closed during the Taconic orogeny in the Northern Appalachians, but that it was not finally destroyed farther south until Permo-Carboniferous time (Alleghanian deformation), when Africa collided with North America (Smith et al., 1973). This does not imply that earlier orogenic episodes did not affect this southern area.

VOLCANIC ISLAND COMPLEXES AND SUBDUCTION EAST OF THE ANCIENT MARGIN

Early to Middle Ordovician volcanic rocks abound to the east of the ancient continental margin. Examples are known all the way from northeastern Newfoundland southward through central New Brunswick and northern Maine to Vermont and probably Connecticut. Many of these volcanic sequences, such as the Snooks Arm and Lushs Bight Groups of northeast Newfoundland, are much too thick to correspond to oceanic crustal layer 2, although some of them are clearly built upon an ophiolite suite. At Long Island of north-central Newfoundland a sequence of predominantly volcanic rocks in excess of 15,000 ft shows a lithological evolution from lowermost mafic dikes, gabbros, and pillow lavas upward through deep-water cherts and turbidites into pyroclastic rocks and volcaniclastic sedimentary rocks capped by limestone and subaerial tuffs. The overall deep to shallow-water lithic change is accompanied by geochemical changes in the volcanic rocks from low-potassium tholeiites at the base to calc-alkaline low-silica andesites toward the top that show progressive enrichment in Al_2O_3 and K_2O and a decrease in CaO and MgO (Kean and Strong, in press). Other nearby volcanic sequences of comparable age, e.g., Mortons Harbour Group (Strong and Payne, 1973) and Roberts Arm Group (Strong, 1973a), show similar lithological and geochemical features, and all are interpreted as ancient analogues of modern island arcs.

Calc-alkaline volcanic accumulations decrease in thickness from east to west across the successive ophiolite belts in northeast Newfoundland, from thicknesses of about 20,000 ft in Notre Dame Bay to only a few thousand feet at the Baie Verte Lineament to a virtual absence of calc-alkaline volcanic products atop the wesward-transported Bay of Islands Complex. The age of the volcanic rocks in northeastern Newfoundland indicates that the island arcs were growing east of the continental margin during emplacement of the west Newfoundland allochthons and furthermore that the evolution of the volcanic islands ceased at the time of final emplacement of the allochthons.

This decrease in calc-alkaline volcanic activity toward the ancient continental margin of eastern North America, coupled with the age of the volcanic rocks and the time of ophiolite obduction, suggests that any subduction zone that existed at the ancient continental margin during this period dipped eastward, at least in Newfoundland. Although contrary to most recent models (e.g., Dewey, 1969; Bird and Dewey, 1970; Dewey and Bird, 1971; Hatcher, 1972; Kennedy, 1973), eastward subduction during the Early and Middle Ordovician would explain the general lack of Ordovician volcanism and intrusion west of the ancient margin and at the same time provide a plausible mechanism for ophiolite obduction, for the structural stacking order of all the transported rocks in western Newfoundland, and for their mode of final emplacement as an already assembled allochthon. Other features that appear to indicate eastward subduction are as follows: (1) a gradual K_2O increase in granitic rocks from central to eastern Newfoundland (Strong et al., 1974); (2) the zonation of mineral deposits throughout the Appalachian System (Strong, 1973b); (3) the petrochemistry of early Paleozoic volcanic rocks southward across Appalachian correlatives in the British Caledonides, which show variation from oceanic to island arc and continental types (Fitton and Hughes, 1970); and (4) seismic-refraction profiles conducted along marine passages through the Appalachians in the north, which show that interfaces between levels of contrasting velocity are consistently inclined southeastward down to the 12-km limit of penetration (Sheridan and Drake, 1968). It is a sobering observation that even such fundamental aspects of the ancient margin as polarity and exact position of a subduction zone are difficult to determine in the Appalachians.

OTHER ANCIENT CONTINENTAL MARGINS AND THE DEFINITION OF THE APPALACHIAN GEOSYNCLINE

At least two other geologically distinct areas, interpreted as ancient continental margins, are recognized in the northern Appalachians of Newfoundland and Nova Scotia (Fig. 5).

East of the volcanic-ophiolite terrane of central Newfoundland an undated crystalline basement complex is overlain by an areally extensive multideformed clastic sequence that displays a structural style and metamorphic facies similar to that of the late Precambrian-Cambrian clastic rocks along the ancient eastern margin of North America to the west. Second-phase isoclines in the metamorphosed clastic sequence face southeast toward the opposing foreland, and this deformation predated nearby Middle Ordovician graywackes that contain fragments of the metamorphic rocks (Kennedy and McGonigal, 1972). The western boundary of this zone is locally marked by shaly mélange that includes volcanic blocks and large blocks of predeformed metamorphic rocks. The western boundary is also marked by a belt of discontinuous mafic-ultramafic complexes and volcanic rocks along the Gander River that either represent dismembered ophiolite or diapiric intrusions (Kean, 1974). Possi-

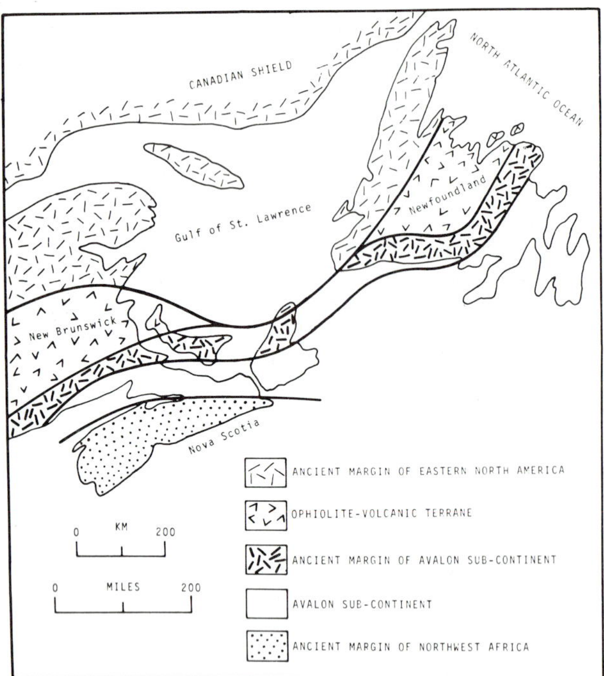

Fig. 5. Ancient continental margins in the northern Appalachians.

ble correlatives of the east Newfoundland basement complex are the George River and Green Head Groups of Nova Scotia and New Brunswick, respectively, the Fitchburg pluton of southern New Hampshire and Massachusetts (E-an Zen, personal communication, 1974), and isotopically dated Grenvillian age rocks of coastal Maine (D. B. Stewart, personal communication, 1972).

Another geologically distinct zone, in a still-more-easterly position, forms the mainland of Nova Scotia and consists of an extremely thick (13 km) and extensive belt of graywacke and slate (Meguma Group) that is exposed along the present Atlantic seaboard. The Meguma Group contains Lower Ordovician graptolites at its top, and provenance studies indicate that it was derived from the southeast, where no discernible source terrane now exists. Schenk (1970) interpreted the Meguma as a continental margin deposit that lay to the west of the African continent at deposition, so that it has arrived in its present position by ocean closing in the Paleozoic and then remained as part of North America during late Mesozoic and Cenozoic opening of the present Atlantic.

Williams (1964) suggested that the Avalon Platform (Kay and Colbert, 1965), which lies east of the central Newfoundland volcanic-ophiolite terrane and which is bounded westward by a zone that itself has the characteristics of a continental margin (Zone G or Gander Zone of Williams et al., 1972, 1974), marked the eastern margin of the Appalachian geosyncline. Rodgers (1972) concluded that the Avalon Zone was neither platformal nor was it a delimiting foreland and that the true eastern margin of the Appalachian geosyncline is the Saharan Shield of north Africa. This new definition of the Appalachian geosyncline implies that the terrane represented by the Avalon Zone of Newfoundland and its equivalents either evolved within the Appalachian geosyncline (Rodgers, 1968, 1972) or represents a separate intervening continent that in the early Paleozoic lay between the ancient continental margin of eastern North America to the west and the Saharan Shield toward the east (Schenk, 1971).

The Cambrian history of the Avalon Zone indicates that it was indeed platformal, if only short lived, and the existence of basement rocks along its western margin in Newfoundland, coupled with accumulating evidence that the whole of the Avalon Zone is underlain by continental basement (Papezik, 1973; Poole, 1973), supports its interpretation as a separate continental block. Possibly it was attached to North Africa during the late Precambrian, but during the early Paleozoic it must have lain to the west of the African continental margin as inferred from the age, position, and mode of origin of the Meguma Group. Certainly the Avalon Zone is part of a much more extensive terrane that can be traced to eastern Massachusetts and probably continues to Florida. Furthermore, its 500-mile width in Newfoundland, as indicated by offshore extensions to Virgin Rocks, Eastern Shoal, and Flemish Cap, makes the Avalon Zone at least twice as wide as the classical cross section of the entire southern Appalachians.

The Appalachian geosyncline and its continental margins may therefore need reappraisal. There was clearly one deep, long depositional trough underlain by oceanic crust and mantle between the ancient eastern margin of North America and the western margin of the Avalon Zone, and probably another during the early Paleozoic farther east, between the Avalon Zone and the Saharan Shield. The situation during the early Paleozoic is summarized in Figure 6, and it is merely a matter of opinion and scale as to how many continental margins or ocean basins one wishes to include in the Appalachian geosyncline. Like so many other concepts in geology, the term "Appalachian geosyncline" came into being at a time when views on this subject were quite different from those held today. The term "Appalachian Orogen" still applies to all the deformed rocks in the Appalachian Mountain System, but many of these rocks are far-traveled, so the sister term, "Appalachian Geosyncline," now has limited intrinsic worth.

CRYPTIC SUTURES

The recognition of the ancient continental margin of eastern North America is based mainly upon an analysis of the stratigraphic record. Where erosion has removed this record or where it is obliterated by intense metamorphism, the recognition of the former margin is much more difficult and more subjective. This condition exists locally in the northern Appalachians of southwest Newfoundland and it may exist regionally in the Piedmont and in other parts of the system.

Two gneissic complexes are juxtaposed in southwestern Newfoundland, whereas toward the north they form the eastern and western margins of the central Newfoundland volcanic-ophiolite terrane. The eastern gneissic belt in the south (Port aux Basques Complex) abuts the western gneissic belt (Cape Ray Complex) at the Cape Ray Fault. The junction is marked by a mylonite zone up to 1 km wide that effectively cuts out the central Newfoundland volcanic-ophiolite terrane (Fig. 5). Accordingly, the Cape Ray Fault has been interpreted as a cryptic suture that juxtaposes two ancient continental margins (Brown, 1973). The Aspy Fault of Cape Breton Island is probably a southwest continuation of the Cape Ray Fault with the same tectonic significance (Wiebe, 1972).

A similar suture may exist within the metamorphic terrane of the Piedmont in North Carolina, possibly between the Inner Piedmont and the Charlotte Belt. The Carolina Slate Belt nearby was deformed and metamorphosed in late Precambrian or Early Cambrian time and possibly sutured to the North American continent at a later time (Glover and Sinha, 1973). Suturing in the Piedmont is also favored by the lack of volcanic products among Valley-and-Ridge Province rocks that are coeval with extensive volcanics in nearby easterly belts (Lynn Glover, personal communication 1973).

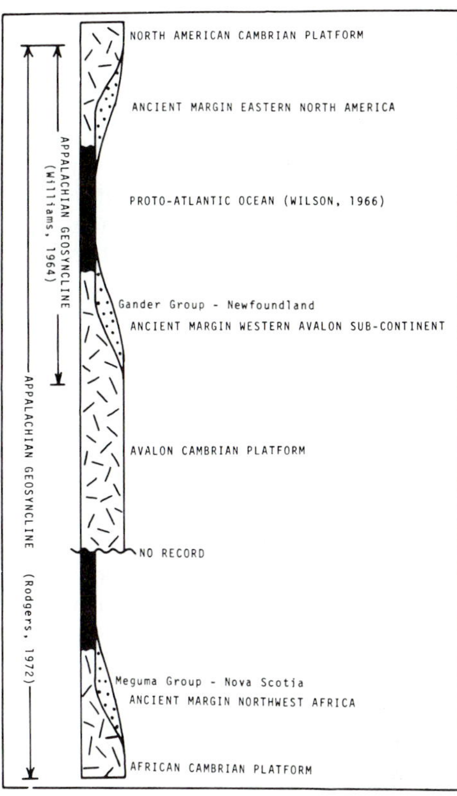

Fig. 6. Tectonic elements of the northern Appalachians and their bearing on the definition of the Appalachian geosyncline.

CONCLUDING REMARKS

Although the idea of an ancient continental margin in eastern North America has been rooted in the geologic literature for more than 100 years, there were few serious attempts to sharply define and locate this margin during its different stages of evolution. The stratigraphically distinct zones that formed the protolith of the Appalachian Orogen were not generally related to the transition between continent and ocean basin; instead they were all viewed as part of one nondescript geosyncline with a myriad of component parts. Even now, much of the terminology persists and is still used in cases where more specific information might be given. It is one thing today to say that a rock group is miogeosynclinal, but quite another to say that is represents a continental terrace wedge. Similarly, it is not sufficient to say that volcanic terranes are eugeosynclinal if it can be shown that one particular volcanic assemblage forms the top of an ophiolite suite, whereas another is calc-alkaline and shows vertical and lateral lithic variation and chemical characteristics that we have come to expect in modern island arcs.

Students of regional tectonics should pay special attention to zones that mark major changes in geological setting. Failure to emphasize such zones in the past has greatly influenced our thinking on such fundamental geologic concepts as the siting of orogens and the permanency of continents and ocean basins. The task before us is to properly delineate ancient continental margins, ocean basins, island arcs, and subduction zones. This is essentially an exercise in identifying ancient analogues of modern phenomena. Even more challenging is the identification of cryptic sutures and the recognition of ancient continental margins and ocean basins where now only the proverbial few fuchsite crystals separate contrasting metamorphic terranes.

SUMMARY

The ancient continental margin of eastern North America is represented by a zone of partially deformed and metamorphosed rocks along the western flank of the Appalachian Orogen. The margin was initiated during late Precambrian as a rift system with accompanying sedimentation and volcanism along its length. A sequence of shallow-water mature sediments, predominantly carbonates, accumulated on the ancient continental shelf during Cambrian and Early Ordovician. The stable margin existed for at least 200 million years until its breakup toward the end of the Early Ordovician. The mechanism of breakup is not clear but it resulted in the formation of a series of offshore island arcs, the obduction of oceanic lithosphere and mantle across the continental margin onto the shelf, the deformation and metamorphism of sediments bordering the continent, the mass transfer of continental-slope sequences westward across the shelf, and a flood of clastic sediments from the continental margin that transgressed toward the continental interior. Taconic orogeny records the destruction of the ancient continental margin and almost complete closing of a proto-Atlantic Ocean in the northern Appalachians. Farther south, the continental margin was strongly modified, but total destruction of the proto-Atlantic did not occur until Permo-Carboniferous time, when Africa collided with North America.

ACKNOWLEDGMENTS

The data presented here were collected by a large number of geologists, only a few of whom are cited in the bibliography or referred to in the text. For more complete references on Appalachian geology the reader is referred to the bibliographies in the following reviews, collections, and syntheses: Neale and Williams (1967), Kay (1969), Poole et al. (1970), Zen et al. (1968), Fisher et al. (1970), Rodgers (1970), Lessing et al. (1972), Zen (1972), and Williams et al. (1972, 1974).

The authors extend special thanks to W. D. Brückner, C. A. Burk, C. L. Drake, E. R. W. Neale, W. H. Poole, John Rodgers, and E-an Zen for critically reading the manuscript, to the Geology 6000 class of 1973 at Memorial University for valuable discussion, and to the National Research Council and Geological Survey of Canada for continued support of our work in Newfoundland.

BIBLIOGRAPHY

Ambrose, J. W., 1964, Exhumed Paleoplains of the Precambrian Shield of North America: Am. Jour. Sci., v. 262, p. 817-857.

Bird, J. M., and Dewey, J. F., 1970, Lithosphere plate-continental margin tectonics and the evolution of the Appalachian orogen: Geol. Soc. America Bull., v. 81, p. 1031-1060.

Bostock, H. H., 1973, Belle Isle, Newfoundland (2 M/4 west): Geol. Survey Can., Paper 73-1, pt. A, p. 2-4.

Brown, P. A., 1973, Possible cryptic suture in southwest Newfoundland: Nature Phys. Sci., v. 245, p. 9-10.

Brown, W. R., 1970, Investigations of the sedimentary record in the Piedmont and Blue Ridge of Virginia, *in* Fisher, G. W., Pettijohn, F. J., Reed, J. C., Jr., and Weaver, K. N., eds., Studies in Appalachian geology: central and southern: New York, Wiley-Interscience, p. 335-349.

Brückner, W. D., 1966, Stratigraphy and Structure of west central Newfoundland, *in* Poole, W. H., ed., Guidebook, geology of parts of Atlantic provinces: Ann. Meeting, Geol. Assoc. Can., Min. Assoc. Can. p. 137-151.

Bullard, E. C., Everett, J. E., and Smith, A. G., 1965, The fit of the continents around the Atlantic, *in* Blackett, P. M. S., Bullard, E. C., and Runcorn, S. K., eds., A symposium on continental drift: Phil. Trans. Roy. Soc. London, ser. A, v. 258, p. 41-51.

Burke, K. and Dewey, J. F., 1973, Plume-generated triple junctions: key indicators in applying Plate Tectonics to old rocks: Jour. Geology, v. 81, p. 406-433.

Chidester, A. H., and Cady, W. M., 1972, Origin and emplacement of alpine-type ultramafic rocks: Nature Phys. Sci., v. 240, p. 27-31.

Church, W. R., and Stevens, R. K., 1971, Early Paleozoic ophiolite complexes of the Newfoundland Appalachians as mantle-oceanic crust sequences. Jour. Geophys. Res., v. 76, p. 1460-1466.

Cloud, P. E., Jr., and Barnes, V. E., 1948, The Ellenburger Group of central Texas: Univ. Texas (Bur. Econ. Geol.) Publ. 4621, 473 p.

Collins, J. A., and Smith, L., 1972, Sphalerite as related to the Tectonic Movements, deposition, diagenesis and karstification of a carbonate platform: Internatl. Geol. Congr., 24th Session, Sect. 6, Stratigraphy and sedimentology, p. 208-215.

Crowley, W. P., 1969, Stratigraphic evidence for a volcanic origin of part of the Bel Air belt of Baltimore Gabbro Complex in Baltimore County, Maryland (abstr.): Geol. Soc. America, pt. 1, p. 10.

Currie, K. L., 1970, An hypothesis on the origin of alkaline rocks suggested by the tectonic setting of the Monteregian Hills: Can. Mineralogist, v. 10, p. 411-420.

Dainty, A. M., Keen, C. E., Keen, M. J., and Blanchard, J. E., 1966, Review of geophysical evidence on crust and upper mantle structure on the Eastern Seaboard of Canada: Am. Geophys. Union Monogr. 10, p. 349-369.

Dewey, J. F., 1969, The evolution of the Appalachian/ Caledonian orogen: Nature, v. 222, p. 124-129.

———, and Bird, J. M., 1971, Origin and emplacement of the ophiolite suite: Appalachian ophiolites in Newfoundland: Jour. Geophys. Res., v. 76, p. 3179-3206.

Dietz, R. S., 1963, Collapsing continental rises: an actualistic concept of geosynclines and mountain building: Jour. Geology, v. 71, p. 314-333.

———, and Holden, J. C., 1966, Miogeoclines (Miogeosynclines) in space and time: Jour. Geology, v. 74, p. 566-583.

Diment, W. H., 1953, A regional gravity survey in Vermont, western Massachusetts and eastern New York [Ph.D. thesis]: Harvard Univ., 176 p.

Doig, R., 1970, An alkaline rock province linking Europe and North America: Can. Jour. Earth Sci., v. 7, p. 22-28.

Drake, C. L., Ewing, M., and Sutton, G. H., 1959, Continental margins and geosynclines: The East Coast of North America north of Cape Hatteras, *in* Ahrens, L. H., Press, F., Rankama, K., and Runcorn, S. K., eds., Physics and chemistry of the earth, v. 3: Pergamon Press, Elmsford, N.Y., p. 110-198.

Enos, P., 1969, Cloridorme Formation, Middle Ordovician flysch, northern Gaspé Peninsula, Quebec: Geol. Soc. America, Spec. Paper 117, 66 p.

Epstein, J. B., Epstein, A. G., and Bergström, S. M., 1972, Significance of Lower Ordovician exotic blocks in the Hamburg klippe, eastern Pennsylvania, *in* Geol. Survey Res. 1972, U.S. Geol. Survey Prof. Paper 800-D, p. D29-D36.

Ewing, G. N., Dainty, A. M., Blanchard, J. E., and Keen, M. J., 1966, Seismic studies on the eastern seaboard of Canada: the Appalachian System: Can. Jour. Earth Sci., v. 3, p. 89-109.

Fisher, G. W., Pettijohn, F. J., Reed, J. C., Jr., and Weaver,

K. N., eds., 1970, Studies in Appalachian geology: central and southern: New York, Wiley-Interscience, 460 p.

Fitton, J. G., and Hughes, D. J., 1970, Volcanism and plate tectonics in the British Ordovician: Earth and Planetary Sci. Letters, v. 8, p. 223-228.

Fitzpatrick, M. M., 1953, Gravity in the eastern townships of Quebec [Ph.D. thesis]: Harvard Univ., 133 p.

Glennie, K. W., Bouef, M. G. A., Hughes-Clark, M. W., Moody-Stuart, M., Pilaar, W. F. H., and Reinhardt, B. M., 1973, Late Cretaceous nappes in Oman Mountains and their geologic evolution: Am. Assoc. Petr. Geol. Bull., v. 57, p. 5-27.

Glover, Lynn, III, and Sinha, A. K., 1973, The Virgilina deformation, a late Precambrian to Early Cambrian (?) orogenic event in the central Piedmont of Virginia and North Carolina: Am. Jour. Sci., Cooper vol., v. 273-A, p. 234-251.

Hatcher, R. D., Jr., 1972, Developmental Model for the southern Appalachians. Geol. Soc. America Bull., v. 83, p. 2735-2760.

Innes, M. J. S., and Argun-Weston, A., 1967, Gravity measurements in Appalachia and their structural implications, *in* Clark, T. H., Appalachian Tectonics: Roy. Soc. Can. Spec. Publ. 10, p. 69-83.

Kay, M., 1951, North American geosynclines: Geol. Soc. America Mem. 48, 143 p.

______, ed., 1969, North Atlantic—geology and continental drift: Am. Assoc. Petr. Geol. Mem. 12, 1082 p.

______, and Colbert, E. H., 1965, Stratigraphy and life history: New York, Wiley, 736 p.

Kean, B. F., 1974, Notes on the geology of the Great Bend and Pipestone Pond ultramafic bodies, *in* Report of activities, 1973: Newfoundland, Dept. Mines and Energy, p. 33-42.

Kennedy, M. J., 1973, Pre-Ordovician polyphase structure in the Burlington Peninsula of the Newfoundland Appalachians: Nature Phys. Sci., v. 241, p. 114-116.

______, and McGonigal, M. H., 1972, The Gander Lake and Davidsville Groups of northeastern Newfoundland: new data and geotectonic implications: Can. Jour. Earth Sci., v. 9, p. 452-459.

King, P. B., 1964, Further thoughts on tectonic framework of southeastern United States: V.P.I. Dept. Geol. Sci. Mem. 1, p. 5-31.

Lessing, P., Hayhurst, R. I., Barlow, J. A., and Woodfork, L. D., eds., 1972, Appalachian structures; origin, evolution, and possible potential for new exploration frontiers: West Virginia Geol. and Econ. Survey, Morgantown, W.V., 322 p.

Lochman-Balk, C., and Wilson, J. L., 1958, Cambrian biostratigraphy in North America: Jour. Paleontol., v. 32, p. 312-350.

Lock, B. E., 1969, Silurian rocks of West White Bay area, Newfoundland, *in* Kay, M., ed., North Atlantic—geology and continental drift: Am. Assoc. Petr. Geol. Mem. 12, p. 433-442.

McBride, E. F., 1962, Flysch and associated beds of the Martinsburg Formation (Ordovician), Central Appalachians: Jour. Sediment. Petr., v. 32, p. 39-91.

McKerrow, W. S., and Ziegler, A. M., 1971, The Lower Silurian Paleogeography of New Brunswick and adjacent areas: Jour. Geol., v. 79, p. 635-646.

Neale, E. R. W., and Williams, H., eds., 1967, Geology of the Atlantic region—the Lilly memorial volume: Geol. Assoc. Can. Spec. Paper 4, 292 p.

Palmer, A. R., 1971, The Cambrian of the Appalachian and Eastern New England regions, eastern United States, *in* Holland, C. M., ed., Cambrian of the New World, New York, Wiley-Interscience, p. 169-217.

Papezik, V. S., 1973, Detrital garnet and muscovite in Late Precambrian sandstone near St. John's, Newfoundland, and their significance: Can. Jour. Earth Sci., v. 10, p. 430-432.

Pitman, W. C., III, and Hayes, J. D., 1973, Upper Cretaceous spreading rates and the Great Transgression (abstr.):Geol. Soc. America Ann. Meeting, v. 5, no. 7.

Poole, W. H., 1973, Detrital garnet and muscovite in late Precambrian sandstone near St. John's, Newfoundland, and their significance: discussion: Can. Jour. Earth Sci., v. 10, p. 1697-1698.

______, Sandford, B. V., Williams, H., and Kelly, D. G., 1970, Geology of southeastern Canada, *in* Douglas, R. J. W., ed., Geology and economic minerals of Canada: Geol. Survey Can. Econ. Geol. Rept. 1, p. 227-304.

Pringle, I. R., Miller, J. A., and Warrell, D. M., 1971, Radiometric age determinations from the Long Range Mountains, Newfoundland: Can. Jour. Earth Sci., v. 8, p. 1325-1330.

Ramberg, H., 1967, Gravity, deformation and the earth's crust: New York, Academic Press, 214 p.

Rankin, D. W., Stern, T. W., Reed, J. C., Jr., and Newell, M. F., 1969, Zircon ages of felsic volcanic rocks in the Upper Precambrian of the Blue Ridge, Appalachian Mountains: Science, v. 166, p. 741-744.

Rodgers, J., 1965, Long Point and Clam Bank Formations, western Newfoundland: Geol. Assoc. Can. Proc., v. 16, p. 83-94.

______, 1968, The eastern edge of the North American Continent during the Cambrian and Early Ordovician, *in* Zen, E-an, White, W. S., Hadley, J. B., and Thompson, J. B., Jr., eds., Studies of Appalachian geology: northern and maritime: Wiley-Interscience, New York, p. 141-149.

______, 1970, The tectonics of the Appalachians: New York, Wiley-Interscience, 271 p.

______, 1971, The Taconic Orogeny: Geol. Soc. Amer. Bull., v. 82, p. 1141-1177.

______, 1972, Latest Precambrian (post-Grenville) rocks of the Appalachian region: Am. Jour. Sci., v. 272, p. 507-520.

Ross, R. J., Jr. and Ingham, J. K., 1970, Distribution of the Toquima-Table Head (Middle Ordovician Whiterock) Faunal realm in the Northern Hemisphere: Geol. Soc. America Bull, v. 81, p. 393-408.

Russell, K. L., 1968, Oceanic ridges and Eustatic changes in sea level: Nature, v. 218, p. 861-862.

Schenk, P. E., 1970, Regional variation of the flysch-like Meguma Group (Lower Paleozoic) of Nova Scotia, compared to recent sedimentation off the Scotian shelf, *in* Lajoie, J., ed., Flysch sedimentology in North America: Geol. Assoc. Can. Spec. Paper 7, p. 127-153.

______, 1971, Southeastern Atlantic Canada, northwestern Africa, and continental drift: Cn. Jour. Earth Sci., v. 8, p. 1218-1251.

Sheridan, R. E., and Drake, C. L., 1968, Seaward extension of the Canadian Appalachians: Can. Jour. Earth Sci., v. 5, p. 337-373.

Smith, G. A., Briden, J. C., and Drewry, G. E., 1973, Phanerozoic world maps, *in* Hughes, N. F., ed., Organisms and continents through time: Palaeontol. Assoc. London Spec. Paper 12, p. 1-42.

Smyth, W. R., 1973, The stratigraphy and structure of the southern part of the Hare Bay Allochthon [Ph.D. thesis]: Memorial Univ. Newfoundland, 172 p.

Stevens, R. K., 1965, Geology of the Humber Arm, west

Newfoundland [M.Sc. thesis]: Memorial Univ. Newfoundland.

———, 1970, Cambro-Ordovician flysch sedimentation and tectonics in west Newfoundland and their possible bearing on a Proto-Atlantic Ocean, *in* Lajoie, J., ed., Flysch sedimentology in North America: Geol. Assoc. Can. Spec. Paper 7, p. 165-177.

Strong, D. F., 1973a, Lushs Bight and Roberts Arm groups of central Newfoundland: possible juxtaposed oceanic and island-arc volcanic suites: Geol. Soc. America Bull., v. 84, p. 3917-3928.

———, 1973b, Plate tectonic setting of Appalachian-Caledonian mineral deposits as indicated by Newfoundland examples: A.I.M.E. Soc. Mining Engr. Preprint 73-1-320, 31 p.

———, (in press), An "off-axis" alkali volcanic suite associated with the Bay of Islands ophiolite, Newfoundland: Earth and Planetary Sci. Letters.

———, and Payne, J. G., 1973, Early Paleozoic volcanism and metamorphism of the Mortons Harbour-Twillingate Area, Newfoundland: Can. Jour. Earth Sci., v. 10, p. 1363-1379.

———, and Williams, H. 1972, Early Paleozoic flood basalts of northwestern Newfoundland: their petrology and tectonic significance: Proc. Geol. Assoc. Can., v. 24, p. 43-54.

———, Dickson, W. L., O'Driscoll, C. F., Kean, B. F., and Stevens, R. K., 1974, Geochemical evidence for an east-dipping Appalachian subduction zone in Newfoundland: Nature (in press).

Whitaker, J. C., 1955, Geology of Catoctin Mountain, Maryland and Virginia: Geol. Soc. America Bull., v. 66, p. 435-462.

Wiebe, R. A., 1972, Igneous and tectonic events in northeastern Cape Breton Island, Nova Scotia: Can. Jour. Earth Sci., v. 9, p. 1262-1277.

Williams, H., 1964, The Appalachians in northeastern Newfoundland—a two-sided symmetrical system: Am. Jour. Sci., v. 262, p. 1137-1158.

———, 1967, Silurian rocks of Newfoundland, *in* Neale, E. R. W., and Williams, H., eds., Geology of the Atlantic region: Geol. Assoc. Can. Spec. Paper 4, p. 93-137.

———, 1971, Mafic-ultramafic complexes in west Newfoundland Appalachians and the evidence for their transportation: a review and interim report: Proc. Geol. Assoc. Can., v. 24, p. 9-25.

———, 1973, Bay of Islands map-area, Newfoundland: Geol. Survey Can. Paper 72-34.

———, and Malpas, J. G., 1972, Sheeted dikes and brecciated dike rocks within transported igneous complexes, Bay of Islands, western Newfoundland: Can. Jour. Earth Sci., v. 9, p. 1216-1229.

———, and Smyth, W. R., 1973, Metamorphic aureoles beneath ophiolite suites and Alpine peridotites: tectonic implications with west Newfoundland examples: Am. Jour. Sci., v. 273, p. 594-621.

———, Kennedy, M. J., and Neale, E. R. W., Coordinators, 1972, The Appalachian structural province, *in* Price, R. A., and Douglas, R. J. W., eds., Variations in tectonic styles in Canada: Geol. Assoc. Can. Spec. Paper 11, p. 181-261.

———, Kennedy, M. J., and Neale, E. R. W., 1974, The northeastward termination of the Appalachian orogen, *in* Nairn, A. E. M., ed., Ocean basins and margins, v. 2, New York, Plenum, p. 79-123.

Wilson, J. T., 1966, Did the Atlantic close and then reopen? Nature, v. 211, p. 676-681.

Zen, E-an, 1967, Time and space relationships of the Taconic allochthon and autochthon: Geol. Soc. America Spec. Paper 97, 107 p.

———, 1968, Nature of Ordovician orogeny in the Taconic area, *in* Zen, E-an, White, W. S., Hadley, J. B., and Thompson, J. B., Jr., eds., Studies of Appalachian geology: northern and maritime: New York, Wiley-Interscience, p. 129-139.

———, 1972, The Taconide zone and the Taconic orogeny in the western part of the northern Appalachian Orogen: Geol. Soc. America Spec. Paper 135, 72 p.

———, White, W. S., Hadley, J. B., Thompson, J. B., Jr., eds., 1968, Studies of Appalachian geology: northern and maritime: New York, Wiley-Interscience, 475 p.

Paleozoic Arctic Margin of North America

K. J. Drummond

INTRODUCTION

Several excellent regional syntheses of the Paleozoic geology of Arctic North America have been published. For a more detailed and fuller account of any area, the reader should consult the reference list and, in particular, Brosgé and Dutro (1973), Churkin (1973), Lathram (1973), D.K. Norris (1973), Douglas et al. (1968), Dawes and Soper (1973), and Trettin et al. (1972). These and other papers are the main foundation for the concepts presented in the present synthesis. This is supplemented by the author's field experience in the Canadian Arctic Islands and regional studies in Arctic and western North America.

REGIONAL SETTING

The major geological regions of Arctic North America are outlined in Figure 1. The Arctic margin of North America consists of the Innuitian orogen, which merges with the East Greenland orogen to the east and the Cordilleran orogen to the west. Along the north, the Innuitian orogen is flanked by deposits of the Arctic Coastal Plain and present Arctic continental shelf and slope. The record of the Paleozoic Arctic margin of North America is now expressed in the Innuitian orogen and the northernmost parts of the East Greenland and Cordilleran orogens.

The nomenclature for many of the tectonic elements comprising Arctic North America is shown in Figure 2. The Innuitian orogen includes the Franklinian fold complex of northern Greenland and the Canadian Arctic Islands and the successor deposits of the Wandel and Sverdrup basins and the Eglinton graben. Southwestward extensions of the Innuitian orogen probably occur in the subsurface of the Mackenzie Delta and along the north coast of Alaska. The Innuitian orogen is interpreted to be present in the Romanzoff British Mountains uplift of northeast Alaska and the northern Yukon. This region was also involved in the later orogenic events of the Cordilleran.

Figure 3 shows the general structural features in the Arctic region of North America. Structural trends within the shield are those of the gneisses and schists comprising the crystalline basement. Faults and folds are shown for all of the Phanerozoic section. Structural trends within the Phanerozoic follow two general directions; (1) a general east-west trend, more or less paralleling the depositional strike of the Franklinian geosyncline and the present shelf margin, and (2) a general northerly trend at more or less right angles to the former, which also parallels the strike of the East Greenland foldbelt. Other structural trends of interest are (1) the series of basement highs along the northern coast of Alaska; (2) northeasterly-trending basement highs in the Mackenzie Delta, representing a northeastern extension of the Dave Lord-Aklavik Arch; (3) the northeasterly trending faults in central Alaska (Lathram, 1973); and (4) a northwesterly-trending high offshore of the Tigara Uplift (Grantz et al., 1970).

EVOLUTION OF THE ANCIENT MARGIN

Basement rocks for the Paleozoic geosynclines bordering Arctic North America are thought to be the Hudsonian crystalline complex (granite gneiss and migmatite), which was formed about 1.7 billion years ago. Figure 4 shows the distribution of Middle and Upper Proterozoic sediments. Middle Proterozoic clastics were deposited as a thick continental terrace wedge in the Cordilleran and East Greenland margins. As indicated, there is no known Middle Proterozoic along the Arctic margin. It is possible that the crystalline Hudsonian shield extended northward into the present site of the Canada basin.

The Middle Proterozoic wedges along the west and the east were deformed by the Racklan and Carolinidian orogenies of possibly Grenvillian age (1,100-800 my). Clastic sedimentation again started in the Upper Proterozoic in these regions with the development of the Cordilleran and East Greenland geosynclines.

EARLY PALEOZOIC FOLDBELT

In the Upper Proterozoic, the Arctic margin is believed to have developed across the general north-south structural grain of the crystalline shield. A thick wedge of clastics derived from the shield was deposited along this ancient margin. An orogeny which formed an early Paleozoic foldbelt ("orogenic welt") is inferred to have occurred in early Cambrian or earlier. The evidence of this ancient Arctic foldbelt is believed to be represented in the various metamorphic complexes; the Cape Columbia and Rens Fiord complexes of northern Ellesmere and northern Axel Heiberg islands (Fig. 2), respectively, and the metamorphic terrains encountered in the subsurface along northern Alaska (Barrow, Cape

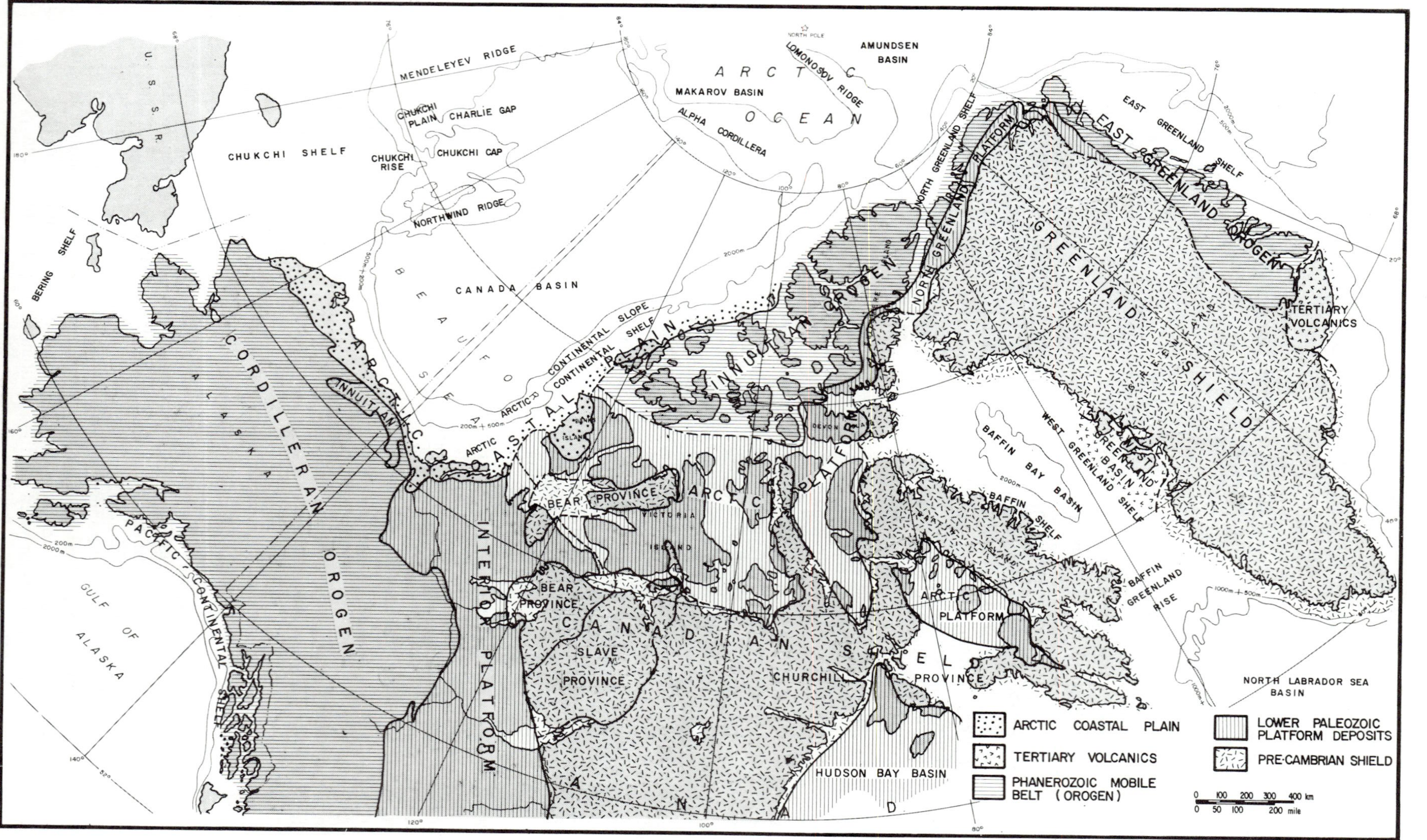

Fig. 1. Geological regions and structural provinces of Arctic North America.

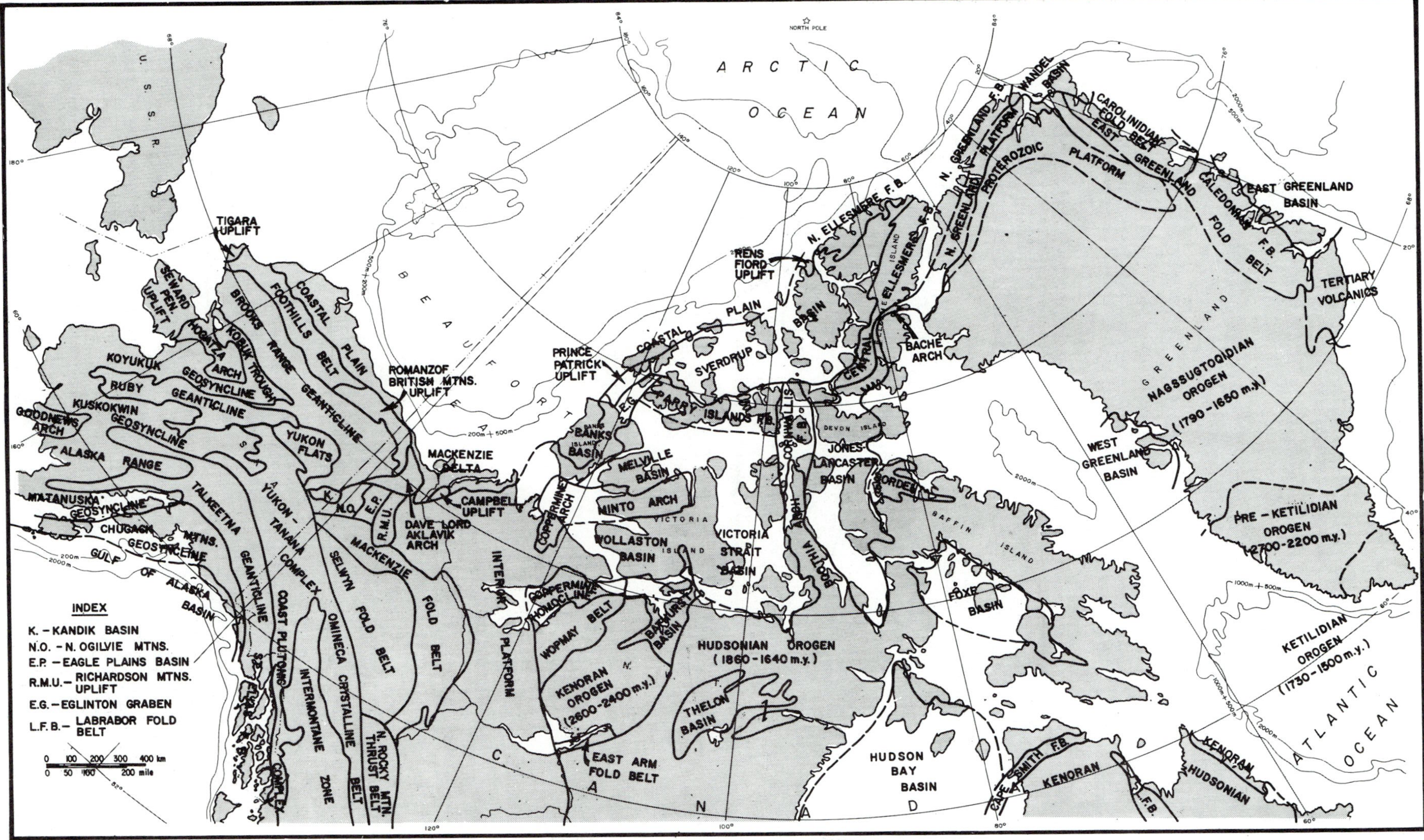

Fig. 2. Index map showing the major regional elements of Arctic North America.

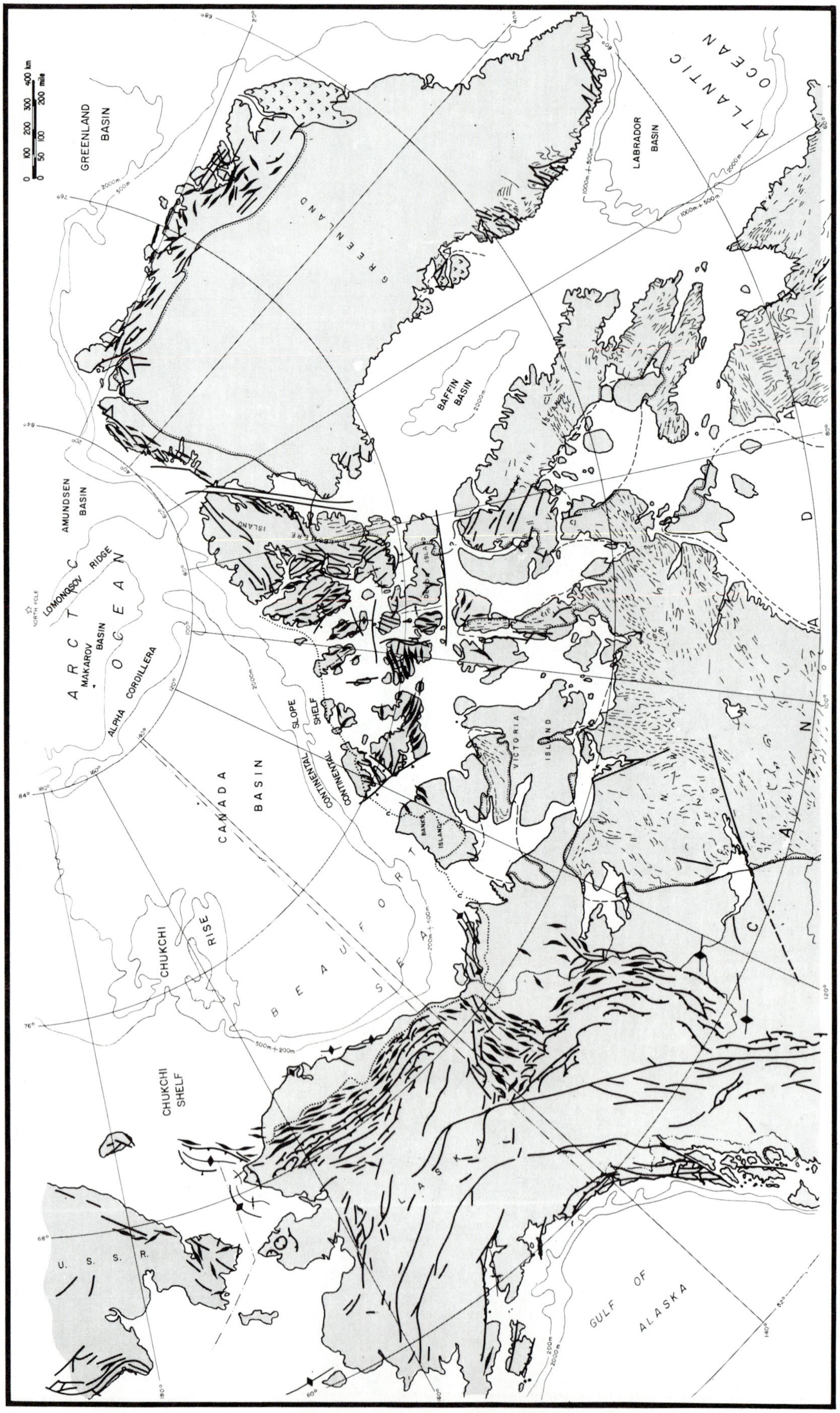

Fig. 3. Simplified tectonic map of Arctic North America. Trends in crystalline shield from Geological Survey of Canada map 1251A (1969).

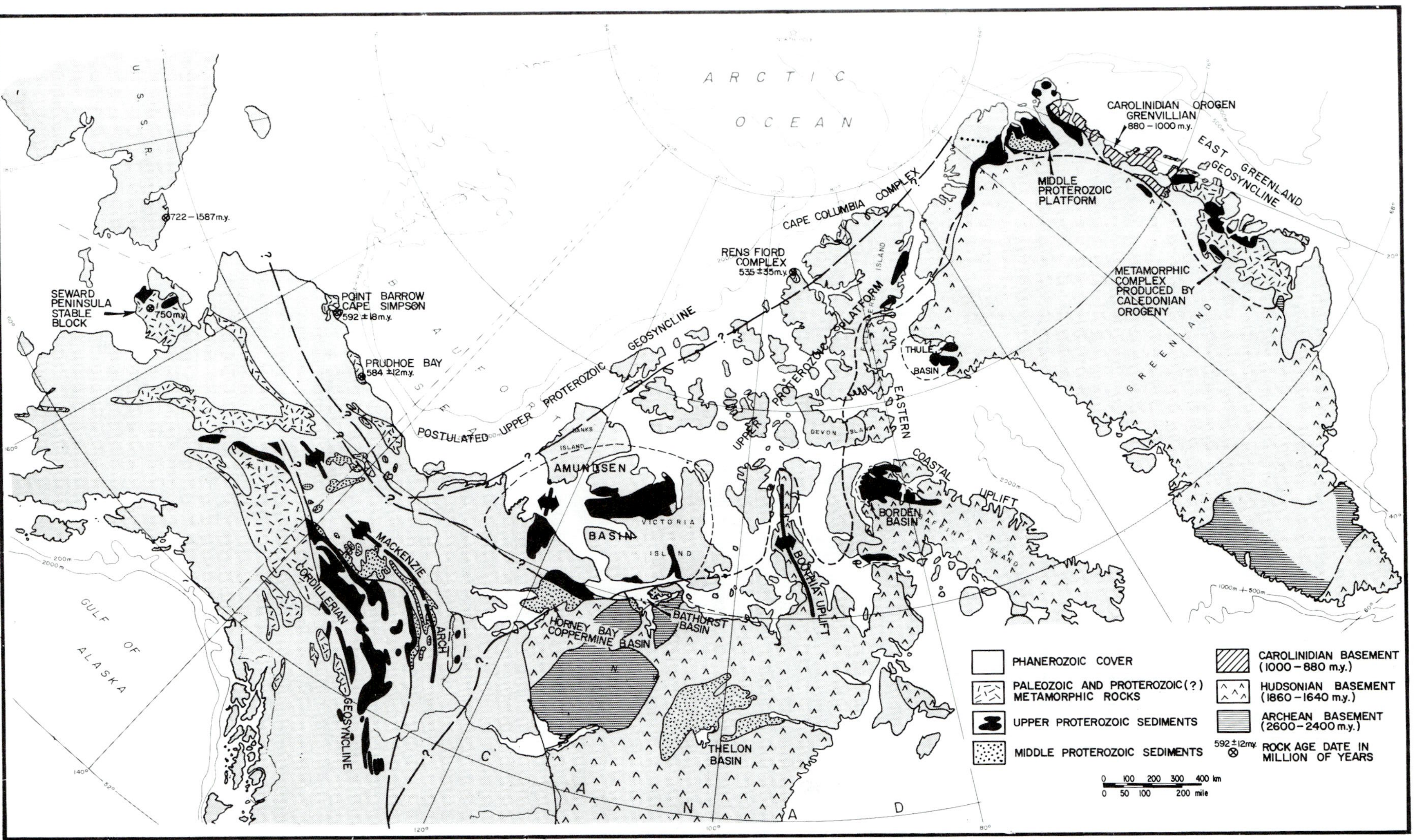

Fig. 4. Distribution of Middle and Upper Proterozoic rocks and metamorphic sequences that may include Proterozoic strata.

Simpson, Prudhoe Bay, etc.). Equivalent rocks are also believed to occur within the Neruokpuk Formation of northeast Alaska and the northwest Yukon of Canada. Published radiometric age data suggest an early Cambrian or earlier regional thermal event rimming the North American Arctic Ocean as follows: Simpson well, Alaska: K-Ar argillite, 592 ± 8 my (Lanphere, 1965); Rens Fiord Complex, Axel Heiberg Island: K-Ar on muscovite from a sheared sandstone, 535 ± 49 my (Wanless et al., 1965). This sandstone unit of the Rens Fiord Complex has been tentatively correlated by H. P. Trettin (1973a) with the Grant Land Formation of northern Ellesmere of inferred Middle to Late Cambrian age. It is suggested that the age determination indicates the age of metamorphism of the source rocks (Trettin, 1969b).

The Cape Columbia complex of northern Ellesmere Island has yielded apparent ages of metamorphism as follows: 403 ± 17 my, 465 ± 19 my, 445 ± 18 my, 389 ± 21 my, 354 ± 15 my, 303 ± 16 my, 189 ± 9 my (see Trettin, 1972, p. 144). It is thus apparent that the northern Ellesmere foldbelt has undergone several orogenies. The oldest date is 465 ± 19 my (Early Ordovician), and "obviously is younger than the first phase of metamorphism and also younger than the uplift that produced the sediments of the Grant Lord Formation" (Trettin, 1971a, p. 32).

Lending support to these ages are K-Ar mica age determinations by R. E. Denison (Mobil Research and Development Corp., personal communication) made on micas of graphitic phyllites from cores of Mobil's West Staines well, on the Alaska North Slope (Sec. 13-9N-23E u.p.m.), as follows: 547 ± 22 my, 564 ± 11 my, 584 ± 12 my, and 553 ± 22 my.

Also adding significantly to indirect evidence of the ancient orogenic welt is the work of H. P. Trettin in northern Ellesmere Island. The provenance for the late Proterozoic northward-thickening wedge of miogeoclinal sediments (Kerr, 1967c) appears to be from the craton to the south; whereas the provenance for an early Paleozoic (Early Ordovician or earlier, possibly Middle to Late Cambrian) deltaic complex is from a postulated metamorphic terrain to the north. These relationships are shown diagrammatically in Figure 5a and b.

The Neruokpuk Formation exposed in the Romanzoff-British Mountains uplift consists of a complex sequence of sedimentary and volcanic rocks, locally metamorphosed, unconformably overlain by Mississippian and younger strata. Recent mapping by the United States Geological Survey in northeast Alaska has resulted in the discovery of five fossil horizons in the Neruokpuk (Dutro et al., 1971). These indicate that much of the Lower Paleozoic is represented, as well as part of the Proterozoic (Dutro et al., 1972). From stratigraphic and structural relationships in northeast Alaska, Dutro et al. (1971) conclude: "Thus it appears that the complex structure of the Neruokpuk results from several Early Paleozoic tectonic episodes rather than from a single Late Devonian-Early Mississippian orogeny." Norris (1972) in the northern Yukon of Canada describes a volcanic conglomerate and limestone unit which "appears, from regional mapping, to overlie progressively lower strata in the upturned Neruokpuk succession toward Alaska." This unit could be correlated with the dated Cambrian unit of Dutro et al. (1972). Norris concludes: "Thus, if the stratigraphic relations between this Cambrian formation and the underlying Neruokpuk have been inferred correctly, the Neruokpuk in Canada is entirely older than the Early Cambrian. Moreover, the acute deformation within it is largely of Precambrian age and is not Ellesmerian in the Romanzoff uplift."

From the foregoing discussion, it appears that a significant orogenic event did affect northern Ellesmere and Axel Heiberg islands and the northern part of Alaska in the Late Proterozoic or Early Cambrian. It is concluded that these are but two parts of an ancient foldbelt that rimmed the Arctic continental margin of North America and possibly extended to northwestern Siberia (Churkin, 1969).

This inferred early Paleozoic foldbelt and the inherited structural grain of the Hudsonian crystalline basement are believed to be the dominant factors controlling the tectonic framework of the Paleozoic margin of Arctic North America. These are the two major structural trends mentioned earlier and shown on Figure 3.

LOWER PALEOZOIC GEOSYNCLINE

By latest Cambrian or Early Ordovician, the ancient foldbelt subsided or was reduced to wave base. In any case, it ceased to be a source of sediment (Fig. 5). A deep-water submarine trough, the Hazen trough, developed by subsidence inland over the deltaic deposits (Trettin, 1973a). The Hazen trough is characterized by a starved basin type of deposition, the Hazen Formation consisting of graptolitic shale, radiolarian chert, and redeposited carbonate. The redeposited carbonate is thought to be derived from the north, where carbonate banks could possibly have developed on an offshore ridge. Facies relationships to the south are not directly known. The Hazen trough was later, beginning in the Silurian, to be filled by a flysch-like succession. Paleo-current directional studies indicate turbidity currents descended from the northwest and deflected along the trough axis to the southwest (Trettin, 1970).

A similar, possibly correlative facies of the Hazen Formation is the basal part of the Ibbett Bay Formation exposed in the Canrobert Hills of northwest Melville Island (Table 1). In this area a thin section of siliceous shale was deposited in deep water in front of a barrier reef complex (Cornwallis Formation). These troughs have been called eugeosynclinal on Figures 5 and 6, although volcanics are only a minor constituent and are restricted to the northern flank adjacent to the ridge.

In the Arctic Islands the elements of the Franklinian geosyncline, which probably existed in the

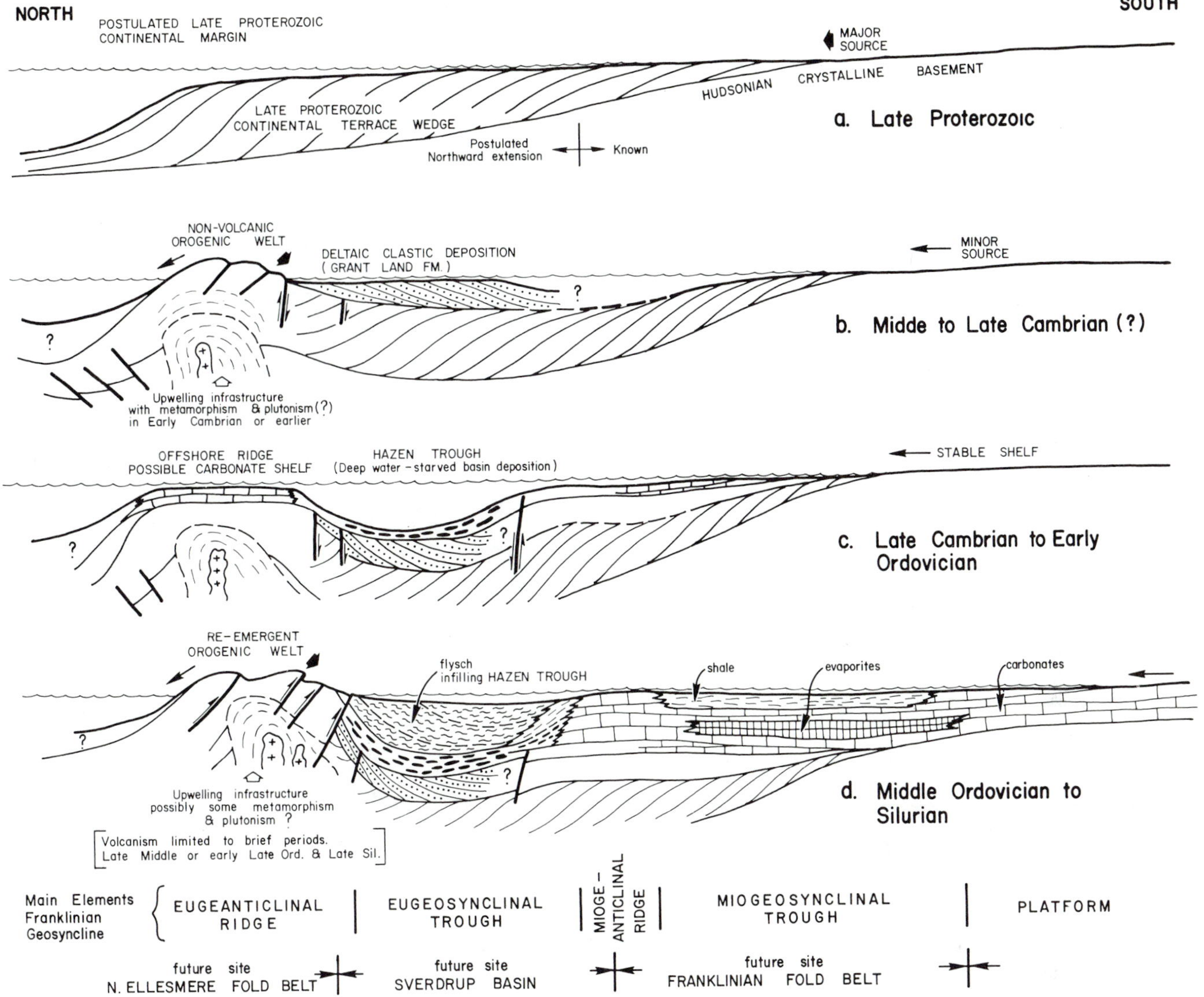

Fig. 5. Schematic sections showing the postulated evolutionary development of the ancient Arctic margin of North America.

Cambrian or earlier, were well defined in the Ordovician. These are illustrated in Figures 5d and 6. In early Ordovician a miogeosynclinal shelf developed in the southern half of the Arctic Islands. This shelf was separated by a hinge line or shelf break from the eugeosynclinal trough in the north. The hinge line is here referred to as a miogeanticlinal ridge. To the close of the Middle Ordovician, the miogeosynclinal area was a broad carbonate shelf. Fringing shelf carbonates developed around the shelf margins and barrier-type carbonate banks along the ridge area. In the central portions, restricted lagoonal environments developed with the deposition of evaporites and/or lagoonal lime muds. At certain times conditions were favorable for the development of bioherms and/or biostromes within the lagoonal environment. Along the northern edge of the miogeanticlinal ridge, shelf carbonates, probably a fore-reef development, rapidly shale-out into the eugeosynclinal trough.

The general geosynclinal elements governing sedimentation in the Middle Ordovician continued into the Upper Ordovician and Silurian. At the beginning, and throughout the Upper Ordovician-Silurian, there was a general deepening of the basin with the development of carbonate banks and shelves and graptolitic shale basins and troughs (Fig. 5). It is postulated that the carbonate banks and shelves developed on the topographically higher areas and the shale basins formed over the lagoonal environment of the Cornwallis Formation. The sedimentary environment of the miogeosyncline from the Middle Ordovician into the Lower Devonian was probably very stable throughout the Arctic. Ordovician-Silurian carbonates attain a maximum thickness of 11,000 ft on Cornwallis Island, generally about 5,000 ft elsewhere. The equivalent Cape Phillips shale section is 8,500 ft thick on Cornwallis Island, 3,000 ft on Melville and Ellesmere islands, and as thin as 1,000 ft on Bathurst Island.

The carbonate platform, which is very broad over the southwestern Arctic Islands, becomes narrower to the northwest. Across northern Greenland

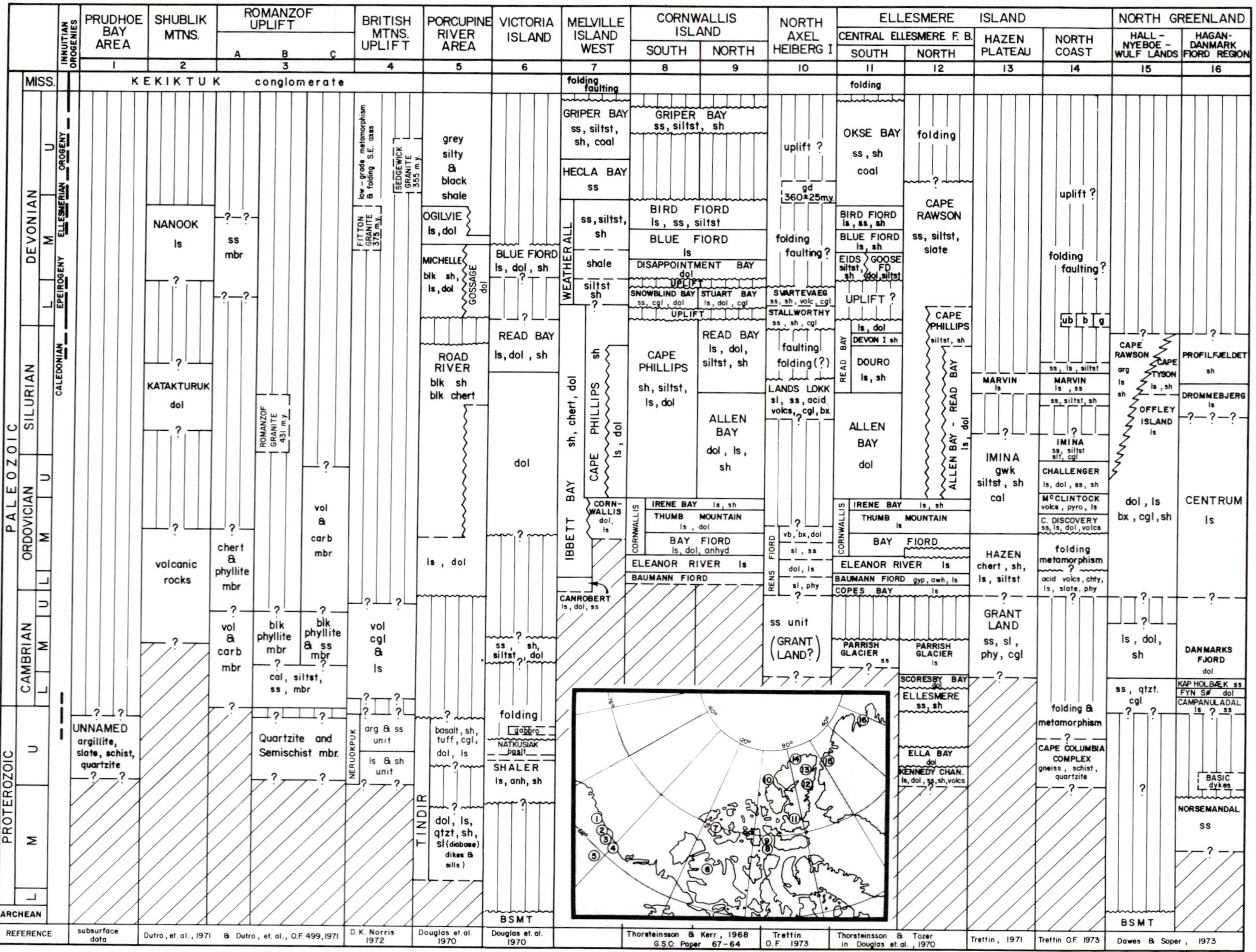

Table 1. Geotectonic correlation chart: Proterozoic and Lower Paleozoic of Arctic North America (chart adapted from Douglas et al., 1970).

the platform is quite narrow and very little development of a miogeosynclinal trough is noted.

To the southwest, platform-carbonate deposition is extensive, with a transition to shale in the north-trending Richardson trough and Selwyn basin. The Ordovician-Silurian record, as shown in Figure 6, is very fragmentary across northern Alaska. Several carbonate shelves are present, possibly developed along a number of arches or ridges, as shown. These carbonate shelves are represented by continuous carbonate deposits of Ordovician to Devonian age. These are more than 12,000 ft thick on Seward Peninsula, 6,000 ft thick in southwest Alaska, and 3,000-4,000 ft thick on the Porcupine platform (Brosgé and Dutro, 1973). Transition to shale is recorded to the southeast of the Porcupine platform and along the Denali fault in central Alaska. It is possible that the metamorphic complexes may represent the former site of deep basinal areas, which lay between the arches shown on Figure 6. Volcanic sequences are represented by andesite within Middle Ordovician carbonate along the eastern margin of Selwyn basin and in northeastern Alaska. Volcanic rocks in the Brooks Range have a scattering of K-Ar dates from 470 to 370 my (Brosgé and Dutro, 1973).

The very distinctive miogeosynclinal phase of the Ordovician-Silurian-Lower Devonian was brought to a close by an orogenic event termed the Caledonian. This event began earliest in East Greenland, possibly as early as Upper Ordovician, and occurred successively later westward in the Franklinian and Cordilleran geosynclines. True orogenic activity possibly occurred only in the East Greenland foldbelt. Activity elsewhere was very widespread, but was mainly epeirogenic, characterized by uplift, mild tilting, and erosion. This event was followed by a return to quiescent conditions and the widespread transgression of Middle Devonian seas.

In the Arctic Islands the most significant uplift of the Caledonian epeirogeny is the Cornwallis foldbelt (Fig. 2), which reactivated the Boothia arch and extended it north across Cornwallis Island and Grinnell Peninsula. Although evidence is poor, Caledonian movements probably affected most of the Arctic Islands. The platform areas and most of Melville Island were generally emergent, with little or no deposition. Caledonian uplifts possibly occurred on the Bache arch and the eugeosyncline, in particular Rens Fiord uplift of northern Axel Heiberg, northern Ellesmere, and northwest of Melville Island. During the Caledonian orogeny, Lower Devonian deposition was mainly confined to the flanks of Caledonian uplifts, the Cornwallis foldbelt and Bache arch, and internally derived sediments from uplifts within the eugeosyncline. Adjacent to the uplifts, these sediments reach thicknesses of 6,000-8,000 ft and thin rapidly away from the uplifts.

Activity within the eugeosyncline is probably a reemergence of the ancient foldbelt ("orogenic welt"), which possibly started in Late Ordovician. These movements were accompanied by volcanism along northern Axel Heiberg and northwestern Ellesmere islands. As mentioned earlier, several apparent ages of metamorphism have been determined for the Cape Columbia complex, suggesting Caledonian orogenic activity. The Rens Fiord uplift originated as a northwest-trending horst-like structure in the Late Silurian-Early Devonian (Trettin, 1973a). This uplift cuts across the predominant depositional and structural pattern.

The Caledonian uplifts, or the Cornwallis foldbelt and the Rens Fiord uplift, are believed to be controlled by inherited structural trends in the underlying crystalline basement.

A widespread epeirogenic uplift possible occurred in latest Silurian and/or Early Devonian over much of Alaska. However, evidence of such a Caledonian event is scattered and meager. Throughout much of northeastern Alaska, a regional unconformity occurs beneath Middle Devonian carbonates (possibly Lower Devonian in some places). The Precambrian quartzite and semischist member of the Neruokpuk in the Romanzoff uplift has been intruded by a granite stock with an apparent K-Ar age of 431 my. This suggests renewed orogenic activity in the ancient foldbelt of northern Alaska.

In the Middle Devonian, tectonic activity of the Caledonian orogeny continued in the East Greenland orogen (Fig. 5). Elsewhere in the Innuitian and Cordilleran orogens, there was a return to stable conditions with the widespread transgression of Middle Devonian seas. In general, carbonate deposits were laid down over wide areas in shallow, well-aerated seas on a slowly subsiding platform. Later in the Middle Devonian and through the Upper Devonian tectonic instability increased within the geosynclines. Only in the southern part of the northern Interior platform and south into Alberta did stable platform conditions prevail throughout the Devonian.

In the Middle Devonian of the Franklinian geosyncline, the sea deepened and transgressed the Caledonian uplifts. During the lower Middle Devonian, carbonate banks (Disappointment Bay and Blue Fiord formations) developed adjacent to, and onlapping the Caledonian uplifts with shale (Eids and Weatherall formations) being deposited in the deeper areas (Table 1). In the upper Middle Devonian, clastic terrigenous deposition became increasingly predominate, indicating tectonic activity within the bordering eugeosyncline, probably the initial activity of the Ellesmerian orogeny. Middle Devonian sediments of the Franklinian miogeosyncline are up to 10,000 ft thick.

The Upper Devonian of the Arctic Islands (Fig. 7) is a thick wedge, up to 10,000 ft thick, of coarse clastics of probable northern provenance. The miogeosynclinal area then became emergent and was subjected to a period of east-west folding in the western islands and north-south folding along the east during the Ellesmerian orogeny in the Late Devonian and/or early Mississippian. The deformation is believed to be primarily gravity slide.

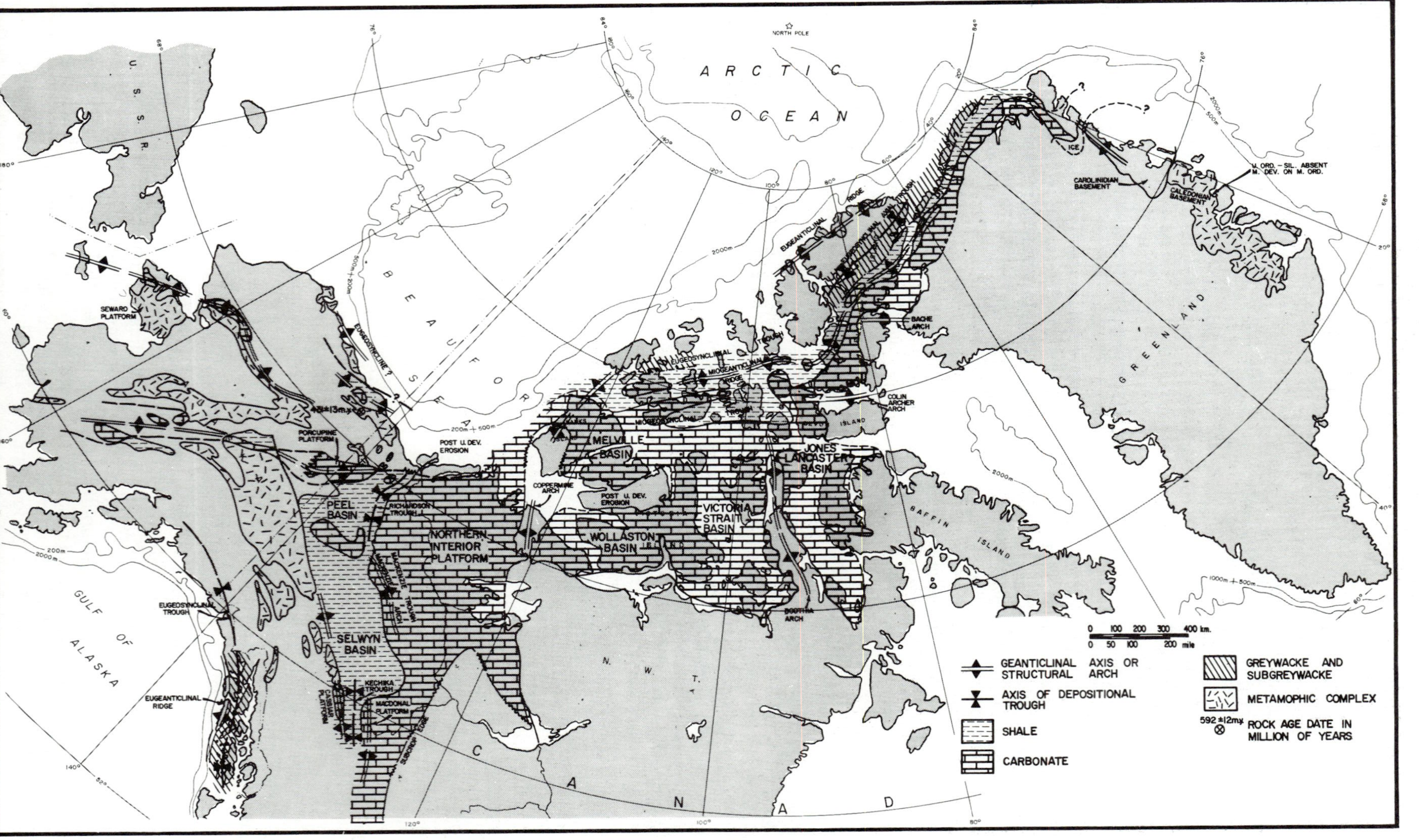

Fig. 6. Distribution of Silurian rocks to illustrate the major elements of the Franklinian geosyncline.

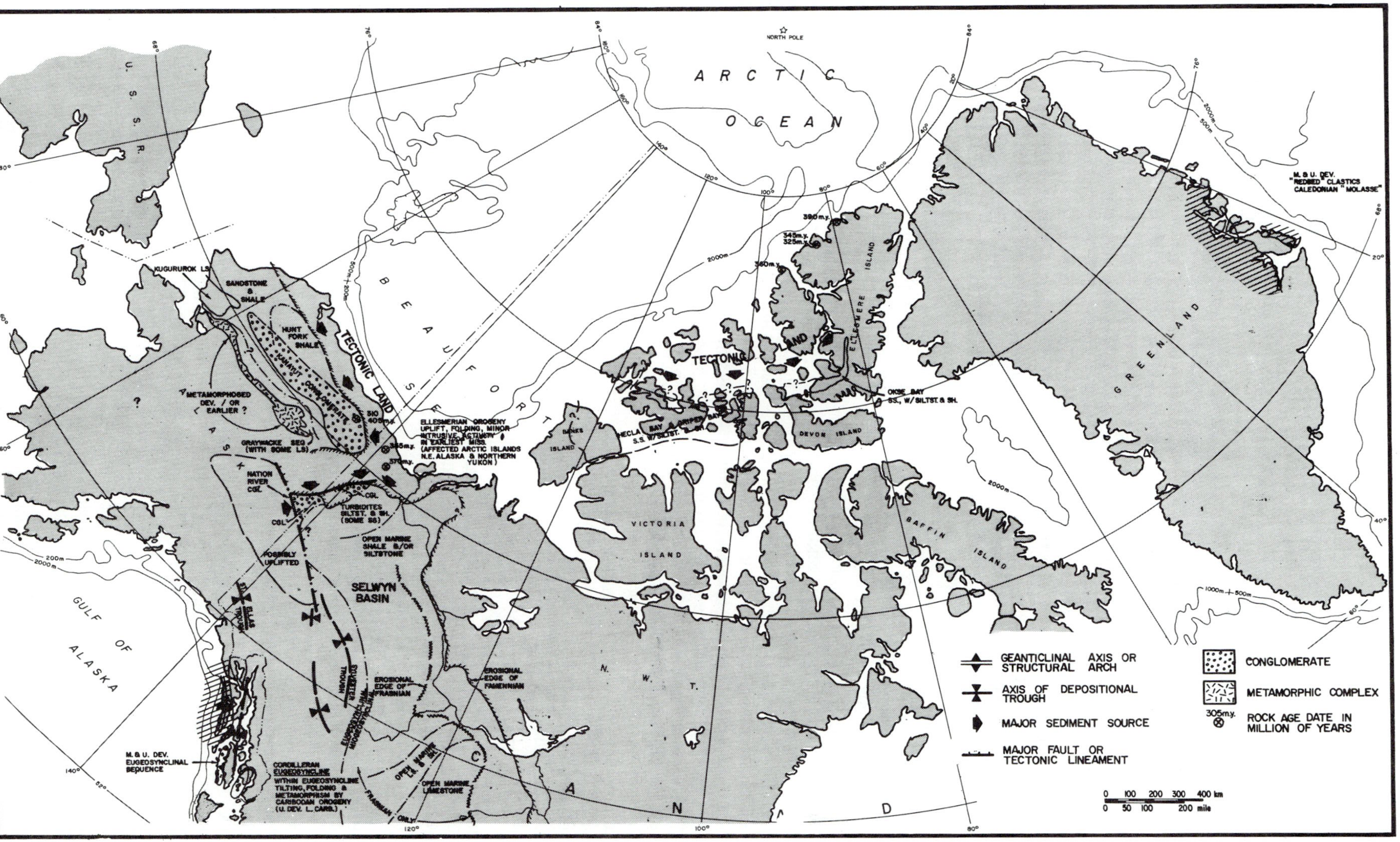

Fig. 7. Distribution of Upper Devonian rocks to show sedimentation and tectonism of the Lower Paleozoic Arctic geosyncline in its final stage.

Ellesmerian orogenic activity within the eugeosynclinal realm is recorded by turbidite and volcanic rocks in the Middle Devonian of northern Axel Heiberg Island. A granitic intrusion in this area gave a K-Ar date of 360 ± 25 my. In northwest Ellesmere Island, a small dioritic intrusion gave a K-Ar date of 390, 325, and 345 my. Basic dykes cutting Silurian strata in this area may be of a similar age.

Devonian deposits are not preserved on northern Ellesmere Island or northern Greenland. In these areas, Carboniferous deposits lie unconformably on folded Silurian strata.

Throughout much of the Lower Paleozoic the tectonic and depositional history of the Franklinian and Cordilleran geosynclines was very similar. However, in the Devonian the history of the two areas became increasingly different. By early Mississippian the Franklinian geosyncline had terminated, whereas geosynclinal deposition and tectonism were to continue in the Cordilleran throughout the Upper Paleozoic and Mesozoic.

The record of the Middle Devonian of Alaska is fragmentary; however, evidence suggests a widespread carbonate deposit over a broad shelf, succeeded by clastic deposition. In general, Middle Devonian rocks lie unconformably on Silurian and older strata.

In late Middle Devonian and throughout the Upper Devonian the Cordilleran was increasingly affected by tectonic activity of the Ellesmerian orogeny. In Late Devonian time the western Cordillera (western Selwyn basin, Cassiar platform) began to subside rapidly and became typically eugeosynclinal. It was the site of extensive tectonic activity, with the accumulation of vast amounts of sedimentary and volcanic material. This type of deposition continued into the early Mississippian.

In northeast Alaska and northern Yukon, the ancient Early Cambrian foldbelt was uplifted in the Upper Devonian, accompanied by plutonic intrusion, deformation, and possibly some metamorphism. Thick clastic-wedge deposits accumulated adjacent to the uplifts in the Brooks Range and northwestern Selwyn basin, grading outward to marine shale and siltstone.

Igneous activity associated with the Ellesmerian orogeny in the northern Cordillera is recorded by granitic intrusion in northeast Alaska and northern Yukon and mafic dykes in the central Brooks Range. K-Ar ages for granites of this area (Romanzoff-British Mountains uplift) are: Romanzoff granite, 310 and 405 my; Mount Fitton (Barn Mountains), 375 my; Mount Sedgewick (British Mountains), 355 my.

Geosynclinal deposition along the Arctic margin of North America came to a close with the Late Devonian-early Mississippian Ellesmerian orogeny. The orogenic welt formed by this orogeny appears to have been a remobilization along the same trend as the earlier late Proterozoic-Early Cambrian orogenic belt. It is significant to note that this belt was to be one of the major controlling features governing Upper Paleozoic and younger geologic history.

SUMMARY

The early Paleozoic continental margin of Arctic North America, throughout its history, is dominated by a marginal ridge or orogenic welt which was probably defined originally by the site of a late Upper Proterozoic-Lower Cambrian foldbelt. Depositional trends and structural deformation of the Ellesmerian orogeny (late Devonian-early Mississippian) are about parallel with the ancient ridge and the present continental margin. In addition, there are anomalous northerly to northwesterly structural and physiographic trends that deviate from depositional trends.

The central part of the Arctic Islands is underlain by Precambrian basement (Hudsonian, 1700 my), with northerly to northwesterly gneissic trends. Mid-Paleozoic, Late Silurian-Early Devonian epeirogenic movements were mainly vertical and occurred parallel to gneissic trends of the older crystalline basement. It was probably this basement-controlled faulting in the Late Silurian-Early Devonian that intensified the basement trends and left such a strong imprint on all subsequent deposition and deformation.

Mid-Tertiary earth movements of the Eurekan orogeny were also controlled by this northerly to northwesterly grain. The Late Tertiary-Pleistocene history and physiographic development of the Arctic Islands, whether by glaciation or vertical faulting, also appears to closely follow this structural grain.

CONCLUSION

Eugeosynclinal realms are thought to represent zones of active interplay between oceanic and continental regimes, and these mobile areas are considered to be the precursors of most fold belts. As such, these Arctic fold belts could represent the sedimentary accumulations of ancient continental margins.

Whatever the history of the very early continental margin of Arctic North America and adjacent Arctic Ocean, it is significant that the ancient margin in the late Proterozoic and through most of the Lower Paleozoic was in much the same location with the same general trend as the present continental margin.

BIBLIOGRAPHY

General

Christie, R. L., et al., 1972, The Canadian Arctic Islands and the Mackenzie region; guidebook excursion A66: XXIV Internatl. Geol. Congr., Montreal, Can.

Douglas, R. J. W., ed., 1970, Geology and economic minerals of Canada: Geol. Survey Can. Econ. Geol. Rept. 1, 5th ed.

King, P. B., 1969: The tectonics of North America: U.S. Geol. Survey Prof. Paper 628 and tectonic map of North America.

Pitcher, M. B., ed., 1973, Arctic geology: Proc. 2nd Internatl. Symp. Arctic Geol., 1971: Am. Assoc. Petr.

Geol. Mem. 19.
Price, R. A., and Douglas, R. J. W., eds., 1972, Variations in tectonic styles in Canada: Geol. Assoc. Can. Spec. Paper 11.
Raasch, G. O., ed., 1961, Geology of the Arctic: Proc. First Internatl. Symp. Arctic Geol., 1960: University of Toronto Press.
Stockwell, C. H., ed., 1969, Tectonic map of Canada; scale 1:5,000,000: Geol. Survey Can., Map 1251A.
Wanless, R. K., ed., 1969, Isotopic age map of Canada; scale 1:5,000,000: Geol. Survey Can., Map 1256A.

Alaska

Adkinson, W. L., and Brosgé, M. M., eds., 1970, Proceedings of the geological seminar on the North Slope of Alaska: Pacific Sect. Am. Assoc. Petr. Geol.
Brosgé, W. P., and Dutro, J. T., Jr., 1973, Paleozoic rocks of northern and central Alaska, *in* Pitcher, M. G., ed., Arctic geology: Am. Assoc. Petr. Geol. Mem. 19, p. 361-375.
———, Brabb, E. E., and King, E. R., 1970, Geologic interpretation of reconnaissance aeromagnetic survey of northeastern Alaska: U.S. Geol. Survey, Bull. 1271-F.
———, and Tailleur, I. L., 1971, Northern Alaska petroleum province, *in* Cram, I. H., ed., Future petroleum provinces of the United States—their geology and potential: Am. Assoc. Petr. Geol. Mem. 15, v. 1, p. 68-99.
Churkin, M., Jr., 1969, Paleozoic tectonic history of the Arctic basin north of Alaska: Science, v. 165, p. 549-555.
———, 1973, Paleozoic and Precambrian rocks of Alaska and their role in its structural evolution: U.S. Geol. Survey Prof. Paper 740.
Dutro, J. T., Jr., et al., 1971, Early Paleozoic fossils in the Neruokpuk Formation, northeast Alaska: U.S. Geol. Survey Open File Report.
———, Brosgé, W. P., and Reiser, H. N., 1972, Significance of recently discovered Cambrian fossils and reinterpretation of Neruokpuk Formation, northeastern Alaska: Am. Assoc. Petrl. Geol. Bull., v. 56, no. 4, p. 808-815.
Gates, G. O., and Gryc, G., 1963: Structure and tectonic history of Alaska, *in* Childs, O. E., and Beebe, B. W., eds., The backbone of the Americas: Am. Assoc. Petr. Geol. Mem. 2, p. 264-277.
Grantz, A., et al., 1970, Reconnaissance geology of the Chukchi Sea as determined by acoustic and magnetic profiles, *in* Adkinson, W. L., and Brosge, W. P., eds., Proceedings of the geologic seminar on the North Slope of Alaska: Pacific Sect. Am. Assoc. Petr. Geol., p. F1-F28.
Gryc, G., et al., 1967, Devonian of Alaska, *in* Oswald, P. H., ed., International symposium on the Devonian system, v. 1: Alta. Soc. Petr. Geol., p. 703-716.
Lanphere, M. A., 1965, Age of Ordovician and Devonian mafic rocks in northern Alaska, *in* Geological survey research 1965: U.S. Geol. Survey Prof. Paper 525-A, p. A101-102.
Lathram, E. H., 1973 Tectonic framework of northern and central Alaska, *in* Pitcher, M. G., ed., Arctic geology: Am. Assoc. Petr. Geol. Mem. 19, p. 351-360.
Morgridge, P. L., and Smith, W. B., Jr., 1972, Geology and discovery of Prudhoe Bay field, eastern Arctic Slope, Alaska, *in* King, R. E., ed., Stratigraphic oil and gas fields: Am. Assoc. Petr. Geol. Mem. 16, and Soc. Expl. Geophysicists Spec. Publ. 10, p. 489.

Yukon and Northwest Territories

Cook, D. G., and Aitken, J. D., 1973, Tectonics of northern Franklin Mountains and Colville Hills, District of Mackenzie, Canada, *in* Pitcher, M. G., ed., Arctic geology: Am. Assoc. Petr. Geol. Mem. 19, p. 13-22.
Donaldson, J. A., Possible correlations between Proterozoic Strata of the Canadian Shield and North American Cordillera: Proc. Belt Symp., Moscow, Idaho, v. 1, p. 61-75.
Douglas, R. J. W., et al., 1970, Geology of western Canada, ch. VIII, p. 367-488; and Geotectonic correlation chart for western Canada, chart III, *in* Douglas, R. J. W., ed., Geology and economic minerals of Canada: Geol. Survey Can. Econ. Geol. Rept. 1, 5th ed.
Gabrielse, H., 1972, Younger Precambrian of the Canadian Cordillera: Am. Jour. Science, v. 272, June, p. 521-536.
Jeletzky, J. A., 1963, Pre-Cretaceous Richardson Mountains trough—its place in the tectonic framework of Arctic Canada and its bearing on some geosynclinal concepts: Roy. Soc. Can. Trans., v. 56, ser. 3, sect. 3, p. 55-84.
Martin, L. J., 1959, Stratigraphy and depositional tectonics of North Yukon—lower Mackenzie river area, Canada: Am. Assoc. Petr. Geol. Bull., v. 43, no. 10, p. 2399-2455.
Norford, B. S., 1964, Reconnaissance of the Ordovician and Silurian rocks of northern Yukon Territory: Geol. Survey Can. Paper 63-39.
Norris, D. K., 1972, Structural and stratigraphic studies in the tectonic complex of northern Yukon Territory, north of Porcupine River, *in* Report of activities: Geol. Survey Can. Paper 72-1B, p. 91-99.
———, 1973, Tectonic styles of northern Yukon Territory and northwestern District of Mackenzie, Canada, *in* Pitcher, M. G., ed., Arctic geology: Am. Assoc. Petr. Geol. Mem. 19, p. 23-40.
Yorath, C. J., 1973, Geology of Beaufort—Mackenzie, Basin and eastern part of northern interior plains, *in* Pitcher, M. G., ed., Arctic geology: Am. Assoc. Petr. Geol. Mem. 19, p. 41-47.
Ziegler, P. A., 1969, The development of sedimentary basins in western and Arctic Canada: Alberta Soc. Petr. Geol.

Arctic Islands

Christie, R. L., 1967, Bache Peninsula, Ellesmere Island, Arctic Archipelago: Geol. Survey Can. Mem. 347.
Douglas, R. J. W., Norris, D. K., Thorsteinsson, R., and Tozer, E. T., 1963, Geology and petroleum potentialities of northern Canada: Geol. Survey Can. Paper 63-31.
Drummond, K. J., 1973: Canadian Arctic Islands, *in* McCrossan, R. G., ed., Future petroleum provinces of Canada: Can. Soc. Petr. Geol. Mem. 1, p. 443-472.
Fortier, Y. O., et al., 1963, Geology of the north-central part of the Arctic Archipelago, northwest territories (Operation Franklin): Geol. Survey Can. Mem. 320.
Kerr, J. W., 1967a, Devonian of the Franklinian miogeosyncline and adjacent central stable region, Arctic Canada, *in* Oswald, D. H., ed., International symposium on the Devonian System: Alberta Soc. Petr. Geol., v. 1, p. 677-692 (1968).
———, 1967b, Nares submarine rift valley and the relative rotation of north Greenland: Bull. Can. Petr. Geol., v. 15, no. 4, p. 483-520.
———, 1967c, Stratigraphy of Proterozoic and Cambrian rocks, central and eastern Ellesmere Island, District

of Franklin: Geol. Survey Can. Paper 67-27, pt. I.
———, 1968, Stratigraphy of central and eastern Ellesmere Island, Arctic Canada, pt. II, Ordovician: Geol. Survey Can. Paper 67-27.
———, and Christie, R. L., 1965, Tectonic history of Boothia Uplift and Cornwallis fold belt, Arctic Canada: Am. Assoc. Petr. Geol. Bull., v. 48, no. 7, p. 905-926.
Klovan, J. E., and Embry, A. F., III, 1970, Upper Devonian stratigraphy and structure, western Canadian Arctic Islands: Am. Assoc. Petr. Geol. Bull., v. 54, p. 2491.
Pelletier, B. R., 1964, Development of submarine physiography in the Canadian Arctic and its relation to crustal movements: Bedford Inst. Oceanography, Dartmouth, N.S., Rept. B.I.O. 64-16 (unpubl. ms.).
Schuchert, C., 1923, Sites and nature of the North American geosynclines: Geol. Soc. America Bull., v. 34, p. 151-229.
Thorsteinsson, R., and Tozer, E. T., 1960, Summary account of structural history of the Canadian Arctic Archipelago since Precambrian time: Geol. Survey Can. Paper 60-7.
———, 1962: Banks, Victoria and Stefansson islands, Arctic Archipelago: Geol. Survey Can. Mem. 330.
———, 1964, Western Queen Elizabeth Islands, Arctic Archipelago: Geol. Survey Can. Mem. 332.
———, 1968, Geology of the Arctic Archipelago, geology and economic minerals of Canada: Geol. Survey Can. Econ. Geol. Rept. 1, 5th ed., ch. X, p. 548-590.
———, 1970, Geotectonic correlation chart for the Arctic Archipelago, *in* Douglas, R. J. W., ed., Geology and economic minerals of Canada: Geol. Survey Can. Econ. Geol. Rept. 1, 5th ed., chart IV.
Thorsteinsson, R., and Kerr, J. W., 1968, Cornwallis Island and adjacent smaller islands, Canadian Arctic Archipalago: Geol. Survey Can. Paper 67-64.
Trettin, H. P., 1967, Devonian of the Franklinian eugeosyncline, *in* Oswald, D. H., ed., International symposium on the Devonian System: Alberta Soc. Petr. Geol., v. 1, p. 693-701.
———, 1969a, Geology of Ordovician to Pennsylvanian rocks, M'Clintock Inlet, north coast of Ellesmere Island, Arctic Archipelago: Geol. Survey Can. Bull., v. 183.
———, 1969b, Pre-Mississippian geology of northern Axel Heiberg and northwestern Ellesmere islands, Arctic Archipelago: Geol. Survey Can. Bull., v. 171.
———, 1970, Ordovician-Silurian flysch sedimentation in the axial trough of the Franklinian geosyncline, northeastern Ellesmere Island, Arctic Canada, *in* Lajoie, J., ed., Flysch sedimentology in North America: Geol. Assoc. Can. Spec. Paper 7, p. 13-35.
———, 1971a, Geology of Lower Paleozoic Formation, Hazen Plateau and Southern Grant Land Mountains, Ellesmere Island, Arctic Archipelago: Geol. Survey Can. Bull., v. 203.
———, 1971b, Reconnaissance of Lower Paleozoic geology, Phillips Inlet region, north coast of Ellesmere Island, District of Franklin: Geol. Survey Can. Paper 71-12, (1972).
———, 1973a, Early Paleozoic evolution of northern parts of Canadian Arctic Archipelago, *in* Pitcher, M. G., ed., Arctic geology; Am. Assoc. Petr. Geol. Mem. 19, p. 57-75.
———, 1973b, Preliminary draft of 1:1,000,000 geological atlas sheets Eureka Sound and Robeson Channel Area, Canadian Arctic Islands (MTS 560, 340 and Canadian Part of 120): Geol. Survey Can. Open File Report.
———, et al., 1972, The Innuitian Province, *in* Price, R. A., and Douglas, R. J. W., eds., Variations in tectonic styles in Canada: Geol. Assoc. Can. Spec. Paper 11, p. 83-179.
Wanless, et al., 1965, Age determinations and geological studies, Geol. Survey Can. Paper 64-17 (pt. 1).
Workum, R. H., 1964, Lower Paleozoic salt, Canadian Arctic Islands: Bull. Can. Petr. Geol., v. 13, no. 1, p. 181-191.

Greenland

Berthelson, A., and Noe-Nygaard, A., 1965, The Precambrian of Greenland, *in* Rankama, K., ed., The Precambrian, v. 2, p. 113-262.
Bridgwater, D., et al., 1973, Development of the Precambrian Shield in West Greenland, Labrador, and Baffin Island, *in* Pitcher, M. G., ed., Arctic geology: Am. Assoc. Petr. Geol. Mem. 19, p. 99-116.
Dawes, P. R., 1971, The North Greenland fold belt and environs: Geol. Survey Greenland, Misc. Paper 89.
———, and Soper, N. J., 1973, Pre-Quaternary history of North Greenland, *in* Pitcher, M. G., ed., Arctic geology: Am. Assoc. Petr. Geol. Mem. 19, p. 117-134.
Escher, A., ed., 1970, Tectonic/geological map of Greenland by the Geological Survey of Greenland, scale 1:2,500,000: Geol. Survey Greenland.
Haller, J., 1971, Geology of the east Greenland Caledonides: New York, Wiley-Interscience.
Henriksen, N., and Jepsen, H. F., 1970, K/Ar age determinations on dolerites from southern Peary Land, *in* Report of activities, 1969: Bull. Geol. Survey Greenland, p. 55-58.

The Ancient Continental Margin of Alaska

J. Casey Moore

INTRODUCTION

The Aleutian Arc-Trench system is considered here to be a response to Tertiary and Quaternary subduction bordering the North Pacific Ocean. Similarly, the Mesozoic geology of this region abounds with evidence of ancient subduction zones and volcano-plutonic arcs. The stratigraphic record of these ancient margins is well exposed and marred by few large-scale tectonic reorganizations. Consequently, the geologic history of this region furnishes some of the most straightforward examples of the evolution of accretionary Pacific margins, which, in turn, may serve as models for more-complex, less-well-exposed areas.

The purpose of this paper is to provide an overview of the Mesozoic history of one of the better known northern Pacific margins, that of southwestern Alaska, including the Bering Sea Shelf. However, it is always difficult to summarize the geology of any large area without utilizing broad generalizations that may obscure the complexities on which they are based. Hopefully, this overview will provide the reader with an entrée into the regional literature which describes the many local problems.

For our purposes, the continental margin of southwestern Alaska can be descriptively characterized by a magmatic arc plus derived sedimentary deposits, and a parallel eugeosynclinal sequence.

MAGMATIC ARC AND ASSOCIATED SEDIMENTARY ROCKS

The Mesozoic volcanic, plutonic, and sedimentary rocks of the Alaska Peninsula and their geophysical extension onto the Bering Shelf define an arcuate terrain at least 1,600 km in length. The oldest rocks of this belt are a Permo-Triassic volcanic-sedimentary sequence which comprises scattered outcrops near the head of the Alaska Peninsula and west of Cook Inlet (Burk, 1965). These Permo-Triassic deposits occur predominantly as bordering country rock or roof pendants within an extensive plutonic terrain. Late Triassic limestones, cherts, and minor mafic volcanic rocks conformably overlie the Permo-Triassic rocks and are similarly restricted in extent (Fig. 1) (Burk, 1965; Detterman and Hartsock, 1966). This areal restriction is due at least in part to limited exposure of these older rocks.

Lower Jurassic volcanic breccia, agglomerate, flows, and related sedimentary rocks crop out over large areas west of Cook Inlet (Detterman and Hartsock, 1966) and are believed to underlie much of the Alaska Peninsula (Burk, 1965). This sequence includes volcanic lithologies of andesitic and basaltic composition and reaches a maximum thickness of about 2,700 m (Detterman and Hartsock, 1966).

The Lower Jurassic volcaniclastic sequence is intruded by plutonic rocks of approximately the same age (176-154 my) as determined by concordant K/Ar dates (Reed and Lanphere, 1973). West of Cook Inlet the plutonic rocks comprise a mass of batholithic dimensions which trends more than 400 km to the southwest, where it is covered by younger deposits. According to Reed and Lanphere these intrusive rocks vary in composition from gabbro to quartz monzonite; however, quartz diorite and granodiorite predominate. The identity in time and space between volcanic and plutonic rocks plus their compositional similarity suggest that they formed as co-magmatic sequences (Reed and Lanphere, 1973). The batholith belt is the eroded remnant of a substantial volcanic chain. Magnetic anomaly patterns of the Bering Shelf (Pratt et al., 1973) strongly suggest an arcuate subsurface continuation of the plutonic rocks to the vicinity of the Pribilof Islands (Fig. 1) (Reed and Lanphere, 1973). It is likely that the coeval volcanic-volcaniclastic suite extends along the same curvilinear trend.

Thick Upper Jurassic to Lower Cretaceous shallow marine arkosic deposits record the rapid uplift and erosion of the Lower Jurassic plutonic rocks and provide independent evidence for their westward subsurface extension (Fig. 1) (Burk, 1965). These Upper Jurassic to Lower Cretaceous rocks range from granitic boulder conglomerates, through arkosic to feldspathic sandstones and siltstones, to mudstones, and routinely attain thicknesses of 1,500-3,000 m (Burk, 1965; Detterman and Hartsock, 1966).

Uplift and mild folding marked the mid-Cretaceous of the Alaska Peninsula. During the Upper Cretaceous (Campanian-Maestrichtian) transgression, about 1,200 m of nonmarine to shallow rocks were deposited on the Alaska Peninsula (Burk, 1965). From 83 to 56 my plutonism resumed west of Cook Inlet (Reed and Lanphere, 1973) and probably elsewhere along the magmatic arc. Although no volcanic equivalents of these plutonic rocks have been recognized on the Alaska Peninsula, a Late Cretaceous flysch sequence of the Shumagin-Kodiak Shelf contains large volumes of volcanic detritus apparently derived from effusive centers to the north (Moore, 1973b). Excepting those areas adjacent to the intrusive masses, sedimentary rocks of the Alaska Peninsula were subject to mild deformation during the Mesozoic (Burk, 1965).

The geology of the Alaska Peninsula and adja-

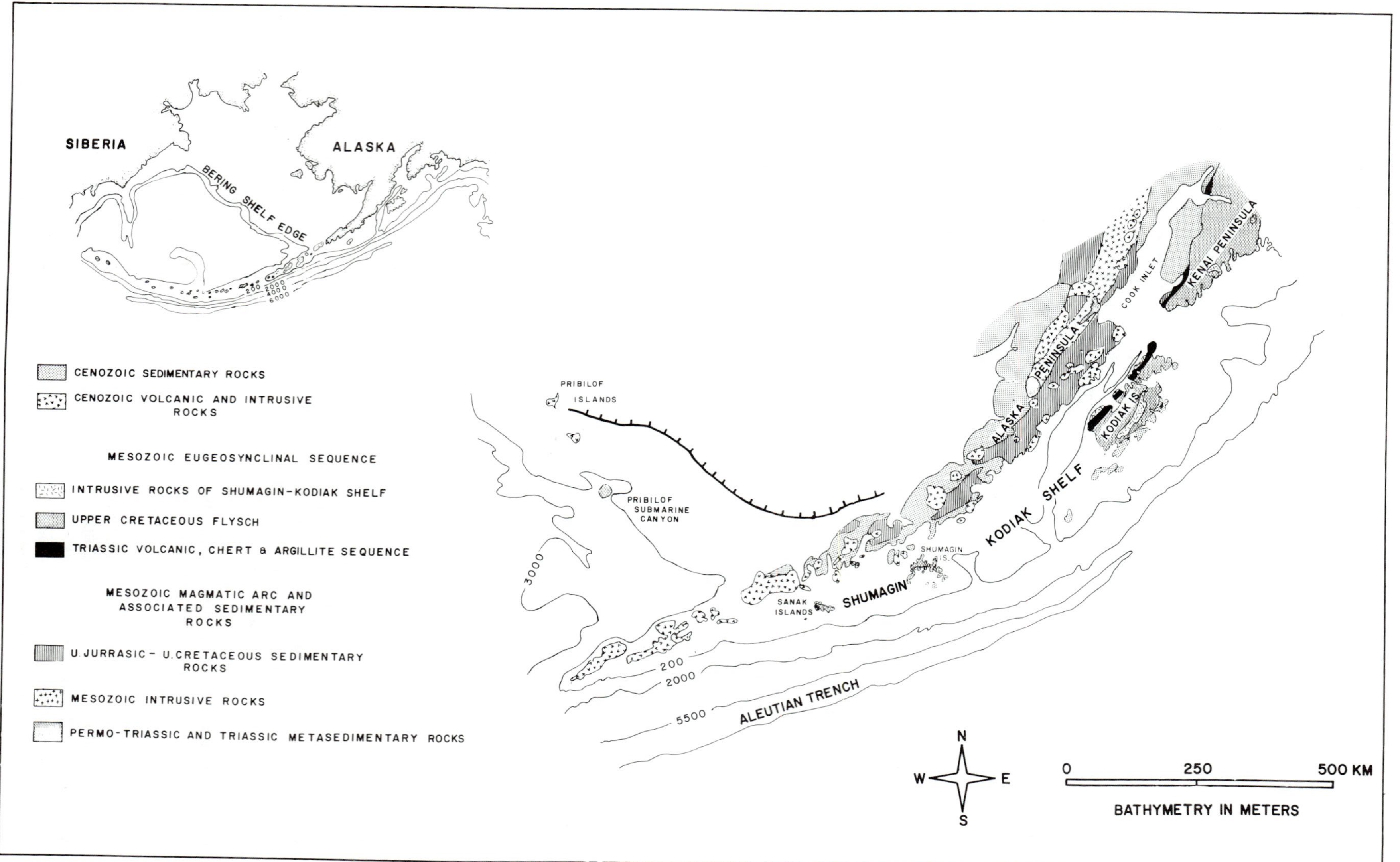

Fig. 1. Regional geology and geophysics of the continental margin of southwestern Alaska (after Burk, 1965; Moore, 1967; Pratt et al., 1973). Hachured line on Bering Shelf separates region of low-frequence, subdued magnetic anomalies (south) from area of high-frequence, large-magnitude anomalies (north).

cent areas to the northeast, plus geophysical data from the Bering Shelf, allows identification of two regional tecto-stratigraphic terrains: (1) a volcano-plutonic arc similar to those bordering modern trenches, and (2) a widespread, mildly deformed, predominantly shallow-marine clastic sedimentary sequence deposited in a forearc setting (arc-trench gap of Dickinson, 1971), primarily in response to volcanic activity plus uplift and erosion of a plutonic terrain.

EUGEOSYNCLINAL SEQUENCE

The continental margin of southern Alaska is underlain by one of the most extensive and complete accretionary eugeosynclinal sequences known anywhere in the world (Burk, 1972). In this region, exposed deep-sea rocks range in age from Triassic to Lower Tertiary (Moore, 1967; Plafker and McNeil, 1966; Clark, 1972; Stoneley, 1967); Miocene deep-sea rocks have been dredged from the shelf edge (von Huene, 1972); and Pleistocene deep-sea deposits have been cored from the lower continental slope (Kulm et al., 1973). In total, these rocks define a sequence that systematically decreases in age seaward. The Mesozoic rocks were accreted along an arcuate zone defined by the Shumagin-Kodiak Shelf and the Bering Shelf edge, whereas the Tertiary and Quaternary rocks were deformed and uplifted during tectonism related to the Aleutian Arc-Trench system.

The Triassic rocks of the Kodiak Islands and Kenai Peninsula are the oldest portion of the eugeosynclinal sequence. These deposits are composed predominantly of pillow lavas, volcanic tuff-breccia, chert, and argillite, with minor sandstone and limestone (Moore, 1969; Capps, 1937; Martin et al., 1915). Crystalline schists and ultramafic bodies are included as exotic tectonic blocks (Forbes and Lanphere, 1973; B. Hill, personal communication, 1973). Forbes and Lanphere (1973) have dated crystalline schists (blue schist and green schist facies) as Late Triassic-Early Jurassic ($\sim$180-190 my). The Triassic sequence is complexly deformed; local exposures may be described as "broken formations" or "mélanges." The total thickness is unknown. Chert and basic volcanic rocks in the Sanak Islands (Moore, 1973a) and a peridotite body in the Pribilof Islands (Barth, 1956) argue for the extension of facies equivalents of the above-described Triassic rocks along an arcuate westerly trend.

Upper Cretaceous flysch ("slate and graywacke belt") underlies a large proportion of the continental shelf seaward of the Triassic pillow lava, chert, and argillite sequence (Fig. 1). The flysch deposits are subaerially exposed on the Kenai Peninsula, and in the Kodiak, Shumagin, and Sanak Islands. Fossils from this terrain indicate a Maestrichtian age (Jones and Clark, 1973). These rocks are thought to be broadly equivalent to lithologically similar rocks in southeastern Alaska, which yield fossils of earliest Cretaceous (Berriasian) and Late Cretaceous (Campanian) ages (Jones and Clark, 1973). Campanian turbidite beds dredged from the acoustic basement of the Bering Shelf edge (Hopkins et al., 1969) are also broadly correlated with the Late Cretaceous flysch of the Shumagin-Kodiak Shelf.

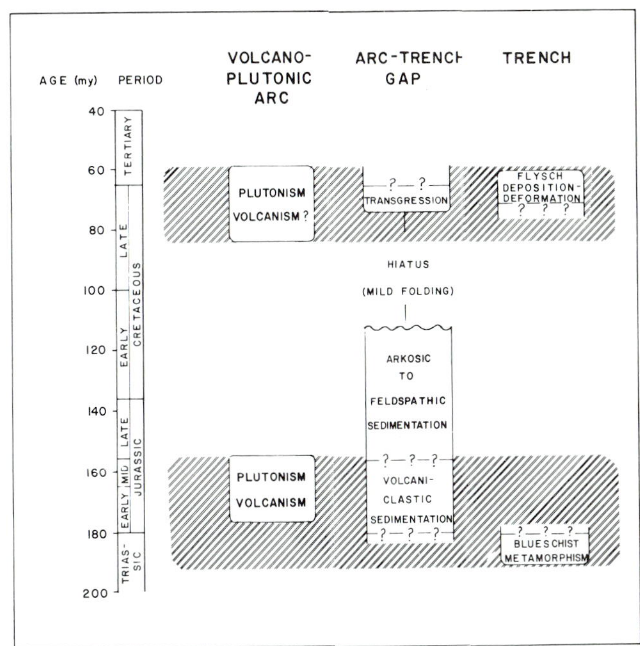

Fig. 2. Correlation of events between magmatic arc and eugeosynclinal sequence, southwestern Alaska (radiometric ages from Reed and Lanphere, 1973; Forbes and Lanphere, 1973).

The Late Cretaceous flysch has been intensively studied near the junction of the Shumagin-Kodiak Shelf and the Bering Shelf edge (Moore, 1973a, 1973b). Here, in the Shumagin and Sanak islands, the sequence consists predominantly of turbidity current deposits. Analysis of more than 500 basal current marks indicates primarily longitudinal flow with lateral feed from the north. Sandstones are composed mostly of primary volcanic detritus. No rocks older than the Cretaceous flysch are found on their seaward side—a southern confining edge to these deposits is unknown.

The flysch sequence was apparently deposited in an elongate trough at the edge of the continental margin. A deep-sea trench is the most probable modern analogue for this basin (Moore, 1972), although some disagreement exists (Scholl and Marlow, 1972). A trench permits the axial confinement of any lateral turbidity flow and would likely accumulate volcanic-rich detritus derived from the associated magmatic arc. Recent mapping of equivalent Cretaceous rocks on the Kenai Peninsula reveals a stratigraphic sequence indicative of trench deposition (Budnik, 1974).

The Cretaceous flysch sequence shows complex deformation with local development of broken formations but no mélanges. In south-central Alaska, equivalent rocks are associated with a chaotic sequence that has been interpreted as a mélange

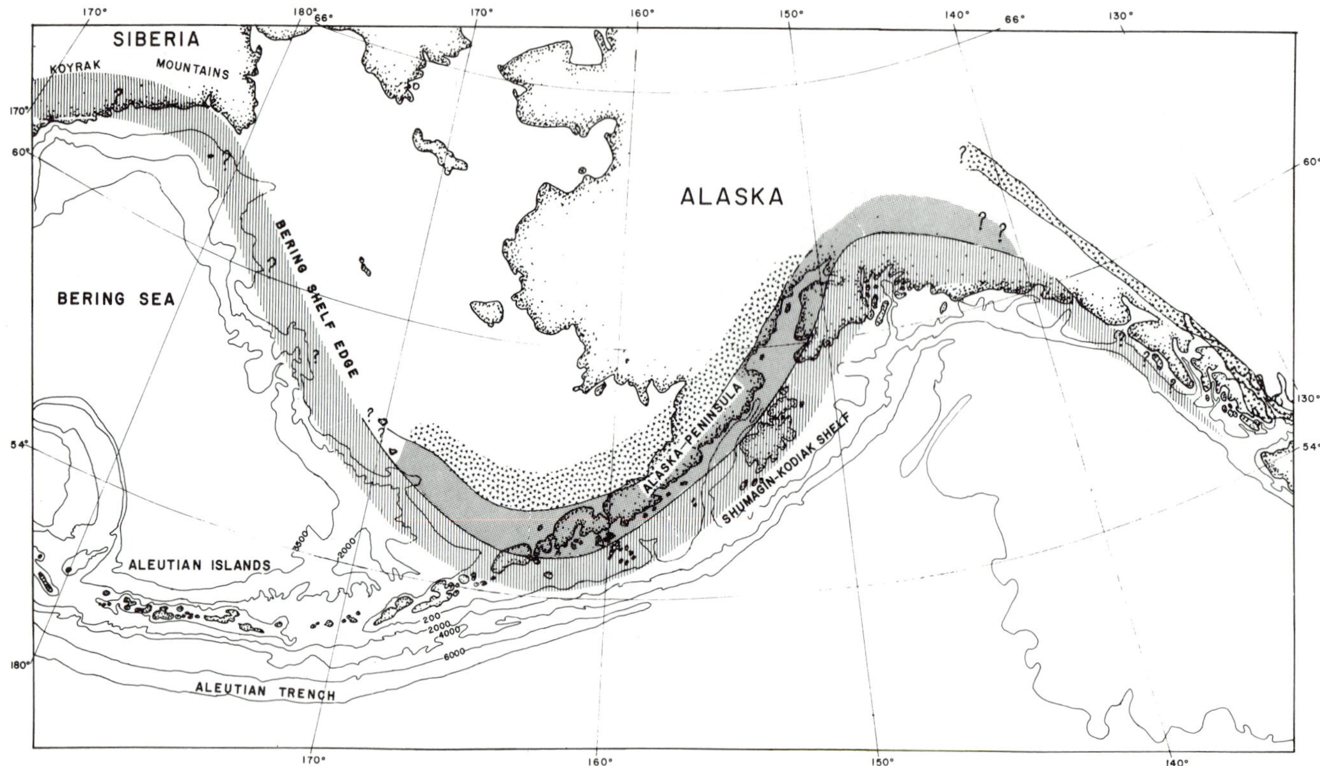

Fig. 3. Proposed extent of Mesozoic Alaska Arc-Trench system (interpretation east of 150° from Berg et al., 1972; of Koryak Mountains from Burk, 1965; Moore, 1972; Scholl et al., in press.) Proposed extent of Mesozoic trench indicated by vertical lines, arc-trench gap by stippling, volcano-plutonic arc by random checks.

(Clark, 1973). In the Shumagin and Sanak islands, fold axes trend along the sedimentary strike and axial surfaces of folds are strongly overturned seaward. The rocks were uplifted and intruded by plutons by 60 my (Moore, 1973a). Both structural and paleocurrent data trend from the Shumagin-Kodiak Shelf and parallel the Bering Shelf edge, suggesting a continuity in sedimentation and structural evolution before the development of the Aleutian Arc-Trench system (Fig. 3).

The geology of the eugeosynclinal sequence reveals at least two phases of subduction-related accretion along the Shumagin-Kodiak Shelf and Bering Shelf edge. One phase may be dated as Late Triassic-Early Jurassic by the metamorphism of the crystalline schists of the Kenai Peninsula. The deposition, deformation, and uplift of the flysch sequence before 60 my brackets the second. It is possible that as work continues in this region, Late Jurassic-Early Cretaceous metamorphism and deformation will be recognized.

CONCLUSIONS

The westward arcuate bend of the Mesozoic marginal orogen of southwestern Alaska is established by the magnetic extension of the volcano-plutonic arc and the continuity of sedimentary structural trends along the composite edge of the Shumagin-Kodiak and Bering shelves. Apparently, the youngest rocks involved in the Mesozoic arc-trench system are latest Cretaceous, whereas the Aleutian Arc-Trench system is thought to have been initiated in the earliest Tertiary (Burk, 1965; Scholl et al., in press). The transition from the Mesozoic arc-trench system then occurred approximately at the Cretaceous-Tertiary boundary. The 60-my plutonic bodies intruding the eugeosynclinal sequence of the Alaska Peninsula represent an anomalous seaward transition of the magmatic front, which may be due, in part, to reorganization of plate boundaries at approximately this time.

A strong correlation exists between phases of magmatism in the arc and accretion-metamorphism-deformation at the trench during the Mesozoic in southwestern Alaska (Fig. 2) (Reed and Lanphere, 1973). The potassium-argon ages of both the blue schist facies metamorphism in the eugeosynclinal sequence and the Jurassic plutons of the magmatic arc overlap within the range of analytical uncertainty. Deposition, deformation, and uplift of the flysch sequence occurred during the second discrete phase of plutonism in the magmatic arc.

The Mesozoic arc-trench couple outlined herein extends for approximately 2,000 km; some data suggest that a similar arc-trench system may have bordered all of the Pacific margin of Alaska at various times during the Mesozoic. Berg et al. (1972) have proposed an Upper Jurassic to Lower Cretaceous arc-trench pair in southeastern Alaska. This system was active during the apparent quiescence of the volcanic arc and trench along the Alaska Peninsula. The possible alternate activity of these nearly orthogonally oriented regions may be accounted for by a change in direction and magnitude of plate motion (assuming that the oroclinal bend existed at that time). Burk (1965), Scholl et al. (in press), and Moore (1972) have speculated that the eugeosynclinal sequence of the Shumagin-Kodiak Shelf and Bering Shelf edge may link with highly deformed Triassic-Cretaceous deep-sea deposits of the Koryak Mountains in eastern Siberia. Thus the entire edge of what is now continental Alaska was bordered by a consuming continental margin during the Mesozoic Era. Accretion of deep-sea deposits and magmatism were not simultaneously active along the composite Alaska Arc-Trench system. Nevertheless, the geometry of this orogenic belt outlines a probable Mesozoic plate boundary, which was significantly different from the modern suture defined by the Aleutian Arc-Trench.

SUMMARY

A review of the Mesozoic geology of southwestern Alaska defines a magmatic arc, a complexly deformed eugeosynclinal (deep-sea) sequence, and a mildly deformed shallow-marine accumulation of the arc-trench gap. These tectono-stratigraphic facies comprise an arcuate orogen extending for approximately 2,000 km along the Alaska Peninsula, Shumagin-Kodiak Shelf, and Bering Shelf. Potassium-argon ages of blueschist metamorphic rocks and of plutonic bodies, and the accretion and uplift of trench deposits define combined phases of magmatism and subduction during the Early Jurassic and Late Cretaceous.

The tecto-stratigraphic facies of southwestern Alaska have lithologic but not exact time equivalents in southeastern Alaska and in the Koryak Mountains of eastern Siberia. The geology of this composite continental margin defines the Alaska Arc-Trench system which was variously active during the Mesozoic. The Alaska Arc-Trench outlines the Mesozoic plate boundary between Alaska and the Pacific Ocean, and diverges significantly from the modern boundary defined by the Aleutian Arc-Trench.

ACKNOWLEDGMENTS

The author has benefited from discussions with Betsy Beyer Hill and from critical reviews by Eli Silver, James Gill, and Roy Budnick. Acknowledgment is made to the Donors of the Petroleum Research Fund, administered by the American Chemical Society, for support during the preparation of this manuscript.

BIBLIOGRAPHY

Barth, T. W. F., 1956, Geology and petrology of the Pribilof Islands, Alaska: U.S. Geol. Survey Bull., v. 1028-F, p. 110-160.

Berg, H. C., Jones, D. L., and Richter, D. H., 1972, Gavina-Nutzotin belt-tectonic significance of an Upper Mesozoic sedimentary and volcanic sequence in southern and southeastern Alaska: U.S. Geol. Survey Prof. Paper 800D, p. D1-D24.

Budnik, R. T., 1974, (in press) Deposition and deformation along an Upper Cretaceous consumptive plate margin, Kenai Peninsula, Alaska (abstr.): Geol. Soc. America, Absts. with Programs, v. 6, no. 1.

Burk, C. A., 1965, Geology of the Alaska Peninsula-island arc and continental margin: Geol. Soc. America Mem. 99, pts. 1,2,3.

———, 1972, Uplifted eugeosynclines and continental margins, *in* Shagan, R., et al., eds., Studies in earth and space science: Geol. Soc. America Mem. 132, p. 75-86.

Capps, S. R., 1937, Kodiak and adjacent islands, Alaska: U.S. Geol. Survey Bull., v. 880C, p. 111-184.

Clark, S. H. B., 1972, Reconnaissance bedrock geologic map of the Chugach Mountains near Anchorage, Alaska: U.S. Geol. Survey Misc. Field Studies Map MF-350.

———, 1973, The McHugh Complex of south-central Alaska: U.S. Geol. Survey Bull., v. 1372-D.

Detterman, R. L., and Hartsock, J. K., 1966, Geology of the Iniskin-Tuxedni region, Alaska, U.S. Geol. Survey Prof. Paper 512: Washington, D.C., U.S. Govt. Printing Office, 78 p.

Dickinson, 1971, Clastic sedimentary sequences deposited in shelf, slope and trough settings between magmatic arcs and associated trenches: Pacific Geology, no. 3, p. 15-30.

Forbes, R. B., and Lanphere, M., 1973, Tectonic significance of mineral ages of blueschists near Seldovia, Alaska: Jour. Geophys. Res., v. 78, p. 1383-1386.

Hopkins, D. M., Scholl, D. W., Addicott, W. O., Pierce, R. L., Smith, P. B., Wolfe, J. A., Gershanovich, D., Kotenev, B., Lohman, K. E., Lipps, J. H., and Obradovich, J., 1969, Cretaceous, Tertiary, and Early Pleistocene rocks from the continental margin in the Bering Sea: Geol. Soc. America Bull., v. 80, p. 1471-1480.

Jones, D. L., and Clark, S. H. B., 1973, Upper Cretaceous (Maastrichtian) fossils from the Kenai-Chugach Mountains, Kodiak, and Shumagin Islands, southern Alaska: U.S. Geol. Survey Jour. Res., v. 1, p. 125-136.

Kulm, L. D., von Huene, R., Duncan, J. R., Ingle, J. C., Kling, S. A., Musich, L. F., Piper, D. J. W., Pratt, R. M., Schrader, H., Weser, O. E., and Wise, S. W., Jr., 1973, Site 181, *in* Musich, L. F., and Weser, O. E., eds., Initial reports of the Deep Sea Drilling Project, v. 18: Washington, D.C., U.S. Govt. Printing Office, ch. 12, p. 407-448.

Martin, G. C., Johnson, B. L., and Grant, U. S., 1915, Resources of the Kenai Peninsula: U.S. Geol. Survey Bull., v. 587, p. 209-238.

Moore, G. W., 1967, Preliminary geologic map of Kodiak Island and vicinity, Alaska: U.S. Geol. Survey Open File Map, Menlo Park, Calif.

———, 1969, New formations on Kodiak and adjacent islands, Alaska: U.S. Geol. Survey Bull., v. 1247A, p. A27-A35.

Moore, J. C., 1972, Uplifted trench sediments: southwestern Alaska Bering Shelf edge: Science, v. 175, p. 1103-1105.

———, 1973a, Complex deformation of Cretaceous trench deposits, southwestern Alaska: Geol. Soc. America Bull., v. 84, p. 2005-2020.

———, 1973b, Cretaceous continental margin sedimentation, southwestern Alaska: Geol. Soc. America Bull., v. 84, p. 595-614.

Plafker, G., and MacNeil, F. S., 1966, Stratigraphic significance of Tertiary fossils from the Orca Group in the Prince William Sound region, Alaska: U.S. Geol. Survey Prof. Paper 550-B, p. B62-B68.

Pratt, R. M., Rutstein, M. S., Walton, F. W., and Buschur, J. A., 1973, Extension of Alaskan structural trends beneath Bristol Bay, Bering Shelf, Alaska: Jour. Geophys. Res., v. 77, p. 4994-4999.

Reed, B. L., and Lanphere, M. A., 1973, Alaska-Aleutian range batholith: geochronology, chemistry, and relation to circum-Pacific plutonism: Geol. Soc. America Bull., v. 84, p. 2583-2610.

Scholl, D. W., and Marlow, M. S., 1972, Sedimentary sequence in modern Pacific trenches and the deformed Pacific eugeosyncline (abstr.): Conf. Modern and Ancient Geosynclinal Sedimentation, Madison, Wisc.

———, Buffington, E. C., and Marlow, M. S., in press, Plate tectonics and the structural evolution of the Aleutian-Bering Sea region—solutions and complications, *in* Forbes, R. B., ed., The geophysics and geology of the Bering Sea region: Geol. Soc. America Spec. Paper.

Stoneley, R., 1967, Structural development of the Gulf of Alaska sedimentary province in southern Alaska: Quart. Jour. Geol. Soc. London, v. 123, p. 25-57.

von Huene, R., 1972, Structure of the continental margin and tectonism at the eastern Aleutian Trench: Geol. Soc. America Bull., v. 83, p. 3613-3626.

The Ancient Continental Margin of Japan

Toshio Kimura

INTRODUCTION

The late Paleozoic Chichibu eugeosyncline in Japan was like a marginal sea but not a subduction zone. Since late Permian, secondary eugeosynclines, similar to small basins between island arcs and trenches, have been formed successively. The present disposition of the Japan Sea, Japanese Islands, Toki and other basins, and Nankai Trench in southwest Japan is quite similar to that of ancient Japan and neighboring sedimentary basins.

FAR EAST CONTINENTAL AREA AND JAPANESE ISLANDS

The geology of the continental area of the Far East is quite different from that of the Japanese Islands. Precambrian and Cambro-Ordovician strata are widely distributed in China and on the Korean Peninsula, but except for some Precambrian rocks in the Hida zone (Fig. 1, H) (Shibata et al., 1972), not in Japan. Silurian, Devonian, and lower Carboniferous strata are almost lacking in north China and the Korean Peninsula, but are widely distributed in Japan. During the Carboniferous and Permian the continental area was land or shallow-water shelf, while Japan was mostly eugeosynclinal.

Ancient Japan has been situated near the continent since Precambrian. The Hida zone at the northwestern corner of Japan yields Precambrian rocks. Silurian coral reef limestones (Takai et al., 1963) and Devonian plant fossils similar to those in China occur in southern Kitakami and in central Shikoku along the southern margin of ancient Japan. The Upper Triassic Miné Group yields the same freshwater estherians as the Taedong (Daido) Group in the Korean Peninsula (Kobayashi, 1954). The geology of the peninsula, however is quite different from that of Japan. The east-trending Median tectonic line in southwest Japan (Fig 1, M) curves strongly to the southwest in west Kyushu. It probably extends to Okinawa, because the Outer Zone of southwest Japan to the south of the Median tectonic line is traceable to Okinawa (Konishi, 1965).

The western part of southwest Japan was probably connected to the Korean Peninsula in the early Cretaceous. The Kanmon (Inkstone) Group is nearly the same in lithofacies, yielding the same freshwater fossils as the Kyŏnsang Group in the peninsula (Kobayashi and Suzuki, 1936). In the late Cretaceous and Paleogene continental deposits and/or rhyolitic volcanics were widely distributed in Japan. In the Paleogene, coal measures were formed in north Kyushu. On the other hand, late Paleogene strata in Tsushima, between the Korean Peninsula and North Kyushu, are shallow marine or deltaic with different lithofacies from those of Kyushu (Matsumoto, 1969). Sole marks are well developed there, showing paleocurrent directions from southwest to northeast (Nagahama, 1967), from the Yellow Sea to the Japan Sea (Fig. 5). A marine basin probably existed in the middle of the present Japan Sea.

Early Miocene "Green Tuff" volcanism on the Japan Sea side of Japan indicates crustal extension there. Early Miocene strata contain cold-climate Aniai-type flora, while in the middle Miocene the warm-climate Daijima type predominated. Fujioka (1972) is of the opinion that invasion of warm currents in the Japan Sea at the Daijima stage formed the warm climate. The Japan Sea was almost completely formed by the middle Miocene. The paleocurrents in Tsushima show that subsidence of the proto-Japan Sea had begun in the late Paleogene.

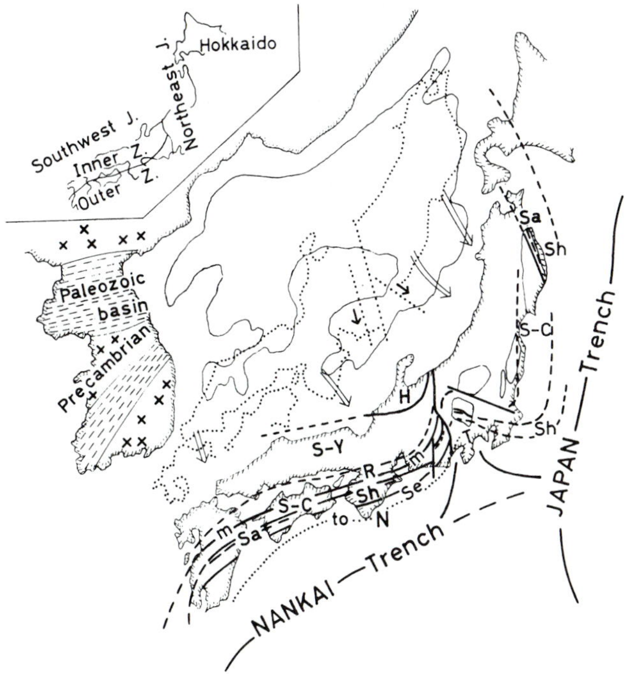

Fig. 1. Tectonic zones in Japan. (H, Hida zone; S-Y, Sangun-Yamaguchi zone; R, Ryoke zone; S-C, Sambagawa-Chichibu zone; S, Sambosan zone; Sh, Shimanto zone; Se, Setogawa zone; N, Neogene eugeosyncline; T, Tanzawa; to, Toki basin; m, median tectonic line. (Dotted lines, position of Japan before opening of Japan Sea.)

Upper Triassic strata in the northwestern corner of Japan are similar to those in Ussuri (Fig. 3) (Kobayashi, 1956). The Cretaceous geosyncline in Hokkaido received abundant "acidic" volcanic materials from the west (Matsumoto and Okada, 1971), probably from Shikhote-Alin. Japan began to separate from the Korean Peninsula in the Paleogene as the result of drifting (Murauchi, 1972). On the other hand, ancient Japan was probably almost straight (Kawai et al., 1961), not curved as it is at present. The configuration of Japan in the Cretaceous was probably as shown in Figure 4.

CHICHIBU EUGEOSYNCLINE AND SIALIC BASEMENT

The Chichibu ("Honshu") geosyncline (Kobayashi, 1941) was a sedimentary basin from Silurian to Cretaceous that cover the main part of Japan. To the north of the geosyncline was the Hida zone, a terrain of Precambrian and late Paleozoic land or shelf. Along the southeastern margin, in the *Kurosegawa terrain*, pre-Silurian basement rocks now crop out along faults within the Kurosegawa tectonic zone (Ichikawa et al., 1956). The late Paleozoic deposits are thinner in the southern part of the Chichibu geosyncline than in the central part and were formed primarily in a shallow-water environment in the Mesozoic (Ogawa, 1974). There were shallow-water swells with pre-Silurian basement below them which occupied a much wider area than the present Kurosegawa tectonic zone. Late Paleozoic eugeosynclinal deposits occur also farther south of the Kurosegawa and are sometimes widespread, as in west Kyushu. Eugeosynclinal conditions prevailed from lower Carboniferous to middle Permian. In late Permian and Triassic, the northern part formed into landmasses after folding of the Akiyoshi series; the main part remained as a miogeosyncline with cherts and other clastic deposits but without basaltic volcanics. In the Cretaceous the entire geosynclinal area, especially the southern half, was intensely folded during the Sakawa series of orogenies.

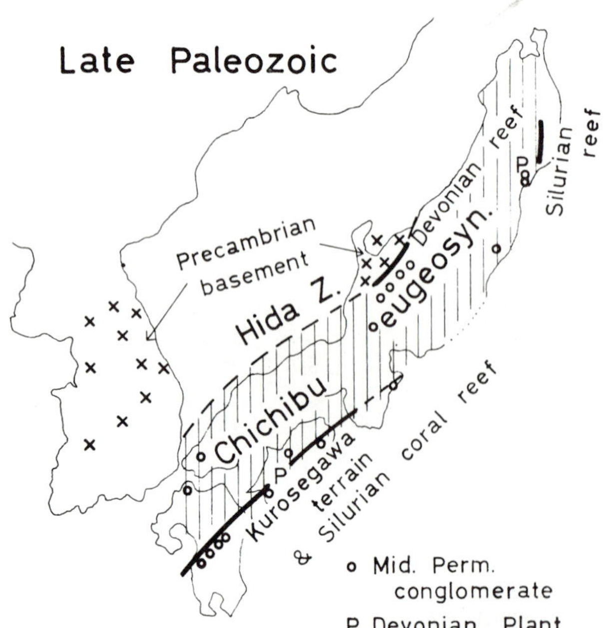

Fig. 2. Eugeosyncline in Japan in late Paleozoic.

In the Chichibu eugeosynclinal area between the Hida zone and the Kurosegawa terrain, pre-Silurian basement occurs in some places (Minato et al., 1965; Kimura, 1973). Coarse clastics in the geosynclinal strata show the presence of sialic islands or submarine swells in the eugeosyncline. To the north of the Ryoke zone in west Chugoku (Fig. 1) the Permo-Carboniferous Ota Group contains thick sandstone. Farther north the Akiyoshi coral reef limestone is of the same age. The sandstone of the Ota Group may have been transported from the south (Kimura, 1960). In the Sambagawa metamorphic zone, south of the Ryoke, conglomerates and very thick quartzofeldspathic Koboke or Oboke sandstones occur, which have been metamorphosed. These clastics have either been transported from landmasses situated in the present Ryoke zone (Ichikawa, 1970) or from the Kurosegawa terrain. Permian strata often contain conglomerates with granite pebbles (Kanmera, 1961). These Carboniferous and Permian quartzofeldspathic sandstones and conglomerates are widely distributed in the Chichibu eugeosynclinal strata.

At present, the granitic layer is rather thick beneath Japan, over 20 km in central Japan (Asada and Asano, 1972), where geosynclinal strata from Silurian to Permian are only about 5 km thick (Yoshida, 1972). There may have been a thick sialic basement during the geosynclinal stage. Geosynclinal strata in Japan form generally intense isoclinal folds at outcrops, which often are accompanied by faults along the fold axes. Therefore, the isoclinal strata have often been erroneously taken to be monoclinal and have tremendous thickness. However, recent studies (Kimura and Tokuyama, 1971) show that the eugeosynclinal strata in Japan are rather thin. The Upper Carboniferous to Upper Triassic strata in the northern Kanto mountains do not exceed 1,000 m. The thick granitic layer and thin eugeosynclinal strata suggest the existence of a rather thick sialic basement beneath the geosynclinal strata. However, the possibility of nappe structures beneath the geosynclinal strata or a thick accumulation of sialic materials beneath the strata carried along the subduction zone raise the question of whether a thick granitic layer beneath Japan means a thick ancient sialic basement.

SEDIMENTARY FACIES OF THE CHICHIBU EUGEOSYNCLINAL STRATA

Silurian and Devonian strata (Fig. 2) are composed of coral reef limestones, rhyolitic and andesitic volcanics, tuffaceous sandstone, and conglomerate and shale with Devonian plant fossils and are not true eugeosynclinal strata. The Hida zone and the Kurosegawa terrain may once have been united or at least close together and later separated to form an eugeosynclinal basin between them much like the present Lau basin near Tonga (Slater et al., 1972). There was a crustal extensional stage during the eugeosynclinal volcanisms (Sugisaki et al., 1971). The Lower-Middle Devonian or pre-Devonian Motai and Matsugadaira groups in Kitakami and Abukuma (Fig. 11) have basaltic volcanics. While they may have been a sialic basement beneath the Chichibu eugeosyncline (Minato et al., 1965), it is possible that the basement consists of eugeosynclinal deposits emplaced at the initial extensional stage of the geosyncline between two zones with rhyolitic and andesitic volcanisms.

Lower Carboniferous strata are characterized by basaltic volcanics, mainly pyroclastics with subordinate pillow lavas, throughout the Chichibu eugeosyncline (Kanmera, 1971; Kimura, 1974). The Mikabu green-rocks in the Sambagawa metamorphic zone in Kanto are probably Lower Carboniferous, and contain within them many sheets of ultramafic rocks.

Visean and Upper Carboniferous strata are not so rich in basaltic volcanics as the Lower Carboniferous. Limestones are well developed instead of the volcanics in the area of limestone facies, and clastics, including turbidites and very thick massive quartzo-feldspathic sandstones in the area of nonlimestone facies. Even in the high-pressure Sambagawa metamorphic zone there are psammitic schists, such as the Koboke or Oboke. These clastics, which are conformably covered by Permian basaltic volcanics, have been deposited close to their sources, some of which are on the oceanic side (Kimura, 1974).

Lower-Middle Permian strata in almost all parts of the Chichibu geosyncline contain abundant basaltic volcanics, especially pyroclastics. The "Mikabu" green schists in Shikoku and Kii are of this age (Fig. 11). The limestone facies in the Inner Zone were formed generally upon shallow-water mounds (Kimura, 1974). The clastic facies are often composed of turbidites and thick massive sandstones, which are intercalated between basaltic volcanics. Visean-middle Permian strata in southern Kitakami are principally composed of limestone of shelf facies and clastics.

The Lower Carboniferous-middle Permian strata in the Chichibu eugeosyncline mentioned above were not deposited on the deep ocean floor but in a marginal sea or basin between the Hida zone and the Kurosegawa terrain. Basaltic volcanics developed between two zones with rhyolitic and andesitic volcanics that are similar to those of interarc basins (Karig, 1971). No subduction zone and no mélange existed in the eugeosynclinal realm (Kimura, 1974), although frequently small-scale submarine slide deposits are found at particular horizons.

TRANSFORMATION OF THE CHICHIBU EUGEOSYNCLINE INTO A MIOGEOSYNCLINE

Basaltic volcanism in the late Paleozoic almost ceased by the late Carboniferous in southern Kitakami and by the late Permian in the main part of the geosynclinal area. Triassic as well as Upper Permian strata have thick turbidites and thick-bedded cherts, which are geosynclinal but without basaltic volcanics (Fig. 3).

In some parts of the limestone terrain in the Inner Zone, as in Akiyoshi, limestone was deposited continuously from Middle to Upper Permian without any depositional gap (Toriyama, 1954). Upper Permian strata are, however, composed mostly of clastics, including turbidites. They cover conformably middle Permian strata containing basaltic volcanics in the Maizuru zone. They also conformably cover middle Permian limestone in central Chugoku (Fig. 11) (Hase, 1964) but are unconformably over the limestone in western Akiyoshi (Fujii and Minakami, 1970). The Upper Permian clastics are well developed in the northern part of the Chichibu geosyncline as in the Maizuru zone (the Maizuru formation) and central Chugoku, and in the southern part of it along the Kurosegawa terrain (the Kuma formation). The Upper Permian clastics are also well developed in southern Kitakami (the Toyoma Group). Conglomerates with granitic rock pebbles (the Yasuba and Usuginu conglomerates) are common especially along the marginal parts of the geosyncline. The uppermost Permian near Maizuru consists of deltaic deposits (the Gujo Formation) (Nakazawa and Nogami, 1958), suggesting the existence of a landmass to the north of the geosyncline.

Lower Triassic (Scythian-Anisian) strata are

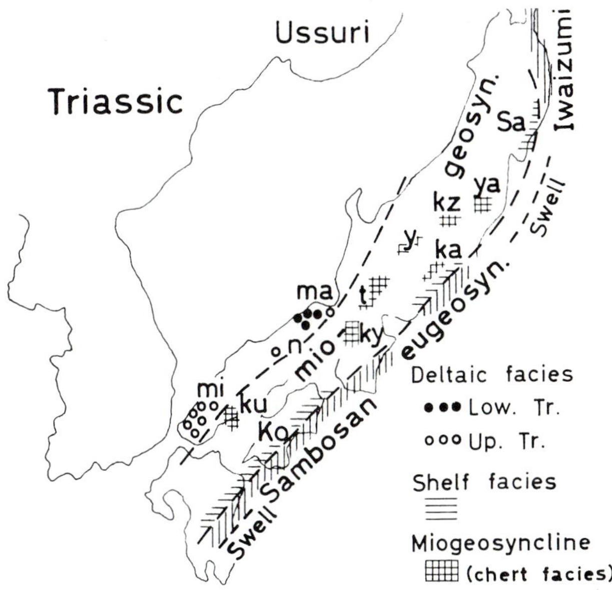

Fig. 3. Geosynclines in Japan in Triassic. (ma, Maizuru; n, Nariwa; mi, Miné; ku, Kuga; ky, to the west of Kyoto; t, Tanigumi and Inuyama; y, Yabuhara; kz, Kuzuu; ya, Yamizo mountains; ka, Kanto mountainous land; Sa, Saragai formation; Ko, Kochigatani formation.)

distributed in nearly the same areas as the Upper Permian: the northern and southern parts of the Chichibu geosyncline and southern Kitakami. Those in the northern part are the Lower Triassic Yakuno group and associated strata in the Maizuru zone (Nakazawa, 1958). Included in these are deltaic facies in the north and offshore facies in the south. The strata of the deltaic facies lie unconformably on intensely folded Paleozoic strata. Lower Triassic strata in the southern part are not well developed. They are of offshore or shelf facies at Kamura in Kyushu, Taho in West Shikoku, and Iwai in Kanto. The Lower Triassic limestone and the conformable middle Permian one form a single rather thin limestone bed at Kamura (Kambe, 1963). In the central part of the geosyncline, Lower Triassic strata are reported to occur to the north of the Kurosegawa terrain at Kurotaki in central Shikoku and at Shionosawa in Kanto and are offshore limestone. The Lower Triassic Inai Group in southern Kitakami is mainly composed of slate with subordinate sandstone. The group covers the Upper Permian Toyoma Slate with slight unconformity.

Upper Triassic (Ladinian-Norian) strata in the northern part of the Chichibu geosynclinal area are generally deltaic. The Molasse-like Miné Group, about 7,000 m thick (Tokuyama, 1962), in western Chugoku covers the intensely folded Sangun metamorphics. Strata composed of thick cherts, thick massive sandstones, turbidites, and other clastics in the southern Sangun-Yamaguchi zone have long been thought to be of Paleozoic age. Recent finds of conodonts (Koike et al., 1971) from the cherts show that they are Upper Triassic. The Chichibu geosyncline was thought by some geologists to have been dried up in Triassic time except for some small inland basins (Minato et al., 1965), but this is not true.

Upper Triassic strata conformably cover middle Permian strata at Yabuhara in central Japan (Kano, 1972), where the Upper Triassic strata were affected by the Ryoke metamorphism (Ono, 1969). Some cherts at Kuzuu in southern northeast Japan cover middle Permian limestone disconformably, with fossil soils and breccia beneath them (Shoji, 1967; Yanagimoto, 1973). The thick-bedded cherts in the area alternate with very thick quartzo-feldspathic sandstone. Some thin beds of chert about 5 cm thick are interbedded with such sandstones at Karasawayama. The Triassic cherts with abundant radiolarian fossils in the central part of the Chichibu geosyncline are not of deep ocean origin but have been formed in a rather shallow water basin.

The Kuga Group near Iwakuni, about 60 km east of the Miné basin, has also been recently reported to be Upper Triassic, and continues into the strata of the Ryoke metamorphics to the south. The group is composed of cherts and thick sandstone beds together with thick submarine slumping deposits containing Permian limestone blocks (Toyohara, 1973). The group also contains sandy mudstone yielding *Monotis* (Hase, 1961). The Kuga Group is of a facies intermediate between the Miné and the strata in the center of the Chichibu geosynclinal area.

Upper Triassic cherts in the Sambagawa-Chichibu zone to the south of the Ryoke zone are found in the northern (the Kamiyoshida formation) and middle (the Ryogami formation) parts of the mountains of Kanto (Koike et al., 1971).

Upper Triassic strata along the southern margin of the Chichibu geosyncline are the Kochigatani Group, which is mainly marine, composed of sandstone. It is very thin compared with the Miné Group, but yields the same molluscan fossils. Therefore, it has been thought to be of offshore-shelf origin, immediately to the south of the Miné or other deltas. The Upper Triassic strata of the chert facies, however, are distributed between them (Fig. 3). The Upper Triassic in southern Kitakami, the Saragai formation, is similar to the Kochigatani in litho- and biofacies. Upper Triassic submarine swells existed in or near the Kurosegawa terrain.

Thus there is an Upper Triassic geosynclinal facies with bedded cherts and turbidites in a central basin between a northern deltaic facies and a southern shelf facies built up on submarine swells. Acidic tuffs occur in the geosynclinal strata, but basic ones have not been reported. This basin is situated in the center of the older Chichibu eugeosyncline, and a miogeosyncline has been produced there. The miogeosynclinal stage may have continued until the Upper Jurassic, at least in the southern Sangun-Yamaguchi and Ryoke zones, because Upper Jurassic ammonites have been reported from the clastics at Inuyama near Nagoya and the southern Yamizo range, the eastern continuation of the southern part of the Sangun-Yamaguchi zone in northeast Japan (Suzuki and Sato, 1972). The Jurassic strata cover conformably or disconformably the Upper Triassic.

SECONDARY EUGEOSYNCLINES

When the basaltic volcanisms waned in the main part of the Chichibu eugeosyncline, similar volcanism began or continued in the Sambosan eugeosyncline (Fig. 3) to the south of the Kurosegawa terrain. Eugeosynclines similar to the Upper Permian-Triassic Sambosan, (Upper Jurassic-) Cretaceous Shimanto (Fig. 4), and Paleogene Setogawa (Fig. 5) (Nakamura) eugeosynclines formed later on the oceanic sides of the older ones (Kimura and Tokuyama, 1971). They are all long, narrow depositional basins and very small in scale or secondary eugeosynclines, compared with the primary Chichibu eugeosyncline. They are similar to small basins between islands arcs and trenches as reported from Okinawa, Tonga, Aleutian, and others (Emery et al., 1969; Karig, 1970; Marlow et al., 1973).

The Toki and other small basins off southwest Japan, the western continuation of the Neogene Tanzawa geosyncline to the south of the Setogawa,

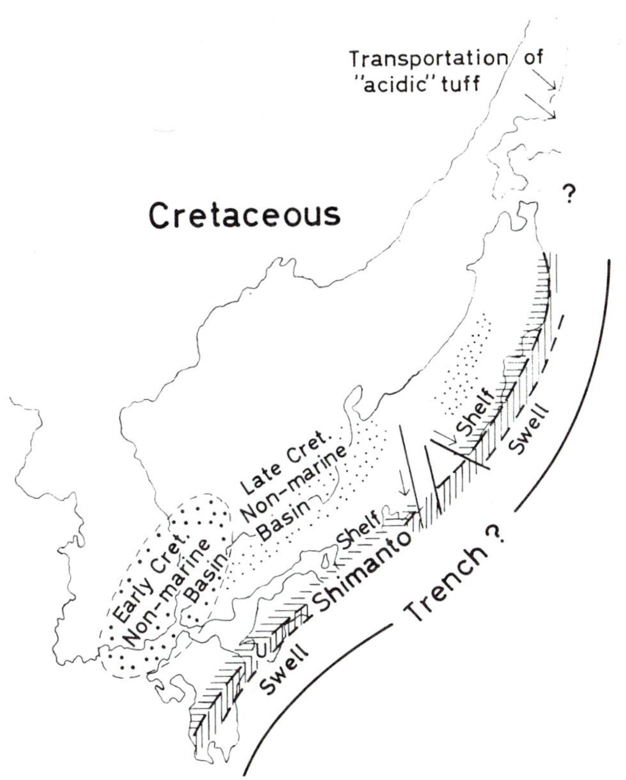

Fig. 4. Geosyncline and other sedimentary basins in Japan in the Cretaceous. (U, Uwajima.)

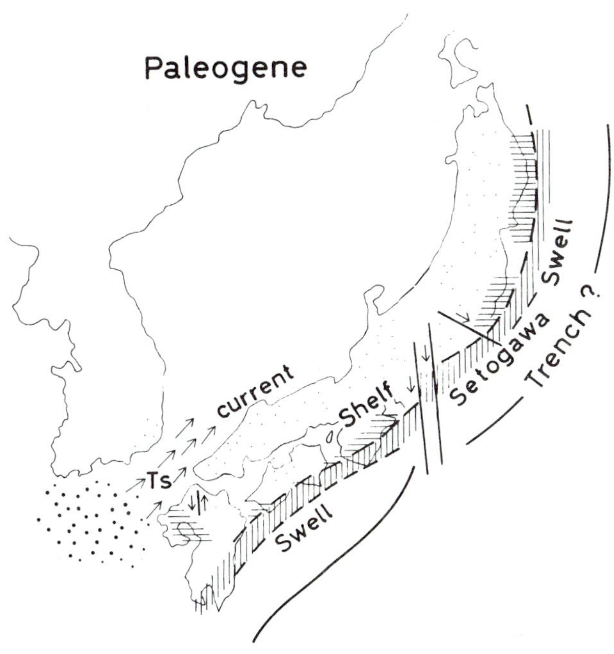

Fig. 5. Geosyncline and other sedimentary basins in Japan in the Paleogene. (Ts, Tsushima.)

are situated to the north of the Nankai trench, with the Tosabae swell between them (Fig. 6). Pliocene turbidites were reported from the basin (Okuda, 1973) and "basic" rocks were dredged from the Tosabae swell (Niino, 1935). The secondary eugeosynclines are quite similar to these recent basins. The Nankai trench to the south marks a subduction zone of the Philippine Sea plate. Ancient trenches and subduction zones must have been similarly situated on the oceanic sides of the ancient secondary eugeosynclines.

North of the Sambosan terrain (Fig. 3), Upper Triassic strata of shelf facies, the Kochigatani Group, are well developed. To the north of the Shimanto terrain (Fig. 4) are the Upper Jurassic Torinosu Group with coral reefs and the Lower Cretaceous paralic Ryoseki Group. To the north of the Setogawa (Fig. 5), Paleogene conglomerate, the Nakaoku Formation of shelf facies was reported (Shiida, 1962) from the Paleozoic as well as a Cretaceous geosynclinal terrain in central Kii (Fig. 6). North of the Toki basin, the Miocene Tanabe shelf deposits and other nonmarine strata, together with the Pliocene Nobori Formation of shelf facies, cover unconformably folded Paleogene geosynclinal strata.

Thus younger eugeosynclines have been formed on the oceanic sides of the shelves which have been produced in the area of the older eugeosynclines. The secondary eugeosynclines migrated in a manner very similar to that in Alaska (Burk, 1972).

To the south of the Shimanto, Setogawa, and Tanzawa geosynclines, there were submarine swells, as already reported (Kimura, 1973). Recently it became clear that the Sambosan eugeosyncline also had submarine swells on its oceanic side. Fossil lateritic soils, which are now changed into "emery" by contact metamorphism with Miocene granite, have been found by Iwao (1972) within the limestone along the southern margin of the Sambosan terrain in east Kyushu.

The Shimanto geosyncline is often united with the Setogawa to form a larger geosyncline. However, the Setogawa is a different geosyncline from the Shimanto in a strict sense, because they have quite

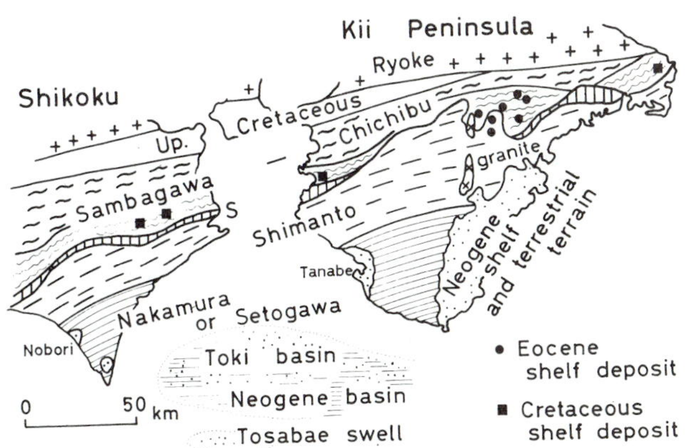

Fig. 6. Eugeosynclinal areas and shelf areas in East Shikoku and Kii. (S, Sambosan eugeosynclinal area.)

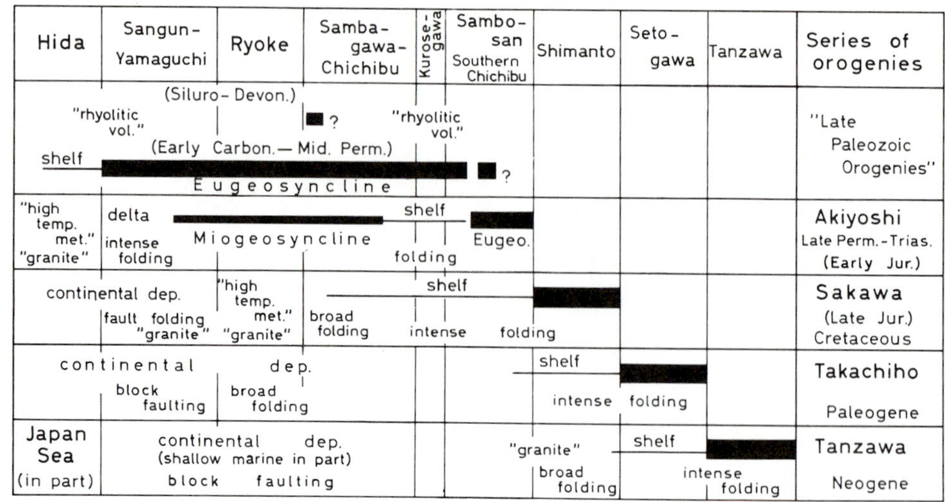

Hida	Sangun-Yamaguchi	Ryoke	Sambagawa-Chichibu	Kurosegawa	Sambosan Southern Chichibu	Shimanto	Setogawa	Tanzawa	Series of orogenies
shelf	"rhyolitic vol." (Siluro-Devon.) (Early Carbon.— Mid. Perm.)	Eugeosyncline	?	"rhyolitic vol."	?				"Late Paleozoic Orogenies"
"high temp. met." "granite"	delta intense folding	Miogeosyncline	shelf folding		Eugeo.				Akiyoshi Late Perm.-Trias. (Early Jur.)
continental dep.	fault folding "granite"	"high temp. met." "granite"	broad folding	shelf intense	folding				Sakawa (Late Jur.) Cretaceous
continental dep.	block faulting	broad folding				shelf intense	folding		Takachiho Paleogene
Japan Sea (in part)	continental dep. (shallow marine in part) block faulting					"granite" broad folding	shelf intense	folding	Tanzawa Neogene

Table 1. Tectonic zones, geosynclines, and earth movements in southwest Japan.

different distributions of strata (Tsuchi et al., 1973), deformation styles, and ages of basaltic volcanism.

The Shimanto geosyncline in a broad sense was thought to be an ancient trench by Matsuda and Uyeda (1971), but this is not true. The Shimanto Group is principally composed of sandstone, shale, and "acidic" tuff with subordinate conglomerates. Basaltic lavas and tuffs occur only in rather small amounts in a lower horizon. "Acidic" tuffs are marked in the middle and upper parts. During Cretaceous time there was extensive subareal rhyolitic volcanism on the northern side of Japan. The Shimanto geosyncline must have been situated near ancient Japan. Turbidites on the southern margin of the Shimanto geosyncline in the southern Oigawa area were transported from southern submarine swells. These were probably triggered by earthquakes near the swells (Kimura, 1966) similar to the Nankai Earthquake of 1946, which occurred near the Tosabae swell south of the present Toki basin. The Shimanto geosyncline and the Toki basin are very similar to one another in sedimentary condition.

The Upper Cretaceous Uwajima Group is developed along a major synclinorial axis plunging westward in west Shikoku (Fig. 4). It is rich in conglomerates and molluscan fossils of shelf facies. The group is underlain conformably by the strata of geosynclinal facies (Tanabe, 1972), so in this area the Shimanto geosyncline was almost filled up by clastics, producing a shelf.

The structures of the Shimanto Group are rather complicated on a small scale, because the group is often intensely deformed to produce lens folds (Kimura, 1968), although those on a large scale are rather simple (Fig. 7). They have not been produced in a subduction zone, and there are no mélanges in the Shimanto terrain. The sedimentary condition of the Shimanto is quite different from the Franciscan in California.

EARTH MOVEMENTS AND MIGRATION OF SECONDARY EUGEOSYNCLINES

The areas of high-temperature metamorphism, granite plutonism, and folding migrated from north to south in Japan (Kimura, 1973) as shown in Table 1, as did the secondary eugeosynclines. Particular deformational features, as well as basaltic volcanisms in the eugeosynclines, do not migrate gradually but in steps. The step migration of the deformation and the secondary eugeosynclines was probably caused by breakdown of the oceanic crust at its bending point and seaward retreat of the subduction zone (Kimura and Tokuyama, 1971). Normal faulting was reported for the Sanriku Earthquake of 1933 (Kanamori, 1971) near the bending point of the Pacific plate. Such normal faulting may be related to the breakdown of an oceanic plate.

In southwest Japan, orogenies in Late Permian to Triassic (the Akiyoshi or "Honshu" orogenies), in Cretaceous (the Sakawa orogenies), in Late Paleo-

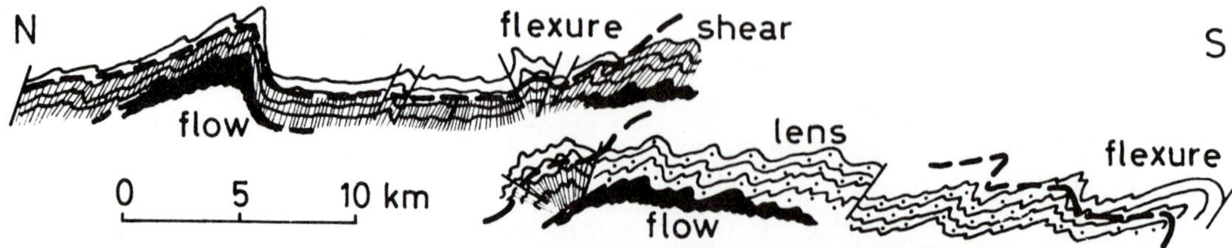

Fig. 7. Profiles of the Shimanto Group and folding styles in the Oigawa area, central Japan.

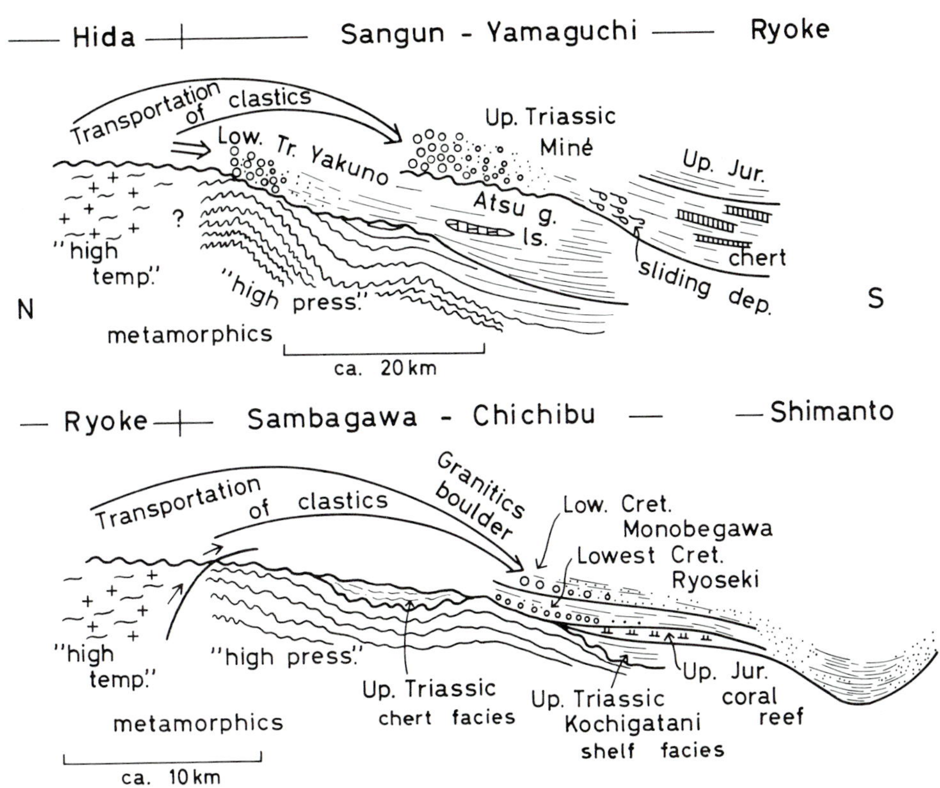

Fig. 8. Stratigraphic profiles, southwest Japan (see the text for a discussion.)

gene (the Takachiho or Amakusa, and "Shimanto" orogenies), and in Neogene and Quaternary (the Tanzawa, Oigawa, Oyashima, "Green Tuff," and other orogenies) are thought to have occurred, in addition to movements in Late Paleozoic. They are all of wider time range than was formerly thought and form a series of orogenies. This series of orogenies is associated, respectively, with the Sambosan, Shimanto, Setogawa, and Tanzawa eugeosynclinal stages or the migrations between them. Each series is a group of earth movements when the Philippine Sea or related plates subducted on the oceanic sides of those eugeosynclines.

Earth movements, however, are not continual even in a series of orogenies. Upper Permian turbidites disconformably overlie middle Permian limestone in western Akiyoshi. Strong earth movements may have occurred to the north of the Chichibu eugeosyncline. Lower Triassic deltaic deposits unconformably cover folded Paleozoic strata near Maizuru (Fig. 8). The Upper Triassic Molasse-like Miné Group unconformably cover the Sangun metamorphics, although the Triassic Atsu group beneath the Miné has some geosynclinal character. The Orogeny before the Upper Triassic was one thought to cover the entire Sangun-Yamaguchi zone or all southwest Japan. However, the Upper Triassic strata of the chert facies in the central part of the Chichibu geosynclinal area cover conformably or disconformably the Permian strata.

The Paleozoic strata near the Kurosegawa terrain were somewhat affected by pre-Upper Triassic deformation. A shift of sedimentary basins from the Upper Triassic Miné basin to the Lower Jurassic Toyora shows an earth movement (Kobayashi, 1941). Thus the Akiyoshi series of orogenies are composed of several earth movements: a subseries of orogenies. The area of intense folding during each subseries was rather narrow, about 30 km wide, although warping or other kinds of deformation occurs in a wider area. The Sakawa series of orogenies is also composed of many subseries: the Oga, Oshima, Sakawa (Kobayashi, 1941), and other orogenies. About 10 unconformities have been reported in the Cretaceous strata in Japan (Matsumoto, 1953). Discontinuities of foldings are sometimes clearly shown in distribution of folding styles (Fig. 8). In the Shimanto Group in the Oigawa area, the folding styles, which depend on the tectonic levels, are nearly parallel to the stratigraphic horizons (Kimura and Tokuyama, 1971), and structural styles are found on opposite sides of an anticlinorium (Fig. 7). This means that the intense folding occurred after the deposition of almost all strata in the geosyncline but not during the deposition of the strata.

In these series or subseries of orogenies, intense folding occurred usually on the oceanic side, broad or fault folding in the middle, and warping or block faulting on the continental side of ancient Japan, forming a set of deformational terrains. The Cretaceous strata in the Shimanto geosynclinal area and southern Chichibu terrain are intensely folded, those in the northern Chichibu are generally broadly folded, and most of those in the Sangun-Yamaguchi

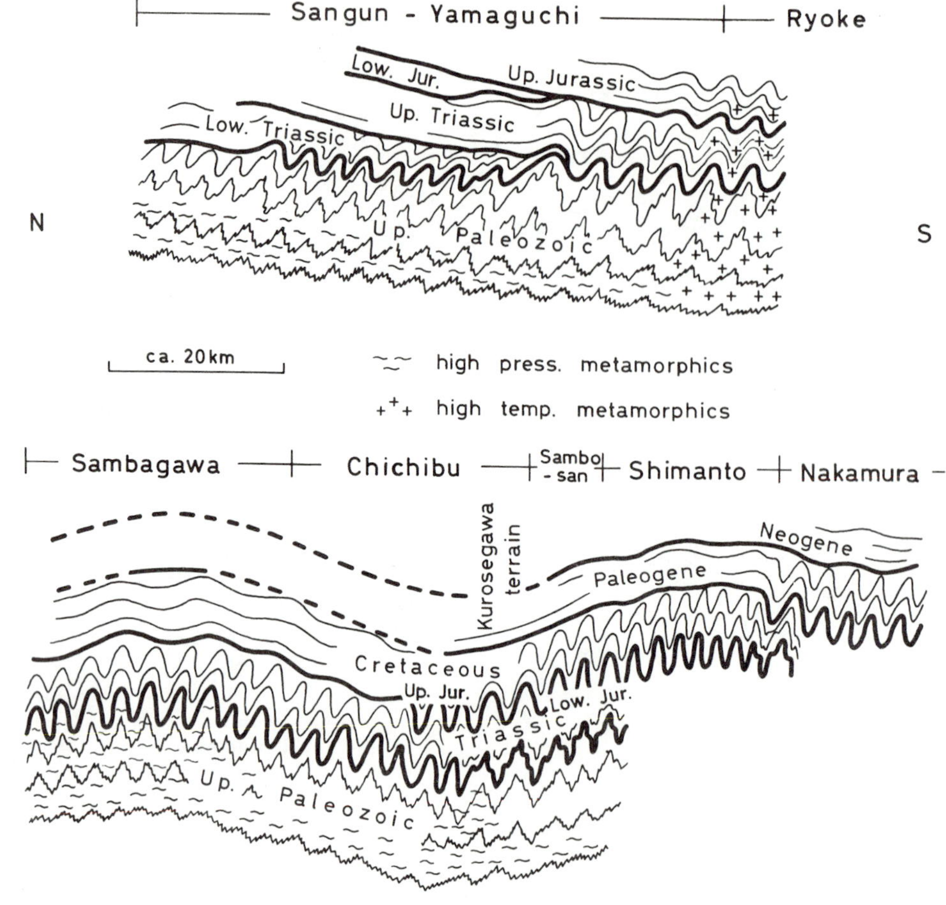

Fig. 9. Schematic profiles of successively folded strata in southwest Japan.

terrain suffered very weak folding except in the areas along faults (Table 1). However, differences of lithologies and older geologic structures produce some irregular distributions of deformation. Pre-Upper Triassic earth movements produced such an irregularity. Folding of that age is negligible in the central part of the Chichibu geosyncline, but considerable near the Kurosegawa terrain in the southern part of it.

The set of deformational terrains migrated with seaward retreat of the subduction zones and the land area spread gradually toward the ocean. Therefore, a particular area, especially the northern part of ancient Japan, suffered intense folding at first, later broad folding (Fig. 9), and still later fault folding. Different kinds of deformation occurred even in a given series of orogenies at different stages. In addition, one set of deformational terrains often contains several different kinds of deformation at the same age. These variations resulted in identification of many independent orogenic zones and orogenies for which many names have been proposed. However, those "orogenies" of the same age must be fundamentally related to regional forces of the same age.

When geosynclinal strata were first intensely folded, the enveloping surface of the folds was generally nearly horizontal and broadly folded later. Paleozoic strata in the northern Chichibu terrain were folded to form isoclinal minor folds with a horizontal-enveloping surface before the deposition of Lower Cretaceous strata, and both were broadly folded later (Ogawa, 1973). Therefore, the Lower Cretaceous strata are clino-unconformable with the Paleozoic on a minor scale, but disconformable on a major scale. This is the reason the orogeny before Lower Cretaceous is thought to be negligible in the Chichibu terrain by Kobayashi (1941), and to be remarkable by Matsumoto and Kanmera (1949) and others. The Shimanto Group in the Oigawa area (Fig. 7) was intensely folded before Paleogene, and broadly folded during the Quaternary.

A superseries or cycle of orogenies in southwest Japan, composed of these series of orogenies mentioned above, began in late Permian and continued until Recent time. This is probably due to the motive forces of orogenies being nearly the same throughout this time. Present secondary geosynclines of the Toki and other basins is closely related to the subduction zone of the Philippine Sea plate. The superseries of orogenies from the late Permian to the present is probably related to this or similar oceanic plates. The end of a series or cycle of orogenies would be distinct where collision of two continents occurred,

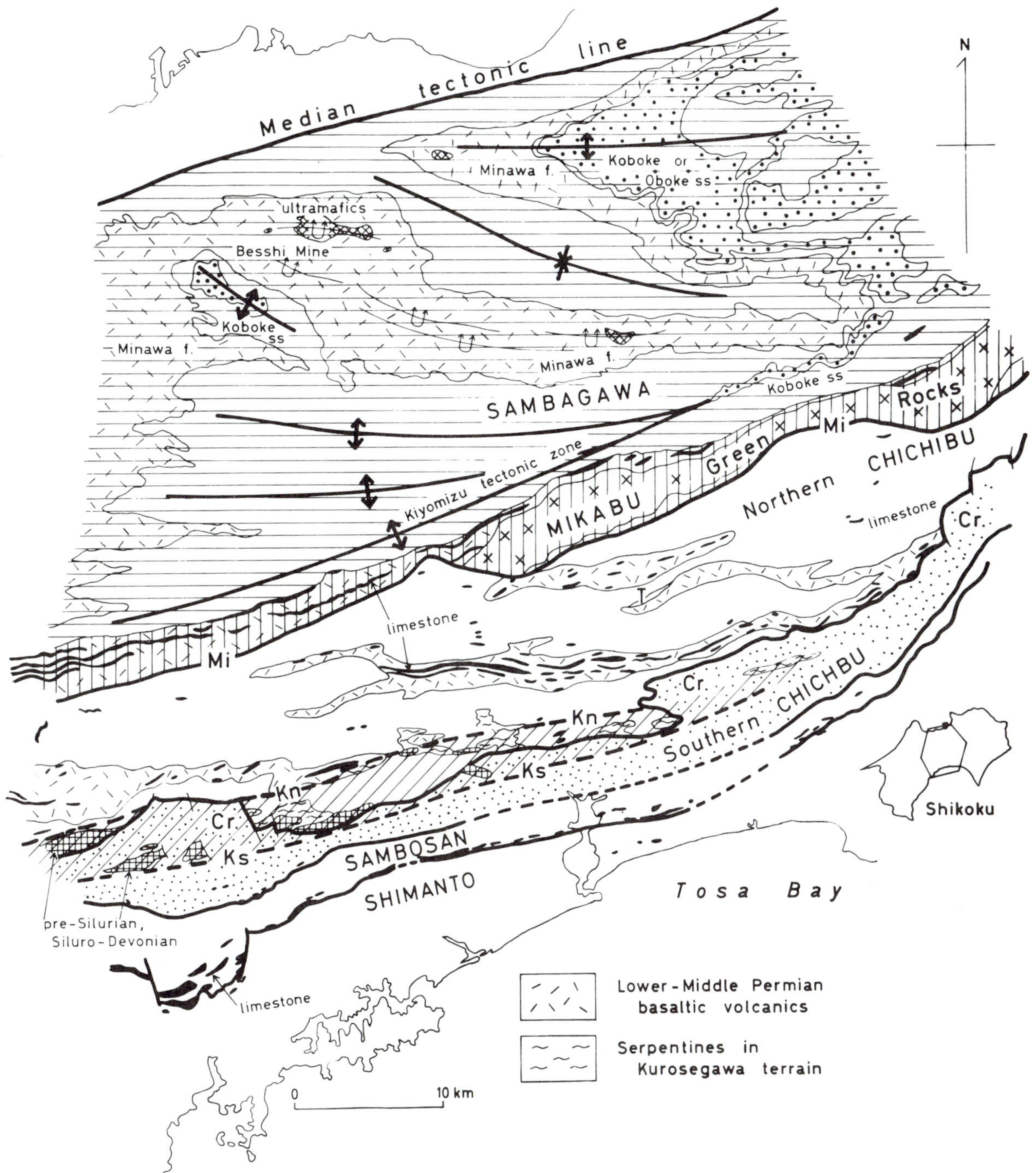

Fig. 10. Tectonic map of central Shikoku. (Mi, Mikabu line; Kn and Ks, approximate northern and southern boundaries of the Kurosegawa tectonic zone, respectively; Cr, Lower Cretaceous sedimentary basins; T, Lower Triassic at Kurotaki.)

as in the cases of the Caledonian, Variscan, and other cycles, but this is not the case in southwest Japan.

In northeast Japan, the migration of secondary eugeosynclines, as well as the structural development, became different from that in southwest Japan from Cretaceous on (Kimura, 1973). A new cycle of superseries began in Cretaceous and became distinct in appearance when the Miocene "Green Tuff" was produced in northeast Japan. This was due to the different movement of the Pacific plate from that of the Philippine Sea plate. In southwest Japan, the Miocene "Green Tuff" occurs on the Japan Sea side. A new primary geosyncline, the Japan Sea, began to be formed in late Paleogene, as already mentioned. A new super-series, or cycle of orogenies, also began at that time in the northern part of southwest Japan, while an older super-series was still going on.

"SUBDUCTION ZONES" AND PAIRED METAMORPHIC BELTS IN JAPAN

Subduction zones as well as mélanges have not yet been observed on the earth's surface in Japan (Kimura, 1974), although they may be concealed deep beneath it. Subduction zones near ancient Japan must always have been on the oceanic sides of the secondary eugeosynclines, as the present subduction zone along the Nankai trench is on the oceanic side of the present Toki and other basins.

The Median tectonic line of the Inner Zone, the Maizuru zone, the Median teconic line, the Kiyomizu tectonic zone, the Mikabu line, and the Kurosegawa tectonic zone are sometimes thought to be fossil benioff zones, or "subduction zones." However, the geology on the two sides of these tectonic zones, except for the Median tectonic line, are similar, and it must be concluded that they are not benioff zones (Kimura, 1974). The Kiyosmizu tectonic zone to the north of the Mikabu zone is nothing less than a shear zone along an anticlinorium axis, as shown in Fig. 10.

There are two paired metamorphic belts in Japan: the Hida-Sangun and the Ryoke-Sambagawa (Miyashiro, 1961). It has been stated that the high-temperature Hida and Ryoke zones were located in a position similar to present northeast Japan, and that the high-pressure Sangun and Sanbagawa type were in a position similar to the Japan trench when the paired belts were produced. However, the geologic histories of the metamorphic terrains show clearly that the opinion is erroneous.

Late Paleozoic strata of the Sangun and Sambagawa metamorphics are not trench deposits. In the metamorphics, quartzo-feldspathic sandstones are common even beneath basaltic volcanics. The geologic structures are very complicated on a small scale, but rather simple on a quite large scale. Lower-Middle Permian basaltic volcanics, principally pyroclastics, can be traced from the Sambagawa metamorphic zone in a narrow belt through the Mikabu zone into the nonmetamorphic Chichibu terrain (Fig. 10). Many overturned folds are found in the Sambagawa terrain. They were formed *before* the metamorphism (Tokuyama, personal communication), because the metamorphic grades are clearly oblique to the structure.

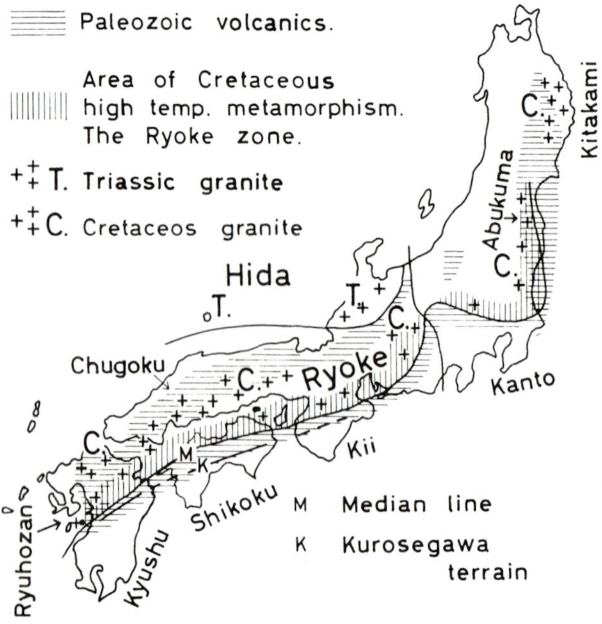

Fig. 11. Terrains of late Paleozoic submarine basaltic volcanism and Cretaceous Ryoke metamorphism in high-temperature type.

Original rocks of the Ryoke are mostly of Triassic and Jurassic age, as already mentioned. Lower Carboniferous and Lower-Middle Permian basaltic volcanics underlie the Ryoke rocks. Lower Carboniferous volcanics in southern Abukuma and Permian ones in the Ryuhozan belt in west Kyushu (Fig. 11) are below the Ryoke rocks. They have been subjected to high-temperature metamorphism, although they are quite similar in original rock facies to the Sambagawa rocks.

In Shikoku there is a broad area with Permian basaltic volcanics between the Median tectonic line and the Kurosegawa terrain (Figs. 10 and 11). The area becomes narrower westward and almost disappears in west Kyushu, where strata similar to the Sambagawa occur in the Ryuhozan belt beyond the Median tectonic line. The belt was probably the western continuation of the Sambagawa-Chichibu zone to the north of the Kurosegawa in Shikoku at the depositional stage. Therefore, the Median tectonic line is diagonal to the original depositional zone. However, there is no evidence of major left lateral strike-slip movement, as suggested by Miyashiro (1972). The sedimentary conditions on the two sides of the Median tectonic line appear not to be greatly different.

In Triassic time when the Hida zone was being subjected to high-temperature metamorphic conditions, the Sangun terrain was one of deltaic deposition (Fig. 7). In Cretaceous time, when the Ryoke was in a high-temperature condition, jadeite-bearing pebbles were carried from the Sambagawa terrain southward to the Lower Cretaceous basin (Seki, 1965), and abundant pebbles and cobbles of granitic rocks were transported from the Ryoke rocks to the basin. When zones were being subjected to high-temperature metamorphism, two high-pressure metamorphic zones were shallow shelf or land. Trenches or subduction zones at those times were farther to the south on the oceanic sides of the Sambosan and Shimanto geosynclines.

How was high-pressure metamorphism produced in Sangun and Sambagawa rocks? This is an important problem not yet solved. In the western Sangun-Yamaguchi zone, the Upper Carboniferous-Permian Akiyoshi limestone forms an overturned fold together with Upper Permian turbidites. Pre-Visean or Visean basaltic volcanics overlying the apparent top of the overturned limestone are not metamorphosed, although volcanics of the same age are generally metamorphosed near the area. The metamorphism occurred after the formation of the nappe structures. The tectonic overburden produced in this manner may have created the metamorphism.

Many large-sized granite boulders in the Lower Cretaceous Monobegawa Group in the Chichibu terrain indicate that the provenance was near the depositional basin. Movement along the Median tectonic line was principally thrusting toward the south in Cretaceous time. Large thrust sheets of Ryoke rocks moved along the Median tectonic line and covered the Sambagawa terrain. Such overthrusting may have produced the metamorphism in the Sambagawa as supposed by Blake et al. (1969). However, the geologic history of Lower Jurassic time is not yet well known. Sedimentary overburden of the Lower Jurassic strata may have caused the metamorphism (Ogawa. 1973), although we cannot find any large quantity of erosion products of Lower Jurassic rocks anywhere in the strata of later age.

SIALIC MASSES AND OCEANIC CRUST IN ANCIENT JAPAN

During the Chichibu eugeosynclinal stage, a part of the continent, the Hida zone, was to the north of it and pre-Silurian sialic rocks of the Kurosegawa terrain were to the south of it. An ancient oceanic crust may have been created in part of the eugeosyncline at its initial stage, as in the present Japan Sea. The area with the oceanic crust may have been changed into one with a sialic basement as the result of later sedimentation and earth movements.

Ancient margins of the Japanese Islands migrated toward the ocean in a stepping style, when new secondary eugeosynclines were formed on the oceanic sides of folded older eugeosynclinal strata. Submarine swells or islands existed on the oceanic sides of these secondary eugeosynclines. How those swells were formed and what they were made of are problems not yet solved. They may be composed of pre-Silurian masses or made of consolidated older strata within or outside the eugeosynclines.

Tokuoka (1970) found widespread orthoquartzite pebbles transported from the south into the Shimanto terrain that may show the presence of a Precambrian continent. If there was a Precambrian mass to the south of the "Shimanto" eugeosyncline, this could indicate oceanward drifting of a part of the Kurosegawa terrain that separated from the main body as new secondary eugeosynclines formed. The older sialic masses on the oceanic side may have been underthrusted beneath Japan as a result of movement along the subduction zone, since there is no such large sialic mass on the oceanic side of the present Toki basin. Alternatively, consolidated older strata within or outside the eugeosynclines may have made the submarine swells or islands outside the geosynclines. The Miocene shelf of the Tanabe Formation once extended to the margin of the Nankai trench (Okuda, 1973), and Miocene clastics were widely distributed upon the shelf. The Pliocene Toki basin was formed within the shelf area, and thus the Tosabae swell is composed at least partly of the Miocene Tanabe Group.

Oceanic crust may be lifted upon the earth surface by compression forces or by isostatic adjustment. In Japan, the presence of old oceanic crust on the earth surface has not yet been proved, but it may be present at depth. Thrusting of oceanic crust onto sialic masses by obduction or by collision of two continents did not occur in Japan.

Isostatic uplift of oceanic crust occurs in Great Valley, California (Bailey and Blake, 1973), where the subduction zone is stationary in position or is advancing, and tremendous accumulations of low-density material have occurred beneath the crust. This is not the case in Japan, where subduction zones retreated and great accumulations of low-density materials did not occur to elevate the oceanic crust or subduction zones to the earth's surface.

SUMMARY

In pre-Silurian time, the Precambrian Hida zone, probably a part of the Far East continent, and the Kurosegawa terrain, the approximate southern margin of the Chichibu eugeosyncline, may have been united or close to one another. The Kurosegawa terrain may subsequently have drifted toward the ocean, leaving the marginal sea-like Chichibu eugeosyncline in which Lower Carboniferous and Lower-Middle Permian submarine basaltic volcanics are prominent. When the Chichibu eugeosyncline changed into a miogeosyncline, a new secondary eugeosyncline was formed on the oceanic side or in the southernmost part of the older primary eugeosyncline. Younger secondary eugeosynclines, which were possibly small ocean basins between island arcs and trenches or subduction zones, were later formed on the oceanic sides of older ones.

The secondary eugeosynclines and subduction zones migrated in a stepping style, in company with the seaward migration of island margins, gradually producing ancient Japan. Ancient oceanic crusts and mélanges have not yet been found on the present earth's surface in those eugeosynclinal areas.

The migration of the secondary eugeosynclines continued until Recent time in southwest Japan, because movements of the Philippine Sea or related oceanic plates did not change fundamentally. In northeast Japan, a great change of structural developments occurred in the Cretaceous. The change was caused by a new movement of the Pacific plate. The Japanese Islands, in which the primary eugeosynclinal area became an almost complete fold mountain by the end of Cretaceous, have migrated toward the ocean since the Paleogene, forming a new primary geosyncline, the Japan Sea.

BIBLIOGRAPHY

Asada, T., and Asano, S., 1972, Crustal structures of Honshu, Japan, The crust and upper mantle of the Japanese area, pt. I, Geophysics: Earthquake Res. Inst. Univ. Tokyo, p. 45-55.

Bailey, E. H., and Blake, M. C., Jr., 1973, Ophiolites of the circum-Pacific belt: Internatl. Symp. "Ophiolites in the earth's crust," Moscow, preprint, p. 95-97.

Blake, M. C., Jr., Irwin, W. D., and Coleman, R. G., 1969, Blueschist-facies metamorphism related to regional thrust faulting: Tectonophysics, v. 8, p. 237-246.

Burk, C. A., 1972, Uplifted eugeosynclines and continental margins: Geol. Soc. America Mem. 132, p. 75-85.

Emery, K. O., et al., 1969, Geological structure and some water characteristics of the East China Sea and the Yellow Sea: Tech. Bull. ECAFE, v. 2, p. 3-43.

Fujii, A., and Minakami, T., 1970, Tsunemori formation—its relation with the Akiyoshi Limestone: Jour. Geol. Soc. Japan, v. 76, p. 547-557 (in Japanese with English résumé).

Fujioka, K., 1972, Geohistorical review on the development of the Japan Sea: Jour. Japan. Assoc. Petr. Technol., v. 37, p. 233-244. (in Japanese).

Hase, A., 1961, A find of *Monotis* (*Entomonotis*) from eastern Yamaguchi Prefecture, Japan: Trans. Proc. Palaeontol. Soc. Japan, N. S., v. 42, p. 79-87.

———, 1964, Geologic history of Hiroshima Prefecture—Paleozoic strata of Hiroshima Prefecture, Explanatory text of Geological Map of Hiroshima Prefecture (in Japanese).

Ichikawa, K., 1970, Some geotectonic problems concerning the Paleozoic-Mesozoic geology of southwest Japan, *in* Hoshino, M., and Aoki, H., eds., Island arc and ocean: Tokai Univ. Press, p. 193-200 (in Japanese with English résumé).

———, et al., 1956, Die Kurosegawa Zone: Jour. Geol. Soc. Japan, v. 62, p. 82-103 (in Japanese with German résumé).

Iwao, S., 1972, On the lateritic textures retained in the "emery" at Shinkiura Mine, Kyushu: Mining Geology, Japan, v. 22, p. 359-369 (in Japanese with English résumé).

Kambe, N., 1963, On the boundary between the Permian and Triassic systems in Japan: Japan. Geol. Survey Rept. 198.

Kanamori, H., 1971, Seismological evidences for a lithospheric normal faulting—the Sanriku earthquake of 1933: Phys. Earth Planet Interiors, v. 4, p. 289-300.

Kanmera, K., 1961, Middle Permian Kozaki formation: Sci. Rept. Fac. Sci. Kyushu Univ. Geology, v. 5, p. 196-215 (in Japanese with English résumé).

———, 1971, Paleozoic and early Mesozoic geosynclinal volcanicity in Japan. Geol. Soc. Japan Mem. 6, p. 97-110 (in Japanese with English résumé).

Kano, K. 1972, Geology of the Yabuhara district in the northern Kiso mountain range, Nagano Pref. [M.S. thesis]: Univ. Tokyo.

Karig, D. E., 1970, Ridges and basins of the Tonga-Kermadec island arc system: Jour. Geophys. Res., v. 75, p. 239-254.

———, 1971, Structural history of the Mariana island arc system: Geol. Soc. America Bull., v. 82, p. 323-344.

Kawai, N., Ito, H., and Kume, S., 1961, Deformation of the Japanese Islands as inferred from rock magnetism: Geophys. Jour. v. 6, p. 124-130.

Kimura, T., 1960, On the geologic structure of the Paleozoic group in Chugoku, West Japan: Sci. Pap. Coll. Gen. Educ. Univ. Tokyo, v. 10, p. 109-124.

———, 1966, Thickness distribution of sandstone beds and cyclic sedimentations in the turbidites sequences at two localities in Japan: Bull. Earthquake Res. Inst. Univ. Tokyo, v. 44, p. 561-607.

———, 1968, Some folded structures and their distribution in Japan: Japan. Jour. Geol. Geography, v. 39, p. 1-26.

———, 1973, The old "inner" arc and its deformation in Japan, The western Pacific: Univ. Western Australia Press, p. 255-273.

———, 1974, 'Ophiolites' and tectonic zones in Japan: Proc. Internatl. Symp. "Ophiolites in the Earth's Crust," Moscow.

———, and Tokuyama, A., 1971, Geosynclinal prisms and tectonics in Japan: Geol. Soc. Japan Mem. 6, p. 9-20.

Kobayashi, T., 1941, The Sakawa orogenic cycles and its bearing on the origin of the Japanese Islands: Jour. Fac. Sci. Imp. Univ. Tokya, v. 5, p. 219-578.

———, 1954, Fossil Estherians and allied fossils: Jour. Fac. Sci. Univ. Tokyo, sect. II, v. 9, p. 1-192.

———, 1956, The insular arc of Japan, its hinter basin and its linking with the peri-Tunghai arc: Proc. 8th Pacific Sci. Congr., v. 2-A, p. 799-807.

———, and Suzuki, K., 1936, Non-marine shells of the Naktong-Wakino series: Japan. Jour. Geol. Geography, v. 13, p. 243-257.

———, et al., 1971, Contribution to the geological history of the Japanese Islands by the conodont biostratigraphy, pt. II: Jour. Geol. Soc. Japan, v. 77, p. 165-168.

Konishi, K., 1965, Geotectonic framework of the Ryukyu Islands: Jour. Geol. Soc. Japan, v. 71, p. 437-457 (in Japanese with English résumé).

Marlow, M. S., et al., 1973, Tectonic history of the central Aleutian Arc: Geol. Soc. America Bull., v. 84, p. 1555-1574.

Matsuda, T., and Uyeda, S., 1971, On the Pacific type orogeny and its model—extension of the paired belts concept and possible origin of marginal seas: Tectonophysics, v. 11, p. 5-27.

Matsumoto, T., 1953, The Cretaceous system in the Japanese Islands: Japan Soc. Promotion Sci., Tokyo.

———, 1969, Geology of Tsushima and relevant problems: Natl. Sci. Museum Japan Mem. 2, p. 5-18 (in Japanese with English résumé).

———, and Kanmera, K., 1949, Contribution to the tectonic history in the Outer Zone of southwest Japan: Mem. Fac. Sci. Kyushu Univ., ser. D, v. 3, p. 77-90.

———, and Okada, H., 1971, Clastic sediments of the Cretaceous Yezo geosyncline: Geol. Soc. Japan Mem. 6, p. 61-74.

Minato, M., Gorai, M., and Hunahashi, M., 1965, The geologic development of the Japanese Islands: Tokyo.

Miyashiro, A., 1961, Evolution of metamorphic belts: Jour. Petr., v. 2, p. 277-311.

———, 1972, Metamorphism and related magmatism in plate tectonic: Am. Jour. Sci., v. 272, p. 629-656.

Murauchi, S., 1972, Crustal structure of the Japan Sea derived by explosion seismology: Kagaku, v. 42, p. 367-407 (in Japanese).

Nagahama, H., 1967, Paleocurrents of the Taishu group, in the Tsushima Island, Kyushu, Japan: Jubilee Publ. Commemom. Prof. Sasa, 60th Birthday, p. 135-144 (in Japanese with English résumé).

Nakazawa, K., 1958, The Triassic system in the Maizuru zone, southwest Japan: Mem. Coll. Sci. Univ. Kyoto, ser. B, v. 24, p. 265-313.

———, and Nogami, Y., 1958, Paleozoic and Mesozoic

formations in the vicinity of Kawanishi, Oe-cho, Kyoto Prefecture, Japan: Jour. Geol. Soc. Japan, v. 64, p. 68-77 (in Japanese with English résumé).

Niino, H., 1935, On the soundings at Tosabae: Geograph. Rev. Japan, v. 11, p. 679-687 (in Japanese with English résumé).

Ogawa, Y., 1973, Tectonic development of the Chichibu terrain in eastern Shikoku, Japan: Jour. Fac. Sci. Univ. Tokyo, sec. II, v. 18 (in press).

———, 1974, Stratigraphy and paleontology of the Chichibu terrain in eastern Shikoku, Japan: Proc. Inst. Natl. Sci. Nihon Univ., v. 9.

Okuda, Y., 1973, Tectonic development of the Toki basin and the origin of the submarine canyons off the Kii strait in relation to sedimentary sequences revealed by electro-sonic profiling [M.S. thesis]: Univ. Tokyo.

Ono, A., 1969, Zoning of the metamorphic rocks in the Takato-Shiojiri area, Nagano Prefecture. Jour. Geol. Soc. Japan, v. 75, p. 521-536 (in Japanese with English résumé).

Seki, Y., 1965, Jadeitic pyroxene found as pebbles in Lower Cretaceous formation of the Kanto Mountains, central Japan: Jour. Japan. Assoc. Min. Petr. and Econ. Geology, v. 53, p. 165-168 (in Japanese with English résumé).

Shibata, K., Adachi, M., and Mizutani, S., 1972, Precambrian chronology of metamorphic rocks in Permian conglomerate from central Japan: Proc. 24th Internatl. Geol. Congr., sec. 1, p. 288-294.

Shiida, I., 1962, Stratigraphic and geotectonic studies of the Paleozoic Chichibu and Mesozoic Hitaka (Shimanto) terrains in the central part of the Kii mountainland, southern Kinki, Japan: Res. Bull. Dept. Gen. Educ. Nagoya Univ., v. 16, p. 1-58 (in Japanese with English résume).

Shoji, R., 1967, Occurrence and petrographical studies of Paleozoic chert of the western Ashio-Mountain, Japan: Jubillee Publ. Commemom. Prof Sasa, 60th Brithday, p. 171-189.

Slater, J. G., et al., 1972, Crustal extension between Tonga and Lau ridges: petrologic and geophysical evidence: Geol. Soc. America Bull., v. 83, p. 505-518.

Sugisaki, R., et al., 1971, Rifting in the Japanese late Paleozoic geosyncline: Nature, v. 233, p. 30-31.

Suzuki, A., and Sato, T., 1972, Discovery of Jurassic ammonite from Toriashi Mountain: Jour. Geol. Soc. Japan, v. 78, p. 213-215 (in Japanese with English résumé).

Takai, F., Matsumoto, T., and Toriyama, R., 1963, Geology of Japan: Univ. Tokyo Press.

Tanabe, K., 1972, Stratigraphy of the Cretaceous formations in the Uwajima district, Ehime Prefecture, Shikoku: Jour. Geol. Soc. Japan, v. 78, p. 177-190 (in Japanese with English résumé).

Tokuoka, T., 1970, Orthoquartzite gravels in the Paleogene Muro group, southwest Japan: Mem. Fac. Sci. Kyoto Univ. Geology and Mining, v. 37, p. 113-132.

Tokuyama, A., 1962, Triassic and some other orogenic sediments of Akiyoshi cycle in Japan: Jour. Fac. Sci. Univ. Tokyo, sec. II, v. 13, p. 379-469.

Toriyama, R., 1954, Geology of Akiyoshi, pt. I, Study of the Akiyoshi limestone group: Mem. Fac. Sci. Kyushu Univ., ser. D, v. 4, p. 39-97.

Toyohara, F., 1973, On the age of the Kuga Group and the Ryoke metamorphic rocks in eastern Yamaguchi Prefecture: Jour. Geol. Soc. Japan (in Japanese).

Tsuchi, F., et al., 1973, Geological map of Shizuoka Prefecture, 1:200,000: Shizuoka Prefecture.

Yanagimoto, Y., 1973, Stratigraphy and geological structure of the Paleozoic and Mesozoic formations in the vicinity of Kuzuu, Tochigi Prefecture: Jour. Geol. Soc. Japan, v. 79, p. 441-451 (in Japanese with English résumé).

Yoshida, S., 1972, Configuration of Yamaguchi zone—analytical study on a fold zone: Jour. Fac. Sci. Univ. Tokyo, sec II, v. 18, p. 371-429.

Early Western Margin of the United States

John C. Maxwell

INTRODUCTION

The development of the western Cordillera of North America perhaps began with the submergence of the western border of the craton in middle to late Proterozoic time and the deposition of the thick Beltian sequence of dominantly shallow-water clastic sediments of presumed marine origin. Windermere and equivalent rocks of latest Proterozoic time, which were deposited unconformably on Beltian rocks, are partly of "eugeosynclinal" facies (abundant graywackes and local marine mafic volcanics) and perhaps mark the beginning of a differentiation of the vast Cordillera into eugeosynclinal and miogeosynclinal belts, which persisted through Paleozoic, Mesozoic, and locally into Cenozoic time. Volcanism—basaltic, andesitic, and rhyolitic—has characterized the Cordillera throughout its history. No geologic period seems to have been free of volcanism (Gilluly, 1965). Extensive plutonic activity, however, began only in middle Mesozoic time, reaching a maximum during Late Cretaceous, but continuing into the late Cenozoic and perhaps ocurring locally even today.

The continental margin of North America has been extended westward some hundreds of kilometers during late Precambrian and Phanerozoic time, by a continuing succession of episodes of marine deposition, volcanism, metamorphism, and plutonism. As will be noted later, there is evidence of the presence from time to time of island chains, of extensive subaerial volcanism, and of large-scale lateral movements along strike-slip faults, all processes going on in the western portion of the mountain system today. It is tempting to suggest that recurrently through its long history of development the western Cordillera looked much as it does today.

King (1969a, 1969b) divides the Cordillera into northern, central, and southern parts and a discontinuous strip, deformed in Cenozoic time, along the coast, which he designates the Pacific Fold Belt. His broad framework will be employed in this paper also. However, the extensive area of old crystalline basement and overlying shelf sediments of the outer Rocky Mountains and Colorado plateau region will not be discussed (Fig. 1).

CORDILLERAN FOLD BELT

Late Precambrian Continental Margin. Development of the Phanerozoic western margin of North America began in Precambrian time, with the deposition of middle Proterozoic Beltian and equivalent strata. Where the base of these rocks is exposed, the rocks can be seen to lie on eroded crystalline and metasedimentary and metavolcanic rocks of the North American craton, and appear to mark the subsiding western edge of that craton. Beltian rocks cut across earlier Precambrian trends and effectively establish the form and location of the Phanerozoic continental margin (Eardley, 1962, pl. 1). If Precambrian North America adjoined other crystalline continental masses, these connections were broken prior to or very early during Beltian sedimentation and have not been reestablished.

It seems to be generally agreed that the Beltian and equivalent strata were deposited on a subsiding continental margin. Price (1964) compared the depositional environment to that of the U.S. Gulf coast; Gabrielse (1972) described them as a continental terrace wedge. Current indicators generally show transport of sediments parallel to the craton margin, perhaps in a marginal trough. Facies changes point to the craton as a source of the vast quantity of generally fine- to medium-grained clastic Beltian sediment. There seems to be no unequivocal evidence of a western source for these sediments. Unconformably above Beltian and equivalent rocks of the Purcell group lies a sequence of late Precambrian to lower Cambrian sedimentary and volcanic rocks of the Windermere group (Fig. 2) (King, 1959; Gabrielse, 1972; Stewart, 1972; Miller et al., 1973). A widespread, thick diamictite "boulder shale" lies at the base, locally containing rounded boulders of granitic and other rocks to several feet in diameter (Fig. 3). In part the boulder beds are related to turbidites, suggesting mud-flow emplacement in relatively deep water; in part they are believed to have been deposited directly from coastal glaciers or dropped from ice rafts. Similar sequences occur in Death Valley. Apparently the western margin of the craton from northern Canada to the southwestern United States was marked by a coastal strip of undeformed to moderately deformed Beltian and equivalent rocks, rimmed on the west by the strip of Windermere and related rocks.

Monger et al. (1972) suggest that the Cordilleran geosyncline was initiated by an episode of rifting in mid-Proterozoic, followed by the deposition of a miogeosynclinal wedge of sediments along the western margin of the Craton (Beltian and Windermere). Stewart (1972), on the other hand, prefers to place the initiation of the geosyncline at the base of the Windermere diamictite sequence, pointing out the lithologic similarity of Windermere rocks and local

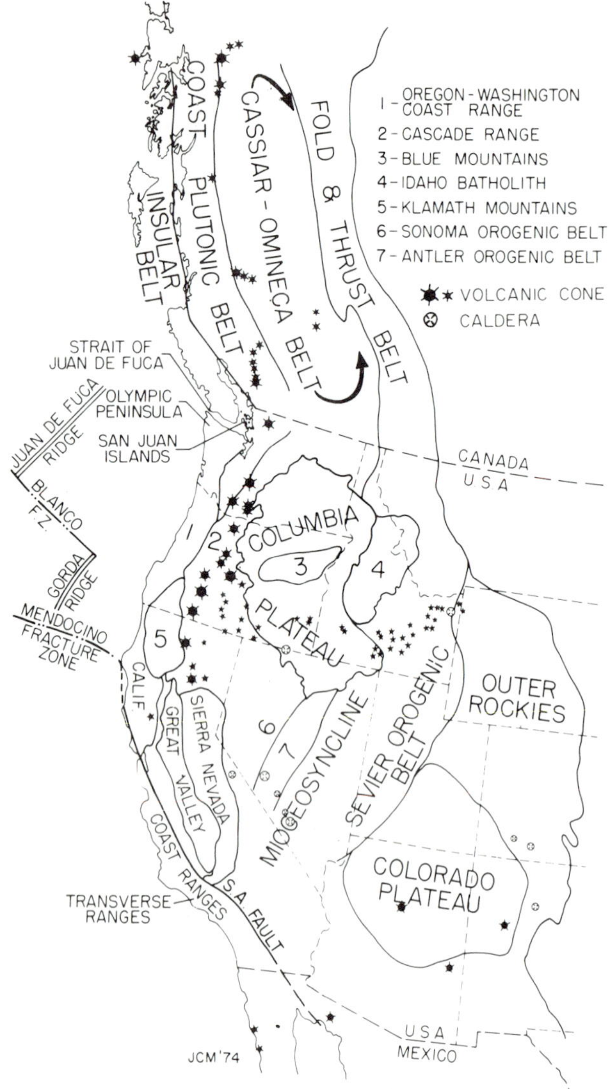

Fig. 1. Location of features mentioned in text (Modified from King, 1969a).

gradation upward into the overlying lower Paleozoic section. The associated volcanics are assumed to be related to the thinning and rifting of the crust during continental separation.

In view of the essential parallelism of Windermere and Beltian strata (even though generally separated by an angular unconformity) and the regional absence of a western cratonic source of sediments following initiation of Beltian sedimentation, it seems reasonable to date the initial continental rifting (if indeed it occurred) as earlier than the beginning of Beltian sedimentation. Reasonable arguments can be made for considering either Windermere or Beltian rocks to be the true base of the Phanerozoic Cordilleran geosyncline.

Paleozoic and Mesozoic Continental Margin.
Antler and Sonoma Trends. In a general way the Beltian-Windermere pattern of westward addition of successively younger belts of thick sediments and associated volcanics continued in the western United States throughout the Paleozoic and Mesozoic, culminating in the early Cenozoic with the youngest Franciscan rocks of the California Coast Ranges (Figs. 5 and 10). Also in a general way, although with less clarity, successively younger periods of emplacement of ultramafic rocks and of orogeny are recognizable westward from the middle or late Proterozoic continental margin, approximately the miogeosynclinal-eugeosynclinal boundary.

The eastern margin of Paleozoic eugeosynclinal rocks, trending north-northeast across central Nevada, is the present extent of klippen of early Paleozoic rocks; these are mostly cherts, shales, and volcanics, lying above a thick carbonate sequence of equivalent age (Roberts et al., 1958; King, 1969a, 1969b). The miogeosynclinal carbonate rocks appear in windows 60 miles (100 km) west of the easternmost klippen. The klippen are remnants of the great Roberts Mountains thrust plate of western assemblage rocks, emplaced at the culmination of the Antler orogeny and accompanied by intense folding and faulting, in Late Devonian or early Mississippian time. Rocks transitional between western eugeosynclinal and eastern miogeosynclinal facies occur in windows beneath the Roberts Mountain thrust plate and as fault slivers within the thrust plate (Roberts et al., 1958; Kay and Crawford, 1964). Figure 4 illustrates diagrammatically the development of the Roberts Mountain thrust and of the overlapping postorogenic clastics of Mississippian age, deposited in a partly continental and partly marine environment east of the thrust.

Although greenstone and chert are abundant constituents of the eugeosynclinal facies of western Nevada, ophiolite assemblages have not been described. The volcanic rocks and cherts appear to occur as interbeds within a marine sedimentary sequence. Ultramafic rocks are sparingly present as tectonic inclusions in strongly deformed sedimentary and volcanic rocks (Poole and Desborough, 1973). No associated gabbroic rocks have been reported (Davis, 1973). Presumably the ultramafic rocks are the only remnants of the oceanic substratum on which lower Paleozoic eugeosynclinal rocks were deposited, prior to the Antler orogeny.

Following the mid-Paleozoic folding and thrusting, the orogenic belt was overlapped by the Antler sequence of Lower Mississippian to Permian, continental and shallow marine clastics and limestones (Silberling and Roberts, 1962). The locus of eugeosynclinal deposition shifted westward to western Nevada and eastern California (Eardley, 1962, plates 5-8). The Sonoma orogeny of late Permian to Early Triassic age again brought eastward thrusting of eugeosynclinal rocks containing small tectonically emplaced slices of serpentinite (Davis, 1973) above rocks of a continental shelf environment. This region was further deformed by folding and thrusting that began in late Early Jurassic and persisted into the Cretaceous (Silberling and Roberts, 1962).

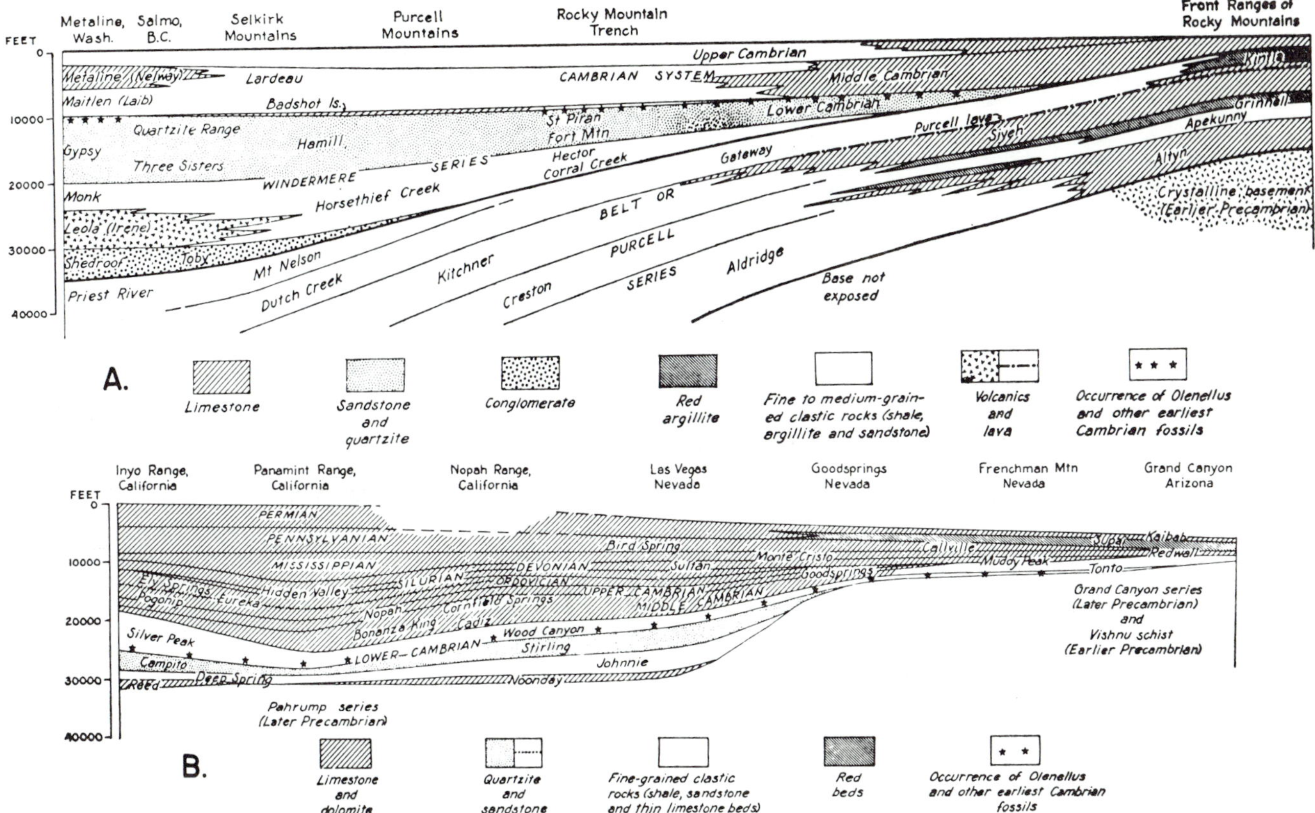

Fig. 2. A. Stratigraphic diagram to show relation of Cambrian of northern Rocky Mountains to Windermere and Belt series. Diagram covers a zone about 200 miles in length immediately north of the International Boundary (after King, 1959, Fig. 75). B. Stratigraphic diagram of Paleozoic rocks of the Cordilleran miogeosyncline across the southern Great Basin from Grand Canyon, Arizona, to the Inyo Mountains, California (after King, 1959, Fig. 77).

Sierra Nevada-Klamath Trends. Westward from the Sonoma orogenic belt lie the Sierra Nevada-Klamath trend and the Coast Ranges. In these also, belts of sediments, of ultramafic rocks, and of orogenic disturbances young progressively westward (Fig. 5), but the genetic relationship of this trend to the Antler-Sonoma belt described above is not at all obvious. The geology of the belts of ultramafic rocks and associated deformed sedimentary and volcanic rocks is best known from the Klamath Mountains (Fig. 6) (Irwin, 1964, 1966, 1972, 1973). Irwin (1973) established *minimum* ages of ultramafic rocks, based on the age of sediments deposited in them, of Ordovician and Devonian in an eastern belt, with successively younger belts to the west of Permian, Permian or Triassic, Late Jurassic Oxfordian or Kimmeridgian, and Late Jurassic Tithonian, the latter occurring in the westernmost or Coast Range sequence (Fig. 5). Hopson and Mattinson (1973) list zircon concordia intercept ages of 455-480 my from mafic-ultramafic rocks near Callahan, California; these rocks are overlain unconformably by sedimentary rocks of late Ordovician and Silurian age.

The map pattern (Weed and Redding sheets, 1:250,000 Geologic Map of California) suggests that the great mass of ultramafic rocks within the Eastern Klamath plate lies in quasi-stratigraphic position between the Ordovician-Silurian rocks of the northern part and the Devonian-Triassic sequence of the southern part of the plate. Whether the ultramafic rocks are all of Ordovician (or earlier) age or also of mid-Paleozoic age comparable to those of the eastern Sierra Nevada belt remains to be demonstrated.

Ordovician ages were also obtained by Hopson and Mattinson (1973) for mafic-ultramafic suites in the Northern Cascades, the Turtleback complex of the San Juan Islands, and possible extensions in southeastern Alaska and the eastern Sierra Nevada (Fig. 5). Other ophiolitic assemblages in the Cascade Mountains of central Washington (Rimrock Lake, Manastash Ridge, Wenatchee Mountains) yield zircon ages clustering near 155 my, close to the 161-my age reported by Hopson and Frano (1973) from the Point Sal ophiolite complex on the coast west of Santa Barbara, California.

Ultramafic rocks of at least two ages occur in the northern, "sedimentary" part of the Sierra Nevada. The largest masses are associated with very thick and intricately deformed Paleozoic eugeosynclinal sedimentary and volcanic rocks, and are thought to lie preferentially along steep faults. A second belt of generally smaller ultramafic bodies is found in the belt of Triassic and Jurassic eugeosynclinal rocks lying along the western margin of the Sierra Nevada and overlapped unconformably by upper Cretaceous sandstones. As mentioned above,

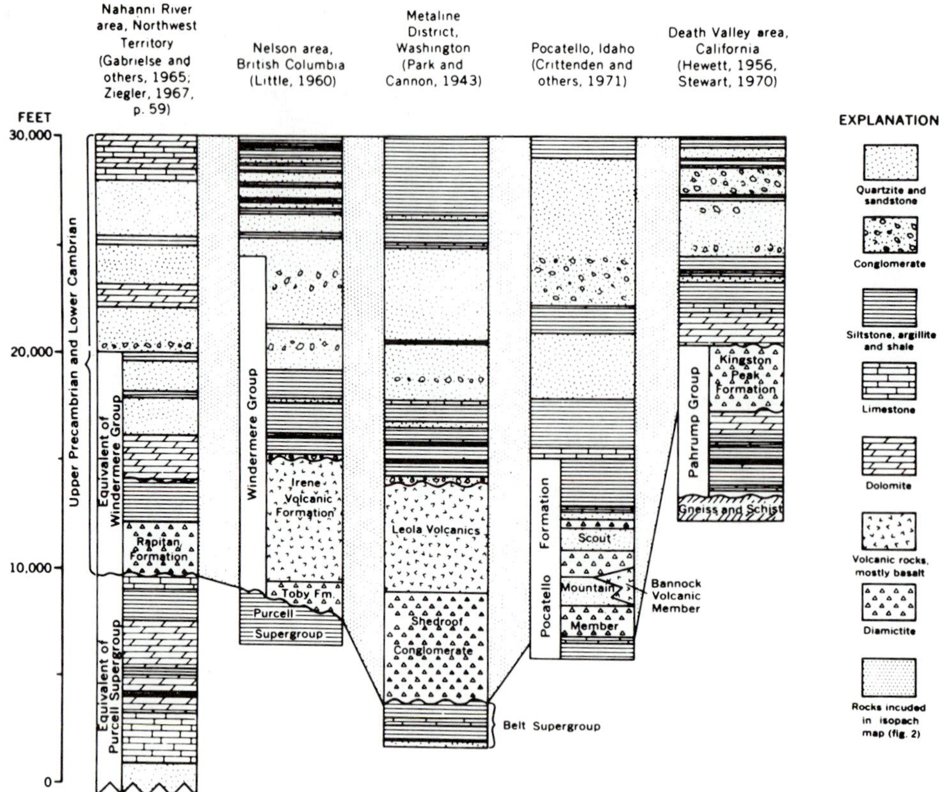

Fig. 3. Precambrian and Lower Cambrian rocks in selected areas in western North America (after Stewart, 1972, Fig. 1).

Hopson and Mattinson (1973) suggest that the eastern belt of ultramafic rocks may correlate with the Ordovician ultramafic rocks of the eastern Klamath Mountains. Eldridge Moores (personal communication) suggests a mid-Paleozoic (post Shoo Fly, pre-Sierra Buttes) age of ophiolite emplacement based on detailed mapping. Ehrenberg (1973) relates the ultramafic rocks and associated amphibolites and metacherts to the Shoo Fly series of Silurian and older slates and cherts. The ophiolites appear to be overlain by a mid to upper Paleozoic, largely volcanic sequence, topped by Permian volcanics. This sequence, in turn, is overlain with angular unconformity by Triassic shelf-type sedimentary rocks (McMath, 1966) and these by a second major outpouring of andesitic and dacitic volcaniclastics and flows of lower to early late Jurassic age. It appears most probable, therefore, that ultramafic rocks in the Paleozoic sequence of the northern Sierras were emplaced near the end of the lower Paleozoic (late Silurian or early Devonian?). Correlation with either the Ordovician ultramafics of the Klamath, or the small serpentinite bodies associated with the Antler and Sonoma orogenic belts to the east is thus in doubt. The possibility of a mid-Paleozoic age for at least part of the Trinity sheet in the Klamath Mountains has been mentioned earlier.

Blue Mountain Trend. At the north end of the Klamath Mountains, structural trends turn from

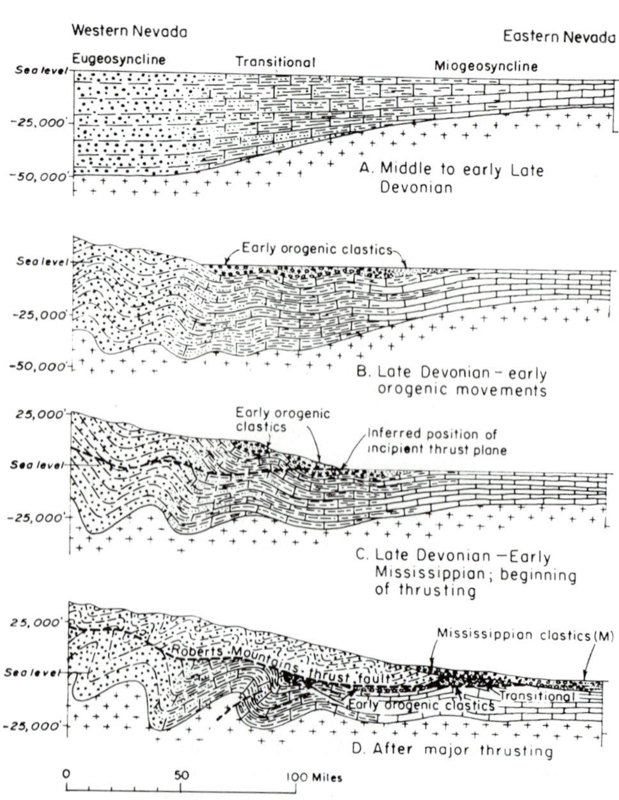

Fig. 4. Inferred sequence of events during Antler orogeny in north-central Nevada (after Roberts et al. 1958, Fig. 10).

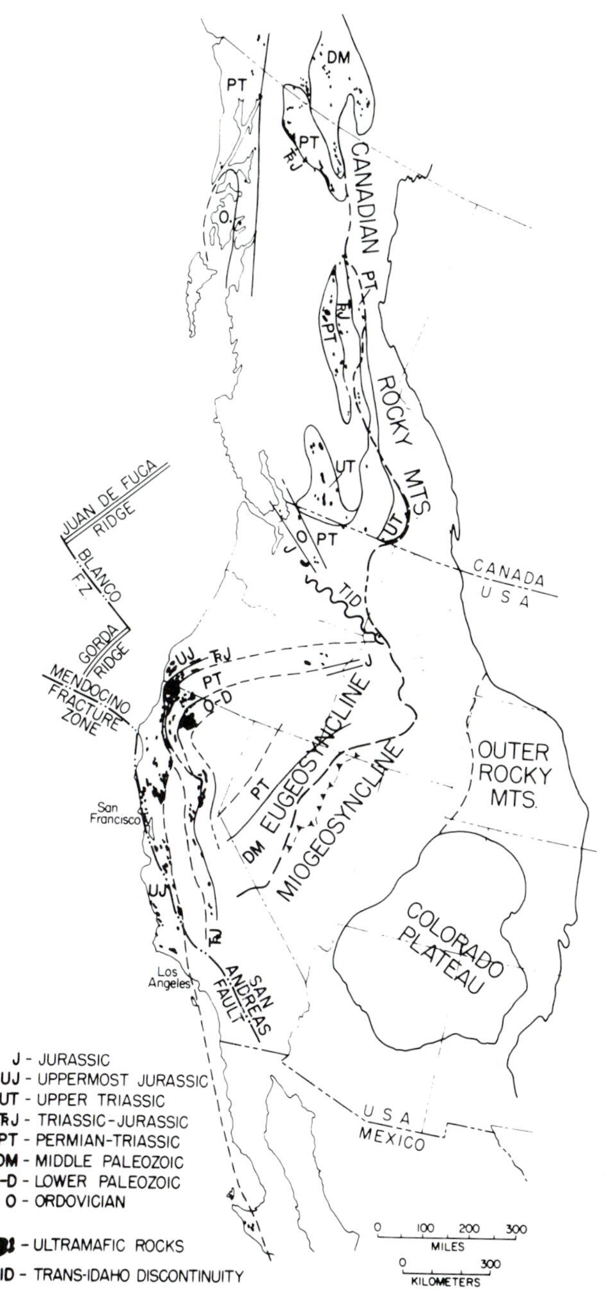

Fig. 5. Apparent ages of ultramafic rocks (based on King, 1969a, plus the following: Canada—Douglas et al., Chart III, 1970; Monger and Preto, 1972; Wheeler, 1966; Campbell, 1966; Souther and Armstrong, 1966; United States—Washington, Oregon, Idaho—Misch, 1966; Thayer and Brown, 1965, 1973; Brown and Thayer, 1966; Hopson and Mattinson, 1973; Vallier, 1973; Brooks, 1973; Peck, 1961; Taubeneck, 1973; Klamath Mountains—Irwin, 1966, 1972, 1973; and Mattinson, 1973; Sierra Nevada—Bateman and Wahrhaftig, 1966; Hopson and Mattinson, 1973; Ehrenberg, 1973; Moores, 1970 and personal communications; Coast Ranges of California—Bailey et al., 1964; Bailey et al., 1970; Blake and Jones, 1974; Pessagno, 1973; Hopson and Frano, 1973; Suppe, 1969, 1972; Dott, 1971; Lanphere, 1971; Rich, 1971; Maxwell, 1974; Raney, 1974; Nevada—Silberling and Roberts, 1962; Poole and Desborough, 1973; Daivs, 1973.

north to northeastward toward the Blue Mountains region of northeastern Oregon and western Idaho. On this trend eugeosynclinal mudstone, chert and metavolcanics of mid-Paleozoic to late Jurassic age emerge through the Columbia Plateau basalts and abut eastward against the Idaho Batholith. Brown and Thayer (1966a) report that ultramafic and mafic rocks of the Canyon Mountain complex intruded the Paleozoic sequence, mainly before the deposition of upper Triassic rocks, but that the latter are also intruded diapirically by serpentinite. Thayer and Brown (1973) interpret the geology, especially the dominance of keratophyre and andesite over basalt in the Permian rocks, to indicate an island arc rather than mid-ocean environment; a possible subduction zone was located along a continental margin at the western limit of the Belt series, the evidence having been obliterated by intrusion of the Idaho batholith in Late Cretaceous time. Brooks (1973) mentions the presence of small tectonic bodies of altered gabbro and serpentinite within a Lower to Middle Jurassic volcanic wacke and siltstone terrain in the southwestern Blue Mountains. The description suggests the Jurassic rocks may be an ophiolite-bearing mélange, of green-schist facies.

Cordillera of Northern Washington and Idaho. Rocks of the Blue Mountains strike generally east-northeasterly into the crystalline rocks of the Idaho batholith. A few miles to the north, metamorphic rocks of Beltian age appear from beneath the covering plateau basalts (Taubeneck, 1973). These and the more continuous outcrops of Beltian rocks to the north and east strike southeasterly, almost at right angles to the Blue Mountain trend. The junction of the two trends, which is not exposed, has been called the Trans-Idaho Discontinuity (Yates, 1968; Taubeneck, 1973) (Fig. 5). To the north of the discontinuity lies the Northern, or Canadian-Southeast Alaskan Cordillera.

In the northern Cordillera, as in the United States to the south, miogeosynclinal rocks of Paleozoic age occur west of the westernmost outcrops of Proterozoic rocks, and these give way westward to Paleozoic and to lower and middle Mesozoic rocks of eugeosynclinal facies (Fig. 7). In the eastern tectonic belt of British Columbia, Wheeler (1966) recognizes a highly complex structural history beginning with folding, metamorphism, and granitic intrusions between Purcell and Windermere times, a second period of folding, metamorphism, and granitic intrusion and emplacement of ultramafic rocks near the Devonian-Mississippian contact, a similar sequence of events about at the Permian-Triassic boundary, and a final episode of strong deformation, metamorphism, and granitic intrusions continued as late as middle Tertiary. The westwardly progressive orderly sequence of emplacement of ultramafic rocks, thick eugeosynclinal deposits, and orogeny, which in general characterizes much of the western Cordillera of the United States, is not apparent in

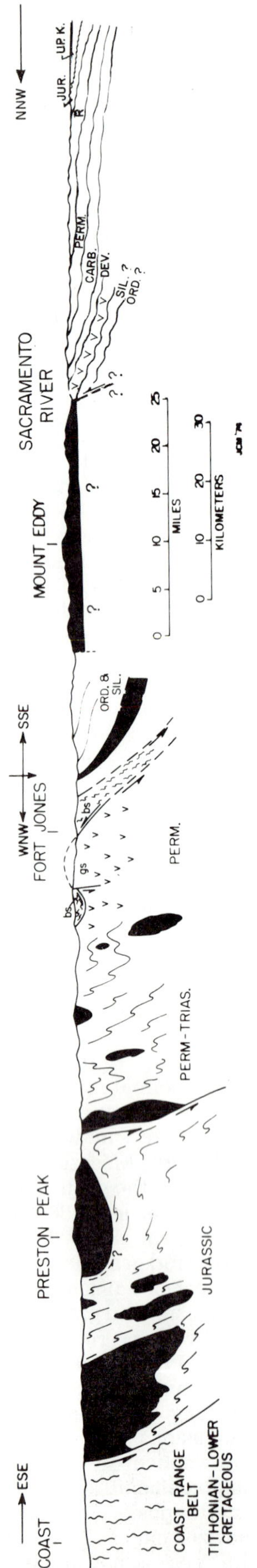

Fig. 6. Northwest-southeast cross section through Klamath Mountains. Cross-hatched pattern, ultramafic rocks (data from the Weed and Redding sheets, Geologic Map of California. Cal. Div. Mines and Geology; Irwin, 1966, 1972, 1973, Hotz, 1973).

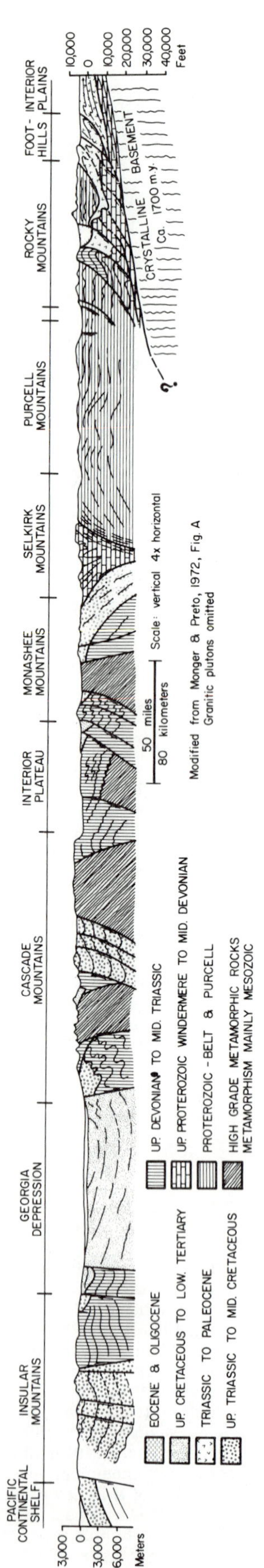

Fig. 7. Cross section along the 49th parallel.

the northern Cordillera, nor in its southern extension into Idaho and Washington.

To complicate the situation further, the Northern Cascades of Washington (Misch, 1966) contain crystalline basement rocks of pre-middle Devonian and possibly even Precambrian age, with associated ultramafic rocks dated, in part at least, as Ordovician (Hopson and Mattinson, 1973). These older rocks, the Yellow Aster series, are overlain by a second metamorphic sequence of probable middle and late Paleozoic age, also with associated ultramafic rocks. The geologic map suggests an eastward, rather than the usual westward polarity, with the thick middle and late Paleozoic eugeosynclinal rocks of the Skagit suite (the metamorphic core of the northern Cascade Mountains) lying mainly to the east of the Yellow Aster series. Wheeler (1966) indicates a possibly similar development between Prince George and Vernon in British Columbia during the lower and middle Silurian and middle Devonian, when sediments were transported easterly from highlands. The culminating period of major deformation, including development of great overthrust sheets, regional metamorphism, and intrusion of granitic rocks, is dated as middle Cretaceous on the west flank of the Cascade Range and Middle or early-late Cretaceous on the east flank. There is no suggestion of a "Nevadan" orogeny in the Northern Cascades.

In considering the probable correlation of pre-Tertiary geology between Canadian and U.S. segments of the Western Cordillera, Monger (1973) proposes the following: The miogeosynclinal Rocky Mountain Thrust and Fold Belt extends southward and correlates with the Sevier orogenic belt of similar stratigraphy and structural style. West of this belt, about nine-tenths of the critical zone is covered by Tertiary rocks. The Canadian Omineca belt of crystalline rocks immediately to the west of the thrust belt (Fig. 1) is comparable to the Idaho batholith and perhaps to the Sierra Nevada in its geologic setting. Farther west, the oldest rocks of the intermontane belt in Canada—Paleozoic and Lower Triassic eugeosynclinal rocks with associated ultramafics—are comparable in lithology and fauna with assemblages in the Blue Mountains, Klamath Mountains, and the west side of the Sierra Nevada. Monger recognizes no counterpart of the Franciscan complex in the Canadian Cordillera, nor any direct counterpart of the granitic Coast Plutonic Complex of British Columbia, the Northern Cascade system of Washington, and the volcanic and sedimentary Insular Belt of British Columbia with rocks farther south in the United States. He suggests that the older terrains of the coast Plutonic complex and insular belts and the northern Cascades may be allochthonous and did not reach their present location until early Mesozoic.

Whatever the explanation for the differences between the Canadian and U.S. Cordillera, it is obvious that they had rather divergent histories, beginning possibly as early as early Paleozoic. The Canadian Cordillera is most simply interpreted as a broad zone of episodic island arc orogenic and intrusive activity (Fig. 8) perhaps with "behind the arc" spreading and emplacement of ultramafic rocks at various times and places (see series of maps, Douglas, 1970, Chap. VIII, especially p. 433; Monger and Preto, 1972, Fig. A).

PACIFIC FOLD BELT

Within the continental United States, exclusive of Alaska, the Pacific Fold Belt may be divided into three segments: the northernmost extending from the Olympic Peninsula to the southwestern corner of Oregon; a central one, the Franciscan province, extending southward from southwestern Oregon to the Transverse Ranges near Santa Barbara; and a third, southward from the Transverse Ranges to the Mexican border and continuing southward along the coast to Baja, California. The last is not discussed.

Oregon-Washington Coast Ranges. This segment of the Pacific Fold Belt extends northward from the Klamath Mountains to the Olympic Peninsula. The Oregon-Washington Coast Ranges are made up of Tertiary, largely Eocene, volcanic-rich clastic rocks, varying from apparently deep water turbidites to deltaic coal-bearing sequences (Baldwin, 1964; Beaulieu, 1971, 1973; Dott, 1971). These rocks are gently to moderately folded and locally faulted. Thick basaltic lava flows, in part submarine, characterize the Eocene sequence and occur sporadically with rocks as young as Pliocene or Pleistocene. In contrast to the other segments of the Pacific Fold Belt, this is an area of relative structural stability, but provided with large volumes of volcanic, mostly basaltic, flows, intrusives, and eroded debris.

At the north and south ends, rocks of the Washington-Oregon Coast Ranges lie in sedimentary contact on older, more highly deformed eugeosynclinal rocks of the Northern and Central Cordilleran belts. To the east, the Eocene rocks pass beneath younger andesitic and basaltic volcanics of the Cascade Range. The great volcanoes that now dominate the Cascade Range are piles of andesitic and smaller amounts of basaltic debris extruded on an uplifted and eroded surface of Middle and Late Tertiary volcanics.

Coast Ranges of California. The Coast Ranges of California (Fig. 9) differ in rock type, age, and deformation style from the above described Oregon-Washington Coast Ranges. In California the Coast Ranges are carved primarily from rocks of the Franciscan complex (Berkland et al., 1972). Characteristically, the Franciscan rocks are massive to thin-bedded graywacke and silty mudstone with locally abundant exotic blocks of pillow lavas, radiolarian cherts, ultramafic rocks, gabbro, dia-

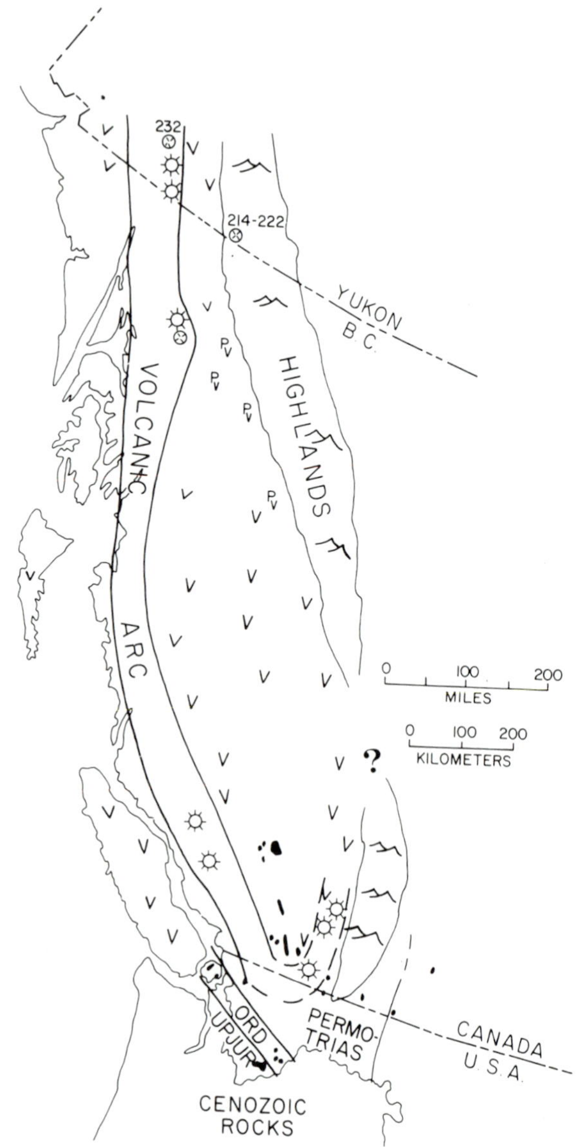

Fig. 8. Possible Upper Triassic volcanic arc and related volcanic, plutonic, and ultramafic rocks. (from Douglas et al., 1970, Fig. VIII-29, p. 429; King, 1969a; and Fig. 5, this paper). black areas, ultramafic rocks; sunbursts, volcanic centers; patterned circles, plutons, with ages in my where determined; v, volcanic rocks; P_V, pillowed.

base and occasional pelagic limestones, and the metamorphic equivalents of these rocks—greenstones, blue-schists and related eclogitic rocks, metacherts, and marbles. Blocks of graywacke, both exotic and locally derived, and great masses and slabs of metagraywacke and metamudstone are also found. Graywackes tend to be rich in volcanic debris. The older graywackes especially are characterized by plagioclase feldspar, mostly albitic, and locally are jadeite and/or lawsonite-bearing. The younger graywackes of the sequence tend to have increasing proportions of potash feldspar. Bailey, Irwin, and Jones (1964) describe these rocks in considerable detail. Locally, large slabs of little deformed graywacke and graywacke-mudstone persist within the Franciscan, but as a whole the complex is highly disturbed to chaotic. Much of it is properly described as a mélange, while other parts appear to be boulder shales and similar products of submarine slumping and mud flows.

Franciscan rocks are bordered on the east by the Great Valley sequence of graywackes and silty mudstones, essentially equivalent in age to rocks of the Franciscan complex. The oldest fossiliferous rocks of both the Franciscan and Great Valley sequences are of Late Jurassic-Tithonian age. The bulk of the Franciscan so far identified is Upper Jurassic and Lower Cretaceous, but scattered fossils of Middle and Upper Cretaceous and Lower Cenozoic age have also been found (Blake and Jones, 1974; Kramer et al., 1974). The youngest rocks of the Great Valley series appear to be Upper Cretaceous.

Within the Franciscan the early Cenozoic fossils (palynomorphs) are restricted to the coastal facies, a sequence of volcanic-rich to feldspathic graywackes and silty mudstones, characterized by discontinuous deformation similar in style to the rest of the Franciscan but much less intensely disturbed in most places (Kleist, 1974). The uppermost beds of the Franciscan complex thus seem to be of the same age or only slightly older than the lowermost beds of the Oregon-Washington Coast Ranges to the north, but are of very different character, lacking the thick basaltic rocks and extensive deltaic and coaly deposits of the Oregon-Washington sequence.

The fundamental structural problem of the California Coast Ranges can be succinctly stated. Fossils and radioactive age determinations demonstrate clearly that the mildly deformed sedimentary rocks of the Great Valley sequence and the adjacent chaotic and partly metamorphosed rocks of the Franciscan complex are largely of the same age. Furthermore, the lithologies of the two are generally similar. Conglomerates with cobbles of granitic and silicic volcanic rocks are widely but sparingly distributed in both Great Valley and Franciscan terranes. And finally, the Franciscan complex seems to lie everywhere structurally beneath the Great Valley sequence. Before fossil ages were well established, the Great Valley sequence was thought to lie in sedimentary contact on the disturbed and metamorphosed rocks of the Franciscan. The preferred explanation now is that the Great Valley sediments have been thrust far to the west over Franciscan rocks (Brown, 1964b; Ernst, 1970; Blake and Jones, 1974). This model involves several structural improbabilities (Raney, 1974; Maxwell, 1974a).

At the base of the Great Valley sequence is a thin-bedded, interbedded mudstone and graded fine graywacke, the Knoxville Formation (Taliaferro, 1943). Close to the basal contact, Knoxville beds stand vertically or are locally overturned toward the east. Basal beds tend to be faulted and locally rubbled within a few feet of the contact. At a few localities, however, it has been demonstrated that Knoxville mudstone lies in sedimentary contact on rocks of the ophiolite suite, which generally lie between Franciscan and Great Valley rocks (Bailey et al., 1970). Turbidite beds rich in serpentine fragments and locally extensive serpentine mud flows occur within the Knoxville (Raney, 1974), and also well up into the Lower Cretaceous, as Rich (1971) has demonstrated in the Wilbur Springs Quadrangle. The distribution of serpentine-rich beds indicates clearly that serpentinite was exposed to the west and was providing sediments locally to flysch deposits of the Great Valley sequence.

The ophiolite zone is generally mapped and characterized as a continuous belt of serpentinite present between Great Valley and Franciscan rocks (see especially Ukiah sheet, 250,000 series, Geologic Map of California). This is, however, an oversimplification (Raney, 1974). The rocks are actually discontinuous, and throughout much of the zone serpentinite is thin, or even absent; when present it is extensively mixed with blocks of extrusive volcanics, radiolarian chert, gabbro, pyroxenite, and igneous breccia. The ophiolite strip at the base of the Great Valley sequence now has the character of a mélange of largely ophiolite blocks.

Stratigraphically beneath and to the west of the Great Valley sequence and ophiolite mélange is a belt of lawsonite-bearing white mica schist with interbedded glaucophane schists, the Southfork Mountain Schist. These schists are very commonly deformed by a secondary chevron-type folding, giving the formation an aspect of intense deformation. However, associated metagraywacke and blue-schist beds rarely show this deformation and tend to remain planer and unfolded. The chevron folding characterizes most of the unit and is not spatially related to the contact between the schists and the overlying ophiolite Great Valley rocks. Blue-schists within the Paskenta Quadrangle originated as graded, volcanic-rich sediments or tuffs, interbedded with argillaceous rocks. Undeformed load casts occur at the base of one blue-schist bed. All blue-schist beds in which grading could be determined are upright and facing east. There is no obvious evidence of an initial folding episode related to generation of the blue schists and associated mica schists. They seemed to have formed in a tectonically quiet environment, presumably characterized by high pressure and relatively low temperature, but little penetrative shearing.

Most authors suggesting thrusting of Great Valley rocks over Franciscan now place the plane of thrusting between the ophiolite rocks and the Southfork Mountain Schists or Franciscan mélange, where the schist is not present. The contact with schist is well exposed at a number of places, however, and shows remarkably little disturbance. There are no associated structural features suggesting large-scale thrust faulting (Raney, 1974) Maxwell, 1974a). The amount of shearing is about that which might occur as the sequence was tilted from essentially horizontal to the present vertical or overturned position. Raney and I postulate, therefore, that the mélange of ophiolite blocks was emplaced by a relatively near surface, probably gravity slide mechanism, and that no fault of regional extent exists at this horizon.

Thus far, twelve tectonic-stratigraphic units have been identified within the Franciscan terrane, structurally beneath the Great Valley sequence ophiolites and Southfork Mountain Schists (Raney, 1974; Lehman, 1974; Jordan, 1974; Gucwa, 1974; Kleist, 1974). These units are distinguished on the basis of gross lithology and especially on the type of contained exotic blocks (Fig. 9). The variation of exotic block content is quite striking between some adjacent units, suggesting a unique origin for the exotic blocks. Considering the possible addition of exotic blocks to a flysch matrix by gravity sliding and slumping or by incorporation from the basement on which the sediments rest, I have suggested tentatively (Maxwell, 1974a) that for the most part the contained exotic blocks are derived from the underlying basement by a process of slicing related to underthrusting of each unit. The mechanism is derived from seismic sections across the Java Trench (Beck, 1972; see also Seely et al. and Kulm and Fowler, this volume).

Fossils collected from the Franciscan are dominantly of Late Jurassic and Early Cretaceous age, but younger fossils appear in the central part of the Franciscan outcrop in the Coast Ranges of northern California and, in general, they become progressively younger westward, so that the youngest Late Cretaceous and Tertiary fossils characterize the westernmost Franciscan-type rocks. This suggests a mechanism of repeated underthrusting of sediments, presumably accumulating in a trench environment that developed to the west of a Sierran metamorphosed basement in Late Jurassic-Tithonian time. The trench was continuously or intermittently active into Lower Tertiary time, apparently terminating during the Eocene. Ages determined isotopically for various units of the Franciscan and contained blocks give rather complicated results (Suppe and Armstrong, 1972; Lanphere, 1971; Gucwa, 1974; Lehman, 1974).

The mechanism of emplacement of the ophiolite rocks, both at the base of the Great Valley sequence and those occurring as exotic blocks within units of the Franciscan complex, is very much in doubt. The fact that they were formed in a deep ocean environment is rather generally accepted. For a number of reasons I have suggested that these ophiolites were originally emplaced diapirically, perhaps in an environment similar to that suggested by Karig (1971). This concept is more fully discussed elsewhere (Maxwell, 1974a, 1974b, 1974c), but, briefly, it involves the observation that ultramafic rocks in a particular orogenic trend seem to have been emplaced only once, as suggested many years ago by Hess (1939). This has been confirmed for the Franciscan by Pessagno (1973), who found that the cherts associated with ophiolites in the Coast Ranges are all of Late Jurassic-Tithonian age. There

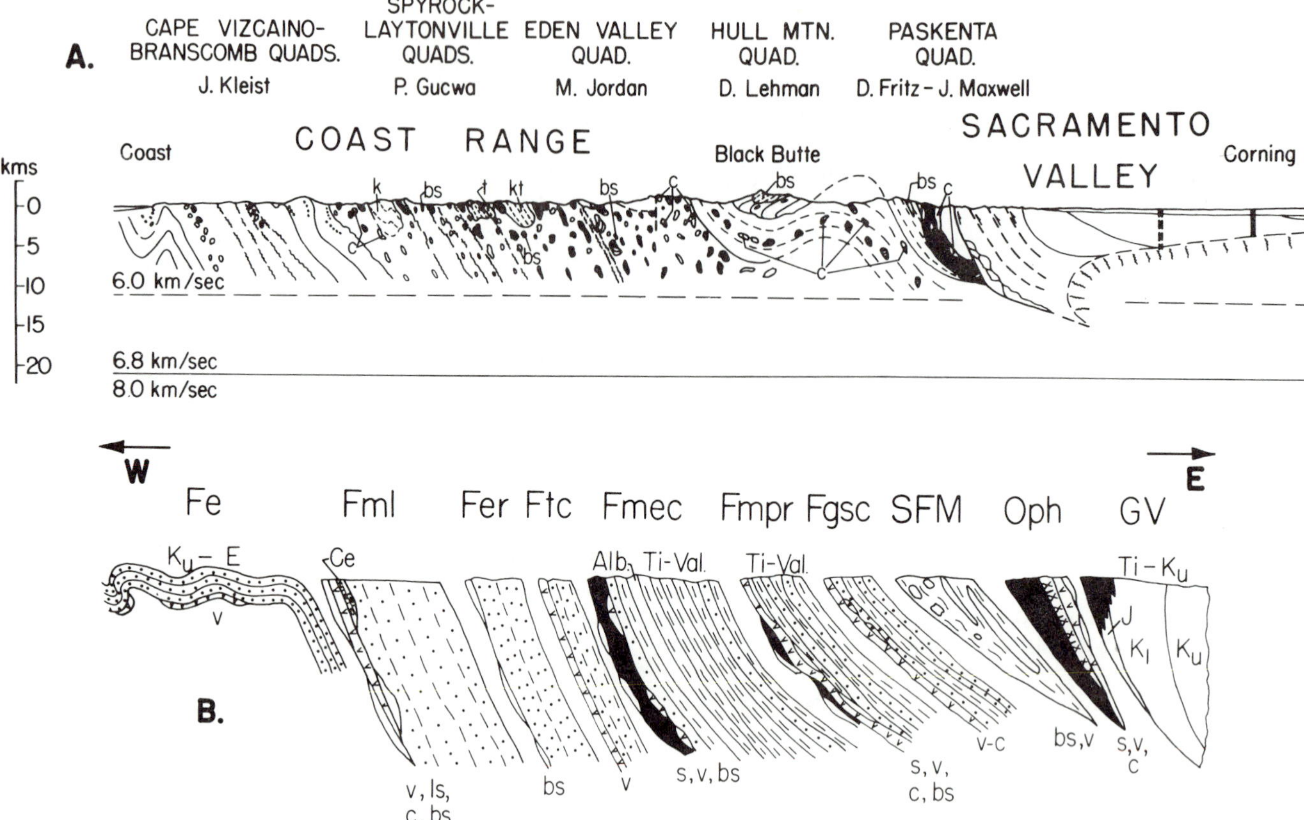

Fig. 9. East-west cross section of northern California Coast Ranges and western part of Sacramento Valley (from Maxwell, 1974a). Line of section extends eastward from the coast 10 miles north of Fort Bragg, south of Covelo, to the vicinity of Corning in the Sacramento Valley. A. Franciscan mélange, exotic blocks: black, ultramafic rocks, mostly serpentinite; bs, blueschist; c, chert; cross-lined and unmarked blocks, largely volcanics or greenstone metavolcanics; k, kt, t, Cretaceous and Tertiary clastic units of the "successor basin" type. B. Pull-apart to illustrate possible tectonically emplaced sedimentary units, with the suggestion that most exotic blocks are derived from basement rocks beneath each unit. F, various units of Franciscan mélange (Maxwell, 1974a); J, Jurassic; Kl, Ku, Lower, Upper Cretaceous; E, Eocene; Ti-Va, Tithonian-Valanginian; Alb, Albian; Ce, Cenomanian. SFM, South Fork Mountain Schist; Oph, Ophiolite unit; GV, Great Valley sedimentary sequence; black, sepentinite; V, Is, c, bs, volcanics, limestone, chert, blue schists of presumed basement beneath sedimentary units. (seismic velocity data from Eaton, 1966).

is some doubt, therefore, that ophiolites represent fragments of normal sea floor, which should have been added continuously throughout the long life (late Triassic to Eocene) of the trench environment. The model I prefer (Maxwell, 1974a) is that of a mantle diapir coming up between Sierra Nevada basement and the developing trench to the west, beneath blue-schists under the inner wall of that trench. The ophiolites and serpentinites so formed provided sedimentary serpentine turbidites, mud flows, and blocks to the Great Valley sediments, which were then being deposited on the seaward-sloping Nevadan basement and also on top of the ophiolite sequence farther seaward. In the meantime, the locus of underthrusting was also displaced westward, thus continuously widening the pile of Franciscan sediments.

The great predominance of Late Jurassic and Early Cretaceous sediments within the Franciscan complex suggest that this was a time of especially vigorous orogeny; activity apparently diminished through upper Cretaceous, dying out during the Lower Cenozoic. A main period of batholithic emplacement in eastern California and western Nevada seems to correspond essentially with this early vigorous stage of the Coast Range orogeny. A later stage of extensive volcanism and magmatic emplacement occurred in Late Cretaceous and Eocene time, coinciding with the dying out of the Coast Range Orogeny. The apparent absence of Late Jurassic and Cretaceous volcanic activity in the area of the present Sierra Nevada is puzzling (Cordell Durrell, personal communication). Abundant andesitic and basaltic volcanic debris in the Franciscan and the lower portion of the Great Valley sequence would seem to require a source in the Sierra Nevada area, but it is hardly credible that all traces of volcanic activity could have been removed by erosion.

Throughout much of the Coast Range orogenic episode, vertical movements seem to have occurred frequently, affecting the surface of the prism of Franciscan sediments and melange. Local basins and fault-bounded troughs and grabens were formed, related to normal faulting, to strike-slip faulting or to diapiric movements within the sedimentary prism (Maxwell, 1974a). Shallow-water deposits such as the Late Cretaceous-Eocene fossiliferous sandstone and gently dipping Miocene coal-bearing sandstones of the Covelo area (Gucwa, 1974) were formed in local, perhaps fault-bounded "successor basins" during and after the waning episode of Coast Range orogeny. Other deposits, however, resemble the Great Valley rocks and have been called outliers; for example, Berkland (1972, 1973) regards the Middle Mountain and Rice Valley graywacke, mudstone, and conglomerate sequences as klippen of Great Valley rocks thrust far to the west over Franciscan. These sections are, however, greatly thinned compared to equivalent rocks of the Great Valley section (Raney, 1974). We suggest that these sediments of Great Valley type were deposited seaward and in continuity with the Great Valley sequence proper, but on a tectonically active shelf made up of Franciscan mélange above actively underthrusting sediments (Raney, 1974; Maxwell, 1974a).

South of San Francisco the simple arrangement of Great Valley sediments to the east and Franciscan complex to the west is complicated by a series of major strike-slip faults such as the San Andreas and other systems. This great fault system has been the subject of much controversy. It has been postulated by some that the fault may have originated as long ago as Late Jurassic and that right lateral strike-slip movements of 600 km or more may have occurred (Hill and Dibblee, 1953; Crowell, 1962; Ross et al., 1973). Crowell (1962) has shown that Oligocene and all older rocks of the Transverse Ranges seem to have been displaced by about the same amount. Subsequently, Crowell (1973), reporting on recent work, indicates that movement on known faults of the San Andreas system of southern California could all be post-Late Miocene in age, with a total slip of about 260 km for the system south of the "big bend" of the Transverse Ranges area.

In summarizing the activity on faults of the San Andreas system north of the "big bend," Crowell (1973) suggests that the faults were active back to about 23 million years ago, preceded by an inactive period between 23 and 49 million years and active again back to about 70 million years, for a total right-slip displacement of about 530 km. Cross (1973) interprets a 20-my gap in igneous activity to indicate a proto-San Andreas fault, active between 65 and 49 my, preceded and followed by subduction. Suppe (1970), likewise addressing the problem of apparent large differences in displacement in the northern and southern parts of the San Andreas system, suggested a two-stage-movement history, in which the northern branch, including the Salinian block, moved twice for a total of approximately 600 km, while the southern branch moved only once for approximately 300 km. The earlier movement of the northern branch presumably took place on a branch of the San Andreas system, including the Newport-Inglewood fault, with the resulting production of a "rhombochasm" recognizable as the present deep-water region of the continental borderland off northern Baja California. A change of direction of block movement by about 25° occurred between the two movement episodes.

EVOLUTION OF THE WESTERN MARGIN OF CONTINENTAL UNITED STATES

The fundamental assumption undergirding this and indeed most other interpretations of ancient, tectonically active continental margins is that rocks of the eugeosynclinal suite—graywacke, bedded chert, basaltic and andesitic volcanics, and ultramafic rocks—represent oceanic crust and associated sediments, which are subsequently attached to a continental craton by compressive (orogenic) mechanisms. Ultramafic rocks of the "alpine" type (those not obviously a product of magmatic differentiation of a gabbroic magma) seem to be found almost exclusively in this environment, whereas other members of the association are not so limited. For that reason the age and distribution of ultramafic rocks receive special emphasis in this study.

The interpretation of the origin and emplacement of ultramafic rocks is fraught with uncertainties. Some with contact metamorphic aureoles are demonstrably intrusive into crustal rocks, but these are relatively uncommon. Most ultramafic rocks in orogenic belts lack evidence of contact metamorphism and are intimately associated, quite often in essentially stratigraphic sequence, with overlying gabbroic, diabasic, and extrusive basaltic rocks. Bedded radiolarian cherts typically occur above or within the extrusive basalts. This sequence of igneous and sedimentary rocks is the ophiolite assemblage of orogenic belts. It is believed to have formed in a deep-sea environment, but the exact mechanism is an enigma (Maxwell, 1969, 1974c). It is now rather generally accepted that ophiolites may form at oceanic ridges, and it is also inferred, although without strong evidence, that similar rocks may form in a "behind-the-arc spreading region" (Karig, 1971). Within orogenic belts ultramafic rocks tend to be distributed in linear zones which also may coincide with the boundary between highly disturbed rock sequences, and are consequently often interpreted as fault zones. The geologic literature is replete with assertions that major zones of faulting have controlled the intrusion of ultramafic masses. In most cases, however, the "intrusive" ultramafics lack evidence of contact metamorphism, tend to be associated with other rocks of the ophiolite suite of hypabyssal and surficial origin, and parallel rather than cross-cut regional and local stratification. The field evidence generally favors surficial emplacement or tectonic slicing rather than fault-controlled intrusions.

The second problem relating to ultramafic rocks is the determination of "age" where evidence of an intrusive origin is lacking. It is now generally agreed that many ultramafic bodies have been emplaced tectonically, long after initial consolidation and cooling. The "age," as used in this paper, refers to the time of emplacement of such ultramafic bodies into the host rock, usually flysch. Since flysch sediments accumulate in a generally unstable environment, they are frequently disrupted by erosional, gravitational, or even tectonic events, and therefore ultramafic rocks often have been reworked into successively younger environments. A minimal absolute age may be established if sediments deposited on ultramafic rocks or associated ophiolitic assemblage can be dated. This is probably the most reliable "age" available. Recently, absolute ages have been obtained from mafic rocks associated with the ultramafics, as described elsewhere in this paper. Efforts to date the ultramafic rocks themselves by isotopic methods seem to give abnormally great, usually Precambrian ages (T. E. Davis, 1973), results that are not clearly understood. Uncertainties also arise from the difficulties of measuring isotopic ages in rocks that are deficient in potassium and other radioactive elements, and because the associated sediments, usually flysch, tend to be sparingly fossiliferous and commonly contain reworked faunas. In addition, as pointed out previously, the ultramafic rocks may be successively reworked into younger flysch formations.

Another possible source of the confusion is the oft-repeated assertion that small bodies of serpentinite have been injected diapirically into overlying rocks. Rarely can the "diapiric" bodies be shown to cut across stratified rocks, however. It should be noted that serpentinite has an apparent viscosity (rheidity) higher than salt (Carey, 1953) and, generally, a smaller density differential with surrounding rocks; and therefore diapiric movements of serpentinite would be expected only with bodies larger than salt diapirs. In summary, it is the opinion of this writer that much of the literature on the dating and mode of emplacement of serpentinite bodies is based on inadequate field data and therefore that any summary of apparent ages, such as those included in this paper, is subject to errors.

With the above caveat, let us then proceed to examine hypotheses relating to the origin of the western continental margin. Most recent literature is cast in the mold of the plate tectonics hypothesis, and specifically, for the west coast of the United States, the reaction through time between a paleo-Pacific plate and the American plate. A detailed analysis of this interaction was prepared by Atwater (1970) and has served as a standard for comparison with onshore geology since that time. Coney (1972) related Cordilleran tectonic events to the relative motions between North America and Africa, based on interpretation of the magnetic pattern within the Atlantic Ocean. In this view the spreading history of the Pacific Ocean is scarcely reflected in Cordilleran geology, except for the relatively recent development of the San Andreas fault system.

Atwater and Molnar (1973) have refined the original Atwater (1970) data, especially in terms of motion along the San Andreas system. In their recent reconstruction the Farallon plate, which presumably existed between a Pacific spreading ridge and the American plate, was progressively overridden by the American plate, begining about 29 my ago, and since that time the North American and Pacific plates have been in contact. The apparent rate of motion along the fault system has varied considerably in magnitude through the last 29 my; indicated movements are approximately twice those inferred from geologic evidence. They point out, however, that the rate postulated from magnetic data approximates the estimate by Hein (1973) for the displacement of Pacific plate sediments past their apparent source regions in North America. They also suggest a total displacement, since 21 my ago of about 600 km with the excess above known movement on the San Andreas perhaps being absorbed by movements both east and west of the San Andreas zone. The Atwater and Molnar reconstruction implies the existence of a consuming plate margin and trench environment along the west coast of the United States north of the Mendocino fracture zone until the present day, and of a trench environment until 29 my ago at the Mendocino escarpment, then gradually eliminated south of that escarpment. This reconstruction is at variance with some geological observations, as will be discussed later.

A geological interpretation of the western continental margin emphasizes the successive episodes of sedimentation, volcanic activity, and orogeny beginning in the Proterozoic and active at least until early Cenozoic time. In plate tectonics terminology, a series of subduction zones marked by trenches and associated volcanic arcs is postulated to have developed successively westward from the Precambrian crystalline craton, with the result that a belt some 600-700 km wide of new continental crust has been added since late Proterozoic time (Fig. 10). The precise mechanisms involved in the addition of various units of new crust are, however, somewhat obscure.

The patterns of distribution of the various ages of ultramafic rocks shown on Fig. 5 indicates a certain degree of complexity within the United States and a very complex picture indeed in the southeast Alaska-British Columbia Cordillera and extension into Washington State. Within the California-Oregon segment, there is an overall orderly westward progression from Old Paleozoic to youngest Jurassic belts of ultramafic rocks, apparently indicating successive additions of oceanic crust and sediments to the continental margin. Abutting against the California trend almost at right angles, however, are the Antler and Sonoma belts of middle Paleozoic and Permo-Triassic ages, respectively. The small bodies of ultramafic rocks within these two belts presumably also are indicative of the initial presence of oceanic crust. Does the abrupt

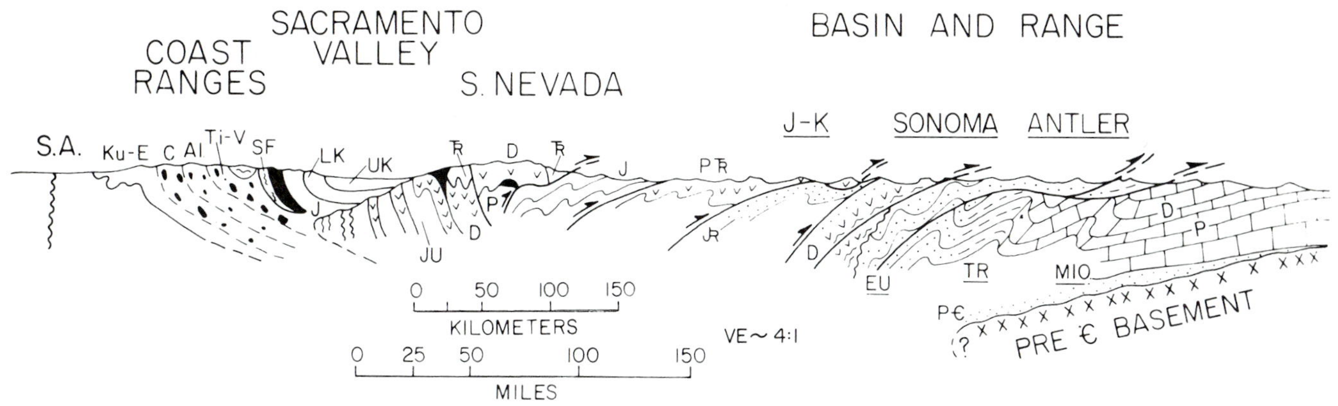

Fig. 10. Schematic geological cross section, coast of northern California eastward to the foreland of the Antler orogen. Notation, west to east: S.A., trace of San Andreas fault; Ku-E, Upper Cretaceous-Eocene; C, Cenomanian; Al, Albian; Ti-V, Tithonian-Valanginian; SF, South Fork Mountain Schist; LK, UK, Lower, Upper Cretaceous; TR, Triassic; D, mid-Paleozoic; J, Jurassic; PTR, Permo-Triassic; J-K, Jurassic-Cretaceous orogenic belt.

discontinuity in trend at the southwest end of the Sonoma and Antler belts (Figs. 5 and 11) signify that a former extension has been sheared off, or do these belts possibly represent areas of thin continental crust or even small ocean basins (G. A. Davis, 1973)? It is noteworthy that the polarity, that is, the directon of accretion of successively younger belts of oceanic rocks, is westerly in both the coastal belt and the Antler-Sonoma region.

Within the southeast Alaska-British Columbia Cordillera the picture is much more complex. In the portion that enters the State of Washington, a belt of Ordovician ultramafic rocks is bordered on the east by ultramafics of presumed Permo-Triassic age, and on the west by occurrences of upper Jurassic ultramafic rocks (Figs. 5 and 8). Polarity was eastward from Ordovician to upper Triassic or Permo-Triassic, and a similar situation exists farther to the north in southeast Alaska and the Coast Ranges of British Columbia (Fig. 8). On the other hand, there is a suggestion of westward polarity from the middle Paleozoic to Permian in the Yukon in northern British Columbia and from Permo-Triassic to lower and middle Mesozoic farther to the south (Fig. 5).

As mapped, the lower and middle Mesozoic belt cuts through or is superimposed on a Permo-Triassic belt, which in turn was apparently cut by the upper Triassic belt. A simpler interpretation would be that all the Permian and lower Mesozoic occurrences represent a single episode of ultramafic emplacement, perhaps of Permian age, with subsequent reworking into younger sediments. Quite obviously, in the United States, and especially in Canada, there is ample room for speculation regarding the scale movements of various elements to bring about the present pattern. The probability that some of the complexity is a reflection of inadequate or misinterpreted data should be kept in mind.

A second problem area regards the structural history of the California-Oregon-Washington Coast Ranges and their relationship to plate motions postulated from magnetic anomalies in the nearby Pacific Ocean. The plate tectonics hypothesis would seem to require the presence of a consuming margin along the entire length of the Coast Ranges until approximately 29 my ago, after which the consuming margin was gradually replaced to the south by a transform fault system, while continuing to be active north of the Mendocino escarpment to the present day. The coastal facies of the Franciscan complex, ranging in age from late Cretaceous to Eocene, is much less deformed than the bulk of the Franciscan to the east, indicating a lessening of orogenic activity during late Cretaceous and early Tertiary and terminating apparently during the Eocene. From the plate tectonics model it seems that evidence of compression should persist here into late Oligocene. The magnetic data require that a consuming plate margin still exists off the Washington and Oregon coasts (E. A. Silver, 1971), yet the volcanic and sedimentary rocks of the Coast Ranges are for the most part only gently warped and faulted. It appears that compressive deformation along the entire Coast Range belt within the United States had essentially ceased by late Cretaceous, and certainly by the beginning of Cenozoic time. It should be noted, however, that the belt of active volcanism along the Cascade Range is essentially coextensive and collinear with the Gorda and Juan de Fuca ridges offshore (Fig. 1), and that this has been cited as evidence of the presence of a consuming margin, even though supporting evidence is not compelling.

Various models have been proposed to account for certain aspects of the evolution of the North American cordillera. One of the most comprehensive, by Burchfiel and Davis (1972), organizes the structural development of the southern part of the cordillera into three periods: (1) late Precambrian through late Paleozoic, (2) early Mesozoic through early Tertiary, and (3) middle Tertiary to recent. The first two represented episodes of accumulation of eugeosynclinal-type sediments, volcanics, and ultramafic rocks, punctuated by periods of orogeny and thrust faulting, while the third period, from middle Tertiary to Recent, was characterized by strike-slip and extensional fault tectonics. During the early period they visualize the development of an offshore Sierran-Klamath island arc above an east-dipping subduction zone with a small ocean basin developed in the present area of the Antler-Sonoma orogenic belts (Figs. 5 and 10). Closure of the small ocean basin caused the Antler and Sonoma orogenies. During the second period of structural development an early Mesozoic volcanic arc of the Andean type arose across the earlier northeast-trending orogenic belts (Fig. 11). They propose that the cross-cutting relationship indicates the truncation and displacement of a portion of the North American continental plate in the California area following Permo-Triassic deformation in the Sonoma area. Early Mesozoic underthrusting then developed, following the older trends north of the truncated margin but continuing southward parallel to the present coastline. The coastal underthrusting continued into early Tertiary time, producing the Nevadan and Coast Range orogenic belts.

Roberts (1972) proposes a somewhat different picture of the evolution of the Cordilleran belt. He likewise recognizes three principal phases during the Phanerozoic time: (1) an initial ortho-geosynclinal phase, (2) an orogenic-late geosynclinal phase, and (3) an orogenic-postgeosynclinal phase. During the first phase extensive sedimentation off the western continental margin produced deep-water (eugeosynclinal), slope (transitional), and shelf (miogeosynclinal) facies. The Antler orogeny in late Devonian through middle Pennsylvanian time terminated phase one and instituted phase two with the development of tectonic highlands during late Paleozoic, which shed sediments both eastward into epicontinental seas and westward into flanking eugeosynclinal troughs. Deposition in the western trough was terminated by the Sonoma orogeny in late Permian and early Triassic. Phase two, the orogenic-late geosynclinal phase, ended with the Nevadan and Sevier orogenies during Jurassic time. Phase three, during Mesozoic and Tertiary time, marked the advent of plate tectonics along the Cordilleran margin and was characterized by orogenic and postorogenic flysch and molasse sedimentation in troughs and basins along the continental margin, and by andesitic volcanism. Roberts suggests that the pre-phase three (preplate tectonics) development of the Western Cordillera was a different nature than that which occurred during the plate tectonics episode.

Another approach to interpretation, consistent with that proposed by Burchfiel and Davis and in part consistent with Roberts' summary, involves the juxtaposition of far-traveled island arcs or continental fragments which have presumably moved long distances, either parallel or perpendicular to the orogenic belt. Hamilton (1969), for example, suggested that ancient island arcs and other structural elements had been swept against continental plates, as for example in the Klamath Mountains, where he recognizes Ordovician ocean floor, Silurian arc debris, Paleozoic oceanic crust and mantle, Devonian-Permian island arc, and a fragment of a Devonian orogenic belt piled together in the eastern Klamaths.

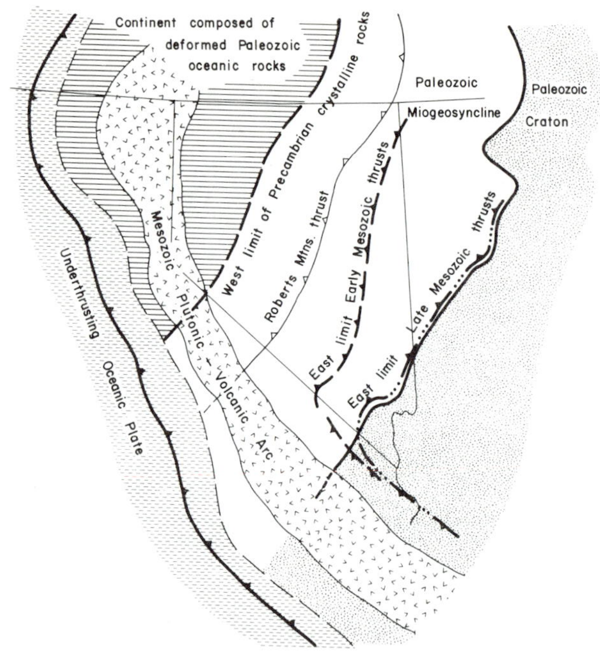

Fig. 11. Sketch map for the Late Mesozoic showing truncation of Paleozoic geosynclinal and deformational trends by a Mesozoic plutonic-volcanic arc of Andean type (after Burchfiel and Davis, 1972, Fig. 7).

Moores (1970) suggests a more-or-less continuous convergence of a paleo-Pacific plate with the American plate along an east-dipping subduction zone beginning in lower Paleozoic and continuing through middle Tertiary. The advancing continental plate three times encountered west-dipping subduction zones, resulting in the emplacement of ultramafic sheets on the continental plate, and expressed structurally as the Antler, Sonoma, and Nevadan orogenies. He further suggested (1972) based on mapping in the northern Sierras, that an eastern block of Mesozoic and Paleozoic sediments and volcanics collided with a western block of island-arc type rocks in upper-late Jurassic time, crushing between them a highly disturbed ophiolite-bearing sequence, possibly containing remnants of subduction zones and old oceanic lithosphere. Hsu (1971) proposed that the Salinia block, west of the San Andreas fault, first rifted from the continental margin in early Mesozoic and then in late Jurassic collided with the continent at a subduction zone as the leading edge of the convergent Pacific plate.

The complex history of the Northern Cordillera, which enters northern Washington, is interpreted (Monger et al., 1972; Monger and Ross, 1971) to be the result of the bringing together of formerly widely separated island-arc-type environments, either by converging plate motion with the elimination of intervening oceanic crust, or by long-distance transcurrent fault movement in early Mesozoic time. The gological evidence supporting such drastic interpretations involves a repetition of Permian fossil zones, an eastern and western zone of Schwagerinids occurring in dirty mixed volcanic facies (arc?), and an intervening belt of Verbeekinids, which occur in clean basaltic (oceanic?) facies. The western Coast Range belt of Verbeekinids possibly was transported from the south along a northwest-trending transcurrent fault in early Mesozoic time from the southwestern United States, where similar faunas occur. Jones et al. (1972) find a similar argument persuasive in deriving a belt of Upper Paleozoic rocks in southeastern Alaska from a California source. The apparent truncation of southwesterly-trending Paleozoic rocks in southern California is perhaps partly explained by long-distance lateral transport of the type postulated by Monger and Ross (1971) and by Jones et al. (1972).

The study of stratigraphic and paleontologic relationships between the various geologic belts provides the most reliable data on which to base interpretation. An example from Monger et al (1972) (Fig. 12) illustrates the interpretation of lithologic assemblages. The existence of stratigraphic units spanning two contiguous belts, or of evidence of detritus shed from one belt to another, provides time links between belts. The absence of such tying elements between two areas now adjacent may indicate the creation of tectonic lands or volcanic arcs, or their arrival from far-distant points. The Antler orogen provides another example (Fig. 4). Eugeosynclinal and miogeosynclinal rocks are tied together by transition beds (Roberts et al., 1958), and therefore for this particular mio-eugeosynclinal boundary no large-scale displacement can be postulated.

Unfortunately, criteria are not ordinarily at hand to determine whether a given belt is autochthonous or allochthonous. The great complexity of geology in the island-studded area surrounding much of the Pacific Ocean suggests that complex orogenic belts may develop close to a continental margin without the need for far-traveled allochthony. The argument of Hamilton (1969) is persuasive, yet no modern island arc or fragment of old orogenic belt is currently disappearing into a trench (subduction zone), although one would expect to see it somewhere in the Pacific if this had been a common phenomenon throughout geologic time. There are points where disaster appears imminent—Australia about to collide with the East Indies, the Caroline Islands about to go into the Marianas Trench, the Fiji islands just skating past the north end of the Tonga Trench, the Emperor Seamounts poised in the bight between the Aleutian and Kurile trenches—but all safe so far.

A related problem is the apparent decoupling of plates at converging plate boundaries, such as that

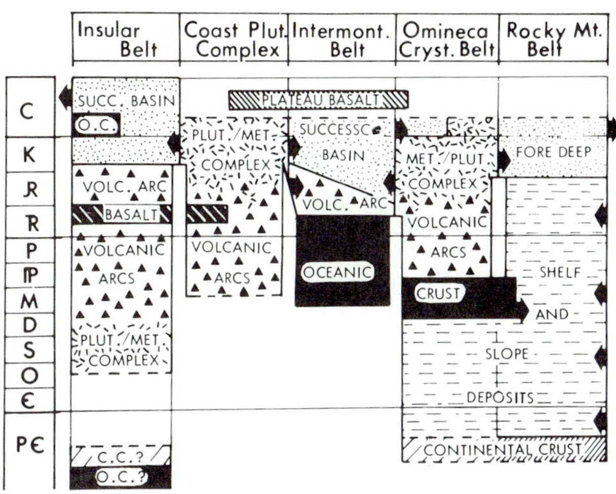

Fig. 12. Space and time distribution of lithological assemblages interpreted largely in actualistic terms within the various physiographic and geological belts of the Canadian Cordillera. Spaces between the vertical columns indicate that no linkages are known between the belts. These linkages take the form of detritus shed from one belt to another, shown by arrows that indicate the direction of clastic movement, or of stratigraphic units spanning two contiguous belts (after Monger et al., 1972, Fig. 2).

postulated to have been present during most of Phanerozoic time at the western margin of the North American plate. Some thousands of kilometers of Pacific ocean crust and associated sediments apparently have disappeared under North America in latest Cretaceous and Cenozoic time alone (Gilluly, 1971; Maxwell, 1974b), and presumably three or four times this amount during the several older orogenic periods if the plate tectonics model is accepted. By contrast, tectonic shortening, where it can be estimated, tends to be measured in tens or hundreds of kilometers, an order of magnitude less than the related plate motions. Volumes of mafic crustal rocks and sediment which apparently have to be "eaten" beneath tectonic continental margins are also much larger than can be easily explained. In short, converging plate motions an order of magnitude smaller than those currently postulated would seem to provide a more satisfactory explanation of the tectonic development of continental margins. A structural geologist is loath to accept extensive (90%?) decoupling between driving mechanism and orogeny, yet at the moment we seem to have no better choice.

SUMMARY

The rocks of the western Cordillera of the United States and adjacent parts of Canada and Mexico record a long history of succeeding episodes of sedimentation, orogeny, and outbuilding of the craton. The development of the present western continental margin perhaps began with deposition of middle Proterozoic Beltian sediments along an apparently continuous western coastline extending from northern Canada to the southwestern United States, truncating trends of earlier Precambrian rocks of the craton. Continental rifting, if it occurred, presumably happened prior to Beltian deposition. Succeeding sequences of thick continental shelf and deep marine deposits have been added since middle Proterozoic, beginning with Windermere beds and related rocks of late Precambrian and early Cambrian time. Windermere rocks are partly of miogeosynclinal and shelf facies, while those to the west of a zone trending through central Nevada are dominantly eugeosynclinal in character.

On the assumption that ultramafic rocks and associated mafic and sedimentary rocks of the ophiolite suite are generated from the upper mantle in a deep-sea environment, westward growth of the continental margin by sedimentation, orogeny, and continental accretion is indicated. In the western United States belts of ultramafic rocks are successively younger westward. The oldest are Ordovician and the youngest latest Jurassic. The ultramafic rocks are located in zones of deformation, now assigned to the Antler, Sonoma, Nevadan, and Coast Range orogenies. If comparable zones of orogeny are associated with the Ordovician and Permo-Triassic belts of ultramafic rocks, they are not well defined in the United States.

The westward accretion of successively younger belts characteristic of the United States is not evident in the Canadian Cordillera. A large-scale lateral transport of crustal elements either along transform faults or at converging plate boundaries, has been suggested to account for the complexity in the Canadian Cordillera and its extension into Washington State. Truncation of Paleozoic and Mesozoic trends in southern California is perhaps related to pre-Nevadan plate movements of this sort.

Plate motions, as interpeted from magnetic patterns in the Pacific and Atlantic oceans, have been correlated here with Coast Range and younger orogenic features of the western continental margin. The correlation is rather good, but striking anomalies remain.

ACKNOWLEDGMENTS

It is a pleasure to acknowledge the support of the National Science Foundation for a continuing program of investigation in the Coast Ranges of northern California. I am particularly indebted to Stephen Etter, Deborah Fritz, Paul Gucwa, Michael Jordan, John Kleist, David Lehman, and Jay Raney for their contributions to this program; to Eldridge Moores and Cordell Durrell for extensive discussions; and to William Muehlburger for criticizing the manuscript.

BIBLIOGRAPHY

Armstrong, R. L., and Higgins, R. E., 1973, K-Ar dating of the beginning of Tertiary volcanism in the Mojave Desert, California: Geol. Soc. America Bull., v. 84, p. 1095-1100.

———, and Suppe, J., 1973, Potassium-argon geochronometry of Mesozoic igneous rocks in Nevada, Utah and Southern California: Geol. Soc. America Bull., v. 84, p. 1375-1392.

Atwater, T., 1970, Implications of plate tectonics for the Cenozoic tectonic evolution of western North America: Geol. Soc. America Bull., v. 81, p. 3513-3536.

———, and Molnar, P., 1973, Relative motion of the Pacific and North American plates deduced from sea-floor spreading in the Atlantic, Indian, and South Pacific oceans, *in* Kovach, R. L., and Nur, A., eds., Proceedings of the conference on Tectonic Problems of San Andreas Fault System: Stanford Univ. Publ. Geol. Sci., V. XIII, p. 136-147.

Bailey, E. H., ed., 1966, Geology of northern California: Calif. Div. Mines and Geol. Bull., v. 190, 507 p.

———, Irwin, W. P., and Jones, D. L., 1964, Franciscan and related rocks: Calif. Div. Mines and Geol. Bull., v. 183, 177 p.

———, Blake, M. C., and Jones, D. L., 1970, On-land Mesozoic oceanic crust in California Coast Ranges: U.S. Geol. Survey Prof. Paper 700C, p. 70-81.

Baldwin, E. M., 1964, Geology of Oregon, 2nd ed.: Univ. Oregon Coop. Book Store, 165 p.

Bally, A. W., Gordy, P. L., and Stewart, G. A., 1966, Structure, seismic data, and orogenic evolution of southern Canadian Rocky Mountains: Bull. Can. Petr. Geol., v. 14, no. 3, p. 337-381.

Barbat, W., 1971, Megatectonics of the Coast Ranges, California: Geol. Soc. America Bull., v. 72, p. 1541-1562.

Bateman, P. C., and Wahrhaftig, C., 1966, Geology of the Sierra Nevada, *in* Bailey, E. H., ed., Geology of Northern California, Calif. Div. Mines and Geol. Bull., v. 190, p. 107-172.

Beaulieu, J. D., ed., 1971, Geologic formations of western Oregon, west of long. 121°30': Oregon Dept. Geol. and Min. Ind. Bull., v. 70, 72 p.

———, ed., 1973, Geologic field trips in northern Oregon and southern Washington: Oregon Dept. Geol. and Min. Ind. Bull., v. 77, 206 p.

Beck, R. H., 1972, The oceans, the new frontier in exploration: APEA Jour. v. 12, pt. 2, p. 5-28.

Berkland, J. O., 1972, Paleogene "frozen" subduction zone in the Coast Ranges of northern California: 24th Internatl. Geol. Congr., sect. 3: Tectonics, p. 99-105.

———, 1973, Rice Valley outlier—new sequence of Cretaceous-Paleocene strata in northern Coast Ranges, California: Geol. Soc. America Bull., v. 84, p. 2389-2406.

———, Raymond, L. A., Kramer, J. C., Moores, E. M., and O'Day, M., 1972, What is Franciscan?: Am. Assoc. Petr. Geol. Bull., v. 12, p. 2295-2302.

Bissell, H. J., 1970, Realms of Permian tectonism and sedimentation in western Utah and eastern Nevada: Am. Assoc. Petr. Geol. Bull., v. 54, p. 285-312.

Blake, M. C., and Jones, D. L., 1974, Origin of Franciscan mélanges in northern California, *in* Dott, R. H., Jr., ed., Modern and ancient geosynclinal sedimentation,

Marshall Kay vol.: Soc. Econ. Paleontologists and Mineralogists Spec. Publ. 19 (in press).

———, Irwin, W. P., and Coleman, R. G., 1967, Upside-down metamorphic zonations, blueschist facies, along a regional thrust in California and Oregon: U.S., Geol. Survey Prof. Paper 575C, p. C1-C9.

Bolt, B. A., and McEvilly, T. V., 1968, Seismological evidence on the tectonics of central and northern California and the Mendocino escarpment; Seismol. Soc. America Bull., v. 58, p. 1725-1767.

———, and Miller, R. D., 1971, Seismicity of northern and central California, 1965-1969; Seismol. Soc. America Bull., v. 61, p. 1831-1847.

Brace, W. F., Ernst, W. G., and Kallberg, R. W., 1970, An experimental study of tectonic overpressure in Franciscan rocks: Geol. Soc. America Bull., v. 81, p. 1325-1338.

Brooks, H. G., 1973, Some Triassic-Jurassic relationships in the southeastern Blue Mountains, Oregon-Idaho (abstr.): Geol. Soc. America, v. 5, no. 1, p. 16.

Brown, C. E., and Thayer, T. P., 1966a, Geologic map of the Canyon City quadrangle, northeastern Oregon: U.S. Geol. Survey Misc. Geol. Inv. Map 1-447.

———, and Thayer, T. P., 1966b, Geologic map of the Mount Vernon quadrangle, Grant County, Oregon: U.S. Geol. Survey Map GQ-548.

Brown, R. D., 1964a, Geologic map of the Stoneford Quadrangle, Glenn, Colusa and Lake counties, California: U.S. Geol. Survey Min. Invest. Map MF-279.

———, 1964b, Thrust-faulting relations in the northern Coast Ranges, California: U.S. Geol. Survey Prof. Paper 475-D, p. D7-D13.

Brown, R. D., Jr., and Hanna, W. F., 1971, Aeromagnetic evidence and geologic structure, northern Olympic Peninsula and Strait of Juan de Fuca, Washington: Am. Assoc. Petr. Geol. Bull., v. 55, p. 1939-1953.

Brown W. H., Fyfe, W. S., and Turner, F. J., 1962, Aragonite in California glaucophane schists and the kinetics of the aragonite-calcite transformation: Jour. Petrology (London), v. 3, p. 566-582.

Burchfiel, B. C., and Davis, G. A., 1972, Structural framework and evolution of the southern part of the Cordilleran orogen, western United States: Am. Jour. Sci., v. 272, p. 97-118.

Burk, C. A., and Moores, E. M., 1968, Problems of major faulting at continental margins, with special reference to the San Andreas fault system, *in* Dickinson, W. R., and Grantz, A., eds., Proceedings of the conference on geologic problems of San Andreas fault system: Stanford Univ. Publ. Geol. Sci., v. XI, p. 358-374.

Byrne, J. V., Fowler, G. A., and Maloney, N. J., 1966, Uplift of the continental margin and possible continental accretion off Oregon: Science, v. 154, p. 1654-1656.

Campbell, R. B., 1966, Tectonics of the South Central Cordillera of British Columbia; *in* Gunning, H. C., ed., Tectonic history and mineral deposits of the Western Cordillera: Can. Inst. Mining and Met., Spec. v. 8, p. 61-71.

Carder, D. S., Qamar, A., and McEvilly, T. V., 1970, Trans-California seismic profile—Pahute Mesa to San Francisco Bay: Seismol. Soc. America Bull., v. 60, p. 1829-1846.

Cardwell, G. T., 1965, Geology and groundwater in Russian river valley areas and in round, Laytonville and little lake valleys, Sonoma and Mendocino counties, California: U.S. Geol. Survey Water Supply Paper 1548, 154 p.

Carey, S. W., 1953, The Rheid concept in geotectonics: Jour. Geol. Soc. Australia, v. 1, no. 1, p. 67-117, (pub. 1954).

Carlson, J. E., and Willden, R., 1968, Transcontinental geophysical survey (35°-39°N) geologic map from 112° W. longitude to the coast of California: U.S. Geol. Survey Map l-532-C.

Chipping, D. H., 1971, Paleoenvironmental significance of chert in the Franciscan formation of western California: Geol. Soc. America Bull., v. 82, p. 1707-1711.

Christensen, M. N., 1965, Late Cenozoic deformation in the central Coast Ranges of California: Geol. Soc. America Bull., v. 76, p. 1105-1124.

Church, S. E., and Tilton, G. R., 1973, Lead and strontium isotopic studies in the Cascade mountains: bearing on andesite genesis: Geol. Soc. America Bull., v. 84, p. 431-454.

Clark, L. D., 1964, Stratigraphy and structure of part of the western Sierra Nevada metamorphic belt, California: U.S. Gol. Survey Prof. Paper 410, 70 p.

Coleman, R. G., 1972, The Colebrook Schist of southwestern Oregon and its relation to the tectonic evolution of the region: U.S. Geol. Survey Bull., v. 1339, 61 p.

———, and Lanphere, M. A., 1971, Distribution and age of highgrade blueschists, associated eclogites, and amphibolites from Oregon and California: Geol. Soc. America Bull., v. 82, p. 2397-2512.

Coney, P. J., 1972, Cordilleran tectonics and North American plate motion: AM. Jour. Sci., v. 272, p. 603-628.

Cross, T. A., 1973, Implications of igneous activity for the early Cenozoic tectonic evolution of western United States (abstr): Geol. Soc. America, v. 5, no. 7, p. 587.

Crowell, J. C., 1962, Displacement along the San Andreas fault, California: Geol. Soc. America Spec. Paper 72, 61 p.

———, 1968, The California coast ranges,Univ. Missouri at Rolla, Jour. no. 1, p. 133-156.

———, 1973, Problems concerning the San Andreas fault system in southern California, *in* Kovach, R. L., and Nur, A., eds., Proceedings of the conference on Tectonic problems of San Andreas fault system: Stanford Univ. Publ. Geol. Sci., v. XIII, p. 125-135.

Curray, J. R., and Nason, R. D., 1967, San Andreas fault north of Point Arena, Calif.: Geol. Soc. America Bull., v. 78, p. 413-418.

Davis, G. A., 1966, Metamorphic and granitic history of the Klamath Mountains, *in* Bailey, E. H., ed., Geology of northern California, Calif. Div. Mines and Geol. Bull., v. 190, p. 39-50.

———, 1973, Subduction-obduction model for the Antler and Sonoma orogenies, western Great Basin area (abstr.): Gol. Soc. America, v. 5, no. 7, p. 592.

———, and Burchfiel, B. C., 1973, Garlock fault: an intracontinental transform structure, Southern California: Geol. Soc. America Bull., v. 84, p. 1407-1422.

Davis, T. E., 1973, Rb-Sr geochronology of ultramafic rocks from Burro Mountain, California (abstr): Geol. Soc. America, v. 5, no. 1, p. 32.

Dehlinger, P., Couch, R. W., McManus, D. A., and Genperle, M., 1970, Northeast Pacific structure, *in* Maxwell, A. E., ed., The sea, v. 4, pt. II: Wiley-Interscience, New York, p. 133-189.

Dickinson, W. R., 1969, Evolution of calc-alkaline rocks in

the geosynclinal system of California and Oregon, *in* Andesite Conference Proceedings: Oregon Dept. Geol. Min. Ind. Bull., no. 65, p. 151-156.

———, 1970, Relations of andesites, granites, and derivative sandstones to arc-trench tectonics: Rev. Geophys. Space Phys., v. 8, p. 813-860.

———, and Rich, E. I., 1972, Petrologic intervals and petrofacies in the Great Valley sequence, Sacramento Valley, California: Geol. Soc. America Bull., v. 83, p. 3007-3024.

Dietz, R. S., 1963, Alpine serpentines as oceanic rind fragments: Geol. Soc. America Bull., v. 74, p. 947-952.

Dole, H. M., ed., 1968, Andesite conference guidebook: Oregon Dept. Geol. and Min. Ind. Bull., v. 62, 107 p.

Dott, R. H., Jr., 1971, Geology of the southwestern Oregon Coast west of the 124 Meridian: Oregon Dept. Geol. and Min. Ind. Bull., v. 69, 63 p.

Douglas, R. J. W., ed., 1970, Geology and economic minerals of Canada: Geol. Survey Can. Econ. Geol. Rept. 1, 838 p.

———, Gabrielse, H., Wheeler, J. O., Stott, D. F., and Belyear, H. R., 1970, Geology of western Canada, Chapt. VIII, *in* Douglas, R. J. W., ed., Geology and economic minerals of Canada: Gol. Survey Can. Econ. Geol. Rept. 1, p. 367-488.

Dudley, P. 1972, Comments on the distribution and age of high-grade blueschists, associated eclogites, and amphibolites from the Tiburon Peninsula, California: Geol. Soc. America Bull., v. 83, p. 3497-3500.

Durrell, C., 1966, Tertiary and Quaternary geology of the northern Sierra Nevada, *in* Bailey, E. H., ed., Geology of northern California: Calif. Div. Mines and Geol. Bull., v. 190, p. 185-197.

Eardley, A. J., 1947, Paleozoic Cordilleran geosyncline and related orogeny: Jour. Geol., v. 55, p. 309-342.

———, 1962, Structural geology of North America, 2nd ed.: New York, Harper & Row, 743 p.

Eaton, J. P., 1966, Crustal structure in northern and central California from seismic evidence, *in* Bailey, E. H., Geology of northern California: Calif. Div. Mines and Geol. Bull., v. 190, p. 419-426.

Ehrenberg, S. N., 1973, Ultramafic and associated rocks near Red Hill, Southwest Almanor quadrangle, northern Sierra Nevada (abstr.): Geol. Soc. America, v. 5, no. 1, p. 37.

Elders, W. A., Rex, R. W., Meidav, T., Robinson, P. T., and Biehler, S., 1972, Crustal spreading in southern California: Science, v. 178, no. 4056, p. 15-24.

Ernst, W. G., 1970, Tectonic contact between the Franciscan mélange and the Great Valley sequence, crustal expression of a late Mesozoic Benioff zone: Jour. Geophys. Res., v. 75, p. 886-902.

———, 1971a, Do mineral parageneses reflect unusually high-pressure conditions of Franciscan metamorphism? Am. Jour. Sci., v. 270, p. 81-108.

———, 1971b, Metamorphic zonations on presumably subducted lithospheric plates from Japan, California and the Alps: Contr. Min. Petr., v. 34, p. 43-59.

———, Seki, Y., Onuki, H., and Gilbert, M. C., 1970, Comparative study of low-grade metamorphism in the California Coast Ranges and the outer metamorphic belt of Japan: Geol. Soc. America Mem. 124, 276 p.

Gabrielse, H., 1972, Younger Precambrian of the Canadian Cordillera: Am. Jour. Sci., v. 272. p. 521-536.

Garfunkel, Z., 1973, History of the San Andreas fault as a plate boundary: Geol. Soc. America Bull., v. 84, p. 2035-2042.

Garrison, L. E., 1972, Geothermal steam in the geysers in Clear Lake region, California: Geol. Soc. America Bull., v. 83, p. 1449-1468.

Garrison, R. E., 1973, Space-time relations of pelagic limestones and volcanic rocks, Olympic Peninsula, Washington: Geol. Soc. America Bull., v. 84, p. 583-594.

Gastil, G., and Phillips, R. P., 1972, The reconstruction of Mesozoic California: 24th Internatl. Geol. Congr. sect. 3, Tectonics, p. 217-229.

Ghent, E. D., and Coleman, R. C., 1973, Eclogites from southwestern Oregon: Geol. Soc. America Bull., v. 84, p. 2471-2488.

Gilluly, J., 1963, Tectonic evolution of the western United States: Geol. Soc. London Quart. Jour., v. 199, p. 133-174.

———, 1965, Volcanism, tectonism, and plutonism in the western United States: Geol. Soc. America Spec. Paper 80, 69 p.

———, 1967, Chronology of tectonic movements in the western United States: Am. Jour. Sci., v. 265, p. 306-331.

———, 1969, Oceanic sediment volumes and continental drift: Science, v. 166, p. 992-993.

———, 1971, Plate tectonics and magmatic evolution; Geol. Soc. America Bull., v. 82, p. 2383-2396.

———, 1972, Tectonics involved in the evolution of mountain ranges, *in* Robertson, E. C., ed., The nature of the solid earth: New York, McGraw-Hill, p. 406-439.

———, 1973, Steady plate motion and episodic orogeny and magmatism: Geol. Soc. America Bull., v. 84, p. 499-514.

———, Reed, J. C., and Cady, W. M., 1970, Sedimentary volumes and their significance: Geol. Soc. America Bull., v. 81, p. 353-376.

Gresens, R. L., 1970, Serpentinites, blueschists and tectonic continental margins: Geol. Soc. America Bull., v. 81, p. 307-310.

Gucwa, P. R., 1974, Geology of the Covelo-Laytonville area, northern California [Ph.D. thesis]: Univ. Texas, Austin.

Gunning, H. C., ed., 1966, Tectonic history and mineral deposits of the Western Cordillera: Can. Inst. Mining and Met., Spec., v. 8, 353 p.

Hamilton, W., 1969, Mesozoic California and the underflow of Pacific mantle: Geol. Soc. America Bull., v. 80, p. 2409-2430.

Hayes, D. E., and Ewing, M., 1970, Pacific boundary structure, *in* Maxwell, A. E., ed., The sea, v. 4, pt. II: New York, Wiley-Interscience, p. 29-72.

Hein, J. R., 1973, Deep-sea sediment source areas: implications of variable rates of movement between California and the Pacific plate: Nature, v. 241, p. 40-41.

Hess, H. H., 1939, Island arcs, gravity anomalies, and serpentinites intrusions: a contribution to the ophiolite problem: Proc. 17th Internatl. Geol. Congr., Moscow, Rept. 2, p. 279-300.

Hietanen, A., 1973, Origin of andesitic and granitic magmas in the northern Sierra Nevada, California: Geol. Soc. America Bull., v. 84, p. 2111-2118.

Hill, D. P., 1972, Crustal and upper mantle structure of the Columbia Plateau from long range seismic-refraction measurements: Geol. Soc. America Bull., v. 83, p. 1639-1648.

———, 1971a, Newport-Inglewood zone and Mesozoic

subduction, California: Geol. Soc. America Bull., v. 82, p. 2957-2962.

———, 1971b, A test of new global tectonics: comparisons of northwest Pacific and California strauctures: Am. Assoc. Petr. Geol. Bull., v. 55, p. 3-9.

———, and Dibblee, T. W., Jr., 1953, San Andreas, Garlock and Big Pine faults, California: Geol. Soc. America Bull., v. 64, p. 443-458.

Himmelberg, G. R., and Loney, R. A., 1973, Petrology of the Vulcan Peak Alpine-type peridotite, southwestern Oregon: Geol. Soc. America Bull., v. 84, p. 1585-1600.

Hopson, C. A., and Frano, C. J., 1973, Late Jurassic ophiolite at Point Sal, Santa Barbara County, California (abstr.): Geol. Soc. America v. 5, no. 1, p. 58.

———, and Mattinson, J. M., 1973, Ordovician and Late Jurassic ophiolitic assemblages in the Pacific Northwest (abstr.): Geol. Soc. America, v. 5, no. 1, p. 57.

Hotz, P. E., 1973, Blueschist metamorphism in the Yreka-Fort Jones area, Klamath Mountains, California: U.S. Geol. Survey Jour. Res., v. 1, no. 1, p. 53-61.

Hsu, K. J., 1968, Principles of mélanges and their bearing on the Franciscan-Knoxville paradox: Geol. Soc. America Bull., v. 79, p. 1063-1074.

———, 1971, Franciscan mélanges as a model for eugeosynclinal sedimentation and underthrusting tectonics, Jour. Geophys. Res., v. 76, p. 1162-1170.

Huffman, O. F., 1972, Lateral displacement of Upper Miocene rocks and the Neogene history of offset along the San Andreas fault in central California: Geol. Soc. America Bull., v. 83, p. 2913-2946.

Huntting, M. T., Bennett, W. A. G., Livingston, V. E., Jr., and Moen, W. S., 1961, Geologic map of Washington: Wash. Div. Mines and Geol., scale 1:500,000.

Irwin, W. P., 1964, Late Mesozoic orogenies in the ultramafic belts of northwestern California and southwestern Oregon: U.S. Geol. Survey Prof. Paper 501-C, p. C1-C9.

———, 1966, Geology of the Klamath Mountains province, *in* Bailey, E. H., ed., Geology of northern California, Calif. Div. Mines and Geol. Bull., v. 190, p. 19-38.

———, 1972, Terranes of the western Paleozoic and Triassic belts in southern Klamath Mountains: U.S. Geol. Survey Res. Prof. Paper 800C, p. C103-C111.

———, 1973, Sequential minimum ages of oceanic crust in accreted tectonic plates of northern California and southern Oregon (abstr): Geol. Soc. America, v. 5, no. 1, p. 62.

———, and Coleman, R. G., 1972, Preliminary map showing global distribution of Alpine-type ultramafic rocks and blueschists: U.S. Geol. Survey Map MF-340.

Jahns, R. H., ed., 1954, Geology of Southern California: Calif. Div. Mines and Geol. Bull., v. 170.

Jones, D. L., and Irwin, W. P., 1971, Structural implications of an offset early Cretaceous shoreline in northern California: Geol. Soc. America Bull., v. 82, p. 815-822.

———, and Irwin, W. P., and Ovenshine, A. T., 1972, Southeastern Alaska—a displaced continental fragment?: Geol. Survey Res., 1972, U.S. Geol. Survey Prof. Paper 800B, p. B211-B217.

Jordan, M. A., 1974, Geology of the Round Valley-Sanhedrin Mountain area, northern California [Ph.D. thesis]: Univ. Texas, Austin.

Karig, D. E., 1971, Origin and development of marginal basins in the western Pacific: Jour. Geophys. Res., v. 76, p. 2542-2561.

Kay, M., and Crawford, J. P., 1964, Paleozoic facies from miogeosynclinal to the eugeosynclinal belt in thrust slices, central Nevada: Geol. Soc. America Bull., v. 75, p. 425-454.

King, P. B., 1959, The evolution of North America: Princeton, N.J., Princeton Univ. Press, 190 p.

———, 1969a, Tectonic map of North America: U.S. Geol. Survey, scale 1:5 million.

———, 1969b, The tectonics of North America—a discussion to accompany the tectonic map of North America: U.S. Geol. Survey Prof. Paper 628, 95 p.

Kleist, J. R., 1974, Geology of coastal belt, Franciscan Complex, near Ft. Bragg, California [Ph.D. thesis]: Univ. Texas, Austin.

Kramer, J. C., Evitt, W. R., and O'Day, M., 1974, Tertiary coastal belt, northern California Coast Ranges: Geol. Soc. America Bull. (in press).

Lanphere, M. A., 1971, Age of the Mesozoic oceanic crust in the California Coast Ranges: Geol. Soc. America Bull., v. 82, p. 8209-3212.

———, Irwin, W. P., and Hotz, P. E., 1968, Isotopic age of the Nevadan orogeny and older plutonic and metamorphic events in the Klamath Mountains, California: Geol. Soc. America Bull., v. 79, p. 1027-1052.

Lehman, D. H., 1974, Structure and petrology of the Hull Mountain area, Northern California [Ph.D. thesis]: Univ. Texas, Austin.

Lipman, P. W., Prostka, H. J., and Christiansen, R. L., 1971, Evolving subduction zones in the western United States, as interpreted from igneous rocks: Science, v. 174, p. 821-825.

Livingston, J. L., 1973, Late Mesozoic continental margin, Southern Diablo Range, California (abstr.): Geol. Soc. America, v. 5, no. 1, p. 74.

Livingston, V. E., Jr., 1969, Geologic history and rocks and minerals of Washington: Wash. Div. Mines and Geol. Inform. Circ. 45.

Lovell, J. P. B., 1969, Tyee formation: undeformed turbidites and their lateral equivalents: Geol. Soc. America Bull., v. 80, p. 9-22.

Marvin, R. V., 1968, Transcontinental geophysical survey (35°-39°N), radiometric age determinations of rocks: U.S. Geol. Survey Map 1-537.

Mason, R. G., and Raff, A. D., 1961, Magnetic survey off the west coast of the United States, 32°N. latitude to 42°N latitude: Geol. Soc. America Bull., v. 72, p. 1259-1266.

Maxwell, J. C., 1969, "Alpine" mafic and ultramafic rocks—the ophiolite suite: a contribution to the discussion of the paper "The origin of ultramafic and ultrabasic rocks" by P. J. Wyllie: Tectonophysics, v. 7, no. 5-6, p. 489-494.

———, 1974a, Anatomy of an Orogen: Geol. Soc. America Bull., (in press).

———, 1974b, The new global tectonics, an assessment, *in* Kahle, C. F., and Meyerhoff, A. A., eds., Continental drift: Am. Assoc. Petr. Geol. Mem. (in press).

———, 1974c, Ophiolites—old oceanic crust or internal diapirs?: Internatl. Symp. Ophiolites in Earth's Crust, Moscow (in press).

McKee, B., 1972, Cascadia: the geologic evolution of the Pacific Northwest: New York, McGraw-Hill, 394 p.

McKee, E. H., 1971, Tertiary igneous chronology of the Great Basin of western United States—implications for tectonic models: Geol. Soc. America Bull., v. 82, p. 3497-3502.

McMath, V. E., 1966, Geology of the Taylorsville area, northern Sierra Nevada, *in* Bailey, E. H., ed., Geology of northern California: Calif. Div. Mines and Geol. Bull., v. 190, p. 173-183.

Medaris, L. G., 1972, High-pressure peridotites in southwestern Oregon: Geol. Sco. America Bull., v. 83, p. 41-58.

———, and Dott, R. H., Jr., 1970, Mantle-derived peridotites in southwestern Oregon: relation to plate tectonics: Science, v. 167, p. 971-974.

Miller, F. K., McKee, E. H., and Yates, R. G., 1973, Age and correlation of the Windermere Group in northeastern Washington: Geol. Soc. America Bull., v. 84, p. 3723-3730.

Misch, P., 1966, Tectonic evolution of the Northern Cascades of Washington State, *in* Gunning, H., ed., Tectonic history and mineral deposits of the Western Cordillera: Can. Inst. Mining and Met., spec. 8, p. 101-148.

Moen, A. D., ed., 1971, Symposium on tectonism of the Pacific Northwest: EOS Trans. Am. Gephys. Union, no. 9, p. 628645.

Monger, J. W. H., 1973, Correlation of pre-Tertiary geology between Canadian and United States' segments of the Cordillera (abstr.): Geol. Soc. Ameica, v. 5, no. 1, p. 84.

———, and Preto, V. A., 1972, Geology of the southern Canadian Cordillera: XXIV Internatl. Geol. Congr. Guidebook Excursions A03-C03, 87 p.

———, Ross, C. A., 1971, Distribution of fusulinaceans in the Western Canadian Cordillera: Can. Jour. Earth Sci., v. 8, p. 259-278.

———, Souther, J. G., and Gabrielse, H., 1972, Evolution of the Canadian Cordillera: a plate-tectonic model: Am. Jour. Sci., v. 272, p. 577-602.

Moody, J., 1966, Crustal shear patterns and orogenesis: Tectonophysics, v. 3, p. 479-522.

Moore, D. G., 1973, Plate-edge deformation and crustal growth, Gulf of California structural province: Geol. Soc. America Bull., v. 84, p. 1883-1906.

Moore, G. W., 1970, Sea-floor spreading at the junction between Gorda Rise and Mendocino Ridge: Geol. Soc. America Bull., v. 81, p. 2817-2824.

Moore, J. G., 1959, The quartz diorite boundary line in the western United States: Jour. Geol., v. 67, p. 198-210.

Moores, E., 1970, Ultramafics and orogeny, with models of the U.S. Cordillera and the Tethys: Nature, v. 228, p. 837-842.

Moores, E. M., 1972, Model for Jurassic island arc—continental margin collision in California (abstr.): Geol. Soc. America, v. 4, no. 3, p. 202.

———, Scott, R. B., and Lumsden, W. W., 1968, Tertiary tectonics of the White Pine-Grant Range region, east-central Nevada, and some regional implications: Geol. Soc. America Bull., v. 79, p. 1703-1726.

Ojakangas, R. W., 1968, Cretaceous sedimentation, Sacramento Valley, California: Geol. Soc. America Bull., v. 79, p. 973-1008.

Page, B. M., 1966, Geology of the Coast Ranges of California, *in* Bailey, E. H., ed., Geology of northern California: Calif. Div. Mines and Geol. Bull., v. 190, p. 255-276.

———, 1970a, Sur-Nacimiento Fault Zone of California: continental margin tectonics: Geol. Soc. America Bull., v. 81, p. 667-690.

———, 1970b, Time of completion of underthrusting of Franciscan beneath Great Valley rocks west of Salinian block, California: Geol. Soc. America Bull., v. 81, p. 2825-2834.

———, 1972, Oceanic crust and mantle fragment in subduction complex near San Luis Obispo, California: Geol. Soc. America Bull., v. 83, p. 957-972.

Peck, D. L., Compiler, 1961, Geologic map of Oregon west of the 121st meridian: U.S. Geol. Survey Misc. Geol. Invest. Map 1-325.

Pessagno, E. A., Jr., 1973, Age and geologic significance of radiolarian cherts in the California Coast Ranges: Geology, v. 1, no. 4, p. 153-156.

Peter, G., Erickson, G. H., Grim, P. J., 1970, Magnetic structure of the Aleutian Trench and northeast Pacific Basin, *in* Maxwell, A. E., ed., The sea, v. 4, pt. II, p. 191-222.

Poole, F. G., and Desborough, G. A., 1973, Alpine-type serpentinites in Nevada and their tectonic significance (abstr.): Geol. Soc. America, v. 5, no. 1, p. 90.

Price, R. A., 1964, The Precambrian Purcell System in the Rocky Mountains of Southern Alberta and British Columbia: Can. Jour. Petr. Geol. Bull., v. 12, p. 399-426.

Raff, A. D., and Mason, R. G., 1961, Magnetic survey off the west coast of North America, 40°N. latitude to 52°N. latitude: Geol. Soc. America Bull., v. 72, p. 1267-1270.

Ragan, D. M., 1967, The Twin Sisters dunite, Washington, *in* Wyllie, P. J., ed., Ultramafic and related rocks: New York, Wiley, p. 160-167.

Raney, J. A., 1974, Geology of the Elk Creek-Stoneyford area, northern California [Ph.D. thesis]: Univ. Texas, Austin.

Rich, E., 1971, Gologic map of the Wilbur Springs quadrangle Colusa and Lake counties, California: U.S. Geol. Survey Misc. Geol. Invest. Map 1-538.

Roberts, R. J., 1972, Evolution of the Cordilleran fold belt: Geol. Soc. America Bull., v. 83, p. 1989-2004.

———, Hotz, P. E., Gilluly, J., and Ferguson, H. G., 1958, Paleozoic rocks of north-central Nevada: Am. Assoc, Petr. Geol. Bull., v. 42, p. 2813-2857.

Rogers, J. J. W., 1969, Tyee formation: undeformed turbidites and their lateral equivalents: discussion: Geol. Soc. America Bull., v. 80, p. 2129-2130.

Ross, D. C., Wentworth, C. M., and McKee, E. H., 1973, Cretaceous mafic conglomerate near Gualala offset 350 miles by San Andreas fault from oceanic crustal source near Eagle Rest Peak, California: U.S. Geol. Survey Jour. Res., v. 1, no. 1, p. 45-52.

Rutland, R. W. R., 1973, On the interpretation of Cordilleran orogenic belts: Am. Jour. Sci., v. 273, p. 811-849.

Savage, J. C., and Buford, R. O., 1973, Geodetic determination of relative plate motion in Central California: Jour. Geophys. Res., v. 78, p. 832-845.

Sbar, M. L., Barazangi, M., Dorman, J., Scholz, C., and Smith, R. B., 1972, Tectonics of the Intermountain seismic belt, western United States: microearthquake seismicity and composite fault plane solutions: Geol. Soc. America Bull., v. 83, p. 13-28.

Scholten, R., 1968, Model for evolution of Rocky Mountains east of Idaho batholith: Tectonophysics, v. 6, no. 2, p. 109-126.

Scholz, C. H., Barazangi, M., and Sbar, M. L., 1971, Late Cenozoic evolution of the Great Basin, western United States, as an ensialic interarc basin: Geol. Soc. America Bull., v. 82, p. 2979-2990.

Silberling, N. J., 1972, Geologic events during Permo-Triassic time along the Pacific margin of the United States; *in* Logan, A., and Hills, L. V., eds., International Permian-Triassic Conference Vol.: Alberta Soc. Petr. Geol. Bull., (in press).

———, and Roberts, R. J., 1962, Pre-Tertiary stratigraphy and structure of northwestern Nevada: Geol. Soc. America Spec. Paper 72, 53 p.

Silver, E. A., 1971, Transitional tectonics and late Cenozoic structure of the continental margin off northernmost California: Geol. Soc. America Bull., v. 82, p. 1-22.

———, 1973, The continental margin off California to Washington: a history of partial coupling between lithospheric plates (abstr.): Geol. Soc. America, v. 5, no. 1, p. 106.

Silver, L. T., 1971, Problems of crystalline rocks of the Transverse Ranges (abstr.): Geol. Soc. America, v. 3, no. 2, p. 193-194.

Snavely, P. D., Jr., MacLeod, N. S., and Wagner, H. C., 1973, Miocene tholeitic basalt of coastal Oregon and Washington and their relations to coeval basalt of the Columbia Plateau: Geol. Soc. America Bull., v. 76, p. 387-424.

Snook, J. R., 1965, Metamorphic and structural history of "Colville Batholith" gneisses, north-central Washington: Geol. Soc. America Bull., v. 76, p. 759-776.

Souther, J. G., and Armstrong, J. E., 1966, North Central Belt of the Cordillera of British Columbia: *in* Gunning, H. C., ed., Tectonic history and mineral deposits of the western Cordillera: Can. Inst. Mining and Met., Spec. v. 8, p. 171-184.

Spall, H., 1972, Paleomagnetism and Precambrian continental drift: 24th Internatl. Geol. Congr., sect. 3: Tectonics, p. 172-179.

Stanley, K. O., Jordan, W. M., and Dott, R. H., Jr., 1971, Early Jurassic paleogeography, western United States: Am. Assoc. Petr. Geol. Bull., v. 55, p. 10-19.

Stevens, C. H., and Ridley, A. P., 1974, Middle Paleozoic off-shelf deposits in southeastern California: evidence for proximity of the Antler orogenic belt?: Geol. Soc. America Bull., v. 85, p. 27-32.

Stewart, J. H., 1972, Initial deposits in the Cordilleran geosyncline: evidence of a late Precambrain (850 my) continental separation: Geol. Soc. America Bull., v. 83, p. 1345-1360.

Suppe, J., 1969, Times of metamorphism in the Franciscan terrain of the northern Coast Ranges, California: Geol. Soc. America Bull., v. 80, p. 135-142.

———, 1970, Offset of late Mesozoic basement terrains by the San Andreas fault system: Geol. Soc. America Bull., v. 81, p. 3253-3258.

———, 1972, Interrelationships of high-pressure metamorphism, deformation and sedimentation in Franciscan tectonics: 24th Internatl. Geol. Congr., sect. 3: Tectonics, p. 552-559.

———, and Armstrong, R. L., 1972, Potassium-argon dating of Franciscan metamorphic rocks: Am. Jour. Sci., v. 272, p. 217-233.

Swanson, D. A., 1969, Lawsonite blueschist from north-central Oregon: Geol. Survey Res., 1969, U.S. Geol. Survey Prof. Paper 650B, p. B8-B11.

Swe, W., and Dickinson, W. R., 1970, Sedimentation and thrusting of late Mesozoic rocks in the Coast Ranges near Clear Lake, California: Geol. Soc. America Bull., v. 81, p. 165-188.

Tabor, R. W., 1972, Age of the Olympic metamorphism, Washington: K-Ar dating of low-grade metamorphic rocks: Geol. Soc. America Bull., v. 83, p. 1805-1816.

Taliaferro, N. L., 1943, Franciscan-Knoxville problem: Am. Assoc. Petr. Geol. Bull., v. 27, p. 109-219.

Taubeneck, W. H., 1966, An evaluation of tectonic rotation in the Pacific Northwest: Jour. Geophys. Res., v. 71, p. 2113-2120.

———, 1973, Field relations near the Trans-Idaho Discontinuity, western Idaho and eastern Wshington (abstr.): Geol. Soc. America, v. 5, no. 1, p. 115.

Thayer, T. P., and Brown, C. E., 1965, Pre-Tertiary orogeny and plutonic intrustive activity in central and northeastern Oregon: Geol. Soc. America Bull., v. 75, p. 1255-1262.

Thiruvathukal, J. V., Berg, J. W., Jr., and Heinrichs, D. F., 1970, Regional gravity of Oregon: Geol. Soc. America Bull., v. 81, p. 725-738.

Thompson, G. A., 1972, Cenozoic Basin Range tectonism in relation to deep structure: 24th Internatl. Geol. Congr., sect. 3: Tectonics, p. 84-90.

———, and Burke, D. B., 1973, Rate and direction of spreading in Dixie Valley, Basin and Range Province, Nevada: Geol. Soc. America Bull., v. 84, p. 627-632.

Tobin, D. G., and Sykes, L. R., 1968, Seismicity and tectonics of the northeastern Pacific Ocean: Jour. Geophys. Res., v. 73, p. 3821-3846.

U.S. Air Force, Compiler, 1968, Transcontinental geophysical survey (35°-39°N). Bouguer gravity maps from 112°W. longitude to the coast of California: U.S. Geol. Survey Map 1-532-B.

Vallier, T. L., 1973, Pre-Tertiary geology of the Snake River Canyon, northeastern Oregon and western Idaho (abstr.): Geol. Soc. America, v. 5, no. 7, p. 846.

Vance, J. A., 1968, Metamorphic aragonite in the prehnite-pumpellyite facies, northwest Washington: Am. Jour. Sci., v. 266, p. 299-315.

Vine, F. J., 1966, Spreading of the ocean floor—new evidence: Science, v. 154, p. 1405-1415.

Walker, G. W., and King, P. B., 1969, Geologic map of Oregon: U.S. Geol. Survey Misc. Geol. Inv. Map 1-595, scale 1:2,500,000.

Wallace, R. E., and Silberling, N. J., 1960, Intrusive rocks of Permian and Triassic age in the Humboldt Range, Nevada: U.S. Geol. Survey Prof. Paper 400-B, p. B291-B293.

Waren, D. H., 1968, Transcontinental geophysical survey (35°-39°N). Seismic refraction profiles of the crust and upper mantle from 112°W. longitude to the coast of California: U.S. Geol. Survey Map 1-532-D.

Waters, A. C., 1955, Volcanic rocks and the tectonic cycle: Geol. Soc. America Spec. Paper 62, p. 703-722.

Wheeler, J. O., 1966, Eastern tectonic belt of western Cordillera in British Columbia, *in* Gunning, H. C., ed., Tectonic history and mineral deposits of the Western Cordillera: Can. Inst. Mining and Met., sp. v. 8, p. 27-45.

Wiebe, R. A., 1970, Pre-Cenozoic tectonic history of the Salinian block western California: Geol. Soc. America Bull., v. 81, p. 1837-1842.

Wilson, J. T., 1965, Transform faults, oceanic ridges and magnetic anomalies southwest of Vancouver Island: Science, v. 150, p. 482-485.

Wise, D. U., 1963, An outrageous hypothesis for the tectonic pattern of the North America Cordillera: Geol. Soc. America Bull., v. 74, p. 357-362.

Wright, T. L., Grolier, M. J., and Swanson, D. A., 1973, Chemical variation related to the stratigraphy of the Columbia River basalt: Geol. Soc. America Bull., v. 84, p. 371-386.

Yates, R. G., 1968, The Trans-Idaho Discontinuity: XXIII Internatl. Geol. Congr. Proc., v. 1, p. 117-123.

Yeats, R. G., 1964, Crystalline klippen in the Index

district, Cascade Range, Washington: Geol. Soc. America Bull., v. 75, p. 549-562.

———, 1968, Southern California structure, sea-floor spreading, and history of the Pacific basin: Geol. Soc. America Bull., v. 79, p. 1693-1702.

Zietz, I., and Kirby, J. R., 1968, Transcontinental geophysical survey (35°-39°N). Magnetic map from 112°W. longitude to the coast of California: U.S. Geol. Survey Map 1-532-A.

Cal

Active Continental Margins: Contrasts Between California and New Zealand

M. C. Blake, Jr., David L. Jones, and C. A. Landis

INTRODUCTION

The margins of the Pacific Ocean in both California and New Zealand each contains two parallel terranes similar in age but very different in lithology, structure, and metamorphic mineral assemblages. In California, the structurally lower terrane is the Franciscan assemblage, deposited in deep marine environments and made up of rarely fossiliferous graywacke and shale with minor amounts of submarine volcanic rocks, radiolarian chert, and foraminiferal limestone. The Franciscan is separated from coeval fossiliferous sedimentary rocks of the Great Valley sequence to the east by a major fault zone along which ultramafic rocks are abundant. In New Zealand, the poorly fossiliferous Alpine assemblage, composed of graywacke with minor submarine volcanics and chert, is separated from coeval sedimentary rocks of the Hokonui assemblage by an ultramafic belt. In both California and New Zealand, these parallel sedimentary belts are adjacent to older crystalline terranes (Sierra Nevada and Tasman or Fiordland belts). These similarities, plus many other features common to both areas, including similar structure, metamorphism, and geophysical anomalies, have led many earlier workers to conclude that they formed under the same conditions, but at different times.

According to a model first proposed by Wellman (1952, 1956) for the New Zealand geosyncline, and later independently proposed for California by Irwin (1957) and Bailey et al. (1964), the "miogeosynclinal" sediments of the Great Valley sequence and the Hokonui assemblage were laid down on the continental shelf and slope or basin margin at the same time that the "eugeosynclinal" sediments and volcanic deposits of the Franciscan and Alpine assemblages were accumulating offshore in the deeper axis of the basin.

Later, this model was somewhat modified to fit the plate tectonic theories related to island arcs and active continental margins (Bailey and Blake, 1969a, 1969b; Dickinson, 1970a, 1971a). In these later models, the source area for the sediments is considered to be an active volcano-plutonic arc developed on or adjacent to older continental crust with the "miogeosynclinal" sediments deposited in a frontal arc or arc-trench gap basin and the "eugeosynclinal" sediments and volcanic rocks deposited in an actively subducting trench (Fig. 1). In New Zealand, sedimentation probably began in late Paleozoic time and culminated in the late Mesozoic. In California, the Franciscan-Great Valley sequence pair was deposited from latest Jurassic to early Tertiary.

Recent studies have pointed out that there are major differences between the two areas and that some data conflict with the simple plate tectonic model. For example, in California, the oldest rocks within the Franciscan are closest to the craton, and the youngest rocks lie to the west (see Maxwell, this volume). In New Zealand, the opposite arrangement exists, in that the youngest rocks in the Torlesse are closest to the cratonic block, and the oldest rocks are farther east. Several recent studies have shown that pronounced differences in clast size and petrography of sandstone for the presumed parallel belts in both California and New Zealand are difficult to reconcile with a single source area providing the detritus for both the arc-trench gap and trench basins (Bradshaw and Andrews, 1973; Force, 1974; Blake and Jones, 1974).

CALIFORNIA COAST RANGE GEOLOGY

Throughout much of western California and extending north into Oregon and south into Baja California, three subparallel linear belts are recognized: from west to east, the Franciscan assemblage, the Great Valley sequence, and granitic and metamorphic rocks of the Sierra Nevada and Klamath mountains (Fig. 2). In the area extending from about Point Arena south to the Transverse Ranges this pattern is complicated by an apparent repetition of the three linear belts west of the San Andreas Fault. This repetition has been explained by several hundred miles of right-lateral movement on the San Andreas fault (Hill and Dibblee, 1953; Page, 1970). The tripartite linear pattern is also broken both east and west of the San Andreas fault where the Great Valley sequence is locally absent and the Franciscan assemblage is juxtaposed directly against the Sierran-Klamath Mountains boundary in northern California, and along parts of the Sur-Nacimiento fault in the southern Coast Ranges.

Franciscan Assemblage

Recent studies in northern California indicate that the Franciscan assemblage can be conveniently subdivided into two distinctive types: coherent, or bedded, units; and mélange, or incoherent, units (Hsü and Ohrbon, 1969; Blake et al., 1971, 1974; Cotton, 1972; Fox et al., 1973; Sims et al., 1973;

which consists of blocks of graywacke, greenstone, chert, and their metamorphosed equivalents, plus serpentinite and high-grade glaucophane schist, eclogite, and amphibolite, set in a highly sheared matrix of shale and minor amounts of sandstone. Individual blocks range in length from less than 1 ft to several miles. Unlike most of the coherent units, the mélange contains megafossils in abundance but their age range is only Tithonian to Valanginian. The mélange matrix is commonly metamorphosed to a very low grade, containing the assemblage quartz-chlorite-white mica-albite-pumpellyite ± calcite ± aragonite (Blake et al., 1971, 1974). Locally, however, lawsonite is present (Cowan, 1973; Gilbert, 1973), and jadeite may be present in graywacke blocks in some mélanges (Ernst, 1971; Raymond, 1973; Suppe, 1969, 1973).

Low-angle thrust faults separating graywacke, mélange, and arkose have been mapped in many places (Fig. 3) (Brown, 1964b; Suppe, 1973; Blake and Jones, 1974; Cowan, 1973; Berkland, 1973; Raymond, 1973). It is not clear, however, whether the graywacke and arkose units were once extensive continuous thrust sheets, subsequently fragmented by later high-angle faulting or formed as discrete small slices within a complex and extensive imbricate thrust system. Few large-scale folds have been mapped, possibly owing to lack of well-defined stratigraphic marker beds. Some large recumbent structures defined by chert beds occur in the Diablo Range (Raymond, 1973), and in the northern Coast Ranges similar cherts have been deformed into folds with amplitudes of tens of kilometers (unpublished mapping by San Jose State University Department of Geology).

Despite these structural complexities, a poorly defined pattern of distribution based on ages of structural units is becoming apparent (Blake and Jones, 1974). In a general way, rock units become younger from east to west, or downward in structurally stacked plates. For example, all uppermost Cretaceous (Campanian) and younger Franciscan rocks appear to be absent in the eastern parts of the Franciscan belt. This relation is best explained by the process of underthrusting, or subducting, successively younger deposits beneath previously deformed sedimentary sequences.

Most fossils found in the Franciscan rocks are species that occur in the adjoining Great Valley sequence. Several species of *Buchia* that occur in Upper Jurassic and lowermost Cretaceous strata are common in both belts. No evidence for deposition in different faunal provinces, postulated for the late Paleozoic of British Columbia (Monger and Ross, 1971), is apparent for the Great Valley sequence and the Franciscan.

One possible exception to this is the extremely limited megafauna associated with the so-called Calera type of pelagic limestones of late Cretaceous (Cenomanian) age. These rocks locally contain abundant debris of hermatypic (reef) corals, echinoids, including the species *Stereocidaris baileyi* Fell, and a nerineaid gastropod. None of these is known in coeval strata of the Great Valley sequence. *Nerinea* is common in warm water (Tethyean) regions (Sohl, 1971), and its presence suggests that the enclosing rocks formed somewhere to the south and were transported to California by plate movements. Significantly, the limestones and intercalated cherts are the only post-Neocomian pelagic sedimentary rocks incorporated within the Franciscan (Pessagno, 1973).

Great Valley Sequence and Development of the Plate Tectonic Model

Structurally overlying all the units of the Franciscan assemblage is the Great Valley sequence, well-bedded mudstone, siltstone, sandstone, and conglomerate that ranges in age from latest Jurassic to latest Cretaceous. The basal strata, which locally include tuffaceous radiolarian chert, lie on an ophiolite sequence composed of volcanic rocks, diabase, gabbro, and ultramafic rocks. The fault contact with the underlying Franciscan assemblage is very steep along most of the length of the Great Valley, but locally it flattens and has been folded into a number of prominent "hooks" (Brown, 1964a; Rich, 1971). Relatively undeformed and unmetamorphosed klippen of the Great Valley sequence and underlying ophiolite are known far to the west of the trace of the main thrust, resting on the deformed and locally metamorphosed Franciscan terrane (Irwin, 1964). To the north, Great Valley strata as old as Early Cretaceous (Valanginian) were deposited unconformably on the granitic basement of the Klamath Mountains; to the east, Turonian strata rest unconformably on the granitic basement of the Sierra Nevada.

Clastic sediments of the Great Valley sequence were derived from the granitic basement of the Sierran and Klamath regions to the east and north. Compositional changes in the sedimentary rocks through time can be correlated with progressively deeper levels of erosion in the source terrane. K-feldspar ranges in abundance from zero to very small amounts in the basal part of the sequence to much larger amounts in the middle and upper parts of the sequence (Bailey and Irwin, 1959; Bailey et al., 1964). Recent studies of sandstone composition have documented these changes (Ojakangas, 1968; Gilbert and Dickinson, 1970; Mansfield, 1972; Dickinson and Rich, 1972), leading to the recognition of five petrologic units based on the amounts and ratios of quartz, feldspar, unstable fragments, and mica (Dickinson and Rich, 1972). The lowest unit, Stony Creek, is low in quartz and mica and has high plagioclase/total feldspar and volcanic/total lithic grain ratios. The uppermost unit, Rumsey, is quartz-rich with high mica and a low plagioclase: total feldspar ratio. The intervening intervals have varying percentages of these components. As earlier workers have indicated (Dickinson and Rich, 1972, p. 3020), this change from volcanic-rich detritus near the base to granite-derived detritus higher in the sequence records the progressive stripping and

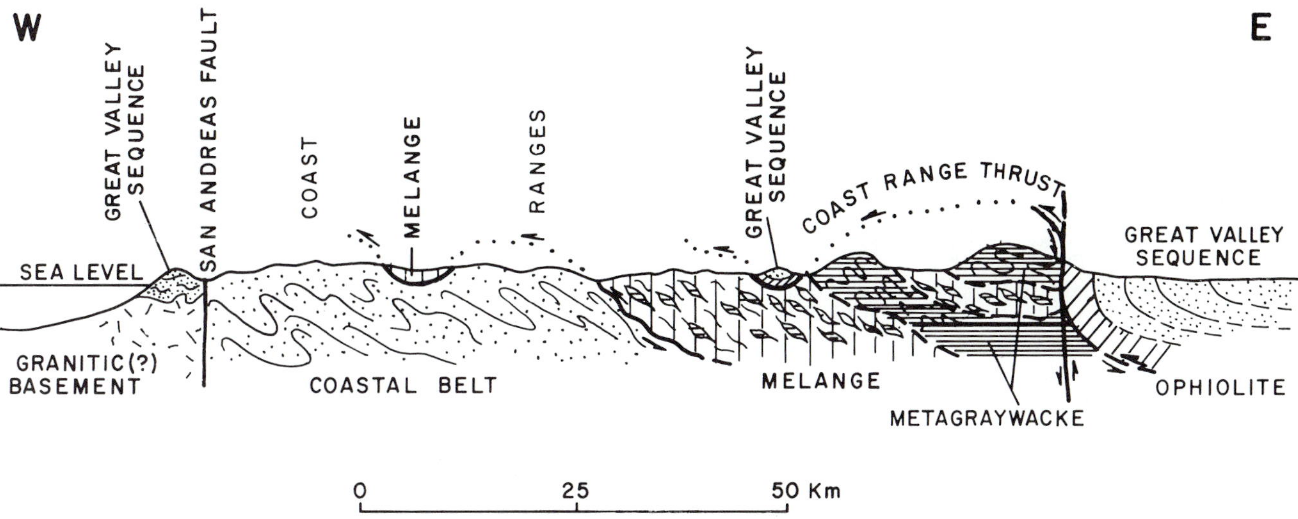

Fig. 3. Schematic cross section drawn across the Coast Ranges of California from Point Arena to Stonyford (location shown in Fig. 2).

erosion of the Sierran-Klamath volcano-plutonic terranes. Not all the volcanic detritus was derived from these sources, however, because locally the basal beds of the Great Valley sequence consist of reworked fragments from the underlying ophiolite (Brown, 1964a; Blake and Jones, 1974).

Much of the Great Valley sequence was deposited by turbidity currents in deep water, except for marginal deposits on the north and east that accumulated in shallow water. Paleocurrent studies along the west side of the Great Valley indicate that transport of sediments was primarily toward the south (Ojakangas, 1968; Mansfield, 1972), with minor transport to the west (Dickinson, 1970b, p. 86; Colburn, 1970, p. 83).

Most of the strata of the Great Valley sequence presumably accumulated as a series of coalescing deep-sea fans near the base of the continental slope. The presence of a submerged western bathymetric barrier lying between the Franciscan trench and the Great Valley trough has been postulated by many workers (e.g., Bailey and Blake, 1969a, b; Bailey et al., 1970; Dickinson and Rich, 1972), but has not been adequately documented. Evidence of axial flow supports, but does not require, the presence of a western barrier. Stratigraphic sequences in outliers to the west of the Great Valley appear to be thinner than correlative sequences to the east in the Great Valley, but most of these outliers are either too poorly known or too incomplete to permit detailed correlations and analyses of stratigraphic trends. Thick Boulder conglomerates, such as the 3,000-m-thick lens at Dry Creek near Healdsburg, suggest derivation from a western source, although conglomerates of comparable thickness in coeval strata do occur at the northern end of the Great Valley, nearly 160 km northeast.

Reconstruction of the original geometry of the Great Valley sequence depositional basin is difficult, particularly for Upper Jurassic and Lower Cretaceous strata, because these rocks are exposed only on the steeply dipping western limb of a large synclinal fold on the west side of the Great Valley. On the gently dipping eastern limb, on the east side of the Great Valley, only Upper Cretaceous rocks are exposed (see Maxwell, this volume).

Clues to the lateral facies relations across the basin can be found at the north end of the Sacramento Valley (Fig. 4), where dissimilar stratigraphic sequences have been juxtaposed across three large left-lateral fault zones (Jones et al., 1969; Jones and Irwin, 1971; Bailey and Jones, 1973; Jones and Bailey, 1973; Jones, 1973). Cumulative displacement on these faults amounts to many tens of kilometers and has resulted in juxtaposition of the Klamath Mountains continental crust on the north against oceanic crust at the base of the Great Valley sequence to the south. These left-lateral faults do not displace Franciscan rocks to the west and are interpreted as tear faults related to the Coast Range thrust (Bailey et al., 1970). Minimum displacement across these three left-lateral faults is 90-100 km, based on an offset Early Cretaceous shoreline (Jones and Irwin, 1971). The present position of these displaced blocks is shown on Figure 4, which is a highly generalized geologic map of the northwestern part of the Sacramento Valley. Because each block has been displaced westward relative to its southern neighbors, a palinspastic reconstruction allows a tentative reconstruction of facies and thickness trends in an east-west direction.

Deposits on the north side of these faults are systematically thicker and contain greater amounts of sandstone and conglomerate than do coeval deposits on the south side. These differences are best seen in the part of the sequence that ranges in age from earliest Cretaceous (Berriasian) to the middle part of the Early Cretaceous (Hauterivian). In the southern fault block these strata comprise 460 m of dominantly mudstone with minor channels filled with conglomerate and coarse-grained sand. In the central block, more than 3,000 m of these beds is present, but the proportion of sandstone is higher.

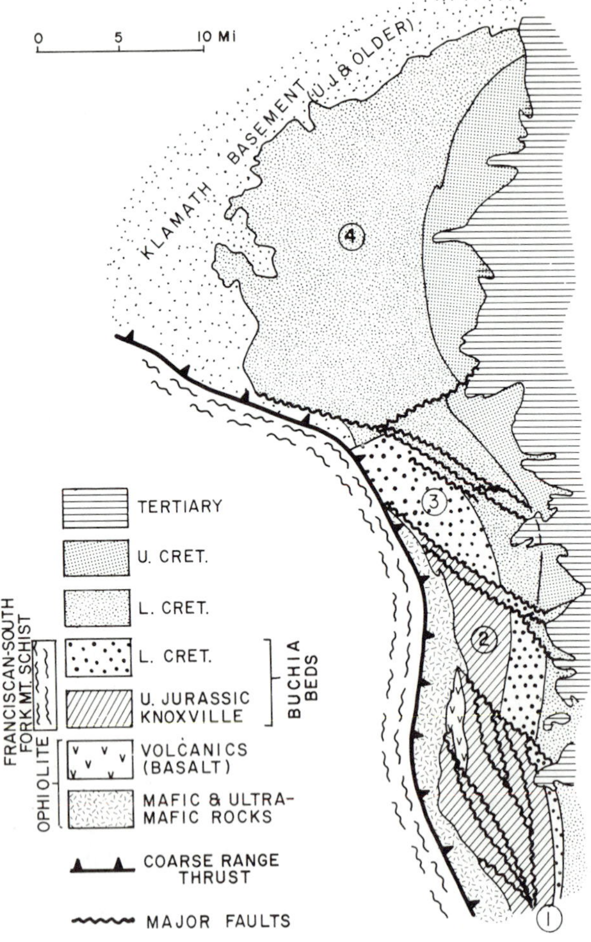

Fig. 4. Generalized geologic map of the northwestern part of the Sacramento Valley, California.

In the northern block, over 4,600 m of mudstone, fine- to coarse-grained sandstone, and grit is present. Associated with these rocks is abundant carbonaceous debris, including plant scraps, leaves, stems, twigs, and rare logs. Molluscan megafossils are rare, and most specimens are displaced from their original dwelling site.

Recognition of the large amount of left-lateral displacement that these rocks have undergone, together with the different facies pattern in each block, permits a tentative reconstruction of original basin geometry by palinspastically restoring the blocks to their original position. In Figure 5, the four stratigraphic sequences shown on Figure 4 have been moved from their present north-south position to their postulated early east-west position.

The resulting depositional pattern (Fig. 5) can be interpreted as a series of thick transgressive fans built on the continental margin, the thickest and coarsest portion of each successive time unit shifted eastward with respect to the underlying strata. The significance of this pattern to the source of Franciscan detritus is discussed in the following section.

Relations Between the Franciscan Assemblage and Great Valley Sequence

Recognition that the Franciscan assemblage and Great Valley sequence were deposited at the same time under different conditions and subjected to different degrees of deformation led Bailey et al. (1964) to propose that the Franciscan assemblage had been juxtaposed with the Great Valley sequence by eastward underthrusting. Other workers, noting the high-pressure minerals and tectonic mélange developed in the lower-plate rocks and drawing on recently developed theories of sea-floor spreading, proposed that the Franciscan rocks had been deposited on an eastward-directed "conveyor belt," underlain by oceanic crust, subsequently carried down beneath the overlying Great Valley sequence and Sierran-Klamath basement along a megathrust or subduction zone (Blake et al., 1967; Hamilton, 1969; Bailey and Blake, 1969a, b; Ernst, 1970; Bailey et al., 1970).

This model was later modified by Dickinson (1970a, 1971a), who proposed that the Great Valley sequence was deposited in an arc-trench gap between the trench (Franciscan) and volcano-plutonic arc (Sierran-Klamath terrane) closely fitting the geometry of many island arcs in the southwest Pacific. Other models were suggested by Hsü (1971), who proposed that an "older" pre-Tithonian Franciscan assemblage had been deformed and metamorphosed prior to the deposition of the Great Valley sequence and subsequently juxtaposed with both these rocks and "younger" Franciscan, and by Moores (1970), who postulated that the east-dipping Coast Range thrust had been preceded by a period of west-dipping subduction.

Objections to all these proposals were made by Blake and Jones (1974), who pointed out that (1) the rocks of the Franciscan assemblage contain no fossils older than Tithonian; (2) all the radiolarian chert dated (Pessagno, 1973) spans a very narrow age range (Tithonian through Neocomian); and (3) the difference in grain size and composition between the Tithonian to Valanginian Franciscan graywacke units (medium- to coarse-grained quartzofeldspathic graywacke) and coeval Great Valley sequence sedimentary rocks (basaltic mudstone and fine- to medium-grained sandstone) make it highly unlikely that all the coarse-grained quartz and feldspar could have been derived from the same Sierran-Klamath terrane that was providing fine-grained volcanic material at the same time to the turbidites of the lower Great Valley sequence. A further complication is seen in the facies trends shown by the palinspastic cross section of the lower part of the Great Valley sequence (Fig. 5).

This reconstruction shows pronounced thinning and shaling-out to the west and poses a problem in locating the source terrane from which the coarser Franciscan detritus was derived. The high silica content of the Franciscan graywackes and the pau-

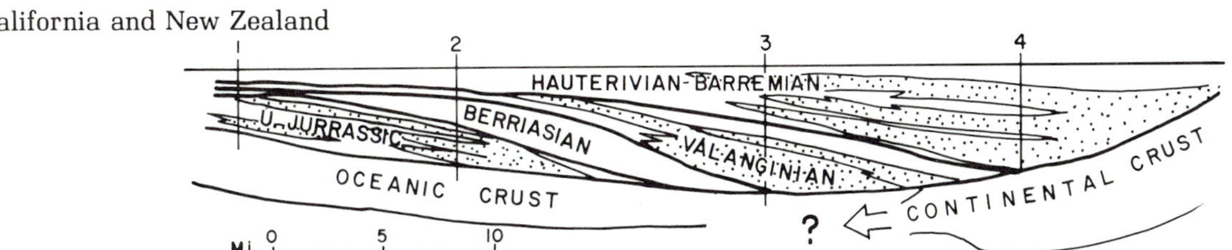

Fig. 5. Reconstruction of the early phases of deposition of the Great Valley sequence of California, by palinspastically restoring the fault-bounded sequences of Figure 4 to near their original position, assuming 15-25 km of horizontal movement between each block. Location of the four stratigraphic sequences shown in Figure 4.

city of pelagic sediments within the Franciscan (Scholl and Marlow, 1974) indicate a continental source. However, the greater proportion of graywacke over shale (estimated sand/shale ratio of >3:1) in the Franciscan as contrasted with coeval strata of the Great Valley sequence (sand/shale ratio of≃1:3) is difficult to explain if the Great Valley sequence was deposited nearer to the only identified source areas. Transportation of large volumes of sand to the Franciscan trench through the Great Valley sequence in submarine canyons is a possibility, but the presence of such canyons has not been substantiated.

Scholl and Marlow (1974) have pointed out that presently active oceanic trenches may not be an actualistic model of the postulated Franciscan "trench," for recent trenches: (1) lack sufficient volume to allow accumulation of such enormous thicknesses of sediments as occur in the Franciscan; and (2) are filled largely with pelagic deposits, not with terrigenous turbidites. Scholl and Marlow calculate that if the Franciscan "trench" had indeed been the site of subduction of vast amounts of Mesozoic oceanic crust, at least one-third of the volume of Franciscan sedimentary rocks, should consist of offscraped pelagic deposits. Pelagic sediments, however, constitute only a few percent of the Franciscan rocks. Scholl and Marlow therefore argue that the Franciscan sediments accumulated not in an active trench but in a basin or series of basins upslope from the trench.

In an attempt to reconcile the various data, we recently proposed (Blake and Jones, 1974) that during Late Jurassic to Early Tertiary time, a remnant volcano-plutonic arc was situated between the Franciscan assemblage and the Great Valley sequence. As this remnant arc was subsequently eroded away and its roots subducted, probably no part of it now remains (Fig. 6). In this hypothesis, the Franciscan graywacke was deposited in an arc-trench gap at the same time that the Great Valley sequence was deposited on ophiolite formed in an interarc basin between the postulated outer arc and the Sierran-Klamath part of the North American plate. During the late Mesozoic and probably extending into the early Tertiary, the mélanges are considered to have formed during subduction by ripping off and shingling parts of the upper plate consisting of ophiolite, overlying chert, and mudstone and sandstone of the basal Great Valley sequence. This helps explain why the mélange belts all appear to be of the same restricted age and so similar lithologically to the rocks of the ophiolite and Great Valley sequence.

NEW ZEALAND

The geologic evolution of New Zealand can be divided into three main sedimentary-orogenic episodes (Carter et al., 1974): (1) Tuhua [Precambrian (?) to early Carboniferous]; (2) Rangitata (late Carboniferous to early Cretaceous); and (3) Kaikoura (middle Cretaceous to Holocene). This review will concentrate on the Rangitata episode.

The Rangitata orogenic zone can be divided into numerous longitudinal provinces (Fig. 7). A fundamental twofold lithologic division distinguishes a western province from an eastern province. The eastern, or New Zealand geosyncline, province may in turn be subdivided into numerous geologic terranes.

Western Province

The western province is underlain largely by the Tuhua sequence, comprising early and middle Paleozoic sedimentary rocks (including quartzite, limestone, flysch-like sandstone and shale, and aluminous shale) that have undergone Devonian and Carboniferous regional metamorphism and plutonism in varying degrees (Aronson, 1968; Grindley, 1971). In the northwestern South Island, Tuhua rocks are overlain unconformably by a thin sequence of metamorphosed fossiliferous, quartzose, aluminous, and carbonaceous Permian sedimentary rocks (Waterhouse and Vella, 1965; Grindley, 1971). Rangitata orogenesis, including plutonism, metamorphism, and uplift, subsequently affected large parts of the province.

The western province extends southeast and northwest beneath the sea onto the Campbell Plateau and the Lord Howe Rise, respectively. Tuhua rocks are considered to have been originally (predrift) in continuity with similar strata and intrusive rocks of southern Australia and west Antarctica (e.g., Wright, 1966; Griffiths, 1971; Halpern, 1968; Harrington et al., 1973). The eastern margin of the province, where it abuts the eastern province, is marked by faulting and intrusion along a tectonically complex zone referred to as the Median Tectonic Line (Landis and Coombs, 1967).

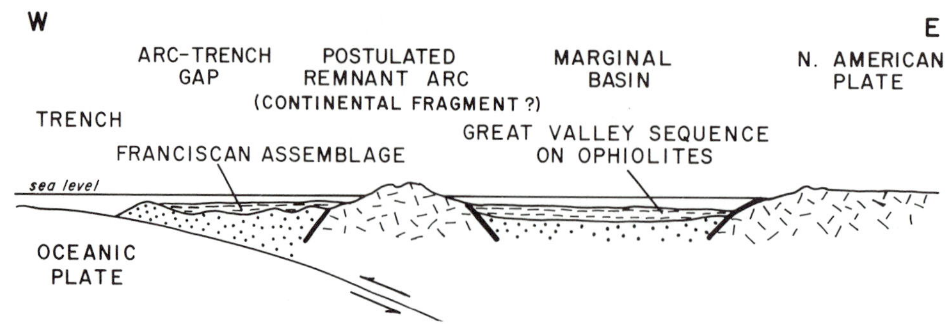

Fig. 6. Hypothetical cross section in northern California during Late Jurassic to Early Tertiary time.

Eastern Province

The eastern province is made up of stratified rocks of the New Zealand geosyncline plus basement ophiolite of the Dun Mountain ophiolite belt (Blake and Landis, 1973; Coombs et al. in press), and upper Paleozoic and Mesozoic hypabyssal and crystalline rocks intrusive into the inner (western) flanks of the geosyncline.

Following Wellman (1952, 1956), the geosyncline can be divided into two major belts, or facies—the Marginal or Hokonui facies along the inner side and the axial or Alpine facies along the outer (eastern) side. The nature of these two belts has been reviewed recently by Fleming (1970) and Landis and Bishop (1972). Carter et al. (1974) propose to supplant the term "facies" with "assemblage."

Each of these assemblages, Hokonui and Alpine, can in turn be subdivided into distinctive geologic terranes (Fig. 7) following proposals of Landis (1969) and Coombs et al. (in press). The characteristics of these terranes are summarized in Table 1. The dividing line between the two assemblages is generally drawn at the Dun Mountain ophiolite belt and the Livingstone-Macpherson fault system (see Fleming, 1970; Landis and Bishop, 1972; E. R. Force, 1974); study of Table 1, however, reveals that the distinction between the two facies is not entirely clear cut.

In general, rocks lying along the inner side of the Dun Mountain belt, that is, the Hokonui assemblage, are andesitic volcanogenic rocks characterized by orderly biostratigraphic and lithostratigraphic patterns, simple and regionally continuous structures, little penetrative deformation and (in post-Permian strata) an absence of chert, limestone, and pillow lava. Sedimentation is inferred to have occurred within and flanking a volcanic arc. Alpine assemblage rocks, along the outer side of the Dun Mountain ophiolite belt, include both volcanogenic and quartzo-feldspathic rocks that are characterized by unresolved and probably complex lithostratigraphy, paucity of fossils, complex structural history including widespread development of schistosity, transposition structures, and mélange, and the presence of minor but conspicuous amounts of chert, pillow lava, and limestone. Numerous environmental facies are represented in Alpine assemblage rocks; these probably include trench, submarine fan, abyssal plain, both arc-flank and oceanic volcanic environments, and nonmarine to shallow marine transgressive and regressive sequences (Fleming, 1970; Landis and Bishop, 1972; Bradshaw and Andrews, 1973). It is noteworthy that the Alpine and Hokonui assemblages each contain distinctive faunal elements that are entirely absent from the other assemblage (Campbell and Warren, 1965; Fleming, 1970; Stevens, 1972).

Origin of the Alpine Assemblage

The most critical problem in reconstructing the New Zealand geosyncline is the origin of contemporaneous sedimentary rocks of the Hokonui and Alpine assemblages.

The Alpine assemblage, including Torlesse, Haast Schist, and Caples-Croiselles-Pelorus terranes, comprises an extremely voluminous and generally unfossiliferous suite of sedimentary rocks in which "graywackes" and argillites predominate. The Caples-Croiselles-Pelorus terrane is characterized by arc-derived quartz-poor volcanogenic detritus, whereas Torlesse sandstones are characterized by abundant quartz and feldspar derived from a plutonic and high-grade metamorphic provenance; representative SiO_2 contents are 65-70% (Reed, 1957; Coombs et al., 1959; Dickinson, 1971b; Landis and Bishop, 1972; E. R. Force, 1974; Coombs et al., in press). The Alpine assemblage is thus divided between a volcanogenic petrofacies on the inner side and a crystalline petrofacies on the outer side. The boundary between the two petrofacies appears to lie within the Haast Schist terrane but has not been precisely located (Landis and Bishop, 1972; Coombs et al., in press). Certain characteristics of the assemblage, however, carry across the petrofacies boundary, that is, the presence of chert, basaltic volcanic rocks, interformational conglomerate, manganiferous rocks, and faunal types (Campbell and Campbell, 1970).

The Alpine assemblage has undergone extensive burial and regional metamorphism. Prehnite-pumpellyite, pumpellyite-actinolite, and greenschist facies assemblages predominate, whereas relatively high-pressure lawsonite and blue amphibole-bearing assemblages are restricted to the western margins of the terrane.

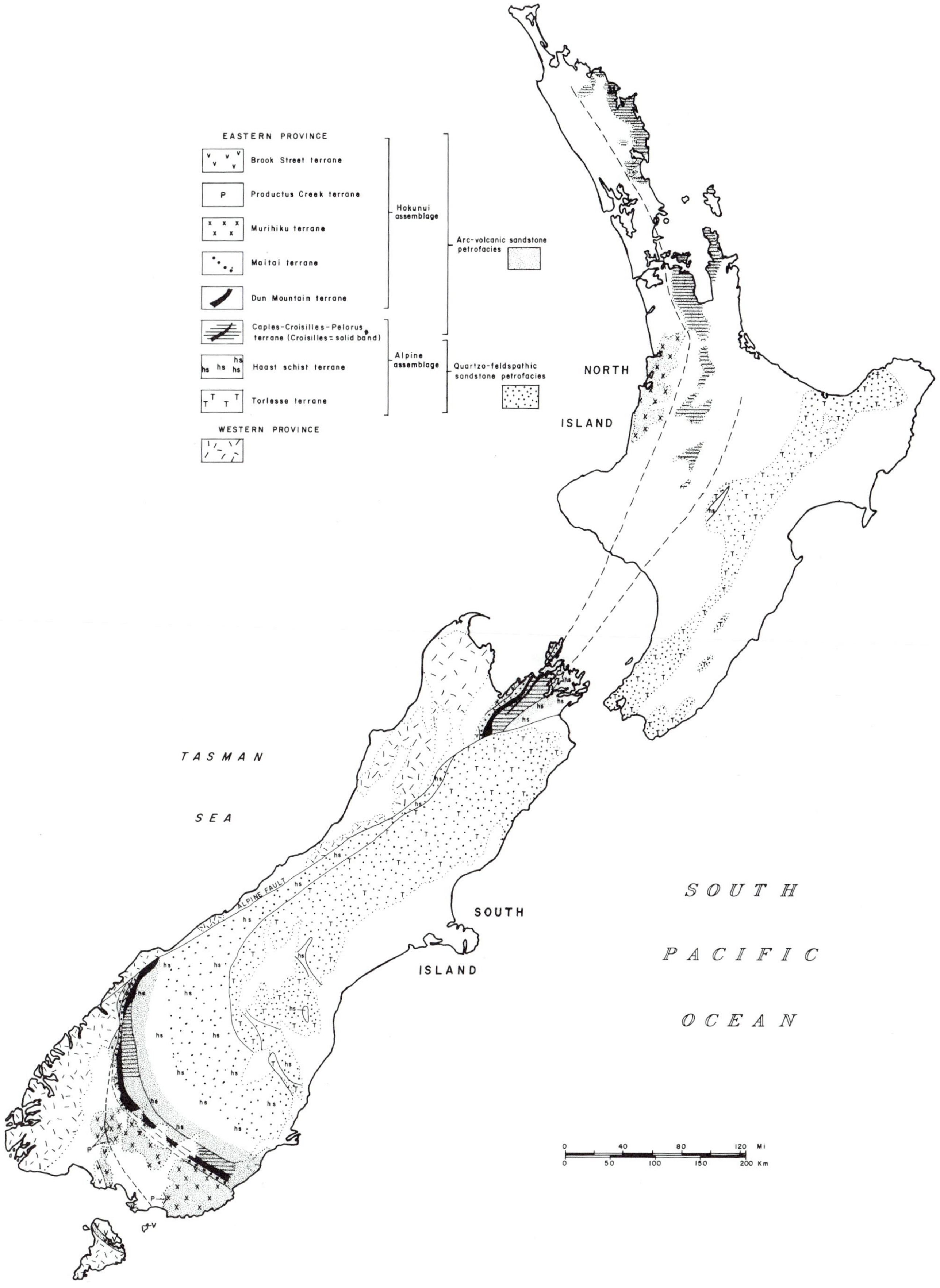

Fig. 7. Major geologic terranes and petrofacies in New Zealand. The South Island region is modified after Coombs et al. (in press); for description of geologic terranes, see Table 1.

Table 1. Description of major lithologic terranes within the Hokonui and Alpine assemblages of New Zealand.

Terrane	HOKONUI ASSEMBLAGE			
	Brook Street	Productus Creek	Murihiku	Maitai
Age of stratified rock	Early Permian	Middle to Late Permian; locally to Triassic (?)	Triassic and Jurassic	Late Permian; locally to Middle Triassic
Marine fossils	Uncommon, mainly brachiopods and bivalves	Abundant, brachiopods, bivalves, corals, gastropods, bryozoans, etc	Locally abundant (especially in rocks lacking plant fossils and turbidites), brachiopods, bivalves, gastropods	Extensive unfossiliferous sequences; local zones rich in bivalves and brachiopods
Plant fossils	None reported	Locally abundant, includes *Glossopterus*	Locally abundant, especially in Jurassic rocks	Locally present
Sandstone and siltstone--volcanogenic	Widespread	Widespread	Predominant rock type	Predominant rock type
Sandstone and siltstone--quartzofeldspathic	None reported	Extremely rare	Uncommon	Restricted to one thin horizon
Turbidites and other redeposited rocks	Present	None reported	Present, especially on outer side	Widely developed
Basaltic and andesitic volcanic rocks	Predominant rock type (largely porphyritic andesite)	Rare	Rare	None reported
Vitric and crystal tuffs (acid to intermediate composition)	Present	Uncommon	Conspicuously well-developed	Uncommon
Chert	None reported	None reported	None reported	Very rare
Limestone	Minor	Predominant rock type	Extremely rare	Characteristic of basal part, elsewhere rare
Conglomerate with rounded exogenous clasts	None reported	Locally conspicuous	Locally conspicuous	Locally conspicuous
Permian-Jurassic intrusive rock	Quartz monzonite, olivine monzonite, soda granite, quartz diorite, diorite, gabbro, peridotite, dunite. Voluminous	Andesite. Minor	Andesite. Minor	None reported
Contact with unit on outer side	Continuous sequence; angular unconformity; fault (Hollyford fault system)	Unconformity	Fault (possibly a faulted unconformity; "Murihiku Tectonic Line")	Continuous sequence; mild unconformity; fault
Geosynclinal autocannibalism: clasts of intrageosynclinal derivation include	Brook Street effusive and sedimentary rock	Brook Street effusive and sedimentary rock	Brook Street; Productus Creek or Maitai; older Murihiku	Brook Street; older Maitai
Structure	Holoclinal sequence up to 20 km thick; moderately to steeply dipping, facing outer side	Homoclinal sequence up to 1.5 km thick; gently to steeply dipping, facing outer side	Regional syncline, axial plane steeply dipping, steep dips on outer side; gentle to steep dips on inner side; up to 10 km thick	Regional syncline, near isoclinal, axial plane dips steeply towards outer side; up to 4 km thick. Becomes complex and possibly mélanged in southern part
Regional and burial metamorphism	Zeolite, prehnite-pumpellyite-actinolite, greenschist; highest grades attained in belt extending 60 km S of Alpine fault	Zeolite	Zeolite	Zeolite, prehnite-pumpellyite, lawsonite, albite-chlorite, pumpellyite-actinolite; highest grades attained in belt 60 km S of Alpine fault
Inferred site of formation	Arc, highly active; mainly shallow marine and subaerial	Flanks of largely inactive Brook Street arc; shallow marine	Frontal Arc Basin flanking active Brook Street arc; during Triassic shallow on inner side, deepening outward; during Jurassic gradually emerging	Frontal Arc Basin or inactive trench; mainly deep marine; flanking largely inactive Brook Street arc

	ALPINE ASSEMBLAGE		
Dun Mountain	Caples-Croiselles-Pelorus	Haast Schist	Torlesse
Early Permian	Early Permian to Jurassic	Triassic, possibly Carboniferous to Jurassic	Carboniferous to Cretaceous
Essentially unfossiliferous (some fossils in mélange blocks)	Fossils rare (some in mélange rocks)	Generally absent, rare in low-grade marginal schist	Extensive unfossiliferous sequences. Siliceous tube fossils (problematica) widespread. Rich fossil assemblages vary locally (including fusulines, ammonites and corals in Permian)
None reported	Uncommon	Some carbonaceous schists	Locally common, especially in outer part
Present	Predominant rock type	Greenschists of volcanoclastic origin	Present
None reported (some present in mélange blocks)	Uncommon	Psammitic schists of quartzofeldspathic origin	Predominant rock type
None reported (some present in mélange)	Common	Probably originally widespread	Widespread and voluminous, especially on inner side
Extensively developed (largely aphyric basalt, including pillow lava)	Present (basalt)	Present	Widespread but not voluminous (largely basalt, including pillow lava and basaltic tuff)
None reported	Uncommon	None reported	Very rare
None reported (some present in mélange blocks)	Uncommon	Siliceous and manganiferous schists	Widespread but not voluminous
None reported (some present in mélange blocks)	Rare (associated with basic volcanic rocks)	Very rare	Widespread but not voluminous (tends to occur in association with basic volcanic rocks and as blocks in mélange)
None reported	Locally conspicuous	Uncommon	Locally conspicuous, especially well-developed on outer side
Gabbro, peridotite, serpentinite, dunite, albite granite. Voluminous	Gabbro, peridotite, albite granite (all as blocks in mélange)	Ultramafite, gabbro. Minor	Rare to absent
Fault (Livingstone-Macpherson fault system)	Metamorphic transition	Metamorphic transition	Unconformably overlain by sedimentary rocks of Kaikoura Sequence
Dun Mountain ophiolite	Caples-Pelorus sediments; possible Brook Street rock	——	Older Torlesse rock
Extensively mélanged, locally homoclinal sequence; up to 4 km thick generally steeply dipping and facing inner side	Complexly folded; more than one phase of deformation; discontinuous central melange zone (Croisilles). Thickness unknown but in excess of 10 km	Complexly deformed, at least three phases of deformation including early isoclinal folding. In Otago, schistosity gently dipping in axial zone; steepens along flanks; in southern Alps, schistosity steepens and steeply plunging folds recognized	Complexly deformed, more than one phase of deformation, melange important locally, especially in outer part. Thickness unknown
Zeolite, prehnite-pumpellyite-actinolite, greenschist. Highest grades attained in belt extending 60 km S of Alpine fault	Zeolite (restricted to North Island), prehnite-pumpellyite, pumpellyite-actinolite, lawsonite-albite-chlorite; highest grades attained in belt extending 60 km S of Alpine fault	Pumpellyite-actinolite, greenschist, albite-epicote amphibolite, amphibolite; highest grades attained in belt adjacent to SE side of Alpine fault	Zeolite, prehnite-pumpellyite, pumpellyite-actinolite (grade increases toward inner side)
Sea floor-oceanic ridge or primitive arc	Frontal Arc and/or trench. Mainly deep marine	Probably mainly deep marine	Continental shelf to abyssal plain. Deep marine prevalent on inner side; shallow water and nonmarine more widespread on outer side

Despite the generally unfossiliferous nature of the Alpine assemblage, more than 450 fossil localities have been recorded (Campbell and Warren, 1965; Stevens, 1972). Major time gaps characterize the Alpine assemblage fossil record, and although these account for more than 50% of the total age span represented, their significance is not fully understood (Fleming, 1970; Landis and Bishop, 1972; Stevens, 1972). A crude biostratigraphic zonation can be recognized within the Torlesse terrane (Campbell and Warren, 1965; Fleming, 1970; Landis and Bishop, 1972). In the southern part, the oldest Torlesse rocks, of late Carboniferous age (Jenkins and Jenkins, 1971), lie nearest the South Pacific Ocean. A large area of Permian strata adjoins the Carboniferous rocks along the inner (western and southern) side, and these Permian rocks are in turn adjoined by Middle and Upper Triassic or younger strata before the transition to the Haast Schist terrane is encountered. A pattern different from the above *eastward-aging* arrangement is recognized in northern South Island, where rocks along the coast of the South Pacific Ocean are of Late Jurassic and Cretaceous age. They are flanked on the inner (western) side by Upper Triassic strata which grade, through a zone of semischistose Upper Triassic or younger strata, into the Haast Schist. Similar *eastward younging* of Torlesse strata appears to characterize eastern North Island (Stevens, 1972).

The sources of Alpine assemblage sediment remain unknown. Numerous aspects of Torlesse rocks, especially their predominantly sandy, quartzo-feldspathic nature, imply deposition near a continental margin. Not all the Torlesse sedimentary rocks were deposited in deep water, but marine conditions appear to have dominated, and flysch-like sediments are widespread. Although fossils are generally rare to absent, at least two small areas of richly fossiliferous Middle Triassic strata occur near the margin of the Haast Schist (Campbell and Warren, 1965; Campbell and Force, 1972; E. R. Force, 1974). These rocks are clearly of shallow marine to nonmarine origin, and in one locality are overlain by a graded-bed facies. Richly fossiliferous limestone bodies, presumably of shallow marine origin, are known from several localities, some of which are in tectonic mélanges (Campbell and Warren, 1965; Hornibrook and Khoon, 1965; Grant-Mackie, 1971; Bradshaw, 1973). In northeastern South Island, Torlesse sandstones tend to be moderately well sorted, thick bedded and carbonaceous, and are probably of shallow marine or nonmarine origin (Bradshaw and Andrews, 1973). No relationship between deep-water and shallow-water Torlesse rocks has been established.

Clearly a tectonically active crystalline terrane must be postulated as a primary Torlesse source, whereas extensive intra-Torlesse erosion and redeposition, or autocannibalism (see Table 1), imply that the Torlesse basin itself was a tectonically active secondary source. The primary source must have been an active continental margin, and it is unlikely that this margin lay above a subduction zone, as volcanic-derived sediment is generally absent from Torlesse rocks.

The inferred provenance of the Caples-Pelorus volcanogenic sandstones along the inner flanks of the Alpine assemblage contrasts markedly with the inferred Torlesse provenance. Caples-Pelorus rocks are characteristically massive, poorly sorted, unfossiliferous, andesitic sandstones interbedded with very fine grained, rhythmically layered sandstone and argillite. They were derived from an eroding volcanic arc, possibly the one that provided sediment to the Hokonui assemblage and have been tectonically juxtaposed with Torlesse rocks.

Origin of the Hokonui Assemblage

The Hokonui assemblage is characterized by lithologically distinctive volcanogenic sedimentary strata of strikingly continuous regional structure and stratigraphy. As such, they contrast markedly with the Alpine assemblage. The Hokonui assemblage sediments were probably derived from Permian to Jurassic volcanic arcs; frontal arc and possibly interarc basins are recognizable (Dickinson, 1971a; Landis and Bishop, 1972). The evolution of this assemblage is divisible into three main episodes: Early Permian, later Permian, and Triassic to Jurassic.

The Hokonui assemblage forms a belt-shaped synclinorium, with oldest rocks lying along the inner and outer margins. Along the inner margin the early Permian Brook Street terrane is an ancient volcano-plutonic arc made up of homoclinal and locally folded sequences of weakly metamorphosed, interlayered volcanogenic sedimentary and andesitic volcanic rocks as much as 20 km thick (Grindley, 1958; Mutch, 1972; Nauman, 1974). Along the outer side of the Hokonui assemblage, a lower Permian Dun Mountain ultramafic belt constitutes an oceanic basement terrane (Blake and Landis, 1973; Coombs et al., in press). The Dun Mountain has been deformed into mélanges and tectonically thinned. Brook Street and Dun Mountain belts comprise contrasting but probably penecontemporaneous belts of volcanic rocks, and it seems likely that an arc-margin and subduction zone lay between the two.

Younger Permian strata that overlie the Brook Street and Dun Mountain terranes comprise, respectively, the Productus Creek (inner) and Maitai (outer) terranes. Both terranes contain conspicuous limestones composed largely of comminuted bivalvian debris (Waterhouse, 1964), and both are dominated by epiclastic volcanic sand; detrital quartz commonly constitutes less than 5% of rocks (Landis, 1969). Silica contents of typical rocks is 50-58%. Productus Creek rocks form homoclinal sequences up to 750 m thick (L. M. Force, 1974). The are coarser grained, more fossiliferous, and more carbonaceous, and in general represent shallower marine deposition (probably intra-arc) than the correlative Maitai rocks. The latter occupy the core of a nearly isoclinal syncline, the Key Summit-Nelson regional syncline, which has been offset 480 km by the Alpine fault. The syncline axis coincides approximately with the depositional strike.

Maitai strata are approximately 4,000 m thick, thinly bedded, sparcely fossiliferous (apart from limestone), fine grained, and commonly flysch-like. They are characterized by remarkable continuity of stratigraphic units, some of which can be traced along a strike for 200 km. They probably accumulated in deep water, possibly filling a frontal arc basin or inactive trench (Landis and Bishop, 1972; Blake and Landis, 1973), but evidence for shallow-water deposition has also been presented (Waterhouse, 1964; Force, 1971). Productus Creek rocks are virtually undeformed and contain large amounts of unaltered volcanic detritus along with authigenic mineral assemblages belonging to the zeolite facies (Landis, 1969). In contrast, Maitai rocks have well-developed slaty cleavage, very few unaltered relict minerals, and are locally metamorphosed to the lawsonite-albite-chlorite (blueschist) metamorphic facies (Landis and Coombs, 1967; Landis, 1969). Permian Hokonui rocks are unknown on North Island.

Triassic and Jurassic rocks of the Murihiku terrane dominate the Hokonui assemblage. The lithologic properties have been described in detail by Coombs (1950, 1954) and more recently by Speden (1971), E. R. Force (1974), and Boles (in press). Murihiku rocks form sequences to 10 km thick within a regional syncline known as the Southland syncline in southern South Island, Heslington syncline in northern South Island across the Alpine fault, and Kawhia syncline in North Island. These synclinal axes appear to lie subparallel to depositional strike. The rocks are moderately fossiliferous, abundantly tuffaceous, and weakly metamorphosed; in some parts of the Triassic sequence (especially on the outer flanks of the regional synclines), they are flysch-like but elsewhere shallow marine and estuarine in character. Chemical and petrographic studies in Southland (Boles, in press) reveal a change from Early Triassic andesitic volcanism to Middle and Late Triassic rhyolitic and dacitic volcanism, then a return to andesitic volcanism in latest Triassic and Early Jurassic times.

Throughout the sequence, plutonic material (metamorphic and granitic clasts) tends to be subordinate, and less than 20% detrital quartz is consistently present. Paleocurrent and sedimentary facies studies suggest that a major volcanic-plutonic source lay along the inner side of the Murihiku basin (Speden, 1971; E. R. Force, 1974; Boles, in press). Some intrusive rocks within the Brook Street terrane have been dated as Triassic and Jurassic (Aronson, 1968; Devereux et al., 1968), and these are regarded as roots of volcanoes that supplied debris to the Murihiku terrane as well as Plutonic detritus. The Permian Hokonui assemblage was also undergoing erosion, as clastic debris from it occurs in Triassic rocks (Wood, 1956; Landis and Coombs, 1967; Boles, in press). Murihiku rocks probably originated in a frontal arc basin (Dickinson, 1971a; Landis and Bishop, 1972). Evidence presented by Boles (in press) suggests a secondary source to the north of the Southland syncline. No suitable source rocks are exposed in that region today; Boles suggests removal by erosion, subduction, or thrusting of a Mesozoic volcanic terrane lying along the outer side of the Hokonui belt.

Summary of Relations between Hokonui and Alpine Assemblages.

In summary, the Hokonui assemblage evolved over a period of approximately 130 my in a region of arc volcanism and active margins. In general it comprised a first-cycle sedimentary system with only minor contributions of plutonic detritus, much of which could have been derived from deeper intrusive parts of the volcanic arc. In contrast, the Alpine assemblage, which evolved more or less contemporaneously, includes a volcanogenic petrofacies on the inner (Tasman) side and a plutonic petrofacies on the outer (Pacific) side. The inner, volcanogenic terrane (Caples-Croiselles-Pelorus) could have been derived from the same volcanic arc that supplied the volcanic material to the Hokonui assemblage (Brook Street terrane). The data available suggest that the outer Torlesse terrane cannot readily be explained as a distal facies of either the Hokonui assemblage or the inner Caples-Croiselles-Pelorus terrane (Bradshaw and Andrews, 1973; E. R. Force, 1974; Coombs et al., in press). More likely the Torlesse rocks were derived from another active continental margin and subsequently rafted into their present position by sea-floor spreading. The Marie Byrd Land-Jones Mountains area of West Antarctica is a possible source area, with appropriate basement geology and inferred predrift position (e.g., Halpern, 1971; Griffiths, 1971). Figure 8 shows how the Hokonui and Alpine assemblages might have formed and subsequently been juxtaposed.

During the Permian, an island arc (Brook Street terrane) probably developed on oceanic crust and supplied volcanic detritus to the Hokonui assemblage and possibly the adjacent Caples-Croiselles-Pelorus terrane (Fig. 8A, B). In latest Triassic time the older Hokonui sediments (Maitai) and the Caples-Croiselles-Pelorus rocks were metamorphosed and deformed along a series of imbricate thrusts related to a west-dipping subduction zone (Fig. 8C). Arc-derived sedimentation continued until near the close of the Jurassic, when it is interpreted to have been terminated by the collision of the continental-derived Torlesse terrane and the island-arc system (Fig. 8D). Concurrently, the Haast Schist may have formed in the suture zone, derived largely from the Torlesse, but in part from the eastern margin of the island-arc system represented by the Caples-Croiselles-Pelorus rocks.

A L. PERMIAN

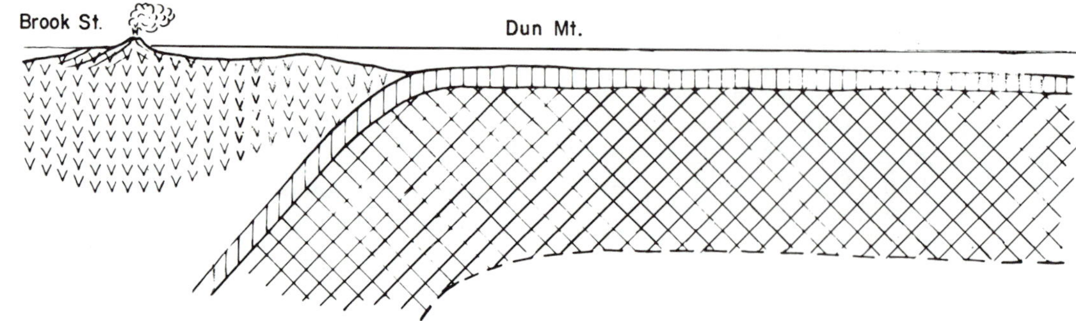

B U. PERMIAN

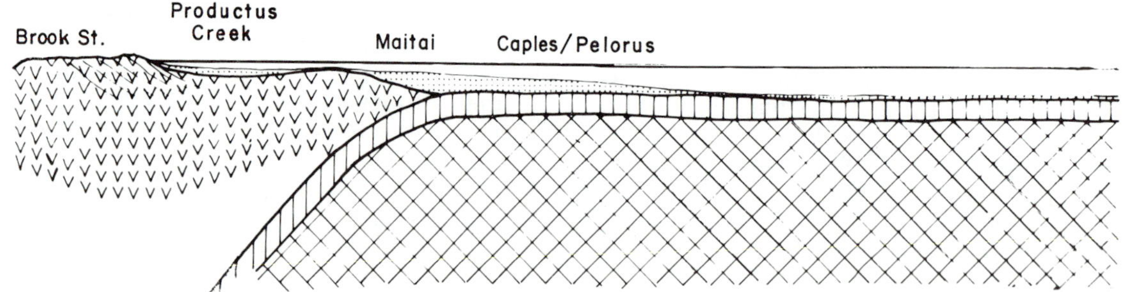

C TRIASSIC

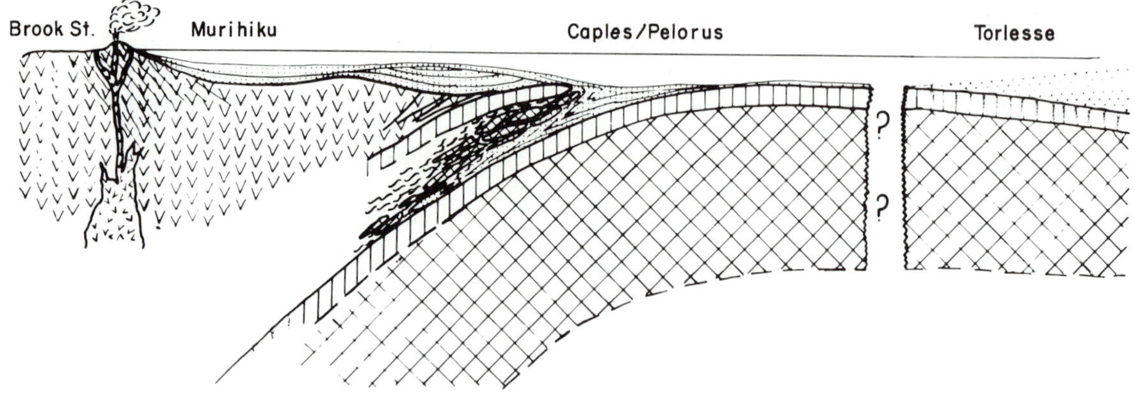

D U.CRETACEOUS- L. JURASSIC

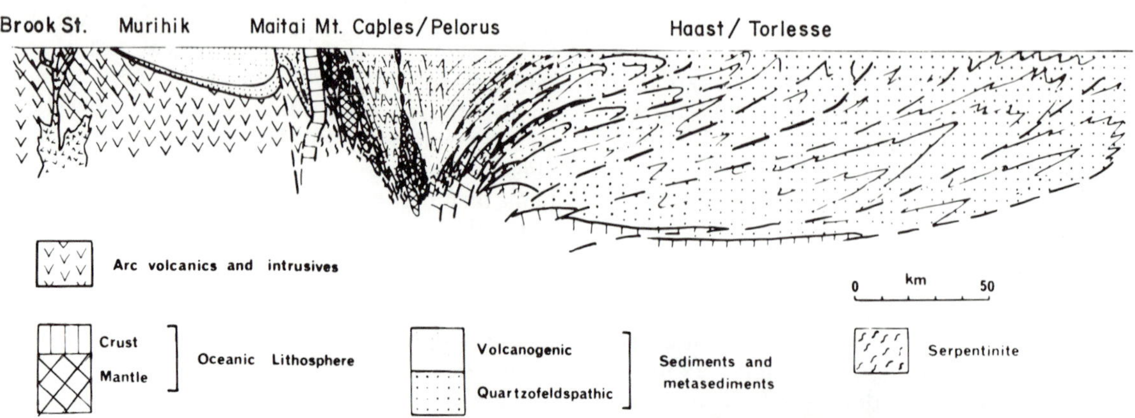

Fig. 8. Proposed model for the evolution of the Hokonui-Alpine assemblages of New Zealand (from Coombs et al., 1974). A. lower Permian; B. upper Permian; C. Triassic; D. Late Jurassic-Early Cretaceous.

Table 2. Comparison of relations between major lithologic assemblages of California and New Zealand.

	NEW ZEALAND		CALIFORNIA	
	Hokonui	Alpine	Great Valley Sequence	Franciscan
Age of sedimentation	Permian to Jurassic	Carboniferous to Cretaceous	Jurassic through Cretaceous	Upper Jurassic to Eocene
Source of sediments	Arc-derived; little or no continentally derived debris	Unidentified continent source lying east of present outcrop belt	Continental crust of Klamath Sierran terranes	Postulated rifted volcano-plutonic arc lying west of Great Valley marginal basin
Contemporaneous volcanic rocks	Entire sequence volcaniclastic, dominantly andesitic, some basaltic	Pillow basalt present but not abundant	Pillow basalt, breccia, & keratophyre present only at base	Pillow basalt, tuff, breccia locally abundant; keratophyre present; rare intrusive rocks of basaltic composition
Basement	Oceanic	Not exposed, but probably oceanic	Oceanic on west side, continental on east side	Not exposed, but probably oceanic
Internal structure and metamorphism	Open folds -- zeolite to lower blueschist facies	Complexly deformed with abundant penetrative structures -- zeolite to amphibolite facies	Homoclinal to gently folded - zeolite-phrenite pumpellyite facies	Complexly deformed penetrative and Mélange structures -- zeolite to upper blueschist facies
Relations to adjacent units	Sutured to older continental block (Fiordland) along median tectonic line	In thrust(?) fault contact with Hokonui assemblage along the Dun Mtn. belt	Depositional on continental crust to east; structurally overlies Franciscan along folded thrust fault to west	Underlies Great Valley sequence along folded thrust
Inferred site of deposition	Basins and aprons on & near active volcanic arc; in part in marginal basin behind arc	Continental rise and slope to shelf	Probably marginal sea west of continental margin, with transgression onto continental platform	Complex trench and borderland west of postulated rifted volcanoplutonic arc: many tectonic islands and locally deep basins

COMPARISONS BETWEEN THE ACTIVE CONTINENTAL MARGINS OF CALIFORNIA AND NEW ZEALAND

From a comparison of significant relations between California and New Zealand, similarities and differences between these Mesozoic continental margins may be summarized as follows (Table 2):

1. In both California and New Zealand, the age of deposition of the paired sedimentary belts is similar, and this fact has been a major factor in the development of most tectono-sedimentary models. Actually in California there is a much closer similarity in age because the use of the term "Great Valley sequence" is usually restricted to Mesozoic strata and does not include the Paleocene and Eocene sedimentary rocks which locally are present and are equivalent in age to the youngest Franciscan rocks in the coastal belt.

2. The provenance of the Hokonui assemblage of New Zealand was clearly a volcanic arc, and while the Great Valley sequence of California includes much first-cycle volcanic detritus, it also contains much plutonic material as well as sedimentary, volcanic, and metamorphic rock fragments derived from the older Sierran-Klamath continental landmass. The Alpine assemblage of New Zealand, unlike the Franciscan, contains both an inner, volcanogenic terrane derived from arc-type volcanism, Caples-Croiselles- Pelorus, and a quartzo-feldspathic terrane, Torlesse, derived from a continental source area. Sandstones from the Torlesse terrane closely resemble those of the Franciscan assemblage.

3. Contemporaneous volcanic rocks (including tuffs), largely of andesitic-dacitic composition, are intercalated within the Hokonui assemblage and support the concept of an island arc, whereas the Great Valley sequence contains volcanic rocks only at the base. Both the Alpine and Franciscan assemblages contain volcanic intercalations, largely pillow lavas, although these appear to be much more abundant in California. Radiolarian cherts, often associated with pillow lavas, are found in both areas.

4. Ophiolite sequences inferred to be oceanic crust underlie the outer side of the Hokonui assemblage as well as the older sedimentary rocks of the Great Valley sequence along the west side of the Great Valley. To the east and north in California, however, the younger Great Valley sequence rocks rest with great unconformity on the older Sierran-Klamath continental basement; these relationships are now known in New Zealand. The basement beneath the Alpine and Franciscan assemblages is not known but is inferred from geophysical data to be oceanic.

5. Structurally, the older, Maitai, portion of the Hokonui assemblage is more deformed that the Great Valley sequence, and locally it contains lawsonite. Parts of the Alpine assemblage, on the other hand, are indistinguishable from metamorphosed Franciscan units, but blue-schist minerals are entirely lacking in the Torlesse terrane, which contains higher T/P assemblages. Tectonic mélange is much more common in the Franciscan assemblage.

6. In New Zealand, the Hokonui assemblage is bounded on both sides by major fault systems separating it from the continental Tasman belt (Fiordland) on the west and the Alpine assemblage on the east. In California, the Great Valley sequence is depositional on continental crust to the east and structurally overlies the Franciscan assemblage to the west along a complexly deformed thrust fault. In some areas, as along much of the Klamath Mountains-Coast Range boundary, the Great Valley sequence is not present; these rocks, like the Franciscan, may have been thrust beneath the crystalline rocks of the Klamath Mountains.

7. The Hokonui assemblage, as suggested earlier, appears to have been derived from an island arc built on oceanic crust. Modern analogues are interarc basins, arc flanks, and perhaps arc-trench gaps seen in the active volcanic arcs of the southwest Pacific. In contrast, most of the Great Valley sequence was probably deposited west of an active continental margin, but whether this was in a marginal sea or an arc-trench gap is not entirely clear; much circumstantial evidence favors a marginal sea. The Alpine assemblage appears to have been deposited in two different sites: (1) the volcanogenic Caples-Croiselles-Pelorus terrane in a series of basins flanking or perhaps within an active volcanic arc-trench system, and (2) the Torlesse on the continental shelf and slope with some actual continental deposition. The Franciscan assemblage appears to have been laid down in a complicated sedimentary environment, including possible trench and borderland west of a postulated dormant arc made up of older volcanic and plutonic material.

This, whereas there are notable similarities in lithology, structural and metamorphic history, and geometry of major lithic belts between the Franciscan assemblage and Great Valley sequence of California and the Hokonui and Alpine assemblages of New Zealand, in detail there are marked differences. This suggests that the two areas followed somewhat different evolutionary paths such that application of the same simplified tectono-sedimentary model to both areas is probably erroneous.

CONCLUSIONS

The Franciscan assemblage and Great Valley sequence of California have been considered to represent subducted trench and overlying arc-trench gap deposits whose sedimentary detritus was derived from an active continental margin represented by the Sierran-Klamath plutonic terrane. A similar model has been proposed for the Alpine and Hokonui assemblages of New Zealand. Recent studies suggest that numerous differences exist between California and New Zealand, and that the trench-arc-trench gap-active continental margin model may not fully apply to either area.

In California, the source of the quartzo-feldspathic sediments of the Franciscan assemblage is uncertain. A Sierran-Klamath source is difficult to envisage, as this implies derivation east of and transportation through the basin presently occupied by the Great Valley sequence. Facies trends within the Great Valley sequence do not support this. Instead, a dormant volcano-plutonic arc may have lain between the two basins and was subsequently eroded away to form the graywacke and shale of the Franciscan assemblage.

In New Zealand, detailed studies suggest that quartzo-feldspathic Torlesse graywackes of the eastern belt were probably derived from an active continental block lying somewhere to the east, where now only deep oceans occur. Derivation from the western continental sources (Fiordlands or Australia) is unlikely, for this would require bypassing enormous volumes of detrital material through the island-arc system that provided the coeval Hokonui volcanogenic sediments that lie between the western continental sources and the Torlesse depositional basin. These two subparallel belts may have been originally separated by a wide area of sea floor and subsequently juxtaposed by ocean-floor spreading.

BIBLIOGRAPHY

Aronson, J. L., 1968, Regional geochronology of New Zealand: Geochim. Cosmochim, Acta, v. 32, p. 669-697.

Bailey, E. H., and Blake, M. C., Jr., 1969a, Tektonicheskoe razvitiye zapadnoy Kalifornii v pozdnem mezozoe, pt. 1 (Tectonic development of western California during the late Mesozoic): Geotektonika, no. 3, p. 17-30 (in Russian).

———, and Blake, M. C., Jr., 1969b, Tektonicheskoe razvitiye zapadnoy Kalifornii v pozdnem mezozoe, pt. 2 (Tectonic development of western California during the late Mesozoic): Geotektonika, no. 4, p. 24-34 (in Russina).

———, and Irwin, W. P., 1959, K-feldspar content of Jurassic and Cretaceous graywackes of the northern Coast Ranges and Sacramento Valley, California: Am. Assoc. Petr. Geol. Bull., v. 43, p. 2797-2809.

———, and Jones, D. L., 1973, Preliminary lithologic map, Colyear Springs quadrangle, California: U.S. Geol. Survey Misc. Field Studies Map MF-516, scale 1:48,000.

———, Irwin, W. P., and Jones, D. L., 1964, Franciscan and related rocks and their significance in the geology of western California: Calif. Div. Mines and Geol. Bull., v. 183, 177 p.

———, Blake, M. C., Jr., and Jones, D. L., 1970, On-land Mesozoic oceanic crust in California Coast Ranges: Geol. Survey Res., 1970, U.S. Geol. Survey Prof. Paper 700-C, p. C70-C81.

Berkland, J. O., 1969, Geology of the Novato quadrangle, Marin County, California [M.S. thesis]: San Jose State College, San Jose, Calif., 146 p.

———, 1972, Paleogene "frozen" subduction zone in the Coast Ranges of northern California, *in* Tectonics: Rept. 24th Internatl. Geol. Congr., Montreal, no.3, p. 99-105.

———, 1973, Rice Valley outlier—new sequence of Cretaceous-Paleocene strata in northern Coast Ranges, California: Geol. Soc. America Bull., v. 84, p. 2389-2406.

———, Raymond, L. A., Kramer, J. C., Moores, E. M., and O'Day, M., 1973, What is Franciscan?: Am. Assoc. Petr. Geol. Bull., v. 56, p. 2295-2302.

Bishop, D. G., 1965, The geology of the Clinton district, South Otago: Trans. Roy. Soc. New Zeland Geol., v. 2, p. 205.

———, 1972, Progressive metamorphism from prehnite-pumpellyite to green-schist facies in the Dansey Pass area, Otago, New Zealand: Geol. Soc. America Bull., v. 83, p. 3163-3176.

———, 1974, Stratigraphic, structural, and metamorphic relationships in the Dansey Pass area: New Zealand Jour. Geol. Geophys. (in press).

Blake, M. C., Jr., 1965, Structure and petrology of low-grade metamorphic rocks, blueschist facies, Yolla Bolly area, northern California: [Ph.D. thesis]: Stanford Univ., Stanford, Calif., 91 p.

———, and Jones, D. L., 1974, Origin of Franciscan mélanges in northern California: Soc. Econ. Paleontologists and Mineralogists Spec. Publ. 19, p. 255-263.

———, and Landis, C. A., 1973, The Dun Mountain ultramafic belt—Permian oceanic crust and upper mantle in New Zealand: U.S. Geol. Survey Jour. Res., v. 1, p. 529-534.

———, Irwin, W. P., and Coleman, R. G., 1967, Upside-down metamorphic zonation, blueschist facies, along a regional thrust in California and Oregon: Geol. Survey Res., 1967, U.S. Geol. Survey Prof. Paper 575-C, p. C1-C9.

———, Smith, J. T., Wentworth, C. M., and Wright, R. H., 1971, Preliminary geologic map of western Sonoma County and northernmost Marin County, California: U.S. Geol. Survey open-file map, scale 1:62,500.

———, Bartow, J. A., Frizzell, V. A., Sorg, D. H., Schlocker, J., Wentworth, C. M., and Wright, R. H., 1974, Preliminary geologic map of Marin, San Francisco, and parts of adjacent counties, California: U.S. Geol. Survey Misc. Field Studies Map MF-573, scale 1:62,500.

Boles, J. R., 1974, Structure, stratigraphy and petrology of mainly Triassic rocks, Hokonui Hills, Southland, New Zealand: New Zealand Jour. Geol. Geophys. (in press).

Bradshaw, J. D., 1972, Stratigraphy and structure of the Torlesse Supergroup (Triassic-Jurassic) in the foothills of the southern Alps near Hawarden (S60-61), Canterbury: New Zealand Jour. Geol. Geophys., v. 15, p. 71-87.

———, 1973, Allochthonous Mesozoic fossil localities in mélange within the Torlesse rocks of north Canterbury: Jour. Roy. Soc. New Zealand, v. 3, p. 161-167.

———, and Andrews, P. B., 1973, Geotectonics and the New Zealand geosyncline: Nature Phys. Sci., v. 241, p. 14-16.

Brown, E. H., 1968, Metamorphic structures in part of the Eastern Otago schists: New Zealand Jour. Geol. Geophys., v. 11, p. 41-65.

Brown, R. D., Jr., 1964a, Geologic map of the Stonyford quadrangle, Glenn, Colusa, and Lake Counties, California: U.S. Geol. Survey Mineral Inv. Field Studies Map MF-279, scale 1:48,000.

———, 1964b, Thrust-fault relations in the northern Coast Ranges, California: U.S. Geol. Survey Prof. Paper 475-D, p. D7-D13.

Campbell, J. D., and Coombs, D. S., 1966, Murihiku Supergroup (Triassic-Jurassic) of Southland and South Otago: New Zealand Jour. Geol. Geophys., v. 9, p. 393-398.

———, and Force, E. R., 1972, Stratigraphy of the Mount Potts Group at Rocky Gully, Rangitata Valley, Canterbury: New Zealand Jour. Geol. Geophys., v. 15, p. 157-167.

———, and Warren, G., 1965, Fossil localities of the Torlesse Group in the South Island: Trans. Roy. Soc. New Zealand Geol., v. 3, p. 99-137.

Campbell, J. K., and Campbell, J. D., 1970, Triassic tube fossils from Tuapeka rocks, Akatore, South Otago: New Zealand Jour. Geol. Geophys., v. 13, p. 392-399.

Carman, M. F., Jr., 1968, A comparison of some Permian rocks on opposite sides of the Alpine fault, South Island: Trans. Roy. Soc. New Zealand Geol., v. 6, p. 91-130.

Carter, R. M., Norris, R. J., Landis, C. A., and Bishop, D. G., 1974, Proposals toward a high-level stratigraphic nomenclature for New Zealand sedimentary rocks: Jour. Roy. Soc. New Zealand, v. 4, no. 1, p. 5-18.

Colburn, I. P., 1970, The trench concept as a model for central California Cretaceous basin of deposition (abstr.): Geol. Soc. America Cordilleran Sect., Hayward, Calif., p. 82-83.

Coombs, D. S., 1950, The geology of the Northern Taringatura Hills, Southland: Trans. Roy. Soc. New Zealand, v. 78, p. 426-448.

———, 1954, The nature and alteration of some Triassic sediments from Southland, New Zealand: Trans. Roy. Soc. New Zealand, v. 82, p. 65-109.

———, Ellis, A. J., Fyfe, W. S., and Taylor, A. M., 1959, The zeolite facies, with comments on the interpretation of hydrothermal syntheses: Geochim. Cosmochim. Acta, v. 17, p. 53-107.

———, Landis, C. A., Norris, R. J., Sinton, J. M., Borns, D., and Nakamura, Y., 1974, The Dun Mountain Ophiolite belt, New Zealand, *in* Ophiolites in the earth's crust: Nauka, Geol. Inst. Acad. Sci. S.S.S.R. and Subcommission for the Tectonic Map of the World (in press).

Cope, R. N., and Reed, J. J., 1967, The Cretaceous palaeogeology of the Taranaki-Cook Strait area: Australasian Inst. Mining and Met. Proc., v. 22, p. 63-72.

Cotton, W. R., 1972, Preliminary geologic map of the Franciscan rocks in the central part of the Diablo Range, Santa Clara and Alameda counties, California: U.S. Geol. Survey Misc. Field Studies Map MF-343, scale 1:62,500.

Cowan, D. S., 1973, Petrology and structure of the Franciscan assemblage southwest of Pacheco Pass, California [Ph.D. thesis]: Stanford Univ., Stanford, Calif., 74 p.

Devereux, I., McDougall, I., and Watters, W. A., 1968, Potassium-argon mineral dates on intrusive rocks from the Foveaux Strait area: New Zealand Jour. Geol. Geophys., v. 11, p. 1230-1235.

Dickinson, W. R., 1970a, Relations of andesites, granites, and derivative sandstones to arc-trench tectonics: Rev. Geophys. Space Phys., v. 8, p. 813-860.

———, 1970b, Tectonic setting and sedimentary petrology of the Great Valley Sequence (abstr.): Geol. Soc. America, Cordilleran Sect. Hayward, Calif., p. 86-87.

———, 1971a, Clastic sedimentary sequences deposited in shelf, slope and trough settings between magmatic arcs and associated trenches: Pacific Geology, v. 3, p. 15-30.

———, 1971b, Detrital modes of New Zealand graywackes: Sediment. Geology, v. 5, p. 37-56.

———, and Rich, E. I., 1972, Petrologic intervals and petrofacies in the Great Valley sequence, Sacramento Valley, California: Geol. Soc. America Bull., v. 83, p. 3007-3024.

Ernst, W. G., 1970, Tectonic contact between the Franciscan mélange and the Great Valley sequence, crustal expression of a late Mesozoic Benioff zone: Jour. Geophys. Res., v. 75, p. 886-902.

———, 1971, Petrologic reconnaissance of Franciscan metagraywackes from the Diablo Range, central California Coast Ranges: Jour. Petrology, v. 12, p. 413-437.

Fleming, C. A., 1970, The Mesozoic of New Zealand—chapters in the history of the circum-Pacific mobile belt: Geol. Soc. London Quart. Jour., v. 125, p. 125-170.

———, and Kear, D., 1960, The Jurassic sequence at Kawhia Harbour, New Zealand: New Zealand Geol. Survey Bull., n.s. 67, 50 p.

Force, E. R., 1974, A comparison of some Triassic rocks in the Hokonui and Alpine belts of South Island, New Zealand: Jour. Geology (in press).

Force, L. M., 1972, *Atomodesma* limestones of the Hokonui belt Permian, South Island, New Zealand [Ph.D thesis]: Lehigh Univ., Bethlehem, Pa., 196 p.

———, 1974, Stratigraphy and palaeoecology of the Productus Creek Group, South Island: New Zealand Jour. Geol. Geophys. (in press).

Fox, K. F., Jr., Sims, J. D., Bartow, J. A., and Helley, E. H., 1973, Preliminary geologic map of eastern Sonoma, Napa, and Solano counties, California: U.S. Geol. Survey Misc. Field Studies Map MF-483, scale 1:62,500.

Gair, H. S., Gregg, D. R., and Speden, I. G., 1962, Triassic fossils from Corbie Creek, North Otago: New Zealand Jour. Geol. Geophys., v. 5, p. 92-113.

Gilbert, W. G., 1973, Franciscan rocks near Sur fault zone, northern Santa Lucia Range, California: Geol. Soc. America Bull., v. 84, p. 3317-3328.

———, and Dickinson, W. R., 1970, Stratigraphic variations in sandstone petrology, late Mesozoic Great Valley sequence, southern Santa Lucia Range, California: Geol. Soc. America Bull., v. 81, p. 949-954.

Grant-Mackie, J. A., 1971, The probably allochthonous nature of Norian and newly-found Carnian rocks of the Orona Valley, western Ruahine Range (abstr.): Symp. on Torlesse Supergroup, Wellington, New Zealand, p. 16-17.

Griffiths, J. R., 1971, Reconstruction of the southwest Pacific margin of Gondwanaland: Nature, v. 234, p. 203-207.

Grindley, G. W., 1958, The geology of the Eglingon Valley, Southland: New Zealand Geol. Survey Bull., n.s. 58, 68 p.

———, 1963, Structure of the Alpine schists of South Westland, southern Alps, New Zealand: New Zealand Jour. Geol. Geophys., v. 6, p. 872-930.

———, 1971, S8 Takaka (1st edition), Geological map of New Zealand, 1:63,360: Dept. Sci. and Ind. Res., Wellington, New Zealand.

Halpern, M., 1968, Ages of Antarctic and Argentine rocks bearing on continental drift: Earth and Planetary Sci. Letters, no. 5, p. 159-167.

———, 1971, Evidence for Gondwanaland from a review of west Antarctic radiometric ages, *in* Research in the Antarctic: Am. Assoc. Advan. Sci. Publ., p. 717-730.

Hamilton, W., 1969, Mesozoic California and the underflow of Pacific mantle: Geol. Soc. America Bull., v. 80, p. 2409-2429.

Harrington, H. J., 1957, Size and abundance of phenocrysts in lavas of terrestrial and geosynclinal environments: Nature, v. 179, p. 271.

———, Burns, K. L., and Thompson, B. R., 1973, Gambier-Beaconsfield and Gambier-Sorell fracture zones and the movement of plates in the Australia-Antarctica-New Zealand region: Nature Phys. Sci., v. 245, p. 109-112.

Hill, M. L., and Dibblee, T. W., Jr., 1953, San Andreas, Garlock, and Big Pine faults, California: Geol. Soc. America Bull., v. 64, p. 443-458.

Hornibrook, N. deB., and Khoon, S. Y., 1965, Fusuline limestone in Torlesse Group near Ben More Dam, Waitaki Valley: Trans. Roy. Soc. New Zealand, v. 3, p. 136-137.

Hsü, K. J., 1971, Franciscan melanges as a model for eugeosynclinal sedimentation and underthrusting tectonics: Jour. Geophys. Res., v. 76, p. 1162-1170.

Hsü, K. J., and Ohrbom, Richard, 1969, Melanges of San Francisco Peninsula—geologic reinterpretation of Type Franciscan: Am. Assoc. Petr. Geol. Bull., v. 53, no. 7, p. 1348-1367.

Irwin, W. P., 1957, Franciscan group in Coast Ranges and its equivalents in Sacramento Valley, California: Am. Assoc. Petr. Geol. Bull., v. 41, p. 2284-2297.

———, 1964, Late Mesozoic orogenies in the ultramafic belts of northwestern California and southwestern Oregon: Geol. Survey Res., 1964, U.S. Geol. Survey Prof. Paper 501-C, p. C1-C9.

———, Wolfe, E. W., Blake, M. C., Jr., and Cunningham, C. G., Jr., 1974, Geologic map of the Pickett Peak quadrangle, Trinity County, California: U.S. Geol. Survey Geol. Quadrangle Map GQ-1111, scale 1:62, 500 (in press).

Jenkins, D. G., and Jenkins, T. B. H., 1971, First diagnostic Carboniferous fossils from New Zealand: Nature, v. 233, p. 117-118.

Jones, D. L., 1973, Structural significance of upper Mesozoic biostratigraphic units in northern California and southwestern Oregon (abstr.): Geol. Soc. America Ann. Meeting, Dallas, Texas, p. 684-685.

———, and Bailey, E. H., 1973, Preliminary biostratigraphic map, Colyear Springs quadrangle, California: U.S. Geol. Survey Misc. Field Studies Map MF-517, scale 1:48,000.

———, and Irwin, W. P., 1971, Structural implications of an offset earth Cretaceous shoreline in northern California: Geol. Soc. America Bull., v. 82, p. 815-822.

———, Bailey, E. H., and Imlay, R. W., 1969, Structural and stratigraphic significance of the *Buchia* zones in the Colyear Spring-Paskenta area, California: U.S. Geol. Survey Prof. Paper 647-A, p. A1-A24.

Kawachi, Y., 1974, Geology and petrochemistry of weakly metamorphosed rocks in the upper Wakatipa district, South New Zealand: New Zealand Jour. Geol. Geophys. (in press).

Landis, C. A., 1969, Upper Permian rocks of South Island, New Zealand: Lithology, stratigraphy, structure, metamorphism and tectonics [Ph.D. thesis]: Univ. Otago, Dunedin, New Zealand, 627 p.

———, and Bishop, D. G., 1972, Plate tectonics and regional stratigraphic-metamorphic relations in the southern part of the New Zealand geosyncline: Geol. Soc. America Bull., v. 83, p. 2267-2284.

———, and Coombs, D. S., 1967, Metamorphic belts and orogenesis in southern New Zealand: Tectonophysics, v. 4, p. 501-518.

Lauder, W. R., 1964, The geology of Pepin Island and part of the adjacent mainland: New Zealand Jour. Geol. Gelphys., v. 7, p. 205-241.

———, 1965, The geology of Dun Mountain, Nelson, New Zealand: New Zealand Jour. Geol. Geophys., v. 8, p. 3-34, 375-504.

Lillie, A. R., 1961, Folds and faults in the New Zealand Alps and their tectonic significance: Proc. Roy. Soc. New Zealand, v. 89, p. 57-85.

Mansfield, C. F., III, 1972, Petrography and sedimentology of the late Mesozoic Great Valley sequence, near Coalinga, California [Ph.D. thesis]: Stanford Univ., Stanford, Calif., 71 p.

Marwick, J., 1953, Divisions and faunas of the Hokonui system (Triassic and Jurassic): New Zealand Geol. Survey Bull. Paleontology, no. 21, 144 p.

Means, W. D., 1963, Mesoscopic structures and multiple deformation in the Otago schist: New Zealand Jour. Geol. Geophys., v. 6, p. 801-816.

Mildenhall, D. C., 1970, Discovery of a New Zealand member of the Permian Glossopteris flora: Australian Jour. Sci., v. 32, p. 474.

Monger, J. W. H., and Ross, C. A., 1971, Distribution of fusulinaceans in the western Canadian Cordillera: Can. Jour. Earth Sci., v. 8, no. 2, p. 259-278.

Moores, E. M., 1970, Ultramafics and orogeny, with models of the U.S. Cordillera and the Tethys: Nature, v. 228, p. 837-842.

Mossman, D. J., 1973, Geology of the Greenhills ultramafic complex, Bluff Peninsula, Southland, New Zealand: Geol. Soc. America Bull., v. 84, p. 39-62.

Mutch, A. R., 1972, Morley Geologic map of New Zealand (Sheet S-159): Wellington, New Zealand Geol. Survey, scale 1:250,000.

———, A. R., 1972, Geology of Morley Subdivision: New Zealand Geol. Survey Bull., v. 78.

Nauman, C. R., 1974, Alabaster Group rocks (Lower Permian) in southern Skippers Range, north-west Otago, New Zealand: Jour. Roy. Soc. New Zealand, v. 3, no. 4, p. 527-544.

Ojakangas, R. W., 1963, Cretaceous sedimentation, Sacramento Valley, California: Geol. Soc. America Bull., v. 79, p. 976-1008.

Page, B. M., 1970, Sur-Nacimiento fault zone of California: continental margin tectonics: Geol. Soc. America Bull., v. 81, p. 667-690.

Pessagno, E. A., Jr., 1973, Age and geologic significance of radiolarian cherts in the California Coast Ranges: Geology, v. 1, no. 4, p. 153-156.

Raymond, L. R., 1973, Franciscan geology of the Mt. Oso area, California [Ph.D. thesis]: California Univ., Davis, 185 p.

Reed, J. J., 1957, Petrology of the Lower Mesozoic rocks of the Wellington district: New Zealand Geol. Survey Bull., n.s. 57, 60 p.

———, 1958, Regional metamorphism in southeast Nelson: New Zealand Geol. Survey Bull., n.s. 60, 64 p.

Rich, E. I., 1971, Geologic map of the Wilbur Springs quadrangle, Colusa and Lake Counties, California:

U.S. Geol. Survey Misc. Geol. Inv. Map I-538, scale 1:48,000.

Scholl, D. W., and Marlow, M. S., 1974, The sedimentary sequences in modern Pacific trenches and the deformed circumpacific eugeosynclines: Soc. Econ. Paleontologists and Mineralogists Spec. Publ. 19 (in press).

Sims, J. D., Fox, K. F., Jr., Bartow, J. A., and Helley, E. J., 1973, Preliminary geologic map of Solano County and parts of Napa, Contra Costa, Marin, and Yolo counties, California: U.S. Geol. Survey Misc. Field Studies Map MF-484, scale 1:62,500.

Skinner, D. N. B., 1972, Subdivision and petrology of the Mesozoic rocks of Coromandel (Manaia Hill Group): New Zealand Jour. Geol. Geophys., v. 15, p. 203-227.

Sohl, N. F., 1971, North American Cretaceous biotic provinces delineated by gastropods: North American Paleontol. Conv. Proc., pt. L, p. 1610-1638.

Speden, I. G., 1971, Geology of Papatowai—subdivision south-east Otago: New Zealand Geol. Survey Bull., n.s. 81, 166 p.

Stevens, G. R., 1972, Paleontology of the Torlesse Supergroup: Rept. New Zealand Geol. Survey, 54.

Suppe, J., 1969, Franciscan geology of the Leech Lake Mountain-Anthony Peak region, northeastern Coast Ranges, California [Ph.D. thesis]: Yale Univ., New Haven, Conn., 99 p.

———, 1973, Geology of the Leech Lake Mountain-Ball Mountain region, California: a cross section of the northeastern Franciscan belt and its tectonic implications: California Univ. Publs. Geol. Sci., v. 107, 82 p.

Vitaliano, C. J., 1968, Petrology and structure of the south-eastern Marlborough Sounds, New Zealand: New Zealand Geol. Survey Bull., n.s. 74, 40 p.

Waterhouse, J. B., 1964, Permian stratigraphy and faunas of New Zealand: New Zealand Geol. Survey Bull., n.s. 72, 101 p.

———, 1966a, The age of the Croisilles Volcanics, Eastern Nelson: Trans. Roy. Soc. New Zealand Geol., v. 3, p. 175-181.

———, 1966b, The Häckel syncline and neighbouring folds of the Upper Tasman Glacier: Trans. Roy. Soc. New Zealand Geol., v. 3, p. 183-195.

———, 1967, Proposal of series and stages for the Permian of New Zealand: Trans. Roy. Soc. New Zealand Geol., v. 5, p. 161-180.

———, and Vella, P., 1965, A Permian fauna from north-west Nelson, New Zealand: Trans. Roy. Soc. New Zealand Geol., v. 3, p. 57-84.

Webby, B. D., 1959, Sedimentation of the alternating greywacke and argillite strata in the Porirua district: New Zealand Jour. Geol. Geophys., v. 2, p. 461-478.

Wellman, H. W., 1952, The Permian-Jurassic stratified rocks: 19th Internatl., Geol. Congr., Algiers Symp. Ser. Gondwana, Proc., p. 13-24.

———, 1956, Structural outline of New Zealand: New Zealand Dept. Sci. Ind. Res. Bull., v. 121, 36 p.

Wood, B. L., 1956, The geology of the Gore Subdivision: New Zealand Geol. Survey Bull., n.s. 53, 128 p.

———, 1963, Structure of the Otago schists: New Zealand Jour. Geol. Geophys., v. 6, p. 641-680.

Wright, J. B., 1966, Convection and continental drift in the southwest Pacific: Tectonophysics, v. 3, p. 69-81.

Possible Ancient Continental Margins in Iran

Jovan Stöcklin

INTRODUCTION

The ophiolites of the Alpine folded region of Iran are examined as an indication of the extent of ancient oceanic realms bordered by ancient continental margins. They are grouped into four geographically and geologically distinct zones, differing from each other in composition, structure, and age. The possibility of these four zones marking former continental margins is then checked against the background of the general structural evolution of Iran. It is concluded that during Paleozoic time Iran was an extension of the Arabian platform, and thus a part of Gondwanaland, possibly bordered by a "Paleo-Tethys" in the north, along the present northern foot of the Alborz Range. Closing of the "Paleo-Tethys," short of a possible modern relict in the South Caspian depression, may have been related to Hercynian orogenic processes in the Scytho-Turanian plate to the north and was completed by Liassic time. A rift in the Arabian-Iranian platform along the "Main Zagros Thrust line" in the early Mesozoic or late Paleozoic was followed by the formation of a "Neo-Tethys" in the south, possibly interrelated and simultaneous with the closing of the "Paleo-Tethys" in the north. Further breakup of Iran led to the formation of several branch troughs of the "Neo-Tethys" and temporary isolation of a "Central-and-East Iranian Microcontinent" in the late Mesozoic. Closing of the "Neo-Tethys" in the early Maestrichtian was followed by reintegration of the "microcontinent" and folding of central and north Iran during the Paleocene paroxysm of the Alpine orogeny.

Iran today forms a solid land-bridge between the old continental masses of Afro-Arabia and Eurasia, i.e., between the former continents of Gondwanaland and Angaraland. In older concepts, such as Argand's (1924), Iran, as part of the Alpine-Himalayan orogenic belt, did not belong in pre-orogenic time to either of the two old continental masses, but to the realm of the Tethys. The latter was the elongated ocean which separated the two continents and from which the Alpine-Himalayan ranges rose when the continents approached each other and compressed the Tethys trough.

With the new concepts of ocean-floor spreading and subduction, introduced in Iran by Takin (1972a), the concept of the Tethys has changed also. Formerly the remnants of the Tethys were seen in those huge piles of marine sediments that make up the greater part of the Alpine-Himalayan ranges. With the new concepts it is thought that most of the Tethys trough has disappeared, "consumed" by subduction. What is left of it within the mountain belt are those narrow scars of ophiolites, which are thought to represent former oceanic crust and to define the sutures along which ancient continents or continental fragments have been welded together—i.e., ancient continental margins.

If these new concepts are valid, Iran with its continental crustal structure must have belonged, not to the realm of the Tethys, but to one or the other or both of the two ancient continents; and if these were ever separated, the line or lines of separation must be found. The possible line of separation, the main Zagros Thrust line, has attracted many investigators during the last few years. It is marked by a narrow zone of ophiolites and separates two realms of fundamentally different geology: the Zagros in the southwest, with its thick, comprehensive, conformable succession of Paleozoic-Mesozoic-Tertiary shelf deposits, warped into gentle folds in a single process in the latest Tertiary, and central Iran in the northeast, with its complex structure and history of repeated folding, magmatism, and metamorphism.

But if the identity of Alpine ophiolite zones with scars separating former continental margins is taken for granted here, other ancient margins possibly could be perceived in Iran.

OPHIOLITES IN IRAN

General Remarks. Ophiolite zones of considerable length and variable (but generally small) width have become known not only in the Zagros, but also in the east, north, and in the very center of Iran (Fig. 1). Most, although not all, of them are associated with radiolarites and other sediments considered to be abyssal, and many display the chaotic structure that has become known as colored mélange (Gansser, 1955), most typically developed in Iranian Baluchestan. Although Gansser (1960) had clearly distinguished the mélanges from ophiolite zones without mélange character, an overgeneralization of the term "colored mélange" in Iran came to be used almost synonymously with ophiolites and radiolarites and, worse, almost as a stratigraphic unit term. This led to confusion and misinterpretations, particularly with regard to the Zagros ophiolites.

Significant Variations. The ophiolite-radiolarite belt of the Zagros appears in two widely separated

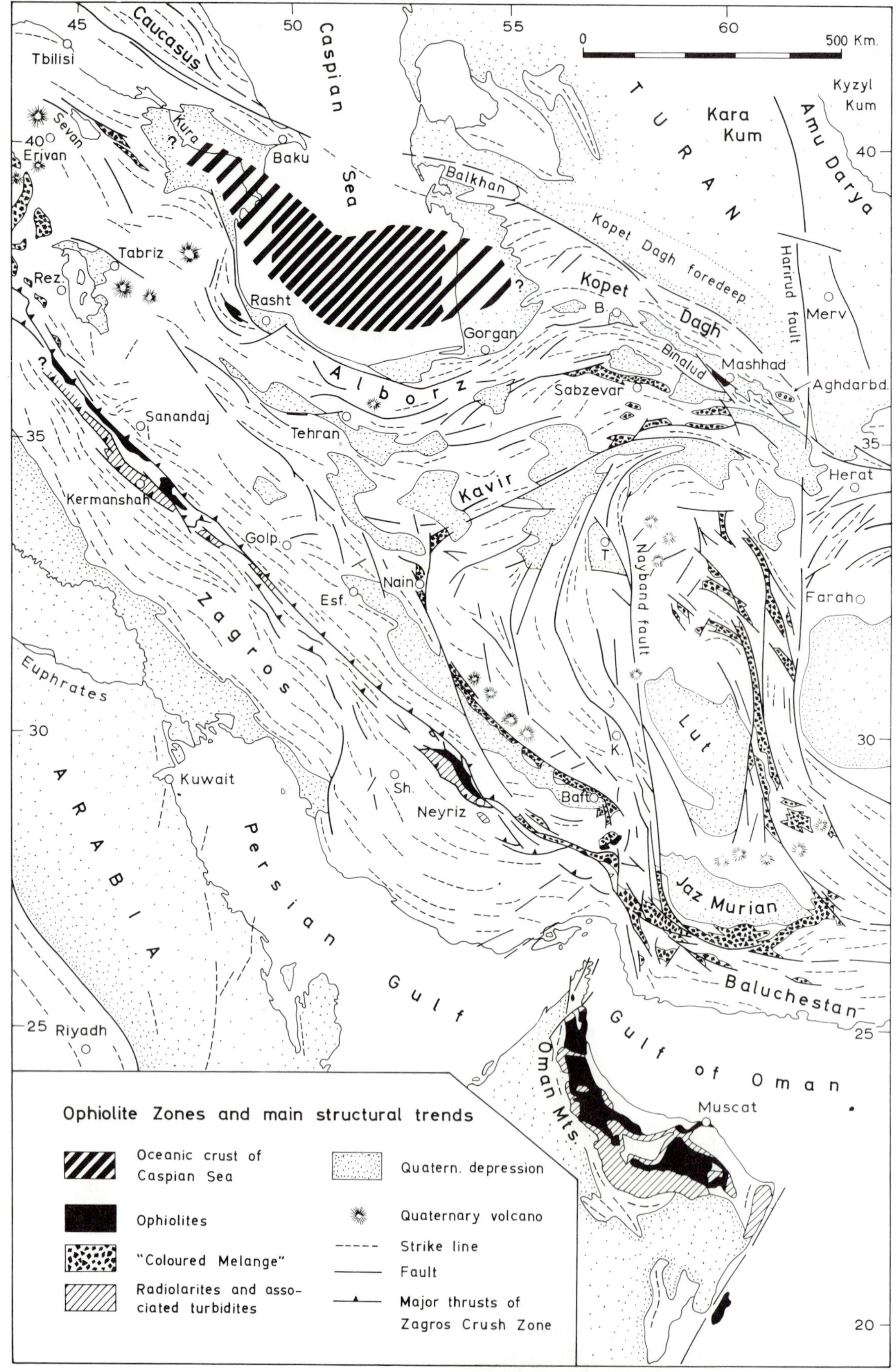

Fig. 1. Ophiolite zones and main structural trends of Iran and adjacent areas.

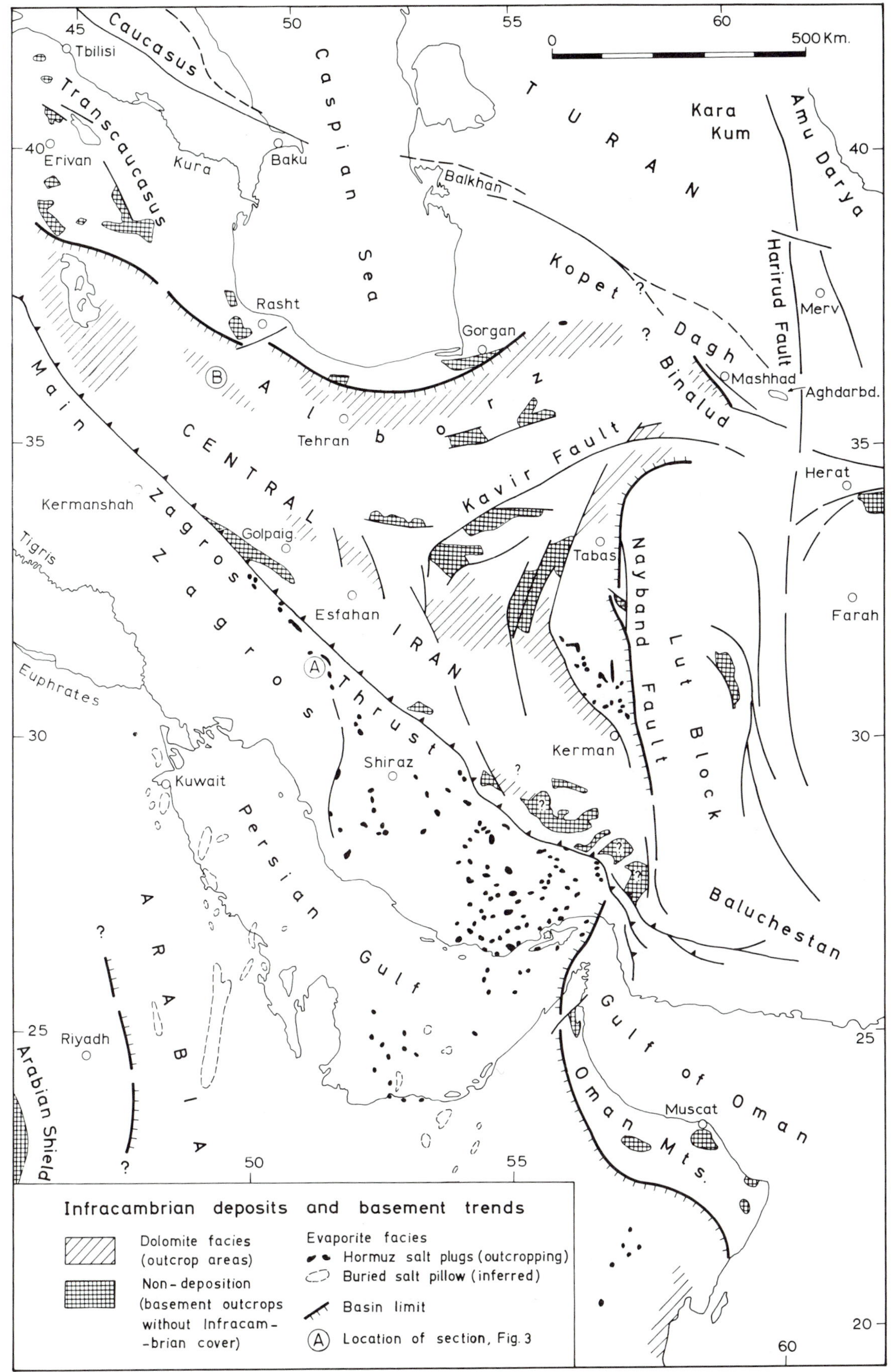

Fig. 2. Infracambrian deposits and basement trends of Iran and adjacent areas.

sectors of the Crush Zone immediately southwest of the Main Thrust: one in the Kermanshah area (Braud, 1970), the other at Neyriz east of Shiraz (Ricou, 1968a, 1968b), each in the focal area of one of the two large, southwest-convex arcs formed by the Zagros Range (the Posht-e-Kuh arc and the Fars arc). Both exposures show very close similarity in composition and structure with the ophiolite-radiolarite complex of the Oman Mountains and together with this form part of Ricou's (1971a) *"croissant ophiolitique péri-arabe."*

When visiting the Neyriz and Oman occurrences, the present writer was struck not by the similarities but by the dissimilarities with the "colored mélanges" of interior Iran and Baluchestan. In the Zagros-Oman belt typical mélanges are, in fact, very subordinate; the rocks show a persistent zonation into a strongly imbricated, but exclusively sedimentary complex in the lower part and an almost exclusively ophiolitic complex on top, with a typical mélange of the two limited to a relatively narrow zone in between, which also includes large exotic limestone blocks (Fig. 4). Also, the sediments associated with the radiolarites in the lower complex have very little in common with those contained in the central Iranian mélanges: to the near exclusion of each other, characteristic detrital and turbiditic limestones abound in the Zagros-Oman belt, flysch-type shaly and tuffaceous deposits and characteristic dense, pelagic limestones in the Iranian mélanges. Spilitic and diabasic rocks play a dominant role in the mélanges but are subordinate in the Zagros and Oman.

The most significant differences, however, have been found by paleontologists: the sediments associated with the radiolarites of the Zagros and Oman contain abundant fossils of Paleozoic and all possible Mesozoic ages but, with very rare exceptions, none younger than Turonian, whereas the Iranian mélanges contain an almost exclusive late Cretaceous, Senonian-Maestrichtian fauna. Moreover, the ages of the cover rocks show clearly that in the Zagros-Oman belt the imbrication and ophiolite emplacement was terminated in pre-Maestrichtian or early Maestrichtian time, whereas these processes extended into late Maestrichtian or Paleocene time in the classical mélange belts to the north and east of the Zagros.

The mélange complexes of Baluchestan and interior Iran can further be grouped into two widely separated geographic zones. One appears as a circular, although somewhat discontinuous belt surrounding the central-and-east Iranian Microcontinent of Takin (1972), extending from Iranian Baluchestan north along the Pakistan and Afghan frontiers, reappearing after a short interruption in several east-trending exposures in the Sabzevar area north of the Kavir fault and again at Nain in central Iran at the western end of this fault, extending from there southeast all along the Nain-Baft fault to join again the Baluchestan occurrences. The junction with Baluchestan, in the area west of the Jaz Murian depression, appears offset along a series of north-trending right-lateral wrench faults (related to the Oman line of Gansser, 1955), and it seems quite possible that here the Baluchestan mélange has been brought into contact with the Zagros-Oman ophiolite belt; if so, a clear discrimination of the two has not been possible as yet in this structurally extremely complicated corner of Iran.

The other mélange zone is situated in the area west of Lake Rezayeh in northwest Iran, extending westward far into Turkish territory. Here again, it seems possible that in its southern part this mélange zone comes into close contact with the ophiolite zone of the Zagros.

In addition to the ophiolite zones mentioned so far, two small and widely separated ophiolite exposures, which may or may not be structurally interrelated, have become known in the far north of Iran, at the northern foot of the Alborz Range. One, only recently discovered by Davies et al. (1972), is situated in the Caspian foothills southwest of Rasht; the other one, recently studied by M. Davoudzadeh but not yet described in any published record, is found at Mashhad at the northern foot of what may be considered as a far eastern extension of the Alborz Range. Both appear associated with metamorphic rocks of Precambrian and/or Paleozoic age, and the emplacement of both is likely to have been related to pre-Jurassic tectonic processes. A connection with the mélange-type ophiolites east of Lake Sevan, in the Transcaucasus, is suspected.

In the following discussion we shall examine to what extent the structural evolution of Iran is in agreement or disagreement with the assumption of ancient continental margins and continental separations being outlined by these four geographically and geologically distinct ophiolite zones in Iran.

EPI-BAIKALIAN CONTINENTAL PLATFORM OF ARABIA-IRAN

Precambrian Basement Consolidation. All available data suggest that in late Precambrian and Paleozoic time Iran was an extension of the Arabian continental platform. As seen in numerous outcrops, a crystalline basement underlies a widespread Infracambrian-Cambrian sedimentary cover of epicontinental character, throughout the greater part of Iran, as it does in Arabia (Stöcklin, 1968b). Consolidation of the basement by metamorphism and partial granitization, and at least in eastern parts of central Iran also by intense folding (Huckriede et al., 1962), took place in late Precambrian time. Recent (and as yet rather tentative) isotopic work on Iranian basement rocks by A. W. Crawford (personal communication) suggests an age of between some 600 and 1,000 my, and a similar range of isotopic data has been obtained for Arabian shield rocks (Schürmann, 1966). Basement consolidation has been identified by various authors with an "Assyntian" or "Baikalian."

Basement Configuration. Certain large-scale basement irregularities are recognized from facies variations in the oldest, Infracambrian and Cambrian cover sediments and show that the fundamental framework of the present Alpine structure of Iran was inherited from Baikalian basement trends (Fig. 2). Thus the Infracambrian-lower Cambrian Hormuz salt basin of south Iran and Arabia was sharply limited in the east by a north-trending basement swell, the "Oman high." A straight northern extension of this seems to appear in the west margin of the Lut Block of east Iran, which forms the eastern limit of the small Infracambrian salt basin north of Kerman (Stöcklin, 1968a), and of a lower Cambrian gypsum facies north of Tabas (Ruttner et al., 1968). The north-south Oman-Lut trend dominates the Alpine structure of east Iran. Similarly, an extensive northwest-southeast basement threshold, coinciding with what was to become the ophiolite-rimmed Main Zagros Thrust, is clearly reflected as a divide between the dominant evaporite (Hormuz) facies in the southwest and a widespread coeval dolomite facies in central Iran. However, local transitions from one into the other facies are known

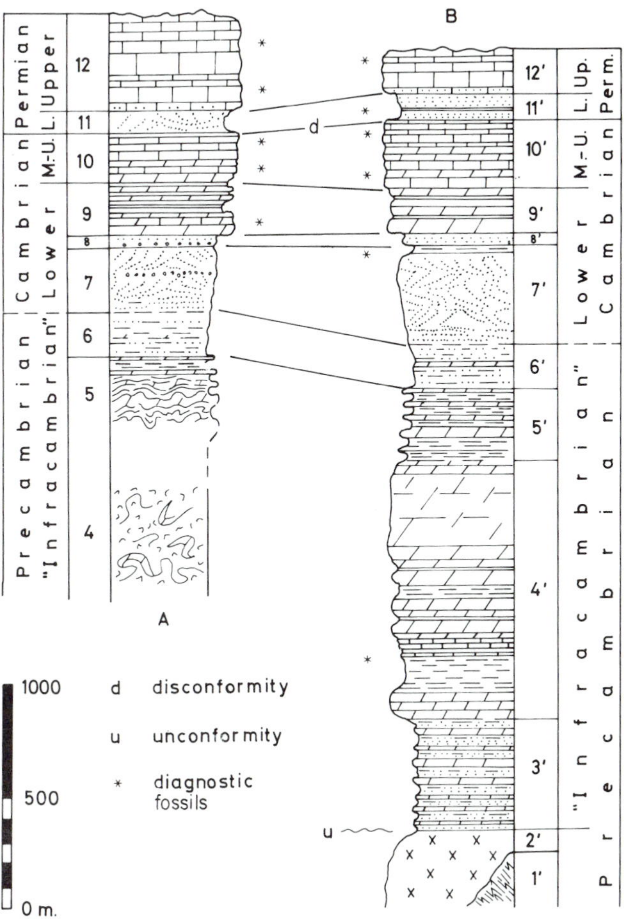

Fig. 3. Comparative sections through epi-Baikalian platform cover on opposite sides of Main Zagros Thrust (see Table 1 for details). A: Bazun Pir, High Zagros; B: Soltanieh Mountains, Northwest Iran (Fig. 2).

on either side of the divide and thus show the essential continuity of early epi-Baikalian environment from Arabia to Iran. In the north of Iran, the Infracambrian-Cambrian deposits, which extend with remarkable facies persistence throughout central Iran, everywhere pinch out against what appears as a formerly elevated Caspian Block; the east-trending, gently arcuate, Alborz Range defines the southern margin of this old block.

Platform Continuity. Except for the partial change from evaporites to dolomite, Infracambrian-Cambrian epicontinental deposition seems to have been perfectly continuous from the Zagros area across the later thrust line to central and north Iran. Figure 3 and Table 1 illustrate this point by confronting two typical sections from opposite sides of the thrust line, 600 km apart.

The platform character of the earliest epi-Baikalian deposits continues through the Paleozoic into the Triassic throughout most of Iran as it does in Arabia (Stöcklin, 1968b). Important sedimentary gaps and regional disconformities are common, but unconformities that would indicate significant crustal deformations do not occur. A widespread Permian transgression characterizes the entire platform. Permian limestones rest with clastic basal beds directly on Devonian or older Paleozoic, or even on basement rocks in most of western Iran (Fig. 3), just as they do in central Arabia; whereas the Paleozoic sequence becomes generally more complete to the east, in eastern central Iran (Ruttner et al., 1968) and, to a lesser extent, also in the eastern Zagros and in eastern Arabia-Oman (Tschopp, 1967).

These facts make us believe that during the Paleozoic, i.e., since at least the Baikalian consolidation of the basement and until the Triassic, Iran formed part of a continuous, undivided Arabian-Iranian platform, and with this, part of Gondwanaland. There is certainly nothing in the geological record to suggest for this time a wide separation of Arabia and Iran such as was implied, e.g., by King (1972).

This conclusion is supported by biogeographical considerations. For the early Cambrian, the distribution of *Archaeocyathids*, *Olenellids*, and *Redlichiids* suggests that Iran belonged to a faunal province that included much of Gondwanaland but was unconnected with northern Asia (Zhuravleva, 1968; Kobayashi, 1972). Similar conclusions were reached at by de Lapparent et al. (1970) with regard to the Carboniferous and Permian of central Afghanistan.

However, the possible post-Paleozoic continental separation indicated by the Zagros ophiolites does follow a line—the Main Zagros Thrust line—which, as we have pointed out, is likely to have been inherited from Baikalian basement structure (Fig. 2), and there is some evidence that this line became tectonically active as early as the Late Paleozoic and perhaps earlier. An unusually shaley and part-

Table 1. Comparative sections through epi-Baikalian platform cover on opposite sides of Main Zagros Thrust (Fig. 3). A: Bazun Pir, High Zagros; B: Soltanieh Mountains, northwest Iran (Figs. 2 and 4).

A	B
12 Dark limestone with fusulinids, corals, brachiopods.	12' Dark limestone with fusulinids, corals, brachiopods.
11 Red-brown quarzitic sandstone and gritstone.	11' Red quartzitic sandstone, shale, some sandy limestone; *Schwagerina, Linoproductus.*
10 Platy dolomite and limestone, passing into flaggy light limestone, partly silty-*glauconitic*, with shale beds, wave marks; layers of sparry limestone. In lower part *Iranoleesia, Anomocare, Lioparella;* in upper part *Chuangia, Briscoia, Idahoia, Billingsella.*	10' Platy dolomite and limestone, passing into yellow, green, and pink thin-bedded limestone, partly silty, with *glauconite*, wave marks; layers of sparry limestone, colored shale. In lower part *Iranoleesia, Anomacarella, Lioparella, Dorypyge;* in upper part *Drepanura, Chuangia, Prochuangia, Briscoia, Idahoia, Kaolishania, Billingsella.*
9 Dark cherty dolomite, some blue-black limestone with *Redlichia*, stromatolites; yellow, green, and pink shales with *salt pseudomorphs*.	9' Dark dolomite with chert, some quartzite, intercalations of yellow, green, and pink shales with *salt pseudomorphs*.
8 Pink to white quartzitic sandstone, chert-pebble beds, cross-bedded.	8' White quartzite, strongly cross-bedded.
7 Purple to pink arkosic sandstone, fine- to coarse-grained, some chert-pebble beds; lower part flaggy, upper part strongly cross-bedded.	7' Purple to pink sandstone, arkosic, fine- to medium-grained, lower part flaggy, upper part strongly cross-bedded. On top shale zone correlative with *Cruziana fascicualta* zone in Alborz and eastern central Iran.
6 Purple shale, partly sandy, strongly micaceous, few thin dolomite bands.	6' Purple shale, silty to fine sandy, strongly micaceous, few thin dolomite bands.
5 Dolomites and dolomitic limestones, laminated, stromatolithic, with chert, solution breccias, and variegated (mainly purple) micaceous shales.	5' Dolomites and dark laminated fetid limestones, with chert, stromatolites, alternating with purple and variegated micaceous shales.
4 Salt with associated dolomites, limestones, shales, (Hormuz Formation), not exposed but inferred from diapiric distortion of overlying beds and presence of salt springs and Hormuz-type salt plugs in vicinity.	4' Upper part massive dolomite with chert, stromatolites. Lower part cherty dolomite with several zones of calcareous, argillaceous, and siliceous shale, *Chuaria circularis.*
	3' Pink sandstone and sandy micaceous shale with intercalations of cherty dolomite.
	________unconformity________
	2' Pink granite intrusive in (1').
	1' Phyllite, mica-schist.

ly volcanic facies of the Permian was recorded by Thiele et al. (1968) and by geologists of the British Petroleum Company (1964), and Permian turbidites were determined by Dimitrijevic (1973), at a few places in the mountain zone just northeast of the Zagros Thrust. Also, an exceptionally strong unconformity between Permian and older Paleozoic beds was noticed by Thiele et al. (1968) near Golpaygan and by A. Haghipour (personal communication) near Rezayeh in the same zone, although the usual conformable relationship appears reestablished in both cases in the near vicinity. These local, but significant, phenomena may perhaps best be explained by differential movements at or near tensional faults that began to form in pre-Permian or early Permian time and may have heralded more important rifting that was to follow. Unfortunately, Alpine metamorphism has obscured much of the evidence in this particular structural zone.

ZAGROS RIFT

Termination of Platform Conditions. In central Iran the long period of tectonic calm that lasted throughout the Paleozoic was abruptly terminated in Triassic time, when important crustal movements initiated a long period of tectonic unrest, culminating in the Alpine orogeny proper in late Cretaceous-Tertiary time. This unrest manifested itself by repeated faulting and folding, magmatism and metamorphism. The initial movements, which can generally be dated as Late Triassic but in places started as early as Ladinian time (Davoudzadeh and Seyed-Emami, 1972), seem to have had primarily disruptive effects, although evidence for a distensional character of these movements is not unambiguous: a slight pressure metamorphism related to this tectonic phase has also been reported (Hushmand-Zadeh, 1973), and volcanic activity was insignificant. However, a sudden change from platform carbonate to paralic and continental clastic deposition, coupled with a distinct disconformity or unconformity, attests to general uplift of central Iran, and its fragmentation into horst- and graben-like blocks is indicated by abrupt thickness changes of the Upper Triassic and Liassic deposits across fault-block boundaries. The most important fracture originated along what is now the Main Zagros Thrust; southwest of it and beyond the ophiolite zone, in the present folded belt of the Zagros and in Arabia, shelf-carbonate deposition continued undisturbedly through the Mesozoic to the Tertiary, placing this region into marked contrast to central Iran from the time of separation in the Triassic.

Main Zagros Thrust Line. Originally, the Main Zagros Thrust was identified with a simple thrust line which had appeared on early geological maps and showed a conspicuous straight alignment; faulting was throught to have been related to downwarping was thought to have been related to downwarp-underthrusting below central Iran (Falcon, 1967; Stöcklin, 1968b). A recent analysis of the thrust line by Braud and Ricou (1971) has shown that we have in fact to deal with two major thrust faults, roughly parallel and sometimes coincident but in places considerably departing from each other, and slightly different in age but both affecting rocks as young as Miocene. The older, southwesterly one is a low-angle thrust, marking the actual southwestward overthrust of central Iran on the Zagros, with a horizontal displacement thought to be at least 40 km. The younger one is a steeply northeast-dipping to subvertical reverse fault with a right-lateral component of unknown magnitude. There is no doubt, however, that the double line of faulting and thrusting is but the reactivation of a much older rift.

Controversy Over Age and Mechanism of Rifting. The dating of the main crustal separation as Triassic from geological evidence in central Iran seems to agree with results of recent studies in the ophiolite-radiolarite zone just southwest of the thrust line by Ricou (1968a, 1968b), and in its continuation in Oman by Glennie et al. (1973). These authors reported Late Triassic fossils as being the oldest autigenous organic remains in the radiolarite suite that is thought to have accumulated in the widening rift. However, while geologists widely agree that the radiolarite-ophiolite complex of the Zagros and Oman were formed in an oceanic trough resulting from some sort of rifting or downbuckling, very different opinions have been expressed as to the age and size of this trough and the mechanism of its formation.

All interpretations depend to a large degree on the question of the true age of the radiolarites. The radiolarite suite (the Pichakun Formation of the Zagros and the corresponding Hawasina Group of Oman) consists of radiolarian chert and associated highly fossiliferous limestone-turbidites and other

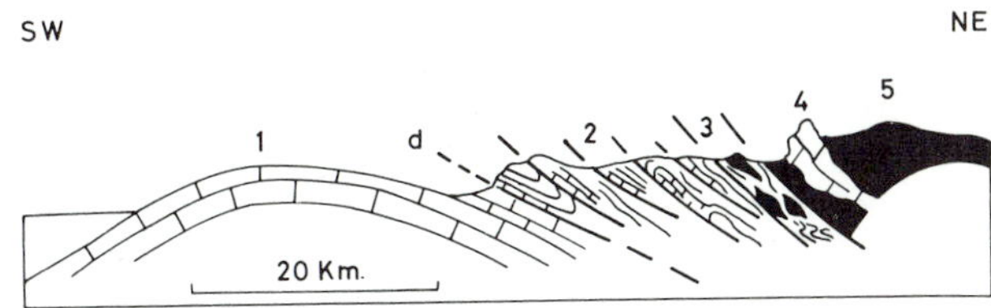

Fig. 4. Schematic section through ophiolite-radiolarite zone at Neyriz, Zagros Crush Zone. 1) Cenomanian-Turonian limestone of Arabian platform cover; 2) limestone turbidites and radiolarites (Pichakun Formation of Ricou, 1968); 3) "Coloured Mélange"; 4) exotic limestone, Triassic; 5) ophiolites; d = disputed contact (? sedimentary transition, ? thrust plane).

sediments, *overlying* with a disputed contact upper Cretaceous shelf carbonates of the Arabian platform (Fig. 4). But while the platform succession ranges into the upper Cretaceous (Turonian at Neyriz, Campanian in Oman), the fauna of the overlying radiolarite suite contains abundant forms of Paleozoic and all possible Mesozoic ages but, as paleontologists insist, has not yielded forms younger than Turonian. And while it is admitted by all that a large portion of the fauna of the turbidites is reworked, a controversy persists over the possible autigenous nature of certain Triassic to middle Cretaceous forms occurring in interbedded micritic limestones.

Accepting their autigenous nature, one also has to accept an allochthonous nature of the radiolarites, which in turn invites construction of nappes involving hundreds of kilometers of horizontal displacement; also, one has, then, to accept the existence of the trough at least as early as Triassic time, and its original position far away from the present location of these trough deposits. If, on the contrary, one believes that *all* organic remains could be reworked, nothing prevents us from assuming that the radiolarites are in normal stratigraphic position and thus have been deposited on the subsiding shelf and margin of Arabia in late Cretaceous time.

Hand in hand with this problem goes that of the age and mode of emplacement of the basic and ultrabasic magmatic rocks (the "Semail ophiolites" of Oman and their counterparts in the Zagros). The bulk of these overlies the radiolarites as an enormous sheet-like mass (Fig. 4).The contact between the sedimentary complex and this ophiolite mass appears usually as a narrow zone of mélange-like intermingling of the two, which in addition includes huge exotic slabs of Permo-Triassic limestone but which obscures rather than clarifies the original relationships between the intermingled rock types. Unambiguous paleontological evidence for the age of the ophiolites is extremely scanty, and such as exists points generally to middle-late Cretaceous age but relates only to the basaltic end members (pillow lavas, tuff breccias, etc.), which in places show normal interbedding with radiolarites, siliceous shales, and pelagic limestones.

According to the writer's own cursory observations in the Zagros and Oman, such synsedimentary associations are occasionally found in the diabasic top part of the ophiolite suite, but otherwise are restricted to the narrow zone of mélange below the main body of the ophiolites; the bulk of the radiolarite suite below the mélange is free of volcanic material. This seems to agree with the observation of Glennie et al. (1973, p. 14) to the extent that volcanic material within the Hawasina (radiolarite) Group appears only in the "highly contorted" units described as the highest Hawasina nappes. According to the same authors (p. 22), "the youngest Semail lavas are dated by middle to late Cretaceous fossils" and (p. 24) "some Semail lavas have been dated by faunas of Coniacian age." Allemann and Peters (1972, p. 675) give a K-Ar age of 96 ± 6 my for a biotite from a Hawasina tuff.

Glennie et al. (1973, p. 14) mention exotic Permo-Triassic limestone blocks that also from Oman "were deposited on a substrate of mafic igneous pillow lavas, now somewhat sheared." From this not wholly convincing evidence, they conclude that "the creation of the Hawasina ocean must have been initiated some unknown time earlier (than Middle Permian)," although they offer (p. 23) an alternative interpretation that would allow dating of this event as "later Permian and Triassic." On the other hand, Ricou (1971b) described a unique case of a thermic contact with formation of skarn minerals between peridotite and a similar exotic limestone in the ophiolite zone of the Zagros.

Allochthony Versus Authochthony. The views of the "autochthonists," which mostly developed before the days of plate tectonics, were more or less explicitly based on conventional "eugeosynclinal" interpretations of the Zagros and Oman ophiolite zones. Thus, for Morton (1959), Falcon (1967), Tschopp (1967), Stöcklin (1968b), Wilson (1969), and others, the radiolarites and ophiolites accumulated in a deep intracratonic trough that developed by strong subsidence of the Arabian plate margin in late Cretaceous time. The ophiolites were thought to have formed in situ as submarine extrusions, as postulated by Dubertret (1955) and Gansser (1955, 1960). With such views, no nappe tectonics were needed to explain the visible structure, although it was implied that an unknown, but supposedly limited, amount of underthrusting of the Arabian basement shield below central Iran had taken place along the Main Zagros Thrust and, accordingly, an unknown, but supposedly limited, width was implied for the marine trough.

The views of the "allochthonists," on the other hand, were in better agreement with, and partly inspired by, the new concepts of plate tectonics and ocean-floor spreading. Reinhardt (1969) developed a model for the origin of the Oman ophiolites from a mid-ocean ridge. For Glennie et al. (1973), the Zagros-Oman trough resulted from continental breakup and subsequent spreading from such a ridge. From palinspatic reconstructions, a width of more than 100 km was estimated by Ricou (1968a) for the Zagros trough, and of more than 1,200 km by Glennie et al (1973) for the Oman trough. Later northeast drift of the Arabian plate with subduction of the oceanic crust (northeast below central Iran in the Zagros, southwest below Arabia in Oman), until collision of the Arabian and Iranian continental margins in the Zagros Crush Zone created the imbrics and nappes in this zone and in Oman. Complex mechanisms were devised by Allemann and Peters (1972) and by Glennie et al. (1973) to explain both partial subduction of oceanic crust below the Arabian continental plate and partial thrusting or gliding upon it. Development in central Iran of a zone of Tertiary acidic magmatism, and related copper mineralization, northeast of and parallel to the hinge line of subduction was seen by Vialon et al. (1972) and Crawford (1972) as the reflection of a northeast-dipping Benioff zone, whereas Takin (1972a) related the volcanicity to zones of weakness and to temporary relaxation of compressional movements after continental collision.

Plate tectonics, evidently, has placed the problem of an ancient Iranian continental margin along the Zagros Thrust in an entirely new light, but is still far from solving it in an unambiguous way. Both the "allochthonists" and the "autochthonists" are faced with puzzling difficulties. The former find it difficult to provide convincing answers to such questions as: Why does the postulated tectonic contact between the "allochthonous" radiolarites and the "autochthonous" substratum have the appearance of a sedimentary transition in many places? Why does the substratum of the nappe complex nowhere appear to be truncated or otherwise deformed in a way that should be expected from rocks overridden by huge nappes? Several supporters of nappe structures in Oman (Allemann and Peters, 1972; Glennie et al., 1973) have tried to resolve these inconsistencies by invoking passive gravity sliding rather than thrusting by active compressional forces as the main mechanism for the emplacement of the nappes; admitting this, however, would mean abandoning one of the strongest geological arguments supporting those tangential forces which are needed to explain continental drift: subduction and continental collision. The adherents of an autochthonous origin of the radiolarites and ophiolites, on the other hand, so far have been unable to disprove a pre-late Cretaceous age of the radiolarite suite as would be necessary to support their case.

Fossil Subduction Zone. On one point, though, all investigators agree: the formation of the radiolarite imbrics or nappes and the emplacement of the ophiolites to their present position on top of the radiolarites, in whatever way these processes are explained, took place in late Cretaceous time and was completed in the Campanian or early Maestrichtian. Shallow-water limestones or Maestrichtian age (late Maestrichtian according to Allemann and Peters, 1972) overlap transgressively and unconformably those rocks and structures. From that time on, the Arabian and Iranian continental plates were joined again, welded together by a rudimentary strip of former oceanic crust, and the Zagros Thrust line had ceased to play its role as a continental margin of Iran. The later history of the Crush Zone does not interest us in the context of this paper.

One question, however, should be briefly touched. Is the suture between the two continental plates still tectonically active? If subduction has occurred, is it still continuing: Records of earthquakes, both historical (Ambraseys, 1971) and recent (Nowroozi, 1971), show that the Zagros is a zone of continuous high seismicity. But Nowroozi's epicenter distribution map can in no way be interpreted as reflecting an active Benioff zone dipping from the Zagros Thrust line, northeast below central Iran; in fact, the great majority of the epicenters lies southwest of the Crush Zone in the broad belt of gentle folding. The ophiolite scar of the Zagros suggests, if any, a stabilized fossil subduction zone.

CENTRAL-AND-EAST IRANIAN MICROCONTINENT

Definition and Structure. The disposition of colored mélange outcrops in central, east, and southeast Iran (Fig. 1) suggests that they belong to one and the same, though perhaps not entirely continuous ophiolite belt. The largest outcrop interruptions in this "ophiolite ring" are found in the northwest and northeast. The northwestern one may be due simply to coverage of the Great Kavir (salt desert) of central Iran by the thick Neogene deposits; the northeastern one lies in an area that is geologically still little explored. In the south, the belt seems to be offset by a bundle of right-lateral faults connected with the important north-trending Nayband fault of east Iran.

The fault-bounded area encircled by this ophiolite belt has been conceived of by Takin (1972a) as a large fragment of the central Iranian continental plate, which for some time in the geological past was isolated from the rest of Iran by a narrow oceanic trough; he named it the central-and-east Iranian Microcontinent.

This roughly circular "microcontinent" measures between 500 and 800 km across and is divided by the Nayband fault into two structurally rather different halves. The eastern half constitutes the north-south-orientated rigid "Lut Block," characterized by a conspicuously low degree of Alpine deformation. Most of the block is covered by Tertiary volcanic and continental deposits, with some widely scattered outcrops of the Mesozoic and Paleozoic substratum, and a nucleus of metamorphic rocks believed to represent the Baikalian basement near the center. The western half shows, from east to west, a succession of fault-bounded, graben-like features, the horsts exposing the Baikalian basement and its Paleozoic cover, the grabens filled with thick and strongly folded Jurassic and Cretaceous sediments. Faults and trough axes are arcuate, with westward convexity increasing as one proceeds to the west, so that the western marginal structures become roughly parallel with the crescent-shaped western boundary of the microcontinent itself.

Basement Trends and Early History. The Paleozoic sequence, well exposed in the numerous horsts, shows the customary succession and facies of the epi-Baikalian platform cover of Iran. The variations that do occur seem to be largely influenced by some of the basement uplifts, which indicates that they are Baikalian relict structures. As in other parts of Iran, strong block faulting, in part reactivating basement faults, occurred in late Triassic and created the arcuate graben pattern, mentioned above, which governed sedimentation in Jurassic time. From this it appears that the large crustal block constituting the "microcontinent" was produced by the same Triassic disruptive movements that created the Zagros rift and the fault-block pattern of all of central Iran. There is, however, no evidence that the circular rift around the microcontinent developed into an oceanic trough at that time: no Triassic and (with an exception discussed below) no Jurassic sediments are found in the trough deposits, and the microcontinental area continued to participate intimately in the further structural development of central Iran. The lower Jurassic shows the same plant- and coal-bearing paralic facies within the microcontinent and outside of it in the remainder of central Iran as well as in the Alborz, and a marine transgression reached the microcontinent from the Alborz region in Toarcian-Bajocian time (Seyed-Emami, 1971).

Distinct folding at about the Jurassic-Cretaceous limit, and related granite intrusions, followed by regional emergence, peneplanation, and widespread deposition of continental redbeds, was followed by renewed marine transgression in Barremian-Aptian time. All these processes affected the microcontinent and the rest of central Iran alike.

Harirud Lineament. Significant differences, however, became apparent when comparing the Jurassic and lower Cretaceous deposits on either side of the north-trending eastern segment of the "ophiolite ring," in the Lut Block of east Iran (Stöcklin et al., 1972) and in the Farah Block of western Afghanistan, south of the Hindukush line (Weippert et al., 1970). The plant-bearing paralic facies of the lower-middle Jurassic, still well developed in the Lut Block is entirely missing in the Farah Block although widespread north of the Hindukush, and the upper Jurassic-lower Cretaceous sequences of the two blocks seem to have little in common. We suspect that these contrasts in facies are related to a southern extension of an old north-south lineament, the "Harirud fault" of Khain (1969), recognized as an important geophysical discontinuity that crosses the Turan plate of central Asia, from the southern tip of the Ural Mountains to the Iran-Afghanistan border (Volvovsky et al., 1966), and which also is apparent in the abrupt change in the general structural trend all along this country border (Figs. 1 and 2). The north-trending eastern segment of our ophiolite belt is apparently related to this important pre-Jurassic lineament.

"Mélange Ring". The ophiolite belt itself reveals a conspicuous uniformity throughout. A stratigraphic succession, or structural zonation such as is found persistently in the Zagros-Oman belt, is nowhere recognized. Larger, elongated ultrabasic bodies up to some 10 km long do occur in places (Davoudzadeh, 1972), but in general the complexes form those chaotic colored mélanges of igneous and sedimentary material, in which original relationships appear hopelessly destroyed. In spite of this disorder, one notices in places clear synsedimentary associations of spilitic lavas (often with pillow structure), diabasic agglomerates and tuffs, radiolarian chert, more-or-less tuffaceous shales, and characteristic white or pink, dense, siliceous or marly limestones which often contain a diagnostic pelagic microfauna of late Cretaceous age, covering the whole range from late Turonian or early Coniacian to late Maestrichtian (Davoudzadeh, 1972; Dimitrijevic, 1973).

The oldest datable rocks found so far in the mélange are *Calpionella*-bearing limestones, associated with radiolarites in the highly fractured area west of the Jaz Murian, where the mélange ring seems to merge with the Zagros-Oman ophiolite belt; conspicuously larger, homogenous, ultrabasic bodies, reminiscent of the Oman ophiolites, appear in the same area, and an unusual admixture of basaltic-spilitic material is found in Jurassic sediments nearby (Vialon et al., 1972). Farther north, however, the only dated pre-Turonian components of the mélange belt, all occurring near its margin and possibly (but not certainly) representing exotic material, are less than half a dozen of limestone blocks containing a Neocomian microfauna. Apart from these, and occasional blocks and slices of metamorphic rocks representing exotic older material, the bulk of the sediments of the mélange ring is of late Cretaceous, essentially Senonian-Maestrichtian age. Significantly, the Senonian was a time of nondeposition through most of this microcontinent!

The upper age limit of the mélange is less evident than that of the ophiolite-radiolarite complex of the Zagros-Oman belt. Occasionally blocks of limestones as young as Lutetian are found intermingled in the mélange, and in Baluchestan and east Iran, the mélange seems to pass without sharp limit into a Lower Tertiary flysch succession, several kilometers thick (Gansser, 1960). But, in other places, particularly at the inner margin of the belt in east Iran (Stöcklin et al., 1972), and also in the Sabzevar area in the north (J. Eftekhar-nezhad, personal communication), shallow-water limestones of late Paleocene to early-middle Eocene overlap the mélange with pronounced unconformity. Evidently, the admixture of such limestones in the mélange was the result of later (post-middle Eocene) tectonic reactivation, a conclusion also reached by Dimitrijevic (1973).

Separation and Isolation of the Microcontinent. With the evidence given above, we may assume that the narrow oceanic trough that surrounded the microcontinent began to form in late Jurassic time as a branch of the Zagros-Oman trough, opening progressively from south to north, and reaching its full development in the late Cretaceous. Rifting along a preexisting arcuate line of faults and subsequent widening of the rift may have been related to the formation of new oceanic (basaltic) crust, possibly from a median ridge, in a way comparable to the Red Sea as suggested by Takin (1972a). This narrow circular or crescent-shaped trough isolated or nearly isolated the central-and-east Iranian Microcontinent from the remainder of the Iranian continental plate. Its isolation, accomplished by Turonian time, was accompanied by its emergence: it became a true, small continental landmass, washed on all sides by ocean water. Deposition of radiolarian chert and other sediments and intermittent effusion of spilitic lava continued until the end of the Cretaceous in the surrounding trough.

Formation of the Colored Mélange. Most investigators agree that synsedimentary slumping and intermingling of unconsolidated, water-saturated material, as well as embedding of consolidated exotic blocks, have played a role in the formation of the mélange. But subsequent narrowing and compression of the trough seem to have been mainly responsible for a tectonic emplacement of the ultrabasic blocks (which nowhere show intrusive relationship with the sediments), and for the final crushing of the complex trough deposits into a huge tectonic breccia. This is particularly evident at the eastern margin of the microcontinent (Stöcklin et al., 1972), where the junction between the rigid Lut Block and the mélange zone is an extremely sharp line of intense faulting, thrusting, and shearing. Here, even a distinct high-pressure metamorphism has affected the rocks on both sides of the junction, with transformation of the mélange into countless imbrics and slices of phyllite, micaschist, metadiabase, amphibole-schist, blueschist, marble, serpentinite, etc. This process of mélange formation may have been continuous during the latest Cretaceous, but probably reached a short climax in a final stage in the early Paleocene, which had its repercussions all over central and north Iran, where the first pronounced folding of the Alpine orogeny took place at that time.

Reintegration of the Microcontinent. These early Paleocene processes were terminated by regional uplifting; but in the broader parts of the mélange belt in the south and east, uplifting and emergence affected only the inner marginal zone of the belt, while subsidence continued in the axial parts of the marine trough, and thick flysch deposits accumulated in the Eocene. Complete reintegration of the microcontinent in the Iranian-Afghanian continental realm was thus achieved only with the closure of the eastern flysch trough in late Eocene or Oligocene time, whereas in the south a remnant of the trough still exists in the present Gulf of Oman.

The microcontinent itself, except for a brief episode of incomplete submergence in the Maestrichtian, was never flooded by the sea since its isolation and emergence in the middle Cretaceous. The widespread Eocene marine transgression encroached on a shelf-like marginal strip all around, but did not penetrate into the interior of the microcontinent, and a last transgression from the west in the late Oligocene and early Miocene just reached its western and southwestern limit, but did not surpass it.

LATE MESOZOIC IRANIAN-ANATOLIAN OCEANIC TROUGH

In northwest Iran, bordering Turkey, ophiolite-radiolarite complexes occur in a number of irregular outcrops extending west into Anatolia. Numerous other, disconnected exposures exist farther west in inner Anatolia (Ankara mélange), but it seems to be difficult to group them into consistent zones. The ages and origins of the Anatolian ophiolite-radiolarite formations have been highly controversial issues (see, e.g., the article of Brinkmann, 1972; Brunn and Monod, 1973), whereas the study of the occurrences in northwest Iran has only just started (Alavi and Bolourchi, 1973). Conclusions bearing on the subject of this paper could, therefore, be only speculative.

The occurrences on the Iranian side have a general resemblence to those of the "mélange ring" in central-east Iran, in that they show similar rock types in a similar disorderly array. Like the latter they contain, in addition to the ubiquitous ultrabasic rocks and radiolarites, abundant diabasic and tuffaceous material, flysch-type shales, pelagic limestones, and occasional conglomerates, and the mélange is overlain by thick Eocene flysch. South of these mélange occurrences, in the wider Sanandaj area, basic and intermediary volcanic rocks are known to be associated with upper Jurassic sediments (Braud and Bellon, no date), with Aptian limestones, and with thick upper Cretaceous calcareous and siliceous shales similar to shales occurring in the mélange. These volcanic rocks may be significant as possibly indicating tensional movements in Late Jurassic and Cretaceous time. However, the mélange itself has not yielded fossils older than late Cretaceous (Coniacian) but includes flysch-type sediments as young as Paleocene (J. Eftekhar-nezhad, personal communication).

As has been mentioned previously, this mélange zone, like the mélange ring in the east, was probably, but not necessarily, connected with the Zagros ophiolite trough in the south. A connection with the mélanges along the important right-lateral North Anatolian fault can only be suspected.

POSSIBLE GONDWANA MARGIN IN THE NORTH

South Caspian Block. The Paleozoic (epi-Baikalian) platform deposits of Iran are still well developed in the Alborz Range in the north, but they show significant changes in the north flank of the range, which faces the Caspian depression. Here the Infracambrian-Cambrian basal part of the sequence rapidly thins and pinches out with approach to metamorphic basement outcrops in the Caspian foothills. These basement rocks are overlain directly by younger Paleozoic or Mesozoic beds. The same is observed farther northwest in the Transcaucases and adjoining parts of northwestern Iran. Here, too, metamorphic basement rocks are covered by sedimentary rocks not older than Devonian. From this we might assume the existence of an old "Transcaucasian-South Caspian Block" (Fig. 2), which played the role of an emerged continental feature in Infracambrian and Cambrian time and which continued to influence the sedimentary and structural history of northern Iran through the Paleozoic and into the Alpine period. An emerged median-mass-like South Caspian Block was also inferred from geophysical data by Malovitski (1968) for late Paleozoic and early-middle Mesozoic time.

All this, however, seems to be in contradiction with the alleged oceanic type of crust underlying the South Caspian depression: deep-seismic sounding in the southern U.S.S.R. suggested that in the Kura depression, west of the southern Caspian, the "granitic" layer is rapidly thinning eastward and is missing altogether below the South Caspian depression (Galperin et al., 1962).

Recently obtained surface and subsurface data have thrown some new light on this puzzling problem. A deep well in the Kura depression reached basalt after passing through a thick Tertiary-Cretaceous sequence (N. K. Kurbanov, personal communication). The deepest well in the Middle East, drilled by the National Iranian Oil Company in the Gorgan Coastal Plain, east of the southern Caspian Sea, was abandoned in Lower Jurassic sediment at a depth of over 5,800 m; if the "oceanic" crust exists there, too, it must be pre-Jurassic. Davies et al. (1972) reported ultrabasic rocks including serpentinites, as being tectonically associated with the greenschists, amphibolites, and gneisses that constitute the Caspian foothills west of Rasht, where they are overlain directly by Jurassic continental deposits. To the southwest these rocks were found faulted against a compressed Ordovician-to-Permian succession, unusually rich in volcanic (basaltic, andesitic) material, with sheared serpentinite lining the fault. Pebbles of ultrabasic rocks were found in Jurassic conglomerate nearby, confirming their pre-Jurassic age. Could this pre-Jurassic ophiolitic material be related to the "oceanic" crust of the southern Caspian?

No other ophiolitic rocks have become known in the densely forested foothills bordering the southern Caspian Sea, and we must go 1,000 km to the east to find the only other known ophiolite exposure that can be compared to those west of Rasht, but its being situated at the faulted northern foot of the far eastern extension of the Alborz Range. But before describing this we have to turn briefly to the structures farther north.

Epi-Hercynian Platform of Turan. North of the eastern Alborz Range, and separated from it by the narrow depression of Bojnurd-Mashhad, follow the gentle folds of the Kopet Dagh Mountains, which consist of a thick, conformable sedimentary sequence ranging from Jurassic to Miocene. The folds fan-out in the west against the South Caspian Block, and border in the north, along a major fault or flexure on the Kopet Dagh foredeep, a marginal depression in the Turan block of central Asia. The Turan plate is described in Russian literature as an epi-Hercynian platform, whereby it is understood that the folded Hercynian substratum includes rocks as young as late Triassic. Hence the platform cover starts with the Rhetian or Liassic, which is known from geophysical and borehole information to overlie Triassic and older rocks with a pronounced unconformity (Volvovsky et al., 1966). A true Hercynian (late Paleozoic) diastrophism is indicated by several unconformities in the Carboniferous-Permian succession, and by a widespread red clastic and partly volcanic "molasse" facies in the Upper Permian-Lower Triassic, as found in outcrops on the Mangyshlak Peninsula of the northeastern Caspian, in the Kyzyl-Kum region, in the western Tien-shan spurs and elsewhere, and in many boreholes in the Kara Kum Desert.

The faulted junction between the Turan block and Kopet Dagh is thus conventionally taken as the boundary between the Hercynian and Alpine realms, a boundary that is drawn westward along the northern foot of the Great Balkhan and across the Caspian Sea (north of the South Caspian Depression and the Mid-Caspian Swell) to the junction of the Great Caucasus Mountains with the Scythian block.

The only exposure of pre-Jurassic rocks in the Kopet Dagh Mountains is found at its eastern extremity, at Aghdarband east of Mashland, where the Kopet Dagh folds die out and pass eastward into the Cretaceous plateau of Afghan Turkestan. Here, at Aghdarband, subhorizontal Liassic beds rest with sharp unconformity on an intricately folded clastic and partly volcanic sequence, dated by rare marine intercalations as middle and later Triassic, and underlain by a great thickness of red conglomerates and sandstones of probably Early Triassic and possibly Permian age, the base not being exposed. This ?Permian-Triassic sequence has nothing in common with the predominantly carbonatic shelf deposits on top of the epi-Baikalian platform succession of Iran, which is still well developed in the Alborz Range, but it resembles closely the above-mentioned late Hercynian "molasse" deposits described from the Turan plate, and also the Trias of northern Afghanistan. From this one should conclude that the junction between the epi-Baikalian platform of Iran and the Hercynian substratum of Turan, passes, not along the northern foot of the Kopet Dagh, but between Kopet Dagh and eastern Alborz, continuing eastward in the Herat-Hindukush line. In this connection the ophiolite occurrence at Mashhad in Iran gains some significance.

These ophiolites appear at the faulted northern foot of the Binalud uplift of the easternmost Alborz Mountains, associated with metamorphic rocks. The following description is based on recent unpublished work by M. Davoudzadeh and coworkers. The metamorphic rocks, which form the northeastern part of the uplift, consist of phyllites and schists with occasional bands of crystalline limestone and quartzite. The ophiolites are partly serpentinized but otherwise rather fresh dunites, peridotites, and pyroxenites (harzburgites) and have the appearance of being interlayered with schists. Both schists and ultrabasics are reworked in lower Jurassic conglomerates occurring nearby, which proves their pre-Jurassic age. The metamorphic rocks have conventionally been regarded as Precambrian, but a Paleozoic age cannot be ruled out. They have, in any case, lithologically nothing in common with a fairly complete, chiefly carbonatic lower-middle Paleozoic sequence exposed in the southern flank of the Binalud uplift (M. Davoudzadeh and M. Shahrabi, personal communication), which has all the characteristics of the epi-Baikalian platform-cover succession of Iran, including Infracambrian and fossiliferous Cambrian beds almost identical with those shown in Figure 3 from outcrops 1,000 km to the west and southwest.

Paleo-Tethys in the North? The Alborz Range, including its Binalud extension in the east, seems thus to mark the northern termination of the epi-Baikalian platform regime of Arabia-Iran. Today, this Paleozoic margin is faulted against two very different tectonic units: the South Caspian "oceanic" block and, farther east, the Kopet Dagh foldbelt, marking the continental Turan plate with its Hercynian substratum.

Could the South Caspian oceanic block represent the relict of a Paleozoic oceanic crust? Could a wedge-shaped eastern extension of this oceanic plate fragment be preserved *below* the Kopet Dagh, buried here as in the Caspian under a thick pile of Mesozoic sediments? Could the ophiolite exposures at Rasht and Mashhad, and even those at Lake Sevan, represent upfaulted slices of such an oceanic plate fragment? Could this oceanic plate fragment, separating the continental plates of Iran and Scythia-Turan, be the remnant of a "Paleo-Tethys" which separated Gondwanaland from Angaraland in Paleozoic time?

Such a separation, and a very wide one, was postulated by Peive (1969) and suspected by Takin (1972b), and from biogeographic considerations Lapparent et al. (1970) placed the Carboniferous-Permian Tethys to the north of the Hindukush. Peive explained the narrowing and disappearance of this ancient ocean by a gigantic obduction of the southern continent (including Iran) until its collision with the epi-Hercynian platform in the north. However, in relating this event to the Alpine mélanges, Peive dated the closing of the ocean along the Caucasus-Kopet Dagh suture as late Mesozoic to early Tertiary, a conclusion that is contradicted by two facts: (1) the absence of any trace of deformation between late Triassic and late Tertiary in the Kopet Dagh; and (2) the face that the lower Jurassic continental deposits of central and north Iran contain a Central Asiatic flora (Assereto et al., 1968), and extend uninterruptedly through the Kopet Dagh to Turan, suggesting that Iran and Turan were a united landmass in early Jurassic time.

The hypothetic Tethys, north of the Alborz Range, would thus have disappeared by Liassic time, be it by simple regression or as a consequence of compression and subduction. A northward subduction could have left its trace in the Hercynian deformation of the Scytho-Turanian continental plate, and a continental collision east of an "unsubduced" Caspian oceanic crustal wedge could have caused the intricate Late Triassic fold and fault structures revealed in the Aghdarband exposure, near the suspected line of collision. The eastern continuation of the old Gondwana continental margin could follow the Herat-Hindukush lineament (Lapparent, 1972). But apart from the meager surface evidence, the subject of pertinent study in the suspected suture zone in northeastern Iran is buried under a thick cover of post-Hercynian sediment, leaving the geologist lost in pure speculation.

CONCLUSIONS

Geological evidence suggests that during the Paleozoic, Iran formed a northern extension of the epi-Baikalian continental platform of Arabia as a part of Gondwanaland. If any substantial separation between Gondwana and Eurasia existed in the Paleozoic, it must have been north of Iran.

The South Caspian depression with its oceanic type of crust may represent a relict of such a northern "Paleo-Tethys." Its closure, short of the Caspian relict, must have been completed in late Triassic time and may have been achieved by northward subduction below the Scytho-Turanian continental plate, leading to the Hercynian deformation of the latter and to the continental collision east of the Caspian relict, along a Binalud-Hindukush suture line.

A rift in the Arabian-Iranian, epi-Baikalian platform along the Main Zagros Thrust line seems to have originated in Triassic or possibly late Paleozoic time and to have led to drifting apart of the Arabian and Iranian continental fragments, with formation of a new oceanic trough between the two. If the formation of this "Neo-Tethys" took place immediately after rifting (which is disputed), closing of the "Paleo-Tethys" in the north and widening of the "Neo-Tethys" in the south could have been interrelated and simultaneous processes, caused by a northward drift of Iran.

A further breakup of the Iranian plate seems to have isolated temporarily, in late Mesozoic time, a Central-and-East Iranian Microcontinent, encircled by a narrow ocean-trough, possibly a branch of the Neo-Tethys. Other temporary troughs, possibly all branches of the Neo-Tethys, are indicated for the late Mesozoic in northwest Iran and Anatolia.

Northward-drift of Afro-Arabia may have led to the narrowing of the Neo-Tethys and collision of the Arabian with the Iranian continental margin along the Zagros Crush Zone in Late Cretaceous time, followed by compression of the Iranian plate, closing of the remaining Neo-Tethys branches in northwest and east Iran, reintegration of the central-and-east Iranian Microcontinent, and folding of central and north Iran during the Paleocene paroxysm of the Alpine orogeny.

These processes can be viewed as the result of a complex interplay among the movements of the surrounding continental masses of Africa, India, and Eurasia as conceived in various ways by Peive (1969), McKenzie and Sclater (1971), Takin (1972), Nowroozi (1972), Crawford (1972), Gansser (1966, 1973), and others. The purpose of the present study was to show how strongly the presence of ancient continental margins in Iran, and their place in time and space as implied in these concepts, are supported or contradicted by Iranian geology. One of the obvious conclusions is that exposed geology alone, while providing much pertinent and exciting information, will be unable to give the final answer.

ACKNOWLEDGMENTS

The writer expresses his gratitude to N. Khadem, Managing Director of the Geological Survey of Iran, for authorizing publication of this paper; to many of his colleagues in the Iranian Geological Survey for discussing pertinent questions, and particularly to J. Eftekhar-nezhad and M. Davoudzadeh for releasing unpublished observations of their own; and to M. Takin for critically reading and improving the manuscript. I also thank Mobil Oil Corp. (J. D.

Moody and C. A. Burk) for the opportunity offered to participate in a geological visit to the Oman Mountains; to the Geological Institutes of the Academy of Sciences of the U.S.S.R. (A. V. Peive), and of the republican Academies of Georgia, Azerbaijan, and Tajikistan, as well as to the Afghan Geological Survey Department (H. E. S.-H. Mirzad), and the French Geological Mission in Afghanistan (Abbe A. F. de Lapparent) for enabling various geological visits to the respective countries, which have most usefully supplemented the writer's Iranian experience, on which this paper is based.

Published by permission of the Government of Iran and the United Nations; the contents of the paper are not necessarily endorsed by either organization.

BIBLIOGRAPHY

Alavi, M., and Bolourchi, M. H., 1973, Maku quadrangle map, scale 1:250,000, with explanatory text: Geol. Survey Iran, Geol. Quadrangle A1.

Alleman, F., and Peters, T., 1972, The ophiolite-radiolarite belt of the North Oman Mountains: Eclogae Geol. Helv., v. 65, no. 3, p. 657-697.

Ambraseys, N. N., 1971, Value of historical records of earthquakes: Nature, v. 232, no. 5310, p. 375-379.

Argand, E., 1924, La tectonique de l'Asie: Compt. Rend. XIIIe Congr. Géol. Internatl., 1922, Liège, p. 171-372.

Assereto, R., Barnard, P. D. W., and Sestine, N. F., 1968, Jurassic stratigraphy of the Central Elburz (Iran): Riv. Ital. Paleontol. Strat., v. 74, no. 1, p. 3-21.

Braud, J., 1970, Les formations du Zagros dans la région de Kermanshah (Iran) et leurs rapports structuraux: Compt. Rend., t. 271, p. 1241-1244.

———, and Bellon, H., no date, Données nouvelles sur le domaine métamorphique du Zagros (Zone de Sanandaj-Sirjan) au niveau de Kermanshah-Hamadan (Iran), nature, âge et interprétation des séries métamorphiques et des intrusions, évolution structurale: Fac. Sci. d'Orsay, Paris, 14 p.

———, and Ricou, L. E., 1971, L'accident du Zagros ou Main Thrust, un charriage et un coulissement: Compt. Rend., t. 272, p. 203-206.

Brinkmann, R., 1972, Mesozoic troughs and crustal structure in Anatolia: Geol. Soc. America Bull., v. 83, p. 819-826.

British Petroleum Company, Ltd., 1964, Geological maps, columns and sections of the High Zagros of southwest Iran: London, Brit. Petr. Co., Ltd.

Brunn, J. H., and Monod, O., 1973, Mesozoic troughs and crustal structure in Anatolia: discussion: Geol. Soc. America Bull., v. 84, p. 1477-1480.

Crawford, A. R., 1972, Iran, continental drift and plate tectonics: 24th Internatl. Geol. Congr., Montreal, sect. 3, p. 106-112.

Davies, R. G., Jones, C. R., Hamzepour, B., and Clark, G. C., 1972, Geology of the Masuleh sheet, 1:100,000, Northwest Iran: Geol. Survey Iran Rept. 24, 110 p.

Davoudzadeh, M., 1972, Geology and petrography of the area north of Nain, central Iran: Geol. Survey Iran Rept. 14, , 89 p.

Davoudzadeh, M., and Seyed-Emami, K., 1972, Stratigraphy of the Triassic Nakhlak Group, Anarak region, central Iran: Geol. Survey Iran Rept. 28, p. 5-28.

Dietz, S., and Holden, J. C., 1970, Reconstruction of Pangaea: breakup and dispersion of continents, Permian to Present: Jour. Geophys. Res., v. 75, no. 26, p. 4939-4956.

Dimitrijevic, M. D., 1973, Geology of Kerman region: Geol. Survey Iran Rept. Yu/52-1973, 334 p.

Dubertret, L, 1955, Geologie des roches vertes du nord-ouest de la Syrie et du Hatay: Notes Mém. Moyen Orient, t. 6, p. 2-179.

Falcon, N. L., 1967, The géology of the north-east margin of the Arabian basement shield: Advan. Sci., Sept. 1967, p. 31-42.

Galperin, E. I., Kosminskaya, I. P., and Krakshina, P. M., 1962, Main characteristics of deep waves registered by deep seismic sounding in the central part of the Caspian Sea, *in* Deep sounding of the earth's crust in the USSR: Moscow (Gostoptekhizdat) (in Russian).

Gansser, A., 1955, New aspects of the geology in central Iran: 4th World Petr. Congr., Rome, Proc., sect. I/A/5, p. 280-300.

———, 1960, Ausseralpine Ophiolithprobleme: Eclogae Geol. Helv., v. 52, no. 2, p. 659-680.

———, 1966, The Indian Ocean and the Himalayas, a geological interpretation: Eclogae Geol. Helv., v. 59, no. 2, p. 831-848.

———, 1973, Orogene Entwicklung in den Anden, im Himalaja und den Alpen, ein Vergleich: Eclogae Geol. Helv., v. 66, no. 1, p. 23-40.

Glennie, K. W., Boeuf, M. G. A., Hughes Clarke, M. W., Moody-Stuart, M., Pilaar, W. F. H., and Reinhardt, B. M., 1973, Late Cretaceous nappes in Oman Mountains and their geologic evolution: Am. Assoc. Petr. Geol. Bull., v. 57, no. 1, p. 5-27.

Huckriede, R., Kursten, M., and Vezlaff, H., 1962, Züir Geologie des Gebietes zwischen Kerman and Sagand (Iran): Geol. Jahrb., Beif. 51, 197 p.

Hushmand-zadeh, A., 1973, Review of problems of metamorphism in the zone of Shar-e-Kord-Sanandaj, *in* The first Iranian geological symposium (abstr.): p. 1: Tehran, Iran Petr. Inst. (in Persian).

Khain, V. E., 1969, Fundamental traits of the structure of of Alpine belt of Eurasia in the region of the Near and Middle East, pt. 2: Moscow Univ. Bull. (Geol.), no. 1, p. 3-25 (in Russian).

King, L. C., 1972, An improved reconstruction of Gondwanaland, *in* NATO Symp. on Continental Drift, Newcastle upon Tyne, Apr. 1972: Scottish Academic Press.

Kobayashi, T., 1972, Three faunal provinces in the Early Cambrian: Proc. Japan Acad., v. 48, p. 242-247.

Kushan, B., 1973, Stratigraphie und Trilobitenfauna in der Mila-Formation (Mittelkambrium-Tremadoc) in Alborz-Gebirge (N-Iran): Palaeontographica, Band 144, Abt. A, p. 113-164.

Lapparent, A. F. de, 1972, L'Afghanistan et la dérive du continent indien: Rev. Géogr. Phys. Géol. Dyn., v. XIV, fasc. 4, p. 449-455.

———, Termier, H., and Termier, G., 1970, Sur la stratigraphie et la paléobiologie de la série permo-carbonifère du Dacht-e-Nawar (Afghanistan): Bull. Soc. Géol. France, ser. 7, t. 12, no. 3, p. 565-572.

Malovitski, Ya. P., 1968, History of the geotectonic development of the depression of the Caspian Sea: Izv. Akad. Nauk S.S.S.R. Ser. Geol., no. 10, p. 103-120 (in Russian).

McKenzie, D., and Sclater, J. G., 1971, The evolution of the Indian Ocean since the Late Cretaceous: Geophys. Jour., v. 25, p. 437-528.

Morton, D. M., 1959, The geology of Oman: 5th World Petr. Congr., New York, Proc., sect. 1, paper 11, 14 p.

National Iranian Oil Company, 1959, Geological map of Iran, scale 1:2,500,000: Tehran, Natl. Iran. Oil Co.

Nowroozi, A. A., 1971, Seismo-tectonics of the Persian plateau, eastern Turkey, Caucasus, and Hindu-Kush regions: Seismol Soc. America Bull., v. 61, no. 2, p. 317-341.

———, 1972, Focal mechanism of earthquakes in Persia, Turkey, West Pakistan, and Afghanistan and plate tectonics of the Middle East: Seismol. Soc. America Bull., v. 62, no. 3, p. 823-850.

Peive, A. V., 1969, The oceanic crust of the geological past: Akad. Nauk S.S.S.R., tonika, no. 4, p. 5-23 (in Russian).

Reinhardt, B. M., 1969, On the genesis and emplacement of ophiolites in the Oman Mountains geosyncline: Schweiz. Mineral. Petrogr. Mitt., v. 49, no. 1, p. 1-30.

Ricou, L. E., 1968a, Sur la mise en place au Crétacé supérieur d'importantes nappes à radiolarites et ophiolites dans les monts Zagros (Iran): Compt. Rend., t. 267, p. 2272-2275.

———, 1968b, Une coupe à travers les séries à radiolarites des monts Pichakun (Zagros, Iran): Bull., Soc. Géol. France, sér. 7, t. 10, p. 478-485.

———, 1971a, Le croissant ophiolitique péri-Arabe, une ceinture de nappes mises en place au Crétacé supérieur: Rev. Géogr. Phys. Géol. Dyn., v. XIII, fasc. 4, p. 327-350.

———, 1971b, Le métamorphisme au contact des péridotites de Neyriz (Zagros interne, Iran): devéloppement de skarns ã pyroxène: Compt. Rend. Som., Soc. Géol. France, fasc. I, p. 43.

Ruttner, A., Nabavi, M. H., and Hajian, J., 1968, Geology of the Shirgesht area (Tabas area, East Iran): Geol. Survey Iran Rept. 4, 133 p.

Schürmann, H. M. E., 1966, The Pre-Cambrian along the Gulf of Suez and the northern part of the Red Sea: Leiden, E. J., Brill, 404 p.

Seyed-Emami, K., 1971, The Jurassic Badamu Formation in the Kerman region, with remarks on the Jurassic stratigraphy of Iran: Geol. Survey Iran Rept. 19, p. 5-79.

Stöcklin, J., 1968a, Salt deposits of the Middle East: Geol. Soc. America Spec. Paper 88, p. 157-181.

———, 1968b, Structural history and tectonics of Iran—a review: Am. Assoc. Petr. Geol. Bull., v. 52, no. 7, p. 1229-1258.

———, and Nabavi, M. H., (compilers), 1973, Tectonic map of Iran, scale 1:2,500,000: Geol. Survey Iran.

———, Eftekhar-nezhad, J., and Hushmand-zadeh, A., 1972, Central Lut Reconnaissance, East Iran: Geol. Survey Iran Rept. 22, 62 p.

Takin, M., 1972a, Iranian geology and continental drift in the Middle East: Nature, v. 235, no. 5334, p. 147-150.

———, 1972b, An outline of the evidence for continental drift in the Iranian region before the early Mesozoic: Geol. Survey Iran, Geol. Note no. 86, 4 p (unpublished).

Thiele, O., Alavi, M., Assefi, R., Hushmand-zadeh, A., Seyed-Emami, K., and Zahedi, M., 1968, Golpaygan quadrangle map, scale 1:250,000, with explanatory text: Geol. Survey Iran, Geol. Quadrangle E7, 24 p.

Tschopp, R. H., 1967, The general geology of Oman: 7th World Petrol. Congr., Mexico, Proc., v. 2, p. 231-241.

Vialong, P. Houchmand-zadeh, A., and Sabzehi, M., 1972, Proposition d'un modèle de l'évolution pétrostructurale de quelques montagnes iraniennes, comme une conséquence de la tectonique de plaques: 24th Internatl. Geol. Congr., Montreal, sect. 3, p. 196-208.

Volvovsky, I. S., Garetzky, R. G., Shlezinger, A. E., and Shreibman, V. I., 1966, Tectonics of the Turanian plate: Tr. Geol. Inst. Akad. Nauk S.S.S.R., v. 165, 287 p. (in Russian).

Weippert, D., Wittekindt, H., and Wolfart, R., 1970, Zur geologischen Entwicklung von Zentral-und Südafghanistan: Geol. Jahrb., Beih. 92, 99 p.

Wilson, H. H., 1969, Late Cretaceous eugeosynclinal sedimentation, gravity tectonics, and ophiolite emplacement in Oman Mountains, southeast Arabia: Am. Assoc. Petr. Geol. Bull., v. 53, no. 3, p. 626-671.

Zhuravleva, I. T., 1968, Early Cambrian biogeography and geochronology according to the Archaeocyathi: 23rd Internatl. Geol. Congr., Prague, Proc. IPU, p. 361-373.

Evolution of the Continental Margins Bounding a Former Southern Tethys

R. Stoneley

INTRODUCTION

Now that the characteristics of modern active and inactive continental margins are being studied, it is appropriate to see whether or not they can be identified in supposed ancient continental margins that have been involved in later orogeny. If so, then the nature of these margins can be determined and the results applied to the study of the pre-orogenic evolution of the region. The present paper is an attempt to apply this procedure to the southern part of the Tethyan orogenic belt, from the Mediterranean Sea eastward to Indonesia.

The suture of the former Tethys Sea has already been identified in the Himalayan region and in Iran (Gansser, 1964, 1966; Falcon 1967; Wells, 1969). By extrapolating from these regions, it is identified (Fig. 1) in the modern Indonesian Trench and it passes thence to the west of the Arakan Yoma of Burma, north of the Himalayas, west of the Baluchistan ranges of Pakistan, between the coasts of Oman and the Makran, along the Zagros Crush Zone, and into the Bitlis Thrust of southeastern Turkey.

Where the geological relationships on either side of this line are examined, it becomes clear that there were consistent differences between the two zones, representing the formerly opposed continental margins, in the Mesozoic and Cenozoic. A coherent pattern of movement prior to the continental collision is proposed, which is reflected also in the mode of orogenic deformation following that collision.

LOCATION OF THE SOUTHERN TETHYS SUTURE

The position of the line of suture between the former Eurasian and Gondwana crustal elements has been discussed at length by Stoneley, 1972. Only those data considered relevant are presented.

Southeast Asia. In the Indonesian region (Fig. 2), the modern representative of the southern Tethys suture is found in the Indonesian Trench, which divides the oceanic crust of the eastern Indian Ocean from continental southeast Asia. The continental margin to the northeast is believed to have been extended, at least since the Jurassic, by the accretion of successively tectonized sedimentary sequences deposited marginally to the continent (Van Bemmelen, 1949; Katili, 1973). The latest of these sequences, comprised of Cenozoic terrigenous sediments deformed largely in the late Miocene-early Pliocene, occupies the outer, nonvolcanic arc. On the continental slope, thence to the present-day trench, later sediments are disturbed by oceanward thrusting presumably related to the subduction of the southern ocean floor (Beck, 1972). This concept is illustrated schematically in Figure 3.

The outer Indonesian arc trends north through the western Nicobar and Andaman Islands (Fig. 2), where its westward bulge may be related to Neogene spreading in the Andaman Sea (Rodolpho, 1969), and into the Arakan Yoma and Naga Hills of the Indo-Burman ranges (Brunnschweiler, 1966). These ranges consist of thick uppermost Cretaceous to Eocene terrigenous deposits resting on Cenomanian to Maestrichtian deep-water sediments and ophiolites: they may, by analogy with Indonesia to the south, be regarded as having accumulated marginally to the Eurasian continent to the east, now overlain by the Tertiary sediments of the central Burma basin, which also includes andesitic volcanoes. The southern Tethys suture, here also marking the edge of the Indian shield, may be expected to lie west of these ranges beneath a thick accumulation of late Cenozoic molassé, postdating the continental collision.

Himalayas. The suture between the Indian and Eurasian continents in the Himalayan region has

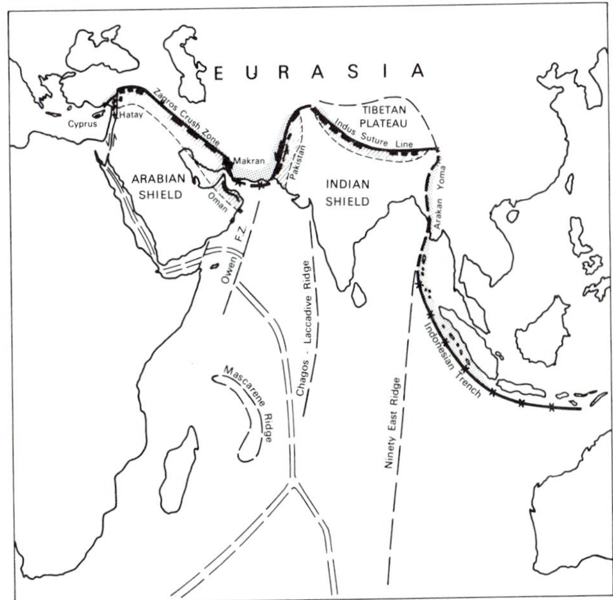

Fig. 1. Regional position of the southern Tethys suture: solid line, suture between Eurasia and Gondwana plates; dots, Upper Cretaceous-Tertiary flysch at margin of Eurasia; hachures, misgeosyncline on margin of Gondwana continent; black, outcrop of north-dipping subduction zone; double line, oceanic spreading ridge.

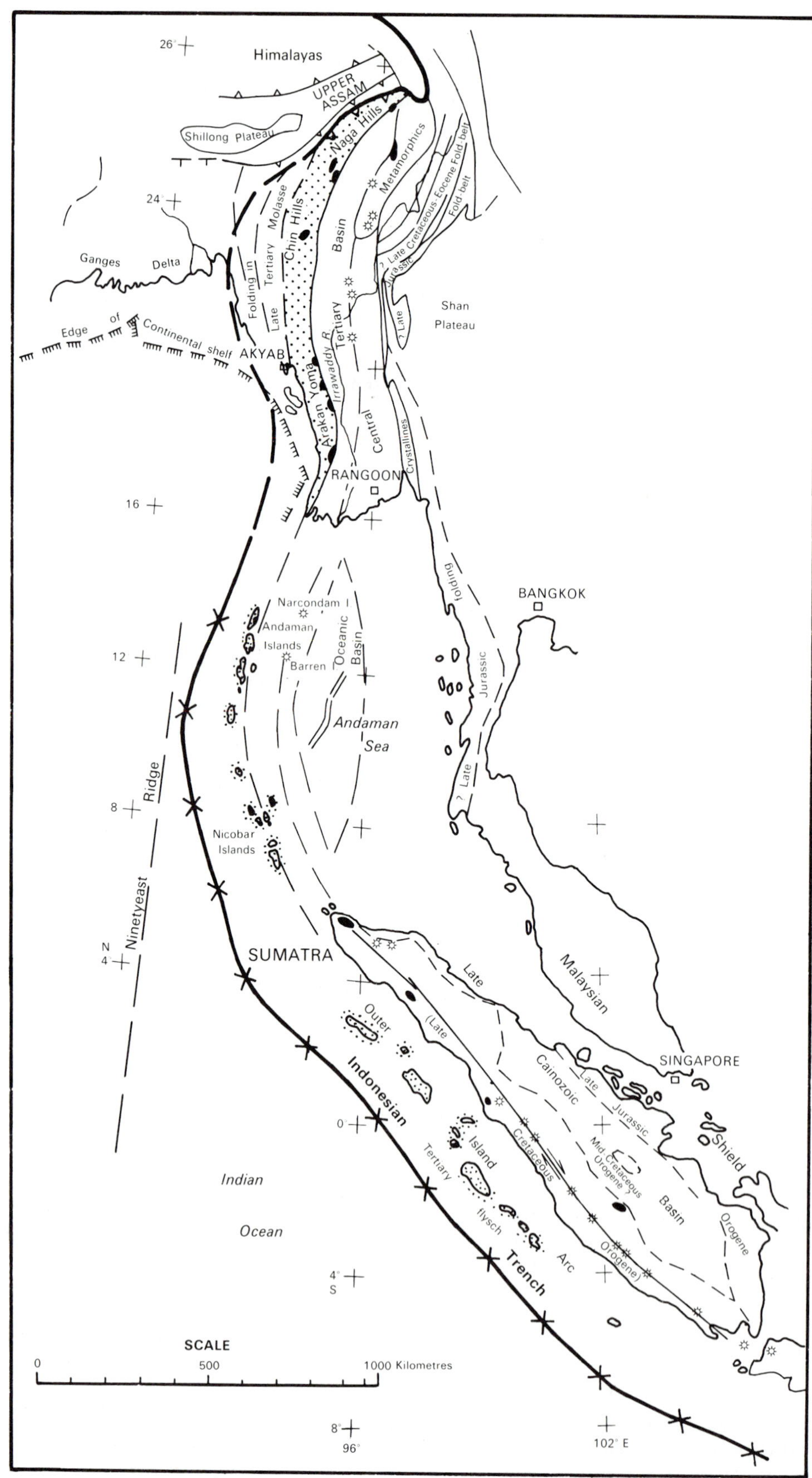

Fig. 2. Principal geological features of South-east Asia: heavy line, southern Tethys suture; dots, flysch accumulations at margin of Eurasia; hachures, Mesozoic-Tertiary miogeosyncline; dashed hachures, miogeosyncline obscured by younger deposits; black, ophiolites; crosses, outcrop of north-dipping subduction; double line, oceanic spreading ridge; starred circle, Quaternary volcano.

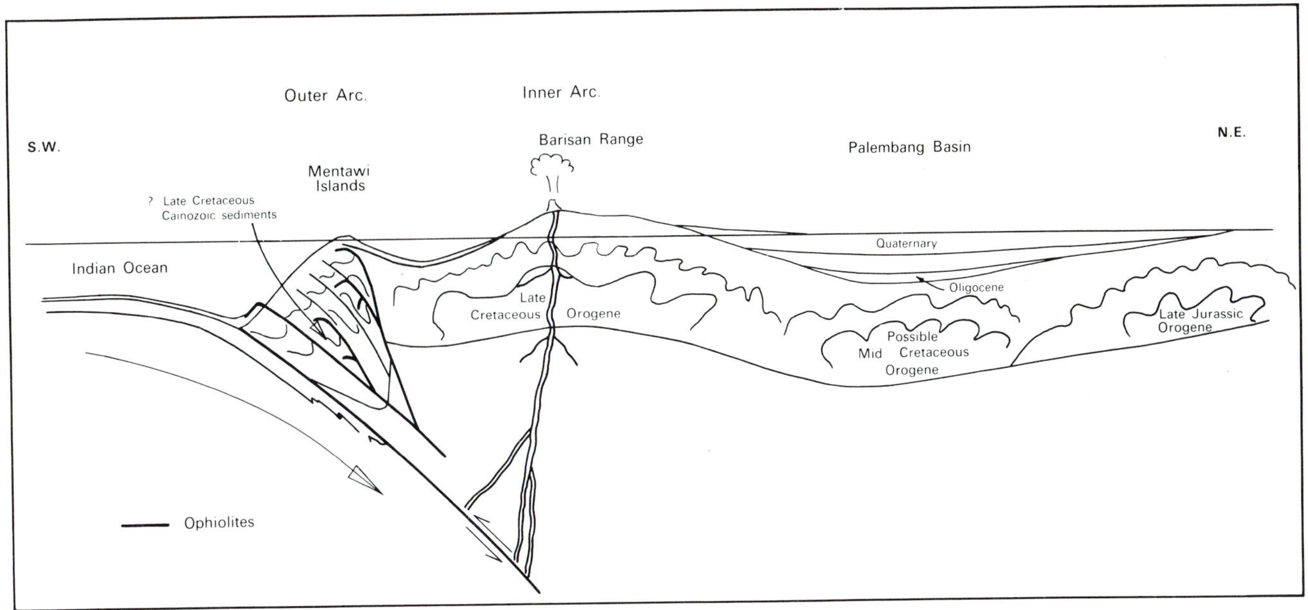

Fig. 3. Schematic section through Sumatra.

been located by Gansser (1964, 1966) along the Indus and Brahmaputra rivers (Fig. 4). It is marked by a strongly compressed belt of the Upper Cretaceous to Eocene Indus flysch, which contains exotic Mesozoic blocks and associated ophiolites. This belt separates the southward thrust Himalayas from the Tibetan Plateau to the north. The Himalayas are interpreted as representing the décollement and southward compression of the upper layers of the Indian shield (Fig. 5).

The uppermost of the nappes thus formed carries a Paleozoic and Tethyan Mesozoic miogeosynclinal sequence, which is terminated by Upper Cretaceous (Senonian) deepwater deposits, olistostromes and ophiolites. The lateral shortening represented by the Himalayan thrusting is estimated by Gansser (1964, 1966) to amount to some 500 km. This figure compares approximately with the width of the Tibetan Plateau (some 700 km), which is suggested by isostatic (Desio and Marussi, 1960; Holmes, 1965) and some seismic (e.g., Gupta and Narain, 1967) considerations to be underlain by nearly double the normal thickness of continental crust.

Karakoram Sector. Westward from the Himalayas, the southern Tethys suture is believed to pass north of the Nanga Parbat-Haramosh massif, which may represent a northerly spur of the Indian shield, and thence south of the Karakoram and Hindu Kush ranges, along the Gardez fault zone of eastern Afghanistan (Gansser, 1964) and into western Pakistan (Fig. 4). Parts of this sector have been affected by Oligocene and younger orogeny, and the suture line itself may be obscured locally by young granites.

The Mesozoic miogeosyncline on the northern margin of the Gondwana block is absent from this sector, and Precambrian and Paleozoic sediments, together with probably uppermost Cretaceous basaltic volcanics (Jan and Kempe, 1973), are affected by Cenozoic metamorphism, granitic intrusion, and locally intense granitization. The latter is particularly prominent at Nanga Parbat (e.g., Gansser, 1964), where Precambrian sediments were granitized in the Oligo-Miocene (Desio, 1973) and uplifted toward the end of the Neogene.

The northward drive of the Indian continent is considered to be absorbed in the Karakoram and Pamir ranges of Eurasia by enormous post-Paleogene northward thrusts. These are spectacularly transformed southeastward into the great dextral Pamir-Karakoram strike-slip fault separating these ranges from the Tibetan Plateau to the east (Peive et al., 1964; Desio, 1973). In the central and southern Pamirs and the northern Karakoram, a Paleozoic-Mesozoic succession of miogeosynclinal proportions represents a second, northern Tethys seaway, which, in the Mesozoic, shows significant differences from the southern, Himalayan Tethys (e.g., Norin, 1946). It is bounded southward by the axial granite zone of the Karakoram.

Still farther south, variably metamorphosed sediments and greenstones border the presumed southern Tethys suture; the mainly terrigenous sediments of Cretaceous and possibly Eocene age (Desio, 1973) might be analogous to the Indus flysch to the east. Associated with the more northern outcrops of these rocks, however, are thick carbonates, at least partly of Permian age (Ivanac et al., 1956), suggesting a more stable sedimentary environment. Thus, while some of the Cretaceous sediments may have accumulated marginally to the continent, some also may have been deposited within the continental margin.

Pakistan. In eastern Afghanistan and western Pakistan (Fig. 4) the suture passes, possibly around another northern spur of the Indian shield in the

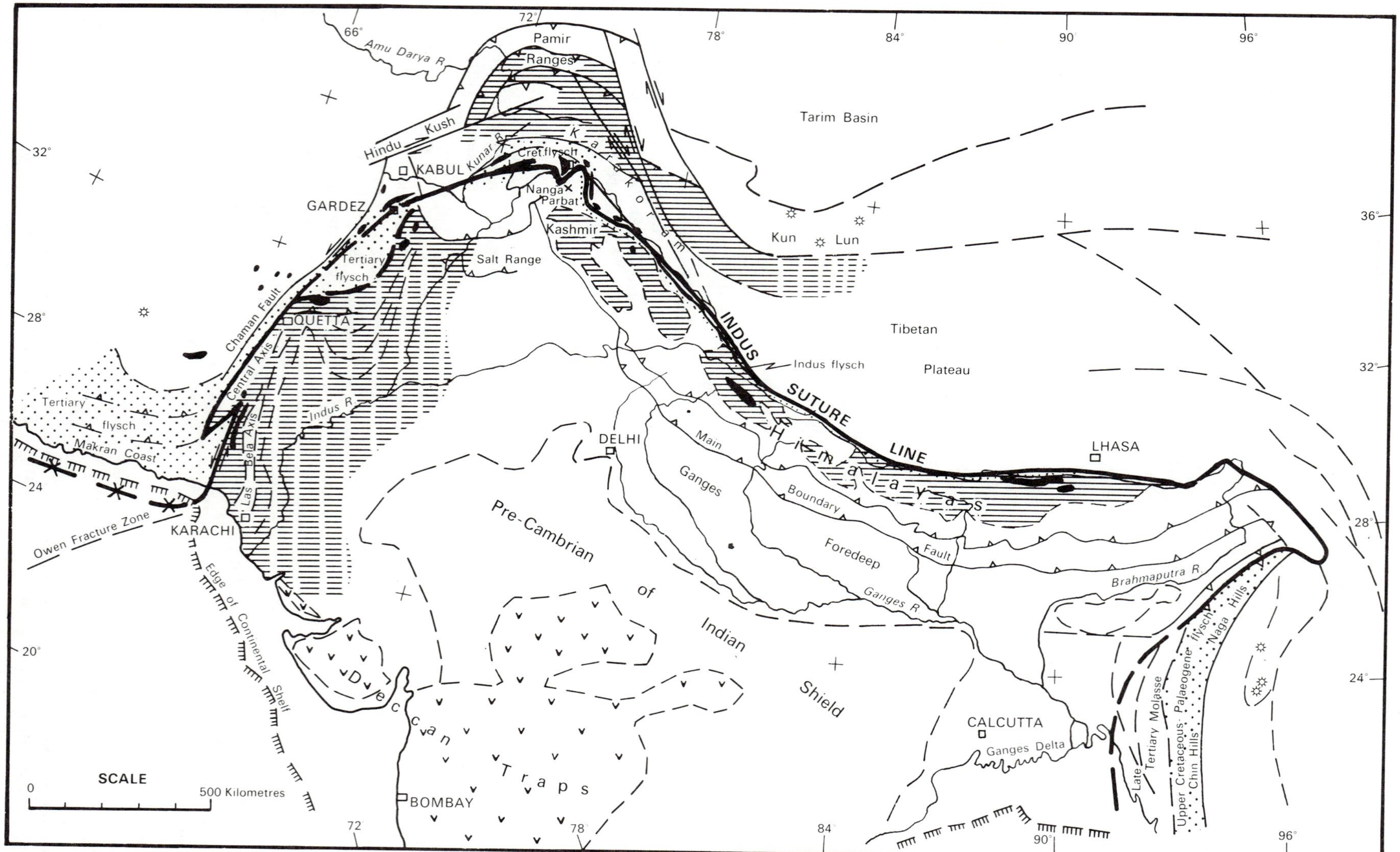

Fig. 4. Principal geological features of the northern Indian shield. (Legend as for Fig. 2.)

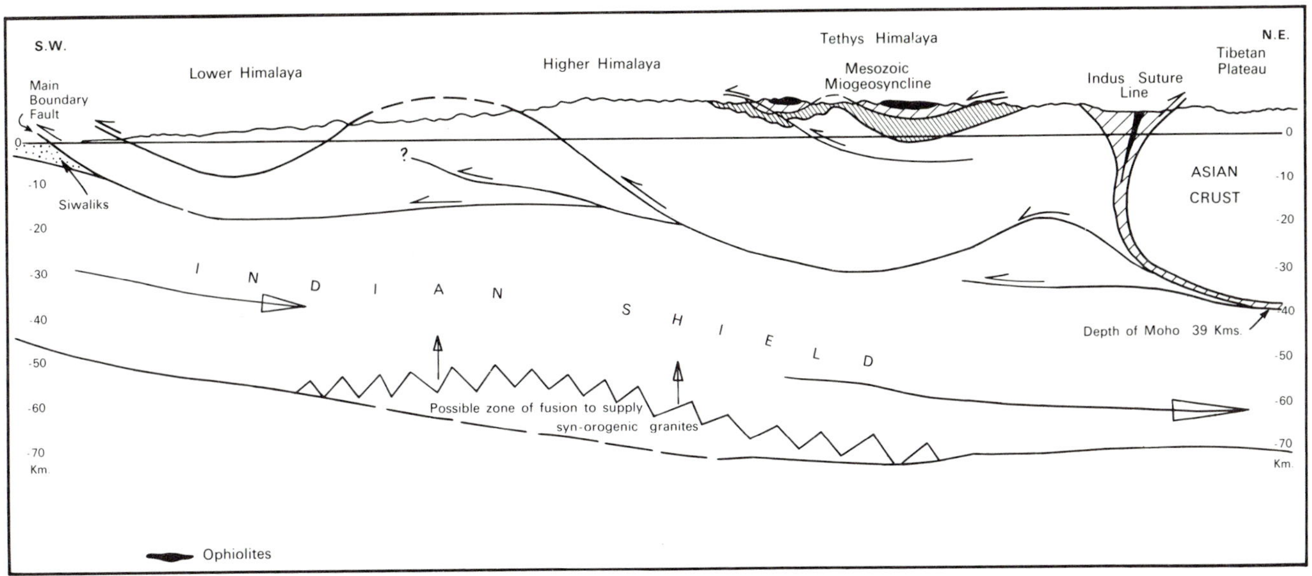

Fig. 5. Schematic section through the Himalayas (based on Gansser, 1964).

vicinity of Kabul (Schreiber et al., 1972), into a trough filled with up to 9 km of Upper Cretaceous to Miocene flysch deposits. This trough, which persists southward for more than 1,000 km into the Makran coastal belt (Jones, 1961; Auden, 1973), separates the Mesozoic-early Tertiary miogeosyncline of Pakistan from more varied and tectonized rocks in Afghanistan. It evidently obscures the precise position of the Indian shield margin; although a still active, left-lateral strike-slip fault (the Chaman Fault), it is clearly related to the suture.

Northwest of the flysch trough, in Afghanistan, relatively stable conditions during the Paleozoic gave place, in the Permo-Triassic, to more varied sedimentation. Tectonic disturbances occurred, particularly in the Late Triassic-Early Jurassic, Early Cretaceous, and Late Cretaceous, culminating in more widespread orogeny in the Oligocene (Schreiber et al., 1972).

The Gondwana miogeosyncline in Pakistan, on the other hand, consists primarily of Mesozoic and Tertiary carbonates (Jones, 1961). It was subdivided during the Cretaceous by a marginal uplift (Central Axis) and a parallel, more internal ridge (Las Bela Axis) with an intervening trough (Jones, 1961). The ridges were associated with littoral deposits, as well as radiolarites and olistostromes, which slid into the adjacent troughs. Ophiolites are reported to have been intruded into the Central Axis in the Late Cretaceous-early Paleocene, whereas the Las Bela Axis displays poorly dated Cretaceous basalts and basic intrusives (Jones, 1961). There is no evidence to suggest contact between the Eurasian and Gondwana continents before the early Eocene: at that time, however, granitic detritus apparently derived from north of the suture flooded onto the northern portion of the miogeosyncline. Compression commenced early in the Oligocene, when a localized flysch trough developed rapidly near Quetta, but the main deformation was Pliocene.

Makran Coast—Oman. In the Makran coastal region (Fig. 6), the flysch trough, which to the northeast separates the Gondwana miogeosyncline of Pakistan from Eurasia, is clearly related to the margin of Eurasia, being bounded to the south by the oceanic depths of the Gulf of Oman.

At its northern edge it rests upon an Upper Cretaceous mélange, including ophiolites (Gansser, 1959). Two sequences of largely terrigenous sediments, of Paleocene to Pliocene age, reach very great thicknesses and are overlain unconformably by more than 1 km of Pliocene (Ahmed, 1969).

They appear to be prograded southward and are affected by southward-directed thrusting (or northward underthrusting), which may have been episodic: lateral shortening in the Neogene alone is estimated by Ahmed (1969) to be 5.5 km. The overall picture is comparable with that seen elsewhere at a continental margin above a subduction zone, even to the presence of andesitic volcanics to the north. The present-day seismic activity (Nowroozi, 1972) is consistent with this. One is therefore led to locate the southern Tethys suture at the foot of the northern continental slope of the Gulf of Oman. The boundary of the opposing Gondwana plate thus being placed in the northern Gulf of Oman, the spur of the Arabian shield extending eastward to the Oman coast may be regarded as a southern continental margin that has not yet reached collision with Eurasia (Fig. 7). Farther to the west, of course, the Arabian shield has indeed collided with Eurasia at the Zagros Crush Zone (see below). In the intervening sector, a right-lateral strike-slip fault (the Zindan Fault, Falcon, 1967) may be regarded as offsetting the suture: it separates the Makran flysch belts to the east from the Gondwana Zagros ranges to the west, and is the counterpart of the left-lateral Chaman fault in Pakistan. Thus it appears that, owing to the areal geometry of the northeastern Arabian margin, the collision that occurred at the

compression within the miogeosyncline, and shallow water sedimentation was resumed above the ophiolites in the Eocene.

Compression in the miogeosyncline dates from the early Miocene and is continuing (e.g., Falcon, 1973); it is reflected in the Zagros and its foothills by a change from carbonate to evaporite and then clastic sedimentation. The thrusting and folding of the Zagros are estimated by Falcon (1973) to represent a shortening of the order of 50 km and probably reflect a décollement of the Phanerozoic section above Eocambrian salt. The process may be analogous to the décollement in the Himalayas, referred to above.

Immediately northeast of the Zagros Crush Zone, the Mesozoic-Cenozoic history was very different. A narrow, probably ensialic, Jurassic flysch trough developed in an area of earlier instability near Isfahan, Iran, and was deformed by southwestward thrusting and local metamorphism in the Late Jurassic or Early Cretaceous (Reyre and Mohafez, 1972). At this time also granites were intruded farther northwest. A further episode of thrusting dates from the Late Cretaceous.

Turkey. The Zagros Crush Zone passes northwest from Iran, through Iraq and into the Bitlis thrust of southeastern Turkey (Fig. 6), which has a relative southward displacement of at least 15-20 km (Rigo de Righi and Cortesini, 1964). This continuation of the southern Tethys suture separates sediments attributable to the northern margin of the Arabian shield from the complex structures of the Taurus foldbelt. Within the shield, the record once more suggests the emplacement during the Maestrichtian of a series of gravitational slides and olistostromes, directed southward into a deep-water trough lying within the marginal zone of the Mesozoic miogeosyncline. Allochthonous ophiolites are suggested by Rigo de Righi and Cortesini (1964) to have been derived from an uplift at the northern margin of the trough. Subsequent sedimentation continued in a series of successor troughs until the overthrusting of the Taurus in the late Miocene and Pliocene.

Westward, toward the Mediterranean, the Late Cretaceous trough becomes narrower and in the Amanos ranges (to the west of the northern prolongation of the 80 km, sinistrally offset Dead Sea Rift; Quennell, 1958), only a localized representative is present, again associated with radiolarites, olistostromes, and ophiolites (Dubertret, 1953). Although the contrary view has been expresssed (Ricou, 1971a), it seems harder in this case to relate the ophiolites to long-distance allochthoneity; an intrusive-extrusive origin, up to within 100 km of the margin of the Arabian shield, appears to be more likely (Dubertret, 1953; Schwan, 1971). Nevertheless, there is possible evidence of compression in the development of a slatey cleavage almost immediately after the emplacement of the ophiolites, still within the Maestrichtian (Schwan, 1971; and noted below).

AGE OF THE SOUTH TETHYAN CONTINENTAL MARGINS

Although there were strong differences between the regions on either side of the southern Tethys suture during most of the Mesozoic and Cenozoic, there is evidence to suggest an earlier continuity between them. If this was the case, the formation of the continental margins they represent should constitute a recognizable event, and we may consider four possible approaches to establishing its date.

Continuity of Facies. It has been suggested (e.g., Wolfart, 1967; Stöcklin, 1968) that, although localized areas of more rapid subsidence may have been present, there was during the Paleozoic a broad continuity of shelf sedimentary facies across the entire region embraced by the Arabian shield, central Iran, and most of Turkey. It was from the Middle Triassic onward that strong differences became apparent between the sediments on either side of the southern Tethys suture (Falcon, 1967). Similarly, Norin (1946) has described an Upper Paleozoic sequence in western Tibet, north of the suture, which is closely comparable both lithologically and paleontologically with Kashmir to the south. Major differences between the regions are a Mesozoic phenomenon.

Rocks Related to Continental Breakup. Continental terrigenous sediments, evaporites, and extrusive basalts related to contemporaneous tensional faulting are features that elsewhere are considered to characterize continental splitting. None of them, however, appears to have been recognized positively in the vicinity of the southern Tethys suture, although spilitic lavas and rare mafic dikes and sills are recorded from a Permo-Triassic to early Jurassic unit within the allochthonous Hawasina rocks of Oman (Glennie et al., 1973).

Rocks Derived from the Southern Tethys. Sediments that must have been deposited in an ocean, which has subsequently disappeared along the line of the southern Tethys suture, are common among the allochthonous masses emplaced along the northern margins of the Gondwana continents in the Late Cretaceous. From Turkey (Rigo de Righi and Cortesini, 1964), Iran (Ricou, 1971a; Stöcklin, this volume), Oman (Allemann and Peters, 1972; Glennie et al., 1973) and the Himalayas (Gansser, 1964), no deepwater sediments older than Triassic, or in some cases Jurassic, have been recorded. Permian, and in one case Ordovician (Allemann and Peters, 1972) exotics are present in Oman and in the Himalayas, but they are apparently of shelf facies and may have been derived from the continental margin itself or, possibly, from shoals within the ocean.

Continental Reconstructions. Paleogeographical reconstructions based on paleomagnetics and the reassembly of the earth's major continents clearly indicate that, at the close of the Paleozoic, an eastward-widening ocean already lay between the Indian and Arabian sections of Gondwanaland and Eurasia (e.g., Kamen-Kaye, 1972; Smith, 1971).

At first sight, this approach conflicts with the first three, which combine to suggest that the ocean, which later closed along the southern Tethys suture, did not exist in the Paleozoic, but began to open during the Triassic. There is, however, a suggestion that a further suture may be found north of the Tibetan Plateau, in the general vicinity of the Kun Lun (Crawford, 1973), and also north of the Iranian plateau along the southern foot of the Kopet Dagh (Crawford, 1972). One may therefore speculate that an intermediate crustal tract, comprising central Iran, Afghanistan, and the Tibetan plateau, possibly extending into southeast Asia, was attached to and formed the northern margin of Gondwanaland in the Paleozoic; that it separated from Gondwanaland in the Triassic and moved northward to collide with Eurasia possibly in the late Jurassic or early Cretaceous; and that it left a southern Tethys in its wake, which was in turn itself closed by collision during the Cenozoic.

The continental margins bounding the southern Tethys ocean are thus considered to date from the Triassic.

DATE OF CONTINENTAL COLLISION

Two separate collisions are involved between the Gondwana continental blocks and Eurasia. They took place to the north of the Indian shield, consequent on Late Cretaceous-Cenozoic spreading from the Carlsberg-southeast Indian Ocean Ridge (McKenzie and Sclater, 1971), and to the north of Arabia, which, until the early Miocene, was joined to Africa. The movement of Afro-Arabia was goverened primarily by Mesozoic and Cenozoic spreading from the Mid-Atlantic Ridge (e.g., Pitman and Talwani, 1972; Smith, 1971).

Indian Collision

We have noted that granitic detritus spread from the north onto the miogeosyncline at the northwestern margin of the Indian shield in the early Eocene. No source for this detritus is available within the Gondwana block, which therefore seems to demand contact with the Afghanistan sector of Eurasia. Compressive folding commenced during the late Eocene. However, farther east, the Indus flysch of the suture line north of the Himalayas contains Eocene limestones, so that the Tethys seaway cannot have been closed completely here until late in that period. Compressive deformation in the Himalayas commenced in the late Oligocene (Gansser, 1973).

The eugeosynclinal rocks east of the Indian shield in the Arakan Yoma of Burma, which because they represent the northward prolongation of the outer Indonesian arc may be regarded as having been deposited at the margin of Eurasia, record a compressive orogeny during the Oligocene. This could have resulted from an eastward deflection of the shield, caused by the oblique collision at its western margin.

We thus have evidence of initial contact between the two continents in Pakistan early in the Eocene, with the remnant of the Tethys farther east being closed toward the end of the Eocene. The relative movement between them, from the Oligocene onward, was expressed in the west by left-lateral movement across the intervening flysch trough, for example on the Chaman Fault, and to the north in underthrusting of the Tibetan Plateau.

Arabian Collision

The date of the collision of Arabia with Eurasia is more difficult to determine. It has been suggested (Wells, 1969) that it was the opening of the Red Sea, commencing in the early Miocene, that caused the closure of the Tethys along the Zagros Crush Zone. The collision had certainly occurred before the late Miocene when the folding in the Zagros commenced, and it may be reflected in the early Miocene change from carbonate to evaporite and redbed sedimentation within the miogeosyncline. The fact that the extension in the Red Sea is of the same order of magnitude as the shortening represented by the structures in the Zagros suggests that Arabia may have been close to Eurasia when the Red Sea spreading commenced.

The events that took place on the northern Arabian margin during the late Cretaceous do not, by comparison with Oman and the Himalayan region, causally demand contact with Eurasia. However, other considerations combine to suggest that the continents were not, in fact, far apart through the early Cenozoic. An Upper Maestrichtian-Paleocene flysch (Amiran Formation) in the Zagros, which at some localities consists almost entirely of radiolarite and some ophiolite debris, is also reported by Falcon (1967) to contain abundant mica. This would require a source northeast of the Zagros Crush Zone. Conversely, Upper Eocene-Oligocene flysch found locally along the Crush Zone is related by Bizon et al. (1972) to the Makran flysch, which is marginal to Eurasia. However, its detritus was derived primarily from the radiolarites, which crop out on the Arabian margin, again perhaps suggesting that a shallow seaway, rather than an ocean, lay between the continents.

Finally, in the northwestern corner of the Arabian shield in Hatay (Schwan, 1971) there is evidence of compressional orogeny immediately following the emplacement of the ophiolites in the Maestrichtian, possibly suggesting collision. Its effects, however, including the development of a slatey cleavage, chevron folding, and an absence of significant horizontal displacement, are different from those attributable to collision elsewhere south of the suture, and it is conceivable that it resulted from pressure due to the emplacement of the ophiolites (cf. Williams and Smyth, 1973).

While we can be sure that the Tethys north of Arabia was closed early in the Miocene, we have to admit some doubt about the relative positions of the two continents during the early Tertiary, there being some evidence to suggest at least temporary contiguity, if not a connection between them. Nevertheless, in the Oman-Makran sector it seems that, owing to the configuration of Arabia, the opposing continental margins have still not yet collided.

CONTINENTAL MARGIN OF EURASIA

Certain features of the regions on opposite sides of the southern Tethys suture indicate that the formerly opposed continental margins underwent very different histories.

Flysch Sedimentation. The characteristic sediments to the north of the suture are thick, rapidly accumulated terrigenous deposits, which were apparently deposited either marginally to Eurasia or in rather short-lived troughs within the continental margin. Among the former, we may cite the Uppermost Cretaceous-Tertiary deposits of the Makran-Pakistan sector, the Indus Flysch north of the Himalayas, the "eugeosynclinal" Upper Cretaceous-Lower Tertiary beds of the Naga Hills-Arakan Yoma, and the sediments of successive troughs in Sumatra. Possibly ensialic troughs are represented by the Jurassic flysch of central Iran and some of the Cretaceous sediments of the Karakoram region.

Carbonates are confined to generally thin beds in the flysch or to relatively short phases of stability, for example in the middle Cretaceous of central Iran.

Orogenic Phases. The areas north of the southern Tethys suture are characterized by discrete orogenic episodes involving thrusting directed generally southward, away from Eurasia, and sometimes metamorphism. These episodes (Table 1) occurred primarily in the late Miocene-Pliocene, late Eocene-Oligocene, late Cretaceous, late Jurassic and/or early Cretaceous, and possibly in the late Triassic-early Jurassic. The effects of impact of the Gondwana continents are expressed in differing ways, but predominantly through vertical movements. The Tibetan Plateau, and to some extent the central Iran-eastern Turkey region, seem to have been uplifted bodily, and it is only in the Karakoram-Pamir sector that the impact has been expressed in vast northward overthrusting and strike-slip faulting.

Andesites. Andesitic volcanic rocks are confined virtually to the northern side of the suture. Most extensive are those in the Eocene of central Iran, which occupy a belt parallel to and some 150 km northeast of the Zagros Crush Zone. Most of the currently active volcanoes are similarly located with respect to the suture line.

Granites. Granites are confined generally to the zones north of the suture and appear to be related to the orogenic phases noted above. Exceptions found south of the suture are the postorogenic granites of the Himalayas and northern Pakistan, and the anatectic granitisation at Nanga Parbat: they were, however, clearly emplaced after any continental collision.

These features are characteristic of an active continental margin and strongly suggest that the southern edge of Eurasia has been the site of northward subduction at least since the middle of the Jurassic. The accumulation of the terrigenous sedimentary series may have taken place during quieter periods, and the orogenic phases may have occurred during episodes of stronger movement of the Tethys ocean floor toward and beneath the Eurasian margin.

MARGINS OF THE GONDWANA CONTINENTS

Quite different phenomena characterize the northern margin of the Indian and Arabian shields, on the south side of the southern Tethys suture.

Mesozoic Miogeosyncline. From the Triassic until the late Cretaceous, predominantly carbonate sedimentation took place in a typical miogeosyncline above the northern edges of the Gondwana continents, in most sectors in continuity with an underlying Paleozoic succession. Granitic detritus is relatively rare in the presumed pre-collision succession and, where present, was derived entirely from the shield areas to the south. The miogeosyncline is notably absent from northern Pakistan, but the reason why this should be so is not clear.

During the late Cretaceous, a deeper-water trough developed within the miogeosyncline, parallel to but apparently within the continental margin, as a precursor to the spectacular events in the Campanian-Maestrichtian, discussed below. From the end of the Cretaceous onward, carbonate sedimentation was resumed, or persisted, until the effects of the continental collision became apparent. Deeper-water basins continued to evolve within the continental margin, notably in southeast Turkey, Iran, Oman, and more locally in Pakistan.

Late Cretaceous Events. A unique series of events followed the development of the elongated trough within the miogeosyncline during the late Cretaceous, and descriptions from all sectors (the Himalayas, Pakistan, Oman, Iran, and Turkey) lead to a similar picture. During the Santonian-Campanian, the trough began to receive turbidites, derived probably from the continental margin itself, and continued to subside to the depth of radiolarian-chert accumulation. It was then invaded by a series of olistostromes, often carrying large masses of material derived from the continental margin and from the Tethys ocean floor to the north.

The olistostromes were covered by extensive allochthonous masses of chromite-bearing ophiolites, again derived from the north and including remnants of diabasic dike swarms together with the full ophiolite sequence, terminating in pillow basalts: the complete sequence may be some thousands of meters thick. This took place more or less synchronously in all sectors in the late Campanian-early Maestrichtian and was followed almost immediately by a resumption of shallow-water carbonate sedimentation.

Structural Deformation. During most of the Mesozoic, there is no evidence to suggest compressive deformation. Rather, only gentle epeirogenic movements, possibly accompanied by normal faulting, led to the development of local unconformities. In the late Cretaceous, however, we have noted that a series of olistostromes and ophiolite "nappes," derived at least partly from the Tethys ocean floor to the north, were emplaced into the rapidly subsiding trough within the continental margin, but we may note that their emplacement took place simultaneously with normal faulting in Oman (Wilson, 1969) and that, in no sector of the south Tethys margin, except possibly Hatay (see above), does there appear to be any evidence that necessarily demands the action of compressive tectonics, as opposed to gravitational gliding.

Indeed, were it not for the presence of the ophiolites, which are now commonly regarded as oceanic floor "obducted" onto the continental margin, it is unlikely that there would be any suggestion that compression was involved at all. As it was, the ophiolites were emplaced while the Indian continent was presumably still some hundreds of kilometers away from Eurasia (Fig. 8).

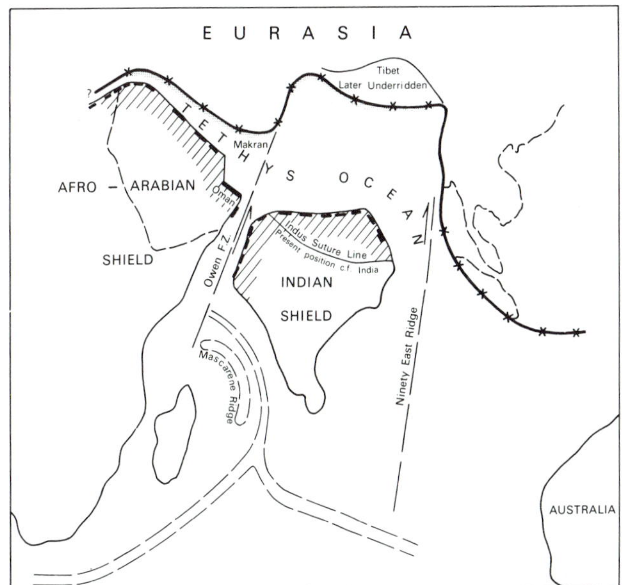

Fig. 8. Tentative reconstruction of the Tethys in the Late Cretaceous. (Legend as for Fig. 1.)

The structures that developed after continental collision represent a stripping off and displacement backward, toward the south, of the upper layers of the continental crust. In the Himalayas, a series of very large thrusts carries sheets involving Precambrian crystalline rocks back toward the south. In the Iranian sector, the presence of Eocambrian salt may have facilitated décollement of the overlying sediments, which, although imbricated close to the suture, are compressed essentially into a series of spectacular folds.

Metamorphism is apparently confined to the north Pakistan sector, for which again the reason is not clear; it may be emphasized, however, that the Mesozoic miogeosyncline is absent here and, furthermore, that only in this sector are the presumed effects of collision prominently displayed on the northern side of the suture. Possibly these exceptional phenomena are to be explained by the former presence of an acute spur at the northern margin of the Gondwana continent.

Igneous Rocks. Apart from the late Cretaceous ophiolite sequences noted above, volcanism in the southern continents was restricted to the extrusion of basalt. This occurred probably in the latest Cretaceous in northern Pakistan (Jan and Kempe, 1973), probably in the late Cretaceous in western Pakistan (Las Bela Axis), slightly later and extending into the Paleocene (Deccan Traps) in India, and in the Neogene in southeastern Turkey. At no time in the Mesozoic or Cenozoic is there any indication that significant andesitic volcanism took place. However, it should be noted that within the olistostromes believed to have been derived from the Tethys, in the Hawasina Series of Oman, occasional andesite flows are encountered in association with basaltic pillow lavas (Allemann and Peters, 1972): they may be a product of local differentiation.

Granites are apparently confined to the post-orogenic stages of deformation in the northern Pakistan-Himalayan region, where they may represent the fusion of the downwarped leading edge of the Indian shield as it tended to underthrust Eurasia.

Apart from the presence of the late Cretaceous ophiolites, all of the evidence suggests that the northern margins of the Gondwana continents were inactive until collision with Eurasia took place. It seems clear that, except possibly during the brief period in the late Campanian-Maestrichtian, these southern continental blocks were being transported passively northward within a lithospheric plate toward subduction at the northern side of the Tethys ocean from at least the middle of the Jurassic onward. Spreading rates from the Indian Ocean give no indication of any interruption to the northward drift of India, even in the Late Cretaceous when, on the contrary, that drift was abnormally rapid (McKenzie and Sclater, 1971).

CONCLUSIONS

The considerations presented here seem to point toward the former presence of an oceanic belt along the line of the southern Tethys suture. This ocean is interpeted as having been closed by a process of subduction at its northern edge—at the southern margin of Eurasia, which shows indications of having been an active continental margin at least since the middle of the Jurassic.

The southern side of the Tethys Ocean appears to have been bounded by inactive continental margins from the Triassic until collision with Eurasia. Two separate continental blocks are involved on the southern side, (Afro-)Arabia and India; possibly since the Triassic or Early Jurassic, they seem to have moved more or less independently, in response to spreading from different mid-ocean ridges to the west and south. Nevertheless, they both moved in a generally northward direction with respect to the margin of Eurasia, and both exhibit similar Mesozoic-Cenozoic geological histories. It is of interest that increases in the rate of northward displacement of India relative to the spreading ridge to the south appear to correlate broadly with major periods of orogeny at the southern margin of Eurasia (Table 1).

This general regime seems to have prevailed in the southern Tethys region, possibly with some lateral positional adjustments of the southern continents, reflected, for example, in the Late Cretaceous ophiolites on the southeastern coast of Arabia at Masira Island, until the southern continents collided individually with Eurasia in the Cenozoic. During this process, the upper layers of the leading margins of Arabia and India may have been stripped off and compressed backward to the south, over the advancing continent itself.

Some of the Gondwana continental crustal section seems to have underridden the Eurasian margin, with little effect except to cause vertical uplift. It is only in the Hindu Kush-Pamir-Karakoram region that the effects on the northern side of the suture reach the surface in spectacular thrusting and related strike-slip faulting. Andesitic volcanism, characteristic of the northern side of the su-

my		EASTERN MED.	EASTERN TURKEY	IRAN	MAKRAN [OMAN]	AFGHANISTAN [PAKISTAN]	KARAKORUM	HIMALAYA	BURMA	ANDAMAN ISLANDS	SUMATRA	INDIAN OCEAN SPREADING
0	QUATERNARY											
	PLIOCENE	UNDER- N. THRUSTING		FOLDING	THRUSTING N.	FOLDING S.		S.	FOLDING & THRUSTING	N. FOLDING	N. FOLDING & N. THRUSTING	SPREADING
10	MIOCENE Late	THRUSTING N.S.	THRUSTING	S.			INTRUSION N.	THRUSTING PROCEEDING	NS. FOLDING	N. FOLDING & THRUSTING	THRUSTING IN OUTER ARC	
20	MIOCENE M.		N.S.			SHEAR ALONG SUTURE		S. N.→S.	NS.		N. OROGENY	INFERRED SPREADING
	MIOCENE Early											
30	OLIGOCENE Late / M. / Early		THRUSTING ↑	OROGENY N.	THRUSTING. N.	OROGENY	OROGENY	N.S.	OROGENY	FOLDING & N. THRUSTING	? OROGENY IN OUTER ARC	
40	EOCENE Late	THRUSTING N.	OROGENY N.	? x		N. FOLDING NS. S.	NS. ? INTRUSION N.	INDUS TROUGH CLOSED	NS.		x x N.	
50	EOCENE M. / Early	GLIDING S.		x x x ?		? GRANITES S.		x ? x			x ? ? N.	
60	PALEOCENE				? N	? S. GLIDING	? INTRUSION N.	x N. ? ?		?		
70	CRETACEOUS Maes.	COMPRESS'N S. x S.	GLIDING S.		x THRUSTING x N.	x N. ? OROGENY	? x OROGENY N	x ? INTRUSION x ? OROGENY N.	THRUSTING x N.	FOLDING x N.	?	SPREADING
	Camp.	x GLIDING S.	x ? x	x FOLDING & x THRUSTING S	x GLIDING S S	x ? x	x x	x GLIDING	N.	x N.	OROGENY	FAST SPREADING
80	Sant.		x OROGENY	NS.		x	x ? x	x S.				
90	Con.		x N. ?			x S. ?	x NS. ?	x N.S. ?				
	Tur.		x S. ?								N. ?	
100	Cen											
	Alb.											
110	Apt.											
120	Neoc.		? OROGENY N.	OROGENY N.							? x x x	?
130							OROGENY		? OROGENY		x x ?	
140	LATE JURASSIC						N.		N. ?		? OROGENY ? N. x ?	
150											x x ?	
160		? OROGENY N.										

TECTONISM... I OPHIOLITES... x NORTH OF THE TETHYS SUTURE... N SOUTH OF THE TETHYS SUTURE... S

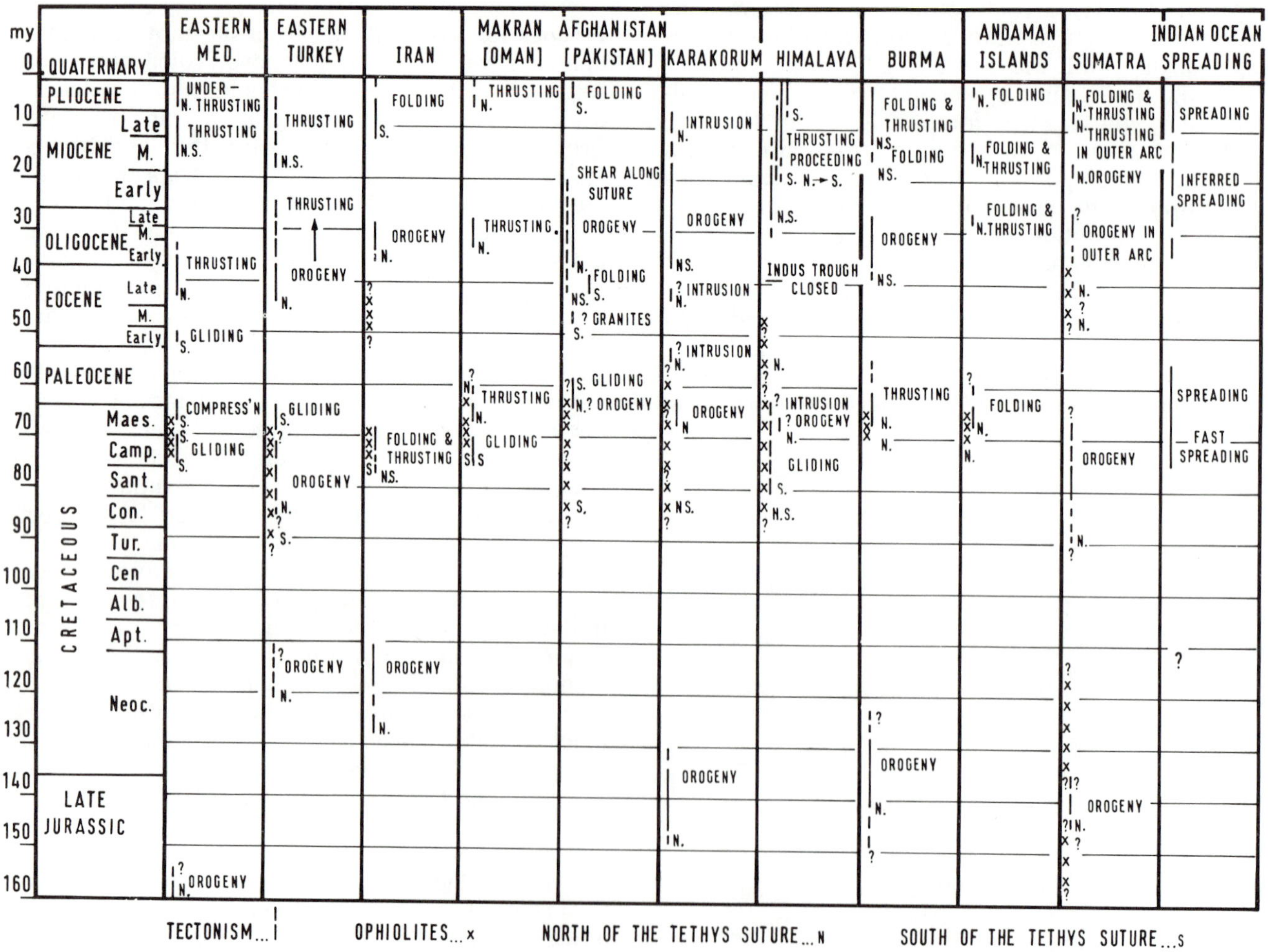

Table 1. Times of orogenic movements and ophiolite emplacement in southern Tethys.

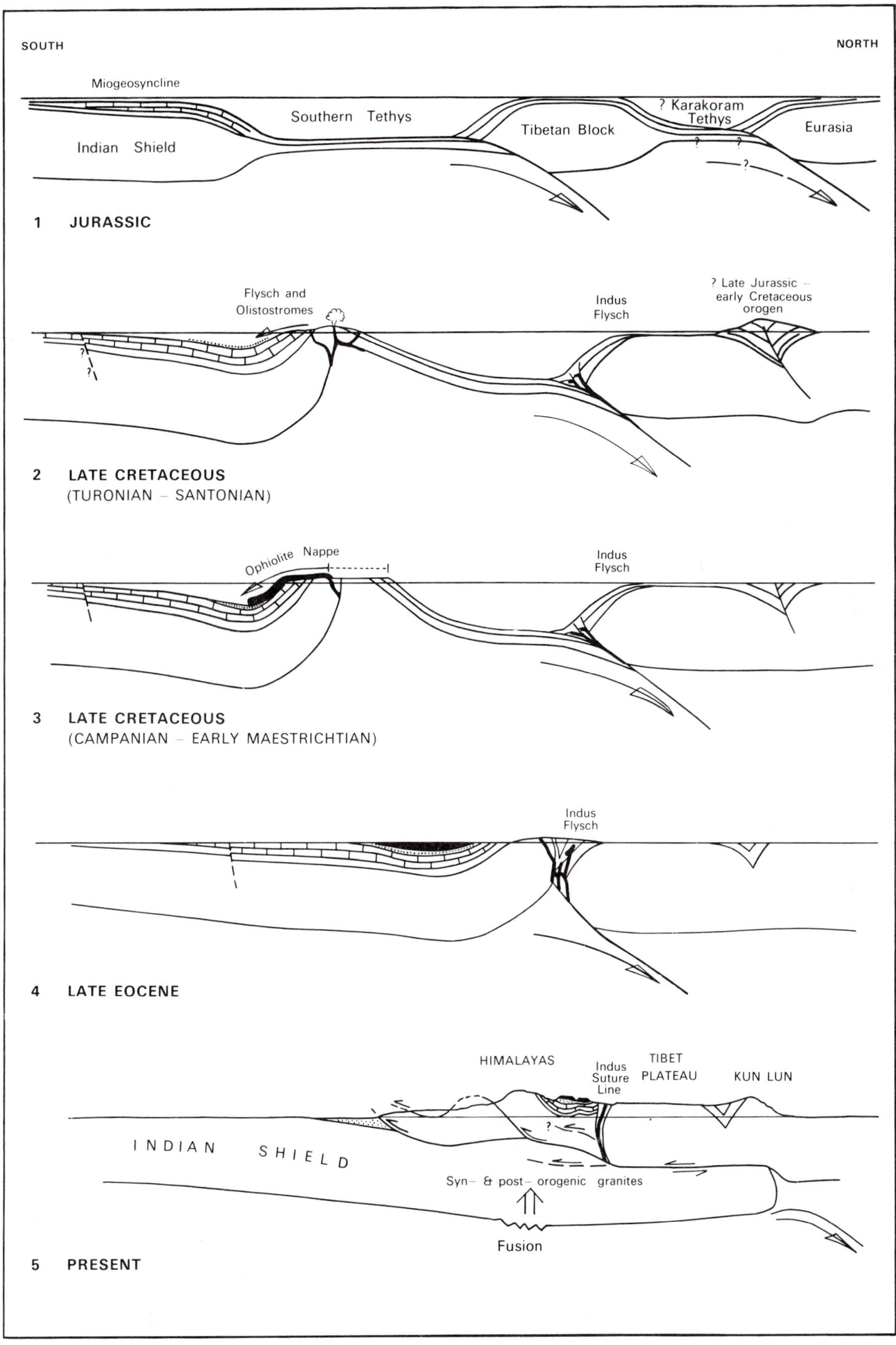

Fig. 9. Hypothetical evolution of the southern Tethys, illustrated by reference to the Himalayan sector.

ture, continues to this day, reflecting a continuation of the tendency toward subduction at the Eurasian margin.

The only complication to this otherwise simple and coherent picture is the emplacement of ophiolite sequences along most sectors of the northern margins of Arabia and India. They occurred more or less synchronously during the Late Cretaceous, and demand a common explanation. It is customary at present to regard all ophiolite sequences as representatives of oceanic lithosphere and to postulate that they were emplaced on continental margins by compressive thrusting (obduction), possibly above a zone of subduction. It is indeed possible that those Late Cretaceous ophiolites associated with the flysch deposits on the northern side of the Tethys suture were emplaced in some such manner, although they can more readily be regarded as scrapings from the sea floor and/or magmatic injections from beneath.

However, in the case of the alpine-type ophiolites found to the south of the suture, on the northern margins of Arabia and India, there is nothing to suggest that subduction was taking place below. There is no indication of marginal flysch sedimentation, of northward directed thrusting, or of andesitic volcanoes. On the contrary, there is evidence that tensional faulting was active nearby during the emplacement of the ophiolites in Oman; and the shape and internal structure of the ophiolite "nappe" in this sector suggest strongly that it reached its present position through gravitational gliding from the north (e.g., Reinhardt, 1969). Additionally, it must be recalled that locally, for example in Iran and the Himalayas, there is evidence of thermal alteration of carbonates in contact with the ophiolites.

Another possibility is that the ophiolites might represent the effects of collision with island arcs. This, however, is unlikely since all the ophiolites were emplaced at the same time in the Late Cretaceous, when the Gondwana continents were at varying distances from collision with Eurasia: indeed it is believed that the Oman sector of Arabia has not, even yet, reached this collision point.

In the apparent absence of evidence necessarily demanding the action of Late Cretaceous compression at the margins of India and Arabia, and in the light of contemporaneous tensional faulting and gravitational gliding, it is not unreasonable to postulate that the ophiolites were emplaced entirely through gravitational gliding from a deep-seated uplift. Some of the ophiolitic material thus emplaced no doubt represents the floor of the Tethys Ocean. Some of it, however, may well comprise mantle material intruded/extruded approximately at the continental margin, partly at least through a crust capable of supporting massive carbonates. There seems to be no factual evidence that such an intrusion under tension of mantle material could not take place under suitable circumstances; nor, if it has, need one expect that the products should differ significantly from those that are intruded/extruded at an active mid-ocean spreading ridge. A further corollary would be that such ophiolites, any more than any other intrusion, do not necessarily mark a former plate boundary.

If, therefore, one does not hold oneself bound by current hypothesis concerning the ophiolites, there is nothing of which the writer is aware to mar the otherwise simple and self-consistent history of the continental margins bounding the former southern Tethys, culminating in the closure of that ocean through collision between the opposing margins. The course of this evolution is illustrated, by reference to the Himalayan sector, in Figure 9.

ACKNOWLEDGMENTS

This study has been made primarily through the literature. The opportunity has nevertheless been taken to discuss the geology with those directly familiar with certain areas.

In particular, the writer has benefited from talks with colleagues in The British Petroleum Company and is especially grateful to Sir Peter Kent for considerable encouragement. It must, however, be emphasized that the writer bears the sole responsibility for the conclusions reached.

Thanks are due to the Chairman and Directors of The British Petroleum Company Limited for permission to publish this paper.

BIBLIOGRAPHY

Ahmed, S. S., 1969, Tertiary geology of part of south Makran, Baluchistan, West Pakistan: Am. Assoc. Petr. Geol. Bull., v. 53, p. 1480-1499.

Allemann, F., and Peters, T., 1972, The ophiolite-radiolarite belt of the North-Oman mountains: Ecologal. Geol. Helv., v. 65, p. 657-698.

Auden, J. B., 1973, Afghanistan-West Pakistan, *in* Spencer, A. M., ed., Mesozoic-Cenozoic orogenic belts: Geol. Soc. London, Spec. Publ. 4, p. 235-253.

Beck, R. H., 1972, The oceans, the new frontier for exploration: Austral. Petr. Expl. Soc. Jour., v. 12, pt. 2, p. 7-28.

Bizon, G., Bizon, J. J., and Ricou, L. E., 1972, Etude tertiaires de la région de Neyriz (Fars interne, Zagros Iranien): Rev. Inst. Franc. Pétrole, t. 27, p. 369-405.

Brunnschweiler, R. O., 1966, On the geology of the Indoburman ranges (Arakan Coast and Yoma, Chin Hills, Naga Hills): Jour. Geol. Soc. Australia, v. 13, p. 137-194.

Crawford, A. R., 1972, Iran, continental drift and plate tectonics: Rept. 24th Internatl. Geol. Congr., Montreal, v. 3, p. 106-112.

———, 1973, A displaced Tibetan massif as a possible source of some Malayan rocks: Geol. Mag., v. 109, p. 483-489.

Desio, A., 1973, Karakoram Mountains, *in* Spencer, A. M., ed., Mesozoic-Cenozoic orogenic belts: Geol. Soc. London Spec. Publ. 4, p. 255-266.

———, and Marussi, A., 1960, On the geotectonics of the granites in the Karakoram and Hindu Kush ranges Central Asia): Rept. 21st Internatl. Geol. Congr., Norden, v. 2, p. 156-167.

Dubertret, L., 1953, Géologie des roches vertes du nord-ouest de la Syrie et du Hatay (Turquie): Notes Mém. Moyen-Orient, t. 6, p. 1-179.

Falcon, N. L., 1967, The geology of the north-east margin of the Arabian basement shield: Advan. Sci., v. 24, p. 1-12.

———, 1973, Southern Iran: Zagros Mountains, *in* Spencer, A. M., ed., Mesozoic-Cenozoic orogenic belts: Geol. Soc. London Spec. Publ. 4, p. 199-211.

Gansser, A., 1959, Ausseralpine Ophiolithprobleme: Eclogal. Geol. Helv., v. 52, p. 659-680.

———, 1964, Geology of the Himalayas: New York, Wiley-Interscience, 289 p.

———, 1966, The Indian Ocean and the Himalayas, a geological interpretation: Eclogal. Geol. Helv., v. 59, p. 831-848.

———, 1973, Himalaya, *in* Spencer, A. M., ed., Mesozoic-Cenozoic orogenic belts: Geol. Soc. London Spec. Publ. 4, p. 267-278.

Glennie, K. W., Boeuf, M. G. A., Hughes Clarke, M. W., Moody-Stuart, M., Pilaar, W. H. P., and Reinhardt, B. M., 1973, Late Cretaceous nappes in Oman Mountains and their geologic evolution: Am. Assoc. Petr. Geol. Bull., v. 57, p. 5-27.

Gupta, H. K., and Narain, H., 1967, Crustal structure in the Himalayas and Tibet Plateau region from surface wave dispersion: Seismol. Soc. Am. Bull., v. 57, p. 235-248.

Holmes, A., 1965, Principles of physical geology (2nd ed.): London, Nelson.

Ivanac, J. F., Traves, D. M., and King, D., 1956, The geology of the northwest portion of Gilgit Agency: Records geol. Survey Pakistan, v. 8, pt. 2, p. 1-27.

Jan, M., and Kempe, D. R. C., 1973, The petrology of the basic and intermediate rocks of Upper Swat, Pakistan: Geol. Mag., v. 110, p. 285-300.

Jones, A. G., ed., 1961, Reconnaissance geology of part of West Pakistan, Oshawa, Ontario, Maracle Press.

Kamen-Kaye, M., 1972, Permian Tethys and Indian Ocean: Am. Assoc. Petr. Geol. Bull., v. 56, p. 1984-1999.

Katili, J. A., 1973, Geochronology of west Indonesia and its implication on plate tectonics: Tectonophysics, v. 19, p. 195-212.

McKenzie, D. P., and Sclater, J. G., 1971, The evolution of the Indian Ocean since the Late Cretaceous: Geophys. Jour., v. 25, p. 437-528.

Morton, D. M., 1959, The geology of Oman: Proc. 5th World Petr. Congr., New York, v. 1, p. 277-294.

Norin, E., 1946, Geological explorations in western Tibet: Rept. Sino-Swedish Exp., Publ. 29, Aktiebolaget Thule, Stockholm.

Nowroozi, A. A., 1972, Focal mechanism of earthquakes in Persia, Turkey, West Pakistan and Afghanistan and plate tectonics of the Middle East: Seismol. Soc. Am. Bull., v. 62, p. 823-850.

Peive, A. V., Burtman, V. S., Ruzhentzev, S. V., and Suvorov, A. I., 1964, Tectonics of the Pamir-Himalayan sector of Asia: Rept. 22nd Internatl. Geol. Congr., New Delhi, v. 11, p. 441-464.

Pitman, W. C., and Talwani, M., 1972, Sea-floor spreading in the North Atlantic: Geol. Soc. America Bull., v. 83, p. 619-646.

Quennell, A. M., 1958, The structural and geomorphic evolution of the Dead Sea Rift: Geol. Soc. London Quart. Jour., v. 114, p. 1-24.

Reinhardt, B. M., 1969, On the genesis and emplacement of ophiolites in the Oman Mountains geosyncline: Schweiz. Mineral. Petrog. Mitt., v. 49, p. 1-30.

Reyre, D., and Mohafez, S., 1972, A first contribution of the NIOC-ERAP agreements to the knowledge of Iranian geology: Paris, Ed. Technip.

Ricou, L. E., 1970, Comments on radiolarite and ophiolite nappes in the Iranian Zagros Mountains: Geol. Mag., v. 107, p. 479-480.

———, 1971a, Le croissant ophiolitique péri-Arabe, une ceinture de nappes mise en place au Crétacé supérieur: Rev. Géogr. Phys. Géol. Dyn. 2e sér., t. 13, p. 327-350.

———, 1971b, Le métamorphism au contact des péridotites de Neyriz (Zagros interne, Iran): développment de skarns à pyroxène: Bull. Soc. Géol. France, 7e sér., t. 13, p. 146-155.

Rigo de Righi, M., and Cortesini, A., 1964, Gravity tectonics in the foothills structure belt of south-east Turkey: Am. Assoc. Petr. Geol. Bull., v. 48, p. 1911-1937.

Rodolpho, K. S., 1969, Bathymetry and marine geology of the Andaman Basin, and tectonic implications for Southeast Asia: Geol. Soc. America Bull., v. 80, p. 1203-1230.

Schreiber, A., Weippert, D., Wittekindt, H-P., and Wolfart, R., 1972, Geology and petroleum potentials of central and south Afghanistan: Am. Assoc. Petr. Geol. Bull., v. 56, p. 1494-1519.

Schwan, W., 1971, Geology and tectonics of the central Amanos mountains of Turkey, *in* Campbell, A. S., ed., Geology and History of Turkey, Petr. Expl. Soc. Libya, Tripoli, p. 283-303.

Smith, A. G., 1971, Alpine deformation and the oceanic areas of the Tethys, Mediterranean and Atlantic: Geol. Soc. America Bull., v. 82, p. 2039-2070.

Stöcklin, J., 1968, Structural history and tectonics of Iran: a review: Am. Assoc. Petr. Geol. Bull., v. 52, p. 1229-1258.

Stoneley, R., 1972, A consideration of the northern margin of parts of Gondwanaland leading to a reconstruction of the drift history of the northwest Indian Ocean region (abstr.): EOS (Trans. Am. Geophys. Union), v. 53, p. 184. (Full text available on microfilm from Am. Geophys. Union.)

Tschopp, R. H., 1967, A general geology of Oman: Proc. 7th World Petr. Congr., Mexico., v. 2, p. 231-242.

Van Bemmelen, R. W., 1949, The geology of Indonesia: The Hague, Govt. Printing Office.

Wells, A. J., 1969, The Crush Zone of the Iranian Zagros Mountains and its implications. Geol. Mag., v. 106, p. 385-394.

Williams, H., and Smyth, W. R., 1973, Metamorphic aureoles beneath ophiolite suites and Alpine peridotites: tectonic implications with west Newfoundland examples: Am. Jour. Sci., v. 273, p. 594-621.

Wilson, H. H., 1969, Late Cretaceous eugeosynclinal sedimentation, gravity tectonics, and ophiolite emplacement in Oman Mountains, south-east Arabia: Am. Assoc. Petr. Geol. Bull., v. 53, p. 626-671.

Wolfart, R., 1967, Zur Entwicklung der paläozoischen Tethys in Vorderasien: Erdoel Kohle, v. 20, p. 168-180.

Part XI

Igneous Activity and Ancient Margins

burgh and Turcotte, 1968, Fig. 11; McKenzie, 1967) either by the close approach of asthenosphere toward the surface, and/or by the emplacement of mafic, (tholeiitic) magma associated with sea-floor spreading; such processes are inferred from the measured high-heat-flow values and from the observed volcanic activity along oceanic ridges.

The situation in the vicinity of a convergent plate junction seems to be much more complex. Isotherms are deflected downward in the subducted slab, because a relatively cool plate is descending into a hotter substrate; the pressure increment accompanying subduction is instantaneous, but because rocks are good insulators, the temperature rises very slowly relative to plate motion. Temperature-depth relationships have been computed by numerous authors, including McKenzie (1969), Oxburgh and Turcotte (1970, 1971), Toksöz et al., (1971, 1973), and Griggs (1972). The extent of downwarping of the isotherms depends on many factors, such as the rate of convergence, the thickness of the slab and its thermal history, the amount of radiogenic heat sources in the material, and the extent of decoupling (Turcotte and Schubert, 1973) between the subducted and nonsubducted masses. Adiabatic heating on compression, and heat liberated by exothermic reactions, are probably of lesser importance in raising the temperature within the subducted slab (Griggs, 1972, p. 382). Downwarp of the isotherms in a downgoing plate is reflected in the low-heat-flow values (i.e., < 1.0 microcal/cm^2/sec) obtained in the vicinity of, and slightly landward from, an oceanic trench—thermal relationships are particularly well studied adjacent to the Japanese arc and trench (Uyeda and Horai, 1964; Uyeda and Vaquier, 1968).

In contrast, the nonsubducted continental margin is the site of high heat flow (i.e., < 2.5 microcal/cm^2/sec) and abundant calc-alkaline activity (Lee and Uyeda, 1965; Sclater and Menard, 1967). Accordingly, isotherms within the earth are bowed upward toward the surface in this region. This high geothermal gradient is thought to be a consequence of heat added to the stable lithospheric slab through frictional dissipation (imperfect decoupling) along the plate junction, and due to the upward bodily flow of mafic and intermediate magmas generated at great depths (= heat transfer). The disposition of the isotherms depends rather critically on the assumed sites of origin and volumes of the melts, as well as on the extent of viscous drag between crustal plates; attempts at calculating the upward bowing of isotherms in the stable plate have been presented by Hasebe et al. (1970) and Turcotte and Oxburgh (1972). Oxburgh and Turcotte (1971, p. 1322-1325) also noted that, where the nonsubducted plate was capped by a thick sialic crust, the decay of radioactive elements would provide a further increment in the upwarp of the isotherms near the continental margin.

An example of the calculated thermal structure near a convergent plate junction is shown in generalized form in Figure 1 (after Turcotte and Oxburgh 1972, Fig. 9); heat transfer accompanying magma ascent and the heat contribution of radioactive elements concentrated in sialic material have been ignored, but the qualitative relationships are shown very adequately. A narrow belt of relatively high-pressure, low-temperature conditions lies at and beneath the locus of the plate junction and seaward of the continental margin. The latter regime, in contrast, is characterized by moderate pressures and much higher temperatures, which are developed over a broad region.

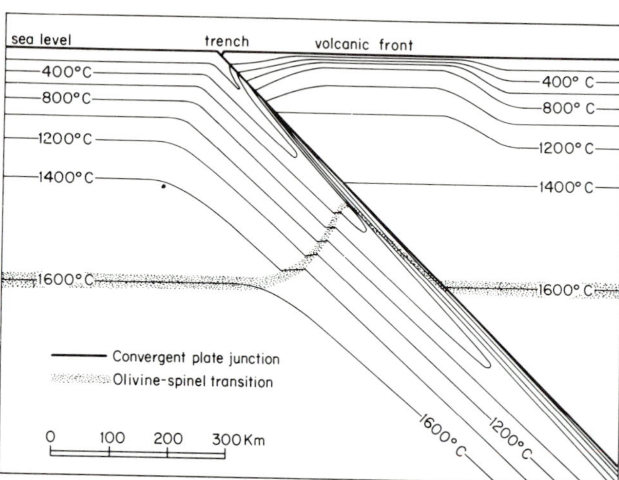

Fig. 1. Computed thermal structure of a convergent plate junction taking frictional heating into account, after Turcotte and Oxburgh (1972, Fig. 9). No vertical exaggeration. Location of the olivine-spinel transition zone is shown, assuming that the attainment of chemical equilibrium is rapid relative to subduction rates.

OVERALL TECTONICS NEAR AN ACTIVE-TYPE CONTINENTAL MARGIN

As pointed out by Dickinson (1970, 1971a, 1971b, 1972), a convergent plate junction is characterized by three principal petrotectonic belts: a trench-subduction zone, an arc-trench gap sequence, and a volcanic-batholithic arc. The farthest seaward of these is the locus of the trench mélange. Here the deposition of immature clastic sediments + deep-sea precipitates (i.e., pelagic sediments) on oceanic crust itself is interpreted as being accompanied by rapid subduction and by the underplating of buoyant, relatively low density material scraped off the downgoing slab along the landward, hanging-wall side of the trench. The arc-trench gap is a region characterized by the deposition of an apron of clastic debris shed onto the submerged portions of the continental margin and adjacent oceanic crust. Like the trench mélange, it presumably has been derived chiefly from the nearby (and landward) volcanic front. The volcanic-plutonic arc represents additions to the sialic crust through the accumulation of calc-alkaline and basaltic volcanogenic materials; thus, even if this lithologic belt begins as an oceanic island arc, the middle and late stages of igneous activity take place in a region characterized by the presence of more-or-less sialic basement. Hatherton and Dickinson (1969) observed that, for a given silica content, the amount of K_2O in a coeval lava suite erupted on the nonsubducted slab tends to increase to higher values proceeding away from the trench; inasmuch as the potash contents of modern volcanic rocks seem to be proportional to the depth of the present-day inclined seismic zone, they postulated that in some way the generation and compositions of melts are related to the depth of the Benioff zone under the island arc or continental margin. In fact, this process seems to represent the major source of addition of sialic material to the continental crust.

Subduction zones apparently are characterized by extreme tectonic imbrication and the formation of nappes and thrust faults—and in some cases by a kind of tectonic erosion of the continental margin (Page, 1970a). In many cases the direction of tectonic transport appears to be compatible with the sense of shear deduced from the inferred subduction direction (e.g., Roeder, 1973), the previously described increase of potash in volcanic rocks disposed across the continental margin (i.e., beyond the volcanic front), and the high-pressure, subduction-zone type of metamorphic zonation—for instance, in the Sanbagawa belt of Japan, the Helvetic + Pennine nappes—the Sesia-Lanzo zone of the Alps, and the Franciscan terrane of central California (Ernst, 1971a).

In contrast, the adjacent arc-trench gap section of rocks is little deformed, probably because, unlike the subduction-zone mélange, these rocks are laid down at the margin of a stable lithospheric slab. The rocks of the volcanic-plutonic arc generally reflect a long and varied history, particularly if old continental crust is involved. Unless subsidiary convergent plate junctions are present, these calc-alkaline igneous + high-temperature metamorphic terranes appear to be dominated by vertical tectonics, with large-scale, broad (to isoclinal) folds and high-angle reverse faults, deformed, reactivated, and offset by the invasion of subjacent plutons (Bateman and Eaton, 1967; Coney, 1971, Fig. 4; Rutland, 1971).

A rather clear example of an active-type continental margin seems to be present in the middle and late Mesozoic geologic record of southern and southeastern Alaska (Burk, 1965, 1966; Berg et al., 1972; Jones et al., 1972; Moore, 1973b). Three lithic belts are present, proceeding landward: (1) the Chugach (Seldovia-McHugh-Valdez) trench mélange, (2) the Matanuska-Wrangell arc-trench gap sequence, and (3) the Gravina-Nutzotin arc series.

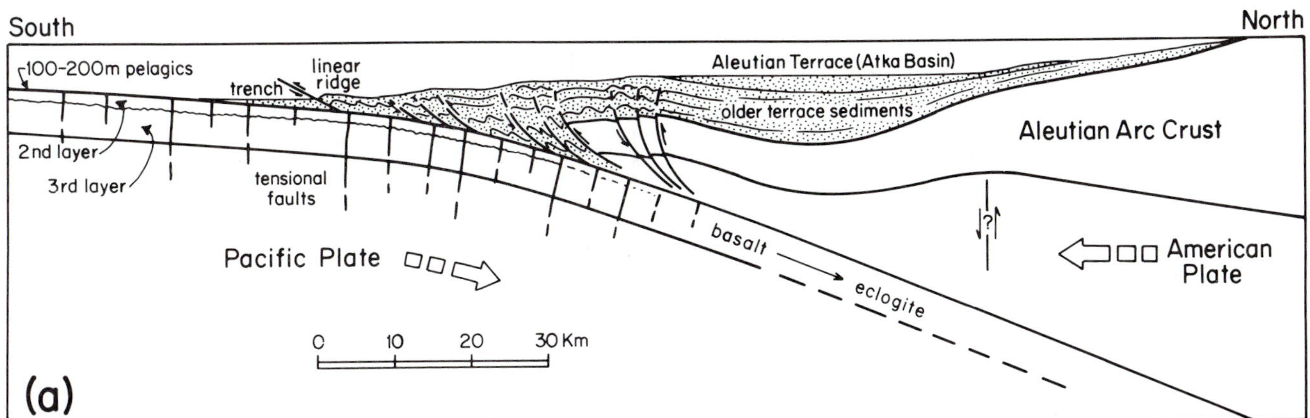

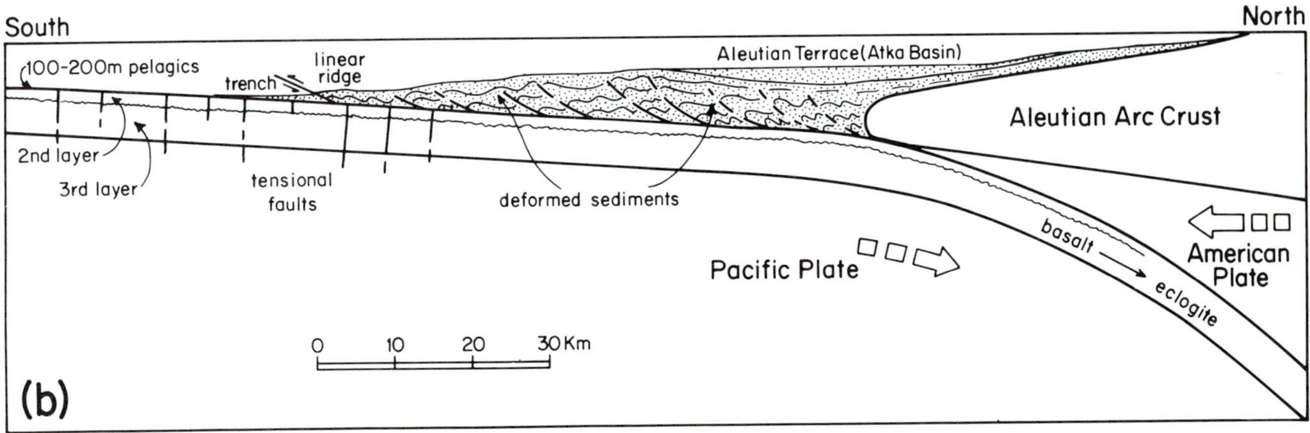

Fig. 2. Interpretive cross sections through the Aleutian arc-trench, after Grow (1973, Fig. 6). No vertical exaggeration. Slight differences in alternative cross sections, models (a) and (b), result from ambiguities in the primary geophysical data. By analogy with Mesozoic rocks exposed in southern and southeastern Alaska, model (a) is preferred (Berg et al., 1972, Fig. 1).

At least locally (e.g., near Anchorage), the subduction zone complex contains fragments of oceanic crust and blue schist (Clark, 1973; Forbes and Lanphere, 1973). Where deeply eroded, the continentward magmatic arc contains abundant plutonic calc-alkaline rocks; K-Ar radiometric data reveal a complex and long-lived sequence of igneous events spanning the time interval from Jurassic to Tertiary (Reed and Lanphere, 1973; J. C. Moore, this volume).

Interpretive structural sections of the analogous present-day Aleutian trench and arc, as deduced by Grow (1973, Fig. 6) from seismic reflections and gravity and dredge-haul information are presented in Figure 2. The angle of Pacific plate underflow beneath the American plate is very low—on the order of 5-20°—within 50-70 km of the trench wall; farther under the nonsubducted slab it steepens only to about 30°. This low angle is similar to parts of the Java Trench (Beck, 1972, Figs. 13 and 14). Inferred vertical tectonics of the volcanic-plutonic terrane are not illustrated in Figure 2, but the relationship between the trench mélange and the sedimentary basin of the arc-trench gap (the Aleutian terrace) is well shown. The figure also shows the direction of tectonic transport, with vergence of folds toward the ocean basin; imbricate thrust surfaces dip toward the continent.

METAMORPHISM NEAR AN ACTIVE-TYPE CONTINENTAL MARGIN

Miyashiro (1961) classified regional metamorphic terranes into five intergradational facies types; he related the paragenetic sequence to contrasting pressure-temperature trajectories during the metamorphic recrystallization. Miyashiro further showed that in the circum-Pacific region, essentially contemporaneous metamorphic belts are paired; a relatively high-temperature, low-pressure (andalusite-sillimanite) type coincides with the volcanic front, whereas the relatively low-temperature, high-pressure (jadeite-glaucophane) type lies oceanward. The former is characterized by the abundance of migmatites and granites, the latter by a profusion of ophiolites. Zwart (1967) also called attention to the duality of these petrotectonic belts, and pointed out that the relatively high-temperature terranes tend to be broad regions typified by telescoped metamorphic zonations, high-angle faults, and more-or-less open folds; whereas the relatively high-pressure terranes tend to be narrow, linear belts, containing broad metamorphic zones, low-angle thrust faults, and abundant nappes. Zwart argued persuasively that the high-pressure metamorphism was favored by intense deformation; in contrast, very high heat flow seemed to be required to account for the high-temperature type of metamorphism.

Other than diagenetic reactions and those attributable simply to burial, regional metamorphism at a continental margin appears to be a response of the preexisting rocks to the pressure-temperature deformational history. For this reason, paired metamorphic belts appear to be confined to Pacific-type continental margins. The overall thermal structure and tectonics of such a margin have been discussed in previous sections. Although coeval dual metamorphic belts are restricted to Pacific-type continental margins, not all such convergent plate junctions contain well-developed paired metamorphic terranes. If subduction involves the closure of only a small sea, the limited amount of underflow of the ocean crust-capped plate beneath the continent may be insufficient to provide the great amounts of calc-alkaline magmas necessary to generate a high-temperature metamorphic-igneous belt in the nonsubducted lithospheric slab [e.g., the closure of late Mesozoic Tethys in central Europe apparently involved small amounts of subduction; hence contemporaneous volcanism in the southern Alps was inappreciable, according to a synthesis by Ernst (1973a, 1973b); however, this plate motion evidently resulted in the continent-continent type of collision and is therefore not classified strictly as of Pacific type]. Or, if the downgoing plate descends relatively slowly, is too thin and/or too hot, the isotherms will be depressed less markedly, resulting in recrystallization at moderate values of pressure and temperature; such a metamorphic terrane would be difficult to distinguish from the higher-temperature metamorphic complex situated near the volcanic front (Miyashiro, 1973).

Clearly, subduction-zone metamorphism should be confined to a narrow body of rock characterized by a high ratio of pressure to temperature, by oceanic basement and by pervasive, imbricate shearing. In contrast, the volcanic-plutonic complex should be a region of very broad, relatively high-temperature recrystallization, depending on the extent of invasion by mantle transecting, practically anhydrous melts, and typified by continental basement and by chiefly vertical tectonics. High heat flow would be expected to promote the partial melting of basal portions of the H_2O-bearing sialic crust. A general model is shown as Figure 3 (see also Miyashiro, 1967, Fig. 25; 1972a, Fig. 2).

Apparently, as portions of the downgoing trench mélange begin to decouple from the subducted lithospheric plate, they rise, and their accumulation results in a building outward of the leading edge of the stable slab, as can be seen in Figure 2. Where this process is operating, the plate junction must move seaward with time, but for simplicity this aspect of the model is not shown in Figure 3.

A 45° dip to the plate junction has been assumed in Figure 3. Obviously the width of each member of an actual paired metamorphic belt complex and that of the intervening arc-trench gap will depend in part on the actual angle of descent and its change through time (e.g., Burk, 1972); this angle will also undoubtedly affect the amount of imbricate thrusting (underplating) within and adjacent to the

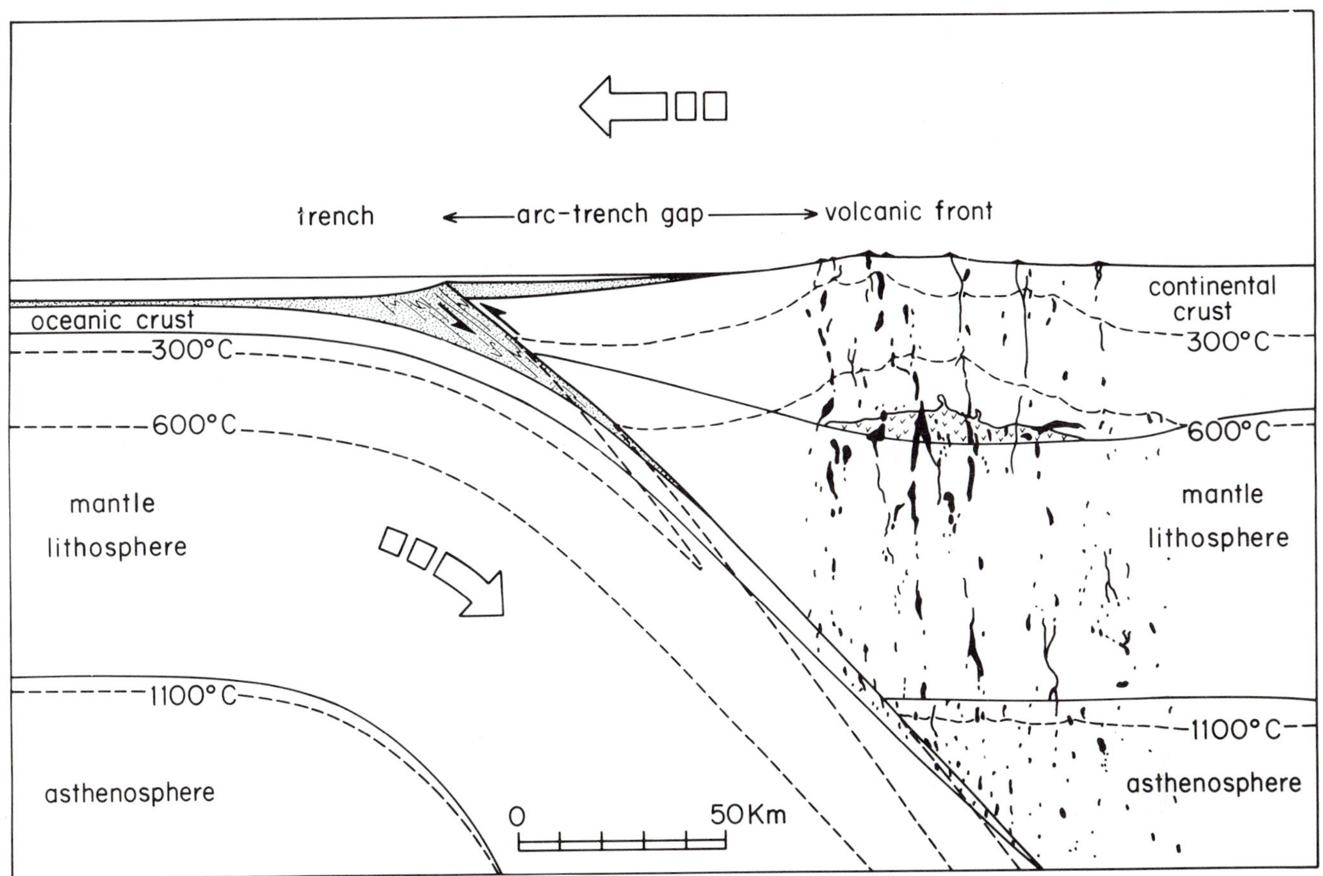

Fig. 3. Generalized tectonic model of a convergent plate junction modified from Ernst (1970, Fig. 3; 1973a, Fig. 4). No vertical exaggeration. Downward deflection of the 300°, 600°, and 1100°C isotherms are largely after Oxburgh and Turcotte (1970, Fig. 14) and Turcotte and Oxburgh (1972, Fig. 9). Partial melting of subducted oceanic crust, previously transformed to amphibolite and/or eclogite, to produce calc-alkaline magmas, and incipient melting of nonsubjected asthenosphere to produce mafic magmas (both shown in black) is modeled after experimental data summarized by Green (1972, 1973), Boettcher (1973), and Wyllie (1973); partial melting of basal, thickened portions of an H_2O-rich sialic crust (shown in checks) is modeled after experimental data of Luth et al. (1964; Fyfe, 1973).

subducted trench melange. Finally, plate thickness and thermal state, rate of convergence, and post-consumption deformation accompanying isostatic readjustment will strongly influence the final nature of the paired belt complex.

Employing metamorphic, petrogenetic relations (Fig. 4), it is possible to illustrate the spatial disposition of metamorphic facies boundaries in the vicinity of the Pacific-type continental margin, modeled in Figure 3. The facies boundaries are shown as pseudounivariant curves of moderate pressure-temperature band width in Figure 4; however, the transition zones are undoubtedly both more extensive and more complicated than indicated. Low-pressure, hornfelsic metamorphic facies and partial melting relations at high fugacities of H_2O have not been shown. The figure was constructed employing experimental phase-equilibrium data on mineral and rock systems, and utilizing oxygen isotopic information for natural analogues wherever possible; oxygen fugacities are assumed to lie well within the field of magnetite stability, and, where present, the metamorphic fluid is presumed to be largely H_2O. Somewhat similar diagrams have been presented by numerous authors (e.g., Winkler, 1967, Fig. 40; 1970; Turner, 1968, Fig. 8-6; Ernst et al., 1970, Fig. 80; Liou, 1971, Fig. 7).

Figure 5 is an attempt to show details of the thermal structure and metamorphic paragenesis presented in Figure 4 for the model continental margin illustrated in Figure 3. The more numerous metamorphic-zone boundaries present in the broad, high-temperature, low-pressure terrane are a consequence both of the greater number of separate facies recognized and the higher geothermal gradient. The accumulation of large volumes of mafic and intermediate magmas, and the great distance (100-300 km) from the trench are factors that evidently are at least partly responsible for the vertical type of orogenic activity. In constrast, the subduction zone is typified by a more simple, broader metamorphic zonation, which is confined to a narrower belt.

The spatial locations of the metamorphic boundaries depicted in Figure 5 depend critically on the correctness of Figures 3 and 4. To the extent that either or both are unrealistic, the distribution of the mineral assemblages will require modification

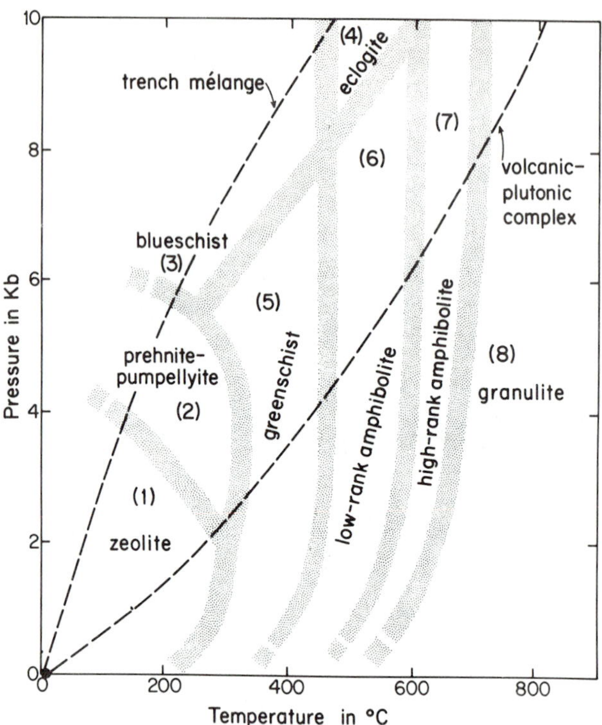

Fig. 4. Schematic metamorphic petrogenetic grid deduced from experimental - oxygen-isotope data for common rock compositions, after Ernst (1973b, Fig. 4). Low-pressure hornfelsic facies are not distinguished in this plot.

from that illustrated. The topology of the deflections of the facies boundaries in the vicinity of abnormally hot or cold regions of the earth is not immediately obvious. Perturbations of these zones are functions of whether the equilibria that define the transitions between the various metamorphic facies possess positive or negative pressure-temperature slopes, and whether the metamorphic pressure-temperature trajectory with increasing depth of burial involves passage from the low-density to the high-density equivalent mineral assemblage, or vice versa (Richardson, 1970; Ernst, 1973a).

Attention is directed to some interesting possible petrologic relationships in Figure 5. Not unexpectedly, (1) zeolite-bearing rocks, (5) green schists, (6) low- and (7) high-rank amphibolites, and finally (8) granulites are found at successively deeper levels of the continental crust at and beneath the volcanic front, whereas in the subducted mass of trench sediments and the underlying oceanic crust, the progression is from (1) zeolite through (2) prehnite-pumpellyite, (3) blue schist, and finally (4) eclogite metamorphic facies. (Whether or not the low-grade metamorphic assemblages are developed near the surface, either on the continent or in the ocean basin, depends crucially on reaction kinetics and the compositional variables, particularly that of the fluid phase.)

Away from the high heat-flow region, in the stable, sial-capped lithospheric slab, a zone of prehnite-pumpellyite facies separates the near-surface zeolites from the intermediate-depth green schists, and at the base of relatively cool continental crust, of at least 35 km thickness, amphibolites evidently may pass laterally into eclogitic rocks. Also, within the typical oceanic realm away from the trench, greenschist mineral assemblages would be expected below about 15 km depth, insofar as appropriate lithologies are present.

Perhaps the most interesting aspect of Figure 5, however, concerns the distribution of metamorphic facies within the subduction zone, but adjacent to the convergent plate junction itself. Because of conductive and frictional heating, the metamorphic progression (1) zeolite → (2) prehnite-pumpellyite → (3) blueschist → (4) eclogite characteristic of axial portions of the trench, passes laterally to the sequence (1) zeolite → (2) prehnite-pumpellyite → (5) greenschist → (6) low-rank amphibolite → (4) eclogite near the "hanging wall"; the latter paragenesis reflects a prograde pressure-temperature trajectory intermediate between the trench mélange and volcanic-plutonic complex geothermal gradients illustrated in Figure 4. All these sequences seem to be in accord with observed parageneses developed among the contrasting metamorphic facies series as recognized by Miyashiro (1961).

It seems clear from the idealized Pacific-type continental margin illustrated in Figure 5, that the metamorphic asymmetry of the linear blueschist belt is a consequence of the progressively deeper underflow of now-resurrected sections of the terrane lying progressively closer to the junction with the unmetamorphosed, non-subducted slab (Ernst, 1971a). In contrast, the much broader high-temperature metamorphic belt exhibits a crude bilateral symmetry, with the highest-grade rocks lying close to the axis of maximum calc-alkaline igneous activity, and flanked on both sides by progressively lower grade metamorphic rocks.

AN EXAMPLE FROM CALIFORNIA

Regional metamorphism along the late Mesozoic continental margin in northern and central California will be described briefly in this section. Here the latest Jurassic, Cretaceous, and Paleocene Franciscan assemblage of rocks is interpreted as representing a trench mélange; the contemporaneous Great Valley sequence, the arc-trench gap deposits; and the Paleozoic-Mesozoic igneous and metamorphic complex of the Klamath, Sierra Nevada, and Salinia provinces, the volcanic-plutonic arc (Dickinson, 1970, p. 844-849). Because relatively few petrologic studies have been made of these terranes, the delineation of metamorphic facies boundaries here must be considered to be only tentative at best. In the volcanic-plutonic arc the situation is even bleaker. Therefore, the present attempt to depict regional metamorphic trends in this complex region is admittedly highly speculative. The approximate distribution of metamorphic assemblages, employing the petrogenetic grid of Figure 4 (and the definitions and phase assemblages as presented by Ernst, 1973b, p. 2056-2057), is illustrated in Figure 6.

In the Franciscan terrane, interpreted as an exhumed subducted complex (Ernst, 1970), the overall metamorphic paragenesis has been presented by Bailey et al. (1970, Fig. 5). In addition, certain parts of this belt have been petrographically investigated in more detailed reconnaissance fashion (e.g.,Ernst, 1971b). The observed metamorphic sequence passes from feebly recrystallized laumontite-bearing metagraywackes in the northwestern coastal belt eastward through a pumpellyite zone to a lawsonite-bearing zone; at least locally, these contrasting mineralogic zones are juxtaposed along faults.

The metamorphic facies sequence is (1) zeolite → (2) prehnite-pumpellyite → (3) blueschist. Although much complicated by later faulting, this late Mesozoic suture is now generally represented by the Coast Range thrust (Bailey et al. 1970; Ernst, 1970; Page, 1970b; Raymond, 1970, 1973).

It also appears that, in general, the ages of deposition of the original rocks and the times of recrystallization, as indicated by contained fossils and by radiometric methods, decrease toward the west (Bailey et al. 1964, p. 105, 115-122; Lee et al., 1963; Keith and Coleman, 1968; Suppe, 1969; Suppe and Armstrong, 1972; Armstrong and Suppe, 1973). Possibly this relationship attests to a series of underplatings of subducted mélanges (i.e., a succession of plate junction "step-outs" oceanward, or a decrease in the angle of the plate boundary and consequent westward migration of the trench). For this reason, Franciscan metamorphic zones depicted in Figure 6 may in part be tectonic contacts. In any case, the Franciscan depositional, metamorphic, and tectonic events took place episodically or continuously during late Mesozoic (and Paleocene) time.

Great Valley strata appear to represent the late Mesozoic arc-trench gap, and as such evidently were laid down along the western margin of the stable American lithospheric plate, both on Jurassic and older portions of the Klamath-Sierra Nevada-Salinia basement and, to the west, on oceanic crust (Hackel, 1966; Bailey et al., 1970; Lanphere, 1971; Page, 1972). The general lack of metamorphism reflects relatively low values of pressure and temperature. Only where the section is very thick, as along the western side of the Sacramento Valley, did burial metamorphism take place, producing laumontite and prehnite (Dickinson et al., 1969;

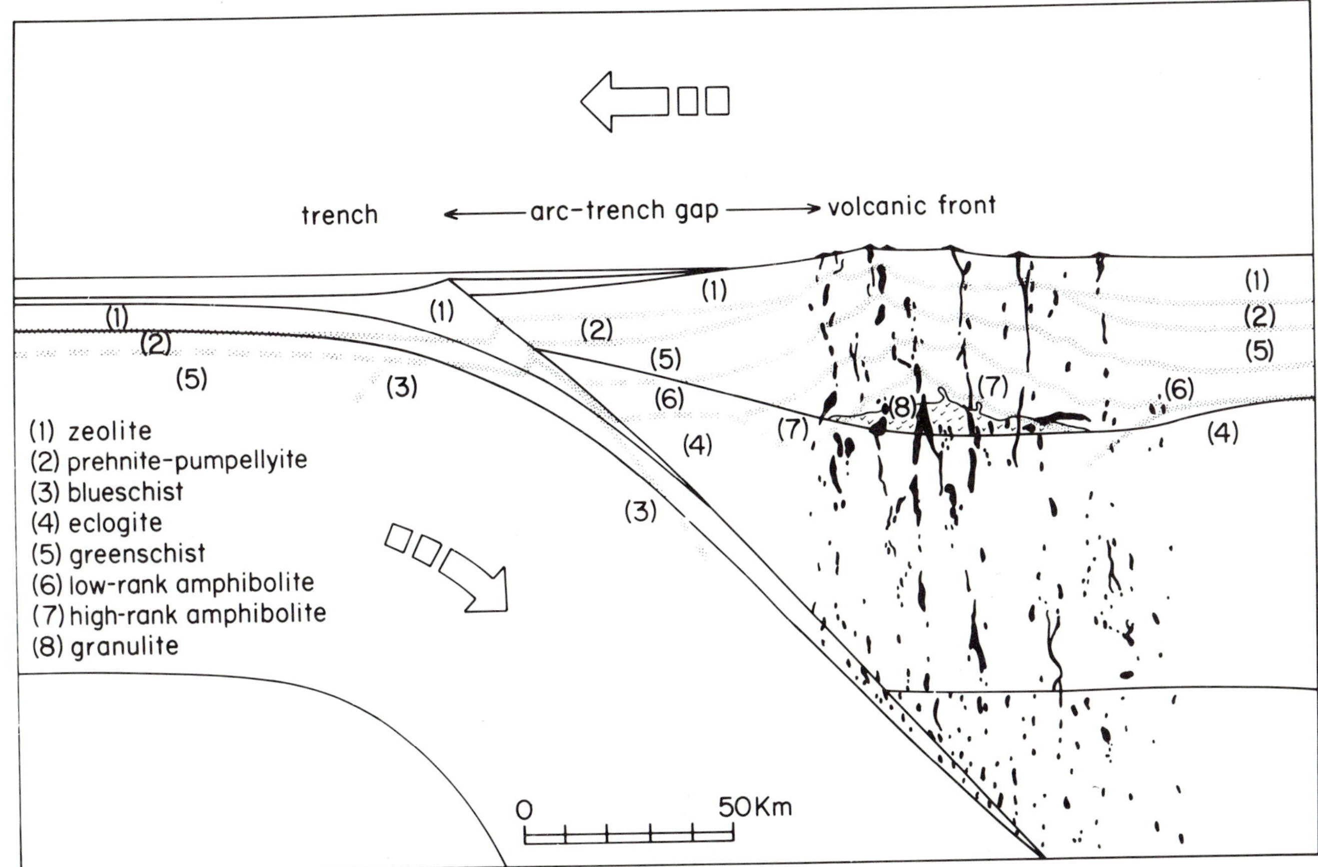

Fig. 5. Approximate spatial distribution of metamporphic facies in crustal rocks near an active convergent plate junction. The petrogenetic grid of Figure 4 has been employed with the thermal structure shown in Figure 3. As in pressure-temperature space, the metamorphic facies boundaries are zones of finite, appreciable volume within the earth. Because of the sluggishness of low-temperature reactions, protoliths may persist metastably in the near-surface environment without the widespread development of the low-grade metamorphic assemblages.

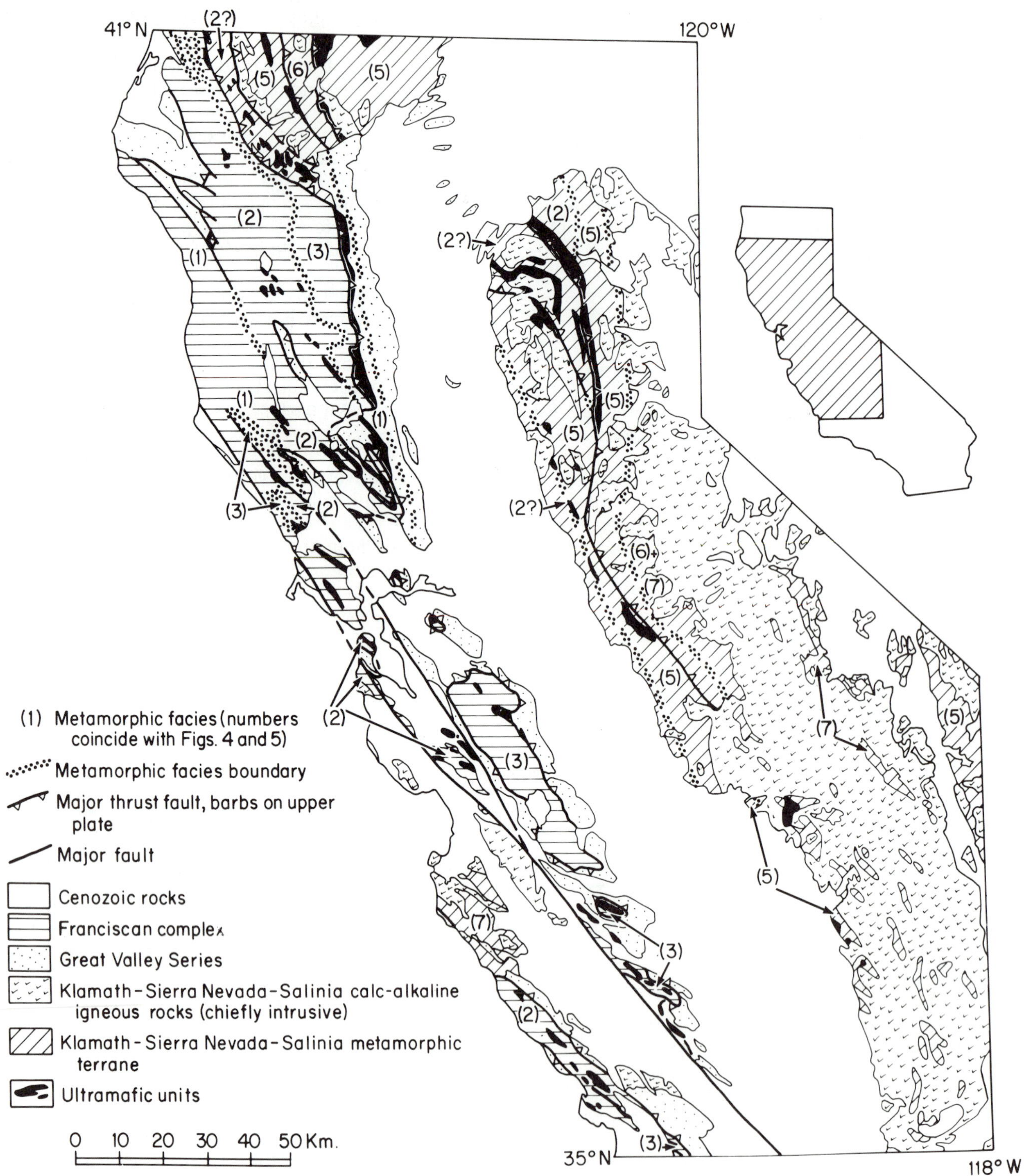

Fig. 6. **Interpretive map of part of northern and central California, showing the present distribution of metamorphic facies of chiefly late Mesozoic age. Metamorphic events of several different ages are juxtaposed; the map, therefore, is a mosaic rather than the product of a single recrystallization episode. Primary paragenetic data have been taken from Durrell (1940), Clark (1954), Compton (1955, 1958, 1960), Hamilton (1956), Parker (1961), Baird (1962), Christensen (1963), Bateman et al. (1963), Best (1963), Best and Weiss (1964), Rinehart and Ross (1964), Kistler (1966), Loomis (1966), McMath (1966), Davis (1969), Dickinson et al. (1969), Ross (1969), Bailey et al. (1970), Bailey and Jones (1973), Ernst (1971b), Kerrick (1970), and Kerrick et al. (1973).**

Bailey and Jones, 1973; Gilbert and Dickinson, 1970; Dickinson and Rich, 1972). Elsewhere, most parts of the Great Valley sequence failed to reach the zeolite grade of metamorphism and recrystallization may be classifed as authigenic.

In contrast to both Franciscan and Great Valley provinces, the Klamath-Sierra Nevada-Salinia terrane has experienced a long and very complicated tectonic and metamorphic history. In the Klamath Mountains, for instance, it is clear that imbricate thrusting has juxtaposed several contrasting terranes of mid-Paleozoic to late Jurassic metamorphic age (Irwin, 1960, 1966; Lanphere et al., 1968; G. A. Davis, personal communication, 1973). Individual sections are representative of subduction zones and volcanic-plutonic arcs (Davis, 1968; Hamilton, 1969; Irwin, 1972; Hotz, 1973). Thus rocks of the central metamorphic belt display (6) low-rank amphibolite assemblages, sutured against (5) greenschists on the east, and on the west (5) greenschist, (3) blueschists, and possibly (4) prehnite-pumpellyite facies rocks. The Klamath complex evidently represents the tectonic welding together of terrane fragments of contrasting petrologies, deformational environments, and recrystallization ages.

Assuming a rather direct correlation between this terrane and the Sierra Nevada (Davis, 1969), it is apparent that remnants of a late Jurassic, but pre-Franciscan, subduction zone mélange, faulted against parts of an older, Paleozoic continental margin are exposed in the Sierran western foothills belt (Moores, 1970; Duffield and Sharp, in press). Elsewhere, the ubiquitous, relatively high-level emplacement of mid- and late-Mesozoic magmatic rocks (Evernden and Kistler, 1970; Crowder et al., 1973) has tended to obliterate the tectonic-petrologic effects of earlier events. Because of much greater abundance and superior exposures of plutonic igneous rocks, most definitive petrologic studies have emphasized the petrology and geochronology of the granitic rocks (e.g., Bateman et al., 1963; Bateman and Wahrhaftig, 1966).

Investigations of the metamorphic rocks in Sierra Nevada roof pendants, which for the most part are small bodies, typically reveal mineral parageneses of the hornblende (and pyroxene) hornfels facies (Fyfe et al., 1958, p. 205-213; Kerrick, 1970), or high-rank amphibolite facies, as used in the present paper. Low-rank amphibolites occur in the easternmost part of the western foothills belt, whereas the central and western zones are characteristically greenschists. Prehnite-pumpellyite facies rocks occur in a northwestern part of the terrane (McMath, 1966, p. 173), and probably also along the eastern margin of the Sacramento and San Joaquin valleys (B.A. Morgan, personal communication, 1973). The paragenetic sequence evidently grades from (2) prehnite-pumpellyite facies west of the volcanic front eastward through (5) greenschists to (6) + (7) low- and high-rank amphibolites (including their hornfelsic equivalents). The eastern side of this broad region lies in Nevada, beyond the area depicted in Figure 6.

It must be cautioned, however, that amphibolite-grade terranes locally occur to the east of the Sierra Nevada batholith. Moreover, the apparent ages of calc-alkaline igneous activity vary widely throughout the entire southwestern United States (Armstrong and Suppe, 1973, Fig. 2). For these reasons the geologic relations are far more complex than shown schematically in Figure 6.

High-rank amphibolites, transitional to charnockites and granulites, are present in the complex but poorly exposed and little studied Salinia block (Compton, 1960, 1966; Wiebe, 1970b, p. 107); the time—or times—of metamorphism of these rocks is uncertain (Wiebe, 1970a, p. 1841). Ages and compositions of the late Mesozoic granitic plutons (Curtis et al., 1958) and the lithologies of the metamorphic rocks suggest affinity with the Sierra Nevada. Salinia is therefore considered to be a southern extension of this granitic terrane; its presence west of the Sierra Nevada and intervening Franciscan + Great Valley units is a consequence of 500-600 km of Cenozoic right-lateral strike slip along the San Andreas fault, which transects the late Mesozoic continental margin at a low angle (Hill and Dibblee, 1953).

The map of California presented in Figure 6 is roughly comparable to the dynamic late Mesozoic cross section of Figure 5, except that approximately 100 my have elapsed between the two highly speculative illustrations.

In summary, late Mesozoic continental margin metamorphism in northern and central California represents one aspect of the comlex thermotectonic response of the American lithospheric slab to Pacific (i.e., Kula or pre-Kula) plate underflow (Atwater, 1970). Detailed elucidation of this paired metamorphic belt will require an integration of more extensive petrologic observations with structural data.

SUMMARY

Rocks bordering ocean basins are subjected to simple burial metamorphism, except where the continental margin roughly coincides with a convergent lithospheric plate junction. In this case, frictional dissipation along the plate boundary and diapiric rise of magmas through the stable slab may cause a marked perturbation in the regional geothermal gradient. Near present-day Pacific-type convergent plate juctions, dynamic interaction of the plates produces a low heat flow in the linear trench, and a broad, high-heat-flow area in the vicinity of the volcanic arc.

A relatively cold regime, oceanward vergence of folds and continentward dipping of thrust faults, tectonic emplacement of Alpine-type peridotites, and blueschistic regional metamorphism (the jadeite-glaucophane type of Miyashiro, 1961) characterize the subduction zone. Continentward, beyond the arc-trench gap, abundant intermediate and felsic calc-alkaline (as well as basaltic) igneous rocks, high-temperature regional + hornfelsic metamorphic rocks (the andalusite-sillimanite type of Miyashiro, 1961), and chiefly vertical tectonics typify the relatively hot volcanic-plutonic complex.

The blueschistic metamorphic belt is developed principally on subducted oceanic crust, and is asymmetric in plan, with highest-grade rocks located adjacent the suture zone. In contrast, the high-temperature metamorphic belt is developed within the confines of nonsubducted continental crust well behind the plate junction; this belt ideally displays bilateral symmetry, and its highest-grade portion coincides roughly with the axis of maximum calc-alkaline igneous activity. The subduction zone terrane reflects a series of events closely related in time, space, and pressure-temperature regime, whereas the volcanic-plutonic arc consists of a complex representing many superposed metamorphic and deformational episodes from various tectonic environments.

The late Mesozoic metamorphic terranes of northern and central California are briefly described as an example. Such paired metamorphic belts are characteristic of continental margins that have experienced Pacific-type underflow. Their complex petrotectonic histories are most readily explained through an understanding of plate tectonic concepts.

ACKNOWLEDGMENTS

I thank C. A. Burk and C. L. Drake for the invitation to a most stimulating GSA Penrose Conference, which dealt with continental margins; G. A. Davis, B. A., Morgan, and John Suppe for critical review of this paper; and the University of California, Los Angeles, for support.

BIBLIOGRAPHY

Armstrong, R. L., and Suppe, J., 1973, Potassium-argon geochemistry of Mesozoic igneous rocks in Nevada, Utah and Southern California: Geol. Soc. America Bull., v. 84, p. 1375-1392.

Atwater, T., 1970, Implications of plate tectonics for the Cenozoic tectonic evolution of western North America: Geol. Soc. America Bull., v. 81, p. 3513-3536.

Bailey, E. H., and Jones, D. L., 1973, Metamorphic facies indicated by vein minerals in basal beds of the Great Valley sequence, northern California: U.S. Geol. Survey Jour. Res., v. 1, p. 383-385.

——, Irwin, W. P., and Jones, D. L., 1964, Franciscan and related rocks and their significance in the geology of western California: Calif. Div. Mines and Geology Bull., v. 183, 177 p.

———, Blake, M. C., Jr., and Jones, D. L., 1970, On-land Mesozoic oceanic crust in California Coast Ranges: U.S. Geol. Survey Prof. Paper 700-C, p. 70-81.

Baird, A. K., 1962, Superposed deformations in the central Sierra Nevada foothills east of the Mother Lode: Univ. Calif. Publ. Geol. Sci., v. 42, p. 1-70.

Bateman, P. C., and Eaton, J. P., 1967, Sierra Nevada batholith: Science, v. 158, p. 1407-1417.

———, and Wahrhaftig, C., 1966, Geology of the Sierra Nevada, *in* Bailey, E. H., ed., Geology of northern California: Calif. Div. Mines and Geology Bull., v. 190, p. 107-172.

———, Clark, L. D., Huber, N. K., Moore, J. G., and Rinehart, C. D., 1963, The Sierra Nevada batholith—a synthesis of recent work across the central part: U.S. Geol. Survey Prof. Paper 414-D, 46 p.

Beck, R. H., 1972, The oceans, the new frontier in exploration: Australian Petr. Expl. Assoc. Jour., v. 12, pt. 2, p. 5-28.

Berg, H. C., Jones, D. L., and Richter, D. H., 1972, Gravina-Nutzotin belt—tectonic significance of an Upper Mesozoic sedimentary and volcanic sequence in southern and southeastern Alaska: U.S. Geol. Survey Prof. Paper 800-D, p. 1-24.

Berkland, J. O., 1972, Paleocene "frozen" subduction zone in the Coast Ranges of northern California: Rept. 24th Internatl. Geol. Congr., Montreal, sect. 3, p. 99-105.

Best, M. G., 1963, Petrology and structural analysis of metamorphic rocks in the southwestern Sierra Nevada foothills, California: Univ. Calif. Publ. Geol. Sci., v. 42, p. 111-157.

———, and Weiss, L. E., 1964, Mineralogical relations in some pelitic hornfelses from the southern Sierra Nevada, California: Am. Mineralogist., v. 49, p. 1240-1266.

Blake, M. C., Jr., Irwin, W. P., and Coleman, R. G., 1967, Upside-down metamorphic zonations, blueschist facies, along a regional thrust in California and Oregon: U.S. Geol. Survey Prof. Paper 575-C, p. 1-9.

———, Irwin, W. P., and Coleman, R. G., 1969, Blueschist-facies metamorphism related to regional thrust-faulting: Tectonophysics, v. 8, p. 237-246.

Boettscher, A. L., 1973, Volcanism and orogenic belts—the origin of andesites: Tectonophysics, v. 17, p. 223-240.

Burk, C. A., 1965, Geology of the Alaska Peninsula—Island Arc and Continental Margin: Geol. Soc. America Mem. 99 (3 pts.), 250 p.

———, 1966, The Aleutian arc and Alaska continental margin, *in* Poole, W. H., ed., Continental margins and island arcs: Geol. Survey Can. Paper 66-15, p. 206-215.

———, 1972, Uplifted eugeosynclines and continental margins: Geol. Soc. America Mem. 132, p. 75-85.

Cann, J. R., and Funnell, B. M., 1967, Palmer Ridge: a section through the upper part of the oceanic crust?: Nature, v. 213, p. 661-664.

Christensen, M. N., 1963, Structure of metamorphic rocks at Mineral King, California: Univ. Calif. Publ. Geol. Sci., v. 42, p. 159-198.

Clark, L. D., 1954, Geology and mineral deposits of the Calaveritas quadrangle, Calaveras County, California: Calif. Div. Mines Spec. Rept. 40, 23 p.

Clark, S. H. B., 1973, The McHugh complex of south-central Alaska: U.S. Geol. Survey Bull. 1372-D, p. 1-10.

Compton, R. R., 1955, Trondhjemite batholith near Bidwell Bar, California: Geol. Soc. America Bull., v. 66, p. 9-44.

______, 1958, Significance of amphibole paragenesis in the Bidwell Bar region, California: Am. Mineralogist, v. 43, p. 890-907.

______, 1960, Charnockitic rocks of the Santa Lucia Range, California: Am. Jour. Sci., v. 258, p. 609-636.

______, 1966, Granitic and metamorphic rocks of the Salinian Block, California coast ranges, *in* Bailey, E. H., ed., Geology of northern California: Calif. Div. Mines and Geology Bull., v. 190, p. 277-287.

Coney, P. J., 1971, Structural evolution of the Cordillera Huayhuash, Andes of Peru: Geol. Soc. America Bull., v. 82, p. 1863-1884.

Crowder, D. F., McKee, E. H., Ross, D. C., and Krauskopf, K. B., 1973, Granitic rocks of the White Mountains area, California-Nevada: age and regional significance: Geol. Soc. America Bull., v. 84, p. 285-296.

Curtis, G. H., Evernden, J. F., and Lipson, J. I., 1958, Age determination of some granitic rocks in California by the postassium-argon method: Calif. Div. Mines and Geology Spec. Rept. 54, 16 p.

Davis, G. A., 1968, Westward thrust faulting in the south-central Klamath Mountains, California: Geol. Soc. America Bull., v. 79, p. 911-934.

______, 1969, Tectonic correlations, Klamath Mountains and western Sierra Nevada, California: Geol. Soc. America Bull., v. 80, p. 1095-1108.

Dewey, J. F., 1969, Continental margins: a model for conversion of Atlantic type to Andean type: Earth and Planetary Sci. Letters, v. 6, p. 189-197.

______, and Bird, J. M., 1970, Mountain belts and the new global tectonics: Jour. Geophys. Res., v. 75, p. 2625-2647.

Dickinson, W. R., 1970, Relation of andesites, granites, and derived sandstones to arc-trench tectonics: Rev. Geophys. Space Phys., v. 8, p. 813-860.

______, 1971a, Clastic sedimentary sequences deposited in shelf, slope and trough settings between magmatic arcs and associated trenches: Pacific Geology, v. 3, p. 1-14.

______, 1971b, Plate tectonic models of geosynclines: Earth and Planetary Sci. Letters, v. 10, p 165-174.

______, 1972, Evidence for plate-tectonic regimes in the rock record: Am. Jour. Sci., v. 272, p. 551-576.

______, and Rich, E. I., 1972, Petrologic intervals and petrofacies in the Great Valley sequence, Sacramento Valley, California: Geol. Soc. America Bull., v. 83, p. 3007-3024.

______, Ojakangas, R. W., and Steward, R. J., 1969, Burial metamorphism of the Late Mesozoic Great Valley sequence, Cache Creek, California: Geol. Soc. America Bull., v. 80, p. 519-526.

Duffield, W. A., and Sharp, R. V., 1974, Geology of the Sierra foothills mélange and adjacent areas, Amador County, California: U.S. Geol. Survey Prof. Paper 827 (in press).

Durrell, C., 1940, Metamorphism in the southern Sierra Nevada northeast of Visalia, California: Univ. Calif. Publ. Geol. Sci., v. 25, p. 1-118.

Ernst, W. G., 1970, Tectonic contact between the Franciscan mélange and the Great Valley sequence—crustal expression of a Late Mesozoic Benioff zone: Jour. Geophys. Res., v. 75, p. 886-901.

______, 1971a, Metamorphic zonations on presumably subducted lithospheric plates from Japan, California and the Alps: Contrib. Mineralogy Petrology, v. 34, p. 43-59.

______, 1971b, Petrologic reconnaissance of Franciscan metagraywackes from the Diablo Range, central California Coast Ranges: Jour. Petrology, v. 12, p. 413-437.

______, 1973a, Blueschist metamorphism and P-T regimes in active subduction zones: Tectonophysics, v. 17, p. 255-272.

______, 1973b, Interpretative synthesis of metamorphism in the Alps: Geol. Soc. America Bull., v. 84, p. 2053-2078.

______, Seki, Y., Onuki, H., and Gilbert, M. C., 1970, Comparative study of low-grade metamorphism in the California Coast Ranges and the Outer Metamorphic Belt of Japan: Geol. Soc. America Mem. 124, 276 p.

Evernden, J. F., and Kistler, R. W., 1970, Chronology of emplacement of Mesozoic batholithic complexes in California and western Nevada: U.S. Geol. Survey Prof. Paper 623, 42 p.

Forbes, R. B., and Lanphere, M. A., 1973, Tectonic significance of mineral ages of blueschists near Seldovia, Alaska: Jour. Geophys. Res., v. 78, p. 1383-1386.

Fyfe, W. S., 1973, The generation of batholiths: Tectonophysics, v. 17, p. 273-283.

______, Turner, F. J., and Verhoogen, J., 1958, Metamorphic reactions and metamorphic facies: Geol. Soc. America Mem. 73, 260 p.

Gilbert, W. G., and Dickinson, W. R., 1970, Stratigraphic variations in sandstone petrology, Great Valley sequence, central California Coast: Geol. Soc. America Bull., v. 81, p. 949-954.

Green, D. H., 1972, Magmatic activity as the major process in the chemical evolution of the earth's crust and mantle: Tectonophysics, v. 13, p. 47-71.

______, 1973, Contrasting melting relations in a pyrolite upper mantle under mid-oceanic ridge, stable crust and island arc environments: Tectonophysics, v. 17, p. 285-297.

Griggs, D. T., 1972, The sinking lithosphere and the focal mechanism of deep earthquakes, *in* Robertson, E. C., Hays, J. F., and Knopoff, L., eds. The Nature of the solid earth: New York, McGraw-Hill, p. 361-384.

Grow, J. A., 1973, Crustal and upper mantle structure of the central Aleutian arc: Geol. Soc. America Bull., v. 84, p. 2169-2192.

Hackel, O., 1966, Summary of the geology of the Great Valley, *in* Bailey, E. H., ed., Geology of northern California: Calif. Div. Mines and Geology Bull., v. 190, p. 217-238.

Hamilton, W. B., 1956, Geology of the Huntington Lake area, Fresno County, California: Calif. Div. Mines Spec. Rept. 46, 25 p.

______, 1969, Mesozoic California and the underflow of Pacific mantle: Geol. Soc. America Bull., v. 80, p. 2409-2430.

Hasebe, K., Fujii, N., and Uyeda, S., 1970, Thermal processes under island arcs: Tectonophysics, v. 10, p. 335-355.

Hatherton, T., and Dickinson, W. R., 1969, The relationship between andesitic volcanism and seismicity in Indonesia, the Lesser Antilles, and other island arcs: Jour. Geophys. Res., v. 74, p. 5301-5310.

Hill, M. L., and Dibblee, T. W., Jr., 1953, San Andreas, Garlock and Big Pine faults, California: Geol. Soc. America Bull., v. 64, p. 443-458.

Hotz, P. E., 1973, Blueschist metamorphism in the Yreka-Fort Jones area, Klamath Mountains, California: U.S. Geol. Survey Jour. Res., v. 1, p. 53-61.

Irwin, W. P., 1960, Geologic reconnaissance of the northern coast ranges and Klamath Mountains, California, with a summary of the mineral resources: Calif. Div. Mines and Geology Bull., v. 179, 80 p.

———, 1966, Geology of the Klamath Mountains Province, *in* Bailey, E. H., ed., Geology of northern California: Calif. Div. Mines and Geology Bull., v. 190, p. 19-38.

———, 1972 Terranes of the western Paleozoic and Triassic belt in the southern Klamath Mountains, California: U.S. Geol. Survey Prof. Paper 800-C, p. 103-111.

Isacks, B., Oliver, J., and Sykes, L. R., 1968, Seismology and the new global tectonics: Jour. Geophys. Res., v. 73, p. 5855-5899.

Jones, D. L., Irwin, W. P., and Ovenshine, A. T., 1972, Southeastern Alaska—a displaced continental fragment?: U.S. Geol. Survey Prof. Paper 800-B, p. 211-217.

Keith, T. E. C., and Coleman, R. G., 1968, Albite-pyroxene-glaucophane schist from Valley Ford, California: U.S. Geol. Survey Prof. Paper 600-C, p. 13-17.

Kerrick D. M., 1970, Contact metamorphism in some areas of the Sierra Nevada, California: Geol. Soc. America Bull., v. 81, p. 2913-2938.

———, Crawford, K. E., and Randazzo, A. F., 1973, Metamorphism of calcareous rocks in three roof pendants in the Sierra Nevada, California: Jour. Petrology, v. 14, p. 303-326.

Kistler, R. W., 1966, Structure and metamorphism in the Mono Craters quadrangle, Sierra Nevada, California: U.S. Geol. Survey Bull., 1221-E, 53 p.

Lanphere, M. A., 1971, Age of the Mesozoic oceanic crust in the California Coast Ranges: Geol. Soc. America Bull., v. 82, p. 3209-3212.

———, Irwin, W. P., and Hotz, P. E., 1968, Isotopic age of the Nevadan orogeny and older plutonic and metamorphic events in the Klamath Mountains, California: Geol. Soc. America Bull., v. 79, p. 1027-1052.

Lee, D. E., Thomas, H. H., Marvin, R. F., and Coleman, R. G., 1963, Isotope ages of glaucophane schists from Cazadero, California: U.S. Geol. Survey Prof. Paper 475-D, p. 105-107.

Lee, W. H. K., and Uyeda, S., 1965, Review of heat flow data, *in* Lee, W. H. K., ed., Terrestrial Heat Flow, Monogr. Am. Geophys. Union, Washington, D.C., p. 87-190.

Liou, J. G., 1971, Synthesis and stability relations of prehnite, $Ca_2Al_2Si_3O_{10}(OH)_2$: Am. Mineralogist, v. 56, p. 507-531.

Loomis, A. A., 1966, Contact metamorphic reactions and processes in the Mt. Tallac roof remnant, Sierra Nevada, California: Jour. Petrology, v. 7, p. 221-245.

Luth, W. C., Jahns, R. H., and Tuttle, O. F., 1964, The granite system at pressures of 4 to 10 kilobars: Jour. Geophys. Res., v. 69, p. 759-773.

McKenzie, D. P., 1967, Some remarks on heat flow and gravity anomalies: Jour. Geophys. Res., v. 72, p. 6261-6274.

———, 1969, Speculations on the consequences and causes of plate motions: Geophys. Jour., v. 18, p. 1-32.

McMath, V. E., 1966, Geology of the Taylorsville area, northern Sierra Nevada, *in* Bailey, E. H., Geology of Northern California: Calif. Div. Mines and Geology Bull., v. 190, p. 173-183.

Melson, W. G., and van Andel, T. H., 1966, Metamorphism in the Mid-Atlantic Ridge 22°N. latitude: Marine Geology, v. 4, p. 165-186.

———, Thompson, G., and van Andel, T. H., 1968, Volcanism and metamorphism in the Mid-Atlantic Ridge, 22°N. latitude: Jour. Geophys. Res., v. 73, p. 5925-5941.

Miyashiro, A., 1961, Evolution of metamorphic belts; Jour. Petrology, v. 2, p. 277-311.

———, 1967, Orogeny, regional metamorphism and magmatism in the Japanese islands: Medd. Dansk Geol. Foren. v. 17, p. 390-446.

———, 1972a, Metamorphism and related magmatism in plate tectonics: Am. Jour. Sci., v. 272, p. 629-656.

———, 1972b, Pressure and temperature conditions and tectonic significance of regional and ocean-floor metamorphism: Tectonophysics, v. 13, p. 141-159.

———, 1973, Paired and unpaired metamorphic belts: Tectonophysics, v. 17, p. 241-254.

———, Shido, F., and Ewing, M., 1970, Petrologic models for the Mid-Atlantic Ridge: Deep-Sea Res., v. 17, p. 109-123.

———, Shido, F., and Ewing, M., 1971, Metamorphism in the Mid-Atlantic Ridge near 24° and 30° N: Phil. Trans. Roy. Soc. London, ser. A, v. 268, p. 589-603.

Moore, J. C., 1973a, Complex deformation of Cretaceous trench deposits, southwestern Alaska: Geol. Soc. America Bull., v. 84, p. 2005-2020.

———, 1973b, Cretaceous continental margin sedimentation, southwestern Alaska: Geol. Soc. America Bull., v. 84, p. 595-614.

Moores, E., 1970, Ultramafics and orogeny, with models of the U.S. Cordillera and the Tethys: Nature, v. 228, p. 837-842.

Oxburgh, E. R., and Turcotte, D. L., 1968, Mid-ocean ridges and geothermal distribution during mantle convection: Jour. Geophys. Res., v. 73, p. 2643-2661.

———, and Turcotte, D. L., 1970, Thermal structure of island arcs: Geol. Soc. America Bull., v. 81, p. 1665-1688.

———, and Turcotte, D. L., 1971, Origin of paired metamorphic belts and crustal dilation in island arc regions: Jour. Geophys. Res., v. 76, p. 1315-1327.

Page, B. M., 1970a, Sur-Nacimiento fault zone of California: continental margin tectonics: Geol. Soc. America Bull., v. 81, p. 667-690.

———, 1970b, Time of completion of underthrusting of Franciscan beneath Great Valley rocks west of Salinia block, California: Geol. Soc. America Bull., v. 81, p. 2825-2834.

———, 1972, Oceanic crust and mantle fragment in subduction complex near San Luis Obispo, California: Geol. Soc. America Bull., v. 83, p. 957-972.

Parker, R. B., 1961, Petrology and structural geometry of pre-granitic rocks in the Sierra Nevada, Alpine county, California: Geol. Soc. America Bull., v. 72, p. 1789-1806.

Raymond, L. A., 1970, Cretaceous sedimentation and regional thrusting, northeastern Diablo Range, California: Geol. Soc. America Bull., v. 81, p. 2123-2128.

———, 1973, Tesla-Ortigalita fault, Coast Range thrust

fault, and Franciscan metamorphism, northeastern Diablo Range, California: Geol. Soc. America Bull., v. 84, p. 3547-3562.

Reed, B. L., and Lanphere, M. A., 1973, Alaska-Aleutian Range batholith: geochronology, chemistry and relation to circum-Pacific plutonism: Geol. Soc. America Bull., v. 84, p. 2583-2610.

Richardson, S. W., 1970, The relation between a petrogenetic grid facies series and the geothermal gradient in metamorphism: Fortschr. Mineral., v. 47, p. 65-76.

Rinehart, C. D., and Ross, D. C., 1964, Geology and mineral deposits of the Mount Morrison quadrangle, Sierra Nevada, California: U.S. Geol. Survey Prof. Paper 385, 106 p.

Roeder, D. H., 1973, Subduction and orogeny: Jour. Geophys. Res., v. 78, p. 5005-5024.

Ross, D. C., 1969, Descriptive petrography of three large granitic bodies in the Inyo Mountains, California: U.S. Geol. Survey Prof. Paper 601, 47 p.

Rutland, R. W. R., 1971, Andean orogeny and ocean floor spreading: Nature, v. 233, p. 252-255.

Sclater, J. G., and Menard, H. W., 1967, Topography and heat flow of the Fiji Plateau: Nature, v. 216, p. 991-993.

Suppe, J., 1969, Times of metamorphism in the Franciscan terrain of the northern Coast Ranges, California: Geol. Soc. America Bull., v. 80, p. 135-142.

———, 1972, Interrelationships of high-pressure metamorphism, deformation and sedimentation in Franciscan tectonics, U.S.A.: Rept. 24th Internatl. Geol. Congr., Montreal, sect. 3, p. 552-559.

———, and Armstrong, R. L., 1972, Potassium-argon dating of Franciscan metamorphic rocks: Am. Jour. Sci., v. 272, p. 217-233.

Takeuchi, H., and Uyeda, S., 1965, A possibility of present-day regional metamorphism: Tectonophysics, v. 2, p. 59-68.

Toksöz, M. N., Minear, J. W., and Julian, B. R., 1971, Temperature field and geophysical effects of a downgoing slab: Jour. Geophys. Res., v. 76, p. 1113-1138.

———, Sleep, N. H., and Smith, A. T., 1973, Evolution of the downgoing lithosphere and the mechanisms of deep focus earthquakes: Geophys. Jour., in press.

Turcotte, D. L., and Oxburgh, E. R., 1972, Mantle convection and the new global tectonics: Ann. Rev. Fluid Mech., v. 4, p. 33-68.

———, and Schubert, G., 1973, Frictional heating of the descending lithosphere, Jour. Geophys. Res., v. 78, p. 5876-5886.

Turner, F. J., 1968, Metamorphic petrology: New York, McGraw-Hill, 403 p.

———, and Verhoogen, J., 1960, Igneous and metamorphic petrology: New York, McGraw-Hill, 694 p.

Uyeda, S., and Horai, K., 1964, Terrestrial heat flow in Japan: Jour. Geophys. Res., v. 69, p. 2121-2141.

———, and Vaquier, V., 1968, Geothermal and geomagnetic data in and around the island arc of Japan, *in* Knopoff, L., Drake, C. L., and Hart, P. J., eds., The crust and upper mantle of the Pacific area, Monogr. 12: Washington, D.C., Am. Geophys. Union, p. 349-366.

Wiebe, R. A., 1970a, Pre-Cenozoic tectonic history of the Salinian block western California: Geol. Soc. America Bull., v. 81, p. 1837-1842.

———, 1970b, Relations of granitic and gabbroic rocks, northern Santa Lucia Range, California: Geol. Soc. America Bull., v. 81, p. 105-116.

Winkler, H. G. F., 1967, Petrogenesis of metamorphic rocks: New York, Springer-Verlag, 237 p.

———, 1970, Abolition of metamorphic facies, introduction of the four divisions of metamorphic stage, and of a classification based on isograds in common rocks: Neues Jahrb. Mineral. Monatsh., p. 189-248.

Wyllie, P. J., 1973, Experimental petrology and global tectonics—a preview: Tectonophysics, v. 17, p. 189-209.

Zwart, H. J., 1967, The duality of orogenic belts: Geol. Mijnbouw, v. 46, p. 283-309.

Ophiolites and Ancient Continental Margins

R. G. Coleman and W. P. Irwin

INTRODUCTION

A consideration of continental margins would not be complete without a discussion of the large tracts of ophiolites, the assemblage of mafic to ultramafic rocks so common along recent and ancient continental margins. Present-day plate tectonic models show ophiolite as oceanic crust being generated at mid-ocean ridges, from whence it slowly migrates toward continental margins—there to be subducted into the mantle. Formation of volcanic arcs and development of calc-alkaline plutonism have been ascribed to this consumption of oceanic crust (Dickinson, 1970). Under some circumstances the slabs of oceanic crust and mantle *overlie* the continental crust, and have apparently overridden (*obducted*) the continental edges (Coleman, 1971b), but an acceptable mechanism for emplacement of slabs of dense oceanic crust on top of lighter continental crust along such continental margins has not been published. Perhaps the continental edge can shear off large sheet-like masses (flakes) of oceanic crust, similar to shavings from a wood plane, the flakes then being driven onto the continental plate during subduction (Oxburgh, 1972). Some geologists (Lockwood, 1971; Maxwell, 1969) have suggested gravity sliding during uplift of the adjacent ocean crust as an emplacement mechanism. Others (Moores, 1970; Dewey and Bird, 1971; Roeder, 1973) have proposed "aborted" subduction.

Although the mechanism of ophiolite emplacement along continental margins doubtless will be debated for a long time, most geologists agree that these slabs of mafic-ultramafic rocks are allochthonous and that they originated in an environment distinctly different from where they occur today. This view contrasts with older debates regarding the ultramafic parts of ophiolites, which centered on the concept of their intrusion as magmas (Hess, 1938). It is now apparent for most examples, where there is a clear absence of high-temperature metamorphic aureoles within adjacent sedimentary rocks, that the ultramafic rocks (peridotites) were crystalline before being tectonically emplaced. Not all the associated basalts, diabases, and gabbros can be explained in a similar manner, however, so some geologists call on submarine volcanic extrusion of the mafic rocks into sedimentary troughs (Brunn, 1960; Aubouin, 1965), to be uplifted and exposed at a later time. Here we summarize data obtained from current studies of ophiolites, which may provide additional evidence bearing on the origin of these rocks and their emplacement along continental margins. The reader is cautioned that these topics are still the subject of active debate.

ANATOMY OF OPHIOLITES

Ophiolites were first studied by European geologists. The name "ophiolite" is derived from the Greek root ophi-, meaning snake or serpent, and was used to refer to serpentinite and associated rocks because sheared serpentinite commonly has green shiny surfaces with a snakeskin-like texture. The ophiolite suite, characterized by the association of red chert, pillow lava, and serpentinite (peridotite), was often referred to as "Steinmann's Trinity." Ophiolites are common in the ancient Tethyan belt from the Mediterranean eastward to the Himalaya Mountains (Gansser, 1959), and were originally considered to represent thick submarine volcanic outpourings developed in the early stages of the Tethyan geosyncline (Steinmann, 1927; Brunn, 1956; Aubouin, 1965). Later orogenic movement often dismembered these layered ophiolite sequences, obscuring their original stratigraphy. This dismemberment produced chaotic mélanges as part of nappes or in major thrust faults (Gansser, 1959; Bailey and McCallien, 1950). Often, as a mapping convenience, rocks as diverse as serpentinite, chert, and pillow lavas were combined into a single unit and called ophiolites (Kaz'min, 1971).

Thus the geologic literature in the first half of the twentieth century contains numerous conflicting and confusing accounts of these rocks. Petrologists of this period generally preferred to work on rocks that had a much less obscure geologic setting, so the term "ophiolite" continued to be used by workers in a broad, ill-defined manner. During the later 1960s and early 1970s, largely as a result of the development of the plate tectonic paradigm, many authors suggested that ophiolites may represent allochthonous fragments of ancient oceanic crust (Gass, 1967; Coleman, 1971b; Dewey and Bird, 1971).

Recent intensive studies of oceanic areas provide a coherent picture of the probable nature of the oceanic crust. These studies, representing a combination of geophysical and petrologic observations, show that the oceanic part of the lithosphere is divisible into three distinct layers (Cann, 1968; Shor and Raitt, 1969; Christensen, 1970) (Fig. 2). Comparison of the oceanic crustal sequences with the established ophiolite sequences of the Tethyan and circum-Pacific belts yields an obvious similarity

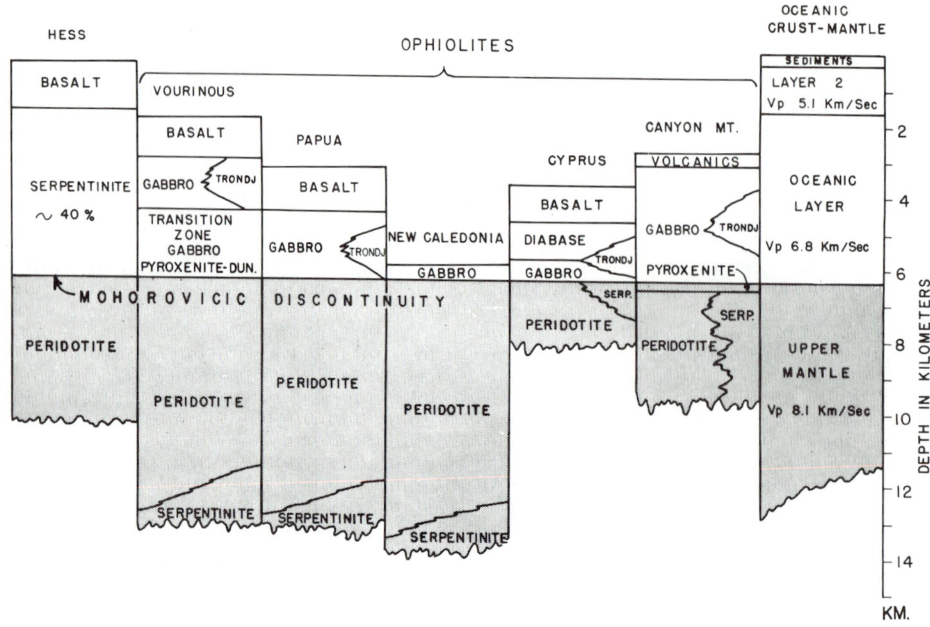

Fig. 1. Comparison of the stratigraphic thickness of igneous units from various ophiolite masses with the geophysical estimate of the oceanic crustal layers (from Coleman, 1971b).

(Bailey et al., 1970; Coleman, 1971b; Dewey and Bird, 1971) (Fig. 2). Thus a major breakthrough was initiated in the study of alpine-ultramafic and associated mafic rocks. The high-temperature history of the peridotites and mafic rocks could be explained as part of their origin at spreading ridges in the oceans (Ave'Lallemant and Carter, 1970), and their present low-temperature tectonic contacts were interpreted to result from tectonic emplacement at convergent plate boundaries (Coleman, 1966, 1967). As these new concepts have been employed toward solution of geologic problems, the fog of ignorance surrounding ophiolites has begun to lift.

A recent conference on ophiolites established a general definition of ophiolite as follows (Geotimes, 1972, p. 25): "Ophiolite as used by those present at the GSA Penrose Conference on ophiolites, refers to a distinctive assemblage of mafic to ultramafic rocks. It should not be used as a rock name or as a lithologic unit in mapping. In a completely developed ophiolite the rock types occur in the following sequence, starting from the bottom and working up:

"Ultramafic complex, consisting of variable proportions of harzburgite, lherzolite and dunite, usually with a metamorphic tectonic fabric (more or less serpentinized).

"Gabbroic complex, ordinarily with cumulus textures commonly containing cumulus peridotites and pyroxenites and usually less deformed than the ultramafic complex.

"Mafic sheeted dike complex.

"Mafic volcanic complex, commonly pillowed.

"Associated rock types include (1) an overlying sedimentary section typically including ribbon cherts, thin shale interbeds, and minor limestones; (2) podiform bodies of chromite generally associated with dunite; (3) sodic felsic intrusive and extrusive rocks.

Faulted contacts between mappable units are common. Whole sections may be missing. An ophiolite may be incomplete, dismembered or metamorphosed, in which case it should be called partial, dismembered or metamorphosed ophiolite. Although ophiolite generally is interpreted to be oceanic crust and upper mantle the use of the term should be independent of its supposed origin."

Within the ultramafic complexes there also occur small masses of orthopyroxenite and clinopyroxenite. The orthopyroxenites form cross-cutting dikes, whereas the clinopyroxenites develop irregular-shaped masses with indistinct boundaries. These pyroxenites appear to be late intrusive rocks within the metamorphic peridotites but are not part of the overlying cumulate peridotite or gabbro.

Perhaps the biggest obstacle to identifying and mapping ophiolites is that commonly they have been modified by partial erosion of formerly complete sequences, tectonic dismemberment, or metamorphism. Thus it is difficult to determine if incomplete or modified ophiolite sequences represent parts of oceanic crust or rocks produced by other geologic processes. Recently, Jackson and Thayer (1972) have distinguished among stratiform, concentric, and alpine peridotite-gabbro complexes. Where only the ultramafic segments of an ophiolite are present, the following characteristics summarized by Jackson and Thayer (1972, p. 289) will provide clues to its origin:

"Alpine complexes are characterized by dunite, harzburgite or lherzolite, olivine gabbro; norite and anorthosite are rare. The rocks generally have tectonite fabrics; gneissic foliation and lineation are characteristic. Layers are lenticular, irregular, commonly discordant and tightly folded. Contact metamorphism is generally slight to obscure; many complexes are fault-bounded. Magmatic deposits consist of high-chromium to high-aluminium chromite. Alpine peridotites may be subdivided into harzburgite and lherzolite types." The subdivision of alpine peridotites into two subtypes (lherzolite and harzburgite) has proved valid for the Mediterranean provinces but not clearly so in other parts of the world (Nicolas and Jackson, 1972).

An important characteristic of ophiolitic peridotites is their metamorphic fabric, the product of subsolidus recrystallization (den Tex, 1971). Experimental deformation of dunites, harzburgites, and lherzolites has shown that most fabrics observed in natural peridotites can be duplicated in the laboratory (Ave'Lallemont and Carter, 1970). Retention of these fabrics in natural peridotites clearly shows deformation under mantle pressure-temperature conditions and also a later deformation resulting from crustal emplacement or metamorphism (Nicolas et al., 1971). Mineral assemblages and bulk composition of peridotites belonging to an ophiolite are more indicative of a history of equilibration and reequilibration under metamorphic conditions (i.e., partial melting or unmixing at various pressure-temperature conditions within the mantle) than of fractional crystallization from a basaltic melt (O'Hara, 1967).

Where ophiolite sequences are relatively complete, such as at Cyprus, Vourinos, and Newfoundland, it now is apparent that the metamorphic peridotites have a different history of deformation and reequilibration than the overlying cumulate ultramafic and mafic rocks, diabase dikes, and pillow lavas (Moores, 1969; Moores and Vine, 1971; Nicolas and Jackson, 1972). Generally there is a sharp break between the metamorphic peridotites and the overlying constructional pile of gabbros, diabase, and pillow lavas, and in some instances they have become separated entirely. This break between the metamorphic peridotites and the overlying constructional pile marks a major discontinuity in time and formation within ophiolite sequences, shown also by isotopic comparisons.

The gabbroic rocks commonly are intimately interlayered with ultramafic rocks near the base of these constructional piles. In some places the cumulate ultramafics may exceed the gabbros and easily can be mistaken for part of the underlying metamorphic peridotites. However, where petrologic details are available, the ultramafic cumulate sequence indicates crystallization at shallow depths. Within the upper parts of the layered gabbros, numerous leucocratic intrusive masses of plagiogranite (tonalite, diorite, trondjhemite, albite granite) are usually developed. These rocks apparently have been produced by differentiation of the mafic magma. These leucocratic rocks commonly enclose brecciated fragments of gabbro or penetrate upward into overlying dike swarms. Coarse-grained gabbro dikes or pegmatitic segregations may also invade the lower parts of the constructional pile. Thus the gabbroic parts of most ophiolites have a complex igneous history of layered rocks cut by dikes or by leucocratic differentiates.

In the ideal ophiolite sequence, swarms of diabase dikes form a zone that separates the underlying gabbro-plagiogranite from the extrusive pillow lavas. Characteristically dike swarms have asymmetric chill zones, and pre-dike rock is completely absent (Moores and Vine, 1971). They are believed to represent the formation of new crust at spreading centers within the oceans. They are probably feeders for the overlying pillow lavas extruded on the ocean floor, and they die out downward into the gabbroic complex from which they presumably came. Structural relations between the dike swarms and cumulate layering in ancient ophiolites commonly indicate that major tectonic movement or rotation occurred between these two events and cannot be reconciled with the simple models of oceanic crust now generally accepted (Thayer, 1973).

Furthermore, the scarcity of dike swarms within known ophiolite complexes of the world is puzzling and perhaps indicates that they are not a universal part of the ocean crust or that not all ophiolites were formed at spreading ridges. Pillow lavas that form the upper parts of ophiolite sequences vary in thickness and may be penetrated by the underlying diabase dikes. Interlayering of pelagic sedimentary rocks near the top of the pillows is evidence of their deep ocean environment. Commonly the pillow basalts are missing and instead keratophyric lavas occur at the top of the ophiolite sequence.

Chemical data on modern oceanic crust in newly formed rifts permit comparison with much new data on the ophiolites (Kay et al., 1970; Miyashiro et al., 1970; Pearce and Cann, 1971; Varne and Rubenach, 1972). Characteristically, the basalts of the deep ocean areas are subalkaline tholeiites, gabbros, and rare plagiogranites and are very similar to members of the ophiolite assemblage. Ultramafic rocks dredged from the ocean crust are also very similar to peridotites of ophiolite sequences (Engel and Fisher, 1969; Bonatti et al., 1971). However, there is some indication that the compositions of mafic rocks from island arcs and marginal basins may also resemble and overlap those of ophiolites (Ewart and Bryan, 1973). The peridotites, gabbros, and basalts of ophiolites have distinctive chemical compositions, for which there are analogues in the oceanic crust (Coleman, 1971b).

Partial melting of mantle peridotites at depths less than 50 km can produce subalkaline basaltic liquid below a spreading center (Green, 1970). Differentiation of some of this liquid in the upper levels of the ridge system could give rise to the plagioclase-rich gabbros often found associated with ophiolites and reported from oceanic ridges.

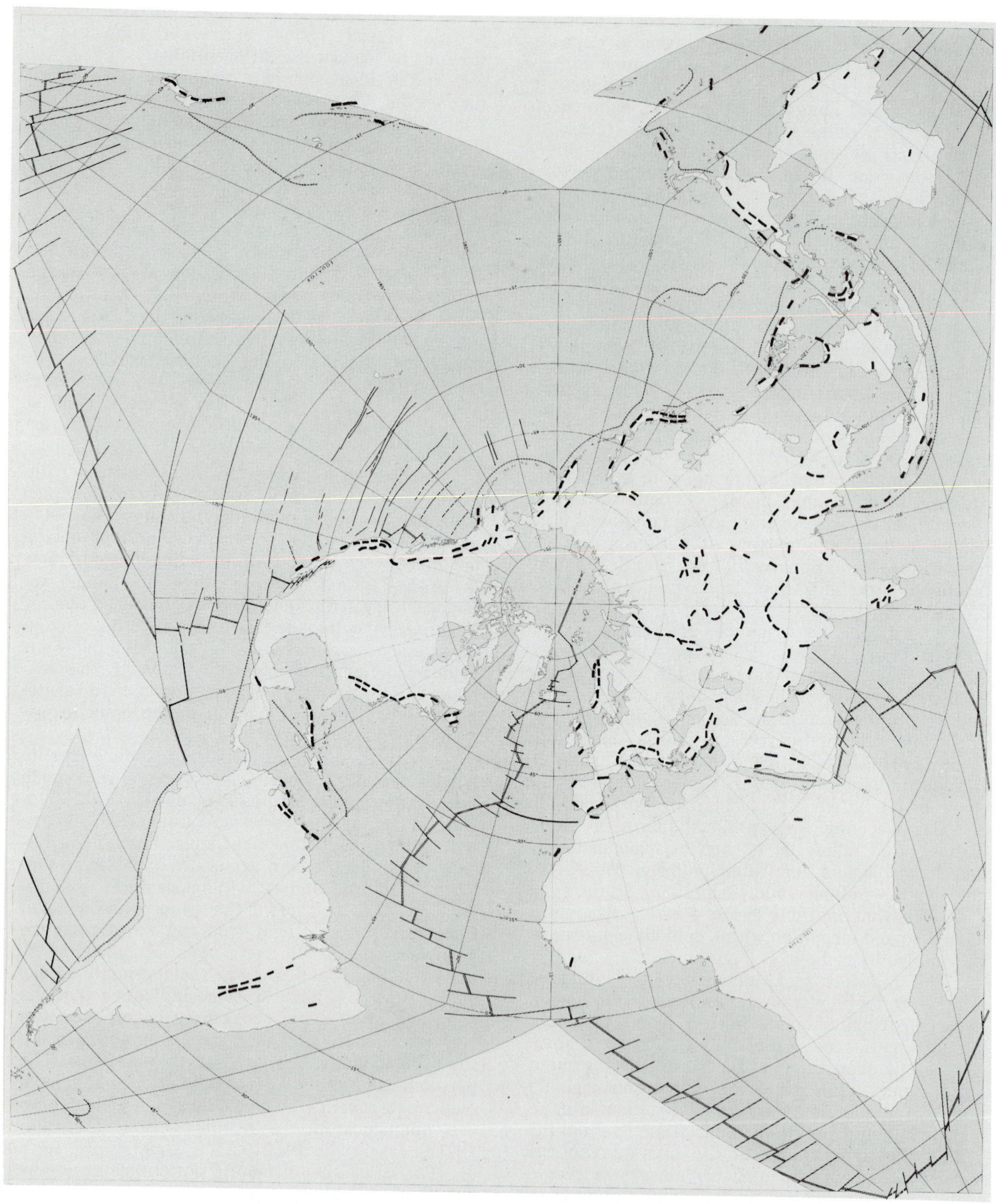

Fig. 2. North polar projection showing the principal ophiolite belts of the world. The trends of the ophiolite belts, shown by heavy dashed lines, are based mainly on the distribution of alpine-type ultramafic rocks (modified from Irwin and Coleman, 1972).

Fractional crystallization of predominately plagioclase plus olivine and/or pyroxene could also account for the small amounts of leucocratic differentiates found at the top of the gabbros. Residues developed by partial melting of primitive mantle peridotites may be represented by the widespread depleted and metamorphic harzburgite-dunite masses that underlie the constructional parts of some ophiolite sequences. Partial melting (20-30%) of lherzolite will leave a depleted harzburgite-dunite residue extremely low in calcium and aluminum and produce a subalkaline basalt liquid whose composition is controlled by fractionation at different pressures and temperatures before extrusion at the ridge axes (O'Hara, 1970).

The cumulate gabbros, unusually high in calcium and aluminum, that rest on the metamorphic peridotites represent crystallization of partially differentiated subalkaline basalt liquids rather than parts of a continuous differentiated series from early ultramafic rocks through gabbro (Coleman, 1971b). As stated earlier, cumulate gabbros and peridotites situated above the metamorphic peridotites in ophiolites may represent shallow magma chambers within the rift zone. Continued disruption of these magma chambers by spreading could explain the very complex features seen within the gabbroic parts of ophiolites.

The depleted metamorphic peridotites have deformational styles that may be related to their movement within and away from the spreading axis of the ridge (den Tex, 1971; Ave'Lallemant and Carter, 1970). The break between metamorphic peridotites and overlying gabbros, dikes, and pillow lavas is also marked by the differences in distribution of the incompatible elements (Rb, Sr, K, Ba). Initial $^{87}Sr/^{86}Sr$ ratios for rocks within the constructional piles are similar and indicate derivation from the same source in the mantle (Peterman et al., 1971). However, initial $^{87}Sr/^{86}Sr$ ratios so far measured in the metamorphic peridotites from ophiolites generally have much higher values than their associated overlying mafic rocks and are much too high to have been the residue of partial melting of modern oceanic basalts (Lanphere, 1973). These chemical differences further confirm the major break within ophiolite sequences and raise the question of the real relationship between these two different rock types.

Assuming that ophiolites represent slabs of ancient oceanic crust, there remain certain unresolved problems of their formation. Conceivably, at least some ophiolites represent island arcs built upon oceanic crust of marginal basins (Ewart and Bryan, 1973) or, as suggested by Nicolas et al. (1972), some lherzolite peridotites may be continental mantle. Nonetheless, the present concept that most ophiolites are fragments of former oceanic lithosphere provides a key to the history of many continental margins and convergent plate boundaries or sutures.

DISTRIBUTION OF OPHIOLITES

Throughout the world, ophiolites are exposed typically along belts of intense tectonism (Fig. 1). They occur along major geosutures and are thought to mark the sites of ancient zones of interaction between oceanic and continental crust, even though many parts of the older belts are now well within the interior of some continental areas.

In general, structures within the suture zones parallel the trend of the zones. Mélanges comprising parts of ophiolite mixed with enclosing sedimentary and metamorphic rocks commonly mark the boundaries of the suture zones. The common occurrence of blueschist metamorphic belts in suture zones containing ophiolites is another significant feature. Amphibolite facies rocks are also present in some areas, but most of the metamorphic rocks of these suture zones seem to have formed under regional dynamothermal conditions rather than contact metamorphism, and their deformational style seems to be related to the tectonic emplacement of the ophiolites. Williams and Smyth (1973) describe metamorphic aureoles attached to the base of ophiolites in western Newfoundland that have been tectonically transported. They consider these metamorphic rocks to have been formed as contact dynamothermal aureoles formed in the early stages of mantle obduction.

The early Paleozoic ophiolites are represented by the Appalachian and Caledonian belts of eastern North America, western Fennoscandia, and the Grampian Highlands of England. Perhaps the largest concentration of Paleozoic ophiolites is in the Ural Mountains, continuing eastward into Mongolia, and includes central Kazakhstan, Tien Shan, Altai-Sayan, western Siberia, and the Mongolia-Okhotsk folded region. Along the Pacific Coast of North America, Paleozoic ophiolites are present in the Klamath Mountains of northern California, in western Canada, and in northern Alaska. In eastern Australia, ophiolites extend from Queensland southward to Tasmania.

Mesozoic and Cenozoic ophiolites are much more abundant and constitute most of the important large exposures of ophiolite. The Tethyan belt extends from the Betic Cordillera and Rif of Spain and Africa eastward through the Alps, the Dinarides in Yugoslavia, through Greece, Turkey, Iran, Oman, Pakistan, the Himalayas, Burma, and Indonesia, where it connects with the circum-Pacific ophiolite belt. Within the Pacific area, ophiolite belts of Mesozoic and Cenozoic age extend from New Zealand northwesterly to New Caledonia, New Guinea, Celebes, Borneo, Philippine Islands, Japan, Sakhalin, Kamchatka, and the Koryak-Chukotka region, and from Alaska and southward along the western Cordillera of North America. Eastward from Guatemala the ophiolite trends through the Greater Antilles, Cuba, and Puerto Rico and perhaps loops into the Caribbean Andes of Venezuela.

The early Paleozoic ophiolites usually occur within folded mountain belts and have undergone repeated deformation and metamorphism, which often transforms the layered fragments into tectonic mixtures or mélanges. Recent work has revealed that the ancient ophiolites are usually large allochthonous tectonic slabs and are now parts of deep-seated nappes in the Urals, Tien Shan, Appalachians, and Klamaths (Peive, 1973; Williams, 1971). Transformation of the metamorphic peridotites (usually dunite-harzburgite) into serpentinite early in the tectonic evolution of ophiolites produces a material whose response to tectonic movements is somewhat analogous to that of salt. Sheared serpentinites are less dense than the rocks surrounding them and commonly provide a plastic medium in which fragments of ophiolites, and sedimentary and metamorphic country rocks, are immersed (Mercier and Vergely, 1972). Once a mélange with a predominately serpentinite matrix is developed, it is vulnerable to any future tectonic event. Because the sheared serpentinite matrix undergoes plastic deformation under stress (Coleman, 1971a), a mélange with a serpentinite matrix responds to tectonism by movement, complicating its inner structure and incorporating blocks of country rock through which it passes.

The complexities of a serpentinite mélange that has undergone both vertical and horizontal movement within a nappe can be bewildering. Nonetheless, with careful mapping and petrologic studies, the original nature of the rock types can often be ascertained. However, extreme dismembering of an ophiolite destroys its original stratigraphy, and Ca-metasomatism of the mafic units can completely obscure their original chemical nature (Coleman, 1966, 1967). Many ophiolites in serpentinite mélanges are controversial because of these complexities, so students of these rocks have turned to the detailed study of less disturbed slabs of ophiolite, such as at Cyprus, Newfoundland, New Guinea, and Oman (Moores and Vine, 1971; Dewey and Bird, 1971; Davies, 1971; Reinhardt, 1969).

The youngest known ophiolites and blueschist are found in New Guinea and New Caledonia. The Papuan ophiolite (Fig. 3), consisting of harzburgite, dunite, gabbro, and basalt, extends 400 km along the eastern part of the island. Davies and Smith (1971, p. 3301-3302) describe the ophiolite as follows: "It consists of three layers (from bottom to top): peridotite (4 to 8 km); gabbro (4 km); basalt (4 to 6 km). Most of the ultramafic rocks are harzburgite, dunite, and enstatite pyroxenite with allotriomorphic metamorphic textures. At the contact between the peridotite and gabbro layers is a discontinuous layer of ultramafic rocks with cumulus textures. Unlike the other ultramafics, these include some clinopyroxene along with the ubiquitous olivine, orthopyroxene and chromite (England and Davies, 1970). The gabbro layer consists of gabbro and norite, some with layering and cumulus textures, and some clearly intrusive into older gabbros. The cumulus gabbroic rocks grade downward into cumulus ultramafic rocks of the ultramafic layer. Elsewhere the contact between the two layers is intrusive, gabbro into peridotite. The basalt layer consists of massive basalt and submarine basalt lavas and pillow lavas; dacitic lavas and pyroclastic rocks are developed locally. The contact between gabbro and basalt layers is in places transitional through a 'high level gabbro' or dolerite phase, but is commonly obscured by Eocene tonalite intrusions."

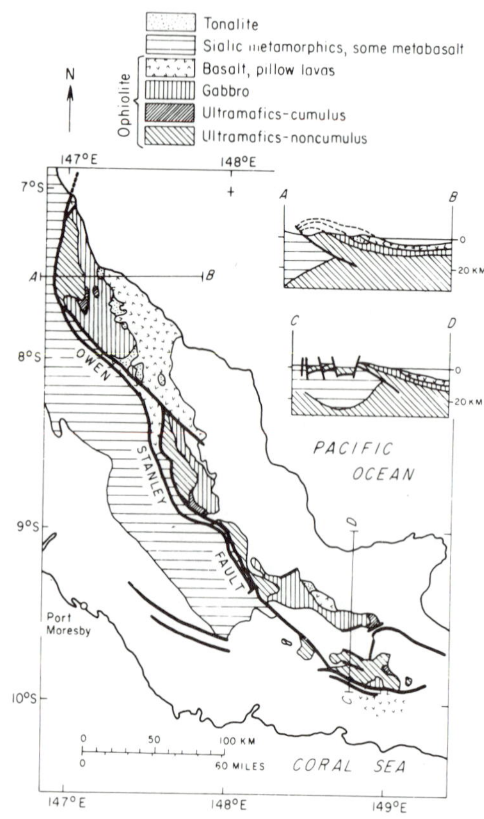

Fig. 3. Papuan ultramafic belt (modified from Davies, 1971).

The gabbro has given K-Ar ages of 147-150 my, and the basalt, 115 my. These rocks are regarded as a slab of oceanic crust emplaced by thrusting during Late Cretaceous or early Tertiary time. The oceanic crust and mantle are thought to have overridden the continental edge or at least to have been underridden by sialic rocks (Milsom, 1973). Blueschist facies metamorphic rocks are sporadically developed below the Owen Stanley fault, along which the Papuan ophiolites are emplaced. A large gravity anomaly associated with the Papuan ophiolites confirms a slab-like structure and further reinforces the idea of overthrusting.

Ultramafic rock consisting principally of dunite and harzburgite forms plates that cover large areas in New Caledonia (Guillon and Routhier, 1971) (Fig. 4). Gravity data on the ultramafic masses indicate that, as in Papua, the ophiolites are slab shaped and

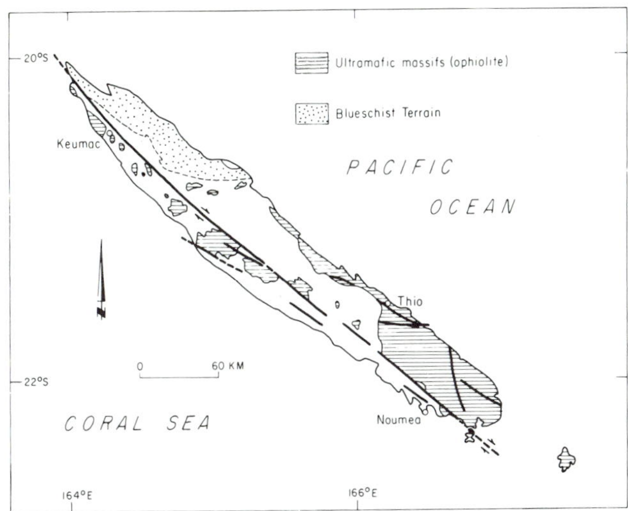

Fig. 4. Tectonic-metamorphic map of New Caledonia showing distribution of ultramafics (modified from Guillon and Routhier, 1971).

emplaced on Cretaceous and Eocene sedimentary rocks. This late Eocene emplacement did not cause any significant metamorphism of the underlying sedimentary rocks (Brothers and Blake, 1973). There are no associated diabase dike swarms or pillow lavas in the ultramafic slab; however, fragments of radiolarian chert, pillow lavas, and serpentinized peridotite are present within zones of Oligocene mélange. The New Caledonia ultramafic rocks lack the complete stratigraphic sequence considered characteristic of oceanic crust, perhaps because erosional or tectonic processes have dismembered or modified the sequence. Oligocene metamorphism on the north part of the island produced a blue-schist terrain and mélange zone 30 km wide, after which northwest-trending transcurrent faulting has offset small klippen of the larger ultramafic slab.

Within the Tethyan belt, perhaps the largest and best exposed ophiolite in the world is the Semail nappe or Semail ophiolite in Oman (Reinhardt, 1969; Glennie et al., 1973; Allemann and Peters, 1972). The Semail ophiolite extends for more than 400 km along the Oman Gulf as a huge sheet (nappe) that forms the highest allochthonous plate in the Oman Mountains (Fig. 5). The plate is about 10 km thick, and is considered to have been thrust 100-200 km westward. It overlies another allochthonous unit, the Hawaŝina complex, which consists of imbricate sheets of a deep-water radiolarite-shale-turbidite sequence ranging in age from Middle Jurassic to earliest Cretaceous. The Semail ophiolite is composed of peridotite (4-5 km) at the base overlain by cumulate peridotite and gabbro (2.5-3 km); the gabbro grades upward into diabase dike swarms and pillow lavas (4 km). Postorogenic sediments of late Maestrichtian and Tertiary age were deposited transgressively on the Semail ophiolite and the Hawasina complex. The Semail ophiolite and Hawasina complex generally are considered to have been thrust onto the Arabian platform.

The preceding descriptions of the younger ophiolites show them generally to be slabs that apparently have overridden the continental edge. Imposition of continued tectonism on such ophiolite slabs tends to destroy and obscure the initial relationships, so that within some older terrains the ophiolites display enigmatic features.

The Pacific Coast region of North America is unusual for the number of subparallel ophiolite belts that are telescoped against the continental margin and illustrate the complexities of older ophiolites. At the latitude of the Klamath Mountains of northern California and southern Oregon, there are as many as four different belts of ophiolite (Fig. 6). Similar to many places elsewhere, the ophiolites consist of rocks of wide variety, commonly including harzburgite, dunite, gabbro, diabase, pillow basalt, and keratophyre, but tectonic disarrangement in most cases has obscured the original sequence and thickness of the lithic components. The ophiolites range in age from Ordovician to Late Jurassic on the basis of the paleontologic age of the oldest strata associated with them. They become sequentially younger oceanward and are considered to be imbricate thrust slices of ocean crust that are accreted to the continental margin (Irwin, 1973).

Estimates of thickness are available for two of these Pacific Coast ophiolites. In California, near the Oregon border (Preston Peak area), the Permian and Triassic ophiolite rests in thrust-fault contact on slaty metasedimentary rocks of the Galice Formation of Late Jurassic (Oxfordian and Kimmeridgian) age. The basal part of the ophiolite consists of approximately 5,000 ft of ultramafic rocks, succeeded upward by 1,500 ft of diabase with xenoliths of gabbro and pyroxenite, and overlain by 500 ft of diabase breccia, siliceous argillite, and chert (Snoke, 1972, Fig. 26). The ophiolite at the base of the Great Valley sequence consists on average of approximately 1 km of serpentinite and minor gabbro succeeded upward by a somewhat lesser thickness of pillow lava, mafic flows and tuff, and minor chert (Bailey et al., 1970, Fig. 2).

Isotopic ages obtained on gabbroic parts of two of the ophiolites are compatible with the oldest paleontologic ages of the overlying strata. The most easterly ophiolite of the Klamath Mountains is overlain by strata as old as Ordovician, and isotopic ages as old as 439 my (Lanphere et al., 1968) and 480 my (Mattinson and Hopson, 1972) were obtained on gabbro associated with the peridotite. In the ophiolite that underlies Late Jurassic (Tithonian) strata at the base of the Great Valley sequence, gabbro that intrudes the peridotite yields isotopic ages of approximately 155 my (Lanphere, 1971). The fairly close coincidence of the isotopic ages of the gabbros and the paleontologic ages of the overlying strata suggests that the span of time between formation of ophiolite and deposition of overlying strata was not long.

The rocks that underlie the ophiolites are metamorphosed, but it is not clear in all cases that the

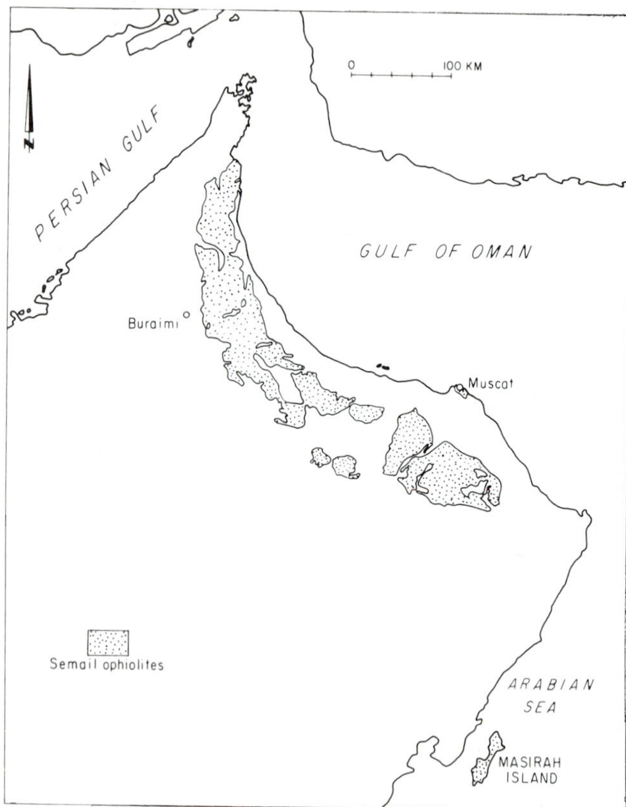

Fig. 5. Distribution of Semail ophiolites in Oman (modified from Glennie et al., 1973)

metamorphism was contemporaneous with the tectonism that emplaced the rocks beneath the ophiolite sheets, as has been suggested by Williams and Smyth (1973) for Newfoundland. The Ordovician ophiolite of the eastern Klamath Mountains is thrust over Abrams Mica Schist, for which a Devonian isotopic (Rb-Sr) age of 380 my was obtained (Lanphere et al., 1968). The ophiolite at the base of the Great Valley sequence overlies blueschist-facies metamorphic rocks that are thought to have developed during overridding by the upper plate (Blake et al., 1967), and for which isotopic ages of approximately 130 my are obtained (Suppe, 1969).

EMPLACEMENT OF OPHIOLITES

If ophiolites do indeed represent oceanic crust, the emplacement of this very dense crust above lighter continental crust is a difficult problem in dynamics. Nonetheless, geologic studies by numerous observers have shown that many ophiolites have been moved tectonically into their present structural situation. Other geologists suggest that many ophiolites form by igneous processes within the early phases of geosynclinal formation, or by oceanization, but none of the arguments satisfactorily explains the internal petrologic evolution of many ophiolites or their allochthonous nature (Brunn, 1973; Maxwell, 1973; Laubscher, 1969; Trumpy, 1971).

Obduction and Subduction

Obduction is emplacement of slabs of oceanic crust over continental crust apparently during convergence of oceanic and continental crust (Coleman, 1971b). Such a convergence is not considered necessarily tied to subduction but could be an indirect product of various types of crustal movements. Where subduction of a slab of oceanic crust was stopped or reversed along a continental edge, at least part of the downgoing slab would not be consumed; later tectonic movements such as thrusting could transport these aborted slabs higher into the crust as nappes. During the translational movement of some ophiolitic slabs and concomittent serpentinization of the peridotite, mélanges could be formed along the zones of movement. In other situations, blue schists and amphibolite facies rocks may develop within the zones of movement.

Reversed Arc Collisions

Karig (1972) has suggested that at least some of the observed ophiolite complexes could have been emplaced by collisions between reversed-arc systems and remnant arcs or larger continental masses. He visualizes the ophiolite as representing oceanic crust beneath the frontal arc and as having formed by marginal basin extension rather than at a spreading oceanic ridge. When the polarity of the arc is reversed (change in direction of subduction), the previous active arc (now a remnant arc) could collide with the active reverse arc and lead to emplacement of oceanic type crust onto the leading edge of the frontal arc. Diagrammatic representations of the various mechanisms of ophiolite emplacement have been given by Dewey and Bird (1971, p. 455, 459).

Continental Root Zones

Within the sutures of some continental zones there are tectonic slices of lherzolite that are not associated with gabbro and basalt. The Lanzo lherzolite mass within the Sesia-Lanzo root zone of the western Alps can be considered a typical occurrence. Here, detailed work combining geophysics, structural analysis, and petrology indicates that this mass was emplaced as part of a deeply rooted thrust (Nicolas et al., 1972; Berckhemer et al., 1968). Gravity and seismic data show that a slab of dense high-velocity mantle has been thrust upward into the continental crust. This has been interpreted as the result of collision between two continental plates, causing the upthrusting of continental mantle that underlay an advancing plate. Even though these lherzolites are not typical ophiolites, their structural situation illustrates the fact that dense mantle material can be transported tectonically upward within the earth's crust along a major crustal suture.

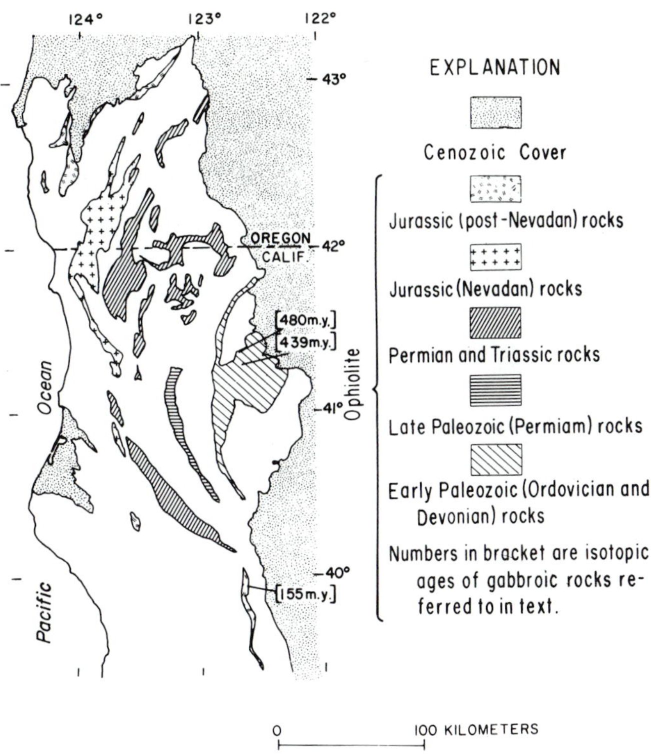

Fig. 6. Distribution of ophiolites in northern California and southern Oregon.

Oceanization

This is a concept of a nebulous process by which continental crust is changed into oceanic crust. Laubscher (1971) demonstrates in rather compelling fashion the usefulness of this concept but does not supply petrologic verification of oceanization. The concept suggests that dense lithosphere may sink into less dense, partially plastic mantle (Van Bemmelen, 1973; Laubscher, 1969). This process is referred to as isostatic subsidence (oceanization), where the crust is replaced by more mobile upper mantle through volcanism. Others call on mantle diapirs or upwellings to produce the same situation. In other words, mafic to ultramafic crust is considered to develop in the absence of a spreading center, while continental crust is consumed and founders into the newly formed oceanic crust. Later gravity sliding of the oceanized material away from the rising area into flanking sedimentary basins provides a model for ophiolite emplacement (Van Bemmelen, 1973).

CONCLUSIONS

The occurrence of ophiolites along continental margins seems well established. The conclusion that some of these ophiolites represent oceanic lithosphere also seems reasonable. There is enough conflicting evidence, however, to suggest that various models should be examined to explain anomalous features of certain ophiolites. The following studies are suggested as an aid in further development of ideas pertaining to ophiolites:

1. Continued detailed mapping of ophiolites to determine their internal igneous relationships and external structural setting.
2. Petrologic studies of as many ophiolites as possible in order to ascertain their conditions of formation. These studies should include chemical analyses for major and minor elements, including isotopic ratios. Comparisons with oceanic and island arc igneous sequences are needed to ascertain possible multiple origins.
3. Geophysical studies that include regional gravity and magnetic measurements on exposed ophiolites. Present comparisons with oceanic crust are based mainly on indirect geophysical measurements.
4. Establishment of tectonic, igneous, and metamorphic ages as related to the development of ophiolites. Such studies should include field evidence based on stratigraphy and paleontology as well as radiometric dating.
5. Metamorphic internal and external history of the ophiolites and associated rocks to provide information on post-formation episodes of deformation and/or emplacement. Establishing metamorphic facies and their relationship to the ophiolite will provide evidence of pressure-temperature conditions of ophiolite emplacement.
6. Continued map compilations on a global and regional scale highlighting distribution of ophiolites and associated sedimentary deposits, metamorphic rocks, and ophiolite-bearing mélanges.
7. Concentrations of nickel, chromium, and copper in ophiolites require fundamental studies regarding their genesis and reserves as ore.

BIBLIOGRAPHY

Allemann, F., and Peters, T., 1972, The ophiolite-radiolarite belt of the North-Oman Mountains: Ecol. Geol. Helv., v. 65, no. 3, p. 657-697.

Aubouin, J., 1965, Geosynclines: New York, Elsevier, 335 p.

Ave'Lallemant, H. G., and Carter, N. L., 1970, Syntectonic recrystallization of olivine and modes of flow in the upper mantle: Geol. Soc. America Bull., v. 81, p. 2181-2202.

Bailey, E. B., and McCallien, W. J., 1950, The Ankara-Melange and the Anatolian thrust: Nature, v. 166, p. 938-940.

———, and McCallien, W. J., 1954, Serpentine lavas, the Ankara-Mélange and the Anatolian thrust: Trans. Roy. Soc. Edinburgh, v. 62, pt. II, p. 403-442.

Bailey, E. H., Blake, M. C., Jr., and Jones, D. L., 1970, On-land Mesozoic oceanic crust in California coast ranges, *in* Geological survey research 1970: U.S. Geol. Survey Prof. Paper 700-C, p. C70-C81.

Blake, M. C., Jr., Irwin, W., P., and Coleman, R. G., 1967, Upside-down metamorphic zonation, blueschist facies, along a regional thrust in California and

Oregon, *in* Geological survey research 1967: U.S. Geol. Survey Prof. Paper 575-C, p. C1-C9.

Berckhemer, M., and the German Research Group for Explosion Seismology, 1968, Topographic des "Ivrea-Körpers" abgeleitet aus seismischen und gravimetrischen Daten: Schweiz, Mineral. Petr., Mitt., v. 48, no. 1, p. 235-246.

Bonatti, E., Honnorez, J., and Ferrara, G., 1971, Peridotite-gabbro-basalt complex from equatorial Mid-Atlantic Ridge: Phil. Trans. Roy. Soc. London, ser. A, v. 268, p. 385-402.

Brothers, R. N., and Blake, M. C., Jr., 1973, Tertiary plate tectonics and high pressure metamorphism in New Caledonia: Tectonophysica, v. 17, p. 337-358.

Brunn, J. H., 1956, Contribution a l'etude géologique du Pinde Serptentrional et d'une partie de la Macedonie occidental: Ann. Geol. Pays Hellenique, v. 7, p. 1-358.

———, 1960, Mise en place et differenciation de l'association pluto-volcanique du cortege ophiolitique: Rev. Geogr. Phys. et Geol. Dynam., v. 3, p. 115-132.

———, 1973, Contribution to the discussion on the emplacement—magmatic or tectonic?— of ophiolites: Internatl. Symp. on Ophiolites in the Earth's Crust (abstr.): Acad. Sci., S.S.S.R., p. 67-68.

Cann, J. R., 1968, Geological processes at mid-ocean ridge crests: Geophys. Jour., v. 15, p. 331-341.

Christensen, N. I., 1970, Composition and evolution of the oceanic crust: Marine Geology, v. 8, p. 139-154.

Coleman, R. G., 1966, New Zealand serpentinites and associated metasomatic rocks: New Zealand Dept. Sci. Ind. Res. Geol. Survey Bull., v. 76, 102 p.

———, 1967, Low-temperature reaction zones and alpine ultramafic rocks of California, Oregon, and Washington: U.S. Geol. Survey Bull., v. 1247, 49 p.

———, 1971a, Petrologic and geophysical nature of serpertinites: Geol. Soc. America Bull., v. 82, p. 897-918.

———, 1971b, Plate tectonic emplacement of upper mantle peridotites along continental edges: Jour. Geophys. Res., v. 76, no. 5, p. 1212-1222.

Davies, H. L., 1971, Peridotite-gabbro-basalt complex in eastern Papua: an overthrust plate of oceanic mantle and crust: Austral. Bur. Mineral Resources Bull., v. 128, 48 p.

———, and Smith, I. E., 1971, Geology of eastern Papua: Geol. Soc. America Bull., v. 82, p. 3299-3312.

Dewey, J. F., and Bird, J. M., 1971, Origin and emplacement of the ophiolite suite: Appalachian ophiolites in Newfoundland: Jour. Geophys. Res., v. 76, no. 14, p. 3179-3206.

Dickinson, W. R., 1970, Relations of andesites, granites, and derivative sandstones to arc-trench tectonics: Rev. Geophys. and Space Phys., v. 8, no. 4, p. 813-860.

Engel, C. G., and Fisher, R. L., 1969, Lherzolite, anorthosite, gabbro and basalt dredged from the mid-Indian Ocean ridge: Science, v. 166, p. 1136.

England, R. N., and Davies, H. L., 1970, Mineralogy of cumulus and noncumulus ultramafic rocks from eastern Papua: Australia Bur. Mineral Resources Rec. 1970/66 (unpubl.).

Ewart, A., and Bryan, W. B., 1973, The petrology and geochemistry of the Tongan Islands: in the western Pacific: *in* Coleman, P. J., ed., Island arcs, marginal seas, geochemistry: Univ. W. Australia Press, p. 503-522.

Gansser, A., 1959, Ausseralpine ophiolith problem: Ecol. Geol. Helv., v. 52, no. 2, p. 659-680.

Gass, I. G., 1967, The ultrabasic volcanic assemblage of the Troodos massif, Cyprus, *in* Wyllie, P. J., ed., Ultramafic and related rocks: New York, Wiley, p. 121-134.

Gastesi, P., 1973, Is the Betancuria Massif, Fuerteventura, Canary Islands, an uplifted piece of oceanic crust?: Nature, v. 246, p. 102-104.

Geotimes, 1972, Penrose field conference on ophiolites: Geotimes, v. 17, no. 12, p. 24-25.

Glennie, K. W., Boeuf, M. G. A., Hughes Clarke, M. W., Moody-Stuart, M., Pilaar, W. F. H., and Reinhardt, B. M., 1973, Late Cretaceous nappes in Oman Mountains and their geologic evolution: Am. Assoc. Petr. Geol. Bull., v. 57, no. 1, p. 5-27.

Green, D. H., 1970, A review of experimental evidence on the origin of basaltic and nephelinitic magmas: Phys. Earth Planetary Interiors, v. 3, p. 221-235.

Guillon, J. H., and Routhier, P., 1971, Les stades d'evolution et de misc en place des massifs ultramafiques de Nouvelle-Caledonie: Bull. B.R.G.M., ser. 2, sect. IV, no. 2, p. 5-38.

Hess, H. H., 1938, A primary peridotite magma: Am. Jour. Sci., ser. 5, v. 35, p. 321-344.

Irwin, W. P., 1973, Sequential minimum ages of oceanic crust in accreted tectonic plates of northern California and southern Oregon (abstr.): Geol. Soc. America, v. 5, no. 1, p. 62-63.

———, and Coleman, R. G., 1972, Preliminary map showing global distribution of alpine-type ultramafic rocks and blueschist: U.S. Geol. Survey Misc. Field Map Studies, Map MF-340.

Jackson, E. D., and Thayer, T. P., 1972, Some criteria for distinguishing between stratiform, concentric and alpine peridotite-gabbro complexes: Proc. 24th Internatl. Geol. Congr., Montreal, v. 2, p. 289-296.

Karig, D. E., 1972, Remnant arcs: Geol. Soc. America Bull., v. 83, p. 1057-1068.

Kay, R. Hubbard, N. J., and Gast, P. W., 1970, Chemical characteristics and origin of oceanic ridge volcanic rocks: Jour. Geophys. Res., v. 75, p. 1585-1614.

Kaz'min, V. G., 1971, The problem of the "Alpine Melange": Geotecktonika, no. 2, p. 19-28.

Lanphere, M. A., 1971, Age of the Mesozoic oceanic crust in the California Coast Ranges: Geol. Soc. America Bull., v. 82, no. 11, p. 3209-3212.

———, 1973, Strontium isotopic relations in the Canyon Mountain, Oregon and Red Mountain, California Ophiolites (abstr.): Am. Geophys. Union Mtg., San Francisco.

———, Irwin, W. P., and Hotz, P. E., 1968, Isotopic age of the Nevadan orogeny and older plutonic and metamorphic events in the Klamath Mountains, California: Geol. Soc. America Bull., v. 79, no. 8, p. 1027-1052.

Laubscher, H. P., 1969, Mountain building: Tectonophysics, v. 7, p. 551-563.

———, 1971, The large-scale kinematics of the western Alps and the northern Apennines and its palinspastic implications: Am. Jour. Sci., v. 271, p. 193-226.

Lockwood, J. P., 1971, Sedimentary and gravity-slide emplacement of serpentine: Geol. Soc. America Bull., v. 82, p. 919-936.

Mattinson, J. M., and Hopson, C. A., 1972, Paleozoic ophiolitic complexes in Washington and northern California: Carnegie Inst. Ann. Rept. Director Geophys. Lab., 1971-1972, p. 578-583.

Maxwell, J. C., 1969, "Alpine" mafic and ultramafic

rocks—the ophiolite suite: a contribution to the discussion of the paper "The origin of ultramafic and ultrabasic rocks," by P. J. Wyllie: Tectonophysics, v. 7, p. 489-494.

———, 1973, Ophiolites—old oceanic crust or internal diapirs?: Internatl. Symp. on Ophiolites in the Earth's Crust, Academy of Sciences, S.S.S.R., p. 71-73.

Melsom, J., 1973, Papuan ultramafic belt: Gravity anomalies and the emplacement of ophiolites: Geol. Soc. America Bull., v, 84, p. 2243-2258.

Mercier, J., and Vergely, P., 1972, Les mélanges ophiolithiques de Macédoine (Grece): Decrochements d'âge anté-crétace superieur: Z. Deut. Geol. Ges., v. 123, p. 469-489.

Miyashiro, A., Shido, F., and Ewing, M., 1970, Crystallization and differentiation in abyssal tholeiites and gabbros from mid-ocean ridges: Earth and Planetary Sci. Letters, v. 7, p. 361-365.

Moores, E. M., 1969, Petrology and structure of the Vourinous ophiolitic complex, northern Greece: Geol. Soc. America Spec. Paper 118, 74 p.

———, 1970, Ultramafics and orogeny, with models of the U.S. Cordillera and the Tethys: Nature, v. 228, p. 837-842.

———, and Vine, F. J., 1971, The Troodos massif, Cyprus and other ophiolites as oceanic crust, evaluation and implications: Phil. Trans. Roy. Soc. London, ser. A, v. 268, p. 443.

Nicolas, A., and Jackson, E. D., 1972, Répartition en deux provinces des péridotites des chaînes alpines longeant La Mediterranée: implications géotectoniques: Bull. Suisse Mineral Petrog., v. 53, p. 385-401.

———, Bouchez, J. L., Boudier, F., and Mercier, J. C., 1971, Textures, structures and fabrics due to solid state flow in some European lherzolites: Tectonophysics, v. 12, p. 55-68.

———, Bouchez, J. L., and Boudier, F., 1972, Interpretation cinematique des deformations plastiques dans le massif de lherzolite de Lanzo (Alpes Piemontaises)—comparaison avec d'autres massifs: Tectonophysics, v. 14, p. 143-171.

O'Hara, M. J., 1967, Mineral facies in ultrabasic rocks, *in* Wyllie, P. J., ed., Ultramafic and related rocks: New York, Wiley, p. 7-18.

———, 1970, Upper mantle composition inferred from laboratory experiments and observation of volcanic products: Phys. Earth Planetary Interiors, v. 3, p. 236-245.

Oxburgh, E. R., 1972, Flake tectonics and continental collision: Nature, v. 239, p. 202-204.

Pearce, J. A., and Cann., J. R., 1971, Ophiolite origin investigated by discriminant analysis using Ti, Zr, and Y: Earth and Planetary Sci. Letters, v. 12, p. 339.

Peive, A. V., 1973, Oceanic crust of the geologic past: Symp. Ophiolites in the Earth's Crust, Acad. Sci., S.S.S.R. Geol. Inst. Moscow, pt. I, p. 3-45.

Peterman, Z. E., Coleman, R. G., Hildreth, R. A., 1971, Sr^{87}/Sr^{86} in mafic rocks of the Troodos Massif, Cyprus, *in* Geological survey research 1971: U.S. Geol. Survey Prof. Paper 750-D, p. D157-D161.

Reinhardt, B. M., 1969, On the genesis and emplacement of ophiolites in the Oman Mountains geosyncline: Schweiz. Mineral. Petrog. Mitt., v. 49, p. 1-30.

Roeder, D. H., 1973, Subduction and orogeny: Jour. Geophy. Res., v. 78, p. 5005-5024.

Roever, W. P., de, 1957, Sind die alpinotypen Peridotit massen vielleicht tektonisch verfrachtete Bruchstücke der Peridotitschale: Geol. Rundschau, v. 46, p. 137-146.

Shor, G. G., Jr., and Raitt, R. W., 1969, Explosion seismic refraction studies of the crust and upper mantle in the Pacific and Indian Oceans, *in* The earth's crust and upper mantle, Geophys. Monogr. 13: Am. Geophys. Union, p. 225-230.

Snoke, A. W., 1972, Petrology and structure of the Preston Peak area, Del Norte and Siskiyou Counties, California [Ph.D. thesis]: Stanford Univ., 274 p.

Steinmann, G., 1927, Die ophiolithischen Zonen in den mediterranean Kellen-gebirgen: Compt. Rend. 14th Internatl. Geol. Congr. Madrid, v. 2, p. 638-667.

Suppe, J., 1969, Times of metamorphism in the Franciscan terrain of northern Coast Ranges, California: Geol. Soc. America Bull., v. 80, no. 1, p. 135-142.

Tex, E., den, 1971, Age, origin, and emplacement of some Alpidic peridotites in the light of recent petrofabric researches: Fortschr. Mineral., v. 48, p. 69-74.

Thayer, T. P., 1973, Some tectonic implications of structural relations between alpine peridotite-gabbro complexes and sheeted dike swarms (abstr.): Symp. Ophiolites in the Earth's Crust, Acad. Sci. S.S.S.R., p. 102-103.

Trumpy, R., 1971, Stratigraphy in mountain belts: Geol. Soc. London Quart. Jour., v. 126, p. 293-318.

Varne, R., and Rubenach, M. J., 1972, Geology of Macquarie Island and its relationship to oceanic crust, *in* Antarctic research series, v. 19: Am. Geophys. Union, p. 251-266.

Van Bemmelen, R. W., 1973, Geodynamic models for the alpine type of orogeny (Test-case II: The alps in central Europe): Tectonophysics, v. 18, p. 33-79.

Williams, H., 1971, Mafic-ultramafic complexes in western Newfoundland Appalachians and the evidence for their transportation: a review and interim report: A Newfoundland Decade, Proc. Geol. Assoc. Canada, v. 24, p. 9-25.

———, and Smyth, W. R., 1973, Metamorphic aureoles beneath ophiolite suites and alpine peridotites: Tectonic implications with west Newfoundland examples: Am. Jour. Sci., v. 273, p. 594-621.

Continental Margins and Ophiolite Obduction: Appalachian Caledonian System

John F. Dewey

INTRODUCTION

Late Cambrian/Early Ordovician ophiolite complexes occur in a variety of tectonic settings in the northwestern orthotectonic belt of the Newfoundland Appalachians and the British Caledonides. The orthotectonic belt is characterized by two major Paleozoic structural stages: the lower stage was deformed and metamorphosed mainly during the Early and Middle Ordovician, and the upper stage was deformed mainly during the Devonian. During the evolution of the lower stage, oceanic crust and mantle were generated by sea-floor spreading, and, shortly after, obducted as ophiolite complexes. One possibility is that all the various ophiolite occurrences belong to a once-continuous single nappe obducted across the orthotectonic belt from the south. Alternatively, the various ophiolite occurrences could have been generated by sea-floor spreading in separate rear-arc, and interarc, oceanic basins and then obducted as thin hot sheets over adjacent continental margins and metamorphic terranes.

In recent years there has been a great revival of interest in Alpine-type mafic/ultramafic complexes mainly as a consequence of the hypothesis that ophiolite complexes represent slices of oceanic crust and mantle generated by sea-floor spreading and subsequently tectonically emplaced into various portions of orogenic belts (Gass, 1968; Temple and Zimmerman, 1969; Moores, 1970; Moores and Vine, 1971; Church and Stevens, 1971; Dewey and Bird, 1971). In this paper the writer takes the view that ophiolite complexes originate as oceanic crust and mantle either at the accreting plate margins of oceanic ridges (Moores and Vine, 1971) or in marginal basins behind or between island arcs (Dewey and Bird, 1971), and are subsequently emplaced at convergent plate margins (subduction zones).

There are many problems related to the origin and emplacement of ophiolite complexes, but four are particularly difficult at present:

1. Sedimentary sequences lying upon the igneous rocks of ophiolite sequences vary greatly in character, from cherts and manganiferous lutites to volcanogenic flysch, and bear a variety of strikingly different relationships to the igneous rocks of the ophiolite suite from conformable to violently unconformable.

2. The time between igneous origin (minimum age determined by the age of sediments lying upon the pillow lavas) and tectonic emplacement may vary considerably but is often very short.

3. In many orogenic belts, ophiolites of a particularly narrow time span are preserved. Thus, although ophiolite complexes provide a key insight to the oceanic origin of many orogenic belts, they do not uniquely determine the life span of the oceans involved (Dewey et al., 1973).

4. There are great variations in the internal sequence of igneous and metamorphic rock types, although the following general sequence, from base to top, applies to all the major well-developed occurrences: dunite, harzburgite, and minor lherzolite, with strong tectonic and metamorphic recrystallization fabrics (these rocks have tectonic relationships with the country rocks); cumulate ultramafic rocks; partly cumulate mafic plutonic rocks; sheeted diabase; pillow lava capped by sediments.

Ophiolite complexes are superbly developed, and exposed, in the Newfoundland Appalachians, and it is useful to describe the internal sequence of rock types developed in these occurrences because it may serve as a "template" for other, less well developed, Appalachian/Caledonian ophiolite sequences. The following description is based upon the Bay of Islands Complex (Smith, 1958; Church and Stevens, 1971; Williams, 1971; Church, 1972; Williams and Malpas, 1972; writer's observations), the Mings Bight Complex (work of J.M. Bird, W. S. F. Kidd, and J. F. Dewey) and the Betts Cove/Tilt Cove Complex (Church and Stevens, 1971; Upadhuay et al., 1971; writer's observations). The location of these complexes is shown in Figure 3. It must be emphasized that the following description is of the *general* sequence of rock types in these ophiolite complexes; there are considerable variations between each, and even within each, complex. Each complex has wholly tectonic external relationships with other rock sequences. Only in the Bay of Islands Complex is an intact ophiolite sequence present. The ophiolite sequences consist, from bottom to top, of the following units:

1. Finely foliated and banded black-green serpentinite-mylonite with pale-weathering bands of a garnet-clinopyroxene-orthopyroxene-kaersutite-phlogopite-ceylonite assemblage (ariégite). The lower contact of this unit is sharply defined against a plagioclase-garnet-diopside amphibolite similar to rocks regarded as thermal-contact aureoles in other Alpine-type ultramafic associations. The "contact" amphibolite "grades" rapidly down through a series of narrow tectonic slices into greenschist facies metavolcanics and pelites. The structural/metamorphic history of the contact zone (serpentinite of

unit 1 and subjacent metamorphic rocks) is complex.

2. Lherzolite and harzburgite with strong tectonic fabrics and a polyphase history of deformation and metamorphism that predates both cross-cutting clinopyroxenite and websterite dikes and the mylonite fabrics of unit 1.

3. Dunite, harzburgite, and minor orthopyroxenite, with strong tectonic foliation folded by tight folds and with a complex polyphase tectonic and metamorphic recrystallization history. The dunite of this unit contains large, deformed, chromite pods and smaller stringers and crystal trains of chromite. Clinopyroxenite, websterite, and occasional bronzitite dikes cross-cut the latest of the folds.

4. Layered dunite, websterite, wehrlite, harzburgite, and minor troctolite and olivine gabbro, with occasional felspathic dunite, having cumulate textures disrupted by intrafolial flowage and tight to isoclinal folding. The banding and foliation are cut by pegmatite clinopyroxenite, rodingite, and anorthositic gabbro dikes. Occasionally, blocks and xenoliths of dunite occur as inclusions in this unit.

5. Dunite, wehrlite, websterite, minor troctolite, gabbro, and harzburgite, with cumulate textures slightly disrupted by foliation and intrafolial folding and cut by clinopyroxenite dikes.

6. Felspathic dunite, wehrlite, websterite, clinopyroxenite, olivine-clinopyroxenite, troctolite, gabbro, with cumulate textures disrupted by flow foliation.

7. A unit of olivine-clinopyroxenite, cliniopyroxenite, troctolite, olivine-gabbro, pegmatitic gabbro, and minor anorthositic gabbro, with exceedingly complicated internal contact relationships. Cumulate, disrupted cumulate, and tectonic foliations are cut by occasional diabase dikes.

8. Cumulate, flow-banded, and flow-foliated, banded gabbro and olivine gabbro with pyroxene commonly altered to brown hornblende. Occasional diabase dikes cross-cut the banding.

9. Massive gabbro and leucogabbro with only occasional banding toward the base. Cross-cut by diabase dikes and occasionally containing diabase xenoliths.

10. Massive gabbro and leucogabbro with diffuse stringers, veins, and pods of quartz-diorite, trondjemite, albite-"granite," and albitite, and occasional diabase xenoliths, cut by diabase dikes.

11. Complex zone of diabase with dikes, veins, and pods of albite-"granite" and minor quartz-keratophyre.

12. Sheeted diabase complex; 100% diabase dikes generally parallel over the area of outcrop but occasionally showing conjugate relationships. This unit is often heavily brecciated.

13. Massive basic lava and pillow lava with diabase dikes and a few sills, and occasional quartz-keratophyre and rhyolite dikes.

14. Pillow lava with minor diabase dikes.

15. Pillow lava, hyaloclastite, and palagonite tuff, interbanded with red chert.

16. Red and green chert and argillite, sometimes manganiferous, with banded palagonite tuffs.

17. Ophiolite-bearing flysch and conglomerate, varying from conformable to unconformable on the igneous rocks and cherts of the ophiolite sequence. Volcanogenic flysch sequences, with gabbro and diabase sills, and interbanded pillow lava units.

Particular problems relate to the mechanisms by which ophiolite complexes are emplaced in orogenic belts and how these mechanisms may be linked to processes at convergent plate boundaries. It is implicit, in the plate tectonics model, that oceanic crust and mantle are mainly subducted in oceanic trenches. It is, therefore, of great interest to understand the mechanisms by which large slabs of dense oceanic crust and mantle are detached from the evolving oceanic lithosphere and tectonically incorporated into orogenic belts. Coleman (1971) has coined the term "obduction" to describe this process and to make the distinction with the usual subduction fate of oceanic crust and mantle.

It is the purpose of this paper to describe some ophiolite occurrences in the Appalachians of Newfoundland and the Caledonides of the British Islands, and to present some possible plate tectonic models for Appalachian/Caledonian ophiolite obduction.

OPHIOLITE NAPPES

Ophiolite complexes vary greatly in size, occur in a wide variety of tectonic positions within orogenic belts, and bear highly variable relationships to sequences against, and onto which, they are emplaced. They occur commonly as blocks in blueschist mélanges, olistostromes, and wildflysch deposits, as highly deformed slices in steep zones of high strain, as well-preserved sequences in faulted synclines, as autochthonous or parautochthonous basement to such sequences as the Great Valley Jurassic assemblage of California, and as giant nappes transported as thin sheets over platform or exogeosynclinal sequences. The last type of occurrence has been studied in great detail in the Bay of Islands Complex, Newfoundland (Stevens, 1970; Church and Stevens; 1971; Williams, 1971), in the Semail Complex in the Oman (Reinhardt, 1969; Glennie et al., 1963), and in the Papuan ophiolites (Davies, 1971). The Bay of Islands Complex and the Semail Complex bear a striking resemblance to one another in tectonic emplacement styles and tectonic/stratigraphic relationships.

Bay Of Islands Ophiolite Complex, Western Newfoundland

This complex occurs as a series of klippen, probably remnants of a single nappe sheet forming the highest unit of a stacked sequence of allochthonous structural slices (Stevens, 1970; Williams, 1971). The general relationships of the Bay of Islands slice to lower allochthonous slices and their subjacent autochthon are shown in Figure 1. The autoch-

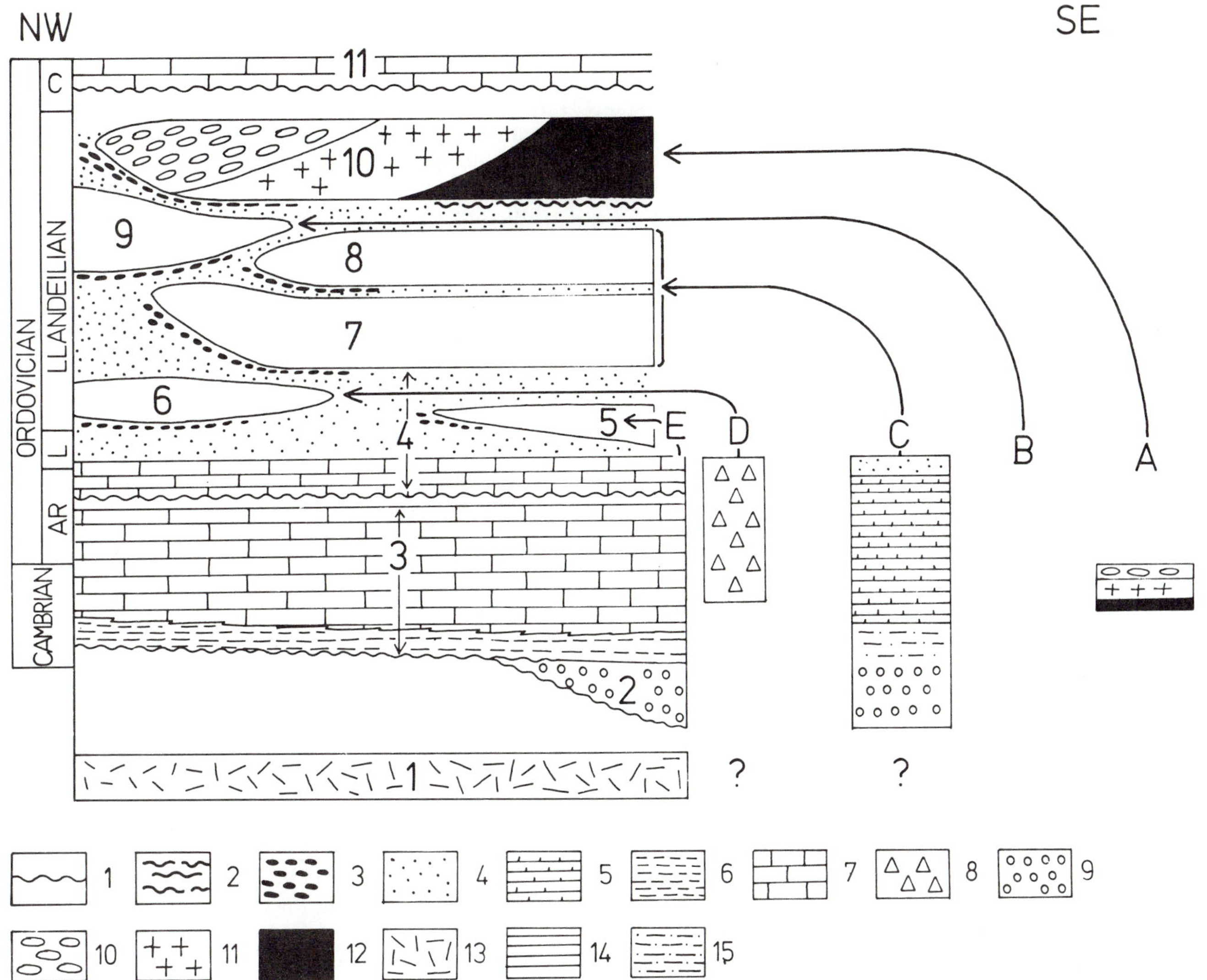

Fig. 1. Schematic illustration of the stratigraphic/structural relationships of the Bay of Islands ophiolite sheet, western Newfoundland (data from Stevens, 1970; Williams, 1971). *Key to symbols:* Large figures refer to stratigraphic/tectonic units described in the text: 1, Precambrian (Grenville) crystalline basement; 2, late Precambrian/Early Cambrian arkosic clastics and basalts; 3, Cambrian/Ordovician clastics and carbonates; 4, Ordovician carbonates and ophiolite-bearing flysch; 5, Penguin Cove allochthon; 6, Cow Head allochthon; 7 and 8, Humber Arm Allochthon; 9, Coastal Complex Allochthon, 10, Bay of Island Allochthon; 11, Ordovician carbonates. Large letters refer to provenance site and order of movement of the allochthons. Small letters in columns at left refer to stratigraphic age: AR, Arenigian; L, Llanvirnian; C, Caradocian. *Key to legend:* 1, unconformity; 2, garnet-amphibolite; 3, ophiolite-bearing melange; 4, ophiolite-bearing flysch; 5, calcareous flysch; 6, shallow marine clastics; 7, shallow marine carbonate; 8, carbonate conglomerate and breccia; 9, red and greed arkose graywacke and shale with basalt; 10, pillow lava; 11, gabbro; 12, ultramafic rocks; 13, crystalline silicic basement rocks; 14, radiolarian chert; 15, deep-water quartz-rich clastics.

thon consists of a Precambrian gneissic basement (unit 1, Fig. 1) overlain unconformably by a platform sequence of clastics followed by carbonates varying in age from Early Cambrian to Middle Ordovician (unit 3, Fig. 1). The highest part of the autochthon consists of Middle Ordovician thin-bedded carbonates, resting with slight unconformity on the earlier carbonates, followed by shale and flysch (unit 4, Fig. 1). The appearance of the shale and flysch sequence marks a major sedimentary polarity reversal, prior to which sediments were derived from the North American continent to the west. The shale and flysch sequence was derived from a region to the east that, until Middle Ordovician times, had been a deep-water "eugeosynclinal belt." The lowest unit of the allochthon is a parautochthonous carbonate slice, or series of slices (unit 5, Fig. 1) probably derived from the eastern edge of the Cambro-Ordovician platform carbonate sequence. The subsequent Cow Head slice (unit 6, Fig. 1) consists dominantly of a sequence of carbonate boulder-slide conglomerates ranging in age from Middle Cambrian to Middle Ordovician. The Cow Head sequence is clearly a proximal sequence marking a major facies change along the eastern edge of the Cambro-Ordovician platform. Allochthonous units 7 and 8 (Humber Arm Allochthon) consists of the following conformable tripartite stratigraphic sequence from base to top: late Precambrian to Early Cambrian red and green pebbly graywacke and slate followed by Early Cambrian dark shale, graywacke, quartzite, and conglomerate (lower flysch division); Middle Cambrian to early Middle Ordovician limestone, calcareous argillite and carbonate conglomerate (calcareous flysch division); Middle Ordovician green graywacke turbidite and shale overlain by massive coarse-grained arkoses (upper flysch division.)

Age and facies considerations indicate that the Humber Arm Allochthon was transported from a provenance site east of the Cow Head Allochthon. The lower flysch division and the calcareous flysch division were derived from the craton and carbonate platform to the west. The upper flysch division, containing an immature clastic suite of potash felspar, quartz, ophiolite debris, and pebbles of the calcareous flysch division, was derived from an easterly source. The next highest slice of the allochthon is the Coastal Complex slice (unit 9, Fig. 1), a poorly understood assemblage of polyphase deformed gabbros, ultramafics, and amphibolites, cut by mafic and silicic igneous bodies. Slices 6, 7, and 8 are essentially encased in flysch. All slices are underlain by mélange units containing ophiolite debris, and debris derived from the immediately superjacent allochthon. The mélange units may be partly avalanche deposits derived from the advancing allochthonous sheets and partly tectonic mélange generated by the overriding sheet. The Bay of Islands slice (unit 10, Fig. 1) forms the highest member of the allochthon. It is a thin sheet in which the ophiolite "stratigraphy" faces westward. Stevens (1970) has argued a latest Cambrian (Tremadocian) age for the Bay of Islands ophiolite complex. On its western margin it is underlain by ophiolitic mélanges, whereas along its eastern contact it is immediately underlain by garnet amphibolites grading rapidly downward into greenschist-facies metavolcanics. The garnet-amphibolites have a complex variety of fabrics, from isotropic to foliated, and a polyphase deformation history.

It is not yet clear whether there is a *continuous* rapid metamorphic gradient from the garnet-amphibolites to the greenschists or whether the garnet amphibolites and greenschists were tectonically juxtaposed during emplacement of the Bay of Islands slice. Also, it is not clear whether these metamorphic rocks form an integral, though transported, part of the subjacent Humber Arm slices or whether they form part of an intervening slice perhaps related to the Coastal Complex metamorphic slice. The emplacement of all allochthonous units onto the platform autochthon was accomplished during the late Middle Ordovician, because a late Middle Ordovician, limestone sequence (unit 11, Fig. 1; rests gently unconformably upon both allochthon and unit 4 of the autochthon.

Facies arguments indicate that the stacking sequence of the allochthonous units (5, 6, 7, 8, 9, 10) is directly related to their paleogeographic provenance site before they were transported westward. Thus unit 5 was derived from the edge of the carbonate shelf, unit 6 from a carbonate boulder slide/shale sequence representing the transition from shelf to basin, and units 7 and 8 from a still more distant site to the east. If this stacking-order argument can be projected to the two higher slices, units 9 and 10 were derived from positions successively farther east of that of units 7 and 8. It is clear, from the presence of ophiolite detritus in the easterly-derived, lower Middle Ordovician, upper flysch division of units 7 and 8, that the Bay of Islands ophiolite slice began to move, and was being eroded, long before the allochthons were finally emplaced on the platform to the west. Also the presence of ophiolite detritus in flysch and mélange units below virgually all the allochthonous units suggests, as argued by Stevens (1970), that the initiation of movement order of the allochthons was the inverse of the stacking order, so that the high-level ophiolite nappe could provide clastic debris to the flysch and mélange units, which were overriden by successive allochthons. This may mean, as argued by Stevens (1970) and Sales (1971), that the allochthons arrived as a stacked, *already assembled*, sequence. This would mean that the allochthons were transported by compressional telescoping progressing from east to west so that the lower slices were successively attached to the base and front of the moving allochthonous assemblage. Alternatively, it is possible that both compressional tectonics and gravity sliding were responsible for the emplacement of the allochthons, the lower slices being emplaced by sliding into the flysch basin, the higher sheets being thrust into place (Bird and Dewey, 1970).

It seems clear that, whatever the mechanism or mechanisms by which the Bay of Islands ophiolite complex was obducted, together with subjacent allochthons, onto the western platform, the pre-Middle Ordovician paleogeography of western Newfoundland consisted of a carbonate shelf (continental shelf) to the west with a continental rise (Cow Head and Humber Arm sequences) and ocean basin (Bay of Islands Ophiolite Complex) to the east (Stevens, 1970; Dewey and Bird, 1971; Williams, 1971) in an arrangement very similar to the Atlantic margin of the southeastern United States at present. It is also clear that, if the Bay of Islands Complex is a fragment of oceanic crust and mantle generated by some form of sea-floor spreading during the Late Cambrian, it is not *diagnostic* for the age of the Appalachian ocean basin because the Cow Head calcareous flysch division of the Humber Arm Allochthon (a continental rise sequence marginal to an ocean basin to the east) was established from latest Early Cambrian times. Furthermore, the marine transgression, registered by the Early Cambrian quartzite sheet at the base of unit 3, was probably related to the continental margin having moved a considerable distance

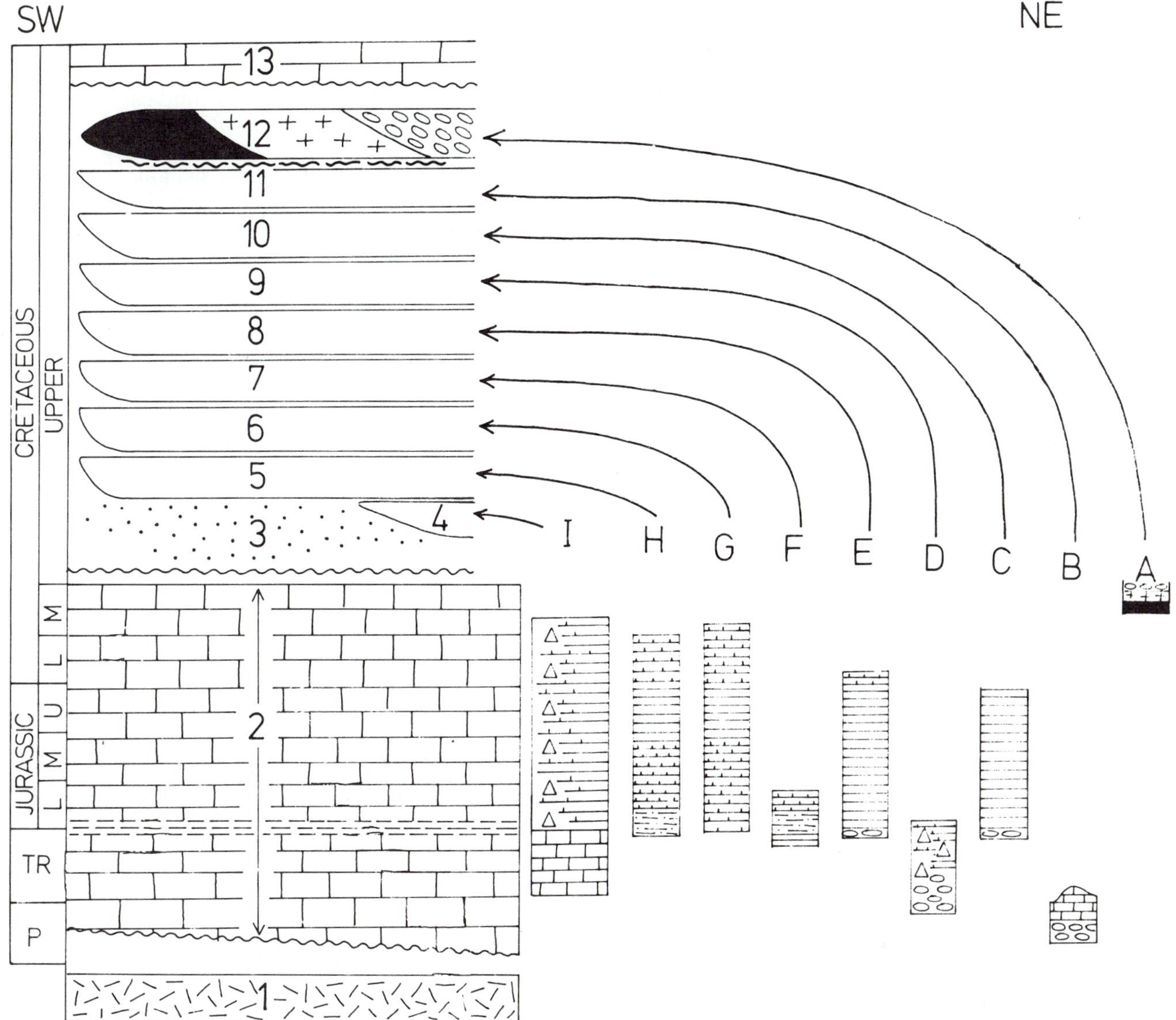

Fig. 2. Schematic illustration of the stratigraphic/structural relationships of the Semail ophiolite sheet, Oman (data from Glennie et al., 1973). *Key to symbols:* numbers refer to stratigraphic/tectonic units described in the text: 1, Precambrian crystalline basement; 2, Permian to Cretaceous carbonates; 3, Cretaceous flysch; 4, 5, 6, 7, 8, 9, 10, 11, Hawasina allochthons: 12, Semail Allochthon; 13, Upper Cretaceous carbonates. Large letters refer to provenance site and movement order of the allochthons. Small letters in columns at left refer to stratigraphic age: P, Permian; TR, Triassic; L, Lower; M, Middle; U, Upper. Key to lithology symbols in Figure 1.

from a spreading oceanic ridge so that it had cooled and subsided (Dewey and Bird, 1970; Sclater et al., 1971). The establishment of the ocean basin by continental rifting is probably registered by the basalts and red beds of unit 2 of the autochthon. If the ocean basin had existed from earliest Cambrian times, it is interesting, and perhaps peculiar, that the only oceanic crust and mantle to be emplaced during the Middle Ordovician obduction phase should be of Late Cambrian age. This suggests that the lower slices were peeled, or stripped, from their older oceanic (?) basement. Considering the east-to-west width of the various allochthonous slices, it is conservatively estimated that the pre-emplacement distance occupied by the successive sheets, in their original provenance sites, was about 175 km.

Semail Ophiolite Complex, Oman

Recent work, and synthesis, by Glennie et al. (1973) has developed a coherent picture of the geometry and timing of emplacement of the Semail Complex, and its relationship to its subjacent allochthonous Hawasina units. These relationships are summarized in Figure 2, a time-span diagram constructed in a way similar to that of Figure 1. The autochthon consists of a pre-Permian crystalline basement (unit 1, Fig. 2) unconformably overlain by a Middle Permian to Middle Cretaceous (Cenomanian) platform carbonate sequence with minor elastics (unit 2, Fig. 2). The highest autochthonous unit is a Late Cretaceous (Coniacian to Campanian) flysch/shale sequence (unit 3, Fig. 2) becoming conglomeratic to the northeast, with ophiolite detritus and carbonate pebbles derived from unit 2. The first allochthonous unit (unit 4, Fig. 2) consists of Early Triassic to Cretaceous stromatolitic dolostones, reefal limestones, carbonate conglomerates, oolitic carbonate turbidites, and cherts, becoming finer northeastward. This unit is similar, in tectonic stacking position and facies, to the Cow Head slice of western Newfoundland. Above this allochthonous unit, six slices comprise the Hawasina Allochthon (units 5-10, Fig. 2). The lower slice (unit 5) consists of Late Triassic to Late Jurassic calcareous conglomerates and turbidites, Late Jurassic/Early Cretaceous red radiolarian cherts, and Early to Middle Cretaceous calcareous turbidites. Unit 6 consists of a Late Triassic to Middle Cretaceous sequence of calcareous turbidites and red radiolarian cherts, the proportion of the latter increasing northeastward. Unit 7 is a sequence of Late Triassic cherts, Late Triassic/Early Jurassic quartz-sandstone turbidites, and Early Jurassic calcareous turbidites. Unit 8 consists of Triassic to Early Cretaceous (?) radiolarian cherts and thin shale interbeds, with pillow basalts near, or at, the base, and manganiferous argillites of Late Jurassic age and calcareous turbidites toward the top. Unit 9 consists of Permian to Early Jurassic carbonate conglomerates, calcareous turbidites, and radiolarian cherts with pillow basalts and mafic sills and dikes toward the base. Unit 10 is a sequence of Early Jurassic to Early Cretaceous radiolarian cherts and calcareous turbidites overlying pillow basalt. Above the Hawasina allochthons is a widespread sheet of giant exotic Permo-Triassic reef carbonate blocks associated stratigraphically with pillow basalts (unit 11, Fig. 2). This sheet is overlain in turn by the Semail ophiolite slice, consisting of a full ophiolite sequence of Upper Cretaceous (Coniacian) age, facing northeastward. Along the base of the Semail slice, garnet-amphibolites similar to those below the Bay of Islands slice are discontinuously developed. All allochthonous slices, except units 11 and 12, are imbricated on northeast-dipping thrust surfaces, and the base of each slice rests variably on different underlying slices.

Emplacement of the whole allochthon (units 4-12) was accomplished during the Campanian Stage of the Late Cretaceous because it rests on Campanian flysch (unit 3) and is unconformably overlain by Maestrichtian limestones (unit 13, Fig. 2). As in western Newfoundland, allochthonous slices consist of partly, or wholly, coeval stratigraphic sequences stacked in a sequence in which each slice has traveled farther than the one below. They can be assembled to form a pre-emplacement paleogeography in which the unit 2 autochthon was deposited as a platform (continental shelf) sequence, at least from Early Jurassic times. The unit 4 slump conglomerates and turbidites probably represent a pre-Coniacian shelf break and upper continental rise. Units 5, 6, and 7 represent upper, middle, and lower continental rise facies, resprectively, and units 8, 9, and 10, an abyssal oceanic realm. Unit 11 may have been a Permo-Triassic seamount complex and unit 12, the farthest-traveled sheet, was generated as Late Cretaceous oceanic crust and upper mantle. Glennie et al. (1973) argue that palinspastic restorations of the allochthonous units to their original provenence sites, prior to Late Cretaceous telescoping and obduction, indicates that the Semail ophiolites lay about 1,200 km from the northern continental margin of Arabia (provenance site of unit 4). Since the Semail ophiolites are of Coniacian age and the whole allochthon was finally emplaced by Maestrichtian times, the rate of overthrusting (? rate of plate convergence) was about 24 cm/yr, a rate about twice that of any known present-day plate convergence. Also, the Semail ophiolite slice was moving for a considerably shorter period of time during the Late Cretaceous than was the Bay of Islands slice during the Middle Ordovician. A further difference between the western Newfoundland and Oman allochthons is that Glennie et al. (1973) do not record any associated flysch and melange units between individual slices of the Hawasina Allochthon and below the Semail Complex. They argue that the dominant mode of emplacement of the Oman allochthons was by gravity sliding, because of the general lack of deformation within the thrust sheets. However, unit 3 flysch sequences contain ophiolite debris, which suggest that the allochthon may have been progressively assembled in reverse sequence from the present stacking order.

A possibly important difference between the Bay of Islands ophiolite slice and the Semail ophiolite slice is that the former ophiolite sequence faces in the direction in which the sheet was transported, so that the basal thrust contact cuts across progressively higher units of the ophiolite sequence in that direction. The reverse is true of the Semail ophiolite slice. This may reflect the initial dip direction of the thrusts along which the ophiolite slices became progressively detached from the oceanic lithosphere.

In both western Newfoundland and in the Oman, obducted ophiolite sheets bear very similar tectonic/stratigraphic relationships to underlying rocks, and in both cases a probable continental shelf, continental rise, ocean basin paleogeography was progressively telescoped into an identical stacking order.

OPHIOLITE OBDUCTION—GENERAL

No single model is likely to account for the emplacement of all obducted ophiolite sheets. Ophiolite obduction is probably a process involving several, if not many, mechanisms. Which was responsible for the emplacement of particular ophiolite occurrences can only be answered by detailed field studies of their structural and stratigraphic relationships. We are faced with the problem of reconstructing complex events from an imcomplete record. Many of these lost relationships may be critical to a full understanding of orogenic evolution. Although plate tectonics has in many ways increased our understanding of orogenic evolution and thereby invigorated our studies of orogenic belts, it has given us fresh insights into the kinds of questions to which we may never know the answers and made us aware of the bewildering array of complexities inherent in orogenic evolution. For example, there has been recent discussion (Cady, 1972) as to whether the Appalachian/Caledonian orogenic belt evolved dominantly on a sialic or an oceanic basement. The present distribution of sialic basement rocks, according to the plate tectonics model, is a poor guide to the nature of the basement during the total evolution of the orogen, because ocean closing must involve subduction and, therefore, the *destruction* of oceanic basement and the *preservation* of continental basement. Furthermore, the age of origin and emplacement of ophiolite complexes may be the worst possible guide to the total oceanic history involved in the evolution of an orogenic belt. It cannot be supposed that the average of the oldest ophiolite complex in an orogenic belt designates the oldest possible age of oceanic tracts involved in the history of that belt, and that the age of ophiolite obduction always marks the total destruction of those oceanic tracts. The possibility of rear-arc and interarc oceanic basins must be considered. Ophiolites generated in such basins will be considerably younger than the oldest oceans involved in the evolution of the orogen, and the time between origin and emplacement may be very short. The oldest oceanic tracts may be completely destroyed and their existence may be argued only from other facies (Bird and Dewey, 1970; Dewey and Bird, 1971) and geometric (Smith, 1971; Dewey et al., 1973) considerations.

The writer suggests that rear-arc basin spreading, with the volcanic arc moving away from the continental margin, followed shortly thereafter by the obduction of marginal basin ophiolites, appears to offer solutions to the following three particular ophiolite problems:

1. As mentioned above, ophiolite origin and emplacement are commonly separated by only a short time span.

2. As Sclater et al. (1971) have shown, there is a clear inverse relationship between age (and heat flow) and elevation of the ocean floor. Obduction of young, high, hot, thin, oceanic lithosphere would be mechanically easier than the detachment and overthrusting of older, low, cool, thick, oceanic crust and mantle.

3. The common direct association of ophiolites with rocks of volcanic arc affinities (andesite flysch, andesites, tonalites, granodiorites) favors a rear-arc basin site for their origin.

APPALACHIAN/CALEDONIAN OPHIOLITE RELATIONSHIPS

The general geology of the western Newfoundland Appalachians and the Scottish and Irish Caledonides (Fig. 3) has been summarized in numerous publications (Bird and Dewey, 1970; Dewey and Bird, 1971; Dewey, 1969, 1971, 1973; Stevens, 1970; Williams, 1964, 1971). The reader is referred to these papers for details beyond the summary below. Several zonal schemes have been proposed for subdividing the Newfoundland Appalachians and the northwestern Caledonides in the British Islands; the writer prefers the zonal scheme shown in Figure 3. In Newfoundland and Scotland, a Northwest Platform (foreland) consists of Precambrian crystalline basement overlain by a Cambro-Ordovician clastic/carbonate sequence. In Newfoundland this sequence is followed by a Middle Ordovician exogeosynclinal shale/flysch sequence enveloping the Humber Arm and Bay of Islands allochthons. The southeastern margin of the foreland zone is a line of major facies change from platform to geosynclinal sequences and a zone of intense thrusting. It is also a line across which metamorphic grade increases sharply into the Western Fleur de Lys Zone. This zone consists of the Fleur de Lys Supergroup, a sequence of polyphase-deformed garnet-amphibolite facies psammites, semipelites, pelites, marbles, and amphibolites (Church, 1969), resting upon a crystalline Precambrian basement with eclogite bodies (M. J. deWit, personal communication). Similar rocks, conformably overlain by a thick sequence of silicic volcanics, occur in the Eastern Fleur de Lys Zone, where they are believed (Dewey and Bird, 1971) to overlie an ophiolitic basement. The Eastern and Western Fleur de Lys zones are separated by a narrow ophiolite belt (Baie Verte/Mings Bight Zone; Bird et al., in preparation). The Eastern Fleur de Lys Zone is bounded to the east by the Betts Cove/Tilt Cove Ophiolite Zone (Upadhuay et al., 1971). The Baie Verte/Mings Bight, and Betts Cove/Tilt Cove, ophiolite complexes with their sedimentary and volcanic sequences are largely greenschist facies and clearly belong to a structural stage younger than the Fleur de Lys metamorphic sequences. The Notre Dame Bay Zone (Fig. 3) is complex and poorly understood but seems to consist of at least two structural stages, an older ophiolite volcanic arc assemblage and a younger ophiolite/volcanic arc assemblage. The Western Fleur de Lys to Notre Dame Bay zones, therefore, constitute an orthotectonic zone assemblage (Fig. 3) consisting of a lower (Fleur de Lys) structural stage, deformed and metamorphosed during Late Cambrian to Early Ordovician times and an upper stage consisting of Ordovician ophiolites and volcanic arc sequences. The Notre Dame Bay Zone is bounded to the southeast by a spectacular mélange (Dunnage mélange: Kay, 1970) to the southeast of which the Fleur de Lys structural stage appears to be absent.

In the British Caledonides (Fig. 3) a similar,

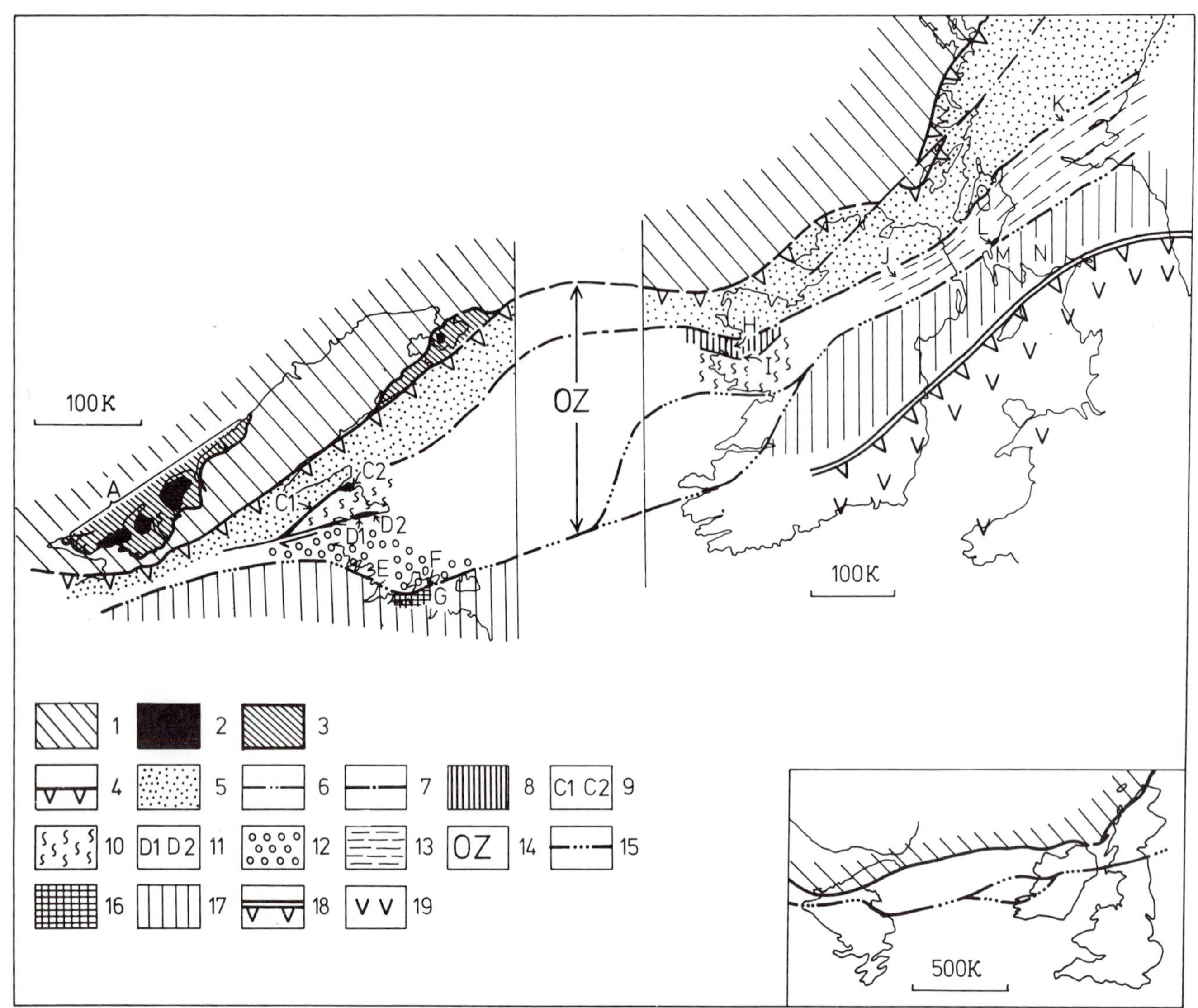

Fig. 3. Schematic tectonic map of the Applachians of western Newfoundland and the Caledonides of the northwestern British Islands (bottom right inset shows the position of Newfoundland relative to the British Islands prior to their separation during the growth of the North Atlantic). *Key to symbols:* 1, Northwest Platform Zone (foreland); 2, ophiolites; 3, Western Newfoundland Allochthon; 4, northwestern thrust margin of the orthotectonic zone assemblage; 5, Grampian Zone of Ireland and Scotland, Western Fleur de Lys Zone of Newfoundland; 6, Great Glen Fault; 7, Highland Boundary Fault Zone of Scotland, Clew Bay Fault Zone of western Ireland; 8, South Mayo Trough Zone; 9, Baie Verte/Mings Bight Ophiolite Zone; 10, Connemara Zone of western Ireland, Eastern Fleur de Lys Zone of Newfoundland; 11, Betts Cove/Tilt ophiolite Zone; 12, Notre Dame Bay Zone; 13, Midland Valley of Scotland zone; 14, orthotectonic zone assemblages; 15, southern margin of the orthotectonic zone assemblage; 16, Dunnage Melange Zone; 17, Central Zone; 18, approximate position of subduction zone associated with Middle Ordovician volcanic arc of southeastern Ireland and northern England; 19, Middle Ordovician volcanic arc. Letters refer to the position of sequences shown in the columns of Figure 4.

though not identical, zonal arrangement occurs. The Grampian Zone, consisting of the Moine/Dalradian metamorphic assemblage, is similar to the Western Fleur de Lys sequence. It was deformed and metamorphosed from Late Cambrian to Middle Ordovician times (Dewey, 1973) and rests, at least in part, on a crystalline Precambrian basement. Dalradian rocks also occur in the Connemara Zone, where M. Max (personal communication) has argued that they rest on an older crystalline basement. The South Mayo Trough Zone consists of a synclinorium of Ordovician and Silurian rocks and may be the Irish equivalent of the Baie Verte/Mings Bight Zone. The northern margin of this zone is the Clew Bay Fault Zone, a probable continuation of the Highland Boundary Fault Zone of Scotland, which forms the southern boundary of the Grampian Zone. The Midland Valley Zone, well defined in Scotland but poorly defined in Ireland, has partly a Dalradian basement (in Tyrone: Cobbing, 1964; Cobbing et al., 1965) and partly an ophiolite basement (Ballantrae Complex: Bird et al., 1971). The Grampian, South Mayo Trough, Connemara, and Midland Valley zones constitute an orthotectonic zone assemblage

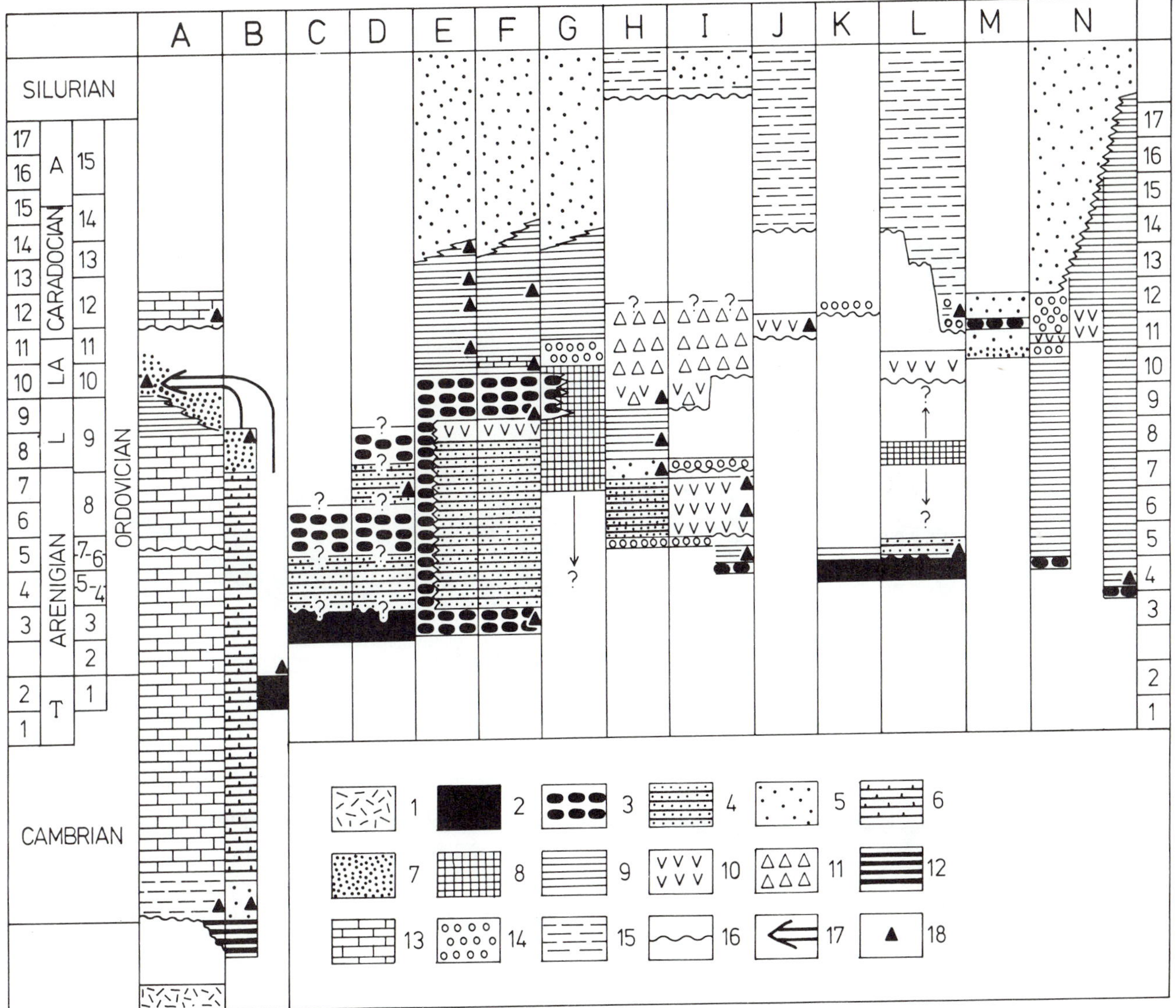

Fig. 4. Correlation chart of some stratigraphic sequences, important to the ophiolite obduction problem, in the Newfoundland Appalachians and British Caledonides. Columns A to N refer to positions shown by those letters in Figure 3. *Key to symbols:* 1, Precambrian crystalline basement; 2, ophiolite; 3, pillow lavas; 4, volcanogenic sediments; 5, flysch derived mainly from metamorphic sources; 6, calcareous flysch; 7, ophiolitic flysch; 8, melanges and olistostromes; 9, argillite; 10, silicic/intermediate volcanics; 11, fluviatile molasse; 12, red and green quartzofeldspathic sediments with basalts; 13, carbonate; 14, conglomerate; 15, shallow marine clastic sediments; 16, unconformity; 17, emplacement of allochthons; 18, fossil assemblages. *Key to stratigraphic ages:* T, Tremadocian; L, Llanvirnian; LA, Llandeilian; A, Ashgillian. 1 - 17 refer to British Cambrian/Ordovician graptolite zones as follows: 1, *Dictyonema flabelliforme;* 2, *Anisograptus;* 3, *Tetragraptus approximatus;* 4, *Didymograptus deflexus;* 5, *Didymograptus nitidus;* 6, *Isograptus gibberulus;* 7, *Didymograptus hirundo;* 8, *Didymograptus "bifidus";* 9, *Didymograptus murchisoni;* 10, *Glyptograptus teretiusculus;* 11, *Nemagraptus gracilis;* 12, *Climacograptus peltifer;* 13, *Climacograptus wilsoni;* 14, *Dicranograptus Clingani;* 15, *Pleurograptus linearis;* 16, *Dicellograptus complanatus;* 17, *Dicellograptus anceps.* 1 - 15 refers to North American Cambrian/Ordovician graptolite zones as follows: 1, *Anisograptus;* 2, *Clonograptus;* 3, *Tetragraptus approximatus;* 4/5, *Tetragraptus fruticosus;* 6, *Didymograptus protobifidus;* 7, *Didymograptus bifidus;* 8, *Isograptus caduceus;* 9, *Paraglossograptus etheridgei;* 10, *Glyptograptus;* 11, *Nemagraptus gracilis;* 12, *Climacograptus bicornis;* 13, *Orthograptus truncatus;* 14, *Orthograptus quadrimucronatus;* 15, *Dicellograptus complanatus.*

(Fig. 3) similar to that of Newfoundland, consisting of two major structural stages, to the south of which, in the Central Zone, the lower stage is absent. In the remainder of this section, the writer briefly describes ophiolite occurrences and their relationships in the orthotectonic zones as a precursor to discussing their mechanism and timing of emplacement. The general relationships of the Bay of Islands ophiolite complex have been discussed earlier and will not be considered further.

Baie Verte/Mings Bight Zone

This zone contains a partially dismembered ophiolite sequence from layered ultramafics to a volcanogenic sediment sequence. The lowest

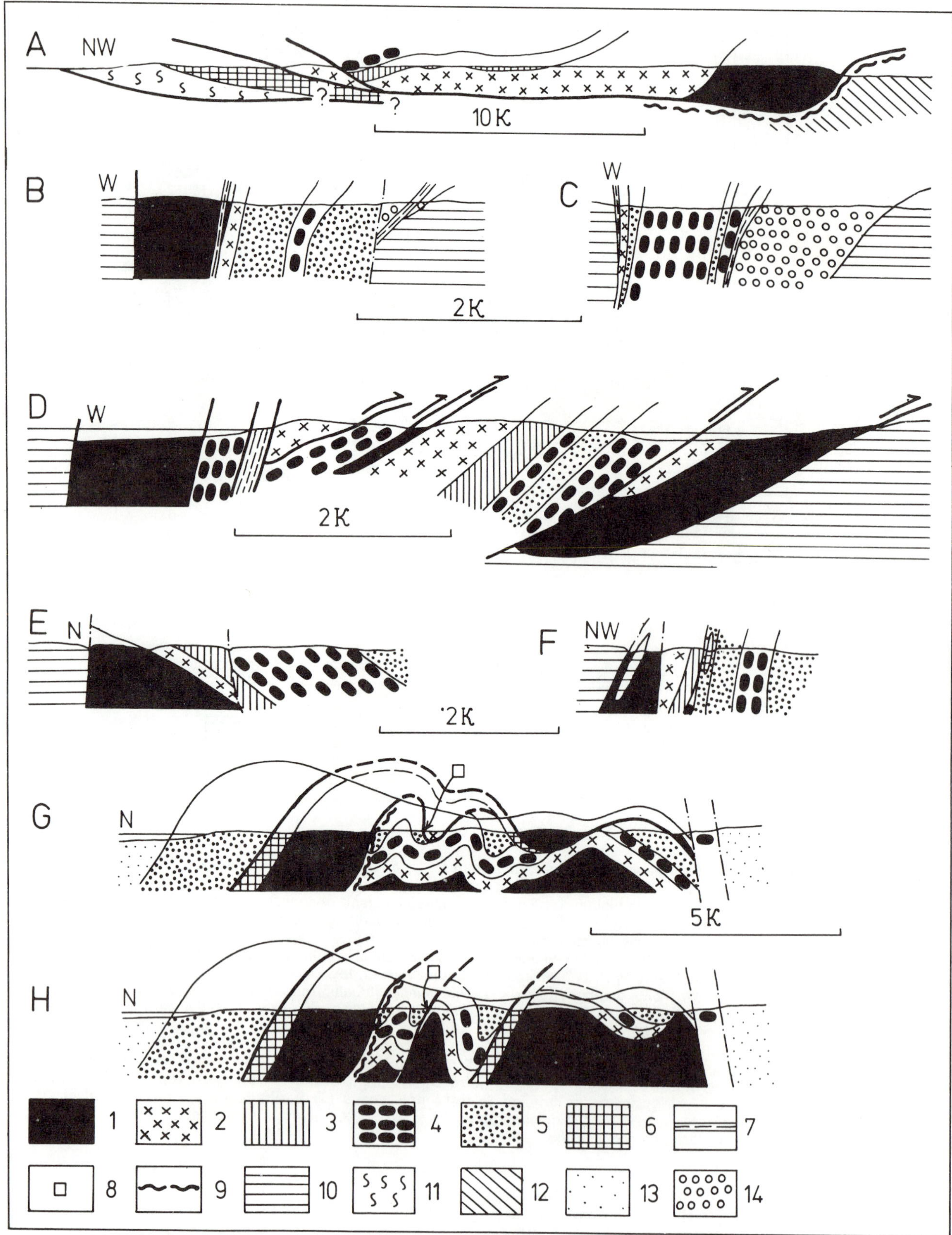

Fig. 5. Sections across several Appalachian/Caledonian ophiolite complexes. *A*, Bay of Island Complex; *B* and *C*, Baie Verte ophiolite complex near C1 (Fig. 3); *D*, Mings Bight ophiolite complex near C2 (Fig. 3); *E*, Betts Cove ophiolite complex near D1 (Fig. 3); *F*, Tilt Cove ophiolite complex near D2 (Fig. 3); *G*, Ballantrae ophiolite complex (section shown in Fig. 6). Several overthrust ophiolite sheets are postulated. *H*, Ballantrae ophiolite complex section shown in Fig. 6. Structure is interpreted here as a single ophiolite cut by high-angle thrusts.

volcanogenic sediments rest both conformably and unconformably on underlaying ophiolite units, occasionally cutting down to gabbro, and have ophiolite debris at the base and andesite debris higher up. The lowest sediments also contain debris from the Eastern Fleur de Lys Zone. Identifiable fossils have not been found, but the Baie Verte ophiolites, by comparison with the Betts Cove/Tilt Cove Zone, are considered to be of Early Ordovician (Arenigian) age (Fig. 4, column C). In most of the zone, the ophiolite sequence is vertical or overturned and faces east, with tectonic contacts against all surrounding rock groups (Fig. 5, B and C). The northern part of the zone is wider and has a synclinal form, a western limb consisting of several overturned westward-dipping thrust slices, and a normal westward-dipping eastern limb with a lower thrust contact against metamorphic rocks of the Eastern Fleur de Lys Zone (Fig. 5, D).

Betts Cove/Tilt Cove Zone

The ophiolite sequence at Betts Cove (Fig. 5, E) has been described briefly by Upadhuay et al., (1971). It consists of a thin conformable ophiolite sequence from cumulate ultramafics to a volcanogenic sediment sequence. The volcanogenic sediments are the lowest formation of the Snooks Arm Group (partly of Arenigian age, Fig. 4, column D). As in the Baie Verte sequence, the lowest sediments have ophiolitic debris, and clastic material derived from the Eastern Fleur de Lys Zone. In the northeastern part of the zone, the lowest sediments cut unconformably down onto gabbros and sheeted complex (Fig. 5, F). The ophiolite sequence and the overlying Snooks Arm Group form the western, partly overturned, limb of a NE-plunging syncline and are entirely faulted against older rocks of the Eastern Fleur de Lys Zone.

Ballantrae Complex, Ayrshire, Scotland

The Ballantrae Complex is a strongly dissected ophiolite suite (Fig. 6) and is of great importance in understanding the origin and emplacement of Appalachian/Caledonian ophiolites because it lies at the southern edge of the orthotectonic zone assemblage. The following description is based on Peach et al. (1926), Balsillie (1932), Anderson (1936), Bloxam and Allen (1960), Bailey and McCallien (1957), Mendum (1968), Bloxam (1968), and the writer's observations. The complex clearly consists of at least two structural stages: a lower stage of ophiolites, shales, cherts, volcanogenic sediments, and mélanges, and an upper stage of volcanics (Mendum, 1968). Shales and cherts in the lower stage contain Arenigian (*D. protobifidus* zone) graptolites (Peach and Horne, 1899; Dewey et al., 1970). The age of the upper stage is unknown but may be Llandeilian (Whittington, 1972). The Ballantrae Complex is unconformably overlain by a shallow marine Llandeilian/Caradocian sequence (Fig. 4 column L). Plutonic, volcanic, and metamorphic rocks of the ophiolite suite are not arranged in an obvious progressive sequence as they are in Newfoundland ophiolite com lexes. Although the Ballantrae Complex is tectoni ally disrupted (Fig. 5, G and H), the following descriptions will illustrate the presence of the characteristic ophiolite units.

1. Ultramafic rocks: Most of the ultramafic rocks are partially, or wholly, serpentinized. Bastite and picotite-bearing serpentinites are common and are probably altered harzburgite and dunite. In the northern ultramafic belt (Fig. 6), interbanded harzburgite and lherzolite are cut by bronzitite, clinopyroxenite, and websterite dikes (Balsillie, 1932). At Knockormal (Bloxam and Allen, 1960), banded wehrlite and clinopyroxenite are associated with ariégite (fassaitic pyroxene-pargasitic amphibole-aluminous garnet-ceylonite).

2. Ultramafic cumulates: In the southern ultramafic belt, cumulate picrites (olivine-clinopyroxene-biotite) are associated with banded wehrlite and clinopyroxenite (Balsillie, 1932).

3. Foliated gabbro: At Fell Hill, Knockdaw Hill, and east of Millenderdale, banded, flasered, and foliated pyroxene-gabbro and brown-hornblende gabbro showing replacement of pyroxene by brown-hornblende and red-brown-biotite, have complex variably orientated zones of ductile to cataclastic deformation and are cut by gray dolerites (Balsillie, 1932).

4. Massive gabbro: This unit is represented by ophitic gabbro, olivine-gabbro, hornblende-gabbro, norite, diallagite pegmatite, occasional leucogabbro, and albitized anorthositic gabbro, with inclusions of bastite serpentinite.

5. Massive gabbro and trondjemite: At Byne Hill (Bloxam, 1968), low-potash olivine-gabbros pass continuously upward into dioritic hornblende-gabbro, diorite/quartz diorite, hornblende-trondjemite, and finally biotite-trondjemite.

6. Sheeted complex: Balsillie (1932) recorded flasered and foliated albite-diabase east of Millenderdale. The writer has briefly visited this locality and it consists of a typical sheeted diabase complex with occasional narrow gabbro screens. Sheeted complex occurs intermittently along the contact between volcanics and serpentinites between Millenderdale and Knockdaw Hill.

7. Mafic volcanics: Pillow lavas, mafic agglomerates, and hyaloclastites, are widespread. Bailey and McCallien (1957) considered that the volcanics belong to two distinct formations separated by an extensive serpentinite sheet, but it seems more likely that the volcanics belong to a single unit forming the topmost unit of an ophiolite sequence.

8. Sediments: Sediments capping the pillow lavas consist mainly of black shales, red and green cherts, tuffs, and agglomerates. The black shales contain an Arenigian (*D. protobifidus* zone) graptolite fauna. At Stockenray (Fig. 6), a distinctive agglomerate contains blocks of porphyritic tachylite, albitized andesite with pale diopside phenocrysts,

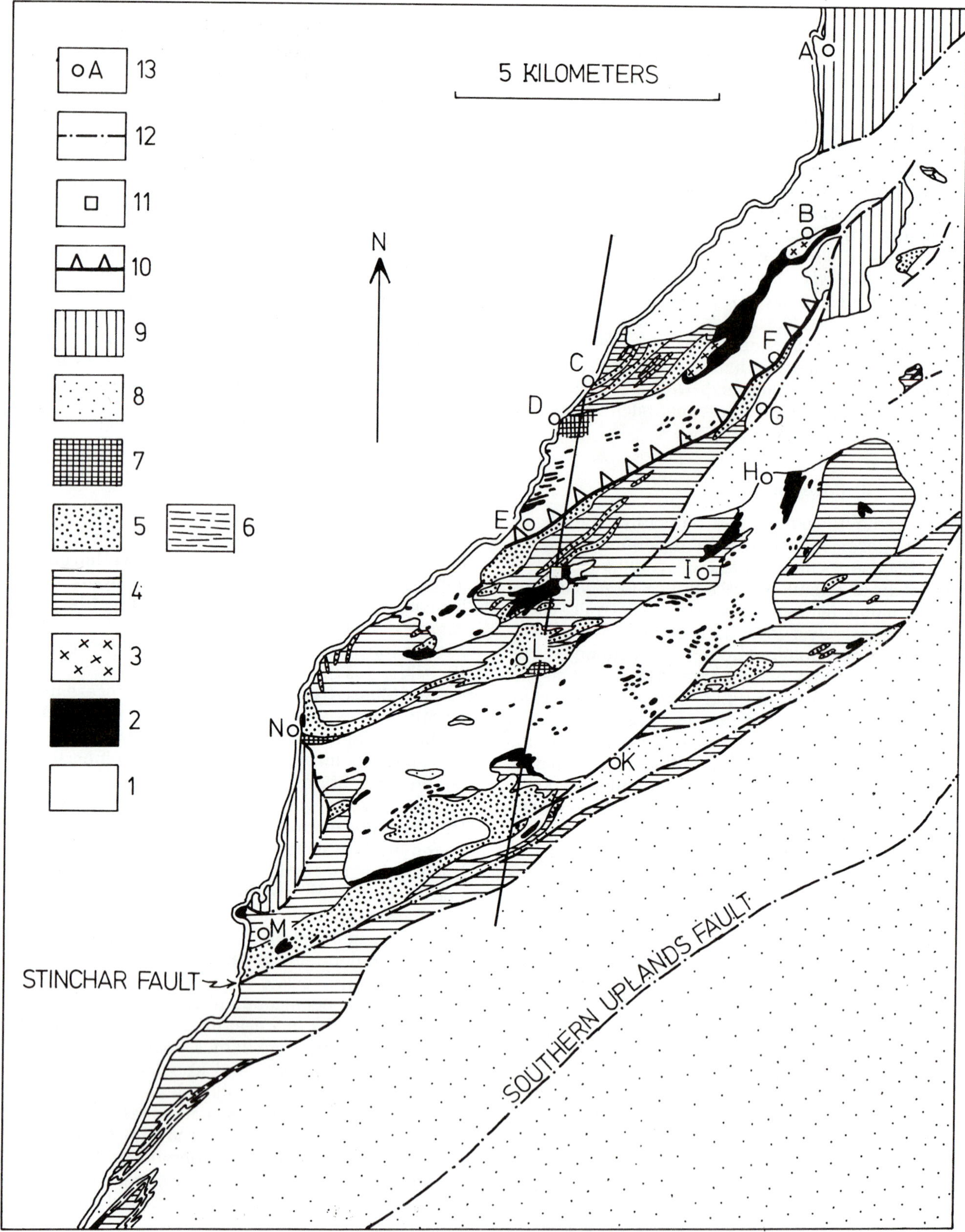

Fig. 6. Geologic map of the Ballantrae ophiolite complex, Ayrshire, Scotland (data from Peach et al., 1926; Anderson, 1936; Bailey and McCallian, 1957). 1, ultramafic rocks; 2, gabbro; 3, quartz diorite and trondjemite; 4, pillow lava; 5, Arenigian argillite, radiolarian chert, limestone, tuff, and agglomerate; 6, Arenigian/Llanvirnian/Llandeilian argillite, radiolarian chert, tuff, and agglomerate; 7, olistostromes and melanges; 8, late Llandeilian to Silurian sediments and volcanics; 9, upper Paleozoic sediments; 10, thrust contact between ultramafic rocks and garnet amphibolites transitional to greenschist facies metavolcanics; 11, glaucophane and crossite schists; 12, faulted contact; 13, localities referred to in the text. *A*, Girvan; *B*, Byne Hill; *C*, Stockenray; *D*, Pinbain Beach; *E*, Lendalfoot; *F*, Loch Lochton; *G*, Laigh Knocklaugh; *H*, Millenderdale; *I*, Knockormal; *J*, Bennane Burn; *K*, Colmonell; *L*, South Ballaird; *M*, Ballantrae; *N*, Bennane Head.

and pink felsitic albite-trachyte.

9. Metamorphic rocks: The Ballantrae Complex contains four distinct metamorphic associations or assemblages.

a. The ariégite, lherzolite, harzburgite, dunite association in the ultramafics was probably generated in the upper mantle during the *origin* of the Ballantrae Complex by some form of sea-floor spreading. Most of the penetrative deformation of the flasered gabbros and the growth of brown hornblende probably occurred at this time, possibly together with the development of albite-analcite-chlorite-epidote-prehnite-zoisite-pectolite assemblages in the volcanics.

b. At Byne Hill, gabbro dikes in the serpentinites are rodingitized (Bloxam, 1968), and albite-diabase schlieren in serpentinites at Lendalfoot are altered, marginally, to porcellanous rodingite (Balsillie, 1932). This rodingitization probably occurred during the serpentinization of the ultramafics as a contact metasomatic effect.

c. Along the NW-dipping contact of volcanics and the northern serpentinite belt between Lendalfoot and Loch Lochton (Fig. 6), both ultramafics and volcanics are affected by metamorphism and penetrative deformation in a zone at varying from about 40 to 80 m wide. The ultramafic rocks are serpentinized and locally reduced to finely banded mylonite and the lherzolites are transformed into tremolite-schists. The volcanics are converted into strongly banded, foliated, hornblende-epidote-felspar-quartz-garnet schists (Anderson, 1936). Near South Ballaird, the contact zone contains foliated scapolite-biotite-amphibolites and, at Lendalfoot, foliated amphibolitic gabbro and diorite contain garnet. Locally, a white rodingite assemblage is superimposed on earlier garnet-amphibolite fabrics. These amphibolites are similar in many respects to those at the base of the Bay of Islands ophiolite sheet in Newfoundland and were presumably generated during the initial tectonic dissection and obduction of the Ballantrae Complex.

d. At Knockormal (Fig. 6) Bloxam and Allen (1960) have recorded the prograde transformation of mafic volcanics into blueschist-facies metamorphics through the following sequence: (1) Andradite/grossularite-bearing spilites with pumpelleyite in veins and replacing plagioclase; (2) actinolite-albite-epidote-chlorite greenschists; (3) glaucophane replacing actinolite and chlorite with distinct narrow bands of crossite and epidote; (4) glaucophane epidote-albite-garnet (almandine-spessartite) assemblages with bands of crossite-pargasite-epidote-albite. These latter blueschist assemblages are partically retrograded to actinolite-prehnite and actinolite-epidote schists.

Three distinct structural/metamorphic episodes have affected the Ballantrae Complex. First there was a period of deformation and metamorphism, during the Early Ordovician sea-floor spreading origin of the complex, varying from zeolite and greenschist in the volcanics, through greenschist/amphibolite in the gabbros, to garnet-spinel mantle assemblages in the ultramafics. Second came a phase of ophiolite obduction during the Early Ordovician during which the ophiolite sequence was disrupted and emplaced as at least two structural slices (Fig. 5, G and H). During the obduction phase, the Lendalfoot/Loch Lochton garnet-amphibolites were formed and a prograde greenschist to blueschist metamorphism developed. The time relations between garnet-amphibolite and blueschist metamorphism is uncertain, but it seems likely that the garnet-amphibolites were developed early in the obduction process while the mantle ultramafics were still hot and that blueschist metamorphism followed. Third came a phase of discontinuous nonpenetrative deformation in Late Silurian times when the complex, together with its unconformable cover of Late Ordovician sediments, was folded and faulted.

10. Mélanges: Distinctive melanges are present near or at the contact of ultramafic and volcanic rocks, at Bennane Head, Pinbain Beach, and South Ballaird (Fig. 6). At South Ballaird, the Knockdolian "agglomerate" contains blocks of sheared dolerite, spilite, and serpentinite. The Bennane Head mélange contains blocks of carbonated picotite/chromite serpentinite, argillite, spilite, and albitized andesite with pale diopside phenocrysts, set in a disrupted contorted matrix of red argillite, chert, and tuff, with slump folds and pull-apart structures. The Pinbain Beach mélange is similar but also contains fragments of ariégite and glaucophane schist. The age of these mélanges cannot be precisely determined but must clearly postdate the blueschist metamorphism of the volcanics. It also seems likely that they predate the eruption of the phenocrystic basalts shown by Mendum (1968) to lie unconformably on deformed Early Ordovician volcanics at Bennane Head. The age of these latter volcanics is uncertain, but they are probably older than Caradocian and may be of Llandeilian age (Fig. 4, column L). The writer suggests that the mélanges were developed at a late stage in the Early Ordovician obduction of the Ballantrae Complex, probably during late Arenigian times.

OBDUCTION OF APPALACHIAN/ CALEDONIAN OPHIOLITES

In Figure 4 the ages, where known, and stratigraphic relationships of ophiolite complexes in the Newfoundland Appalachians and British Caledonides are shown. Also included in this chart are stratigraphic sequences that are relevant to, and place certain constraints on, the timing and nature of ophiolite obduction in this part of the Appalachian/Caledonian orogenic belt. The columns (A, B, C, etc.) summarize the characteristics of stratigraphic sequences whose position is shown in Figure 3. The ophiolites all appear to be of Late Cambrian/Early Ordovician age and were obducted to their present tectonic positions at various times mainly during the

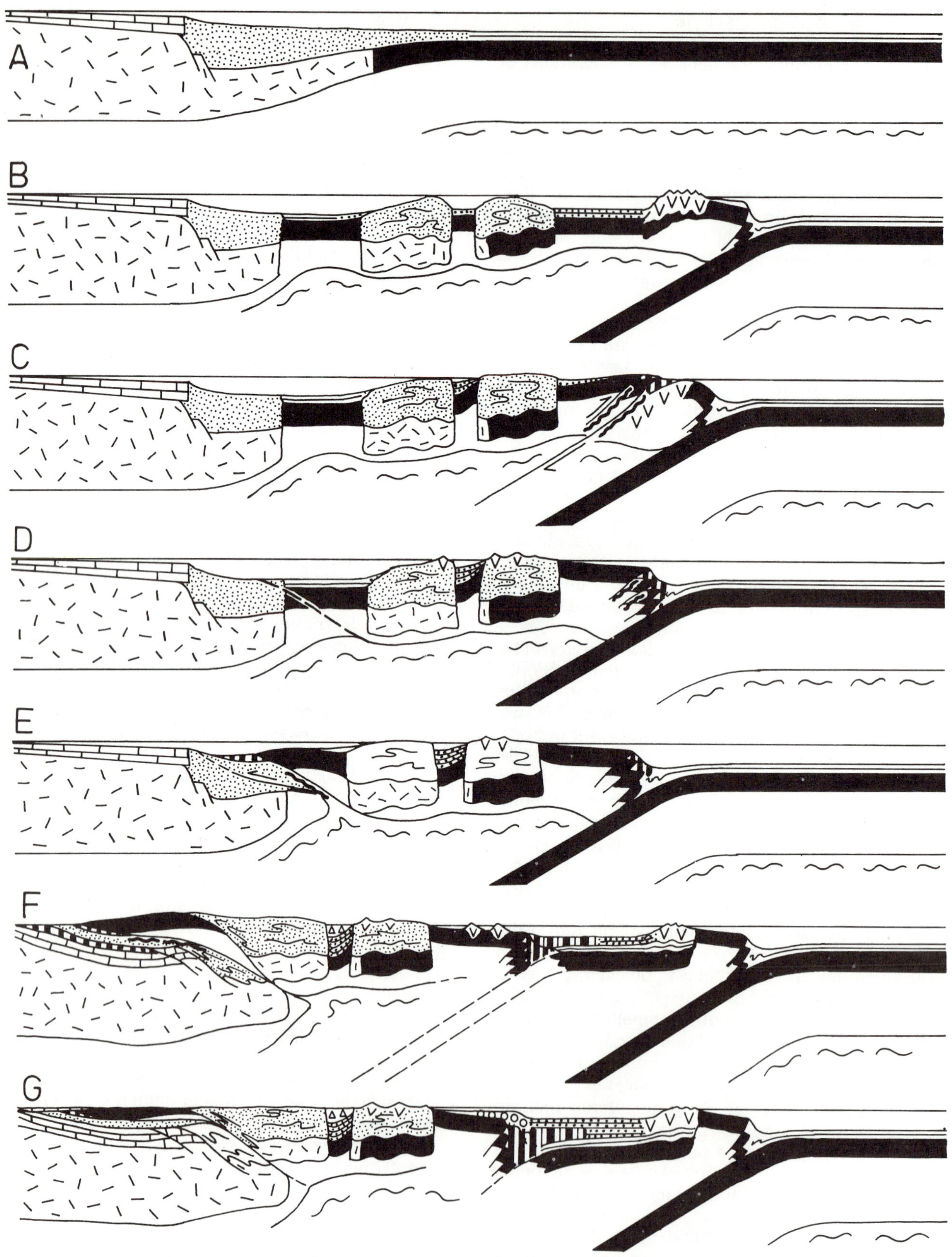

Fig. 7. Schematic illustration of the evolution of the orthotectonic zone assemblage of the Newfoundland Appalachians and British Caledonides, involving the generation of ophiolite sequences in rear-arc and interarc oceanic basins. *A*, Middle Cambrian; *B*, Late Cambrian/Early Ordovician; *C*, early Arenigian; *D*, middle Arenigian; *E*, late Arenigian/early Llanvirnian; *F*, Llandeilian; *G*, early Caradocian.

Early and Middle Ordovician. It is of considerable importance to correctly evaluate the timing of the origin and emplacement, and stratigraphic/structural relationships, of the ophiolites because not only do they provide a link between the two main structural stages of the orthotectonic zone, but they determine whether the ophiolites were expelled from rear-arc/interarc/intraarc oceanic basins above subduction zones or whether they were obducted from larger oceanic basins unrelated to volcanic arcs.

The basic question to be answered is: (can a single model explain the origin and emplacement of all the ophiolite occurrences? Several views have been expressed on this matter, and basically they fall into two divergent hypotheses. Stevens and Church (1969), Stevens (1970), Church and Stevens (1971), and Church (1972) have argued that all the ophiolite occurrences originally formed part of a single sheet obducted during the Middle Ordovician. These workers consider the short time between origin and emplacement to indicate the existence of a short-lived narrow Ordovician ocean. This hypothesis involves the abduction of a continuous high-level ophiolite nappe, across the whole of the orthotectonic zone, that was subsequently tectonically dissected and preserved as erosional remnants. This further involves the idea that Fleur de Lys/Grampian deformation and metamorphism was a pre-oceanic intracontinental event unrelated to the subduction process. An alternative view has been expressed by Dewey and Bird (1971), who consider the ophiolites to be remnants of rear-arc and intraarc basins generated during the subduction of an older adjacent ocean that had been in existence since the latest Precambrian or Early Cambrian. This hypothesis involves separate sites of origin and different times and modes of obduction for the Bay of Islands, Baie Verte, Betts Cove, and Ballantrae ophiolite complexes.

The writer prefers a model like the one of Dewey and Bird (1971), perhaps slightly modified, in which the ophiolite sequences with their thick overlying volcanogenic flysch sequences are viewed as a series of several small oceanic basins lying behind and between distending volcanic arcs that lay marginal to a larger ocean that had been in existence since Early Cambrian times. One possible model of this kind is illustrated in Figure 7. The model is essentially a schematic collage built from Newfoundland, Irish, and Scottish evidence and it will be appreciated that it cannot represent a definitive evolution for any one cross section of the orogenic belt. Furthermore, even if the different ophiolite sequences originated and rooted in separate oceanic basins, there are several ways in which their evolution can be linked to the growth of the orthotectonic terrane.

By Middle Cambrian times, a continental shelf carbonate platform lay to the west of a continental rise that lay marginal to an ocean of unknown dimensions (Fig. 7, A). Under the upper continental rise, a thick sequence of quartzofelspathic clastics and calcareous flysch passed oceanward into similar sediments of the Fleur de Lys and Grampian terranes. Older continental rocks underlie at least part of these sequences in Newfoundland, Ireland, and Scotland, but may have formed part of a thinned and distended continental crust of the kind that Talwani and Eldholm (1972) have shown to exist beneath the continental rise of the eastern North Atlantic. It is also possible that an oceanic basement underlay part of the continental rise west of the Fleur de Lys and Grampian terranes. Such an oceanic basement would have been of Cambrian age and could have eventually been obducted as the Bay of Islands ophiolite sheet. During latest Cambrian/earliest Ordovician times the oceanic lithosphere decoupled so that oceanic basement began to slide beneath the continental rise (Fig. 7, B). During this phase of subduction, the thick stratigraphic sequences of the continental rise were deformed, metamorphosed, and injected by granodioritic magmas (Dewey, 1969; Bird and Dewey, 1970). At this time, sediments continued to accumulate on the upper continental rise, this fact perhaps supporting the idea that an oceanic tract separated the continental margin from the Fleur de Lys terrane. At this time, or possibly slightly later, the Fleur de Lys terrane began to fragment and move away from the continental margin by the growth of a series of rear-arc and interarc basins (Fig. 7, B). The evolving basement of these basins consisted of oceanic crust and mantle (now obducted ophiolite complexes) that became progressively buried by volcanogenic flysch sequences derived from locally rising blocks of Fleur de Lys and Grampian metamorphic rocks. Possibly the Coastal Complex of western Newfoundland was a small sliver of Fleur de Lys metamorphics, perhaps founded on Cambrian oceanic crust, stranded in the wake of the retreating Fleur de Lys block. Such an origin would result in the eventual obduction of the Coastal Complex in its present position in the stacking order of the western Newfoundland allochthons. It would also explain the basic dikes that cut earlier metamorphic rocks in the Coastal Complex. During Arenigian times, volcanogenic flysch sequences accumulated in the South Mayo Trough, but the Ballantrae marginal basin began to collapse by the shearing and telescoping of its thin hot lithosphere. This is viewed as a process similar to that which Blake and Jones (1973) have argued was responsible for the late Jurassic obduction of the Great Valley ophiolites of California. Slices of oceanic crust and mantle started moving, as a volcanic arc was progressively subducted beneath them (Fig. 7, C and D). During the initial stages of obduction, garnet-amphibolites developed in the mafic volcanics as hot mantle ultramafics rode over them. As the sheets cooled and became involved in subduction, blueschist metamorphism began, and olistostrome mélanges developed. During latest Arenigian times, subduction continued just south of the Ballantrae Complex and sedimentation continued in the South Mayo Trough (Fig. 7, D). West of the

Fleur de Lys metamorphic terrane, a hot slab of oceanic crust and mantle began to detach and move westward behind a frontal wave of ophiolite-bearing flysch (Fig. 7, E). During Llandeilian times, the assembled western Newfoundland allochthon was obducted onto the continental shelf, while a molasse sequence, derived from adjacent rising metamorphic/volcanic cordillera, filled the South Mayo Trough (Fig. 7, F). During early Caradocian times, an oceanic tract immediately south of the Ballantrae Complex became the site of deposition of thick clastic sequences (Fig. 4, column G) derived from Grampian metamorphic, Ballantrae ophiolite, and volcanic arc sources. This ocean tract may have been partly inherited from earliest Ordovician times but may also have been the site of generation of oceanic crust and mantle behind a volcanic arc evidenced by the late Llandeilian/early Caradocian andesites of the Central Zone (Fig. 4, column N). Thus the site of subduction appears to have moved southward in Middle Ordovician times. At this time a subduction zone with opposite polarity was active to the south (Fig. 3), behind which were generated the Borrowdale silicic/intermediate volcanic sequences of the English Lake District (Dewey, 1969; Fitton and Hughes, 1970).

This kind of model accounts for many of the problems and apparent paradoxes of the Fleur de Lys/Grampian deformation and metamorphism, in particular its complex relationships to Early and Middle Ordovician ophiolite/volcanogenic clastic sequences, its protracted polyphase history, and its progressive, probably diachronous, development. The writer views the Fleur de Lys/Grampian "event" as resulting from thermal and mechanical processes involving subduction in a complex volcanic arc terrane, in which ophiolite origin and emplacement resulted from the opening and closing of rear-arc and interarc oceanic basins. The width of these supposed interarc oceanic basins is unknown, as also is the width of the ocean whose irregular closure led to the Late Silurian/Devonian deformation of the Appalachian/Caledonian orogenic belt (Dewey, 1969).

BIBLIOGRAPHY

Anderson, J. G. C., 1936, Age of the Girvan-Ballantrae-Serpentine: Geol. Mag., v. 73, p. 535-545.

Bailey, E. B., and McCallien, W. J., 1957, The Ballantrae serpentine, Ayrshire: Trans. Edinburgh Geol. Soc., v. 17, p. 33-53.

Bailey, E. H., Blake, M. C., and Jones, D. L., 1970, On-land Mesozoic oceanic crust in California Coast Ranges: U.S. Geol. Surv. Prof. Paper 700-C, p. 70-81.

Balsillie, D., 1932, The Ballantrae igneous complex, South Ayrshire: Geol. Mag., v. 69, p. 107-131.

Bird, J. M., and Dewey, J. F., 1970, Lithosphere plate-continental margin tectonics and the evolution of the Appalachian orogen: Geol. Soc. America Bull., v. 81, p. 1031-1060.

———, Dewey, J. R., and Kidd, W. S. F., 1971, Proto-Atlantic oceanic crust and mantle: Appalachian/Caledonian ophiolites: Nature, v. 231, p. 28-31.

———, Dewey, J. F., and Kidd, W. S. F., in preparation, The Mings Bight Ophiolite Complex, Newfoundland: Jour. Geoph. Research.

Blake, M. C., and Jones, D. L., 1973, Origin of Franciscan mélange in northern California, *in* Dott, R. H., ed., Geosynclinal Sedimentation (in press).

Bloxam, T. W., 1968, The petrology of Byne Hill, Ayrshire: Trans. Roy. Soc. Edinburgh, v. 68, p. 105-123.

Bloxam, T. W., and Allen, J. B., 1960, Glaucophane-schist, eclogite, and associated rocks from Knockormal in the Girvan-Ballantrae Complex, South Ayrshire: Trans. Roy. Soc. Edinburgh, v. 64, p. 1-29.

Cady, W. M., 1972, Are the Ordovician Northern Appalachians and the Mesozoic Cordilleran System homologous? Jour. Geophys. Res., v. 77, p. 3806-3815.

Church, W. R., 1969, Metamorphic rocks of Burlington Peninsula and adjoining areas of Newfoundland, and their bearing on continental drift in North Atlantic: Am. Assoc. Petrol. Geol. Mem. 12, p. 212-233.

———, 1972, Ophiolite: its definition, origin as oceanic crust, and mode of emplacement in orogenic belts, with special reference to the Appalachians: Dept. Energy, Mines and Resources of Canada Publ., v. 42, p. 71-85.

———, and Stevens, R. K., 1971, Early Paleozoic ophiolite complexes of the Newfoundland Appalachians as mantle-oceanic crust sequences: Jour. Geophys. Res., v. 76, p. 1460-1466.

Cobbing, E. J., 1964, The Highland boundary fault in East Tyrone: Geol. Mag., v. 101, p. 496-501.

———, Manning, P. I., and Griffith, A. E., 1965, Ordovician-Dalradian unconformity in Tyrone: Geol. Mag., v. 206, p. 1132-1135.

Coleman, R. G., 1967, Low-temperature reaction zones and Alpine ultramafic rocks of California, Oregon, and Washington: U.S. Geol. Surv. Bull., v. 1247, p. 1-49.

———, 1971, Plate tectonic emplacement of upper mantle peridotites along continental edges: Jour. Geophys. Res., v. 76, p. 1212-1222.

Davies, H. L., 1971, Peridotite-gabbro-basalt complex in eastern Papua: an overthrust plate of oceanic crust and mantle: Bureau Min. Res. Geol. Geophys. Austral. Bull., v. 128, p. 1-48.

Dewey, J. F., 1962, The provenance and emplacement of Upper Arenigian turbidites in Co. Mayo, Eire: Geol. Mag., v. 119, p. 238-252.

———, 1963, The Lower Palaeozoic stratigraphy of central Murrisk, Co. Mayo, Ireland, and the evolution of the South Mayo Trough: Quart. Jour. Geol. Soc. London, v. 119, p. 313-344.

———, 1969, Evolution of the Appalachian/Caledonian orogen: Nature, v. 222, p. 124-129.

———, 1971, A model for the Lower Palaeozoic evolution of the southern margin of the early Caledonides of Scotland and Ireland: Scot. Jour. Geol., v. 7, p. 219-240.

———, 1973, The geology of the southern termination of the Caledonides, *in* Nairn, A., ed., The geology of the margins of the North Atlantic: New York, Plenum, (in press), New York, Plenum Press.

———, 1974, Some aspects of finite plate evolution: their implications for the evolution of transforms, triple junctions, and orogenic belts: Am. Jour. Sci. (in press).

———, and Bird, J. M., 1970, Mountain belts and the new global tectonics: Jour. Geophys. Res., v. 75, p. 2625-2647.

———, and Bird, J. M., 1971, Origin and emplacement of the ophiolite suite: Appalachian ophiolites in Newfoundland: Jour. Geophys. Res., v. 76, p. 3179-3206.

———, and Burke, K. C. A., 1973, Tibetan, Variscan, and Precambrian, basement reactivation: products of continental collision: Jour. Geol. (in press).

———, and Pankhurst, R. J., 1970, The evolution of the Scottish Caledonides in relation to their isotopic age pattern: Trans. Roy. Soc. Edinburgh, v. 68, p. 361-389.

———, McKerrow, W. S., and Moorbath, S., 1970, The relationship between isotopic ages, uplift, and sedimentation during Ordovician times in western Ireland: Scot. Jour. Geol., v. 6, p. 133-145.

———, Rickards, R. B., and Skevington, D., 1970, New light on the age of Dalradian deformation and metamorphism in western Ireland: Norsk. Geol. Tiddskr., v. 50, p. 19-44.

———, Pitman, W. C., Ryan, W. B. F., and Bonnin, J., 1973, Plate tectonics and the evolution of the Alpine System: Geol. Soc. America Bull., v. 84, p. 3137-3180.

Dickinson, W. R., 1971, Plate tectonic models of geosynclines: Earth and Planetary Sci. Letters, v. 10, p. 165-174.

Dietrich, V., 1969, Die Ophiolithe des Oberhalbsteins (Graubünden) und das Ophiolithmaterial der ostschweizerischen Molasseablagerungen, ein petrographischer Vergleich: Europäische Hochschulschriften, v. 17, 179 p.

Fitton, J. F., and Highes, D. J., 1970, Volcanism and plate tectonics in the British Ordovician: Earth and Planetary Sci. Letters, v. 8, p. 223-228.

Gass, I. G., 1968, Is the Troodos massif of Cyprus a fragment of Mesozoic ocean floor?: Nature, v. 220, p. 39.

Glennie, K. W., Bouef, M. G. A., Hughes Clark, M. W., Moody-Stuart, M., Pilaar, W. F. H., and Reinhardt, B. M., 1973. Late Cretaceous nappes in Oman mountains and their geologic evolution: Amer. Assoc. Petrol. Geol. Bull., 57, 5-27

Hamilton, W., 1969, Mesozoic California and the underflow of the Pacific mantle: Geol. Soc. America Bull., v. 80, p. 2409-2430.

Helwig, J., 1969, Redefinition of exploits group, Lower Paleozoic, northeast Newfoundland: Am. Assoc. Petrol. Geol. Mem. v. 12, p. 408-413.

Horne, G. S., and Helwig, J., 1969, Ordovician stratigraphy of Notre Dame Bay, Newfoundland: Am. Assoc. Petrol. Geol. Mem. v. 12, p. 388-407.

Isacks, B., and Molnar, P., 1969, Mantle earthquake mechanisms and the sinking of the lithosphere: Nature, v. 223, p. 1121-1124.

Jehu, T. J., and Campbell, R., 1917, The Highland Border rocks of the Aberfoyle district: Trans. Roy. Soc. Edinburgh, v. 52, p. 175-212.

Johnson, M. R. W., and Harris, A. L., 1967, Dalradian-Arenig relations in parts of the Highland Border, Scotland, and their significance in the chronology of the Caledonian orogeny: Scot. Jour. Geol., v. 3, p. 1-16.

Karig, D. E., 1971, Origin and development of marginal basins in the western Pacific: Jour. Geophys. Res., v. 76, p. 2542-2560.

Kay, M., 1970, Flysch and bouldery mudstone in northeast Newfoundland: Geol. Assoc. Can. Spec. Paper 7, 155-164.

McKenzie, D. P., 1969, Speculations on the consequences and causes of plate motion: Geophys. Jour. Roy. Astron. Soc., v. 18, p. 1-32.

McKerrow, W. S., and Campbell, C. J., 1960, The stratigraphy and structure of the Lower Palaeozoic rocks of north-west Galway: Proc. Roy. Dublin Soc., v. 1, p. 27-51.

McManus, J., 1967, Sedimentology of the Partry Series in the Partry Mountains, Co. Mayo, Eire: Geol. Mag., v. 104, p. 585-607.

Mendum, J. R., 1968, Unconformities in the Ballantrae volcanic sequence: Trans. Leeds Geol. Assoc., v. 7, p. 261-264.

Mitchell, A. H., and Reading, H. G., 1969, Continental margins, geosynclines, and ocean-floor spreading: Jour. Geol., v. 77, p. 629-646.

Moores, E. M., 1970, Ultramafics and orogeny, with models of the U.S. Cordillera and the Tethys: Nature, v. 228, p. 837-842.

———, and Vine, F. J., 1971, The Troodos Massif, Cyprus, and other ophiolites as oceanic crust: evaluation and implications: Phil. Trans. Roy. Soc. London, ser. A, v. 268, p. 443-466.

Packham, G. H., and Falvey, D. A., 1971, An hypothesis for the formation of marginal seas in the western Pacific: Tectonophysics, v. 11, p. 79-109.

Peach, B. N., and Horne, J., 1899, The Silurian rocks of Great Britain, v. 1, Scotland: Mem. Geol. Surv. Great Britain, v. 18, 749 p.

———, Horne, J., and Geikie, A., 1926, Girvan 3rd ed.: Geol. Surv. Great Britain, 1 Inch: 1 Mile Geol. Map Ser., Sheet 7.

Rabinowitz, P. D., and Ryan, W. B. F., 1970, Gravity anomalies and crustal shortening in the eastern Mediterranean: Tectonophysics, v. 10, p. 585-608.

Reinhardt, B. M., 1969, On the genesis and emplacement of ophiolites in the Oman Mountains geosyncline: Schweiz. Mineral. Petrog. Mitt., v. 49, p. 1-30.

Ritchie, M., and Eckford, R. J. A., 1931, The lavas of Tweeddale and their position in the Caradocian Sequence: Geol. Suv. Great Britain, Summary of Progress for 1930, p. 46-57.

Sales, J. K., 1971, The Taconic allochthon—not a detached gravity slide (abs): Geol. Soc. America Ann. Meeting, Washington, p. 693.

Sclater, J. G., Anderson, R. N., and Bell, M. L., 1971, Elevation of ridges and the evolution of the central eastern Pacific: Jour. Geophys. Res., v. 76, p. 7888-7915.

Smith, A. G., 1971, Alpine deformation and the oceanic areas of the Tethys: Geol. Soc. America Bull., v. 82, p. 2039-2070.

Smith, C. H., 1958, Bay of Islands igneous complex, western Newfoundland: Geol. Surv. Can. Mem. v. 290, 132 p.

Stanton, W. I., 1960, The Lower Palaeozoic rocks of southwest Murrisk, Ireland: Quart. Jour. Geol. Soc. London, v. 116, p. 269-291.

Stevens, R. K., 1970, Cambro-Ordovician flysch sedimentation and tectonics in west Newfoundland and their possible bearing on a proto-Atlantic Ocean: Geol. Assoc. Can. Spec. Paper 7, p. 165-177.

———, and Church, W. R., 1969, Age of ultramafic rocks in the northwestern Appalachians (abs.): Geol. Soc. America Ann. Meeting, Atlantic City, p. 215-216.

Talwani, M., and Eldholm, O., 1972, Continental margin off Norway: A geophysical study: Geol. Soc. America Bull., v. 83, p. 3575-3606.

Temple, P. G., and Zimmerman, J., 1969, Tectonic significance of Alpine ophiolites in Greece and Turkey (abs.): Geol. Soc. America Ann. Meeting, Atlantic City, p. 221-222.

Upadhuay, H. D., Dewey, J. F., and Neale, E. R. W., 1971, The Betts Cove Ophiolite Complex, Newfoundland: Appalachian oceanic crust and mantle: Geol. Assoc. Can. Proc., v. 24, p. 27-34.

Whittington, H. B., 1972, Scotland: Geol. Soc. London Spec. Rept. 3, p. 49-53.

Williams, A., 1972, Ireland: Geol. Soc. London Spec. Rept. 3, p. 53-59.

Williams, H., 1964, The Appalachians in Newfoundland—a two-sided symmetrical system: Am. Jour. Sci., v. 262, p. 1137-1158.

———, 1971, Mafic-ultramafic complexes in western Newfoundland Appalachians and the evidence for their transportation: a review and interim report: Geol. Assoc. Can. Proc., v. 24, p. 9-25.

———, and Malpas, J., 1972, Sheeted dikes and brecciated dike rocks within transported igneous complexes, Bay of Islands, western Newfoundland: Can. Jour. Earth Sci., v. 9, p. 1216-1229.

Part XII

Resources at Continental Margins

Petroleum Resources Potential of Continental Margins

Lewis G. Weeks

INTRODUCTION

This paper had its origin in an assessment of the potential petroleum resources of all marine areas of the earth, prepared for the United Nations at the request of the Head of the U.N. Ocean Economics and Technology branch, to whom a document covering the study was presented in April 1973. The writer is grateful to the United Nations for permission to publish this revised version.

The purpose of the initial study was to provide one of the necessary bases for an agreement among the nations on fair and acceptable boundaries in the oceans between national and international jurisdiction for purposes of development of the natural resources. The 1958 Geneva Convention was one of several negotiated at the United Nations Conference on the Law of the Seas; the Convention was signed by the United States June 10, 1964. The agreement stipulates that the signatories may extend exploration and exploitation of the sea-bottom and subbottom resources out to 200 m of water depth, and beyond that limit as far as those resources profitably can be exploited. Resulting complications were largely due to the complex nature of so many coastlines and offshore topographies, and also to the lack of established offshore boundaries between neighboring nations.

Quantitative assessments of the petroleum prospects of the world were initiated by the writer in the 1930s for the purpose of providing meaningful analyses of prospects throughout the world. These assessments have continually been upgraded and revised with the worldwide development of new geologic understanding and improved industry exploration experience and capabilities. The writer has long called attention to the relatively limited life potential of the world's petroleum resources in terms of national and world history, a fact that had been made evident by the figures themselves.

ECONOMIC AND POLITICAL RESTRICTIONS TO RESOURCE ESTIMATES

Anyone who has sat in on discussions of this subject in the United Nations is impressed by the diversity of views concerning where the boundary between national and international responsibilities should be drawn, as well as matters related to general jurisdiction. Among the numerous boundary proposals, four seem to be most favored. Two of these are based on water depth, one along the 200-m isobath and the other along the 3,000-m isobath. The other two are, respectively, 40 and 200 nautical miles from the coast.

The concept of a boundary based on water depth derives from the fact that 100 fathoms, 600 ft, or 200 m has commonly been considered as the approximate depth of the outer edge of the continental shelf. Its depth actually averages, however, about 150 m, and it varies from as little as 50 m to more than 500 m. The width of the shelf varies from less than 10 to several hundred miles, and is not, in itself, an index of favorability for petroleum or other resources.

Beyond the shelf edge, the slope of the sea floor increases rapidly and extends to the deep floor of the ocean basin. The upper, steeper part of this entire slope is known as the "continental slope"; the gentler deeper part, from the foot of the continental slope to the abyssal plains, is generally referred to as the "continental rise." The foot of the continental slope, which occurs at a variable depth of about 2,000-3,000 m, might be considered as the continental edge, for it is the approximate boundary between the kind of rocks that are typical of the continental crust and the distinctively different kind of rocks that make up the earth's crust under the deep, main ocean bottoms.

Thus the tying of the boundary between national and international jurisdiction to either the 200 or the 3,000-m isobath has merit from the standpoint that each marks approximately the outer edge of a prominent geomorphic feature of the ocean bottom, features that are present off most continental and insular lands. The use of the approximately 3,000-m depth has particular merit in several respects: the continental slope is the most conspicuous and earth-embracing feature of the surface of the solid earth. In consequence, the foot of the slope, besides being a morphologically and geologically prominent and natural (not an artificial) boundary, is about the most clearly delineated boundary on the ocean bottom; but exact definition of this depth must still be determined.

There are yet other reasons why a water depth of 3,000 m would seem to be a most logical jurisdiction boundary. Both of the two proposed boundaries that are based on distances from the coastline (40 and 200 nautical miles) present difficult problems of acceptable delineation because of the very great irregularity and complexity of the continental, insular, and archipelagic coasts. Also, the commercial interest of many nations in the subsea has already advanced well beyond 40 miles from the coast, and even beyond 200 miles in the case

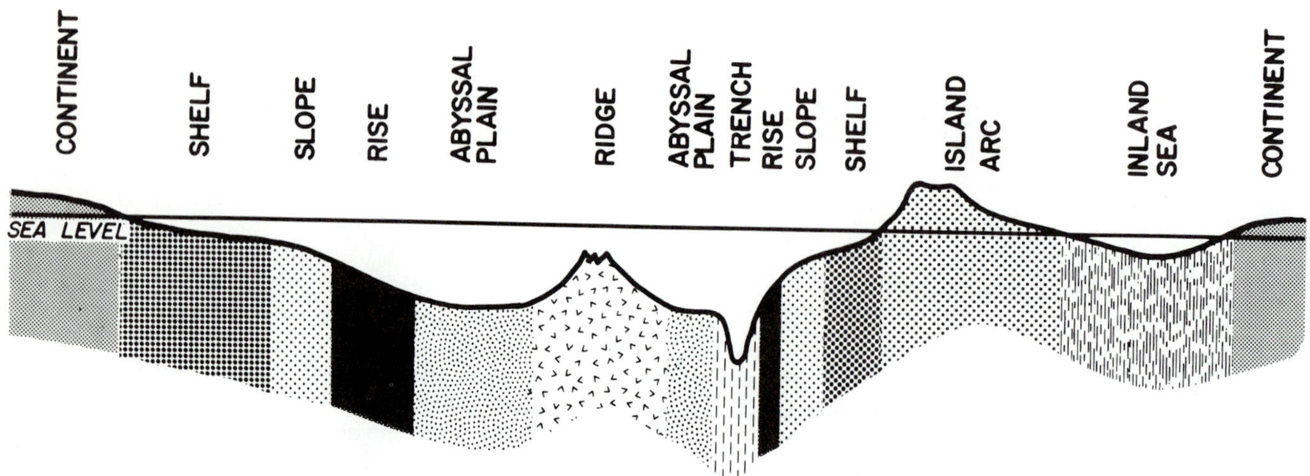

Fig. 1. Physiographic zones of the world's oceans as used in this study.

of several whose shelves extend out to as much as 300-400 miles. Furthermore, operations beyond the edge of the shelf are becoming increasingly widespread. Active interest in the petroleum possibilities has, in fact, already extended to the 3,000-m isobath off the coasts of several countries.

It would seem geologically reasonable that a boundary between jurisdictions drawn along the foot of the continental slope would, in addition to being both natural and reasonable, be the most equitable and the most likely to be acceptable by every coastal nation.

The area of the world's oceans, including their connecting seas and major gulf and bays, are shown as follows:

Ocean	Statute Square Miles	Nautical Square Miles
Atlantic	41,147,000	31,101,000
Pacific	70,018,000	52,924,000
Indian	28,617,000	21,630,000
	139,782,000	105,655,000

Subsea Production History and Resource Assessment Summary

The following figures give the 1972 and the estimated 1980 oil and products demand of the 14 leading consumer nations:

Country	1972	1980 est.	Country	1972	1980 est.
United States	16,054.0	24,606.0	Canada	1,803.0	2,340.0
U.S.S.R.	6,070.0	9,600.0	Brazil	633.0	1,220.0
Japan	4,744.0	10,134.1	China	675.0	985.0
W. Germany	2,964.1	4,721.9	Australia	578.9	970.0
U.K.	2,273.4	3,211.2	Mexico	560.0	845.0
France	2,332.1	4,249.2	Argentina	501.0	583.0
Italy	2,020.8	3,810.3	India	486.0	862.0

As is commonly done, the Arctic Ocean's 4,732,000 square statute miles (3,577,000 square nautical miles) is included in this study with the Atlantic, with which it directly connects.

The following gives the major ocean subdivisions as used in this report (fig. 2).

1. Western North Atlantic (north of Equator and west of Mid-Atlantic Ridge).
2. Eastern North Atlantic (north of Equator and east of Mid-Atlantic Ridge).
3. Western South Atlantic (south of Equator and west of Mid-Atlantic Ridge).
4. Eastern South Atlantic (south of Equator and east of Mid-Atlantic Ridge).
5. Western North Pacific (north of Equator and east of Mid-Atlantic Ridge).
6. Eastern North Pacific (north of Equator and east of mid-ocean line).
7. Western South Pacific (south of Equator and west of mid-ocean line).
8. Eastern South Pacific (south of Equator and east of mid-ocean line).
9. Western Indian Ocean (west of 75°E longitude).
10. Eastern Indian Ocean (east of 75°E longitude).

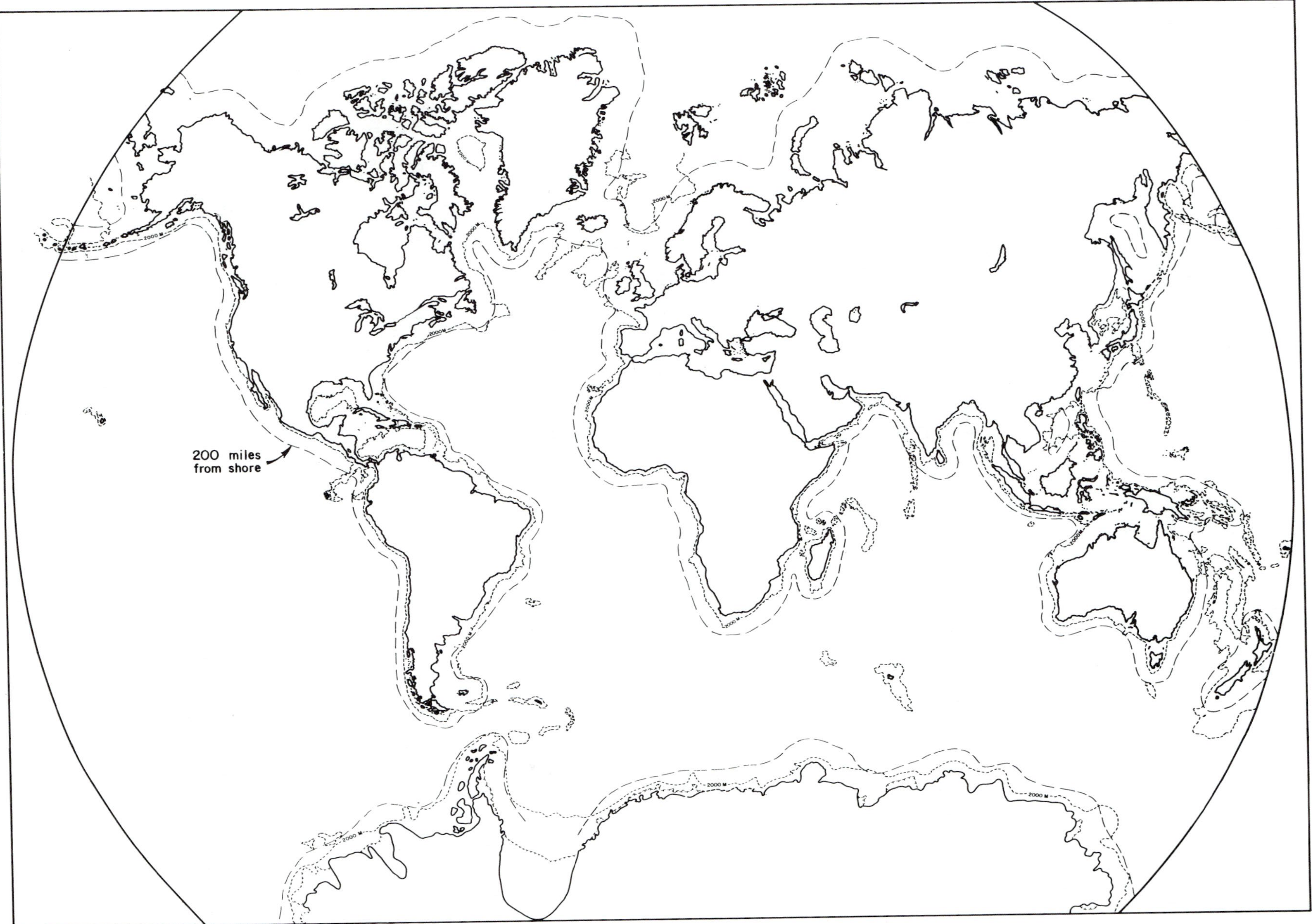

Fig. 2. Generalized map of the world's continental margins as used in this summary: 2,000 m isobath in short-dashed line.

The foregoing tables, and the tables that follow, set forth the steps by which the estimates of potential ultimate recoverable petroleum shown in Table 2 (which add up to 2,272 billion barrels of oil and oil equivalent of the gas) were calculated and distributed among the 10 major ocean subdivision defined above.

Table 1 gives the areas of the world's subsea physiographic provinces and an estimate of the potentially prospective petroleum-bearing area of each. This table leads to Table 2, which concludes the setting out of a worldwide framework of this nature. As shown on Table 2, the study arrived at a world subsea total of 2,272 billion barrels of estimated potential ultimate recoverable barrels of oil, plus oil equivalent of the gas. It should be noted that *in making this estimate it was assumed that all of this petroleum will be economically recoverable.* This may not be found possible, particularly with respect to deep-water areas of lower-than-average productivity.

It should perhaps be repeated here that the incidence of petroleum occurrence varies widely. The factors that control its occurrence are purely and entirely geological. Size of area in itself may mean little or nothing. For instance, continental shelf area A, with 10 times the area of continental shelf in area B may conceivably carry less than one-tenth the recoverable petroleum of area B. The percent prospective in Table 1 and the percentage of prospective area estimated to be productive in Table 2 are approximate world averages for geologies corresponding to that indicated for the particular physiographic province, or part thereof.

Tables 3 to 11 give the total area, the physiographic province areas, the petroleum production, the current proved reserves, and the total amount of petroleum discovered to January 1, 1973, in each of the 10 major subdivisions into which the oceans of the world are divided above. The numbers also show the distribution of area, production, reserves, and total discoveries, both landward and seaward of four limiting boundaries: 200-m isobath, 3,000-m isobath, 40 miles from the coast, and 200 miles from the coast. It should be noted that the shelf, slope, and rise areas of the small ocean basins of Tables 1 and 2 are included together with those of the main open oceans in the figures for these physiographic provinces in Tables 3 through 13.

Tables 3 to 13 present production for 1972 and cumulative to January 1, 1973, for each of the 10 major world ocean subdivisions. Published production and reserve figures do not vary greatly, in contrast to estimates of potential resources. Although variations of the latter may be expected, it is important that they be of the correct order of magnitude.

In addition, Tables 3 through 12 give an estimate of the total production from the particular ocean subdivisions for the year 1980. The world total, 1980, subsea production is estimated at 5.7 billion barrels of oil, plus gas equivalent. This is 74% greater than the approximately 3.288 billion barrels produced in 1972. The general level of prosperity over the next 8 years in the major consuming areas will have a definite bearing on the size of the demand for petroleum energy in 1980. Although the ratio of world offshore to onshore proved reserves is of the order of 1:5, the unproved potential offshore and onshore resources are more nearly equal. Thus the future will bring a progressive increase in the ratio of offshore discoveries, proved reserves, and production over those existing onshore.

Finally, Table 13 shows the world subsea distribution of area, of total petroleum discovered to 1/1/73, of potential estimated but undiscovered, and of ultimate potential of 2,272 billion barrels of petroleum for each of the major designated provinces. The distribution is shown for each of the provinces of the oceans, both landward and seaward of each of the same four limiting boundaries: the 200-m isobath, 3,000-m isobath, 40 miles from the coast, and 200 miles from the coast.

PROSPECTIVE SUBSEA PHYSIOGRAPHIC PROVINCES

The slope and rise sediments appear to be comprised of about 60-80% mud, 10-20% sand and silt, and 5-15% coarser clastics, shells, etc. The sediments of the slope vary in thickness from a few thousand to as much as 25,000 ft. Deposits of the rise are not present everywhere; their thickness varies from negligible to as much as 40,000 ft. They are generally thickest opposite deltas and other major sediment-discharge areas. Thus a considerable part of the rise sediments probably represents delta and other relatively young deposits of the Pleistocene, for the Pleistocene glacial epoch was a time of considerably lowered sea level.

The continental rise is wide off much of the eastern coast of North and South America, the Arabian Sea, Bay of Bengal, eastern Africa, and off much of western Africa. In other words, the rise is generally widest off the stable coasts, where it may extend as much as 900 miles from shore. On the other hand, off mobile coasts, as in the circum-Pacific, much of the rise and slope sediments have been carried into trenches, and presumably incorporated by metamorphism into the mountain roots and continental crust. Rise sediments extend across the entire bottom of many of the small, partially land-locked, ocean basins.

The sediments beneath the slopes and rises, and beneath the small oceans or seas, may in many places attain a greater maximum thickness and volume than those under the neighboring shelves.

The rise and lower-slope sediments are now mainly being deposited in a rather oxidizing environment, which is not generally favorable for the preservation of organic matter and as a possible source of petroleum. This may not always have been so. There are good geological reasons for questioning the adequacy of good porosity and favorable structure.

This is part of the explanation for the very great variation in the incidence of oil by geologic ages—why, for instance, the main part of the Upper Cretaceous accounts for but a minute percentage of the world's petroleum? Without expanding into a lengthy geological side issue, there appear to be good geological reasons why the presence of important quantities of petroleum in the sediments of the deep slope and rise seems unlikely, short of the Lower Cretaceous-Upper Jurassic, and this will be true only in limited parts of certain of the oceans. To recover such petroleum will entail serious problems of engineering and economics. To the writer's mind, there is far more to the question of petroleum prospects than the *volume of sediments* so often mentioned. Other global conditions had modifying effects. The general crustal stability or mobility at different periods is important. The circumferential balance between sea-floor spreading on the one hand and subduction on the other is important.

There are large differences in the area of watershed facing the three oceans today. The total area of land now draining into the Atlantic (nearly 17 million $miles^2$ or 43 million km^2) is about four times that draining into the Pacific, and it is also nearly four times that which drains into the Indian Ocean. However, historically, this difference varied greatly from that of the present. The magnitude of the area and the degree of tectonic activity in past geologic history have had an important bearing on the nature, the significance, and the areal and age distribution of petroleum off the respective coasts.

Another factor overlooked by those who lightly forecast large volumes of petroleum generation and accumulation in the deeper sediments of the oceans is the temperature gradient in those sediments. There is convincing worldwide evidence that temperature has had a primary influence on the genesis of petroleum. The optimum temperature has not been reached everywhere in sedimentary basins, particularly in areas of low temperature gradient. In deltas, for instance, which commonly have a rather low temperature gradient, a greater-than-usual depth of overburden is needed for the generation of petroleum Unfavorably low temperature conditions probably existed beneath the greater part of the long, stable marginal platforms, following the deposition of the major volume of their sediments. Under such conditions petroleum may be present only at depth—if and where thick sediments exist. In the thousands of miles of subduction trenches, such as lie off the shores of the circum-Pacific, the sediments are largely of young Tertiary and Quarternary age, and have generally low temperature gradients.

The foregoing are some examples of the many factors that may be much more important than total sediment volume, and they most certainly need to be taken into account in evaluation of hydrocarbon potential.

Interesting are the numerous shallower seas, whose sediments may overlie a continental or intermediate-type crust, particularly where there has been contemporaneous tectonic activity. Many of these small, partially land-locked sea basins have thick sediment sections. Thus, while their total area is only about 2% of the major oceans, they contain a volume of sediments of the order of at least 10% of that beneath all sea floors. They include such seas as the Gulf of Mexico, North, Mediterranean, Adriatic, Black, Red, South and East China, Okhotsk, Bering, and various Arctic seas. Many of these have very good petroleum prospects.

SUBSEA PHYSIOGRAPHIC PROVINCES OF LITTLE PROSPECTS

The abyssal plains have not, in general, had the kind of deposition and postdeposition deformational history favorable for petroleum generation and accumulation, even in most quantities, much less in the quantities that would be required by any form of economic recovery. Deposition has been at a very slow rate of only about 20-50 ft/million years. This fact, together with the sediment composition, the environment of deposition, and the great water depth, are factors that do not favor the deep oceans as a potential source of important commercial supplies of petroleum.

The abyssal plains of individual Pacific basins range in depth from 16,000 to 23,000 ft; depths in the trenches range up to over 26,000 ft (five of the Western Pacific trenches have been sounded to depths exceeding 33,000 ft). Even with the considerable exploration accomplished during the International Indian Ocean Year, 1965, large parts of the area remain little explored. The Mid-Atlantic Ridge, which occupies about one-third of that ocean's width, is part of a world-girdling ridge, some 45,000 miles in length, which attains heights of 15,000-20,000 ft above the neighboring abyssal plains. The Arctic Ocean, which we have included with the Atlantic in the accompanying tables, is divided by the Lomonosov Ridge (still depth 5,000 ft) into two smaller basins. The geology of the Pacific islands (which comprise a major part of the "oceanic ridges and rises" and of the "volcanic ridges and cones" of Tables 1 and 2) is not considered to be generally favorable for petroleum occurrence.

RANGE OF PETROLEUM OCCURRENCE CONTROLLED BY GEOLOGY

Any worthwhile evaluation of potential petroleum resources must consider two basic factors: (1) the areal and volumetric distribution of the sedimentary rocks, and (2) the composition and deformation of the sediments. The area or volume of sediments may have little bearing on the size of the petroleum yield. Worldwide experience of the petroleum industry clearly shows that the incidence of ultimate hydrocarbon yield varies from as little as several hundred to as much as several million barrels of oil per cubic mile of sediments.

To be sure, petroleum is a normal and widespread constituent of sediments, just as numerous minerals occur widely in the rocks of the earth's crust. However, just as is the case with minerals, economic concentrations of petroleum occur in but a relatively small percentage of the total world sedimentary area or sediment volume. Much the greater percentage of the area and volume will ultimately produce very little commercial petroleum. And if we consider *only* those sedimentary basins which prove to be commercially productive, the percentage of their total areal extent beneath which commercial petroleum will ultimately be found will, in only very limited instances, approach 10%; on a worldwide basis it will not average more than about 3%.

In several of the many productive basins of the world, the percentage of successful wildcat wells has been as high as 50-85% during some of the early exploration years; much more often it is as low as 10%. The world average incidence of petroleum occurrence per unit of area (or volume) of all sediments is of course well below economic limits.

As has so often been stated, petroleum occurrence depends wholly on the geology and on nothing else. The only sound basis for estimating the potential of an area, and this applies most particularly to the new areas of the world, must be a sound geologic understanding of the factors that control local occurrence. These cannot then be evaluated except from a worldwide background of understanding, under all manner of conditions. These geological factors are very real, just as are those which control the occurrence of iron ores, coal, gold, zinc, and so on, whether in the sea or on land. On land, well over 80% of the world's petroleum reserves occur scattered through basins that comprise less than 20% of the total sedimentary basin area. The situation offshore appears to be little different. To illustrate, the total world offshore area out to a water depth of say 1,000 ft (310 m) is about 10,763,000 miles2 (27,876,000 km2). Of this total, about 42% is essentially nonbasin area; and thus it is devoid of any value for petroleum. This leaves an area equal to 58% of the total shelf that is underlain by sedimentary basin. About 38% of the 58% of total shelf is probably economically submarginal; this leaves about 20% as commercially attractive. Of this 20%, about 4% may be considered as prime area, the remaining 16% as having a varying, lesser degree of attractiveness.

As on land, forecasts of the size of subsea resources are sometimes unduly optimistic; imaginations become unrestrained and the basis of calculation becomes questionable. Large volumes of sediments are cited, too often with very exaggerated assumptions as to their unit-hydrocarbon potential. Mention is made of shows of oil logged in exploratory wells off numerous coasts, implying that potential, economically recoverable, supplies probably occur off the shores of most coastal countries and that these foretell some vaguely dimensioned vast total potential. Hydrocarbons are, of course, a normal constituent of sediments, just as all minerals occur widely in the earth's crust. However, the facts of petroleum occurrence, based on the worldwide experience of the petroleum industry, as well as on geology, also show that attractive occurrences have a limited distribution.

With regard to the lower slope and rise, it appears that excessively optimistic forecasts have been made, for not only is the water depth great, but geologic experience indicates that any accumulations of important volume probably lie deep within the lower part of only some of these thicker sediment sequences. However, there are certain geologic situations where important petroleum accumulations may possibly occur below water depths of 1-2 km; and industry, sensing the serious imminent shortages and the need for rapid progress of engineering research, seems to be actively preparing for these possible opportunities.

Seismic surveys are now being conducted routinely in water depths of up to as much as 6,000 ft, and expensive detailed surveys in more than 3,000 ft of water are now common. Subsea fixed platforms have been placed at depths of 600 ft, and plans go forward for their use at water depths of 1,000 ft. Other advances include efficient hole-reentry systems, submerged petroleum storage tanks, deep-water pipelines, manned and unmanned vehicles for subsea operations, and remote-control robot devices for various deep-water tasks, such as subsea well completions. This technology will obviously modify evaluations of future hydrocarbon resource potential.

PETROLEUM'S BASICALLY UNIQUE STATUS

Energy is the most basic element in any nation's economy. This is true whether the energy is derived from the fossil fuels, or from the labor of human bodies. The progress of living standards has paralleled with remarkable precision the growth of economical and productive energy use. Studies show that if all forms of energy used are expressed in terms of equivalent gallons of oil, gross national products vary quite uniformly from country to country on the basis of approximately $1 of income for each gallon equivalent of oil energy used. Next to food, petroleum is probably the most important resource in raising local standards of living.

PROBLEMS IMPOSED BY GROWING PETROLEUM DEMAND

In the 1970s the world will consume more petroleum than in the prior 112 years of petroleum history. In Europe and Japan, the consumption of energy has been growing at three to five times the rate of pupulation increase. World oil production attained a rate of 55 million barrels per day in the latter part of 1973, just over 20 billion barrels annually. About 3.6 billion barrels of the 1973 oil production, or 18%, came from the subsea. Of the 1973 world gas production of about 53 billion thousand cubic feet, about 14% was produced offshore.

Cumulative world oil and gas production to January 1, 1974, was approximately 270 billion barrels of oil and 550 trillion ft3 of natural gas. Of these amounts, about 6.0% and 4.5%, respectively, came from beneath the oceans. World proved reserves of oil and gas are estimated at 640 billion barrels of oil and 550 trillion ft^3 of natural gas. Of these reserves, the offshore now accounts for about 18% of the oil and 9.5% of the natural gas. Thus the supply-reserve ratios of the world for oil and gas are about 33 and 38, respectively.

The world annual percentage increase in oil production from 1950 to 1960 was approximately 7%; and 8% from 1960 to 1970. However, in the 1971-1972 period it was about 6.5%. In the past the annual increase in consumption of petroleum has varied rather closely with the degree of general prosperity in the principal consuming nations.

If continued at the current rate of consumption growth, the world will consume something like 730 billion barrels of oil and 1,000 trillion ft^3 of natural gas over the next 20 years. These amounts are about 110% and 120%, respectively, of the current proved reserves of these sources of energy. At recent growth rates, the oil and natural gas demand 20 years hence will be over three times that of today. To supply this the petroleum industry would be called upon to find on the order of 1,000 billion barrels of oil to replace the amount consumed and maintain a safe inventory. This would call for finding, in the short span of 20 years, about 50% more oil and gas than was found in the past 114 years of the petroleum industry!

As a result of recent worldwide exploration, relatively large new petroleum reserves have been found. These include the Alaskan North Slope and Canadian Arctic, the North Sea, West Siberian Lowlands, northern Africa, the west Africa offshore, the Persian Gulf, and the offshore of Indonesia and Australia. Whereas on a world basis these will provide large additional supplies of energy, the sources will not be adequate to fill the burgeoning needs of the future. In more and more of even the newly developing basins of the world, all or most of the east-to-get-at, inexpensive energy has been found. This is true even in the Middle East.

Much of the future energy resources of the world lie in the world's continental margins. The extent to which we can develop these depends upon international and national politics (a factor over which the international oil corporation has no control), as well as the technological competence to explore successfully for these resources, and the engineering ability to produce them economically.

Perhaps the situation can best be put in perspective by stating that something like 70% of the world's ultimate conventional petroleum resources will have been produced within the normal lifetime of individuals now in their twenties.

World oil production will probably peak early in the twenty first century; however, it will not fulfill demand beginning quite long before that. After its peak, petroleum production will begin a very slow decline, lasting over 100 years, contributing less and less to the world's energy needs. Only at rising prices and/or lower economically supportable costs can an ever-diminishing fraction of the growing petroleum energy requirements be supplied.

The Energy Economics Division of the Chase Manhattan Bank forecasts $1 trillion in total capital investment requirements for the world petroleum industry from 1970 to 1985, predicting that $400 billion of this will be financed externally, if the industry grows at an 8% annual rate as in the previous 15-year period. This $400 billion is approximately eight times larger than the amount raised in the previous period, and much larger than the gross national product of most nations of the world. Most of these risk expenditures will be spent offshore. World offshore expenditures for the year 1972, of approximately $4 billion, will certainly have to be more than doubled by the year 1980 if the annual potential demands are to be supplied.

Mineral Resources of the Sea, Report E/4973, April 26, 1971, of the U.N. Secretary General gives the total value of the world production of marine mineral resources in 1969 as $7.1 billion, of which petroleum accounted for $6.1 billion. On the basis of the rate of increase of marine petroleum production, and allowing for the recent increases in price to a range of, say, $6-12 per barrel, the value of offshore petroleum production appears currently to be running at the rate of about $20-40 billion annually. On the basis of present prices, and in terms of current dollars, this value of offshore oil production seems destined to rise several times before the advent, some three decades or so hence, of an inevitable waning of production.

Table 1. Areas of worldwide marine physiographic provinces (all figures in millions).

Physiographic Province	Miles 2	Prospective Areas % of World	Prospective Areas Group %	Total %	Miles 2
Continental shelf (open ocean)	8.0	5.7	25.8	25.0	2.00
Continental slope (open ocean)	12.0	8.6		15.0	1.80
Continental rise (open ocean)	6.50	4.7		10.0	0.65
Small ocean basins	7.33	5.2		15.0	1.10
Trenches and associated ridges	2.32	1.6		10.0	0.23
Abyssal plains	53.43	38.2	74.2	0.0	0.0
Ocean ridges and rises	45.95	32.9		0.0	0.0
Volcanic ridges and cones	4.25	3.1		0.0	0.0
	139.78	100.0	100.0	4.1	5.78

Table 2. Worldwide marine resources (all figures in millions).

Physiographic province	Prospective Area (miles 2)	Productive Areas % of prospective	Productive Areas Acres	Bbl/acre	Potential Ultimate Recoverable (bbl)
Continental shelf (open ocean)	2.00	3	38.40	35,000	1,344,000
Continental slope (open ocean)	1.80	3-1	23.04	20,000	460,800
Continental rise (open ocean)	0.65	1	4.48	20,000	89,600
Small ocean basins	1.10	4-1	14.08	25,000	352,000
Trenches and associated ridges	0.23	1	1.28	20,000	25,600
Abyssal plains	0.0	0	0.0	0	0
Oceanic ridges and rises	0.0	0	0.0	0	0
Volcanic ridges and cones	0.0	0	0.0	0	0
	5.78	2.2	81.28	—	2,272,000

Table 3. Western North Atlantic (resources) (12,600,000 square statute miles).

Proposed Jurisdictional Boundary Meters Depth	Area Landward (L) Seaward (S) (Miles2)	Production 1972	Production 1980 (est.)	Production Cumulative	Proved Reserves	Discovered to 1/1/1973
		Barrels of Oil and Equivalent Gas (millions)				
200	2,018,000 (L)	1,324	1,500	9,650	15,745	25,395
	10,582,000 (S)	—		—	—	—
3,000	3,554,000 (L)	1,324		9,650	15,745	25,395
	9,046,000 (S)	—		—	—	—
Miles Out						
40	1,598,000 (L)	1,324		9,650	15,745	25,395
	11,102,000 (S)	—		—	—	—
200	6,103,600 (L)	1,324		9,650	15,745	25,395
	6,496,400 (S)	—		—	—	—

Table 4. Eastern North Atlantic (resources) (9,840,000 square statute miles).

Proposed Jurisdictional Boundary Meters Depth	Area Landward (L) Seaward (S) (Miles^2)	Barrels of Oil and Equivalent Gas (millions)				
		1972	Production 1980 (est.)	Cumulative	Proved Reserves	Discovered to 1/1/1973
200	1,302,600 (L)	300	950	1,000	24,200	25,200
	8,538,000 (S)	—		—	—	—
3,000	3,805,000 (L)	300		1,000	24,200	25,200
	6,635,000 (S)	—		—	—	—
Miles Out						
40	1,455,300 (L)	270		800	7,000	7,800
	8,384,700 (S)	30		200	17,200	17,400
200	4,645,300 (L)	300		1,000	24,200	25,200
	5,195,000 (S)	—		—	—	—

Table 5. Western South Atlantic (resources) (10,500,000 square statute miles).

Proposed Jurisdictional Boundary Meters Depth	Area Landward (L) Seaward (S) (Miles^2)	Barrels of Oil and Equivalent Gas (millions)				
		1972	Production 1980 (est.)	Cumulative	Proved Reserves	Discovered to 1/1/1973
200	580,100 (L)	—	35	—	25	25
	10,019,800 (S)	—		—	—	—
3,000	1,181,900 (L)	—		—	25	25
	9,318,100 (S)	—		—	—	—
Miles Out						
40	298,000 (L)	—		—	25	25
	10,302,000 (S)	—		—	—	—
200	1,805,000 (L)	—		—	25	25
	9,695,000 (S)	—		—	—	—

Table 6. Eastern South Atlantic (resources) (8,207,000 square statute miles).

Proposed Jurisdictional Boundary Meters Depth	Area Landward (L) Seaward (S) (Miles^2)	Barrels of Oil and Equivalent Gas (millions)				
		1972	Production 1980 (est.)	Cumulative	Proved Reserves	Discovered to 1/1/1973
200	81,500 (L)	64.3	140	197	5,068	5,265
	8,125,500 (S)	—		—	—	—
3,000	416,700 (L)	64.3		197	5,068	5,265
	7,791,300 (S)	—		—	—	—
Miles Out						
40	99,000 (L)	64.3		197	5,068	5,265
	8,108,000 (S)	—		—	—	—
200	428,400 (L)	64.3		197	5,068	5,265
	7,778,600 (S)	—		—	—	—

Table 7. Western North Pacific (resources) (16,900,000 square statute miles).

Proposed Jurisdictional Boundary Meters Depth	Area Landward (L) Seaward (S) (Miles2)	Barrels of Oil and Equivalent Gas (millions)				
		Production				
		1972	1980 (est.)	Cumulative	Proved Reserves	Discovered to 1/1/1973
200	1,951,600 (L) 14,948,400 (S)	36.33 —	140	101 —	3,235 —	3,336 —
3,000	3,171,400 (L) 13,728,600 (S)	36.33 —		101 —	3,235 —	3,336 —
Miles Out						
40	357,000 (L) 16,543,000 (S)	36.33 —		101 —	3,235 —	3,336 —
200	5,528,600 (L) 11,471,400 (S)	36.33 —		101 —	3,235 —	3,336 —

Table 8. Eastern North Pacific (resources) (16,960,000 square statute miles).

Proposed Jurisdictional Boundary Meters Depth	Area Landward (L) Seaward (S) (Miles2)	Barrels of Oil and Equivalent Gas (millions)				
		Production				
		1972	1980 (est.)	Cumulative	Proved Reserves	Discovered to 1/1/1973
200	331,200 (L) 16,568,000 (S)	150 —	200	400 —	600 —	1,000 —
3,000	544,000 (L) 16,355,900 (S)	150 —		400 —	600 —	1,000 —
Miles Out						
40	232,000 (L) 16,668,000 (S)	150 —		400 —	600 —	1,000 —
200	1,113,300 (L) 15,786,700 (S)	150 —		400 —	600 —	1,000 —

Table 9. Western South Pacific (resources) (18,109,000 square statute miles).

Proposed Jurisdictional Boundary Meters Depth	Area Landward (L) Seaward (S) (Miles2)	Barrels of Oil and Equivalent Gas (millions)				
		Production				
		1972	1980 (est.)	Cumulative	Proved Reserves	Discovered to 1/1/1973
200	648,300 (L) 17,460,700 (S)	117 —	180	247 —	6,000 —	6,247 —
3,000	1,723,000 (L) 16,386,000 (S)	117 —		247 —	6,000 —	6,247 —
Miles Out						
40	740,000 (L) 17,369,000 (S)	117 —		247 —	6,000 —	6,247 —
200	3,692,000 (L) 14,417,000 (S)	117 —		247 —	6,000 —	6,247 —

Table 10. Eastern South Pacific (resources) (18,109,000 square statute miles).

Proposed Jurisdictional Boundary	Area	Barrels of Oil and Equivalent Gas (millions)				
Meters Depth	Landward (L) Seaward (S) (Miles2)	1972	Production 1980 (est.)	Cumulative	Proved Reserves	Discovered to 1/1/1973
200	55,400 (L)	12.17	15	70	183	253
	18,053,600 (S)	—		—	—	—
3,000	339,200 (L)	12.17		70	183	253
	17,769,800 (S)	—		—	—	—
Miles Out						
40	264,000 (L)	12.17		70	183	253
	17,845,000 (S)	—		—	—	—
200	1,396,700 (L)	12.17		70	183	253
	16,712,300 (S)	—		—	—	—

Table 11. Western Indian Ocean (resources) (14,000,000 square statute miles).

Proposed Jurisdictional Boundary	Area	Barrels of Oil and Equivalent Gas (millions)				
Meters Depth	Landward (L) Seaward (S) (Miles2)	1972	Production 1980 (est.)	Cumulative	Proved Reserves	Discovered to 1/1/1973
200	700,000 (L)	1,285	2,500	9,300	88,600	97,733
	13,300,000 (S)	—		—	—	—
3,000	1,286,000 (L)	1,285		—	—	—
	12,713,400 (S)					
Miles Out						
40	529,000 (L)	1,285		9,300	88,600	97,733
	13,471,000 (S)	—		—	—	—
200	2,806,000 (L)	1,285		9,300	88,600	97,733
	11,294,000 (S)	—		—	—	—

Table 12. Eastern Indian Ocean (resources) (14,617,000 square statute miles).

Proposed Jurisdictional Boundary	Area	Barrels of Oil and Equivalent Gas (millions)				
Meters Depth	Landward (L) Seaward (S) (Miles2)	1972	Production 1980 (est.)	Cumulative	Proved Reserves	Discovered to 1/1/1973
200	793,800 (L)	—	75	40	2,580	2,620
	13,823,200 (S)	—		—	—	—
3,000	1,521,500 (L)	—		40	2,580	2,620
	13,095,500 (S)	—		—	—	—
Miles Out						
40	476,000 (L)	—		—	—	—
	14,141,000 (S)	—		40	2,560	2,620
200	2,249,100 (L)	—		40	2,560	2,620
	12,367,900 (S)	—		—	—	—

Table 13. World (marine) resources (139,782 square statute miles).

Proposed Jurisdictional Boundary Meters Depth	Area Landward (L) Seaward (S) (Miles^2)	Designated Provinces or Areas	Oil and Equivalent Gas (millions of bbls) Discovered to 1/1/1973	Estimated Potential Undiscovered	Ultimate
200	8,462,500 (L)	Shelf	167,481	1,376,519	1,544,000
	131,419,200 (S)	Seaward of shelf	—	728,000	728,000
3,000	17,543,300 (L)	Shelf and slope	167,481	1,937,319	2,104,800
	122,839,600 (S)	Seaward of slope	—	167,200	167,200
Miles Out					
40	7,847,000 (L)	To 40 miles out	147,454	1,202,546	1,350,000
	117,615,000 (S)	Seaward of 40 miles	20,027	901,973	922,000
200	24,768,000 (L)	To 200 miles out	167,481	1,820,519	1,988,000
	111,214,300 (S)	Seaward of 200 miles	—	284,000	284,000

Note: Most of this oil potential occurs in the Persian Gulf. Small amounts occur in the Gulf of Mexico, North Sea, and northwest Australia shelf, but the largest marine potential is not at a true continental margin.

ANNOTATED BIBLIOGRAPHY

(The geological and geophysical references to resource potential of continental margins is endless. It is perhaps more important here to note recent generalizations that consider the ultimate extent of petroleum development in continental margins.)

A discussion of potential oil reserves: paper presented at United Nations Conference on Conservation and Utilization of Resources, 1949: United Nations, Proc. Plenary Sessions, v. 1, 1950.

Estimates of potential oil reserves, in Panorama of Science (Ann. Suppl. Smithsonian Series 1951): New York, The Series Publishers, Inc., 1952.

Factors of sedimentary basin development that control oil occurrence: Am. Assoc. Petr. Geol. Bull., Nov. 1952.

Habitat of oil: a symposium containing 56 papers on oil occurrence worldwide: Am. Assoc. Petr. Geol., 1958, 1384 p.

The next hundred years energy demand and sources of supply, with particular reference to Canada: Jour. Alberta Soc. Petr. Geol., May 1961.

Origin, migration and occurrence of petroleum, in Petroleum Exploration Handbook: New York, McGraw-Hill, 1961, 836 p.

World gas reserves, production, occurrence: a comprehensive and pioneering study of 62 countries in which gas has been developed: Institute of Economics of the Gas Industry, Southwestern Legal Foundation, Matthew Bender, New York, 1962.

World petroleum exploration review, and exploration principles: World Petr. Congr., Plenary Session, Frankfurt, Germany, 1963.

World offshore petroleum resources: Am. Assoc. Petr. Geol. Bull., Oct., 1965.

Marine geology: economic problems and prospects: Ann. N.Y. Acad. Sci., Apr. 1967.

Offshore petroleum developments and resources: 2nd Ann. Conf. Oceanogr. Investment, Overseas Press Club, New York, Dec. 1968.

Marine geology and petroleum resources: Proc. 8th World Petr. Congr., Moscow, June 1971.

Mineral resources of the sea: Rept. Secretary-General, United Nations Economic and Social Council, E/4973, Apr. 1971.

Critical interrelated geologic, economic and political problems facing the geologist, petroleum industry and the nation: Am. Assoc. Petr. Geol. Bull., v. 56, no. 10, Oct. 1972, p. 1919-1930.

Trends and prospects in crude petroleum production, with particular reference to the period 1971-1980: Bureau d'Etudes Industrielles et de Cooperation de l'Institut Francais du Petrole, Mar. 1971.

Summary petroleum and selected mineral statistics for 120 countries, including offshore areas: U.S. Geol. Survey Prof. Paper 817, 1973.

Mineral Resources Potential of Continental Margins

Michael J. Cruickshank

INTRODUCTION

This report discusses the mineral resources of the continental margins, excluding fluid hydrocarbons, which are described elsewhere. For the sake of completeness, all classes of minerals, including the dissolved elements in seawater, are discussed.

The continental margins extend over an area of approximately 74.7 million km^2, or 50% of existing land areas. According to Menard and Smith (1966), the water depth ranges from 0 to 5.0 km, but more generally the seaward limit may be taken to the 2,000- or 2,500-m isobath (Fig. 1).

On the basis of depth, the sea may be divided into four provinces: the littoral, 0-200 m; the bathyal, 200-2,500 m; abyssal, 2,500-6,000; and the hadal, 6,000-12,000 meters. Although strictly speaking the edge of the continent might be considered in some instances to extend to the hadal depth of adjacent trenches, the continental margin will be considered here as encompassing only the littoral and bathyal zones.

Zonation strictly on the basis of depth does not specifically delineate the margin since the shallow waters of the world overlie a geological environment as complex and as varied as its terrestrial counterpart. Many workers have classified these regions (Kossinna, 1921; Heezen and Menard, 1963; Menard and Smith, 1966; McKelvey et al., 1969a; McKelvey and Wang, 1969), and it appears generally agreed that the sea floor may be divided into at least eight significant provinces having distinctive topography and usually characteristic structure and relations to other provinces (Figs. 2, 3). Marginal provinces include the continental shelf and slope, and the continental rise and small ocean basins.

The continental shelf and slope, which comprise the greater part of the continental margins, the sea floor from the shoreline to the base of the steep continental slope, together are sometimes termed the Continental Terrace. The major part of the submerged edge of the continental blocks is thus included. Depths of the bottom edge of the slope, as recorded, vary considerably between 1.5 and 5.5 km, depending on the adjacent seaward physiography. Isolated areas of continental blocks such as the Kerguelen Islands and the Seychelles, are included here with neighboring continents (Fig. 4).

Mineral deposits of the continental terrace will for the most part approximate those found in equivalent or adjacent terrestrial areas and will include unconsolidated deposits of heavy minerals mostly close inshore or in estuarine or drowned river valleys; sands, gravels, shells, and similar nonmetallic deposits laid down under shallow water or subareal conditions; and in local areas authigenic deposits of phosphorite glauconite and ferromanganese oxides with associated minerals. Consolidated deposits within the bedrock should occur on the average in equal proportion to those underlying terrestrial areas and will be as varied in size and mineral content.

The major deposits of the continental rise and small ocean basins appear to be gently sloping thick layers of sediments eroded from the adjacent continent and overlying the typical oceanic crust. Where the edge of the continent forms a closed or partially closed marine basin, such as in the Gulf of Mexico or the Western Mediterranean, sediment accumulation may be many kilometers thick. The province is possibly favorable for sulfur and potash, in areas underlain by saline deposits, but it is not thought to be favorable for other minerals.

Many of the theories of plate tectonics may provide a basis for the forecasting of minerals associated with continental margins. The origin of some of the large ocean basins is considered to be related to a process of sea-floor spreading that brings basic igneous rocks to the surface along the axial zones of the large mid-oceanic rise-rift systems, and carries new ocean floor away from the mid-ocean ridges at rates of as much as 15 cm/yr. The old crust, far from the mid-oceanic ridges, may be thrust back within the mantle beneath the adjacent continental or oceanic plate, forming deep ocean trenches, island arcs, and deep-seated secondary tectonic activity and vulcanism; or it may collide with, override, or extrude from beneath, the rock plate, depending on the juxtaposition of the two.

Where an oceanic rise or rift system is associated closely with the edge of a continental plate, such as in the Red Sea, or the Gulf of California, the potential for both acid and basic hydrothermal mineralization is quite profound. The prime examples of this type of concentration are the Red Sea metalliferous muds and the Salton Sea metalliferous brines both discovered within the last decade. As pointed out previously (Cruickshank, 1970; Blissenbach, 1972), if emanations of this type are associated with young continental rifts, the residual deposits so formed should be sought on the outer edge of the continental shelf or slope rather than in contemporary mid-ocean ridges. Other forms of mineralization associated with a subduc-

Fig. 1. Distribution of known mineral deposits in the littoral and bathyal regions within the 2,500 m isobath. (Au, gold; Ba, barite; c, coal; Fe, iron; g, gravel; Hg, mercury; Ls, limestone; M, monazite; Pt, platinum; sh, shell; Sn, tin; Ti, titanium, Zr, zirconium)

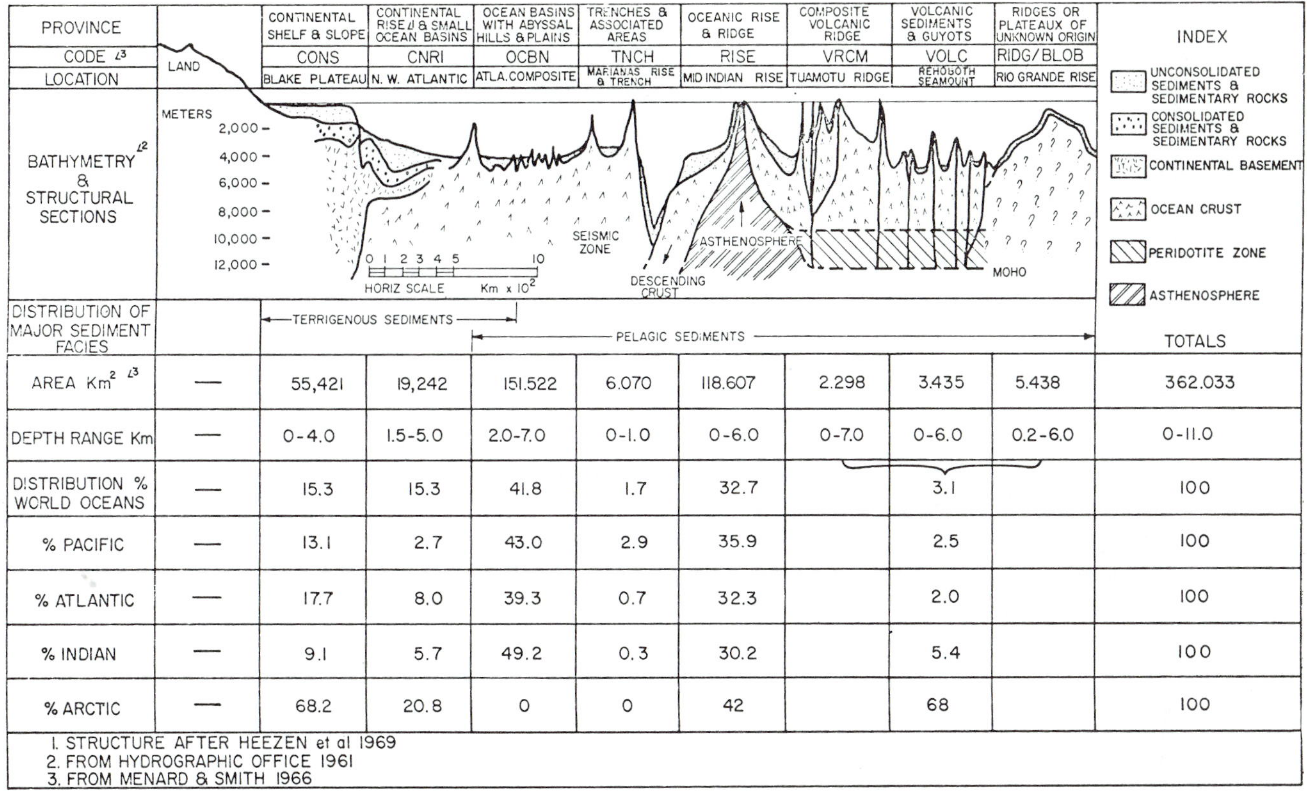

PROVINCE		CONTINENTAL SHELF & SLOPE	CONTINENTAL RISE & SMALL OCEAN BASINS	OCEAN BASINS WITH ABYSSAL HILLS & PLAINS	TRENCHES & ASSOCIATED AREAS	OCEANIC RISE & RIDGE	COMPOSITE VOLCANIC RIDGE	VOLCANIC SEDIMENTS & GUYOTS	RIDGES OR PLATEAUX OF UNKNOWN ORIGIN	
CODE [1,3]	LAND	CONS	CNRI	OCBN	TNCH	RISE	VRCM	VOLC	RIDG/BLOB	
LOCATION		BLAKE PLATEAU	N. W. ATLANTIC	ATLA. COMPOSITE	MARIANAS RISE & TRENCH	MID INDIAN RISE	TUAMOTU RIDGE	REHOBOTH SEAMOUNT	RIO GRANDE RISE	
BATHYMETRY & STRUCTURAL SECTIONS [1,2]										
DISTRIBUTION OF MAJOR SEDIMENT FACIES		←TERRIGENOUS SEDIMENTS→		←PELAGIC SEDIMENTS→						TOTALS
AREA Km² [1,3]	—	55,421	19,242	151.522	6.070	118.607	2.298	3.435	5.438	362.033
DEPTH RANGE Km	—	0-4.0	1.5-5.0	2.0-7.0	0-1.0	0-6.0	0-7.0	0-6.0	0.2-6.0	0-11.0
DISTRIBUTION % WORLD OCEANS	—	15.3	15.3	41.8	1.7	32.7		3.1		100
% PACIFIC	—	13.1	2.7	43.0	2.9	35.9		2.5		100
% ATLANTIC	—	17.7	8.0	39.3	0.7	32.3		2.0		100
% INDIAN	—	9.1	5.7	49.2	0.3	30.2		5.4		100
% ARCTIC	—	68.2	20.8	0	0	42		68		100

1. STRUCTURE AFTER HEEZEN et al 1969
2. FROM HYDROGRAPHIC OFFICE 1961
3. FROM MENARD & SMITH 1966

Fig. 2. Physiographic provinces of the world's oceans and seas showing areas, depth ranges, and distributions.

tion or collision zone may be yet located, and may present additional indicators for paleomarine deposits.

For example, it seems likely that the metallogenic belts which parallel the Peru-Chile trench, the Java trench, and the Japan trench are due to intrusion of metallic concentrations formed by melting of the subducted lithospheric plates at depths of 120-200 km beneath the surface. Inference made from such new theories of ore genesis suggests (Wang, 1973) that many undiscovered ore bodies may possibly occur along other parts of the island arcs in the western and northern Pacific, such as the Aleutians, Kuriles, Ryukyus, and behind the Mariana trench and Tonga-Kermadac trench. Most segments of these island arc-ocean boundaries are within the continental margin. Further, the recognition of paleosubduction zones (Mitchel and Garson, 1971) may lead to the recognition of additional submerged metallogenic target areas.

MARINE MINERAL RESOURCES ESTIMATES

The economics of mineral resources are largely tied to the needs of consumers. Shortages can increase the value of a mineral commodity to the extent that it moves resources up to the category of reserves. Surpluses can reduce commodities from the reserve category to that of resources (Fig. 5). McKelvey (1968) has classified the mineral resources to differentiate between the situations prevailing at the time of categorization. He states that in order to give economic and geologic perspective to estimates of resources, it is desirable to view them in a framework that takes account of the degree of certainty of knowledge about their existence and character, and the feasibility of their recovery and sale. In the classification in Fig. 6, the degree of certainty of knowledge of the dimensions and quality of mineral deposits is shown on the abscissa and the feasibility of recovery and marketing on the ordinate. The classification of individual deposits shifts with progress in exploration, advance in technology, and changes in economic conditions. Recoverable reserves are marketable materials that are producible under locally prevailing economic and technologic conditions. Paramarginal resources are prospectively marketable materials that are recoverable at prices as much as 1.5 times those prevailing now or with a comparable advance in technology. Submarginal resources are materials recoverable at prices higher than 1.5 times those prevailing now but that have some foreseeable use and prospective value.

Seen in this framework, the presently recoverable proved reserves of most minerals are relatively small compared to the resources that may eventually be found by exploration or become recoverable as a result of technologic advances or changes in economic conditions. This is particularly true for subsea resources, because only a small part of the

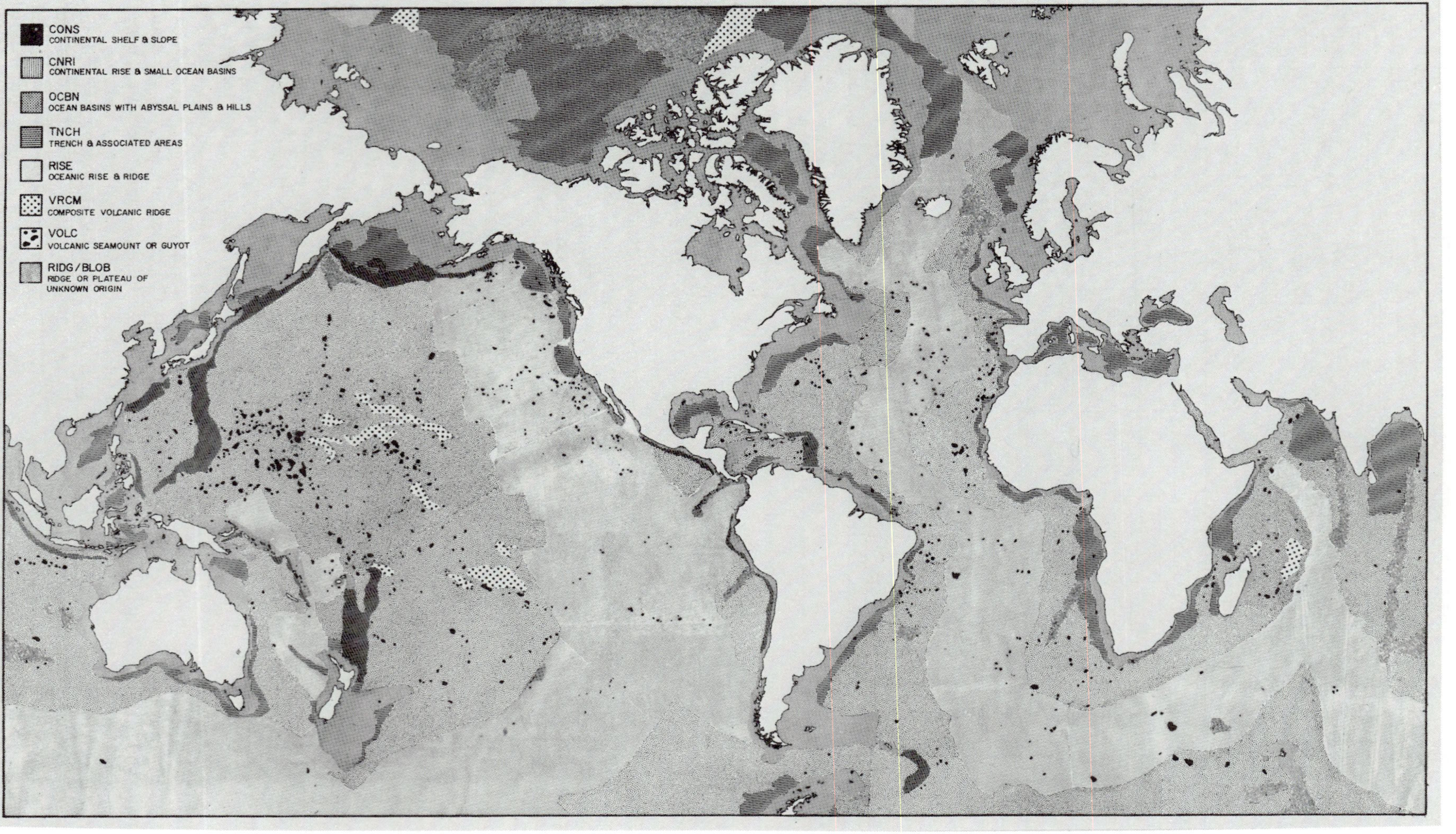

Fig. 3. Physiographic provinces of the world's oceans and seas (after McKelvey and Wang, 1970).

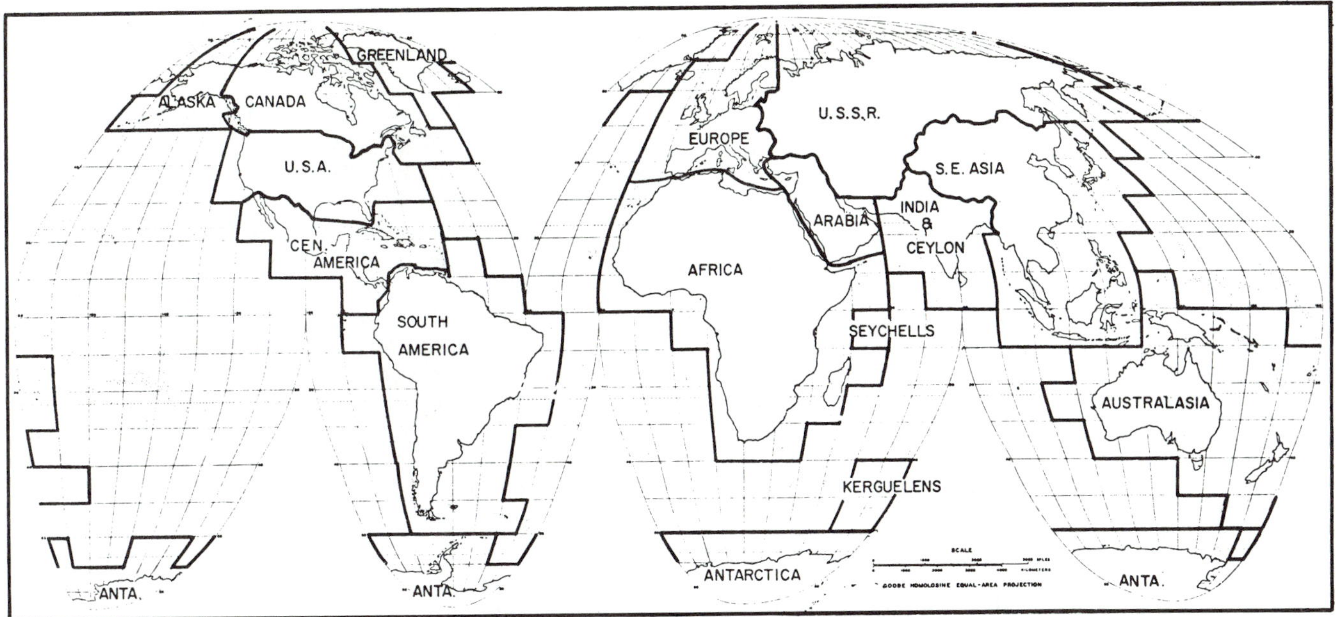

Fig. 4. Geopolitical subdivisions of the continental land masses of the world.

seabed has been explored, and most of the resources it contains are not yet economically or technically recoverable.

In this study, resources are considered in their entirety, and the figures given include all classes of reserves and resources without differentiation as to class. These are termed Apparent Resources. Three major classes of resources are known: the dissolved minerals in seawater, the minerals in unconsolidated deposits that occur in a variety of locations from coastal beaches to the deep-sea floor, and the minerals in consolidated deposits in the underlying bedrock. To assess their economic potential, it is necessary to understand their environment and the many factors that influence their origin and distribution. Although our present knowledge on these matters in the ocean compared to our knowledge on the land is unfortunately lacking, sufficient statistical data are available to make an assessment of apparent resources (within orders of magnitude, at least).

Apparent Resources are presented in terms of Order of Magnitude Dollars, where \$OM 9.345 equals \$0.345 x 10^9. These numbers represent, within order of magnitude, the gross value of contained metals which would be available to the resource market at 1970 prices.

SEAWATER

Mineral commodities in seawater are basically fresh water and minerals dissolved as salts or in elemental form. Only a few of these can be currently classed in the reserve category, but all are potentially extractable given the technological means. Other resources associated with seawater are freshwater springs, particulate matter, and biogeochemical concentrations, each of which is considered a potential resource. For the purpose of resource estimates, a limiting quantity of 1% of contained elements is assumed to be potentially available.

Dissolved Minerals

More than 77 dissolved elements have been isolated, to date, from seawater, and others can be inferred by their presence in marine authigenes or biota. Forchhammer (1865) made the classic statement that the quantity of the different elements in seawater is not proportional to the quantity of elements that river water pours into the sea, but inversely proportional to the facility with which the elements in seawater are made insoluble by general chemical or organochemical actions in the sea. It could be further propounded that all the elements in the earth's crust are involved in a continuous temporal cycle, of which their residence time in the ocean is but a fraction. In the search for mineral resources, that time when the element is in its most concentrated or most easily isolated form is the most sought after. For many elements, the residence time in seawater is optimum, and for others it is the insoluble state that follows.

Reported abundances of major elements are quite well established, but those of the minor and trace elements show wide variations, up to 2.4 orders of magnitude for copper, for example, and five orders for gold (Table 1). This is probably indicative of actual variation in content, which could be due to a variety of environmental factors and would tend to suggest that for certain elements there may be high-grade areas of ocean water. This

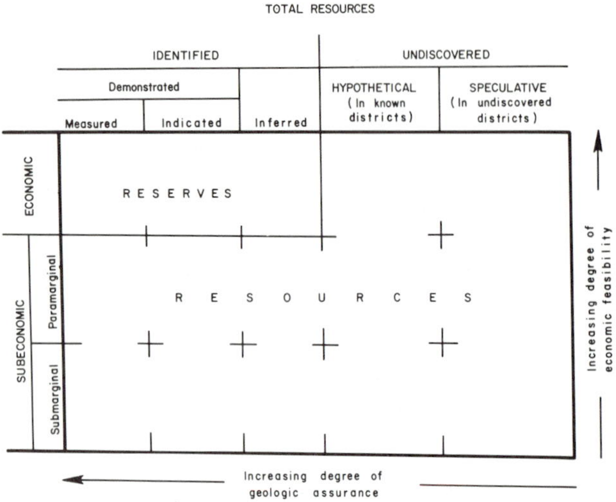

Fig. 5. Classification of reserves and resources according to McKelvey (1968).

is substantiated by the analysis of the hot brines in the Red Sea (Table 2), which shows total salinity of one order and concentrations of trace metals up to four orders of magnitude greater than for normal seawater. The role of brines in ore-forming processes is being widely studied (Bush, 1970; Beales and Onasick, 1970) following the discovery of the Red Sea deposits, and the possibility of hydrothermal ore fluids being associated with oceanic rises is being investigated by a number of researchers (Bostrom and Peterson, 1966).

Of the concentration of 45 dissolved elements reported by various authors, those for the major elements do not differ significantly, and for the minor elements, differences are generally less than one order of magnitude. Few analyses for the trace elements are available and these may differ by several orders of magnitude particularly in coastal waters. Where differences are apparent, the concentration used in comparing resources has been selected from the most recent or authoritative source. The potential resources for dissolved minerals in the world oceans are presented in Table 3 in terms of present value in order-of-magnitude dollars (where \$OM 13.36 = \$0.36 x 10^{13}).

No account is made here of metalliferous brine concentrates such as in the Red Sea; and, although they are obvious targets for early exploitation, it is unlikely that their inclusion would significantly alter the figures presented.

The resources of fresh water in the oceans are virtually unlimited. One percent of the contained fresh water would represent 1.36 x 10^{16} metric tons, valued at \$3.6 x 10^{15} at \$1 per 1,000 gal.

Freshwater Springs

Fresh groundwater occurs within some aquifers off the southeastern coast of the United States, mostly landward of the 200-m isobath on the continental shelf (McKelvey and Wang, 1969). These deposits were reported by Manheim and Horn (1968) in their description of the composition of deeper subsurface waters along the Atlantic continental margin. Kohout (1966) refers to a number of scientific investigations of submarine springs and reports known occurrences off the coasts of 19 countries. A spring currently being utilized in Argolis Bay in Greece is reported (New Scientist, 1970) to be emitting 10 m^3/sec, or almost 230 million gallons, of fresh water each day. Submarine springs may be closely related to coastal groundwater supply problems as well as to geothermal and hydrothermal activity. Their study is a somewhat neglected field well worthy of additional attention.

Values of these sources of water are not included, although they may be of local significance. The Argolis Bay Spring alone (New Scientist, 1970) represents an annual value of \$84 million dollars on the same cost basis described for fresh water.

Particulate Matter

The ocean waters at all times contain a significant amount of particulate matter, but quantitative data on the distribution of this material is inadequate for any generalizations (El Wardani, 1966). It is apparent that filter-feeding animals are important in the vertical distribution of particulate matter, which has been recorded in relatively large quantities in deep-ocean waters by Ewing and Thorndyke (1965) and Groot and Ewing (1963). Its role in trace-element concentration may be important, as the bulk of the material seems to be, in the form of charged colloidal particles, which according to Goldberg (1957) are effective scavengers of the oppositely charged ions in solution. According to Mero (1965) most of the gold in seawater is thought to be there in the form of a colloidal suspension, or as particles adhering to the surfaces of clay minerals in suspension. Lisitzin (1972) reports an average suspension concentration in the water column at about 1 mg/liter. From this he has calculated the total amount of suspended sediment of the world oceans to be 1.37 x 10^{12} tons. As with other forms of distributed trace elements, it is likely that the concentration at any one point will be dependent upon the character of the local biogeochemical environment.

No accounting has been made for minerals from this source, since no data are available on which to base even the grossest estimates.

Biogeochemical Concentration

It is now well known that many organisms living in the ocean are able to enrich selectively certain elements within their mass. Widely quoted is the enrichment by tunicates of vanadium by a factor of 28 x 10^5 (Mero, 1965). Numerous other data have been presented, and enrichment factors for 37 elements are compared in Table 4 for several sources. According to Vinogradova (1964), there is an important and delicately balanced relationship between the biomass and elemental concentration in seawater, and it is doubtful if the two are ever long in equilibrium. Vinogradova refers to his theory of biogeochemical provinces on land as being quite applicable also to marine waters. With regard to quantitative distributions, it should be pointed out that most of the values used in Table 4 are based on dry weights of organisms. On the basis of bulk or wet weights of organisms, corresponding to a water content of 90% for the biota, the enrichment factors would be reduced approximately by two orders of magnitude.

Elements concentrated in marine biota, although offering potentially economic resources in selected areas, are considered to be transient in the geochemical cycle of the oceans and therefore would be included in the figures given for dissolved minerals. Were sufficient quantitative data available to justify estimates, it is unlikely that they would form a significant fraction of the total resource, although they might well be significant on a regional basis.

UNCONSOLIDATED DEPOSITS

For the purpose of this discussion unconsolidated deposits are defined as naturally occurring concentrations of mineral in the marine environment which are not indurated and which may be a potential mineral resource. They are amenable to dredging.

The classification of mineral-bearing sediments has been admirably accomplished by many authorities from the point of view of the sedimentologist (Twenhofel, 1932), the economic geologist (Lindgren, 1933), and the oceanographer (Sverdrup et al., 1942; Kuenen, 1950; Arrhenius, 1961; Lisitzin, 1972), and many alternative approaches are available.

Classifications are based upon mode of occurrence, origin, physical appearance, or chemical composition. Two major divisions of marine sediments are commonly made which cause confusion because of common usage of the same terminology: Terrigenous/Hemipelagic/Pelagic or Terrigenous/Authigenic/Biotic. Terrigenous in the first example usually refers to either the location or the origin of the sediment, whereas in the second it refers strictly to the origin (Table 5). Red clay is probably the greatest enigma in classification terminology. Named in the Challenger reports from the color of the earliest samples, it nevertheless may be brick red, brown, or even blue (Sverdrup et al., 1942). It is a pelagic sediment and yet it is made up of greater than 70% terrigenous material. Using origin as the basis for classification, red clay would be a terrigenous sediment in the Terrigenous/Pelagic grouping. Using location, there would be no terrigenous deep-sea sediments. The influence of transport on the ultimate sediment type is very significant, and Figure 6 illustrates this as well as

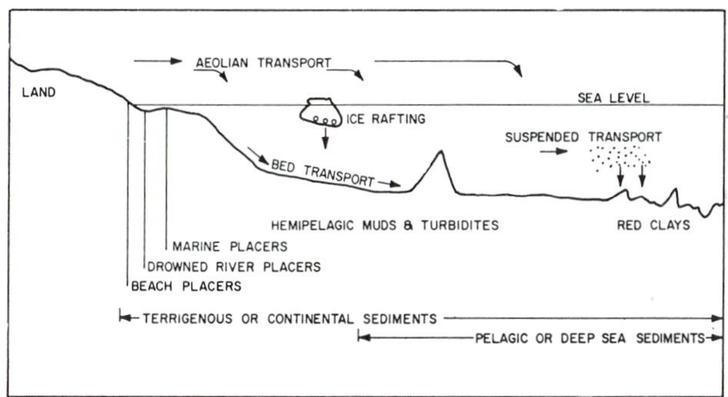

Fig. 6. The role of transport mechanisms in the differentiation of Terrigenous Recent Marine Sediments of the Continental Margin and beyond.

Table 6. Many other contradictions and confusions occur due to the casual use of terminology and lack of definition.

In an earlier study by Cruickshank et al. (1968), it was pointed out that the major change in systems components affecting the technology of mineral exploitation by dredging were the increase in water depth and the accompanying distance from land. For this reason, a classification based on water depth is considered to be most relevant. In this discussion, the deposits, listed in Table 7, are classified first on the basis of their depositional region, and second on the basis of the origin of their principal mineral constituents. Common names for deposit types or mineral species are used and where multiple terminology occurs in the literature, this will be clarified. Table 8 defines the environmental classification used in this study.

The areas of the sea floor covered by specific marine sediments for each of the three depth environments are listed in Table 9. Biogenous sediments are now widespread in the margins.

Resources of mineral commodities from unconsolidated deposits are computed separately for littoral and bathyal regions. They are totaled in Table 10 by commodity, but where breakdown in this detail is not possible, the inclusion or exclusion of any commodity from the total is indicated.

Littoral Deposits

Littoral deposits include deposits of nonmetallics, heavy minerals, and native elements and gemstones from the continental shelves and slopes. Phosphorite and glauconite are considered in this text to be bathyal deposits, although they commonly occur as shelf deposits. Manganese nodules from the Blake Plateau are included also. Many of the data used in these calculations have been extracted from an excellent compilation by McKelvey, et al. (1968) on the potential mineral resources of the United States outer continental shelf. Where data were not available, or supplementary data were used, this is indicated. Figures for world resources were extrapolated from those for the United States (Table 11) on the basis of dollar value per unit area for each of the continental plates. Some anomalies are created; for example, the United States has no reserves of diamonds and very little tin. However, the subtotals for commodity types are considered to be within reasonable limits. If data for production from each continental plate for each commodity could be obtained, it is likely that a more detailed breakdown could then be made.

Nonmetallic Materials. These bulk materials are widely distributed and have been investigated and mined in many areas over the years. The most prolific deposits are probably common sands and gravels, which occur in reworked deposits of generally medium- to high-energy environments in most areas of the continental shelves. Those on the American continent are well described in the literature by Campbell et al. (1970), Duane (1967, 1969), Emery (1965), Schlee (1964, 1968), Manheim (1972), and McKelvey et al. (1969a).

Estimates for sand and gravel resources have been made by a number of persons and organizations for different areas of the U.S. shelf (Table 12). Estimates for each region have been made on the basis of these figures and are given in Table 13. The present value of sand and gravel on the U.S. Continental Terrace is estimated to amount to an average of \$551,000 per km^2.

Calcareous sands, shell sands, corals, and aragonite are also widely distributed in shallow waters, generally in warmer climates. Occasionally in shallow coastal waters *phosphoritic sands* are deposited over large areas (d'Anglejan-Chatillon, 1964). Those so far reported have been low grade, 3-5% P_2O_5, the higher-grade phosphorite being generally found in slightly deeper water.

Refractory muds containing a high percentage of organics have been recently described in deposits off New England (NEMRIP, 1971). They are extensive in low-energy coastal environments where anoxic conditions exist. However, no value has been assigned to them here.

Heavy Minerals. These are mostly made up of sands containing a low percentage of heavy minerals of specific gravity generally around 4-8. The minerals may be derived from various sources (Table 14) and their concentration depends on the sequential existence of a source rock, a mechanism for weathering, a means of transportation, and some form of stratigraphic trap. The most important economic deposits contain oxides of *tin, titanium, iron, thorium, chromium,* or *yttrium,* and they are widely distributed on the continental shelf in existing and submerged beaches, river valleys, and bars. Emery and Noakes (1968) review thoroughly the environment of distribution and the minerals of economic importance on a world scale and they are described by many other authors on a regional basis.

Values of heavy mineral placers, which include those for tin, iron, titanium, chromium, zircon, thorium, and yttrium, have been extrapolated from resource data in McKelvey et al. (1968) for the United States (Table 11). Unit prices for each commodity are from the U.S. Bureau of Mines (1970). The data have been extrapolated to the world shelves on the basis of value of commodity per unit area.

Native Elements and Gemstones. These are confined mainly to *gold* and *platinum* group metals of high specific gravity which do not travel far from their source. However, extensive deposits of very fine gold are indicated to be widespread in the Bering Sea (McKelvey et al., 1968). Placer deposits of native *silver* have been reported on land as well as *copper* (Cruickshank, 1965, unpublished notes). It can be said that such deposits offshore will be as widely distributed as on land, where to date placer gold has been found on every continent. Of the gemstones, only *diamonds* (Webb, 1965) have been located in economic deposits offshore, in South West Africa, but the likelihood of finding deposits of other gemstones is quite high. *Precious corals* and *pearls* might be considered as gem minerals, and their distribution is also quite extensive.

Figures are given for *gold* resources in Alaska which were developed for the U.S. Government Heavy Metals program prior to 1968 and reported by McKelvey et al. (1969a) as a potential range. The lower figures given have been used at a price of \$50 per ounce. Much of the gold is assumed to be very fine, or "flour," gold. Up to this time no methods have been developed to concentrate this size range. Estimates for *platinum* are from the same source. No estimates are made for diamonds, but they are assumed to be included in extrapolation to other areas.

BATHYAL DEPOSITS

The bathyus is defined here as a middepth area, between 200 and 2,500 m. Two types of deposit are designated: those which lie upon the surface of the sea floor and those which form, or occur, within the substrata. The minerals of terrigenous origin are of lesser importance in this region, which is not only deeper than the littoral, but farther from land. Some data are available for specific deposits and specific regions, but extrapolation to world resources is not possible on an area basis because of a paucity of data characterizing the environment of deposition. Increases of one order of magnitude have been assumed in these cases. Table 15 summarizes the estimated resources for surficial and substratal deposits.

Surficial Deposits

Most important are the authigenic deposits of *phosphorite*, which have been reported at depths varying from 60 to 3,000 m (Mero, 1965), mostly in areas of upwelling. McKelvey and Wang (1969) have located 19 areas of potential deposition throughout the world oceans (Fig. 9) and the known areas off California (Barnes, 1970; Inderbitzen et al., 1970; Mero, 1961), Mexico (d'Anglejan-Chatillon, 1964), North Africa (Summerhayes et al., 1970), South Africa (Parker and Siesser, 1972), Australia (van Andel, 1965), and New Zealand (Williams, 1965) have been described in varying detail. The deposits are nodular for the most part, apparently accreted from solution but also formed from diagenesis or replacement of existing calcareous rocks. Analysis of phosphorites may vary (Table 16) between 20 and 30% P_2O_5, and their distribution is widespread. The literature on marine phosphorite is quite extensive and has been amply reviewed in a number of publications (McKelvey et al., 1969a; Mero, 1961; Gulbrandsen, 1969). Marine phosphorite deposits off Southern California have been intensively examined and a number of resource estimates made. McKelvey et al. (1968) have discussed the potential for the Atlantic coast and also made estimates of *glauconite* resources on both coasts. These data are summarized in Table 17.

With the exception of Fe/Mn oxides, glauconite is the only iron mineral which is known to be precipitating in the ocean at the present time. The mineral is typically related to areas of high redox potential. The periodic catastrophic deaths of innumerable sea creatures results in ammonia production, and therefore in high pH values, favorable for the leaching of SiO_2 from bottom sediments. Iron is taken into solution under the reducing conditions resulting from the decomposition of organic matter. Regions of recent glauconite formation are typically coasts, lacking important rivers. Extrapolation on the basis of area has not been made, and world resources of phosphorite and glauconite are assumed to be at least one order of magnitude greater than those of the United States.

Concretions from the bathyal sea floor containing between 50 and 85% *barite* have been reported in the literature from Ceylon (Jones, 1887, in Clark, 1924, p. 138), California (Revelle and Emery, 1951), and Indonesia (Mero, 1965, p. 76). So far their existence has been of academic interest only, but as Bonatti (1972) points out, it is a common authigenic component of deep-sea sediments in the Pacific, and concentrates of greater than 8% by weight have been recorded in sediments close to the East Pacific Rise. Origin is ascribed variously to the sinking of the remains of barium-rich planktonic organisms, precipitation from seawater, or the introduction of hydrothermal solutions through the sea floor (Arrhenius and Bonatti, 1965). The depth of recorded occurrence has ranged from 304 to 1235 m, and its distribution in the Pacific is shown in Figure 7. Analysis of nodules from various locations are given in Table 18. McKelvey et al. (1968) discuss the hypothesis advanced by Shane et al. (1968) that barite is a normal marine chemical sediment that will be found to be far more common than previously suspected. Although reported in both bathyal and abyssal regions, resource estimates are reported here as bathyal. Table 19 indicates a concentration range of 0.15-3.0% barium across the East Pacific Rise. Assuming about 50,000 km^2 of 2% barite-rich sediments in the East Pacific area, 1 m thick, a tonnage of 4.5 x 10^9 barite could be assessed. A minimum of 10 x 10^9 might safely be assumed for worldwide oceanic occurrences, valued at \$239 x 10^9 at 1970 prices.

Arrhenium (1961) reports dolomitization of calcite and aragonite at shallow and intermediate depths, where the rate of dissolution of calcium carbonate is low. Outcrops of *dolomitized limestone* occur on the Nasca Ridge in the South Pacific, and intensive dolomitization has been observed at basaltic intrusion interfaces with carbonate sediment. Wind appears to be a significant transport mechanism for detrital carbonate and in areas with a heavy aeolian fallout and a high preservation of accumulating carbonates, Arrhenius suggests that aeolian dolomite might be quantitatively important.

Substratal Deposits

The major deposits classically termed hemipelagic sediments are widespread but, with two exceptions are of little economic interest. They are mostly associated with continental rocks as terrigenous sediments, but also derive from admixtures of organic materials and authigenes. Some general analyses of hemipelagic deposits are presented in Table 20. One of the more common types of sediment is *carbonaceous mud*, described by various authors as blue mud (Kuenen, 1950), organic sediment (Mero, 1965), basin sediment (Emery, 1960), or carbonaceous mud (McKelvey et al., 1969b). Essentially, the muds contain fine terrigenous material with an admixture of planktonic foraminifera. In the Timor Trough, at 600 m depth (Kuenen, 1950), showed 9.8% terrigenous materials greater than 20 μm, 30% lutite less than 2 μm, and 17% $CaCO_3$. Trending to deeper water increases the ratio of calcareous to terrigenous fractions. In most cases the terrigenous fraction can be correlated with a nearby continental rock. However, the most significant constituent is the organic material, generally preserved by the reducing environment. The muds appear to be the progenitors of the black or metalliferous shales, well known on land, and commonly containing 5-20 gal of oil equivalent per ton (McKelvey et al., 1969b), may also contain varying amounts of minor metals such as vanadium, chromium, zinc, nickel, molybdenium, uranium, and sulfur. Emery (1960) has calculated that the Santa Barbara Basin alone contains upward of 3 x 10^{12} tons of these sediments.

Similar reducing environments occur in many parts of the world oceans, and as Mero (1965) notes, metallic sulfides are frequently formed, most notably pyrite. He suggests that if springs containing copper, nickel, cobalt, or other metals should empty into the floor of the basins, deposits of the sulfides of these metals would likely form. Similar

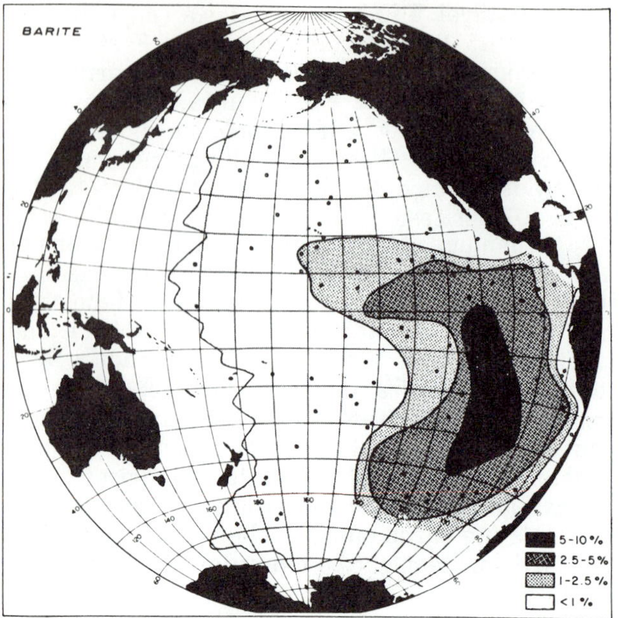

Fig. 7. Distribution of barite in Pacific sediments (after Arrhenius and Bonatti, 1965).

deposition might be found at the interface of highly biotic sedimentation on the continental shelf within the zone of minimum oxygen, as observed in the Gulf of Mexico. Assuming that the Santa Barbara Basin might be typical of the size of these deposits, and assuming world deposits for each of the contained commodities (Table 21) to be one order of magnitude larger, a world total value of $OM16.20 may be assigned to these muds.

Red and yellow muds are merely special types of blue mud that owe their exceptional color to the climatic condition of the adjoining land, such as loess deposited from the Yangtze River or iron-rich lateritic soils in other areas. The aereal distribution is not great.

In *glacial marine sediments*, formed mainly around the Antarctic continent, the place of clay minerals is taken by rock flour from glacial activity, and lime content, despite a rich benthonic life, is practically nil.

Coral mud and sand is composed of detrital material from coral reefs which is broken down to form a fine, light-colored lime mud, or near the reefs, a coral sand. Coral mud will grade into other types of mud or ooze with change of environment.

Calcareous mud is found generally in enclosed seas and has a high lime content, but no coral debris. Composed largely of globegerina skeletal remains, it has been suggested the term "globigerina" mud be used for this type of sediment to distinguish it from the pelagic ooze. Unlike the ooze, the terrigenous fraction is relatively coarse grained.

Green mud is normally colored by the presence of glauconite (K, Fe silicate), in which case it may be termed "glauconite mud" (or occasionally colored by chlorophyll or green terrigenous material).

Generally darker and coarser grained than blue muds, *volcanic muds* result from volcanic activity, particularly explosive eruptions.

Recent discoveries in the Red Sea of highly enriched brines and sedimentary mud precipitates containing iron, lead, zinc, copper, silver, and gold suggest that potentially economic mineral deposits might be found under similar tectonic circumstances in other parts of the world oceans. The composition of these *metalliferous muds* is highly variable (Tooms, 1969), and existing samples evince metal contents varying by as much as 200 times. Table 22 presents analyses for copper, zinc, silver, and gold, recorded by different workers. An origin for the source brines has been proposed on the basis of major element composition, and oxygen and hydrogen isotope ratios, which suggest that they were formed as a result of solution of evaporates by seawater of similar composition to that at the southern end of the Red Sea. The brines have been interpreted as having moved great distances through the fractured rocks underlying the Red Sea, becoming heated and extracting metals from the country rock through which they passed. On reemergence, precipitation of the metals is being caused either by reduction of dissolved sulfates at the brine-seawater interface by bacterial action and reaction of the resultant metal-poor H_2S solutions with the metal chloride complexes; or alternatively by H_2S introduced into the brine pools in metal-poor solutions which are then reacting with the metal chlorides. All the evidence is against formation of the sulfides as a result of simple cooling. There is still considerable speculation and controversy as to the genesis of the contained metals, and considerable discrepancies in the quantitative data reported. Until more field data are acquired, the true nature of the deposits will not be certain.

Gross estimates of value in place have been made from data presented in Table 23. Again an increase of one order of magnitude for world resources has been assumed, using the lower figure.

CONSOLIDATED DEPOSITS

It can be stated, with little fear of contradiction, that the continental shelf and slope are, in fact, submerged parts of the continental land masses. The mineral deposits contained in the continental rocks beneath the sea floor will be similar to those presently known on land, and their origins will be common. Unlike land deposits, however, are some of the authigenic minerals, such as phosphorite and iron/manganese oxides, which form coatings on submerged rock surfaces in the bathyal and abyssal regions. Other deposits are formed by diagenesis from seawater, and hydrothermal fluids emanating from mid-ocean tectonic activity may form deposits of types as yet undiscovered.

The major classification of consolidated deposits is based on technological factors (Table 24) which may sometimes be related indirectly to the mineral origins. The designation is according to amenability to mining by surficial or open-pit methods, such as quarrying or ripping; or by underground methods requiring the maintenance of a 1-atm environment for life support; or by solution or fluid mining, requiring only the exchange of fluid solvents or energy through boreholes.

Some hard-rock deposits on land are amenable to mining by open-pit methods, and two such deposits are known to occur offshore. A *barite* deposit in Alaska crops out on the sea floor with exposures also on an adjacent island. It is a typical vein-type hydrothermal deposit and may be associated at depth with sulfide or other mineralization.

In Australia, extensive *bauxite* deposits have been located in the Gulf of Carpenteria, and are indicated to be exposed over large areas of the sea floor. Similar deposits will no doubt be uncovered elsewhere. In warm, shallow water, *coral* forms extensive deposits of high-purity calcite in many parts of the world, and *phosphorite* and *ferromanganese oxide* crusts have been mapped in several areas, notably the Blake Plateau for both minerals, and off Hawaii for manganese. Dolomitization has been reported on outcropping indurated *calcareous sediments* in the Pacific (Arrhenius, 1961), although its extent is not known.

Most subsurface mineral deposits on the continental shelf which are or have been exploited have been worked by underground methods. Extensions of *coal* fields are widely known in the coastal waters of Great Britain (Armstrong, 1965), Turkey, Japan (Beki, 1970), Canada (Buckham, 1947), and Chile.

Bedded *iron ore* has been located offshore of Canada (Pierce, 1957) and Finland (Cruickshank, 1962), and *limestone* is mined in Finland. Vein, massive, and disseminated deposits of most minerals known today might be expected offshore. The rocks of the continental terrace are potential sources of nearly all the minerals now produced on land.

Certain minerals exist in hard rocks in a fluid state. These include *hydrocarbons*, which are not considered here; *helium*, which occurs isolated or as a constituent in natural gases; and *geothermal energy*, which although not a mineral, is certainly one of the nonliving resources. As an alternative to mineral fuels, submarine geothermal deposits might be logically included. Austin (1966) lists over 600 references to geothermal deposits and points out they are the only apparent alternative indigenous power source to fossil fuels in the continental shelf and slope. Large resources may be present in the Gulf of Mexico and offshore of Alaska, Italy, New Zealand, Mexico, Japan, U.S.S.R., Iceland, Chile, El Salvador, Guadaloupe, Martinique, Nicaragua Philippines, Taiwan, and Turkey, all of which have known geothermal areas (Homan, 1972). They may occur in areas of thick geosynclinal accumulations of young sediments, in zones of high heat flow, associated with rift and fracture zones, and in areas of volcanic activity (McKelvey and Wang, 1969). Theobold et al. (1970) estimate the U.S. reserves at greater than 10^{22} cal, equivalent to 6.98 x 10^{12} barrels of crude oil. Offshore reserves will likely be found in areas favorable locally for petroleum (McKelvey and Wang, 1969).

Other deposits amenable to borehole extraction include *sulfur* (Grindrod, 1960), which may be found associated with oil shales, organic-rich muds, gypsum, anhydrite and salt domes. The latter are quite widespread, particularly in enclosed basin environments, which also act as sources for the formation of the elemental sulfur. *Salt* is itself a resource of wide extent in basin environments, and bedded *potash* deposits (Investors' Guardian, 1969) are also associated with such evaporite basins (McKelvey et al., 1968). Other amenable deposits for which no marine resources are yet known are *carbonates*, *iodates*, *borates*, and *sulfides* of copper (Hansen et al., 1968) and other metals. *Hydrothermal fluids* might also be included, and although no resources are defined, there are indications of occurrences in New Zealand (Weissberg, 1969), California (White, 1963), East Africa (Degens et al., 1972), and the Red Sea (Tooms, 1969).

The estimation of mineral resources is by necessity a somewhat nebulous undertaking. However, some attempt must be made to assess the unknown, and we would be justified in assuming, therefore, that resource estimates which are applicable to continental land masses will be applicable to the submerged parts of continents also. Valid methods exist (Allais, 1957; Blondel, 1956; Slichter, 1960) for the extrapolation of mineral resource potentials to unexplored areas; using known data, supplemented by statistical methods, values may be derived for recoverable minerals in place. McKelvey (1972), in reviewing the problem of mineral resources estimates, outlined the various applicable methods that had then been used.

Of the two principal approaches taken to date, one is to extrapolate observations related to industrial activity, such as annual rate of production or annual demand (Hubbert, 1969), which may be adequate for short-term projections in terms of a specific number of years; the other is the extrapolation of observations relating to abundance of the mineral in any particular geologic environment. This quasi-statistical approach was first developed by Nolan (1936) using data on the mineral production of the region around Boulder Dam to develop statistical inferences that could be applied to new districts. This was further developed by Blondel (1956), and Allais (1957) presented a case study of the Sahara region, where inferences were made on the basis of a Poisson distribution of mineral occurrences and a log-normal distribution of mineral values. These figures were projected for base metals and appear to be low on a unit-area basis when compared to total resource projections (Table 25), but it should be noted that they refer only to *presently* exploitable resources.

A third approach is to project the resources in relation to the crustal abundance of the elements. According to McKelvey (1960) the mineable reserves of common industrial materials in elemental form in the United States bear a direct relationship to their crustal abundance: $R = A \times 10^{10} - 10^{11}$, where R equals the reserves in short tons and A equals the crustal abundance in percent. He suggested that for extrapolation to world reserves, where the land area would be 17.3 times greater, (excluding Antarctica), a figure of $A \times 10^{11}$ would probably give the right order of magnitude for total resources. This work was confirmed by Sekine (1963), who made similar computations using reserve data for 22 elements in Japan. Using data for 56 elements, the ratio of dollar values calculated from $R = A \times 10^{6}$, to dollar values calculated from estimates made individually by commodity specialists, was compared (Table 26). For 30 elements the estimates were within one order or magnitude and for 45 elements within two orders of magnitude. Of the nine elements where the ratios were three orders of magnitude or greater, seven of them were rare metals, underestimated by the Bureau of Mines (Rh, Ce, Ga, Ge, Sc, Th, and Y), and the other two, iodine and nitrogen compounds, were overestimated, by the inclusion for the former of reserves in brines and for the latter of nitrogen fixed from the atmosphere in coal.

These results tend to confirm the validity of the relationship propounded by McKelvey (1960) and would appear to provide an acceptable means of estimating resources, at least on a world scale, where numbers within orders of magnitude are sufficiently meaningful. As further confirmation, a least-squares analysis run for correlation between crustal abundance of the elements and projected world resources of the elements in order-of-magnitude dollars gave a coefficient of 0.762.

In the estimation of resources for the consolidated deposits, it is assumed that the relationship with crustal abundance holds for minerals of the continental rocks and does not include minerals derived from seawater or oceanic crust. Placer deposits of terrestrial origin are assumed to be included in estimates made from crustal abundance and the value of those previously estimated for unconsolidated deposits (Table 27) has been subtracted from the total. Values for the three classes of deposit amenable to surface, underground, or fluid-mining have been separated where possible, but for the most part the values have been assigned to the underground class (Table 28). Transfers of resources between the classes could be expected as a result of more detailed analyses. It must be understood, however, that the figures given are gross estimates and should be used only with the greatest caution. Much additional work in mineral commodity analysis and minerals exploration potential will be required!

No estimates have been made for authigenic deposits of phosphorite and ferromanganese crusts; but as Mero (1965) has pointed out for manganese, their inclusion would greatly increase the tonnage estimates. A 5-cm crust of ferromanganese oxide with a specific gravity of 2.4, for example, would contain 12.0 g/cm^2 or 120,000 $tons/km^2$.

Estimates for geothermal energy based on figures given by Theobold et al. (1970), extrapolated to the area of the U.S. continental shelf alone, amount to over 10^{13} dollars in equivalent petroleum values.

Table 1. Concentration of dissolved mineral commodities in seawater, reported by various authorities.

	Symbol	McIlhenny and Ballard (1963) (ppm)	Mero (1965) (mg/1)	Goldberg (1963) (mg/1)	Marine Commiss (1970) (ppm)	Hood (1963) (mg/1)	Sverdrup et al. (1942) (ppm)	Harvey (1969) (mg/1)	Chem. Eng. News (1964) (ppm)	Reported Range (ppm)
Energy Resources										
Anthracite	C									
Bituminous coal and lignite	C									
Carbon	C	28	28	28	—	—	28	—	28	n.s.
Deuterium	—	34	—	—	—	—	—	16,000		16-34
Geothermal	—									
Helium	He									
Hydrogen	H									
Natural gas	—									
Peat	C									
Petroleum	HC									
Shale oil	HC									
Thorium	Th	0.0005	0.00005	—	—	0.05	0.0005	0.01-0.001	0.0007	0.000001-0.0005
Uranium	U	0.002	0.003	—	0.003	3.0	0.0015	—	0.003	0.0015 -0.003
Ferrous Minerals										
Chromium	Cr	0.001	0.00005	—	0.00005	0.13- 0.25	—	1 2.5	—	0.00005-0.002
Cobalt	Co	0.0001	0.0005	0.0005	0.0005	0.2- 0.7	—	0.1	0.0005	0.0001 -0.0007
Columbium	—									
Iron	Fe	0.002	0.01	—	0.01	1.7-150	—	1-60	0.01	0.001 -0.15
Manganese	Mn	0.01	0.002	0.002	0.002	0.1- 8.0	0.001 -0.001	0.7-10	0.002	0.0007 -0.01
Molybdenum	Mo	0.0005	0.01	—	0.01	4.0-12.0	0.0005	0.3-16	0.01	0.0005 -0.016
Nickel	Ni	0.0001	0.002	—	0.002	2.0	0.0001	0.1-6	0.0005	0.0001 -0.0002
Rhenium	Re									
Silicon	Si	4.0	3.0	3.0	—	3.0	0.2 -4.0	10-3,000	3.0	0.01 -4.0
Tantalum	Ta									
Tungsten	W									
Vanadium	V	0.0003	0.002	—	—	—	0.0003	0.2- 0.7	0.002	0.0002 -0.002
Nonferrous Minerals										
Aluminum	Al	1.0	0.01	—	0.01	1.0 -10.0	0.5	0 -1,900	0.01	0-1.9
Antimony	Sb	0.0002	0.0005	—	—	—	—	0.2	0.0005	0.0002 -0.0005
Arsenic	As	0.02	0.003	—	—	—	0.01 -0.02	3 -35	0.003	0.003 -0.035
Beryllium	Be									
Bismuth	Bi	0.0002	0.00002	—	—	—	—	0.2	0.03	0.00002 -0.0002
Cadmium	Cd	0.0003	0.0001	0.0001	—	—	—	0.032-0.057	0.0001	0.00003 -0.0003
Cesium	Cs	0.002	0.0005	—	—	—	0.002	2	0.0005	0.002 -0.0005
Copper	Cu	0.001	0.003	0.003	0.003	0.5 - 3.5	0.001-0.01	1 -25	0.003	0.001 -0.25
Gallium	Ga	0.0005	0.00003	—	—	—	—	0.0005	0.0005	0.000003 -0.0005
Germanium	Ge									
Gold	Au	0.000006	0.000004	—	0.000004	0.015- 0.4	0.000006	0.004-0.008	0.000004	0.000004-0.4
Hafnium	Hf									
Indium	In									
Lead	Pb	0.004	0.00003	0.00003	0.0001	0.6 - 1.5	0.004	4 -5	0.003	0.0003 -0.005
Magnesium	Mg	1,298	1,350	1,350	—		1,272	—	1,300	n.s.
Mercury	Hg	0.00003	0.00003	—	—	0.15 - 0.27	0.00003	0.03	0.00003	0.00003 -0.00027
Platinum-group metals	—									

Table 1 (cont.)

	Symbol	McIlhenny and Ballard (1963) (ppm)	Mero (1965) (mg/1)	Goldberg (1963) (mg/1)	Marine Commiss. (1970) (ppm)	Hood (1963) (mg/1)	Sverdrup et al. (1942) (ppm)	Harvey (1969) (mg/1)	Chem. Eng. News (1964) (ppm)	Reported Range (ppm)
Nonferrous Minerals										
Radium	Ra									
Rare-earth elements	—									
Rubidium	Rb	0.2	0.12	—	—	120	0.2	200	0.2	0.12 -0.2
Scandium	Sc									
Selenium	Se	0.004	0.004	—	—	—	0.004	4	0.004	n.s.
Silver	Ag	0.0003	0.0003	—	0.0003	0.145	0.0003	0.15-0.3	0.0003	0.00015 -0.0003
Tellurium	Te									
Thallium	Tl									
Tin	Sn	0.003	0.003	—	0.003	0.3	—	3	0.003	0.003 -0.0003
Titanium	Ti	0.001	0.001	—	0.001	—	—	1 -9	0.001	0.001 -0.009
Yttrium	Y	0.0003	0.0003	—	—	—	0.0003	0.3	0.0003	n.s.
Zinc	Zn	0.009	0.01	0.01	0.01	—	0.005	9 -21	0.001	0.001 -0.021
Zirconium	Zr									
Nonmetallic Minerals										
Argon	Ar									
Asbestos	—									
Barium	Ba	0.05	0.03	0.03	—	—	0.05	30- 90	0.03	0.03-0.05
Boron	B	4.6	4.6	4.6	—	—	4.6	—	4.8	n.s.
Bromine	Br	66	65	65	—	—	65	—	65	n.s.
Calcium	Ca	408	400	400	—	—	400	—	400	n.s.
Chlorine	Cl	19,361	19,000	19,000	—	—	18,980	—	19,000	n.s.
Clays	—									
Corundum and emery	—									
Diamond	—									
Diatomite	—									
Feldspar	—									
Flourine	—	1.3	1.3	1.3	—	1.3	1.4	1,300-1,400	1.3	1.3-1.4
Garnet	—									
Gem stones	—									
Graphite (natural)	—									
Gypsum	—									
Iodine	I	0.05	0.06	—	—	—	0.05	50	0.05	0.05-0.06
Kyanite and related minerals	—									
Lithium	Li	0.1	0.17	—	—	—	0.1	100	0.2	0.1-0.2
Mica	—									
Nitrogen	N	1.0	0.5	—	—	0.5	0.01 - 0.7	0.1- 600	0.8	0.0001-1.0
Oxygen	O									
Perlite	—									
Phosphorus	P	0.07	0.07	—	0.07	—	0.001- 0.10	0- 60	0.07	0-0.1
Potassium	K	388	380	380	380	—	380	—	400	n.s.
Pumice	—									
Quartz crystal	—									
Sand and gravel	—									
Sodium	Na	10,768	10,500	10,500	—	—	10,556	—	10,600	n.s.
Stone	—									
Strontium	Sr	13	80	8.0	—	8.0	13	—	8.0	8.0-13
Sulfur	S	880	885	885	885	—	884	—	900	n.s.
Talc, soapstone, and pyrophyllite	—									
Vermiculite	—									

Table 2. Enrichment factors in Red Sea metalliferous brines relative to normal seawater of 38.2‰ salinity (Craig, 1969).

Component	Brine Analysis	Concentrations (g/kg) Brine	Seawater	Enrichment Factor
Pb	Atl.	6.3×10^{-4}	3×10^{-8}	27,200
Mn	Atl.	8.2×10^{-2}	4.2×10^{-6}	25,300
Fe	Atl.	8.1×10^{-2}	$< 2 \times 10^{-5}$	>5,200
Zn	Atl.	5.4×10^{-3}	$< 5 \times 10^{-6}$	>1,400
Ba	Atl.	9×10^{-4}	16.7×10^{-6}	70
Cu	Atl.	2.6×10^{-4}	5.5×10^{-6}	61
Ca	Atl.	5.15	0.450	15
Si	Atl.	2.76×10^{-2}	2.4×10^{-3}	15
Na	Atl.	92.85	11.75	10
Cl	Atl.	155.5	21.13	9.5
Sr	Atl.	4.8×10^{-2}	8.9×10^{-3}	7
K	Atl.	1.87	0.423	5.7
Br	Atl.	0.128	7.36×10^{-2}	2.2
B	Disc.	7.8×10^{-3}	5.1×10^{-3}	2.0
Li	Disc.	2.62×10^{-4}	2.07×10^{-4}	1.6
HCO_3	Atl.	0.143	0.125	1.5
H_2O	Atl.	742.50	961.65	1.0
Mg	Atl.	0.764	1.413	0.70
I	Disc.	3.0×10^{-5}	6.5×10^{-5}	0.60
SO_4	Atl.	0.84	2.96	0.37
NO_3	Disc.	4.4×10^{-5}	7.5×10^{-4}	0.08
F	Disc.	5.12×10^{-5}	1.4×10^{-3}	0.05

Table 3. Value of dissolved mineral commodities in seawater assuming a recovery potential of 1% of the total.

	Symbol	Concentration[a] (ppm)	World Oceans Total[b] (10^9 metric tons)	Present Value[c] ($/metric ton)	Potential Resource (1% total $$10^9$)	Value of Potential Resource (1% total $)	Value in OM Dollars[d] (1% total $OM)
Energy Resources							
Anthracite	C						
Bituminous coal and lignite	C						
Carbon	C	28	38.9 X 10^3	9.34	3.63 X 10^3	3.6 X 10^{12}	13.36
Deuterium	—	16	22.2 X 10^3	111 X $$0^6$	24.6 X 10^9	2.5 X 10^{19}	20.25
Geothermal	—						
Helium	He						
Hydrogen	H						
Natural gas	—						
Peat	C						
Petroleum	HC						
Shale oil	HC						
Thorium	Th	0.0001	0.139	15.035.3	20.89	2.1 X 10^{10}	11.21
Uranium	U	0.002	2.78	20.789.5	577.56	5.8 X 10^{11}	12.58
Ferrous Minerals							
Chromium	Cr	0.0001	0.139	3.086.5	4.291	4.3 X 10^9	10.43
Cobalt	Co	0.0004	0.556	4.078.5	22.66	2.3 X 10^{10}	11.23
Columbium	—						
Iron	Fe	0.01	13.9	16.9	2.357	2.4 X 10^9	10.24
Manganese	Mn	0.002	2.78	60.98	1.69	1.7 X 10^9	10.17
Molybdenum	Mo	0.01	13.9	3.571.5	496.5	5.0 X 10^{11}	12.50
Nickel	Ni	0.002	2.78	2.070.9	57.57	5.8 X 10^{10}	11.58
Rhenium	Re						
Silicon	Si	3.0	4.17 X 10^3	316.3	13.18 X 10^3	1.3 X 10^{13}	14.13
Tantalum	Ta						
Tungsten	W						
Vanadium	V	0.0003	0.417	3.745.3	15.62	1.6 X 10^{10}	11.16
Nonferrous Minerals							
Aluminum	Al	0.05	695	561.8	3.90 X 10^3	3.9.10^{12}	13.39
Antimony	Sb	0.0002	0.278	1.007.9	28.02	2.8.10^{10}	11.28
Arsenic	As	0.02	27.8	176.3	49.01	4.9.10^{10}	11.49
Beryllium	Be						
Bismuth	Bi	0.0002	0.278	8.812.5	24.50	2.5.10^{10}	11.25
Cadmium	Cd	0.0001	0.139	5.838.9	8.06	8.1.10^9	10.81
Cesium	Cs	0.0005	0.695	1.321.8	9.187	9.2.10^9	10.92
Copper	Cu	0.001	1.39	930.4	12.93	1.3.10^{10}	11.13
Gallium	Ga	0.0005	0.695	1.2 X 10^6	8.34 X 10^3	8.3.10^{12}	13.83
Germanium	Ge						
Gold	Au	0.000011[f]	0.0153	1.6 X 10^6	246.02	2.5.10^{11}	12.25
Hafnium	Hf						
Indium	In						
Lead	Pb	0.001	1.39	297.4	41.34	4.1.10^{10}	11.41
Mangesium	Mg	1300	1.81 X 10^6	776.6	1.41 X 10^7	1.4.10^{16}	17.14
Mercury	Hg	0.00003	0.0417	15.210.0	6.34	6.3.10^9	10.63
Platinum-group metals	—						
Radium	Ra						
Rare-earth elements	—						

Rubidium	Rb	0.12	167	991.4	1.656 X 10^{3}	1.7.10^{12}	13.17
Scandium	Sc						
Selenium	Se	0.004	5.56	11,235.9	627.72	6.2.10^{11}	12.62
Silver	Ag	0.0003	0.417	68,802.6	286.91	2.9.10^{11}	12.29
Tellurium	Te						
Thallium	Tl						
Tin	Sn	0.003	4.17	3,262.8	136.06	1.4.10^{10}	11.14
Titanium	Ti	0.001	1.39	2,908.1	40.42	4.0.10^{9}	10.40
Yttrium	Y	0.0003	0.417	123,370	514.45	5.1.10^{11}	12.51
Zinc	Zn	0.01	13.9	297.4	41.34	4.1.10^{10}	11.41
Zirconium	Zr						
Nonmetallic Minerals							
Argon	Ar						
Asbestos	—						
Barium	Ba	0.03	41.7	29.08	12.13	1.2.10^{10}	11.12
Boron	B	4.6	6.40 X 10^{3}	520.7	33.3 X 10^{3}	3.3.10^{13}	14.33
Bromine	Br	66	91.7 X 10^{3}	627.9	576 X 10^{3}	5.8.10^{14}	15.58
Calcium	Ca	400	556 X 10^{3}	4.2	235 X 10^{3}	2.4.10^{14}	15.24
Chlorine	Cl	19,000	26.4 X 10^{6}	79.31	20.94 X 10^{6}	2.1.10^{16}	17.21
Clays	—						
Corundum and emery	—						
Diamond	—						
Diatomite	—						
Feldspar	—						
Fluorine	—	1.3	1.81 X 10^{3}	110.2	1.99 X 10^{3}	2.0.10^{12}	13.20
Garnet	—						
Gem stones	—						
Graphite (natural)	—						
Gypsum	—						
Iodine	I	0.05	69.5	2599.6	18.07	1.8.10^{10}	11.18
Kyanite and related minerals	—						
Lithium	Li	0.1	139	2026.8	2.817 X 10^{3}	2.8.10^{12}	13.28
Mica	—						
Nitrogen	N	0.5	695	25.33	176.0	1.76 X 10^{11}	12.18
Oxygen	O						
Perlite	—						
Phosphorus	P	0.07	97.3	49.57	48.23	4.8.10^{10}	11.48
Potassium	K	380	528 X 10^{3}	36.90	140 X 10^{3}	1.4.10^{14}	15.14
Pumice	—						
Quartz crystal	—						
Sand and gravel	8						
Sodium	Na	10,600	14.7 X 10^{6}	20.46	3.01 X 10^{6}	3.0.10^{15}	16.30
Stone	—						
Strontium	Sr	13	18.1 X 10^{3}	169.8	9.77 X 10^{3}	9.8.10^{12}	13.98
Sulfur	S	880	1.22 X 10^{6}	41.34	504 X 10^{3}	5.0.10^{14}	15.50
Talc, soapstone, and pyrophyllite							
Vermiculite	—						
Water (fresh)	—	981,000	1.36 X 10^{9}	0.40[b]	5.44 X 10^{6}	5.4.10^{15}	16.54

[a]Table 1, most recent or authoritative source.
[b]Volume of oceans, 1.35 X 10^{9} km^3 (Menard and Smith, 1966); density of oceans, 1.027 (Clark, 1924); mass, 1.39 X 10^{18} metric tons.
[c]Mostly from U.S. Bureau of Mines, 1970: market value of contained metal.
[d]Order of magnitude dollars where $OM 13.36 = $0.36 X 10^{13}.
[e]Equivalent energy/bbl crude oil (Hubbert, 1969).
[f]Rosenbaum et al., 1969.
[g]$50/oz.
[h]Based on McIlhenny (1967).

Table 4. Enrichment factors in orders of magnitude for 37 elements in marine organisms compared to seawater.

	Symbol	Seawater Concentration (ppm)	Biotic Enrichment Factor				
			Brooks and Kaplan (1972)[a]	Terry (1961)[b]	Abbot (1971)[c]	Vinogradov (1964)[d]	Kukal (1971)[e]
Energy Resources							
Anthracite	C						
Bituminous coal and lignite	C						
Carbon	C	2.8 X 10'	3	—	—	—	4.1
Deuterium	—						
Geothermal	—						
Helium	He						
Hydrogen	H						
Natural gas	—						
Peat	C						
Petroleum	HC						
Shale oil	HC						
Thorium	Th						
Uranium	U	2×10^{-3}	—	—	4	—	—
Ferrous Minerals							
Chromium	Cr	1×10^{-4}	—	3	6	—	—
Cobalt	Co	4×10^{-4}	—	4	4	—	—
Columbium	—						
Iron	Fe	1×10^{-2}	3	5	6	4.5	4.2
Manganese	Mn	2×10^{-3}	4	4	4	2.3	3.4
Molybdenum	Mo	1×10^{-2}	2	4	—	—	—
Nickel	Ni	2×10^{-3}	3	4	4		
Rhenium	Re						
Silicon	Si	3.0	1	—	—	4.5	5.2
Tantalum	Ta						
Tungsten	W						
Vanadium	V	3×10^{-4}	4	5	6	—	4.1
Nonferrous Minerals							
Aluminum	Al	6.0×10^{-1}	—	—	5	4.5	1.9
Antimony	Sb	2×10^{-4}	—	2	—	—	—
Arsenic	As	2×10^{-2}	—	3	—	—	1.7
Beryllium	Be						
Bismuth	Bi	2×10^{-4}	—	3	3	—	—
Cadmium	Cd	1×10^{-4}	5	3	5	—	—
Cesium	Cs						
Copper	Cu	1×10^{-3}	4	4	4	3.4	3.5
Gallium	Ga	5×10^{-4}	—	3	—	—	—
Germanium	Ge		—	4	—	—	—
Gold	Au	1.1×10^{-5}	—	3	3	—	—
Hafnium	Hf						
Indium	In						
Lead	Pb	1×10^{-3}	3	3	—	—	
Magnesium	Mg	1.3×10^{3}	-1	—	—	0.1	0.7

Mercury	Hg	3 X 10^{-5}	—	—	5	—	
Platinum-group metals	—						
Radium	Ra						
Rare-earth elements	—						
Rubidium	Rb						
Scandium	Sc						
Selenium	Se						
Silver	Ag	3 X 10^{-4}	3	4	5	4.5	—
Tellurium	Te						
Thallium	Tl						
Tin	Sn	3 X 10^{-3}	—	3	—	—	—
Titanium	Ti	1 X 10^{-3}	—	3	6	—	—
Yttrium	Y						
Zinc	Zn	1 X 10^{-2}	4	4	6	3.4	4.4
Zirconium	Zr						
<u>Nonmetallic Minerals</u>							
Argon	Ar						
Asbestos	—						
Barium	Ba	3.0 X 10^{-2}	—	—	5	3.4	—
Boron	B	4.6	0	—	—	—	-1.4
Bromine	Br						
Calcium	Ca	4.0 X 10^{2}	0	—	—	0.1	1.1
Chlorine	Cl	1.9 X 10^{4}	0	—	—	—	0.2
Clays	—						
Corundum and emery	—						
Diamond	—						
Diatomite	—						
Feldspar	—						
Fluorine	—	1.3	—	—	—	—	0.7
Garnet	—						
Gem stones	—						
Graphite (natural)	—						
Gypsum	—						
Iodine	I	5.0 X 10^{-2}	2	—	—	—	—
Kyanite and related minerals	—						
Lithium	Li	1.0 X 10^{-1}	—	—	—	2.3	—
Mica	—						
Nitrogen	N	6.0 X 10^{-1}	5	—	—	—	5.2
Oxygen	O						
Perlite	—						
Phosphorous	P	7.0 X 10^{-2}	5	—	—	4.5	5.2
Potassium	K	3.8 X 10^{2}	1	—	—	—	1.3
Pumice	—						
Quartz crystal	—						
Sand and gravel	—						
Sodium	Na	1.1 X 10^{4}	-1	—	—	—	0.3
Stone	—						
Strontium	Sr	1.3 X 10^{1}	.	—	—	0.1	—
Sulfur	S	8.8 X 10^{2}	0	—	—	—	1.1
Talc, soapstone, and pyrophyllite	—						
Vermiculite	—						

[a]Data from Mason (1958) and others. Marine biosphere based on copepods and lamellibrachs.
[b]Data from Noddack and Noddack (1939) and Goldberg (1957). Algae, diatoms, and other.
[c]Tunicates, diatoms, and other.
[d]Diatoms and copepods in Black Sea.
[e]From Kukal (1971).

Table 5. Classification of unconsolidated deposits according to origin.

Type	Definition
Terrigenous	Products of mechanical breakdown of continental rock
Biogenic	Products of living organisms
Authigenic	Products of chemical deposition in place
Diagenic	Products of chemical replacement
Volcanogenic	Detrital products from volcanic activity
Cosmic	Meteoritic and other material from outer space

Table 6. Transport time for common marine sediments.

Type	Nature of Sediment	Transport Time
Placers	Alluvial deposit containing a valuable constituent.	Several seasons
Hemipelagic muds	Pelagic sediment in bathyal depths in which the terrigenous content is greater than 3% and may be related by analysis to adjacent land mass (after Kuenen, (1950).	Fast, may be single event
Turbidites	Outer shelf coarse sediment transported to continental slope or abyssal plain by turbidity current.	Very fast single event
Red clays	Deep-sea pelagic sediments with maximum deposition rates for the terrigenous component in the range of 5 X 10^{5}-5 X 10^{-4} cm/yr (Arrhenius, 1961); contains less than 30% biotic material.	Very slow, hundreds of years

Table 7. Unconsolidated mineral deposits of the continental margins.

	Littoral			Bathyal	
	Nonmetallics	Heavy Minerals	Native Elements	Surficial	Substratal
Terrigenous	Sand and gravel: SiO_2 Silica sand: SiO_2 Industrial sand: kyanite Sillimanite: Al_2SiO_5 Staurolite Garnet Refractory muds	Heavy mineral sands Magnetite: Fe_3O_4 Ilmenite: $FeTiO_3$ Rutite: TiO_2 Monaite: (Ce, La, Th) PO_4 Chromite: $FeCr_2O_4$ Zircon: $ZrSiO_4$ Cassiterite: SnO_2 Xenotime: YPO_4 Beryl Columbite Cinnabar	Native metals Gold, Au Platinum: Pt Copper: Cu Gemstones Diamond		Hemipelagics Blue mud Volcanic mud Red and yellow mud Gravel sediments
Biogenic	Calc. sands: $(CaCO_3)$ Shell beds: $CaCO_3$ Corals: $CaCO_3$		Precious corals Pearls		Organic muds Calcareous Coral muds
Authigenic	Phosphorite: $Ca_5\ (PO4)_3$ Aragonite: $CaCO_3$ Glauconite			Phosphorite Barium sulfate concretions Glauconite: K (MgFe)Al Silicate Mn nodules and crusts Mn, Fe, Ni, Co, Cu	Metalliferous muds Au/Ag/Cu/Pb/Zn Green muds
Diagenic				Phosphorite Dolomite	

Table 8. Environmental classification of unconsolidated marine mineral deposits in the continental margin.

Classification	Definition
Littoral	Occurring in the continental shelf and slope at a depth in the range approximately 0-300 m (Hedgpeth, 1957, p. 18). Included are *nonmetallics* of specific gravity 2-4 and mineral content 15-100%, *heavy minerals* of specific gravity 4-8 and mineral content 0.1-15%, and *native elements and gems* of specific gravity 8-20 (except gems, gravity 3-3.5) and mineral content less than 1ppm.
Bathyal	Occurring largely on the continental shelf and slope or in small ocean basins at depths between 300 and 2,500 m. Included are surficial deposits occurring at the sea water-sea floor interface, and substratal deposits occurring at depth below the sea floor. Synonymous with hemipelagic.

[a]As with most natural phenomena, the definition is an approximation only. There are overlaps between the classes.

Table 9. Areas of Continental Margin Covered by Specific Marine Sediments (after Kuenen 1960, p. 362 & 346).

Sediment Type	Ave. Depth M.	Area $km^2 X 10^6$	% Seafloor
Littoral	0-200	30	8
Bathyal	200-2500	73	18
Blue Mud	2560	5	13
Glacial Marine	ND[2]	ND	
Red & Yellow Mud	1130	0.5	0.2
Coral Mud and Sand	1350	320	10
Calcareous Mud	ND		
Green Mud	935	4	1
Volcanic Mud	1880	2	0.5
Abyssal & Hadal	2500-12000	268	74

[1]Depths of individual types after Clark (1924) p. 516
[2]No data.

Table 10. Estimated world resources of unconsolidated marine minerals on the continental margin.

	Symbol	Value $/km² U.S. Continental Shelf (from Table 11)	Value $OM[a] World: Littoral[b]	Value $OM[a] World: Bathyal[c]	Total Value $OM[a] Continental Margin
Energy Resources					
Anthracite	C				
Bituminous coal and lignite	C				
Carbon	C				
Deuterium	—				
Geothermal	—				
Helium	He				
Hydrogen	H				
Natural gas	—				
Peat	C				
Petroleum	HC			15.36[d]	15.36
Shale oil	PC				
Thorium	Th	326	11.18		11.18
Uranium	U		n.a.[e]	14.24[d]	14.24
Ferrous Minerals					
Chromium	Cr	391	11.22	14.14[d]	14.14
Cobalt	Co	84[f]	10.47		10.47
Columbium	—				
Iron	Fe	1.842	12.10		12.10
Manganese	Mn	44[f]	10.24		10.24
Molybdenum	Mo			14.83[d]	14.83
Nickel	Ni	66[f]	10.37	14.48[d]	14.48
Rhenium	Re				
Silicon	Si				
Tantalum	Ta				
Tungsten	W				
Vanadium	V			15.87[d]	15.87
Nonferrous Minerals					
Aluminum	Al				
Antimony	Sb				
Arsenic	As				
Beryllium	Be				
Bismuth	Bi				
Cadmium	Cd				
Cesium	Cs				
Copper	Cu	5[f]	9.28	9.87[g]	10.12
Gallium	Ga				
Germanium	Ge				

Table 10 (cont.)

	Symbol	Value $/km² U.S. Continental Shelf (from Table 11)	Value $OM[a] World: Littoral[b]	Value $OM[a] World: Bathyal[c]	Total Value $OM[a] Continental Margin
Nonferrous Minerals					
Gold	Au	163,300	13.91		13.91
Hafnium	Hf				
Indium	In				
Lead	Pb				
Magnesium	Mg				
Mercury	Hg				
Platinum-group metals	—	30	10.17		10.17
Radium	Ra				
Rare-earth elements	—				
Rubidium	Rb				
Scandium	Sc				
Selenium	Se				
Silver	Ag			9.24[g]	9.24
Tellurium	Te				
Thallium	Tl				
Tin	Sn	125,000	12.69		12.69
Titanium	Ti	13,286	12.74		12.74
Yttrium, Xenotime	Y	3,650	12.20		12.20
Zinc	Zn			14.91[d,g]	14.91
Zirconium	Zr	342	11.19		
Nonmetallic Minerals					
Argon	Ar				
Asbestos	—				
Barium	Ba			11.10	11.10
Boron	B				
Bromine	Br				
Calcium	Ca	6,000	12.33		12.33
Chlorine	Cl				
Clays	—				
Corundum and emery	—				
Diamond	—				
Diatomite	—				
Feldspar	—				
Flourine	—				
Garnet	—	137	10.76		10.76
Gem stones	—				
Graphite (natural)	—				
Gypsum	—				
Iodine	I				
Kyanite and related minerals	—	2,210	12.12		12.12
Lithium	Li				
Mica	—				
Nitrogen	N				
Oxygen	O				
Perlite	—				
Phosphorus	P			11.52	11.52
Potassium	K			11.20	11.20
Pumice	—				
Quartz crystal	—				
Sand and gravel	—	551,000	14.31		14.31
Sodium	Na				
Stone	—				
Strontium	Sr				
Sulfur	S			14.32	14.32
Talc, soapstone, and pyrophyllite	—				
Vermiculite	—				

[a]Order of magnitude dollars where $OM 11.18 represents $\$0.18 \times 10^{11}$.
[b]Value of littoral deposits (from Table 12) extrapolated on basis of area of world continental shelf 55.42×10^6 km².
[c]Value of bathyal deposits from Table 15.
[d]Carbonaceous mud. See Table 21 for details.
[e]n.a., not available.
[f]Includes metals from Mn nodules on continental shelf.
[g]Metalliferous mud. See Table 23 for details.

Table 11. Unconsolidated mineral deposits of the U.S. continental shelf and slope. Estimated resources,

	Nonmetallics					
Mineral:	Sand and Gravel[b]	Kyanite Sillimanite	Staurolite	Garnet	Shells and Aragonite	Mn Oxide Nodules (Blake Plateau):
Commodity:	SiO_2				Ca	FeMn CoNiCu
Alaska	8.0 $\times 10^5$	—	—	—	—	—
Pacific	2.4 $\times 10^4$	—	—	—	500	—
Atlantic	4.8 $\times 10^5$	44	154	—	5,000[3]	25
Gulf	9.0 $\times 10^4$	44	18	4	5,000	—
Total U.S.	14 $\times 10^5$	88	172	4	10,500	25
Value \$/m.t.[f]	1.21	77	70	105	1.75	29.1[g]
Total value	1.69 $\times 10^{12}$	6.78 $\times 10^9$	1.204 $\times 10^9$	4.2 $\times 10^8$	1.84 $\times 10^{10}$	7.28 $\times 10^8$
Total \$OM	13.169	10.678	10.120	9.420	11.184	9.728

[a]Data from McKelvey et al. (1968) except where indicated. Where range is given, mean is used.

[b]From Table 14.

[c]Lower figure in range used.

[d]m.t., metrictons.

Table 12. Estimated regional resources of sand and gravel on U.S. continental terrace by various workers.

Location	Area (yds^3 X 10^6)	Source
New Jersey	3,043	Duane (1969)
Connecticut	130	Duane (1969)
Rhode Island	141	Duane (1969)
Massachusetts	137	Duane (1969)
Maine	123	Duane (1969)
Florida Coast	600	Duane (1969)
East Coast total	4,174	
New Jersey (gravel)	10-30 X 10^3	McKelvey et al. (1969b)
New England Coast		
sand	450 X 10^3 m.t.[a]	Manheim (1972)
gravel	31 X 10^3 m.t.	Manheim (1972)
	481 X 10^3 m.t.	
California, Russian River	100 m.t.	MMTC unpublished
California, Redondo Beach	5	Fisher (1969)
California	Considerably less than Atlantic	McKelvey et al. (1969b)
Southeastern Alaska	Large quantities	McKelvey et al. (1969b)
Hawaii, Oahu	370	Campbell et al. (1970)

[a]m.t., metric tons

Table 13. Estimated regional resources of sand and gravel on U.S. continental terrace.

Location	Area (km^2 X 10)	Resources (metric tons X 10^6)	Basis for Estimate
Hawaii	10	—	Not continental shelf[a]
Alaska	2,000	800 X 10^3	Equivalent to 50%
Pacific Coast	237	24 X 10^3	10% Atlantic Coast/unit area
Atlantic Coast	497	481 X 10^3	Manheim (1972)
Gulf Coast	581	90 X 10^3	15% Atlantic Coast/unit area
Total United States	3,325	1,395 X 10^3	
Value		\$1.69 X 10^{12}	\$1.21 per metric ton

[a]Under Menard's definition Hawaii is volcanic and is not included in the area of continental shelf.

by region (in metrictons, except for gold and platinum in ounces and diamonds in carats).

Heavy Mineral Sands								Native Elements		
Cassiterite	Magnetite	Ilmenite	Chromite	Zircon	Rutile	Monazite	Xenotime	Gold[c]	Platinum	Diamonds
Sn	Fe	Ti	Cr	Zr	Ti	Th		Au	Pt	
0.013	500	500	—	—	—	—	—	10.010	0.13	—
—	52	21	40	5	4	—	—	11	0.35	—
—	—	1,109	—	154	15	1	0.1	—	—	—
—	13	90	—	16	14	4.3		—	—	—
0.013	565	1,720	40	175	33	5.3	0.1	10,021	0.48	—
2,962	10	20	30	60	193	190	112 X 10^3	50/oz	194/oz	6.80/caret
3.85 X 10^{10}	5.65 X 10^9	3.44 X 10^{10}	1.2 X 10^9	1.05 X 10^9	6.37 X 10^9	1.00 X 10^9	1.12 X 10^{10}	5.01 X 10^{11}	9.3 X 10^7	
11.385	10.565	11.344	10.120	10.105	10.637	10.106	11.112	12.501	8.931	—

[e]Author's estimate.

[f]U.S. Bureau of Mines (1970) converted to metric tons.

[g]Computed from Mero (1965) and U.S. Bureau of Mines (1970).

Table 14. Characteristic assemblages of the more common detrital heavy minerals from different source rocks (after Baker, 1962).

Source rock	Characteristic heavy mineral assemblage
Acid igneous rocks	Apatite, biotite, hornblende, magnetite, monazite, sphene, tourmaline, zircon
Basic igneous rocks	Anatase, augite, brookite, chromite, hypersthene, ilmenite, leucoxene, magnetite, olivine, rutile, spinel, zircon
Pegmatites	Fluorite, garnet, monazite, topaz, tourmaline, zircon
Low-rank metamorphic rocks	Biotite, chlorite (if clastic), leucoxene, tourmaline
High-rank metamorphic rocks	Andalusite, biotite, clinozoisite, epidote, garnet, hornblende, kyanite, magnetite, sillimanite, staurolite, zoisite
Metalliferous veins	Cassiterite, pyrite, wolframite, gold
Reworked sediments	Garnet, leucoxene, rutile, tourmaline, zircon
Chemical precipitates	Hematite, limonite, pyrite
Sediments containing authigenetically precipitated minerals	Dolomite (almost wholly authigenic); occasionally chlorites; sometimes tourmaline, anatase, fluorite, rutile, brookite, pyrite, zircon (overgrowths); rarely garnet, staurolite, zoisite, apatite, sphene

Table 15. Estimated world resources of unconsolidated minerals from bathyal deposits in order of magnitude dollars.

Locale:	Surficial			Substratal	
	Phosphorite[a]	Barite[b]	Glauconite	Organic Mud[c]	Metalliferous Mud[d]
Element:	P	Ba	K	V, Cr, Zn, Ni, Mo, U.S. petr.	Zn, Ag, Pb, Cu, Au
United States	10.32			15.15	
Pacific		11.24	11.20		
American					
African					
Arabian					10.18
Indian					
Eurasian					
Antarctic					
	11.32	12.24	12.20[e]	16.15[e]	11.18[e]

[a]McKelvey and Wang (1969); "hundreds" assumed to be 300. Also Mero (1965, p. 73).
[b]Author's estimate; see the text.
[c]Based on McKelvey et al. (1969a), Emery (1961).
[d]Based on Degens and Ross (1969).
[e]Assumed 10 times greater for world.

Table 16. Chemical composition of marine phosphorite from various locations.

	S. California[a]	S. Africa[a]	Chile[b]
CaO	47.4	37.3	31.04
P_2O_5	29.6	22.7	25.62
CO_2	3.9	7.1	3.04
F	3.3	—	2.55
Me oxides	0.43	9.4	6.70
Organic	0.1	—	0.60
	84.7	76.5	69.55

[a]After Mero (1965).
[b]After Baturin (1971).

Table 17. Estimated U.S. resources of marine phosphorite and glauconite[a].

	Phosphorite, P (metric tons X 10^6)	Glauconite, K (metric tons X 10^6)
Alaska	—	—
Pacific	63[b]	5,000
Atlantic	260	5,000
Gulf	—	—
Total United States	323	10,000
Value $/m.t.[c]	10	2
Total $	3.23 X 10^9	2 X 10^{10}
Total $OM	10.323	11.20

[a]Data from McKelvey et al. (1969a) except where indicated.
[b]Inderbitzen et al. (1970).
[c]U.S. Bureau of Mines (1970); converted to metric tons.

Table 18. Chemical analyses (in %) of barite concretions (from Revelle and Emery 1951, p. 712).

	U.S.S.R., Kertch Nodules	East Indies, Kai Island Nodules	California, Sea-floor Concretions
SiO_2	9.41	6.42	6.95
Al_2O_3	5.56	2.32	3.76
Fe_2O_3	1.06	1.67	2.26
MgO	0.65	0.42	0.48
CaO		2.01	4.21
BaO	53.22	53.85	50.96
Na_2O			n.d.[a]
K_2O			n.d.
SO_3	28.12	28.56	26.42
P_2O_5	0.25		trace
NaCl			n.d.
F			n.d.
CO_2			1.78
H_2O	1.39	2.94	2.97
Organic matter	n.d.		
MnO	0.20		
TiO_2	trace		
Total	99.86	98.19	99.79
BaSO[b]	81.88	81.85	77.54

[a]n.d., no data.
[b]BaSO, computed from % BaO.

Table 19. Analyses of samples across the East Pacific Rise (after Bostrom and Petersen, 1966).[a] [a]There is a general positive correlation with heat flow and, excepting barium and strontium, a negative correlation with water depth.

Element	Range along traverse (ppm)
Fe	10^4 - 0.105×10^6
Mn	0.5×10^4 - 0.8×10^5
Cu	80 - 1,150
Ni	100 - 800
Co	40 - 280
Cr	20 - 200
Pb	20 - 140
Ba	1,500 - 30,000
Sr	4,000 - 3,500

Table 20. Analyses of hemipelagic sediments (from Clark, 1924, p. 517).

	Blue Mud	Red Mud	Green Mud[a]	Green Sand[a]	Volcanic Mud
Ignition	5.60	6.02	3.30	9.10	6.22
SiO_2	64.20	31.66	31.27	29.70	34.12
Al_2O_3	13.55	9.21	4.08	3.25	9.22
Fe_2O_3	8.38	4.52	12.72	5.05	15.46
MnO_2	—	—	—	—	trace
CaO	2.51	25.68	0.30	0.22	1.44
MgO	0.25	2.07	0.12	0.13	0.22
Na_2O	—	1.63	—	—	—
K_2O	—	1.33	—	—	—
$CaCO_3$	2.94	—	46.36	49.46	32.22
$Ca_3P_2O_8$	1.39	—	0.70	trace	trace
$CaSO_4$	0.42	—	0.58	1.07	0.27
$MgCO_3$	0.76	—	0.57	2.02	0.83
SO_3	—	0.27	—	—	—
CO_2	—	17.13	—	—	—
Cl	—	2.46	—	—	—
	100.00	101.98	100.00	100.00	100.00

[a]Not glauconitic.

Table 21. Gross value of contained minerals in carbonaceous muds.

Commodity	Units[a]	$/Ton	World Total $OM[b]
Petroleum	5 - 20 gal/ton	0.50	14.15
Uranium	0.005-0.01%	0.94	14.24
Chromium	1.0 %	28.06	15.84
Molybdenum	0.1-0.3 %	3.24	14.83
Nickel	0.1-0.3 %	1.88	14.48
Vanadium	1.0 %	34.00	15.87
Zinc	1.0 %	2,70	14.69
Sulfur[c]	3.0 %	1.26	14.32
			16.20

[a]From McKelvey et al. (1968).
[b]$OM 14.15 = $0.15 X 10^{14}.
[c]Given as "several" percent.

Table 22. Analyses of selected elements in Red Sea metalliferous muds by various workers (from Tooms, 1970).

Cu	0.19%	0.74%	1.3%
Zn	0.93%	2.1 %[b]	3.4%
Ag	12 dwt[a]		34 dwt
Au	0.2 dwt		0.3 dwt

[a]dwt, dry weight tons.
[b]salt-free basis

Table 23. Gross value estimates for elements in metalliferous muds from Atlantis II Deep (data from Degens and Ross, 1969).

		(82 X 10^6 dry tons)		(2 X 10^9 dry tons)	
	Value $/T[a]	Assay % (averaged)	$ X 10^6 Value	Assay % (averaged)	$ X 10^6 Value
Fe		29	—	12	—
Zn	300	34	860	0.93	3,580
Cu	930	13	991	0.19	3,534
Pb	300	0.1	20	—	—
Ag	68,800	0 0054	305	0.0021	2,890
Au	1.6 X 10^6	0.5 ppm	66	0.3 ppm	960
			2,242		12,964

[a]From Table 3.

Table 24. Consolidated mineral desposits in the continental margin.

Environment	Recovery Techniques		
	Surficial	Underground	Fluid
Littoral	Coral	Coal	Coal
	Barite	Iron ore	Shale oil Fluid hydrocarbon[a]
	Bauxite	Limestone	Geothermal energy
		Lode and vein deposits (all elements)	Helium Sulfur Sulfides (Cu) Chlorides (Na, Mg, Ca) Nitrates (K) Carbonates Iodates Borates
Bathyal	Phosphorite Fe/Mn oxide	Indurated metalliferous muds	Hydrothermal fluids

[a]Excluded from this study.

Table 25. Value of mineral resources projected by different authors.

Author		Number of Commodities Considered	$/km^2
Blondel	(1956)	14	259
Allais	(1957)	NK[a]	487
Blondel and Lasky	(1956)	NK[a]	1,783 4 18
McKelvey	(1960)[b]	11	9,444
U.S. Bureau of Mines	(1970)[c]	11	716,000
Theobold et al.	(1970)[d]	1	20.6×10^6
This study	(1972)[e]	60	350,000

[a]Not known.
[b]Based on tonnages reported in Sekine (1962).
[c]For energy resources excluding geothermal.
[d]U.S. geothermal energy.
[e]Continental terrace, excluding energy resources.

Table 26. Comparison of world mineral resource estimates by extrapolation from data on crustal abundance of the elements with estimates made by commodity specialists of the U.S. Bureau of Mines.

	Symbol	Crustal Abundance, A^a (ppm)	Value, V^b ($OM)	World Resource, $R_A{}^c$ ($OM)	World Resources $R_B{}^d$ ($OM)	Ratio, R_A/R_B	OM Deviation, R_A/R_B
Energy Resources							
Anthracite	C						
Bituminous coal and lignite	C		9.34		13.30		
Carbon	C	0.003	5.15		14.23		
Deuterium	—		24.24		14.31		
Geothermal	—						
Helium	He	14,000					
Hydrogen	H						
Natural gas	—						
Peat	C				13.34		
Petroleum	Hc		12.87				
Shale oil	HC				15.95		
Thorium	Th	10	15,031.28	12.15	11.24	6.3	1
Uranium	U	2	20,783.72	11.42	11.38	1.1	1
Ferrous Minerals							
Chromium	Cr	200	58.52	11.12	11.41	0.3	1
Cobalt	Co	23	4,077.40	11.94	10.89	10.5	2
Columbium	—	80	2,997.44	12.24	11.28	8.7	1
Iron	Fe	50,500	16.91	12.85	13.17	0.5	1
Manganese	Mn	1,000	61.01	11.61	13.10	0.05	2
Molybdenum	Mo	1	3,570.48	10.36	11.18	0.2	1
Nickel	Ni	80	2,071.76	12.17	12.14	1.2	1
Rhenium	Re	0.001	1,278,320.00	10.13	12.61	0.002	3
Silicon	Si	276,900	316.27	14.88	13.14	60.8	2
Tantalum	Ta	2	24,244.00	11.49	10.68	7.1	1
Tungsten	W	1	5,972.84	10.60	10.88	0.7	1
Vanadium	V	110	3,746.80	12.41	11.43	9.7	1
Nonferrous Minerals							
Aluminum	Al	80,700	562.02	11.45	11.68	0.7	1
Antimony	Sb	0.2	1,008.33	9.20	10.39	0.05	2
Arsenic	As	2	176.32	9.35	9.67	0.5	1
Beryllium	Be	2	136,648.00	12.27	12.19	1.4	1
Bismuth	Bi	0.2	8,816.00	10.18	9.87	2.2	1
Cadmium	Cd	0.2	5,406.00	10.11	10.38	0.3	1
Cesium	Cs	46	1,322.40	11.61	9.13	460.6	3
Copper	Cu	45	930.09	11.42	12.29	0.1	1
Gallium	Ga	15	1,199,681.28	14.18	11.30	600	3
Germanium	Ge	2	176,320.00	12.35	9.24	1,470	4
Gold	Au	0.005	126.22×10^4	10.63	11.47	0.1	1
Hafnium	Hf	5	187,340.00	12.94	11.53	17.7	2
Indium	In	0.1	8.52×10^4	10.85	9.20	43.5	2
Lead	Pb	15	297.54	10.45	11:28	0.2	1
Magnesium	Mg	20,800	776.91	14.16	13.18	8.9	1
Mercury	Hg	0.5	15,511.59	10.78	10.80	0.9	1

Platinum-group metals	—	0.005	624.64 X 10^4	11.31	11.82	0.4	1
Radium	Ra	1.3 X 10^{-6}	26.00 X 10^9	11.33	11.12	2.9	1
Rare-earth elements	—	2	3,306.00	12.66	11.31	21.6	2
Rubidium	Rb	0.001	991.80	6.99	8.29	0.003	2
Scandium	Sc	5	46.06 X 10^5	14.233	10.35	6,638	4
Selenium	Se	0.009	11,240.40	10.10	9.96	1.0	1
Silver	Ag	0.1	6.88 X 10^4	10.68	11.23	0.3	1
Tellurium	Te	0.0018	13,224.00	8.24	9.74	0.03	2
Thallium	Tl	10	16,530.00	12.17	8.11	15,000	5
Tin	Sn	3	3,264.12	10.98	11.30	0.3	1
Titanium	Ti	4,400	2,909.28	14.13	12.39	32.9	2
Yttrium	Y	40	123,424.00	13.49	11.20	411.6	3
Zinc	Zn	65	297.54	11.19	11.34	0.6	1
Zirconium	Zr	160	6,612.00	13.11	12:10	10.2	2
Nonmetallic Minerals							
Argon	Ar	0.04	257.87	8.10	00		
Asbestos	—		95.01		9.86		
Barium	Ba	400	29.09	11.12	10.22	5.2	1
Boron	B	3	520.92	10.16	11.34	0.05	2
Bromine	Br	3	628.14	10.19	00		
Calcium	Ca	36,500	4.23	12.15	00		
Chlorine	Cl	200	79.34	11.16	00		
Clays	—		4.74				
Corundum and emery	—		88.16				
Diamond	—		13.60 X 10^5		n.d.		
Diatomite	—		6,391				
Feldspar	—		12.20		11.12		
Flourine	—		110.20	11.88	10.39	22.7	2
Garnet	—		115.71				
Gem stones	—						
Graphite (natural)	—		55.10		10.32		
Gypsum	—		4.04		12.37		
Iodine	I	0.3	2,600.72	9.78	11.95	0.008	3
Kyanite and related minerals	—		84.85		10.89		
Lithium	Li	30	2,027.68	11.61	11.11	5.5	1
Mica	—		1,564.84			n.d.	
Nitrogen	N	46	66.12	10.30	12.37	0.008	3
Oxygen	O	467,100	13.22	13.62	00		
Perlite	—		10.88				
Phosphorus	P	1,180	49.59	11.59	12.98	0.06	2
Potassium	K	25,800	36.92	12.95	13.37	0.3	1
Pumice	—		1.74				
Quartz crystal	—		2,622.76				
Sand and gravel	—		1.22				
Sodium	Na	27,500	20.46	12.56	00		
Stone	—		1.64				
Strontium	Sr	450	53.99	11.24	9.25	99.2	2
Sulfur	S	520	41.33	11.22	12.10	2.1	1
Talc, soapstone, and pyrophyllite	—						
Vermiculite	—						

[a]From Barth (1962).
[b]From U.S. Bureau of Mines (1970); adjusted for metric tons.
[c]From $RA = A \times 10^{-4} \times V \times 10^{10}$ (McKelvey, 1972), where \$OM 12.15 = \$0.15 x 10^{12}.
[d]From U.S. Bureau of Mines (1970) projections.

Table 27. Unconsolidated mineral production as a percentage of world production for selected commodities.

Commodity	Symbol	World Production[a] (1880-1965 X 10^6)	% Total
Sand and gravel	—	n.d.[b]	n.d.
Lime sands	Ca	n.d.	n.d.
Magnetite	Fe	n.d.	n.d.
Ilmenite	Ti	195	45
Rutile	Ti	177	—
Monazite	Th	2.6	n.d.
Chromite	Cr	n.d.	n.d.
Zircon	Zr	111.6	100
Cassiterite	Sn	738.5	
Gold	Au	5,600	14
Platinum	Pt	3,230	18
Diamonds	C	3,366	90
Garnet	—	n.d.	100

[a]From Emery and Noakes (1968).
[b]n.d., no data.

Table 28. Estimated world resources[a] of minerals in consolidated deposits in the continental margins, for deposits amenable to surface, underground, and fluid-mining methods.

	Symbol	Underground ($OM)	Fluid ($OM)	Total Resources ($OM)
Energy Resources				
Anthracite	C	13.111[b]		13.111
Bituminous coal and lignite	C	14.874[b]		13.874
Carbon	C		14.114[c]	n.a.[d]
Deuterium	—			
Geothermal	—			14.114
Helium	He			
Hydrogen	H			
Natural gas	—			
Peat	C		15.355[b]	
Petroleum	HC			
Shale oil	HC			15.355
Thorium	Th	11.112		11.112
Uranium	U	11.155		11.155
Ferrous Minerals				
Chromium	Cr	10.437		10.437
Cobalt	Co	11.350		11.350
Columbium	—	11.897		11.897
Iron	Fe	12.319		12.319
Manganese	Mn	11.228		11.228
Molybdenum	Mo	10.133		10.133
Nickel	Ni	11.616		11.616
Rhenium	Re	9.478		9.478
Silicon	Si	14.327		14.327
Tantalum	Ta	11.181		11.181
Tungsten	W	10.223		10.223
Vanadium	V	12.154		12.154
Nonferrous Minerals				
Aluminum	Al	(11.770 surface)		11.770
Antimony	Sb	8.755		8.755
Arsenic	As	9.132		9.132
Beryllium	Be	12.102		12.102
Bismuth	Bi	9.658		9.658
Cadmium	Cd	9.403		9.403
Cesium	Cs	11.227		11.227
Copper	Cu		11.156	11.156

Table 28 (cont.)

Nonferrous Minerals	Symbol	Underground ($OM)	Fluid ($OM)	Total Resources ($OM)
Gallium	Ga	13.672		13.672
Germanium	Ge	12.132		12.132
Gold	Au	10.203		10.203
Hafnium	Hf	12.350		12.350
Indium	In	10.318		10.318
Lead	Pb	10.167		10.167
Magnesium	Mg	13.605		13.605
Mercury	Hg		10.290	10.290
Platinum-group metals	—	10.956[e]		10.956
Radium	Ra	11.126		11.126
Rare-earth elements	—	12.247		12.247
Rubidium	Rb	6.371		6.371
Scandium	Sc	13.810[f]		13.810
Selenium	Se	9.377		9.377
Silver	Ag	10.257		10.257
Tellurium	Te	7.889		7.889
Thallium	Tl	11.616[e]		11.616
Tin	Sn	9.951[e]		9.951
Titanium	Ti	13.263		13.263
Yttrium	Y	13.185		13.185
Zinc	Zn	10.721		10.721
Zirconium	Zr	—[e]		
Nonmetallic Minerals				
Argon	Ar			n.a.
Asbestos	—	9.322[b]		9.322
Barium	Ba	10.433		10.433
Boron	B		9.583	9.583
Bromine	Br		9.702	9.702
Calcium	Ca	11.575		11.575
Chlorine	Cl		10.594	10.594
Clays	—			
Corundum and emery	—			
Diamond	—	n.d.[g]		
Diatomite	—			n.d.
Feldspar	—	10.463		
Flourine	—	11.330		10.463[b]
Garnet	—	—[e]		11.330
Gem stones	—	—[e]		
Graphite (natural)	—	10.194[b]		
Gypsum	—	12.137[b]		10.194
Iodine	I	9.299		12.137
Kyanite and related minerals	—	10.331[b]		9.299
Lithium	Li	11.227		10.331
Mica	—			11.227
Nitrogen	N	10.114		
Oxygen	O	n.a.		10.114
Perlite	—			n.a.
Phosphorus	P	11.219[a]		
Potassium	K			11.219
Pumice	—		12.356	12.356
Quartz crystal	—			
Sand and gravel	—			
Sodium	Na			
Stone	—		12.210	12.210
Strontium	Sr	10.908		
Sulfur	S			10.908
Talc, soapstone, and pyrophyllite	—		10.805	10.805
Vermiculite	—			

[a]In order of magnitude dollars, where $\$OM\ 13.111 = \$0.111 \times 10^{13}$; computed from crustal abundance except where noted.
[b]From U.S. Bureau of Mines (1970); estimates adjusted.
[c]From Theobold et al. (1970).
[d]n.a., not available.
[e]Reduced by amount in unconsolidated form. See Table 27.
[f]Unstable price.
[g]n.d., no data.

BIBLIOGRAPHY

Abbott, W., 1971, Metallurgical mariculture: Ocean Ind., June 1971, p. 43.

Allais, M., 1957, Method of appraising economic prospects of mining exploration over large territories: Algerian Sahara case study: Management Sci., v. 3, no. 4, p. 285-319.

Armstrong, G., 1965, Undersea mining of coal in the United Kingdom: National Coal Board, unpubl. memo.

Arrhenius, G., 1961, Pelagic sediments, *in* Hill, M. N., ed., The sea, v. 3: New York, Wiley p. 655-727.

———, and Bonatti, E., 1965, Neptunism and volcanism in the ocean, *in* Sears, ed., Progress of oceanography, v. 3, pp. 7-22.

Austin, C. F., 1966, Undersea geothermal deposits—their selection and potential use: U.S. N.O.T.S., Tech. Paper 4122, p. 71.

Baker, G., 1962, Detrital heavy minerals in natural accumulates: Austral. Inst. Mining and Met., 146 pp.

Barnes, B. B., 1970, Marine phosphorite deposit delineation techniques tested on the Coronado Bank, Southern California: Offshore Tech. Conf., Houston, May, OTC 1259, 36 p.

Barth, T. F. W., 1962, Theoretical petrology: New York, Wiley, 416 p.

Baturin, G. N., 1971, Stages of phosphorite formation on the ocean floor: Nature Phys. Sci., v. 232, July, p. 61-62.

Beales, F. W., and Onasick, E. P., 1970, Stratigraphic habitat of Mississippi type orebodies: Trans. Inst. Mining and Met., v. 79.

Beki, T., 1970, Short note on the geology of the Miike coal field: Geol. Survey Japan, Mar.

Blissenbach, E., 1972, Continental drift and metalliferous sediments: Conf. Oceanol. Internatl., Brighton, England, p. 412-416.

Blondel, F., 1956, Les lois statistiques de la répartition géographique des productions minières: Revue Ind. Minérale, p. 319-328.

———, and Lasky, S. F., 1956, Mineral reserves and mineral resource: Econ. Geology, v. 60, 6Y6-97.

Bonatti, E., 1972, Authigenesis of minerals—marine, *in* Hill, M. N., ed., Encyclopedia of geochemical and environmental sciences, v. IVA, p. 48-56.

Bostrom, K., and Peterson, M. N. A., 1966, Precipitates from hydrothermal exhalations on the East Pacific Rise: Econ. Geology, v. 61, p. 1258-1265.

Brooks, R. R., and Kaplan, I. R., 1972, Biogeochemistry, *in* Encyclopedia of geochemistry, New York, Van Nostrand Reinhold, p. 74-82.

Buckham, A. F., 1947, The Nanaimo coal field: Trans. Can. Inst. Mining Met., v. 50, p. 460-472.

Bush, P. R., 1970, Chloride—rich brines from Sabkha sediments and their possible role in ore formation: Trans. Inst. Mining and Met., v. 79, p. B137-144.

Chemical and Engineering News, 1964, Chemistry and the oceans: Chem. and Eng. News Spec. Rept., no. 22, June.

Campbell, J. P., et al., 1970, Reconnaissance sand inventory off leward Oahu: Hawaii Inst. Geophys. Univ. Hawaii, June 1970, p. 1-14.

Clark, F. W., 1924, The data on geochemistry: U.S. Geol. Survey Bull., v. 770.

Commission on Marine Science, Engineering, and Resources, 1970, Our nation and the sea: a plan for national action: Washington, D. C.

Craig, H., 1969, Geochemistry and origin of the Red Sea brines, *in* Degens, E. T., and Ross, D. A., eds., Hot brines and recent heavy metal deposits in the Red Sea: New York, Springer-Verlag, 600 p.

Cruickshank, M. J., 1962, Exploration and exploitation of offshore mineral deposits [M.Sc. thesis 969]: Colorado School of Mines, 185 p.

———, 1970, Mining and mineral recovery 1969, *in* Undersea technology handbook, directory, 1970, p. A 11-21.

———, Romanowitz, C. M., Overall, M. P., 1968, Offshore mining—present and future: Eng. Mining Jour., Jan.

d'Anglejan-Chatillon, B. F., 1964, The marine phosphorite deposit of Baja, California, Mexico: present environment and recent history: dissertation: La Jolla, Calif., Library Scripps Inst. Oceanogr. Univ. Calif.

Degens, E. T., and Ross, D. A., eds., 1969, Hot brines and recent heavy metal deposits in the Red Sea: New York, Springer-Verlag, 600 p.

———, Okada, H., and Hathaway, J. C., 1972, Microcrystalline sphalerite in resin globules suspended in Lake Kivu, East Africa: Mineral. Deposita (Berlin), v. 7, p. 1-12.

Duane, D. B., 1967, Sand deposits on the continental shelf: a presently exploitable resource: Proc. Marine Tech. Soc.

———, 1969, Sand inventory program—a study of New Jersey and northern New England coastal waters: Shore and Beach, Oct., p. 12-17.

El Wardani, S. A., 1966, Pelagic biogeochemistry, *in* Fairbridge, R. W., ed., Encyclopedia of oceanography: New York, Van Nostrand Reinhold.

Emery, K. O., 1960, The sea off southern California: New York, Wiley, 366 p.

———, 1965, Some potential resources of the Atlantic Continental Margin: Woods Hold Oceanographic Inst. Contr. 1615, U.S. Geol. Survey Prof. Paper 525-C, p. C157-C160.

———, and Noakes, L. C., 1968, Economic placer deposits of the Continental Shelf: Tech. Bull. ECAFE, v. 1, p. 95-111.

———, and Noakes, L. C., 1969, Economic deposits of heavy-mineral placer's on the world's continental shelves: Conf. Oceanol. Internatl., Brighton, England.

Ewing, M. and Thorndyke, 1965, Suspended matter in deep ocean water: Science, v. 147, p. 1291-1294.

Fisher, C. H., 1969, Mining the ocean for beach sand, *in* Civil engineering in the oceans II, Am. Soc. Civil Engr., Dec., p. 717-723.

Forchhammer, G., 1865, Chemical composition of seawater: Trans. Phil. Soc., v. 155, p. 203-262.

Goldberg, E. D., 1957, Biogeochemistry of trace elements: Geol. Soc. America Mem. 67, v. 1, p. 345-358.

———, 1963, The oceans as a chemical system, *in* Hill, M. N., ed., The sea, v. 2, New York, Wiley-Interscience, p. 3-25.

Grinrod, J., 1960, An offshore sulphur mine: Mining Mag., Sept., p. 142.

Groot, J. J., and Ewing, M., 1963, Suspended clay in a water sample from the deep ocean: Science, v. 142, p. 579-580.

Gulbrandsen, R. A., 1969, Physical and chemical factors in the formation of marine apatite: Econ. Geology, v. 64, no. 4, p. 365-380.

Hansen, M., and Rabb, D. D., 1968, Seek profitability answers to nuclear in-situ copper leaching at

Safford: World Mining.

Harvey, H. W., 1969, The chemistry and fertility of sea water: New York, Cambridge Univ. Press.

Hedgpeth, J. W., eds., 1957, Treatise on marine ecology and paleoecology: v. I., Geol. Soc. America Mem. 67, 1296 p.

Heezen, B. C., and Menard, H. W., 1963, Topography at the deep sea floor, *in* Hill, M. N., ed., The sea, v. 3: New York, Wiley, p. 233-280.

Homan, F., 1972, Energy from the earth: Alaska Construction and Oil Rept., Apr., p. 64-71.

Hood, 1963, Reference not available.

Hubbert, M. K., 1969, Energy resources, *in* Resources and man: Natl. Acad. Sci.—Natl. Res. Council, San Francisco, Freeman, W. H.

Inderbitzen, A. L., Carsola, A. J., and Everhart, D. L., 1970, The submarine phosphate deposits off Southern California: Offshore Tech. Conf. Texas, Paper OTC 1257.

Investors' Guardian, 1969, English potash: the giants join battle on the Yorkshire moors.

Jones, E. J., 1887, Barite in nodules: Jour. Asiatic Soc. Bengal, v. 56, p.E. 2, p. 209.

Kohout, F. A., 1966, Submarine springs, *in* Fairbridge, R. W., ed., Encyclopedia of oceanography: New York, Van Nostrand Reinhold, p. 878-883.

Kossinna, E., 1921, Die Tiefendes Iveltemeeres: Inst. Meereskunde, Verolx, Geogr-nature, U.S., 70 pp.

Kuenen, P. H., 1950, Marine geology: New York, Wiley, 558 p.

Kukal, Z., 1971, Geology of recent sediments: New York, Academic Press, 490 p.

Lindgren, W., 1933, Mineral deposits: New York, McGraw-Hill, 930 p.

Lisitzin, A. P., 1972, Sedimentation in the world ocean: Soc. Econ. Paleontologists and Mineralogists Spec. Publ. 17, p. 218.

Mamen, C., 1959, Goderick mine goes into production: Can. Mining Jour., Dec., p. 59-63.

Manheim, F. T., and Horn, D., 1968, Composition of deeper subsurface waters along the Atlantic continental margin: Southeastern Geology, v. 9, p. 215-236.

Mason, B., 1958, Principles of geochemistry: New York, Wiley.

McIlhenny, W. F., 1967, Ocean raw materials, *in* Encyclopedia of chemical technology: New York, Wiley-Interscience.

______, and Ballard, D. A., 1963, The sea as a source of dissolved chemicals: Symposium on economic importance of chemicals from the sea: Natl. Am. Chem. Soc., Los Angeles, Calif., April.

McKelvey, V. E., 1960, Relations of reserves of the elements to their crustal abundance: Am. Jour. Sci., v. 258-A (Bradley Vol.), p. 234-241.

______, 1968, Mineral potential of the submerged parts of the continents: Proceedings of a symposium on mineral resources of the world ocean: Univ. Rhode Island, U.S.N., U.S. Geol. Survey Publ. 4.

______, 1972, Mineral resource estimates and public policy: Am. Scientist, v. 60, Jan.-Feb., 1972, p. 32-40.

______, and Wang, F. F. H., 1969, Preliminary maps, world sub sea mineral resources: Washington, D.C., U.S. Geol. Survey Misc. Geol. Invest., Map I-632.

______, and Wang, F. F. H., 1970, World subsea mineral resources, maps: rev. USGS MGI Map I-632.

______, Wang, F. H., Schweinfurth, S. P., and Overstreet, W. C., 1968, Potential mineral resources of the U.S. outer continental shelf: Public Land Law Review Commission; Study of OCS lands of U.S., v. IV, app. 5-A: U.S. Dept. Com. P.B. Rept. 188717, 117 p.

______, Tracey, J. I., Stoertz, G. E., and Vedder, J. G., 1969a, Subsea mineral resources and problems related to their development: U.S. Geol. Survey Circ. 619.

______, Stoertz, G. B., and Vedder, J. G., 1969b, Subsea physiographic provinces and their mineral potential: U.S. Geol. Survey Circ. 619.

Menard, H. W., and Smith, S. M., 1966, Hypsometry of ocean basin provinces: Jour. Geophys. Res., v. 71, no. 18, p. 4305-4325.

Mero, J. L., 1961, Sea floor phosphorite: Calif. Min. Inform. Service, Div. Mines and Geology, v. 14, no. 11, p. 1-12.

______, 1965, The mineral resources of the sea: New York, American Elsevier, 312 p.

Mitchel, A. H. G., and Garson, M. S., 1971, Relationship of porphyry copper and circum-Pacific tin deposits to palaeo-Benioff zones: Inst. Mining and Met., Aug.

NEMRIP, 1971, Use found for mud lying off New England coast: New England Marine Resources Inform., Publ. 31, Univ. Rhode Island, dec.

New Scientist, 1970, Tapping the ocean's fresh water: New Scientist, Sept., p. 583.

Noddack, I., and Noddack, W., 1939, Die Haufigkeiten der Schwermetalle in Meeres Tieren: Arkiv Zool., v. 32A, p. 1-35.

Nolan, T. B., 1936, Non ferrous metal deposits, *in* Mineral resources of the region around Boulder Dam: U.S. Geol. Survey Bull., v. 871, p. 5-77.

Parker, R. J., and Siesser, W. G., Petrology and origin of some phosphorites from the South African continental margin: Jour. Sedimentary Petrology, v. 42, no. 2, p. 434-440, Figs. 1-4.

Pierce, U. S. 1957, Mining iron ore under the sea: Compressed Air Mag., Feb., p. 30-34.

Revelle, R. and Emery, K. O., 1951, Barite concretions from the ocean floor: Geol. Soc. America Bull., v. 62, p. 707-724.

Schlee, J., 1964, New Jersey offshore gravel deposit: Rept. Pit and Quarry, Woods Hold Oceanogr. Contr. 1514.

______, 1968, Sand and gravel on the continental shelf off the northeastern United States: Geol. Survey Circ. 602.

Sekine, Y., 1963, On the concept of concentration of ore forming elements and the relationship of their frequency in the earth's crust: Internatl. Geology Rev., v. 5, no. 5, p. 505.

Slichter, L. B., 1960, The need of a new philosophy of prospecting: Mining Eng., v. 12, p. 570-575.

Summerhayes, C. P., et al., 1970, Phosphorite prospecting using a submersible scintillation counter: Econ. Geology, v. 65, p. 718-723.

Sverdrup, H. U., Johnson, M. W., and Fleming, R. H., 1942, The oceans, their physics, chemistry, and general biology: Englewood Cliffs, N.J., Prentice-Hall.

Terry, R. D., 1961, Oceanography, its tools, methods, resources and applications: Autonetics, 347 pages.

Theobold, P. K., et al., 1970, Energy resources of the U.S.: Geol. Survey Circ. 650.

Tooms, J. S., 1969, Metal deposits in the Red Sea—their nature, origin, and economic worth: Imperial Coll. Sci. and Technology, Roy. School Mines, London.

______, 1970, Review of knowledge of metalliferrous

brines and related deposits: Trans. Inst. Mining and Met. (London), v. 79, p. B116-126.

Twenhofel, W. H., 1932, Treatise on sedimentation: Baltimore, Md., Williams & Wilkins, 926 p.

U.S. Bureau of Mines, 1970, Mineral Facts and problems: U.S. Bur. Mines Bull., v. 650, 1291 p.

van Andel, T. H., 1965, Marine phosphorite prospects in Australia and marine geology in the BMR: Dept. Natl. Develop., Bur. Mineral Resources Geology and Geophys. Record 1965/188.

Vinogradova, Z. A., 1964, Certain biochemical aspects of a comparative study of the Black Sea, Sea of Azov and Caspian Sea: Oceanology (Moscow), v. 4, no. 2, p. 232-242.

Wang, F. F. H., 1973, Solid mineral deposits and mineral fluids within bedrock: U.N. publication, in press.

Webb, B., 1965, Technology of sea diamond mining: Proc. 1st Ann. Conf. Marine Tech. Soc., p. 8-23.

Weissberg, B. G., 1969, Gold-silver ore-grade precipitates from New Zealand thermal waters: Econ. Geology, v. 64, p. 95-108.

White, D. E., 1963, Econ. Geology, v. 63, p. 301-335.

Williams, J. G., 1965, Phosphorites *in* Exploration and mining congress, v. 2, 8th Commonwealth Mining and Met. Congr., Austral. Inst. Mining and Met., p. 262.

Part XIII

Continental Margins in Perspective

Continental Margins in Perspective

C. A. Burk and C. L. Drake

INTRODUCTION

The basic objective of this volume is to provide a broad inventory of our present state of knowledge of continental margins, not only for its own value, but also as a basis for future studies, research, and exploration. The chapters in the volume deal with the characteristics and history of these margins, ancient and modern, with examples from many parts of the world. From the vantage point of editors of the volume, it seems appropriate to comment briefly here on some of the more apparent and broadly important aspects of the continental margins of the world and to suggest directions of research or important problems that could significantly improve our understanding of these important features. All references cited will be to contributions in this volume and locations of areas referenced are given in the frontispiece.

SHAPE OF CONTINENTAL MARGINS

As noted repeatedly, continental margins are obviously important academically, economically, and politically. They constitute the complex transition between the two great hypsometric surfaces of the earth, the ocean floors and the continental platforms. These margins have long been divided into active and inactive types, according to whether or not they were associated with young deformation, volcanism, or active seismicity (Heezen; Fisher).

However, it should be emphasized that inactive margins are not undeformed, and that the deformation of active margins is very poorly understood. Concepts of sea-floor spreading and the plate tectonics model provide useful frameworks for interpreting the evolution of continental margins, but many problems remain to be solved. The study of continental margins was based initially simply on bathymetry. Indeed, the shape and configuration of margins does reflect the geological history and underlying structure, modified to varying degrees by sedimentation, and bathymetric studies will continue to be an important tool.

DEEP STRUCTURE OF CONTINENTAL MARGINS

In deciphering the history and structure of the continents, geologists worked to the edge of the oceans, where further work was impossible, except on a few scattered islands, exposing bits of the continental shelf. Consequently, determination of the structure of margins has been largely dependent of the indirect methods of geophysics. These received some early attention, but the vast ocean floors demanded most of the intitial efforts of marine geophysicists, and it is only very recently that concerted attention has been devoted to the margins.

In fact, geophysical studies are extremely difficult at continental margins. Structural complexity and density variations create major problems in computing and interpreting reasonable models from the earth's gravity field. Seismic-refraction studies are complicated by the surface and subsurface topography, difficulties in interpreting velocity variations, and refracting horizons commonly dipping in opposite directions. Seismic-reflection studies require a relatively large energy source, and deep data are typically obscured by numerous multiple reflections. However, new studies with modern equipment have been given impetus not only by our lack of knowledge, but also by the potential for future resource development. Drilling on continental margins will continue to be of extremely great value, especially in deeper waters. We are also learning now how to extrapolate the exposed rocks of ancient margins into better interpretations of young continental margins.

The exact position of the edge of the continental crust is extremely difficult to determine (Rabinowitz), even using all available geophysical techniques (Mayhew). The general transition between the thick continental crust and the thin oceanic crust is well known (Worzel), as are the broad acoustic differences between them, attributed largely to the lack of granitic rocks in the oceanic crust. Details of the geological and geophysical changes across continental margins have been studied in a preliminary fashion only.

In some areas of the world, it is not entirely clear that such a bathymetric boundary is associated with a true continental margin—in the Bahama Platform, for example, it is not certain whether these thick carbonates are underlain by oceanic or continental crust (see Meyerhoff and Hatten), a problem not limited to modern margins. Because of the high acoustic velocity of these carbonates and the large hypsometric variations, any type of geophysical study is extremely difficult on the Bahama Platform. Nevertheless, any improved knowledge of this important area would greatly increase our understanding of continental margins and our ability to interpret the geology of former marginal areas.

The margins of many small ocean basins may or may not have a deep crustal structure characteristic of either the active or passive type of margin (for example, Gulf of Mexico, Baffin Bay, parts of the Mediterranean Sea, the Black and Caspian seas, Sea of Japan, Sea of Okhotsk, the Lau Basin). Frequently, they show evidence of massive vertical movements. The margins of these small basins deserve continued and careful study. Similarly, small continental masses, commonly with anomalous crustal characteristics, must be studied further (for example, Orphan Knoll, Jan Mayen and Rockall ridges, possibly the Canary and Cape Verde islands, New Caledonia and Fiji islands, Lord Howe Rise, Seychelles and Mascarene ridges).

Seismicity is much more related to the margins of plates than to continental margins—as is to be expected, since seismicity is one of the basic tenants of any plate tectonics model (Oliver et al.). This model helps to explain the much earlier classification of "active" and "passive" continental margins, but there are many problems yet to be solved. The entire western margin of South America would be classified geologically as the same (active) type, yet the absence of seismicity in southern Chile would ideally remove it from the grouping of other active margins (see Hatherton).

Structures transverse to the strike of active margins are apparent in exposed geology and in volcanic alignments, and this segmented character has now been shown in seismicity (Carr et al.). Such observations open new opportunities for studying the nature of the deep crustal and mantle transition at active margins.

The geophysical characteristics of the deep continent-ocean crustal transition must depend in large measure on the geological history of these continental margins. There has been much past discussion regarding continental accretion through time; assuming that this necessarily requires lateral extension of continental crust, accretion is now occurring by sedimentation at both active and passive margins. However, it seems likely that the total volume of the ocean basins has been constant throughout much of geological history (Wise), in which case crustal thickening must occur to accommodate this lateral expansion. This, in turn, must require changing geological and geophysical characteristics within these margins as they evolve tectonically. Are these geophysical characteristics apparent in the different and presently evolving continental margins and how are they manifested in ancient margins?

It is worth noting also that the coincidence of continental margins and belts of active seismicity must be more than fortuitous, particularly around the Pacific Ocean. If so, this implies some crustal or upper mantle control (at continental margins) over processes presumably involving the entire lithosphere and possibly much of the upper mantle.

SEDIMENTATION AT CONTINENTAL MARGINS

The record of much of the geological history of continental margins is preserved in the accumulated sediments or represented by unconformities. It is important in understanding the evolution of any margin to separate the eustatic and tectonic effects (Wise). Intercontinental chronostratigraphic correlation thus becomes obviously important.

Studies of the world's deep oceans (especially the JOIDES Deep Sea Drilling Project) have greatly opened the perspective of marine sedimentology within the last half decade (Berger). Marine sediments in the past have been studied largely in terms of detailed analyses of small variations in environments and lithology largely within inner-shelf depths. The abundant new sediment samples from the deep oceans has spurred new and truly global thinking about the framework of sedimentation, particularly the complex sediments of the entire continental margin (Swift; Ginsburg and James; Urien and Ewing; Seibold and Hinz; Flood and Hollister; Curray and Moore; von Huene; Berger; Edgar).

New geophysical surveys have revealed unconformities (some even folded) within the sediments of the generally stable inactive margins. Where these unconformities can be dated, they become powerful tools in understanding the evolution of continental margins (Seibold and Hinz). It has long been known that sediments of the shelf moved oceanward, commonly through spectacular canyons eroded into the continental margin, or prograded over a subsiding shelf edge to form the world's great deltas and vast submarine fans, the largest of which is fed by sediment of the Ganges River (Curray and Moore). Studies here have progressed to the point where 12 km of continental rise sediments can be recognized as underlying a 4-km cover of the Bengal Fan deposits. However, a delicate balance exists between the volume of such sediment supply and the nature of deep-ocean currents in the development of continental rises (Flood and Hollister). Much has yet to be learned about the importance of these deep currents in controlling sediment distribution and in shaping the sea-floor morphology.

The place of continental rises in active continental margins is taken by deep oceanic trenches. Most trenches contain only a small quantity of sediments (Scholl); however, the continental margins facing trenches seem to be composed of vast volumes of sediments and deeper crustal material presumably scraped from the subducting ocean plate, along with whatever sediments had accumulated in trench depressions (Seely et al.; Kulm and Fowler; Moore). Studies of ancient active-margin deposits (Moore; Kimura) suggest that the bulk of the "eugeosynclinal" marginal sequence is turbidites, accumulated in a trench, rather than pelagic deposits of the open ocean. Careful study of these deepest of sediments is important (von Huene; Scholl); even though their instantaneous importance may seem small volumetrically, their cumulative importance may be exceptionally great. Successor sediments accumulating in tectonic troughs of both the slope and shelf of active margins must also be of fundamental importance (Scholl; Kulm and Fowler; Shor).

The stratigraphy of deep-ocean sediments was first established on the geophysically mapable basis of their acoustic characteristics. The sediments themselves were studied from numerous shallow-penetration cores (to 30 m). These descriptions could then be tied into newly understood sedimentary processes beyond the continental margins (Berger). In just the last few years, as a result of the Deep Sea Drilling Project, deep-ocean stratigraphy has come around full cycle (Edgar), so that known sediment types, ages, and processes can now be extrapolated with seismic-reflection data into the adjacent continental margins.

TECTONICS OF CONTINENTAL MARGINS

It is natural to look first at active-type continental margins to study major deformation, and indeed the tectonics here are impressive but very poorly understood. Modern seismic-reflection techniques have indicated that crustal thickening landward of trenches may occur as a result of imbricate underthrusting (Seely et al.; Kulm and Fowler). This helps to explain the complex structure and apparently excessive thicknesses of "eugeosynclinal" sequences, but the model is not completely satisfying in terms of the vertical movements required to uplift crustal ophiolites (Coleman and Irwin) and blueschist-facies metamorphic rocks (Ernst), or subsequently to depress the inner margin to accumulate great thicknesses of superimposed sediments confined behind a marginal ridge (Seely et al.; Kulm and Fowler; Scholl). There are also geological problems that must be resolved in many of the recently deformed but now uplifted margins.

The asymmetry of active continental margins (Hatherton) seems particularly important to us—there are no present subduction zones dipping seaward, beneath oceanic crust—which again implies some crustal control over upper mantle and asthenosphere processes. This configuration is sometimes used to explain the geological features of more ancient margins. Active margins are thus a fruitful area for both detailed and regional structural studies.

The passive continental margins are passive only in the sense of earthquakes and volcanic activity. The enormous vertical movements that have taken place in these regions, particularly in small ocean basins, indicate that a process of oceanization cannot be ignored (see Kent).

The plate tectonics model has suggested a possible new type of passive continental margin, one that results from shear strain, including possibly the Barents Sea margin, the northern Gulf of Guinea, northern Faulkland (Malvinas) Plateau, and southeastern Africa (Talwani and Eldholm; Delteil et al.; Dalziel; Siesser et al.). If these and others actually have a different origin than other types of passive margins, it should be apparent in their internal structure (see also Renard and Mascle).

It is important to emphasize that the "Pacific Plate" is bounded everywhere by an active continental margin—except in the Chatham Rise and Campbell Plateau of New Zealand (Katz). This is the one passive margin entirely within the "Pacific Plate."

A particularly important passive continental margin of the nonsheared(?) type is eastern Africa (Kent). This margin, entirely within a lithospheric plate, has been affected by major epeirogenic movements related to those of the African continent, but bearing only questionable relationships to the evolution of adjacent plate boundaries and deep oceans as interpreted from major magnetic data. Further studies of the tectonics of this margin may also relate to such "small continents" as Madagascar, Seychelles and Mascarene plateaus, possibly the Roset Islands, Kerguelen Plateau, and eventually to Antarctica.

A particularly unique tectonic problem is the presence of oceanic ophiolites and cherts overlying continental crust at a margin. In some instances this may be related to oceanic closing and collision at an active margin (see Dewey; Williams and Stevens), but in many areas there has been no apparent "collision" (Coleman and Irwin), such as Oman, New Guinea and New Caledonia, and the detailed processes which emplaced these rocks at continental margins is not yet understood.

NATURE OF MODERN CONTINENTAL MARGINS

The present must indeed be the key to the past regarding continental margins, since their definition is based on the present configuration and structure of the earth. Detailed and regional investigations of particular margins provides the basis for our present state of knowledge. Obviously, not all these continental margins are completely young—many have a known history that may date beyond the early Mesozoic. Thus it is impossible to separate the study of present and ancient margins. We have tried in this volume to combine the approaches of tectonics, sedimentology, time correlations, volcanism, plutonism, and metamorphism to studies of specific continental margins of every type, hopefully providing a representative sample of our present state of knowledge.

Atlantic Region

The western North Atlantic has long been the type locality of passive continental margins and is one of the best studied (Keen and Keen; Mayhew; Rabinowitz; Flood and Hollister; Sheridan; Meyerhoff and Hatten). Not only can the recent margin be extended back to the early Mesozoic, but interpretations reach back to early Paleozoic (Dewey; Williams and Stevens; Sheridan). The evolution of all North Atlantic is important in this regard (Keen and Keen; Talwani and Eldholm; Roberts; Montadert et al.). The evolution and deep structure of Iceland, atop the Mid-Atlantic Ridge, and its margins, is an equally important development of this area (Pálmason), particularly the great thickness of Neogene subaerial plateau basalts (to 12 km) covering most of the island and now uplifted, tilted, and eroded.

The eastern south Atlantic contains complex fault and diapiric structures (Seibold and Hinz; Delteil et al.; Driver and Pardo; Siesser et al.) which are only just now beginning to be understood. The Canary Islands, which expose ophiolites and appear to reflect the ocean-continent transition, and the Cape Verde Islands, which contain carbonatites, will unquestionably become more important in the future. The western margins of the south Atlantic are in some ways similar and in others dissimilar to the eastern south Atlantic margin (Dalziel; Zambrano and Urien; Campos et al.). The margins bordering the Scotia Sea (Dalziel) may contain both active and passive margins, as well as that resulting from shear strain.

Throughout all of this large Atlantic region, the presence of latitudinally controlled evaporite and carbonate sediments apparently have a major affect on the structure and marine geophysical data (e.g., Mayhew). Perhaps someday we can map facies transitions within continental margins with seismic refraction.

In each of the regions reported, new aspects are described that should be considered possibly applicable to other parts of the world's continental margins (great variations in sedimentation, possible importance of wrench faulting, significance of marginal ridges, large and differential vertical movements, volcanic and diapiric activity).

Indian Ocean Region

As noted earlier, careful studies of the continental margins bordering the Indian Ocean may be of particular significance, especially if the important Antarctica margin (Houtz) and the various "mini-continents" scattered throughout its area are included. Southern and eastern Africa show unusual tectonic configurations, large and differential vertical movements, and submerged marginal crust with transitional characteristics (Siesser et al.; Kent).

In general, this is also true of the margins of the Indian Peninsula—great subsidence and large vertical tectonics—except that these features have been overprinted by great masses of sediments supplied by the major rivers draining the Himalayan Mountains (Closs et al.; Curray and Moore). Deep-sea drilling into such related areas as the Ninety-East Ridge and the northern Laccadive Islands has greatly changed our perspective on the evolution of this region. What is the timing and structural relationships among the Bengal Fan, the Ninety-East Ridge, the Andaman-Timor Trench, and the complex continental margin of western Australia (Veevers)? These have typically been considered separately, in terms of a small continent (Ceylon), a large deltaic wedge (Bengal), a subducting and accreting margin (Andaman-Java), a complexly thrust margin (Timor), a stable subsided margin (since Hercynian—northwest Australia), and a possibly sheared margin in western Australia and eastern Africa.

In terms of ancient continental margins, the northern Indian subcontinent may be particularly important in enclosing the old seaway of the Tethys Ocean (cf. Stöcklin; Stoneley), especially since the tectonics of this ancient belt seem to extend to the present continental margins of Java and Sumatra to the east, and to Turkey in the west (Biju-Duval et al.). An extension of the Indian Ocean (the Red Sea); also contains what is generally considered to be one of the youngest of the world's continental margins (Coleman).

Pacific Region

The Pacific is the largest ocean of the world and consequently is bordered by the greatest length of continental margins, and it is considered to be characterized by active-type margins (Fisher; Hatherton). Indeed, its margins mark the "rim-of-fire" and have been shown to contain the bulk of the world's intermediate and deep earthquakes (Oliver et al.). Analogies between continental margins have been made at extremely great distances across this ocean (cf. Blake et al.).

The New Zealand-Tonga Trench represents one of the few which extends from continental into oceanic crust (Katz), others include the Kamchatka-Kurile Trench, the Scotia Arc (Dalziel), and the Aleutian Trench (Moore). Each of these deserves careful future study to consider how the same tectonic stresses affect both continental and oceanic crust, and to what extent their location is controlled by the edges of continents.

Bordering the western Pacific Ocean are crustal areas that are neither clearly oceanic nor continental, such as the Lord Howe Rise, Lau Basin, New Caledonia, Ryukyu Trough, seas of Japan and southern Okhotsk (Bentz; Hawkins; Dubois et al.; Uyeda). The evolution of these "intermediate" regions may provide important clues to understanding the history of more typical continental margins. The data of the Deep Sea Drilling Project are invaluable in this regard, but much more advanced and careful geophysical studies are needed on these intermediate margins.

The continental margins of the eastern Pacific have been studied extensively, primarily because they are easily accessible to the major research institutions (cf. Kulm and Fowler; Seely et al.; Shor; Hayes; Mordojovich). The deep Pacific is rapidly becoming the most studied of the deep-ocean basins, especially within the model of plate tectonics. This tectonic framework has become increasingly complex (including numerous former plates that have now disappeared). All these complexities should be apparent in the evolution of the adjacent continental margins.

The history of most of the continental margins of the Pacific region can be traced back into the Mesozoic, but it seems possible that very ancient continental margins, geologically continuous with present margins, can be recognized in Japan (Kimura; Uyeda), in Alaska (Moore), California (Maxwell; Blake et al.), and in southern Chile (Dalziel). The "back-arc" basin of the Sea of Japan can be interpreted as a young extensional feature (Uyeda), but this cannot be the case in the deep Bering Sea (Ludwig). Such sinuous features as Bowers and Lomonosov ridges in the Bering Sea suggest that deformation other than brittle plate movement must have occurred.

Arctic and Antarctic Margins

The linear extent of continental margins in the Arctic (Ostenso) closely approximates that in Antarctica (Houtz), but the former area is largely deep ocean and the latter entirely continental. The ancient history of the Arctic margins is much better known from land geology (e.g., Drummond), whereas the largest speculations about oceanic history are from the same region. Antarctic geology and margins are poorly known, but the apparent history of the adjacent circum-Antarctic oceans have been much more thoroughly studied (Houtz; Dalziel; Katz). Both these regions show large vertical movements at the continental margins.

MARGINS OF SMALL OCEAN BASINS

The Bering Sea consists of a very wide continental shelf off Alaska and a deep ocean basin north of the Aleutian Islands (Ludwig). The data presently available indicate that the margin of the North Pacific Ocean was at the western shelf edge of Alaska until earliest Tertiary (Moore), at which time the Aleutian Trench and Island Arc isolated the deep Bering Sea from the Pacific Ocean, and the western Alaska and northern Kamchatka margins were converted from active to passive types. The continental margins of the Bering Sea may thus have a somewhat unique history; and careful investigations, combining both marine geophysics and exposed geology, might be particularly rewarding in this region. The history of the Sea of Okhotsk may be quite similar to that of the Bering Sea, and perhaps the Phillipine Sea, and possibly the Andaman Sea as well (Curray and Moore).

Unique small ocean basins are represented by the Caspian Sea, the Black Sea, and the Gulf of Mexico (Ross; Stöcklin; Antoine et al.; Worzel). These are unquestionably of deep-ocean character in part, but are variously considered to be "oceanized" crustal depressions, superimposed on the tectonic grain of continental crust, or to be remnants of a much older ocean basin that has not yet become involved with the tectonics of the adjacent continental crust. Such small ocean basins have much yet to teach us about the evolution of continental margins, particularly since they are surrounded by exposed rocks of known geological history. To what extent is the history of the Mediterranean margins similar to those of these very very small ocean basins (Auzende and Olivet; Biju-Duval et al.)?

The Sea of Japan (Kimura; Uyeda), the Philippine Sea, and the Lau Basin of the southwest Pacific (Hawkins) may yet be even different small ocean basins, and the evolution of their continental margins might be quite distinct from most others.

The Caribbean Sea (Case) has long been compared with other areas containing active or ancient volcanic arcs, such as Southeast Asia and the Mediterranean. The Sunda Seas are known to overlie slightly submerged continental crust, and analogies with the Mediterranean depend upon the evolution of the old "Tethyan Sea." New studies are yielding data that seem to show a very complex pattern of evolution within the Mediterranean by both vertical and horizontal tectonic movements (Biju-Duval et al.; Auzende and Olivet; Stoneley).

The Arctic Ocean is a small ocean basin with a very complex and varied history (Ostenso; Drummond; Talwani and Eldholm; Keen and Keen). A large amount of data has been collected in this region, considering its remoteness and climatic inhospitality ("the only ocean where refraction work can be done with trucks"). The Arctic Ocean appears to contain all the multiple features characteristic of the other deep oceans of the world (Ostenso), yet its small and closed perimeter should permit closer definition of the evolution of its continental margins (e.g., Drummond).

Slivers of oceanic crust, determined by seismic-refraction velocities, are present in the Red Sea (Coleman), Gulf of California, and possibly Lake Bakail in the USSR. Lake Bakail seems to have a history at least as old as Late Cretaceous, but the other two areas appear to be relatively new features. The evolution of these margins should be further examined in relation to the better-known history of adjacent exposed rocks.

ANCIENT CONTINENTAL MARGINS

As noted earlier, the study of present continental margins commonly requires evaluation of their history well into the past. Our best interpretations indicate that many old continental margins are now completely encased in continental crust, the old oceanic basin having disappeared. The recognition of such ancient margins now entirely within a continent lends strong support to subduction, seafloor spreading, drifting continents, oceanic consumption, and the plate tectonics model.

One of the classic ancient oceans is the Tethys Sea, which may possibly have had more than one closing (Stoneley). When investigated in detail, Tethyan history becomes increasingly complex. It seems possible that there were two Tethys seaways in Iran (Stöcklin), and another ocean may have encircled the Lut Block of eastern Iran, a possible "mini-continent." One seaway in Iran seems to have opened simultaneously as another closed. The ancient Atlantic Ocean appears to have a similarly complex history (Dewey; Williams and Stevens). We have much yet to learn from continental studies of such ancient margins.

The history of western North America is equally complex and renewed interpretations there are equally rewarding (Maxwell). The younger part of this margin has been compared in many respects to the history of Japan and New Zealand (Blake et al.). There are many similarities, indeed, but also some prominent dissimilarities. Studies in New Zealand are especially important, particularly where the continental crust of the adjacent Chatham Rise and Campbell Plateau are considered as well (Katz).

An impressive aspect of the evolution of ancient Japanese continental margins is the apparent seaward migration through time of the margin and the associated trench (Kimura). The confident recognition of ancient trench deposits is very important, as well as their distinction from sediments accumulated in troughs within the slope or shelf (von Huene; Scholl). On the basis of trench sedimentation, the Mesozoic margin of the North Pacific Ocean can be interpreted to lie along the northern margins of the Bering Sea (Moore).

It is possible that we can recognize ancient continental margins at least 1 billion years old (Goodwin). Certainly this offers vast opportunities for understanding the total crustal evolution of the earth. There are also strong indications of major changes in the tectonic regime of the earth; for example, at the start of the Phanerozoic and during the Hercynian.

IGNEOUS AND METAMORPHIC IMPLICATIONS

The concept that most ophiolites are fragments of oceanic crust provides an important key to the recognition of ancient continental margins (Coleman and Irwin). Where these rocks now overlie continental crust, there can be no doubt about the former existence of a continental margin. Particularly important occurrences include northern California, New Guinea, New Caledonia, Oman (Coleman and Irwin), Newfoundland and the Appalachians (Williams and Stevens), Appalachian-Caledonian system (Dewey), Iran (Stöcklin), and much of Tethys (Stoneley). The exact methods of emplacement of these ophiolites is still debatable, but their significance to ancient continental margins is not. Stöcklin distinguishes between two different kinds of ophiolite-mélange belts in Iran, which may be useful elsewhere in the world. Also, ophiolites do occur in areas not easily related to active margins (e.g., the eastern Canary Islands).

Active continental margins also appear to be characterized by paired metamorphic belts (Ernst). A high-pressure blueschist facies commonly affects the sediments associated with ophiolites, while a high-temperature facies is developed farther away from the ocean. These relationships are most readily explained in terms of subduction, and many problems have yet to be solved. By what process, for example, are these deeply buried blueschist rocks uplifted and exposed at continental margins along with the associated ophiolites of the oceanic crust and the eclogites of continental crust?

Both volcanism and seismicity are characteristic of presently active continental margins. Preliminary attempts are now being made to recognize old volcanic arcs associated with ancient margins (e.g., Japan; Kimura). Williams and Stevens appropriately noted that "it is not sufficient to say that volcanic terranes are eugeosynclinal if it can be shown that one particular volcanic assemblage forms the top of an ophiolite suite, whereas another is calc-alkaline and shows vertical and lateral lithic variations and chemical characteristics that we have come to expect in modern island arcs."

Plateau basalts are an important constituent of many passive margins (e.g. Iceland, Greenland, Faroes, Britain, Brazil, southern Africa, the traps of India Yemen, Ethiopia), and Kent even suggests that the plateau basalts of Mozambique and Madigascar may be continuous with layer 2 of parts of the western Indian Ocean, and are possibly underlain by sediments.

CONTINENTAL MARGINS AND NATURAL RESOURCES

The study of continental margins has been given new impetus with the recognition that resources there are available with the technology of today or the near future. These resources can be grouped by their related geology and water depth, which also largely controls the potential for economic recovery, or they can be classed by commodity, independent of the costs of recovery (Weeks; Cruickshank).

The annual value of offshore petroleum production now exceeds that of all other marine resources combined, including fisheries and seawater chemicals. However, only a very small part of this currently comes from true continental margins (Drake and Burk), the largest part being produced from such epicontinental seas as the Persian Gulf and Lake Maracaibo. It seems certain that petroleum will continue to dominate the total value of marine resources over the next several decades (Weeks).

Particularly important to further understanding of continental margins will be the drilling and exploration carried out with greatly advanced geophysical techniques (see Seely et al.; Driver and Pardo; Montadert et al.; Biju-Duval et al.; Renard and Mascle; Katz; Dubois et al.; Bentz; Auzende and Olivet; Mordojovich; Delteil et al.).

An inventory of the nonpetroleum resources is much more difficult (Cruickshank) because of their great variety and multiple geological environments. There is no reason to expect that all the mineral resources known on land are not present within the various continental margins of the world, but even a small amount of overlying water restricts the availability and the economics of mining on continental margins. However, certain other resources are unique to the oceans, such as some forms of phosphate rock, manganese nodules, and particularly the minerals in seawater itself. It seems likely that manganese nodules may eventually yield large resources, simply because of their vast extent and large total volume. Another important contribution may eventually be from localized, metal-enriched sediments and brines.

CONCLUDING REMARKS

The collection of reports that make up this volume was assembled in the spirit of the International Geodynamics Project and was strongly encouraged by the U.S. Geodynamics Committee. The significance of continental margins was recognized early by the International Upper Mantle Committee, which established a commission on continental margins and island arcs, devoted to studies of present active and inactive margins, and another on the world rift system, to study active rifts and their relationship to continental margins.

These activities were continued and expanded by the Inter-Union Commission on Geodynamics through both regional and topical working groups. In fact, it is difficult to find an activity in Geodynamics to which continental margins are not relevant, or vice versa.

The difficulties of working in these areas are manifold, as illustrated by the chapters in this volume. But the scientific, economic, and social benefits to be gained from understanding them completely warrant an increased expenditure of scientific effort and resources. Many surprises have turned up during the last decade and, although much has been learned, it is not unreasonable to conclude that many additional surprises await us.

BIBLIOGRAPHY

All references cited are to reports appearing in this volume.

ARCTIC OCEAN
OSTENSO

DRUMMOND
LUDWIG
MOORE
SCHOLL
KEENS
DEWEY
GOODWIN
KULM & FOWLER
WILLIAMS & STEVENS
ERNST
SHERADAN, MAYHEW
MAXWELL
RABINOWITZ, FLOOD & HOLLISTER
BLAKE, et al.
WORZEL
ANTOINE, et al
MEYERHOFF & HATTEN
CARR, et al
CASE
SEELY, et al
SHOR
DUBOIS et al
HAWKINS
HAYES
ZAMBRANO & URIEN
CAMPOS et al
BENTZ
MORDOJOVICH
URIEN & EWING
BLAKE, et al.
KATZ
DALZIEL

ANTARCTICA–
HOUTZ

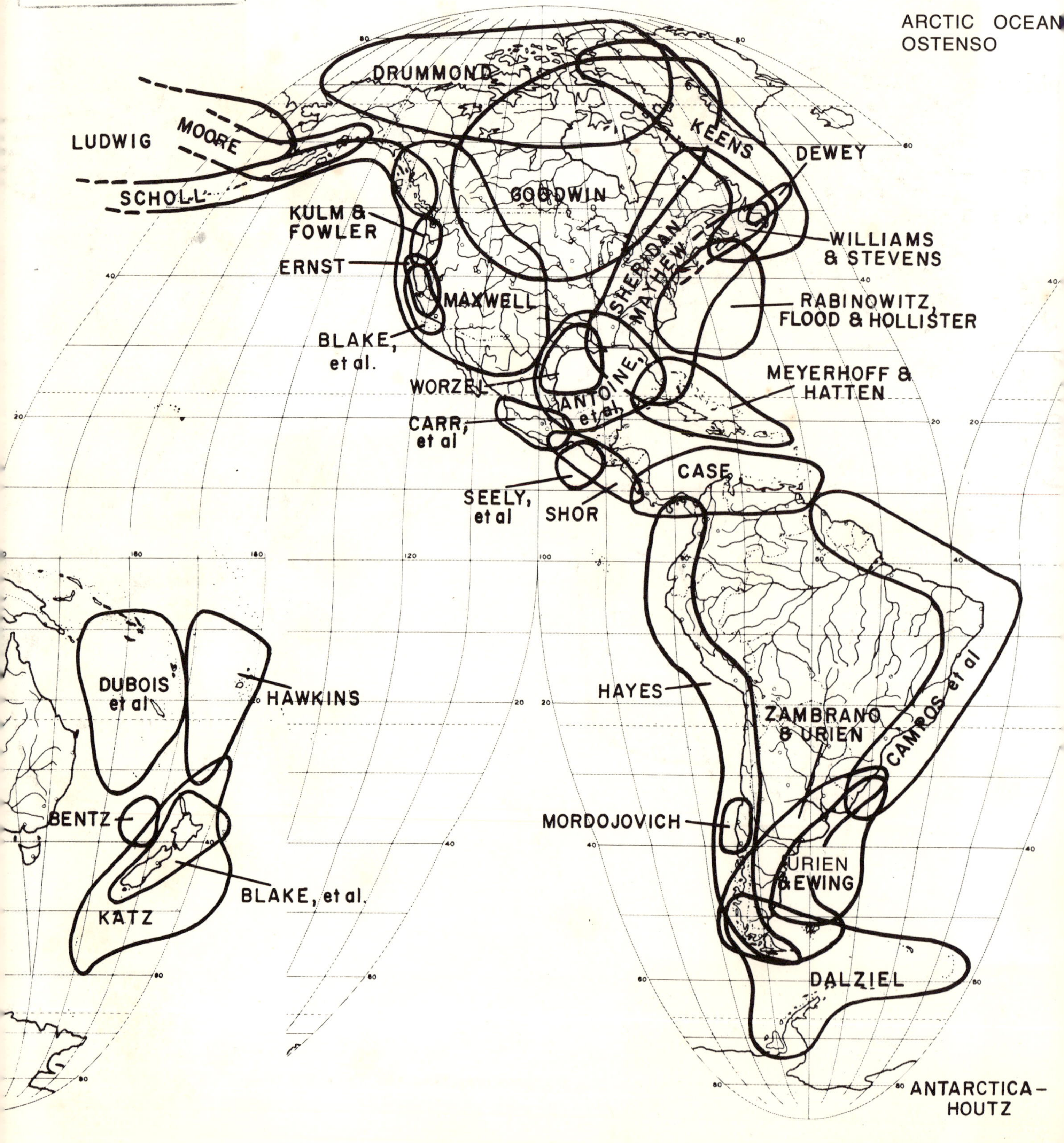